INTERNATIONAL RESIDENTIAL CODE®
for One- and Two-Family Dwellings

A Member of the International Code Family®

2021

INCLUDES
Residential requirements from
NFPA 70: National Electrical Code® 2020
The electrical code designated for use with the I-Codes®

INTERNATIONAL CODE COUNCIL®

2021 International Residential Code®

Date of First Publication: January 29, 2021

First Printing: January 2021

ISBN: 978-1-60983-957-4 (soft-cover edition)
ISBN: 978-1-60983-958-1 (loose-leaf edition)

T025919

PRINTED IN THE USA

PREFACE

Introduction

The *International Residential Code*® (IRC®) establishes minimum requirements for one- and two family dwellings and townhouses using prescriptive provisions. It is founded on broad-based principles that make possible the use of new materials and new building designs. This 2021 edition is fully compatible with all of the International Codes® (I-Codes®) published by the International Code Council (ICC), including the *International Building Code*® (IBC®), *International Energy Conservation Code*® (IECC®), *International Existing Building Code*® (IEBC®), *International Fire Code*® (IFC®), *International Fuel Gas Code*® (IFGC®), *International Green Construction Code*® (IgCC®), *International Mechanical Code*® (IMC®), *International Plumbing Code*® (IPC®), *International Private Sewage Disposal Code*® (IPSDC®), *International Property Maintenance Code*® (IPMC®), *International Swimming Pool and Spa Code*® (ISPSC®), *International Wildland-Urban Interface Code*® (IWUIC®), *International Zoning Code*® (IZC®) and *International Code Council Performance Code*® (ICCPC®).

The I-Codes, including the IRC, are used in a variety of ways in both the public and private sectors. Most industry professionals are familiar with the I-Codes as the basis of laws and regulations in communities across the US and in other countries. However, the impact of the codes extends well beyond the regulatory arena, as they are used in a variety of nonregulatory settings, including:

- Voluntary compliance programs such as those promoting sustainability, energy efficiency and disaster resistance.
- The insurance industry, to estimate and manage risk, and as a tool in underwriting and rate decisions.
- Certification and credentialing of individuals involved in the fields of building design, construction and safety.
- Certification of building and construction-related products.
- US federal agencies, to guide construction in an array of government-owned properties.
- Facilities management.
- "Best practices" benchmarks for designers and builders, including those who are engaged in projects in jurisdictions that do not have a formal regulatory system or a governmental enforcement mechanism.
- College, university and professional school textbooks and curricula.
- Reference works related to building design and construction.

In addition to the codes themselves, the code development process brings together building professionals on a regular basis. It provides an international forum for discussion and deliberation about building design, construction methods, safety, performance requirements, technological advances and innovative products.

Development

This 2021 edition presents the code as originally issued, with changes reflected in the 2003 through 2018 editions and further changes approved by the ICC Code Development Process through 2019. Residential electrical provisions are based on the 2020 National Electrical Code® (NFPA 70). A new edition such as this is promulgated every 3 years.

Fuel gas provisions have been included through an agreement with the American Gas Association (AGA). Electrical provisions have been included through an agreement with the NFPA.

This code is founded on principles intended to establish provisions consistent with the scope of a residential code that adequately protects public health, safety and welfare; provisions that do not unnecessarily increase construction costs; provisions that do not restrict the use of new materials, products or methods of construction; and provisions that do not give preferential treatment to particular types or classes of materials, products or methods of construction.

Maintenance

The IRC is kept up to date through the review of proposed changes submitted by code enforcement officials, industry representatives, design professionals and other interested parties. Proposed changes are carefully considered through an open code development process in which all interested and affected parties may participate.

The ICC Code Development Process reflects principles of openness, transparency, balance, due process and consensus, the principles embodied in OMB Circular A-119, which governs the federal government's use of private-sector standards. The ICC process is open to anyone; there is no cost to participate, and people can participate without travel cost through the ICC's cloud-based app, cdpAccess®. A broad cross section of interests are represented in the ICC Code Development Process. The codes, which are updated regularly, include safeguards that allow for emergency action when required for health and safety reasons.

In order to ensure that organizations with a direct and material interest in the codes have a voice in the process, the ICC has developed partnerships with key industry segments that support the ICC's important public safety mission. Some code development committee members were nominated by the following industry partners and approved by the ICC Board:

- National Association of Home Builders (NAHB)
- National Council of Structural Engineers Association (NCSEA)

The code development committees evaluate and make recommendations regarding proposed changes to the codes. Their recommendations are then subject to public comment and council-wide votes. The ICC's governmental members—public safety officials who have no financial or business interest in the outcome—cast the final votes on proposed changes.

The contents of this work are subject to change through the code development cycles and by any governmental entity that enacts the code into law. For more information regarding the code development process, contact the Codes and Standards Development Department of the ICC.

The maintenance process for the fuel gas provisions is based on the process used to maintain the IFGC, in conjunction with the AGA. The maintenance process for the electrical provisions is undertaken by the NFPA.

While the I-Code development procedure is thorough and comprehensive, the ICC, its members and those participating in the development of the codes disclaim any liability resulting from the publication or use of the I-Codes, or from compliance or noncompliance with their provisions. The ICC does not have the power or authority to police or enforce compliance with the contents of this code.

Code Development Committee Responsibilities (Letter Designations in Front of Section Numbers)

In each code development cycle, proposed changes to this code are considered at the Committee Action Hearings by the applicable International Code Development Committee as follows:

[RB] = IRC—Building Code Development Committee

[RE] = International Residential Energy Conservation Code Development Committee;

[MP] = IRC—Mechanical/Plumbing Code Development Committee

The [RE] committee is also responsible for the IECC—Residential Provisions.

For the development of the 2024 edition of the I-Codes, there will be two groups of code development committees and they will meet in separate years, as shown in the following Code Development Hearings Table.

Code change proposals submitted for IRC Chapters 1 and 3 through 10, Appendices AE, AF, AH, AJ, AK, AL, AM, AN, AO, AQ, AR, AS, AT, AU, AV and AW, and definitions designated [RB] are heard by the IRC—Building Committee during the Group B (2022) cycle code development hearing. Code change proposals submitted for Chapter 11 are heard by the International Energy Conservation Code Development Committee during the Group B (2022) cycle code development hearing. Proposed changes to all other chapters are heard by the IRC Plumbing and Mechanical Committee during the Group A (2021) code development cycle.

It is very important that anyone submitting code change proposals understand which code development committee is responsible for the section of the code that is the subject of the code change proposal. For further information on the code development committee responsibilities, please visit the ICC website at www.iccsafe.org/current-code-development-cycle.

CODE DEVELOPMENT HEARINGS

Group A Codes (Heard in 2021, Code Change Proposals Deadline: January 11, 2021)	Group B Codes (Heard in 2022, Code Change Proposals Deadline: January 10, 2022)
International Building Code – Egress (Chapters 10, 11, Appendix E) – Fire Safety (Chapters 7, 8, 9, 14, 26) – General (Chapters 2–6, 12, 27–33, Appendices A, B, C, D, K, N)	Administrative Provisions (Chapter 1 of all codes except IECC, IRC and IgCC; IBC Appendix O; the appendices titled "Board of Appeals" for all codes except IECC, IRC, IgCC, ICCPC and IZC; administrative updates to currently referenced standards; and designated definitions)
International Fire Code	**International Building Code**– Structural (Chapters 15–25, Appendices F, G, H, I, J, L, M)
International Fuel Gas Code	**International Existing Building Code**
International Mechanical Code	**International Energy Conservation Code—Commercial**
International Plumbing Code	**International Energy Conservation Code—Residential** – IECC—Residential – IRC—Energy (Chapter 11)
International Property Maintenance Code	**International Green Construction Code** (Chapter 1)
International Private Sewage Disposal Code	**International Residential Code** – IRC—Building (Chapters 1–10, Appendices AE, AF, AH, AJ, AK, AL, AM, AO, AQ, AR, AS, AT, AU, AV, AW)
International Residential Code – IRC—Mechanical (Chapters 12–23) – IRC—Plumbing (Chapters 25–33, Appendices AG, AI, AN, AP)	
International Swimming Pool and Spa Code	
International Wildland-Urban Interface Code	
International Zoning Code	

Note: Proposed changes to the ICCPC will be heard by the code development committee noted in brackets [ ] in the text of the ICCPC.

Marginal Markings

Solid vertical lines in the margins within the body of the code indicate a technical change from the requirements of the 2018 edition. Deletion indicators in the form of an arrow (➡) are provided in the margin where an entire section, paragraph, exception or table has been deleted or an item in a list of items or a row of a table has been deleted.

A single asterisk [*] placed in the margin indicates that text or a table has been relocated within the code. A double asterisk [**] placed in the margin indicates that the text or table immediately following it has been relocated there from elsewhere in the code. The following table indicates such relocations in the 2021 edition of the IRC.

RELOCATIONS

2021 LOCATION	2018 LOCATION
Table N1102.1.2	Table N1102.1.4
Table N1102.1.3	Table N1102.1.2
N1103.3.2	N1103.3.7
N1103.3.3	N1103.3.6
N1103.3.3.1	N1103.3.6.1
N1103.3.4	N1103.3.2
N1103.3.4.1	N1103.3.2.1
N1103.3.5	N1103.3.3
N1103.3.6	N1103.3.4
N1103.3.7	N1103.3.5
N1107.2	N1101.13.1

Coordination of the International Codes

The coordination of technical provisions is one of the strengths of the ICC family of model codes. The codes can be used as a complete set of complementary documents, which will provide users with full integration and coordination of technical provisions. Individual codes can also be used in subsets or as stand-alone documents. To make sure that each individual code is as complete as possible, some technical provisions that are relevant to more than one subject area are duplicated in some of the model codes. This allows users maximum flexibility in their application of the I-Codes.

Italicized Terms

Terms italicized in code text, other than document titles, are defined in Chapter 2. The terms selected to be italicized have definitions that the user should read carefully to better understand the code. Where italicized, the Chapter 2 definition applies. If not italicized, common-use definitions apply.

Adoption

The ICC maintains a copyright in all of its codes and standards. Maintaining copyright allows the ICC to fund its mission through sales of books, in both print and electronic formats. The ICC welcomes adoption of its codes by jurisdictions that recognize and acknowledge the ICC's copyright in the code, and further acknowledge the substantial shared value of the public/private partnership for code development between jurisdictions and the ICC.

The ICC also recognizes the need for jurisdictions to make laws available to the public. All I-Codes and I-Standards, along with the laws of many jurisdictions, are available for free in a nondownloadable form on the ICC's website. Jurisdictions should contact the ICC at adoptions@iccsafe.org to learn how to adopt and distribute laws based on the IRC in a manner that provides necessary access, while maintaining the ICC's copyright.

To facilitate adoption, several sections of this code contain blanks for fill-in information that needs to be supplied by the adopting jurisdiction as part of the adoption legislation. For this code, please see:

Section R101.1. Insert: **[NAME OF JURISDICTION]**

Table R301.2. Jurisdictions to fill in details as directed by provisions of the code.

Section P2603.5.1. Insert: **[NUMBER OF INCHES IN TWO LOCATIONS]**

Effective Use of the International Residential Code

The IRC was created to serve as a complete, comprehensive code regulating the construction of single-family houses, two-family houses (duplexes) and buildings consisting of three or more townhouse units. All buildings within the scope of the IRC are limited to three stories above grade plane. For example, a four-story single-family house would fall within the scope of the IBC, not the IRC. The benefits of devoting a separate code to residential construction include the fact that the user need not navigate through a multitude of code provisions that do not apply to residential construction in order to locate that which is applicable. A separate code also allows for residential and nonresidential code provisions to be distinct and tailored to the structures that fall within the appropriate code's scopes.

The IRC contains coverage for all components of a house or townhouse, including structural components, fireplaces and chimneys, thermal insulation, mechanical systems, fuel gas systems, plumbing systems and electrical systems.

The IRC is a prescriptive-oriented (specification) code with some examples of performance code language. It has been said that the IRC is the complete cookbook for residential construction. Section R301.1, for example, is written in performance language, but states that the prescriptive requirements of the code will achieve such performance.

It is important to understand that the IRC contains coverage for what is conventional and common in residential construction practice. While the IRC will provide all of the needed coverage for most residential construction, it might not address construction practices and systems that are atypical or rarely encountered in the industry. Sections such as R301.1.3, R301.2.2.1.1, R320.1, M1301.1, G2401.1 and P2601.1 refer to other codes either as an alternative to the provisions of the IRC or where the IRC lacks coverage for a particular type of structure, design, system, appliance or method of construction. In other words, the IRC is meant to be all inclusive for typical residential construction and it relies on other codes only where alternatives are desired or where the code lacks coverage for the uncommon aspect of residential construction. Of course, the IRC constantly evolves to address new technologies and construction practices that were once uncommon, but are now common.

The IRC is unique in that much of it, including Chapters 3 through 9 and Chapters 34 through 43, is presented in an ordered format that is consistent with the normal progression of construction, starting with the design phase and continuing through the final trim-out phase. This is consistent with the "cookbook" philosophy of the IRC.

ARRANGEMENT AND FORMAT OF THE 2021 IRC

The IRC is divided into nine main parts, specifically: Part I—Administrative, Part II—Definitions, Part III—Building Planning and Construction, Part IV—Energy Conservation, Part V—Mechanical, Part VI—Fuel Gas, Part VII—Plumbing, Part VIII—Electrical and Part IX—Referenced Standards.

The following provides a brief description of the content of each chapter and appendix of the IRC:

Chapter 1 Scope and Administration

This chapter contains provisions for the application, enforcement and administration of subsequent requirements of the code. In addition to establishing the scope of the code, Chapter 1 identifies which buildings and structures come under its purview. Chapter 1 is largely concerned with maintaining "due process of law" in enforcing the building criteria contained in the body of the code. Only through careful observation of the administrative provisions can the building official reasonably expect to demonstrate that "equal protection under the law" has been provided.

Chapter 2 Definitions

Terms defined in the code are listed alphabetically in Chapter 2. It is important to note that three chapters have their own definitions sections: Chapter 11 for the defined terms unique to energy conservation, Chapter 24 for the defined terms unique to fuel gas and Chapter 35 for the terms applicable to electrical Chapters 34 through 43. Where Chapter 24 or 35 defines a term differently than it is defined in Chapter 2, the definition applies in that chapter only. Chapter 2 definitions apply in all other locations in the code.

Where understanding a term's definition is key to or necessary for understanding a particular code provision, the term is shown in italics where it appears in the code. This is true only for those terms that have a meaning that is unique to the code. In other words, the generally understood meaning of a term or phrase might not be sufficient or consistent with the meaning prescribed by the code; therefore, it is essential that the code-defined meaning be known.

Guidance regarding not only tense, gender and plurality of defined terms, but also terms not defined in this code, is provided.

Chapter 3 Building Planning

Chapter 3 provides guidelines for a minimum level of structural integrity, life safety, fire safety and livability for inhabitants of dwelling units regulated by this code. Chapter 3 is a compilation of the code requirements specific to the building planning sector of the design and construction process. This chapter sets forth code requirements dealing with light, ventilation, sanitation, minimum room size, ceiling height and environmental comfort. Chapter 3 establishes life-safety provisions including limitations on glazing used in hazardous areas, specifications on stairways, use of guards at elevated surfaces, window and fall protection, and rules for means of egress. Snow, wind and seismic design live and dead loads and flood-resistant construction, as well as solar energy systems, and swimming pools, spas and hot tubs, are addressed in this chapter.

Chapter 4 Foundations

Chapter 4 provides the requirements for the design and construction of foundation systems for buildings regulated by this code. Provisions for seismic load, flood load and frost protection are contained in this chapter. A foundation system consists of two interdependent components: the foundation structure itself and the supporting soil.

The prescriptive provisions of this chapter provide requirements for constructing footings and walls for foundations of wood, masonry, concrete and precast concrete. In addition to a foundation's ability to support the required design loads, this chapter addresses several other factors that can affect foundation performance. These include controlling surface water and subsurface drainage, requiring soil tests where conditions warrant and evaluating proximity to slopes and minimum depth requirements. The chapter also provides requirements to minimize adverse effects of moisture, decay and pests in basements and crawl spaces.

Chapter 5 Floors

Chapter 5 provides the requirements for the design and construction of floor systems that will be capable of supporting minimum required design loads. This chapter covers four different types: wood floor framing, wood floors on the ground, cold-formed steel floor framing and concrete slabs on the ground. Allowable span tables are provided that greatly simplify the determination of joist, girder and sheathing sizes for raised floor systems of wood framing and cold-formed steel framing. This chapter also contains prescriptive requirements for wood-framed exterior decks and their attachment to the main building.

Chapter 6 Wall Construction

Chapter 6 contains provisions that regulate the design and construction of walls. The wall construction covered in Chapter 6 consists of five different types: wood framed, cold-formed steel framed, masonry, concrete and structural insulated panel (SIP). The primary concern of this chapter is the structural integrity of wall construction and transfer of all imposed loads to the supporting structure. This chapter provides the requirements for the design and construction of wall systems that are capable of supporting the minimum design vertical loads (dead, live and snow loads) and lateral loads (wind or seismic loads). This chapter contains the prescriptive requirements for wall bracing and/or shear walls to resist the imposed lateral loads due to wind and seismic activity.

Chapter 6 also regulates exterior windows and doors installed in walls. This chapter contains criteria for the performance of exterior windows and doors and includes provisions for testing and labeling, garage doors, windborne debris protection and anchorage details.

Chapter 7 Wall Covering

Chapter 7 contains provisions for the design and construction of interior and exterior wall coverings. This chapter establishes the various types of materials, materials standards and methods of application permitted for use as interior coverings, including interior plaster, gypsum board, ceramic tile, wood veneer paneling, hardboard paneling, wood shakes and wood shingles. Chapter 7 also contains requirements for the use of vapor retarders for moisture control in walls.

Exterior wall coverings provide the weather-resistant exterior envelope that protects the building's interior from the elements. Chapter 7 provides the requirements for wind resistance and water-resistive barrier for exterior wall coverings. This chapter prescribes the exterior wall coverings as well as the water-resistive barrier required beneath the exterior materials. Exterior wall coverings regulated by this section include aluminum, stone and masonry veneer, wood, hardboard, particleboard, wood structural panel siding, wood shakes and shingles, exterior plaster, steel, vinyl, fiber cement and exterior insulation finish systems.

Chapter 8 Roof-ceiling Construction

Chapter 8 regulates the design and construction of roof-ceiling systems. This chapter contains two roof-ceiling framing systems: wood framing and cold-formed steel framing. Allowable span tables are provided to simplify the selection of rafter and ceiling joist size for wood roof framing and cold-formed steel framing. Chapter 8 also provides requirements for the application of ceiling finishes, the proper ventilation of concealed spaces in roofs (e.g., enclosed attics and rafter spaces), unvented attic assemblies and attic access.

Chapter 9 Roof Assemblies

Chapter 9 regulates the design and construction of roof assemblies. A roof assembly includes the roof deck, vapor retarder, substrate or thermal barrier, insulation, vapor retarder and roof covering. This chapter provides the requirement for wind resistance of roof coverings.

The types of roof covering materials and installation regulated by Chapter 9 are: asphalt shingles, clay and concrete tile, metal roof shingles, mineral-surfaced roll roofing, slate and slate-type shingles, wood shakes and shingles, built-up roofs, metal roof panels, modified bitumen roofing, thermoset and thermoplastic single-ply roofing, sprayed polyurethane foam roofing, liquid applied coatings and photovoltaic shingles. Chapter 9 also provides requirements for roof drainage, flashing, above deck thermal insulation, rooftop-mounted photovoltaic systems and recovering or replacing an existing roof covering.

Chapter 10 Chimneys and Fireplaces

Chapter 10 contains requirements for the safe construction of masonry chimneys and fireplaces and establishes the standards for the use and installation of factory-built chimneys, fireplaces and masonry heaters. Chimneys and fireplaces constructed of masonry rely on prescriptive requirements for the details of their construction; the factory-built type relies on the listing and labeling method of approval. Chapter 10 provides the requirements for seismic reinforcing and anchorage of masonry fireplaces and chimneys.

Chapter 11 [RE] Energy Efficiency

The purpose of Chapter 11 [RE] is to provide minimum design requirements that will promote efficient utilization of energy in buildings. The requirements are directed toward the design of building envelopes with adequate thermal resistance and low air leakage, and toward the design and selection of mechanical, water heating, electrical and illumination systems that promote effective use of depletable energy resources. The provisions of Chapter 11 [RE] are duplicated from the *International Energy Conservation Code—Residential Provisions*, as applicable for buildings which fall under the scope of the IRC.

For ease of use and coordination of provisions, the corresponding IECC—Residential Provisions section number is indicated following the IRC section number [e.g., N1102.1 (R402.1)].

Chapter 12 Mechanical Administration

Chapter 12 establishes the limits of applicability of the code and describes how the code is to be applied and enforced. A mechanical code, like any other code, is intended to be adopted as a legally enforceable document and it cannot be effective without adequate provisions for its administration and enforcement. The provisions of Chapter 12 establish the authority and duties of the code official appointed by the jurisdiction having authority and also establish the rights and privileges of the design professional, contractor and property owner. It also relates this chapter to the administrative provisions in Chapter 1.

Chapter 13 General Mechanical System Requirements

Chapter 13 contains broadly applicable requirements related to appliance listing and labeling, appliance location and installation, appliance and systems access, protection of structural elements and clearances to combustibles, among others.

Chapter 14 Heating and Cooling Equipment and Appliances

Chapter 14 is a collection of requirements for various heating and cooling appliances, dedicated to single topics by section. The common theme is that all of these types of appliances use energy in one form or another, and the improper installation of such appliances would present a hazard to the occupants of the dwellings, due to either the potential for fire or the accidental release of refrigerants. Both situations are undesirable in dwellings that are covered by this code.

Chapter 15 Exhaust Systems

Chapter 15 is a compilation of code requirements related to residential exhaust systems, including kitchens and bathrooms, clothes dryers and range hoods. The code regulates the materials used for constructing and installing such duct systems. Air brought into the building for ventilation, combustion or makeup purposes is protected from contamination by the provisions found in this chapter.

Chapter 16 Duct Systems

Chapter 16 provides requirements for the installation of ducts for supply, return and exhaust air systems. This chapter contains no information on the design of these systems from the standpoint of air movement, but is concerned with the structural integrity of the systems and the overall impact of the systems on the fire-safety performance of the building. This chapter regulates the materials and methods of construction which affect the performance of the entire air distribution system.

Chapter 17 Combustion Air

Complete combustion of solid and liquid fuel is essential for the proper operation of appliances, control of harmful emissions and achieving maximum fuel efficiency. If insufficient quantities of oxygen are supplied, the combustion process will be incomplete, creating dangerous byproducts and wasting energy in the form of unburned fuel (hydrocarbons). The byproducts of incomplete combustion are poisonous, corrosive and combustible, and can cause serious appliance or equipment malfunctions that pose fire or explosion hazards.

The combustion air provisions in this code from previous editions have been deleted from Chapter 17 in favor of a single section that directs the user to NFPA 31 for oil-fired appliance combustion air requirements and the manufacturer's installation instructions for solid fuel-burning appliances. If fuel gas appliances are used, the provisions of Chapter 24 must be followed.

Chapter 18 Chimneys and Vents

Chapter 18 regulates the design, construction, installation, maintenance, repair and approval of chimneys, vents and their connections to fuel-burning appliances. A properly designed chimney or vent system is needed to conduct the flue gases produced by a fuel-burning appliance to the outdoors. The provisions of this chapter are intended to minimize the hazards associated with high temperatures and potentially toxic and corrosive combustion gases. This chapter addresses factory-built and masonry chimneys, vents and venting systems used to vent oil-fired and solid fuel-burning appliances.

Chapter 19 Special Appliances, Equipment and Systems

Chapter 19 regulates the installation of fuel-burning appliances that are not covered in other chapters, such as ranges and ovens, sauna heaters, fuel cell power plants and hydrogen systems. Because the subjects in this chapter do not contain the volume of text necessary to warrant individual chapters, they have been combined into a single chapter. The only commonality is that the subjects use energy to perform some task or function. The intent is to provide a reasonable level of protection for the occupants of the dwelling.

Chapter 20 Boilers and Water Heaters

Chapter 20 regulates the installation of boilers and water heaters. Its purpose is to protect the occupants of the dwelling from the potential hazards associated with such appliances. A water heater is any appliance that heats potable water and supplies it to the plumbing hot water distribution system. A boiler either heats water or generates steam for space heating and is generally a closed system.

Chapter 21 Hydronic Piping

Hydronic piping includes piping, fittings and valves used in building space conditioning systems. Applications include hot water, chilled water, steam, steam condensate, brines and water/antifreeze mixtures. Chapter 21 regulates installation, alteration and repair of all hydronic piping systems to ensure the reliability, serviceability, energy efficiency and safety of such systems.

Chapter 22 Special Piping and Storage Systems

Chapter 22 regulates the design and installation of fuel oil storage and piping systems. The regulations include reference to construction standards for above-ground and underground storage tanks, material standards for piping systems (both above-ground and underground) and extensive requirements for the proper assembly of system piping and components. The purpose of this chapter is to prevent fires, leaks and spills involving fuel oil storage and piping systems, whether inside or outside structures and above or underground.

Chapter 23 Solar Thermal Energy Systems

Chapter 23 contains requirements for the construction, alteration and repair of all systems and components of solar thermal energy systems used for space heating or cooling, and domestic hot water heating or processing. The provisions of this chapter are limited to those necessary to achieve installations that are relatively hazard free.

A solar thermal energy system can be designed to handle 100 percent of the energy load of a building, although this is rarely accomplished. Because solar energy is a low-intensity energy source and dependent on the weather, it is usually necessary to supplement a solar thermal energy system with traditional energy sources.

As our world strives to find alternate means of producing power for the future, the requirements of this chapter will become more and more important over time.

Chapter 24 Fuel Gas

Chapter 24 regulates the design and installation of fuel gas distribution piping and systems, appliances, appliance venting systems and combustion air provisions. The definition of "Fuel gas" includes natural, liquefied petroleum and manufactured gases and mixtures of these gases.

The purposes of this chapter are to establish the minimum acceptable level of safety and to protect life and property from the potential dangers associated with the storage, distribution and use of fuel gases and the byproducts of combustion of such fuels. This code also protects the personnel who install, maintain, service and replace the systems and appliances addressed herein.

Chapter 24 is composed entirely of text extracted from the IFGC; therefore, whether using the IFGC or the IRC, the fuel gas provisions will be identical. Note that to avoid the potential for confusion and conflicting definitions, Chapter 24 has its own definition section.

Chapter 25 Plumbing Administration

The requirements of Chapter 25 do not supersede the administrative provisions of Chapter 1. Rather, the administrative guidelines of Chapter 25 pertain to plumbing installations that are best referenced and located within the plumbing chapters. This chapter addresses how to apply the plumbing provisions of this code to specific types or phases of construction. This chapter also outlines the responsibilities of the applicant, installer and inspector with regard to testing plumbing installations.

Chapter 26 General Plumbing Requirements

The content of Chapter 26 is often referred to as "miscellaneous," rather than general plumbing requirements. This is the only chapter of the plumbing chapters of the code whose requirements do not interrelate. If a requirement cannot be located in another plumbing chapter, it should be located in this chapter. Chapter 26 contains safety requirements for the installation of plumbing systems and includes requirements for the identification of pipe, pipe fittings, traps, fixtures, materials and devices used in plumbing systems. If specific provisions do not demand that a requirement be located in another chapter, the requirement is located in this chapter.

Chapter 27 Plumbing Fixtures

Chapter 27 requires fixtures to be of the proper type, approved for the purpose intended and installed properly to promote usability and safe, sanitary conditions. This chapter regulates the quality of fixtures and faucets by requiring those items to comply with nationally recognized standards. Because fixtures must be properly installed so that they are usable by the occupants of the building, this chapter contains the requirements for the installation of fixtures.

Chapter 28 Water Heaters

Chapter 28 regulates the design, approval and installation of water heaters and related safety devices. The intent is to minimize the hazards associated with the installation and operation of water heaters. Although this chapter does not regulate the size of a water heater, it does regulate all other aspects of the water heater installation such as temperature and pressure relief valves, safety drip pans and connections. Where a water heater also supplies water for space heating, this chapter regulates the maximum water temperature supplied to the water distribution system.

Chapter 29 Water Supply and Distribution

This chapter regulates the supply of potable water from both public and individual sources to every fixture and outlet so that it remains potable and uncontaminated by cross connections. Chapter 29 also regulates the design of the water distribution system, which will allow fixtures to function properly. Because it is critical that the potable water supply system remain free of actual or potential sanitary hazards, this chapter has the requirements for providing backflow protection devices.

Chapter 30 Sanitary Drainage

The purpose of Chapter 30 is to regulate the materials, design and installation of sanitary drainage piping systems as well as the connections made to the system. The intent is to design and install sanitary drainage systems that will function reliably, are neither undersized nor oversized and are constructed from materials, fittings and connections whose quality is regulated by this section. This chapter addresses the proper use of fittings for directing the flow into and within the sanitary drain piping system. Materials and provisions necessary for servicing the drainage system are also included in this chapter.

Chapter 31 Vents

Venting protects the trap seal of each trap. The vents are designed to limit differential pressures at each trap to 1 inch of water column (249 Pa). Because waste flow in the drainage system creates pressure fluctuations that can negatively affect traps, the sanitary drainage system must have a properly designed venting system. Chapter 31 covers the requirements for vents and venting. All of the provisions set forth in this chapter are intended to limit the pressure differentials in the drainage system to a maximum of 1 inch of water column (249 Pa) above or below atmospheric pressure (i.e., positive or negative pressures).

Chapter 32 Traps

Traps prevent sewer gas from escaping from the drainage piping into the building. Water seal traps are the simplest and most reliable means of preventing sewer gas from entering the interior environment. This chapter lists prohibited trap types and specifies the minimum trap size for each type of fixture.

Chapter 33 Storm Drainage

Rainwater infiltration into the ground adjacent to a building can cause the interior of foundation walls to become wet. The installation of a subsoil drainage system prevents the buildup of rainwater on the exterior of the foundation walls. This chapter provides the specifications for subsoil drain piping. Where the discharge of the subsoil drain system is to a sump, this chapter also provides coverage for sump construction, pumps and discharge piping.

Chapter 34 General Requirements

This chapter contains broadly applicable, general and miscellaneous requirements including scope, listing and labeling, equipment locations and clearances for conductor materials and connections and conductor identification.

Chapter 35 Electrical Definitions

Chapter 35 is the repository of the definitions of terms used in the body of Part VIII of the code. To avoid the potential for confusion and conflicting definitions, Part VIII, Electrical, has its own definition chapter.

Codes are technical documents and every word, term and punctuation mark can add to or change the meaning of a technical requirement. The code often uses terms that have a unique meaning in the code, which can differ substantially from the ordinarily understood meaning of the term as used outside of the code.

The terms defined in Chapter 35 are deemed to be of prime importance in establishing the meaning and intent of the electrical code text that uses the terms. The user of the code should be familiar with and consult this chapter because the definitions are essential to the correct interpretation of the code and because the user may not be aware that a term is defined.

Chapter 36 Services

This chapter covers the design, sizing and installation of the building's electrical service equipment and grounding electrode system. It includes an easy-to-use load calculation method and service conductor sizing table. The electrical service is generally the first part of the electrical system to be designed and installed.

Chapter 37 Branch Circuit and Feeder Requirements

Chapter 37 addresses the requirements for designing the power distribution system, which consists of feeders and branch circuits emanating from the service equipment. This chapter dictates the ratings of circuits and the allowable loads, the number and types of branch circuits required, the wire sizing for such branch circuits and feeders and the requirements for protection from overcurrent for conductors. A load calculation method specific to feeders is also included. This chapter is used to design the electrical system on the load side of the service.

Chapter 38 Wiring Methods

Chapter 38 specifies the allowable wiring methods, such as cable, conduit and raceway systems, and provides the installation requirements for the wiring methods. This chapter is primarily applicable to the "rough-in" phase of construction.

Chapter 39 Power and Lighting Distribution

This chapter mostly contains installation requirements for the wiring that serves the lighting outlets, receptacle outlets, appliances and switches located throughout the building. The required distribution and spacing of receptacle outlets and lighting outlets is prescribed in this chapter, as well as the requirements for ground-fault and arc-fault circuit-interrupter protection.

Chapter 40 Devices and Luminaires

This chapter focuses on the devices, including switches and receptacles, and lighting fixtures that are typically installed during the final phase of construction.

Chapter 41 Appliance Installation

Chapter 41 addresses the installation of appliances including HVAC appliances, water heaters, fixed space-heating equipment, dishwashers, garbage disposals, range hoods and suspended paddle fans.

Chapter 42 Swimming Pools

This chapter covers the electrical installation requirements for swimming pools, storable swimming pools, wading pools, decorative pools, fountains, hot tubs, spas and hydromassage bathtubs. The allowable wiring methods are specified along with the required clearances between electrical system components and pools, spas and tubs. This chapter includes the special grounding requirements related to pools, spas and tubs, and also prescribes the equipotential bonding requirements that are unique to pools, spas and tubs.

Chapter 43 Class 2 Remote-control, Signaling and Power-limited Circuits

This chapter covers the power supplies, wiring methods and installation requirements for the Class 2 circuits found in dwellings. Such circuits include thermostat wiring, alarm systems, security systems, automated control systems and doorbell systems.

Chapter 44 Referenced Standards

The code contains numerous references to standards that are used to regulate materials and methods of construction. Chapter 44 contains a comprehensive list of all standards that are referenced in the code. The standards are part of the code to the extent of the reference to the standard. Compliance with the referenced standard is necessary for compliance with this code. By providing specifically adopted standards, the construction and installation requirements necessary for compliance with the code can be readily determined. The basis for code compliance is, therefore, established and available on an equal basis to the code official, contractor, designer and owner.

Chapter 44 is organized in a manner that makes it easy to locate specific standards. It lists all of the referenced standards, alphabetically, by acronym of the promulgating agency of the standard. Each agency's standards are then listed in either alphabetical or numeric order based upon the standard identification. The list also contains the title of the standard; the edition (date) of the standard referenced; any addenda included as part of the ICC adoption; and the section or sections of this code that reference the standard.

Appendix AA Sizing and Capacities of Gas Piping

This appendix is informative and not part of the code. It provides design guidance, useful facts and data and multiple examples of how to apply the sizing tables and sizing methodologies of Chapter 24.

Appendix AB Sizing of Venting Systems Serving Appliances Equipped with Draft Hoods, Category I Appliances, and Appliances Listed for Use with Type B Vents

This appendix is informative and not part of the code. It contains multiple examples of how to apply the vent and chimney tables and methodologies of Chapter 24.

Appendix AC Exit Terminals of Mechanical Draft and Direct-vent Venting Systems

This appendix is informative and not part of the code. It consists of a figure and notes that visually depict code requirements from Chapter 24 for vent terminals with respect to the openings found in building exterior walls.

Appendix AD Recommended Procedure for Safety Inspection of an Existing Appliance Installation

This appendix is informative and not part of the code. It provides recommended procedures for testing and inspecting an appliance installation to determine if the installation is operating safely and if the appliance is in a safe condition.

Appendix AE Manufactured Housing Used as Dwellings

The criteria for the construction of manufactured homes are governed by the National Manufactured Housing Construction and Safety Act. While this act may seem to cover the bulk of the construction of manufactured housing, it does not cover those areas related to the placement of the housing on the property. The provisions of Appendix AE are not applicable to the design and construction of manufactured homes. Appendix AE provides a complete set of regulations in conjunction with federal law for the installation of manufactured housing. This appendix also contains provisions for existing manufactured home installations.

Appendix AF Radon Control Methods

Radon comes from the natural (radioactive) decay of the element radium in soil, rock and water and finds its way into the air. Appendix AF contains requirements to mitigate the transfer of radon gases from the soil into the dwelling. The provisions of this appendix regulate the design and construction of radon-resistant measures intended to reduce the entry of radon gases into the living space of residential buildings.

Appendix AG Piping Standards for Various Applications

Appendix AG provides standards for various types of plastic piping products. This appendix is informative and is not part of the code.

Appendix AH Patio Covers

Appendix AH sets forth the regulations and limitations for patio covers. The provisions address those uses permitted in patio cover structures, the minimum design loads to be assigned for structural purposes, and the effect of the patio cover on egress and emergency escape or rescue from sleeping rooms. This appendix also contains the special provisions for aluminum screen enclosures in hurricane-prone regions.

Appendix AI Private Sewage Disposal

Appendix AI simply provides the opportunity to utilize the International Private Sewage Disposal Code for the design and installation of private sewage disposal in one- and two-family dwellings.

Appendix AJ Existing Buildings and Structures

Appendix AJ contains the provisions for the repair, renovation, alteration and reconstruction of existing buildings and structures that are within the scope of this code. To accomplish this objective and to make the rehabilitation process more available, this appendix allows for a controlled departure from full code compliance without compromising minimum life safety, fire safety, structural and environmental features of the rehabilitated existing building or structure.

Appendix AK Sound Transmission

Appendix AK regulates the sound transmission of wall and floor-ceiling assemblies separating dwelling units and townhouse units. Airborne sound insulation is required for walls. Airborne sound insulation and impact sound insulation are required for floor-ceiling assemblies. The provisions in Appendix AK set forth a minimum Sound Transmission Class (STC) rating for common walls and floor-ceiling assemblies between dwelling units. In addition, a minimum Impact Insulation Class (IIC) rating is also established to limit structureborne sound through common floor-ceiling assemblies separating dwelling units.

Appendix AL Permit Fees

Appendix AL provides guidance to jurisdictions for setting appropriate permit fees. This appendix will aid many jurisdictions to assess permit fees that will assist to fairly and properly administer the code. This appendix can be used for informational purposes only or may be adopted when specifically referenced in the adopting ordinance.

Appendix AM Home Day Care—R-3 Occupancy

Appendix AM provides means of egress and smoke detection requirements for a Group R-3 Occupancy that is to be used as a home day care for more than five children who receive custodial care for less than 24 hours. This appendix is strictly for guidance and/or adoption by those jurisdictions that have Licensed Home Care Provider laws and statutes that allow more than five children to be cared for in a person's home. When a jurisdiction adopts this appendix, the provisions for day care and child care facilities in the IBC should be considered also.

Appendix AN Venting Methods

Because venting of sanitary drainage systems is a difficult concept to understand, and Chapter 31 uses only words to describe venting requirements, illustrations can offer greater insight into what the words mean. Appendix AN has a number of illustrations for commonly installed sanitary drainage systems in order for the reader to gain a better understanding of this code's venting requirements.

Appendix AO Automatic Vehicular Gates

Appendix AO provides the requirements for the design and construction of automatic vehicular gates. The provisions are for where automatic gates are installed for use at a vehicular entrance or exit on the lot of a one- or two-family dwelling. The requirements provide protection for individuals from potential entrapment between an automatic gate and a stationary object or surface.

Appendix AP Sizing of Water Piping System

Appendix AP provides two recognized methods for sizing the water service and water distribution piping for a building. The method under Section AP103 provides friction loss diagrams that require the user to "plot" points and read values from the diagrams in order to perform the required calculations and necessary checks. This method is the most accurate of the two presented in this appendix. The method under Section AP201 is known to be conservative; however, very few calculations are necessary in order to determine a pipe size that satisfies the flow requirements of any application.

Appendix AQ Tiny Houses

For dwelling units that are 400 square feet (37 m^2) or less in floor area, excluding lofts, Appendix AQ provides relaxed provisions as compared to those in the body of the code. These provisions primarily address reduced ceiling heights for loft areas and specific stair and ladder detail requirements that allow for more compact designs where accessing lofts.

Appendix AR Light Straw-clay Construction

This appendix regulates the use of light straw-clay as a construction material. It is limited in application to nonbearing wall infill systems.

Appendix AS Strawbale Construction

This appendix provides prescriptive requirements for the use of strawbale as a construction material. It is limited in application to the walls of one-story structures, except where additional engineering is provided.

Appendix AT Solar-ready Provisions—Detached One- and Two-family Dwellings and Townhouses

This appendix provides requirements for preparation of a house for future installation of solar equipment for electrical power or heating. Given the growing popularity of solar power and the possible need for the equipment in the future, this appendix, if adopted, would require an area be provided on the building roof that would accommodate solar equipment. In addition, pathways for routing of plumbing and conduit need to be provided.

Appendix AU Cob Construction (Monolithic Adobe)

This appendix provides prescriptive requirements for the use of natural cob (monolithic adobe) as a construction material. It is limited in application to the walls of one-story structures, except where additional engineering is provided.

Appendix AV Board of Appeals

This appendix provides criteria for Board of Appeals members and procedures by which the Board of Appeals should conduct its business.

Appendix AW 3D-printed Building Construction

Appendix AW provides for the design, construction and inspection of buildings, structures and building elements fabricated by 3D-printed construction techniques.

TABLE OF CONTENTS

Part I—Administrative

CHAPTER 1
SCOPE AND ADMINISTRATION

User note:

About this chapter: *Chapter 1 establishes the limits of applicability of this code and describes how the code is to be applied and enforced. Chapter 1 is in two parts: Part 1—Scope and Application (Sections R101–R102) and Part 2—Administration and Enforcement (Sections R103–R114). Section R101 identifies which buildings and structures come under its purview and references other I-Codes as applicable. Standards and codes are scoped to the extent referenced (see Section R102.4).*

The one- and two-family dwelling code is intended to be adopted as a legally enforceable document, and it cannot be effective without adequate provisions for its administration and enforcement. The provisions of Chapter 1 establish the authority and duties of the building official appointed by the authority having jurisdiction and also establish the rights and privileges of the design professional, contractor and property owner.

PART 1—SCOPE AND APPLICATION

SECTION R101
SCOPE AND GENERAL REQUIREMENTS

R101.1 Title. These provisions shall be known as the *Residential Code for One- and Two-family Dwellings* of [NAME OF JURISDICTION], and shall be cited as such and will be referred to herein as "this code."

R101.2 Scope. The provisions of this code shall apply to the construction, *alteration*, movement, enlargement, replacement, *repair*, equipment, use and occupancy, location, removal and demolition of detached one- and two-family dwellings and *townhouses* not more than three stories above *grade plane* in height with a separate means of egress and their *accessory structures* not more than three stories above *grade plane* in height.

Exception: The following shall be permitted to be constructed in accordance with this code where provided with an automatic spinkler system complying with Section P2904:

1. Live/work units located in townhouses and complying with the requirements of Section 508.5 of the *International Building Code.*
2. Owner-occupied *lodging houses* with five or fewer guestrooms.
3. A care facility with five or fewer persons receiving custodial care within a *dwelling unit.*
4. A care facility with five or fewer persons receiving medical care within a *dwelling unit.*
5. A care facility for five or fewer persons receiving care that are within a single-family dwelling.

R101.3 Purpose. The purpose of this code is to establish minimum requirements to provide a reasonable level of safety, health and general welfare through affordability, structural strength, means of egress, stability, sanitation, light and ventilation, energy conservation and safety to life and property from fire and other hazards and to provide a reasonable level of safety to fire fighters and emergency responders during emergency operations.

SECTION R102
APPLICABILITY

R102.1 General. Where there is a conflict between a general requirement and a specific requirement, the specific requirement shall be applicable. Where, in any specific case, different sections of this code specify different materials, methods of construction or other requirements, the most restrictive shall govern.

R102.2 Other laws. The provisions of this code shall not be deemed to nullify any provisions of local, state or federal law.

R102.3 Application of references. References to chapter or section numbers, or to provisions not specifically identified by number, shall be construed to refer to such chapter, section or provision of this code.

R102.4 Referenced codes and standards. The codes and standards referenced in this code shall be considered part of the requirements of this code to the prescribed extent of each such reference and as further regulated in Sections R102.4.1 and R102.4.2.

Exception: Where enforcement of a code provision would violate the conditions of the *listing* of the *equipment* or *appliance*, the conditions of the *listing* and manufacturer's instructions shall apply.

R102.4.1 Conflicts. Where conflicts occur between provisions of this code and referenced codes and standards, the provisions of this code shall apply.

R102.4.2 Provisions in referenced codes and standards. Where the extent of the reference to a referenced code or standard includes subject matter that is within the scope of this code, the provisions of this code, as applicable, shall take precedence over the provisions in the referenced code or standard.

R102.5 Appendices. Provisions in the appendices shall not apply unless specifically referenced in the adopting ordinance.

R102.6 Partial invalidity. In the event any part or provision of this code is held to be illegal or void, this shall not have the effect of making void or illegal any of the other parts or provisions.

R102.7 Existing structures. The legal occupancy of any structure existing on the date of adoption of this code shall be permitted to continue without change, except as is specifically covered in this code, the *International Property Maintenance Code* or the *International Fire Code*, or as is deemed necessary by the *building official* for the general safety and welfare of the occupants and the public.

R102.7.1 Additions, alterations or repairs. *Additions, alterations* or repairs to any structure shall conform to the requirements for a new structure without requiring the existing structure to comply with the requirements of this code, unless otherwise stated. *Additions, alterations,* repairs and relocations shall not cause an existing structure to become less compliant with the provisions of this code than the existing building or structure was prior to the *addition, alteration* or *repair.* An existing building together with its *additions* shall comply with the height limits of this code. Where the *alteration* causes the use or occupancy to be changed to one not within the scope of this code, the provisions of the *International Existing Building Code* shall apply.

PART 2—ADMINISTRATION AND ENFORCEMENT

SECTION R103 DEPARTMENT OF BUILDING SAFETY

R103.1 Creation of enforcement agency. The department of building safety is hereby created and the official in charge thereof shall be known as the *building official.*

R103.2 Appointment. The *building official* shall be appointed by the *jurisdiction.*

R103.3 Deputies. In accordance with the prescribed procedures of this *jurisdiction* and with the concurrence of the appointing authority, the *building official* shall have the authority to appoint a deputy *building official*, the related technical officers, inspectors, plan examiners and other employees. Such employees shall have powers as delegated by the *building official.*

SECTION R104 DUTIES AND POWERS OF THE BUILDING OFFICIAL

R104.1 General. The *building official* is hereby authorized and directed to enforce the provisions of this code. The *building official* shall have the authority to render interpretations of this code and to adopt policies and procedures in order to clarify the application of its provisions. Such interpretations, policies and procedures shall be in compliance with the intent and purpose of this code. Such policies and procedures shall not have the effect of waiving requirements specifically provided for in this code.

R104.2 Applications and permits. The *building official* shall receive applications, review *construction documents* and issue *permits* for the erection and *alteration* of buildings and structures, inspect the premises for which such permits have been issued and enforce compliance with the provisions of this code.

R104.3 Notices and orders. The *building official* shall issue necessary notices or orders to ensure compliance with this code.

R104.4 Inspections. The *building official* shall make the required inspections, or the *building official* shall have the authority to accept reports of inspection by *approved agencies* or individuals. Reports of such inspections shall be in writing and be certified by a responsible officer of such *approved agency* or by the responsible individual. The *building official* is authorized to engage such expert opinion as deemed necessary to report on unusual technical issues that arise, subject to the approval of the appointing authority.

R104.5 Identification. The *building official* shall carry proper identification when inspecting structures or premises in the performance of duties under this code.

R104.6 Right of entry. Where it is necessary to make an inspection to enforce the provisions of this code, or where the *building official* has reasonable cause to believe that there exists in a structure or upon a premises a condition that is contrary to or in violation of this code that makes the structure or premises unsafe, dangerous or hazardous, the *building official* or designee is authorized to enter the structure or premises at reasonable times to inspect or to perform the duties imposed by this code, provided that if such structure or premises be occupied that credentials be presented to the occupant and entry requested. If such structure or premises is unoccupied, the *building official* shall first make a reasonable effort to locate the *owner*, the owner's authorized agent, or other *person* having charge or control of the structure or premises and request entry. If entry is refused, the *building official* shall have recourse to the remedies provided by law to secure entry.

R104.7 Department records. The *building official* shall keep official records of applications received, *permits* and certificates issued, fees collected, reports of inspections, and notices and orders issued. Such records shall be retained in the official records for the period required for the retention of public records.

R104.8 Liability. The *building official*, member of the board of appeals or employee charged with the enforcement of this code, while acting for the *jurisdiction* in good faith and without malice in the discharge of the duties required by this code or other pertinent law or ordinance, shall not thereby be rendered civilly or criminally liable personally and is hereby relieved from personal liability for any damage accruing to persons or property as a result of any act or by reason of an act or omission in the discharge of official duties.

R104.8.1 Legal defense. Any suit or criminal complaint instituted against an officer or employee because of an act performed by that officer or employee in the lawful discharge of duties and under the provisions of this code shall be defended by legal representatives of the *jurisdiction* until the final termination of the proceedings. The *building official* or any subordinate shall not be liable for cost in any action, suit or proceeding that is instituted in pursuance of the provisions of this code.

R104.9 Approved materials and equipment. Materials, *equipment* and devices *approved* by the *building official* shall be constructed and installed in accordance with such approval.

R104.9.1 Used materials and equipment. Used materials, *equipment* and devices shall not be reused unless *approved* by the *building official.*

R104.10 Modifications. Where there are practical difficulties involved in carrying out the provisions of this code, the *building official* shall have the authority to grant modifications for individual cases, provided the *building official* shall first find that special individual reason makes the strict letter of this code impractical and the modification is in compliance with the intent and purpose of this code and that such modification does not lessen health, life and fire safety or structural requirements. The details of action granting modifications shall be recorded and entered in the files of the department of building safety.

R104.10.1 Flood hazard areas. The *building official* shall not grant modifications to any provisions required in flood hazard areas as established by Table R301.2 unless a determination has been made that:

1. There is good and sufficient cause showing that the unique characteristics of the size, configuration or topography of the site render the elevation standards of Section R322 inappropriate.
2. Failure to grant the modification would result in exceptional hardship by rendering the lot undevelopable.
3. The granting of modification will not result in increased flood heights, additional threats to public safety, extraordinary public expense, cause fraud on or victimization of the public, or conflict with existing laws or ordinances.
4. The modification is the minimum necessary to afford relief, considering the flood hazard.
5. Written notice specifying the difference between the design flood elevation and the elevation to which the building is to be built, stating that the cost of flood insurance will be commensurate with the increased risk resulting from the reduced floor elevation and stating that construction below the design flood elevation increases risks to life and property, has been submitted to the applicant.

R104.11 Alternative materials, design and methods of construction and equipment. The provisions of this code are not intended to prevent the installation of any material or to prohibit any design or method of construction not specifically prescribed by this code. The *building official* shall have the authority to approve an alternative material, design or method of construction upon application of the *owner* or the owner's authorized agent. The *building official* shall first find that the proposed design is satisfactory and complies with the intent of the provisions of this code, and that the material, method or work offered is, for the purpose intended, not less than the equivalent of that prescribed in this code in quality, strength, effectiveness, fire resistance, durability and safety. Compliance with the specific performance-based provisions of the International Codes shall be an alternative to the specific requirements of this code. Where the alternative material, design or method of construction is not *approved,* the *building official* shall respond in writing, stating the reasons why the alternative was not *approved.*

R104.11.1 Tests. Where there is insufficient evidence of compliance with the provisions of this code, or evidence that a material or method does not conform to the requirements of this code, or in order to substantiate claims for alternative materials or methods, the *building official* shall have the authority to require tests as evidence of compliance to be made at no expense to the *jurisdiction.* Test methods shall be as specified in this code or by other recognized test standards. In the absence of recognized and accepted test methods, the *building official* shall approve the testing procedures. Tests shall be performed by an *approved* agency. Reports of such tests shall be retained by the *building official* for the period required for retention of public records.

SECTION R105
PERMITS

R105.1 Required. Any *owner* or owner's authorized agent who intends to construct, enlarge, alter, *repair,* move, demolish or change the occupancy of a building or structure, or to erect, install, enlarge, alter, *repair,* remove, convert or replace any electrical, gas, mechanical or plumbing system, the installation of which is regulated by this code, or to cause any such work to be performed, shall first make application to the *building official* and obtain the required *permit.*

R105.2 Work exempt from permit. Exemption from *permit* requirements of this code shall not be deemed to grant authorization for any work to be done in any manner in violation of the provisions of this code or any other laws or ordinances of this *jurisdiction. Permits* shall not be required for the following:

Building:

1. Other than *storm shelters,* one-*story* detached *accessory structures,* provided that the floor area does not exceed 200 square feet (18.58 m^2).
2. Fences not over 7 feet (2134 mm) high.
3. Retaining walls that are not over 4 feet (1219 mm) in height measured from the bottom of the footing

to the top of the wall, unless supporting a surcharge.

4. Water tanks supported directly upon *grade* if the capacity does not exceed 5,000 gallons (18 927 L) and the ratio of height to diameter or width does not exceed 2 to 1.
5. Sidewalks and driveways.
6. Painting, papering, tiling, carpeting, cabinets, counter tops and similar finish work.
7. Prefabricated swimming pools that are less than 24 inches (610 mm) deep.
8. Swings and other playground equipment.
9. Window awnings supported by an exterior wall that do not project more than 54 inches (1372 mm) from the exterior wall and do not require additional support.
10. Decks not exceeding 200 square feet (18.58 m^2) in area, that are not more than 30 inches (762 mm) above *grade* at any point, are not attached to a dwelling and do not serve the exit door required by Section R311.4.

Electrical:

1. *Listed* cord-and-plug connected temporary decorative lighting.
2. Reinstallation of attachment plug receptacles but not the outlets therefor.
3. Replacement of branch circuit overcurrent devices of the required capacity in the same location.
4. Electrical wiring, devices, *appliances,* apparatus or *equipment* operating at less than 25 volts and not capable of supplying more than 50 watts of energy.
5. Minor repair work, including the replacement of lamps or the connection of *approved* portable electrical equipment to *approved* permanently installed receptacles.

Gas:

1. Portable heating, cooking or clothes drying *appliances.*
2. Replacement of any minor part that does not alter approval of *equipment* or make such *equipment* unsafe.
3. Portable-fuel-cell *appliances* that are not connected to a fixed piping system and are not interconnected to a power grid.

Mechanical:

1. Portable heating *appliances.*
2. Portable ventilation *appliances.*
3. Portable cooling units.
4. Steam, hot- or chilled-water piping within any heating or cooling *equipment* regulated by this code.
5. Replacement of any minor part that does not alter approval of *equipment* or make such *equipment* unsafe.
6. Portable evaporative coolers.
7. Self-contained refrigeration systems containing 10 pounds (4.54 kg) or less of refrigerant or that are actuated by motors of 1 horsepower (746 W) or less.
8. Portable-fuel-cell *appliances* that are not connected to a fixed piping system and are not interconnected to a power grid.

Plumbing:

1. The stopping of leaks in drains, water, soil, waste or vent pipe; provided, however, that if any concealed trap, drainpipe, water, soil, waste or vent pipe becomes defective and it becomes necessary to remove and replace the same with new material, such work shall be considered as new work and a *permit* shall be obtained and inspection made as provided in this code.
2. The clearing of stoppages or the repairing of leaks in pipes, valves or fixtures, and the removal and reinstallation of water closets, provided such repairs do not involve or require the replacement or rearrangement of valves, pipes or fixtures.

R105.2.1 Emergency repairs. Where *equipment* replacements and repairs must be performed in an emergency situation, the *permit* application shall be submitted within the next working business day to the *building official.*

R105.2.2 Repairs. Application or notice to the *building official* is not required for ordinary repairs to structures, replacement of lamps or the connection of *approved* portable electrical equipment to *approved* permanently installed receptacles. Such repairs shall not include the cutting away of any wall, partition or portion thereof, the removal or cutting of any structural beam or load-bearing support, or the removal or change of any required means of egress, or rearrangement of parts of a structure affecting the egress requirements; nor shall ordinary repairs include *addition* to, *alteration* of, replacement or relocation of any water supply, sewer, drainage, drain leader, gas, soil, waste, vent or similar piping, electric wiring or mechanical or other work affecting public health or general safety.

R105.2.3 Public service agencies. A *permit* shall not be required for the installation, *alteration* or *repair* of generation, transmission, distribution, metering or other related equipment that is under the ownership and control of public service agencies by established right.

R105.3 Application for permit. To obtain a *permit,* the applicant shall first file an application therefor in writing on a form furnished by the department of building safety for that purpose. Such application shall:

1. Identify and describe the work to be covered by the *permit* for which application is made.

2. Describe the land on which the proposed work is to be done by legal description, street address or similar description that will readily identify and definitely locate the proposed building or work.
3. Indicate the use and occupancy for which the proposed work is intended.
4. Be accompanied by *construction documents* and other information as required in Section R106.1.
5. State the valuation of the proposed work.
6. Be signed by the applicant or the applicant's authorized agent.
7. Give such other data and information as required by the *building official*.

R105.3.1 Action on application. The *building official* shall examine or cause to be examined applications for *permits* and amendments thereto within a reasonable time after filing. If the application or the *construction documents* do not conform to the requirements of pertinent laws, the *building official* shall reject such application in writing stating the reasons therefor. If the *building official* is satisfied that the proposed work conforms to the requirements of this code and laws and ordinances applicable thereto, the *building official* shall issue a *permit* therefor as soon as practicable.

R105.3.1.1 Determination of substantially improved or substantially damaged existing buildings in flood hazard areas. For applications for reconstruction, rehabilitation, *addition, alteration, repair* or other improvement of existing buildings or structures located in a flood hazard area as established by Table R301.2, the *building official* shall examine or cause to be examined the *construction documents* and shall make a determination with regard to the value of the proposed work. For buildings that have sustained damage of any origin, the value of the proposed work shall include the cost to repair the building or structure to its predamaged condition. If the *building official* finds that the value of proposed work equals or exceeds 50 percent of the market value of the building or structure before the damage has occurred or the improvement is started, the proposed work is a substantial improvement or *repair* of substantial damage and the building official shall require existing portions of the entire building or structure to meet the requirements of Section R322.

For the purpose of this determination, a substantial improvement shall mean any *repair*, reconstruction, rehabilitation, addition or improvement of a building or structure, the cost of which equals or exceeds 50 percent of the market value of the building or structure before the improvement or *repair* is started. Where the building or structure has sustained substantial damage, repairs necessary to restore the building or structure to its predamaged condition shall be considered substantial improvements regardless of the actual repair work performed. The term shall not include either of the following:

1. Improvements to a building or structure that are required to correct existing health, sanitary or safety code violations identified by the building official and that are the minimum necessary to ensure safe living conditions.
2. Any *alteration* of a *historic building* or structure, provided that the *alteration* will not preclude the continued designation as a *historic building* or structure. For the purposes of this exclusion, a *historic building* shall be any of the following:
    2.1. Listed or preliminarily determined to be eligible for listing in the National Register of Historic Places.
    2.2. Determined by the Secretary of the US Department of Interior as contributing to the historical significance of a registered historic district or a district preliminarily determined to qualify as an historic district.
    2.3. Designated as historic under a state or local historic preservation program that is approved by the Department of Interior.

R105.3.2 Time limitation of application. An application for a *permit* for any proposed work shall be deemed to have been abandoned 180 days after the date of filing unless such application has been pursued in good faith or a *permit* has been issued; except that the *building official* is authorized to grant one or more extensions of time for additional periods not exceeding 180 days each. The extension shall be requested in writing and justifiable cause demonstrated.

R105.4 Validity of permit. The issuance or granting of a *permit* shall not be construed to be a *permit* for, or an *approval* of, any violation of any of the provisions of this code or of any other ordinance of the *jurisdiction. Permits* presuming to give authority to violate or cancel the provisions of this code or other ordinances of the *jurisdiction* shall not be valid. The issuance of a *permit* based on *construction documents* and other data shall not prevent the *building official* from requiring the correction of errors in the *construction documents* and other data. The *building official* is authorized to prevent occupancy or use of a structure where in violation of this code or of any other ordinances of this *jurisdiction*.

R105.5 Expiration. Every *permit* issued shall become invalid unless the work authorized by such *permit* is commenced within 180 days after its issuance or after commencement of work if more than 180 days pass between inspections. The building official is authorized to grant, in writing, one or more extensions of time, for periods not more than 180 days each. The extension shall be requested in writing and justifiable cause demonstrated.

R105.6 Suspension or revocation. The *building official* is authorized to suspend or revoke a *permit* issued under the provisions of this code wherever the *permit* is issued in error or on the basis of incorrect, inaccurate or incomplete information, or in violation of any ordinance or regulation or any of the provisions of this code.

R105.7 Placement of permit. The building *permit* or a copy shall be kept on the site of the work until the completion of the project.

R105.8 Responsibility. It shall be the duty of every *person* who performs work for the installation or *repair* of building, structure, electrical, gas, mechanical or plumbing systems, for which this code is applicable, to comply with this code.

R105.9 Preliminary inspection. Before issuing a *permit,* the *building official* is authorized to examine or cause to be examined buildings, structures and sites for which an application has been filed.

SECTION R106 CONSTRUCTION DOCUMENTS

R106.1 Submittal documents. Submittal documents consisting of *construction documents*, and other data shall be submitted in two or more sets, or in a digital format where allowed by the *building official*, with each application for a *permit.* The *construction documents* shall be prepared by a *registered design professional* where required by the statutes of the *jurisdiction* in which the project is to be constructed. Where special conditions exist, the *building official* is authorized to require additional *construction documents* to be prepared by a *registered design professional.*

Exception: The *building official* is authorized to waive the submission of *construction documents* and other data not required to be prepared by a *registered design professional* if it is found that the nature of the work applied for is such that reviewing of *construction documents* is not necessary to obtain compliance with this code.

R106.1.1 Information on construction documents. *Construction documents* shall be drawn upon suitable material. Electronic media documents are permitted to be submitted where *approved* by the *building official. Construction documents* shall be of sufficient clarity to indicate the location, nature and extent of the work proposed and show in detail that it will conform to the provisions of this code and relevant laws, ordinances, rules and regulations, as determined by the *building official.*

R106.1.2 Manufacturer's installation instructions. Manufacturer's installation instructions, as required by this code, shall be available on the job site at the time of inspection.

R106.1.3 Information on braced wall design. For buildings and structures utilizing braced wall design, and where required by the *building official*, *braced wall lines* shall be identified on the *construction documents.* Pertinent information including, but not limited to, bracing methods, location and length of *braced wall panels* and foundation requirements of *braced wall panels* at top and bottom shall be provided.

R106.1.4 Information for construction in flood hazard areas. For buildings and structures located in whole or in part in flood hazard areas as established by Table R301.2, *construction documents* shall include:

1. Delineation of flood hazard areas, floodway boundaries and flood zones and the design flood elevation, as appropriate.
2. The elevation of the proposed lowest floor, including *basement*; in areas of shallow flooding (AO Zones), the height of the proposed lowest floor, including *basement*, above the highest adjacent *grade.*
3. The elevation of the bottom of the lowest horizontal structural member in coastal high-hazard areas (V Zone) and in Coastal A Zones where such zones are delineated on flood hazard maps identified in Table R301.2 or otherwise delineated by the *jurisdiction.*
4. If design flood elevations are not included on the community's Flood Insurance Rate Map (FIRM), the *building official* and the applicant shall obtain and reasonably utilize any design flood elevation and floodway data available from other sources.

R106.1.5 Information on storm shelters. *Construction documents* for *storm shelters* shall include the information required in ICC 500.

R106.2 Site plan or plot plan. The *construction documents* submitted with the application for *permit* shall be accompanied by a site plan showing the size and location of new construction and existing structures on the site and distances from *lot lines*. In the case of demolition, the site plan shall show construction to be demolished and the location and size of existing structures and construction that are to remain on the site or plot. The *building official* is authorized to waive or modify the requirement for a site plan where the application for *permit* is for *alteration* or *repair* or where otherwise warranted.

R106.3 Examination of documents. The *building official* shall examine or cause to be examined *construction documents* for code compliance.

R106.3.1 Approval of construction documents. Where the *building official* issues a *permit*, the *construction documents* shall be *approved* in writing or by a stamp that states "REVIEWED FOR CODE COMPLIANCE." One set of *construction documents* so reviewed shall be retained by the *building official.* The other set shall be returned to the applicant, shall be kept at the site of work and shall be open to inspection by the *building official* or a duly authorized representative.

R106.3.2 Previous approvals. This code shall not require changes in the *construction documents*, construction or designated occupancy of a structure for which a lawful *permit* has been heretofore issued or otherwise lawfully authorized, and the construction of which has been pursued in good faith within 180 days after the effective date of this code and has not been abandoned.

R106.3.3 Phased approval. The *building official* is authorized to issue a *permit* for the construction of foundations or any other part of a building or structure before the *construction documents* for the whole building or structure have been submitted, provided that adequate information and detailed statements have been filed complying with pertinent requirements of this code. The holder of such *permit* for the foundation or other parts of a building or structure shall proceed at the holder's own risk with the building operation and without assurance that a *permit* for the entire structure will be granted.

R106.4 Amended construction documents. Work shall be installed in accordance with the *approved construction documents*, and any changes made during construction that are not in compliance with the *approved construction documents* shall be resubmitted for approval as an amended set of *construction documents*.

R106.5 Retention of construction documents. One set of *approved construction documents* shall be retained by the *building official* for a period of not less than 180 days from date of completion of the permitted work, or as required by state or local laws.

SECTION R107 TEMPORARY STRUCTURES AND USES

R107.1 General. The *building official* is authorized to issue a *permit* for temporary structures and temporary uses. Such *permits* shall be limited as to time of service, but shall not be permitted for more than 180 days. The *building official* is authorized to grant extensions for demonstrated cause.

R107.2 Conformance. Temporary structures and uses shall conform to the structural strength, fire safety, means of egress, light, *ventilation* and sanitary requirements of this code as necessary to ensure the public health, safety and general welfare.

R107.3 Temporary power. The *building official* is authorized to give permission to temporarily supply and use power in part of an electric installation before such installation has been fully completed and the final certificate of completion has been issued. The part covered by the temporary certificate shall comply with the requirements specified for temporary lighting, heat or power in NFPA 70.

R107.4 Termination of approval. The *building official* is authorized to terminate such *permit* for a temporary structure or use and to order the temporary structure or use to be discontinued.

SECTION R108 FEES

R108.1 Payment of fees. A *permit* shall not be valid until the fees prescribed by law have been paid, nor shall an amendment to a *permit* be released until the additional fee, if any, has been paid.

R108.2 Schedule of permit fees. On buildings, structures, electrical, gas, mechanical and plumbing systems or *alterations* requiring a *permit*, a fee for each *permit* shall be paid as required, in accordance with the schedule as established by the applicable governing authority.

R108.3 Building permit valuations. Building *permit* valuation shall include total value of the work for which a *permit* is being issued, such as electrical, gas, mechanical, plumbing equipment and other permanent systems, including materials and labor.

R108.4 Related fees. The payment of the fee for the construction, *alteration*, removal or demolition for work done in connection to or concurrently with the work authorized by a building *permit* shall not relieve the applicant or holder of the *permit* from the payment of other fees that are prescribed by law.

R108.5 Refunds. The *building official* is authorized to establish a refund policy.

R108.6 Work commencing before permit issuance. Any *person* who commences work requiring a *permit* on a building, structure, electrical, gas, mechanical or plumbing system before obtaining the necessary *permits* shall be subject to a fee established by the applicable governing authority that shall be in addition to the required *permit* fees.

SECTION R109 INSPECTIONS

R109.1 Types of inspections. For on-site construction, from time to time the *building official*, upon notification from the *permit* holder or his agent, shall make or cause to be made any necessary inspections and shall either approve that portion of the construction as completed or shall notify the *permit* holder or his or her agent wherein the same fails to comply with this code.

R109.1.1 Foundation inspection. Inspection of the foundation shall be made after poles or piers are set or trenches or *basement* areas are excavated and any required forms erected and any required reinforcing steel is in place and supported prior to the placing of concrete. The foundation inspection shall include excavations for thickened slabs intended for the support of bearing walls, partitions, structural supports, or equipment and special requirements for wood foundations.

R109.1.2 Plumbing, mechanical, gas and electrical systems inspection. Rough inspection of plumbing, mechanical, gas and electrical systems shall be made prior to covering or concealment, before fixtures or *appliances* are set or installed, and prior to framing inspection.

Exception: Backfilling of ground-source heat pump loop systems tested in accordance with Section M2105.28 prior to inspection shall be permitted.

R109.1.3 Floodplain inspections. For construction in flood hazard areas as established by Table R301.2, upon placement of the lowest floor, including *basement*, and prior to further vertical construction, the *building official* shall require submission of documentation, prepared and sealed by a *registered design professional*, of the elevation of the lowest floor, including *basement*, required in Section R322.

R109.1.4 Frame and masonry inspection. Inspection of framing and masonry construction shall be made after the roof, masonry, framing, firestopping, draftstopping and bracing are in place and after the plumbing, mechanical and electrical rough inspections are *approved*.

R109.1.5 Other inspections. In addition to inspections in Sections R109.1.1 through R109.1.4, the *building official* shall have the authority to make or require any other inspections to ascertain compliance with this code and other laws enforced by the *building official*.

R109.1.5.1 Fire-resistance-rated construction inspection. Where fire-resistance-rated construction is required between *dwelling units* or due to location on property, the *building official* shall require an inspection of such construction after lathing or gypsum board or gypsum panel products are in place, but before any plaster is applied, or before board or panel joints and fasteners are taped and finished.

R109.1.6 Final inspection. Final inspection shall be made after the permitted work is complete and prior to occupancy.

R109.1.6.1 Elevation documentation. If located in a flood hazard area, the documentation of elevations required in Section R322.1.10 shall be submitted to the *building official* prior to the final inspection.

R109.2 Inspection agencies. The *building official* is authorized to accept reports of *approved* agencies, provided such agencies satisfy the requirements as to qualifications and reliability.

R109.3 Inspection requests. It shall be the duty of the *permit* holder or their agent to notify the *building official* that such work is ready for inspection. It shall be the duty of the *person* requesting any inspections required by this code to provide access to and means for inspection of such work.

R109.4 Approval required. Work shall not be done beyond the point indicated in each successive inspection without first obtaining the approval of the *building official*. The *building official*, upon notification, shall make the requested inspections and shall either indicate the portion of the construction that is satisfactory as completed, or shall notify the *permit* holder or an agent of the *permit* holder wherein the same fails to comply with this code. Any portions that do not comply shall be corrected and such portion shall not be covered or concealed until authorized by the *building official*.

SECTION R110 CERTIFICATE OF OCCUPANCY

R110.1 Use and change of occupancy. A building or structure shall not be used or occupied in whole or in part, and a *change of occupancy* of a building or structure or portion thereof shall not be made, until the *building official* has issued a certificate of occupancy therefor as provided herein. Issuance of a certificate of occupancy shall not be construed as an approval of a violation of the provisions of this code or of other ordinances of the *jurisdiction*. Certificates presuming to give authority to violate or cancel the provisions of this code or other ordinances of the *jurisdiction* shall not be valid.

Exceptions:

1. Certificates of occupancy are not required for work exempt from *permits* under Section R105.2.
2. Accessory buildings or structures.

R110.2 Change in use. Changes in the character or use of an existing structure shall not be made except as specified in Sections 506 and 507 of the *International Existing Building Code*.

R110.3 Certificate issued. After the *building official* inspects the building or structure and does not find violations of the provisions of this code or other laws that are enforced by the department, the *building official* shall issue a certificate of occupancy containing the following:

1. The *permit* number.
2. The address of the structure.
3. The name and address of the *owner* or the owner's authorized agent.
4. A description of that portion of the structure for which the certificate is issued.
5. A statement that the described portion of the structure has been inspected for compliance with the requirements of this code.
6. The name of the *building official*.
7. The edition of the code under which the *permit* was issued.
8. Where an automatic sprinkler system is provided and whether the sprinkler system is required.
9. Any special stipulations and conditions of the building *permit*.

R110.4 Temporary occupancy. The *building official* is authorized to issue a temporary certificate of occupancy before the completion of the entire work covered by the *permit*, provided that such portion or portions shall be occupied safely. The *building official* shall set a time period during which the temporary certificate of occupancy is valid.

R110.5 Revocation. The *building official* is authorized to suspend or revoke a certificate of occupancy issued under the provisions of this code, in writing, wherever the certificate is issued in error, or on the basis of incorrect information supplied, or where it is determined that the building or structure or portion thereof is in violation of the provisions of this code or other ordinance of the *jurisdiction*.

SECTION R111
SERVICE UTILITIES

R111.1 Connection of service utilities. A *person* shall not make connections from a utility, a source of energy, fuel, or power to any building or system that is regulated by this code for which a *permit* is required, until *approved* by the *building official*.

R111.2 Temporary connection. The *building official* shall have the authority to authorize the temporary connection of the building or system to the utility, source of energy, fuel or power.

R111.3 Authority to disconnect service utilities. The *building official* shall have the authority to authorize disconnection of utility service to the building, structure or system regulated by this code and the referenced codes and standards set forth in Section R102.4 in case of emergency where necessary to eliminate an immediate hazard to life or property or where such utility connection has been made without the approval required by Section R111.1 or R111.2. The *building official* shall notify the serving utility and where possible the *owner* or the owner's authorized agent and occupant of the building, structure or service system of the decision to disconnect prior to taking such action. If not notified prior to disconnection, the *owner*, the owner's authorized agent or occupant of the building, structure or service system shall be notified in writing as soon as practical thereafter.

SECTION R112
BOARD OF APPEALS

R112.1 General. In order to hear and decide appeals of orders, decisions or determinations made by the *building official* relative to the application and interpretation of this code, there shall be and is hereby created a board of appeals. The *building official* shall be an ex officio member of said board but shall not have a vote on any matter before the board. The board of appeals shall be appointed by the governing body and shall hold office at its pleasure. The board shall adopt rules of procedure for conducting its business, and shall render decisions and findings in writing to the appellant with a duplicate copy to the *building official*.

R112.2 Limitations on authority. An application for appeal shall be based on a claim that the true intent of this code or the rules legally adopted thereunder have been incorrectly interpreted, the provisions of this code do not fully apply or an equally good or better form of construction is proposed. The board shall not have authority to waive requirements of this code.

R112.3 Qualifications. The board of appeals shall consist of members who are qualified by experience and training to pass judgment on matters pertaining to building construction and are not employees of the *jurisdiction*.

R112.4 Administration. The *building official* shall take immediate action in accordance with the decision of the board.

SECTION R113
VIOLATIONS

R113.1 Unlawful acts. It shall be unlawful for any *person*, firm or corporation to erect, construct, alter, extend, *repair*, move, remove, demolish or occupy any building, structure or equipment regulated by this code, or cause same to be done, in conflict with or in violation of any of the provisions of this code.

R113.2 Notice of violation. The *building official* is authorized to serve a notice of violation or order on the *person* responsible for the erection, construction, *alteration*, extension, *repair*, moving, removal, demolition or occupancy of a building or structure in violation of the provisions of this code, or in violation of a detail statement or a plan *approved* thereunder, or in violation of a *permit* or certificate issued under the provisions of this code. Such order shall direct the discontinuance of the illegal action or condition and the abatement of the violation.

R113.3 Prosecution of violation. If the notice of violation is not complied with in the time prescribed by such notice, the *building official* is authorized to request the legal counsel of the *jurisdiction* to institute the appropriate proceeding at law or in equity to restrain, correct or abate such violation, or to require the removal or termination of the unlawful occupancy of the building or structure in violation of the provisions of this code or of the order or direction made pursuant thereto.

R113.4 Violation penalties. Any *person* who violates a provision of this code or fails to comply with any of the requirements thereof or who erects, constructs, alters or repairs a building or structure in violation of the *approved construction documents* or directive of the *building official*, or of a *permit* or certificate issued under the provisions of this code, shall be subject to penalties as prescribed by law.

SECTION R114
STOP WORK ORDER

R114.1 Authority. Where the *building official* finds any work regulated by this code being performed in a manner contrary to the provisions of this code or in a dangerous or unsafe manner, the *building official* is authorized to issue a stop work order.

R114.2 Issuance. The stop work order shall be in writing and shall be given to the *owner* of the property, the owner's authorized agent or the *person* performing the work. Upon issuance of a stop work order, the cited work shall immediately cease. The stop work order shall state the reason for the order and the conditions under which the cited work is authorized to resume.

R114.3 Emergencies. Where an emergency exists, the building official shall not be required to give a written notice prior to stopping the work.

R114.4 Failure to comply. Any *person* who shall continue any work after having been served with a stop work order, except such work as that *person* is directed to perform to remove a violation or unsafe condition, shall be subject to fines established by the authority having jurisdiction.

Part II—Definitions

CHAPTER 2
DEFINITIONS

User notes:

About this chapter: Codes, by their very nature, are technical documents. Every word, term and punctuation mark can add to or change the meaning of a technical requirement. It is necessary to maintain a consensus on the specific meaning of each term contained in the code. Chapter 2 performs this function by stating clearly what specific terms mean for the purpose of the code.

Code development reminder: Code change proposals to definitions in this chapter preceded by a bracketed letter are considered by the IRC—Building Code Development Committee [RB], the IRC—Mechanical/Plumbing Code Development Committee [MP] or the IECC—Residential Code Development Committee [RE] during the Group B (2022) Code Development Cycle.

SECTION R201
GENERAL

R201.1 Scope. Unless otherwise expressly stated, the following words and terms shall, for the purposes of this code, have the meanings indicated in this chapter.

R201.2 Interchangeability. Words used in the present tense include the future; words in the masculine gender include the feminine and neuter; the singular number includes the plural and the plural, the singular.

R201.3 Terms defined in other codes. Where terms are not defined in this code such terms shall have the meanings ascribed in other code publications of the International Code Council.

R201.4 Terms not defined. Where terms are not defined through the methods authorized by this section, such terms shall have ordinarily accepted meanings such as the context implies.

SECTION R202
DEFINITIONS

[RE] ABOVE-GRADE WALL. For the definition applicable in Chapter 11, see Section N1101.6.

[RB] ACCESS (TO). That which enables a device, an *appliance* or equipment to be reached by *ready access* or by a means that first requires the removal or movement of a panel, door or similar obstruction.

➡ **[RB] ACCESSORY STRUCTURE.** A structure that is accessory to and incidental to that of the *dwelling(s)* and that is located on the same *lot.*

[RB] ADDITION. An extension or increase in floor area, number of stories or height of a building or structure. For the definition applicable in Chapter 11, see Section N1101.6.

[RB] ADHERED STONE OR MASONRY VENEER. Stone or masonry veneer secured and supported through the adhesion of an *approved* bonding material applied to an *approved* backing.

[MP] AIR ADMITTANCE VALVE. A one-way valve designed to allow air into the plumbing drainage system where a negative pressure develops in the piping. This device shall close by gravity and seal the terminal under conditions of zero differential pressure (no flow conditions) and under positive internal pressure.

[RE] AIR BARRIER. For the definition applicable in Chapter 11, see Section N1101.6.

[MP] AIR BREAK (DRAINAGE SYSTEM). An arrangement where a discharge pipe from a fixture, *appliance* or device drains indirectly into a receptor below the flood-level rim of the receptor and above the trap seal.

[MP] AIR CIRCULATION, FORCED. A means of providing space conditioning utilizing movement of air through ducts or plenums by mechanical means.

[MP] AIR GAP, DRAINAGE SYSTEM. The unobstructed vertical distance through free atmosphere between the outlet of a waste pipe and the flood-level rim of the fixture or receptor into which it is discharging.

[MP] AIR GAP, WATER-DISTRIBUTION SYSTEM. The unobstructed vertical distance through free atmosphere between the lowest opening from a water supply discharge to the flood-level rim of a plumbing fixture.

[MP] AIR-CONDITIONING SYSTEM. A system that consists of heat exchangers, blowers, filters, supply, exhaust and return-air systems, and shall include any apparatus installed in connection therewith.

[RB] AIR-IMPERMEABLE INSULATION. An insulation having an air permanence equal to or less than 0.02 L/s-m^2 at 75 Pa pressure differential as tested in accordance with ASTM E283 or E2178.

[RB] ALTERATION. Any construction, retrofit or renovation to an existing structure other than *repair* or *addition* that requires a *permit.* Also, a change in a building, electrical, gas, mechanical or plumbing system that involves an extension, *addition* or change to the arrangement, type or purpose of the original installation that requires a *permit.* For the definition applicable in Chapter 11, see Section N1101.6.

[RB] ALTERNATING TREAD DEVICE. A device that has a series of steps between 50 and 70 degrees (0.87 and 1.22 rad) from horizontal, usually attached to a center support rail in an alternating manner so that the user does not have both feet on the same level at the same time.

[RB] ANCHORED STONE OR MASONRY VENEER. Stone or masonry veneer secured with *approved* mechanical fasteners to an *approved* backing.

[MP] ANCHORS. See "*Supports.*"

[MP] ANTISIPHON. A term applied to valves or mechanical devices that eliminate siphonage.

[MP] APPLIANCE. A device or apparatus that is manufactured and designed to utilize energy and for which this code provides specific requirements.

[RB] APPROVED. Acceptable to the *building official*.

[RB] APPROVED AGENCY. An established and recognized agency that is regularly engaged in conducting tests, furnishing inspection services or furnishing product certification, and has been *approved* by the building official.

[MP] APPROVED SOURCE. An independent person, firm or corporation, *approved* by the *building official*, who is competent and experienced in the application of engineering principles to materials, methods or systems analyses.

[RB] ASPECT RATIO. The ratio of longest to shortest perpendicular dimensions, or for wall sections, the ratio of height to length.

[RB] ATTIC. The unfinished space between the ceiling assembly and the *roof assembly*.

[RB] ATTIC, HABITABLE. A finished or unfinished *habitable space* within an attic.

[RE] AUTOMATIC. For the definition applicable in Chapter 11, see Section N1101.6.

[MP] BACKFLOW, DRAINAGE. A reversal of flow in the drainage system.

[MP] BACKFLOW, WATER DISTRIBUTION. The flow of water or other liquids into the potable water-supply piping from any sources other than its intended source. Backsiphonage is one type of backflow.

[MP] BACKFLOW PREVENTER. A backflow prevention assembly, a backflow prevention device or other means or method to prevent backflow into the potable water supply.

[MP] BACKFLOW PREVENTER, REDUCED-PRESSURE-ZONE TYPE. A backflow-prevention device consisting of two independently acting check valves, internally force loaded to a normally closed position and separated by an intermediate chamber (or zone) in which there is an automatic relief means of venting to atmosphere internally loaded to a normally open position between two tightly closing shutoff valves and with means for testing for tightness of the checks and opening of relief means.

[MP] BACKPRESSURE. Pressure created by any means in the water distribution system that by being in excess of the pressure in the water supply mains causes a potential backflow condition.

[MP] BACKPRESSURE, LOW HEAD. A pressure less than or equal to 4.33 psi (29.88 kPa) or the pressure exerted by a 10-foot (3048 mm) column of water.

[MP] BACKSIPHONAGE. The flowing back of used or contaminated water from piping into a potable water-supply pipe due to a negative pressure in such pipe.

[MP] BACKWATER VALVE. A device installed in a drain or pipe to prevent backflow of sewage.

[MP] BALANCED VENTILATION. Any combination of concurrently operating mechanical exhaust and mechanical supply whereby the total mechanical exhaust airflow rate is within 10 percent of the total mechanical supply airflow rate.

[MP] BALANCED VENTILATION SYSTEM. A ventilation system where the total supply airflow and total exhaust airflow are simultaneously within 10 percent of their averages. The balanced ventilation system airflow is the average of the supply and exhaust airflows.

[RB] BASEMENT. A *story* that is not a *story above grade plane* (see "*Story above grade plane*").

[RE] BASEMENT WALL. For the definition applicable in Chapter 11, see Section N1101.6.

[RB] BASIC WIND SPEED. Three-second gust speed at 33 feet (10 058 mm) above the ground in Exposure C (see Section R301.2.1) as given in Figure R301.2(5)A.

[MP] BATHROOM GROUP. A group of fixtures, including or excluding a bidet, consisting of a water closet, lavatory, and bathtub or shower. Such fixtures are located together on the same floor level.

[MP] BEND. A drainage fitting, designed to provide a change in direction of a drain pipe of less than the angle specified by the amount necessary to establish the desired slope of the line (see "*Elbow*" and "*Sweep*").

[MP] BOILER. A self-contained *appliance* from which hot water is circulated for heating purposes and then returned to the boiler, and that operates at water pressures not exceeding 160 pounds per square inch gage (psig) (1102 kPa gauge) and at water temperatures not exceeding 250°F (121°C).

[RB] BOND BEAM. A horizontal grouted element within masonry in which reinforcement is embedded.

[RB] BRACED WALL LINE. A straight line through the building plan that represents the location of the lateral resistance provided by the wall bracing.

[RB] BRACED WALL LINE, CONTINUOUSLY SHEATHED. A *braced wall line* with structural sheathing applied to all sheathable surfaces including the areas above and below openings.

[RB] BRACED WALL PANEL. A full-height section of wall constructed to resist in-plane shear loads through interaction of framing members, sheathing material and anchors. The panel's length meets the requirements of its particular bracing method, and contributes toward the total amount of bracing required along its *braced wall line* in accordance with Section R602.10.1.

[MP] BRANCH. Any part of the piping system other than a riser, main or stack.

[MP] BRANCH, FIXTURE. See "*Fixture branch, drainage.*"

[MP] BRANCH, HORIZONTAL. See "*Horizontal branch, drainage.*"

[MP] BRANCH, MAIN. A water-distribution pipe that extends horizontally off a main or riser to convey water to branches or fixture groups.

[MP] BRANCH, VENT. A vent connecting two or more individual vents with a vent stack or stack vent.

[MP] BRANCH INTERVAL. A vertical measurement of distance, 8 feet (2438 mm) or more in *developed length*, between the connections of *horizontal* branches to a drainage stack. Measurements are taken down the stack from the highest *horizontal* branch connection.

[MP] BTU/H. The *listed* maximum capacity of an *appliance*, absorption unit or burner expressed in British thermal units input per hour.

[RB] BUILDING. Any one- or two-family dwelling or *townhouse*, or portion thereof, used or intended to be used for human habitation, for living, sleeping, cooking or eating purposes, or any combination thereof, or any *accessory structure*. For the definition applicable in Chapter 11, see Section N1101.6.

[RB] BUILDING, EXISTING. Existing building is a building erected prior to the adoption of this code, or one for which a legal building *permit* has been issued.

[MP] BUILDING DRAIN. The lowest piping that collects the discharge from all other drainage piping inside the house and extends 30 inches (762 mm) in *developed length* of pipe, beyond the exterior walls and conveys the drainage to the *building sewer*.

[RB] BUILDING LINE. The line established by law, beyond which a building shall not extend, except as specifically provided by law.

[RB] BUILDING OFFICIAL. The officer or other designated authority charged with the administration and enforcement of this code, or a duly authorized representative. For the definition applicable in Chapter 11, see Section N1101.6.

[MP] BUILDING SEWER. That part of the drainage system that extends from the end of the *building drain* and conveys its discharge to a public sewer, private sewer, individual sewage-disposal system or other point of disposal.

[RE] BUILDING SITE. For the definition applicable in Chapter 11, see Section N1101.6.

[RE] BUILDING THERMAL ENVELOPE. For the definition applicable in Chapter 11, see Section N1101.6.

[RB] BUILDING-INTEGRATED PHOTOVOLTAIC PRODUCT. A building product that incorporates *photovoltaic modules* and functions as a component of the building envelope.

[RB] BUILDING-INTEGRATED PHOTOVOLTAIC ROOF PANEL (BIPV Roof Panel). A *photovoltaic panel* that functions as a component of the building envelope.

[RB] BUILT-UP ROOF COVERING. Two or more layers of felt cemented together and surfaced with a cap sheet, mineral aggregate, smooth coating or similar surfacing material.

[RB] CAP PLATE. The top plate of the double top plates used in *structural insulated panel* (SIP) construction. The cap plate is cut to match the *panel thickness* such that it overlaps the wood structural panel facing on both sides.

[RB] CARBON MONOXIDE ALARM. A single- or multiple-station alarm intended to detect carbon monoxide gas and alert occupants by a distinct audible signal. It incorporates a sensor, control components and an alarm notification appliance in a single unit.

[RB] CARBON MONOXIDE DETECTOR. A device with an integral sensor to detect carbon monoxide gas and transmit an alarm signal to a connected alarm control unit.

[RB] CEILING HEIGHT. The clear vertical distance from the finished floor to the finished ceiling.

[RB] CEMENT PLASTER. A mixture of Portland or blended cement, Portland cement or blended cement and hydrated lime, masonry cement or plastic cement and aggregate and other *approved* materials as specified in this code.

[RB] CHANGE OF OCCUPANCY. A change in the use of a building or portion of a building that involves a change in the application of the requirements of this code.

[MP] CHIMNEY. A primary vertical structure containing one or more flues, for the purpose of carrying gaseous products of combustion and air from a fuel-burning *appliance* to the outside atmosphere.

[MP] CHIMNEY CONNECTOR. A pipe that connects a fuel-burning *appliance* to a chimney.

[MP] CHIMNEY TYPES.

Residential-type appliance. An *approved* chimney for removing the products of combustion from fuel-burning, residential-type *appliances* producing combustion gases not in excess of 1,000°F (538°C) under normal operating conditions, and capable of producing combustion gases of 1,400°F (760°C) during intermittent forces firing for periods up to 1 hour. All temperatures shall be measured at the *appliance* flue outlet. Residential-type *appliance* chimneys include masonry and factory-built types.

[MP] CIRCUIT VENT. A vent that connects to a horizontal drainage branch and vents two traps to not more than eight traps or trapped fixtures connected into a battery.

[MP] CIRCULATING HOT WATER SYSTEM. A specifically designed water distribution system where one or more pumps are operated in the service hot water piping to circulate heated water from the water-heating equipment to fixtures and back to the water-heating equipment. For the definition applicable in Chapter 11, see Section N1101.6.

[RB] CLADDING. The exterior materials that cover the surface of the building envelope that is directly loaded by the wind.

[MP] CLEANOUT. An access opening in the drainage system utilized for the removal of obstructions. Types of cleanouts include a removable plug or cap, and a removable fixture or fixture trap.

[RE] CLIMATE ZONE. A geographical region based on climatic criteria as specified in this code. For the definition applicable in Chapter 11, see Section N1101.6.

[RB] CLOSET. A small room or chamber used for storage.

[RB] COLLAPSIBLE SOILS. Soils that exhibit volumetric reduction in response to partial or full wetting under load.

[MP] COLLECTION PIPE. Unpressurized pipe used within the collection system that drains on-site nonpotable water or rainwater to a storage tank by gravity.

[MP] COMBINATION WASTE AND VENT SYSTEM. A specially designed system of waste piping embodying the horizontal wet venting of one or more sinks, lavatories or floor drains by means of a common waste and vent pipe adequately sized to provide free movement of air above the flow line of the drain.

[RB] COMBUSTIBLE MATERIAL. Any material not defined as noncombustible.

[MP] COMBUSTION AIR. The air provided to fuel-burning equipment including air for fuel combustion, draft hood dilution and *ventilation* of the equipment enclosure.

[MP] COMMON VENT. A single pipe venting two trap arms within the same *branch interval*, either back-to-back or one above the other.

[RB] COMPRESSIBLE SOILS. Soils that exhibit volumetric reduction in response to the application of load even in the absence of wetting or drying.

[MP] CONDENSATE. The liquid that separates from a gas due to a reduction in temperature; for example, water that condenses from flue gases and water that condenses from air circulating through the cooling coil in air conditioning equipment.

[MP] CONDENSING APPLIANCE. An *appliance* that condenses water generated by the burning of fuels.

[RB] CONDITIONED AIR. Air treated to control its temperature, relative humidity or quality.

[RE] CONDITIONED FLOOR AREA. For the definition applicable in Chapter 11, see Section N1101.6.

[RE] CONDITIONED SPACE. For the definition applicable in Chapter 11, see Section N1101.6.

[RB] CONSTRUCTION DOCUMENTS. Written, graphic and pictorial documents prepared or assembled for describing the design, location and physical characteristics of the elements of a project necessary for obtaining a building *permit*. Construction drawings shall be drawn to an appropriate scale.

[MP] CONTAMINATION. A high-hazard or health-hazard impairment of the quality of the potable water that creates an actual hazard to the public health through poisoning or through the spread of disease by sewage, industrial fluids or waste.

[RE] CONTINUOUS AIR BARRIER. For the definition applicable in Chapter 11, see Section N1101.6.

[RE] CONTINUOUS INSULATION (ci). For the definition applicable in Chapter 11, see Section N1101.6.

[MP] CONTINUOUS WASTE. A drain from two or more similar adjacent fixtures connected to a single trap.

[MP] CONTROL, LIMIT. An automatic control responsive to changes in liquid flow or level, pressure, or temperature for limiting the operation of an *appliance*.

[MP] CONTROL, PRIMARY SAFETY. A safety control responsive directly to flame properties that senses the presence or absence of flame and, in event of ignition failure or unintentional flame extinguishment, automatically causes shutdown of mechanical equipment.

[MP] CONVECTOR. A system incorporating a heating element in an enclosure in which air enters an opening below the heating element, is heated and leaves the enclosure through an opening located above the heating element.

[RB] CORE. The lightweight middle section of a *structural insulated panel*, composed of foam plastic insulation, that provides the link between the two facing shells.

[RB] CORROSION RESISTANCE. The ability of a material to withstand deterioration of its surface or its properties where exposed to its environment.

[RB] COURT. A space, open and unobstructed to the sky, located at or above *grade* level on a *lot* and bounded on three or more sides by walls or a building.

[RB] CRAWL SPACE. An underfloor space that is not a *basement*.

[RE] CRAWL SPACE WALL. For the definition applicable in Chapter 11, see Section N1101.6.

[RB] CRIPPLE WALL. A framed wall extending from the top of the foundation to the underside of the floor framing of the first *story above grade plane*.

[RB] CRIPPLE WALL CLEAR HEIGHT. The vertical height of a *cripple wall* from the top of the foundation to the underside of floor framing above.

[MP] CROSS CONNECTION. Any connection between two otherwise separate piping systems that allows a flow from one system to the other.

[RB] CROSS-LAMINATED TIMBER. A prefabricated engineered wood product consisting of not less than three layers of solid-sawn lumber or *structural composite lumber* where the adjacent layers are cross-oriented and bonded with structural adhesive to form a solid wood element.

[RE] CURTAIN WALL. For the definition applicable in Chapter 11, see Section N1101.6.

[RB] DALLE GLASS. A decorative composite glazing material made of individual pieces of glass that are embedded in a cast matrix of concrete or epoxy.

[MP] DAMPER, VOLUME. A device that will restrict, retard or direct the flow of air in any duct, or the products of combustion of heat-producing equipment, vent connector, vent or chimney.

[RB] DEAD LOADS. The weight of the materials of construction incorporated into the building, including but not limited to walls, floors, roofs, ceilings, *stairways*, built-in partitions, finishes, cladding, and other similarly incorporated architectural and structural items, and fixed service equipment.

[RB] DECORATIVE GLASS. A carved, leaded or Dalle glass or glazing material with a purpose that is decorative or artistic, not functional; with coloring, texture or other design qualities or components that cannot be removed without destroying the glazing material; and with a surface, or assembly into which it is incorporated, that is divided into segments.

[RE] DEMAND RECIRCULATION WATER SYSTEM. For the definition applicable in Chapter 11, see Section N1101.6.

[MP] DESIGN PROFESSIONAL. See "*Registered design professional.*"

[MP] DEVELOPED LENGTH. The length of a pipeline measured along the center line of the pipe and fittings.

[MP] DIAMETER. Unless specifically stated, the term "diameter" is the nominal diameter as designated by the *approved* material standard.

[RB] DIAPHRAGM. A horizontal or nearly horizontal system acting to transmit lateral forces to the vertical resisting elements. Where the term "*diaphragm*" is used, it includes horizontal bracing systems.

[MP] DILUTION AIR. Air that enters a draft hood or draft regulator and mixes with flue gases.

[MP] DIRECT SYSTEM. A solar thermal system in which the gas or liquid in the solar collector loop is not separated from the load.

[MP] DIRECT-VENT APPLIANCE. A fuel-burning *appliance* with a sealed combustion system that draws all air for combustion from the outside atmosphere and discharges all flue gases to the outside atmosphere.

[MP] DRAFT. The pressure difference existing between the *appliance* or any component part and the atmosphere, that causes a continuous flow of air and products of combustion through the gas passages of the *appliance* to the atmosphere.

Induced draft. The pressure difference created by the action of a fan, blower or ejector, that is located between the *appliance* and the chimney or vent termination.

Natural draft. The pressure difference created by a vent or chimney because of its height, and the temperature difference between the flue gases and the atmosphere.

[MP] DRAFT HOOD. A device built into an *appliance*, or a part of the vent connector from an *appliance*, that is designed to provide for the ready escape of the flue gases from the *appliance* in the event of no draft, backdraft or stoppage beyond the draft hood; prevent a backdraft from entering the *appliance*; and neutralize the effect of stack action of the chimney or gas vent on the operation of the *appliance*.

[MP] DRAFT REGULATOR. A device that functions to maintain a desired draft in the *appliance* by automatically reducing the draft to the desired value.

[RB] DRAFT STOP. A material, device or construction installed to restrict the movement of air within open spaces of concealed areas of building components such as crawl spaces, floor-ceiling assemblies, roof-ceiling assemblies and *attics*.

[MP] DRAIN. Any pipe that carries soil and waterborne wastes in a building drainage system.

[MP] DRAINAGE FITTING. A pipe fitting designed to provide connections in the drainage system that have provisions for establishing the desired slope in the system. These fittings are made from a variety of both metals and plastics. The methods of coupling provide for required slope in the system.

[MP] DRAIN-BACK SYSTEM. A solar thermal system in which the fluid in the solar collector loop is drained from the collector into a holding tank under prescribed circumstances.

[RE] DUCT. For the definition applicable in Chapter 11, see Section N1101.6.

[MP] DUCT SYSTEM. A continuous passageway for the transmission of air that, in addition to ducts, includes duct fittings, dampers, plenums, fans and accessory air-handling *equipment* and *appliances*.

For the definition applicable in Chapter 11, see Section N1101.6.

[RB] DWELLING. Any building that contains one or two *dwelling units* used, intended, or designed to be built, used, rented, leased, let or hired out to be occupied, or that are occupied for living purposes.

[RB] DWELLING UNIT. A single unit providing complete independent living facilities for one or more persons, including permanent provisions for living, sleeping, eating, cooking and sanitation. For the definition applicable in Chapter 11, see Section N1101.6.

[MP] DWV. Abbreviated term for drain, waste and vent piping as used in common plumbing practice.

[MP] EFFECTIVE OPENING. The minimum cross-sectional area at the point of water-supply discharge, measured or expressed in terms of diameter of a circle and if the opening is not circular, the diameter of a circle of equivalent cross-sectional area. (This is applicable to *air gap*.)

[MP] ELBOW. A pressure pipe fitting designed to provide an exact change in direction of a pipe run. An elbow provides a sharp turn in the flow path (see "Bend" and "Sweep").

[RB] EMERGENCY ESCAPE AND RESCUE OPENING. An operable exterior window, door or other similar device that provides for a means of escape and access for rescue in the event of an emergency. (See also "*Grade floor emergency escape and rescue opening.*")

[RE] ENERGY ANALYSIS. For the definition applicable in Chapter 11, see Section N1101.6.

[RE] ENERGY COST. For the definition applicable in Chapter 11, see Section N1101.6.

[RE] ENERGY SIMULATION TOOL. For the definition applicable in Chapter 11, see Section N1101.6.

[RB] ENERGY STORAGE SYSTEMS (ESS). One device or multiple devices, assembled together, capable of storing electrical energy to be supplied at a future time.

[RB] ENGINEERED WOOD RIM BOARD. A full-depth *structural composite lumber*, wood structural panel, structural glued laminated timber or prefabricated wood I-joist member designed to transfer horizontal (shear) and vertical (compression) loads, provide attachment for *diaphragm* sheathing, siding and exterior deck ledgers and provide lateral support at the ends of floor or roof joists or rafters.

[MP] EQUIPMENT. Piping, ducts, vents, control devices and other components of systems other than *appliances* that are permanently installed and integrated to provide control of environmental conditions for buildings. This definition shall also include other systems specifically regulated in this code.

[MP] EQUIVALENT LENGTH. For determining friction losses in a piping system, the effect of a particular fitting equal to the friction loss through a straight piping length of the same nominal diameter.

[RE] ERI REFERENCE DESIGN. For the definition applicable in Chapter 11, see Section N1101.6.

[RB] ESCARPMENT. With respect to topographic wind effects, a cliff or steep slope generally separating two levels or gently sloping areas.

[MP] ESSENTIALLY NONTOXIC TRANSFER FLUIDS. Fluids having a Gosselin rating of 1, including propylene glycol; mineral oil; polydimethy oil oxane; hydrochlorofluorocarbon, chlorofluorocarbon and hydrofluorocarbon refrigerants; and FDA-approved boiler water additives for steam boilers.

[MP] ESSENTIALLY TOXIC TRANSFER FLUIDS. Soil, water or graywater and fluids having a Gosselin rating of 2 or more including ethylene glycol, hydrocarbon oils, ammonia refrigerants and hydrazine.

[MP] EVAPORATIVE COOLER. A device used for reducing air temperature by the process of evaporating water into an airstream.

[MP] EXCESS AIR. Air that passes through the combustion chamber and the *appliance* flue in excess of what is theoretically required for complete combustion.

[MP] EXHAUST HOOD, FULL OPENING. An exhaust hood with an opening not less than the diameter of the connecting vent.

[MP] EXISTING INSTALLATIONS. Any plumbing system regulated by this code that was legally installed prior to the effective date of this code, or for which a *permit* to install has been issued.

[RB] EXPANSIVE SOILS. Soils that exhibit volumetric increase or decrease (swelling or shrinking) in response to partial or full wetting or drying under load.

[RB] EXTERIOR INSULATION AND FINISH SYSTEMS (EIFS). EIFS are nonstructural, nonload-bearing exterior wall cladding systems that consist of an insulation board attached either adhesively or mechanically, or both, to the substrate; an integrally reinforced base coat; and a textured protective finish coat.

[RB] EXTERIOR INSULATION AND FINISH SYSTEMS (EIFS) WITH DRAINAGE. An EIFS that incorporates a means of drainage applied over a *water-resistive barrier*.

[RE] EXTERIOR WALL. For the definition applicable in Chapter 11, see Section N1101.6.

[RB] EXTERIOR WALL COVERING. A material or assembly of materials applied on the exterior side of exterior walls for the purpose of providing a weather-resistive barrier, insulation or for aesthetics, including but not limited to, veneers, siding, exterior insulation and finish systems, architectural *trim* and embellishments such as cornices, soffits, and fascias.

[RB] FACING. The wood structural panel facings that form the two outmost rigid layers of the *structural insulated panel*.

[MP] FACTORY-BUILT CHIMNEY. A *listed* and *labeled* chimney composed of factory-made components assembled in the field in accordance with the manufacturer's instructions and the conditions of the *listing*.

[MP] FACTORY-MADE AIR DUCT. A *listed and labeled* duct manufactured in a factory and assembled in the field in accordance with the manufacturer's instructions and conditions of the *listing*.

[RE] FENESTRATION. Products classified as either vertical fenestration or *skylights and sloped glazing*, installed in such a manner as to preserve the weather-resistant barrier of the wall or roof in which they are installed. Fenestration includes products with glass or other transparent or translucent materials.

For the definition applicable in Chapter 11, see Section N1101.6.

[RE] FENESTRATION, VERTICAL. Windows that are fixed or movable, opaque doors, glazed doors, glazed block and combination opaque and glazed doors installed in a wall at less than 15 degrees (0.26 rad) from vertical.

For the definition applicable in Chapter 11, see Section N1101.6.

[RE] FENESTRATION PRODUCT, SITE-BUILT. For the definition applicable in Chapter 11, see Section N1101.6.

[RB] FIBER-CEMENT (BACKERBOARD, SIDING, SOFFIT, TRIM AND UNDERLAYMENT) PRODUCTS. Manufactured thin section composites of hydraulic cementitious matrices and discrete nonasbestos fibers.

[RB] FIRE SEPARATION DISTANCE. The distance measured from the building face to one of the following:

1. To the closest interior *lot line*.
2. To the centerline of a street, an alley or public way.
3. To an imaginary line between two buildings on the *lot*.

The distance shall be measured at a right angle from the face of the wall.

[RB] FIREBLOCKING. Building materials or materials *approved* for use as fireblocking, installed to resist the free passage of flame to other areas of the building through concealed spaces.

[RB] FIREPLACE. An assembly consisting of a hearth and fire chamber of *noncombustible material* and provided with a chimney, for use with solid fuels.

[MP] FIREPLACE STOVE. A free-standing, chimney-connected solid-fuel-burning heater designed to be operated with the fire chamber doors in either the open or closed position.

[RB] FIREPLACE THROAT. The opening between the top of the firebox and the smoke chamber.

[RB] FIRE-RETARDANT-TREATED WOOD. Wood products that, when impregnated with chemicals by a pressure process or other means during manufacture, exhibit reduced surface burning characteristics and resist propagation of fire.

Other means during manufacture. A process where the wood raw material is treated with a fire-retardant formulation while undergoing creation as a finished product.

Pressure process. A process for treating wood using an initial vacuum followed by the introduction of pressure above atmospheric.

[MP] FIXTURE. See "*Plumbing fixture.*"

[MP] FIXTURE BRANCH, DRAINAGE. A drain serving two or more fixtures that discharges into another portion of the drainage system.

[MP] FIXTURE BRANCH, WATER-SUPPLY. A water-supply pipe between the fixture supply and a main water-distribution pipe or fixture group main.

[MP] FIXTURE DRAIN. The drain from the trap of a fixture to the junction of that drain with any other drain pipe.

[MP] FIXTURE FITTING.

Supply fitting. A fitting that controls the volume or directional flow or both of water and that is either attached to or accessed from a fixture or is used with an open or atmospheric discharge.

Waste fitting. A combination of components that conveys the sanitary waste from the outlet of a fixture to the connection of the sanitary drainage system.

[MP] FIXTURE GROUP, MAIN. The main water-distribution pipe (or secondary branch) serving a plumbing fixture grouping such as a bath, kitchen or laundry area to which two or more individual fixture branch pipes are connected.

[MP] FIXTURE SUPPLY. The water-supply pipe connecting a fixture or fixture fitting to a fixture branch.

[MP] FIXTURE UNIT, DRAINAGE (d.f.u.). A measure of probable discharge into the drainage system by various types of plumbing fixtures, used to size DWV piping systems. The drainage fixture-unit value for a particular fixture depends on its volume rate of drainage discharge, on the time duration of a single drainage operation and on the average time between successive operations.

[MP] FIXTURE UNIT, WATER-SUPPLY (w.s.f.u.). A measure of the probable hydraulic demand on the water supply by various types of plumbing fixtures used to size water-piping systems. The water-supply fixture-unit value for a particular fixture depends on its volume rate of supply, on the time duration of a single supply operation and on the average time between successive operations.

[RB] FLAME SPREAD. The propagation of flame over a surface.

[RB] FLAME SPREAD INDEX. A comparative measure, expressed as a dimensionless number, derived from visual measurements of the spread of flame versus time for a material tested in accordance with ASTM E84 or UL 723.

[MP] FLEXIBLE AIR CONNECTOR. A conduit for transferring air between an air duct or plenum and an air terminal unit, an air inlet or an air outlet. Such conduit is limited in its use, length and location.

[RB] FLIGHT. A continuous run of rectangular treads or *winders* or combination thereof from one landing to another.

[MP] FLOOD-LEVEL RIM. The edge of the receptor or fixture from which water overflows.

[MP] FLOOR DRAIN. A plumbing fixture for recess in the floor having a floor-level strainer intended for the purpose of the collection and disposal of wastewater used in cleaning the floor and for the collection and disposal of accidental spillage to the floor.

[MP] FLOOR FURNACE. A self-contained furnace suspended from the floor of the space being heated, taking air for combustion from outside such space, and with means for lighting the *appliance* from such space.

[MP] FLOW PRESSURE. The static pressure reading in the water-supply pipe near the faucet or water outlet while the faucet or water outlet is open and flowing at capacity.

[MP] FLUE. See "*Vent.*"

[MP] FLUE, APPLIANCE. The passages within an *appliance* through which combustion products pass from the combustion chamber to the flue collar.

[MP] FLUE COLLAR. The portion of a fuel-burning *appliance* designed for the attachment of a draft hood, vent connector or venting system.

[MP] FLUE GASES. Products of combustion plus excess air in *appliance* flues or heat exchangers.

[MP] FLUSH VALVE. A device located at the bottom of a flush tank that is operated to flush water closets.

[MP] FLUSHOMETER TANK. A device integrated within an air accumulator vessel that is designed to discharge a predetermined quantity of water to fixtures for flushing purposes.

[MP] FLUSHOMETER VALVE. A flushometer valve is a device that discharges a predetermined quantity of water to fixtures for flushing purposes and is actuated by direct water pressure.

[RB] FOAM BACKER BOARD. Foam plastic used in siding applications where the foam plastic is a component of the siding.

[RB] FOAM PLASTIC INSULATION. A plastic that is intentionally expanded by the use of a foaming agent to produce a reduced-density plastic containing voids consisting of open or closed cells distributed throughout the plastic for thermal insulating or acoustic purposes and that has a density less than 20 pounds per cubic foot (320 kg/m^3) unless it is used as interior *trim*.

[RB] FOAM PLASTIC INTERIOR TRIM. Exposed foam plastic used as picture molds, chair rails, crown moldings, baseboards, *handrails*, ceiling beams, door *trim* and window *trim* and similar decorative or protective materials used in fixed applications.

[RB] FUEL CELL POWER SYSTEM, STATIONARY. A stationary energy generation system that converts the chemical energy of a fuel and oxidant to electric energy (DC or AC electricity) by an electrochemical process.

Field-fabricated fuel cell power system. A *stationary fuel cell power system* that is assembled at the job site and is not a preengineered or prepackaged factory-assembled fuel cell power system.

Preengineered fuel cell power system. A *stationary fuel cell power system* consisting of components and modules that are produced in a factory, and shipped to the job site for assembly.

Prepackaged fuel cell power system. A *stationary fuel cell power system* that is factory assembled as a single, complete unit and shipped as a complete unit for installation at the job site.

[MP] FUEL-PIPING SYSTEM. All piping, tubing, valves and fittings used to connect fuel utilization equipment to the point of fuel delivery.

[MP] FULL-OPEN VALVE. A water control or shutoff component in the water supply system piping that, where adjusted for maximum flow, the flow path through the component's closure member is not a restriction in the component's through-flow area.

[MP] FULLWAY VALVE. A valve that in the full open position has an opening cross-sectional area that is not less than 85 percent of the cross-sectional area of the connecting pipe.

[MP] FURNACE. A vented heating *appliance* designed or arranged to discharge heated air into a *conditioned space* or through a duct or ducts.

[RB] GLASS MAT GYPSUM PANEL. A gypsum panel consisting of a noncombustible core primarily of gypsum, surfaced with glass mat partially or completely embedded in the core.

[RB] GLAZING AREA. The interior surface area of all glazed fenestration, including the area of sash, curbing or other framing elements, that enclose *conditioned space*. Includes the area of glazed fenestration assemblies in walls bounding conditioned *basements*.

[RB] GRADE. The finished ground level adjoining the building at all exterior walls.

[MP] GRADE, PIPING. See "*Slope*."

[RB] GRADE FLOOR EMERGENCY ESCAPE AND RESCUE OPENING. An emergency escape and rescue opening located such that the bottom of the clear opening is not more than 44 inches (1118 mm) above or below the finished ground level adjacent to the opening. (See also "*Emergency escape and rescue opening.*")

[RB] GRADE PLANE. A reference plane representing the average of the finished ground level adjoining the building at all exterior walls. Where the finished ground level slopes away from the exterior walls, the reference plane shall be established by the lowest points within the area between the building and the *lot line* or, where the *lot line* is more than 6 feet (1829 mm) from the building between the structure and a point 6 feet (1829 mm) from the building.

[MP] GRAYWATER. Waste discharged from lavatories, bathtubs, showers, clothes washers and laundry trays.

[MP] GRIDDED WATER DISTRIBUTION SYSTEM. A water distribution system where every water distribution pipe is interconnected so as to provide two or more paths to each fixture supply pipe.

[RB] GROSS AREA OF EXTERIOR WALLS. The normal projection of all *exterior walls*, including the area of all windows and doors installed therein.

[MP] GROUND-SOURCE HEAT PUMP LOOP SYSTEM. Piping buried in horizontal or vertical excavations or placed in a body of water for the purpose of transporting heat transfer liquid to and from a heat pump. Included in this definition are closed loop systems in which the liquid is recirculated and open loop systems in which the liquid is drawn from a well or other source.

[RB] GUARD. A building component or a system of building components located near the open sides of elevated walking surfaces that minimizes the possibility of a fall from the walking surface to the lower level.

[RB] GUESTROOM. Any room or rooms used or intended to be used by one or more guests for living or sleeping purposes.

[RB] GYPSUM BOARD. The generic name for a family of sheet products consisting of a noncombustible core primarily of gypsum with paper surfacing. Gypsum wallboard, gypsum sheathing, gypsum base for gypsum *veneer* plaster, exterior gypsum soffit board, predecorated gypsum board and water-resistant gypsum backing board complying with the standards listed in Section R702.3 and Part IX of this code are types of gypsum board.

[RB] GYPSUM PANEL PRODUCT. The general name for a family of sheet products consisting essentially of gypsum.

[RB] GYPSUM SHEATHING. Gypsum panel products specifically manufactured with enhanced water resistance for use as a substrate for exterior surface materials.

[RB] GYPSUM WALLBOARD. A gypsum board used primarily as interior surfacing for building structures.

[RB] HABITABLE SPACE. A space in a building for living, sleeping, eating or cooking. Bathrooms, toilet rooms, closets, halls, storage or utility spaces and similar areas are not considered *habitable spaces.*

[RB] HANDRAIL. A horizontal or sloping rail intended for grasping by the hand for guidance or support.

[MP] HANGERS. See "*Supports.*"

[MP] HAZARDOUS LOCATION. Any location considered to be a fire hazard for flammable vapors, dust, combustible fibers or other highly combustible substances.

[MP] HEAT PUMP. An *appliance* having heating or heating and cooling capability and that uses refrigerants to extract heat from air, liquid or other sources.

[RE] HEATED SLAB. For the definition applicable in Chapter 11, see Section N1101.6.

[RB] HEIGHT, BUILDING. The vertical distance from *grade plane* to the average height of the highest roof surface.

[RB] HEIGHT, STORY. The vertical distance from top to top of two successive tiers of beams or finished floor surfaces; and, for the topmost *story,* from the top of the floor finish to the top of the ceiling joists or, where there is not a ceiling, to the top of the roof rafters.

[RE] HIGH-EFFICACY LIGHT SOURCES. For the definition applicable in Chapter 11, see Section N1101.6.

[MP] HIGH-TEMPERATURE (H.T.) CHIMNEY. A high-temperature chimney complying with the requirements of UL 103. A Type H.T. chimney is identifiable by the markings "Type H.T." on each chimney pipe section.

[RB] HILL. With respect to topographic wind effects, a land surface characterized by strong relief in any horizontal direction.

[RB] HISTORIC BUILDING. A building or structure that is one or more of the following:

1. Listed, or certified as eligible for listing, by the State Historic Preservation Officer or the Keeper of the National Register of Historic Places in the National Register of Historic Places.
2. Designated as historic under an applicable state or local law.
3. Certified as a contributing resource within a National Register-listed, or a state-designated or locally designated historic district.

For the definition applicable in Chapter 11, see Section N1101.6.

[MP] HORIZONTAL BRANCH, DRAINAGE. A drain pipe extending laterally from a soil or waste stack or *building drain,* that receives the discharge from one or more *fixture drains.*

[MP] HORIZONTAL PIPE. Any pipe or fitting that makes an angle of less than 45 degrees (0.79 rad) with the horizontal.

[MP] HOT WATER. Water at a temperature greater than 120°F (49°C).

[RB] HURRICANE-PRONE REGIONS. Areas vulnerable to hurricanes, defined as the US Atlantic Ocean and Gulf of Mexico coasts where the ultimate design wind speed, V_{ult}, is greater than 115 miles per hour (51 m/s), and Hawaii, Puerto Rico, Guam, Virgin Islands and America Samoa.

[MP] HYDROGEN-GENERATING APPLIANCE. A self-contained package or factory-matched packages of integrated systems for generating gaseous hydrogen. Hydrogen-generating *appliances* utilize electrolysis, reformation, chemical or other processes to generate hydrogen.

[MP] IGNITION SOURCE. A flame, spark or hot surface capable of igniting flammable vapors or fumes. Such sources include *appliance* burners, burner ignitions and electrical switching devices.

[RB] IMPACT PROTECTIVE SYSTEM. Construction that has been shown by testing to withstand the impact of test missiles and that is applied, attached, or locked over exterior glazing.

[MP] INDIRECT SYSTEM. A solar thermal system in which the gas or liquid in the solar collector loop circulates between the solar collector and a heat exchanger and such gas or liquid is not drained from the system or supplied to the load during normal operation.

[MP] INDIRECT WASTE PIPE. A waste pipe that discharges into the drainage system through an *air gap* into a trap, fixture or receptor.

[MP] INDIVIDUAL SEWAGE DISPOSAL SYSTEM. A system for disposal of sewage by means of a septic tank or mechanical treatment, designed for use apart from a public sewer to serve a single establishment or building.

[MP] INDIVIDUAL VENT. A pipe installed to vent a single *fixture drain* that connects with the vent system above or terminates independently outside the building.

[MP] INDIVIDUAL WATER SUPPLY. A supply other than an *approved* public water supply that serves one or more families.

[RE] INFILTRATION. For the definition applicable in Chapter 11, see Section N1101.6.

[RB] INSULATED SIDING. A type of continuous insulation, with manufacturer-installed insulating material as an integral part of the cladding product, having a minimum *R*-value of R-2. For the definition applicable in Chapter 11, see Section N1101.6.

[RB] INSULATED VINYL SIDING. A vinyl cladding product, with manufacturer-installed foam plastic insulating material as an integral part of the cladding product, having a thermal resistance of not less than R-2.

[RB] INSULATING CONCRETE FORM (ICF). A concrete forming system using stay-in-place forms of rigid foam plastic insulation, a hybrid of cement and foam insulation, a hybrid of cement and wood chips, or other insulating material for constructing cast-in-place concrete walls.

[RB] INSULATING SHEATHING. A rigid panel or board insulation material having a thermal resistance of not less than R-2 of the core material with properties suitable for use on walls, floors, roofs or foundations.

For the definition applicable in Chapter 11, see Section N1101.6.

[RB] INTERMODAL SHIPPING CONTAINER. A six-sided steel unit originally constructed as a general cargo container used for the transport of goods and materials.

[RB] JURISDICTION. The governmental unit that has adopted this code.

[RB] KITCHEN. An area used, or designated to be used, for the preparation of food.

[RB] LABEL. An identification applied on a product by the manufacturer that contains the name of the manufacturer, the function and performance characteristics of the product or material, and the name and identification of an *approved agency* and that indicates that the representative sample of the product or material has been tested and evaluated by an *approved agency*. (See also "*Manufacturer's designation*" and "*Mark*.")

[RB] LABELED. Equipment, materials or products to which have been affixed a *label*, seal, symbol or other identifying *mark* of a nationally recognized testing laboratory, *approved* agency or other organization concerned with product evaluation that maintains periodic inspection of the production of such *labeled* items and whose labeling indicates either that the *equipment*, material or product meets identified standards or has been tested and found suitable for a specified purpose. For the definition applicable in Chapter 11, see Section N1101.6.

[RB] LIGHT-FRAME CONSTRUCTION. Construction whose vertical and horizontal structural elements are primarily formed by a system of repetitive wood or cold-formed steel framing members.

[RB] LISTED. Equipment, materials, products or services included in a list published by an organization acceptable to the code official and concerned with evaluation of products or services that maintains periodic inspection of production of *listed equipment* or materials or periodic evaluation of services and whose listing states either that the *equipment*, material, product or service meets identified standards or has been tested and found suitable for a specified purpose. For the definition applicable in Chapter 11, see Section N1101.6.

[RB] LIVE LOADS. Those loads produced by the use and occupancy of the building or other structure and do not include construction or environmental loads such as wind load, snow load, rain load, earthquake load, flood load or dead load.

[RB] LIVE/WORK UNIT. A *dwelling unit* or sleeping unit in which a significant portion of the space includes a nonresidential use that is operated by the tenant.

[MP] LIVING SPACE. Space within a *dwelling unit* utilized for living, sleeping, eating, cooking, bathing, washing and sanitation purposes.

[MP] LOCAL EXHAUST. An exhaust system that uses one or more fans to exhaust air from a specific room or rooms within a dwelling.

[MP] LOCKING-TYPE TAMPER-RESISTANT CAP. A cap designed to be unlocked by a specially designed tool or key to prevent removal of the cap by means of hand-loosening or by commonly available tools.

[RB] LODGING HOUSE. A one-family dwelling where one or more occupants are primarily permanent in nature, and rent is paid for guestrooms.

[RB] LOT. A measured portion or parcel of land considered as a unit having fixed boundaries.

[RB] LOT LINE. The line that bounds a plot of ground described as a lot in the title to the property.

[RE] LOW-VOLTAGE LIGHTING. For the definition applicable in Chapter 11, see Section N1101.6.

[MP] MACERATING TOILET SYSTEMS. A system comprised of a sump with macerating pump and with connections for a water closet and other plumbing fixtures, that is designed to accept, grind and pump wastes to an *approved* point of discharge.

[MP] MAIN. The principal pipe artery to which branches may be connected.

[MP] MAIN SEWER. See "*Public sewer.*"

[MP] MANIFOLD WATER DISTRIBUTION SYSTEMS. A fabricated piping arrangement in which a large supply main is fitted with multiple branches in close proximity in which water is distributed separately to fixtures from each branch.

[RE] MANUAL. For the definition applicable in Chapter 11, see Section N1101.6.

[RB] MANUFACTURED HOME. A structure, transportable in one or more sections, that in the traveling mode is 8 body feet (2438 body mm) or more in width or 40 body feet (12 192 body mm) or more in length, or, where erected on site, is 320 square feet (30 m^2) or more, and that is built on a permanent chassis and designed to be used as a *dwelling* with or without a permanent foundation where connected to the required utilities, and includes the plumbing, heating, air-conditioning and electrical systems contained therein; except that such term shall include any structure that meets all the requirements of this paragraph except the size requirements and with respect to which the manufacturer voluntarily files a certification required by the secretary (HUD) and complies with the standards established under this title. For mobile homes built prior to June 15, 1976, a *label* certifying compliance to the Standard for Mobile Homes, NFPA 501, in effect at the time of manufacture is required. For the purpose of these provisions, a mobile home shall be considered to be a *manufactured home*.

[RB] MANUFACTURER'S DESIGNATION. An identification applied on a product by the manufacturer indicating that a product or material complies with a specified standard or set of rules. (See also "*Mark*" and "*Label*.")

[RB] MANUFACTURER'S INSTALLATION INSTRUCTIONS. Printed instructions included with equipment as part of the conditions of their *listing* and *labeling*.

[RB] MARK. An identification applied on a product by the manufacturer indicating the name of the manufacturer and the function of a product or material. (See also "*Manufacturer's designation*" and "*Label.*")

[RB] MASONRY, SOLID. Masonry consisting of *solid masonry* units laid contiguously with the joints between the units filled with mortar.

[RB] MASONRY CHIMNEY. A field-constructed chimney composed of *solid masonry* units, bricks, stones or concrete.

[RB] MASONRY HEATER. A masonry heater is a solid fuel burning heating *appliance* constructed predominantly of concrete or *solid masonry* having a mass of not less than 1,100 pounds (500 kg), excluding the chimney and foundation. It is designed to absorb and store a substantial portion of heat from a fire built in the firebox by routing exhaust gases through internal heat exchange channels in which the flow path downstream of the firebox includes not less than one 180-degree (3.14-rad) change in flow direction before entering the chimney and that deliver heat by radiation through the masonry surface of the heater.

[RB] MASONRY UNIT. Brick, tile, stone, architectural cast stone, glass block or concrete block conforming to the requirements specified in Section 2103 of the *International Building Code.*

Clay. A building unit larger in size than a brick, composed of burned clay, shale, fire clay or mixtures thereof.

Concrete. A building unit or block larger in size than 12 inches by 4 inches by 4 inches (305 mm by 102 mm by 102 mm) made of cement and suitable aggregates.

Glass. Nonload-bearing masonry composed of glass units bonded by mortar.

Hollow. A *masonry unit* with a net cross-sectional area in any plane parallel to the loadbearing surface that is less than 75 percent of its gross cross-sectional area measured in the same plane.

Solid. A *masonry unit* with a net cross-sectional area in every plane parallel to the loadbearing surface that is 75 percent or more of its cross-sectional area measured in the same plane.

[RB] MEAN ROOF HEIGHT. The average of the roof eave height and the height to the highest point on the roof surface, except that eave height shall be used for roof angle of less than or equal to 10 degrees (0.18 rad).

[MP] MECHANICAL DRAFT SYSTEM. A venting system designed to remove flue or vent gases by mechanical means, that consists of an induced draft portion under nonpositive static pressure or a forced draft portion under positive static pressure.

Forced draft venting system. A portion of a venting system using a fan or other mechanical means to cause the removal of flue or vent gases under positive static pressure.

Induced draft venting system. A portion of a venting system using a fan or other mechanical means to cause the removal of flue or vent gases under nonpositive static vent pressure.

Power venting system. A portion of a venting system using a fan or other mechanical means to cause the removal of flue or vent gases under positive static vent pressure.

[MP] MECHANICAL EXHAUST SYSTEM. A system for removing air from a room or space by mechanical means.

[MP] MECHANICAL JOINT.

1. A connection between pipes, fittings or pipes and fittings that is not welded, brazed, caulked, soldered, solvent cemented or heat fused.
2. A general form of gastight or liquid-tight connections obtained by the joining of parts through a positive holding mechanical construction such as, but not limited to, flanged, screwed, clamped or flared connections.

[MP] MECHANICAL SYSTEM. A system specifically addressed and regulated in this code and composed of components, devices, *appliances* and *equipment.*

[RB] METAL ROOF PANEL. An interlocking metal sheet having an installed weather exposure of not less than 3 square feet (0.28 m^2) per sheet.

[RB] METAL ROOF SHINGLE. An interlocking metal sheet having an installed weather exposure less than 3 square feet (0.28 m^2) per sheet.

[RB] MEZZANINE. An intermediate level or levels between the floor and ceiling of any *story.*

[RB] MODIFIED BITUMEN ROOF COVERING. One or more layers of polymer modified asphalt sheets. The sheet materials shall be fully adhered or mechanically attached to the substrate or held in place with an *approved* ballast layer.

[RB] MULTIPLE-STATION SMOKE ALARM. Two or more single-station alarm devices that are capable of interconnection such that actuation of one causes all integral or separate audible alarms to operate.

[RB] NAILABLE SUBSTRATE. A product or material such as framing, sheathing or furring, composed of wood or wood-based materials, or other materials and fasteners providing equivalent fastener withdrawal resistance.

[MP] NATURAL DRAFT SYSTEM. A venting system designed to remove flue or vent gases under nonpositive static vent pressure entirely by natural draft.

[RB] NATURALLY DURABLE WOOD. The heartwood of the following species with the exception that an occasional piece with corner sapwood is permitted if 90 percent or more of the width of each side on which it occurs is heartwood.

Decay resistant. Redwood, cedar, black locust and black walnut.

Termite resistant. Alaska yellow cedar, redwood, Eastern red cedar and Western red cedar including all sapwood of Western red cedar.

[RB] NONCOMBUSTIBLE MATERIAL. A material that passes ASTM E136.

[RB] NOSING. The leading edge of treads of stairs and of landings at the top of *stairway* flights.

[RB] OCCUPIED SPACE. The total area of all buildings or structures on any *lot* or parcel of ground projected on a horizontal plane, excluding permitted projections as allowed by this code.

[MP] OFFSET. A combination of fittings that makes two changes in direction, bringing one section of the pipe out of line and into a line parallel with the other section.

[MP] ON-SITE NONPOTABLE WATER REUSE SYSTEMS. Water systems for the collection, treatment, storage, distribution, and reuse of nonpotable water generated on site, including but not limited to graywater systems. This definition does not include rainwater harvesting systems.

[RE] OPAQUE DOOR. For the definition applicable in Chapter 11, see Section N1101.6.

[RB] OWNER. Any person, agent, firm or corporation having a legal or equitable interest in the property.

[RB] PAN FLASHING. Corrosion-resistant flashing at the base of an opening that is integrated into the building exterior wall to direct water to the exterior and is premanufactured, fabricated, formed or applied at the job site.

[RB] PANEL THICKNESS. Thickness of core plus two layers of structural wood panel facings.

[MP] PELLET FUEL-BURNING APPLIANCE. A closed combustion, vented *appliance* equipped with a fuel feed mechanism for burning processed pellets of solid fuel of a specified size and composition.

[MP] PELLET VENT. A vent *listed* and *labeled* for use with a *listed* pellet fuel-burning *appliance.*

[RB] PERFORMANCE CATEGORY. A designation of wood structural panels as related to the panel performance used in Chapters 4, 5, 6 and 8.

[RB] PERMIT. An official document or certificate issued by the *building official* that authorizes performance of a specified activity.

[RB] PERSON. An individual, heirs, executors, administrators or assigns, and a firm, partnership or corporation, its or their successors or assigns, or the agent of any of the aforesaid.

[RB] PHOTOVOLTAIC MODULE. A complete, environmentally protected unit consisting of solar cells, optics and other components, exclusive of a tracker, designed to generate DC power where exposed to sunlight.

[RB] PHOTOVOLTAIC PANEL. A collection of *photovoltaic modules* mechanically fastened together, wired, and designed to provide a field-installable unit.

[RB] PHOTOVOLTAIC PANEL SYSTEM. A system that incorporates discrete photovoltaic panels that convert solar radiation into electricity, including rack support systems.

[RB] PHOTOVOLTAIC SHINGLES. A *roof covering* that resembles shingles and that incorporates *photovoltaic modules.*

[MP] PITCH. See "*Slope.*"

[RB] PLASTIC COMPOSITE. A generic designation that refers to wood-plastic composites and plastic lumber.

[RB] PLATFORM CONSTRUCTION. A method of construction by which floor framing bears on load bearing walls that are not continuous through the *story* levels or floor framing.

[MP] PLENUM. A chamber that forms part of an air-circulation system other than the *occupied space* being conditioned.

[MP] PLUMBING. For the purpose of this code, plumbing refers to those installations, repairs, maintenance and *alterations* regulated by Chapters 25 through 33.

[MP] PLUMBING APPLIANCE. An energized household *appliance* with plumbing connections, such as a dishwasher, food waste disposer, clothes washer or water heater.

[MP] PLUMBING APPURTENANCE. A device or assembly that is an adjunct to the basic plumbing system and does not demand additional water supply or add any discharge load to the system. It is presumed that it performs some useful function in the operation, maintenance, servicing, economy or safety of the plumbing system. Examples include filters, relief valves and aerators.

[MP] PLUMBING FIXTURE. A receptacle or device that is connected to a water supply system or discharges to a drainage system or both. Such receptacles or devices require a supply of water; or discharge liquid waste or liquidborne solid waste; or require a supply of water and discharge waste to a drainage system.

[MP] PLUMBING SYSTEMS. Includes the water distribution pipes; plumbing fixtures and traps; water-treating or water-using equipment; soil, waste and vent pipes; and building drains; in addition to their respective connections, devices and appurtenances within a structure or premises; and the water service, building sewer and building storm sewer serving such structure or premises.

[MP] POLLUTION. A low-hazard or nonhealth-hazard impairment of the quality of the potable water to a degree that does not create a hazard to the public health and that does adversely and unreasonably affect the aesthetic qualities of such potable water for domestic use.

[RB] POLYPROPYLENE SIDING. A shaped material, made principally from polypropylene homopolymer, or copolymer, that in some cases contains fillers or reinforcements, that is used to clad exterior walls or buildings.

[MP] PORTABLE-FUEL-CELL APPLIANCE. A fuel cell generator of electricity that is not fixed in place. A portable-fuel-cell *appliance* utilizes a cord and plug connection to a grid-isolated load and has an integral fuel supply.

[RB] POSITIVE ROOF DRAINAGE. The drainage condition in which consideration has been made for the loading deflections of the *roof deck*, and additional slope has been provided to ensure drainage of the roof within 48 hours of precipitation.

[MP] POTABLE WATER. Water free from impurities present in amounts sufficient to cause disease or harmful physiological effects and conforming in bacteriological and chemical quality to the requirements of the public health authority having *jurisdiction*.

[RB] PRECAST CONCRETE. A structural concrete element cast elsewhere than its final position in the structure.

[RB] PRECAST CONCRETE FOUNDATION WALLS. Preengineered, *precast concrete* wall panels that are designed to withstand specified stresses and used to build below-*grade* foundations.

[MP] PRESS-CONNECT JOINT. A permanent mechanical joint incorporating an elastomeric seal or an elastomeric seal and corrosion-resistant grip or bite ring. The joint is made with a pressing tool and jaw or ring approved by the fitting manufacturer.

[MP] PRESSURE-RELIEF VALVE. A pressure-actuated valve held closed by a spring or other means and designed to automatically relieve pressure at the pressure at which it is set.

[RE] PROPOSED DESIGN. For the definition applicable in Chapter 11, see Section N1101.6.

[MP] PUBLIC SEWER. A common sewer directly controlled by public authority.

[MP] PUBLIC WATER MAIN. A water-supply pipe for public use controlled by public authority.

[RB] PUBLIC WAY. Any street, alley or other parcel of land open to the outside air leading to a public street, that has been deeded, dedicated or otherwise permanently appropriated to the public for public use and that has a clear width and height of not less than 10 feet (3048 mm).

[MP] PURGE. To clear of air, gas or other foreign substances.

[MP] PUSH-FIT FITTING. A mechanical fitting that joins pipes or tubes and achieves a seal by mating the pipe or tube into the fitting.

[MP] QUICK-CLOSING VALVE. A valve or faucet that closes automatically where released manually or controlled by mechanical means for fast-action closing.

[RB] RAMP. A walking surface that has a running slope steeper than 1 unit vertical in 20 units horizontal (5-percent slope).

[RE] RATED DESIGN. For the definition applicable in Chapter 11, see Section N1101.6.

[RB] READY ACCESS (TO). That which enables a device, *appliance* or equipment to be directly reached, without requiring the removal or movement of any panel, door or similar obstruction.

[MP] RECEPTOR. A fixture or device that receives the discharge from indirect waste pipes.

[MP] RECLAIMED WATER. Nonpotable water that has been derived from the treatment of wastewater by a facility or system licensed or permitted to produce water meeting the *jurisdiction's* water requirements for its intended uses. Also known as "recycled water."

[MP] REFRIGERANT. A substance used to produce refrigeration by its expansion or evaporation.

[MP] REFRIGERANT COMPRESSOR. A specific machine, with or without accessories, for compressing a given refrigerant vapor.

[MP] REFRIGERATING SYSTEM. A combination of interconnected parts forming a closed circuit in which refrigerant is circulated for the purpose of extracting, then rejecting, heat. A direct refrigerating system is one in which the evaporator or condenser of the refrigerating system is in direct contact with the air or other substances to be cooled or heated. An indirect refrigerating system is one in which a secondary coolant cooled or heated by the refrigerating system is circulated to the air or other substance to be cooled or heated.

[RB] REGISTERED DESIGN PROFESSIONAL. An individual who is registered or licensed to practice their respective design profession as defined by the statutory requirements of the professional registration laws of the state or *jurisdiction* in which the project is to be constructed.

[MP] RELIEF VALVE, VACUUM. A device to prevent excessive buildup of vacuum in a pressure vessel.

[RB] REPAIR. The reconstruction, replacement or renewal of any part of an existing building for the purpose of its maintenance or to correct damage.

For the definition applicable in Chapter 11, see Section N1101.6.

[RB] REROOFING. The process of recovering or replacing an existing *roof covering*. See "*Roof recover*."

For the definition applicable in Chapter 11, see Section N1101.6.

[RE] RESIDENTIAL BUILDING. For the definition applicable in Chapter 11, see Section N1101.6.

[MP] RETURN AIR. Air removed from an *approved conditioned space* or location and recirculated or exhausted.

[RB] RIDGE. With respect to topographic wind effects, an elongated crest of a *hill* characterized by strong relief in two directions.

[MP] RISER (PLUMBING). A water pipe that extends vertically one full *story* or more to convey water to branches or to a group of fixtures.

[RB] RISER (STAIR). The vertical component of a step or *stair*.

[RB] ROOF ASSEMBLY. A system designed to provide weather protection and resistance to design loads. The system consists of a *roof covering* and *roof deck* or a single component serving as both the *roof covering* and the *roof deck*. A *roof assembly* can include an underlayment, thermal barrier, ignition barrier, insulation or a vapor retarder. For the definition applicable in Chapter 11, see Section N1101.6.

[RB] ROOF COATING. A fluid-applied, adhered coating used for roof maintenance or *roof repair*, or as a component of a *roof covering* system or *roof assembly*.

[RB] ROOF COVERING. The covering applied to the *roof deck* for weather resistance, fire classification or appearance.

[RB] ROOF COVERING SYSTEM. See "*Roof assembly*."

[RB] ROOF DECK. The flat or sloped surface not including its supporting members or vertical supports.

[RB] ROOF RECOVER. The process of installing an additional *roof covering* over an existing roof covering without removing the existing roof covering. For the definition applicable in Chapter 11, see Section N1101.6.

[RB] ROOF REPAIR. Reconstruction or renewal of any part of an existing roof for the purposes of its maintenance. For the definition applicable in Chapter 11, see Section N1101.6.

[RB] ROOF REPLACEMENT. The process of removing the existing *roof covering*, repairing any damaged substrate and installing a new *roof covering*. For the definition applicable in Chapter 11, see Section N1101.6.

[MP] ROOM HEATER. A free-standing heating *appliance* installed in the space being heated and not connected to ducts.

[MP] ROUGH-IN. The installation of the parts of the plumbing system that must be completed prior to the installation of fixtures. This includes DWV, water supply and built-in fixture supports.

[RB] RUNNING BOND. The placement of *masonry units* such that head joints in successive courses are horizontally offset not less than one-quarter the unit length.

[RE] *R*-VALUE (THERMAL RESISTANCE). For the definition applicable in Chapter 11, see Section N1101.6.

[MP] SANITARY SEWER. A sewer that carries sewage and excludes storm, surface and groundwater.

[RB] SCUPPER. An opening in a wall or parapet that allows water to drain from a roof.

[RB] SEISMIC DESIGN CATEGORY (SDC). A classification assigned to a structure based on its occupancy category and the severity of the design earthquake ground motion at the site.

[MP] SEPTIC TANK. A watertight receptor that receives the discharge of a building sanitary drainage system and is constructed so as to separate solids from the liquid, digest organic matter through a period of detention, and allow the liquids to discharge into the soil outside of the tank through a system of open joint or perforated piping or a seepage pit.

[RE] SERVICE WATER HEATING. For the definition applicable in Chapter 11, see Section N1101.6.

[MP] SEWAGE. Any liquid waste containing animal matter, vegetable matter or other impurity in suspension or solution.

[MP] SEWAGE PUMP. A permanently installed mechanical device for removing sewage or liquid waste from a sump.

[RB] SHALL. The term, where used in the code, is construed as mandatory.

[RB] SHEAR WALL. A general term for walls that are designed and constructed to resist racking from seismic and wind by use of masonry, concrete, cold-formed steel or wood framing in accordance with Chapter 6 of this code and the associated limitations in Section R301.2 of this code.

[RB] SHINGLE FASHION. A method of installing roof or wall coverings, *water-resistive barriers*, flashing or other building components such that upper layers of material are placed overlapping lower layers of material to provide drainage and protect against water intrusion at unsealed penetrations and joints or in combination with sealed joints.

[RB] SINGLE-PLY MEMBRANE. A roofing membrane that is field applied using one layer of membrane material (either homogeneous or composite) rather than multiple layers.

[RB] SINGLE-STATION SMOKE ALARM. An assembly incorporating the detector, control equipment and alarm sounding device in one unit that is operated from a power supply either in the unit or obtained at the point of installation.

[RE] SKYLIGHT. For the definition applicable in Chapter 11, see Section N1101.6.

[RB] SKYLIGHT, UNIT. A factory assembled, glazed fenestration unit, containing one panel of glazing material, that allows for natural daylighting through an opening in the *roof assembly* while preserving the weather-resistant barrier of the roof.

[RB] SKYLIGHTS AND SLOPED GLAZING. Glass or other transparent or translucent glazing material installed at a slope of 15 degrees (0.26 rad) or more from vertical. *Unit skylights*, *tubular daylighting devices* and glazing materials in solariums, *sunrooms*, roofs and sloped walls are included in this definition. For the definition applicable in Chapter 11, see Section N1101.6.

[RB] SLEEPING UNIT. A single unit that provides rooms or spaces for one or more persons, includes permanent provisions for sleeping and can include provisions for living, eating and either sanitation or kitchen facilities but not both. Such rooms and spaces that are also part of a *dwelling unit* are not sleeping units.

[MP] SLIP JOINT. A mechanical-type joint used primarily on fixture traps. The joint tightness is obtained by compressing a friction-type washer such as rubber, nylon, neoprene, lead or special packing material against the pipe by the tightening of a (slip) nut.

[MP] SLOPE. The fall (pitch) of a line of pipe in reference to a horizontal plane. In drainage, the slope is expressed as the fall in units vertical per units horizontal (percent) for a length of pipe.

[RB] SMOKE-DEVELOPED INDEX. A comparative measure, expressed as a dimensionless number, derived from measurements of smoke obscuration versus time for a material tested in accordance with ASTM E84 or UL 723.

[MP] SOIL STACK OR PIPE. A pipe that conveys sewage containing fecal material.

[RB] SOLAR ENERGY SYSTEM. A system that converts solar radiation to usable energy, including *photovoltaic panel systems* and *solar thermal systems.*

[RE] SOLAR HEAT GAIN COEFFICIENT (SHGC). For the definition applicable in Chapter 11, see Section N1101.6.

[MP] SOLAR THERMAL COLLECTOR. Components in a *solar thermal system* that collect and convert solar radiation to thermal energy.

[MP] SOLAR THERMAL SYSTEM. A system that converts solar radiation to thermal energy for use in heating or cooling.

[RB] SOLID MASONRY. Load-bearing or nonload-bearing construction using *masonry units* where the net cross-sectional area of each unit in any plane parallel to the bearing surface is not less than 75 percent of its gross cross-sectional area. *Solid masonry* units shall conform to ASTM C55, C62, C73, C145 or C216.

[RB] SPLINE. A strip of wood structural panel cut from the same material used for the panel facings, used to connect two structural insulated panels. The strip (spline) fits into a groove cut into the vertical edges of the two structural insulated panels to be joined. Splines are used behind each facing of the structural insulated panels being connected as shown in Figure R610.8.

[MP] STACK. Any main vertical DWV line, including offsets, that extends one or more stories as directly as possible to its vent terminal.

[RB] STACK BOND. The placement of *masonry units* in a bond pattern is such that head joints in successive courses are vertically aligned. For the purpose of this code, requirements for stack bond shall apply to all masonry laid in other than *running bond.*

[MP] STACK VENT. The extension of soil or waste stack above the highest horizontal drain connected.

[RB] STAIR. A change in elevation, consisting of one or more *risers.*

[RB] STAIRWAY. One or more flights of stairs, either interior or exterior, with the necessary landings and connecting platforms to form a continuous and uninterrupted passage from one level to another.

[RB] STAIRWAY, SPIRAL. A stairway with a plan view of closed circular form and uniform section-shaped treads radiating from a minimum-diameter circle.

[RE] STANDARD REFERENCE DESIGN. For the definition applicable in Chapter 11, see Section N1101.6.

[RB] STANDARD TRUSS. Any construction that does not permit the roof-ceiling insulation to achieve the required *R*-value over the exterior walls.

[MP] STATIONARY FUEL CELL POWER PLANT. A self-contained package or factory-matched packages that constitute an automatically operated assembly of integrated systems for generating useful electrical energy and recoverable thermal energy that is permanently connected and fixed in place.

[MP] STORM SEWER, DRAIN. A pipe used for conveying rainwater, surface water, subsurface water and similar liquid waste.

[RB] STORM SHELTER. A building, structure or portion thereof, constructed in accordance with ICC 500 and designated for use during a severe wind storm event, such as a hurricane or tornado.

[RB] STORY. That portion of a building included between the upper surface of a floor and the upper surface of the floor or roof next above.

[RB] STORY ABOVE GRADE PLANE. Any *story* having its finished floor surface entirely above *grade plane,* or in which the finished surface of the floor next above is either of the following:

1. More than 6 feet (1829 mm) above *grade plane.*
2. More than 12 feet (3658 mm) above the finished ground level at any point.

[RB] STRUCTURAL COMPOSITE LUMBER. Structural members manufactured using wood elements bonded together with exterior adhesives.

Examples of structural composite lumber are:

Laminated strand lumber (LSL). A composite of wood strand elements with wood fibers primarily oriented along the length of the member, where the least dimension of the wood strand elements is 0.10 inch (2.54 mm) or less and their average lengths are not less than 150 times the least dimension of the wood strand elements.

Laminated veneer lumber (LVL). A composite of wood veneer elements with wood fibers primarily oriented along the length of the member, where the veneer element thicknesses are 0.25 inch (6.4 mm) or less.

Oriented strand lumber (OSL). A composite of wood strand elements with wood fibers primarily oriented along the length of the member, where the least dimension of the wood strand elements is 0.10 inch (2.54 mm) or less and their average lengths are not less than 75 times and less than 150 times the least dimension of the wood strand elements.

Parallel strand lumber (PSL). A composite of wood strand elements with wood fibers primarily oriented along the length of the member, where the least dimension of the wood strand elements is 0.25 inch (6.4 mm) or less and their average lengths are not less than 300 times the least dimension of the wood strand elements.

[RB] STRUCTURAL INSULATED PANEL (SIP). A structural sandwich panel that consists of a lightweight foam plastic core securely laminated between two thin, rigid wood structural panel facings.

[RB] STRUCTURE. That which is built or constructed.

[RB] SUBSOIL DRAIN. A drain that collects subsurface water or seepage water and conveys such water to a place of disposal.

[MP] SUMP. A tank or pit that receives sewage or waste, located below the normal *grade* of the gravity system and that must be emptied by mechanical means.

[MP] SUMP PUMP. A pump installed to empty a sump. These pumps are used for removing storm water only. The pump is selected for the specific head and volume of the load and is usually operated by level controllers.

[RB] SUNROOM. A one-*story* structure attached to a *dwelling* with a *glazing area* in excess of 40 percent of the gross area of the structure's exterior walls and roof.

For the definition applicable in Chapter 11, see Section N1101.6.

[MP] SUPPLY AIR. Air delivered to a *conditioned space* through ducts or plenums from the heat exchanger of a heating, cooling or ventilating system.

[MP] SUPPORTS. Devices for supporting, hanging and securing pipes, fixtures and equipment.

[MP] SWEEP. A drainage fitting designed to provide a change in direction of a drain pipe of less than the angle specified by the amount necessary to establish the desired slope of the line. Sweeps provide a longer turning radius than bends and a less turbulent flow pattern (see "*Bend*" and "*Elbow*").

[MP] TEMPERATURE- AND PRESSURE-RELIEF (T AND P) VALVE. A combination relief valve designed to function as both a temperature-relief and pressure-relief valve.

[MP] TEMPERATURE-RELIEF VALVE. A temperature-actuated valve designed to discharge automatically at the temperature at which it is set.

[RB] TERMITE-RESISTANT MATERIAL. Pressure-preservative-treated wood in accordance with the AWPA standards in Section R317.1, naturally durable termite-resistant wood, steel, concrete, masonry or other *approved* material.

[RE] THERMAL ISOLATION. For the definition applicable in Chapter 11, see Section N1101.6.

[RE] THERMAL RESISTANCE, *R*-VALUE. See "*R-value.*"

[RE] THERMAL TRANSMITTANCE, *U*-FACTOR. See "*U-factor.*"

[RE] THERMOSTAT. For the definition applicable in Chapter 11, see Section N1101.6.

[MP] THIRD-PARTY CERTIFICATION AGENCY. An *approved* agency operating a product or material certification system that incorporates initial product testing, assessment and surveillance of a manufacturer's quality control system.

[MP] THIRD-PARTY CERTIFIED. Certification obtained by the manufacturer indicating that the function and performance characteristics of a product or material have been determined by testing and ongoing surveillance by an *approved* third-party certification agency. Assertion of certification is in the form of identification in accordance with the requirements of the third-party certification agency.

[RB] TOWNHOUSE. A *building* that contains three or more attached *townhouse units.*

[RB] TOWNHOUSE UNIT. A single-family *dwelling unit* in a *townhouse* that extends from foundation to roof and that has a *yard* or *public way* on not less than two sides.

[MP] TRAP. A fitting, either separate or built into a fixture, that provides a liquid seal to prevent the emission of sewer gases without materially affecting the flow of sewage or wastewater through it.

[MP] TRAP ARM. That portion of a *fixture drain* between a trap weir and the vent fitting.

[MP] TRAP PRIMER. A device or system of piping to maintain a water seal in a trap, typically installed where infrequent use of the trap would result in evaporation of the trap seal, such as floor drains.

[MP] TRAP SEAL. The trap seal is the maximum vertical depth of liquid that a trap will retain, measured between the crown weir and the top of the dip of the trap.

[RB] TRIM. Picture molds, chair rails, baseboards, *handrails,* door and window frames, and similar decorative or protective materials used in fixed applications.

[RB] TRUSS DESIGN DRAWING. The graphic depiction of an individual truss, that describes the design and physical characteristics of the truss.

[RB] TUBULAR DAYLIGHTING DEVICE (TDD). A nonoperable fenestration unit primarily designed to transmit daylight from a roof surface to an interior ceiling via a tubular conduit. The basic unit consists of an exterior glazed weathering surface, a light-transmitting tube with a reflective interior surface, and an interior-sealing device such as a translucent ceiling panel. The unit may be factory assembled, or field assembled from a manufactured kit.

[MP] TYPE L VENT. A *listed* and *labeled* vent conforming to UL 641 for venting oil-burning *appliances listed* for use with Type L vents or with gas *appliances listed* for use with Type B vents.

[RE] *U*-FACTOR (THERMAL TRANSMITTANCE). For the definition applicable in Chapter 11, see Section N1101.6.

[RB] UNDERLAYMENT. One or more layers of felt, sheathing paper, nonbituminous saturated felt, or other *approved* material over which a roof covering, with a slope of 2 units vertical in 12 units horizontal (17-percent slope) or greater, is applied.

[MP] VACUUM BREAKER. A device that prevents backsiphonage of water by admitting atmospheric pressure through ports to the discharge side of the device.

[RB] VAPOR DIFFUSION PORT. An assembly constructed or installed within a *roof assembly* at an opening in the *roof deck* to convey water vapor from an unvented attic to the outside atmosphere.

[RB] VAPOR PERMEABLE. The property of having a moisture vapor permeance rating of 5 perms (2.9×10^{-10} kg/Pa $\times$ s $\times$ m^2) or greater, where tested in accordance with Procedure A or Procedure B of ASTM E96. A vapor permeable material permits the passage of moisture vapor.

[RB] VAPOR RETARDER CLASS. A measure of the ability of a material or assembly to limit the amount of moisture that passes through that material or assembly. Vapor retarder class shall be defined using the desiccant method with Procedure A of ASTM E96 as follows:

Class I: $\leq$ 0.1 perm rating

Class II: > 0.1 to $\leq$ 1.0 perm rating

Class III: > 1.0 to $\leq$ 10 perm rating

[MP] VENT. A passageway for conveying flue gases from fuel-fired *appliances*, or their vent connectors, to the outside atmosphere.

[MP] VENT COLLAR. See "*Flue collar.*"

[MP] VENT CONNECTOR. That portion of a venting system that connects the flue collar or draft hood of an *appliance* to a vent.

[MP] VENT DAMPER DEVICE, AUTOMATIC. A device intended for installation in the venting system, in the outlet of an individual, automatically operated fuel-burning *appliance* and that is designed to open the venting system automatically where the *appliance* is in operation and to close off the venting system automatically where the *appliance* is in a standby or shutdown condition.

[MP] VENT GASES. Products of combustion from fuel-burning *appliances*, plus excess air and dilution air, in the venting system above the draft hood or draft regulator.

[MP] VENT STACK. A vertical vent pipe installed to provide circulation of air to and from the drainage system and that extends through one or more stories.

[MP] VENT SYSTEM. Piping installed to equalize pneumatic pressure in a drainage system to prevent trap seal loss or blowback due to siphonage or back pressure.

[RB] VENTILATION. The natural or mechanical process of supplying conditioned or unconditioned air to, or removing such air from, any space.

For the definition applicable in Chapter 11, see Section N1101.6.

[RE] VENTILATION AIR. For the definition applicable in Chapter 11, see Section N1101.6.

[MP] VENTING. Removal of combustion products to the outdoors.

[MP] VENTING SYSTEM. A continuous open passageway from the flue collar of an *appliance* to the outside atmosphere for the purpose of removing flue or vent gases. A venting system is usually composed of a vent or a chimney and vent connector, if used, assembled to form the open passageway.

[MP] VERTICAL PIPE. Any pipe or fitting that makes an angle of 45 degrees (0.79 rad) or more with the horizontal.

[RB] VINYL SIDING. A shaped material, made principally from rigid polyvinyl chloride (PVC), that is used to cover exterior walls of buildings.

[RE] VISIBLE TRANSMITTANCE (VT). For the definition applicable in Chapter 11, see Section N1101.6.

[RB] WALL, RETAINING. A wall not laterally supported at the top, that resists lateral soil load and other imposed loads.

[RB] WALLS. Walls shall be defined as follows:

Load-bearing wall. A wall supporting any vertical load in addition to its own weight.

Nonbearing wall. A wall which does not support vertical loads other than its own weight.

[MP] WASTE. Liquidborne waste that is free of fecal matter.

[MP] WASTE PIPE OR STACK. Piping that conveys only liquid sewage not containing fecal material.

[MP] WASTE RECEPTOR. A floor sink, standpipe, hub drain or a floor drain that receives the discharge of one or more indirect waste pipes.

[MP] WATER DISTRIBUTION SYSTEM. Piping that conveys water from the service to the plumbing fixtures, *appliances*, appurtenances, equipment, devices or other systems served, including fittings and control valves.

[MP] WATER HEATER. Any heating *appliance* or equipment that heats potable water and supplies such water to the potable hot water distribution system.

[MP] WATER MAIN. A water supply pipe for public use.

[MP] WATER OUTLET. A valved discharge opening, including a hose bibb, through which water is removed from the potable water system supplying water to a plumbing fixture or plumbing *appliance* that requires either an *air gap* or backflow prevention device for protection of the supply system.

[MP] WATER SERVICE PIPE. The outside pipe from the water main or other source of potable water supply to the water distribution system inside the building, terminating at the service valve.

[MP] WATER SUPPLY SYSTEM. The water service pipe, the water-distributing pipes and the necessary connecting pipes, fittings, control valves and appurtenances in or adjacent to the building or premises.

[RB] WATER-RESISTIVE BARRIER. A material behind an exterior wall covering that is intended to resist liquid water that has penetrated behind the exterior covering from further intruding into the exterior wall assembly.

[MP] WET VENT. A vent that receives the discharge of wastes from other fixtures.

[MP] WHOLE-HOUSE MECHANICAL VENTILATION SYSTEM. An exhaust system, supply system, or combination thereof that is designed to mechanically exchange indoor air for outdoor air where operating continuously or through a programmed intermittent schedule to satisfy the whole-house ventilation rate.

For the definition applicable in Chapter 11, see Section N1101.6.

[RB] WINDBORNE DEBRIS REGION. Areas within *hurricane-prone regions* located in accordance with one of the following:

1. Within 1 mile (1.61 km) of the coastal mean high-water line where the ultimate design wind speed, V_{ult}, is 130 mph (58 m/s) or greater.
2. In areas where an Exposure D condition exists upwind at the waterlne and the ultimate design wind speed, V_{ult}, is 140 mph (63 m/s) or greater; or Hawaii.

[RB] WINDER. A tread with nonparallel edges.

[RB] WOOD STRUCTURAL PANEL. A panel manufactured from veneers; or wood strands or wafers; bonded together with waterproof synthetic resins or other suitable bonding systems. Examples of wood structural panels are plywood, orientated strand board (OSB) or composite panels.

[RB] YARD. An open space, other than a court, unobstructed from the ground to the sky, except where specifically provided by this code, on the *lot* on which a building is situated.

[RE] ZONE. For the definition applicable in Chapter 11, see Section N1101.6.

Part III—Building Planning and Construction

CHAPTER 3
BUILDING PLANNING

User note:

About this chapter: *Chapter 3 contains a wide array of building planning requirements that are critical to designing a safe and usable building. This includes, but is not limited to, requirements related to general structural design, fire-resistant construction, light, ventilation, sanitation, plumbing fixture clearances, minimum room area and ceiling height, safety glazing, means of egress, automatic fire sprinkler systems, smoke and carbon monoxide alarm systems, accessibility, solar energy systems, swimming pools, spas and hot tubs.*

SECTION R301
DESIGN CRITERIA

R301.1 Application. Buildings and structures, and parts thereof, shall be constructed to safely support all loads, including dead loads, *live loads*, roof loads, flood loads, snow loads, wind loads and seismic loads as prescribed by this code. The construction of buildings and structures in accordance with the provisions of this code shall result in a system that provides a complete load path that meets the requirements for the transfer of loads from their point of origin through the load-resisting elements to the foundation. Buildings and structures constructed as prescribed by this code are deemed to comply with the requirements of this section.

R301.1.1 Alternative provisions. As an alternative to the requirements in Section R301.1, the following standards are permitted subject to the limitations of this code and the limitations therein. Where engineered design is used in conjunction with these standards, the design shall comply with the *International Building Code.*

1. AWC *Wood Frame Construction Manual* (WFCM).
2. AISI *Standard for Cold-Formed Steel Framing—Prescriptive Method for One- and Two-Family Dwellings* (AISI S230).
3. ICC *Standard on the Design and Construction of Log Structures* (ICC 400).

R301.1.2 Construction systems. The requirements of this code are based on platform and balloon-frame construction for light-frame buildings. The requirements for concrete and masonry buildings are based on a balloon framing system. Other framing systems must have equivalent detailing to ensure force transfer, continuity and compatible deformations.

R301.1.3 Engineered design. Where a building of otherwise conventional construction contains structural elements exceeding the limits of Section R301 or otherwise not conforming to this code, these elements shall be designed in accordance with accepted engineering practice. The extent of such design need only demonstrate compliance of nonconventional elements with other applicable provisions and shall be compatible with the performance of the conventional framed system. Engineered design in accordance with the *International Building Code* is permitted for buildings and structures, and parts thereof, included in the scope of this code.

R301.1.4 Intermodal shipping containers. Intermodal shipping containers that are repurposed for use as buildings or structures shall be designed in accordance with the structural provisions in Section 3115 of the *International Building Code.*

R301.2 Climatic and geographic design criteria. Buildings shall be constructed in accordance with the provisions of this code as limited by the provisions of this section. Additional criteria shall be established by the local *jurisdiction* and set forth in Table R301.2.

R301.2.1 Wind design criteria. Buildings and portions thereof shall be constructed in accordance with the wind provisions of this code using the ultimate design wind speed in Table R301.2 as determined from Figure R301.2(2). The structural provisions of this code for wind loads are not permitted where wind design is required as specified in Section R301.2.1.1. Where different construction methods and structural materials are used for various portions of a building, the applicable requirements of this section for each portion shall apply. Where not otherwise specified, the wind loads listed in Table R301.2.1(1) adjusted for height and exposure using Table R301.2.1(2) shall be used to determine design load performance requirements for wall coverings, curtain walls, roof coverings, exterior windows, skylights, garage doors and exterior doors. Asphalt shingles shall be designed for wind speeds in accordance with Section R905.2.4. *Metal roof shingles* shall be designed for wind speeds in accordance with Section R905.4.4. A continuous load path shall be provided to transmit the applicable uplift forces in Section R802.11 from the *roof assembly* to the foundation. Where ultimate design wind speeds in Figure R301.2(2) are less than the lowest wind speed indicated in the prescriptive provisions of this code, the lowest wind speed indicated in the prescriptive provisions of this code shall be used.

R301.2.1.1 Wind limitations and wind design required. The wind provisions of this code shall not apply to the design of buildings where wind design is required in accordance with Figure R301.2.1.1, or where the ultimate design wind speed, V_{ult}, in Figure R301.2(2) equals or exceeds 140 miles per hour (225 kph) in a special wind region.

Exceptions:

1. For concrete construction, the wind provisions of this code shall apply in accordance with the limitations of Sections R404 and R608.
2. For structural insulated panels, the wind provisions of this code shall apply in accordance with the limitations of Section R610.
3. For cold-formed steel *light-frame construction*, the wind provisions of this code shall apply in accordance with the limitations of Sections R505, R603 and R804.

In regions where wind design is required in accordance with Figure R301.2.1.1 or where the ultimate design wind speed, V_{ult}, in Figure R301.2(2) equals or exceeds 140 miles per hour (225 kph) in a special wind region, the design of buildings for wind loads shall be in accordance with one or more of the following methods:

1. AWC *Wood Frame Construction Manual* (WFCM).
2. ICC *Standard for Residential Construction in High-Wind Regions* (ICC 600).
3. ASCE *Minimum Design Loads for Buildings and Other Structures* (ASCE 7).
4. AISI *Standard for Cold-Formed Steel Framing—Prescriptive Method for One- and Two-Family Dwellings* (AISI S230).
5. *International Building Code.*

The elements of design not addressed by the methods in Items 1 through 5 shall be in accordance with the provisions of this code.

Where ASCE 7 or the *International Building Code* is used for the design of the building, the wind speed map and exposure category requirements as specified in ASCE 7 and the *International Building Code* shall be used.

R301.2.1.1.1 Sunrooms. *Sunrooms* shall comply with AAMA/NPEA/NSA 2100. For the purpose of applying the criteria of AAMA/NPEA/NSA 2100 based on the intended use, *sunrooms* shall be identified as one of the following categories by the permit applicant, *design professional* or the property *owner* or owner's agent in the *construction documents.* Component and cladding pressures shall be used for the design of elements that do not qualify as main windforce-resisting systems. Main windforce-resisting system pressures shall be used for the design of elements assigned to provide support and stability for the overall *sunroom.*

Category I: A thermally isolated *sunroom* with walls that are open or enclosed with insect screening or 0.5 mm (20 mil) maximum thickness plastic film. The space is nonhabitable and unconditioned.

Category II: A thermally isolated *sunroom* with enclosed walls. The openings are enclosed with translucent or transparent plastic or glass. The space is nonhabitable and unconditioned.

Category III: A thermally isolated *sunroom* with enclosed walls. The openings are enclosed with translucent or transparent plastic or glass. The *sunroom* fenestration complies with additional requirements for air infiltration resistance and water penetration resistance. The space is nonhabitable and unconditioned.

Category IV: A thermally isolated *sunroom* with enclosed walls. The *sunroom* is designed to be heated or cooled by a separate temperature control or system and is thermally isolated from the primary structure. The *sunroom* fenestration complies with additional requirements for water penetration resistance, air infiltration resistance and thermal performance. The space is nonhabitable and conditioned.

Category V: A *sunroom* with enclosed walls. The *sunroom* is designed to be heated or cooled and is open to the main structure. The *sunroom* fenestration complies with additional requirements for water penetration resistance, air infiltration resistance and thermal performance. The space is habitable and conditioned.

R301.2.1.2 Protection of openings. Exterior glazing in buildings located in *windborne debris regions* shall be protected from windborne debris. Glazed opening protection for windborne debris shall meet the requirements of the Large Missile Test of ASTM E1886 and ASTM E1996 as modified in Section 301.2.1.2.1. Garage door glazed opening protection for windborne debris shall meet the requirements of an *approved* impact-resisting standard or ANSI/DASMA 115.

Exception: *Wood structural panels* with a thickness of not less than $^7/_{16}$ inch (11 mm) and a span of not more than 8 feet (2438 mm) shall be permitted for opening protection. Panels shall be precut and attached to the framing surrounding the opening containing the product with the glazed opening. Panels shall be predrilled as required for the anchorage method and shall be secured with the attachment hardware provided. Attachments shall be designed to resist the component and cladding loads determined in accordance with either Table R301.2.1(1) or ASCE 7, with the permanent corrosion-resistant attachment hardware provided and anchors permanently installed on the building. Attachment in accordance with Table R301.2.1.2 is permitted for buildings with a mean roof height of 45 feet (13 728 mm) or less where the ultimate design wind speed, V_{ult}, is 180 mph (290 kph) or less.

TABLE R301.2 CLIMATIC AND GEOGRAPHIC DESIGN CRITERIA

GROUND SNOW LOAD[o]	WIND DESIGN				SEISMIC DESIGN CATEGORY[f]	SUBJECT TO DAMAGE FROM			ICE BARRIER UNDERLAYMENT REQUIRED[h]	FLOOD HAZARDS[g]	AIR FREEZING INDEX[i]	MEANANNU-ALTEMP[j]
	Speed[d] (mph)	Topographic effects[k]	Special wind region[l]	Windborne debris zone[m]		Weathering[a]	Frost line depth[b]	Termite[c]				
—	—	—	—	—	—	—	—	—	—	—	—	—
MANUAL J DESIGN CRITERIA[n]												
Elevation			Altitude correction factor[e]	Coincident wet bulb	Indoor winter design dry-bulb temperature	Indoor winter design dry-bulb temperature			Outdoor winter design dry-bulb temperature		Heating temperature difference	
—			—	—	—	—			—		—	
Latitude			Daily range	Indoor summer design relative humidity	Indoor summer design relative humidity	Indoor summer design dry-bulb temperature			Outdoor summer design dry-bulb temperature		Cooling temperature difference	
—			—	—	—	—			—		—	

For SI: 1 pound per square foot = 0.0479 kPa, 1 mile per hour = 0.447 m/s.

a. Where weathering requires a higher strength concrete or grade of masonry than necessary to satisfy the structural requirements of this code, the frost line depth strength required for weathering shall govern. The weathering column shall be filled in with the weathering index, "negligible," "moderate" or "severe" for concrete as determined from Figure R301.2(1). The grade of masonry units shall be determined from ASTM C34, ASTM C55, ASTM C62, ASTM C73, ASTM C90, ASTM C129, ASTM C145, ASTM C216 or ASTM C652.

b. Where the frost line depth requires deeper footings than indicated in Figure R403.1(1), the frost line depth strength required for weathering shall govern. The jurisdiction shall fill in the frost line depth column with the minimum depth of footing below finish grade.

c. The jurisdiction shall fill in this part of the table to indicate the need for protection depending on whether there has been a history of local subterranean termite damage.

d. The jurisdiction shall fill in this part of the table with the wind speed from the basic wind speed map [Figure R301.2(2). Wind exposure category shall be determined on a site-specific basis in accordance with Section R301.2.1.4.

e. The jurisdiction shall fill in this section of the table to establish the design criteria using Table 10A from ACCA Manual J or established criteria determined by the jurisdiction.

f. The jurisdiction shall fill in this part of the table with the seismic design category determined from Section R301.2.2.1.

g. The jurisdiction shall fill in this part of the table with: the date of the jurisdiction's entry into the National Flood Insurance Program (date of adoption of the first code or ordinance for management of flood hazard areas); and the title and date of the currently effective Flood Insurance Study or other flood hazard study

h. In accordance with Sections R905.1.2, R905.4.3.1, R905.5.3.1, R905.6.3.1, R905.7.3.1 and R905.8.3.1, where there has been a history of local damage from the effects of ice damming, the jurisdiction shall fill in this part of the table with "YES." Otherwise, the jurisdiction shall fill in this part of the table with "NO."

i. The jurisdiction shall fill in this part of the table with the 100-year return period air freezing index (BF-days) from Figure R403.3(2) or from the 100-year (99 percent) value on the National Climatic Data Center data table "Air Freezing Index-USA Method (Base 32°F)."

j. The jurisdiction shall fill in this part of the table with the mean annual temperature from the National Climatic Data Center data table "Air Freezing Index-USA Method (Base 32°F)."

k. In accordance with Section R301.2.1.5, where there is local historical data documenting structural damage to buildings due to topographic wind speed-up effects, the jurisdiction shall fill in this part of the table with "YES." Otherwise, the jurisdiction shall indicate "NO" in this part of the table.

l. In accordance with Figure R301.2(2), where there is local historical data documenting unusual wind conditions, the jurisdiction shall fill in this part of the table with "YES" and identify any specific requirements. Otherwise, the jurisdiction shall indicate "NO" in this part of the table.

m. In accordance with Section R301.2.1.2 the jurisdiction shall indicate the wind-borne debris wind zone(s). Otherwise, the jurisdiction shall indicate "NO" in this part of the table.

n. The jurisdiction shall fill in these sections of the table to establish the design criteria using Table 1a or 1b from ACCA Manual J or established criteria determined by the jurisdiction.

o. The jurisdiction shall fill in this section of the table using the Ground Snow Loads in Figures R301.2(3) and R301.2(4).

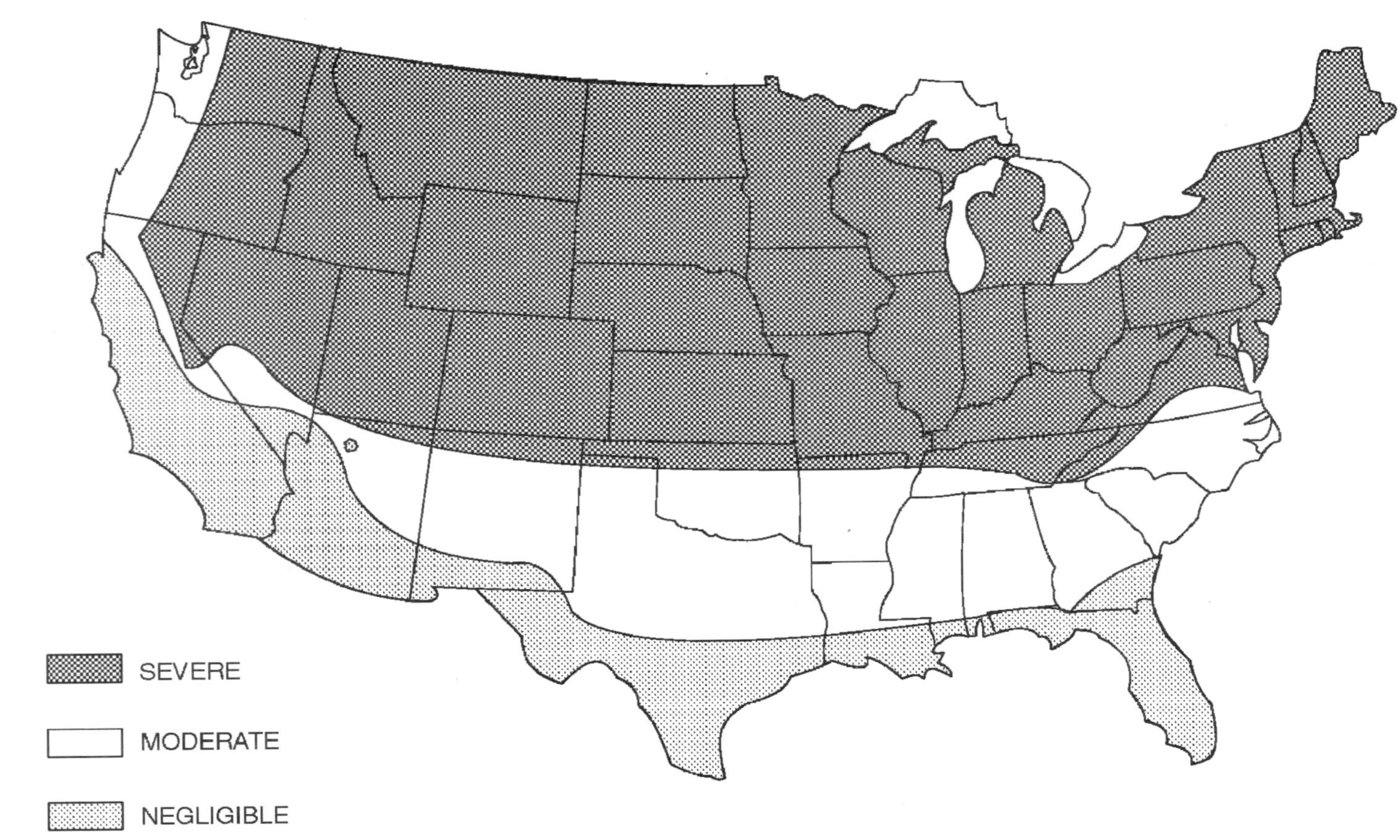

a. Alaska and Hawaii are classified as severe and negligible, respectively.

b. Lines defining areas are approximate only. Local conditions may be more or less severe than indicated by region classification. A sever classification is where weather conditions result in significant snowfall combined with extended periods during which there is little or no natural thawing, causing deicing salts to be used extensively.

FIGURE R301.2(1)
WEATHERING PROBABILITY MAP FOR CONCRETE[a, b]

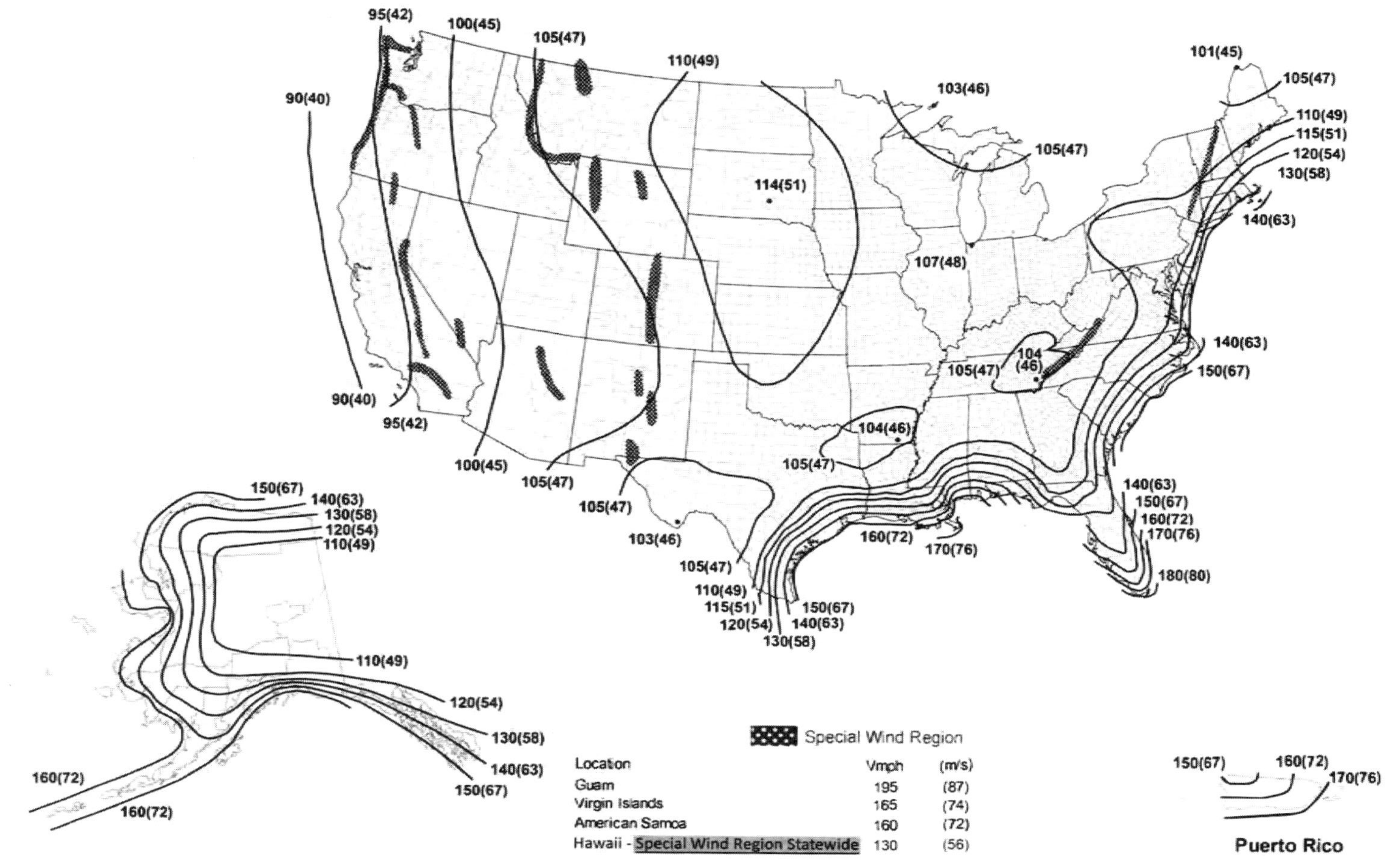

Notes:

1. Values are nominal design 3-second gust wind speeds in miles per hour (m/s) at 33 ft (10m) above ground for Exposure C category.
2. Linear interpolation is permitted between contours. Point values are provided to aid with interpolation.
3. Islands, coastal areas, and land boundaries outside the last contour shall use the last wind speed contour.
4. Mountainous terrain, gorges, ocean promontories, and special wind regions shall be examined for unusual wind conditions.
5. Wind speeds correspond to approximately a 7% probability of exceedance in 50 years (Annual Exceedance Probability = 0.00143, MRI = 700 Years).
6. Location-specific basic wind speeds shall be permitted to be determined using www.atcouncil.org/windspeed

FIGURE R301.2(2)
ULTIMATE DESIGN WIND SPEEDS

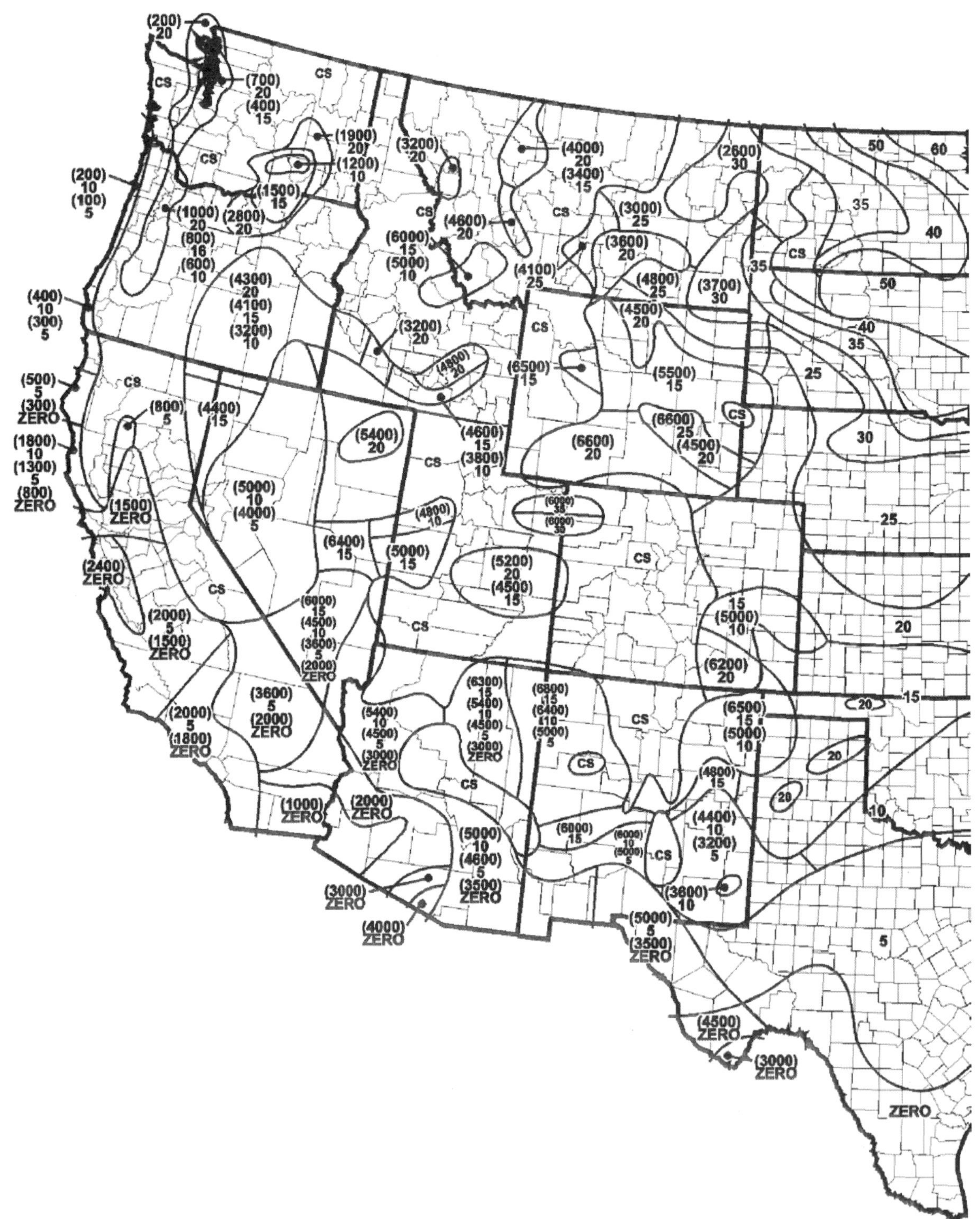

For SI:1 foot = 34.8 mm, 1 pound per square foot =0.0479 kPa, 1 mile =1.61 km.

a. In CS areas, site-specific case studies are required to establish ground snow loads. Extreme local variations in ground snow loads in these areas preclude mapping at this scale.

b. Numbers in parentheses represent the upper elevation limits in feet for the ground snow load values presented below. Site-specific case studies are required to establish ground snow loads at elevations not covered.

FIGURE R301.2(3)
GROUND SNOW LOADS, P_g, FOR THE UNITED STATES (lb/ft^2)

For SI: 1 foot = 304.8 mm, 1 pound per square foot = 0.0479 kPa, 1 mile = 1.61 km.

a. In CS areas, site-specific case studies are required to establish ground snow loads. Extreme local variations in ground snow loads in these areas preclude mapping at this scale.

b. Numbers in parentheses represent the upper elevation limits in feet for the ground snow load values presented below. Site-specific case studies are required to establish ground snow loads at elevations not covered.

FIGURE R301.2(4)
GROUND SNOW LOADS, P_g, FOR THE UNITED STATES (lb/ft^2)

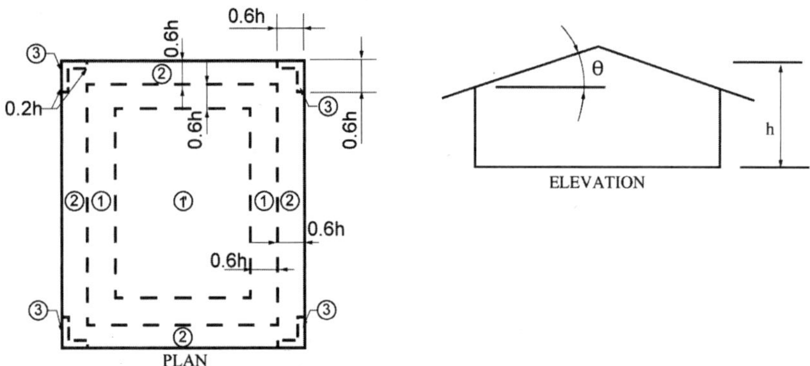

Gable and Flat Roofs $\theta \leq 7°$

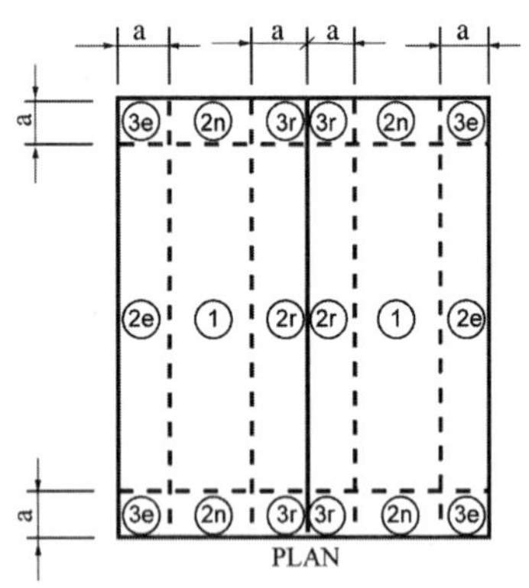

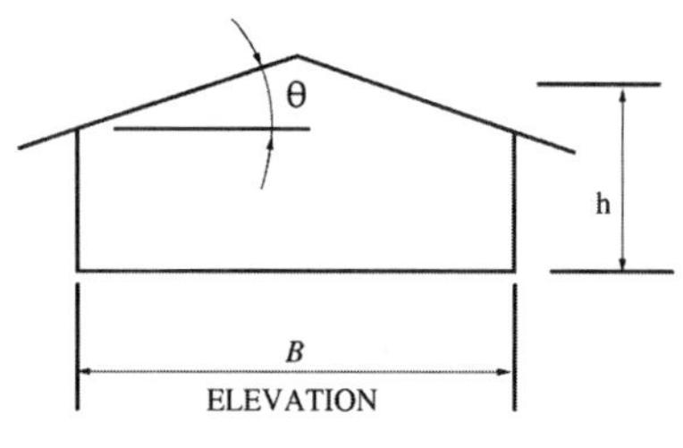

Gable and Flat Roofs $7° < \theta \leq 45°$

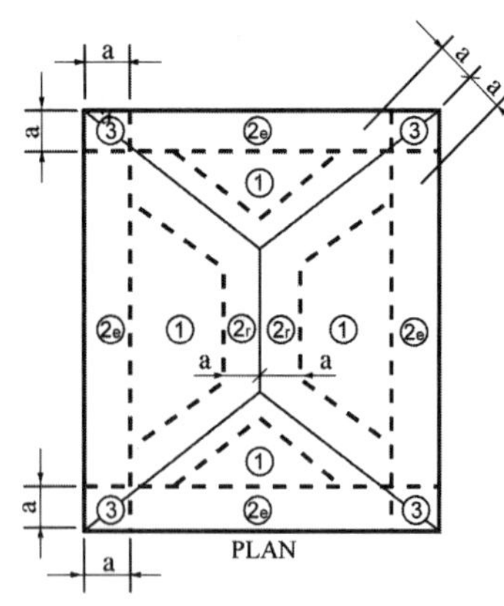

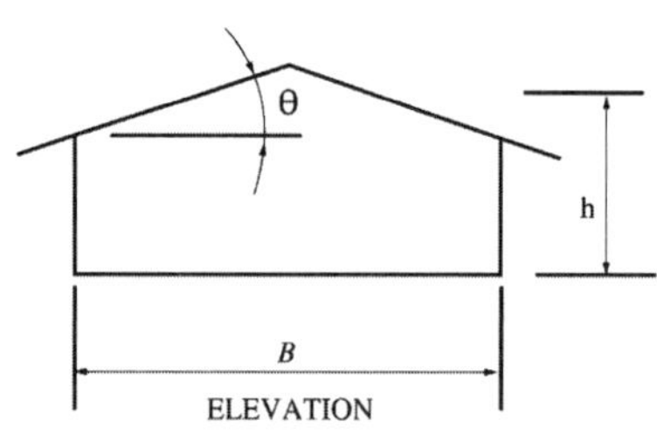

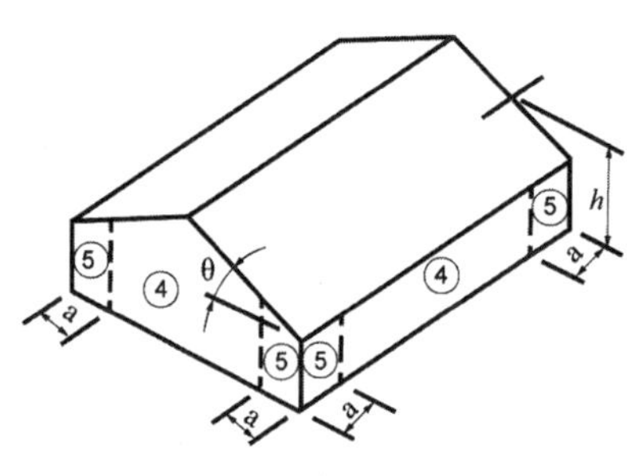

Hip Roofs $7° < \theta \leq 45°$

Walls

For SI: 1 foot = 304.8 mm, 1 degree = 0.0175 rad.
Note: a = 4 feet in all cases.

FIGURE R301.2.1
COMPONENT AND CLADDING PRESSURE ZONES

TABLE R301.2.1(1) COMPONENT AND CLADDING LOADS FOR A BUILDING WITH A MEAN ROOF HEIGHT OF 30 FEET LOCATED IN EXPOSURE B (ASD) (psf)[a, b, c, d, e, f, g]

	ZONE	EFFECTIVE WIND AREAS (square feet)	ULTIMATE DESIGN WIND SPEED, V_{ult}																									
			90.0		95.0		105.0		115.0		130.0		150.0		170.0		95.0		105.0		115.0		130.0		150.0		170.0	
			Pos	Neg	Pos	Neg	Pos	Neg	Pos	Neg	Pos	Neg	Pos	Neg	Pos	Neg	Pos	Neg	Pos	Neg	Pos	Neg	Pos	Neg	Pos	Neg	Pos	Neg
Flat and gable roof 0 to 7 degrees	1	10.0	3.6	-13.9	4.0	-15.5	4.4	-17.2	4.8	-19.0	5.3	-20.8	5.8	-22.7	6.3	-24.8	7.4	-29.1	8.6	-33.7	9.9	-38.7	11.2	-44.0	12.7	-49.7	14.2	-55.7
	1	20.0	3.3	-13.0	3.7	-14.5	4.1	-16.0	4.5	-17.7	5.0	-19.4	5.4	-21.2	5.9	-23.1	7.0	-27.1	8.1	-31.4	9.3	-36.1	10.5	-41.1	11.9	-46.4	13.3	-52.0
	1	50.0	3.0	-11.8	3.4	-13.1	3.8	-14.5	4.1	-16.0	4.5	-17.6	5.0	-19.2	5.4	-20.9	6.3	-24.5	7.4	-28.4	8.4	-32.6	9.6	-37.1	10.8	-41.9	12.2	-47.0
	1	100.0	2.8	-10.8	3.1	-12.1	3.5	-13.4	3.8	-14.7	4.2	-16.2	4.6	-17.7	5.0	-19.2	5.9	-22.6	6.8	-26.2	7.8	-30.0	8.9	-34.2	10.0	-38.6	11.3	-43.3
	2	10.0	3.6	-18.4	4.0	-20.5	4.4	-22.7	4.8	-25.0	5.3	-27.4	5.8	-30.0	6.3	-32.7	7.4	-38.3	8.6	-44.5	9.9	-51.0	11.2	-28.1	12.7	-65.6	14.2	-73.5
	2	20.0	3.3	-17.2	3.7	-19.2	4.1	-21.2	4.5	-23.4	5.0	-25.7	5.4	-28.1	5.9	-30.6	7.0	-35.9	8.1	-41.6	9.3	-47.8	10.5	-54.3	11.9	-61.4	13.3	-68.8
	2	50.0	3.0	-15.6	3.4	-17.4	3.8	-19.3	4.1	-21.3	4.5	-23.3	5.0	-25.5	5.4	-27.8	6.3	-32.6	7.4	-37.8	8.4	-43.4	9.6	-49.4	10.8	-55.8	12.2	-62.5
	2	100.0	2.8	-14.4	3.1	-16.1	3.5	-17.8	3.8	-19.7	4.2	-21.6	4.6	-23.6	5.0	-25.7	5.9	-30.1	6.8	-35.0	7.8	-40.1	8.9	-45.7	10.0	-51.5	11.3	-57.8
	3	10.0	3.6	-25.0	4.0	-27.9	4.4	-30.9	4.8	-34.1	5.3	-37.4	5.8	-40.9	6.3	-44.5	7.4	-52.2	8.6	-60.6	9.9	-69.6	11.2	-79.1	12.7	-89.4	14.2	-100.2
	3	20.0	3.3	-22.6	3.7	-25.2	4.1	-28.0	4.5	-30.8	5.0	-33.8	5.4	-37.0	5.9	-40.3	7.0	-47.2	8.1	-54.8	9.3	-62.9	10.5	-71.6	11.9	-80.8	13.3	-90.6
	3	50.0	3.0	-19.4	3.4	-21.7	3.8	-24.0	4.1	-26.5	4.5	-29.0	5.0	-31.7	5.4	-34.6	6.3	-40.6	7.4	-47.0	8.4	-54.0	9.6	-61.4	10.8	-69.4	12.2	-77.8
	3	100.0	2.8	-17.4	3.1	-19.0	3.5	-21.0	3.8	-23.2	4.2	-25.5	4.6	-27.8	5.0	-30.3	5.9	-35.6	6.8	-41.2	7.8	-47.3	8.9	-53.9	10.0	-60.8	11.3	-68.2
Gable roof > 7 to 20 degrees	1, 2e	10.0	5.4	-16.2	6.0	-18.0	6.7	-19.9	7.4	-22.0	8.1	-24.1	8.8	-26.4	9.6	-28.7	11.3	-33.7	13.1	39.1	15.0	44.9	17.1	-51.0	19.3	-57.6	21.6	-64.6
	1, 2e	20.0	4.9	-16.2	5.4	-18.0	6.0	-19.9	6.6	-22.0	7.2	-24.1	7.9	-26.4	8.6	-28.7	10.1	-33.7	11.7	-39.1	13.5	-44.9	15.3	-51.0	17.3	-57.6	19.4	-64.6
	1, 2e	50.0	4.1	-9.9	4.6	-11.0	5.1	-12.2	5.6	-13.4	6.1	-14.7	6.7	-16.1	7.3	-17.5	8.6	-20.6	10.0	-23.8	11.4	-27.4	13.0	-31.1	14.7	-35.2	16.4	-39.4
	1, 2e	100.0	3.6	-5.0	4.0	-5.6	4.4	-6.2	4.8	-6.9	5.3	-7.5	5.8	-8.2	6.3	-9.0	7.4	-10.5	8.6	-12.2	9.9	-14.0	11.2	-15.9	12.7	-18.0	14.2	-20.2
	2n, 2r, 3e	10.0	5.4	-23.6	6.0	-26.3	6.7	-29.1	7.4	-32.1	8.1	-35.2	8.8	-38.5	9.6	-41.9	11.3	-49.2	13.1	-57.0	15.0	-65.4	17.1	-74.5	19.3	-84.1	21.6	-94.2
	2n, 2r, 3e	20.0	4.9	-20.3	5.4	-22.7	6.0	-25.1	6.6	-27.7	7.2	-30.4	7.9	-33.2	8.6	-36.2	10.1	-42.4	11.7	-49.2	13.5	-56.5	15.3	-64.3	17.3	-72.6	19.4	-81.4
	2n, 2r, 3e	50.0	4.1	-16.0	4.6	-17.9	5.1	-19.8	5.6	-21.8	6.1	-24.0	6.7	-26.2	7.3	-28.5	8.6	-33.5	10.0	-38.8	11.4	-44.6	13.0	-50.7	14.7	-57.2	16.4	-64.2
	2n, 2r, 3e	100.0	3.6	-12.8	4.0	-14.3	4.4	-15.8	4.8	-17.4	5.3	-19.1	5.8	-20.9	6.3	-22.8	7.4	-26.7	8.6	-31.0	9.9	-35.6	11.2	-40.5	12.7	-45.7	14.2	-51.3
	3r	10.0	5.4	-28.0	6.0	-30.2	6.7	-34.6	7.4	-38.1	8.1	-41.8	8.8	-45.7	9.6	-49.8	11.3	-58.4	13.1	-67.8	15.0	-77.8	17.1	-88.5	19.3	-99.9	21.6	-112.0
	3r	20.0	4.9	-24.0	5.4	-26.7	6.0	-29.6	6.6	-32.7	7.2	-35.9	7.9	-39.2	8.6	-42.7	10.1	-50.1	11.7	-58.1	13.5	-66.7	15.3	-75.9	17.3	-85.6	19.4	-96.0
	3r	50.0	4.1	-18.7	4.6	-20.8	5.1	-23.1	5.6	-25.4	6.1	-27.9	6.7	-30.5	7.3	-33.2	8.6	-39.0	10.0	-45.2	11.4	-51.9	13.0	-59.0	14.7	-66.6	16.4	-74.7
	3r	100.0	3.6	-14.7	4.0	-16.3	4.4	-18.1	4.8	-20.0	5.3	-21.9	5.8	-24.0	6.3	-26.1	7.4	-30.6	8.6	-35.5	9.9	-40.8	11.2	-46.4	12.7	-52.3	14.2	-58.7
Gable roof > 20 to 27 degrees	1, 2e	10.0	6.5	-12.4	7.3	-13.9	8.0	-15.4	8.9	-16.9	9.7	-18.6	10.6	-20.3	11.6	-22.1	13.6	-26.0	15.8	-30.1	18.1	-34.6	20.6	-39.3	23.3	-44.4	26.1	-49.9
	1, 2e	20.0	5.6	-12.4	6.3	-13.9	7.0	-15.4	7.7	-16.9	8.4	-18.6	9.2	-20.3	10.0	-22.1	11.7	-26.0	13.6	-30.1	15.6	-34.6	17.8	-39.3	20.1	-44.4	22.5	-49.8
	1, 2e	50.0	4.4	-10.6	5.0	-11.8	5.5	-3.1	6.1	-14.4	6.6	-15.8	7.3	-17.3	7.9	-18.8	9.3	-22.1	10.8	-25.6	12.3	-29.4	14.0	-33.5	15.9	-37.8	17.8	-42.4
	1, 2e	100.0	3.6	-9.1	4.0	-10.2	4.4	-11.3	4.8	-12.4	5.3	-13.6	5.8	-14.9	6.3	-16.2	7.4	-19.0	8.6	-22.1	9.9	-25.3	11.2	-28.8	12.7	-32.5	14.2	-36.5
	2n, 2r, 3e	10.0	6.5	-19.9	7.3	-22.1	8.0	-24.5	8.9	-27.0	9.7	-29.7	10.6	-32.4	11.6	-35.3	13.6	-41.4	15.8	-48.0	18.1	-55.2	20.6	-62.8	23.3	-70.8	26.1	-79.4
	2n, 2r, 3e	20.0	5.6	-17.4	6.3	-19.4	7.0	-21.5	7.7	-23.7	8.4	-26.0	9.2	-28.4	10.0	-31.0	11.7	-36.3	13.6	-42.1	15.6	-48.4	17.8	-55.0	20.1	-62.1	22.5	-69.6
	2n, 2r, 3e	50.0	4.4	-14.2	5.0	-15.8	5.5	-17.5	6.1	-19.3	6.6	-21.1	7.3	-23.1	7.9	-25.2	9.3	-29.5	10.8	-34.2	12.3	-39.3	14.0	-44.7	15.9	-50.5	17.8	-56.6
	2n, 2r, 3e	100.0	3.6	-11.7	4.0	-13.0	4.4	-14.5	4.8	-15.9	5.3	-17.5	5.8	-19.1	6.3	-20.8	7.4	-24.4	8.6	-28.3	9.9	-32.5	11.2	-37.0	12.7	-41.8	14.2	-46.8
	3r	10.0	6.5	-23.6	7.3	-26.3	8.0	-29.1	8.9	-32.1	9.7	-35.2	10.6	-38.5	11.6	-41.9	13.6	-49.2	15.8	-57.0	18.1	-65.4	20.6	-74.5	23.3	-84.1	26.1	-94.2
	3r	20.0	5.6	-19.9	6.3	-22.1	7.0	-24.5	7.7	-27.0	8.4	-29.7	9.2	-32.4	10.0	-35.3	11.7	-41.4	13.6	-48.0	15.6	-55.2	17.8	-62.8	20.1	-70.8	22.5	-79.4
	3r	50.0	4.4	-14.7	5.0	-16.3	5.5	-18.1	6.1	-20.0	6.6	-21.9	7.3	-24.0	7.9	-26.1	9.3	-30.6	10.8	-35.5	12.3	-40.8	14.0	-46.4	15.9	-52.3	17.8	-58.7
	3r	100.0	3.6	-14.7	4.0	-16.3	4.4	-18.1	4.8	-20.0	5.3	-21.9	5.8	-24.0	6.3	-26.1	7.4	-30.6	8.6	-35.5	9.9	-40.8	11.2	-46.4	12.7	-52.3	14.2	-58.7

(continued)

TABLE R301.2.1(1)—continued
COMPONENT AND CLADDING LOADS FOR A BUILDING WITH A MEAN ROOF HEIGHT OF 30 FEET LOCATED IN EXPOSURE B (ASD) (psf) [a, b, c, d, e, f, g]

	ZONE	EFFECTIVE WIND AREAS (square feet)	ULTIMATE DESIGN WIND SPEED, V_{ult}																									
			90.0		95.0		105.0		115.0		130.0		150.0		170.0		95.0		105.0		115.0		130.0		150.0		170.0	
			Pos	Neg	Pos	Neg	Pos	Neg	Pos	Neg	Pos	Neg	Pos	Neg	Pos	Neg	Pos	Neg	Pos	Neg	Pos	Neg	Pos	Neg	Pos	Neg	Pos	Neg
Gable roof > 27 to 45 degrees	1, 2e, 2r	10.0	8.0	-14.7	8.9	-16.3	9.9	-18.1	10.9	-20.0	12.0	-21.9	13.1	-24.0	14.2	-26.1	16.7	-30.6	19.4	-35.5	22.2	-40.8	25.3	-16.4	28.5	-52.3	32.0	-58.7
	1, 2e, 2r	20.0	7.1	-12.4	7.9	-13.9	8.8	-15.4	9.7	-16.9	10.6	-18.6	11.6	-20.3	12.6	-22.1	14.8	-26.0	17.2	-30.1	19.8	-34.6	22.5	-39.3	25.4	-44.4	28.5	-49.8
	1, 2e, 2r	50.0	5.9	-9.5	6.6	-10.6	7.3	-11.7	8.1	-12.9	8.9	-14.2	9.7	-15.5	10.5	-16.9	12.4	-19.8	14.3	-22.9	16.5	-26.3	18.7	-30.0	21.1	-33.8	23.7	-37.9
	1, 2e, 2r	100.0	5.0	-7.3	5.6	-8.1	6.2	-9.0	6.9	-9.9	7.5	-10.8	8.2	-11.9	9.0	-12.9	10.5	-15.1	12.2	-17.6	14.0	-20.2	15.9	-22.9	18.0	-25.9	20.2	-29.0
	2n, 3r	10.0	8.0	-16.2	8.9	-18.0	9.9	-19.9	10.9	-22.0	12.0	-24.1	13.1	-26.4	-14.2	-28.7	16.7	-33.7	19.4	-39.1	22.2	-44.9	25.3	-51.0	28.5	-57.6	32.0	-64.6
	2n, 3r	20.0	7.1	-14.4	7.9	-16.1	8.8	-17.8	9.7	-19.7	10.6	-21.6	11.6	-23.6	12.6	-25.7	14.8	-30.1	17.2	-34.9	19.8	-40.1	22.5	-45.6	25.4	-51.5	28.5	-57.8
	2n, 3r	50.0	5.9	-12.2	6.6	-13.5	7.3	-15.0	8.1	-16.5	8.9	-18.2	9.7	-19.9	10.5	-21.6	12.4	-25.4	14.3	-29.4	16.5	-33.8	18.7	-38.4	21.1	-43.4	23.7	-48.6
	2n, 3r	100.0	5.0	-10.4	5.6	-11.6	6.2	-12.9	6.9	-14.2	7.5	-15.6	8.2	-17.1	9.0	-18.6	10.5	-21.8	12.2	-25.3	14.0	-29.0	15.9	-33.0	18.0	-37.3	20.0	-41.8
	3e	10.0	8.0	-19.9	8.9	-22.1	9.9	-24.5	10.9	-27.0	12.0	-29.7	13.1	-32.4	14.2	-35.3	16.7	-41.4	19.4	-48.0	22.2	-55.2	25.3	-62.8	28.8	-70.8	32.0	-79.4
	3e	20.0	7.1	-17.6	7.9	-19.6	8.8	-21.8	9.7	-24.0	10.6	-26.3	11.6	-28.8	12.6	-31.3	14.8	-36.8	17.2	-42.7	19.8	-49.0	22.5	-55.7	25.4	-62.9	28.5	-70.5
	3e	50.0	5.9	-14.7	6.6	-16.3	7.3	-18.1	8.1	-20.0	8.9	-21.9	9.7	-24.0	10.5	-26.1	12.4	-30.6	14.3	-35.5	16.6	-40.8	18.7	-46.4	21.1	-52.3	23.7	-58.7
	3e	100.0	5.0	-12.4	5.6	-13.9	6.2	-15.4	6.9	-16.9	7.5	-18.6	8.2	-20.3	9.0	-22.1	10.5	-26.0	12.2	-30.1	14.0	-34.6	15.9	-39.3	18.0	-44.4	20.2	-49.8
Hipped roof > 7 to 20 degrees[g]	1	10.0	6.5	-14.7	7.3	-16.3	8.0	-18.1	8.9	-20.0	9.7	-21.9	10.6	-24.0	11.6	-26.1	13.6	-30.6	15.8	-35.5	18.1	-40.8	20.6	-46.4	23.3	-52.3	26.1	-58.7
	1	20.0	5.6	-14.7	6.3	-16.3	7.0	-18.1	7.7	-20.0	8.4	-21.9	9.2	-24.0	10.0	-26.1	11.7	-30.6	13.6	-35.5	15.6	-40.8	17.8	-46.4	20.1	-52.3	22.5	-58.7
	1	50.0	4.4	-11.3	5.0	-12.6	5.5	-14.0	6.1	-15.4	6.6	-16.9	7.3	-18.5	7.9	-20.2	9.3	-23.7	10.8	-27.4	12.3	-31.5	14.0	-35.8	15.9	-40.4	17.8	-45.3
	1	100.0	3.6	-8.7	4.0	-9.7	4.4	-10.8	4.8	-11.9	5.3	-13.1	5.8	-14.3	6.3	-15.5	7.4	-18.2	8.6	-21.2	9.9	-24.3	11.2	-27.6	12.7	-31.2	14.2	-35.0
	2r	10.0	6.5	-19.1	7.3	-21.3	8.0	-23.6	8.9	-26.0	9.7	-28.6	10.6	-31.2	11.6	-34.0	13.6	-39.9	15.8	-46.3	18.1	-53.1	20.6	-60.4	23.3	-68.2	26.1	-76.5
	2r	20.0	5.6	-17.2	6.3	-19.2	7.0	-21.3	7.7	-23.4	8.4	-25.7	9.2	-28.1	10.0	-30.6	11.7	-35.9	13.6	-41.7	15.6	-47.9	17.8	-54.4	20.1	-61.5	22.5	-68.9
	2r	50.0	4.4	-14.7	5.0	-16.4	5.5	-18.2	6.1	-20.0	6.6	-22.0	7.3	-24.0	7.9	-26.1	9.3	-30.7	10.8	-35.6	12.3	-40.9	14.0	-46.5	15.9	-52.5	17.8	-58.8
	2r	100.0	3.6	-12.8	4.0	-14.3	4.4	-15.8	4.8	-17.4	5.3	-19.1	5.8	-20.9	6.3	-22.8	7.4	-26.7	8.6	-31.0	9.9	-35.6	11.2	-40.5	12.7	-45.7	14.2	-51.3
	2e, 3	10.0	6.5	-20.6	7.3	-22.9	8.0	-25.4	8.9	-28.0	9.7	-30.8	10.6	-33.6	11.6	-36.6	13.6	-43.0	15.8	-49.8	18.1	-57.2	20.6	-65.1	23.3	-73.5	26.1	-82.4
	2e, 3	20.0	5.6	-18.5	6.3	-20.6	7.0	-22.9	7.7	-25.2	8.4	-27.7	9.2	-30.3	10.0	-32.9	11.7	-38.7	13.6	-44.8	15.6	-51.5	17.8	-58.6	20.1	-66.1	22.5	-74.1
	2e, 3	50.0	4.4	-15.8	5.0	-17.6	5.5	-19.5	6.1	-21.5	6.6	-23.6	7.3	-25.8	7.9	-28.0	9.3	-32.9	10.8	-38.2	12.3	-43.8	14.0	-49.9	15.9	-56.3	17.8	-63.1
	2e, 3	100.0	3.6	-13.7	4.0	-15.3	4.0	-16.9	4.8	-18.7	5.3	-20.5	5.8	-22.4	6.3	-24.4	7.4	-28.6	8.6	-33.2	9.9	-38.1	11.2	-43.3	12.7	-48.9	14.2	-54.8
Hipped roof > 20 to 27 degrees	1	10.0	6.5	-11.7	7.3	-13.0	8.0	-14.5	8.9	-15.9	9.7	-17.5	10.6	-19.1	11.6	-20.8	13.6	-24.4	15.8	-28.3	18.1	-32.5	20.6	-37.0	23.3	-41.8	26.1	-46.8
	1	20.0	5.6	-10.4	6.3	-11.6	7.0	-12.8	7.7	-14.1	8.4	-15.5	9.2	-16.9	10.0	-18.4	11.7	-21.6	13.6	-25.1	15.6	-28.8	17.8	-32.8	20.1	-37.0	22.5	-41.5
	1	50.0	4.4	-8.6	5.0	-9.6	5.5	-10.6	6.1	-11.7	6.6	-12.8	7.3	-14.0	7.9	-15.3	9.3	-17.9	10.8	-20.8	12.3	-23.9	14.0	-27.2	15.9	-30.7	17.8	-34.4
	1	100.0	3.6	-7.3	4.0	-8.1	4.4	-9.0	4.8	-9.9	5.3	-10.8	5.8	-11.9	6.3	-12.9	7.4	-15.1	8.6	-17.6	9.9	-20.2	11.2	-22.9	12.7	-25.9	14.2	-29.0
	2e, 2r, 3	10.0	6.5	-16.2	7.3	-18.0	8.0	-19.9	8.9	-22.0	9.7	-24.1	10.6	-26.4	11.6	-28.7	13.6	-33.7	15.8	-39.1	18.1	-44.9	20.6	-51.0	23.3	-57.6	26.1	-64.6
	2e, 2r, 3	20.0	5.6	-14.4	6.3	-16.1	7.0	-17.8	7.7	-19.7	8.4	-21.6	9.2	-23.6	10.0	-25.7	11.7	-30.1	13.6	-34.9	15.6	-40.1	17.8	-45.6	20.1	-51.5	22.5	-57.8
	2e, 2r, 3	50.0	4.4	-12.2	5.0	-13.5	5.5	-15.0	6.1	-16.5	6.6	-18.2	7.3	-19.9	7.9	-21.6	9.3	-25.4	10.8	-29.4	12.3	-33.8	14.0	-38.4	15.9	-43.4	17.8	-48.6
	2e, 2r, 3	100.0	3.6	-10.4	4.0	-11.6	4.4	-12.9	4.8	-14.2	5.3	-15.6	5.8	-17.1	6.3	-18.6	7.4	-21.8	8.6	-25.3	9.9	-29.0	11.2	-33.0	12.7	-37.3	14.2	-41.8

(continued)

TABLE R301.2.1(1)—continued
COMPONENT AND CLADDING LOADS FOR A BUILDING WITH A MEAN ROOF HEIGHT OF 30 FEET LOCATED IN EXPOSURE B (ASD) (psf) [a, b, c, d, e, f, g]

	ZONE	EFFECTIVE WIND AREAS (square feet)	ULTIMATE DESIGN WIND SPEED, V_{ult}																									
			90.0		95.0		105.0		115.0		130.0		150.0		170.0		95.0		105.0		115.0		130.0		150.0		170.0	
			Pos	Neg	Pos	Neg	Pos	Neg	Pos	Neg	Pos	Neg	Pos	Neg	Pos	Neg	Pos	Neg	Pos	Neg	Pos	Neg	Pos	Neg	Pos	Neg	Pos	Neg
Hipped roof > 27 to 45 degrees	1	10.0	6.2	-12.4	6.9	-13.9	7.7	-15.4	8.5	-16.9	9.3	-18.6	10.2	-20.3	11.1	-22.1	13.0	-26.0	15.1	-30.1	17.3	-34.6	19.7	-39.3	22.2	-44.4	24.9	-49.8
	1	20.0	5.4	-11.0	6.0	-12.3	6.7	-13.6	7.4	-15.0	8.1	-16.5	8.9	-18.0	9.6	-19.6	11.3	-23.0	13.1	-26.7	15.1	-30.7	17.1	-34.9	19.4	-39.4	21.7	-44.2
	1	50.0	4.4	-9.2	4.9	-10.2	5.4	-11.3	5.9	-12.5	6.5	-13.7	7.1	-15.0	7.7	-16.3	9.1	-19.2	10.5	-22.2	12.1	-25.5	13.8	-29.0	15.5	-32.8	17.4	-36.7
	1	100.0	3.6	-7.8	4.0	-8.7	4.4	-9.6	4.8	-10.6	5.3	-11.6	5.8	-12.7	6.3	-13.8	7.4	-16.2	8.6	-18.8	9.9	-21.6	11.2	-24.6	12.7	-27.8	14.2	-31.1
	2e	10.0	6.2	-14.8	6.9	-16.5	7.7	-18.3	8.5	-20.2	9.3	-22.1	10.2	-24.2	11.1	-26.3	13.0	-30.9	15.1	-35.9	17.3	-41.2	19.7	-46.8	22.2	-52.9	24.9	-59.3
	2e	20.0	5.4	-11.7	6.0	-13.0	6.7	-14.5	7.4	-15.9	8.1	-17.5	8.9	-19.1	9.6	-20.8	11.3	-24.4	13.1	-28.3	15.1	-32.5	17.1	-37.0	19.4	-41.8	21.7	-46.8
	2e	50.0	4.4	-7.3	4.9	-8.1	5.4	-9.0	5.9	-9.9	6.5	-10.8	7.1	-11.9	7.7	-12.9	9.1	-15.1	10.5	-17.6	12.1	-20.2	13.8	-22.9	15.5	-25.9	17.4	-29.0
	2e	100.0	3.6	-7.3	4.0	-8.1	4.4	-9.0	4.8	-9.9	5.3	-10.8	5.8	-11.9	6.3	-12.9	7.4	-15.1	8.6	-17.6	9.9	-20.2	11.2	-22.9	12.7	-25.9	14.2	-29.0
	2r	10.0	6.2	-18.7	6.9	-20.9	7.7	-23.1	8.5	-25.5	9.3	-28.0	10.2	-30.6	11.1	-33.3	13.0	-39.1	15.1	-45.4	17.3	-52.1	19.7	-59.2	22.2	-66.9	24.9	-75.0
	2r	20.0	5.4	-15.7	6.0	-17.5	6.7	-19.4	7.4	-21.4	8.1	-23.5	8.9	-25.7	9.6	-28.0	11.3	-32.8	13.1	-38.1	15.1	-43.7	17.1	-49.8	19.4	-56.2	21.7	-63.0
	2r	50.0	4.4	-11.7	4.9	-13.1	5.4	-14.5	5.9	-16.0	6.5	-17.5	7.1	-19.2	7.7	-20.9	9.1	-24.5	10.5	-28.4	12.1	-32.6	13.8	-37.1	15.5	-41.9	17.4	-47.0
	2r	100.0	3.6	-8.7	4.0	-9.7	4.4	-10.8	4.8	-11.9	5.3	-13.1	5.8	-14.3	6.3	-15.5	7.4	-18.2	8.6	-21.2	9.9	-24.3	11.2	-27.6	12.7	-31.2	14.2	-35.0
	3	10.0	6.2	-20.0	6.9	-22.3	7.7	-24.7	8.5	-27.2	9.3	-29.9	10.2	-32.7	11.1	-35.6	13.0	-41.7	15.1	-48.4	17.3	-55.6	19.7	-63.2	22.2	-71.4	24.9	-80.0
	3	20.0	5.4	-15.0	6.0	-16.8	6.7	-18.6	7.4	-20.5	8.1	-22.5	8.9	-24.6	9.6	-26.7	11.3	-31.4	13.1	-36.4	15.1	-41.8	17.1	-47.5	19.4	-53.7	21.7	-60.2
	3	50.0	4.4	-8.7	4.9	-9.7	5.4	-10.8	5.9	-11.9	6.5	-13.1	7.1	-14.3	7.7	-15.5	9.1	-18.2	10.5	-21.2	12.1	-24.3	13.8	-27.6	15.5	-31.2	17.4	-35.0
	3	100.0	3.6	-8.7	4.0	-9.7	4.4	-10.8	4.8	-11.9	5.3	-13.1	5.8	-14.3	6.3	-15.5	7.4	-18.2	8.6	-21.2	9.9	-24.3	11.2	-27.6	12.7	-31.2	14.2	-35.0
Wall	4	10.0	8.7	-9.5	9.7	-10.6	10.8	-11.7	11.9	-12.9	13.1	-14.2	14.3	-15.5	15.5	-16.9	18.2	-19.8	21.2	-22.9	24.3	-26.3	27.6	-30.0	31.2	-33.8	35.0	-37.9
	4	20.0	8.3	-9.1	9.3	-10.1	10.3	-11.2	11.4	-12.4	12.5	-13.6	13.6	-14.8	14.8	-16.2	17.4	-19.0	20.2	-22.0	23.2	-25.3	26.4	-28.7	29.8	-32.4	33.4	-36.4
	4	50.0	7.8	-8.6	8.7	-9.5	9.7	-10.6	10.7	-11.7	11.7	-12.8	12.8	-14.0	13.9	-15.2	16.3	-17.9	18.9	-20.7	21.7	-23.8	24.7	-27.1	27.9	-30.6	31.3	-34.3
	4	100.0	7.4	-8.2	8.3	-9.1	9.2	-10.1	10.1	-11.1	11.1	-12.2	12.1	-13.3	13.2	-14.5	15.5	-17.1	18.0	-19.8	20.6	-22.7	23.5	-25.8	26.5	-29.2	29.7	-32.7
	4	500.0	6.5	-7.3	7.3	-8.1	8.0	-9.0	8.9	-9.9	9.7	-10.8	10.6	-11.9	11.6	-12.9	13.5	-15.1	15.8	-17.6	18.1	-20.2	20.6	-22.9	23.3	-25.9	26.1	-29.0
	5	10.0	8.7	-11.7	9.7	-13.0	10.8	-14.5	11.9	-15.9	13.1	-17.5	14.3	-19.1	15.5	-20.8	18.2	-24.4	21.2	-28.3	24.3	-32.5	27.6	-37.0	31.2	-41.8	35.0	-46.8
	5	20.0	8.3	-10.9	9.3	-12.2	10.3	-13.5	11.4	-14.9	12.5	-16.3	13.6	-17.8	14.8	-19.4	17.4	-22.8	20.2	-26.4	23.2	-30.3	26.4	-34.5	29.8	-39.0	33.4	-43.7
	5	50.0	7.8	-9.9	8.7	-11.0	9.7	-12.2	10.7	-13.4	11.7	-14.7	12.8	-16.1	13.9	-17.5	16.3	-20.6	18.9	-23.9	21.7	-27.4	24.7	-31.2	27.9	-35.2	31.3	-39.5
	5	100.0	7.4	-9.1	8.3	-10.1	9.2	-11.2	10.1	-12.4	11.1	-13.6	12.1	-14.8	13.2	-16.1	15.5	-19.0	18.0	-22.0	20.6	-25.2	23.5	-28.7	26.5	-32.4	29.7	-36.3
	5	500.0	6.5	-7.3	7.3	-8.1	8.0	-9.0	8.9	-9.9	9.7	-10.8	10.6	-11.9	11.6	-12.9	13.6	-15.1	15.8	-17.6	18.1	-20.2	20.6	-22.9	23.3	-25.9	26.1	-29.0

For SI: 1 foot = 304.8 mm, 1 square foot = 0.0929 m², 1 mile per hour = 0.447 m/s, 1 pound per square foot = 0.0479 kPa.

a. The effective wind area shall be equal to the span length multiplied by an effective width. This width shall be not less than one-third the span length. For cladding fasteners, the effective wind areas shall not be greater than the area that is tributary to an individual fastener.

b. For effective areas between those given, the load shall be interpolated or the load associated with the lower effective areas shall be used.

c. Table values shall be adjusted for height and exposure by multiplying by the adjustment coefficient in Table R301.2.1(2).

d. See Figure R318.4 for locations of termite infestation probability zones.

e. Plus and minus signs signify pressures acting toward and away from the building surfaces.

f. Positive and negative design wind pressures shall not be less than 10 psf.

g. Where the ratio of the building mean roof height to the building length or width is less than 0.8, uplift loads shall be permitted to be calculated in accordance with ASCE 7.

R301.2.1.2.1 Application of ASTM E1996. The text of Section 2.2 of ASTM E1996 shall be substituted as follows:

2.2 ASCE Standard:

ASCE 7-10 American Society of Civil Engineers *Minimum Design Loads for Buildings and Other Structures*

The text of Section 6.2.2 of ASTM E1996 shall be substituted as follows:

6.2.2 Unless otherwise specified, select the wind zone based on the ultimate design wind speed, V_{ult}, as follows:

6.2.2.1 Wind Zone 1–130 mph ≤ ultimate design wind speed, V_{ult} < 140 mph.

6.2.2.2 Wind Zone 2–140 mph ≤ ultimate design wind speed, V_{ult} < 150 mph at greater than 1 mile (1.6 km) from the coastline. The coastline shall be measured from the mean high-water mark.

6.2.2.3 Wind Zone 3–150 mph (67 m/s) ≤ ultimate design wind speed, V_{ult} ≤ 170 mph (76 m/s), or 140 mph (54 m/s) ≤ ultimate design wind speed, V_{ult} ≤ 170 mph (76 m/s) and within 1 mile (1.6 km) of the coastline. The coastline shall be measured from the mean high-water mark.

6.2.2.4 Wind Zone 4–ultimate design wind speed, V_{ult} > 170 mph (76 m/s).

R301.2.1.3 Wind speed conversion. Where referenced documents are based on nominal design wind speeds and do not provide the means for conversion between ultimate design wind speeds and nominal design wind speeds, the ultimate design wind speeds, V_{ult}, of Figure R301.2(2) shall be converted to nominal design wind speeds, V_{asd}, using Table R301.2.1.3.

TABLE R301.2.1(2)
HEIGHT AND EXPOSURE ADJUSTMENT COEFFICIENTS FOR Table R301.2.1(1)

MEAN ROOF HEIGHT	EXPOSURE		
	B	C	D
15	0.82	1.21	1.47
20	0.89	1.29	1.55
25	0.94	1.35	1.61
30	1.00	1.40	1.66
35	1.05	1.45	1.70
40	1.09	1.49	1.74
45	1.12	1.53	1.78
50	1.16	1.56	1.81
55	1.19	1.59	1.84
60	1.22	1.62	1.87

TABLE R301.2.1.2
WINDBORNE DEBRIS PROTECTION FASTENING SCHEDULE FOR WOOD STRUCTURAL PANELS[a, b, c, d]

FASTENER TYPE	FASTENER SPACING (inches)[a, b]		
	Panel span ≤ 4 feet	4 feet < panel span ≤ 6 feet	6 feet < panel span ≤ 8 feet
No. 8 wood-screw-based anchor with 2-inch embedment length	16	10	8
No. 10 wood-screw-based anchor with 2-inch embedment length	16	12	9
$^1/_4$-inch lag-screw-based anchor with 2-inch embedment length	16	16	16

For SI: 1 inch = 25.4 mm, 1 foot = 304.8 mm, 1 pound = 4.448 N, 1 mile per hour = 0.447 m/s.

a. This table is based on 180 mph ultimate design wind speeds, V_{ult}, and a 45-foot mean roof height.

b. Fasteners shall be installed at opposing ends of the wood structural panel. Fasteners shall be located not less than 1 inch from the edge of the panel.

c. Anchors shall penetrate through the exterior wall covering with an embedment length of not less than 2 inches into the building frame. Fasteners shall be located not less than $2^1/_2$ inches from the edge of concrete block or concrete.

d. Panels attached to masonry or masonry/stucco shall be attached using vibration-resistant anchors having an ultimate withdrawal capacity of not less than 1,500 pounds.

TABLE R301.2.1.3
WIND SPEED CONVERSIONS[a]

V_{ult}	110	115	120	130	140	150	160	170	180	190	200
V_{asd}	85	89	93	101	108	116	124	132	139	147	155

For SI: 1 mile per hour = 0.447 m/s.

a. Linear interpolation is permitted.

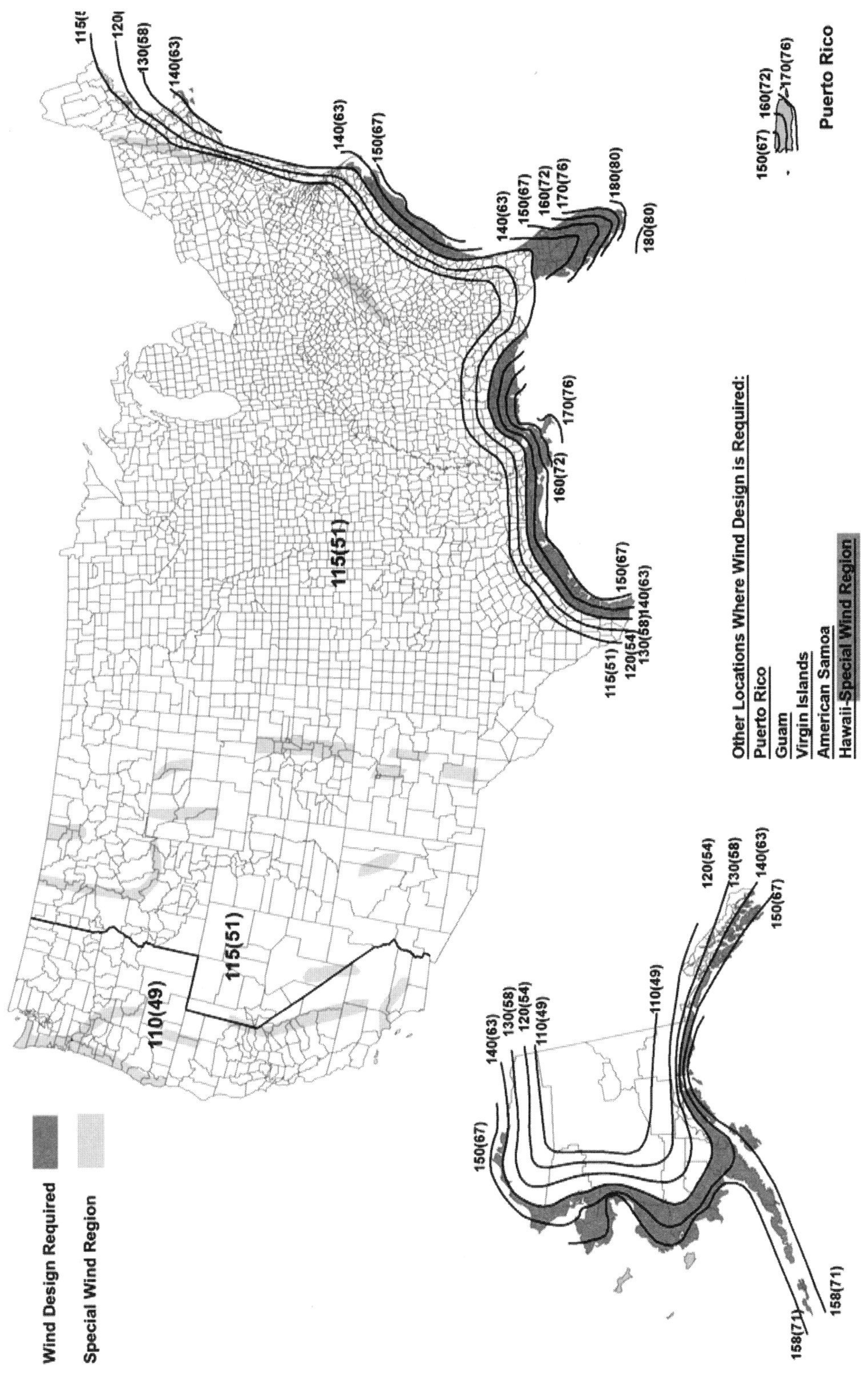

FIGURE R301.2.1.1
REGIONS WHERE WIND DESIGN IS REQUIRED

R301.2.1.4 Exposure category. For each wind direction considered, an exposure category that adequately reflects the characteristics of ground surface irregularities shall be determined for the site at which the building or structure is to be constructed. For a site located in the transition zone between categories, the category resulting in the largest wind forces shall apply. Account shall be taken of variations in ground surface roughness that arise from natural topography and vegetation as well as from constructed features. For a site where multiple detached one- and two-family *dwellings*, townhouses or other structures are to be constructed as part of a subdivision or master-planned community, or are otherwise designated as a developed area by the authority having *jurisdiction*, the exposure category for an individual structure shall be based on the site conditions that will exist at the time when all adjacent structures on the site have been constructed, provided that their construction is expected to begin within 1 year of the start of construction for the structure for which the exposure category is determined. For any given wind direction, the exposure in which a specific building or other structure is sited shall be assessed as being one of the following categories:

1. Exposure B. Urban and suburban areas, wooded areas or other terrain with numerous closely spaced obstructions having the size of single-family *dwellings* or larger. Exposure B shall be assumed unless the site meets the definition of another type exposure.
2. Exposure C. Open terrain with scattered obstructions, including surface undulations or other irregularities, having heights generally less than 30 feet (9144 mm) extending more than 1,500 feet (457 m) from the building site in any quadrant. This exposure shall apply to any building located within Exposure B type terrain where the building is directly adjacent to open areas of Exposure C type terrain in any quadrant for a distance of more than 600 feet (183 m). This category includes flat, open country and grasslands.
3. Exposure D. Flat, unobstructed areas exposed to wind flowing over open water, smooth mud flats, salt flats and unbroken ice for a distance of not less than 5,000 feet (1524 m). This exposure shall apply only to those buildings and other structures exposed to the wind coming from over the unobstructed area. Exposure D extends downwind from the edge of the unobstructed area a distance of 600 feet (183 m) or 20 times the height of the building or structure, whichever is greater.

R301.2.1.5 Topographic wind effects. In areas designated in Table R301.2 as having local historical data documenting structural damage to buildings caused by wind speed-up at isolated *hills*, ridges and escarpments that are abrupt changes from the general topography of the area, topographic wind effects shall be considered in the design of the building in accordance with Section R301.2.1.5.1 or in accordance with the provisions of ASCE 7. See Figure R301.2.1.5.1(1) for topographic features for wind speed-up effect.

In these designated areas, topographic wind effects shall apply only to buildings sited on the top half of an isolated *hill, ridge* or escarpment where all of the following conditions exist:

1. The average slope of the top half of the *hill, ridge* or escarpment is 10 percent or greater.
2. The *hill, ridge* or escarpment is 60 feet (18 288 mm) or greater in height for Exposure B, 30 feet (9144 mm) or greater in height for Exposure C, and 15 feet (4572 mm) or greater in height for Exposure D.
3. The *hill, ridge* or escarpment is isolated or unobstructed by other topographic features of similar height in the upwind direction for a distance measured from its high point of 100 times its height or 2 miles (3.2 km), whichever is less. See Figure R301.2.1.5.1(3) for upwind obstruction.
4. The *hill, ridge* or escarpment protrudes by a factor of two or more above the height of other upwind topographic features located in any quadrant within a radius of 2 miles (3.2 km) measured from its high point.

R301.2.1.5.1 Simplified topographic wind speed-up method. As an alternative to the ASCE 7 topographic wind provisions, the provisions of Section R301.2.1.5.1 shall be permitted to be used to design for wind speed-up effects, where required by Section R301.2.1.5.

Structures located on the top half of isolated *hills*, ridges or escarpments meeting the conditions of Section R301.2.1.5 shall be designed for an increased basic wind speed as determined by Table R301.2.1.5.1. On the high side of an escarpment, the increased basic wind speed shall extend horizontally downwind from the edge of the escarpment 1.5 times the horizontal length of the upwind slope (1.5L) or 6 times the height of the escarpment (6H), whichever is greater. See Figure R301.2.1.5.1(2) for where wind speed increase is applied.

TABLE R301.2.1.5.1
ULTIMATE DESIGN WIND SPEED MODIFICATION FOR TOPOGRAPHIC WIND EFFECT[a, b]

ULTIMATE DESIGN WIND SPEED FROM FIGURE R301.2(2) (mph)	AVERAGE SLOPE OF THE TOP HALF OF HILL, RIDGE OR ESCARPMENT (percent)						
	0.10	0.125	0.15	0.175	0.20	0.23	0.25
	Required ultimate design wind speed-up, modified for topographic wind speed-up (mph)						
95	114	119	123	127	131	137	140
100	120	125	129	134	138	144	147
105	126	131	135	141	145	151	154
110	132	137	142	147	152	158	162
115	138	143	148	154	159	165	169
120	144	149	155	160	166	172	176
130	156	162	168	174	179	NA	NA
140	168	174	181	NA	NA	NA	NA
150	180	NA	NA	NA	NA	NA	NA

For SI: 1 mile per hour = 0.447 m/s, 1 foot = 304.8 mm.
NA = Not Applicable.

a. Table applies to a feature height of 500 feet or less and dwellings sited a distance equal or greater than half the feature height.

b. Where the ultimate design wind speed as modified by Table R301.2.1.5.1 equals or exceeds 140 miles per hour, the building shall be considered as "wind design required" in accordance with Section R301.2.1.1.

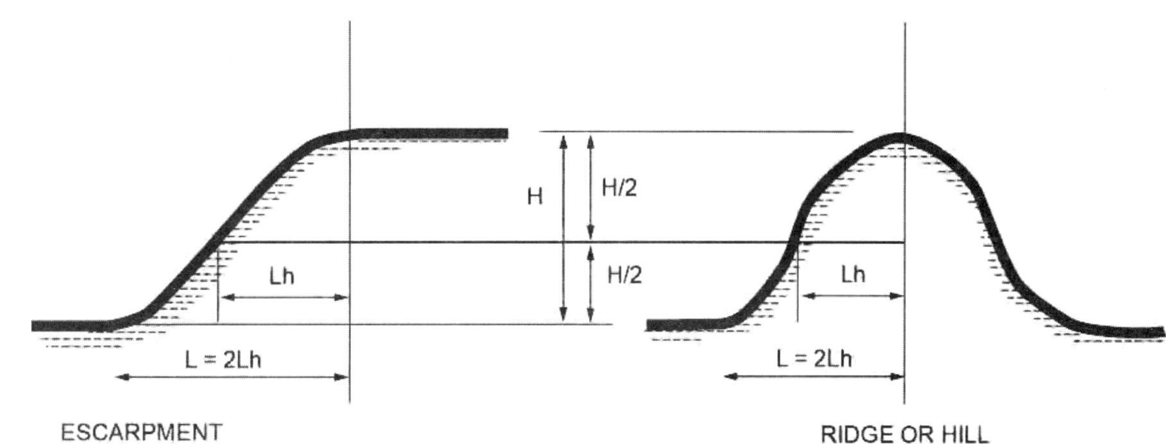

FIGURE R301.2.1.5.1(1)
TOPOGRAPHIC FEATURES FOR WIND SPEED-UP EFFECT

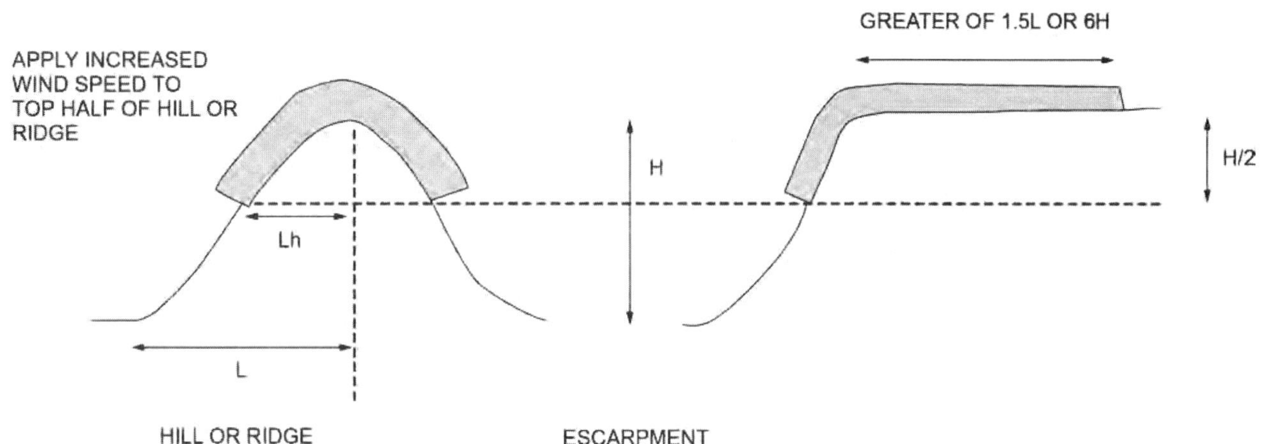

FIGURE R301.2.1.5.1(2)
ILLUSTRATION OF WHERE ON A TOPOGRAPHIC FEATURE, WIND SPEED INCREASE IS APPLIED

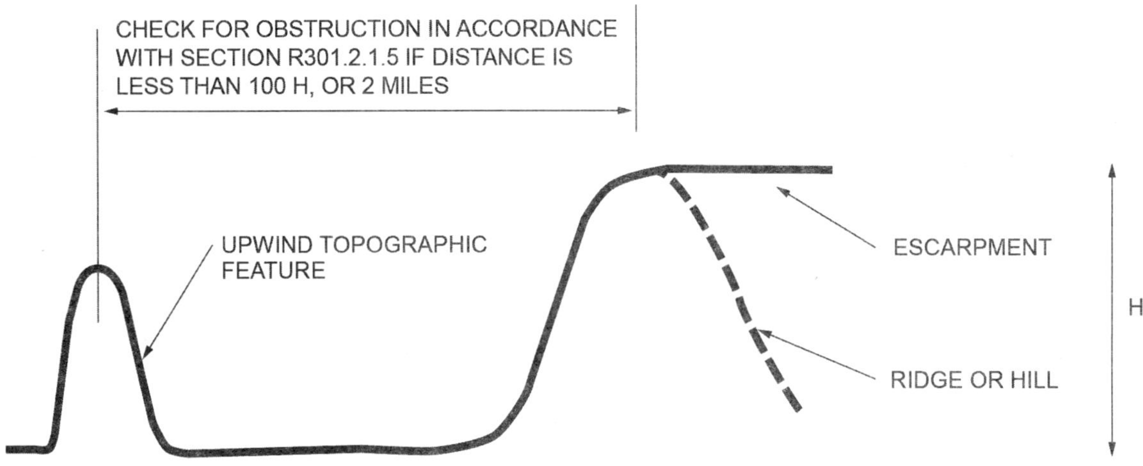

FIGURE R301.2.1.5.1(3)
UPWIND OBSTRUCTION

R301.2.2 Seismic provisions. Buildings in *Seismic Design Categories* C, D_0, D_1 and D_2 shall be constructed in accordance with the requirements of this section and other seismic requirements of this code. The seismic provisions of this code shall apply as follows:

1. *Townhouses* in *Seismic Design Categories* C, D_0, D_1 and D_2.
2. Detached one- and two-family *dwellings* in *Seismic Design Categories*, D_0, D_1 and D_2.

Buildings in Seismic Design Category E shall be designed to resist seismic loads in accordance with the *International Building Code*, except where the seismic design category is reclassified to a lower seismic design category in accordance with Section R301.2.2.1. Components of buildings not required to be designed to resist seismic loads shall be constructed in accordance with the provisions of this code.

R301.2.2.1 Determination of seismic design category. Buildings shall be assigned a seismic design category in accordance with Figures R301.2.2.1(1) through R301.2.2.1(6).

R301.2.2.1.1 Alternate determination of seismic design category. If soil conditions are determined by the building official to be Site Class A, B, or D, the seismic design category and short-period design spectral response accelerations, S_{DS}, for a site shall be allowed to be determined in accordance with Figures R301.2.2.1.1(1) through R301.2.2.1.1(6), or Section 1613.2 of the *International Building Code*. The value of S_{DS} determined in accordance with Section 1613.2 of the *International Building Code* is permitted to be used to set the seismic design category in accordance with Table R301.2.2.1.1, and to interpolate between values in Tables R602.10.3(3) and R603.9.2(1) and other seismic design requirements of this code.

R301.2.2.1.2 Alternative determination of Seismic Design Category E. Buildings located in Seismic Design Category E in accordance with Figures R301.2.2.1(1) through R301.2.2.1(6), or Figures R301.2.2.1.1(1) through R301.2.2.1.1(6) where applicable, are permitted to be reclassified as being in Seismic Design Category D_2 provided that one of the following is done:

1. A more detailed evaluation of the seismic design category is made in accordance with the provisions and maps of the *International Building Code*. Buildings located in Seismic Design Category E in accordance with Table R301.2.2.1.1, but located in Seismic Design Category D in accordance with the *International Building Code*, shall be permitted to be designed using the Seismic Design Category D_2 requirements of this code.
2. Buildings located in Seismic Design Category E that conform to the following additional restrictions are permitted to be constructed in accordance with the provisions for Seismic Design Category D_2 of this code:
   - 2.1. All exterior shear wall lines or *braced wall panels* are in one plane vertically from the foundation to the uppermost *story*.
   - 2.2. Floors shall not cantilever past the *exterior walls*.
   - 2.3. The building is within the requirements of Section R301.2.2.6 for being considered as regular.

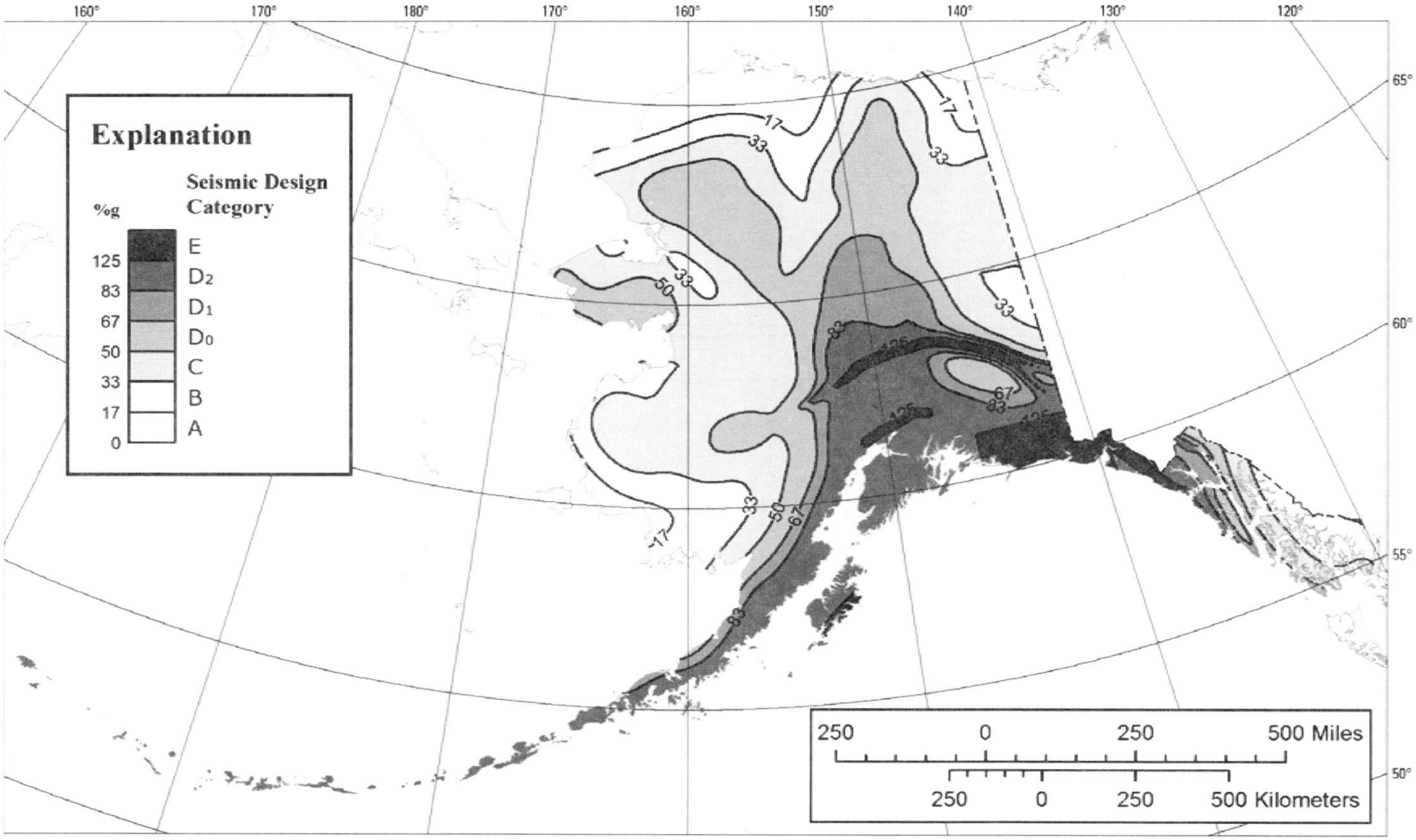

Explanation

%g		Seismic Design Category
125		E
83		D_2
67		D_1
50		D_0
33		C
17		B
0		A

250 0 250 500 Miles

250 0 250 500 Kilometers

Map prepared by U.S. Geological Survey in collaboration with the Federal Emergency Management Agency (FEMA)-funded Building Seismic Safety Council's (BSSC) Code Resource Support Committee (CRSC).

REFERENCES

Building Seismic Safety Council, 2015, NEHRP Recommended Seismic Provisions for New Buildings and Other Structures: FEMA P-1050, Federal Emergency Management Agency, Washington, DC.

Huhng, Yin-Nan, Whittaker, A.S., and Luco, Nicolas, 2008, Maximum spectral demands in the near-fault region, Earthquake Spectra Volume 24, Issue 1, pp. 319-341.

Luco, Nicolas, Ellingwood, B.R., Hamburger, R.O., Hooper, J.D., Kimball, J.K., and Kircher, C.A., 2007, Risk-Targeted versus Current Seismic Design Maps for the Conterminous United States, Structural Engineers Association of California 2007 Convention Proceedings, pp. 163-175.

Wesson, Robert L., Boyd, Oliver S., Mueller, Charles S., Bufe, Charles G., Frankel, Arthur D., Petersen, Mark D., 2007, Revision of time-Independent probabilistic seismic hazard maps for Alaska: U.S. Geological Survey Open-File Report 2007-1043.

a. The seismic design categories and corresponding short-period design spectral response accelerations, S_{DS}, shown in Figures R301.2.2.1(1) through 301.2.2.1(6) are based on soil Site Class D, used as an assumed default, as defined in Section 1613.2.2 of the *International Building Code*.

FIGURE R301.2.2.1(1)
SEISMIC DESIGN CATEGORIES—ALASKA[a]

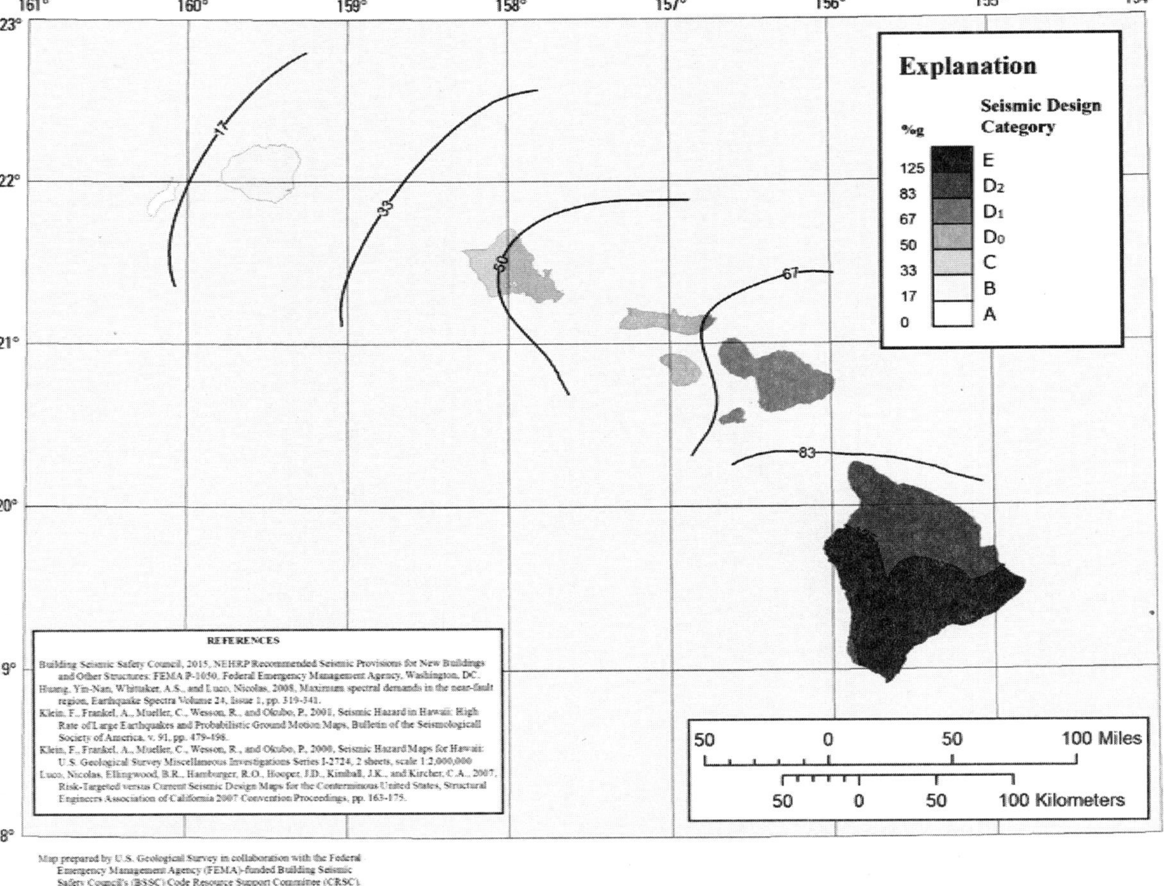

a. The seismic design categories and corresponding short-period design spectral response accelerations, S_{DS}, shown in Figures R301.2.2.1(1) through 301.2.2.1(6) are based on soil Site Class D, used as an assumed default, as defined in Section 1613.2.2 of the *International Building Code*.

FIGURE R301.2.2.1(2)
SEISMIC DESIGN CATEGORIES—HAWAII[a]

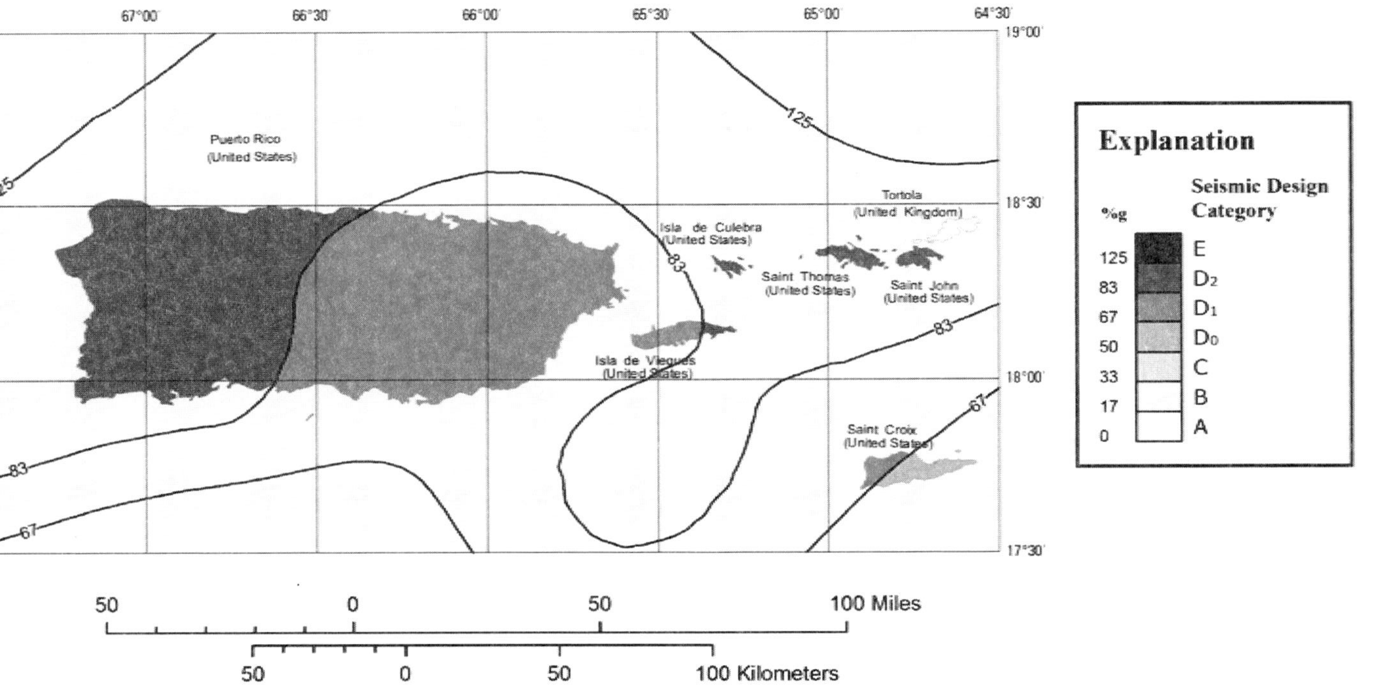

REFERENCES

Building Seismic Safety Council, 2015, NEHRP Recommended Seismic Provisions for New Buildings and Other Structures: FEMA P-1050, Federal Emergency Management Agency, Washington, DC.

Huang, Yin-Nan, Whittaker, A.S., and Luco, Nicolas, 2008, Maximum spectral demands in the near-fault region, Earthquake Spectra Volume 24, Issue 1, pp. 319-341.

Luco, Nicolas, Ellingwood, B.R., Hamburger, R.O., Hooper, J.D., Kimball, J.K., and Kircher, C.A., 2007, Risk-Targeted versus Current Seismic Design Maps for the Conterminous United States, Structural Engineers Association of California 2007 Convention Proceedings, pp. 163-175.

Mueller, C., Frankel, A., Petersen, M., and Leyendecker, E., 2003, Documentation for 2003 USGS Seismic Hazard Maps for Puerto Rico and the U.S. Virgin Islands, U.S. Geological Survey Open-File Report 03-379.

Map prepared by U.S. Geological Survey in collaboration with the Federal Emergency Management Agency (FEMA)-funded Building Seismic Safety Council's (BSSC) Code Resource Support Committee (CRSC).

a. The seismic design categories and corresponding short-period design spectral response accelerations, S_{DS}, shown in Figures R301.2.2.1(1) through 301.2.2.1(6) are based on soil Site Class D, used as an assumed default, as defined in Section 1613.2.2 of the *International Building Code*.

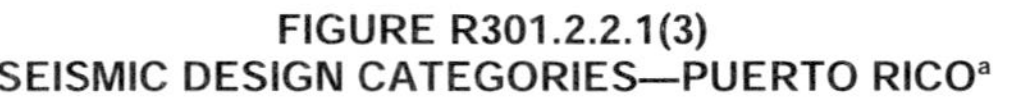

FIGURE R301.2.2.1(3)
SEISMIC DESIGN CATEGORIES—PUERTO RICO[a]

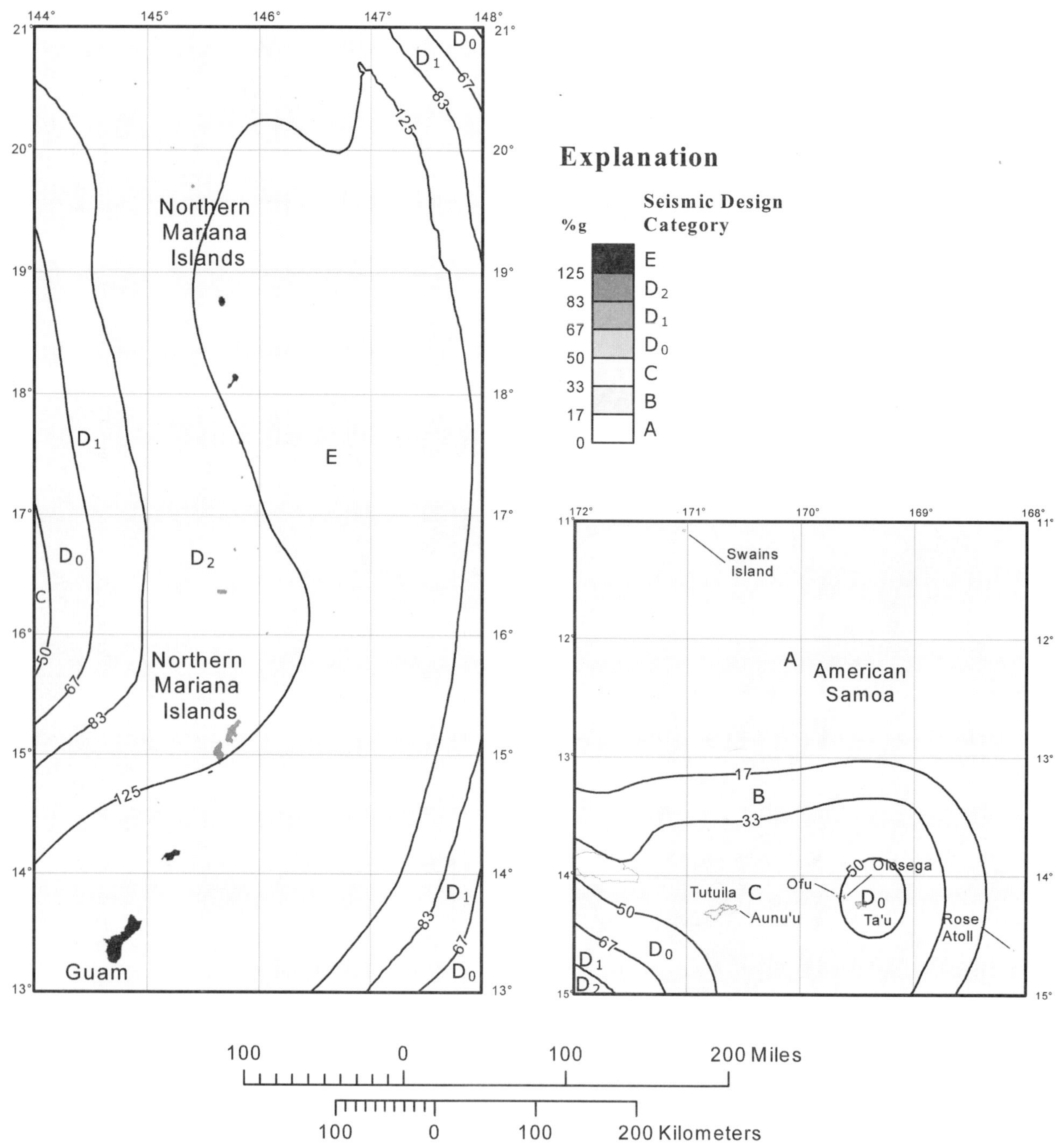

a. The seismic design categories and corresponding short-period design spectral response accelerations, S_{DS}, shown in Figures R301.2.2.1(1) through 301.2.2.1(6) are based on soil Site Class D, used as an assumed default, as defined in Section 1613.2.2 of the *International Building Code*.

FIGURE R301.2.2.1(4)
SEISMIC DESIGN CATEGORIES—NORTHERN MARIANA ISLANDS AND AMERICAN SAMOA[a]

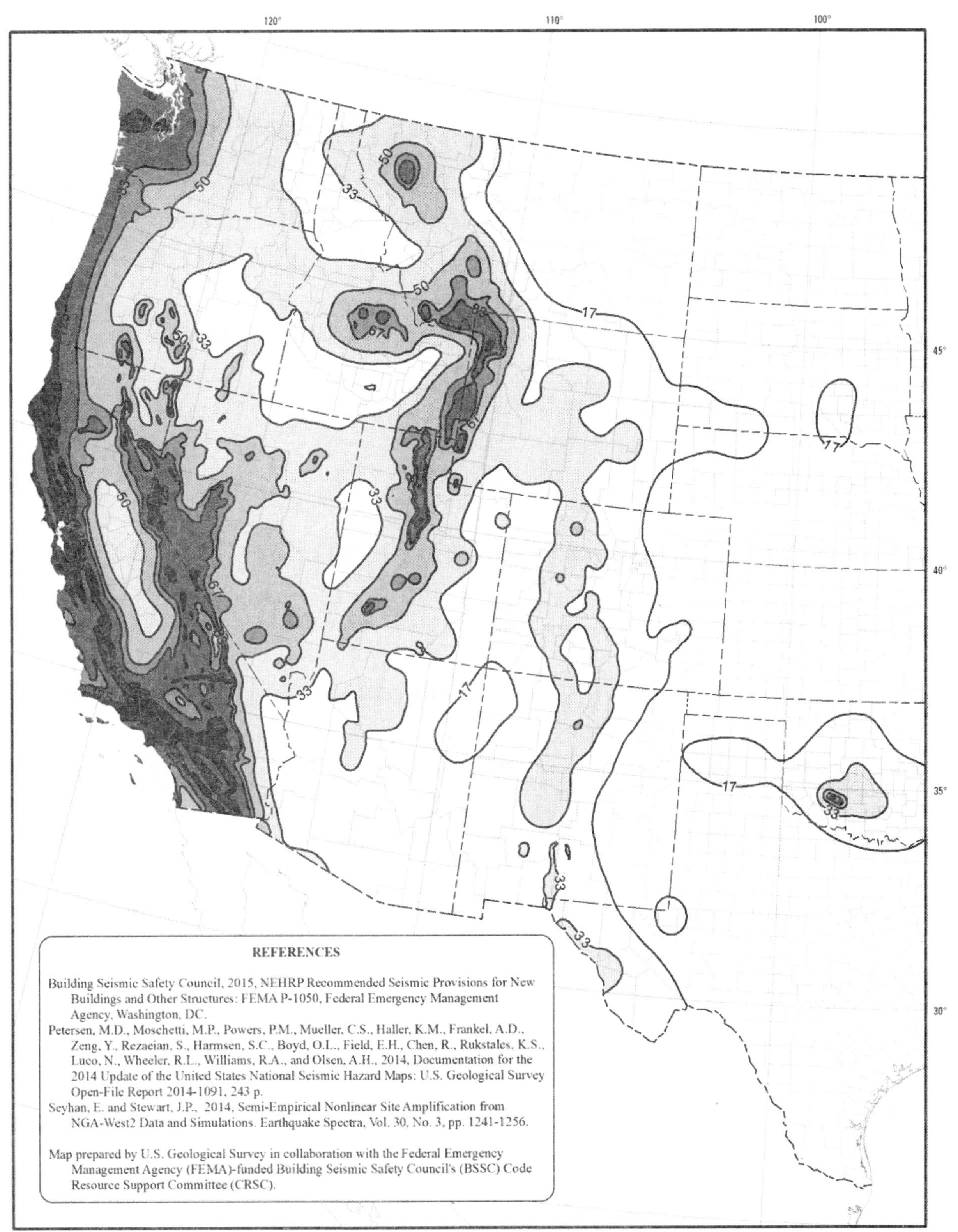

a. The seismic design categories and corresponding short-period design spectral response accelerations, S_{DS}, shown in Figures R301.2.2.1(1) through 301.2.2.1(6) are based on soil Site Class D, used as an assumed default, as defined in Section 1613.2.2 of the *International Building Code*.

FIGURE R301.2.2.1(5)
SEISMIC DESIGN CATEGORIES—UNITED STATES[a]

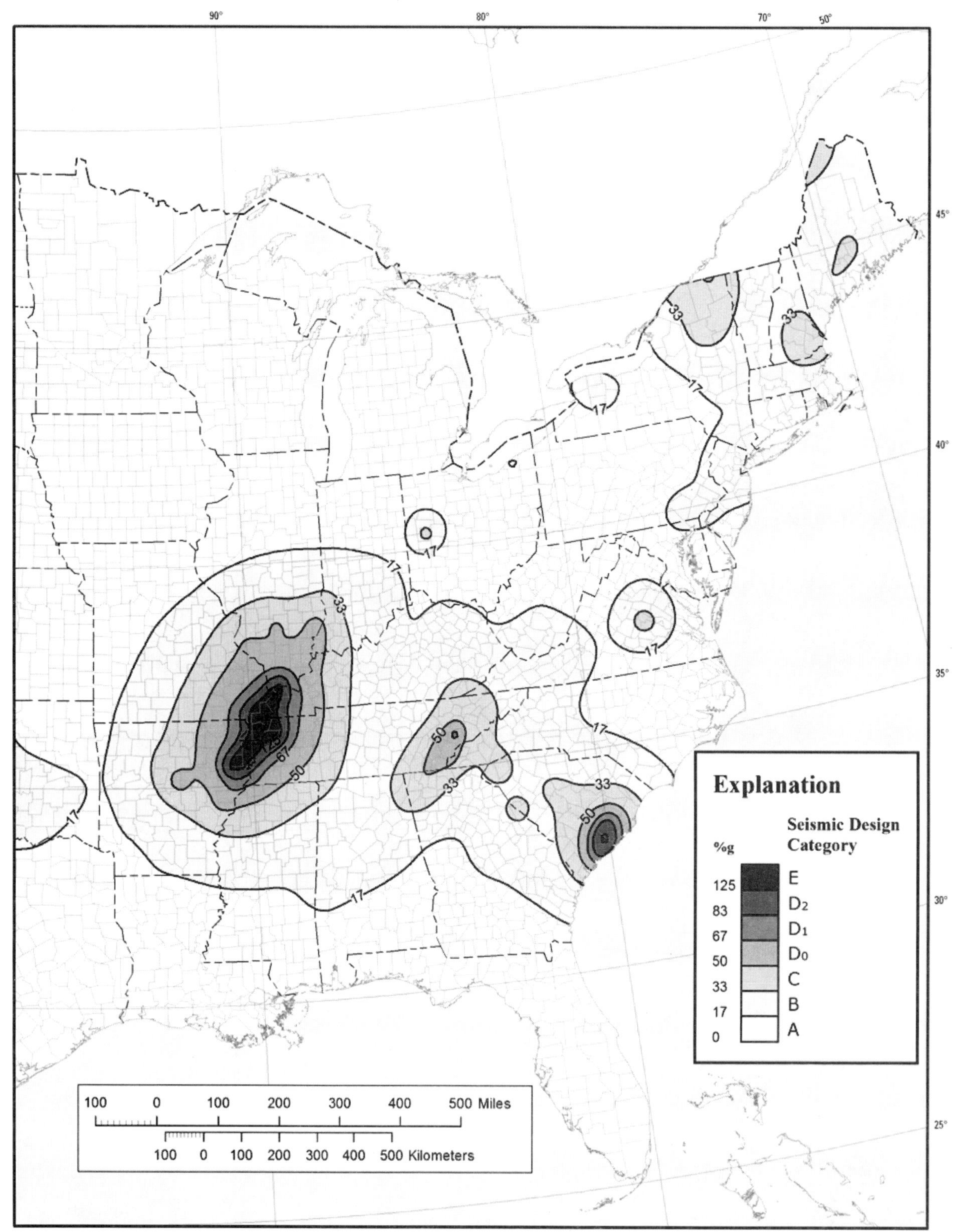

a. The seismic design categories and corresponding short-period design spectral response accelerations, S_{DS}, shown in Figures R301.2.2.1(1) through 301.2.2.1(6), are based on soil Site Class D, used as an assumed default, as defined in Section 1613.2.2 of the *International Building Code*.

FIGURE R301.2.2.1(6)
SEISMIC DESIGN CATEGORIES—UNITED STATES[a]

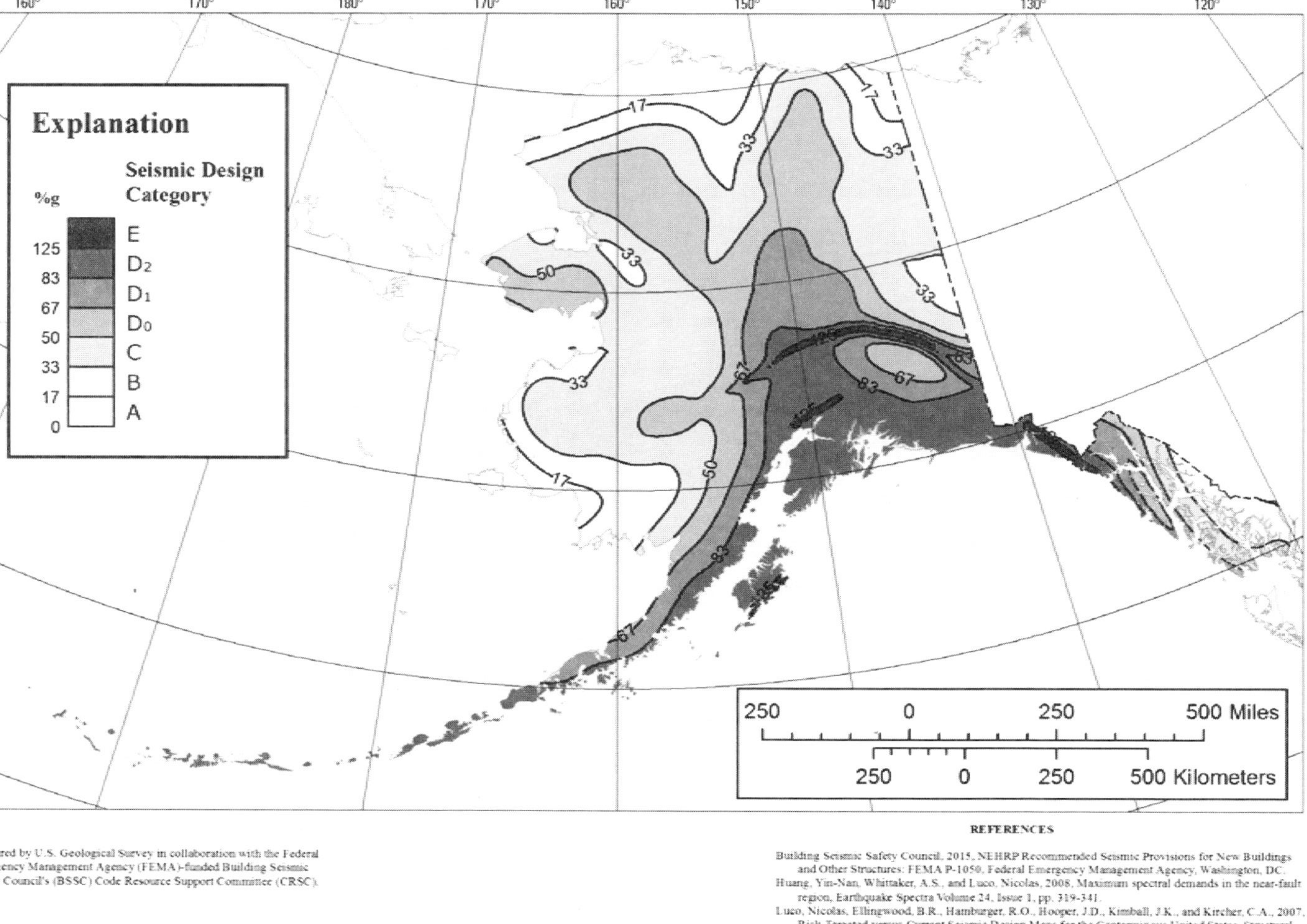

a. The seismic design categories and corresponding short-period design spectral response accelerations, S_{DS}, shown in Figures R301.2.2.1.1(1) through 301.2.2.1.1(6) are permitted to be used where soil conditions are determined by the building official to be Site Class A, B or D.

FIGURE R301.2.2.1.1(1)
ALTERNATE SEISMIC DESIGN CATEGORIES—ALASKA[a]

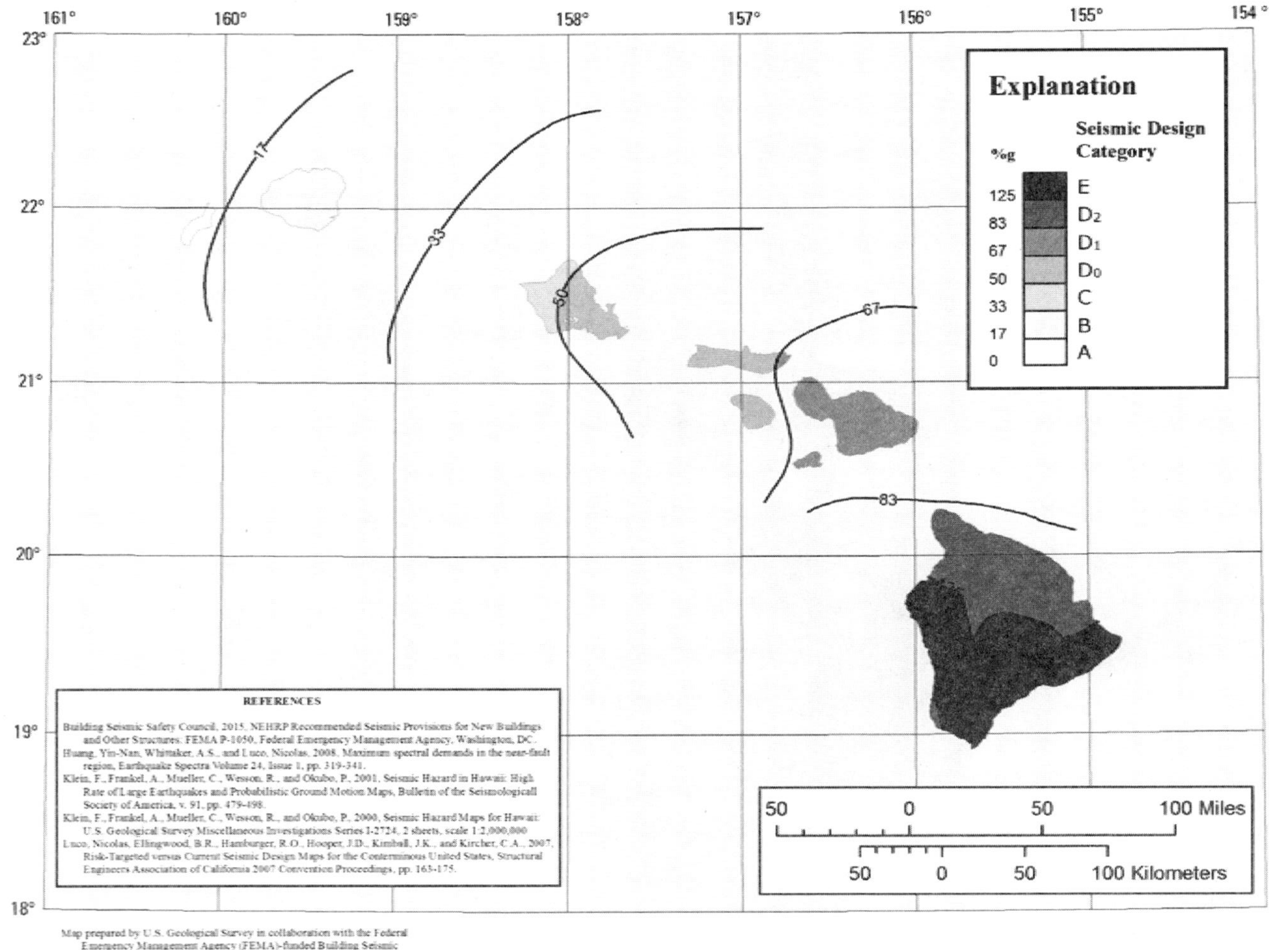

a. The seismic design categories and corresponding short-period design spectral response accelerations, S_{DS}, shown in Figures R301.2.2.1.1(1) through 301.2.2.1.1(6) are permitted to be used where soil conditions are determined by the building official to be Site Class A, B or D.

FIGURE R301.2.2.1.1(2)
ALTERNATE SEISMIC DESIGN CATEGORIES—HAWAII[a]

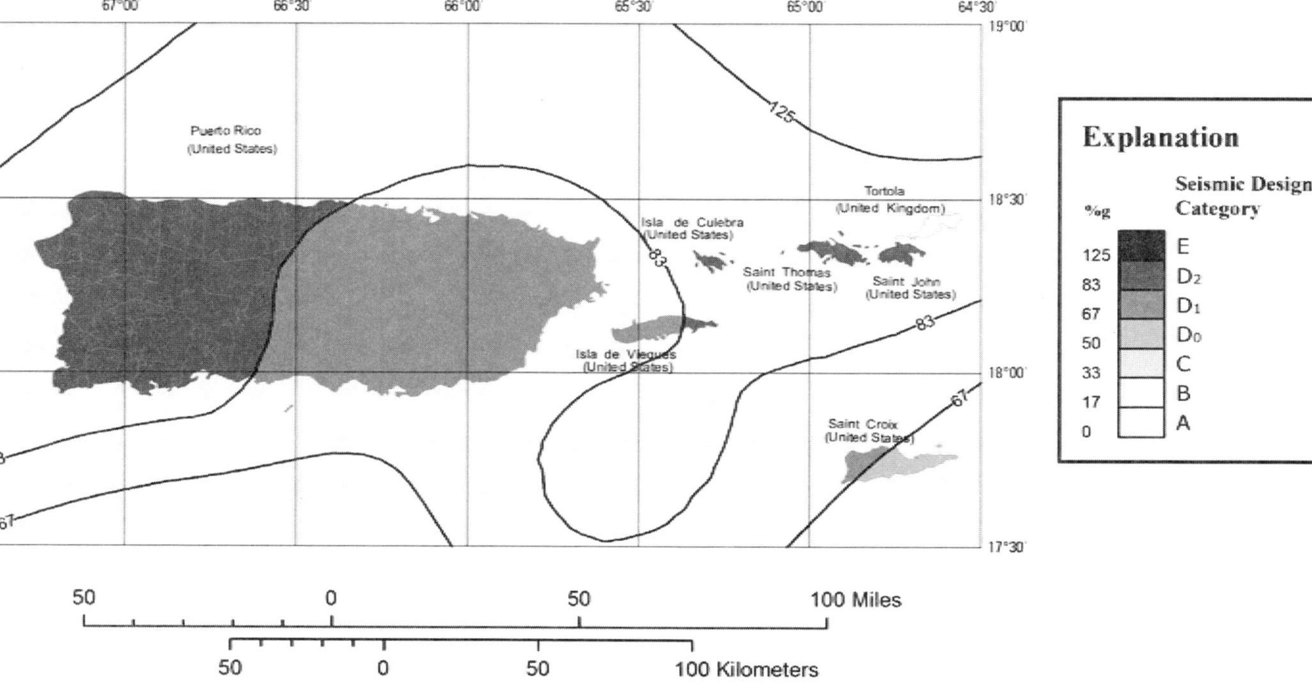

Explanation

%g	Seismic Design Category
125	E
83	D_2
67	D_1
50	D_0
33	C
17	B
0	A

REFERENCES

Building Seismic Safety Council, 2015, NEHRP Recommended Seismic Provisions for New Buildings and Other Structures: FEMA P-1050, Federal Emergency Management Agency, Washington, DC.

Huang, Yin-Nan, Whittaker, A.S., and Luco, Nicolas, 2008, Maximum spectral demands in the near-fault region, Earthquake Spectra Volume 24, Issue 1, pp. 319-341.

Luco, Nicolas, Ellingwood, B.R., Hamburger, R.O., Hooper, J.D., Kimball, J.K., and Kircher, C.A., 2007, Risk-Targeted versus Current Seismic Design Maps for the Conterminous United States, Structural Engineers Association of California 2007 Convention Proceedings, pp. 163-175.

Mueller, C., Frankel, A., Petersen, M., and Leyendecker, E., 2003, Documentation for 2003 USGS Seismic Hazard Maps for Puerto Rico and the U.S. Virgin Islands, U.S. Geological Survey Open-File Report 03-379.

Map prepared by U.S. Geological Survey in collaboration with the Federal Emergency Management Agency (FEMA)-funded Building Seismic Safety Council's (BSSC) Code Resource Support Committee (CRSC).

a. The seismic design categories and corresponding short-period design spectral response accelerations, S_{DS}, shown in Figures R301.2.2.1.1(1) through 301.2.2.1.1(6) are permitted to be used where soil conditions are determined by the building official to be Site Class A, B or D.

FIGURE R301.2.2.1.1(3)
ALTERNATE SEISMIC DESIGN CATEGORIES—PUERTO RICO[a]

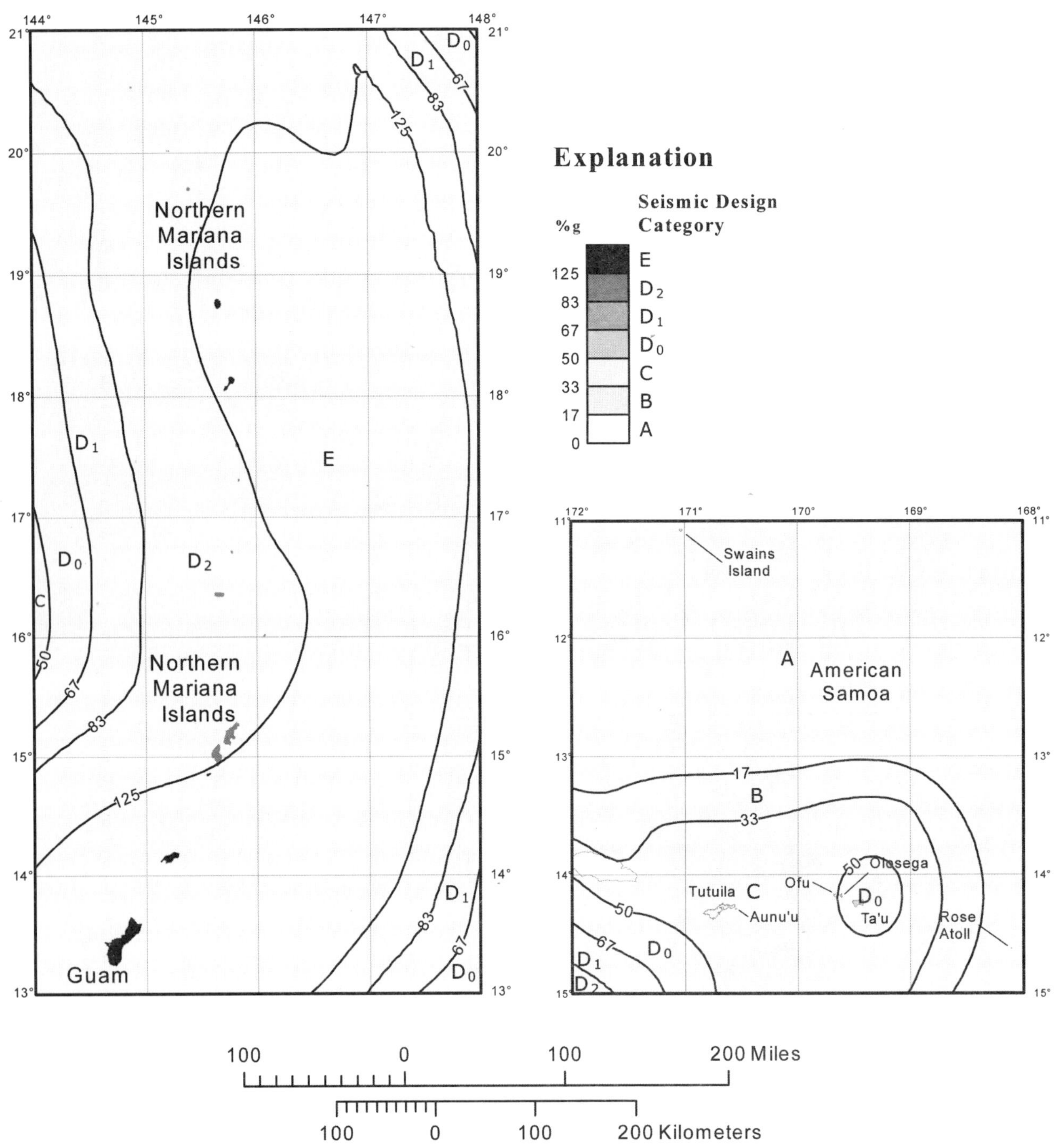

a. The seismic design categories and corresponding short-period design spectral response accelerations, S_{DS}, shown in Figures R301.2.2.1.1(1) through 301.2.2.1.1(6) are permitted to be used where soil conditions are determined by the building official to be Site Class A, B or D.

FIGURE R301.2.2.1.1(4)
ALTERNATE SEISMIC DESIGN CATEGORIES—NORTHERN MARIANA ISLANDS AND AMERICAN SAMOA[a]

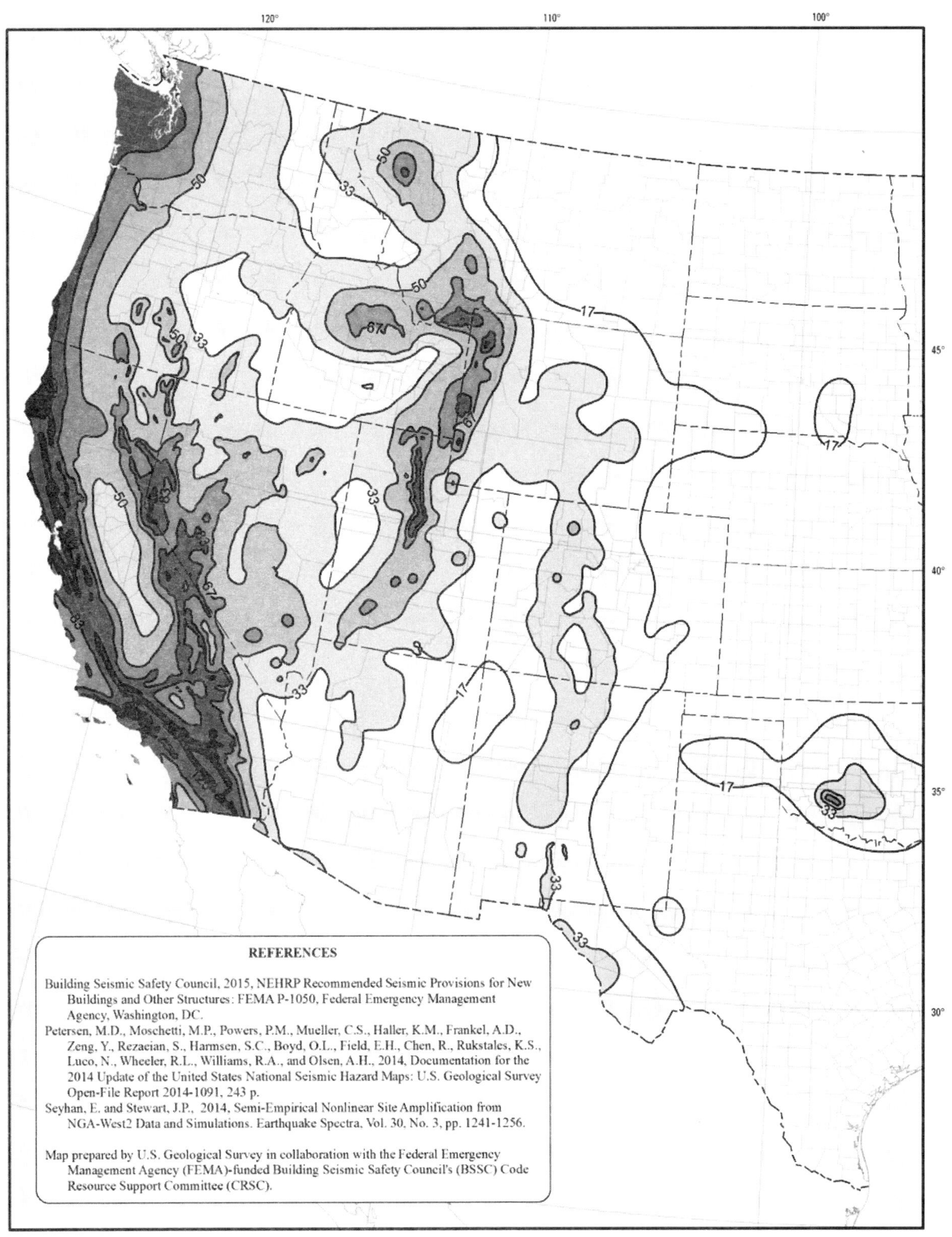

a. The seismic design categories and corresponding short-period design spectral response accelerations, S_{DS}, shown in Figures R301.2.2.1.1(1) through 301.2.2.1.1(6) are permitted to be used where soil conditions are determined by the building official to be Site Class A, B or D.

FIGURE R301.2.2.1.1(5)
ALTERNATE SEISMIC DESIGN CATEGORIES—UNITED STATES[a]

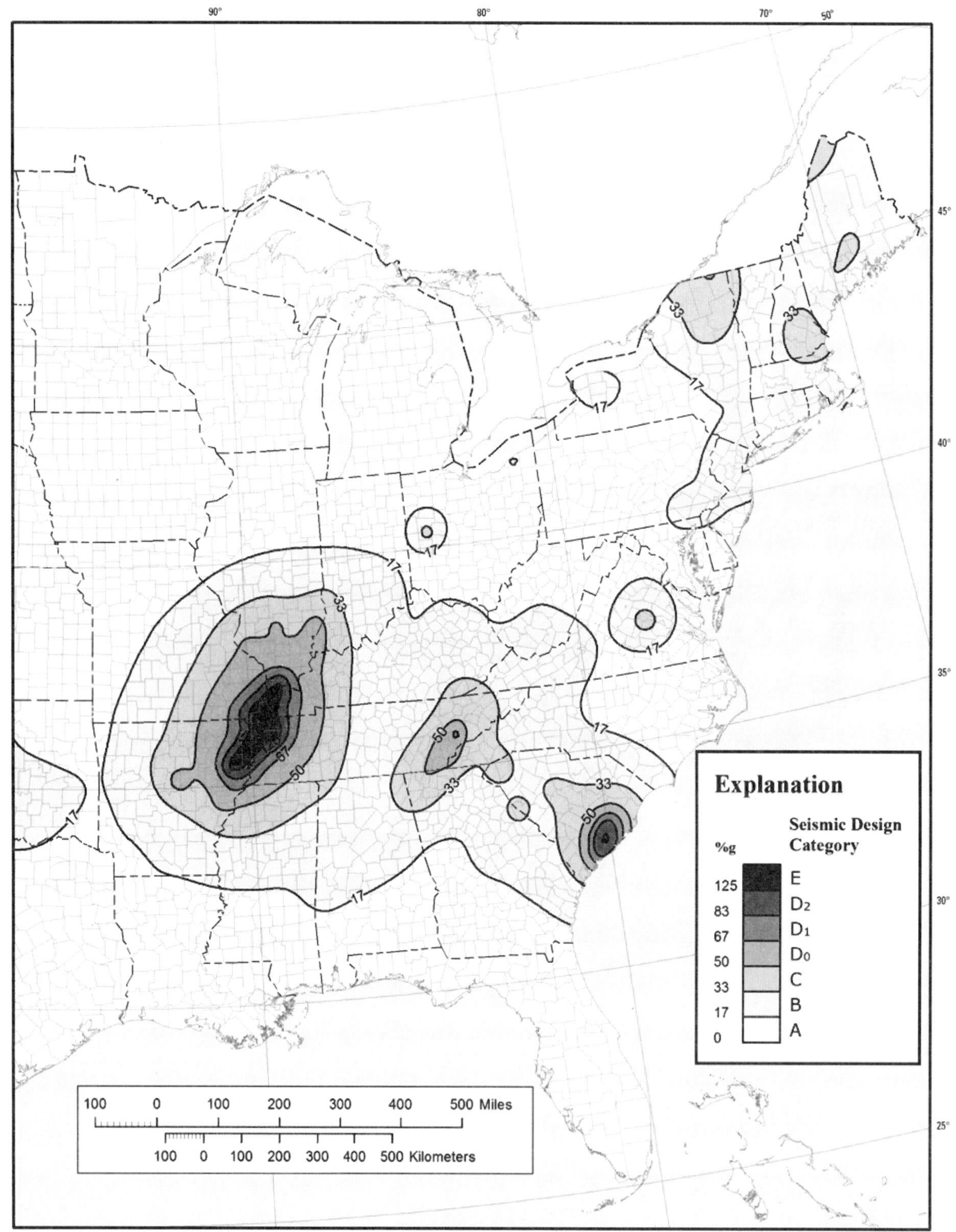

a. The seismic design categories and corresponding short-period design spectral response accelerations, S_{DS}, shown in Figures R301.2.2.1.1(1) through 301.2.2.1.1(6) are permitted to be used where soil conditions are determined by the building official to be Site Class A, B or D.

FIGURE R301.2.2.1.1(6)
ALTERNATE SEISMIC DESIGN CATEGORIES—UNITED STATES[a]

TABLE R301.2.2.1.1
SEISMIC DESIGN CATEGORY DETERMINATION

CALCULATED S_{DS}	SEISMIC DESIGN CATEGORY
$S_{DS} \leq 0.17g$	A
$0.17g < S_{DS} \leq 0.33g$	B
$0.33g < S_{DS} \leq 0.50g$	C
$0.50g < S_{DS} \leq 0.67g$	D_0
$0.67g < S_{DS} \leq 0.83g$	D_1
$0.83g < S_{DS} \leq 1.25g$	D_2
$1.25g < S_{DS}$	E

R301.2.2.2 Weights of materials. Average dead loads shall not exceed 15 pounds per square foot (720 Pa) for the combined roof and ceiling assemblies (on a horizontal projection) or 10 pounds per square foot (480 Pa) for floor assemblies, except as further limited by Section R301.2.2. Dead loads for walls above *grade* shall not exceed:

1. Fifteen pounds per square foot (720 Pa) for exterior light-frame wood walls.
2. Fourteen pounds per square foot (670 Pa) for exterior light-frame cold-formed steel walls.
3. Ten pounds per square foot (480 Pa) for interior light-frame wood walls.
4. Five pounds per square foot (240 Pa) for interior light-frame cold-formed steel walls.
5. Eighty pounds per square foot (3830 Pa) for 8-inch-thick (203 mm) masonry walls.
6. Eighty-five pounds per square foot (4070 Pa) for 6-inch-thick (152 mm) concrete walls.
7. Ten pounds per square foot (480 Pa) for SIP walls.

Exceptions:

1. Roof and ceiling dead loads not exceeding 25 pounds per square foot (1190 Pa) shall be permitted provided that the wall bracing amounts in Section R602.10.3 are increased in accordance with Table R602.10.3(4).
2. Light-frame walls with stone or masonry veneer shall be permitted in accordance with the provisions of Sections R702.1 and R703.
3. Fireplaces and chimneys shall be permitted in accordance with Chapter 10.

R301.2.2.3 Stone and masonry veneer. Anchored stone and masonry veneer shall comply with the requirements of Sections R702.1 and R703.

R301.2.2.4 Masonry construction. Masonry construction in *Seismic Design Categories* D_0 and D_1 shall comply with the requirements of Section R606.12.1. Masonry construction in Seismic Design Category D_2 shall comply with the requirements of Section R606.12.4.

R301.2.2.5 Concrete construction. Buildings with exterior above-*grade* concrete walls shall comply with PCA 100 or shall be designed in accordance with ACI 318.

Exception: Detached one- and two-family dwellings in Seismic Design Category C with exterior above-*grade* concrete walls are allowed to comply with the requirements of Section R608.

R301.2.2.6 Irregular buildings. The seismic provisions of this code shall not be used for structures, or portions thereof, located in *Seismic Design Categories* C, D_0, D_1 and D_2 and considered to be irregular in accordance with this section. A building or portion of a building shall be considered to be irregular where one or more of the conditions defined in Items 1 through 8 occur. Irregular structures, or irregular portions of structures, shall be designed in accordance with accepted engineering practice to the extent the irregular features affect the performance of the remaining structural system. Where the forces associated with the irregularity are resisted by a structural system designed in accordance with accepted engineering practice, the remainder of the building shall be permitted to be designed using the provisions of this code.

1. ***Shear wall* or braced wall offsets out of plane.** Conditions where exterior *shear wall* lines or *braced wall panels* are not in one plane vertically from the foundation to the uppermost story in which they are required.

   Exception: For wood *light-frame construction*, floors with cantilevers or setbacks not exceeding four times the nominal depth of the wood floor joists are permitted to support *braced wall panels* that are out of plane with *braced wall panels* below provided that all of the following are satisfied:

   1. Floor joists are nominal 2 inches by 10 inches (51 mm by 254 mm) or larger and spaced not more than 16 inches (406 mm) on center.
   2. The ratio of the back span to the cantilever is not less than 2 to 1.
   3. Floor joists at ends of *braced wall panels* are doubled.

4. For wood-frame construction, a continuous rim joist is connected to ends of cantilever joists. Where spliced, the rim joists shall be spliced using a galvanized metal tie not less than 0.058 inch (1.5 mm) (16 gage) and $1^1/_2$ inches (38 mm) wide fastened with six 16d nails on each side of the splice; or a block of the same size as the rim joist and of sufficient length to fit securely between the joist space at which the splice occurs, fastened with eight 16d nails on each side of the splice.
5. Gravity loads carried at the end of cantilevered joists are limited to uniform wall and roof loads and the reactions from headers having a span of 8 feet (2438 mm) or less.

2. **Lateral support of roofs and floors.** Conditions where a section of floor or roof is not laterally supported by *shear walls* or *braced wall lines* on all edges.

   Exception: Portions of floors that do not support *shear walls, braced wall panels* above, or roofs shall be permitted to extend not more than 6 feet (1829 mm) beyond a *shear wall* or *braced wall line.*

3. ***Shear wall* or braced wall offsets in plane.** Conditions where the end of a *braced wall panel* occurs over an opening in the wall below and extends more than 1 foot (305 mm) horizontally past the edge of the opening. This provision is applicable to *shear walls* and *braced wall panels* offset in plane and to *braced wall panels* offset out of plane in accordance with the exception to Item 1.

   Exception: For wood light-frame wall construction, one end of a *braced wall panel* shall be permitted to extend more than 1 foot (305 mm) over an opening not more than 8 feet (2438 mm) in width in the wall below provided that the opening includes a header in accordance with all of the following:

   1. The building width, loading condition and framing member species limitations of Table R602.7(1) shall apply.
   2. The header is composed of:

      2.1. Not less than one 2 × 12 or two 2 × 10 for an opening not more than 4 feet (1219 mm) wide.

      2.2. Not less than two 2 × 12 or three 2 × 10 for an opening not more than 6 feet (1829 mm) in width.

      2.3. Not less than three 2 × 12 or four 2 × 10 for an opening not more than 8 feet (2438 mm) in width.
   3. The entire length of the *braced wall panel* does not occur over an opening in the wall below.

4. **Floor and roof opening.** Conditions where an opening in a floor or roof exceeds the lesser of 12 feet (3658 mm) or 50 percent of the least floor or roof dimension.

5. **Floor level offset.** Conditions where portions of a floor level are vertically offset.

   Exceptions:

   1. Framing supported directly by continuous foundations at the perimeter of the building.
   2. For wood *light-frame construction,* floors shall be permitted to be vertically offset where the floor framing is lapped or tied together as required by Section R502.6.1.

6. **Perpendicular *shear wall* and wall bracing.** Conditions where *shear walls* and *braced wall lines* do not occur in two perpendicular directions.

7. **Wall bracing in stories containing masonry or concrete construction.** Conditions where stories above *grade plane* are partially or completely braced by wood wall framing in accordance with Section R602 or cold-formed steel wall framing in accordance with Section R603 include masonry or concrete construction. Where this irregularity applies, the entire story shall be designed in accordance with accepted engineering practice.

   Exceptions: Fireplaces, chimneys and masonry veneer in accordance with this code.

8. **Hillside *light-frame construction.*** Conditions in which all of the following apply:

   8.1. The grade slope exceeds 1 unit vertical in 5 units horizontal where averaged across the full length of any side of the dwelling.

   8.2. The tallest cripple wall clear height exceeds 7 feet (2134 mm), or where a post and beam system occurs at the dwelling perimeter, the post and beam system tallest post clear height exceeds 7 feet (2134 mm).

   8.3. Of the total plan area below the lowest framed floor, whether open or enclosed, less than 50 percent is living space having interior wall finishes conforming to Section R702.

Where Item 8 is applicable, design in accordance with accepted engineering practice shall be provided for the floor immediately above the cripple walls or post and beam system and all structural elements and connections from this diaphragm down to and including connections to the foundation and design of the foundation to transfer lateral loads from the framing above.

Exception: *Light-frame construction* in which the lowest framed floor is supported directly on concrete or masonry walls over the full length of all sides except the downhill side of the dwelling need not be considered an irregular dwelling under Item 8.

R301.2.2.7 Height limitations. Wood-framed buildings shall be limited to three *stories* above *grade plane* or the limits given in Table R602.10.3(3). Wood-framed buildings in Seismic Design Category D_2 exceeding two *stories* shall be designed for wind and seismic loads in accordance with accepted engineering practice. Cold-formed steel-framed buildings shall be limited to less than or equal to three *stories* above *grade plane* in accordance with AISI S230. *Mezzanines* as defined in Section R202 that comply with Section R325 shall not be considered as *stories*. *Structural insulated panel* buildings shall be limited to two *stories* above *grade plane*.

R301.2.2.8 Cold-formed steel framing in Seismic Design Categories D_0, D_1 and D_2. In *Seismic Design Categories* D_0, D_1 and D_2 in addition to the requirements of this code, cold-formed steel framing shall comply with the requirements of AISI S230.

R301.2.2.9 Masonry chimneys. In *Seismic Design Categories* D_0, D_1 and D_2, masonry chimneys shall be reinforced and anchored to the building in accordance with Sections R1003.3 and R1003.4.

R301.2.2.10 Anchorage of water heaters. In *Seismic Design Categories* D_0, D_1 and D_2, and in *townhouses* in Seismic Design Category C, water heaters and thermal storage units shall be anchored against movement and overturning in accordance with Section M1307.2 or P2801.8.

R301.2.3 Snow loads. Wood-framed construction, cold-formed, steel-framed construction and masonry and concrete construction, and *structural insulated panel* construction in regions with ground snow loads 70 pounds per square foot (3.35 kPa) or less, shall be in accordance with Chapters 5, 6 and 8. Buildings in regions with ground snow loads greater than 70 pounds per square foot (3.35 kPa) shall be designed in accordance with accepted engineering practice.

R301.2.4 Floodplain construction. Buildings and structures constructed in whole or in part in flood hazard areas (including A or V Zones) as established in Table R301.2, and substantial improvement and *repair* of substantial damage of buildings and structures in flood hazard areas, shall be designed and constructed in accordance with Section R322. Buildings and structures that are located in more than one flood hazard area shall comply with the provisions associated with the most restrictive flood hazard area. Buildings and structures located in whole or in part in identified floodways shall be designed and constructed in accordance with ASCE 24.

R301.2.4.1 Alternative provisions. As an alternative to the requirements in Section R322, ASCE 24 is permitted subject to the limitations of this code and the limitations therein.

R301.3 Story height. The wind and seismic provisions of this code shall apply to buildings with *story heights* not exceeding the following:

1. For wood wall framing, the *story height* shall not exceed 11 feet 7 inches (3531 mm) and the laterally unsupported bearing wall stud height permitted by Table R602.3(5).

   Exception: A *story height* not exceeding 13 feet 7 inches (4140 mm) is permitted provided that the maximum wall stud clear height does not exceed 12 feet (3658 mm), the wall studs are in accordance with Exception 2 or 3 of Section R602.3.1 or an engineered design is provided for the wall framing members, and wall bracing for the building is in accordance with Section R602.10. Studs shall be laterally supported at the top and bottom plate in accordance with Section R602.3.

2. For cold-formed steel wall framing, the *story height* shall be not more than 11 feet 7 inches (3531 mm) and the unsupported bearing wall stud height shall be not more than 10 feet (3048 mm).
3. For masonry walls, the *story height* shall be not more than 13 feet 7 inches (4140 mm) and the bearing wall clear height shall be not more than 12 feet (3658 mm).

   Exception: An additional 8 feet (2438 mm) of bearing wall clear height is permitted for gable end walls.

4. For insulating concrete form walls, the maximum *story height* shall not exceed 11 feet 7 inches (3531 mm) and the maximum unsupported wall height per *story* as permitted by Section R608 tables shall not exceed 10 feet (3048 mm).
5. For structural insulated panel (SIP) walls, the *story height* shall be not more than 11 feet 7 inches (3531 mm) and the bearing wall height per *story* as permitted by Section R610 tables shall not exceed 10 feet (3048 mm).

For walls other than wood-framed walls, individual walls or wall studs shall be permitted to exceed these limits as permitted by Chapter 6, provided that the *story heights* of this section are not exceeded. An engineered design shall be provided for the wall or wall framing members where the limits of Chapter 6 are exceeded. Where the *story height* limits of this section are exceeded, the design of the building, or the noncompliant portions thereof, to resist wind and seismic loads shall be in accordance with the *International Building Code*.

R301.4 Dead load. The actual weights of materials and construction shall be used for determining dead load with consideration for the dead load of fixed service equipment.

R301.5 Live load. The minimum uniformly distributed *live load* shall be as provided in Table R301.5.

R301.6 Roof load. The roof shall be designed for the *live load* indicated in Table R301.6 or the ground snow load indicated in Table R301.2, whichever is greater.

R301.7 Deflection. The allowable deflection of any structural member under the *live load* listed in Sections R301.5 and R301.6 or wind loads determined by Section R301.2.1 shall not exceed the values in Table R301.7.

R301.8 Nominal sizes. For the purposes of this code, dimensions of lumber specified shall be deemed to be nominal dimensions unless specifically designated as actual dimensions.

TABLE R301.5
MINIMUM UNIFORMLY DISTRIBUTED LIVE LOADS (in pounds per square foot)

USE	UNIFORM LOAD (psf)	CONCENTRATED LOAD (lb)
Uninhabitable attics without storage[b]	10	—
Uninhabitable attics with limited storage[b, g]	20	—
Habitable attics and attics served with fixed stairs	30	—
Balconies (exterior) and decks[e]	40	—
Fire escapes	40	—
Guards	—	200[h, i]
Guard in-fill components[f]	—	50[h]
Handrail[d]	200[h]	—
Passenger vehicle garages[a]	50[a]	2,000[h]
Areas other than sleeping areas	40	—
Sleeping areas	30	—
Stairs	40[c]	300[c]

For SI: 1 inch = 25.4 mm, 1 pound per square foot = 0.0479 kPa, 1 square inch = 645 mm^2, 1 pound = 4.45 N.

a. Elevated garage floors shall be capable of supporting the uniformly distributed live load or a 2,000-pound concentrated load applied on an area of $4^1/_2$ inches by $4^1/_2$ inches, whichever produces the greater stresses.

b. Uninhabitable attics without storage are those where the clear height between joists and rafters is not more than 42 inches, or where there are not two or more adjacent trusses with web configurations capable of accommodating an assumed rectangle 42 inches in height by 24 inches in width, or greater, within the plane of the trusses. This live load need not be assumed to act concurrently with any other live load requirements.

c. Individual stair treads shall be capable of supporting the uniformly distributed live load or a 300-pound concentrated load applied on an area of 2 inches by 2 inches, whichever produces the greater stresses.

d. A single concentrated load applied in any direction at any point along the top. For a guard not required to serve as a handrail, the load need not be applied to the top element of the guard in a direction parallel to such element.

e. See Section R507.1 for decks attached to exterior walls.

f. Guard in-fill components (all those except the handrail), balusters and panel fillers shall be designed to withstand a horizontally applied normal load of 50 pounds on an area equal to 1 square foot. This load need not be assumed to act concurrently with any other live load requirement.

g. Uninhabitable attics with limited storage are those where the clear height between joists and rafters is 42 inches or greater, or where there are two or more adjacent trusses with web configurations capable of accommodating an assumed rectangle 42 inches in height by 24 inches in width, or greater, within the plane of the trusses.

The live load need only be applied to those portions of the joists or truss bottom chords where all of the following conditions are met:

1. The attic area is accessed from an opening not less than 20 inches in width by 30 inches in length that is located where the clear height in the attic is not less than 30 inches.
2. The slopes of the joists or truss bottom chords are not greater than 2 units vertical in 12 units horizontal.
3. Required insulation depth is less than the joist or truss bottom chord member depth.

The remaining portions of the joists or truss bottom chords shall be designed for a uniformly distributed concurrent live load of not less than 10 pounds per square foot.

h. Glazing used in handrail assemblies and guards shall be designed with a load adjustment factor of 4. The load adjustment factor shall be applied to each of the concentrated loads applied to the top of the rail, and to the load on the in-fill components. These loads shall be determined independent of one another, and loads are assumed not to occur with any other live load.

i. Where the top of a guard system is not required to serve as a handrail, the single concentrated load shall be applied at any point along the top, in the vertical downward direction and in the horizontal direction away from the walking surface. Where the top of a guard is also serving as the handrail, a single concentrated load shall be applied in any direction at any point along the top. Concentrated loads shall not be applied concurrently.

TABLE R301.6
MINIMUM ROOF LIVE LOADS IN POUNDS-FORCE PER SQUARE FOOT OF HORIZONTAL PROJECTION

ROOF SLOPE	TRIBUTARY LOADED AREA IN SQUARE FEET FOR ANY STRUCTURAL MEMBER		
	0 to 200	201 to 600	Over 600
Flat or rise less than 4 inches per foot (1:3)	20	16	12
Rise 4 inches per foot (1:3) to less than 12 inches per foot (1:1)	16	14	12
Rise 12 inches per foot (1:1) and greater	12	12	12

For SI: 1 square foot = 0.0929 m^2, 1 pound per square foot = 0.0479 kPa, 1 inch per foot = 83.3 mm/m.

TABLE R301.7
ALLOWABLE DEFLECTION OF STRUCTURAL MEMBERS[b, c]

STRUCTURAL MEMBER	ALLOWABLE DEFLECTION
Rafters having slopes greater than 3:12 with finished ceiling not attached to rafters	*L*/180
Interior walls and partitions	*H*/180
Floors	*L*/360
Ceilings with brittle finishes (including plaster and stucco)	*L*/360
Ceilings with flexible finishes (including gypsum board)	*L*/240
All other structural members	*L*/240
Exterior walls—wind loads[a] with plaster or stucco finish	*H*/360
Exterior walls—wind loads[a] with other brittle finishes	*H*/240
Exterior walls—wind loads[a] with flexible finishes	*H*/120[d]
Lintels supporting masonry veneer walls[e]	*L*/600

Note: *L* = span length, *H* = span height.

a. For the purpose of the determining deflection limits herein, the wind load shall be permitted to be taken as 0.7 times the component and cladding (ASD) loads obtained from Table R301.2.1(1).

b. For cantilever members, *L* shall be taken as twice the length of the cantilever.

c. For aluminum structural members or panels used in roofs or walls of sunroom additions or patio covers, not supporting edge of glass or sandwich panels, the total load deflection shall not exceed *L*/60. For continuous aluminum structural members supporting edge of glass, the total load deflection shall not exceed *L*/175 for each glass lite or *L*/60 for the entire length of the member, whichever is more stringent. For sandwich panels used in roofs or walls of sunroom additions or patio covers, the total load deflection shall not exceed *L*/120.

d. Deflection for exterior walls with interior gypsum board finish shall be limited to an allowable deflection of *H*/180.

e. Refer to Section R703.8.2. The dead load of supported materials shall be included when calculating the deflection of these members.

SECTION R302
FIRE-RESISTANT CONSTRUCTION

R302.1 Exterior walls. Construction, projections, openings and penetrations of exterior walls of *dwellings* and accessory buildings shall comply with Table R302.1(1); or *dwellings* equipped throughout with an *automatic sprinkler system* installed in accordance with Section P2904 shall comply with Table R302.1(2).

Exceptions:

1. Walls, projections, openings or penetrations in walls perpendicular to the line used to determine the *fire separation distance.*
2. Walls of *individual dwelling units* and their *accessory structures* located on the same *lot.*
3. Detached tool sheds and storage sheds, playhouses and similar structures exempted from *permits* are not required to provide wall protection based on location on the *lot.* Projections beyond the exterior wall shall not extend over the *lot line.*
4. Detached garages accessory to a *dwelling* located within 2 feet (610 mm) of a *lot line* are permitted to have roof eave projections not exceeding 4 inches (102 mm).
5. Foundation vents installed in compliance with this code are permitted.

R302.2 Townhouses. Walls separating *townhouse units* shall be constructed in accordance with Section R302.2.1 or R302.2.2 and shall comply with Sections 302.2.3 through 302.2.5.

R302.2.1 Double walls. Each *townhouse unit* shall be separated from other *townhouse units* by two 1-hour fire-resistance-rated wall assemblies tested in accordance with ASTM E119, UL 263 or Section 703.2.2 of the *International Building Code.*

R302.2.2 Common walls. Common walls separating *townhouse units* shall be assigned a fire-resistance rating in accordance with Item 1 or 2 and shall be rated for fire exposure from both sides. Common walls shall extend to and be tight against the exterior sheathing of the exterior walls, or the inside face of exterior walls without stud cavities, and the underside of the roof sheathing. The common wall shared by two *townhouse units* shall be constructed without plumbing or mechanical equipment, ducts or vents, other than water-filled fire sprinkler piping in the cavity of the common wall. Electrical installations shall be in accordance with Chapters 34 through 43. Penetrations of the membrane of common walls for electrical outlet boxes shall be in accordance with Section R302.4.

1. Where an automatic sprinkler system in accordance with Section P2904 is provided, the common wall shall be not less than a 1-hour fire-resistance-rated wall assembly tested in accordance with ASTM E119, UL 263 or Section 703.2.2 of the *International Building Code.*
2. Where an automatic sprinkler system in accordance with Section P2904 is not provided, the common wall shall be not less than a 2-hour fire-resistance-rated wall assembly tested in accordance with ASTM E119, UL 263 or Section 703.2.2 of the *International Building Code.*

Exception: Common walls are permitted to extend to and be tight against the inside of the exterior walls if the cavity between the end of the common wall and the

TABLE R302.1(1)
EXTERIOR WALLS

EXTERIOR WALL ELEMENT		MINIMUM FIRE-RESISTANCE RATING	MINIMUM FIRE SEPARATION DISTANCE
Walls	Fire-resistance rated	1 hour—tested in accordance with ASTM E119, UL 263 or Section 703.3 of the *International Building Code* with exposure from both sides	0 feet
	Not fire-resistance rated	0 hours	≥ 5 feet
Projections	Not allowed	NA	< 2 feet
	Fire-resistance rated	1 hour on the underside, or heavy timber, or fire-retardant-treated wood[a, b]	≥ 2 feet to < 5 feet
	Not fire-resistance rated	0 hours	≥ 5 feet
Openings in walls	Not allowed	NA	< 3 feet
	25% maximum of wall area	0 hours	3 feet
	Unlimited	0 hours	5 feet
Penetrations	All	Comply with Section R302.4	< 3 feet
		None required	3 feet

For SI: 1 foot = 304.8 mm.
NA = Not Applicable.

a. The fire-resistance rating shall be permitted to be reduced to 0 hours on the underside of the eave overhang if fireblocking is provided from the wall top plate to the underside of the roof sheathing.
b. The fire-resistance rating shall be permitted to be reduced to 0 hours on the underside of the rake overhang where gable vent openings are not installed.

TABLE R302.1(2)
EXTERIOR WALLS—DWELLINGS WITH FIRE SPRINKLERS

EXTERIOR WALL ELEMENT		MINIMUM FIRE-RESISTANCE RATING	MINIMUM FIRE SEPARATION DISTANCE
Walls	Fire-resistance rated	1 hour—tested in accordance with ASTM E119, UL 263 or Section 703.2.2 of the *International Building Code* with exposure from the outside	0 feet
	Not fire-resistance rated	0 hours	3 feet[a]
Projections	Not allowed	NA	< 2 feet
	Fire-resistance rated	1 hour on the underside, or heavy timber, or fire-retardant-treated wood[b, c]	2 feet[a]
	Not fire-resistance rated	0 hours	3 feet
Openings in walls	Not allowed	NA	< 3 feet
	Unlimited	0 hours	3 feet[a]
Penetrations	All	Comply with Section R302.4	< 3 feet
		None required	3 feet[a]

For SI: 1 foot = 304.8 mm.
NA = Not Applicable.

a. For residential subdivisions where all dwellings are equipped throughout with an automatic sprinkler system installed in accordance with Section P2904, the fire separation distance for exterior walls not fire-resistance rated and for fire-resistance-rated projections shall be permitted to be reduced to 0 feet, and unlimited unprotected openings and penetrations shall be permitted, where the adjoining lot provides an open setback yard that is 6 feet or more in width on the opposite side of the property line.
b. The fire-resistance rating shall be permitted to be reduced to 0 hours on the underside of the eave overhang if fireblocking is provided from the wall top plate to the underside of the roof sheathing.
c. The fire-resistance rating shall be permitted to be reduced to 0 hours on the underside of the rake overhang where gable vent openings are not installed.

exterior sheathing is filled with a minimum of two 2-inch nominal thickness wood studs.

R302.2.3 Continuity. The fire-resistance-rated wall or assembly separating *townhouse units* shall be continuous from the foundation to the underside of the roof sheathing, deck or slab. The fire-resistance rating shall extend the full length of the wall or assembly, including wall extensions through and separating attached enclosed *accessory structures.*

R302.2.4 Parapets for townhouses. Parapets constructed in accordance with Section R302.2.5 shall be constructed for *townhouses* as an extension of exterior walls or common walls separating *townhouse units* in accordance with the following:

1. Where roof surfaces adjacent to the wall or walls are at the same elevation, the parapet shall extend not less than 30 inches (762 mm) above the roof surfaces.
2. Where roof surfaces adjacent to the wall or walls are at different elevations and the higher roof is not more than 30 inches (762 mm) above the lower roof, the parapet shall extend not less than 30 inches (762 mm) above the lower roof surface.

   Exception: A parapet is not required in the preceding two cases where the roof covering complies with a minimum Class C rating as tested in accordance with ASTM E108 or UL 790 and the roof decking or sheathing is of *noncombustible materials* or fire-retardant-treated wood for a distance of 4 feet (1219 mm) on each side of the wall or walls, or one layer of $^5/_8$-inch (15.9 mm) Type X gypsum board is installed directly beneath the roof decking or sheathing, supported by not less than nominal 2-inch (51 mm) ledgers attached to the sides of the roof framing members, for a distance of not less than 4 feet (1219 mm) on each side of the wall or walls and any openings or penetrations in the roof are not within 4 feet (1219 mm) of the common walls. Fire-retardant-treated wood shall meet the requirements of Sections R802.1.5 and R803.2.1.2.
3. A parapet is not required where roof surfaces adjacent to the wall or walls are at different elevations and the higher roof is more than 30 inches (762 mm) above the lower roof. The common wall construction from the lower roof to the underside of the higher *roof deck* shall have not less than a 1-hour fire-resistance rating. The wall shall be rated for exposure from both sides.

R302.2.5 Parapet construction. Parapets shall have the same fire-resistance rating as that required for the supporting wall or walls. On any side adjacent to a roof surface, the parapet shall have noncombustible faces for the uppermost 18 inches (457 mm), to include counter-flashing and coping materials. Where the roof slopes toward a parapet at slopes greater than 2 units vertical in 12 units horizontal (16.7-percent slope), the parapet shall extend to the same height as any portion of the roof within a distance of 3 feet (914 mm), and the height shall be not less than 30 inches (762 mm).

R302.2.6 Structural independence. Each *townhouse unit* shall be structurally independent.

Exceptions:

1. Foundations supporting exterior walls or common walls.
2. Structural roof and wall sheathing from each unit fastened to the common wall framing.
3. Nonstructural wall and roof coverings.
4. Flashing at termination of roof covering over common wall.
5. *Townhouse units* separated by a common wall as provided in Section R302.2.2, Item 1 or 2.
6. *Townhouse units* protected by a fire sprinkler system complying with Section P2904 or NFPA 13D.

R302.3 Two-family dwellings. *Dwelling units* in two-family dwellings shall be separated from each other by wall and floor assemblies having not less than a 1-hour fire-resistance rating where tested in accordance with ASTM E119, UL 263 or Section 703.2.2 of the *International Building Code.* Such separation shall be provided regardless of whether a *lot line* exists between the two *dwelling units* or not. Fire-resistance-rated floor/ceiling and wall assemblies shall extend to and be tight against the exterior wall, and wall assemblies shall extend from the foundation to the underside of the roof sheathing.

Exceptions:

1. A fire-resistance rating of $^1/_2$ hour shall be permitted in buildings equipped throughout with an automatic sprinkler system installed in accordance with Section P2904.
2. Wall assemblies need not extend through attic spaces where the ceiling is protected by not less than $^5/_8$-inch (15.9 mm) Type X gypsum board, an attic draft stop constructed as specified in Section R302.12.1 is provided above and along the wall assembly separating the *dwellings* and the structural framing supporting the ceiling is protected by not less than $^1/_2$-inch (12.7 mm) gypsum board or equivalent.

R302.3.1 Supporting construction. Where floor assemblies are required to be fire-resistance rated by Section R302.3, the supporting construction of such assemblies shall have an equal or greater fire-resistance rating.

R302.4 Dwelling unit rated penetrations. Penetrations of wall or floor-ceiling assemblies required to be fire-resistance rated in accordance with Section R302.2 or R302.3 shall be protected in accordance with this section.

R302.4.1 Through penetrations. Through penetrations of fire-resistance-rated wall or floor assemblies shall comply with Section R302.4.1.1 or R302.4.1.2.

Exceptions:

1. Where the penetrating items are steel, ferrous or copper pipes, tubes or conduits, the annular space shall be protected as follows:
   1.1. In concrete or masonry wall or floor assemblies, concrete, grout or mortar shall be permitted where installed to the full thickness of the wall or floor assembly or the thickness required to maintain the fire-resistance rating, provided that both of the following are complied with:
      1.1.1. The nominal diameter of the penetrating item is not more than 6 inches (152 mm).
      1.1.2. The area of the opening through the wall does not exceed 144 square inches (92 900 mm^2).
   1.2. The material used to fill the annular space shall prevent the passage of flame and hot gases sufficient to ignite cotton waste where subjected to ASTM E119 or UL 263 time temperature fire conditions under a positive pressure differential of not less than 0.01 inch of water (3 Pa) at the location of the penetration for the time period equivalent to the fire-resistance rating of the construction penetrated.
2. The annular space created by the penetration of water-filled fire sprinkler piping, provided that the annular space is filled using a material complying with Item 1.2 of Exception 1.

R302.4.1.1 Fire-resistance-rated assembly. Penetrations shall be installed as tested in the *approved* fire-resistance-rated assembly.

R302.4.1.2 Penetration firestop system. Penetrations shall be protected by an *approved* penetration firestop system installed as tested in accordance with ASTM E814 or UL 1479, with a positive pressure differential of not less than 0.01 inch of water (3 Pa) and shall have an F rating of not less than the required fire-resistance rating of the wall or floor-ceiling assembly penetrated.

R302.4.2 Membrane penetrations. Membrane penetrations shall comply with Section R302.4.1. Where walls are required to have a fire-resistance rating, recessed fixtures shall be installed so that the required fire-resistance rating will not be reduced.

Exceptions:

1. Membrane penetrations of not more than 2-hour fire-resistance-rated walls and partitions by steel electrical boxes that do not exceed 16 square inches (0.0103 m^2) in area provided that the aggregate area of the openings through the membrane does not exceed 100 square inches (0.0645 m^2) in any 100 square feet (9.29 m^2) of wall area. The annular space between the wall membrane and the box shall not exceed $^1/_8$ inch (3.1 mm). Such boxes on opposite sides of the wall shall be separated by one of the following:
   1.1. By a horizontal distance of not less than 24 inches (610 mm) where the wall or partition is constructed with individual noncommunicating stud cavities.
   1.2. By a horizontal distance of not less than the depth of the wall cavity where the wall cavity is filled with cellulose loose-fill, rockwool or slag mineral wool insulation.
   1.3. By solid fireblocking in accordance with Section R302.11.
   1.4. By protecting both boxes with *listed* putty pads.
   1.5. By other *listed* materials and methods.
2. Membrane penetrations by *listed* electrical boxes of any materials provided that the boxes have been tested for use in fire-resistance-rated assemblies and are installed in accordance with the instructions included in the *listing*. The annular space between the wall membrane and the box shall not exceed $^1/_8$ inch (3.1 mm) unless *listed* otherwise. Such boxes on opposite sides of the wall shall be separated by one of the following:
   2.1. By the horizontal distance specified in the *listing* of the electrical boxes.
   2.2. By solid fireblocking in accordance with Section R302.11.
   2.3. By protecting both boxes with *listed* putty pads.
   2.4. By other *listed* materials and methods.
3. The annular space created by the penetration of a fire sprinkler or water-filled fire sprinkler piping, provided that the annular space is covered by a metal escutcheon plate.
4. Ceiling membrane penetrations by *listed* luminaires or by luminaires protected with *listed* materials that have been tested for use in fire-resistance-rated assemblies and are installed in accordance with the instructions included in the *listing*.

R302.5 Dwelling-garage opening and penetration protection. Openings and penetrations through the walls or ceilings separating the *dwelling* from the garage shall be in accordance with Sections R302.5.1 through R302.5.3.

R302.5.1 Opening protection. Openings from a private garage directly into a room used for sleeping purposes shall not be permitted. Other openings between the garage and residence shall be equipped with solid wood doors not less than $1^3/_8$ inches (35 mm) in thickness, solid or honeycomb-core steel doors not less than $1^3/_8$ inches (35 mm) thick, or 20-minute fire-rated doors. Doors shall be self-latching and equipped with a self-closing or automatic-closing device.

R302.5.2 Duct penetration. Ducts in the garage and ducts penetrating the walls or ceilings separating the *dwelling* from the garage shall be constructed of a minimum No. 26 gage (0.48 mm) sheet steel or other *approved* material and shall not have openings into the garage.

R302.5.3 Other penetrations. Penetrations through the separation required in Section R302.6 shall be protected as required by Section R302.11, Item 4.

R302.6 Dwelling-garage fire separation. The garage shall be separated as required by Table R302.6. Openings in garage walls shall comply with Section R302.5. Attachment of gypsum board shall comply with Table R702.3.5. The wall separation provisions of Table R302.6 shall not apply to garage walls that are perpendicular to the adjacent *dwelling unit* wall.

R302.7 Under-stair protection. Enclosed space under stairs that is *accessed* by a door or access panel shall have walls, under-stair surface and any soffits protected on the enclosed side with $^1/_2$-inch (12.7 mm) gypsum board.

R302.8 Foam plastics. For requirements for foam plastics, see Section R316.

R302.8.1 Interior finish. Foam plastics used as interior finishes shall comply with Section R316.5.10.

R302.9 Flame spread index and smoke-developed index for wall and ceiling finishes. Flame spread and smoke-developed indices for wall and ceiling finishes shall be in accordance with Sections R302.9.1 through R302.9.4.

R302.9.1 Flame spread index. Wall and ceiling finishes shall have a flame spread index of not greater than 200.

Exception: Flame spread index requirements for finishes shall not apply to *trim* defined as picture molds, chair rails, baseboards and *handrails*; to doors and windows or their frames; or to materials that are less than $^1/_{28}$ inch (0.91 mm) in thickness cemented to the surface of walls or ceilings if these materials exhibit flame spread index values not greater than those of paper of this thickness cemented to a noncombustible backing.

R302.9.2 Smoke-developed index. Wall and ceiling finishes shall have a *smoke-developed index* of not greater than 450.

R302.9.3 Testing. Tests shall be made in accordance with ASTM E84 or UL 723.

R302.9.4 Alternative test method. As an alternative to having a flame spread index of not greater than 200 and a *smoke-developed index* of not greater than 450 where tested in accordance with ASTM E84 or UL 723, wall and ceiling finishes shall be permitted to be tested in accordance with NFPA 286. Materials tested in accordance with NFPA 286 shall meet the following criteria:

The interior finish shall comply with the following:

1. During the 40 kW exposure, flames shall not spread to the ceiling.
2. The flame shall not spread to the outer extremity of the sample on any wall or ceiling.
3. Flashover, as defined in NFPA 286, shall not occur.
4. The peak heat release rate throughout the test shall not exceed 800 kW.
5. The total smoke released throughout the test shall not exceed 1,000 m^2.

R302.9.5 High-density polyethylene (HDPE) and polypropylene (PP). Where high-density polyethylene or polypropylene is used as an interior finish material, it shall be tested in accordance with NFPA 286 and comply with the criteria in Section R302.9.4.

R302.10 Flame spread index and smoke-developed index for insulation. Flame spread and *smoke-developed index* for insulation shall be in accordance with Sections R302.10.1 through R302.10.5.

R302.10.1 Insulation. Insulating materials installed within floor-ceiling assemblies, roof-ceiling assemblies, wall assemblies, crawl spaces and *attics* shall comply with the requirements of this section. They shall exhibit a flame spread index not to exceed 25 and a *smoke-developed index* not to exceed 450 where tested in accordance with ASTM E84 or UL 723. Insulating materials, where

TABLE R302.6
DWELLING-GARAGE SEPARATION

SEPARATION	MATERIAL
From the residence and attics	Not less than $^1/_2$-inch gypsum board or equivalent applied to the garage side
From habitable rooms above the garage	Not less than $^5/_8$-inch Type X gypsum board or equivalent
Structure(s) supporting floor/ceiling assemblies used for separation required by this section	Not less than $^1/_2$-inch gypsum board or equivalent
Garages located less than 3 feet from a dwelling unit on the same lot	Not less than $^1/_2$-inch gypsum board or equivalent applied to the interior side of exterior walls that are within this area

For SI: 1 inch = 25.4 mm, 1 foot = 304.8 mm.

tested in accordance with the requirements of this section, shall include facings, where used, such as vapor retarders, *vapor permeable* membranes and similar coverings.

Exceptions:

1. Where such materials are installed in concealed spaces, the flame spread index and *smoke-developed index* limitations do not apply to the facings, provided that the facing is installed in substantial contact with the unexposed surface of the ceiling, floor or wall finish.
2. Cellulose fiber loose-fill insulation that is not spray applied and that complies with the requirements of Section R302.10.3 shall not be required to meet the flame spread index requirements but shall be required to meet a *smoke-developed index* of not more than 450 where tested in accordance with CAN/ULC S102.2.
3. Foam plastic insulation shall comply with Section R316.

R302.10.2 Loose-fill insulation. Loose-fill insulation materials that cannot be mounted in the ASTM E84 or UL 723 apparatus without a screen or artificial supports shall comply with the flame spread and smoke-developed limits of Section R302.10.1 where tested in accordance with CAN/ULC S102.2.

Exception: Cellulosic fiber loose-fill insulation shall not be required to be tested in accordance with CAN/ULC S102.2, provided that such insulation complies with the requirements of Sections R302.10.1 and R302.10.3.

R302.10.3 Cellulosic fiber loose-fill insulation. Cellulosic fiber loose-fill insulation shall comply with CPSC 16 CFR, Parts 1209 and 1404. Each package of such insulating material shall be clearly *labeled* in accordance with CPSC 16 CFR, Parts 1209 and 1404.

R302.10.4 Exposed attic insulation. Exposed insulation materials installed on attic floors shall have a critical radiant flux of not less than 0.12 watt per square centimeter.

R302.10.5 Testing. Tests for critical radiant flux shall be made in accordance with ASTM E970.

R302.11 Fireblocking. In combustible construction, fireblocking shall be provided to cut off both vertical and horizontal concealed draft openings and to form an effective fire barrier between stories, and between a top story and the roof space.

Fireblocking shall be provided in wood-framed construction in the following locations:

1. In concealed spaces of stud walls and partitions, including furred spaces and parallel rows of studs or staggered studs, as follows:
   1.1. Vertically at the ceiling and floor levels.
   1.2. Horizontally at intervals not exceeding 10 feet (3048 mm).
2. At interconnections between concealed vertical and horizontal spaces such as occur at soffits, drop ceilings and cove ceilings.
3. In concealed spaces between stair stringers at the top and bottom of the run. Enclosed spaces under stairs shall comply with Section R302.7.
4. At openings around vents, pipes, ducts, cables and wires at ceiling and floor level, with an *approved* material to resist the free passage of flame and products of combustion. The material filling this annular space shall not be required to meet the ASTM E136 requirements.
5. For the fireblocking of chimneys and fireplaces, see Section R1003.19.
6. Fireblocking of cornices of a two-family *dwelling* is required at the line of *dwelling unit* separation.

R302.11.1 Fireblocking materials. Except as provided in Section R302.11, Item 4, fireblocking shall consist of the following materials.

1. Two-inch (51 mm) nominal lumber.
2. Two thicknesses of 1-inch (25.4 mm) nominal lumber with broken lap joints.
3. One thickness of $^{23}/_{32}$-inch (18.3 mm) *wood structural panels* with joints backed by $^{23}/_{32}$-inch (18.3 mm) *wood structural panels.*
4. One thickness of $^{3}/_{4}$-inch (19.1 mm) particleboard with joints backed by $^{3}/_{4}$-inch (19.1 mm) particleboard.
5. One-half-inch (12.7 mm) gypsum board.
6. One-quarter-inch (6.4 mm) cement-based millboard.
7. Batts or blankets of mineral wool or glass fiber or other *approved* materials installed in such a manner as to be securely retained in place.
8. Cellulose insulation installed as tested in accordance with ASTM E119 or UL 263, for the specific application.

R302.11.1.1 Batts or blankets of mineral or glass fiber. Batts or blankets of mineral or glass fiber or other *approved* nonrigid materials shall be permitted for compliance with the 10-foot (3048 mm) horizontal fireblocking in walls constructed using parallel rows of studs or staggered studs.

R302.11.1.2 Unfaced fiberglass. Unfaced fiberglass batt insulation used as fireblocking shall fill the entire cross section of the wall cavity to a height of not less than 16 inches (406 mm) measured vertically. Where piping, conduit or similar obstructions are encountered, the insulation shall be packed tightly around the obstruction.

R302.11.1.3 Loose-fill insulation material. Loose-fill insulation material shall not be used as a fireblock unless specifically tested in the form and manner intended for use to demonstrate its ability to remain in place and to retard the spread of fire and hot gases.

R302.11.2 Fireblocking integrity. The integrity of fireblocks shall be maintained.

R302.12 Draftstopping. In combustible construction where there is usable space both above and below the concealed space of a floor-ceiling assembly, draftstops shall be installed so that the area of the concealed space does not exceed 1,000 square feet (92.9 m^2). Draftstopping shall divide the concealed space into approximately equal areas. Where the assembly is enclosed by a floor membrane above and a ceiling membrane below, draftstopping shall be provided in floor-ceiling assemblies under the following circumstances:

1. Ceiling is suspended under the floor framing.
2. Floor framing is constructed of truss-type open-web or perforated members.

R302.12.1 Materials. Draftstopping materials shall be not less than $^1/_2$-inch (12.7 mm) gypsum board, $^3/_8$-inch (9.5 mm) *wood structural panels* or other *approved* materials adequately supported. Draftstopping shall be installed parallel to the floor framing members unless otherwise *approved* by the *building official.* The integrity of the draftstops shall be maintained.

R302.13 Fire protection of floors. Floor assemblies that are not required elsewhere in this code to be fire-resistance rated, shall be provided with a $^1/_2$-inch (12.7 mm) gypsum wallboard membrane, $^5/_8$-inch (16 mm) *wood structural panel* membrane, or equivalent on the underside of the floor framing member. Penetrations or openings for ducts, vents, electrical outlets, lighting, devices, luminaires, wires, speakers, drainage, piping and similar openings or penetrations shall be permitted.

Exceptions:

1. Floor assemblies located directly over a space protected by an automatic sprinkler system in accordance with Section P2904, NFPA 13D, or other *approved* equivalent sprinkler system.
2. Floor assemblies located directly over a *crawl space* not intended for storage or for the installation of fuel-fired or electric-powered heating *appliances.*
3. Portions of floor assemblies shall be permitted to be unprotected where complying with the following:
   - 3.1. The aggregate area of the unprotected portions does not exceed 80 square feet (7.4 m^2) per story.
   - 3.2. Fireblocking in accordance with Section R302.11.1 is installed along the perimeter of the unprotected portion to separate the unprotected portion from the remainder of the floor assembly.
4. Wood floor assemblies using dimension lumber or *structural composite lumber* equal to or greater than 2-inch by 10-inch (50.8 mm by 254 mm) nominal dimension, or other *approved* floor assemblies demonstrating equivalent fire performance.

R302.14 Combustible insulation clearance. Combustible insulation shall be separated not less than 3 inches (76 mm) from recessed luminaires, fan motors and other heat-producing devices.

Exception: Where heat-producing devices are *listed* for lesser clearances, combustible insulation complying with the listing requirements shall be separated in accordance with the conditions stipulated in the listing.

Recessed luminaires installed in the *building thermal envelope* shall meet the requirements of Section N1102.4.5 of this code.

SECTION R303
LIGHT, VENTILATION AND HEATING

R303.1 Habitable rooms. Habitable rooms shall have an aggregate glazing area of not less than 8 percent of the floor area of such rooms. Natural *ventilation* shall be through windows, skylights, doors, louvers or other *approved* openings to the outdoor air. Such openings shall be provided with ready access or shall otherwise be readily controllable by the building occupants. The openable area to the outdoors shall be not less than 4 percent of the floor area being ventilated.

Exceptions:

1. For habitable rooms other than kitchens, the glazed areas need not be openable where the opening is not required by Section R310 and a whole-house mechanical *ventilation* system or a mechanical ventilation system capable of producing 0.35 air changes per hour in the habitable rooms is installed in accordance with Section M1505.
2. For kitchens, the glazed areas need not be openable where the opening is not required by Section R310 and a local exhaust system is installed in accordance with Section M1505.
3. The glazed areas need not be installed in rooms where Exception 1 is satisfied and artificial light is provided that is capable of producing an average illumination of 6 footcandles (65 lux) over the area of the room at a height of 30 inches (762 mm) above the floor level.
4. Use of *sunroom* and patio covers, as defined in Section R202, shall be permitted for natural *ventilation* if in excess of 40 percent of the exterior *sunroom* walls are open, or are enclosed only by insect screening.

R303.2 Adjoining rooms. For the purpose of determining light and *ventilation* requirements, rooms shall be considered to be a portion of an adjoining room where not less than one-half of the area of the common wall is open and unobstructed and provides an opening of not less than one-tenth of the floor area of the interior room and not less than 25 square feet (2.3 m^2).

Exception: Openings required for light or *ventilation* shall be permitted to open into a *sunroom* with thermal isolation or a patio cover, provided that there is an openable area between the adjoining room and the *sunroom* or patio cover of not less than one-tenth of the floor area of the interior room and not less than 20 square feet (2 m^2). The minimum openable area to the outdoors shall be based on the total floor area being ventilated.

R303.3 Bathrooms. Bathrooms, water closet compartments and other similar rooms shall be provided with aggregate glazing area in windows of not less than 3 square feet (0.3 m^2), one-half of which shall be openable.

Exception: The glazed areas shall not be required where artificial light and a local exhaust system are provided. The minimum local exhaust rates shall be determined in accordance with Section M1505. Exhaust air from the space shall be exhausted directly to the outdoors.

R303.4 Mechanical ventilation. Buildings and *dwelling units* complying with Section N1102.4.1 shall be provided with mechanical ventilation in accordance with Section M1505, or with other *approved* means of ventilation.

R303.5 Opening location. Outdoor intake and exhaust openings shall be located in accordance with Sections R303.5.1 and R303.5.2.

R303.5.1 Intake openings. Mechanical and gravity outdoor air intake openings shall be located not less than 10 feet (3048 mm) from any hazardous or noxious contaminant, such as vents, chimneys, plumbing vents, streets, alleys, parking lots and loading docks.

For the purpose of this section, the exhaust from *dwelling unit* toilet rooms, bathrooms and *kitchens* shall not be considered as hazardous or noxious.

Exceptions:

1. The 10-foot (3048 mm) separation is not required where the intake opening is located 3 feet (914 mm) or greater below the contaminant source.
2. Vents and chimneys serving fuel-burning *appliances* shall be terminated in accordance with the applicable provisions of Chapters 18 and 24.
3. Clothes dryer exhaust ducts shall be terminated in accordance with Section M1502.3.

R303.5.2 Exhaust openings. Exhaust air shall not be directed onto walkways.

R303.6 Outside opening protection. Air exhaust and intake openings that terminate outdoors shall be protected with corrosion-resistant screens, louvers or grilles having an opening size of not less than $^1/_4$ inch (6 mm) and a maximum opening size of $^1/_2$ inch (13 mm), in any dimension. Openings shall be protected against local weather conditions. Outdoor air exhaust and intake openings shall meet the provisions for exterior wall opening protectives in accordance with this code.

R303.7 Interior stairway illumination. Interior *stairways* shall be provided with an artificial light source to illuminate the landings and treads. The light source shall be capable of illuminating treads and landings to levels of not less than 1 footcandle (11 lux) as measured at the center of treads and landings. There shall be a wall switch at each floor level to control the light source where the *stairway* has six or more *risers.*

Exception: A switch is not required where remote, central or automatic control of lighting is provided.

R303.8 Exterior stairway illumination. Exterior *stairways* shall be provided with an artificial light source located at the top landing of the *stairway.* Exterior *stairways* providing access to a *basement* from the outdoor *grade* level shall be provided with an artificial light source located at the bottom landing of the *stairway.*

R303.9 Required glazed openings. Required glazed openings shall open directly onto a street or public alley, or a *yard* or court located on the same *lot* as the building.

Exceptions:

1. Required glazed openings that face into a roofed porch where the porch abuts a street, *yard* or court and the longer side of the porch is not less than 65 percent unobstructed and the ceiling height is not less than 7 feet (2134 mm).
2. Eave projections shall not be considered as obstructing the clear open space of a *yard* or court.
3. Required glazed openings that face into the area under a deck, balcony, bay or floor cantilever where a clear vertical space not less than 36 inches (914 mm) in height is provided.

R303.9.1 Sunroom additions. Required glazed openings shall be permitted to open into *sunroom additions* or patio covers that abut a street, *yard* or court if in excess of 40 percent of the exterior *sunroom* walls are open, or are enclosed only by insect screening, and the ceiling height of the *sunroom* is not less than 7 feet (2134 mm).

R303.10 Required heating. Where the winter design temperature in Table R301.2 is below 60°F (16°C), every *dwelling unit* shall be provided with heating facilities capable of maintaining a room temperature of not less than 68°F (20°C) at a point 3 feet (914 mm) above the floor and 2 feet (610 mm) from exterior walls in habitable rooms at the design temperature. The installation of one or more portable space heaters shall not be used to achieve compliance with this section.

SECTION R304
MINIMUM ROOM AREAS

R304.1 Minimum area. Habitable rooms shall have a floor area of not less than 70 square feet (6.5 m^2).

Exception: Kitchens.

R304.2 Minimum dimensions. Habitable rooms shall be not less than 7 feet (2134 mm) in any horizontal dimension.

Exception: Kitchens.

R304.3 Height effect on room area. Portions of a room with a sloping ceiling measuring less than 5 feet (1524 mm) or a furred ceiling measuring less than 7 feet (2134 mm) from the finished floor to the finished ceiling shall not be considered as contributing to the minimum required habitable area for that room.

SECTION R305 CEILING HEIGHT

R305.1 Minimum height. *Habitable space*, hallways and portions of *basements* containing these spaces shall have a ceiling height of not less than 7 feet (2134 mm). Bathrooms, toilet rooms and laundry rooms shall have a ceiling height of not less than 6 feet 8 inches (2032 mm).

Exceptions:

1. For rooms with sloped ceilings, the required floor area of the room shall have a ceiling height of not less than 5 feet (1524 mm) and not less than 50 percent of the required floor area shall have a ceiling height of not less than 7 feet (2134 mm).
2. The ceiling height above bathroom and toilet room fixtures shall be such that the fixture is capable of being used for its intended purpose. A shower or tub equipped with a showerhead shall have a ceiling height of not less than 6 feet 8 inches (2032 mm) above an area of not less than 30 inches (762 mm) by 30 inches (762 mm) at the showerhead.
3. Beams, girders, ducts or other obstructions in *basements* containing *habitable space* shall be permitted to project to within 6 feet 4 inches (1931 mm) of the finished floor.
4. Beams and girders spaced apart not less than 36 inches (914 mm) in clear finished width shall project not more than 78 inches (1981 mm) from the finished floor.

R305.1.1 Basements. Portions of *basements* that do not contain *habitable space* or hallways shall have a ceiling height of not less than 6 feet 8 inches (2032 mm).

Exception: At beams, girders, ducts or other obstructions, the ceiling height shall be not less than 6 feet 4 inches (1931 mm) from the finished floor.

SECTION R306 SANITATION

R306.1 Toilet facilities. Every *dwelling unit* shall be provided with a water closet, lavatory, and a bathtub or shower.

R306.2 Kitchen. Each *dwelling unit* shall be provided with a kitchen area and every kitchen area shall be provided with a sink.

R306.3 Sewage disposal. Plumbing fixtures shall be connected to a sanitary sewer or to an *approved* private sewage disposal system.

R306.4 Water supply to fixtures. Plumbing fixtures shall be connected to an *approved* water supply. Kitchen sinks, lavatories, bathtubs, showers, bidets, laundry tubs and washing machine outlets shall be provided with hot and cold water.

SECTION R307 TOILET, BATH AND SHOWER SPACES

R307.1 Space required. Fixtures shall be spaced in accordance with Figure R307.1, and in accordance with the requirements of Section P2705.1.

R307.2 Bathtub and shower spaces. Bathtub and shower floors and walls above bathtubs with installed shower heads and in shower compartments shall be finished with a nonabsorbent surface. Such wall surfaces shall extend to a height of not less than 6 feet (1829 mm) above the floor.

SECTION R308 GLAZING

R308.1 Identification. Except as indicated in Section R308.1.1 each pane of glazing installed in hazardous locations as defined in Section R308.4 shall be provided with a manufacturer's designation specifying who applied the designation, the type of glass and the safety glazing standard with which it complies, and that is visible in the final installation. The designation shall be acid etched, sandblasted, ceramic-fired, laser etched, embossed, or be of a type that once applied cannot be removed without being destroyed. A *label* shall be permitted in lieu of the manufacturer's designation.

Exceptions:

1. For other than tempered glass, manufacturer's designations are not required provided that the *building official* approves the use of a certificate, affidavit or other evidence confirming compliance with this code.
2. Tempered spandrel glass is permitted to be identified by the manufacturer with a removable paper designation.

R308.1.1 Identification of multiple assemblies. Multipane assemblies having individual panes not exceeding 1 square foot (0.09 m^2) in exposed area shall have not less than one pane in the assembly identified in accordance with Section R308.1. Other panes in the assembly shall be *labeled* "CPSC 16 CFR 1201" or "ANSI Z97.1" as appropriate.

R308.2 Louvered windows or jalousies. Regular, float, wired or patterned glass in jalousies and louvered windows shall be not less than nominal $^3/_{16}$ inch (5 mm) thick and not more than 48 inches (1219 mm) in length. Exposed glass edges shall be smooth.

R308.2.1 Wired glass prohibited. Wired glass with wire exposed on longitudinal edges shall not be used in jalousies or louvered windows.

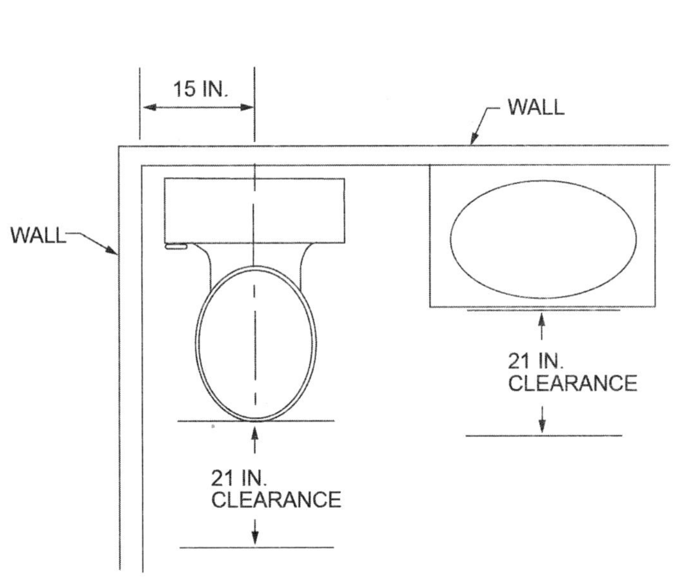

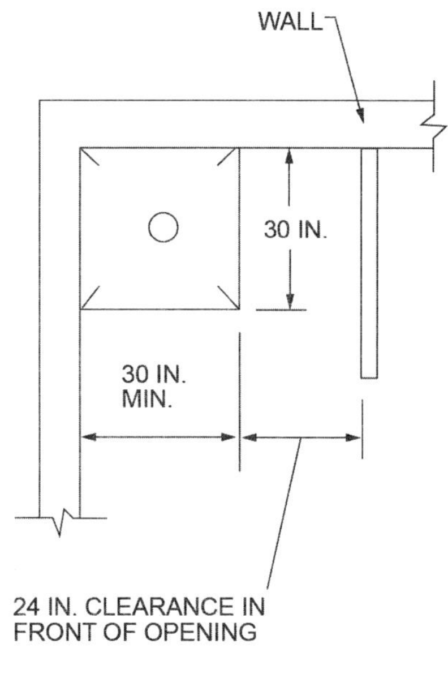

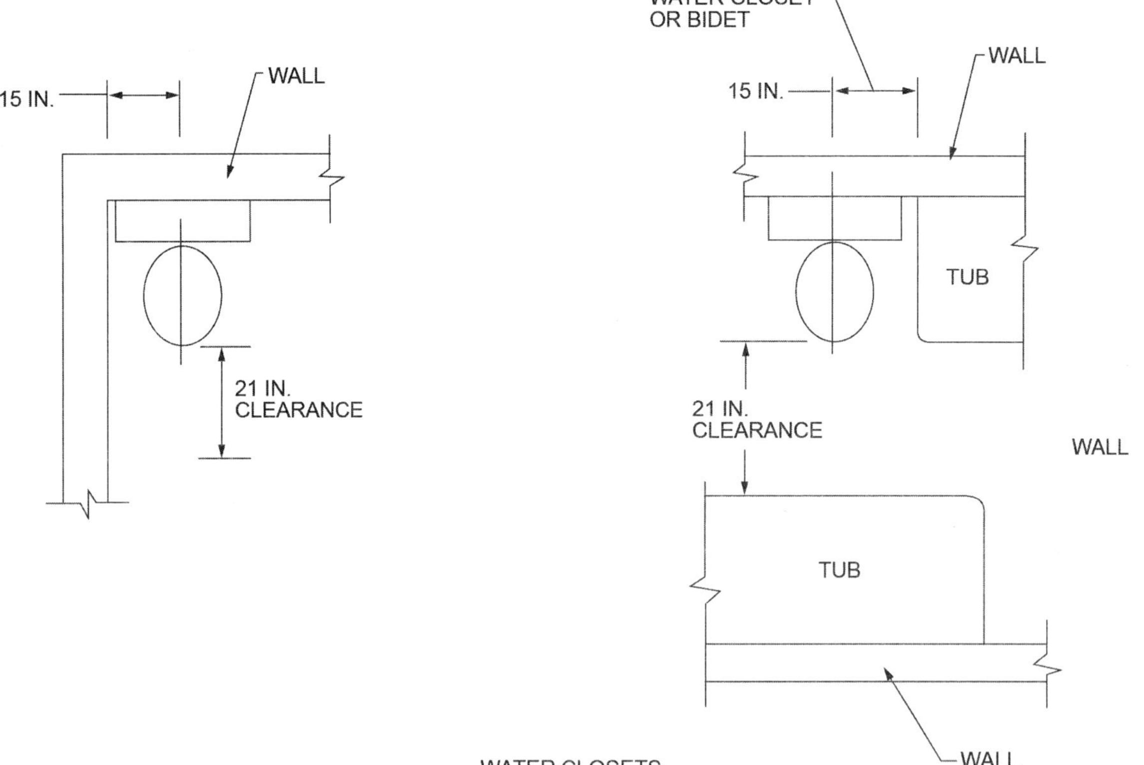

For SI: 1 inch = 25.4 mm.

FIGURE R307.1
MINIMUM FIXTURE CLEARANCES

R308.3 Human impact loads. Individual glazed areas, including glass mirrors in hazardous locations such as those indicated as defined in Section R308.4, shall pass the test requirements of Section R308.3.1.

Exceptions:

1. Louvered windows and jalousies shall comply with Section R308.2.
2. Mirrors and other glass panels mounted or hung on a surface that provides a continuous backing support.
3. Glass unit masonry complying with Section R607.

R308.3.1 Impact test. Where required by other sections of the code, glazing shall be tested in accordance with CPSC 16 CFR 1201. Glazing shall comply with the test criteria for Category II unless otherwise indicated in Table R308.3.1(1).

Exception: Glazing not in doors or enclosures for hot tubs, whirlpools, saunas, steam rooms, bathtubs and showers shall be permitted to be tested in accordance with ANSI Z97.1. Glazing shall comply with the test criteria for Class A unless otherwise indicated in Table R308.3.1(2).

R308.4 Hazardous locations. The locations specified in Sections R308.4.1 through R308.4.7 shall be considered to be specific hazardous locations for the purposes of glazing.

R308.4.1 Glazing in doors. Glazing in fixed and operable panels of swinging, sliding and bifold doors shall be considered to be a hazardous location.

Exceptions:

1. Glazed openings of a size through which a 3-inch-diameter (76 mm) sphere is unable to pass.
2. Decorative glazing.

R308.4.2 Glazing adjacent to doors. Glazing in an individual fixed or operable panel adjacent to a door shall be considered to be a hazardous location where the bottom exposed edge of the glazing is less than 60 inches (1524 mm) above the floor or walking surface and it meets either of the following conditions:

1. Where the glazing is within 24 inches (610 mm) of either side of the door in the plane of the door in a closed position.
2. Where the glazing is on a wall less than 180 degrees (3.14 rad) from the plane of the door in a closed position and within 24 inches (610 mm) of the hinge side of an in-swinging door.

Exceptions:

1. Decorative glazing.
2. Where there is an intervening wall or other permanent barrier between the door and the glazing.
3. Where access through the door is to a closet or storage area 3 feet (914 mm) or less in depth. Glazing in this application shall comply with Section R308.4.3.
4. Glazing that is adjacent to the fixed panel of patio doors.

R308.4.3 Glazing in windows. Glazing in an individual fixed or operable panel that meets all of the following conditions shall be considered to be a hazardous location:

1. The exposed area of an individual pane is larger than 9 square feet (0.836 m^2).
2. The bottom edge of the glazing is less than 18 inches (457 mm) above the floor.
3. The top edge of the glazing is more than 36 inches (914 mm) above the floor.

TABLE R308.3.1(1)
MINIMUM CATEGORY CLASSIFICATION OF GLAZING USING CPSC 16 CFR 1201

EXPOSED SURFACE AREA OF ONE SIDE OF ONE LITE	GLAZING IN STORM OR COMBINATION DOORS (Category Class)	GLAZING INDOORS (Category Class)	GLAZED PANELS REGULATED BY SECTION R308.4.3 (Category Class)	GLAZED PANELS REGULATED BY SECTION R308.4.2 (Category Class)	GLAZING INDOORS AND ENCLOSURES REGULATED BY SECTION 308.4.5 (Category Class)	SLIDING GLASS DOORS PATIO TYPE (Category Class)
9 square feet or less	I	I	NR	I	II	II
More than 9 square feet	II	II	II	II	II	II

For SI: 1 square foot = 0.0929 m^2.
NR = No Requirement.

TABLE R308.3.1(2)
MINIMUM CATEGORY CLASSIFICATION OF GLAZING USING ANSI Z97.1

EXPOSED SURFACE AREA OF ONE SIDE OF ONE LITE	GLAZED PANELS REGULATED BY SECTION R308.4.3 (Category Class)	GLAZED PANELS REGULATED BY SECTION R308.4.2 (Category Class)	DOORS AND ENCLOSURES REGULATED BY SECTION R308.4.5[a] (Category Class)
9 square feet or less	No requirement	B	A
More than 9 square feet	A	A	A

For SI: 1 square foot = 0.0929 m^2.
a. Use is permitted only by the exception to Section R308.3.1.

4. One or more walking surfaces are within 36 inches (914 mm), measured horizontally and in a straight line, of the glazing.

Exceptions:

1. Decorative glazing.
2. Where glazing is adjacent to a walking surface and a horizontal rail is installed 34 to 38 inches (864 to 965 mm) above the walking surface. The rail shall be capable of withstanding a horizontal load of 50 pounds per linear foot (730 N/m) without contacting the glass and have a cross-sectional height of not less than $1^1/_2$ inches (38 mm).
3. Outboard panes in insulating glass units and other multiple glazed panels where the bottom edge of the glass is 25 feet (7620 mm) or more above *grade*, a roof, walking surfaces or other horizontal [within 45 degrees (0.79 rad) of horizontal] surface adjacent to the glass exterior.

R308.4.4 Glazing in guards and railings. Glazing in *guards* and railings, including structural baluster panels and nonstructural in-fill panels, regardless of area or height above a walking surface shall be considered to be a hazardous location.

R308.4.4.1 Structural glass baluster panels. *Guards* with structural glass baluster panels shall be installed with an attached top rail or *handrail*. The top rail or *handrail* shall be supported by not less than three glass baluster panels, or shall be otherwise supported to remain in place should one glass baluster panel fail.

Exception: An attached top rail or *handrail* is not required where the glass baluster panels are laminated glass with two or more glass plies of equal thickness and of the same glass type.

R308.4.5 Glazing and wet surfaces. Glazing in walls, enclosures or fences containing or adjacent to hot tubs, spas, whirlpools, saunas, steam rooms, bathtubs, showers and indoor or outdoor swimming pools where the bottom exposed edge of the glazing is less than 60 inches (1524 mm) measured vertically above any standing or walking surface shall be considered to be a hazardous location. This shall apply to single glazing and each pane in multiple glazing.

Exception: Glazing that is more than 60 inches (1524 mm), measured horizontally, from the water's edge of a bathtub, hot tub, spa, whirlpool or swimming pool or from the edge of a shower, sauna or steam room.

R308.4.6 Glazing adjacent to stairs and ramps. Glazing where the bottom exposed edge of the glazing is less than 36 inches (914 mm) above the plane of the adjacent walking surface of *stairways*, landings between flights of stairs and *ramps* shall be considered to be a hazardous location.

Exceptions:

1. Where glazing is adjacent to a walking surface and a horizontal rail is installed at 34 to 38 inches (864 to 965 mm) above the walking surface. The rail shall be capable of withstanding a horizontal load of 50 pounds per linear foot (730 N/m) without contacting the glass and have a cross-sectional height of not less than $1^1/_2$ inches (38 mm).
2. Glazing 36 inches (914 mm) or more measured horizontally from the walking surface.

R308.4.7 Glazing adjacent to the bottom stair landing. Glazing adjacent to the landing at the bottom of a *stairway* where the glazing is less than 36 inches (914 mm) above the landing and within a 60-inch (1524 mm) horizontal arc less than 180 degrees (3.14 rad) from the bottom tread *nosing* shall be considered to be a hazardous location. (See Figure R308.4.7.)

Exception: Where the glazing is protected by a *guard* complying with Section R312 and the plane of the glass is more than 18 inches (457 mm) from the *guard*.

R308.5 Site-built windows. Site-built windows shall comply with Section 2404 of the *International Building Code*.

R308.6 Skylights and sloped glazing. *Skylights and sloped glazing* shall comply with the following sections.

R308.6.1 Definitions. The following terms are defined in Chapter 2:

SKYLIGHT, UNIT.

SKYLIGHTS AND SLOPED GLAZING.

TUBULAR DAYLIGHTING DEVICE (TDD).

R308.6.2 Materials. Glazing materials shall be limited to the following:

1. Laminated glass with not less than a 0.015-inch (0.38 mm) polyvinyl butyral interlayer for glass panes 16 square feet (1.5 m^2) or less in area located such that the highest point of the glass is not more than 12 feet (3658 mm) above a walking surface; for higher or larger sizes, the interlayer thickness shall be not less than 0.030 inch (0.76 mm).
2. Fully tempered glass.
3. Heat-strengthened glass.
4. Wired glass.
5. *Approved* rigid plastics.

R308.6.3 Screens, general. For fully tempered or heat-strengthened glass, a broken glass retention screen meeting the requirements of Section R308.6.7 shall be installed below the full area of the glass, except for fully tempered glass that meets Condition 1 or 2 listed in Section R308.6.5.

R308.6.4 Screens with multiple glazing. Where the inboard pane is fully tempered, heat-strengthened or wired glass, a broken glass retention screen meeting the requirements of Section R308.6.7 shall be installed below the full area of the glass, except for Condition 1 or 2 listed in Section R308.6.5. Other panes in the multiple glazing shall be of any type listed in Section R308.6.2.

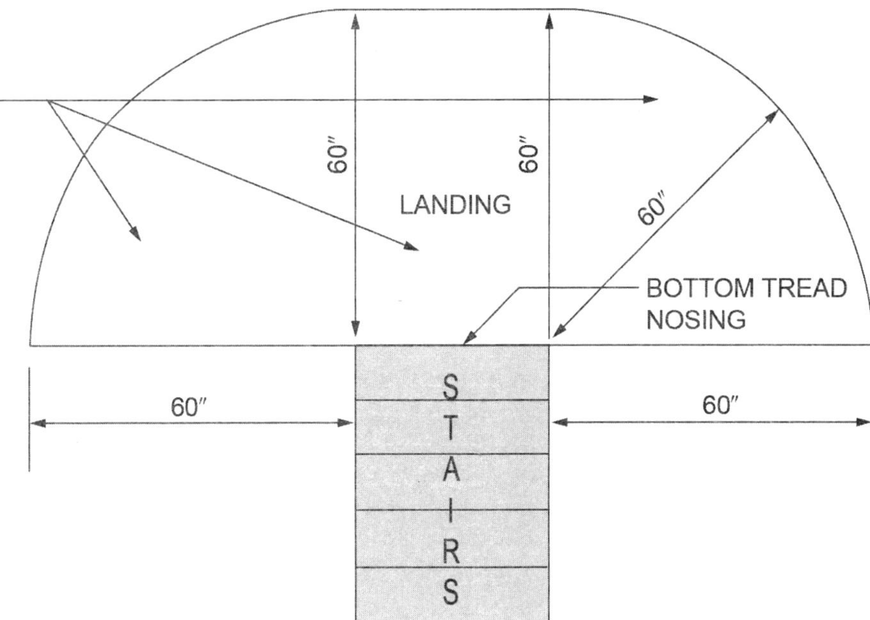

For SI: 1 inch = 25.4 mm.

FIGURE R308.4.7
HAZARDOUS GLAZING LOCATIONS AT BOTTOM STAIR LANDINGS

R308.6.5 Screens not required. Screens shall not be required where laminated glass complying with Item 1 of Section R308.6.2 is used as single glazing or the inboard pane in multiple glazing. Screens shall not be required where fully tempered glass is used as single glazing or the inboard pane in multiple glazing and either of the following conditions is met:

1. The glass area is 16 square feet (1.49 m^2) or less; the highest point of glass is not more than 12 feet (3658 mm) above a walking surface; the nominal glass thickness is not more than $^3/_{16}$ inch (4.8 mm); and for multiple glazing only the other pane or panes are fully tempered, laminated or wired glass.
2. The glass area is greater than 16 square feet (1.49 m^2); the glass is sloped 30 degrees (0.52 rad) or less from vertical; and the highest point of glass is not more than 10 feet (3048 mm) above a walking surface.

R308.6.6 Glass in greenhouses. Any glazing material is permitted to be installed without screening in the sloped areas of greenhouses, provided that the greenhouse height at the ridge does not exceed 20 feet (6096 mm) above *grade*.

R308.6.7 Screen characteristics. The screen and its fastenings shall: be capable of supporting twice the weight of the glazing; be firmly and substantially fastened to the framing members; be installed within 4 inches (102 mm) of the glass; and have a mesh opening of not greater than 1 inch by 1 inch (25 mm by 25 mm).

R308.6.8 Curbs for skylights. *Unit skylights* installed in a roof with a pitch of less than three units vertical in 12 units horizontal (25-percent slope) shall be mounted on a curb extending not less than 4 inches (102 mm) above the plane of the roof, unless otherwise specified in the manufacturer's installation instructions.

R308.6.9 Testing and labeling. *Unit skylights* and *tubular daylighting devices* shall be tested by an *approved* independent laboratory, and bear a *label* identifying manufacturer, performance grade rating and *approved* inspection agency to indicate compliance with the requirements of AAMA/WDMA/CSA 101/I.S.2/A440.

R308.6.9.1 Comparative analysis for glass-glazed unit skylights. Structural wind load design pressures for glass-glazed *unit skylights* different than the size tested in accordance with Section R308.6.9 shall be permitted to be different than the design value of the tested unit where determined in accordance with one of the following comparative analysis methods:

1. Structural wind load design pressures for glass-glazed *unit skylights* smaller than the size tested in accordance with Section R308.6.9 shall be permitted to be higher than the design value of the tested unit provided that such higher pressures are determined by accepted engineering analysis. Components of the smaller unit shall be the same as those of the tested unit. Such calculated design pressures shall be validated by an additional test of the glass-glazed *unit skylight* having the highest allowable design pressure.
2. In accordance with WDMA I.S.11.

SECTION R309
GARAGES AND CARPORTS

R309.1 Floor surface. Garage floor surfaces shall be of *approved noncombustible material.*

The area of floor used for parking of automobiles or other vehicles shall be sloped to facilitate the movement of liquids to a drain or toward the main vehicle entry doorway.

R309.2 Carports. Carports shall be open on not less than two sides. Carport floor surfaces shall be of *approved noncombustible material.* Carports not open on two or more sides shall be considered to be a garage and shall comply with the provisions of this section for garages.

The area of floor used for parking of automobiles or other vehicles shall be sloped to facilitate the movement of liquids to a drain or toward the main vehicle entry doorway.

Exception: Asphalt surfaces shall be permitted at ground level in carports.

R309.3 Flood hazard areas. Garages and carports located in flood hazard areas as established by Table R301.2 shall be constructed in accordance with Section R322.

R309.4 Automatic garage door openers. Automatic garage door openers, if provided, shall be *listed* and *labeled* in accordance with UL 325.

R309.5 Fire sprinklers. Private garages shall be protected by fire sprinklers where the garage wall has been designed based on Table R302.1(2), Note a. Sprinklers in garages shall be connected to an automatic sprinkler system that complies with Section P2904. Garage sprinklers shall be residential sprinklers or quick-response sprinklers, designed to provide a density of 0.05 gpm/ft^2. Garage doors shall not be considered obstructions with respect to sprinkler placement.

SECTION R310
EMERGENCY ESCAPE AND RESCUE OPENINGS

R310.1 Emergency escape and rescue opening required. *Basements, habitable attics* and every sleeping room shall have not less than one operable *emergency escape and rescue opening.* Where *basements* contain one or more sleeping rooms, an *emergency escape and rescue opening* shall be required in each sleeping room. *Emergency escape and rescue openings* shall open directly into a *public way,* or to a *yard* or court having a minimum width of 36 inches (914 mm) that opens to a *public way.*

Exceptions:

1. *Storm shelters* and *basements* used only to house mechanical *equipment* not exceeding a total floor area of 200 square feet (18.58 m^2).
2. Where the *dwelling unit* or *townhouse unit* is equipped with an automatic sprinkler system installed in accordance with Section P2904, sleeping rooms in *basements* shall not be required to have *emergency escape and rescue openings* provided that the *basement* has one of the following:
   2.1. One means of egress complying with Section R311 and one *emergency escape and rescue opening.*
   2.2. Two means of egress complying with Section R311.
3. A *yard* shall not be required to open directly into a *public way* where the *yard* opens to an unobstructed path from the *yard* to the *public way.* Such path shall have a width of not less than 36 inches (914 mm).

R310.1.1 Operational constraints and opening control devices. *Emergency escape and rescue openings* shall be operational from the inside of the room without the use of keys, tools or special knowledge. Window opening control devices and fall prevention devices complying with ASTM F2090 shall be permitted for use on windows serving as a required *emergency escape and rescue opening* and shall be not more than 70 inches (178 cm) above the finished floor.

R310.2 Emergency escape and rescue openings. *Emergency escape and rescue openings* shall have minimum dimensions in accordance with Sections R310.2.1 through R310.2.4.

R310.2.1 Minimum size. *Emergency escape and rescue openings* shall have a net clear opening of not less than 5.7 square feet (0.530 m^2).

Exception: The minimum net clear opening for *grade-floor emergency escape and rescue openings* shall be 5 square feet (0.465 m^2).

R310.2.2 Minimum dimensions. The minimum net clear opening height dimension shall be 24 inches (610 mm). The minimum net clear opening width dimension shall be 20 inches (508 mm). The net clear opening dimensions shall be the result of normal operation of the opening.

R310.2.3 Maximum height from floor. Emergency escape and rescue openings shall have the bottom of the clear opening not greater than 44 inches (1118 mm) above the floor.

R310.2.4 Emergency escape and rescue openings under decks, porches and cantilevers. *Emergency escape and rescue openings* installed under decks, porches and cantilevers shall be fully openable and provide a path not less than 36 inches (914 mm) in height and 36 inches (914 mm) in width to a *yard* or court.

R310.3 Emergency escape and rescue doors. Where a door is provided as the required *emergency escape and rescue opening,* it shall be a side-hinged door or a sliding door.

R310.4 Area wells. An *emergency escape and rescue opening* where the bottom of the clear opening is below the adjacent grade shall be provided with an area well in accordance with Sections R310.4.1 through R310.4.4.

R310.4.1 Minimum size. The horizontal area of the area well shall be not less than 9 square feet (0.9 m^2), with a horizontal projection and width of not less than 36 inches (914 mm). The size of the area well shall allow the *emergency escape and rescue opening* to be fully opened.

Exception: The ladder or steps required by Section R310.4.2 shall be permitted to encroach not more than 6 inches (152 mm) into the required dimensions of the area well.

R310.4.2 Ladder and steps. Area wells with a vertical depth greater than 44 inches (1118 mm) shall be equipped with an *approved*, permanently affixed ladder or steps. The ladder or steps shall not be obstructed by the *emergency escape and rescue opening* where the window or door is in the open position. Ladders or steps required by this section shall not be required to comply with Section R311.7.

R310.4.2.1 Ladders. Ladders and rungs shall have an inside width of not less than 12 inches (305 mm), shall project not less than 3 inches (76 mm) from the wall and shall be spaced not more than 18 inches (457 mm) on center vertically for the full height of the area well.

R310.4.2.2 Steps. Steps shall have an inside width of not less than 12 inches (305 mm), a minimum tread depth of 5 inches (127 mm) and a maximum *riser* height of 18 inches (457 mm) for the full height of the area well.

R310.4.3 Drainage. Area wells shall be designed for proper drainage by connecting to the building's foundation drainage system required by Section R405.1.

Exception: A drainage system for area wells is not required where the foundation is on well-drained soil or sand-gravel mixture soils in accordance with the United Soil Classification System, Group I Soils, as detailed in Table R405.1.

R310.4.4 Bars, grilles, covers and screens. Where bars, grilles, covers, screens or similar devices are placed over *emergency escape and rescue openings*, bulkhead enclosures or area wells that serve such openings, the minimum net clear opening size shall comply with Sections R310.2 through R310.2.2 and R310.4.1. Such devices shall be releasable or removable from the inside without the use of a key or tool or force greater than that required for the normal operation of the escape and rescue opening.

R310.5 Replacement windows for emergency escape and rescue openings. Replacement windows installed in buildings meeting the scope of this code shall be exempt from Sections R310.2 and R310.4.4, provided that the replacement window meets the following conditions:

1. The replacement window is the manufacturer's largest standard size window that will fit within the existing frame or existing rough opening. The replacement window is of the same operating style as the existing window or a style that provides for an equal or greater window opening area than the existing window.
2. The replacement window is not part of a change of occupancy.

R310.6 Dwelling additions. Where *dwelling additions* contain sleeping rooms, an *emergency escape and rescue opening* shall be provided in each new sleeping room. Where *dwelling additions* have *basements*, an *emergency escape and rescue opening* shall be provided in the new *basement*.

Exceptions:

1. An *emergency escape and rescue opening* is not required in a new *basement* that contains a sleeping room with an *emergency escape and rescue opening*.
2. An *emergency escape and rescue opening* is not required in a new *basement* where there is an *emergency escape and rescue opening* in an existing *basement* that is *accessed* from the new *basement*.
3. An operable window complying with Section 310.7.1 shall be acceptable as an *emergency escape and rescue opening*.

R310.7 Alterations or repairs of existing basements. New sleeping rooms created in an existing *basement* shall be provided with *emergency escape and rescue openings* in accordance with Section R310.1. Other than new sleeping rooms, where existing basements undergo alterations or repairs, an *emergency escape and rescue opening* is not required.

Exception: An operable window complying with Section 310.7.1 shall be acceptable as an *emergency escape and rescue opening*.

R310.7.1 Existing emergency escape and rescue openings. Where a *change of occupancy* would require an *emergency escape and rescue opening* in accordance with Section 310.1, operable windows serving as the *emergency escape and rescue opening* shall comply with the following:

1. An existing operable window shall provide a minimum net clear opening of 4 square feet (0.38 m^2) with a minimum net clear opening height of 22 inches (559 mm) and a minimum net clear opening width of 20 inches (508 mm).
2. A replacement window where such window complies with both of the following:
    2.1. The replacement window meets the size requirements in Item 1.
    2.2. The replacement window is the manufacturer's largest standard-size window that will fit within the existing frame or existing rough opening. The replacement window shall be permitted to be of the same operating style as the existing window or a style that provides for an equal or greater window opening area than the existing window.

SECTION R311
MEANS OF EGRESS

R311.1 Means of egress. *Dwellings* shall be provided with a means of egress in accordance with this section. The means of egress shall provide a continuous and unobstructed path of vertical and horizontal egress travel from all portions of the *dwelling* to the required egress door without requiring travel through a garage. The required egress door shall open directly into a *public way* or to a *yard* or court that opens to a *public way*.

R311.2 Egress door. Not less than one egress door shall be provided for each *dwelling unit*. The egress door shall be side-hinged, and shall provide a clear width of not less than 32 inches (813 mm) where measured between the face of the door and the stop, with the door open 90 degrees (1.57 rad). The clear height of the door opening shall be not less than 78 inches (1981 mm) in height measured from the top of the threshold to the bottom of the stop. Other doors shall not be required to comply with these minimum dimensions. Egress doors shall be readily openable from inside the *dwelling* without the use of a key or special knowledge or effort.

R311.3 Floors and landings at exterior doors. There shall be a landing or floor on each side of each exterior door. The width of each landing shall be not less than the door served. Landings shall have a dimension of not less than 36 inches (914 mm) measured in the direction of travel. The slope at exterior landings shall not exceed $^1/_4$ unit vertical in 12 units horizontal (2 percent).

> **Exception:** Exterior balconies less than 60 square feet (5.6 m^2) and only *accessed* from a door are permitted to have a landing that is less than 36 inches (914 mm) measured in the direction of travel.

R311.3.1 Floor elevations at the required egress doors. Landings or finished floors at the required egress door shall be not more than $1^1/_2$ inches (38 mm) lower than the top of the threshold.

> **Exception:** The landing or floor on the exterior side shall be not more than $7^3/_4$ inches (196 mm) below the top of the threshold provided that the door does not swing over the landing or floor.

Where exterior landings or floors serving the required egress door are not at *grade*, they shall be provided with access to *grade* by means of a *ramp* in accordance with Section R311.8 or a *stairway* in accordance with Section R311.7.

R311.3.2 Floor elevations at other exterior doors. Doors other than the required egress door shall be provided with landings or floors not more than $7^3/_4$ inches (196 mm) below the top of the threshold.

> **Exception:** A top landing is not required where a *stairway* of not more than two *risers* is located on the exterior side of the door, provided that the door does not swing over the *stairway*.

R311.3.3 Storm and screen doors. Storm and screen doors shall be permitted to swing over exterior stairs and landings.

R311.4 Vertical egress. Egress from habitable levels including habitable attics and *basements* that are not provided with an egress door in accordance with Section R311.2 shall be by a *ramp* in accordance with Section R311.8 or a *stairway* in accordance with Section R311.7.

R311.5 Landing, deck, balcony and stair construction and attachment. Exterior landings, decks, balconies, stairs and similar facilities shall be positively anchored to the primary structure to resist both vertical and lateral forces or shall be designed to be self-supporting. Attachment shall not be accomplished by use of toenails or nails subject to withdrawal.

R311.6 Hallways. The width of a hallway shall be not less than 3 feet (914 mm).

R311.7 Stairways. Where required by this code or provided, *stairways* shall comply with this section.

Exceptions:

1. Stairways not within or serving a building, porch or deck.
2. Stairways leading to nonhabitable attics.
3. Stairways leading to *crawl spaces*.

R311.7.1 Width. *Stairways* shall be not less than 36 inches (914 mm) in clear width at all points above the permitted *handrail* height and below the required headroom height. The clear width of *stairways* at and below the *handrail* height, including treads and landings, shall be not less than $31^1/_2$ inches (787 mm) where a *handrail* is installed on one side and 27 inches (698 mm) where *handrails* are installed on both sides.

> **Exception:** The width of *spiral stairways* shall be in accordance with Section R311.7.10.1.

R311.7.2 Headroom. The headroom in *stairways* shall be not less than 6 feet 8 inches (2032 mm) measured vertically from the sloped line adjoining the tread *nosing* or from the floor surface of the landing or platform on that portion of the *stairway*.

Exceptions:

1. Where the *nosings* of treads at the side of a flight extend under the edge of a floor opening through which the *stair* passes, the floor opening shall not project horizontally into the required headroom more than $4^3/_4$ inches (121 mm).
2. The headroom for spiral *stairways* shall be in accordance with Section R311.7.10.1.

R311.7.3 Vertical rise. A flight of stairs shall not have a vertical rise greater than 12 feet 7 inches (3835 mm) between floor levels or landings.

R311.7.4 Walkline. The walkline across *winder* treads and landings shall be concentric to the turn and parallel to the direction of travel entering and exiting the turn. The walkline shall be located 12 inches (305 mm) from the inside of the turn. The 12-inch (305 mm) dimension shall be measured from the widest point of the clear stair width at the walking surface. Where *winders* are adjacent within a flight, the point of the widest clear stair width of the adjacent *winders* shall be used.

R311.7.5 Stair treads and risers. *Stair* treads and *risers* shall meet the requirements of this section. For the purposes of this section, dimensions and dimensioned surfaces shall be exclusive of carpets, rugs or runners.

R311.7.5.1 Risers. The *riser* height shall be not more than $7^3/_4$ inches (196 mm). The *riser* height shall be measured vertically between leading edges of the adjacent treads. The greatest *riser* height within any flight of stairs shall not exceed the smallest by more than $^3/_8$ inch (9.5 mm). *Risers* shall be vertical or sloped from the underside of the *nosing* of the tread above at an angle not more than 30 degrees (0.51 rad) from the vertical. At open *risers*, openings located more than 30 inches (762 mm), as measured vertically, to the floor or *grade* below shall not permit the passage of a 4-inch-diameter (102 mm) sphere.

Exceptions:

1. The opening between adjacent treads is not limited on *spiral stairways*.
2. The *riser* height of *spiral stairways* shall be in accordance with Section R311.7.10.1.

R311.7.5.2 Treads. The tread depth shall be not less than 10 inches (254 mm). The tread depth shall be measured horizontally between the vertical planes of the foremost projection of adjacent treads and at a right angle to the tread's leading edge. The greatest tread depth within any flight of stairs shall not exceed the smallest by more than $^3/_8$ inch (9.5 mm).

R311.7.5.2.1 Winder treads. *Winder* treads shall have a tread depth of not less than 10 inches (254 mm) measured between the vertical planes of the foremost projection of adjacent treads at the intersections with the walkline. *Winder* treads shall have a tread depth of not less than 6 inches (152 mm) at any point within the clear width of the *stair*. Within any flight of stairs, the largest *winder* tread depth at the walkline shall not exceed the smallest *winder* tread by more than $^3/_8$ inch (9.5 mm). Consistently shaped *winders* at the walkline shall be allowed within the same flight of stairs as rectangular treads and shall not be required to be within $^3/_8$ inch (9.5 mm) of the rectangular tread depth.

Exception: The tread depth at *spiral stairways* shall be in accordance with Section R311.7.10.1.

R311.7.5.3 Nosings. *Nosings* at treads, landings and floors of *stairways* shall have a radius of curvature at the *nosing* not greater than $^9/_{16}$ inch (14 mm) or a bevel not greater than $^1/_2$ inch (12.7 mm). A *nosing* projection not less than $^3/_4$ inch (19 mm) and not more than $1^1/_4$ inches (32 mm) shall be provided on *stairways*. The greatest *nosing* projection shall not exceed the smallest *nosing* projection by more than $^3/_8$ inch (9.5 mm) within a *stairway*.

Exception: A *nosing* projection is not required where the tread depth is not less than 11 inches (279 mm).

R311.7.5.4 Exterior plastic composite stair treads. *Plastic composite* exterior stair treads shall comply with the provisions of this section and Section R507.2.2.

R311.7.6 Landings for stairways. There shall be a floor or landing at the top and bottom of each *stairway*. The width perpendicular to the direction of travel shall be not less than the width of the flight served. For landings of shapes other than square or rectangular, the depth at the walk line and the total area shall be not less than that of a quarter circle with a radius equal to the required landing width. Where the *stairway* has a straight run, the depth in the direction of travel shall be not less than 36 inches (914 mm).

Exception: A floor or landing is not required at the top of an interior flight of stairs, including stairs in an enclosed garage, provided that a door does not swing over the stairs.

R311.7.7 Stairway walking surface. The walking surface of treads and landings of *stairways* shall be sloped not steeper than 1 unit vertical in 48 units horizontal (2-percent slope).

Exception: Where the surface of a landing is required elsewhere in the code to drain surface water, the walking surface of the landing shall be sloped not steeper than 1 unit vertical in 20 units horizontal (5-percent slope) in the direction of travel.

R311.7.8 Handrails. *Handrails* shall be provided on not less than one side of each flight of stairs with four or more *risers*.

R311.7.8.1 Height. *Handrail* height, measured vertically from the sloped plane adjoining the tread *nosing*, or finish surface of ramp slope, shall be not less than 34 inches (864 mm) and not more than 38 inches (965 mm).

Exceptions:

1. The use of a volute, turnout or starting easing shall be allowed over the lowest tread.
2. Where *handrail* fittings or bendings are used to provide continuous transition between flights, transitions at *winder* treads, the transition from *handrail* to *guard*, or used at the start of a flight, the *handrail* height at the fittings or bendings shall be permitted to exceed 38 inches (965 mm).

R311.7.8.2 Handrail projection. *Handrails* shall not project more than $4^1/_2$ inches (114 mm) on either side of the *stairway*.

Exception: Where *nosings* of landings, floors or passing flights project into the *stairway* reducing the clearance at passing *handrails*, *handrails* shall project not more than $6^1/_2$ inches (165 mm) into the *stairway*, provided that the stair width and *handrail* clearance are not reduced to less than that required.

R311.7.8.3 Handrail clearance. *Handrails* adjacent to a wall shall have a space of not less than $1^1/_2$ inches (38 mm) between the wall and the *handrails*.

R311.7.8.4 Continuity. *Handrails* shall be continuous for the full length of the flight, from a point directly above the top riser of the flight to a point directly above the lowest riser of the flight. *Handrail* ends shall be returned toward a wall, guard walking surface continuous to itself, or terminate to a post.

Exceptions:

1. *Handrail* continuity shall be permitted to be interrupted by a newel post at a turn in a flight with winders, at a landing, or over the lowest tread.
2. A volute, turnout or starting easing shall be allowed to terminate over the lowest tread and over the top landing.

R311.7.8.5 Grip size. Required *handrails* shall be of one of the following types or provide equivalent graspability.

1. Type I. *Handrails* with a circular cross section shall have an outside diameter of not less than $1^1/_4$ inches (32 mm) and not greater than 2 inches (51 mm). If the *handrail* is not circular, it shall have a perimeter of not less than 4 inches (102 mm) and not greater than $6^1/_4$ inches (160 mm) and a cross section of not more than $2^1/_4$ inches (57 mm). Edges shall have a radius of not less than 0.01 inch (0.25 mm).
2. Type II. *Handrails* with a perimeter greater than $6^1/_4$ inches (160 mm) shall have a graspable finger recess area on both sides of the profile. The finger recess shall begin within $^3/_4$ inch (19 mm) measured vertically from the tallest portion of the profile and have a depth of not less than $^5/_{16}$ inch (8 mm) within $^7/_8$ inch (22 mm) below the widest portion of the profile. This required depth shall continue for not less than $^3/_8$ inch (10 mm) to a level that is not less than $1^3/_4$ inches (45 mm) below the tallest portion of the profile. The width of the *handrail* above the recess shall be not less than $1^1/_4$ inches (32 mm) and not more than $2^3/_4$ inches (70 mm). Edges shall have a radius of not less than 0.01 inch (0.25 mm).

R311.7.8.6 Exterior plastic composite handrails. *Plastic composite* exterior *handrails* shall comply with the requirements of Section R507.2.2.

R311.7.9 Illumination. *Stairways* shall be provided with illumination in accordance with Sections R303.7 and R303.8.

R311.7.10 Special stairways. *Spiral stairways* and bulkhead enclosure *stairways* shall comply with the requirements of Section R311.7 except as specified in Sections R311.7.10.1 and R311.7.10.2.

R311.7.10.1 Spiral stairways. The clear width at and below the *handrails* at *spiral stairways* shall be not less than 26 inches (660 mm) and the walkline radius shall be not greater than $24^1/_2$ inches (622 mm). Each tread shall have a depth of not less than $6^3/_4$ inches (171 mm) at the walkline. Treads shall be identical, and the rise shall be not more than $9^1/_2$ inches (241 mm). Headroom shall be not less than 6 feet 6 inches (1982 mm).

R311.7.10.2 Bulkhead enclosure stairways. *Stairways* serving bulkhead enclosures, not part of the required building egress, providing access from the outside *grade* level to the *basement* shall be exempt from the requirements of Sections R311.3 and R311.7 where the height from the *basement* finished floor level to *grade* adjacent to the *stairway* is not more than 8 feet (2438 mm) and the *grade* level opening to the *stairway* is covered by a bulkhead enclosure with hinged doors or other *approved* means.

R311.7.11 Alternating tread devices. Alternating tread devices shall not be used as an element of a means of egress. Alternating tread devices shall be permitted provided that a required means of egress *stairway* or *ramp* serves the same space at each adjoining level or where a means of egress is not required. The clear width at and below the *handrails* shall be not less than 20 inches (508 mm).

Exception: Alternating tread devices are allowed to be used as an element of a means of egress for lofts, *mezzanines* and similar areas of 200 gross square feet (18.6 m^2) or less where such devices do not provide exclusive access to a kitchen or bathroom.

R311.7.11.1 Treads of alternating tread devices. Alternating tread devices shall have a tread depth of not less than 5 inches (127 mm), a projected tread depth of not less than $8^1/_2$ inches (216 mm), a tread width of not less than 7 inches (178 mm) and a *riser* height of not more than $9^1/_2$ inches (241 mm). The tread depth shall be measured horizontally between the vertical planes of the foremost projections of adjacent treads. The *riser* height shall be measured vertically between the leading edges of adjacent treads. The *riser* height and tread depth provided shall result in an angle of ascent from the horizontal of between 50 and 70 degrees (0.87 and 1.22 rad). The initial tread of the device shall begin at the same elevation as the platform, landing or floor surface.

R311.7.11.2 Handrails of alternating tread devices. *Handrails* shall be provided on both sides of alternating tread devices and shall comply with Sections R311.7.8.2 through R311.7.8.6. *Handrail* height shall be uniform, not less than 30 inches (762 mm) and not more than 34 inches (864 mm).

R311.7.12 Ship's ladders. Ship's ladders shall not be used as an element of a means of egress. Ship's ladders shall be permitted provided that a required means of egress *stairway* or *ramp* serves the same space at each adjoining level or where a means of egress is not required. The clear width at and below the *handrails* shall be not less than 20 inches (508 mm).

Exception: Ship's ladders are allowed to be used as an element of a means of egress for lofts, *mezzanines* and similar areas of 200 gross square feet (18.6 m^2) or less that do not provide exclusive access to a kitchen or bathroom.

R311.7.12.1 Treads of ship's ladders. Treads shall have a depth of not less than 5 inches (127 mm). The tread shall be projected such that the total of the tread depth plus the *nosing* projection is not less than $8^1/_2$ inches (216 mm). The *riser* height shall be not more than $9^1/_2$ inches (241 mm).

R311.7.12.2 Handrails of ship's ladders. *Handrails* shall be provided on both sides of ship's ladders and shall comply with Sections R311.7.8.2 through R311.7.8.6. *Handrail* height shall be uniform, not less than 30 inches (762 mm) and not more than 34 inches (864 mm).

R311.8 Ramps. Where required by this code or provided, *ramps* shall comply with this section.

Exception: Ramps not within or serving a building, porch or deck.

R311.8.1 Maximum slope. *Ramps* serving the egress door required by Section R311.2 shall have a slope of not more than 1 unit vertical in 12 units horizontal (8.3-percent slope).

Other *ramps* shall have a maximum slope of 1 unit vertical in 8 units horizontal (12.5 percent).

Exception: Where it is technically infeasible to comply because of site constraints, *ramps* shall have a slope of not more than 1 unit vertical in 8 units horizontal (12.5 percent).

R311.8.2 Landings required. There shall be a floor or landing at the top and bottom of each *ramp*, where doors open onto *ramps*, and where *ramps* change directions. The width of the landing perpendicular to the *ramp* slope shall be not less than the width of the *ramp*. The depth of the landing in the direction of the ramp slope shall be not less than 36 inches (914 mm).

R311.8.3 Handrails required. *Handrails* shall be provided on not less than one side of *ramps* exceeding a slope of 1 unit vertical in 12 units horizontal (8.33-percent slope).

R311.8.3.1 Height. *Handrail* height, measured above the finished surface of the *ramp* slope, shall be not less than 34 inches (864 mm) and not more than 38 inches (965 mm).

R311.8.3.2 Grip size. *Handrails* on *ramps* shall comply with Section R311.7.8.5.

R311.8.3.3 Continuity. *Handrails* where required on *ramps* shall be continuous for the full length of the *ramp*. *Handrail* ends shall be returned or shall terminate in newel posts or safety terminals. *Handrails* adjacent to a wall shall have a space of not less than $1^1/_2$ inches (38 mm) between the wall and the *handrails*.

SECTION R312
GUARDS AND WINDOW FALL PROTECTION

R312.1 Guards. *Guards* shall be provided in accordance with Sections R312.1.1 through R312.1.4.

R312.1.1 Where required. *Guards* shall be provided for those portions of open-sided walking surfaces, including floors, stairs, *ramps* and landings that are located more than 30 inches (762 mm) measured vertically to the floor or *grade* below at any point within 36 inches (914 mm) horizontally to the edge of the open side. Insect screening shall not be considered as a *guard*.

R312.1.2 Height. Required *guards* at open-sided walking surfaces, including stairs, porches, balconies or landings, shall be not less than 36 inches (914 mm) in height as measured vertically above the adjacent walking surface or the line connecting the *nosings*.

Exceptions:

1. *Guards* on the open sides of stairs shall have a height of not less than 34 inches (864 mm) measured vertically from a line connecting the *nosings*.
2. Where the top of the *guard* serves as a *handrail* on the open sides of stairs, the top of the *guard* shall be not less than 34 inches (864 mm) and not more than 38 inches (965 mm) as measured vertically from a line connecting the *nosings*.

R312.1.3 Opening limitations. Required *guards* shall not have openings from the walking surface to the required *guard* height that allow passage of a sphere 4 inches (102 mm) in diameter.

Exceptions:

1. The triangular openings at the open side of *stair*, formed by the *riser*, tread and bottom rail of a *guard*, shall not allow passage of a sphere 6 inches (153 mm) in diameter.
2. *Guards* on the open side of stairs shall not have openings that allow passage of a sphere $4^3/_8$ inches (111 mm) in diameter.

R312.1.4 Exterior plastic composite guards. *Plastic composite* exterior *guards* shall comply with the requirements of Section R317.4.

R312.2 Window fall protection. Window fall protection shall be provided in accordance with Sections R312.2.1 and R312.2.2.

R312.2.1 Window opening height. In *dwelling units*, where the bottom of the clear opening of an operable window opening is located less than 24 inches (610 mm) above the finished floor and greater than 72 inches (1829 mm) above the finished *grade* or other surface below on the exterior of the building, the operable window shall comply with one of the following:

1. Operable window openings will not allow a 4-inch-diameter (102 mm) sphere to pass through where the openings are in their largest opened position.
2. Operable windows are provided with window opening control devices or fall prevention devices that comply with ASTM F2090.

R312.2.2 Emergency escape and rescue openings. Where an operable window serves as an *emergency escape and rescue opening*, a window opening control device or fall prevention device, after operation to release the control device or fall prevention device allowing the window to fully open, shall not reduce the net clear opening area of the window unit to less than the area required by Sections R310.2.1 and R310.2.2.

SECTION R313 AUTOMATIC FIRE SPRINKLER SYSTEMS

R313.1 Townhouse automatic fire sprinkler systems. An automatic sprinkler system shall be installed in *townhouses*.

Exception: An automatic sprinkler system shall not be required where *additions* or *alterations* are made to existing *townhouses* that do not have an automatic sprinkler system installed.

R313.1.1 Design and installation. Automatic sprinkler systems for *townhouses* shall be designed and installed in accordance with Section P2904 or NFPA 13D.

R313.2 One- and two-family dwellings automatic sprinkler systems. An automatic sprinkler system shall be installed in one- and two-family *dwellings*.

Exception: An automatic sprinkler system shall not be required for *additions* or *alterations* to existing buildings that are not already provided with a sprinkler system.

R313.2.1 Design and installation. Automatic sprinkler systems shall be designed and installed in accordance with Section P2904 or NFPA 13D.

SECTION R314 SMOKE ALARMS

R314.1 General. Smoke alarms shall comply with NFPA 72 and Section R314.

R314.1.1 Listings. Smoke alarms shall be *listed* in accordance with UL 217. Combination smoke and carbon monoxide alarms shall be *listed* in accordance with UL 217 and UL 2034.

R314.2 Where required. Smoke alarms shall be provided in accordance with this section.

R314.2.1 New construction. Smoke alarms shall be provided in *dwelling units*.

R314.2.2 Alterations, repairs and additions. Where *alterations*, *repairs* or *additions* requiring a *permit* occur, the individual *dwelling unit* shall be equipped with smoke alarms located as required for new *dwellings*.

Exceptions:

1. Work involving the exterior surfaces of *dwellings*, such as the replacement of roofing or siding, the addition or replacement of windows or doors, or the addition of a porch or deck.
2. Installation, *alteration* or repairs of plumbing or mechanical systems.

R314.3 Location. Smoke alarms shall be installed in the following locations:

1. In each sleeping room.
2. Outside each separate sleeping area in the immediate vicinity of the bedrooms.
3. On each additional story of the *dwelling*, including *basements* and *habitable attics* and not including crawl spaces and uninhabitable *attics*. In *dwellings* or *dwelling units* with split levels and without an intervening door between the adjacent levels, a smoke alarm installed on the upper level shall suffice for the adjacent lower level provided that the lower level is less than one full *story* below the upper level.
4. Not less than 3 feet (914 mm) horizontally from the door or opening of a bathroom that contains a bathtub or shower unless this would prevent placement of a smoke alarm required by this section.
5. In the hallway and in the room open to the hallway in *dwelling units* where the ceiling height of a room open to a hallway serving bedrooms exceeds that of the hallway by 24 inches (610 mm) or more.

R314.3.1 Installation near cooking appliances. Smoke alarms shall not be installed in the following locations unless this would prevent placement of a smoke alarm in a location required by Section R314.3.

1. Ionization smoke alarms shall not be installed less than 20 feet (6096 mm) horizontally from a permanently installed cooking *appliance*.
2. Ionization smoke alarms with an alarm-silencing switch shall not be installed less than 10 feet (3048 mm) horizontally from a permanently installed cooking *appliance*.
3. Photoelectric smoke alarms shall not be installed less than 6 feet (1828 mm) horizontally from a permanently installed cooking *appliance*.
4. Smoke alarms *listed* and marked "helps reduce cooking nuisance alarms" shall not be installed

less than 6 feet (1828 mm) horizontally from a permanently installed cooking *appliance.*

R314.4 Interconnection. Where more than one smoke alarm is required to be installed within an individual *dwelling unit* in accordance with Section R314.3, the alarm devices shall be interconnected in such a manner that the actuation of one alarm will activate all of the alarms in the individual *dwelling unit.* Physical interconnection of smoke alarms shall not be required where *listed* wireless alarms are installed and all alarms sound upon activation of one alarm.

R314.5 Combination alarms. Combination smoke and carbon monoxide alarms shall be permitted to be used in lieu of smoke alarms.

R314.6 Power source. Smoke alarms shall receive their primary power from the building wiring where such wiring is served from a commercial source and, where primary power is interrupted, shall receive power from a battery. Wiring shall be permanent and without a disconnecting switch other than those required for overcurrent protection.

Exceptions:

1. Smoke alarms shall be permitted to be battery operated where installed in buildings without commercial power.
2. Smoke alarms installed in accordance with Section R314.2.2 shall be permitted to be battery powered.

R314.7 Fire alarm systems. Fire alarm systems shall be permitted to be used in lieu of smoke alarms and shall comply with Sections R314.7.1 through R314.7.4.

R314.7.1 General. Fire alarm systems shall comply with the provisions of this code and the household fire warning equipment provisions of NFPA 72. Smoke detectors shall be *listed* in accordance with UL 268.

R314.7.2 Location. Smoke detectors shall be installed in the locations specified in Section R314.3.

R314.7.3 Permanent fixture. Where a household fire alarm system is installed, it shall become a permanent fixture of the occupancy, owned by the homeowner.

R314.7.4 Combination detectors. Combination smoke and carbon monoxide detectors shall be permitted to be installed in fire alarm systems in lieu of smoke detectors, provided that they are *listed* in accordance with UL 268 and UL 2075.

SECTION R315
CARBON MONOXIDE ALARMS

R315.1 General. Carbon monoxide alarms shall comply with Section R315.

R315.1.1 Listings. Carbon monoxide alarms shall be *listed* in accordance with UL 2034. Combination carbon monoxide and smoke alarms shall be *listed* in accordance with UL 217 and UL 2034.

R315.2 Where required. Carbon monoxide alarms shall be provided in accordance with Sections R315.2.1 and R315.2.2.

R315.2.1 New construction. For new construction, carbon monoxide alarms shall be provided in *dwelling units* where either or both of the following conditions exist.

1. The *dwelling unit* contains a fuel-fired *appliance.*
2. The *dwelling unit* has an attached garage with an opening that communicates with the *dwelling unit.*

R315.2.2 Alterations, repairs and additions. Where *alterations, repairs* or *additions* requiring a *permit* occur, the individual *dwelling unit* shall be equipped with carbon monoxide alarms located as required for new *dwellings.*

Exceptions:

1. Work involving the exterior surfaces of *dwellings,* such as the replacement of roofing or siding, or the addition or replacement of windows or doors, or the addition of a porch or deck.
2. Installation, *alteration* or repairs of plumbing systems.
3. Installation, alteration or repairs of mechanical systems that are not fuel fired.

R315.3 Location. Carbon monoxide alarms in *dwelling units* shall be installed outside of each separate sleeping area in the immediate vicinity of the bedrooms. Where a fuel-burning *appliance* is located within a bedroom or its attached bathroom, a carbon monoxide alarm shall be installed within the bedroom.

R315.4 Combination alarms. Combination carbon monoxide and smoke alarms shall be permitted to be used in lieu of carbon monoxide alarms.

R315.5 Interconnectivity. Where more than one carbon monoxide alarm is required to be installed within an individual *dwelling unit* in accordance with Section R315.3, the alarm devices shall be interconnected in such a manner that the actuation of one alarm will activate all of the alarms in the individual *dwelling unit.* Physical interconnection of carbon monoxide alarms shall not be required where *listed* wireless alarms are installed and all alarms sound upon activation of one alarm.

Exception: Interconnection of carbon monoxide alarms in existing areas shall not be required where *alterations* or *repairs* do not result in removal of interior wall or ceiling finishes exposing the structure, unless there is an attic, *crawl space* or *basement* available that could provide access for interconnection without the removal of interior finishes.

R315.6 Power source. Carbon monoxide alarms shall receive their primary power from the building wiring where such wiring is served from a commercial source and, where primary power is interrupted, shall receive power from a battery. Wiring shall be permanent and without a disconnecting switch other than those required for overcurrent protection.

Exceptions:

1. Carbon monoxide alarms shall be permitted to be battery operated where installed in buildings without commercial power.
2. Carbon monoxide alarms installed in accordance with Section R315.2.2 shall be permitted to be battery powered.

R315.7 Carbon monoxide detection systems. Carbon monoxide detection systems shall be permitted to be used in lieu of carbon monoxide alarms and shall comply with Sections R315.7.1 through R315.7.4.

R315.7.1 General. Household carbon monoxide detection systems shall comply with NFPA 720. Carbon monoxide detectors shall be *listed* in accordance with UL 2075.

R315.7.2 Location. Carbon monoxide detectors shall be installed in the locations specified in Section R315.3. These locations supersede the locations specified in NFPA 720.

R315.7.3 Permanent fixture. Where a household carbon monoxide detection system is installed, it shall become a permanent fixture of the occupancy and owned by the homeowner.

R315.7.4 Combination detectors. Combination carbon monoxide and smoke detectors installed in carbon monoxide detection systems in lieu of carbon monoxide detectors shall be *listed* in accordance with UL 268 and UL 2075 .

SECTION R316 FOAM PLASTIC

R316.1 General. The provisions of this section shall govern the materials, design, application, construction and installation of foam plastic materials.

R316.2 Labeling and identification. Packages and containers of foam plastic insulation and foam plastic insulation components delivered to the job site shall bear the *label* of an *approved agency* showing the manufacturer's name, the product listing, product identification and information sufficient to determine that the end use will comply with the requirements.

R316.3 Surface burning characteristics. Unless otherwise allowed in Section R316.5, foam plastic, or foam plastic cores used as a component in manufactured assemblies, used in building construction shall comply with Section R316.3.1 or R316.3.2. Loose-fill-type foam plastic insulation shall be tested as board stock for the flame spread index and *smoke-developed index.*

Exception: Spray foam plastic insulation more than 4 inches (102 mm) in thickness shall have a flame spread index of not more than 25 and a *smoke-developed index* of not more than 450 where tested at a thickness of 4 inches (102 mm) and at the density intended for use. Such spray foam plastic shall be separated from the interior of a building by $^1/_2$-inch (12.7 mm) gypsum wallboard or by a material that has been tested in accordance with NFPA 275, and shall meet the acceptance criteria of both the Temperature Transmission Fire Test and the Integrity Fire Test.

R316.3.1 Foam plastic insulation 4 inches thick or less. Foam plastic insulation installed at 4 inches (102 mm) in thickness or less shall have a flame spread index of not more than 75 and a *smoke-developed index* of not more than 450 where tested in the maximum thickness and density intended for use in accordance with ASTM E84 or UL 723.

R316.3.2 Foam plastic insulation more than 4 inches thick. Foam plastic insulation installed at more than 4 inches (102 mm) in thickness shall have a flame spread index of not more than 75 and a *smoke-developed index* of not more than 450 where tested at a thickness of 4 inches (102 mm) in accordance with ASTM E84 or UL 723, provided that the end use is *approved* in accordance with Section R316.6 using the thickness and density intended for use.

R316.4 Thermal barrier. Unless otherwise allowed in Section R316.5, foam plastic shall be separated from the interior of a building by an *approved* thermal barrier of not less than $^1/_2$-inch (12.7 mm) gypsum wallboard, $^{23}/_{32}$-inch (18.2 mm) *wood structural panel* or a material that is tested in accordance with and meets the acceptance criteria of both the Temperature Transmission Fire Test and the Integrity Fire Test of NFPA 275.

R316.5 Specific requirements. The following requirements shall apply to these uses of foam plastic unless specifically *approved* in accordance with Section R316.6 or by other sections of the code or the requirements of Sections R316.2 through R316.4 have been met.

R316.5.1 Masonry or concrete construction. The thermal barrier specified in Section R316.4 is not required in a masonry or concrete wall, floor or roof where the foam plastic insulation is separated from the interior of the building by not less than a 1-inch (25 mm) thickness of masonry or concrete.

R316.5.2 Roofing. The thermal barrier specified in Section R316.4 is not required where the foam plastic in a *roof assembly* or under a roof covering is installed in accordance with the code and the manufacturer's instructions and is separated from the interior of the building by tongue-and-groove wood planks or *wood structural panel* sheathing, in accordance with Section R803, that is not less than $^{15}/_{32}$ inch (11.9 mm) thick bonded with exterior glue, identified as Exposure 1 and with edges supported by blocking or tongue-and-groove joints or an equivalent material. The *smoke-developed index* for roof applications shall not be limited.

R316.5.3 Attics. The thermal barrier specified in Section R316.4 is not required where all of the following apply:

1. *Attic* access is required by Section R807.1.

2. The space is entered only for purposes of repairs or maintenance.
3. The foam plastic insulation has been tested in accordance with Section R316.6 or the foam plastic insulation is protected against ignition using one of the following ignition barrier materials:
   - 3.1. $1^1/_2$-inch-thick (38 mm) mineral fiber insulation.
   - 3.2. $^1/_4$-inch-thick (6.4 mm) *wood structural panels*.
   - 3.3. $^3/_8$-inch (9.5 mm) particleboard.
   - 3.4. $^1/_4$-inch (6.4 mm) hardboard.
   - 3.5. $^3/_8$-inch (9.5 mm) gypsum board.
   - 3.6. Corrosion-resistant steel having a base metal thickness of 0.016 inch (0.406 mm).
   - 3.7. $1^1/_2$-inch-thick (38 mm) cellulose insulation.
   - 3.8. $^1/_4$-inch (6.4 mm) fiber-cement panel, soffit or backer board.

The ignition barrier is not required where the foam plastic insulation has been tested in accordance with Section R316.6.

R316.5.4 Crawl spaces. The thermal barrier specified in Section R316.4 is not required where all of the following apply:

1. *Crawl space* access is required by Section R408.4.
2. Entry is made only for purposes of repairs or maintenance.
3. The foam plastic insulation has been tested in accordance with Section R316.6 or the foam plastic insulation is protected against ignition using one of the following ignition barrier materials:
   - 3.1. $1^1/_2$-inch-thick (38 mm) mineral fiber insulation.
   - 3.2. $^1/_4$-inch-thick (6.4 mm) *wood structural panels*.
   - 3.3. $^3/_8$-inch (9.5 mm) particleboard.
   - 3.4. $^1/_4$-inch (6.4 mm) hardboard.
   - 3.5. $^3/_8$-inch (9.5 mm) gypsum board.
   - 3.6. Corrosion-resistant steel having a base metal thickness of 0.016 inch (0.406 mm).
   - 3.7. $^1/_4$-inch (6.4 mm) fiber-cement panel, soffit or backer board.

R316.5.5 Foam-filled exterior doors. Foam-filled exterior doors are exempt from the requirements of Sections R316.3 and R316.4.

R316.5.6 Foam-filled garage doors. Foam-filled garage doors in attached or detached garages are exempt from the requirements of Sections R316.3 and R316.4.

R316.5.7 Foam backer board. The thermal barrier specified in Section R316.4 is not required where siding backer board foam plastic insulation has a thickness of not more than 0.5 inch (12.7 mm) and a potential heat of not more than 2000 Btu per square foot (22 720 kJ/m^2) when tested in accordance with NFPA 259 and it complies with one or more of the following:

1. The foam plastic insulation is separated from the interior of the building by not less than 2 inches (51 mm) of mineral fiber insulation.
2. The foam plastic insulation is installed over existing exterior wall finish in conjunction with re-siding.
3. The foam plastic insulation has been tested in accordance with Section R316.6.

R316.5.8 Re-siding. The thermal barrier specified in Section R316.4 is not required where the foam plastic insulation is installed over existing exterior wall finish in conjunction with re-siding provided that the foam plastic has a thickness of not more than 0.5 inch (12.7 mm) and a potential heat of not more than 2000 Btu per square foot (22 720 kJ/m^2) when tested in accordance with NFPA 259.

R316.5.9 Interior trim. The thermal barrier specified in Section R316.4 is not required for exposed foam plastic interior *trim*, provided that all of the following are met:

1. The density is not less than 20 pounds per cubic foot (320 kg/m^3).
2. The thickness of the *trim* is not more than 0.5 inch (12.7 mm) and the width is not more than 8 inches (204 mm).
3. The interior *trim* shall not constitute more than 10 percent of the aggregate wall and ceiling area of any room or space.
4. The flame spread index does not exceed 75 when tested in accordance with ASTM E84 or UL 723. The *smoke-developed index* is not limited.

R316.5.10 Interior finish. Foam plastics used as interior finishes shall comply with Section R316.6 and shall meet the flame spread index and *smoke-developed index* requirements of Sections R302.9.1 and R302.9.2.

R316.5.11 Sill plates and headers. Foam plastic spray applied to sill plates and headers or installed in the perimeter joist space without the thermal barrier specified in Section R316.4 shall comply with all of the following:

1. The thickness of the foam plastic shall be not more than $3^1/_4$ inches (83 mm).
2. The density of the foam plastic shall be in the range of 0.5 to 2.0 pounds per cubic foot (8 to 32 kg/m^3).
3. The foam plastic shall have a flame spread index of 25 or less and an accompanying *smoke-developed index* of 450 or less when tested in accordance with ASTM E84 or UL 723.

R316.5.12 Sheathing. Foam plastic insulation used as sheathing shall comply with Section R316.3 and Section R316.4. Where the foam plastic sheathing is exposed to the *attic* space at a gable or kneewall, the provisions of Section R316.5.3 shall apply. Where foam plastic insulation is used as exterior wall sheathing on framed wall assemblies, it shall comply with Section R316.8.

R316.5.13 Floors. The thermal barrier specified in Section R316.4 is not required to be installed on the walking surface of a structural floor system that contains foam plastic insulation where the foam plastic is covered by not less than a nominal $^1/_2$-inch-thick (12.7 mm) *wood structural panel* or equivalent. The thermal barrier specified in Section R316.4 is required on the underside of the structural floor system that contains foam plastic insulation where the underside of the structural floor system is exposed to the interior of the building.

R316.6 Specific approval. Foam plastic not meeting the requirements of Sections R316.3 through R316.5 shall be specifically *approved* on the basis of one of the following *approved* tests: NFPA 286 with the acceptance criteria of Section R302.9.4, FM 4880, UL 1040 or UL 1715, or fire tests related to actual end-use configurations. Approval shall be based on the actual end-use configuration and shall be performed on the finished foam plastic assembly in the maximum thickness intended for use. Assemblies tested shall include seams, joints and other typical details used in the installation of the assembly and shall be tested in the manner intended for use.

R316.7 Termite damage. The use of foam plastics in areas of "very heavy" termite infestation probability shall be in accordance with Section R318.4.

R316.8 Wind resistance. Foam plastic insulation complying with ASTM C578 and ASTM C1289 and used as exterior wall sheathing on framed wall assemblies shall comply with SBCA FS 100 for wind pressure resistance unless installed directly over a sheathing material that is separately capable of resisting the wind load or otherwise exempted from the scope of SBCA FS 100.

SECTION R317
PROTECTION OF WOOD AND WOOD-BASED PRODUCTS AGAINST DECAY

R317.1 Location required. Protection of wood and wood-based products from decay shall be provided in the following locations by the use of *naturally durable wood* or wood that is preservative-treated in accordance with AWPA U1.

1. In crawl spaces or unexcavated areas located within the periphery of the building foundation, wood joists or the bottom of a wood structural floor where closer than 18 inches (457 mm) to exposed ground, wood girders where closer than 12 inches (305 mm) to exposed ground, and wood columns where closer than 8 inches (204 mm) to exposed ground.
2. Wood framing members, including columns, that rest directly on concrete or masonry exterior foundation walls and are less than 8 inches (203 mm) from the exposed ground.
3. Sills and sleepers on a concrete or masonry slab that is in direct contact with the ground unless separated from such slab by an impervious moisture barrier.
4. The ends of wood girders entering exterior masonry or concrete walls having clearances of less than $^1/_2$ inch (12.7 mm) on tops, sides and ends.
5. Wood siding, sheathing and wall framing on the exterior of a building having a clearance of less than 6 inches (152 mm) from the ground or less than 2 inches (51 mm) measured vertically from concrete steps, porch slabs, patio slabs and similar horizontal surfaces exposed to the weather.
6. Wood structural members supporting moisture-permeable floors or roofs that are exposed to the weather, such as concrete or masonry slabs, unless separated from such floors or roofs by an impervious moisture barrier.
7. Wood furring strips or other wood framing members attached directly to the interior of exterior masonry walls or concrete walls below *grade* except where an *approved* vapor retarder is applied between the wall and the furring strips or framing members.
8. Portions of wood structural members that form the structural supports of buildings, balconies, porches or similar permanent building appurtenances where those members are exposed to the weather without adequate protection from a roof, eave, overhang or other covering that would prevent moisture or water accumulation on the surface or at joints between members.

    Exception: Sawn lumber used in buildings located in a geographical region where experience has demonstrated that climatic conditions preclude the need to use naturally durable or preservative-treated wood where the structure is exposed to the weather.

9. Wood columns in contact with *basement* floor slabs unless supported by concrete piers or metal pedestals projecting not less than 1 inch (25 mm) above the concrete floor and separated from the concrete pier by an impervious moisture barrier.

R317.1.1 Field treatment. Field-cut ends, notches and drilled holes of preservative-treated wood shall be treated in the field in accordance with AWPA M4.

R317.1.2 Ground contact. All wood in contact with the ground, embedded in concrete in direct contact with the ground or embedded in concrete exposed to the weather that supports permanent structures intended for human occupancy shall be *approved* pressure-preservative-treated wood suitable for ground contact use, except that untreated wood used entirely below groundwater level or continuously submerged in fresh water shall not be required to be pressure-preservative treated.

R317.2 Quality mark. Lumber and plywood required to be pressure-preservative treated in accordance with Section R318.1 shall bear the quality *mark* of an *approved* inspection agency that maintains continuing supervision, testing and inspection over the quality of the product and that has been *approved* by an accreditation body that complies with the requirements of the American Lumber Standard Committee treated wood program.

R317.2.1 Required information. The required quality *mark* on each piece of pressure-preservative-treated lumber or plywood shall contain the following information:

1. Identification of the treating plant.
2. Type of preservative.
3. The minimum preservative retention.
4. End use for which the product was treated.
5. Standard to which the product was treated.
6. Identity of the *approved* inspection agency.
7. The designation "Dry," if applicable.

Exception: Quality *marks* on lumber less than 1 inch (25 mm) nominal thickness, or lumber less than nominal 1 inch by 5 inches (25 mm by 127 mm) or 2 inches by 4 inches (51 mm by 102 mm) or lumber 36 inches (914 mm) or less in length shall be applied by stamping the faces of exterior pieces or by end labeling not less than 25 percent of the pieces of a bundled unit.

R317.3 Fasteners and connectors in contact with preservative-treated and fire-retardant-treated wood. Fasteners, including nuts and washers, and connectors in contact with preservative-treated wood and fire-retardant-treated wood shall be in accordance with this section. The coating weights for zinc-coated fasteners shall be in accordance with ASTM A153. Stainless steel driven fasteners shall be in accordance with the material requirements of ASTM F1667.

R317.3.1 Fasteners for preservative-treated wood. Fasteners, including nuts and washers, for preservative-treated wood shall be of hot-dipped, zinc-coated galvanized steel, stainless steel, silicon bronze or copper. Staples shall be of stainless steel. Coating types and weights for connectors in contact with preservative-treated wood shall be in accordance with the connector manufacturer's recommendations. In the absence of manufacturer's recommendations, not less than ASTM A653 type G185 zinc-coated galvanized steel, or equivalent, shall be used.

Exceptions:

1. $^{1}/_{2}$-inch-diameter (12.7 mm) or greater steel bolts.
2. Fasteners other than nails, staples and timber rivets shall be permitted to be of mechanically deposited zinc-coated steel with coating weights in accordance with ASTM B695, Class 55 minimum.
3. Plain carbon steel fasteners in SBX/DOT and zinc borate preservative-treated wood in an interior, dry environment shall be permitted.

R317.3.2 Fastenings for wood foundations. Fastenings, including nuts and washers, for wood foundations shall be as required in AWC PWF.

R317.3.3 Fasteners for fire-retardant-treated wood used in exterior applications or wet or damp locations. Fasteners, including nuts and washers, for fire-retardant-treated wood used in exterior applications or wet or damp locations shall be of hot-dipped, zinc-coated galvanized steel, stainless steel, silicon bronze or copper. Fasteners other than nails, staples and timber rivets shall be permitted to be of mechanically deposited zinc-coated steel with coating weights in accordance with ASTM B695, Class 55 minimum.

R317.3.4 Fasteners for fire-retardant-treated wood used in interior applications. Fasteners, including nuts and washers, for fire-retardant-treated wood used in interior locations shall be in accordance with the manufacturer's recommendations. In the absence of the manufacturer's recommendations, Section R317.3.3 shall apply.

R317.4 Plastic composites. *Plastic composite* exterior deck boards, stair treads, *guards* and *handrails* containing wood, cellulosic or other biodegradable materials shall comply with the requirements of Section R507.2.2.

SECTION R318
PROTECTION AGAINST SUBTERRANEAN TERMITES

R318.1 Subterranean termite control methods. In areas subject to damage from termites as indicated by Table R301.2, protection shall be by one, or a combination, of the following methods:

1. Chemical termiticide treatment in accordance with Section R318.2.
2. Termite-baiting system installed and maintained in accordance with the *label*.
3. Pressure-preservative-treated wood in accordance with the provisions of Section R317.1.
4. Naturally durable termite-resistant wood.
5. Physical barriers in accordance with Section R318.3 and used in locations as specified in Section R317.1.
6. Cold-formed steel framing in accordance with Sections R505.2.1 and R603.2.1.

R318.1.1 Quality mark. Lumber and plywood required to be pressure-preservative treated in accordance with Section R318.1 shall bear the quality *mark* of an *approved* inspection agency that maintains continuing supervision, testing and inspection over the quality of the product and that has been *approved* by an accreditation body that complies with the requirements of the American Lumber Standard Committee treated wood program.

R318.1.2 Field treatment. Field-cut ends, notches and drilled holes of pressure-preservative-treated wood shall be retreated in the field in accordance with AWPA M4.

R318.2 Chemical termiticide treatment. Chemical termiticide treatment shall include soil treatment or field-applied wood treatment. The concentration, rate of application and method of treatment of the chemical termiticide shall be in strict accordance with the termiticide *label*.

R318.3 Barriers. *Approved* physical barriers, such as metal or plastic sheeting or collars specifically designed for termite prevention, shall be installed in a manner to prevent termites from entering the structure. Shields placed on top of an exterior foundation wall shall be used only if in combination with another method of protection.

R318.4 Foam plastic protection. In areas where the probability of termite infestation is "very heavy" as indicated in Figure R318.4, extruded and expanded polystyrene, polyisocyanurate and other foam plastics shall not be installed on the exterior face or under interior or exterior foundation walls or slab foundations located below *grade*. The clearance between foam plastics installed above *grade* and exposed earth shall be not less than 6 inches (152 mm).

Exceptions:

1. Buildings where the structural members of walls, floors, ceilings and roofs are entirely of *noncombustible materials* or pressure-preservative-treated wood.
2. Where in addition to the requirements of Section R318.1, an *approved* method of protecting the foam plastic and structure from subterranean termite damage is used.
3. On the interior side of basement *walls*.

SECTION R319 SITE ADDRESS

R319.1 Address identification. Buildings shall be provided with *approved* address identification. The address identification shall be legible and placed in a position that is visible from the street or road fronting the property. Address identification characters shall contrast with their background. Address numbers shall be Arabic numbers or alphabetical letters. Numbers shall not be spelled out. Each character shall be not less than 4 inches (102 mm) in height with a stroke width of not less than 0.5 inch (12.7 mm). Where required by the fire code official, address identification shall be provided in additional *approved* locations to facilitate emergency response. Where access is by means of a private road and the building address cannot be viewed from the *public way*, a monument, pole or other sign or means shall be used to identify the structure. Address identification shall be maintained.

SECTION R320 ACCESSIBILITY

R320.1 Scope. Where there are four or more *dwelling units* or *sleeping units* in a single structure, the provisions of Chapter 11 of the *International Building Code* for Group R-3 shall apply.

Exception: Owner-occupied *lodging houses* with five or fewer guestrooms are not required to be accessible.

R320.2 Live/work units. In *live/work units,* the nonresidential portion shall be accessible in accordance with Sections 508.5.9 and 508.5.11 of the *International Building Code*. In a structure where there are four or more *live/work units*, the dwelling portion of the *live/work unit* shall comply with Section 1108.6.2.1 of the *International Building Code*.

SECTION R321 ELEVATORS AND PLATFORM LIFTS

R321.1 Elevators. Where provided, passenger elevators, limited-use and limited-application elevators or private residence elevators shall comply with ASME A17.1/CSA B44.

R321.2 Platform lifts. Where provided, platform lifts shall comply with ASME A18.1.

R321.3 Accessibility. Elevators or platform lifts that are part of an accessible route required by Chapter 11 of the *International Building Code*, shall comply with ICC A117.1.

SECTION R322 FLOOD-RESISTANT CONSTRUCTION

R322.1 General. Buildings and structures constructed in whole or in part in flood hazard areas, including A or V Zones and Coastal A Zones, as established in Table R301.2, and substantial improvement and *repair* of substantial damage of buildings and structures in flood hazard areas, shall be designed and constructed in accordance with the provisions contained in this section. Buildings and structures that are located in more than one flood hazard area shall comply with the provisions associated with the most restrictive flood hazard area. Buildings and structures located in whole or in part in identified floodways shall be designed and constructed in accordance with ASCE 24.

R322.1.1 Alternative provisions. As an alternative to the requirements in Section R322, ASCE 24 is permitted subject to the limitations of this code and the limitations therein.

R322.1.2 Structural systems. Structural systems of buildings and structures shall be designed, connected and anchored to resist flotation, collapse or permanent lateral movement due to structural loads and stresses from flooding equal to the design flood elevation.

R322.1.3 Flood-resistant construction. Buildings and structures erected in areas prone to flooding shall be constructed by methods and practices that minimize flood damage.

R322.1.4 Establishing the design flood elevation. The design flood elevation shall be used to define flood hazard areas. At a minimum, the design flood elevation shall be the higher of the following:

1. The base flood elevation at the depth of peak elevation of flooding, including wave height, that has a 1-percent (100-year flood) or greater chance of being equaled or exceeded in any given year.
2. The elevation of the design flood associated with the area designated on a flood hazard map adopted by the community, or otherwise legally designated.

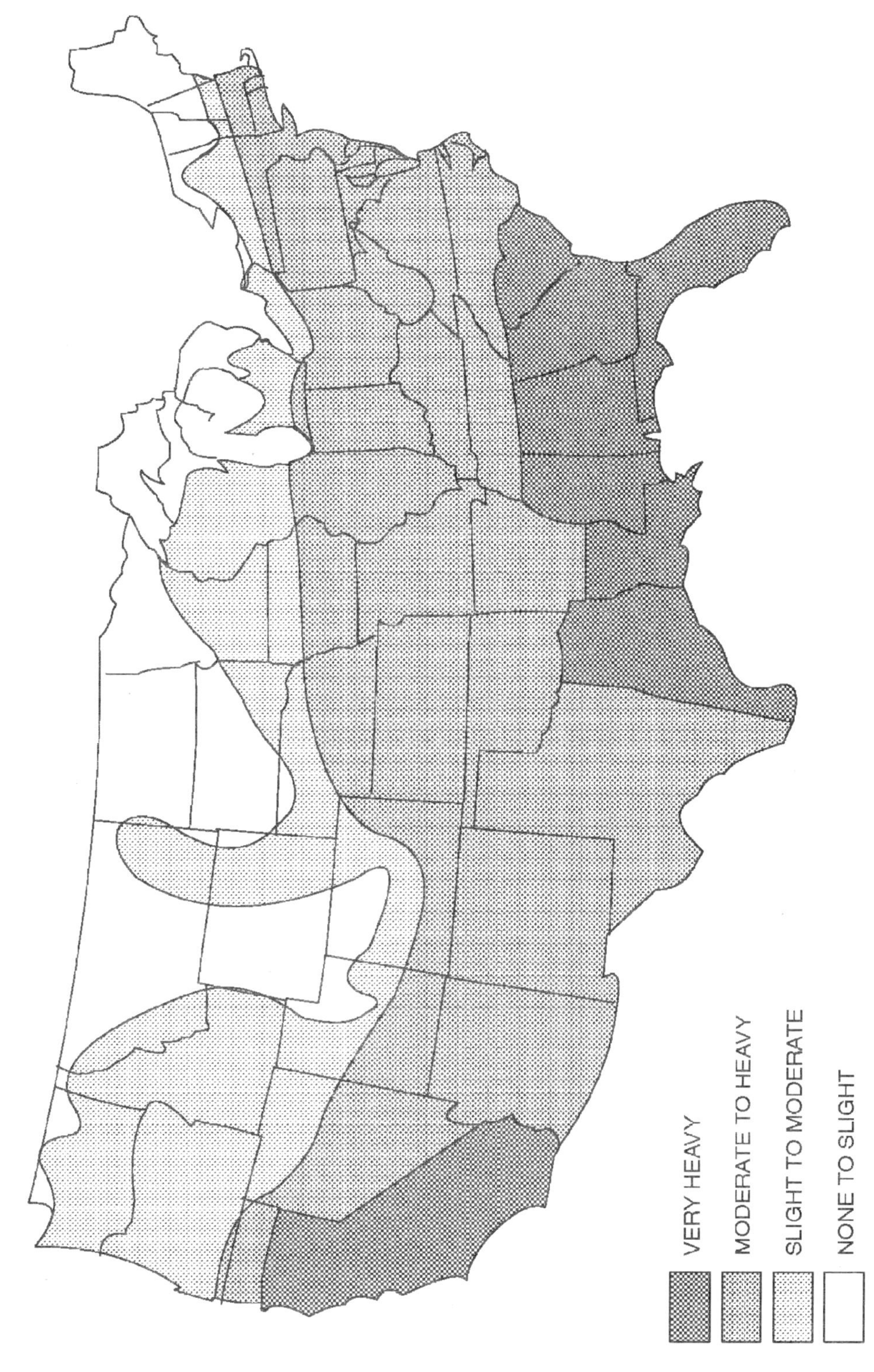

Note: Lines defining areas are approximate only. Local conditions may be more or less severe than indicated by the region classification.

**FIGURE R318.4
TERMITE INFESTATION PROBABILITY MAP**

R322.1.4.1 Determination of design flood elevations. If design flood elevations are not specified, the *building official* is authorized to require the applicant to comply with either of the following:

1. Obtain and reasonably use data available from a federal, state or other source.
2. Determine the design flood elevation in accordance with accepted hydrologic and hydraulic engineering practices used to define special flood hazard areas. Determinations shall be undertaken by a *registered design professional* who shall document that the technical methods used reflect currently accepted engineering practice. Studies, analyses and computations shall be submitted in sufficient detail to allow thorough review and *approval*.

R322.1.4.2 Determination of impacts. In riverine flood hazard areas where design flood elevations are specified but floodways have not been designated, the applicant shall demonstrate that the effect of the proposed buildings and structures on design flood elevations, including fill, when combined with other existing and anticipated flood hazard area encroachments, will not increase the design flood elevation more than 1 foot (305 mm) at any point within the *jurisdiction*.

R322.1.5 Lowest floor. The lowest floor shall be the lowest floor of the lowest enclosed area, including *basement*, and excluding any unfinished flood-resistant enclosure that is useable solely for vehicle parking, building access or limited storage provided that such enclosure is not built so as to render the building or structure in violation of this section.

R322.1.6 Protection of mechanical, plumbing and electrical systems. Electrical systems, *equipment* and components; heating, ventilating, air-conditioning; plumbing *appliances* and plumbing fixtures; *duct systems*; and other service *equipment* shall be located at or above the elevation required in Section R322.2 or R322.3. If replaced as part of a substantial improvement, electrical systems, *equipment* and components; heating, ventilating, air-conditioning and plumbing *appliances* and plumbing fixtures; *duct systems*; and other service *equipment* shall meet the requirements of this section. Systems, fixtures, and *equipment* and components shall not be mounted on or penetrate through walls intended to break away under flood loads.

Exception: Locating electrical systems, *equipment* and components; heating, ventilating, air-conditioning; plumbing *appliances* and plumbing fixtures; *duct systems*; and other service *equipment* is permitted below the elevation required in Section R322.2 or R322.3 provided that they are designed and installed to prevent water from entering or accumulating within the components and to resist hydrostatic and hydrodynamic loads and stresses, including the effects of buoyancy, during the occurrence of flooding to the required elevation in accordance with ASCE 24. Electrical wiring systems are permitted to be located below the required elevation provided that they conform to the provisions of the electrical part of this code for wet locations.

R322.1.7 Protection of water supply and sanitary sewage systems. New and replacement water supply systems shall be designed to minimize or eliminate infiltration of flood waters into the systems in accordance with the plumbing provisions of this code. New and replacement sanitary sewage systems shall be designed to minimize or eliminate infiltration of floodwaters into systems and discharges from systems into floodwaters in accordance with the plumbing provisions of this code and Chapter 3 of the *International Private Sewage Disposal Code*.

R322.1.8 Flood-resistant materials. Building materials and installation methods used for flooring and interior and exterior walls and wall coverings below the elevation required in Section R322.2 or R322.3 shall be flood damage-resistant materials that conform to the provisions of FEMA TB-2.

R322.1.9 Manufactured homes. The bottom of the frame of new and replacement *manufactured homes* on foundations that conform to the requirements of Section R322.2 or R322.3, as applicable, shall be elevated to or above the elevations specified in Section R322.2 (flood hazard areas including A Zones) or R322.3 in coastal high-hazard areas (V Zones and Coastal A Zones). The anchor and tie-down requirements of the applicable state or federal requirements shall apply. The foundation and anchorage of *manufactured homes* to be located in identified floodways shall be designed and constructed in accordance with ASCE 24.

R322.1.10 As-built elevation documentation. A *registered design professional* shall prepare and seal documentation of the elevations specified in Section R322.2 or R322.3.

R322.2 Flood hazard areas (including A Zones). Areas that have been determined to be prone to flooding and that are not subject to high-velocity wave action shall be designated as flood hazard areas. Flood hazard areas that have been delineated as subject to wave heights between $1^1/_2$ feet (457 mm) and 3 feet (914 mm) or otherwise designated by the *jurisdiction* shall be designated as Coastal A Zones and are subject to the requirements of Section R322.3. Buildings and structures constructed in whole or in part in flood hazard areas shall be designed and constructed in accordance with Sections R322.2.1 through R322.2.4.

R322.2.1 Elevation requirements.

1. Buildings and structures in flood hazard areas, not including flood hazard areas designated as Coastal A Zones, shall have the lowest floors elevated to or above the base flood elevation plus 1 foot (305 mm), or the design flood elevation, whichever is higher.
2. In areas of shallow flooding (AO Zones), buildings and structures shall have the lowest floor

(including *basement*) elevated to a height above the highest adjacent *grade* of not less than the depth number specified in feet (mm) on the FIRM plus 1 foot (305 mm), or not less than 3 feet (915 mm) if a depth number is not specified.

3. *Basement* floors that are below *grade* on all sides shall be elevated to or above base flood elevation plus 1 foot (305 mm), or the design flood elevation, whichever is higher.
4. Garage and carport floors shall comply with one of the following:
    4.1. They shall be elevated to or above the elevations required in Item 1 or Item 2, as applicable.
    4.2. They shall be at or above *grade* on not less than one side. Where a garage or carport is enclosed by walls, the garage or carport shall be used solely for parking, building access or storage.

Exception: Enclosed areas below the elevation required in this section, including *basements* with floors that are not below *grade* on all sides, shall meet the requirements of Section R322.2.2.

R322.2.2 Enclosed area below required elevation. Enclosed areas, including *crawl spaces*, that are below the elevation required in Section R322.2.1 shall:

1. Be used solely for parking of vehicles, building access or storage.
2. Be provided with flood openings that meet the following criteria and are installed in accordance with Section R322.2.2.1:
    2.1. The total net area of nonengineered openings shall be not less than 1 square inch (645 mm^2) for each square foot (0.093 m^2) of enclosed area where the enclosed area is measured on the exterior of the enclosure walls, or the openings shall be designed as engineered openings and the *construction documents* shall include a statement by a *registered design professional* that the design of the openings will provide for equalization of hydrostatic flood forces on exterior walls by allowing for the automatic entry and exit of floodwaters as specified in Section 2.7.2.2 of ASCE 24.
    2.2. Openings shall be not less than 3 inches (76 mm) in any direction in the plane of the wall.
    2.3. The presence of louvers, blades, screens and faceplates or other covers and devices shall allow the automatic flow of floodwater into and out of the enclosed areas and shall be accounted for in the determination of the net open area.

R322.2.2.1 Installation of openings. The walls of enclosed areas shall have openings installed such that:

1. There shall be not less than two openings on different sides of each enclosed area; if a building has more than one enclosed area, each area shall have openings.
2. The bottom of each opening shall be not more than 1 foot (305 mm) above the higher of the final interior grade or floor and the finished exterior grade immediately under each opening.
3. Openings shall be permitted to be installed in doors and windows; doors and windows without installed openings do not meet the requirements of this section.

R322.2.3 Foundation design and construction. Foundation walls for buildings and structures erected in flood hazard areas shall meet the requirements of Chapter 4.

Exception: Unless designed in accordance with Section R404:

1. The unsupported height of 6-inch (152 mm) plain masonry walls shall be not more than 3 feet (914 mm).
2. The unsupported height of 8-inch (203 mm) plain masonry walls shall be not more than 4 feet (1219 mm).
3. The unsupported height of 8-inch (203 mm) reinforced masonry walls shall be not more than 8 feet (2438 mm).

For the purpose of this exception, unsupported height is the distance from the finished *grade* of the under-floor space to the top of the wall.

R322.2.4 Tanks. Underground tanks shall be anchored to prevent flotation, collapse and lateral movement under conditions of the base flood. Above-ground tanks shall be installed at or above the elevation required in Section R322.2.1 or shall be anchored to prevent flotation, collapse and lateral movement under conditions of the base flood.

R322.3 Coastal high-hazard areas (including V Zones and Coastal A Zones, where designated). Areas that have been determined to be subject to wave heights in excess of 3 feet (914 mm) or subject to high-velocity wave action or wave-induced erosion shall be designated as coastal high-hazard areas. Flood hazard areas that have been designated as subject to wave heights between $1^1/_2$ feet (457 mm) and 3 feet (914 mm) or otherwise designated by the *jurisdiction* shall be designated as Coastal A Zones. Buildings and structures constructed in whole or in part in coastal high-hazard areas and Coastal A Zones, where designated, shall be designed and constructed in accordance with Sections R322.3.1 through R322.3.10.

R322.3.1 Location and site preparation.

1. New buildings and buildings that are determined to be substantially improved pursuant to Section

R105.3.1.1 shall be located landward of the reach of mean high tide.

2. For any alteration of sand dunes and mangrove stands, the *building official* shall require submission of an engineering analysis that demonstrates that the proposed alteration will not increase the potential for flood damage.

R322.3.2 Elevation requirements.

1. Buildings and structures erected within coastal high-hazard areas and Coastal A Zones, shall be elevated so that the bottom of the lowest horizontal structural members supporting the lowest floor, with the exception of piling, pile caps, columns, grade beams and bracing, is elevated to or above the base flood elevation plus 1 foot (305 mm) or the design flood elevation, whichever is higher.
2. *Basement* floors that are below *grade* on all sides are prohibited.
3. Garages used solely for parking, building access or storage, and carports shall comply with Item 1 or shall be at or above *grade* on not less than one side and, if enclosed with walls, such walls shall comply with Item 6.
4. The use of fill for structural support is prohibited.
5. Minor grading, and the placement of minor quantities of fill, shall be permitted for landscaping and for drainage purposes under and around buildings and for support of parking slabs, pool decks, patios and walkways.
6. Walls and partitions enclosing areas below the elevation required in this section shall meet the requirements of Sections R322.3.5 and R322.3.6.

R322.3.3 Foundations. Buildings and structures erected in coastal high-hazard areas and Coastal A Zones shall be supported on pilings or columns and shall be adequately anchored to such pilings or columns and shall comply with the following:

1. The space below the elevated building shall be either free of obstruction or, if enclosed with walls, the walls shall meet the requirements of Section R322.3.5.
2. Pilings shall have adequate soil penetrations to resist the combined wave and wind loads (lateral and uplift) and pile embedment shall include consideration of decreased resistance capacity caused by scour of soil strata surrounding the piling.
3. Columns and their supporting foundations shall be designed to resist combined wave and wind loads, lateral and uplift, and shall include consideration of decreased resistance capacity caused by scour of soil strata surrounding the columns. Spread footing, mat, raft or other foundations that support columns shall not be permitted where soil investigations that are required in accordance with Section R401.4 indicate that soil material under the spread footing, mat, raft or other foundation is subject to scour or erosion from wave-velocity flow conditions. If permitted, spread footing, mat, raft or other foundations that support columns shall be designed in accordance with ASCE 24.
4. Flood and wave loads shall be those associated with the design flood. Wind loads shall be those required by this code.
5. Foundation designs and *construction documents* shall be prepared and sealed in accordance with Section R322.3.9.

Exception: In Coastal A Zones, stem wall foundations supporting a floor system above and backfilled with soil or gravel to the underside of the floor system shall be permitted provided that the foundations are designed to account for wave action, debris impact, erosion and local scour. Where soils are susceptible to erosion and local scour, stem wall foundations shall have deep footings to account for the loss of soil.

R322.3.4 Concrete slabs. Concrete slabs used for parking, floors of enclosures, landings, decks, walkways, patios and similar uses that are located beneath structures, or slabs that are located such that if undermined or displaced during base flood conditions could cause structural damage to the building foundation, shall be designed and constructed in accordance with one of the following:

1. To be structurally independent of the foundation system of the structure, to not transfer flood loads to the main structure, and to be frangible and break away under flood conditions prior to base flood conditions. Slabs shall be a maximum of 4 inches (102 mm) thick, shall not have turned-down edges, shall not contain reinforcing, shall have isolation joints at pilings and columns, and shall have control or construction joints in both directions spaced not more than 4 feet (1219 mm) apart.
2. To be self-supporting, structural slabs capable of remaining intact and functional under base flood conditions, including erosion and local scour, and the main structure shall be capable of resisting any added flood loads and effects of local scour caused by the presence of the slabs.

R322.3.5 Walls below required elevation. Walls and partitions are permitted below the elevation required in Section R322.3.2, provided that such walls and partitions are not part of the structural support of the building or structure and:

1. Electrical, mechanical and plumbing system components are not to be mounted on or penetrate through walls that are designed to break away under flood loads; and
2. Are constructed with insect screening or open lattice; or
3. Are designed to break away or collapse without causing collapse, displacement or other structural damage to the elevated portion of the building or

supporting foundation system. Such walls, framing and connections shall have a resistance of not less than 10 (479 Pa) and not more than 20 pounds per square foot (958 Pa) as determined using allowable stress design; or

4. Where wind loading values of this code exceed 20 pounds per square foot (958 Pa), as determined using allowable stress design, the *construction documents* shall include documentation prepared and sealed by a *registered design professional* that:
   4.1. The walls and partitions below the required elevation have been designed to collapse from a water load less than that which would occur during the base flood.
   4.2. The elevated portion of the building and supporting foundation system have been designed to withstand the effects of wind and flood loads acting simultaneously on structural and nonstructural building components. Water-loading values used shall be those associated with the design flood. Wind-loading values shall be those required by this code.
5. Walls intended to break away under flood loads as specified in Item 3 or 4 have flood openings that meet the criteria in Section R322.2.2, Item 2.

R322.3.6 Enclosed areas below required elevation. Enclosed areas below the elevation required in Section R322.3.2 shall be used solely for parking of vehicles, building access or storage.

R322.3.6.1 Protection of building envelope. An exterior door that meets the requirements of Section R609 shall be installed at the top of stairs that provide access to the building and that are enclosed with walls designed to break away in accordance with Section R322.3.5.

R322.3.7 Stairways and ramps. *Stairways* and *ramps* that are located below the lowest floor elevations specified in Section R322.3.2 shall comply with one or more of the following:

1. Be designed and constructed with open or partially open *risers* and *guards.*
2. *Stairways* and *ramps* not part of the required means of egress shall be designed and constructed to break away during design flood conditions without causing damage to the building or structure, including foundation.
3. Be retractable, or able to be raised to or above the lowest floor elevation, provided that the ability to be retracted or raised prior to the onset of flooding is not contrary to the means of egress requirements of the code.
4. Be designed and constructed to resist flood loads and minimize transfer of flood loads to the building or structure, including foundation.

Areas below *stairways* and *ramps* shall not be enclosed with walls below the required in Section R322.3.2 elevation unless such walls are constructed in accordance with Section R322.3.5.

R322.3.8 Decks and porches. Attached decks and porches shall meet the elevation requirements of Section R322.3.2 and shall either meet the foundation requirements of this section or shall be cantilevered from or knee braced to the building or structure. Self-supporting decks and porches that are below the elevation required in Section R322.3.2 shall not be enclosed by solid, rigid walls, including walls designed to break away. Self-supporting decks and porches shall be designed and constructed to remain in place during base flood conditions or shall be frangible and break away under base flood conditions.

R322.3.9 Construction documents. The *construction documents* shall include documentation that is prepared and sealed by a *registered design professional* that the design and methods of construction to be used meet the applicable criteria of this section.

R322.3.10 Tanks. Underground tanks shall be anchored to prevent flotation, collapse and lateral movement under conditions of the base flood. Above-ground tanks shall be installed at or above the elevation required in Section R322.3.2. Where elevated on platforms, the platforms shall be cantilevered from or knee braced to the building or shall be supported on foundations that conform to the requirements of Section R322.3.

SECTION R323
STORM SHELTERS

R323.1 General. This section applies to *storm shelters* where constructed as separate detached buildings or where constructed as safe rooms within buildings for the purpose of providing refuge from storms that produce high winds, such as tornados and hurricanes. In addition to other applicable requirements in this code, storm shelters shall be constructed in accordance with ICC 500.

R323.1.1 Sealed documentation. The *construction documents* for all structural components and *impact protective systems* of the *storm shelter* shall be prepared and sealed by a *registered design professional* indicating that the design meets the criteria of ICC 500.

Exception: *Storm shelters,* structural components and impact-protective systems that are *listed* and *labeled* to indicate compliance with ICC 500.

SECTION R324
SOLAR ENERGY SYSTEMS

R324.1 General. Solar energy systems shall comply with the provisions of this section.

R324.2 Solar thermal systems. Solar thermal systems shall be designed and installed in accordance with Chapter 23.

R324.3 Photovoltaic systems. Photovoltaic (PV) systems shall be designed and installed in accordance with Sections R324.3.1 through R324.7.1 and the manufacturer's installation instructions. The electrical portion of solar PV systems shall be designed and installed in accordance with NFPA 70.

R324.3.1 Equipment listings. *Photovoltaic panels* and modules shall be *listed* and *labeled* in accordance with UL 1703 or with both UL 61730-1 and UL 61730-2. Inverters shall be *listed* and *labeled* in accordance with UL 1741. Systems connected to the utility grid shall use inverters *listed* for utility interaction. Mounting systems *listed* and *labeled* in accordance with UL 2703 shall be installed in accordance with the manufacturer's installation instructions and their listings.

R324.4 Rooftop-mounted photovoltaic systems. Rooftop-mounted *photovoltaic panel systems* installed on or above the roof covering shall be designed and installed in accordance with this section.

R324.4.1 Structural requirements. Rooftop-mounted *photovoltaic panel systems* shall be designed to structurally support the system and withstand applicable gravity loads in accordance with Chapter 3. The roof on which these systems are installed shall be designed and constructed to support the loads imposed by such systems in accordance with Chapter 8.

R324.4.1.1 Roof load. Portions of roof structures not covered with *photovoltaic panel systems* shall be designed for dead loads and roof loads in accordance with Sections R301.4 and R301.6. Portions of roof structures covered with *photovoltaic panel systems* shall be designed for the following load cases:

1. Dead load (including *photovoltaic panel* weight) plus snow load in accordance with Table R301.2.
2. Dead load (excluding *photovoltaic panel* weight) plus roof *live load* or snow load, whichever is greater, in accordance with Section R301.6.

R324.4.1.2 Wind load. Rooftop-mounted *photovoltaic panel* or *module* systems and their supports shall be designed and installed to resist the component and cladding loads specified in Table R301.2.1(1), adjusted for height and exposure in accordance with Table R301.2.1(2).

R324.4.2 Fire classification. Rooftop-mounted *photovoltaic panel systems* shall have the same fire classification as the *roof assembly* required in Section R902.

R324.4.3 Roof penetrations. Roof penetrations shall be flashed and sealed in accordance with Chapter 9.

R324.5 Building-integrated photovoltaic systems. Building-integrated photovoltaic (BIPV) systems that serve as roof coverings shall be designed and installed in accordance with Section R905.

R324.5.1 Photovoltaic shingles. Photovoltaic shingles shall comply with Section R905.16.

R324.5.2 Fire classification. *Building-integrated photovoltaic systems* shall have a fire classification in accordance with Section R902.3.

R324.5.3 BIPV roof panels. BIPV roof panels shall comply with Section R905.17.

R324.6 Roof access and pathways. Roof access, pathways and setback requirements shall be provided in accordance with Sections R324.6.1 through R324.6.2.1. Access and minimum spacing shall be required to provide emergency access to the roof, to provide pathways to specific areas of the roof, provide for smoke ventilation opportunity areas, and to provide emergency egress from the roof.

Exceptions:

1. Detached, nonhabitable structures, including but not limited to detached garages, parking shade structures, carports, solar trellises and similar structures, shall not be required to provide roof access.
2. Roof access, pathways and setbacks need not be provided where the code official has determined that rooftop operations will not be employed.
3. These requirements shall not apply to roofs with slopes of 2 units vertical in 12 units horizontal (17-percent slope) or less.
4. BIPV systems *listed* in accordance with Section 690.12(B)(2) of NFPA 70, where the removal or cutting away of portions of the BIPV system during fire-fighting operations has been determined to not expose a fire fighter to electrical shock hazards.

R324.6.1 Pathways. Not fewer than two pathways, on separate roof planes from lowest roof edge to ridge and not less than 36 inches (914 mm) wide, shall be provided on all buildings. Not fewer than one pathway shall be provided on the street or driveway side of the roof. For each roof plane with a photovoltaic array, a pathway not less than 36 inches wide (914 mm) shall be provided from the lowest roof edge to ridge on the same roof plane as the photovoltaic array, on an adjacent roof plane, or straddling the same and adjacent roof planes. Pathways shall be over areas capable of supporting fire fighters accessing the roof. Pathways shall be located in areas with minimal obstructions such as vent pipes, conduit, or mechanical equipment.

R324.6.2 Setback at ridge. For photovoltaic arrays occupying not more than 33 percent of the plan view total roof area, not less than an 18-inch (457 mm) clear setback is required on both sides of a horizontal ridge. For photovoltaic arrays occupying more than 33 percent of the plan view total roof area, not less than a 36-inch (914 mm) clear setback is required on both sides of a horizontal ridge.

R324.6.2.1 Alternative setback at ridge. Where an automatic sprinkler system is installed within the dwelling in accordance with NFPA 13D or Section P2904, setbacks at ridges shall comply with one of the following:

1. For photovoltaic arrays occupying not more than 66 percent of the plan view total roof area, not less than an 18-inch (457 mm) clear setback is required on both sides of a horizontal ridge.
2. For photovoltaic arrays occupying more than 66 percent of the plan view total roof area, not less than a 36-inch (914 mm) clear setback is required on both sides of a horizontal ridge.

R324.6.3 Emergency escape and rescue openings. Panels and modules installed on dwellings shall not be placed on the portion of a roof that is below an *emergency escape and rescue opening.* A pathway not less than 36 inches (914 mm) wide shall be provided to the emergency escape and rescue opening.

Exception: BIPV systems *listed* in accordance with Section 690.12(B)(2) of NFPA 70, where the removal or cutting away of portions of the BIPV system during fire-fighting operations has been determined to not expose a fire fighter to electrical shock hazards.

R324.7 Ground-mounted photovoltaic systems. Ground-mounted photovoltaic systems shall be designed and installed in accordance with Section R301.

R324.7.1 Fire separation distances. Ground-mounted photovoltaic systems shall be subject to the *fire separation distance* requirements determined by the local *jurisdiction.*

SECTION R325 MEZZANINES

R325.1 General. *Mezzanines* shall comply with Sections R325 through R325.5.

R325.2 Mezzanines. The clear height above and below *mezzanine* floor construction shall be not less than 7 feet (2134 mm).

R325.3 Area limitation. The aggregate area of a *mezzanine* or *mezzanines* shall be not greater than one-third of the floor area of the room or space in which they are located. The enclosed portion of a room shall not be included in a determination of the floor area of the room in which the *mezzanine* is located.

Exception: The aggregate area of a *mezzanine* located within a *dwelling unit* equipped with an automatic sprinkler system in accordance with Section P2904 shall not be greater than one-half of the floor area of the room, provided that the *mezzanine* meets all of the following requirements:

1. Except for enclosed closets and bathrooms, the *mezzanine* is open to the room in which such *mezzanine* is located.
2. The opening to the room is unobstructed except for walls not more than 42 inches (1067 mm) in height, columns and posts.
3. The exceptions to Section R325.5 are not applied.

R325.4 Means of egress. The means of egress for *mezzanines* shall comply with the applicable provisions of Section R311.

R325.5 Openness. *Mezzanines* shall be open and unobstructed to the room in which they are located except for walls not more than 36 inches (914 mm) in height, columns and posts.

Exceptions:

1. *Mezzanines* or portions thereof are not required to be open to the room in which they are located, provided that the aggregate floor area of the enclosed space is not greater than 10 percent of the *mezzanine* area.
2. In buildings that are not more than two stories above *grade plane* and equipped throughout with an automatic sprinkler system in accordance with Section R313, a *mezzanine* shall not be required to be open to the room in which the *mezzanine* is located.

SECTION R326 HABITABLE ATTICS

R326.1 General. Habitable attics shall comply with Sections R326.2 and R326.3.

R326.2 Minimum dimensions. A habitable attic shall have a floor area in accordance with Section R304 and a ceiling height in accordance with Section R305.

R326.3 Story above grade plane. A habitable attic shall be considered a story above *grade plane.*

Exceptions: A habitable attic shall not be considered to be a story above *grade plane* provided that the habitable attic meets all the following:

1. The aggregate area of the habitable attic is either of the following:
    - 1.1. Not greater than one-third of the floor area of the story below.
    - 1.2. Not greater than one-half of the floor area of the story below where the habitable attic is located within a dwelling unit equipped with a fire sprinkler system in accordance with Section P2904.
2. The occupiable space is enclosed by the roof assembly above, knee walls, if applicable, on the sides and the floor-ceiling assembly below.
3. The floor of the habitable attic does not extend beyond the exterior walls of the story below.
4. Where a habitable attic is located above a third story, the dwelling unit or townhouse unit shall be equipped with a fire sprinkler system in accordance with Section P2904.

R326.4 Means of egress. The means of egress for habitable attics shall comply with the applicable provisions of Section R311.

SECTION R327 SWIMMING POOLS, SPAS AND HOT TUBS

R327.1 General. The design and construction of pools and spas shall comply with the *International Swimming Pool and Spa Code.*

SECTION R328 ENERGY STORAGE SYSTEMS

R328.1 General. *Energy storage systems (ESS)* shall comply with the provisions of this section.

Exceptions:

1. *ESS listed* and *labeled* in accordance with UL 9540 and marked "For use in residential dwelling units" where installed in accordance with the manufacturer's instructions and NFPA 70.
2. ESS less than 1 kWh (3.6 megajoules).

R328.2 Equipment listings. *Energy storage systems (ESS)* shall be *listed* and *labeled* in accordance with UL 9540.

Exception: Where *approved*, repurposed unlisted battery systems from electric vehicles are allowed to be installed outdoors or in detached sheds located not less than 5 feet (1524 mm) from exterior walls, property lines and public ways.

R328.3 Installation. *ESS* shall be installed in accordance with the manufacturer's instructions and their *listing.*

R328.3.1 Spacing. Individual units shall be separated from each other by not less than 3 feet (914 mm) except where smaller separation distances are documented to be adequate based on large-scale fire testing complying with Section 1207.1.5 of the *International Fire Code.*

R328.4 Locations. *ESS* shall be installed only in the following locations:

1. Detached garages and detached accessory structures.
2. Attached garages separated from the *dwelling unit* living space in accordance with Section R302.6.
3. Outdoors or on the exterior side of exterior walls located not less than 3 feet (914 mm) from doors and windows directly entering the *dwelling unit.*
4. Enclosed utility closets, basements, storage or utility spaces within *dwelling units* with finished or noncombustible walls and ceilings. Walls and ceilings of unfinished wood-framed construction shall be provided with not less than $^5/_8$-inch (15.9 mm) Type X gypsum wallboard.

ESS shall not be installed in sleeping rooms, or closets or spaces opening directly into sleeping rooms.

R328.5 Energy ratings. Individual *ESS* units shall have a maximum rating of 20 kWh. The aggregate rating of the *ESS* shall not exceed:

1. 40 kWh within utility closets, basements and storage or utility spaces.
2. 80 kWh in attached or detached garages and detached accessory structures.
3. 80 kWh on exterior walls.
4. 80 kWh outdoors on the ground.

ESS installations exceeding the permitted individual or aggregate ratings shall be installed in accordance with Section 1207 of the *International Fire Code.*

R328.6 Electrical installation. *ESS* shall be installed in accordance with NFPA 70. Inverters shall be *listed* and *labeled* in accordance with UL 1741 or provided as part of the UL 9540 listing. Systems connected to the utility grid shall use inverters *listed* for utility interaction.

R328.7 Fire detection. Rooms and areas within *dwelling units,* basements and attached garages in which *ESS* are installed shall be protected by smoke alarms in accordance with Section R314. A heat detector, *listed* and interconnected to the smoke alarms, shall be installed in locations within *dwelling units* and attached garages where smoke alarms cannot be installed based on their listing.

R328.8 Protection from impact. *ESS* installed in a location subject to vehicle damage shall be protected by *approved* barriers.

R328.9 Ventilation. Indoor installations of *ESS* that produce hydrogen or other flammable gases during charging shall be provided with mechanical *ventilation* in accordance with Section M1307.4.

R328.10 Electric vehicle use. The temporary use of an *owner* or occupant's electric-powered vehicle to power a *dwelling unit* while parked in an attached or detached garage or outdoors shall comply with the vehicle manufacturer's instructions and NFPA 70.

R328.11 Documentation and labeling. The following information shall be provided:

1. A copy of the manufacturer's installation, operation, maintenance and decommissioning instructions shall be provided to the owner or placed in a conspicuous location near the *ESS* equipment.
2. A label on the installed system containing the contact information for the qualified maintenance and service providers.

SECTION R329 STATIONARY ENGINE GENERATORS

R329.1 General. Stationary engine generators shall be *listed* and *labeled* in accordance with UL 2200 and shall comply with this section. The connection of stationary engine generators to the premise wiring system shall be by means of a *listed* transfer switch.

R329.2 Installation. The installation of stationary engine generators shall be in an *approved* location and in accordance with the listing, the manufacturer's installation instructions and Chapters 34 through 43.

SECTION R330 STATIONARY FUEL CELL POWER SYSTEMS

R330.1 General. *Stationary fuel cell power systems* in new and existing buildings and structures shall comply with Section 1206 of the *International Fire Code.*

CHAPTER 4
FOUNDATIONS

User note:

About this chapter: Chapter 4 provides requirements for constructing footings and walls for foundations of wood, masonry, concrete and precast concrete. In addition to a foundation's ability to support the required design loads, this chapter addresses several other factors that can affect foundation performance. These include controlling surface water and subsurface drainage, requiring soil tests where conditions warrant and evaluating proximity to slopes and minimum depth requirements. This chapter also provides requirements to minimize adverse effects of moisture, decay and pests in basements and crawl spaces.

SECTION R401
GENERAL

R401.1 Application. The provisions of this chapter shall control the design and construction of the foundation and foundation spaces for buildings. In addition to the provisions of this chapter, the design and construction of foundations in flood hazard areas as established by Table R301.2 shall meet the provisions of Section R322. Wood foundations shall be designed and installed in accordance with AWC PWF.

Exception: The provisions of this chapter shall be permitted to be used for wood foundations only in the following situations:

1. In buildings that have not more than two floors and a roof.
2. Where interior *basement* and foundation walls are constructed at intervals not exceeding 50 feet (15 240 mm).

Wood foundations in Seismic Design Category D_0, D_1 or D_2 shall be designed in accordance with accepted engineering practice.

R401.2 Requirements. Foundation construction shall be capable of accommodating all loads in accordance with Section R301 and of transmitting the resulting loads to the supporting soil. Fill soils that support footings and foundations shall be designed, installed and tested in accordance with accepted engineering practice.

R401.3 Drainage. Surface drainage shall be diverted to a storm sewer conveyance or other *approved* point of collection that does not create a hazard. *Lots* shall be graded to drain surface water away from foundation walls. The *grade* shall fall not fewer than 6 inches (152 mm) within the first 10 feet (3048 mm).

Exception: Where *lot lines*, walls, slopes or other physical barriers prohibit 6 inches (152 mm) of fall within 10 feet (3048 mm), drains or swales shall be constructed to ensure drainage away from the structure. Impervious surfaces within 10 feet (3048 mm) of the building foundation shall be sloped not less than 2 percent away from the building.

R401.4 Soil tests. Where quantifiable data created by accepted soil science methodologies indicate *expansive soils, compressible soils*, shifting soils or other questionable soil characteristics are likely to be present, the *building official* shall determine whether to require a soil test to determine the soil's characteristics at a particular location. This test shall be done by an *approved agency* using an *approved* method.

R401.4.1 Geotechnical evaluation. In lieu of a complete geotechnical evaluation, the load-bearing values in Table R401.4.1 shall be assumed.

TABLE R401.4.1
PRESUMPTIVE LOAD-BEARING VALUES OF FOUNDATION MATERIALS[a]

CLASS OF MATERIAL	LOAD-BEARING PRESSURE (pounds per square foot)
Crystalline bedrock	12,000
Sedimentary and foliated rock	4,000
Sandy gravel and/or gravel (GW and GP)	3,000
Sand, silty sand, clayey sand, silty gravel and clayey gravel (SW, SP, SM, SC, GM and GC)	2,000
Clay, sandy, silty clay, clayey silt, silt and sandy siltclay (CL, ML, MH and CH)	1,500[b]

For SI: 1 pound per square foot = 0.0479 kPa.

a. Where soil tests are required by Section R401.4, the allowable bearing capacities of the soil shall be part of the recommendations.

b. Where the building official determines that in-place soils with an allowable bearing capacity of less than 1,500 psf are likely to be present at the site, the allowable bearing capacity shall be determined by a soils investigation.

R401.4.2 Compressible or shifting soil. Instead of a complete geotechnical evaluation, where top or subsoils are compressible or shifting, they shall be removed to a depth and width sufficient to ensure stable moisture content in each active zone and shall not be used as fill or stabilized within each active zone by chemical, dewatering or presaturation.

SECTION R402
MATERIALS

R402.1 Wood foundations. Wood foundation systems shall be designed and installed in accordance with the provisions of this code.

R402.1.1 Fasteners. Fasteners used below *grade* to attach plywood to the exterior side of exterior *basement* or crawl-space wall studs, or fasteners used in knee wall construction, shall be of Type 304 or 316 stainless steel. Fasteners used above *grade* to attach plywood and all lumber-to-lumber fasteners except those used in knee wall construction shall be of Type 304 or 316 stainless steel, silicon bronze, copper, hot-dipped galvanized (zinc coated) steel nails, or hot-tumbled galvanized (zinc coated) steel nails. Electro-galvanized steel nails and galvanized (zinc coated) steel staples shall not be permitted.

R402.1.2 Wood treatment. Lumber and plywood shall be pressure-preservative treated and dried after treatment in accordance with AWPA U1 (Commodity Specification A, Special Requirement 4.2), and shall bear the *label* of an accredited agency. Where lumber or plywood is cut or drilled after treatment, the treated surface shall be field treated with copper naphthenate, the concentration of which shall contain not less than 2-percent copper metal, by repeated brushing, dipping or soaking until the wood cannot absorb more preservative.

R402.2 Concrete. Concrete shall have a minimum specified compressive strength of f'_c, as shown in Table R402.2. Concrete subject to moderate or severe weathering as indicated in Table R301.2 shall be air entrained as specified in Table R402.2. The maximum weight of fly ash, other pozzolans, silica fume, slag or blended cements that is included in concrete mixtures for garage floor slabs and for exterior porches, carport slabs and steps that will be exposed to deicing chemicals shall not exceed the percentages of the total weight of cementitious materials specified in Section 19.3.3.4 of ACI 318. Materials used to produce concrete and testing thereof shall comply with the applicable standards listed in Chapters 19 and 20 of ACI 318 or ACI 332.

R402.2.1 Materials for concrete. Materials for concrete shall comply with the requirements of Section R608.5.1.

R402.3 Precast concrete. *Precast concrete* foundations shall be designed in accordance with Section R404.5 and shall be installed in accordance with the provisions of this code and the manufacturer's instructions.

R402.3.1 Precast concrete foundation materials. Materials used to produce *precast concrete* foundations shall meet the following requirements:

1. All concrete used in the manufacture of *precast concrete* foundations shall have a minimum compressive strength of 5,000 psi (34 470 kPa) at 28 days. Concrete exposed to a freezing and thawing environment shall be air entrained with a minimum total air content of 5 percent.
2. Structural reinforcing steel shall meet the requirements of ASTM A615, A706M or A996M. The minimum yield strength of reinforcing steel shall be 40,000 psi (Grade 40) (276 MPa). Steel reinforcement for *precast concrete foundation walls* shall have a minimum concrete cover of $^3/_4$ inch (19.1 mm).
3. Panel-to-panel connections shall be made with Grade II steel fasteners.
4. The use of nonstructural fibers shall conform to ASTM C1116.
5. Grout used for bedding precast foundations placed on concrete footings shall meet ASTM C1107.

TABLE R402.2
MINIMUM SPECIFIED COMPRESSIVE STRENGTH OF CONCRETE

TYPE OR LOCATION OF CONCRETE CONSTRUCTION	MINIMUM SPECIFIED COMPRESSIVE STRENGTH[a] (f'_c)		
	Weathering Potential[b]		
	Negligible	Moderate	Severe
Basement walls, foundations and other concrete not exposed to the weather	2,500	2,500	2,500[c]
Basement slabs and interior slabs on grade, except garage floor slabs	2,500	2,500	2,500[c]
Basement walls, foundation walls, exterior walls and other vertical concrete work exposed to the weather	2,500	3,000[d]	3,000[d]
Porches, carport slabs and steps exposed to the weather, and garage floor slabs	2,500	3,000[d, e, f]	3,500[d, e, f]

For SI: 1 pound per square inch = 6.895 kPa.

a. Strength at 28 days psi.

b. See Table R301.2 for weathering potential.

c. Concrete in these locations that is subject to freezing and thawing during construction shall be air-entrained concrete in accordance with Note d.

d. Concrete shall be air-entrained. Total air content (percent by volume of concrete) shall be not less than 5 percent or more than 7 percent.

e. See Section R402.2 for maximum cementitious materials content.

f. For garage floors with a steel-troweled finish, reduction of the total air content (percent by volume of concrete) to not less than 3 percent is permitted if the specified compressive strength of the concrete is increased to not less than 4,000 psi.

R402.4 Masonry. Masonry systems shall be designed and installed in accordance with this chapter and shall have a minimum specified compressive strength of 1,500 psi (10.3 MPa).

SECTION R403
FOOTINGS

R403.1 General. All exterior walls shall be supported on continuous solid or fully grouted masonry or concrete footings, crushed stone footings, wood foundations, or other *approved* structural systems that shall be of sufficient design to accommodate all loads according to Section R301 and to transmit the resulting loads to the soil within the limitations as determined from the character of the soil. Footings shall be supported on undisturbed natural soils or engineered fill. Concrete footing shall be designed and constructed in accordance with the provisions of Section R403 or in accordance with ACI 332.

TABLE R403.1(1)
MINIMUM WIDTH AND THICKNESS FOR CONCRETE FOOTINGS FOR LIGHT-FRAME CONSTRUCTION (inches)[a, b, c, d]

GROUND SNOW LOAD OR ROOF LIVE LOAD	STORY AND TYPE OF STRUCTURE WITH LIGHT FRAME	LOAD-BEARING VALUE OF SOIL (psf)					
		1,500	2,000	2,500	3,000	3,500	4,000
20 psf roof live load or 25 psf ground snow load	1 story—slab-on-grade	12 × 6	12 × 6	12 × 6	12 × 6	12 × 6	12 × 6
	1 story—with crawl space	12 × 6	12 × 6	12 × 6	12 × 6	12 × 6	12 × 6
	1 story—plus basement	16 × 6	12 × 6	12 × 6	12 × 6	12 × 6	12 × 6
	2 story—slab-on-grade	13 × 6	12 × 6	12 × 6	12 × 6	12 × 6	12 × 6
	2 story—with crawl space	15 × 6	12 × 6	12 × 6	12 × 6	12 × 6	12 × 6
	2 story—plus basement	19 × 6	14 × 6	12 × 6	12 × 6	12 × 6	12 × 6
	3 story—slab-on-grade	16 × 6	12 × 6	12 × 6	12 × 6	12 × 6	12 × 6
	3 story—with crawl space	18 × 6	14 × 6	12 × 6	12 × 6	12 × 6	12 × 6
	3 story—plus basement	22 × 7	16 × 6	13 × 6	12 × 6	12 × 6	12 × 6
30 psf	1 story—slab-on-grade	12 × 6	12 × 6	12 × 6	12 × 6	12 × 6	12 × 6
	1 story—with crawl space	13 × 6	12 × 6	12 × 6	12 × 6	12 × 6	12 × 6
	1 story—plus basement	16 × 6	12 × 6	12 × 6	12 × 6	12 × 6	12 × 6
	2 story—slab-on-grade	13 × 6	12 × 6	12 × 6	12 × 6	12 × 6	12 × 6
	2 story—with crawl space	16 × 6	12 × 6	12 × 6	12 × 6	12 × 6	12 × 6
	2 story—plus basement	19 × 6	14 × 6	12 × 6	12 × 6	12 × 6	12 × 6
	3 story—slab-on-grade	16 × 6	14 × 6	12 × 6	12 × 6	12 × 6	12 × 6
	3 story—with crawl space	19 × 6	14 × 6	12 × 6	12 × 6	12 × 6	12 × 6
	3 story—plus basement	22 × 7	16 × 6	13 × 6	12 × 6	12 × 6	12 × 6
50 psf	1 story—slab-on-grade	12 × 6	12 × 6	12 × 6	12 × 6	12 × 6	12 × 6
	1 story—with crawl space	14 × 6	12 × 6	12 × 6	12 × 6	12 × 6	12 × 6
	1 story—plus basement	18 × 6	13 × 6	12 × 6	12 × 6	12 × 6	12 × 6
	2 story—slab-on-grade	15 × 6	13 × 6	12 × 6	12 × 6	12 × 6	12 × 6
	2 story—with crawl space	17 × 6	13 × 6	12 × 6	12 × 6	12 × 6	12 × 6
	2 story—plus basement	21 × 7	15 × 6	12 × 6	12 × 6	12 × 6	12 × 6
	3 story—slab-on-grade	18 × 6	13 × 6	12 × 6	12 × 6	12 × 6	12 × 6
	3 story—with crawl space	20 × 6	15 × 6	12 × 6	12 × 6	12 × 6	12 × 6
	3 story—plus basement	24 × 8	18 × 6	14 × 6	12 × 6	12 × 6	12 × 6
70 psf	1 story—slab-on-grade	14 × 6	12 × 6	12 × 6	12 × 6	12 × 6	12 × 6
	1 story—with crawl space	16 × 6	12 × 6	12 × 6	12 × 6	12 × 6	12 × 6
	1 story—plus basement	19 × 6	14 × 6	12 × 6	12 × 6	12 × 6	12 × 6
	2 story—slab-on-grade	17 × 6	12 × 6	12 × 6	12 × 6	12 × 6	12 × 6
	2 story—with crawl space	19 × 6	14 × 6	12 × 6	12 × 6	12 × 6	12 × 6
	2 story—plus basement	22 × 7	17 × 6	13 × 6	12 × 6	12 × 6	12 × 6
	3 story—slab-on-grade	20 × 6	15 × 6	12 × 6	12 × 6	12 × 6	12 × 6
	3 story—with crawl space	22 × 7	16 × 6	13 × 6	12 × 6	12 × 6	12 × 6
	3 story—plus basement	24 × 8	19 × 6	15 × 6	13 × 6	12 × 6	12 × 6

For SI: 1 inch = 25.4 mm, 1 foot = 304.8 mm, 1 pound per square foot = 47.9 N/m^2.

a. Linear interpolation of footing width is permitted between the soil bearing pressures in the table. Extrapolation is not permitted.

b. The table is based on the following conditions and loads: building width, 32 feet; wall height, 9 feet; basement wall height, 8 feet; dead loads, 15 psf roof and ceiling assembly, 10 psf floor assembly, 12 psf wall assembly; live loads, roof and ground snow loads as listed, 40 psf first floor, 30 psf second and third floors. Footing sizes are calculated assuming a clear span roof/ceiling assembly and an interior bearing wall or beam at each floor.

c. Where the building width perpendicular to the wall footing is greater than 32 feet, the footing width shall be increased by 2 inches and footing depth shall be increased by 1 inch for every 4 feet of increase in building width.

d. Where the building width perpendicular to the wall footing is less than 32 feet, a 2-inch decrease in footing width and 1-inch decrease in footing depth is permitted for every 4 feet of decrease in building width provided that the minimum width is 12 inches and minimum depth is 6 inches.

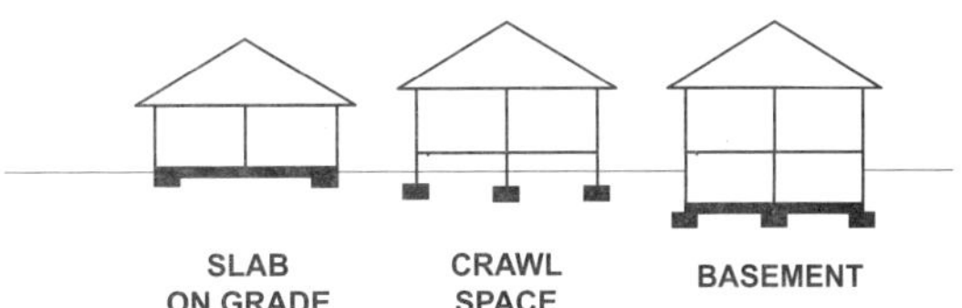

TABLE R403.1(2)
MINIMUM WIDTH AND THICKNESS FOR CONCRETE FOOTINGS FOR LIGHT-FRAME CONSTRUCTION WITH BRICK VENEER OR LATH AND PLASTER (inches)[a, b, c, d]

GROUND SNOW LOAD OR ROOF LIVE LOAD	STORY AND TYPE OF STRUCTURE WITH BRICK VENEER	LOAD-BEARING VALUE OF SOIL (psf)					
		1,500	2,000	2,500	3,000	3,500	4,000
20 psf roof live load or 25 psf ground snow load	1 story—slab-on-grade	12 × 6	12 × 6	12 × 6	12 × 6	12 × 6	12 × 6
	1 story—with crawl space	15 × 6	12 × 6	12 × 6	12 × 6	12 × 6	12 × 6
	1 story—plus basement	18 × 6	14 × 6	12 × 6	12 × 6	12 × 6	12 × 6
	2 story—slab-on-grade	18 × 6	13 × 6	12 × 6	12 × 6	12 × 6	12 × 6
	2 story—with crawl space	20 × 6	15 × 6	12 × 6	12 × 6	12 × 6	12 × 6
	2 story—plus basement	23 × 8	17 × 6	14 × 6	12 × 6	12 × 6	12 × 6
	3 story—slab-on-grade	23 × 8	17 × 6	14 × 6	12 × 6	12 × 6	12 × 6
	3 story—with crawl space	25 × 9	19 × 6	15 × 6	13 × 6	12 × 6	12 × 6
	3 story—plus basement	29 × 11	21 × 7	17 × 6	14 × 6	12 × 6	12 × 6
30 psf	1 story—slab-on-grade	13 × 6	12 × 6	12 × 6	12 × 6	12 × 6	12 × 6
	1 story—with crawl space	15 × 6	12 × 6	12 × 6	12 × 6	12 × 6	12 × 6
	1 story—plus basement	18 × 6	14 × 6	12 × 6	12 × 6	12 × 6	12 × 6
	2 story—slab-on-grade	18 × 6	14 × 6	12 × 6	12 × 6	12 × 6	12 × 6
	2 story—with crawl space	20 × 6	15 × 6	12 × 6	12 × 6	12 × 6	12 × 6
	2 story—plus basement	24 × 8	18 × 6	14 × 6	12 × 6	12 × 6	12 × 6
	3 story—slab-on-grade	23 × 8	18 × 6	14 × 6	12 × 6	12 × 6	12 × 6
	3 story—with crawl space	26 × 9	19 × 6	15 × 6	13 × 6	12 × 6	12 × 6
	3 story—plus basement	29 × 11	22 × 7	17 × 6	14 × 6	12 × 6	12 × 6
50 psf	1 story—slab-on-grade	14 × 6	12 × 6	12 × 6	12 × 6	12 × 6	12 × 6
	1 story—with crawl space	17 × 6	13 × 6	12 × 6	12 × 6	12 × 6	12 × 6
	1 story—plus basement	20 × 60	15 × 6	12 × 6	12 × 6	12 × 6	12 × 6
	2 story—slab-on-grade	20 × 6	15 × 6	12 × 6	12 × 6	12 × 6	12 × 6
	2 story—with crawl space	22 × 7	17 × 6	13 × 6	12 × 6	12 × 6	12 × 6
	2 story—plus basement	25 × 9	19 × 6	15 × 6	13 × 6	12 × 6	12 × 6
	3 story—slab-on-grade	25 × 9	19 × 6	15 × 6	13 × 6	12 × 6	12 × 6
	3 story—with crawl space	27 × 10	21 × 7	16 × 6	14 × 6	12 × 6	12 × 6
	3 story—plus basement	31 × 12	23 × 8	18 × 6	15 × 6	13 × 6	12 × 6
70 psf	1 story—slab-on-grade	14 × 6	12 × 6	12 × 6	12 × 6	12 × 6	12 × 6
	1 story—with crawl space	17 × 6	14 × 6	12 × 6	12 × 6	12 × 6	12 × 6
	1 story—plus basement	22 × 7	16 × 6	13 × 6	12 × 6	12 × 6	12 × 6
	2 story—slab-on grade	21 × 7	16 × 6	13 × 6	12 × 6	12 × 6	12 × 6
	2 story—with crawl space	24 × 8	18 × 6	14 × 6	12 × 6	12 × 6	12 × 6
	2 story—plus basement	27 × 10	20 × 6	16 × 6	13 × 6	12 × 6	12 × 6
	3 story—slab-on-grade	27 × 10	20 × 6	16 × 6	13 × 6	12 × 6	12 × 6
	3 story—with crawl space	29 × 11	22 × 7	17 × 6	15 × 6	12 × 6	12 × 6
	3 story—plus basement	32 × 12	24 × 8	19 × 6	16 × 6	14 × 6	12 × 6

For SI: 1 inch = 25.4 mm, 1 foot = 304.8 mm, 1 pound per square foot = 47.9 N/m^2.

a. Linear interpolation of footing width is permitted between the soil bearing pressures in the table. Extrapolation is not permitted.

b. The table is based on the following conditions and loads: building width, 32 feet; wall height, 9 feet; basement wall height, 8 feet; dead loads, 15 psf roof and ceiling assembly, 10 psf floor assembly, 12 psf wall assembly; live loads, roof and ground snow loads as listed, 40 psf first floor, 30 psf second and third floors. Footing sizes are calculated assuming a clear span roof/ceiling assembly and an interior bearing wall or beam at each floor.

c. Where the building width perpendicular to the wall footing is greater than 32 feet, the footing width shall be increased by 2 inches and footing depth shall be increased by 1 inch for every 4 feet of increase in building width.

d. Where the building width perpendicular to the wall footing is less than 32 feet, a 2-inch decrease in footing width and 1-inch decrease in footing depth is permitted for every 4 feet of decrease in building width provided that the minimum width is 12 inches and minimum depth is 6 inches.

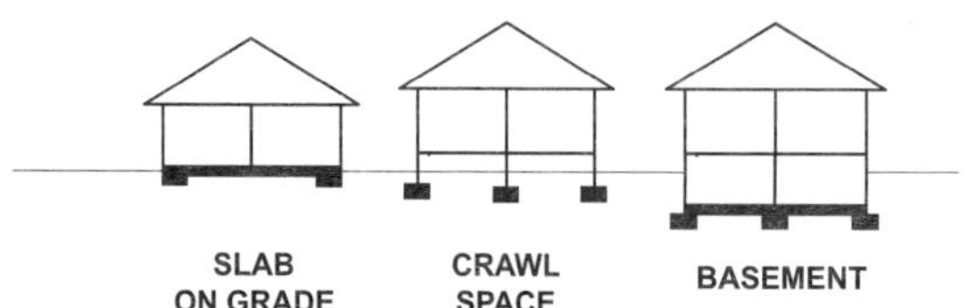

TABLE R403.1(3)
MINIMUM WIDTH AND THICKNESS FOR CONCRETE FOOTINGS WITH CAST-IN-PLACE CONCRETE OR PARTIALLY GROUTED MASONRY WALL CONSTRUCTION (inches)[a, b, c, d]

GROUND SNOW LOAD OR ROOF LIVE LOAD	STORY AND TYPE OF STRUCTURE WITH CMU OR CONCRETE	LOAD-BEARING VALUE OF SOIL (psf)					
		1,500	2,000	2,500	3,000	3,500	4,000
20 psf roof live load or 25 psf ground snow load	1 story—slab-on-grade	13 × 6	12 × 6	12 × 6	12 × 6	12 × 6	12 × 6
	1 story—with crawl space	16 × 6	12 × 6	12 × 6	12 × 6	12 × 6	12 × 6
	1 story—plus basement	19 × 6	14 × 6	12 × 6	12 × 6	12 × 6	12 × 6
	2 story—slab-on-grade	19 × 6	14 × 6	12 × 6	12 × 6	12 × 6	12 × 6
	2 story—with crawl space	22 × 7	16 × 6	13 × 6	12 × 6	12 × 6	12 × 6
	2 story—plus basement	25 × 9	19 × 6	15 × 6	12 × 6	12 × 6	12 × 6
	3 story—slab-on-grade	25 × 9	19 × 6	15 × 6	13 × 6	12 × 6	12 × 6
	3 story—with crawl space	28 × 10	21 × 7	17 × 6	14 × 6	12 × 6	12 × 6
	3 story—plus basement	31 × 12	23 × 8	18 × 6	15 × 6	13 × 6	12 × 6
30 psf	1 story—slab-on-grade	13 × 6	12 × 6	12 × 6	12 × 6	12 × 6	12 × 6
	1 story—with crawl space	16 × 6	12 × 6	12 × 6	12 × 6	12 × 6	12 × 6
	1 story—plus basement	19 × 7	14 × 6	12 × 6	12 × 6	12 × 6	12 × 6
	2 story—slab-on-grade	19 × 6	15 × 6	12 × 6	12 × 6	12 × 6	12 × 6
	2 story—with crawl space	22 × 7	16 × 6	13 × 6	12 × 6	12 × 6	12 × 6
	2 story—plus basement	25 × 9	19 × 6	15 × 6	13 × 6	12 × 6	12 × 6
	3 story—slab-on-grade	26 × 9	19 × 6	15 × 6	13 × 6	12 × 6	12 × 6
	3 story—with crawl space	28 × 10	21 × 7	17 × 6	14 × 6	12 × 6	12 × 6
	3 story—plus basement	31 × 12	23 × 8	19 × 6	16 × 6	13 × 6	12 × 6
50 psf	1 story—slab-on-grade	15 × 6	12 × 6	12 × 6	12 × 6	12 × 6	12 × 6
	1 story—with crawl space	18 × 6	13 × 6	12 × 6	12 × 6	12 × 6	12 × 6
	1 story—plus basement	21× 7	16 × 6	12 × 6	12 × 6	12 × 6	12 × 6
	2 story—slab-on-grade	21 × 7	16 × 6	13 × 6	12 × 6	12 × 6	12 × 6
	2 story—with crawl space	24 × 8	18 × 6	14 × 6	12 × 6	12 × 6	12 × 6
	2 story—plus basement	27 × 10	20 × 6	16 × 6	13 × 6	12 × 6	12 × 6
	3 story—slab-on-grade	27 × 10	20 × 6	16 × 6	14 × 6	12 × 6	12 × 6
	3 story—with crawl space	30 × 11	22 × 7	18 × 6	15 × 6	13 × 6	12 × 6
	3 story—plus basement	33 × 13	25 × 9	20 × 6	16 × 6	14 × 6	12 × 6
70 psf	1 story—slab-on-grade	17 × 6	13 × 6	12 × 6	12 × 6	12 × 6	12 × 6
	1 story—with crawl space	19 × 6	14 × 6	12 × 6	12 × 6	12 × 6	12 × 6
	1 story—plus basement	22 × 7	17 × 6	13 × 6	12 × 6	12 × 6	12 × 6
	2 story—slab-on-grade	23 × 8	17 × 6	14 × 6	12 × 6	12 × 6	12 × 6
	2 story—with crawl space	25 × 9	19 × 6	15 × 6	12× 6	12 × 6	12 × 6
	2 story—plus basement	28 × 10	21 × 7	17 × 6	14 × 6	12 × 6	12 × 6
	3 story—slab-on-grade	29 × 11	22 × 7	17 × 6	14 × 6	12 × 6	12 × 6
	3 story—with crawl space	31 × 12	23 × 8	19 × 6	16 × 6	13 × 6	12 × 6
	3 story—plus basement	34 × 13	26 × 9	21 × 7	17 × 6	15 × 6	13 × 6

For SI: 1 inch = 25.4 mm, 1 foot = 304.8 mm, 1 pound per square foot = 47.9 N/m^2.

a. Linear interpolation of footing width is permitted between the soil bearing pressures in the table. Extrapolation is not permitted.

b. The table is based on the following conditions and loads: building width, 32 feet; wall height, 9 feet; basement wall height, 8 feet; dead loads, 15 psf roof and ceiling assembly, 10 psf floor assembly, 12 psf wall assembly; live loads, roof and ground snow loads as listed, 40 psf first floor, 30 psf second and third floors. Footing sizes are calculated assuming a clear span roof/ceiling assembly and an interior bearing wall or beam at each floor.

c. Where the building width perpendicular to the wall footing is greater than 32 feet, the footing width shall be increased by 2 inches and footing depth shall be increased by 1 inch for every 4 feet of increase in building width.

d. Where the building width perpendicular to the wall footing is less than 32 feet, a 2-inch decrease in footing width and 1-inch decrease in footing depth is permitted for every 4 feet of decrease in building width provided that the minimum width is 12 inches and minimum depth is 6 inches.

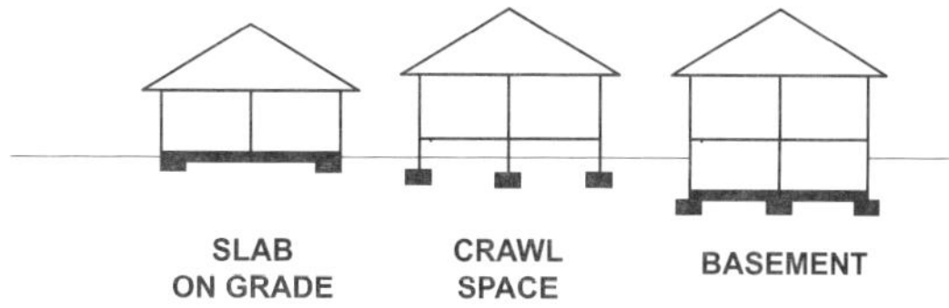

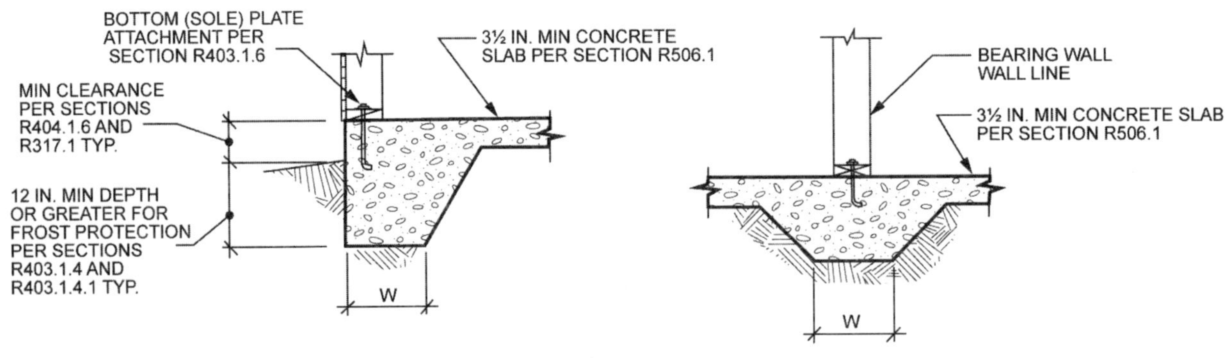

1 MONOLITHIC SLAB-ON-GROUND WITH TURNED-DOWN FOOTING
SCALE: NOT TO SCALE

2 THICKENED SLAB-ON-GROUND FOOTING AT BEARING WALLS OR BRACED WALL LINES
SCALE: NOT TO SCALE

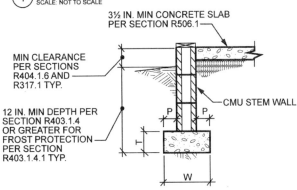

3½ IN. MIN CONCRETE SLAB PER SECTION R506.1
MIN CLEARANCE PER SECTIONS R404.1.6 AND R317.1 TYP.
12 IN. MIN DEPTH PER SECTION R403.1.4 OR GREATER FOR FROST PROTECTION PER SECTION R403.1.4.1 TYP.
CMU STEM WALL
P P T W

3 SLAB-ON-GROUND WITH MASONRY STEM WALL AND SPREAD FOOTING
SCALE: NOT TO SCALE

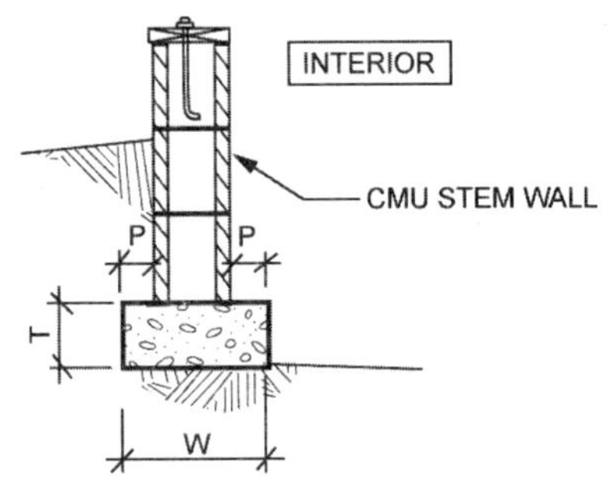

4 BASEMENT OR CRAWL SPACE WITH MASONRY WALL AND SPREAD FOOTING
SCALE: NOT TO SCALE

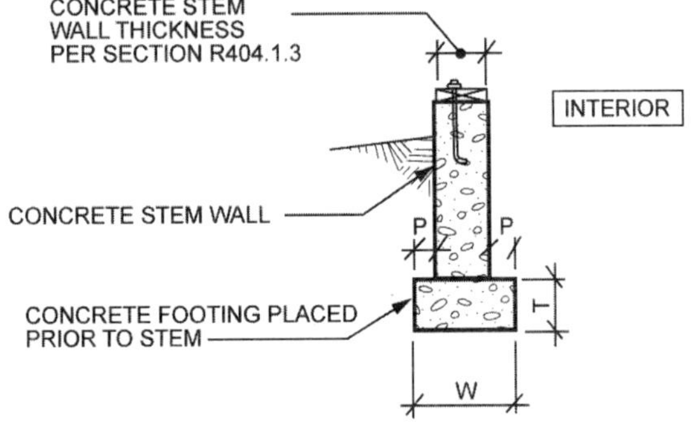

5 BASEMENT OR CRAWL SPACE WITH CONCRETE WALL AND SPREAD FOOTING
SCALE: NOT TO SCALE

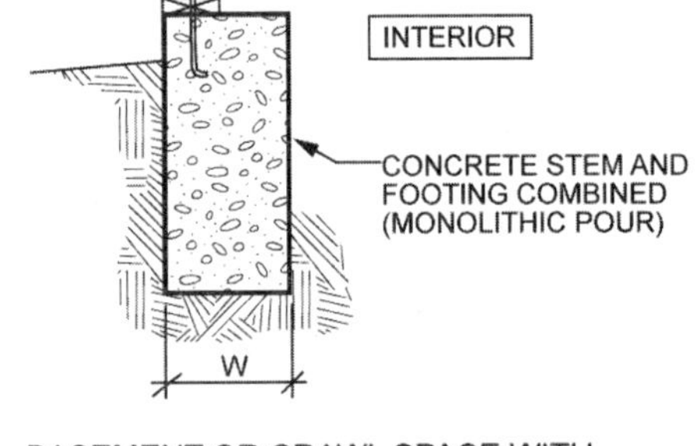

6 BASEMENT OR CRAWL SPACE WITH FOUNDATION WALL BEARING DIRECTLY ON SOIL
SCALE: NOT TO SCALE

For SI: 1 inch = 25.4 mm.
W = Width of footing, T = Thickness of footing and P = Projection per Section R403.1.1.

NOTES:

a. See Section R404.3 for sill requirements.
b. See Section R403.1.6 for sill attachment.
c. See Section R506.2.3 for vapor barrier requirements.
d. See Section R403.1 for base.
e. See Figure R403.1.3 for additional footing requirements for structures in Seismic Design Categories D_0, D_1 and D_2 and townhouses in Seismic Design Category C.
f. See Section R408 for under-floor ventilation and access requirements.

FIGURE R403.1(1)
PLAIN CONCRETE FOOTINGS WITH MASONRY AND CONCRETE STEM WALLS IN SEISMIC DESIGN CATEGORIES A, B AND C[a, b, c, d, e, f]

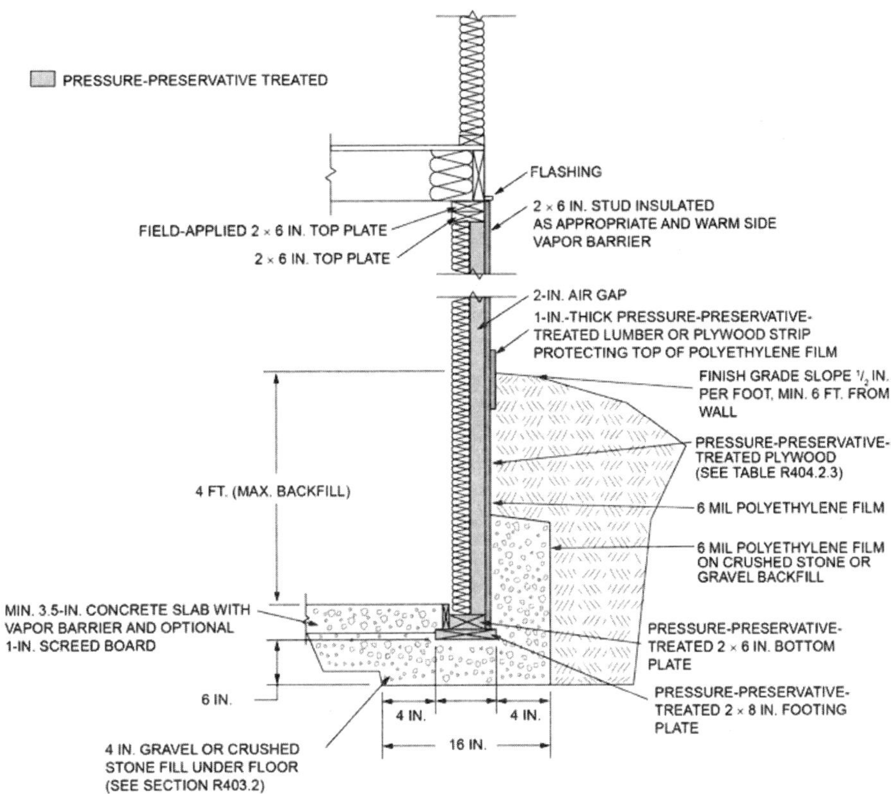

For SI: 1 inch = 25.4 mm, 1 foot = 304.8 mm, 1 mil = 0.0254 mm.

FIGURE R403.1(2)
PERMANENT WOOD FOUNDATION BASEMENT WALL SECTION

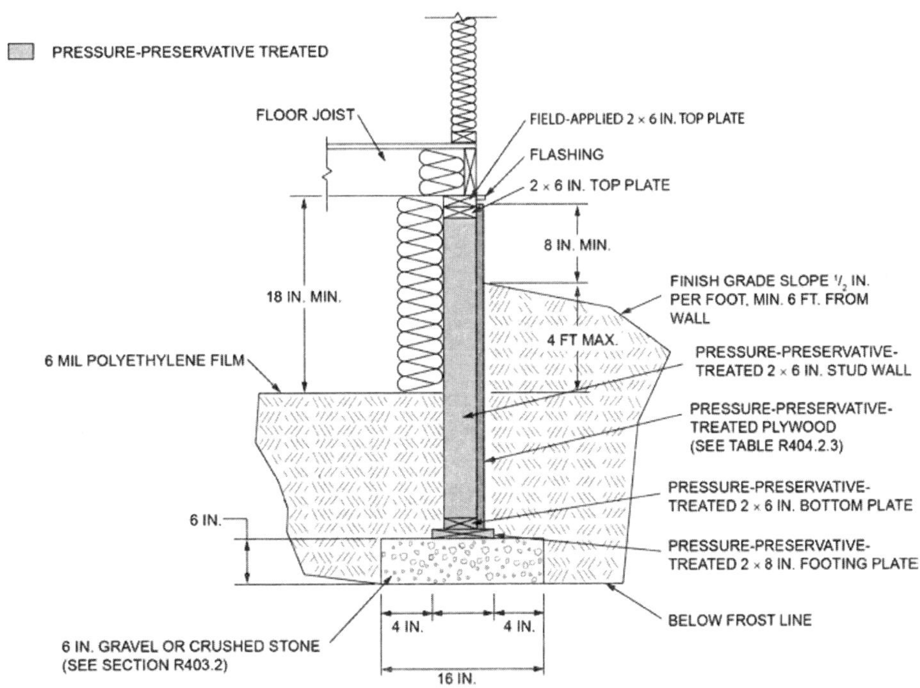

For SI: 1 inch = 25.4 mm, 1 foot = 304.8 mm, 1 mil = 0.0254 mm.

FIGURE R403.1(3)
PERMANENT WOOD FOUNDATION CRAWL SPACE SECTION

R403.1.1 Minimum size. The minimum width, W, and thickness, T, for concrete footings shall be in accordance with Tables R403.1(1) through R403.1(3) and Figure R403.1(1) or R403.1.3, as applicable, but not less than 12 inches (305 mm) in width and 6 inches (152 mm) in depth. The footing width shall be based on the load-bearing value of the soil in accordance with Table R401.4.1. Footing projections, P, shall be not less than 2 inches (51 mm) and shall not exceed the thickness of the footing. Footing thickness and projection for fireplaces shall be in accordance with Section R1001.2. The size of footings supporting piers and columns shall be based on the tributary load and allowable soil pressure in accordance with Table R401.4.1. Footings for wood foundations shall be in accordance with the details set forth in Section R403.2, and Figures R403.1(2) and R403.1(3). Footings for precast foundations shall be in accordance with the details set forth in Section R403.4, Table R403.4, and Figures R403.4(1) and R403.4(2).

R403.1.2 Continuous footing in Seismic Design Categories D_0, D_1 and D_2. Exterior walls of buildings located in *Seismic Design Categories* D_0, D_1 and D_2 shall be supported by continuous solid or fully grouted masonry or concrete footings. Other footing materials or systems shall be designed in accordance with accepted engineering practice. Required interior *braced wall panels* in buildings located in *Seismic Design Categories* D_0, D_1 and D_2 with plan dimensions greater than 50 feet (15 240 mm) shall be supported by continuous solid or fully grouted masonry or concrete footings in accordance with Section R403.1.3.4, except for two-story buildings in Seismic Design Category D_2, in which all *braced wall panels*, interior and exterior, shall be supported on continuous foundations.

Exception: Two-story buildings shall be permitted to have interior *braced wall panels* supported on continuous foundations at intervals not exceeding 50 feet (15 240 mm) provided that:

1. The height of cripple walls does not exceed 4 feet (1219 mm).
2. First-floor *braced wall panels* are supported on doubled floor joists, continuous blocking or floor beams.
3. The distance between bracing lines does not exceed twice the building width measured parallel to the *braced wall line*.

R403.1.3 Footing and stem wall reinforcing in Seismic Design Categories D_0, D_1 and D_2. Concrete footings located in *Seismic Design Categories* D_0, D_1 and D_2, as established in Table R301.2, shall have minimum reinforcement in accordance with this section and Figure R403.1.3. Reinforcement shall be installed with support and cover in accordance with Section R403.1.3.5.

R403.1.3.1 Concrete stem walls with concrete footings. In *Seismic Design Categories* D_0, D_1 and D_2 where a construction joint is created between a concrete footing and a concrete stem wall, not fewer than one No. 4 vertical bar shall be installed at not more than 4 feet (1219 mm) on center. The vertical bar shall have a standard hook and extend to the bottom of the footing and shall have support and cover as specified in Section R403.1.3.5.3 and extend not less than 14 inches (357 mm) into the stem wall. Standard hooks shall comply with Section R608.5.4.5. Not fewer than one No. 4 horizontal bar shall be installed within 12 inches (305 mm) of the top of the stem wall and one No. 4 horizontal bar shall be located 3 to 4 inches (76 mm to 102 mm) from the bottom of the footing.

R403.1.3.2 Masonry stem walls with concrete footings. In *Seismic Design Categories* D_0, D_1 and D_2 where a masonry stem wall is supported on a concrete footing, not fewer than one No. 4 vertical bar shall be installed at not more than 4 feet (1219 mm) on center. The vertical bar shall have a standard hook and extend to the bottom of the footing and shall have support and cover as specified in Section R403.1.3.5.3 and extend not less than 14 inches (357 mm) into the stem wall. Standard hooks shall comply with Section R608.5.4.5. Not fewer than one No. 4 horizontal bar shall be installed within 12 inches (305 mm) of the top of the wall and one No. 4 horizontal bar shall be located 3 to 4 inches (76 mm to 102 mm) from the bottom of the footing. Masonry stem walls shall be solid grouted.

R403.1.3.3 Slabs-on-ground with turned-down footings. In *Seismic Design Categories* D_0, D_1 and D_2, slabs-on-ground cast monolithically with turned-down footings shall have not fewer than one No. 4 bar at the top and the bottom of the footing or one No. 5 bar or two No. 4 bars in the middle third of the footing depth.

Where the slab is not cast monolithically with the footing, No. 3 or larger vertical dowels with standard hooks on each end shall be installed at not more than 4 feet (1219 mm) on center in accordance with Figure R403.1.3, Detail 2. Standard hooks shall comply with Section R608.5.4.5.

R403.1.3.4 Interior bearing and *braced wall panel* footings in Seismic Design Categories D_0, D_1 and D_2. In *Seismic Design Categories* D_0, D_1 and D_2, interior footings supporting bearing walls or *braced wall panels*, and cast monolithically with a slab on *grade*, shall extend to a depth of not less than 12 inches (305 mm) below the top of the slab.

R403.1.3.5 Reinforcement. Footing and stem wall reinforcement shall comply with Sections R403.1.3.5.1 through R403.1.3.5.4.

R403.1.3.5.1 Steel reinforcement. Steel reinforcement shall comply with the requirements of ASTM A615, A706M or A996M. ASTM A996 bars produced from rail steel shall be Type R. The minimum yield strength of reinforcing steel shall be 40,000 psi (Grade 40) (276 MPa).

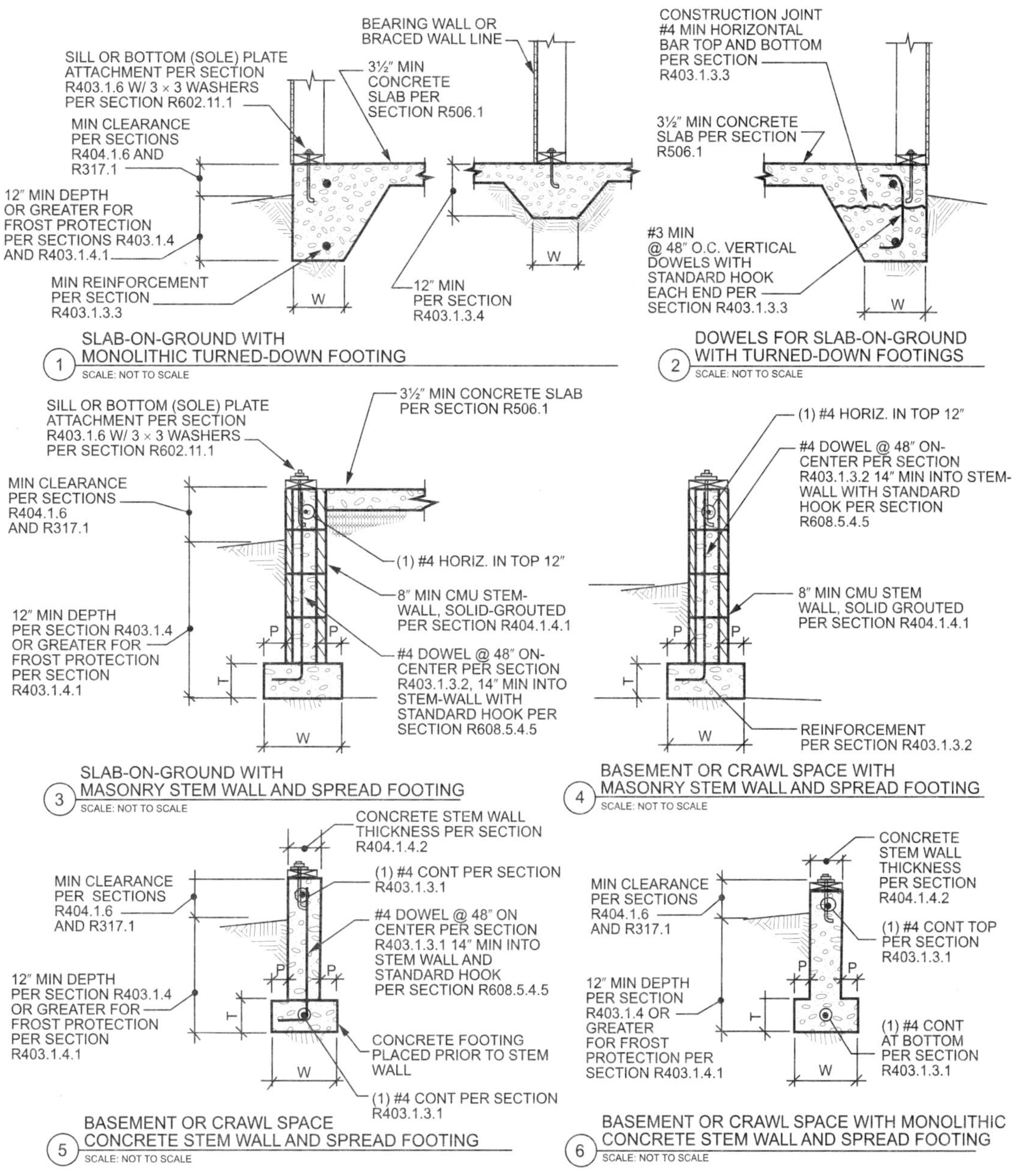

For SI: 1 inch = 25.4 mm.

W = Width of footing, T = Thickness of footing and P = Projection per Section R403.1.1.

NOTES:

a. See Section R404.3 for sill requirements.

b. See Section R403.1.6 for sill attachment.

c. See Section R506.2.3 for vapor barrier requirements.

d. See Section R403.1 for base.

e. See Section R408 for under-floor ventilation and access requirements.

f. See Section R403.1.3.5 for reinforcement requirements.

FIGURE R403.1.3
REINFORCED CONCRETE FOOTINGS AND MASONRY AND CONCRETE STEM WALLS IN SDC D_0, D_1 AND D_2[a, b, c, d, e, f]

R403.1.3.5.2 Location of reinforcement in wall. The center of vertical reinforcement in stem walls shall be located at the centerline of the wall. Horizontal and vertical reinforcement shall be located in footings and stem walls to provide the minimum cover required by Section R403.1.3.5.3.

R403.1.3.5.3 Support and cover. Reinforcement shall be secured in the proper location in the forms with tie wire or other bar support system to prevent displacement during the concrete placement operation. Steel reinforcement in concrete cast against the earth shall have a minimum cover of 3 inches

(75 mm). Minimum cover for reinforcement in concrete cast in removable forms that will be exposed to the earth or weather shall be $1^1/_2$ inches (38 mm) for No. 5 bars and smaller, and 2 inches (50 mm) for No. 6 bars and larger. For concrete cast in removable forms that will not be exposed to the earth or weather, and for concrete cast in stay-in-place forms, minimum cover shall be $^3/_4$ inch (19 mm).

R403.1.3.5.4 Lap splices. Vertical and horizontal reinforcement shall be the longest lengths practical. Where splices are necessary in reinforcement, the length of lap splice shall be in accordance with Table R608.5.4(1) and Figure R608.5.4(1). The maximum gap between noncontact parallel bars at a lap splice shall not exceed the smaller of one-fifth the required lap length and 6 inches (152 mm) [see Figure R608.5.4(1)].

R403.1.3.6 Isolated concrete footings. In detached one- and two-family dwellings that are three *stories* or less in height and constructed with stud bearing walls, isolated plain concrete footings supporting columns or pedestals are permitted.

R403.1.4 Minimum depth. Exterior footings shall be placed not less than 12 inches (305 mm) below the undisturbed ground surface. Where applicable, the depth of footings shall also conform to Section R403.1.4.1. Deck footings shall be in accordance with Section R507.3.

R403.1.4.1 Frost protection. Except where otherwise protected from frost, foundation walls, piers and other permanent supports of buildings and structures shall be protected from frost by one or more of the following methods:

1. Extended below the frost line specified in Table R301.2.
2. Constructed in accordance with Section R403.3.
3. Constructed in accordance with ASCE 32.
4. Erected on solid rock.

Footings shall not bear on frozen soil unless the frozen condition is permanent.

Exceptions:

1. Protection of free-standing *accessory structures* with an area of 600 square feet (56 m^2) or less, of *light-frame construction*, with an eave height of 10 feet (3048 mm) or less shall not be required.
2. Protection of free-standing *accessory structures* with an area of 400 square feet (37 m^2) or less, of other than *light-frame construction*, with an eave height of 10 feet (3048 mm) or less shall not be required.

R403.1.5 Slope. The top surface of footings shall be level. The bottom surface of footings shall not have a slope exceeding 1 unit vertical in 10 units horizontal (10-percent slope). Footings shall be stepped where it is necessary to change the elevation of the top surface of the footings or where the slope of the bottom surface of the footings will exceed 1 unit vertical in 10 units horizontal (10-percent slope).

R403.1.6 Foundation anchorage. Wood sill plates and wood walls supported directly on continuous foundations shall be anchored to the foundation in accordance with this section.

Cold-formed steel framing shall be anchored directly to the foundation or fastened to wood sill plates in accordance with Section R505.3.1 or R603.3.1, as applicable. Wood sill plates supporting cold-formed steel framing shall be anchored to the foundation in accordance with this section.

Wood sole plates at all exterior walls on monolithic slabs, wood sole plates of *braced wall panels* at building interiors on monolithic slabs and all wood sill plates shall be anchored to the foundation with minimum $^1/_2$-inch-diameter (12.7 mm) anchor bolts spaced not greater than 6 feet (1829 mm) on center or *approved* anchors or anchor straps spaced as required to provide equivalent anchorage to $^1/_2$-inch-diameter (12.7 mm) anchor bolts. Bolts shall extend not less than 7 inches (178 mm) into concrete or grouted cells of *concrete masonry units*. The bolts shall be located in the middle third of the width of the plate. A nut and washer shall be tightened on each anchor bolt. There shall be not fewer than two bolts per plate section with one bolt located not more than 12 inches (305 mm) or less than seven bolt diameters from each end of the plate section. Interior bearing wall sole plates on monolithic slab foundation that are not part of a *braced wall panel* shall be positively anchored with *approved* fasteners. Sill plates and sole plates shall be protected against decay and termites where required by Sections R317 and R318. Anchor bolts shall be permitted to be located while concrete is still plastic and before it has set. Where anchor bolts resist placement or the consolidation of concrete around anchor bolts is impeded, the concrete shall be vibrated to ensure full contact between the anchor bolts and concrete.

Exceptions:

1. Walls 24 inches (610 mm) total length or shorter connecting offset *braced wall panels* shall be anchored to the foundation with not fewer than one anchor bolt located in the center third of the plate section and shall be attached to adjacent *braced wall panels* at corners as shown in Item 9 of Table R602.3(1).
2. Connection of walls 12 inches (305 mm) total length or shorter connecting offset *braced wall panels* to the foundation without anchor bolts shall be permitted. The wall shall be attached to adjacent *braced wall panels* at corners as shown in Item 9 of Table R602.3(1).

R403.1.6.1 Foundation anchorage in Seismic Design Categories C, D_0, D_1 and D_2. In addition to the requirements of Section R403.1.6, the following requirements shall apply to wood light-frame structures in *Seismic Design Categories* D_0, D_1 and D_2 and wood light-frame *townhouses* in Seismic Design Category C.

1. Plate washers conforming to Section R602.11.1 shall be provided for all anchor bolts over the full length of required *braced wall lines* except where *approved* anchor straps are used. Properly sized cut washers shall be permitted for anchor bolts in wall lines not containing *braced wall panels*.
2. Interior braced wall plates shall have anchor bolts spaced at not more than 6 feet (1829 mm) on center and located within 12 inches (305 mm) of the ends of each plate section where supported on a continuous foundation.
3. Interior bearing wall sole plates shall have anchor bolts spaced at not more than 6 feet (1829 mm) on center and located within 12 inches (305 mm) of the ends of each plate section where supported on a continuous foundation.
4. The maximum anchor bolt spacing shall be 4 feet (1219 mm) for buildings over two *stories* in height.
5. Stepped cripple walls shall conform to Section R602.11.2.
6. Where continuous wood foundations in accordance with Section R404.2 are used, the force transfer shall have a capacity equal to or greater than the connections required by Section R602.11.1 or the *braced wall panel* shall be connected to the wood foundations in accordance with the *braced wall panel*-to-floor fastening requirements of Table R602.3(1).

R403.1.7 Footings on or adjacent to slopes. The placement of buildings and structures on or adjacent to slopes steeper than 1 unit vertical in 3 units horizontal (33.3-percent slope) shall conform to Sections R403.1.7.1 through R403.1.7.4.

R403.1.7.1 Building clearances from ascending slopes. In general, buildings below slopes shall be set a sufficient distance from the slope to provide protection from slope drainage, erosion and shallow failures. Except as provided in Section R403.1.7.4 and Figure R403.1.7.1, the following criteria will be assumed to provide this protection. Where the existing slope is steeper than 1 unit vertical in 1 unit horizontal (100-percent slope), the toe of the slope shall be assumed to be at the intersection of a horizontal plane drawn from the top of the foundation and a plane drawn tangent to the slope at an angle of 45 degrees (0.79 rad) to the horizontal. Where a retaining wall is constructed at the toe of the slope, the height of the slope shall be measured from the top of the wall to the top of the slope.

R403.1.7.2 Footing setback from descending slope surfaces. Footings on or adjacent to slope surfaces shall be founded in material with an embedment and setback from the slope surface sufficient to provide vertical and lateral support for the footing without detrimental settlement. Except as provided for in Section R403.1.7.4 and Figure R403.1.7.1, the following setback is deemed adequate to meet the criteria. Where the slope is steeper than 1 unit vertical in 1 unit horizontal (100-percent slope), the required setback shall be measured from an imaginary plane 45 degrees (0.79 rad) to the horizontal, projected upward from the toe of the slope.

R403.1.7.3 Foundation elevation. On graded sites, the top of any exterior foundation shall extend above the elevation of the street gutter at point of discharge or the inlet of an *approved* drainage device not less than 12 inches (305 mm) plus 2 percent. Alternate elevations are permitted subject to the approval of the *building official*, provided that it can be demonstrated that required drainage to the point of discharge and away from the structure is provided at all locations on the site.

R403.1.7.4 Alternate setbacks and clearances. Alternate setbacks and clearances are permitted, subject to the approval of the *building official*. The *building official* is permitted to require an investigation and recommendation of a qualified engineer to demonstrate that the intent of this section has been satisfied. Such an investigation shall include consideration of material, height of slope, slope gradient, load intensity and erosion characteristics of slope material.

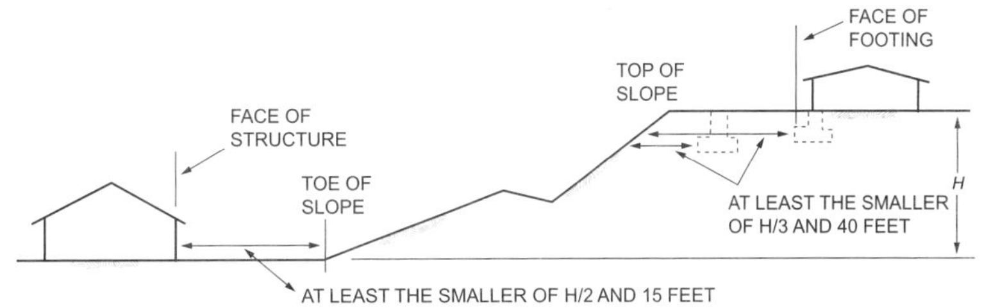

For SI: 1 foot = 304.8 mm.

FIGURE R403.1.7.1
FOUNDATION CLEARANCE FROM SLOPES

R403.1.8 Foundations on expansive soils. Foundation and floor slabs for buildings located on *expansive soils* shall be designed in accordance with Section 1808.6 of the *International Building Code*.

Exception: Slab-on-ground and other foundation systems that have performed adequately in soil conditions similar to those encountered at the building site are permitted subject to the approval of the *building official*.

R403.1.8.1 Expansive soils classifications. Soils meeting all of the following provisions shall be considered to be expansive, except that tests to show compliance with Items 1, 2 and 3 shall not be required if the test prescribed in Item 4 is conducted:

1. Plasticity Index (PI) of 15 or greater, determined in accordance with ASTM D4318.
2. More than 10 percent of the soil particles pass a No. 200 sieve (75 μm), determined in accordance with ASTM D422.
3. More than 10 percent of the soil particles are less than 5 micrometers in size, determined in accordance with ASTM D422.
4. Expansion Index greater than 20, determined in accordance with ASTM D4829.

R403.2 Footings for wood foundations. Footings for wood foundations shall be in accordance with Figures R403.1(2) and R403.1(3). Gravel shall be washed and well graded. The maximum size stone shall not exceed $^3/_4$ inch (19.1 mm). Gravel shall be free from organic, clayey or silty soils. Sand shall be coarse, not smaller than $^1/_{16}$-inch (1.6 mm) grains and shall be free from organic, clayey or silty soils. Crushed stone shall have a maximum size of $^1/_2$ inch (12.7 mm).

R403.3 Frost-protected shallow foundations. For buildings where the monthly mean temperature of the building is maintained at not less than 64°F (18°C), footings are not required to extend below the frost line where protected from frost by insulation in accordance with Figure R403.3(1) and Table R403.3(1). Foundations protected from frost in accordance with Figure R403.3(1) and Table R403.3(1) shall not be used for unheated spaces such as porches, utility rooms, garages and carports, and shall not be attached to *basements* or *crawl spaces* that are not maintained at a minimum monthly mean temperature of 64°F (18°C).

Materials used below *grade* for the purpose of insulating footings against frost shall be *labeled* as complying with ASTM C578.

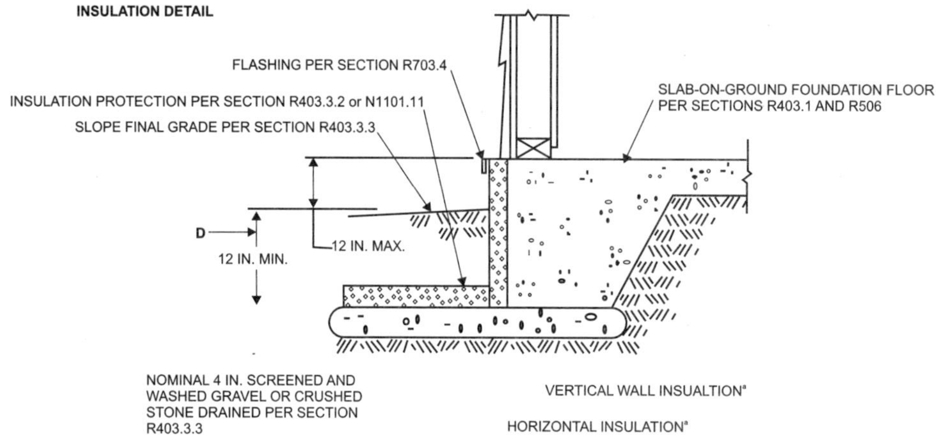

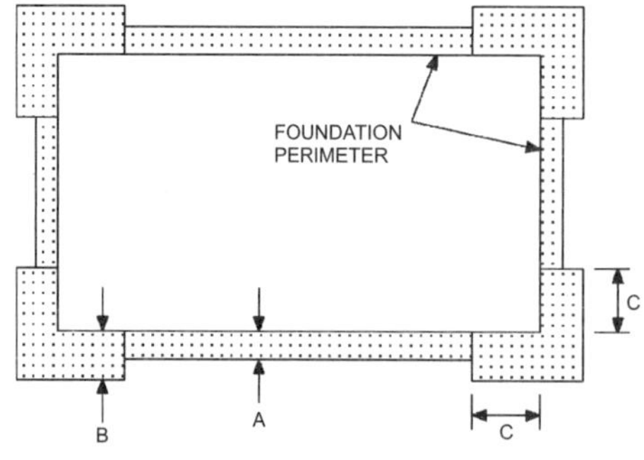

For SI: 1 inch = 25.4 mm.

a. See Table R403.3(1) for required dimensions and *R*-values for vertical and horizontal insulation and minimum footing depth.

FIGURE R403.3(1)
INSULATION PLACEMENT FOR FROST-PROTECTED FOOTINGS IN HEATED BUILDINGS

For SI: °C = [(°F) – 32]/1.8.

Note: The air-freezing index is defined as cumulative degree days below 32°F. It is used as a measure of the combined magnitude and duration of air temperature below freezing. The index was computed over a 12-month period (July–June) for each of the 3,044 stations used in the above analysis. Dates from the 1951–80 period were fitted to a Weibull probability distribution to produce an estimate of the 100-year return period.

FIGURE R403.3(2)
AIR-FREEZING INDEX AN ESTIMATE OF THE 100-YEAR RETURN PERIOD

TABLE R403.3(1)
MINIMUM FOOTING DEPTH AND INSULATION REQUIREMENTS FOR FROST-PROTECTED FOOTINGS IN HEATED BUILDINGS[a]

AIR-FREEZING INDEX (°F days)[b]	MINIMUM FOOTING DEPTH, *D* (inches)	VERTICAL INSULATION *R*-VALUE[c, d]	HORIZONTAL INSULATION *R*-VALUE[c, e]		HORIZONTAL INSULATION DIMENSIONS PER Figure R403.3(1) (inches)		
			Along walls	At corners	A	B	C
1,500 or less	12	4.5	Not required	Not required	Not required	Not required	Not required
2,000	14	5.6	Not required	Not required	Not required	Not required	Not required
2,500	16	6.7	1.7	4.9	12	24	40
3,000	16	7.8	6.5	8.6	12	24	40
3,500	16	9.0	8.0	11.2	24	30	60
4,000	16	10.1	10.5	13.1	24	36	60

For SI: 1 inch = 25.4 mm, °C = [(°F) – 32]/1.8.

a. Insulation requirements are for protection against frost damage in heated buildings. Greater values could be required to meet energy conservation standards.

b. See Figure R403.3(2) or Table R403.3(2) for Air-Freezing Index values.

c. Insulation materials shall provide the stated minimum *R*-values under long-term exposure to moist, below-ground conditions in freezing climates. The following *R*-values shall be used to determine insulation thicknesses required for this application: Type II expanded polystyrene (EPS)-3.2 R per inch for vertical insulation and 2.6 R per inch for horizontal insulation; Type IX expanded polystyrene (EPS)-3.4 R per inch for vertical insulation and 2.8 R per inch for horizontal insulation; Types IV, V, VI, VII, and X extruded polystyrene (XPS)-4.5 R per inch for vertical insulation and 4.0 R per inch for horizontal insulation.

d. Vertical insulation shall be expanded polystyrene insulation or extruded polystyrene insulation.

e. Horizontal insulation shall be expanded polystyrene insulation or extruded polystyrene insulation.

TABLE R403.3(2)
AIR-FREEZING INDEX FOR US LOCATIONS BY COUNTY

STATE	AIR-FREEZING INDEX					
	1,500 or less	2,000	2,500	3,000	3,500	4,000
Alabama	All counties	—	—	—	—	—
Alaska	Ketchikan Gateway, Prince of Wales-Outer Ketchikan (CA), Sitka, Wrangell-Petersburg (CA)	—	Aleutians West (CA), Haines, Juneau, Skagway-Hoonah-Angoon (CA), Yakutat	—	—	All counties not listed
Arizona	All counties	—	—	—	—	—
Arkansas	All counties	—	—	—	—	—
California	All counties not listed	Nevada, Sierra	—	—	—	—
Colorado	All counties not listed	Archuleta, Custer, Fremont, Huerfano, Las Animas, Ouray, Pitkin, San Miguel	Clear Creek, Conejos, Costilla, Dolores, Eagle, La Plata, Park, Routt, San Juan, Summit	Alamosa, Grand, Jackson, Larimer, Moffat, Rio Blanco, Rio Grande	Chaffee, Gunnison, Lake, Saguache	Hinsdale, Mineral
Connecticut	All counties not listed	Hartford, Litchfield	—	—	—	—
Delaware	All counties	—	—	—	—	—
District of Columbia	All counties	—	—	—	—	—
Florida	All counties	—	—	—	—	—
Georgia	All counties	—	—	—	—	—
Hawaii	All counties	—	—	—	—	—
Idaho	All counties not listed	Adams, Bannock, Blaine, Clearwater, Idaho, Lincoln, Oneida, Power, Valley, Washington	Bingham, Bonneville, Camas, Caribou, Elmore, Franklin, Jefferson, Madison, Teton	Bear Lake, Butte, Custer, Fremont, Lemhi	Clark	—

(continued)

TABLE R403.3(2)—continued
AIR-FREEZING INDEX FOR US LOCATIONS BY COUNTY

STATE	AIR-FREEZING INDEX					
	1,500 or less	2,000	2,500	3,000	3,500	4,000
Illinois	All counties not listed	Boone, Bureau, Cook, Dekalb, DuPage, Fulton, Grundy, Henderson, Henry, Iroquois, Jo Daviess, Kane, Kankakee, Kendall, Knox, La Salle, Lake, Lee, Livingston, Marshall, Mason, McHenry, McLean, Mercer, Peoria, Putnam, Rock Island, Stark, Tazewell, Warren, Whiteside, Will, Woodford	Carroll, Ogle, Stephenson, Winnebago	—	—	—
Indiana	All counties not listed	Allen, Benton, Cass, Fountain, Fulton, Howard, Jasper, Kosciusko, La Porte, Lake, Marshall, Miami, Newton, Porter, Pulaski, Starke, Steuben, Tippecanoe, Tipton, Wabash, Warren, White	—	—	—	—
Iowa	Appanoose, Davis, Fremont, Lee, Van Buren	All counties not listed	Allamakee, Black Hawk, Boone, Bremer, Buchanan, Buena Vista, Butler, Calhoun, Cerro Gordo, Cherokee, Chickasaw, Clay, Clayton, Delaware, Dubuque, Fayette, Floyd, Franklin, Grundy, Hamilton, Hancock, Hardin, Humboldt, Ida, Jackson, Jasper, Jones, Linn, Marshall, Palo Alto, Plymouth, Pocahontas, Poweshiek, Sac, Sioux, Story, Tama, Webster, Winnebago, Woodbury, Worth, Wright	Dickinson, Emmet, Howard, Kossuth, Lyon, Mitchell, O'Brien, Osceola, Winneshiek	—	—
Kansas	All counties	—	—	—	—	—
Kentucky	All counties	—	—	—	—	—

(continued)

TABLE R403.3(2)—continued
AIR-FREEZING INDEX FOR US LOCATIONS BY COUNTY

STATE	AIR-FREEZING INDEX					
	1,500 or less	2,000	2,500	3,000	3,500	4,000
Louisiana	All counties	—	—	—	—	—
Maine	York	Knox, Lincoln, Sagadahoc	Androscoggin, Cumberland, Hancock, Kennebec, Waldo, Washington	Aroostook, Franklin, Oxford, Penobscot, Piscataquis, Somerset	—	—
Maryland	All counties	—	—	—	—	—
Massachusetts	All counties not listed	Berkshire, Franklin, Hampden, Worcester	—	—	—	—
Michigan	Berrien, Branch, Cass, Kalamazoo, Macomb, Ottawa, St. Clair, St. Joseph	All counties not listed	Alger, Charlevoix, Cheboygan, Chippewa, Crawford, Delta, Emmet, Iosco, Kalkaska, Lake, Luce, Mackinac, Menominee, Missaukee, Montmorency, Ogemaw, Osceola, Otsego, Roscommon, Schoolcraft, Wexford	Baraga, Dickinson, Iron, Keweenaw, Marquette	Gogebic, Houghton, Ontonagon	—
Minnesota	—	—	Houston, Winona	All counties not listed	Aitkin, Big Stone, Carlton, Crow Wing, Douglas, Itasca, Kanabec, Lake, Morrison, Pine, Pope, Stearns, Stevens, Swift, Todd, Wadena	Becker, Beltrami, Cass, Clay, Clearwater, Grant, Hubbard, Kittson, Koochiching, Lake of the Woods, Mahnomen, Marshall, Norman, Otter Tail, Pennington, Polk, Red Lake, Roseau, St. Louis, Traverse, Wilkin
Mississippi	All counties	—	—	—	—	—
Missouri	All counties not listed	Atchison, Mercer, Nodaway, Putnam	—	—	—	—
Montana	Mineral	Broadwater, Golden Valley, Granite, Lake, Lincoln, Missoula, Ravalli, Sanders, Sweet Grass	Big Horn, Carbon, Jefferson, Judith Basin, Lewis and Clark, Meagher, Musselshell, Powder River, Powell, Silver Bow, Stillwater, Westland	Carter, Cascade, Deer Lodge, Falcon, Fergus, Flathead, Gallatin, Glacier, Madison, Park, Petroleum, Ponder, Rosebud, Teton, Treasure, Yellowstone	Beaverhead, Blaine, Chouteau, Custer, Dawson, Garfield, Liberty, McCone, Prairie, Toole, Wibaux	Daniels, Hill, Phillips, Richland, Roosevelt, Sheridan, Valley

(continued)

TABLE R403.3(2)—continued
AIR-FREEZING INDEX FOR US LOCATIONS BY COUNTY

STATE	AIR-FREEZING INDEX					
	1,500 or less	2,000	2,500	3,000	3,500	4,000
Nebraska	Adams, Banner, Chase, Cheyenne, Clay, Deuel, Dundy, Fillmore, Franklin, Frontier, Furnas, Gage, Garden, Gosper, Harlan, Hayes, Hitchcock, Jefferson, Kimball, Morrill, Nemaha, Nuckolls, Pawnee, Perkins, Phelps, Red Willow, Richardson, Saline, Scotts Bluff, Seward, Thayer, Webster	All counties not listed	Boyd, Burt, Cedar, Cuming, Dakota, Dixon, Dodge, Knox, Thurston	—	—	—
Nevada	All counties not listed	Elko, Eureka, Nye, Washoe, White Pine	—	—	—	—
New Hampshire	—	All counties not listed	—	—	—	Carroll, Coos, Grafton
New Jersey	All counties	—	—	—	—	—
New Mexico	All counties not listed	Rio Arriba	Colfax, Mora, Taos	—	—	—
New York	Albany, Bronx, Cayuga, Columbia, Cortland, Dutchess, Genessee, Kings, Livingston, Monroe, Nassau, New York, Niagara, Onondaga, Ontario, Orange, Orleans, Putnam, Queens, Richmond, Rockland, Seneca, Suffolk, Wayne, Westchester, Yates	All counties not listed	Clinton, Essex, Franklin, Hamilton, Herkimer, Jefferson, Lewis, St. Lawrence, Warren	—	—	—
North Carolina	All counties	—	—	—	—	—
North Dakota	—	—	—	Billings, Bowman	Adams, Dickey, Golden Valley, Hettinger, LaMoure, Oliver, Ransom, Sargent, Sioux, Slope, Stark	All counties not listed

(continued)

TABLE R403.3(2)—continued
AIR-FREEZING INDEX FOR US LOCATIONS BY COUNTY

STATE	AIR-FREEZING INDEX					
	1,500 or less	2,000	2,500	3,000	3,500	4,000
Ohio	All counties not listed	Ashland, Crawford, Defiance, Holmes, Huron, Knox, Licking, Morrow, Paulding, Putnam, Richland, Seneca, Williams	—	—	—	—
Oklahoma	All counties	—	—	—	—	—
Oregon	All counties not listed	Baker, Crook, Grant, Harney	—	—	—	—
Pennsylvania	All counties not listed	Berks, Blair, Bradford, Cambria, Cameron, Centre, Clarion, Clearfield, Clinton, Crawford, Elk, Forest, Huntingdon, Indiana, Jefferson, Lackawanna, Lycoming, McKean, Pike, Potter, Susquehanna, Tioga, Venango, Warren, Wayne, Wyoming	—	—	—	—
Rhode Island	All counties	—	—	—	—	—
South Carolina	All counties	—	—	—	—	—
South Dakota	—	Bennett, Custer, Fall River, Lawrence, Mellette, Shannon, Todd, Tripp	Bon Homme, Charles Mix, Davison, Douglas, Gregory, Jackson, Jones, Lyman	All counties not listed	Beadle, Brookings, Brown, Campbell, Codington, Corson, Day, Deuel, Edmunds, Faulk, Grant, Hamlin, Kingsbury, Marshall, McPherson, Perkins, Roberts, Spink, Walworth	—
Tennessee	All counties	—	—	—	—	—
Texas	All counties	—	—	—	—	—
Utah	All counties not listed	Box Elder, Morgan, Weber	Garfield, Salt Lake, Summit	Carbon, Daggett, Duchesne, Rich, Sanpete, Uintah, Wasatch	—	—
Vermont	—	Bennington, Grand Isle, Rutland, Windham	Addison, Chittenden, Franklin, Orange, Washington, Windsor	Caledonia, Essex, Lamoille, Orleans	—	—
Virginia	All counties	—	—	—	—	—

(continued)

TABLE R403.3(2)—continued
AIR-FREEZING INDEX FOR US LOCATIONS BY COUNTY

STATE	AIR-FREEZING INDEX					
	1,500 or less	2,000	2,500	3,000	3,500	4,000
Washington	All counties not listed	Chelan, Douglas, Ferry, Okanogan	—	—	—	—
West Virginia	All counties	—	—	—	—	—
Wisconsin	—	Kenosha, Kewaunee, Racine, Sheboygan, Walworth	All counties not listed	Ashland, Barron, Burnett, Chippewa, Clark, Dunn, Eau Claire, Florence, Forest, Iron, Jackson, La Crosse, Langlade, Marathon, Monroe, Pepin, Polk, Portage, Price, Rust, St. Croix, Taylor, Trempealeau, Vilas, Wood	Bayfield, Douglas, Lincoln, Oneida, Sawyer, Washburn	—
Wyoming	Goshen, Platte	Converse, Crook, Laramie, Niobrara	Campbell, Carbon, Hot Springs, Johnson, Natrona, Sheridan, Uinta, Weston	Albany, Big Horn, Park, Washakie	Fremont, Teton	Lincoln, Sublette, Sweetwater

INSULATION DETAIL

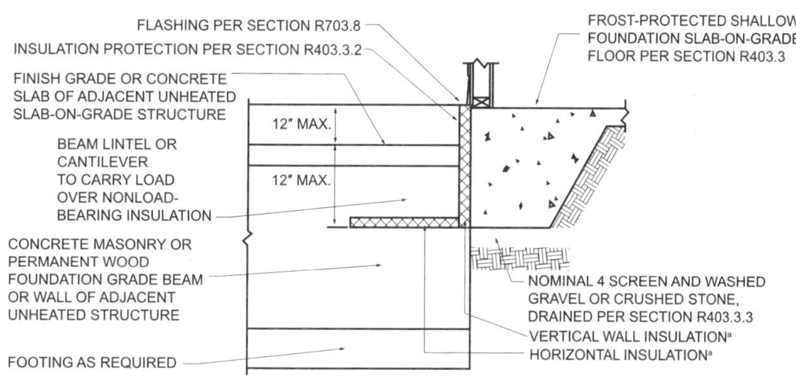

HORIZONTAL INSULATION PLAN

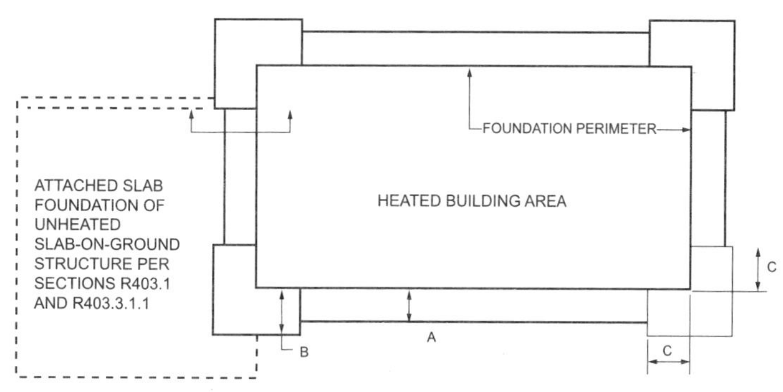

For SI: 1 inch = 25.4 mm.

a. See Table R403.3(1) for required dimensions and *R*-values for vertical and horizontal insulation.

FIGURE R403.3(3)
INSULATION PLACEMENT FOR FROST-PROTECTED FOOTINGS ADJACENT TO UNHEATED SLAB-ON-GROUND STRUCTURE

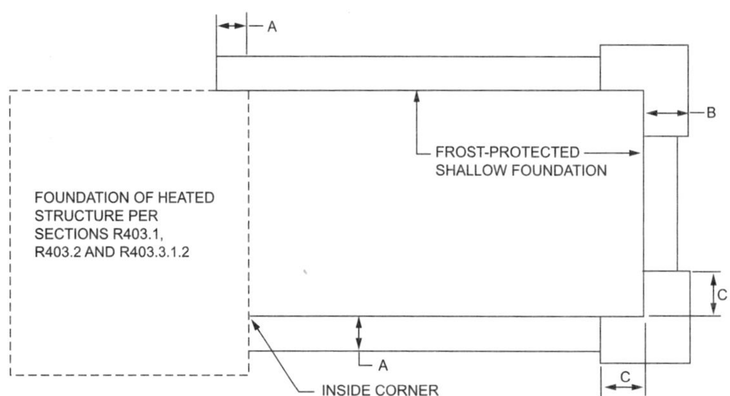

FIGURE R403.3(4)
INSULATION PLACEMENT FOR FROST-PROTECTED FOOTINGS ADJACENT TO HEATED STRUCTURE

R403.3.1 Foundations adjoining frost-protected shallow foundations. Foundations that adjoin frost-protected shallow foundations shall be protected from frost in accordance with Section R403.1.4.

R403.3.1.1 Attachment to unheated slab-on-ground structure. Vertical wall insulation and horizontal insulation of frost-protected shallow foundations that adjoin a slab-on-ground foundation that does not have a monthly mean temperature maintained at not less than 64°F (18°C) shall be in accordance with Figure R403.3(3) and Table R403.3(1). Vertical wall insulation shall extend between the frost-protected shallow foundation and the adjoining slab foundation. Required horizontal insulation shall be continuous under the adjoining slab foundation and through any foundation walls adjoining the frost- protected shallow foundation. Where insulation passes through a foundation wall, it shall be either of a type complying with this section and having bearing capacity equal to or greater than the structural loads imposed by the building, or the building shall be designed and constructed using beams, lintels, cantilevers or other means of transferring building loads such that the structural loads of the building do not bear on the insulation.

R403.3.1.2 Attachment to heated structure. Where a frost-protected shallow foundation abuts a structure that has a monthly mean temperature maintained at not less than 64°F (18°C), horizontal insulation and vertical wall insulation shall not be required between the frost-protected shallow foundation and the adjoining structure. Where the frost-protected shallow foundation abuts the heated structure, the horizontal insulation and vertical wall insulation shall extend along the adjoining foundation in accordance with Figure R403.3(4) a distance of not less than Dimension A in Table R403.3(1).

Exception: Where the frost-protected shallow foundation abuts the heated structure to form an inside corner, vertical insulation extending along the adjoining foundation is not required.

R403.3.2 Protection of horizontal insulation below ground. Horizontal insulation placed less than 12 inches (305 mm) below the ground surface or that portion of horizontal insulation extending outward more than 24 inches (610 mm) from the foundation edge shall be protected against damage by use of a concrete slab or asphalt paving on the ground surface directly above the insulation or by cementitious board, plywood rated for below-ground use, or other *approved* materials placed below ground, directly above the top surface of the insulation.

R403.3.3 Drainage. Final *grade* shall be sloped in accordance with Section R401.3. In other than Group I Soils, as detailed in Table R405.1, gravel or crushed stone beneath horizontal insulation below ground shall drain to daylight or into an *approved* sewer system.

R403.3.4 Termite protection. The use of foam plastic in areas of "very heavy" termite infestation probability shall be in accordance with Section R318.4.

R403.4 Footings for precast concrete foundations. Footings for *precast concrete* foundations shall comply with Section R403.4.

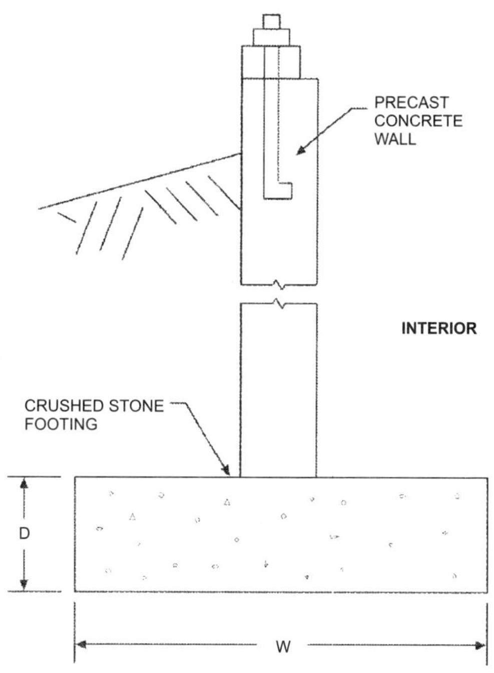

FIGURE R403.4(1)
BASEMENT OR CRAWL SPACE WITH PRECAST FOUNDATION WALL BEARING ON CRUSHED STONE

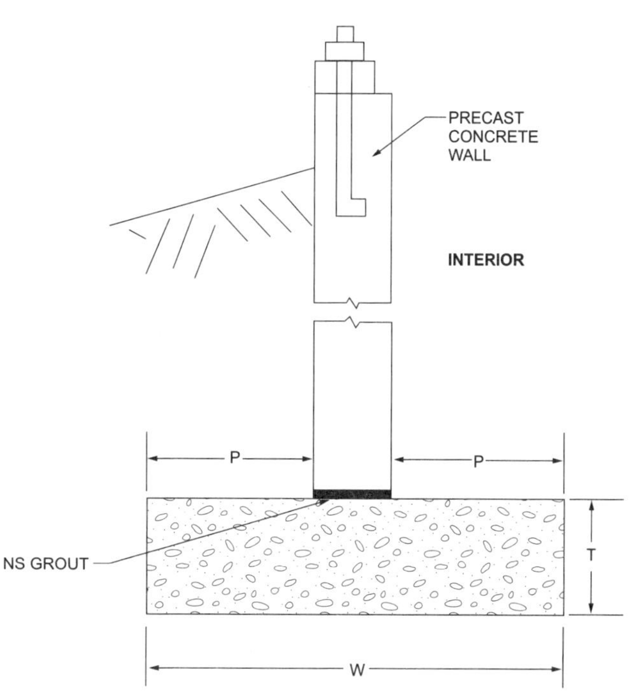

FIGURE R403.4(2)
BASEMENT OR CRAWL SPACE WITH PRECAST FOUNDATION WALL ON SPREAD FOOTING

TABLE R403.4
MINIMUM DEPTH (D) AND WIDTH (W) OF CRUSHED STONE FOOTINGS[a, b] (inches)

NUMBER OF STORIES	UNIFORM WALL LOAD	DEPTH (D) AND WIDTH (W)	LOAD-BEARING VALUE OF SOIL (psf)																	
			1500			2000			2500			3000			3500			4000		
			MH, CH, CL, ML[c]			SC, GC, SM, GM, SP, SW[c]						GP, GW[c]								
			Wall width (inches)			Wall width (inches)			Wall width (inches)			Wall width (inches)			Wall width (inches)			Wall width (inches)		
			8	10	12	8	10	12	8	10	12	8	10	12	8	10	12	8	10	12
Conventional light-frame construction																				
1-story	1,100 plf	D	4	4	4	4	4	4	4	4	4	4	4	4	4	4	4	4	4	4
		W	13	15	17	13	15	17	13	15	17	13	15	17	13	15	17	13	15	17
2-story	1,800 plf	D	6	4	4	4	4	4	4	4	4	4	4	4	4	4	4	4	4	4
		W	15	15	17	13	15	17	13	15	17	13	15	17	13	15	17	13	15	17
3-story	2,900 plf	D	14	12	10	9	7	5	6	4	4	4	4	4	4	4	4	4	4	4
		W	25	24	24	19	19	18	15	15	17	13	15	17	13	15	17	13	15	17
4-inch brick veneer over light-frame or 8-inch hollow concrete masonry																				
1-story	1,500 plf	D	4	4	4	4	4	4	4	4	4	4	4	4	4	4	4	4	4	4
		W	13	15	17	13	15	17	13	15	17	13	15	17	13	15	17	13	15	17
2-story	2,700 plf	D	12	11	9	8	6	4	5	4	4	4	4	4	4	4	4	4	4	4
		W	22	23	23	18	17	17	14	15	17	13	15	17	13	15	17	13	15	17
3-story	4,000 plf	D	21	20	18	14	13	11	10	8	7	7	6	4	5	4	4	4	4	4
		W	33	34	33	25	26	25	20	20	21	17	17	17	14	15	17	13	15	17
8-inch solid or fully grouted masonry																				
1-story	2,000 plf	D	7	6	4	4	4	4	4	4	4	4	4	4	4	4	4	4	4	4
		W	17	17	17	13	15	17	13	15	17	13	15	17	13	15	17	13	15	17
2-story	3,600 plf	D	19	17	15	12	11	9	9	7	5	6	4	4	4	4	4	4	4	4
		W	30	30	30	22	23	23	19	19	18	15	15	17	13	15	17	13	15	17
3-story	5,300 plf	D	30	29	27	21	19	18	16	14	12	12	10	8	9	8	6	7	6	4
		W	43	44	44	33	32	33	27	27	26	22	22	22	19	20	19	17	17	17

For SI: 1 inch = 25.4 mm, 1 plf = 14.6 N/m, 1 pound per square foot = 47.9 N/m².

a. Linear interpolation of stone depth between wall widths is permitted within each Load-Bearing Value of Soil (psf).

b. Crushed stone must be consolidated in 8-inch lifts with a plate vibrator.

c. Soil classes are in accordance with the Unified Soil Classification System. Refer to Table R405.1.

R403.4.1 Crushed stone footings. Clean crushed stone shall be free from organic, clayey or silty soils. Crushed stone shall be angular in nature and meet ASTM C33, with the maximum size stone not to exceed $^1/_2$ inch (12.7 mm) and the minimum stone size not to be smaller than $^1/_{16}$ inch (1.6 mm). Crushed stone footings for precast foundations shall be installed in accordance with Figure R403.4(1) and Table R403.4. Crushed stone footings shall be consolidated using a vibratory plate in not greater than 8-inch (203 mm) lifts. Crushed stone footings shall be limited to *Seismic Design Categories* A, B and C.

R403.4.2 Concrete footings. Concrete footings shall be installed in accordance with Section R403.1 and Figure R403.4(2).

SECTION R404
FOUNDATION AND RETAINING WALLS

R404.1 Concrete and masonry foundation walls. Concrete foundation walls shall be selected and constructed in accordance with the provisions of Section R404.1.3. Masonry foundation walls shall be selected and constructed in accordance with the provisions of Section R404.1.2.

R404.1.1 Design required. Concrete or masonry foundation walls shall be designed in accordance with accepted engineering practice where either of the following conditions exists:

1. Walls are subject to hydrostatic pressure from ground water.
2. Walls supporting more than 48 inches (1219 mm) of unbalanced backfill that do not have permanent lateral support at the top or bottom.

TABLE R404.1.1(1)
PLAIN MASONRY FOUNDATION WALLS[f]

MAXIMUM UNSUPPORTED WALL HEIGHT (feet)	MAXIMUM UNBALANCED BACKFILL HEIGHT[c] (feet)	PLAIN MASONRY[a] MINIMUM NOMINAL WALL THICKNESS (inches)		
		Soil classes[b]		
		GW, GP, SW and SP	GM, GC, SM, SM-SC and ML	SC, MH, ML-CL and inorganic CL
5	4	6 solid[d] or 8	6 solid[d] or 8	6 solid[d] or 8
	5	6 solid[d] or 8	8	10
6	4	6 solid[d] or 8	6 solid[d] or 8	6 solid[d] or 8
	5	6 solid[d] or 8	8	10
	6	8	10	12
7	4	6 solid[d] or 8	8	8
	5	6 solid[d] or 8	10	10
	6	10	12	10 solid[d]
	7	12	10 solid[d]	12 solid[d]
8	4	6 solid[d] or 8	6 solid[d] or 8	8
	5	6 solid[d] or 8	10	12
	6	10	12	12 solid[d]
	7	12	12 solid[d]	Note e
	8	10 grout[d]	12 grout[d]	Note e
9	4	6 grout[d] or 8 solid[d] or 12	6 grout[d] or 8 solid[d]	8 grout[d] or 10 solid[d]
	5	6 grout[d] or 10 solid[d]	8 grout[d] or 12 solid[d]	8 grout[d]
	6	8 grout[d] or 12 solid[d]	10 grout[d]	10 grout[d]
	7	10 grout[d]	10 grout[d]	12 grout
	8	10 grout[d]	12 grout	Note e
	9	12 grout	Note e	Note e

For SI: 1 inch = 25.4 mm, 1 foot = 304.8 mm.

a. Mortar shall be Type M or S and masonry shall be laid in running bond. Ungrouted hollow masonry units are permitted except where otherwise indicated.

b. Soil classes are in accordance with the Unified Soil Classification System. Refer to Table R405.1.

c. Unbalanced backfill height is the difference in height between the exterior finish ground level and the lower of the top of the concrete footing that supports the foundation wall or the interior finish ground level. Where an interior concrete slab-on-grade is provided and is in contact with the interior surface of the foundation wall, measurement of the unbalanced backfill height from the exterior finish ground level to the top of the interior concrete slab is permitted.

d. Solid indicates solid masonry unit; grout indicates grouted hollow units.

e. Wall construction shall be in accordance with Table R404.1.1(2), R404.1.1(3) or R404.1.1(4), or a design shall be provided.

f. The use of this table shall be prohibited for soil classifications not shown.

TABLE R404.1.1(2)
8-INCH MASONRY FOUNDATION WALLS WITH REINFORCING WHERE d ≥ 5 INCHES[a, c, f]

MAXIMUM UNSUPPORTED WALL HEIGHT	HEIGHT OF UNBALANCED BACKFILL[e]	MINIMUM VERTICAL REINFORCEMENT AND SPACING (INCHES)[b, c] Soil classes and lateral soil load[d] (psf per foot below grade)		
		GW, GP, SW and SP soils 30	GM, GC, SM, SM-SC and ML soils 45	SC, ML-CL and inorganic CL soils 60
6 feet 8 inches	4 feet (or less)	#4 at 48	#4 at 48	#4 at 48
	5 feet	#4 at 48	#4 at 48	#4 at 48
	6 feet 8 inches	#4 at 48	#5 at 48	#6 at 48
7 feet 4 inches	4 feet (or less)	#4 at 48	#4 at 48	#4 at 48
	5 feet	#4 at 48	#4 at 48	#4 at 48
	6 feet	#4 at 48	#5 at 48	#5 at 48
	7 feet 4 inches	#5 at 48	#6 at 48	#6 at 40
8 feet	4 feet (or less)	#4 at 48	#4 at 48	#4 at 48
	5 feet	#4 at 48	#4 at 48	#4 at 48
	6 feet	#4 at 48	#5 at 48	#5 at 48
	7 feet	#5 at 48	#6 at 48	#6 at 40
	8 feet	#5 at 48	#6 at 48	#6 at 32
8 feet 8 inches	4 feet (or less)	#4 at 48	#4 at 48	#4 at 48
	5 feet	#4 at 48	#4 at 48	#5 at 48
	6 feet	#4 at 48	#5 at 48	#6 at 48
	7 feet	#5 at 48	#6 at 48	#6 at 40
	8 feet 8 inches	#6 at 48	#6 at 32	#6 at 24
9 feet 4 inches	4 feet (or less)	#4 at 48	#4 at 48	#4 at 48
	5 feet	#4 at 48	#4 at 48	#5 at 48
	6 feet	#4 at 48	#5 at 48	#6 at 48
	7 feet	#5 at 48	#6 at 48	#6 at 40
	8 feet	#6 at 48	#6 at 40	#6 at 24
	9 feet 4 inches	#6 at 40	#6 at 24	#6 at 16
10 feet	4 feet (or less)	#4 at 48	#4 at 48	#4 at 48
	5 feet	#4 at 48	#4 at 48	#5 at 48
	6 feet	#4 at 48	#5 at 48	#6 at 48
	7 feet	#5 at 48	#6 at 48	#6 at 32
	8 feet	#6 at 48	#6 at 32	#6 at 24
	9 feet	#6 at 40	#6 at 24	#6 at 16
	10 feet	#6 at 32	#6 at 16	#6 at 16

For SI: 1 inch = 25.4 mm, 1 foot = 304.8 mm, 1 pound per square foot per foot = 0.157 kPa/mm.

a. Mortar shall be Type M or S and masonry shall be laid in running bond.

b. Alternative reinforcing bar sizes and spacings having an equivalent cross-sectional area of reinforcement per lineal foot of wall shall be permitted provided the spacing of the reinforcement does not exceed 72 inches in Seismic Design Categories A, B and C, and 48 inches in Seismic Design Categories D_0, D_1 and D_2.

c. Vertical reinforcement shall be Grade 60 minimum. The distance, *d*, from the face of the soil side of the wall to the center of vertical reinforcement shall be not less than 5 inches.

d. Soil classes are in accordance with the Unified Soil Classification System and design lateral soil loads are for moist conditions without hydrostatic pressure. Refer to Table R405.1.

e. Unbalanced backfill height is the difference in height between the exterior finish ground level and the lower of the top of the concrete footing that supports the foundation wall or the interior finish ground level. Where an interior concrete slab-on-grade is provided and is in contact with the interior surface of the foundation wall, measurement of the unbalanced backfill height from the exterior finish ground level to the top of the interior concrete slab is permitted.

f. The use of this table shall be prohibited for soil classifications not shown.

TABLE R404.1.1(3)
10-INCH MASONRY FOUNDATION WALLS WITH REINFORCING WHERE d ≥ 6.75 INCHES[a, c, f]

MAXIMUM UNSUPPORTED WALL HEIGHT	HEIGHT OF UNBALANCED BACKFILL[e]	MINIMUM VERTICAL REINFORCEMENT AND SPACING (INCHES)[b, c]		
		Soil classes and later soil load[d] (psf per foot below grade)		
		GW, GP, SW and SP soils 30	GM, GC, SM, SM-SC and ML soils 45	SC, ML-CL and inorganic CL soils 60
6 feet 8 inches	4 feet (or less)	#4 at 56	#4 at 56	#4 at 56
	5 feet	#4 at 56	#4 at 56	#4 at 56
	6 feet 8 inches	#4 at 56	#5 at 56	#5 at 56
7 feet 4 inches	4 feet (or less)	#4 at 56	#4 at 56	#4 at 56
	5 feet	#4 at 56	#4 at 56	#4 at 56
	6 feet	#4 at 56	#4 at 56	#5 at 56
	7 feet 4 inches	#4 at 56	#5 at 56	#6 at 56
8 feet	4 feet (or less)	#4 at 56	#4 at 56	#4 at 56
	5 feet	#4 at 56	#4 at 56	#4 at 56
	6 feet	#4 at 56	#4 at 56	#5 at 56
	7 feet	#4 at 56	#5 at 56	#6 at 56
	8 feet	#5 at 56	#6 at 56	#6 at 48
8 feet 8 inches	4 feet (or less)	#4 at 56	#4 at 56	#4 at 56
	5 feet	#4 at 56	#4 at 56	#4 at 56
	6 feet	#4 at 56	#4 at 56	#5 at 56
	7 feet	#4 at 56	#5 at 56	#6 at 56
	8 feet 8 inches	#5 at 56	#6 at 48	#6 at 32
9 feet 4 inches	4 feet (or less)	#4 at 56	#4 at 56	#4 at 56
	5 feet	#4 at 56	#4 at 56	#4 at 56
	6 feet	#4 at 56	#5 at 56	#5 at 56
	7 feet	#4 at 56	#5 at 56	#6 at 56
	8 feet	#5 at 56	#6 at 56	#6 at 40
	9 feet 4 inches	#6 at 56	#6 at 40	#6 at 24
10 feet	4 feet (or less)	#4 at 56	#4 at 56	#4 at 56
	5 feet	#4 at 56	#4 at 56	#4 at 56
	6 feet	#4 at 56	#5 at 56	#5 at 56
	7 feet	#5 at 56	#6 at 56	#6 at 48
	8 feet	#5 at 56	#6 at 48	#6 at 40
	9 feet	#6 at 56	#6 at 40	#6 at 24
	10 feet	#6 at 48	#6 at 32	#6 at 24

For SI: 1 inch = 25.4 mm, 1 foot = 304.8 mm, 1 pound per square foot per foot = 0.157 kPa/mm.

a. Mortar shall be Type M or S and masonry shall be laid in running bond.

b. Alternative reinforcing bar sizes and spacings having an equivalent cross-sectional area of reinforcement per lineal foot of wall shall be permitted provided the spacing of the reinforcement does not exceed 72 inches in Seismic Design Categories A, B and C, and 48 inches in Seismic Design Categories D_0, D_1 and D_2.

c. Vertical reinforcement shall be Grade 60 minimum. The distance, *d*, from the face of the soil side of the wall to the center of vertical reinforcement shall be not less than 6.75 inches.

d. Soil classes are in accordance with the Unified Soil Classification System and design lateral soil loads are for moist conditions without hydrostatic pressure. Refer to Table R405.1.

e. Unbalanced backfill height is the difference in height between the exterior finish ground level and the lower of the top of the concrete footing that supports the foundation wall or the interior finish ground level. Where an interior concrete slab-on-grade is provided and is in contact with the interior surface of the foundation wall, measurement of the unbalanced backfill height from the exterior finish ground level to the top of the interior concrete slab is permitted.

f. The use of this table shall be prohibited for soil classifications not shown.

TABLE R404.1.1(4)
12-INCH MASONRY FOUNDATION WALLS WITH REINFORCING WHERE d ≥ 8.75 INCHES[a, c, f]

MAXIMUM UNSUPPORTED WALL HEIGHT	HEIGHT OF UNBALANCED BACKFILL[e]	MINIMUM VERTICAL REINFORCEMENT AND SPACING (INCHES)[b, c]		
		Soil classes and lateral soil load[d] (psf per foot below grade)		
		GW, GP, SW and SP soils 30	GM, GC, SM, SM-SC and ML soils 45	SC, ML-CL and inorganic CL soils 60
6 feet 8 inches	4 feet (or less)	#4 at 72	#4 at 72	#4 at 72
	5 feet	#4 at 72	#4 at 72	#4 at 72
	6 feet 8 inches	#4 at 72	#4 at 72	#5 at 72
7 feet 4 inches	4 feet (or less)	#4 at 72	#4 at 72	#4 at 72
	5 feet	#4 at 72	#4 at 72	#4 at 72
	6 feet	#4 at 72	#4 at 72	#5 at 72
	7 feet 4 inches	#4 at 72	#5 at 72	#6 at 72
8 feet	4 feet (or less)	#4 at 72	#4 at 72	#4 at 72
	5 feet	#4 at 72	#4 at 72	#4 at 72
	6 feet	#4 at 72	#4 at 72	#5 at 72
	7 feet	#4 at 72	#5 at 72	#6 at 72
	8 feet	#5 at 72	#6 at 72	#6 at 64
8 feet 8 inches	4 feet (or less)	#4 at 72	#4 at 72	#4 at 72
	5 feet	#4 at 72	#4 at 72	#4 at 72
	6 feet	#4 at 72	#4 at 72	#5 at 72
	7 feet	#4 at 72	#5 at 72	#6 at 72
	8 feet 8 inches	#5 at 72	#7 at 72	#6 at 48
9 feet 4 inches	4 feet (or less)	#4 at 72	#4 at 72	#4 at 72
	5 feet	#4 at 72	#4 at 72	#4 at 72
	6 feet	#4 at 72	#5 at 72	#5 at 72
	7 feet	#4 at 72	#5 at 72	#6 at 72
	8 feet	#5 at 72	#6 at 72	#6 at 56
	9 feet 4 inches	#6 at 72	#6 at 48	#6 at 40
10 feet	4 feet (or less)	#4 at 72	#4 at 72	#4 at 72
	5 feet	#4 at 72	#4 at 72	#4 at 72
	6 feet	#4 at 72	#5 at 72	#5 at 72
	7 feet	#4 at 72	#6 at 72	#6 at 72
	8 feet	#5 at 72	#6 at 72	#6 at 48
	9 feet	#6 at 72	#6 at 56	#6 at 40
	10 feet	#6 at 64	#6 at 40	#6 at 32

For SI: 1 inch = 25.4 mm, 1 foot = 304.8 mm, 1 pound per square foot per foot = 0.157 kPa/mm.

a. Mortar shall be Type M or S and masonry shall be laid in running bond.

b. Alternative reinforcing bar sizes and spacings having an equivalent cross-sectional area of reinforcement per lineal foot of wall shall be permitted provided the spacing of the reinforcement does not exceed 72 inches in Seismic Design Categories A, B and C, and 48 inches in Seismic Design Categories D_0, D_1 and D_2.

c. Vertical reinforcement shall be Grade 60 minimum. The distance, *d*, from the face of the soil side of the wall to the center of vertical reinforcement shall be not less than 8.75 inches.

d. Soil classes are in accordance with the Unified Soil Classification System and design lateral soil loads are for moist conditions without hydrostatic pressure. Refer to Table R405.1.

e. Unbalanced backfill height is the difference in height between the exterior finish ground level and the lower of the top of the concrete footing that supports the foundation wall or the interior finish ground levels. Where an interior concrete slab-on-grade is provided and in contact with the interior surface of the foundation wall, measurement of the unbalanced backfill height is permitted to be measured from the exterior finish ground level to the top of the interior concrete slab is permitted.

f. The use of this table shall be prohibited for soil classifications not shown.

R404.1.2 Design of masonry foundation walls. Masonry foundation walls shall be designed and constructed in accordance with the provisions of this section or in accordance with the provisions of TMS 402. Where TMS 402 or the provisions of this section are used to design masonry foundation walls, project drawings, typical details and specifications are not required to bear the seal of the architect or engineer responsible for design, unless otherwise required by the state law of the *jurisdiction* having authority.

TABLE R404.1.2(1)
MINIMUM HORIZONTAL REINFORCEMENT FOR CONCRETE BASEMENT WALLS[a, b]

MAXIMUM UNSUPPORTED WALL HEIGHT (feet)	LOCATION OF HORIZONTAL REINFORCEMENT
≤ 8	One No. 4 bar within 12 inches of the top of the wall story and one No. 4 bar near mid-height of the wall story.
> 8	One No. 4 bar within 12 inches of the top of the wall story and one No. 4 bar near third points in the wall story.

For SI: 1 inch = 25.4 mm, 1 foot = 304.8 mm, 1 pound per square inch = 6.895 kPa.

a. Horizontal reinforcement requirements are for reinforcing bars with a minimum yield strength of 40,000 psi and concrete with a minimum concrete compressive strength of 2,500 psi.

b. See Section R404.1.3.2 for minimum reinforcement required for foundation walls supporting above-grade concrete walls.

TABLE R404.1.2(2)
MINIMUM VERTICAL REINFORCEMENT FOR 6-INCH NOMINAL FLAT CONCRETE BASEMENT WALLS[b, c, d, e, g, h, i, j, k]

MAXIMUM UNSUPPORTED WALL HEIGHT (feet)	MAXIMUM UNBALANCED BACKFILL HEIGHT[f] (feet)	MINIMUM VERTICAL REINFORCEMENT-BAR SIZE AND SPACING (inches)		
		Soil classes[a] and design lateral soil (psf per foot of depth)		
		GW, GP, SW, SP 30	GM, GC, SM, SM-SC and ML 45	SC, ML-CL and inorganic CL 60
8	4	NR	NR	NR
	5	NR	6 @ 39	6 @ 48
	6	5 @ 39	6 @ 48	6 @ 35
	7	6 @ 48	6 @ 34	6 @ 25
	8	6 @ 39	6 @ 25	6 @ 18
9	4	NR	NR	NR
	5	NR	5 @ 37	6 @ 48
	6	5 @ 36	6 @ 44	6 @ 32
	7	6 @ 47	6 @ 30	6 @ 22
	8	6 @ 34	6 @ 22	6 @ 16
	9	6 @ 27	6 @ 17	DR
10	4	NR	NR	NR
	5	NR	5 @ 35	6 @ 48
	6	6 @ 48	6 @ 41	6 @ 30
	7	6 @ 43	6 @ 28	6 @ 20
	8	6 @ 31	6 @ 20	DR
	9	6 @ 24	6 @ 15	DR
	10	6 @ 19	DR	DR

For SI: 1 inch = 25.4 mm, 1 foot = 304.8 mm, 1 pound per square foot per foot = 0.1571 kPa^2/m, 1 pound per square inch = 6.895 kPa.

NR = Not Required.

DR = Design Required.

a. Soil classes are in accordance with the Unified Soil Classification System. Refer to Table R405.1.

b. Table values are based on reinforcing bars with a minimum yield strength of 60,000 psi concrete with a minimum specified compressive strength of 2,500 psi and vertical reinforcement being located at the centerline of the wall. See Section R404.1.3.3.7.2.

c. Vertical reinforcement with a yield strength of less than 60,000 psi and bars of a different size than specified in the table are permitted in accordance with Section R404.1.3.3.7.6 and Table R404.1.2(9).

d. Deflection criterion is $L/240$, where L is the height of the basement wall in inches.

e. Interpolation is not permitted.

f. Where walls will retain 4 feet or more of unbalanced backfill, they shall be laterally supported at the top and bottom before backfilling.

g. NR indicates vertical wall reinforcement is not required, except for 6-inch-nominal walls formed with stay-in-place forming systems in which case vertical reinforcement shall be No. 4@48 inches on center.

h. See Section R404.1.3.2 for minimum reinforcement required for basement walls supporting above-grade concrete walls.

i. See Table R608.3 for tolerance from nominal thickness permitted for flat walls.

j. DR means design is required in accordance with the applicable building code, or in the absence of a code, in accordance with ACI 318.

k. The use of this table shall be prohibited for soil classifications not shown.

TABLE R404.1.2(3) MINIMUM VERTICAL REINFORCEMENT FOR 8-INCH (203 mm) NOMINAL FLAT CONCRETE BASEMENT WALLS[b, c, d, e, f, h, i, j]

MAXIMUM UNSUPPORTED WALL HEIGHT (feet)	MAXIMUM UNBALANCED BACKFILL HEIGHT[g] (feet)	MINIMUM VERTICAL REINFORCEMENT-BAR SIZE AND SPACING (inches)		
		Soil classes[a] and design lateral soil (psf per foot of depth)		
		GW, GP, SW, SP 30	GM, GC, SM, SM-SC and ML 45	SC, ML-CL and inorganic CL 60
8	4	NR	NR	NR
	5	NR	NR	NR
	6	NR	NR	6 @ 37
	7	NR	6 @ 36	6 @ 35
	8	6 @ 41	6 @ 35	6 @ 26
9	4	NR	NR	NR
	5	NR	NR	NR
	6	NR	NR	6 @ 35
	7	NR	6 @ 35	6 @ 32
	8	6 @ 36	6 @ 32	6 @ 23
	9	6 @ 35	6 @ 25	6 @ 18
10	4	NR	NR	NR
	5	NR	NR	NR
	6	NR	NR	6 @ 35
	7	NR	6 @ 35	6 @ 29
	8	6 @ 35	6 @ 29	6 @ 21
	9	6 @ 34	6 @ 22	6 @ 16
	10	6 @ 27	6 @ 17	6 @ 13

For SI: 1 inch = 25.4 mm, 1 foot = 304.8 mm, 1 pound per square foot per foot = 0.1571 kPa^2/m, 1 pound per square inch = 6.895 kPa.

NR = Not Required.

a. Soil classes are in accordance with the Unified Soil Classification System. Refer to Table R405.1.

b. Table values are based on reinforcing bars with a minimum yield strength of 60,000 psi, concrete with a minimum specified compressive strength of 2,500 psi and vertical reinforcement being located at the centerline of the wall. See Section R404.1.3.3.7.2.

c. Vertical reinforcement with a yield strength of less than 60,000 psi and bars of a different size than specified in the table are permitted in accordance with Section R404.1.3.3.7.6 and Table R404.1.2(9).

d. NR indicates vertical reinforcement is not required.

e. Deflection criterion is $L/240$, where L is the height of the basement wall in inches.

f. Interpolation is not permitted.

g. Where walls will retain 4 feet or more of unbalanced backfill, they shall be laterally supported at the top and bottom before backfilling.

h. See Section R404.1.3.2 for minimum reinforcement required for basement walls supporting above-grade concrete walls.

i. See Table R608.3 for tolerance from nominal thickness permitted for flat walls.

j. The use of this table shall be prohibited for soil classifications not shown.

TABLE R404.1.2(4)
MINIMUM VERTICAL REINFORCEMENT FOR 10-INCH NOMINAL FLAT CONCRETE BASEMENT WALLS[b, c, d, e, f, h, i, j]

MAXIMUM UNSUPPORTED WALL HEIGHT (feet)	MAXIMUM UNBALANCED BACKFILL HEIGHT[g] (feet)	MINIMUM VERTICAL REINFORCEMENT-BAR SIZE AND SPACING (inches)		
		Soil classes[a] and design lateral soil (psf per foot of depth)		
		GW, GP, SW, SP 30	GM, GC, SM, SM-SC and ML 45	SC, ML-CL and inorganic CL 60
8	4	NR	NR	NR
	5	NR	NR	NR
	6	NR	NR	NR
	7	NR	NR	NR
	8	6 @ 48	6 @ 35	6 @ 28
9	4	NR	NR	NR
	5	NR	NR	NR
	6	NR	NR	NR
	7	NR	NR	6 @ 31
	8	NR	6 @ 31	6 @ 28
	9	6 @ 37	6 @ 28	6 @ 24
10	4	NR	NR	NR
	5	NR	NR	NR
	6	NR	NR	NR
	7	NR	NR	6 @ 28
	8	NR	6 @ 28	6 @ 28
	9	6 @ 33	6 @ 28	6 @ 21
	10	6 @ 28	6 @ 23	6 @ 17

For SI: 1 inch = 25.4 mm, 1 foot = 304.8 mm, 1 pound per square foot per foot = 0.1571 kPa^2/m, 1 pound per square inch = 6.895 kPa.

NR = Not Required.

a. Soil classes are in accordance with the Unified Soil Classification System. Refer to Table R405.1.

b. Table values are based on reinforcing bars with a minimum yield strength of 60,000 psi concrete with a minimum specified compressive strength of 2,500 psi and vertical reinforcement being located at the centerline of the wall. See Section R404.1.3.3.7.2.

c. Vertical reinforcement with a yield strength of less than 60,000 psi and bars of a different size than specified in the table are permitted in accordance with Section R404.1.3.3.7.6 and Table R404.1.2(9).

d. NR indicates vertical reinforcement is not required.

e. Deflection criterion is $L/240$, where L is the height of the basement wall in inches.

f. Interpolation is not permitted.

g. Where walls will retain 4 feet or more of unbalanced backfill, they shall be laterally supported at the top and bottom before backfilling.

h. See Section R404.1.3.2 for minimum reinforcement required for basement walls supporting above-grade concrete walls.

i. See Table R608.3 for tolerance from nominal thickness permitted for flat walls.

j. The use of this table shall be prohibited for soil classifications not shown.

TABLE R404.1.2(5)
MINIMUM VERTICAL WALL REINFORCEMENT FOR 6-INCH WAFFLE-GRID BASEMENT WALLS[b, c, d, e, g, h, i, j]

MAXIMUM UNSUPPORTED WALL HEIGHT (feet)	MAXIMUM UNBALANCED BACKFILL HEIGHT[f] (feet)	MINIMUM VERTICAL REINFORCEMENT-BAR SIZE AND SPACING (inches)		
		Soil classes[a] and design lateral soil (psf per foot of depth)		
		GW, GP, SW, SP 30	GM, GC, SM, SM-SC and ML 45	SC, ML-CL and inorganic CL 60
8	4	4 @ 48	4 @ 46	6 @ 39
	5	4 @ 45	5 @ 46	6 @ 47
	6	5 @ 45	6 @ 40	DR
	7	6 @ 44	DR	DR
	8	6 @ 32	DR	DR
9	4	4 @ 48	4 @ 46	4 @ 37
	5	4 @ 42	5 @ 43	6 @ 44
	6	5 @ 41	6 @ 37	DR
	7	6 @ 39	DR	DR
	> 8	DR[i]	DR	DR
10	4	4 @ 48	4 @ 46	4 @ 35
	5	4 @ 40	5 @ 40	6 @ 41
	6	5 @ 38	6 @ 34	DR
	7	6 @ 36	DR	DR
	> 8	DR	DR	DR

For SI: 1 inch = 25.4 mm, 1 foot = 304.8 mm, 1 pound per square foot per foot = 0.1571 kPa^2/m, 1 pound per square inch = 6.895 kPa.

DR = Design Required.

a. Soil classes are in accordance with the Unified Soil Classification System. Refer to Table R405.1.

b. Table values are based on reinforcing bars with a minimum yield strength of 60,000 psi concrete with a minimum specified compressive strength of 2,500 psi and vertical reinforcement being located at the centerline of the wall. See Section R404.1.3.3.7.2.

c. Maximum spacings shown are the values calculated for the specified bar size. Where the bar used is Grade 60 and the size specified in the table, the actual spacing in the wall shall not exceed a whole-number multiple of 12 inches (12, 24, 36 and 48) that is less than or equal to the tabulated spacing. Vertical reinforcement with a yield strength of less than 60,000 psi and bars of a different size than specified in the table are permitted in accordance with Section R404.1.3.3.7.6 and Table R404.1.2(9).

d. Deflection criterion is $L/240$, where L is the height of the basement wall in inches.

e. Interpolation is not permitted.

f. Where walls will retain 4 feet or more of unbalanced backfill, they shall be laterally supported at the top and bottom before backfilling.

g. See Section R404.1.3.2 for minimum reinforcement required for basement walls supporting above-grade concrete walls.

h. See Table R608.3 for thicknesses and dimensions of waffle-grid walls.

i. DR means design is required in accordance with the applicable building code, or in the absence of a code, in accordance with ACI 318.

j. The use of this table shall be prohibited for soil classifications not shown.

TABLE R404.1.2(6)
MINIMUM VERTICAL REINFORCEMENT FOR 8-INCH WAFFLE-GRID BASEMENT WALLS[b, c, d, e, f, h, i, j, k]

MAXIMUM UNSUPPORTED WALL HEIGHT (feet)	MAXIMUM UNBALANCED BACKFILL HEIGHT[g] (feet)	MINIMUM VERTICAL REINFORCEMENT-BAR SIZE AND SPACING (inches)		
		Soil classes[a] and design lateral soil (psf per foot of depth)		
		GW, GP, SW, SP 30	GM, GC, SM, SM-SC and ML 45	SC, ML-CL and inorganic CL 60
8	4	NR	NR	NR
	5	NR	5 @ 48	5 @ 46
	6	5 @ 48	5 @ 43	6 @ 45
	7	5 @ 46	6 @ 43	6 @ 31
	8	6 @ 48	6 @ 32	6 @ 23
9	4	NR	NR	NR
	5	NR	5 @ 47	5 @ 46
	6	5 @ 46	5 @ 39	6 @ 41
	7	5 @ 42	6 @ 38	6 @ 28
	8	6 @ 44	6 @ 28	6 @ 20
	9	6 @ 34	6 @ 21	DR
10	4	NR	NR	NR
	5	NR	5 @ 46	5 @ 44
	6	5 @ 46	5 @ 37	6 @ 38
	7	5 @ 38	6 @ 35	6 @ 25
	8	6 @ 39	6 @ 25	DR
	9	6 @ 30	DR	DR
	10	6 @ 24	DR	DR

For SI: 1 inch = 25.4 mm, 1 foot = 304.8 mm, 1 pound per square foot per foot = 0.1571 kPa^2/m, 1 pound per square inch = 6.895 kPa.

NR = Not Required.

DR = Design Required.

a. Soil classes are in accordance with the Unified Soil Classification System. Refer to Table R405.1.

b. Table values are based on reinforcing bars with a minimum yield strength of 60,000 psi concrete with a minimum specified compressive strength of 2,500 psi and vertical reinforcement being located at the centerline of the wall. See Section R404.1.3.3.7.2.

c. Maximum spacings shown are the values calculated for the specified bar size. Where the bar used is Grade 60 (420 MPa) and the size specified in the table, the actual spacing in the wall shall not exceed a whole-number multiple of 12 inches (12, 24, 36 and 48) that is less than or equal to the tabulated spacing. Vertical reinforcement with a yield strength of less than 60,000 psi and bars of a different size than specified in the table are permitted in accordance with Section R404.1.3.3.7.6 and Table R404.1.2(9).

d. NR indicates vertical reinforcement is not required.

e. Deflection criterion is $L/240$, where L is the height of the basement wall in inches.

f. Interpolation shall not be permitted.

g. Where walls will retain 4 feet or more of unbalanced backfill, they shall be laterally supported at the top and bottom before backfilling.

h. See Section R404.1.3.2 for minimum reinforcement required for basement walls supporting above-grade concrete walls.

i. See Table R608.3 for thicknesses and dimensions of waffle-grid walls.

j. DR means design is required in accordance with the applicable building code, or in the absence of a code, in accordance with ACI 318.

k. The use of this table shall be prohibited for soil classifications not shown.

TABLE R404.1.2(7) MINIMUM VERTICAL REINFORCEMENT FOR 6-INCH (152 mm) SCREEN-GRID BASEMENT WALLS[b, c, d, e, g, h, i, j]

MAXIMUM UNSUPPORTED WALL HEIGHT (feet)	MAXIMUM UNBALANCED BACKFILL HEIGHT[f] (feet)	MINIMUM VERTICAL REINFORCEMENT-BAR SIZE AND SPACING (inches)		
		Soil classes[a] and design lateral soil (psf per foot of depth)		
		GW, GP, SW, SP 30	GM, GC, SM, SM-SC and ML 45	SC, ML-CL and inorganic CL 60
8	4	4 @ 48	4 @ 48	5 @ 43
	5	4 @ 48	5 @ 48	5 @ 37
	6	5 @ 48	6 @ 45	6 @ 32
	7	6 @ 48	DR	DR
	8	6 @ 36	DR	DR
9	4	4 @ 48	4 @ 48	4 @ 41
	5	4 @ 48	5 @ 48	6 @ 48
	6	5 @ 45	6 @ 41	DR
	7	6 @ 43	DR	DR
	> 8	DR	DR	DR
10	4	4 @ 48	4 @ 48	4 @ 39
	5	4 @ 44	5 @ 44	6 @ 46
	6	5 @ 42	6 @ 38	DR
	7	6 @ 40	DR	DR
	> 8	DR	DR	DR

For SI: 1 inch = 25.4 mm, 1 foot = 304.8 mm, 1 pound per square foot per foot = 0.1571 kPa^2/m, 1 pound per square inch = 6.895 kPa.

DR = Design Required.

a. Soil classes are in accordance with the Unified Soil Classification System. Refer to Table R405.1.

b. Table values are based on reinforcing bars with a minimum yield strength of 60,000 psi, concrete with a minimum specified compressive strength of 2,500 psi and vertical reinforcement being located at the centerline of the wall. See Section R404.1.3.3.7.2.

c. Maximum spacings shown are the values calculated for the specified bar size. Where the bar used is Grade 60 and the size specified in the table, the actual spacing in the wall shall not exceed a whole-number multiple of 12 inches (12, 24, 36 and 48) that is less than or equal to the tabulated spacing. Vertical reinforcement with a yield strength of less than 60,000 psi and bars of a different size than specified in the table are permitted in accordance with Section R404.1.3.3.7.6 and Table R404.1.2(9).

d. Deflection criterion is $L/240$, where L is the height of the basement wall in inches.

e. Interpolation is not permitted.

f. Where walls will retain 4 feet or more of unbalanced backfill, they shall be laterally supported at the top and bottom before backfilling.

g. See Sections R404.1.3.2 for minimum reinforcement required for basement walls supporting above-grade concrete walls.

h. See Table R608.3 for thicknesses and dimensions of screen-grid walls.

i. DR means design is required in accordance with the applicable building code, or in the absence of a code, in accordance with ACI 318.

j. The use of this table shall be prohibited for soil classifications not shown.

TABLE R404.1.2(8)
MINIMUM VERTICAL REINFORCEMENT FOR 6-, 8-, 10- AND 12-INCH NOMINAL FLAT BASEMENT WALLS[b, c, d, e, f, h, i, k, n, o]

MAXIMUM UNSUPPORTED WALL HEIGHT (feet)	MAXIMUM UNBALANCED BACKFILL HEIGHT[g] (feet)	MINIMUM VERTICAL REINFORCEMENT-BAR SIZE AND SPACING (inches)											
		Soil classes[a] and design lateral soil (psf per foot of depth)											
		GW, GP, SW, SP 30				GM, GC, SM, SM-SC and ML 45				SC, ML-CL and inorganic CL 60			
		Minimum nominal wall thickness (inches)											
		6	8	10	12	6	8	10	12	6	8	10	12
5	4	NR	NR	NR	NR	NR	NR	NR	NR	NR	NR	NR	NR
	5	NR	NR	NR	NR	NR	NR	NR	NR	NR	NR	NR	NR
6	4	NR	NR	NR	NR	NR	NR	NR	NR	NR	NR	NR	NR
	5	NR	NR	NR	NR	NR	NR[l]	NR	NR	4 @ 35	NR[l]	NR	NR
	6	NR	NR	NR	NR	5 @ 48	NR	NR	NR	5 @ 36	NR	NR	NR
7	4	NR	NR	NR	NR	NR	NR	NR	NR	NR	NR	NR	NR
	5	NR	NR	NR	NR	NR	NR	NR	NR	5 @ 47	NR	NR	NR
	6	NR	NR	NR	NR	5 @ 42	NR	NR	NR	6 @ 43	5 @ 48	NR[l]	NR
	7	5 @ 46	NR	NR	NR	6 @ 42	5 @ 46	NR[l]	NR	6 @ 34	6 @ 48	NR	NR
8	4	NR	NR	NR	NR	NR	NR	NR	NR	NR	NR	NR	NR
	5	NR	NR	NR	NR	4 @ 38	NR[l]	NR	NR	5 @ 43	NR	NR	NR
	6	4 @ 37	NR[l]	NR	NR	5 @ 37	NR	NR	NR	6 @ 37	5 @ 43	NR[l]	NR
	7	5 @ 40	NR	NR	NR	6 @ 37	5 @ 41	NR[l]	NR	6 @ 34	6 @ 43	NR	NR
	8	6 @ 43	5 @ 47	NR[l]	NR	6 @ 34	6 @ 43	NR	NR	6 @ 27	6 @ 32	6 @ 44	NR
9	4	NR	NR	NR	NR	NR	NR	NR	NR	NR	NR	NR	NR
	5	NR	NR	NR	NR	4 @ 35	NR[l]	NR	NR	5 @ 40	NR	NR	NR
	6	4 @ 34	NR[l]	NR	NR	6 @ 48	NR	NR	NR	6 @ 36	6 @ 39	NR[l]	NR
	7	5 @ 36	NR	NR	NR	6 @ 34	5 @ 37	NR	NR	6 @ 33	6 @ 38	5 @ 37	NR[l]
	8	6 @ 38	5 @ 41	NR[l]	NR	6 @ 33	6 @ 38	5 @ 37	NR[l]	6 @ 24	6 @ 29	6 @ 39	4 @ 48[m]
	9	6 @ 34	6 @ 46	NR	NR	6 @ 26	6 @ 30	6 @ 41	NR	6 @ 19	6 @ 23	6 @ 30	6 @ 39
10	4	NR	NR	NR	NR	NR	NR	NR	NR	NR	NR	NR	NR
	5	NR	NR	NR	NR	4 @ 33	NR[l]	NR	NR	5 @ 38	NR	NR	NR
	6	5 @ 48	NR[l]	NR	NR	6 @ 45	NR	NR	NR	6 @ 34	5 @ 37	NR	NR
	7	6 @ 47	NR	NR	NR	6 @ 34	6 @ 48	NR	NR	6 @ 30	6 @ 35	6 @ 48	NR[l]
	8	6 @ 34	5 @ 38	NR	NR	6 @ 30	6 @ 34	6 @ 47	NR[l]	6 @ 22	6 @ 26	6 @ 35	6 @ 45[m]
	9	6 @ 34	6 @ 41	4 @ 48	NR[l]	6 @ 23	6 @ 27	6 @ 35	4 @ 48[m]	DR	6 @ 22	6 @ 27	6 @ 34
	10	6 @ 28	6 @ 33	6 @ 45	NR	DR[j]	6 @ 23	6 @ 29	6 @ 38	DR	6 @ 22	6 @ 22	6 @ 28

For SI: 1 inch = 25.4 mm, 1 foot = 304.8 mm, 1 pound per square foot per foot = 0.1571 kPa^2/m, 1 pound per square inch = 6.895 kPa.

NR = Not Required.

DR = Design Required.

a. Soil classes are in accordance with the Unified Soil Classification System. Refer to Table R405.1.

b. Table values are based on reinforcing bars with a minimum yield strength of 60,000 psi.

c. Vertical reinforcement with a yield strength of less than 60,000 psi and bars of a different size than specified in the table are permitted in accordance with Section R404.1.3.3.7.6 and Table R404.1.2(9).

d. NR indicates vertical wall reinforcement is not required, except for 6-inch nominal walls formed with stay-in-place forming systems in which case vertical reinforcement shall be No. 4@48 inches on center.

e. Allowable deflection criterion is $L/240$, where L is the unsupported height of the basement wall in inches.

f. Interpolation is not permitted.

g. Where walls will retain 4 feet or more of unbalanced backfill, they shall be laterally supported at the top and bottom before backfilling.

h. Vertical reinforcement shall be located to provide a cover of $1^1/_4$ inches measured from the inside face of the wall. The center of the steel shall not vary from the specified location by more than the greater of 10 percent of the wall thickness or $^3/_8$ inch.

i. Concrete cover for reinforcement measured from the inside face of the wall shall be not less than $^3/_4$ inch. Concrete cover for reinforcement measured from the outside face of the wall shall be not less than $1^1/_2$ inches for No. 5 bars and smaller, and not less than 2 inches for larger bars.

j. DR means design is required in accordance with the applicable building code, or in the absence of a code, in accordance with ACI 318.

k. Concrete shall have a specified compressive strength, f'_c, of not less than 2,500 psi at 28 days, unless a higher strength is required by Note l or m.

l. The minimum thickness is permitted to be reduced 2 inches, provided that the minimum specified compressive strength of concrete, f'_c, is 4,000 psi.

m. A plain concrete wall with a minimum nominal thickness of 12 inches is permitted, provided that the minimum specified compressive strength of concrete, f'_c, is 3,500 psi.

n. See Table R608.3 for tolerance from nominal thickness permitted for flat walls.

o. The use of this table shall be prohibited for soil classifications not shown.

TABLE R404.1.2(9)
MINIMUM SPACING FOR ALTERNATE BAR SIZE AND ALTERNATE GRADE OF STEEL[a, b, c]

BAR SPACING FROM APPLICABLE TABLE IN SECTION R404.1.3.2 (inches)	BAR SIZE FROM APPLICABLE TABLE IN SECTION R404.1.3.2														
	#4					#5					#6				
	Alternate bar size and alternate grade of steel desired														
	Grade 60		Grade 40			Grade 60		Grade 40			Grade 60		Grade 40		
	#5	#6	#4	#5	#6	#4	#6	#4	#5	#6	#4	#5	#4	#5	#6
	Maximum spacing for alternate bar size and alternate grade of steel (inches)														
8	12	18	5	8	12	5	11	3	5	8	4	6	2	4	5
9	14	20	6	9	13	6	13	4	6	9	4	6	3	4	6
10	16	22	7	10	15	6	14	4	7	9	5	7	3	5	7
11	17	24	7	11	16	7	16	5	7	10	5	8	3	5	7
12	19	26	8	12	18	8	17	5	8	11	5	8	4	6	8
13	20	29	9	13	19	8	18	6	9	12	6	9	4	6	9
14	22	31	9	14	21	9	20	6	9	13	6	10	4	7	9
15	23	33	10	16	22	10	21	6	10	14	7	11	5	7	10
16	25	35	11	17	23	10	23	7	11	15	7	11	5	8	11
17	26	37	11	18	25	11	24	7	11	16	8	12	5	8	11
18	28	40	12	19	26	12	26	8	12	17	8	13	5	8	12
19	29	42	13	20	28	12	27	8	13	18	9	13	6	9	13
20	31	44	13	21	29	13	28	9	13	19	9	14	6	9	13
21	33	46	14	22	31	14	30	9	14	20	10	15	6	10	14
22	34	48	15	23	32	14	31	9	15	21	10	16	7	10	15
23	36	48	15	24	34	15	33	10	15	22	10	16	7	11	15
24	37	48	16	25	35	15	34	10	16	23	11	17	7	11	16
25	39	48	17	26	37	16	35	11	17	24	11	18	8	12	17
26	40	48	17	27	38	17	37	11	17	25	12	18	8	12	17
27	42	48	18	28	40	17	38	12	18	26	12	19	8	13	18
28	43	48	19	29	41	18	40	12	19	26	13	20	8	13	19
29	45	48	19	30	43	19	41	12	19	27	13	20	9	14	19
30	47	48	20	31	44	19	43	13	20	28	14	21	9	14	20
31	48	48	21	32	45	20	44	13	21	29	14	22	9	15	21
32	48	48	21	33	47	21	45	14	21	30	15	23	10	15	21
33	48	48	22	34	48	21	47	14	22	31	15	23	10	16	22
34	48	48	23	35	48	22	48	15	23	32	15	24	10	16	23
35	48	48	23	36	48	23	48	15	23	33	16	25	11	16	23
36	48	48	24	37	48	23	48	15	24	34	16	25	11	17	24
37	48	48	25	38	48	24	48	16	25	35	17	26	11	17	25
38	48	48	25	39	48	25	48	16	25	36	17	27	12	18	25
39	48	48	26	40	48	25	48	17	26	37	18	27	12	18	26
40	48	48	27	41	48	26	48	17	27	38	18	28	12	19	27
41	48	48	27	42	48	26	48	18	27	39	19	29	12	19	27
42	48	48	28	43	48	27	48	18	28	40	19	30	13	20	28
43	48	48	29	44	48	28	48	18	29	41	20	30	13	20	29
44	48	48	29	45	48	28	48	19	29	42	20	31	13	21	29
45	48	48	30	47	48	29	48	19	30	43	20	32	14	21	30
46	48	48	31	48	48	30	48	20	31	44	21	32	14	22	31
47	48	48	31	48	48	30	48	20	31	44	21	33	14	22	31
48	48	48	32	48	48	31	48	21	32	45	22	34	15	23	32

For SI: 1 inch = 25.4 mm.

a. This table is for use with tables in Section R404.1.3.2 that specify the minimum bar size and maximum spacing of vertical wall reinforcement for foundation walls and above-grade walls. Reinforcement specified in tables in Section R404.1.3.2 is based on Grade 60 steel reinforcement.

b. Bar spacing shall not exceed 48 inches on center and shall be not less than one-half the nominal wall thickness.

c. For Grade 50 steel bars (ASTM A996, Type R), use spacing for Grade 40 bars or interpolate between Grades 40 and 60.

R404.1.2.1 Masonry foundation walls. *Concrete masonry* and *clay masonry* foundation walls shall be constructed as set forth in Table R404.1.1(1), R404.1.1(2), R404.1.1(3) or R404.1.1(4) and shall comply with applicable provisions of Section R606. In buildings assigned to *Seismic Design Categories* D_0, D_1 and D_2, *concrete masonry* and *clay masonry* foundation walls shall also comply with Section R404.1.4.1. Rubble stone masonry foundation walls shall be constructed in accordance with Sections R404.1.8 and R606.4.2. Rubble stone masonry walls shall not be used in *Seismic Design Categories* D_0, D_1 and D_2, or in *townhouses* in Seismic Design Category C.

R404.1.3 Concrete foundation walls. Concrete foundation walls that support light-frame walls shall be designed and constructed in accordance with the provisions of this section, ACI 318, ACI 332 or PCA 100. Concrete foundation walls that support above-grade concrete walls that are within the applicability limits of Section R608.2 shall be designed and constructed in accordance with the provisions of this section, ACI 318, ACI 332 or PCA 100. Concrete foundation walls that support above-grade concrete walls that are not within the applicability limits of Section R608.2 shall be designed and constructed in accordance with the provisions of ACI 318, ACI 332 or PCA 100. Where ACI 318, ACI 332, PCA 100 or the provisions of this section are used to design concrete foundation walls, project drawings, typical details and specifications are not required to bear the seal of the architect or engineer responsible for design, unless otherwise required by the state law of the *jurisdiction* having authority.

R404.1.3.1 Concrete cross section. Concrete walls constructed in accordance with this code shall comply with the shapes and minimum concrete cross-sectional dimensions required by Table R608.3. Other types of forming systems resulting in concrete walls not in compliance with this section and Table R608.3 shall be designed in accordance with ACI 318.

R404.1.3.2 Reinforcement for foundation walls. Concrete foundation walls shall be laterally supported at the top and bottom. Horizontal reinforcement shall be provided in accordance with Table R404.1.2(1). Vertical reinforcement shall be provided in accordance with Table R404.1.2(2), R404.1.2(3), R404.1.2(4), R404.1.2(5), R404.1.2(6), R404.1.2(7) or R404.1.2(8). Vertical reinforcement for flat *basement* walls retaining 4 feet (1219 mm) or more of unbalanced backfill is permitted to be determined in accordance with Table R404.1.2(9). For *basement* walls supporting above-grade concrete walls, vertical reinforcement shall be the greater of that required by Tables R404.1.2(2) through R404.1.2(8) or by Section R608.6 for the above-grade wall. In buildings assigned to Seismic Design Category D_0, D_1 or D_2, concrete foundation walls shall also comply with Section R404.1.4.2.

R404.1.3.2.1 Concrete foundation stem walls supporting above-grade concrete walls. Foundation stem walls that support above-grade concrete walls shall be designed and constructed in accordance with this section.

1. Stem walls not laterally supported at top. Concrete stem walls that are not monolithic with slabs-on-ground or are not otherwise laterally supported by slabs-on-ground shall comply with this section. Where unbalanced backfill retained by the stem wall is less than or equal to 18 inches (457 mm), the stem wall and above-grade wall it supports shall be provided with vertical reinforcement in accordance with Section R608.6 and Table R608.6(1), R608.6(2) or R608.6(3) for above-grade walls. Where unbalanced backfill retained by the stem wall is greater than 18 inches (457 mm), the stem wall and above-grade wall it supports shall be provided with vertical reinforcement in accordance with Section R608.6 and Table R608.6(1).
2. Stem walls laterally supported at top. Concrete stem walls that are monolithic with slabs-on-ground or are otherwise laterally supported by slabs-on-ground shall be vertically reinforced in accordance with Section R608.6 and Table R608.6(1), R608.6(2) or R608.6(3) for above-grade walls. Where the unbalanced backfill retained by the stem wall is greater than 18 inches (457 mm), the connection between the stem wall and the slab-on-ground, and the portion of the slab-on-ground providing lateral support for the wall shall be designed in accordance with PCA 100 or with accepted engineering practice. Where the unbalanced backfill retained by the stem wall is greater than 18 inches (457 mm), the minimum nominal thickness of the wall shall be 6 inches (152 mm).

R404.1.3.2.2 Concrete foundation stem walls supporting light-frame above-grade walls. Concrete foundation stem walls that support light-frame above-grade walls shall be designed and constructed in accordance with this section.

1. Stem walls not laterally supported at top. Concrete stem walls that are not monolithic with slabs-on-ground or are not otherwise laterally supported by slabs-on-ground and retain 48 inches (1219 mm) or less of unbalanced fill, measured from the top of the wall, shall be constructed in accordance with Section R404.1.3. Foundation stem walls that retain more than 48 inches (1219 mm) of unbalanced fill, measured from the top of the wall, shall be designed in accordance with Sections R404.1.1 and R404.4.

2. Stem walls laterally supported at top. Concrete stem walls that are monolithic with slabs-on-ground or are otherwise laterally supported by slabs-on-ground shall be constructed in accordance with Section R404.1.3. Where the unbalanced backfill retained by the stem wall is greater than 48 inches (1219 mm), the connection between the stem wall and the slab-on-ground, and the portion of the slab-on-ground providing lateral support for the wall, shall be designed in accordance with PCA 100 or in accordance with accepted engineering practice.

R404.1.3.3 Concrete, materials for concrete, and forms. Materials used in concrete, the concrete itself and forms shall conform to requirements of this section or ACI 318.

R404.1.3.3.1 Compressive strength. The minimum specified compressive strength of concrete, f'_c, shall comply with Section R402.2 and shall be not less than 2,500 psi (17.2 MPa) at 28 days in buildings assigned to Seismic Design Category A, B or C and 3,000 psi (20.5 MPa) in buildings assigned to Seismic Design Category D_0, D_1 or D_2.

R404.1.3.3.2 Concrete mixing and delivery. Mixing and delivery of concrete shall comply with ASTM C94 or ASTM C685.

R404.1.3.3.3 Maximum aggregate size. The nominal maximum size of coarse aggregate shall not exceed one-fifth the narrowest distance between sides of forms, or three-fourths the clear spacing between reinforcing bars or between a bar and the side of the form.

Exception: Where *approved*, these limitations shall not apply where removable forms are used and workability and methods of consolidation permit concrete to be placed without honeycombs or voids.

R404.1.3.3.4 Proportioning and slump of concrete. Proportions of materials for concrete shall be established to provide workability and consistency to permit concrete to be worked readily into forms and around reinforcement under conditions of placement to be employed, without segregation or excessive bleeding. Slump of concrete placed in removable forms shall not exceed 6 inches (152 mm).

Exception: Where *approved*, the slump is permitted to exceed 6 inches (152 mm) for concrete mixtures that are resistant to segregation, and are in accordance with the form manufacturer's recommendations.

Slump of concrete placed in stay-in-place forms shall exceed 6 inches (152 mm). Slump of concrete shall be determined in accordance with ASTM C143.

R404.1.3.3.5 Consolidation of concrete. Concrete shall be consolidated by suitable means during placement and shall be worked around embedded items and reinforcement and into corners of forms. Where stay-in-place forms are used, concrete shall be consolidated by internal vibration.

Exception: Where *approved* for concrete to be placed in stay-in-place forms, self-consolidating concrete mixtures with slumps equal to or greater than 8 inches (203 mm) that are specifically designed for placement without internal vibration need not be internally vibrated.

R404.1.3.3.6 Form materials and form ties. Forms shall be made of wood, steel, aluminum, plastic, a composite of cement and foam insulation, a composite of cement and wood chips, or other *approved* material suitable for supporting and containing concrete. Forms shall be accurately positioned and secured before placing concrete and shall provide sufficient strength to contain concrete during the concrete placement operation.

Form ties shall be steel, solid plastic, foam plastic, a composite of cement and wood chips, a composite of cement and foam plastic, or other suitable material capable of resisting the forces created by fluid pressure of fresh concrete.

R404.1.3.3.6.1 Stay-in-place forms. Stay-in-place concrete forms shall comply with this section.

1. Surface burning characteristics. The flame-spread index and *smoke-developed index* of forming material, other than foam plastic, left exposed on the interior shall comply with Section R302. The surface burning characteristics of foam plastic used in *insulating concrete forms* shall comply with Section R316.3.
2. Interior covering. Stay-in-place forms constructed of rigid foam plastic shall be protected on the interior of the building as required by Section R316. Where gypsum board is used to protect the foam plastic, it shall be installed with a mechanical fastening system. Use of adhesives in addition to mechanical fasteners is permitted.
3. Exterior wall covering. Stay-in-place forms constructed of rigid foam plastics shall be protected from sunlight and physical damage by the application of an *approved* exterior wall covering complying with this code. Exterior surfaces of other stay-in-place forming systems shall be protected in accordance with this code.
4. Termite protection. In areas where the probability of termite infestation is "very heavy" as indicated by Table R301.2 or

Figure R301.2.1, foam plastic insulation shall be permitted below grade on foundation walls in accordance with Section R318.4.

5. Flat ICF wall system forms shall conform to ASTM E2634.

R404.1.3.3.7 Reinforcement.

R404.1.3.3.7.1 Steel reinforcement. Steel reinforcement shall comply with the requirements of ASTM A615, A706M or A996. ASTM A996 bars produced from rail steel shall be Type R. In buildings assigned to Seismic Design Category A, B or C, the minimum yield strength of reinforcing steel shall be 40,000 psi (Grade 40) (276 MPa). In buildings assigned to Seismic Design Category D_0, D_1 or D_2, the minimum yield strength shall be 60,000 psi (Grade 60) (414 MPa).

R404.1.3.3.7.2 Location of reinforcement in wall. The center of vertical reinforcement in *basement* walls determined from Tables R404.1.2(2) through R404.1.2(7) shall be located at the centerline of the wall. Vertical reinforcement in *basement* walls determined from Table R404.1.2(8) shall be located to provide a maximum cover of $1^1/_4$ inches (32 mm) measured from the inside face of the wall. Regardless of the table used to determine vertical wall reinforcement, the center of the steel shall not vary from the specified location by more than the greater of 10 percent of the wall thickness and $^3/_8$ inch (10 mm). Horizontal and vertical reinforcement shall be located in foundation walls to provide the minimum cover required by Section R404.1.3.3.7.4.

R404.1.3.3.7.3 Wall openings. Vertical wall reinforcement required by Section R404.1.3.2 that is interrupted by wall openings shall have additional vertical reinforcement of the same size placed within 12 inches (305 mm) of each side of the opening.

R404.1.3.3.7.4 Support and cover. Reinforcement shall be secured in the proper location in the forms with tie wire or other bar support system to prevent displacement during the concrete placement operation. Steel reinforcement in concrete cast against the earth shall have a minimum cover of 3 inches (75 mm). Minimum cover for reinforcement in concrete cast in removable forms that will be exposed to the earth or weather shall be $1^1/_2$ inches (38 mm) for No. 5 bars and smaller, and 2 inches (50 mm) for No. 6 bars and larger. For concrete cast in removable forms that will not be exposed to the earth or weather, and for concrete cast in stay-in-place forms, minimum cover shall be $^3/_4$ inch (19 mm). The minus tolerance for cover shall not exceed the smaller of one-third the required cover or $^3/_8$ inch (10 mm).

R404.1.3.3.7.5 Lap splices. Vertical and horizontal wall reinforcement shall be the longest lengths practical. Where splices are necessary in reinforcement, the length of lap splice shall be in accordance with Table R608.5.4(1) and Figure R608.5.4(1). The maximum gap between noncontact parallel bars at a lap splice shall not exceed the smaller of one-fifth the required lap length and 6 inches (152 mm) [See Figure R608.5.4(1)].

R404.1.3.3.7.6 Alternate grade of reinforcement and spacing. Where tables in Section R404.1.3.2 specify vertical wall reinforcement based on minimum bar size and maximum spacing, which are based on Grade 60 (414 MPa) steel reinforcement, different size bars or bars made from a different grade of steel are permitted provided that an equivalent area of steel per linear foot of wall is provided. Use of Table R404.1.2(9) is permitted to determine the maximum bar spacing for different bar sizes than specified in the tables or bars made from a different grade of steel. Bars shall not be spaced less than one-half the wall thickness, or more than 48 inches (1219 mm) on center.

R404.1.3.3.7.7 Standard hooks. Where reinforcement is required by this code to terminate with a standard hook, the hook shall comply with Section R608.5.4.5 and Figure R608.5.4(3).

R404.1.3.3.7.8 Construction joint reinforcement. Construction joints in foundation walls shall be made and located to not impair the strength of the wall. Construction joints in plain concrete walls, including walls required to have not less than No. 4 bars at 48 inches (1219 mm) on center by Sections R404.1.3.2 and R404.1.4.2, shall be located at points of lateral support, and not fewer than one No. 4 bar shall extend across the construction joint at a spacing not to exceed 24 inches (610 mm) on center. Construction joint reinforcement shall have not less than 12 inches (305 mm) embedment on both sides of the joint. Construction joints in reinforced concrete walls shall be located in the middle third of the span between lateral supports, or located and constructed as required for joints in plain concrete walls.

Exception: Use of vertical wall reinforcement required by this code is permitted in lieu of construction joint reinforcement provided that the spacing does not exceed 24 inches (610 mm), or the combination of wall reinforcement and No. 4 bars described in this section does not exceed 24 inches (610 mm).

R404.1.3.3.8 Exterior wall coverings. Requirements for installation of masonry veneer, stucco and other wall coverings on the exterior of concrete walls and other construction details not covered in this section shall comply with the requirements of this code.

R404.1.3.4 Requirements for Seismic Design Category C. Concrete foundation walls supporting above-grade concrete walls in *townhouses* assigned to Seismic Design Category C shall comply with ACI 318, ACI 332 or PCA 100 (see Section R404.1.3).

R404.1.4 Seismic Design Category D_0, D_1 or D_2.

R404.1.4.1 Masonry foundation walls. In buildings assigned to Seismic Design Category D_0, D_1 or D_2, as established in Table R301.2, masonry foundation walls shall comply with this section. In addition to the requirements of Table R404.1.1(1), plain masonry foundation walls shall comply with the following:

1. Wall height shall not exceed 8 feet (2438 mm).
2. Unbalanced backfill height shall not exceed 4 feet (1219 mm).
3. Minimum nominal thickness for plain masonry foundation walls shall be 8 inches (203 mm).
4. Masonry stem walls shall have a minimum vertical reinforcement of one No. 4 (No. 13) bar located not greater than 4 feet (1219 mm) on center in grouted cells. Vertical reinforcement shall be tied to the horizontal reinforcement in the footings.

Foundation walls, supporting more than 4 feet (1219 mm) of unbalanced backfill or exceeding 8 feet (2438 mm) in height shall be constructed in accordance with Table R404.1.1(2), R404.1.1(3) or R404.1.1(4). Masonry foundation walls shall have two No. 4 (No. 13) horizontal bars located in the upper 12 inches (305 mm) of the wall.

R404.1.4.2 Concrete foundation walls. In buildings assigned to Seismic Design Category D_0, D_1 or D_2, as established in Table R301.2, concrete foundation walls that support light-frame walls shall comply with this section, and concrete foundation walls that support above-grade concrete walls shall comply with ACI 318, ACI 332 or PCA 100 (see Section R404.1.3). In addition to the horizontal reinforcement required by Table R404.1.2(1), plain concrete walls supporting light-frame walls shall comply with the following:

1. Wall height shall not exceed 8 feet (2438 mm).
2. Unbalanced backfill height shall not exceed 4 feet (1219 mm).
3. Minimum thickness for plain concrete foundation walls shall be 7.5 inches (191 mm) except that 6 inches (152 mm) is permitted where the maximum wall height is 4 feet, 6 inches (1372 mm).

Foundation walls less than 7.5 inches (191 mm) in thickness, supporting more than 4 feet (1219 mm) of unbalanced backfill or exceeding 8 feet (2438 mm) in height shall be provided with horizontal reinforcement in accordance with Table R404.1.2(1), and vertical reinforcement in accordance with Table R404.1.2(2), R404.1.2(3), R404.1.2(4), R404.1.2(5), R404.1.2(6), R404.1.2(7) or R404.1.2(8). Where Tables R404.1.2(2) through R404.1.2(8) permit plain concrete walls, not less than No. 4 (No. 13) vertical bars at a spacing not exceeding 48 inches (1219 mm) shall be provided.

R404.1.5 Foundation wall thickness based on walls supported. The thickness of masonry or concrete foundation walls shall be not less than that required by Section R404.1.5.1 or R404.1.5.2, respectively.

R404.1.5.1 Masonry wall thickness. Masonry foundation walls shall be not less than the thickness of the wall supported, except that masonry foundation walls of not less than 8-inch (203 mm) nominal thickness shall be permitted under brick veneered frame walls and under 10-inch-wide (254 mm) cavity walls where the total height of the wall supported, including gables, is not more than 20 feet (6096 mm), provided that the requirements of Section R404.1.1 are met.

R404.1.5.2 Concrete wall thickness. The thickness of concrete foundation walls shall be equal to or greater than the thickness of the wall in the story above. Concrete foundation walls with corbels, brackets or other projections built into the wall for support of masonry veneer or other purposes are not within the scope of the tables in this section.

Where a concrete foundation wall is reduced in thickness to provide a shelf for the support of masonry veneer, the reduced thickness shall be equal to or greater than the thickness of the wall in the story above. Vertical reinforcement for the foundation wall shall be based on Table R404.1.2(8) and located in the wall as required by Section R404.1.3.3.7.2 where that table is used. Vertical reinforcement shall be based on the thickness of the thinner portion of the wall.

Exception: Where the height of the reduced thickness portion measured to the underside of the floor assembly or sill plate above is less than or equal to 24 inches (610 mm) and the reduction in thickness does not exceed 4 inches (102 mm), the vertical reinforcement is permitted to be based on the thicker portion of the wall.

R404.1.5.3 Pier and curtain wall foundations. Use of pier and curtain wall foundations shall be permitted to support *light-frame construction* not more than two *stories* in height, provided that the following requirements are met:

1. All *load-bearing walls* shall be placed on continuous concrete footings placed integrally with the exterior wall footings.
2. The minimum actual thickness of a load-bearing masonry wall shall be not less than 4 inches (102 mm) nominal or $3^3/_8$ inches (92 mm) actual thickness, and shall be bonded integrally

with piers spaced in accordance with Section R606.6.4.

3. Piers shall be constructed in accordance with Sections R606.7 and R606.7.1, and shall be bonded into the load-bearing masonry wall in accordance with Section R606.13.1 or R606.13.1.1.

4. The maximum height of a 4-inch (102 mm) load-bearing masonry foundation wall supporting wood-frame walls and floors shall be not more than 4 feet (1219 mm).

5. Anchorage shall be in accordance with Section R403.1.6, Figure R404.1.5.3, or as specified by engineered design accepted by the *building official*.

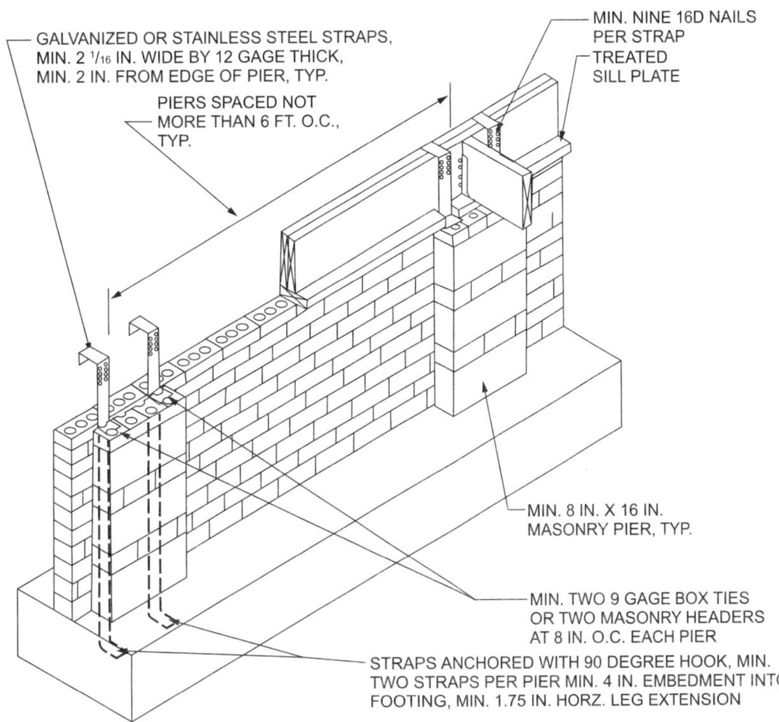

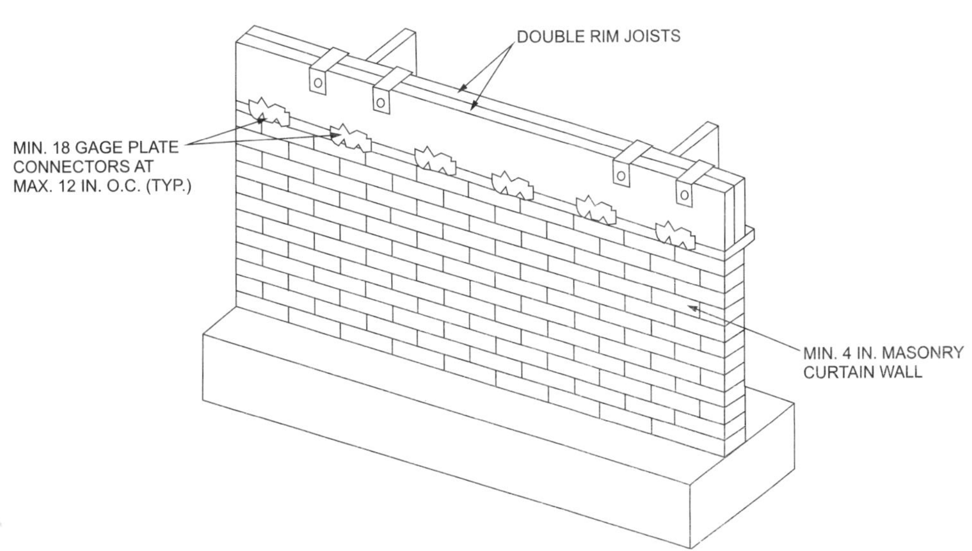

For SI: 1 inch = 25.4 mm, 1 foot = 304.8 mm, 1 degree = 0.0175 rad.

FIGURE R404.1.5.3
FOUNDATION WALL CLAY MASONRY CURTAIN WALL WITH CONCRETE MASONRY PIERS

6. The unbalanced fill for 4-inch (102 mm) foundation walls shall not exceed 24 inches (610 mm) for *solid masonry* or 12 inches (305 mm) for *hollow masonry.*
7. In *Seismic Design Categories* D_0, D_1 and D_2, prescriptive reinforcement shall be provided in the horizontal and vertical direction. Provide minimum horizontal joint reinforcement of two No. 9 gage wires spaced not less than 6 inches (152 mm) or one $^1/_4$-inch-diameter (6.4 mm) wire at 10 inches (254 mm) on center vertically. Provide minimum vertical reinforcement of one No. 4 bar at 48 inches (1220 mm) on center horizontally grouted in place.

R404.1.6 Height above finished grade. Concrete and masonry foundation walls shall extend above the finished *grade* adjacent to the foundation at all points not less than 4 inches (102 mm) where masonry veneer is used and not less than 6 inches (152 mm) elsewhere.

R404.1.7 Backfill placement. Backfill shall not be placed against the wall until the wall has sufficient strength and has been anchored to the floor above, or has been sufficiently braced to prevent damage by the backfill.

Exception: Bracing is not required for walls supporting less than 4 feet (1219 mm) of unbalanced backfill.

R404.1.8 Rubble stone masonry. Rubble stone masonry foundation walls shall have a minimum thickness of 16 inches (406 mm), shall not support an unbalanced backfill exceeding 8 feet (2438 mm) in height, shall not support a soil pressure greater than 30 pounds per square foot per foot (4.71 kPa/m), and shall not be constructed in *Seismic Design Categories* D_0, D_1, D_2 or *townhouses* in Seismic Design Category C, as established in Figure R301.2(2).

R404.1.9 Isolated masonry piers. Isolated masonry piers shall be constructed in accordance with this section and the general masonry construction requirements of Section R606. *Hollow masonry* piers shall have a minimum nominal thickness of 8 inches (203 mm), with a nominal height not exceeding four times the nominal thickness and a nominal length not exceeding three times the nominal thickness. Where *hollow masonry units* are solidly filled with concrete or grout, piers shall be permitted to have a nominal height not exceeding ten times the nominal thickness. Footings for isolated masonry piers shall be sized in accordance with Section R403.1.1.

R404.1.9.1 Pier cap. *Hollow masonry* piers shall be capped with 4 inches (102 mm) of *solid masonry* or concrete, a masonry cap block, or shall have cavities of the top course filled with concrete or grout. Where required, termite protection for the pier cap shall be provided in accordance with Section R318.

R404.1.9.2 Masonry piers supporting floor girders. Masonry piers supporting wood girders sized in accordance with Tables R602.7(1) and R602.7(2) shall be permitted in accordance with this section. Piers supporting girders for interior bearing walls shall have a minimum nominal dimension of 12 inches (305 mm) and a maximum height of 10 feet (3048 mm) from top of footing to bottom of sill plate or girder. Piers supporting girders for exterior bearing walls shall have a minimum nominal dimension of 12 inches (305 mm) and a maximum height of 4 feet (1220 mm) from top of footing to bottom of sill plate or girder. Girders and sill plates shall be anchored to the pier or footing in accordance with Section R403.1.6 or Figure R404.1.5.3. Floor girder bearing shall be in accordance with Section R502.6.

R404.1.9.3 Masonry piers supporting braced wall panels. Masonry piers supporting *braced wall panels* shall be designed in accordance with accepted engineering practice.

R404.1.9.4 Seismic design of masonry piers. Masonry piers in *dwellings* located in Seismic Design Category D_0, D_1 or D_2, and *townhouses* in Seismic Design Category C, shall be designed in accordance with accepted engineering practice.

R404.1.9.5 Masonry piers in flood hazard areas. Masonry piers for *dwellings* in flood hazard areas shall be designed in accordance with Section R322.

R404.2 Wood foundation walls. Wood foundation walls shall be constructed in accordance with the provisions of Sections R404.2.1 through R404.2.6 and with the details shown in Figures R403.1(2) and R403.1(3).

R404.2.1 Identification. Load-bearing lumber shall be identified by the grade *mark* of a lumber grading or inspection agency that has been *approved* by an accreditation body that complies with DOC PS 20. In lieu of a grade *mark*, a certificate of inspection issued by a lumber grading or inspection agency meeting the requirements of this section shall be accepted. *Wood structural panels* shall conform to DOC PS 1 or DOC PS 2 and shall be identified by a grade *mark* or certificate of inspection issued by an *approved agency.*

R404.2.2 Stud size. The studs used in foundation walls shall be 2-inch by 6-inch (51 mm by 152 mm) members. Where spaced 16 inches (406 mm) on center, a wood species with an F_b value of not less than 1,250 pounds per square inch (8619 kPa) as *listed* in ANSI AWC NDS shall be used. Where spaced 12 inches (305 mm) on center, an F_b of not less than 875 psi (6033 kPa) shall be required.

R404.2.3 Height of backfill. For wood foundations that are not designed and installed in accordance with AWC PWF, the height of backfill against a foundation wall shall not exceed 4 feet (1219 mm). Where the height of fill is more than 12 inches (305 mm) above the interior grade of a *crawl space* or floor of a *basement,* the thickness of the plywood sheathing shall meet the requirements of Table R404.2.3.

R404.2.4 Backfilling. Wood foundation walls shall not be backfilled until the *basement* floor and first floor have been constructed or the walls have been braced. For *crawl space* construction, backfill or bracing shall be installed on the interior of the walls prior to placing backfill on the exterior.

TABLE R404.2.3
PLYWOOD GRADE AND THICKNESS FOR WOOD FOUNDATION CONSTRUCTION (30 pcf equivalent-fluid weight soil pressure)

HEIGHT OF FILL (inches)	STUD SPACING (inches)	FACE GRAIN ACROSS STUDS			FACE GRAIN PARALLEL TO STUDS		
		Grade[a]	Minimum thickness (inches)	Span rating	Grade[a]	Minimum thickness (inches)[b, c]	Span rating
24	12	B	15/32	32/16	A	15/32	32/16
					B	15/32[c]	32/16
	16	B	15/32	32/16	A	15/32[c]	32/16
					B	19/32[c] (4, 5 ply)	40/20
36	12	B	15/32	32/16	A	15/32	32/16
					B	15/32[c] (4, 5 ply)	32/16
					B	19/32 (4, 5 ply)	40/20
	16	B	15/32[c]	32/16	A	19/32	40/20
					B	23/32	48/24
48	12	B	15/32	32/16	A	15/32[c]	32/16
					B	19/32[c] (4, 5 ply)	40/20
	16	B	19/32	40/20	A	19/32[c]	40/20
					A	23/32	48/24

For SI: 1 inch = 25.4 mm, 1 foot = 304.8 mm.

a. Plywood shall be of the following minimum grades in accordance with DOC PS 1 or DOC PS 2:
  1. DOC PS 1 Plywood grades marked:
    1.1. Structural I C-D (Exposure 1).
    1.2. C-D (Exposure 1).
  2. DOC PS 2 Plywood grades marked:
    2.1. Structural I Sheathing (Exposure 1).
    2.2. Sheathing (Exposure 1).
  3. Where a major portion of the wall is exposed above ground and a better appearance is desired, the following plywood grades marked exterior are suitable:
    3.1. Structural I A-C, Structural I B-C or Structural I C-C (Plugged) in accordance with DOC PS 1.
    3.2. A-C Group 1, B-C Group 1, C-C (Plugged) Group 1 or MDO Group 1 in accordance with DOC PS 1.
    3.3. Single Floor in accordance with DOC PS 1 or DOC PS 2.

b. Minimum thickness 15/32 inch, except crawl space sheathing shall have not less than 3/8 inch for face grain across studs 16 inches on center and maximum 2-foot depth of unequal fill.

c. For this fill height, thickness and grade combination, panels that are continuous over less than three spans (across less than three stud spacings) require blocking 16 inches above the bottom plate. Offset adjacent blocks and fasten through studs with two 16d corrosion-resistant nails at each end.

R404.2.5 Drainage and dampproofing. Wood foundation *basements* shall be drained and dampproofed in accordance with Sections R405 and R406, respectively.

R404.2.6 Fastening. *Wood structural panel* foundation wall sheathing shall be attached to framing in accordance with Table R602.3(1) and Section R402.1.1.

R404.3 Wood sill plates. Wood sill plates shall be not less than 2-inch by 4-inch (51 mm by 102 mm) nominal lumber. Sill plate anchorage shall be in accordance with Sections R403.1.6 and R602.11.

R404.4 Retaining walls. Retaining walls that are not laterally supported at the top and that retain in excess of 48 inches (1219 mm) of unbalanced fill, or retaining walls exceeding 24 inches (610 mm) in height that resist lateral loads in addition to soil, shall be designed in accordance with accepted engineering practice to ensure stability against overturning, sliding, excessive foundation pressure and water uplift. Retaining walls shall be designed for a safety factor of 1.5 against lateral sliding and overturning. This section shall not apply to foundation walls supporting buildings.

R404.5 Precast concrete foundation walls.

R404.5.1 Design. *Precast concrete* foundation walls shall be designed in accordance with accepted engineering practice. The design and manufacture of *precast concrete* foundation wall panels shall comply with the materials requirements of Section R402.3 or ACI 318. The panel design drawings shall be prepared by a *registered design professional* where required by the statutes of the *jurisdiction* in which the project is to be constructed in accordance with Section R106.1.

R404.5.2 Precast concrete foundation design drawings. *Precast concrete* foundation wall design drawings shall be submitted to the *building official* and *approved* prior to installation. Drawings shall include, at a minimum, the following information:

1. Design loading as applicable.
2. Footing design and material.

3. Concentrated loads and their points of application.
4. Soil bearing capacity.
5. Maximum allowable total uniform load.
6. Seismic design category.
7. Basic wind speed.

R404.5.3 Identification. *Precast concrete* foundation wall panels shall be identified by a certificate of inspection *label* issued by an *approved* third-party inspection agency.

SECTION R405 FOUNDATION DRAINAGE

R405.1 Concrete or masonry foundations. Drains shall be provided around concrete or masonry foundations that retain earth and enclose habitable or usable spaces located below *grade*. Drainage tiles, gravel or crushed stone drains, perforated pipe or other *approved* systems or materials shall be installed at or below the top of the footing or below the bottom of the slab and shall discharge by gravity or mechanical means into an *approved* drainage system. Gravel or crushed stone drains shall extend not less than 1 foot (305 mm) beyond the outside edge of the footing and 6 inches (152 mm) above the top of the footing and be covered with an *approved* filter membrane material. The top of open joints of drain tiles shall be protected with strips of building paper. Except where otherwise recommended by the drain manufacturer, perforated drains shall be surrounded with an *approved* filter membrane or the filter membrane shall cover the washed gravel or crushed rock covering the drain. Drainage tiles or perforated pipe shall be placed on not less than 2 inches (51 mm) of washed gravel or crushed rock not less than one sieve size larger than the tile joint opening or perforation and covered with not less than 6 inches (152 mm) of the same material.

Exception: A drainage system is not required where the foundation is installed on well-drained ground or sand-gravel mixture soils according to the Unified Soil Classification System, Group I soils, as detailed in Table R405.1.

R405.1.1 Precast concrete foundation. *Precast concrete* walls that retain earth and enclose habitable or useable space located below-*grade* that rest on crushed stone

TABLE R405.1
PROPERTIES OF SOILS CLASSIFIED ACCORDING TO THE UNIFIED SOIL CLASSIFICATION SYSTEM

SOIL GROUP	UNIFIED SOIL CLASSIFICATION SYSTEM SYMBOL	SOIL DESCRIPTION	DRAINAGE CHARACTERISTICS[a]	FROST HEAVE POTENTIAL	VOLUME CHANGE POTENTIAL EXPANSION[b]
Group I	GW	Well-graded gravels, gravel sand mixtures, little or no fines	Good	Low	Low
	GP	Poorly graded gravels or gravel sand mixtures, little or no fines	Good	Low	Low
	SW	Well-graded sands, gravelly sands, little or no fines	Good	Low	Low
	SP	Poorly graded sands or gravelly sands, little or no fines	Good	Low	Low
	GM	Silty gravels, gravel-sand-silt mixtures	Good	Medium	Low
	SM	Silty sand, sand-silt mixtures	Good	Medium	Low
Group II	GC	Clayey gravels, gravel-sand-clay mixtures	Medium	Medium	Low
	SC	Clayey sands, sand-clay mixture	Medium	Medium	Low
	ML	Inorganic silts and very fine sands, rock flour, silty or clayey fine sands or clayey silts with slight plasticity	Medium	High	Low
	CL	Inorganic clays of low to medium plasticity, gravelly clays, sandy clays, silty clays, lean clays	Medium	Medium	Medium to Low
Group III	CH	Inorganic clays of high plasticity, fat clays	Poor	Medium	High
	MH	Inorganic silts, micaceous or diatomaceous fine sandy or silty soils, elastic silts	Poor	High	High
Group IV	OL	Organic silts and organic silty clays of low plasticity	Poor	Medium	Medium
	OH	Organic clays of medium to high plasticity, organic silts	Unsatisfactory	Medium	High
	Pt	Peat and other highly organic soils	Unsatisfactory	Medium	High

For SI: 1 inch = 25.4 mm.

a. The percolation rate for good drainage is over 4 inches per hour, medium drainage is 2 inches to 4 inches per hour, and poor is less than 2 inches per hour.

b. Soils with a low potential expansion typically have a plasticity index (PI) of 0 to 15, soils with a medium potential expansion have a PI of 10 to 35 and soils with a high potential expansion have a PI greater than 20.

footings shall have a perforated drainage pipe installed below the base of the wall on either the interior or exterior side of the wall, not less than 1 foot (305 mm) beyond the edge of the wall. If the exterior drainage pipe is used, an *approved* filter membrane material shall cover the pipe. The drainage system shall discharge into an *approved* sewer system or to daylight.

R405.2 Wood foundations. Wood foundations enclosing habitable or usable spaces located below *grade* shall be adequately drained in accordance with Sections R405.2.1 through R405.2.3.

R405.2.1 Base. A porous layer of gravel, crushed stone or coarse sand shall be placed to a minimum thickness of 4 inches (102 mm) under the *basement* floor. Provision shall be made for automatic draining of this layer and the gravel or crushed stone wall footings.

R405.2.2 Vapor retarder. A 6-mil-thick (0.15 mm) polyethylene vapor retarder shall be applied over the porous layer with the *basement* floor constructed over the polyethylene.

R405.2.3 Drainage system. In other than Group I soils, a sump shall be provided to drain the porous layer and footings. The sump shall be not less than 24 inches (610 mm) in diameter or 20 inches square (0.0129 m^2), shall extend not less than 24 inches (610 mm) below the bottom of the *basement* floor and shall be capable of positive gravity or mechanical drainage to remove any accumulated water. The drainage system shall discharge into an *approved* sewer system or to daylight.

SECTION R406 FOUNDATION WATERPROOFING AND DAMPPROOFING

R406.1 Concrete and masonry foundation dampproofing. Except where required by Section R406.2 to be waterproofed, foundation walls that retain earth and enclose interior spaces and floors below *grade* shall be dampproofed from the finished grade to the higher of the top of the footing or 6 inches (152 mm) below the top of the basement floor. Masonry walls shall have not less than $^3/_8$-inch (9.5 mm) Portland cement parging applied to the exterior of the wall. The parging shall be dampproofed in accordance with one of the following:

1. Bituminous coating.
2. Three pounds per square yard (1.63 kg/m^2) of acrylic modified cement.
3. One-eighth-inch (3.2 mm) coat of surface-bonding cement complying with ASTM C887.
4. Any material permitted for waterproofing in Section R406.2.
5. Other *approved* methods or materials.

Exception: Parging of unit masonry walls is not required where a material is *approved* for direct application to the masonry.

Concrete walls shall be dampproofed by applying any one of the listed dampproofing materials or any one of the waterproofing materials listed in Section R406.2 to the exterior of the wall.

R406.2 Concrete and masonry foundation waterproofing. In areas where a high water table or other severe soil-water conditions are known to exist, exterior foundation walls that retain earth and enclose interior spaces and floors below *grade* shall be waterproofed from the finished *grade* to the higher of the top of the footing or 6 inches (152 mm) below the top of the basement floor. Walls shall be waterproofed in accordance with one of the following:

1. Two-ply hot-mopped felts.
2. Fifty-five-pound (25 kg) roll roofing.
3. Forty-mil (1 mm) polymer-modified asphalt.
4. Sixty-mil (1.5 mm) flexible polymer cement.
5. One-eighth-inch (3 mm) cement-based, fiber-reinforced, waterproof coating.
6. Sixty-mil (1.5 mm) solvent-free liquid-applied synthetic rubber.

All joints in membrane waterproofing shall be lapped and sealed with an adhesive compatible with the membrane.

Exception: Organic-solvent-based products such as hydrocarbons, chlorinated hydrocarbons, ketones and esters shall not be used for ICF walls with expanded polystyrene form material. Use of plastic roofing cements, acrylic coatings, latex coatings, mortars and pargings to seal ICF walls is permitted. Cold-setting asphalt or hot asphalt shall conform to Type C of ASTM D449. Hot asphalt shall be applied at a temperature of less than 200°F (93°C).

R406.3 Dampproofing for wood foundations. Wood foundations enclosing habitable or usable spaces located below *grade* shall be dampproofed in accordance with Sections R406.3.1 through R406.3.4.

R406.3.1 Panel joint sealed. Plywood panel joints in the foundation walls shall be sealed full length with a caulking compound capable of producing a moistureproof seal under the conditions of temperature and moisture content at which it will be applied and used.

R406.3.2 Below-grade moisture barrier. A 6-mil-thick (0.15 mm) polyethylene film shall be applied over the below-*grade* portion of exterior foundation walls prior to backfilling. Joints in the polyethylene film shall be lapped 6 inches (152 mm) and sealed with adhesive. The top edge of the polyethylene film shall be bonded to the sheathing to form a seal. Film areas at *grade* level shall be protected from mechanical damage and exposure by a pressure-preservative treated lumber or plywood strip attached to the wall several inches above finished *grade* level and extending approximately 9 inches (229 mm) below *grade*. The joint between the strip and the wall shall be caulked full length prior to fastening the strip to the wall. Where *approved*, other coverings appropriate to the architectural treatment shall be permitted to be used. The polyethylene film shall extend down to the bottom of the wood footing plate but shall not overlap or extend into the gravel or crushed stone footing.

R406.3.3 Porous fill. The space between the excavation and the foundation wall shall be backfilled with the same material used for footings, up to a height of 1 foot (305 mm) above the footing for well-drained sites, or one-half the total backfill height for poorly drained sites. The porous fill shall be covered with strips of 30-pound (13.6 kg) asphalt paper or 6-mil (0.15 mm) polyethylene to permit water seepage while avoiding infiltration of fine soils.

R406.3.4 Backfill. The remainder of the excavated area shall be backfilled with the same type of soil as was removed during the excavation.

R406.4 Precast concrete foundation system dampproofing. Except where required by Section R406.2 to be waterproofed, *precast concrete* foundation walls enclosing habitable or useable spaces located below *grade* shall be dampproofed in accordance with Section R406.1.

R406.4.1 Panel joints sealed. *Precast concrete* foundation panel joints shall be sealed full height with a sealant meeting ASTM C920, Type S or M, Grade NS, Class 25, Use NT, M or A. Joint sealant shall be installed in accordance with the manufacturer's instructions.

SECTION R407 COLUMNS

R407.1 Wood column protection. Wood columns shall be protected against decay as set forth in Section R317.

R407.2 Steel column protection. All surfaces (inside and outside) of steel columns shall be given a shop coat of rust-inhibitive paint, except for corrosion-resistant steel and steel treated with coatings to provide corrosion resistance.

R407.3 Structural requirements. The columns shall be restrained to prevent lateral displacement at the bottom end. Wood columns shall be not less in nominal size than 4 inches by 4 inches (102 mm by 102 mm). Steel columns shall be not less than 3-inch-diameter (76 mm) Schedule 40 pipe manufactured in accordance with ASTM A53/A53M Grade B or *approved* equivalent.

Exception: In *Seismic Design Categories* A, B and C, columns not more than 48 inches (1219 mm) in height on a pier or footing are exempt from the bottom end lateral displacement requirement within under-floor areas enclosed by a continuous foundation.

SECTION R408 UNDER-FLOOR SPACE

R408.1 Moisture control. The under-floor space between the bottom of the floor joists and the earth under any building (except space occupied by a *basement*) shall comply with Section R408.2 or R408.3.

R408.2 Openings for under-floor ventilation. Ventilation openings through foundation or exterior walls surrounding the under-floor space shall be provided in accordance with this section. The minimum net area of ventilation openings shall be not less than 1 square foot (0.0929 m^2) for each 150 square feet (14 m^2) of under-floor area. One ventilation opening shall be within 3 feet (915 mm) of each external corner of the under-floor space. Ventilation openings shall be covered for their height and width with any of the following materials provided that the least dimension of the covering shall not exceed $^1/_4$ inch (6.4 mm), and operational louvers are permitted:

1. Perforated sheet metal plates not less than 0.070 inch (1.8 mm) thick.
2. Expanded sheet metal plates not less than 0.047 inch (1.2 mm) thick.
3. Cast-iron grill or grating.
4. Extruded load-bearing brick vents.
5. Hardware cloth of 0.035 inch (0.89 mm) wire or heavier.
6. Corrosion-resistant wire mesh, with the least dimension being $^1/_8$ inch (3.2 mm) thick.

Exceptions:

1. The total area of ventilation openings shall be permitted to be reduced to $^1/_{1,500}$ of the under-floor area where the ground surface is covered with an *approved* Class I vapor retarder material.
2. Where the ground surface is covered with an *approved* Class 1 vapor retarder material, ventilation openings are not required to be within 3 feet (915 mm) of each external corner of the under-floor space provided that the openings are placed to provide cross ventilation of the space.

R408.3 Unvented crawl space. For unvented under-floor spaces, the following items shall be provided:

1. Exposed earth shall be covered with a continuous Class I vapor retarder. Joints of the vapor retarder shall overlap by 6 inches (152 mm) and shall be sealed or taped. The edges of the vapor retarder shall extend not less than 6 inches (152 mm) up the stem wall and shall be attached and sealed to the stem wall or insulation.
2. One of the following shall be provided for the under-floor space:
    2.1. Continuously operated mechanical exhaust ventilation at a rate equal to 1 cubic foot per minute (0.47 L/s) for each 50 square feet (4.7 m^2) of *crawl space* floor area, including an air pathway to the common area (such as a duct or transfer grille), and perimeter walls insulated in accordance with Section N1102.2.10.1 of this code.
    2.2. *Conditioned air* supply sized to deliver at a rate equal to 1 cubic foot per minute (0.47 L/s) for each 50 square feet (4.7 m^2) of under-floor area, including a return air pathway to the common area (such as a duct or transfer grille), and perimeter walls insulated in

accordance with Section N1102.2.10.1 of this code.

2.3. Plenum in existing structures complying with Section M1601.5, if under-floor space is used as a plenum.

2.4. Dehumidification sized in accordance with manufacturer's specifications.

R408.4 Access. Access shall be provided to all under-floor spaces. Access openings through the floor shall be not smaller than 18 inches by 24 inches (457 mm by 610 mm). Openings through a perimeter wall shall be not less than 16 inches by 24 inches (407 mm by 610 mm). Where any portion of the through-wall access is below *grade*, an areaway not less than 16 inches by 24 inches (407 mm by 610 mm) shall be provided. The bottom of the areaway shall be below the threshold of the access opening. Through wall access openings shall not be located under a door to the residence. See Section M1305.1.3 for access requirements where mechanical *equipment* is located under floors.

R408.5 Removal of debris. The under-floor *grade* shall be cleaned of all vegetation and organic material. Wood forms used for placing concrete shall be removed before a building is occupied or used for any purpose. Construction materials shall be removed before a building is occupied or used for any purpose.

R408.6 Finished grade. The finished *grade* of under-floor surface shall be permitted to be located at the bottom of the footings; however, where there is evidence that the groundwater table can rise to within 6 inches (152 mm) of the finished floor at the building perimeter or where there is evidence that the surface water does not readily drain from the building site, the *grade* in the under-floor space shall be as high as the outside finished *grade*, unless an *approved* drainage system is provided.

R408.7 Flood resistance. For buildings located in flood hazard areas as established in Table R301.2:

1. Walls enclosing the under-floor space shall be provided with flood openings in accordance with Section R322.2.2.
2. The finished ground level of the under-floor space shall be equal to or higher than the outside finished ground level on at least one side.

Exception: Under-floor spaces that meet the requirements of FEMA TB 11-1.

R408.8 Under-floor vapor retarder. In Climate Zones 1A, 2A and 3A below the warm-humid line, a continuous Class I or II vapor retarder shall be provided on the exposed face of air-permeable insulation installed between the floor joists and exposed to the grade in the under-floor space. The vapor retarder shall have a maximum water vapor permeance of 1.5 perms when tested in accordance with Procedure B of ASTM E96.

Exception: The vapor retarder shall not be required in unvented *crawl spaces* constructed in accordance with Section R408.3.

CHAPTER 5
FLOORS

User note:

About this chapter: *Chapter 5 provides the requirements for the design and construction of floor systems that will be capable of supporting minimum required design loads. This chapter covers wood floor framing, wood floors on the ground, cold-formed steel floor framing and concrete slabs on the ground. Allowable span tables are provided that greatly simplify the determination of joist, girder and sheathing sizes for raised floor systems of wood framing and cold-formed steel framing. This chapter also contains prescriptive requirements for wood-framed exterior decks and their attachment to the main building.*

SECTION R501
GENERAL

R501.1 Application. The provisions of this chapter shall control the design and construction of the floors for buildings, including the floors of attic spaces used to house mechanical or plumbing fixtures and *equipment.*

R501.2 Requirements. Floor construction shall be capable of accommodating all loads in accordance with Section R301 and of transmitting the resulting loads to the supporting structural elements.

SECTION R502
WOOD FLOOR FRAMING

R502.1 General. Wood and wood-based products used for load-supporting purposes shall conform to the applicable provisions of this section.

R502.1.1 Sawn lumber. Sawn lumber shall be identified by a grade *mark* of an accredited lumber grading or inspection agency and have design values certified by an accreditation body that complies with DOC PS 20. In lieu of a grade *mark,* a certificate of inspection issued by a lumber grading or inspection agency meeting the requirements of this section shall be accepted.

R502.1.1.1 Preservative-treated lumber. Preservative treated dimension lumber shall be identified as required by Section R317.2.

R502.1.1.2 End-jointed lumber. *Approved* end-jointed lumber identified by a grade *mark* conforming to Section R502.1.1 shall be permitted to be used interchangeably with solid-sawn members of the same species and grade. End-jointed lumber used in an assembly required elsewhere in this code to have a fire-resistance rating shall have the designation "Heat-Resistant Adhesive" or "HRA" included in its grade *mark.*

R502.1.2 Prefabricated wood I-joists. Structural capacities and design provisions for prefabricated wood I-joists shall be established and monitored in accordance with ASTM D5055.

R502.1.3 Structural glued laminated timbers. Glued laminated timbers shall be manufactured and identified as required in ANSI A190.1, ANSI 117 and ASTM D3737.

R502.1.4 Structural log members. Structural log members shall comply with the provisions of ICC 400.

R502.1.5 Structural composite lumber. Structural capacities for *structural composite lumber* shall be established and monitored in accordance with ASTM D5456.

R502.1.6 Cross-laminated timber. Cross-laminated timber shall be manufactured and identified as required by ANSI/APA PRG 320.

R502.1.7 Engineered wood rim board. Engineered wood rim boards shall conform to ANSI/APA PRR 410 or shall be evaluated in accordance with ASTM D7672. Structural capacities shall be in accordance with ANSI/APA PRR 410 or established in accordance with ASTM D7672. Rim boards conforming to ANSI/APA PRR 410 shall be marked in accordance with that standard.

R502.2 Design and construction. Floors shall be designed and constructed in accordance with the provisions of this chapter, Figure R502.2 and Sections R317 and R318 or in accordance with ANSI AWC NDS.

R502.2.1 Framing at braced wall lines. A load path for lateral forces shall be provided between floor framing and *braced wall panels* located above or below a floor, as specified in Section R602.10.8.

R502.2.2 Blocking and subflooring. Blocking for fastening panel edges or fixtures shall be not less than utility grade lumber. Subflooring shall be not less than utility grade lumber, No. 4 common grade boards or *wood structural panels* as specified in Section R503.2. Fireblocking shall be of any grade lumber.

R502.3 Allowable joist spans. Spans for floor joists shall be in accordance with Tables R502.3.1(1) and R502.3.1(2). For other grades and species and for other loading conditions, refer to the AWC STJR.

R502.3.1 Sleeping areas and attic joists. Table R502.3.1(1) shall be used to determine the maximum allowable span of floor joists that support sleeping areas and *attics* that are accessed by means of a fixed *stairway* in accordance with Section R311.7 provided that the design *live load* does not exceed 30 pounds per square foot (1.44 kPa) and the design dead load does not exceed 20 pounds per square foot (0.96 kPa). The allowable span of ceiling joists that support *attics* used for limited storage or no storage shall be determined in accordance with Section R802.5.

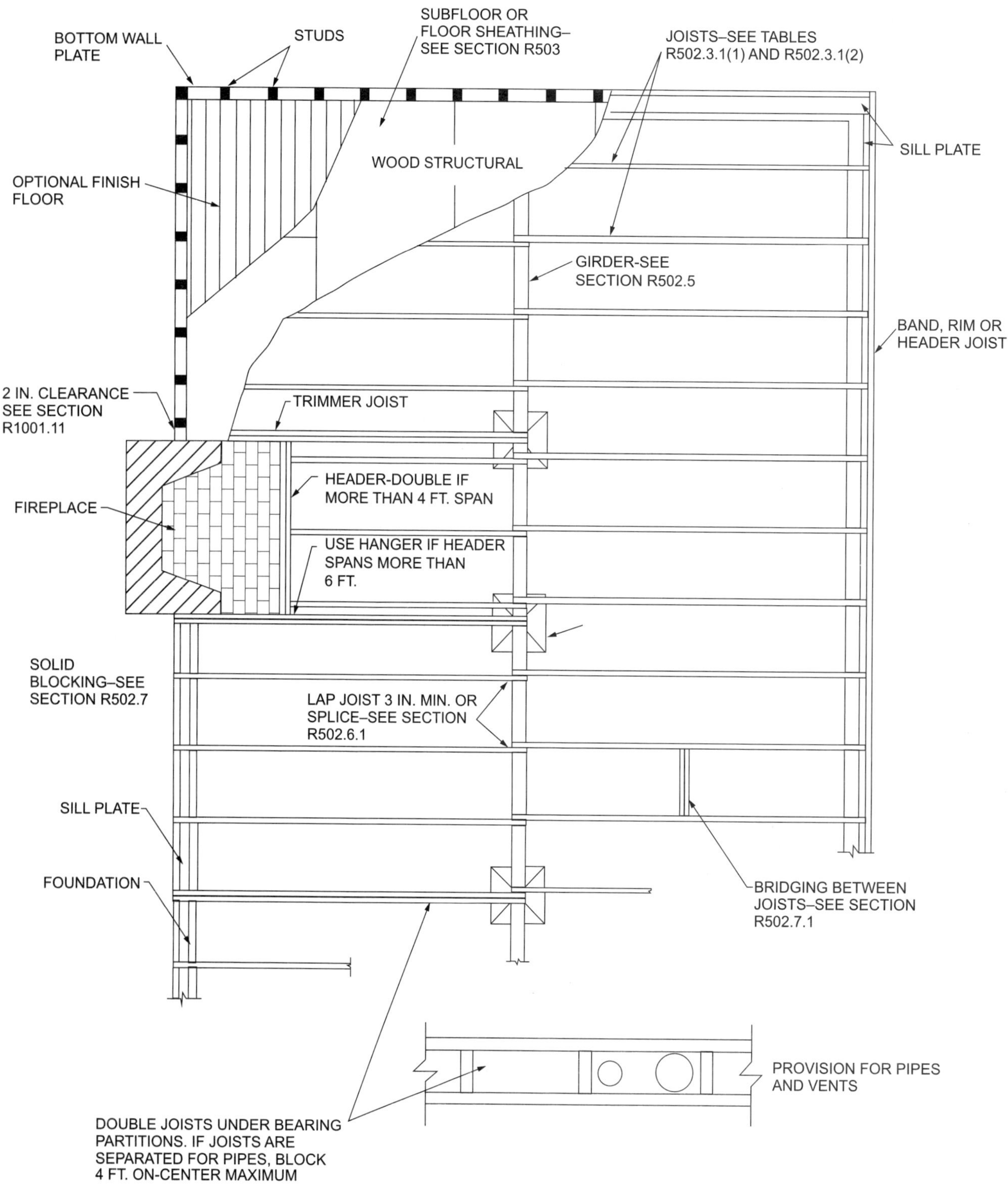
BOTTOM WALL PLATE
STUDS
SUBFLOOR OR FLOOR SHEATHING–SEE SECTION R503
JOISTS–SEE TABLES R502.3.1(1) AND R502.3.1(2)
SILL PLATE
OPTIONAL FINISH FLOOR
WOOD STRUCTURAL
GIRDER-SEE SECTION R502.5
BAND, RIM OR HEADER JOIST
2 IN. CLEARANCE SEE SECTION R1001.11
TRIMMER JOIST
FIREPLACE
HEADER-DOUBLE IF MORE THAN 4 FT. SPAN
USE HANGER IF HEADER SPANS MORE THAN 6 FT.
SOLID BLOCKING–SEE SECTION R502.7
LAP JOIST 3 IN. MIN. OR SPLICE–SEE SECTION R502.6.1
SILL PLATE
FOUNDATION
BRIDGING BETWEEN JOISTS–SEE SECTION R502.7.1
PROVISION FOR PIPES AND VENTS
DOUBLE JOISTS UNDER BEARING PARTITIONS. IF JOISTS ARE SEPARATED FOR PIPES, BLOCK 4 FT. ON-CENTER MAXIMUM

For SI: 1 inch = 25.4 mm, 1 foot = 304.8 mm.

FIGURE R502.2
FLOOR CONSTRUCTION

TABLE R502.3.1(1)
FLOOR JOIST SPANS FOR COMMON LUMBER SPECIES (Residential sleeping areas, live load = 30 psf, L/Δ = 360)[a]

JOIST SPACING (inches)	SPECIES AND GRADE		DEAD LOAD = 10 psf				DEAD LOAD = 20 psf			
			2 × 6	2 × 8	2 × 10	2 × 12	2 × 6	2 × 8	2 × 10	2 × 12
			Maximum floor joist spans							
			(ft-in)	(ft-in)	(ft-in)	(ft-in)	(ft-in)	(ft-in)	(ft-in)	(ft-in)
12	Douglas fir-larch	SS	12-6	16-6	21-0	25-7	12-6	16-6	21-0	25-7
	Douglas fir-larch	#1	12-0	15-10	20-3	24-8	12-0	15-7	19-0	22-0
	Douglas fir-larch	#2	11-10	15-7	19-10	23-4	11-8	14-9	18-0	20-11
	Douglas fir-larch	#3	9-11	12-7	15-5	17-10	8-11	11-3	13-9	16-0
	Hem-fir	SS	11-10	15-7	19-10	24-2	11-10	15-7	19-10	24-2
	Hem-fir	#1	11-7	15-3	19-5	23-7	11-7	15-3	18-9	21-9
	Hem-fir	#2	11-0	14-6	18-6	22-6	11-0	14-4	17-6	20-4
	Hem-fir	#3	9-8	12-4	15-0	17-5	8-8	11-0	13-5	15-7
	Southern pine	SS	12-3	16-2	20-8	25-1	12-3	16-2	20-8	25-1
	Southern pine	#1	11-10	15-7	19-10	24-2	11-10	15-7	18-7	22-0
	Southern pine	#2	11-3	14-11	18-1	21-4	10-9	13-8	16-2	19-1
	Southern pine	#3	9-2	11-6	14-0	16-6	8-2	10-3	12-6	14-9
	Spruce-pine-fir	SS	11-7	15-3	19-5	23-7	11-7	15-3	19-5	23-7
	Spruce-pine-fir	#1	11-3	14-11	19-0	23-0	11-3	14-7	17-9	20-7
	Spruce-pine-fir	#2	11-3	14-11	19-0	23-0	11-3	14-7	17-9	20-7
	Spruce-pine-fir	#3	9-8	12-4	15-0	17-5	8-8	11-0	13-5	15-7
16	Douglas fir-larch	SS	11-4	15-0	19-1	23-3	11-4	15-0	19-1	23-3
	Douglas fir-larch	#1	10-11	14-5	18-5	21-4	10-8	13-6	16-5	19-1
	Douglas fir-larch	#2	10-9	14-2	17-5	20-3	10-1	12-9	15-7	18-1
	Douglas fir-larch	#3	8-7	10-11	13-4	15-5	7-8	9-9	11-11	13-10
	Hem-fir	SS	10-9	14-2	18-0	21-11	10-9	14-2	18-0	21-11
	Hem-fir	#1	10-6	13-10	17-8	21-1	10-6	13-4	16-3	18-10
	Hem-fir	#2	10-0	13-2	16-10	19-8	9-10	12-5	15-2	17-7
	Hem-fir	#3	8-5	10-8	13-0	15-1	7-6	9-6	11-8	13-6
	Southern pine	SS	11-2	14-8	18-9	22-10	11-2	14-8	18-9	22-10
	Southern pine	#1	10-9	14-2	18-0	21-4	10-9	13-9	16-1	19-1
	Southern pine	#2	10-3	13-3	15-8	18-6	9-4	11-10	14-0	16-6
	Southern pine	#3	7-11	10-0	11-1	14-4	7-1	8-11	10-10	12-10
	Spruce-pine-fir	SS	10-6	13-10	17-8	21-6	10-6	13-10	17-8	21-4
	Spruce-pine-fir	#1	10-3	13-6	17-2	19-11	9-11	12-7	15-5	17-10
	Spruce-pine-fir	#2	10-3	13-6	17-2	19-11	9-11	12-7	15-5	17-10
	Spruce-pine-fir	#3	8-5	10-8	13-0	15-1	7-6	9-6	11-8	13-6

(continued)

TABLE R502.3.1(1)—continued
FLOOR JOIST SPANS FOR COMMON LUMBER SPECIES (Residential sleeping areas, live load = 30 psf, L/Δ = 360)[a]

JOIST SPACING (inches)	SPECIES AND GRADE		DEAD LOAD = 10 psf				DEAD LOAD = 20 psf			
			2 × 6	2 × 8	2 × 10	2 × 12	2 × 6	2 × 8	2 × 10	2 × 12
			Maximum floor joist spans							
			(ft-in)	(ft-in)	(ft-in)	(ft-in)	(ft-in)	(ft-in)	(ft-in)	(ft-in)
19.2	Douglas fir-larch	SS	10-8	14-1	18-0	21-10	10-8	14-1	18-0	21-4
	Douglas fir-larch	#1	10-4	13-7	16-9	19-6	9-8	12-4	15-0	17-5
	Douglas fir-larch	#2	10-1	13-0	15-11	18-6	9-3	11-8	14-3	16-6
	Douglas fir-larch	#3	7-10	10-0	12-2	14-1	7-0	8-11	10-11	12-7
	Hem-fir	SS	10-1	13-4	17-0	20-8	10-1	13-4	17-0	20-7
	Hem-fir	#1	9-10	13-0	16-7	19-3	9-7	12-2	14-10	17-2
	Hem-fir	#2	9-5	12-5	15-6	17-1	8-11	11-4	13-10	16-1
	Hem-fir	#3	7-8	9-9	11-10	13-9	6-10	8-8	10-7	12-4
	Southern pine	SS	10-6	13-10	17-8	21-6	10-6	13-10	17-8	21-6
	Southern pine	#1	10-1	13-4	16-5	19-6	9-11	12-7	14-8	17-5
	Southern pine	#2	9-6	12-1	14-4	16-10	8-6	10-10	12-10	15-1
	Southern pine	#3	7-3	9-1	11-0	13-1	6-5	8-2	9-10	11-8
	Spruce-pine-fir	SS	9-10	13-0	16-7	20-2	9-10	13-0	16-7	19-6
	Spruce-pine-fir	#1	9-8	12-9	15-8	18-3	9-1	11-6	14-1	16-3
	Spruce-pine-fir	#2	9-8	12-9	15-8	18-3	9-1	11-6	14-1	16-3
	Spruce-pine-fir	#3	7-8	9-9	11-10	13-9	6-10	8-8	10-7	12-4
24	Douglas fir-larch	SS	9-11	13-1	16-8	20-3	9-11	13-1	16-5	19-1
	Douglas fir-larch	#1	9-7	12-4	15-0	17-5	8-8	11-0	13-5	15-7
	Douglas fir-larch	#2	9-3	11-8	14-3	16-6	8-3	10-5	12-9	14-9
	Douglas fir-larch	#3	7-0	8-11	10-11	12-7	6-3	8-0	9-9	11-3
	Hem-fir	SS	9-4	12-4	15-9	19-2	9-4	12-4	15-9	18-5
	Hem-fir	#1	9-2	12-1	14-10	17-2	8-7	10-10	13-3	15-5
	Hem-fir	#2	8-9	11-4	13-10	16-1	8-0	10-2	12-5	14-4
	Hem-fir	#3	6-10	8-8	10-7	12-4	6-2	7-9	9-6	11-0
	Southern pine	SS	9-9	12-10	16-5	19-11	9-9	12-10	16-5	19-8
	Southern pine	#1	9-4	12-4	14-8	17-5	8-10	11-3	13-1	15-7
	Southern pine	#2	8-6	10-10	12-10	15-1	7-7	9-8	11-5	13-6
	Southern pine	#3	6-5	8-2	9-10	11-8	5-9	7-3	8-10	10-5
	Spruce-pine-fir	SS	9-2	12-1	15-5	18-9	9-2	12-1	15-0	17-5
	Spruce-pine-fir	#1	8-11	11-6	14-1	16-3	8-1	10-3	12-7	14-7
	Spruce-pine-fir	#2	8-11	11-6	14-1	16-3	8-1	10-3	12-7	14-7
	Spruce-pine-fir	#3	6-10	8-8	10-7	12-4	6-2	7-9	9-6	11-0

For SI: 1 inch = 25.4 mm, 1 foot = 304.8 mm, 1 pound per square foot = 0.0479 kPa.

Note: Check sources for availability of lumber in lengths greater than 20 feet.

a. Dead load limits for townhouses in Seismic Design Category C and all structures in Seismic Design Categories D_0, D_1 and D_2 shall be determined in accordance with Section R301.2.2.2.

TABLE R502.3.1(2)
FLOOR JOIST SPANS FOR COMMON LUMBER SPECIES (Residential living areas, live load = 40 psf, L/Δ = 360)[b]

JOIST SPACING (inches)	SPECIES AND GRADE		DEAD LOAD = 10 psf				DEAD LOAD = 20 psf			
			2 × 6	2 × 8	2 × 10	2 × 12	2 × 6	2 × 8	2 × 10	2 × 12
			Maximum floor joist spans							
			(ft-in)	(ft-in)	(ft-in)	(ft-in)	(ft-in)	(ft-in)	(ft-in)	(ft-in)
12	Douglas fir-larch	SS	11-4	15-0	19-1	23-3	11-4	15-0	19-1	23-3
	Douglas fir-larch	#1	10-11	14-5	18-5	22-0	10-11	14-2	17-4	20-1
	Douglas fir-larch	#2	10-9	14-2	18-0	20-11	10-8	13-6	16-5	19-1
	Douglas fir-larch	#3	8-11	11-3	13-9	16-0	8-1	10-3	12-7	14-7
	Hem-fir	SS	10-9	14-2	18-0	21-11	10-9	14-2	18-0	21-11
	Hem-fir	#1	10-6	13-10	17-8	21-6	10-6	13-10	17-1	19-10
	Hem-fir	#2	10-0	13-2	16-10	20-4	10-0	13-1	16-0	18-6
	Hem-fir	#3	8-8	11-0	13-5	15-7	7-11	10-0	12-3	14-3
	Southern pine	SS	11-2	14-8	18-9	22-10	11-2	14-8	18-9	22-10
	Southern pine	#1	10-9	14-2	18-0	21-11	10-9	14-2	16-11	20-1
	Southern pine	#2	10-3	13-6	16-2	19-1	9-10	12-6	14-9	17-5
	Southern pine	#3	8-2	10-3	12-6	14-9	7-5	9-5	11-5	13-6
	Spruce-pine-fir	SS	10-6	13-10	17-8	21-6	10-6	13-10	17-8	21-6
	Spruce-pine-fir	#1	10-3	13-6	17-3	20-7	10-3	13-3	16-3	18-10
	Spruce-pine-fir	#2	10-3	13-6	17-3	20-7	10-3	13-3	16-3	18-10
	Spruce-pine-fir	#3	8-8	11-0	13-5	15-7	7-11	10-0	12-3	14-3
16	Douglas fir-larch	SS	10-4	13-7	17-4	21-1	10-4	13-7	17-4	21-1
	Douglas fir-larch	#1	9-11	13-1	16-5	19-1	9-8	12-4	15-0	17-5
	Douglas fir-larch	#2	9-9	12-9	15-7	18-1	9-3	11-8	14-3	16-6
	Douglas fir-larch	#3	7-8	9-9	11-11	13-10	7-0	8-11	10-11	12-7
	Hem-fir	SS	9-9	12-10	16-5	19-11	9-9	12-10	16-5	19-11
	Hem-fir	#1	9-6	12-7	16-0	18-10	9-6	12-2	14-10	17-2
	Hem-fir	#2	9-1	12-0	15-2	17-7	8-11	11-4	13-10	16-1
	Hem-fir	#3	7-6	9-6	11-8	13-6	6-10	8-8	10-7	12-4
	Southern pine	SS	10-2	13-4	17-0	20-9	10-2	13-4	17-0	20-9
	Southern pine	#1	9-9	12-10	16-1	19-1	9-9	12-7	14-8	17-5
	Southern pine	#2	9-4	11-10	14-0	16-6	8-6	10-10	12-10	15-1
	Southern pine	#3	7-1	8-11	10-10	12-10	6-5	8-2	9-10	11-8
	Spruce-pine-fir	SS	9-6	12-7	16-0	19-6	9-6	12-7	16-0	19-6
	Spruce-pine-fir	#1	9-4	12-3	15-5	17-10	9-1	11-6	14-1	16-3
	Spruce-pine-fir	#2	9-4	12-3	15-5	17-10	9-1	11-6	14-1	16-3
	Spruce-pine-fir	#3	7-6	9-6	11-8	13-6	6-10	8-8	10-7	12-4

(continued)

TABLE R502.3.1(2)—continued
FLOOR JOIST SPANS FOR COMMON LUMBER SPECIES (Residential living areas, live load = 40 psf, L/Δ = 360)[b]

JOIST SPACING (inches)	SPECIES AND GRADE		DEAD LOAD = 10 psf				DEAD LOAD = 20 psf			
			2 × 6	2 × 8	2 × 10	2 × 12	2 × 6	2 × 8	2 × 10	2 × 12
			Maximum floor joist spans							
			(ft-in)	(ft-in)	(ft-in)	(ft-in)	(ft-in)	(ft-in)	(ft-in)	(ft-in)
19.2	Douglas fir-larch	SS	9-8	12-10	16-4	19-10	9-8	12-10	16-4	19-6
	Douglas fir-larch	#1	9-4	12-4	15-0	17-5	8-10	11-3	13-8	15-11
	Douglas fir-larch	#2	9-2	11-8	14-3	16-6	8-5	10-8	13-0	15-1
	Douglas fir-larch	#3	7-0	8-11	10-11	12-7	6-5	8-2	9-11	11-6
	Hem-fir	SS	9-2	12-1	15-5	18-9	9-2	12-1	15-5	18-9
	Hem-fir	#1	9-0	11-10	14-10	17-2	8-9	11-1	13-6	15-8
	Hem-fir	#2	8-7	11-3	13-10	16-1	8-2	10-4	12-8	14-8
	Hem-fir	#3	6-10	8-8	10-7	12-4	6-3	7-11	9-8	11-3
	Southern pine	SS	9-6	12-7	16-0	19-6	9-6	12-7	16-0	19-6
	Southern pine	#1	9-2	12-1	14-8	17-5	9-0	11-5	13-5	15-11
	Southern pine	#2	8-6	10-10	12-10	15-1	7-9	9-10	11-8	13-9
	Southern pine	#3	6-5	8-2	9-10	11-8	5-11	7-5	9-0	10-8
	Spruce-pine-fir	SS	9-0	11-10	15-1	18-4	9-0	11-10	15-1	17-9
	Spruce-pine-fir	#1	8-9	11-6	14-1	16-3	8-3	10-6	12-10	14-10
	Spruce-pine-fir	#2	8-9	11-6	14-1	16-3	8-3	10-6	12-10	14-10
	Spruce-pine-fir	#3	6-10	8-8	10-7	12-4	6-3	7-11	9-8	11-3
24	Douglas fir-larch	SS	9-0	11-11	15-2	18-5	9-0	11-11	15-0	17-5
	Douglas fir-larch	#1	8-8	11-0	13-5	15-7	7-11	10-0	12-3	14-3
	Douglas fir-larch	#2	8-3	10-5	12-9	14-9	7-6	9-6	11-8	13-6
	Douglas fir-larch	#3	6-3	8-0	9-9	11-3	5-9	7-3	8-11	10-4
	Hem-fir	SS	8-6	11-3	14-4	17-5	8-6	11-3	14-4	16-10[a]
	Hem-fir	#1	8-4	10-10	13-3	15-5	7-10	9-11	12-1	14-0
	Hem-fir	#2	7-11	10-2	12-5	14-4	7-4	9-3	11-4	13-1
	Hem-fir	#3	6-2	7-9	9-6	11-0	5-7	7-1	8-8	10-1
	Southern pine	SS	8-10	11-8	14-11	18-1	8-10	11-8	14-11	18-0
	Southern pine	#1	8-6	11-3	13-1	15-7	8-1	10-3	12-0	14-3
	Southern pine	#2	7-7	9-8	11-5	13-6	7-0	8-10	10-5	12-4
	Southern pine	#3	5-9	7-3	8-10	10-5	5-3	6-8	8-1	9-6
	Spruce-pine-fir	SS	8-4	11-0	14-0	17-0	8-4	11-0	13-8	15-11
	Spruce-pine-fir	#1	8-1	10-3	12-7	14-7	7-5	9-5	11-6	13-4
	Spruce-pine-fir	#2	8-1	10-3	12-7	14-7	7-5	9-5	11-6	13-4
	Spruce-pine-fir	#3	6-2	7-9	9-6	11-0	5-7	7-1	8-8	10-1

For SI: 1 inch = 25.4 mm, 1 foot = 304.8 mm, 1 pound per square foot = 0.0479 kPa.

Note: Check sources for availability of lumber in lengths greater than 20 feet.

a. End bearing length shall be increased to 2 inches.

b. Dead load limits for townhouses in Seismic Design Category C and all structures in Seismic Design Categories D_0, D_1, and D_2 shall be determined in accordance with Section R301.2.2.2.

R502.3.2 Other floor joists. Table R502.3.1(2) shall be used to determine the maximum allowable span of floor joists that support other areas of the building, other than sleeping areas and *attics*, provided that the design *live load* does not exceed 40 pounds per square foot (1.92 kPa) and the design dead load does not exceed 20 pounds per square foot (0.96 kPa).

R502.3.3 Floor cantilevers. Floor cantilever spans shall not exceed the nominal depth of the wood floor joist. Floor cantilevers constructed in accordance with Table R502.3.3(1) shall be permitted where supporting a light-frame bearing wall and roof only. Floor cantilevers supporting an exterior balcony are permitted to be constructed in accordance with Table R502.3.3(2).

R502.4 Joists under bearing partitions. Joists under parallel bearing partitions shall be of adequate size to support the load. Double joists, sized to adequately support the load, that are separated to permit the installation of piping or vents shall be full-depth solid blocked with lumber not less than 2 inches (51 mm) in nominal thickness spaced not more than 4 feet (1219 mm) on center. Bearing partitions perpendicular to joists shall not be offset from supporting girders, walls or partitions more than the joist depth unless such joists are of sufficient size to carry the additional load.

R502.5 Allowable girder and header spans. The allowable spans of girders and headers fabricated of dimension lumber shall not exceed the values set forth in Tables R602.7(1), R602.7(2) and R602.7(3).

R502.6 Bearing. The ends of each joist, beam or girder shall have not less than $1^1/_2$ inches (38 mm) of bearing on wood or metal, have not less than 3 inches of bearing (76 mm) on masonry or concrete or be supported by *approved* joist hangers. Alternatively, the ends of joists shall be supported on a 1-inch by 4-inch (25 mm by 102 mm) ribbon strip and shall be nailed to the adjacent stud. The bearing on masonry or concrete shall be direct, or a sill plate of 2-inch-minimum (51 mm) nominal thickness shall be provided under the joist, beam or girder. The sill plate shall provide a minimum nominal bearing area of 48 square inches (30 865 mm^2).

R502.6.1 Floor systems. Joists framing from opposite sides over a bearing support shall lap not less than 3 inches (76 mm) and shall be nailed together with a minimum three 10d face nails. A wood or metal splice with strength equal to or greater than that provided by the nailed lap is permitted.

R502.6.2 Joist framing. Joists framing into the side of a wood girder shall be supported by *approved* framing anchors or on ledger strips not less than nominal 2 inches by 2 inches (51 mm by 51 mm).

TABLE R502.3.3(1)
CANTILEVER SPANS FOR FLOOR JOISTS SUPPORTING LIGHT-FRAME EXTERIOR BEARING WALL AND ROOF ONLY[a, b, c, f, g, h]
(Floor live load ≤ 40 psf, roof live load ≤ 20 psf)

MEMBER & SPACING	MAXIMUM CANTILEVER SPAN (uplift force at backspan support in lb)[d, e]											
	Ground Snow Load											
	≤ 20 psf			30 psf			50 psf			70 psf		
	Roof Width			Roof Width			Roof Width			Roof Width		
	24 ft	32 ft	40 ft	24 ft	32 ft	40 ft	24 ft	32 ft	40 ft	24 ft	32 ft	40 ft
2 × 8 @ 12″	20″ (177)	15″ (227)	—	18″ (209)	—	—	—	—	—	—	—	—
2 × 10 @ 16″	29″ (228)	21″ (297)	16″ (364)	26″ (271)	18″ (354)	—	20″ (375)	—	—	—	—	—
2 × 10 @ 12″	36″ (166)	26″ (219)	20″ (270)	34″ (198)	22″ (263)	16″ (324)	26″ (277)	—	—	19″ (356)	—	—
2 × 12 @ 16″	—	32″ (287)	25″ (356)	36″ (263)	29″ (345)	21″ (428)	29″ (367)	20″ (484)	—	23″ (471)	—	—
2 × 12 @ 12″	—	42″ (209)	31″ (263)	—	37″ (253)	27″ (317)	36″ (271)	27″ (358)	17″ (447)	31″ (348)	19″ (462)	—
2 × 12 @ 8″	—	48″ (136)	45″ (169)	—	48″ (164)	38″ (206)	—	40″ (233)	26″ (294)	36″ (230)	29″ (304)	18″ (379)

For SI: 1 inch = 25.4 mm, 1 foot = 304.8 mm, 1 pound per square foot = 0.0479 kPa.

a. Tabulated values are for clear-span roof supported solely by exterior bearing walls.

b. Spans are based on No. 2 Grade lumber of Douglas fir-larch, Southern pine, hem-fir and spruce-pine-fir for repetitive (three or more) members.

c. Ratio of backspan to cantilever span shall be not less than 3:1.

d. Connections capable of resisting the indicated uplift force shall be provided at the backspan support.

e. Uplift force is for a backspan to cantilever span ratio of 3:1. Tabulated uplift values are permitted to be reduced by multiplying by a factor equal to 3 divided by the actual backspan ratio provided (3/backspan ratio).

f. See Section R301.2.2.6, Item 1, for additional limitations on cantilevered floor joists for detached one- and two-family dwellings in Seismic Design Category D_0, D_1 or D_2 and townhouses in Seismic Design Category C, D_0, D_1 or D_2.

g. A full-depth rim joist shall be provided at the unsupported end of the cantilever joists. Solid blocking shall be provided at the supported end. Where the cantilever length is 24 inches or less and the building is assigned to Seismic Design Category A, B or C, solid blocking at the support for the cantilever shall not be required.

h. Linear interpolation shall be permitted for building widths and ground snow loads other than shown.

TABLE R502.3.3(2)
CANTILEVER SPANS FOR FLOOR JOISTS SUPPORTING EXTERIOR BALCONY[a, b, e, f]

MEMBER SIZE	SPACING	MAXIMUM CANTILEVER SPAN (uplift force at backspan support in lb)[c, d]		
		Ground Snow Load		
		≤ 30 psf	50 psf	70 psf
2 × 8	12″	42″ (139)	39″ (156)	34″ (165)
2 × 8	16″	36″ (151)	34″ (171)	29″ (180)
2 × 10	12″	61″ (164)	57″ (189)	49″ (201)
2 × 10	16″	53″ (180)	49″ (208)	42″ (220)
2 × 10	24″	43″ (212)	40″ (241)	34″ (255)
2 × 12	16″	72″ (228)	67″ (260)	57″ (268)
2 × 12	24″	58″ (279)	54″ (319)	47″ (330)

For SI: 1 inch = 25.4 mm, 1 pound per square foot = 0.0479 kPa.

a. Spans are based on No. 2 Grade lumber of Douglas fir-larch, Southern pine, hem-fir, and spruce-pine-fir for repetitive (three or more) members.

b. Ratio of backspan to cantilever span shall be not less than 2:1.

c. Connections capable of resisting the indicated uplift force shall be provided at the backspan support.

d. Uplift force is for a backspan to cantilever span ratio of 2:1. Tabulated uplift values are permitted to be reduced by multiplying by a factor equal to 2 divided by the actual backspan ratio provided (2/backspan ratio).

e. A full-depth rim joist shall be provided at the unsupported end of the cantilever joists. Solid blocking shall be provided at the supported end. Where the cantilever length is 24 inches or less and the building is assigned to Seismic Design Category A, B or C, solid blocking at the support for the cantilever shall not be required.

f. Linear interpolation shall be permitted for ground snow loads other than shown.

R502.7 Lateral restraint at supports. Joists shall be supported laterally at the ends by full-depth solid blocking not less than 2 inches (51 mm) nominal in thickness; or by attachment to a full-depth header, band or rim joist, or to an adjoining stud or shall be otherwise provided with lateral support to prevent rotation.

Exceptions:

1. Trusses, *structural composite lumber*, structural glued-laminated members and I-joists shall be supported laterally as required by the manufacturer's recommendations.
2. In *Seismic Design Categories* D_0, D_1 and D_2, lateral restraint shall be provided at each intermediate support.

R502.7.1 Bridging. Joists exceeding a nominal 2 inches by 12 inches (51 mm by 305 mm) shall be supported laterally by solid blocking, diagonal bridging (wood or metal), or a continuous 1-inch by 3-inch (25 mm by 76 mm) strip nailed across the bottom of joists perpendicular to joists at intervals not exceeding 8 feet (2438 mm).

Exception: Trusses, *structural composite lumber*, structural glued-laminated members and I-joists shall be supported laterally as required by the manufacturer's recommendations.

R502.8 Cutting, drilling and notching. Structural floor members shall not be cut, bored or notched in excess of the limitations specified in this section. See Figure R502.8.

R502.8.1 Sawn lumber. Notches in solid lumber joists, rafters and beams shall not exceed one-sixth of the depth of the member, shall not be longer than one-third of the depth of the member and shall not be located in the middle one-third of the span. Notches at the ends of the member shall not exceed one-fourth the depth of the member. The tension side of members 4 inches (102 mm) or greater in nominal thickness shall not be notched except at the ends of the members. The diameter of holes bored or cut into members shall not exceed one-third the depth of the member. Holes shall not be closer than 2 inches (51 mm) to the top or bottom of the member, or to any other hole located in the member. Where the member is notched, the hole shall not be closer than 2 inches (51 mm) to the notch.

R502.8.2 Engineered wood products. Cuts, notches and holes bored in trusses, *structural composite lumber*, structural glue-laminated members, cross-laminated timber members or I-joists are prohibited except where permitted by the manufacturer's recommendations or where the effects of such alterations are specifically considered in the design of the member by a *registered design professional*.

R502.9 Fastening. Floor framing shall be nailed in accordance with Table R602.3(1). Where posts and beam or girder construction is used to support floor framing, positive connections shall be provided to ensure against uplift and lateral displacement.

R502.10 Framing of openings. Openings in floor framing shall be framed with header and trimmer joists. Where the header joist span does not exceed 4 feet (1219 mm), the header joist shall be a single member the same size as the floor joist. Single trimmer joists shall be used to carry a single header joist that is located within 3 feet (914 mm) of the trimmer joist bearing. Where the header joist span exceeds 4 feet (1219 mm), the trimmer joists and the header joist shall be doubled and of sufficient cross section to support the floor joists framing into the header.

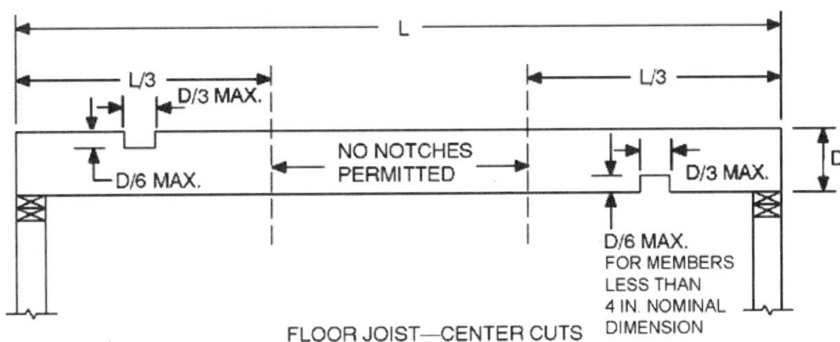

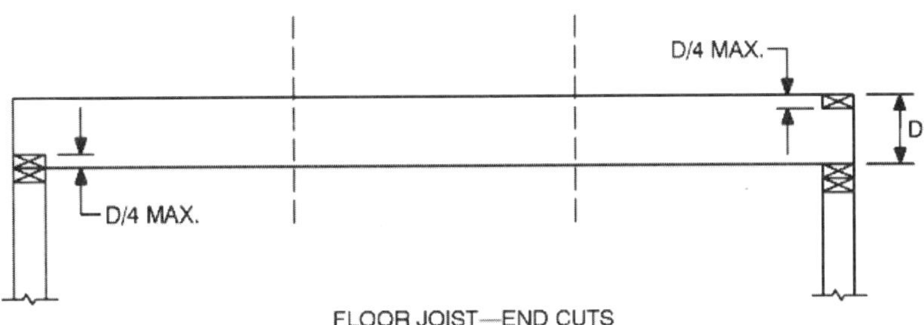

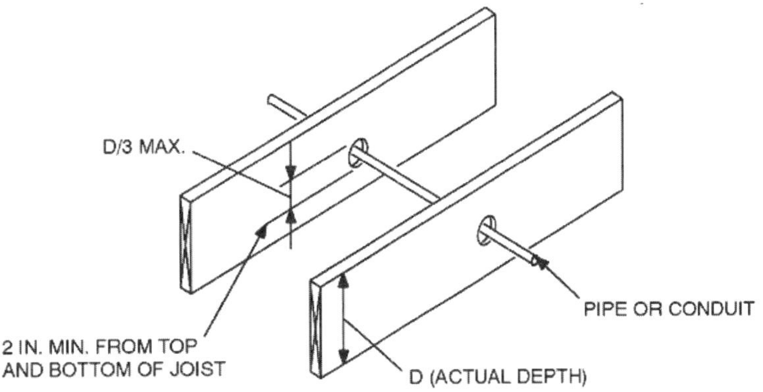

For SI: 1 inch = 25.4 mm.

FIGURE R502.8
CUTTING, NOTCHING AND DRILLING

R502.11 Wood trusses.

R502.11.1 Design. Wood trusses shall be designed in accordance with *approved* engineering practice. The design and manufacture of metal-plate-connected wood trusses shall comply with ANSI/TPI 1. The *truss design drawings* shall be prepared by a *registered design professional* where required by the statutes of the *jurisdiction* in which the project is to be constructed in accordance with Section R106.1.

R502.11.2 Bracing. Trusses shall be braced to prevent rotation and provide lateral stability in accordance with the requirements specified in the *construction documents* for the building and on the individual *truss design drawings*. In the absence of specific bracing requirements, trusses shall be braced in accordance with accepted industry practices, such as the SBCA *Building Component Safety Information (BCSI) Guide to Good Practice for Handling, Installing & Bracing of Metal Plate Connected Wood Trusses.*

R502.11.3 Alterations to trusses. Truss members and components shall not be cut, notched, spliced or otherwise altered in any way without the approval of a *registered design professional. Alterations* resulting in the addition of load that exceeds the design load for the truss, shall not be permitted without verification that the truss is capable of supporting the additional loading.

R502.11.4 Truss design drawings. *Truss design drawings*, prepared in compliance with Section R502.11.1, shall be submitted to the *building official* and *approved* prior to installation. *Truss design drawings* shall be provided with the shipment of trusses delivered to the job site. *Truss design drawings* shall include, at a minimum, the information specified as follows:

1. Slope or depth, span and spacing.
2. Location of all joints.
3. Required bearing widths.
4. Design loads as applicable:
    - 4.1. Top chord *live load*.
    - 4.2. Top chord dead load.
    - 4.3. Bottom chord *live load*.
    - 4.4. Bottom chord dead load.
    - 4.5. Concentrated loads and their points of application.
    - 4.6. Controlling wind and earthquake loads.
5. Adjustments to lumber and joint connector design values for conditions of use.
6. Each reaction force and direction.
7. Joint connector type and description, such as size, thickness or gage, and the dimensioned location of each joint connector except where symmetrically located relative to the joint interface.
8. Lumber size, species and grade for each member.
9. Connection requirements for:
    - 9.1. Truss-to-girder-truss.
    - 9.2. Truss ply-to-ply.
    - 9.3. Field splices.
10. Calculated deflection ratio, maximum description for live and total load, or both.
11. Maximum axial compression forces in the truss members to enable the building designer to design the size, connections and anchorage of the permanent continuous lateral bracing. Forces shall be shown on the truss drawing or on supplemental documents.
12. Required permanent truss member bracing location.

R502.12 Draftstopping required. Draftstopping shall be provided in accordance with Section R302.12.

R502.13 Fireblocking required. Fireblocking shall be provided in accordance with Section R302.11.

SECTION R503
FLOOR SHEATHING

R503.1 Lumber sheathing. Maximum allowable spans for lumber used as floor sheathing shall conform to Tables R503.1, R503.2.1.1(1) and R503.2.1.1(2).

TABLE R503.1
MINIMUM THICKNESS OF LUMBER FLOOR SHEATHING

JOIST OR BEAM SPACING (inches)	MINIMUM NET THICKNESS	
	Perpendicular to joist	Diagonal to joist
24	$^{11}/_{16}$	$^{3}/_{4}$
16	$^{5}/_{8}$	$^{5}/_{8}$
48[a]	$1^{1}/_{2}$ T & G	N/A
54[b]		
60[c]		

For SI: 1 inch = 25.4 mm, 1 pound per square inch = 6.895 kPa.
N/A = Not Applicable.

a. For this support spacing, lumber sheathing shall have a minimum F_b of 675 and minimum E of 1,100,000 (see ANSI AWC NDS).

b. For this support spacing, lumber sheathing shall have a minimum F_b of 765 and minimum E of 1,400,000 (see ANSI AWC NDS).

c. For this support spacing, lumber sheathing shall have a minimum F_b of 855 and minimum E of 1,700,000 (see ANSI AWC NDS).

R503.1.1 End joints. End joints in lumber used as subflooring shall occur over supports unless end-matched lumber is used, in which case each piece shall bear on not less than two joists. Subflooring shall be permitted to be omitted where joist spacing does not exceed 16 inches (406 mm) and a 1-inch (25 mm) nominal tongue-and-groove wood strip flooring is applied perpendicular to the joists.

R503.2 Wood structural panel sheathing.

R503.2.1 Identification and grade. *Wood structural panel* sheathing used for structural purposes shall conform to CSA O325, CSA O437 DOC PS 1 or DOC PS 2. Panels shall be identified for grade, bond classification and Performance Category by a grade *mark* or certificate of inspection issued by an *approved* agency. The Performance Category value shall be used as the "nominal *panel thickness*" or "*panel thickness*" wherever referenced in this code.

R503.2.1.1 Subfloor and combined subfloor underlayment. Where used as subflooring or combination subfloor underlayment, *wood structural panels* shall be of one of the grades specified in Table R503.2.1.1(1). Where sanded plywood is used as combination subfloor underlayment, the grade, bond classification, and Performance Category shall be as specified in Table R503.2.1.1(2).

TABLE R503.2.1.1(1) ALLOWABLE SPANS AND LOADS FOR WOOD STRUCTURAL PANELS FOR ROOF AND SUBFLOOR SHEATHING AND COMBINATION SUBFLOOR UNDERLAYMENT[a, b, c]

SPAN RATING	MINIMUM NOMINAL PANEL THICKNESS (inch)	ALLOWABLE LIVE LOAD (psf)[h, l]		MAXIMUM SPAN (inches)		LOAD (pounds per square foot, at maximum span)		MAXIMUM SPAN (inches)
		SPAN @ 16" o.c.	SPAN @ 24" o.c.	With edge support[d]	Without edge support	Total load	Live load	
Sheathing[e]				Roof[f]				Subfloor[j]
16/0	$^{3}/_{8}$	30	—	16	16	40	30	0
20/0	$^{3}/_{8}$	50	—	20	20	40	30	0
24/0	$^{3}/_{8}$	100	30	24	20[g]	40	30	0
24/16	$^{7}/_{16}$	100	40	24	24	50	40	16
32/16	$^{15}/_{32}$, $^{1}/_{2}$	180	70	32	28	40	30	16[h]
40/20	$^{19}/_{32}$, $^{5}/_{8}$	305	130	40	32	40	30	20[h, i]
48/24	$^{23}/_{32}$, $^{3}/_{4}$	—	175	48	36	45	35	24
60/32	$^{7}/_{8}$	—	305	60	48	45	35	32
Underlayment, C-C plugged, single floor[e]				Roof[f]				Combination subfloor underlayment[k]
16 o.c.	$^{19}/_{32}$, $^{5}/_{8}$	100	40	24	24	50	40	16[i]
20 o.c.	$^{19}/_{32}$, $^{5}/_{8}$	150	60	32	32	40	30	20[i, j]
24 o.c.	$^{23}/_{32}$, $^{3}/_{4}$	240	100	48	36	35	25	24
32 o.c.	$^{7}/_{8}$	—	185	48	40	50	40	32
48 o.c.	$1^{3}/_{32}$, $1^{1}/_{8}$	—	290	60	48	50	40	48

For SI: 1 inch = 25.4 mm, 1 pound per square foot = 0.0479 kPa.

a. The allowable total loads were determined using a dead load of 10 psf. If the dead load exceeds 10 psf, then the live load shall be reduced accordingly.

b. Panels continuous over two or more spans with long dimension (strength axis) perpendicular to supports. Spans shall be limited to values shown because of possible effect of concentrated loads.

c. Applies to panels 24 inches or wider.

d. Lumber blocking, panel edge clips (one midway between each support, except two equally spaced between supports where span is 48 inches), tongue-and-groove panel edges, or other approved type of edge support.

e. Includes Structural I panels in these grades.

f. Uniform load deflection limitation: $^{1}/_{180}$ of span under live load plus dead load, $^{1}/_{240}$ of span under live load only.

g. Maximum span 24 inches for $^{15}/_{32}$- and $^{1}/_{2}$-inch panels.

h. Maximum span 24 inches where $^{3}/_{4}$-inch wood finish flooring is installed at right angles to joists.

i. Maximum span 24 inches where 1.5 inches of lightweight concrete or approved cellular concrete is placed over the subfloor.

j. Unsupported edges shall have tongue-and-groove joints or shall be supported with blocking unless minimum nominal $^{1}/_{4}$-inch-thick wood panel-type underlayment, fiber-cement underlayment with end and edge joints offset not less than 2 inches or $1^{1}/_{2}$ inches of lightweight concrete or approved cellular concrete is placed over the subfloor, or $^{3}/_{4}$-inch wood finish flooring is installed at right angles to the supports. Fiber-cement underlayment shall comply with ASTM C1288 or ISO 8336 Category C. Allowable uniform live load at maximum span, based on deflection of $^{1}/_{360}$ of span, is 100 psf.

k. Unsupported edges shall have tongue-and-groove joints or shall be supported by blocking unless nominal $^{1}/_{4}$-inch-thick wood panel-type underlayment, fiber-cement underlayment with end and edge joints offset not less than 2 inches or $^{3}/_{4}$-inch wood finish flooring is installed at right angles to the supports. Fiber-cement underlayment shall comply with ASTM C1288 or ISO 8336 Category C. Allowable uniform live load at maximum span, based on deflection of $^{1}/_{360}$ of span, is 100 psf, except panels with a span rating of 48 on center are limited to 65 psf total uniform load at maximum span.

l. Allowable live load values at spans of 16 inches on center and 24 inches on center taken from referenced standard APA E30, *APA Engineered Wood Construction Guide*. Refer to referenced standard for allowable spans not listed in the table.

TABLE R503.2.1.1(2) ALLOWABLE SPANS FOR SANDED PLYWOOD COMBINATION SUBFLOOR UNDERLAYMENT[a]

IDENTIFICATION	SPACING OF JOISTS (inches)		
	16	20	24
Species group[b]	—	—	—
1	$^{1}/_{2}$	$^{5}/_{8}$	$^{3}/_{4}$
2, 3	$^{5}/_{8}$	$^{3}/_{4}$	$^{7}/_{8}$
4	$^{3}/_{4}$	$^{7}/_{8}$	1

For SI: 1 inch = 25.4 mm, 1 pound per square foot = 0.0479 kPa.

a. Plywood continuous over two or more spans and face grain perpendicular to supports. Unsupported edges shall be tongue-and-groove or blocked except where nominal $^{1}/_{4}$-inch-thick wood panel-type underlayment, fiber-cement underlayment or $^{3}/_{4}$-inch wood finish floor is used. Fiber-cement underlayment shall comply with ASTM C1288 or ISO 8336 Category C. Allowable uniform live load at maximum span based on deflection of $^{1}/_{360}$ of span is 100 psf.

b. Applicable to all grades of sanded exterior-type plywood.

R503.2.2 Allowable spans. The maximum allowable span for *wood structural panels* used as subfloor or combination subfloor underlayment shall be as set forth in Table R503.2.1.1(1), or APA E30. The maximum span for sanded plywood combination subfloor underlayment shall be as set forth in Table R503.2.1.1(2).

R503.2.3 Installation. *Wood structural panels* used as subfloor or combination subfloor underlayment shall be attached to wood framing in accordance with Table R602.3(1) and shall be attached to cold-formed steel framing in accordance with Table R505.3.1(2).

R503.3 Particleboard.

R503.3.1 Identification and grade. Particleboard shall conform to ANSI A208.1 and shall be so identified by a grade *mark* or certificate of inspection issued by an *approved agency*.

R503.3.2 Floor underlayment. Particleboard floor underlayment shall conform to Type PBU and shall be not less than $^1/_4$ inch (6.4 mm) in thickness.

R503.3.3 Installation. Particleboard underlayment shall be installed in accordance with the recommendations of the manufacturer and attached to framing in accordance with Table R602.3(1).

SECTION R504 PRESSURE PRESERVATIVE-TREATED WOOD FLOORS (ON GROUND)

R504.1 General. Pressure preservative-treated wood *basement* floors and floors on ground shall be designed to withstand axial forces and bending moments resulting from lateral soil pressures at the base of the exterior walls and floor live and dead loads. Floor framing shall be designed to meet joist deflection requirements in accordance with Section R301.

R504.1.1 Unbalanced soil loads. Unless special provision is made to resist sliding caused by unbalanced lateral soil loads, wood *basement* floors shall be limited to applications where the differential depth of fill on opposite exterior foundation walls is 2 feet (610 mm) or less.

R504.1.2 Construction. Joists in wood *basement* floors shall bear tightly against the narrow face of studs in the foundation wall or directly against a band joist that bears on the studs. Plywood subfloor shall be continuous over lapped joists or over butt joints between in-line joists. Sufficient blocking shall be provided between joists to transfer lateral forces at the base of the end walls into the floor system.

R504.1.3 Uplift and buckling. Where required, resistance to uplift or restraint against buckling shall be provided by interior bearing walls or properly designed stub walls anchored in the supporting soil below.

R504.2 Site preparation. The area within the foundation walls shall have all vegetation, topsoil and foreign material removed, and any fill material that is added shall be free of vegetation and foreign material. The fill shall be compacted to ensure uniform support of the pressure preservative-treated wood floor sleepers.

R504.2.1 Base. A minimum 4-inch-thick (102 mm) granular base of gravel having a maximum size of $^3/_4$ inch (19.1 mm) or crushed stone having a maximum size of $^1/_2$ inch (12.7 mm) shall be placed over the compacted earth.

R504.2.2 Moisture barrier. Polyethylene sheeting of minimum 6-mil (0.15 mm) thickness shall be placed over the granular base. Joints shall be lapped 6 inches (152 mm) and left unsealed. The polyethylene membrane shall be placed over the pressure preservative-treated wood sleepers and shall not extend beneath the footing plates of the exterior walls.

R504.3 Materials. Framing materials, including sleepers, joists, blocking and plywood subflooring, shall be pressure-preservative treated and dried after treatment in accordance with AWPA U1 (Commodity Specification A, Special Requirement 4.2), and shall bear the *label* of an accredited agency.

SECTION R505 COLD-FORMED STEEL FLOOR FRAMING

R505.1 Cold-formed steel floor framing. Elements shall be straight and free of any defects that would significantly affect structural performance. Cold-formed steel floor framing members shall be in accordance with the requirements of this section.

R505.1.1 Applicability limits. The provisions of this section shall control the construction of cold-formed steel floor framing for buildings not greater than 60 feet (18 288 mm) in length perpendicular to the joist span, not greater than 40 feet (12 192 mm) in width parallel to the joist span and less than or equal to three *stories* above *grade plane*. Cold-formed steel floor framing constructed in accordance with the provisions of this section shall be limited to sites where the ultimate design wind speed is less than 140 miles per hour (63 m/s), Exposure Category B or C, and the ground snow load is less than or equal to 70 pounds per square foot (3.35 kPa).

R505.1.1.1 Alternate applications. Cold-formed steel floor framing for buildings exceeding the applicability limits of Section R505.1.1 is permitted to be designed and constructed in accordance with AISI S230, subject to the limits therein.

R505.1.2 In-line framing. Where supported by cold-formed steel-framed walls in accordance with Section R603, cold-formed steel floor framing shall be constructed with floor joists located in-line with load-bearing studs located below the joists in accordance with the tolerances specified in AISI S240, Section B1.2.3.

R505.1.3 Floor trusses. Cold-formed steel trusses shall be designed, braced and installed in accordance with AISI S230, Section D8. In the absence of specific bracing requirements, trusses shall be braced in accordance with accepted industry practices, such as the SBCA *Cold-Formed Steel Building Component Safety Information (CFSBCSI), Guide to Good Practice for Handling, Installing & Bracing of Cold-Formed Steel Trusses.* Truss members shall not be notched, cut or altered in any manner without an *approved* design.

R505.2 Structural framing. Load-bearing cold-formed steel floor framing members shall be in accordance with this section.

R505.2.1 Material. Load-bearing cold-formed steel framing members shall be cold formed to shape from structural quality sheet steel complying with the requirements of AISI S240, Section A3.

R505.2.2 Corrosion protection. Load-bearing cold-formed steel framing shall have a metallic coating complying with AISI S240, Section A4.

R505.2.3 Dimension, thickness and material grade. Load-bearing cold-formed steel floor framing members shall comply with AISI S230, Section A4.3 and material grade requirements as specified in AISI S230, Section A4.4.

R505.2.4 Identification. Load-bearing cold-formed steel framing members shall meet the product identification requirements of AISI S240, Section A5.5.

R505.2.5 Fastening. Screws for steel-to-steel connections shall be installed with a minimum edge distance and center-to-center spacing of $^1/_2$ inch (12.7 mm), shall be self-drilling tapping, and shall conform to ASTM C1513. Floor sheathing shall be attached to cold-formed steel joists with minimum No. 8 self-drilling tapping screws that conform to ASTM C1513. Screws attaching floor sheathing to cold-formed steel joists shall have a minimum head diameter of 0.292 inch (7.4 mm) with countersunk heads and shall be installed with a minimum edge distance of $^3/_8$ inch (9.5 mm). Gypsum board ceilings shall be attached to cold-formed steel joists with minimum No. 6 screws conforming to ASTM C954 or ASTM C1513 with a bugle-head style and shall be installed in accordance with Section R702. For all connections, screws shall extend through the steel not fewer than three exposed threads. Fasteners shall have a rust-inhibitive coating suitable for the installation in which they are being used, or be manufactured from material not susceptible to corrosion.

R505.2.6 Web holes, web hole reinforcing and web hole patching. Web holes in floor framing members shall comply with the conditions as prescribed in AISI S230, Section A4.5. Web holes not in compliance with the conditions as prescribed in AISI S230, Section A4.5 shall be reinforced in accordance with the provisions of AISI S230, Section A4.6 or patched in accordance with the provisions of AISI S230, Section A4.7.

R505.3 Floor construction. Cold-formed steel floors shall be constructed in accordance with this section.

R505.3.1 Floor-to-foundation or load-bearing wall connections. Cold-formed steel-framed floors shall be anchored to foundations, wood sills or *load-bearing walls* in accordance with Table R505.3.1(1) and Figure R505.3.1(1), R505.3.1(2), R505.3.1(3), R505.3.1(4), R505.3.1(5) or R505.3.1(6). Anchor bolts shall be located not more than 12 inches (305 mm) from corners or the termination of bottom tracks. Continuous cold-formed steel joists supported by interior *load-bearing walls* shall be constructed in accordance with Figure R505.3.1(7). Lapped cold-formed steel joists shall be constructed in accordance with Figure R505.3.1(8). End floor joists constructed on foundation walls parallel to the joist span shall be doubled unless a C-shaped bearing stiffener, sized in accordance with Section R505.3.4, is installed web-to-web with the floor joist beneath each supported wall stud, as shown in Figure R505.3.1(9). Fastening of cold-formed steel joists to other framing members shall be in accordance with Section R505.2.5 and Table R505.3.1(2).

TABLE R505.3.1(1)
FLOOR-TO-FOUNDATION OR BEARING WALL CONNECTION REQUIREMENTS[a, b]

FRAMING CONDITION	BASIC ULTIMATE WIND SPEED (mph) AND EXPOSURE	
	110 mph Exposure Category C or less than 139 mph Exposure Category B	Less than 139 mph Exposure Category C
Floor joist to wall track of exterior wall in accordance with Figure R505.3.1(1)	2-No. 8 screws	3-No. 8 screws
Rim track or end joist to load-bearing wall top track in accordance with Figure R505.3.1(1)	1-No. 8 screw at 24 inches o.c.	1-No. 8 screw at 24 inches o.c.
Rim track or end joist to wood sill in accordance with Figure R505.3.1(2)	Steel plate spaced at 4 feet o.c. with 4-No. 8 screws and 4-10d or 6-8d common nails	Steel plate spaced at 2 feet o.c. with 4-No. 8 screws and 4-10d or 6-8d common nails
Rim track or end joist to foundation in accordance with Figure R505.3.1(3)	$^1/_2$-inch minimum diameter anchor bolt and clip angle spaced at 6 feet o.c. with 8-No. 8 screws	$^1/_2$-inch minimum diameter anchor bolt and clip angle spaced at 4 feet o.c. with 8-No. 8 screws
Cantilevered joist to foundation in accordance with Figure R505.3.1(4)	$^1/_2$-inch minimum diameter anchor bolt and clip angle spaced at 6 feet o.c. with 8-No. 8 screws	$^1/_2$-inch minimum diameter anchor bolt and clip angle spaced at 4 feet o.c. with 8-No. 8 screws
Cantilevered joist to wood sill in accordance with Figure R505.3.1(5)	Steel plate spaced at 4 feet o.c. with 4-No. 8 screws and 4-10d or 6-8d common nails	Steel plate spaced at 2 feet o.c. with 4-No. 8 screws and 4-10d or 6-8d common nails
Cantilevered joist to exterior load-bearing wall track in accordance with Figure R505.3.1(6)	2-No. 8 screws	3-No. 8 screws

For SI: 1 inch = 25.4 mm, 1 mile per hour = 0.447 m/s, 1 foot = 304.8 mm.

a. Anchor bolts are to be located not more than 12 inches from corners or the termination of bottom tracks such as at door openings or corners. Bolts extend not less than 15 inches into masonry or 7 inches into concrete. Anchor bolts connecting cold-formed steel framing to the foundation structure are to be installed so that the distance from the center of the bolt hole to the edge of the connected member is not less than one and one-half bolt diameters.

b. All screw sizes shown are minimum.

TABLE R505.3.1(2)
FLOOR FASTENING SCHEDULE[a]

DESCRIPTION OF BUILDING ELEMENTS	NUMBER AND SIZE OF FASTENERS	SPACING OF FASTENERS
Floor joist to track of an interior load-bearing wall in accordance with Figures R505.3.1(7) and R505.3.1(8)	2-No. 8 screws	Each joist
Floor joist to track at end of joist	2-No. 8 screws	One per flange or two per bearing stiffener
Subfloor to floor joists	No. 8 screws	6 in. o.c. on edges and 12 in. o.c. at intermediate supports

For SI: 1 inch = 25.4 mm.
a. All screw sizes shown are minimum.

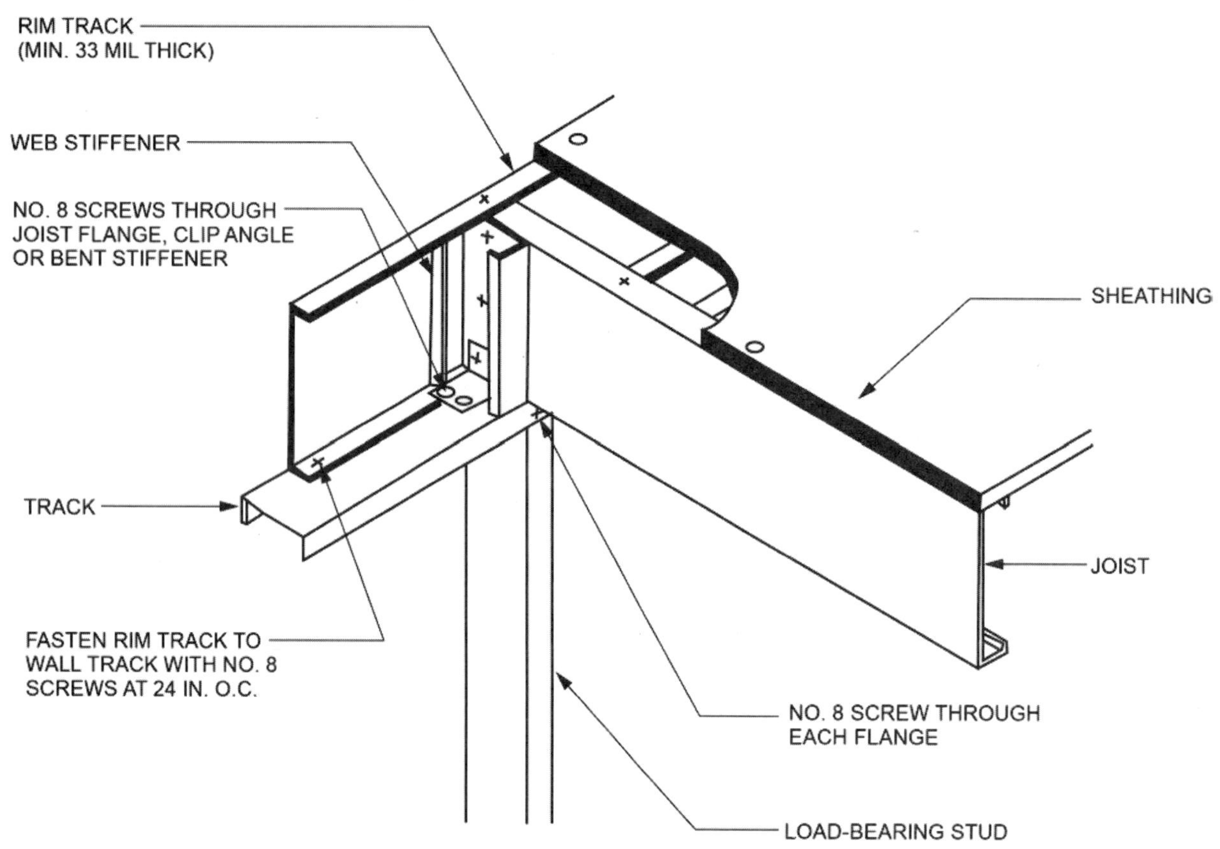

For SI: 1 mil = 0.0254 mm, 1 inch = 25.4 mm.

FIGURE R505.3.1(1)
FLOOR-TO-EXTERIOR LOAD-BEARING WALL STUD CONNECTION

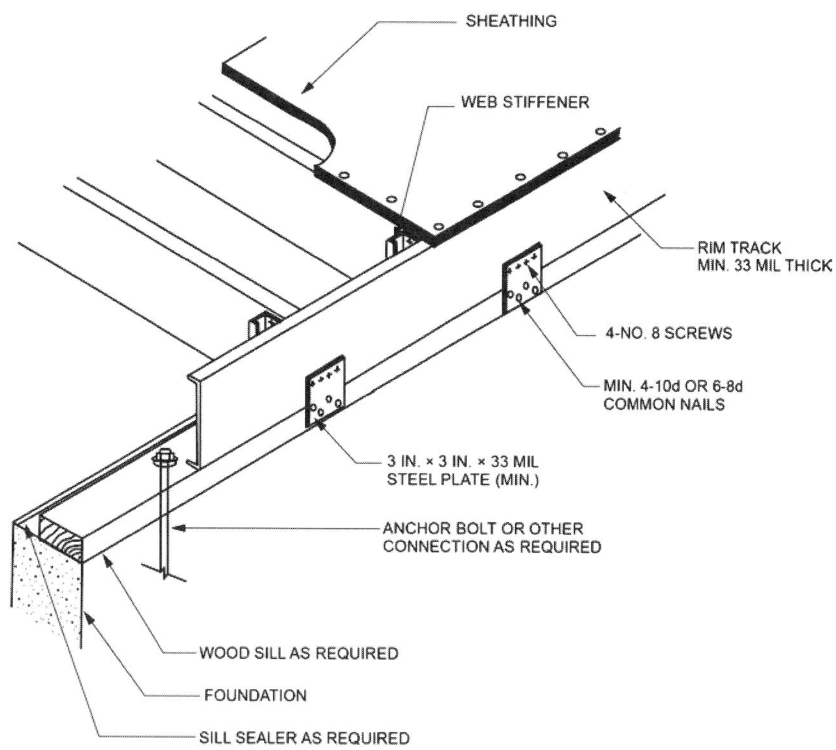

For SI: 1 mil = 0.0254 mm, 1 inch = 25.4 mm.

FIGURE R505.3.1(2)
FLOOR-TO-WOOD-SILL CONNECTION

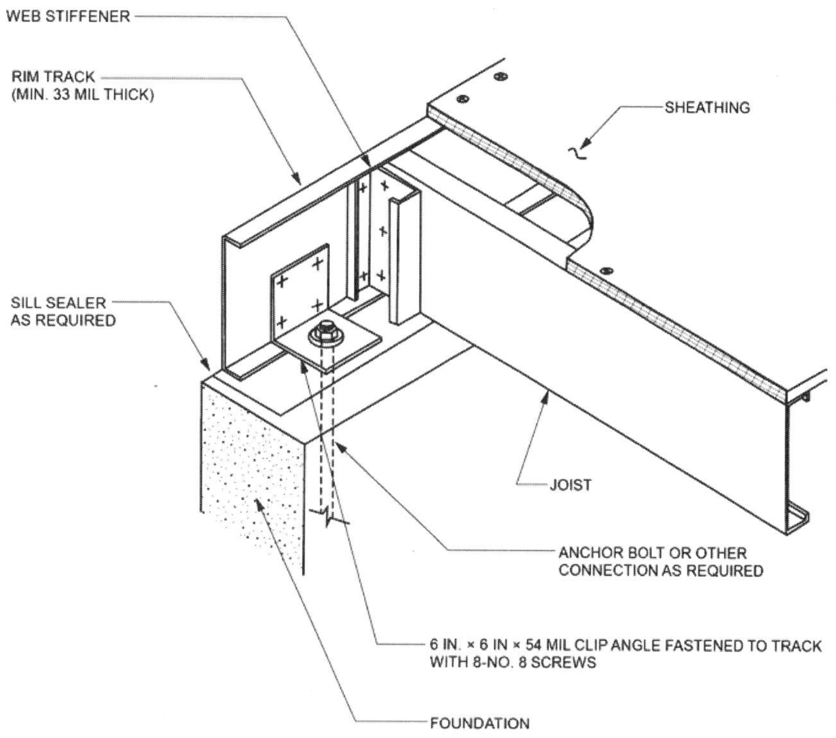

For SI: 1 mil = 0.0254 mm, 1 inch = 25.4 mm.

FIGURE R505.3.1(3)
FLOOR-TO-FOUNDATION CONNECTION

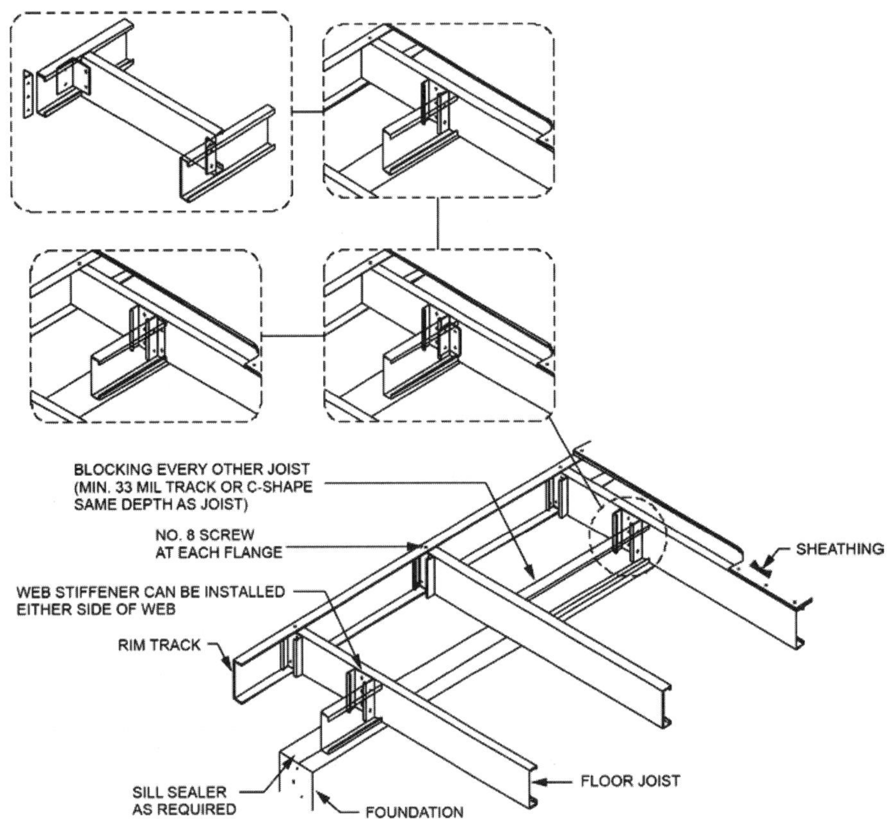

For SI: 1 mil = 0.0254 mm.

FIGURE R505.3.1(4)
CANTILEVERED FLOOR-TO-FOUNDATION CONNECTION

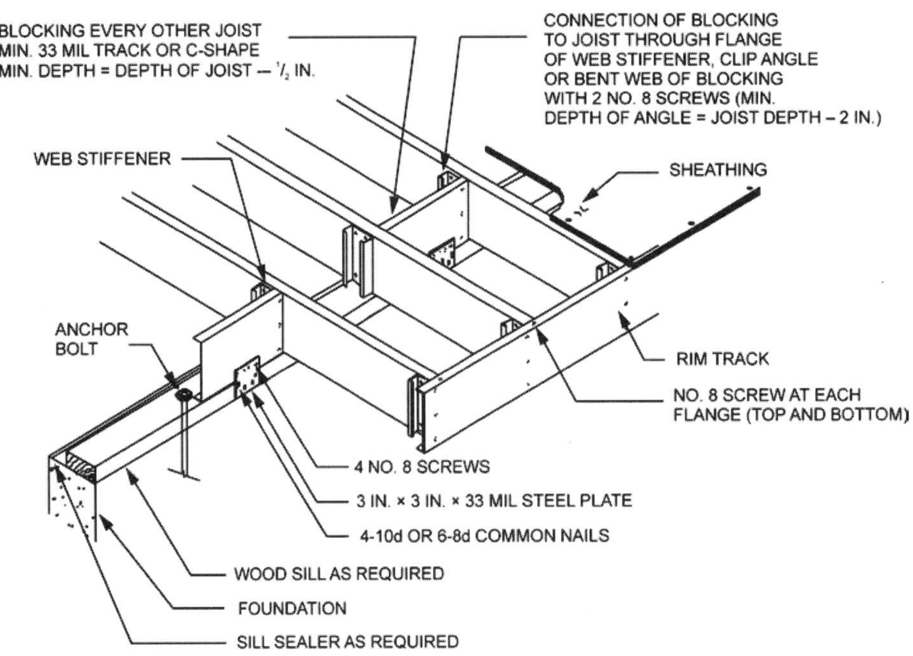

For SI: 1 mil = 0.0254 mm, 1 inch = 25.4 mm.

FIGURE R505.3.1(5)
CANTILEVERED FLOOR-TO-WOOD-SILL CONNECTION

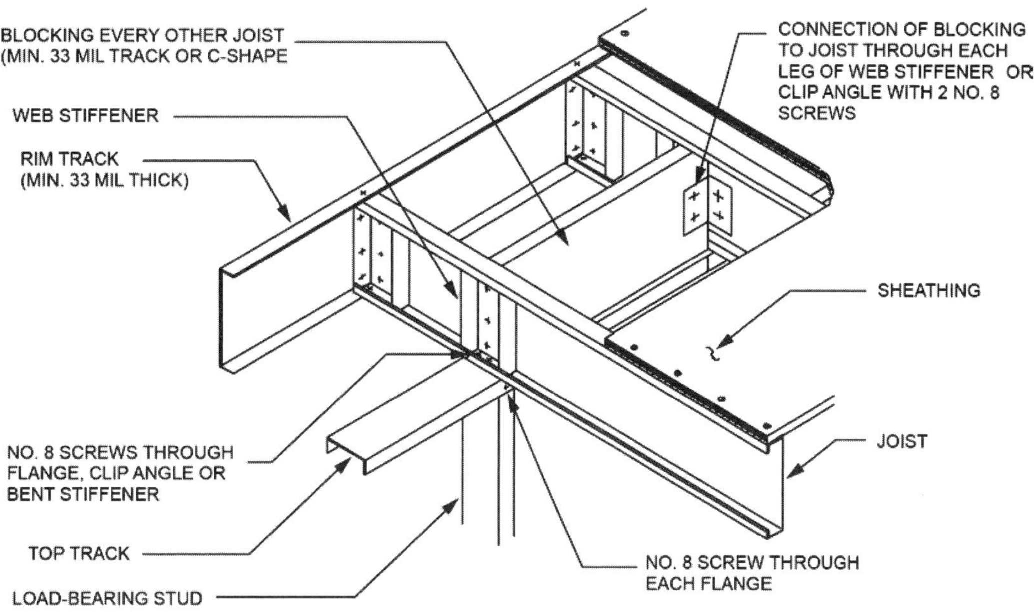

For SI: 1 mil = 0.0254 mm.

FIGURE R505.3.1(6)
CANTILEVERED FLOOR TO EXTERIOR LOAD-BEARING WALL CONNECTION

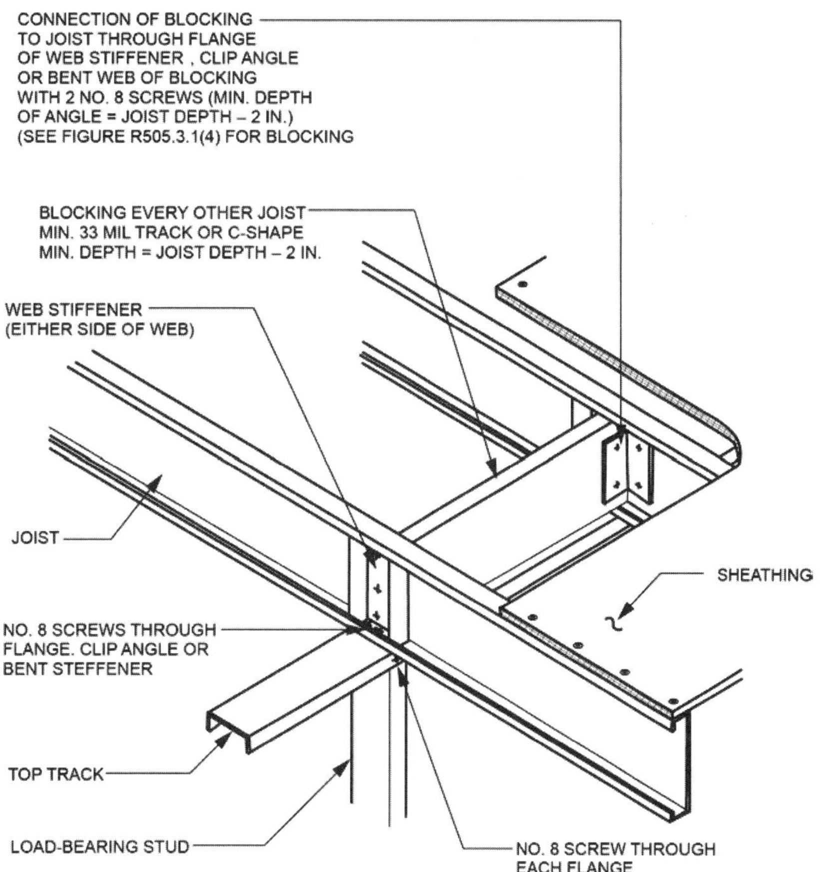

For SI: 1 mil = 0.0254 mm, 1 inch = 25.4 mm.

FIGURE R505.3.1(7)
CONTINUOUS SPAN JOIST SUPPORTED ON INTERIOR LOAD-BEARING WALL

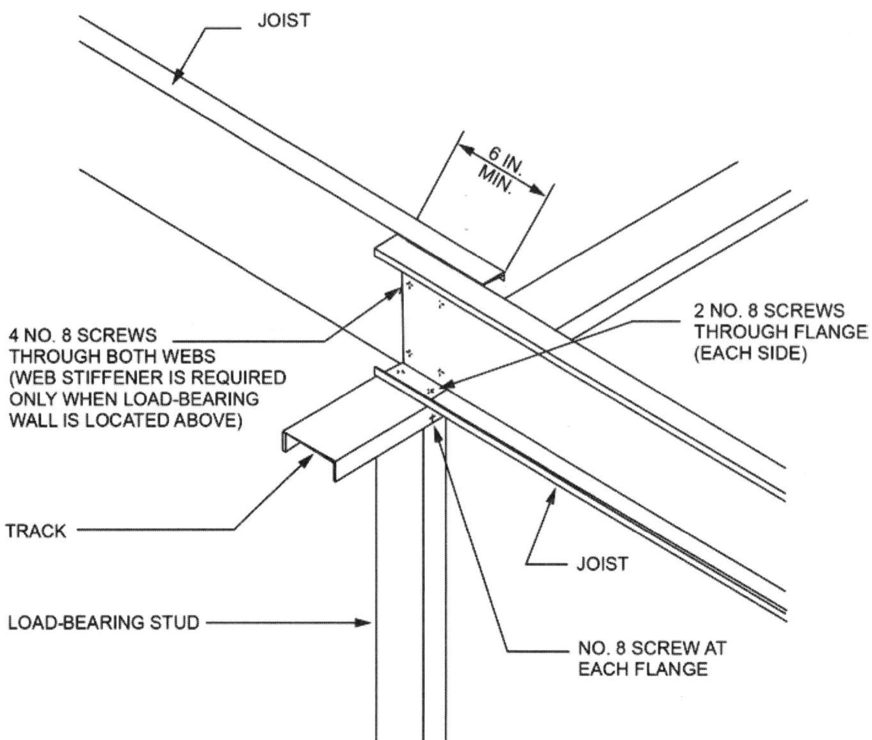

For SI: 1 inch = 25.4 mm.

FIGURE R505.3.1(8)
LAPPED JOISTS SUPPORTED ON INTERIOR LOAD-BEARING WALL

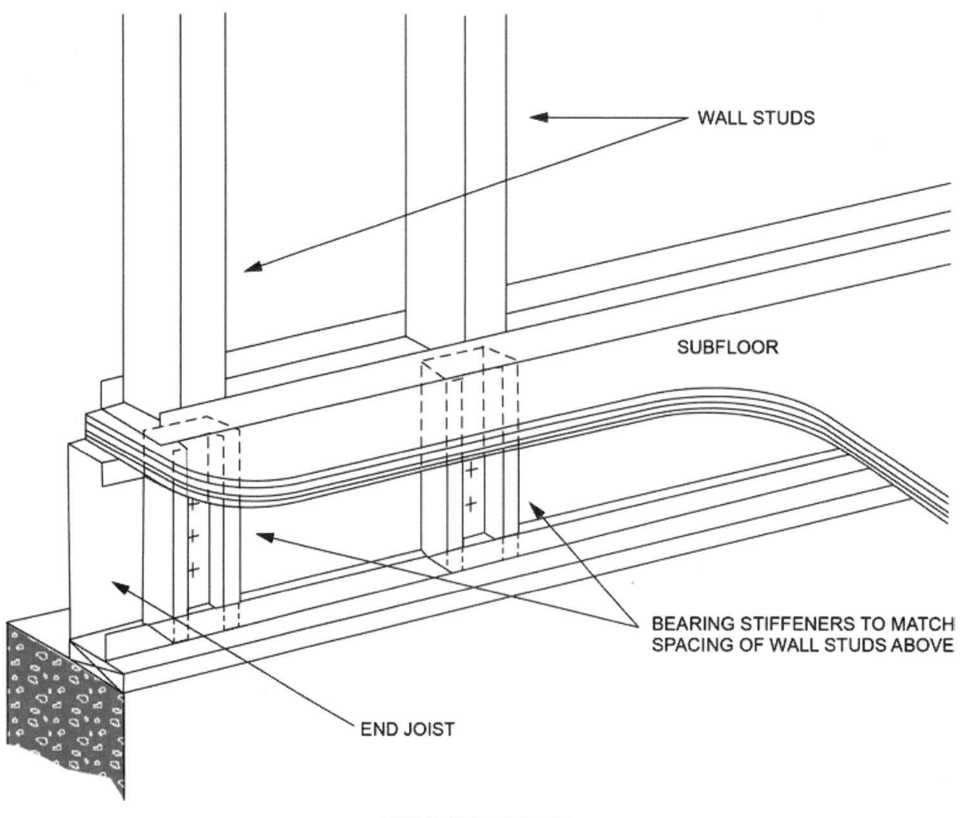

FIGURE R505.3.1(9)
BEARING STIFFENERS FOR END JOISTS

R505.3.2 Minimum floor joist sizes. Floor joist size and thickness shall be determined in accordance with the limits set forth in Table R505.3.2 for single or continuous spans. Where continuous joist members are used, the interior bearing supports shall be located within 2 feet (610 mm) of midspan of the cold-formed steel joists, and the individual spans shall not exceed the spans in Table R505.3.2. Floor joists shall have a bearing support length of not less than $1^1/_2$ inches (38 mm) for exterior wall supports and $3^1/_2$ inches (89 mm) for interior wall supports. Tracks shall be not less than 33 mils (0.84 mm) thick except where used as part of a floor header or trimmer in accordance with Section R505.3.8. Bearing stiffeners shall be installed in accordance with Section R505.3.4.

R505.3.3 Joist bracing and blocking. Joist bracing and blocking shall be in accordance with this section.

R505.3.3.1 Joist top flange bracing. The top flanges of cold-formed steel joists shall be laterally braced by the application of floor sheathing fastened to the joists in accordance with Section R505.2.5 and Table R505.3.1(2).

R505.3.3.2 Joist bottom flange bracing/blocking. Floor joists with spans that exceed 12 feet (3658 mm) shall have the bottom flanges laterally braced in accordance with one of the following:

1. Gypsum board installed with minimum No. 6 screws in accordance with Section R702.
2. Continuous steel straps installed in accordance with Figure R505.3.3.2(1). Steel straps shall be spaced at not greater than 12 feet (3658 mm) on center and shall be not less than $1^1/_2$ inches (38 mm) in width and 33 mils (0.84 mm) in thickness. Straps shall be fastened to the bottom flange of each joist with one No. 8 screw, fastened to blocking with two No. 8 screws, and fastened at each end (of strap) with two No. 8 screws. Blocking in accordance with Figure R505.3.3.2(1) or R505.3.3.2(2) shall be installed between joists at each end of the continuous strapping and at a maximum spacing of 12 feet (3658 mm) measured along the continuous strapping (perpendicular to the joist run). Blocking shall also be located at the termination of all straps. As an alternative to blocking at the ends, anchoring the strap to a stable building component with two No. 8 screws shall be permitted.

R505.3.3.3 Blocking at interior bearing supports. Blocking is not required for continuous back-to-back floor joists at bearing supports. Blocking shall be installed between every other joist for single continuous floor joists across bearing supports in accordance with Figure R505.3.1(7). Blocking shall consist of C-shaped or track section with a minimum thickness of 33 mils (0.84 mm). Blocking shall be fastened to each adjacent joist through a 33-mil (0.84 mm) clip angle, bent web of blocking or flanges of web stiffeners with two No. 8 screws on each side. The minimum depth of the blocking shall be equal to the depth of the joist minus 2 inches (51 mm). The minimum length of the angle shall be equal to the depth of the joist minus 2 inches (51 mm).

TABLE R505.3.2
ALLOWABLE SPANS FOR COLD-FORMED STEEL JOISTS—SINGLE OR CONTINUOUS SPANS[a, b, c, d, e, f]

JOIST DESIGNATION	30 PSF LIVE LOAD				40 PSF LIVE LOAD			
	Spacing (inches)				Spacing (inches)			
	12	16	19.2	24	12	16	19.2	24
550S162-33	11′-8″	10′-4″	9′-5″	8′-5″	10′-7″	9′-2″	8′-5″	7′-6″
550S162-43	12′-8″	11′-6″	10′-8″	10′-5″	11′-6″	10′-4″	9′-10″	9′-3″
550S162-54	13′-7″	12′-4″	11′-7″	10′-9″	12′-4″	11′-3″	10′-7″	9′-10″
550S162-68	14′-7″	13′-3″	12′-6″	11′-7″	13′-3″	12′-0″	11′-4″	10′-6″
800S162-33	14′-6″	12′-6″	11′-5″	10′-3″	12′-10″	11′-1″	10′-2″	9′-1″
800S162-43	17′-0″	15′-1″	13′-9″	12′-4″	15′-5″	13′-5″	12′-3″	10′-11″
800S162-54	18′-3″	16′-7″	15′-8″	14′-6″	16′-7″	15′-1″	14′-2″	13′-2″
800S162-68	19′-9″	17′-11″	16′-11″	15′-8″	17′-11″	16′-3″	15′-4″	14′-3″
1000S162-43	19′-4″	16′-9″	15′-3″	13′-8″	17′-2″	14′-10″	13′-7″	12′-2″
1000S162-54	21′-9″	19′-9″	18′-7″	17′-3″	19′-9″	18′-0″	16′-11″	15′-8″
1000S162-68	23′-7″	21′-5″	20′-2″	18′-9″	21′-5″	19′-6″	18′-4″	17′-0″
1200S162-54	25′-1″	22′-10″	21′-6″	19′-9″	22′-10″	20′-9″	19′-6″	17′-6″
1200S162-68	27′-3″	24′-9″	23′-4″	21′-8″	24′-9″	22′-6″	21′-2″	19′-8″

For SI: 1 inch = 25.4 mm, 1 foot = 304.8 mm, 1 pound per square foot = 0.0479 kPa, 1 mil = 0.0254 mm.

a. Deflection criteria: *L*/480 for live loads, *L*/240 for total loads.

b. Floor dead load = 10 psf.

c. Table provides the maximum clear span in feet and inches.

d. Bearing stiffeners are to be installed at all support points and concentrated loads.

e. Minimum Grade 33 ksi steel shall be used for 33 mil and 43 mil thickness. Minimum Grade 50 ksi steel shall be used for 54 and 68 mil thickness.

f. Table R505.3.2 is not applicable for 800S162-33 and 1000S162-43 continuous joist members.

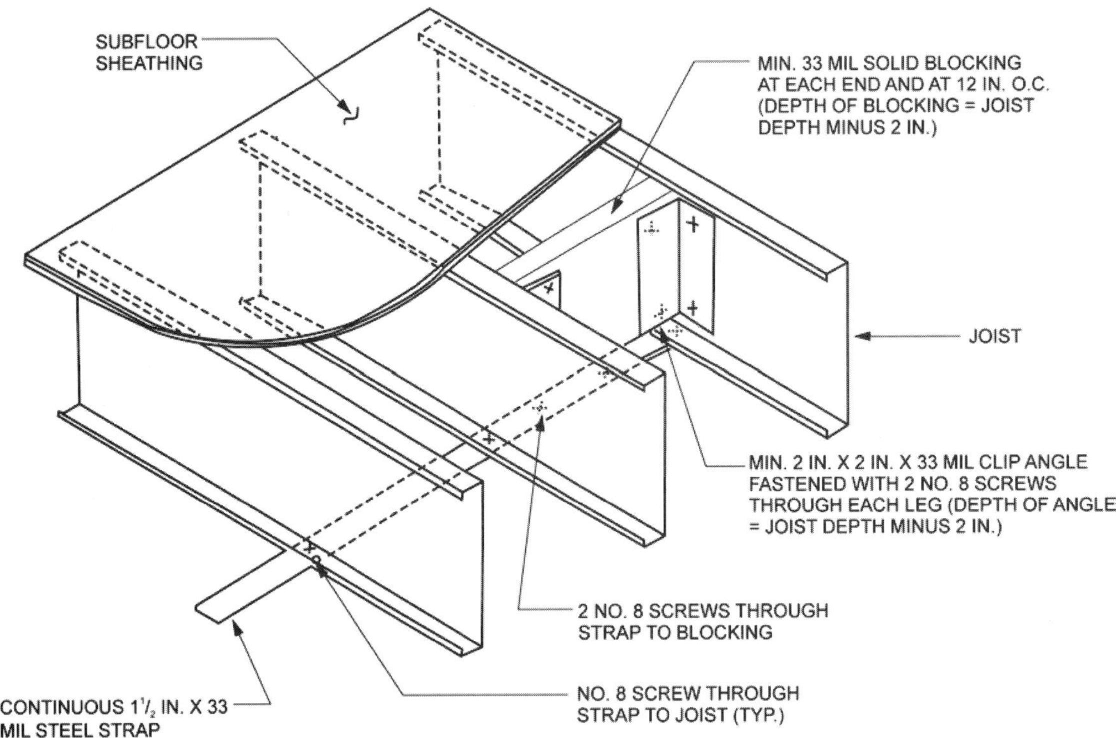

For SI: 1 mil = 0.0254 mm, 1 inch = 25.4 mm.

FIGURE R505.3.3.2(1)
JOIST BLOCKING (SOLID)

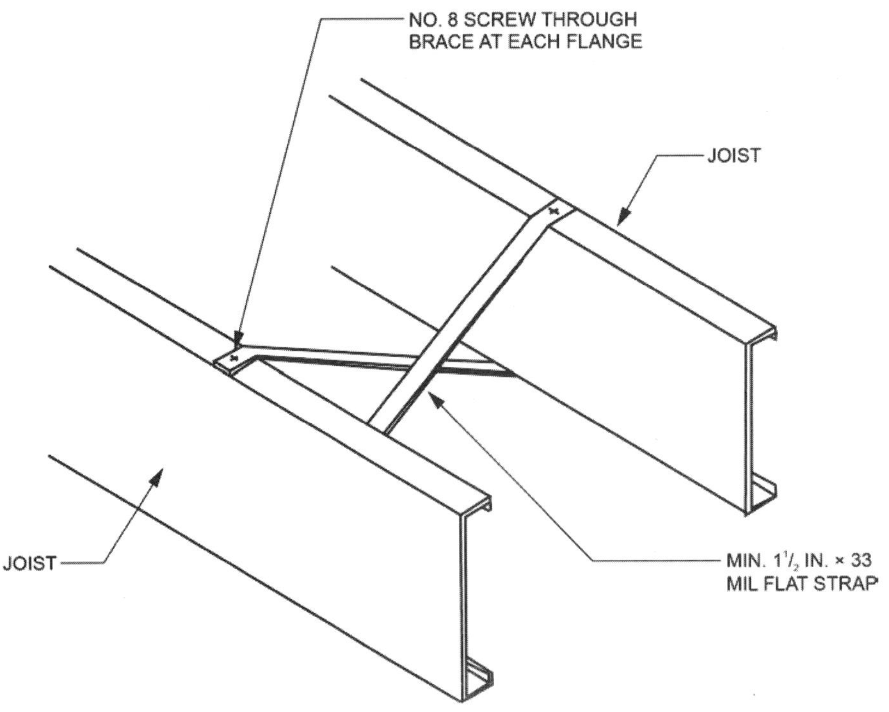

For SI: 1 mil = 0.0254 mm, 1 inch = 25.4 mm.

FIGURE R505.3.3.2(2)
JOIST BLOCKING (STRAP)

R505.3.3.4 Blocking at cantilevers. Blocking shall be installed between every other joist over cantilever bearing supports in accordance with Figure R505.3.1(4), R505.3.1(5) or R505.3.1(6). Blocking shall consist of C-shaped or track section with minimum thickness of 33 mils (0.84 mm). Blocking shall be fastened to each adjacent joist through bent web of blocking, 33 mil clip angle or flange of web stiffener with two No. 8 screws at each end. The depth of the blocking shall be equal to the depth of the joist. The minimum length of the angle shall be equal to the depth of the joist minus 2 inches (51 mm). Blocking shall be fastened through the floor sheathing and to the support with three No. 8 screws (top and bottom).

R505.3.4 Bearing stiffeners. Bearing stiffeners shall be installed at each joist bearing location in accordance with this section, except for joists lapped over an interior support not carrying a *load-bearing wall* above. Floor joists supporting jamb studs with multiple members shall have two bearing stiffeners in accordance with Figure R505.3.4(1). Bearing stiffeners shall be fabricated from a C-shaped, track or clip angle member in accordance with one of the following:

1. C-shaped bearing stiffeners:
   1.1. Where the joist is not carrying a *load-bearing wall* above, the bearing stiffener shall be a minimum 33 mil (0.84 mm) thickness.
   1.2. Where the joist is carrying a *load-bearing wall* above, the bearing stiffener shall be not less than the same designation thickness as the wall stud above.
2. Track bearing stiffeners:
   2.1. Where the joist is not carrying a *load-bearing wall* above, the bearing stiffener shall be a minimum 43 mil (1.09 mm) thickness.
   2.2. Where the joist is carrying a *load-bearing wall* above, the bearing stiffener shall be not less than one designation thickness greater than the wall stud above.

The minimum length of a bearing stiffener shall be the depth of member being stiffened minus $^3/_8$ inch (9.5 mm). Each bearing stiffener shall be fastened to the web of the member it is stiffening as shown in Figure R505.3.4(2).

R505.3.5 Cutting and notching. Flanges and lips of load-bearing cold-formed steel floor framing members shall not be cut or notched.

R505.3.6 Floor cantilevers. Floor cantilevers for the top floor of a two- or three-story building or the first floor of a one-story building shall not exceed 24 inches (610 mm). Cantilevers, not exceeding 24 inches (610 mm) and supporting two stories and roof (first floor of a two-story building), shall be permitted provided that all cantilevered joists are doubled (nested or back-to-back). The doubled cantilevered joists shall extend not less than 6 feet (1829 mm) toward the inside and shall be fastened with not less than two No. 8 screws spaced at 24 inches (610 mm) on center through the webs (for back-to-back) or flanges (for nested joists).

R505.3.7 Splicing. Joists and other structural members shall not be spliced without an *approved* design. Splicing of tracks shall conform to Figure R505.3.7.

R505.3.8 Framing of floor openings. Openings in floors shall be framed with header and trimmer joists. Header joist spans shall not exceed 6 feet (1829 mm) or 8 feet (2438 mm) in length in accordance with Figure R505.3.8(1) or R505.3.8(2), respectively. Header and trimmer joists shall be fabricated from joist and track members, having a minimum size and thickness at least equivalent to the adjacent floor joists, and shall be installed in accordance with Figures R505.3.8(1), R505.3.8(2), R505.3.8(3) and R505.3.8(4). Each header joist shall be connected to trimmer joists with four 2-inch by 2-inch (51-mm by 51-mm) clip angles. Each clip angle shall be fastened to both the header and trimmer joists with four No. 8 screws, evenly spaced, through each leg of the clip angle. The clip angles shall have a thickness not less than that of the floor joist. Each track section for a built-up header or trimmer joist shall extend the full length of the joist (continuous).

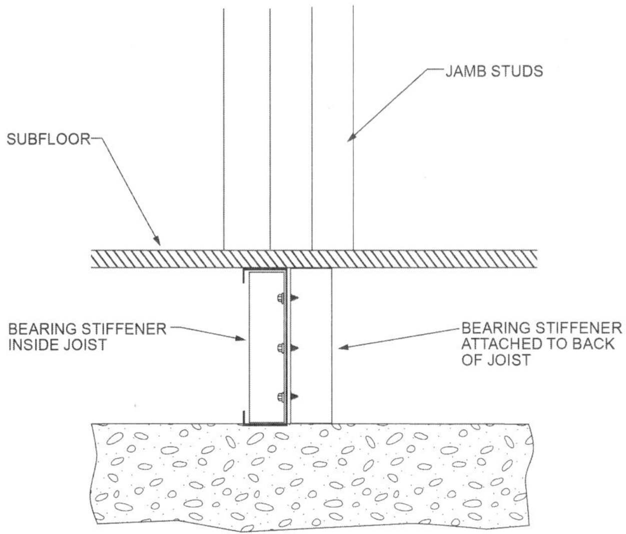

FIGURE R505.3.4(1)
BEARING STIFFENERS UNDER JAMB STUDS

SECTION R506
CONCRETE FLOORS (ON GROUND)

R506.1 General. Concrete slab-on-ground floors shall be designed and constructed in accordance with the provisions of this section or ACI 332. Floors shall be a minimum $3^1/_2$ inches (89 mm) thick (for *expansive soils*, see Section R403.1.8). The specified compressive strength of concrete shall be as set forth in Section R402.2.

R506.2 Site preparation. The area within the foundation walls shall have all vegetation, top soil and foreign material removed.

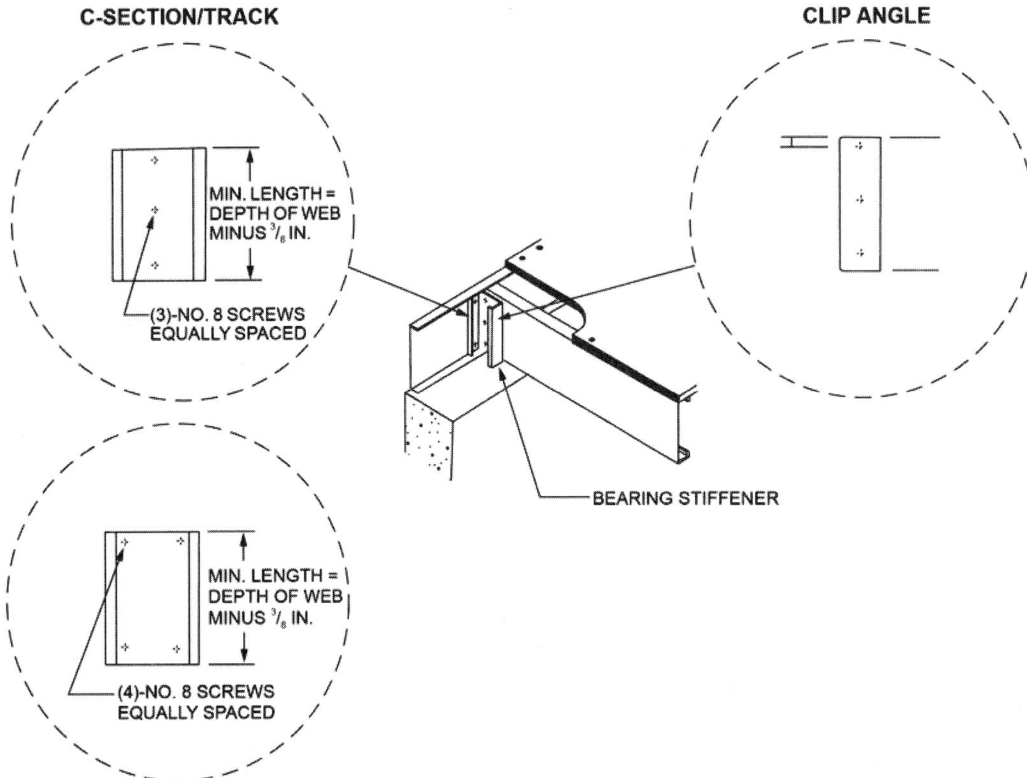

For SI: 1 inch = 25.4 mm.

FIGURE R505.3.4(2)
BEARING STIFFENER

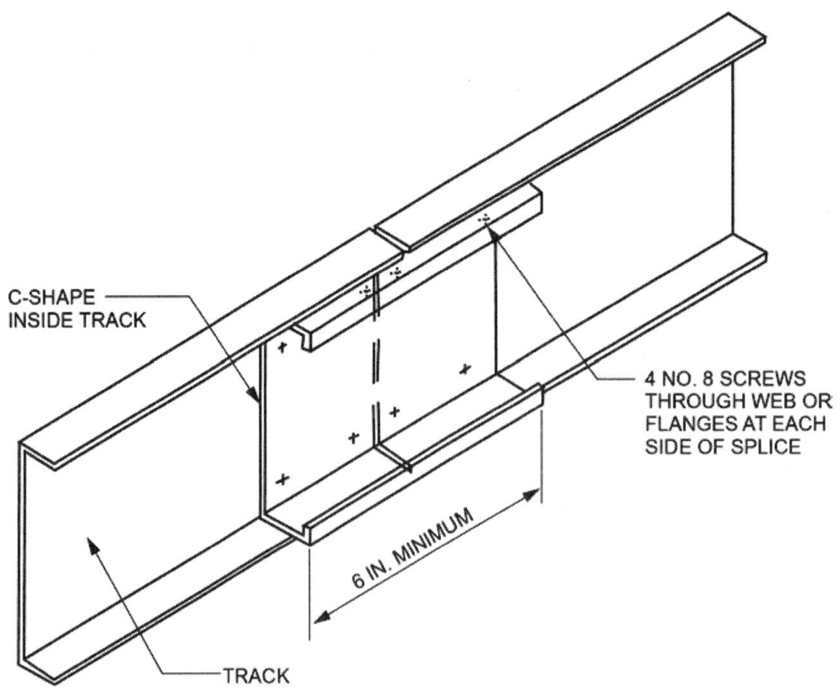

For SI: 1 inch = 25.4 mm.

FIGURE R505.3.7
TRACK SPLICE

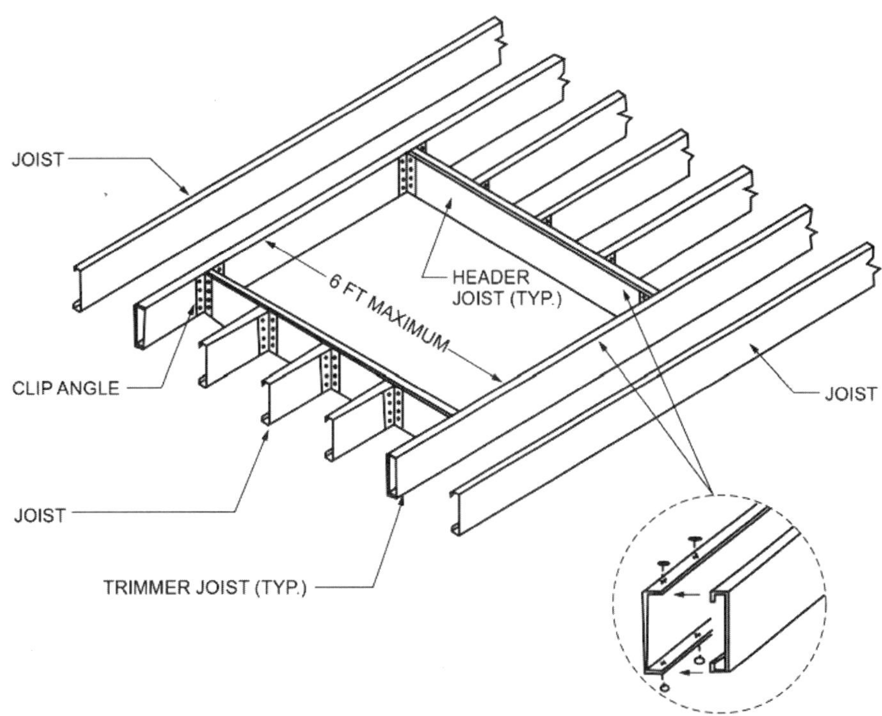

For SI: 1 foot = 304.8 mm.

FIGURE R505.3.8(1)
COLD-FORMED STEEL FLOOR CONSTRUCTION—6-FOOT FLOOR OPENING

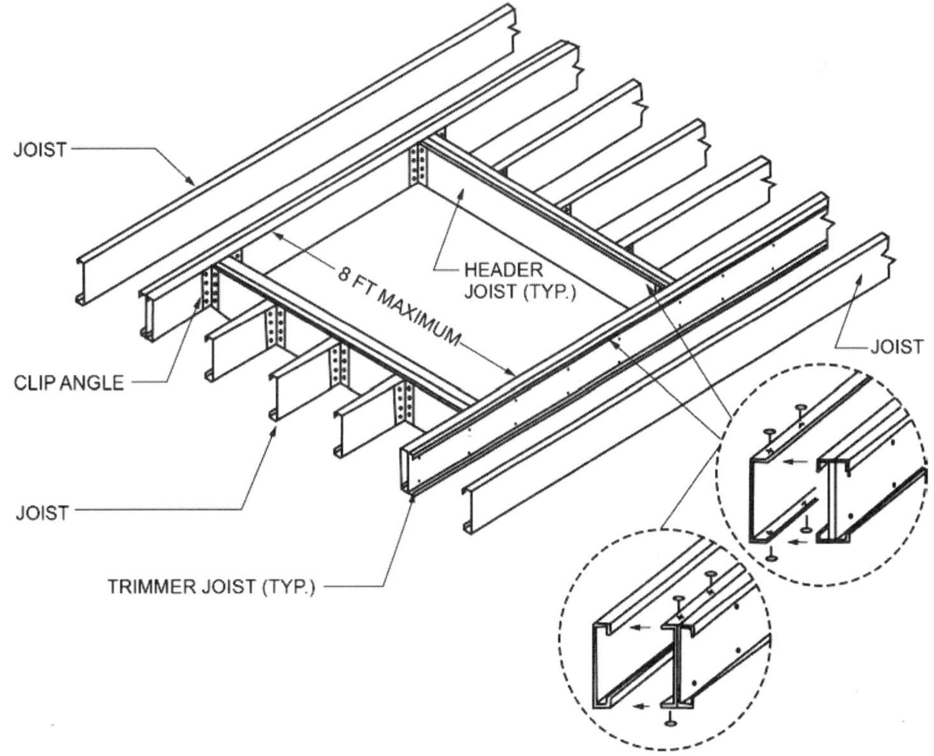

For SI: 1 foot = 304.8 mm.

FIGURE R505.3.8(2)
COLD-FORMED STEEL FLOOR CONSTRUCTION—8-FOOT FLOOR OPENING

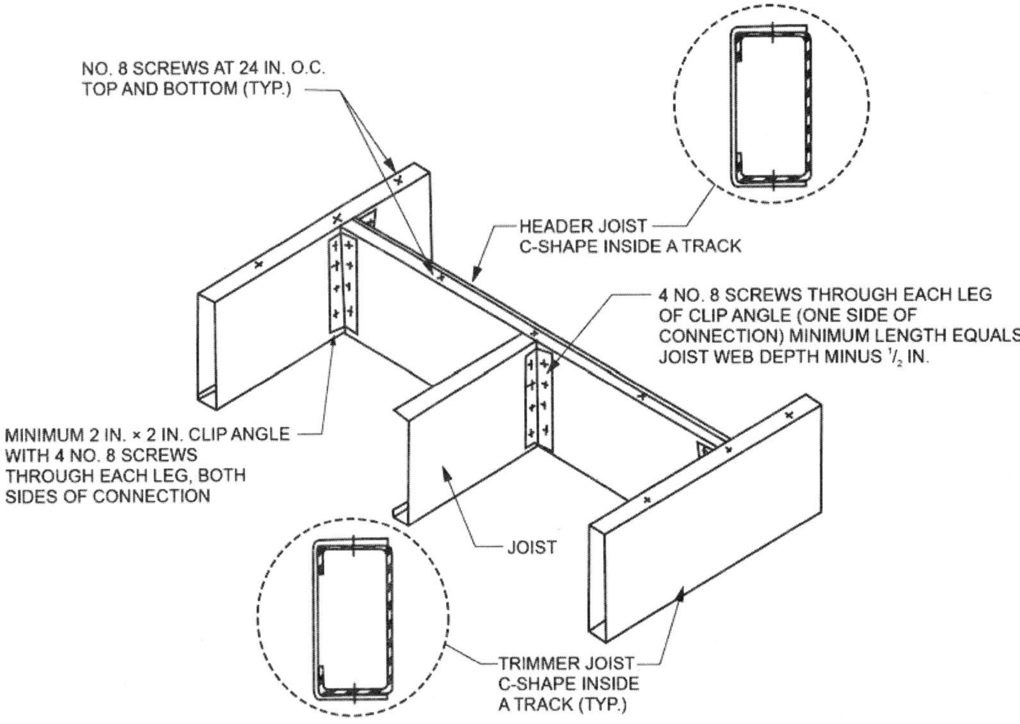

For SI: 1 inch = 25.4 mm, 1 foot = 304.8 mm.

FIGURE R505.3.8(3)
COLD-FORMED STEEL FLOOR CONSTRUCTION: FLOOR HEADER TO TRIMMER CONNECTION—6-FOOT OPENING

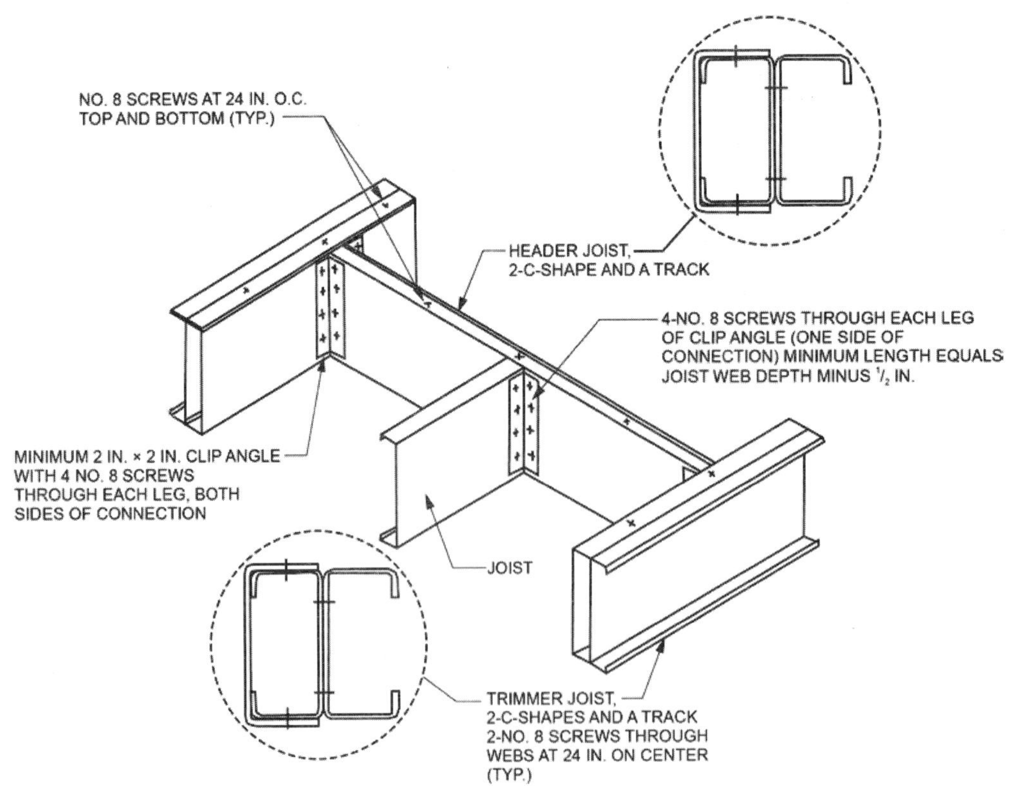

For SI: 1 inch = 25.4 mm, 1 foot = 304.8 mm.

FIGURE R505.3.8(4)
COLD-FORMED STEEL FLOOR CONSTRUCTION: FLOOR HEADER TO TRIMMER CONNECTION—8-FOOT OPENING

R506.2.1 Fill. Fill material shall be free of vegetation and foreign material. The fill shall be compacted to ensure uniform support of the slab, and except where *approved*, the fill depths shall not exceed 24 inches (610 mm) for clean sand or gravel and 8 inches (203 mm) for earth.

R506.2.2 Base. A 4-inch-thick (102 mm) base course consisting of clean graded sand, gravel, crushed stone, crushed concrete or crushed blast-furnace slag passing a 2-inch (51 mm) sieve shall be placed on the prepared subgrade where the slab is below *grade*.

Exception: A base course is not required where the concrete slab is installed on well-drained or sand-gravel mixture soils classified as Group I according to the United Soil Classification System in accordance with Table R405.1.

R506.2.3 Vapor retarder. A minimum 10-mil (0.010 inch; 0.254 mm) vapor retarder conforming to ASTM E1745 Class A requirements with joints lapped not less than 6 inches (152 mm) shall be placed between the concrete floor slab and the base course or the prepared subgrade where a base course does not exist.

Exception: The vapor retarder is not required for the following:

1. Garages, utility buildings and other unheated *accessory structures*.
2. For unheated storage rooms having an area of less than 70 square feet (6.5 m^2) and carports.
3. Driveways, walks, patios and other flatwork not likely to be enclosed and heated at a later date.
4. Where *approved* by the *building official*, based on local site conditions.

R506.2.4 Reinforcement support. Where provided in slabs-on-ground, reinforcement shall be supported to remain in place from the center to upper one-third of the slab for the duration of the concrete placement.

SECTION R507
EXTERIOR DECKS

R507.1 Decks. Wood-framed decks shall be in accordance with this section. Decks shall be designed for the *live load* required in Section R301.5 or the ground snow load indicated in Table R301.2, whichever is greater. For decks using materials and conditions not prescribed in this section, refer to Section R301.

R507.2 Materials. Materials used for the construction of decks shall comply with this section.

R507.2.1 Wood materials. Wood materials shall be No. 2 grade or better lumber, preservative-treated in accordance with Section R317, or *approved*, naturally durable lumber, and termite protected where required in accordance with Section R318. Where design in accordance with Section R301 is provided, wood structural members shall be designed using the wet service factor defined in AWC NDS. Cuts, notches and drilled holes of preservative-treated wood members shall be treated in accordance with Section R317.1.1. All preservative-treated wood products in contact with the ground shall be *labeled* for such usage.

R507.2.1.1 Engineered wood products. Engineered wood products shall be in accordance with Section R502.

R507.2.2 Plastic composite deck boards, stair treads, guards or handrails. *Plastic composite* exterior deck boards, stair treads, *guards* and *handrails* shall comply with the requirements of ASTM D7032 and this section.

R507.2.2.1 Labeling. *Plastic composite* deck boards and stair treads, or their packaging, shall bear a *label* that indicates compliance with ASTM D7032 and includes the allowable load and maximum allowable span determined in accordance with ASTM D7032. Plastic or composite *handrails* and *guards*, or their packaging, shall bear a *label* that indicates compliance with ASTM D7032 and includes the maximum allowable span determined in accordance with ASTM D7032.

R507.2.2.2 Flame spread index. *Plastic composite* deck boards, stair treads, *guards*, and *handrails* shall exhibit a flame spread index not exceeding 200 when tested in accordance with ASTM E84 or UL 723 with the test specimen remaining in place during the test.

Exception: *Plastic composites* determined to be noncombustible.

R507.2.2.3 Decay resistance. *Plastic composite* deck boards, stair treads, *guards* and *handrails* containing wood, cellulosic or other biodegradable materials shall be decay resistant in accordance with ASTM D7032.

R507.2.2.4 Termite resistance. Where required by Section 318, *plastic composite* deck boards, stair treads, *guards* and *handrails* containing wood, cellulosic or other biodegradable materials shall be termite resistant in accordance with ASTM D7032.

R507.2.2.5 Installation of plastic composites. *Plastic composite* deck boards, stair treads, *guards* and *handrails* shall be installed in accordance with this code and the manufacturer's instructions.

R507.2.3 Fasteners and connectors. Metal fasteners and connectors used for all decks shall be in accordance with Section R317.3 and Table R507.2.3.

R507.2.4 Flashing. Flashing shall be corrosion-resistant metal of nominal thickness not less than 0.019 inch (0.48 mm) or *approved* nonmetallic material that is compatible with the substrate of the structure and the decking materials.

R507.2.5 Alternate materials. Alternative materials, including glass and metals, shall be permitted.

TABLE R507.2.3
FASTENER AND CONNECTOR SPECIFICATIONS FOR DECKS[a, b]

ITEM	MATERIAL	MINIMUM FINISH/COATING	ALTERNATE FINISH/COATING[e]
Nails and glulam rivets	In accordance with ASTM F1667	Hot-dipped galvanized per ASTM A153, Class D for $^{3}/_{8}$-inch diameter and less	Stainless steel, silicon bronze or copper
Bolts[c] Lag screws[d] (including nuts and washers)	In accordance with ASTM A307 (bolts), ASTM A563 (nuts), ASTM F844 (washers)	Hot-dipped galvanized per ASTM A153, Class C (Class D for $^{3}/_{8}$-inch diameter and less) or mechanically galvanized per ASTM B695, Class 55 or 410 stainless steel	Stainless steel, silicon bronze or copper
Metal connectors	Per manufacturer's specification	ASTM A653 type G185 zinc-coated galvanized steel or post hot-dipped galvanized per ASTM A123 providing a minimum average coating weight of 2.0 oz./ft^2 (total both sides)	Stainless steel

For SI: 1 inch = 25.4 mm, 1 foot = 304.8 mm.

a. Equivalent materials, coatings and finishes shall be permitted.

b. Fasteners and connectors exposed to salt water or located within 300 feet of a salt water shoreline shall be stainless steel.

c. Holes for bolts shall be drilled a minimum $^{1}/_{32}$ inch and a maximum $^{1}/_{16}$ inch larger than the bolt.

d. Lag screws $^{1}/_{2}$ inch and larger shall be predrilled to avoid wood splitting per the *National Design Specification (NDS) for Wood Construction.*

e. Stainless-steel-driven fasteners shall be in accordance with ASTM F1667.

R507.3 Footings. Decks shall be supported on concrete footings or other *approved* structural systems designed to accommodate all loads in accordance with Section R301. Deck footings shall be sized to carry the imposed loads from the deck structure to the ground as shown in Figure R507.3.

Exceptions:

1. Footings shall not be required for free-standing decks consisting of joists directly supported on grade over their entire length.
2. Footings shall not be required for free-standing decks that meet all of the following criteria:
   2.1. The joists bear directly on *precast concrete* pier blocks at grade without support by beams or posts.
   2.2. The area of the deck does not exceed 200 square feet (18.6 m^2).
   2.3. The walking surface is not more than 20 inches (508 mm) above grade at any point within 36 inches (914 mm) measured horizontally from the edge.

R507.3.1 Minimum size. The minimum size of concrete footings shall be in accordance with Table R507.3.1, based on the tributary area and allowable soil-bearing pressure in accordance with Table R401.4.1.

R507.3.2 Minimum depth. Deck footings shall be placed not less than 12 inches (305 mm) below the undisturbed ground surface.

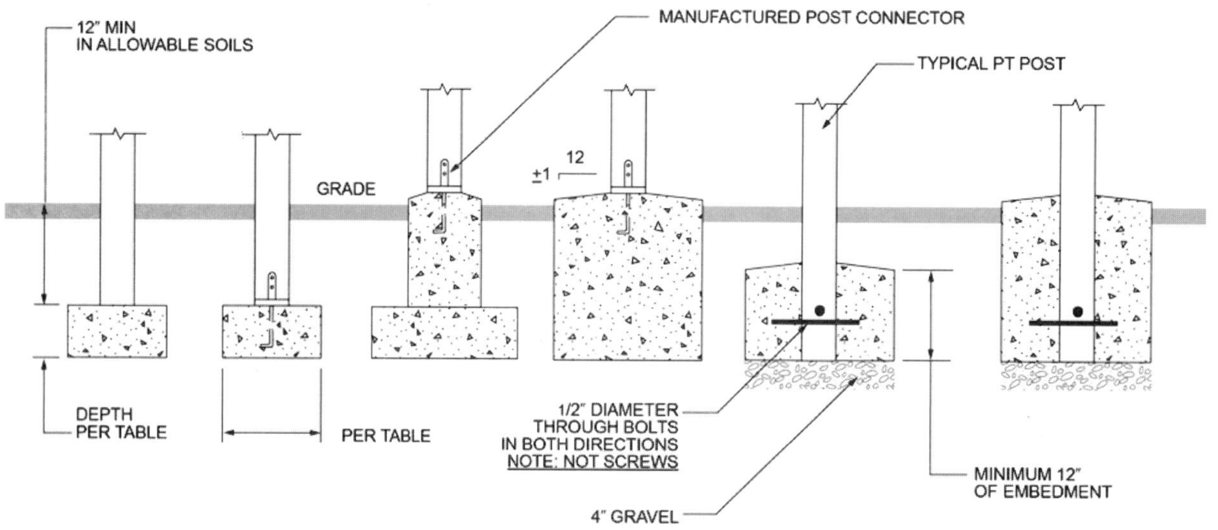

For SI: 1 inch = 25.4 mm.

FIGURE R507.3
DECK POSTS TO DECK FOOTING CONNECTION

TABLE R507.3.1
MINIMUM FOOTING SIZE FOR DECKS

LIVE OR GROUND SNOW LOAD[b] (psf)	TRIBUTARY AREA (ft^2)	LOAD-BEARING VALUE OF SOILS[a, c, d] (psf)								
		1,500[e]			2,000[e]			≥ 3,000[e]		
		Side of a square footing (inches)	Diameter of a round footing (inches)	Thickness (inches)[f]	Side of a square footing (inches)	Diameter of a round footing (inches)	Thickness (inches)[f]	Side of a square footing (inches)	Diameter of a round footing (inches)	Thickness (inches)[f]
40	5	7	8	6	7	8	6	7	8	6
	20	10	12	6	9	9	6	7	8	6
	40	14	16	6	12	14	6	10	12	6
	60	17	19	6	15	17	6	12	14	6
	80	20	22	7	17	19	6	14	16	6
	100	22	25	8	19	21	6	15	17	6
	120	24	27	9	21	23	7	17	19	6
	140	26	29	10	22	25	8	18	21	6
	160	28	31	11	24	27	9	20	22	7
50	5	7	8	6	7	8	6	7	8	6
	20	11	13	6	10	11	6	8	9	6
	40	15	17	6	13	15	6	11	13	6
	60	19	21	6	16	18	6	13	15	6
	80	21	24	8	19	21	6	15	17	6
	100	24	27	9	21	23	7	17	19	6
	120	26	30	10	23	26	8	19	21	6
	140	28	32	11	25	28	9	20	23	7
	160	30	34	12	26	30	10	21	24	8
60	5	7	8	6	7	8	6	7	8	6
	20	12	14	6	11	12	6	9	10	6
	40	16	19	6	14	16	8	12	14	6
	60	20	23	7	17	20	6	14	16	6
	80	23	26	9	20	23	7	16	19	6
	100	26	29	10	22	25	8	18	21	6
	120	28	32	11	25	28	9	20	23	7
	140	31	35	12	27	30	10	22	24	8
	160	33	37	13	28	32	11	23	26	9
70	5	7	8	6	7	8	6	7	8	6
	20	12	14	6	11	13	6	9	10	6
	40	18	20	6	15	17	6	12	14	6
	60	21	24	8	19	21	6	15	17	6
	80	25	28	9	21	24	8	18	20	6
	100	28	31	11	24	27	9	20	22	7
	120	30	34	12	26	30	10	21	24	8
	140	33	37	13	28	32	11	23	26	9
	160	35	40	15	30	34	12	25	28	9

For SI: 1 inch = 25.4 mm, 1 square foot = 0.0929 m^2, 1 pound per square foot = 0.0479 kPa.

a. Interpolation permitted, extrapolation not permitted.

b. Based on highest load case: Dead + Live or Dead + Snow.

c. Footing dimensions shall allow complete bearing of the post.

d. If the support is a brick or CMU pier, the footing shall have a minimum 2-inch projection on all sides.

e. Area, in square feet, of deck surface supported by post and footings.

f. Minimum thickness shall only apply to plain concrete footings.

R507.3.3 Frost protection. Where decks are attached to a frost-protected structure, deck footings shall be protected from frost by one or more of the following methods:

1. Extending below the frost line specified in Table R301.2.
2. Erecting on solid rock.
3. Other *approved* methods of frost protection.

R507.4 Deck posts. For single-level decks, wood post size shall be in accordance with Table R507.4.

TABLE R507.4
DECK POST HEIGHT

LOADS (psf)[b]	POST SPECIES[c]	POST SIZE[d]	TRIBUTARY AREA (ft²)[g, h]							
			20	40	60	80	100	120	140	160
			MAXIMUM DECK POST HEIGHT[a] (feet-inches)							
40 live load	Southern pine	4 × 4	14-0	13-8	11-0	9-5	8-4	7-5	6-9	6-2
		4× 6	14-0	14-0	13-11	12-0	10-8	9-8	8-10	8-2
		6 × 6	14-0	14-0	14-0	14-0	14-0	14-0	14-0	14-0
		8 × 8	14-0	14-0	14-0	14-0	14-0	14-0	14-0	14-0
	Douglas fir[e] Hem-fir[e] Spruce-pine-fir[e]	4 × 4	14-0	13-6	10-10	9-3	8-0	7-0	6-2	5-3
		4 × 6	14-0	14-0	13-10	11-10	10-6	9-5	8-7	7-10
		6 × 6	14-0	14-0	14-0	14-0	14-0	14-0	14-0	14-0
		8 × 8	14-0	14-0	14-0	14-0	14-0	14-0	14-0	14-0
	Redwood[f] Western cedars[f] Ponderosa pine[f] Red pine[f]	4 × 4	14-0	13-2	10-3	8-1	5-8	NP	NP	NP
		4 × 6	14-0	14-0	13-6	11-4	9-9	8-4	6-9	4-7
		6 × 6	14-0	14-0	14-0	14-0	14-0	14-0	13-7	9-7
		8 × 8	14-0	14-0	14-0	14-0	14-0	14-0	14-0	14-0
50 ground snow load	Southern pine	4 × 4	14-0	12-2	9-10	8-5	7-5	6-7	5-11	5-4
		4 × 6	14-0	14-0	12-6	10-9	9-6	8-7	7-10	7-3
		6 × 6	14-0	14-0	14-0	14-0	14-0	14-0	14-0	13-4
		8 × 8	14-0	14-0	14-0	14-0	14-0	14-0	14-0	14-0
	Douglas fir[e] Hem-fir[e] Spruce-pine-fir[e]	4 × 4	14-0	12-1	9-8	8-2	7-1	6-2	5-3	4-2
		4 × 6	14-0	14-0	12-4	10-7	9-4	8-4	7-7	6-11
		6 × 6	14-0	14-0	14-0	14-0	14-0	14-0	14-0	12-10
		8 × 8	14-0	14-0	14-0	14-0	14-0	14-0	14-0	14-0
	Redwood[f] Western cedars[f] Ponderosa pine[f] Red pine[f]	4 × 4	14-0	11-8	9-0	6-10	3-7	NP	NP	NP
		4 × 6	14-0	14-0	12-0	10-0	8-6	7-0	5-3	NP
		6 × 6	14-0	14-0	14-0	14-0	14-0	14-0	10-8	2-4
		8 × 8	14-0	14-0	14-0	14-0	14-0	14-0	14-0	14-0
60 ground snow load	Southern pine	4 × 4	14-0	11-1	8-11	7-7	6-7	5-10	5-2	4-6
		4 × 6	14-0	14-0	11-4	9-9	8-7	7-9	7-1	6-6
		6 × 6	14-0	14-0	14-0	14-0	14-0	14-0	12-9	11-2
		8 × 8	14-0	14-0	14-0	14-0	14-0	14-0	14-0	14-0
	Douglas fir[e] Hem-fir[e] Spruce-pine-fir[e]	4 × 4	14-0	10-11	8-8	7-3	6-2	5-0	3-7	NP
		4 × 6	14-0	13-11	11-2	9-7	8-4	7-5	6-8	5-11
		6 × 6	14-0	14-0	14-0	14-0	14-0	14-0	12-2	10-2
		8 × 8	14-0	14-0	14-0	14-0	14-0	14-0	14-0	14-0
	Redwood[f] Western cedars[f] Ponderosa pine[f] Red pine[f]	4 × 4	14-0	10-6	7-9	4-7	NP	NP	NP	NP
		4 × 6	14-0	13-7	10-9	8-9	7-0	4-9	NP	NP
		6 × 6	14-0	14-0	14-0	14-0	14-0	9-9	NP	NP
		8 × 8	14-0	14-0	14-0	14-0	14-0	14-0	14-0	14-0

(continued)

TABLE R507.4—continued
DECK POST HEIGHT

LOADS (psf)[b]	POST SPECIES[c]	POST SIZE[d]	TRIBUTARY AREA (ft²)[g, h]							
			20	40	60	80	100	120	140	160
			MAXIMUM DECK POST HEIGHT[a] (feet-inches)							
70 ground snow load	Southern pine	4 × 4	14-0	10-2	8-2	6-11	5-11	5-2	4-4	3-4
		4 × 6	14-0	12-11	10-5	8-11	7-10	7-1	6-5	5-10
		6 × 6	14-0	14-0	14-0	14-0	14-0	12-9	10-11	8-7
		8 × 8	14-0	14-0	14-0	14-0	14-0	14-0	14-0	14-0
	Douglas fir[e] Hem-fir[e] Spruce-pine-fir[e]	4 × 4	14-0	10-1	7-11	6-6	5-3	3-7	NP	NP
		4 × 6	14-0	12-10	10-3	8-9	7-7	6-8	5-10	4-11
		6 × 6	14-0	14-0	14-0	14-0	14-0	12-2	9-9	5-9
		8 × 8	14-0	14-0	14-0	14-0	14-0	14-0	14-0	14-0
	Redwood[f] Western cedars[f] Ponderosa pine[f] Red pine[f]	4 × 4	14-0	9-5	6-5	NP	NP	NP	NP	NP
		4 × 6	14-0	12-6	9-8	7-7	5-3	NP	NP	NP
		6 × 6	14-0	14-0	14-0	14-0	10-8	NP	NP	NP
		8 × 8	14-0	14-0	14-0	14-0	14-0	14-0	14-0	14-0

For SI: 1 inch = 25.4 mm, 1 foot = 304.8 mm, 1 pound per square foot = 0.0479 kPa.
NP = Not Permitted.
a. Measured from the underside of the beam to the top of footing or pier.
b. 10 psf dead load. Snow load not assumed to be concurrent with live load.
c. No. 2 grade, wet service factor included.
d. Notched deck posts shall be sized to accommodate beam size in accordance with Section R507.5.2.
e. Includes incising factor.
f. Incising factor not included.
g. Area, in square feet, of deck surface supported by post and footings.
h. Interpolation permitted. Extrapolation not permitted.

R507.4.1 Deck post to deck footing connection. Where posts bear on concrete footings in accordance with Section R403 and Figure R507.3, lateral restraint shall be provided by manufactured connectors or a minimum post embedment of 12 inches (305 mm) in surrounding soils or concrete piers. Other footing systems shall be permitted.

Exception: Where expansive, compressible, shifting or other questionable soils are present, surrounding soils shall not be relied on for lateral support.

R507.5 Deck beams. Maximum allowable spans for wood deck beams, as shown in Figure R507.5, shall be in accordance with Tables R507.5(1) through R507.5(4). Beam plies shall be fastened together with two rows of 10d (3-inch × 0.128-inch) nails minimum at 16 inches (406 mm) on center along each edge. Beams shall be permitted to cantilever at each end up to one-fourth of the actual beam span. Deck beams of other materials shall be permitted where designed in accordance with accepted engineering practices.

R507.5.1 Deck beam bearing. The ends of beams shall have not less than $1^1/_2$ inches (38 mm) of bearing on wood or metal and not less than 3 inches (76 mm) of bearing on concrete or masonry for the entire width of the beam. Where multiple-span beams bear on intermediate posts, each ply must have full bearing on the post in accordance with Figures R507.5.1(1) and R507.5.1(2).

R507.5.2 Deck beam connection to supports. Deck beams shall be attached to supports in a manner capable of transferring vertical loads and resisting horizontal displacement. Deck beam connections to wood posts shall be in accordance with Figures R507.5.1(1) and R507.5.1(2). Manufactured post-to-beam connectors shall be sized for the post and beam sizes. Bolts shall have washers under the head and nut.

R507.6 Deck joists. Maximum allowable spans for wood deck joists, as shown in Figure R507.6, shall be in accordance with Table R507.6. The maximum joist spacing shall be limited by the decking materials in accordance with Table R507.7.

R507.6.1 Deck joist bearing. The ends of joists shall have not less than $1^1/_2$ inches (38 mm) of bearing on wood or metal and not less than 3 inches (76 mm) of bearing on concrete or masonry over its entire width. Joists bearing on top of a multiple-ply beam or ledger shall be fastened in accordance with Table R602.3(1). Joists bearing on top of a single-ply beam or ledger shall be attached by a mechanical connector. Joist framing into the side of a beam or ledger board shall be supported by *approved* joist hangers.

R507.6.2 Deck joist lateral restraint. Joist ends and bearing locations shall be provided with lateral resistance to prevent rotation. Where lateral restraint is provided by joist hangers or blocking between joists, their depth shall

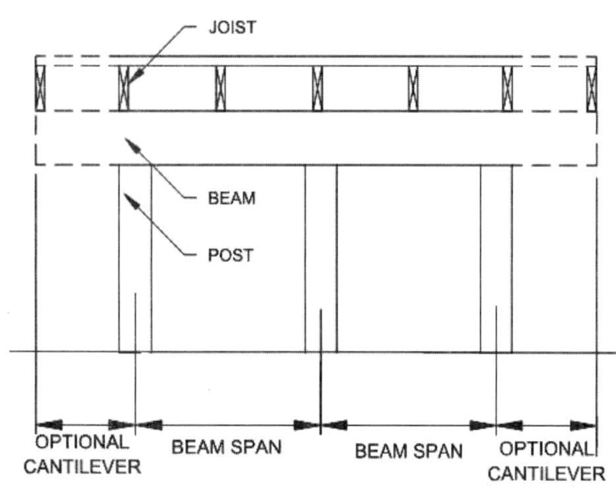

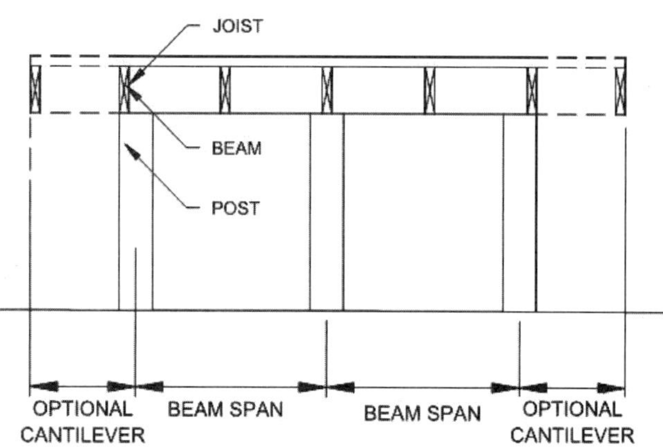

FIGURE R507.5
TYPICAL DECK JOIST SPANS

equal not less than 60 percent of the joist depth. Where lateral restraint is provided by rim joists, they shall be secured to the end of each joist with not fewer than three 10d (3-inch by 0.128-inch) (76 mm by 3.3 mm) nails or three No. 10 x 3-inch-long (76 mm) wood screws.

R507.7 Decking. Maximum allowable spacing for joists supporting wood decking, excluding *stairways*, shall be in accordance with Table R507.7. Wood decking shall be attached to each supporting member with not less than two 8d threaded nails or two No. 8 wood screws. Maximum allowable spacing for joists supporting *plastic composite* decking shall be in accordance with Section R507.2. Other *approved* decking or fastener systems shall be installed in accordance with the manufacturer's installation requirements.

R507.8 Vertical and lateral supports. Where supported by attachment to an exterior wall, decks shall be positively anchored to the primary structure and designed for both vertical and lateral loads. Such attachment shall not be accomplished by the use of toenails or nails subject to withdrawal. For decks with cantilevered framing members, connection to exterior walls or other framing members shall be designed and constructed to resist uplift resulting from the full *live load* specified in Table R301.5 acting on the cantilevered portion of the deck. Where positive connection to the primary building structure cannot be verified during inspection, decks shall be self-supporting.

R507.9 Vertical and lateral supports at band joist. Vertical and lateral supports for decks shall comply with this section.

R507.9.1 Vertical supports. Vertical loads shall be transferred to band joists with ledgers in accordance with this section.

R507.9.1.1 Ledger details. Deck ledgers shall be a minimum 2-inch by 8-inch (51 mm by 203 mm) nominal, pressure-preservative-treated Southern pine, incised pressure-preservative-treated hem-fir, or *approved*, naturally durable, No. 2 grade or better lumber. Deck ledgers shall not support concentrated loads from beams or girders. Deck ledgers shall not be supported on stone or masonry veneer.

R507.9.1.2 Band joist details. Band joists supporting a ledger shall be a minimum 2-inch-nominal (51 mm), solid-sawn, spruce-pine-fir or better lumber or a minimum 1-inch (25 mm) nominal engineered wood rim boards in accordance with Section R502.1.7. Band joists shall bear fully on the primary structure capable of supporting all required loads.

TABLE R507.5(1) MAXIMUM DECK BEAM SPAN—40 PSF LIVE LOAD[c]

BEAM SPECIES[d]	BEAM SIZE[e]	EFFECTIVE DECK JOIST SPAN LENGTH[a, i, j] (feet)						
		6	8	10	12	14	16	18
		MAXIMUM DECK BEAM SPAN LENGTH (feet-inches)[a, b, f]						
Southern pine	1 – 2 × 6	4-7	4-0	3-7	3-3	3-0	2-10	2-8
	1 – 2 × 8	5-11	5-1	4-7	4-2	3-10	3-7	3-5
	1 – 2 × 10	7-0	6-0	5-5	4-11	4-7	4-3	4-0
	1 – 2 × 12	8-3	7-1	6-4	5-10	5-5	5-0	4-9
	2 – 2 × 6	6-11	5-11	5-4	4-10	4-6	4-3	4-0
	2 – 2 × 8	8-9	7-7	6-9	6-2	5-9	5-4	5-0
	2 – 2 × 10	10-4	9-0	8-0	7-4	6-9	6-4	6-0
	2 – 2 × 12	12-2	10-7	9-5	8-7	8-0	7-5	7-0
	3 – 2 × 6	8-6	7-5	6-8	6-1	5-8	5-3	4-11
	3 – 2 × 8	10-11	9-6	8-6	7-9	7-2	6-8	6-4
	3 – 2 × 10	13-0	11-2	10-0	9-2	8-6	7-11	7-6
	3 – 2 × 12	15-3	13-3	11-10	10-9	10-0	9-4	8-10
Douglas fir-larch[g] Hem-fir[g] Spruce-pine-fir	1 – 2 x 6	4-1	3-6	3-0	2-8	2-5	2-3	2-1
	1 – 2 × 8	5-6	4-8	4-0	3-6	3-2	2-11	2-9
	1 – 2 × 10	6-8	5-10	5-1	4-6	4-1	3-9	3-6
	1 – 2 × 12	7-9	6-9	6-0	5-6	5-0	3-9	3-6
	2 – 2 × 6	6-1	5-3	4-9	4-4	3-11	3-7	3-3
	2 – 2 × 8	8-2	7-1	6-4	5-9	5-2	4-8	4-4
	2 – 2 × 10	10-0	8-7	7-9	7-0	6-6	6-0	5-6
	2 – 2 × 12	11-7	10-0	8-11	8-2	7-7	7-1	6-8
	3 – 2 × 6	7-8	6-8	6-0	5-6	5-1	4-9	4-6
	3 – 2 × 8	10-3	8-10	7-11	7-3	6-8	6-3	5-11
	3 – 2 × 10	12-6	10-10	9-8	8-10	8-2	7-8	7-2
	3 – 2 × 12	14-6	12-7	11-3	10-3	9-6	8-11	8-5
Redwood[h] Western cedars[h] Ponderosa pine[h] Red pine[h]	1 – 2 × 6	4-2	3-7	3-1	2-9	2-6	2-3	2-2
	1 – 2 × 8	5-4	4-7	4-1	3-7	3-3	3-0	2-10
	1 – 2 × 10	6-6	5-7	5-0	4-7	4-2	3-10	3-7
	1 – 2 × 12	7-6	6-6	5-10	5-4	4-11	4-7	4-4
	2 – 2 × 6	6-2	5-4	4-10	4-5	4-0	3-8	3-4
	2 – 2 × 8	7-10	6-10	6-1	5-7	5-2	4-10	4-5
	2 – 2 × 10	9-7	8-4	7-5	6-9	6-3	5-10	5-6
	2 – 2 × 12	11-1	9-8	8-7	7-10	7-3	6-10	6-5
	3 – 2 × 6	7-8	6-9	6-0	5-6	5-1	4-9	4-6
	3 – 2 × 8	9-10	8-6	7-7	6-11	6-5	6-0	5-8
	3 – 2 × 10	12-0	10-5	9-4	8-6	7-10	7-4	6-11
	3 – 2 × 12	13-11	12-1	10-9	9-10	9-1	8-6	8-1

For SI: 1 inch = 25.4 mm, 1 foot = 304.8 mm, 1 pound per square foot = 0.0479 kPa, 1 pound = 0.454 kg.

a. Interpolation permitted. Extrapolation not permitted.

b. Beams supporting a single span of joists with or without cantilever.

c. Dead load = 10 psf, L/Δ = 360 at main span, L/Δ = 180 at cantilever. Snow load is not assumed to be concurrent with live load.

d. No. 2 grade, wet service factor included.

e. Beam depth shall be equal to or greater than the depth of intersecting joist for a flush beam connection.

f. Beam cantilevers are limited to the adjacent beam's span divided by 4.

g. Includes incising factor.

h. Incising factor not included.

i. Deck joist span as shown in Figure R507.5.

j. For calculation of effective deck joist span, the actual joist span length shall be multiplied by the joist span factor in accordance with Table R507.5(5).

TABLE R507.5(2)
MAXIMUM DECK BEAM SPAN—50 PSF GROUND SNOW LOAD[c]

BEAM SPECIES[d]	BEAM SIZE[e]	EFFECTIVE DECK JOIST SPAN LENGTH (feet)[a, i, j]						
		6	8	10	12	14	16	18
		MAXIMUM DECK BEAM SPAN LENGTH (feet-inches)[a, b, f]						
Southern pine	1 – 2 × 6	4-6	3-11	3-6	3-2	2-11	2-9	2-7
	1 – 2 × 8	5-9	4-11	4-5	4-0	3-9	3-6	3-3
	1 – 2 × 10	6-9	5-10	5-3	4-9	4-5	4-2	3-11
	1 – 2 × 12	8-0	6-11	6-2	5-8	5-3	4-11	4-7
	2 – 2 × 6	6-8	5-9	5-2	4-9	4-4	4-1	3-10
	2 – 2 × 8	8-6	7-4	6-7	6-0	5-7	5-2	4-11
	2 – 2 × 10	10-1	8-9	7-10	7-1	6-7	6-2	5-10
	2 – 2 × 12	11-11	10-3	9-2	8-5	7-9	7-3	6-10
	3 – 2 × 6	7-11	7-2	6-6	5-11	5-6	5-1	4-10
	3 – 2 × 8	10-5	9-3	8-3	7-6	6-11	6-6	6-2
	3 – 2 × 10	12-8	10-11	9-9	8-11	8-3	7-9	7-3
	3 – 2 × 12	14-11	12-11	11-6	10-6	9-9	9-1	8-7
Douglas fir-larch[g] Hem-fir[g] Spruce-pine-fir[g]	1 – 2 × 6	4-0	3-5	2-11	2-7	2-4	2-2	2-0
	1 – 2 × 8	5-4	4-7	3-11	3-5	3-1	2-10	2-8
	1 – 2 × 10	6-7	5-8	4-11	4-5	4-0	3-8	3-5
	1 – 2 × 12	7-7	6-7	5-11	5-4	4-10	4-6	4-2
	2 – 2 × 6	6-0	5-2	4-7	4-2	3-10	3-5	3-2
	2 – 2 × 8	8-0	6-11	6-2	5-8	5-0	4-7	4-2
	2 – 2 × 10	9-9	8-5	7-7	6-11	6-4	5-10	5-4
	2 – 2 × 12	11-4	9-10	8-9	8-0	7-5	6-11	6-6
	3 – 2 × 6	7-6	6-6	5-9	5-3	4-11	4-7	4-4
	3 – 2 × 8	10-0	8-8	7-9	7-1	6-6	6-1	5-8
	3 – 2 × 10	12-3	10-7	9-6	8-8	8-0	7-6	7-0
	3 – 2 × 12	14-3	12-4	11-0	10-1	9-4	8-9	8-3
Redwood[h] Western cedars[h] Ponderosa pine[h] Red pine[h]	1 – 2 × 6	4-1	3-6	3-0	2-8	2-5	2-3	2-1
	1 – 2 × 8	5-2	4-6	4-0	3-6	3-2	2-11	2-9
	1 – 2 × 10	6-4	5-6	4-11	4-6	4-1	3-9	3-6
	1 – 2 × 12	7-4	6-4	5-8	5-2	4-10	4-6	4-3
	2 – 2 × 6	6-1	5-3	4-8	4-4	3-11	3-6	3-3
	2 – 2 × 8	7-8	6-8	5-11	5-5	5-0	4-8	4-3
	2 – 2 × 10	9-5	8-2	7-3	6-8	6-2	5-9	5-5
	2 – 2 × 12	10-11	9-5	8-5	7-8	7-2	6-8	6-3
	3 – 2 × 6	7-1	6-5	5-11	5-5	5-0	4-8	4-5
	3 – 2 × 8	9-4	8-4	7-5	6-10	604	5-11	5-7
	3 – 2 × 10	11-9	10-2	9-1	8-4	7-8	7-2	6-9
	3 – 2 × 12	13-8	11-10	10-7	9-8	8-11	8-4	7-10

For SI: 25.4 mm, 1 foot = 304.8 mm, 1 pound per square foot = 0.0479 kPa, 1 pound = 0.454 kg.

a. Interpolation allowed. Extrapolation is not allowed.

b. Beams supporting a single span of joists with or without cantilever.

c. Dead load = 10 psf, L/Δ = 360 at main span, L/Δ = 180 at cantilever. Snow load not assumed to be concurrent with live load.

d. No. 2 grade, wet service factor included.

e. Beam depth shall be equal to or greater than the depth of intersecting joist for a flush beam connection.

f. Beam cantilevers are limited to the adjacent beam's span divided by 4.

g. Includes incising factor.

h. Incising factor not included.

i. Deck joist span as shown in Figure R507.5.

j. For calculation of effective deck joist span, the actual joist span length shall be multiplied by the joist span factor in accordance with Table R507.5(5).

TABLE R507.5(3)
MAXIMUM DECK BEAM SPAN—60 PSF GROUND SNOW LOAD[c]

BEAM SPECIES[d]	BEAM SIZE[e]	EFFECTIVE DECK JOIST SPAN LENGTH[a, i, j] (feet)						
		6	8	10	12	14	16	18
		MAXIMUM DECK BEAM SPAN LENGTH (feet-inches)[a, b, f]						
Southern pine	1 – 2 × 6	4-2	3-7	3-3	2-11	2-9	2-6	2-5
	1 – 2 × 8	5-3	4-7	4-1	3-9	3-5	3-3	3-0
	1 – 2 × 10	6-3	5-5	4-10	4-5	4-1	3-10	3-7
	1 – 2 × 12	7-5	6-5	5-9	5-3	4-10	4-6	4-3
	2 – 2 × 6	6-2	5-4	4-9	4-4	4-0	3-9	3-7
	2 – 2 × 8	7-10	6-10	6-1	5-7	5-2	4-10	4-6
	2 – 2 × 10	9-4	8-1	7-3	6-7	6-1	5-8	5-4
	2 – 2 × 12	11-0	9-6	8-6	7-9	7-2	6-9	6-4
	3 – 2 × 6	7-5	6-9	6-0	5-6	5-1	4-9	4-6
	3 – 2 × 8	9-9	8-6	7-8	6-11	6-5	6-0	5-8
	3 – 2 × 10	11-8	10-2	9-1	8-3	7-8	7-2	6-9
	3 – 2 × 12	13-9	11-11	10-8	9-9	9-0	8-5	7-11
Douglas fir-larch[g] Hem-fir[g] Spuce-pine-fir[g]	1 – 2 × 6	3-8	3-1	2-8	2-4	2-2	2-0	1-10
	1 – 2 × 8	5-0	4-1	3-6	3-1	2-10	2-7	2-5
	1 – 2 × 10	6-1	5-2	4-6	4-0	3-7	3-4	3-2
	1 – 2 × 12	7-1	6-1	5-5	4-10	4-5	4-1	3-10
	2 – 2 × 6	5-6	4-9	4-3	3-10	3-5	3-1	2-10
	2 – 2 × 8	7-5	6-5	5-9	5-0	4-6	4-1	3-9
	2 – 2 × 10	9-0	7-10	7-0	6-4	5-9	5-2	4-10
	2 – 2 × 12	10-6	9-1	8-1	7-5	6-10	6-4	5-10
	3 – 2 × 6	6-11	6-0	5-4	4-11	4-6	4-2	3-10
	3 – 2 × 8	9-3	8-0	7-2	6-6	6-1	5-6	5-0
	3 – 2 × 10	11-4	9-10	8-9	8-0	7-5	6-11	6-5
	3 – 2 × 12	13-2	11-5	10-2	9-4	8-7	8-1	7-7
Redwood[h] Western cedars[h] Ponderosa pine[h] Red pine[h]	1 – 2 × 6	3-9	3-2	2-9	2-5	2-2	2-0	1-11
	1 – 2 × 8	4-10	4-2	3-7	3-2	2-11	2-8	2-6
	1 – 2 × 10	5-10	5-1	4-6	4-1	3-8	3-5	3-3
	1 – 2 × 12	6-10	5-11	5-3	4-10	4-5	4-2	3-11
	2 – 2 × 6	5-7	4-10	4-4	3-11	3-6	3-2	2-11
	2 – 2 × 8	7-1	6-2	5-6	5-0	4-7	4-2	3-10
	2 – 2 × 10	8-8	7-6	6-9	6-2	5-8	5-4	4-11
	2 – 2 × 12	10-1	8-9	7-10	7-2	6-7	6-2	5-10
	3 – 2 × 6	6-8	6-1	5-5	5-0	4-7	4-3	3-11
	3 – 2 × 8	8-9	7-9	6-22	6-4	5-20	5-5	5-3
	3 – 2 × 10	10-11	9-5	8-5	7-8	7-3	6-8	6-3
	3 – 2 × 12	12-8	10-11	9-9	8-11	8-3	7-9	7-3

For SI: 1 inch = 25.4 mm, 1 foot = 304.8 mm, 1 pound per square foot = 0.0479 kPa, 1 pound = 0.454 kg.

a. Interpolation allowed. Extrapolation is not allowed.

b. Beams supporting a single span of joists with or without cantilever.

c. Dead load = 10 psf, L/Δ = 360 at main span, L/Δ = 180 at cantilever. Snow load not assumed to be concurrent with live load.

d. No. 2 grade, wet service factor included.

e. Beam depth shall be equal to or greater than the depth of intersecting joist for a flush beam connection.

f. Beam cantilevers are limited to the adjacent beam's span divided by 4.

g. Includes incising factor.

h. Incising factor not included.

i. Deck joist span as shown in Figure R507.5.

j. For calculation of effective deck joist span, the actual joist span length shall be multiplied by the joist span factor in accordance with Table R507.5(5).

TABLE R507.5(4)
MAXIMUM DECK BEAM SPAN—70 PSF GROUND SNOW LOAD[c]

BEAM SPECIES[d]	BEAM SIZE[e]	EFFECTIVE DECK JOIST SPAN LENGTH (feet)[a, i, j]						
		6	8	10	12	14	16	18
		MAXIMUM DECK BEAM SPAN LENGTH (feet-inches)[a, b, f]						
Southern pine	1 – 2 × 6	3-11	3-4	3-0	2-9	2-6	2-4	2-3
	1 – 2 × 8	4-11	4-3	3-10	3-6	3-3	3-0	2-10
	1 – 2 × 10	5-10	5-1	4-6	4-2	3-10	3-7	3-4
	1 – 2 × 12	6-11	6-0	5-4	4-11	4-6	4-3	4-0
	2 – 2 × 6	5-9	5-0	4-6	4-1	3-9	3-6	3-4
	2 – 2 × 8	7-4	6-4	5-8	5-2	4-10	4-6	4-3
	2 – 2 × 10	8-9	7-7	6-9	6-2	5-8	5-4	5-0
	2 – 2 × 12	10-3	8-11	8-0	7-3	6-9	6-3	5-11
	3 – 2 × 6	7-0	6-3	5-7	5-1	4-9	4-5	4-2
	3 – 2 × 8	9-3	8-0	7-2	6-6	6-0	5-8	5-4
	3 – 2 × 10	10-11	9-6	8-6	7-9	7-2	6-8	6-4
	3 – 2 × 12	12-11	11-2	10-0	9-1	8-5	7-11	7-5
Douglas fir-larch[g] Hem-fir[g] Spruce-pine-fir[g]	1 – 2 × 6	3-5	2-10	2-5	2-2	2-0	1-10	1-9
	1 – 2 × 8	4-7	3-8	3-2	2-10	2-7	2-5	2-4
	1 – 2 × 10	5-8	4-9	4-1	3-8	3-4	3-1	2-11
	1 – 2 × 12	6-7	5-8	5-0	4-6	4-1	3-10	3-7
	2 – 2 × 6	5-2	4-6	4-0	3-5	3-1	2-10	2-7
	2 – 2 × 8	6-11	6-0	5-3	4-7	4-1	3-8	3-5
	2 – 2 × 10	8-5	7-4	6-6	5-10	5-2	4-9	4-5
	2 – 2 × 12	9-10	8-6	7-7	6-11	6-4	5-9	5-4
	3 – 2 × 6	6-6	5-7	5-0	4-7	4-2	3-9	3-5
	3 – 2 × 8	8-8	7-6	6-8	6-1	5-6	5-0	4-7
	3 – 2 × 10	10-7	9-2	8-2	7-6	6-11	6-4	5-10
	3 – 2 × 12	12-4	10-8	9-7	8-9	8-1	7-7	7-1
Redwood[h] Western cedars[h] Ponderosa pine[h] Red pine[h]	1 – 2 × 6	3-6	2-11	2-6	2-3	2-0	1-11	1-9
	1 – 2 × 8	4-6	3-10	3-3	2-11	2-8	2-6	2-4
	1 – 2 × 10	5-6	4-9	4-2	3-9	3-5	3-2	3-0
	1 – 2 × 12	6-4	5-6	4-11	4-6	4-2	3-11	3-8
	2 – 2 × 6	5-3	4-7	4-1	3-6	3-2	2-11	2-8
	2 – 2 × 8	6-8	5-9	5-2	4-8	4-2	3-10	3-6
	2 – 2 × 10	8-2	7-1	6-4	5-9	5-4	4-10	4-6
	2 – 2 × 12	9-5	8-2	7-4	6-8	6-2	5-9	5-5
	3 – 2 × 6	6-4	5-8	5-1	4-8	4-3	3-10	3-6
	3 – 2 × 8	8-4	7-3	6-5	5-11	5-5	5-1	4-8
	3 – 2 × 10	10-2	8-10	7-11	7-2	6-8	6-3	5-11
	3 – 2 × 12	11-10	10-3	9-2	8-4	7-9	7-3	6-10

For SI: 1 inch = 25.4 mm, 1 foot = 304.8 mm, 1 pound per square foot = 0.0479 kPa, 1 pound = 0.454 kg.

a. Interpolation allowed. Extrapolation is not allowed.

b. Beams supporting a single span of joists with or without cantilever.

c. Dead load = 10 psf, L/Δ = 360 at main span, L/Δ = 180 at cantilever. Snow load not assumed to be concurrent with live load.

d. No. 2 grade, wet service factor included.

e. Beam depth shall be equal to or greater than the depth of intersecting joist for a flush beam connection.

f. Beam cantilevers are limited to the adjacent beam's span divided by 4.

g. Includes incising factor.

h. Incising factor not included.

i. Deck joist span as shown in Figure R507.5.

j. For calculation of effective deck joist span, the actual joist span length shall be multiplied by the joist span factor in accordance with Table R507.5(5).

TABLE R507.5(5)
JOIST SPAN FACTORS FOR CALCULATING EFFECTIVE DECK JOIST SPAN
[for use with Note j in Tables R507.5(1), R507.5(2), R507.5(3) and R507.5(4)]

C/J[a]	JOIST SPAN FACTOR
0 (no cantilever)	0.66
1/12 (0.87)	0.72
1/10 (0.10)	0.80
1/8 (0.125)	0.84
1/6 (0.167)	0.90
1/4 (0.250)	1.00

For SI: 1 foot = 304.8 mm.
a. C = actual joist cantilever length (feet); J = actual joist span length (feet).

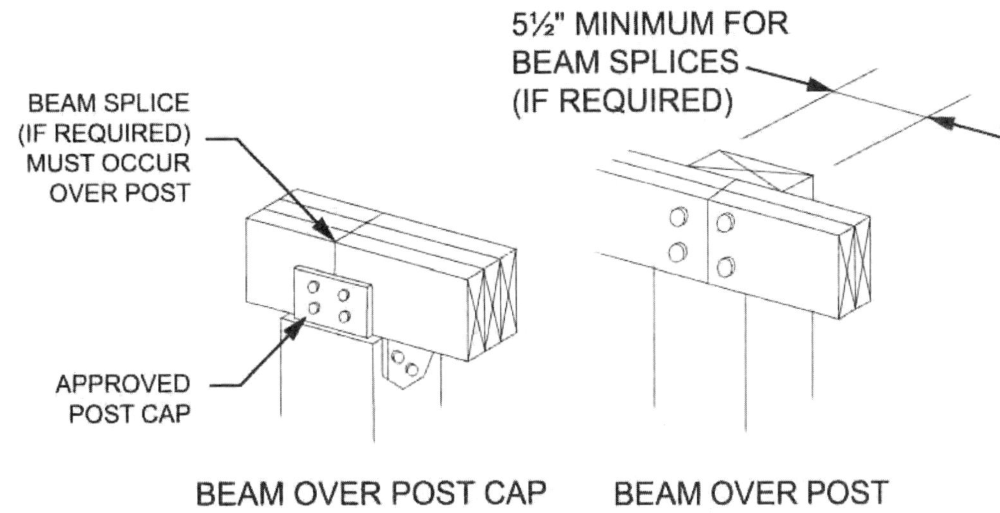

For SI: 1 inch = 25.4 mm.

FIGURE R507.5.1(1)
DECK BEAM TO DECK POST

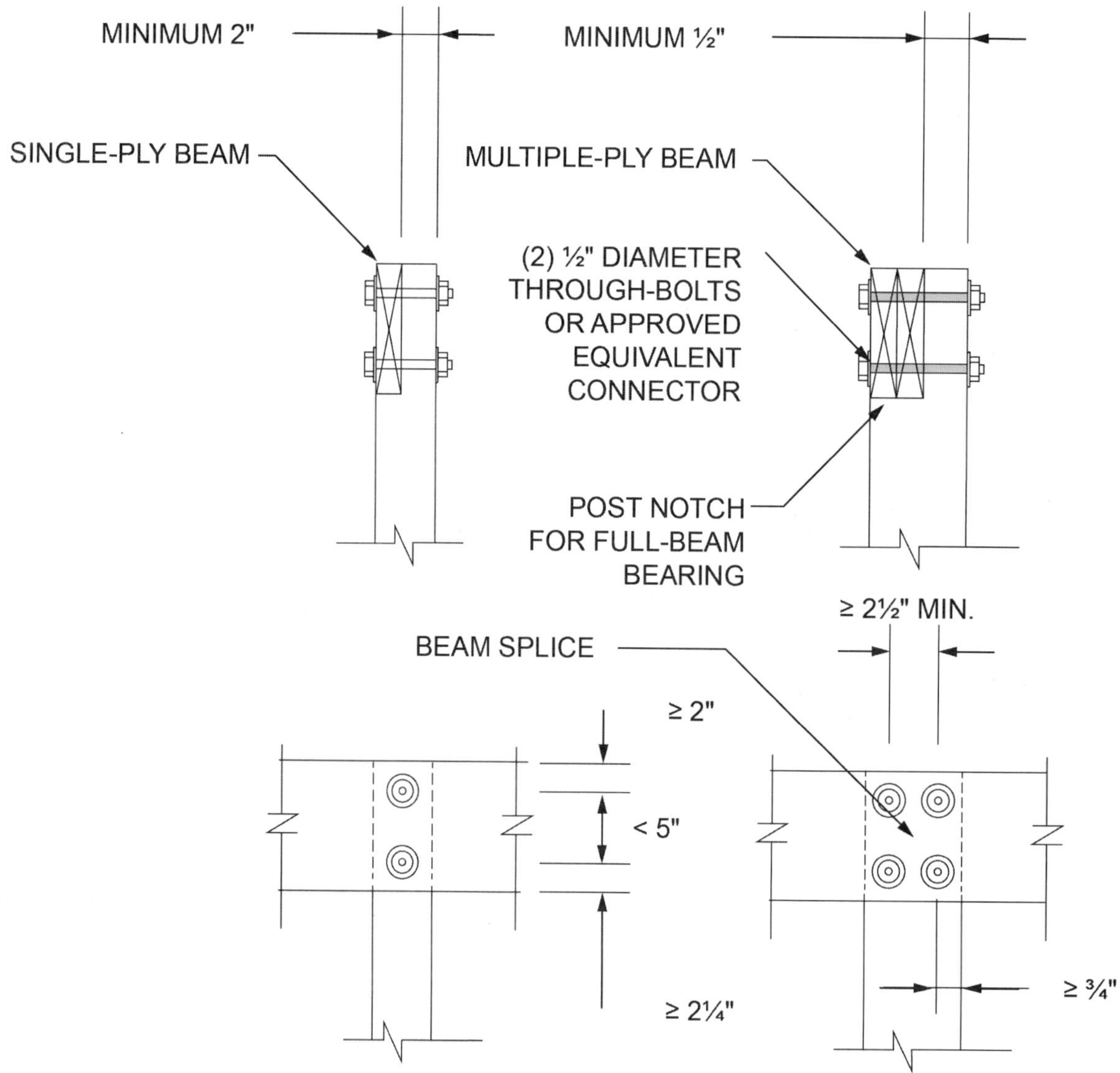

For SI: 1 inch = 25.4 mm.

FIGURE R507.5.1(2)
NOTCHED POST-TO-BEAM CONNECTION

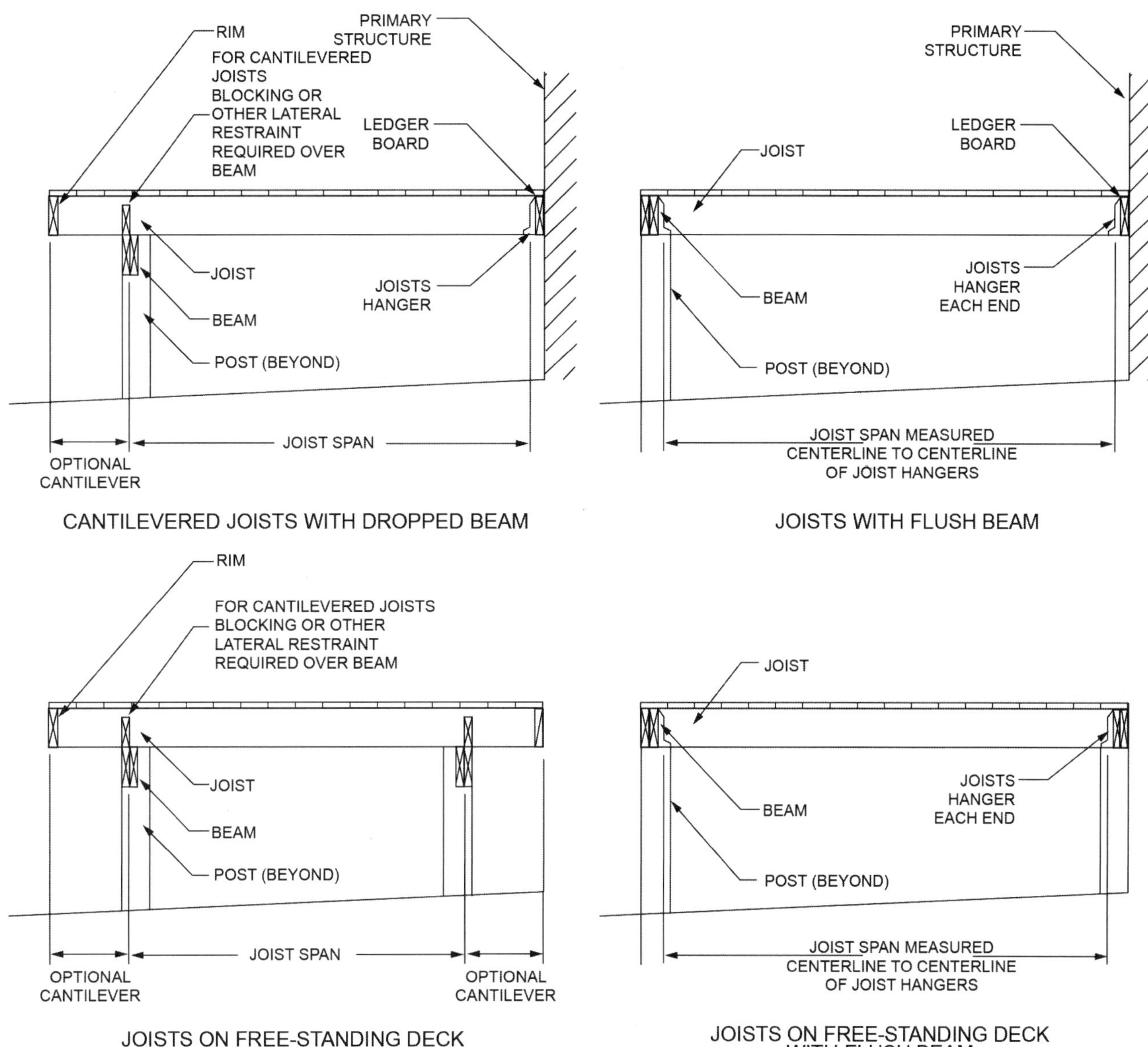

FIGURE R507.6
TYPICAL DECK JOIST SPANS

**TABLE R507.6
MAXIMUM DECK JOIST SPANS**

LOAD[a] (psf)	JOIST SPECIES[b]	JOIST SIZE	ALLOWABLE JOIST SPAN[b, c] (feet-inches)			MAXIMUM CANTILEVER[d,f] (feet-inches)							
			Joist spacing (inches)			Joist back span[g] (feet)							
			12	16	24	4	6	8	10	12	14	16	18
40 live load	Southern pine	2 × 6	9-11	9-0	7-7	1-0	1-6	1-5	NP	NP	NP	NP	NP
		2 × 8	13-1	11-10	9-8	1-0	1-6	2-0	2-6	2-3	NP	NP	NP
		2 × 10	16-2	14-0	11-5	1-0	1-6	2-0	2-6	3-0	3-4	3-4	NP
		2 × 12	18-0	16-6	13-6	1-0	1-6	2-0	2-6	3-0	3-6	4-0	4-1
	Douglas fir-larch[e] Hem-fir[e] Spruce-pine-fir[e]	2 × 6	9-6	8-4	6-10	1-0	1-6	1-4	NP	NP	NP	NP	NP
		2 × 8	12-6	11-1	9-1	1-0	1-6	2-0	2-3	2-0	NP	NP	NP
		2 × 10	15-8	13-7	11-1	1-0	1-6	2-0	2-6	3-0	3-3	NP	NP
		2 × 12	18-0	15-9	12-10	1-0	1-6	2-0	2-6	3-0	3-6	3-11	3-11
	Redwood[f] Western cedars[f] Ponderosa pine[f] Red pine[f]	2 × 6	8-10	8-0	6-10	1-0	1-4	1-1	NP	NP	NP	NP	NP
		2 × 8	11-8	10-7	8-8	1-0	1-6	2-0	1-11	NP	NP	NP	NP
		2 × 10	14-11	13-0	10-7	1-0	1-6	2-0	2-6	3-0	2-9	NP	NP
		2 × 12	17-5	15-1	12-4	1-0	1-6	2-0	2-6	3-0	3-6	3-8	NP
50 ground snow load	Southern pine	2 × 6	9-2	8-4	7-4	1-0	1-6	1-5	NP	NP	NP	NP	NP
		2 × 8	12-1	11-0	9-5	1-0	1-6	2-0	2-5	2-3	NP	NP	NP
		2 × 10	15-5	13-9	11-3	1-0	1-6	2-0	2-6	3-0	3-1	NP	NP
		2 × 12	18-0	16-2	13-2	1-0	1-6	2-0	2-6	3-0	3-6	3-10	3-10
	Douglas fir-larch[e] Hem-fir[e] Spruce-pine-fir[e]	2 × 6	8-10	8-0	6-8	1-0	1-6	1-4	NP	NP	NP	NP	NP
		2 × 8	11-7	10-7	8-11	1-0	1-6	2-0	2-3	NP	NP	NP	NP
		2 × 10	14-10	13-3	10-10	1-0	1-6	2-0	2-6	3-0	3-0	NP	NP
		2 × 12	17-9	15-5	12-7	1-0	1-6	2-0	2-6	3-0	3-6	3-8	NP
	Redwood[f] Western cedars[f] Ponderosa pine[f] Red pine[f]	2 × 6	8-3	7-6	6-6	1-0	1-4	1-1	NP	NP	NP	NP	NP
		2 × 8	10-10	9-10	8-6	1-0	1-6	2-0	1-11	NP	NP	NP	NP
		2 × 10	13-10	12-7	10-5	1-0	1-6	2-0	2-6	2-9	NP	NP	NP
		2 × 12	16-10	14-9	12-1	1-0	1-6	2-0	2-6	3-0	3-5	3-5	NP
60 ground snow load	Southern pine	2 × 6	8-8	7-10	6-10	1-0	1-6	1-5	NP	NP	NP	NP	NP
		2 × 8	11-5	10-4	8-9	1-0	1-6	2-0	2-4	NP	NP	NP	NP
		2 × 10	14-7	12-9	10-5	1-0	1-6	2-0	2-6	2-11	2-11	NP	NP
		2 × 12	17-3	15-0	12-3	1-0	1-6	2-0	2-6	3-0	3-6	3-7	NP
	Douglas fir-larch[e] Hem-fir[e] Spruce-pine-fir[e]	2 × 6	8-4	7-6	6-2	1-0	1-6	1-4	NP	NP	NP	NP	NP
		2 × 8	10-11	9-11	8-3	1-0	1-6	2-0	2-2	NP	NP	NP	NP
		2 × 10	13-11	12-4	10-0	1-0	1-6	2-0	2-6	2-10	NP	NP	NP
		2 × 12	16-6	14-3	11-8	1-0	1-6	2-0	2-6	3-0	3-5	3-5	NP
	Redwood[f] Western cedars[f] Ponderosa pine[f] Red pine[f]	2 × 6	7-9	7-0	6-2	1-0	1-4	NP	NP	NP	NP	NP	NP
		2 × 8	10-2	9-3	7-11	1-0	1-6	2-0	1-11	NP	NP	NP	NP
		2 × 10	13-0	11-9	9-7	1-0	1-6	2-0	2-6	2-7	NP	NP	NP
		2 × 12	15-9	13-8	11-2	1-0	1-6	2-0	2-6	3-0	3-2	NP	NP

(continued)

TABLE R507.6—continued
MAXIMUM DECK JOIST SPANS

LOAD[a] (psf)	JOIST SPECIES[b]	JOIST SIZE	ALLOWABLE JOIST SPAN [b, c] (feet-inches)			MAXIMUM CANTILEVER[d,f] (feet-inches)							
			Joist spacing (inches)			Joist back span[g] (feet)							
			12	16	24	4	6	8	10	12	14	16	18
70 ground snow load	Southern pine	2 × 6	8-3	7-6	6-5	1-0	1-6	1-5	NP	NP	NP	NP	NP
		2 × 8	10-10	9-10	8-2	1-0	1-6	2-0	2-2	NP	NP	NP	NP
		2 × 10	13-9	11-11	9-9	1-0	1-6	2-0	2-6	2-9	NP	NP	NP
		2 × 12	16-2	14-0	11-5	1-0	1-6	2-0	2-6	3-0	3-5	3-5	NP
	Douglas fir-larch[e] Hem-fir[e] Spruce-pine-fir[e]	2 × 6	7-11	7-1	5-9	1-0	1-6	NP	NP	NP	NP	NP	NP
		2 × 8	10-5	9-5	7-8	1-0	1-6	2-0	2-1	NP	NP	NP	NP
		2 × 10	13-3	11-6	9-5	1-0	1-6	2-0	2-6	2-8	NP	NP	NP
		2 × 12	15-5	13-4	10-11	1-0	1-6	2-0	2-6	3-0	3-3	NP	NP
	Redwood[f] Western cedars[f] Ponderosa pine[f] Red pine[f]	2 × 6	7-4	6-8	5-10	1-0	1-4	NP	NP	NP	NP	NP	NP
		2 × 8	9-8	8-10	7-4	1-0	1-6	1-11	NP	NP	NP	NP	NP
		2 × 10	12-4	11-0	9-0	1-0	1-6	2-0	2-6	2-6	NP	NP	NP
		2 × 12	14-9	12-9	10-5	1-0	1-6	2-0	2-6	3-0	3-0	NP	NP

For SI: 1 inch = 25.4 mm, 1 foot = 304.8 mm, 1 pound per square foot = 0.0479 kPa, 1 pound = 0.454 kg.

NP = Not Permitted.

a. Dead load = 10 psf. Snow load not assumed to be concurrent with live load.

b. No. 2 grade, wet service factor included.

c. L/Δ = 360 at main span.

d. L/Δ = 180 at cantilever with a 220-pound point load applied to end.

e. Includes incising factor.

f. Incising factor not included.

g. Interpolation allowed. Extrapolation is not allowed.

TABLE R507.7
MAXIMUM JOIST SPACING FOR WOOD DECKING

DECKING MATERIAL TYPE AND NOMINAL SIZE	DECKING PERPENDICULAR TO JOIST		DECKING DIAGONAL TO JOIST[a]	
	Single span[c]	Multiple span[c]	Single span[c]	Multiple span[c]
	Maximum on-center joist spacing (inches)			
$1^1/_4$-inch-thick wood[b]	12	16	8	12
2-inch-thick wood	24	24	18	24

For SI: 1 inch = 25.4 mm, 1 foot = 304.8 mm, 1 degree = 0.01745 rad.

a. Maximum angle of 45 degrees from perpendicular for wood deck boards.

b. Other maximum span provided by an accredited lumber grading or inspection agency also allowed.

c. Individual wood deck boards supported by two joists shall be considered single span and three or more joists shall be considered multiple span.

R507.9.1.3 Ledger to band joist details. Fasteners used in deck ledger connections in accordance with Table R507.9.1.3(1) shall be hot-dipped galvanized or stainless steel and shall be installed in accordance with Table R507.9.1.3(2) and Figures R507.9.1.3(1) and R507.9.1.3(2).

R507.9.1.4 Alternate ledger details. Alternate framing configurations supporting a ledger constructed to meet the load requirements of Section R301.5 shall be permitted.

TABLE R507.9.1.3(1)
DECK LEDGER CONNECTION TO BAND JOIST

LOAD[c] (psf)	JOIST SPAN[a] (feet)	ON-CENTER SPACING OF FASTENERS[b] (inches)		
		$^1/_2$-inch diameter lag screw with $^1/_2$-inch maximum sheathing[d, e]	$^1/_2$-inch diameter bolt with $^1/_2$-inch maximum sheathing[e]	$^1/_2$-inch diameter bolt with 1-inch maximum sheathing[f]
40 live load	6	30	36	36
	8	23	36	36
	10	18	34	29
	12	15	29	24
	14	13	24	21
	16	11	21	18
	18	10	19	16
50 ground snow load	6	29	36	36
	8	22	36	35
	10	17	33	28
	12	14	27	23
	14	12	23	20
	16	11	20	17
	18	9	18	15
60 ground snow load	6	25	36	36
	8	18	35	30
	10	15	28	24
	12	12	23	20
	14	10	20	17
	16	9	17	15
	18	8	15	13
70 ground snow load	6	22	36	35
	8	16	31	26
	10	13	25	21
	12	11	20	17
	14	9	17	15
	16	8	15	13
	18	7	13	11

For SI: 1 inch = 25.4 mm, 1 foot = 304.8 mm, 1 pound per square foot = 0.0479 kPa.

a. Interpolation permitted. Extrapolation is not permitted.

b. Ledgers shall be flashed in accordance with Section R703.4 to prevent water from contacting the house band joist.

c. Dead Load = 10 psf. Snow load shall not be assumed to act concurrently with live load.

d. The tip of the lag screw shall fully extend beyond the inside face of the band joist.

e. Sheathing shall be wood structural panel or solid sawn lumber.

f. Sheathing shall be permitted to be wood structural panel, gypsum board, fiberboard, lumber or foam sheathing. Up to $^1/_2$-inch thickness of stacked washers shall be permitted to substitute for up to $^1/_2$ inch of allowable sheathing thickness where combined with wood structural panel or lumber sheathing.

TABLE R507.9.1.3(2)
PLACEMENT OF LAG SCREWS AND BOLTS IN DECK LEDGERS AND BAND JOISTS

	MINIMUM END AND EDGE DISTANCES AND SPACING BETWEEN ROWS			
	TOP EDGE	BOTTOM EDGE	ENDS	ROW SPACING
Ledger[a]	2 inches[d]	$^3/_4$ inch	2 inches[b]	$1^5/_8$ inches[b]
Band Joist[c]	$^3/_4$ inch	2 inches	2 inches[b]	$1^5/_8$ inches[b]

For SI: 1 inch = 25.4 mm.

a. Lag screws or bolts shall be staggered from the top to the bottom along the horizontal run of the deck ledger in accordance with Figure R507.9.1.3(1).

b. Maximum 5 inches.

c. For engineered rim joists, the manufacturer's recommendations shall govern.

d. The minimum distance from bottom row of lag screws or bolts to the top edge of the ledger shall be in accordance with Figure R507.9.1.3(1).

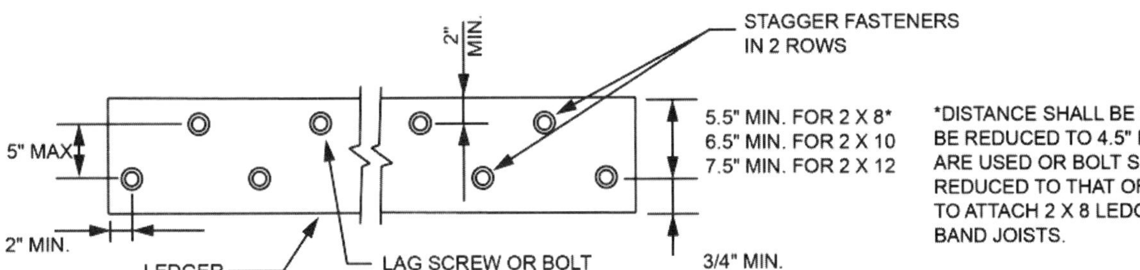

For SI: 1 inch = 25.4 mm.

FIGURE R507.9.1.3(1)
PLACEMENT OF LAG SCREWS AND BOLTS IN LEDGERS

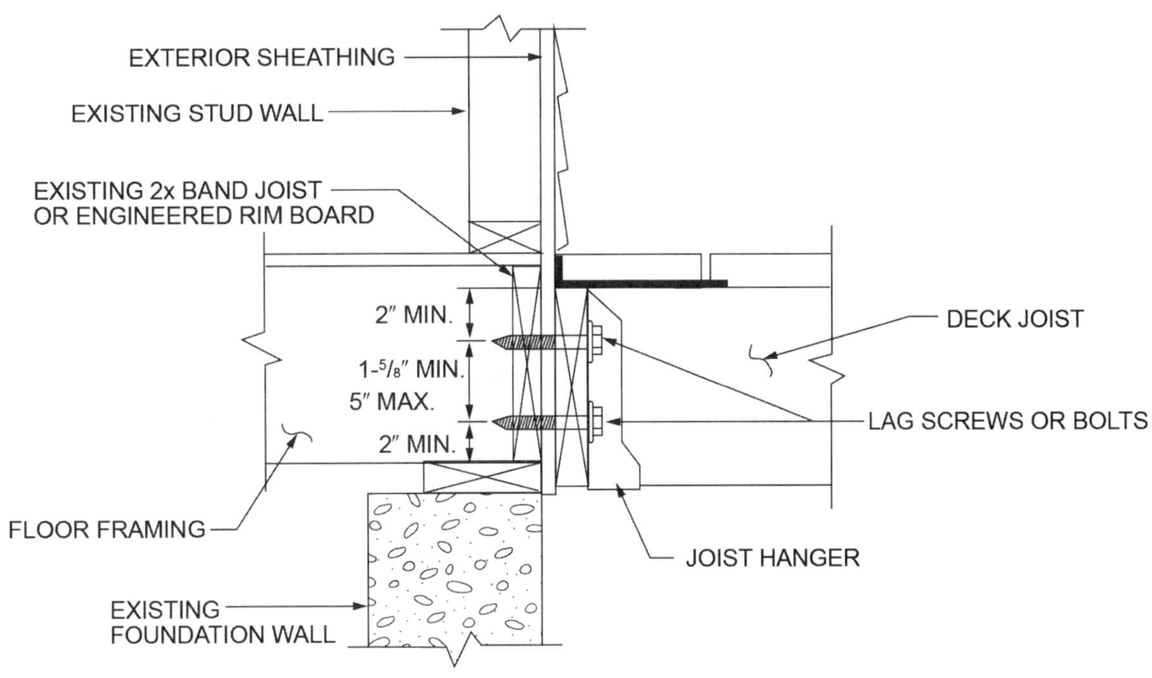

For SI: 1 inch = 25.4 mm.

FIGURE R507.9.1.3(2)
PLACEMENT OF LAG SCREWS AND BOLTS IN BAND JOISTS

R507.9.2 Lateral connection. Lateral loads shall be transferred to the ground or to a structure capable of transmitting them to the ground. Where the lateral load connection is provided in accordance with Figure R507.9.2(1), hold-down tension devices shall be installed in not less than two locations per deck, within 24 inches (610 mm) of each end of the deck. Each device shall have an allowable stress design capacity of not less than 1,500 pounds (6672 N). Where the lateral load connections are provided in accordance with Figure R507.9.2(2), the hold-down tension devices shall be installed in not less than four locations per deck, and each device shall have an allowable stress design capacity of not less than 750 pounds (3336 N).

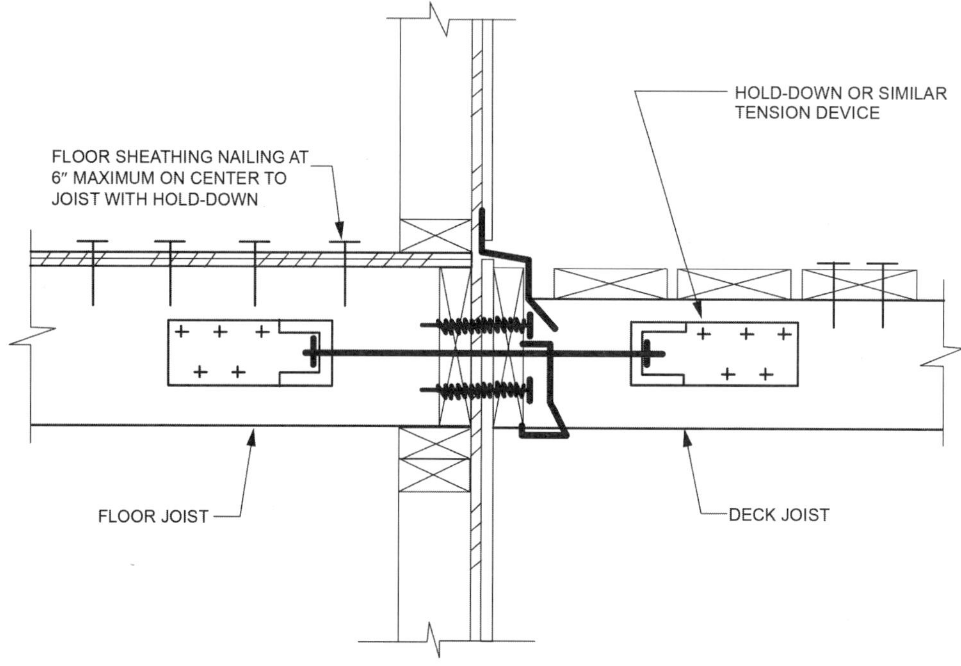

For SI: 1 inch = 25.4 mm.

FIGURE R507.9.2(1)
DECK ATTACHMENT FOR LATERAL LOADS

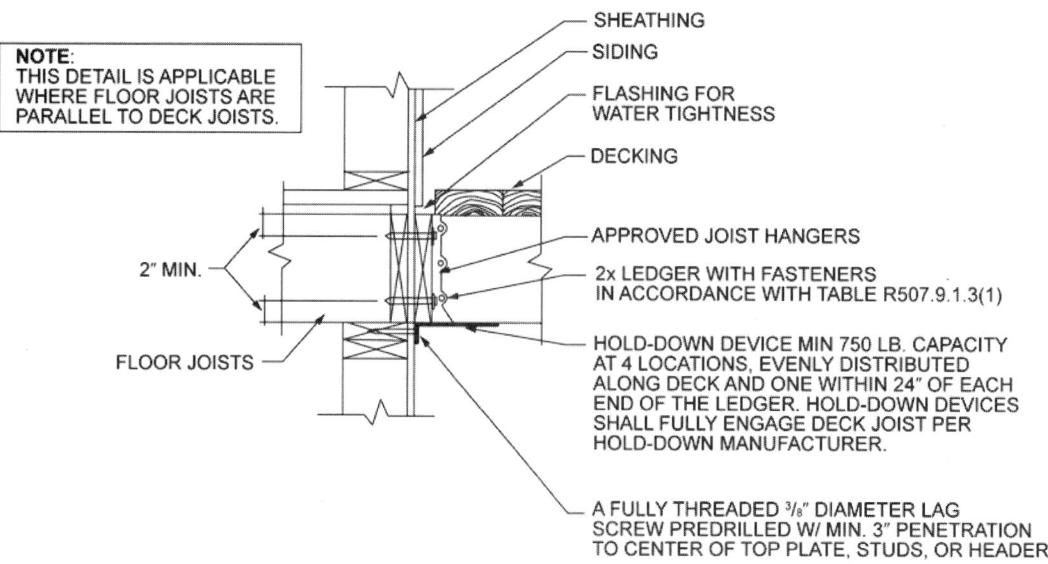

For SI: 1 inch = 25.4 mm, 1 foot = 304.8 mm.

FIGURE R507.9.2(2)
DECK ATTACHMENT FOR LATERAL LOADS

R507.10 Exterior guards. *Guards* shall be constructed to meet the requirements of Sections R301.5 and R312, and this section.

R507.10.1 Support of guards. Where *guards* are supported on deck framing, *guard* loads shall be transferred to the deck framing with a continuous load path to the deck joists.

R507.10.1.1 Guards supported by side of deck framing. Where *guards* are connected to the interior or exterior side of a deck joist or beam, the joist or beam shall be connected to the adjacent joists to prevent rotation of the joist or beam. Connections relying only on fasteners in end grain withdrawal are not permitted.

R507.10.1.2 Guards supported on top of deck framing. Where *guards* are mounted on top of the decking, the *guards* shall be connected to the deck framing or blocking and installed in accordance with manufacturer's instructions to transfer the *guard* loads to the adjacent joists.

R507.10.2 Wood posts at deck guards. Where 4-inch by 4-inch (102 mm by 102 mm) wood posts support guard loads applied to the top of the guard, such posts shall not be notched at the connection to the supporting structure.

R507.10.3 Plastic composite guards. *Plastic composite guards* shall comply with the provisions of Section R507.2.2.

R507.10.4 Other guards. Other *guards* shall be in accordance with either manufacturer's instructions or accepted engineering principles.

CHAPTER 6
WALL CONSTRUCTION

User note:

About this chapter: Chapter 6 contains prescriptive provisions for the design and construction of walls. The wall construction covered in Chapter 6 consists of five different types: wood framed, cold-formed steel framed, masonry, concrete and structural insulated panel (SIP). The primary concern of this chapter is the structural integrity of wall construction and transfer of all imposed loads to the supporting structure.

SECTION R601
GENERAL

R601.1 Application. The provisions of this chapter shall control the design and construction of walls and partitions for buildings.

R601.2 Requirements. Wall construction shall be capable of accommodating all loads imposed in accordance with Section R301 and of transmitting the resulting loads to the supporting structural elements.

R601.2.1 Compressible floor-covering materials. Compressible floor-covering materials that compress more than $^1/_{32}$ inch (0.8 mm) when subjected to 50 pounds (23 kg) applied over 1 inch square (645 mm) of material and are greater than $^1/_8$ inch (3.2 mm) in thickness in the uncompressed state shall not extend beneath walls, partitions or columns, which are fastened to the floor.

SECTION R602
WOOD WALL FRAMING

R602.1 General. Wood and wood-based products used for load-supporting purposes shall conform to the applicable provisions of this section.

R602.1.1 Sawn lumber. Sawn lumber shall be identified by a grade *mark* of an accredited lumber grading or inspection agency and have design values certified by an accreditation body that complies with DOC PS 20. In lieu of a grade *mark*, a certification of inspection issued by a lumber grading or inspection agency meeting the requirements of this section shall be accepted.

R602.1.2 End-jointed lumber. *Approved* end-jointed lumber identified by a grade *mark* conforming to Section R602.1 shall be permitted to be used interchangeably with solid-sawn members of the same species and grade. End-jointed lumber used in an assembly required elsewhere in this code to have a fire-resistance rating shall have the designation "Heat Resistant Adhesive" or "HRA" included in its grade *mark*.

R602.1.3 Structural glued-laminated timbers. Glued-laminated timbers shall be manufactured and identified as required in ANSI A190.1, ANSI 117 and ASTM D3737.

R602.1.4 Structural log members. Structural log members shall comply with the provisions of ICC 400.

R602.1.5 Structural composite lumber. Structural capacities for *structural composite lumber* shall be established and monitored in accordance with ASTM D5456.

R602.1.6 Cross-laminated timber. Cross-laminated timber shall be manufactured and identified as required by ANSI/APA PRG 320.

R602.1.7 Engineered wood rim board. Engineered wood rim boards shall conform to ANSI/APA PRR 410 or shall be evaluated in accordance with ASTM D7672. Structural capacities shall be in accordance with either ANSI/APA PRR 410 or established in accordance with ASTM D7672. Rim boards conforming to ANSI/APA PRR 410 shall be marked in accordance with that standard.

R602.1.8 Wood structural panels. *Wood structural panel* sheathing shall conform to DOC PS 1, DOC PS 2 or, when manufactured in Canada, CSA 0325 or CSA 0437. Panels shall be identified for grade, bond classification, and performance category by a grade *mark* or certificate of inspection issued by an *approved* agency.

R602.1.9 Particleboard. Particleboard shall conform to ANSI A208.1. Particleboard shall be identified by the grade *mark* or certificate of inspection issued by an *approved* agency.

R602.1.10 Fiberboard. Fiberboard shall conform to ASTM C208. Fiberboard sheathing, where used structurally, shall be identified by an *approved* agency as conforming to ASTM C208.

R602.1.11 Structural insulated panels. *Structural insulated panels* shall be manufactured and identified in accordance with ANSI/APA PRS 610.1.

R602.2 Grade. Studs shall be a minimum No. 3, standard or stud grade lumber.

Exception: Bearing studs not supporting floors and nonbearing studs shall be permitted to be utility grade lumber, provided that the studs are spaced in accordance with Table R602.3(5).

R602.3 Design and construction. Exterior walls of wood-frame construction shall be designed and constructed in accordance with the provisions of this chapter and Figures R602.3(1) and R602.3(2), or in accordance with AWC NDS. Components of exterior walls shall be fastened in accordance with Tables R602.3(1) through R602.3(4). Wall sheathing shall be fastened directly to framing members and, where placed on the exterior side of an exterior wall, shall be capable of resisting the wind pressures listed in Table R301.2.1(1) adjusted for height and exposure using Table R301.2.1(2) and shall conform to the requirements of Table R602.3(3). Wall sheathing used only for exterior wall covering purposes shall comply with Section R703.

Studs shall be continuous from support at the sole plate to a support at the top plate to resist loads perpendicular to the wall. The support shall be a foundation or floor, ceiling or roof *diaphragm* or shall be designed in accordance with accepted engineering practice.

Exception: Jack studs, trimmer studs and cripple studs at openings in walls that comply with Tables R602.7(1) and R602.7(2).

R602.3.1 Stud size, height and spacing. The size, height and spacing of studs shall be in accordance with Table R602.3(5).

Exceptions:

1. Utility grade studs shall not be spaced more than 16 inches (406 mm) on center, shall not support more than a roof and ceiling, and shall not exceed 8 feet (2438 mm) in height for exterior walls and *load-bearing walls* or 10 feet (3048 mm) for interior nonload-bearing walls.
2. Where ground snow loads are less than or equal to 25 pounds per square foot (1.2 kPa), and the ultimate design wind speed is less than or equal to 130 mph (58.1 m/s), 2-inch by 6-inch (38 mm by 140 mm) studs supporting a roof load with not more than 6 feet (1829 mm) of tributary length shall have a maximum height of 18 feet (5486 mm) where spaced at 16 inches (406 mm) on center, or 20 feet (6096 mm) where spaced at 12 inches (305 mm) on center. Studs shall be No. 2 grade lumber or better.
3. Exterior load-bearing studs not exceeding 12 feet (3658 mm) in height provided in accordance with Table R602.3(6). The minimum number of full-height studs adjacent to openings shall be in accordance with Section R602.7.5. The building shall be located in Exposure B, the roof *live load* shall not exceed 20 psf (0.96 kPa), and the ground snow load shall not exceed 30 psf (1.4 kPa). Studs and plates shall be No. 2 grade lumber or better.

R602.3.2 Top plate. Wood stud walls shall be capped with a double top plate installed to provide overlapping at corners and intersections with bearing partitions. End joints in top plates shall be offset not less than 24 inches (610 mm). Joints in plates need not occur over studs. Plates shall be not less than 2-inches (51 mm) nominal thickness and have a width not less than the width of the studs.

Exception: A single top plate used as an alternative to a double top plate shall comply with the following:

1. The single top plate shall be tied at corners, intersecting walls, and at in-line splices in straight wall lines in accordance with Table R602.3.2.
2. The rafters or joists shall be centered over the studs with a tolerance of not more than 1 inch (25 mm).
3. Omission of the top plate is permitted over headers where the headers are adequately tied to adjacent wall sections in accordance with Table R602.3.2.

R602.3.3 Bearing studs. Where joists, trusses or rafters are spaced more than 16 inches (406 mm) on center and the bearing studs below are spaced 24 inches (610 mm) on center, such members shall bear within 5 inches (127 mm) of the studs beneath.

Exceptions:

1. The top plates are two 2-inch by 6-inch (38 mm by 140 mm) or two 3-inch by 4-inch (64 mm by 89 mm) members.
2. A third top plate is installed.
3. Solid blocking equal in size to the studs is installed to reinforce the double top plate.

R602.3.4 Bottom (sole) plate. Studs shall have full bearing on a nominal 2-by (51 mm) or larger plate or sill having a width not less than to the width of the studs.

R602.3.5 Braced wall panel uplift load path. Braced wall panels located at exterior walls that support roof rafters or trusses (including stories below top story) shall have the framing members connected in accordance with one of the following:

1. Fastening in accordance with Table R602.3(1) where:
   1.1. The ultimate design wind speed does not exceed 115 mph (51 m/s), the wind exposure category is B, the roof pitch is 5:12 or greater, and the roof span is 32 feet (9754 mm) or less.
   1.2. The net uplift value at the top of a wall does not exceed 100 plf (146 N/mm). The net uplift value shall be determined in accordance with Section R802.11 and shall be permitted to be reduced by 60 plf (86 N/mm) for each full wall above.
2. Where the net uplift value at the top of a wall exceeds 100 plf (146 N/mm), installing *approved* uplift framing connectors to provide a continuous load path from the top of the wall to the foundation or to a point where the uplift force is 100 plf (146 N/mm) or less. The net uplift value shall be as determined in Item 1.2.
3. Wall sheathing and fasteners designed to resist combined uplift and shear forces in accordance with accepted engineering practice.

R602.4 Interior load-bearing walls. Interior *load-bearing walls* shall be constructed, framed and fireblocked as specified for exterior walls.

TABLE R602.3(1)
FASTENING SCHEDULE

ITEM	DESCRIPTION OF BUILDING ELEMENTS	NUMBER AND TYPE OF FASTENER[a, b, c]	SPACING AND LOCATION
Roof			
1	Blocking between ceiling joists, rafters or trusses to top plate or other framing below	4-8d box ($2^1/_2$″ × 0.113″); or 3-8d common ($2^1/_2$″ × 0.131″); or 3-10d box (3″ × 0.128″); or 3-3″ × 0.131″ nails	Toe nail
	Blocking between rafters or truss not at the wall top plates, to rafter or truss	2-8d common ($2^1/_2$″ × 0.131″); or 2-3″ × 0.131″ nails	Each end toe nail
		2-16d common ($3^1/_2$″ × 0.162″); or 3-3″ × 0.131″ nails	End nail
	Flat blocking to truss and web filler	16d common ($3^1/_2$″ × 0.162″); or 3″ × 0.131″ nails	6″ o.c. face nail
2	Ceiling joists to top plate	4-8d box ($2^1/_2$″ × 0.113″); or 3-8d common ($2^1/_2$″ × 0.131″); or 3-10d box (3″ × 0.128″); or 3-3″ × 0.131″ nails	Per joist, toe nail
3	Ceiling joist not attached to parallel rafter, laps over partitions [see Section R802.5.2 and Table R802.5.2(1)]	4-10d box (3″ × 0.128″); or 3-16d common ($3^1/_2$″ × 0.162″); or 4-3″ × 0.131″ nails	Face nail
4	Ceiling joist attached to parallel rafter (heel joint) [see Section R802.5.2 and Table R802.5.2(1)]	Table R802.5.2(1)	Face nail
5	Collar tie to rafter, face nail	4-10d box (3″ × 0.128″); or 3-10d common (3″ × 0.148″); or 4-3″ × 0.131″ nails	Face nail each rafter
6	Rafter or roof truss to plate	3-16d box ($3^1/_2$″ × 0.135″); or 3-10d common (3″ × 0.148″); or 4-10d box (3″ × 0.128″); or 4-3″ × 0.131″ nails	2 toe nails on one side and 1 toe nail on opposite side of each rafter or truss[i]
7	Roof rafters to ridge, valley or hip rafters or roof rafter to minimum 2″ ridge beam	4-16d box ($3^1/_2$″ × 0.135″); or 3-10d common (3″ × 0.148″); or 4-10d box (3″ × 0.128″); or 4-3″ × 0.131″ nails	Toe nail
		3-16d box ($3^1/_2$″ × 0.135″); or 2-16d common ($3^1/_2$″ × 0.162″); or 3-10d box (3″ × 0.128″); or 3-3″ × 0.131″ nails	End nail
Wall			
8	Stud to stud (not at braced wall panels)	16d common ($3^1/_2$″ × 0.162″)	24″ o.c. face nail
		10d box (3″ × 0.128″); or 3″ × 0.131″ nails	16″ o.c. face nail
9	Stud to stud and abutting studs at intersecting wall corners (at braced wall panels)	16d box ($3^1/_2$″ × 0.135″); or 3″ × 0.131″ nails	12″ o.c. face nail
		16d common ($3^1/_2$″ × 0.162″)	16″ o.c. face nail
10	Built-up header (2″ to 2″ header with $^1/_2$″ spacer)	16d common ($3^1/_2$″ × 0.162″)	16″ o.c. each edge face nail
		16d box ($3^1/_2$″ × 0.135″)	12″ o.c. each edge face nail
11	Continuous header to stud	5-8d box ($2^1/_2$″ × 0.113″); or 4-8d common ($2^1/_2$″ × 0.131″); or 4-10d box (3″ × 0.128″)	Toe nail

(continued)

TABLE R602.3(1)—continued
FASTENING SCHEDULE

ITEM	DESCRIPTION OF BUILDING ELEMENTS	NUMBER AND TYPE OF FASTENER[a, b, c]	SPACING AND LOCATION
Wall			
12	Adjacent full-height stud to end of header	4-16d box (3$^{1}/_{2}$″× 0.135″); or 3-16d common (3$^{1}/_{2}$″ × 0.162″); or 4-10d box (3″ × 0.128″); or 4-3″ × 0.131″ nails	End nail
13	Top plate to top plate	16d common (3$^{1}/_{2}$″ × 0.162″)	16″ o.c. face nail
		10d box (3″ × 0.128″); or 3″ × 0.131″ nails	12″ o.c. face nail
14	Double top plate splice	8-16d common (3$^{1}/_{2}$″ × 0.162″); or 12-16d box (3$^{1}/_{2}$″ × 0.135″); or 12-10d box (3″ × 0.128″); or 12-3″ × 0.131″ nails	Face nail on each side of end joint (minimum 24″ lap splice length each side of end joint)
15	Bottom plate to joist, rim joist, band joist or blocking (not at braced wall panels)	16d common (3$^{1}/_{2}$″ × 0.162″)	16″ o.c. face nail
		16d box (3$^{1}/_{2}$″ × 0.135″); or 3″ × 0.131″ nails	12″ o.c. face nail
Roof			
16	Bottom plate to joist, rim joist, band joist or blocking (at braced wall panel)	3-16d box (3$^{1}/_{2}$″ × 0.135″); or 2-16d common (3$^{1}/_{2}$″ × 0.162″); or 4-3″ × 0.131″ nails	16″ o.c. face nail
17	Top or bottom plate to stud	4-8d box (2$^{1}/_{2}$″ × 0.113″); or 3-16d box (3$^{1}/_{2}$″ × 0.135″); or 4-8d common (2$^{1}/_{2}$″ × 0.131″); or 4-10d box (3″ × 0.128″); or 4-3″ × 0.131″ nails	Toe nail
		3-16d box (3$^{1}/_{2}$″ × 0.135″); or 2-16d common (3$^{1}/_{2}$″ × 0.162″); or 3-10d box (3″ × 0.128″); or 3-3″ × 0.131″ nails	End nail
18	Top plates, laps at corners and intersections	3-10d box (3″ × 0.128″); or 2-16d common (3$^{1}/_{2}$″ × 0.162″); or 3-3″ × 0.131″ nails	Face nail
19	1″ brace to each stud and plate	3-8d box (2$^{1}/_{2}$″ × 0.113″); or 2-8d common (2$^{1}/_{2}$″ × 0.131″); or 2-10d box (3″ × 0.128″); or 2 staples 1$^{3}/_{4}$″	Face nail
20	1″ × 6″ sheathing to each bearing	3-8d box (2$^{1}/_{2}$″ × 0.113″); or 2-8d common (2$^{1}/_{2}$″ × 0.131″); or 2-10d box (3″ × 0.128″); or 2 staples, 1″ crown, 16 ga., 1$^{3}/_{4}$″ long	Face nail
21	1″ × 8″ and wider sheathing to each bearing	3-8d box (2$^{1}/_{2}$″ × 0.113″); or 3-8d common (2$^{1}/_{2}$″ × 0.131″); or 3-10d box (3″ × 0.128″); or 3 staples, 1″ crown, 16 ga., 1$^{3}/_{4}$″ long	Face nail
		Wider than 1″ × 8″ 4-8d box (2$^{1}/_{2}$″ × 0.113″); or 3-8d common (2$^{1}/_{2}$″ × 0.131″); or 3-10d box (3″ × 0.128″); or 4 staples, 1″ crown, 16 ga., 1$^{3}/_{4}$″ long	

(continued)

TABLE R602.3(1)—continued
FASTENING SCHEDULE

ITEM	DESCRIPTION OF BUILDING ELEMENTS	NUMBER AND TYPE OF FASTENER[a, b, c]	SPACING AND LOCATION
		Floor	
22	Joist to sill, top plate or girder	4-8d box ($2^1/_2$″ × 0.113″); or 3-8d common ($2^1/_2$″ × 0.131″); or 3-10d box (3″ × 0.128″); or 3-3″ × 0.131″ nails	Toe nail
23	Rim joist, band joist or blocking to sill or top plate (roof applications also)	8d box ($2^1/_2$″ × 0.113″)	4″ o.c. toe nail
		8d common ($2^1/_2$″ × 0.131″); or 10d box (3″ × 0.128″); or 3″ × 0.131″ nails	6″ o.c. toe nail
24	1″ × 6″ subfloor or less to each joist	3-8d box ($2^1/_2$″ × 0.113″); or 2-8d common ($2^1/_2$″ × 0.131″); or 3-10d box (3″ × 0.128″); or 2 staples, 1″ crown, 16 ga., $1^3/_4$″ long	Face nail
25	2″ subfloor to joist or girder	3-16d box ($3^1/_2$″ × 0.135″); or 2-16d common ($3^1/_2$″ × 0.162″)	Blind and face nail
26	2″ planks (plank & beam—floor & roof)	3-16d box ($3^1/_2$″ × 0.135″); or 2-16d common ($3^1/_2$″ × 0.162″)	At each bearing, face nail
27	Band or rim joist to joist	3-16d common ($3^1/_2$″ × 0.162″); or 4-10 box (3″ × 0.128″); or 4-3″ × 0.131″ nails; or 4-3″ × 14 ga. staples, $^7/_{16}$″ crown	End nail
28	Built-up girders and beams, 2-inch lumber layers	20d common (4″ × 0.192″); or	Nail each layer as follows: 32″ o.c. at top and bottom and staggered.
		10d box (3″ × 0.128″); or 3″ × 0.131″ nails	24″ o.c. face nail at top and bottom staggered on opposite sides
		And: 2-20d common (4″ × 0.192″); or 3-10d box (3″ × 0.128″); or 3-3″ × 0.131″ nails	Face nail at ends and at each splice
29	Ledger strip supporting joists or rafters	4-16d box ($3^1/_2$″ × 0.135″); or 3-16d common ($3^1/_2$″ × 0.162″); or 4-10d box (3″ × 0.128″); or 4-3″ × 0.131″ nails	At each joist or rafter, face nail
30	Bridging or blocking to joist, rafter or truss	2-10d box (3″ × 0.128″); or 2-8d common ($2^1/_2$″ × 0.131″); or 3″ × 0.131″ nails	Each end, toe nail

ITEM	DESCRIPTION OF BUILDING ELEMENTS	NUMBER AND TYPE OF FASTENER[a, b, c]	SPACING OF FASTENERS	
			Edges[h] (inches)	Intermediate supports[c, e] (inches)
	Wood structural panels, subfloor, roof and interior wall sheathing to framing and particleboard wall sheathing to framing [see Table R602.3(3) for wood structural panel exterior wall sheathing to wall framing]			
31	$^3/_8$″ – $^1/_2$″	6d common or deformed (2″ × 0.113″ × 0.266″ head); or $2^3/_8$″ × 0.113″ × 0.266″ head nail (subfloor, wall)[i]	6	6[f]
		8d common ($2^1/_2$″ × 0.131″) nail (roof); or RSRS-01 ($2^3/_8$″ × 0.113″) nail (roof)[b]	6	6[f]
32	$^{19}/_{32}$″ – $^3/_4$″	8d common (2-$2^1/_2$″ × 0.131″) nail (subfloor, wall)	6	12
		8d common ($2^1/_2$″ × 0.131″) nail (roof); or RSRS-01; ($2^3/_8$″ × 0.113″) nail (roof)[b]	6	6[f]
		Deformed $2^3/_8$″ × 0.113″ × 0.266″ head (wall or subfloor)	6	12
33	$^7/_8$″ – $1^1/_4$″	10d common (3″ × 0.148″) nail; or ($2^1/_2$″ × 0.131 × 0.281″ head) deformed nail	6	12

(continued)

TABLE R602.3(1)—continued
FASTENING SCHEDULE

ITEM	DESCRIPTION OF BUILDING ELEMENTS	NUMBER AND TYPE OF FASTENER[a, b, c]	SPACING AND LOCATION	
		Other wall sheathing[g]		
34	$^1/_2$" structural cellulosic fiberboard sheathing	$1^1/_2$" × 0.120" galvanized roofing nail, $^7/_{16}$" head diameter; or $1^1/_4$" long 16 ga. staple with $^7/_{16}$" or 1" crown	3	6
35	$^{25}/_{32}$" structural cellulosic fiberboard sheathing	$1^3/_4$" × 0.120" galvanized roofing nail, $^7/_{16}$" head diameter; or $1^1/_4$" long 16 ga. staple with $^7/_{16}$" or 1" crown	3	6
36	$^1/_2$" gypsum sheathing[d]	$1^1/_2$" × 0.120" galvanized roofing nail, $^7/_{16}$" head diameter, or $1^1/_4$" long 16 ga.; staple galvanized, $1^1/_2$" long; $^7/_{16}$" or 1" crown or $1^1/_4$" screws, Type W or S	7	7
37	$^5/_8$" gypsum sheathing[d]	$1^3/_4$" × 0.120" galvanized roofing nail, $^7/_{16}$" head diameter, or $1^1/_4$" long 16 ga.; staple galvanized, $1^1/_2$" long; $^7/_{16}$" or 1" crown or $1^1/_4$" screws, Type W or S	7	7
		Wood structural panels, combination subfloor underlayment to framing		
38	$^3/_4$" and less	Deformed (2" × 0.113") or Deformed (2" × 0.120") nail; or 8d common ($2^1/_2$" × 0.131") nail	6	12
39	$^7/_8$" - 1"	8d common ($2^1/_2$" × 0.131") nail; or Deformed (2" × 0.113"); or Deformed ($2^1/_2$" × 0.120") nail	6	12
40	$1^1/_8$" - $1^1/_4$"	10d common (3" × 0.148") nail; or Deformed (2" × 0.113"); or Deformed ($2^1/_2$" × 0.120") nail	6	12

For SI: 1 inch = 25.4 mm, 1 foot = 304.8 mm, 1 mile per hour = 0.447 m/s; 1 ksi = 6.895 MPa.

a. Nails are smooth-common, box or deformed shanks except where otherwise stated. Nails used for framing and sheathing connections are carbon steel and shall have minimum average bending yield strengths as shown: 80 ksi for shank diameter of 0.192 inch (20d common nail), 90 ksi for shank diameters larger than 0.142 inch but not larger than 0.177 inch, and 100 ksi for shank diameters of 0.142 inch or less. Connections using nails and staples of other materials, such as stainless steel, shall be designed by accepted engineering practice or approved under Section R104.11.

b. RSRS-01 is a Roof Sheathing Ring Shank nail meeting the specifications in ASTM F1667.

c. Nails shall be spaced at not more than 6 inches on center at all supports where spans are 48 inches or greater.

d. Four-foot by 8-foot or 4-foot by 9-foot panels shall be applied vertically.

e. Spacing of fasteners not included in this table shall be based on Table R602.3(2).

f. For wood structural panel roof sheathing attached to gable end roof framing and to intermediate supports within 48 inches of roof edges and ridges, nails shall be spaced at 4 inches on center where the ultimate design wind speed is greater than 130 mph in Exposure B or greater than 110 mph in Exposure C.

g. Gypsum sheathing shall conform to ASTM C1396 and shall be installed in accordance with ASTM C1280 or GA 253. Fiberboard sheathing shall conform to ASTM C208.

h. Spacing of fasteners on floor sheathing panel edges applies to panel edges supported by framing members and required blocking and at floor perimeters only. Spacing of fasteners on roof sheathing panel edges applies to panel edges supported by framing members and required blocking. Blocking of roof or floor sheathing panel edges perpendicular to the framing members need not be provided except as required by other provisions of this code. Floor perimeter shall be supported by framing members or solid blocking.

i. Where a rafter is fastened to an adjacent parallel ceiling joist in accordance with this schedule, provide two toe nails on one side of the rafter and toe nails from the ceiling joist to top plate in accordance with this schedule. The toe nail on the opposite side of the rafter shall not be required.

**TABLE R602.3(2)
ALTERNATE ATTACHMENTS TO TABLE R602.3(1)**

NOMINAL MATERIAL THICKNESS (inches)	DESCRIPTION[a, b] OF FASTENER AND LENGTH (inches)	SPACING[c] OF FASTENERS	
		Edges (inches)	Intermediate supports (inches)
Wood structural panels subfloor, roof[g] and wall sheathing to framing and particleboard wall sheathing to framing[f]			
Up to $^1/_2$	Staple 15 ga. $1^3/_4$	4	8
	0.097–0.099 Nail $2^1/_4$	3	6
	Staple 16 ga. $1^3/_4$	3	6
$^{19}/_{32}$ and $^5/_8$	0.113 Nail 2	3	6
	Staple 15 and 16 ga. 2	4	8
	0.097–0.099 Nail $2^1/_4$	4	8
$^{23}/_{32}$ and $^3/_4$	Staple 14 ga. 2	4	8
	Staple 15 ga. $1^3/_4$	3	6
	0.097–0.099 Nail $2^1/_4$	4	8
	Staple 16 ga. 2	4	8
1	Staple 14 ga. $2^1/_4$	4	8
	0.113 Nail $2^1/_4$	3	6
	Staple 15 ga. $2^1/_4$	4	8
	0.097–0.099 Nail $2^1/_2$	4	8
NOMINAL MATERIAL THICKNESS (inches)	DESCRIPTION[a, b] OF FASTENER AND LENGTH (inches)	SPACING[c] OF FASTENERS	
		Edges (inches)	Body of panel[d] (inches)
Floor underlayment; plywood-hardboard-particleboard[f]-fiber-cement[h]			
Fiber-cement			
$^1/_4$	$1^1/_4$ long × 0.099″ corrosion-resistant, ring shank nails (finished flooring other than tile)	3	6
	Staple 18 ga., $^7/_8$ long, $^1/_4$ crown (finished flooring other than tile)	3	6
	$1^1/_4$ long × .121 shank × .375 head diameter corrosion-resistant (galvanized or stainless steel) roofing nails (for tile finish)	8	8
	$1^1/_4$ long, No. 8 × .375 head diameter, ribbed wafer-head screws (for tile finish)	8	8
Plywood			
$^1/_4$ and $^5/_{16}$	$1^1/_4$ ring or screw shank nail-minimum $12^1/_2$ ga. (0.099″) shank diameter	3	6
	Staple 18 ga., $^7/_8$, $^3/_{16}$ crown width	2	5
$^{11}/_{32}$, $^3/_8$, $^{15}/_{32}$ and $^1/_2$	$1^1/_4$ ring or screw shank nail-minimum $12^1/_2$ ga. (0.099″) shank diameter	6	8[e]
$^{19}/_{32}$, $^5/_8$, $^{23}/_{32}$ and $^3/_4$	$1^1/_2$ ring or screw shank nail-minimum $12^1/_2$ ga. (0.099″) shank diameter	6	8
	Staple 16 ga. $1^1/_2$	6	8
Hardboard[f]			
0.200	$1^1/_2$ long × 0.080″ ring-grooved shank underlayment nail	6	6
	$1^3/_8$ long × 0.080″ polymer cement-coated sinker nail	6	6
	Staple 18 ga., $^7/_8$ long (plastic coated)	3	6
Particleboard			
$^1/_4$	$1^1/_2$ long × 0.099″ ring-grooved shank underlayment nail	3	6
	Staple 18 ga., $^7/_8$ long, $^3/_{16}$ crown	3	6
$^3/_8$	2 long × 0.120″ ring-grooved shank underlayment nail	6	10
	Staple 16 ga., $1^1/_8$ long, $^3/_8$ crown	3	6
$^1/_2$, $^5/_8$	2 long × 0.120″ ring-grooved shank underlayment nail	6	10
	Staple 16 ga., $1^5/_8$ long, $^3/_8$ crown	3	6

(continued)

TABLE R602.3(2)—continued
ALTERNATE ATTACHMENTS TO TABLE R602.3(1)

For SI: 1 inch = 25.4 mm.

a. Nail is a general description and shall be permitted to be T-head, modified round head or round head.

b. Staples shall have a minimum crown width of $^{7}/_{16}$-inch except as noted.

c. Nails or staples shall be spaced at not more than 6 inches on center at all supports where spans are 48 inches or greater. Nails or staples shall be spaced at not more than 12 inches on center at intermediate supports for floors.

d. Fasteners shall be placed in a grid pattern throughout the body of the panel.

e. For 5-ply panels, intermediate nails shall be spaced not more than 12 inches on center each way.

f. Hardboard underlayment shall conform to CPA/ANSI A135.4.

g. Alternate fastening is only permitted for roof sheathing where the ultimate design wind speed is less than or equal to 110 mph, and where fasteners are installed 3 inches on center at all supports.

h. Fiber-cement underlayment shall conform to ASTM C1288 or ISO 8336, Category C.

TABLE R602.3(3)
REQUIREMENTS FOR WOOD STRUCTURAL PANEL WALL SHEATHING USED TO RESIST WIND PRESSURES[a, b, c]

MINIMUM NAIL		MINIMUM WOOD STRUCTURAL PANEL SPAN RATING	MINIMUM NOMINAL PANEL THICKNESS (inches)	MAXIMUM WALL STUD SPACING (inches)	PANEL NAIL SPACING		ULTIMATE DESIGN WIND SPEED V_{ult} (mph)		
Size	Penetration (inches)				Edges (inches o.c.)	Field (inches o.c.)	Wind exposure category		
							B	C	D
6d Common (2.0″ × 0.113″)	1.5	24/0	$^{3}/_{8}$	16	6	12	140	115	110
8d Common (2.5″ × 0.131″)	1.75	24/16	$^{7}/_{16}$	16	6	12	170	140	135
				24	6	12	140	115	110

For SI: 1 inch = 25.4 mm, 1 mile per hour = 0.447 m/s.

a. Panel strength axis parallel or perpendicular to supports. Three-ply plywood sheathing with studs spaced more than 16 inches on center shall be applied with panel strength axis perpendicular to supports.

b. Table is based on wind pressures acting toward and away from building surfaces in accordance with Section R301.2. Lateral bracing requirements shall be in accordance with Section R602.10.

c. Wood structural panels with span ratings of Wall-16 or Wall-24 shall be permitted as an alternate to panels with a 24/0 span rating. Plywood siding rated 16 o.c. or 24 o.c. shall be permitted as an alternate to panels with a 24/16 span rating. Wall-16 and Plywood siding 16 o.c. shall be used with studs spaced not more than 16 inches on center.

TABLE R602.3(4)
ALLOWABLE SPANS FOR PARTICLEBOARD WALL SHEATHING[a]

THICKNESS (inch)	GRADE	STUD SPACING (inches)	
		Where siding is nailed to studs	Where siding is nailed to sheathing
$^{3}/_{8}$	M-1 Exterior glue	16	—
$^{1}/_{2}$	M-2 Exterior glue	16	16

For SI: 1 inch = 25.4 mm.

a. Wall sheathing not exposed to the weather. If the panels are applied horizontally, the end joints of the panel shall be offset so that four panel corners will not meet. Panel edges must be supported. Leave a $^{1}/_{16}$-inch gap between panels and nail not less than $^{3}/_{8}$ inch from panel edges.

TABLE R602.3(5)
SIZE, HEIGHT AND SPACING OF WOOD STUDS[a]

STUD SIZE (inches)	BEARING WALLS					NONBEARING WALLS	
	Laterally unsupported stud height[a] (feet)	Maximum spacing where supporting a roof-ceiling assembly or a habitable attic assembly, only (inches)	Maximum spacing where supporting one floor, plus a roof-ceiling assembly or a habitable attic assembly (inches)	Maximum spacing where supporting two floors, plus a roof-ceiling assembly or a habitable attic assembly (inches)	Maximum spacing where supporting one floor height[a] (inches)	Laterally unsupported stud height[a] (feet)	Maximum spacing (inches)
2 × 3[b]	—	—	—	—	—	10	16
2 × 4	10	24[c]	16[c]	—	24	14	24
3 × 4	10	24	24	16	24	14	24
2 × 5	10	24	24	—	24	16	24
2 × 6	10	24	24	16	24	20	24

For SI: 1 inch = 25.4 mm, 1 foot = 304.8 mm.

a. Listed heights are distances between points of lateral support placed perpendicular to the plane of the wall. Bearing walls shall be sheathed on not less than one side or bridging shall be installed not greater than 4 feet apart measured vertically from either end of the stud. Increases in unsupported height are permitted where in compliance with Exception 2 of Section R602.3.1 or designed in accordance with accepted engineering practice.

b. Shall not be used in exterior walls.

c. A habitable attic assembly supported by 2 × 4 studs is limited to a roof span of 32 feet. Where the roof span exceeds 32 feet, the wall studs shall be increased to 2 × 6 or the studs shall be designed in accordance with accepted engineering practice.

TABLE R602.3(6)
ALTERNATE WOOD BEARING WALL STUD SIZE, HEIGHT AND SPACING

STUD HEIGHT	SUPPORTING	STUD SPACING[a]	ULTIMATE DESIGN WIND SPEED					
			115 mph		130 mph[b]		140 mph[b]	
			Maximum roof/floor span[c]		Maximum roof/floor span[c]		Maximum roof/floor span[c]	
			12 ft	24 ft	12 ft	24 ft	12 ft	24 ft
11 ft	Roof only	12 in	2 × 4	2 × 4	2 × 4	2 × 4	2 × 4	2 × 4
		16 in	2 × 4	2 × 4	2 × 4	2 × 6	2 × 4	2 × 6
		24 in	2 × 6	2 × 6	2 × 6	2 × 6	2 × 6	2 × 6
	Roof and one floor	12 in	2 × 4	2 × 6	2 × 4	2 × 6	2 × 4	2 × 6
		16 in	2 × 6	2 × 6	2 × 6	2 × 6	2 × 6	2 × 6
		24 in	2 × 6	2 × 6	2 × 6	2 × 6	2 × 6	2 × 6
12 ft	Roof only	12 in	2 × 4	2 × 4	2 × 4	2 × 6	2 × 4	2 × 6
		16 in	2 × 4	2 × 6	2 × 6	2 × 6	2 × 6	2 × 6
		24 in	2 × 6	2 × 6	2 × 6	2 × 6	2 × 6	2 × 6
	Roof and one floor	12 in	2 × 4	2 × 6	2 × 6	2 × 6	2 × 6	2 × 6
		16 in	2 × 6	2 × 6	2 × 6	2 × 6	2 × 6	2 × 6
		24 in	2 × 6	2 × 6	2 × 6	2 × 6	2 × 6	DR

For SI: 1 inch = 25.4 mm, 1 foot = 304.8 mm, 1 mph = 0.447 m/s, 1 pound = 4.448 N.

DR = Design Required.

a. Wall studs not exceeding 16 inches on center shall be sheathed with minimum $^1/_2$-inch gypsum board on the interior and $^3/_8$-inch wood structural panel sheathing on the exterior. Wood structural panel sheathing shall be attached with 8d (2.5″ × 0.131″) nails not greater than 6 inches on center along panel edges and 12 inches on center at intermediate supports, and all panel joints shall occur over studs or blocking.

b. Where the ultimate design wind speed exceeds 115 mph, studs shall be attached to top and bottom plates with connectors having a minimum 300-pound lateral capacity.

c. The maximum span is applicable to both single- and multiple-span roof and floor conditions. The *roof assembly* shall not contain a habitable attic.

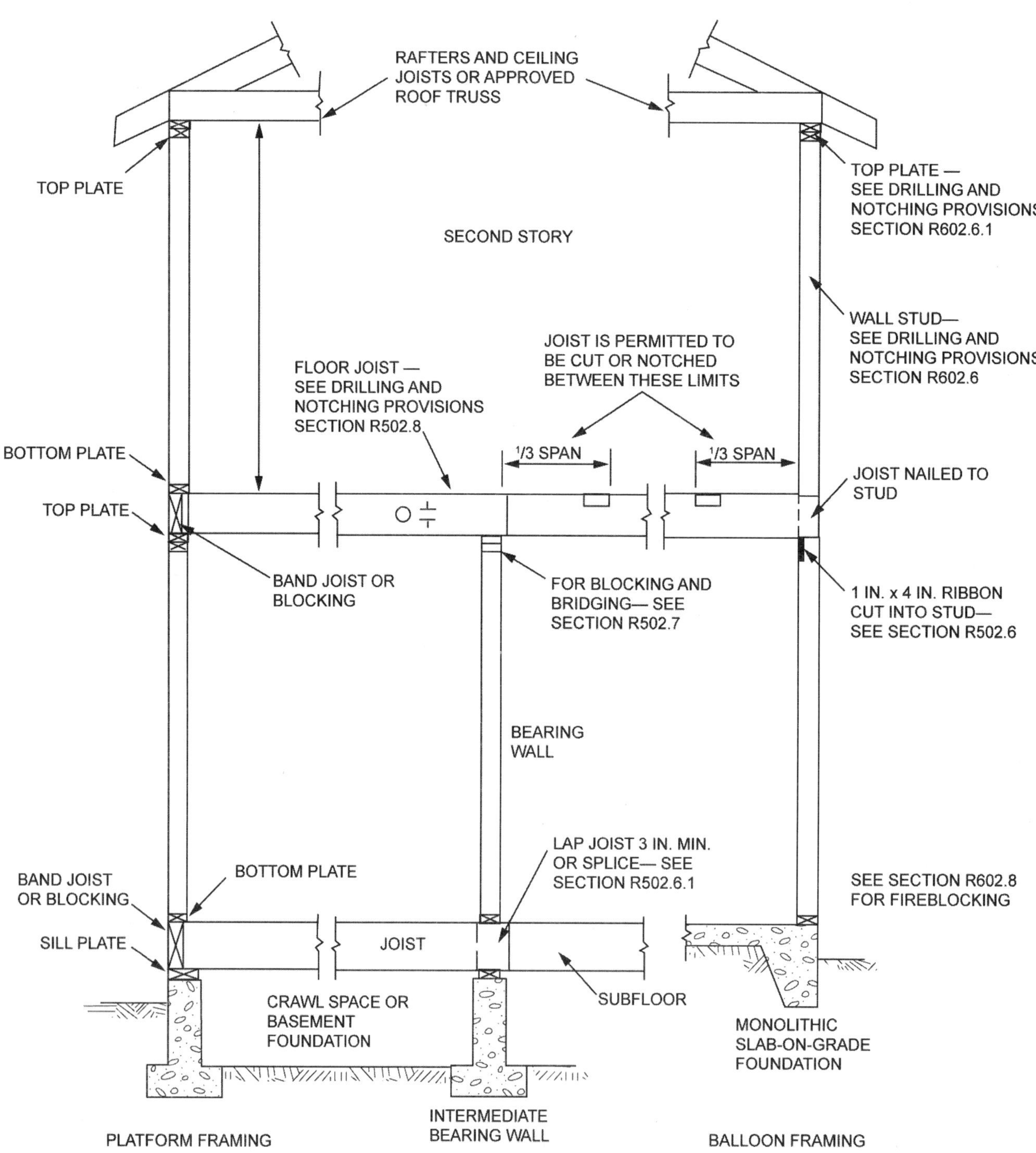
RAFTERS AND CEILING JOISTS OR APPROVED ROOF TRUSS
TOP PLATE
TOP PLATE — SEE DRILLING AND NOTCHING PROVISIONS SECTION R602.6.1
SECOND STORY
WALL STUD— SEE DRILLING AND NOTCHING PROVISIONS SECTION R602.6
JOIST IS PERMITTED TO BE CUT OR NOTCHED BETWEEN THESE LIMITS
FLOOR JOIST — SEE DRILLING AND NOTCHING PROVISIONS SECTION R502.8
BOTTOM PLATE
1/3 SPAN
1/3 SPAN
JOIST NAILED TO STUD
TOP PLATE
BAND JOIST OR BLOCKING
FOR BLOCKING AND BRIDGING— SEE SECTION R502.7
1 IN. x 4 IN. RIBBON CUT INTO STUD— SEE SECTION R502.6
BEARING WALL
LAP JOIST 3 IN. MIN. OR SPLICE— SEE SECTION R502.6.1
BAND JOIST OR BLOCKING
BOTTOM PLATE
SEE SECTION R602.8 FOR FIREBLOCKING
SILL PLATE
JOIST
SUBFLOOR
CRAWL SPACE OR BASEMENT FOUNDATION
MONOLITHIC SLAB-ON-GRADE FOUNDATION
INTERMEDIATE BEARING WALL
PLATFORM FRAMING
BALLOON FRAMING

For SI: 1 inch = 25.4 mm.

FIGURE R602.3(1)
TYPICAL WALL, FLOOR AND ROOF FRAMING

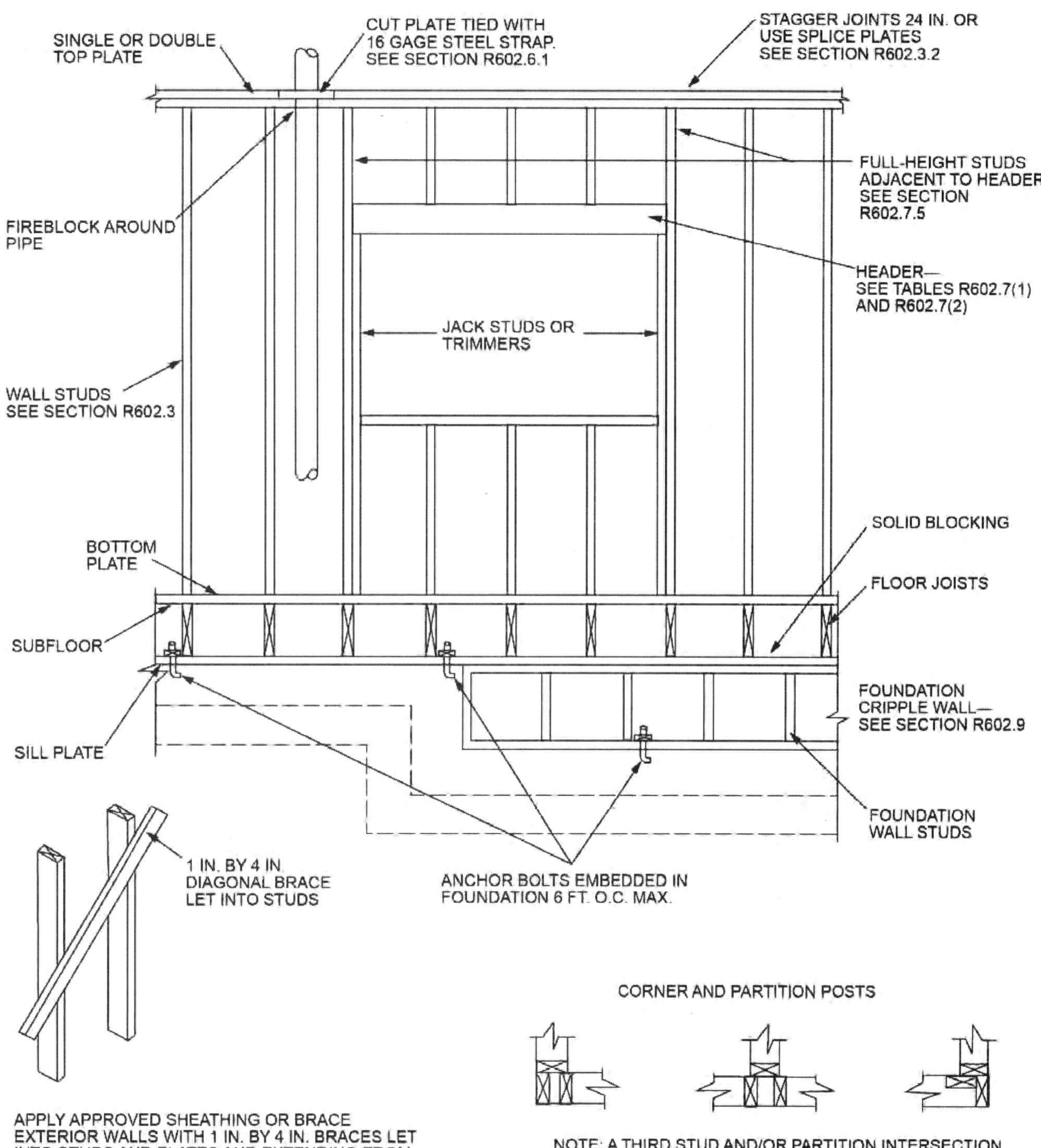
SINGLE OR DOUBLE TOP PLATE
CUT PLATE TIED WITH 16 GAGE STEEL STRAP. SEE SECTION R602.6.1
STAGGER JOINTS 24 IN. OR USE SPLICE PLATES SEE SECTION R602.3.2
FULL-HEIGHT STUDS ADJACENT TO HEADER SEE SECTION R602.7.5
FIREBLOCK AROUND PIPE
HEADER— SEE TABLES R602.7(1) AND R602.7(2)
JACK STUDS OR TRIMMERS
WALL STUDS SEE SECTION R602.3
SOLID BLOCKING
BOTTOM PLATE
FLOOR JOISTS
SUBFLOOR
FOUNDATION CRIPPLE WALL— SEE SECTION R602.9
SILL PLATE
FOUNDATION WALL STUDS
1 IN. BY 4 IN. DIAGONAL BRACE LET INTO STUDS
ANCHOR BOLTS EMBEDDED IN FOUNDATION 6 FT. O.C. MAX.
CORNER AND PARTITION POSTS
APPLY APPROVED SHEATHING OR BRACE EXTERIOR WALLS WITH 1 IN. BY 4 IN. BRACES LET INTO STUDS AND PLATES AND EXTENDING FROM BOTTOM PLATE TO TOP PLATE, OR OTHER APPROVED METAL STRAP DEVICES INSTALLED IN ACCORDANCE WITH THE MANUFACTURER'S SPECIFICATIONS. SEE SECTION R602.10.
NOTE: A THIRD STUD AND/OR PARTITION INTERSECTION BACKING STUDS SHALL BE PERMITTED TO BE OMITTED THROUGH THE USE OF WOOD BACKUP CLEATS, METAL DRYWALL CLIPS OR OTHER APPROVED DEVICES THAT WILL SERVE AS ADEQUATE BACKING FOR THE FACING MATERIALS.

For SI: 1 inch = 25.4 mm, 1 foot = 304.8 mm.

FIGURE R602.3(2)
FRAMING DETAILS

TABLE R602.3.2
SINGLE TOP-PLATE SPLICE CONNECTION DETAILS

CONDITION	TOP-PLATE SPLICE LOCATION			
	Corners and intersecting walls		Butt joints in straight walls	
	Splice plate size	Minimum nails each side of joint	Splice plate size	Minimum nails each side of joint
Structures in SDC A-C; and in SDC D_0, D_1 and D_2 with braced wall line spacing less than 25 feet	3″ × 6″ × 0.036″ galvanized steel plate or equivalent	(6) 8d box ($2^1/_2$″ × 0.113″) nails	3″× 12″ × 0.036″ galvanized steel plate or equivalent	(12) 8d box ($2^1/_2$″ × 0.113″) nails
Structures in SDC D_0, D_1 and D_2, with braced wall line spacing greater than or equal to 25 feet	3″ × 8″ × 0.036″ galvanized steel plate or equivalent	(9) 8d box ($2^1/_2$″ × 0.113″) nails	3″ × 16″ × 0.036″ galvanized steel plate or equivalent	(18) 8dbox ($2^1/_2$″ × 0.113″) nails

For SI: 1 inch = 25.4 mm, 1 foot = 304.8 mm.

R602.5 Interior nonbearing walls. Interior *nonbearing walls* shall be permitted to be constructed with 2-inch by 3-inch (51 mm by 76 mm) studs spaced 24 inches (610 mm) on center or, where not part of a *braced wall line*, 2-inch by 4-inch (51 mm by 102 mm) flat studs spaced at 16 inches (406 mm) on center. Interior *nonbearing walls* shall be capped with not less than a single top plate. Interior *nonbearing walls* shall be fireblocked in accordance with Section R602.8.

R602.6 Drilling and notching of studs. Drilling and notching of studs shall be in accordance with the following:

1. Notching. A stud in an exterior wall or bearing partition shall not be cut or notched to a depth exceeding 25 percent of its depth. Studs in nonbearing partitions shall not be notched to a depth exceeding 40 percent of a single stud depth.
2. Boring. The diameter of bored holes in studs shall not exceed 60 percent of the stud depth, the edge of the hole shall not be less than $^5/_8$ inch (16 mm) from the edge of the stud, and the hole shall not be located in the same section as a cut or notch. Where the diameter of a bored hole in a stud located in exterior walls or bearing partitions is over 40 percent, such stud shall be doubled and not more than two successive doubled studs shall be so bored. See Figures R602.6(1) and R602.6(2).

Exception: Where approved, stud shoes are installed in accordance with the manufacturer's instructions.

R602.6.1 Drilling and notching of top plate. Where piping or ductwork is placed in or partly in an exterior wall or interior *load-bearing wall*, necessitating cutting, drilling or notching of the top plate by more than 50 percent of its width, a galvanized metal tie not less than 0.054 inch thick (1.37 mm) (16 ga) and $1^1/_2$ inches (38 mm) wide shall be fastened across and to the plate at each side of the opening with not less than eight 10d (0.148 inch diameter) nails having a minimum length of $1^1/_2$ inches (38 mm) at each side or equivalent. The metal tie must extend not less than 6 inches past the opening. See Figure R602.6.1.

Exception: Where the entire side of the wall with the notch or cut is covered by wood structural panel sheathing.

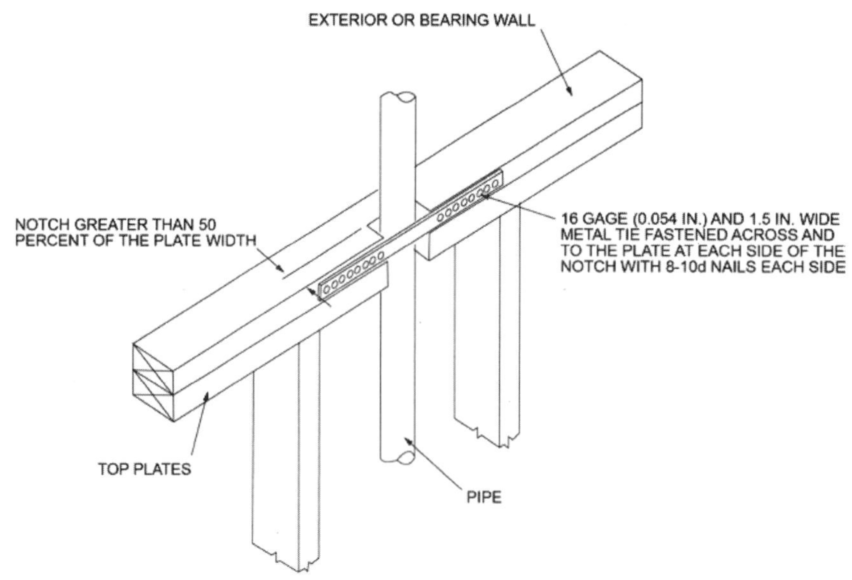

For SI: 1 inch = 25.4 mm.

FIGURE R602.6.1
TOP PLATE FRAMING TO ACCOMMODATE PIPING

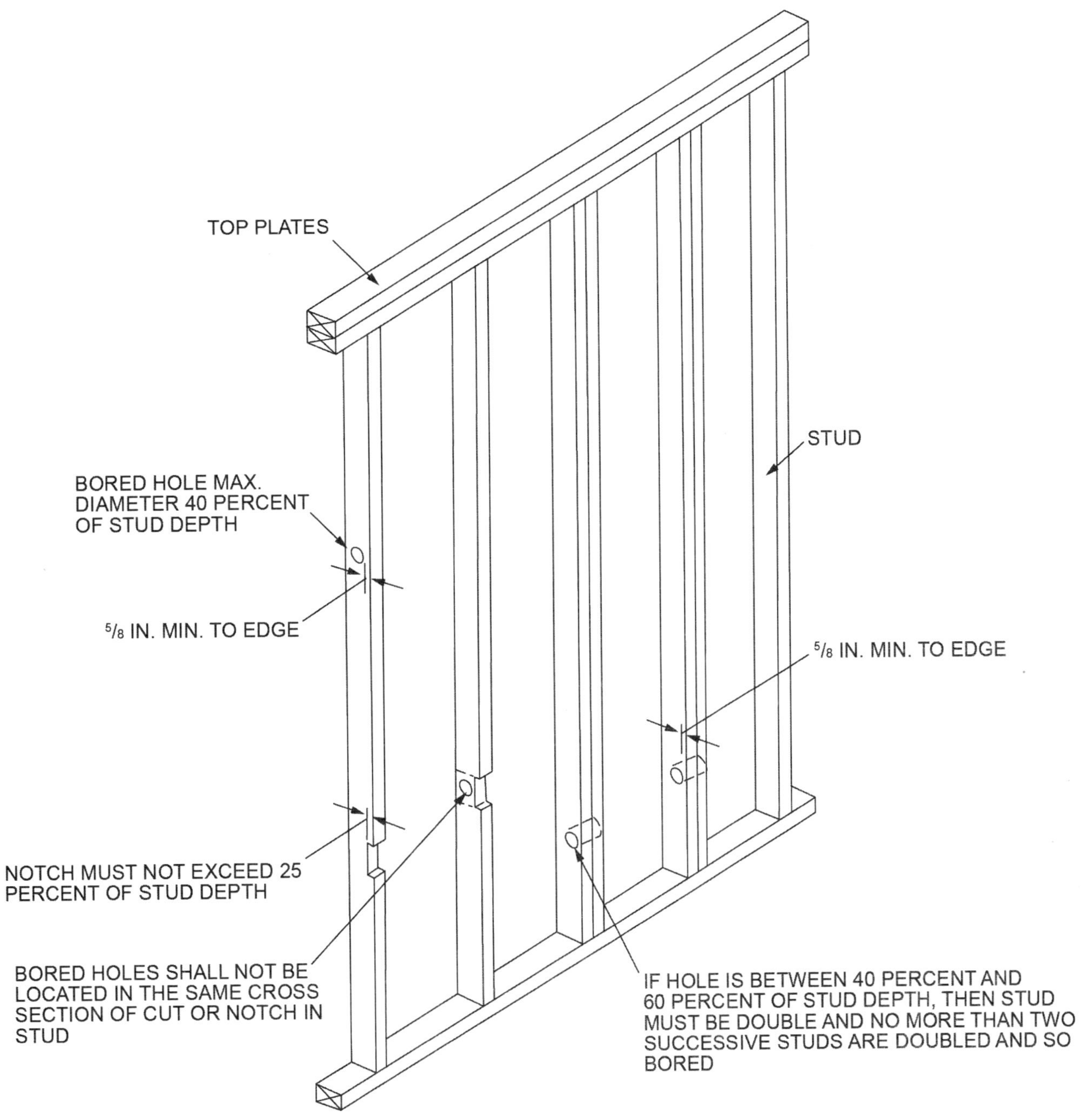

For SI: 1 inch = 25.4 mm.

Note: Condition for exterior and bearing walls.

FIGURE R602.6(1)
NOTCHING AND BORED HOLE LIMITATIONS FOR EXTERIOR WALLS AND BEARING WALLS

R602.7 Headers. For header spans, see Tables R602.7(1), R602.7(2) and R602.7(3).

R602.7.1 Single member headers. Single headers shall be framed with a single flat 2-inch-nominal (51 mm) member or wall plate not less in width than the wall studs on the top and bottom of the header in accordance with Figures R602.7.1(1) and R602.7.1(2) and face nailed to the top and bottom of the header with 10d box nails (3 inches × 0.128 inches) spaced 12 inches on center.

R602.7.2 Rim board headers. Rim board header size, material and span shall be in accordance with Table R602.7(1). Rim board headers shall be constructed in accordance with Figure R602.7.2 and shall be supported at each end by full-height studs. The number of full-

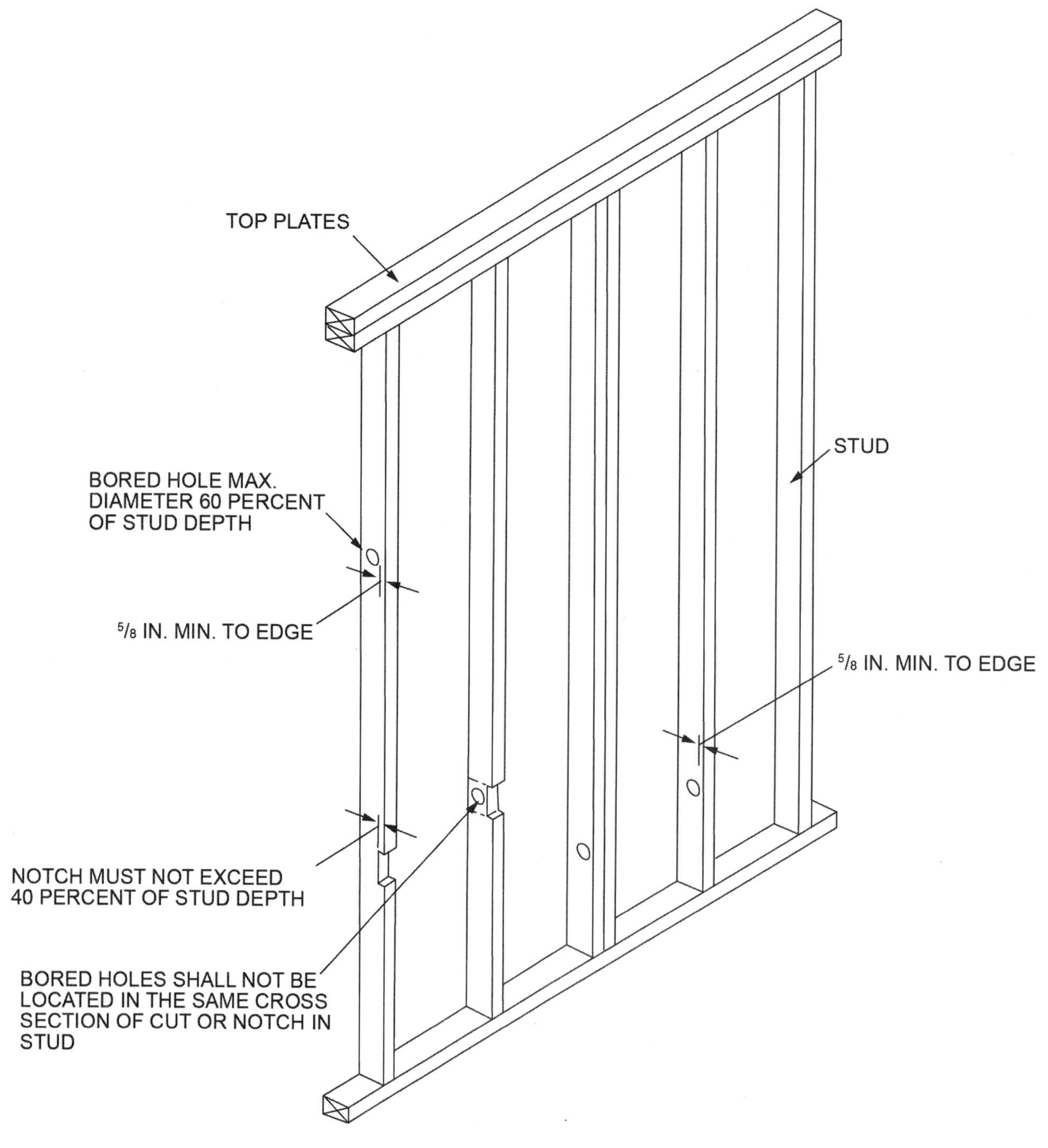

For SI: 1 inch = 25.4 mm.

FIGURE R602.6(2)
NOTCHING AND BORED HOLE LIMITATIONS FOR INTERIOR NONBEARING WALLS

height studs at each end shall be not less than the number of studs displaced by half of the header span based on the maximum stud spacing in accordance with Table R602.3(5). Rim board headers supporting concentrated loads shall be designed in accordance with accepted engineering practice.

R602.7.3 Wood structural panel box headers. Wood structural panel box headers shall be constructed in accordance with Figure R602.7.3 and Table R602.7.3.

R602.7.4 Nonbearing walls. Load-bearing headers are not required in interior or exterior *nonbearing walls.* A single flat 2-inch by 4-inch (51 mm by 102 mm) member shall be permitted to be used as a header in interior or exterior *nonbearing walls* for openings up to 8 feet (2438 mm) in width if the vertical distance to the parallel nailing surface above is not more than 24 inches (610 mm). For such nonbearing headers, cripples or blocking are not required above the header.

R602.7.5 Supports for headers. Headers shall be supported on each end with one or more jack studs or with *approved* framing anchors in accordance with Table R602.7(1) or R602.7(2). The full-height stud adjacent to each end of the header shall be end nailed to each end of the header in accordance with Table R602.3(1). The minimum number of full-height studs at each end of a header shall be in accordance with Table R602.7.5.

TABLE R602.7.5
MINIMUM NUMBER OF FULL-HEIGHT STUDS AT EACH END OF HEADERS IN EXTERIOR WALLS[a]

MAXIMUM HEADER SPAN (feet)	ULTIMATE DESIGN WIND SPEED AND EXPOSURE CATEGORY	
	< 140 mph, Exposure B or < 130 mph, Exposure C	≤ 115 mph, Exposure B[b]
4	1	1
6	2	1
8	2	1
10	3	2
12	3	2
14	3	2
16	4	2
18	4	2

For SI: 1 foot = 304.8 mm, 1 mile per hour = 0.447 m/s.

a. For header spans between those given, use the minimum number of full-height studs associated with the larger header span.

b. The tabulated minimum number of full-height studs is applicable where jack studs are provided to support the header at each end in accordance with Table R602.7(1). Where a framing anchor is used to support the header in lieu of a jack stud in accordance with Note d of Table R602.7(1), the minimum number of full-height studs at each end of a header shall be in accordance with requirements for wind speed < 140 mph, Exposure B.

R602.8 Fireblocking required. Fireblocking shall be provided in accordance with Section R302.11.

R602.9 Cripple walls. Foundation cripple walls shall be framed of studs not smaller than the studding above. Where exceeding 4 feet (1219 mm) in height, such walls shall be framed of studs having the size required for an additional story.

Exterior cripple walls with a stud height less than 14 inches (356 mm) shall be continuously sheathed on one side with wood structural panels fastened to both the top and bottom plates in accordance with Table R602.3(1), or the cripple walls shall be constructed of solid blocking.

Cripple walls shall be supported on continuous foundations.

R602.10 Wall bracing. Buildings shall be braced in accordance with this section or, when applicable, Section R602.12. Where a building, or portion thereof, does not comply with one or more of the bracing requirements in this section, those portions shall be designed and constructed in accordance with Section R301.1.

R602.10.1 Braced wall lines. For the purpose of determining the amount and location of bracing required in each story level of a building, *braced wall lines* shall be designated as straight lines in the building plan placed in accordance with this section.

R602.10.1.1 Length of a braced wall line. The length of a *braced wall line* shall be the distance between its ends. The end of a *braced wall line* shall be the intersection with a perpendicular *braced wall line*, an angled *braced wall line* as permitted in Section R602.10.1.4 or an exterior wall as shown in Figure R602.10.1.1.

R602.10.1.2 Location of braced wall lines and permitted offsets. Each *braced wall line* shall be located such that no more than two-thirds of the required *braced wall panel* length is located to one side of the *braced wall line.* Braced wall panels shall be permitted to be offset up to 4 feet (1219 mm) from the designated *braced wall line.* Braced wall panels parallel to a *braced wall line* shall be offset not more than 4 feet (1219 mm) from the designated *braced wall line* location as shown in Figure R602.10.1.1.

Exterior walls parallel to a *braced wall line* shall be offset not more than 4 feet (1219 mm) from the designated *braced wall line* location as shown in Figure R602.10.1.1.

Interior walls used as bracing shall be offset not more than 4 feet (1219 mm) from a *braced wall line* through the interior of the building as shown in Figure R602.10.1.1.

TABLE R602.7(1)
GIRDER SPANS[a] AND HEADER SPANS[a] FOR EXTERIOR BEARING WALLS
(Maximum spans for Douglas fir-larch, hem-fir, Southern pine and spruce-pine-fir[b] and required number of jack studs)

GIRDERS AND HEADERS SUPPORTING	SIZE	GROUND SNOW LOAD (psf)[e]																	
		30						50						70					
		Building width[c] (feet)																	
		12		24		36		12		24		36		12		24		36	
		Span[f]	NJ[d]	Span[f]	NJ[d]	Span[f]	NJ[d]	Span[f]	NJ[d]	Span[f]	NJ[d]	Span[f]	NJ[d]	Span[f]	NJ[d]	Span[f]	NJ[d]	Span[f]	NJ[d]
Roof and ceiling HEADER TYP ROOF AND CEILING	1-2 × 6	4-0	1	3-1	2	2-7	2	3-5	1	2-8	2	2-3	2	3-0	2	2-4	2	2-0	2
	1-2 × 8	5-1	2	3-11	2	3-3	2	4-4	2	3-4	2	2-10	2	3-10	2	3-0	2	2-6	3
	1-2 × 10	6-0	2	4-8	2	3-11	2	5-2	2	4-0	2	3-4	3	4-7	2	3-6	3	3-0	3
	1-2 × 12	7-1	2	5-5	2	4-7	3	6-1	2	4-8	3	3-11	3	5-5	2	4-2	3	3-6	3
	2-2 × 4	4-0	1	3-1	1	2-7	1	3-5	1	2-7	1	2-2	1	3-0	1	2-4	1	2-0	1
	2-2 × 6	6-0	1	4-7	1	3-10	1	5-1	1	3-11	1	3-3	2	4-6	1	3-6	2	2-11	2
	2-2 × 8	7-7	1	5-9	1	4-10	2	6-5	1	5-0	2	4-2	2	5-9	1	4-5	2	3-9	2
	2-2 × 10	9-0	1	6-10	2	5-9	2	7-8	2	5-11	2	4-11	2	6-9	2	5-3	2	4-5	2
	2-2 × 12	10-7	2	8-1	2	6-10	2	9-0	2	6-11	2	5-10	2	8-0	2	6-2	2	5-2	3
	3-2 × 8	9-5	1	7-3	1	6-1	1	8-1	1	6-3	1	5-3	2	7-2	1	5-6	2	4-8	2
	3-2 × 10	11-3	1	8-7	1	7-3	2	9-7	1	7-4	2	6-2	2	8-6	1	6-7	2	5-6	2
	3-2 × 12	13-2	1	10-1	2	8-6	2	11-3	2	8-8	2	7-4	2	10-0	2	7-9	2	6-6	2
	4-2 × 8	10-11	1	8-4	1	7-0	1	9-4	1	7-2	1	6-0	1	8-3	1	6-4	1	5-4	2
	4-2 × 10	12-11	1	9-11	1	8-4	1	11-1	1	8-6	1	7-2	2	9-10	1	7-7	2	6-4	2
	4-2 × 12	15-3	1	11-8	1	9-10	2	13-0	1	10-0	2	8-5	2	11-7	1	8-11	2	7-6	2
Roof, ceiling and one center-bearing floor ROOF, CEILING AND ONE FLOOR (CENTER BEARING)	1-2 × 6	3-3	1	2-7	2	2-2	2	3-0	2	2-4	2	2-0	2	2-9	2	2-2	2	1-10	2
	1-2 × 8	4-1	2	3-3	2	2-9	2	3-9	2	3-0	2	2-6	3	3-6	2	2-9	2	2-4	3
	1-2 × 10	4-11	2	3-10	2	3-3	3	4-6	2	3-6	3	3-0	3	4-1	2	3-3	3	2-9	3
	1-2 × 12	5-9	2	4-6	3	3-10	3	5-3	2	4-2	3	3-6	3	4-10	3	3-10	3	3-3	4
	2-2 × 4	3-3	1	2-6	1	2-2	1	3-0	1	2-4	1	2-0	1	2-8	1	2-2	1	1-10	1
	2-2 × 6	4-10	1	3-9	1	3-3	2	4-5	1	3-6	2	3-0	2	4-1	1	3-3	2	2-9	2
	2-2 × 8	6-1	1	4-10	2	4-1	2	5-7	2	4-5	2	3-9	2	5-2	2	4-1	2	3-6	2
	2-2 × 10	7-3	2	5-8	2	4-10	2	6-8	2	5-3	2	4-5	2	6-1	2	4-10	2	4-1	2
	2-2 × 12	8-6	2	6-8	2	5-8	2	7-10	2	6-2	2	5-3	3	7-2	2	5-8	2	4-10	3
	3-2 × 8	7-8	1	6-0	1	5-1	2	7-0	1	5-6	2	4-8	2	6-5	1	5-1	2	4-4	2
	3-2 × 10	9-1	1	7-2	2	6-1	2	8-4	1	6-7	2	5-7	2	7-8	2	6-1	2	5-2	2
	3-2 × 12	10-8	2	8-5	2	7-2	2	9-10	2	7-8	2	6-7	2	9-0	2	7-1	2	6-1	2
	4-2 × 8	8-10	1	6-11	1	5-11	1	8-1	1	6-4	1	5-5	2	7-5	1	5-11	1	5-0	2
	4-2 × 10	10-6	1	8-3	2	7-0	2	9-8	1	7-7	2	6-5	2	8-10	1	7-0	2	6-0	2
	4-2 × 12	12-4	1	9-8	2	8-3	2	11-4	2	8-11	2	7-7	2	10-4	2	8-3	2	7-0	2
Roof, ceiling and one clear-span floor ROOF, CEILING AND ONE FLOOR (CLEAR SPAN)	1-2 × 6	2-11	2	2-3	2	1-11	2	2-9	2	2-1	2	1-9	2	2-7	2	2-0	2	1-8	2
	1-2 × 8	3-9	2	2-10	2	2-5	3	3-6	2	2-8	2	2-3	3	3-3	2	2-6	3	2-2	3
	1-2 × 10	4-5	2	3-5	3	2-10	3	4-2	2	3-2	3	2-8	3	3-11	2	3-0	3	2-6	3
	1-2 × 12	5-2	2	4-0	3	3-4	3	4-10	3	3-9	3	3-2	4	4-7	3	3-6	3	3-0	4
	2-2 × 4	2-11	1	2-3	1	1-10	1	2-9	1	2-1	1	1-9	1	2-7	1	2-0	1	1-8	1
	2-2 × 6	4-4	1	3-4	2	2-10	2	4-1	1	3-2	2	2-8	2	3-10	1	3-0	2	2-6	2
	2-2 × 8	5-6	2	4-3	2	3-7	2	5-2	2	4-0	2	3-4	2	4-10	2	3-9	2	3-2	2
	2-2 × 10	6-7	2	5-0	2	4-2	2	6-1	2	4-9	2	4-0	2	5-9	2	4-5	2	3-9	3
	2-2 × 12	7-9	2	5-11	2	4-11	3	7-2	2	5-7	2	4-8	3	6-9	2	5-3	3	4-5	3
	3-2 × 8	6-11	1	5-3	2	4-5	2	6-5	1	5-0	2	4-2	2	6-1	1	4-8	2	4-0	2
	3-2 × 10	8-3	2	6-3	2	5-3	2	7-8	2	5-11	2	5-0	2	7-3	2	5-7	2	4-8	2
	3-2 × 12	9-8	2	7-5	2	6-2	2	9-0	2	7-0	2	5-10	2	8-6	2	6-7	2	5-6	3
	4-2 × 8	8-0	1	6-1	1	5-1	2	7-5	1	5-9	2	4-10	2	7-0	1	5-5	2	4-7	2
	4-2 × 10	9-6	1	7-3	2	6-1	2	8-10	1	6-10	2	5-9	2	8-4	1	6-5	2	5-5	2
	4-2 × 12	11-2	2	8-6	2	7-2	2	10-5	2	8-0	2	6-9	2	9-10	2	7-7	2	6-5	2

(continued)

TABLE R602.7(1)—continued
GIRDER SPANS[a] AND HEADER SPANS[a] FOR EXTERIOR BEARING WALLS
(Maximum spans for Douglas fir-larch, hem-fir, Southern pine and spruce-pine-fir[b] and required number of jack studs)

GIRDERS AND HEADERS SUPPORTING	SIZE	GROUND SNOW LOAD (psf)[e]																	
		30						50						70					
		Building width[c] (feet)																	
		12		24		36		12		24		36		12		24		36	
		Span[f]	NJ[d]	Span[f]	NJ[d]	Span[f]	NJ[d]	Span[f]	NJ[d]	Span[f]	NJ[d]	Span[f]	NJ[d]	Span[f]	NJ[d]	Span[f]	NJ[d]	Span[f]	NJ[d]
Roof, ceiling and two center-bearing floors (ROOF, CEILING AND TWO FLOORS (CENTER BEARING))	1-2 × 6	2-8	2	2-1	2	1-10	2	2-7	2	2-0	2	1-9	2	2-5	2	1-11	2	1-8	2
	1-2 × 8	3-5	2	2-8	2	2-4	3	3-3	2	2-7	2	2-2	3	3-1	2	2-5	3	2-1	3
	1-2 × 10	4-0	2	3-2	3	2-9	3	3-10	2	3-1	3	2-7	3	3-8	2	2-11	3	2-5	3
	1-2 × 12	4-9	3	3-9	3	3-2	4	4-6	3	3-7	3	3-1	4	4-3	3	3-5	3	2-11	4
	2-2 × 4	2-8	1	2-1	1	1-9	1	2-6	1	2-0	1	1-8	1	2-5	1	1-11	1	1-7	1
	2-2 × 6	4-0	1	3-2	2	2-8	2	3-9	1	3-0	2	2-7	2	3-7	1	2-10	2	2-5	2
	2-2 × 8	5-0	2	4-0	2	3-5	2	4-10	2	3-10	2	3-3	2	4-7	2	3-7	2	3-1	2
	2-2 × 10	6-0	2	4-9	2	4-0	2	5-8	2	4-6	2	3-10	3	5-5	2	4-3	2	3-8	3
	2-2 × 12	7-0	2	5-7	2	4-9	3	6-8	2	5-4	3	4-6	3	6-4	2	5-0	3	4-3	3
	3-2 × 8	6-4	1	5-0	2	4-3	2	6-0	1	4-9	2	4-1	2	5-8	2	4-6	2	3-10	2
	3-2 × 10	7-6	2	5-11	2	5-1	2	7-1	2	5-8	2	4-10	2	6-9	2	5-4	2	4-7	2
	3-2 × 12	8-10	2	7-0	2	5-11	2	8-5	2	6-8	2	5-8	3	8-0	2	6-4	2	5-4	3
	4-2 × 8	7-3	1	5-9	1	4-11	2	6-11	1	5-6	2	4-8	2	6-7	1	5-2	2	4-5	2
	4-2 × 10	8-8	1	6-10	2	5-10	2	8-3	2	6-6	2	5-7	2	7-10	2	6-2	2	5-3	2
	4-2 × 12	10-2	2	8-1	2	6-10	2	9-8	2	7-8	2	6-7	2	9-2	2	7-3	2	6-2	2
Roof, ceiling, and two clear-span floors (ROOF, CEILING AND TWO FLOORS (CLEAR SPAN))	1-2 × 6	2-3	2	1-9	2	1-5	2	2-3	2	1-9	2	1-5	3	2-2	2	1-8	2	1-5	3
	1-2 × 8	2-10	2	2-2	3	1-10	3	2-10	2	2-2	3	1-10	3	2-9	2	2-1	3	1-10	3
	1-2 × 10	3-4	2	2-7	3	2-2	3	3-4	3	2-7	3	2-2	4	3-3	3	2-6	3	2-2	4
	1-2 × 12	4-0	3	3-0	3	2-7	4	4-0	3	3-0	4	2-7	4	3-10	3	3-0	4	2-6	4
	2-2 × 4	2-3	1	1-8	1	1-4	1	2-3	1	1-8	1	1-4	1	2-2	1	1-8	1	1-4	2
	2-2 × 6	3-4	1	2-6	2	2-2	2	3-4	2	2-6	2	2-2	2	3-3	2	2-6	2	2-1	2
	2-2 × 8	4-3	2	3-3	2	2-8	2	4-3	2	3-3	2	2-8	2	4-1	2	3-2	2	2-8	3
	2-2 × 10	5-0	2	3-10	2	3-2	3	5-0	2	3-10	2	3-2	3	4-10	2	3-9	3	3-2	3
	2-2 × 12	5-11	2	4-6	3	3-9	3	5-11	2	4-6	3	3-9	3	5-8	2	4-5	3	3-9	3
	3-2 × 8	5-3	1	4-0	2	3-5	2	5-3	2	4-0	2	3-5	2	5-1	2	3-11	2	3-4	2
	3-2 × 10	6-3	2	4-9	2	4-0	2	6-3	2	4-9	2	4-0	2	6-1	2	4-8	2	4-0	3
	3-2 × 12	7-5	2	5-8	2	4-9	3	7-5	2	5-8	2	4-9	3	7-2	2	5-6	3	4-8	3
	4-2 × 8	6-1	1	4-8	2	3-11	2	6-1	1	4-8	2	3-11	2	5-11	1	4-7	2	3-10	2
	4-2 × 10	7-3	2	5-6	2	4-8	2	7-3	2	5-6	2	4-8	2	7-0	2	5-5	2	4-7	2
	4-2 × 12	8-6	2	6-6	2	5-6	2	8-6	2	6-6	2	5-6	2	8-3	2	6-4	2	5-4	3

For SI: 1 inch = 25.4 mm, 1 foot = 304.8 mm, 1 pound per square foot = 0.0479 kPa.

a. Spans are given in feet and inches.

b. Spans are based on minimum design properties for No. 2 grade lumber of Douglas fir-larch, hem-fir, Southern pine, and spruce-pine-fir.

c. Building width is measured perpendicular to the ridge. For widths between those shown, spans are permitted to be interpolated.

d. NJ = Number of jack studs required to support each end. Where the number of required jack studs equals one, the header is permitted to be supported by an approved framing anchor attached to the full-height wall stud and to the header.

e. Use 30 psf ground snow load for cases in which ground snow load is less than 30 psf and the roof live load is equal to or less than 20 psf.

f. Spans are calculated assuming the top of the header or girder is laterally braced by perpendicular framing. Where the top of the header or girder is not laterally braced (for example, cripple studs bearing on the header), tabulated spans for headers consisting of 2 × 8, 2 × 10, or 2 × 12 sizes shall be multiplied by 0.70 or the header or girder shall be designed.

TABLE R602.7(2)
GIRDER SPANS[a] AND HEADER SPANS[a] FOR INTERIOR BEARING WALLS
(Maximum spans for Douglas fir-larch, hem-fir, southern pine and spruce-pine-fir[b] and required number of jack studs)

HEADERS AND GIRDERS SUPPORTING	SIZE	BUILDING Width[c] (feet)					
		12		24		36	
		Span[e]	NJ[d]	Span[e]	NJ[d]	Span[e]	NJ[d]
One floor only	2-2 × 4	4-1	1	2-10	1	2-4	1
	2-2 × 6	6-1	1	4-4	1	3-6	1
	2-2 × 8	7-9	1	5-5	1	4-5	2
	2-2 × 10	9-2	1	6-6	2	5-3	2
	2-2 × 12	10-9	1	7-7	2	6-3	2
	3-2 × 8	9-8	1	6-10	1	5-7	1
	3-2 × 10	11-5	1	8-1	1	6-7	2
	3-2 × 12	13-6	1	9-6	2	7-9	2
	4-2 × 8	11-2	1	7-11	1	6-5	1
	4-2 × 10	13-3	1	9-4	1	7-8	1
	4-2 × 12	15-7	1	11-0	1	9-0	2
Two floors	2-2 × 4	2-7	1	1-11	1	1-7	1
	2-2 × 6	3-11	1	2-11	2	2-5	2
	2-2 × 8	5-0	1	3-8	2	3-1	2
	2-2 × 10	5-11	2	4-4	2	3-7	2
	2-2 × 12	6-11	2	5-2	2	4-3	3
	3-2 × 8	6-3	1	4-7	2	3-10	2
	3-2 × 10	7-5	1	5-6	2	4-6	2
	3-2 × 12	8-8	2	6-5	2	5-4	2
	4-2 × 8	7-2	1	5-4	1	4-5	2
	4-2 × 10	8-6	1	6-4	2	5-3	2
	4-2 × 12	10-1	1	7-5	2	6-2	2

For SI: 1 inch = 25.4 mm, 1 foot = 304.8 mm.

a. Spans are given in feet and inches.

b. Spans are based on minimum design properties for No. 2 grade lumber of Douglas fir-larch, hem-fir, Southern pine, and spruce-pine-fir.

c. Building width is measured perpendicular to the ridge. For widths between those shown, spans are permitted to be interpolated.

d. NJ = Number of jack studs required to support each end. Where the number of required jack studs equals one, the header is permitted to be supported by an approved framing anchor attached to the full-height wall stud and to the header.

e. Spans are calculated assuming the top of the header or girder is laterally braced by perpendicular framing. Where the top of the header or girder is not laterally braced (for example, cripple studs bearing on the header), tabulated spans for headers consisting of 2 × 8, 2 × 10, or 2 × 12 sizes shall be multiplied by 0.70 or the header or girder shall be designed.

TABLE R602.7(3)
GIRDER AND HEADER SPANS[a] FOR OPEN PORCHES
(Maximum span for Douglas fir-larch, hem-fir, Southern pine and spruce-pine-fir[b])

SIZE	SUPPORTING ROOF						SUPPORTING FLOOR	
	Ground Snow Load (psf)							
	30		50		70			
	Depth of Porch[c] (feet)							
	8	14	8	14	8	14	8	14
2-2 × 6	7-6	5-8	6-2	4-8	5-4	4-0	6-4	4-9
2-2 × 8	10-1	7-7	8-3	6-2	7-1	5-4	8-5	6-4
2-2 × 10	12-4	9-4	10-1	7-7	8-9	6-7	10-4	7-9
2-2 × 12	14-4	10-10	11-8	8-10	10-1	7-8	11-11	9-0

For SI: 1 inch = 25.4 mm, 1 foot = 304.8 mm, 1 pound per square foot = 0.0479 kPa.

a. Spans are given in feet and inches.

b. Tabulated values assume No. 2 grade lumber, wet service and incising for refractory species. Use 30 psf ground snow load for cases in which ground snow load is less than 30 psf and the roof live load is equal to or less than 20 psf.

c. Porch depth is measured horizontally from building face to centerline of the header. For depths between those shown, spans are permitted to be interpolated.

TOP PLATE
CRIPPLE
PLATE TO HEADER CONNECTION IN ACCORDANCE WITH SECTION R602.7.1
JACK STUD

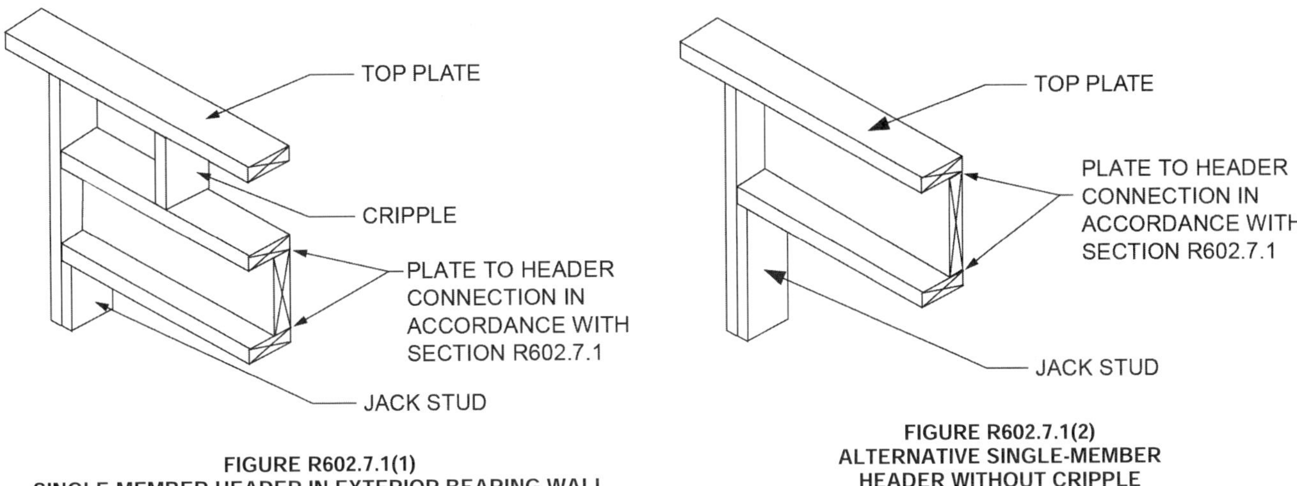

FIGURE R602.7.1(1)
SINGLE-MEMBER HEADER IN EXTERIOR BEARING WALL

TOP PLATE
PLATE TO HEADER CONNECTION IN ACCORDANCE WITH SECTION R602.7.1
JACK STUD

FIGURE R602.7.1(2)
ALTERNATIVE SINGLE-MEMBER HEADER WITHOUT CRIPPLE

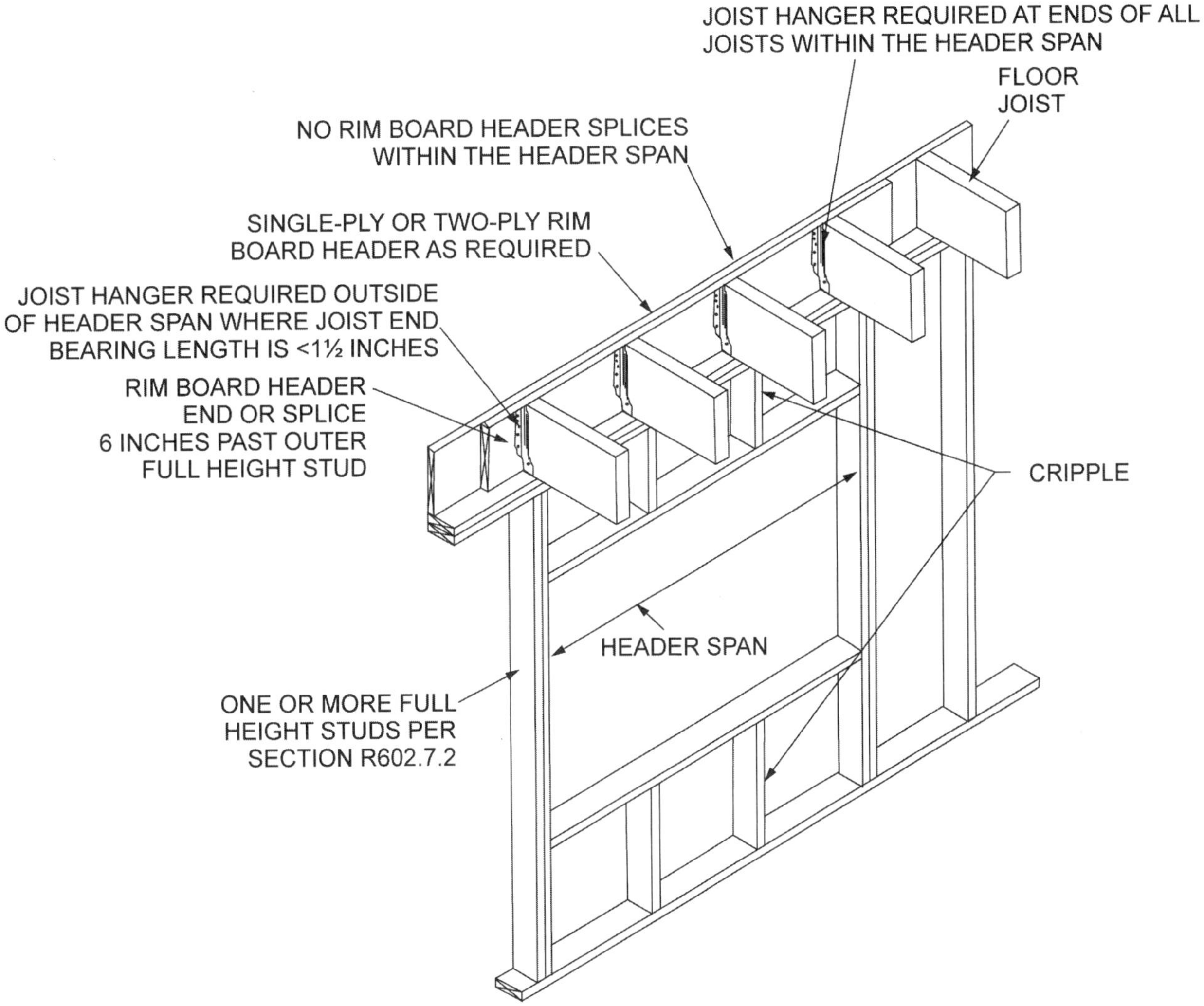

JOIST HANGER REQUIRED AT ENDS OF ALL JOISTS WITHIN THE HEADER SPAN
FLOOR JOIST
NO RIM BOARD HEADER SPLICES WITHIN THE HEADER SPAN
SINGLE-PLY OR TWO-PLY RIM BOARD HEADER AS REQUIRED
JOIST HANGER REQUIRED OUTSIDE OF HEADER SPAN WHERE JOIST END BEARING LENGTH IS <1½ INCHES
RIM BOARD HEADER END OR SPLICE 6 INCHES PAST OUTER FULL HEIGHT STUD
CRIPPLE
HEADER SPAN
ONE OR MORE FULL HEIGHT STUDS PER SECTION R602.7.2

For SI: 25.4 mm = 1 inch.

FIGURE R602.7.2
RIM BOARD HEADER CONSTRUCTION

TABLE R602.7.3
MAXIMUM SPANS FOR WOOD STRUCTURAL PANEL BOX HEADERS[a]

HEADER CONSTRUCTION[b]	HEADER DEPTH (inches)	HOUSE DEPTH (feet)				
		24	26	28	30	32
Wood structural panel—one side	9	4	4	3	3	—
	15	5	5	4	3	3
Wood structural panel—both sides	9	7	5	5	4	3
	15	8	8	7	7	6

For SI: 1 inch = 25.4 mm, 1 foot = 304.8 mm.

a. Spans are based on single story with clear-span trussed roof or two story with floor and roof supported by interior-bearing walls.

b. See Figure R602.7.3 for construction details.

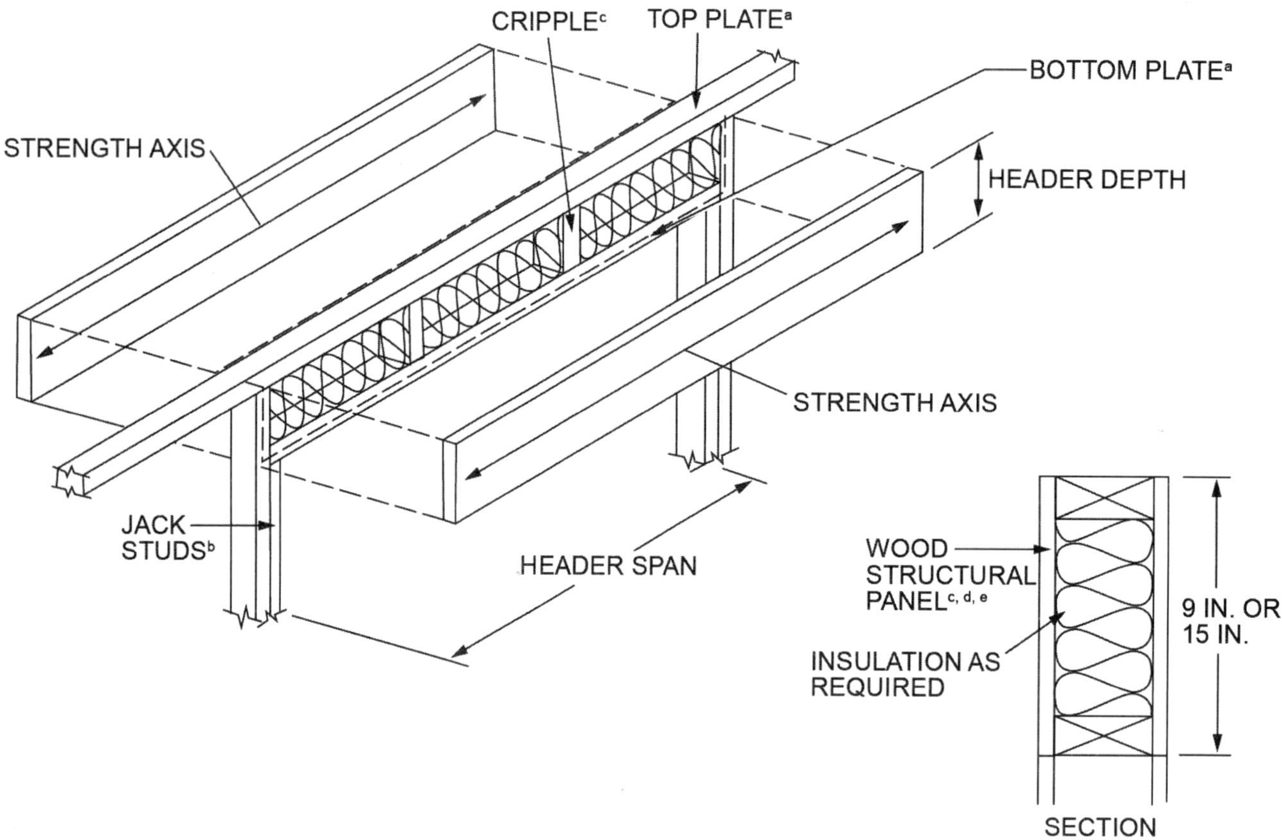

For SI: 1 inch = 25.4 mm, 1 foot = 304.8 mm.

Notes:

a. The top and bottom plates shall be continuous at header location.

b. Jack studs shall be used for spans over 4 feet.

c. Cripple spacing shall be the same as for studs.

d. Wood structural panel faces shall be single pieces of $^{15}/_{32}$-inch-thick Exposure 1 (exterior glue) or thicker, installed on the interior or exterior or both sides of the header.

e. Wood structural panel faces shall be nailed to framing and cripples with 8d common or galvanized box nails spaced 3 inches on center, staggering alternate nails $^{1}/_{2}$ inch. Galvanized nails shall be hot-dipped or tumbled.

FIGURE R602.7.3
TYPICAL WOOD STRUCTURAL PANEL BOX HEADER CONSTRUCTION

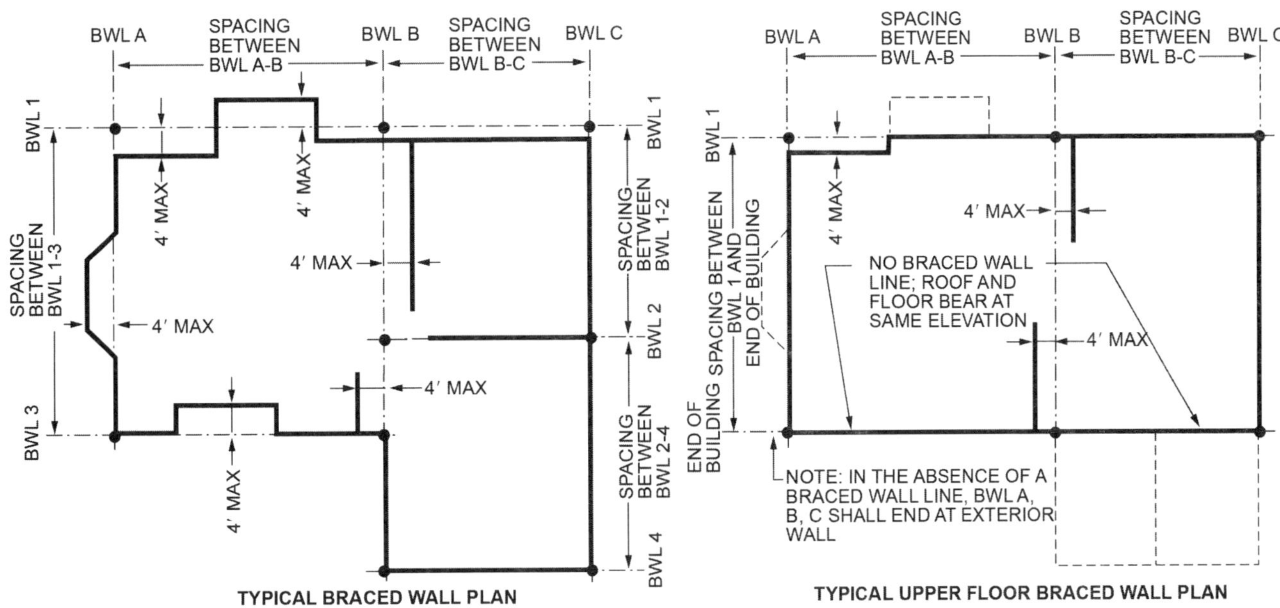

For SI: 1 foot = 304.8 mm.

FIGURE R602.10.1.1
BRACED WALL LINES

R602.10.1.3 Spacing of braced wall lines. The spacing between parallel *braced wall lines* shall be in accordance with Table R602.10.1.3. Intermediate *braced wall lines* through the interior of the building shall be permitted.

R602.10.1.4 Angled walls. Any portion of a wall along a *braced wall line* shall be permitted to angle out of plane for a maximum diagonal length of 8 feet (2438 mm). Where the angled wall occurs at a corner, the length of the *braced wall line* shall be measured from the projected corner as shown in Figure R602.10.1.4. Where the diagonal length is greater than 8 feet (2438 mm), it shall be considered to be a separate *braced wall line* and shall be braced in accordance with Section R602.10.1.

R602.10.2 Braced wall panels. *Braced wall panels* shall be full-height sections of wall that shall not have vertical or horizontal offsets. *Braced wall panels* shall be constructed and placed along a *braced wall line* in accordance with this section and the bracing methods specified in Section R602.10.4.

R602.10.2.1 Braced wall panel uplift load path. The bracing lengths in Table R602.10.3(1) apply only when uplift loads are resisted in accordance with Section R602.3.5.

TABLE R602.10.1.3
BRACED WALL LINE SPACING

APPLICATION	CONDITION	BUILDING TYPE	BRACED WALL LINE SPACING CRITERIA	
			Maximum Spacing	Exception to Maximum Spacing
Wind bracing	Ultimate design wind speed < 140 mph	Detached, townhouse	60 feet	None
Seismic bracing	SDC A – C	Detached	Use wind bracing	
	SDC A – B	Townhouse	Use wind bracing	
	SDC C	Townhouse	35 feet	Up to 50 feet when length of required bracing per Table R602.10.3(3) is adjusted in accordance with Table R602.10.3(4).
	SDC D_0, D_1, D_2	Detached, townhouses, one- and two-story only	25 feet	Up to 35 feet to allow for a single room not to exceed 900 square feet. Spacing of all other braced wall lines shall not exceed 25 feet.
	SDC D_0, D_1, D_2	Detached, townhouse	25 feet	Up to 35 feet when length of required bracing per Table R602.10.3(3) is adjusted in accordance with Table R602.10.3(4).

For SI: 1 foot = 304.8 mm, 1 square foot = 0.0929 m^2, 1 mile per hour = 0.447 m/s.

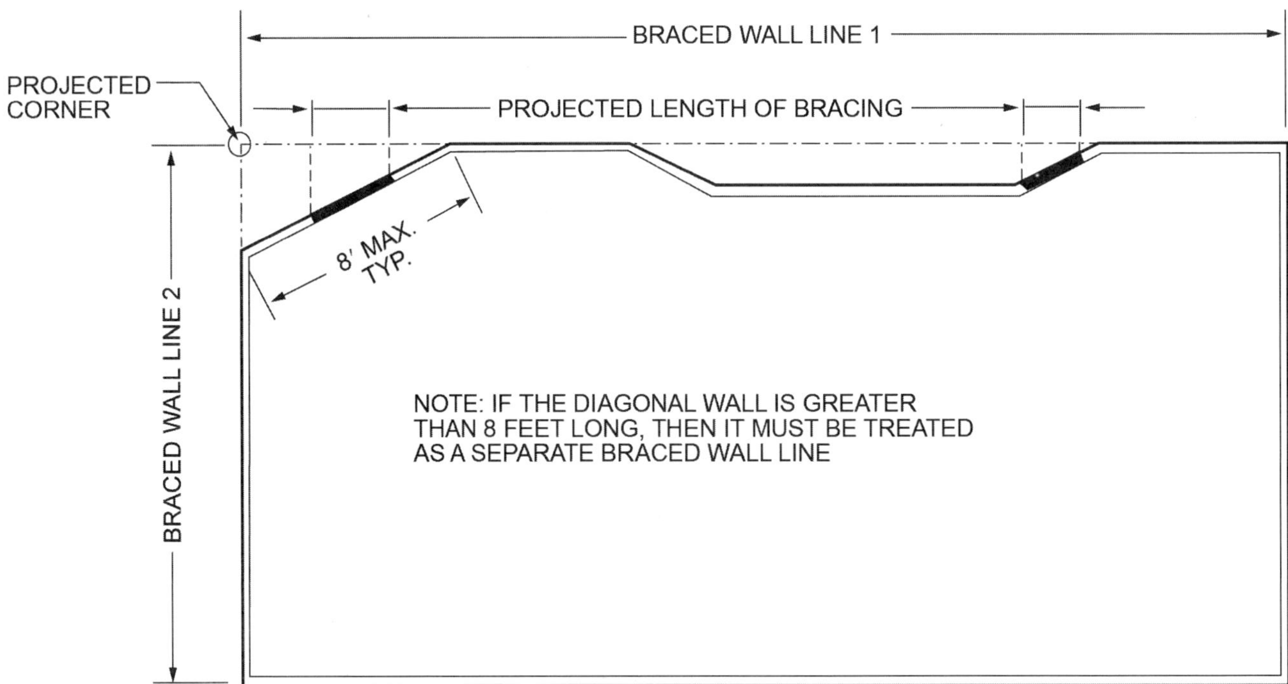

For SI: 1 foot = 304.8 mm.

FIGURE R602.10.1.4
ANGLED WALLS

R602.10.2.2 Locations of braced wall panels. A *braced wall panel* shall begin within 10 feet (3810 mm) from each end of a *braced wall line* as determined in Section R602.10.1.1. The distance between adjacent edges of braced wall panels along a *braced wall line* shall be not greater than 20 feet (6096 mm) as shown in Figure R602.10.2.2.

Exceptions:

1. *Braced wall panels* in *Seismic Design Categories* D_0, D_1 and D_2 shall comply with Section R602.10.2.2.1.
2. *Braced wall panels* with continuous sheathing in *Seismic Design Categories* A, B and C shall comply with Section R602.10.7.

R602.10.2.2.1 Location of braced wall panels in Seismic Design Categories D_0, D_1 and D_2. Braced wall panels shall be located at each end of a *braced wall line*.

Exceptions:

1. Braced wall panels constructed of Method WSP or BV-WSP and continuous sheathing methods as specified in Section R602.10.4 shall be permitted to begin not more than 10 feet (3048 mm) from each end of a *braced wall line* provided that each end complies with one of the following:
   - 1.1. A minimum 24-inch-wide (610 mm) panel for Methods WSP, CS-WSP, CS-G and CS-PF is applied to each side of the building corner as shown in End Condition 4 of Figure R602.10.7.
   - 1.2. The end of each *braced wall panel* closest to the end of the *braced wall line* shall have an 1,800-pound (8 kN) hold-down device fastened to the stud at the edge of the *braced wall panel* closest to the corner and to the foundation or framing below as shown in End Condition 5 of Figure R602.10.7.
2. Braced wall panels constructed of Method PFH or ABW, or of Method BV-WSP where a hold-down is provided in accordance with Table R602.10.6.5.4, shall be permitted to begin not more than 10 feet (3048 mm) from each end of a *braced wall line*.

R602.10.2.3 Minimum number of braced wall panels. *Braced wall lines* with a length of 16 feet (4877 mm) or less shall have not less than two *braced wall panels* of any length or one *braced wall panel* equal to 48 inches (1219 mm) or more. *Braced wall lines* greater than 16 feet (4877 mm) shall have not less than two *braced wall panels*.

R602.10.3 Required length of bracing. The required length of bracing along each *braced wall line* shall be determined as follows:

1. All buildings in *Seismic Design Categories* A and B shall use Table R602.10.3(1) and the applicable adjustment factors in Table R602.10.3(2).
2. Detached buildings in *Seismic Design Category* C shall use Table R602.10.3(1) and the applicable adjustment factors in Table R602.10.3(2).
3. Townhouses in *Seismic Design Category* C shall use the greater value determined from Table R602.10.3(1) or R602.10.3(3) and the applicable adjustment factors in Table R602.10.3(2) or R602.10.3(4), respectively.
4. All buildings in *Seismic Design Categories* D_0, D_1 and D_2 shall use the greater value determined from Table R602.10.3(1) or R602.10.3(3) and the applicable adjustment factors in Table R602.10.3(2) or R602.10.3(4), respectively.

Only *braced wall panels* parallel to the *braced wall line* shall contribute toward the required length of bracing of that *braced wall line*. *Braced wall panels* along an angled wall meeting the minimum length requirements of Tables R602.10.5 and R602.10.5.2 shall be permitted to contribute its projected length toward the minimum required length of bracing for the *braced wall line* as shown in Figure R602.10.1.4. Any *braced wall panel* on an angled wall at the end of a *braced wall line* shall contribute its projected length for only one of the *braced wall lines* at the projected corner.

> **Exception:** The length of wall bracing for dwellings in *Seismic Design Categories* D_0, D_1 and D_2 with stone or masonry veneer installed in accordance with Section R703.8 and exceeding the first-story height shall be in accordance with Section R602.10.6.5.

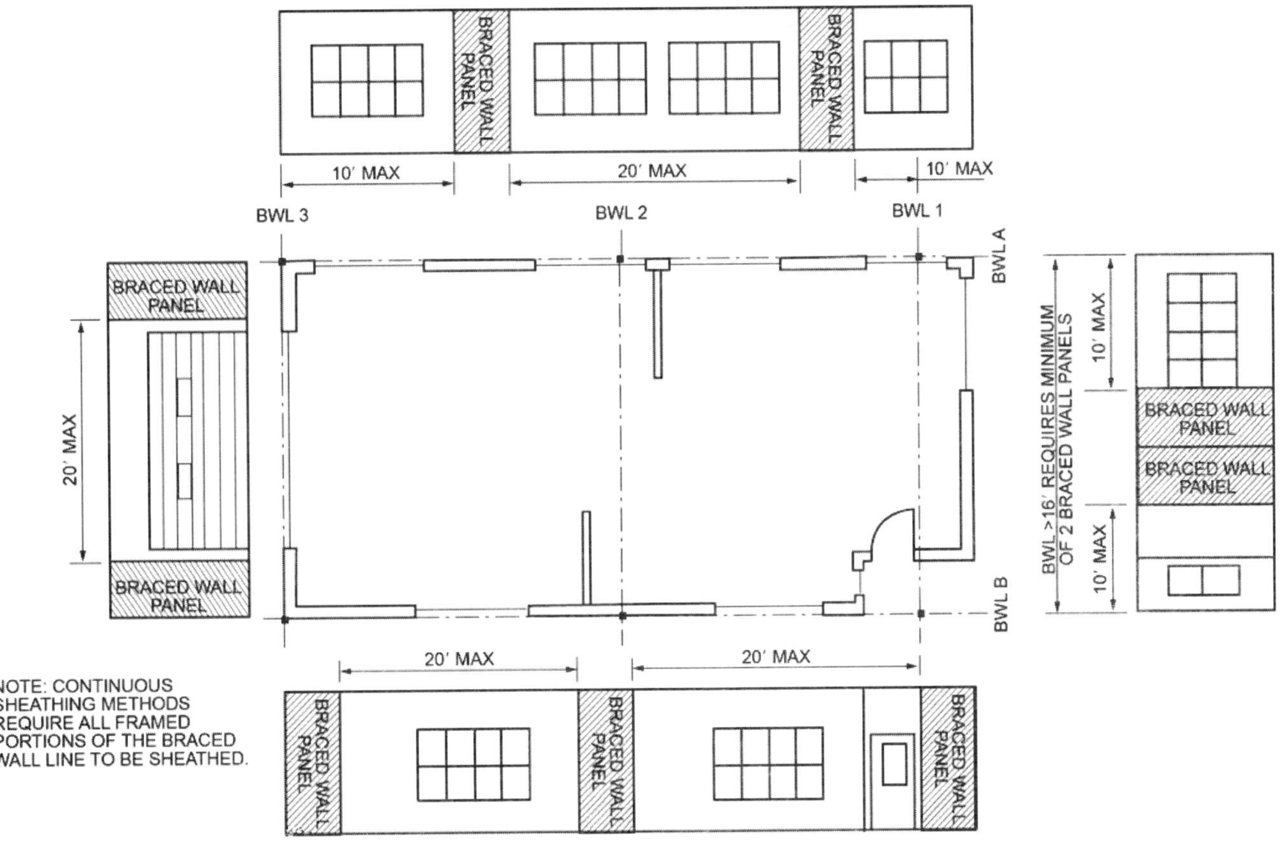

For SI: 1 foot = 304.8 mm.

FIGURE R602.10.2.2
LOCATION OF BRACED WALL PANELS

TABLE R602.10.3(1)
BRACING REQUIREMENTS BASED ON WIND SPEED

• EXPOSURE CATEGORY B • 30-FOOT MEAN ROOF HEIGHT • 10-FOOT WALL HEIGHT • 2 BRACED WALL LINES			MINIMUM TOTAL LENGTH (FEET) OF BRACED WALL PANELS REQUIRED ALONG EACH BRACED WALL LINE[a]			
Ultimate Design Wind Speed (mph)	Story Location	Braced Wall Line Spacing[c] (feet)	Method LIB[b]	Method GB	Methods DWB, WSP, SFB, PBS, PCP, HPS, BV-WSP, ABW, PFH, PFC, CS-SFB	Methods CS-WSP, CS-G, CS-PF
< 95 mph		10	2.5	2.5	1.5	1.5
		20	4.5	4.5	2.5	2.5
		30	6.5	6.5	4.0	3.5
		40	8.5	8.5	5.0	4.0
		50	10.5	10.5	6.0	5.0
		60	12.5	12.5	7.0	6.0
		10	5.0	5.0	3.0	2.5
		20	8.5	8.5	5.0	4.5
		30	12.5	12.5	7.0	6.0
		40	16.0	16.0	9.5	8.0
		50	20.0	20.0	11.5	10.0
		60	23.5	23.5	13.5	11.5
		10	NP	7.0	4.0	3.5
		20	NP	13.0	7.5	6.5
		30	NP	18.5	10.5	9.0
		40	NP	24.0	13.5	11.5
		50	NP	29.5	17.0	14.5
		60	NP	35.0	20.0	17.0
≤ 110		10	3.5	3.5	2.0	1.5
		20	6.0	6.0	3.5	3.0
		30	8.5	8.5	5.0	4.5
		40	11.5	11.5	6.5	5.5
		50	14.0	14.0	8.0	7.0
		60	16.5	16.5	9.5	8.0
		10	6.5	6.5	3.5	3.0
		20	11.5	11.5	6.5	5.5
		30	16.5	16.5	9.5	8.0
		40	21.5	21.5	12.5	10.5
		50	26.5	26.5	15.5	13.0
		60	31.5	31.5	18.0	15.5
		10	NP	9.5	5.5	4.5
		20	NP	17.0	10.0	8.5
		30	NP	24.5	14.0	12.0
		40	NP	32.0	18.5	15.5
		50	NP	39.5	22.5	19.0
		60	NP	46.5	26.5	23.0

(continued)

TABLE R602.10.3(1)—continued
BRACING REQUIREMENTS BASED ON WIND SPEED

• EXPOSURE CATEGORY B • 30-FOOT MEAN ROOF HEIGHT • 10-FOOT WALL HEIGHT • 2 BRACED WALL LINES			MINIMUM TOTAL LENGTH (FEET) OF BRACED WALL PANELS REQUIRED ALONG EACH BRACED WALL LINE[a]			
Ultimate Design Wind Speed (mph)	Story Location	Braced Wall Line Spacing[c] (feet)	Method LIB[b]	Method GB	Methods DWB, WSP, SFB, PBS, PCP, HPS, BV-WSP, ABW, PFH, PFC, CS-SFB	Methods CS-WSP, CS-G, CS-PF
≤ 115	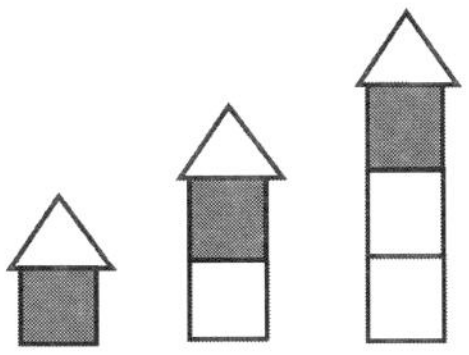	10	3.5	3.5	2.0	2.0
		20	6.5	6.5	3.5	3.5
		30	9.5	9.5	5.5	4.5
		40	12.5	12.5	7.0	6.0
		50	15.0	15.0	9.0	7.5
		60	18.0	18.0	10.5	9.0
	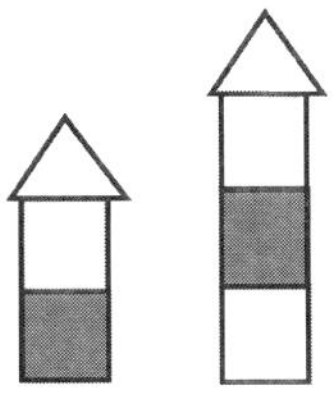	10	7.0	7.0	4.0	3.5
		20	12.5	12.5	7.5	6.5
		30	18.0	18.0	10.5	9.0
		40	23.5	23.5	13.5	11.5
		50	29.0	29.0	16.5	14.0
		60	34.5	34.5	20.0	17.0
	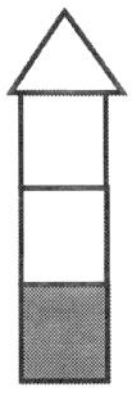	10	NP	10.0	6.0	5.0
		20	NP	18.5	11.0	9.0
		30	NP	27.0	15.5	13.0
		40	NP	35.0	20.0	17.0
		50	NP	43.0	24.5	21.0
		60	NP	51.0	29.0	25.0
≤ 120	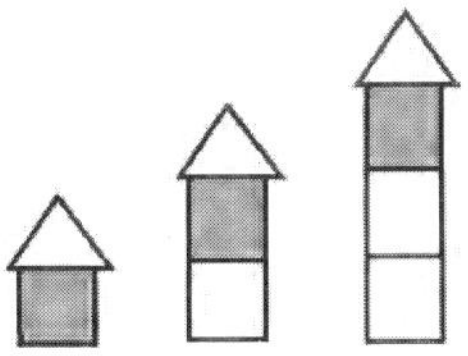	10	4.0	4.0	2.5	2.0
		20	7.0	7.0	4.0	3.5
		30	10.5	10.5	6.0	5.0
		40	13.5	13.5	8.0	6.5
		50	16.5	16.5	9.5	8.0
		60	19.5	19.5	11.5	9.5
	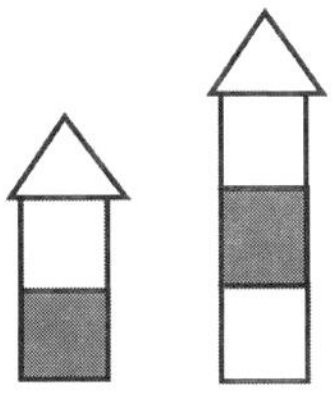	10	7.5	7.5	4.5	3.5
		20	14.0	14.0	8.0	7.0
		30	20.0	20.0	11.5	9.5
		40	25.5	25.5	15.0	12.5
		50	31.5	31.5	18.0	15.5
		60	37.5	37.5	21.5	18.5
	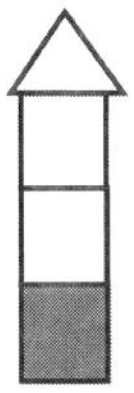	10	NP	11.0	6.5	5.5
		20	NP	20.5	11.5	10.0
		30	NP	29.0	17.0	14.5
		40	NP	38.0	22.0	18.5
		50	NP	47.0	27.0	23.0
		60	NP	55.5	32.0	27.0

(continued)

TABLE R602.10.3(1)—continued
BRACING REQUIREMENTS BASED ON WIND SPEED

• EXPOSURE CATEGORY B • 30-FOOT MEAN ROOF HEIGHT • 10-FOOT WALL HEIGHT • 2 BRACED WALL LINES			MINIMUM TOTAL LENGTH (FEET) OF BRACED WALL PANELS REQUIRED ALONG EACH BRACED WALL LINE[a]			
Ultimate Design Wind Speed (mph)	Story Location	Braced Wall Line Spacing[c] (feet)	Method LIB[b]	Method GB	Methods DWB, WSP, SFB, PBS, PCP, HPS, BV-WSP, ABW, PFH, PFC, CS-SFB	Methods CS-WSP, CS-G, CS-PF
≤ 130	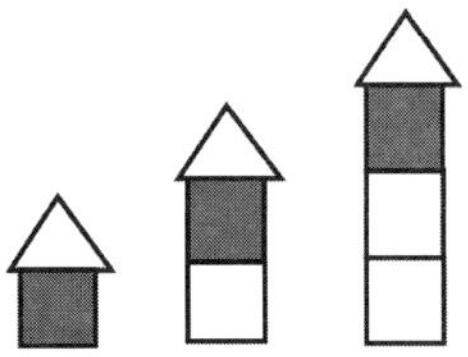	10	4.5	4.5	2.5	2.5
		20	8.5	8.5	5.0	4.0
		30	12.0	12.0	7.0	6.0
		40	15.5	15.5	9.0	7.5
		50	19.5	19.5	11.0	9.5
		60	23.0	23.0	13.0	11.0
	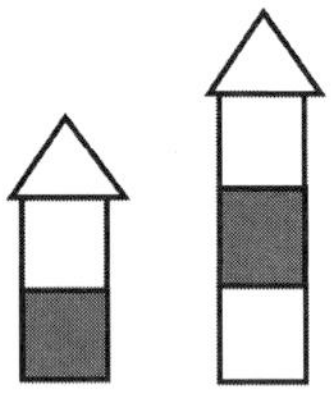	10	8.5	8.5	5.0	4.5
		20	16.0	16.0	9.5	8.0
		30	23.0	23.0	13.5	11.5
		40	30.0	30.0	17.5	15.0
		50	37.0	37.0	21.5	18.0
		60	44.0	44.0	25.0	21.5
		10	NP	13.0	7.5	6.5
		20	NP	24.0	13.5	11.5
		30	NP	34.5	19.5	17.0
		40	NP	44.5	25.5	22.0
		50	NP	55.0	31.5	26.5
		60	NP	65.0	37.5	31.5
< 140	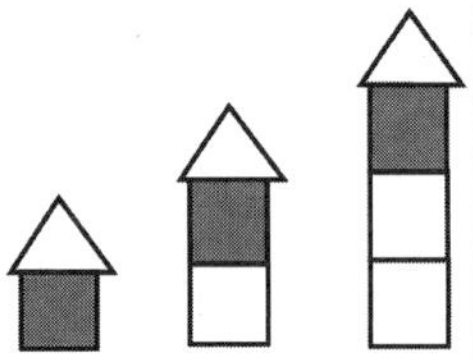	10	5.5	5.5	3.0	2.5
		20	10.0	10.0	5.5	5.0
		30	14.0	14.0	8.0	7.0
		40	18.0	18.0	10.5	9.0
		50	22.5	22.5	13.0	11.0
		60	26.5	26.5	15.0	13.0
	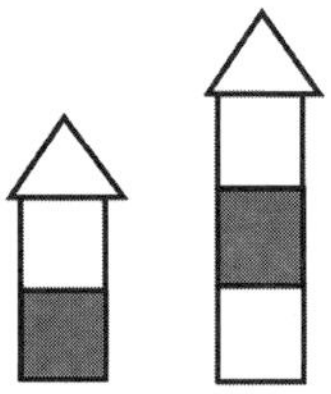	10	10.0	10.0	6.0	5.0
		20	18.5	18.5	11.0	9.0
		30	27.0	27.0	15.5	13.0
		40	35.0	35.0	20.0	17.0
		50	43.0	43.0	24.5	21.0
		60	51.0	51.0	29.0	25.0
	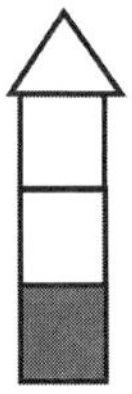	10	NP	15.0	8.5	7.5
		20	NP	27.5	16.0	13.5
		30	NP	39.5	23.0	19.5
		40	NP	51.5	29.5	25.0
		50	NP	63.5	36.5	31.0
		60	NP	75.5	43.0	36.5

For SI: 1 inch = 25.4 mm, 1 foot = 304.8 mm, 1 mile per hour = 0.447 m/s.

NP = Not Permitted.

a. Linear interpolation shall be permitted.

b. Method LIB shall have gypsum board fastened to not less than one side with nails or screws in accordance with Table R602.3(1) for exterior sheathing or Table R702.3.5 for interior gypsum board. Spacing of fasteners at panel edges shall not exceed 8 inches.

c. Where three or more parallel braced wall lines are present and the distances between adjacent braced wall lines are different, the average dimension shall be permitted to be used for braced wall line spacing.

TABLE R602.10.3(2)
WIND ADJUSTMENT FACTORS TO THE REQUIRED LENGTH OF WALL BRACING

ITEM NUMBER	ADJUSTMENT BASED ON	STORY/SUPPORTING	CONDITION	ADJUSTMENT FACTOR[a, b] [multiply length from Table R602.10.3(1) by this factor]	APPLICABLE METHODS
1	Exposure category[d]	One-story structure	B	1.00	All methods
			C	1.20	
			D	1.50	
		Two-story structure	B	1.00	
			C	1.30	
			D	1.60	
		Three-story structure	B	1.00	
			C	1.40	
			D	1.70	
2	Roof eave-to-ridge height	Roof only	≤ 5 feet	0.70	
			10 feet	1.00	
			15 feet	1.30	
			20 feet	1.60	
		Roof + 1 floor	≤ 5 feet	0.85	
			10 feet	1.00	
			15 feet	1.15	
			20 feet	1.30	
		Roof + 2 floors	≤ 5 feet	0.90	
			10 feet	1.00	
			15 feet	1.10	
			20 feet	Not permitted	
3	Story height (Section R301.3)	Any story	8 feet	0.90	
			9 feet	0.95	
			10 feet	1.00	
			11 feet	1.05	
			12 feet	1.10	
4	Number of braced wall lines (per plan direction)[c]	Any story	2	1.00	
			3	1.30	
			4	1.45	
			≥ 5	1.60	
5	Additional 800-pound hold-down device	Top story only	Fastened to the end studs of each braced wall panel and to the foundation or framing below	0.80	DWB, WSP, SFB, PBS, PCP, HPS
6	Interior gypsum board finish (or equivalent)	Any story	Omitted from inside face of braced wall panels	1.40	DWB, WSP, SFB, PBS, PCP, HPS, CS-WSP, CS-G, CS-SFB
7	Gypsum board fastening	Any story	4 inches o.c. at panel edges, including top and bottom plates, and all horizontal joints blocked	0.7	GB
8	Horizontal blocking	Any story	Horizontal block is omitted	2.0	WSP, PBS, CS-WSP

For SI: 1 inch = 25.4 mm, 1 foot = 304.8 mm, 1 pound = 4.48 N.

a. Linear interpolation shall be permitted.

b. The total adjustment factor is the product of all applicable adjustment factors.

c. The adjustment factor is permitted to be 1.0 when determining bracing amounts for intermediate braced wall lines provided the bracing amounts on adjacent braced wall lines are based on a spacing and number that neglects the intermediate braced wall line.

d. The same adjustment factor shall be applied to all braced wall lines on all floors of the structure, based on the worst-case exposure category.

TABLE R602.10.3(3)
BRACING REQUIREMENTS BASED ON SEISMIC DESIGN CATEGORY

• WALL HEIGHT = 10 FEET • 10 PSF FLOOR DEAD LOAD • 15 PSF ROOF/CEILING DEAD LOAD • BRACED WALL LINE SPACING ≤ 25 FEET			MINIMUM TOTAL LENGTH (FEET) OF BRACED WALL PANELS REQUIRED ALONG EACH BRACED WALL LINE[a, g]				
Seismic Design Category[b]	Story Location	Braced Wall Line Length (feet)[c]	Method LIB[d]	Method GB	Methods DWB, SFB, PBS, PCP, HPS, CS-SFB[e]	Methods WSP, ABW[f], PFH[f] and PFG[e, f]	Methods CS-WSP, CS-G, CS-PF
C (townhouses only)		10	2.5	2.5	2.5	1.6	1.4
		20	5.0	5.0	5.0	3.2	2.7
		30	7.5	7.5	7.5	4.8	4.1
		40	10.0	10.0	10.0	6.4	5.4
		50	12.5	12.5	12.5	8.0	6.8
		10	NP	4.5	4.5	3.0	2.6
		20	NP	9.0	9.0	6.0	5.1
		30	NP	13.5	13.5	9.0	7.7
		40	NP	18.0	18.0	12.0	10.2
		50	NP	22.5	22.5	15.0	12.8
		10	NP	6.0	6.0	4.5	3.8
		20	NP	12.0	12.0	9.0	7.7
		30	NP	18.0	18.0	13.5	11.5
		40	NP	24.0	24.0	18.0	15.3
		50	NP	30.0	30.0	22.5	19.1
D_0		10	NP	2.8	2.8	1.8	1.6
		20	NP	5.5	5.5	3.6	3.1
		30	NP	8.3	8.3	5.4	4.6
		40	NP	11.0	11.0	7.2	6.1
		50	NP	13.8	13.8	9.0	7.7
		10	NP	5.3	5.3	3.8	3.2
		20	NP	10.5	10.5	7.5	6.4
		30	NP	15.8	15.8	11.3	9.6
		40	NP	21.0	21.0	15.0	12.8
		50	NP	26.3	26.3	18.8	16.0
		10	NP	7.3	7.3	5.3	4.5
		20	NP	14.5	14.5	10.5	9.0
		30	NP	21.8	21.8	15.8	13.4
		40	NP	29.0	29.0	21.0	17.9
		50	NP	36.3	36.3	26.3	22.3

(continued)

TABLE R602.10.3(3)—continued
BRACING REQUIREMENTS BASED ON SEISMIC DESIGN CATEGORY

• WALL HEIGHT = 10 FEET • 10 PSF FLOOR DEAD LOAD • 15 PSF ROOF/CEILING DEAD LOAD • BRACED WALL LINE SPACING ≤ 25 FEET			MINIMUM TOTAL LENGTH (FEET) OF BRACED WALL PANELS REQUIRED ALONG EACH BRACED WALL LINE[a, g]					
Seismic Design Category[b]	Story Location	Braced Wall Line Length (feet)[c]	Method LIB[d]	Method GB	Methods DWB, SFB, PBS, PCP, HPS, CS-SFB[e]	Methods WSP, ABW[f], PFH[f] and PFG[e, f]	Methods CS-WSP, CS-G, CS-PF
D_1	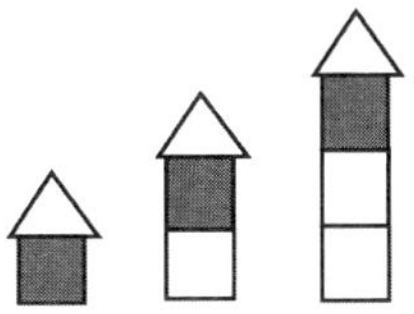	10	NP	3.0	3.0	2.0	1.7
		20	NP	6.0	6.0	4.0	3.4
		30	NP	9.0	9.0	6.0	5.1
		40	NP	12.0	12.0	8.0	6.8
		50	NP	15.0	15.0	10.0	8.5
	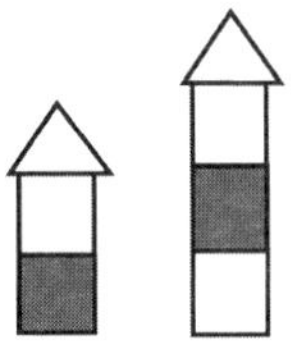	10	NP	6.0	6.0	4.5	3.8
		20	NP	12.0	12.0	9.0	7.7
		30	NP	18.0	18.0	13.5	11.5
		40	NP	24.0	24.0	18.0	15.3
		50	NP	30.0	30.0	22.5	19.1
	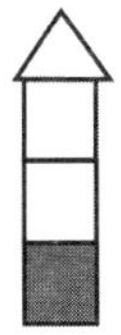	10	NP	8.5	8.5	6.0	5.1
		20	NP	17.0	17.0	12.0	10.2
		30	NP	25.5	25.5	18.0	15.3
		40	NP	34.0	34.0	24.0	20.4
		50	NP	42.5	42.5	30.0	25.5
D_2[h]	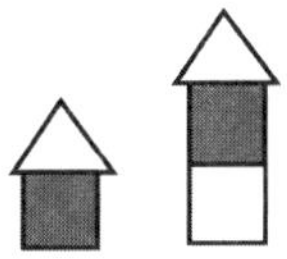	10	NP	4.0	4.0	2.5	2.1
		20	NP	8.0	8.0	5.0	4.3
		30	NP	12.0	12.0	7.5	6.4
		40	NP	16.0	16.0	10.0	8.5
		50	NP	20.0	20.0	12.5	10.6
	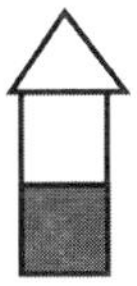	10	NP	7.5	7.5	5.5	4.7
		20	NP	15.0	15.0	11.0	9.4
		30	NP	22.5	22.5	16.5	14.0
		40	NP	30.0	30.0	22.0	18.7
		50	NP	37.5	37.5	27.5	23.4
	Three-story dwelling	10	NP	NP	NP	NP	NP
		20	NP[d]	NP	NP	NP	NP
		30	NP	NP	NP	NP	NP
		40	NP	NP	NP	NP	NP
		50	NP	NP	NP	NP	NP
	Cripple wall below one- or two-story dwelling	10	NP	NP	NP	7.5	6.4
		20	NP	NP	NP	15.0	12.8
		30	NP	NP	NP	22.5	19.1
		40	NP	NP	NP	30.0	25.5
		50	NP	NP	NP	37.5	31.9

(continued)

TABLE R602.10.3(3)—continued
BRACING REQUIREMENTS BASED ON SEISMIC DESIGN CATEGORY

For SI: 1 inch = 25.4 mm, 1 foot = 304.8 mm, 1 pound per square foot = 0.0479 kPa.

NP = Not Permitted.

a. Linear interpolation shall be permitted.

b. Interpolation of bracing length between the S_{ds} values associated with the seismic design categories shall be permitted when a site-specific S_{ds} value is determined in accordance with Section 1613.2 of the *International Building Code.*

c. Where the braced wall line length is greater than 50 feet, braced wall lines shall be permitted to be divided into shorter segments having lengths of 50 feet or less, and the amount of bracing within each segment shall be in accordance with this table.

d. Method LIB shall have gypsum board fastened to not less than one side with nails or screws in accordance with Table R602.3(1) for exterior sheathing or Table R702.3.5 for interior gypsum board. Spacing of fasteners at panel edges shall not exceed 8 inches.

e. Methods PFG and CS-SFB do not apply in Seismic Design Categories D_0, D_1 and D_2.

f. Methods PFH, PFG and ABW are only permitted on a single story or a first of two stories.

g. Where more than one bracing method is used, mixing methods shall be in accordance with Section R602.10.4.1.

h. One- and two-family dwellings in Seismic Design Category D_2 exceeding two stories shall be designed in accordance with accepted engineering practice.

TABLE R602.10.3(4)
SEISMIC ADJUSTMENT FACTORS TO THE REQUIRED LENGTH OF WALL BRACING

ITEM NUMBER	ADJUSTMENT BASED ON	STORY[g]	CONDITION	ADJUSTMENT FACTOR[a, b] [Multiply length from Table R602.10.3(3) by this factor]	APPLICABLE METHODS
1	Story height (Section 301.3)	Any story	≤ 10 feet	1.0	All methods
			> 10 feet and ≤ 12 feet	1.2	
2	Braced wall line spacing, townhouses in SDC C	Any story	≤ 35 feet	1.0	
			> 35 feet and ≤ 50 feet	1.43	
3	Braced wall line spacing, in SDC D_0, D_1, D_2[c]	Any story	> 25 feet and ≤ 30 feet	1.2	
			> 30 feet and ≤ 35 feet	1.4	
4	Wall dead load	Any story	> 8 psf and < 15 psf	1.0	
			< 8psf	0.85	
5	Roof/ceiling dead load for wall supporting	1-, 2- or 3-story building	≤15 psf	1.0	
		2- or 3-story building	> 15 psf and ≤ 25 psf	1.1	
		1-story building or top story	> 15 psf and ≤ 25 psf	1.2	
6	Walls with stone or masonry veneer, townhouses in SDC C[d, e]	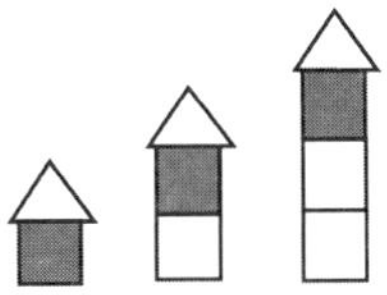		1.0	All methods
		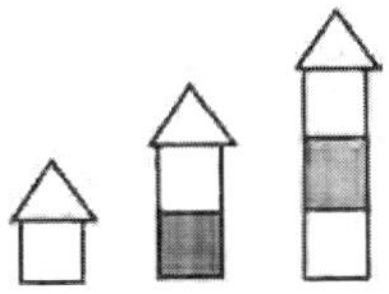		1.5	
		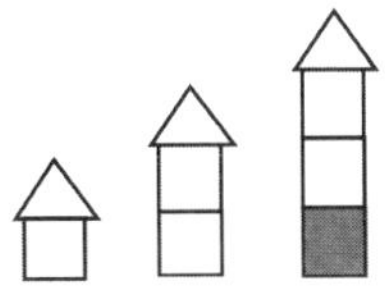		1.5	
7	Walls with stone or masonry veneer, detached one- and two-family dwellings in SDC $D_0 - D_2$[d, f]	Any story	See Section R602.10.6.5.4		BV-WSP
8	Walls with stone or masonry veneer, detached one- and two-family dwellings in SDC $D_0 - D_2$[d, f]	First and second story of two-story dwelling	Limited brick veneer on second story. See Section R602.10.6.5.3.	1.2	WSP, CS-WSP
9	Interior gypsum board finish (or equivalent)	Any story	Omitted from inside face of braced wall panels	1.5	DWB, WSP, SFB, PBS, PCP, HPS, CS-WSP, CS-G, CS-SFB
10	Horizontal blocking	Any story	Horizontal blocking omitted	2.0	WSP, PBS, CS-WSP

For SI: 1 foot = 304.8 mm, 1 pound per square foot = 0.0479 kPa.

a. Linear interpolation shall be permitted.

b. The total length of bracing required for a given wall line is the product of all applicable adjustment factors.

c. The length-to-width ratio for the floor/roof diaphragm shall not exceed 3:1.

d. Applies to stone or masonry veneer exceeding the first story height.

e. The adjustment factor for stone or masonry veneer shall be applied to all exterior braced wall lines and all braced wall lines on the interior of the building, backing or perpendicular to and laterally supporting veneered walls.

f. See Section R602.10.6.5 for requirements where stone or masonry veneer does not exceed the first-story height.

g. One- and two-family dwellings in Seismic Design Category D_2 exceeding two stories shall be designed in accordance with accepted engineering practice.

R602.10.4 Construction methods for braced wall panels. Intermittent and continuously sheathed *braced wall panels* shall be constructed in accordance with this section and the methods listed in Table R602.10.4.

R602.10.4.1 Mixing methods. Mixing of bracing methods shall be permitted as follows:

1. Mixing intermittent bracing and continuous sheathing methods from story to story shall be permitted.
2. Mixing intermittent bracing methods from *braced wall line* to *braced wall line* within a story shall be permitted. In regions within *Seismic Design Categories* A, B and C where the ultimate design wind speed is less than or equal to 130 mph (58m/s), mixing of intermittent bracing and continuous sheathing methods from *braced wall line* to *braced wall line* within a story shall be permitted.
3. Mixing intermittent bracing methods along a *braced wall line* shall be permitted in *Seismic Design Categories* A and B, and detached dwellings in *Seismic Design Category* C, provided that the length of required bracing in accordance with Table R602.10.3(1) or R602.10.3(3) is the highest value of all intermittent bracing methods used.
4. Mixing of continuous sheathing methods CS-WSP, CS-G and CS-PF along a *braced wall line* shall be permitted. Intermittent methods ABW, PFH and PFG shall be permitted to be used along a *braced wall line* with continuous sheathed methods, provided that the length of required bracing for that *braced wall line* is determined in accordance with Table R602.10.3(1) or R602.10.3(3) using the highest value of the bracing methods used.
5. In *Seismic Design Categories* A and B, and for detached one- and two-family dwellings in *Seismic Design Category* C, mixing of intermittent bracing methods along the interior portion of a *braced wall line* with continuous sheathing methods CS-WSP, CS-G and CS-PF along the exterior portion of the same *braced wall line* shall be permitted. The length of required bracing shall be the highest value of all intermittent bracing methods used in accordance with Table R602.10.3(1) or R602.10.3(3) as adjusted by Tables R602.10.3(2) and R602.10.3(4), respectively. The requirements of Section R602.10.7 shall apply to each end of the continuously sheathed portion of the *braced wall line.*

R602.10.4.2 Continuous sheathing methods. Continuous sheathing methods require structural panel sheathing to be used on all sheathable surfaces on one side of a *braced wall line* including areas above and below openings and gable end walls and shall meet the requirements of Section R602.10.7.

R602.10.4.3 Braced wall panel interior finish material. *Braced wall panels* shall have gypsum wall board installed on the side of the wall opposite the bracing material. Gypsum wall board shall be not less than $^1/_2$ inch (12.7 mm) in thickness and be fastened with nails or screws in accordance with Table R602.3(1) for exterior sheathing or Table R702.3.5 for interior gypsum wall board. Spacing of fasteners at panel edges for gypsum wall board opposite Method LIB bracing shall not exceed 8 inches (203 mm). Interior finish material shall not be glued in *Seismic Design Categories* D_0, D_1 and D_2.

Exceptions:

1. Interior finish material is not required opposite wall panels that are braced in accordance with Methods GB, BV-WSP, ABW, PFH, PFG and CS-PF, unless otherwise required by Section R302.6.
2. An *approved* interior finish material with an in-plane shear resistance equivalent to gypsum board shall be permitted to be substituted, unless otherwise required by Section R302.6.
3. Except for Method LIB, gypsum wall board is permitted to be omitted provided that the required length of bracing in Tables R602.10.3(1) and R602.10.3(3) is multiplied by the appropriate adjustment factor in Tables R602.10.3(2) and R602.10.3(4), respectively, unless otherwise required by Section R302.6.

R602.10.4.4 Panel joints. Vertical joints of panel sheathing shall occur over and be fastened to common studs. Horizontal joints of panel sheathing in *braced wall panels* shall occur over and be fastened to common blocking of a thickness of $1^1/_2$ inches (38 mm) or greater.

Exceptions:

1. For methods WSP and CS-WSP, blocking of horizontal joints is permitted to be omitted when adjustment factor No. 8 of Table R602.10.3(2) or No. 9 of Table R602.10.3(4) is applied.
2. Vertical joints of panel sheathing shall be permitted to occur over double studs, where adjoining panel edges are attached to separate studs with the required panel edge fastening schedule, and the adjacent studs are attached together with two rows of 10d box nails [3 inches by 0.128 inch (76.2 mm by 3.25 mm)] at 10 inches o.c. (254 mm).
3. Blocking at horizontal joints shall not be required in wall segments that are not counted as *braced wall panels.*
4. Where Method GB panels are installed horizontally, blocking of horizontal joints is not required.

TABLE R602.10.4
BRACING METHODS

METHODS, MATERIAL		MINIMUM THICKNESS	FIGURE	CONNECTION CRITERIA[a]	
				Fasteners	Spacing
Intermittent Bracing Methods	**LIB** Let-in-bracing	1 × 4 wood or approved metal straps at 45° to 60° angles for maximum 16″ stud spacing	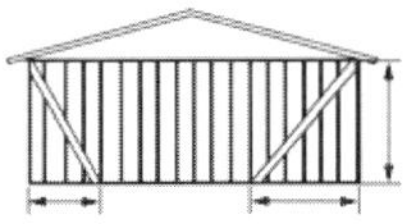	Wood: 2-8d common nails or 3-8d ($2^1/_2$″ long × 0.113″ dia.) nails	Wood: per stud and top and bottom plates
				Metal strap: per manufacturer	Metal: per manufacturer
	DWB Diagonal wood boards	$^3/_4$″ (1″ nominal) for maximum 24″ stud spacing	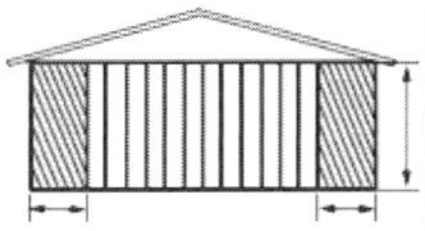	2-8d ($2^1/_2$″ long × 0.113″ dia.) nails or 2-$1^3/_4$″ long staples	Per stud
	WSP Wood structural panel (See Section R604)	$^3/_8$″	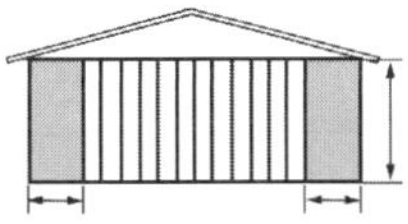	Exterior sheathing per Table R602.3(3)	6″ edges 12″ field
				Interior sheathing per Table R602.3(1) or R602.3(2)	Varies by fastener
	BV-WSP[e] Wood structural panels with stone or masonry veneer (See Section R602.10.6.5)	$^7/_{16}$″	See Figure R602.10.6.5.2	8d common ($2^1/_2$″ × 0.131) nails	4″ at panel edges 12″ at intermediate supports 4″ at braced wall panel end posts
	SFB Structural fiberboard sheathing	$^1/_2$″ or $^{25}/_{32}$″ for maximum 16″ stud spacing	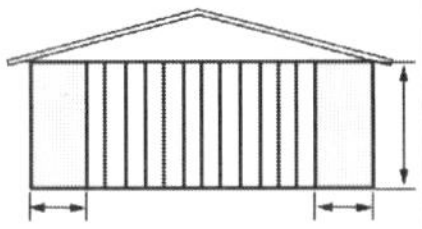	$1^1/_2$″ long × 0.12″ dia. (for $^1/_2$″ thick sheathing) $1^3/_4$″ long × 0.12″ dia. (for $^{25}/_{32}$″ thick sheathing) galvanized roofing nails	3″ edges 6″ field
	GB Gypsum board	$^1/_2$″	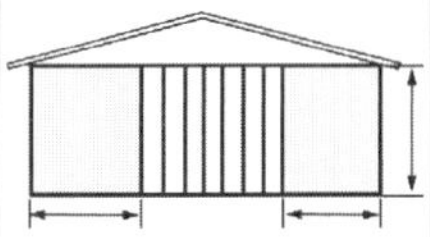	Nails or screws per Table R602.3(1) for exterior locations	For all braced wall panel locations: 7″edges (including top and bottom plates) 7″ field
				Nails or screws per Table R702.3.5 for interior locations	
	PBS Particleboard sheathing (See Section R605)	$^3/_8$″ or $^1/_2$″ for maximum 16″stud spacing	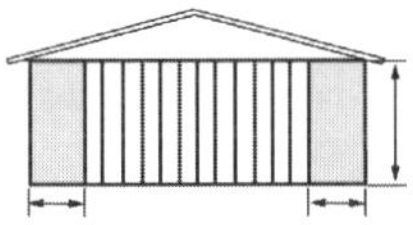	For $^3/_8$″, 6d common (2″ long × 0.113″ dia.) nails; For $^1/_2$″, 8d common ($2^1/_2$″ long × 0.131″ dia.) nails	3″ edges 6″ field
	PCP Portland cement plaster	See Section R703.6 for maximum 16″ stud spacing	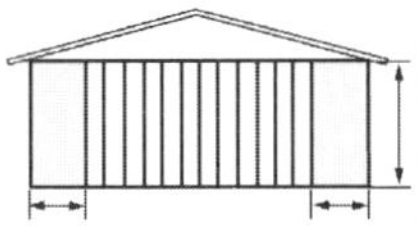	$1^1/_2$″ long, 11 gage, 0.120" dia., $^7/_{16}$″ dia. head nails or $^7/_8$″ long, 16 gage staples	6″ o.c. on all framing members
	HPS Hardboard panel siding	$^7/_{16}$″ for maximum 16″ stud spacing	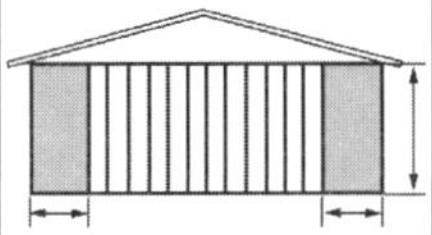	0.092″ dia., 0.225″ dia. head nails with length to accommodate $1^1/_2$″ penetration into studs	4″ edges 8″ field
	ABW Alternate braced wall	$^3/_8$″	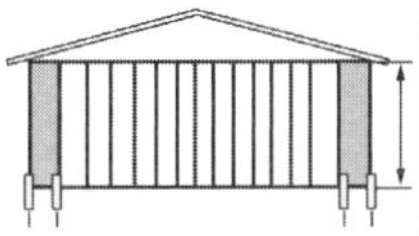	See Section R602.10.6.1	See Section R602.10.6.1

(continued)

TABLE R602.10.4—continued
BRACING METHODS

	METHODS, MATERIAL	MINIMUM THICKNESS	FIGURE	CONNECTION CRITERIA[a]	
				Fasteners	Spacing
Intermittent Bracing Methods	**PFH** Portal frame with hold-downs	$^{3}/_{8}$″		See Section R602.10.6.2	See Section R602.10.6.2
Intermittent Bracing Methods	**PFG** Portal frame at garage	$^{7}/_{16}$″		See Section R602.10.6.3	See Section R602.10.6.3
Continuous Sheathing Methods	**CS-WSP** Continuously sheathed wood structural panel	$^{3}/_{8}$″		Exterior sheathing per Table R602.3(3)	6″ edges 12″ field
Continuous Sheathing Methods				Interior sheathing per Table R602.3(1) or R602.3(2)	Varies by fastener
Continuous Sheathing Methods	**CS-G**[b, c] Continuously sheathed wood structural panel adjacent to garage openings	$^{3}/_{8}$″		See Method CS-WSP	See Method CS-WSP
Continuous Sheathing Methods	**CS-PF** Continuously sheathed portal frame	$^{7}/_{16}$″		See Section R602.10.6.4	See Section R602.10.6.4
Continuous Sheathing Methods	**CS-SFB**[d] Continuously sheathed structural fiberboard	$^{1}/_{2}$″ or $^{25}/_{32}$″ for maximum 16″ stud spacing		$1^{1}/_{2}$″ long × 0.12″ dia. (for $^{1}/_{2}$″ thick sheathing) $1^{3}/_{4}$″ long × 0.12″ dia. (for $^{25}/_{32}$″ thick sheathing) galvanized roofing nails	3″ edges 6″ field

For SI: 1 inch = 25.4 mm, 1 foot = 304.8 mm, 1 degree = 0.0175 rad, 1 pound per square foot = 47.8 N/m^2, 1 mile per hour = 0.447 m/s.

a. Adhesive attachment of wall sheathing, including Method GB, shall not be permitted in Seismic Design Categories C, D_0, D_1 and D_2.

b. Applies to panels next to garage door opening where supporting gable end wall or roof load only. Shall only be used on one wall of the garage. In Seismic Design Categories D_0, D_1 and D_2, roof covering dead load shall not exceed 3 psf.

c. Garage openings adjacent to a Method CS-G panel shall be provided with a header in accordance with Table R602.7(1). A full-height clear opening shall not be permitted adjacent to a Method CS-G panel.

d. Method CS-SFB does not apply in Seismic Design Categories D_0, D_1 and D_2.

e. Method applies to detached one- and two-family dwellings in Seismic Design Categories D_0 through D_2 only.

R602.10.5 Minimum length of a braced wall panel. The minimum length of a *braced wall panel* shall comply with Table R602.10.5. For Methods CS-WSP and CS-SFB, the minimum panel length shall be based on the adjacent clear opening height in accordance with Table R602.10.5 and Figure R602.10.5. Where a panel has an opening on either side of differing heights, the taller opening height shall be used to determine the panel length.

R602.10.5.1 Contributing length. For purposes of computing the required length of bracing in Tables R602.10.3(1) and R602.10.3(3), the contributing length of each *braced wall panel* shall be as specified in Table R602.10.5.

R602.10.5.2 Partial credit. For Methods DWB, WSP, SFB, PBS, PCP and HPS in *Seismic Design Categories* A, B and C, panels between 36 inches and 48 inches (914 mm and 1219 mm) in length shall be considered a *braced wall panel* and shall be permitted to partially contribute toward the required length of bracing in Tables R602.10.3(1) and R602.10.3(3), and the contributing length shall be determined from Table R602.10.5.2.

R602.10.6 Construction of Methods ABW, PFH, PFG, CS-PF and BV-WSP. Methods ABW, PFH, PFG, CS-PF and BV-WSP shall be constructed as specified in Sections R602.10.6.1 through R602.10.6.5.

R602.10.6.1 Method ABW: Alternate braced wall panels. Method ABW *braced wall panels* shall be constructed in accordance with Figure R602.10.6.1. The hold-down force shall be in accordance with Table R602.10.6.1.

R602.10.6.2 Method PFH: Portal frame with hold-downs. Method PFH *braced wall panels* shall be constructed in accordance with Figure R602.10.6.2.

R602.10.6.3 Method PFG: Portal frame at garage door openings in Seismic Design Categories A, B and C. Where supporting a roof or one story and a roof, a Method PFG *braced wall panel* constructed in accordance with Figure R602.10.6.3 shall be permitted on either side of garage door openings.

R602.10.6.4 Method CS-PF: Continuously sheathed portal frame. Continuously sheathed portal frame *braced wall panels* shall be constructed in accordance with Figure R602.10.6.4 and Table R602.10.6.4.

R602.10.6.5 Wall bracing for dwellings with stone and masonry veneer in Seismic Design Categories D_0, D_1 and D_2. Townhouses in *Seismic Design Categories* D_0, D_1 and D_2 with stone or masonry veneer exceeding the first-story height shall be designed in accordance with accepted engineering practice.

One- and two-family dwellings in *Seismic Design Category* D_2 exceeding two stories and having stone or masonry veneer shall be designed in accordance with accepted engineering practice.

Where stone and masonry veneer are installed in accordance with Section R703.8, wall bracing on exterior *braced wall lines* and *braced wall lines* on the interior of the building, backing or perpendicular to and laterally supporting veneered walls shall comply with this section.

R602.10.6.5.1 Veneer on first story only. Where dwellings in *Seismic Design Categories* D_0, D_1 and D_2 have stone or masonry veneer installed in accordance with Section R703.8 and the veneer does not exceed the first-story height, wall bracing shall be in accordance with Section R602.10, exclusive of Section R602.10.6.5.

R602.10.6.5.2 Veneer exceeding first-story height. Where detached one- or two-family dwellings in *Seismic Design Categories* D_0, D_1 and D_2 have stone or masonry veneer installed in accordance with Section R703.8, and the veneer exceeds the first-*story height*, wall bracing at exterior *braced wall lines* and *braced wall lines* on the interior of the building shall be constructed using Method BV-WSP in accordance with this section and Figure R602.10.6.5.2. Cripple walls shall not be permitted, and required interior *braced wall lines* shall be supported on continuous foundations.

R602.10.6.5.3 Limited veneer exceeding first-story height. Where detached one- or two-family dwellings in *Seismic Design Categories* D_0, D_1 and D_2 have exterior veneer installed in accordance with Section R703.8 and where brick veneer installed above the first-story height meets the following limitations, bracing in accordance with Method WSP or CS-WSP shall be permitted provided that the total length of *braced wall panels* specified by Table R602.10.3(3) is multiplied by 1.2 for each first- and second-story *braced wall line.*

1. The dwelling does not extend more than two stories above *grade plane.*
2. The veneer does not exceed 5 inches (127 mm) in thickness.
3. The height of veneer on gable-end walls does not extend more than 8 feet (2438 mm) above the bearing wall top plate elevation.
4. Where veneer is installed on multiple walls above the first story, the total area of the veneer on the second-story exterior walls shall not exceed 25 percent of the occupied second floor area.
5. Where the veneer is installed on one entire second-story exterior wall, including walls on bay windows and similar appurtenances, brick veneer shall not be installed on any of the other walls on that floor.

TABLE R602.10.5
MINIMUM LENGTH OF BRACED WALL PANELS

METHOD (See Table R602.10.4)		MINIMUM LENGTH[a] (inches)					CONTRIBUTING LENGTH (inches)
		Wall Height					
		8 feet	9 feet	10 feet	11 feet	12 feet	
DWB, WSP, SFB, PBS, PCP, HPS, BV-WSP		48	48	48	53	58	Actual[b]
GB		48	48	48	53	58	Double sided = Actual Single sided = 0.5 × Actual
LIB		55	62	69	NP	NP	Actual[b]
ABW	SDC A, B and C, ultimate design wind speed $<$ 140 mph	28	32	34	38	42	48
	SDC D_0, D_1 and D_2, ultimate design wind speed $<$ 140 mph	32	32	34	NP	NP	
CS-G		24	27	30	33	36	Actual[b]
CS-WSP, CS-SFB	Adjacent clear opening height (inches)						
	$\leq$ 64	24	27	30	33	36	Actual[b]
	68	26	27	30	33	36	
	72	27	27	30	33	36	
	76	30	29	30	33	36	
	80	32	30	30	33	36	
	84	35	32	32	33	36	
	88	38	35	33	33	36	
	92	43	37	35	35	36	
	96	48	41	38	36	36	
	100	—	44	40	38	38	
	104	—	49	43	40	39	
	108	—	54	46	43	41	
	112	—	—	50	45	43	
	116	—	—	55	48	45	
	120	—	—	60	52	48	
	124	—	—	—	56	51	
	128	—	—	—	61	54	
	132	—	—	—	66	58	
	136	—	—	—	—	62	
	140	—	—	—	—	66	
	144	—	—	—	—	72	
METHOD (See Table R602.10.4)		Portal header height					
		8 feet	9 feet	10 feet	11 feet	12 feet	
PFH	Supporting roof only	16	16	16	Note c	Note c	48
	Supporting one story and roof	24	24	24	Note c	Note c	
PFG		24	27	30	Note d	Note d	1.5 × Actual[b]
CS-PF	SDC A, B and C	16	18	20	Note e	Note e	1.5 × Actual[b]
	SDC D_0, D_1 and D_2	16	18	20	Note e	Note e	Actual[b]

For SI: 1 inch = 25.4 mm, 1 foot = 304.8 mm, 1 mile per hour = 0.447 m/s.

NP = Not Permitted.

a. Linear interpolation shall be permitted.

b. Use the actual length where it is greater than or equal to the minimum length.

c. Maximum header height for PFH is 10 feet in accordance with Figure R602.10.6.2, but wall height shall be permitted to be increased to 12 feet with pony wall.

d. Maximum header height for PFG is 10 feet in accordance with Figure R602.10.6.3, but wall height shall be permitted to be increased to 12 feet with pony wall.

e. Maximum header height for CS-PF is 10 feet in accordance with Figure R602.10.6.4, but wall height shall be permitted to be increased to 12 feet with pony wall.

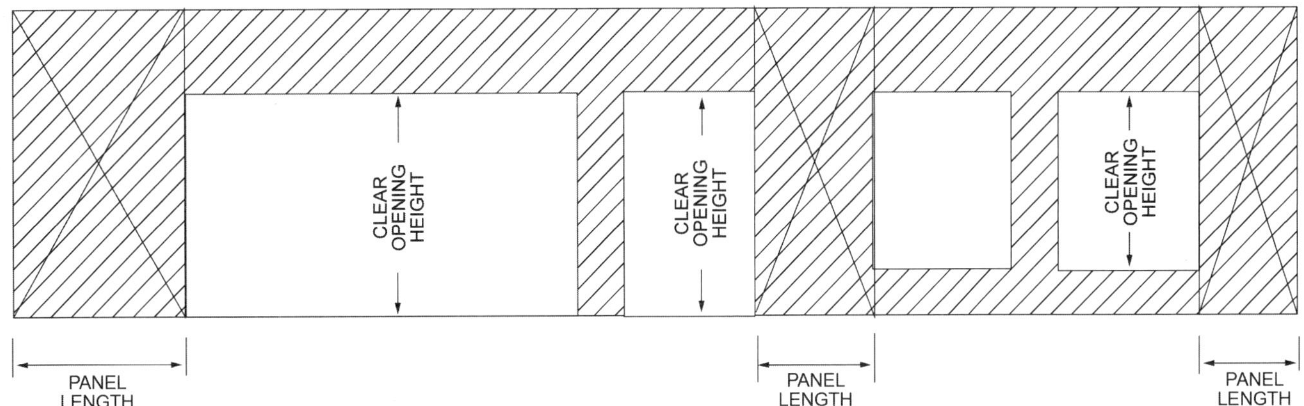

FIGURE R602.10.5
BRACED WALL PANELS WITH CONTINUOUS SHEATHING

TABLE R602.10.5.2
PARTIAL CREDIT FOR BRACED WALL PANELS LESS THAN 48 INCHES IN ACTUAL LENGTH

ACTUAL LENGTH OF BRACED WALL PANEL (inches)	CONTRIBUTING LENGTH OF BRACED WALL PANEL (inches)[a]	
	8-foot Wall Height	9-foot Wall Height
48	48	48
42	36	36
36	27	NA

For SI: 1 inch = 25.4 mm, 1 foot = 304.8 mm.
NA = Not Applicable.
a. Linear interpolation shall be permitted.

TABLE R602.10.6.1
MINIMUM HOLD-DOWN FORCES FOR METHOD ABW BRACED WALL PANELS

SEISMIC DESIGN CATEGORY AND WIND SPEED	SUPPORTING/STORY	HOLD-DOWN FORCE (pounds)				
		Height of Braced Wall Panel				
		8 feet	9 feet	10 feet	11 feet	12 feet
SDC A, B and C Ultimate design wind speed < 140 mph	One story	1,800	1,800	1,800	2,000	2,200
	First of two stories	3,000	3,000	3,000	3,300	3,600
SDC D_0, D_1 and D_2 Ultimate design wind speed <140 mph	One story	1,800	1,800	1,800	NP	NP
	First of two stories	3,000	3,000	3,000	NP	NP

For SI: 1 foot = 304.8 mm, 1 pound = 4.45 N, 1 mile per hour = 0.447 m/s.
NP = Not Permitted.

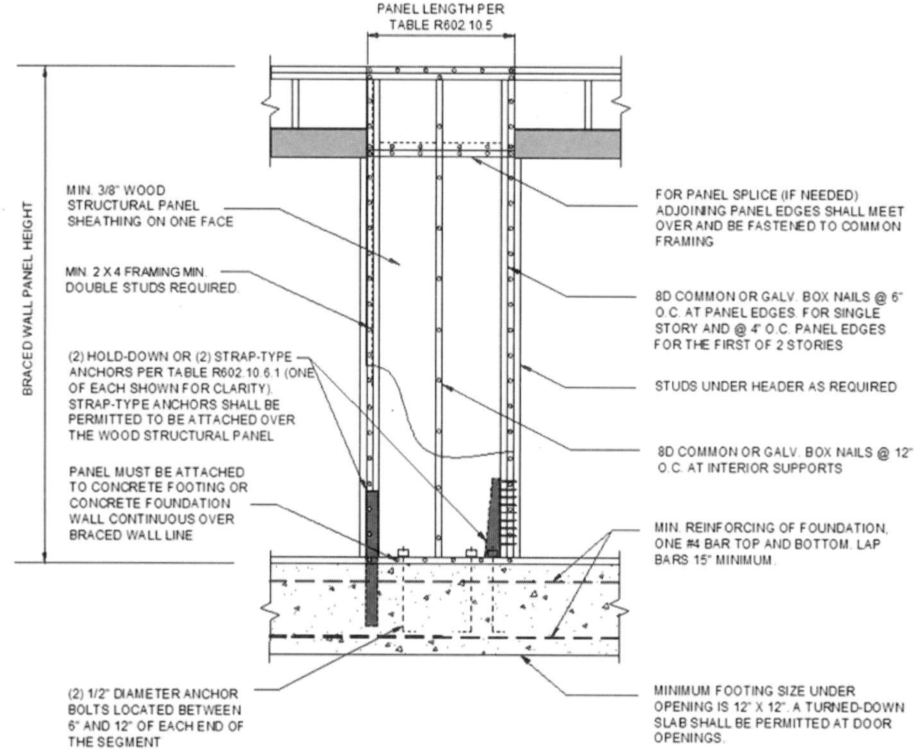
PANEL LENGTH PER TABLE R602.10.5
BRACED WALL PANEL HEIGHT
MIN. 3/8" WOOD STRUCTURAL PANEL SHEATHING ON ONE FACE
MIN. 2 X 4 FRAMING MIN. DOUBLE STUDS REQUIRED.
(2) HOLD-DOWN OR (2) STRAP-TYPE ANCHORS PER TABLE R602.10.6.1 (ONE OF EACH SHOWN FOR CLARITY). STRAP-TYPE ANCHORS SHALL BE PERMITTED TO BE ATTACHED OVER THE WOOD STRUCTURAL PANEL
PANEL MUST BE ATTACHED TO CONCRETE FOOTING OR CONCRETE FOUNDATION WALL CONTINUOUS OVER BRACED WALL LINE
(2) 1/2" DIAMETER ANCHOR BOLTS LOCATED BETWEEN 6" AND 12" OF EACH END OF THE SEGMENT
FOR PANEL SPLICE (IF NEEDED) ADJOINING PANEL EDGES SHALL MEET OVER AND BE FASTENED TO COMMON FRAMING
8D COMMON OR GALV. BOX NAILS @ 6" O.C. AT PANEL EDGES FOR SINGLE STORY AND @ 4" O.C. PANEL EDGES FOR THE FIRST OF 2 STORIES
STUDS UNDER HEADER AS REQUIRED
8D COMMON OR GALV. BOX NAILS @ 12" O.C. AT INTERIOR SUPPORTS
MIN. REINFORCING OF FOUNDATION, ONE #4 BAR TOP AND BOTTOM. LAP BARS 15" MINIMUM.
MINIMUM FOOTING SIZE UNDER OPENING IS 12" X 12". A TURNED-DOWN SLAB SHALL BE PERMITTED AT DOOR OPENINGS.

For SI: 1 inch = 25.4 mm.

FIGURE R602.10.6.1
METHOD ABW—ALTERNATE BRACED WALL PANEL

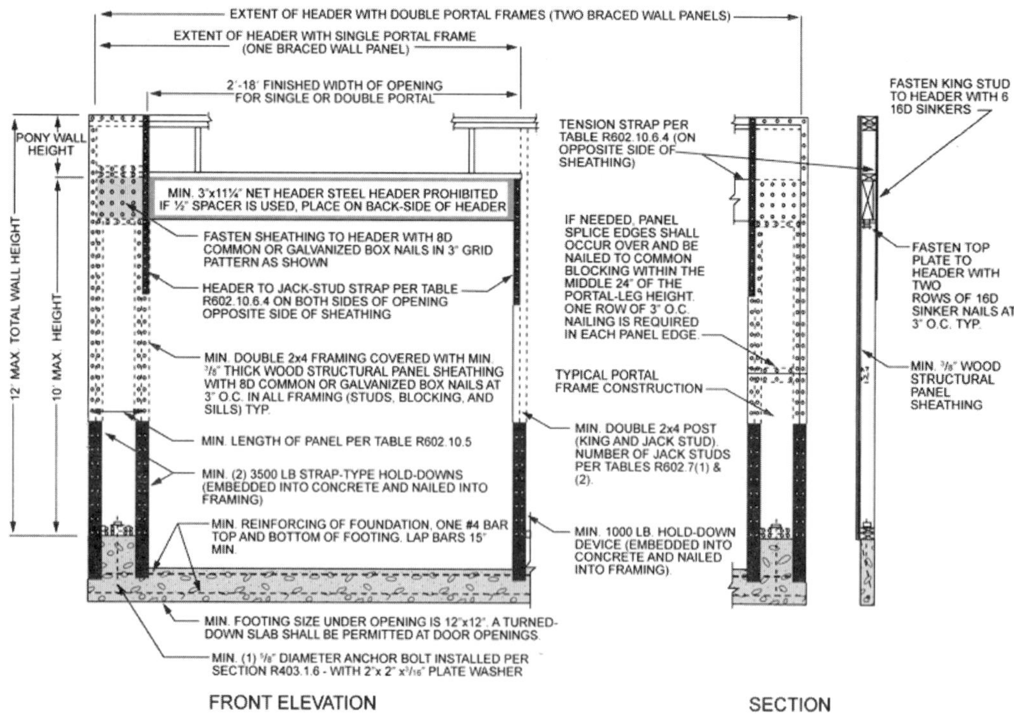
EXTENT OF HEADER WITH DOUBLE PORTAL FRAMES (TWO BRACED WALL PANELS)
EXTENT OF HEADER WITH SINGLE PORTAL FRAME (ONE BRACED WALL PANEL)
2'-18' FINISHED WIDTH OF OPENING FOR SINGLE OR DOUBLE PORTAL
PONY WALL HEIGHT
12' MAX. TOTAL WALL HEIGHT
10' MAX. HEIGHT
MIN. 3"x11¼" NET HEADER STEEL HEADER PROHIBITED IF ½" SPACER IS USED, PLACE ON BACK-SIDE OF HEADER
FASTEN SHEATHING TO HEADER WITH 8D COMMON OR GALVANIZED BOX NAILS IN 3" GRID PATTERN AS SHOWN
HEADER TO JACK-STUD STRAP PER TABLE R602.10.6.4 ON BOTH SIDES OF OPENING OPPOSITE SIDE OF SHEATHING
MIN. DOUBLE 2x4 FRAMING COVERED WITH MIN. ⅜" THICK WOOD STRUCTURAL PANEL SHEATHING WITH 8D COMMON OR GALVANIZED BOX NAILS AT 3" O.C. IN ALL FRAMING (STUDS, BLOCKING, AND SILLS) TYP.
MIN. LENGTH OF PANEL PER TABLE R602.10.5
MIN. (2) 3500 LB STRAP-TYPE HOLD-DOWNS (EMBEDDED INTO CONCRETE AND NAILED INTO FRAMING)
MIN. REINFORCING OF FOUNDATION, ONE #4 BAR TOP AND BOTTOM OF FOOTING. LAP BARS 15" MIN.
MIN. FOOTING SIZE UNDER OPENING IS 12"x12". A TURNED-DOWN SLAB SHALL BE PERMITTED AT DOOR OPENINGS.
MIN. (1) ⅝" DIAMETER ANCHOR BOLT INSTALLED PER SECTION R403.1.6 - WITH 2"x 2" x³/₁₆" PLATE WASHER
TENSION STRAP PER TABLE R602.10.6.4 (ON OPPOSITE SIDE OF SHEATHING)
IF NEEDED, PANEL SPLICE EDGES SHALL OCCUR OVER AND BE NAILED TO COMMON BLOCKING WITHIN THE MIDDLE 24" OF THE PORTAL-LEG HEIGHT. ONE ROW OF 3" O.C. NAILING IS REQUIRED IN EACH PANEL EDGE.
TYPICAL PORTAL FRAME CONSTRUCTION
MIN. DOUBLE 2x4 POST (KING AND JACK STUD). NUMBER OF JACK STUDS PER TABLES R602.7(1) & (2).
MIN. 1000 LB. HOLD-DOWN DEVICE (EMBEDDED INTO CONCRETE AND NAILED INTO FRAMING).
FASTEN KING STUD TO HEADER WITH 6 16D SINKERS
FASTEN TOP PLATE TO HEADER WITH TWO ROWS OF 16D SINKER NAILS AT 3" O.C. TYP.
MIN. ⅜" WOOD STRUCTURAL PANEL SHEATHING
FRONT ELEVATION
SECTION

For SI: 1 inch = 25.4 mm, 1 foot = 304.8 mm.

FIGURE R602.10.6.2
METHOD PFH—PORTAL FRAME WITH HOLD-DOWNS

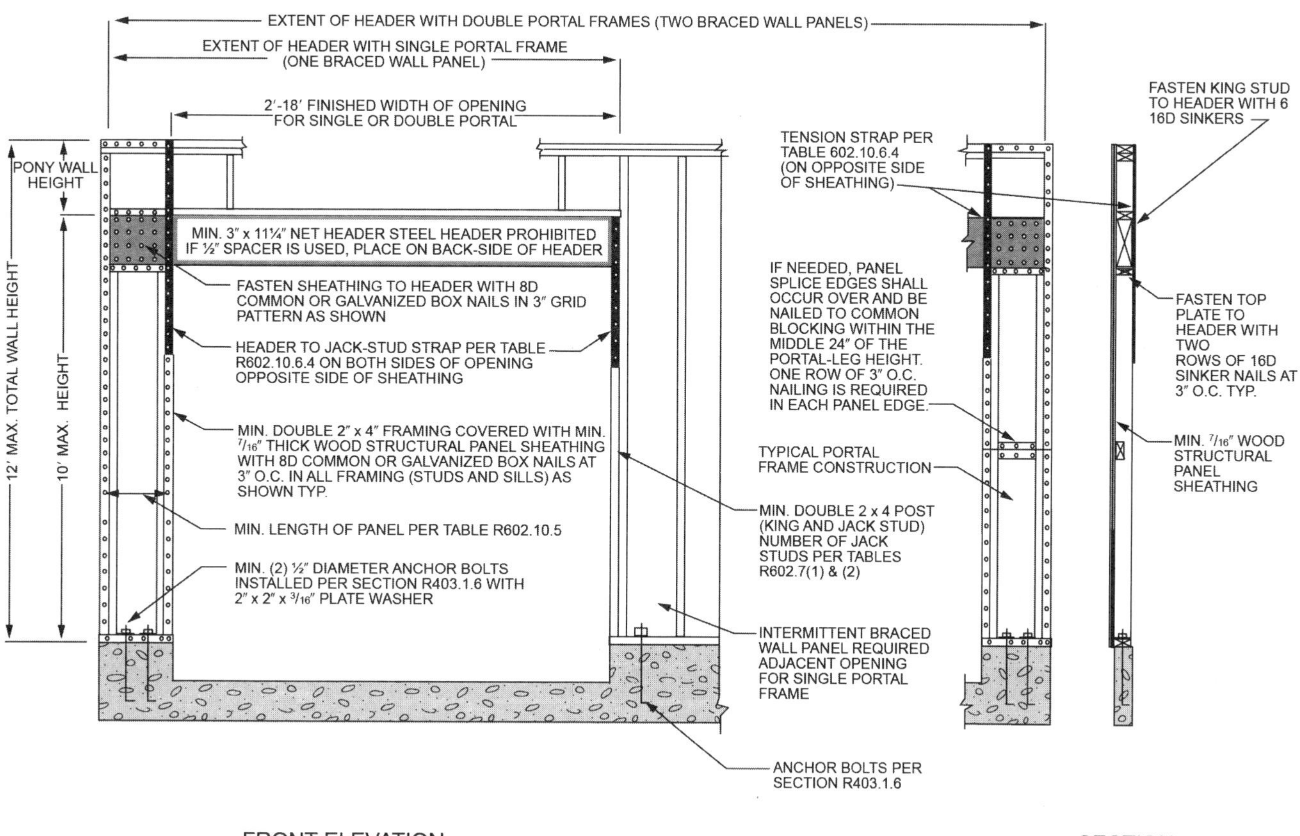
EXTENT OF HEADER WITH DOUBLE PORTAL FRAMES (TWO BRACED WALL PANELS)
EXTENT OF HEADER WITH SINGLE PORTAL FRAME (ONE BRACED WALL PANEL)
2′-18′ FINISHED WIDTH OF OPENING FOR SINGLE OR DOUBLE PORTAL
PONY WALL HEIGHT
12′ MAX. TOTAL WALL HEIGHT
10′ MAX. HEIGHT
MIN. 3″ x 11¼″ NET HEADER STEEL HEADER PROHIBITED IF ½″ SPACER IS USED, PLACE ON BACK-SIDE OF HEADER
FASTEN SHEATHING TO HEADER WITH 8D COMMON OR GALVANIZED BOX NAILS IN 3″ GRID PATTERN AS SHOWN
HEADER TO JACK-STUD STRAP PER TABLE R602.10.6.4 ON BOTH SIDES OF OPENING OPPOSITE SIDE OF SHEATHING
MIN. DOUBLE 2″ x 4″ FRAMING COVERED WITH MIN. 7/16″ THICK WOOD STRUCTURAL PANEL SHEATHING WITH 8D COMMON OR GALVANIZED BOX NAILS AT 3″ O.C. IN ALL FRAMING (STUDS AND SILLS) AS SHOWN TYP.
MIN. LENGTH OF PANEL PER TABLE R602.10.5
MIN. (2) ½″ DIAMETER ANCHOR BOLTS INSTALLED PER SECTION R403.1.6 WITH 2″ x 2″ x 3/16″ PLATE WASHER
TENSION STRAP PER TABLE 602.10.6.4 (ON OPPOSITE SIDE OF SHEATHING)
IF NEEDED, PANEL SPLICE EDGES SHALL OCCUR OVER AND BE NAILED TO COMMON BLOCKING WITHIN THE MIDDLE 24″ OF THE PORTAL-LEG HEIGHT. ONE ROW OF 3″ O.C. NAILING IS REQUIRED IN EACH PANEL EDGE.
TYPICAL PORTAL FRAME CONSTRUCTION
MIN. DOUBLE 2 x 4 POST (KING AND JACK STUD) NUMBER OF JACK STUDS PER TABLES R602.7(1) & (2)
INTERMITTENT BRACED WALL PANEL REQUIRED ADJACENT OPENING FOR SINGLE PORTAL FRAME
ANCHOR BOLTS PER SECTION R403.1.6
FASTEN KING STUD TO HEADER WITH 6 16D SINKERS
FASTEN TOP PLATE TO HEADER WITH TWO ROWS OF 16D SINKER NAILS AT 3″ O.C. TYP.
MIN. 7/16″ WOOD STRUCTURAL PANEL SHEATHING
FRONT ELEVATION
SECTION

For SI: 1 inch = 25.4 mm, 1 foot = 304.8 mm.

FIGURE R602.10.6.3
METHOD PFG—PORTAL FRAME AT GARAGE DOOR OPENINGS IN SEISMIC DESIGN CATEGORIES A, B AND C

TABLE R602.10.6.4
TENSION STRAP CAPACITY FOR RESISTING WIND PRESSURES
PERPENDICULAR TO METHODS PFH, PFG AND CS-PF BRACED WALL PANELS[a]

MINIMUM WALL STUD FRAMING NOMINAL SIZE AND GRADE	MAXIMUM PONY WALL HEIGHT (feet)	MAXIMUM TOTAL WALL HEIGHT (feet)	MAXIMUM OPENING WIDTH (feet)	TENSION STRAP CAPACITY REQUIRED (pounds)[a]					
				Ultimate Design Wind Speed V_{ult} (mph)					
				≤ 110	115	130	≤ 110	115	130
				Exposure B			Exposure C		
2 × 4 No. 2 Grade	0	10	18	1,000	1,000	1,000	1,000	1,000	1,050
	1	10	9	1,000	1,000	1,000	1,000	1,000	1,750
			16	1,000	1,025	2,050	2,075	2,500	3,950
			18	1,000	1,275	2,375	2,400	2,850	DR
	2	10	9	1,000	1,000	1,475	1,500	1,875	3,125
			16	1,775	2,175	3,525	3,550	4,125	DR
			18	2,075	2,500	3,950	3,975	DR	DR
	2	12	9	1,150	1,500	2,650	2,675	3,175	DR
			16	2,875	3,375	DR	DR	DR	DR
			18	3,425	3,975	DR	DR	DR	DR
	4	12	9	2,275	2,750	DR	DR	DR	DR
			12	3,225	3,775	DR	DR	DR	DR
2 × 6 Stud Grade	2	12	9	1,000	1,000	1,700	1,700	2,025	3,050
			16	1,825	2,150	3,225	3,225	3,675	DR
			18	2,200	2,550	3,725	3,750	DR	DR
	4	12	9	1,450	1,750	2,700	2,725	3,125	DR
			16	2,050	2,400	DR	DR	DR	DR
			18	3,350	3,800	DR	DR	DR	DR

For SI: 1 foot = 304.8 mm, 1 pound = 4.45 N, 1 mile per hour = 0.447 m/s.
DR = Design Required.
a. Straps shall be installed in accordance with manufacturer's recommendations.

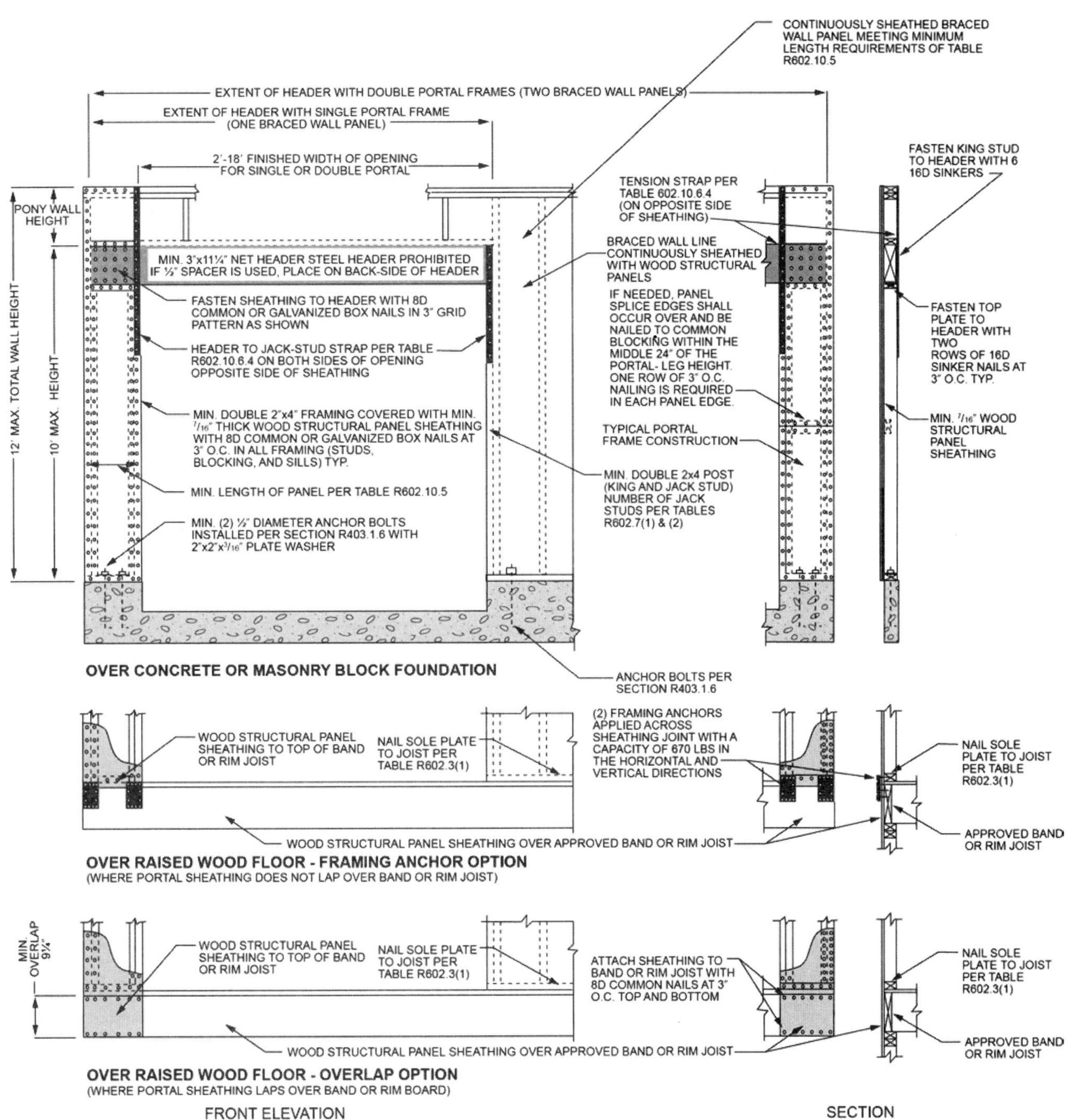
CONTINUOUSLY SHEATHED BRACED WALL PANEL MEETING MINIMUM LENGTH REQUIREMENTS OF TABLE R602.10.5
EXTENT OF HEADER WITH DOUBLE PORTAL FRAMES (TWO BRACED WALL PANELS)
EXTENT OF HEADER WITH SINGLE PORTAL FRAME (ONE BRACED WALL PANEL)
2'-18' FINISHED WIDTH OF OPENING FOR SINGLE OR DOUBLE PORTAL
PONY WALL HEIGHT
12' MAX. TOTAL WALL HEIGHT
10' MAX. HEIGHT
MIN. 3"x11¼" NET HEADER STEEL HEADER PROHIBITED IF ½" SPACER IS USED, PLACE ON BACK-SIDE OF HEADER
FASTEN SHEATHING TO HEADER WITH 8D COMMON OR GALVANIZED BOX NAILS IN 3" GRID PATTERN AS SHOWN
HEADER TO JACK-STUD STRAP PER TABLE R602.10.6.4 ON BOTH SIDES OF OPENING OPPOSITE SIDE OF SHEATHING
MIN. DOUBLE 2"x4" FRAMING COVERED WITH MIN. 7/16" THICK WOOD STRUCTURAL PANEL SHEATHING WITH 8D COMMON OR GALVANIZED BOX NAILS AT 3" O.C. IN ALL FRAMING (STUDS, BLOCKING, AND SILLS) TYP.
MIN. LENGTH OF PANEL PER TABLE R602.10.5
MIN. (2) ½" DIAMETER ANCHOR BOLTS INSTALLED PER SECTION R403.1.6 WITH 2"x2"x3/16" PLATE WASHER
TENSION STRAP PER TABLE 602.10.6.4 (ON OPPOSITE SIDE OF SHEATHING)
BRACED WALL LINE CONTINUOUSLY SHEATHED WITH WOOD STRUCTURAL PANELS
IF NEEDED, PANEL SPLICE EDGES SHALL OCCUR OVER AND BE NAILED TO COMMON BLOCKING WITHIN THE MIDDLE 24" OF THE PORTAL- LEG HEIGHT. ONE ROW OF 3" O.C. NAILING IS REQUIRED IN EACH PANEL EDGE.
TYPICAL PORTAL FRAME CONSTRUCTION
MIN. DOUBLE 2x4 POST (KING AND JACK STUD) NUMBER OF JACK STUDS PER TABLES R602.7(1) & (2)
FASTEN KING STUD TO HEADER WITH 6 16D SINKERS
FASTEN TOP PLATE TO HEADER WITH TWO ROWS OF 16D SINKER NAILS AT 3" O.C. TYP.
MIN. 7/16" WOOD STRUCTURAL PANEL SHEATHING
OVER CONCRETE OR MASONRY BLOCK FOUNDATION
ANCHOR BOLTS PER SECTION R403.1.6
WOOD STRUCTURAL PANEL SHEATHING TO TOP OF BAND OR RIM JOIST
NAIL SOLE PLATE TO JOIST PER TABLE R602.3(1)
(2) FRAMING ANCHORS APPLIED ACROSS SHEATHING JOINT WITH A CAPACITY OF 670 LBS IN THE HORIZONTAL AND VERTICAL DIRECTIONS
NAIL SOLE PLATE TO JOIST PER TABLE R602.3(1)
APPROVED BAND OR RIM JOIST
WOOD STRUCTURAL PANEL SHEATHING OVER APPROVED BAND OR RIM JOIST
OVER RAISED WOOD FLOOR - FRAMING ANCHOR OPTION
(WHERE PORTAL SHEATHING DOES NOT LAP OVER BAND OR RIM JOIST)
MIN. 9¼" OVERLAP
WOOD STRUCTURAL PANEL SHEATHING TO TOP OF BAND OR RIM JOIST
NAIL SOLE PLATE TO JOIST PER TABLE R602.3(1)
ATTACH SHEATHING TO BAND OR RIM JOIST WITH 8D COMMON NAILS AT 3" O.C. TOP AND BOTTOM
NAIL SOLE PLATE TO JOIST PER TABLE R602.3(1)
APPROVED BAND OR RIM JOIST
WOOD STRUCTURAL PANEL SHEATHING OVER APPROVED BAND OR RIM JOIST
OVER RAISED WOOD FLOOR - OVERLAP OPTION
(WHERE PORTAL SHEATHING LAPS OVER BAND OR RIM BOARD)
FRONT ELEVATION
SECTION

For SI: 1 inch = 25.4 mm, 1 foot = 304.8 mm.

FIGURE R602.10.6.4
METHOD CS-PF—CONTINUOUSLY SHEATHED PORTAL FRAME PANEL CONSTRUCTION

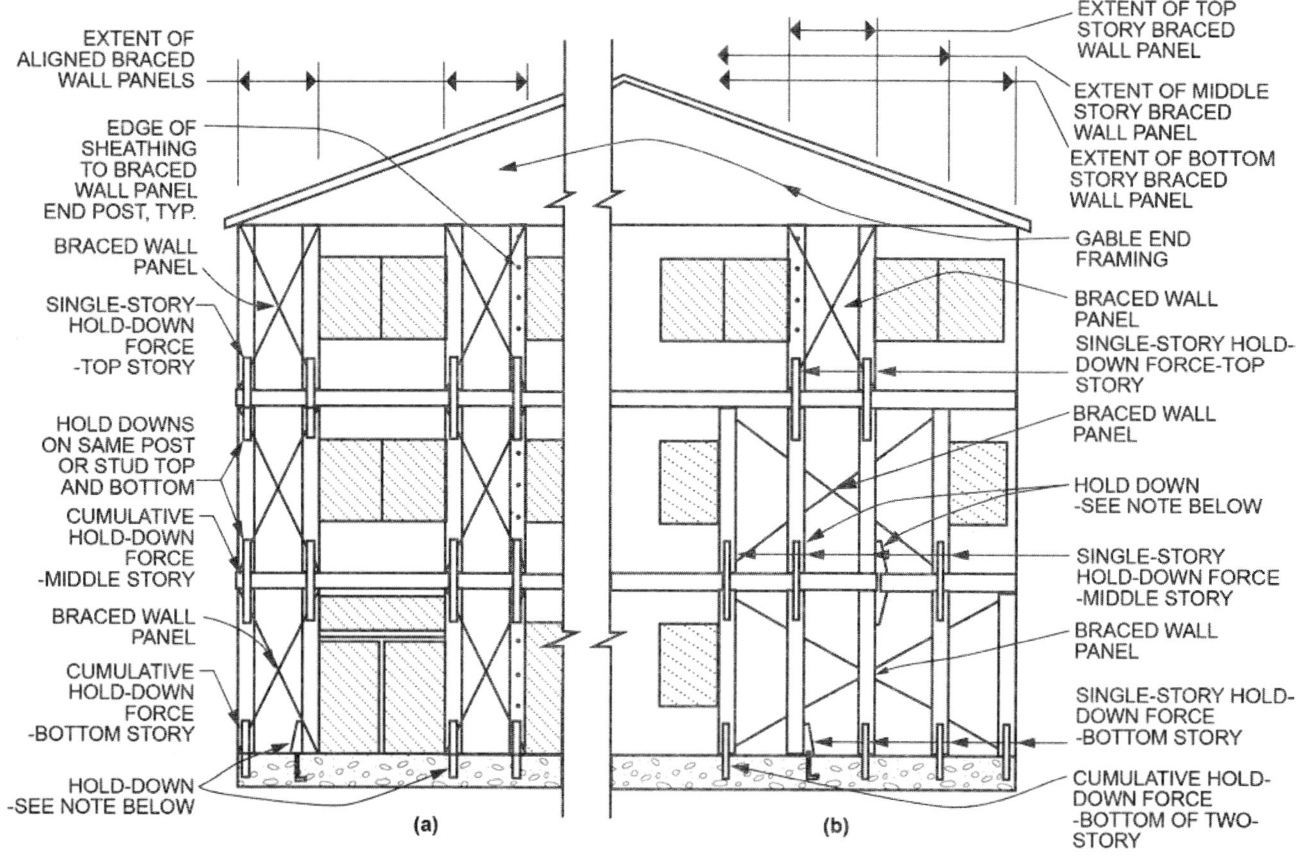

(a) Braced wall panels stacked (aligned story to story). Use cumulative hold-down force.

(b) Braced wall panels mixed stacked and not stacked. Use hold-down force as noted.

Note: Hold downs should be strap ties, tension ties, or other approved hold-down devices and shall be installed in accordance with the manufacturer's instructions.

FIGURE R602.10.6.5.2
METHOD BV-WSP—WALL BRACING FOR DWELLINGS WITH STONE AND MASONRY VENEER IN SEISMIC DESIGN CATEGORIES D_0, D_1 and D_2

R602.10.6.5.4 Length of bracing. The length of bracing along each *braced wall line* shall be the greater of that required by the ultimate design wind speed and *braced wall line* spacing in accordance with Table R602.10.3(1) as adjusted by the factors in Table R602.10.3(2) or the seismic design category and *braced wall line* length in accordance either with Table R602.10.6.5.4 when using Method BV-WSP, or Table R602.10.3(3) as adjusted by the factors in Table R602.10.3(4) when using Method WSP or CS-WSP. Angled walls shall be permitted to be counted in accordance with Section R602.10.1.4, and *braced wall panel* location shall be in accordance with Section R602.10.2.2. Spacing between *braced wall lines* shall be in accordance with Table R602.10.1.3. The seismic adjustment factors in Table R602.10.3(4) shall not be applied to the length of bracing determined using Table R602.10.6.5.4, except that the bracing amount increase for *braced wall line* spacing greater than 25 feet (7620 mm) in accordance with Table R602.10.1.3 shall be required. The minimum total length of bracing in a *braced wall line,* after all adjustments have been taken, shall be not less than 48 inches (1219 mm) total.

R602.10.7 Ends of braced wall lines with continuous sheathing. Each end of a *braced wall line* with continuous sheathing shall have one of the conditions shown in Figure R602.10.7.

TABLE R602.10.6.5.4
METHOD BV-WSP WALL BRACING REQUIREMENTS[d]

SEISMIC DESIGN CATEGORY	STORY	BRACED WALL LINE LENGTH (FEET)					SINGLE-STORY HOLD-DOWN FORCE (pounds)[b]	CUMULATIVE HOLD-DOWN FORCE (pounds)[c]
		10	20	30	40	50		
		Minimum Total Length (feet) of Braced Wall Panels Required Along each Braced Wall Line						
D_0		4.0	7.0	10.5	14.0	17.5	NA	—
		4.0	7.0	10.5	14.0	17.5	1,900	—
		4.5	9.0	13.5	18.0	22.5	3,500	5,400
		6.0	12.0	18.0	24.0	30.0	3,500	8,900
D_1		4.5	9.0	13.5	18.0	22.5	2,100	—
		4.5	9.0	13.5	18.0	22.5	3,700	5,800
		6.0	12.0	18.0	24.0	30.0	3,700	9,500

(continued)

TABLE R602.10.6.5.4—continued
METHOD BV-WSP WALL BRACING REQUIREMENTS[d]

SEISMIC DESIGN CATEGORY	STORY	BRACED WALL LINE LENGTH (FEET)					SINGLE-STORY HOLD-DOWN FORCE (pounds)[b]	CUMULATIVE HOLD-DOWN FORCE (pounds)[c]
		10	20	30	40	50		
		Minimum Total Length (feet) of Braced Wall Panels Required Along each Braced Wall Line						
D_2[a]		5.5	11.0	16.5	22.0	27.5	2,300	—
		5.5	11.0	16.5	22.0	27.5	3,900	6,200
	Three-story dwelling	NP	NP	NP	NP	NP	NA	NA

For SI: 1 inch = 25.4 mm, 1 foot = 304.8 mm, 1 pound per square foot = 0.479 kPa, 1 pound-force = 4.448 N.

NP = Not Permitted.

NA = Not Applicable.

a. One- and two-family dwellings in Seismic Design Category D_2 exceeding two stories shall be designed in accordance with accepted engineering practices.

b. Hold-down force is minimum allowable stress design load for connector providing uplift tie from wall framing at end of braced wall panel at the noted story to wall framing at end of braced wall panel at the story below, or to foundation or foundation wall. Use single-story hold-down force where edges of braced wall panels do not align; a continuous load path to the foundation shall be maintained.

c. Where hold-down connectors from stories above align with stories below, use cumulative hold-down force to size middle- and bottom-story hold-down connectors.

d. Interpolation between braced wall lengths is permitted.

R602.10.8 Braced wall panel connections. *Braced wall panels* shall be connected to floor framing or foundations as follows:

1. Where joists are perpendicular to a *braced wall panel* above or below, a rim joist, band joist or blocking shall be provided along the entire length of the *braced wall panel* in accordance with Figure R602.10.8(1). Fastening of top and bottom wall plates to framing, rim joist, band joist or blocking shall be in accordance with Table R602.3(1).
2. Where joists are parallel to a *braced wall panel* above or below, a rim joist, end joist or other parallel framing member shall be provided directly above and below the *braced wall panel* in accordance with Figure R602.10.8(2). Where a parallel framing member cannot be located directly above and below the panel, full-depth blocking at 16-inch (406 mm) spacing shall be provided between the parallel framing members to each side of the *braced wall panel* in accordance with Figure R602.10.8(2). Fastening of blocking and wall plates shall be in accordance with Table R602.3(1) and Figure R602.10.8(2).
3. Connections of *braced wall panels* to concrete or masonry shall be in accordance with Section R403.1.6.

R602.10.8.1 Braced wall panel connections for Seismic Design Categories D_0, D_1 and D_2. *Braced wall panels* shall be fastened to required foundations in accordance with Section R602.11.1, and top plate lap splices shall be face-nailed with not less than eight 16d nails on each side of the splice.

R602.10.8.2 Connections to roof framing. Top plates of exterior *braced wall panels* shall be attached to rafters or roof trusses above in accordance with Table R602.3(1) and this section. Where required by this section, blocking between rafters or roof trusses shall be attached to top plates of *braced wall panels* and to rafters and roof trusses in accordance with Table R602.3(1). A continuous band, rim or header joist or roof truss parallel to the *braced wall panels* shall be permitted to replace the blocking required by this section. Blocking shall not be required over openings in continuously sheathed *braced wall lines*. In addition to the requirements of this section, lateral support shall be provided for rafters and ceiling joists in accordance with Section R802.8 and for trusses in accordance with Section R802.10.3. Roof *ventilation* shall be provided in accordance with Section R806.1.

1. For *Seismic Design Categories* A, B and C where the distance from the top of the *braced wall panel* to the top of the rafters or roof trusses above is $9^1/_4$ inches (235 mm) or less,

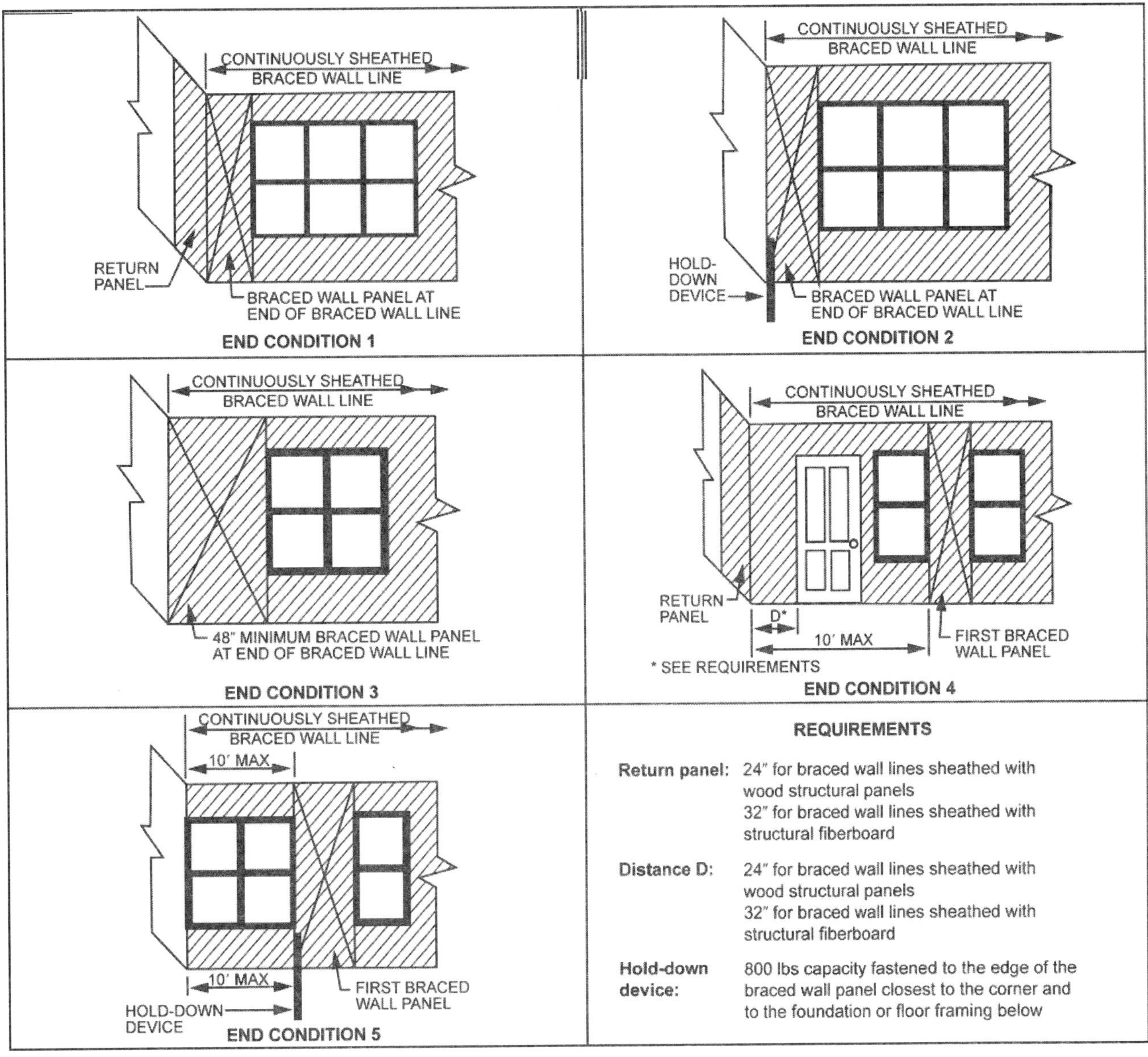

For SI: 1 inch = 25.4 mm, 1 foot = 304.8 mm, 1 pound = 4.45 N.

FIGURE R602.10.7
END CONDITIONS FOR BRACED WALL LINES WITH CONTINUOUS SHEATHING

blocking between rafters or roof trusses shall not be required. Where the distance from the top of the *braced wall panel* to the top of the rafters or roof trusses above is between $9^1/_4$ inches (235 mm) and $15^1/_4$ inches (387 mm), blocking between rafters or roof trusses shall be provided above the *braced wall panel* in accordance with Figure R602.10.8.2(1).

Exception: Where the outside edge of truss vertical web members aligns with the outside face of the wall studs below, wood structural panel sheathing extending above the top plate as shown in Figure R602.10.8.2(3) shall be permitted to be fastened to each truss web with three-8d nails ($2^1/_2$ inches × 0.131 inch) and blocking between the trusses shall not be required.

2. For *Seismic Design Categories* D_0, D_1 and D_2, where the distance from the top of the *braced wall panel* to the top of the rafters or roof trusses is $15^1/_4$ inches (387 mm) or less, blocking between rafters or roof trusses shall be provided above the *braced wall panel* in accordance with Figure R602.10.8.2(1).

3. Where the distance from the top of the *braced wall panel* to the top of rafters or roof trusses exceeds $15^1/_4$ inches (387 mm), the top plates of the *braced wall panel* shall be connected to perpendicular rafters or roof trusses above in accordance with one or more of the following methods:
   3.1. Soffit blocking panels constructed in accordance with Figure R602.10.8.2(2).
   3.2. Vertical blocking panels constructed in accordance with Figure R602.10.8.2(3).
   3.3. Blocking panels provided by the roof truss manufacturer and designed in accordance with Section R802.
   3.4. Blocking, blocking panels or other methods of lateral load transfer designed in accordance with the AWC WFCM or accepted engineering practice.

R602.10.9 Braced wall panel support. *Braced wall panel* support shall be provided as follows:

1. Cantilevered floor joists complying with Section R502.3.3 shall be permitted to support *braced wall panels*.
2. Raised floor system post or pier foundations supporting *braced wall panels* shall be designed in accordance with accepted engineering practice.
3. Masonry stem walls with a length of 48 inches (1219 mm) or less supporting *braced wall panels* shall be reinforced in accordance with Figure R602.10.9. Masonry stem walls with a length greater than 48 inches (1219 mm) supporting *braced wall panels* shall be constructed in accordance with Section R403.1 Methods ABW and PFH shall not be permitted to attach to masonry stem walls.
4. Concrete stem walls with a length of 48 inches (1219 mm) or less, greater than 12 inches (305 mm) tall and less than 6 inches (152 mm) thick shall have reinforcement sized and located in accordance with Figure R602.10.9.

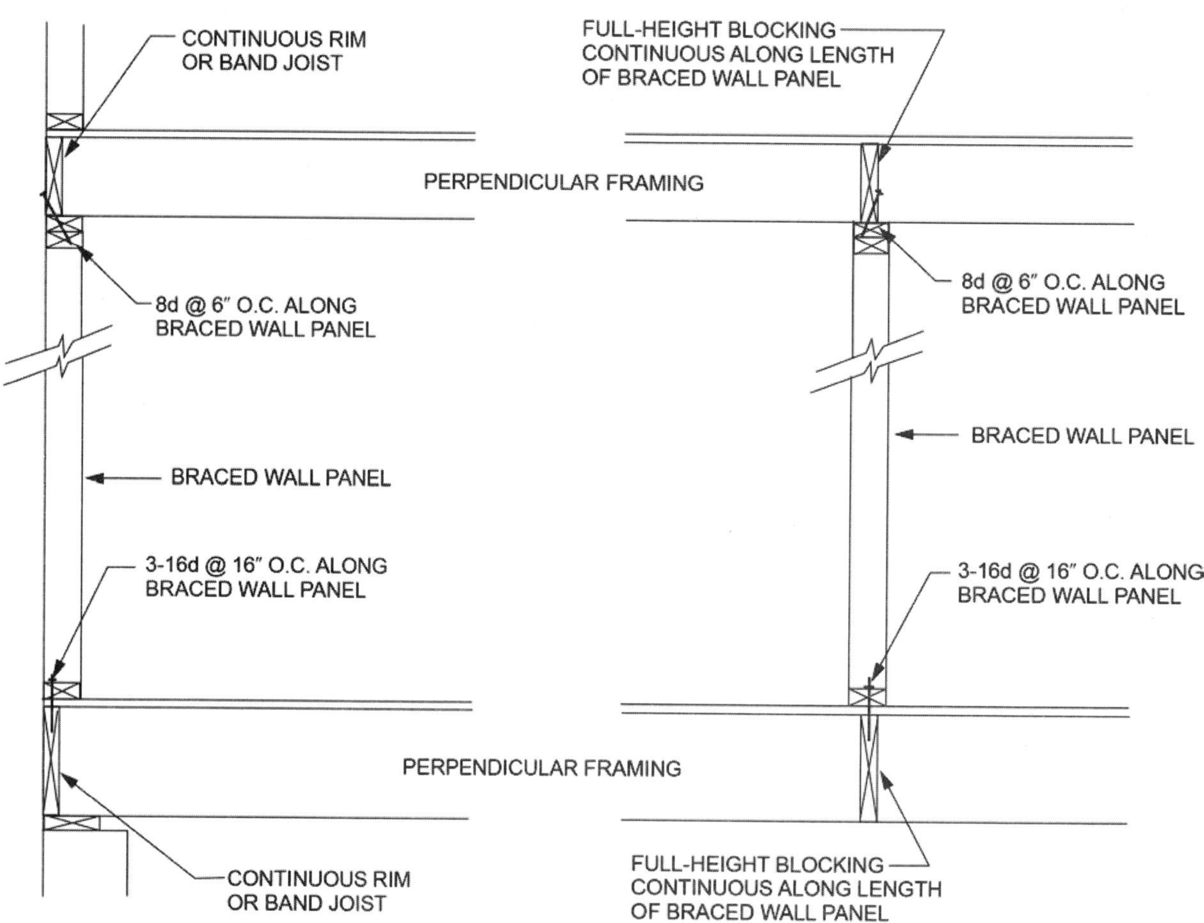

For SI: 1 inch = 25.4 mm.

FIGURE R602.10.8(1)
BRACED WALL PANEL CONNECTION WHEN PERPENDICULAR TO FLOOR/CEILING FRAMING

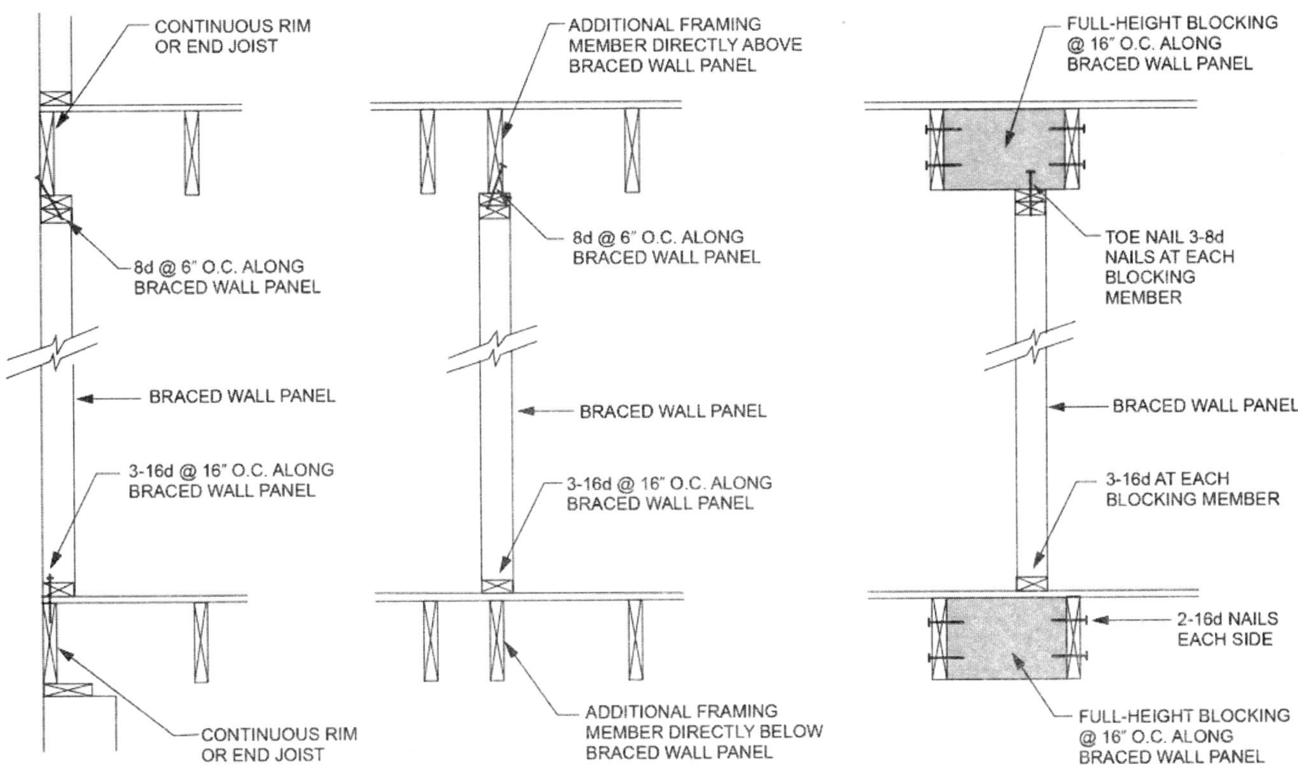

For SI: 1 inch = 25.4 mm.

FIGURE R602.10.8(2)
BRACED WALL PANEL CONNECTION WHEN PARALLEL TO FLOOR/CEILING FRAMING

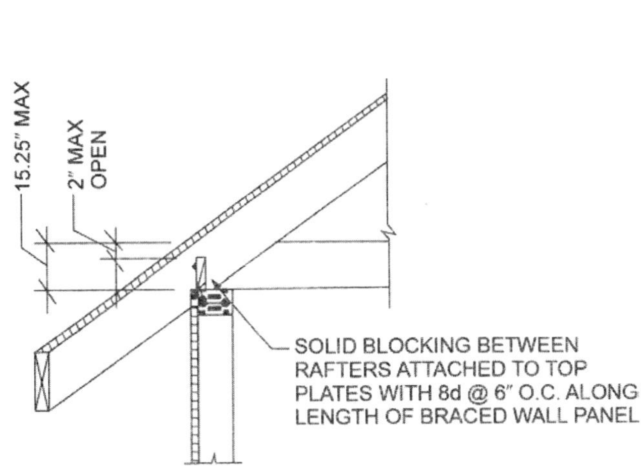

For SI: 1 inch = 25.4 mm.

FIGURE R602.10.8.2(1)
BRACED WALL PANEL CONNECTION TO PERPENDICULAR RAFTERS

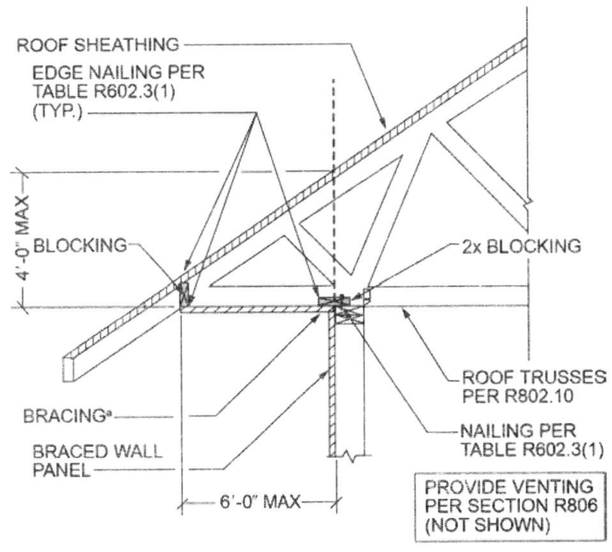

For SI: 1 inch = 25.4 mm, 1 foot = 304.8 mm.

a. Methods of bracing shall be as described in Section R602.10.4.

FIGURE R602.10.8.2(2)
BRACED WALL PANEL CONNECTION OPTION TO PERPENDICULAR RAFTERS OR ROOF TRUSSES

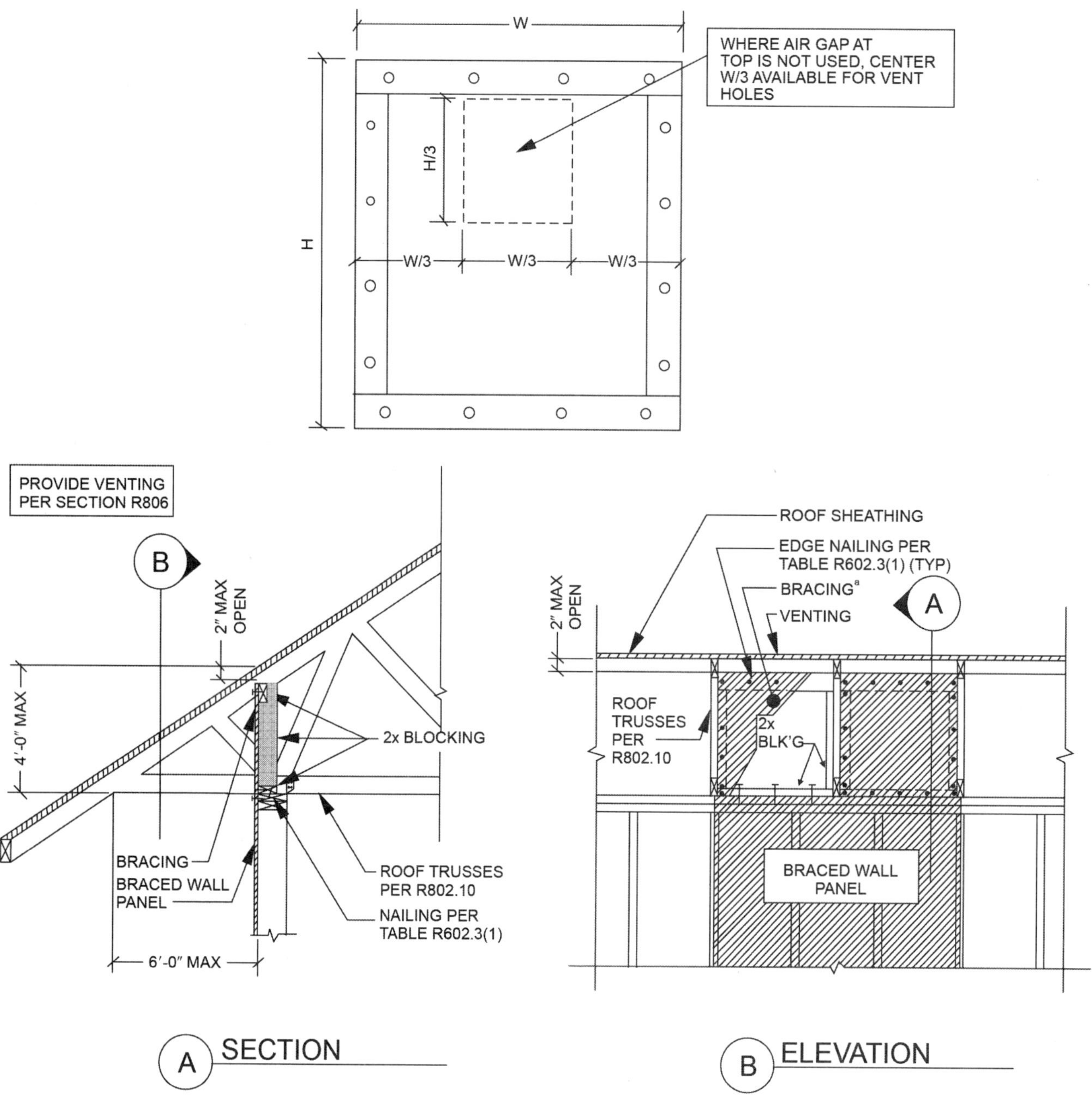

For SI: 1 inch = 25.4 mm, 1 foot =304.8 mm

a. Methods of bracing shall be as described in Section R602.10.4.

FIGURE R602.10.8.2(3)
BRACED WALL PANEL CONNECTION OPTION TO PERPENDICULAR RAFTERS OR ROOF TRUSSES

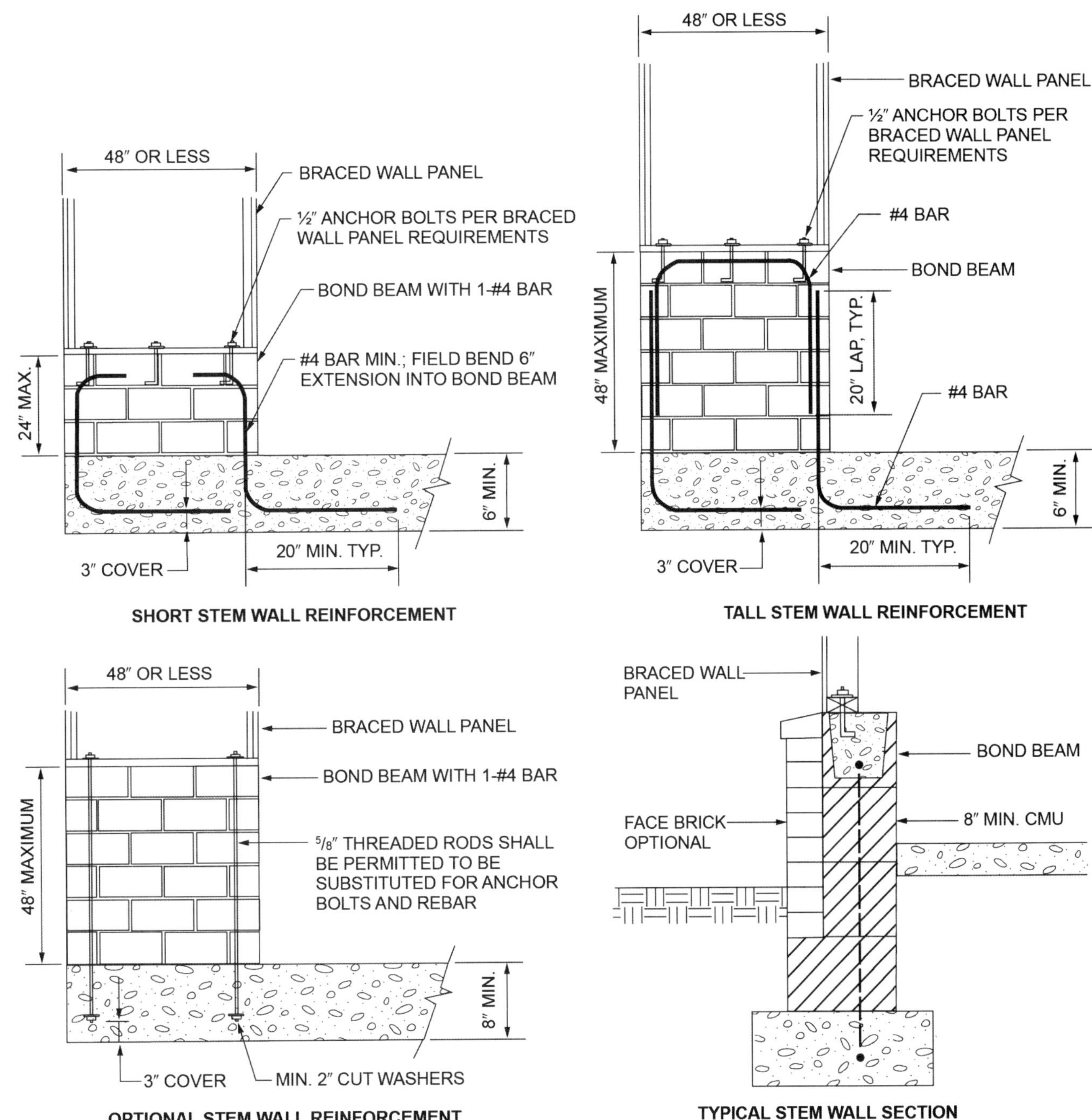

For SI: 1 inch = 25.4 mm.

FIGURE R602.10.9
MASONRY STEM WALLS SUPPORTING BRACED WALL PANELS

R602.10.9.1 Braced wall panel support for Seismic Design Categories D_0, D_1 and D_2. In *Seismic Design Categories* D_0, D_1 and D_2, *braced wall panel* footings shall be as specified in Section R403.1.2.

R602.10.10 Cripple wall bracing. Cripple walls shall be constructed in accordance with Section R602.9 and braced in accordance with this section. Cripple walls shall be braced with the length and method of bracing used for the wall above in accordance with Tables R602.10.3(1) and R602.10.3(3), and the applicable adjustment factors in Table R602.10.3(2) or R602.10.3(4), respectively, except that the length of cripple wall bracing shall be multiplied by a factor of 1.15. Where gypsum wall board is not used on the inside of the cripple wall bracing, the length adjustments for the elimination of the gypsum wallboard, or equivalent, shall be applied as directed in Tables R602.10.3(2) and R602.10.3(4) to the length of cripple wall bracing required. This adjustment shall be taken in addition to the 1.15 increase.

R602.10.10.1 Cripple wall bracing for Seismic Design Categories D_0 and D_1 and townhouses in Seismic Design Category C. In addition to the requirements in Section R602.10.10, cripple wall bracing shall be limited to methods WSP and CS-WSP, and the distance between adjacent edges of *braced wall panels* for cripple walls along a *braced wall line* shall be 14 feet (4267 mm) maximum.

Where *braced wall lines* at interior walls are not supported on a continuous foundation below, the adjacent parallel cripple walls, where provided, shall be braced with Method WSP or Method CS-WSP in accordance with Section R602.10.4. The length of bracing required in accordance with Table R602.10.3(3) for the cripple walls shall be multiplied by 1.5. Where the cripple walls do not have sufficient length to provide the required bracing, the spacing of panel edge fasteners shall be reduced to 4 inches (102 mm) on center and the required bracing length adjusted by 0.7. If the required length can still not be provided, the cripple wall shall be designed in accordance with accepted engineering practice.

R602.10.10.2 Cripple wall bracing for Seismic Design Category D_2. In *Seismic Design Category* D_2, cripple walls shall be braced in accordance with Tables R602.10.3(3) and R602.10.3(4).

R602.10.10.3 Redesignation of cripple walls. Where all cripple wall segments along a *braced wall line* do not exceed 48 inches (1219 mm) in height, the cripple walls shall be permitted to be redesignated as a first-*story* wall for purposes of determining wall bracing requirements. Where any cripple wall segment in a *braced wall line* exceeds 48 inches (1219 mm) in height, the entire cripple wall shall be counted as an additional *story*. If the cripple walls are redesignated, the stories above the redesignated *story* shall be counted as the second and third stories, respectively.

R602.11 Wall anchorage. *Braced wall line* sills shall be anchored to concrete or masonry foundations in accordance with Sections R403.1.6 and R602.11.1.

R602.11.1 Wall anchorage for all buildings in Seismic Design Categories D_0, D_1 and D_2 and townhouses in Seismic Design Category C. Plate washers, not less than 0.229 inch by 3 inches by 3 inches (5.8 mm by 76 mm by 76 mm) in size, shall be provided between the foundation sill plate and the nut except where *approved* anchor straps are used. The hole in the plate washer is permitted to be diagonally slotted with a width of up to $^3/_{16}$ inch (5 mm) larger than the bolt diameter and a slot length not to exceed $1^3/_4$ inches (44 mm), provided a standard cut washer is placed between the plate washer and the nut.

R602.11.2 Stepped foundations in Seismic Design Categories D_0, D_1 and D_2. In all buildings located in *Seismic Design Categories* D_0, D_1 or D_2, where the height of a required *braced wall line* that extends from foundation to floor above varies more than 4 feet (1219 mm), the *braced wall line* shall be constructed in accordance with the following:

1. Where the lowest floor framing rests directly on a sill bolted to a foundation not less than 8 feet (2440 mm) in length along a line of bracing, the line shall be considered as braced. The double plate of the cripple stud wall beyond the segment of footing that extends to the lowest framed floor shall be spliced by extending the upper top plate not less than 4 feet (1219 mm) along the foundation. Anchor bolts shall be located not more than 1 foot and 3 feet (305 and 914 mm) from the step in the foundation. See Figure R602.11.2.
2. Where cripple walls occur between the top of the foundation and the lowest floor framing, the bracing requirements of Sections R602.10.10, R602.10.10.1 and R602.10.10.2 shall apply.
3. Where only the bottom of the foundation is stepped and the lowest floor framing rests directly on a sill bolted to the foundations, the requirements of Sections R403.1.6 and R602.11.1 shall apply.

R602.12 Simplified wall bracing. Buildings meeting all of the following conditions shall be permitted to be braced in accordance with this section as an alternative to the requirements of Section R602.10. The entire building shall be braced in accordance with this section; the use of other bracing provisions of Section R602.10, except as specified herein, shall not be permitted.

1. There shall be not more than three stories above the top of a concrete or masonry foundation or basement wall. Permanent wood foundations shall not be permitted.
2. Floors shall not cantilever more than 24 inches (607 mm) beyond the foundation or bearing wall below.
3. Wall height shall not be greater than 10 feet (3048 mm).

4. The building shall have a roof eave-to-ridge height of 15 feet (4572 mm) or less.
5. Exterior walls shall have gypsum board with a minimum thickness of $^1/_2$ inch (12.7 mm) installed on the interior side fastened in accordance with Table R702.3.5.
6. The structure shall be located where the ultimate design wind speed is less than or equal to 130 mph (58 m/s), and the exposure category is B or C.
7. The structure shall be located in *Seismic Design Category* A, B or C for detached one- and two-family dwellings or *Seismic Design Category* A or B for townhouses.
8. Cripple walls shall not be permitted in three-story buildings.

R602.12.1 Circumscribed rectangle. The bracing required for each building shall be determined by circumscribing a rectangle around the entire building on each floor as shown in Figure R602.12.1. The rectangle shall surround all enclosed offsets and projections such as *sunrooms* and attached garages. Open structures, such as carports and decks, shall be permitted to be excluded. The rectangle shall not have a side greater than 60 feet (18 288 mm), and the ratio between the long side and short side shall be not greater than 3:1.

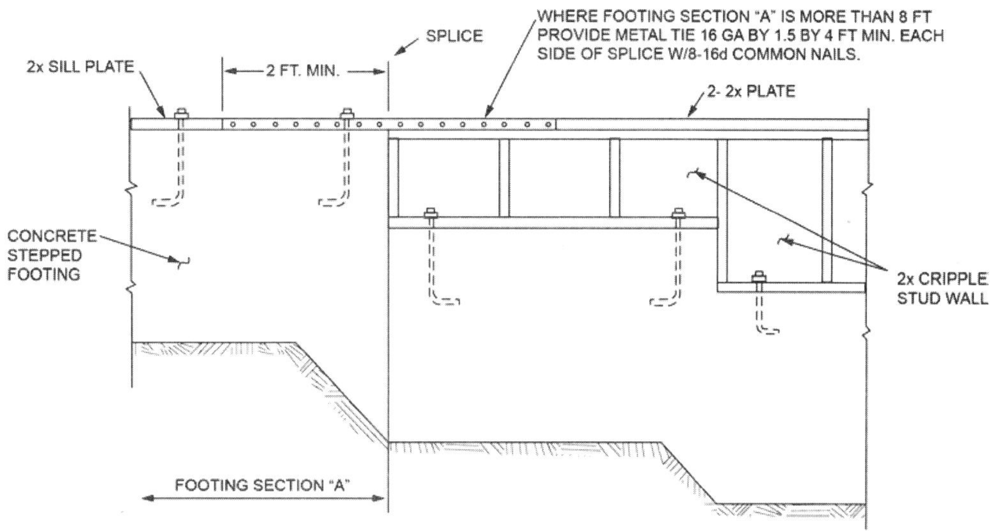

For SI: 1 foot = 304.8 mm.

Note: Where footing Section "A" is less than 8 feet long in a 25-foot-long wall, install bracing at cripple stud wall.

FIGURE R602.11.2
STEPPED FOUNDATION CONSTRUCTION

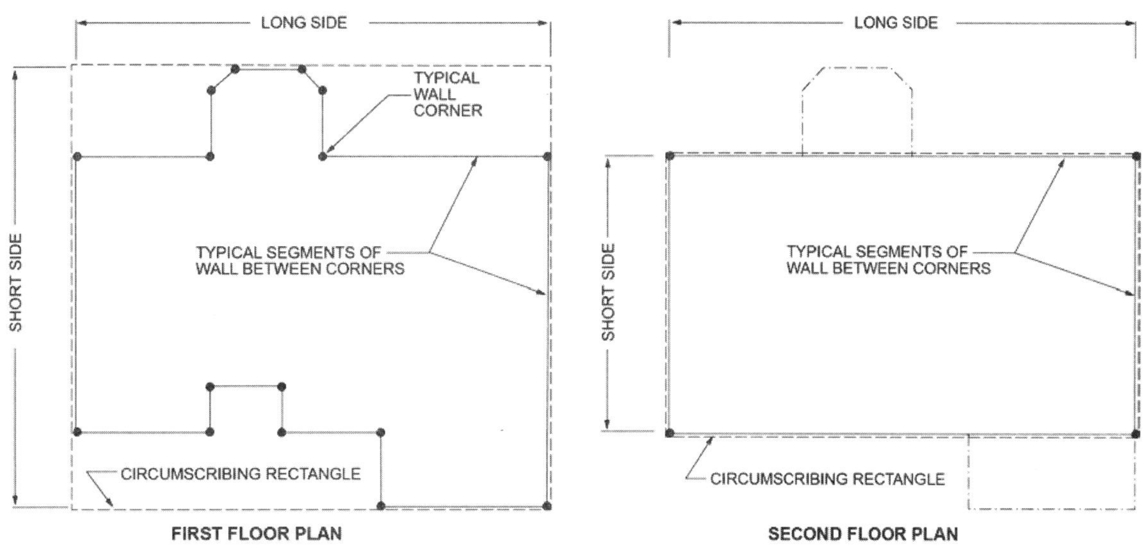

FIGURE R602.12.1
RECTANGLE CIRCUMSCRIBING AN ENCLOSED BUILDING

R602.12.2 Sheathing materials. The following sheathing materials installed on the exterior side of exterior walls shall be used to construct a bracing unit as defined in Section R602.12.3. Mixing materials is prohibited.

1. Wood structural panels with a minimum thickness of $^3/_8$ inch (9.5 mm) fastened in accordance with Table R602.3(3).
2. Structural fiberboard sheathing with a minimum thickness of $^1/_2$ inch (12.7 mm) fastened in accordance with Table R602.3(1).

R602.12.3 Bracing unit. A bracing unit shall be a full-height sheathed segment of the exterior wall without openings or vertical or horizontal offsets and a minimum length as specified herein. Interior walls shall not contribute toward the amount of required bracing. Mixing of Items 1 and 2 is prohibited on the same story.

1. Where all framed portions of all exterior walls are sheathed in accordance with Section R602.12.2, including wall areas between bracing units, above and below openings and on gable end walls, the minimum length of a bracing unit shall be 3 feet (914 mm).
2. Where the exterior walls are braced with sheathing panels in accordance with Section R602.12.2 and areas between bracing units are covered with other materials, the minimum length of a bracing unit shall be 4 feet (1219 mm).

R602.12.3.1 Multiple bracing units. Segments of wall compliant with Section R602.12.3 and longer than the minimum bracing unit length shall be considered as multiple bracing units. The number of bracing units shall be determined by dividing the wall segment length by the minimum bracing unit length. Full-height sheathed segments of wall narrower than the minimum bracing unit length shall not contribute toward a bracing unit except as specified in Section R602.12.6.

R602.12.4 Number of bracing units. Each side of the circumscribed rectangle, as shown in Figure R602.12.1, shall have, at a minimum, the number of bracing units in accordance with Table R602.12.4 placed on the parallel exterior walls facing the side of the rectangle. Bracing units shall then be placed using the distribution requirements specified in Section R602.12.5.

R602.12.5 Distribution of bracing units. The placement of bracing units on exterior walls shall meet all of the following requirements as shown in Figure R602.12.5.

1. A bracing unit shall begin not more than 12 feet (3658 mm) from any wall corner.
2. The distance between adjacent edges of bracing units shall be not greater than 20 feet (6096 mm).
3. Segments of wall greater than 8 feet (2438 mm) in length shall have not less than one bracing unit.

R602.12.6 Narrow panels. The bracing methods referenced in Section R602.10 and specified in Sections R602.12.6.1 through R602.12.6.3 shall be permitted where using simplified wall bracing.

R602.12.6.1 Method CS-G. *Braced wall panels* constructed as Method CS-G in accordance with Tables R602.10.4 and R602.10.5 shall be permitted for one-story garages where all framed portions of all exterior walls are sheathed with *wood structural panels*. Each CS-G panel shall be equivalent to 0.5 of a bracing unit. Segments of wall that include a Method CS-G panel shall meet the requirements of Section R602.10.4.2.

R602.12.6.2 Method CS-PF. *Braced wall panels* constructed as Method CS-PF in accordance with Section R602.10.6.4 shall be permitted where all framed portions of all exterior walls are sheathed with *wood structural panels*. Each CS-PF panel shall equal 0.75 bracing units. Segments of wall that include a Method CS-PF panel shall meet the requirements of Section R602.10.4.2.

R602.12.6.3 Methods ABW, PFH and PFG. *Braced wall panels* constructed as Method ABW, PFH and PFG shall be permitted where bracing units are constructed using *wood structural panels* applied either continuously or intermittently. Each ABW and PFH panel shall equal one bracing unit and each PFG panel shall be equal to 0.75 bracing unit.

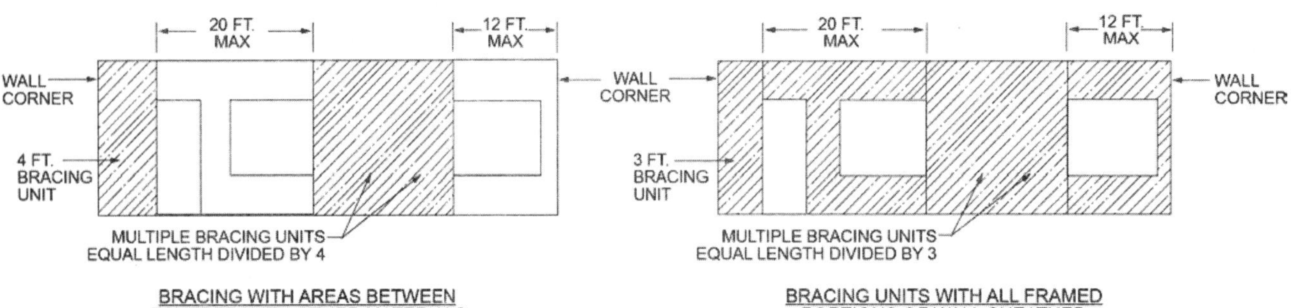

For SI: 1 foot = 304.8 mm.

FIGURE R602.12.5
BRACING UNIT DISTRIBUTION

TABLE R602.12.4
MINIMUM NUMBER OF BRACING UNITS ON EACH SIDE OF THE CIRCUMSCRIBED RECTANGLE

ULTIMATE DESIGN WIND SPEED (mph)	STORY LEVEL	EAVE-TO-RIDGE HEIGHT (feet)	MINIMUM NUMBER OF BRACING UNITS ON EACH LONG SIDE[a, b, d]						MINIMUM NUMBER OF BRACING UNITS ON EACH SHORT SIDE[a, b, d]					
			Length of short side (feet)[c]						Length of long side (feet)[c]					
			10	20	30	40	50	60	10	20	30	40	50	60
115		10	1	2	2	2	3	3	1	2	2	2	3	3
			2	3	3	4	5	6	2	3	3	4	5	6
			2	3	4	6	7	8	2	3	4	6	7	8
		15	1	2	3	3	4	4	1	2	3	3	4	4
			2	3	4	5	6	7	2	3	4	5	6	7
			2	4	5	6	7	9	2	4	5	6	7	9
130		10	1	2	2	3	3	4	1	2	2	3	3	4
			2	3	4	5	6	7	2	3	4	5	6	7
			2	4	5	7	8	10	2	4	5	7	8	10
		15	2	3	3	4	4	6	2	3	3	4	4	6
			3	4	6	7	8	10	3	4	6	7	8	10
			3	6	7	10	11	13	3	6	7	10	11	13

For SI: 1 inch = 25.4 mm, 1 foot = 304.8 mm, 1 mile per hour = 0.447m/s.

a. Interpolation shall not be permitted.

b. Cripple walls or wood-framed basement walls in a walk-out condition shall be designated as the first story and the stories above shall be redesignated as the second and third stories, respectively, and shall be prohibited in a three-story structure.

c. Actual lengths of the sides of the circumscribed rectangle shall be rounded to the next highest unit of 10 when using this table.

d. For Exposure Category C, multiply bracing units by a factor of 1.20 for a one-story building, 1.30 for a two-story building and 1.40 for a three-story building.

R602.12.7 Lateral support. For bracing units located along the eaves, the vertical distance from the outside edge of the top wall plate to the roof sheathing above shall not exceed 9.25 inches (235 mm) at the location of a bracing unit unless lateral support is provided in accordance with Section R602.10.8.2.

R602.12.8 Stem walls. Masonry stem walls with a height and length of 48 inches (1219 mm) or less supporting a bracing unit or a Method CS-G, CS-PF or PFG *braced wall panel* shall be constructed in accordance with Figure R602.10.9. Concrete stem walls with a length of 48 inches (1219 mm) or less, greater than 12 inches (305 mm) tall and less than 6 inches (152 mm) thick shall be reinforced sized and located in accordance with Figure R602.10.9.

SECTION R603
COLD-FORMED STEEL WALL FRAMING

R603.1 General. Elements shall be straight and free of any defects that would significantly affect structural performance. Cold-formed steel wall framing members shall be in accordance with the requirements of this section.

R603.1.1 Applicability limits. The provisions of this section shall control the construction of exterior cold-formed steel wall framing and interior load-bearing cold-formed steel wall framing for buildings not more than 60 feet (18 288 mm) long perpendicular to the joist or truss span, not more than 40 feet (12 192 mm) wide parallel to the joist or truss span, and less than or equal to three stories above *grade plane*. Exterior walls installed in accordance with the provisions of this section shall be considered as *load-bearing walls*. Cold-formed steel walls constructed in accordance with the provisions of this section shall be limited to sites where the ultimate design wind speed is less than 140 miles per hour (63 m/s), Exposure Category B or C, and the ground snow load is less than or equal to 70 pounds per square foot (3.35 kPa).

R603.1.1.1 Alternate applications. Cold-formed steel wall framing for buildings exceeding the applicability limits of Section R603.1.1 are permitted to be designed and constructed in accordance with AISI S230, subject to the limits therein.

R603.1.2 In-line framing. Load-bearing cold-formed steel studs constructed in accordance with Section R603 shall be located in-line with joists, trusses and rafters in accordance with the tolerances specified in AISI S240, Section B1.2.3.

R603.2 Structural framing. Load-bearing cold-formed steel wall framing members shall be in accordance with this section.

R603.2.1 Material. Load-bearing cold-formed steel framing members shall be cold formed to shape from structural-quality sheet steel complying with the requirements of AISI 240, Section A3.

R603.2.2 Corrosion protection. Load-bearing cold-formed steel framing shall have a protective coating complying with AISI S240, Section A4.

R603.2.3 Dimension, thickness and material grade. Load-bearing cold-formed steel wall framing members shall comply with the dimensional and thickness requirements specified in AISI S230, Section A4.3 and material grade requirements as specified in AISI S230, Section A4.4.

R603.2.4 Identification. Load-bearing cold-formed steel framing members shall meet the product identification requirements of AISI S240, Section A5.5.

R603.2.5 Fastening. Screws for steel-to-steel connections shall be installed with a minimum edge distance and center-to-center spacing of $^1/_2$ inch (12.7 mm), shall be self-drilling tapping and shall conform to ASTM C1513. Structural sheathing shall be attached to cold-formed steel studs with minimum No. 8 self-drilling tapping screws that conform to ASTM C1513. Screws for attaching structural sheathing to cold-formed steel wall framing shall have a minimum head diameter of 0.292 inch (7.4 mm) with countersunk heads and shall be installed with a minimum edge distance of $^3/_8$ inch (9.5 mm). Gypsum board shall be attached to cold-formed steel wall framing with minimum No. 6 screws conforming to ASTM C954 or ASTM C1513 with a bugle-head style and shall be installed in accordance with Section R702. For connections, screws shall extend through the steel not fewer than three exposed threads. Fasteners shall have rust-inhibitive coating suitable for the installation in which they are being used, or be manufactured from material not susceptible to corrosion.

R603.2.6 Web holes, web hole reinforcing and web hole patching. Web holes in wall studs shall comply with the conditions as prescribed in AISI S230, Section A4.5. Web holes not in conformance to the conditions as prescribed in AISI S230, Section A4.5 shall be reinforced in accordance with the provisions of AISI S230, Section A4.6 or patched in accordance with the provisions of AISI S230, Section A4.7.

R603.3 Wall construction. Exterior cold-formed steel framed walls and interior load-bearing cold-formed steel framed walls shall be constructed in accordance with the provisions of this section.

R603.3.1 Wall to foundation or floor connection. Cold-formed steel framed walls shall be anchored to foundations or floors in accordance with Table R603.3.1 and Figure R603.3.1(1), R603.3.1(2), R603.3.1(3) or R603.3.1(4). Anchor bolts shall be located not more than 12 inches (305 mm) from corners or the termination of bottom tracks. Anchor bolts shall extend not less than 15 inches (381 mm) into masonry or 7 inches (178 mm) into concrete. Foundation anchor straps shall be permitted, in lieu of anchor bolts, if spaced as required to provide equivalent anchorage to the required anchor bolts and installed in accordance with manufacturer's requirements.

TABLE R603.3.1 WALL TO FOUNDATION OR FLOOR CONNECTION REQUIREMENTS[a, b]

FRAMING CONDITION			ULTIMATE WIND SPEED AND EXPOSURE CATEGORY (mph)					
			115 B	120 B	130 B or 115 C	< 140 B or 120 C	130 C	< 140 C
Wall bottom track to floor per Figure R603.3.1(1)			1-No. 8 screw at 12″ o.c.	1-No. 8 screw at 8″ o.c.	2-No. 8 screws at 8″ o.c.	2-No. 8 screws at 6″ o.c.	3-No. 8 screws at 8″ o.c.	3-No. 8 screws at 6″ o.c.
Wall bottom track to foundation per Figure R603.3.1(2)[d]			$^{1}/_{2}$″ minimum diameter anchor bolt at 6′ o.c.	$^{1}/_{2}$″ minimum diameter anchor bolt at 6′ o.c.	$^{1}/_{2}$″ minimum diameter anchor bolt at 4′ o.c.	$^{1}/_{2}$″ minimum diameter anchor bolt at 4′ o.c.	$^{1}/_{2}$″ minimum diameter anchor bolt at 3′-4″ o.c.	$^{1}/_{2}$″ minimum diameter anchor bolt at 2′-8″ o.c.
Wall bottom track to wood sill per Figure R603.3.1(3)			Steel plate spaced at 4′ o.c., with 4-No. 8 screws and 4-10d or 6-8d common nails	Steel plate spaced at 4′ o.c., with 4-No. 8 screws and 4-10d or 6-8d common nails	Steel plate spaced at 3′ o.c., with 4-No. 8 screws and 4-10d or 6-8d common nails	Steel plate spaced at 3′ o.c., with 4-No. 8 screws and 4-10d or 6-8d common nails	Steel plate spaced at 2′ o.c., with 4-No. 8 screws and 4-10d or 6-8d common nails	Steel plate spaced at 1′-4″ o.c., with 4-No. 8 screws and 4-10d or 6-8d common nails
Wind uplift connector strength (lb)[c, e]	Stud Spacing (inches)	Roof Span (feet)						
	16	24	NR	NR	NR	NR	NR	NR
		28	NR	NR	NR	NR	NR	339
		32	NR	NR	NR	NR	NR	382
		36	NR	NR	NR	NR	333	426
		40	NR	NR	NR	NR	368	470
	24	24	NR	NR	NR	NR	343	443
		28	NR	NR	NR	NR	395	508
		32	NR	NR	NR	330	447	573
		36	NR	NR	NR	371	500	639
		40	NR	NR	345	411	552	704

For SI: 1 inch = 25.4 mm, 1 mile per hour = 0.447 m/s, 1 foot = 304.8 mm, 1 pound = 4.45 N.

a. Anchor bolts are to be located not more than 12 inches from corners or the termination of bottom tracks, such as at door openings or corners. Bolts are to extend not less than 15 inches into masonry or 7 inches into concrete.

b. All screw sizes shown are minimum.

c. NR = Uplift connector not required.

d. Foundation anchor straps are permitted in place of anchor bolts, if spaced as required to provide equivalent anchorage to the required anchor bolts and installed in accordance with manufacturer's requirements.

e. See Figure R603.3.1(4) for details.

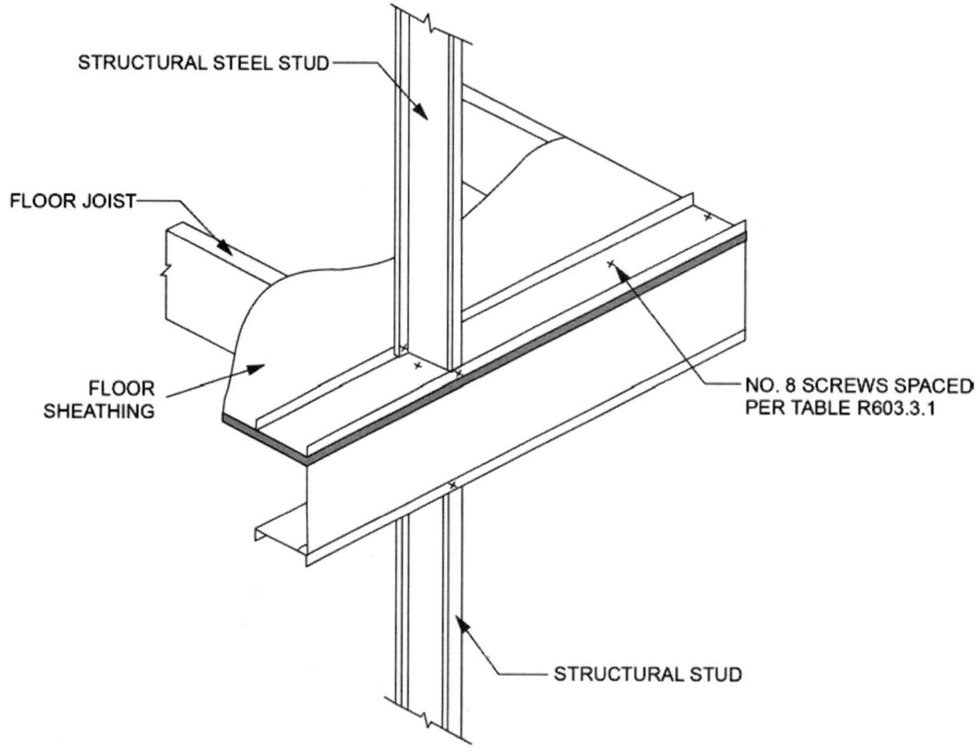

FIGURE R603.3.1(1)
WALL TO FLOOR CONNECTION

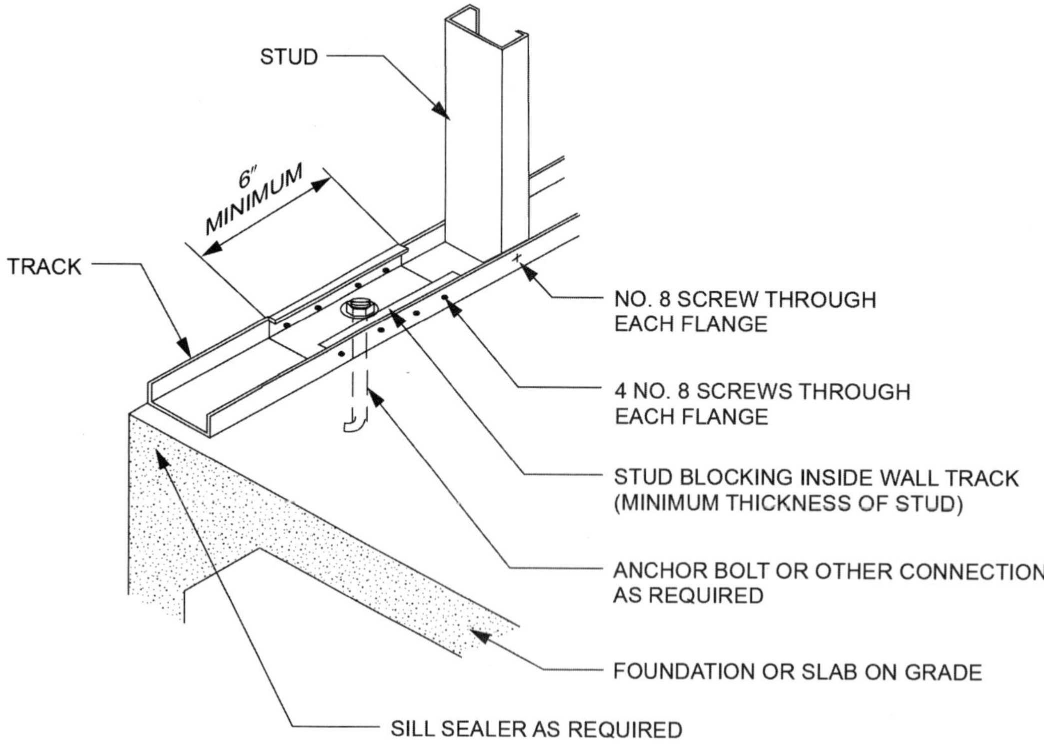

For SI: 1 inch = 25.4 mm.

FIGURE R603.3.1(2)
WALL TO FOUNDATION CONNECTION

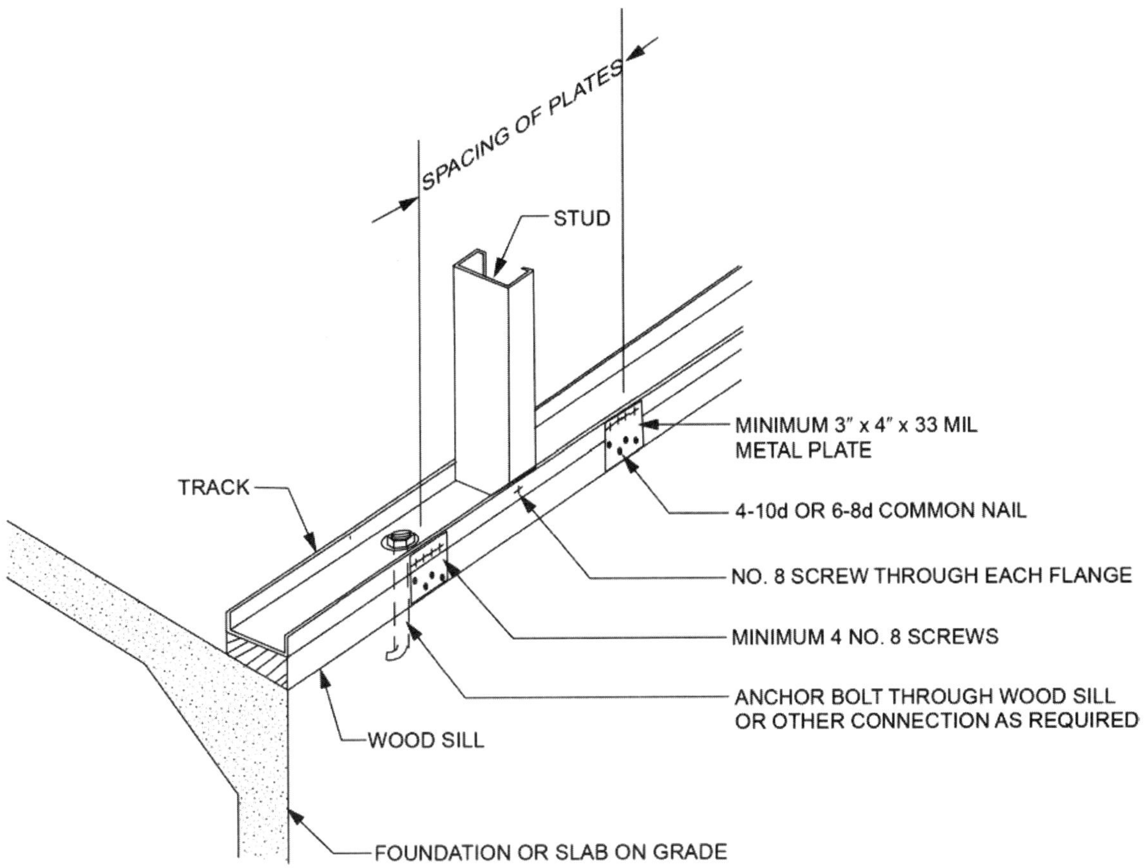

For SI: 1 mil = 0.0254 mm, 1 inch = 25.4 mm.

FIGURE R603.3.1(3)
WALL TO WOOD SILL CONNECTION

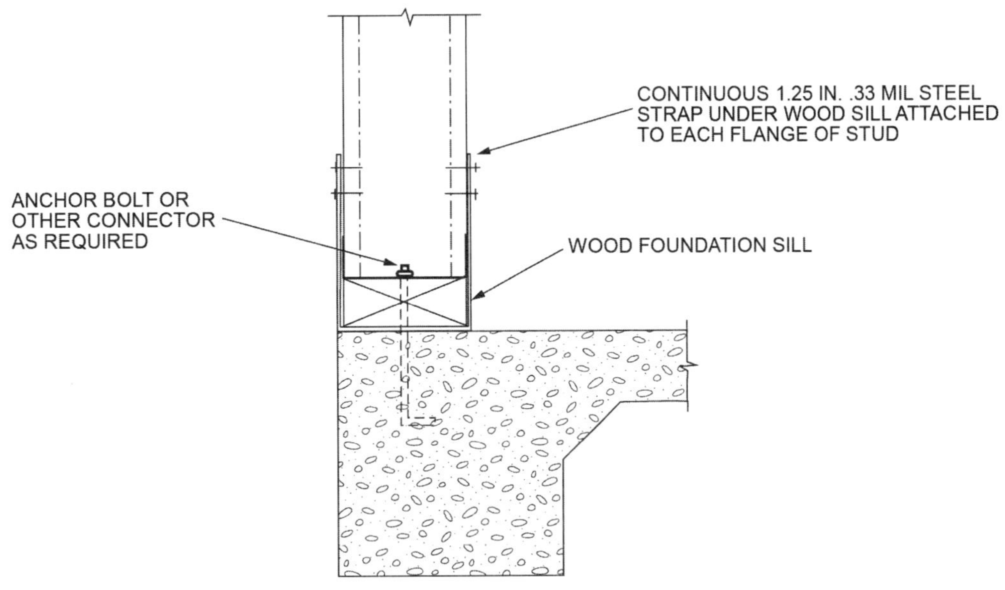

For SI: 1 mil = 0.0254 mm, 1 inch = 25.4 mm.

FIGURE R603.3.1(4)
WIND UPLIFT CONNECTOR

R603.3.1.1 Gable endwalls. Gable endwalls with heights greater than 10 feet (3048 mm) shall be anchored to foundations or floors in accordance with Table R603.3.1.1(1) or R603.3.1.1(2).

R603.3.2 Minimum stud sizes. Cold-formed steel walls shall be constructed in accordance with Figure R603.3.1(1), R603.3.1(2) or R603.3.1(3), as applicable. Exterior wall stud size and thickness shall be determined in accordance with the limits set forth in Tables R603.3.2(2) through R603.3.2(16). Interior *load-bearing wall* stud size and thickness shall be determined in accordance with the limits set forth in Tables R603.3.2(2) through R603.3.2(16) based on an ultimate design wind speed of 115 miles per hour (51 m/s), Exposure Category B, and the building width, stud spacing and ground snow load, as appropriate. Fastening requirements shall be in accordance with Section R603.2.5 and Table R603.3.2(1). Top and bottom tracks shall have the same minimum thickness as the wall studs.

Exterior wall studs shall be permitted to be reduced to the next thinner size, as shown in Tables R603.3.2(2) through R603.3.2(16), but not less than 33 mils (0.84 mm), where both of the following conditions exist:

1. Minimum of 1/2-inch (12.7 mm) gypsum board is installed and fastened on the interior surface in accordance with Section R702.
2. Wood structural sheathing panels of minimum 7/16-inch-thick (11.1 mm) oriented strand board or 15/32-inch-thick (12 mm) plywood are installed and fastened in accordance with Section R603.9.1 and Table R603.3.2(1) on the outside surface.

Interior load-bearing walls shall be permitted to be reduced to the next thinner size, as shown in Tables R603.3.2(2) through R603.3.2(16), but not less than 33 mils (0.84 mm), where not less than 1/2-inch (12.7 mm) gypsum board is installed and fastened in accordance with Section R702 on both sides of the wall. The tabulated stud thickness for *load-bearing walls* shall be used where the attic load is 10 pounds per square foot (480 Pa) or less. A limited attic storage load of 20 pounds per square foot (960 Pa) shall be permitted provided that the next higher snow load column is used to select the stud size from Tables R603.3.2(2) through R603.3.2(16).

TABLE R603.3.1.1(1)
GABLE ENDWALL TO FLOOR CONNECTION REQUIREMENTS[a, b, c]

ULTIMATE WIND SPEED (mph)		WALL BOTTOM TRACK TO FLOOR JOIST OR TRACK CONNECTION		
Exposure Category		Stud height, *h* (feet)		
B	C	10 < *h* ≤ 14	14 < *h* ≤ 18	18 < *h* ≤ 22
115	—	1-No. 8 screw @ 12″ o.c.	1-No. 8 screw @ 12″ o.c.	1-No. 8 screw @ 12″ o.c.
120	—	1-No. 8 screw @ 12″ o.c.	1-No. 8 screw @ 12″ o.c.	1-No. 8 screw @ 12″ o.c.
130	115	1-No. 8 screw @ 12″ o.c.	1-No. 8 screw @ 12″ o.c.	2-No. 8 screws @ 12″ o.c.
< 140	120	1-No. 8 screw @ 12″ o.c.	1-No. 8 screw @ 12″ o.c.	2-No. 8 screws @ 12″ o.c.
—	130	2-No. 8 screws @ 12″ o.c.	1-No. 8 screw @ 8″ o.c.	2-No. 8 screws @ 8″ o.c.
—	< 140	2-No. 8 screws @ 12″ o.c.	1-No. 8 screw @ 8″ o.c.	2-No. 8 screws @ 8″ o.c.

For SI: 1 inch = 25.4 mm, 1 mile per hour = 0.447 m/s, 1 foot = 304.8 mm.

a. Refer to Table R603.3.1.1(2) for gable endwall bottom track to foundation connections.

b. Where attachment is not given, special design is required.

c. Stud height, *h*, is measured from wall bottom track to wall top track or brace connection height.

TABLE R603.3.1.1(2)
GABLE ENDWALL BOTTOM TRACK TO FOUNDATION CONNECTION REQUIREMENTS[a, b, c]

ULTIMATE WIND SPEED (mph)		MINIMUM SPACING FOR 1/2-INCH-DIAMETER ANCHOR BOLTS[d]		
Exposure Category		Stud height, *h* (feet)		
B	C	10 < *h* ≤ 14	14 < *h* ≤ 18	18 < *h* ≤ 22
115	—	6′- 0″ o.c.	6′- 0″ o.c.	6′- 0″ o.c.
120	—	6′- 0″ o.c.	5′- 7″ o.c.	6′- 0″ o.c.
130	115	5′- 0″ o.c.	6′- 0″ o.c.	6′- 0″ o.c.
< 140	120	6′- 0″ o.c.	5′- 6″ o.c.	6′- 0″ o.c.
—	130	5′- 3″ o.c.	6′- 0″ o.c.	6′- 0″ o.c.
—	< 140	3′- 0″ o.c.	3′- 0″ o.c.	3′- 0″ o.c.

For SI: 1 inch = 25.4 mm, 1 mile per hour = 0.447 m/s, 1 foot = 304.8 mm.

a. Refer to Table R603.3.1.1(1) for gable endwall bottom track to floor joist or track connection connections.

b. Where attachment is not given, special design is required.

c. Stud height, *h*, is measured from wall bottom track to wall top track or brace connection height.

d. Foundation anchor straps are permitted in place of anchor bolts if spaced as required to provide equivalent anchorage to the required anchor bolts and installed in accordance with manufacturer's requirements.

For two-story buildings, the tabulated stud thickness for walls supporting one floor, roof and ceiling shall be used where the second-floor *live load* is 30 pounds per square foot (1440 Pa). Second-floor *live loads* of 40 psf (1920 Pa) shall be permitted provided that the next higher snow load column is used to select the stud size from Tables R603.3.2(2) through R603.3.2(11).

For three-story buildings, the tabulated stud thickness for walls supporting one or two floors, roof and ceiling shall be used where the third-floor *live load* is 30 pounds per square foot (1440 Pa). Third-floor *live loads* of 40 pounds per square foot (1920 Pa) shall be permitted provided that the next higher snow load column is used to select the stud size from Tables R603.3.2(12) through R603.3.2(16).

TABLE R603.3.2(1)
WALL FASTENING SCHEDULE[a]

DESCRIPTION OF BUILDING ELEMENT	NUMBER AND SIZE OF FASTENERS[a]	SPACING OF FASTENERS
Wall stud to top or bottom track	2-No. 8 screws	Each end of stud, one per flange
Structural sheathing to wall studs	No. 8 screws[b]	6″ o.c. on edges and 12″ o.c. at intermediate supports
$^1/_2$″ gypsum board to framing	No. 6 screws	12″ o.c.

For SI: 1 inch = 25.4 mm.

a. All screw sizes shown are minimum.

b. Screws for attachment of structural sheathing panels are to be bugle-head, flat-head, or similar head styles with a minimum head diameter of 0.29 inch.

TABLE R603.3.2(2)
24-FOOT-WIDE BUILDING SUPPORTING ROOF AND CEILING ONLY[a, b, c, d]

ULTIMATE WIND SPEED AND EXPOSURE CATEGORY (mph)		MEMBER SIZE	STUD SPACING (inches)	MINIMUM STUD THICKNESS (mils)											
				8-foot Studs				9-foot Studs				10-foot Studs			
				Ground Snow Load (psf)											
Exp. B	Exp. C			20	30	50	70	20	30	50	70	20	30	50	70
115	—	350S162	16	33	33	33	33	33	33	33	33	33	33	33	33
			24	33	33	33	43	33	33	33	43	33	33	43	43
		550S162	16	33	33	33	33	33	33	33	33	33	33	33	33
			24	33	33	33	33	33	33	33	33	33	33	33	43
120	—	350S162	16	33	33	33	33	33	33	33	33	33	33	33	33
			24	33	33	33	43	33	33	33	43	43	43	43	43
		550S162	16	33	33	33	33	33	33	33	33	33	33	33	33
			24	33	33	33	43	33	33	33	33	33	33	33	43
130	115	350S162	16	33	33	33	33	33	33	33	33	33	33	33	33
			24	33	33	43	43	43	43	43	43	43	43	43	54
		550S162	16	33	33	33	33	33	33	33	33	33	33	33	33
			24	33	33	33	43	33	33	33	43	33	33	33	43
< 140	120	350S162	16	33	33	33	33	33	33	33	33	33	33	33	43
			24	33	33	43	43	43	43	43	43	54	54	54	54
		550S162	16	33	33	33	33	33	33	33	33	33	33	33	33
			24	33	33	33	43	33	33	33	43	43	43	43	43
—	130	350S162	16	33	33	33	33	33	33	33	33	43	43	43	43
			24	43	43	43	43	54	54	54	54	54	54	54	54
		550S162	16	33	33	33	33	33	33	33	33	33	33	33	33
			24	33	33	33	43	43	43	43	43	43	43	43	43
—	< 140	350S162	16	33	33	33	33	43	43	43	43	43	43	43	43
			24	43	43	43	54	54	54	54	54	54	54	54	54
		550S162	16	33	33	33	33	33	33	33	33	33	33	33	33
			24	43	43	43	43	43	43	43	43	43	43	43	43

For SI: 1 inch = 25.4 mm, 1 foot = 304.8 mm, 1 mil = 0.0254 mm, 1 mile per hour = 0.447 m/s, 1 pound per square foot = 0.0479 kPa, 1 ksi = 1,000 psi = 6.895 MPa.

a. Deflection criterion: *L*/240.

b. Design load assumptions:
  Second-floor dead load is 10 psf.
  Second-floor live load is 30 psf.
  Roof/ceiling dead load is 12 psf.
  Attic live load is 10 psf.

c. Building width is in the direction of horizontal framing members supported by the wall studs.

d. Minimum Grade 33 ksi steel shall be used for 33 mil and 43 mil thicknesses. Minimum Grade 50 ksi steel shall be used for 54 and 68 mil thicknesses.

TABLE R603.3.2(3)
28-FOOT-WIDE BUILDING SUPPORTING ROOF AND CEILING ONLY[a, b, c, d]

ULTIMATE WIND SPEED AND EXPOSURE CATEGORY (mph)		MEMBER SIZE	STUD SPACING (inches)	MINIMUM STUD THICKNESS (mils)											
				8-foot Studs				9-foot Studs				10-foot Studs			
				Ground Snow Load (psf)											
Exp. B	Exp. C			20	30	50	70	20	30	50	70	20	30	50	70
115	—	350S162	16	33	33	33	33	33	33	33	33	33	33	33	33
			24	33	33	43	43	33	33	43	43	33	33	43	54
		550S162	16	33	33	33	33	33	33	33	33	33	33	33	33
			24	33	33	33	43	33	33	33	43	33	33	33	43
120	—	350S162	16	33	33	33	33	33	33	33	33	33	33	33	33
			24	33	33	43	43	33	33	43	43	43	43	43	54
		550S162	16	33	33	33	33	33	33	33	33	33	33	33	33
			24	33	33	33	43	33	33	33	43	33	33	33	43
130	115	350S162	16	33	33	33	33	33	33	33	33	33	33	33	43
			24	33	33	43	54	43	43	43	54	43	43	43	54
		550S162	16	33	33	33	33	33	33	33	33	33	33	33	33
			24	33	33	33	43	33	33	33	43	33	33	33	43
< 140	120	350S162	16	33	33	33	33	33	33	33	33	33	33	33	43
			24	33	33	43	54	43	43	43	54	54	54	54	54
		550S162	16	33	33	33	33	33	33	33	33	33	33	33	33
			24	33	33	33	43	33	33	33	43	43	43	43	43
—	130	350S162	16	33	33	33	33	33	33	33	43	43	43	43	43
			24	43	43	43	54	54	54	54	54	54	54	54	54
		550S162	16	33	33	33	33	33	33	33	33	33	33	33	33
			24	33	33	33	43	43	43	43	43	43	43	43	43
—	< 140	350S162	16	33	33	33	43	43	43	43	43	43	43	43	43
			24	43	43	43	54	54	54	54	54	54	54	54	54
		550S162	16	33	33	33	33	33	33	33	33	33	33	33	33
			24	43	43	43	43	43	43	43	43	43	43	43	43

For SI: 1 inch = 25.4 mm, 1 foot = 304.8 mm, 1 mil = 0.0254 mm, 1 mile per hour = 0.447 m/s, 1 pound per square foot = 0.0479 kPa, 1 ksi = 1,000 psi = 6.895 MPa.

a. Deflection criterion: *L*/240.

b. Design load assumptions:
Second-floor dead load is 10 psf.
Second-floor live load is 30 psf.
Roof/ceiling dead load is 12 psf.
Attic live load is 10 psf.

c. Building width is in the direction of horizontal framing members supported by the wall studs.

d. Minimum Grade 33 ksi steel shall be used for 33 mil and 43 mil thicknesses. Minimum Grade 50 ksi steel shall be used for 54 and 68 mil thicknesses.

TABLE R603.3.2(4)
32-FOOT-WIDE BUILDING SUPPORTING ROOF AND CEILING ONLY[a, b, c, d]

ULTIMATE WIND SPEED AND EXPOSURE CATEGORY (mph)		MEMBER SIZE	STUD SPACING (inches)	MINIMUM STUD THICKNESS (mils)											
				8-foot Studs				9-foot Studs				10-foot Studs			
				Ground Snow Load (psf)											
Exp. B	Exp. C			20	30	50	70	20	30	50	70	20	30	50	70
115	—	350S162	16	33	33	33	33	33	33	33	33	33	33	33	43
			24	33	33	43	54	33	33	43	54	43	43	43	54
		550S162	16	33	33	33	33	33	33	33	33	33	33	33	33
			24	33	33	33	43	33	33	33	43	33	33	33	43
120	—	350S162	16	33	33	33	33	33	33	33	33	33	33	33	43
			24	33	33	43	54	33	33	43	54	43	43	43	54
		550S162	16	33	33	33	33	33	33	33	33	33	33	33	33
			24	33	33	33	43	33	33	33	43	33	33	43	43
130	115	350S162	16	33	33	33	43	33	33	33	43	33	33	33	43
			24	33	33	43	54	43	43	43	54	43	43	54	54
		550S162	16	33	33	33	33	33	33	33	33	33	33	33	33
			24	33	33	43	43	33	33	33	43	33	33	43	43
< 140	120	350S162	16	33	33	33	43	33	33	33	43	33	33	33	43
			24	33	33	43	54	43	43	43	54	54	54	54	54
		550S162	16	33	33	33	33	33	33	33	33	33	33	33	33
			24	33	33	43	43	33	33	33	43	43	43	43	43
—	130	350S162	16	33	33	33	43	33	33	33	43	43	43	43	43
			24	43	43	43	54	54	54	54	54	54	54	54	54
		550S162	16	33	33	33	33	33	33	33	33	33	33	33	33
			24	33	33	43	43	43	43	43	43	43	43	43	43
—	< 140	350S162	16	33	33	33	43	43	43	43	43	43	43	43	43
			24	43	43	54	54	54	54	54	54	54	54	54	54
		550S162	16	33	33	33	33	33	33	33	33	33	33	33	33
			24	43	43	43	43	43	43	43	43	43	43	43	43

For SI: 1 inch = 25.4 mm, 1 foot = 304.8 mm, 1 mil = 0.0254 mm, 1 mile per hour = 0.447 m/s, 1 pound per square foot = 0.0479 kPa, 1 ksi = 1,000 psi = 6.895 MPa.

a. Deflection criterion: *L*/240.

b. Design load assumptions:

Second-floor dead load is 10 psf.

Second-floor live load is 30 psf.

Roof/ceiling dead load is 12 psf.

Attic live load is 10 psf.

c. Building width is in the direction of horizontal framing members supported by the wall studs.

d. Minimum Grade 33 ksi steel shall be used for 33 mil and 43 mil thicknesses. Minimum Grade 50 ksi steel shall be used for 54 and 68 mil thicknesses.

TABLE R603.3.2(5)
36-FOOT-WIDE BUILDING SUPPORTING ROOF AND CEILING ONLY[a, b, c, d]

ULTIMATE WIND SPEED AND EXPOSURE CATEGORY (mph)		MEMBER SIZE	STUD SPACING (inches)	MINIMUM STUD THICKNESS (mils)											
				8-foot Studs				9-foot Studs				10-foot Studs			
				Ground Snow Load (psf)											
Exp. B	Exp. C			20	30	50	70	20	30	50	70	20	30	50	70
115	—	350S162	16	33	33	33	43	33	33	33	43	33	33	33	43
			24	33	33	43	54	33	33	43	54	43	43	54	54
		550S162	16	33	33	33	33	33	33	33	33	33	33	33	33
			24	33	33	43	43	33	33	43	43	33	33	43	43
120	—	350S162	16	33	33	33	43	33	33	33	43	33	33	33	43
			24	33	33	43	54	33	33	43	54	43	43	54	54
		550S162	16	33	33	33	33	33	33	33	33	33	33	33	33
			24	33	33	43	43	33	33	43	43	33	33	43	43
130	115	350S162	16	33	33	33	43	33	33	33	43	33	33	43	43
			24	33	43	43	54	43	43	43	54	43	43	54	54
		550S162	16	33	33	33	33	33	33	33	33	33	33	33	33
			24	33	33	43	43	33	33	43	43	33	33	43	43
< 140	120	350S162	16	33	33	33	43	33	33	33	33	33	33	43	43
			24	43	43	43	54	43	43	43	54	54	54	54	54
		550S162	16	33	33	33	33	33	33	33	33	33	33	33	33
			24	33	33	43	43	33	33	43	43	43	43	43	54
—	130	350S162	16	33	33	33	43	33	33	33	43	43	43	43	43
			24	43	43	54	54	54	54	54	54	54	54	54	54
		550S162	16	33	33	33	33	33	33	33	33	33	33	33	43
			24	33	33	43	54	43	43	43	43	43	43	43	54
—	< 140	350S162	16	33	33	33	43	43	43	43	43	43	43	43	54
			24	43	43	54	54	54	54	54	54	54	54	54	68
		550S162	16	33	33	33	33	33	33	33	33	33	33	33	43
			24	43	43	43	54	43	43	43	43	43	43	43	54

For SI: 1 inch = 25.4 mm, 1 foot = 304.8 mm, 1 mil = 0.0254 mm, 1 mile per hour = 0.447 m/s, 1 pound per square foot = 0.0479 kPa, 1 ksi = 1,000 psi = 6.895 MPa.

a. Deflection criterion: *L*/240.

b. Design load assumptions:
   Second-floor dead load is 10 psf.
   Second-floor live load is 30 psf.
   Roof/ceiling dead load is 12 psf.
   Attic live load is 10 psf.

c. Building width is in the direction of horizontal framing members supported by the wall studs.

d. Minimum Grade 33 ksi steel shall be used for 33 mil and 43 mil thicknesses. Minimum Grade 50 ksi steel shall be used for 54 and 68 mil thicknesses.

TABLE R603.3.2(6)
40-FOOT-WIDE BUILDING SUPPORTING ROOF AND CEILING ONLY[a, b, c, d]

ULTIMATE WIND SPEED AND EXPOSURE CATEGORY (mph)		MEMBER SIZE	STUD SPACING (inches)	MINIMUM STUD THICKNESS (mils)											
				8-foot Studs				9-foot Studs				10-foot Studs			
				Ground Snow Load (psf)											
Exp. B	Exp. C			20	30	50	70	20	30	50	70	20	30	50	70
115	—	350S162	16	33	33	33	43	33	33	33	43	33	33	43	43
			24	33	33	43	54	33	43	43	54	43	43	54	54
		550S162	16	33	33	33	43	33	33	33	33	33	33	33	33
			24	33	33	43	54	33	33	43	43	33	33	43	54
120	—	350S162	16	33	33	33	43	33	33	33	43	33	33	43	43
			24	33	43	43	54	33	43	43	54	43	43	54	54
		550S162	16	33	33	33	43	33	33	33	33	33	33	33	43
			24	33	33	43	54	33	33	43	43	33	33	43	54
130	115	350S162	16	33	33	33	43	33	33	33	43	33	33	43	43
			24	43	43	54	54	43	43	54	54	43	54	54	54
		550S162	16	33	33	33	43	33	33	33	33	33	33	33	43
			24	33	33	43	54	33	33	43	54	33	33	43	54
< 140	120	350S162	16	33	33	33	43	33	33	33	43	33	33	43	43
			24	43	43	54	54	43	43	54	54	54	54	54	54
		550S162	16	33	33	33	43	33	33	33	33	33	33	33	43
			24	33	33	43	54	33	33	43	54	43	43	43	54
—	130	350S162	16	33	33	43	43	33	33	43	43	43	43	43	54
			24	43	43	54	54	54	54	54	54	54	54	54	68
		550S162	16	33	33	33	43	33	33	33	43	33	33	33	43
			24	33	33	43	54	43	43	43	54	43	43	43	54
—	< 140	350S162	16	33	33	43	43	43	43	43	43	43	43	43	54
			24	43	43	54	54	54	54	54	54	54	54	54	68
		550S162	16	33	33	33	43	33	33	33	43	33	33	33	43
			24	43	43	43	54	43	43	43	54	43	43	43	54

For SI: 1 inch = 25.4 mm, 1 foot = 304.8 mm, 1 mil = 0.0254 mm, 1 mile per hour = 0.447 m/s, 1 pound per square foot = 0.0479 kPa, 1 ksi = 1,000 psi = 6.895 MPa.

a. Deflection criterion: *L*/240.

b. Design load assumptions:
   Second-floor dead load is 10 psf.
   Second-floor live load is 30 psf.
   Roof/ceiling dead load is 12 psf.
   Attic live load is 10 psf.

c. Building width is in the direction of horizontal framing members supported by the wall studs.

d. Minimum Grade 33 ksi steel shall be used for 33 mil and 43 mil thicknesses. Minimum Grade 50 ksi steel shall be used for 54 and 68 mil thicknesses.

TABLE R603.3.2(7)
24-FOOT-WIDE BUILDING SUPPORTING ONE FLOOR, ROOF AND CEILING[a, b, c, d]

ULTIMATE WIND SPEED AND EXPOSURE CATEGORY (mph)		MEMBER SIZE	STUD SPACING (inches)	MINIMUM STUD THICKNESS (mils)											
				8-foot Studs				9-foot Studs				10-foot Studs			
				Ground Snow Load (psf)											
Exp. B	Exp. C			20	30	50	70	20	30	50	70	20	30	50	70
115	—	350S162	16	33	33	33	33	33	33	33	33	33	33	33	43
			24	33	33	43	43	43	43	43	43	43	43	43	54
		550S162	16	33	33	33	33	33	33	33	33	33	33	33	33
			24	33	33	33	43	33	33	33	43	33	33	33	43
120	—	350S162	16	33	33	33	33	33	33	33	33	33	33	33	43
			24	43	43	43	43	43	43	43	43	43	43	54	54
		550S162	16	33	33	33	33	33	33	33	33	33	33	33	33
			24	33	33	33	43	33	33	33	43	33	33	33	43
130	115	350S162	16	33	33	33	43	33	33	33	43	43	43	43	43
			24	43	43	43	54	43	43	54	54	54	54	54	54
		550S162	16	33	33	33	33	33	33	33	33	33	33	33	33
			24	33	33	33	43	33	33	33	43	33	33	43	43
< 140	120	350S162	16	33	33	33	43	33	33	43	43	43	43	43	43
			24	43	43	43	54	43	54	54	54	54	54	54	54
		550S162	16	33	33	33	33	33	33	33	33	33	33	33	33
			24	33	33	43	43	33	33	33	43	43	43	43	43
—	130	350S162	16	33	33	33	43	43	43	43	43	43	43	43	54
			24	43	43	54	54	54	54	54	54	54	54	54	54
		550S162	16	33	33	33	33	33	33	33	33	33	33	33	33
			24	33	33	43	43	43	43	43	43	43	43	43	43
—	< 140	350S162	16	43	43	43	43	43	43	43	43	54	54	54	54
			24	54	54	54	54	54	54	54	54	54	54	54	54
		550S162	16	33	33	33	33	33	33	33	33	33	33	33	33
			24	43	43	43	43	43	43	43	43	43	43	43	43

For SI: 1 inch = 25.4 mm, 1 foot = 304.8 mm, 1 mil = 0.0254 mm, 1 mile per hour = 0.447 m/s, 1 pound per square foot = 0.0479 kPa,
1 ksi = 1,000 psi = 6.895 MPa.

a. Deflection criterion: $L/240$.

b. Design load assumptions:
   Second-floor dead load is 10 psf.
   Second-floor live load is 30 psf.
   Roof/ceiling dead load is 12 psf.
   Attic live load is 10 psf.

c. Building width is in the direction of horizontal framing members supported by the wall studs.

d. Minimum Grade 33 ksi steel shall be used for 33 mil and 43 mil thicknesses. Minimum Grade 50 ksi steel shall be used for 54 and 68 mil thicknesses.

TABLE R603.3.2(8)
28-FOOT-WIDE BUILDING SUPPORTING ONE FLOOR, ROOF AND CEILING[a, b, c, d]

ULTIMATE WIND SPEED AND EXPOSURE CATEGORY (mph)		MEMBER SIZE	STUD SPACING (inches)	MINIMUM STUD THICKNESS (mils)											
				8-foot Studs				9-foot Studs				10-foot Studs			
				Ground Snow Load (psf)											
Exp. B	Exp. C			20	30	50	70	20	30	50	70	20	30	50	70
115	—	350S162	16	33	33	33	43	33	33	33	43	33	33	43	43
			24	43	43	43	54	43	43	43	54	43	43	54	54
		550S162	16	33	33	33	33	33	33	33	33	33	33	33	33
			24	33	33	43	43	33	33	43	43	33	33	43	43
120	—	350S162	16	33	33	33	43	33	33	33	43	33	33	43	43
			24	43	43	43	54	43	43	43	54	54	54	54	54
		550S162	16	33	33	33	33	33	33	33	33	33	33	33	33
			24	33	33	43	43	33	33	43	43	33	33	43	43
130	115	350S162	16	33	33	33	43	33	33	43	43	43	43	43	43
			24	43	43	43	54	43	54	54	54	54	54	54	54
		550S162	16	33	33	33	33	33	33	33	33	33	33	33	33
			24	33	33	43	43	33	33	43	43	43	43	43	43
< 140	120	350S162	16	33	33	33	43	43	43	43	43	43	43	43	43
			24	43	43	54	54	54	54	54	54	54	54	54	54
		550S162	16	33	33	33	33	33	33	33	33	33	33	33	33
			24	33	33	43	43	33	33	43	43	43	43	43	43
—	130	350S162	16	33	33	43	43	43	43	43	43	43	43	54	54
			24	54	54	54	54	54	54	54	54	54	54	54	54
		550S162	16	33	33	33	33	33	33	33	33	33	33	33	43
			24	33	33	43	43	43	43	43	43	43	43	43	43
—	< 140	350S162	16	43	43	43	43	43	43	43	43	54	54	54	54
			24	54	54	54	54	54	54	54	54	54	54	54	54
		550S162	16	33	33	33	33	33	33	33	33	33	33	33	43
			24	43	43	43	43	43	43	43	43	43	43	43	54

For SI: 1 inch = 25.4 mm, 1 foot = 304.8 mm, 1 mil = 0.0254 mm, 1 mile per hour = 0.447 m/s, 1 pound per square foot = 0.0479 kPa, 1 ksi = 1,000 psi = 6.895 MPa.

a. Deflection criterion: *L*/240.

b. Design load assumptions:
   Second-floor dead load is 10 psf.
   Second-floor live load is 30 psf.
   Roof/ceiling dead load is 12 psf.
   Attic live load is 10 psf.

c. Building width is in the direction of horizontal framing members supported by the wall studs.

d. Minimum Grade 33 ksi steel shall be used for 33 mil and 43 mil thicknesses. Minimum Grade 50 ksi steel shall be used for 54 and 68 mil thicknesses.

TABLE R603.3.2(9)
32-FOOT-WIDE BUILDING SUPPORTING ONE FLOOR, ROOF AND CEILING[a, b, c, d]

ULTIMATE WIND SPEED AND EXPOSURE CATEGORY (mph)		MEMBER SIZE	STUD SPACING (inches)	MINIMUM STUD THICKNESS (mils)											
				8-foot Studs				9-foot Studs				10-foot Studs			
				Ground Snow Load (psf)											
Exp. B	Exp. C			20	30	50	70	20	30	50	70	20	30	50	70
115	—	350S162	16	33	33	33	43	33	33	33	43	33	43	43	43
			24	43	43	43	54	43	43	43	54	54	54	54	54
		550S162	16	33	33	33	43	33	33	33	33	33	33	33	43
			24	33	43	43	54	33	33	43	43	33	33	43	43
120	—	350S162	16	33	33	33	43	33	33	33	43	43	43	43	43
			24	43	43	43	54	43	43	43	54	54	54	54	54
		550S162	16	33	33	33	43	33	33	33	33	33	33	33	43
			24	33	43	43	54	33	33	43	43	33	33	43	54
130	115	350S162	16	33	33	43	43	43	43	43	43	43	43	43	43
			24	43	43	54	54	54	54	54	54	54	54	54	54
		550S162	16	33	33	33	43	33	33	33	33	33	33	33	43
			24	33	43	43	54	33	33	43	43	43	43	43	54
< 140	120	350S162	16	33	33	43	43	43	43	43	43	43	43	43	54
			24	43	54	54	54	54	54	54	54	54	54	54	54
		550S162	16	33	33	33	43	33	33	33	43	33	33	33	43
			24	33	43	43	54	33	43	43	43	43	43	43	54
—	130	350S162	16	43	43	43	43	43	43	43	43	43	54	54	54
			24	54	54	54	54	54	54	54	54	54	54	54	54
		550S162	16	33	33	33	43	33	33	33	43	33	33	33	43
			24	43	43	43	54	43	43	43	54	43	43	43	54
—	< 140	350S162	16	43	43	43	43	43	43	43	54	54	54	54	54
			24	54	54	54	54	54	54	54	54	54	54	54	68
		550S162	16	33	33	33	43	33	33	33	43	33	33	33	43
			24	43	43	43	54	43	43	43	54	43	43	43	54

For SI: 1 inch = 25.4 mm, 1 foot = 304.8 mm, 1 mil = 0.0254 mm, 1 mile per hour = 0.447 m/s, 1 pound per square foot = 0.0479 kPa, 1 ksi = 1,000 psi = 6.895 MPa.

a. Deflection criterion: *L*/240.

b. Design load assumptions:
Second-floor dead load is 10 psf.
Second-floor live load is 30 psf.
Roof/ceiling dead load is 12 psf.
Attic live load is 10 psf.

c. Building width is in the direction of horizontal framing members supported by the wall studs.

d. Minimum Grade 33 ksi steel shall be used for 33 mil and 43 mil thicknesses. Minimum Grade 50 ksi steel shall be used for 54 and 68 mil thicknesses.

TABLE R603.3.2(10)
36-FOOT-WIDE BUILDING SUPPORTING ONE FLOOR, ROOF AND CEILING[a, b, c, d]

ULTIMATE WIND SPEED AND EXPOSURE CATEGORY (mph)		MEMBER SIZE	STUD SPACING (inches)	MINIMUM STUD THICKNESS (mils)											
				8-foot Studs				9-foot Studs				10-foot Studs			
				Ground Snow Load (psf)											
Exp. B	Exp. C			20	30	50	70	20	30	50	70	20	30	50	70
115	—	350S162	16	33	33	43	43	33	33	43	43	43	43	43	43
			24	43	43	54	54	43	43	54	54	54	54	54	54
		550S162	16	33	33	33	43	33	33	33	43	33	33	33	43
			24	43	43	43	54	43	43	43	54	43	43	43	54
120	—	350S162	16	33	33	43	43	33	33	43	43	43	43	43	43
			24	43	43	54	54	43	43	54	54	54	54	54	54
		550S162	16	33	33	33	43	33	33	33	43	33	33	33	43
			24	43	43	43	54	43	43	43	54	43	43	43	54
130	115	350S162	16	33	33	43	43	43	43	43	43	43	43	43	54
			24	43	54	54	54	54	54	54	54	54	54	54	68
		550S162	16	33	33	33	43	33	33	33	43	33	33	33	43
			24	43	43	43	54	43	43	43	54	43	43	43	54
< 140	120	350S162	16	43	43	43	43	43	43	43	43	43	43	54	54
			24	54	54	54	54	54	54	54	54	54	54	54	68
		550S162	16	33	33	33	43	33	33	33	43	33	33	33	43
			24	43	43	43	54	43	43	43	54	43	43	43	54
—	130	350S162	16	43	43	43	43	43	43	43	43	54	54	54	54
			24	54	54	54	54	54	54	54	54	54	54	54	68
		550S162	16	33	33	33	43	33	33	33	43	33	33	33	43
			24	43	43	43	54	43	43	43	54	43	43	43	54
—	< 140	350S162	16	43	43	43	54	43	43	54	54	54	54	54	54
			24	54	54	54	54	54	54	54	54	54	54	54	68
		550S162	16	33	33	33	43	33	33	33	43	33	33	43	43
			24	43	43	43	54	43	43	43	54	43	43	54	54

For SI: 1 inch = 25.4 mm, 1 foot = 304.8 mm, 1 mil = 0.0254 mm, 1 mile per hour = 0.447 m/s, 1 pound per square foot = 0.0479 kPa, 1 ksi = 1,000 psi = 6.895 MPa.

a. Deflection criterion: *L*/240.

b. Design load assumptions:

Second-floor dead load is 10 psf.

Second-floor live load is 30 psf.

Roof/ceiling dead load is 12 psf.

Attic live load is 10 psf.

c. Building width is in the direction of horizontal framing members supported by the wall studs.

d. Minimum Grade 33 ksi steel shall be used for 33 mil and 43 mil thicknesses. Minimum Grade 50 ksi steel shall be used for 54 and 68 mil thicknesses.

TABLE R603.3.2(11)
40-FOOT-WIDE BUILDING SUPPORTING ONE FLOOR, ROOF AND CEILING[a, b, c, d]

ULTIMATE WIND SPEED AND EXPOSURE CATEGORY (mph)		MEMBER SIZE	STUD SPACING (inches)	MINIMUM STUD THICKNESS (mils)											
				8-foot Studs				9-foot Studs				10-foot Studs			
				Ground Snow Load (psf)											
Exp. B	Exp. C			20	30	50	70	20	30	50	70	20	30	50	70
115	—	350S162	16	33	33	43	43	33	33	43	43	43	43	43	54
			24	43	43	54	54	43	43	54	54	54	54	54	68
		550S162	16	33	33	33	43	33	33	33	43	33	33	33	43
			24	43	43	54	54	43	43	43	54	43	43	43	54
120	—	350S162	16	33	33	43	43	33	33	43	43	43	43	43	54
			24	43	43	54	54	54	54	54	54	54	54	54	68
		550S162	16	33	33	33	43	33	33	33	43	33	33	33	43
			24	43	43	54	54	43	43	43	54	43	43	43	54
130	115	350S162	16	43	43	43	54	43	43	43	43	43	43	54	54
			24	54	54	54	54	54	54	54	54	54	54	54	68
		550S162	16	33	33	43	43	33	33	33	43	33	33	43	43
			24	43	43	54	54	43	43	43	54	43	43	54	54
< 140	120	350S162	16	43	43	43	54	43	43	43	54	43	43	54	54
			24	54	54	54	54	54	54	54	54	54	54	54	68
		550S162	16	33	33	43	43	33	33	33	43	33	33	43	43
			24	43	43	54	54	43	43	43	54	43	43	54	54
—	130	350S162	16	43	43	43	54	43	43	43	54	54	54	54	54
			24	54	54	54	68	54	54	54	54	54	54	68	68
		550S162	16	33	33	43	43	33	33	33	43	33	33	43	43
			24	43	43	54	54	43	43	43	54	43	43	54	54
—	< 140	350S162	16	43	43	43	54	43	43	54	54	54	54	54	54
			24	54	54	54	68	54	54	54	68	54	54	68	68
		550S162	16	33	33	43	43	33	33	43	43	33	43	43	43
			24	43	43	54	54	43	43	43	54	43	43	54	54

For SI: 1 inch = 25.4 mm, 1 foot = 304.8 mm, 1 mil = 0.0254 mm, 1 mile per hour = 0.447 m/s, 1 pound per square foot = 0.0479 kPa, 1 ksi = 1,000 psi = 6.895 MPa.

a. Deflection criterion: *L*/240.

b. Design load assumptions:

Second-floor dead load is 10 psf.

Second-floor live load is 30 psf.

Roof/ceiling dead load is 12 psf.

Attic live load is 10 psf.

c. Building width is in the direction of horizontal framing members supported by the wall studs.

d. Minimum Grade 33 ksi steel shall be used for 33 mil and 43 mil thicknesses. Minimum Grade 50 ksi steel shall be used for 54 and 68 mil thicknesses.

TABLE R603.3.2(12)
24-FOOT-WIDE BUILDING SUPPORTING TWO FLOORS, ROOF AND CEILING[a, b, c, d]

ULTIMATE WIND SPEED AND EXPOSURE CATEGORY (mph)		MEMBER SIZE	STUD SPACING (inches)	MINIMUM STUD THICKNESS (mils)											
				8-foot Studs				9-foot Studs				10-foot Studs			
				Ground Snow Load (psf)											
Exp. B	Exp. C			20	30	50	70	20	30	50	70	20	30	50	70
115	—	350S162	16	43	43	43	43	33	33	33	43	43	43	43	43
			24	54	54	54	54	54	54	54	54	54	54	54	54
		550S162	16	33	33	43	43	33	33	33	33	33	33	33	43
			24	43	43	54	54	43	43	43	43	43	43	43	54
120	—	350S162	16	43	43	43	43	33	33	43	43	43	43	43	43
			24	54	54	54	54	54	54	54	54	54	54	54	54
		550S162	16	33	33	43	43	33	33	33	33	33	33	33	43
			24	43	43	54	54	43	43	43	43	43	43	43	54
130	115	350S162	16	43	43	43	43	43	43	43	43	43	43	43	54
			24	54	54	54	54	54	54	54	54	54	54	54	54
		550S162	16	33	33	43	43	33	33	33	33	33	33	33	43
			24	43	43	54	54	43	43	43	43	43	43	43	54
< 140	120	350S162	16	43	43	43	43	43	43	43	43	43	43	54	54
			24	54	54	54	54	54	54	54	54	54	54	54	54
		550S162	16	33	33	43	43	33	33	33	33	33	33	33	43
			24	43	43	54	54	43	43	43	43	43	43	43	54
—	130	350S162	16	43	43	43	43	43	43	43	43	54	54	54	54
			24	54	54	54	54	54	54	54	54	54	54	68	68
		550S162	16	33	33	43	43	33	33	33	33	33	33	33	43
			24	43	43	54	54	43	43	43	43	43	43	43	54
—	< 140	350S162	16	43	43	43	43	43	43	54	54	54	54	54	54
			24	54	54	54	54	54	54	54	54	54	54	68	68
		550S162	16	33	33	43	43	33	33	33	33	33	33	43	43
			24	43	43	54	54	43	43	43	43	54	54	54	54

For SI: 1 inch = 25.4 mm, 1 foot = 304.8 mm, 1 mil = 0.0254 mm, 1 mile per hour = 0.447 m/s, 1 pound per square foot = 0.0479 kPa, 1 ksi = 1,000 psi = 6.895 MPa.

a. Deflection criterion: *L*/240.

b. Design load assumptions:
   - Top- and middle-floor dead load is 10 psf.
   - Top-floor live load is 30 psf.
   - Middle-floor live load is 40 psf.
   - Roof/ceiling dead load is 12 psf.
   - Attic live load is 10 psf.

c. Building width is in the direction of horizontal framing members supported by the wall studs.

d. Minimum Grade 33 ksi steel shall be used for 33 mil and 43 mil thicknesses. Minimum Grade 50 ksi steel shall be used for 54 and 68 mil thicknesses.

TABLE R603.3.2(13)
28-FOOT-WIDE BUILDING SUPPORTING TWO FLOORS, ROOF AND CEILING[a, b, c, d]

ULTIMATE WIND SPEED AND EXPOSURE CATEGORY (mph)		MEMBER SIZE	STUD SPACING (inches)	MINIMUM STUD THICKNESS (mils)											
				8-foot Studs				9-foot Studs				10-foot Studs			
				Ground Snow Load (psf)											
Exp. B	Exp. C			20	30	50	70	20	30	50	70	20	30	50	70
115	—	350S162	16	43	43	43	43	43	43	43	43	43	43	43	43
			24	54	54	54	54	54	54	54	54	54	54	54	54
		550S162	16	43	43	43	43	43	43	43	43	43	43	43	43
			24	54	54	54	54	54	54	54	54	54	54	54	54
120	—	350S162	16	43	43	43	43	43	43	43	43	43	43	43	43
			24	54	54	54	54	54	54	54	54	54	54	54	54
		550S162	16	43	43	43	43	43	43	43	43	43	43	43	43
			24	54	54	54	54	54	54	54	54	54	54	54	54
130	115	350S162	16	43	43	43	43	43	43	43	43	43	43	54	54
			24	54	54	54	54	54	54	54	54	54	54	54	68
		550S162	16	43	43	43	43	43	43	43	43	43	43	43	43
			24	54	54	54	54	54	54	54	54	54	54	54	54
< 140	120	350S162	16	43	43	43	43	43	43	43	43	54	54	54	54
			24	54	54	54	54	54	54	54	54	54	54	68	68
		550S162	16	43	43	43	43	43	43	43	43	43	43	43	43
			24	54	54	54	54	54	54	54	54	54	54	54	54
—	130	350S162	16	43	43	43	43	43	43	43	54	54	54	54	54
			24	54	54	54	54	54	54	54	54	68	68	68	68
		550S162	16	43	43	43	43	43	43	43	43	43	43	43	43
			24	54	54	54	54	54	54	54	54	54	54	54	54
—	< 140	350S162	16	43	43	43	54	54	54	54	54	54	54	54	54
			24	54	54	54	54	54	54	54	68	68	68	68	68
		550S162	16	43	43	43	43	43	43	43	43	43	43	43	43
			24	54	54	54	54	54	54	54	54	54	54	54	54

For SI: 1 inch = 25.4 mm, 1 foot = 304.8 mm, 1 mil = 0.0254 mm, 1 mile per hour = 0.447 m/s, 1 pound per square foot = 0.0479 kPa, 1 ksi = 1,000 psi = 6.895 MPa.

a. Deflection criterion: *L*/240.

b. Design load assumptions:

Top- and middle-floor dead load is 10 psf.

Top-floor live load is 30 psf.

Middle-floor live load is 40 psf.

Roof/ceiling dead load is 12 psf.

Attic live load is 10 psf.

c. Building width is in the direction of horizontal framing members supported by the wall studs.

d. Minimum Grade 33 ksi steel shall be used for 33 mil and 43 mil thicknesses. Minimum Grade 50 ksi steel shall be used for 54 and 68 mil thicknesses.

TABLE R603.3.2(14)
32-FOOT-WIDE BUILDING SUPPORTING TWO FLOORS, ROOF AND CEILING[a, b, c, d]

ULTIMATE WIND SPEED AND EXPOSURE CATEGORY (mph)		MEMBER SIZE	STUD SPACING (inches)	MINIMUM STUD THICKNESS (mils)											
				8-foot Studs				9-foot Studs				10-foot Studs			
				Ground Snow Load (psf)											
Exp. B	Exp. C			20	30	50	70	20	30	50	70	20	30	50	70
115	—	350S162	16	43	43	43	54	43	43	43	43	43	43	43	54
			24	54	54	54	68	54	54	54	54	54	54	54	68
		550S162	16	43	43	43	43	43	43	43	43	43	43	43	43
			24	54	54	54	54	54	54	54	54	54	54	54	54
120	—	350S162	16	43	43	43	54	43	43	43	43	43	43	43	54
			24	54	54	54	68	54	54	54	54	54	54	54	68
		550S162	16	43	43	43	43	43	43	43	43	43	43	43	43
			24	54	54	54	54	54	54	54	54	54	54	54	54
130	115	350S162	16	43	43	43	54	43	43	43	43	54	54	54	54
			24	54	54	54	68	54	54	54	54	54	68	68	68
		550S162	16	43	43	43	43	43	43	43	43	43	43	43	43
			24	54	54	54	54	54	54	54	54	54	54	54	54
< 140	120	350S162	16	43	43	43	54	43	43	43	54	54	54	54	54
			24	54	54	54	68	54	54	54	54	68	68	68	68
		550S162	16	43	43	43	43	43	43	43	43	43	43	43	43
			24	54	54	54	54	54	54	54	54	54	54	54	54
—	130	350S162	16	43	43	43	54	43	54	54	54	54	54	54	54
			24	54	54	54	68	54	54	54	68	68	68	68	68
		550S162	16	43	43	43	43	43	43	43	43	43	43	43	43
			24	54	54	54	54	54	54	54	54	54	54	54	54
—	< 140	350S162	16	43	43	54	54	54	54	54	54	54	54	54	54
			24	54	54	54	68	54	68	68	68	68	68	68	68
		550S162	16	43	43	43	43	43	43	43	43	43	43	43	43
			24	54	54	54	54	54	54	54	54	54	54	54	54

For SI: 1 inch = 25.4 mm, 1 foot = 304.8 mm, 1 mil = 0.0254 mm, 1 mile per hour = 0.447 m/s, 1 pound per square foot = 0.0479 kPa, 1 ksi = 1,000 psi = 6.895 MPa.

a. Deflection criterion: *L*/240.

b. Design load assumptions:

Top- and middle-floor dead load is 10 psf.

Top-floor live load is 30 psf.

Middle-floor live load is 40 psf.

Roof/ceiling dead load is 12 psf.

Attic live load is 10 psf.

c. Building width is in the direction of horizontal framing members supported by the wall studs.

d. Minimum Grade 33 ksi steel shall be used for 33 mil and 43 mil thicknesses. Minimum Grade 50 ksi steel shall be used for 54 and 68 mil thicknesses.

TABLE R603.3.2(15)
36-FOOT-WIDE BUILDING SUPPORTING TWO FLOORS, ROOF AND CEILING[a, b, c, d]

ULTIMATE WIND SPEED AND EXPOSURE CATEGORY (mph)		MEMBER SIZE	STUD SPACING (inches)	MINIMUM STUD THICKNESS (mils)											
				8-foot Studs				9-foot Studs				10-foot Studs			
				Ground Snow Load (psf)											
Exp. B	Exp. C			20	30	50	70	20	30	50	70	20	30	50	70
115	—	350S162	16	54	54	54	54	43	43	43	54	54	54	54	54
			24	68	68	68	68	54	54	54	68	68	68	68	68
		550S162	16	43	43	43	54	43	43	43	43	43	43	43	43
			24	54	54	54	54	54	54	54	54	54	54	54	54
120	—	350S162	16	54	54	54	54	43	43	43	54	54	54	54	54
			24	68	68	68	68	54	54	54	68	68	68	68	68
		550S162	16	43	43	43	54	43	43	43	43	43	43	43	43
			24	54	54	54	54	54	54	54	54	54	54	54	54
130	115	350S162	16	54	54	54	54	43	43	43	54	54	54	54	54
			24	68	68	68	68	54	54	54	68	68	68	68	68
		550S162	16	43	43	43	54	43	43	43	43	43	43	43	43
			24	54	54	54	54	54	54	54	54	54	54	54	54
< 140	120	350S162	16	54	54	54	54	43	43	54	54	54	54	54	54
			24	68	68	68	68	54	54	54	68	68	68	68	68
		550S162	16	43	43	43	54	43	43	43	43	43	43	43	43
			24	54	54	54	54	54	54	54	54	54	54	54	54
—	130	350S162	16	54	54	54	54	54	54	54	54	54	54	54	54
			24	68	68	68	68	54	54	68	68	68	68	68	68
		550S162	16	43	43	43	54	43	43	43	43	43	43	43	43
			24	54	54	54	54	54	54	54	54	54	54	54	54
—	< 140	350S162	16	54	54	54	54	54	54	54	54	54	54	54	54
			24	68	68	68	68	68	68	68	68	68	68	68	68
		550S162	16	43	43	43	54	43	43	43	43	43	43	43	43
			24	54	54	54	54	54	54	54	54	54	54	54	54

For SI: 1 inch = 25.4 mm, 1 foot = 304.8 mm, 1 mil = 0.0254 mm, 1 mile per hour = 0.447 m/s, 1 pound per square foot = 0.0479 kPa, 1 ksi = 1,000 psi = 6.895 MPa.

a. Deflection criterion: *L*/240.

b. Design load assumptions:

Top- and middle-floor dead load is 10 psf.

Top-floor live load is 30 psf.

Middle-floor live load is 40 psf.

Roof/ceiling dead load is 12 psf.

Attic live load is 10 psf.

c. Building width is in the direction of horizontal framing members supported by the wall studs.

d. Minimum Grade 33 ksi steel shall be used for 33 mil and 43 mil thicknesses. Minimum Grade 50 ksi steel shall be used for 54 and 68 mil thicknesses.

TABLE R603.3.2(16) 40-FOOT-WIDE BUILDING SUPPORTING TWO FLOORS, ROOF AND CEILING[a, b, c, d]

ULTIMATE WIND SPEED AND EXPOSURE CATEGORY (mph)		MEMBER SIZE	STUD SPACING (inches)	MINIMUM STUD THICKNESS (mils)											
				8-foot Studs				9-foot Studs				10-foot Studs			
				Ground Snow Load (psf)											
Exp. B	Exp. C			20	30	50	70	20	30	50	70	20	30	50	70
115	—	350S162	16	54	54	54	54	54	54	54	54	54	54	54	54
			24	68	68	68	68	68	68	68	68	68	68	68	68
		550S162	16	54	54	54	54	43	43	54	54	43	43	54	54
			24	54	54	54	68	54	54	54	54	54	54	54	54
120	—	350S162	16	54	54	54	54	54	54	54	54	54	54	54	54
			24	68	68	68	68	68	68	68	68	68	68	68	68
		550S162	16	54	54	54	54	43	43	54	54	43	43	54	54
			24	54	54	54	68	54	54	54	54	54	54	54	54
130	115	350S162	16	54	54	54	54	54	54	54	54	54	54	54	54
			24	68	68	68	68	68	68	68	68	68	68	68	68
		550S162	16	54	54	54	54	43	43	54	54	43	43	54	54
			24	54	54	54	68	54	54	54	54	54	54	54	54
< 140	120	350S162	16	54	54	54	54	54	54	54	54	54	54	54	54
			24	68	68	68	68	68	68	68	68	68	68	68	68
		550S162	16	54	54	54	54	43	43	54	54	43	43	54	54
			24	54	54	54	68	54	54	54	54	54	54	54	54
—	130	350S162	16	54	54	54	54	54	54	54	54	54	54	54	54
			24	68	68	68	68	68	68	68	68	68	68	68	68
		550S162	16	54	54	54	54	43	43	54	54	43	43	54	54
			24	54	54	54	68	54	54	54	54	54	54	54	54
—	< 140	350S162	16	54	54	54	54	54	54	54	54	54	54	54	54
			24	68	68	68	68	68	68	68	68	68	68	68	68
		550S162	16	54	54	54	54	43	43	54	54	43	43	54	54
			24	54	54	54	68	54	54	54	54	54	54	54	54

For SI: 1 inch = 25.4 mm, 1 foot = 304.8 mm, 1 mil = 0.0254 mm, 1 mile per hour = 0.447 m/s, 1 pound per square foot = 0.0479 kPa, 1 ksi = 1,000 psi = 6.895 MPa.

a. Deflection criterion: *L*/240.

b. Design load assumptions:

Top- and middle-floor dead load is 10 psf.

Top-floor live load is 30 psf.

Middle-floor live load is 40 psf.

Roof/ceiling dead load is 12 psf.

Attic live load is 10 psf.

c. Building width is in the direction of horizontal framing members supported by the wall studs.

d. Minimum Grade 33 ksi steel shall be used for 33 mil and 43 mil thicknesses. Minimum Grade 50 ksi steel shall be used for 54 and 68 mil thicknesses.

R603.3.2.1 Gable endwalls. The size and thickness of gable endwall studs with heights less than or equal to 10 feet (3048 mm) shall be permitted in accordance with the limits set forth in Table R603.3.2.1(1). The size and thickness of gable endwall studs with heights greater than 10 feet (3048 mm) shall be determined in accordance with the limits set forth in Table R603.3.2.1(2).

R603.3.3 Stud bracing. The flanges of cold-formed steel studs shall be laterally braced in accordance with one of the following:

1. Gypsum board on both sides, structural sheathing on both sides, or gypsum board on one side and structural sheathing on the other side of *load-bearing walls* with gypsum board installed with minimum No. 6 screws in accordance with Section R702 and structural sheathing installed in accordance with Section R603.9 and Table R603.3.2(1).
2. Horizontal steel straps fastened in accordance with Figure R603.3.3(1) on both sides at mid-height for 8-foot (2438 mm) walls, and at one-third points for 9-foot and 10-foot (2743 mm and 3048 mm) walls. Horizontal steel straps shall be not less than $1^1/_2$ inches in width and 33 mils in thickness (38 mm by 0.84 mm). Straps shall be

TABLE R603.3.2.1(1)
ALL BUILDING WIDTHS GABLE ENDWALLS 8, 9 OR 10 FEET IN HEIGHT[a, b, c, d]

ULTIMATE WIND SPEED AND EXPOSURE CATEGORY (mph)		MEMBER SIZE	STUD SPACING (inches)	MINIMUM STUD THICKNESS (mils)		
Exp. B	Exp. C			8-foot Studs	9-foot Studs	10-foot Studs
115	—	350S162	16	33	33	33
			24	33	33	33
		550S162	16	33	33	33
			24	33	33	33
120	—	350S162	16	33	33	33
			24	33	33	43
		550S162	16	33	33	33
			24	33	33	33
130	115	350S162	16	33	33	33
			24	33	43	43
		550S162	16	33	33	33
			24	33	33	33
< 140	120	350S162	16	33	33	43
			24	33	43	54
		550S162	16	33	33	33
			24	33	33	33
—	130	350S162	16	33	33	43
			24	43	43	54
		550S162	16	33	33	33
			24	33	43	43
—	< 140	350S162	16	33	43	43
			24	43	54	54
		550S162	16	33	33	33
			24	43	43	43

For SI: 1 inch = 25.4 mm, 1 foot = 304.8 mm, 1 mil = 0.0254 mm, 1 mile per hour = 0.447 m/s, 1 pound per square foot = 0.0479 kPa, 1 ksi = 1,000 psi = 6.895 MPa.

a. Deflection criterion *L*/240.

b. Design load assumptions:
Ground snow load is 70 psf.
Roof/ceiling dead load is 12 psf.
Floor dead load is 10 psf.
Floor live load is 40 psf.
Attic dead load is 10 psf.

c. Building width is in the direction of horizontal framing members supported by the wall studs.

d. Minimum Grade 33 ksi steel shall be used for 33 mil and 43 mil thicknesses. Minimum Grade 50 ksi steel shall be used for 54 and 68 mil thicknesses.

attached to the flanges of studs with one No. 8 screw. In-line blocking shall be installed between studs at the termination of straps and at 12-foot (3658 mm) intervals along the strap. Straps shall be fastened to the blocking with two No. 8 screws.

3. Sheathing on one side and strapping on the other side fastened in accordance with Figure R603.3.3(2). Sheathing shall be installed in accordance with Item 1. Steel straps shall be installed in accordance with Item 2.

R603.3.4 Cutting and notching. Flanges and lips of cold-formed steel studs and headers shall not be cut or notched.

R603.3.5 Splicing. Steel studs and other structural members shall not be spliced without an *approved* design. Tracks shall be spliced in accordance with Figure R603.3.5.

R603.4 Corner framing. In exterior walls, corner studs and the top tracks shall be installed in accordance with Figure R603.4.

TABLE R603.3.2.1(2)
ALL BUILDING WIDTHS GABLE ENDWALLS OVER 10 FEET IN HEIGHT[a, b, c, d]

ULTIMATE WIND SPEED AND EXPOSURE CATEGORY (mph)		MEMBER SIZE	STUD SPACING (inches)	MINIMUM STUD THICKNESS (mils)					
				Stud Height, *h* (feet)					
Exp. B	Exp. C			$10 < h \le 12$	$12 < h \le 14$	$14 < h \le 16$	$16 < h \le 18$	$18 < h \le 20$	$20 < h \le 22$
115	—	350S162	16	33	43	68	97	—	—
			24	43	68	—	—	—	—
		550S162	16	33	33	33	43	43	54
			24	33	43	43	54	68	97
120	—	350S162	16	43	54	97	—	—	—
			24	54	97	—	—	—	—
		550S162	16	33	33	43	43	54	68
			24	33	43	54	54	68	97
130	115	350S162	16	43	54	97	—	—	—
			24	54	97	—	—	—	—
		550S162	16	33	33	43	54	54	97
			24	43	43	54	68	97	97
< 140	120	350S162	16	43	68	—	—	—	—
			24	68	—	—	—	—	—
		550S162	16	33	43	43	54	68	97
			24	43	54	54	68	97	—
—	130	350S162	16	54	97	—	—	—	—
			24	97	—	—	—	—	—
		550S162	16	33	43	54	68	97	—
			24	43	54	54	97	—	—
—	< 140	350S162	16	54	97	—	—	—	—
			24	97	—	—	—	—	—
		550S162	16	43	43	54	97	97	—
			24	54	54	68	—	—	—

For SI: 1 inch = 25.4 mm, 1 foot = 304.8 mm, 1 mil = 0.0254 mm, 1 mile per hour = 0.447 m/s, 1 pound per square foot = 0.0479 kPa, 1 ksi = 1,000 psi = 6.895 MPa.

a. Deflection criterion *L*/240.

b. Design load assumptions:
Ground snow load is 70 psf.
Roof/ceiling dead load is 12 psf.
Floor dead load is 10 psf.
Floor live load is 40 psf.
Attic dead load is 10 psf.

c. Building width is in the direction of horizontal framing members supported by the wall studs.

d. Minimum Grade 33 ksi steel shall be used for 33 mil and 43 mil thicknesses. Minimum Grade 50 ksi steel shall be used for 54 and 68 mil thicknesses.

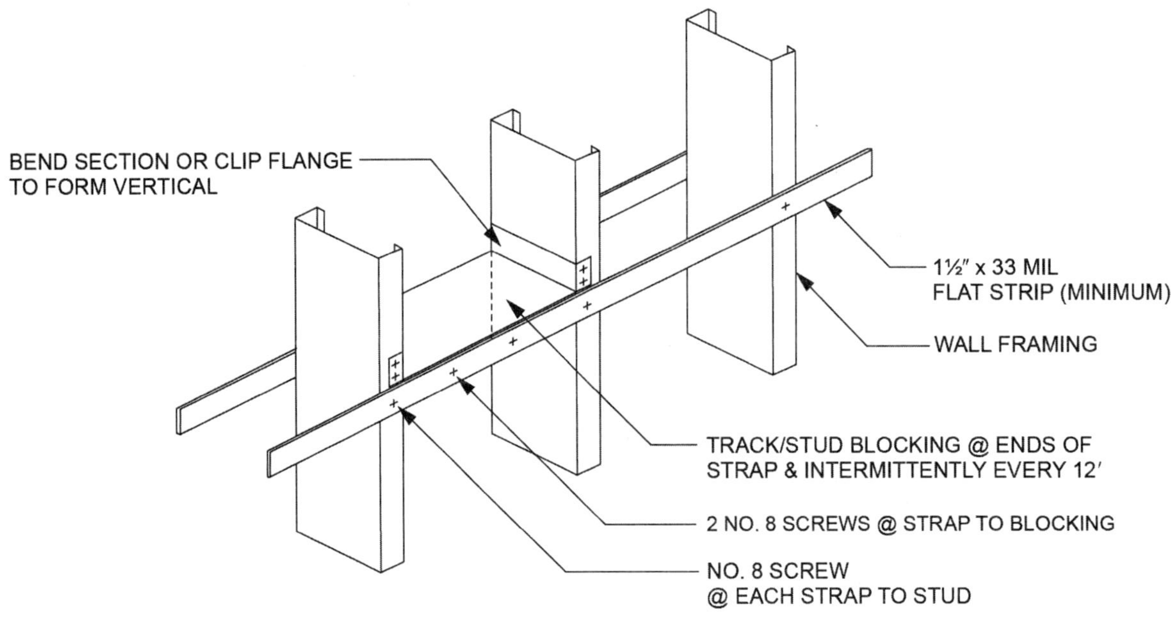

For SI: 1 mil = 0.0254 mm, 1 inch = 25.4 mm, 1 foot = 304.8 mm.

FIGURE R603.3.3(1)
STUD BRACING WITH STRAPPING ONLY

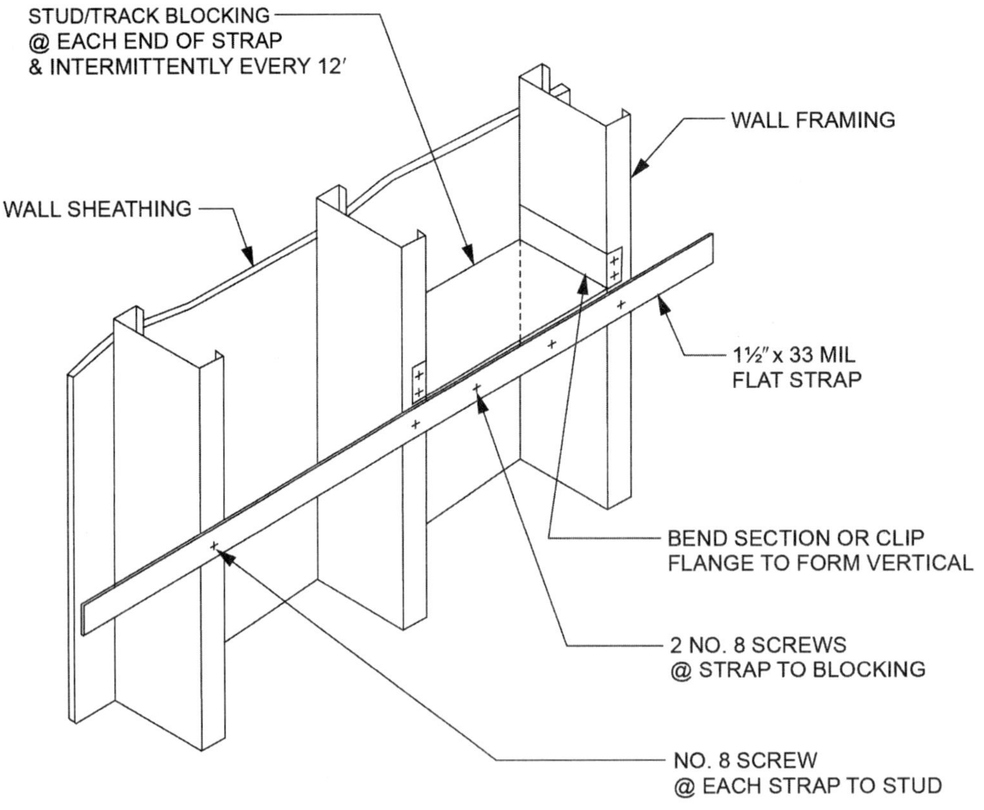

For SI: 1 mil = 0.0254 mm, 1 inch = 25.4 mm, 1 foot = 304.8 mm.

FIGURE R603.3.3(2)
STUD BRACING WITH STRAPPING AND SHEATHING MATERIAL

R603.5 Exterior wall covering. The method of attachment of exterior wall covering materials to cold-formed steel stud wall framing shall conform to the manufacturer's installation instructions.

R603.6 Headers. Headers shall be installed above all wall openings in exterior walls and interior *load-bearing walls*. Box beam headers and back-to-back headers each shall be

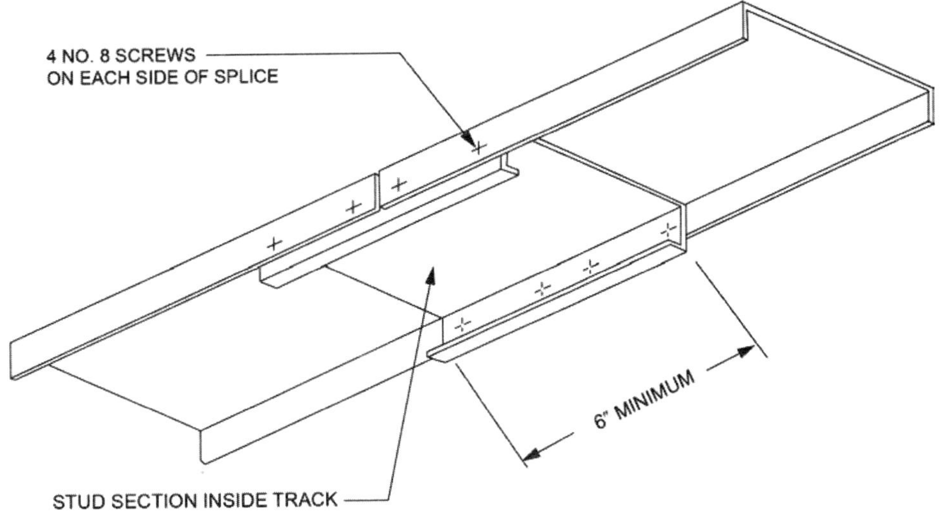

For SI: 1 inch = 25.4 mm.

FIGURE R603.3.5
TRACK SPLICE

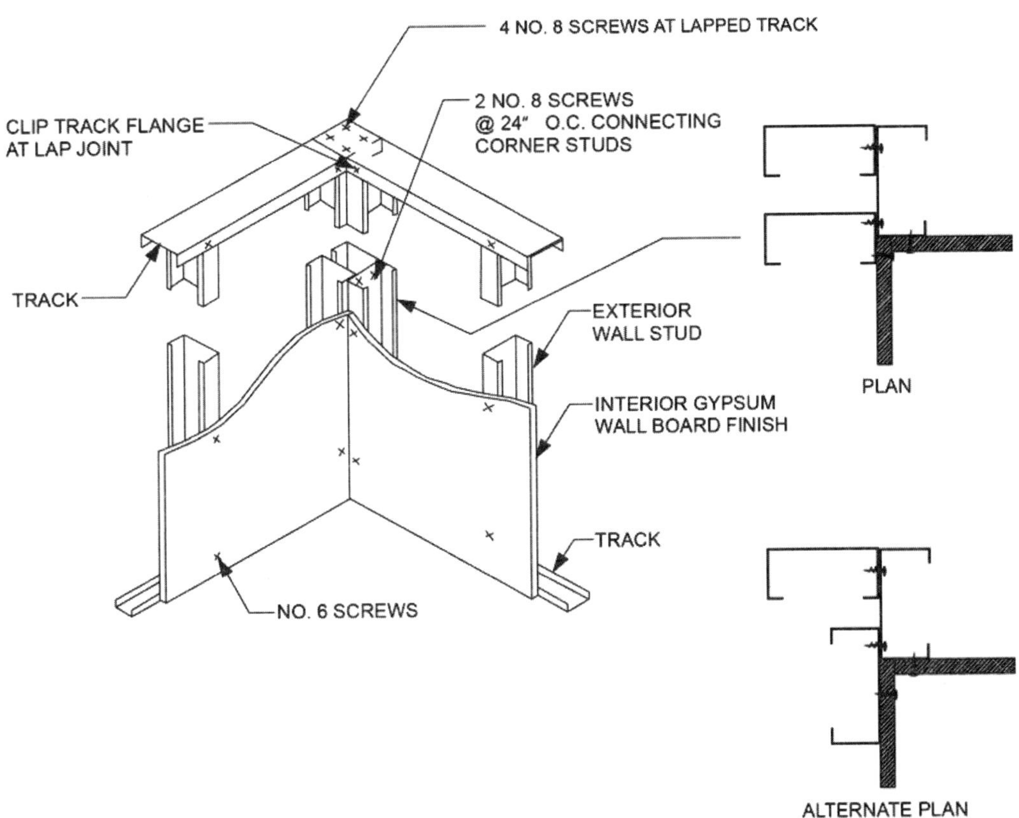

For SI: 1 inch = 25.4 mm.

FIGURE R603.4
CORNER FRAMING

formed from two equal sized C-shaped members in accordance with Figures R603.6(1) and R603.6(2), respectively, and Tables R603.6(1) through R603.6(6). L-shaped headers shall be permitted to be constructed in accordance with AISI S230. Alternately, headers shall be permitted to be designed and constructed in accordance with AISI S240.

R603.6.1 Headers in gable endwalls. Box beam and back-to-back headers in gable endwalls shall be permitted to be constructed in accordance with Section R603.6 or with the header directly above the opening in accordance with Figures R603.6.1(1) and R603.6.1(2) and the following provisions:

1. Two 362S162-33 for openings less than or equal to 4 feet (1219 mm).
2. Two 600S162-43 for openings greater than 4 feet (1219 mm) but less than or equal to 6 feet (1829 mm).
3. Two 800S162-54 for openings greater than 6 feet (1829 mm) but less than or equal to 9 feet (2743 mm).

R603.7 Jack and king studs. The number of jack and king studs installed on each side of a header shall comply with Table R603.7(1). King, jack and cripple studs shall be of the same dimension and thickness as the adjacent wall studs. Headers shall be connected to king studs in accordance with Table R603.7(2) and the following provisions:

1. For box beam headers, one-half of the total number of required screws shall be applied to the header and one- half to the king stud by use of C-shaped or track member in accordance with Figure R603.6(1). The track or C-shaped sections shall extend the depth of the header minus $^1/_2$ inch (12.7 mm) and shall have a minimum thickness not less than that of the wall studs.
2. For back-to-back headers, one-half the total number of screws shall be applied to the header and one-half to the king stud by use of a minimum 2-inch by 2-inch (51 mm by 51 mm) clip angle in accordance with Figure R603.6(2). The clip angle shall extend the depth of the header minus $^1/_2$ inch (12.7 mm) and shall have a minimum thickness not less than that of the wall studs. Jack and king studs shall be interconnected with structural sheathing in accordance with Figures R603.6(1) and R603.6(2).

TABLE R603.6(1)
BOX-BEAM AND BACK-TO-BACK HEADER SPANS
Headers Supporting Roof and Ceiling Only[a, b, d]

MEMBER DESIGNATION	GROUND SNOW LOAD (20 psf)					GROUND SNOW LOAD (30 psf)				
	Building width[c] (feet)					Building width[c] (feet)				
	24	28	32	36	40	24	28	32	36	40
2-350S162-33	3′-3″	2′-8″	2′-2″	—	—	2′-8″	2′-2″	—	—	—
2-350S162-43	4′-2″	3′-9″	3′-4″	2′-11″	2′-7″	3′-9″	3′-4″	2′-11″	2′-7″	2′-2″
2-350S162-54	6′-2″	5′-10″	5′-8″	5′-3″	4′-10″	5′-11″	5′-8″	5′-2″	4′-10″	4′-6″
2-350S162-68	6′-7″	6′-3″	6′-0″	5′-10″	5′-8″	6′-4″	6′-1″	5′-10″	5′-8″	5′-6″
2-550S162-33	4′-8″	4′-0″	3′-6″	3′-0″	2′-6″	4′-1″	3′-6″	3′-0″	2′-6″	—
2-550S162-43	6′-0″	5′-4″	4′-10″	4′-4″	3′-11″	5′-5″	4′-10″	4′-4″	3′-10″	3′-5″
2-550S162-54	8′-9″	8′-5″	8′-1″	7′-9″	7′-3″	8′-6″	8′-1″	7′-8″	7′-2″	6′-8″
2-550S162-68	9′-5″	9′-0″	8′-8″	8′-4″	8′-1″	9′-1″	8′-8″	8′-4″	8′-1″	7′-10″
2-800S162-33	4′-5″	3′-11″	3′-5″	3′-1″	2′-10″	3′-11″	3′-6″	3′-1″	2′-9″	2′-3″
2-800S162-43	7′-3″	6′-7″	5′-11″	5′-4″	4′-10″	6′-7″	5′-11″	5′-4″	4′-9″	4′-3″
2-800S162-54	10′-10″	10′-2″	9′-7″	9′-0″	8′-5″	10′-2″	9′-7″	8′-11″	8′-4″	7′-9″
2-800S162-68	12′-8″	11′-10″	11′-2″	10′-7″	10′-1″	11′-11″	11′-2″	10′-7″	10′-0″	9′-6″
2-1000S162-43	7′-10″	6′-10″	6′-1″	5′-6″	5′-0″	6′-11″	6′-1″	5′-5″	4′-11″	4′-6″
2-1000S162-54	12′-3″	11′-5″	10′-9″	10′-2″	9′-6″	11′-6″	10′-9″	10′-1″	9′-5″	8′-9″
2-1000S162-68	14′-5″	13′-5″	12′-8″	12′-0″	11′-6″	13′-6″	12′-8″	12′-0″	11′-5″	10′-10″
2-1200S162-54	12′-11″	11′-3″	10′-0″	9′-0″	8′-2″	11′-5″	10′-0″	9′-0″	8′-1″	7′-4″
2-1200S162-68	15′-11″	14′-10″	14′-0″	13′-4″	12′-8″	15′-0″	14′-0″	13′-3″	12′-7″	11′-11″

For SI: 1 mil = 0.0254 mm, 1 inch = 25.4 mm, 1 foot = 304.8 mm, 1 pound per square foot = 0.0479 kPa, 1 pound per square inch = 6.895 kPa, 1 ksi = 1,000 psi = 6.895 MPa.

a. Deflection criteria: *L*/360 for live loads, *L*/240 for total loads.

b. Design load assumptions:
Roof/ceiling dead load is 12 psf.
Attic dead load is 10 psf.

c. Building width is in the direction of horizontal framing members supported by the header.

d. Minimum Grade 33 ksi steel shall be used for 33 mil and 43 mil thicknesses. Minimum Grade 50 ksi steel shall be used for 54 and 68 mil thicknesses.

TABLE R603.6(2)
BOX-BEAM AND BACK-TO-BACK HEADER SPANS
Headers Supporting Roof and Ceiling Only[a, b, d]

MEMBER DESIGNATION	GROUND SNOW LOAD (50 psf)					GROUND SNOW LOAD (70 psf)				
	Building width[c] (feet)					Building width[c] (feet)				
	24	28	32	36	40	24	28	32	36	40
2-350S162-33	—	—	—	—	—	—	—	—	—	—
2-350S162-43	2′-4″	—	—	—	—	—	—	—	—	—
2-350S162-54	4′-8″	4′-2″	3′-9″	3′-5″	3′-1″	3′-7″	3′-2″	2′-9″	2′-5″	2′-0″
2-350S162-68	5′-7″	5′-2″	4′-9″	4′-4″	3′-11″	4′-7″	4′-1″	3′-7″	3′-2″	2′-10″
2-550S162-33	2′-2″	—	—	—	—	—	—	—	—	—
2-550S162-43	3′-8″	3′-1″	2′-6″	—	—	2′-3″	—	—	—	—
2-550S162-54	6′-11″	6′-3″	5′-9″	5′-3″	4′-9″	5′-6″	4′-11″	4′-5″	3′-11″	3′-5″
2-550S162-68	8′-0″	7′-6″	6′-11″	6′-5″	5′-11″	6′-9″	6′-1″	5′-6″	5′-0″	4′-7″
2-800S162-33	2′-7″	—	—	—	—	—	—	—	—	—
2-800S162-43	4′-6″	3′-9″	3′-1″	2′-5″	—	2′-10″	—	—	—	—
2-800S162-54	8′-0″	7′-3″	6′-8″	6′-1″	5′-7″	6′-5″	5′-9″	5′-1″	4′-7″	4′-0″
2-800S162-68	9′-9″	9′-0″	8′-3″	7′-8″	7′-1″	8′-0″	7′-3″	6′-7″	6′-0″	5′-6″
2-1000S162-43	4′-8″	4′-1″	3′-6″	2′-9″	—	3′-3″	2′-2″	—	—	—
2-1000S162-54	9′-1″	8′-2″	7′-3″	6′-7″	6′-0″	7′-0″	6′-2″	5′-6″	5′-0″	4′-6″
2-1000S162-68	11′-1″	10′-2″	9′-5″	8′-8″	8′-1″	9′-1″	8′-3″	7′-6″	6′-10″	6′-3″
2-1200S162-54	7′-8″	6′-9″	6′-1″	5′-6″	5′-0″	5′-10″	5′-1″	4′-7″	4′-1″	3′-9″
2-1200S162-68	12′-3″	11′-3″	10′-4″	9′-7″	8′-11″	10′-1″	9′-1″	8′-3″	7′-6″	6′-10″

For SI: 1 mil = 0.0254 mm, 1 inch = 25.4 mm, 1 foot = 304.8 mm, 1 pound per square foot = 0.0479 kPa, 1 pound per square inch = 6.895 kPa, 1 ksi = 1,000 psi = 6.895 MPa.

a. Deflection criteria: *L*/360 for live loads, *L*/240 for total loads.

b. Design load assumptions:
   Roof/ceiling dead load is 12 psf.
   Attic dead load is 10 psf.

c. Building width is in the direction of horizontal framing members supported by the header.

d. Minimum Grade 33 ksi steel shall be used for 33 mil and 43 mil thicknesses. Minimum Grade 50 ksi steel shall be used for 54 and 68 mil thicknesses.

TABLE R603.6(3)
BOX-BEAM AND BACK-TO-BACK HEADER SPANS
Headers Supporting One Floor, Roof and Ceiling[a, b, d]

MEMBER DESIGNATION	GROUND SNOW LOAD (20 psf)					GROUND SNOW LOAD (30 psf)				
	Building width[c] (feet)					Building width[c] (feet)				
	24	28	32	36	40	24	28	32	36	40
2-350S162-33	—	—	—	—	—	—	—	—	—	—
2-350S162-43	2′-2″	—	—	—	—	2′-1″	—	—	—	—
2-350S162-54	4′-4″	3′-10″	3′-5″	3′-1″	2′-9″	4′-3″	2′-9″	3′-4″	3′-0″	2′-8″
2-350S162-68	5′-0″	4′-9″	4′-7″	4′-2″	3′-9″	4′-11″	4′-8″	4′-6″	4′-1″	3′-9″
2-550S162-33	—	—	—	—	—	—	—	—	—	—
2-550S162-43	3′-5″	2′-9″	2′-1″	—	—	3′-3″	2′-7″	—	—	—
2-550S162-54	6′-6″	5′-10″	5′-3″	4′-9″	4′-4″	6′-4″	5′-9″	5′-2″	4′-8″	4′-3″
2-550S162-68	7′-2″	6′-10″	6′-5″	5′-11″	5′-6″	7′-0″	6′-9″	6′-4″	5′-10″	5′-4″
2-800S162-33	2′-1″	—	—	—	—	—	—	—	—	—
2-800S162-43	4′-2″	3′-4″	2′-7″	—	—	4′-0″	3′-3″	2′-5″	—	—
2-800S162-54	7′-6″	6′-9″	6′-2″	5′-7″	5′-0″	7′-5″	6′-8″	6′-0″	5′-5″	4′-11″
2-800S162-68	9′-3″	8′-5″	7′-8″	7′-1″	6′-6″	9′-1″	8′-3″	7′-7″	7′-0″	6′-5″
2-1000S162-43	4′-4″	3′-9″	2′-11″	—	—	4′-3″	3′-8″	2′-9″	—	—
2-1000S162-54	8′-6″	7′-6″	6′-8″	6′-0″	5′-5″	8′-4″	7′-4″	6′-6″	5′-10″	5′-4″
2-1000S162-68	10′-6″	9′-7″	8′-9″	8′-0″	7′-5″	10′-4″	9′-5″	8′-7″	7′-11″	7′-3″
2-1200S162-54	7′-1″	6′-2″	5′-6″	5′-0″	4′-6″	6′-11″	6′-1″	5′-5″	4′-10″	4′-5″
2-1200S162-68	11′-7″	10′-7″	9′-8″	8′-11″	8′-2″	11′-5″	10′-5″	9′-6″	8′-9″	8′-0″

For SI: 1 inch = 25.4 mm, 1 foot = 304.8 mm, 1 pound per square foot = 0.0479 kPa, 1 pound per square inch = 6.895 kPa.

a. Deflection criteria: *L*/360 for live loads, *L*/240 for total loads.

b. Design load assumptions:

Second-floor dead load is 10 psf.

Roof/ceiling dead load is 12 psf.

Second-floor live load is 30 psf.

Attic dead load is 10 psf.

c. Building width is in the direction of horizontal framing members supported by the header.

d. Minimum Grade 33 ksi steel shall be used for 33 mil and 43 mil thicknesses. Minimum Grade 50 ksi steel shall be used for 54 and 68 mil thicknesses.

TABLE R603.6(4) BOX-BEAM AND BACK-TO-BACK HEADER SPANS Headers Supporting One Floor, Roof and Ceiling[a, b, d]

MEMBER DESIGNATION	GROUND SNOW LOAD (50 psf)					GROUND SNOW LOAD (70 psf)				
	Building width[c] (feet)					Building width[c] (feet)				
	24	28	32	36	40	24	28	32	36	40
2-350S162-33	—	—	—	—	—	—	—	—	—	—
2-350S162-43	—	—	—	—	—	—	—	—	—	—
2-350S162-54	3′-5″	3′-0″	2′-7″	2′-2″	—	2′-8″	2′-2″	—	—	—
2-350S162-68	4′-6″	4′-1″	3′-8″	3′-3″	2′-11″	3′-9″	3′-3″	2′-10″	2′-5″	2′-1″
2-550S162-33	—	—	—	—	—	—	—	—	—	—
2-550S162-43	2′-0″	—	—	—	—	—	—	—	—	—
2-550S162-54	5′-3″	3′-8″	4′-1″	3′-8″	3′-2″	4′-3″	3′-8″	3′-1″	2′-7″	2′-0″
2-550S162-68	6′-5″	5′-10″	5′-3″	4′-9″	4′-4″	5′-5″	4′-9″	4′-3″	3′-9″	3′-4″
2-800S162-33	—	—	—	—	—	—	—	—	—	—
2-800S162-43	2′-6″	—	—	—	—	—	—	—	—	—
2-800S162-54	6′-1″	5′-5″	4′-10″	4′-3″	3′-9″	4′-11″	4′-3″	3′-8″	3′-0″	2′-5″
2-800S162-68	7′-8″	6′-11″	6′-3″	5′-9″	5′-2″	6′-5″	5′-9″	5′-1″	4′-6″	4′-0″
2-1000S162-43	2′-10″	—	—	—	—	—	—	—	—	—
2-1000S162-54	6′-7″	5′-10″	5′-3″	4′-9″	4′-3″	5′-4″	4′-9″	4′-1″	3′-5″	2′-9″
2-1000S162-68	8′-8″	7′-10″	7′-2″	6′-6″	5′-11″	7′-4″	6′-6″	5′-9″	5′-1″	4′-6″
2-1200S162-54	5′-6″	4′-10″	4′-4″	3′-11″	3′-7″	4′-5″	3′-11″	3′-6″	3′-2″	2′-11″
2-1200S162-68	9′-7″	8′-8″	7′-11″	7′-2″	6′-6″	8′-1″	7′-2″	6′-4″	5′-8″	5′-0″

For SI: 1 mil = 0.0254 mm, 1 inch = 25.4 mm, 1 foot = 304.8 mm, 1 pound per square foot = 0.0479 kPa, 1 pound per square inch = 6.895 kPa, 1 ksi = 1,000 psi = 6.895 MPa.

a. Deflection criteria: *L*/360 for live loads, *L*/240 for total loads.

b. Design load assumptions:

Second-floor dead load is 10 psf.

Roof/ceiling dead load is 12 psf.

Second-floor live load is 30 psf.

Attic dead load is 10 psf.

c. Building width is in the direction of horizontal framing members supported by the header.

d. Minimum Grade 33 ksi steel shall be used for 33 mil and 43 mil thicknesses. Minimum Grade 50 ksi steel shall be used for 54 and 68 mil thicknesses.

TABLE R603.6(5)
BOX-BEAM AND BACK-TO-BACK HEADER SPANS
Headers Supporting Two Floors, Roof and Ceiling[a, b, d]

MEMBER DESIGNATION	GROUND SNOW LOAD (20 psf)					GROUND SNOW LOAD (30 psf)				
	Building width[c] (feet)					Building width[c] (feet)				
	24	28	32	36	40	24	28	32	36	40
2-350S162-33	—	—	—	—	—	—	—	—	—	—
2-350S162-43	—	—	—	—	—	—	—	—	—	—
2-350S162-54	2′-5″	—	—	—	—	2′-4″	—	—	—	—
2-350S162-68	3′-6″	3′-0″	2′-6″	2′-1″	—	3′-5″	2′-11″	2′-6″	2′-0″	—
2-550S162-33	—	—	—	—	—	—	—	—	—	—
2-550S162-43	—	—	—	—	—	—	—	—	—	—
2-550S162-54	3′-11″	3′-3″	2′-8″	2′-0″	—	3′-10″	3′-3″	2′-7″	—	—
2-550S162-68	5′-1″	4′-5″	3′-10″	3′-3″	2′-9″	5′-0″	4′-4″	3′-9″	3′-3″	2′-9″
2-800S162-33	—	—	—	—	—	—	—	—	—	—
2-800S162-43	—	—	—	—	—	—	—	—	—	—
2-800S162-54	4′-7″	3′-10″	3′-1″	2′-5″	—	4′-6″	3′-9″	3′-0″	2′-4″	—
2-800S162-68	6′-0″	5′-3″	4′-7″	3′-11″	3′-4″	6′-0″	5′-2″	4′-6″	3′-11″	3′-3″
2-1000S162-43	—	—	—	—	—	—	—	—	—	—
2-1000S162-54	5′-0″	4′-4″	3′-6″	2′-9″	—	4′-11″	4′-3″	3′-5″	2′-7″	—
2-1000S162-68	6′-10″	6′-0″	5′-3″	4′-6″	3′-10″	6′-9″	5′-11″	5′-2″	4′-5″	3′-9″
2-1200S162-54	4′-2″	3′-7″	3′-3″	2′-11″	—	4′-1″	3′-7″	3′-2″	2′-10″	—
2-1200S162-68	7′-7″	6′-7″	5′-9″	5′-0″	4′-2″	7′-6″	6′-6″	5′-8″	4′-10″	4′-1″

For SI: 1 mil = 0.0254 mm, 1 inch = 25.4 mm, 1 foot = 304.8 mm, 1 pound per square foot = 0.0479 kPa, 1 pound per square inch = 6.895 kPa, 1 ksi = 1,000 psi = 6.895 MPa.

a. Deflection criteria: *L*/360 for live loads, *L*/240 for total loads.

b. Design load assumptions:
   Second-floor dead load is 10 psf.
   Roof/ceiling dead load is 12 psf.
   Second-floor live load is 40 psf.
   Third floor live load is 30 psf.
   Attic live load is 10 psf.

c. Building width is in the direction of horizontal framing members supported by the header.

d. Minimum Grade 33 ksi steel shall be used for 33 mil and 43 mil thicknesses. Minimum Grade 50 ksi steel shall be used for 54 and 68 mil thicknesses.

TABLE R603.6(6)
BOX-BEAM AND BACK-TO-BACK HEADER SPANS
Headers Supporting Two Floors, Roof and Ceiling[a, b, d]

MEMBER DESIGNATION	GROUND SNOW LOAD (50 psf)					GROUND SNOW LOAD (70 psf)				
	Building width[c] (feet)					Building width[c] (feet)				
	24	28	32	36	40	24	28	32	36	40
2-350S162-33	—	—	—	—	—	—	—	—	—	—
2-350S162-43	—	—	—	—	—	—	—	—	—	—
2-350S162-54	2′-2″	—	—	—	—	—	—	—	—	—
2-350S162-68	3′-3″	2′-9″	2′-3″	—	—	2′-11″	2′-5″	—	—	—
2-550S162-33	—	—	—	—	—	—	—	—	—	—
2-550S162-43	—	—	—	—	—	—	—	—	—	—
2-550S162-54	3′-7″	2′-11″	2′-3″	—	—	3′-3″	2′-7″	—	—	—
2-550S162-68	4′-9″	2′-1″	3′-6″	3′-0″	2′-5″	4′-4″	3′-9″	3′-2″	2′-8″	2′-1″
2-800S162-33	—	—	—	—	—	—	—	—	—	—
2-800S162-43	—	—	—	—	—	—	—	—	—	—
2-800S162-54	4′-3″	3′-5″	2′-8″	—	—	3′-9″	3′-0″	2′-3″	—	—
2-800S162-68	5′-8″	4′-11″	4′-2″	3′-7″	2′-11″	5′-3″	4′-6″	3′-10″	3′-3″	2′-7″
2-1000S162-43	—	—	—	—	—	—	—	—	—	—
2-1000S162-54	4′-8″	3′-11″	3′-1″	2′-2″	—	4′-3″	3′-5″	2′-7″	—	—
2-1000S162-68	6′-5″	5′-7″	4′-9″	4′-1″	3′-4″	5′-11″	5′-1″	4′-5″	3′-8″	2′-11″
2-1200S162-54	3′-11″	3′-5″	3′-0″	2′-4″	—	3′-7″	3′-2″	2′-10″	—	—
2-1200S162-68	7′-1″	6′-2″	5′-3″	4′-6″	3′-8″	6′-6″	5′-8″	4′-10″	4′-0″	3′-3″

For SI: 1 mil = 0.0254 mm, 1 inch = 25.4 mm, 1 foot = 304.8 mm, 1 pound per square foot = 0.0479 kPa, 1 pound per square inch = 6.895 kPa, 1 ksi = 1,000 psi = 6.895 MPa.

a. Deflection criteria: *L*/360 for live loads, *L*/240 for total loads.

b. Design load assumptions:
Second-floor dead load is 10 psf.
Roof/ceiling dead load is 12 psf.
Second-floor live load is 40 psf.
Third floor live load is 30 psf.
Attic live load is 10 psf.

c. Building width is in the direction of horizontal framing members supported by the header.

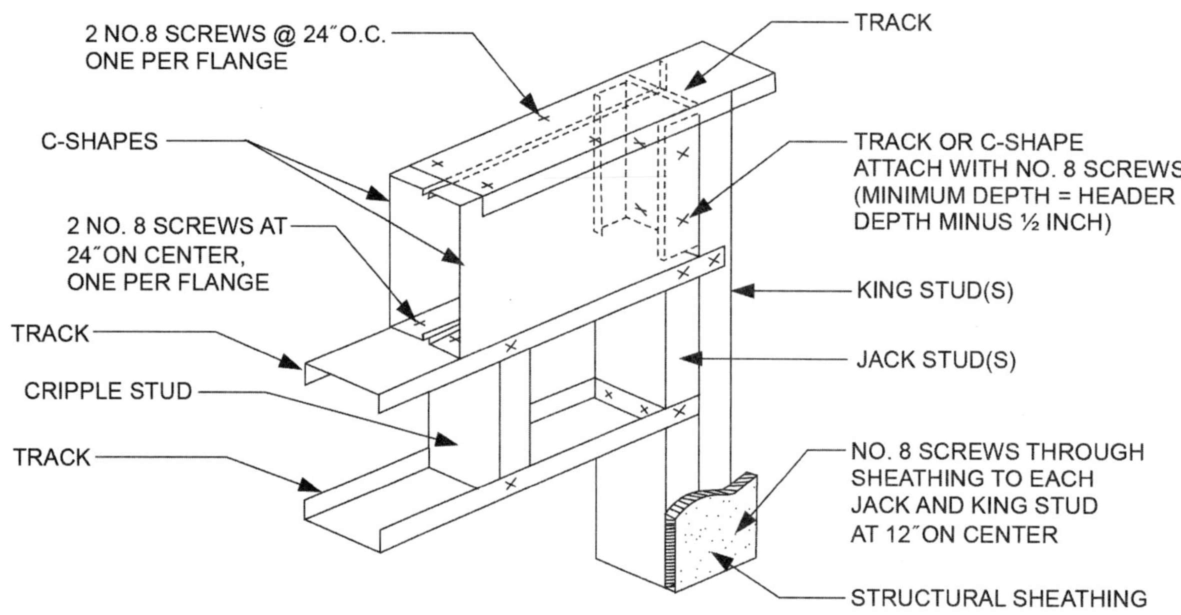

For SI: 1 inch = 25.4 mm.

FIGURE R603.6(1)
BOX BEAM HEADER

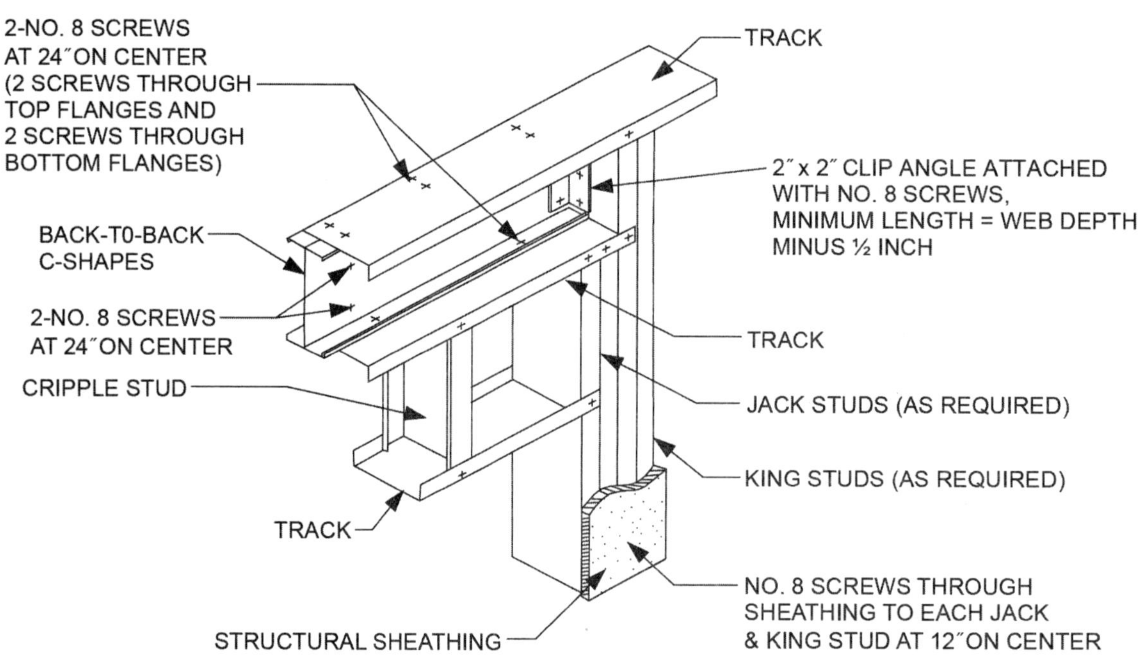

For SI: 1 inch = 25.4 mm.

FIGURE R603.6(2)
BACK-TO-BACK HEADER

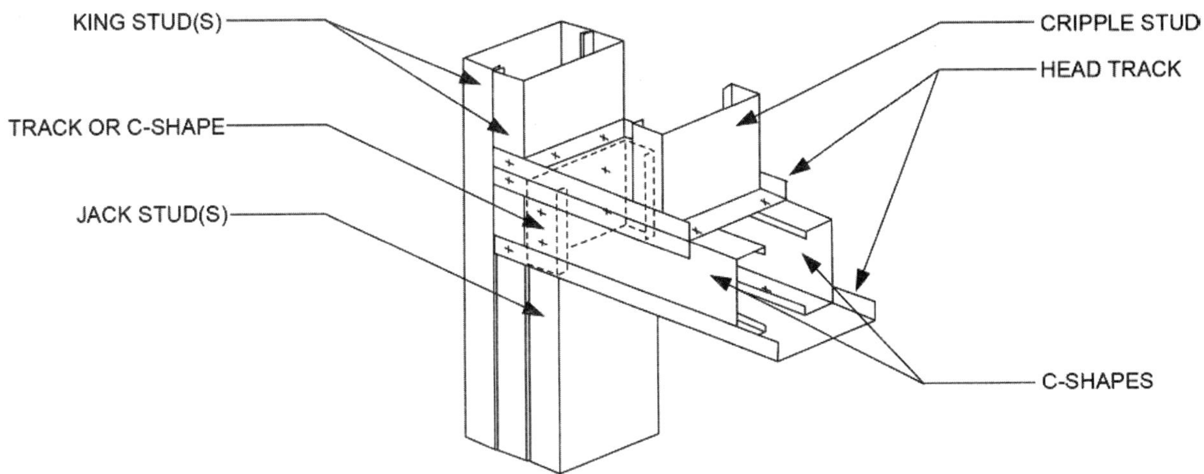

FIGURE R603.6.1(1)
BOX BEAM HEADER IN GABLE ENDWALL

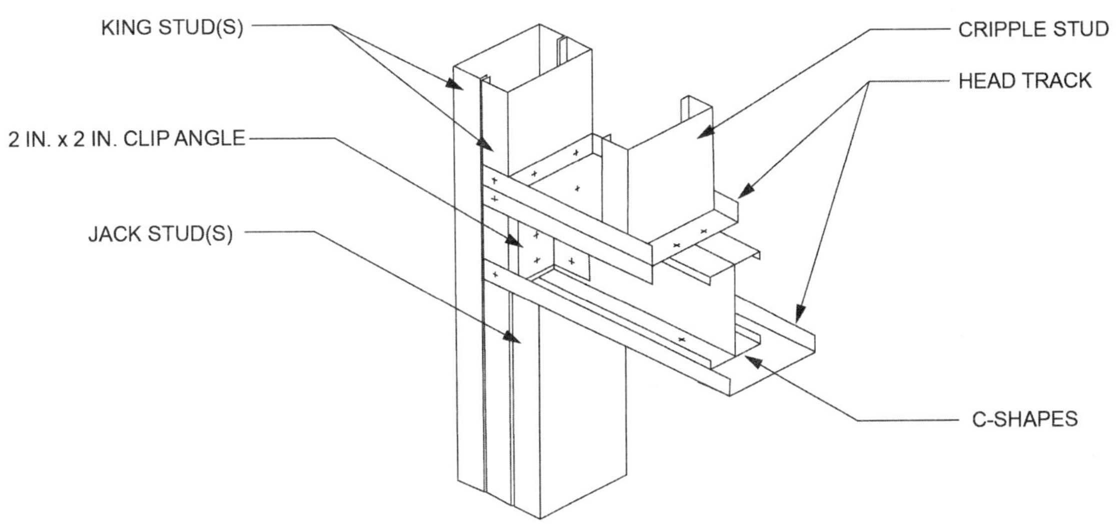

For SI: 1 inch = 25.4 mm.

FIGURE R603.6.1(2)
BACK-TO-BACK HEADER IN GABLE ENDWALL

R603.8 Head and sill track. Head track spans above door and window openings and sill track spans beneath window openings shall comply with Table R603.8. For openings less than 4 feet (1219 mm) in height that have both a head track and a sill track, multiplying the spans by 1.75 shall be permitted in Table R603.8. For openings less than or equal to 6 feet (1829 mm) in height that have both a head track and a sill track, multiplying the spans in Table R603.8 by 1.50 shall be permitted.

TABLE R603.7(1)
TOTAL NUMBER OF JACK AND KING STUDS REQUIRED AT EACH END OF AN OPENING

SIZE OF OPENING (feet-inches)	24-INCH O.C. STUD SPACING		16-INCH O.C. STUD SPACING	
	No. of jack studs	No. of king studs	No. of jack studs	No. of king studs
Up to 3′-6″	1	1	1	1
> 3′-6″ to 5′-0″	1	2	1	2
> 5′-0″ to 5′-6″	1	2	2	2
> 5′-6″ to 8′-0″	1	2	2	2
> 8′-0″ to 10′-6″	2	2	2	3
> 10′-6″ to 12′-0″	2	2	3	3
> 12′-0″ to 13′-0″	2	3	3	3
> 13′-0″ to 14′-0″	2	3	3	4
> 14′-0″ to 16′-0″	2	3	3	4
> 16′-0″ to 18′-0″	3	3	4	4

For SI: 1 inch = 25.4 mm, 1 foot = 304.8 mm.

TABLE R603.7(2)
HEADER TO KING STUD CONNECTION REQUIREMENTS[a, b, c, d]

HEADER SPAN (feet)	ULTIMATE WIND SPEED (mph), EXPOSURE CATEGORY					
	115 B	120 B	130 B	< 140 B	130 C	<140 C
			115 C	120 C		
≤ 4	4-No. 8 screws	4-No. 8 screws	4-No. 8 screws	4-No. 8 screws	6-No. 8 screws	6-No. 8 screws
> 4 to 8	4-No. 8 screws	4-No. 8 screws	4-No. 8 screws	6-No. 8 screws	8-No. 8 screws	8-No. 8 screws
> 8 to 12	4-No. 8 screws	6-No. 8 screws	6-No. 8 screws	8-No. 8 screws	10-No. 8 screws	12-No. 8 screws
> 12 to 16	4-No. 8 screws	6-No. 8 screws	8-No. 8 screws	10-No. 8 screws	12-No. 8 screws	14-No. 8 screws

For SI: 1 foot = 304.8 mm, 1 mile per hour = 0.447 m/s, 1 pound = 4.448 N.

a. All screw sizes shown are minimum.

b. For headers located on the first floor of a two-story building or the first or second floor of a three-story building, the total number of screws is permitted to be reduced by 2 screws, but the total number of screws shall be not less than four.

c. For roof slopes of 6:12 or greater, the required number of screws shall be permitted to be reduced by half, but the total number of screws shall be not less than four.

d. Screws can be replaced by an uplift connector that has a capacity of the number of screws multiplied by 164 pounds.

TABLE R603.8
HEAD AND SILL TRACK SPAN

ULTIMATE WIND SPEED AND EXPOSURE CATEGORY (mph)		ALLOWABLE HEAD AND SILL TRACK SPAN[a, b, c] (feet-inches)					
		TRACK DESIGNATION[d]					
B	C	350T125-33	350T125-43	350T125-54	550T125-33	550T125-43	550T125-54
115	—	5′-9″	6′-9″	9′-3″	7′-3″	9′-1″	12′-5″
120	—	5′-6″	6′-6″	8′-11″	7′-0″	8′-9″	11′-11″
130	115	4′-10″	5′-9″	7′-10″	6′-2″	7′-8″	10′-6″
< 140	120	4′-8″	5′-6″	7′-6″	5′-11″	7′-4″	10′-1″
—	130	4′-3″	5′-1″	6′-11″	5′-6″	6′-9″	9′-4″
—	< 140	4′-0″	4′-9″	6′-5″	5′-1″	6′-4″	8′-8″

For SI: 1 mil = 0.0254 mm, 1 inch = 25.4 mm, 1 foot = 304.8 mm, 1 mile per hour = 0.447 m/s, 1 ksi = 1,000 psi = 6.895 MPa.

a. Deflection limit: *L*/240.

b. Head and sill track spans are based on components and cladding wind pressures and 48-inch tributary span.

c. For openings less than 4 feet in height that have both a head track and sill track, the spans are permitted to be multiplied by 1.75. For openings less than or equal to 6 feet in height that have both a head track and a sill track, the spans are permitted to be multiplied by a factor of 1.5.

d. Minimum Grade 33 ksi steel shall be used for 33 mil and 43 mil thicknesses. Minimum Grade 50 ksi steel shall be used for 54 and 68 mil thicknesses.

R603.9 Structural sheathing. Structural sheathing shall be installed in accordance with Figure R603.9 and this section on all sheathable exterior wall surfaces, including areas above and below openings.

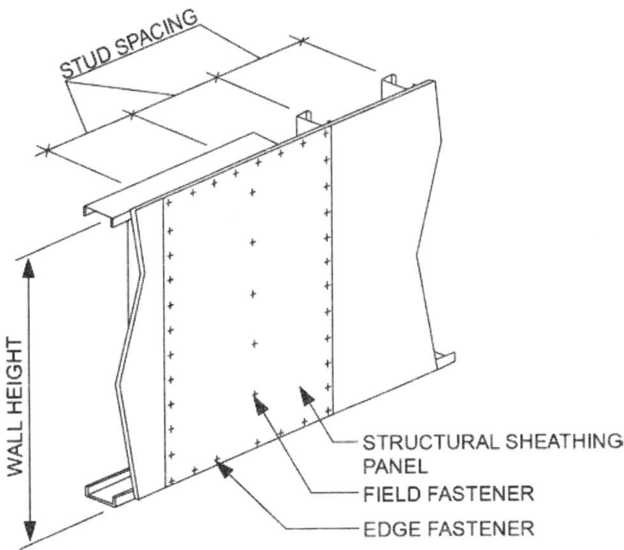

FIGURE R603.9
STRUCTURAL SHEATHING FASTENING PATTERN

R603.9.1 Sheathing materials. Structural sheathing panels shall consist of minimum $^{7}/_{16}$-inch-thick (11 mm) oriented strand board or $^{15}/_{32}$-inch-thick (12 mm) plywood.

R603.9.2 Determination of minimum length of full-height sheathing. The minimum length of full-height sheathing on each *braced wall line* shall be determined by multiplying the length of the *braced wall line* by the percentage obtained from Table R603.9.2(1) and by the plan aspect-ratio adjustment factors obtained from Table R603.9.2(2). The minimum length of full-height sheathing shall be not less than 20 percent of the *braced wall line* length.

To be considered full-height sheathing, structural sheathing shall extend from the bottom to the top of the wall without interruption by openings. Only sheathed, full-height wall sections, uninterrupted by openings, which are not less than 48 inches (1219 mm) wide, shall be counted toward meeting the minimum percentages in Table R603.9.2(1). In addition, structural sheathing shall comply with all of the following requirements:

1. Be installed with the long dimension parallel to the stud framing and shall cover the full vertical height of wall from the bottom of the bottom track to the top of the top track of each *story*. Installing the long dimension perpendicular to the stud framing or using shorter segments shall be permitted provided that the horizontal joint is blocked as described in Item 2.
2. Be blocked where the long dimension is installed perpendicular to the stud framing. Blocking shall be not less than 33 mil (0.84 mm) thickness. Each horizontal structural sheathing panel shall be fastened with No. 8 screws spaced at 6 inches (152 mm) on center to the blocking at the joint.
3. Be applied to each end (corners) of each of the exterior walls with a minimum 48-inch-wide (1219 mm) panel.

Exception: Where stone or masonry veneer is installed, the required length of full-height sheathing and overturning anchorage required shall be determined in accordance with Section R603.9.5.

TABLE R603.9.2(1)
MINIMUM PERCENTAGE OF FULL-HEIGHT STRUCTURAL SHEATHING ON EXTERIOR WALLS[a, b]

WALL SUPPORTING	ROOF SLOPE	ULTIMATE WIND SPEED AND EXPOSURE (mph)					
		115 B	120 B	130 B / 115 C	< 140 B / 120 C	< 130 C	< 140 C
Roof and ceiling only (one story or top floor of two- or three-story building)	3:12	9	11	11	13	17	20
	6:12	13	15	17	22	28	35
	9:12	23	27	29	33	53	59
	12:12	32	39	40	44	70	76
One story, roof and ceiling (first floor of a two-story building or second floor of a three-story building)	3:12	26	32	34	39	53	67
	6:12	27	33	34	44	61	75
	9:12	38	45	46	61	78	92
	12:12	43	53	57	72	106	116
Two stories, roof and ceiling (first floor of a three-story building)	3:12	43	53	57	64	89	113
	6:12	41	51	51	67	95	114
	9:12	53	63	63	89	104	126
	12:12	54	67	74	100	142	157

For SI: 1 mph = 0.447 m/s.

a. Linear interpolation is permitted.

b. For hip-roofed homes the minimum percentage of full-height sheathing, based on wind, is permitted to be multiplied by a factor of 0.95 for roof slopes not exceeding 7:12 and a factor of 0.9 for roof slopes greater than 7:12.

TABLE R603.9.2(2)
FULL-HEIGHT SHEATHING LENGTH ADJUSTMENT FACTORS

PLAN ASPECT RATIO	LENGTH ADJUSTMENT FACTORS	
	Short wall	Long wall
1:1	1.0	1.0
1.5:1	1.5	0.67
2:1	2.0	0.50
3:1	3.0	0.33
4:1	4.0	0.25

R603.9.2.1 Full-height sheathing. The minimum percentage of full-height structural sheathing shall be multiplied by 1.10 for 9-foot-high (2743 mm) walls and multiplied by 1.20 for 10-foot-high (3048 mm) walls.

R603.9.2.2 Full-height sheathing in lowest story. In the lowest *story* of a *dwelling*, multiplying the percentage of full-height sheathing required in Table R603.9.2(1) by 0.6 shall be permitted where hold-down anchors are provided in accordance with Section R603.9.4.2.

R603.9.3 Structural sheathing fastening. Edges and interior areas of structural sheathing panels shall be fastened to framing members and tracks in accordance with Figure R603.9 and Table R603.3.2(1). Screws for attachment of structural sheathing panels shall be bugle-head, flat-head, or similar head style with a minimum head diameter of 0.29 inch (8 mm).

For continuously sheathed *braced wall lines* using *wood structural panels* installed with No. 8 screws spaced 4 inches (102 mm) on center at all panel edges and 12 inches (304.8 mm) on center on intermediate framing members, the following shall apply:

1. Multiplying the percentages of full-height sheathing in Table R603.9.2(1) by 0.72 shall be permitted.
2. For bottom track attached to foundations or framing below, the bottom track anchor or screw connection spacing in Tables R505.3.1(1) and R603.3.1 shall be multiplied by two-thirds.

R603.9.4 Uplift connection requirements. Uplift connections shall be provided in accordance with this section.

R603.9.4.1 Ultimate design wind speeds greater than 130 mph. Where ultimate design wind speeds exceed 130 miles per hour (58 m/s), Exposure Category C walls shall be provided with direct uplift connections in accordance with AISI S230, Section E13.3, and AISI S230, Section F8.2, as required for 140 miles per hour (63 m/s), Exposure Category C.

R603.9.4.2 Hold-down anchor. Where the percentage of full-height sheathing is adjusted in accordance with Section R603.9.2.2, a hold-down anchor, with a strength of 4,300 pounds (19 kN), shall be provided at each end of each full-height sheathed wall section used to meet the minimum percent sheathing requirements of Section R603.9.2. Hold-down anchors shall be attached to back-to-back studs; structural sheathing panels shall have edge fastening to the studs, in accordance with Section R603.9.3 and AISI S230, Table E11-1.

A single hold-down anchor, installed in accordance with Figure R603.9.4.2, shall be permitted at the corners of buildings.

R603.9.5 Structural sheathing for stone and masonry veneer. Where stone and masonry veneer are installed in accordance with Section R703.8, the length of full-height sheathing for exterior and interior wall lines backing or perpendicular to and laterally supporting walls with veneer shall comply with this section.

R603.9.5.1 Seismic Design Category C. In *Seismic Design Category* C, the length of structural sheathing for walls supporting one *story*, roof and ceiling shall be the greater of the amounts required by Section R603.9.2, except Section R603.9.2.2 shall be permitted.

R603.9.5.2 Seismic Design Categories D_0, D_1 and D_2. In *Seismic Design Categories* D_0, D_1 and D_2, the required length of structural sheathing and overturning anchorage shall be determined in accordance with Tables R603.9.5(1), R603.9.5(2), R603.9.5(3), and R603.9.5(4). Overturning anchorage shall be installed on the doubled studs at the end of each full-height wall segment.

SECTION R604 WOOD STRUCTURAL PANELS

R604.1 Identification and grade. *Wood structural panels* shall conform to DOC PS 1, DOC PS 2, ANSI/APA PRP 210, CSA O325 or CSA O437. Panels shall be identified by a grade *mark* or certificate of inspection issued by an *approved* agency.

R604.2 Allowable spans. The maximum allowable spans for wood structural panel wall sheathing shall not exceed the values set forth in Table R602.3(3).

R604.3 Installation. Wood structural panel wall sheathing shall be attached to framing in accordance with Table R602.3(1) or R602.3(3).

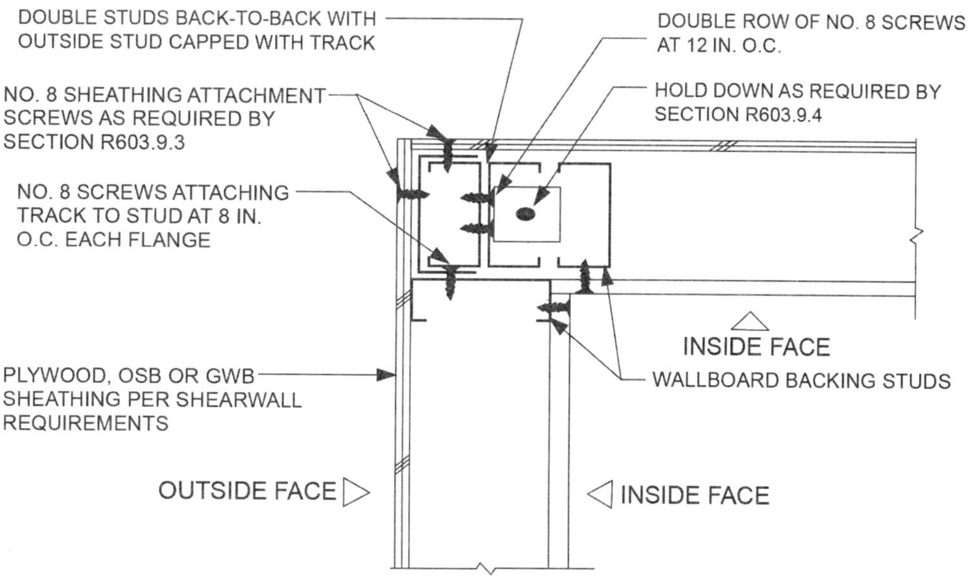

For SI: 1 inch = 25.4 mm.

FIGURE R603.9.4.2
CORNER STUD HOLD-DOWN DETAIL

SECTION R605
PARTICLEBOARD

R605.1 Identification and grade. Particleboard shall conform to ANSI A208.1 and shall be so identified by a grade *mark* or certificate of inspection issued by an *approved* agency. Particleboard shall comply with the grades specified in Table R602.3(4).

SECTION R606
GENERAL MASONRY CONSTRUCTION

R606.1 General. Masonry construction shall be designed and constructed in accordance with the provisions of this section, TMS 402, TMS 403 or TMS 404.

R606.1.1 Professional registration not required. Where the empirical design provisions of Appendix A of TMS 402, the provisions of TMS 403, or the provisions of this section are used to design masonry, project drawings, typical details and specifications are not required to bear the seal of the architect or engineer responsible for design, unless otherwise required by the state law of the *jurisdiction* having authority.

R606.2 Masonry construction materials.

R606.2.1 Concrete masonry units. *Concrete masonry units* shall conform to the following standards: ASTM C55 for concrete brick; ASTM C73 for calcium silicate face brick; ASTM C90 for load-bearing *concrete masonry units*; ASTM C744 for prefaced concrete and calcium silicate *masonry units*; or ASTM C1634 for concrete facing brick.

R606.2.2 Clay or shale masonry units. Clay or shale *masonry units* shall conform to the following standards: ASTM C34 for structural clay *load-bearing wall* tile; ASTM C56 for structural clay nonload-bearing wall tile; ASTM C62 for building brick (*solid masonry* units made from clay or shale); ASTM C126 for ceramic-glazed structural clay facing tile, facing brick and *solid masonry* units; ASTM C212 for structural clay facing tile; ASTM C216 for facing brick (*solid masonry* units made from clay or shale); ASTM C652 for hollow brick (*hollow masonry units* made from clay or shale); ASTM C1088 for solid units of thin veneer brick; or ASTM C1405 for glazed brick (single-fired solid brick units).

Exception: Structural clay tile for nonstructural use in fireproofing of structural members and in wall furring shall not be required to meet the compressive strength specifications. The fire-resistance rating shall be determined in accordance with ASTM E119 or UL 263 and shall comply with the requirements of Section R302.

R606.2.3 AAC masonry. AAC *masonry units* shall conform to ASTM C1691 and ASTM C1693 for the strength class specified.

R606.2.4 Stone masonry units. Stone *masonry units* shall conform to the following standards: ASTM C503 for marble building stone (exterior); ASTM C568 for limestone building stone; ASTM C615 for granite building stone; ASTM C616 for sandstone building stone; or ASTM C629 for slate building stone.

R606.2.5 Architectural cast stone. Architectural cast stone shall conform to ASTM C1364.

**TABLE R603.9.5(1)
REQUIRED LENGTH OF FULL-HEIGHT SHEATHING AND ASSOCIATED OVERTURNING ANCHORAGE FOR WALLS SUPPORTING WALLS WITH STONE OR MASONRY VENEER AND USING 33-MIL COLD-FORMED STEEL FRAMING AND 6-INCH SCREW SPACING ON THE PERIMETER OF EACH PANEL OF STRUCTURAL SHEATHING**

SEISMIC DESIGN CATEGORY	STORY	BRACED WALL LINE LENGTH (feet)						SINGLE-STORY HOLD-DOWN FORCE (pounds)	CUMULATIVE HOLD-DOWN FORCE (pounds)
		10	20	30	40	50	60		
		Minimum total length of braced wall panels required along each braced wall line (feet)							
D_0		3.3	4.7	6.1	7.4	8.8	10.2	3,360	—
		5.3	8.7	12.1	15.4	18.8	22.2	3,360	6,720
		7.3	12.7	18.0	23.4	28.8	34.2	3,360	10,080
D_1		4.1	5.8	7.5	9.2	10.9	12.7	3,360	—
		6.6	10.7	14.9	19.1	23.3	27.5	3,360	6,720
		9.0	15.7	22.4	29.0	35.7	42.2	3,360	10,080
D_2		5.7	8.2	10.6	13.0	15.4	17.8	3,360	—
		9.2	15.1	21.1	27.0	32.9	38.8	3,360	6,720
		12.7	22.1	31.5	40.9	50.3	59.7	3,360	10,080

For SI: 1 mil = 0.0254 mm, 1 inch = 25.4 mm, 1 foot = 304.8 mm, 1 pound-force = 4.448 N.

TABLE R603.9.5(2)
REQUIRED LENGTH OF FULL-HEIGHT SHEATHING AND ASSOCIATED OVERTURNING ANCHORAGE FOR WALLS SUPPORTING WALLS WITH STONE OR MASONRY VENEER AND USING 43-MIL COLD-FORMED STEEL FRAMING AND 6-INCH SCREW SPACING ON THE PERIMETER OF EACH PANEL OF STRUCTURAL SHEATHING

SEISMIC DESIGN CATEGORY	STORY	BRACED WALL LINE LENGTH (feet)						SINGLE-STORY HOLD-DOWN FORCE (pounds)	CUMULATIVE HOLD-DOWN FORCE (pounds)
		10	20	30	40	50	60		
		Minimum total length of braced wall panels required along each braced wall line (feet)							
D_0		2.8	4.0	5.1	6.3	7.5	8.7	3,960	—
		4.5	7.4	10.2	13.1	16.0	18.8	3,960	7,920
		6.2	10.7	15.3	19.9	24.4	29.0	3,960	11, 880
D_1		3.5	4.9	6.4	7.8	9.3	10.7	3,960	—
		5.6	9.1	12.7	16.2	19.8	23.3	3,960	7,920
		7.7	13.3	19.0	24.6	30.3	35.9	3,960	11, 880
D_2		4.9	6.9	9.0	11.0	13.1	15.1	3,960	—
		7.8	12.9	17.9	22.9	27.9	32.9	3,960	7,920
		10.8	18.8	26.7	34.7	42.7	50.7	3,960	11, 880

For SI: 1 mil = 0.0254 mm, 1 inch = 25.4 mm, 1 foot = 304.8 mm, 1 pound-force = 4.448 N.

TABLE R603.9.5(3)
REQUIRED LENGTH OF FULL-HEIGHT SHEATHING AND ASSOCIATED OVERTURNING ANCHORAGE FOR WALLS SUPPORTING WALLS WITH STONE OR MASONRY VENEER AND USING 33-MIL COLD-FORMED STEEL FRAMING AND 4-INCH SCREW SPACING ON THE PERIMETER OF EACH PANEL OF STRUCTURAL SHEATHING

SEISMIC DESIGN CATEGORY	STORY	BRACED WALL LINE LENGTH (feet)						SINGLE-STORY HOLD-DOWN FORCE (pounds)	CUMULATIVE HOLD-DOWN FORCE (pounds)
		10	20	30	40	50	60		
		Minimum total length of braced wall panels required along each braced wall line (feet)							
D_0		2.5	3.6	4.6	5.7	6.8	7.8	4,392	—
		4.0	6.6	9.2	11.8	14.4	17.0	4,392	8,784
		5.6	9.7	13.8	17.9	22.0	26.2	4,392	13,176
D_1		3.1	4.4	5.7	7.1	8.4	9.7	4,392	—
		5.0	8.2	11.4	14.6	17.8	21.0	4,392	8,784
		6.9	12.0	17.1	22.2	27.3	32.4	4,392	13,176
D_2		4.4	6.2	8.1	10.0	11.8	13.7	4,392	—
		7.1	11.6	16.1	20.6	25.1	29.7	4,392	8,784
		9.7	16.9	24.1	31.3	38.5	45.7	4,392	13,176

For SI: 1 mil = 0.0254 mm, 1 inch = 25.4 mm, 1 foot = 304.8 mm, 1 pound-force = 4.448 N.

TABLE R603.9.5(4)
REQUIRED LENGTH OF FULL-HEIGHT SHEATHING AND ASSOCIATED OVERTURNING ANCHORAGE FOR WALLS SUPPORTING WALLS WITH STONE OR MASONRY VENEER AND USING 43-MIL COLD-FORMED STEEL FRAMING AND 4-INCH SCREW SPACING ON THE PERIMETER OF EACH PANEL OF STRUCTURAL SHEATHING

SEISMIC DESIGN CATEGORY	STORY	BRACED WALL LINE LENGTH (feet)						SINGLE-STORY HOLD-DOWN FORCE (pounds)	CUMULATIVE HOLD-DOWN FORCE (pounds)
		10	20	30	40	50	60		
		Minimum total length of braced wall panels required along each braced wall line (feet)							
D_0		1.9	2.7	3.4	4.2	5.0	5.8	5,928	—
		3.0	4.9	6.8	8.8	10.7	12.6	5,928	11,856
D_1		2.3	3.3	4.3	5.2	6.2	7.2	5,928	—
		3.7	6.1	8.5	10.8	13.2	15.6	5,928	11,856
D_2		3.3	4.6	6.0	7.4	8.7	10.1	5,928	—
		5.2	8.6	11.9	15.3	18.6	22.0	5,928	11,856

For SI: 1 mil = 0.0254 mm, 1 inch = 25.4 mm, 1 foot = 304.8 mm, 1 pound-force = 4.448 N.

R606.2.6 Adhered manufactured stone masonry veneer units. Adhered manufactured stone masonry veneer units shall conform to ASTM C1670.

R606.2.7 Second-hand units. Second-hand *masonry units* shall not be reused unless they conform to the requirements of new units. The units shall be of whole, sound materials and free from cracks and other defects that will interfere with proper laying or use. Old mortar shall be cleaned from the unit before reuse.

R606.2.8 Mortar. Except for mortars listed in Sections R606.2.9, R606.2.10 and R606.2.11, mortar for use in masonry construction shall meet the proportion specifications of Table R606.2.8 or the property specifications of ASTM C270. The type of mortar shall be in accordance with Sections R606.2.8.1, R606.2.8.2 and R606.2.8.3.

R606.2.8.1 Foundation walls. Mortar for masonry foundation walls constructed as set forth in Tables R404.1.1(1) through R404.1.1(4) shall be Type M or S mortar.

R606.2.8.2 Masonry in Seismic Design Categories A, B and C. Mortar for masonry serving as the lateral force-resisting system in *Seismic Design Categories* A, B and C shall be Type M, S or N mortar.

R606.2.8.3 Masonry in Seismic Design Categories D_0, D_1 and D_2. Mortar for masonry serving as the lateral-force-resisting system in *Seismic Design Categories* D_0, D_1 and D_2 shall be Type M or S Portland cement-lime or mortar cement.

R606.2.9 Surface-bonding mortar. Surface-bonding mortar shall comply with ASTM C887. Surface bonding of *concrete masonry units* shall comply with ASTM C946.

R606.2.10 Mortar for AAC masonry. Thin-bed mortar for AAC masonry shall comply with Article 2.1 C.1 of TMS 602. Mortar used for the leveling courses of AAC masonry shall comply with Article 2.1 C.2 of TMS 602.

R606.2.11 Mortar for adhered masonry veneer. Mortar for use with adhered masonry veneer shall conform to ASTM C270 Type S or Type N or shall comply with ANSI A118.4 for latex-modified Portland cement mortar.

R606.2.12 Grout. Grout shall consist of cementitious material and aggregate in accordance with ASTM C476 or the proportion specifications of Table R606.2.12. Type M or Type S mortar to which sufficient water has been added to produce pouring consistency shall be permitted to be used as grout.

R606.2.13 Metal reinforcement and accessories. Metal reinforcement and accessories shall conform to Article 2.4 of TMS 602.

R606.3 Construction requirements.

R606.3.1 Bed and head joints. Unless otherwise required or indicated on the project drawings, head and bed joints shall be $^{3}/_{8}$ inch (9.5 mm) thick, except that the thickness of the bed joint of the starting course placed over foundations shall be not less than $^{1}/_{4}$ inch (6.4 mm) and not more than $^{3}/_{4}$ inch (19.1 mm). Mortar joint thick-

TABLE R606.2.8
MORTAR PROPORTIONS[a, b]

MORTAR	TYPE	PROPORTIONS BY VOLUME (cementitious materials)								
		Portland cement or blended cement	Mortar cement			Masonry cement			Hydrated lime[c] or lime putty	Aggregate ratio (measured in damp, loose conditions)
			M	S	N	M	S	N		
Cement-lime	M	1	—	—	—	—	—	—	$^{1}/_{4}$	Not less than $2^{1}/_{4}$ and not more than 3 times the sum of separate volumes of lime, if used, and cement
	S	1	—	—	—	—	—	—	over $^{1}/_{4}$ to $^{1}/_{2}$	
	N	1	—	—	—	—	—	—	over $^{1}/_{2}$ to $1^{1}/_{4}$	
	O	1	—	—	—	—	—	—	over $1^{1}/_{4}$ to $2^{1}/_{2}$	
Mortar cement	M	1	—	—	1	—	—	—	—	
	M	—	1	—	—	—	—	—		
	S	$^{1}/_{2}$	—	—	1	—	—	—		
	S	—	—	1	—	—	—	—		
	N	—	—	—	1	—	—	—		
	O	—	—	—	1	—	—	—		
Masonry cement	M	1	—	—	—	—	—	1	—	
	M	—	—	—	—	1	—	—		
	S	$^{1}/_{2}$	—	—	—	—	—	1		
	S	—	—	—	—	—	1	—		
	N	—	—	—	—	—	—	1		
	O	—	—	—	—	—	—	1		

For SI: 1 cubic foot = 0.0283 m^3, 1 pound = 0.454 kg.

a. For the purpose of these specifications, the weight of 1 cubic foot of the respective materials shall be considered to be as follows:

Hydrated lime = 40 pounds
Lime putty (Quicklime) = 80 pounds
Masonry cement = Weight printed on bag
Mortar cement = Weight printed on bag
Portland cement = 94 pounds
Sand, damp and loose = 80 pounds of dry sand

b. Two air-entraining materials shall not be combined in mortar.

c. Hydrated lime conforming to the requirements of ASTM C207.

TABLE R606.2.12
GROUT PROPORTIONS BY VOLUME FOR MASONRY CONSTRUCTION

TYPE	PORTLAND CEMENT OR BLENDED CEMENT SLAG CEMENT	HYDRATED LIME OR LIME PUTTY	AGGREGATE MEASURED IN A DAMP, LOOSE CONDITION	
			Fine	Coarse
Fine	1	0 to 1/10	$2^{1}/_{4}$ to 3 times the sum of the volumes of the cementitious materials	—
Coarse	1	0 to 1/10	$2^{1}/_{4}$ to 3 times the sum of the volumes of the cementitious materials	1 to 2 times the sum of the volumes of the cementitious materials

ness for load-bearing masonry shall be within the following tolerances from the specified dimensions:

1. Bed joint: + $^1/_8$ inch (3.2 mm).
2. Head joint: - $^1/_4$ inch (6.4 mm), + $^3/_8$ inch (9.5 mm).
3. Collar joints: - $^1/_4$ inch (6.4 mm), + $^3/_8$ inch (9.5 mm).

R606.3.2 Masonry unit placement. The mortar shall be sufficiently plastic and units shall be placed with sufficient pressure to extrude mortar from the joint and produce a tight joint. Deep furrowing of bed joints that produces voids shall not be permitted. Any units disturbed to the extent that initial bond is broken after initial placement shall be removed and relaid in fresh mortar. Surfaces to be in contact with mortar shall be clean and free of deleterious materials.

R606.3.2.1 Solid masonry. *Solid masonry* units shall be laid with full head and bed joints and all interior vertical joints that are designed to receive mortar shall be filled.

R606.3.2.2 Hollow masonry. For *hollow masonry units*, head and bed joints shall be filled solidly with mortar for a distance in from the face of the unit not less than the thickness of the face shell.

R606.3.3 Installation of wall ties. The installation of wall ties shall be as follows:

1. The ends of wall ties shall be embedded in mortar joints. Wall ties shall have not less than $^5/_8$-inch (15.9 mm) mortar coverage from the exposed face.
2. Wall ties shall not be bent after being embedded in grout or mortar.
3. For *solid masonry* units, solid grouted hollow units, or hollow units in anchored masonry veneer, wall ties shall be embedded in mortar bed not less than $1^1/_2$ inches (38 mm).
4. For *hollow masonry units* in other than anchored masonry veneer, wall ties shall engage outer face shells by not less than $^1/_2$ inch (13 mm).

R606.3.4 Protection for reinforcement. Bars shall be completely embedded in mortar or grout. Joint reinforcement embedded in horizontal mortar joints shall not have less than $^5/_8$-inch (15.9 mm) mortar coverage from the exposed face. Other reinforcement shall have a minimum coverage of one bar diameter over all bars, but not less than $^3/_4$ inch (19 mm), except where exposed to weather or soil, in which case the minimum coverage shall be 2 inches (51 mm).

R606.3.4.1 Corrosion protection. Minimum corrosion protection of joint reinforcement, anchor ties and wire fabric for use in masonry wall construction shall conform to Table R606.3.4.1.

R606.3.5 Grouting requirements.

R606.3.5.1 Grout placement. Grout shall be a plastic mix suitable for pumping without segregation of the constituents and shall be mixed thoroughly. Grout shall be placed by pumping or by an *approved* alternate method and shall be placed before any initial set occurs and not more than $1^1/_2$ hours after water has been added. Grout shall be consolidated by puddling or mechanical vibrating during placing and reconsolidated after excess moisture has been absorbed but before plasticity is lost. Grout shall not be pumped through aluminum pipes.

Maximum pour heights and the minimum dimensions of spaces provided for grout placement shall conform to Table R606.3.5.1. Grout shall be poured in lifts with a maximum height of 8 feet (2438 mm). Where a total grout pour exceeds 8 feet (2438 mm) in height, the grout shall be placed in lifts not exceeding 64 inches (1626 mm) and special inspection during grouting shall be required. If the work is stopped for 1 hour or longer, the horizontal construction joints shall be formed by stopping all tiers at the same elevation and with the grout 1 inch (25 mm) below the top.

R606.3.5.2 Cleanouts. Provisions shall be made for cleaning the space to be grouted. Mortar that projects more than $^1/_2$ inch (12.7 mm) into the grout space and any other foreign matter shall be removed from the grout space prior to inspection and grouting. Where required by the building official, cleanouts shall be provided in the bottom course of masonry for each grout pour where the grout pour height exceeds 64 inches (1626 mm). In solid grouted masonry, cleanouts shall be spaced horizontally not more than 32 inches (813 mm) on center. The cleanouts shall be sealed before grouting and after inspection.

R606.3.5.3 Construction. Requirements for grouted masonry construction shall be as follows:

1. Masonry shall be built to preserve the unobstructed vertical continuity of the cells or spaces to be filled. In partially grouted

TABLE R606.3.4.1
MINIMUM CORROSION PROTECTION

MASONRY METAL ACCESSORY	STANDARD
Joint reinforcement, interior walls	ASTM A641, Class 1
Wire ties or anchors in exterior walls completely embedded in mortar or grout	ASTM A641, Class 3
Wire ties or anchors in exterior walls not completely embedded in mortar or grout	ASTM A153, Class B-2
Joint reinforcement in exterior walls or interior walls exposed to moist environment	ASTM A153, Class B-2
Sheet metal ties or anchors exposed to weather	ASTM A153, Class B-2
Sheet metal ties or anchors completely embedded in mortar or grout	ASTM A653, Coating Designation G60
Stainless steel hardware for any exposure	ASTM A167, Type 304

construction, cross webs forming cells to be filled shall be full-bedded in mortar to prevent leakage of grout. Head and end joints shall be solidly filled with mortar for a distance in from the face of the wall or unit not less than the thickness of the longitudinal face shells.

2. Vertical reinforcement shall be held in position at top and bottom and at intervals not exceeding 200 diameters of the reinforcement.
3. Cells containing reinforcement shall be filled solidly with grout.
4. The thickness of grout or mortar between *masonry units* and reinforcement shall be not less than $^1/_4$ inch (6.4 mm), except that $^1/_4$-inch (6.4 mm) bars shall be permitted to be laid in horizontal mortar joints not less than $^1/_2$ inch (12.7 mm) thick, and steel wire reinforcement shall be permitted to be laid in horizontal mortar joints not less than twice the thickness of the wire diameter.

R606.3.6 Grouted multiple-wythe masonry. Grouted multiple-wythe masonry shall conform to all the requirements specified in Section R606.3.5 and the requirements of this section.

R606.3.6.1 Bonding of backup wythe. Where all interior vertical spaces are filled with grout in multiple-wythe construction, masonry headers shall not be permitted. Metal wall ties shall be used in accordance with Section R606.13.2 to prevent spreading of the wythes and to maintain the vertical alignment of the wall. Wall ties shall be installed in accordance with Section R606.13.2 where the backup wythe in multiple-wythe construction is fully grouted.

R606.3.6.2 Grout barriers. Vertical grout barriers or dams shall be built of *solid masonry* across the grout space the entire height of the wall to control the flow of the grout horizontally. Grout barriers shall be not more than 25 feet (7620 mm) apart. The grouting of any section of a wall between control barriers shall be completed in one day without interruptions greater than 1 hour.

R606.3.7 Masonry bonding pattern. Masonry laid in running and *stack bond* shall conform to Sections R606.3.7.1 and R606.3.7.2.

R606.3.7.1 Masonry laid in running bond. In each wythe of masonry laid in *running bond*, head joints in successive courses shall be offset by not less than one-fourth the unit length, or the masonry walls shall be reinforced longitudinally as required in Section R606.3.7.2.

R606.3.7.2 Masonry laid in stack bond. Where unit masonry is laid with less head joint offset than in Section R606.3.7.1, the minimum area of horizontal reinforcement placed in mortar bed joints or in bond beams spaced not more than 48 inches (1219 mm) apart shall be 0.0007 times the vertical cross-sectional area of the wall.

R606.4 Thickness of masonry. The nominal thickness of masonry walls shall conform to the requirements of Sections R606.4.1 through R606.4.4.

R606.4.1 Minimum thickness. The minimum thickness of masonry bearing walls more than one story high shall be 8 inches (203 mm). *Solid masonry* walls of one-story dwellings and garages shall be not less than 6 inches (152 mm) in thickness where not greater than 9 feet (2743 mm) in height, provided that where gable construction is used, an additional 6 feet (1829 mm) is permitted to the peak of the gable. Masonry walls shall be laterally supported in either the horizontal or vertical direction at intervals as required by Section R606.6.4.

R606.4.2 Rubble stone masonry wall. The minimum thickness of rough, random or coursed rubble stone masonry walls shall be 16 inches (406 mm).

R606.4.3 Change in thickness. Where walls of masonry of hollow units or masonry-bonded hollow walls are decreased in thickness, a course of *solid masonry* or

TABLE R606.3.5.1
GROUT SPACE DIMENSIONS AND POUR HEIGHTS

GROUT TYPE	GROUT POUR MAXIMUM HEIGHT (feet)	MINIMUM WIDTH OF GROUT SPACES[a, b] (inches)	MINIMUM GROUT[b, c] SPACE DIMENSIONS FOR GROUTING CELLS OF HOLLOW UNITS (inches × inches)
Fine	1	0.75	1.5 × 2
	5	2	2 × 3
	12	2.5	2.5 × 3
	24	3	3 × 3
Coarse	1	1.5	1.5 × 3
	5	2	2.5 × 3
	12	2.5	3 × 3
	24	3	3 × 4

For SI: 1 inch = 25.4 mm, 1 foot = 304.8 mm.

a. For grouting between masonry wythes.

b. Grout space dimension is the clear dimension between any masonry protrusion and shall be increased by the horizontal projection of the diameters of the horizontal bars within the cross section of the grout space.

c. Area of vertical reinforcement shall not exceed 6 percent of the area of the grout space.

masonry units filled with mortar or grout shall be constructed between the wall below and the thinner wall above, or special units or construction shall be used to transmit the loads from face shells or wythes above to those below.

R606.4.4 Parapet walls. Unreinforced *solid masonry* parapet walls shall be not less than 8 inches (203 mm) thick and their height shall not exceed four times their thickness. Unreinforced hollow unit masonry parapet walls shall be not less than 8 inches (203 mm) thick, and their height shall not exceed three times their thickness. Masonry parapet walls in areas subject to wind loads of 30 pounds per square foot (1.44 kPa) located in *Seismic Design Category* D_0, D_1 or D_2, or on townhouses in *Seismic Design Category* C shall be reinforced in accordance with Section R606.12.

R606.5 Corbeled masonry. Corbeled masonry shall be in accordance with Sections R606.5.1 through R606.5.3.

R606.5.1 Units. *Solid masonry* units or masonry units filled with mortar or grout shall be used for corbeling.

R606.5.2 Corbel projection. The maximum projection of one unit shall not exceed one-half the height of the unit or one-third the thickness at right angles to the wall. The maximum corbeled projection beyond the face of the wall shall not exceed:

1. One-half of the wall thickness for multiple-wythe walls bonded by mortar or grout and wall ties or masonry headers.
2. One-half the wythe thickness for single wythe walls, masonry-bonded hollow walls, multiple-wythe walls with open collar joints and veneer walls.

R606.5.3 Corbeled masonry supporting floor or roof-framing members. Where corbeled masonry is used to support floor or roof-framing members, the top course of the corbel shall be a header course or the top course bed joint shall have ties to the vertical wall.

R606.6 Support conditions. Bearing and support conditions shall be in accordance with Sections R606.6.1 through R606.6.4.

R606.6.1 Bearing on support. Each masonry wythe shall be supported by not less than two-thirds of the wythe thickness.

R606.6.2 Support at foundation. Cavity wall or masonry veneer construction shall be permitted to be supported on an 8-inch (203 mm) foundation wall, provided the 8-inch (203 mm) wall is corbeled to the width of the wall system above with masonry constructed of *solid masonry* units or masonry units filled with mortar or grout. The total horizontal projection of the corbel shall not exceed 2 inches (51 mm) with individual corbels projecting not more than one-third the thickness of the unit or one-half the height of the unit. The hollow space behind the corbeled masonry shall be filled with mortar or grout.

R606.6.3 Beam supports. Beams, girders or other concentrated loads supported by a wall or column shall have a bearing of not less than 3 inches (76 mm) in length measured parallel to the beam on *solid masonry* not less than 4 inches (102 mm) in thickness, or on a metal bearing plate of adequate design and dimensions to distribute the load safely, or on a continuous reinforced masonry member projecting not less than 4 inches (102 mm) from the face of the wall.

R606.6.3.1 Joist bearing. Joists shall have a bearing of not less than $1^1/_2$ inches (38 mm), except as provided in Section R606.6.3, and shall be supported in accordance with Figure R606.11(1).

R606.6.4 Lateral support. Masonry walls shall be laterally supported in either the horizontal or the vertical direction. The maximum spacing between lateral supports shall not exceed the distances in Table R606.6.4. Lateral support shall be provided by cross walls, pilasters, buttresses or structural frame members where the limiting distance is taken horizontally, or by floors or roofs where the limiting distance is taken vertically.

TABLE R606.6.4
SPACING OF LATERAL SUPPORT FOR MASONRY WALLS

CONSTRUCTION	MAXIMUM WALL LENGTH TO THICKNESS OR WALL HEIGHT TO THICKNESS[a, b]
Bearing walls:	
Solid or solid grouted	20
All other	18
Nonbearing walls:	
Exterior	18
Interior	36

For SI: 1 foot = 304.8 mm.

a. Except for cavity walls and cantilevered walls, the thickness of a wall shall be its nominal thickness measured perpendicular to the face of the wall. For cavity walls, the thickness shall be determined as the sum of the nominal thicknesses of the individual wythes. For cantilever walls, except for parapets, the ratio of height to nominal thickness shall not exceed 6 for solid masonry, or 4 for hollow masonry. For parapets, see Section R606.4.4.

b. An additional unsupported height of 6 feet is permitted for gable end walls.

R606.6.4.1 Horizontal lateral support. Lateral support in the horizontal direction provided by intersecting masonry walls shall be provided by one of the methods in Section R606.6.4.1.1 or R606.6.4.1.2.

R606.6.4.1.1 Bonding pattern. Fifty percent of the units at the intersection shall be laid in an overlapping masonry bonding pattern, with alternate units having a bearing of not less than 3 inches (76 mm) on the unit below.

R606.6.4.1.2 Metal reinforcement. Interior nonload-bearing walls shall be anchored at their intersections, at vertical intervals of not more than 16 inches (406 mm) with joint reinforcement of not less than 9 gage [0.148 inch (4 mm)], or $^1/_4$-inch (6 mm) galvanized mesh hardware cloth. Intersecting masonry walls, other than interior nonload-bearing walls, shall be anchored at vertical intervals of not more than 8 inches (203 mm) with joint reinforcement of not less than 9 gage (4 mm) and shall extend not less than 30 inches (762 mm) in each direction at the intersection. Other metal ties, joint reinforcement or anchors, if used, shall be spaced to provide equivalent area of anchorage to that required by this section.

R606.6.4.2 Vertical lateral support. Vertical lateral support of masonry walls in *Seismic Design Category* A, B or C shall be provided in accordance with one of the methods in Section R606.6.4.2.1 or R606.6.4.2.2.

R606.6.4.2.1 Roof structures. Masonry walls shall be anchored to roof structures with metal strap anchors spaced in accordance with the manufacturer's instructions, $^1/_2$-inch (13 mm) bolts spaced not more than 6 feet (1829 mm) on center, or other *approved* anchors. Anchors shall be embedded not less than 16 inches (406 mm) into the masonry, or be hooked or welded to bond beam reinforcement placed not less than 6 inches (152 mm) from the top of the wall.

R606.6.4.2.2 Floor diaphragms. Masonry walls shall be anchored to floor *diaphragm* framing by metal strap anchors spaced in accordance with the manufacturer's instructions, $^1/_2$-inch-diameter (13 mm) bolts spaced at intervals not to exceed 6 feet (1829 mm) and installed as shown in Figure R606.11(1), or by other *approved* methods.

R606.7 Piers. The unsupported height of masonry piers shall not exceed 10 times their least dimension. Where structural clay tile or hollow *concrete masonry units* are used for isolated piers to support beams and girders, the cellular spaces shall be filled solidly with grout or Type M or S mortar, except that unfilled hollow piers shall be permitted to be used if their unsupported height is not more than four times their least dimension. Where *hollow masonry units* are solidly filled with grout or Type M, S or N mortar, the allowable compressive stress shall be permitted to be increased as provided in Table R606.9.

R606.7.1 Pier cap. Hollow piers shall be capped with 4 inches (102 mm) of *solid masonry* or concrete, a masonry cap block, or shall have cavities of the top course filled with concrete or grout.

R606.8 Chases. Chases and recesses in masonry walls shall not be deeper than one-third the wall thickness. The maximum length of a horizontal chase or horizontal projection shall not exceed 4 feet (1219 mm) and shall have not less than 8 inches (203 mm) of masonry in back of the chases and recesses and between adjacent chases or recesses and the jambs of openings. Chases and recesses in masonry walls shall be designed and constructed so as not to reduce the required strength or required fire resistance of the wall and shall not be permitted within the required area of a pier. Masonry directly above chases or recesses wider than 12 inches (305 mm) shall be supported on noncombustible lintels.

R606.9 Allowable stresses. Allowable compressive stresses in masonry shall not exceed the values prescribed in Table R606.9. In determining the stresses in masonry, the effects of all loads and conditions of loading and the influence of all forces affecting the design and strength of the several parts shall be taken into account.

R606.9.1 Combined units. In walls or other structural members composed of different kinds or grades of units, materials or mortars, the maximum stress shall not exceed the allowable stress for the weakest of the combination of units, materials and mortars of which the member is composed. The net thickness of any facing unit that is used to resist stress shall be not less than $1^1/_2$ inches (38 mm).

R606.10 Lintels. Masonry over openings shall be supported by steel lintels, reinforced concrete or masonry lintels or masonry arches, designed to support load imposed.

R606.11 Anchorage. Masonry walls shall be anchored to floor and roof systems in accordance with the details shown in Figure R606.11(1), R606.11(2) or R606.11(3). Footings shall be permitted to be considered as points of lateral support.

R606.12 Seismic requirements. The seismic requirements of this section shall apply to the design of masonry and the construction of masonry building elements located in *Seismic Design Category* D_0, D_1 or D_2. Townhouses in *Seismic Design Category* C shall comply with the requirements of Section R606.12.2. These requirements shall not apply to glass unit masonry conforming to Section R610, anchored masonry veneer conforming to Section R703.8 or adhered masonry veneer conforming to Section R703.12.

R606.12.1 General. Masonry structures and masonry elements shall comply with the requirements of Sections R606.12.2 through R606.12.4 based on the seismic design category established in Table R301.2.1(1). Masonry structures and masonry elements shall comply with the requirements of Section R606.12 and Figures R606.11(1), R606.11(2) and R606.11(3) or shall be designed in accordance with TMS 402 or TMS 403.

R606.12.1.1 Floor and roof diaphragm construction. Floor and roof *diaphragms* shall be constructed of *wood structural panels* attached to wood framing in accordance with Table R602.3(1) or to cold-formed steel floor framing in accordance with Table R505.3.1(2) or to cold-formed steel roof framing in accordance with Table R804.3. Additionally, sheathing panel edges perpendicular to framing members shall be backed by blocking, and sheathing shall be connected to the blocking with fasteners at the edge spacing. For *Seismic Design Categories* C, D_0, D_1 and D_2, where the width-to-thickness dimension of the *diaphragm* exceeds 2-to-1, edge spacing of fasteners shall be 4 inches (102 mm) on center.

TABLE R606.9
ALLOWABLE COMPRESSIVE STRESSES FOR EMPIRICAL DESIGN OF MASONRY

CONSTRUCTION; COMPRESSIVE STRENGTH OF UNIT, GROSS AREA	ALLOWABLE COMPRESSIVE STRESSES[a] GROSS CROSS-SECTIONAL AREA[b]	
	Type M or S mortar	Type N mortar
Solid masonry of brick and other solid units of clay or shale; sand-lime or concrete brick:		
8,000 + psi	350	300
4,500 psi	225	200
2,500 psi	160	140
1,500 psi	115	100
Grouted[c] masonry, of clay or shale; sand-lime or concrete:		
4,500 + psi	225	200
2,500 psi	160	140
1,500 psi	115	100
Solid masonry of solid concrete masonry units:		
3,000 + psi	225	200
2,000 psi	160	140
1,200 psi	115	100
Masonry of hollow load-bearing units:		
2,000 + psi	140	120
1,500 psi	115	100
1,000 psi	75	70
700 psi	60	55
Hollow walls (cavity or masonry bonded[d]) solid units:		
2,500 + psi	160	140
1,500 psi	115	100
Hollow units	75	70
Stone ashlar masonry:		
Granite	720	640
Limestone or marble	450	400
Sandstone or cast stone	360	320
Rubble stone masonry:		
Coarse, rough or random	120	100

For SI: 1 pound per square inch = 6.895 kPa.

a. Linear interpolation shall be used for determining allowable stresses for masonry units having compressive strengths that are intermediate between those given in the table.

b. Gross cross-sectional area shall be calculated on the actual rather than nominal dimensions.

c. See Section R606.13.

d. Where floor and roof loads are carried on one wythe, the gross cross-sectional area is that of the wythe under load; if both wythes are loaded, the gross cross-sectional area is that of the wall minus the area of the cavity between the wythes. Walls bonded with metal ties shall be considered as cavity walls unless the collar joints are filled with mortar or grout.

R606.12.2 Seismic Design Category C. Townhouses located in *Seismic Design Category* C shall comply with the requirements of this section.

R606.12.2.1 Minimum length of wall without openings. Table R606.12.2.1 shall be used to determine the minimum required solid wall length without openings at each masonry exterior wall. The provided percentage of solid wall length shall include only those wall segments that are 3 feet (914 mm) or longer. The maximum clear distance between wall segments included in determining the solid wall length shall not exceed 18 feet (5486 mm). *Shear wall* segments required to meet the minimum wall length shall be in accordance with Section R606.12.2.2.3.

R606.12.2.2 Design of elements not part of the lateral force-resisting system.

R606.12.2.2.1 Load-bearing frames or columns. Elements not part of the lateral force-resisting system shall be analyzed to determine their effect on the response of the system. The frames or columns shall be adequate for vertical load-carrying capacity and induced moment caused by the design *story* drift.

R606.12.2.2.2 Masonry partition walls. Masonry partition walls, masonry screen walls and other masonry elements that are not designed to resist vertical or lateral loads, other than those induced by their own weight, shall be isolated from the structure so that vertical and lateral forces are not imparted to these elements. Isolation joints and connectors between these elements and the structure shall be designed to accommodate the design *story* drift.

R606.12.2.2.3 Reinforcement requirements for masonry elements. Masonry elements listed in Section R606.12.2.2.2 shall be reinforced in either the horizontal or vertical direction as shown in Figure R606.11(2) and in accordance with the following:

1. Horizontal reinforcement. Horizontal joint reinforcement shall consist of not less than two longitudinal W1.7 wires spaced not more than 16 inches (406 mm) for walls greater than 4 inches (102 mm) in width and not less than one longitudinal W1.7 wire spaced not more than 16 inches (406 mm) for walls not exceeding 4 inches (102 mm) in width; or not less than one No. 4 bar spaced not more than 48 inches (1219 mm). Where two longitudinal wires of joint reinforcement are used, the space between these wires shall be the widest that the mortar joint will accommodate. Horizontal reinforcement shall be provided within 16 inches (406 mm) of the top and bottom of these masonry elements.
2. Vertical reinforcement. Vertical reinforcement shall consist of not less than one No. 4 bar spaced not more than 48 inches (1219 mm). Vertical reinforcement shall be located within 16 inches (406 mm) of the ends of masonry walls.

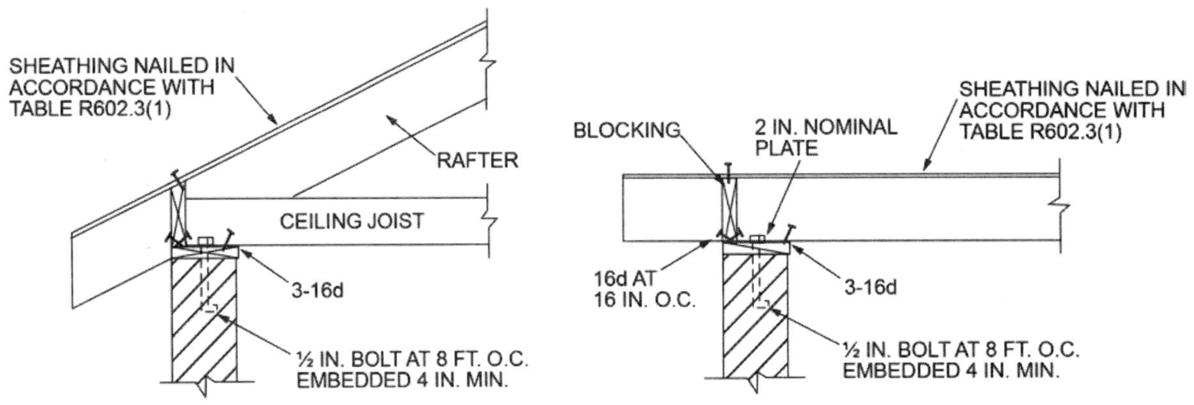

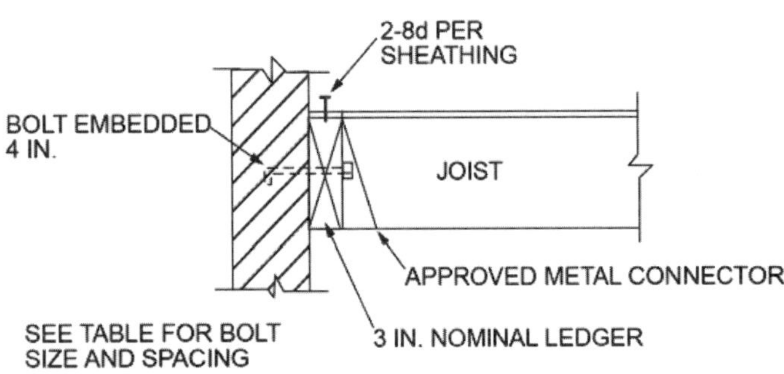

LEDGER BOLT SIZE AND SPACING

JOIST SPAN	BOLT SIZE AND SPACING	
	ROOF	FLOOR
10 FT.	$^1/_2$ AT 2 FT. 6 IN. $^7/_8$ AT 3 FT. 6 IN.	$^1/_2$ AT 2 FT. 0 IN. $^7/_8$ AT 2 FT. 9 IN.
10—15 FT.	$^1/_2$ AT 1 FT. 9 IN. $^7/_8$ AT 2 FT. 6 IN.	$^1/_2$ AT 1 FT. 4 IN. $^7/_8$ AT 2 FT. 0 IN.
15—20 FT.	$^1/_2$ AT 1 FT. 3 IN. $^7/_8$ AT 2 FT. 0 IN.	$^1/_2$ AT 1 FT. 0 IN. $^7/_8$ AT 1 FT. 6 IN.

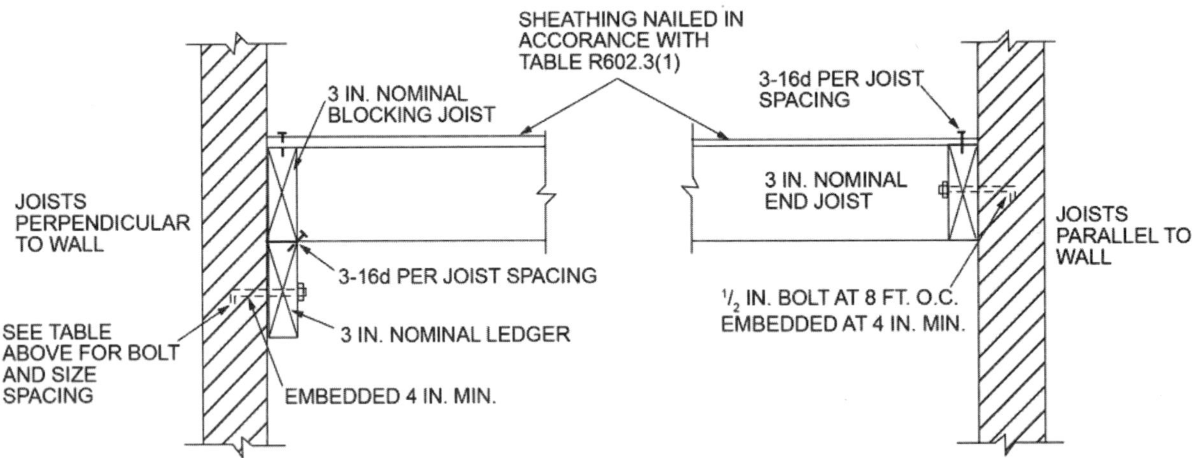

For SI: 1 inch = 25.4 mm, 1 foot = 304.8 mm, 1 pound per square foot = 0.0479 kPa.

Note: Where bolts are located in hollow masonry, the cells in the courses receiving the bolt shall be grouted solid.

FIGURE R606.11(1)
ANCHORAGE REQUIREMENTS FOR MASONRY WALLS LOCATED IN SEISMIC DESIGN CATEGORY A, B OR C AND WHERE WIND LOADS ARE LESS THAN 30 PSF

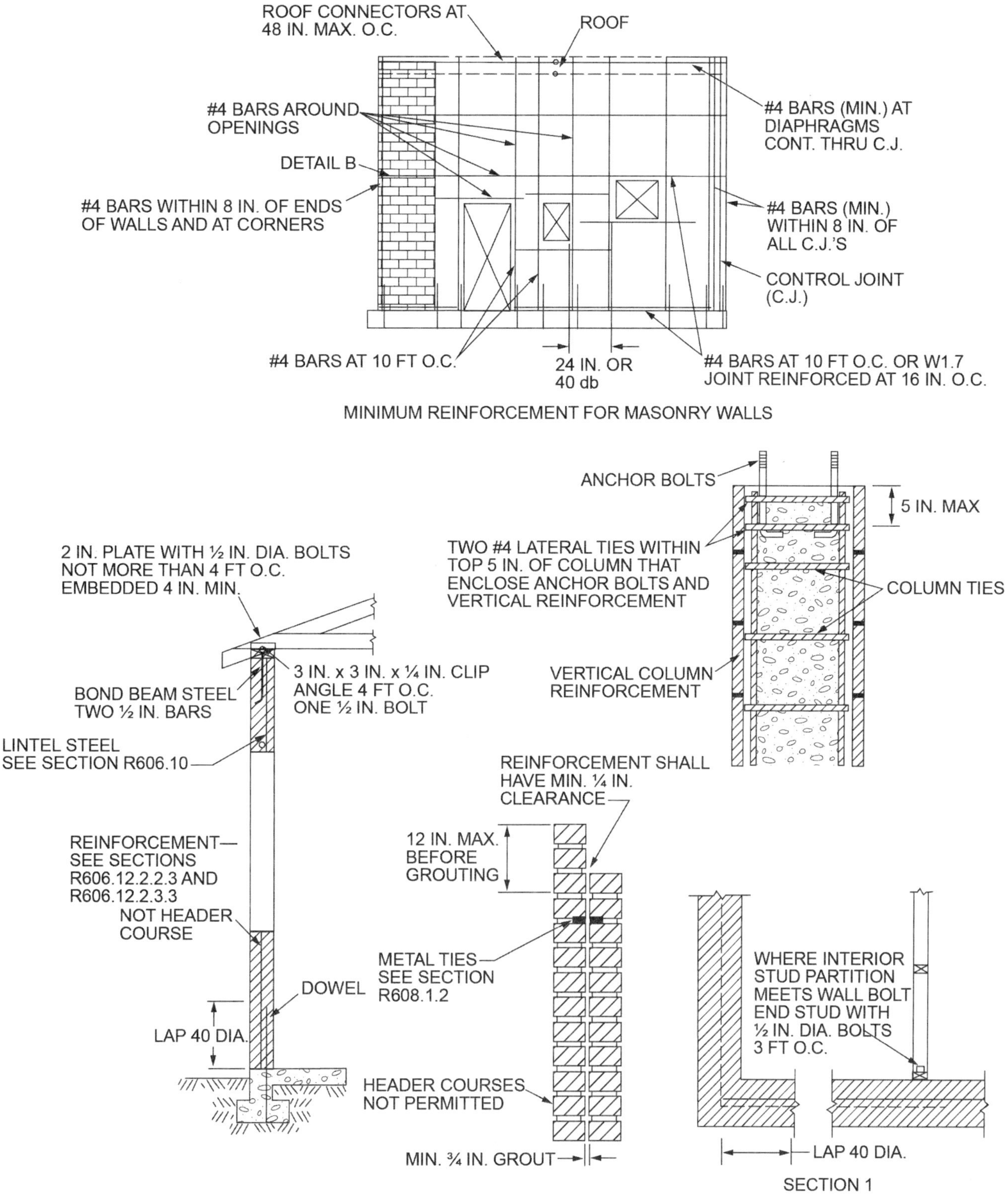

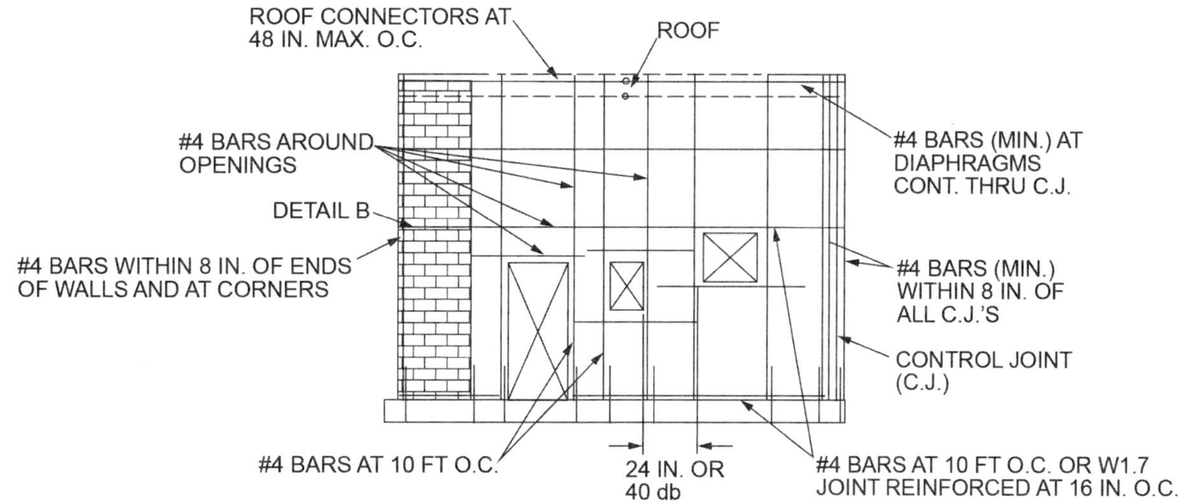

ROOF CONNECTORS AT 48 IN. MAX. O.C.
ROOF
#4 BARS AROUND OPENINGS
#4 BARS (MIN.) AT DIAPHRAGMS CONT. THRU C.J.
DETAIL B
#4 BARS WITHIN 8 IN. OF ENDS OF WALLS AND AT CORNERS
#4 BARS (MIN.) WITHIN 8 IN. OF ALL C.J.'S
CONTROL JOINT (C.J.)
#4 BARS AT 10 FT O.C.
24 IN. OR 40 db
#4 BARS AT 10 FT O.C. OR W1.7 JOINT REINFORCED AT 16 IN. O.C.
MINIMUM REINFORCEMENT FOR MASONRY WALLS
ANCHOR BOLTS
5 IN. MAX
TWO #4 LATERAL TIES WITHIN TOP 5 IN. OF COLUMN THAT ENCLOSE ANCHOR BOLTS AND VERTICAL REINFORCEMENT
COLUMN TIES
VERTICAL COLUMN REINFORCEMENT
2 IN. PLATE WITH ½ IN. DIA. BOLTS NOT MORE THAN 4 FT O.C. EMBEDDED 4 IN. MIN.
3 IN. x 3 IN. x ¼ IN. CLIP ANGLE 4 FT O.C. ONE ½ IN. BOLT
BOND BEAM STEEL TWO ½ IN. BARS
LINTEL STEEL SEE SECTION R606.10
REINFORCEMENT SHALL HAVE MIN. ¼ IN. CLEARANCE
12 IN. MAX. BEFORE GROUTING
REINFORCEMENT— SEE SECTIONS R606.12.2.2.3 AND R606.12.2.3.3
NOT HEADER COURSE
METAL TIES SEE SECTION R608.1.2
DOWEL
LAP 40 DIA.
WHERE INTERIOR STUD PARTITION MEETS WALL BOLT END STUD WITH ½ IN. DIA. BOLTS 3 FT O.C.
HEADER COURSES NOT PERMITTED
MIN. ¾ IN. GROUT
LAP 40 DIA.
SECTION 1

For SI: 1 inch = 25.4 mm, 1 foot = 304.8 mm.

FIGURE R606.11(2)
REQUIREMENTS FOR REINFORCED GROUTED MASONRY CONSTRUCTION IN SEISMIC DESIGN CATEGORY C

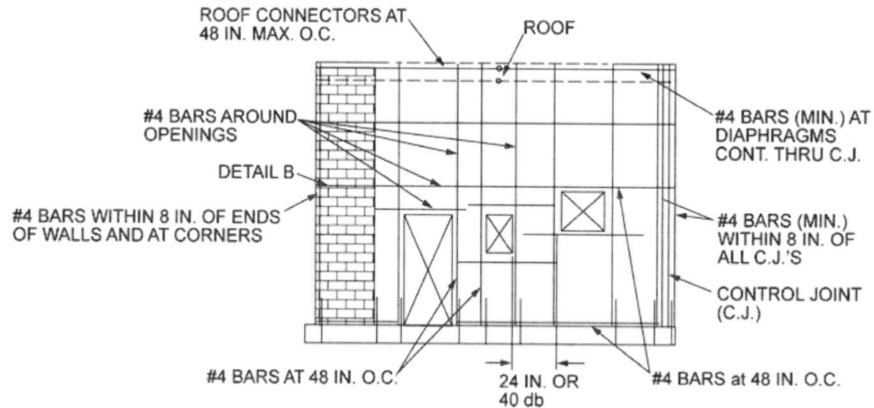

MINIMUM REINFORCEMENT FOR MASONRY WALLS

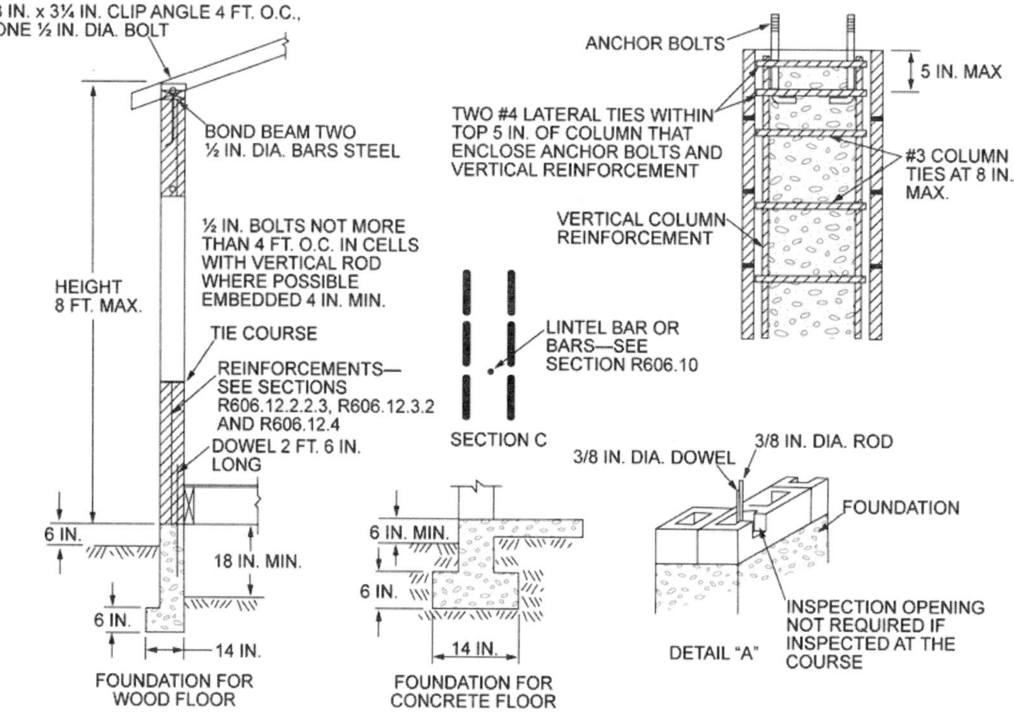

For SI: 1 inch = 25.4 mm, 1 foot = 304.8 mm.

Note: A full bed joint must be provided. Cells containing vertical bars are to be filled to the top of wall and provide inspection opening as shown on detail "A." Horizontal bars are to be laid as shown on detail "B." Lintel bars are to be laid as shown on Section C.

FIGURE R606.11(3)
REQUIREMENTS FOR REINFORCED MASONRY CONSTRUCTION IN SEISMIC DESIGN CATEGORY D_0, D_1 OR D_2

TABLE R606.12.2.1
MINIMUM SOLID WALL LENGTH ALONG EXTERIOR WALL LINES

SESIMIC DESIGN CATEGORY	MINIMUM SOLID WALL LENGTH (percent)[a]		
	One story or top story of two story	Wall supporting light-frame second story and roof	Wall supporting masonry second story and roof
Townhouses in C	20	25	35
D_0 or D_1	25	NP	NP
D_2	30	NP	NP

NP = Not Permitted, except with design in accordance with the *International Building Code*.

a. For all walls, the minimum required length of solid walls shall be based on the table percent multiplied by the dimension, parallel to the wall direction under consideration, of a rectangle inscribing the overall building plan.

R606.12.2.3 Design of elements part of the lateral force-resisting system.

R606.12.2.3.1 Connections to masonry shear walls. Connectors shall be provided to transfer forces between masonry walls and horizontal elements in accordance with the requirements of Section 4.1.4 of TMS 402. Connectors shall be designed to transfer horizontal design forces acting either perpendicular or parallel to the wall, but not less than 200 pounds per linear foot (2919 N/m) of wall. The maximum spacing between connectors shall be 4 feet (1219 mm). Such anchorage mechanisms shall not induce tension stresses perpendicular to grain in ledgers or nailers.

R606.12.2.3.2 Connections to masonry columns. Connectors shall be provided to transfer forces between masonry columns and horizontal elements in accordance with the requirements of Section 4.1.4 of TMS 402. Where anchor bolts are used to connect horizontal elements to the tops of columns, the bolts shall be placed within lateral ties. Lateral ties shall enclose both the vertical bars in the column and the anchor bolts. There shall be not less than two No. 4 lateral ties provided in the top 5 inches (127 mm) of the column.

R606.12.2.3.3 Minimum reinforcement requirements for masonry shear walls. Vertical reinforcement of not less than one No. 4 bar shall be provided at corners, within 16 inches (406 mm) of each side of openings, within 8 inches (203 mm) of each side of movement joints, within 8 inches (203 mm) of the ends of walls, and at a maximum spacing of 10 feet (3048 mm).

Horizontal joint reinforcement shall consist of not less than two wires of W1.7 spaced not more than 16 inches (406 mm); or bond beam reinforcement of not less than one No. 4 bar spaced not more than 10 feet (3048 mm) shall be provided. Horizontal reinforcement shall be provided at the bottom and top of wall openings and shall extend not less than 24 inches (610 mm) nor less than 40 bar diameters past the opening; continuously at structurally connected roof and floor levels; and within 16 inches (406 mm) of the top of walls.

R606.12.3 Seismic Design Category D_0 or D_1. Structures in *Seismic Design Category* D_0 or D_1 shall comply with the requirements of *Seismic Design Category* C and the additional requirements of this section. AAC masonry shall not be used for the design of masonry elements that are part of the lateral force-resisting system.

R606.12.3.1 Design requirements. Masonry elements other than those covered by Section R606.12.2.2.2 shall be designed in accordance with the requirements of Chapters 1 through 7 and Sections 8.1 and 8.3 of TMS 402 and shall meet the minimum reinforcement requirements contained in Sections R606.12.3.2 and R606.12.3.2.1. Otherwise, masonry shall be designed in accordance with TMS 403.

Exception: Masonry walls limited to one *story* in height and 9 feet (2743 mm) between lateral supports need not be designed provided they comply with the minimum reinforcement requirements of Sections R606.12.3.2 and R606.12.3.2.1.

R606.12.3.2 Minimum reinforcement requirements for masonry walls. Masonry walls other than those covered by Section R606.12.2.2.3 shall be reinforced in both the vertical and horizontal direction. The sum of the cross-sectional area of horizontal and vertical reinforcement shall be not less than 0.002 times the gross cross-sectional area of the wall, and the minimum cross-sectional area in each direction shall be not less than 0.0007 times the gross cross-sectional area of the wall. Reinforcement shall be uniformly distributed. Table R606.12.3.2 shows the minimum reinforcing bar sizes required for varying thicknesses of masonry walls. The maximum spacing of reinforcement shall be 48 inches (1219 mm) provided that the walls are solid grouted and constructed of hollow open-end units, hollow units laid with full head joints or two wythes of solid units. The maximum spacing of reinforcement shall be 24 inches (610 mm) for all other masonry.

R606.12.3.2.1 Shear wall reinforcement requirements. The maximum spacing of vertical and horizontal reinforcement shall be the smaller of one-third the length of the *shear wall*, one-third the height of the *shear wall*, or 48 inches (1219 mm). The minimum cross-sectional area of vertical reinforcement

TABLE R606.12.3.2
MINIMUM DISTRIBUTED WALL REINFORCEMENT FOR BUILDINGS ASSIGNED TO SEISMIC DESIGN CATEGORY D_0 or D_1

NOMINAL WALL THICKNESS (inches)	MINIMUM SUM OF THE VERTICAL AND HORIZONTAL REINFORCEMENT AREAS[a] (square inches per foot)	MINIMUM REINFORCEMENT AS DISTRIBUTED IN BOTH HORIZONTAL AND VERTICAL DIRECTIONS[b] (square inches per foot)	MINUMUM BAR SIZE FOR REINFORCEMENT SPACED AT 48 INCHES
6	0.135	0.047	#4
8	0.183	0.064	#5
10	0.231	0.081	#6
12	0.279	0.098	#6

For SI: 1 inch = 25.4 mm, 1 foot = 304.8 mm, 1 square inch per foot = 2064 mm^2/m.

a. Based on the minimum reinforcing ratio of 0.002 times the gross cross-sectional area of the wall.

b. Based on the minimum reinforcing ratio each direction of 0.0007 times the gross cross-sectional area of the wall.

shall be one-third of the required shear reinforcement. Shear reinforcement shall be anchored around vertical reinforcing bars with a standard hook.

R606.12.3.3 Minimum reinforcement for masonry columns. Lateral ties in masonry columns shall be spaced not more than 8 inches (203 mm) on center and shall be not less than $^{3}/_{8}$-inch (9.5 mm) diameter. Lateral ties shall be embedded in grout.

R606.12.3.4 Material restrictions. Type N mortar or masonry cement shall not be used as part of the lateral force-resisting system.

R606.12.3.5 Lateral tie anchorage. Standard hooks for lateral tie anchorage shall be either a 135-degree (2.4 rad) standard hook or a 180-degree (3.2 rad) standard hook.

R606.12.4 Seismic Design Category D_2. Structures in *Seismic Design Category* D_2 shall comply with the requirements of *Seismic Design Category* D_1 and to the additional requirements of this section.

R606.12.4.1 Design of elements not part of the lateral force-resisting system. *Stack bond* masonry that is not part of the lateral force-resisting system shall have a horizontal cross-sectional area of reinforcement of not less than 0.0015 times the gross cross-sectional area of masonry. Table R606.12.4.1 shows minimum reinforcing bar sizes for masonry walls. The maximum spacing of horizontal reinforcement shall be 24 inches (610 mm). These elements shall be solidly grouted and shall be constructed of hollow open-end units or two wythes of solid units.

TABLE R606.12.4.1
MINIMUM REINFORCING FOR STACKED BONDED MASONRY WALLS IN SEISMIC DESIGN CATEGORY D_2

NOMINAL WALL THICKNESS (inches)	MINIMUM BAR SIZE SPACED AT 24 INCHES
6	#4
8	#5
10	#5
12	#6

For SI: 1 inch = 25.4 mm.

R606.12.4.2 Design of elements part of the lateral force-resisting system. *Stack bond* masonry that is part of the lateral force-resisting system shall have a horizontal cross-sectional area of reinforcement of not less than 0.0025 times the gross cross-sectional area of masonry. Table R606.12.4.2 shows minimum reinforcing bar sizes for masonry walls. The maximum spacing of horizontal reinforcement shall be 16 inches (406 mm). These elements shall be solidly grouted and shall be constructed of hollow open-end units or two wythes of solid units.

TABLE R606.12.4.2
MINIMUM REINFORCING FOR STACKED BONDED MASONRY WALLS IN SEISMIC DESIGN CATEGORY D_2

NOMINAL WALL THICKNESS (inches)	MINIMUM BAR SIZE SPACED AT 16 INCHES
6	#4
8	#5
10	#5
12	#6

For SI: 1 inch = 25.4 mm.

R606.13 Multiple-wythe masonry. The facing and backing of multiple-wythe masonry walls shall be bonded in accordance with Section R606.13.1, R606.13.2 or R606.13.3. In cavity walls, neither the facing nor the backing shall be less than 3 inches (76 mm) nominal in thickness and the cavity shall be not more than 4 inches (102 mm) nominal in width. The backing shall not be less than the thickness of the facing.

Exception: Cavities shall be permitted to exceed the 4-inch (102 mm) nominal dimension provided that tie size and tie spacing have been established by calculation.

R606.13.1 Bonding with masonry headers. Bonding with solid or *hollow masonry* headers shall comply with Sections R606.13.1.1 and R606.13.1.2.

R606.13.1.1 Solid units. Where the facing and backing (adjacent wythes) of *solid masonry* construction are bonded by means of masonry headers, not less than 4 percent of the wall surface of each face shall be composed of headers extending not less than 3 inches (76 mm) into the backing. The distance between adjacent full-length headers shall not exceed 24 inches (610 mm) either vertically or horizontally. In walls in which a single header does not extend through the wall, headers from the opposite sides shall overlap not less than 3 inches (76 mm), or headers from opposite sides shall be covered with another header course overlapping the header below not less than 3 inches (76 mm).

R606.13.1.2 Hollow units. Where two or more hollow units are used to make up the thickness of a wall, the stretcher courses shall be bonded at vertical intervals not exceeding 34 inches (864 mm) by lapping not less than 3 inches (76 mm) over the unit below, or by lapping at vertical intervals not exceeding 17 inches (432 mm) with units that are not less than 50 percent thicker than the units below.

R606.13.2 Bonding with wall ties or joint reinforcement. Bonding with wall ties or joint reinforcement shall comply with Section R606.13.2.3.

R606.13.2.1 Bonding with wall ties. Bonding with wall ties, except as required by Section R607, where the facing and backing (adjacent wythes) of masonry walls are bonded with $^{3}/_{16}$-inch-diameter (5 mm) wall ties embedded in the horizontal mortar joints, there shall be not less than one metal tie for each $4^1/_2$ square feet (0.418 m^2) of wall area. Ties in alternate courses shall be staggered. The maximum vertical distance between ties shall not exceed 24 inches (610 mm), and the maximum horizontal distance shall not exceed 36 inches (914 mm). Rods or ties bent to rectangular shape shall be used with *hollow masonry units* laid with the cells vertical. In other walls, the ends of ties shall be bent to 90-degree (0.79 rad) angles to provide hooks not less than 2 inches (51 mm) long. Additional bonding ties shall be provided at all openings, spaced not more than 3 feet (914 mm) apart around the perimeter and within 12 inches (305 mm) of the opening.

R606.13.2.2 Bonding with adjustable wall ties. Where the facing and backing (adjacent wythes) of masonry are bonded with adjustable wall ties, there shall be not less than one tie for each 2.67 square feet (0.248 m^2) of wall area. Neither the vertical nor the horizontal spacing of the adjustable wall ties shall exceed 24 inches (610 mm). The maximum vertical offset of bed joints from one wythe to the other shall be 1.25 inches (32 mm). The maximum clearance between connecting parts of the ties shall be $^1/_{16}$ inch (2 mm). Where pintle legs are used, ties shall have not less than two $^3/_{16}$-inch-diameter (5 mm) legs.

R606.13.2.3 Bonding with prefabricated joint reinforcement. Where the facing and backing (adjacent wythes) of masonry are bonded with prefabricated joint reinforcement, there shall be not less than one cross wire serving as a tie for each 2.67 square feet (0.248 m^2) of wall area. The vertical spacing of the joint reinforcement shall not exceed 16 inches (406 mm). Cross wires on prefabricated joint reinforcement shall not be smaller than No. 9 gage. The longitudinal wires shall be embedded in the mortar.

R606.13.3 Bonding with natural or cast stone. Bonding with natural and cast stone shall conform to Sections R606.13.3.1 and R606.13.3.2.

R606.13.3.1 Ashlar masonry. In ashlar masonry, bonder units, uniformly distributed, shall be provided to the extent of not less than 10 percent of the wall area. Such bonder units shall extend not less than 4 inches (102 mm) into the backing wall.

R606.13.3.2 Rubble stone masonry. Rubble stone masonry 24 inches (610 mm) or less in thickness shall have bonder units with a maximum spacing of 3 feet (914 mm) vertically and 3 feet (914 mm) horizontally, and if the masonry is of greater thickness than 24 inches (610 mm), shall have one bonder unit for each 6 square feet (0.557 m^2) of wall surface on both sides.

R606.14 Anchored and adhered masonry veneer.

R606.14.1 Anchored veneer. Anchored masonry veneer installed over a backing of wood or cold-formed steel shall meet the requirements of Section R703.8.

R606.14.2 Adhered veneer. Adhered masonry veneer shall be installed in accordance with the requirements of Section R703.12.

SECTION R607
GLASS UNIT MASONRY

R607.1 General. Panels of glass unit masonry located in load-bearing and nonload-bearing exterior and interior walls shall be constructed in accordance with this section.

R607.2 Materials. Hollow glass units shall be partially evacuated and have a minimum average glass face thickness of $^3/_{16}$ inch (5 mm). The surface of units in contact with mortar shall be treated with a polyvinyl butyral coating or latex-based paint. The use of reclaimed units is prohibited.

R607.3 Units. Hollow or solid glass block units shall be standard or thin units.

R607.3.1 Standard units. The specified thickness of standard units shall be not less than $3^7/_8$ inches (98 mm).

R607.3.2 Thin units. The specified thickness of thin units shall be not less than $3^1/_8$ inches (79 mm) for hollow units and not less than 3 inches (76 mm) for solid units.

R607.4 Isolated panels. Isolated panels of glass unit masonry shall conform to the requirements of this section.

R607.4.1 Exterior standard-unit panels. The maximum area of each individual standard-unit panel shall be 144 square feet (13.4 m^2) where the design wind pressure is 20 pounds per square foot (958 Pa). The maximum area of such panels subjected to design wind pressures other than 20 pounds per square foot (958 Pa) shall be in accordance with Figure R607.4.1. The maximum panel dimension between structural supports shall be 25 feet (7620 mm) in width or 20 feet (6096 mm) in height.

R607.4.2 Exterior thin-unit panels. The maximum area of each individual thin-unit panel shall be 85 square feet (7.9 m^2). The maximum dimension between structural supports shall be 15 feet (4572 mm) in width or 10 feet (3048 mm) in height. Thin units shall not be used in applications where the design wind pressure as stated in Table R301.2.1(1) exceeds 20 pounds per square foot (958 Pa).

R607.4.3 Interior panels. The maximum area of each individual standard-unit panel shall be 250 square feet (23.2 m^2). The maximum area of each thin-unit panel shall be 150 square feet (13.9 m^2). The maximum dimension between structural supports shall be 25 feet (7620 mm) in width or 20 feet (6096 mm) in height.

R607.4.4 Curved panels. The width of curved panels shall conform to the requirements of Sections R607.4.1, R607.4.2 and R607.4.3, except additional structural supports shall be provided at locations where a curved section joins a straight section, and at inflection points in multiple-curve walls.

R607.5 Panel support. Glass unit masonry panels shall conform to the support requirements of this section.

R607.5.1 Deflection. The maximum total deflection of structural members that support glass unit masonry shall not exceed $^{l}/_{600}$.

R607.5.2 Lateral support. Glass unit masonry panels shall be laterally supported along the top and sides of the panel. Lateral supports for glass unit masonry panels shall be designed to resist not less than 200 pounds per lineal feet (2918 N/m) of panel, or the actual applied loads, whichever is greater. Except for single-unit panels, lateral support shall be provided by panel anchors along the top and sides spaced not greater than 16 inches (406 mm) on center or by channel-type restraints. Single-unit panels shall be supported by channel-type restraints.

Exceptions:

1. Lateral support is not required at the top of panels that are one unit wide.
2. Lateral support is not required at the sides of panels that are one unit high.

R607.5.2.1 Panel anchor restraints. Panel anchors shall be spaced not greater than 16 inches (406 mm) on center in both jambs and across the head. Panel anchors shall be embedded not less than 12 inches (305 mm) and shall be provided with two fasteners so as to resist the loads specified in Section R607.5.2.

R607.5.2.2 Channel-type restraints. Glass unit masonry panels shall be recessed not less than 1 inch (25 mm) within channels and chases. Channel-type restraints shall be oversized to accommodate expansion material in the opening, packing and sealant between the framing restraints, and the glass unit masonry perimeter units.

R607.6 Sills. Before the bedding of glass units, the sill area shall be covered with a water-base asphaltic emulsion coating. The coating shall be not less than $^{1}/_{8}$ inch (3 mm) thick.

R607.7 Expansion joints. Glass unit masonry panels shall be provided with expansion joints along the top and sides at all structural supports. Expansion joints shall be not less than $^{3}/_{8}$ inch (10 mm) in thickness and shall have sufficient thickness to accommodate displacements of the supporting structure. Expansion joints shall be entirely free of mortar and other debris and shall be filled with resilient material.

R607.8 Mortar. Glass unit masonry shall be laid with Type S or N mortar. Mortar shall not be retempered after initial set. Mortar unused within $1^{1}/_{2}$ hours after initial mixing shall be discarded.

R607.9 Reinforcement. Glass unit masonry panels shall have horizontal joint reinforcement spaced not greater than 16 inches (406 mm) on center located in the mortar bed joint. Horizontal joint reinforcement shall extend the entire length of the panel but shall not extend across expansion joints.

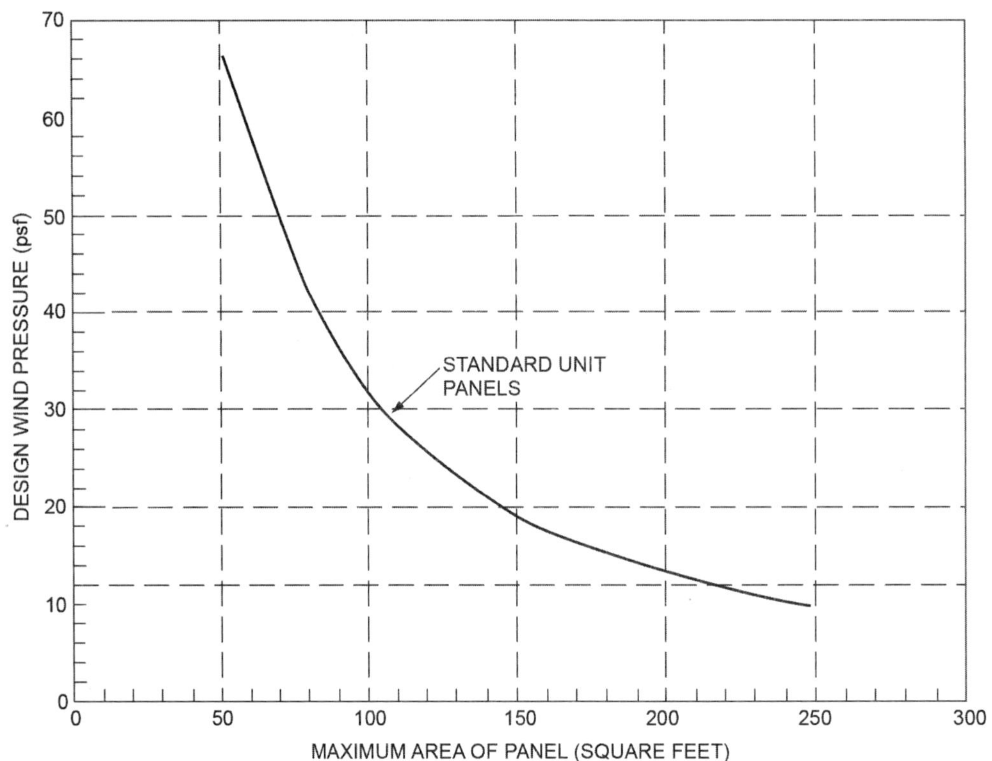

For SI: 1 square foot = 0.0929 m^2, 1 pound per square foot = 0.0479 kPa.

FIGURE R607.4.1
GLASS UNIT MASONRY DESIGN WIND LOAD RESISTANCE

Longitudinal wires shall be lapped not less than 6 inches (152 mm) at splices. Joint reinforcement shall be placed in the bed joint immediately below and above openings in the panel. The reinforcement shall have not less than two parallel longitudinal wires of size W1.7 or greater, and have welded cross wires of size W1.7 or greater.

R607.10 Placement. Glass units shall be placed so head and bed joints are filled solidly. Mortar shall not be furrowed. Head and bed joints of glass unit masonry shall be $^1/_4$ inch (6.4 mm) thick, except that vertical joint thickness of radial panels shall be not less than $^1/_8$ inch (3 mm) or greater than $^5/_8$ inch (16 mm). The bed joint thickness tolerance shall be minus $^1/_{16}$ inch (1.6 mm) and plus $^1/_8$ inch (3 mm). The head joint thickness tolerance shall be plus or minus $^1/_8$ inch (3 mm).

SECTION R608 EXTERIOR CONCRETE WALL CONSTRUCTION

R608.1 General. Exterior concrete walls shall be designed and constructed in accordance with the provisions of this section or in accordance with the provisions of PCA 100, ACI 318 or ACI 332. Where PCA 100, ACI 318, ACI 332 or the provisions of this section are used to design concrete walls, project drawings, typical details and specifications are not required to bear the seal of the architect or engineer responsible for design, unless otherwise required by the state law of the *jurisdiction* having authority.

R608.1.1 Interior construction. These provisions are based on the assumption that interior walls and partitions, both load-bearing and nonload-bearing, floors and roof/ceiling assemblies are constructed of *light-frame construction* complying with the limitations of this code and the additional limitations of Section R608.2. Design and construction of light-frame assemblies shall be in accordance with the applicable provisions of this code. Where second-story exterior walls are of *light-frame construction*, they shall be designed and constructed as required by this code.

Aspects of concrete construction not specifically addressed by this code, including interior concrete walls, shall comply with ACI 318.

R608.1.2 Other concrete walls. Exterior concrete walls constructed in accordance with this code shall comply with the shapes and minimum concrete cross-sectional dimensions of Table R608.3. Other types of forming systems resulting in concrete walls not in compliance with this section shall be designed in accordance with ACI 318.

R608.2 Applicability limits. The provisions of this section shall apply to the construction of exterior concrete walls for buildings not greater than 60 feet (18 288 mm) in plan dimensions, floors with clear spans not greater than 32 feet (9754 mm) and roofs with clear spans not greater than 40 feet (12 192 mm). Buildings shall not exceed 35 feet (10 668 mm) in mean roof height or two stories in height above grade. Floor/ceiling dead loads shall not exceed 10 pounds per square foot (479 Pa), roof/ceiling dead loads shall not exceed 15 pounds per square foot (718 Pa) and attic live loads shall not exceed 20 pounds per square foot (958 Pa). Roof overhangs shall not exceed 2 feet (610 mm) of horizontal projection beyond the exterior wall and the dead load of the overhangs shall not exceed 8 pounds per square foot (383 Pa).

Walls constructed in accordance with the provisions of this section shall be limited to buildings subjected to a maximum design wind speed of 160 mph (72 m/s) Exposure B, 136 mph (61 m/s) Exposure C and 125 mph (56 m/s) Exposure D. Walls constructed in accordance with the provisions of this section shall be limited to detached one- and two-family *dwellings* and townhouses assigned to *Seismic Design Category* A or B, and detached one- and two-family *dwellings* assigned to *Seismic Design Category* C.

Buildings that are not within the scope of this section shall be designed in accordance with PCA 100 or ACI 318.

R608.3 Concrete wall systems. Concrete walls constructed in accordance with these provisions shall comply with the shapes and minimum concrete cross-sectional dimensions of Table R608.3.

R608.3.1 Flat wall systems. Flat concrete wall systems shall comply with Table R608.3 and Figure R608.3(1) and have a minimum nominal thickness of 4 inches (102 mm).

R608.3.2 Waffle-grid wall systems. Waffle-grid wall systems shall comply with Table R608.3 and Figure R608.3(2) and shall have a minimum nominal thickness of 6 inches (152 mm) for the horizontal and vertical concrete members (cores). The core and web dimensions shall comply with Table R608.3. The maximum weight of waffle-grid walls shall comply with Table R608.3.

R608.3.3 Screen-grid wall systems. Screen-grid wall systems shall comply with Table R608.3 and Figure R608.3(3) and shall have a minimum nominal thickness of 6 inches (152 mm) for the horizontal and vertical concrete members (cores). The core dimensions shall comply with Table R608.3. The maximum weight of screen-grid walls shall comply with Table R608.3.

R608.4 Stay-in-place forms. Stay-in-place concrete forms shall comply with this section.

R608.4.1 Surface burning characteristics. The flame spread index and *smoke-developed index* of forming material, other than foam plastic, left exposed on the interior shall comply with Section R302.9. The surface burning characteristics of foam plastic used in *insulating concrete forms* shall comply with Section R316.3.

R608.4.2 Interior covering. Stay-in-place forms constructed of rigid foam plastic shall be protected on the interior of the building as required by Sections R316.4 and R702.3.4. Where gypsum board is used to protect the foam plastic, it shall be installed with a mechanical fastening system. Use of adhesives is permitted in addition to mechanical fasteners.

TABLE R608.3
DIMENSIONAL REQUIREMENTS FOR WALLS[a]

WALL TYPE AND NOMINAL THICKNESS	MAXIMUM WALL WEIGHT[b] (psf)	MINIMUM WIDTH, W, OF VERTICAL CORES (inches)	MINIMUM THICKNESS, T, OF VERTICAL CORES (inches)	MAXIMUM SPACING OF VERTICAL CORES (inches)	MAXIMUM SPACING OF HORIZONTAL CORES (inches)	MINIMUM WEB THICKNESS (inches)
4″ Flat[c]	50	NA	NA	NA	NA	NA
6″ Flat[c]	75	NA	NA	NA	NA	NA
8″ Flat[c]	100	NA	NA	NA	NA	NA
10″ Flat[c]	125	NA	NA	NA	NA	NA
6″ Waffle-grid	56	8[d]	5.5[d]	12	16	2
8″ Waffle-grid	76	8[e]	8[e]	12	16	2
6″ Screen-grid	53	6.25[f]	6.25[f]	12	12	NA

For SI: 1 inch = 25.4 mm, 1 pound per square foot = 0.0479 kPa, 1 pound per cubic foot = 2402.77 kg/m^3, 1 square inch = 645.16 mm^2, 1 inch^4 = 42 cm^4.
NA = Not Applicable.

a. Width "W," thickness "T," spacing and web thickness, refer to Figures R608.3(2) and R608.3(3).

b. Wall weight is based on a unit weight of concrete of 150 pcf. For flat walls the weight is based on the nominal thickness. The tabulated values do not include any allowance for interior and exterior finishes.

c. Nominal wall thickness. The actual as-built thickness of a flat wall shall not be more than $^1/_2$ inch less or more than $^1/_4$ inch more than the nominal dimension indicated.

d. Vertical core is assumed to be elliptical-shaped. Another shape of core is permitted provided the minimum thickness is 5 inches, the moment of inertia, *I*, about the centerline of the wall (ignoring the web) is not less than 65 inch^4, and the area, *A*, is not less than 31.25 square inches. The width used to calculate *A* and *I* shall not exceed 8 inches.

e. Vertical core is assumed to be circular. Another shape of core is permitted provided the minimum thickness is 7 inches, the moment of inertia, *I*, about the centerline of the wall (ignoring the web) is not less than 200 inch^4, and the area, *A*, is not less than 49 square inches. The width used to calculate *A* and *I* shall not exceed 8 inches.

f. Vertical core is assumed to be circular. Another shape of core is permitted provided the minimum thickness is 5.5 inches, the moment of inertia, *I*, about the centerline of the wall is not less than 76 inch^4, and the area, *A*, is not less than 30.25 square inches. The width used to calculate *A* and *I* shall not exceed 6.25 inches.

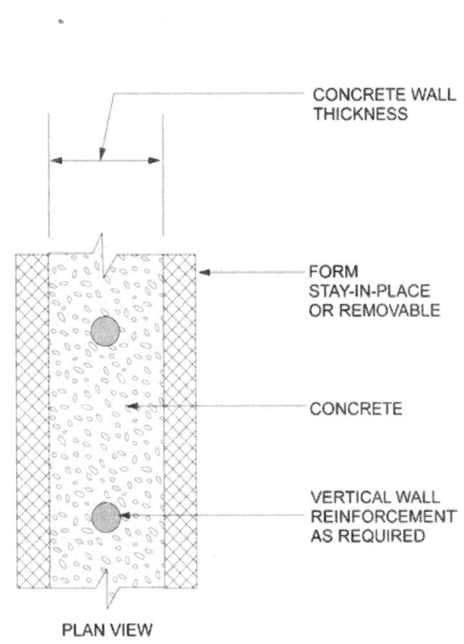

FIGURE R608.3(1)
FLAT WALL SYSTEM

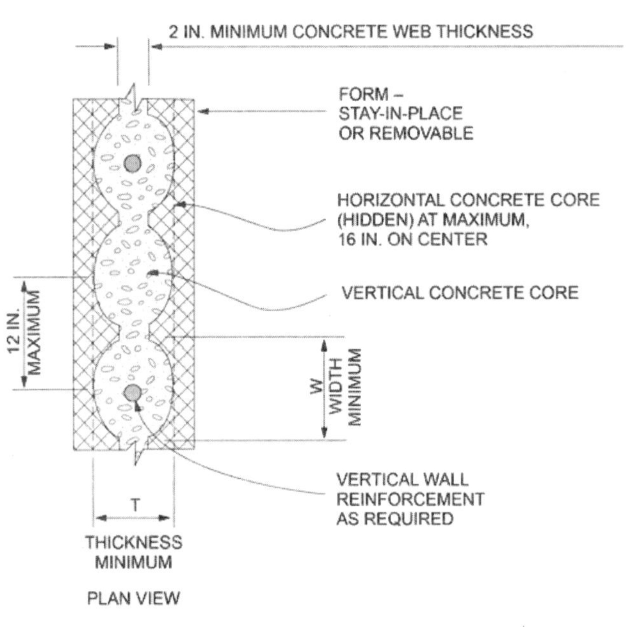

For SI: 1 inch = 25.4 mm.

FIGURE R608.3(2)
WAFFLE-GRID WALL SYSTEM

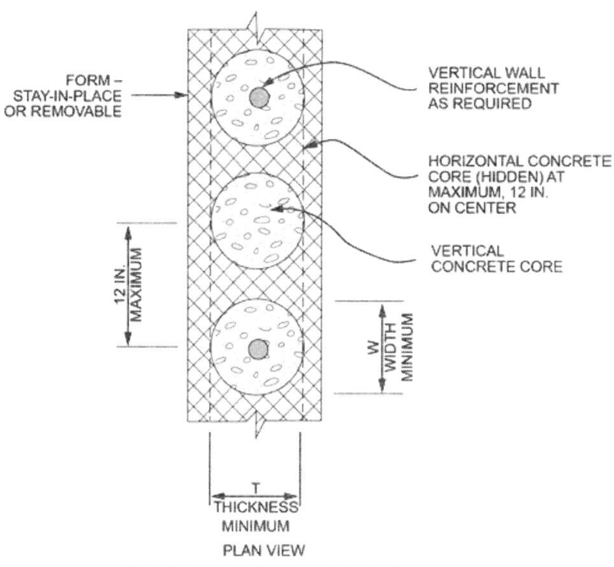

For SI: 1 inch = 25.4 mm.

FIGURE R608.3(3)
SCREEN-GRID WALL SYSTEM

R608.4.3 Exterior wall covering. Stay-in-place forms constructed of rigid foam plastics shall be protected from sunlight and physical damage by the application of an *approved* exterior wall covering complying with this code. Exterior surfaces of other stay-in-place forming systems shall be protected in accordance with this code.

Requirements for installation of masonry veneer, stucco and other finishes on the exterior of concrete walls and other construction details not covered in this section shall comply with the requirements of this code.

R608.4.4 Flat ICF wall systems. Flat ICF wall system forms shall conform to ASTM E2634.

R608.5 Materials. Materials used in the construction of concrete walls shall comply with this section.

R608.5.1 Concrete and materials for concrete. Materials used in concrete, and the concrete itself, shall conform to requirements of this section, PCA 100, ACI 318 or ACI 332.

R608.5.1.1 Cements. The following standards as referenced in Chapter 44 shall be permitted to be used:

1. ASTM C150
2. ASTM C595
3. ASTM C1157

R608.5.1.2 Concrete mixing and delivery. Mixing and delivery of concrete shall comply with ASTM C94 or ASTM C685.

R608.5.1.3 Maximum aggregate size. The nominal maximum size of coarse aggregate shall not exceed one-fifth the narrowest distance between sides of forms, or three-fourths the clear spacing between reinforcing bars or between a bar and the side of the form.

Exception: When *approved,* these limitations shall not apply where removable forms are used and workability and methods of consolidation permit concrete to be placed without honeycombs or voids.

R608.5.1.4 Proportioning and slump of concrete. Proportions of materials for concrete shall be established to provide workability and consistency to permit concrete to be worked readily into forms and around reinforcement under conditions of placement to be employed, without segregation or excessive bleeding. Slump of concrete placed in removable forms shall not exceed 6 inches (152 mm).

Exception: When *approved,* the slump is permitted to exceed 6 inches (152 mm) for concrete mixtures that are resistant to segregation, and are in accordance with the form manufacturer's recommendations.

Slump of concrete placed in stay-in-place forms shall exceed 6 inches (152 mm). Slump of concrete shall be determined in accordance with ASTM C143.

R608.5.1.5 Compressive strength. The minimum specified compressive strength of concrete, f'_c, shall comply with Section R402.2 and shall be not less than 2,500 pounds per square inch (17.2 MPa) at 28 days.

R608.5.1.6 Consolidation of concrete. Concrete shall be consolidated by suitable means during placement and shall be worked around embedded items and reinforcement and into corners of forms. Where stay-in-place forms are used, concrete shall be consolidated by internal vibration.

Exception: When *approved,* self-consolidating concrete mixtures with slumps equal to or greater than 8 inches (203 mm) that are specifically designed for placement without internal vibration need not be internally vibrated.

R608.5.2 Steel reinforcement and anchor bolts.

R608.5.2.1 Steel reinforcement. Steel reinforcement shall comply with ASTM A615, ASTM A706, or ASTM A996. ASTM A996 bars produced from rail steel shall be Type R.

R608.5.2.2 Anchor bolts. Anchor bolts for use with connection details in accordance with Figures R608.9(1) through R608.9(12) shall be bolts with heads complying with ASTM A307 or ASTM F1554. ASTM A307 bolts shall be Grade A with heads. ASTM F1554 bolts shall be Grade 36 minimum. Instead of bolts with heads, it is permissible to use rods with threads on both ends fabricated from steel complying with ASTM A36. The threaded end of the rod to be embedded in the concrete shall be provided with a hex or square nut.

R608.5.2.3 Sheet steel angles and tension tie straps. Angles and tension tie straps for use with connection details in accordance with Figures R608.9(1) through R608.9(12) shall be fabricated from sheet steel complying with ASTM A653 SS, ASTM A792 SS, or ASTM A875 SS. The steel shall be minimum Grade 33 unless a higher grade is required by the applicable figure.

R608.5.3 Form materials and form ties. Forms shall be made of wood, steel, aluminum, plastic, a composite of cement and foam insulation, a composite of cement and wood chips, or other *approved* material suitable for supporting and containing concrete. Forms shall provide sufficient strength to contain concrete during the concrete placement operation.

Form ties shall be steel, solid plastic, foam plastic, a composite of cement and wood chips, a composite of cement and foam plastic, or other suitable material capable of resisting the forces created by fluid pressure of fresh concrete.

R608.5.4 Reinforcement installation details.

R608.5.4.1 Support and cover. Reinforcement shall be secured in the proper location in the forms with tie wire or other bar support system such that displacement will not occur during the concrete placement operation. Steel reinforcement in concrete cast against the earth shall have a minimum cover of 3 inches (76 mm). Minimum cover for reinforcement in concrete cast in removable forms that will be exposed to the earth or weather shall be $1^1/_2$ inches (38 mm) for No. 5 bars and smaller, and 2 inches (50 mm) for No. 6 bars and larger. For concrete cast in removable forms that will not be exposed to the earth or weather, and for concrete cast in stay-in-place forms, minimum cover shall be $^3/_4$ inch (19 mm). The minus tolerance for cover shall not exceed the smaller of one-third the required cover and $^3/_8$ inch (10 mm). See Section R608.5.4.4 for cover requirements for hooks of bars developed in tension.

R608.5.4.2 Location of reinforcement in walls. For location of reinforcement in foundation walls and above-grade walls, see Sections R404.1.3.3.7.2 and R608.6.5, respectively.

R608.5.4.3 Lap splices. Vertical and horizontal wall reinforcement required by Sections R608.6 and R608.7 shall be the longest lengths practical. Where splices are necessary in reinforcement, the length of lap splices shall be in accordance with Table R608.5.4(1) and Figure R608.5.4(1). The maximum gap between noncontact parallel bars at a lap splice shall not exceed the smaller of one-fifth the required lap length and 6 inches (152 mm). See Figure R608.5.4(1).

R608.5.4.4 Development of bars in tension. Where bars are required to be developed in tension by other provisions of this code, development lengths and cover for hooks and bar extensions shall comply with Table R608.5.4(1) and Figure R608.5.4(2). The development lengths shown in Table R608.5.4(1) shall apply to bundled bars in lintels installed in accordance with Section R608.8.2.2.

R608.5.4.5 Standard hooks. Where reinforcement is required by this code to terminate with a standard hook, the hook shall comply with Figure R608.5.4(3).

R608.5.4.6 Webs of waffle-grid walls. Reinforcement, including stirrups, shall not be placed in webs of waffle-grid walls, including lintels. Webs are permitted to have form ties.

TABLE R608.5.4(1)
LAP SPLICE AND TENSION DEVELOPMENT LENGTHS

	BAR SIZE NO.	YIELD STRENGTH OF STEEL, f_y psi (MPa)	
		40,000 (280)	60,000 (420)
		Splice length or tension development length (inches)	
Lap splice length-tension	4	20	30
	5	25	38
	6	30	45
Tension development length for straight bar	4	15	23
	5	19	28
	6	23	34
Tension development length for: a. 90-degree and 180-degree standard hooks with not less than $2^1/_2$ inches of side cover perpendicular to plane of hook. b. 90-degree standard hooks with not less than 2 inches of cover on the bar extension beyond the hook.	4	6	9
	5	7	11
	6	8	13
Tension development length for bar with 90-degree or 180-degree standard hook having less cover than required in Items a and b.	4	8	12
	5	10	15
	6	12	18

For SI: 1 inch = 25.4 mm, 1 degree = 0.0175 rad, 1 pound per square inch = 6.895 kPa.

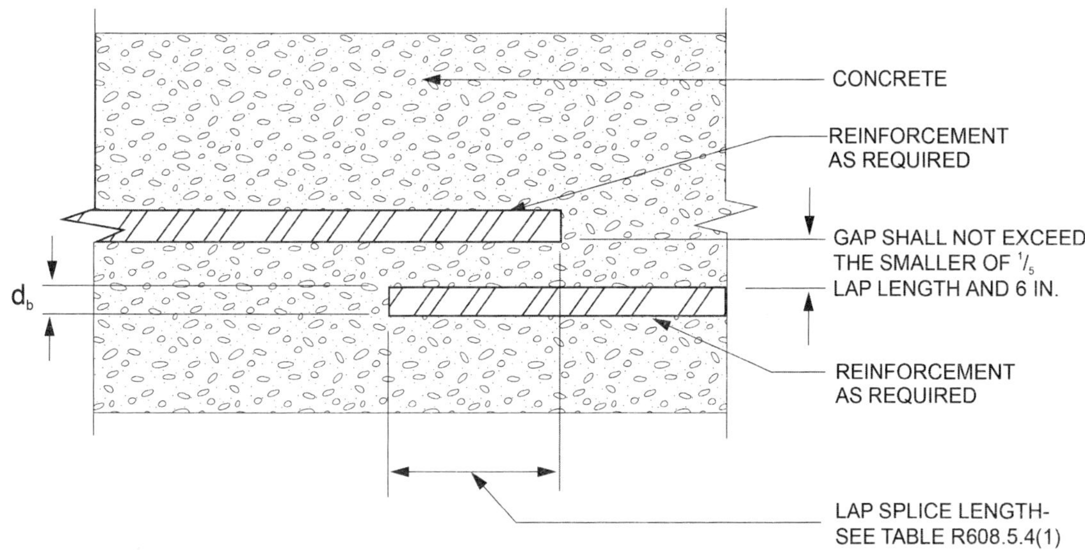

For SI: 1 inch = 25.4 mm.

FIGURE R608.5.4(1)
LAP SPLICES

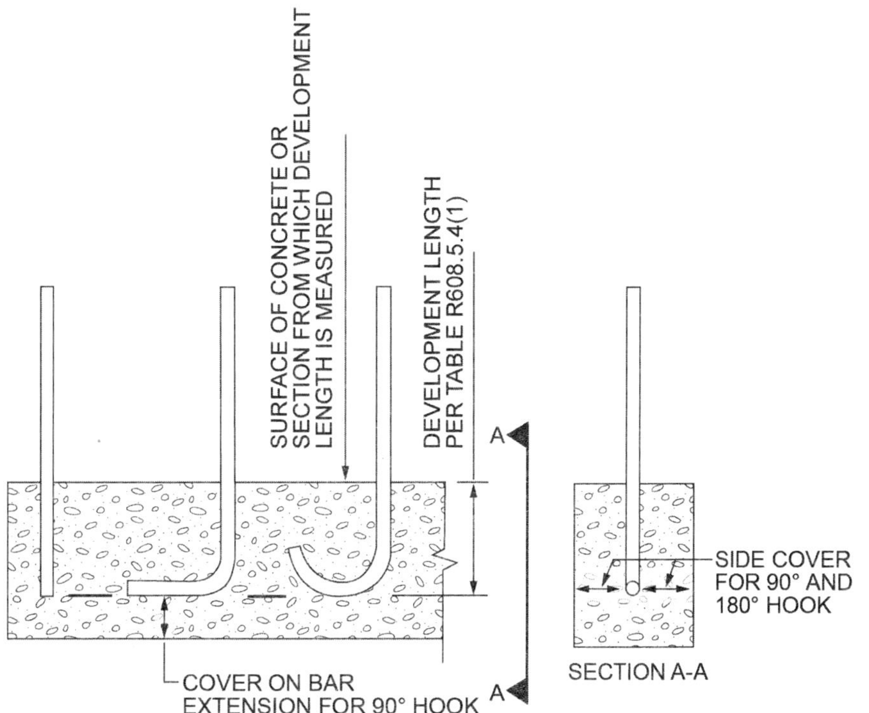

For SI: 1 degree = 0.0175 rad.

FIGURE R608.5.4(2)
DEVELOPMENT LENGTH AND COVER FOR HOOKS AND BAR EXTENSION

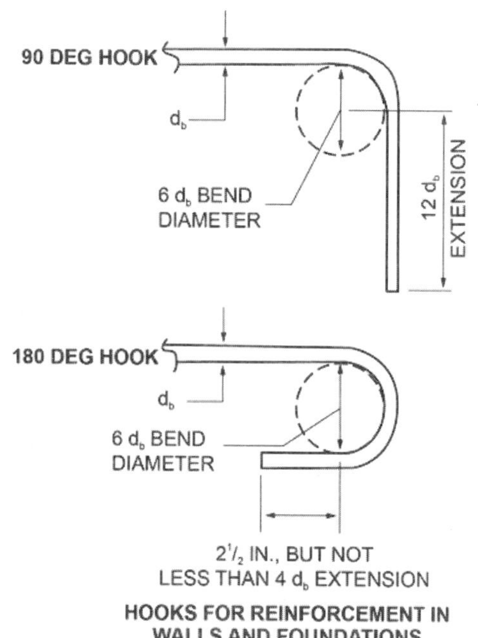

HOOKS FOR REINFORCEMENT IN WALLS AND FOUNDATIONS

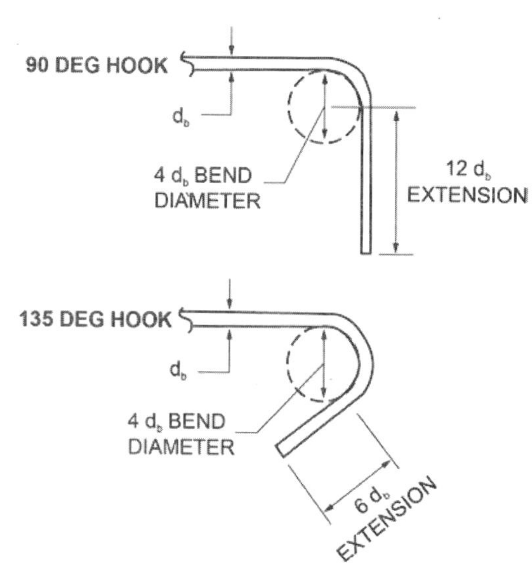

HOOKS FOR STIRRUPS IN LINTELS

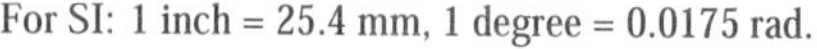
For SI: 1 inch = 25.4 mm, 1 degree = 0.0175 rad.

FIGURE R608.5.4(3)
STANDARD HOOKS

R608.5.4.7 Alternate grade of reinforcement and spacing. Where tables in Sections R404.1.3 and R608.6 specify vertical wall reinforcement based on minimum bar size and maximum spacing, which are based on Grade 60 (420 MPa) steel reinforcement, different size bars or bars made from a different grade of steel are permitted provided an equivalent area of steel per linear foot of wall is provided. Use of Table R608.5.4(2) is permitted to determine the maximum bar spacing for different bar sizes than specified in the tables and bars made from a different grade of steel. Bars shall not be spaced less than one-half the wall thickness, or more than 48 inches (1219 mm) on center.

R608.5.5 Construction joints in walls. Construction joints shall be made and located to not impair the strength of the wall. Construction joints in plain concrete walls, including walls required to have not less than No. 4 bars at 48 inches (1219 mm) on center by Section R608.6, shall be located at points of lateral support, and not less than one No. 4 bar shall extend across the construction joint at a spacing not to exceed 24 inches (610 mm) on center. Construction joint reinforcement shall have not less than 12 inches (305 mm) of embedment on both sides of the joint. Construction joints in reinforced concrete walls shall be located in the middle third of the span between lateral supports, or located and constructed as required for joints in plain concrete walls.

Exception: Vertical wall reinforcement required by this code is permitted to be used in lieu of construction joint reinforcement, provided the spacing does not exceed 24 inches (610 mm), or the combination of wall reinforcement and No. 4 bars described in Section R608.5.5 does not exceed 24 inches (610 mm).

R608.6 Above-grade wall requirements.

R608.6.1 General. The minimum thickness of load-bearing and nonload-bearing above-grade walls and reinforcement shall be as set forth in the appropriate table in this section based on the type of wall form to be used. The wall shall be designed in accordance with ACI 318 where the wall or building is not within the limitations of Section R608.2, where design is required by the tables in this section or where the wall is not within the scope of the tables in this section.

Above-grade concrete walls shall be constructed in accordance with this section and Figure R608.6(1), R608.6(2), R608.6(3) or R608.6(4). Above-grade concrete walls that are continuous with stem walls and not laterally supported by the slab-on-ground shall be designed and constructed in accordance with this section. Concrete walls shall be supported on continuous foundation walls or slabs-on-ground that are monolithic with the footing in accordance with Section R403. The minimum length of solid wall without openings shall be in accordance with Section R608.7. Reinforcement around openings, including lintels, shall be in accordance with Section R608.8. Lateral support for above-grade walls in the out-of-plane direction shall be provided by connections to the floor framing system, if applicable, and to ceiling and roof framing systems in accordance with Section R608.9. The wall thickness shall be equal to or greater than the thickness of the wall in the *story* above.

TABLE R608.5.4(2)
MAXIMUM SPACING FOR ALTERNATIVE BAR SIZE AND ALTERNATIVE GRADE OF STEEL[a, b, c]

BAR SPACING FROM APPLICABLE TABLE IN SECTION R608.6 (inches)	BAR SIZE FROM APPLICABLE TABLE IN SECTION R608.6														
	#4					#5					#6				
	Alternative bar size and alternative grade of steel desired														
	Grade 60		Grade 40			Grade 60		Grade 40			Grade 60		Grade 40		
	#5	#6	#4	#5	#6	#4	#6	#4	#5	#6	#4	#5	#4	#5	#6
	Maximum spacing for alternative bar size and alternative grade of steel (inches)														
8	12	18	5	8	12	5	11	3	5	8	4	6	2	4	5
9	14	20	6	9	13	6	13	4	6	9	4	6	3	4	6
10	16	22	7	10	15	6	14	4	7	9	5	7	3	5	7
11	17	24	7	11	16	7	16	5	7	10	5	8	3	5	7
12	19	26	8	12	18	8	17	5	8	11	5	8	4	6	8
13	20	29	9	13	19	8	18	6	9	12	6	9	4	6	9
14	22	31	9	14	21	9	20	6	9	13	6	10	4	7	9
15	23	33	10	16	22	10	21	6	10	14	7	11	5	7	10
16	25	35	11	17	23	10	23	7	11	15	7	11	5	8	11
17	26	37	11	18	25	11	24	7	11	16	8	12	5	8	11
18	28	40	12	19	26	12	26	8	12	17	8	13	5	8	12
19	29	42	13	20	28	12	27	8	13	18	9	13	6	9	13
20	31	44	13	21	29	13	28	9	13	19	9	14	6	9	13
21	33	46	14	22	31	14	30	9	14	20	10	15	6	10	14
22	34	48	15	23	32	14	31	9	15	21	10	16	7	10	15
23	36	48	15	24	34	15	33	10	15	22	10	16	7	11	15
24	37	48	16	25	35	15	34	10	16	23	11	17	7	11	16
25	39	48	17	26	37	16	35	11	17	24	11	18	8	12	17
26	40	48	17	27	38	17	37	11	17	25	12	18	8	12	17
27	42	48	18	28	40	17	38	12	18	26	12	19	8	13	18
28	43	48	19	29	41	18	40	12	19	26	13	20	8	13	19
29	45	48	19	30	43	19	41	12	19	27	13	20	9	14	19
30	47	48	20	31	44	19	43	13	20	28	14	21	9	14	20
31	48	48	21	32	45	20	44	13	21	29	14	22	9	15	21
32	48	48	21	33	47	21	45	14	21	30	15	23	10	15	21
33	48	48	22	34	48	21	47	14	22	31	15	23	10	16	22
34	48	48	23	35	48	22	48	15	23	32	15	24	10	16	23
35	48	48	23	36	48	23	48	15	23	33	16	25	11	16	23
36	48	48	24	37	48	23	48	15	24	34	16	25	11	17	24
37	48	48	25	38	48	24	48	16	25	35	17	26	11	17	25
38	48	48	25	39	48	25	48	16	25	36	17	27	12	18	25
39	48	48	26	40	48	25	48	17	26	37	18	27	12	18	26
40	48	48	27	41	48	26	48	17	27	38	18	28	12	19	27
41	48	48	27	42	48	26	48	18	27	39	19	29	12	19	27
42	48	48	28	43	48	27	48	18	28	40	19	30	13	20	28
43	48	48	29	44	48	28	48	18	29	41	20	30	13	20	29
44	48	48	29	45	48	28	48	19	29	42	20	31	13	21	29
45	48	48	30	47	48	29	48	19	30	43	20	32	14	21	30
46	48	48	31	48	48	30	48	20	31	44	21	32	14	22	31
47	48	48	31	48	48	30	48	20	31	44	21	33	14	22	31
48	48	48	32	48	48	31	48	21	32	45	22	34	15	23	32

For SI: 1 inch = 25.4 mm.

a. This table is for use with tables in Section R608.6 that specify the minimum bar size and maximum spacing of vertical wall reinforcement for foundation walls and above-grade walls. Reinforcement specified in tables in Section R608.6 is based on Grade 60 (420 MPa) steel reinforcement.

b. Bar spacing shall not exceed 48 inches on center and shall be not less than one-half the nominal wall thickness.

c. For Grade 50 (350 MPa) steel bars (ASTM A996, Type R), use spacing for Grade 40 (280 MPa) bars or interpolate between Grade 40 (280 MPa) and Grade 60 (420 MPa).

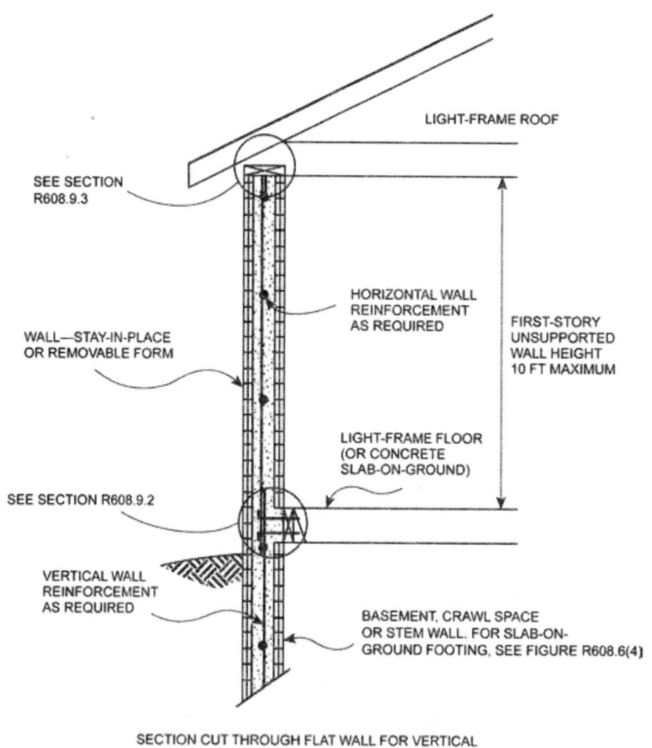

For SI: 1 foot = 304.8 mm.

FIGURE R608.6(1)
ABOVE-GRADE CONCRETE WALL CONSTRUCTION ONE STORY

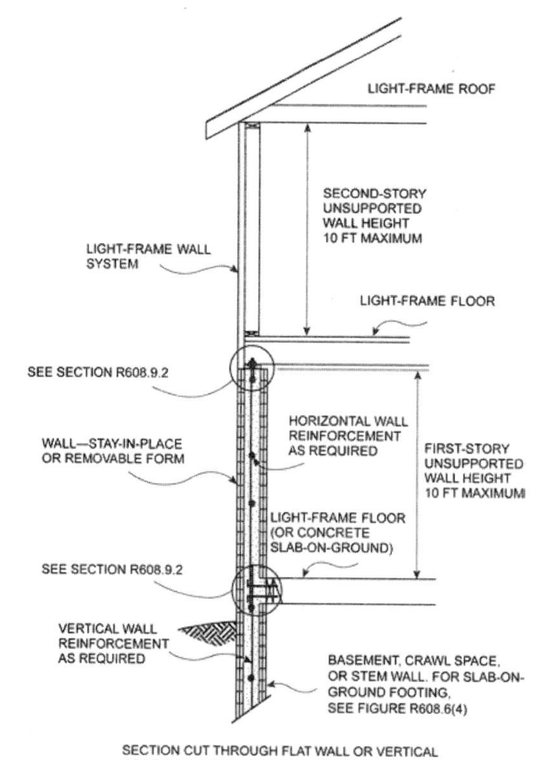

For SI: 1 foot = 304.8 mm.

FIGURE R608.6(2)
ABOVE-GRADE CONCRETE WALL CONSTRUCTION CONCRETE FIRST STORY AND LIGHT-FRAME SECOND STORY

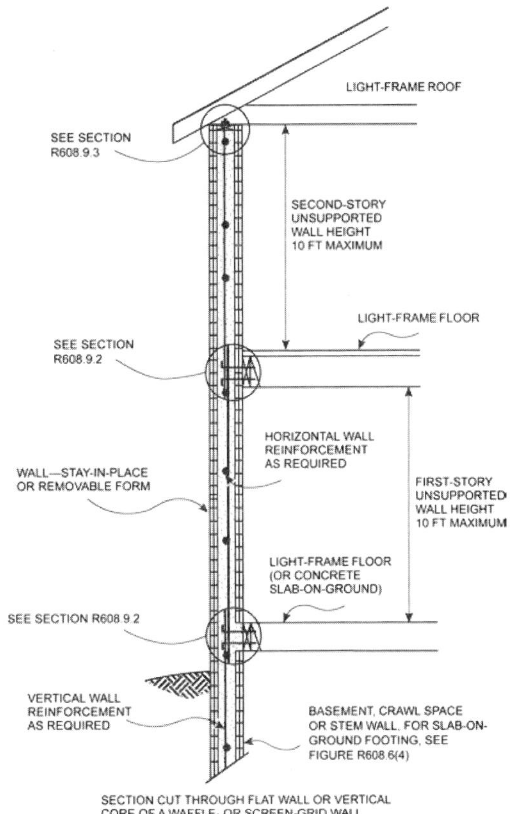

For SI: 1 foot = 304.8 mm.

FIGURE R608.6(3)
ABOVE-GRADE CONCRETE WALL CONSTRUCTION TWO-STORY

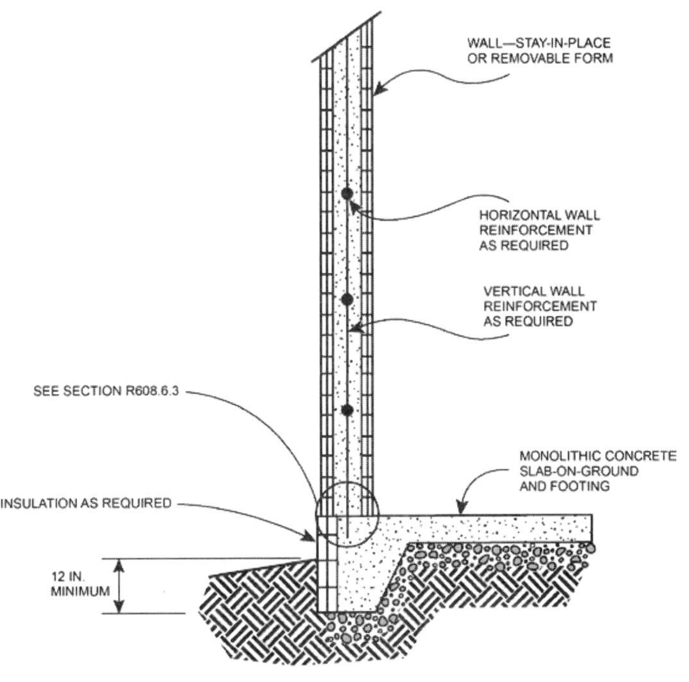

For SI: 1 foot = 304.8 mm.

FIGURE R608.6(4)
ABOVE-GRADE CONCRETE WALL SUPPORTED ON MONOLITHIC SLAB-ON-GROUND FOOTING

TABLE R608.6(1)
MINIMUM VERTICAL REINFORCEMENT FOR FLAT ABOVE-GRADE WALLS[a, b, c, d, e]

MAXIMUM WIND SPEED (mph)			MAXIMUM UNSUPPORTED WALL HEIGHT PER STORY (feet)	MINIMUM VERTICAL REINFORCEMENT-BAR SIZE AND SPACING (inches)[f, g]							
				Nominal[h] wall thickness (inches)							
Exposure Category				4		6		8		10	
B	C	D		Top[i]	Side[i]	Top[i]	Side[i]	Top[i]	Side[i]	Top[i]	Side[i]
115	—	—	8	4@48	4@48	4@48	4@48	4@48	4@48	4@48	4@48
			9	4@48	4@39	4@48	4@48	4@48	4@48	4@48	4@48
			10	4@41	4@34	4@48	4@48	4@48	4@48	4@48	4@48
120	—	—	8	4@48	4@43	4@48	4@48	4@48	4@48	4@48	4@48
			9	4@48	4@36	4@48	4@48	4@48	4@48	4@48	4@48
			10	4@37	4@34	4@48	4@48	4@48	4@48	4@48	4@48
130	110	—	8	4@48	4@38	4@48	4@48	4@48	4@48	4@48	4@48
			9	4@39	4@34	4@48	4@48	4@48	4@48	4@48	4@48
			10	4@34	4@34	4@48	4@48	4@48	4@48	4@48	4@48
140	119	110	8	4@43	4@34	4@48	4@48	4@48	4@48	4@48	4@48
			9	4@34	4@34	4@48	4@48	4@48	4@48	4@48	4@48
			10	4@34	4@31	4@48	4@48	4@48	4@48	4@48	4@48
150	127	117	8	4@37	4@34	4@48	4@48	4@48	4@48	4@48	4@48
			9	4@34	4@33	4@48	4@48	4@48	4@48	4@48	4@48
			10	4@31	4@27	4@48	4@48	4@48	4@48	4@48	4@48
160	136	125	8	4@34	4@34	4@48	4@48	4@48	4@48	4@48	4@48
			9	4@34	4@29	4@48	4@48	4@48	4@48	4@48	4@48
			10	4@27	4@24	4@48	4@48	4@48	4@48	4@48	4@48

For SI: 1 inch = 25.4 mm, 1 foot = 304.8 mm, 1 mile per hour = 0.447 m/s, 1 pound per square inch = 1.895 kPa, 1 square foot = 0.0929 m^2.

a. Table is based on ASCE 7 components and cladding wind pressures for an enclosed building using a mean roof height of 35 feet, interior wall area 4, an effective wind area of 10 square feet, topographic factor, K_{zt}, equal to 1.0, and Risk Category II.

b. Table is based on concrete with a minimum specified compressive strength of 2,500 psi.

c. See Section R608.6.5 for location of reinforcement in wall.

d. Deflection criterion is $L/240$, where L is the unsupported height of the wall in inches.

e. Interpolation is not permitted.

f. Where No. 4 reinforcing bars at a spacing of 48 inches are specified in the table as indicated by shaded cells, use of bars with a minimum yield strength of 40,000 psi or 60,000 psi is permitted.

g. Other than for No. 4 bars spaced at 48 inches on center, table values are based on reinforcing bars with a minimum yield strength of 60,000 psi. Vertical reinforcement with a yield strength of less than 60,000 psi or bars of a different size than specified in the table are permitted in accordance with Section R608.5.4.7 and Table R608.5.4(2).

h. See Table R608.3 for tolerances on nominal thicknesses.

i. "Top" means gravity load from roof or floor construction bears on top of wall. "Side" means gravity load from floor construction is transferred to wall from a wood ledger or cold-formed steel track bolted to side of wall. For nonload-bearing walls where floor framing members span parallel to the wall, use of the "Top" bearing condition is permitted.

TABLE R608.6(2)
MINIMUM VERTICAL REINFORCEMENT FOR WAFFLE-GRID ABOVE-GRADE WALLS[a, b, c, d, e]

MAXIMUM WIND SPEED (mph)			MAXIMUM UNSUPPORTED WALL HEIGHT PER STORY (feet)	MINIMUM VERTICAL REINFORCEMENT-BAR SIZE AND SPACING (inches)[f, g]			
				Nominal[h] wall thickness (inches)			
Exposure Category				6		8	
B	C	D		Top[i]	Side[i]	Top[i]	Side[i]
115	—	—	8	4@48	4@48	4@48	4@48
			9	4@48	5@43	4@48	4@48
			10	5@47	5@37	4@48	4@48
120	—	—	8	4@48	5@48	4@48	4@48
			9	4@48	5@40	4@48	4@48
			10	5@43	5@37	4@48	4@48
130	110	—	8	4@48	5@42	4@48	4@48
			9	5@45	5@37	4@48	4@48
			10	5@37	5@37	4@48	4@48
140	119	110	8	4@48	5@38	4@48	4@48
			9	5@39	5@37	4@48	4@48
			10	5@37	5@35	4@48	4@48
150	127	117	8	5@43	5@37	4@48	4@48
			9	5@37	5@37	4@48	4@48
			10	5@36	6@44	4@48	4@48
160	136	125	8	5@38	5@37	4@48	4@48
			9	5@37	6@47	4@48	4@48
			10	6@45	6@39	4@48	6@46

For SI: 1 inch = 25.4 mm, 1 foot = 304.8 mm, 1 mile per hour = 0.447 m/s, 1 pound per square inch = 6.895 kPa, 1 square foot = 0.0929 m^2.

a. Table is based on ASCE 7 components and cladding wind pressures for an enclosed building using a mean roof height of 35 feet, interior wall area 4, an effective wind area of 10 square feet, topographic factor, K_{zt} equal to 1.0, and Risk Category II.

b. Table is based on concrete with a minimum specified compressive strength of 2,500 psi.

c. See Section R608.6.5 for location of reinforcement in wall.

d. Deflection criterion is $L/240$, where L is the unsupported height of the wall in inches.

e. Interpolation is not permitted.

f. Where No. 4 reinforcing bars at a spacing of 48 inches are specified in the table as indicated by shaded cells, use of bars with a minimum yield strength of 40,000 psi or 60,000 psi is permitted.

g. Other than for No. 4 bars spaced at 48 inches on center, table values are based on reinforcing bars with a minimum yield strength of 60,000 psi. Maximum spacings shown are the values calculated for the specified bar size. Where the bar used is Grade 60 and the size specified in the table, the actual spacing in the wall shall not exceed a whole-number multiple of 12 inches such as, 12, 24, 36 and 48, that is less than or equal to the tabulated spacing. Vertical reinforcement with a yield strength of less than 60,000 psi or bars of a different size than specified in the table are permitted in accordance with Section R608.5.4.7 and Table R608.5.4(2).

h. See Table R608.3 for minimum core dimensions and maximum spacing of horizontal and vertical cores.

i. "Top" means gravity load from roof or floor construction bears on top of wall. "Side" means gravity load from floor construction is transferred to wall from a wood ledger or cold-formed steel track bolted to side of wall. For nonload-bearing walls and where floor framing members span parallel to the wall, the "top" bearing condition is permitted to be used.

TABLE R608.6(3)
MINIMUM VERTICAL REINFORCEMENT FOR 6-INCH SCREEN-GRID ABOVE-GRADE WALLS[a, b, c, d, e]

MAXIMUM WIND SPEED (mph)			MAXIMUM UNSUPPORTED WALL HEIGHT PER STORY (feet)	MINIMUM VERTICAL REINFORCEMENT-BAR SIZE AND SPACING (inches)[f, g]	
Exposure Category				Nominal[h] wall thickness (inches)	
				6	
B	C	D		Top[i]	Side[i]
115	—	—	8	4@48	4@48
			9	4@48	5@41
			10	4@48	6@48
120	—	—	8	4@48	4@48
			9	4@48	5@38
			10	5@42	6@48
130	110	—	8	4@48	5@41
			9	5@44	6@48
			10	5@35	6@48
140	119	110	8	4@48	5@36
			9	5@38	6@48
			10	6@48	6@48
150	127	117	8	5@42	6@48
			9	6@48	6@48
			10	6@48	6@42
160	136	125	8	5@37	6@48
			9	6@48	6@45
			10	6@44	6@38

For SI: 1 inch = 25.4 mm, 1 foot = 304.8 mm, 1 mile per hour = 0.447 m/s, 1 pound per square inch = 6.895 kPa, 1 square foot = 0.0929 m^2.

a. Table is based on ASCE 7 components and cladding wind pressures for an enclosed building using a mean roof height of 35 feet, interior wall area 4, an effective wind area of 10 square feet, topographic factor, K_{zt}, equal to 1.0, and Risk Category II.

b. Table is based on concrete with a minimum specified compressive strength of 2,500 psi.

c. See Section R608.6.5 for location of reinforcement in wall.

d. Deflection criterion is *L*/240, where *L* is the unsupported height of the wall in inches.

e. Interpolation is not permitted.

f. Where No. 4 reinforcing bars at a spacing of 48 inches are specified in the table as indicated by shaded cells, use of bars with a minimum yield strength of 40,000 psi or 60,000 psi is permitted.

g. Other than for No. 4 bars spaced at 48 inches on center, table values are based on reinforcing bars with a minimum yield strength of 60,000 psi. Maximum spacings shown are the values calculated for the specified bar size. Where the bar used is Grade 60 and the size specified in the table, the actual spacing in the wall shall not exceed a whole-number multiple of 12 inches such as, 12, 24, 36 and 48, that is less than or equal to the tabulated spacing. Vertical reinforcement with a yield strength of less than 60,000 psi or bars of a different size than specified in the table are permitted in accordance with Section R608.5.4.7 and Table R608.5.4(2).

h. See Table R608.3 for minimum core dimensions and maximum spacing of horizontal and vertical cores.

i. "Top" means gravity load from roof or floor construction bears on top of wall. "Side" means gravity load from floor construction is transferred to wall from a wood ledger or cold-formed steel track bolted to side of wall. For nonload-bearing wall and where floor framing members span parallel to the wall, use of the "Top" bearing condition is permitted.

TABLE R608.6(4) MINIMUM VERTICAL REINFORCEMENT FOR FLAT, WAFFLE- AND SCREEN-GRID ABOVE-GRADE WALLS DESIGNED CONTINUOUS WITH FOUNDATION STEM WALLS[a, b, c, d, e, k]

MAXIMUM WIND SPEED (mph)			HEIGHT OF STEM WALL[h, i] (feet)	MAXIMUM DESIGN LATERAL SOIL LOAD (psf/ft)	MAXIMUM UNSUPPORTED HEIGHT OF ABOVE-GRADE WALL (feet)	MINIMUM VERTICAL REINFORCEMENT-BAR SIZE AND SPACING (inches)[f, g]						
Exposure Category						Wall type and nominal thickness[j] (inches)						
						Flat				Waffle		Screen
B	C	D				4	6	8	10	6	8	6
115	—	—	3	30	8	4@30	4@48	4@48	4@48	4@22	4@26	4@21
					10	4@23	5@43	4@48	4@48	4@17	4@20	4@16
				60	10	4@19	5@37	4@48	4@48	4@14	4@17	4@14
			6	30	10	DR	5@21	6@35	4@48	DR	4@10	DR
				60	10	DR	5@12	6@25	6@28	DR	DR	DR
120	—	—	3	30	8	4@28	4@48	4@48	4@48	4@21	4@48	4@20
					10	4@22	5@41	4@48	4@48	4@16	4@19	4@15
				60	10	4@18	5@35	4@48	4@48	4@14	4@17	4@13
			6	30	10	DR	5@21	6@35	4@48	DR	4@10	DR
				60	10	DR	5@12	6@25	6@28	DR	DR	DR
130	110	—	3	30	8	4@25	4@48	4@48	4@48	4@18	4@22	4@18
					10	4@19	5@36	4@48	4@48	4@14	4@17	4@13
			6	60	10	4@16	5@34	4@48	4@48	4@12	4@17	4@12
				30	10	DR	5@19	6@35	4@48	DR	4@9	DR
				60	10	DR	5@12	6@24	6@28	DR	DR	DR
140	119	110	3	30	8	4@22	5@42	4@48	4@48	4@16	4@20	4@16
					10	4@17	5@34	4@48	4@48	4@21	4@17	4@12
				60	10	4@15	5@34	4@48	4@48	4@11	4@17	4@10
			6	30	10	DR	5@18	6@35	6@35	DR	4@48	DR
				60	10	DR	5@11	6@23	6@28	DR	DR	DR
150	127	117	3	30	8	4@20	5@37	4@48	4@48	4@15	4@18	4@14
					10	4@15	5@34	4@48	4@48	4@11	4@17	4@11
				60	10	4@13	5@34	4@48	4@48	4@10	4@16	4@9
			6	30	10	DR	5@17	6@33	6@32	DR	4@8	DR
				60	10	DR	DR	6@22	6@28	DR	DR	DR
160	136	125	3	30	8	4@18	5@34	4@48	4@48	4@13	4@17	4@13
					10	4@13	5@34	4@48	4@48	4@10	4@16	4@9
				60	10	4@11	5@31	6@45	4@48	4@9	4@14	4@8
			6	30	10	DR	5@15	6@31	6@30	DR	4@7	DR
				60	10	DR	DR	6@21	6@27	DR	DR	DR

(continued)

TABLE R608.6(4)—continued
MINIMUM VERTICAL REINFORCEMENT FOR FLAT, WAFFLE- AND SCREEN-GRID ABOVE-GRADE WALLS DESIGNED CONTINUOUS WITH FOUNDATION STEM WALLS[a, b, c, d, e, k]

For SI: 1 inch = 25.4 mm, 1 foot = 304.8 mm, 1 mile per hour = 0.447 m/s, 1 pound per square inch = 6.895 kPa, 1 square foot = 0.0929 m^2.

DR = Design Required.

a. Table is based on ASCE 7 components and cladding wind pressures for an enclosed building using a mean roof height of 35 feet, interior wall area 4, an effective wind area of 10 square feet, topographic factor, K_{zt}, equal to 1.0, and Risk Category II.

b. Table is based on concrete with a minimum specified compressive strength of 2,500 psi.

c. See Section R608.6.5 for location of reinforcement in wall.

d. Deflection criterion is *L*/240, where *L* is the height of the wall in inches from the exterior finish ground level to the top of the above-grade wall.

e. Interpolation is not permitted. For intermediate values of basic wind speed, heights of stem wall and above-grade wall, and design lateral soil load, use next higher value.

f. Where No. 4 reinforcing bars at a spacing of 48 inches are specified in the table as indicated by shaded cells, use of bars with a minimum yield strength of 40,000 psi or 60,000 psi is permitted.

g. Other than for No. 4 bars spaced at 48 inches on center, table values are based on reinforcing bars with a minimum yield strength of 60,000 psi. Maximum spacings shown are the values calculated for the specified bar size. In waffle and screen-grid walls where the bar used is Grade 60 and the size specified in the table, the actual spacing in the wall shall not exceed a whole-number multiple of 12 inches such as, 12, 24, 36 and 48, that is less than or equal to the tabulated spacing. Vertical reinforcement with a yield strength of less than 60,000 psi and bars of a different size than specified in the table are permitted in accordance with Section R608.5.4.7 and Table R608.5.4(2).

h. Height of stem wall is the distance from the exterior finish ground level to the top of the slab-on-ground.

i. Where the distance from the exterior finish ground level to the top of the slab-on-ground is equal to or greater than 4 feet, the stem wall shall be laterally supported at the top and bottom before backfilling. Where the wall is designed and constructed to be continuous with the above-grade wall, temporary supports bracing the top of the stem wall shall remain in place until the above-grade wall is laterally supported at the top by floor or roof construction.

j. See Table R608.3 for tolerances on nominal thicknesses, and minimum core dimensions and maximum spacing of horizontal and vertical cores for waffle- and screen-grid walls.

k. Tabulated values are applicable to construction where gravity loads bear on top of wall, and conditions where gravity loads from floor construction are transferred to wall from a wood ledger or cold-formed steel track bolted to side of wall. See Tables R608.6(1), R608.6(2) and R608.6(3).

R608.6.2 Wall reinforcement for wind. Vertical wall reinforcement for resistance to out-of-plane wind forces shall be determined from Table R608.6(1), R608.6(2), R608.6(3) or R608.6(4). For the design of nonload-bearing walls, in Tables R608.6(1), R608.6(2) and R608.6(3) use the appropriate column labeled "Top." (see Sections R608.7.2.2.2 and R608.7.2.2.3). There shall be a vertical bar at corners of exterior walls. Unless more horizontal reinforcement is required by Section R608.7.2.2.1, the minimum horizontal reinforcement shall be four No. 4 bars [Grade 40 (280 MPa)] placed as follows: top bar within 12 inches (305 mm) of the top of the wall, bottom bar within 12 inches (305 mm) of the finish floor and one bar each at approximately one-third and two-thirds of the wall height.

R608.6.3 Continuity of wall reinforcement between stories. Vertical reinforcement required by this section shall be continuous between elements providing lateral support for the wall. Reinforcement in the wall of the *story* above shall be continuous with the reinforcement in the wall of the *story* below, or the foundation wall, if applicable. Lap splices, where required, shall comply with Section R608.5.4.3 and Figure R608.5.4(1). Where the above-grade wall is supported by a monolithic slab-on-ground and footing, dowel bars with a size and spacing to match the vertical above-grade concrete wall reinforcement shall be embedded in the monolithic slab-on-ground and footing the distance required to develop the dowel bar in tension in accordance with Section R608.5.4.4 and Figure R608.5.4(2) and lap-spliced with the above-grade wall reinforcement in accordance with Section R608.5.4.3 and Figure R608.5.4(1).

Where a construction joint in the wall is located below the level of the floor and less than the distance required to develop the bar in tension, the distance required to develop the bar in tension shall be measured from the top of the concrete below the joint. See Section R608.5.5.

Exception: Where reinforcement in the wall above cannot be made continuous with the reinforcement in the wall below, the bottom of the reinforcement in the wall above shall be terminated in accordance with one of the following:

1. Extend below the top of the floor the distance required to develop the bar in tension in accordance with Section R608.5.4.4 and Figure R608.5.4(2).
2. Lap-spliced in accordance with Section R608.5.4.3 and Figure R608.5.4(1) with a dowel bar that extends into the wall below the distance required to develop the bar in tension in accordance with Section R608.5.4.4 and Figure R608.5.4(2).

R608.6.4 Termination of reinforcement. Where indicated in Items 1 through 3, vertical wall reinforcement in the top-most *story* with concrete walls shall be terminated with a 90-degree (1.57 rad) standard hook complying with Section R608.5.4.5 and Figure R608.5.4(3).

1. Vertical bars adjacent to door and window openings required by Section R608.8.1.2.
2. Vertical bars at the ends of required solid wall segments (see Section R608.7.2.2.2).
3. Vertical bars (other than end bars, see Item 2) used as shear reinforcement in required solid wall segments where the reduction factor for design strength, R_3, used is based on the wall having horizontal and vertical shear reinforcement (see Section R608.7.2.2.3).

The bar extension of the hook shall be oriented parallel to the horizontal wall reinforcement and be within 4 inches (102 mm) of the top of the wall.

Horizontal reinforcement shall be continuous around the building corners by bending one of the bars and lap-splicing it with the bar in the other wall in accordance with Section R608.5.4.3 and Figure R608.5.4(1).

In required solid wall segments where the reduction factor for design strength, R_3, is based on the wall having horizontal and vertical shear reinforcement in accordance with Section R608.7.2.2.1, horizontal wall reinforcement shall be terminated with a standard hook complying with Section R608.5.4.5 and Figure R608.5.4(3) or in a lap-splice, except at corners where the reinforcement shall be continuous as required.

Exception: In lieu of bending horizontal reinforcement at corners, separate bent reinforcing bars shall be permitted provided that the bent bar is lap-spliced with the horizontal reinforcement in both walls in accordance with Section R608.5.4.3 and Figure R608.5.4(1).

R608.6.5 Location of reinforcement in wall. Except for vertical reinforcement at the ends of required solid wall segments, which shall be located as required by Section R608.7.2.2.2, the location of the vertical reinforcement shall not vary from the center of the wall by more than the greater of 10 percent of the wall thickness and $^3/_8$-inch (10 mm). Horizontal and vertical reinforcement shall be located to provide not less than the minimum cover required by Section R608.5.4.1.

R608.7 Solid walls for resistance to lateral forces.

R608.7.1 Length of solid wall. Each exterior wall line in each *story* shall have a total length of solid wall required by Section R608.7.1.1. A solid wall is a section of flat, waffle-grid or screen-grid wall, extending the full *story height* without openings or penetrations, except those permitted by Section R608.7.2. Solid wall segments that contribute to the total length of solid wall shall comply with Section R608.7.2.

R608.7.1.1 Length of solid wall for wind. Buildings shall have solid walls in each exterior endwall line (the side of a building that is parallel to the span of the roof or floor framing) and sidewall line (the side of a building that is perpendicular to the span of the roof or floor framing) to resist lateral in-plane wind forces. The site-appropriate basic wind speed and exposure category shall be used in Tables R608.7.1.1(1) through (3) to determine the unreduced total length, *UR*, of solid wall required in each exterior endwall line and sidewall line. For buildings with a mean roof height of less than 35 feet (10 668 mm), the unreduced values determined from Tables R608.7.1.1(1) through (3) are permitted to be reduced by multiplying by the applicable factor, R_1, from Table R608.7.1.1(4); however, reduced values shall be not less than the minimum values in Tables R608.7.1.1(1) through (3). Where the floor-to-ceiling height of a *story* is less than 10 feet (3048 mm), the unreduced values determined from Tables R608.7.1.1(1) through (3), including minimum values, are permitted to be reduced by multiplying by the applicable factor, R_2, from Table R608.7.1.1(5). To account for different design strengths than assumed in determining the values in Tables R608.7.1.1(1) through (3), the unreduced lengths determined from Tables R608.7.1.1(1) through (3), including minimum values, are permitted to be reduced by multiplying by the applicable factor, R_3, from Table R608.7.1.1(6). The reductions permitted by Tables R608.7.1.1(4), R608.7.1.1(5) and R608.7.1.1(6) are cumulative.

The total length of solid wall segments, *TL*, in a wall line that comply with the minimum length requirements of Section R608.7.2.1 [see Figure R608.7.1.1(1)] shall be equal to or greater than the product of the unreduced length of solid wall from Tables R608.7.1.1(1) through (3), *UR* and the applicable reduction factors, if any, from Tables R608.7.1.1(4), R608.7.1.1(5) and R608.7.1.1(6) as indicated by Equation R6-1.

$$TL \geq R_1 \times R_2 \times R_3 \times UR \qquad \textbf{(Equation R6-1)}$$

where:

TL = Total length of solid wall segments in a wall line that comply with Section R608.7.2.1 [see Figure R608.7.1.1(1)].

R_1 = 1.0 or reduction factor for mean roof height from Table R608.7.1.1(4).

R_2 = 1.0 or reduction factor for floor-to-ceiling wall height from Table R608.7.1.1(5).

R_3 = 1.0 or reduction factor for design strength from Table R608.7.1.1(6).

UR = Unreduced length of solid wall from Tables R608.7.1.1(1) through (3).

The total length of solid wall in a wall line, *TL*, shall be not less than that provided by two solid wall segments complying with the minimum length requirements of Section R608.7.2.1.

To facilitate determining the required wall thickness, wall type, number and *grade* of vertical bars at each end of each solid wall segment, and whether shear reinforcement is required, use of Equation R6-2 is permitted.

$$R_3 \leq \frac{TL}{R_1 \times R_2 \times UR} \qquad \textbf{(Equation R6-2)}$$

After determining the maximum permitted value of the reduction factor for design strength, R_3, in accordance with Equation R6-2, select a wall type from Table R608.7.1.1(6) with R_3 less than or equal to the value calculated.

TABLE R608.7.1.1(1)
UNREDUCED LENGTH, *UR*, OF SOLID WALL REQUIRED IN EACH EXTERIOR ENDWALL FOR WIND PERPENDICULAR TO RIDGE ONE STORY OR TOP STORY OF TWO STORY[a, c, d, e, f, g]

SIDEWALL LENGTH (feet)	ENDWALL LENGTH (feet)	ROOF SLOPE	UNREDUCED LENGTH, *UR*, OF SOLID WALL REQUIRED IN ENDWALLS FOR WIND PERPENDICULAR TO RIDGE (feet)						
			Basic Wind Speed (mph) Exposure						
			115B	120B	130B	140B	150B	160B	Minimum[b]
			—	—	110C	119C	127C	136C	
			—	—	—	110D	117D	125D	
15	15	< 1:12	1.03	1.12	1.32	1.53	1.76	2.00	0.92
		5:12	1.43	1.56	1.83	2.12	2.43	2.77	1.15
		7:12	2.00	2.18	2.56	2.97	3.41	3.88	1.25
		12:12	3.20	3.48	4.09	4.74	5.44	6.19	1.54
	30	< 1:12	1.03	1.12	1.32	1.53	1.76	2.00	0.98
		5:12	1.43	1.56	1.83	2.12	2.43	2.77	1.43
		7:12	2.78	3.03	3.56	4.13	4.74	5.39	1.64
		12:12	5.17	5.63	6.61	7.67	8.80	10.01	2.21
	45	< 1:12	1.03	1.12	1.32	1.53	1.76	2.00	1.04
		5:12	1.43	1.56	1.83	2.12	2.43	2.77	1.72
		7:12	3.57	3.88	4.56	5.28	6.07	6.90	2.03
		12:12	7.15	7.78	9.13	10.59	12.16	13.84	2.89
	60	< 1:12	1.03	1.12	1.32	1.53	1.76	2.00	1.09
		5:12	1.43	1.56	1.83	2.12	2.43	2.77	2.01
		7:12	4.35	4.73	5.55	6.44	7.39	8.41	2.42
		12:12	9.12	9.93	11.66	13.52	15.52	17.66	3.57
30	15	< 1:12	1.84	2.01	2.35	2.73	3.13	3.57	1.82
		5:12	2.56	2.78	3.27	3.79	4.35	4.95	2.23
		7:12	3.61	3.93	4.61	5.34	6.13	6.98	2.42
		12:12	5.61	6.10	7.16	8.31	9.54	10.85	2.93
	30	< 1:12	1.84	2.01	2.35	2.73	3.13	3.57	1.93
		5:12	2.56	2.78	3.27	3.79	4.35	4.95	2.75
		7:12	4.92	5.35	6.28	7.29	8.37	9.52	3.12
		12:12	8.92	9.71	11.39	13.22	15.17	17.26	4.14
	45	< 1:12	1.84	2.01	2.35	2.73	3.13	3.57	2.03
		5:12	2.56	2.78	3.27	3.79	4.35	4.95	3.26
		7:12	6.23	6.78	7.96	9.23	10.60	12.06	3.82
		12:12	12.23	13.31	15.63	18.12	20.80	23.67	5.36
	60	< 1:12	1.84	2.01	2.35	2.73	3.13	3.57	2.14
		5:12	2.56	2.78	3.27	3.79	4.35	4.95	3.78
		7:12	7.54	8.21	9.64	11.17	12.83	14.60	4.52
		12:12	15.54	16.92	19.86	23.03	26.44	30.08	6.57

(continued)

TABLE R608.7.1.1(1)—continued
UNREDUCED LENGTH, *UR*, OF SOLID WALL REQUIRED IN EACH EXTERIOR ENDWALL FOR WIND PERPENDICULAR TO RIDGE ONE STORY OR TOP STORY OF TWO STORY[a, c, d, e, f, g]

SIDEWALL LENGTH (feet)	ENDWALL LENGTH (feet)	ROOF SLOPE	UNREDUCED LENGTH, *UR*, OF SOLID WALL REQUIRED IN ENDWALLS FOR WIND PERPENDICULAR TO RIDGE (feet)						
			Basic Wind Speed (mph) Exposure						
			115B	120B	130B	140B	150B	160B	Minimum[b]
			—	—	110C	119C	127C	136C	
			—	—	—	110D	117D	125D	
60	15	< 1:12	3.42	3.72	4.36	5.06	5.81	6.61	3.63
		5:12	4.75	5.17	6.06	7.03	8.07	9.19	4.40
		7:12	6.76	7.36	8.64	10.02	11.51	13.09	4.75
		12:12	10.35	11.27	13.23	15.34	17.61	20.04	5.71
	30	< 1:12	3.42	3.72	4.36	5.06	5.81	6.61	3.83
		5:12	4.75	5.17	6.06	7.03	8.07	9.19	5.37
		7:12	9.12	9.93	11.66	13.52	15.52	17.66	6.07
		12:12	16.30	17.75	20.83	24.16	27.73	31.55	8.00
	45	< 1:12	3.55	3.87	4.54	5.27	6.05	6.88	4.03
		5:12	4.94	5.37	6.31	7.31	8.40	9.55	6.34
		7:12	11.71	12.75	14.97	17.36	19.93	22.67	7.39
		12:12	22.70	24.71	29.00	33.64	38.62	43.94	10.29
	60	< 1:12	3.68	4.01	4.71	5.46	6.27	7.13	4.23
		5:12	5.11	5.57	6.54	7.58	8.70	9.90	7.31
		7:12	14.38	15.66	18.37	21.31	24.46	27.83	8.71
		12:12	29.30	31.90	37.44	43.42	49.85	56.72	12.57

For SI: 1 inch = 25.4 mm, 1 foot = 304.8 mm, 1 mile per hour = 0.447 m/s, 1 pound-force per linear foot = 0.146 kN/m, 1 pound per square foot = 47.88 Pa.

a. Tabulated lengths were derived by calculating design wind pressures in accordance with Figure 28.4-1 of ASCE 7 for a building with a mean roof height of 35 feet, topographic factor, K_{zt} equal to 1.0, and Risk Category II. For wind perpendicular to the ridge, the effects of a 2-foot overhang on each endwall are included. The design pressures were used to calculate forces to be resisted by solid wall segments in each. The forces to be resisted by each wall line were then divided by the default design strength of 840 pounds per linear foot of length to determine the unreduced length, *UR*, of solid wall length required in each endwall. The actual mean roof height of the building shall not exceed the least horizontal dimension of the building.

b. Tabulated lengths in the "minimum" column are based on the requirement of Section 28.4.4 of ASCE 7 that the main windforce-resisting system be designed for a minimum pressure of 16 psf multiplied by the wall area of the building and 8 psf multiplied by the roof area of the building projected onto a vertical plane normal to the assumed wind direction. Tabulated lengths in shaded cells are less than the "minimum" value. Where the minimum controls, it is permitted to be reduced in accordance with Notes c, d and e. See Section R608.7.1.1.

c. For buildings with a mean roof height of less than 35 feet, tabulated lengths are permitted to be reduced by multiplying by the appropriate factor, R_1, from Table R608.7.1.1(4). The reduced length shall be not less than the "minimum" value shown in the table.

d. Tabulated lengths for "one story or top story of two story" are based on a floor-to-ceiling height of 10 feet. Tabulated lengths for "first story of two story" are based on floor-to-ceiling heights of 10 feet each for the first and second story. For floor-to-ceiling heights less than assumed, use the lengths in this table or Table R608.1.1(2) or (3), or multiply the value in the table by the reduction factor, R_2, from Table R608.7.1.1(5).

e. Tabulated lengths are based on the default design shear strength of 840 pounds per linear foot of solid wall segment. The tabulated lengths are permitted to be reduced by multiplying by the applicable reduction factor for design strength, R_3, from Table R608.7.1.1(6).

f. The reduction factors, R_1, R_2 and R_3, in Tables R608.7.1.1(4), R608.7.1.1(5), and R608.7.1.1(6), respectively, are permitted to be compounded, subject to the limitations of Note b. However, the minimum number and minimum length of solid wall segments in each wall line shall comply with Sections R608.7.1 and R608.7.2.1, respectively.

g. For intermediate values of sidewall length, endwall length, roof slope and basic wind speed, use the next higher value, or determine by interpolation.

TABLE R608.7.1.1(2)
UNREDUCED LENGTH, *UR*, OF SOLID WALL REQUIRED IN EACH EXTERIOR ENDWALL FOR WIND PERPENDICULAR TO RIDGE FIRST STORY OF TWO STORY[a, c, d, e, f, g]

SIDEWALL LENGTH (feet)	ENDWALL LENGTH (feet)	ROOF SLOPE	UNREDUCED LENGTH, *UR*, OF SOLID WALL REQUIRED IN ENDWALLS FOR WIND PERPENDICULAR TO RIDGE (feet)						
			Basic Wind Speed (mph) Exposure						
			115B	120B	130B	140B	150B	160B	Minimum[b]
			—	—	110C	119C	127C	136C	
			—	—	—	110D	117D	125D	
15	15	< 1:12	2.98	3.25	3.81	4.42	5.07	5.77	2.54
		5:12	4.13	4.50	5.28	6.12	7.03	8.00	2.76
		7:12	4.31	4.70	5.51	6.39	7.34	8.35	2.87
		12:12	5.51	6.00	7.04	8.16	9.37	10.66	3.15
	30	< 1:12	2.98	3.25	3.81	4.42	5.07	5.77	2.59
		5:12	4.13	4.50	5.28	6.12	7.03	8.00	3.05
		7:12	5.09	5.55	6.51	7.55	8.67	9.86	3.26
		12:12	7.48	8.15	9.56	11.09	12.73	14.49	3.83
	45	< 1:12	2.98	3.25	3.81	4.42	5.07	5.77	2.65
		5:12	4.13	4.50	5.28	6.12	7.03	8.00	3.34
		7:12	5.88	6.40	7.51	8.71	10.00	11.37	3.65
		12:12	9.46	10.30	12.09	14.02	16.09	18.31	4.51
	60	< 1:12	2.98	3.25	3.81	4.42	5.07	5.77	2.71
		5:12	4.13	4.50	5.28	6.12	7.03	8.00	3.63
		7:12	6.66	7.25	8.51	9.87	11.32	12.89	4.04
		12:12	11.43	12.45	14.61	16.94	19.45	22.13	5.19
30	15	< 1:12	5.32	5.79	6.80	7.89	9.05	10.30	5.06
		5:12	7.39	8.04	9.44	10.95	12.57	14.30	5.47
		7:12	7.94	8.65	10.15	11.77	13.51	15.37	5.65
		12:12	9.94	10.82	12.70	14.73	16.91	19.24	6.17
	30	< 1:12	5.32	5.79	6.80	7.89	9.05	10.30	5.16
		5:12	7.39	8.04	9.44	10.95	12.57	14.30	5.98
		7:12	9.25	10.07	11.82	13.71	15.74	17.91	6.35
		12:12	13.25	14.43	16.93	19.64	22.54	25.65	7.38
	45	< 1:12	5.32	5.79	6.80	7.89	9.05	10.30	5.27
		5:12	7.39	8.04	9.44	10.95	12.57	14.30	6.50
		7:12	10.56	11.50	13.50	15.65	17.97	20.45	7.06
		12:12	16.56	18.03	21.16	24.55	28.18	32.06	8.60
	60	< 1:12	5.32	5.79	6.80	7.89	9.05	10.30	5.38
		5:12	7.39	8.04	9.44	10.95	12.57	14.30	7.01
		7:12	11.87	12.93	15.17	17.60	20.20	22.98	7.76
		12:12	19.87	21.64	25.40	29.45	33.81	38.47	9.81

(continued)

TABLE R608.7.1.1(2)—continued
UNREDUCED LENGTH, *UR*, OF SOLID WALL REQUIRED IN EACH EXTERIOR ENDWALL FOR WIND PERPENDICULAR TO RIDGE FIRST STORY OF TWO STORY[a, c, d, e, f, g]

SIDEWALL LENGTH (feet)	ENDWALL LENGTH (feet)	ROOF SLOPE	UNREDUCED LENGTH, *UR*, OF SOLID WALL REQUIRED IN ENDWALLS FOR WIND PERPENDICULAR TO RIDGE (feet)						
			Basic Wind Speed (mph) Exposure						
			115B	120B	130B	140B	150B	160B	Minimum[b]
			—	—	110C	119C	127C	136C	
			—	—	—	110D	117D	125D	
60	15	< 1:12	9.87	10.74	12.61	14.62	16.79	19.10	10.10
		5:12	13.71	14.93	17.52	20.32	23.33	26.54	10.87
		7:12	15.08	16.42	19.27	22.35	25.66	29.20	11.22
		12:12	18.67	20.33	23.86	27.67	31.77	36.14	12.19
	30	< 1:12	9.87	10.74	12.61	14.62	16.79	19.10	10.30
		5:12	13.71	14.93	17.52	20.32	23.33	26.54	11.85
		7:12	17.44	18.99	22.29	25.85	29.67	33.76	12.54
		12:12	24.62	26.81	31.46	36.49	41.89	47.66	14.48
	45	< 1:12	10.27	11.18	13.12	15.21	17.47	19.87	10.50
		5:12	14.26	15.52	18.22	21.13	24.26	27.60	12.82
		7:12	20.21	22.01	25.83	29.95	34.39	39.12	13.86
		12:12	31.20	33.97	39.87	46.23	53.07	60.39	16.76
	60	< 1:12	10.64	11.59	13.60	15.77	18.11	20.60	10.70
		5:12	14.77	16.09	18.88	21.90	25.14	28.60	13.79
		7:12	23.05	25.09	29.45	34.15	39.21	44.61	15.18
		12:12	37.97	41.34	48.52	56.27	64.60	73.49	19.05

For SI: 1 inch = 25.4 mm, 1 foot = 304.8 mm, 1 mile per hour = 0.447 m/s, 1 pound force per linear foot = 0.146 kN/m, 1 pound per square foot = 47.88 Pa.

a. Tabulated lengths were derived by calculating design wind pressures in accordance with Figure 28.4-1 of ASCE 7 for a building with a mean roof height of 35 feet, topographic factor, K_{zt} equal to 1.0, and Risk Category II. For wind perpendicular to the ridge, the effects of a 2-foot overhang on each endwall are included. The design pressures were used to calculate forces to be resisted by solid wall segments in each endwall. The forces to be resisted by each wall line were then divided by the default design strength of 840 pounds per linear foot of length to determine the unreduced length, *UR*, of solid wall length required in each endwall. The actual mean roof height of the building shall not exceed the least horizontal dimension of the building.

b. Tabulated lengths in the "minimum" column are based on the requirement of Section 28.4.4 of ASCE 7 that the main windforce-resisting system be designed for a minimum pressure of 1016 psf multiplied by the wall area of the building and 8 psf multiplied by the roof area of the building projected onto a vertical plane normal to the assumed wind direction. Tabulated lengths in shaded cells are less than the "minimum" value. Where the minimum controls, it is permitted to be reduced in accordance with Notes c, d and e. See Section R608.7.1.1.

c. For buildings with a mean roof height of less than 35 feet, tabulated lengths are permitted to be reduced by multiplying by the appropriate factor, R_1, from Table R608.7.1.1(4). The reduced length shall be not less than the "minimum" value shown in the table.

d. Tabulated lengths for "one story or top story of two story" are based on a floor-to-ceiling height of 10 feet. Tabulated lengths for "first story of two story" are based on floor-to-ceiling heights of 10 feet each for the first and second story. For floor-to-ceiling heights less than assumed, use the lengths in this table or Table R608.7.1.1(1) or R608.7.1.1(3), or multiply the value in the table by the reduction factor, R_2, from Table R608.7.1.1(5).

e. Tabulated lengths are based on the default design shear strength of 840 pounds per linear foot of solid wall segment. The tabulated lengths are permitted to be reduced by multiplying by the applicable reduction factor for design strength, R_3, from Table R608.7.1.1(6).

f. The reduction factors, R_1, R_2 and R_3, in Tables R608.7.1.1(4), R608.7.1.1(5), and R608.7.1.1(6), respectively, are permitted to be compounded, subject to the limitations of Note b. However, the minimum number and minimum length of solid wall segments in each wall line shall comply with Sections R608.7.1 and R608.7.2.1, respectively.

g. For intermediate values of sidewall length, endwall length, roof slope and basic wind speed, use the next higher value, or determine by interpolation.

TABLE R608.7.1.1(3)
UNREDUCED LENGTH, *UR*, OF SOLID WALL REQUIRED IN EACH EXTERIOR SIDEWALL FOR WIND PARALLEL TO RIDGE[a, c, d, e, f, g]

SIDEWALL LENGTH (feet)	ENDWALL LENGTH (feet)	ROOF SLOPE	UNREDUCED LENGTH, *UR*, OF SOLID WALL REQUIRED IN SIDEWALLS FOR WIND PARALLEL TO RIDGE (feet)						
			Basic Wind Speed (mph) Exposure						
			115B	120B	130B	140B	150B	160B	Minimum[b]
			—	—	110C	119C	127C	136C	
			—	—	—	110D	117D	125D	
			One story or top story of two story						
< 30	15	< 1:12	1.08	1.18	1.39	161	1.84	2.10	0.90
		5:12	1.29	1.40	1.65	1.91	2.19	2.49	1.08
		7:12	1.38	1.50	1.76	2.04	2.35	2.67	1.17
		12:12	1.63	1.78	2.09	2.42	2.78	3.16	1.39
	30	< 1:12	2.02	2.20	2.59	3.00	3.44	3.92	1.90
		5:12	2.73	2.97	3.48	4.04	4.64	5.28	2.62
		7:12	3.05	3.32	3.89	4.51	5.18	5.89	2.95
		12:12	3.93	4.27	5.02	5.82	6.68	7.60	3.86
	45	< 1:12	3.03	3.30	3.87	4.49	5.15	5.86	2.99
		5:12	4.55	4.96	5.82	6.75	7.74	8.81	4.62
		7:12	5.24	5.71	6.70	7.77	8.92	10.15	5.36
		12:12	7.16	7.79	9.14	10.61	12.17	13.85	7.39
	60	< 1:12	4.11	4.47	5.25	6.09	6.99	7.96	4.18
		5:12	6.78	7.39	8.67	10.05	11.54	13.13	7.07
		7:12	8.00	8.71	10.22	11.85	13.61	15.48	8.38
		12:12	11.35	12.36	14.51	16.82	19.31	21.97	12.00
60	45	< 1:12	3.17	3.46	4.06	4.70	5.40	6.14	2.99
		5:12	4.75	5.18	6.07	7.04	8.09	9.20	4.62
		7:12	5.47	5.96	6.99	8.11	9.31	10.59	5.36
		12:12	7.45	8.11	9.52	11.04	12.68	14.43	7.39
	60	< 1:12	4.41	4.81	5.64	6.54	7.51	8.54	4.18
		5:12	7.22	7.86	9.23	10.70	12.29	13.98	7.07
		7:12	8.50	9.25	10.86	12.59	14.46	16.45	8.38
		12:12	12.02	13.09	15.36	17.81	20.45	23.27	12.00

(continued)

TABLE R608.7.1.1(3)—continued
UNREDUCED LENGTH, *UR*, OF SOLID WALL REQUIRED IN EACH EXTERIOR SIDEWALL FOR WIND PARALLEL TO RIDGE[a, c, d, e, f, g]

SIDEWALL LENGTH (feet)	ENDWALL LENGTH (feet)	ROOF SLOPE	UNREDUCED LENGTH, *UR*, OF SOLID WALL REQUIRED IN SIDEWALLS FOR WIND PARALLEL TO RIDGE (feet)						
			Basic Wind Speed (mph) Exposure						Minimum[b]
			115B	120B	130B	140B	150B	160B	
			—	—	110C	119C	127C	136C	
			—	—	—	110D	117D	125D	
			One story or top story of two story						
< 30	15	< 1:12	3.03	3.30	3.88	4.49	5.16	5.87	2.52
		5:12	3.24	3.52	4.14	4.80	5.51	6.26	2.70
		7:12	3.33	3.62	4.25	4.93	5.66	6.44	2.79
		12:12	3.58	3.90	4.58	5.31	6.10	6.94	3.01
	30	< 1:12	5.50	5.99	7.03	8.16	9.36	10.65	5.14
		5:12	6.21	6.76	7.93	9.20	10.56	12.01	5.86
		7:12	6.52	7.10	8.34	9.67	11.10	12.63	6.19
		12:12	7.41	8.06	9.46	10.97	12.60	14.33	7.10
	45	< 1:12	8.00	8.71	10.22	11.85	13.61	15.48	7.85
		5:12	9.52	10.37	12.17	14.11	16.20	18.43	9.48
		7:12	10.21	11.12	13.05	15.14	17.38	19.77	10.21
		12:12	12.13	13.20	15.50	17.97	20.63	23.47	12.25
	60	< 1:12	10.56	11.50	13.50	15.65	17.97	20.44	10.65
		5:12	13.24	14.41	16.91	19.62	22.52	25.62	13.54
		7:12	14.45	15.73	18.46	21.41	24.58	27.97	14.85
		12:12	17.80	19.38	22.75	26.38	30.29	34.46	18.48
60	45	< 1:12	8.39	9.14	10.72	12.44	14.28	16.25	7.85
		5:12	9.97	10.86	12.74	14.78	16.97	19.30	9.48
		7:12	10.69	11.64	13.66	15.84	18.19	20.69	10.21
		12:12	12.67	13.80	16.19	18.78	21.56	24.53	12.25
	60	< 1:12	11.37	12.38	14.53	16.85	19.35	22.01	10.65
		5:12	14.18	15.44	18.12	21.02	24.13	27.45	13.54
		7:12	15.46	16.83	19.75	22.91	26.29	29.92	14.85
		12:12	18.98	20.66	24.25	28.13	32.29	36.74	18.48

For SI: 1 inch = 25.4 mm, 1 foot = 304.8 mm, 1 mile per hour = 0.447 m/s, 1 pound force per linear foot = 0.146 kN/m, 1 pound per square foot = 47.88 Pa.

a. Tabulated lengths were derived by calculating design wind pressures in accordance with Figure 28.4-1 of ASCE 7 for a building with a mean roof height of 35 feet, topographic factor, K_{zt} equal to 1.0, and Risk Category II. The design pressures were used to calculate forces to be resisted by solid wall segments in each sidewall. The forces to be resisted by each wall line were then divided by the default design strength of 840 pounds per linear foot of length to determine the unreduced length, *UR*, of solid wall length required in each sidewall. The actual mean roof height of the building shall not exceed the least horizontal dimension of the building.

b. Tabulated lengths in the "minimum" column are based on the requirement of Section 28.4.4 of ASCE 7 that the main windforce-resisting system be designed for a minimum pressure of 16 psf multiplied by the wall area of the building and 8 psf multiplied by the roof area of the building projected onto a vertical plane normal to the assumed wind direction. Tabulated lengths in shaded cells are less than the "minimum" value. Where the minimum controls, it is permitted to be reduced in accordance with Notes c, d and e. See Section R608.7.1.1.

c. For buildings with a mean roof height of less than 35 feet, tabulated lengths are permitted to be reduced by multiplying by the appropriate factor, R_1, from Table R608.7.1.1(4). The reduced length shall be not less than the "minimum" value shown in the table.

d. Tabulated lengths for "one story or top story of two story" are based on a floor-to-ceiling height of 10 feet. Tabulated lengths for "first story of two story" are based on floor-to-ceiling heights of 10 feet each for the first and second story. For floor-to-ceiling heights less than assumed, use the lengths in this table or Table R608.7.1.1(1) or Table R608.7.1.1(2), or multiply the value in the table by the reduction factor, R_2, from Table R608.7.1.1(5).

e. Tabulated lengths are based on the default design shear strength of 840 pounds per linear foot of solid wall segment. The tabulated lengths are permitted to be reduced by multiplying by the applicable reduction factor for design strength, R_3, from Table R608.7.1.1(6).

f. The reduction factors, R_1, R_2 and R_3, in Table R608.7.1.1(4), Table R608.7.1.1(5), and Table R608.7.1.1(6), respectively, are permitted to be compounded, subject to the limitations of Note b. However, the minimum number and minimum length of solid walls segments in each wall line shall comply with Sections R608.7.1 and R608.7.2.1, respectively.

g. For intermediate values of sidewall length, endwall length, roof slope and basic wind speed, use the next higher value, or determine by interpolation.

TABLE R608.7.1.1(4)
REDUCTION FACTOR, R_1, FOR BUILDINGS WITH MEAN ROOF HEIGHT LESS THAN 35 FEET[a]

MEAN ROOF HEIGHT[b, c] (feet)	REDUCTION FACTOR R_1, FOR MEAN ROOF HEIGHT		
	Exposure category		
	B	C	D
< 15	0.96	0.84	0.87
20	0.96	0.89	0.91
25	0.96	0.93	0.94
30	0.96	0.97	0.98
35	1.00	1.00	1.00

For SI: 1 foot = 304.8 mm, 1 degree = 0.0175 rad.

a. See Section R608.7.1.1 and Note c to Table R608.7.1.1(1) for application of reduction factors in this table. This reduction is not permitted for "minimum" values.

b. For intermediate values of mean roof height, use the factor for the next greater height, or determine by interpolation.

c. Mean roof height is the average of the roof eave height and height of the highest point on the roof surface, except that for roof slopes of less than or equal to $2^1/_8$:12 (10 degrees), the mean roof height is permitted to be taken as the roof eave height.

TABLE R608.7.1.1(5)
REDUCTION FACTOR, R_2, FOR FLOOR-TO-CEILING WALL HEIGHTS LESS THAN 10 FEET[a, b]

STORY UNDER CONSIDERATION	FLOOR-TO-CEILING HEIGHT[c] (feet)	ENDWALL LENGTH (feet)	ROOF SLOPE	REDUCTION FACTOR, R_2
Endwalls—for wind perpendicular to ridge				
One story or top story of two story	8	15	< 5:12	0.83
			7:12	0.90
			12:12	0.94
		60	< 5:12	0.83
			7:12	0.95
			12:12	0.98
First story of two story	16 combined first and second story	15	< 5:12	0.83
			7:12	0.86
			12:12	0.89
		60	< 5:12	0.83
			7:12	0.91
			12:12	0.95
Sidewalls—for wind parallel to ridge				
One story or top story of two story	8	15	< 1:12	0.84
			5:12	0.87
			7:12	0.88
			12:12	0.89
		60	< 1:12	0.86
			5:12	0.92
			7:12	0.93
			12:12	0.95
First story of two story	16 combined first and second story	15	< 1:12	0.83
			5:12	0.84
			7:12	0.85
			12:12	0.86
		60	< 1:12	0.84
			5:12	0.87
			7:12	0.88
			12:12	0.90

For SI: 1 foot = 304.8 mm.

a. See Section R608.7.1.1 and Note d to Table R608.7.1.1(1) for application of reduction factors in this table.

b. For intermediate values of endwall length and roof slope, use the next higher value or determine by interpolation.

c. Tabulated values in Tables R608.7.1.1(1) and R608.7.1.1(3) for "one story or top story of two story" are based on a floor-to-ceiling height of 10 feet. Tabulated values in Tables R608.7.1.1(2) and R608.7.1.1(3) for "first story of two story" are based on floor-to-ceiling heights of 10 feet each for the first and second story. For floor to ceiling heights between those shown in this table and those assumed in Table R608.7.1.1(1), R608.7.1.1(2) or R608.7.1.1(3), use the solid wall lengths in Table R608.7.1.1(1), R608.7.1.1(2) or R608.7.1.1(3), or determine the reduction factor by interpolating between 1.0 and the factor shown in this table.

TABLE R608.7.1.1(6)
REDUCTION FACTOR FOR DESIGN STRENGTH, R_3, FOR FLAT, WAFFLE- AND SCREEN-GRID WALLS[a, c]

NOMINAL THICKNESS OF WALL (inches)	VERTICAL BARS AT EACH END OF SOLID WALL SEGMENT		VERTICAL REINFORCEMENT LAYOUT DETAIL [see Figure R608.7.1.1(2)]	REDUCTION FACTOR, R_3, FOR LENGTH OF SOLID WALL			
				Horizontal and vertical shear reinforcement provided			
				No		Yes[d]	
	Number of bars	Bar size		40,000[b]	60,000[b]	40,000[b]	60,000[b]
Flat walls							
4	2	4	1	0.74	0.61	0.74	0.50
	3	4	2	0.61	0.61	0.52	0.27
	2	5	1	0.61	0.61	0.48	0.25
	3	5	2	0.61	0.61	0.26	0.18
6	2	4	3	0.70	0.48	0.70	0.48
	3	4	4	0.49	0.38	0.49	0.33
	2	5	3	0.46	0.38	0.46	0.31
	3	5	4	0.38	0.38	0.32	0.16
8	2	4	3	0.70	0.47	0.70	0.47
	3	4	5	0.47	0.32	0.47	0.32
	2	5	3	0.45	0.31	0.45	0.31
	4	4	6	0.36	0.28	0.36	0.25
	3	5	5	0.31	0.28	0.31	0.16
	4	5	6	0.28	0.28	0.24	0.12
10	2	4	3	0.70	0.47	0.70	0.47
	2	5	3	0.45	0.30	0.45	0.30
	4	4	7	0.36	0.25	0.36	0.25
	6	4	8	0.25	0.22	0.25	0.13
	4	5	7	0.24	0.22	0.24	0.12
	6	5	8	0.22	0.22	0.12	0.08
Waffle-grid walls[e]							
6	2	4	3	0.78	0.78	0.70	0.48
	3	4	4	0.78	0.78	0.49	0.25
	2	5	3	0.78	0.78	0.46	0.23
	3	5	4	0.78	0.78	0.24	0.16
8	2	4	3	0.78	0.78	0.70	0.47
	3	4	5	0.78	0.78	0.47	0.24
	2	5	3	0.78	0.78	0.45	0.23
	4	4	6	0.78	0.78	0.36	0.18
	3	5	5	0.78	0.78	0.23	0.16
	4	5	6	0.78	0.78	0.18	0.13
Screen-grid walls[e]							
6	2	4	3	0.93	0.93	0.70	0.48
	3	4	4	0.93	0.93	0.49	0.25
	2	5	3	0.93	0.93	0.46	0.23
	3	5	4	0.93	0.93	0.24	0.16

For SI: 1 inch = 25.4 mm, 1,000 pounds per square inch = 6.895 MPa.

a. See Note e to Table R608.7.1.1(1) for application of adjustment factors in this table.

b. Yield strength in pounds per square inch of vertical wall reinforcement at ends of solid wall segments.

c. Values are based on concrete with a specified compressive strength, f'_c, of 2,500 psi. Where concrete with f'_c of not less than 3,000 psi is used, values in shaded cells are permitted to be decreased by multiplying by 0.91.

d. Horizontal and vertical shear reinforcement shall be provided in accordance with Section R608.7.2.2.

e. Each end of each solid wall segment shall have rectangular flanges. In the through-the-wall dimension, the flange shall be not less than $5^1/_2$ inches for 6-inch-nominal waffle- and screen-grid walls, and not less than $7^1/_2$ inches for 8-inch-nominal waffle-grid walls. In the in-plane dimension, flanges shall be long enough to accommodate the vertical reinforcement required by the layout detail selected from Figure R608.7.1.1(2) and provide the cover required by Section R608.5.4.1. If necessary to achieve the required dimensions, form material shall be removed or use of flat wall forms is permitted.

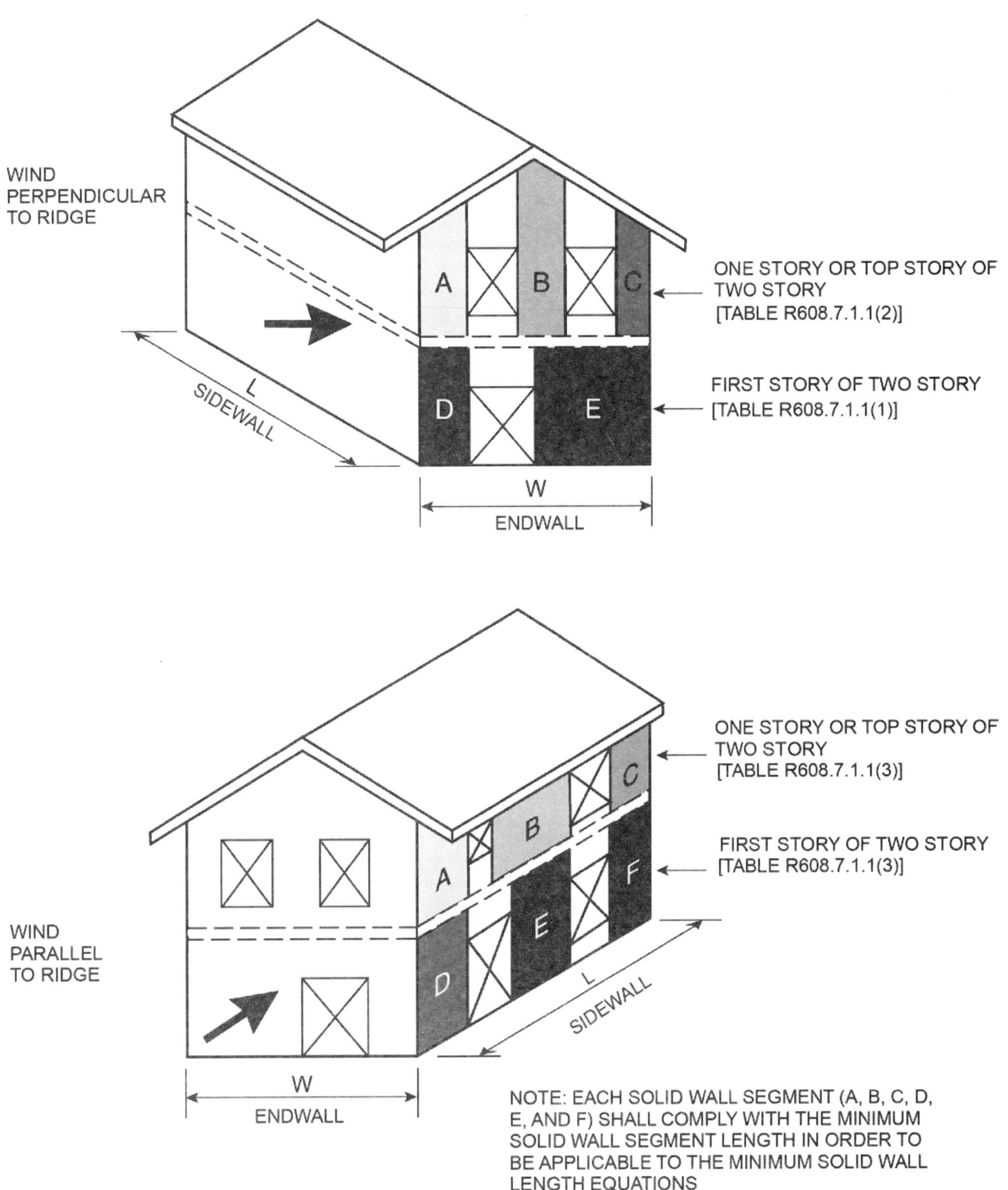

FIGURE R608.7.1.1(1)
MINIMUM SOLID WALL LENGTH

DETAIL NO.	NOM. WALL THICKNESS, IN.	REINFORCEMENT LAYOUT AT ENDS OF SOLID WALL SEGMENTS	NOTES
1	4	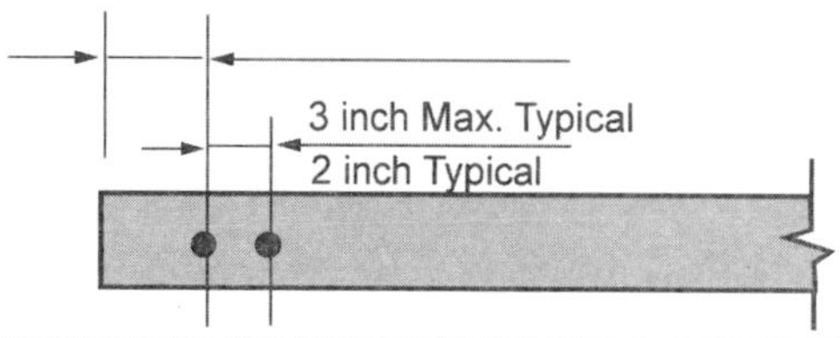 3 inch Max. Typical 2 inch Typical	For SI: 1 inch = 25.4 mm. 1. See Table R608.7(4) for use of details.
2	4	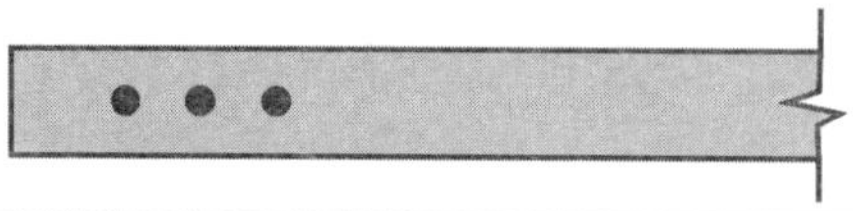	2. Minimum length of solid wall segment and size and grade of reinforcement in each end of each solid wall segment shall be determined from Table R608.7.1.1(6).
3	6 8 10	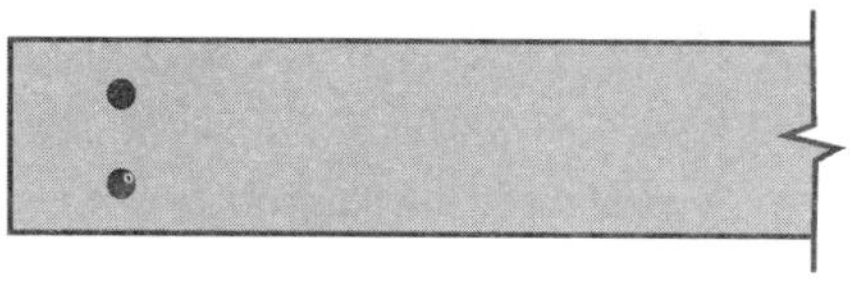	3. For minimum cover requirements, see Section R608.5.4.1.
4	6		4. For details 3 - 8 where two or more bars are in the same row parallel to the end of the segment, place bars so that corner bars are as close to the sides of the wall segments as minimum cover requirements of Section R608.5.4.1 will permit.
5	8	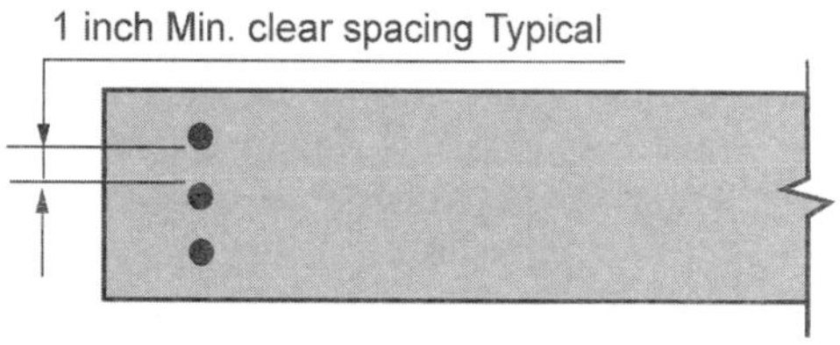 1 inch Min. clear spacing Typical	
6	8		5. For waffle- and screen-grid walls, each end of each solid wall segment shall have rectangular flanges. In the through-the-wall dimension, the flange shall be not less than 5½ inches for 6-inch nominal waffle- and screen-grid forms, and not less than 7½ inches for 8-inch nominal waffle-grid forms. In the in-plane dimension, flanges shall be long enough to accommodate the vertical reinforcement required by the layout detail selected and provide the cover required by Section R608.5.4.1. If necessary to achieve the required dimensions, form material shall be removed or flat wall forms are permitted. See Table R608.7.1.1(6), Note e.
7	10		
8	10	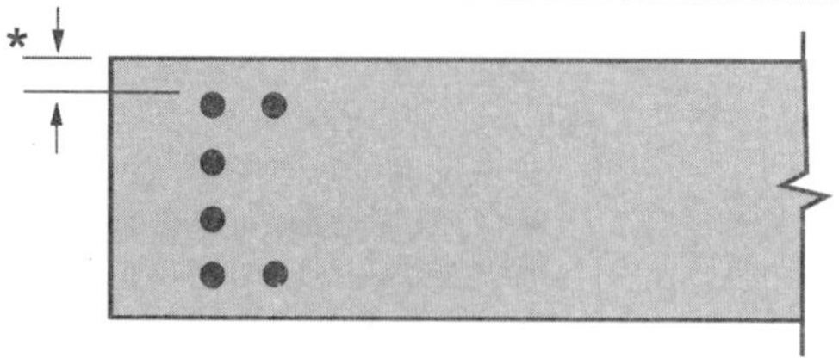 * * For minimum cover see Section R608.5.4.1	

FIGURE R608.7.1.1(2)
VERTICAL REINFORCEMENT LAYOUT DETAIL

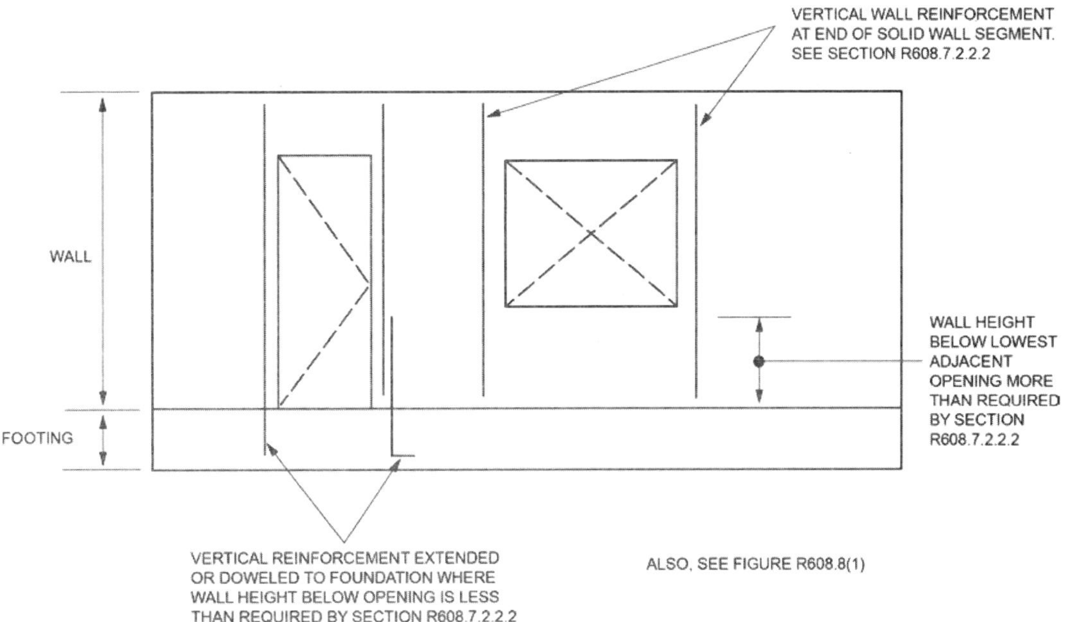

FIGURE R608.7.1.1(3)
VERTICAL WALL REINFORCEMENT ADJACENT TO WALL OPENINGS

R608.7.2 Solid wall segments. Solid wall segments that contribute to the required length of solid wall shall comply with this section. Reinforcement shall be provided in accordance with Section R608.7.2.2 and Table R608.7.1.1(6). Solid wall segments shall extend the full story-height without openings, other than openings for the utilities and other building services passing through the wall. In flat walls and waffle-grid walls, such openings shall have an area of less than 30 square inches (19 355 mm^2) without any dimension exceeding 6$^1/_4$ inches (159 mm), and shall not be located within 6 inches (152 mm) of the side edges of the solid wall segment. In screen-grid walls, such openings shall be located in the portion of the solid wall segment between horizontal and vertical cores of concrete and opening size and location are not restricted provided there is not any concrete removed.

R608.7.2.1 Minimum length of solid wall segment and maximum spacing. Only solid wall segments equal to or greater than 24 inches (610 mm) in length shall be included in the total length of solid wall required by Section R608.7.1. In addition, not more than two solid wall segments equal to or greater than 24 inches (610 mm) in length and less than 48 inches (1219 mm) in length shall be included in the required total length of solid wall. The maximum clear opening width shall be 18 feet (5486 mm). See Figure R608.7.1.1(1).

R608.7.2.2 Reinforcement in solid wall segments.

R608.7.2.2.1 Horizontal shear reinforcement. Where reduction factors for design strength, R_3, from Table R608.7.1.1(6) based on horizontal and vertical shear reinforcement being provided are used, solid wall segments shall have horizontal reinforcement consisting of minimum No. 4 bars. Horizontal shear reinforcement shall be the same grade of steel required for the vertical reinforcement at the ends of solid wall segments by Section R608.7.2.2.2.

The spacing of horizontal reinforcement shall not exceed the smaller of one-half the length of the solid wall segment, minus 2 inches (51 mm), and 18 inches (457 mm). Horizontal shear reinforcement shall terminate in accordance with Section R608.6.4.

R608.7.2.2.2 Vertical reinforcement. Vertical reinforcement applicable to the reduction factor(s) for design strength, R_3, from Table R608.7.1.1(6) that is used, shall be located at each end of each solid wall segment in accordance with the applicable detail in Figure R608.7.1.1(2). The No. 4 vertical bar required on each side of an opening by Section R608.8.1.2 is permitted to be used as reinforcement at the ends of solid wall segments where installed in accordance with the applicable detail in Figure R608.7.1.1(2). There shall be not less than two No. 4 bars at each end of solid wall segments located as required by the applicable detail in Figure R608.7.1.1(2). One of the bars at each end of solid wall segments shall be deemed to meet the requirements for vertical wall reinforcement required by Section R608.6.

The vertical wall reinforcement at each end of each solid wall segment shall be developed below the bottom of the adjacent wall opening [see Figure R608.7.1.1(3)] by one of the following methods:

1. Where the wall height below the bottom of the adjacent opening is equal to or greater than 22 inches (559 mm) for No. 4 or 28

inches (711 mm) for No. 5 vertical wall reinforcement, reinforcement around openings in accordance with Section R608.8.1 shall be sufficient.

2. Where the wall height below the bottom of the adjacent opening is less than required by Item 1, the vertical wall reinforcement adjacent to the opening shall extend into the footing far enough to develop the bar in tension in accordance with Section R608.5.4.4 and Figure R608.5.4(2), or shall be lap-spliced with a dowel that is embedded in the footing far enough to develop the dowel-bar in tension.

R608.7.2.2.3 Vertical shear reinforcement. Where reduction factors for design strength, R_3, from Table R608.7.1.1(6) based on horizontal and vertical shear reinforcement being provided are used, solid wall segments shall have vertical reinforcement consisting of minimum No. 4 bars. Vertical shear reinforcement shall be the same grade of steel required by Section R608.7.2.2.2 for the vertical reinforcement at the ends of solid wall segments. The spacing of vertical reinforcement throughout the length of the segment shall not exceed the smaller of one third the length of the segment, and 18 inches (457 mm). Vertical shear reinforcement shall be continuous between stories in accordance with Section R608.6.3, and shall terminate in accordance with Section R608.6.4. Vertical shear reinforcement required by this section is permitted to be used for vertical reinforcement required by Table R608.6(1), R608.6(2), R608.6(3) or R608.6(4), whichever is applicable.

R608.7.2.3 Solid wall segments at corners. At all interior and exterior corners of exterior walls, a solid wall segment shall extend the full height of each wall *story.* The segment shall have the length required to develop the horizontal reinforcement above and below the adjacent opening in tension in accordance with Section R608.5.4.4. For an exterior corner, the limiting dimension is measured on the outside of the wall, and for an interior corner the limiting dimension is measured on the inside of the wall. See Section R608.8.1. The length of a segment contributing to the required length of solid wall shall comply with Section R608.7.2.1.

The end of a solid wall segment complying with the minimum length requirements of Section R608.7.2.1 shall be located not more than 6 feet (1829 mm) from each corner.

R608.8 Requirements for lintels and reinforcement around openings.

R608.8.1 Reinforcement around openings. Reinforcement shall be provided around openings in walls equal to or greater than 2 feet (610 mm) in width in accordance with this section and Figure R608.8(1), in addition to the minimum wall reinforcement required by Sections R404.1.3, R608.6 and R608.7. Vertical wall reinforcement required by this section is permitted to be used as reinforcement at the ends of solid wall segments required by Section R608.7.2.2.2 provided it is located in accordance with Section R608.8.1.2. Wall openings shall have a minimum depth of concrete over the width of the opening of 8 inches (203 mm) in flat walls and waffle-grid walls, and 12 inches (305 mm) in screen-grid walls. Wall openings in waffle-grid and screen-grid walls shall be located such that not less than one-half of a vertical core occurs along each side of the opening.

R608.8.1.1 Horizontal reinforcement. Lintels complying with Section R608.8.2 shall be provided above wall openings equal to or greater than 2 feet (610 mm) in width.

Openings equal to or greater than 2 feet (610 mm) in width shall have not less than one No. 4 bar placed within 12 inches (305 mm) of the bottom of the opening. See Figure R608.8(1).

Horizontal reinforcement placed above and below an opening shall extend beyond the edges of the opening the dimension required to develop the bar in tension in accordance with Section R608.5.4.4.

Exception: Continuous horizontal wall reinforcement placed within 12 inches (305 mm) of the top of the wall *story* as required in Sections R404.1.3.2 and R608.6.2 is permitted in lieu of top or bottom lintel reinforcement required by Section R608.8.2 provided that the continuous horizontal wall reinforcement meets the location requirements specified in Figures R608.8(2), R608.8(3), and R608.8(4) and the size requirements specified in Tables R608.8(2) through R608.8(10).

R608.8.1.2 Vertical reinforcement. Not less than one No. 4 bar [Grade 40 (280 MPa)] shall be provided on each side of openings equal to or greater than 2 feet (610 mm) in width. The vertical reinforcement required by this section shall extend the full height of the wall story and shall be located within 12 inches (305 mm) of each side of the opening. The vertical reinforcement required on each side of an opening by this section is permitted to serve as reinforcement at the ends of solid wall segments in accordance with Section R608.7.2.2.2, provided it is located as required by the applicable detail in Figure R608.7.1.1(2). Where the vertical reinforcement required by this section is used to satisfy the requirements of Section R608.7.2.2.2 in waffle- and screen-grid walls, a concrete flange shall be created at the ends of the solid wall segments in accordance with Table R608.7.1.1(6), Note e. In the topmost story, the reinforcement shall terminate in accordance with Section R608.6.4.

R608.8.2 Lintels. Lintels shall be provided over all openings equal to or greater than 2 feet (610 mm) in width. Lintels with uniform loading shall conform to Sections R608.8.2.1 and R608.8.2.2, or Section R608.8.2.3. Lintels supporting concentrated loads, such as from roof or floor beams or girders, shall be designed in accordance with ACI 318.

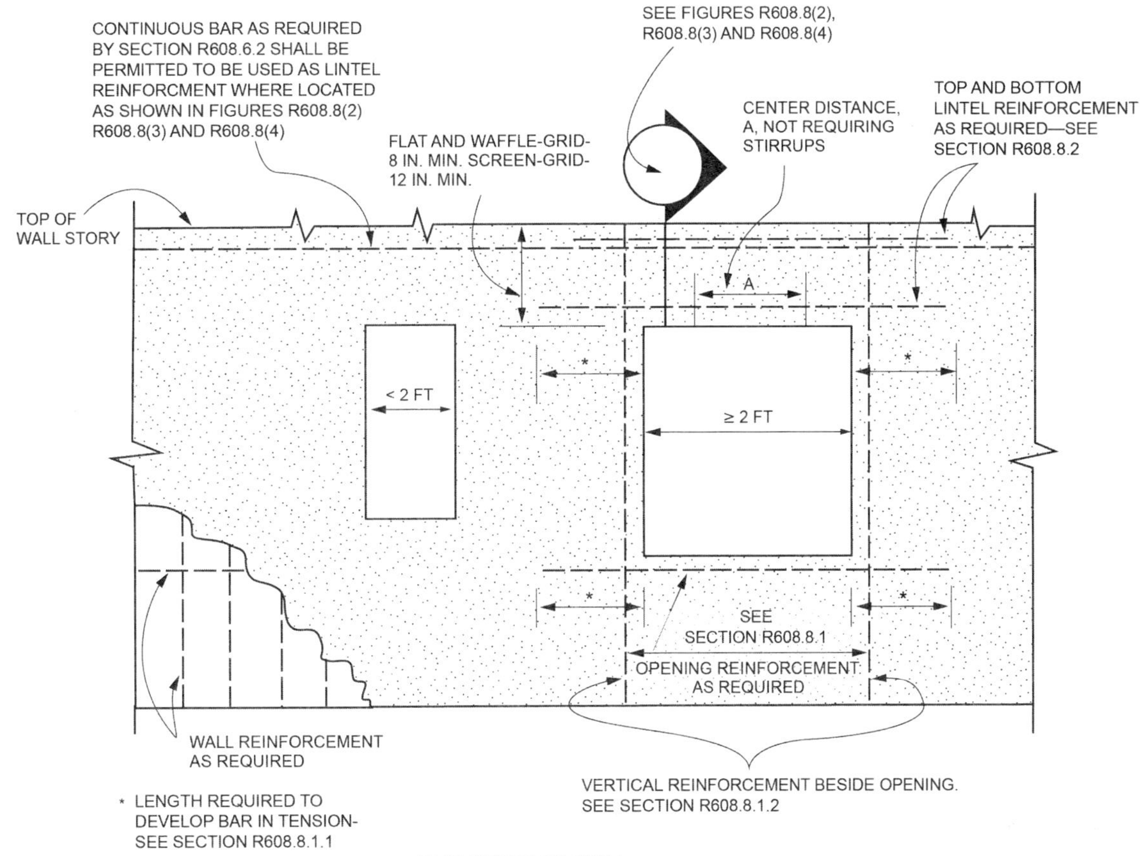

For SI: 1 inch = 25.4 mm, 1 foot = 304.8 mm.

FIGURE R608.8(1)
REINFORCEMENT OF OPENINGS

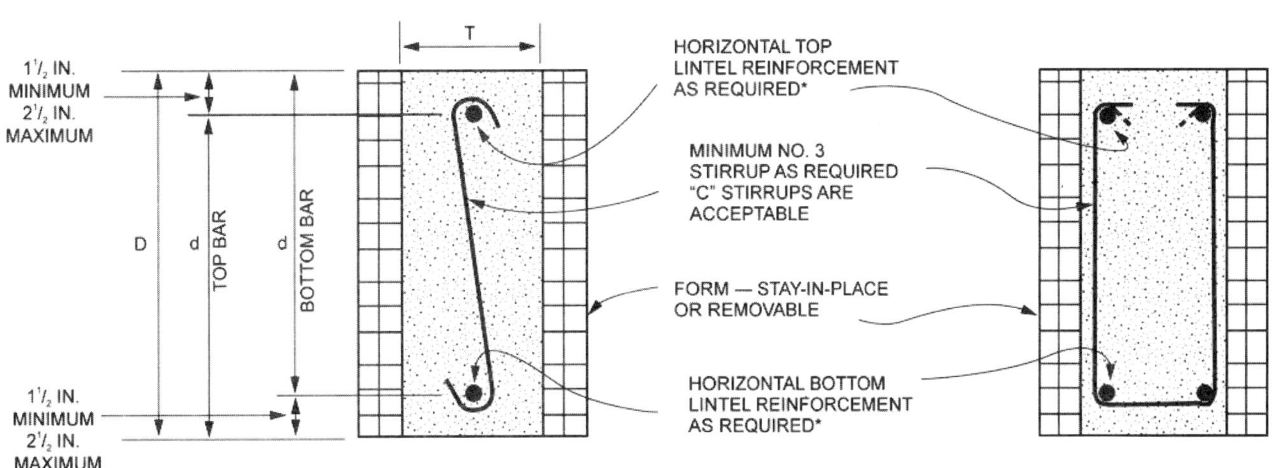

For SI: 1 inch = 25.4 mm.

FIGURE R608.8(2)
LINTEL FOR FLAT WALLS

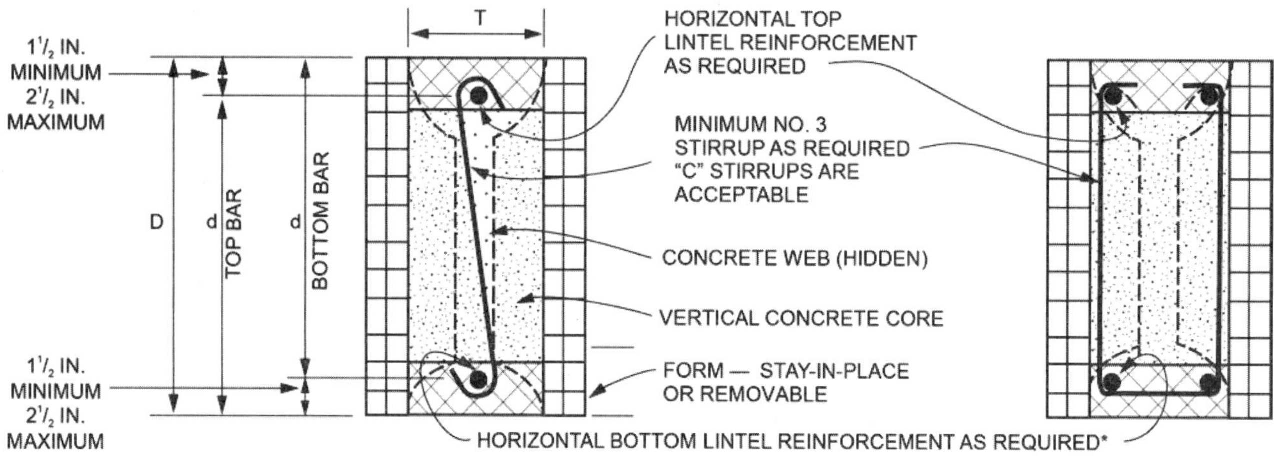

(a) SINGLE FORM HEIGHT SECTION CUT THROUGH VERTICAL CORE OF A WAFFLE-GRID LINTEL

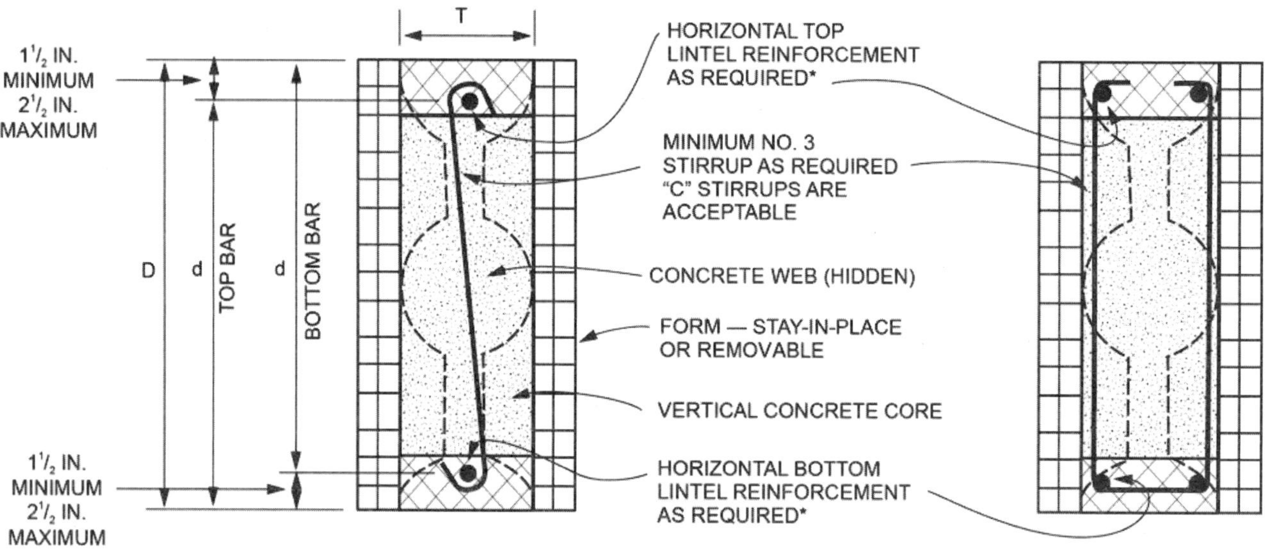

(b) DOUBLE FORM HEIGHT SECTION CUT THROUGH VERTICAL CORE OF A WAFFLE-GRID LINTEL

*FOR BUNDLED BARS, SEE SECTION R608.8.2.2.

NOTE: CROSS HATCHING REPRESENTS THE AREA IN WHICH FORM MATERIAL SHALL BE REMOVED, IF NECESSARY, TO CREATE FLANGES CONTINUOUS THE LENGTH OF THE LINTEL. FLANGES SHALL HAVE A MINIMUM THICKNESS OF 3 IN., AND A MINIMUM WIDTH OF 5 IN. AND 7 IN. IN 6 IN. NOMINAL AND 8 IN. NOMINAL WAFFLE-GRID WALLS, RESPECTIVELY. SEE NOTE a TO TABLES R608.8(6) AND R608.8(10).

For SI: 1 inch = 25.4 mm.

FIGURE R608.8(3)
LINTELS FOR WAFFLE-GRID WALLS

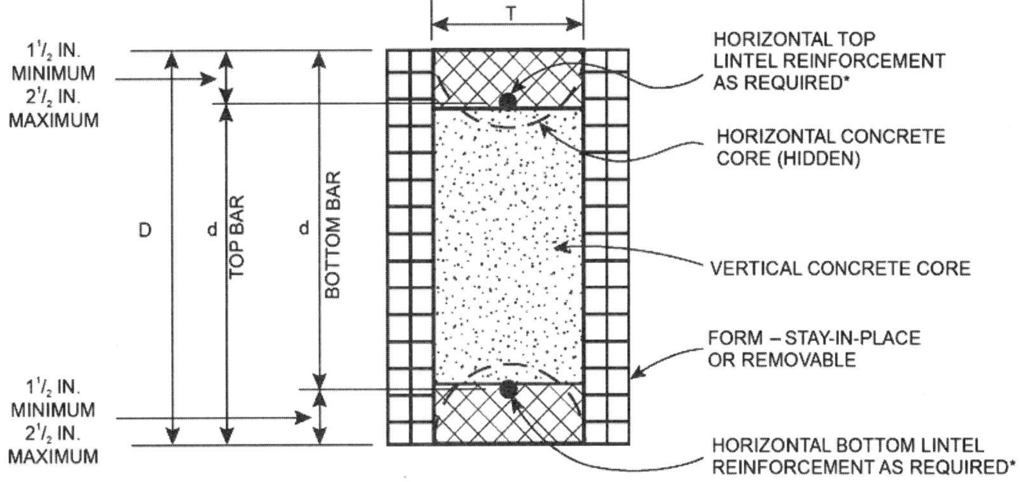

(a) SINGLE FORM HEIGHT SECTION CUT THROUGH VERTICAL CORE OF A SCREEN-GRID LINTEL

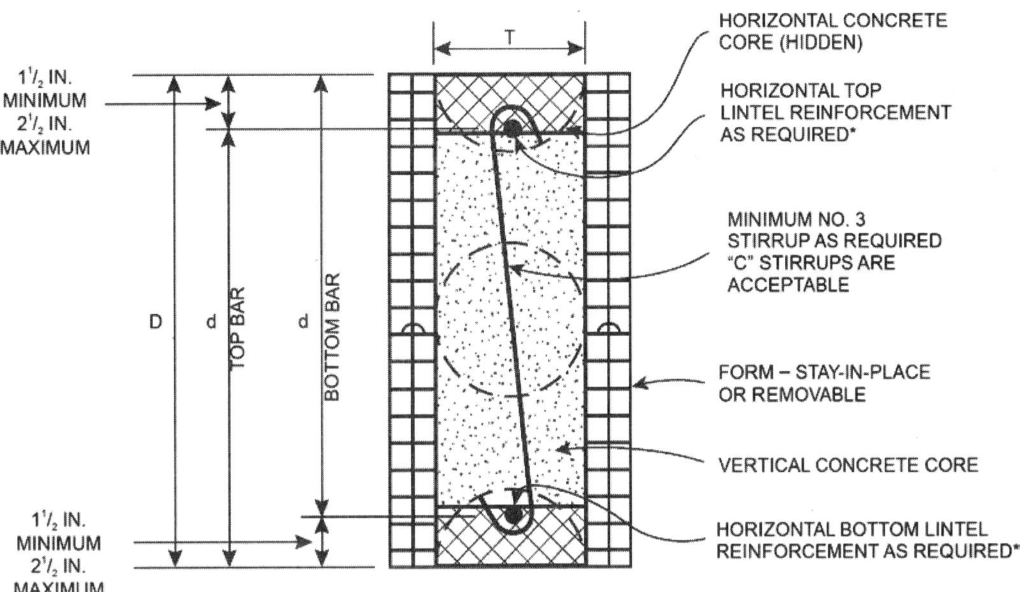

(b) DOUBLE FORM HEIGHT SECTION CUT THROUGH VERTICAL CORE OF A SCREEN-GRID LINTEL

*FOR BUNDLED BARS, SEE SECTION R608.8.2.2

NOTE: CROSS HATCHING REPRESENTS THE AREA IN WHICH FORM MATERIAL SHALL BE REMOVED, IF NECESSARY, TO CREATE FLANGES CONTINUOUS THE LENGTH OF THE LINTEL. FLANGES SHALL HAVE A MINIMUM THICKNESS OF 2.5 IN. AND A MINIMUM WIDTH OF 5 IN. SEE NOTE a TO TABLES R608.8(8) AND R608.8(10).

For SI: 1 inch = 25.4 mm.

FIGURE R608.8(4)
LINTELS FOR SCREEN-GRID WALLS

TABLE R608.8(1)
LINTEL DESIGN LOADING CONDITIONS[a, b, d]

<table>
<tr><th colspan="3">DESCRIPTION OF LOADS AND OPENINGS ABOVE INFLUENCING DESIGN OF LINTEL</th><th>DESIGN LOAD CONDITION[c]</th></tr>
<tr><td colspan="4">Opening in wall of top story of two-story building, or first story of one-story building</td></tr>
<tr><td rowspan="2">Wall supporting loads from roof, including attic floor, if applicable, and</td><td colspan="2">Top of lintel equal to or less than W/2 below top of wall</td><td>2</td></tr>
<tr><td colspan="2">Top of lintel greater than W/2 below top of wall</td><td>NLB</td></tr>
<tr><td colspan="3">Wall not supporting loads from roof or attic floor</td><td>NLB</td></tr>
<tr><td colspan="4">Opening in wall of first story of two-story building where wall immediately above is of concrete construction, or opening in basement wall of one-story building where wall immediately above is of concrete construction</td></tr>
<tr><td rowspan="3">LB ledger board mounted to side of wall with bottom of ledger less than or equal to W/2 above top of lintel, and</td><td colspan="2">Top of lintel greater than W/2 below bottom of opening in story above</td><td>1</td></tr>
<tr><td rowspan="2">Top of lintel less than or equal to W/2 below bottom of opening in story above, and</td><td>Opening is entirely within the footprint of the opening in the story above</td><td>1</td></tr>
<tr><td>Opening is partially within the footprint of the opening in the story above</td><td>4</td></tr>
<tr><td colspan="3">LB ledger board mounted to side of wall with bottom of ledger more than W/2 above top of lintel</td><td>NLB</td></tr>
<tr><td rowspan="3">NLB ledger board mounted to side of wall with bottom of ledger less than or equal to W/2 above top of lintel, or no ledger board, and</td><td colspan="2">Top of lintel greater than W/2 below bottom of opening in story above</td><td>NLB</td></tr>
<tr><td rowspan="2">Top of lintel less than or equal to W/2 below bottom of opening in story above, and</td><td>Opening is entirely within the footprint of the opening in the story above</td><td>NLB</td></tr>
<tr><td>Opening is partially within the footprint of the opening in the story above</td><td>1</td></tr>
<tr><td colspan="4">Opening in basement wall of two-story building where walls of two stories above are of concrete construction</td></tr>
<tr><td rowspan="3">LB ledger board mounted to side of wall with bottom of ledger less than or equal toW/2 above top of lintel, and</td><td colspan="2">Top of lintel greater than W/2 below bottom of opening in story above</td><td>1</td></tr>
<tr><td rowspan="2">Top of lintel less than or equal to W/2 below bottom of opening in story above, and</td><td>Opening is entirely within the footprint of the opening in the story above</td><td>1</td></tr>
<tr><td>Opening is partially within the footprint of the opening in the story above</td><td>5</td></tr>
<tr><td colspan="3">LB ledger board mounted to side of wall with bottom of ledger more than W/2 above top of lintel</td><td>NLB</td></tr>
<tr><td rowspan="3">NLB ledger board mounted to side of wall with bottom of ledger less than or equal to W/2 above top of lintel, or no ledger board, and</td><td colspan="2">Top of lintel greater than W/2 below bottom of opening in story above</td><td>NLB</td></tr>
<tr><td rowspan="2">Top of lintel less than or equal to W/2 below bottom of opening in story above, and</td><td>Opening is entirely within the footprint of the opening in the story above</td><td>NLB</td></tr>
<tr><td>Opening is partially within the footprint of the opening in the story above</td><td>1</td></tr>
<tr><td colspan="4">Opening in wall of first story of two-story building where wall immediately above is of light-frame construction, or opening in basement wall of one-story building where wall immediately above is of light-frame construction</td></tr>
<tr><td rowspan="2">Wall supporting loads from roof, second floor and top-story wall of light-frame construction, and</td><td colspan="2">Top of lintel equal to or less than W/2 below top of wall</td><td>3</td></tr>
<tr><td colspan="2">Top of lintel greater than W/2 below top of wall</td><td>NLB</td></tr>
<tr><td colspan="3">Wall not supporting loads from roof or second floor</td><td>NLB</td></tr>
</table>

a. LB means load bearing, NLB means nonload bearing, and W means width of opening.

b. Footprint is the area of the wall below an opening in the story above, bounded by the bottom of the opening and vertical lines extending downward from the edges of the opening.

c. For design loading condition "NLB" see Tables R608.8(9) and R608.8(10). For all other design loading conditions, see Tables R608.8(2) through R608.8(8).

d. An NLB ledger board is a ledger attached to a wall that is parallel to the span of the floor, roof or ceiling framing that supports the edge of the floor, ceiling or roof.

TABLE R608.8(2)
MAXIMUM ALLOWABLE CLEAR SPANS FOR 4-INCH-NOMINAL THICK FLAT LINTELS IN LOAD-BEARING WALLS[a, b, c, d, e, f, m]
ROOF CLEAR SPAN 40 FEET AND FLOOR CLEAR SPAN 32 FEET

LINTEL DEPTH, D[g] (inches)	NUMBER OF BARS AND BAR SIZE IN TOP AND BOTTOM OF LINTEL	STEEL YIELD STRENGTH[h], f_y (psi)	DESIGN LOADING CONDITION DETERMINED FROM Table R608.8(1)								
			1	2		3		4		5	
			Maximum ground snow load (psf)								
			—	30	70	30	70	30	70	30	70
			Maximum clear span of lintel (feet-inches)								
8	Span without stirrups[i, j]		3-2	3-4	2-4	2-6	2-2	2-1	2-0	2-0	2-0
	1-#4	40,000	5-2	5-5	4-1	4-3	3-10	3-7	3-4	2-9	2-9
		60,000	6-2	6-5	4-11	5-1	4-6	4-2	3-8	2-11	2-10
	1-#5	40,000	6-3	6-7	5-0	5-2	4-6	4-2	3-8	2-11	2-10
		60,000	DR	DR	DR	DR	DR	DR	DR	DR	DR
	Center distance A[k, l]		1-1	1-2	0-8	0-9	0-7	0-6	0-5	0-4	0-4
12	Span without stirrups[i, j]		3-4	3-7	2-9	2-11	2-8	2-6	2-5	2-2	2-2
	1-#4	40,000	6-7	7-0	5-4	5-7	5-0	4-9	4-4	3-8	3-7
		60,000	7-11	8-6	6-6	6-9	6-0	5-9	5-3	4-5	4-4
	1-#5	40,000	8-1	8-8	6-7	6-10	6-2	5-10	5-4	4-6	4-5
		60,000	9-8	10-4	7-11	8-2	7-4	6-11	6-2	4-10	4-8
	2-#4	40,000	9-1	9-8	7-4	7-8	6-10	6-6	6-0	4-10	4-8
	1-#6	60,000	DR	DR	DR	DR	DR	DR	DR	DR	DR
	Center distance A[k, l]		1-8	1-11	1-1	1-3	1-0	0-11	0-9	0-6	0-6
16	Span without stirrups[i, j]		4-7	5-0	3-11	4-0	3-8	3-7	3-4	3-1	3-0
	1-#4	40,000	6-8	7-3	5-6	5-9	5-2	4-11	4-6	3-10	3-8
		60,000	9-3	10-1	7-9	8-0	7-2	6-10	6-3	5-4	5-2
	1-#4	40,000	9-6	10-4	7-10	8-2	7-4	6-11	6-5	5-5	5-3
		60,000	11-5	12-5	9-6	9-10	8-10	8-4	7-9	6-6	6-4
	2-#4	40,000	10-7	11-7	8-10	9-2	8-3	7-9	7-2	6-1	5-11
	1-#6	60,000	12-9	13-10	10-7	11-0	9-10	9-4	8-7	6-9	6-6
	2-#5	40,000	13-0	14-1	10-9	11-2	9-11	9-2	8-2	6-6	6-3
		60,000	DR	DR	DR	DR	DR	DR	DR	DR	DR
	Center distance A[k, l]		2-3	2-8	1-7	1-8	1-4	1-3	1-0	0-9	0-8
20	Span without stirrups A[i, j]		5-9	6-5	5-0	5-2	4-9	4-7	4-4	3-11	3-11
	1-#4	40,000	7-5	8-2	6-3	6-6	5-10	5-7	5-1	4-4	4-2
		60,000	9-0	10-0	7-8	7-11	7-1	6-9	6-3	5-3	5-1
	1-#5	40,000	9-2	10-2	7-9	8-1	7-3	6-11	6-4	5-4	5-2
		60,000	12-9	14-2	10-10	11-3	10-1	9-7	8-10	7-5	7-3
	2-#4	40,000	11-10	13-2	10-1	10-5	9-4	8-11	8-2	6-11	6-9
	1-#6	60,000	14-4	15-10	12-1	12-7	11-3	10-9	9-11	8-4	8-1
	2-#5	40,000	14-7	16-2	12-4	12-9	11-4	10-6	9-5	7-7	7-3
		60,000	17-5	19-2	14-9	15-3	13-5	12-4	11-0	8-8	8-4
	2-#6	40,000	16-4	18-11	12-7	13-3	11-4	10-6	9-5	7-7	7-3
		60,000	DR	DR	DR	DR	DR	DR	DR	DR	DR
	Center distance A[k, l]		2-9	3-5	2-0	2-2	1-9	1-7	1-4	0-11	0-11

(continued)

TABLE R608.8(2)—continued
MAXIMUM ALLOWABLE CLEAR SPANS FOR 4-INCH-NOMINAL THICK FLAT LINTELS IN LOAD-BEARING WALLS[a, b, c, d, e, f, m]
ROOF CLEAR SPAN 40 FEET AND FLOOR CLEAR SPAN 32 FEET

LINTEL DEPTH, D[g] (inches)	NUMBER OF BARS AND BAR SIZE IN TOP AND BOTTOM OF LINTEL	STEEL YIELD STRENGTH[h], f_y (psi)	DESIGN LOADING CONDITION DETERMINED FROM Table R608.8(1)								
			1	2		3		4		5	
			Maximum ground snow load (psf)								
			—	30	70	30	70	30	70	30	70
			Maximum clear span of lintel (feet-inches)								
24	Span without stirrups[i, j]		6-11	7-9	6-1	6-3	5-9	5-7	5-3	4-9	4-8
	1-#4	40,000	8-0	9-0	6-11	7-2	6-5	6-2	5-8	4-9	4-8
		60,000	9-9	11-0	8-5	8-9	7-10	7-6	6-11	5-10	5-8
	1-#5	40,000	10-0	11-3	8-7	8-11	8-0	7-7	7-0	5-11	5-9
		60,000	13-11	15-8	12-0	12-5	11-2	10-7	9-10	8-3	8-0
	2-#4	40,000	12-11	14-6	11-2	11-6	10-5	9-10	9-1	7-8	7-5
	1-#6	60,000	15-7	17-7	13-6	13-11	12-7	11-11	11-0	9-3	9-0
	2-#5	40,000	15-11	17-11	13-7	14-3	12-8	11-9	10-8	8-7	8-4
		60,000	19-1	21-6	16-5	17-1	15-1	14-0	12-6	9-11	9-7
	2-#6	40,000	17-7	21-1	14-1	14-10	12-8	11-9	10-8	8-7	8-4
		60,000	DR	DR	DR	DR	DR	DR	DR	DR	DR
	Center distance A[k, l]		3-3	4-1	2-5	2-7	2-1	1-11	1-7	1-2	1-1

For SI: 1 inch = 25.4 mm, 1 foot = 304.8 mm, 1 pound per square inch = 6.895 kPa, 1 pound per square foot = 0.0479 kPa, Grade 40 = 280 MPa, Grade 60 = 420 MPa.

a. See Table R608.3 for tolerances permitted from nominal thickness.

b. Table values are based on concrete with a minimum specified compressive strength of 2,500 psi. See Note j.

c. Table values are based on uniform loading. See Section R608.8.2 for lintels supporting concentrated loads.

d. Deflection criterion is $L/240$, where L is the clear span of the lintel in inches, or $^1/_2$-inch, whichever is less.

e. Linear interpolation is permitted between ground snow loads and between lintel depths.

f. DR indicates design required.

g. Lintel depth, D, is permitted to include the available height of wall located directly above the lintel, provided that the increased lintel depth spans the entire length of the lintel.

h. Stirrups shall be fabricated from reinforcing bars with the same yield strength as that used for the main longitudinal reinforcement.

i. Allowable clear span without stirrups applicable to all lintels of the same depth, D. Top and bottom reinforcement for lintels without stirrups shall be not less than the least amount of reinforcement required for a lintel of the same depth and loading condition with stirrups. All other spans require stirrups spaced at not more than $d/2$.

j. Where concrete with a minimum specified compressive strength of 3,000 psi is used, clear spans for lintels without stirrups shall be permitted to be multiplied by 1.05. If the increased span exceeds the allowable clear span for a lintel of the same depth and loading condition with stirrups, the top and bottom reinforcement shall be equal to or greater than that required for a lintel of the same depth and loading condition that has an allowable clear span that is equal to or greater than that of the lintel without stirrups that has been increased.

k. Center distance, A, is the center portion of the clear span where stirrups are not required. This is applicable to all longitudinal bar sizes and steel yield strengths.

l. Where concrete with a minimum specified compressive strength of 3,000 psi is used, center distance, A, shall be permitted to be multiplied by 1.10.

m. The maximum clear opening width between two solid wall segments shall be 18 feet. See Section R608.7.2.1. Lintel clear spans in the table greater than 18 feet are shown for interpolation and information only.

TABLE R608.8(3)
MAXIMUM ALLOWABLE CLEAR SPANS FOR 6-INCH-NOMINAL THICK FLAT LINTELS IN LOAD-BEARING WALLS[a, b, c, d, e, f, m]
ROOF CLEAR SPAN 40 FEET AND FLOOR CLEAR SPAN 32 FEET

LINTEL DEPTH, D^g (inches)	NUMBER OF BARS AND BAR SIZE IN TOP AND BOTTOM OF LINTEL	STEEL YIELD STRENGTH[h], f_y (psi)	DESIGN LOADING CONDITION DETERMINED FROM Table R608.8(1)								
			1	2		3		4		5	
			Maximum ground snow load (psf)								
			—	30	70	30	70	30	70	30	70
			Maximum clear span of lintel (feet-inches)								
8	Span without stirrups[i, j]		4-2	4-8	3-1	3-3	2-10	2-6	2-3	2-0	2-0
	1-#4	40,000	5-1	5-5	4-2	4-3	3-10	3-6	3-3	2-8	2-7
		60,000	6-2	6-7	5-0	5-2	4-8	4-2	3-11	3-3	3-2
	1-#5	40,000	6-3	6-8	5-1	5-3	4-9	4-3	4-0	3-3	3-2
		60,000	7-6	8-0	6-1	6-4	5-8	5-1	4-9	3-8	3-6
	2-#4	40,000	7-0	7-6	5-8	5-11	5-3	4-9	4-5	3-8	3-6
	1-#6	60,000	DR	DR	DR	DR	DR	DR	DR	DR	DR
	Center distance $A^{k, l}$		1-7	1-10	1-1	1-2	0-11	0-9	0-8	0-5	0-5
12	Span without stirrups[i, j]		4-2	4-8	3-5	3-6	3-2	2-11	2-9	2-5	2-4
	1-#4	40,000	5-7	6-1	4-8	4-10	4-4	3-11	3-8	3-0	2-11
		60,000	7-9	8-6	6-6	6-9	6-1	5-6	5-1	4-3	4-1
	1-#5	40,000	7-11	8-8	6-8	6-11	6-2	5-7	5-2	4-4	4-2
		60,000	9-7	10-6	8-0	8-4	7-6	6-9	6-3	5-2	5-1
	2-#4	40,000	8-11	9-9	7-6	7-9	6-11	6-3	5-10	4-10	4-8
	1-#6	60,000	10-8	11-9	8-12	9-4	8-4	7-6	7-0	5-10	5-8
	2-#5	40,000	10-11	12-0	9-2	9-6	8-6	7-8	7-2	5-6	5-3
		60,000	12-11	14-3	10-10	11-3	10-1	9-0	8-1	6-1	5-10
	2-#6	40,000	12-9	14-0	10-8	11-1	9-7	8-1	7-3	5-6	5-3
		60,000	DR	DR	DR	DR	DR	DR	DR	DR	DR
	Center distance $A^{k, l}$		2-6	3-0	1-9	1-10	1-6	1-3	1-1	0-9	0-8
16	Span without stirrups[i, j]		5-7	6-5	4-9	4-11	4-5	4-0	3-10	3-4	3-4
	1-#4	40,000	6-5	7-2	5-6	5-9	5-2	4-8	4-4	3-7	3-6
		60,000	7-10	8-9	6-9	7-0	6-3	5-8	5-3	4-4	4-3
	1-#5	40,000	7-11	8-11	6-10	7-1	6-5	5-9	5-4	4-5	4-4
		60,000	11-1	12-6	9-7	9-11	8-11	8-0	7-6	6-2	6-0
	2-#4	40,000	10-3	11-7	8-10	9-2	8-3	7-6	6-11	5-9	5-7
	1-#6	60,000	12-5	14-0	10-9	11-1	10-0	9-0	8-5	7-0	6-9
	2-#5	40,000	12-8	14-3	10-11	11-4	10-2	9-2	8-7	6-9	6-6
		60,000	15-2	17-1	13-1	13-7	12-3	11-0	10-3	7-11	7-7
	2-#6	40,000	14-11	16-9	12-8	13-4	11-4	9-8	8-8	6-9	6-6
		60,000	DR	DR	DR	DR	DR	DR	DR	DR	DR
	Center distance $A^{k, l}$		3-3	4-1	2-5	2-7	2-1	1-9	1-6	1-0	1-0

(continued)

TABLE R608.8(3)—continued
MAXIMUM ALLOWABLE CLEAR SPANS FOR 6-INCH-NOMINAL THICK FLAT LINTELS IN LOAD-BEARING WALLS[a, b, c, d, e, f, m]
ROOF CLEAR SPAN 40 FEET AND FLOOR CLEAR SPAN 32 FEET

LINTEL DEPTH, D^g (inches)	NUMBER OF BARS AND BAR SIZE IN TOP AND BOTTOM OF LINTEL	STEEL YIELD STRENGTH[h], f_y (psi)	DESIGN LOADING CONDITION DETERMINED FROM Table R608.8(1)								
			1	2		3		4		5	
			Maximum ground snow load (psf)								
			—	30	70	30	70	30	70	30	70
			Maximum clear span of lintel (feet-inches)								
20	Span without stirrups[i, j]		6-11	8-2	6-1	6-3	5-8	5-2	4-11	4-4	4-3
	1-#5	40,000	8-9	10-1	7-9	8-0	7-3	6-6	6-1	5-1	4-11
		60,000	10-8	12-3	9-5	9-9	8-10	8-0	7-5	6-2	6-0
	2-#4 1-#6	40,000	9-11	11-4	8-9	9-1	8-2	7-4	6-10	5-8	5-7
		60,000	13-9	15-10	12-2	12-8	11-5	10-3	9-7	7-11	7-9
	2-#5	40,000	14-0	16-2	12-5	12-11	11-7	10-6	9-9	7-11	7-8
		60,000	16-11	19-6	15-0	15-6	14-0	12-7	11-9	9-1	8-9
	2-#6	40,000	16-7	19-1	14-7	15-3	13-1	11-3	10-2	7-11	7-8
		60,000	19-11	22-10	17-4	18-3	15-6	13-2	11-10	9-1	8-9
	Center distance $A^{k, l}$		3-11	5-2	3-1	3-3	2-8	2-2	1-11	1-4	1-3
24	Span without stirrups[i, j]		8-2	9-10	7-4	7-8	6-11	6-4	5-11	5-3	5-2
	1-#5	40,000	9-5	11-1	8-7	8-10	8-0	7-3	6-9	5-7	5-5
		60,000	11-6	13-6	10-5	10-9	9-9	8-9	8-2	6-10	6-8
	2-#4 1-#6	40,000	10-8	12-6	9-8	10-0	9-0	8-2	7-7	6-4	6-2
		60,000	12-11	15-2	11-9	12-2	11-0	9-11	9-3	7-8	7-6
	2-#5	40,000	15-2	17-9	13-9	14-3	12-10	11-7	10-10	9-0	8-9
		60,000	18-4	21-6	16-7	17-3	15-6	14-0	13-1	10-4	10-0
	2-#6	40,000	18-0	21-1	16-4	16-11	14-10	12-9	11-8	9-2	8-11
		60,000	21-7	25-4	19-2	20-4	17-2	14-9	13-4	10-4	10-0
	Center distance $A^{k, l}$		4-6	6-2	3-8	4-0	3-3	2-8	2-3	1-7	1-6

For SI: 1 inch = 25.4 mm, 1 foot = 304.8 mm, 1 pound per square inch = 6.895 kPa, 1 pound per square foot = 0.0479 kPa, Grade 40 = 280 MPa, Grade 60 = 420 MPa.

a. See Table R608.3 for tolerances permitted from nominal thickness.

b. Table values are based on concrete with a minimum specified compressive strength of 2,500 psi. See Note j.

c. Table values are based on uniform loading. See Section R608.8.2 for lintels supporting concentrated loads.

d. Deflection criterion is $L/240$, where L is the clear span of the lintel in inches, or $^1/_2$ inch, whichever is less.

e. Linear interpolation is permitted between ground snow loads and between lintel depths.

f. DR indicates design required.

g. Lintel depth, D, is permitted to include the available height of wall located directly above the lintel, provided that the increased lintel depth spans the entire length of the lintel.

h. Stirrups shall be fabricated from reinforcing bars with the same yield strength as that used for the main longitudinal reinforcement.

i. Allowable clear span without stirrups applicable to all lintels of the same depth, D. Top and bottom reinforcement for lintels without stirrups shall be not less than the least amount of reinforcement required for a lintel of the same depth and loading condition with stirrups. All other spans require stirrups spaced at not more than $d/2$.

j. Where concrete with a minimum specified compressive strength of 3,000 psi is used, clear spans for lintels without stirrups shall be permitted to be multiplied by 1.05. If the increased span exceeds the allowable clear span for a lintel of the same depth and loading condition with stirrups, the top and bottom reinforcement shall be equal to or greater than that required for a lintel of the same depth and loading condition that has an allowable clear span that is equal to or greater than that of the lintel without stirrups that has been increased.

k. Center distance, A, is the center portion of the clear span where stirrups are not required. This is applicable to all longitudinal bar sizes and steel yield strengths.

l. Where concrete with a minimum specified compressive strength of 3,000 psi is used, center distance, A, shall be permitted to be multiplied by 1.10.

m. The maximum clear opening width between two solid wall segments shall be 18 feet. See Section R608.7.2.1. Lintel clear spans in the table greater than 18 feet are shown for interpolation and information only.

TABLE R608.8(4)
MAXIMUM ALLOWABLE CLEAR SPANS FOR 8-INCH-NOMINAL THICK FLAT LINTELS IN LOAD-BEARING WALLS[a, b, c, d, e, f, m]
ROOF CLEAR SPAN 40 FEET AND FLOOR CLEAR SPAN 32 FEET

LINTEL DEPTH, D^g (inches)	NUMBER OF BARS AND BAR SIZE IN TOP AND BOTTOM OF LINTEL	STEEL YIELD STRENGTH[h], f_y (psi)	DESIGN LOADING CONDITION DETERMINED FROM Table R608.8(1)								
			1	2		3		4		5	
			Maximum ground snow load (psf)								
			—	30	70	30	70	30	70	30	70
			Maximum clear span of lintel (feet-inches)								
8	Span without stirrups[i, j]		4-4	4-9	3-7	3-9	3-4	2-10	2-7	2-1	2-0
	1-#4	40,000	4-4	4-9	3-7	3-9	3-4	2-11	2-9	2-3	2-2
		60,000	6-1	6-7	5-0	5-3	4-8	4-0	3-9	3-1	3-0
	1-#5	40,000	6-2	6-9	5-2	5-4	4-9	4-1	3-10	3-2	3-1
		60,000	7-5	8-1	6-2	6-5	5-9	4-11	4-7	3-9	3-8
	2-#4	40,000	6-11	7-6	5-9	6-0	5-4	4-7	4-4	3-6	3-5
	1-#6	60,000	8-3	9-0	6-11	7-2	6-5	5-6	5-2	4-2	4-1
	2-#5	40,000	8-5	9-2	7-0	7-3	6-6	5-7	5-3	4-2	4-0
		60,000	DR	DR	DR	DR	DR	DR	DR	DR	DR
	Center distance $A^{k, l}$		2-1	2-6	1-5	1-6	1-3	0-11	0-10	0-6	0-6
12	Span without stirrups[i, j]		4-10	5-8	4-0	4-2	3-9	3-2	3-0	2-7	2-6
	1-#4	40,000	5-5	6-1	4-8	4-10	4-4	3-9	3-6	2-10	2-10
		60,000	6-7	7-5	5-8	5-11	5-4	4-7	4-3	3-6	3-5
	1-#5	40,000	6-9	7-7	5-9	6-0	5-5	4-8	4-4	3-7	3-6
		60,000	9-4	10-6	8-1	8-4	7-6	6-6	6-1	5-0	4-10
	2-#4	40,000	8-8	9-9	7-6	7-9	7-0	6-0	5-8	4-7	4-6
	1-#6	60,000	10-6	11-9	9-1	9-5	8-5	7-3	6-10	5-7	5-5
	2-#5	40,000	10-8	12-0	9-3	9-7	8-7	7-5	6-11	5-6	5-4
		60,000	12-10	14-5	11-1	11-6	10-4	8-11	8-4	6-7	6-4
	2-#6	40,000	12-7	14-2	10-10	11-3	10-2	8-3	7-6	5-6	5-4
		60,000	DR	DR	DR	DR	DR	DR	DR	DR	DR
	Center distance $A^{k, l}$		3-2	4-0	2-4	2-6	2-0	1-6	1-4	0-11	0-10
16	Span without stirrups[i, j]		6-5	7-9	5-7	5-10	5-2	4-5	4-2	3-7	3-6
	1-#4	40,000	6-2	7-1	5-6	5-8	5-1	4-5	4-2	3-5	3-4
		60,000	7-6	8-8	6-8	6-11	6-3	5-5	5-1	4-2	4-0
	1-#5	40,000	7-8	8-10	6-10	7-1	6-4	5-6	5-2	4-3	4-1
		60,000	9-4	10-9	8-4	8-7	7-9	6-8	6-3	5-2	5-0
	2-#4	40,000	8-8	10-0	7-8	8-0	7-2	6-2	5-10	4-9	4-8
	1-#6	60,000	12-0	13-11	10-9	11-2	10-0	8-8	8-1	6-8	6-6
	2-#5	40,000	12-3	14-2	11-0	11-4	10-3	8-10	8-3	6-9	6-7
		60,000	14-10	17-2	13-3	13-8	12-4	10-8	10-0	7-11	7-8
	2-#6	40,000	14-6	16-10	13-0	13-5	12-1	10-1	9-2	6-11	6-8
		60,000	17-5	20-2	15-7	16-1	14-6	11-10	10-8	7-11	7-8
	Center distance $A^{k, l}$		4-1	5-5	3-3	3-6	2-10	2-1	1-10	1-3	1-2

(continued)

TABLE R608.8(4)—continued
MAXIMUM ALLOWABLE CLEAR SPANS FOR 8-INCH-NOMINAL THICK FLAT LINTELS IN LOAD-BEARING WALLS[a, b, c, d, e, f, m]
ROOF CLEAR SPAN 40 FEET AND FLOOR CLEAR SPAN 32 FEET

LINTEL DEPTH, D[g] (inches)	NUMBER OF BARS AND BAR SIZE IN TOP AND BOTTOM OF LINTEL	STEEL YIELD STRENGTH[h], f_y (psi)	DESIGN LOADING CONDITION DETERMINED FROM Table R608.8(1)								
			1	2		3		4		5	
			Maximum ground snow load (psf)								
			—	30	70	30	70	30	70	30	70
			Maximum clear span of lintel (feet-inches)								
20	Span without stirrups[i, j]		7-10	9-10	7-1	7-5	6-7	5-8	5-4	4-7	4-6
	1-#5	40,000	8-4	9-11	7-8	8-0	7-2	6-3	5-10	4-9	4-8
		60,000	10-2	12-1	9-5	9-9	8-9	7-7	7-1	5-10	5-8
	2-#4	40,000	9-5	11-3	8-8	9-0	8-1	7-0	6-7	5-5	5-3
	1-#6	60,000	11-6	13-8	10-7	11-0	9-11	8-7	8-0	6-7	6-5
	2-#5	40,000	11-9	13-11	10-10	11-2	10-1	8-9	8-2	6-8	6-7
		60,000	16-4	19-5	15-0	15-7	14-0	12-2	11-4	9-3	9-0
	2-#6	40,000	16-0	19-0	14-9	15-3	13-9	11-10	10-10	8-3	8-0
		60,000	19-3	22-11	17-9	18-5	16-7	13-7	12-4	9-3	9-0
	Center distance A[k, l]		4-10	6-10	4-1	4-5	3-7	2-8	2-4	1-7	1-6
24	Span without stirrups[i, j]		9-2	11-9	8-7	8-11	8-0	6-11	6-6	5-7	5-6
	1-#5	40,000	8-11	10-10	8-6	8-9	7-11	6-10	6-5	5-3	5-2
		60,000	10-11	13-3	10-4	10-8	9-8	8-4	7-10	6-5	6-3
	2-#4	40,000	10-1	12-3	9-7	9-11	8-11	7-9	7-3	6-0	5-10
	1-#6	60,000	12-3	15-0	11-8	12-1	10-11	9-5	8-10	7-3	7-1
	2-#5	40,000	12-6	15-3	11-11	12-4	11-1	9-7	9-0	7-5	7-3
		60,000	17-6	21-3	16-7	17-2	15-6	13-5	12-7	10-4	10-1
	2-#6	40,000	17-2	20-11	16-3	16-10	15-3	13-2	12-4	9-7	9-4
		60,000	20-9	25-3	19-8	20-4	18-5	15-4	14-0	10-7	10-3
	Center distance A[k, l]		5-6	8-1	4-11	5-3	4-4	3-3	2-10	1-11	1-10

For SI: 1 inch = 25.4 mm, 1 foot = 304.8 mm, 1 pound per square inch = 6.895 kPa, 1 pound per square foot = 0.0479 kPa, Grade 40 = 280 MPa, Grade 60 = 420 MPa.

Note: Top and bottom reinforcement for lintels without stirrups, as shown in shaded cells, shall be equal to or greater than that required for lintel of the same depth and loading condition that has an allowable clear span that is equal to or greater than that of the lintel without stirrups.

a. See Table R608.3 for tolerances permitted from nominal thickness.

b. Table values are based on concrete with a minimum specified compressive strength of 2,500 psi. See Note j.

c. Table values are based on uniform loading. See Section R608.8.2 for lintels supporting concentrated loads.

d. Deflection criterion is *L*/240, where *L* is the clear span of the lintel in inches, or $^1/_2$ inch, whichever is less.

e. Linear interpolation is permitted between ground snow loads and between lintel depths.

f. DR indicates design required.

g. Lintel depth, *D*, is permitted to include the available height of wall located directly above the lintel, provided that the increased lintel depth spans the entire length of the lintel.

h. Stirrups shall be fabricated from reinforcing bars with the same yield strength as that used for the main longitudinal reinforcement.

i. Allowable clear span without stirrups applicable to all lintels of the same depth, *D*. Top and bottom reinforcement for lintels without stirrups shall be not less than the least amount of reinforcement required for a lintel of the same depth and loading condition with stirrups. All other spans require stirrups spaced at not more than *d*/2.

j. Where concrete with a minimum specified compressive strength of 3,000 psi is used, clear spans for lintels without stirrups shall be permitted to be multiplied by 1.05. If the increased span exceeds the allowable clear span for a lintel of the same depth and loading condition with stirrups, the top and bottom reinforcement shall be equal to or greater than that required for a lintel of the same depth and loading condition that has an allowable clear span that is equal to or greater than that of the lintel without stirrups that has been increased.

k. Center distance, A, is the center portion of the clear span where stirrups are not required. This is applicable to all longitudinal bar sizes and steel yield strengths.

l. Where concrete with a minimum specified compressive strength of 3,000 psi is used, center distance, *A*, shall be permitted to be multiplied by 1.10.

m. The maximum clear opening width between two solid wall segments shall be 18 feet. See Section R608.7.2.1. Lintel clear spans in the table greater than 18 feet are shown for interpolation and information only.

TABLE R608.8(5)
MAXIMUM ALLOWABLE CLEAR SPANS FOR 10-INCH-NOMINAL THICK FLAT LINTELS IN LOAD-BEARING WALLS[a, b, c, d, e, f, m]
ROOF CLEAR SPAN 40 FEET AND FLOOR CLEAR SPAN 32 FEET

LINTEL DEPTH, D^g (inches)	NUMBER OF BARS AND BAR SIZE IN TOP AND BOTTOM OF LINTEL	STEEL YIELD STRENGTH[h], f_y (psi)	DESIGN LOADING CONDITION DETERMINED FROM Table R608.8(1)								
			1	2		3		4		5	
			Maximum ground snow load (psf)								
			—	30	70	30	70	30	70	30	70
			Maximum clear span of lintel (feet-inches)								
8	Span without stirrups[i, j]		6-0	7-2	4-7	4-10	4-1	3-1	2-11	2-3	2-2
	1-#4	40,000	4-3	4-9	3-7	3-9	3-4	2-9	2-7	2-1	2-1
		60,000	5-11	6-7	5-0	5-3	4-8	3-10	3-8	2-11	2-11
	1-#5	40,000	6-1	6-9	5-2	5-4	4-9	3-11	3-9	3-0	2-11
		60,000	7-4	8-1	6-3	6-5	5-9	4-9	4-6	3-7	3-7
	2-#4	40,000	6-10	7-6	5-9	6-0	5-5	4-5	4-2	3-4	3-4
	1-#6	60,000	8-2	9-1	6-11	7-2	6-6	5-4	5-0	4-1	4-0
	2-#5	40,000	8-4	9-3	7-1	7-4	6-7	5-5	5-1	4-1	4-0
		60,000	9-11	11-0	8-5	8-9	7-10	6-6	6-1	4-8	4-6
	2-#6	40,000	9-9	10-10	8-3	8-7	7-9	6-4	5-10	4-1	4-0
		60,000	DR	DR	DR	DR	DR	DR	DR	DR	DR
	Center distance $A^{k, l}$		2-6	3-1	1-10	1-11	1-7	1-1	0-11	0-7	0-7
12	Span without stirrups[i, j]		5-5	6-7	4-7	4-10	4-3	3-5	3-3	2-8	2-8
	1-#4	40,000	5-3	6-0	4-8	4-10	4-4	3-7	3-4	2-9	2-8
		60,000	6-5	7-4	5-8	5-10	5-3	4-4	4-1	3-4	3-3
	1-#5	40,000	6-6	7-6	5-9	6-0	5-5	4-5	4-2	3-5	3-4
		60,000	7-11	9-1	7-0	7-3	6-7	5-5	5-1	4-2	4-0
	2-#4	40,000	7-4	8-5	6-6	6-9	6-1	5-0	4-9	3-10	3-9
	1-#6	60,000	10-3	11-9	9-1	9-5	8-6	7-0	6-7	5-4	5-3
	2-#5	40,000	10-5	12-0	9-3	9-7	8-8	7-2	6-9	5-5	5-4
		60,000	12-7	14-5	11-2	11-6	10-5	8-7	8-1	6-6	6-4
	2-#6	40,000	12-4	14-2	10-11	11-4	10-2	8-5	7-8	5-7	5-5
		60,000	14-9	17-0	13-1	13-6	12-2	10-0	9-1	6-6	6-4
	Center distance $A^{k, l}$		3-9	4-11	2-11	3-2	2-7	1-9	1-7	1-0	1-0
16	Span without stirrups[i, j]		7-1	9-0	6-4	6-8	5-10	4-9	4-6	3-9	3-8
	1-#4	40,000	5-11	7-0	5-5	5-8	5-1	4-3	4-0	3-3	3-2
		60,000	7-3	8-7	6-8	6-11	6-3	5-2	4-10	3-11	3-10
	1-#5	40,000	7-4	8-9	6-9	7-0	6-4	5-3	4-11	4-0	3-11
		60,000	9-0	10-8	8-3	8-7	7-9	6-5	6-0	4-11	4-9
	2-#4	40,000	8-4	9-11	7-8	7-11	7-2	5-11	5-7	4-6	4-5
	1-#6	60,000	10-2	12-0	9-4	9-8	8-9	7-3	6-10	5-6	5-5
	2-#5	40,000	10-4	12-3	9-6	9-10	8-11	7-4	6-11	5-8	5-6
		60,000	14-4	17-1	13-3	13-8	12-4	10-3	9-8	7-10	7-8
	2-#6	40,000	14-1	16-9	13-0	13-5	12-2	10-1	9-6	7-0	6-10
		60,000	17-0	20-2	15-8	16-2	14-7	12-0	10-11	8-0	7-9
	Center distance $A^{k, l}$		4-9	6-8	4-0	4-4	3-6	2-5	2-2	1-5	1-4

(continued)

TABLE R608.8(5)—continued
MAXIMUM ALLOWABLE CLEAR SPANS FOR 10-INCH-NOMINAL THICK FLAT LINTELS IN LOAD-BEARING WALLS[a, b, c, d, e, f, m]
ROOF CLEAR SPAN 40 FEET AND FLOOR CLEAR SPAN 32 FEET

LINTEL DEPTH, D^g (inches)	NUMBER OF BARS AND BAR SIZE IN TOP AND BOTTOM OF LINTEL	STEEL YIELD STRENGTH[h], f_y (psi)	DESIGN LOADING CONDITION DETERMINED FROM Table R608.8(1)								
			1	2		3		4		5	
			Maximum ground snow load (psf)								
			—	30	70	30	70	30	70	30	70
			Maximum clear span of lintel (feet-inches)								
20	Span without stirrups[i, j]		8-7	11-4	8-1	8-5	7-5	6-1	5-9	4-10	4-9
	1-#4	40,000	6-5	7-10	6-2	6-4	5-9	4-9	4-6	3-8	3-7
		60,000	7-10	9-7	7-6	7-9	7-0	5-10	5-6	4-5	4-4
	1-#5	40,000	8-0	9-9	7-8	7-11	7-2	5-11	5-7	4-6	4-5
		60,000	9-9	11-11	9-4	9-8	8-9	7-3	6-10	5-6	5-5
	2-#4	40,000	9-0	11-1	8-8	8-11	8-1	6-9	6-4	5-2	5-0
	1-#6	60,000	11-0	13-6	10-6	10-11	9-10	8-2	7-9	6-3	6-2
	2-#5	40,000	11-3	13-9	10-9	11-1	10-0	8-4	7-10	6-5	6-3
		60,000	15-8	19-2	15-0	15-6	14-0	11-8	11-0	8-11	8-9
	2-#6	40,000	15-5	18-10	14-8	15-2	13-9	11-5	10-9	8-6	8-3
		60,000	18-7	22-9	17-9	18-5	16-7	13-10	12-9	9-5	9-2
	Center distance A[k, l]		5-7	8-4	5-1	5-5	4-5	3-1	2-9	1-10	1-9
24	Span without stirrups[i, j]		9-11	13-7	9-9	10-2	9-0	7-5	7-0	5-10	5-9
	1-#5	40,000	8-6	10-8	8-5	8-8	7-10	6-6	6-2	5-0	4-11
		60,000	10-5	13-0	10-3	10-7	9-7	8-0	7-6	6-1	6-0
	2-#4	40,000	9-7	12-1	9-6	9-9	8-10	7-5	7-0	5-8	5-6
	1-#6	60,000	11-9	14-9	11-7	11-11	10-10	9-0	8-6	6-11	6-9
	2-#5	40,000	12-0	15-0	11-9	12-2	11-0	9-2	8-8	7-1	6-11
		60,000	14-7	18-3	14-4	14-10	13-5	11-2	10-7	8-7	8-5
	2-#6	40,000	14-3	17-11	14-1	14-7	13-2	11-0	10-4	8-5	8-3
		60,000	19-11	25-0	19-7	20-3	18-4	15-3	14-5	10-10	10-7
	Center distance A[k, l]		6-3	9-11	6-1	6-6	5-4	3-9	3-4	2-2	2-1

For SI: 1 inch = 25.4 mm, 1 foot = 304.8 mm, 1 pound per square inch = 6.895 kPa, 1 pound per square foot = 0.0479 kPa, Grade 40 = 280 MPa, Grade 60 = 420 MPa.

Note: Top and bottom reinforcement for lintels without stirrups, as shown in shaded cells, shall be equal to or greater than that required for lintel of the same depth and loading condition that has an allowable clear span that is equal to or greater than that of the lintel without stirrups.

a. See Table R608.3 for tolerances permitted from nominal thickness.

b. Table values are based on concrete with a minimum specified compressive strength of 2,500 psi. See Note j.

c. Table values are based on uniform loading. See Section R608.8.2 for lintels supporting concentrated loads.

d. Deflection criterion is $L/240$, where L is the clear span of the lintel in inches, or $^1/_2$ inch, whichever is less.

e. Linear interpolation is permitted between ground snow loads and between lintel depths.

f. DR indicates design required.

g. Lintel depth, D, is permitted to include the available height of wall located directly above the lintel, provided that the increased lintel depth spans the entire length of the lintel.

h. Stirrups shall be fabricated from reinforcing bars with the same yield strength as that used for the main longitudinal reinforcement.

i. Allowable clear span without stirrups applicable to all lintels of the same depth, D. Top and bottom reinforcement for lintels without stirrups shall be not less than the least amount of reinforcement required for a lintel of the same depth and loading condition with stirrups. All other spans require stirrups spaced at not more than $d/2$.

j. Where concrete with a minimum specified compressive strength of 3,000 psi is used, clear spans for lintels without stirrups shall be permitted to be multiplied by 1.05. If the increased span exceeds the allowable clear span for a lintel of the same depth and loading condition with stirrups, the top and bottom reinforcement shall be equal to or greater than that required for a lintel of the same depth and loading condition that has an allowable clear span that is equal to or greater than that of the lintel without stirrups that has been increased.

k. Center distance, A, is the center portion of the clear span where stirrups are not required. This is applicable to all longitudinal bar sizes and steel yield strengths.

l. Where concrete with a minimum specified compressive strength of 3,000 psi is used, center distance, A, shall be permitted to be multiplied by 1.10.

m. The maximum clear opening width between two solid wall segments shall be 18 feet. See Section R608.7.2.1. Lintel clear spans in the table greater than 18 feet are shown for interpolation and information only.

TABLE R608.8(6)
MAXIMUM ALLOWABLE CLEAR SPANS FOR 6-INCH-THICK WAFFLE-GRID LINTELS IN LOAD-BEARING WALLS[a, b, c, d, e, f, o]
MAXIMUM ROOF CLEAR SPAN 40 FEET AND MAXIMUM FLOOR SPAN 32 FEET

LINTEL DEPTH, D[g] (inches)	NUMBER OF BARS AND BAR SIZE IN TOP AND BOTTOM OF LINTEL	STEEL YIELD STRENGTH[h], f_y (psi)	DESIGN LOADING CONDITION DETERMINED FROM Table R608.8(1)								
			1	2		3		4		5	
			Maximum ground snow load (psf)								
			—	30	70	30	70	30	70	30	70
			Maximum clear span of lintel (feet-inches)								
8[i]	Span without stirrups[k, l]		2-7	2-9	2-0	2-1	2-0	2-0	2-0	2-0	2-0
	1-#4	40,000	5-2	5-5	4-0	4-3	3-7	3-3	2-11	2-4	2-3
		60,000	5-9	6-3	4-0	4-3	3-7	3-3	2-11	2-4	2-3
	1-#5	40,000	5-9	6-3	4-0	4-3	3-7	3-3	2-11	2-4	2-3
		60,000	5-9	6-3	4-0	4-3	3-7	3-3	2-11	2-4	2-3
	2-#4	40,000	5-9	6-3	4-0	4-3	3-7	3-3	2-11	2-4	2-3
	1-#6	60,000	DR	DR	DR	DR	DR	DR	DR	DR	DR
	Center distance A[m, n]		0-9	0-10	0-6	0-6	0-5	0-5	0-4	STL	STL
12[i]	Span without stirrups[k, l]		2-11	3-1	2-6	2-7	2-5	2-4	2-3	2-1	2-0
	1-#4	40,000	5-9	6-2	4-8	4-10	4-4	4-1	3-9	3-2	3-1
		60,000	8-0	8-7	6-6	6-9	6-0	5-5	4-11	3-11	3-10
	1-#5	40,000	8-1	8-9	6-8	6-11	6-0	5-5	4-11	3-11	3-10
		60,000	9-1	10-3	6-8	7-0	6-0	5-5	4-11	3-11	3-10
	2-#4 1-#6	40,000	9-1	9-9	6-8	7-0	6-0	5-5	4-11	3-11	3-10
	Center distance A[m, n]		1-3	1-5	0-10	0-11	0-9	0-8	0-6	STL	STL
16[i]	Span without stirrups[k, l]		4-0	4-4	3-6	3-7	3-4	3-3	3-1	2-10	2-10
	1-#4	40,000	6-7	7-3	5-6	5-9	5-2	4-10	4-6	3-9	3-8
		60,000	8-0	8-10	6-9	7-0	6-3	5-11	5-5	4-7	4-5
	1-#5	40,000	8-2	9-0	6-11	7-2	6-5	6-0	5-7	4-8	4-6
		60,000	11-5	12-6	9-3	9-9	8-4	7-7	6-10	5-6	5-4
	2-#4	40,000	10-7	11-7	8-11	9-3	8-3	7-7	6-10	5-6	5-4
	1-#6	60,000	12-2	14-0	9-3	9-9	8-4	7-7	6-10	5-6	5-4
	2-#5	40,000	12-2	14-2	9-3	9-9	8-4	7-7	6-10	5-6	5-4
		60,000	DR	DR	DR	DR	DR	DR	DR	DR	DR
	Center distance A[m, n]		1-8	2-0	1-2	1-3	1-0	0-11	0-9	STL	STL
20[i]	Span without stirrups[k, l]		5-0	5-6	4-6	4-7	4-3	4-1	4-0	3-8	3-8
	1-#4	40,000	7-2	8-2	6-3	6-6	5-10	5-6	5-1	4-3	4-2
		60,000	8-11	9-11	7-8	7-11	7-1	6-8	6-2	5-2	5-0
	1-#5	40,000	9-1	10-2	7-9	8-1	7-3	6-10	6-4	5-4	5-2
		60,000	12-8	14-2	10-11	11-3	10-2	9-6	8-9	7-1	6-10
	2-#4	40,000	10-3	11-5	8-9	9-1	8-2	7-8	7-1	6-0	5-10
	1-#6	60,000	14-3	15-11	11-9	12-5	10-8	9-9	8-9	7-1	6-10
	2-#5	40,000	14-6	16-3	11-6	12-1	10-4	9-6	8-6	6-11	6-8
		60,000	DR	DR	DR	DR	DR	DR	DR	DR	DR
	Center distance A[m, n]		2-0	2-6	1-6	1-7	1-3	1-1	1-0	STL	STL

(continued)

TABLE R608.8(6)—continued
MAXIMUM ALLOWABLE CLEAR SPANS FOR 6-INCH-THICK WAFFLE-GRID LINTELS IN LOAD-BEARING WALLS[a, b, c, d, e, f, o]
MAXIMUM ROOF CLEAR SPAN 40 FEET AND MAXIMUM FLOOR SPAN 32 FEET

LINTEL DEPTH, D^g (inches)	NUMBER OF BARS AND BAR SIZE IN TOP AND BOTTOM OF LINTEL	STEEL YIELD STRENGTH[h], f_y (psi)	DESIGN LOADING CONDITION DETERMINED FROM Table R608.8(1)								
			1	2		3		4		5	
			Maximum ground snow load (psf)								
			—	30	70	30	70	30	70	30	70
			Maximum clear span of lintel (feet-inches)								
24w[j]	Span without stirrups[k, l]		6-0	6-8	5-5	5-7	5-3	5-0	4-10	4-6	4-5
	1-#4	40,000	7-11	9-0	6-11	7-2	6-5	6-0	5-7	4-8	4-7
		60,000	9-8	10-11	8-5	8-9	7-10	7-4	6-10	5-9	5-7
	1-#5	40,000	9-10	11-2	8-7	8-11	8-0	7-6	7-0	5-10	5-8
		60,000	12-0	13-7	10-6	10-10	9-9	9-2	8-6	7-2	6-11
	2-#4	40,000	11-1	12-7	9-8	10-1	9-1	8-6	7-10	6-7	6-5
	1-#6	60,000	15-6	17-7	13-6	14-0	12-8	11-10	10-8	8-7	8-4
	2-#5	40,000	15-6	17-11	12-8	13-4	11-6	10-7	9-7	7-10	7-7
		60,000	DR	DR	DR	DR	DR	DR	DR	DR	DR
	Center distance $A^{m, n}$		2-4	3-0	1-9	1-11	1-6	1-4	1-2	STL	STL

For SI: 1 inch = 25.4 mm, 1 foot = 304.8 mm, 1 pound per square inch = 6.895 kPa, 1 pound per square foot = 0.0479 kPa, Grade 40 = 280 MPa, Grade 60 = 420 MPa.

a. Where lintels are formed with waffle-grid forms, form material shall be removed, if necessary, to create top and bottom flanges of the lintel that are not less than 3 inches in depth (in the vertical direction), are not less than 5 inches in width for 6-inch-nominal waffle-grid forms and not less than 7 inches in width for 8-inch-nominal waffle-grid forms. See Figure R608.8(3). Flat form lintels shall be permitted in place of waffle-grid lintels. See Tables R608.8(2) through R608.8(5).

b. See Table R608.3 for tolerances permitted from nominal thicknesses and minimum dimensions and spacing of cores.

c. Table values are based on concrete with a minimum specified compressive strength of 2,500 psi. See Notes l and n. Table values are based on uniform loading. See Section R608.8.2 for lintels supporting concentrated loads.

d. Deflection criterion is $L/240$, where L is the clear span of the lintel in inches, or $^1/_2$ inch, whichever is less.

e. Linear interpolation is permitted between ground snow loads.

f. DR indicates design required. STL indicates stirrups required throughout lintel.

g. Lintel depth, D, is permitted to include the available height of wall located directly above the lintel, provided that the increased lintel depth spans the entire length of the lintel.

h. Stirrups shall be fabricated from reinforcing bars with the same yield strength as that used for the main longitudinal reinforcement.

i. Lintels less than 24 inches in depth with stirrups shall be formed from flat-wall forms [see Tables R608.8(2) through R608.8(5)], or, if necessary, form material shall be removed from waffle-grid forms so as to provide the required cover for stirrups. Allowable spans for lintels formed with flat-wall forms shall be determined from Tables R608.8(2) through R608.8(5).

j. Where stirrups are required for 24-inch-deep lintels, the spacing shall not exceed 12 inches on center.

k. Allowable clear span without stirrups applicable to all lintels of the same depth, D. Top and bottom reinforcement for lintels without stirrups shall be not less than the least amount of reinforcement required for a lintel of the same depth and loading condition with stirrups. All other spans require stirrups spaced at not more than $d/2$.

l. Where concrete with a minimum specified compressive strength of 3,000 psi is used, clear spans for lintels without stirrups shall be permitted to be multiplied by 1.05. If the increased span exceeds the allowable clear span for a lintel of the same depth and loading condition with stirrups, the top and bottom reinforcement shall be equal to or greater than that required for a lintel of the same depth and loading condition that has an allowable clear span that is equal to or greater than that of the lintel without stirrups that has been increased.

m. Center distance, A, is the center portion of the span where stirrups are not required. This is applicable to all longitudinal bar sizes and steel yield strengths.

n. Where concrete with a minimum specified compressive strength of 3,000 psi is used, center distance, A, shall be permitted to be multiplied by 1.10.

o. The maximum clear opening width between two solid wall segments shall be 18 feet. See Section R608.7.2.1. Lintel spans in the table greater than 18 feet are shown for interpolation and information only.

TABLE R608.8(7)
MAXIMUM ALLOWABLE CLEAR SPANS FOR 8-INCH-THICK WAFFLE-GRID LINTELS IN LOAD-BEARING WALLS[a, b, c, d, e, f, o] MAXIMUM ROOF CLEAR SPAN 40 FEET AND MAXIMUM FLOOR CLEAR SPAN 32 FEET

LINTEL DEPTH, D[g] (inches)	NUMBER OF BARS AND BAR SIZE IN TOP AND BOTTOM OF LINTEL	STEEL YIELD STRENGTH[h], f_y (psi)	DESIGN LOADING CONDITION DETERMINED FROM Table R608.8(1)								
			1	2		3		4		5	
			Maximum ground snow load (psf)								
			—	30	70	30	70	30	70	30	70
			Maximum clear span of lintel (feet-inches)								
8[i]	Span without stirrups[k, l]		2-6	2-9	2-0	2-1	2-0	2-0	2-0	2-0	2-0
	1-#4	40,000	4-5	4-9	3-7	3-9	3-4	3-0	2-10	2-3	2-2
		60,000	5-6	6-2	4-0	4-3	3-7	3-1	2-10	2-3	2-2
	1-#5	40,000	5-6	6-2	4-0	4-3	3-7	3-1	2-10	2-3	2-2
	Center distance A[m, n]		0-9	0-10	0-6	0-6	0-5	0-4	0-4	STL	STL
12[i]	Span without stirrups[k, l]		2-10	3-1	2-6	2-7	2-5	2-3	2-2	2-0	2-0
	1-#4	40,000	5-7	6-1	4-8	4-10	4-4	3-11	3-8	3-0	2-11
		60,000	6-9	7-5	5-8	5-11	5-4	4-9	4-5	3-8	3-7
	1-#5	40,000	6-11	7-7	5-10	6-0	5-5	4-10	4-6	3-9	3-7
		60,000	8-8	10-1	6-7	7-0	5-11	5-2	4-8	3-9	3-7
	2-#4	40,000	8-8	9-10	6-7	7-0	5-11	5-2	4-8	3-9	3-7
	1-#6	60,000	8-8	10-1	6-7	7-0	5-11	5-2	4-8	3-9	3-7
	Center distance A[m, n]		1-2	1-5	0-10	0-11	0-9	0-7	0-6	STL	STL
16[i]	Span without stirrups[k, l]		3-10	4-3	3-6	3-7	3-4	3-2	3-0	2-10	2-9
	1-#4	40,000	6-5	7-2	5-6	5-9	5-2	4-8	4-4	3-7	3-6
		60,000	7-9	8-9	6-9	7-0	6-3	5-8	5-3	4-4	4-3
	1-#5	40,000	7-11	8-11	6-10	7-1	6-5	5-9	5-4	4-5	4-4
		60,000	9-8	10-11	8-4	8-8	7-10	7-0	6-6	5-2	5-1
	2-#4	40,000	9-0	10-1	7-9	8-0	7-3	6-6	6-1	5-0	4-11
	1-#6	60,000	11-5	13-10	9-2	9-8	8-3	7-2	6-6	5-2	5-1
	Center distance A[m, n]		1-6	1-11	1-2	1-3	1-0	0-10	0-8	STL	STL
20[i]	Span without stirrups[k, l]		4-10	5-5	4-5	4-7	4-3	4-0	3-11	3-7	3-7
	1-#4	40,000	7-0	8-1	6-3	6-5	5-10	5-3	4-11	4-1	3-11
		60,000	8-7	9-10	7-7	7-10	7-1	6-5	6-0	4-11	4-10
	1-#5	40,000	8-9	10-1	7-9	8-0	7-3	6-6	6-1	5-1	4-11
		60,000	10-8	12-3	9-6	9-10	8-10	8-0	7-5	6-2	6-0
	2-#4	40,000	9-10	11-4	8-9	9-1	8-2	7-4	6-10	5-8	5-7
	1-#6	60,000	12-0	13-10	10-8	11-0	9-11	9-0	8-4	6-8	6-6
	2-#5	40,000	12-3	14-1	10-10	11-3	10-2	8-11	8-1	6-6	6-4
		60,000	14-0	17-6	11-8	12-3	10-6	9-1	8-4	6-8	6-6
	Center distance A[m, n]		1-10	2-5	1-5	1-7	1-3	1-0	0-11	STL	STL

(continued)

TABLE R608.8(7)—continued
MAXIMUM ALLOWABLE CLEAR SPANS FOR 8-INCH-THICK WAFFLE-GRID LINTELS IN LOAD-BEARING WALLS[a, b, c, d, e, f, o]
MAXIMUM ROOF CLEAR SPAN 40 FEET AND MAXIMUM FLOOR CLEAR SPAN 32 FEET

LINTEL DEPTH, D[g] (inches)	NUMBER OF BARS AND BAR SIZE IN TOP AND BOTTOM OF LINTEL	STEEL YIELD STRENGTH[h], f_y (psi)	DESIGN LOADING CONDITION DETERMINED FROM Table R608.8(1)								
			1	2		3		4		5	
			Maximum ground snow load (psf)								
			—	30	70	30	70	30	70	30	70
			Maximum clear span of lintel (feet-inches)								
24[j]	Span without stirrups[k, l]		5-9	6-7	5-5	5-6	5-2	4-11	4-9	4-5	4-4
	1-#4	40,000	7-6	8-10	6-10	7-1	6-5	5-9	5-5	4-6	4-4
		60,000	9-2	10-9	8-4	8-8	7-10	7-1	6-7	5-6	5-4
	1-#5	40,000	9-5	11-0	8-6	8-10	8-0	7-2	6-8	5-7	5-5
		60,000	11-5	13-5	10-5	10-9	9-9	8-9	8-2	6-10	6-8
	2-#4	40,000	10-7	12-5	9-8	10-0	9-0	8-1	7-7	6-3	6-2
	1-#6	60,000	12-11	15-2	11-9	12-2	11-0	9-11	9-3	7-8	7-6
	2-#5	40,000	13-2	15-6	12-0	12-5	11-2	9-11	9-2	7-5	7-3
		60,000	16-3	21-0	14-1	14-10	12-9	11-1	10-1	8-1	7-11
	2-#6	40,000	14-4	18-5	12-6	13-2	11-5	9-11	9-2	7-5	7-3
	Center distance A[m, n]		2-1	2-11	1-9	1-10	1-6	1-3	1-1	STL	STL

For SI: 1 inch = 25.4 mm, 1 foot = 304.8 mm, 1 pound per square inch = 6.895 kPa, 1 pound per square foot = 0.0479 kPa, Grade 40 = 280 MPa, Grade 60 = 420 MPa.

a. Where lintels are formed with waffle-grid forms, form material shall be removed, if necessary, to create top and bottom flanges of the lintel that are not less than 3 inches in depth (in the vertical direction), are not less than 5 inches in width for 6-inch-nominal waffle-grid forms and not less than 7 inches in width for 8-inch-nominal waffle-grid forms. See Figure R608.8(3). Flat-form lintels shall be permitted in lieu of waffle-grid lintels. See Tables R608.8(2) through R608.8(5).

b. See Table R608.3 for tolerances permitted from nominal thicknesses and minimum dimensions and spacing of cores.

c. Table values are based on concrete with a minimum specified compressive strength of 2,500 psi. See Notes l and n. Table values are based on uniform loading. See Section R608.8.2 for lintels supporting concentrated loads.

d. Deflection criterion is $L/240$, where L is the clear span of the lintel in inches, or $^1/_2$ inch, whichever is less.

e. Linear interpolation is permitted between ground snow loads.

f. STL indicates stirrups required throughout lintel.

g. Lintel depth, D, is permitted to include the available height of wall located directly above the lintel, provided that the increased lintel depth spans the entire length of the lintel.

h. Stirrups shall be fabricated from reinforcing bars with the same yield strength as that used for the main longitudinal reinforcement.

i. Lintels less than 24 inches in depth with stirrups shall be formed from flat-wall forms [see Tables R608.8(2) through R608.8(5)], or, if necessary, form material shall be removed from waffle-grid forms so as to provide the required cover for stirrups. Allowable spans for lintels formed with flat-wall forms shall be determined from Tables R608.8(2) through R608.8(5).

j. Where stirrups are required for 24-inch-deep lintels, the spacing shall not exceed 12 inches on center.

k. Allowable clear span without stirrups applicable to all lintels of the same depth, D. Top and bottom reinforcement for lintels without stirrups shall be not less than the least amount of reinforcement required for a lintel of the same depth and loading condition with stirrups. All other spans require stirrups spaced at not more than $d/2$.

l. Where concrete with a minimum specified compressive strength of 3,000 psi is used, clear spans for lintels without stirrups shall be permitted to be multiplied by 1.05. If the increased span exceeds the allowable clear span for a lintel of the same depth and loading condition with stirrups, the top and bottom reinforcement shall be equal to or greater than that required for a lintel of the same depth and loading condition that has an allowable clear span that is equal to or greater than that of the lintel without stirrups that has been increased.

m. Center distance, A, is the center portion of the span where stirrups are not required. This is applicable to all longitudinal bar sizes and steel yield strengths.

n. Where concrete with a minimum specified compressive strength of 3,000 psi is used, center distance, A, shall be permitted to be multiplied by 1.10.

o. The maximum clear opening width between two solid wall segments shall be 18 feet. See Section R608.7.2.1. Lintel spans in the table greater than 18 feet are shown for interpolation and information only.

TABLE R608.8(8)
MAXIMUM ALLOWABLE CLEAR SPANS FOR 6-INCH-THICK SCREEN-GRID LINTELS IN LOAD-BEARING WALLS[a, b, c, d, e, f, p]
ROOF CLEAR SPAN 40 FEET AND FLOOR CLEAR SPAN 32 FEET

LINTEL DEPTH, D[g] (inches)	NUMBER OF BARS AND BAR SIZE IN TOP AND BOTTOM OF LINTEL	STEEL YIELD STRENGTH[h], f_y (psi)	DESIGN LOADING CONDITION DETERMINED FROM Table R608.8(1)								
			1	2		3		4		5	
			Maximum ground snow load (psf)								
			—	30	70	30	70	30	70	30	70
			Maximum clear span of lintel (feet-inches)								
12[i, j]	Span without stirrups		2-9	2-11	2-4	2-5	2-3	2-3	2-2	2-0	2-0
16[i, j]	Span without stirrups		3-9	4-0	3-4	3-5	3-2	3-1	3-0	2-9	2-9
20[i, j]	Span without stirrups		4-9	5-1	4-3	4-4	4-1	4-0	3-10	3-7	3-7
24[k]	Span without stirrups[l, m]		5-8	6-3	5-2	5-3	5-0	4-10	4-8	4-4	4-4
	1-#4	40,000	7-11	9-0	6-11	7-2	6-5	6-1	5-8	4-9	4-7
		60,000	9-9	11-0	8-5	8-9	7-10	7-5	6-10	5-9	5-7
	1-#5	40,000	9-11	11-2	8-7	8-11	8-0	7-7	7-0	5-11	5-9
		60,000	12-1	13-8	10-6	10-10	9-9	9-3	8-6	7-2	7-0
	2-#4	40,000	11-2	12-8	9-9	10-1	9-1	8-7	7-11	6-8	6-6
	1-#6	60,000	15-7	17-7	12-8	13-4	11-6	10-8	9-8	7-11	7-8
	2-#5	40,000	14-11	18-0	12-2	12-10	11-1	10-3	9-4	7-8	7-5
		60,000	DR	DR	DR	DR	DR	DR	DR	DR	DR
	Center distance A[n, o]		2-0	2-6	1-6	1-7	1-4	1-2	1-0	STL	STL

For SI: 1 inch = 25.4 mm, 1 foot = 304.8 mm, 1 pound per square inch = 6.895 kPa, 1 pound per square foot = 0.0479 kPa, Grade 40 = 280 MPa, Grade 60 = 420 MPa.

a. Where lintels are formed with screen-grid forms, form material shall be removed if necessary to create top and bottom flanges of the lintel that are not less than 5 inches in width and not less than 2.5 inches in depth (in the vertical direction). See Figure R608.8(4). Flat-form lintels shall be permitted in lieu of screen-grid lintels. See Tables R608.8(2) through R608.8(5).

b. See Table R608.3 for tolerances permitted from nominal thickness and minimum dimensions and spacings of cores.

c. Table values are based on concrete with a minimum specified compressive strength of 2,500 psi. See Notes m and o. Table values are based on uniform loading. See Section R608.7.2.1 for lintels supporting concentrated loads.

d. Deflection criterion is $L/240$, where L is the clear span of the lintel in inches, or $^1/_2$ inch, whichever is less.

e. Linear interpolation is permitted between ground snow loads.

f. DR indicates design required. STL indicates stirrups required throughout lintel.

g. Lintel depth, D, is permitted to include the available height of wall located directly above the lintel, provided that the increased lintel depth spans the entire length of the lintel.

h. Stirrups shall be fabricated from reinforcing bars with the same yield strength as that used for the main longitudinal reinforcement.

i. Stirrups are not required for lintels less than 24 inches in depth fabricated from screen-grid forms. Top and bottom reinforcement shall consist of a No. 4 bar having a yield strength of 40,000 psi or 60,000 psi.

j. Lintels between 12 and 24 inches in depth with stirrups shall be formed from flat-wall forms [see Tables R608.8(2) through R608.8(5)], or form material shall be removed from screen-grid forms to provide a concrete section comparable to that required for a flat wall. Allowable spans for flat lintels with stirrups shall be determined from Tables R608.8(2) through R608.8(5).

k. Where stirrups are required for 24-inch-deep lintels, the spacing shall not exceed 12 inches on center.

l. Allowable clear span without stirrups applicable to all lintels of the same depth, D. Top and bottom reinforcement for lintels without stirrups shall be not less than the least amount of reinforcement required for a lintel of the same depth and loading condition with stirrups. All other spans require stirrups spaced at not more than 12 inches.

m. Where concrete with a minimum specified compressive strength of 3,000 psi is used, clear spans for lintels without stirrups shall be permitted to be multiplied by 1.05. If the increased span exceeds the allowable clear span for a lintel of the same depth and loading condition with stirrups, the top and bottom reinforcement shall be equal to or greater than that required for a lintel of the same depth and loading condition that has an allowable clear span that is equal to or greater than that of the lintel without stirrups that has been increased.

n. Center distance, A, is the center portion of the span where stirrups are not required. This is applicable to all longitudinal bar sizes and steel yield strengths.

o. Where concrete with a minimum specified compressive strength of 3,000 psi is used, center distance, A, shall be permitted to be multiplied by 1.10.

p. The maximum clear opening width between two solid wall segments shall be 18 feet. See Section R608.7.2.1. Lintel spans in the table greater than 18 feet are shown for interpolation and information only.

TABLE R608.8(9)
MAXIMUM ALLOWABLE CLEAR SPANS FOR FLAT LINTELS WITHOUT STIRRUPS IN NONLOAD-BEARING WALLS[a, b, c, d, e, g]

LINTEL DEPTH, D (inches)	NUMBER OF BARS AND BAR SIZE	STEEL YIELD STRENGTH, f_y (psi)	NOMINAL WALL THICKNESS (inches)							
			4		6		8		10	
			Lintel Supporting							
			Concrete Wall	Light-frame Gable	Concrete Wall	Light-frame Gable	Concrete Wall	Light-frame Gable	Concrete Wall	Light-frame Gable
			Maximum Clear Span of Lintel (feet-inches)							
8	1-#4	40,000	10-11	11-5	9-7	11-2	7-10	9-5	7-3	9-2
		60,000	12-5	11-7	10-11	13-5	9-11	13-2	9-3	12-10
	1-#5	40,000	12-7	11-7	11-1	13-8	10-1	13-5	9-4	13-1
		60,000	DR	DR	12-7	16-4	11-6	14-7	10-9	14-6
	2-#4 1-#6	40,000	DR	DR	12-0	15-3	10-11	15-0	10-2	14-8
		60,000	DR	DR	DR	DR	12-2	15-3	11-7	15-3
	2-#5	40,000	DR	DR	DR	DR	12-7	16-7	11-9	16-7
		60,000	DR	DR	DR	DR	DR	DR	13-3	16-7
	2-#6	40,000	DR	DR	DR	DR	DR	DR	13-2	17-8
		60,000	DR	DR	DR	DR	DR	DR	DR	DR
12	1-#4	40,000	11-5	9-10	10-6	12-0	9-6	11-6	8-9	11-1
		60,000	11-5	9-10	11-8	13-3	10-11	14-0	10-1	13-6
	1-#5	40,000	11-5	9-10	11-8	13-3	11-1	14-4	10-3	13-9
		60,000	11-5	9-10	11-8	13-3	11-10	16-0	11-9	16-9
	2-#4 1-#6	40,000	DR	DR	11-8	13-3	11-10	16-0	11-2	15-6
		60,000	DR	DR	11-8	13-3	11-10	16-0	11-11	18-4
	2-#5	40,000	DR	DR	11-8	13-3	11-10	16-0	11-11	18-4
		60,000	DR	DR	11-8	13-3	11-10	16-0	11-11	18-4
16	1-#4	40,000	13-6	13-0	11-10	13-8	10-7	12-11	9-11	12-4
		60,000	13-6	13-0	13-8	16-7	12-4	15-9	11-5	15-0
	1-#5	40,000	13-6	13-0	13-10	17-0	12-6	16-1	11-7	15-4
		60,000	13-6	13-0	13-10	17-1	14-0	19-7	13-4	18-8
	2-#4 1-#6	40,000	13-6	13-0	13-10	17-1	13-8	18-2	12-8	17-4
		60,000	13-6	13-0	13-10	17-1	14-0	20-3	14-1	—
	2-#5	40,000	13-6	13-0	13-10	17-1	14-0	20-3	14-1	—
		60,000	DR	DR	13-10	17-1	14-0	20-3	14-1	—
20	1-#4	40,000	14-11	15-10	13-0	14-10	11-9	13-11	10-10	13-2
		60,000	15-3	15-10	14-11	18-1	13-6	17-0	12-6	16-2
	1-#5	40,000	15-3	15-10	15-2	18-6	13-9	17-5	12-8	16-6
		60,000	15-3	15-10	15-8	20-5	15-9	—	14-7	20-1
	2-#4 1-#6	40,000	15-3	15-10	15-8	20-5	14-11	—	13-10	—
		60,000	15-3	15-10	15-8	20-5	15-10	—	15-11	—
	2-#5	40,000	15-3	15-10	15-8	20-5	15-10	—	15-11	—
		60,000	15-3	15-10	15-8	20-5	15-10	—	15-11	—

(continued)

TABLE R608.8(9)—continued
MAXIMUM ALLOWABLE CLEAR SPANS FOR FLAT LINTELS WITHOUT STIRRUPS IN NONLOAD-BEARING WALLS[a, b, c, d, e, g]

LINTEL DEPTH, D[f] (inches)	NUMBER OF BARS AND BAR SIZE	STEEL YIELD STRENGTH, f_y (psi)	NOMINAL WALL THICKNESS (inches)							
			4		6		8		10	
			Lintel Supporting							
			Concrete Wall	Light-frame Gable	Concrete Wall	Light-frame Gable	Concrete Wall	Light-frame Gable	Concrete Wall	Light-frame Gable
			Maximum Clear Span of Lintel (feet-inches)							
24	1-#4	40,000	16-1	17-1	13-11	15-10	12-7	14-9	11-8	13-10
		60,000	16-11	18-5	16-1	19-3	14-6	18-0	13-5	17-0
	1-#5	40,000	16-11	18-5	16-3	19-8	14-9	18-5	13-8	17-4
		60,000	16-11	18-5	17-4	—	17-0	—	15-8	—
	2-#4	40,000	16-11	18-5	17-4	—	16-1	—	14-10	—
	1-#6	60,000	16-11	18-5	17-4	—	17-6	—	17-1	—
	2-#5	40,000	16-11	18-5	17-4	—	17-6	—	17-4	—
		60,000	16-11	18-5	17-4	—	17-6	—	17-8	—

For SI: 1 inch = 25.4 mm, 1 foot = 304.8 mm, 1 pound per square inch = 6.895 kPa, Grade 40 = 280 MPa, Grade 60 = 420 MPa.

DR = Design Required.

a. See Table R608.3 for tolerances permitted from nominal thickness.

b. Table values are based on concrete with a minimum specified compressive strength of 2,500 psi. See Note e.

c. Deflection criterion is $L/240$, where L is the clear span of the lintel in inches, or $^1/_2$ inch, whichever is less.

d. Linear interpolation between lintels depths, D, is permitted provided the two cells being used to interpolate are shaded.

e. Where concrete with a minimum specified compressive strength of 3,000 psi is used, spans in cells that are shaded shall be permitted to be multiplied by 1.05.

f. Lintel depth, D, is permitted to include the available height of wall located directly above the lintel, provided that the increased lintel depth spans the entire length of the lintel.

g. The maximum clear opening width between two solid wall segments shall be 18 feet. See Section R608.7.2.1. Lintel spans in the table greater than 18 feet are shown for interpolation and information purposes only.

TABLE R608.8(10)
MAXIMUM ALLOWABLE CLEAR SPANS FOR WAFFLE-GRID AND SCREEN-GRID LINTELS WITHOUT STIRRUPS IN NONLOAD-BEARING WALLS[c, d, e, f, g]

LINTEL DEPTH[h], D (inches)	FORM TYPE AND NOMINAL WALL THICKNESS (inches)					
	6-inch Waffle-grid[a]		8-inch Waffle-grid[a]		6-inch Screen-grid[b]	
	Lintel supporting					
	Concrete Wall	Light-frame Gable	Concrete Wall	Light-frame Gable	Concrete Wall	Light-frame Gable
	Maximum Clear Span of Lintel (feet-inches)					
8	10-3	8-8	8-8	8-3	—	—
12	9-2	7-6	7-10	7-1	8-8	6-9
16	10-11	10-0	9-4	9-3	—	—
20	12-5	12-2	10-7	11-2	—	—
24	13-9	14-2	11-10	12-11	13-0	12-9

For SI: 1 inch = 25.4 mm, 1 foot = 304.8 mm, Grade 40 = 280 MPa, Grade 60 = 420 MPa.

a. Where lintels are formed with waffle-grid forms, form material shall be removed, if necessary, to create top and bottom flanges of the lintel that are not less than 3 inches in depth (in the vertical direction), are not less than 5 inches in width for 6-inch waffle-grid forms and not less than 7 inches in width for 8-inch waffle-grid forms. See Figure R608.8(3). Flat-form lintels shall be permitted in lieu of waffle-grid lintels. See Tables R608.8(2) through R608.8(5).

b. Where lintels are formed with screen-grid forms, form material shall be removed if necessary to create top and bottom flanges of the lintel that are not less than 5 inches in width and not less than 2.5 inches in depth (in the vertical direction). See Figure R608.8(4). Flat-form lintels shall be permitted in lieu of screen-grid lintels. See Tables R608.8(2) through R608.8(5).

c. See Table R608.3 for tolerances permitted from nominal thickness and minimum dimensions and spacing of cores.

d. Table values are based on concrete with a minimum specified compressive strength of 2,500 psi. See Note g.

e. Deflection criterion is $L/240$, where L is the clear span of the lintel in inches, or $^1/_2$ inch, whichever is less.

f. Top and bottom reinforcement shall consist of a No. 4 bar having a minimum yield strength of 40,000 psi.

g. Where concrete with a minimum specified compressive strength of 3,000 psi is used, spans in shaded cells shall be permitted to be multiplied by 1.05.

h. Lintel depth, D, is permitted to include the available height of wall located directly above the lintel, provided that the increased lintel depth spans the entire length of the lintel.

R608.8.2.1 Lintels designed for gravity load-bearing conditions. Where a lintel will be subjected to gravity load conditions 1 through 5 of Table R608.8(1), the clear span of the lintel shall not exceed that permitted by Tables R608.8(2) through R608.8(8). The maximum clear span of lintels with and without stirrups in flat walls shall be determined in accordance with Tables R608.8(2) through R608.8(5), and constructed in accordance with Figure R608.8(2). The maximum clear span of lintels with and without stirrups in waffle-grid walls shall be determined in accordance with Tables R608.8(6) and R608.8(7), and constructed in accordance with Figure R608.8(3). The maximum clear span of lintels with and without stirrups in screen-grid walls shall be determined in accordance with Table R608.8(8), and constructed in accordance with Figure R608.8(4).

Where required by the applicable table, No. 3 stirrups shall be installed in lintels at a maximum spacing of *d/2* where *d* equals the depth of the lintel, *D*, less the cover of the concrete as shown in Figures R608.8(2) through R608.8(4). The smaller value of *d* computed for the top and bottom bar shall be used to determine the maximum stirrup spacing. Where stirrups are required in a lintel with a single bar or two bundled bars in the top and bottom, they shall be fabricated like the letter "c" or "s" with 135-degree (2.36 rad) standard hooks at each end that comply with Section R608.5.4.5 and Figure R608.5.4(3) and installed as shown in Figures R608.8(2) through R608.8(4). Where two bars are required in the top and bottom of the lintel and the bars are not bundled, the bars shall be separated by not less than 1 inch (25 mm). The free end of the stirrups shall be fabricated with 90- or 135-degree (1.57 or 2.36 rad) standard hooks that comply with Section R608.5.4.5 and Figure R608.5.4(3) and installed as shown in Figures R608.8(2) and R608.8(3). For flat, waffle-grid and screen-grid lintels, stirrups are not required in the center distance, *A*, portion of spans in accordance with Figure R608.8(1) and Tables R608.8(2) through R608.8(8). See Section R608.8.2.2, Item 5, for requirement for stirrups through out lintels with bundled bars.

R608.8.2.2 Bundled bars in lintels. It is permitted to bundle two bars in contact with each other in lintels if all of the following are observed:

1. Bars equal to or less than No. 6 are bundled.
2. Where the wall thickness is not sufficient to provide not less than 3 inches (76 mm) of clear space beside bars (total on both sides) oriented horizontally in a bundle, the bundled bars shall be oriented in a vertical plane.
3. Where vertically oriented bundled bars terminate with standard hooks to develop the bars in tension beyond the support (see Section R608.5.4.4), the hook extensions shall be staggered to provide not less than 1 inch (25 mm) clear spacing between the extensions.
4. Bundled bars shall not be lap spliced within the lintel span and the length on each end of the lintel that is required to develop the bars in tension.
5. Bundled bars shall be enclosed within stirrups throughout the length of the lintel. Stirrups and the installation thereof shall comply with Section R608.8.2.1.

R608.8.2.3 Lintels without stirrups designed for nonload-bearing conditions. The maximum clear span of lintels without stirrups designed for nonload-bearing conditions of Table R608.8(1) shall be determined in accordance with this section. The maximum clear span of lintels without stirrups in flat walls shall be determined in accordance with Table R608.8(9), and the maximum clear span of lintels without stirrups in walls of waffle-grid or screen-grid construction shall be determined in accordance with Table R608.8(10).

R608.9 Requirements for connections—general. Concrete walls shall be connected to footings, floors, ceilings and roofs in accordance with this section.

R608.9.1 Connections between concrete walls and light-frame floor, ceiling and roof systems. Connections between concrete walls and light-frame floor, ceiling and roof systems using the prescriptive details of Figures R608.9(1) through R608.9(12) shall comply with this section and Sections R608.9.2 and R608.9.3.

R608.9.1.1 Anchor bolts. Anchor bolts used to connect light-frame floor, ceiling and roof systems to concrete walls in accordance with Figures R608.9(1) through R608.9(12) shall have heads, or shall be rods with threads on both ends with a hex or square nut on the end embedded in the concrete. Bolts and threaded rods shall comply with Section R608.5.2.2. Anchor bolts with J- or L-hooks shall not be used where the connection details in these figures are used.

R608.9.1.2 Removal of stay-in-place form material at bolts. Holes in stay-in-place forms for installing bolts for attaching face-mounted wood ledger boards to the wall shall be not less than 4 inches (102 mm) in diameter for forms not greater than $1^1/_2$ inches (38 mm) in thickness, and increased 1 inch (25 mm) in diameter for each $^1/_2$-inch (12.7 mm) increase in form thickness. Holes in stay-in-place forms for installing bolts for attaching face-mounted cold-formed steel tracks to the wall shall be not less than 4 inches (102 mm) square. The wood ledger board or steel track shall be in direct contact with the concrete at each bolt location.

Exception: A vapor retarder or other material less than or equal to $^1/_{16}$ inch (1.6 mm) in thickness is permitted to be installed between the wood ledger or cold-formed track and the concrete.

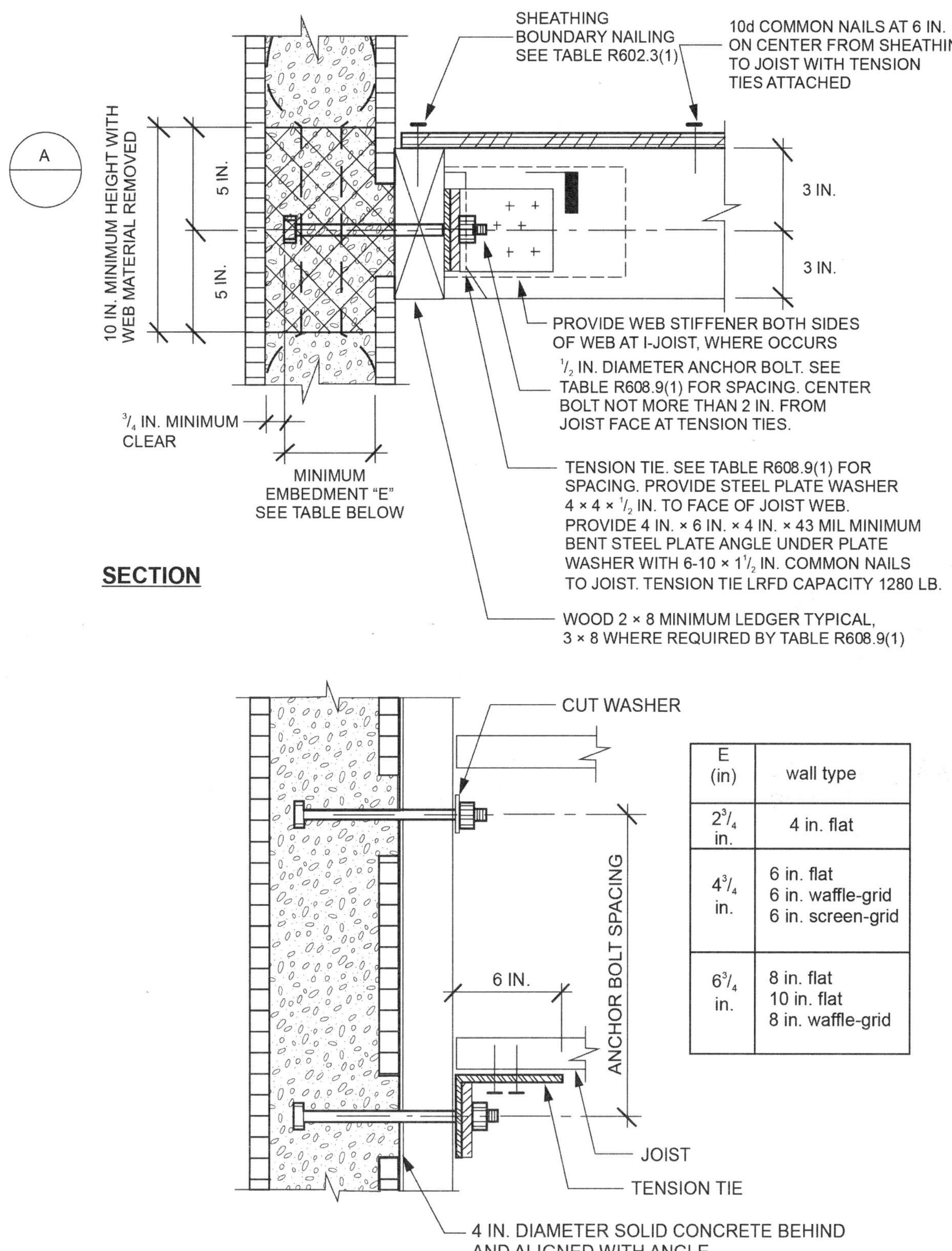

E (in)	wall type
$2^3/_4$ in.	4 in. flat
$4^3/_4$ in.	6 in. flat 6 in. waffle-grid 6 in. screen-grid
$6^3/_4$ in.	8 in. flat 10 in. flat 8 in. waffle-grid

For SI: 1 mil = 0.0254 mm, 1 inch = 25.4 mm, 1 pound-force = 4.448 N.

FIGURE R608.9(1)
WOOD-FRAMED FLOOR TO SIDE OF CONCRETE WALL, FRAMING PERPENDICULAR

TABLE R608.9(1)
WOOD-FRAMED FLOOR TO SIDE OF CONCRETE WALL, FRAMING PERPENDICULAR[a, b]

ANCHOR BOLT SPACING (inches)	TENSION TIE SPACING (inches)	BASIC WIND SPEED (mph)					
		115B	120B	130B	140B	150B	160B
		—	—	110C	119C	127C	136C
		—	—	—	110D	117D	125D
12	12						
12	24						
12	36						
12	48						
16	16						
16	32						
16	48						
19.2	19.2						
19.2	38.4						

For SI:1 inch = 25.4 mm, 1 mile per hour = 0.447 m/s.

a. This table is for use with the detail in Figure R608.9(1). Use of this detail is permitted where a cell is not shaded and prohibited where shaded.

b. Wall design per other provisions of Section R608 is required.

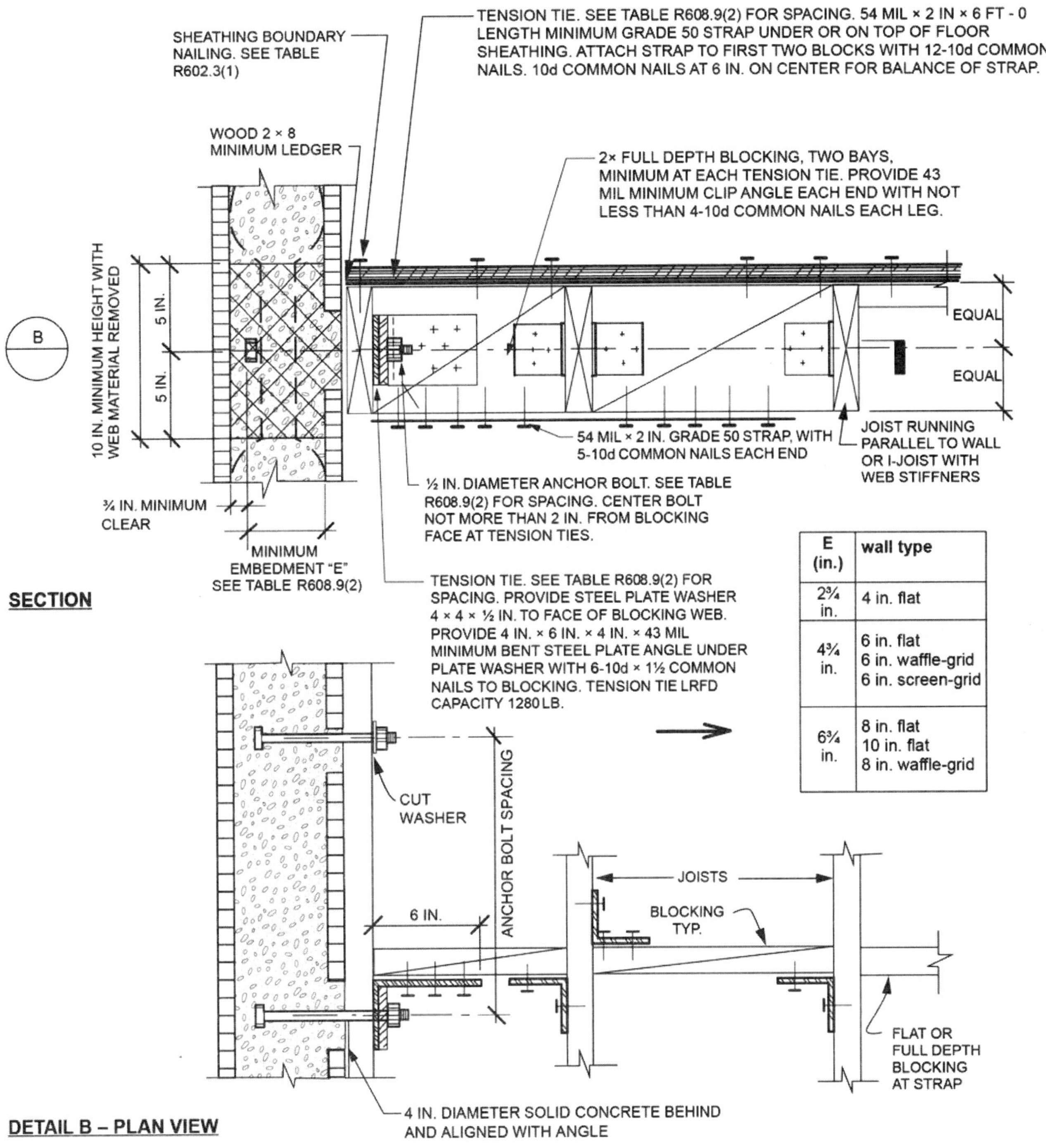

E (in.)	wall type
2¾ in.	4 in. flat
4¾ in.	6 in. flat 6 in. waffle-grid 6 in. screen-grid
6¾ in.	8 in. flat 10 in. flat 8 in. waffle-grid

For SI: 1 mil = 0.0254 mm, 1 inch = 25.4 mm, 1 foot = 304.8 mm, 1 pound-force = 4.448 N.

FIGURE R608.9(2)
WOOD-FRAMED FLOOR TO SIDE OF CONCRETE WALL, FRAMING PARALLEL

TABLE R608.9(2)
WOOD-FRAMED FLOOR TO SIDE OF CONCRETE WALL, FRAMING PARALLEL[a, b]

ANCHOR BOLT SPACING (inches)	TENSION TIE SPACING (inches)	BASIC WIND SPEED (mph) AND WIND EXPOSURE CATEGORY					
		115B	120B	130B	140B	150B	160B
		—	—	110C	119C	127C	136C
		—	—	—	110D	117D	125D
12	12						
12	24						
12	36						
12	48						
16	16						
16	32						
16	48						
19.2	19.2						
19.2	38.4						
24	24						
24	48						

For SI: 1 inch = 25.4 mm, 1 mile per hour = 0.447 m/s.

a. This table is for use with the detail in Figure R608.9(2). Use of this detail is permitted where a cell is not shaded and prohibited where shaded.

b. Wall design per other provisions of Section R608 is required.

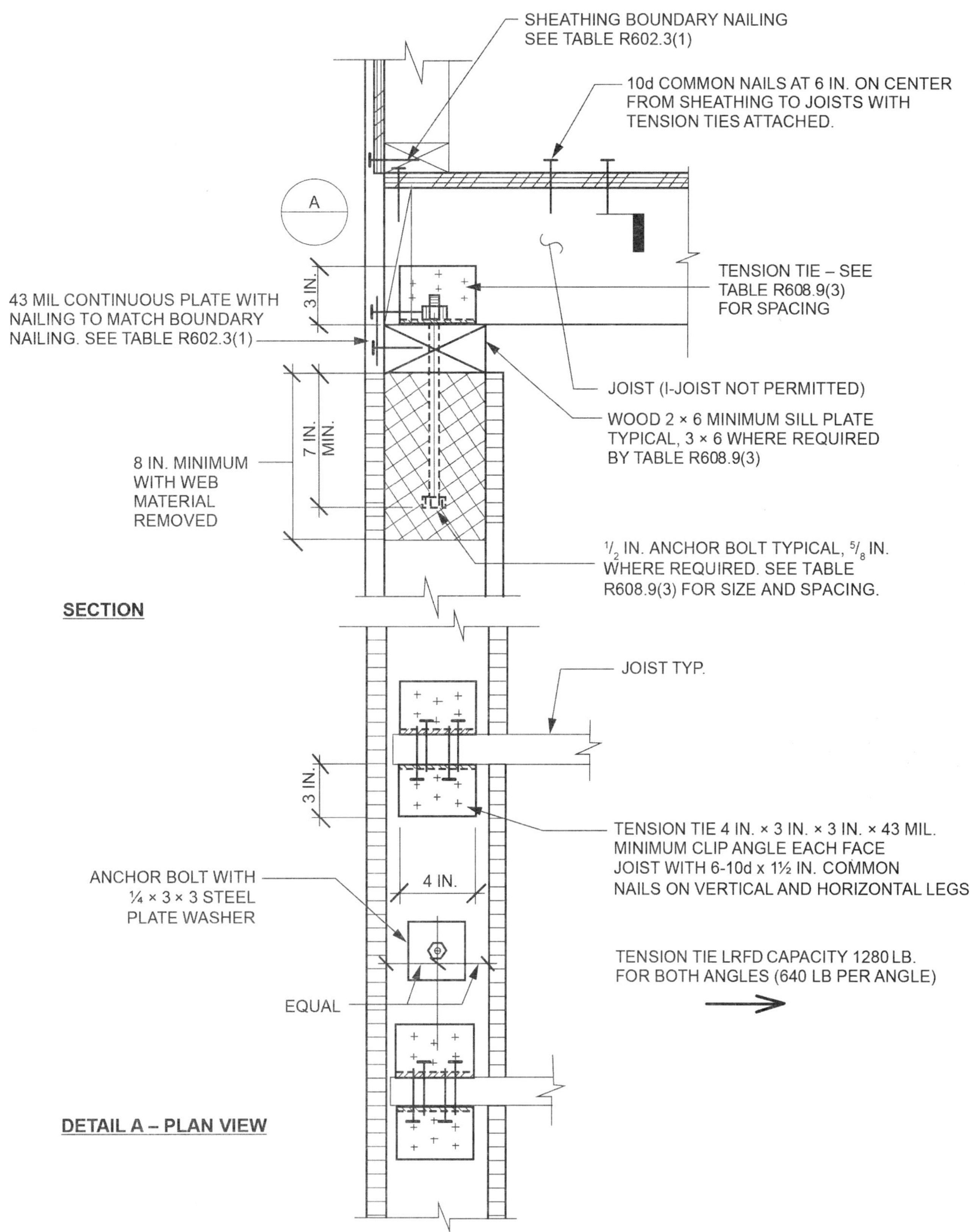

For SI: 1 mil = 0.0254 mm, 1 inch = 25.4 mm, 1 pound-force = 4.448 N.

FIGURE R608.9(3)
WOOD-FRAMED FLOOR TO TOP OF CONCRETE WALL, FRAMING PERPENDICULAR

TABLE R608.9(3)
WOOD-FRAMED FLOOR TO TOP OF CONCRETE WALL, FRAMING PERPENDICULAR[a, b, c, d, e]

ANCHOR BOLT SPACING (inches)	TENSION TIE SPACING (inches)	BASIC WIND SPEED (mph) AND WIND EXPOSURE CATEGORY					
		115B	120B	130B	140B	150B	160B
		—	—	110C	119C	127C	136C
		—	—	—	110D	117D	125D
12	12						6
12	24					6	6
12	36					6	6
12	48				6	6	6
16	16					6	6A
16	32				6	6	6A
16	48			6	6	6	6A
19.2	19.2				6A	6A	6B
19.2	38.4			6	6A	6A	6B
24	24			6A	6B	6B	6B
24	48		6	6A	6B	6B	8B

For SI: 1 inch = 25.4 mm, 1 mile per hour = 0.447 m/s.

a. This table is for use with the detail in Figure R608.9(3). Use of this detail is permitted where cell is not shaded.

b. Wall design per other provisions in Section R608 is required.

c. For wind design, minimum 4-inch-nominal wall is permitted in unshaded cells that do not contain a number.

d. Numbers 6 and 8 indicate minimum permitted nominal wall thickness in inches necessary to develop required strength (capacity) of connection. As a minimum, this nominal thickness shall occur in the portion of the wall indicated by the cross hatching in Figure R608.9(3). For the remainder of the wall, see Note b.

e. Letter "A" indicates that a minimum nominal 3 × 6 sill plate is required. Letter "B" indicates that a $^5/_8$-inch-diameter anchor bolt and a minimum nominal 3 × 6 sill plate are required.

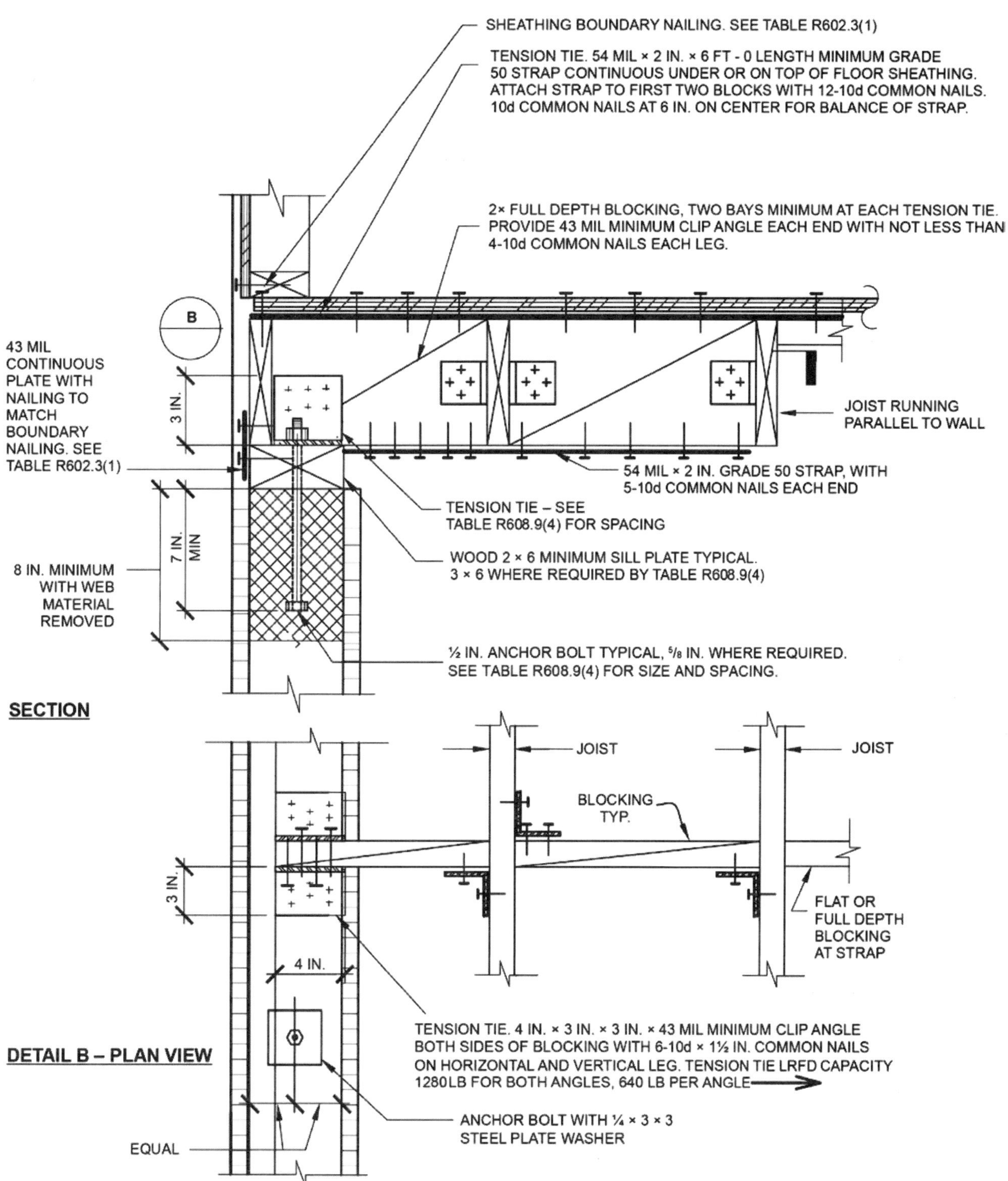

For SI: 1 mil = 0.0254 mm, 1 inch = 25.4 mm, 1 foot = 304.8 mm, 1 pound-force = 4.448 N.

FIGURE R608.9(4)
WOOD-FRAMED FLOOR TO TOP OF CONCRETE WALL, FRAMING PARALLEL

TABLE R608.9(4)
WOOD-FRAMED FLOOR TO TOP OF CONCRETE WALL, FRAMING PARALLEL[a, b, c, d, e]

ANCHOR BOLT SPACING (inches)	TENSION TIE SPACING (inches)	BASIC WIND SPEED (mph) AND WIND EXPOSURE CATEGORY					
		115B	120B	130B	140B	150B	160B
		—	—	110C	119C	127C	136C
		—	—	—	110D	117D	125D
12	12						6
12	24					6	6
12	36					6	6
12	48				6	6	6
16	16					6	6A
16	32				6	6	6A
16	48			6	6	6	6A
19.2	19.2				6A	6A	6B
19.2	38.4			6	6A	6A	6B
24	24			6A	6B	6B	6B
24	48		6	6A	6B	6B	8B

For SI: 1 inch = 25.4 mm, 1 mile per hour = 0.447 m/s.

a. This table is for use with the detail in Figure R608.9(4). Use of this detail is permitted where a cell is not shaded.

b. Wall design per other provisions of Section R608 is required.

c. For wind design, minimum 4-inch-nominal wall is permitted in unshaded cells that do not contain a number.

d. Numbers 6 and 8 indicate minimum permitted nominal wall thickness in inches necessary to develop required strength (capacity) of connection. As a minimum, this nominal thickness shall occur in the portion of the wall indicated by the cross hatching in Figure R608.9(4). For the remainder of the wall, see Note b.

e. Letter "A" indicates that a minimum nominal 3 × 6 sill plate is required. Letter "B" indicates that a $^5/_8$-inch-diameter anchor bolt and a minimum nominal 3 × 6 sill plate are required.

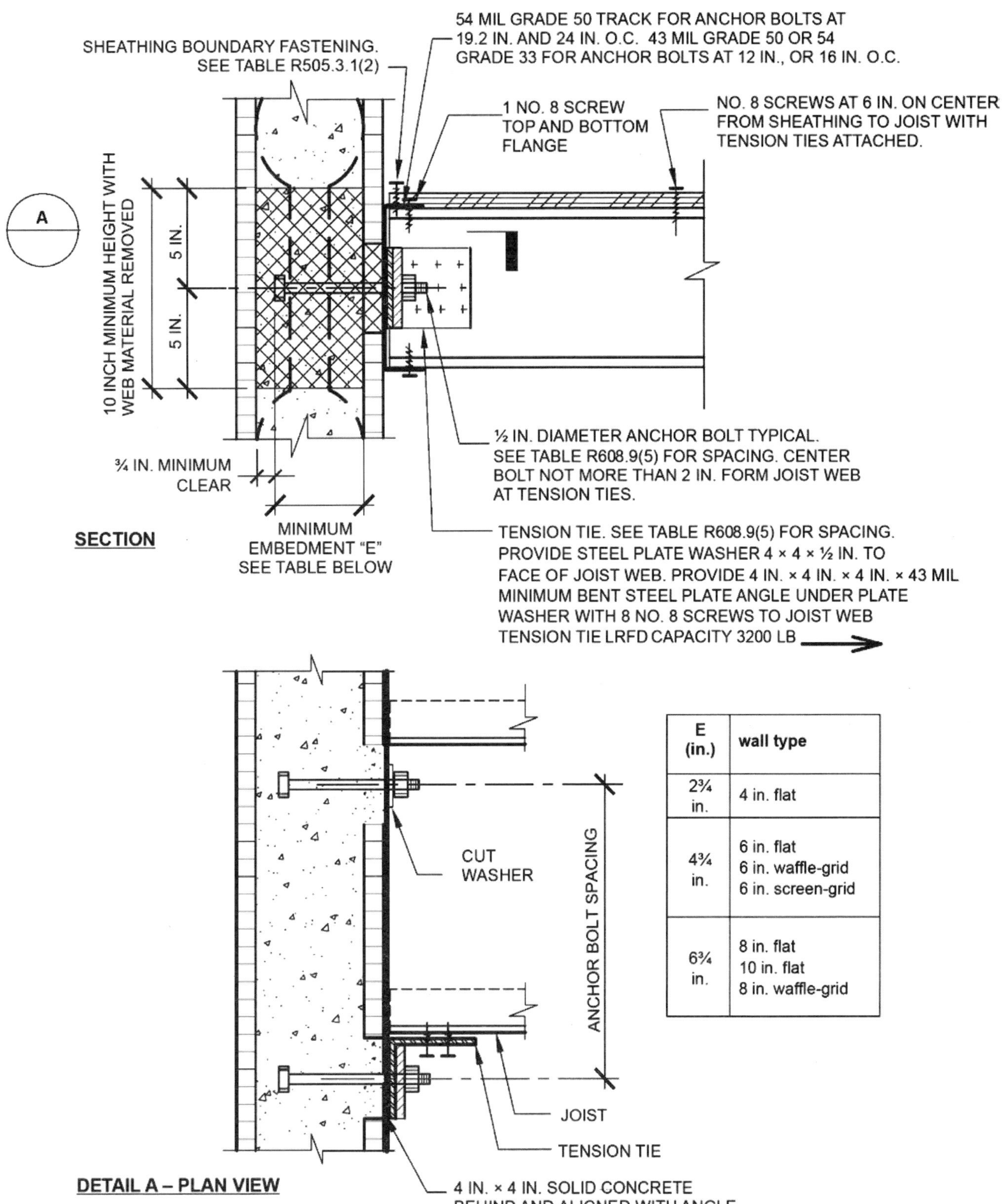

E (in.)	wall type
2¾ in.	4 in. flat
4¾ in.	6 in. flat 6 in. waffle-grid 6 in. screen-grid
6¾ in.	8 in. flat 10 in. flat 8 in. waffle-grid

For SI: 1 mil = 0.0254 mm, 1 inch = 25.4 mm, 1 pound-force = 4.448 N.

FIGURE R608.9(5)
COLD-FORMED STEEL FLOOR TO SIDE OF CONCRETE WALL, FRAMING PERPENDICULAR

TABLE R608.9(5)
COLD-FORMED STEEL-FRAMED FLOOR TO SIDE OF CONCRETE WALL, FRAMING PERPENDICULAR[a, b, c]

ANCHOR BOLT SPACING (inches)	TENSION TIE SPACING (inches)	BASIC WIND SPEED (mph) AND WIND EXPOSURE CATEGORY					
		115B	120B	130B	140B	150B	160B
		—	—	110C	119C	127C	136C
		—	—	—	110D	117D	125D
12	12						
12	24						
12	36						
12	48						
16	16						
16	32						
16	48						
19.2	19.2						
19.2	38.4						
24	24						
24	48						

For SI:1 inch = 25.4 mm, 1 mile per hour = 0.4470 m/s.

a. This table is for use with the detail in Figure R608.9(5). Use of this detail is permitted where a cell is not shaded.

b. Wall design per other provisions of Section R608 is required.

c. For wind design, minimum 4-inch-nominal wall is permitted in unshaded cells that do not contain a number.

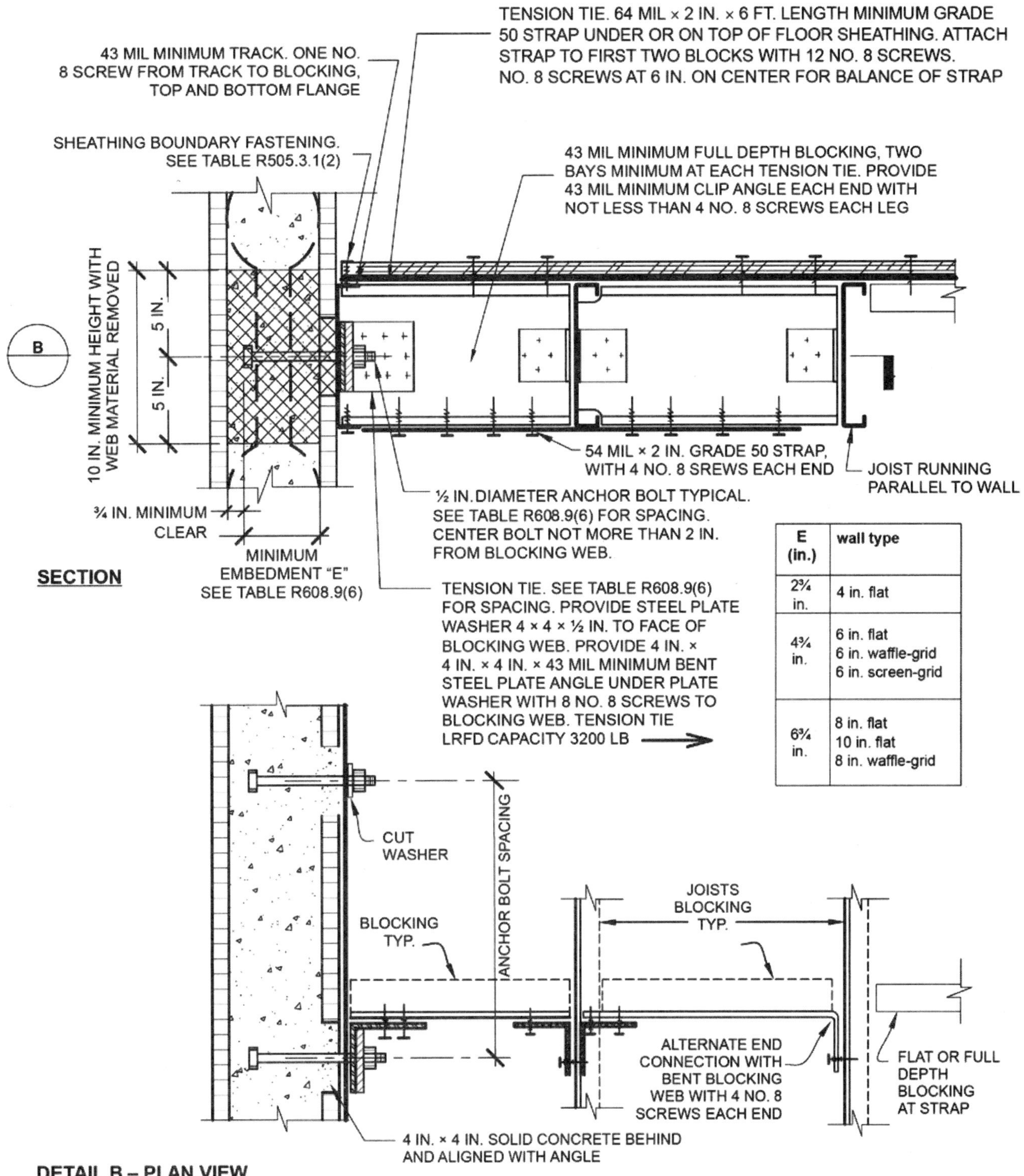

E (in.)	wall type
2¾ in.	4 in. flat
4¾ in.	6 in. flat 6 in. waffle-grid 6 in. screen-grid
6¾ in.	8 in. flat 10 in. flat 8 in. waffle-grid

For SI: 1 mil = 0.0254 mm, 1 inch = 25.4 mm, 1 pound-force = 4.448 N.

FIGURE R608.9(6)
COLD-FORMED STEEL FLOOR TO SIDE OF CONCRETE WALL, FRAMING PARALLEL

TABLE R608.9(6)
COLD-FORMED STEEL-FRAMED FLOOR TO SIDE OF CONCRETE WALL, FRAMING PARALLEL[a, b, c]

ANCHOR BOLT SPACING (inches)	TENSION TIE SPACING (inches)	BASIC WIND SPEED (mph) AND WIND EXPOSURE CATEGORY					
		115B	120B	130B	140B	150B	160B
		—	—	110C	119C	127C	136C
		—	—	—	110D	117D	125D
12	12						
12	24						
12	36						
12	48						
16	16						
16	32						
16	48						
19.2	19.2						
19.2	38.4						
24	24						
24	48						

For SI: 1 inch = 25.4 mm, 1 mile per hour = 0.447 m/s.

a. This table is for use with the detail in Figure R608.9(6). Use of this detail is permitted where a cell is not shaded.

b. Wall design per other provisions of Section R608 is required.

c. For wind design, minimum 4-inch-nominal wall is permitted in unshaded cells that do not contain a number.

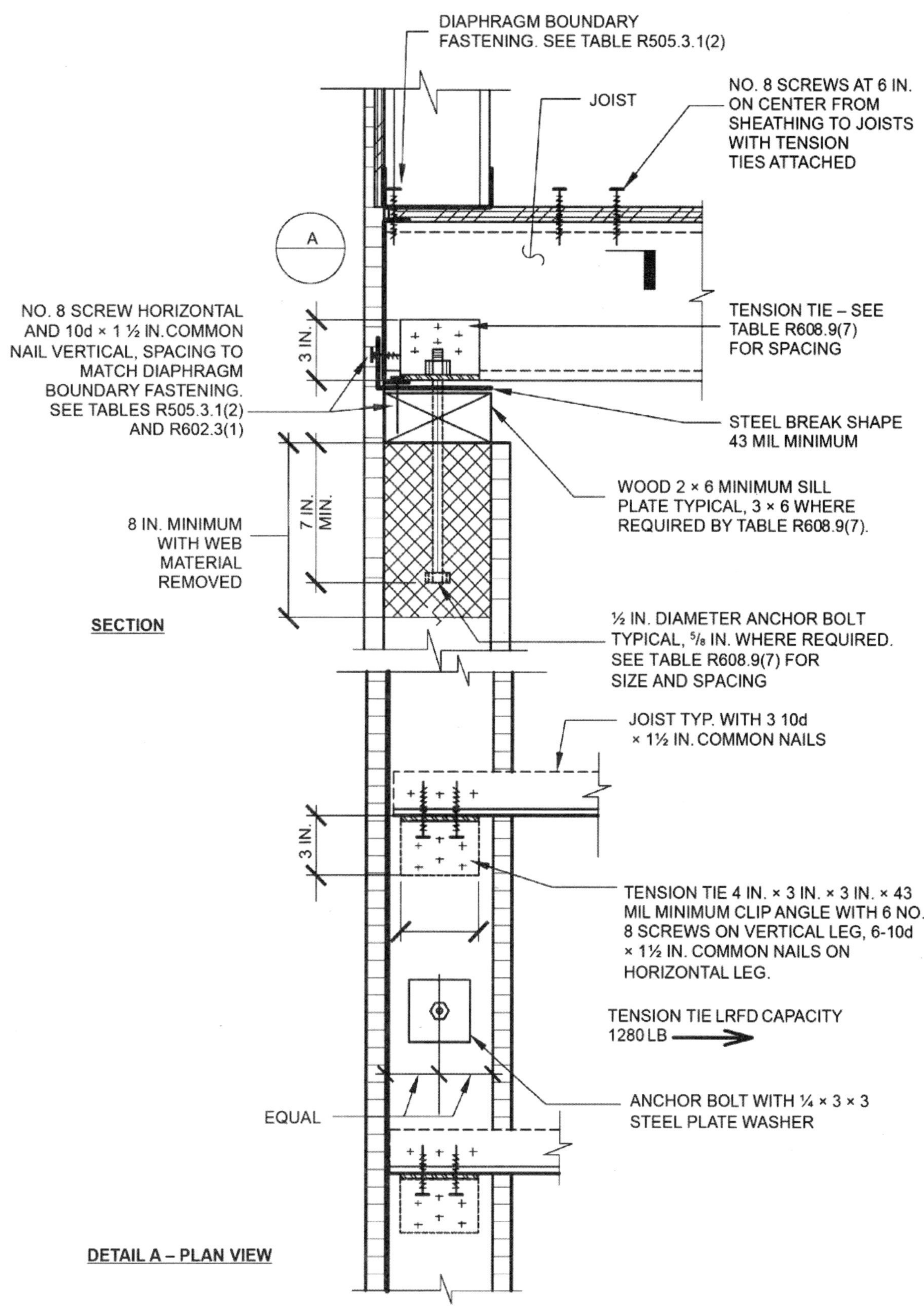

For SI: 1 mil = 0.0254 mm, 1 inch = 25.4 mm, 1 pound-force = 4.448 N.

FIGURE R608.9(7)
COLD-FORMED STEEL FLOOR TO TOP OF CONCRETE WALL, FRAMING PERPENDICULAR

TABLE R608.9(7)
COLD-FORMED STEEL-FRAMED FLOOR TO TOP OF CONCRETE WALL, FRAMING PERPENDICULAR[a, b, c, d, e]

ANCHOR BOLT SPACING (inches)	TENSION TIE SPACING (inches)	BASIC WIND SPEED AND WIND EXPOSURE CATEGORY (mph)					
		115B	120B	130B	140B	150B	160B
		—	—	110C	119C	127C	136C
		—	—	—	110D	117D	125D
12	12						6
12	24					6	6
16	16					6	6A
16	32				6	6	6A
19.2	19.2				6A	6A	6B
19.2	38.4			6	6A	6A	6B
24	24			6A	6B	6B	6B

For SI: 1 inch = 25.4 mm, 1 mile per hour = 0.447 m/s.

a. This table is for use with the detail in Figure R608.9(7). Use of this detail is permitted where a cell is not shaded.

b. Wall design per other provisions of Section R608 is required.

c. For wind design, minimum 4-inch-nominal wall is permitted in unshaded cells that do not contain a number.

d. Number 6 indicates minimum permitted nominal wall thickness in inches necessary to develop required strength (capacity) of connection. As a minimum, this nominal thickness shall occur in the portion of the wall indicated by the cross hatching in Figure R608.9(7). For the remainder of the wall, see Note b.

e. Letter "A" indicates that a minimum nominal 3 × 6 sill plate is required. Letter "B" indicates that a $^{5}/_{8}$-inch-diameter anchor bolt and a minimum nominal 3 × 6 sill plate are required.

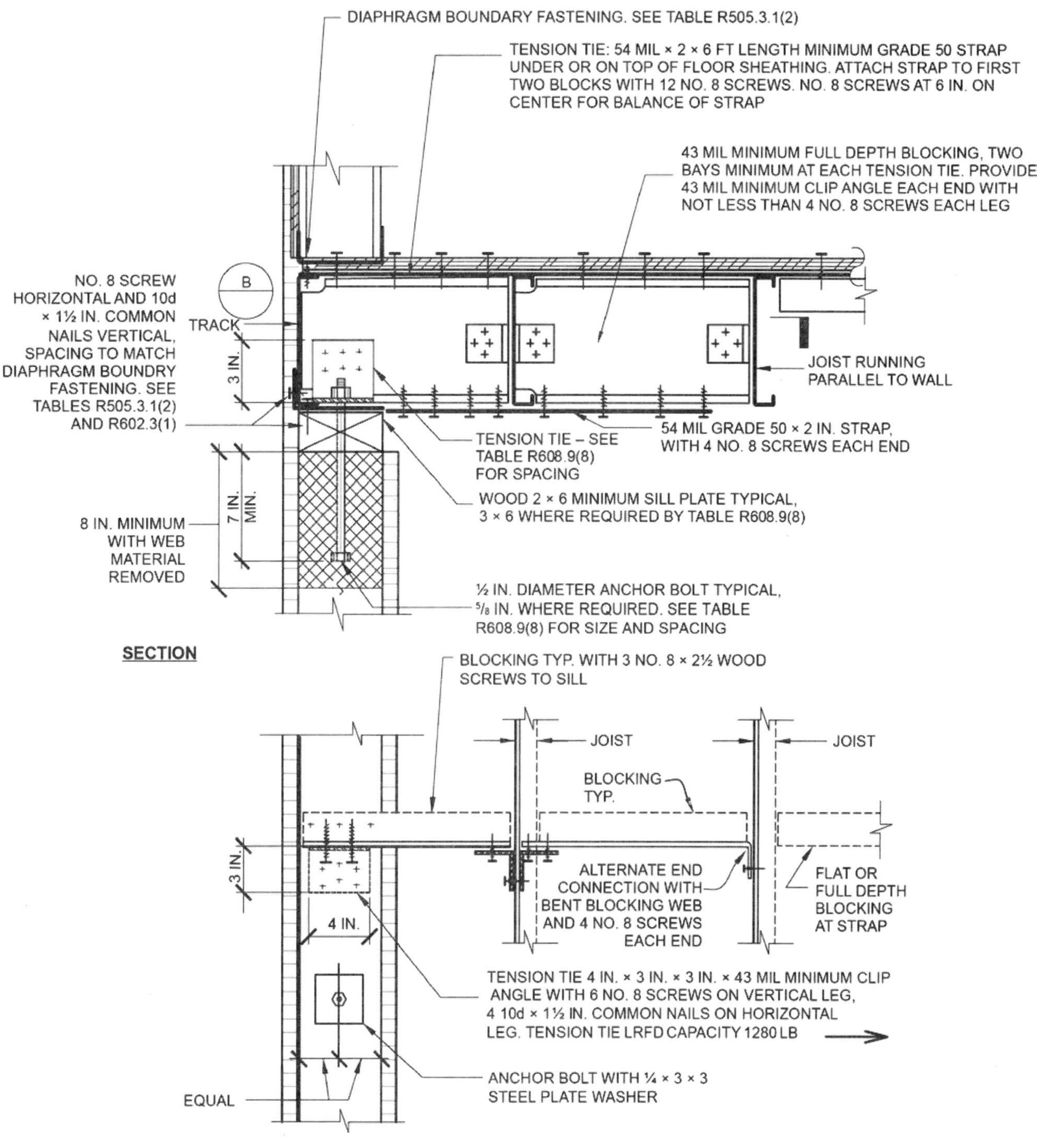

For SI: 1 mil = 0.0254 mm, 1 inch = 25.4 mm, 1 pound-force = 4.448 N.

FIGURE R608.9(8)
COLD-FORMED STEEL FLOOR TO TOP OF CONCRETE WALL, FRAMING PARALLEL

TABLE R608.9(8) COLD-FORMED STEEL-FRAMED FLOOR TO TOP OF CONCRETE WALL, FRAMING PARALLEL[a, b, c, d, e]

ANCHOR BOLT SPACING (inches)	TENSION TIE SPACING (inches)	BASIC WIND SPEED AND WIND EXPOSURE CATEGORY (mph)					
		115B	120B	130B	140B	150B	160B
		—	—	110C	119C	127C	136C
		—	—	—	110D	117D	125D
12	12						6
12	24					6	6
16	16					6	6A
16	32				6	6	6A
19.2	19.2				6A	6A	6B
19.2	38.4			6	6A	6A	6B
24	24			6A	6B	6B	6B

For SI: 1 inch = 25.4 mm, 1 mile per hour = 0.447 m/s.

a. This table is for use with the detail in Figure R608.9(8). Use of this detail is permitted where a cell is not shaded.

b. Wall design per other provisions of Section R608 is required.

c. For wind design, minimum 4-inch-nominal wall is permitted in unshaded cells that do not contain a number.

d. Number 6 indicates minimum permitted nominal wall thickness in inches necessary to develop required strength (capacity) of connection. As a minimum, this nominal thickness shall occur in the portion of the wall indicated by the cross hatching in Figure R608.9(8). For the remainder of the wall, see Note b.

e. Letter "A" indicates that a minimum nominal 3 × 6 sill plate is required. Letter "B" indicates that a $^{5}/_{8}$-inch-diameter anchor bolt and a minimum nominal 3 × 6 sill plate are required.

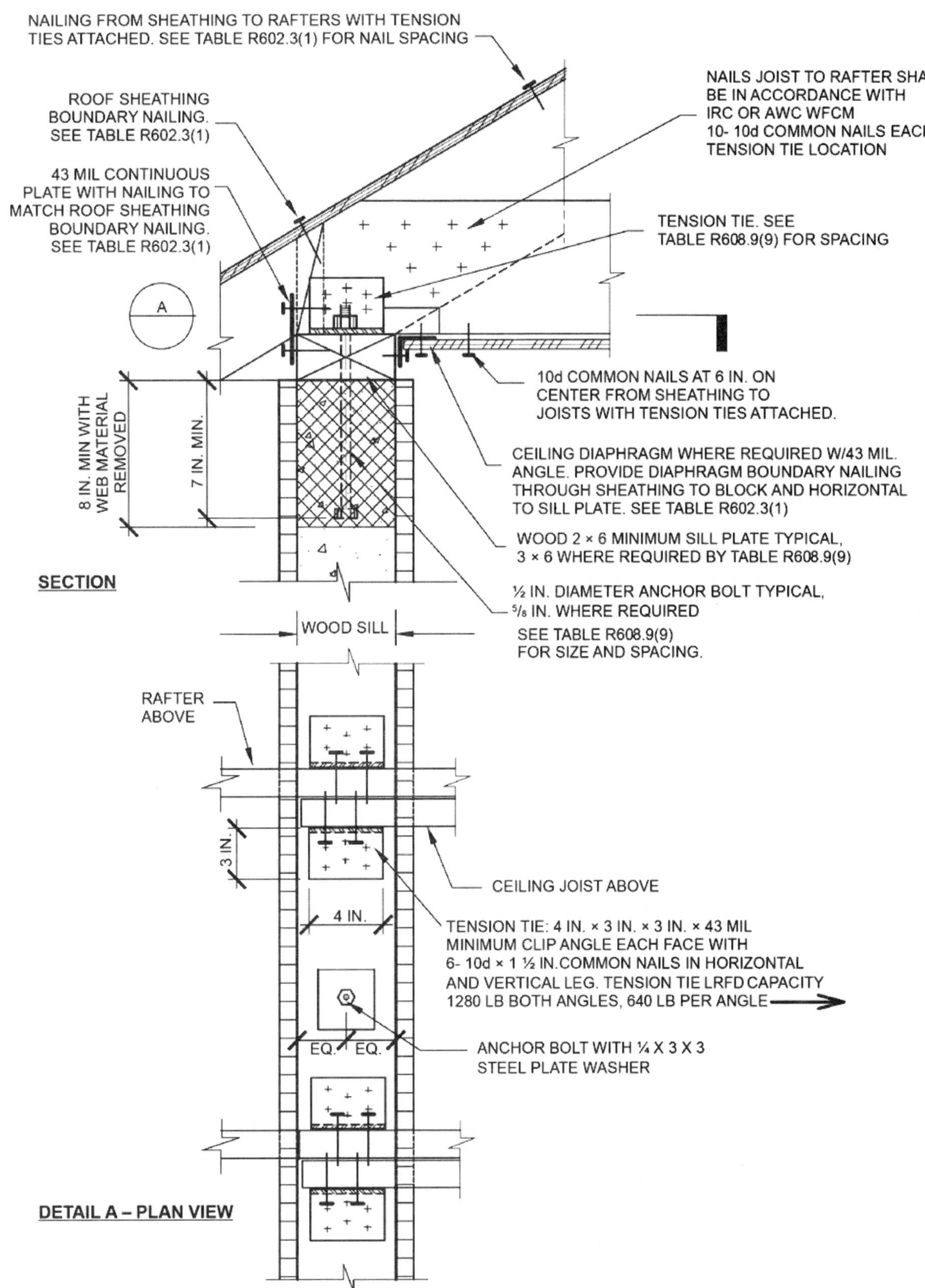

For SI: 1 mil = 0.0254 mm, 1 inch = 25.4 mm, 1 pound-force = 4.448 N.

FIGURE R608.9(9)
WOOD-FRAMED ROOF TO TOP OF CONCRETE WALL, FRAMING PERPENDICULAR

TABLE R608.9(9)
WOOD-FRAMED ROOF TO TOP OF CONCRETE WALL, FRAMING PERPENDICULAR[a, b, c, d, e]

ANCHOR BOLT SPACING (inches)	TENSION TIE SPACING (inches)	BASIC WIND SPEED (mph) AND WIND EXPOSURE CATEGORY					
		115B	120B	130B	140B	150B	160B
		—	—	110C	119C	127C	136C
		—	—	—	110D	117D	125D
12	12						6
12	24						6
12	36					6	6
12	48				6	6	6
16	16					6	6
16	32					6	6
16	48				6	6	6
19.2	19.2					6	6
19.2	38.4				6	6	
24	24				6		
24	48			6	8B		

For SI: 1 inch = 25.4 mm, 1 mile per hour = 0.447 m/s.

a. This table is for use with the detail in Figure R608.9(9). Use of this detail is permitted where a cell is not shaded, and prohibited where shaded.

b. Wall design per other provisions of Section R608 is required.

c. For wind design, minimum 4-inch-nominal wall is permitted in unshaded cells that do not contain a number.

d. Numbers 6 and 8 indicate minimum permitted nominal wall thickness in inches necessary to develop required strength (capacity) of connection. As a minimum, this nominal thickness shall occur in the portion of the wall indicated by the cross hatching in Figure R608.9(9). For the remainder of the wall, see Note b.

e. Letter "B" indicates that a $^{5}/_{8}$-inch-diameter anchor bolt and a minimum nominal 3 × 6 sill plate are required.

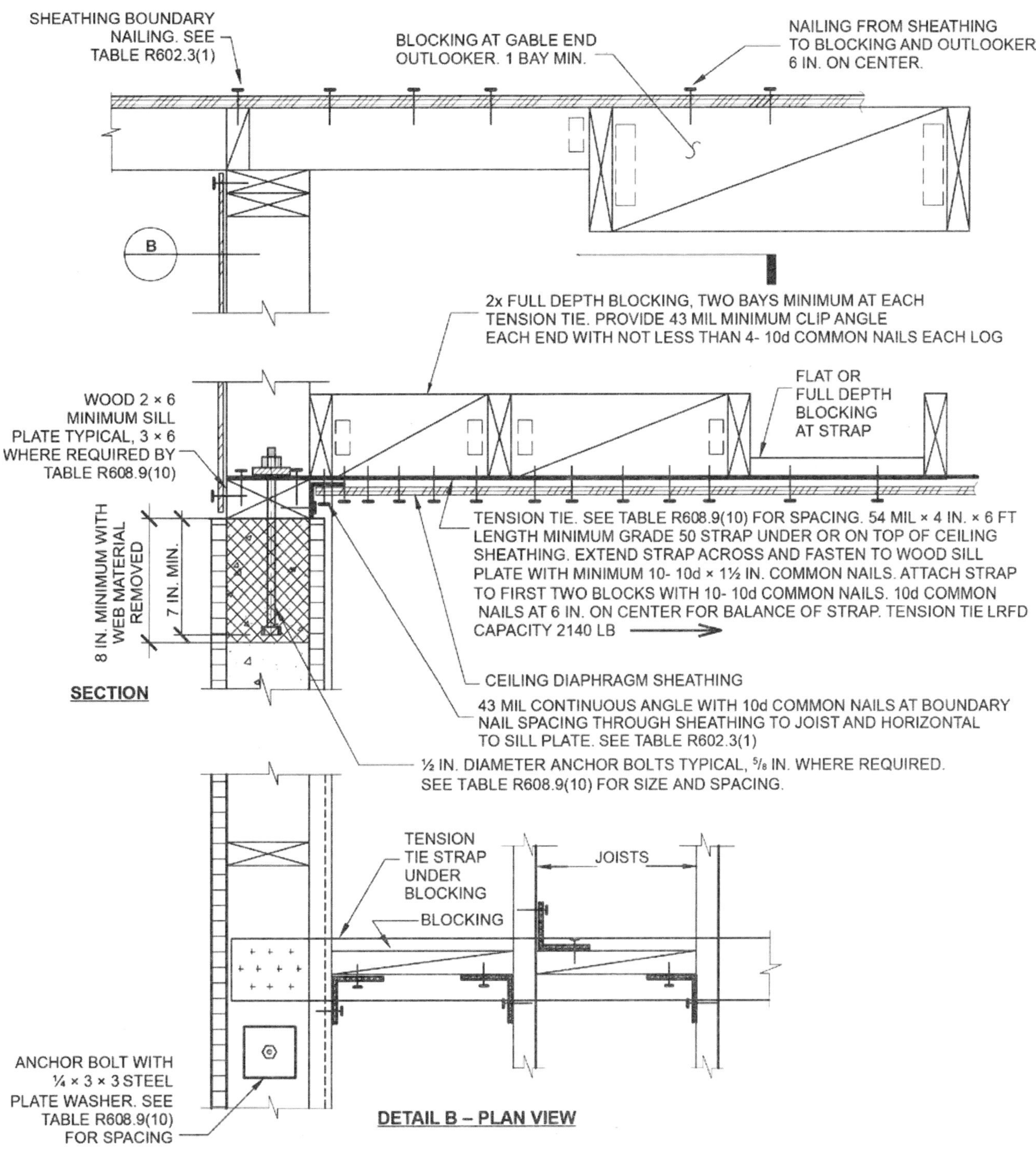
SHEATHING BOUNDARY NAILING. SEE TABLE R602.3(1)
BLOCKING AT GABLE END OUTLOOKER. 1 BAY MIN.
NAILING FROM SHEATHING TO BLOCKING AND OUTLOOKER 6 IN. ON CENTER.
B
2x FULL DEPTH BLOCKING, TWO BAYS MINIMUM AT EACH TENSION TIE. PROVIDE 43 MIL MINIMUM CLIP ANGLE EACH END WITH NOT LESS THAN 4- 10d COMMON NAILS EACH LOG
FLAT OR FULL DEPTH BLOCKING AT STRAP
WOOD 2 × 6 MINIMUM SILL PLATE TYPICAL, 3 × 6 WHERE REQUIRED BY TABLE R608.9(10)
8 IN. MINIMUM WITH WEB MATERIAL REMOVED
7 IN. MIN.
TENSION TIE. SEE TABLE R608.9(10) FOR SPACING. 54 MIL × 4 IN. × 6 FT LENGTH MINIMUM GRADE 50 STRAP UNDER OR ON TOP OF CEILING SHEATHING. EXTEND STRAP ACROSS AND FASTEN TO WOOD SILL PLATE WITH MINIMUM 10- 10d × 1½ IN. COMMON NAILS. ATTACH STRAP TO FIRST TWO BLOCKS WITH 10- 10d COMMON NAILS. 10d COMMON NAILS AT 6 IN. ON CENTER FOR BALANCE OF STRAP. TENSION TIE LRFD CAPACITY 2140 LB
SECTION
CEILING DIAPHRAGM SHEATHING
43 MIL CONTINUOUS ANGLE WITH 10d COMMON NAILS AT BOUNDARY NAIL SPACING THROUGH SHEATHING TO JOIST AND HORIZONTAL TO SILL PLATE. SEE TABLE R602.3(1)
½ IN. DIAMETER ANCHOR BOLTS TYPICAL, ⅝ IN. WHERE REQUIRED. SEE TABLE R608.9(10) FOR SIZE AND SPACING.
TENSION TIE STRAP UNDER BLOCKING
JOISTS
BLOCKING
ANCHOR BOLT WITH ¼ × 3 × 3 STEEL PLATE WASHER. SEE TABLE R608.9(10) FOR SPACING
DETAIL B – PLAN VIEW

For SI: 1 mil = 0.0254 mm, 1 inch = 25.4 mm, 1 pound-force = 4.448 N.

FIGURE R608.9(10)
WOOD-FRAMED ROOF TO TOP OF CONCRETE WALL, FRAMING PARALLEL

TABLE R608.9(10)
WOOD-FRAMED ROOF TO TOP OF CONCRETE WALL, FRAMING PARALLEL[a, b, c, d, e]

ANCHOR BOLT SPACING (inches)	TENSION TIE SPACING (inches)	BASIC WIND SPEED (mph) AND WIND EXPOSURE CATEGORY					
		115B	120B	130B	140B	150B	160B
		—	—	110C	119C	127C	136C
		—	—	—	110D	117D	125D
12	12						6
12	24						6
12	36					6	6
12	48				6	6	6
16	16					6	6
16	32					6	6
16	48				6	6	6
19.2	19.2					6	6
19.2	38.4				6	6	
24	24				6		
24	48			6	8B		

For SI: 1 inch = 25.4 mm, 1 mile per hour = 0.447 m/s.

a. This table is for use with the detail in Figure R608.9(10). Use of this detail is permitted where a cell is not shaded, and prohibited where shaded.

b. Wall design per other provisions of Section R608 is required.

c. For wind design, minimum 4-inch-nominal wall is permitted in cells that do not contain a number.

d. Numbers 6 and 8 indicate minimum permitted nominal wall thickness in inches necessary to develop required strength (capacity) of connection. As a minimum, this nominal thickness shall occur in the portion of the wall indicated by the cross hatching in Figure R608.9(10). For the remainder of the wall, see Note b.

e. Letter "B" indicates that a $^{5}/_{8}$-inch-diameter anchor bolt and a minimum nominal 3 × 6 sill plate are required.

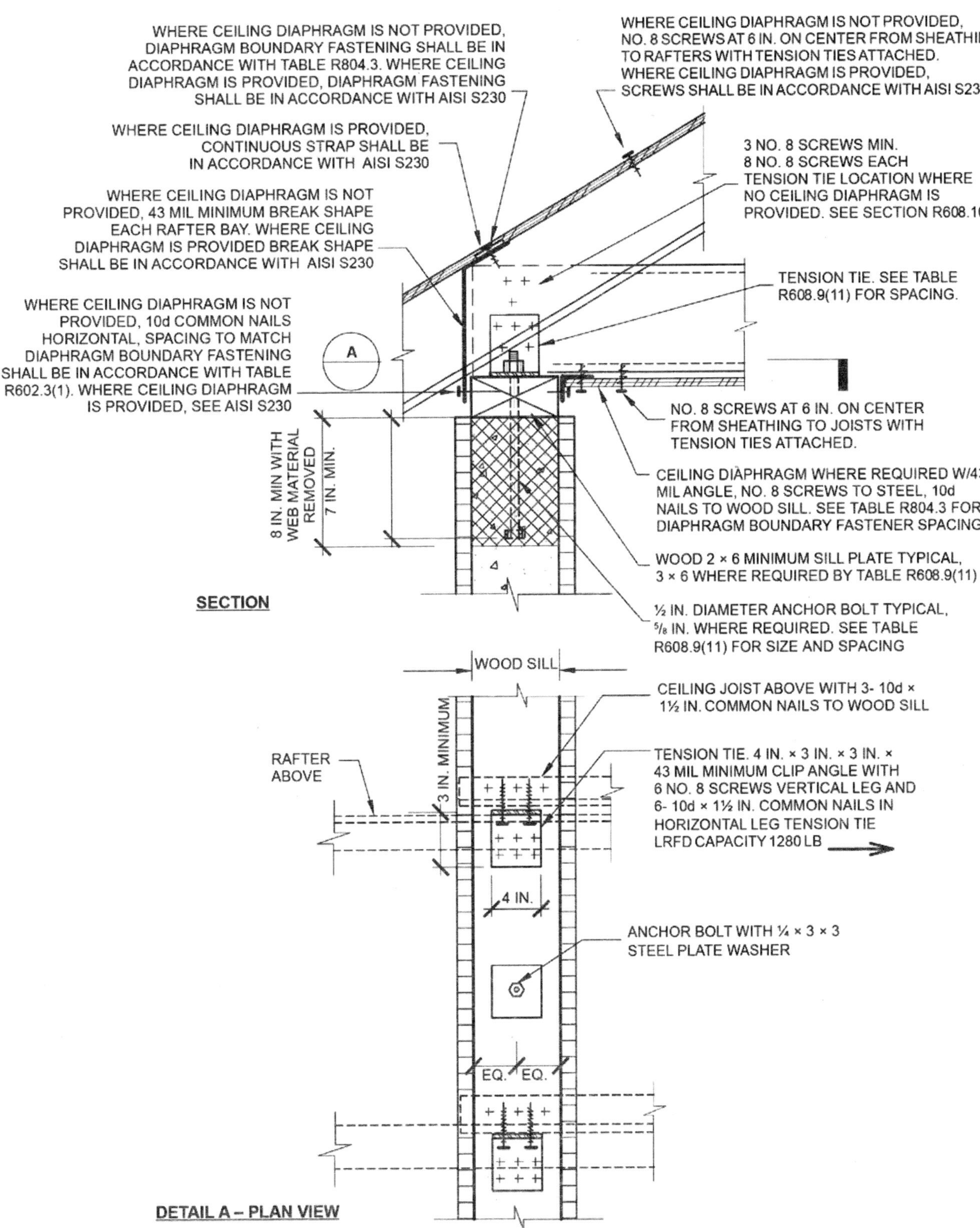
WHERE CEILING DIAPHRAGM IS NOT PROVIDED, DIAPHRAGM BOUNDARY FASTENING SHALL BE IN ACCORDANCE WITH TABLE R804.3. WHERE CEILING DIAPHRAGM IS PROVIDED, DIAPHRAGM FASTENING SHALL BE IN ACCORDANCE WITH AISI S230
WHERE CEILING DIAPHRAGM IS NOT PROVIDED, NO. 8 SCREWS AT 6 IN. ON CENTER FROM SHEATHING TO RAFTERS WITH TENSION TIES ATTACHED. WHERE CEILING DIAPHRAGM IS PROVIDED, SCREWS SHALL BE IN ACCORDANCE WITH AISI S230.
WHERE CEILING DIAPHRAGM IS PROVIDED, CONTINUOUS STRAP SHALL BE IN ACCORDANCE WITH AISI S230
3 NO. 8 SCREWS MIN. 8 NO. 8 SCREWS EACH TENSION TIE LOCATION WHERE NO CEILING DIAPHRAGM IS PROVIDED. SEE SECTION R608.10
WHERE CEILING DIAPHRAGM IS NOT PROVIDED, 43 MIL MINIMUM BREAK SHAPE EACH RAFTER BAY. WHERE CEILING DIAPHRAGM IS PROVIDED BREAK SHAPE SHALL BE IN ACCORDANCE WITH AISI S230
TENSION TIE. SEE TABLE R608.9(11) FOR SPACING.
WHERE CEILING DIAPHRAGM IS NOT PROVIDED, 10d COMMON NAILS HORIZONTAL, SPACING TO MATCH DIAPHRAGM BOUNDARY FASTENING SHALL BE IN ACCORDANCE WITH TABLE R602.3(1). WHERE CEILING DIAPHRAGM IS PROVIDED, SEE AISI S230
A
NO. 8 SCREWS AT 6 IN. ON CENTER FROM SHEATHING TO JOISTS WITH TENSION TIES ATTACHED.
8 IN. MIN WITH WEB MATERIAL REMOVED
7 IN. MIN.
CEILING DIAPHRAGM WHERE REQUIRED W/43 MIL ANGLE, NO. 8 SCREWS TO STEEL, 10d NAILS TO WOOD SILL. SEE TABLE R804.3 FOR DIAPHRAGM BOUNDARY FASTENER SPACING
WOOD 2 × 6 MINIMUM SILL PLATE TYPICAL, 3 × 6 WHERE REQUIRED BY TABLE R608.9(11)
SECTION
½ IN. DIAMETER ANCHOR BOLT TYPICAL, ⅝ IN. WHERE REQUIRED. SEE TABLE R608.9(11) FOR SIZE AND SPACING
WOOD SILL
CEILING JOIST ABOVE WITH 3- 10d × 1½ IN. COMMON NAILS TO WOOD SILL
3 IN. MINIMUM
RAFTER ABOVE
TENSION TIE. 4 IN. × 3 IN. × 3 IN. × 43 MIL MINIMUM CLIP ANGLE WITH 6 NO. 8 SCREWS VERTICAL LEG AND 6- 10d × 1½ IN. COMMON NAILS IN HORIZONTAL LEG TENSION TIE LRFD CAPACITY 1280 LB
4 IN.
ANCHOR BOLT WITH ¼ × 3 × 3 STEEL PLATE WASHER
EQ.
EQ.
DETAIL A – PLAN VIEW

For SI: 1 mil = 0.0254 mm, 1 inch = 25.4 mm, 1 foot = 304.8 mm, 1 pound-force = 4.448 N.

FIGURE R608.9(11)
COLD-FORMED STEEL ROOF TO TOP OF CONCRETE WALL, FRAMING PERPENDICULAR

TABLE R608.9(11)
COLD-FORMED STEEL ROOF TO TOP OF CONCRETE WALL, FRAMING PERPENDICULAR[a, b, c, d, e]

ANCHOR BOLT SPACING (inches)	TENSION TIE SPACING (inches)	BASIC WIND SPEED (mph) AND WIND EXPOSURE CATEGORY					
		115B	120B	130B	140B	150B	160B
		—	—	110C	119C	127C	136C
		—	—	—	110D	117D	125D
12	12						6
12	24						6
16	16					6	6
16	32					6	6
19.2	19.2					6	6
19.2	38.4				6	6	6
24	24				6	6A	6B

For SI: 1 inch = 25.4 mm, 1 mile per hour = 0.447 m/s.

a. This table is for use with the detail in Figure R608.9(11). Use of this detail is permitted where a cell is not shaded.

b. Wall design per other provisions of Section R608 is required.

c. For wind design, minimum 4-inch-nominal wall is permitted in unshaded cells that do not contain a number.

d. Number 6 indicates minimum permitted nominal wall thickness in inches necessary to develop required strength (capacity) of connection. As a minimum, this nominal thickness shall occur in the portion of the wall indicated by the cross hatching in Figure R608.9(11). For the remainder of the wall, see Note b.

e. Letter "A" indicates that a minimum nominal 3 × 6 sill plate is required. Letter "B" indicates that a $^5/_8$-inch-diameter anchor bolt and a minimum nominal 3 × 6 sill plate are required.

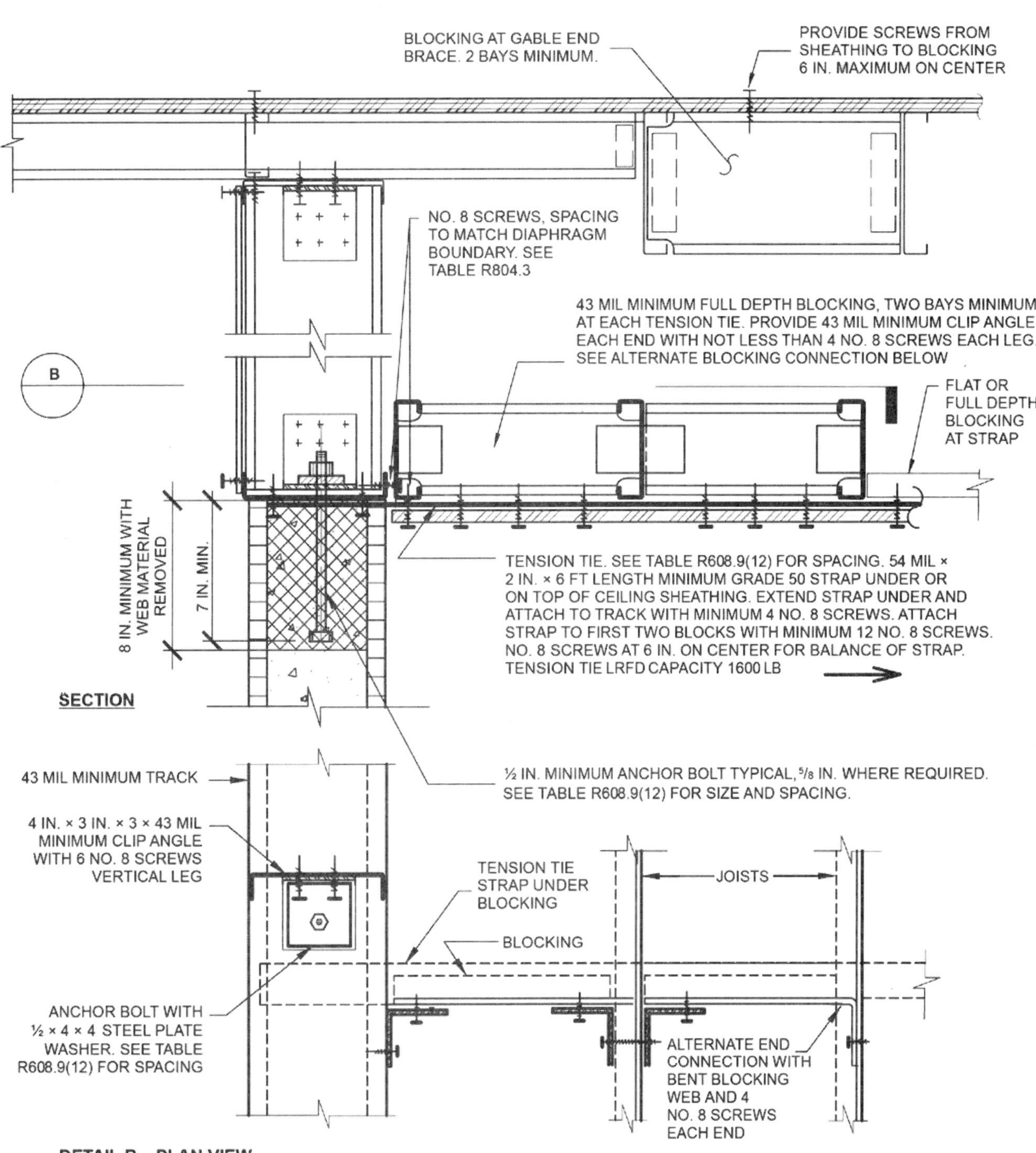

For SI: 1 mil = 0.0254 mm, 1 inch = 25.4 mm, 1 pound-force = 4.448 N.

FIGURE R608.9(12)
COLD-FORMED STEEL ROOF TO TOP OF CONCRETE WALL, FRAMING PARALLEL

TABLE R608.9(12)
COLD-FORMED STEEL ROOF TO TOP OF CONCRETE WALL, FRAMING PARALLEL[a, b, c, d, e]

ANCHOR BOLT SPACING (inches)	TENSION TIE SPACING (inches)	BASIC WIND SPEED (mph) AND WIND EXPOSURE CATEGORY					
		115B	120B	130B	140B	150B	160B
		—	—	110C	119C	127C	136C
		—	—	—	110D	117D	125D
12	12						6
12	24						6
16	16					6	6
16	32					6	6
19.2	19.2					6	6
19.2	38.4				6	6	6
24	24				6	6	6B

For SI: 1 inch = 25.4 mm, 1 mile per hour = 0.447 m/s.

a. This table is for use with the detail in Figure R608.9(12). Use of this detail is permitted where a cell is not shaded.

b. Wall design per other provisions of Section R608 is required.

c. For wind design, minimum 4-inch-nominal wall is permitted in cells that do not contain a number.

d. Number 6 indicates minimum permitted nominal wall thickness in inches necessary to develop required strength (capacity) of connection. As a minimum, this nominal thickness shall occur in the portion of the wall indicated by the cross hatching in Figure R608.9(12). For the remainder of the wall, see Note b.

e. Letter "B" indicates that a $^5/_8$-inch-diameter anchor bolt is required.

R608.9.2 Connections between concrete walls and light-frame floor systems. Connections between concrete walls and light-frame floor systems shall be in accordance with one of the following:

1. For floor systems of wood-framed construction, the provisions of Section R608.9.1 and the prescriptive details of Figures R608.9(1) through R608.9(4), where permitted by the tables accompanying those figures. Portions of connections of wood-framed floor systems not noted in the figures shall be in accordance with Section R502, or AWC WFCM, if applicable. Wood framing members shall be of a species having a specific gravity equal to or greater than 0.42.
2. For floor systems of cold-formed steel construction, the provisions of Section R608.9.1 and the prescriptive details of Figures R608.9(5) through R608.9(8), where permitted by the tables accompanying those figures. Portions of connections of cold-formed steel-framed floor systems not noted in the figures shall be in accordance with Section R505, or AISI S230, if applicable.
3. Proprietary connectors selected to resist loads and load combinations in accordance with Appendix A (ASD) or Appendix B (LRFD) of PCA 100.
4. An engineered design using loads and load combinations in accordance with Appendix A (ASD) or Appendix B (LRFD) of PCA 100.
5. An engineered design using loads and material design provisions in accordance with this code, or in accordance with ASCE 7, ACI 318, and AWC NDS for wood-framed construction or AISI S100 for cold-formed steel frame construction.

R608.9.3 Connections between concrete walls and light-frame ceiling and roof systems. Connections between concrete walls and light-frame ceiling and roof systems shall be in accordance with one of the following:

1. For ceiling and roof systems of wood-framed construction, the provisions of Section R608.9.1 and the prescriptive details of Figures R608.9(9) and R608.9(10), where permitted by the tables accompanying those figures. Portions of connections of wood-framed ceiling and roof systems not noted in the figures shall be in accordance with Section R802, or AWC WFCM, if applicable. Wood framing members shall be of a species having a specific gravity equal to or greater than 0.42.
2. For ceiling and roof systems of cold-formed steel construction, the provisions of Section R608.9.1 and the prescriptive details of Figures R608.9(11) and R608.9(12), where permitted by the tables accompanying those figures. Portions of connections of cold-formed steel-framed ceiling and roof systems not noted in the figures shall be in accordance with Section R804, or AISI S230, if applicable.
3. Proprietary connectors selected to resist loads and load combinations in accordance with Appendix A (ASD) or Appendix B (LRFD) of PCA 100.
4. An engineered design using loads and load combinations in accordance with Appendix A (ASD) or Appendix B (LRFD) of PCA 100.
5. An engineered design using loads and material design provisions in accordance with this code, or in accordance with ASCE 7, ACI 318, and AWC NDS for wood-framed construction or AISI S100 for cold-formed steel-framed construction.

R608.10 Floor, roof and ceiling diaphragms. Floors and roofs in buildings with exterior walls of concrete shall be designed and constructed as diaphragms. Where gable-end walls occur, ceilings shall be designed and constructed as diaphragms. The design and construction of floors, roofs and ceilings of wood framing or cold-formed-steel framing serving as diaphragms shall comply with the applicable requirements of this code, or AWC WFCM or AISI S230, if applicable. Wood framing members shall be of a species having a specific gravity equal to or greater than 0.42.

SECTION R609 EXTERIOR WINDOWS AND DOORS

R609.1 General. This section prescribes performance and construction requirements for exterior windows and doors installed in walls. Windows and doors shall be installed in accordance with the fenestration manufacturer's written instructions. Window and door openings shall be flashed in accordance with Section R703.4. Written installation instructions shall be provided by the fenestration manufacturer for each window or door.

R609.2 Performance. Exterior windows and doors shall be capable of resisting the design wind loads specified in Table R301.2.1(1) adjusted for height and exposure in accordance with Table R301.2.1(2) or determined in accordance with ASCE 7 using the allowable stress design load combinations of ASCE 7. For exterior windows and doors tested in accordance with Sections R609.3 and R609.5, required design wind pressures determined from ASCE 7 using the ultimate strength design (USD) are permitted to be multiplied by 0.6. Design wind loads for exterior glazing not part of a labeled assembly shall be permitted to be determined in accordance with Chapter 24 of the *International Building Code*. Design wind loads for exterior glazing not part of a labeled assembly shall be permitted to be determined in accordance with Chapter 24 of the *International Building Code*.

R609.3 Testing and labeling. Exterior windows and sliding doors shall be tested by an *approved* independent laboratory, and bear a *label* identifying manufacturer, performance characteristics and *approved* inspection agency to indicate compliance with AAMA/WDMA/CSA 101/I.S.2/A440. Exterior side-hinged doors shall be tested and *labeled* as conforming to AAMA/WDMA/CSA 101/I.S.2/A440 or AMD 100, or comply with Section R609.5.

Exception: Decorative glazed openings.

R609.3.1 Comparative analysis. Structural wind load design pressures for window and door units different than the size tested in accordance with Section R609.3 shall be permitted to be different than the design value of the tested unit where determined in accordance with one of the following comparative analysis methods:

1. Structural wind load design pressures for window and door units smaller than the size tested in accordance with Section R609.3 shall be permitted to be higher than the design value of the tested unit provided such higher pressures are determined by accepted engineering analysis. Components of the smaller unit shall be the same as those of the tested unit. Where such calculated design pressures are used, they shall be validated by an additional test of the window or door unit having the highest allowable design pressure.
2. In accordance with WDMA I.S.11.

R609.4 Garage doors. Garage doors shall be tested in accordance with either ASTM E330 or ANSI/DASMA 108, and shall meet the pass/fail criteria of ANSI/DASMA 108.

R609.4.1 Garage door labeling. Garage doors shall be *labeled* with a permanent *label* provided by the garage door manufacturer. The *label* shall identify the garage door manufacturer, the garage door model/series number, the positive and negative design wind pressure rating, the installation instruction drawing reference number, and the applicable test standard.

R609.5 Other exterior window and door assemblies. Exterior windows and door assemblies not included within the scope of Section R609.3 or R609.4 shall be tested in accordance with ASTM E330. Glass in assemblies covered by this section shall comply with Section R308.5.

R609.6 Windborne debris protection. Protection of exterior windows, glass doors and doors with glass in buildings located in *windborne debris regions* shall be in accordance with Section R301.2.1.2.

R609.6.1 Fenestration testing and labeling. *Fenestration* shall be tested by an *approved* independent laboratory, *listed* by an *approved* entity, and bear a *label* identifying the manufacturer, performance characteristics and an *approved* inspection agency to indicate compliance with the requirements of the following specification(s):

1. ASTM E1886 and ASTM E1996; or
2. AAMA 506.

R609.6.2 Impact protective systems testing and labeling. *Impact protective systems* shall be tested for impact resistance by an *approved* independent laboratory for compliance with ASTM E1886 and ASTM E1996. *Impact protective systems* shall be tested for design wind pressure by an *approved* independent laboratory for compliance with ASTM E330. Required design wind pressures shall be determined in accordance with Table R301.2.1(1), adjusted for height and exposure in accordance with Table R301.2.1(2) or determined in accordance with ASCE 7. For the purposes of this section, design wind pressures determined in accordance with ASCE 7 are permitted to be multiplied by 0.6.

Impact protective systems bear a *label* identifying the manufacturer, performance characteristics and an *approved* inspection agency. *Impact protective systems* shall have a permanent *label* providing traceability to the manufacturer, product designation and performance characteristics. The permanent *label* shall be acid etched, sand blasted, ceramic fired, laser etched, embossed or of a type that, once applied, cannot be removed without being destroyed.

R609.7 Anchorage methods. The methods cited in this section apply only to anchorage of window and glass door assemblies to the main force-resisting system.

R609.7.1 Anchoring requirements. Window and glass door assemblies shall be anchored in accordance with the published manufacturer's recommendations to achieve the design pressure specified. Substitute anchoring systems used for substrates not specified by the fenestration manufacturer shall provide equal or greater anchoring performance as demonstrated by accepted engineering practice.

R609.7.2 Anchorage details. Products shall be anchored in accordance with the minimum requirements illustrated in Figures R609.7.2(1), R609.7.2(2), R609.7.2(3), R609.7.2(4), R609.7.2(5), R609.7.2(6), R609.7.2(7) and R609.7.2(8).

R609.7.2.1 Masonry, concrete or other structural substrate. Where the wood shim or buck thickness is less than $1^1/_2$ inches (38 mm), window and glass door assemblies shall be anchored through the jamb, or by jamb clip and anchors shall be embedded directly into the masonry, concrete or other substantial substrate material. Anchors shall adequately transfer load from the window or door frame into the rough opening substrate [see Figures R609.7.2(1) and R609.7.2(2)].

Where the wood shim or buck thickness is $1^1/_2$ inches (38 mm) or more, the buck is securely fastened to the masonry, concrete or other substantial substrate, and the buck extends beyond the interior face of the window or door frame, window and glass door assemblies shall be anchored through the jamb, or by jamb clip, or through the flange to the secured wood buck. Anchors shall be embedded into the secured wood buck to adequately transfer load from the window or door frame assembly [see Figures R609.7.2(3), R609.7.2(4) and R609.7.2(5)].

R609.7.2.2 Wood or other approved framing material. Where the framing material is wood or other *approved* framing material, window and glass door assemblies shall be anchored through the frame, or by frame clip, or through the flange. Anchors shall be embedded into the frame construction to adequately transfer load [see Figures R609.7.2(6), R609.7.2(7) and R609.7.2(8)].

R609.8 Mullions. Mullions shall be tested by an *approved* testing laboratory in accordance with AAMA 450, or be engineered in accordance with accepted engineering practice. Mullions tested as stand-alone units or qualified by engineering shall use performance criteria cited in Sections R609.8.1, R609.8.2 and R609.8.3. Mullions qualified by an actual test of an entire assembly shall comply with Sections R609.8.1 and R609.8.3.

R609.8.1 Load transfer. Mullions shall be designed to transfer the design pressure loads applied by the window and door assemblies to the rough opening substrate.

R609.8.2 Deflection. Mullions shall be capable of resisting the design pressure loads applied by the window and door assemblies to be supported without deflecting more than $L/175$, where L is the span of the mullion in inches.

R609.8.3 Structural safety factor. Mullions shall be capable of resisting a load of 1.5 times the design pressure loads applied by the window and door assemblies to be supported without exceeding the appropriate material stress levels. If tested by an *approved* laboratory, the 1.5 times the design pressure load shall be sustained for 10 seconds, and the permanent deformation shall not exceed 0.4 percent of the mullion span after the 1.5 times design pressure load is removed.

SECTION R610 STRUCTURAL INSULATED PANEL WALL CONSTRUCTION

R610.1 General. Structural insulated panel (SIP) walls shall be designed in accordance with the provisions of this section. Where the provisions of this section are used to design *structural insulated panel* walls, project drawings, typical details and specifications are not required to bear the seal of the architect or engineer responsible for design, unless otherwise required by the state law of the *jurisdiction* having authority.

R610.2 Applicability limits. The provisions of this section shall control the construction of exterior *structural insulated panel* walls and interior load-bearing *structural insulated panel* walls for buildings not greater than 60 feet (18 288 mm) in length perpendicular to the joist or truss span, not greater than 40 feet (12 192 mm) in width parallel to the joist or truss span and not greater than two stories in height with each wall not greater than 10 feet (3048 mm) high. Exterior walls installed in accordance with the provisions of this section shall be considered as *load-bearing walls. Structural insulated panel* walls constructed in accordance with the provisions of this section shall be limited to sites where the ultimate design wind speed (V_{ult}) is not greater than 155 miles per hour (69 m/s) in Exposure B or 140 miles per hour (63 m/s) in Exposure C, the ground snow load is not greater than 70 pounds per square foot (3.35 kPa), and the seismic design category is A, B or C.

R610.3 Materials. SIPs shall comply with the requirements of ANSI/APA PRS 610.1.

R610.3.1 Lumber. The minimum lumber framing material used for SIPs prescribed in this document is NLGA graded No. 2 Spruce-pine-fir. Substitution of other wood species/grades that meet or exceed the mechanical properties and specific gravity of No. 2 Spruce-pine-fir shall be permitted.

R610.3.2 SIP screws. Screws used for the erection of SIPs as specified in Section R610.5 shall be fabricated from steel, shall be provided by the SIP manufacturer and shall be sized to penetrate the wood member to which the assembly is being attached by not less than 1 inch (25 mm). The screws shall be corrosion resistant and have a minimum shank diameter of 0.188 inch (4.7 mm) and a minimum head diameter of 0.620 inch (15.5 mm).

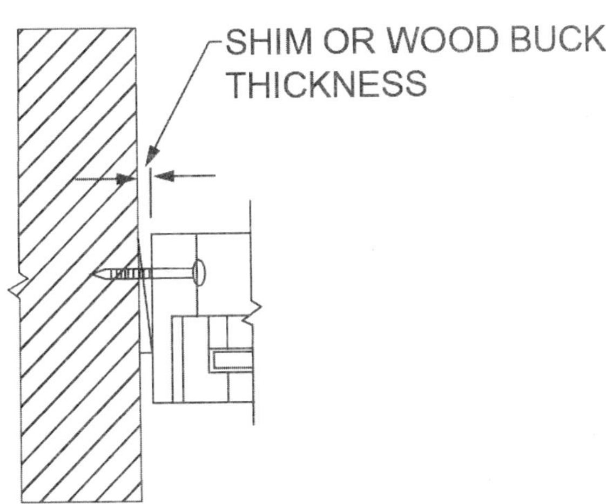

FIGURE R609.7.2(1)
THROUGH THE FRAME

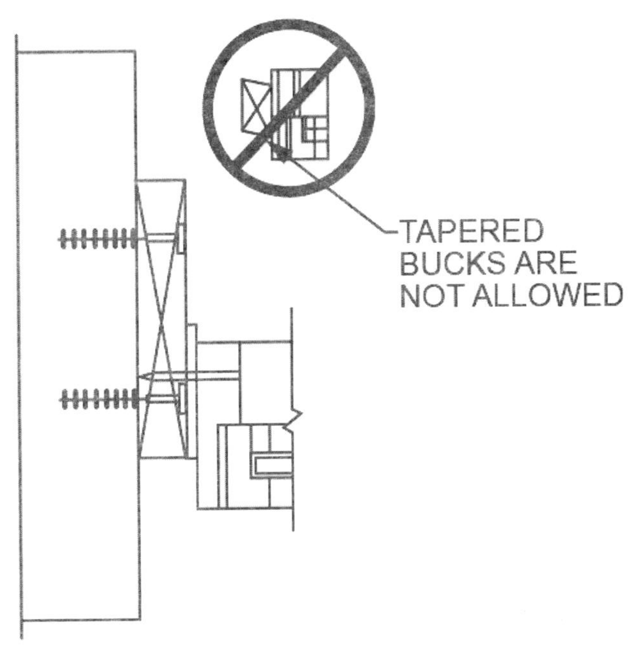

FIGURE R609.7.2(3)
THROUGH THE FRAME

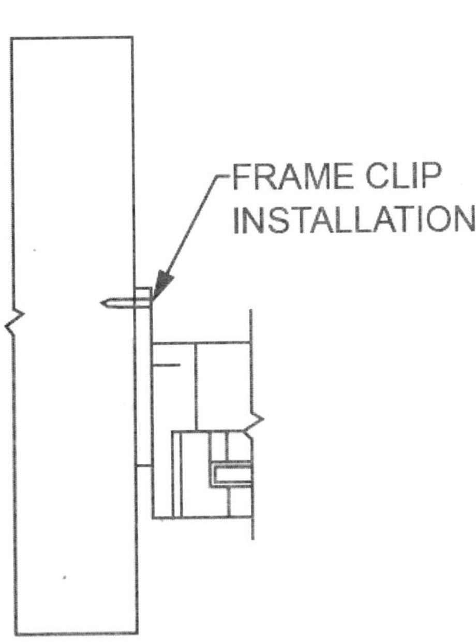

FIGURE R609.7.2(2)
FRAME CLIP

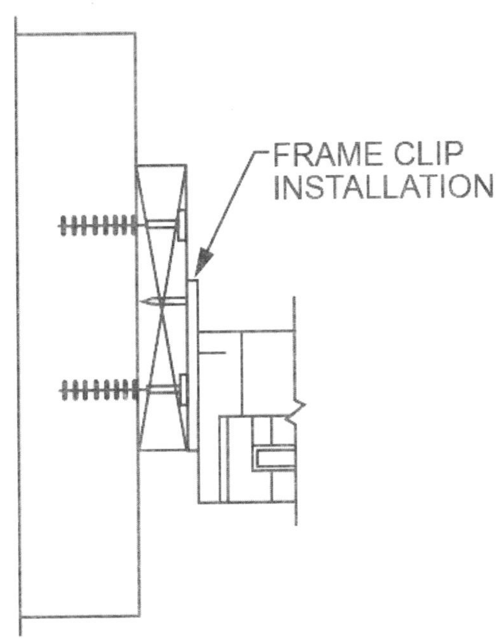

FIGURE R609.7.2(4)
FRAME CLIP

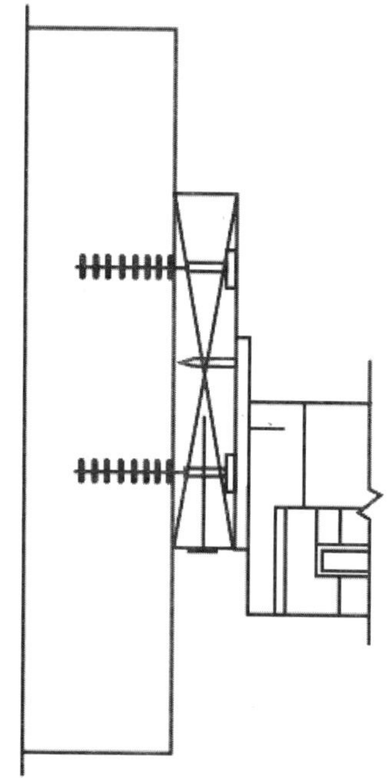

FIGURE R609.7.2(5)
THROUGH THE FLANGE

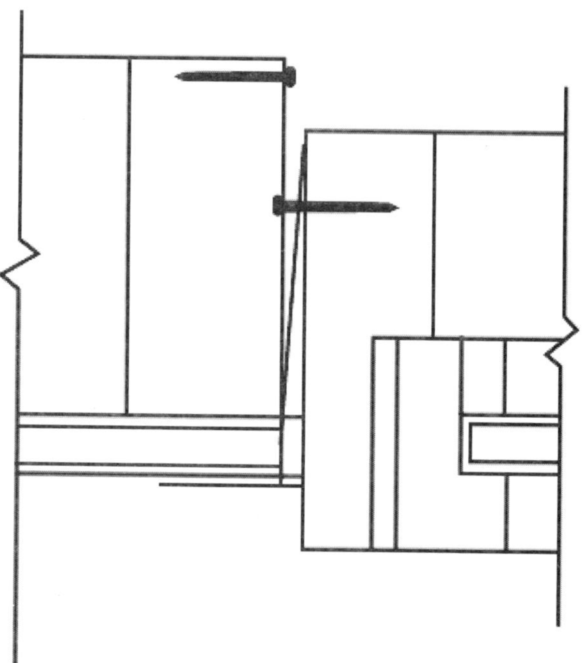

FIGURE R609.7.2(7)
FRAME CLIP

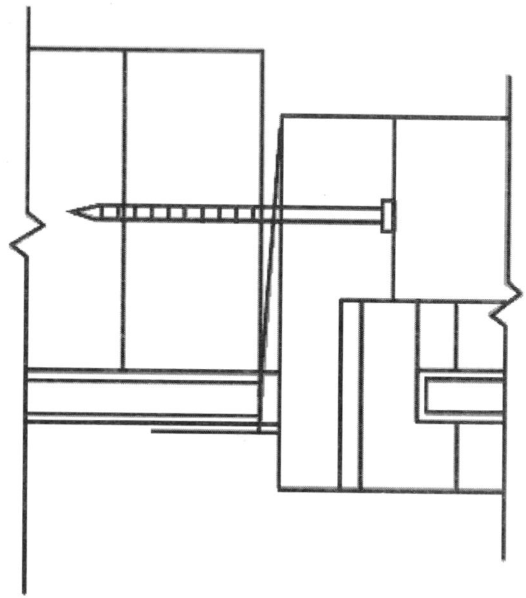

FIGURE R609.7.2(6)
THROUGH THE FLANGE

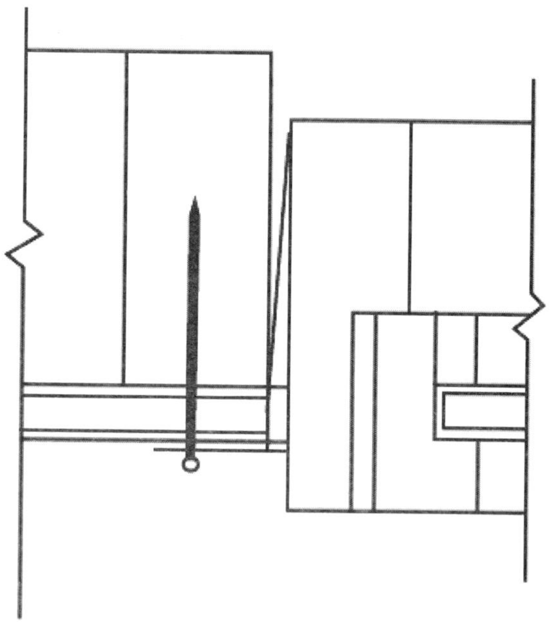

FIGURE R609.7.2(8)
THROUGH THE FLANGE

R610.3.3 Nails. Nails specified in Section R610 shall be common or galvanized box unless otherwise stated.

R610.4 SIP wall panels. SIPs shall comply with Figure R610.4 and shall have minimum *panel thickness* in accordance with Tables R610.5(1) and R610.5(2) for above-grade walls. SIPs shall be identified by grade *mark* or certificate of inspection issued by an *approved* agency in accordance with ANSI/APA PRS 610.1.

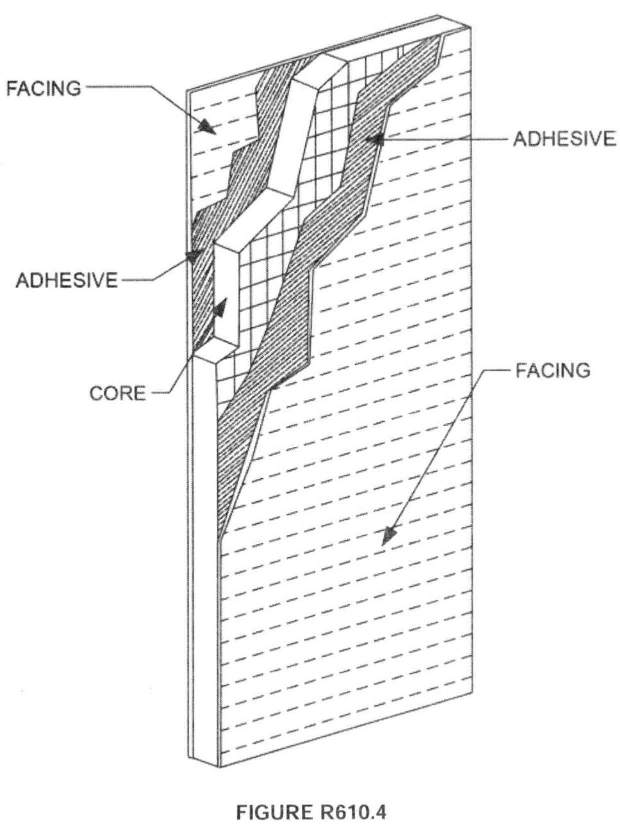

FIGURE R610.4
SIP WALL PANEL

R610.5 Wall construction. Exterior walls of SIP construction shall be designed and constructed in accordance with the provisions of this section and Tables R610.5(1) and R610.5(2) and Figures R610.5(1) through R610.5(5). SIP walls shall be fastened to other wood building components in accordance with Tables R602.3(1) through R602.3(4).

Framing shall be attached in accordance with Table R602.3(1) unless otherwise provided for in Section R610.

R610.5.1 Top plate connection. SIP walls shall be capped with a double top plate installed to provide overlapping at corner, intersections and *splines* in accordance with Figure R610.5.1. The double top plates shall be made up of a single 2-by (nominal 2-inch) top plate having a width equal to the width of the panel core, and shall be recessed into the SIP below. Over this top plate a cap plate shall be placed. The cap plate width shall match the SIP thickness and overlap the facers on both sides of the panel. End joints in top plates shall be offset not less than 24 inches (610 mm).

R610.5.2 Bottom (sole) plate connection. SIP walls shall have full bearing on a sole plate having a width equal to the nominal width of the foam core. Where SIP walls are supported directly on continuous foundations, the wall wood sill plate shall be anchored to the foundation in accordance with Figure R610.5.2 and Section R403.1.

R610.5.3 Panel-to-panel connection. SIPs shall be connected at vertical in-plane joints in accordance with Figure R610.8 or by other *approved* methods.

R610.5.4 Corner framing. Corner framing of SIP walls shall be constructed in accordance with Figure R610.5.4.

R610.5.5 Wall bracing. SIP walls shall be braced in accordance with Section R602.10. SIP walls shall be considered continuous wood structural panel sheathing (bracing Method CS-WSP) for purposes of computing required bracing. SIP walls shall meet the requirements of Section R602.10.4.2 except that SIP corners shall be fabricated as shown in Figure R610.8. Where SIP walls are used for wall bracing, the SIP bottom plate shall be attached to wood framing below in accordance with Table R602.3(1).

R610.5.6 Thermal barrier. SIP walls shall be separated from the interior of a building by an *approved* thermal barrier in accordance with Section R316.4.

R610.6 Interior load-bearing walls. Interior *load-bearing walls* shall be constructed as specified for exterior walls.

R610.7 Drilling and notching. The maximum vertical chase penetration in SIPs shall have a maximum side dimension of 2 inches (51 mm) centered in the panel. Vertical chases shall have a minimum spacing of 24 inches (610 mm) on center. Not more than two horizontal chases shall be permitted in each wall panel, one at 14 inches (360 mm) plus or minus 2 inches (51 mm) from the bottom of the panel and one at 48 inches (1220 mm) plus or minus 2 inches (51 mm) from the bottom edge of the SIP's panel. Additional penetrations are permitted where justified by analysis.

R610.8 Headers. SIP headers shall be designed and constructed in accordance with Table R610.8 and Figure R610.5.1. SIP headers shall be continuous sections without *splines*. Headers shall be not less than $11^7/_8$ inches (302 mm) deep. Headers longer than 4 feet (1219 mm) shall be constructed in accordance with Section R602.7. The strength axis of the factors on the header shall be oriented horizontally.

R610.8.1 Wood structural panel box headers. Wood structural panel box headers shall be allowed where SIP headers are not applicable. Wood structural panel box headers shall be constructed in accordance with Figure R602.7.3 and Table R602.7.3.

TABLE R610.5(1)
MINIMUM THICKNESS FOR SIP WALL SUPPORTING SIP OR LIGHT-FRAME ROOF ONLY (inches)[a]

ULTIMATE DESIGN WIND SPEED V_{ult} (mph)		GROUND SNOW LOAD (psf)	BUILDING WIDTH (ft)														
			24			28			32			36			40		
			Wall Height (feet)			Wall Height (feet)			Wall Height (feet)			Wall Height (feet)			Wall Height (feet)		
Exp. B	Exp. C		8	9	10	8	9	10	8	9	10	8	9	10	8	9	10
110	—	20	4.5	4.5	4.5	4.5	4.5	4.5	4.5	4.5	4.5	4.5	4.5	4.5	4.5	4.5	4.5
		30	4.5	4.5	4.5	4.5	4.5	4.5	4.5	4.5	4.5	4.5	4.5	4.5	4.5	4.5	4.5
		50	4.5	4.5	4.5	4.5	4.5	4.5	4.5	4.5	4.5	4.5	4.5	4.5	4.5	4.5	4.5
		70	4.5	4.5	4.5	4.5	4.5	4.5	4.5	4.5	4.5	4.5	4.5	6.5	4.5	4.5	6.5
115	—	20	4.5	4.5	4.5	4.5	4.5	4.5	4.5	4.5	4.5	4.5	4.5	4.5	4.5	4.5	4.5
		30	4.5	4.5	4.5	4.5	4.5	4.5	4.5	4.5	4.5	4.5	4.5	4.5	4.5	4.5	4.5
		50	4.5	4.5	4.5	4.5	4.5	4.5	4.5	4.5	4.5	4.5	4.5	4.5	4.5	4.5	6.5
		70	4.5	4.5	4.5	4.5	4.5	4.5	4.5	4.5	6.5	4.5	4.5	DR	4.5	4.5	DR
130	110	20	4.5	4.5	6.5	4.5	4.5	6.5	4.5	4.5	6.5	4.5	4.5	DR	4.5	4.5	DR
		30	4.5	4.5	6.5	4.5	4.5	6.5	4.5	4.5	DR	4.5	4.5	DR	4.5	4.5	DR
		50	4.5	4.5	DR	4.5	4.5	DR	4.5	4.5	DR	4.5	6.5	DR	4.5	DR	DR
		70	4.5	4.5	DR	4.5	DR	DR	4.5	DR	DR	4.5	DR	DR	DR	DR	DR
140	120	20	4.5	6.5	DR	4.5	6.5	DR	4.5	DR	DR	4.5	DR	DR	4.5	DR	DR
		30	4.5	6.5	DR	4.5	DR	DR	4.5	DR	DR	4.5	DR	DR	4.5	DR	DR
		50	4.5	DR	DR	4.5	DR	DR	DR	DR	DR	DR	DR	DR	DR	DR	DR
		70	4.5	DR	DR	DR	DR	DR	DR	DR	DR	DR	DR	DR	DR	DR	DR

For SI: 1 inch = 25.4 mm, 1 foot = 304.8 mm, 1 pound per square foot = 0.0479 kPa, 1 mile per hour = 0.447 m/s.

DR = Design Required.

a. Design assumptions:
   Maximum deflection criteria: L/240.
   Maximum roof dead load: 10 psf.
   Maximum roof live load: 70 psf.
   Maximum ceiling dead load: 5 psf.
   Maximum ceiling live load: 20 psf.
   Wind loads based on Table R301.2.1(1).
   Strength axis of facing material applied vertically.

TABLE R610.5(2)
MINIMUM THICKNESS FOR SIP WALL SUPPORTING SIP OR LIGHT-FRAME ONE STORY AND ROOF ONLY (inches)[a]

ULTIMATE DESIGN WIND SPEED V_{ult} (mph)		GROUND SNOW LOAD (psf)	BUILDING WIDTH (ft)														
			24			28			32			36			40		
			Wall Height (feet)			Wall Height (feet)			Wall Height (feet)			Wall Height (feet)			Wall Height (feet)		
Exp. B	Exp. C		8	9	10	8	9	10	8	9	10	8	9	10	8	9	10
110	—	20	4.5	4.5	4.5	4.5	4.5	4.5	4.5	4.5	6.5	4.5	4.5	DR	4.5	4.5	DR
		30	4.5	4.5	4.5	4.5	4.5	4.5	4.5	4.5	6.5	4.5	4.5	DR	4.5	6.5	DR
		50	4.5	4.5	4.5	4.5	4.5	6.5	4.5	4.5	DR	4.5	DR	DR	DR	DR	DR
		70	4.5	4.5	6.5	4.5	4.5	DR	4.5	DR	DR	DR	DR	DR	DR	DR	DR
115	—	20	4.5	4.5	4.5	4.5	4.5	6.5	4.5	4.5	DR	4.5	4.5	DR	4.5	DR	DR
		30	4.5	4.5	4.5	4.5	4.5	6.5	4.5	4.5	DR	4.5	6.5	DR	4.5	DR	DR
		50	4.5	4.5	6.5	4.5	4.5	DR	4.5	DR	DR	4.5	DR	DR	DR	DR	DR
		70	4.5	4.5	DR	4.5	DR	DR	DR	DR	DR	DR	DR	DR	DR	DR	DR
120	—	20	4.5	4.5	6.5	4.5	4.5	DR	4.5	4.5	DR	4.5	DR	DR	4.5	DR	DR
		30	4.5	4.5	DR	4.5	4.5	DR	4.5	6.5	DR	4.5	DR	DR	DR	DR	DR
		50	4.5	4.5	DR	4.5	DR	DR	4.5	DR	DR	DR	DR	DR	DR	DR	DR
		70	4.5	DR	DR	4.5	DR	DR	DR	DR	DR	DR	DR	DR	DR	DR	DR
130	110	20	4.5	6.5	DR	4.5	DR	DR	4.5	DR	DR	DR	DR	DR	DR	DR	DR
		30	4.5	DR	DR	4.5	DR	DR	DR	DR	DR	DR	DR	DR	DR	DR	DR
		50	4.5	DR	DR	DR	DR	DR	DR	DR	DR	DR	DR	DR	DR	DR	DR
		70	DR	DR	DR	DR	DR	DR	DR	DR	DR	DR	DR	DR	DR	DR	DR

For SI: 1 inch = 25.4 mm, 1 foot = 304.8 mm, 1 pound per square foot = 0.0479 kPa, 1 mile per hour = 0.447 m/s.
DR = Design Required.

a. Design assumptions:
Maximum deflection criteria: L/240.
Maximum roof dead load: 10 psf.
Maximum roof live load: 70 psf.
Maximum ceiling dead load: 5 psf.
Maximum ceiling live load: 20 psf.
Maximum second-floor dead load: 10 psf.
Maximum second-floor live load: 30 psf.
Maximum second-floor dead load from walls: 10 psf.
Maximum first-floor dead load: 10 psf.
Maximum first-floor live load: 40 psf.
Wind loads based on Table R301.2.1(1).
Strength axis of facing material applied vertically.

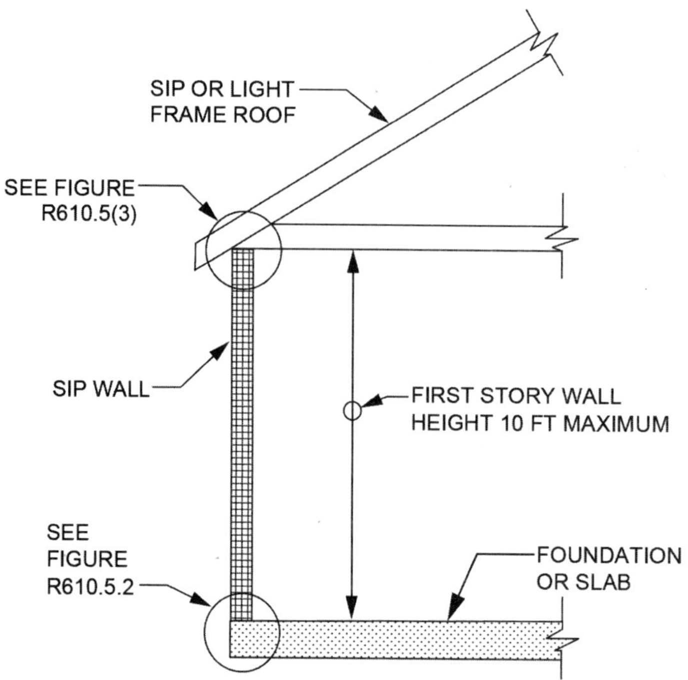

For SI: 1 foot = 304.8 mm.

Note: Figure illustrates SIP-specific attachment requirements. Other connections shall be made in accordance with Tables R602.3(1) and R602.3(2), as appropriate.

FIGURE R610.5(1)
MAXIMUM ALLOWABLE HEIGHT OF SIP WALLS

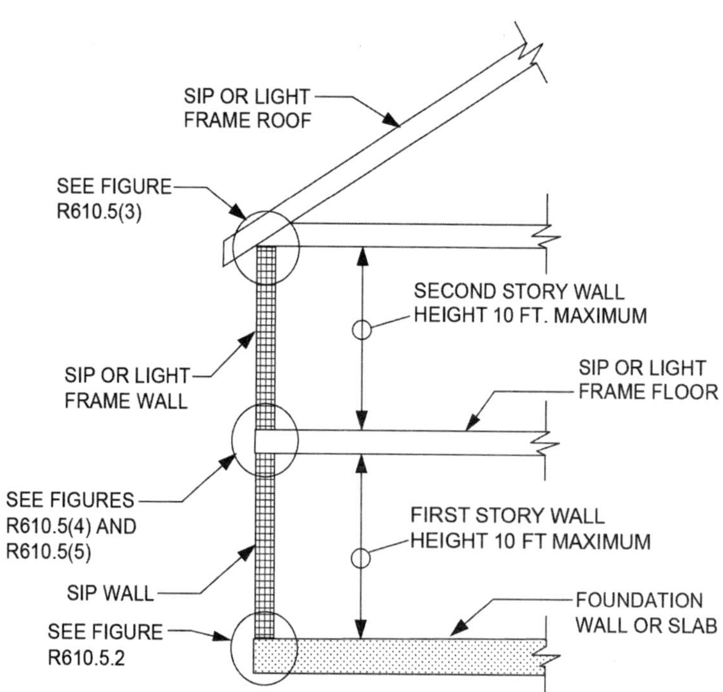

For SI: 1 foot = 304.8 mm.

Note: Figure illustrates SIP-specific attachment requirements. Other connections shall be made in accordance with Tables R602.3(1) and R602.3(2), as appropriate.

FIGURE R610.5(2)
MAXIMUM ALLOWABLE HEIGHT OF SIP WALLS

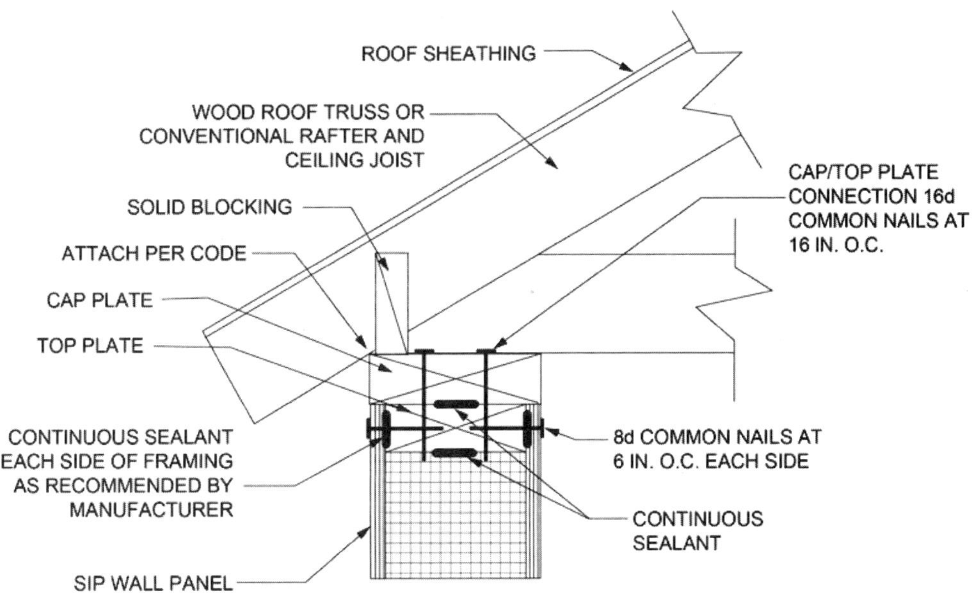

For SI: 1 inch = 25.4 mm.

Note: Figure illustrates SIP-specific attachment requirements. Other connections shall be made in accordance with Tables R602.3(1) and R602.3(2), as appropriate.

FIGURE R610.5(3)
TRUSSED ROOF TO TOP PLATE CONNECTION

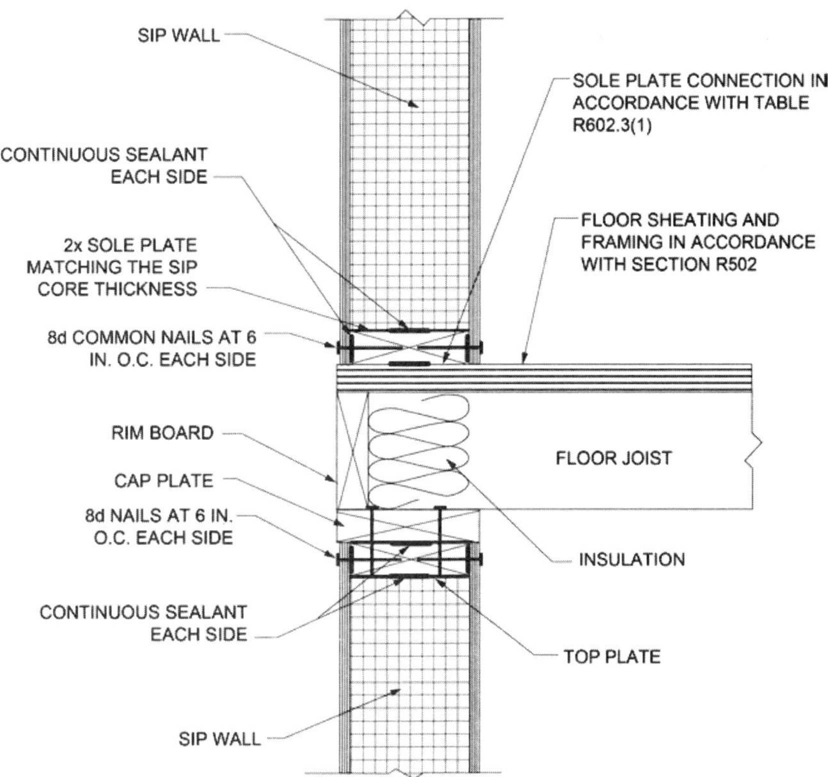

For SI: 1 inch = 25.4 mm.

Note: Figure illustrates SIP-specific attachment requirements. Other connections shall be made in accordance with Tables R602.3(1) and R602.3(2), as appropriate.

FIGURE R610.5(4)
SIP WALL-TO-WALL PLATFORM FRAME CONNECTION

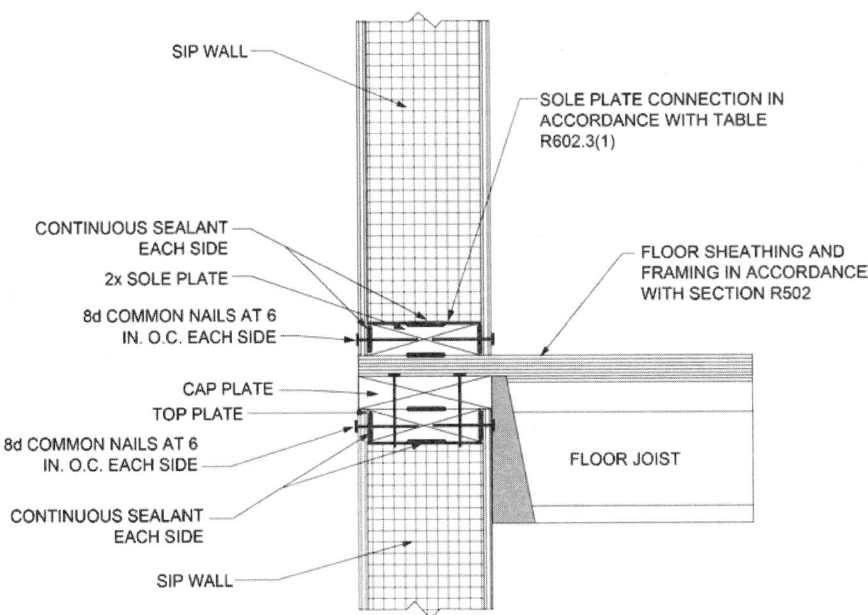

For SI: 1 inch = 25.4 mm.

Note: Figure illustrates SIP-specific attachment requirements. Other connections shall be made in accordance with Tables R602.3(1) and R602.3(2), as appropriate.

FIGURE R610.5(5)
SIP WALL-TO-WALL HANGING FLOOR FRAME CONNECTION
(I-Joist floor shown for Illustration only)

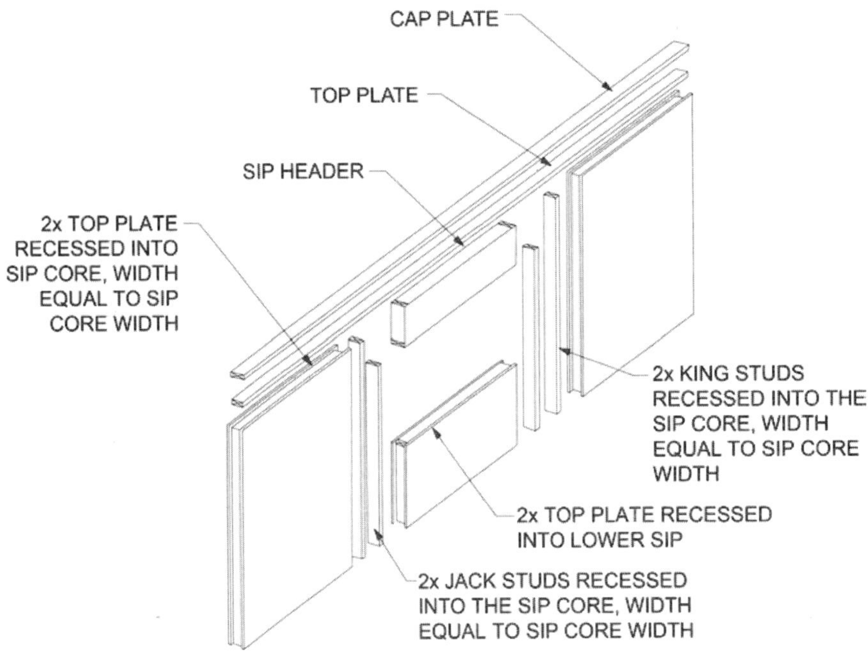

For SI: 1 inch = 25.4 mm.

Notes:

1. Top plates shall be continuous over header.
2. Lower 2x top plate shall have a width equal to the SIP core width and shall be recessed into the top edge of the panel. Cap plate shall be placed over the recessed top plate and shall have a width equal to the SIP's width.
3. SIP facing surfaces shall be nailed to framing and cripples with 8d common or galvanized box nails spaced 6 inches on center.

FIGURE R610.5.1
SIP WALL FRAMING CONFIGURATION

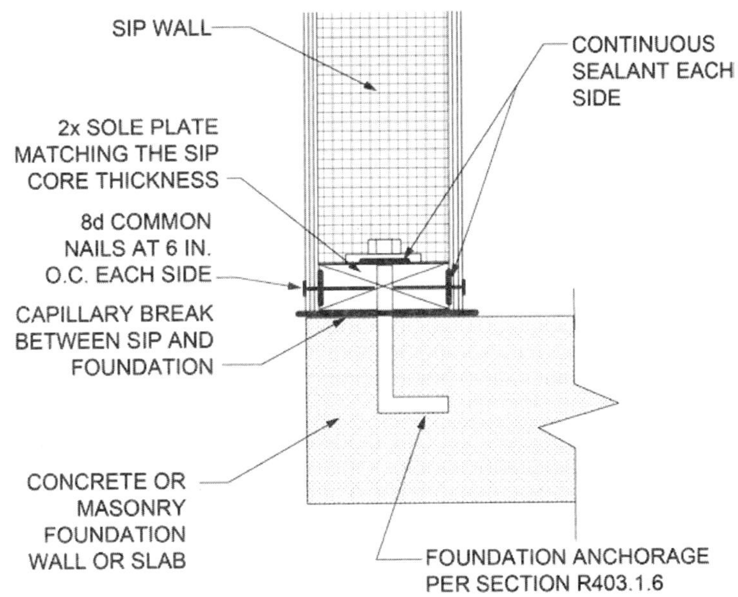

For SI: 1 inch = 25.4 mm.

FIGURE R610.5.2
SIP WALL TO CONCRETE SLAB FOR FOUNDATION WALL ATTACHMENT

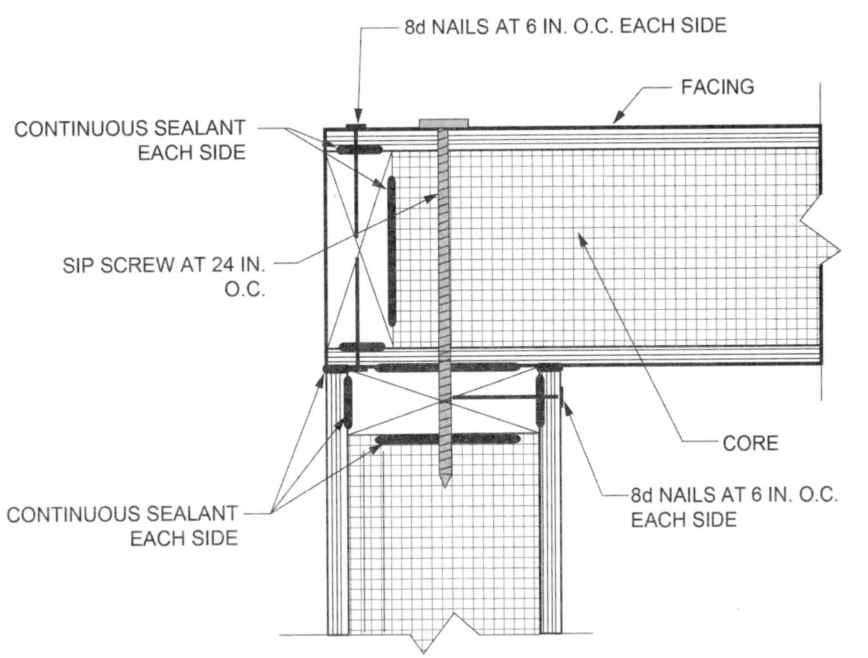

For SI: 1 inch = 25.4 mm.

FIGURE R610.5.4
SIP CORNER FRAMING DETAIL

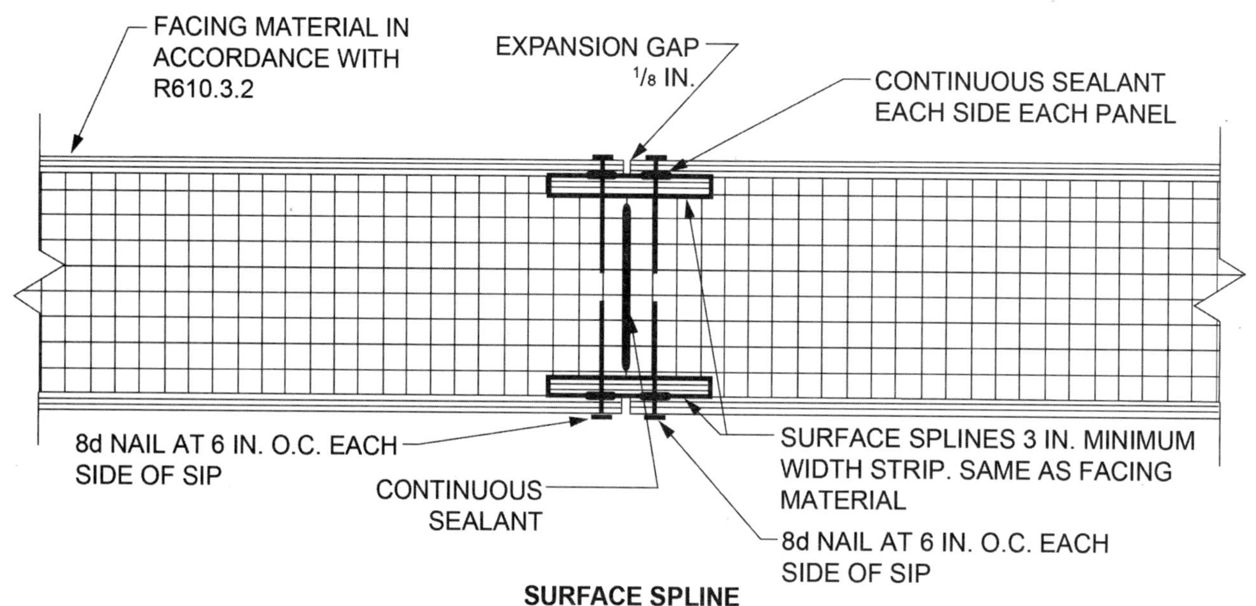

SURFACE SPLINE

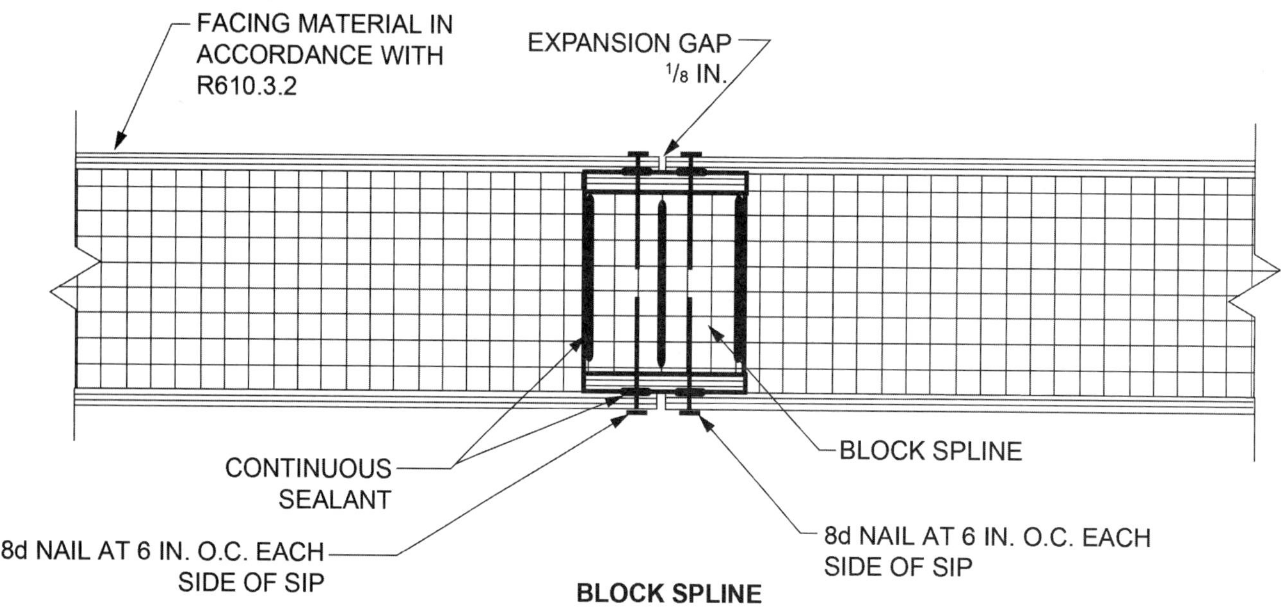

BLOCK SPLINE

For SI: 1 inch = 25.4 mm.

FIGURE R610.8
TYPICAL SIP WALL PANEL-TO-PANEL CONNECTION DETAILS

**TABLE R610.8
MAXIMUM SPANS FOR 11$^7/_8$-INCH OR DEEPER SIP HEADERS (feet)[a, c, d]**

LOAD CONDITION	GROUND SNOW LOAD (psf)	BUILDING[b] width (feet)				
		24	28	32	36	40
Supporting roof only	20	4	4	4	4	2
	30	4	4	4	2	2
	50	2	2	2	2	2
	70	2	2	2	DR	DR
Supporting roof and one-story	20	2	2	DR	DR	DR
	30	2	2	DR	DR	DR
	50	2	DR	DR	DR	DR
	70	DR	DR	DR	DR	DR

For SI: 1 inch = 25.4 mm, 1 foot = 304.8 mm, 1 pound per square foot = 0.0479 kPa.

DR = Design Required.

a. Design assumptions:
   Maximum deflection criterion: L/240.
   Maximum roof dead load: 10 psf.
   Maximum ceiling load: 5 psf.
   Maximum ceiling live load: 20 psf
   Maximum second-floor live load: 30 psf.
   Maximum second-floor dead load: 10 psf.
   Maximum second-floor dead load from walls: 10 psf.
   Maximum first floor dead load: 10 psf.
   Wind loads based on Table R301.2.1(1).
   Strength axis of facing material applied horizontally.

b. Building width is in the direction of horizontal framing members supported by the header.

c. The table provides for roof slopes between 3:12 and 12:12.

d. The maximum roof overhang is 24 inches (610 mm).

CHAPTER 7
WALL COVERING

User note:

About this chapter: *Chapter 7 establishes the various types of materials, materials standards and methods of application permitted as interior and exterior wall coverings. Interior coverings include interior plaster, gypsum board, ceramic tile, wood veneer paneling, hardboard paneling, wood shakes and wood shingles. Exterior wall coverings regulated by this section include aluminum, stone and masonry veneer, wood, hardboard, particleboard, wood structural panel siding, wood shakes and shingles, exterior plaster, steel, vinyl, fiber cement and exterior insulation finish systems. This chapter also contains requirements for the use of vapor retarders for moisture control in walls; wind resistance and water-resistive barriers for exterior wall coverings; and the water-resistive barrier required beneath exterior materials.*

SECTION R701 GENERAL

R701.1 Application. The provisions of this chapter shall control the design and construction of the interior and exterior wall covering for buildings.

R701.2 Installation. Products sensitive to adverse weather shall not be installed until adequate weather protection for the installation is provided. Exterior sheathing shall be dry before applying exterior cover.

SECTION R702 INTERIOR COVERING

R702.1 General. Interior coverings or wall finishes shall be installed in accordance with this chapter and Tables R702.1(1), R702.1(2), R702.1(3) and R702.3.5. Interior masonry veneer shall comply with the requirements of Section R703.7.1 for support and Section R703.7.4 for anchorage, except an airspace is not required. Interior finishes and materials shall conform to the flame spread and smoke-development requirements of Section R302.9.

R702.2 Interior plaster.

R702.2.1 Gypsum plaster. Gypsum plaster materials shall conform to ASTM C5, C22, C28, C35, C59, C61, C587, C631, C847, C933, C1032 and C1047, and shall be installed or applied in compliance with ASTM C841, C842 and C843. Gypsum lath or gypsum base for veneer plaster shall conform to ASTM C1396 and shall be installed in compliance with ASTM C844. Plaster shall be not less than three coats where applied over metal lath and not less than two coats where applied over other bases permitted by this section, except that veneer plaster shall be applied in one coat not to exceed $^{3}/_{16}$ inch (4.76 mm) thickness, provided the total thickness is in accordance with Table R702.1(1).

TABLE R702.1(1)
THICKNESS OF PLASTER

PLASTER BASE	FINISHED THICKNESS OF PLASTER FROM FACE OF LATH, MASONRY, CONCRETE (inches)	
	Gypsum Plaster	Cement Plaster
Expanded metal lath	$^{5}/_{8}$, minimum[a]	$^{5}/_{8}$, minimum[a]
Wire lath	$^{5}/_{8}$, minimum[a]	$^{3}/_{4}$, minimum (interior)[b]
		$^{7}/_{8}$, minimum (exterior)[b]
Gypsum lath[g]	$^{1}/_{2}$, minimum	$^{3}/_{4}$, minimum (interior)[b]
Masonry walls[c]	$^{1}/_{2}$, minimum	$^{1}/_{2}$, minimum
Monolithic concrete walls[c, d]	$^{5}/_{8}$, maximum	$^{7}/_{8}$, maximum
Monolithic concrete ceilings[c, d]	$^{3}/_{8}$, maximum[e]	$^{1}/_{2}$, maximum
Gypsum veneer base[f, g]	$^{1}/_{16}$, minimum	$^{3}/_{4}$, minimum (interior)[b]
Gypsum sheathing[g]	—	$^{3}/_{4}$, minimum (interior)[b]
		$^{7}/_{8}$, minimum (exterior)[b]

For SI: 1 inch = 25.4 mm.

a. Where measured from back plane of expanded metal lath, exclusive of ribs, or self-furring lath, plaster thickness shall be $^{3}/_{4}$ inch minimum.

b. Where measured from face of support or backing.

c. Because masonry and concrete surfaces vary in plane, thickness of plaster need not be uniform.

d. Where applied over a liquid bonding agent, finish coat shall be permitted to be applied directly to concrete surface.

e. Approved acoustical plaster shall be permitted to be applied directly to concrete or over base coat plaster, beyond the maximum plaster thickness shown.

f. Attachment shall be in accordance with Table R702.3.5.

g. Where gypsum board is used as a base for cement plaster, a water-resistive barrier complying with Section R703.2 shall be provided.

TABLE R702.1(2)
GYPSUM PLASTER PROPORTIONS[a]

NUMBER	COAT	PLASTER BASE OR LATH	MAXIMUM VOLUME AGGREGATE PER 100 POUNDS NEAT PLASTER[b] (cubic feet)	
			Damp Loose Sand[a]	Perlite or Vermiculite[c]
Two-coat work	Base coat	Gypsum lath	2.5	2
	Base coat	Masonry	3	3
Three-coat work	First coat	Lath	2[d]	2
	Second coat	Lath	3[d]	2[e]
	First and second coats	Masonry	3	3

For SI: 1 inch = 25.4 mm, 1 cubic foot = 0.0283 m³, 1 pound = 0.454 kg.

a. Wood-fibered gypsum plaster shall be mixed in the proportions of 100 pounds of gypsum to not more than 1 cubic foot of sand where applied on masonry or concrete.

b. Where determining the amount of aggregate in set plaster, a tolerance of 10 percent shall be allowed.

c. Combinations of sand and lightweight aggregate shall be permitted to be used, provided the volume and weight relationship of the combined aggregate to gypsum plaster is maintained.

d. If used for both first and second coats, the volume of aggregate shall be permitted to be 2.5 cubic feet.

e. Where plaster is 1 inch or more in total thickness, the proportions for the second coat may be increased to 3 cubic feet.

TABLE R702.1(3)
CEMENT PLASTER PROPORTIONS, PARTS BY VOLUME

COAT	CEMENT PLASTER TYPE	CEMENTITIOUS MATERIALS				VOLUME OF AGGREGATE PER SUM OF SEPARATE VOLUMES OF CEMENTITIOUS MATERIALS[b]
		Portland Cement Type I, II or III; Blended Hydraulic Cement Type IP, I (S < 70), IL, or IT (S < 70); or Hydraulic Cement Type GU, HE, MS, HS or MH	Plastic Cement	Masonry Cement Type M, S or N	Lime	
First	Portland or blended	1	—	—	$^3/_4$–$1^1/_2$[a]	$2^1/_2$–4
	Masonry	—	—	1	—	$2^1/_2$–4
	Plastic	—	1	—	—	$2^1/_2$–4
Second	Portland or blended	1	—	—	$^3/_4$–$1^1/_2$	3–5
	Masonry	—	—	1	—	3–5
	Plastic	—	1	—	—	3–5
Finish	Portland or blended	1	—	—	$1^1/_2$–2	$1^1/_2$–3
	Masonry	—	—	1	—	$1^1/_2$–3
	Plastic	—	1	—	—	$1^1/_2$–3

a. Lime by volume of 0 to $^3/_4$ shall be used where the plaster will be placed over low-absorption surfaces such as dense clay tile or brick.

b. The same or greater sand proportion shall be used in the second coat than used in the first coat.

R702.2.2 Cement plaster. Cement plaster materials shall conform to ASTM C91 (Type M, S or N), C150 (Types I, II and III), C595 [Types IP, I (PM), IS and I (SM)], C847, C897, C933, C1032, C1047 and C1328, and shall be installed or applied in compliance with ASTM C926 and C1063. Gypsum lath shall conform to ASTM C1396. Plaster shall be not less than three coats where applied over metal lath and not less than two coats where applied over other bases permitted by this section.

R702.2.2.1 Application. Each coat shall be kept in a moist condition for not less than 24 hours prior to application of the next coat.

Exception: Applications installed in accordance with ASTM C926.

R702.2.2.2 Curing. The finish coat for two-coat cement plaster shall not be applied sooner than 48 hours after application of the first coat. For three-coat cement plaster, the second coat shall not be applied sooner than 24 hours after application of the first coat. The finish coat for three-coat cement plaster shall not be applied sooner than 48 hours after application of the second coat.

R702.2.3 Support. Support spacing for gypsum or metal lath on walls or ceilings shall not exceed 16 inches (406 mm) for $^3/_8$-inch-thick (9.5 mm) or 24 inches (610 mm) for $^1/_2$-inch-thick (12.7 mm) plain gypsum lath. Gypsum lath shall be installed at right angles to support framing with end joints in adjacent courses staggered by not less than one framing space.

R702.3 Gypsum board and gypsum panel products.

R702.3.1 Materials. Gypsum board and gypsum panel product materials and accessories shall conform to ASTM C22, C475, C514, C1002, C1047, C1177, C1178, C1278, C1396, C1658 or C1766 and shall be installed in accordance with the provisions of this section. Adhesives for the installation of gypsum board and gypsum panel products shall conform to ASTM C557.

R702.3.1.1 Adhesives. Expandable foam adhesives for the installation of gypsum board and gypsum panel products shall conform to ASTM D6464. Other adhesives for the installation of gypsum board and gypsum panel products shall conform to ASTM C557. Supports and fasteners used to attach gypsum board and gypsum panel products shall comply with Table R702.3.5 or other *approved* method.

R702.3.2 Wood framing. Wood framing supporting gypsum board and gypsum panel products shall be not less than 2 inches (51 mm) nominal thickness in the least dimension except that wood furring strips not less than 1-inch by 2-inch (25 mm by 51 mm) nominal dimension shall be permitted to be used over solid backing or framing spaced not more than 24 inches (610 mm) on center.

R702.3.3 Cold-formed steel framing. Cold-formed steel framing supporting gypsum board and gypsum panel products shall be not less than $1^1/_4$ inches (32 mm) wide in the least dimension. Nonload-bearing cold-formed steel framing shall comply with AISI S220. Load-bearing cold-formed steel framing shall comply with AISI S240.

R702.3.4 Insulating concrete form walls. Foam plastics for insulating concrete form walls constructed in accordance with Sections R404.1.2 and R608 on the interior of *habitable spaces* shall be protected in accordance with Section R316.4. Use of adhesives in conjunction with mechanical fasteners is permitted. Adhesives used for interior and exterior finishes shall be compatible with the insulating form materials.

R702.3.5 Application. Supports and fasteners used to attach gypsum board and gypsum panel products shall comply with Table R702.3.5. Gypsum sheathing shall be attached to exterior walls in accordance with Table R602.3(1). Gypsum board and gypsum panel products shall be applied at right angles or parallel to framing members. All edges and ends of gypsum board and gypsum panel products shall occur on the framing members, except those edges and ends that are perpendicular to the framing members. Interior gypsum board shall not be installed where it is directly exposed to the weather or to water.

R702.3.5.1 Screw fastening. Screws for attaching gypsum board and gypsum panel products to wood framing shall be Type W or Type S in accordance with ASTM C1002 and shall penetrate the wood not less than $^5/_8$ inch (15.9 mm). Gypsum board and gypsum panel products shall be attached to cold-formed steel framing with minimum No. 6 screws. Screws for attaching gypsum board and gypsum panel products to cold-formed steel framing less than 0.033 inch (1 mm) thick shall be Type S in accordance with ASTM C1002 or bugle head style in accordance with ASTM C1513 and shall penetrate the steel not less than $^3/_8$ inch (9.5 mm). Screws for attaching gypsum board and gypsum panel products to cold-formed steel framing 0.033 inch to 0.112 inch (1 mm to 3 mm) thick shall be in accordance with ASTM C954 or bugle head style in accordance with ASTM C1513. Screws for attaching gypsum board and gypsum panel products to structural insulated panels shall penetrate the wood structural panel facing not less than $^7/_{16}$ inch (11.1 mm).

R702.3.6 Horizontal gypsum board diaphragm ceilings. Gypsum board and gypsum panel products shall be permitted on wood joists to create a horizontal *diaphragm* in accordance with Table R702.3.6. Gypsum board and gypsum panel products shall be installed perpendicular to ceiling framing members. End joints of adjacent courses of board and panels shall not occur on the same joist. The maximum allowable *diaphragm* proportions shall be $1^1/_2$:1 between shear resisting elements. Rotation or cantilever conditions shall not be permitted. Gypsum board or gypsum panel products shall not be used in *diaphragm* ceilings to resist lateral forces imposed by masonry or concrete construction. Perimeter edges shall be blocked using wood members not less than 2-inch by 6-inch (51 mm by 152 mm) nominal dimension. Blocking material shall be installed flat over the top plate of the wall to provide a nailing surface not less than 2 inches (51 mm) in width for the attachment of the gypsum board or gypsum panel product.

R702.3.7 Water-resistant gypsum backing board. Gypsum board used as the base or backer for adhesive application of ceramic tile or other required nonabsorbent finish material shall conform to ASTM C1178, C1278 or C1396. Use of water-resistant gypsum backing board shall be permitted on ceilings. Water-resistant gypsum board shall not be installed over a Class I or II vapor retarder in a shower or tub compartment. Cut or exposed edges, including those at wall intersections, shall be sealed as recommended by the manufacturer.

R702.3.7.1 Limitations. Water-resistant gypsum backing board shall not be used where there will be direct exposure to water, or in areas subject to continuous high humidity.

R702.4 Ceramic tile.

R702.4.1 General. Ceramic tile surfaces shall be installed in accordance with ANSI A108.1, A108.4, A108.5, A108.6, A108.11, A118.1, A118.3, A136.1 and A137.1.

TABLE R702.3.5
MINIMUM THICKNESS AND APPLICATION OF GYPSUM BOARD AND GYPSUM PANEL PRODUCTS

THICKNESS OF GYPSUM BOARD OR GYPSUM PANEL PRODUCTS (inches)	APPLICATION	ORIENTATION OF GYPSUM BOARD OR GYPSUM PANEL PRODUCTS TO FRAMING	MAXIMUM SPACING OF FRAMING MEMBERS (inches o.c.)	MAXIMUM SPACING OF FASTENERS (inches)		SIZE OF NAILS FOR APPLICATION TO WOOD FRAMING[c]
				Nails[a]	Screws[b]	
Application without adhesive						
$^3/_8$	Ceiling[d]	Perpendicular	16	7	12	13 gage, $1^1/_4$″ long, $^{19}/_{64}$″ head; 0.098″ diameter, $1^1/_4$″ long, ring shank; or 4d cooler nail, 0.080″ diameter, $1^3/_8$″ long, $^7/_{32}$″ head.
	Wall	Either direction	16	8	16	
$^1/_2$	Ceiling	Either direction	16	7	12	13 gage, $1^3/_8$″ long, $^{19}/_{64}$″ head; 0.098″ diameter, $1^1/_4$″ long, ring shank; 5d cooler nail, 0.086″ diameter, $1^5/_8$″ long, $^{15}/_{64}$″ head; or gypsum board nail, 0.086″ diameter, 1 $^5/_8$″ long, $^9/_{32}$″ head.
	Ceiling[d]	Perpendicular	24	7	12	
	Wall	Either direction	24	8	12	
	Wall	Either direction	16	8	16	
$^5/_8$	Ceiling	Either direction	16	7	12	13 gage, $1^5/_8$″ long, $^{19}/_{64}$″ head; 0.098″ diameter, $1^3/_8$″ long, ring shank; 6d cooler nail, 0.092″ diameter, $1^7/_8$″ long, $^1/_4$″ head; or gypsum board nail, 0.0915″ diameter, $1^7/_8$″ long, $^{19}/_{64}$″ head.
	Ceiling	Perpendicular	24	7	12	
	Type X at garage ceiling beneath habitable rooms	Perpendicular	24	6	6	$1^7/_8$″ long 0.099″ diameter galvanized nails or equivalent drywall screws. Screws shall comply with Section R702.3.5.1.
	Wall	Either direction	24	8	12	13 gage, $1^5/_8$″ long, $^{19}/_{64}$″ head; 0.098″ diameter, $1^3/_8$″ long, ring shank; 6d cooler nail, 0.092″ diameter, $1^7/_8$″ long, $^1/_4$″ head; or gypsum board nail, 0.0915″ diameter, $1^7/_8$″ long, $^{19}/_{64}$″ head.
	Wall	Either direction	16	8	16	
Application with adhesive						
$^3/_8$	Ceiling[d]	Perpendicular	16	16	16	Same as above for $^3/_8$″ gypsum board and gypsum panel products.
	Wall	Either direction	16	16	24	
$^1/_2$ or $^5/_8$	Ceiling	Either direction	16	16	16	Same as above for $^1/_2$″ and $^5/_8$″ gypsum board and gypsum panel products, respectively.
	Ceiling[d]	Perpendicular	24	12	16	
	Wall	Either direction	24	16	24	
Two $^3/_8$ layers	Ceiling	Perpendicular	16	16	16	Base ply nailed as above for $^1/_2$″ gypsum board and gypsum panel products; face ply installed with adhesive.
	Wall	Either direction	24	24	24	

For SI: 1 inch = 25.4 mm.

a. For application without adhesive, a pair of nails spaced not less than 2 inches apart or more than $2^1/_2$ inches apart shall be permitted to be used with the pair of nails spaced 12 inches on center.

b. Screws shall be in accordance with Section R702.3.5.1. Screws for attaching gypsum board or gypsum panel products to structural insulated panels shall penetrate the wood structural panel facing not less than $^7/_{16}$ inch.

c. Where cold-formed steel framing is used with a clinching design to receive nails by two edges of metal, the nails shall be not less than $^5/_8$ inch longer than the gypsum board or gypsum panel product thickness and shall have ringed shanks. Where the cold-formed steel framing has a nailing groove formed to receive the nails, the nails shall have barbed shanks or be 0.086-inch diameter, $1^5/_8$ inches long, $^{15}/_{64}$-inch head for $^1/_2$-inch gypsum board or gypsum panel product; and 0.099-inch diameter, $1^7/_8$ inches long, $^{15}/_{64}$-inch head for $^5/_8$-inch gypsum board or gypsum panel product.

d. Three-eighths-inch-thick single-ply gypsum board or gypsum panel product shall not be used on a ceiling where a water-based textured finish is to be applied, or where it will be required to support insulation above a ceiling. On ceiling applications to receive a water-based texture material, either hand or spray applied, the gypsum board or gypsum panel product shall be applied perpendicular to framing. Where applying a water-based texture material, the minimum gypsum board thickness shall be increased from $^3/_8$ inch to $^1/_2$ inch for 16-inch on center framing, and from $^1/_2$ inch to $^5/_8$ inch for 24-inch on center framing or $^1/_2$-inch sag-resistant gypsum ceiling board shall be used.

TABLE R702.3.6
SHEAR CAPACITY FOR HORIZONTAL WOOD-FRAMED GYPSUM BOARD DIAPHRAGM CEILING ASSEMBLIES

MATERIAL	THICKNESS OF MATERIAL (min.) (inch)	SPACING OF FRAMING MEMBERS (max.) (inch)	SHEAR VALUE[a, b] (plf of ceiling)	MINIMUM FASTENER SIZE[c, d]
Gypsum board or gypsum panel product	$^{1}/_{2}$	16 o.c.	90	5d cooler or wallboard nail; $1^{5}/_{8}$-inch long; 0.086-inch shank; $^{15}/_{64}$-inch head
Gypsum board or gypsum panel product	$^{1}/_{2}$	24 o.c.	70	5d cooler or wallboard nail; $1^{5}/_{8}$-inch long; 0.086-inch shank; $^{15}/_{64}$-inch head

For SI: 1 inch = 25.4 mm, 1 pound per linear foot = 1.488 kg/m.

a. Values are not cumulative with other horizontal diaphragm values and are for short-term loading caused by wind or seismic loading. Values shall be reduced 25 percent for normal loading.

b. Values shall be reduced 50 percent in *Seismic Design Categories* D_0, D_1, D_2 and E.

c. $1^{1}/_{4}$-inch, No. 6 Type S or W screws shall be permitted to be substituted for the listed nails.

d. Fasteners shall be spaced not more than 7 inches on center at all supports, including perimeter blocking, and not less than $^{3}/_{8}$ inch from the edges and ends of the gypsum board.

R702.4.2 Backer boards. Materials used as backers for wall tile in tub and shower areas and wall panels in shower areas shall be of materials listed in Table R702.4.2, and installed in accordance with the manufacturer's recommendations.

TABLE R702.4.2
BACKER BOARD MATERIALS

MATERIAL	STANDARD
Glass mat gypsum backing panel	ASTM C1178
Fiber-reinforced gypsum panels	ASTM C1278
Nonasbestos fiber-cement backer board	ASTM C1288 or ISO 8336, Category C
Nonasbestos fiber mat-reinforced cementitious backer units	ASTM C1325

R702.5 Other finishes. Wood veneer paneling and hardboard paneling shall be placed on wood or cold-formed steel framing spaced not more than 16 inches (406 mm) on center. Wood veneer and hard board paneling less than $^{1}/_{4}$-inch (6 mm) nominal thickness shall not have less than a $^{3}/_{8}$-inch (10 mm) gypsum board or gypsum panel product backer. Wood veneer paneling not less than $^{1}/_{4}$-inch (6 mm) nominal thickness shall conform to ANSI/HPVA HP-1. Hardboard paneling shall conform to CPA/ANSI A135.5.

R702.6 Wood shakes and shingles. Wood shakes and shingles shall conform to CSSB *Grading Rules for Wood Shakes and Shingles* and shall be permitted to be installed directly to the studs with maximum 24 inches (610 mm) on-center spacing.

R702.6.1 Attachment. Nails, staples or glue are permitted for attaching shakes or shingles to the wall, and attachment of the shakes or shingles directly to the surface shall be permitted provided the fasteners are appropriate for the type of wall surface material. Where nails or staples are used, two fasteners shall be provided and shall be placed so that they are covered by the course above.

R702.6.2 Furring strips. Where furring strips are used, they shall be 1 inch by 2 inches or 1 inch by 3 inches (25 mm by 51 mm or 25 mm by 76 mm), spaced a distance on center equal to the desired exposure, and shall be attached to the wall by nailing through other wall material into the studs.

R702.7 Vapor retarders. Vapor retarder materials shall be classified in accordance with Table R702.7(1). A vapor retarder shall be provided on the interior side of frame walls of the class indicated in Table R702.7(2), including compliance with Table R702.7(3) or R702.7(4) where applicable. An *approved* design using accepted engineering practice for hygrothermal analysis shall be permitted as an alternative. The climate zone shall be determined in accordance with Section N1101.7.

Exceptions:

1. *Basement walls.*
2. Below-grade portion of any wall.
3. Construction where accumulation, condensation or freezing of moisture will not damage the materials.
4. A vapor retarder shall not be required in Climate Zones 1, 2 and 3.

R702.7.1 Spray foam plastic insulation for moisture control with Class II and III vapor retarders. For purposes of compliance with Tables R702.7(3) and R702.7(4), spray foam with a maximum permeance of 1.5 perms at the installed thickness applied to the interior side of wood structural panels, fiberboard, *insulating sheathing* or gypsum shall be deemed to meet the continuous insulation moisture control requirement in accordance with one of the following conditions:

1. The spray foam *R*-value is equal to or greater than the specified continuous insulation *R*-value.
2. The combined *R*-value of the spray foam and continuous insulation is equal to or greater than the specified continuous insulation *R*-value.

TABLE R702.7(1)
VAPOR RETARDER MATERIALS AND CLASSES

CLASS	ACCEPTABLE MATERIALS
I	Sheet polyethylene, nonperforated aluminum foil or other approved materials with a perm rating less than or equal to 0.1.
II	Kraft-faced fiberglass batts, vapor retarder paint or other approved materials applied in accordance with the manufacturer's installation instructions for a perm rating greater than 0.1 and less than or equal to 1.0.
III	Latex paint, enamel paint or other approved materials applied in accordance with the manufacturer's installation instructions for a perm rating greater than 1.0 and less than or equal to 10.0.

TABLE R702.7(2)
VAPOR RETARDER OPTIONS

CLIMATE ZONE	VAPOR RETARDER CLASS		
	CLASS I[a]	CLASS II[a]	CLASS III
1, 2	Not Permitted	Not Permitted	Permitted
3, 4 (except Marine 4)	Not Permitted	Permitted[c]	Permitted
Marine 4, 5, 6, 7, 8	Permitted[b]	Permitted[c]	See Table R702.7(3)

a. Class I and II vapor retarders with vapor permeance greater than 1 perm when measured by ASTM E96 water method (Procedure B) shall be allowed on the interior side of any frame wall in all climate zones.

b. Use of a Class I interior vapor retarder in frame walls with a Class I vapor retarder on the exterior side shall require an approved design.

c. Where a Class II vapor retarder is used in combination with foam plastic insulating sheathing installed as continuous insulation on the exterior side of frame walls, the continuous insulation shall comply with Table R702.7(4) and the Class II vapor retarder shall have a vapor permeance greater than 1 perm when measured by ASTM E96 water method (Procedure B).

TABLE R702.7(3)
CLASS III VAPOR RETARDERS

CLIMATE ZONE	CLASS III VAPOR RETARDERS PERMITTED FOR:[a, b]
Marine 4	Vented cladding over wood structural panels.
	Vented cladding over fiberboard.
	Vented cladding over gypsum.
	Continuous insulation with *R*-value ≥ 2.5 over 2 × 4 wall.
	Continuous insulation with *R*-value ≥ 3.75 over 2 × 6 wall.
5	Vented cladding over wood structural panels.
	Vented cladding over fiberboard.
	Vented cladding over gypsum.
	Continuous insulation with *R*-value ≥ 5 over 2 × 4 wall.
	Continuous insulation with *R*-value ≥ 7.5 over 2 × 6 wall.
6	Vented cladding over fiberboard.
	Vented cladding over gypsum.
	Continuous insulation with *R*-value ≥ 7.5 over 2 × 4 wall.
	Continuous insulation with *R*-value ≥ 11.25 over 2 × 6 wall.
7	Continuous insulation with *R*-value ≥ 10 over 2 × 4 wall.
	Continuous insulation with *R*-value ≥ 15 over 2 × 6 wall.
8	Continuous insulation with *R*-value ≥ 12.5 over 2 × 4 wall. Continuous insulation with *R*-value ≥ 20 over 2 × 6 wall.

a. Vented cladding shall include vinyl, polypropylene, or horizontal aluminum siding, brick veneer with a clear airspace as specified in Table R703.8.4(1), and other approved vented claddings.

b. The requirements in this table apply only to insulation used to control moisture in order to permit the use of Class III vapor retarders. The insulation materials used to satisfy this option also contribute to but do not supersede the thermal envelope requirements of Chapter 11.

**TABLE R702.7(4)
CONTINUOUS INSULATION WITH CLASS II VAPOR RETARDER**

CLIMATE ZONE	CLASS II VAPOR RETARDERS PERMITTED FOR:[a]
3	Continuous insulation with R-value ≥ 2.
4, 5 and 6	Continuous insulation with R-value ≥ 3 over 2 × 4 wall. Continuous insulation with R-value ≥ 5 over 2 × 6 wall.
7	Continuous insulation with R-value ≥ 5 over 2 × 4 wall. Continuous insulation with R-value ≥ 7.5 over 2 × 6 wall.
8	Continuous insulation with R-value ≥ 7.5 over 2 × 4 wall. Continuous insulation with R-value ≥ 10 over 2 × 6 wall.

a. The requirements in this table apply only to insulation used to control moisture in order to permit the use of Class II vapor retarders. The insulation materials used to satisfy this option also contribute to but do not supersede the thermal envelope requirements of Chapter 11.

SECTION R703
EXTERIOR COVERING

R703.1 General. Exterior walls shall provide the building with a weather-resistant exterior wall envelope. The exterior wall envelope shall include flashing as described in Section R703.4.

Exception: Log walls designed and constructed in accordance with the provisions of ICC 400.

R703.1.1 Water resistance. The exterior wall envelope shall be designed and constructed in a manner that prevents the accumulation of water within the wall assembly by providing a water-resistant barrier behind the exterior cladding as required by Section R703.2 and a means of draining to the exterior water that penetrates the exterior cladding.

Exceptions:

1. A weather-resistant exterior wall envelope shall not be required over concrete or masonry walls designed in accordance with Chapter 6 and flashed in accordance with Section R703.4 or R703.8.
2. Compliance with the requirements for a means of drainage, and the requirements of Sections R703.2 and R703.4, shall not be required for an exterior wall envelope that has been demonstrated to resist wind-driven rain through testing of the exterior wall envelope, including joints, penetrations and intersections with dissimilar materials, in accordance with ASTM E331 under the following conditions:
   - 2.1. Exterior wall envelope test assemblies shall include at least one opening, one control joint, one wall/eave interface and one wall sill. All tested openings and penetrations shall be representative of the intended end-use configuration.
   - 2.2. Exterior wall envelope test assemblies shall be at least 4 feet by 8 feet (1219 mm by 2438 mm) in size.
   - 2.3. Exterior wall assemblies shall be tested at a minimum differential pressure of 6.24 pounds per square foot (299 Pa).
   - 2.4. Exterior wall envelope assemblies shall be subjected to the minimum test exposure for a minimum of 2 hours.

The exterior wall envelope design shall be considered to resist wind-driven rain where the results of testing indicate that water did not penetrate control joints in the exterior wall envelope, joints at the perimeter of openings penetration or intersections of terminations with dissimilar materials.

R703.1.2 Wind resistance. Wall coverings, backing materials and their attachments shall be capable of resisting wind loads in accordance with Tables R301.2.1(1) and R301.2.1(2). Wind-pressure resistance of the siding, soffit and backing materials shall be determined by ASTM E330 or other applicable standard test methods. Where wind-pressure resistance is determined by design analysis, data from *approved* design standards and analysis conforming to generally accepted engineering practice shall be used to evaluate the siding, soffit and backing material and its fastening. All applicable failure modes including bending rupture of siding, fastener withdrawal and fastener head pull-through shall be considered in the testing or design analysis. Where the wall covering, soffit and backing material resist wind load as an assembly, use of the design capacity of the assembly shall be permitted.

R703.2 Water-resistive barrier. Not fewer than one layer of *water-resistive barrier* shall be applied over studs or sheathing of all exterior walls with flashing as indicated in Section R703.4, in such a manner as to provide a continuous water-resistive barrier behind the exterior wall veneer. The water-resistive barrier material shall be continuous to the top of walls and terminated at penetrations and building appendages in a manner to meet the requirements of the exterior wall envelope as described in Section R703.1. Water-resistive barrier materials shall comply with one of the following:

1. No. 15 felt complying with ASTM D226, Type 1.
2. ASTM E2568, Type 1 or 2.

3. ASTM E331 in accordance with Section R703.1.1.
4. Other approved materials in accordance with the manufacturer's installation instructions.

No.15 asphalt felt and *water-resistive barriers* complying with ASTM E2556 shall be applied horizontally, with the upper layer lapped over the lower layer not less than 2 inches (51 mm), and where joints occur, shall be lapped not less than 6 inches (152 mm).

R703.3 Wall covering nominal thickness and attachments. The nominal thickness and attachment of exterior wall coverings shall be in accordance with Table R703.3(1), the wall covering material requirements of this section, and the wall covering manufacturer's installation instructions. Cladding attachment over foam sheathing shall comply with the additional requirements and limitations of Sections R703.15 through R703.17. Nominal material thicknesses in Table R703.3(1) are based on a maximum stud spacing of 16 inches (406 mm) on center. Where specified by the siding manufacturer's instructions and supported by a test report or other documentation, attachment to studs with greater spacing is permitted. Fasteners for exterior wall coverings attached to wood framing shall be in accordance with Section R703.3.3 and Table R703.3(1). Exterior wall coverings shall be attached to cold-formed steel light frame construction in accordance with the cladding manufacturer's installation instructions, the requirements of Table R703.3(1) using screw fasteners substituted for the nails specified in accordance with Table R703.3(2), or an *approved* design.

R703.3.1 Soffit installation. Soffits shall comply with Section R704.

R703.3.2 Wind limitations. Where the design wind pressure exceeds 30 psf or where the limits of Table R703.3.2 are exceeded, the attachment of wall coverings and soffits shall be designed to resist the component and cladding loads specified in Table R301.2.1(1) for walls, adjusted for height and exposure in accordance with Table R301.2.1(2). For the determination of wall covering and soffit attachment, component and cladding loads shall be determined using an effective wind area of 10 square feet (0.93 m^2).

R703.3.3 Fasteners. Exterior wall coverings and roof overhang soffits shall be securely fastened with aluminum, galvanized, stainless steel or rust-preventative coated nails or staples in accordance with Table R703.3(1) or with other *approved* corrosion-resistant fasteners in accordance with the wall covering manufacturer's installation instructions. Nails and staples shall comply with ASTM F1667. Nails shall be T-head, modified round head, or round head with smooth or deformed shanks. Staples shall have a minimum crown width of $^7/_{16}$ inch (11.1 mm) outside diameter and be manufactured of minimum 16-gage wire. Where fiberboard, gypsum, or foam plastic sheathing backing is used, nails or staples shall be driven into the studs. Where wood or wood structural panel sheathing is used, fasteners shall be driven into studs unless otherwise permitted to be driven into sheathing in accordance with either the siding manufacturer's installation instructions or Table R703.3.3.

R703.3.4 Minimum fastener length and penetration. Fasteners shall have the greater of the minimum length specified in Table R703.3(1) or as required to provide a minimum penetration into framing as follows:

1. Fasteners for horizontal aluminum siding, steel siding, particleboard panel siding, wood structural panel siding in accordance with ANSI/APA-PRP 210, fiber-cement panel siding and fiber-cement lap siding installed over foam plastic sheathing shall penetrate not less than $1^1/_2$ inches (38 mm) into framing or shall be in accordance with the manufacturer's installation instructions.
2. Fasteners for hardboard panel and lap siding shall penetrate not less than $1^1/_2$ inches (38 mm) into framing.
3. Fasteners for vinyl siding and insulated vinyl siding installed over wood or wood structural panel sheathing shall penetrate not less than $1^1/_4$ inches (32 mm) into sheathing and framing combined. Vinyl siding and insulated vinyl siding shall be permitted to be installed with fasteners penetrating into or through wood or wood structural sheathing of minimum thickness as specified by the manufacturer's instructions or test report, with or without penetration into the framing. Where the fastener penetrates fully through the sheathing, the end of the fastener shall extend not less than $^1/_4$ inch (6.4 mm) beyond the opposite face of the sheathing. Fasteners for vinyl siding and insulated vinyl siding installed over foam plastic sheathing shall be in accordance with Section R703.11.2. Fasteners for vinyl siding and insulated vinyl siding installed over fiberboard or gypsum sheathing shall penetrate not less than $1^1/_4$ inches (32 mm) into framing.
4. Fasteners for vertical or horizontal wood siding shall penetrate not less than $1^1/_2$ inches (38 mm) into studs, studs and wood sheathing combined, or blocking.
5. Fasteners for siding material installed over foam plastic sheathing shall have sufficient length to accommodate foam plastic sheathing thickness and to penetrate framing or sheathing and framing combined, as specified in Items 1 through 4.

TABLE R703.3(1)
SIDING MINIMUM ATTACHMENT AND MINIMUM THICKNESS

SIDING MATERIAL		NOMINAL THICKNESS (inches)	JOINT TREATMENT	TYPE OF SUPPORTS FOR THE SIDING MATERIAL AND FASTENERS					
				Wood or wood structural panel sheathing into stud	Fiberboard sheathing into stud	Gypsum sheathing into stud	Foam plastic sheathing into stud[l]	Direct to studs	Number or spacing of fasteners
Anchored veneer: brick, concrete, masonry or stone (see Section R703.8)		2	Section R703.8	Section R703.8					
Adhered veneer: concrete, stone or masonry (see Section R703.12)		—	Section R703.12	Section R703.12					
Fiber cement siding	Panel siding (see Section R703.10.1)	$^{5}/_{16}$	Section R703.10.1	6d common (2″ × 0.113″)	6d common (2″ × 0.113″)	6d common (2″ × 0.113″)	6d common (2″ × 0.113″)	4d common ($1^{1}/_{2}$″ × 0.099″)	6″ panel edges 12″ inter. sup.
	Lap siding (see Section R703.10.2)	$^{5}/_{16}$	Section R703.10.2	6d common (2″ × 0.113″)	6d common (2″ × 0.113″)	6d common (2″ × 0.113″)	6d common (2″ × 0.113″)	6d common (2″ × 0.113″) or 11 gage roofing nail	Note f
Hardboard panel siding (see Section R703.5)		$^{7}/_{16}$	—	0.120″ nail (shank) with 0.225″ head	0.120″ nail (shank) with 0.225″ head	0.120″ nail (shank) with 0.225″ head	0.120″ nail (shank) with 0.225″ head	0.120″ nail (shank) with 0.225″ head	6″ panel edges 12″ inter. sup.[d]
Hardboard lap siding (see Section R703.5)		$^{7}/_{16}$	Note e	0.099″ nail (shank) with 0.240″ head	0.099″ nail (shank) with 0.240″ head	0.099″ nail (shank) with 0.240″ head	0.099″ nail (shank) with 0.240″ head	0.099″ nail (shank) with 0.240″ head	Same as stud spacing 2 per bearing
Horizontal aluminum[a]	Without insulation	0.019[b]	Lap	Siding nail $1^{1}/_{2}$″ × 0.120″	Siding nail 2″ × 0.120″	Siding nail 2″ × 0.120″	Siding nail[h] $1^{1}/_{2}$″ × 0.120″	Not allowed	Same as stud spacing
		0.024	Lap	Siding nail $1^{1}/_{2}$″ × 0.120″	Siding nail 2″ × 0.120″	Siding nail 2″ × 0.120″	Siding nail[h] $1^{1}/_{2}$″ × 0.120″	Not allowed	
	With insulation	0.019	Lap	Siding nail $1^{1}/_{2}$″ × 0.120″	Siding nail $2^{1}/_{2}$″ × 0.120″	Siding nail $2^{1}/_{2}$″ × 0.120″	Siding nail[h] $1^{1}/_{2}$″ × 0.120″	Siding nail $1^{1}/_{2}$″ × 0.120″	
Insulated vinyl siding[j]		0.035 (vinyl siding layer only)	Lap	0.120 nail (shank) with a 0.313 head or 16-gage crown[h, i]	0.120 nail (shank) with a 0.313 head or 16-gage crown[h]	0.120 nail (shank) with a 0.313 head or 16-gage crown[h]	0.120 nail (shank) with a 0.313 head Section R703.11.2	Not allowed	16 inches on center or specified by manufacturer instructions, test report or other sections of this code
Particleboard panels		$^{3}/_{8}$	—	6d box nail (2″ × 0.099″)	6d box nail (2″ × 0.099″)	6d box nail (2″ × 0.099″)	6d box nail (2″ × 0.099″)	Not allowed	6″ panel edges 12″ inter. sup.
		$^{1}/_{2}$	—	6d box nail (2″ × 0.099″)	6d box nail (2″ × 0.099″)	6d box nail (2″ × 0.099″)	6d box nail (2″ × 0.099″)	6d box nail (2″ × 0.099″)	
		$^{5}/_{8}$	—	6d box nail (2″ × 0.099″)	8d box nail ($2^{1}/_{2}$″ × 0.113″)	8d box nail ($2^{1}/_{2}$″ × 0.113″)	6d box nail (2″ × 0.099″)	6d box nail (2″ × 0.099″)	
Polypropylene siding[k]		Not applicable	Lap	Section 703.14.1	Section 703.14.1	Section 703.14.1	Section 703.14.1	Not allowed	As specified by the manufacturer instructions, test report or other sections of this code

(continued)

TABLE R703.3(1)—continued
SIDING MINIMUM ATTACHMENT AND MINIMUM THICKNESS

SIDING MATERIAL		NOMINAL THICKNESS (inches)	JOINT TREATMENT	TYPE OF SUPPORTS FOR THE SIDING MATERIAL AND FASTENERS					
				Wood or wood structural panel sheathing into stud	Fiberboard sheathing into stud	Gypsum sheathing into stud	Foam plastic sheathing into stud[l]	Direct to studs	Number or spacing of fasteners
Steel[c]		29 ga.	Lap	Siding nail ($1^{3}/_{4}$″ × 0.113″) Staple–$1^{3}/_{4}$	Siding nail ($2^{3}/_{4}$″ × 0.113″) Staple–$2^{1}/_{2}$	Siding nail ($2^{1}/_{2}$″ × 0.113″) Staple–$2^{1}/_{4}$	Siding nail ($1^{3}/_{4}$″ × 0.113″) Staple–$1^{3}/_{4}$	Not allowed	Same as stud spacing
Vinyl siding (see Section R703.11)		0.035	Lap	0.120″ nail (shank) with a 0.313″ head or 16-gage staple with $^{3}/_{8}$- to $^{1}/_{2}$-inch crown[h, i]	0.120″ nail (shank) with a 0.313″ head or 16-gage staple with $^{3}/_{8}$- to $^{1}/_{2}$-inch crown[h]	0.120″ nail (shank) with a 0.313″ head or 16-gage staple with $^{3}/_{8}$- to $^{1}/_{2}$-inch crown[h]	0.120″ nail (shank) with a 0.313 head Section R703.11.2	Not allowed	16 inches on center or as specified by the manufacturer instructions or test report
Wood siding (see Section R703.5)	Wood rustic, drop	$^{3}/_{8}$ min.	Lap	6d box or siding nail (2″ × 0.099″)	6d box or siding nail (2″ ×0.099″)	6d box or siding nail (2″ × 0.099″)	6d box or siding nail (2″ × 0.099″)	8d box or siding nail ($2^{1}/_{2}$″ × 0.113″) Staple–2″	Face nailing up to 6″ widths, 1 nail per bearing; 8″ width sand over, 2 nails per bearing
	Shiplap	$^{19}/_{32}$ average	Lap						
	Bevel	$^{7}/_{16}$							
	Butt tip	$^{3}/_{16}$	Lap						
Wood structural panel ANSI/APA PRP-210 siding (exterior grade) (see Section R703.5)		$^{3}/_{8}$ – $^{1}/_{2}$	Note e	2″ × 0.099″ siding nail	$2^{1}/_{2}$″ × 0.113″ siding nail	$2^{1}/_{2}$″ × 0.113″ siding nail	$2^{1}/_{2}$″ × 0.113″ siding nail	2″ × 0.099″ siding nail	6″ panel edges 12″ inter. sup.
Wood structural panel lap siding (see Section R703.5)		$^{3}/_{8}$ – $^{1}/_{2}$	Note e Note g	2″ × 0.099″ siding nail	$2^{1}/_{2}$″ × 0.113″ siding nail	$2^{1}/_{2}$″ × 0.113″ siding nail	$2^{1}/_{2}$″ × 0.113″ siding nail	2″ × 0.099″ siding nail	8″ along bottom edge

For SI: 1 inch = 25.4 mm.

a. Aluminum nails shall be used to attach aluminum siding.

b. Aluminum (0.019 inch) shall be unbacked only where the maximum panel width is 10 inches and the maximum flat area is 8 inches. The tolerance for aluminum siding shall be +0.002 inch of the nominal dimension.

c. Shall be of approved type.

d. Where used to resist shear forces, the spacing must be 4 inches at panel edges and 8 inches on interior supports.

e. Vertical end joints shall occur at studs and shall be covered with a joint cover or shall be caulked.

f. Face nailing: one 6d common nail through the overlapping planks at each stud. Concealed nailing: one 11-gage $1^{1}/_{2}$-inch-long galv. roofing nail through the top edge of each plank at each stud in accordance with the manufacturer's installation instructions.

g. Vertical joints, if staggered, shall be permitted to be away from studs if applied over wood structural panel sheathing.

h. Minimum fastener length must be sufficient to penetrate sheathing other nailable substrate and framing a total of a minimum of $1^{1}/_{4}$ inches or in accordance with the manufacturer's installation instructions.

i. Where specified by the manufacturer's instructions and supported by a test report, fasteners are permitted to penetrate into or fully through nailable sheathing or other nailable substrate of minimum thickness specified by the instructions or test report, without penetrating into framing.

j. Insulated vinyl siding shall comply with ASTM D7793.

k. Polypropylene siding shall comply with ASTM D7254.

l. Cladding attachment over foam sheathing shall comply with the additional requirements and limitations of Sections R703.15, R703.16 and R703.17.

TABLE R703.3(2)
SCREW FASTENER SUBSTITUTION FOR SIDING ATTACHMENT TO COLD-FORMED STEEL LIGHT FRAME CONSTRUCTION[a, b, c, d, e]

NAIL DIAMETER PER TABLE R703.3(1)	MINIMUM SCREW FASTENER SIZE
0.099″	No. 6
0.113″	No. 7
0.120″	No. 8

For SI: 1 inch = 25.4 mm.

a. Screws shall comply with ASTM C1513 and shall penetrate a minimum of three threads through minimum 33 mil (20 gage) cold-formed steel frame construction.

b. Screw head diameter shall be not less than the nail head diameter required by Table R703.3(1).

c. Number and spacing of screw fasteners shall comply with Table R703.3(1).

d. Pan head, hex washer head, modified truss head or other screw head types with a flat attachment surface under the head shall be used for vinyl siding attachment.

e. Aluminum siding shall not be fastened directly to cold-formed steel light frame construction.

TABLE R703.3.2
LIMITS FOR ATTACHMENT PER Table R703.3(1)

MAXIMUM MEAN ROOF HEIGHT			
Ultimate Wind Speed (mph 3-second gust)	Exposure		
	B	C	D
95	NL	NL	NL
100	NL	NL	NL
105	NL	NL	NL
110	NL	NL	40′
115	NL	50′	20′
120	NL	30′	DR
130	60′	15′	DR
140	35′	DR	DR

For SI: 1 foot = 304.8 mm, 1 mile per hour = 0.447 m/s.
NL = Not Limited by Table R703.3.2, DR = Design Required.

R703.4 Flashing. *Approved* corrosion-resistant flashing shall be applied *shingle-fashion* in a manner to prevent entry of water into the wall cavity or penetration of water to the building structural framing components. Self-adhered membranes used as flashing shall comply with AAMA 711. Fluid-applied membranes used as flashing in exterior walls shall comply with AAMA 714. The flashing shall extend to the surface of the exterior wall finish. *Approved* corrosion-resistant flashings shall be installed at the following locations:

1. Exterior window and door openings. Flashing at exterior window and door openings shall be installed in accordance with Section R703.4.1.
2. At the intersection of chimneys or other masonry construction with frame or stucco walls, with projecting lips on both sides under stucco copings.
3. Under and at the ends of masonry, wood or metal copings and sills.
4. Continuously above all projecting wood *trim*.
5. Where exterior porches, decks or stairs attach to a wall or floor assembly of wood-frame construction.
6. At wall and roof intersections.
7. At built-in gutters.

R703.4.1 Flashing installation at exterior window and door openings. Flashing at exterior window and door openings shall extend to the surface of the exterior wall finish or to a *water-resistive barrier* complying with Section 703.2 for subsequent drainage. Air sealing shall be installed around all window and door openings on the interior side of the rough opening gap. Mechanically attached flexible flashings shall comply with AAMA 712. Flashing at exterior window and door openings shall be installed in accordance with one or more of the following:

1. The fenestration manufacturer's installation and flashing instructions, or for applications not addressed in the fenestration manufacturer's instructions, in accordance with the flashing manufacturer's instructions. Where flashing instructions or details are not provided, *pan flashing* shall be installed at the sill of exterior window and door openings. *Pan flashing* shall be sealed or sloped in such a manner as to direct water to the surface of the exterior wall finish or to the water-resistive barrier for subsequent drainage. Openings using *pan flashing* shall incorporate flashing or protection at the head and sides.
2. In accordance with the flashing design or method of a *registered design professional.*
3. In accordance with other *approved* methods.

TABLE R703.3.3
OPTIONAL SIDING ATTACHMENT SCHEDULE FOR FASTENERS WHERE NO STUD PENETRATION NECESSARY

APPLICATION	NUMBER AND TYPE OF FASTENER	SPACING OF FASTENERS[b]
Exterior wall covering (weighing 3 psf or less) attachment to wood structural panel sheathing, either direct or over foam sheathing a maximum of 2 inches thick.[a] Note: Does not apply to vertical siding.	Ring shank roofing nail (0.120″ min. dia.)	12″ o.c.
	Ring shank nail (0.148″ min. dia.)	15″ o.c.
	No. 6 screw (0.138″ min. dia.)	12″ o.c.
	No. 8 screw (0.164″ min. dia.)	16″ o.c.

For SI: 1 inch = 25.4 mm, 1 pound per square foot = 0.479 kPa.

a. Fastener length shall be sufficient to penetrate the back side of the wood structural panel sheathing by at least $^1/_4$ inch. The wood structural panel sheathing shall be not less than $^7/_{16}$ inch in thickness.

b. Spacing of fasteners is per 12 inches of siding width. For other siding widths, multiply "Spacing of Fasteners" above by a factor of 12/s, where "s" is the siding width in inches. Fastener spacing shall never be greater than the manufacturer's minimum recommendations.

R703.5 Wood, hardboard and wood structural panel siding. Wood, hardboard and wood structural panel siding shall be installed in accordance with this section and Table R703.3(1). Hardboard siding shall comply with ANSI A135.5.

R703.5.1 Vertical wood siding. Wood siding applied vertically shall be nailed to horizontal nailing strips or blocking set not more than 24 inches (610 mm) on center.

R703.5.2 Panel siding. Three-eighths-inch (9.5 mm) wood structural panel siding shall not be applied directly to studs spaced more than 16 inches (406 mm) on center where long dimension is parallel to studs. Wood structural panel siding $^7/_{16}$ inch (11.1 mm) or thinner shall not be applied directly to studs spaced more than 24 inches (610 mm) on center. The stud spacing shall not exceed the panel span rating provided by the manufacturer unless the panels are installed with the face grain perpendicular to the studs or over sheathing *approved* for that stud spacing.

Joints in wood, hardboard or wood structural panel siding shall be made as follows unless otherwise *approved.* Vertical joints in panel siding shall occur over framing members, unless wood or wood structural panel sheathing is used, and shall be shiplapped or covered with a batten. Horizontal joints in panel siding shall be lapped not less than 1 inch (25 mm) or shall be shiplapped or flashed with Z-flashing and occur over solid blocking, wood or wood structural panel sheathing.

R703.5.3 Horizontal wood siding. Horizontal lap siding shall be installed in accordance with the manufacturer's recommendations. Where there are no recommendations the siding shall be lapped not less than 1 inch (25 mm), or $^1/_2$ inch (12.7 mm) if rabbeted, and shall have the ends caulked, covered with a batten or sealed and installed over a strip of flashing.

R703.6 Wood shakes and shingles. Wood shakes and shingles shall conform to CSSB.

R703.6.1 Application. Wood shakes or shingles shall be applied either single course or double course over nominal $^1/_2$-inch (12.7 mm) wood-based sheathing or to furring strips over $^1/_2$-inch (12.7 mm) nominal nonwood sheathing. A *water-resistive barrier* shall be provided over all sheathing, with horizontal overlaps in the membrane of not less than 2 inches (51 mm) and vertical overlaps of not less than 6 inches (152 mm). Where horizontal furring strips are used, they shall be 1 inch by 3 inches or 1 inch by 4 inches (25 mm by 76 mm or 25 mm by 102 mm) and shall be fastened to the studs with minimum 7d or 8d box nails and shall be spaced a distance on center equal to the actual weather exposure of the shakes or shingles, not to exceed the maximum exposure specified in Table R703.6.1. When installing shakes or shingles over a nonpermeable *water-resistive barrier,* furring strips shall be placed first vertically over the barrier and in addition, horizontal furring strips shall be fastened to the vertical furring strips prior to attaching the shakes or shingles to the horizontal furring strips. The spacing between adjacent shingles to allow for expansion shall be $^1/_8$ inch (3.2 mm) to $^1/_4$ inch (6.4 mm) apart, and between adjacent shakes shall be $^3/_8$ inch (9.5 mm) to $^1/_2$ inch (12.7 mm) apart. The offset spacing between joints in adjacent courses shall be not less than $1^1/_2$ inches (38 mm).

TABLE R703.6.1
MAXIMUM WEATHER EXPOSURE FOR WOOD SHAKES AND SHINGLES ON EXTERIOR WALLS[a, b, c]
(Dimensions are in inches)

LENGTH	EXPOSURE FOR SINGLE COURSE	EXPOSURE FOR DOUBLE COURSE
Shingles[a]		
16	7	12[b]
18	8	14[c]
24	$10^1/_2$	16[d]
Shakes[a]		
18	8	14
24	$10^1/_2$	18

For SI: 1 inch = 25.4 mm.

a. Dimensions given are for No. 1 grade.

b. A maximum 9-inch exposure is permitted for No. 2 grade.

c. A maximum 10-inch exposure is permitted for No. 2 grade.

d. A maximum 14-inch exposure is permitted for No. 2 grade.

R703.6.2 Weather exposure. The maximum weather exposure for shakes and shingles shall not exceed that specified in Table R703.6.1.

R703.6.3 Attachment. Wood shakes or shingles shall be installed according to this chapter and the manufacturer's instructions. Each shake or shingle shall be held in place by two stainless steel Type 304, Type 316 or hot-dipped zinc-coated galvanized corrosion-resistant box nails in accordance with Table R703.6.3(1) or R703.6.3(2). The hot-dipped zinc-coated galvanizing shall be in compliance with ASTM A153, 1.0 ounce per square foot. Alternatively, 16-gage stainless steel Type 304 or Type 316 staples with crown widths $^7/_{16}$ inch (11 mm) minimum, $^3/_4$ inch (19 mm) maximum, shall be used and the crown of the staple shall be placed parallel with the butt of the shake or the shingle. In single-course application, the fasteners shall be concealed by the course above and shall be driven approximately 1 inch (25 mm) above the butt line of the succeeding course and $^3/_4$ inch (19 mm) from the edge. In double-course applications, the exposed shake or shingle shall be face-nailed with two fasteners, driven approximately 2 inches (51 mm) above the butt line and $^3/_4$ inch (19 mm) from each edge. Fasteners installed within 15 miles (24 km) of saltwater coastal areas shall be stainless steel Type 316. Fasteners for fire-retardant-treated shakes or shingles in accordance with Section R902 or pressure-impregnated-preservative-treated shakes or shingles in accordance with AWPA U1 shall be stainless steel Type 316. The fasteners shall penetrate the sheathing or furring strips by not less than $^1/_2$ inch (13 mm) and shall not be overdriven. Fasteners for untreated (natural) and treated products shall comply with ASTM F1667.

R703.6.4 Bottom courses. The bottom courses shall be doubled.

TABLE R703.6.3(1)
SINGLE-COURSE SIDEWALL FASTENERS

PRODUCT TYPE	NAIL TYPE, MINIMUM LENGTH AND SHANK DIAMETER (inches)
R & R and sanded shingles	
16″ and 18″ shingles	3d box 1 $^{1}/_{4}$ × 0.076
24″ shingles	4d box 1$^{1}/_{2}$ × 0.076
Grooved shingles	
16″ and 18″ shingles	3d box 1$^{1}/_{4}$ × 0.076
24″ shingles	4d box 1$^{1}/_{2}$ × 0.076
Split and sawn shakes	
18″ straight-split shakes	5d box 1$^{3}/_{4}$ × 0.080
18″ and 24″ handsplit shakes	6d box 2 × 0.099
24″ tapersplit shakes	5d box 1$^{3}/_{4}$ × 0.080
18″ and 24″ tapersawn shakes	6d box 2 × 0.099

For SI: 1 inch = 25.4 mm.

TABLE R703.6.3(2)
DOUBLE-COURSE SIDEWALL FASTENERS

PRODUCT TYPE	NAIL TYPE, MINIMUM LENGTH AND SHANK DIAMETER (inches)
R & R and sanded shingles	
16″, 8″ and 24″ shingles	5d box 1$^{3}/_{4}$ × 0.08 or 5d casing nails 1$^{3}/_{4}$ × 0.080
Grooved shingles	
16″, 18″ and 24″shingles	5d box 1$^{3}/_{4}$ × 0.080
Split and sawn shakes	
18″ straight-split shakes	7d box 2$^{1}/_{4}$ × 0.099 or 8d box 2$^{1}/_{2}$ × 0.113
18″ and 24″ handsplit shakes	7d box 2$^{1}/_{4}$ × 0.099 or 8d box 2$^{1}/_{2}$ × 0.113
24″ tapersplit shakes	7d box 2$^{1}/_{4}$ × 0.099 or 8d box 2$^{1}/_{2}$ × 0.113
18″ and 24″ tapersawn shakes	7d box 2$^{1}/_{4}$ × 0.099 or 8d box 2$^{1}/_{2}$ × 0.113

For SI: 1 inch = 25.4 mm.

R703.7 Exterior plaster (stucco). Installation of exterior plaster shall be in compliance with ASTM C926, ASTM C1063 and the provisions of this code.

R703.7.1 Lath. Lath and lath attachments shall be of corrosion-resistant materials in accordance with ASTM C1063. Expanded metal, welded wire, or woven wire lath shall be attached to wood framing members or furring. Where the exterior plaster is serving as wall bracing in accordance with Table R602.10.4, the lath shall be attached directly to framing. The lath shall be attached with 1$^{1}/_{2}$-inch-long (38 mm), 11-gage nails having a $^{7}/_{16}$-inch (11.1 mm) head, or $^{7}/_{8}$-inch-long (22.2 mm), 16-gage staples, spaced not more than 7 inches (178 mm) on center along framing members or furring and not more than 24 inches (610 mm) on center between framing members or furring, or as otherwise *approved*. Additional fastening between wood framing members shall not be prohibited. Lath attachments to cold-formed steel framing or to masonry, stone, or concrete substrates shall be in accordance with ASTM C1063. Where lath is installed directly over foam sheathing, lath connections shall also be in accordance with Section R703.15, R703.16 or R703.17. Where lath is attached to furring installed over foam sheathing, the furring connections shall be in accordance with Section R703.15, R703.16 or R703.17.

Exception: Lath is not required over masonry, cast-in-place concrete, *precast concrete* or stone substrates prepared in accordance with ASTM C1063.

703.7.1.1 Furring. Where provided, furring shall consist of wood furring strips not less than 1 inch by 2 inches (25 mm by 51 mm), minimum $^{3}/_{4}$-inch (19 mm) metal channels, or self-furring lath, and shall be installed in accordance with ASTM C1063. Furring shall be spaced not greater than 24 inches (600 mm) on center and, where installed over wood or cold-formed steel framing, shall be fastened into framing members.

R703.7.2 Plaster. Plastering with cement plaster shall be in accordance with ASTM C926. Cement materials shall be in accordance with one of the following:

1. Masonry cement conforming to ASTM C91, Type M, S or N.
2. Portland cement conforming to ASTM C150, Type I, II or III.
3. Blended hydraulic cement conforming to ASTM C595, Type IP, IS (< 70), IL, or IT (S < 70).

4. Hydraulic cement conforming to ASTM C1157, Type GU, HE, MS, HS or MH.
5. Plastic (stucco) cement conforming to ASTM C1328.

Plaster shall be not less than three coats where applied over metal lath or wire lath and shall be not less than two coats where applied over masonry, concrete, pressure-preservative-treated wood or decay-resistant wood as specified in Section R317.1 or gypsum backing. If the plaster surface is completely covered by veneer or other facing material or is completely concealed, plaster application need be only two coats, provided the total thickness is as set forth in Table R702.1(1).

On wood-frame construction with an on-grade floor slab system, exterior plaster shall be applied to cover, but not extend below, lath, paper and screed.

The proportion of aggregate to cementitious materials shall be as set forth in Table R702.1(3).

R703.7.2.1 Weep screeds. A minimum 0.019-inch (0.5 mm) (No. 26 galvanized sheet gage), corrosion-resistant weep screed or plastic weep screed, with a minimum vertical attachment flange of $3^1/_2$ inches (89 mm), shall be provided at or below the foundation plate line on exterior stud walls in accordance with ASTM C926. The weep screed shall be placed not less than 4 inches (102 mm) above the earth or 2 inches (51 mm) above paved areas and shall be of a type that will allow trapped water to drain to the exterior of the building. The weather-resistant barrier shall lap the attachment flange. The exterior lath shall cover and terminate on the attachment flange of the weep screed.

R703.7.3 Water-resistive barriers. Water-resistive barriers shall be installed as required in Section R703.2 and, where applied over wood-based sheathing, shall comply with Section R703.7.3.1 or R703.7.3.2.

R703.7.3.1 Dry climates. In Dry (B) climate zones indicated in Figure N1101.7, *water-resistive barriers* shall comply with one of the following:

1. The *water-resistive barrier* shall be two layers of 10-minute Grade D paper or have a water resistance equal to or greater than two layers of a *water-resistive barrier* complying with ASTM E2556, Type I. The individual layers shall be installed independently such that each layer provides a separate continuous plane. Flashing installed in accordance with Section R703.4 and intended to drain to the *water-resistive barrier* shall be directed between the layers.
2. The *water-resistive barrier* shall be 60-minute Grade D paper or have a water resistance equal to or greater than one layer of a water-resistive barrier complying with ASTM E2556, Type II. The *water-resistive barrier* shall be separated from the stucco by a layer of foam plastic *insulating sheathing* or other non-water-absorbing layer, or a designed drainage space.

R703.7.3.2 Moist or marine climates. In the Moist (A) or Marine (C) climate zones indicated in Figure N1101.7, *water-resistive barriers* shall comply with one of the following:

1. In addition to complying with Section R703.7.3.1, a space or drainage material not less than $^3/_{16}$ inch (5 mm) in depth shall be added to the exterior side of the *water-resistive barrier*.
2. In addition to complying with Section R703.7.3.1, Item 2, drainage on the exterior of the *water-resistive barrier* shall have a drainage efficiency of not less than 90 percent, as measured in accordance with ASTM E2273 or Annex A2 of ASTM E2925.

R703.7.4 Application. Each coat shall be kept in a moist condition for at least 48 hours prior to application of the next coat.

Exception: Applications installed in accordance with ASTM C926.

R703.7.5 Curing. The finish coat for two-coat cement plaster shall not be applied sooner than seven days after application of the first coat. For three-coat cement plaster, the second coat shall not be applied sooner than 48 hours after application of the first coat. The finish coat for three-coat cement plaster shall not be applied sooner than seven days after application of the second coat.

R703.8 Anchored stone and masonry veneer, general. Anchored stone and masonry veneer shall be installed in accordance with this chapter, Table R703.3(1) and Figures R703.8(1) and R703.8(2). These veneers installed over a backing of wood or cold-formed steel shall be limited to the first *story above grade plane* and shall not exceed 5 inches (127 mm) in thickness. See Section R602.10 for wall bracing requirements for masonry veneer for wood-framed construction and Section R603.9.5 for wall bracing requirements for masonry veneer for cold-formed steel construction.

Exceptions:

1. For buildings in *Seismic Design Categories* A, B and C, exterior stone or masonry veneer, as specified in Table R703.8(1), with a backing of wood or steel framing shall be permitted to the height specified in Table R703.8(1) above a noncombustible foundation.
2. For detached one- or two-family dwellings in *Seismic Design Categories* D_0, D_1 and D_2, exterior stone or masonry veneer, as specified in Table R703.8(2), with a backing of wood framing shall be permitted to the height specified in Table R703.8(2) above a noncombustible foundation.

R703.8.1 Interior veneer support. Veneers used as interior wall finishes shall be permitted to be supported on wood or cold-formed steel floors that are designed to support the loads imposed.

R703.8.2 Exterior veneer support. Except in *Seismic Design Categories* D_0, D_1 and D_2, exterior masonry

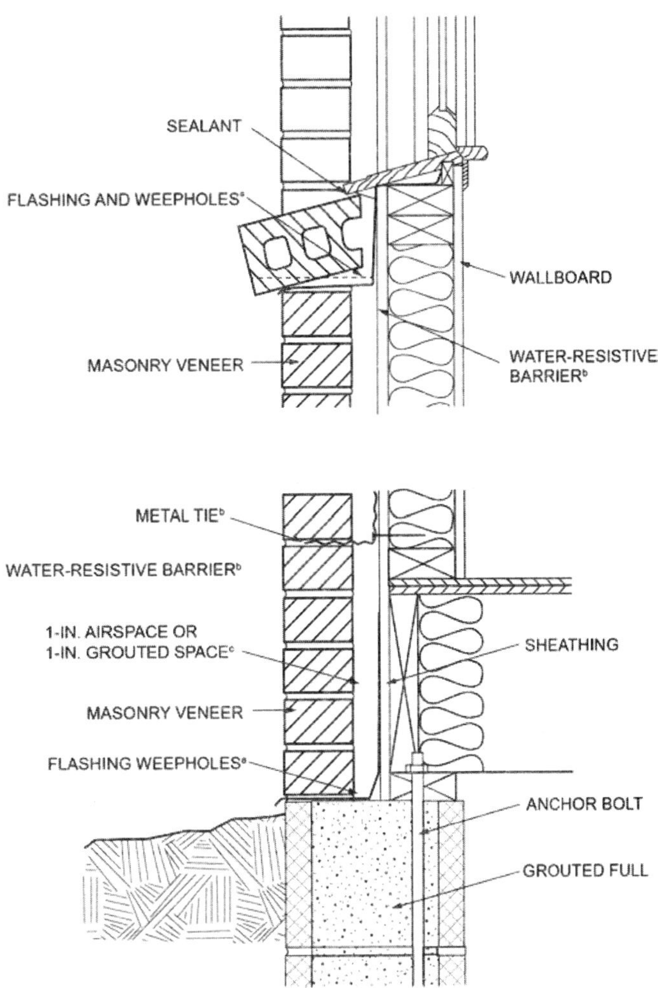

For SI: 1 inch = 25.4 mm.

a. See Sections R703.4, R703.8.5 and R703.8.6.

b. See Sections R703.2 and R703.8.4.

c. See Table R703.8.4(1) and Section R703.8.4.2.

d. Figures R703.8(1) and R703.8(2) illustrate typical construction details for a masonry veneer wall. For the actual mandatory requirements of this code, see the indicated sections of text. Other details of masonry veneer wall construction shall be permitted provided the requirements of the indicated sections of text are met.

FIGURE R703.8(1)
TYPICAL MASONRY VENEER WALL DETAILS[d]

veneers having an installed weight of 40 pounds per square foot (195 kg/m^2) or less shall be permitted to be supported on wood or cold-formed steel construction. Where masonry veneer supported by wood or cold-formed steel construction adjoins masonry veneer supported by the foundation, there shall be a movement joint between the veneer supported by the wood or cold-formed steel construction and the veneer supported by the foundation. The wood or cold-formed steel construction supporting the masonry veneer shall be designed to limit the deflection to $^1/_{600}$ of the span for the supporting members. The design of the wood or cold-formed steel construction shall consider the weight of the veneer and any other loads.

R703.8.2.1 Support by steel angle. A minimum 6-inch by 4-inch by $^5/_{16}$-inch (152 mm by 102 mm by 8 mm) steel angle, with the long leg placed vertically, shall be anchored to double 2-inch by 4-inch (51 mm by 102 mm) wood studs or double 350S162 cold-formed steel studs at a maximum on-center spacing of 16 inches (406 mm). Anchorage of the steel angle at every double stud spacing shall be not less than two $^7/_{16}$-inch-diameter (11 mm) by 4-inch (102 mm) lag screws for wood construction or two $^7/_{16}$-inch (11.1 mm) bolts with washers for cold-formed steel construction. The steel angle shall have a minimum clearance to underlying construction of $^1/_{16}$ inch (1.6 mm). Not less than two-thirds the width of the masonry veneer thickness shall bear on the steel angle. Flashing and weep holes shall be located in the

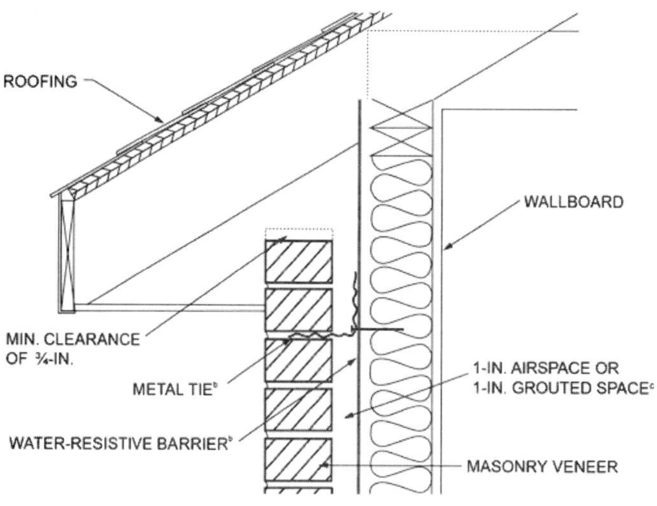

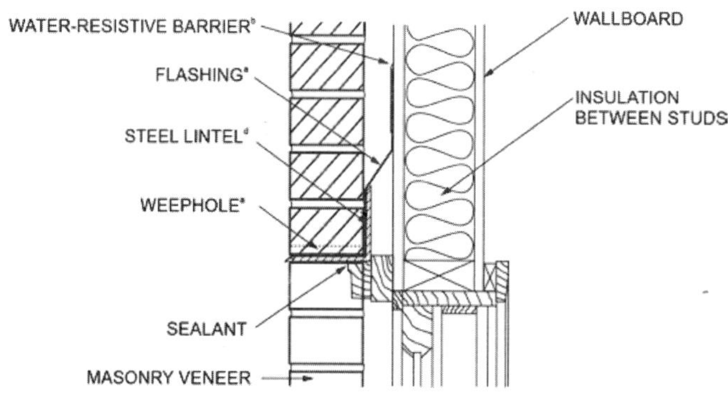

For SI: 1 inch = 25.4 mm.

a. See Sections R703.4, R703.8.5 and R703.8.6.

b. See Sections R703.2 and R703.8.4.

c. See Table R703.8.4(1) and Section R703.8.4.2.

d. See Section R703.8.3.

e. Figures R703.8(1) and R703.8(2) illustrate typical construction details for a masonry veneer wall. For the actual mandatory requirements of this code, see the indicated sections of text. Other details of masonry veneer wall construction shall be permitted provided that the requirements of the indicated sections of text are met.

FIGURE R703.8(2)
TYPICAL MASONRY VENEER WALL DETAILS[e]

masonry veneer in accordance with Figure R703.8.2.1. The maximum height of masonry veneer above the steel angle support shall be 12 feet 8 inches (3861 mm). The airspace separating the masonry veneer from the wood backing shall be in accordance with Sections R703.8.4 and R703.8.4.2. The method of support for the masonry veneer on wood construction shall be constructed in accordance with Figure R703.8.2.1.

The maximum slope of the roof construction without stops shall be 7:12. Roof construction with slopes greater than 7:12 but not more than 12:12 shall have stops of a minimum 3-inch by 3-inch by $^{1}/_{4}$-inch (76 mm by 76 mm by 6.4 mm) steel plate welded to the angle at 24 inches (610 mm) on center along the angle or as *approved* by the *building official.*

R703.8.2.2 Support by roof construction. A steel angle shall be placed directly on top of the roof construction. The roof supporting construction for the steel angle shall consist of not fewer than three 2-inch by 6-inch (51 mm by 152 mm) wood members for wood construction or three 550S162 cold-formed steel members for cold-formed steel light frame construction. A wood member abutting the vertical wall stud construction shall be anchored with not fewer than three $^{5}/_{8}$-inch (15.9 mm) diameter by 5-inch (127 mm) lag screws to every wood stud spacing. Each additional wood roof member shall be anchored by the use of two 10d nails at every wood stud spacing. A cold-formed steel member abutting the vertical wall stud shall be anchored with not fewer than nine No. 8 screws to every cold-formed

TABLE R703.8(1)
STONE OR MASONRY VENEER LIMITATIONS AND REQUIREMENTS, WOOD OR STEEL FRAMING, SEISMIC DESIGN CATEGORIES A, B AND C

SEISMIC DESIGN CATEGORY	NUMBER OF WOOD- OR STEEL-FRAMED STORIES	MAXIMUM HEIGHT OF VENEER ABOVE NONCOMBUSTIBLE FOUNDATION[a] (feet)	MAXIMUM NOMINAL THICKNESS OF VENEER (inches)	MAXIMUM WEIGHT OF VENEER (psf)[b]	WOOD- OR STEEL-FRAMED STORY
A or B	Steel: 1 or 2 Wood: 1, 2 or 3	30	5	50	all
C	1	30	5	50	1 only
	2	30	5	50	top
					bottom
	Wood only: 3	30	5	50	top
					middle
					bottom

For SI: 1 inch = 25.4 mm, 1 foot = 304.8 mm, 1 pound per square foot = 0.479 kPa.

a. An additional 8 feet is permitted for gable end walls. See also story height limitations of Section R301.3.

b. Maximum weight is installed weight and includes weight of mortar, grout, lath and other materials used for installation. Where veneer is placed on both faces of a wall, the combined weight shall not exceed that specified in this table.

TABLE R703.8(2)
STONE OR MASONRY VENEER LIMITATIONS AND REQUIREMENTS, ONE- AND TWO-FAMILY DETACHED DWELLINGS, SEISMIC DESIGN CATEGORIES D_0, D_1 AND D_2

SEISMIC DESIGN CATEGORY	NUMBER OF WOOD-FRAMED STORIES[a]	MAXIMUM HEIGHT OF VENEER ABOVE NONCOMBUSTIBLE FOUNDATION OR FOUNDATION WALL (feet)	MAXIMUM NOMINAL THICKNESS OF VENEER (inches)	MAXIMUM WEIGHT OF VENEER (psf)[b]
D_0	1	20[c]	4	40
	2	20[c]	4	40
	3	30[d]	4	40
D_1	1	20[c]	4	40
	2	20[c]	4	40
	3	20[c]	4	40
D_2	1	20[c]	3	30
	2	20[c]	3	30

For SI: 1 inch = 25.4 mm, 1 foot = 304.8 mm, 1 pound per square foot = 0.479 kPa, 1 pound-force = 4.448 N.

a. Cripple walls are not permitted in *Seismic Design Categories* D_0, D_1 and D_2.

b. Maximum weight is installed weight and includes weight of mortar, grout and lath, and other materials used for installation.

c. The veneer shall not exceed 20 feet in height above a noncombustible foundation, with an additional 8 feet permitted for gable end walls, or 30 feet in height with an additional 8 feet for gable end walls where the lower 10 feet have a backing of concrete or masonry wall. See story height limitations of Section R301.3.

d. The veneer shall not exceed 30 feet in height above a noncombustible foundation, with an additional 8 feet permitted for gable end walls. See story height limitations of Section R301.3.

steel stud. Each additional cold-formed steel roof member shall be anchored to the adjoining roof member using two No. 8 screws at every stud spacing. Not less than two-thirds the width of the masonry veneer thickness shall bear on the steel angle. Flashing and weep holes shall be located in the masonry veneer wythe in accordance with Figure R703.8.2.2. The maximum height of the masonry veneer above the steel angle support shall be 12 feet 8 inches (3861 mm). The airspace separating the masonry veneer from the wood backing shall be in accordance with Sections R703.8.4 and R703.8.4.2. The support for the masonry veneer shall be constructed in accordance with Figure R703.8.2.2.

The maximum slope of the roof construction without stops shall be 7:12. Roof construction with slopes greater than 7:12 but not more than 12:12 shall have stops of a minimum 3-inch by 3-inch by $^1/_4$-inch (76 mm by 76 mm by 6.4 mm) steel plate welded to the angle at 24 inches (610 mm) on center along the angle or as *approved* by the *building official.*

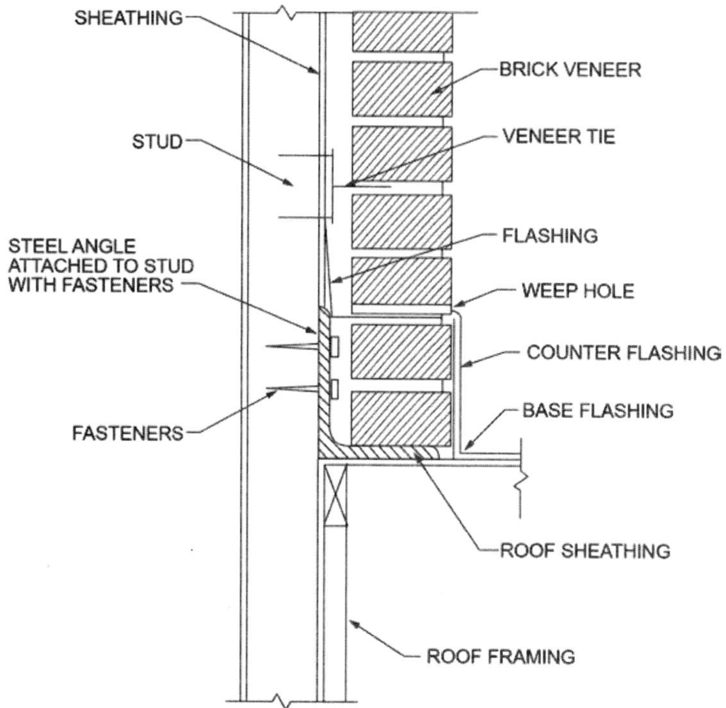

SUPPORT BY STEEL ANGLE

FIGURE R703.8.2.1
EXTERIOR MASONRY VENEER SUPPORT BY STEEL ANGLES

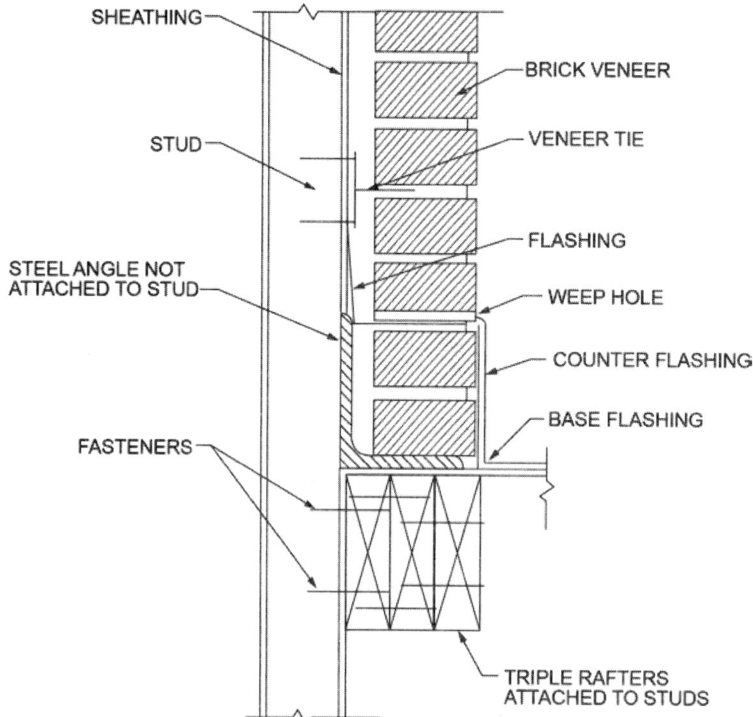

SUPPORT BY ROOF MEMBERS

FIGURE R703.8.2.2
EXTERIOR MASONRY VENEER SUPPORT BY ROOF MEMBERS

R703.8.3 Lintels. Masonry veneer shall not support any vertical load other than the dead load of the veneer above. Veneer above openings shall be supported on lintels of *noncombustible materials*. The lintels shall have a length of bearing not less than 4 inches (102 mm). Steel lintels shall be shop coated with a rust-inhibitive paint, except for lintels made of corrosion-resistant steel or steel treated with coatings to provide corrosion resistance. Construction of openings shall comply with either Section R703.8.3.1 or 703.8.3.2.

R703.8.3.1 Allowable span. The allowable span shall not exceed the values set forth in Table R703.8.3.1.

R703.8.3.2 Maximum span. The allowable span shall not exceed 18 feet 3 inches (5562 mm) and shall be constructed to comply with Figure R703.8.3.2 and the following:

1. Provide a minimum length of 18 inches (457 mm) of masonry veneer on each side of opening as shown in Figure R703.8.3.2.
2. Provide a minimum 5-inch by $3^1/_2$-inch by $^5/_{16}$-inch (127 mm by 89 mm by 7.9 mm) steel angle above the opening and shore for a minimum of 7 days after installation.
3. Provide double-wire joint reinforcement extending 12 inches (305 mm) beyond each side of the opening. Lap splices of joint reinforcement not less than 12 inches (305 mm). Comply with one of the following:
   3.1. Double-wire joint reinforcement shall be $^3/_{16}$-inch (4.8 mm) diameter and shall be placed in the first two bed joints above the opening.
   3.2. Double-wire joint reinforcement shall be 9 gauge (0.144 inch or 3.66 mm diameter) and shall be placed in the first three bed joints above the opening.
4. Provide the height of masonry veneer above opening, in accordance with Table R703.8.3.2.

TABLE R703.8.3.2
HEIGHT OF MASONRY VENEER ABOVE OPENING

MINIMUM HEIGHT OF MASONRY VENEER ABOVE OPENING (inches)	MAXIMUM HEIGHT OF MASONRY VENEER ABOVE OPENING (feet)
13	< 5
24	5 to < 12
60	12 to height above support allowed by Section R703.8

For SI: 1 inch = 25.4 mm, 1 foot = 304.8 mm.

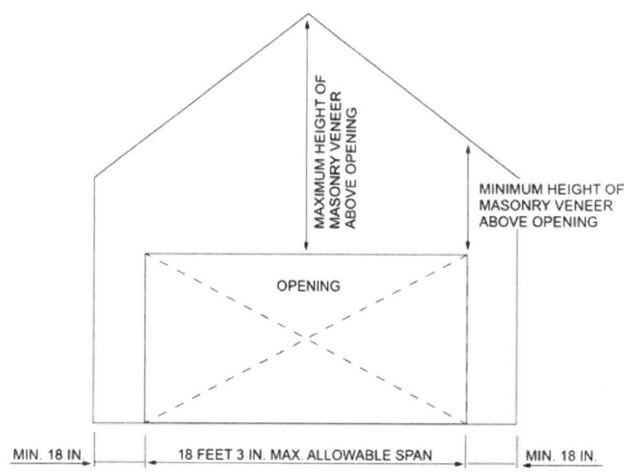

For SI: 1 inch = 25.4 mm, 1 foot = 304.8 mm.

FIGURE R703.8.3.2
MASONRY VENEER OPENING

R703.8.4 Anchorage. Masonry veneer shall be anchored to the supporting wall studs with corrosion-resistant metal ties embedded in mortar or grout and extending into the veneer a minimum of $1^1/_2$ inches (38 mm), with not less than $^5/_8$-inch (15.9 mm) mortar or grout cover to outside face. Masonry veneer shall conform to Table R703.8.4(1). Where the masonry veneer tie attachment is

TABLE R703.8.3.1
ALLOWABLE SPANS FOR LINTELS SUPPORTING MASONRY VENEER[a, b, c, d]

SIZE OF STEEL ANGLE[a, c, d] (inches)	NO STORY ABOVE	ONE STORY ABOVE	TWO STORIES ABOVE	NO. OF $^1/_2$-INCH OR EQUIVALENT REINFORCING BARS IN REINFORCED LINTEL[b, d]
$3 \times 3 \times {}^1/_4$	6′-0″	4′-6″	3′-0″	1
$4 \times 3 \times {}^1/_4$	8′-0″	6′-0″	4′-6″	1
$5 \times 3^1/_2 \times {}^5/_{16}$	10′-0″	8′-0″	6′-0″	2
$6 \times 3^1/_2 \times {}^5/_{16}$	14′-0″	9′-6″	7′-0″	2
$2\text{-}6 \times 3^1/_2 \times {}^5/_{16}$	20′-0″	12′-0″	9′-6″	4

For SI: 1 inch = 25.4 mm, 1 foot = 304.8 mm.

a. Long leg of the angle shall be placed in a vertical position.

b. Depth of reinforced lintels shall be not less than 8 inches and all cells of hollow masonry lintels shall be grouted solid. Reinforcing bars shall extend not less than 8 inches into the support.

c. Steel members indicated are adequate typical examples; other steel members meeting structural design requirements shall be permitted to be used.

d. Either steel angle or reinforced lintel shall span opening.

fastened to wood structural panel not less than 7/16 performance category through insulating sheathing not greater than 2 inches (51 mm) in thickness, see Table R703.8.4(2). Where Table R703.8.4(2) is used, attachment to the studs behind the sheathing is not required.

R703.8.4.1 Size and spacing. Veneer ties, if strand wire, shall be not less in thickness than No. 9 U.S. gage [(0.148 inch) (4 mm)] wire and shall have a hook embedded in the mortar joint, or if sheet metal, shall be not less than No. 22 U.S. gage by [(0.0299 inch) (0.76 mm)] $^7/_8$ inch (22 mm) corrugated. Each tie shall support not more than 2.67 square feet (0.25 m^2) of wall area and shall be spaced not more than 32 inches (813 mm) on center horizontally and 24 inches (635 mm) on center vertically.

Exception: In Seismic Design Category D_0, D_1 or D_2 or townhouses in Seismic Design Category C or in wind areas of more than 30 pounds per square foot pressure (1.44 kPa), each tie shall support not more than 2 square feet (0.2 m^2) of wall area.

R703.8.4.1.1 Veneer ties around wall openings. Additional metal ties shall be provided around wall openings greater than 16 inches (406 mm) in either dimension. Metal ties around the perimeter of openings shall be spaced not more than 3 feet (9144 mm) on center and placed within 12 inches (305 mm) of the wall opening.

R703.8.4.2 Grout fill. As an alternative to the airspace required by Table R703.8.4(1), grout shall be permitted to fill the airspace. Where the airspace is filled with grout, a *water-resistive barrier* is required over studs or sheathing. Where the airspace is filled, replacing the sheathing and *water-resistive barrier* with a wire mesh and *approved water-resistive barrier* or an *approved water-resistive barrier*-backed reinforcement attached directly to the studs is permitted.

R703.8.5 Flashing. Flashing shall be located beneath the first course of masonry above finished ground level above the foundation wall or slab and at other points of support, including structural floors, shelf angles and lintels where masonry veneers are designed in accordance with Section R703.8. See Section R703.4 for additional requirements.

R703.8.6 Weepholes. Weepholes shall be provided in the outside wythe of masonry walls at a maximum spacing of 33 inches (838 mm) on center. Weepholes shall be not less than $^3/_{16}$ inch (5 mm) in diameter. Weepholes shall be located immediately above the flashing.

R703.9 Exterior insulation and finish system (EIFS)/EIFS with drainage. Exterior insulation and finish systems (EIFS) shall comply with this chapter and Section R703.9.1. EIFS with drainage shall comply with this chapter and Section R703.9.2.

TABLE R703.8.4(1)
TIE ATTACHMENT AND AIRSPACE REQUIREMENTS

BACKING AND TIE	MINIMUM TIE	MINIMUM TIE FASTENER[a]	AIRSPACE[b]	
Wood stud backing with corrugated sheet metal	22 U.S. gage (0.0299 in.) × $^7/_8$ in. wide	8d common nail[c] ($2^1/_2$ in. × 0.131 in.)	Nominal 1 in. between sheathing and veneer	
Wood stud backing with adjustable metal strand wire	W1.7 (No. 9 U.S. gage; 0.148 in. dia.) with hook embedded in mortar joint[d]	8d common nail[c] ($2^1/_2$ in. × 0.131 in.)	Minimum nominal 1 in. between sheathing and veneer	Maximum $4^5/_8$ in. between backing and veneer
Wood stud backing with adjustable metal strand wire	W2.8 (0.187 in. dia.) with hook embedded in mortar joint[e, f]	8d common nail[c] ($2^1/_2$ in. × 0.131 in.)	Greater than $4^5/_8$ in. between backing and veneer	Maximum $6^5/_8$ in. between backing and veneer
Cold-formed steel stud backing with adjustable metal strand wire	W1.7 (No. 9 U.S. gage; 0.148 in. dia.) with hook embedded in mortar joint[d]	No. 10 screw extending through the steel framing a minimum of three exposed threads	Minimum nominal 1 in. between sheathing and veneer	Maximum $4^5/_8$ in. between backing and veneer
Cold-formed steel stud backing with adjustable metal strand wire	W2.8 (0.187 in. dia.) with hook embedded in mortar joint[e, f]	No. 10 screw extending through the steel framing a minimum of three exposed threads	Greater than $4^5/_8$ in. between backing and veneer	Maximum $6^5/_8$ in. between backing and veneer

For SI: 1 inch = 25.4 mm.

a. All fasteners shall have rust-inhibitive coating suitable for the installation in which they are being used, or be manufactured from material not susceptible to corrosion.

b. An airspace that provides drainage shall be permitted to contain mortar from construction.

c. In Seismic Design Category D_0, D_1 or D_2, the minimum tie fastener shall be an 8d ring-shank nail ($2^1/_2$ in. × 0.131 in.).

d. Adjustable tie pintles shall include not fewer than 1 pintle leg of wire size W2.8 (MW18) with a maximum offset of $1^1/_4$ inches.

e. Adjustable tie pintles shall include not fewer than 2 pintle legs with a maximum offset of $1^1/_4$ inches. Distance between inside face of brick and end of pintle shall be a maximum of 2 inches.

f. Adjustable tie backing attachment components shall consist of one of the following: eyes with minimum wire W2.8 (MW18), barrel with minimum $^1/_4$-inch outside diameter, or plate with minimum thickness of 0.074 inch and minimum width of $1^1/_4$ inches.

TABLE R703.8.4(2)
REQUIRED BRICK TIE SPACING FOR DIRECT APPLICATION TOWOOD STRUCTURAL PANEL SHEATHING[a, b, c]

FASTENER TYPE[d]	SIZE (DIA. OR SCREW #)	REQUIRED BRICK-TIE SPACING (VERTICAL-TIE SPACING/HORIZONTAL-TIE SPACING) (inches/inches)											
		110 mph V_{ult}			115 mph V_{ult}			130 mph V_{ult}			140 mph V_{ult}		
		Zone 5, Exposure B	Zone 5, Exposure C	Zone 5, Exposure D	Zone 5, Exposure B	Zone 5, Exposure C	Zone 5, Exposure D	Zone 5, Exposure B	Zone 5, Exposure C	Zone 5, Exposure D	Zone 5, Exposure B	Zone 5, Exposure C	Zone 5, Exposure D
Ring Shank Nails	0.091	16/16, 16/12, 12/16, 12/12	16/12, 12/16, 12/12	12/12	16/16, 16/12, 12/16, 12/12	16/12, 12/16, 12/12	12/12	16/12, 12/16, 12/12	12/12	—	12/12	—	—
	0.148	24/16, 16/24, 16/16, 16/12, 12/16, 12/12	16/16, 16/12, 12/16, 12/12	16/16, 16/12, 12/16, 12/12	24/16, 16/24, 16/16, 16/12, 12/16, 12/12	16/16, 16/12, 12/16, 12/12	16/16, 16/12, 12/16, 12/12	16/16, 16/12, 12/16, 12/12	16/12, 12/16, 12/12	16/12, 12/16, 12/12	16/16, 16/12, 12/16, 12/12	16/12, 12/16, 12/12	12/12
Screws	#6	24/16, 16/24, 16/16, 16/12, 12/16, 12/12	16/16, 16/12, 12/16, 12/12	16/16, 16/12, 12/16, 12/12	24/16, 16/24, 16/16, 16/12, 12/16, 12/12	16/16, 16/12, 12/16, 12/12	16/16, 16/12, 12/16, 12/12	16/16, 16/12, 12/16, 12/12	16/12, 12/16, 12/12	16/12, 12/16, 12/12	16/16, 16/12, 12/16, 12/12	16/12, 12/16, 12/12	12/12
	#8	24/16, 16/24, 16/16, 16/12, 12/16, 12/12	24/16, 16/24, 16/16, 16/12, 12/16, 12/12	16/16, 16/12, 12/16, 12/12	24/16, 16/24, 16/16, 16/12, 12/16, 12/12	16/16, 16/12, 12/16, 12/12	16/16, 16/12, 12/16, 12/12	24/16, 16/24, 16/16, 16/12, 12/16, 12/12	16/16, 16/12, 12/16, 12/12	16/12, 12/16, 12/12	16/16, 16/12, 12/16, 12/12	16/12, 12/16, 12/12	16/12, 12/16, 12/12
	#10	24/16, 16/24, 16/16, 16/12, 12/16, 12/12	24/16, 16/24, 16/16, 16/12, 12/16, 12/12	24/16, 16/24, 16/16, 16/12, 12/16, 12/12	24/16, 16/24, 16/16, 16/12, 12/16, 12/12	24/16, 16/24, 16/16, 16/12, 12/16, 12/12	16/16, 16/12, 12/16, 12/12	24/16, 16/24, 16/16, 16/12, 12/16, 12/12	16/16, 16/12, 12/16, 12/12	16/16, 16/12, 12/16, 12/12	24/16, 16/24, 16/16, 16/12, 12/16, 12/12	16/16, 16/12, 12/16, 12/12	16/12, 12/16, 12/12
	#14	24/16, 16/24, 16/16, 16/12, 12/16, 12/12	24/16, 16/24, 16/16, 16/12, 12/16, 12/12	24/16, 16/24, 16/16, 16/12, 12/16, 12/12	24/16, 16/24, 16/16, 16/12, 12/16, 12/12	24/16, 16/24, 16/16, 16/12, 12/16, 12/12	24/16, 16/24, 16/16, 16/12, 12/16, 12/12	24/16, 16/24, 16/16, 16/12, 12/16, 12/12	24/16, 16/24, 16/16, 16/12, 12/16, 12/12	16/16, 16/12, 12/16, 12/12	24/16, 16/24, 16/16, 16/12, 12/16, 12/12	16/16, 16/12, 12/16, 12/12	16/16, 16/12, 12/16, 12/12

For SI: 1 inch = 25.4 mm, 1 mph = 0.447 m/s.

a. This table is based on attachment of brick ties directly to wood structural panel sheathing only. Additional attachment of the brick tie to lumber framing is not required. The brick ties shall be permitted to be placed over any insulating sheathing, not to exceed 2 inches in thickness. Wood structural panel sheathing shall be a minimum 7/16 performance category. The table is based on a building height of 30 feet or less.

b. Wood structural panels shall have a specific gravity of 0.42 or greater in accordance with NDS.

c. Foam sheathing shall have a minimum compressive strength of 15 psi in accordance with ASTM C578 or ASTM C1289.

d. Fasteners shall be sized such that the tip of the fastener passes completely through the wood structural panel sheathing by not less than $^1/_4$ inch.

R703.9.1 Exterior insulation and finish systems (EIFS). EIFS shall comply with the following:

1. ASTM E2568.
2. EIFS shall be limited to applications over substrates of concrete or masonry wall assemblies.
3. Flashing of EIFS shall be provided in accordance with the requirements of Section R703.4.
4. EIFS shall be installed in accordance with the manufacturer's instructions.
5. EIFS shall terminate not less than 6 inches (152 mm) above the finished ground level.
6. Decorative *trim* shall not be face-nailed through the EIFS.

R703.9.2 Exterior insulation and finish system (EIFS) with drainage. EIFS with drainage shall comply with the following:

1. ASTM E2568.
2. EIFS with drainage shall be required over all wall assemblies with the exception of substrates of concrete or masonry wall assemblies.
3. EIFS with drainage shall have an average minimum drainage efficiency of 90 percent when tested in accordance with ASTM E2273.
4. The *water-resistive barrier* shall comply with Section R703.2 or ASTM E2570.
5. The *water-resistive barrier* shall be applied between the EIFS and the wall sheathing.
6. Flashing of EIFS with drainage shall be provided in accordance with the requirements of Section R703.4.
7. EIFS with drainage shall be installed in accordance with the manufacturer's instructions.
8. EIFS with drainage shall terminate not less than 6 inches (152 mm) above the finished ground level.
9. Decorative *trim* shall not be face-nailed through the EIFS with drainage.

R703.10 Fiber cement siding.

R703.10.1 Panel siding. Fiber-cement panels shall comply with the requirements of ASTM C1186, Type A, minimum Grade II or ISO 8336, Category A, minimum Class 2. Panels shall be installed with the long dimension either parallel or perpendicular to framing. Vertical and horizontal joints shall occur over framing members and shall be protected with caulking, or with battens or flashing, or be vertical or horizontal shiplap, or otherwise designed to comply with Section R703.1. Panel siding shall be installed with fasteners in accordance with Table R703.3(1) or the approved manufacturer's instructions.

R703.10.2 Lap siding. Fiber-cement lap siding having a maximum width of 12 inches (305 mm) shall comply with the requirements of ASTM C1186, Type A, minimum Grade II or ISO 8336, Category A, minimum Class 2. Lap siding shall be lapped a minimum of $1^1/_4$ inches (32 mm) and lap siding not having tongue-and-groove end joints shall have the ends protected with caulking, covered with an H-section joint cover, located over a strip of flashing, or shall be designed to comply with Section R703.1. Lap siding courses shall be installed with the fastener heads exposed or concealed, in accordance with Table R703.3(1) or approved manufacturer's instructions.

R703.11 Vinyl siding. Vinyl siding shall be certified and *labeled* as conforming to the requirements of ASTM D3679 by an *approved* quality control agency.

R703.11.1 Installation. Vinyl siding, soffit and accessories shall be installed in accordance with the manufacturer's instructions.

R703.11.1.1 Fasteners. Unless specified otherwise by the manufacturer's instructions, fasteners for vinyl siding shall be 0.120-inch (3 mm) shank diameter nail with a 0.313-inch (8 mm) head or 16-gage staple with a $^3/_8$-inch (9.5 mm) to $^1/_2$-inch (12.7 mm) crown.

R703.11.1.2 Penetration depth. Unless specified otherwise by the manufacturer's instructions, fasteners shall penetrate into building framing. The total penetration into sheathing, furring framing or other *nailable substrate* shall be a minimum $1^1/_4$ inches (32 mm). Where specified by the manufacturer's instructions and supported by a test report, fasteners are permitted to penetrate into or fully through nailable sheathing or other *nailable substrate* of minimum thickness specified by the instructions or test report without penetrating into framing. Where the fastener penetrates fully through the sheathing, the end of the fastener shall extend a minimum of $^1/_4$ inch (6.4 mm) beyond the opposite face of the sheathing or *nailable substrate.*

R703.11.1.3 Spacing. Unless specified otherwise by the manufacturer's instructions, the maximum spacing between fasteners for horizontal siding shall be 16 inches (406 mm), and for vertical siding 12 inches (305 mm) both horizontally and vertically. Where specified by the manufacturer's instructions and supported by a test report, greater fastener spacing is permitted.

R703.11.2 Installation over foam plastic sheathing. Where vinyl siding or *insulated vinyl siding* is installed over foam plastic sheathing, the vinyl siding shall comply with Section R703.11 and shall have a wind load design pressure rating in accordance with Table R703.11.2.

Exceptions:

1. Where the foam plastic sheathing is applied directly over *wood structural panels,* fiberboard, gypsum sheathing or other *approved* backing capable of independently resisting the design wind pressure, the vinyl siding shall be installed in accordance with Sections R703.3.3 and R703.11.1.
2. Where the vinyl siding manufacturer's product specifications provide an *approved* wind load design pressure rating for installation over foam plastic sheathing, use of this wind load

design pressure rating shall be permitted and the siding shall be installed in accordance with the *manufacturer's installation instructions.*

3. Where the foam plastic sheathing and its attachment have a design wind pressure resistance complying with Sections R316.8 and R301.2.1, the vinyl siding shall be installed in accordance with Sections R703.3.3 and R703.11.1.

R703.12 Adhered masonry veneer installation. Adhered masonry veneer shall comply with the requirements of Section R703.7.3 and the requirements in Sections 12.1 and 12.3 of TMS 402. Adhered masonry veneer shall be installed in accordance with Section R703.7.1, Article 3.3C of TMS 602 or the manufacturer's instructions.

R703.12.1 Clearances. On exterior stud walls, adhered masonry veneer shall be installed:

1. Minimum of 4 inches (102 mm) above the earth;
2. Minimum of 2 inches (51 mm) above paved areas; or
3. Minimum of $^1/_2$ inch (12.7 mm) above exterior walking surfaces that are supported by the same foundation that supports the exterior wall.

R703.12.2 Flashing at foundation. A corrosion-resistant screed or flashing of a minimum 0.019-inch (0.48 mm) or 26-gage galvanized or plastic with a minimum vertical attachment flange of $3^1/_2$ inches (89 mm) shall be installed to extend a minimum of 1 inch (25 mm) below the foundation plate line on exterior stud walls in accordance with Section R703.4.

R703.12.3 Water-resistive barrier. A *water-resistive barrier* shall be installed as required by Section R703.2 and shall comply with the requirements of Section R703.7.3. The *water-resistive barrier* shall lap over the exterior of the attachment flange of the screed or flashing provided in accordance with Section R703.12.2.

R703.13 Insulated vinyl siding. *Insulated vinyl siding* shall be certified and *labeled* as conforming to the requirements of ASTM D7793 by an *approved* quality control agency.

R703.13.1 Insulated vinyl siding and accessories. *Insulated vinyl siding* and accessories shall be installed in accordance with the manufacturer's installation instructions.

R703.14 Polypropylene siding. *Polypropylene siding* shall be certified and *labeled* as conforming to the requirements of ASTM D7254, and those of Section R703.14.2 or Section R703.14.3, by an *approved* quality control agency.

R703.14.1 Polypropylene siding and accessories. *Polypropylene siding* and accessories shall be installed in accordance with manufacturer's installation instructions.

R703.14.1.1 Installation. *Polypropylene siding* shall be installed over and attached to wood structural panel sheathing with minimum thickness of $^7/_{16}$ inch (11.1 mm), or other substrate, composed of wood or wood-based material and fasteners having equivalent withdrawal resistance.

R703.14.1.2 Fastener requirements. Unless otherwise specified in the approved manufacturer's instructions, nails shall be corrosion resistant, with a minimum 0.120-inch (3 mm) shank and minimum

TABLE R703.11.2
REQUIRED MINIMUM WIND LOAD DESIGN PRESSURE RATING FOR VINYL SIDING INSTALLED OVER FOAM PLASTIC SHEATHING ALONE

ULTIMATE DESIGN WIND SPEED (MPH)	ADJUSTED MINIMUM DESIGN WIND PRESSURE (ASD) (PSF)[a, b]					
	Case 1: With interior gypsum wallboard[c]			Case 2: Without interior gypsum wallboard[c]		
	Exposure			Exposure		
	B	C	D	B	C	D
≤ 95	-30.0	-33.2	-39.4	-33.9	-47.4	-56.2
100	-30.0	-36.8	-43.6	-37.2	-52.5	-62.2
105	-30.0	-40.5	-48.1	-41.4	-57.9	-68.6
110	-31.8	-44.5	-52.8	-45.4	-63.5	-75.3
115	-35.5	-49.7	-59.0	-50.7	-71.0	-84.2
120	-37.4	-52.4	-62.1	-53.4	-74.8	-88.6
130	-44.9	-62.8	-74.5	-64.1	-89.7	-106
> 130	See Note d					

For SI: 1 inch = 25.4 mm, 1 foot = 304.8 mm, 1 square foot = 0.0929 m^2, 1 mile per hour = 0.447 m/s, 1 pound per square foot = 0.0479 kPa.

a. Linear interpolation is permitted.

b. The table values are based on a maximum 30-foot mean roof height, and effective wind area of 10 square feet Wall Zone 5 (corner), and the ASD design component and cladding wind pressure from Table R301.2.1(1), adjusted for exposure in accordance with Table R301.2.1(2), multiplied by the following adjustment factors: 1.87 (Case 1) and 2.67 (Case 2).

c. Gypsum wallboard, gypsum panel product or equivalent.

d. For the indicated wind speed condition and where foam sheathing is the only sheathing on the exterior of a frame wall with vinyl siding, the wall assembly shall be capable of resisting an impact without puncture at least equivalent to that of a wood frame wall with minimum $^7/_{16}$-inch OSB sheathing as tested in accordance with ASTM E1886. The vinyl siding shall comply with an adjusted design wind pressure requirement in accordance with Note b, using an adjustment factor of 2.67.

0.313-inch (8 mm) head diameter. Nails shall be a minimum of $1^1/_4$ inches (32 mm) long or as necessary to penetrate sheathing or substrate not less than $^3/_4$ inch (19.1 mm). Where the nail fully penetrates the sheathing or *nailable substrate*, the end of the fastener shall extend not less than $^1/_4$ inch (6.4 mm) beyond the opposite face of the sheathing or substrate. Staples are not permitted.

R703.14.2 Fire separation. *Polypropylene siding* shall not be installed on walls with a *fire separation distance* of less than 5 feet (1524 mm) and walls closer than 10 feet (3048 mm) to a building on another lot.

Exception: Walls perpendicular to the line used to determine the *fire separation distance*.

R703.14.3 Flame spread index. The certification of the flame spread index shall be accompanied by a test report stating that all portions of the test specimen ahead of the flame front remained in position during the test in accordance with ASTM E84 or UL 723.

R703.15 Cladding attachment over foam sheathing to wood framing. Cladding shall be specified and installed in accordance with Section R703, the cladding manufacturer's approved instructions, including any limitations for use over foam plastic sheathing, or an *approved* design. In addition, the cladding or furring attachments through foam sheathing to framing shall meet or exceed the minimum fastening requirements of Section R703.15.1, Section R703.15.2, or an *approved* design for support of cladding weight.

Exceptions:

1. Where the cladding manufacturer has provided *approved* installation instructions for application over foam sheathing, those requirements shall apply.
2. For exterior insulation and finish systems, refer to Section R703.9.
3. For anchored masonry or stone veneer installed over foam sheathing, refer to Section R703.8.

R703.15.1 Direct attachment. Where cladding is installed directly over foam sheathing without the use of furring, cladding minimum fastening requirements to support the cladding weight shall be as specified in Table R703.15.1.

R703.15.2 Furred cladding attachment. Where wood furring is used to attach cladding over foam sheathing, furring minimum fastening requirements to support the cladding weight shall be as specified in Table R703.15.2. Where placed horizontally, wood furring shall be preservative-treated wood in accordance with Section R317.1 or *naturally durable wood* and fasteners shall be corrosion resistant in accordance Section R317.3.

TABLE R703.15.1
CLADDING MINIMUM FASTENING REQUIREMENTS FOR DIRECT ATTACHMENT OVER FOAM PLASTIC SHEATHING TO SUPPORT CLADDING WEIGHT[a]

CLADDING FASTENER THROUGH FOAM SHEATHING[b]	CLADDING FASTENER TYPE AND MINIMUM SIZE[c]	CLADDING FASTENER VERTICAL SPACING (inches)	MAXIMUM THICKNESS OF FOAM SHEATHING[d] (inches)									
			16″ o.c. Fastener Horizontal Spacing					24″ o.c. Fastener Horizontal Spacing				
			Cladding Weight:					Cladding Weight:				
			3 psf	11 psf	15 psf	18 psf	25 psf	3 psf	11 psf	15 psf	18 psf	25 psf
Wood framing (minimum $1^1/_4$-inch penetration)	0.113″ diameter nail	6	2.00	1.45	1.00	0.75	DR	2.00	0.85	0.55	DR	DR
		8	2.00	1.00	0.65	DR	DR	2.00	0.55	DR	DR	DR
		12	2.00	0.55	DR	DR	DR	1.85	DR	DR	DR	DR
	0.120″ diamete rnail	6	3.00	1.70	1.15	0.90	0.55	3.00	1.05	0.65	0.50	DR
		8	3.00	1.20	0.80	0.60	DR	3.00	0.70	DR	DR	DR
		12	3.00	0.70	DR	DR	DR	2.15	DR	DR	DR	DR
	0.131″ diameter nail	6	4.00	2.15	1.50	1.20	0.75	4.00	1.35	0.90	0.70	DR
		8	4.00	1.55	1.05	0.80	DR	4.00	0.90	0.55	DR	DR
		12	4.00	0.90	0.55	DR	DR	2.70	0.50	DR	DR	DR
	0.162″ diameter nail	6	4.00	3.55	2.50	2.05	1.40	4.00	2.25	1.55	1.25	0.80
		8	4.00	2.55	1.80	1.45	0.95	4.00	1.60	1.10	0.85	0.50
		12	4.00	1.60	1.10	0.85	0.50	4.00	0.95	0.60	DR	DR

For SI: 1 inch = 25.4 mm, 1 pound per square foot = 0.0479 kPa, 1 pound per square inch = 6.895 kPa.

DR = Design Required.

o.c. = On Center.

a. Wood framing shall be Spruce-pine-fir or any wood species with a specific gravity of 0.42 or greater in accordance with AWC NDS.

b. The thickness of wood structural panels complying with the specific gravity requirement of Note a shall be permitted to be included in satisfying the minimum penetration into framing. For cladding connections to wood structural panels, refer to Table R703.3.3. For brick veneer tie connections to wood structural panels, refer to Table R703.8.4(2).

c. Nail fasteners shall comply with ASTM F1667, except nail length shall be permitted to exceed ASTM F1667 standard lengths.

d. Foam sheathing shall have a minimum compressive strength of 15 psi in accordance with ASTM C578 or ASTM C1289.

TABLE R703.15.2
FURRING MINIMUM FASTENING REQUIREMENTS FOR APPLICATION OVER FOAM PLASTIC SHEATHING TO SUPPORT CLADDING WEIGHT[a, b]

FURRING MATERIAL	FRAMING MEMBER	FASTENER TYPE AND MINIMUM SIZE	MINIMUM PENETRATION INTO WALL FRAMING (inches)[c]	FASTENER SPACING IN FURRING (inches)	MAXIMUM THICKNESS OF FOAM SHEATHING[e] (inches)									
					16″ o.c. Furring[f]					24″ o.c. Furring[f]				
					Siding Weight:					Siding Weight:				
					3 psf	11 psf	15 psf	18 psf	25 psf	3 psf	11 psf	15 psf	18 psf	25 psf
Minimum 1× wood furring[d]	Minimum 2× wood stud	0.131″ diameter nail	$1^{1}/_{4}$	8	4.00	2.45	1.75	1.45	0.95	4.00	1.60	1.10	0.85	DR
				12	4.00	1.60	1.10	0.85	DR	4.00	0.95	0.55	DR	DR
				16	4.00	1.10	0.70	DR	DR	3.05	0.60	DR	DR	DR
		0.162″ diameter nail	$1^{1}/_{4}$	8	4.00	4.00	3.05	2.45	1.60	4.00	2.75	1.85	1.45	0.85
				12	4.00	2.75	1.85	1.45	0.85	4.00	1.65	1.05	0.75	DR
				16	4.00	1.90	1.25	0.95	DR	4.00	1.05	0.60	DR	DR
		No.10 wood screw	1	12	4.00	2.30	1.60	1.20	0.70	4.00	1.40	0.85	0.60	DR
				16	4.00	1.65	1.05	0.75	DR	4.00	0.90	DR	DR	DR
				24	4.00	0.90	DR	DR	DR	2.85	DR	DR	DR	DR
		$^{1}/_{4}$″ lag screw	$1^{1}/_{2}$	12	4.00	2.65	1.90	1.50	0.90	4.00	1.65	1.05	0.80	DR
				16	4.00	1.95	1.25	0.95	0.50	4.00	1.10	0.65	DR	DR
				24	4.00	1.10	0.65	DR	DR	3.25	0.50	DR	DR	DR

For SI: 1 inch = 25.4 mm, 1 pound per square foot = 0.0479 kPa, 1 pound per square inch = 6.895 kPa.

DR = Design Required.

o.c. = On Center.

a. Wood framing and furring shall be Spruce-pine-fir or any wood species with a specific gravity of 0.42 or greater in accordance with AWC NDS.

b. Nail fasteners shall comply with ASTM F1667, except nail length shall be permitted to exceed ASTM F1667 standard lengths.

c. The thickness of wood structural panels complying with the specific gravity requirements of Note a shall be permitted to be included in satisfying the minimum required penetration into framing.

d. Where the required cladding fastener penetration into wood material exceeds $^{3}/_{4}$ inch and is not more than $1^{1}/_{2}$ inches, a minimum 2× wood furring or an approved design shall be used.

e. Foam sheathing shall have a minimum compressive strength of 15 psi in accordance with ASTM C578 or ASTM C1289.

f. Furring shall be spaced not more than 24 inches on center, in a vertical or horizontal orientation. In a vertical orientation, furring shall be located over wall studs and attached with the required fastener spacing. In a horizontal orientation, the indicated 8-inch and 12-inch fastener spacing in furring shall be achieved by use of two fasteners into studs at 16 inches and 24 inches on center, respectively.

R703.16 Cladding attachment over foam sheathing to cold-formed steel framing. Cladding shall be specified and installed in accordance with Section R703, the cladding manufacturer's approved instructions, including any limitations for use over foam plastic sheathing, or an *approved* design. In addition, the cladding or furring attachments through foam sheathing to framing shall meet or exceed the minimum fastening requirements of Section R703.16.1, Section R703.16.2 or an *approved* design for support of cladding weight.

Exceptions:

1. Where the cladding manufacturer has provided approved installation instructions for application over foam sheathing, those requirements shall apply.
2. For exterior insulation and finish systems, refer to Section R703.9.
3. For anchored masonry or stone veneer installed over foam sheathing, refer to Section R703.8.

R703.16.1 Direct attachment. Where cladding is installed directly over foam sheathing without the use of furring, cladding minimum fastening requirements to support the cladding weight shall be as specified in Table R703.16.1.

R703.16.2 Furred cladding attachment. Where steel or wood furring is used to attach cladding over foam sheathing, furring minimum fastening requirements to support the cladding weight shall be as specified in Table R703.16.2. Where placed horizontally, wood furring shall be preservative-treated wood in accordance with Section R317.1 or *naturally durable wood* and fasteners shall be corrosion resistant in accordance with Section R317.3. Steel furring shall have a minimum G60 galvanized coating.

TABLE R703.16.1
CLADDING MINIMUM FASTENING REQUIREMENTS FOR DIRECT ATTACHMENT OVER FOAM PLASTIC SHEATHING TO SUPPORT CLADDING WEIGHT[a, b]

CLADDING FASTENER THROUGH FOAM SHEATHING INTO:	CLADDING FASTENER TYPE AND MINIMUM SIZE[c]	CLADDING FASTENER VERTICAL SPACING (inches)	MAXIMUM THICKNESS OF FOAM SHEATHING[d] (inches)									
			16" o.c. Fastener Horizontal Spacing					24" o.c. Fastener Horizontal Spacing				
			Cladding Weight:					Cladding Weight:				
			3 psf	11 psf	15 psf	18 psf	25 psf	3 psf	11 psf	15 psf	18 psf	25 psf
Steel framing (minimum penetration of steel thickness + 3 threads)	No. 8 screw into 33-mil steel or thicker	6	3.00	2.95	2.50	2.20	1.45	3.00	2.35	1.75	1.25	DR
		8	3.00	2.55	2.00	1.60	0.60	3.00	1.80	0.90	DR	DR
		12	3.00	1.80	0.95	DR	DR	3.00	0.65	DR	DR	DR
	No. 10 screw into 33-mil steel	6	4.00	3.50	3.05	2.70	1.95	4.00	2.90	2.20	1.70	0.55
		8	4.00	3.10	2.50	2.05	1.00	4.00	2.25	1.35	0.70	DR
		12	4.00	2.25	1.35	0.70	DR	3.70	1.05	DR	DR	DR
	No. 10 screw into 43-mil steel or thicker	6	4.00	4.00	4.00	4.00	3.60	4.00	4.00	3.80	3.45	2.70
		8	4.00	4.00	4.00	3.70	3.00	4.00	3.85	3.25	2.80	1.80
		12	4.00	3.85	3.25	2.80	1.80	4.00	3.05	2.15	1.50	DR

For SI: 1 inch = 25.4 mm, 1 mil = 0.0254 mm, 1 pound per square foot = 0.0479 kPa, 1 pound per square inch = 6.895 kPa.
DR = Design Required.
o.c. = On Center.

a. Steel framing shall be minimum 33-ksi steel for 33-mil and 43-mil steel, and 50-ksi steel for 54-mil steel or thicker.
b. Where cladding is attached to wood structural panel sheathing only, fastening requirements shall be in accordance with Table R703.3.3.
c. Screws shall comply with the requirements of ASTM C1513.
d. Foam sheathing shall have a minimum compressive strength of 15 psi in accordance with ASTM C578 or ASTM C1289.

TABLE R703.16.2
FURRING MINIMUM FASTENING REQUIREMENTS FOR APPLICATION OVER FOAM PLASTIC SHEATHING TO SUPPORT CLADDING WEIGHT[a]

FURRING MATERIAL	FRAMING MEMBER	FASTENER TYPE AND MINIMUM SIZE[b]	MINIMUM PENETRATION INTO WALL FRAMING (inches)	FASTENER SPACING IN FURRING (inches)	MAXIMUM THICKNESS OF FOAM SHEATHING[d] (inches)									
					16" o.c. Furring[e]					24" o.c. Furring[e]				
					Cladding Weight:					Cladding Weight:				
					3 psf	11 psf	15 psf	18 psf	25 psf	3 psf	11 psf	15 psf	18 psf	25 psf
Minimum 33-mil steel furring or minimum 1× wood furring[c]	33-mil steel stud	No. 8 screw	Steel thickness + 3 threads	12	3.00	1.80	0.95	DR	DR	3.00	0.65	DR	DR	DR
				16	3.00	1.00	DR	DR	DR	2.85	DR	DR	DR	DR
				24	2.85	DR	DR	DR	DR	2.20	DR	DR	DR	DR
		No. 10 screw	Steel thickness + 3 threads	12	4.00	2.25	1.35	0.70	DR	3.70	1.05	DR	DR	DR
				16	3.85	1.45	DR	DR	DR	3.40	DR	DR	DR	DR
				24	3.40	DR	DR	DR	DR	2.70	DR	DR	DR	DR
	43-mil or thicker steel stud	No. 8 Screw	Steel thickness + 3 threads	12	3.00	1.80	0.95	DR	DR	3.00	0.65	DR	DR	DR
				16	3.00	1.00	DR	DR	DR	2.85	DR	DR	DR	DR
				24	2.85	DR	DR	DR	DR	2.20	DR	DR	DR	DR
		No. 10 screw	Steel thickness + 3 threads	12	4.00	3.85	3.25	2.80	1.80	4.00	3.05	2.15	1.50	DR
				16	4.00	3.30	2.55	1.95	0.60	4.00	2.25	1.05	DR	DR
				24	4.00	2.25	1.05	DR	DR	4.00	0.65	DR	DR	DR

For SI: 1 inch = 25.4 mm, 1 mil = 0.0254 mm, 1 pound per square foot = 0.0479 kPa, 1 pound per square inch = 6.895 kPa.
DR = Design Required.
o.c. = On Center.

a. Wood furring shall be Spruce-pine-fir or any softwood species with a specific gravity of 0.42 or greater. Steel furring shall be minimum 33-ksi steel. Steel studs shall be minimum 33-ksi steel for 33-mil and 43-mil thickness, and 50-ksi steel for 54-mil steel or thicker.
b. Screws shall comply with the requirements of ASTM C1513.
c. Where the required cladding fastener penetration into wood material exceeds $^3/_4$ inch and is not more than $1^1/_2$ inches, a minimum 2-inch nominal wood furring or an approved design shall be used.
d. Foam sheathing shall have a minimum compressive strength of 15 psi in accordance with ASTM C578 or ASTM C1289.
e. Furring shall be spaced not more than 24 inches (610 mm) on center, in a vertical or horizontal orientation. In a vertical orientation, furring shall be located over wall studs and attached with the required fastener spacing. In a horizontal orientation, the indicated 8-inch and 12-inch fastener spacing in furring shall be achieved by use of two fasteners into studs at 16 inches and 24 inches on center, respectively.

R703.17 Cladding attachment over foam sheathing to masonry or concrete wall construction. Cladding shall be specified and installed in accordance with Section 703.3 and the cladding manufacturer's instructions or an *approved* design. Foam sheathing shall be attached to masonry or concrete construction in accordance with the insulation manufacturer's installation instructions or an *approved* design. Furring and furring attachments through foam sheathing into concrete or masonry substrate shall be designed to resist design loads determined in accordance with Section R301, including support of cladding weight as applicable. Fasteners used to attach cladding or furring through foam sheathing to masonry or concrete substrates shall be *approved* for application into masonry or concrete material and shall be installed in accordance with the fastener manufacturer's instructions.

Exceptions:

1. Where the cladding manufacturer has provided approved installation instructions for application over foam sheathing and connection to a masonry or concrete substrate, those requirements shall apply.
2. For exterior insulation and finish systems, refer to Section R703.9.
3. For anchored masonry or stone veneer installed over foam sheathing, refer to Section R703.8.

SECTION R704
SOFFITS

R704.1 General wind limitations. Where the design wind pressure is 30 pounds per square foot (1.44 kPa) or less, soffits shall comply with Section R704.2. Where the design wind pressure exceeds 30 pounds per square foot (1.44 kPa), soffits shall comply with Section R704.3. The design wind pressure on soffits shall be determined using the component and cladding loads specified in Table R301.2.1(1) for walls using an effective wind area of 10 square feet (0.93 m^2) and adjusted for height and exposure in accordance with Table R301.2.1(2).

R704.2 Soffit installation where the design wind pressure is 30 psf or less. Where the design wind pressure is 30 pounds per square foot (1.44 kPa) or less, soffit installation shall comply with Section R704.2.1, R704.2.2, R704.2.3 or R704.2.4. Soffit materials not addressed in Sections R704.2.1 through R704.2.4 shall be in accordance with the manufacturer's installation instructions.

R704.2.1 Vinyl soffit panels. Vinyl soffit panels shall be installed using fasteners specified by the manufacturer and shall be fastened at both ends to a supporting component such as a nailing strip, fascia or subfascia component in accordance with Figure R704.2.1(1). Where the unsupported span of soffit panels is greater than 16 inches (406 mm), intermediate nailing strips shall be provided in accordance with Figure R704.2.1(2). Vinyl soffit panels shall be installed in accordance with the manufacturer's installation instructions. Fascia covers shall be installed in accordance with the manufacturer's installation instructions.

R704.2.2 Fiber-cement soffit panels. Fiber-cement soffit panels shall be a minimum of $^{1}/_{4}$ inch (6.4 mm) in thickness and shall comply with the requirements of ASTM C1186, Type A, minimum Grade II, or ISO 8336, Category A, minimum Class 2. Panel joints shall occur

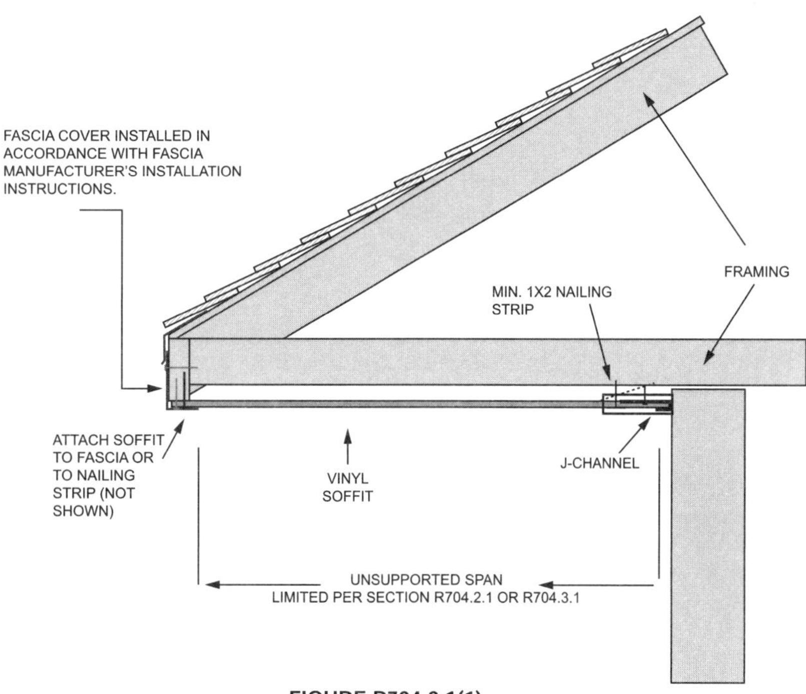

FIGURE R704.2.1(1)
TYPICAL SINGLE-SPAN VINYL SOFFIT PANEL SUPPORT

over framing or over wood structural panel sheathing. Soffit panels shall be installed with spans and fasteners in accordance with the manufacturer's installation instructions.

R704.2.3 Hardboard soffit panels. Hardboard soffit panels shall be not less than $^{7}/_{16}$ inch (11.11 mm) in thickness and shall be fastened to framing or nailing strips with $2^{1}/_{2}$-inch by 0.113-inch (64 mm by 2.9 mm) siding nails spaced not more than 6 inches (152 mm) on center at panel edges and 12 inches (305 mm) on center at intermediate supports.

R704.2.4 Wood structural panel soffit. The minimum nominal thickness for wood structural panel soffits shall be $^{3}/_{8}$ inch (9.5 mm) and shall be fastened to framing or nailing strips with 2-inch by 0.099-inch (51 mm by 2.5 mm) nails. Fasteners shall be spaced not less than 6 inches (152 mm) on center at panel edges and 12 inches (305 mm) on center at intermediate supports.

R704.3 Soffit installation where the design wind pressure exceeds 30 psf. Where the design wind pressure is greater than 30 psf, soffit installation shall comply with Section R704.3.1, R704.3.2, R704.3.3 or R704.3.4. Soffit materials not addressed in Sections R704.3.1 through R704.3.4 shall be in accordance with the manufacturer's installation instructions.

R704.3.1 Vinyl soffit panels. Vinyl soffit panels and their attachments shall be capable of resisting wind loads specified in Table R301.2.1(1) for walls using an effective wind area of 10 square feet (0.929 m^2) and adjusted for height and exposure in accordance with Table R301.2.1(2). Vinyl soffit panels shall be installed using fasteners specified by the manufacturer and shall be fastened at both ends to a supporting component such as a nailing strip, fascia or subfascia component in accordance with Figure R704.2.1(1). Where the unsupported span of soffit panels is greater than 12 inches (305 mm), intermediate nailing strips shall be provided in accordance with Figure R704.2.1(2). Vinyl soffit panels shall be installed in accordance with the manufacturer's installation instructions. Fascia covers shall be installed in accordance with the manufacturer's installation instructions.

R704.3.2 Fiber-cement soffit panels. Fiber-cement soffit panels shall comply with Section R704.2.2 and shall be capable of resisting wind loads specified in Table R301.2.1(1) for walls using an effective wind area of 10 square feet (0.929 m^2) and adjusted for height and exposure in accordance with Table R301.2.1(2).

R704.3.3 Hardboard soffit panels. Hardboard soffit panels shall comply with the manufacturer's installation instructions and shall be capable of resisting wind loads specified in Table R301.2.1(1) for walls using an effective wind area of 10 square feet (0.929 m^2) and adjusted for height and exposure in accordance with Table R301.2.1(2).

R704.3.4 Wood structural panel soffit. Wood structural panel soffits shall be capable of resisting wind loads specified in Table R301.2.1(1) for walls using an effective wind area of 10 square feet (0.929 m^2) and adjusted for height and exposure in accordance with Table R301.2.1(2). Alternatively, wood structural panel soffits shall be installed in accordance with Table R704.3.4.

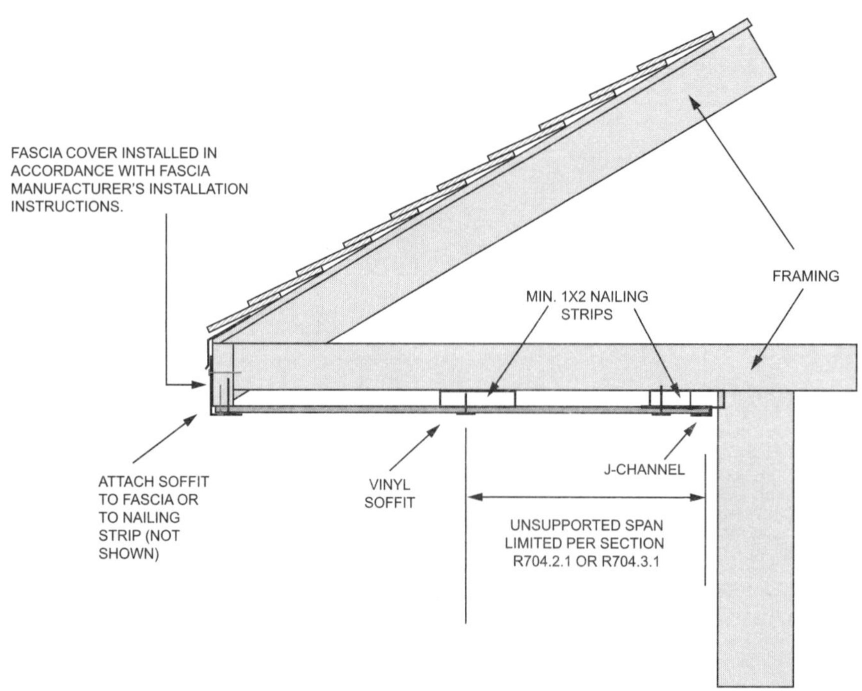

**FIGURE R704.2.1(2)
TYPICAL DOUBLE-SPAN VINYL SOFFIT PANEL SUPPORT**

TABLE R704.3.4
PRESCRIPTIVE ALTERNATIVE FOR WOOD STRUCTURAL PANEL SOFFIT [b, c, d, e]

MAXIMUM DESIGN PRESSURE (+ or - psf)	MINIMUM PANEL SPAN RATING	MINIMUM PANEL PERFORMANCE CATEGORY	NAIL TYPE AND SIZE	FASTENER[a] SPACING ALONG EDGES AND INTERMEDIATE SUPPORTS	
				Galvanized Steel	Stainless Steel
30	24/0	3/8	6d box (2 × 0.099 × 0.266 head diameter)	6[f]	4
40	24/0	3/8	6d box (2 × 0.099 × 0.266 head diameter)	6	4
50	24/0	3/8	6d box (2 × 0.099 × 0.266 head diameter)	4	4
			8d common ($2^1/_2$ × 0.131 × 0.281 head diameter)	6	6
60	24/0	3/8	6d box (2 × 0.099 × 0.266 head diameter)	4	3
			8d common ($2^1/_2$ × 0.131 × 0.281 head diameter)	6	4
70	24/16	7/16	8d common ($2^1/_2$ × 0.131 × 0.281 head diameter)	4	4
			10d box (3 × 0.128 × 0.312 head diameter)	6	4
80	24/16	7/16	8d common ($2^1/_2$ × 0.131 × 0.281 head diameter)	4	4
			10d box (3 × 0.128 × 0.312 head diameter)	6	4
90	32/16	15/32	8d common ($2^1/_2$ × 0.131 × 0.281 head diameter)	4	3
			10d box (3 × 0.128 × 0.312 head diameter)	6	4

For SI: 1 inch = 25.4 mm, 1 pound per square foot = 0.0479 kPa.

a. Fasteners shall comply with Sections R703.3.2 and R703.3.3.

b. Maximum spacing of soffit framing members shall not exceed 24 inches.

c. Wood structural panels shall be of an exterior exposure grade.

d. Wood structural panels shall be installed with strength axis perpendicular to supports with not fewer than two continuous spans.

e. Wood structural panels shall be attached to soffit framing members with specific gravity of at least 0.42. Framing members shall be minimum 2 × 3 nominal with the larger dimension in the cross section aligning with the length of fasteners to provide sufficient embedment depths.

f. Spacing at intermediate supports shall be not greater than 12 inches on center.

CHAPTER 8
ROOF-CEILING CONSTRUCTION

User note:

***About this chapter:** Chapter 8 addresses the design and construction of roof-ceiling systems. This chapter contains two roof-ceiling framing systems: wood framing and cold-formed steel framing. Allowable span tables are provided to simplify the selection of rafter and ceiling joist size for wood roof framing and cold-formed steel framing. Chapter 8 also provides requirements for the application of ceiling finishes, the proper ventilation of concealed spaces in roofs (for example, enclosed attics and rafter spaces), unvented attic assemblies and attic access.*

SECTION R801
GENERAL

R801.1 Application. The provisions of this chapter shall control the design and construction of the roof-ceiling system for buildings.

R801.2 Requirements. Roof and ceiling construction shall be capable of accommodating all loads imposed in accordance with Section R301 and of transmitting the resulting loads to the supporting structural elements.

R801.3 Roof drainage. In areas where *expansive soils* or *collapsible soils* are known to exist, all *dwellings* shall have a controlled method of water disposal from roofs that will collect and discharge roof drainage to the ground surface not less than 5 feet (1524 mm) from foundation walls or to an *approved* drainage system.

SECTION R802
WOOD ROOF FRAMING

R802.1 General. Wood and wood-based products used for load-supporting purposes shall conform to the applicable provisions of this section.

R802.1.1 Sawn lumber. Sawn lumber shall be identified by a grade *mark* of an accredited lumber grading or inspection agency and have design values certified by an accreditation body that complies with DOC PS 20. In lieu of a grade mark, a certificate of inspection issued by a lumber grading or inspection agency meeting the requirements of this section shall be accepted.

R802.1.1.1 End-jointed lumber. *Approved* end-jointed lumber identified by a grade *mark* conforming to Section R802.1.1 shall be permitted to be used interchangeably with solid-sawn members of the same species and grade. End-jointed lumber used in an assembly required elsewhere in this code to have a fire-resistance rating shall have the designation "Heat-Resistant Adhesive" or "HRA" included in its grade *mark*.

R802.1.2 Structural glued-laminated timbers. Glued-laminated timbers shall be manufactured and identified as required in ANSI A190.1, ANSI 117 and ASTM D3737.

R802.1.3 Structural log members. Structural log members shall comply with the provisions of ICC 400.

R802.1.4 Structural composite lumber. Structural capacities for *structural composite lumber* shall be established and monitored in accordance with ASTM D5456.

R802.1.5 Fire-retardant-treated wood. Fire-retardant-treated wood (FRTW) is any wood product that, when impregnated with chemicals by a pressure process or other means during manufacture, shall have, when tested in accordance with ASTM E84 or UL 723, a listed flame spread index of 25 or less. In addition, the ASTM E84 or UL 723 test shall be continued for an additional 20-minute period and the flame front shall not progress more than 10.5 feet (3200 mm) beyond the center line of the burners at any time during the test.

R802.1.5.1 Pressure process. For wood products impregnated with chemicals by a pressure process, the process shall be performed in closed vessels under pressures not less than 50 pounds per square inch gauge (psig) (344.7 kPa).

R802.1.5.2 Other means during manufacture. For wood products impregnated with chemicals by other means during manufacture, the treatment shall be an integral part of the manufacturing process of the wood product. The treatment shall provide permanent protection to all surfaces of the wood product. The use of paints, coating, stains or other surface treatments is not an *approved* method of protection as required by this section.

R802.1.5.3 Testing. For fire-retardant-treated wood products, the front and back faces of the wood product shall be tested in accordance with and produce the results required in Section R802.1.5.

R802.1.5.3.1 Fire testing of wood structural panels. *Wood structural panels* shall be tested with a ripped or cut longitudinal gap of $^{1}/_{8}$ inch (3.2 mm).

R802.1.5.4 Labeling. In addition to the *labels* required by Section 802.1.1 for sawn lumber and Section 803.2.1 for *wood structural panels*, each piece of *fire-retardant-treated* lumber and *wood structural panel* shall be *labeled*. The *label* shall contain:

1. The identification *mark* of an *approved agency* in accordance with Section 1703.5 of the *International Building Code*.
2. Identification of the treating manufacturer.

3. The name of the fire-retardant treatment.
4. The species of wood treated.
5. Flame spread index and *smoke-developed index.*
6. Method of drying after treatment.
7. Conformance to applicable standards in accordance with Sections R802.1.5.5 through R802.1.5.10.
8. For FRTW exposed to weather, or a damp or wet location, the words "No increase in the listed classification when subjected to the Standard Rain Test" (ASTM D2898).

R802.1.5.5 Strength adjustments. Design values for untreated lumber and *wood structural panels* as specified in Section R802.1 shall be adjusted for fire-retardant-treated wood. Adjustments to design values shall be based on an *approved* method of investigation that takes into consideration the effects of the anticipated temperature and humidity to which the fire-retardant-treated wood will be subjected, the type of treatment and redrying procedures.

R802.1.5.6 Wood structural panels. The effect of treatment and the method of redrying after treatment, and exposure to high temperatures and high humidities on the flexure properties of fire-retardant-treated softwood plywood shall be determined in accordance with ASTM D5516. The test data developed by ASTM D5516 shall be used to develop adjustment factors, maximum loads and spans, or both for untreated plywood design values in accordance with ASTM D6305. Each manufacturer shall publish the allowable maximum loads and spans for service as floor and roof sheathing for their treatment.

R802.1.5.7 Lumber. For each species of wood treated, the effect of the treatment and the method of redrying after treatment and exposure to high temperatures and high humidities on the allowable design properties of fire-retardant-treated lumber shall be determined in accordance with ASTM D5664. The test data developed by ASTM D5664 shall be used to develop modification factors for use at or near room temperature and at elevated temperatures and humidity in accordance with ASTM D6841. Each manufacturer shall publish the modification factors for service at temperatures of not less than 80°F (27°C) and for roof framing. The roof framing modification factors shall take into consideration the climatological location.

R802.1.5.8 Exposure to weather. Where fire-retardant-treated wood is exposed to weather or damp or wet locations, it shall be identified as "Exterior" to indicate there is not an increase in the *listed* flame spread index as defined in Section R802.1.5 when subjected to ASTM D2898.

R802.1.5.9 Interior applications. Interior fire-retardant-treated wood shall have a moisture content of not over 28 percent when tested in accordance with ASTM D3201 procedures at 92-percent relative humidity. Interior fire-retardant-treated wood shall be tested in accordance with Section R802.1.5.6 or R802.1.5.7. Interior fire-retardant-treated wood designated as Type A shall be tested in accordance with the provisions of this section.

R802.1.5.10 Moisture content. Fire-retardant-treated wood shall be dried to a moisture content of 19 percent or less for lumber and 15 percent or less for *wood structural panels* before use. For wood kiln dried after treatment (KDAT) the kiln temperatures shall not exceed those used in kiln drying the lumber and plywood submitted for the tests described in Section R802.1.5.6 for plywood and R802.1.5.7 for lumber.

R802.1.6 Cross-laminated timber. Cross-laminated timber shall be manufactured and identified as required by ANSI/APA PRG 320.

R802.1.7 Engineered wood rim board. Engineered wood rim boards shall conform to ANSI/APA PRR 410 or shall be evaluated in accordance with ASTM D7672. Structural capacities shall be in accordance with ANSI/APA PRR 410 or established in accordance with ASTM D7672. Rim boards conforming to ANSI/APA PRR 410 shall be marked in accordance with that standard.

R802.1.8 Prefabricated wood I-joists. Structural capacities and design provisions for prefabricated wood I-joists shall be established and monitored in accordance with ASTM D5055.

R802.2 Design and construction. The roof and ceiling assembly shall provide continuous ties across the structure to prevent roof thrust from being applied to the supporting walls. The assembly shall be designed and constructed in accordance with the provisions of this chapter and Figures R606.11(1), R606.11(2) and R606.11(3) or in accordance with AWC NDS.

R802.3 Ridge. A ridge board used to connect opposing rafters shall be not less than 1 inch (25 mm) nominal thickness and not less in depth than the cut end of the rafter. Where ceiling joist or rafter ties do not provide continuous ties across the structure as required by Section R802.5.2, the ridge shall be supported by a wall or ridge beam designed in accordance with accepted engineering practice and supported on each end by a wall or column.

R802.4 Rafters. Rafters shall be in accordance with this section.

R802.4.1 Rafter size. Rafters shall be sized based on the rafter spans in Tables R802.4.1(1) through R802.4.1(8). Rafter spans shall be measured along the horizontal projection of the rafter. For other grades and species and for other loading conditions, refer to the AWC STJR.

TABLE R802.4.1(1)
RAFTER SPANS FOR COMMON LUMBER SPECIES (Roof live load = 20 psf, ceiling not attached to rafters, L/Δ = 180)

RAFTER SPACING (inches)	SPECIES AND GRADE		DEAD LOAD = 10 psf					DEAD LOAD = 20 psf				
			2 × 4	2 × 6	2 × 8	2 × 10	2 × 12	2 × 4	2 × 6	2 × 8	2 × 10	2 × 12
			Maximum rafter spans[a]									
			(feet-inches)	(feet-inches)	(feet-inches)	(feet-inches)	(feet-inches)	(feet-inches)	(feet-inches)	(feet-inches)	(feet-inches)	(feet-inches)
12	Douglas fir-larch	SS	11-6	18-0	23-9	Note b	Note b	11-6	18-0	23-9	Note b	Note b
	Douglas fir-larch	#1	11-1	17-4	22-5	Note b	Note b	10-6	15-4	19-5	23-9	Note b
	Douglas fir-larch	#2	10-10	16-10	21-4	26-0	Note b	10-0	14-7	18-5	22-6	26-0
	Douglas fir-larch	#3	8-9	12-10	16-3	19-10	23-0	7-7	11-1	14-1	17-2	19-11
	Hem-fir	SS	10-10	17-0	22-5	Note b	Note b	10-10	17-0	22-5	Note b	Note b
	Hem-fir	#1	10 -7	16-8	22-0	Note b	Note b	10-4	15-2	19-2	23-5	Note b
	Hem-fir	#2	10-1	15-11	20-8	25-3	Note b	9-8	14-2	17-11	21-11	25-5
	Hem-fir	#3	8-7	12-6	15-10	19-5	22-6	7-5	10-10	13-9	16-9	19-6
	Southern pine	SS	11-3	17-8	23-4	Note b	Note b	11-3	17-8	23-4	Note b	Note b
	Southern pine	#1	10-10	17-0	22-5	Note b	Note b	10-6	15-8	19-10	23-2	Note b
	Southern pine	#2	10-4	15-7	19-8	23-5	Note b	9-0	13-6	17-1	20-3	23-10
	Southern pine	#3	8-0	11-9	14-10	18-0	21-4	6-11	10-2	12-10	15-7	18-6
	Spruce-pine-fir	SS	10-7	16-8	21-11	Note b	Note b	10-7	16-8	21-9	Note b	Note b
	Spruce-pine-fir	#1	10-4	16-3	21-0	25-8	Note b	9-10	14-4	18-2	22-3	25-9
	Spruce-pine-fir	#2	10-4	16-3	21-0	25-8	Note b	9-10	14-4	18-2	22-3	25-9
	Spruce-pine-fir	#3	8-7	12-6	15-10	19-5	22-6	7-5	10-10	13-9	16-9	19-6
16	Douglas fir-larch	SS	10-5	16-4	21-7	Note b	Note b	10-5	16-3	20-7	25-2	Note b
	Douglas fir-larch	#1	10-0	15-4	19-5	23-9	Note b	9-1	13-3	16-10	20-7	23-10
	Douglas fir-larch	#2	9-10	14-7	18-5	22-6	26-0	8-7	12-7	16-0	19-6	22-7
	Douglas fir-larch	#3	7-7	11-1	14-1	17-2	19-11	6-7	9-8	12-12	14-11	17-3
	Hem-fir	SS	9-10	15-6	20-5	Note b	Note b	9-10	15-6	19-11	24-4	Note b
	Hem-fir	#1	9-8	15-2	19-2	23-5	Note b	9-0	13-1	16-7	20-4	23-7
	Hem-fir	#2	9-2	14-2	17-11	21-11	25-5	8-5	12-3	15-6	18-11	22-0
	Hem-fir	#3	7-5	10-10	13-9	16-9	19-6	6-5	9-5	11-11	14-6	16-10
	Southern pine	SS	10-3	16-1	21-2	Note b	Note b	10-3	16-1	21-2	25-7	Note b
	Southern pine	#1	9-10	15-6	19-10	23-2	Note b	9-1	13-7	17-2	20-1	23-10
	Southern pine	#2	9-0	13-6	17-1	20-3	23-10	7-9	11-8	14-9	17-6	20-8
	Southern pine	#3	6-11	10-2	12-10	15-7	18-6	6-0	8-10	11-2	13-6	16-0
	Spruce-pine-fir	SS	9-8	15-2	19-11	25-5	Note b	9-8	14-10	18-10	23-0	Note b
	Spruce-pine-fir	#1	9-5	14-4	18-2	22-3	25-9	8-6	12-5	15-9	19-3	22-4
	Spruce-pine-fir	#2	9-5	14-4	18-2	22-3	25-9	8-6	12-5	15-9	19-3	22-4
	Spruce-pine-fir	#3	7-5	10-10	13-9	16-9	19-6	6-5	9-5	11-11	14-6	16-10

(continued)

TABLE R802.4.1(1)—continued
RAFTER SPANS FOR COMMON LUMBER SPECIES (Roof live load = 20 psf, ceiling not attached to rafters, L/Δ = 180)

RAFTER SPACING (inches)	SPECIES AND GRADE		DEAD LOAD = 10 psf					DEAD LOAD = 20 psf				
			2 × 4	2 × 6	2 × 8	2 × 10	2 × 12	2 × 4	2 × 6	2 × 8	2 × 10	2 × 12
			Maximum rafter spans[a]									
			(feet-inches)	(feet-inches)	(feet-inches)	(feet-inches)	(feet-inches)	(feet-inches)	(feet-inches)	(feet-inches)	(feet-inches)	(feet-inches)
19.2	Douglas fir-larch	SS	9-10	15-5	20-4	25-11	Note b	9-10	14-10	18-10	23-0	Note b
	Douglas fir-larch	#1	9-5	14-0	17-9	21-8	25-2	8-4	12-2	15-4	18-9	21-9
	Douglas fir-larch	#2	9-1	13-3	16-10	20-7	23-10	7-10	11-6	14-7	17-10	20-8
	Douglas fir-larch	#3	6-11	10-2	12-10	15-8	18-3	6-0	8-9	11-2	12-7	15-9
	Hem-fir	SS	9-3	14-7	19-2	24-6	Note b	9-3	14-4	18-2	22-3	25-9
	Hem-fir	#1	9-1	13-10	17-6	21-5	24-10	8-2	12-0	15-2	18-6	21-6
	Hem-fir	#2	8-8	12-11	16-4	20-0	23-2	7-8	11-2	14-2	17-4	20-1
	Hem-fir	#3	6-9	9-11	12-7	15-4	17-9	5-10	8-7	10-10	13-3	15-5
	Southern pine	SS	9-8	15-2	19-11	25-5	Note b	9-8	15-2	19-7	23-4	Note b
	Southern pine	#1	9-3	14-3	18-1	21-2	25-2	8-4	12-4	15-8	18-4	21-9
	Southern pine	#2	8-2	12-3	15-7	18-6	21-9	7-1	10-8	13-6	16-0	18-10
	Southern pine	#3	6-4	9-4	11-9	14-3	16-10	5-6	8-1	10-2	12-4	14-7
	Spruce-pine-fir	SS	9-1	14-3	18-9	23-11	Note b	9-1	13-7	17-2	21-0	24-4
	Spruce-pine-fir	#1	8-10	13-1	16-7	20-3	23-6	7-9	11-4	14-4	17-7	20-4
	Spruce-pine-fir	#2	8-10	13-1	16-7	20-3	23-6	7-9	11-4	14-4	17-7	20-4
	Spruce-pine-fir	#3	6-9	9-11	12-7	15-4	17-9	5-10	8-7	10-10	13-3	15-5
24	Douglas fir-larch	SS	9-1	14-4	18-10	23-9	Note b	9-1	13-3	16-10	20-7	23-10
	Douglas fir-larch	#1	8-7	12-6	15-10	19-5	22-6	7-5	10-10	13-9	16-9	19-6
	Douglas fir-larch	#2	8-2	11-11	15-1	18-5	21-4	7-0	10-4	13-0	15-11	18-6
	Douglas fir-larch	#3	6-2	9-1	11-6	14-1	16-3	5-4	7-10	10-0	12-2	14-1
	Hem-fir	SS	8-7	13-6	17-10	22-9	Note b	8-7	12-10	16-3	19-10	23-0
	Hem-fir	#1	8-5	12-4	15-8	19-2	22-2	7-4	10-9	13-7	16-7	19-3
	Hem-fir	#2	7-11	11-7	14-8	17-10	20-9	6-10	10-0	12-8	15-6	17-11
	Hem-fir	#3	6-1	8-10	11-3	13-8	15-11	5-3	7-8	9-9	11-10	13-9
	Southern pine	SS	8-11	14-1	18-6	23-8	Note b	8-11	13-10	17-6	20-10	24-8
	Southern pine	#1	8-7	12-9	16-2	18-11	22-6	7-5	11-1	14-0	16-5	19-6
	Southern pine	#2	7-4	11-0	13-11	16-6	19-6	6-4	9-6	12-1	14-4	16-10
	Southern pine	#3	5-8	8-4	10-6	12-9	15-1	4-11	7-3	9-1	11-0	13-1
	Spruce-pine-fir	SS	8-5	13-3	17-5	21-8	25-2	8-4	12-2	15-4	18-9	21-9
	Spruce-pine-fir	#1	8-0	11-9	14-10	18-2	21-0	6-11	10-2	12-10	15-8	18-3
	Spruce-pine-fir	#2	8-0	11-9	14-10	18-2	21-0	6-11	10-2	12-10	15-8	18-3
	Spruce-pine-fir	#3	6-1	8-10	11-3	13-8	15-11	5-3	7-8	9-9	11-10	13-9

Check sources for availability of lumber in lengths greater than 20 feet.

For SI: 1 inch = 25.4 mm, 1 foot = 304.8 mm, 1 pound per square foot = 0.0479 kPa.

a. The tabulated rafter spans assume that ceiling joists are located at the bottom of the attic space or that some other method of resisting the outward push of the rafters on the bearing walls, such as rafter ties, is provided at that location. Where ceiling joists or rafter ties are located higher in the attic space, the rafter spans shall be multiplied by the adjustment factors in Table R802.4.1(9).

b. Span exceeds 26 feet in length.

TABLE R802.4.1(2)
RAFTER SPANS FOR COMMON LUMBER SPECIES (Roof live load = 20 psf, ceiling attached to rafters, $L/\Delta = 240$)

RAFTER SPACING (inches)	SPECIES AND GRADE		DEAD LOAD = 10 psf					DEAD LOAD = 20 psf				
			2 × 4	2 × 6	2 × 8	2 × 10	2 × 12	2 × 4	2 × 6	2 × 8	2 × 10	2 × 12
			Maximum rafter spans[a]									
			(feet-inches)	(feet-inches)	(feet-inches)	(feet-inches)	(feet-inches)	(feet-inches)	(feet-inches)	(feet-inches)	(feet-inches)	(feet-inches)
12	Douglas fir-larch	SS	10-5	16-4	21-7	Note b	Note b	10-5	16-4	21-7	Note b	Note b
	Douglas fir-larch	#1	10-0	15-9	20-10	Note b	Note b	10-0	15-4	19-5	23-9	Note b
	Douglas fir-larch	#2	9-10	15-6	20-5	26-0	Note b	9-10	14-7	18-5	22-6	26-0
	Douglas fir-larch	#3	8-9	12-10	16-3	19-10	23-0	7-7	11-1	14-1	17-2	19-11
	Hem-fir	SS	9-10	15-6	20-5	Note b	Note b	9-10	15-6	20-5	Note b	Note b
	Hem-fir	#1	9-8	15-2	19-11	25-5	Note b	9-8	15-2	19-2	23-5	Note b
	Hem-fir	#2	9-2	14-5	19-0	24-3	Note b	9-2	14-2	17-11	21-11	25-5
	Hem-fir	#3	8-7	12-6	15-10	19-5	22-6	7-5	10-10	13-9	16-9	19-6
	Southern pine	SS	10-3	16-1	21-2	Note b	Note b	10-3	16-1	21-2	Note b	Note b
	Southern pine	#1	9-10	15-6	20-5	Note b	Note b	9-10	15-6	19-10	23-2	Note b
	Southern pine	#2	9-5	14-9	19-6	23-5	Note b	9-0	13-6	17-1	20-3	23-10
	Southern pine	#3	8-0	11-9	14-10	18-0	21-4	6-11	10-2	12-10	15-7	18-6
	Spruce-pine-fir	SS	9-8	15-2	19-11	25-5	Note b	9-8	15-2	19-11	25-5	Note b
	Spruce-pine-fir	#1	9-5	14-9	19-6	24-10	Note b	9-5	14-4	18-2	22-3	25-9
	Spruce-pine-fir	#2	9-5	14-9	19-6	24-10	Note b	9-5	14-4	18-2	22-3	25-9
	Spruce-pine-fir	#3	8-7	12-6	15-10	19-5	22-6	7-5	10-10	13-9	16-9	19-6
16	Douglas fir-larch	SS	9-6	14-11	19-7	25-0	Note b	9-6	14-11	19-7	25-0	Note b
	Douglas fir-larch	#1	9-1	14-4	18-11	23-9	Note b	9-1	13-3	16-10	20-7	23-10
	Douglas fir-larch	#2	8-11	14-1	18-5	22-6	26-0	8-7	12-7	16-0	19-6	22-7
	Douglas fir-larch	#3	7-7	11-1	14-1	17-2	19-11	6-7	9-8	12-2	14-11	17-3
	Hem-fir	SS	8-11	14-1	18-6	23-8	Note b	8-11	14-1	18-6	23-8	Note b
	Hem-fir	#1	8-9	13-9	18-1	23-1	Note b	8-9	13-1	16-7	20-4	23-7
	Hem-fir	#2	8-4	13-1	17-3	21-11	25-5	8-4	12-3	15-6	18-11	22-0
	Hem-fir	#3	7-5	10-10	13-9	16-9	19-6	6-5	9-5	11-11	14-6	16-10
	Southern pine	SS	9-4	14-7	19-3	24-7	Note b	9-4	14-7	19-3	24-7	Note b
	Southern pine	#1	8-11	14-1	18-6	23-2	Note b	8-11	13-7	17-2	20-1	23-10
	Southern pine	#2	8-7	13-5	17-1	20-3	23-10	7-9	11-8	14-9	17-6	20-8
	Southern pine	#3	6-11	10-2	12-10	15-7	18-6	6-0	8-10	11-2	13-6	16-0
	Spruce-pine-fir	SS	8-9	13-9	18-1	23-1	Note b	8-9	13-9	18-1	23-0	Note b
	Spruce-pine-fir	#1	8-7	13-5	17-9	22-3	25-9	8-6	12-5	15-9	19-3	22-4
	Spruce-pine-fir	#2	8-7	13-5	17-9	22-3	25-9	8-6	12-5	15-9	19-3	22-4
	Spruce-pine-fir	#3	7-5	10-10	13-9	16-9	19-6	6-5	9-5	11-11	14-6	16-10

(continued)

TABLE R802.4.1(2)—continued
RAFTER SPANS FOR COMMON LUMBER SPECIES (Roof live load = 20 psf, ceiling attached to rafters, *L*/Δ = 240)

RAFTER SPACING (inches)	SPECIES AND GRADE		DEAD LOAD = 10 psf					DEAD LOAD = 20 psf				
			2 × 4	2 × 6	2 × 8	2 × 10	2 × 12	2 × 4	2 × 6	2 × 8	2 × 10	2 × 12
			Maximum rafter spans[a]									
			(feet-inches)	(feet-inches)	(feet-inches)	(feet-inches)	(feet-inches)	(feet-inches)	(feet-inches)	(feet-inches)	(feet-inches)	(feet-inches)
19.2	Douglas fir-larch	SS	8-11	14-0	18-5	23-7	Note b	8-11	14-0	18-5	23-0	Note b
	Douglas fir-larch	#1	8-7	13-6	17-9	21-8	25-2	8-4	12-2	15-4	18-9	21-9
	Douglas fir-larch	#2	8-5	13-3	16-10	20-7	23-10	7-10	11-6	14-7	17-10	20-8
	Douglas fir-larch	#3	6-11	10-2	12-10	15-8	18-3	6-0	8-9	11-2	13-7	15-9
	Hem-fir	SS	8-5	13-3	17-5	22-3	Note b	8-5	13-3	17-5	22-3	25-9
	Hem-fir	#1	8-3	12-11	17-1	21-5	24-10	8-2	12-0	15-2	18-6	21-6
	Hem-fir	#2	7-10	12-4	16-3	20-0	23-2	7-8	11-2	14-2	17-4	20-1
	Hem-fir	#3	6-9	9-11	12-7	15-4	17-9	5-10	8-7	10-10	13-3	15-5
	Southern pine	SS	8-9	13-9	18-2	23-1	Note b	8-9	13-9	18-2	23-1	Note b
	Southern pine	#1	8-5	13-3	17-5	21-2	25-2	8-4	12-4	15-8	18-4	21-9
	Southern pine	#2	8-1	12-3	15-7	18-6	21-9	7-1	10-8	13-6	16-0	18-10
	Southern pine	#3	6-4	9-4	11-9	14-3	16-10	5-6	8-1	10-2	12-4	14-7
	Spruce-pine-fir	SS	8-3	12-11	17-1	21-9	Note b	8-3	12-11	17-1	21-0	24-4
	Spruce-pine-fir	#1	8-1	12-8	16-7	20-3	23-6	7-9	11-4	14-4	17-7	20-4
	Spruce-pine-fir	#2	8-1	12-8	16-7	20-3	23-6	7-9	11-4	14-4	17-7	20-4
	Spruce-pine-fir	#3	6-9	9-11	12-7	15-4	17-9	5-10	8-7	10-10	13-3	15-5
24	Douglas fir-larch	SS	8-3	13-0	17-2	21-10	Note b	8-3	13-0	16-10	20-7	23-10
	Douglas fir-larch	#1	8-0	12-6	15-10	19-5	22-6	7-5	10-10	13-9	16-9	19-6
	Douglas fir-larch	#2	7-10	11-11	15-1	18-5	21-4	7-0	10-4	13-0	15-11	18-6
	Douglas fir-larch	#3	6-2	9-1	11-6	14-1	16-3	5-4	7-10	10-0	12-2	14-1
	Hem-fir	SS	7-10	12-3	16-2	20-8	25-1	7-10	12-3	16-2	19-10	23-0
	Hem-fir	#1	7-8	12-0	15-8	19-2	22-2	7-4	10-9	13-7	16-7	19-3
	Hem-fir	#2	7-3	11-5	14-8	17-10	20-9	6-10	10-0	12-8	15-6	17-11
	Hem-fir	#3	6-1	8-10	11-3	13-8	15-11	5-3	7-8	9-9	11-10	13-9
	Southern pine	SS	8-1	12-9	16-10	21-6	Note b	8-1	12-9	16-10	20-10	24-8
	Southern pine	#1	7-10	12-3	16-2	18-11	22-6	7-5	11-1	14-0	16-5	19-6
	Southern pine	#2	7-4	11-0	13-11	16-6	19-6	6-4	9-6	12-1	14-4	16-10
	Southern pine	#3	5-8	8-4	10-6	12-9	15-1	4-11	7-3	9-1	11-0	13-1
	Spruce-pine-fir	SS	7-8	12-0	15-10	20-2	24-7	7-8	12-0	15-4	18-9	21-9
	Spruce-pine-fir	#1	7-6	11-9	14-10	18-2	21-0	6-11	10-2	12-10	15-8	18-3
	Spruce-pine-fir	#2	7-6	11-9	14-10	18-2	21-0	6-11	10-2	12-10	15-8	18-3
	Spruce-pine-fir	#3	6-1	8-10	11-3	13-8	15-11	5-3	7-8	9-9	11-10	13-9

Check sources for availability of lumber in lengths greater than 20 feet.

For SI: 1 inch = 25.4 mm, 1 foot = 304.8 mm, 1 pound per square foot = 0.0479 kPa.

a. The tabulated rafter spans assume that ceiling joists are located at the bottom of the attic space or that some other method of resisting the outward push of the rafters on the bearing walls, such as rafter ties, is provided at that location. Where ceiling joists or rafter ties are located higher in the attic space, the rafter spans shall be multiplied by the adjustment factors in Table R802.4.1(9).

b. Span exceeds 26 feet in length.

TABLE R802.4.1(3)
RAFTER SPANS FOR COMMON LUMBER SPECIES (Ground snow load = 30 psf, ceiling not attached to rafters, L/Δ = 180)

RAFTER SPACING (inches)	SPECIES AND GRADE		DEAD LOAD = 10 psf					DEAD LOAD = 20 psf				
			2 × 4	2 × 6	2 × 8	2 × 10	2 × 12	2 × 4	2 × 6	2 × 8	2 × 10	2 × 12
			Maximum rafter spans[a]									
			(feet-inches)	(feet-inches)	(feet-inches)	(feet-inches)	(feet-inches)	(feet-inches)	(feet-inches)	(feet-inches)	(feet-inches)	(feet-inches)
12	Douglas fir-larch	SS	10-0	15-9	20-9	Note b	Note b	10-0	15-9	20-5	24-11	Note b
	Douglas fir-larch	#1	9-8	14-9	18-8	22-9	Note b	9-0	13-2	16-8	20-4	23-7
	Douglas fir-larch	#2	9-6	14-0	17-8	21-7	25-1	8-6	12-6	15-10	19-4	22-5
	Douglas fir-larch	#3	7-3	10-8	13-6	16-6	19-2	6-6	9-6	12-1	14-9	17-1
	Hem-fir	SS	9-6	14-10	19-7	25-0	Note b	9-6	14-10	19-7	24-1	Note b
	Hem-fir	#1	9-3	14-6	18-5	22-6	26-0	8-11	13-0	16-6	20-1	23-4
	Hem-fir	#2	8-10	13-7	17-2	21-0	24-4	8-4	12-2	15-4	18-9	21-9
	Hem-fir	#3	7-1	10-5	13-2	16-1	18-8	6-4	9-4	11-9	14-5	16-8
	Southern pine	SS	9-10	15-6	20-5	Note b	Note b	9-10	15-6	20-5	25-4	Note b
	Southern pine	#1	9-6	14-10	19-0	22-3	Note b	9-0	13-5	17-0	19-11	23-7
	Southern pine	#2	8-7	12-11	16-4	19-5	22-10	7-8	11-7	14-8	17-4	20-5
	Southern pine	#3	6-7	9-9	12-4	15-0	17-9	5-11	8-9	11-0	13-5	15-10
	Spruce-pine-fir	SS	9-3	14-7	19-2	24-6	Note b	9-3	14-7	18-8	22-9	Note b
	Spruce-pine-fir	#1	9-1	13-9	17-5	21-4	24-8	8-5	12-4	15-7	19-1	22-1
	Spruce-pine-fir	#2	9-1	13-9	17-5	21-4	24-8	8-5	12-4	15-7	19-1	22-1
	Spruce-pine-fir	#3	7-1	10-5	13-2	16-1	18-8	6-4	9-4	11-9	14-5	16-8
16	Douglas fir-larch	SS	9-1	14-4	18-10	24-1	Note b	9-1	14-0	17-8	21-7	25-1
	Douglas fir-larch	#1	8-9	12-9	16-2	19-9	22-10	7-10	11-5	14-5	17-8	20-5
	Douglas fir-larch	#2	8-3	12-1	15-4	18-9	21-8	7-5	10-10	13-8	16-9	19-5
	Douglas fir-larch	#3	6-4	9-3	11-8	14-3	16-7	5-8	8-3	10-6	12-9	14-10
	Hem-fir	SS	8-7	13-6	17-10	22-9	Note b	8-7	13-6	17-1	20-10	24-2
	Hem-fir	#1	8-5	12-7	15-11	19-6	22-7	7-8	11-3	14-3	17-5	20-2
	Hem-fir	#2	8-0	11-9	14-11	18-2	21-1	7-2	10-6	13-4	16-3	18-10
	Hem-fir	#3	6-2	9-0	11-5	13-11	16-2	5-6	8-1	10-3	12-6	14-6
	Southern pine	SS	8-11	14-1	18-6	23-8	Note b	8-11	14-1	18-5	1-11	25-11
	Southern pine	#1	8-7	13-0	16-6	19-3	22-10	7-10	11-7	14-9	17-3	20-5
	Southern pine	#2	7-6	11-2	14-2	16-10	19-10	6-8	10-0	12-8	15-1	17-9
	Southern pine	#3	5-9	8-6	10-8	13-0	15-4	5-2	7-7	9-7	11-7	13-9
	Spruce-pine-fir	SS	8-5	13-3	17-5	22-1	25-7	8-5	12-9	16-2	19-9	22-10
	Spruce-pine-fir	#1	8-2	11-11	15-1	18-5	21-5	7-3	10-8	13-6	16-6	19-2
	Spruce-pine-fir	#2	8-2	11-11	15-1	18-5	21-5	7-3	10-8	13-6	16-6	19-2
	Spruce-pine-fir	#3	6-2	9-0	11-5	13-11	16-2	5-6	8-1	10-3	12-6	14-6

(continued)

TABLE R802.4.1(3)—continued
RAFTER SPANS FOR COMMON LUMBER SPECIES (Ground snow load = 30 psf, ceiling not attached to rafters, L/Δ = 180)

RAFTER SPACING (inches)	SPECIES AND GRADE		DEAD LOAD = 10 psf					DEAD LOAD = 20 psf				
			2 × 4	2 × 6	2 × 8	2 × 10	2 × 12	2 × 4	2 × 6	2 × 8	2 × 10	2 × 12
			Maximum rafter spans[a]									
			(feet-inches)	(feet-inches)	(feet-inches)	(feet-inches)	(feet-inches)	(feet-inches)	(feet-inches)	(feet-inches)	(feet-inches)	(feet-inches)
19.2	Douglas fir-larch	SS	8-7	13-6	17-9	22-1	25-7	8-7	12-9	16-2	19-9	22-10
	Douglas fir-larch	#1	7-11	11-8	14-9	18-0	20-11	7-1	10-5	13-2	16-1	18-8
	Douglas fir-larch	#2	7-7	11-0	14-0	17-1	19-10	6-9	9-10	12-6	15-3	17-9
	Douglas fir-larch	#3	5-9	8-5	10-8	13-1	15-2	5-2	7-7	9-7	11-8	13-6
	Hem-fir	SS	8-1	12-9	16-9	21-4	24-8	8-1	12-4	15-7	19-1	22-1
	Hem-fir	#1	7-10	11-6	14-7	17-9	20-7	7-0	10-3	13-0	15-11	18-5
	Hem-fir	#2	7-4	10-9	13-7	16-7	19-3	6-7	9-7	12-2	14-10	17-3
	Hem-fir	#3	5-7	8-3	10-5	12-9	14-9	5-0	7-4	9-4	11-5	13-2
	Southern pine	SS	8-5	13-3	17-5	22-3	Note b	8-5	13-3	16-10	20-0	23-7
	Southern pine	#1	8-0	11-10	15-1	17-7	20-11	7-1	10-7	13-5	15-9	18-8
	Southern pine	#2	6-10	10-2	12-11	15-4	18-1	6-1	9-2	11-7	13-9	16-2
	Southern pine	#3	5-3	7-9	9-9	11-10	14-0	4-8	6-11	8-9	10-7	12-6
	Spruce-pine-fir	SS	7-11	12-5	16-5	20-2	23-4	7-11	11-8	14-9	18-0	20-11
	Spruce-pine-fir	#1	7-5	10-11	13-9	16-10	19-6	6-8	9-9	12-4	15-1	17-6
	Spruce-pine-fir	#2	7-5	10-11	13-9	16-10	19-6	6-8	9-9	12-4	15-1	17-6
	Spruce-pine-fir	#3	5-7	8-3	10-5	12-9	14-9	5-0	7-4	9-4	11-5	13-2
24	Douglas fir-larch	SS	8-0	12-6	16-2	19-9	22-10	7-10	11-5	14-5	17-8	20-5
	Douglas fir-larch	#1	7-1	10-5	13-2	16-1	18-8	6-4	9-4	11-9	14-5	16-8
	Douglas fir-larch	#2	6-9	9-10	12-6	15-3	17-9	6-0	8-10	11-2	13-8	15-10
	Douglas fir-larch	#3	5-2	7-7	9-7	11-8	13-6	4-7	6-9	8-7	10-5	12-1
	Hem-fir	SS	7-6	11-10	15-7	19-1	22-1	7-6	11-0	13-11	17-0	19-9
	Hem-fir	#1	7-0	10-3	13-0	15-11	18-5	6-3	9-2	11-8	14-3	16-6
	Hem-fir	#2	6-7	9-7	12-2	14-10	17-3	5-10	8-7	10-10	13-3	15-5
	Hem-fir	#3	5-0	7-4	9-4	11-5	13-2	4-6	6-7	8-4	10-2	11-10
	Southern pine	SS	7-10	12-3	16-2	20-0	23-7	7-10	11-10	15-0	17-11	21-2
	Southern pine	#1	7-1	10-7	13-5	15-9	18-8	6-4	9-6	12-0	14-1	16-8
	Southern pine	#2	6-1	9-2	11-7	13-9	16-2	5-5	8-2	10-4	12-3	14-6
	Southern pine	#3	4-8	6-11	8-9	10-7	12-6	4-2	6-2	7-10	9-6	11-2
	Spruce-pine-fir	SS	7-4	11-7	14-9	18-0	20-11	7-1	10-5	13-2	16-1	18-8
	Spruce-pine-fir	#1	6-8	9-9	12-4	15-1	17-6	5-11	8-8	11-0	13-6	15-7
	Spruce-pine-fir	#2	6-8	9-9	12-4	15-1	17-6	5-11	8-8	11-0	13-6	15-7
	Spruce-pine-fir	#3	5-0	7-4	9-4	11-5	13-2	4-6	6-7	8-4	10-2	11-10

Check sources for availability of lumber in lengths greater than 20 feet.

For SI: 1 inch = 25.4 mm, 1 foot = 304.8 mm, 1 pound per square foot = 0.0479 kPa.

a. The tabulated rafter spans assume that ceiling joists are located at the bottom of the attic space or that some other method of resisting the outward push of the rafters on the bearing walls, such as rafter ties, is provided at that location. Where ceiling joists or rafter ties are located higher in the attic space, the rafter spans shall be multiplied by the adjustment factors in Table R802.4.1(9).

b. Span exceeds 26 feet in length.

TABLE R802.4.1(4)
RAFTER SPANS FOR COMMON LUMBER SPECIES (Ground snow load = 30 psf, ceiling attached to rafters, L/Δ = 240)

RAFTER SPACING (inches)	SPECIES AND GRADE		DEAD LOAD = 10 psf					DEAD LOAD = 20 psf				
			2 × 4	2 × 6	2 × 8	2 × 10	2 × 12	2 × 4	2 × 6	2 × 8	2 × 10	2 × 12
			Maximum rafter spans[a]									
			(feet-inches)	(feet-inches)	(feet-inches)	(feet-inches)	(feet-inches)	(feet-inches)	(feet-inches)	(feet-inches)	(feet-inches)	(feet-inches)
12	Douglas fir-larch	SS	9-1	14-4	18-10	24-1	Note b	9-1	14-4	18-10	24-1	Note b
	Douglas fir-larch	#1	8-9	13-9	18-2	22-9	Note b	8-9	13-2	16-8	20-4	23-7
	Douglas fir-larch	#2	8-7	13-6	17-8	21-7	25-1	8-6	12-6	15-10	19-4	22-5
	Douglas fir-larch	#3	7-3	10-8	13-6	16-6	19-2	6-6	9-6	12-1	14-9	17-1
	Hem-fir	SS	8-7	13-6	17-10	22-9	Note b	8-7	13-6	17-10	22-9	Note b
	Hem-fir	#1	8-5	13-3	17-5	22-3	26-0	8-5	13-0	16-6	20-1	23-4
	Hem-fir	#2	8-0	12-7	16-7	21-0	24-4	8-0	12-2	15-4	18-9	21-9
	Hem-fir	#3	7-1	10-5	13-2	16-1	18-8	6-4	9-4	11-9	14-5	16-8
	Southern pine	SS	8-11	14-1	18-6	23-8	Note b	8-11	14-1	18-6	23-8	Note b
	Southern pine	#1	8-7	13-6	17-10	22-3	Note b	8-7	13-5	17-0	19-11	23-7
	Southern pine	#2	8-3	12-11	16-4	19-5	22-10	7-8	11-7	14-8	17-4	20-5
	Southern pine	#3	6-7	9-9	12-4	15-0	17-9	5-11	8-9	11-0	13-5	15-10
	Spruce-pine-fir	SS	8-5	13-3	17-5	22-3	Note b	8-5	13-3	17-5	22-3	Note b
	Spruce-pine-fir	#1	8-3	12-11	17-0	21-4	24-8	8-3	12-4	15-7	19-1	22-1
	Spruce-pine-fir	#2	8-3	12-11	17-0	21-4	24-8	8-3	12-4	15-7	19-1	22-1
	Spruce-pine-fir	#3	7-1	10-5	13-2	16-1	18-8	6-4	9-4	11-9	14-5	16-8
16	Douglas fir-larch	SS	8-3	13-0	17-2	21-10	Note b	8-3	13-0	17-2	21-7	25-1
	Douglas fir-larch	#1	8-0	12-6	16-2	19-9	22-10	7-10	11-5	14-5	17-8	20-5
	Douglas fir-larch	#2	7-10	12-1	15-4	18-9	21-8	7-5	10-10	13-8	16-9	19-5
	Douglas fir-larch	#3	6-4	9-3	11-8	14-3	16-7	5-8	8-3	10-6	12-9	14-10
	Hem-fir	SS	7-10	12-3	16-2	20-8	25-1	7-10	12-3	16-2	20-8	24-2
	Hem-fir	#1	7-8	12-0	15-10	19-6	22-7	7-8	11-3	14-3	17-5	20-2
	Hem-fir	#2	7-3	11-5	14-11	18-2	21-1	7-2	10-6	13-4	16-3	18-10
	Hem-fir	#3	6-2	9-0	11-5	13-11	16-2	5-6	8-1	10-3	12-6	14-6
	Southern pine	SS	8-1	12-9	16-10	21-6	Note b	8-1	12-9	16-10	21-6	25-11
	Southern pine	#1	7-10	12-3	16-2	19-3	22-10	7-10	11-7	14-9	17-3	20-5
	Southern pine	#2	7-6	11-2	14-2	16-10	19-10	6-8	10-0	12-8	15-1	17-9
	Southern pine	#3	5-9	8-6	10-8	13-0	15-4	5-2	7-7	9-7	11-7	13-9
	Spruce-pine-fir	SS	7-8	12-0	15-10	20-2	24-7	7-8	12-0	15-10	19-9	22-10
	Spruce-pine-fir	#1	7-6	11-9	15-1	18-5	21-5	7-3	10-8	13-6	16-6	19-2
	Spruce-pine-fir	#2	7-6	11-9	15-1	18-5	21-5	7-3	10-8	13-6	16-6	19-2
	Spruce-pine-fir	#3	6-2	9-0	11-5	13-11	16-2	5-6	8-1	10-3	12-6	14-6

(continued)

TABLE R802.4.1(4)—continued
RAFTER SPANS FOR COMMON LUMBER SPECIES (Ground snow load = 30 psf, ceiling attached to rafters, L/Δ = 240)

RAFTER SPACING (inches)	SPECIES AND GRADE		DEAD LOAD = 10 psf					DEAD LOAD = 20 psf				
			2 × 4	2 × 6	2 × 8	2 × 10	2 × 12	2 × 4	2 × 6	2 × 8	2 × 10	2 × 12
			Maximum rafter spans[a]									
			(feet-inches)	(feet-inches)	(feet-inches)	(feet-inches)	(feet-inches)	(feet-inches)	(feet-inches)	(feet-inches)	(feet-inches)	(feet-inches)
19.2	Douglas fir-larch	SS	7-9	12-3	16-1	20-7	25-0	7-9	12-3	16-1	19-9	22-10
	Douglas fir-larch	#1	7-6	11-8	14-9	18-0	20-11	7-1	10-5	13-2	16-1	18-8
	Douglas fir-larch	#2	7-4	11-0	14-0	17-1	19-10	6-9	9-1	12-6	15-3	17-9
	Douglas fir-larch	#3	5-9	8-5	10-8	13-1	15-2	5-2	7-7	9-7	11-8	13-6
	Hem-fir	SS	7-4	11-7	15-3	19-5	23-7	7-4	11-7	15-3	19-1	22-1
	Hem-fir	#1	7-2	11-4	14-7	17-9	20-7	7-0	16-3	13-0	15-11	18-5
	Hem-fir	#2	6-10	10-9	13-7	16-7	19-3	6-7	9-7	12-2	14-10	17-3
	Hem-fir	#3	5-7	8-3	10-5	12-9	14-9	5-0	7-4	9-4	11-5	13-2
	Southern pine	SS	7-8	12-0	15-10	20-2	24-7	7-8	12-0	15-10	20-0	23-7
	Southern pine	#1	7-4	11-7	15-1	17-7	20-11	7-1	10-7	13-5	15-9	18-8
	Southern pine	#2	6-10	10-2	12-11	15-4	18-1	6-1	9-2	11-7	13-9	16-2
	Southern pine	#3	5-3	7-9	9-9	11-10	14-0	4-8	6-11	8-9	10-7	12-6
	Spruce-pine-fir	SS	7-2	11-4	14-11	19-0	23-1	7-2	11-4	14-9	18-0	20-11
	Spruce-pine-fir	#1	7-0	10-11	13-9	16-10	19-6	6-8	9-9	12-4	15-1	17-6
	Spruce-pine-fir	#2	7-0	10-11	13-9	16-10	19-6	6-8	9-9	12-4	15-1	17-6
	Spruce-pine-fir	#3	5-7	8-3	10-5	12-9	14-9	5-0	7-4	9-4	11-5	13-2
24	Douglas fir-larch	SS	7-3	11-4	15-0	19-1	22-10	7-3	11-4	14-5	17-8	20-5
	Douglas fir-larch	#1	7-0	10-5	13-2	16-1	18-8	6-4	9-4	11-9	14-5	16-8
	Douglas fir-larch	#2	6-9	9-10	12-6	15-3	17-9	6-0	8-10	11-2	13-8	15-10
	Douglas fir-larch	#3	5-2	7-7	9-7	11-8	13-6	4-7	6-9	8-7	10-5	12-1
	Hem-fir	SS	6-10	10-9	14-2	18-0	21-11	6-10	10-9	13-11	17-0	19-9
	Hem-fir	#1	6-8	10-3	13-0	15-11	18-5	6-3	9-2	11-8	14-3	16-6
	Hem-fir	#2	6-4	9-7	12-2	14-10	17-3	5-10	8-7	10-10	13-3	15-5
	Hem-fir	#3	5-0	7-4	9-4	11-5	13-2	4-6	6-7	8-4	10-2	11-10
	Southern pine	SS	7-1	11-2	14-8	18-9	22-10	7-1	11-2	14-8	17-11	21-2
	Southern pine	#1	6-10	10-7	13-5	15-9	18-8	6-4	9-6	12-0	14-1	16-8
	Southern pine	#2	6-1	9-2	11-7	13-9	16-2	5-5	8-2	10-4	12-3	14-6
	Southern pine	#3	4-8	6-11	8-9	10-7	12-6	4-2	6-2	7-10	9-6	11-2
	Spruce-pine-fir	SS	6-8	10-6	13-10	17-8	20-11	6-8	10-5	13-2	16-1	18-8
	Spruce-pine-fir	#1	6-6	9-9	12-4	15-1	17-6	5-11	8-8	11-0	13-6	15-7
	Spruce-pine-fir	#2	6-6	9-9	12-4	15-1	17-6	5-11	8-8	11-0	13-6	15-7
	Spruce-pine-fir	#3	5-0	7-4	9-4	11-5	13-2	4-6	6-7	8-4	10-2	11-10

Check sources for availability of lumber in lengths greater than 20 feet.

For SI: 1 inch = 25.4 mm, 1 foot = 304.8 mm, 1 pound per square foot = 0.0479 kPa.

a. The tabulated rafter spans assume that ceiling joists are located at the bottom of the attic space or that some other method of resisting the outward push of the rafters on the bearing walls, such as rafter ties, is provided at that location. Where ceiling joists or rafter ties are located higher in the attic space, the rafter spans shall be multiplied by the adjustment factors in Table R802.4.1(9).

b. Span exceeds 26 feet in length.

TABLE R802.4.1(5)
RAFTER SPANS FOR COMMON LUMBER SPECIES (Ground snow load = 50 psf, ceiling not attached to rafters, $L/\Delta = 180$)

RAFTER SPACING (inches)	SPECIES AND GRADE		DEAD LOAD = 10 psf					DEAD LOAD = 20 psf				
			2 × 4	2 × 6	2 × 8	2 × 10	2 × 12	2 × 4	2 × 6	2 × 8	2 × 10	2 × 12
			Maximum rafter spans[a]									
			(feet-inches)	(feet-inches)	(feet-inches)	(feet-inches)	(feet-inches)	(feet-inches)	(feet-inches)	(feet-inches)	(feet-inches)	(feet-inches)
12	Douglas fir-larch	SS	8-5	13-3	17-6	22-4	26-0	8-5	13-3	17-3	21-1	24-5
	Douglas fir-larch	#1	8-2	12-0	15-3	18-7	21-7	7-7	11-2	14-1	17-3	20-0
	Douglas fir-larch	#2	7-10	11-5	14-5	17-8	20-5	7-3	10-7	13-4	16-4	18-11
	Douglas fir-larch	#3	6-0	8-9	11-0	13-6	15-7	5-6	8-1	10-3	12-6	14-6
	Hem-fir	SS	8-0	12-6	16-6	21-1	25-6	8-0	12-6	16-6	20-4	23-7
	Hem-fir	#1	7-10	11-10	15-0	18-4	21-3	7-6	11-0	13-11	17-0	19-9
	Hem-fir	#2	7-5	11-1	14-0	17-2	19-11	7-0	10-3	13-0	15-10	18-5
	Hem-fir	#3	5-10	8-6	10-9	13-2	15-3	5-5	7-10	10-0	12-2	14-1
	Southern pine	SS	8-4	13-1	17-2	21-11	Note b	8-4	13-1	17-2	21-5	25-3
	Southern pine	#1	8-0	12-3	15-6	18-2	21-7	7-7	11-4	14-5	16-10	20-0
	Southern pine	#2	7-0	10-6	13-4	15-10	18-8	6-6	9-9	12-4	14-8	17-3
	Southern pine	#3	5-5	8-0	10-1	12-3	14-6	5-0	7-5	9-4	11-4	13-5
	Spruce-pine-fir	SS	7-10	12-3	16-2	20-8	24-1	7-10	12-3	15-9	19-3	22-4
	Spruce-pine-fir	#1	7-8	11-3	14-3	17-5	20-2	7-1	10-5	13-2	16-1	18-8
	Spruce-pine-fir	#2	7-8	11-3	14-3	17-5	20-2	7-1	10-5	13-2	16-1	18-8
	Spruce-pine-fir	#3	5-10	8-6	10-9	13-2	15-3	5-5	7-10	10-0	12-2	14-1
16	Douglas fir-larch	SS	7-8	12-1	15-11	19-9	22-10	7-8	11-10	14-11	18-3	21-2
	Douglas fir-larch	#1	7-1	10-5	13-2	16-1	18-8	6-7	9-8	12-2	14-11	17-3
	Douglas fir-larch	#2	6-9	9-10	12-6	15-3	17-9	6-3	9-2	11-7	14-2	16-5
	Douglas fir-larch	#3	5-2	7-7	9-7	11-18	13-6	4-9	7-0	8-10	10-10	12-6
	Hem-fir	SS	7-3	11-5	15-0	19-1	22-1	7-3	11-5	14-5	17-8	20-5
	Hem-fir	#1	7-0	10-3	13-0	15-11	18-5	6-6	9-6	12-1	14-9	17-1
	Hem-fir	#2	6-7	9-7	12-2	14-10	17-3	6-1	8-11	11-3	13-9	15-11
	Hem-fir	#3	5-0	7-4	9-4	11-5	13-2	4-8	6-10	8-8	10-6	12-3
	Southern pine	SS	7-6	11-10	15-7	19-11	23-7	7-6	11-10	15-7	18-6	21-10
	Southern pine	#1	7-1	10-7	13-5	15-9	18-8	6-7	9-10	12-5	14-7	17-3
	Southern pine	#2	6-1	9-2	11-7	13-9	16-2	5-8	8-5	10-9	12-9	15-0
	Southern pine	#3	4-8	6-11	8-9	10-7	12-6	4-4	6-5	8-1	9-10	11-7
	Spruce-pine-fir	SS	7-1	11-2	14-8	18-0	20-11	7-1	10-9	13-8	15-11	19-4
	Spruce-pine-fir	#1	6-8	9-9	12-4	15-1	17-6	6-2	9-0	11-5	13-11	16-2
	Spruce-pine-fir	#2	6-8	9-9	12-4	15-1	17-6	6-2	9-0	11-5	13-11	16-2
	Spruce-pine-fir	#3	5-0	7-4	9-4	11-5	13-2	4-8	6-10	8-8	10-6	12-3

(continued)

TABLE R802.4.1(5)—continued
RAFTER SPANS FOR COMMON LUMBER SPECIES (Ground snow load = 50 psf, ceiling not attached to rafters, L/Δ = 180)

RAFTER SPACING (inches)	SPECIES AND GRADE		DEAD LOAD = 10 psf					DEAD LOAD = 20 psf				
			2 × 4	2 × 6	2 × 8	2 × 10	2 × 12	2 × 4	2 × 6	2 × 8	2 × 10	2 × 12
			Maximum rafter spans[a]									
			(feet-inches)	(feet-inches)	(feet-inches)	(feet-inches)	(feet-inches)	(feet-inches)	(feet-inches)	(feet-inches)	(feet-inches)	(feet-inches)
19.2	Douglas fir-larch	SS	7-3	11-4	14-9	18-0	20-11	7-3	10-9	13-8	16-8	19-4
	Douglas fir-larch	#1	6-6	9-6	12-0	14-8	17-1	6-0	8-10	11-2	13-7	15-9
	Douglas fir-larch	#2	6-2	9-0	11-5	13-11	16-2	5-8	8-4	10-9	12-11	15-0
	Douglas fir-larch	#3	4-8	6-11	8-9	10-8	12-4	4-4	6-4	8-1	9-10	11-5
	Hem-fir	SS	6-10	10-9	14-2	17-5	20-2	6-10	10-5	13-2	16-1	18-8
	Hem-fir	#1	6-5	9-5	11-11	14-6	16-10	8-11	8-8	11-0	13-5	15-7
	Hem-fir	#2	6-0	8-9	11-1	13-7	15-9	5-7	8-1	10-3	12-7	14-7
	Hem-fir	#3	4-7	6-9	8-6	10-5	12-1	4-3	6-3	7-11	9-7	11-2
	Southern pine	SS	7-1	11-2	14-8	18-3	21-7	7-1	11-2	14-2	16-11	20-0
	Southern pine	#1	6-6	9-8	12-3	14-4	17-1	6-0	9-0	11-4	13-4	15-9
	Southern pine	#2	5-7	8-4	10-7	12-6	14-9	5-2	7-9	9-9	11-7	13-8
	Southern pine	#3	4-3	6-4	8-0	9-8	11-5	4-0	5-10	7-4	8-11	10-7
	Spruce-pine-fir	SS	6-8	10-6	13-5	16-5	19-1	6-8	9-10	12-5	15-3	17-8
	Spruce-pine-fir	#1	6-1	8-11	11-3	13-9	15-11	5-7	8-3	10-5	12-9	14-9
	Spruce-pine-fir	#2	6-1	8-11	11-3	13-9	15-11	5-7	8-3	10-5	12-9	14-9
	Spruce-pine-fir	#3	4-7	6-9	8-6	10-5	12-1	4-3	6-3	7-11	9-7	11-2
24	Douglas fir-larch	SS	6-8	10-5	13-2	16-1	18-8	6-7	9-8	12-2	14-11	17-3
	Douglas fir-larch	#1	5-10	8-6	10-9	13-2	15-3	5-5	7-10	10-0	12-2	14-1
	Douglas fir-larch	#2	5-6	8-1	10-3	12-6	14-6	5-1	7-6	9-5	11-7	13-5
	Douglas fir-larch	#3	4-3	6-2	7-10	9-6	11-1	3-11	5-8	7-3	8-10	10-3
	Hem-fir	SS	6-4	9-11	12-9	15-7	18-0	6-4	9-4	11-9	14-5	16-8
	Hem-fir	#1	5-9	8-5	10-8	13-0	15-1	8-4	7-9	9-10	12-0	13-11
	Hem-fir	#2	5-4	7-10	9-11	12-1	14-1	4-11	7-3	9-2	11-3	13-0
	Hem-fir	#3	4-1	6-0	7-7	9-4	10-9	3-10	5-7	7-1	8-7	10-0
	Southern pine	SS	6-7	10-4	13-8	16-4	19-3	6-7	10-0	12-8	15-2	17-10
	Southern pine	#1	5-10	8-8	11-0	12-10	15-3	5-5	8-0	10-2	11-11	14-1
	Southern pine	#2	5-0	7-5	9-5	11-3	13-2	4-7	6-11	8-9	10-5	12-3
	Southern pine	#3	3-10	5-8	7-1	8-8	10-3	3-6	5-3	6-7	8-0	9-6
	Spruce-pine-fir	SS	6-2	9-6	12-0	14-8	17-1	6-0	8-10	11-2	13-7	15-9
	Spruce-pine-fir	#1	5-5	7-11	10-1	12-4	14-3	5-0	7-4	9-4	11-5	13-2
	Spruce-pine-fir	#2	5-5	7-11	10-1	12-4	14-3	5-0	7-4	9-4	11-5	13-2
	Spruce-pine-fir	#3	4-1	6-0	7-7	9-4	10-9	3-10	5-7	7-1	8-7	10-0

Check sources for availability of lumber in lengths greater than 20 feet.

For SI: 1 inch = 25.4 mm, 1 foot = 304.8 mm, 1 pound per square foot = 0.0479 kPa.

a. The tabulated rafter spans assume that ceiling joists are located at the bottom of the attic space or that some other method of resisting the outward push of the rafters on the bearing walls, such as rafter ties, is provided at that location. Where ceiling joists or rafter ties are located higher in the attic space, the rafter spans shall be multiplied by the adjustment factors in Table R802.4.1(9).

b. Span exceeds 26 feet in length.

TABLE R802.4.1(6)
RAFTER SPANS FOR COMMON LUMBER SPECIES (Ground snow load = 50 psf, ceiling attached to rafters, L/Δ = 240)

RAFTER SPACING (inches)	SPECIES AND GRADE		DEAD LOAD = 10 psf					DEAD LOAD = 20 psf				
			2 × 4	2 × 6	2 × 8	2 × 10	2 × 12	2 × 4	2 × 6	2 × 8	2 × 10	2 × 12
			Maximum rafter spans[a]									
			(feet-inches)	(feet-inches)	(feet-inches)	(feet-inches)	(feet-inches)	(feet-inches)	(feet-inches)	(feet-inches)	(feet-inches)	(feet-inches)
12	Douglas fir-larch	SS	7-8	12-1	15-11	20-3	24-8	7-8	12-1	15-11	20-3	24-5
	Douglas fir-larch	#1	7-5	11-7	15-3	18-7	21-7	7-5	11-2	14-1	17-3	20-0
	Douglas fir-larch	#2	7-3	11-5	14-5	17-8	20-5	7-3	10-7	13-4	16-4	18-11
	Douglas fir-larch	#3	6-0	8-9	11-0	13-6	15-7	5-6	8-1	10-3	12-6	14-6
	Hem-fir	SS	7-3	11-5	15-0	19-2	23-4	7-3	11-5	15-0	19-2	23-4
	Hem-fir	#1	7-1	11-2	14-8	18-4	21-3	7-1	11-0	13-11	17-0	19-9
	Hem-fir	#2	6-9	10-8	14-0	17-2	19-11	6-9	10-3	13-0	15-10	18-5
	Hem-fir	#3	5-10	8-6	10-9	13-2	15-3	5-5	7-10	10-0	12-2	14-1
	Southern pine	SS	7-6	11-10	15-7	19-11	24-3	7-6	11-10	15-7	19-11	24-3
	Southern pine	#1	7-3	11-5	15-0	18-2	21-7	7-3	11-4	14-5	16-10	20-0
	Southern pine	#2	6-11	10-6	13-4	15-10	18-8	6-6	9-9	12-4	14-8	17-3
	Southern pine	#3	5-5	8-0	10-1	12-3	14-6	5-0	7-5	9-4	11-4	13-5
	Spruce-pine-fir	SS	7-1	11-2	14-8	18-9	22-10	7-1	11-2	14-8	18-9	22-4
	Spruce-pine-fir	#1	6-11	10-11	14-3	17-5	20-2	6-11	10-5	13-2	16-1	18-8
	Spruce-pine-fir	#2	6-11	10-11	14-3	17-5	20-2	6-11	10-5	13-2	16-1	18-8
	Spruce-pine-fir	#3	5-10	8-6	10-9	13-2	15-3	5-5	7-10	10-0	12-2	14-1
16	Douglas fir-larch	SS	7-0	11-0	14-5	18-5	22-5	7-0	11-0	14-5	18-3	21-2
	Douglas fir-larch	#1	6-9	10-5	13-2	16-1	18-8	6-7	9-8	12-2	14-11	17-3
	Douglas fir-larch	#2	6-7	9-10	12-6	15-3	17-9	6-3	9-2	11-7	14-2	16-5
	Douglas fir-larch	#3	5-2	7-7	9-7	11-8	13-6	4-9	7-0	8-10	10-10	12-6
	Hem-fir	SS	6-7	10-4	13-8	17-5	21-2	6-7	10-4	13-8	17-5	20-5
	Hem-fir	#1	6-5	10-2	13-0	15-11	18-5	6-5	9-6	12-1	14-9	17-1
	Hem-fir	#2	6-2	9-7	12-2	14-10	17-3	6-1	8-11	11-3	13-9	15-11
	Hem-fir	#3	5-0	7-4	9-4	11-5	13-2	4-8	6-10	8-8	10-6	12-3
	Southern pine	SS	6-10	10-9	14-2	18-1	22-0	6-10	10-9	14-2	18-1	21-10
	Southern pine	#1	6-7	10-4	13-5	15-9	18-8	6-7	9-10	12-5	14-7	17-3
	Southern pine	#2	6-1	9-2	11-7	13-9	16-2	5-8	8-5	10-9	12-9	15-0
	Southern pine	#3	4-8	6-11	8-9	10-7	12-6	4-4	6-5	8-1	9-10	11-7
	Spruce-pine-fir	SS	6-5	10-2	13-4	17-0	20-9	6-5	10-2	13-4	16-8	19-4
	Spruce-pine-fir	#1	6-4	9-9	12-4	15-1	17-6	6-2	9-0	11-5	13-11	16-2
	Spruce-pine-fir	#2	6-4	9-9	12-4	15-1	17-6	6-2	9-0	11-5	13-11	16-2
	Spruce-pine-fir	#3	5-0	7-4	9-4	11-5	13-2	4-8	6-10	8-8	10-6	12-3

(continued)

TABLE R802.4.1(6)—continued
RAFTER SPANS FOR COMMON LUMBER SPECIES (Ground snow load = 50 psf, ceiling attached to rafters, L/Δ = 240)

RAFTER SPACING (inches)	SPECIES AND GRADE		DEAD LOAD = 10 psf					DEAD LOAD = 20 psf				
			2 × 4	2 × 6	2 × 8	2 × 10	2 × 12	2 × 4	2 × 6	2 × 8	2 × 10	2 × 12
			Maximum rafter spans[a]									
			(feet-inches)	(feet-inches)	(feet-inches)	(feet-inches)	(feet-inches)	(feet-inches)	(feet-inches)	(feet-inches)	(feet-inches)	(feet-inches)
19.2	Douglas fir-larch	SS	6-7	10-4	13-7	17-4	20-11	6-7	10-4	13-7	16-8	19-4
	Douglas fir-larch	#1	6-4	9-6	12-0	14-8	17-1	6-0	8-10	11-2	13-7	15-9
	Douglas fir-larch	#2	6-2	9-0	11-5	13-11	16-2	5-8	8-4	10-7	12-11	15-0
	Douglas fir-larch	#3	4-8	6-11	8-9	10-8	12-4	4-4	6-4	8-1	9-10	11-5
	Hem-fir	SS	6-2	9-9	12-10	16-5	19-11	6-2	9-9	12-10	16-1	18-8
	Hem-fir	#1	6-1	9-5	11-11	14-6	16-10	5-11	8-8	11-0	13-5	15-7
	Hem-fir	#2	5-9	8-9	11-1	13-7	15-9	5-7	8-1	10-3	12-7	14-7
	Hem-fir	#3	4-7	6-9	8-6	10-5	12-1	4-3	6-3	7-11	9-7	11-2
	Southern pine	SS	6-5	10-2	13-4	17-0	20-9	6-5	10-2	13-4	16-11	20-0
	Southern pine	#1	6-2	9-8	12-3	14-4	17-1	6-0	9-0	11-4	13-4	15-9
	Southern pine	#2	5-7	8-4	10-7	12-6	14-9	5-2	7-9	9-9	11-7	13-8
	Southern pine	#3	4-3	6-4	8-0	9-8	11-5	4-0	5-10	7-4	8-11	10-7
	Spruce-pine-fir	SS	6-1	9-6	12-7	16-0	19-1	6-1	9-6	12-5	15-3	17-8
	Spruce-pine-fir	#1	5-11	8-11	11-3	13-9	15-11	5-7	8-3	10-5	12-9	14-9
	Spruce-pine-fir	#2	5-11	8-11	11-3	13-9	15-11	5-7	8-3	10-5	12-9	14-9
	Spruce-pine-fir	#3	4-7	6-9	8-6	10-5	12-1	4-3	6-3	7-11	9-7	11-2
24	Douglas fir-larch	SS	6-1	9-7	12-7	16-1	18-8	6-1	9-7	12-2	14-11	17-3
	Douglas fir-larch	#1	5-10	8-6	10-9	13-2	15-3	5-5	7-10	10-0	12-2	14-1
	Douglas fir-larch	#2	5-6	8-1	10-3	12-6	14-6	5-1	7-6	9-5	11-7	13-5
	Douglas fir-larch	#3	4-3	6-2	7-10	9-6	11-1	3-11	5-8	7-3	8-10	10-3
	Hem-fir	SS	5-9	9-1	11-11	15-2	18-0	5-9	9-1	11-9	14-5	15-11
	Hem-fir	#1	5-8	8-5	10-8	13-0	15-1	5-4	7-9	9-10	12-0	13-11
	Hem-fir	#2	5-4	7-10	9-11	12-1	14-1	4-11	7-3	9-2	11-3	13-0
	Hem-fir	#3	4-1	6-0	7-7	9-4	10-9	3-10	5-7	7-1	8-7	10-0
	Southern pine	SS	6-0	9-5	12-5	15-10	19-3	6-0	9-5	12-5	15-2	17-10
	Southern pine	#1	5-9	8-8	11-0	12-10	15-3	5-5	8-0	10-2	11-11	14-1
	Southern pine	#2	5-0	7-5	9-5	11-3	13-2	4-7	6-11	8-9	10-5	12-3
	Southern pine	#3	3-10	5-8	7-1	8-8	10-3	3-6	5-3	6-7	8-0	9-6
	Spruce-pine-fir	SS	5-8	8-10	11-8	14-8	17-1	5-8	8-10	11-2	13-7	15-9
	Spruce-pine-fir	#1	5-5	7-11	10-1	12-4	14-3	5-0	7-4	9-4	11-5	13-2
	Spruce-pine-fir	#2	5-5	7-11	10-1	12-4	14-3	5-0	7-4	9-4	11-5	13-2
	Spruce-pine-fir	#3	4-1	6-0	7-7	9-4	10-9	3-10	5-7	7-1	8-7	10-0

Check sources for availability of lumber in lengths greater than 20 feet.

For SI: 1 inch = 25.4 mm, 1 foot = 304.8 mm, 1 pound per square foot = 0.0479 kPa.

a. The tabulated rafter spans assume that ceiling joists are located at the bottom of the attic space or that some other method of resisting the outward push of the rafters on the bearing walls, such as rafter ties, is provided at that location. Where ceiling joists or rafter ties are located higher in the attic space, the rafter spans shall be multiplied by the adjustment factors in Table R802.4.1(9).

TABLE R802.4.1(7)
RAFTER SPANS FOR COMMON LUMBER SPECIES (Ground snow load = 70 psf, ceiling not attached to rafters, L/Δ = 180)

RAFTER SPACING (inches)	SPECIES AND GRADE		DEAD LOAD = 10 psf					DEAD LOAD = 20 psf				
			2 × 4	2 × 6	2 × 8	2 × 10	2 × 12	2 × 4	2 × 6	2 × 8	2 × 10	2 × 12
			Maximum Rafter Spans[a]									
			(feet-inches)	(feet-inches)	(feet-inches)	(feet-inches)	(feet-inches)	(feet-inches)	(feet-inches)	(feet-inches)	(feet-inches)	(feet-inches)
12	Douglas fir-larch	SS	7-7	11-10	15-8	19-9	22-10	7-7	11-10	15-3	18-7	21-7
	Douglas fir-larch	#1	7-1	10-5	13-2	16-1	18-8	6-8	9-10	12-5	15-2	17-7
	Douglas fir-larch	#2	6-9	9-10	12-6	15-3	17-9	6-4	9-4	11-9	14-5	16-8
	Douglas fir-larch	#3	5-2	7-7	9-7	11-8	13-6	4-10	7-1	9-0	11-0	12-9
	Hem-fir	SS	7-2	11-3	14-9	18-10	22-1	7-2	11-3	14-8	18-0	20-10
	Hem-fir	#1	7-0	10-3	13-0	15-11	18-5	6-7	9-8	12-3	15-0	17-5
	Hem-fir	#2	6-7	9-7	12-2	14-10	17-3	6-2	9-1	11-5	14-0	16-3
	Hem-fir	#3	5-0	7-4	9-4	11-5	13-2	4-9	6-11	8-9	10-9	12-5
	Southern pine	SS	7-5	11-8	15-4	19-7	23-7	7-5	11-8	15-4	18-10	22-3
	Southern pine	#1	7-1	10-7	13-5	15-9	18-8	6-9	10-0	12-8	14-10	17-7
	Southern pine	#2	6-1	9-2	11-7	13-9	16-2	5-9	8-7	10-11	12-11	15-3
	Southern pine	#3	4-8	6-11	8-9	10-7	12-6	4-5	6-6	8-3	10-0	11-10
	Spruce-pine-fir	SS	7-0	11-0	14-6	18-0	20-11	7-0	11-0	13-11	17-0	19-8
	Spruce-pine-fir	#1	6-8	9-9	12-4	15-1	17-6	6-3	9-2	11-8	14-2	16-6
	Spruce-pine-fir	#2	6-8	9-9	12-4	15-1	17-6	6-3	9-2	11-8	14-2	16-6
	Spruce-pine-fir	#3	5-0	7-4	9-4	11-5	13-2	4-9	6-11	8-9	10-9	12-5
16	Douglas fir-larch	SS	6-10	10-9	14-0	17-1	19-10	6-10	10-5	13-2	16-1	18-8
	Douglas fir-larch	#1	6-2	9-0	11-5	13-11	16-2	5-10	8-6	10-9	13-2	15-3
	Douglas fir-larch	#2	5-10	8-7	10-10	13-3	15-4	5-6	8-1	10-3	12-6	14-6
	Douglas fir-larch	#3	4-6	6-6	8-3	10-1	11-9	4-3	6-2	7-10	9-6	11-1
	Hem-fir	SS	6-6	10-2	13-5	16-6	19-2	6-6	10-1	12-9	15-7	18-0
	Hem-fir	#1	6-1	8-11	11-3	13-9	16-0	5-9	8-5	10-8	13-0	15-1
	Hem-fir	#2	5-8	8-4	10-6	12-10	14-11	5-4	7-10	9-11	12-1	14-1
	Hem-fir	#3	4-4	6-4	8-1	9-10	11-5	4-1	6-0	7-7	9-4	10-9
	Southern pine	SS	6-9	10-7	14-0	17-4	20-5	6-9	10-7	13-9	16-4	19-3
	Southern pine	#1	6-2	9-2	11-8	13-8	16-2	5-10	8-8	11-0	12-10	15-3
	Southern pine	#2	5-3	7-11	10-0	11-11	14-0	5-0	7-5	9-5	11-3	13-2
	Southern pine	#3	4-1	6-0	7-7	9-2	10-10	3-10	5-8	7-1	8-8	10-3
	Spruce-pine-fir	SS	6-4	10-0	12-9	15-7	18-1	6-4	9-6	12-0	14-8	17-1
	Spruce-pine-fir	#1	5-9	8-5	10-8	13-1	15-2	5-5	7-11	10-1	12-4	14-3
	Spruce-pine-fir	#2	5-9	8-5	10-8	13-1	15-2	5-5	7-11	10-1	12-4	14-3
	Spruce-pine-fir	#3	4-4	6-4	8-1	9-10	11-5	4-1	6-0	7-7	9-4	10-9

(continued)

TABLE R802.4.1(7)—continued
RAFTER SPANS FOR COMMON LUMBER SPECIES (Ground snow load = 70 psf, ceiling not attached to rafters, L/Δ = 180)

RAFTER SPACING (inches)	SPECIES AND GRADE		DEAD LOAD = 10 psf					DEAD LOAD = 20 psf				
			2 × 4	2 × 6	2 × 8	2 × 10	2 × 12	2 × 4	2 × 6	2 × 8	2 × 10	2 × 12
			Maximum Rafter Spans[a]									
			(feet-inches)	(feet-inches)	(feet-inches)	(feet-inches)	(feet-inches)	(feet-inches)	(feet-inches)	(feet-inches)	(feet-inches)	(feet-inches)
19.2	Douglas fir-larch	SS	6-6	10-1	12-9	15-7	18-1	6-6	9-6	12-0	14-8	17-1
	Douglas fir-larch	#1	5-7	8-3	10-5	12-9	14-9	5-4	7-9	9-10	12-0	13-11
	Douglas fir-larch	#2	5-4	7-10	9-11	12-1	14-0	5-0	7-4	9-4	11-5	13-2
	Douglas fir-larch	#3	4-1	6-0	7-7	9-3	10-8	3-10	5-7	7-1	8-8	10-1
	Hem-fir	SS	6-1	9-7	12-4	15-1	17-4	6-1	9-2	11-8	14-2	15-5
	Hem-fir	#1	5-7	8-2	10-3	12-7	14-7	5-3	7-8	9-8	11-10	13-9
	Hem-fir	#2	5-2	7-7	9-7	11-9	13-7	4-11	7-2	9-1	11-1	12-10
	Hem-fir	#3	4-0	5-10	7-4	9-0	10-5	3-9	5-6	6-11	8-6	9-10
	Southern pine	SS	6-4	10-0	13-2	15-10	18-8	6-4	9-10	12-6	14-11	17-7
	Southern pine	#1	5-8	8-5	10-8	12-5	14-9	5-4	7-11	10-0	11-9	13-11
	Southern pine	#2	4-10	7-3	9-2	10-10	12-9	4-6	6-10	8-8	10-3	12-1
	Southern pine	#3	3-8	5-6	6-11	8-4	9-11	3-6	5-2	6-6	7-11	9-4
	Spruce-pine-fir	SS	6-0	9-2	11-8	14-3	16-6	5-11	8-8	11-0	13-5	15-7
	Spruce-pine-fir	#1	5-3	7-8	9-9	11-11	13-10	5-0	7-3	9-2	11-3	13-0
	Spruce-pine-fir	#2	5-3	7-8	9-9	11-11	13-10	5-0	7-3	9-2	11-3	13-0
	Spruce-pine-fir	#3	4-0	5-10	7-4	9-0	10-5	3-9	5-6	6-11	8-6	9-10
24	Douglas fir-larch	SS	6-0	9-0	11-5	13-11	16-2	5-10	8-6	10-9	13-2	15-3
	Douglas fir-larch	#1	5-0	7-4	9-4	11-5	13-2	4-9	6-11	8-9	10-9	12-5
	Douglas fir-larch	#2	4-9	7-0	8-10	10-10	12-6	4-6	6-7	8-4	10-2	11-10
	Douglas fir-larch	#3	3-8	5-4	6-9	8-3	9-7	3-5	5-0	6-4	7-9	9-10
	Hem-fir	SS	5-8	8-8	11-0	13-6	13-11	5-7	8-3	10-5	12-4	12-4
	Hem-fir	#1	5-0	7-3	9-2	11-3	13-0	4-8	6-10	8-8	10-7	12-4
	Hem-fir	#2	4-8	6-9	8-7	10-6	12-2	4-4	6-5	8-1	9-11	11-6
	Hem-fir	#3	3-7	5-2	6-7	8-1	9-4	3-4	4-11	6-3	7-7	8-10
	Southern pine	SS	5-11	9-3	11-11	14-2	16-8	5-11	8-10	11-2	13-4	15-9
	Southern pine	#1	5-0	7-6	9-6	11-1	13-2	4-9	7-1	9-0	10-6	12-5
	Southern pine	#2	4-4	6-5	8-2	9-9	11-5	4-1	6-1	7-9	9-2	10-9
	Southern pine	#3	3-4	4-11	6-2	7-6	8-10	3-1	4-7	5-10	7-1	8-4
	Spruce-pine-fir	SS	5-6	8-3	10-5	12-9	14-9	5-4	7-9	9-10	12-0	12-11
	Spruce-pine-fir	#1	4-8	6-11	8-9	10-8	12-4	4-5	6-6	8-3	10-0	11-8
	Spruce-pine-fir	#2	4-8	6-11	8-9	10-8	12-4	4-5	6-6	8-3	10-0	11-8
	Spruce-pine-fir	#3	3-7	5-2	6-7	8-1	9-4	3-4	4-11	6-3	7-7	8-10

Check sources for availability of lumber in lengths greater than 20 feet.

For SI: 1 inch = 25.4 mm, 1 foot = 304.8 mm, 1 pound per square foot = 0.0479 kPa.

a. The tabulated rafter spans assume that ceiling joists are located at the bottom of the attic space or that some other method of resisting the outward push of the rafters on the bearing walls, such as rafter ties, is provided at that location. Where ceiling joists or rafter ties are located higher in the attic space, the rafter spans shall be multiplied by the adjustment factors in Table R802.4.1(9).

TABLE R802.4.1(8)
RAFTER SPANS FOR COMMON LUMBER SPECIES (Ground snow load = 70 psf, ceiling attached to rafters, L/Δ = 240)

RAFTER SPACING (inches)	SPECIES AND GRADE		DEAD LOAD = 10 psf					DEAD LOAD = 20 psf				
			2 × 4	2 × 6	2 × 8	2 × 10	2 × 12	2 × 4	2 × 6	2 × 8	2 × 10	2 × 12
			Maximum rafter spans[a]									
			(feet-inches)	(feet-inches)	(feet-inches)	(feet-inches)	(feet-inches)	(feet-inches)	(feet-inches)	(feet-inches)	(feet-inches)	(feet-inches)
12	Douglas fir-larch	SS	6-10	10-9	14-3	18-2	22-1	6-10	10-9	14-3	18-2	21-7
	Douglas fir-larch	#1	6-7	10-5	13-2	16-1	18-8	6-7	9-10	12-5	15-2	17-7
	Douglas fir-larch	#2	6-6	9-10	12-6	15-3	17-9	6-4	9-4	11-9	14-5	16-8
	Douglas fir-larch	#3	5-2	7-7	9-7	11-8	13-6	4-10	7-1	9-0	11-0	12-9
	Hem-fir	SS	6-6	10-2	13-5	17-2	20-10	6-6	10-2	13-5	17-2	20-10
	Hem-fir	#1	6-4	10-0	13-0	15-11	18-5	6-4	9-8	12-3	15-0	17-5
	Hem-fir	#2	6-1	9-6	12-2	14-10	17-3	6-1	9-1	11-5	14-0	16-3
	Hem-fir	#3	5-0	7-4	9-4	11-5	13-2	4-9	6-11	8-9	10-9	12-5
	Southern pine	SS	6-9	10-7	14-0	17-10	21-8	6-9	10-7	14-0	17-10	21-8
	Southern pine	#1	6-6	10-2	13-5	15-9	18-8	6-6	10-0	12-8	14-10	17-7
	Southern pine	#2	6-1	9-2	11-7	13-9	16-2	5-9	8-7	10-11	12-11	15-3
	Southern pine	#3	4-8	6-11	8-9	10-7	12-6	4-5	6-6	8-3	10-0	11-10
	Spruce-pine-fir	SS	6-4	10-0	13-2	16-9	20-5	6-4	10-0	13-2	16-9	19-8
	Spruce-pine-fir	#1	6-2	9-9	12-4	15-1	17-6	6-2	9-2	11-8	14-2	16-6
	Spruce-pine-fir	#2	6-2	9-9	12-4	15-1	17-6	6-2	9-2	11-8	14-2	16-6
	Spruce-pine-fir	#3	5-0	7-4	9-4	11-5	13-2	4-9	6-11	8-9	10-9	12-5
16	Douglas fir-larch	SS	6-3	9-10	12-11	16-6	19-10	6-3	9-10	12-11	16-1	18-8
	Douglas fir-larch	#1	6-0	9-0	11-5	13-11	16-2	5-10	8-6	10-9	13-2	15-3
	Douglas fir-larch	#2	5-10	8-7	10-10	13-3	15-4	5-6	8-1	10-3	12-6	14-6
	Douglas fir-larch	#3	4-6	6-6	8-3	10-1	11-9	4-3	6-2	7-10	9-6	11-1
	Hem-fir	SS	5-11	9-3	12-2	15-7	18-11	5-11	9-3	12-2	15-7	18-0
	Hem-fir	#1	5-9	8-11	11-3	13-9	16-0	5-9	8-5	10-8	13-0	15-1
	Hem-fir	#2	5-6	8-4	10-6	12-10	14-11	5-4	7-10	9-11	12-1	14-1
	Hem-fir	#3	4-4	6-4	8-1	9-10	11-5	4-1	6-0	7-7	9-4	10-9
	Southern pine	SS	6-1	9-7	12-8	16-2	19-8	6-1	9-7	12-8	16-2	19-3
	Southern pine	#1	5-11	9-2	11-8	13-8	16-2	5-10	8-8	11-0	12-10	15-3
	Southern pine	#2	5-3	7-11	10-0	11-11	14-0	5-0	7-5	9-5	11-3	13-2
	Southern pine	#3	4-1	6-0	7-7	9-2	10-10	3-10	5-8	7-1	8-8	10-3
	Spruce-pine-fir	SS	5-9	9-1	11-11	15-3	18-1	5-9	9-1	11-11	14-8	17-1
	Spruce-pine-fir	#1	5-8	8-5	10-8	13-1	15-2	5-5	7-11	10-1	12-4	14-3
	Spruce-pine-fir	#2	5-8	8-5	10-8	13-1	15-2	5-5	7-11	10-1	12-4	14-3
	Spruce-pine-fir	#3	4-4	6-4	8-1	9-10	11-5	4-1	6-0	7-7	9-4	10-9

(continued)

TABLE R802.4.1(8)—continued
RAFTER SPANS FOR COMMON LUMBER SPECIES (Ground snow load = 70 psf, ceiling attached to rafters, L/Δ = 240)

RAFTER SPACING (inches)	SPECIES AND GRADE		DEAD LOAD = 10 psf					DEAD LOAD = 20 psf				
			2 × 4	2 × 6	2 × 8	2 × 10	2 × 12	2 × 4	2 × 6	2 × 8	2 × 10	2 × 12
			Maximum rafter spans[a]									
			(feet-inches)	(feet-inches)	(feet-inches)	(feet-inches)	(feet-inches)	(feet-inches)	(feet-inches)	(feet-inches)	(feet-inches)	(feet-inches)
19.2	Douglas fir-larch	SS	5-10	9-3	12-2	15-6	18-1	5-10	9-3	12-0	14-8	17-1
	Douglas fir-larch	#1	5-7	8-3	10-5	12-9	14-9	5-4	7-9	9-10	12-0	13-11
	Douglas fir-larch	#2	5-4	7-10	9-11	12-1	14-0	5-0	7-4	9-4	11-5	13-2
	Douglas fir-larch	#3	4-1	6-0	7-7	9-3	10-8	3-10	5-7	7-1	8-8	10-1
	Hem-fir	SS	5-6	8-8	11-6	14-8	17-4	5-6	8-8	11-6	14-2	15-5
	Hem-fir	#1	5-5	8-2	10-3	12-7	14-7	5-3	7-8	9-8	11-10	13-9
	Hem-fir	#2	5-2	7-7	9-7	11-9	13-7	4-11	7-2	9-1	11-1	12-10
	Hem-fir	#3	4-0	5-10	7-4	9-0	10-5	3-9	5-6	6-11	8-6	9-10
	Southern pine	SS	5-9	9-1	11-11	15-3	18-6	5-9	9-1	11-11	14-11	17-7
	Southern pine	#1	5-6	8-5	10-8	12-5	14-9	5-4	7-11	10-0	11-9	13-11
	Southern pine	#2	4-10	7-3	9-2	10-10	12-9	4-6	6-10	8-8	10-3	12-1
	Southern pine	#3	3-8	5-6	6-11	8-4	9-11	3-6	5-2	6-6	7-11	9-4
	Spruce-pine-fir	SS	5-5	8-6	11-3	14-3	16-6	5-5	8-6	11-0	13-5	15-7
	Spruce-pine-fir	#1	5-3	7-8	9-9	11-11	13-10	5-0	7-3	9-2	11-3	13-0
	Spruce-pine-fir	#2	5-3	7-8	9-9	11-11	13-10	5-0	7-3	9-2	11-3	13-0
	Spruce-pine-fir	#3	4-0	5-10	7-4	9-0	10-5	3-9	5-6	6-11	8-6	9-10
24	Douglas fir-larch	SS	5-5	8-7	11-3	13-11	16-2	5-5	8-6	10-9	13-2	15-3
	Douglas fir-larch	#1	5-0	7-4	9-4	11-5	13-2	4-9	6-11	8-9	10-9	12-5
	Douglas fir-larch	#2	4-9	7-0	8-10	10-10	12-6	4-6	6-7	8-4	10-2	11-10
	Douglas fir-larch	#3	3-8	5-4	6-9	8-3	9-7	3-5	5-0	6-4	7-9	9-0
	Hem-fir	SS	5-2	8-1	10-8	13-6	13-11	5-2	8-1	10-5	12-4	12-4
	Hem-fir	#1	5-0	7-3	9-2	11-3	13-0	4-8	6-10	8-8	10-7	12-4
	Hem-fir	#2	4-8	6-9	8-7	10-6	12-2	4-4	6-5	8-1	9-11	11-6
	Hem-fir	#3	3-7	5-2	6-7	8-1	9-4	3-4	4-11	6-3	7-7	8-10
	Southern pine	SS	5-4	8-5	11-1	14-2	16-8	5-4	8-5	11-1	13-4	15-9
	Southern pine	#1	5-0	7-6	9-6	11-1	13-2	4-9	7-1	9-0	10-6	12-5
	Southern pine	#2	4-4	6-5	8-2	9-9	11-5	4-1	6-1	7-9	9-2	10-9
	Southern pine	#3	3-4	4-11	6-2	7-6	8-10	3-1	4-7	5-10	7-1	8-4
	Spruce-pine-fir	SS	5-0	7-11	10-5	12-9	14-9	5-0	7-9	9-10	12-0	12-11
	Spruce-pine-fir	#1	4-8	6-11	8-9	10-8	12-4	4-5	6-6	8-3	10-0	11-8
	Spruce-pine-fir	#2	4-8	6-11	8-9	10-8	12-4	4-5	6-6	8-3	10-0	11-8
	Spruce-pine-fir	#3	3-7	5-2	6-7	8-1	9-4	3-4	4-11	6-3	7-7	8-10

Check sources for availability of lumber in lengths greater than 20 feet.

For SI: 1 inch = 25.4 mm, 1 foot = 304.8 mm, 1 pound per square foot = 0.0479 kPa.

a. The tabulated rafter spans assume that ceiling joists are located at the bottom of the attic space or that some other method of resisting the outward push of the rafters on the bearing walls, such as rafter ties, is provided at that location. Where ceiling joists or rafter ties are located higher in the attic space, the rafter spans shall be multiplied by the adjustment factors in Table R802.4.1(9).

TABLE R802.4.1(9)
RAFTER SPAN ADJUSTMENT FACTOR

H_C/H_R[a]	RAFTER SPAN ADJUSTMENT FACTOR
1/3	0.67
1/4	0.76
1/5	0.83
1/6	0.90
1/7.5 or less	1.00

a. H_C = Height of ceiling joists or rafter ties measured vertically above the top of the rafter support walls; H_R = Height of roof ridge measured vertically above the top of the rafter support walls.

R802.4.2 Framing details. Rafters shall be framed opposite from each other to a ridge board, shall not be offset more than $1^1/_2$ inches (38 mm) from each other and shall be connected with a collar tie or ridge strap in accordance with Section R802.4.6 or directly opposite from each other to a gusset plate in accordance with Table R602.3(1). Rafters shall be nailed to the top wall plates in accordance with Table R602.3(1) unless the *roof assembly* is required to comply with the uplift requirements of Section R802.11.

R802.4.3 Hips and valleys. Hip and valley rafters shall be not less than 2 inches (51 mm) nominal in thickness and not less in depth than the cut end of the rafter. Hip and valley rafters shall be supported at the ridge by a brace to a bearing partition or be designed to carry and distribute the specific load at that point.

R802.4.4 Rafter supports. Where the roof pitch is less than 3:12 (25-percent slope), structural members that support rafters, such as ridges, hips and valleys, shall be designed as beams, and bearing shall be provided for rafters in accordance with Section R802.6.

R802.4.5 Purlins. Installation of purlins to reduce the span of rafters is permitted as shown in Figure R802.4.5. Purlins shall be sized not less than the required size of the rafters that they support. Purlins shall be continuous and shall be supported by 2-inch by 4-inch (51 mm by 102 mm) braces installed to bearing walls at a slope not less than 45 degrees (0.79 rad) from the horizontal. The braces shall be spaced not more than 4 feet (1219 mm) on center and the unbraced length of braces shall not exceed 8 feet (2438 mm).

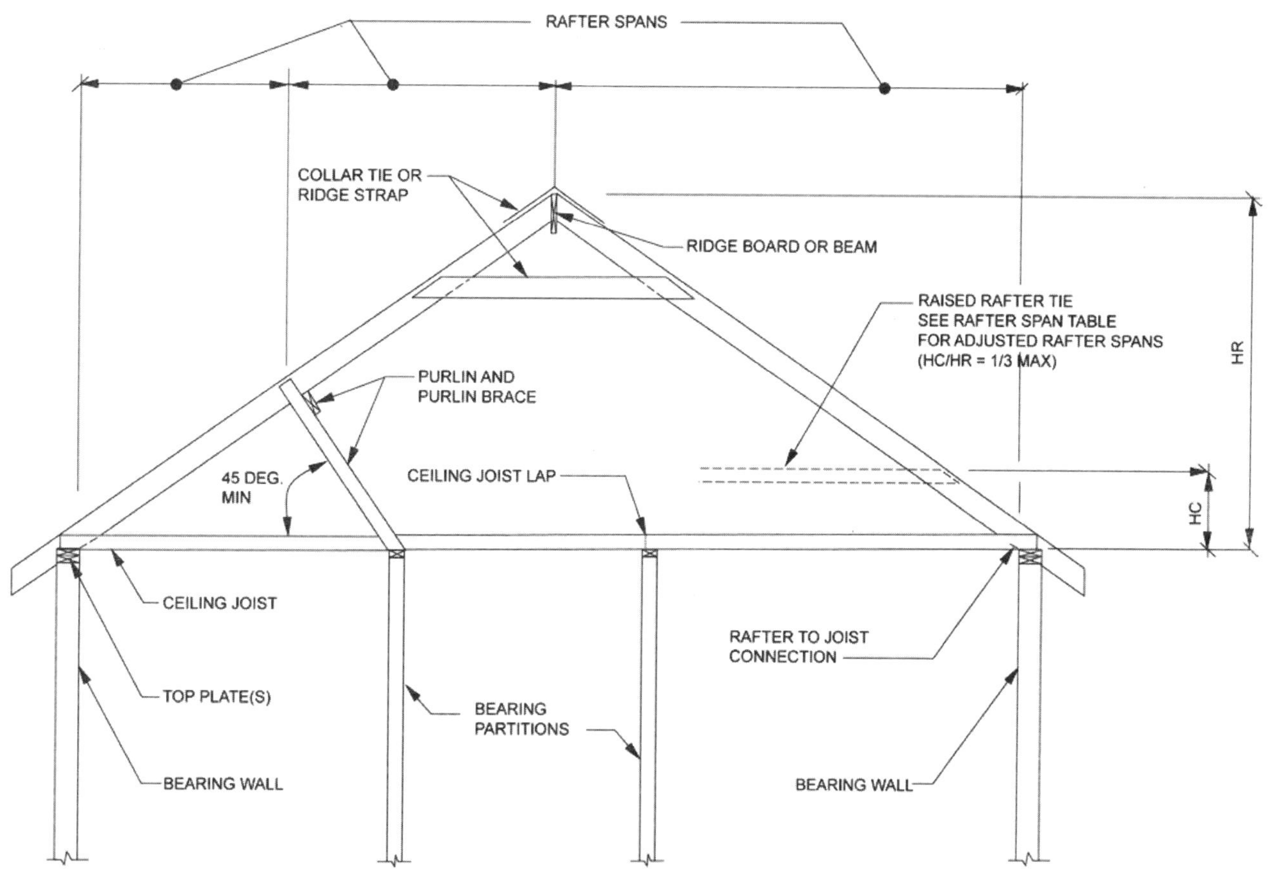

For SI: 1 degree = 0.018 rad.
H_C = Height of ceiling joists or rafter ties measured vertically above the top of rafter support walls.
H_R = Height of roof ridge measured vertically above the top of the rafter support walls.

FIGURE R802.4.5
BRACED RAFTER CONSTRUCTION

R802.4.6 Collar ties. Where collar ties are used to connect opposing rafters, they shall be located in the upper third of the attic space and fastened in accordance with Table R602.3(1). Collar ties shall be not less than 1 inch by 4 inches (25 mm × 102 mm) nominal, spaced not more than 4 feet (1219 mm) on center. Ridge straps shall be permitted to replace collar ties. Ridge straps shall be not less than $1^1/_4$-inch (32 mm) × 20 gage and shall be nailed to the top edge of each rafter with not fewer than three 10d common (3″ × 0.148″) nails with the closest nail not closer than $2^3/_8$ inches (60.3 mm) from the end of the rafter.

R802.5 Ceiling joists. Ceiling joists shall be continuous across the structure or securely joined where they meet over interior partitions in accordance with Section R802.5.2.1. Ceiling joists shall be fastened to the top plate in accordance with Table R602.3(1).

R802.5.1 Ceiling joist size. Ceiling joists shall be sized based on the joist spans in Tables R802.5.1(1) and R802.5.1(2). For other grades and species and for other loading conditions, refer to the AWC STJR.

TABLE R802.5.1(1)
CEILING JOIST SPANS FOR COMMON LUMBER SPECIES (Uninhabitable attics without storage, live load = 10 psf, L/Δ = 240)

CEILING JOIST SPACING (inches)	SPECIES AND GRADE		DEAD LOAD = 5 psf			
			2 × 4	2 × 6	2 × 8	2 × 10
			Maximum ceiling joist spans			
			(feet-inches)	(feet-inches)	(feet-inches)	(feet-inches)
12	Douglas fir-larch	SS	13-2	20-8	Note a	Note a
	Douglas fir-larch	#1	12-8	19-11	Note a	Note a
	Douglas fir-larch	#2	12-5	19-6	25-8	Note a
	Douglas fir-larch	#3	11-1	16-3	20-7	25-2
	Hem-fir	SS	12-5	19-6	25-8	Note a
	Hem-fir	#1	12-2	19-1	25-2	Note a
	Hem-fir	#2	11-7	18-2	24-0	Note a
	Hem-fir	#3	10-10	15-10	20-1	24-6
	Southern pine	SS	12-11	20-3	Note a	Note a
	Southern pine	#1	12-5	19-6	25-8	Note a
	Southern pine	#2	11-10	18-8	24-7	Note a
	Southern pine	#3	10-1	14-11	18-9	22-9
	Spruce-pine-fir	SS	12-2	19-1	25-2	Note a
	Spruce-pine-fir	#1	11-10	18-8	24-7	Note a
	Spruce-pine-fir	#2	11-10	18-8	24-7	Note a
	Spruce-pine-fir	#3	10-10	15-10	20-1	24-6
16	Douglas fir-larch	SS	11-11	18-9	24-8	Note a
	Douglas fir-larch	#1	11-6	18-1	23-10	Note a
	Douglas fir-larch	#2	11-3	17-8	23-4	Note a
	Douglas fir-larch	#3	9-7	14-1	17-10	21-9
	Hem-fir	SS	11-3	17-8	23-4	Note a
	Hem-fir	#1	11-0	17-4	22-10	Note a
	Hem-fir	#2	10-6	16-6	21-9	Note a
	Hem-fir	#3	9-5	13-9	17-5	21-3
	Southern pine	SS	11-9	18-5	24-3	Note a
	Southern pine	#1	11-3	17-8	23-10	Note a
	Southern pine	#2	10-9	16-11	21-7	25-7
	Southern pine	#3	8-9	12-11	16-3	19-9
	Spruce-pine-fir	SS	11-0	17-4	22-10	Note a
	Spruce-pine-fir	#1	10-9	16-11	22-4	Note a
	Spruce-pine-fir	#2	10-9	16-11	22-4	Note a
	Spruce-pine-fir	#3	9-5	13-9	17-5	21-3

(continued)

TABLE R802.5.1(1)—continued
CEILING JOIST SPANS FOR COMMON LUMBER SPECIES (Uninhabitable attics without storage, live load = 10 psf, L/Δ = 240)

CEILING JOIST SPACING (inches)	SPECIES AND GRADE		DEAD LOAD = 5 psf			
			2 × 4	2 × 6	2 × 8	2 × 10
			Maximum ceiling joist spans			
			(feet-inches)	(feet-inches)	(feet-inches)	(feet-inches)
19.2	Douglas fir-larch	SS	11-3	17-8	23-3	Note a
	Douglas fir-larch	#1	10-10	17-0	22-5	Note a
	Douglas fir-larch	#2	10-7	16-8	21-4	26-0
	Douglas fir-larch	#3	8-9	12-10	16-3	19-10
	Hem-fir	SS	10-7	16-8	21-11	Note a
	Hem-fir	#1	10-4	16-4	21-6	Note a
	Hem-fir	#2	9-11	15-7	20-6	25-3
	Hem-fir	#3	8-7	12-6	15-10	19-5
	Southern -pine	SS	11-0	17-4	22-10	Note a
	Southern pine	#1	10-7	16-8	22-0	Note a
	Southern pine	#2	10-2	15-7	19-8	23-5
	Southern pine	#3	8-0	11-9	14-10	18-0
	Spruce-pine-fir	SS	10-4	16-4	21-6	Note a
	Spruce-pine-fir	#1	10-2	15-11	21-0	25-8
	Spruce-pine-fir	#2	10-2	15-11	21-0	25-8
	Spruce-pine-fir	#3	8-7	12-6	15-10	19-5
24	Douglas fir-larch	SS	10-5	16-4	21-7	Note a
	Douglas fir-larch	#1	10-0	15-9	20-1	24-6
	Douglas fir-larch	#2	9-10	15-0	19-1	23-3
	Douglas fir-larch	#3	7-10	11-6	14-7	17-9
	Hem-fir	SS	9-10	15-6	20-5	Note a
	Hem-fir	#1	9-8	15-2	19-10	24-3
	Hem-fir	#2	9-2	14-5	18-6	22-7
	Hem-fir	#3	7-8	11-2	14-2	17-4
	Southern pine	SS	10-3	16-1	21-2	Note a
	Southern pine	#1	9-10	15-6	20-5	24-0
	Southern pine	#2	9-3	13-11	17-7	20-11
	Southern pine	#3	7-2	10-6	13-3	16-1
	Spruce-pine-fir	SS	9-8	15-2	19-11	25-5
	Spruce-pine-fir	#1	9-5	14-9	18-9	22-11
	Spruce-pine-fir	#2	9-5	14-9	18-9	22-11
	Spruce-pine-fir	#3	7-8	11-2	14-2	17-4

Check sources for availability of lumber in lengths greater than 20 feet.

For SI: 1 inch = 25.4 mm, 1 foot = 304.8 mm, 1 pound per square foot = 0.0479 kPa.

a. Span exceeds 26 feet in length.

TABLE R802.5.1(2)
CEILING JOIST SPANS FOR COMMON LUMBER SPECIES (Uninhabitable attics with limited storage, live load = 20 psf, L/Δ = 240)

CEILING JOIST SPACING (inches)	SPECIES AND GRADE		DEAD LOAD = 10 psf			
			2 × 4	2 × 6	2 × 8	2 × 10
			Maximum ceiling joist spans			
			(feet-inches)	(feet-inches)	(feet-inches)	(feet-inches)
12	Douglas fir-larch	SS	10-5	16-4	21-7	Note a
	Douglas fir-larch	#1	10-0	15-9	20-1	24-6
	Douglas fir-larch	#2	9-10	15-0	19-1	23-3
	Douglas fir-larch	#3	7-10	11-6	14-7	17-9
	Hem-fir	SS	9-10	15-6	20-5	Note a
	Hem-fir	#1	9-8	15-2	19-10	24-3
	Hem-fir	#2	9-2	14-5	18-6	22-7
	Hem-fir	#3	7-8	11-2	14-2	17-4
	Southern pine	SS	10-3	16-1	21-2	Note a
	Southern pine	#1	9-10	15-6	20-5	24-0
	Southern pine	#2	9-3	13-11	17-7	20-11
	Southern pine	#3	7-2	10-6	13-3	16-1
	Spruce-pine-fir	SS	9-8	15-2	19-11	25-5
	Spruce-pine-fir	#1	9-5	14-9	18-9	22-11
	Spruce-pine-fir	#2	9-5	14-9	18-9	22-11
	Spruce-pine-fir	#3	7-8	11-2	14-2	17-4
16	Douglas fir-larch	SS	9-6	14-11	19-7	25-0
	Douglas fir-larch	#1	9-1	13-9	17-5	21-3
	Douglas fir-larch	#2	8-11	13-0	16-6	20-2
	Douglas fir-larch	#3	6-10	9-11	12-7	15-5
	Hem-fir	SS	8-11	14-1	18-6	23-8
	Hem-fir	#1	8-9	13-7	17-2	21-0
	Hem-fir	#2	8-4	12-8	16-0	19-7
	Hem-fir	#3	6-8	9-8	12-4	15-0
	Southern pine	SS	9-4	14-7	19-3	24-7
	Southern pine	#1	8-11	14-0	17-9	20-9
	Southern pine	#2	8-0	12-0	15-3	18-1
	Southern pine	#3	6-2	9-2	11-6	14-0
	Spruce-pine-fir	SS	8-9	13-9	18-1	23-1
	Spruce-pine-fir	#1	8-7	12-10	16-3	19-10
	Spruce-pine-fir	#2	8-7	12-10	16-3	19-10
	Spruce-pine-fir	#3	6-8	9-8	12-4	15-0

(continued)

TABLE R802.5.1(2)—continued
CEILING JOIST SPANS FOR COMMON LUMBER SPECIES (Uninhabitable attics with limited storage, live load = 20 psf, L/Δ = 240)

CEILING JOIST SPACING (inches)	SPECIES AND GRADE		DEAD LOAD = 10 psf			
			2 × 4	2 × 6	2 × 8	2 × 10
			Maximum ceiling joist spans			
			(feet-inches)	(feet-inches)	(feet-inches)	(feet-inches)
19.2	Douglas fir-larch	SS	8-11	14-0	18-5	23-7
	Douglas fir-larch	#1	8-7	12-6	15-10	19-5
	Douglas fir-larch	#2	8-2	11-11	15-1	18-5
	Douglas fir-larch	#3	6-2	9-1	11-6	14-1
	Hem-fir	SS	8-5	13-3	17-5	22-3
	Hem-fir	#1	8-3	12-4	15-8	19-2
	Hem-fir	#2	7-10	11-7	14-8	17-10
	Hem-fir	#3	6-1	8-10	11-3	13-8
	Southern pine	SS	8-9	13-9	18-2	23-1
	Southern pine	#1	8-5	12-9	16-2	18-11
	Southern pine	#2	7-4	11-0	13-11	16-6
	Southern pine	#3	5-8	8-4	10-6	12-9
	Spruce-pine-fir	SS	8-3	12-11	17-1	21-8
	Spruce-pine-fir	#1	8-0	11-9	14-10	18-2
	Spruce-pine-fir	#2	8-0	11-9	14-10	18-2
	Spruce-pine-fir	#3	6-1	8-10	11-3	13-8
24	Douglas fir-larch	SS	8-3	13-0	17-2	21-3
	Douglas fir-larch	#1	7-8	11-2	14-2	17-4
	Douglas fir-larch	#2	7-3	10-8	13-6	16-5
	Douglas fir-larch	#3	5-7	8-1	10-3	12-7
	Hem-fir	SS	7-10	12-3	16-2	20-6
	Hem-fir	#1	7-7	11-1	14-0	17-1
	Hem-fir	#2	7-1	10-4	13-1	16-0
	Hem-fir	#3	5-5	7-11	10-0	12-3
	Southern pine	SS	8-1	12-9	16-10	21-6
	Southern pine	#1	7-8	11-5	14-6	16-11
	Southern pine	#2	6-7	9-10	12-6	14-9
	Southern pine	#3	5-1	7-5	9-5	11-5
	Spruce-pine-fir	SS	7-8	12-0	15-10	19-5
	Spruce-pine-fir	#1	7-2	10-6	13-3	16-3
	Spruce-pine-fir	#2	7-2	10-6	13-3	16-3
	Spruce-pine-fir	#3	5-5	7-11	10-0	12-3

Check sources for availability of lumber in lengths greater than 20 feet.

For SI: 1 inch = 25.4 mm, 1 foot = 304.8 mm, 1 pound per square foot = 0.0479 kPa.

a. Span exceeds 26 feet in length.

R802.5.2 Ceiling joist and rafter connections. Where ceiling joists run parallel to rafters and are located in the bottom third of the rafter height, they shall be installed in accordance with Figure R802.4.5 and fastened to rafters in accordance with Table R802.5.2(1). Where the ceiling joists are installed above the bottom third of the rafter height, the ridge shall be designed as a beam in accordance with Section R802.3. Where ceiling joists do not run parallel to rafters, rafters shall be tied across the structure with a rafter tie in accordance with Section R802.5.2.2, or the ridge shall be designed as a beam in accordance with Section R802.3.

TABLE R802.5.2(2)
HEEL JOINT CONNECTION ADJUSTMENT FACTORS

H_C/H_R [a, b]	HEEL JOINT CONNECTION ADJUSTMENT FACTOR
1/3	1.5
1/4	1.33
1/5	1.25
1/6	1.2
1/10 or less	1.11

a. H_C = Height of ceiling joists or rafter ties measured vertically from the top of the rafter support walls to the bottom of the ceiling joists or rafter ties; H_R = Height of roof ridge measured vertically from the top of the rafter support walls to the bottom of the roof ridge.

b. Where H_C/H_R exceeds 1/3, connections shall be designed in accordance with accepted engineering practice.

TABLE R802.5.2(1)
RAFTER/CEILING JOIST HEEL JOINT CONNECTIONS[g]

RAFTER SLOPE	RAFTER SPACING (inches)	GROUND SNOW LOAD (psf)											
		20[e]			30			50			70		
		Roof span (feet)											
		12	24	36	12	24	36	12	24	36	12	24	36
		Required number of 16d common nails per heel joint splices [a, b, c, d, f]											
3:12	12	3	5	8	3	6	9	5	9	13	6	12	17
	16	4	7	10	4	8	12	6	12	17	8	15	23
	19.2	4	8	12	5	10	14	7	14	21	9	18	27
	24	5	10	15	6	12	18	9	17	26	12	23	34
4:12	12	3	4	6	3	5	7	4	7	10	5	9	13
	16	3	5	8	3	6	9	5	9	13	6	12	17
	19.2	3	6	9	4	7	11	6	11	16	7	14	21
	24	4	8	11	5	9	13	7	13	19	9	17	26
5:12	12	3	3	5	3	4	6	3	6	8	4	7	11
	16	3	4	6	3	5	7	4	7	11	5	9	14
	19.2	3	5	7	3	6	9	5	9	13	6	11	17
	24	3	6	9	4	7	11	6	11	16	7	14	21
7:12	12	3	3	4	3	3	4	3	4	6	3	5	8
	16	3	3	5	3	4	5	3	5	8	4	7	10
	19.2	3	4	5	3	4	6	3	6	9	4	8	12
	24	3	5	7	3	5	8	4	8	11	5	10	15
9:12	12	3	3	3	3	3	3	3	3	5	3	4	6
	16	3	3	4	3	3	4	3	4	6	3	5	8
	19.2	3	3	4	3	4	5	3	5	7	3	6	9
	24	3	4	5	3	4	6	3	6	9	4	8	12
12:12	12	3	3	3	3	3	3	3	3	4	3	3	5
	16	3	3	3	3	3	3	3	3	5	3	4	6
	19.2	3	3	3	3	3	4	3	4	6	3	5	7
	24	3	3	4	3	3	5	3	5	7	3	6	9

For SI: 1 inch = 25.4 mm, 1 foot = 304.8 mm, 1 pound per square foot = 0.0479 kPa.

a. 10d common (3″ × 0.148″) nails shall be permitted to be substituted for 16d common ($3^1/_2$″ × 0.162″) nails where the required number of nails is taken as 1.2 times the required number of 16d common nails, rounded up to the next full nail.

b. Heel joint connections are not required where the ridge is supported by a load-bearing wall, header or ridge beam.

c. Where intermediate support of the rafter is provided by vertical struts or purlins to a load-bearing wall, the tabulated heel joint connection requirements shall be permitted to be reduced proportionally to the reduction in span.

d. Equivalent nailing patterns are required for ceiling joist to ceiling joist lap splices.

e. Applies to roof live load of 20 psf or less.

f. Tabulated heel joint connection requirements assume that ceiling joists or rafter ties are located at the bottom of the attic space. Where ceiling joists or rafter ties are located higher in the attic, heel joint connection requirements shall be increased by the adjustment factors in Table 802.5.2(2).

g. Tabulated requirements are based on 10 psf roof dead load in combination with the specified roof snow load and roof live load.

R802.5.2.1 Ceiling joists lapped. Ends of ceiling joists shall be lapped not less than 3 inches (76 mm) or butted over bearing partitions or beams and toenailed to the bearing member. Where ceiling joists are used to provide the continuous tie across the building, lapped joists shall be nailed together in accordance with Table R802.5.2(1) and butted joists shall be tied together with a connection of equivalent capacity. Laps in joists that do not provide the continuous tie across the building shall be permitted to be nailed in accordance with Table R602.3(1).

R802.5.2.2 Rafter ties. Wood rafter ties shall be not less than 2 inches by 4 inches (51 mm × 102 mm) installed in accordance with Table R802.5.2(1) at a maximum of 24 inches (610 mm) on center. Other *approved* rafter tie methods shall be permitted.

R802.5.2.3 Blocking. Blocking shall be not less than utility grade lumber.

R802.6 Bearing. The ends of each rafter or ceiling joist shall have not less than $1^1/_2$ inches (38 mm) of bearing on wood or metal and not less than 3 inches (76 mm) on masonry or concrete. The bearing on masonry or concrete shall be direct, or a sill plate of 2-inch (51 mm) minimum nominal thickness shall be provided under the rafter or ceiling joist. The sill plate shall provide a minimum nominal bearing area of 48 square inches (30 968 mm^2). Where the roof pitch is greater than or equal to 3 units vertical in 12 units horizontal (25-percent slope), and ceiling joists or rafter ties are connected to rafters to provide a continuous tension tie in accordance with Section R802.5.2, vertical bearing of the top of the rafter against the ridge board shall satisfy this bearing requirement.

R802.6.1 Finished ceiling material. If the finished ceiling material is installed on the ceiling prior to the attachment of the ceiling to the walls, such as in construction at a factory, a compression strip of the same thickness as the finished ceiling material shall be installed directly above the top plate of bearing walls if the compressive strength of the finished ceiling material is less than the loads it will be required to withstand. The compression strip shall cover the entire length of such top plate and shall be not less than one-half the width of the top plate. It shall be of material capable of transmitting the loads transferred through it.

R802.7 Cutting, drilling and notching. Structural roof members shall not be cut, bored or notched in excess of the limitations specified in this section.

R802.7.1 Sawn lumber. Cuts, notches and holes in solid lumber joists, rafters, blocking and beams shall comply with the provisions of Section R502.8.1 except that cantilevered portions of rafters shall be permitted in accordance with Section R802.7.1.1.

R802.7.1.1 Cantilevered portions of rafters. Notches on cantilevered portions of rafters are permitted provided the dimension of the remaining portion of the rafter is not less than $3^1/_2$ inches (89 mm) and the length of the cantilever does not exceed 24 inches (610 mm) in accordance with Figure R802.7.1.1.

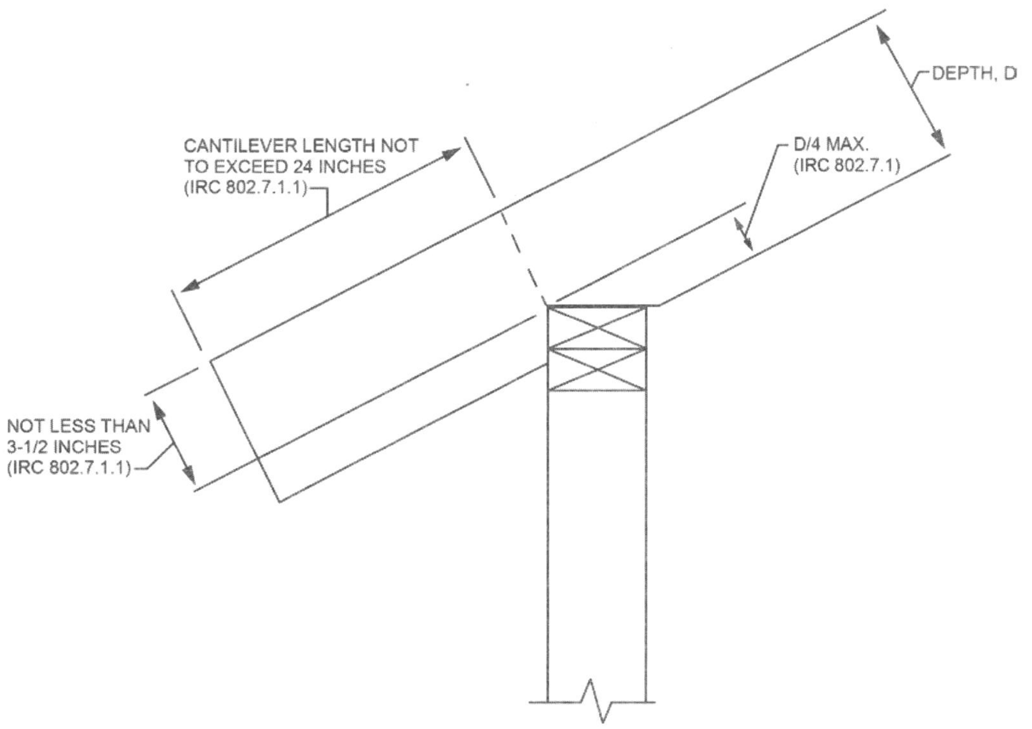

For SI: 1 inch = 25.4 mm.

FIGURE R802.7.1.1
RAFTER NOTCH

R802.7.1.2 Ceiling joist taper cut. Taper cuts at the ends of the ceiling joist shall not exceed one-fourth the depth of the member in accordance with Figure R802.7.1.2.

R802.7.2 Engineered wood products. Cuts, notches and holes bored in trusses, *structural composite lumber*, structural glue-laminated members, cross-laminated timber members or I-joists are prohibited except where permitted by the manufacturer's recommendations or where the effects of such alterations are specifically considered in the design of the member by a *registered design professional*.

R802.8 Lateral support. Roof framing members and ceiling joists having a depth-to-thickness ratio exceeding 5 to 1 based on nominal dimensions shall be provided with lateral support at points of bearing to prevent rotation. For roof rafters with ceiling joists attached in accordance with Table R602.3(1), the depth-to-thickness ratio for the total assembly shall be determined using the combined thickness of the rafter plus the attached ceiling joist.

Exception: Roof trusses shall be braced in accordance with Section R802.10.3.

R802.8.1 Bridging. Rafters and ceiling joists having a depth-to-thickness ratio exceeding 6 to 1 based on nominal dimensions shall be supported laterally by solid blocking, diagonal bridging (wood or metal) or a continuous 1-inch by 3-inch (25 mm by 76 mm) wood strip nailed across the rafters or ceiling joists at intervals not exceeding 8 feet (2438 mm).

R802.9 Framing of openings. Openings in roof and ceiling framing shall be framed with header and trimmer joists. Where the header joist span does not exceed 4 feet (1219 mm), the header joist shall be permitted to be a single member the same size as the ceiling joist or rafter. Single trimmer joists shall be permitted to be used to carry a single header joist that is located within 3 feet (914 mm) of the trimmer joist bearing. Where the header joist span exceeds 4 feet (1219 mm), the trimmer joists and the header joist shall be doubled and of sufficient cross section to support the ceiling joists or rafter framing into the header. *Approved* hangers shall be used for the header joist to trimmer joist connections where the header joist span exceeds 6 feet (1829 mm). Tail joists over 12 feet (3658 mm) long shall be supported at the header by framing anchors or on ledger strips not less than 2 inches by 2 inches (51 mm by 51 mm).

R802.10 Wood trusses.

R802.10.1 Truss design drawings. *Truss design drawings*, prepared in conformance to Section R802.10.1, shall be provided to the *building official* and *approved* prior to installation. *Truss design drawings* shall be provided with the shipment of trusses delivered to the job site. *Truss design drawings* shall include, at a minimum, the following information:

1. Slope or depth, span and spacing.
2. Location of all joints.
3. Required bearing widths.

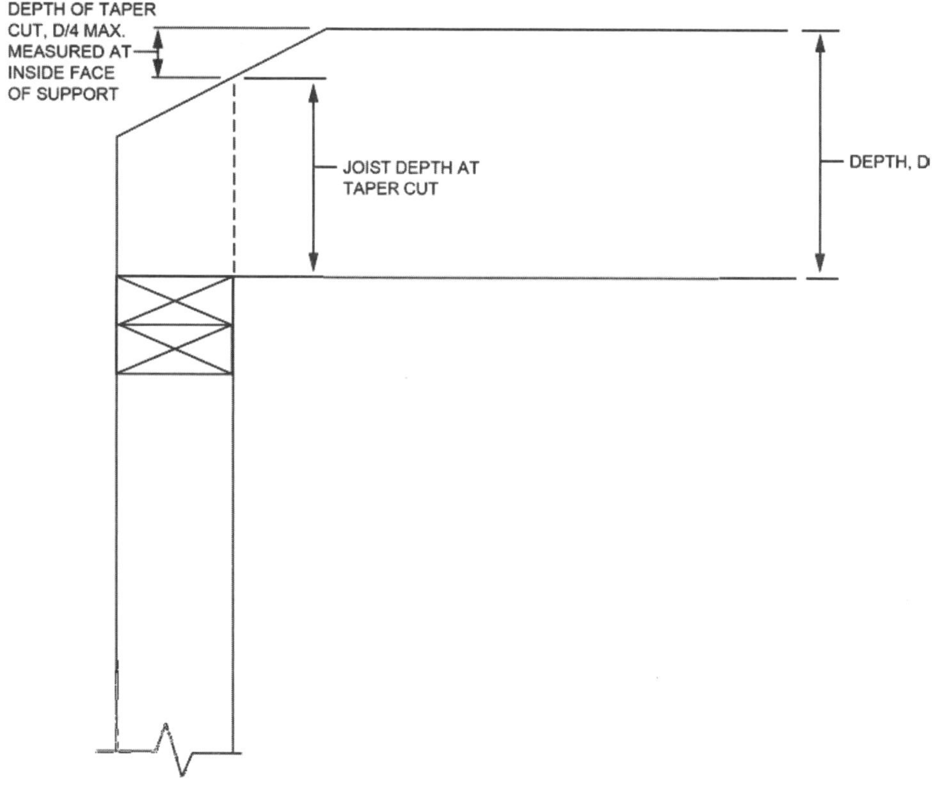

FIGURE R802.7.1.2
CEILING JOIST TAPER CUT

4. Design loads as applicable.
   4.1. Top chord *live load* (as determined from Section R301.6).
   4.2. Top chord dead load.
   4.3. Bottom chord *live load.*
   4.4. Bottom chord dead load.
   4.5. Concentrated loads and their points of application.
   4.6. Controlling wind and earthquake loads.
5. Adjustments to lumber and joint connector design values for conditions of use.
6. Each reaction force and direction.
7. Joint connector type and description such as size, thickness or gage and the dimensioned location of each joint connector except where symmetrically located relative to the joint interface.
8. Lumber size, species and *grade for each member.*
9. Connection requirements for:
   9.1. Truss to girder-truss.
   9.2. Truss ply to ply.
   9.3. Field splices.
10. Calculated deflection ratio or maximum description for live and total load.
11. Maximum axial compression forces in the truss members to enable the building designer to design the size, connections and anchorage of the permanent continuous lateral bracing. Forces shall be shown on the *truss design drawing* or on supplemental documents.
12. Required permanent truss member bracing location.

R802.10.2 Design. Wood trusses shall be designed in accordance with accepted engineering practice. The design and manufacture of metal-plate-connected wood trusses shall comply with ANSI/TPI 1. The *truss design drawings* shall be prepared by a *registered design professional* where required by the statutes of the *jurisdiction* in which the project is to be constructed in accordance with Section R106.1.

R802.10.2.1 Applicability limits. The provisions of this section shall control the design of truss roof framing where snow controls for buildings that are not greater than 60 feet (18 288 mm) in length perpendicular to the joist, rafter or truss span, not greater than 36 feet (10 973 mm) in width parallel to the joist, rafter or truss span, not more than three stories above *grade plane* in height, and have roof slopes not smaller than 3:12 (25-percent slope) or greater than 12:12 (100-percent slope). Truss roof framing constructed in accordance with the provisions of this section shall be limited to sites subjected to a maximum design wind speed of 140 miles per hour (63 m/s), Exposure B or C, and a maximum ground snow load of 70 psf (3352 Pa). For consistent loading of all truss types, roof snow load is to be computed as: 0.7 p_g.

R802.10.3 Bracing. Trusses shall be braced to prevent rotation and provide lateral stability in accordance with the requirements specified in the *construction documents* for the building and on the individual *truss design drawings.* In the absence of specific bracing requirements, trusses shall be braced in accordance with accepted industry practice such as the SBCA *Building Component Safety Information (BCSI) Guide to Good Practice for Handling, Installing & Bracing of Metal Plate Connected Wood Trusses.*

R802.10.4 Alterations to trusses. Truss members shall not be cut, notched, drilled, spliced or otherwise altered in any way without the approval of a registered *design professional. Alterations* resulting in the addition of load such as HVAC equipment water heater that exceeds the design load for the truss shall not be permitted without verification that the truss is capable of supporting such additional loading.

R802.11 Roof tie uplift resistance. *Roof assemblies* shall have uplift resistance in accordance with Sections R802.11.1 and R802.11.2.

Exceptions: Rafters or trusses shall be permitted to be attached to their supporting wall assemblies in accordance with Table R602.3(1) where either of the following occur:

1. Where the uplift force per rafter or truss does not exceed 200 pounds (90.8 kg) as determined by Table R802.11.
2. Where the basic wind speed does not exceed 115 miles per hour (51.4 m/s), the wind exposure category is B, the roof pitch is 5 units vertical in 12 units horizontal (42-percent slope) or greater, the roof span is 32 feet (9754 mm) or less, and rafters and trusses are spaced not more than 24 inches (610 mm) on center.

R802.11.1 Truss uplift resistance. Trusses shall be attached to supporting wall assemblies by connections capable of resisting uplift forces as specified on the *truss design drawings* for the ultimate design wind speed as determined by Figure R301.2(2) and listed in Table R301.2 or as shown on the *construction documents.* Uplift forces shall be permitted to be determined as specified by Table R802.11, if applicable, or as determined by accepted engineering practice.

R802.11.2 Rafter uplift resistance. Individual rafters shall be attached to supporting wall assemblies by connections capable of resisting uplift forces as determined by Table R802.11 or as determined by accepted engineering practice. Connections for beams used in a roof system shall be designed in accordance with accepted engineering practice.

TABLE R802.11
RAFTER OR TRUSS UPLIFT CONNECTION FORCES FROM WIND (ASD) (POUNDS PER CONNECTION)[a, b, c, d, e, f, g, h]

RAFTER OR TRUSS SPACING	ROOF SPAN (feet)	EXPOSURE B									
		Ultimate Design Wind Speed V_{ULT} (mph)									
		110		115		120		130		140	
		Roof Pitch		Roof Pitch		Roof Pitch		Roof Pitch		Roof Pitch	
		< 5:12	≥ 5:12	< 5:12	≥ 5:12	< 5:12	≥ 5:12	< 5:12	≥ 5:12	< 5:12	≥ 5:12
12″ o.c.	12	48	43	59	53	70	64	95	88	122	113
	18	59	52	74	66	89	81	122	112	157	146
	24	71	62	89	79	108	98	149	137	192	178
	28	79	69	99	88	121	109	167	153	216	200
	32	86	75	109	97	134	120	185	170	240	222
	36	94	82	120	106	146	132	203	186	264	244
	42	106	92	135	120	166	149	230	211	300	278
	48	118	102	151	134	185	166	258	236	336	311
16″ o.c.	12	64	57	78	70	93	85	126	117	162	150
	18	78	69	98	88	118	108	162	149	209	194
	24	94	82	118	105	144	130	198	182	255	237
	28	105	92	132	117	161	145	222	203	287	266
	32	114	100	145	129	178	160	246	226	319	295
	36	125	109	160	141	194	176	270	247	351	325
	42	141	122	180	160	221	198	306	281	399	370
	48	157	136	201	178	246	221	343	314	447	414
24″ o.c.	12	96	86	118	106	140	128	190	176	244	226
	18	118	104	148	132	178	162	244	224	314	292
	24	142	124	178	158	216	196	298	274	384	356
	28	158	138	198	176	242	218	334	306	432	400
	32	172	150	218	194	268	240	370	340	480	444
	36	188	164	240	212	292	264	406	372	528	488
	42	212	184	270	240	332	298	460	422	600	556
	48	236	204	302	268	370	332	516	472	672	622
		EXPOSURE C									
12″ o.c.	12	95	88	110	102	126	118	161	151	198	186
	18	121	111	141	131	163	151	208	195	257	242
	24	148	136	173	160	200	185	256	239	317	298
	28	166	152	195	179	225	208	289	269	358	335
	32	184	168	216	199	249	231	321	299	398	373
	36	202	185	237	219	274	254	353	329	438	411
	42	229	210	269	248	312	289	402	375	499	468
	48	256	234	302	278	349	323	450	420	560	524
16″ o.c.	12	126	117	146	136	168	157	214	201	263	247
	18	161	148	188	174	217	201	277	259	342	322
	24	197	181	230	213	266	246	340	318	422	396
	28	221	202	259	238	299	277	384	358	476	446
	32	245	223	287	265	331	307	427	398	529	496
	36	269	246	315	291	364	338	469	438	583	547
	42	305	279	358	330	415	384	535	499	664	622
	48	340	311	402	370	464	430	599	559	745	697

(continued)

TABLE R802.11—continued
RAFTER OR TRUSS UPLIFT CONNECTION FORCES FROM WIND (ASD) (POUNDS PER CONNECTION)[a, b, c, d, e, f, g, h]

RAFTER OR TRUSS SPACING	ROOF SPAN (feet)	EXPOSURE B									
		Ultimate Design Wind Speed V_{ULT} (mph)									
		110		115		120		130		140	
		Roof Pitch		Roof Pitch		Roof Pitch		Roof Pitch		Roof Pitch	
		< 5:12	≥ 5:12	< 5:12	≥ 5:12	< 5:12	≥ 5:12	< 5:12	≥ 5:12	< 5:12	≥ 5:12
24″ o.c.	12	190	176	220	204	252	236	322	302	396	372
	18	242	222	282	262	326	302	416	390	514	484
	24	296	272	346	320	400	370	512	478	634	596
	28	332	304	390	358	450	416	578	538	716	670
	32	368	336	432	398	498	462	642	598	796	746
	36	404	370	474	438	548	508	706	658	876	822
	42	458	420	538	496	624	578	804	750	998	936
	48	512	468	604	556	698	646	900	840	1,120	1,048

For SI: 1 inch = 25.4 mm, 1 foot = 304.8 mm, 1 mile per hour = 0.447 m/s, 1 pound = 0.454 kg, 1 pound per square foot = 47.9 N/m^2, 1 pound per linear foot = 14.6 N/m.

a. The uplift connection forces are based on a maximum 33-foot mean roof height and Wind Exposure Category B or C. For Exposure D, the uplift connection force shall be selected from the Exposure C portion of the table using the next highest tabulated ultimate design wind speed. The adjustment coefficients in Table R301.2.1(2) shall not be used to multiply the tabulated forces for Exposures C and D or for other mean roof heights.

b. The uplift connection forces include an allowance for roof and ceiling assembly dead load of 15 psf.

c. The tabulated uplift connection forces are limited to a maximum roof overhang of 24 inches.

d. The tabulated uplift connection forces shall be permitted to be multiplied by 0.75 for connections not located within 8 feet of building corners.

e. For buildings with hip roofs with 5:12 and greater pitch, the tabulated uplift connection forces shall be permitted to be multiplied by 0.70. This reduction shall not be combined with any other reduction in tabulated forces.

f. For wall-to-wall and wall-to-foundation connections, the uplift connection force shall be permitted to be reduced by 60 pounds per linear foot for each full wall above.

g. Linear interpolation between tabulated roof spans and wind speeds shall be permitted.

h. The tabulated forces for a 12-inch on-center spacing shall be permitted to be used to determine the uplift load in pounds per linear foot.

SECTION R803
ROOF SHEATHING

R803.1 Lumber sheathing. Allowable spans for lumber used as roof sheathing shall conform to Table R803.1. Spaced lumber sheathing for wood shingle and shake roofing shall conform to the requirements of Sections R905.7 and R905.8. Spaced lumber sheathing is not allowed in Seismic Design Category D_2.

TABLE R803.1
MINIMUM THICKNESS OF LUMBER ROOF SHEATHING

RAFTER OR BEAM SPACING (inches)	MINIMUM NET THICKNESS (inches)
24	$^5/_8$
48[a]	$1^1/_2$ T & G
60[b]	
72[c]	

For SI: 1 inch = 25.4 mm.

a. Minimum 270F_b, 340,000E.

b. Minimum 420F_b, 660,000E.

c. Minimum 600F_b, 1,150,000E.

R803.2 Wood structural panel sheathing.

R803.2.1 Identification and grade. *Wood structural panels* shall conform to DOC PS 1, DOC PS 2, CSA 0325 or CSA 0437, and shall be identified for grade, bond classification and performance category by a grade mark or certificate of inspection issued by an *approved* agency. *Wood structural panels* shall comply with the grades specified in Table R503.2.1.1(1).

R803.2.1.1 Exposure durability. *Wood structural panels*, when designed to be permanently exposed in outdoor applications, shall be of an exterior exposure durability. *Wood structural panel* roof sheathing exposed to the underside shall be permitted to be of interior type bonded with exterior glue, identified as Exposure 1.

R803.2.1.2 Fire-retardant-treated plywood. The allowable unit stresses for fire-retardant-treated plywood, including fastener values, shall be developed from an *approved* method of investigation that considers the effects of anticipated temperature and humidity to which the fire-retardant-treated plywood will be subjected, the type of treatment and redrying process. The fire-retardant-treated plywood shall be graded by an *approved agency*.

R803.2.2 Allowable spans. The maximum allowable spans for *wood structural panel* roof sheathing shall not exceed the values set forth in Table R503.2.1.1(1) or APA E30.

R803.2.3 Installation. *Wood structural panel* used as roof sheathing shall be installed with joints staggered or not staggered in accordance with Table R602.3(1), APA E30 for wood roof framing or with Table R804.3 for cold-formed steel roof framing. Wood structural panel roof sheathing in accordance with Table R503.2.1.1(1) shall not cantilever more than 9 inches (229 mm) beyond the gable endwall unless supported by gable overhang framing.

SECTION R804
COLD-FORMED STEEL ROOF FRAMING

R804.1 General. Elements shall be straight and free of any defects that would significantly affect their structural performance. Cold-formed steel roof framing members shall be in accordance with the requirements of this section.

R804.1.1 Applicability limits. The provisions of this section shall control the construction of cold-formed steel roof framing for buildings not greater than 60 feet (18 288 mm) perpendicular to the joist, rafter or truss span, not greater than 40 feet (12 192 mm) in width parallel to the joist span or truss, less than or equal to three stories above *grade* plane and with roof slopes not less than 3:12 (25-percent slope) or greater than 12:12 (100-percent slope). Cold-formed steel roof framing constructed in accordance with the provisions of this section shall be limited to sites where the ultimate design wind speed is less than 140 miles per hour (63 m/s), Exposure Category B or C, and the ground snow load is less than or equal to 70 pounds per square foot (3350 Pa).

R804.1.1.1 Alternate applications. Cold-formed steel roof and ceiling framing for buildings exceeding the applicability limits of Section R804.1.1 is permitted to be designed and constructed in accordance with AISI S230, subject to the limits therein.

R804.1.2 In-line framing. Cold-formed steel roof framing constructed in accordance with Section R804 shall be located in line with the tolerances specified in AISI S240, Section B1.2.3.

R804.2 Structural framing. Load-bearing, cold-formed steel roof framing members shall be in accordance with this section.

R804.2.1 Material. Load-bearing, cold-formed steel framing members shall be cold formed to shape from structural quality sheet steel complying with the requirements of AISI S240, Section A3.

R804.2.2 Corrosion protection. Load-bearing, cold-formed steel framing shall have a protective coating complying with AISI S240, Section A4.

R804.2.3 Dimension, thickness and material grade. Load-bearing, cold-formed steel roof framing members shall comply with AISI S230, Section A4.3 and material grade requirements as specified in AISI S230, Section A4.4.

R804.2.4 Identification. Load-bearing, cold-formed steel framing members shall meet the product identification requirements of AISI S240, Section A5.5.

R804.2.5 Fastening requirements. Screws for steel-to-steel connections shall be installed with a minimum edge distance and center-to-center spacing of $^{1}/_{2}$ inch (12.7 mm), shall be self-drilling tapping and shall conform to ASTM C1513. Structural sheathing shall be attached to cold-formed steel roof rafters with minimum No. 8 self-drilling tapping screws that conform to ASTM C1513. Screws for attaching structural sheathing to cold-formed steel roof framing shall have a minimum head diameter of 0.292 inch (7.4 mm) with countersunk heads and shall be installed with a minimum edge distance of $^{3}/_{8}$ inch (9.5 mm). Gypsum board ceilings shall be attached to cold-formed steel joists with minimum No. 6 screws conforming to ASTM C954 or ASTM C1513 with a bugle-head style and shall be installed in accordance with Section R805. For all connections, screws shall extend through the steel not fewer than three exposed threads. Fasteners shall have rust-inhibitive coating suitable for the installation in which they are being used, or be manufactured from material not susceptible to corrosion.

R804.2.6 Web holes, web hole reinforcing and web hole patching. Web holes in roof or ceiling joists shall comply with the conditions as prescribed in AISI S230, Section A4.5. Web holes not in conformance to the conditions of AISI S230, Section A4.5 shall be reinforced in accordance with the provisions of AISI S230, Section A4.6 or patched in accordance with the provisions of AISI S230, Section A4.7.

R804.3 Roof construction. Cold-formed steel roof systems constructed in accordance with the provisions of this section shall consist of both ceiling joists and rafters in accordance with Figure R804.3 and fastened in accordance with Table R804.3.

R804.3.1 Ceiling joists. Cold-formed steel ceiling joists shall be in accordance with this section.

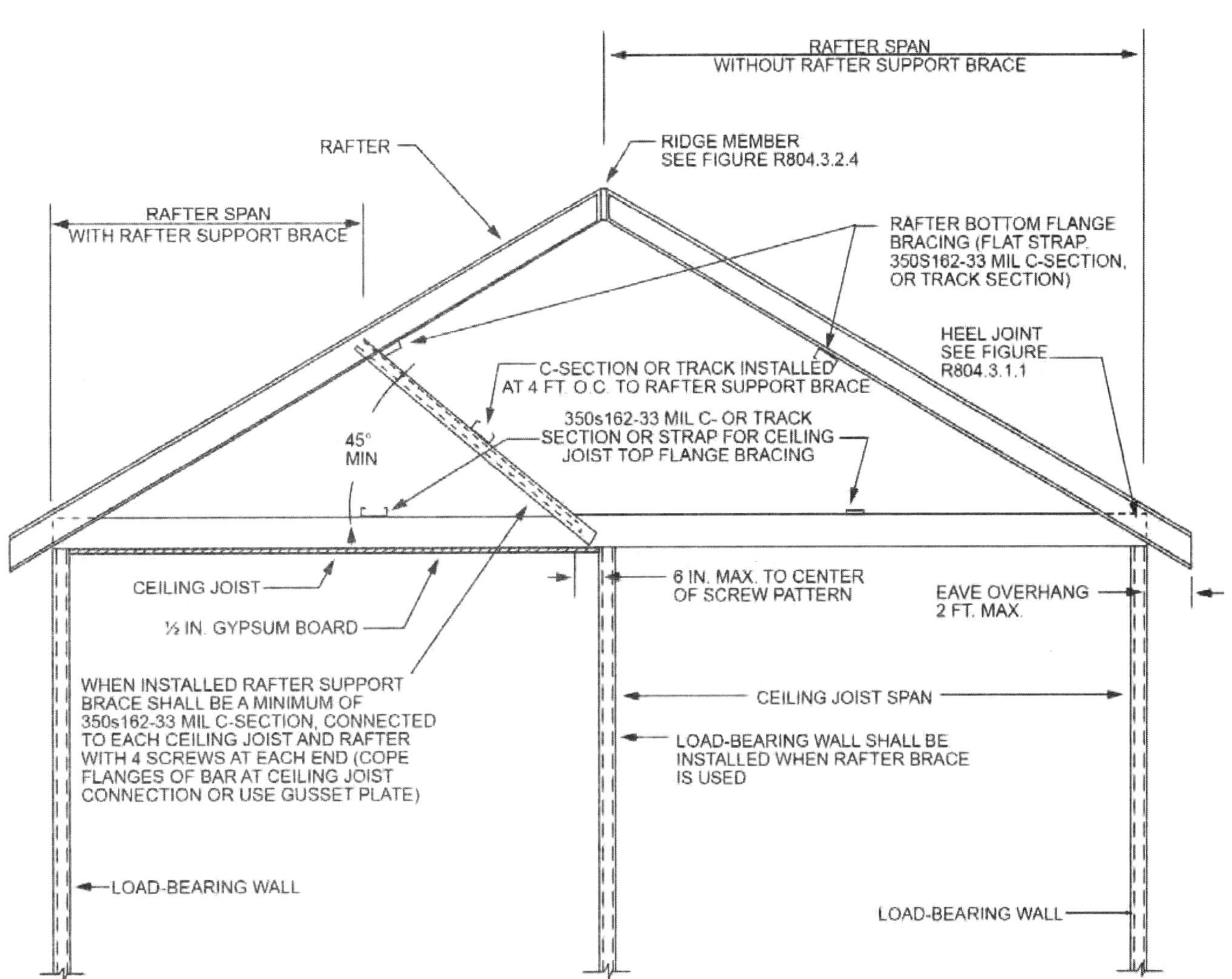

For SI: 1 inch = 25.4 mm, 1 foot = 304.8 mm, 1 mil = 0.0254 mm.

FIGURE R804.3
COLD-FORMED STEEL ROOF CONSTRUCTION

**TABLE R804.3
ROOF FRAMING FASTENING SCHEDULE[a, b]**

DESCRIPTION OF BUILDING ELEMENTS			NUMBER AND SIZE OF FASTENERS[a]				SPACING OF FASTENERS
Roof sheathing (oriented strand board or plywood) to rafter			No. 8 screws				6″ o.c. on edges and 12″ o.c. at interior supports. 6″ o.c. at gable end truss.
Gypsum board to ceiling joists			No. 6 screws				12″ o.c.
Gable end truss to endwall top track			No. 10 screws				12″ o.c.
Rafter to ceiling joist and to ridge member			Minimum No. 10 screws, in accordance with Table R804.3.1.1(3)				Evenly spaced, not less than $^1/_2$″ from all edges.
Ceiling joist or roof truss to top track of bearing wall[b]	Ceiling joist or truss spacing (in.)	Roof Span (ft)	Ultimate Design Wind Speed (mph) and Exposure Category				Each ceiling joist or roof truss.
			130 B 115 C	< 139 B 120 C	130 C	< 139 C	
	16	24	3	3	4	5	
		28	3	3	4	5	
		32	3	4	5	6	
		36	4	4	5	6	
		40	4	4	6	7	
	24	24	4	5	6	7	
		28	4	5	6	8	
		32	4	6	7	8	
		36	4	6	8	9	
		40	6	6	8	10	

For SI: 1 inch = 25.4 mm, 1 foot = 304.8 mm, 1 pound per square foot = 0.0479 kPa, 1 mil = 0.0254 mm.

a. Screws are a minimum No. 10 unless noted otherwise.

b. Indicated number of screws shall be applied through the flanges of the truss or ceiling joist or through each leg of a 54 mil clip angle. See Section R804.3.8 for additional requirements to resist uplift forces.

R804.3.1.1 Minimum ceiling joist size. Ceiling joist size and thickness shall be determined in accordance with the limits set forth in Tables R804.3.1.1(1) and R804.3.1.1(2). When determining the size of ceiling joists, the lateral support of the top flange shall be classified as unbraced, braced at midspan or braced at third points in accordance with Section R804.3.1.3. Where sheathing material is attached to the top flange of ceiling joists or where the bracing is spaced closer than at third points of the joists, the "third point" values from Tables R804.3.1.1(1) and R804.3.1.1(2) shall be used.

Ceiling joists shall have a bearing support length of not less than $1^1/_2$ inches (38 mm) and shall be connected to roof rafters (heel joint) with No. 10 screws in accordance with Figure R804.3.1.1 and Table R804.3.1.1(3).

Where continuous joists are framed across interior bearing supports, the interior bearing supports shall be located within 24 inches (610 mm) of midspan of the ceiling joist, and the individual spans shall not exceed the applicable spans in Tables R804.3.1.1(1) and R804.3.1.1(2).

Where the attic is to be used as an *occupied space*, the ceiling joists shall be designed in accordance with Section R505.

R804.3.1.2 Ceiling joist bottom flange bracing. The bottom flanges of ceiling joists shall be laterally braced by the application of gypsum board or continuous steel straps installed perpendicular to the joist run in accordance with one of the following:

1. Gypsum board shall be fastened with No. 6 screws in accordance with Section R702.
2. Steel straps with a minimum size of $1^1/_2$ inches by 33 mils (38 mm by 0.84 mm) shall be installed at a maximum spacing of 4 feet (1219 mm). Straps shall be fastened to the bottom flange at each joist with one No. 8 screw and shall be fastened to blocking with two No. 8 screws. Blocking shall be installed between joists at a maximum spacing of 12 feet (3658 mm) measured along a line of continuous strapping (perpendicular to the joist run), and at the termination of all straps.

TABLE R804.3.1.1(1)
CEILING JOIST SPANS 10 PSF LIVE LOAD (NO ATTIC STORAGE)[a, b, c, d]

MEMBER DESIGNATION	ALLOWABLE SPAN (feet-inches)					
	Lateral Support of Top (Compression) Flange					
	Unbraced		Midspan Bracing		Third-point Bracing	
	Ceiling Joist Spacing (inches)					
	16	24	16	24	16	24
350S162-33	9′-6″	8′-6″	11′-10″	9′-10″	11′-10″	10′-4″
350S162-43	10′-4″	9′-3″	12′-10″	11′-3″	12′-10″	11′-3″
350S162-54	11′-1″	9′-11″	13′-9″	12′-0″	13′-9″	12′-0″
350S162-68	12′-2″	10′-10″	14′-9″	12′-10″	14′-9″	12′-10″
550S162-33	10′-11″	9′-10″	15′-7″	12′-0″	16′-10″	12′-0″
550S162-43	11′-8″	10′-6″	16′-10″	14′-10″	18′-4″	16′-0″
550S162-54	12′-7″	11′-3″	18′-0″	16′-2″	19′-4″	17′-2″
550S162-68	13′-7″	12′-1″	19′-3″	17′-3″	20′-6″	18′-5″
800S162-33	—	—	—	—	—	—
800S162-43	13′-1″	11′-9″	18′-9″	16′-9″	21′-2″	18′-7″
800S162-54	13′-11″	12′-6″	20′-1″	18′-1″	21′-5″	20′-5″
800S162-68	14′-11″	13′-4″	21′-4″	19′-2″	22′-9″	21′-9″
1000S162-43	—	—	—	—	—	—
1000S162-54	14′-10″	13′-4″	21′-4″	19′-2″	22′-8″	21′-8″
1000S162-68	15′-10″	14′-3″	22′-9″	20′-5″	24′-3″	23′-3″
1200S162-43	—	—	—	—	—	—
1200S162-54	—	—	—	—	—	—
1200S162-68	16′-8″	14′-11″	23′-11″	21′-7″	25′-5″	24′-5″

For SI: 1 inch = 25.4 mm, 1 foot = 304.8 mm, 1 mil = 0.0254 mm, 1 pound per square foot = 0.0479 kPa.

a. Deflection criterion: *L*/240 for total loads.

b. Ceiling dead load = 5 psf.

c. Minimum Grade 33 ksi steel shall be used for 33 mil and 43 mil thicknesses. Minimum Grade 50 ksi steel shall be used for 54 and 68 mil thicknesses.

d. Listed allowable spans are not applicable for 350S162-33, 550S162-33, 550S162-43 and 800S162-43 continuous joist members.

**TABLE R804.3.1.1(2)
CEILING JOIST SPANS 20 PSF LIVE LOAD (LIMITED ATTIC STORAGE)[a, b, c, d]**

MEMBER DESIGNATION	ALLOWABLE SPAN (feet-inches)					
	Lateral Support of Top (Compression) Flange					
	Unbraced		Midspan Bracing		Third-point Bracing	
	Ceiling Joist Spacing (inches)					
	16	24	16	24	16	24
350S162-33	8′-0″	6′-5″	9′-2″	7′-5″	9′-11″	7′-5″
350S162-43	8′-11″	7′-8″	10′-9″	8′-9″	10′-0″	9′-6″
350S162-54	9′-7″	8′-7″	11′-7″	10′-2″	11′-7″	10′-2″
350S162-68	10′-4″	9′-3″	12′-5″	10′-10″	12′-5″	10′-10″
550S162-33	9′-5″	6′-11″	10′-5″	6′-11″	10′-5″	6′-11″
550S162-43	10′-2″	9′-2″	14′-2″	11′-8″	15′-2″	11′-8″
550S162-54	10′-10″	9′-9″	15′-7″	14′-0″	16′-7″	14′-5″
550S162-68	11′-8″	10′-5″	16′-7″	14′-10″	17′-9″	15′-6″
800S162-33	—	—	—	—	—	—
800S162-43	11′-4″	10′-2″	16′-1″	11′-0″	16′-6″	11′-0″
800S162-54	12′-0″	10′-10″	17′-4″	15′-7″	18′-7″	17′-7″
800S162-68	12′-10″	11′-6″	18′-6″	16′-7″	19′-11″	18′-11″
1000S162-43	—	—	—	—	—	—
1000S162-54	12′-10″	11′-7″	18′-5″	16′-6″	19′-8″	18′-8″
1000S162-68	13′-8″	12′-3″	19′-8″	17′-9″	21′-1″	20′-1″
1200S162-43	—	—	—	—	—	—
1200S162-54	—	—	—	—	—	—
1200S162-68	14′-5″	12′-11″	20′-9″	18′-7″	22′-0″	21′-0″

For SI: 1 inch = 25.4 mm, 1 foot = 304.8 mm, 1 mil = 0.0254 mm, 1 pound per square foot = 0.0479 kPa.

a. Deflection criterion: *L*/240 for total loads.

b. Ceiling deal load = 5 psf.

c. Minimum Grade 33 ksi steel shall be used for 33 mil and 43 mil thicknesses. Minimum Grade 50 ksi steel shall be used for 54 and 68 mil thicknesses.

d. Listed allowable spans are not applicable for 350S162-33, 350S162-43, 550S162-33, 550S162-43 and 800S162-43 continuous joist members.

TABLE R804.3.1.1(3)
NUMBER OF SCREWS REQUIRED FOR CEILING JOIST TO ROOF RAFTER CONNECTION[a]

ROOF SLOPE	NUMBER OF SCREWS																			
	Building width (feet)																			
	24				28				32				36				40			
	Ground snow load (psf)																			
	20	30	50	70	20	30	50	70	20	30	50	70	20	30	50	70	20	30	50	70
3/12	5	6	9	11	5	7	10	13	6	8	11	15	7	8	13	17	8	9	14	19
4/12	4	5	7	9	4	5	8	10	5	6	9	12	5	7	10	13	6	7	11	14
5/12	3	4	6	7	4	4	6	8	4	5	7	10	5	5	8	11	5	6	9	12
6/12	3	3	5	6	3	4	6	7	4	4	6	8	4	5	7	9	4	5	8	10
7/12	3	3	4	6	3	3	5	7	3	4	6	7	4	4	6	8	4	5	7	9
8/12	2	3	4	5	3	3	5	6	3	4	5	7	3	4	6	8	4	4	6	8
9/12	2	3	4	5	3	3	4	6	3	3	5	6	3	4	5	7	3	4	6	8
10/12	2	2	4	5	2	3	4	5	3	3	5	6	3	3	5	7	3	4	6	7
11/12	2	2	3	4	2	3	4	5	3	3	4	6	3	3	5	6	3	4	5	7
12/12	2	2	3	4	2	3	4	5	2	3	4	5	3	3	5	6	3	4	5	7

For SI: 1 foot = 304.8 mm, 1 pound per square foot = 0.0479kPa.
a. Screws shall be No. 10.

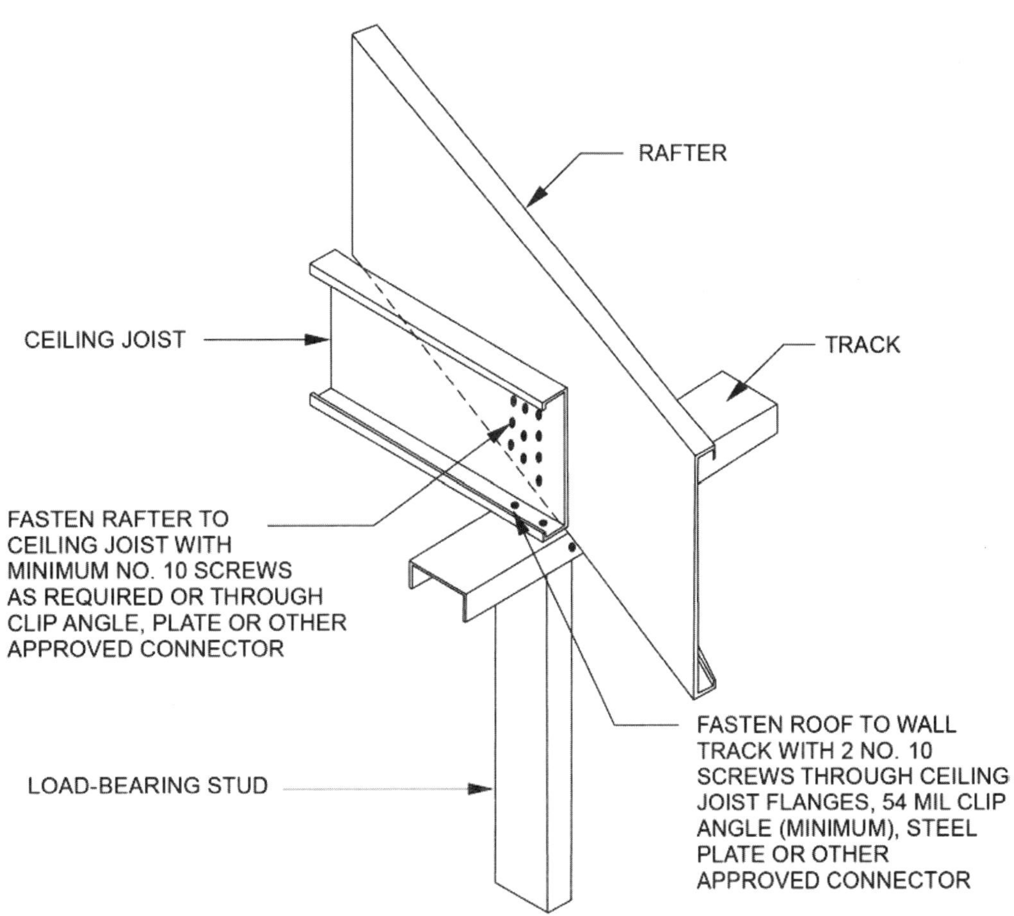

For SI: 1 mil = 0.0254 mm.

FIGURE R804.3.1.1
JOIST TO RAFTER CONNECTION

R804.3.1.3 Ceiling joist top flange bracing. The top flanges of ceiling joists shall be laterally braced as required by Tables R804.3.1.1(1) and R804.3.1.1(2), in accordance with one of the following:

1. Minimum 33-mil (0.84 mm) C-shaped member in accordance with Figure R804.3.1.3(1).
2. Minimum 33-mil (0.84 mm) track section in accordance with Figure R804.3.1.3(1).
3. Minimum 33-mil (0.84 mm) hat section in accordance with Figure R804.3.1.3(1).
4. Minimum 54-mil (1.37 mm) $1^1/_2$-inch (38 mm) cold-rolled channel section in accordance with Figure R804.3.1.3(1).

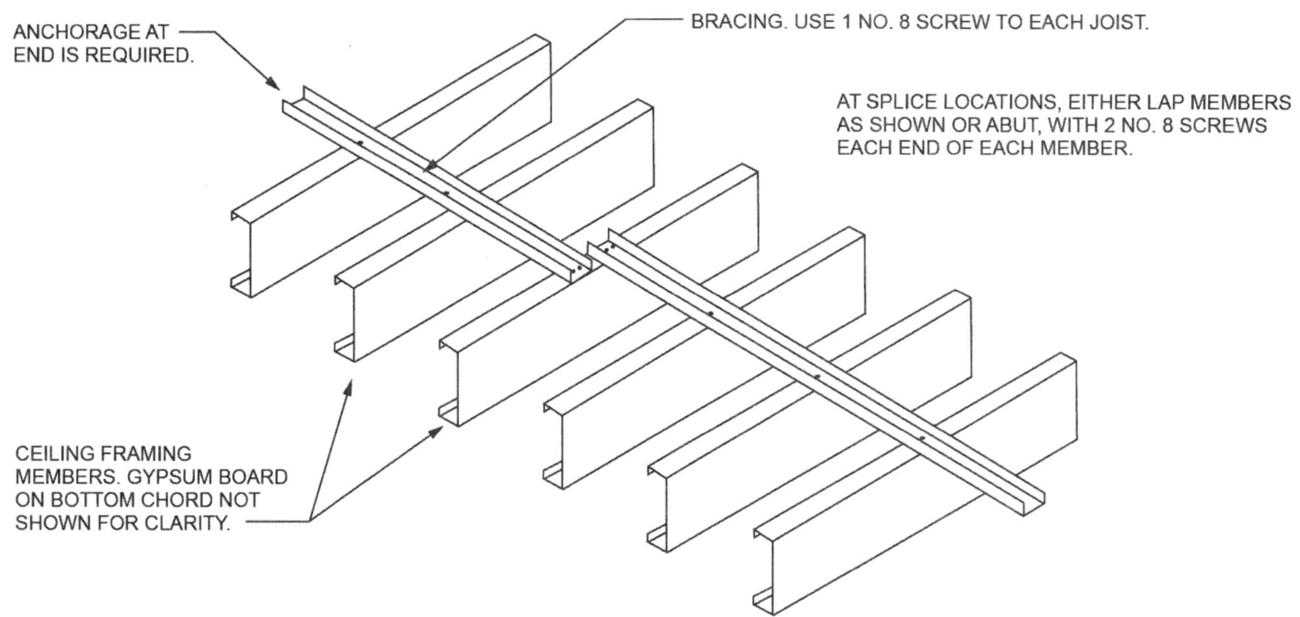

FIGURE R804.3.1.3(1)
CEILING JOIST TOP FLANGE BRACING WITH C-SHAPED, TRACK OR COLD-ROLLED CHANNEL

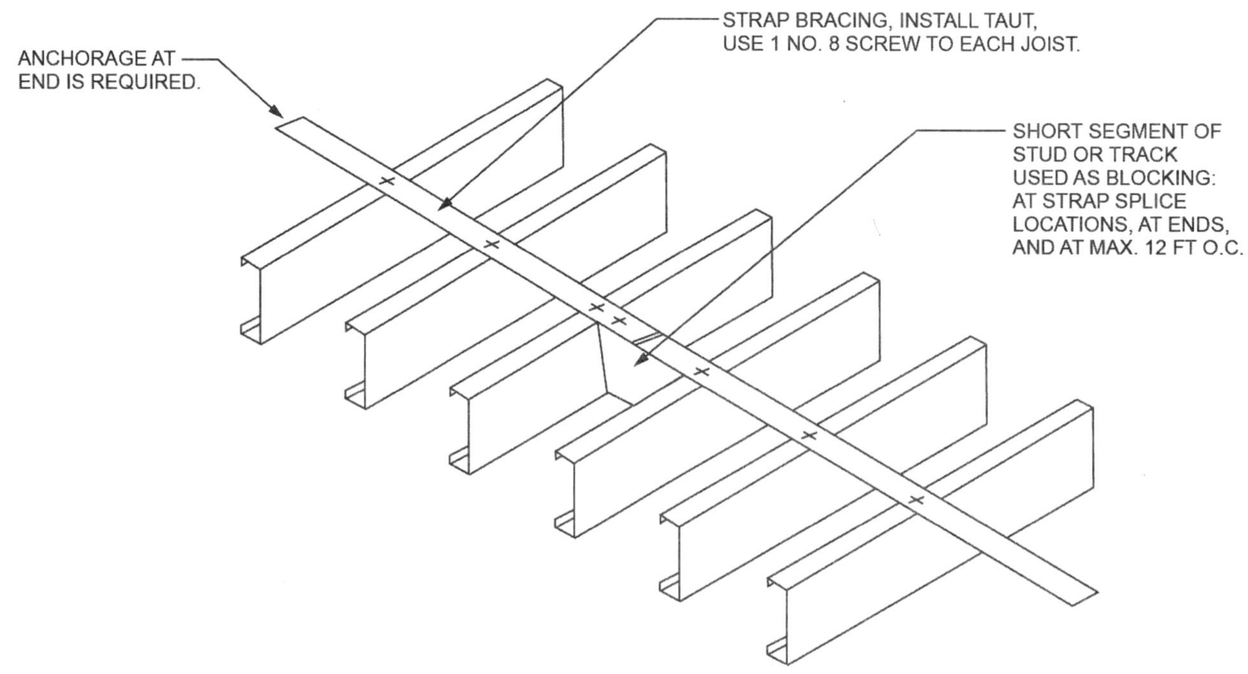

For SI: 1 foot = 304.8 mm.

FIGURE R804.3.1.3(2)
CEILING JOIST TOP FLANGE BRACING WITH CONTINUOUS STEEL STRAP AND BLOCKING

5. Minimum $1^1/_2$-inch by 33-mil (38 mm by 0.84 mm) continuous steel strap in accordance with Figure R804.3.1.3(2).

Lateral bracing shall be installed perpendicular to the ceiling joists and shall be fastened to the top flange of each joist with one No. 8 screw. Blocking shall be installed between joists in line with bracing at a maximum spacing of 12 feet (3658 mm) measured perpendicular to the joists. Ends of lateral bracing shall be attached to blocking or anchored to a stable building component with two No. 8 screws.

R804.3.1.4 Ceiling joist splicing. Splices in ceiling joists shall be permitted, if ceiling joist splices are supported at interior bearing points and are constructed in accordance with Figure R804.3.1.4. The number of screws on each side of the splice shall be the same as required for the heel joint connection in Table R804.3.1.1(3).

R804.3.2 Roof rafters. Cold-formed steel roof rafters shall be in accordance with this section.

R804.3.2.1 Minimum roof rafter sizes. Roof rafter size and thickness shall be determined in accordance with the limits set forth in Table R804.3.2.1(1) based on the horizontal projection of the roof rafter span. For determination of roof rafter sizes, reduction of roof spans shall be permitted where a roof rafter support brace is installed in accordance with Section R804.3.2.2. The reduced roof rafter span shall be taken as the larger of the distances from the roof rafter support brace to the ridge or to the heel measured horizontally.

For the purpose of determining roof rafter sizes in Table R804.3.2.1(1), ultimate design wind speeds shall be converted to equivalent ground snow loads in accordance with Table R804.3.2.1(2). Roof rafter sizes shall be based on the higher of the ground snow load or the equivalent snow load converted from the ultimate design wind speed.

R804.3.2.1.1 Eave overhang. Eave overhangs shall not exceed 24 inches (610 mm) measured horizontally.

R804.3.2.1.2 Rake overhangs. Rake overhangs shall not exceed the limitations provided for Option 1 or 2 in Figure R804.3.2.1.2. Outlookers at gable endwalls shall be installed in accordance with Figure R804.3.2.1.2. The required strength for uplift connectors required for Option 1 shall be determined in accordance with AISI S230, Table F3-4.

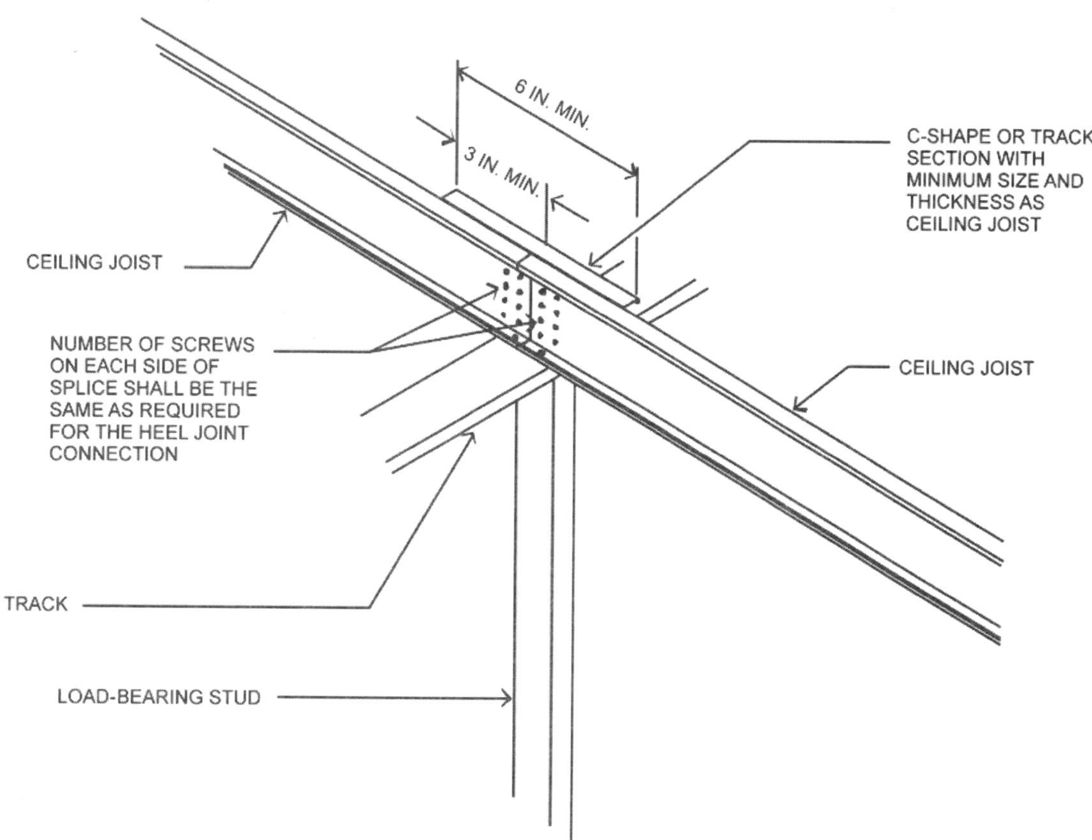

For SI: 1 foot = 304.8 mm.

FIGURE R804.3.1.4
SPLICED CEILING JOISTS

TABLE R804.3.2.1(1)
ROOF RAFTER SPANS[a, b, c, d]

MEMBER DESIGNATION	ALLOWABLE SPAN MEASURED HORIZONTALLY (feet-inches)							
	Ground snow load (psf)							
	20		30		50		70	
	Rafter spacing (inches)							
	16	24	16	24	16	24	16	24
550S162-33	13′-11″	11′-4″	11′-9″	9′-7″	9′-5″	7′-8″	8′-1″	6′-7″
550S162-43	15′-9″	13′-8″	14′-3″	11′-8″	11′-4″	9′-3″	9′-9″	7′-11″
550S162-54	16′-11″	14′-10″	15′-3″	13′-4″	13′-3″	11′-7″	12′-0″	10′-6″
550S162-68	18′-2″	15′-10″	16′-5″	14′-4″	14′-3″	12′-5″	12′-11″	11′-3″
800S162-33	16′-4″	13′-4″	13′-11″	11′-4″	11′-1″	9′-0″	9′-6″	6′-7″
800S162-43	19′-7″	16′-0″	16′-8″	13′-7″	13′-4″	10′-10″	11′-5″	9′-4″
800S162-54	22′-9″	19′-11″	20′-7″	17′-11″	17′-10″	4′-9″	15′-6″	12′-7″
800S162-68	24′-7″	21′-6″	22′-2″	19′-5″	19′-3″	16′-10″	17′-5″	14′-8″
1000S162-43	22′-2″	18′-1″	18′-10″	15′-4″	15′-1″	12′-4″	12′-11″	10′-7″
1000S162-54	27′-1″	23′-8″	24′-6″	20′-9″	20′-5″	16′-8″	17′-6″	14′-3″
1000S162-68	29′-5″	25′-8″	26′-6″	23′-2″	23′-0″	19′-6″	20′-6″	16′-9″
1200S162-54	31′-3″	27′-0″	28′-1″	22′-11″	22′-6″	18′-4″	19′-4″	15′-9″
1200S162-68	34′-0″	29′-8″	30′-8″	26′-9″	26′-6″	21′-7″	22′-8″	18′-6″

For SI: 1 inch = 25.4 mm, 1 foot = 304.8 mm, 1 pound per square foot = 0.0479 kPa.

a. Table provides maximum horizontal rafter spans in feet and inches for slopes between 3:12 and 12:12.

b. Deflection criteria: *L*/240 for live loads and *L*/180 for total loads.

c. Roof dead load = 12 psf.

d. Grade 33 ksi steel is permitted to be used for 33 mil and 43 mil thicknesses. Grade 50 ksi steel shall be used for 54 and 68 mil thicknesses.

TABLE R804.3.2.1(2)
ULTIMATE DESIGN WIND SPEED TO EQUIVALENT SNOW LOAD CONVERSION

ULTIMATE WIND SPEED AND EXPOSURE		EQUIVALENT GROUND SNOW LOAD (psf)									
		Roof slope									
Exposure	Wind speed (mph)	3:12	4:12	5:12	6:12	7:12	8:12	9:12	10:12	11:12	12:12
B	115	20	20	20	20	20	20	20	20	20	20
	120	20	20	20	20	20	20	20	20	20	20
	130	20	20	20	20	20	20	20	20	20	20
	< 140	20	20	20	20	20	20	20	30	30	30
C	115	20	20	20	20	20	20	20	20	30	30
	120	20	20	20	20	20	20	20	30	30	50
	130	20	20	30	30	30	30	30	30	50	50
	< 140	30	30	50	50	30	30	50	50	50	50

For SI: 1 mile per hour = 0.447 m/s, 1 pound per square foot = 0.0479 kPa.

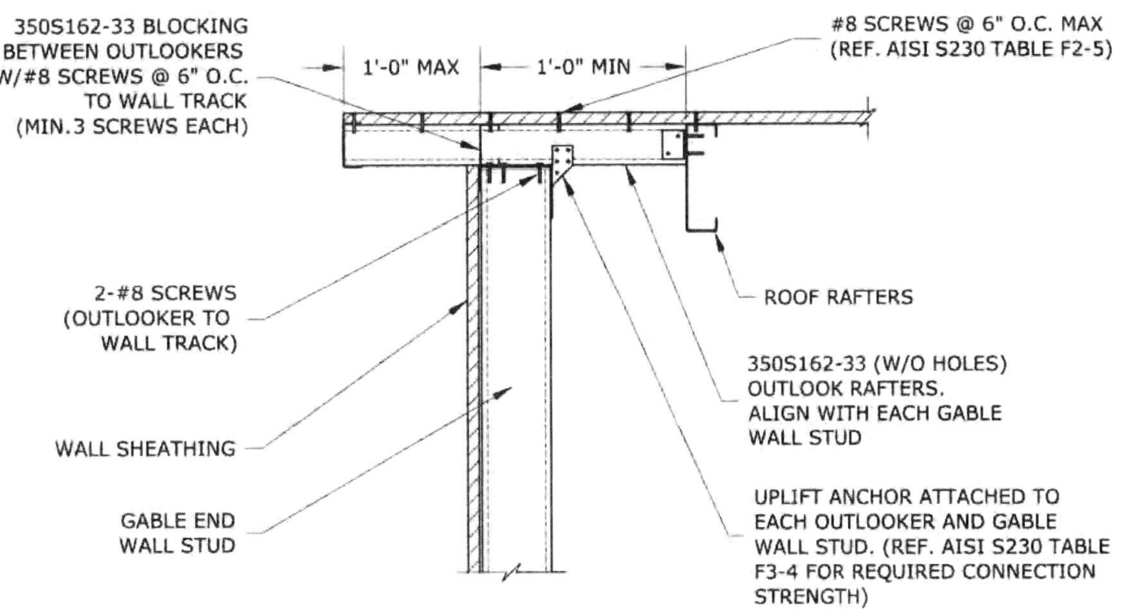

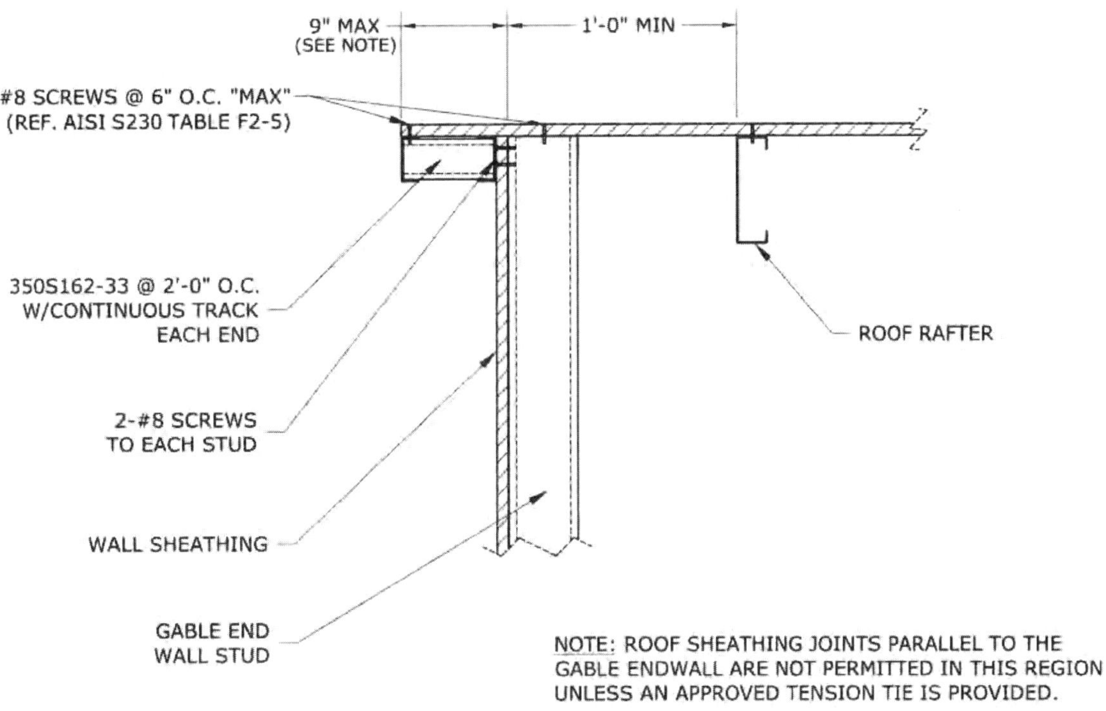

For SI: 1 inch = 25.4 mm, 1 foot = 304.8 mm, 1 mile per hour = 1.6 kph.

FIGURE R804.3.2.1.2
GABLE ENDWALL OVERHANG DETAILS

R804.3.2.2 Roof rafter support brace. Where used to reduce roof rafter spans in determining roof rafter sizes, a roof rafter support brace shall meet all of the following conditions:

1. Minimum 350S162-33 C-shaped brace member with maximum length of 8 feet (2438 mm).
2. Minimum brace member slope of 45 degrees (0.785 rad) to the horizontal.
3. Minimum connection of brace to a roof rafter and ceiling joist with four No.10 screws at each end.
4. Maximum 6 inches (152 mm) between brace/ceiling joist connection and *load-bearing wall* below.
5. Each roof rafter support brace greater than 4 feet (1219 mm) in length, shall be braced with a supplemental brace having a minimum size of 350S162-33 or 350T162-33 such that the maximum unsupported length of the roof rafter support brace is 4 feet (1219 mm). The supplemental brace shall be continuous and shall be connected to each roof rafter support brace using two No. 8 screws.

R804.3.2.3 Roof rafter splice. Roof rafters shall not be spliced.

R804.3.2.4 Roof rafter to ceiling joist and ridge member connection. Roof rafters shall be connected to a parallel ceiling joist to form a continuous tie between exterior walls in accordance with Figure R804.3.1.1 and Table R804.3.1.1(3). Ceiling joists shall be connected to the top track of the *load-bearing wall* in accordance with Table R804.3, either with the required number of No. 10 screws applied through the flange of the ceiling joist or by using a 54-mil (1.37 mm) clip angle with the required number of No.10 screws in each leg. Roof rafters shall be connected to a ridge member with a minimum 2-inch by 2-inch (51 mm by 51 mm) clip angle fastened with No. 10 screws to the ridge member in accordance with Figure R804.3.2.4 and Table R804.3.2.4. The clip angle shall have a steel thickness equivalent to or greater than the roof rafter thickness and shall extend the depth of the roof rafter member to the extent possible. The ridge member shall be fabricated from a C-shaped member and a track section that shall have a minimum size and steel thickness equivalent to or greater than that of adjacent roof rafters and shall be installed in accordance with Figure R804.3.2.4. The ridge member shall extend the full depth of the sloped roof rafter cut.

R804.3.2.5 Roof rafter bottom flange bracing. The bottom flanges of roof rafters shall be continuously braced, at a maximum spacing of 4 feet (1219 mm) as measured parallel to the roof rafters, with one of the following members:

1. Minimum 33-mil (0.84 mm) C-shaped member.
2. Minimum 33-mil (0.84 mm) track section.
3. Minimum $1^{1}/_{2}$-inch by 33-mil (38 mm by 0.84 mm) steel strap.

The bracing element shall be fastened to the bottom flange of each roof rafter with one No. 8 screw and shall be fastened to blocking with two No. 8 screws.

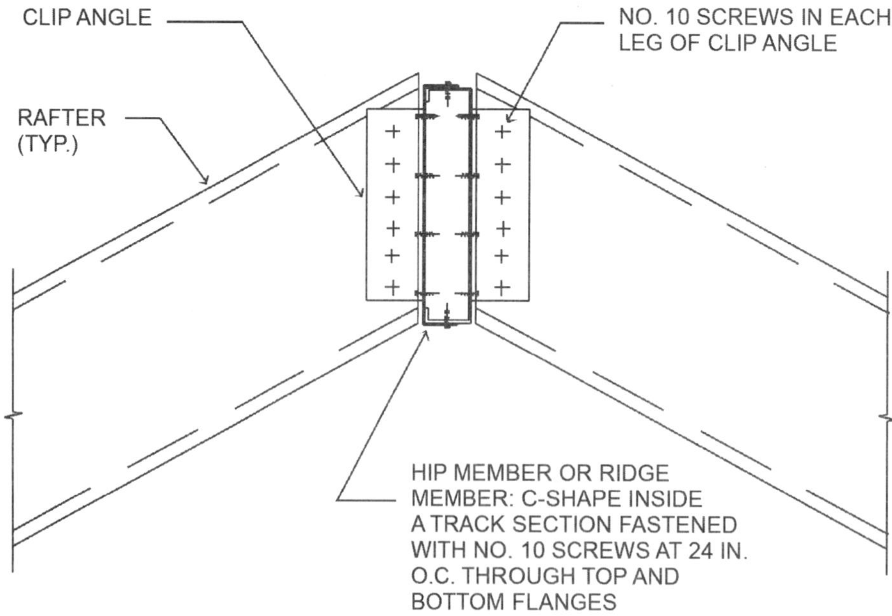

For SI: 1 inch = 25.4 mm.

FIGURE R804.3.2.4
RIDGE MEMBER CONNECTION

Blocking shall be installed between roof rafters in-line with the continuous bracing at a maximum spacing of 12 feet (3658 mm) measured perpendicular to the roof rafters. The ends of continuous bracing shall be fastened to blocking or anchored to a stable building component with two No. 8 screws.

R804.3.3 Cutting and notching. Flanges and lips of load-bearing, cold-formed steel roof framing members shall not be cut or notched.

R804.3.4 Headers. Roof-ceiling framing above wall openings shall be supported on headers. The allowable spans for headers in *load-bearing walls* shall not exceed the values set forth in Section R603.6 and Tables R603.6(1) through R603.6(6).

R804.3.5 Framing of openings in roofs and ceilings. Openings in roofs and ceilings shall be framed with header and trimmer joists. Header joist spans shall not exceed 4 feet (1219 mm) in length. Header and trimmer joists shall be fabricated from joist and track members having a minimum size and thickness equivalent to the adjacent ceiling joists or roof rafters and shall be installed in accordance with Figures R804.3.5(1) and R804.3.5(2). Each header joist shall be connected to trimmer joists with not less than four 2-inch by 2-inch (51 by 51 mm) clip angles. Each clip angle shall be fastened to both the header and trimmer joists with four No. 8 screws, evenly spaced, through each leg of the clip angle. The steel thickness of the clip angles shall be not less than that of the ceiling joist or roof rafter. Each track section for a built-up header or trimmer joist shall extend the full length of the joist (continuous).

R804.3.6 Roof trusses. Cold-formed steel trusses shall be designed and installed in accordance with AISI S230, Section F6. In the absence of specific bracing requirements, trusses shall be braced in accordance with accepted industry practices, such as the SBCA *Cold-Formed Steel Building Component Safety Information (CFSBCSI) Guide to Good Practice for Handling, Installing & Bracing of Cold-Formed Steel Trusses.* Trusses shall be connected to the top track of the *load-bearing wall* in accordance with Table R804.3, either with the required number of No. 10 screws applied through the flange of the truss or by using a 54-mil (1.37 mm) clip angle with the required number of No. 10 screws in each leg.

R804.3.7 Ceiling and roof diaphragms. Ceiling and roof diaphragms shall be in accordance with this section.

R804.3.7.1 Ceiling diaphragms. At gable endwalls a ceiling *diaphragm* shall be provided by attaching a minimum $^1/_2$-inch (12.7 mm) gypsum board or a minimum $^3/_8$-inch (9.5 mm) wood structural panel sheathing, that complies with Section R803, to the bottom of ceiling joists or roof trusses and connected to wall framing in accordance with Figures R804.3.7.1(1) and R804.3.7.1(2), unless studs are designed as full height without bracing at the ceiling. Flat blocking shall consist of C-shaped or track section with a minimum thickness of 33 mils (0.84 mm). For a gypsum board sheathed ceiling, the *diaphragm* length shall be in accordance with Table R804.3.7.1. For a wood structural panel sheathed ceiling, the *diaphragm* length shall be not less than 12 feet (3658 mm) for building widths less than 36 feet (10 973 mm), or not less than 14 feet (4267 mm) for building widths greater than or equal to 36 feet (10 973 mm).

The ceiling *diaphragm* shall be secured with screws spaced at a maximum 6 inches (152 mm) o.c. at panel edges and a maximum 12 inches (305 mm) o.c. in the field. The required lengths in Table R804.3.7.1 for gypsum board sheathed ceiling diaphragms shall be permitted to be multiplied by 0.35 if all panel edges are blocked. Multiplying the required lengths in Table R804.3.7.1 for gypsum board sheathed ceiling diaphragms by 0.9 shall be permitted if all panel edges are secured with screws spaced at 4 inches (102 mm) o.c.

R804.3.7.2 Roof diaphragm. A roof *diaphragm* shall be provided by attaching not less than $^3/_8$-inch (9.5 mm) *wood structural panel* that complies with Section R803 to roof rafters or truss top chords in accordance with Table R804.3. Buildings with 3:1 or larger plan *aspect ratio* and with roof rafter slope (pitch) of 9:12 or larger shall have the roof rafters and ceiling joists blocked in accordance with Figure R804.3.7.2.

R804.3.8 Roof tie-down. *Roof assemblies* shall be connected to walls below in accordance with Table R804.3. A continuous load path shall be provided to transfer uplift loads to the foundation.

TABLE R804.3.2.4
SCREWS REQUIRED AT EACH LEG OF CLIP ANGLE FOR ROOF RAFTER TO RIDGE MEMBER CONNECTION[a]

BUILDING WIDTH (feet)	NUMBER OF SCREWS			
	Ground snow load (psf)			
	0 to 20	21 to 30	31 to 50	51 to 70
24	2	2	3	4
28	2	3	4	5
32	2	3	4	5
36	3	3	5	6
40	3	4	5	7

For SI: 1 foot = 304.8 mm, 1 pound per square foot = 0.0479 kPa.

a. Screws shall be No. 10 minimum.

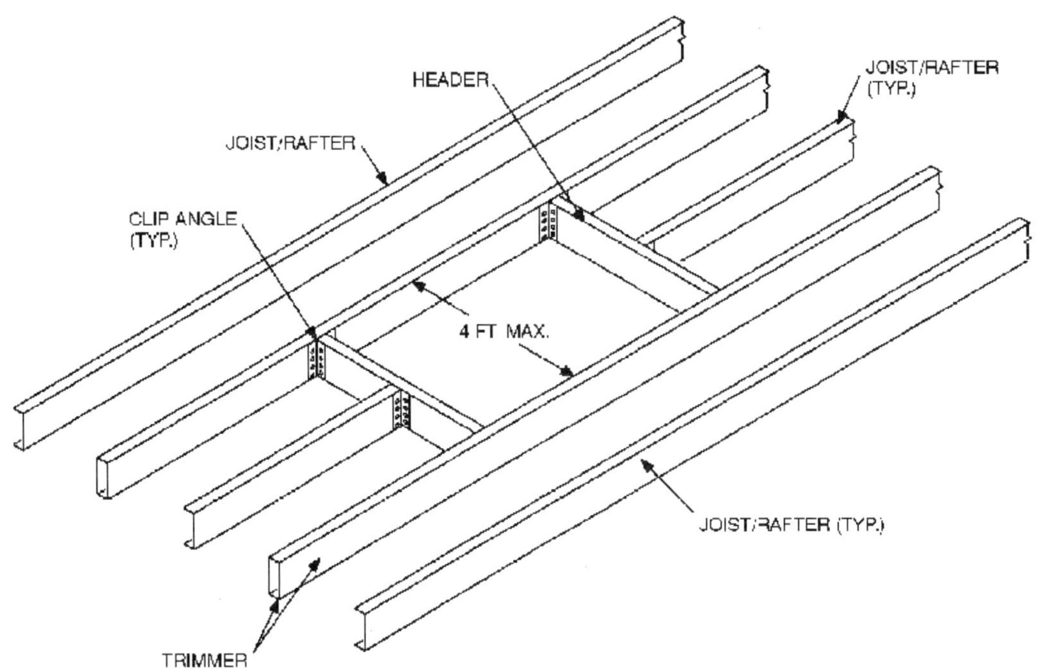

For SI: 1 foot = 304.8 mm.

FIGURE R804.3.5(1)
ROOF OR CEILING OPENING

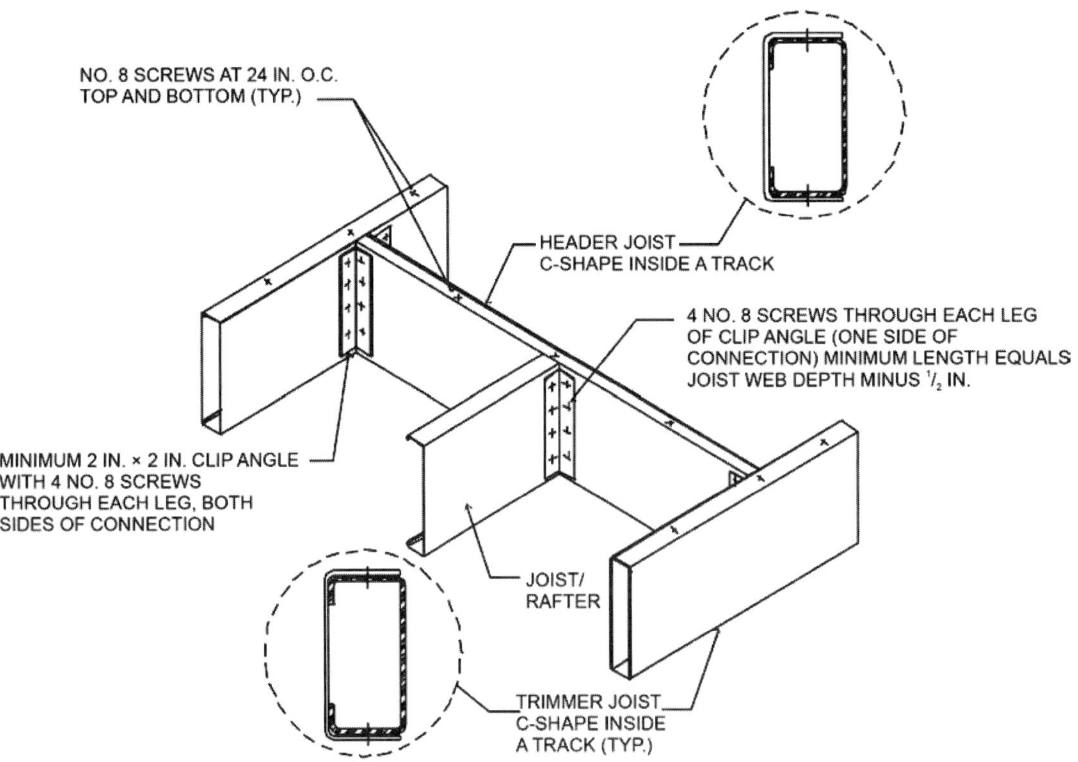

For SI: 1 inch = 25.4 mm.

FIGURE R804.3.5(2)
HEADER TO TRIMMER CONNECTION

**TABLE R804.3.7.1
REQUIRED LENGTHS FOR CEILING DIAPHRAGMS AT GABLE
ENDWALLS GYPSUM BOARD SHEATHED, CEILING HEIGHT = 8 FEET[a, b, c, d, e, f, g]**

EXPOSURE CATEGORY		ULTIMATE DESIGN WIND SPEED (mph)					
B		115	120	130	< 140	—	—
C		—	—	115	120	130	< 140
Roof pitch	Building endwall width (feet)	Minimum diaphragm length (feet)					
3:12 to 6:12	24–28	16	18	24	26	30	34
	> 28–32	20	20	26	32	34	40
	> 32–36	24	26	30	36	42	46
	> 36–40	26	28	36	40	48	52
6:12 to 9:12	> 24–28	20	20	26	30	34	38
	> 28–32	24	26	30	36	42	46
	> 32–36	26	30	38	42	48	54
	> 36–40	30	34	40	50	56	62
9:12 to 12:12	> 24–28	22	24	30	34	38	44
	> 28–32	26	28	36	40	46	52
	> 32–36	30	32	40	48	54	62
	> 36–40	36	38	48	56	64	72

For SI: 1 inch = 25.4 mm, 1 mile per hour = 0.447 m/s, 1 foot = 304.8 mm, 1 mil = 0.0254 mm.

a. Ceiling diaphragm is composed of $^{1}/_{2}$-inch gypsum board (min. thickness) secured with screws spaced at 6 inches o.c. at panel edges and 12 inches o.c. infield. Use No. 8 screws (min.) where framing members have a designation thickness of 54 mils or less and No. 10 screws (min.) where framing members have a designation thickness greater than 54 mils.

b. Maximum aspect ratio (length/width) of diaphragms is 2:1.

c. Building width is in the direction of horizontal framing members supported by the wall studs.

d. Required diaphragm lengths are to be provided at each end of the structure.

e. Multiplying required diaphragm lengths by 0.35 is permitted if all panel edges are blocked.

f. Multiplying required diaphragm lengths by 0.9 is permitted if all panel edges are secured with screws spaced at 4 inches o.c.

g. To determine the minimum diaphragm length for buildings with ceiling heights of 9 feet or 10 feet values in this table shall be multiplied by 1.15.

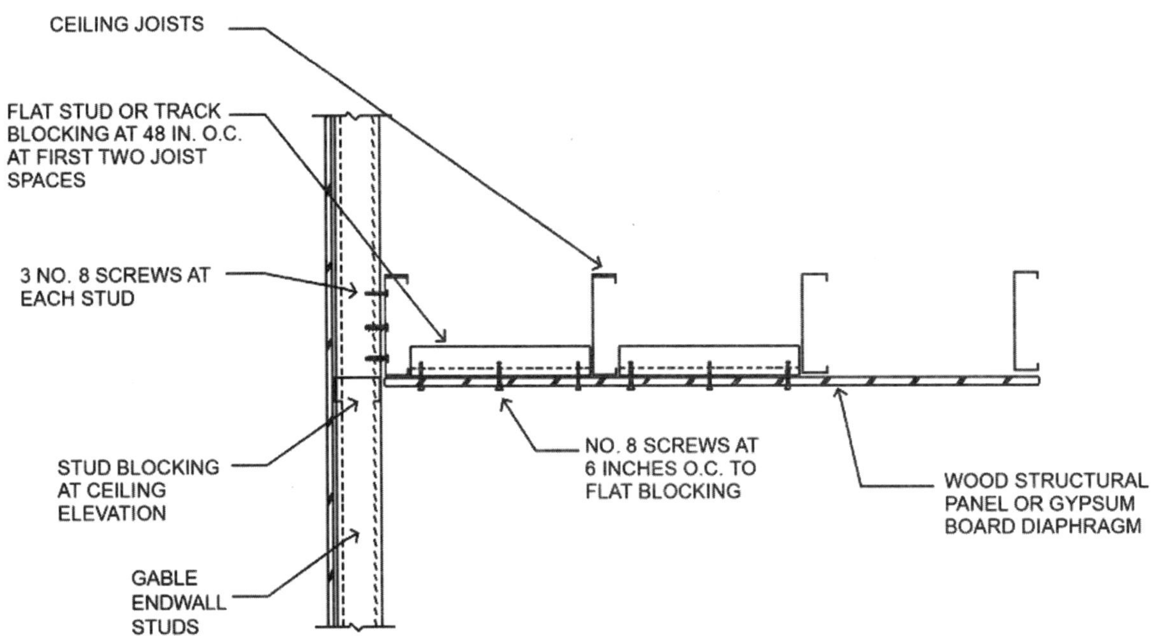

For SI: 1 inch = 25.4 mm.

FIGURE R804.3.7.1(1)
CEILING DIAPHRAGM TO GABLE ENDWALL DETAIL

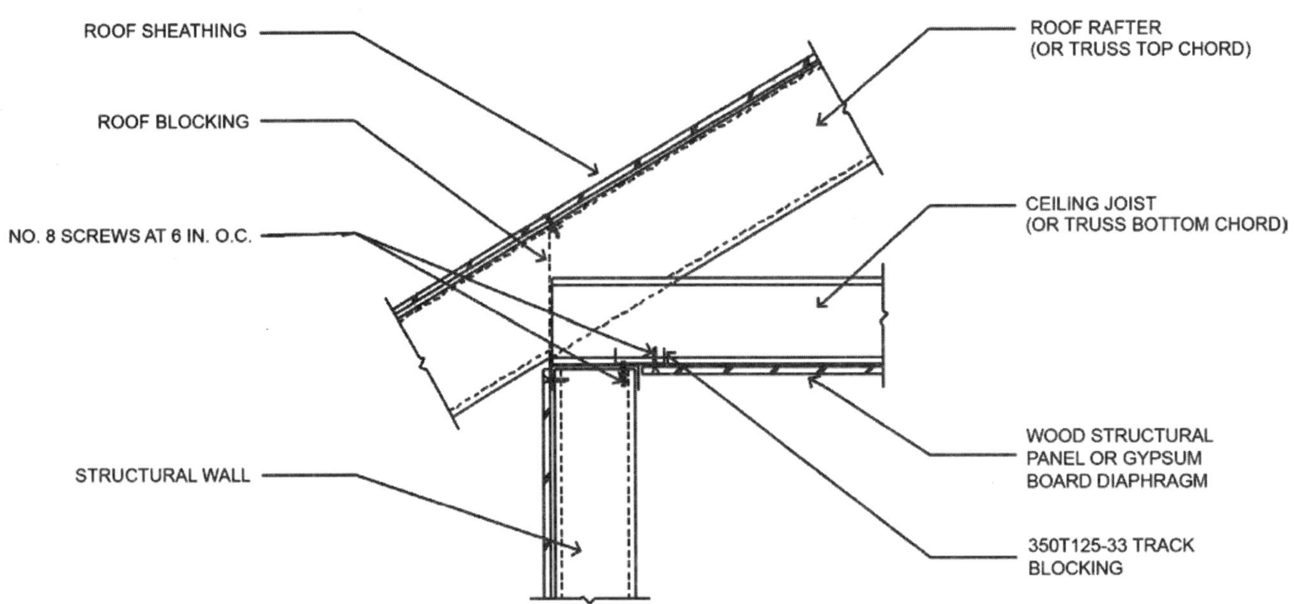

For SI: 1 inch = 25.4 mm.

FIGURE R804.3.7.1(2)
CEILING DIAPHRAGM TO SIDEWALL DETAIL

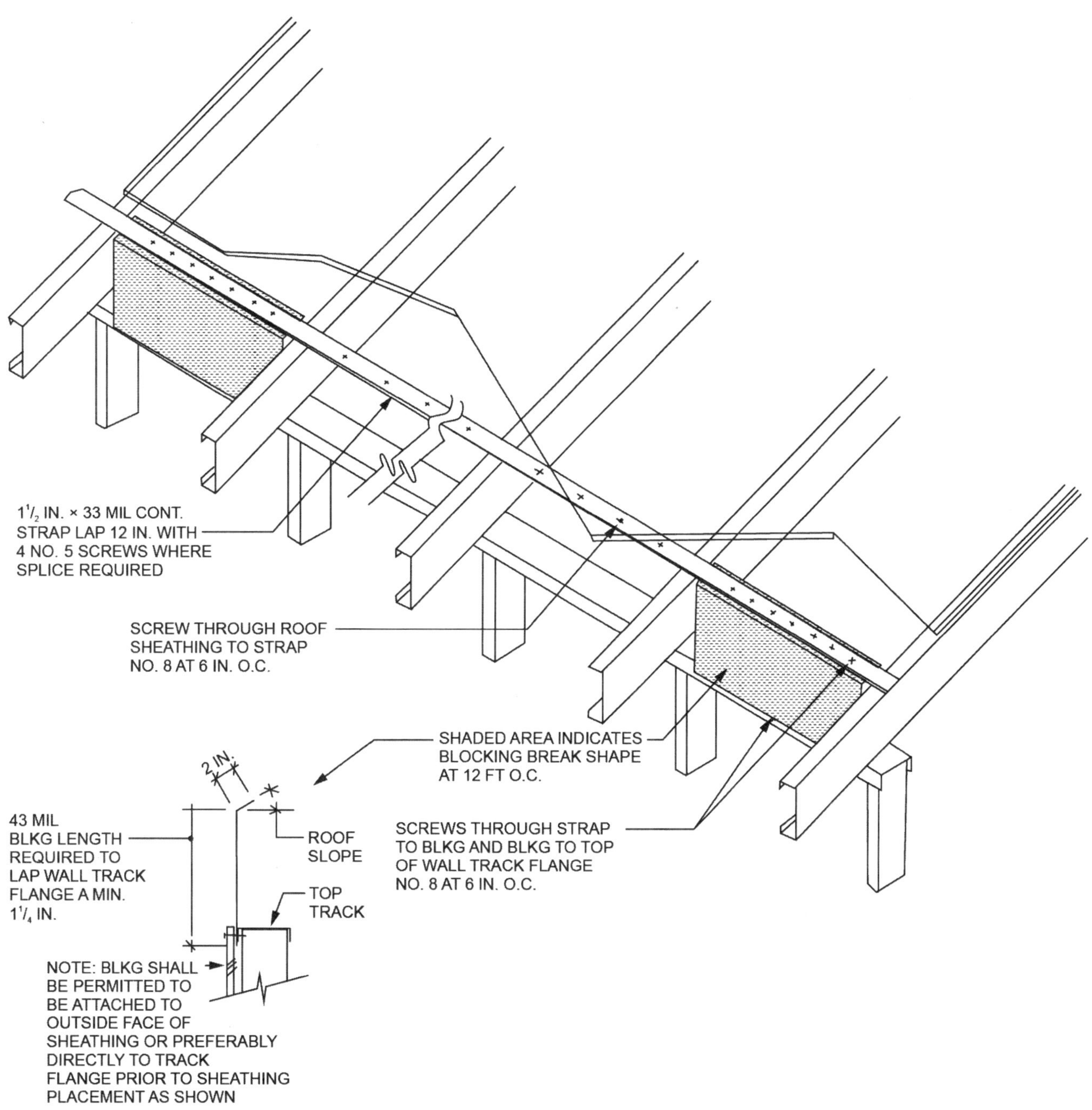

For SI: 1 mil = 0.0254 mm, 1 inch = 25.4 mm, 1 foot = 304.8 mm.

FIGURE R804.3.7.2
ROOF BLOCKING DETAIL

SECTION R805 CEILING FINISHES

R805.1 Ceiling installation. Ceilings shall be installed in accordance with the requirements for interior wall finishes as provided in Sections R702.1 through R702.6.

SECTION R806 ROOF VENTILATION

R806.1 Ventilation required. Enclosed attics and enclosed rafter spaces formed where ceilings are applied directly to the underside of roof rafters shall have cross *ventilation* for each separate space by ventilating openings protected against the entrance of rain or snow. Ventilation openings shall have a least dimension of $^1/_{16}$ inch (1.6 mm) minimum and $^1/_4$ inch (6.4 mm) maximum. Ventilation openings having a least dimension larger than $^1/_4$ inch (6.4 mm) shall be provided with corrosion-resistant wire cloth screening, hardware cloth, perforated vinyl or similar material with openings having a least dimension of $^1/_{16}$ inch (1.6 mm) minimum and $^1/_4$ inch (6.4 mm) maximum. Openings in roof framing members shall conform to the requirements of Section R802.7. Required ventilation openings shall open directly to the outside air and shall be protected to prevent the entry of birds, rodents, snakes and other similar creatures.

R806.2 Minimum vent area. The minimum net free ventilating area shall be $^1/_{150}$ of the area of the vented space.

Exception: The minimum net free ventilation area shall be $^1/_{300}$ of the vented space provided both of the following conditions are met:

1. In Climate Zones 6, 7 and 8, a Class I or II vapor retarder is installed on the warm-in-winter side of the ceiling.
2. Not less than 40 percent and not more than 50 percent of the required ventilating area is provided by ventilators located in the upper portion of the attic or rafter space. Upper ventilators shall be located not more than 3 feet (914 mm) below the ridge or highest point of the space, measured vertically. The balance of the required *ventilation* provided shall be located in the bottom one-third of the attic space. Where the location of wall or roof framing members conflicts with the installation of upper ventilators, installation more than 3 feet (914 mm) below the ridge or highest point of the space shall be permitted.

R806.3 Vent and insulation clearance. Where eave or cornice vents are installed, blocking, bridging and insulation shall not block the free flow of air. Not less than a 1-inch (25 mm) space shall be provided between the insulation and the roof sheathing and at the location of the vent.

R806.4 Installation and weather protection. Ventilators shall be installed in accordance with manufacturer's instructions. Installation of ventilators in roof systems shall be in accordance with the requirements of Section R903. Installation of ventilators in wall systems shall be in accordance with the requirements of Section R703.1.

R806.5 Unvented attic and unvented enclosed rafter assemblies. Unvented *attics* and unvented enclosed roof framing assemblies created by ceilings that are applied directly to the underside of the roof framing members and structural roof sheathing applied directly to the top of the roof framing members/rafters, shall be permitted where all the following conditions are met:

1. The unvented *attic* space is completely within the *building thermal envelope.*
2. Interior Class I vapor retarders are not installed on the ceiling side (*attic* floor) of the unvented *attic* assembly or on the ceiling side of the unvented enclosed roof framing assembly.
3. Where wood shingles or shakes are used, a minimum $^1/_4$-inch (6.4 mm) vented airspace separates the shingles or shakes and the roofing underlayment above the structural sheathing.
4. In Climate Zones 5, 6, 7 and 8, any *air-impermeable insulation* shall be a Class II vapor retarder, or shall have a Class II vapor retarder coating or covering in direct contact with the underside of the insulation.
5. Insulation shall comply with Item 5.3 and either Item 5.1 or 5.2:
    - 5.1. Item 5.1.1, 5.1.2, 5.1.3 or 5.1.4 shall be met, depending on the air permeability of the insulation directly under the structural roof sheathing.
        - 5.1.1. Where only *air-impermeable insulation* is provided, it shall be applied in direct contact with the underside of the structural roof sheathing.
        - 5.1.2. Where *air-permeable insulation* is installed directly below the structural sheathing, rigid board or sheet insulation shall be installed directly above the structural roof sheathing in accordance with the *R*-values in Table R806.5 for condensation control.
        - 5.1.3. Where both *air-impermeable* and *air-permeable insulation* are provided, the *air-impermeable insulation* shall be applied in direct contact with the underside of the structural roof sheathing in accordance with Item 5.1.1 and shall be in accordance with the *R*-values in Table R806.5 for condensation control. The *air-permeable insulation* shall be installed directly under the *air-impermeable insulation.*

5.1.4. Alternatively, sufficient rigid board or sheet insulation shall be installed directly above the structural roof sheathing to maintain the monthly average temperature of the underside of the structural roof sheathing above 45°F (7°C). For calculation purposes, an interior air temperature of 68°F (20°C) is assumed and the exterior air temperature is assumed to be the monthly average outside air temperature of the three coldest months.

5.2. In Climate Zones 1, 2 and 3, air-permeable insulation installed in unvented *attics* shall meet the following requirements:

5.2.1. An approved *vapor diffusion port* shall be installed not more than 12 inches (305 mm) from the highest point of the roof, measured vertically from the highest point of the roof to the lower edge of the port.

5.2.2. The port area shall be greater than or equal to 1:600 of the ceiling area. Where there are multiple ports in the attic, the sum of the port areas shall be greater than or equal to the area requirement.

5.2.3. The vapor-permeable membrane in the *vapor diffusion port* shall have a vapor permeance rating of greater than or equal to 20 perms when tested in accordance with Procedure A of ASTM E96.

5.2.4. The *vapor diffusion port* shall serve as an air barrier between the *attic* and the exterior of the building.

5.2.5. The *vapor diffusion port* shall protect the *attic* against the entrance of rain and snow.

5.2.6. Framing members and blocking shall not block the free flow of water vapor to the port. Not less than a 2-inch (51 mm) space shall be provided between any blocking and the roof sheathing. Air-permeable insulation shall be permitted within that space.

5.2.7. The roof slope shall be greater than or equal to 3:12 (vertical/horizontal).

5.2.8. Where only air-permeable insulation is used, it shall be installed directly below the structural roof sheathing, on top of the attic floor, or on top of the ceiling.

5.2.9. *Air-impermeable insulation*, where used in conjunction with air-permeable insulation, shall be directly above or below the structural roof sheathing and is not required to meet the *R*-value in Table R806.5. Where directly below the structural roof sheathing, there shall be no space between the *air-impermeable insulation* and air-permeable insulation.

5.2.10. Where air-permeable insulation is used and is installed directly below the roof structural sheathing, air shall be supplied at a flow rate greater than or equal to 50 CFM (23.6 L/s) per 1,000 square feet (93 m^2) of ceiling. The air shall be supplied from ductwork providing supply air to the occupiable space when the conditioning system is operating. Alternatively, the air shall be supplied by a supply fan when the conditioning system is operating.

Exceptions:

1. Where both air-impermeable and air-permeable insulation are used, and the *R*-value in Table 806.5 is met, air supply to the attic is not required.
2. Where only air-permeable insulation is used and is installed on top of the attic floor, or on top of the ceiling, air supply to the attic is not required.

5.3. Where preformed insulation board is used as the air-impermeable insulation layer, it shall be sealed at the perimeter of each individual sheet interior surface to form a continuous layer.

TABLE R806.5
INSULATION FOR CONDENSATION CONTROL

CLIMATE ZONE	MINIMUM RIGID BOARD ON AIR-IMPERMEABLE INSULATION *R*-VALUE[a, b]
2B and 3B tile roof only	0 (none required)
1, 2A, 2B, 3A, 3B, 3C	R-5
4C	R-10
4A, 4B	R-15
5	R-20
6	R-25
7	R-30
8	R-35

a. Contributes to but does not supersede the requirements in Section N1102.

b. Alternatively, sufficient continuous insulation shall be installed directly above the structural roof sheathing to maintain the monthly average temperature of the underside of the structural roof sheathing above 45°F (7°C). For calculation purposes, an interior air temperature of 68°F (20°C) is assumed and the exterior air temperature is assumed to be the monthly average outside air temperature of the three coldest months.

SECTION R807
ATTIC ACCESS

R807.1 Attic access. Buildings with combustible ceiling or roof construction shall have an attic access opening to attic areas that have a vertical height of 30 inches (762 mm) or greater over an area of not less than 30 square feet (2.8 m^2). The vertical height shall be measured from the top of the ceiling framing members to the underside of the roof framing members.

The rough-framed opening shall be not less than 22 inches by 30 inches (559 mm by 762 mm) and shall be located in a hallway or other location with *ready access.* Where located in a wall, the opening shall be not less than 22 inches wide by 30 inches high (559 mm wide by 762 mm high). Where the access is located in a ceiling, minimum unobstructed headroom in the attic space shall be 30 inches (762 mm) at some point above the access measured vertically from the bottom of ceiling framing members. See Section M1305.1.2 for access requirements where mechanical *equipment* is located in *attics.*

CHAPTER 9
ROOF ASSEMBLIES

User note:

About this chapter: *Chapter 9 addresses the design and construction of roof assemblies. A roof assembly includes the roof deck, substrate or thermal barrier, insulation, vapor retarder and roof covering. This chapter provides the requirement for wind resistance of roof coverings. The types of roof covering materials and installation addressed by Chapter 9 are: asphalt shingles, clay and concrete tile, metal roof shingles, mineral-surfaced roll roofing, slate and slate-type shingles, wood shakes and shingles, built-up roofs, metal roof panels, modified bitumen roofing, thermoset and thermoplastic single-ply roofing, sprayed polyurethane foam roofing, liquid applied coatings and photovoltaic shingles. Chapter 9 also provides requirements for roof drainage, flashing, above-deck thermal insulation, rooftop-mounted photovoltaic systems and recovering or replacing an existing roof covering.*

SECTION R901
GENERAL

R901.1 Scope. The provisions of this chapter shall govern the design, materials, construction and quality of *roof assemblies.*

SECTION R902
FIRE CLASSIFICATION

R902.1 Roof covering materials. Roofs shall be covered with materials as set forth in Sections R904 and R905. Class A, B or C roofing shall be installed in *jurisdictions* designated by law as requiring their use or where the edge of the roof is less than 3 feet (914 mm) from a *lot line.* Class A, B and C roofing required by this section to be *listed* shall be tested in accordance with ASTM E108 or UL 790.

Exceptions:

1. Class A *roof assemblies* include those with coverings of brick, masonry and exposed concrete *roof deck.*
2. Class A *roof assemblies* include ferrous or copper shingles or sheets, metal sheets and shingles, clay or concrete roof tile, or slate installed on noncombustible decks.
3. Class A *roof assemblies* include minimum 16 ounces per square foot (4.882 kg/m^2) copper sheets installed over combustible decks.
4. Class A *roof assemblies* include slate installed over *underlayment* over combustible decks.

R902.2 Fire-retardant-treated shingles and shakes. Fire-retardant-treated wood shakes and shingles shall be treated by impregnation with chemicals by the full-cell vacuum-pressure process, in accordance with AWPA C1. Each bundle shall be marked to identify the manufactured unit and the manufacturer, and shall be *labeled* to identify the classification of the material in accordance with the testing required in Section R902.1, the treating company and the quality control agency.

R902.3 Building-integrated photovoltaic product. *Building-integrated photovoltaic* (BIPV) *products* installed as the roof covering shall be tested, *listed* and *labeled* for fire classification in accordance with UL 7103. Class A, B or C BIPV products shall be installed where the edge of the roof is less than 3 feet (914 mm) from a *lot line.*

R902.4 Rooftop-mounted photovoltaic panel systems. Rooftop-mounted *photovoltaic panel systems* installed on or above the roof covering shall be tested, *listed* and identified with a fire classification in accordance with UL 2703. Class A, B or C *photovoltaic panel systems* and modules shall be installed in *jurisdictions* designated by law as requiring their use or where the edge of the roof is less than 3 feet (914 mm) from a *lot line.*

SECTION R903
WEATHER PROTECTION

R903.1 General. *Roof decks* shall be covered with *approved* roof coverings secured to the building or structure in accordance with the provisions of this chapter. *Roof assemblies* shall be designed and installed in accordance with this code and the *approved* manufacturer's instructions such that the *roof assembly* shall serve to protect the building or structure.

R903.2 Flashing. Flashings shall be installed in a manner that prevents moisture from entering the wall and roof through joints in copings, through moisture permeable materials and at intersections with parapet walls and other penetrations through the roof plane.

R903.2.1 Locations. Flashings shall be installed at wall and roof intersections, wherever there is a change in roof slope or direction and around roof openings. A flashing shall be installed to divert the water away from where the eave of a sloped roof intersects a vertical sidewall. Where flashing is of metal, the metal shall be corrosion resistant with a thickness of not less than 0.019 inch (0.5 mm) (No. 26 galvanized sheet).

R903.2.2 Crickets and saddles. A cricket or saddle shall be installed on the ridge side of any chimney or penetration more than 30 inches (762 mm) wide as measured perpendicular to the slope. Cricket or saddle coverings shall be sheet metal or of the same material as the roof covering.

Exception: *Unit skylights* installed in accordance with Section R308.6 and flashed in accordance with the manufacturer's instructions shall be permitted to be installed without a cricket or saddle.

R903.3 Coping. Parapet walls shall be properly coped with noncombustible, weatherproof materials of a width not less than the thickness of the parapet wall.

R903.4 Roof drainage. Unless roofs are sloped to drain over roof edges, roof drains shall be installed at each low point of the roof.

R903.4.1 Secondary (emergency overflow) drains or scuppers. Where roof drains are required, secondary emergency overflow roof drains or *scuppers* shall be provided where the roof perimeter construction extends above the roof in such a manner that water will be entrapped if the primary drains allow buildup for any reason. Overflow drains having the same size as the roof drains shall be installed with the inlet flow line located 2 inches (51 mm) above the low point of the roof, or overflow *scuppers* having three times the size of the roof drains and having a minimum opening height of 4 inches (102 mm) shall be installed in the adjacent parapet walls with the inlet flow located 2 inches (51 mm) above the low point of the roof served. The installation and sizing of overflow drains, leaders and conductors shall comply with Sections 1106 and 1108 of the *International Plumbing Code*, as applicable.

Overflow drains shall discharge to an *approved* location and shall not be connected to roof drain lines.

SECTION R904 MATERIALS

R904.1 Scope. The requirements set forth in this section shall apply to the application of roof covering materials specified herein. *Roof assemblies* shall be applied in accordance with this chapter and the manufacturer's installation instructions. Installation of *roof assemblies* shall comply with the applicable provisions of Section R905.

R904.2 Compatibility of materials. *Roof assemblies* shall be of materials that are compatible with each other and with the building or structure to which the materials are applied.

R904.3 Material specifications and physical characteristics. Roof covering materials shall conform to the applicable standards listed in this chapter.

R904.4 Product identification. Roof covering materials shall be delivered in packages bearing the manufacturer's identifying marks and *approved* testing agency *labels* required. Bulk shipments of materials shall be accompanied by the same information issued in the form of a certificate or on a bill of lading by the manufacturer.

SECTION R905 REQUIREMENTS FOR ROOF COVERINGS

R905.1 Roof covering application. Roof coverings shall be applied in accordance with the applicable provisions of this section and the manufacturer's installation instructions. Unless otherwise specified in this section, roof coverings shall be installed to resist the component and cladding loads specified in Table R301.2.1(1), adjusted for height and exposure in accordance with Table R301.2.1(2).

R905.1.1 Underlayment. *Underlayment* for asphalt shingles, clay and concrete tile, *metal roof shingles*, mineral-surfaced roll roofing, slate and slate-type shingles, wood shingles, wood shakes, *metal roof panels* and *photovoltaic shingles* shall conform to the applicable standards listed in this chapter. *Underlayment* materials required to comply with ASTM D226, D1970, D4869 and D6757 shall bear a *label* indicating compliance to the standard designation and, if applicable, type classification indicated in Table R905.1.1(1). *Underlayment* shall be applied in accordance with Table R905.1.1(2). *Underlayment* shall be attached in accordance with Table R905.1.1(3).

Exceptions:

1. As an alternative, self-adhering polymer-modified bitumen underlayment bearing a label indicating compliance with ASTM D1970
2. As an alternative, a minimum 4-inch-wide (102 mm) strip of self-adhering polymer-modified bitumen membrane bearing a *label* indicating compliance with ASTM D1970, installed in accordance with the *manufacturer's installation instructions* for the deck material, shall be applied over all joints in the roof decking. An *approved underlayment* complying with Table R905.1.1(1) for the applicable roof covering for areas where wind design is not required in accordance with Figure R301.2.1.1 shall be applied over the entire roof over the 4-inch-wide (102 mm) membrane strips. Underlayment shall be applied in accordance with Table R905.1.1(2) using the application requirements for areas where wind design is not required in accordance with Figure R301.2.1.1. Underlayment shall be attached in accordance with Table R905.1.1(3).

R905.1.2 Ice barriers. In areas where there has been a history of ice forming along the eaves causing a backup of water as designated in Table R301.2, an ice barrier shall be installed for asphalt shingles, metal roof shingles, mineral-surfaced roll roofing, slate and slate-type shingles, wood shingles and wood shakes. The ice barrier shall consist of not fewer than two layers of *underlayment* cemented together, or a self-adhering polymer-modified bitumen sheet shall be used in place of normal *underlayment* and extend from the lowest edges of all roof surfaces to a point not less than 24 inches (610 mm) inside the exterior wall line of the building. On roofs with slope equal to or greater than 8 units vertical in 12 units horizontal (67-percent slope), the ice barrier shall also be applied not less than 36 inches (914 mm) measured along the roof slope from the eave edge of the building.

Exception: Detached *accessory structures* not containing conditioned floor area.

R905.2 Asphalt shingles. The installation of asphalt shingles shall comply with the provisions of this section.

R905.2.1 Sheathing requirements. Asphalt shingles shall be fastened to solidly sheathed decks.

R905.2.2 Slope. Asphalt shingles shall be used only on roof slopes of 2 units vertical in 12 units horizontal (17-percent slope) or greater. For roof slopes from 2 units vertical in 12 units horizontal (17-percent slope) up to 4 units vertical in 12 units horizontal (33-percent slope), double *underlayment* application is required in accordance with Section R905.1.1.

R905.2.3 Underlayment. *Underlayment* shall comply with Section R905.1.1.

R905.2.4 Asphalt shingles. Asphalt shingles shall comply with ASTM D3462.

R905.2.4.1 Wind resistance of asphalt shingles. Asphalt shingles shall be tested in accordance with ASTM D7158. Asphalt shingles shall meet the classification requirements of Table R905.2.4.1 for the appropriate ultimate design wind speed. Asphalt shingle packaging shall bear a *label* to indicate compliance with ASTM D7158 and the required classification in Table R905.2.4.1.

Exception: Asphalt shingles not included in the scope of ASTM D7158 shall be tested and *labeled* in accordance with ASTM D3161. Asphalt shingle packaging shall bear a *label* to indicate compliance with ASTM D3161 and the required classification in Table R905.2.4.1.

R905.2.5 Fasteners. Fasteners for asphalt shingles shall be galvanized steel, stainless steel, aluminum or copper roofing nails, minimum 12-gage [0.105 inch (3 mm)] shank with a minimum $^{3}/_{8}$-inch-diameter (9.5 mm) head, complying with ASTM F1667, of a length to penetrate through the roofing materials and not less than $^{3}/_{4}$ inch (19.1 mm) into the roof sheathing. Where the roof sheathing is less than $^{3}/_{4}$ inch (19.1 mm) thick, the fasteners shall penetrate through the sheathing.

R905.2.6 Attachment. Asphalt shingles shall have the minimum number of fasteners required by the manufacturer's *approved* installation instructions, but not less than four fasteners per strip shingle or two fasteners per individual shingle. Where the roof slope exceeds 21 units vertical in 12 units horizontal (21:12, 175-percent slope), shingles shall be installed in accordance with the manufacturer's *approved* installation instructions.

TABLE R905.1.1(1)
UNDERLAYMENT TYPES

ROOF COVERING	SECTION	AREAS WHERE WIND DESIGN IS NOT REQUIRED IN ACCORDANCE WITH FIGURE R301.2.1.1	AREAS WHERE WIND DESIGN IS REQUIRED IN ACCORDANCE WITH FIGURE R301.2.1.1
Asphalt shingles	R905.2	ASTM D226 Type I or II ASTM D48696 Type I, II, III or IV ASTM D6757	ASTM D226 Type II ASTM D4869 Type III or Type IV
Clay and concrete tile	R905.3	ASTM D226 Type II ASTM D2626 Type I ASTM D6380 Class M mineral-surfaced roll roofing	ASTM D226 Type II
Metal roof shingles	R905.4	ASTM D226 Type I or II ASTM D4869 Type I, II, III or IV	ASTM D226 Type II ASTM D4869 Type III or Type IV
Mineral-surfaced roll roofing	R905.5	ASTM D226 Type I or II ASTM D4869 Type I, II, III or IV	ASTM D226 Type II ASTM D4869 Type III or Type IV
Slate and slate-type shingles	R905.6	ASTM D226 Type I ASTM D4869 Type I, II, III or IV	ASTM D226 Type II ASTM D4869 Type III or Type IV
Wood shingles	R905.7	ASTM D226 Type I or II ASTM D4869 Type I, II, III or IV	ASTM D226 Type II ASTM D4869 Type III or Type IV
Wood shakes	R905.8	ASTM D226 Type I or II ASTM D4869 Type I, II, III or IV	ASTM D226 Type II ASTM D4869 Type III or Type IV
Metal panels	R905.10	Manufacturer's instructions	ASTM D226 Type II ASTM D4869 Type III or Type IV
Photovoltaic shingles	R905.16	ASTM D4869 Type I, II, III or IV ASTM D6757	ASTM D4869 Type III or Type IV

For SI: 1 mile per hour = 0.447 m/s.

**TABLE R905.1.1(2)
UNDERLAYMENT APPLICATION**

ROOF COVERING	SECTION	AREAS WHERE WIND DESIGN IS NOT REQUIRED IN ACCORDANCE WITH FIGURE R301.2.1.1	AREAS WHERE WIND DESIGN IS REQUIRED IN ACCORDANCE WITH FIGURE R301.2.1.1
Asphalt shingles	R905.2	For roof slopes from 2 units vertical in 12 units horizontal (2:12), up to 4 units vertical in 12 units horizontal (4:12), underlayment shall be two layers applied in the following manner: apply a 19-inch strip of underlayment felt parallel to and starting at the eaves. Starting at the eave, apply 36-inch-wide sheets of underlayment, overlapping successive sheets 19 inches. Distortions in the underlayment shall not interfere with the ability of the shingles to seal. End laps shall be 4 inches and shall be offset by 6 feet. For roof slopes of 4 units vertical in 12 units horizontal (4:12) or greater, underlayment shall be one layer applied in the following manner: underlayment shall be applied shingle fashion, parallel to and starting from the eave and lapped 2 inches. Distortions in the underlayment shall not interfere with the ability of the shingles to seal. End laps shall be 4 inches and shall be offset by 6 feet.	Underlayment shall be two layers applied in the following manner: apply a 19-inch strip of underlayment felt parallel to and starting at the eaves. Starting at the eave, apply 36-inch-wide sheets of underlayment, overlapping successive sheets 19 inches. Distortions in the underlayment shall not interfere with the ability of the shingles to seal. End laps shall be 4 inches and shall be offset by 6 feet.
Clay and concrete tile	R905.3	For roof slopes from $2^1/_2$ units vertical in 12 units horizontal ($2^1/_2$:12), up to 4 units vertical in 12 units horizontal (4:12), underlayment shall be not fewer than two layers applied as follows: starting at the eave, apply a 19-inch strip of underlayment parallel with the eave. Starting at the eave, apply 36-inch-wide strips of underlayment felt, overlapping successive sheets 19 inches. End laps shall be 4 inches and shall be offset by 6 feet. For roof slopes of 4 units vertical in 12 units horizontal (4:12) or greater, underlayment shall be not fewer than one layer of underlayment felt applied shingle fashion, parallel to and starting from the eaves and lapped 2 inches. End laps shall be 4 inches and shall be offset by 6 feet.	Underlayment shall be two layers applied in the following manner: apply a 19-inch strip of underlayment felt parallel to and starting at the eaves. Starting at the eave, apply 36-inch-wide sheets of underlayment, overlapping successive sheets 19 inches. Distortions in the underlayment shall not interfere with the ability of the shingles to seal. End laps shall be 4 inches and shall be offset by 6 feet.
Metal roof shingles	R905.4	Apply in accordance with the manufacturer's installation instructions.	Underlayment shall be two layers applied in the following manner: apply a 19-inch strip of underlayment felt parallel to and starting at the eaves. Starting at the eave, apply 36-inch-wide sheets of underlayment, overlapping successive sheets 19 inches. End laps shall be 4 inches and shall be offset by 6 feet.
Mineral-surfaced roll roofing	R905.5		
Slate and slate-type shingles	R905.6		
Wood shingles	R905.7		
Wood shakes	R905.8		
Metal panels	R905.10		
Photovoltaic shingles	R905.16	For roof slopes from 2 units vertical in 12 units horizontal (2:12), up to 4 units vertical in 12 units horizontal (4:12), underlayment shall be two layers applied in the following manner: apply a 19-inch strip of underlayment felt parallel to and starting at the eaves. Starting at the eave, apply 36-inch-wide sheets of underlayment, overlapping successive sheets 19 inches. Distortions in the underlayment shall not interfere with the ability of the shingles to seal. End laps shall be 4 inches and shall be offset by 6 feet. For roof slopes of 4 units vertical in 12 units horizontal (4:12) or greater, underlayment shall be one layer applied in the following manner: underlayment shall be applied shingle fashion, parallel to and starting from the eave and lapped 2 inches. Distortions in the underlayment shall not interfere with the ability of the shingles to seal. End laps shall be 4 inches and shall be offset by 6 feet.	Underlayment shall be two layers applied in the following manner: apply a 19-inch strip of underlayment felt parallel to and starting at the eaves. Starting at the eave, apply 36-inch-wide sheets of underlayment, overlapping successive sheets 19 inches. Distortions in the underlayment shall not interfere with the ability of the shingles to seal. End laps shall be 4 inches and shall be offset by 6 feet.

For SI: 1 inch = 25.4 mm, 1 foot = 304.8 mm, 1 mile per hour = 0.447 m/s.

TABLE R905.1.1(3)
UNDERLAYMENT APPLICATION

ROOF COVERING	SECTION	AREAS WHERE WIND DESIGN IS NOT REQUIRED IN ACCORDANCE WITH FIGURE R301.2.1.1	AREAS WHERE WIND DESIGN IS REQUIRED IN ACCORDANCE WITH FIGURE R301.2.1.1
Asphalt shingles	R905.2	Fastened sufficiently to hold in place	The underlayment shall be attached with corrosion-resistant fasteners in a grid pattern of 12 inches between side laps with a 6-inch spacing at side and end laps. Underlayment shall be attached using annular ring or deformed shank nails with 1-inch-diameter metal or plastic caps. Metal caps shall have a thickness of not less than 32-gage sheet metal. Power-driven metal caps shall have a minimum thickness of 0.010 inch. Minimum thickness of the outside edge of plastic caps shall be 0.035 inch. The cap nail shank shall be not less than 0.083 inch. The cap nail shank shall have a length sufficient to penetrate through the roof sheathing or not less than $^{3}/_{4}$ inch into the roof sheathing.
Clay and concrete tile	R905.3		
Photovoltaic	R905.16		
Metal roof shingles	R905.4	Manufacturer's installation instructions.	The underlayment shall be attached with corrosion-resistant fasteners in a grid pattern of 12 inches between side laps with a 6-inch spacing at side and end laps. Underlayment shall be attached using annular ring or deformed shank nails with 1-inch-diameter metal or plastic caps. Metal caps shall have a thickness of not less than 32-gage sheet metal. Power-driven metal caps shall have a minimum thickness of 0.010 inch. Minimum thickness of the outside edge of plastic caps shall be 0.035 inch. The cap nail shank shall be not less than 0.083 inch. The cap nail shank shall have a length sufficient to penetrate through the roof sheathing or not less than $^{3}/_{4}$ inch into the roof sheathing.
Mineral-surfaced roll roofing	R905.5		
Slate and slate-type shingles	R905.6		
Wood shingles	R905.7		
Wood shakes	R905.8		
Metal panels	R905.10		

For SI: 1 inch = 25.4 mm, 1 mile per hour = 0.447 m/s.

TABLE R905.2.4.1
CLASSIFICATION OF ASPHALT ROOF SHINGLES

MAXIMUM ULTIMATE DESIGN WIND SPEED, V_{ult} FROM Figure R301.2(2) (mph)	MAXIMUM BASIC WIND SPEED, V_{ASD} FROM TABLE R301.2.1.3 (mph)	ASTM D7158[a] SHINGLE CLASSIFICATION	ASTM D3161 SHINGLE CLASSIFICATION
110	85	D, G or H	A, D or F
116	90	D, G or H	A, D or F
129	100	G or H	A, D or F
142	110	G or H	F
155	120	G or H	F
168	130	H	F
181	140	H	F
194	150	H	F

For SI: 1 foot = 304.8 mm, 1 mile per hour = 0.447 m/s.

a. The standard calculations contained in ASTM D7158 assume Exposure Category B or C and a building height of 60 feet or less. Additional calculations are required for conditions outside of these assumptions.

R905.2.7 Ice barrier. Where required, ice barriers shall comply with Section R905.1.2.

R905.2.8 Flashing. Flashing for asphalt shingles shall comply with this section and the asphalt shingle manufacturer's *approved* installation instructions.

R905.2.8.1 Base and cap flashing. Base and cap flashing shall be installed in accordance with manufacturer's instructions. Base flashing shall be of either corrosion-resistant metal of minimum nominal 0.019-inch (0.5 mm) thickness or mineral-surfaced roll roofing weighing not less than 77 pounds per 100 square feet (4 kg/m^2). Cap flashing shall be corrosion-resistant metal of minimum nominal 0.019-inch (0.5 mm) thickness.

R905.2.8.2 Valleys. Valley linings shall be installed in accordance with the manufacturer's instructions before applying shingles. Valley linings of the following types shall be permitted:

1. For open valleys (valley lining exposed) lined with metal, the valley lining shall be not less than 24 inches (610 mm) wide and of any of the corrosion-resistant metals in Table R905.2.8.2.
2. For open valleys, valley lining of two plies of mineral-surfaced roll roofing, complying with ASTM D3909 or ASTM D6380 Class M, shall be permitted. The bottom layer shall be 18 inches (457 mm) and the top layer not less than 36 inches (914 mm) wide.
3. For closed valleys (valley covered with shingles), valley lining of one ply of smooth roll roofing complying with ASTM D6380 and not less than 36 inches wide (914 mm) or valley lining as described in Item 1 or 2 shall be permitted. Self-adhering polymer-modified bitumen *underlayment* complying with ASTM D1970 shall be permitted in lieu of the lining material.

R905.2.8.3 Sidewall flashing. Base flashing against a vertical sidewall shall be continuous or step flashing and shall be not less than 4 inches (102 mm) in height and 4 inches (102 mm) in width and shall direct water away from the vertical sidewall onto the roof or into the gutter. Where siding is provided on the vertical sidewall, the vertical leg of the flashing shall be continuous under the siding. Where anchored masonry veneer is provided on the vertical sidewall, the base flashing shall be provided in accordance with this section and counterflashing shall be provided in accordance with Section R703.8.2.2. Where exterior plaster or adhered masonry veneer is provided on the vertical sidewall, the base flashing shall be provided in accordance with this section and Section R703.6.3.

R905.2.8.4 Other flashing. Flashing against a vertical front wall, as well as soil stack, vent pipe and chimney flashing, shall be applied in accordance with the asphalt shingle manufacturer's printed instructions.

R905.2.8.5 Drip edge. A drip edge shall be provided at eaves and rake edges of shingle roofs. Adjacent segments of drip edge shall be overlapped not less than 2 inches (51 mm). Drip edges shall extend not less than $^1/_4$ inch (6.4 mm) below the roof sheathing and extend up back onto the *roof deck* not less than 2 inches (51 mm). Drip edges shall be mechanically fastened to the *roof deck* at not more than 12 inches (305 mm) o.c. with fasteners as specified in Section R905.2.5. *Underlayment* shall be installed over the drip edge along eaves and under the drip edge along rake edges.

R905.3 Clay and concrete tile. The installation of clay and concrete tile shall comply with the provisions of this section.

R905.3.1 Deck requirements. Concrete and clay tile shall be installed only over solid sheathing.

Exception: Spaced lumber sheathing in accordance with Section R803.1 shall be permitted in *Seismic Design Categories* A, B and C.

R905.3.2 Deck slope. Clay and concrete roof tile shall be installed on roof slopes of $2^1/_2$ units vertical in 12 units horizontal (25-percent slope) or greater. For roof slopes from $2^1/_2$ units vertical in 12 units horizontal (25-percent

TABLE R905.2.8.2
VALLEY LINING MATERIAL

MATERIAL	MINIMUM THICKNESS (inches)	GAGE	WEIGHT (pounds)
Aluminum	0.024	—	—
Cold-rolled copper	0.0216 nominal	—	ASTM B370, 16 oz. per square foot
Galvanized steel	0.0179	26 (zinc coated G90)	—
High-yield copper	0.0162 nominal	—	ASTM B370, 12 oz. per square foot
Lead	—	—	$2^1/_2$
Lead-coated copper	0.0216 nominal	—	ASTM B101, 16 oz. per square foot
Lead-coated high-yield copper	0.0162 nominal	—	ASTM B101, 12 oz. per square foot
Painted terne	—	—	20
Stainless steel	—	28	—
Zinc alloy	0.027	—	—

For SI: 1 inch = 25.4 mm, 1 pound = 0.454 kg, 1 square foot = 0.93 m^2.

slope) to 4 units vertical in 12 units horizontal (33-percent slope), double *underlayment* application is required in accordance with Section R905.3.3.

R905.3.3 Underlayment. *Underlayment* shall comply with Section R905.1.1.

R905.3.4 Clay tile. Clay roof tile shall comply with ASTM C1167.

R905.3.5 Concrete tile. Concrete roof tile shall comply with ASTM C1492.

R905.3.6 Fasteners. Nails shall be corrosion resistant and not less than 11-gage [0.120 inch (3 mm)], $^{5}/_{16}$-inch (11 mm) head, and of sufficient length to penetrate the deck not less than $^{3}/_{4}$ inch (19 mm) or through the thickness of the deck, whichever is less. Attaching wire for clay or concrete tile shall not be smaller than 0.083 inch (2 mm). Perimeter fastening areas include three tile courses but not less than 36 inches (914 mm) from either side of hips or ridges and edges of eaves and gable rakes.

R905.3.7 Application. Tile shall be applied in accordance with this chapter and the manufacturer's installation instructions, based on the following:

1. Climatic conditions.
2. Roof slope.
3. *Underlayment* system.
4. Type of tile being installed.

Clay and concrete roof tiles shall be fastened in accordance with this section and the manufacturer's installation instructions. Perimeter tiles shall be fastened with not less than one fastener per tile. Tiles with installed weight less than 9 pounds per square foot (0.4 kg/m^2) require not less than one fastener per tile regardless of roof slope. Clay and concrete roof tile attachment shall be in accordance with the manufacturer's installation instructions where applied in areas where the ultimate design wind speed exceeds 130 miles per hour (58 m/s) and on buildings where the roof is located more than 40 feet (12 192 mm) above grade. In areas subject to snow, not less than two fasteners per tile are required. In other areas, clay and concrete roof tiles shall be attached in accordance with Table R905.3.7.

TABLE R905.3.7
CLAY AND CONCRETE TILE ATTACHMENT

SHEATHING	ROOF SLOPE	NUMBER OF FASTENERS
Solid without battens	All	One per tile
Spaced or solid with battens and slope < 5:12	Fasteners not required	—
Spaced sheathing without battens	5:12 ≤ slope < 12:12	One per tile/ every other row
	12:12 ≤ slope < 24:12	One per tile

R905.3.8 Flashing. At the juncture of roof vertical surfaces, flashing and counterflashing shall be provided in accordance with this chapter and the manufacturer's installation instructions and, where of metal, shall be not less than 0.019 inch (0.5 mm) (No. 26 galvanized sheet gage) corrosion-resistant metal. The valley flashing shall extend not less than 11 inches (279 mm) from the centerline each way and have a splash diverter rib not less than 1 inch (25 mm) in height at the flow line formed as part of the flashing. Sections of flashing shall have an end lap of not less than 4 inches (102 mm). For roof slopes of 3 units vertical in 12 units horizontal (25-percent slope) and greater, valley flashing shall have a 36-inch-wide (914 mm) *underlayment* of one layer of Type I *underlayment* running the full length of the valley, in addition to other required *underlayment*. In areas where the average daily temperature in January is 25°F (-4°C) or less, metal valley flashing *underlayment* shall be solid-cemented to the roofing *underlayment* for slopes less than 7 units vertical in 12 units horizontal (58-percent slope) or be of self-adhering polymer-modified bitumen sheet.

R905.4 Metal roof shingles. The installation of metal roof shingles shall comply with the provisions of this section.

R905.4.1 Deck requirements. *Metal roof shingles* shall be applied to a solid or closely fitted deck, except where the roof covering is specifically designed to be applied to spaced sheathing.

R905.4.2 Deck slope. *Metal roof shingles* shall not be installed on roof slopes below 3 units vertical in 12 units horizontal (25-percent slope).

R905.4.3 Underlayment. *Underlayment* shall comply with Section R905.1.1.

R905.4.3.1 Ice barrier. Where required, ice barriers shall comply with Section R905.1.2.

R905.4.4 Material standards. *Metal roof shingle* roof coverings shall comply with Table R905.10.3(1). The materials used for *metal roof shingle* roof coverings shall be naturally corrosion resistant or be made corrosion resistant in accordance with the standards and minimum thicknesses listed in Table R905.10.3(2).

R905.4.4.1 Wind resistance of metal roof shingles. *Metal roof shingles* applied to a solid or closely fitted deck shall be tested in accordance with ASTM D3161, FM 4474, UL 580 or UL 1897. *Metal roof shingles* tested in accordance with ASTM D3161 shall meet the classification requirements of Table R905.4.4.1 for the appropriate maximum basic wind speed and the metal shingle packaging shall bear a *label* to indicate compliance with ASTM D3161 and the required classification in Table R905.2.4.1.

R905.4.5 Application. *Metal roof shingles* shall be secured to the roof in accordance with this chapter and the *approved* manufacturer's installation instructions.

R905.4.6 Flashing. Roof valley flashing shall be of corrosion-resistant metal of the same material as the roof covering or shall comply with the standards in Table R905.10.3(1). The valley flashing shall extend not less than 8 inches (203 mm) from the centerline each way and shall have a splash diverter rib not less than $^{3}/_{4}$ inch (19 mm) in height at the flow line formed as part of the flashing. Sections of flashing shall have an end lap of not less than 4 inches (102 mm). The metal valley flashing shall have a 36-inch-wide (914 mm) *underlayment* directly under it consisting of one layer of *underlayment* running the full length of the valley, in addition to *underlayment* required for *metal roof shingles*. In areas where the average daily temperature in January is 25°F (-4°C) or less, the metal valley flashing *underlayment* shall be solid-cemented to the roofing *underlayment* for roof slopes under 7 units vertical in 12 units horizontal (58-percent slope) or self-adhering polymer-modified bitumen sheet.

R905.5 Mineral-surfaced roll roofing. The installation of mineral-surfaced roll roofing shall comply with this section.

R905.5.1 Deck requirements. Mineral-surfaced roll roofing shall be fastened to solidly sheathed roofs.

R905.5.2 Deck slope. Mineral-surfaced roll roofing shall not be applied on roof slopes below 1 unit vertical in 12 units horizontal (8-percent slope).

R905.5.3 Underlayment. *Underlayment* shall comply with Section R905.1.1.

R905.5.3.1 Ice barrier. Where required, ice barriers shall comply with Section R905.1.2.

R905.5.4 Material standards. Mineral-surfaced roll roofing shall conform to ASTM D3909 or ASTM D6380, Class M.

R905.5.5 Application. Mineral-surfaced roll roofing shall be installed in accordance with this chapter and the manufacturer's instructions.

R905.6 Slate shingles. The installation of slate shingles shall comply with the provisions of this section.

R905.6.1 Deck requirements. Slate shingles shall be fastened to solidly sheathed roofs.

R905.6.2 Deck slope. Slate shingles shall be used only on slopes of 4 units vertical in 12 units horizontal (33-percent slope) or greater.

R905.6.3 Underlayment. *Underlayment* shall comply with Section R905.1.1.

R905.6.3.1 Ice barrier. Where required, ice barriers shall comply with Section R905.1.2.

R905.6.4 Material standards. Slate shingles shall comply with ASTM C406.

R905.6.5 Application. Minimum headlap for slate shingles shall be in accordance with Table R905.6.5. Slate shingles shall be secured to the roof with two fasteners per slate. Slate shingles shall be installed in accordance with this chapter and the manufacturer's instructions.

**TABLE R905.6.5
SLATE SHINGLE HEADLAP**

SLOPE	HEADLAP (inches)
4:12 ≤ slope < 8:12	4
8:12 ≤ slope < 20:12	3
Slope ≥ 20:12	2

For SI: 1 inch = 25.4 mm.

R905.6.6 Flashing. Flashing and counterflashing shall be made with sheet metal. Valley flashing shall be not less than 15 inches (381 mm) wide. Valley and flashing metal shall be a minimum uncoated thickness of 0.0179-inch (0.5 mm) zinc coated G90. Chimneys, stucco or brick walls shall have not less than two plies of felt for a cap flashing consisting of a 4-inch-wide (102 mm) strip of felt set in plastic cement and extending 1 inch (25 mm) above the first felt and a top coating of plastic cement. The felt shall extend 2 inches (51 mm) over the base flashing.

R905.7 Wood shingles. The installation of wood shingles shall comply with the provisions of this section.

R905.7.1 Deck requirements. Wood shingles shall be installed on solid or spaced sheathing. Where spaced sheathing is used, sheathing boards shall be not less than 1-inch by 4-inch (25 mm by 102 mm) nominal dimensions and shall be spaced on centers equal to the weather exposure to coincide with the placement of fasteners.

R905.7.1.1 Solid sheathing required. In areas where the average daily temperature in January is 25°F (-4°C) or less, solid sheathing is required on that portion of the roof requiring the application of an ice barrier.

R905.7.2 Deck slope. Wood shingles shall be installed on slopes of 3 units vertical in 12 units horizontal (25-percent slope) or greater.

**TABLE R905.4.4.1
CLASSIFICATION OF STEEP SLOPE METAL ROOF SHINGLES TESTED IN ACCORDANCE WITH ASTM D3161**

MAXIMUM ULTIMATE DESIGN WIND SPEED, V_{ult}, FROM FIGURE R301.2(2) (mph)	MAXIMUM BASIC WIND SPEED, V_{ASD}, FROM TABLE R301.2.1.3 (mph)	ASTM D3161 SHINGLE CLASSIFICATION
110	85	A, D or F
116	90	A, D or F
129	100	A, D or F
142	110	F
155	120	F
168	130	F
181	140	F
194	150	F

For SI: 1 mile per hour = 1.609 kph.

R905.7.3 Underlayment. *Underlayment* shall comply with Section R905.1.1.

R905.7.3.1 Ice barrier. Where required, ice barriers shall comply with Section R905.1.2.

R905.7.4 Material standards. Wood shingles shall be of *naturally durable wood* and comply with the requirements of Table R905.7.4.

TABLE R905.7.4
WOOD SHINGLE MATERIAL REQUIREMENTS

MATERIAL	MINIMUM GRADES	APPLICABLE GRADING RULES
Wood shingles of naturally durable wood	1, 2 or 3	CSSB

R905.7.5 Application. Wood shingles shall be installed in accordance with this chapter and the manufacturer's instructions. Wood shingles shall be laid with a side lap not less than $1^1/_2$ inches (38 mm) between joints in courses, and two joints shall not be in direct alignment in any three adjacent courses. Spacing between shingles shall be not less than $^1/_4$ inch to $^3/_8$ inch (6.4 mm to 9.5 mm). Weather exposure for wood shingles shall not exceed those set in Table R905.7.5(1). Fasteners for untreated (naturally durable) wood shingles shall be box nails in accordance with Table R905.7.5(2). Nails shall be stainless steel Type 304 or 316 or hot-dipped galvanized with a coating weight of ASTM A153 Class D (1.0 oz/ft^2). Alternatively, two 16-gage stainless steel Type 304 or 316 staples with crown widths $^7/_{16}$ inch (11.1 mm) minimum, $^3/_4$ inch (19.1 mm) maximum, shall be used. Fasteners installed within 15 miles (24 km) of saltwater coastal areas shall be stainless steel Type 316. Fasteners for fire-retardant-treated shingles in accordance with Section R902 or pressure-impregnated-preservative-treated shingles of naturally durable wood in accordance with AWPA U1 shall be stainless steel Type 316. Fasteners shall have a minimum penetration into the sheathing of $^3/_4$ inch (19.1 mm). For sheathing less than $^3/_4$ inch in (19.1 mm) thickness, each fastener shall penetrate through the sheathing. Wood shingles shall be attached to the roof with two fasteners per shingle, positioned in accordance with the manufacturer's installation instructions. Fastener packaging shall bear a *label* indicating the appropriate grade material or coating weight.

TABLE R905.7.5(1)
WOOD SHINGLE WEATHER EXPOSURE AND ROOF SLOPE

ROOFING MATERIAL	LENGTH (inches)	GRADE	EXPOSURE (inches)	
			3:12 pitch to < 4:12	4:12 pitch or steeper
Shingles of naturally durable wood	16	No. 1	$3^3/_4$	5
		No. 2	$3^1/_2$	4
		No. 3	3	$3^1/_2$
	18	No. 1	$4^1/_4$	$5^1/_2$
		No. 2	4	$4^1/_2$
		No. 3	$3^1/_2$	4
	24	No. 1	$5^3/_4$	$7^1/_2$
		No. 2	$5^1/_2$	$6^1/_2$
		No. 3	5	$5^1/_2$

For SI: 1 inch = 25.4 mm.

TABLE R905.7.5(2)
NAIL REQUIREMENTS FOR WOOD SHAKES AND WOOD SHINGLES

PRODUCT TYPE	NAIL TYPE, MINIMUM LENGTH AND SHANK DIAMETER (inches)
Shakes	
18″ straight-split	5d box $1^3/_4$″ × 0.080
18″ and 24″ handsplit and resawn	6d box 2″ × 0.099
24″ taper-split	5d box $1^3/_4$″ × 0.080
18″ and 24″ tapersawn	6d box 2″ × 0.099
Shingles	
16″ and 18″	3d box $1^1/_4$″ × 0.076
24″	4d box $1^1/_2$″ × 0.076

For SI: 1 inch = 25.4 mm.

R905.7.6 Valley flashing. Roof flashing shall be not less than No. 26 gage [0.019 inches (0.5 mm)] corrosion-resistant sheet metal and shall extend 10 inches (254 mm) from the centerline each way for roofs having slopes less than 12 units vertical in 12 units horizontal (100-percent slope), and 7 inches (178 mm) from the centerline each way for slopes of 12 units vertical in 12 units horizontal (100-percent slope) and greater. Sections of flashing shall have an end lap of not less than 4 inches (102 mm).

R905.7.7 Label required. Each bundle of shingles shall be identified by a *label* of an *approved* grading or inspection bureau or agency.

R905.8 Wood shakes. The installation of wood shakes shall comply with the provisions of this section.

R905.8.1 Deck requirements. Wood shakes shall be used only on solid or spaced sheathing. Where spaced sheathing is used, sheathing boards shall be not less than 1-inch by 4-inch (25 mm by 102 mm) nominal dimensions and shall be spaced on centers equal to the weather exposure to coincide with the placement of fasteners. Where 1-inch by 4-inch (25 mm by 102 mm) spaced sheathing is installed at 10 inches (254 mm) on center, additional 1-inch by 4-inch (25 mm by 102 mm) boards shall be installed between the sheathing boards.

R905.8.1.1 Solid sheathing required. In areas where the average daily temperature in January is 25°F (-4°C) or less, solid sheathing is required on that portion of the roof requiring an ice barrier.

R905.8.2 Deck slope. Wood shakes shall only be used on slopes of 3 units vertical in 12 units horizontal (25-percent slope) or greater.

R905.8.3 Underlayment. *Underlayment* shall comply with Section R905.1.1.

R905.8.3.1 Ice barrier. Where required, ice barriers shall comply with Section R905.1.2.

R905.8.4 Interlayment. Interlayment shall comply with ASTM D226, Type I.

R905.8.5 Material standards. Wood shakes shall comply with the requirements of Table R905.8.5.

R905.8.6 Application. Wood shakes shall be installed in accordance with this chapter and the manufacturer's installation instructions. Wood shakes shall be laid with a side lap not less than $1^1/_2$ inches (38 mm) between joints in adjacent courses. Spacing between shakes in the same course shall be $^3/_8$ inch to $^5/_8$ inch (9.5 mm to 15.9 mm) including tapersawn shakes. Weather exposures for wood shakes shall not exceed those set in Table R905.8.6. Fasteners for untreated (naturally durable) wood shakes shall be box nails in accordance with Table R905.7.5(2). Nails shall be stainless steel Type 304, or Type 316 or hot-dipped with a coating weight of ASTM A153 Class D (1.0 oz/ft^2). Alternatively, two 16-gage Type 304 or Type 316 stainless steel staples, with crown widths $^7/_{16}$ inch (11.1 mm) minimum, $^3/_4$ inch (19.1 mm) maximum, shall be used. Fasteners installed within 15 miles (24 km) of saltwater coastal areas shall be stainless steel Type 316. Wood shakes shall be attached to the roof with two fasteners per shake positioned in accordance with the manufacturer's installation instructions. Fasteners for fire-retardant-treated (as defined in Section R902) shakes or pressure-impregnated-preservative-treated shakes of *naturally durable wood* in accordance with AWPA U1 shall be stainless steel Type 316. Fasteners shall have a minimum penetration into the sheathing of $^3/_4$ inch (19.1 mm). Where the sheathing is less than $^3/_4$ inch (19.1 mm) thick, each fastener shall penetrate through the sheathing. Fastener packaging shall bear a *label* indicating the appropriate grade material or coating weight.

R905.8.7 Shake placement. The starter course at the eaves shall be doubled and the bottom layer shall be either 15-inch (381 mm), 18-inch (457 mm) or 24-inch (610 mm) wood shakes or wood shingles. Fifteen-inch (381 mm) or 18-inch (457 mm) wood shakes shall be permitted to be used for the final course at the ridge. Shakes shall be interlaid with 18-inch-wide (457 mm) strips of not less than No. 30 felt shingled between each course in such a manner that felt is not exposed to the weather by positioning the lower edge of each felt strip above the butt end of the shake it covers a distance equal to twice the weather exposure.

R905.8.8 Valley flashing. Roof valley flashing shall be not less than No. 26 gage [0.019 inch (0.5 mm)] corrosion-resistant sheet metal and shall extend not less than 11 inches (279 mm) from the centerline each way. Sections of flashing shall have an end lap of not less than 4 inches (102 mm).

R905.8.9 Label required. Each bundle of shakes shall be identified by a *label* of an *approved* grading or inspection bureau or agency.

TABLE R905.8.5
WOOD SHAKE MATERIAL REQUIREMENTS

MATERIAL	MINIMUM GRADES	APPLICABLE GRADING RULES
Wood shakes of naturally durable wood	1	Cedar Shake and Shingle Bureau
Tapersawn shakes of naturally durable wood	1 or 2	Cedar Shake and Shingle Bureau
Preservative-treated shakes and shingles of naturally durable wood	1	Cedar Shake and Shingle Bureau
Fire-retardant-treated shakes and shingles of naturally durable wood	1	Cedar Shake and Shingle Bureau
Preservative-treated tapersawn shakes of Southern pine treated in accordance with AWPA Standard U1 (Commodity Specification A, Special Requirement 4.6)	1 or 2	Forest Products Laboratory of the Texas Forest Services

TABLE R905.8.6
WOOD SHAKE WEATHER EXPOSURE AND ROOF SLOPE

ROOFING MATERIAL	LENGTH (inches)	GRADE	EXPOSURE (inches)
			4:12 pitch or steeper
Shakes of naturally durable wood	18	No. 1	$7^1/_2$
	24	No. 1	10[a]
Preservative-treated tapersawn shakes of Southern Yellow Pine	18	No. 1	$7^1/_2$
	24	No. 1	10
	18	No. 2	$5^1/_2$
	24	No. 2	$7^1/_2$
Taper-sawn shakes of naturally durable wood	18	No. 1	$7^1/_2$
	24	No. 1	10
	18	No. 2	$5^1/_2$
	24	No. 2	$7^1/_2$

For SI: 1 inch = 25.4 mm.

a. For 24-inch by $^3/_8$-inch handsplit shakes, the maximum exposure is $7^1/_2$ inches.

R905.9 Built-up roofs. The installation of built-up roofs shall comply with the provisions of this section and the manufacturer's *approved* installation instructions.

R905.9.1 Slope. Built-up roofs shall have a design slope of not less than $^{1}/_{4}$ unit vertical in 12 units horizontal (2-percent slope) for drainage, except for coal-tar built-up roofs, which shall have a design slope of a minimum $^{1}/_{8}$ unit vertical in 12 units horizontal (1-percent slope).

R905.9.2 Material standards. *Built-up roof covering* materials shall comply with the standards in Table R905.9.2 or UL 55A.

R905.9.3 Application. Built-up roofs shall be installed in accordance with this chapter and the manufacturer's instructions.

R905.10 Metal roof panels. The installation of *metal roof panels* shall comply with the provisions of this section.

R905.10.1 Deck requirements. *Metal roof panel* roof coverings shall be applied to solid or spaced sheathing, except where the roof covering is specifically designed to be applied to spaced supports.

R905.10.2 Slope. Minimum slopes for *metal roof panels* shall comply with the following:

1. The minimum slope for lapped, nonsoldered-seam metal roofs without applied lap sealant shall be 3 units vertical in 12 units horizontal (25-percent slope).
2. The minimum slope for lapped, nonsoldered-seam metal roofs with applied lap sealant shall be $^{1}/_{2}$ unit vertical in 12 units horizontal (4-percent slope). Lap sealants shall be applied in accordance with the *approved* manufacturer's installation instructions.
3. The minimum slope for standing-seam roof systems shall be $^{1}/_{4}$ unit vertical in 12 units horizontal (2-percent slope).

R905.10.3 Material standards. Metal-sheet roof covering systems that incorporate supporting structural members shall be designed in accordance with the *International Building Code*. Metal-sheet roof coverings installed over structural decking shall comply with Table R905.10.3(1). The materials used for metal-sheet roof coverings shall be naturally corrosion resistant or provided with corrosion resistance in accordance with the standards and minimum thicknesses shown in Table R905.10.3(2).

R905.10.4 Attachment. *Metal roof panels* shall be secured to the supports in accordance with this chapter and the manufacturer's installation instructions. In the absence of manufacturer's installation instructions, the following fasteners shall be used:

1. Galvanized fasteners shall be used for steel roofs.
2. Copper, brass, bronze, copper alloy and 300-series stainless steel fasteners shall be used for copper roofs.
3. Stainless steel fasteners are acceptable for metal roofs.

R905.10.5 Underlayment. *Underlayment* shall comply with Section R905.1.1.

TABLE R905.9.2
BUILT-UP ROOFING MATERIAL STANDARDS

MATERIAL STANDARD	STANDARD
Acrylic coatings used in roofing	ASTM D6083
Aggregate surfacing	ASTM D1863
Asphalt adhesive used in roofing	ASTM D3747
Asphalt cements used in roofing	ASTM D2822; D3019; D4586
Asphalt-coated glass fiber base sheet	ASTM D4601
Asphalt coatings used in roofing	ASTM D1227; D2823; D2824; D4479
Asphalt glass felt	ASTM D2178
Asphalt primer used in roofing	ASTM D41
Asphalt-saturated and asphalt-coated organic felt base sheet	ASTM D2626
Asphalt-saturated organic felt (perforated)	ASTM D2626
Asphalt used in roofing	ASTM D312
Coal-tar cements used in roofing	ASTM D4022; D5643
Coal-tar primer used in roofing, dampproofing and waterproofing	ASTM D43
Coal-tar saturated organic felt	ASTM D227
Coal-tar used in roofing	ASTM D450, Type I or II
Glass mat, coal tar	ASTM D4990
Glass mat, venting type	ASTM D4897
Mineral-surfaced inorganic cap sheet	ASTM D3909
Thermoplastic fabrics used in roofing	ASTM D5665; D5726

R905.11 Modified bitumen roofing. The installation of modified bitumen roofing shall comply with the provisions of this section and the manufacturer's *approved* installation instructions.

R905.11.1 Slope. Modified bitumen roofing shall have a design slope of not less than one-fourth unit vertical in 12 units horizontal (2-percent slope) for drainage.

R905.11.2 Material standards. Modified bitumen roofing shall comply with the standards in Table R905.11.2.

R905.11.2.1 Base sheet. A base sheet that complies with the requirements of Section 1507.11.2 of the *International Building Code*, ASTM D1970 or ASTM D4601 shall be permitted to be used with a modified bitumen cap sheet.

R905.11.3 Application. Modified bitumen roofs shall be installed in accordance with this chapter and the manufacturer's instructions.

R905.12 Thermoset single-ply roofing. The installation of thermoset single-ply roofing shall comply with the provisions of this section.

R905.12.1 Slope. Thermoset *single-ply membrane* roofs shall have a design slope of not less than $^{1}/_{4}$ unit vertical in 12 units horizontal (2-percent slope) for drainage.

R905.12.2 Material standards. Thermoset single-ply roof coverings shall comply with ASTM D4637 or ASTM D5019.

TABLE R905.10.3(1)
METAL ROOF COVERING STANDARDS

ROOF COVERING TYPE	STANDARD APPLICATION RATE/THICKNESS
Aluminum	ASTM B209, 0.024 minimum thickness for roll-formed panels and 0.019-inch minimum thickness for press-formed shingles.
Cold-rolled copper	ASTM B370 minimum 16 oz/sq ft and 12 oz/sq ft high-yield copper for metal-sheet roof-covering systems; 12 oz/sq ft for preformed metal shingle systems.
Galvanized steel	ASTM A653 G90 Zinc coated
Hard lead	2 lb/sq ft
Lead-coated copper	ASTM B101
Soft lead	3 lb/sq ft
Stainless steel	ASTM A240, 300 Series alloys
Steel	ASTM A924
Terne (tin) and terne-coated stainless	Terne coating of 40 lb per double base box, field painted where applicable in accordance with manufacturer's installation instructions.
Zinc	0.027 inch minimum thickness: 99.995% electrolytic high-grade zinc with alloy additives of copper (0.08–0.20%), titanium (0.07%–0.12%) and aluminum (0.015%).

For SI: 1 ounce per square foot = 0.305 kg/m^2, 1 pound per square foot = 4.214 kg/m^2, 1 inch = 25.4 mm, 1 pound = 0.454 kg.

TABLE R905.10.3(2)
MINIMUM CORROSION RESISTANCE

55% aluminum-zinc-alloy-coated steel	ASTM A792 AZ 50
5% aluminum alloy-coated steel	ASTM A875 GF60
Aluminum-coated steel	ASTM A463 T2 65
Galvanized steel	ASTM A653 G-90
Prepainted steel	ASTM A755[a]

a. Paint systems in accordance with ASTM A755 shall be applied over steel products with corrosion-resistant coatings complying with ASTM A792, ASTM A875, ASTM A463 or ASTM A653.

TABLE R905.11.2
MODIFIED BITUMEN ROOFING MATERIAL STANDARDS

MATERIAL	STANDARD
Acrylic coating	ASTM D6083
Asphalt adhesive	ASTM D3747
Asphalt cement	ASTM D3019
Asphalt coating	ASTM D1227; D2824
Asphalt primer	ASTM D41
Modified bitumen roof membrane	ASTM D6162; D6163; D6164; D6222; D6223; D6298

R905.12.3 Application. Thermoset single-ply roofs shall be installed in accordance with this chapter and the manufacturer's instructions.

R905.13 Thermoplastic single-ply roofing. The installation of thermoplastic single-ply roofing shall comply with the provisions of this section.

R905.13.1 Slope. Thermoplastic *single-ply membrane* roofs shall have a design slope of not less than $^1/_4$ unit vertical in 12 units horizontal (2-percent slope).

R905.13.2 Material standards. Thermoplastic single-ply roof coverings shall comply with ASTM D4434, D6754 or D6878.

R905.13.3 Application. Thermoplastic single-ply roofs shall be installed in accordance with this chapter and the manufacturer's instructions.

R905.14 Sprayed polyurethane foam roofing. The installation of sprayed polyurethane foam roofing shall comply with the provisions of this section.

R905.14.1 Slope. Sprayed polyurethane foam roofs shall have a design slope of not less than $^1/_4$ unit vertical in 12 units horizontal (2-percent slope) for drainage.

R905.14.2 Material standards. Spray-applied polyurethane foam insulation shall comply with ASTM C1029, Type III or IV or ASTM D7425.

R905.14.3 Application. Foamed-in-place roof insulation shall be installed in accordance with this chapter and the manufacturer's instructions. A liquid-applied protective coating that complies with Table R905.14.3 shall be applied not less than 2 hours nor more than 72 hours following the application of the foam.

TABLE R905.14.3
PROTECTIVE COATING MATERIAL STANDARDS

MATERIAL	STANDARD
Acrylic coating	ASTM D6083
Moisture-cured polyurethane coating	ASTM D6947
Silicone coating	ASTM D6694

R905.14.4 Foam plastics. Foam plastic materials and installation shall comply with Section R316.

R905.15 Liquid-applied roofing. The installation of liquid-applied roofing shall comply with the provisions of this section.

R905.15.1 Slope. Liquid-applied roofing shall have a design slope of not less than $^1/_4$ unit vertical in 12 units horizontal (2-percent slope).

R905.15.2 Material standards. Liquid-applied roofing shall comply with ASTM C836, C957, D1227, D3468, D6083, D6694 or D6947.

R905.15.3 Application. Liquid-applied roofing shall be installed in accordance with this chapter and the manufacturer's installation instructions.

R905.16 Photovoltaic shingles. The installation of *photovoltaic shingles* shall comply with the provisions of this section, Section R324 and NFPA 70.

R905.16.1 Deck requirements. *Photovoltaic shingles* shall be applied to a solid or closely-fitted deck, except where the roof covering is specifically designed to be applied over spaced sheathing.

R905.16.2 Deck slope. *Photovoltaic shingles* shall be used only on roof slopes of 2 units vertical in 12 units horizontal (2:12) or greater.

R905.16.3 Underlayment. *Underlayment* shall comply with Section R905.1.1.

R905.16.3.1 Ice barrier. Where required, ice barriers shall comply with Section R905.1.2.

R905.16.4 Material standards. *Photovoltaic shingles* shall be *listed* and *labeled* in accordance with UL 7103 or with both UL 61730-1 and UL 61730-2.

R905.16.5 Attachment. *Photovoltaic shingles* shall be attached in accordance with the manufacturer's installation instructions.

R905.16.6 Wind resistance. *Photovoltaic shingles* shall comply with the classification requirements of Table R905.16.6 for the appropriate maximum basic wind speed.

R905.17 Building-integrated photovoltaic (BIPV) roof panels applied directly to the roof deck. The installation of *BIPV roof panels* shall comply with the provisions of this section, Section R324 and NFPA 70.

TABLE R905.16.6
CLASSIFICATION OF PHOTOVOLTAIC SHINGLES

MAXIMUM ULTIMATE DESIGN WIND SPEED, V_{ult}, FROM FIGURE R301.2(2) (mph)	MAXIMUM BASIC WIND SPEED, V_{ASD}, FROM TABLE R301.2.1.3 (mph)	UL 7103 SHINGLE CLASSIFICATION
110	85	A, D or F
116	90	A, D or F
129	100	A, D or F
142	110	F
155	120	F
168	130	F
181	140	F
194	150	F

For SI:1 mile per hour = 1.609 kph.

R905.17.1 Deck requirements. *BIPV roof panels* shall be applied to a solid or closely-fitted deck, except where the roof covering is specifically designed to be applied over spaced sheathing.

R905.17.2 Deck slope. *BIPV roof panels* shall be used only on roof slopes of 2 units vertical in 12 units horizontal (17-percent slope) or greater.

R905.17.3 Underlayment. *Underlayment* shall comply with Section 905.1.1.

R905.17.3.1 Ice barrier. Where required, an ice barrier shall comply with Section R905.1.2.

R905.17.4 Ice barrier. In areas where there has been a history of ice forming along the eaves causing a backup of water, as designated in Table R301.2, an ice barrier that consists of not less than two layers of *underlayment* cemented together or of a self-adhering polymer-modified bitumen sheet shall be used in lieu of normal *underlayment* and extend from the lowest edges of all roof surfaces to a point not less than 24 inches (610 mm) inside the exterior wall line of the building.

Exception: Detached *accessory structures* that do not contain conditioned floor area.

R905.17.5 Material standards. *BIPV roof panels* shall be *listed* and *labeled* in accordance with UL 7103 or with both UL 61730-1 and UL 61730-2.

R905.17.6 Attachment. *BIPV roof panels* shall be attached in accordance with the manufacturer's installation instructions.

SECTION R906
ROOF INSULATION

R906.1 General. Where above-deck thermal insulation is installed, such insulation shall be covered with an *approved* roof covering and shall comply with NFPA 276 or UL 1256.

R906.2 Material standards. Above-deck thermal insulation board shall comply with the standards in Table R906.2.

TABLE R906.2
MATERIAL STANDARDS FOR ROOF INSULATION

Cellular glass board	ASTM C552
Composite boards	ASTM C1289, Type III, IV, V or VI
Expanded polystyrene	ASTM C578
Extruded polystyrene board	ASTM C578
Fiber-reinforced gypsum board	ASTM C1278
Glass-faced gypsum board	ASTM C1177
Mineral wool board	ASTM C726
Perlite board	ASTM C728
Polyisocyanurate board	ASTM C1289, Type I or II
Wood fiberboard	ASTM C208

SECTION R907
ROOFTOP-MOUNTED PHOTOVOLTAIC PANEL SYSTEMS

R907.1 Rooftop-mounted photovoltaic panel systems. Rooftop-mounted *photovoltaic panel systems* shall be designed and installed in accordance with Section R324 and NFPA 70.

SECTION R908
REROOFING

R908.1 General. Materials and methods of application used for recovering or replacing an existing roof covering shall comply with the requirements of Chapter 9.

Exceptions:

1. *Reroofing* shall not be required to meet the minimum design slope requirement of one-quarter unit vertical in 12 units horizontal (2-percent slope) in Section R905 for roofs that provide *positive roof drainage.*
2. For roofs that provide positive drainage, recovering or replacing an existing roof covering shall not require the secondary (emergency overflow) drains or *scuppers* of Section R903.4.1 to be added to an existing roof.

R908.2 Structural and construction loads. The structural roof components shall be capable of supporting the roof covering system and the material and equipment loads that will be encountered during installation of the roof covering system.

R908.3 Roof replacement. *Roof replacement* shall include the removal of existing layers of roof coverings down to the *roof deck.*

Exception: Where the existing *roof assembly* includes an ice barrier membrane that is adhered to the *roof deck*, the existing ice barrier membrane shall be permitted to remain in place and covered with an additional layer of ice barrier membrane in accordance with Section R905.

R908.3.1 Roof recover. The installation of a new roof covering over an existing roof covering shall be permitted where any of the following conditions occur:

1. Where the new roof covering is installed in accordance with the roof covering manufacturer's approved instructions
2. Complete and separate roofing systems, such as standing-seam metal roof systems, that are designed to transmit the roof loads directly to the building's structural system and do not rely on existing roofs and roof coverings for support, shall not require the removal of existing roof coverings.
3. Metal panel, metal shingle and concrete and clay tile roof coverings shall be permitted to be installed over existing wood shake roofs where applied in accordance with Section R908.4.

4. The application of a new protective *roof coating* over an existing protective *roof coating, metal roof panel, metal roof shingle*, mineral surfaced roll roofing, built-up roof, modified bitumen roofing, thermoset and thermoplastic single-ply roofing and spray polyurethane foam roofing system shall be permitted without tear-off of existing roof coverings.

R908.3.1.1 Roof recover not allowed. A *roof recover* shall not be permitted where any of the following conditions occur:

1. Where the existing roof or roof covering is water soaked or has deteriorated to the point that the existing roof or roof covering is not adequate as a base for additional roofing.
2. Where the existing roof covering is slate, clay, cement or asbestos-cement tile.
3. Where the existing roof has two or more applications of any type of roof covering.

R908.4 Roof recovering. Where the application of a new roof covering over wood shingle or shake roofs creates a combustible concealed space, the entire existing surface shall be covered with gypsum board, mineral fiber, glass fiber or other *approved* materials securely fastened in place.

R908.5 Reinstallation of materials. Existing slate, clay or cement tile shall be permitted for reinstallation, except that damaged, cracked or broken slate or tile shall not be reinstalled. Any existing flashings, edgings, outlets, vents or similar devices that are a part of the assembly shall be replaced where rusted, damaged or deteriorated. Aggregate surfacing materials shall not be reinstalled.

R908.6 Flashings. Flashings shall be reconstructed in accordance with *approved* manufacturer's installation instructions. Metal flashing to which bituminous materials are to be adhered shall be primed prior to installation.

CHAPTER 10

CHIMNEYS AND FIREPLACES

User note:

About this chapter: Chapter 10 contains requirements for the construction, seismic reinforcing and anchorage of masonry chimneys and fireplaces; and establishes standards for the use and installation of factory-built chimneys, fireplaces and masonry heaters. Chimneys and fireplaces constructed of masonry rely on prescriptive requirements for the details of their construction; factory-built versions rely on the listing and labeling method of approval.

SECTION R1001 MASONRY FIREPLACES

R1001.1 General. Masonry fireplaces shall be constructed in accordance with this section and the applicable provisions of Chapters 3 and 4.

R1001.2 Footings and foundations. Footings for masonry fireplaces and their chimneys shall be constructed of concrete or *solid masonry* not less than 12 inches (305 mm) thick and shall extend not less than 6 inches (152 mm) beyond the face of the fireplace or foundation wall on all sides. Footings shall be founded on natural, undisturbed earth or engineered fill below frost depth. In areas not subjected to freezing, footings shall be not less than 12 inches (305 mm) below finished *grade*.

R1001.2.1 Ash dump cleanout. Cleanout openings located within foundation walls below fireboxes, where provided, shall be equipped with ferrous metal or masonry doors and frames constructed to remain tightly closed except when in use. Cleanouts shall be located to allow *access* so that ash removal will not create a hazard to *combustible materials*.

R1001.3 Seismic reinforcing. Masonry or concrete chimneys in Seismic Design Category D_0, D_1 or D_2 shall be reinforced. Reinforcing shall conform to the requirements set forth in Table R1001.1 and Section R606.

R1001.3.1 Vertical reinforcing. For chimneys up to 40 inches (1016 mm) wide, four No. 4 continuous vertical bars shall be placed between wythes of *solid masonry* or within the cells of hollow unit masonry and grouted in accordance with Section R606. Grout shall be prevented from bonding with the flue liner so that the flue liner is free to move with thermal expansion. For chimneys more than 40 inches (1016 mm) wide, two additional No. 4 vertical bars shall be provided for each additional flue incorporated into the chimney or for each additional 40 inches (1016 mm) in width or fraction thereof.

R1001.3.2 Horizontal reinforcing. Vertical reinforcement shall be placed within $^1/_4$-inch (6.4 mm) ties, or other reinforcing of equivalent net cross-sectional area, placed in the bed joints in accordance with Section R606 at not less than every 18 inches (457 mm) of vertical height. Two such ties shall be installed at each bend in the vertical bars.

R1001.4 Seismic anchorage. Masonry or concrete chimneys in Seismic Design Category D_0, D_1 or D_2 shall be anchored at each floor, ceiling or roof line more than 6 feet (1829 mm) above *grade*, except where constructed completely within the exterior walls. Anchorage shall conform to the requirements of Section R1001.4.1.

R1001.4.1 Anchorage. Two $^3/_{16}$-inch by 1-inch (5 mm by 25 mm) straps shall be embedded not less than 12 inches (305 mm) into the chimney. Straps shall be hooked around the outer bars and extend 6 inches (152 mm) beyond the bend. Each strap shall be fastened to not less than four floor ceiling or floor joists or rafters with two $^1/_2$-inch (12.7 mm) bolts.

R1001.4.1.1 Cold-formed steel framing. Where cold-formed steel framing is used, the location where the $^1/_2$-inch (12.7 mm) bolts are used to attach the straps to the framing shall be reinforced with not less than a 3-inch × 3-inch × 0.229-inch (76 mm × 76 mm × 5.8 mm) steel plate on top of the strap that is screwed to the framing with not fewer than seven No. 6 screws for each bolt.

R1001.5 Firebox walls. Masonry fireboxes shall be constructed of *solid masonry* units, *hollow masonry units* grouted solid, stone or concrete. Where a lining of firebrick not less than 2 inches (51 mm) thick or other *approved* lining is provided, the minimum thickness of back and sidewalls shall each be 8 inches (203 mm) of *solid masonry*, including the lining. The width of joints between firebricks shall not be greater than $^1/_4$ inch (6.4 mm). Where a lining is not provided, the total minimum thickness of back and side walls shall be 10 inches (254 mm) of *solid masonry*. Firebrick shall conform to ASTM C27 or C1261 and shall be laid with medium- duty refractory mortar conforming to ASTM C199.

R1001.5.1 Steel fireplace units. Installation of steel fireplace units with *solid masonry* to form a masonry fireplace is permitted where installed either in accordance with the requirements of their listing or the requirements of this section. Steel fireplace units incorporating a steel firebox lining shall be constructed with steel not less than $^1/_4$ inch (6.4 mm) thick, and an air-circulating chamber that is ducted to the interior of the building. The firebox lining shall be encased with *solid masonry* to provide a total thickness at the back and sides of not less than 8 inches (203 mm), of which not less than 4 inches (102 mm) shall be of *solid masonry* or concrete. Circulating air ducts used with steel fireplace units shall be constructed of metal or masonry.

TABLE R1001.1
SUMMARY OF REQUIREMENTS FOR MASONRY FIREPLACES AND CHIMNEYS

ITEM	LETTER[a]	REQUIREMENTS
Hearth slab thickness	A	4 inches
Hearth extension (each side of opening)	B	8-inch fireplace opening < 6 square feet. 12-inch fireplace opening ≥ 6 square feet.
Hearth extension (front of opening)	C	16-inch fireplace opening < 6 square feet. 20-inch fireplace opening ≥ 6 square feet.
Hearth slab reinforcing	D	Reinforced to carry its own weight and all imposed loads.
Thickness of wall of firebox	E	10-inch solid brick or 8 inches where a firebrick lining is used. Joints in firebrick $^{1}/_{4}$-inch maximum.
Distance from top of opening to throat	F	8 inches
Smoke chamber wall thickness Unlined walls	G	6 inches 8 inches
Chimney Vertical reinforcing[b]	H	Four No. 4 full-length bars for chimney up to 40 inches wide. Add two No. 4 bars for each additional 40 inches or fraction of width or each additional flue.
Horizontal reinforcing	J	$^{1}/_{4}$-inch ties at 18 inches and two ties at each bend in vertical steel.
Bond beams	K	No specified requirements.
Fireplace lintel	L	Noncombustible material.
Chimney walls with flue lining	M	Solid masonry units or hollow masonry units grouted solid with not less than 4-inch nominal thickness.
Distances between adjacent flues	—	See Section R1003.13.
Effective flue area (based on area of fireplace opening)	P	See Section R1003.15.
Clearances Combustible material Mantel and trim Above roof	R	 See Sections R1001.11 and R1003.18. See Section R1001.11, Exception 4. 3 feet at roofline and 2 feet at 10 feet.
Anchorage[b] Strap Number Embedment into chimney Fasten to Bolts	S	 $^{3}/_{16}$-inch × 1-inch Two 12 inches hooked around outer bar with 6-inch extension. 4 joists Two $^{1}/_{2}$-inch diameter.
Footing Thickness Width	T	 12 inches min. 6 inches each side of fireplace wall.

For SI: 1 inch = 25.4 mm, 1 foot = 304.8 mm, 1 square foot = 0.0929 m^2.

Note: This table provides a summary of major requirements for the construction of masonry chimneys and fireplaces. Letter references are to Figure R1001.1, which shows examples of typical construction. This table does not cover all requirements, nor does it cover all aspects of the indicated requirements. For the actual mandatory requirements of the code, see the indicated section of text.

a. The letters refer to Figure R1001.1.

b. Not required in Seismic Design Category A, B or C.

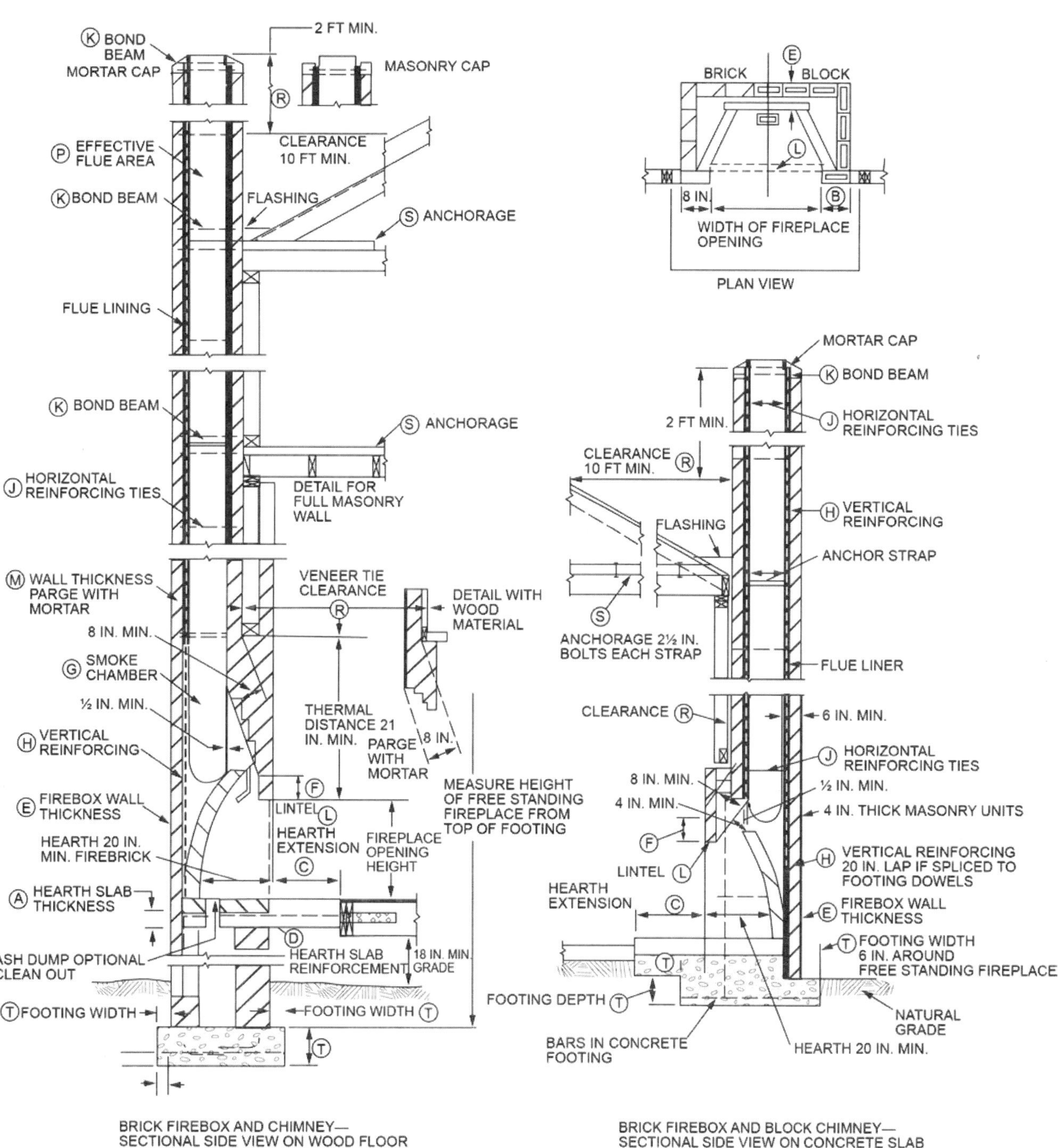

For SI: 1 inch = 25.4 mm, 1 foot = 304.8 mm.

FIGURE R1001.1
FIREPLACE AND CHIMNEY DETAILS

R1001.6 Firebox dimensions. The firebox of a concrete or masonry fireplace shall have a depth of not less than 20 inches (508 mm). The throat shall be not less than 8 inches (203 mm) above the fireplace opening. The throat opening shall be not less than 4 inches (102 mm) deep. The cross-sectional area of the passageway above the firebox, including the throat, damper and smoke chamber, shall be not less than the cross-sectional area of the flue.

Exception: Rumford fireplaces shall be permitted provided that the depth of the fireplace is not less than 12 inches (305 mm) and not less than one-third of the width of the fireplace opening, that the throat is not less than 12 inches (305 mm) above the lintel and is not less than one-twentieth the cross-sectional area of the fireplace opening.

R1001.7 Lintel and throat. Masonry over a fireplace opening shall be supported by a lintel of *noncombustible material.* The minimum required bearing length on each end of the fireplace opening shall be 4 inches (102 mm). The fireplace throat or damper shall be located not less than 8 inches (203 mm) above the lintel.

R1001.7.1 Damper. Masonry fireplaces shall be equipped with a ferrous metal damper located not less than 8 inches (203 mm) above the top of the fireplace opening. Dampers shall be installed in the fireplace or the chimney venting the fireplace, and shall be operable from the room containing the fireplace.

R1001.8 Smoke chamber. Smoke chamber walls shall be constructed of *solid masonry* units, *hollow masonry units* grouted solid, stone or concrete. The total minimum thickness of front, back and side walls shall be 8 inches (203 mm) of *solid masonry.* The inside surface shall be parged smooth with refractory mortar conforming to ASTM C199. Where a lining of firebrick not less than 2 inches (51 mm) thick, or a lining of vitrified clay not less than $^5/_8$ inch (16 mm) thick, is provided, the total minimum thickness of front, back and side walls shall be 6 inches (152 mm) of *solid masonry,* including the lining. Firebrick shall conform to ASTM C1261 and shall be laid with medium-duty refractory mortar conforming to ASTM C199. Vitrified clay linings shall conform to ASTM C315.

R1001.8.1 Smoke chamber dimensions. The inside height of the smoke chamber from the fireplace throat to the beginning of the flue shall not be greater than the inside width of the fireplace opening. The inside surface of the smoke chamber shall not be inclined more than 45 degrees (0.79 rad) from vertical where prefabricated smoke chamber linings are used or where the smoke chamber walls are rolled or sloped rather than corbeled. Where the inside surface of the smoke chamber is formed by corbeled masonry, the walls shall not be corbeled more than 30 degrees (0.52 rad) from vertical.

R1001.9 Hearth and hearth extension. Masonry fireplace hearths and hearth extensions shall be constructed of concrete or masonry, supported by *noncombustible materials,* and reinforced to carry their own weight and all imposed loads. *Combustible material* shall not remain against the underside of hearths and hearth extensions after construction.

R1001.9.1 Hearth thickness. The minimum thickness of fireplace hearths shall be 4 inches (102 mm).

R1001.9.2 Hearth extension thickness. The minimum thickness of hearth extensions shall be 2 inches (51 mm).

Exception: Where the bottom of the firebox opening is raised not less than 8 inches (203 mm) above the top of the hearth extension, a hearth extension of not less than $^3/_8$-inch-thick (10 mm) brick, concrete, stone, tile or other *approved noncombustible material* is permitted.

R1001.10 Hearth extension dimensions. Hearth extensions shall extend not less than 16 inches (406 mm) in front of and not less than 8 inches (203 mm) beyond each side of the fireplace opening. Where the fireplace opening is 6 square feet (0.6 m^2) or larger, the hearth extension shall extend not less than 20 inches (508 mm) in front of and not less than 12 inches (305 mm) beyond each side of the fireplace opening.

R1001.11 Fireplace clearance. Wood beams, joists, studs and other *combustible material* shall have a clearance of not less than 2 inches (51 mm) from the front faces and sides of masonry fireplaces and not less than 4 inches (102 mm) from the back faces of masonry fireplaces. The airspace shall not be filled, except to provide fireblocking in accordance with Section R1001.12.

Exceptions:

1. Masonry fireplaces *listed* and *labeled* for use in contact with combustibles in accordance with UL 127 and installed in accordance with the manufacturer's instructions are permitted to have *combustible material* in contact with their exterior surfaces.
2. Where masonry fireplaces are part of masonry or concrete walls, *combustible materials* shall not be in contact with the masonry or concrete walls less than 12 inches (306 mm) from the inside surface of the nearest firebox lining.
3. Exposed combustible *trim* and the edges of sheathing materials such as wood siding, flooring and gypsum board shall be permitted to abut the masonry fireplace sidewalls and hearth extension in accordance with Figure R1001.11, provided such combustible *trim* or sheathing is not less than 12 inches (305 mm) from the inside surface of the nearest firebox lining.
4. Exposed combustible mantels or *trim* is permitted to be placed directly on the masonry fireplace front surrounding the fireplace opening providing such *combustible materials* are not placed within 6 inches (152 mm) of a fireplace opening. *Combustible material* within 12 inches (306 mm) of the fireplace opening shall not project more than $^1/_8$ inch (3 mm) for each 1-inch (25 mm) distance from such an opening.

R1001.12 Fireplace fireblocking. Fireplace fireblocking shall comply with the provisions of Section R602.8.

R1001.13 Fireplace accessories. *Listed* and *labeled* fireplace accessories shall be installed in accordance with the conditions of the listing and the manufacturer's instructions. Fireplace accessories shall comply with UL 907.

SECTION R1002 MASONRY HEATERS

R1002.1 Definition. A *masonry heater* is a heating *appliance* constructed of concrete or *solid masonry*, hereinafter referred to as masonry, that is designed to absorb and store heat from a solid-fuel fire built in the firebox by routing the exhaust gases through internal heat exchange channels in which the flow path downstream of the firebox includes flow in a horizontal or downward direction before entering the chimney and that delivers heat by radiation from the masonry surface of the heater.

R1002.2 Installation. *Masonry heaters* shall be installed in accordance with this section and comply with one of the following:

1. *Masonry heaters* shall comply with the requirements of ASTM E1602.
2. *Masonry heaters* shall be *listed* and *labeled* in accordance with UL 1482 or CEN 15250 and installed in accordance with the manufacturer's instructions.

R1002.3 Footings and foundation. The firebox floor of a *masonry heater* shall be a minimum thickness of 4 inches (102 mm) of *noncombustible material* and be supported on a noncombustible footing and foundation in accordance with Section R1003.2.

R1002.4 Seismic reinforcing. In *Seismic Design Categories* D_0, D_1 and D_2, *masonry heaters* shall be anchored to the masonry foundation in accordance with Section R1003.3. Seismic reinforcing shall not be required within the body of a *masonry heater* whose height is equal to or less than 3.5 times its body width and where the *masonry chimney* serving the heater is not supported by the body of the heater. Where the *masonry chimney* shares a common wall with the facing of the *masonry heater*, the chimney portion of the structure shall be reinforced in accordance with Section R1003.

R1002.5 Masonry heater clearance. *Combustible materials* shall not be placed within 36 inches (914 mm) of the outside surface of a *masonry heater* in accordance with NFPA 211 Section 8-7 (clearances for solid-fuel-burning *appliances*), and the required space between the heater and *combustible material* shall be fully vented to permit the free flow of air around all heater surfaces.

Exceptions:

1. Where the *masonry heater* wall is not less than 8 inches (203 mm) thick of *solid masonry* and the wall of the heat exchange channels is not less than 5 inches (127 mm) thick of *solid masonry*, *combustible materials* shall not be placed within 4 inches (102 mm) of the outside surface of a *masonry heater*. A clearance of not less than 8 inches (203 mm) shall be provided between the gas-tight capping slab of the heater and a combustible ceiling.
2. *Masonry heaters listed* and *labeled* in accordance with UL 1482 or CEN 15250 shall be installed in accordance with the listing specifications and the manufacturer's written instructions.

SECTION R1003 MASONRY CHIMNEYS

R1003.1 Definition. A *masonry chimney* is a chimney constructed of *solid masonry* units, *hollow masonry units* grouted solid, stone or concrete, hereinafter referred to as masonry. *Masonry chimneys* shall be constructed, anchored, supported and reinforced as required in this chapter.

R1003.2 Footings and foundations. Footings for *masonry chimneys* shall be constructed of concrete or *solid masonry* not less than 12 inches (305 mm) thick and shall extend not

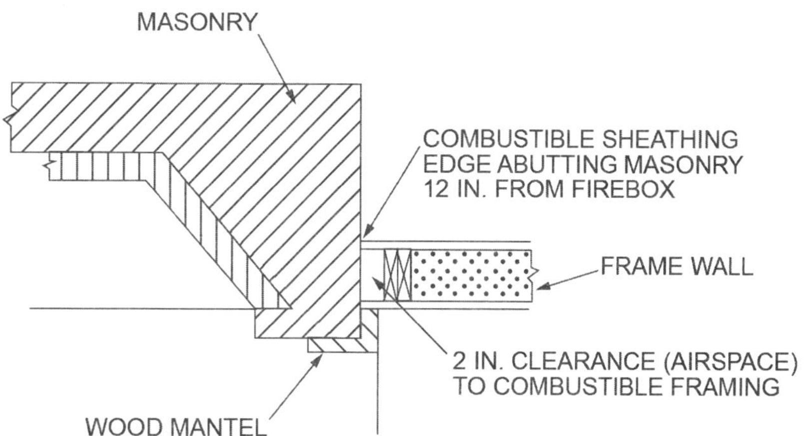

For SI: 1 inch = 25.4 mm.

FIGURE R1001.11 CLEARANCE FROM COMBUSTIBLES

less than 6 inches (152 mm) beyond the face of the foundation or support wall on all sides. Footings shall be founded on natural undisturbed earth or engineered fill below frost depth. In areas not subjected to freezing, footings shall be not less than 12 inches (305 mm) below finished *grade*.

R1003.3 Seismic reinforcing. Masonry or concrete chimneys shall be constructed, anchored, supported and reinforced as required in this chapter. In Seismic Design Category D_0, D_1 or D_2 masonry and concrete chimneys shall be reinforced and anchored as detailed in Sections R1003.3.1, R1003.3.2 and R1003.4. In Seismic Design Category A, B or C, reinforcement and seismic anchorage are not required.

R1003.3.1 Vertical reinforcing. For chimneys up to 40 inches (1016 mm) wide, four No. 4 continuous vertical bars, anchored in the foundation, shall be placed in the concrete, or between wythes of *solid masonry*, or within the cells of hollow unit masonry, and grouted in accordance with Section R608.1.1. Grout shall be prevented from bonding with the flue liner so that the flue liner is free to move with thermal expansion. For chimneys more than 40 inches (1016 mm) wide, two additional No. 4 vertical bars shall be installed for each additional 40 inches (1016 mm) in width or fraction thereof.

R1003.3.2 Horizontal reinforcing. Vertical reinforcement shall be placed enclosed within $^1/_4$-inch (6.4 mm) ties, or other reinforcing of equivalent net cross-sectional area, spaced not to exceed 18 inches (457 mm) on center in concrete, or placed in the bed joints of unit masonry, at not less than every 18 inches (457 mm) of vertical height. Two such ties shall be installed at each bend in the vertical bars.

R1003.4 Seismic anchorage. Masonry and concrete chimneys and foundations in Seismic Design Category D_0, D_1 or D_2 shall be anchored at each floor, ceiling or roof line more than 6 feet (1829 mm) above *grade*, except where constructed completely within the exterior walls. Anchorage shall conform to the requirements in Section R1003.4.1.

R1003.4.1 Anchorage. Two $^3/_{16}$-inch by 1-inch (5 mm by 25 mm) straps shall be embedded not less than 12 inches (305 mm) into the chimney. Straps shall be hooked around the outer bars and extend 6 inches (152 mm) beyond the bend. Each strap shall be fastened to not less than four floor joists with two $^1/_2$-inch (12.7 mm) bolts.

R1003.4.1.1 Cold-formed steel framing. Where cold-formed steel framing is used, the location where the $^1/_2$-inch (12.7 mm) bolts are used to attach the straps to the framing shall be reinforced with not less than a 3-inch × 3-inch × 0.229-inch (76 mm × 76 mm × 5.8 mm) steel plate on top of a strap that is screwed to the framing with not fewer than seven No. 6 screws for each bolt.

R1003.5 Corbeling. *Masonry chimneys* shall not be corbeled more than one-half of the chimney's wall thickness from a wall or foundation, nor shall a chimney be corbeled from a wall or foundation that is less than 12 inches (305 mm) thick unless it projects equally on each side of the wall, except that on the second *story* of a two-story *dwelling*, corbeling of chimneys on the exterior of the enclosing walls shall be permitted to be equal to the wall thickness. The projection of a single course shall not exceed one-half the unit height or one-third of the unit bed depth, whichever is less.

R1003.6 Changes in dimension. The chimney wall or chimney flue lining shall not change in size or shape within 6 inches (152 mm) above or below where the chimney passes through floor components, ceiling components or roof components.

R1003.7 Offsets. Where a *masonry chimney* is constructed with a fireclay flue liner surrounded by one wythe of masonry, the maximum offset shall be such that the centerline of the flue above the offset does not extend beyond the center of the chimney wall below the offset. Where the chimney offset is supported by masonry below the offset in an *approved* manner, the maximum offset limitations shall not apply. Each individual corbeled masonry course of the offset shall not exceed the projection limitations specified in Section R1003.5.

R1003.8 Additional load. Chimneys shall not support loads other than their own weight unless they are designed and constructed to support the additional load. Construction of *masonry chimneys* as part of the masonry walls or reinforced concrete walls of the building shall be permitted.

R1003.9 Termination. Chimneys shall extend not less than 2 feet (610 mm) higher than any portion of a building within 10 feet (3048 mm), but shall be not less than 3 feet (914 mm) above the highest point where the chimney passes through the roof.

R1003.9.1 Chimney caps. *Masonry chimneys* shall have a concrete, metal or stone cap, a drip edge and a caulked bond break around any flue liners in accordance with ASTM C1283. The concrete, metal or stone cap shall be sloped to shed water.

R1003.9.2 Spark arrestors. Where a spark arrestor is installed on a *masonry chimney*, the spark arrestor shall meet all of the following requirements:

1. The net free area of the arrestor shall be not less than four times the net free area of the outlet of the chimney flue it serves.
2. The arrestor screen shall have heat and corrosion resistance equivalent to 19-gage galvanized steel or 24-gage stainless steel.
3. Openings shall not permit the passage of spheres having a diameter greater than $^1/_2$ inch (12.7 mm) nor block the passage of spheres having a diameter less than $^3/_8$ inch (9.5 mm).
4. The spark arrestor shall be located with *access* for cleaning and the screen or chimney cap shall be removable to allow for cleaning of the chimney flue.

R1003.9.3 Rain caps. Where a masonry or metal rain cap is installed on a *masonry chimney*, the net free area under the cap shall be not less than four times the net free area of the outlet of the chimney flue it serves.

R1003.10 Wall thickness. *Masonry chimney* walls shall be constructed of *solid masonry* units or *hollow masonry units* grouted solid with not less than a 4-inch (102 mm) nominal thickness.

R1003.10.1 Masonry veneer chimneys. Where masonry is used to veneer a frame chimney, through-flashing and weep holes shall be installed as required by Section R703.

R1003.11 Flue lining (material). *Masonry chimneys* shall be lined. The lining material shall be appropriate for the type of *appliance* connected, in accordance with the terms of the *appliance* listing and manufacturer's instructions.

R1003.11.1 Residential-type appliances (general). Flue lining systems shall comply with one of the following:

1. Clay flue lining complying with the requirements of ASTM C315.
2. *Listed* and *labeled* chimney lining systems complying with UL 1777.
3. Factory-built chimneys or chimney units *listed* for installation within *masonry chimneys.*
4. Other *approved* materials that will resist corrosion, erosion, softening or cracking from flue gases and condensate at temperatures up to 1,800°F (982°C).

R1003.11.2 Flue linings for specific appliances. Flue linings other than these covered in Section R1003.11.1, intended for use with specific types of *appliances,* shall comply with Sections R1003.11.3 through R1003.11.6.

R1003.11.3 Gas appliances. Flue lining systems for gas *appliances* shall be in accordance with Chapter 24.

R1003.11.4 Pellet fuel-burning appliances. Flue lining and vent systems for use in *masonry chimneys* with pellet fuel-burning *appliances* shall be limited to the following:

1. Flue lining systems complying with Section R1003.11.1.
2. Pellet vents *listed* for installation within *masonry chimneys* (see Section R1003.11.6 for marking).

R1003.11.5 Oil-fired appliances approved for use with Type L vent. Flue lining and vent systems for use in *masonry chimneys* with oil-fired *appliances approved* for use with Type L vent shall be limited to the following:

1. Flue lining systems complying with Section R1003.11.1.
2. *Listed* chimney liners complying with UL 641 (see Section R1003.11.6 for marking).

R1003.11.6 Notice of usage. Where a flue is relined with a material not complying with Section R1003.11.1, the chimney shall be plainly and permanently identified by a *label* attached to a wall, ceiling or other conspicuous location adjacent to where the connector enters the chimney. The *label* shall include the following message or equivalent language:

THIS CHIMNEY FLUE IS FOR USE ONLY WITH [TYPE OR CATEGORY OF APPLIANCE] APPLIANCES THAT BURN [TYPE OF FUEL]. DO NOT CONNECT OTHER TYPES OF APPLIANCES.

R1003.12 Clay flue lining (installation). Clay flue liners shall be installed in accordance with ASTM C1283 and extend from a point not less than 8 inches (203 mm) below the lowest inlet or, in the case of fireplaces, from the top of the smoke chamber to a point above the enclosing walls. The lining shall be carried up vertically, with a slope not greater than 30 degrees (0.52 rad) from the vertical.

Clay flue liners shall be laid in medium-duty water insoluble refractory mortar conforming to ASTM C199 with tight mortar joints left smooth on the inside and installed to maintain an airspace or insulation not to exceed the thickness of the flue liner separating the flue liners from the interior face of the chimney masonry walls. Flue liners shall be supported on all sides. Only enough mortar shall be placed to make the joint and hold the liners in position.

R1003.12.1 Listed materials. *Listed* materials used as flue linings shall be installed in accordance with the terms of their listings and manufacturer's instructions.

R1003.12.2 Space around lining. The space surrounding a chimney lining system or vent installed within a *masonry chimney* shall not be used to vent any other *appliance.*

Exception: This shall not prevent the installation of a separate flue lining in accordance with the manufacturer's instructions.

R1003.13 Multiple flues. Where two or more flues are located in the same chimney, masonry wythes shall be built between adjacent flue linings. The masonry wythes shall be not less than 4 inches (102 mm) thick and bonded into the walls of the chimney.

Exception: Where venting only one *appliance,* two flues shall be permitted to adjoin each other in the same chimney with only the flue lining separation between them. The joints of the adjacent flue linings shall be staggered not less than 4 inches (102 mm).

R1003.14 Flue area (appliance). Chimney flues shall not be smaller in area than that of the area of the connector from the *appliance* [see Tables R1003.14(1) and R1003.14(2)]. The sizing of a chimney flue to which multiple *appliance* venting systems are connected shall be in accordance with Section M1805.3.

TABLE R1003.14(1)
NET CROSS-SECTIONAL AREA OF ROUND FLUE SIZES[a]

FLUE SIZE, INSIDE DIAMETER (inches)	CROSS-SECTIONAL AREA (square inches)
6	28
7	38
8	50
10	78
$10^3/_4$	90
12	113
15	176
18	254

For SI: 1 inch = 25.4 mm, 1 square inch = 645.16 mm^2.
a. Flue sizes are based on ASTM C315.

TABLE R1003.14(2)
NET CROSS-SECTIONAL AREA OF SQUARE AND RECTANGULAR FLUE SIZES

FLUE SIZE, OUTSIDE NOMINAL DIMENSIONS (inches)	CROSS-SECTIONAL AREA (square inches)
4.5 × 8.5	23
4.5 × 13	34
8 × 8	42
8.5 × 8.5	49
8 × 12	67
8.5 × 13	76
12 × 12	102
8.5 × 18	101
13 × 13	127
12 × 16	131
13 × 18	173
16 × 16	181
16 × 20	222
18 × 18	233
20 × 20	298
20 × 24	335
24 × 24	431

For SI: 1 inch = 25.4 mm, 1 square inch = 645.16 mm^2.

R1003.15 Flue area (masonry fireplace). Flue sizing for chimneys serving fireplaces shall be in accordance with Section R1003.15.1 or R1003.15.2.

R1003.15.1 Option 1. Round chimney flues shall have a minimum net cross-sectional area of not less than one-twelfth of the fireplace opening. Square chimney flues shall have a minimum net cross-sectional area of one-tenth of the fireplace opening. Rectangular chimney flues with an *aspect ratio* less than 2 to 1 shall have a minimum net cross-sectional area of one-tenth of the fireplace opening. Rectangular chimney flues with an *aspect ratio* of 2 to 1 or more shall have a minimum net cross-sectional area of one-eighth of the fireplace opening. Cross-sectional areas of clay flue linings are shown in Tables R1003.14(1) and R1003.14(2) or as provided by the manufacturer or as measured in the field.

R1003.15.2 Option 2. The minimum net cross-sectional area of the chimney flue shall be determined in accordance with Figure R1003.15.2. A flue size providing not less than the equivalent net cross-sectional area shall be used. Cross-sectional areas of clay flue linings are shown in Tables R1003.14(1) and R1003.14(2) or as provided by the manufacturer or as measured in the field. The height of the chimney shall be measured from the firebox floor to the top of the chimney flue.

R1003.16 Inlet. Inlets to *masonry chimneys* shall enter from the side. Inlets shall have a thimble of fireclay, rigid refractory material or metal that will prevent the connector from pulling out of the inlet or from extending beyond the wall of the liner.

R1003.17 Masonry chimney cleanout openings. Cleanout openings shall be provided within 6 inches (152 mm) of the base of each flue within every *masonry chimney*. The upper edge of the cleanout shall be located not less than 6 inches (152 mm) below the lowest chimney inlet opening. The height of the opening shall be not less than 6 inches (152 mm). The cleanout shall be provided with a noncombustible cover.

Exception: Chimney flues serving masonry fireplaces where cleaning is possible through the fireplace opening.

R1003.18 Chimney clearances. Any portion of a *masonry chimney* located in the interior of the building or within the exterior wall of the building shall have a minimum airspace clearance to combustibles of 2 inches (51 mm). Chimneys located entirely outside the exterior walls of the building, including chimneys that pass through the soffit or cornice, shall have a minimum airspace clearance of 1 inch (25 mm). The airspace shall not be filled, except to provide fire blocking in accordance with Section R1003.19.

Exceptions:

1. *Masonry chimneys* equipped with a chimney lining system *listed* and *labeled* for use in chimneys in contact with combustibles in accordance with UL 1777 and installed in accordance with the manufacturer's instructions are permitted to have *combustible material* in contact with their exterior surfaces.
2. Where *masonry chimneys* are constructed as part of masonry or concrete walls, *combustible materials* shall not be in contact with the masonry or concrete wall less than 12 inches (305 mm) from the inside surface of the nearest flue lining.
3. Exposed combustible *trim* and the edges of sheathing materials, such as wood siding and flooring, shall be permitted to abut the *masonry chimney* side walls, in accordance with Figure R1003.18, provided such combustible trim or sheathing is not less than 8 inches (203 mm) from the inside surface of the nearest flue lining.

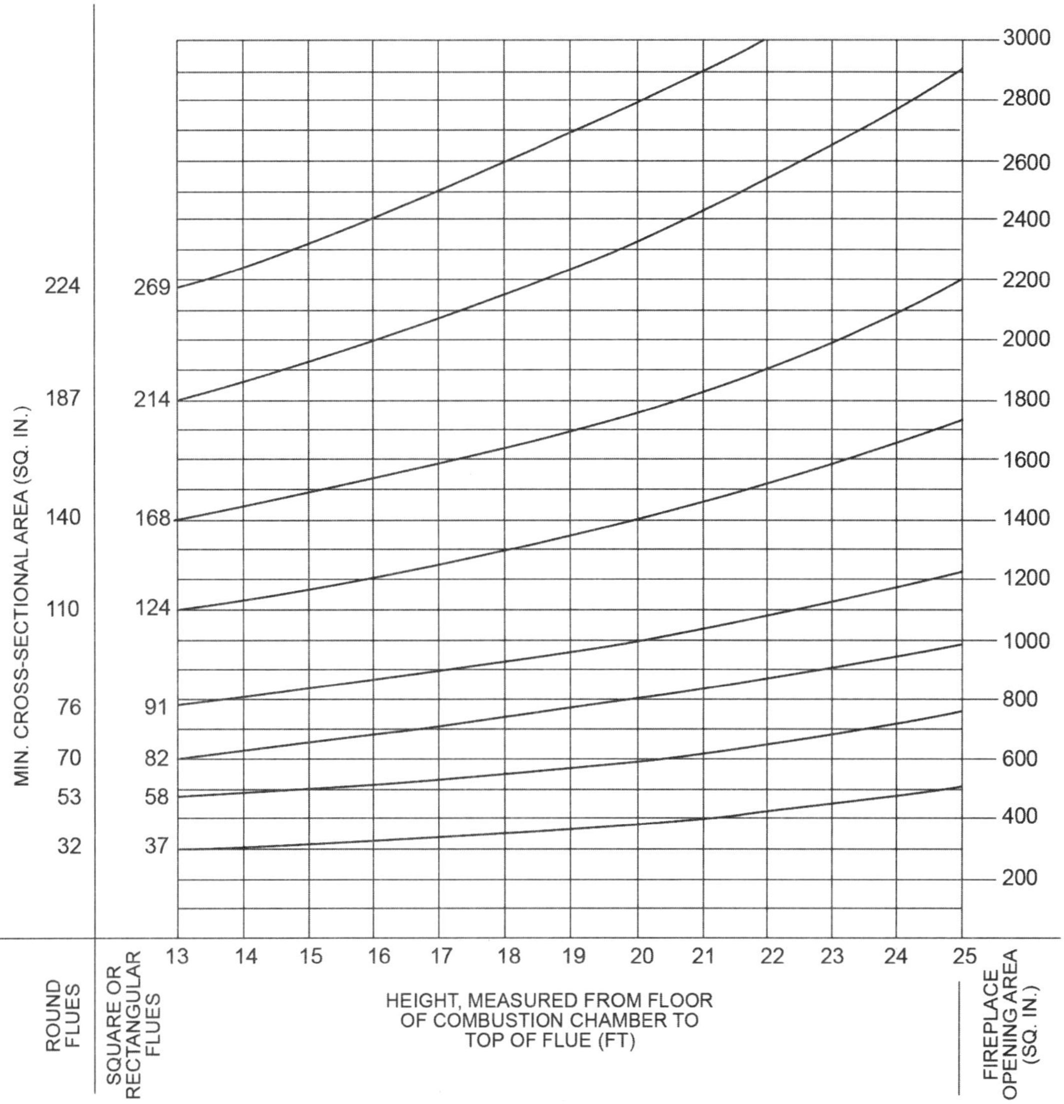

MIN. CROSS-SECTIONAL AREA (SQ. IN.)

ROUND FLUES	SQUARE OR RECTANGULAR FLUES
224	269
187	214
140	168
110	124
76	91
70	82
53	58
32	37

FIREPLACE OPENING AREA (SQ. IN.): 3000, 2800, 2600, 2400, 2200, 2000, 1800, 1600, 1400, 1200, 1000, 800, 600, 400, 200

HEIGHT, MEASURED FROM FLOOR OF COMBUSTION CHAMBER TO TOP OF FLUE (FT): 13, 14, 15, 16, 17, 18, 19, 20, 21, 22, 23, 24, 25

For SI: 1 foot = 304.8 mm, 1 square inch = 645.16 mm^2.

FIGURE R1003.15.2
FLUE SIZES FOR MASONRY CHIMNEYS

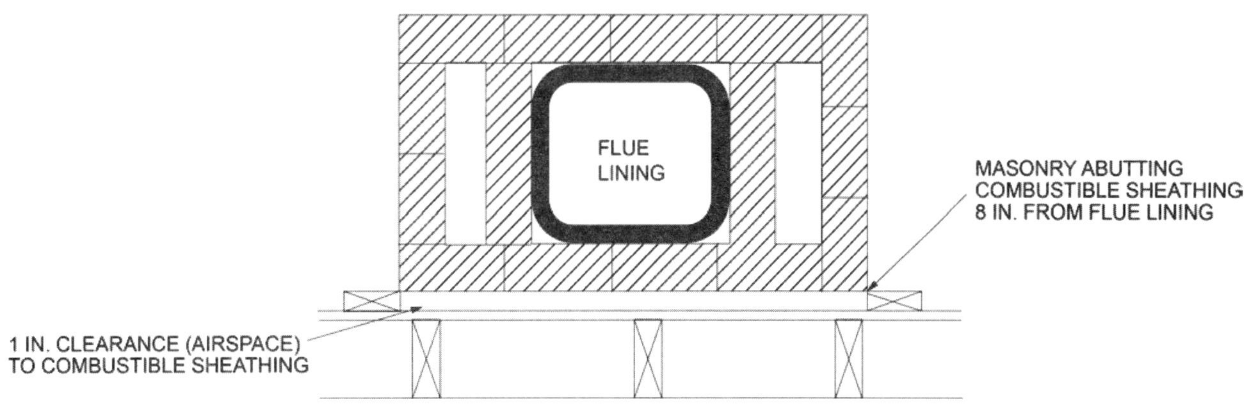

For SI: 1 inch = 25.4 mm.

FIGURE R1003.18
CLEARANCE FROM COMBUSTIBLES

R1003.19 Chimney fireblocking. Spaces between chimneys and floors and ceilings through which chimneys pass shall be fireblocked with *noncombustible material* securely fastened in place. The fireblocking of spaces between chimneys and wood joists, beams or headers shall be self-supporting or be placed on strips of metal or metal lath laid across the spaces between *combustible material* and the chimney.

R1003.20 Chimney crickets. Chimneys shall be provided with crickets where the dimension parallel to the ridgeline is greater than 30 inches (762 mm) and does not intersect the ridgeline. The intersection of the cricket and the chimney shall be flashed and counterflashed in the same manner as normal roof-chimney intersections. Crickets shall be constructed in compliance with Figure R1003.20 and Table R1003.20.

TABLE R1003.20
CRICKET DIMENSIONS

ROOF SLOPE	H
12:12	$^{1}/_{2}$ of W
8:12	$^{1}/_{3}$ of W
6:12	$^{1}/_{4}$ of W
4:12	$^{1}/_{6}$ of W
3:12	$^{1}/_{8}$ of W

SECTION R1004
FACTORY-BUILT FIREPLACES

R1004.1 General. Factory-built fireplaces shall be *listed* and *labeled* and shall be installed in accordance with the conditions of the *listing*. Factory-built fireplaces shall be tested in accordance with UL 127.

R1004.2 Hearth extensions. Hearth extensions of *approved* factory-built fireplaces shall be installed in accordance with the *listing* of the fireplace. The hearth extension shall be readily distinguishable from the surrounding floor area. *Listed* and *labeled* hearth extensions shall comply with UL 1618.

R1004.3 Decorative shrouds. Decorative shrouds shall not be installed at the termination of chimneys for factory-built fireplaces except where the shrouds are *listed* and *labeled* for use with the specific factory-built fireplace system and installed in accordance with the manufacturer's instructions.

R1004.4 Unvented gas log heaters. An unvented gas log heater shall not be installed in a factory-built fireplace unless the fireplace system has been specifically tested, *listed* and *labeled* for such use in accordance with UL 127.

R1004.5 Gasketed fireplace doors. A gasketed fireplace door shall not be installed on a factory-built fireplace except where the fireplace system has been specifically tested, *listed* and *labeled* for such use in accordance with UL 127.

SECTION R1005
FACTORY-BUILT CHIMNEYS

R1005.1 Listing. *Factory-built chimneys* shall be *listed* and *labeled* and shall be installed and terminated in accordance with the *manufacturer's installation instructions.*

R1005.2 Decorative shrouds. Decorative shrouds shall not be installed at the termination of *factory-built chimneys* except where the shrouds are *listed* and *labeled* for use with the specific *factory-built chimney* system and installed in accordance with the *manufacturer's installation instructions.*

R1005.3 Solid-fuel appliances. *Factory-built chimneys* installed in *dwelling units* with solid-fuel-burning *appliances* shall comply with the Type HT requirements of UL 103 and shall be marked "Type HT" and "Residential Type and Building Heating Appliance Chimney."

> **Exception:** *Chimneys* for use with open combustion chamber fireplaces shall comply with the requirements of

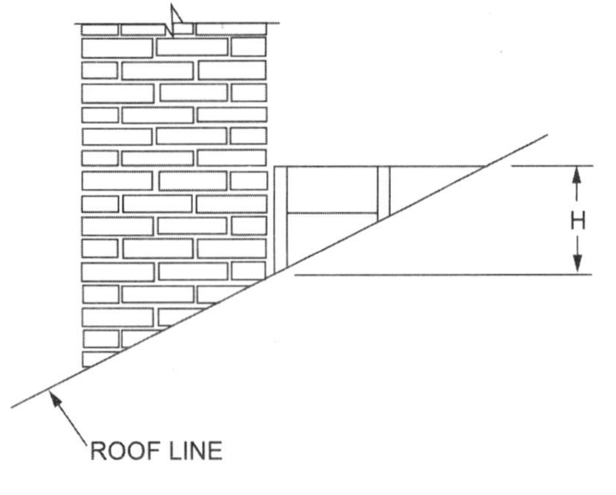

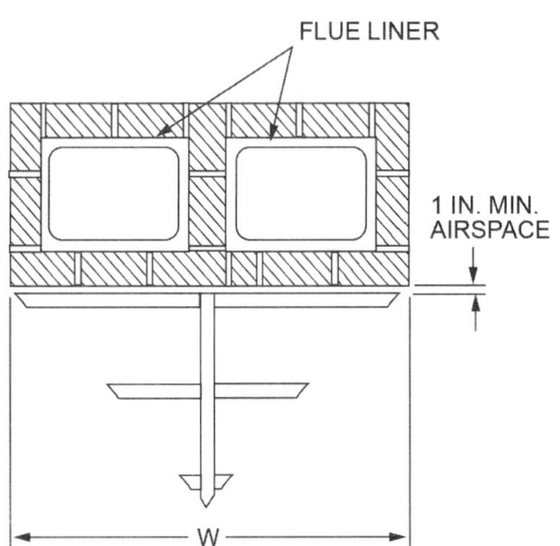

For SI: 1 inch = 25.4 mm.

FIGURE R1003.20
CHIMNEY CRICKET

UL 103 and shall be marked "Residential Type and Building Heating Appliance Chimney."

Chimneys for use with open combustion chamber *appliances* installed in buildings other than *dwelling units* shall comply with the requirements of UL 103 and shall be marked "Building Heating Appliance Chimney" or "Residential Type and Building Heating Appliance Chimney."

R1005.4 Factory-built fireplaces. *Chimneys* for use with factory-built fireplaces shall comply with the requirements of UL 127.

R1005.5 Support. Where *factory-built chimneys* are supported by structural members, such as joists and rafters, those members shall be designed to support the additional load.

R1005.6 Medium-heat appliances. *Factory-built chimneys* for medium-heat *appliances* producing flue gases having a temperature above 1,000°F (538°C), measured at the entrance to the *chimney*, shall comply with UL 959.

R1005.7 Factory-built chimney offsets. Where a *factory-built chimney* assembly incorporates offsets, no part of the *chimney* shall be at an angle of more than 30 degrees (0.52 rad) from vertical at any point in the assembly and the chimney assembly shall not include more than four elbows.

R1005.8 Insulation shield. Where *factory-built chimneys* pass through insulated assemblies, an insulation shield constructed of steel having a thickness of not less than 0.0187 inch (0.4712 mm) (No. 26 gage) shall be installed to provide clearance between the *chimney* and the insulation material. The clearance shall be not less than the clearance to combustibles specified by the chimney *manufacturer's installation instructions*. Where *chimneys* pass through attic space, the shield shall terminate not less than 2 inches (51 mm) above the insulation materials and shall be secured in place to prevent displacement. Insulation shields provided as part of a *listed* chimney system shall be installed in accordance with the manufacturer's installation instructions.

SECTION R1006 EXTERIOR AIR SUPPLY

R1006.1 Exterior air. Factory-built or masonry fireplaces covered in this chapter shall be equipped with an exterior air supply to ensure proper fuel combustion unless the room is mechanically ventilated and controlled so that the indoor pressure is neutral or positive.

R1006.1.1 Factory-built fireplaces. Exterior *combustion air* ducts for factory-built fireplaces shall be a *listed* component of the fireplace and shall be installed in accordance with the fireplace manufacturer's instructions.

R1006.1.2 Masonry fireplaces. *Listed combustion air* ducts for masonry fireplaces shall be installed in accordance with the terms of their *listing* and the manufacturer's instructions.

R1006.2 Exterior air intake. The exterior air intake shall be capable of supplying all *combustion air* from the exterior of the *dwelling* or from spaces within the *dwelling* ventilated with outdoor air such as nonmechanically ventilated crawl or attic spaces. The exterior air intake shall not be located within the garage or *basement* of the dwelling. The exterior air intake, for other than *listed* factory-built fireplaces, shall not be located at an elevation higher than the firebox. The exterior air intake shall be covered with a corrosion-resistant screen of $^1/_4$-inch (6.4 mm) mesh.

R1006.3 Clearance. Unlisted *combustion air* ducts shall be installed with a minimum 1-inch (25 mm) clearance to combustibles for all parts of the duct within 5 feet (1524 mm) of the duct outlet.

R1006.4 Passageway. The *combustion air* passageway shall be not less than 6 square inches (3870 mm^2) and not more than 55 square inches (0.035 m^2), except that *combustion air* systems for *listed* fireplaces shall be constructed in accordance with the fireplace manufacturer's instructions.

R1006.5 Outlet. The exterior air outlet shall be located in the back or side of the firebox chamber or shall be located outside of the firebox, at the level of the hearth and not greater than 24 inches (610 mm) from the firebox opening. The outlet shall be closable and designed to prevent burning material from dropping into concealed combustible spaces.

Part IV—Energy Conservation

CHAPTER 11 [RE]
ENERGY EFFICIENCY

User note:

About this chapter: *The purpose of Chapter 11 [RE] is to provide minimum design requirements that will promote efficient utilization of energy in buildings. The requirements are directed toward the design of building envelopes with adequate thermal resistance and low air leakage, and toward the design and selection of mechanical, water heating, electrical and illumination systems that promote effective use of depletable energy resources.*

SECTION N1101
GENERAL

N1101.1 Scope. This chapter regulates the energy efficiency for the design and construction of buildings regulated by this code.

Note: The text of Sections N1101.2 through N1113 parallels the text of the 2021 edition of the International Energy Conservation Code—Residential Provisions (IECC-R). *The section numbers appearing in parenthesis after each section number are the section numbers of the corresponding text in the* IECC-R. *If a section does not have a section number in parenthesis after it, then there is no corresponding text in the* IECC-R.

N1101.2 (R101.3) Intent. This chapter shall regulate the design and construction of *buildings* for the effective use and conservation of energy over the useful life of each *building.* This chapter is intended to provide flexibility to permit the use of innovative approaches and techniques to achieve this objective. This chapter is not intended to abridge safety, health or environmental requirements contained in other applicable codes or ordinances.

N1101.3 (R101.5.1) Compliance materials. The *code official* shall be permitted to approve specific computer software, worksheets, compliance manuals and other similar materials that meet the intent of this chapter.

N1101.4 (R102.1.1) Above code programs. The *code official* or other authority having *jurisdiction* shall be permitted to deem a national, state or local energy-efficiency program to exceed the energy efficiency required by this code. *Buildings approved* in writing by such an energy-efficiency program shall be considered to be in compliance with this code. The requirements identified in Table N1105.2, as applicable, shall be met.

N1101.5 (R103.2) Information on construction documents. Construction documents shall be drawn to scale on suitable material. Electronic media documents are permitted to be submitted when *approved* by the *code official.* Construction documents shall be of sufficient clarity to indicate the location, nature and extent of the work proposed, and show in sufficient detail pertinent data and features of the building, systems and equipment as herein governed. Details shall include the following as applicable:

1. Energy compliance path.
2. Insulation materials and their *R*-values.
3. Fenestration *U*-factors and solar heat gain coefficients (SHGC).
4. Area-weighted *U*-factor and solar heat gain coefficient (SHGC) calculations.
5. Mechanical system design criteria.
6. Mechanical and service water heating systems and equipment types, sizes and efficiencies.
7. Equipment and system controls.
8. Duct sealing, duct and pipe insulation and location.
9. Air sealing details.

N1101.5.1 (R103.2.1) Building thermal envelope depiction. The *building thermal envelope* shall be represented on the *construction documents.*

N1101.6 (R202) Defined terms. The following words and terms shall, for the purposes of this chapter, have the meanings shown herein.

ABOVE-GRADE WALL. A wall more than 50 percent above grade and enclosing *conditioned space.* This includes between-floor spandrels, peripheral edges of floors, roof and *basement* knee walls, dormer walls, gable end walls, walls enclosing a mansard roof and *skylight* shafts.

ACCESS (TO). That which enables a device, appliance or equipment to be reached by *ready access* or by a means that first requires the removal or movement of a panel or similar obstruction.

ADDITION. An extension or increase in the *conditioned space* floor area, number of stories or height of a building or structure.

AIR BARRIER. One or more materials joined together in a continuous manner to restrict or prevent the passage of air through the *building thermal envelope* and its assemblies.

ALTERATION. Any construction, retrofit or renovation to an existing structure other than *repair* or *addition*. Also, a change in a building, electrical, gas, mechanical or plumbing system that involves an extension, *addition* or change to the arrangement, type or purpose of the original installation.

AUTOMATIC. Self-acting, operating by its own mechanism when actuated by some impersonal influence, as, for example, a change in current strength, pressure, temperature or mechanical configuration (see "*Manual*").

BASEMENT WALL. A wall 50 percent or more below grade and enclosing *conditioned space*.

BUILDING. Any structure used or intended for supporting or sheltering any use or occupancy, including any mechanical systems, service water-heating systems and electric power and lighting systems located on the *building site* and supporting the building.

BUILDING SITE. A contiguous area of land that is under the ownership or control of one entity.

BUILDING THERMAL ENVELOPE. The *basement walls*, *exterior walls*, floors, ceilings, roofs and any other *building* element assemblies that enclose *conditioned space* or provide a boundary between *conditioned space* and exempt or *unconditioned space*.

CAVITY INSULATION. Insulating material located between framing members.

CIRCULATING HOT WATER SYSTEM. A specifically designed water distribution system where one or more pumps are operated in the service hot water piping to circulate heated water from the water-heating equipment to fixtures and back to the water-heating equipment.

CLIMATE ZONE. A geographical region based on climatic criteria as specified in this code.

CONDITIONED FLOOR AREA. The horizontal projection of the floors associated with the *conditioned space*.

CONDITIONED SPACE. An area, room or space that is enclosed within the *building thermal envelope* and that is directly heated or cooled or indirectly heated or cooled. Spaces are indirectly heated or cooled where they communicate through openings with *conditioned spaces*, where they are separated from *conditioned spaces* by uninsulated walls, floors or ceilings, or where they contain uninsulated ducts, piping or other sources of heating or cooling.

CONTINUOUS AIR BARRIER. A combination of materials and assemblies that restrict or prevent the passage of air through the *building thermal envelope*.

CONTINUOUS INSULATION (ci). Insulating material that is continuous across all structural members without thermal bridges other than fasteners and service openings. It is installed on the interior or exterior, or is integral to any opaque surface, of the *building* envelope.

CRAWL SPACE WALL. The opaque portion of a wall that encloses a *crawl space* and is partially or totally below grade.

CURTAIN WALL. Fenestration products used to create an external nonload-bearing wall that is designed to separate the exterior and interior environments.

DEMAND RECIRCULATION WATER SYSTEM. A water distribution system where one or more pumps prime the service hot water piping with heated water on demand for hot water.

DIMMER. A control device that is capable of continuously varying the light output and energy use of light sources.

DUCT. A tube or conduit utilized for conveying air. The air passages of self-contained systems are not to be construed as air ducts.

DUCT SYSTEM. A continuous passageway for the transmission of air that, in addition to ducts, includes duct fittings, dampers, plenums, fans and accessory air-handling *equipment* and *appliances*.

DWELLING UNIT. A single unit providing complete independent living facilities for one or more persons, including permanent provisions for living, sleeping, eating, cooking and sanitation.

DWELLING UNIT ENCLOSURE AREA. The sum of the area of ceiling, floors and walls separating a *dwelling unit's* conditioned space from the exterior or from adjacent conditioned or unconditioned spaces. Wall height shall be measured from the finished floor of the *dwelling unit* to the underside of the floor above.

ENERGY ANALYSIS. A method for estimating the annual energy use of the *proposed design* and *standard reference design* based on estimates of energy use.

ENERGY COST. The total estimated annual cost for purchased energy for the building functions regulated by this code, including applicable demand charges.

ENERGY SIMULATION TOOL. An *approved* software program or calculation-based methodology that projects the annual energy use of a *building*.

ERI REFERENCE DESIGN. A version of the *rated design* that meets the minimum requirements of the 2006 *International Energy Conservation Code*.

EXTERIOR WALL. Walls including both above-grade walls and *basement walls*.

FENESTRATION. Products classified as either *vertical fenestration* or *skylights*.

Skylights. Glass or other transparent or translucent glazing material installed at a slope of less than 60 degrees (1.05 rad) from horizontal, including *unit skylights*, *tubular daylighting devices*, and glazing materials in solariums, *sunrooms*, roofs and sloped walls.

Vertical fenestration. Windows that are fixed or operable, opaque doors, glazed doors, glazed block and combination opaque/glazed doors composed of glass or other transparent or translucent glazing materials and installed at a slope of not less than 60 degrees (1.05 rad) from horizontal.

FENESTRATION PRODUCT, SITE-BUILT. A fenestration designed to be made up of field-glazed or field-assembled units using specific factory cut or otherwise factory-formed framing and glazing units. Examples of site-built fenestration include storefront systems, curtain walls, and atrium roof systems.

HEATED SLAB. Slab-on-grade construction in which the heating elements, hydronic tubing, or hot air distribution system is in contact with, or placed within or under, the slab.

HIGH-EFFICACY LIGHT SOURCES. Compact fluorescent lamps, light-emitting diode (LED) lamps, T-8 or smaller diameter linear fluorescent lamps, other lamps with an efficacy of not less than 65 lumens per watt, or luminaires with an efficacy of not less than 45 lumens per watt.

HISTORIC BUILDING. Any building or structure that is one or more of the following:

1. Listed, or certified as eligible for listing by the State Historic Preservation Officer or the Keeper of the National Register of Historic Places, in the National Register of Historic Places.
2. Designated as historic under an applicable state or local law.
3. Certified as a contributing resource within a National Register-listed, state-designated or locally designated historic district.

INFILTRATION. The uncontrolled inward air leakage into a building caused by the pressure effects of wind or the effect of differences in the indoor and outdoor air density or both.

INSULATED SIDING. A type of continuous insulation with manufacturer-installed insulating material as an integral part of the cladding product having an *R*-value of not less than R-2.

INSULATING SHEATHING. An insulating board with a core material having an *R*-value of not less than R-2.

LABELED. Equipment, materials or products to which have been affixed a *label*, seal, symbol or other identifying *mark* of a nationally recognized testing laboratory, *approved* agency or other organization concerned with product evaluation that maintains periodic inspection of the production of such *labeled* items and whose labeling indicates either that the equipment, material or product meets identified standards or has been tested and found suitable for a specified purpose.

LISTED. Equipment, materials, products or services included in a list published by an organization acceptable to the *code official* and concerned with evaluation of products or services that maintains periodic inspection of production of *listed* equipment or materials or periodic evaluation of services and where the listing states either that the equipment, material, product or service meets identified standards or has been tested and found suitable for a specified purpose.

MANUAL. Capable of being operated by personal intervention (see "*Automatic*").

OCCUPANT SENSOR CONTROL. An automatic control device that detects the presence or absence of people within an area and causes lighting, equipment or appliances to be regulated accordingly.

ON-SITE RENEWABLE ENERGY. Energy from renewable energy resources harvested at the building site.

OPAQUE DOOR. A door that is not less than 50-percent opaque in surface area.

PROPOSED DESIGN. A description of the proposed building used to estimate annual energy use for determining compliance based on total building performance.

RATED DESIGN. A description of the proposed *building* used to determine the energy rating index.

READY ACCESS (TO). That which enables a device, appliance or equipment to be directly reached without requiring the removal or movement of any panel or similar obstruction.

RENEWABLE ENERGY CERTIFICATE (REC). An instrument that represents the environmental attributes of one megawatt hour of renewable energy; also known as an energy attribute certificate (EAC).

RENEWABLE ENERGY RESOURCES. Energy derived from solar radiation, wind, waves, tides, landfill gas, biogas, biomass or extracted from hot fluid or steam heated within the earth.

REPAIR. The reconstruction or renewal of any part of an existing *building* for the purpose of its maintenance or to correct damage.

REROOFING. The process of recovering or replacing an existing roof covering. See "*Roof recover*" and "*Roof replacement.*"

RESIDENTIAL BUILDING. For this chapter, includes detached one- and two-family dwellings and townhouses as well as Group R-2, R-3 and R-4 buildings three stories or less in height above *grade plane.*

ROOF ASSEMBLY. A system designed to provide weather protection and resistance to design loads. The system consists of a roof covering and *roof deck* or a single component serving as both the roof covering and the *roof deck.* A *roof assembly* includes the roof covering, underlayment, and *roof deck*, and can also include a thermal barrier, ignition barrier, insulation or a vapor retarder.

ROOF RECOVER. The process of installing an additional roof covering over an existing roof covering without removing the existing roof covering.

ROOF REPAIR. Reconstruction or renewal of any part of an existing roof for the purposes of its maintenance.

ROOF REPLACEMENT. The process of removing the existing roof covering, repairing any damaged substrate and installing a new roof covering.

***R*-VALUE (THERMAL RESISTANCE).** The inverse of the time rate of heat flow through a body from one of its bounding surfaces to the other surface for a unit temperature difference between the two surfaces, under steady state conditions, per unit area ($h \times ft^2 \times °F/Btu$) [$(m^2 \times K)/W$].

SERVICE WATER HEATING. Supply of hot water for purposes other than comfort heating.

SOLAR HEAT GAIN COEFFICIENT (SHGC). The ratio of the solar heat gain entering the space through the fenestration assembly to the incident solar radiation. Solar heat gain includes directly transmitted solar heat and absorbed solar radiation that is then reradiated, conducted or convected into the space.

STANDARD REFERENCE DESIGN. A version of the *proposed design* that meets the minimum requirements of this code and is used to determine the maximum annual energy use requirement for compliance based on total building performance.

SUNROOM. A one-story structure attached to a dwelling with a glazing area in excess of 40 percent of the gross area of the structure's *exterior walls* and roof.

THERMAL DISTRIBUTION EFFICIENCY (TDE). The resistance to changes in air heat as air is conveyed through a distance of air duct. TDE is a heat-loss calculation evaluating the difference in the heat of the air between the air duct inlet and outlet caused by differences in temperatures between the air in the duct and the duct material. TDE is expressed as a percent difference between the inlet and outlet heat in the duct.

THERMAL ISOLATION. Physical and space conditioning separation from *conditioned spaces.* The *conditioned spaces* shall be controlled as separate *zones* for heating and cooling or conditioned by separate equipment.

THERMOSTAT. An automatic control device used to maintain temperature at a fixed or adjustable set point.

***U*-FACTOR (THERMAL TRANSMITTANCE).** The coefficient of heat transmission (air to air) through a building component or assembly, equal to the time rate of heat flow per unit area and unit temperature difference between the warm side and cold side air films (Btu/h $\times$ ft^2 $\times$ °F) [W/(m^2 $\times$ K)].

VENTILATION. The natural or mechanical process of supplying conditioned or unconditioned air to, or removing such air from, any space.

VENTILATION AIR. That portion of supply air that comes from outside (outdoors) plus any recirculated air that has been treated to maintain the desired quality of air within a designated space.

VISIBLE TRANSMITTANCE [VT]. The ratio of visible light entering the space through the fenestration product assembly to the incident visible light. Visible Transmittance includes the effects of glazing material and frame and is expressed as a number between 0 and 1.

WHOLE HOUSE MECHANICAL VENTILATION SYSTEM. An exhaust system, supply system, or combination thereof that is designed to mechanically exchange indoor air with outdoor air when operating continuously or through a programmed intermittent schedule to satisfy the whole house ventilation rates.

ZONE. A space or group of spaces within a *building* with heating or cooling requirements that are sufficiently similar so that desired conditions can be maintained throughout using a single controlling device.

N1101.7 (R301.1) Climate zones. Climate zones from Figure N1101.7 or Table N1101.7 shall be used for determining the applicable requirements in Sections N1101 through N1113. Locations not indicated in Table N1101.7 shall be assigned a climate zone in accordance with Section N1101.7.2.

N1101.7.1 (R301.2) Warm Humid counties. In Table N1101.7, Warm Humid counties are identified by an asterisk.

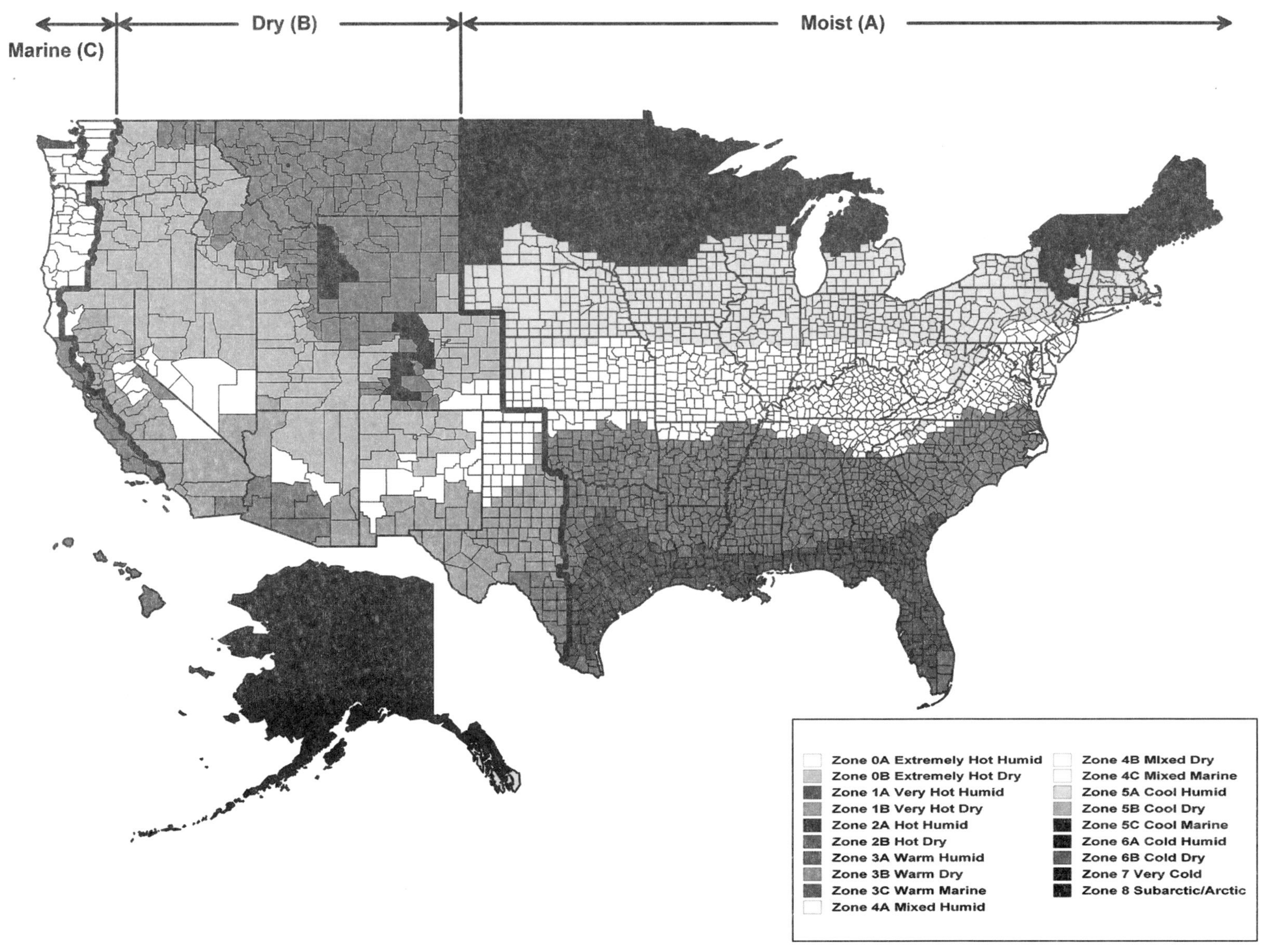

FIGURE N1101.7 (R301.1)
CLIMATE ZONES

TABLE N1101.7 (R301.1)
CLIMATE ZONES, MOISTURE REGIMES, AND WARM HUMID DESIGNATIONS BY STATE, COUNTY AND TERRITORY[a]

US STATES
ALABAMA
3A Autauga*
2A Baldwin*
3A Barbour*
3A Bibb
3A Blount
3A Bullock*
3A Butler*
3A Calhoun
3A Chambers
3A Cherokee
3A Chilton
3A Choctaw*
3A Clarke*
3A Clay
3A Cleburne
2A Coffee*
3A Colbert
3A Conecuh*
3A Coosa
2A Covington*
3A Crenshaw*
3A Cullman
2A Dale*
3A Dallas*
3A DeKalb
3A Elmore*
2A Escambia*
3A Etowah
3A Fayette
3A Franklin
2A Geneva*
3A Greene
3A Hale
2A Henry*
2A Houston*
3A Jackson
3A Jefferson
3A Lamar
3A Lauderdale
3A Lawrence
3A Lee
3A Limestone
3A Lowndes*
3A Macon*
3A Madison
3A Marengo*
3A Marion
3A Marshall
2A Mobile*
3A Monroe*
3A Montgomery*
3A Morgan
3A Perry*
3A Pickens
3A Pike*
3A Randolph
3A Russell*
3A Shelby
3A St. Clair
3A Sumter
3A Talladega
3A Tallapoosa
3A Tuscaloosa
3A Walker
3A Washington*
3A Wilcox*
3A Winston
ALASKA
7 Aleutians East
7 Aleutians West
7 Anchorage
7 Bethel
7 Bristol Bay
8 Denali
7 Dillingham
8 Fairbanks North Star
6A Haines
6A Juneau
7 Kenai Peninsula
5C Ketchikan Gateway
6A Kodiak Island
7 Lake and Peninsula
7 Matanuska-Susitna
8 Nome
8 North Slope
8 Northwest Arctic

(continued)

TABLE N1101.7 (R301.1)—continued
CLIMATE ZONES, MOISTURE REGIMES, AND WARM HUMID DESIGNATIONS BY STATE, COUNTY AND TERRITORY[a]

US STATES—continued
ALASKA *(continued)*
5C Prince of Wales Outer Ketchikan
5C Sitka
6A Skagway-Hoonah-Angoon
8 Southeast Fairbanks
7 Valdez-Cordova
8 Wade Hampton
6A Wrangell-Petersburg
7 Yakutat
8 Yukon-Koyukuk
ARIZONA
5B Apache
3B Cochise
5B Coconino
4B Gila
3B Graham
3B Greenlee
2B La Paz
2B Maricopa
3B Mohave
5B Navajo
2B Pima
2B Pinal
3B Santa Cruz
4B Yavapai
2B Yuma
ARKANSAS
3A Arkansas
3A Ashley
4A Baxter
4A Benton
4A Boone
3A Bradley
3A Calhoun
4A Carroll
3A Chicot
3A Clark
3A Clay
3A Cleburne
3A Cleveland
3A Columbia*
3A Conway
3A Craighead
3A Crawford
3A Crittenden
3A Cross
3A Dallas
3A Desha
3A Drew
3A Faulkner
3A Franklin
4A Fulton
3A Garland
3A Grant
3A Greene
3A Hempstead*
3A Hot Spring
3A Howard
3A Independence
4A Izard
3A Jackson
3A Jefferson
3A Johnson
3A Lafayette*
3A Lawrence
3A Lee
3A Lincoln
3A Little River*
3A Logan
3A Lonoke
4A Madison
4A Marion
3A Miller*
3A Mississippi
3A Monroe
3A Montgomery
3A Nevada
4A Newton
3A Ouachita
3A Perry
3A Phillips
3A Pike
3A Poinsett
3A Polk
3A Pope
3A Prairie
3A Pulaski
3A Randolph

(continued)

TABLE N1101.7 (R301.1)—continued
CLIMATE ZONES, MOISTURE REGIMES, AND WARM HUMID DESIGNATIONS BY STATE, COUNTY AND TERRITORY[a]

US STATES—continued
ARKANSAS *(continued)*
3A Saline
3A Scott
4A Searcy
3A Sebastian
3A Sevier*
3A Sharp
3A St. Francis
4A Stone
3A Union*
3A Van Buren
4A Washington
3A White
3A Woodruff
3A Yell
CALIFORNIA
3C Alameda
6B Alpine
4B Amador
3B Butte
4B Calaveras
3B Colusa
3B Contra Costa
4C Del Norte
4B El Dorado
3B Fresno
3B Glenn
4C Humboldt
2B Imperial
4B Inyo
3B Kern
3B Kings
4B Lake
5B Lassen
3B Los Angeles
3B Madera
3C Marin
4B Mariposa
3C Mendocino
3B Merced
5B Modoc
6B Mono
3C Monterey
3C Napa
5B Nevada
3B Orange
3B Placer
5B Plumas
3B Riverside
3B Sacramento
3C San Benito
3B San Bernardino
3B San Diego
3C San Francisco
3B San Joaquin
3C San Luis Obispo
3C San Mateo
3C Santa Barbara
3C Santa Clara
3C Santa Cruz
3B Shasta
5B Sierra
5B Siskiyou
3B Solano
3C Sonoma
3B Stanislaus
3B Sutter
3B Tehama
4B Trinity
3B Tulare
4B Tuolumne
3C Ventura
3B Yolo
3B Yuba
COLORADO
5B Adams
6B Alamosa
5B Arapahoe
6B Archuleta
4B Baca
4B Bent
5B Boulder
5B Broomfield
6B Chaffee
5B Cheyenne
7 Clear Creek
6B Conejos
6B Costilla
5B Crowley

(continued)

TABLE N1101.7 (R301.1)—continued
CLIMATE ZONES, MOISTURE REGIMES, AND WARM HUMID DESIGNATIONS BY STATE, COUNTY AND TERRITORY[a]

US STATES—continued
COLORADO *(continued)*
5B Custer
5B Delta
5B Denver
6B Dolores
5B Douglas
6B Eagle
5B Elbert
5B El Paso
5B Fremont
5B Garfield
5B Gilpin
7 Grand
7 Gunnison
7 Hinsdale
5B Huerfano
7 Jackson
5B Jefferson
5B Kiowa
5B Kit Carson
7 Lake
5B La Plata
5B Larimer
4B Las Animas
5B Lincoln
5B Logan
5B Mesa
7 Mineral
6B Moffat
5B Montezuma
5B Montrose
5B Morgan
4B Otero
6B Ouray
7 Park
5B Phillips
7 Pitkin
4B Prowers
5B Pueblo
6B Rio Blanco
7 Rio Grande
7 Routt
6B Saguache
7 San Juan
6B San Miguel
5B Sedgwick
7 Summit
5B Teller
5B Washington
5B Weld
5B Yuma
CONNECTICUT
5A (all)
DELAWARE
4A (all)
DISTRICT OF COLUMBIA
4A (all)
FLORIDA
2A Alachua*
2A Baker*
2A Bay*
2A Bradford*
2A Brevard*
1A Broward*
2A Calhoun*
2A Charlotte*
2A Citrus*
2A Clay*
2A Collier*
2A Columbia*
2A DeSoto*
2A Dixie*
2A Duval*
2A Escambia*
2A Flagler*
2A Franklin*
2A Gadsden*
2A Gilchrist*
2A Glades*
2A Gulf*
2A Hamilton*
2A Hardee*
2A Hendry*
2A Hernando*
2A Highlands*
2A Hillsborough*
2A Holmes*
2A Indian River*
2A Jackson*

(continued)

TABLE N1101.7 (R301.1)—continued
CLIMATE ZONES, MOISTURE REGIMES, AND WARM HUMID DESIGNATIONS BY STATE, COUNTY AND TERRITORY[a]

US STATES—continued
FLORIDA *(continued)*
2A Jefferson*
2A Lafayette*
2A Lake*
2A Lee*
2A Leon*
2A Levy*
2A Liberty*
2A Madison*
2A Manatee*
2A Marion*
2A Martin*
1A Miami-Dade*
1A Monroe*
2A Nassau*
2A Okaloosa*
2A Okeechobee*
2A Orange*
2A Osceola*
1A Palm Beach*
2A Pasco*
2A Pinellas*
2A Polk*
2A Putnam*
2A Santa Rosa*
2A Sarasota*
2A Seminole*
2A St. Johns*
2A St. Lucie*
2A Sumter*
2A Suwannee*
2A Taylor*
2A Union*
2A Volusia*
2A Wakulla*
2A Walton*
2A Washington*
GEORGIA
2A Appling*
2A Atkinson*
2A Bacon*
2A Baker*
3A Baldwin
3A Banks
3A Barrow
3A Bartow
3A Ben Hill*
2A Berrien*
3A Bibb
3A Bleckley*
2A Brantley*
2A Brooks*
2A Bryan*
3A Bulloch*
3A Burke
3A Butts
2A Calhoun*
2A Camden*
3A Candler*
3A Carroll
3A Catoosa
2A Charlton*
2A Chatham*
3A Chattahoochee*
3A Chattooga
3A Cherokee
3A Clarke
3A Clay*
3A Clayton
2A Clinch*
3A Cobb
2A Coffee*
2A Colquitt*
3A Columbia
2A Cook*
3A Coweta
3A Crawford
3A Crisp*
3A Dade
3A Dawson
2A Decatur*
3A DeKalb
3A Dodge*
3A Dooly*
2A Dougherty*
3A Douglas
2A Early*
2A Echols*
2A Effingham*

(continued)

TABLE N1101.7 (R301.1)—continued
CLIMATE ZONES, MOISTURE REGIMES, AND WARM HUMID DESIGNATIONS BY STATE, COUNTY AND TERRITORY[a]

US STATES—continued
GEORGIA *(continued)*
3A Elbert
3A Emanuel*
2A Evans*
3A Fannin
3A Fayette
3A Floyd
3A Forsyth
3A Franklin
3A Fulton
3A Gilmer
3A Glascock
2A Glynn*
3A Gordon
2A Grady*
3A Greene
3A Gwinnett
3A Habersham
3A Hall
3A Hancock
3A Haralson
3A Harris
3A Hart
3A Heard
3A Henry
3A Houston*
3A Irwin*
3A Jackson
3A Jasper
2A Jeff Davis*
3A Jefferson
3A Jenkins*
3A Johnson*
3A Jones
3A Lamar
2A Lanier*
3A Laurens*
3A Lee*
2A Liberty*
3A Lincoln
2A Long*
2A Lowndes*
3A Lumpkin
3A Macon*
3A Madison
3A Marion*
3A McDuffie
2A McIntosh*
3A Meriwether
2A Miller*
2A Mitchell*
3A Monroe
3A Montgomery*
3A Morgan
3A Murray
3A Muscogee
3A Newton
3A Oconee
3A Oglethorpe
3A Paulding
3A Peach*
3A Pickens
2A Pierce*
3A Pike
3A Polk
3A Pulaski*
3A Putnam
3A Quitman*
3A Rabun
3A Randolph*
3A Richmond
3A Rockdale
3A Schley*
3A Screven*
2A Seminole*
3A Spalding
3A Stephens
3A Stewart*
3A Sumter*
3A Talbot
3A Taliaferro
2A Tattnall*
3A Taylor*
3A Telfair*
3A Terrell*
2A Thomas*
2A Tift*
2A Toombs*
3A Towns

(continued)

TABLE N1101.7 (R301.1)—continued
CLIMATE ZONES, MOISTURE REGIMES, AND WARM HUMID DESIGNATIONS BY STATE, COUNTY AND TERRITORY[a]

US STATES—continued
GEORGIA *(continued)*
3A Treutlen*
3A Troup
3A Turner*
3A Twiggs*
3A Union
3A Upson
3A Walker
3A Walton
2A Ware*
3A Warren
3A Washington
2A Wayne*
3A Webster*
3A Wheeler*
3A White
3A Whitfield
3A Wilcox*
3A Wilkes
3A Wilkinson
2A Worth*
HAWAII
1A (all)*
IDAHO
5B Ada
6B Adams
6B Bannock
6B Bear Lake
5B Benewah
6B Bingham
6B Blaine
6B Boise
6B Bonner
6B Bonneville
6B Boundary
6B Butte
6B Camas
5B Canyon
6B Caribou
5B Cassia
6B Clark
5B Clearwater
6B Custer
5B Elmore
6B Franklin
6B Fremont
5B Gem
5B Gooding
5B Idaho
6B Jefferson
5B Jerome
5B Kootenai
5B Latah
6B Lemhi
5B Lewis
5B Lincoln
6B Madison
5B Minidoka
5B Nez Perce
6B Oneida
5B Owyhee
5B Payette
5B Power
5B Shoshone
6B Teton
5B Twin Falls
6B Valley
5B Washington
ILLINOIS
5A Adams
4A Alexander
4A Bond
5A Boone
5A Brown
5A Bureau
4A Calhoun
5A Carroll
5A Cass
5A Champaign
4A Christian
4A Clark
4A Clay
4A Clinton
4A Coles
5A Cook
4A Crawford
4A Cumberland
5A DeKalb
5A De Witt

(continued)

TABLE N1101.7 (R301.1)—continued
CLIMATE ZONES, MOISTURE REGIMES, AND WARM HUMID DESIGNATIONS BY STATE, COUNTY AND TERRITORY[a]

US STATES—continued
ILLINOIS *(continued)*
5A Douglas
5A DuPage
5A Edgar
4A Edwards
4A Effingham
4A Fayette
5A Ford
4A Franklin
5A Fulton
4A Gallatin
4A Greene
5A Grundy
4A Hamilton
5A Hancock
4A Hardin
5A Henderson
5A Henry
5A Iroquois
4A Jackson
4A Jasper
4A Jefferson
4A Jersey
5A Jo Daviess
4A Johnson
5A Kane
5A Kankakee
5A Kendall
5A Knox
5A Lake
5A La Salle
4A Lawrence
5A Lee
5A Livingston
5A Logan
5A Macon
4A Macoupin
4A Madison
4A Marion
5A Marshall
5A Mason
4A Massac
5A McDonough
5A McHenry
5A McLean
5A Menard
5A Mercer
4A Monroe
4A Montgomery
5A Morgan
5A Moultrie
5A Ogle
5A Peoria
4A Perry
5A Piatt
5A Pike
4A Pope
4A Pulaski
5A Putnam
4A Randolph
4A Richland
5A Rock Island
4A Saline
5A Sangamon
5A Schuyler
5A Scott
4A Shelby
5A Stark
4A St. Clair
5A Stephenson
5A Tazewell
4A Union
5A Vermilion
4A Wabash
5A Warren
4A Washington
4A Wayne
4A White
5A Whiteside
5A Will
4A Williamson
5A Winnebago
5A Woodford
INDIANA
5A Adams
5A Allen
4A Bartholomew
5A Benton
5A Blackford
5A Boone
4A Brown

(continued)

TABLE N1101.7 (R301.1)—continued
CLIMATE ZONES, MOISTURE REGIMES, AND WARM HUMID DESIGNATIONS BY STATE, COUNTY AND TERRITORY[a]

US STATES—continued
INDIANA *(continued)*
5A Carroll
5A Cass
4A Clark
4A Clay
5A Clinton
4A Crawford
4A Daviess
4A Dearborn
4A Decatur
5A De Kalb
5A Delaware
4A Dubois
5A Elkhart
4A Fayette
4A Floyd
5A Fountain
4A Franklin
5A Fulton
4A Gibson
5A Grant
4A Greene
5A Hamilton
5A Hancock
4A Harrison
4A Hendricks
5A Henry
5A Howard
5A Huntington
4A Jackson
5A Jasper
5A Jay
4A Jefferson
4A Jennings
4A Johnson
4A Knox
5A Kosciusko
5A LaGrange
5A Lake
5A LaPorte
4A Lawrence
5A Madison
4A Marion
5A Marshall
4A Martin
5A Miami
4A Monroe
5A Montgomery
4A Morgan
5A Newton
5A Noble
4A Ohio
4A Orange
4A Owen
5A Parke
4A Perry
4A Pike
5A Porter
4A Posey
5A Pulaski
4A Putnam
5A Randolph
4A Ripley
4A Rush
4A Scott
4A Shelby
4A Spencer
5A Starke
5A Steuben
5A St. Joseph
4A Sullivan
4A Switzerland
5A Tippecanoe
5A Tipton
4A Union
4A Vanderburgh
5A Vermillion
4A Vigo
5A Wabash
5A Warren
4A Warrick
4A Washington
5A Wayne
5A Wells
5A White
5A Whitley
IOWA
5A Adair
5A Adams

(continued)

TABLE N1101.7 (R301.1)—continued
CLIMATE ZONES, MOISTURE REGIMES, AND WARM HUMID DESIGNATIONS BY STATE, COUNTY AND TERRITORY[a]

US STATES—continued
IOWA *(continued)*
5A Allamakee
5A Appanoose
5A Audubon
5A Benton
6A Black Hawk
5A Boone
5A Bremer
5A Buchanan
5A Buena Vista
5A Butler
5A Calhoun
5A Carroll
5A Cass
5A Cedar
6A Cerro Gordo
5A Cherokee
5A Chickasaw
5A Clarke
6A Clay
5A Clayton
5A Clinton
5A Crawford
5A Dallas
5A Davis
5A Decatur
5A Delaware
5A Des Moines
6A Dickinson
5A Dubuque
6A Emmet
5A Fayette
5A Floyd
5A Franklin
5A Fremont
5A Greene
5A Grundy
5A Guthrie
5A Hamilton
6A Hancock
5A Hardin
5A Harrison
5A Henry
5A Howard
5A Humboldt
5A Ida
5A Iowa
5A Jackson
5A Jasper
5A Jefferson
5A Johnson
5A Jones
5A Keokuk
6A Kossuth
5A Lee
5A Linn
5A Louisa
5A Lucas
6A Lyon
5A Madison
5A Mahaska
5A Marion
5A Marshall
5A Mills
6A Mitchell
5A Monona
5A Monroe
5A Montgomery
5A Muscatine
6A O'Brien
6A Osceola
5A Page
6A Palo Alto
5A Plymouth
5A Pocahontas
5A Polk
5A Pottawattamie
5A Poweshiek
5A Ringgold
5A Sac
5A Scott
5A Shelby
6A Sioux
5A Story
5A Tama
5A Taylor
5A Union
5A Van Buren
5A Wapello

(continued)

TABLE N1101.7 (R301.1)—continued
CLIMATE ZONES, MOISTURE REGIMES, AND WARM HUMID DESIGNATIONS BY STATE, COUNTY AND TERRITORY[a]

US STATES—continued
IOWA *(continued)*
5A Warren
5A Washington
5A Wayne
5A Webster
6A Winnebago
5A Winneshiek
5A Woodbury
6A Worth
5A Wright
KANSAS
4A Allen
4A Anderson
4A Atchison
4A Barber
4A Barton
4A Bourbon
4A Brown
4A Butler
4A Chase
4A Chautauqua
4A Cherokee
5A Cheyenne
4A Clark
4A Clay
4A Cloud
4A Coffey
4A Comanche
4A Cowley
4A Crawford
5A Decatur
4A Dickinson
4A Doniphan
4A Douglas
4A Edwards
4A Elk
4A Ellis
4A Ellsworth
4A Finney
4A Ford
4A Franklin
4A Geary
5A Gove
4A Graham
4A Grant
4A Gray
5A Greeley
4A Greenwood
4A Hamilton
4A Harper
4A Harvey
4A Haskell
4A Hodgeman
4A Jackson
4A Jefferson
5A Jewell
4A Johnson
4A Kearny
4A Kingman
4A Kiowa
4A Labette
4A Lane
4A Leavenworth
4A Lincoln
4A Linn
5A Logan
4A Lyon
4A Marion
4A Marshall
4A McPherson
4A Meade
4A Miami
4A Mitchell
4A Montgomery
4A Morris
4A Morton
4A Nemaha
4A Neosho
4A Ness
5A Norton
4A Osage
4A Osborne
4A Ottawa
4A Pawnee
5A Phillips
4A Pottawatomie
4A Pratt
5A Rawlins
4A Reno

(continued)

TABLE N1101.7 (R301.1)—continued
CLIMATE ZONES, MOISTURE REGIMES, AND WARM HUMID DESIGNATIONS BY STATE, COUNTY AND TERRITORY[a]

US STATES—continued
KANSAS *(continued)*
5A Republic
4A Rice
4A Riley
4A Rooks
4A Rush
4A Russell
4A Saline
5A Scott
4A Sedgwick
4A Seward
4A Shawnee
5A Sheridan
5A Sherman
5A Smith
4A Stafford
4A Stanton
4A Stevens
4A Sumner
5A Thomas
4A Trego
4A Wabaunsee
5A Wallace
4A Washington
5A Wichita
4A Wilson
4A Woodson
4A Wyandotte
KENTUCKY
4A (all)
LOUISIANA
2A Acadia*
2A Allen*
2A Ascension*
2A Assumption*
2A Avoyelles*
2A Beauregard*
3A Bienville*
3A Bossier*
3A Caddo*
2A Calcasieu*
3A Caldwell*
2A Cameron*
3A Catahoula*
3A Claiborne*
3A Concordia*
3A De Soto*
2A East Baton Rouge*
3A East Carroll
2A East Feliciana*
2A Evangeline*
3A Franklin*
3A Grant*
2A Iberia*
2A Iberville*
3A Jackson*
2A Jefferson*
2A Jefferson Davis*
2A Lafayette*
2A Lafourche*
3A La Salle*
3A Lincoln*
2A Livingston*
3A Madison*
3A Morehouse
3A Natchitoches*
2A Orleans*
3A Ouachita*
2A Plaquemines*
2A Pointe Coupee*
2A Rapides*
3A Red River*
3A Richland*
3A Sabine*
2A St. Bernard*
2A St. Charles*
2A St. Helena*
2A St. James*
2A St. John the Baptist*
2A St. Landry*
2A St. Martin*
2A St. Mary*
2A St. Tammany*
2A Tangipahoa*
3A Tensas*
2A Terrebonne*
3A Union*
2A Vermilion*
3A Vernon*

(continued)

TABLE N1101.7 (R301.1)—continued
CLIMATE ZONES, MOISTURE REGIMES, AND WARM HUMID DESIGNATIONS BY STATE, COUNTY AND TERRITORY[a]

US STATES—continued
LOUISIANA *(continued)*
2A Washington*
3A Webster*
2A West Baton Rouge*
3A West Carroll
2A West Feliciana*
3A Winn*
MAINE
6A Androscoggin
7 Aroostook
6A Cumberland
6A Franklin
6A Hancock
6A Kennebec
6A Knox
6A Lincoln
6A Oxford
6A Penobscot
6A Piscataquis
6A Sagadahoc
6A Somerset
6A Waldo
6A Washington
6A York
MARYLAND
5A Allegany
4A Anne Arundel
4A Baltimore
4A Baltimore (city)
4A Calvert
4A Caroline
4A Carroll
4A Cecil
4A Charles
4A Dorchester
4A Frederick
5A Garrett
4A Harford
4A Howard
4A Kent
4A Montgomery
4A Prince George's
4A Queen Anne's
4A Somerset
4A St. Mary's
4A Talbot
4A Washington
4A Wicomico
4A Worcester
MASSACHUSETTS
5A (all)
MICHIGAN
6A Alcona
6A Alger
5A Allegan
6A Alpena
6A Antrim
6A Arenac
6A Baraga
5A Barry
5A Bay
6A Benzie
5A Berrien
5A Branch
5A Calhoun
5A Cass
6A Charlevoix
6A Cheboygan
6A Chippewa
6A Clare
5A Clinton
6A Crawford
6A Delta
6A Dickinson
5A Eaton
6A Emmet
5A Genesee
6A Gladwin
6A Gogebic
6A Grand Traverse
5A Gratiot
5A Hillsdale
6A Houghton
5A Huron
5A Ingham
5A Ionia
6A Iosco
6A Iron
6A Isabella

(continued)

TABLE N1101.7 (R301.1)—continued
CLIMATE ZONES, MOISTURE REGIMES, AND WARM HUMID DESIGNATIONS BY STATE, COUNTY AND TERRITORY[a]

US STATES—continued
MICHIGAN *(continued)*
5A Jackson
5A Kalamazoo
6A Kalkaska
5A Kent
7 Keweenaw
6A Lake
5A Lapeer
6A Leelanau
5A Lenawee
5A Livingston
6A Luce
6A Mackinac
5A Macomb
6A Manistee
7 Marquette
6A Mason
6A Mecosta
6A Menominee
5A Midland
6A Missaukee
5A Monroe
5A Montcalm
6A Montmorency
5A Muskegon
6A Newaygo
5A Oakland
6A Oceana
6A Ogemaw
6A Ontonagon
6A Osceola
6A Oscoda
6A Otsego
5A Ottawa
6A Presque Isle
6A Roscommon
5A Saginaw
5A Sanilac
6A Schoolcraft
5A Shiawassee
5A St. Clair
5A St. Joseph
5A Tuscola
5A Van Buren
5A Washtenaw
5A Wayne
6A Wexford
MINNESOTA
7 Aitkin
6A Anoka
6A Becker
7 Beltrami
6A Benton
6A Big Stone
6A Blue Earth
6A Brown
7 Carlton
6A Carver
7 Cass
6A Chippewa
6A Chisago
6A Clay
7 Clearwater
7 Cook
6A Cottonwood
7 Crow Wing
6A Dakota
6A Dodge
6A Douglas
6A Faribault
5A Fillmore
6A Freeborn
6A Goodhue
6A Grant
6A Hennepin
5A Houston
7 Hubbard
6A Isanti
7 Itasca
6A Jackson
6A Kanabec
6A Kandiyohi
7 Kittson
7 Koochiching
6A Lac qui Parle
7 Lake
7 Lake of the Woods
6A Le Sueur
6A Lincoln

(continued)

TABLE N1101.7 (R301.1)—continued
CLIMATE ZONES, MOISTURE REGIMES, AND WARM HUMID DESIGNATIONS BY STATE, COUNTY AND TERRITORY[a]

US STATES—continued
MINNESOTA *(continued)*
6A Lyon
7 Mahnomen
7 Marshall
6A Martin
6A McLeod
6A Meeker
6A Mille Lacs
6A Morrison
6A Mower
6A Murray
6A Nicollet
6A Nobles
7 Norman
6A Olmsted
6A Otter Tail
7 Pennington
7 Pine
6A Pipestone
7 Polk
6A Pope
6A Ramsey
7 Red Lake
6A Redwood
6A Renville
6A Rice
6A Rock
7 Roseau
6A Scott
6A Sherburne
6A Sibley
6A Stearns
6A Steele
6A Stevens
7 St. Louis
6A Swift
6A Todd
6A Traverse
6A Wabasha
7 Wadena
6A Waseca
6A Washington
6A Watonwan
6A Wilkin
5A Winona
6A Wright
6A Yellow Medicine
MISSISSIPPI
3A Adams*
3A Alcorn
3A Amite*
3A Attala
3A Benton
3A Bolivar
3A Calhoun
3A Carroll
3A Chickasaw
3A Choctaw
3A Claiborne*
3A Clarke
3A Clay
3A Coahoma
3A Copiah*
3A Covington*
3A DeSoto
3A Forrest*
3A Franklin*
2A George*
3A Greene*
3A Grenada
2A Hancock*
2A Harrison*
3A Hinds*
3A Holmes
3A Humphreys
3A Issaquena
3A Itawamba
2A Jackson*
3A Jasper
3A Jefferson*
3A Jefferson Davis*
3A Jones*
3A Kemper
3A Lafayette
3A Lamar*
3A Lauderdale
3A Lawrence*
3A Leake
3A Lee

(continued)

TABLE N1101.7 (R301.1)—continued
CLIMATE ZONES, MOISTURE REGIMES, AND WARM HUMID DESIGNATIONS BY STATE, COUNTY AND TERRITORY[a]

US STATES—continued
MISSISSIPPI *(continued)*
3A Leflore
3A Lincoln*
3A Lowndes
3A Madison
3A Marion*
3A Marshall
3A Monroe
3A Montgomery
3A Neshoba
3A Newton
3A Noxubee
3A Oktibbeha
3A Panola
2A Pearl River*
3A Perry*
3A Pike*
3A Pontotoc
3A Prentiss
3A Quitman
3A Rankin*
3A Scott
3A Sharkey
3A Simpson*
3A Smith*
2A Stone*
3A Sunflower
3A Tallahatchie
3A Tate
3A Tippah
3A Tishomingo
3A Tunica
3A Union
3A Walthall*
3A Warren*
3A Washington
3A Wayne*
3A Webster
3A Wilkinson*
3A Winston
3A Yalobusha
3A Yazoo
MISSOURI
5A Adair
5A Andrew
5A Atchison
4A Audrain
4A Barry
4A Barton
4A Bates
4A Benton
4A Bollinger
4A Boone
4A Buchanan
4A Butler
4A Caldwell
4A Callaway
4A Camden
4A Cape Girardeau
4A Carroll
4A Carter
4A Cass
4A Cedar
4A Chariton
4A Christian
5A Clark
4A Clay
4A Clinton
4A Cole
4A Cooper
4A Crawford
4A Dade
4A Dallas
5A Daviess
5A DeKalb
4A Dent
4A Douglas
3A Dunklin
4A Franklin
4A Gasconade
5A Gentry
4A Greene
5A Grundy
5A Harrison
4A Henry
4A Hickory
5A Holt
4A Howard
4A Howell

(continued)

TABLE N1101.7 (R301.1)—continued
CLIMATE ZONES, MOISTURE REGIMES, AND WARM HUMID DESIGNATIONS BY STATE, COUNTY AND TERRITORY[a]

US STATES—continued
MISSOURI *(continued)*
4A Iron
4A Jackson
4A Jasper
4A Jefferson
4A Johnson
5A Knox
4A Laclede
4A Lafayette
4A Lawrence
5A Lewis
4A Lincoln
5A Linn
5A Livingston
5A Macon
4A Madison
4A Maries
5A Marion
4A McDonald
5A Mercer
4A Miller
4A Mississippi
4A Moniteau
4A Monroe
4A Montgomery
4A Morgan
4A New Madrid
4A Newton
5A Nodaway
4A Oregon
4A Osage
4A Ozark
3A Pemiscot
4A Perry
4A Pettis
4A Phelps
5A Pike
4A Platte
4A Polk
4A Pulaski
5A Putnam
5A Ralls
4A Randolph
4A Ray
4A Reynolds
4A Ripley
4A Saline
5A Schuyler
5A Scotland
4A Scott
4A Shannon
5A Shelby
4A St. Charles
4A St. Clair
4A St. Francois
4A St. Louis
4A St. Louis (city)
4A Ste. Genevieve
4A Stoddard
4A Stone
5A Sullivan
4A Taney
4A Texas
4A Vernon
4A Warren
4A Washington
4A Wayne
4A Webster
5A Worth
4A Wright
MONTANA
6B (all)
NEBRASKA
5A (all)
NEVADA
4B Carson City (city)
5B Churchill
3B Clark
4B Douglas
5B Elko
4B Esmeralda
5B Eureka
5B Humboldt
5B Lander
4B Lincoln
4B Lyon
4B Mineral
4B Nye
5B Pershing

(continued)

TABLE N1101.7 (R301.1)—continued
CLIMATE ZONES, MOISTURE REGIMES, AND WARM HUMID DESIGNATIONS BY STATE, COUNTY AND TERRITORY[a]

US STATES—continued
NEVADA *(continued)*
5B Storey
5B Washoe
5B White Pine
NEW HAMPSHIRE
6A Belknap
6A Carroll
5A Cheshire
6A Coos
6A Grafton
5A Hillsborough
5A Merrimack
5A Rockingham
5A Strafford
6A Sullivan
NEW JERSEY
4A Atlantic
5A Bergen
4A Burlington
4A Camden
4A Cape May
4A Cumberland
4A Essex
4A Gloucester
4A Hudson
5A Hunterdon
4A Mercer
4A Middlesex
4A Monmouth
5A Morris
4A Ocean
5A Passaic
4A Salem
5A Somerset
5A Sussex
4A Union
5A Warren
NEW MEXICO
4B Bernalillo
4A Catron
3B Chaves
4B Cibola
5B Colfax
4B Curry
4B DeBaca
3B Doña Ana
3B Eddy
4B Grant
4B Guadalupe
5B Harding
3B Hidalgo
3B Lea
4B Lincoln
5B Los Alamos
3B Luna
5B McKinley
5B Mora
3B Otero
4B Quay
5B Rio Arriba
4B Roosevelt
5B Sandoval
5B San Juan
5B San Miguel
5B Santa Fe
3B Sierra
4B Socorro
5B Taos
5B Torrance
4B Union
4B Valencia
NEW YORK
5A Albany
5A Allegany
4A Bronx
5A Broome
5A Cattaraugus
5A Cayuga
5A Chautauqua
5A Chemung
6A Chenango
6A Clinton
5A Columbia
5A Cortland
6A Delaware
5A Dutchess
5A Erie
6A Essex
6A Franklin

(continued)

TABLE N1101.7 (R301.1)—continued
CLIMATE ZONES, MOISTURE REGIMES, AND WARM HUMID DESIGNATIONS BY STATE, COUNTY AND TERRITORY[a]

US STATES—continued
NEW YORK *(continued)*
6A Fulton
5A Genesee
5A Greene
6A Hamilton
6A Herkimer
6A Jefferson
4A Kings
6A Lewis
5A Livingston
6A Madison
5A Monroe
6A Montgomery
4A Nassau
4A New York
5A Niagara
6A Oneida
5A Onondaga
5A Ontario
5A Orange
5A Orleans
5A Oswego
6A Otsego
5A Putnam
4A Queens
5A Rensselaer
4A Richmond
5A Rockland
5A Saratoga
5A Schenectady
5A Schoharie
5A Schuyler
5A Seneca
5A Steuben
6A St. Lawrence
4A Suffolk
6A Sullivan
5A Tioga
5A Tompkins
6A Ulster
6A Warren
5A Washington
5A Wayne
4A Westchester
5A Wyoming
5A Yates
NORTH CAROLINA
3A Alamance
3A Alexander
5A Alleghany
3A Anson
5A Ashe
5A Avery
3A Beaufort
3A Bertie
3A Bladen
3A Brunswick*
4A Buncombe
4A Burke
3A Cabarrus
4A Caldwell
3A Camden
3A Carteret*
3A Caswell
3A Catawba
3A Chatham
3A Cherokee
3A Chowan
3A Clay
3A Cleveland
3A Columbus*
3A Craven
3A Cumberland
3A Currituck
3A Dare
3A Davidson
3A Davie
3A Duplin
3A Durham
3A Edgecombe
3A Forsyth
3A Franklin
3A Gaston
3A Gates
4A Graham
3A Granville
3A Greene
3A Guilford
3A Halifax

(continued)

TABLE N1101.7 (R301.1)—continued
CLIMATE ZONES, MOISTURE REGIMES, AND WARM HUMID DESIGNATIONS BY STATE, COUNTY AND TERRITORY[a]

US STATES—continued
NORTH CAROLINA *(continued)*
3A Harnett
4A Haywood
4A Henderson
3A Hertford
3A Hoke
3A Hyde
3A Iredell
4A Jackson
3A Johnston
3A Jones
3A Lee
3A Lenoir
3A Lincoln
4A Macon
4A Madison
3A Martin
4A McDowell
3A Mecklenburg
4A Mitchell
3A Montgomery
3A Moore
3A Nash
3A New Hanover*
3A Northampton
3A Onslow*
3A Orange
3A Pamlico
3A Pasquotank
3A Pender*
3A Perquimans
3A Person
3A Pitt
3A Polk
3A Randolph
3A Richmond
3A Robeson
3A Rockingham
3A Rowan
3A Rutherford
3A Sampson
3A Scotland
3A Stanly
4A Stokes
4A Surry
4A Swain
4A Transylvania
3A Tyrrell
3A Union
3A Vance
3A Wake
3A Warren
3A Washington
5A Watauga
3A Wayne
3A Wilkes
3A Wilson
4A Yadkin
5A Yancey
NORTH DAKOTA
6A Adams
6A Barnes
7 Benson
6A Billings
7 Bottineau
6A Bowman
7 Burke
6A Burleigh
6A Cass
7 Cavalier
6A Dickey
7 Divide
6A Dunn
6A Eddy
6A Emmons
6A Foster
6A Golden Valley
7 Grand Forks
6A Grant
6A Griggs
6A Hettinger
6A Kidder
6A LaMoure
6A Logan
7 McHenry
6A McIntosh
6A McKenzie
6A McLean
6A Mercer

(continued)

TABLE N1101.7 (R301.1)—continued
CLIMATE ZONES, MOISTURE REGIMES, AND WARM HUMID DESIGNATIONS BY STATE, COUNTY AND TERRITORY[a]

US STATES—continued
NORTH DAKOTA *(continued)*
6A Morton
6A Mountrail
7 Nelson
6A Oliver
7 Pembina
7 Pierce
7 Ramsey
6A Ransom
7 Renville
6A Richland
7 Rolette
6A Sargent
6A Sheridan
6A Sioux
6A Slope
6A Stark
6A Steele
6A Stutsman
7 Towner
6A Traill
7 Walsh
7 Ward
6A Wells
6A Williams
OHIO
4A Adams
5A Allen
5A Ashland
5A Ashtabula
4A Athens
5A Auglaize
5A Belmont
4A Brown
4A Butler
5A Carroll
5A Champaign
5A Clark
4A Clermont
4A Clinton
5A Columbiana
5A Coshocton
5A Crawford
5A Cuyahoga
5A Darke
5A Defiance
5A Delaware
5A Erie
5A Fairfield
4A Fayette
4A Franklin
5A Fulton
4A Gallia
5A Geauga
4A Greene
5A Guernsey
4A Hamilton
5A Hancock
5A Hardin
5A Harrison
5A Henry
4A Highland
4A Hocking
5A Holmes
5A Huron
4A Jackson
5A Jefferson
5A Knox
5A Lake
4A Lawrence
5A Licking
5A Logan
5A Lorain
5A Lucas
4A Madison
5A Mahoning
5A Marion
5A Medina
4A Meigs
5A Mercer
5A Miami
5A Monroe
5A Montgomery
5A Morgan
5A Morrow
5A Muskingum
5A Noble
5A Ottawa
5A Paulding

(continued)

TABLE N1101.7 (R301.1)—continued
CLIMATE ZONES, MOISTURE REGIMES, AND WARM HUMID DESIGNATIONS BY STATE, COUNTY AND TERRITORY[a]

US STATES—continued
OHIO *(continued)*
5A Perry
4A Pickaway
4A Pike
5A Portage
5A Preble
5A Putnam
5A Richland
4A Ross
5A Sandusky
4A Scioto
5A Seneca
5A Shelby
5A Stark
5A Summit
5A Trumbull
5A Tuscarawas
5A Union
5A Van Wert
4A Vinton
4A Warren
4A Washington
5A Wayne
5A Williams
5A Wood
5A Wyandot
OKLAHOMA
3A Adair
4A Alfalfa
3A Atoka
4B Beaver
3A Beckham
3A Blaine
3A Bryan
3A Caddo
3A Canadian
3A Carter
3A Cherokee
3A Choctaw
4B Cimarron
3A Cleveland
3A Coal
3A Comanche
3A Cotton
4A Craig
3A Creek
3A Custer
4A Delaware
3A Dewey
4A Ellis
4A Garfield
3A Garvin
3A Grady
4A Grant
3A Greer
3A Harmon
4A Harper
3A Haskell
3A Hughes
3A Jackson
3A Jefferson
3A Johnston
4A Kay
3A Kingfisher
3A Kiowa
3A Latimer
3A Le Flore
3A Lincoln
3A Logan
3A Love
4A Major
3A Marshall
3A Mayes
3A McClain
3A McCurtain
3A McIntosh
3A Murray
3A Muskogee
3A Noble
4A Nowata
3A Okfuskee
3A Oklahoma
3A Okmulgee
4A Osage
4A Ottawa
3A Pawnee
3A Payne
3A Pittsburg
3A Pontotoc

(continued)

TABLE N1101.7 (R301.1)—continued
CLIMATE ZONES, MOISTURE REGIMES, AND WARM HUMID DESIGNATIONS BY STATE, COUNTY AND TERRITORY[a]

US STATES—continued
OKLAHOMA *(continued)*
3A Pottawatomie
3A Pushmataha
3A Roger Mills
3A Rogers
3A Seminole
3A Sequoyah
3A Stephens
4B Texas
3A Tillman
3A Tulsa
3A Wagoner
4A Washington
3A Washita
4A Woods
4A Woodward
OREGON
5B Baker
4C Benton
4C Clackamas
4C Clatsop
4C Columbia
4C Coos
5B Crook
4C Curry
5B Deschutes
4C Douglas
5B Gilliam
5B Grant
5B Harney
5B Hood River
4C Jackson
5B Jefferson
4C Josephine
5B Klamath
5B Lake
4C Lane
4C Lincoln
4C Linn
5B Malheur
4C Marion
5B Morrow
4C Multnomah
4C Polk
5B Sherman
4C Tillamook
5B Umatilla
5B Union
5B Wallowa
5B Wasco
4C Washington
5B Wheeler
4C Yamhill
PENNSYLVANIA
4A Adams
5A Allegheny
5A Armstrong
5A Beaver
5A Bedford
4A Berks
5A Blair
5A Bradford
4A Bucks
5A Butler
5A Cambria
5A Cameron
5A Carbon
5A Centre
4A Chester
5A Clarion
5A Clearfield
5A Clinton
5A Columbia
5A Crawford
4A Cumberland
4A Dauphin
4A Delaware
5A Elk
5A Erie
5A Fayette
5A Forest
4A Franklin
5A Fulton
5A Greene
5A Huntingdon
5A Indiana
5A Jefferson
5A Juniata
5A Lackawanna

(continued)

TABLE N1101.7 (R301.1)—continued
CLIMATE ZONES, MOISTURE REGIMES, AND WARM HUMID DESIGNATIONS BY STATE, COUNTY AND TERRITORY[a]

US STATES—continued
PENNSYLVANIA *(continued)*
4A Lancaster
5A Lawrence
4A Lebanon
5A Lehigh
5A Luzerne
5A Lycoming
5A McKean
5A Mercer
5A Mifflin
5A Monroe
4A Montgomery
5A Montour
5A Northampton
5A Northumberland
4A Perry
4A Philadelphia
5A Pike
5A Potter
5A Schuylkill
5A Snyder
5A Somerset
5A Sullivan
5A Susquehanna
5A Tioga
5A Union
5A Venango
5A Warren
5A Washington
5A Wayne
5A Westmoreland
5A Wyoming
4A York
RHODE ISLAND
5A (all)
SOUTH CAROLINA
3A Abbeville
3A Aiken
3A Allendale*
3A Anderson
3A Bamberg*
3A Barnwell*
2A Beaufort*
3A Berkeley*
3A Calhoun
3A Charleston*
3A Cherokee
3A Chester
3A Chesterfield
3A Clarendon
3A Colleton*
3A Darlington
3A Dillon
3A Dorchester*
3A Edgefield
3A Fairfield
3A Florence
3A Georgetown*
3A Greenville
3A Greenwood
3A Hampton*
3A Horry*
2A Jasper*
3A Kershaw
3A Lancaster
3A Laurens
3A Lee
3A Lexington
3A Marion
3A Marlboro
3A McCormick
3A Newberry
3A Oconee
3A Orangeburg
3A Pickens
3A Richland
3A Saluda
3A Spartanburg
3A Sumter
3A Union
3A Williamsburg
3A York
SOUTH DAKOTA
6A Aurora
6A Beadle
5A Bennett
5A Bon Homme
6A Brookings
6A Brown

(continued)

TABLE N1101.7 (R301.1)—continued
CLIMATE ZONES, MOISTURE REGIMES, AND WARM HUMID DESIGNATIONS BY STATE, COUNTY AND TERRITORY[a]

US STATES—continued
SOUTH DAKOTA *(continued)*
5A Brule
6A Buffalo
6A Butte
6A Campbell
5A Charles Mix
6A Clark
5A Clay
6A Codington
6A Corson
6A Custer
6A Davison
6A Day
6A Deuel
6A Dewey
5A Douglas
6A Edmunds
6A Fall River
6A Faulk
6A Grant
5A Gregory
5A Haakon
6A Hamlin
6A Hand
6A Hanson
6A Harding
6A Hughes
5A Hutchinson
6A Hyde
5A Jackson
6A Jerauld
5A Jones
6A Kingsbury
6A Lake
6A Lawrence
6A Lincoln
5A Lyman
6A Marshall
6A McCook
6A McPherson
6A Meade
5A Mellette
6A Miner
6A Minnehaha
6A Moody
6A Pennington
6A Perkins
6A Potter
6A Roberts
6A Sanborn
6A Shannon
6A Spink
5A Stanley
6A Sully
5A Todd
5A Tripp
6A Turner
5A Union
6A Walworth
5A Yankton
6A Ziebach
TENNESSEE
4A Anderson
3A Bedford
4A Benton
4A Bledsoe
4A Blount
4A Bradley
4A Campbell
4A Cannon
4A Carroll
4A Carter
4A Cheatham
3A Chester
4A Claiborne
4A Clay
4A Cocke
3A Coffee
3A Crockett
4A Cumberland
3A Davidson
3A Decatur
4A DeKalb
4A Dickson
3A Dyer
3A Fayette
4A Fentress
3A Franklin
3A Gibson

(continued)

TABLE N1101.7 (R301.1)—continued
CLIMATE ZONES, MOISTURE REGIMES, AND WARM HUMID DESIGNATIONS BY STATE, COUNTY AND TERRITORY[a]

US STATES—continued
TENNESSEE *(continued)*
3A Giles
4A Grainger
4A Greene
3A Grundy
4A Hamblen
3A Hamilton
4A Hancock
3A Hardeman
3A Hardin
4A Hawkins
3A Haywood
3A Henderson
4A Henry
3A Hickman
4A Houston
4A Humphreys
4A Jackson
4A Jefferson
4A Johnson
4A Knox
4A Lake
3A Lauderdale
3A Lawrence
3A Lewis
3A Lincoln
4A Loudon
4A Macon
3A Madison
3A Marion
3A Marshall
3A Maury
4A McMinn
3A McNairy
4A Meigs
4A Monroe
4A Montgomery
3A Moore
4A Morgan
4A Obion
4A Overton
3A Perry
4A Pickett
4A Polk
4A Putnam
4A Rhea
4A Roane
4A Robertson
3A Rutherford
4A Scott
4A Sequatchie
4A Sevier
3A Shelby
4A Smith
4A Stewart
4A Sullivan
4A Sumner
3A Tipton
4A Trousdale
4A Unicoi
4A Union
4A Van Buren
4A Warren
4A Washington
3A Wayne
4A Weakley
4A White
3A Williamson
4A Wilson
TEXAS
2A Anderson*
3B Andrews
2A Angelina*
2A Aransas*
3A Archer
4B Armstrong
2A Atascosa*
2A Austin*
4B Bailey
2B Bandera
2A Bastrop*
3B Baylor
2A Bee*
2A Bell*
2A Bexar*
3A Blanco*
3B Borden
2A Bosque*
3A Bowie*

(continued)

TABLE N1101.7 (R301.1)—continued
CLIMATE ZONES, MOISTURE REGIMES, AND WARM HUMID DESIGNATIONS BY STATE, COUNTY AND TERRITORY[a]

US STATES—continued
TEXAS *(continued)*
2A Brazoria*
2A Brazos*
3B Brewster
4B Briscoe
2A Brooks*
3A Brown*
2A Burleson*
3A Burnet*
2A Caldwell*
2A Calhoun*
3B Callahan
1A Cameron*
3A Camp*
4B Carson
3A Cass*
4B Castro
2A Chambers*
2A Cherokee*
3B Childress
3A Clay
4B Cochran
3B Coke
3B Coleman
3A Collin*
3B Collingsworth
2A Colorado*
2A Comal*
3A Comanche*
3B Concho
3A Cooke
2A Coryell*
3B Cottle
3B Crane
3B Crockett
3B Crosby
3B Culberson
4B Dallam
2A Dallas*
3B Dawson
4B Deaf Smith
3A Delta
3A Denton*
2A DeWitt*
3B Dickens
2B Dimmit
4B Donley
2A Duval*
3A Eastland
3B Ector
2B Edwards
2A Ellis*
3B El Paso
3A Erath*
2A Falls*
3A Fannin
2A Fayette*
3B Fisher
4B Floyd
3B Foard
2A Fort Bend*
3A Franklin*
2A Freestone*
2B Frio
3B Gaines
2A Galveston*
3B Garza
3A Gillespie*
3B Glasscock
2A Goliad*
2A Gonzales*
4B Gray
3A Grayson
3A Gregg*
2A Grimes*
2A Guadalupe*
4B Hale
3B Hall
3A Hamilton*
4B Hansford
3B Hardeman
2A Hardin*
2A Harris*
3A Harrison*
4B Hartley
3B Haskell
2A Hays*
3B Hemphill
3A Henderson*

(continued)

TABLE N1101.7 (R301.1)—continued
CLIMATE ZONES, MOISTURE REGIMES, AND WARM HUMID DESIGNATIONS BY STATE, COUNTY AND TERRITORY[a]

US STATES—continued
TEXAS *(continued)*
1A Hidalgo*
2A Hill*
4B Hockley
3A Hood*
3A Hopkins*
2A Houston*
3B Howard
3B Hudspeth
3A Hunt*
4B Hutchinson
3B Irion
3A Jack
2A Jackson*
2A Jasper*
3B Jeff Davis
2A Jefferson*
2A Jim Hogg*
2A Jim Wells*
2A Johnson*
3B Jones
2A Karnes*
3A Kaufman*
3A Kendall*
2A Kenedy*
3B Kent
3B Kerr
3B Kimble
3B King
2B Kinney
2A Kleberg*
3B Knox
3A Lamar*
4B Lamb
3A Lampasas*
2B La Salle
2A Lavaca*
2A Lee*
2A Leon*
2A Liberty*
2A Limestone*
4B Lipscomb
2A Live Oak*
3A Llano*
3B Loving
3B Lubbock
3B Lynn
2A Madison*
3A Marion*
3B Martin
3B Mason
2A Matagorda*
2B Maverick
3B McCulloch
2A McLennan*
2A McMullen*
2B Medina
3B Menard
3B Midland
2A Milam*
3A Mills*
3B Mitchell
3A Montague
2A Montgomery*
4B Moore
3A Morris*
3B Motley
3A Nacogdoches*
2A Navarro*
2A Newton*
3B Nolan
2A Nueces*
4B Ochiltree
4B Oldham
2A Orange*
3A Palo Pinto*
3A Panola*
3A Parker*
4B Parmer
3B Pecos
2A Polk*
4B Potter
3B Presidio
3A Rains*
4B Randall
3B Reagan
2B Real
3A Red River*
3B Reeves

(continued)

TABLE N1101.7 (R301.1)—continued
CLIMATE ZONES, MOISTURE REGIMES, AND WARM HUMID DESIGNATIONS BY STATE, COUNTY AND TERRITORY[a]

US STATES—continued
TEXAS *(continued)*
2A Refugio*
4B Roberts
2A Robertson*
3A Rockwall*
3B Runnels
3A Rusk*
3A Sabine*
3A San Augustine*
2A San Jacinto*
2A San Patricio*
3A San Saba*
3B Schleicher
3B Scurry
3B Shackelford
3A Shelby*
4B Sherman
3A Smith*
3A Somervell*
2A Starr*
3A Stephens
3B Sterling
3B Stonewall
3B Sutton
4B Swisher
2A Tarrant*
3B Taylor
3B Terrell
3B Terry
3B Throckmorton
3A Titus*
3B Tom Green
2A Travis*
2A Trinity*
2A Tyler*
3A Upshur*
3B Upton
2B Uvalde
2B Val Verde
3A Van Zandt*
2A Victoria*
2A Walker*
2A Waller*
3B Ward
2A Washington*
2B Webb
2A Wharton*
3B Wheeler
3A Wichita
3B Wilbarger
1A Willacy*
2A Williamson*
2A Wilson*
3B Winkler
3A Wise
3A Wood*
4B Yoakum
3A Young
2B Zapata
2B Zavala
UTAH
5B Beaver
5B Box Elder
5B Cache
5B Carbon
6B Daggett
5B Davis
6B Duchesne
5B Emery
5B Garfield
5B Grand
5B Iron
5B Juab
5B Kane
5B Millard
6B Morgan
5B Piute
6B Rich
5B Salt Lake
5B San Juan
5B Sanpete
5B Sevier
6B Summit
5B Tooele
6B Uintah
5B Utah
6B Wasatch
3B Washington
5B Wayne

(continued)

TABLE N1101.7 (R301.1)—continued
CLIMATE ZONES, MOISTURE REGIMES, AND WARM HUMID DESIGNATIONS BY STATE, COUNTY AND TERRITORY[a]

US STATES—continued
UTAH *(continued)*
5B Weber
VERMONT
6A (all)
VIRGINIA
4A (all except as follows:)
5A Alleghany
5A Bath
3A Brunswick
3A Chesapeake
5A Clifton Forge
5A Covington
3A Emporia
3A Franklin
3A Greensville
3A Halifax
3A Hampton
5A Highland
3A Isle of Wight
3A Mecklenburg
3A Newport News
3A Norfolk
3A Pittsylvania
3A Portsmouth
3A South Boston
3A Southampton
3A Suffolk
3A Surry
3A Sussex
3A Virginia Beach
WASHINGTON
5B Adams
5B Asotin
5B Benton
5B Chelan
5C Clallam
4C Clark
5B Columbia
4C Cowlitz
5B Douglas
6B Ferry
5B Franklin
5B Garfield
5B Grant
4C Grays Harbor
5C Island
4C Jefferson
4C King
5C Kitsap
5B Kittitas
5B Klickitat
4C Lewis
5B Lincoln
4C Mason
5B Okanogan
4C Pacific
6B Pend Oreille
4C Pierce
5C San Juan
4C Skagit
5B Skamania
4C Snohomish
5B Spokane
6B Stevens
4C Thurston
4C Wahkiakum
5B Walla Walla
4C Whatcom
5B Whitman
5B Yakima
WEST VIRGINIA
5A Barbour
4A Berkeley
4A Boone
4A Braxton
5A Brooke
4A Cabell
4A Calhoun
4A Clay
4A Doddridge
4A Fayette
4A Gilmer
5A Grant
4A Greenbrier
5A Hampshire
5A Hancock
5A Hardy
5A Harrison
4A Jackson

(continued)

TABLE N1101.7 (R301.1)—continued
CLIMATE ZONES, MOISTURE REGIMES, AND WARM HUMID DESIGNATIONS BY STATE, COUNTY AND TERRITORY[a]

US STATES—continued
WEST VIRGINIA *(continued)*
4A Jefferson
4A Kanawha
4A Lewis
4A Lincoln
4A Logan
5A Marion
5A Marshall
4A Mason
4A McDowell
4A Mercer
5A Mineral
4A Mingo
5A Monongalia
4A Monroe
4A Morgan
4A Nicholas
5A Ohio
5A Pendleton
4A Pleasants
5A Pocahontas
5A Preston
4A Putnam
4A Raleigh
5A Randolph
4A Ritchie
4A Roane
4A Summers
5A Taylor
5A Tucker
4A Tyler
4A Upshur
4A Wayne
4A Webster
5A Wetzel
4A Wirt
4A Wood
4A Wyoming
WISCONSIN
5A Adams
6A Ashland
6A Barron
6A Bayfield
6A Brown
6A Buffalo
6A Burnett
5A Calumet
6A Chippewa
6A Clark
5A Columbia
5A Crawford
5A Dane
5A Dodge
6A Door
6A Douglas
6A Dunn
6A Eau Claire
6A Florence
5A Fond du Lac
6A Forest
5A Grant
5A Green
5A Green Lake
5A Iowa
6A Iron
6A Jackson
5A Jefferson
5A Juneau
5A Kenosha
6A Kewaunee
5A La Crosse
5A Lafayette
6A Langlade
6A Lincoln
6A Manitowoc
6A Marathon
6A Marinette
6A Marquette
6A Menominee
5A Milwaukee
5A Monroe
6A Oconto
6A Oneida
5A Outagamie
5A Ozaukee
6A Pepin
6A Pierce
6A Polk
6A Portage

(continued)

TABLE N1101.7 (R301.1)—continued
CLIMATE ZONES, MOISTURE REGIMES, AND WARM HUMID DESIGNATIONS BY STATE, COUNTY AND TERRITORY[a]

US STATES—continued
WISCONSIN *(continued)*
6A Price
5A Racine
5A Richland
5A Rock
6A Rusk
5A Sauk
6A Sawyer
6A Shawano
6A Sheboygan
6A St. Croix
6A Taylor
6A Trempealeau
5A Vernon
6A Vilas
5A Walworth
6A Washburn
5A Washington
5A Waukesha
6A Waupaca
5A Waushara
5A Winnebago
6A Wood
WYOMING
6B Albany
6B Big Horn
6B Campbell
6B Carbon
6B Converse
6B Crook
6B Fremont
5B Goshen
6B Hot Springs
6B Johnson
5B Laramie
7 Lincoln
6B Natrona
6B Niobrara
6B Park
5B Platte
6B Sheridan
7 Sublette
6B Sweetwater
7 Teton
6B Uinta
6B Washakie
6B Weston
US TERRITORIES
AMERICAN SAMOA
1A (all)*
GUAM
1A (all)*
NORTHERN MARIANA ISLANDS
1A (all)*
PUERTO RICO
1A (all except as follows:)*
2B Barraquitas
2B Cayey
VIRGIN ISLANDS
1A (all)*

a. Key: A – Moist, B – Dry, C – Marine. Absence of moisture designation indicates moisture regime is irrelevant. Asterisk (*) indicates a Warm Humid location.

N1101.7.2 (R301.3) Climate zone definitions. To determine the climate zones for locations not listed in this code, use the following information to determine climate zone numbers and letters in accordance with Items 1 through 5.

1. Determine the thermal climate zone, 0 through 8, from Table N1101.7.2 using the heating (HDD) and cooling degree-days (CDD) for the location.
2. Determine the moisture zone (Marine, Dry or Humid) in accordance with Items 2.1 through 2.3.
   - 2.1. If monthly average temperature and precipitation data are available, use the Marine, Dry and Humid definitions to determine the moisture zone (C, B or A).
   - 2.2. If annual average temperature information (including degree-days) and annual precipitation (i.e., annual mean) are available, use Items 2.2.1 through 2.2.3 to determine the moisture zone. If the moisture zone is not Marine, then use the Dry definition to determine whether Dry or Humid.
     - 2.2.1. If thermal climate zone is 3 and CDD50°F ≤ 4,500 (CDD10°C ≤ 2500), climate zone is Marine (3C).
     - 2.2.2. If thermal climate zone is 4 and CDD50°F ≤ 2,700 (CDD10°C ≤ 1500), climate zone is Marine (4C).
     - 2.2.3. If thermal climate zone is 5 and CDD50°F ≤ 1,800 (CDD10°C ≤ 1000), climate zone is Marine (5C).
   - 2.3. If only degree-day information is available, use Items 2.3.1 through 2.3.3 to determine the moisture zone. If the moisture zone is not Marine, then it is not possible to assign Humid or Dry moisture zone for this location.
     - 2.3.1. If thermal climate zone is 3 and CDD50°F ≤ 4,500 (CDD10°C ≤ 2500), climate zone is Marine (3C).
     - 2.3.2. If thermal climate zone is 4 and CDD50°F ≤ 2,700 (CDD10°C ≤ 1500), climate zone is Marine (4C).
     - 2.3.3. If thermal climate zone is 5 and CDD50°F ≤ 1,800 (CDD10°C ≤ 1000), climate zone is Marine (5C).
3. Marine (C) Zone definition: Locations meeting all of the criteria in Items 3.1 through 3.4.
   - 3.1. Mean temperature of coldest month between 27°F (-3°C) and 65°F (18°C).
   - 3.2. Warmest month mean < 72°F (22°C).
   - 3.3. Not fewer than four months with mean temperatures over 50°F (10°C).
   - 3.4. Dry season in summer. The month with the heaviest precipitation in the cold season has at least three times as much precipitation as the month with the least precipitation in the rest of the year. The cold season is October through March in the Northern Hemisphere and April through September in the Southern Hemisphere.
4. Dry (B) definition: Locations meeting the criteria in Items 4.1 through 4.4.
   - 4.1. Not Marine (C).
   - 4.2. If 70 percent or more of the precipitation, *P*, occurs during the high sun period, defined as April through September in the Northern Hemisphere and October

TABLE N1101.7.2 (R301.3)
THERMAL CLIMATE ZONE DEFINITIONS

ZONE NUMBER	THERMAL CRITERIA	
	IP Units	SI Units
0	10,800 < CDD50ºF	6000 < CDD10ºC
1	9,000 < CDD50ºF < 10,800	5000 < CDD10°C < 6000
2	6,300 < CDD50°F ≤ 9,000	3500 < CDD10°C ≤ 5000
3	CDD50°F ≤ 6,300 AND HDD65°F ≤ 3,600	CDD10°C ≤ 3500 AND HDD18°C ≤ 2000
4	CDD50°F ≤ 6,300 AND 3,600 < HDD65°F ≤ 5,400	CDD10°C ≤ 3500 AND 2000 < HDD18°C ≤ 3000
5	CDD50°F < 6,300 AND 5,400 < HDD65°F ≤ 7,200	CDD10°C < 3500 AND 3000 < HDD18°C ≤ 4000
6	7,200 < HDD65°F ≤ 9,000	4000 < HDD18°C ≤ 5000
7	9,000 < HDD65°F ≤ 12,600	5000 < HDD18°C ≤ 7000
8	12,600 < HDD65°F	7000 < HDD18°C

For SI: °C = [(°F) − 32]/1.8.

through March in the Southern Hemisphere, then the dry/humid threshold is in accordance with Equation 11-1.

$P < 0.44 \times (T - 7)$
$[P < 20.0 \times (T + 14)$ in SI units]

(Equation 11-1)

where:

P = Annual precipitation, inches (mm).

T = Annual mean temperature, °F (°C).

4.3. If between 30 and 70 percent of the precipitation, P, occurs during the high sun period, defined as April through September in the Northern Hemisphere and October through March in the Southern Hemisphere, then the dry/humid threshold is in accordance with Equation 11-2.

$P < 0.44 \times (T - 19.5)$
$[P < 20.0 \times (T + 7)$ in SI units]

(Equation 11-2)

where:

P = Annual precipitation, inches (mm).

T = Annual mean temperature, °F (°C).

4.4. If 30 percent or less of the precipitation, P, occurs during the high sun period, defined as April through September in the Northern Hemisphere and October through March in the Southern Hemisphere, then the dry/humid threshold is in accordance with Equation 11-3.

$P < 0.44 \times (T - 32)$
$[P < 20.0 \times T$ in SI units]

(Equation 11-3)

where:

P = Annual precipitation, inches (mm).

T = Annual mean temperature, °F (°C).

5. Humid (A) definition: Locations that are not Marine (C) or Dry (B).

N1101.8 (R301.4) Tropical climate region. The tropical climate region shall be defined as:

1. Hawaii, Puerto Rico, Guam, American Samoa, US Virgin Islands, Commonwealth of Northern Mariana Islands; and
2. Islands in the area between the Tropic of Cancer and the Tropic of Capricorn.

N1101.9 (R302.1) Interior design conditions. The interior design temperatures used for heating and cooling load calculations shall be a maximum of 72°F (22°C) for heating and minimum of 75°F (24°C) for cooling.

N1101.10 (R303.1) Identification. Materials, systems and equipment shall be identified in a manner that will allow a determination of compliance with the applicable provisions of this code.

N1101.10.1 (R303.1.1) Building thermal envelope insulation. An R-value identification mark shall be applied by the manufacturer to each piece of *building thermal envelope* insulation that is 12 inches (305 mm) or greater in width. Alternatively, the insulation installers shall provide a certification that indicates the type, manufacturer and R-value of insulation installed in each element of the *building thermal envelope*. For blown-in or sprayed fiberglass and cellulose insulation, the initial installed thickness, settled thickness, settled R-value, installed density, coverage area and number of bags installed shall be indicated on the certification. For sprayed polyurethane foam (SPF) insulation, the installed thickness of the areas covered and the R-value of the installed thickness shall be indicated on the certification. For *insulated siding*, the R-value shall be on a label on the product's package and shall be indicated on the certification. The insulation installer shall sign, date and post the certification in a conspicuous location on the job site.

Exception: For roof insulation installed above the deck, the R-value shall be *labeled* as required by the material standards specified in Table R906.2.

N1101.10.1.1 (R303.1.1.1) Blown-in or sprayed roof and ceiling insulation. The thickness of blown-in or sprayed fiberglass and cellulose roof and ceiling insulation shall be written in inches (mm) on markers that are installed at not less than one for every 300 square feet (28 m^2) throughout the attic space. The markers shall be affixed to the trusses or joists and marked with the minimum initial installed thickness with numbers not less than 1 inch (25 mm) in height. Each marker shall face the attic access opening. The thickness and installed R-value of sprayed polyurethane foam insulation shall be indicated on the certification provided by the insulation installer.

N1101.10.2 (R303.1.2) Insulation mark installation. Insulating materials shall be installed such that the manufacturer's R-value *mark* is readily observable at inspection. For insulation materials that are installed without an observable manufacturer's R-value mark, such as blown or draped products, an insulation certificate complying with Section N1101.10.1 shall be left immediately after installation by the installer, in a conspicuous location within the building, to certify the installed R-value of the insulation material.

N1101.10.3 (R303.1.3) Fenestration product rating. U-factors of fenestration products such as windows, doors and *skylights* shall be determined in accordance with NFRC 100.

Exception: Where required, garage door U-factors shall be determined in accordance with either NFRC 100 or ANSI/DASMA 105.

U-factors shall be determined by an accredited, independent laboratory, and *labeled* and certified by the manufacturer.

Products lacking such a *labeled* U-factor shall be assigned a default U-factor from Table N1101.10.3(1) or N1101.10.3(2). The *solar heat gain coefficient* (SHGC) and visible transmittance (VT) of glazed fenestration products such as windows, glazed doors and *skylights* shall be determined in accordance with NFRC 200 by an accredited, independent laboratory, and *labeled* and certified by the manufacturer. Products lacking such a *labeled* SHGC or VT shall be assigned a default SHGC or VT from Table N1101.10.3(3).

**TABLE N1101.10.3(1) [R303.1.3(1)]
DEFAULT GLAZED WINDOW, GLASS DOOR AND SKYLIGHT *U*-FACTORS**

FRAME TYPE	WINDOW AND GLASS DOOR		SKYLIGHT	
	Single pane	Double pane	Single	Double
Metal	1.20	0.80	2.00	1.30
Metal with Thermal Break	1.10	0.65	1.90	1.10
Nonmetal or Metal Clad	0.95	0.55	1.75	1.05
Glazed Block	0.60			

**TABLE N1101.10.3(2) [R303.1.3(2)]
DEFAULT OPAQUE DOOR *U*-FACTORS**

DOOR TYPE	OPAQUE *U*-FACTOR
Uninsulated Metal	1.20
Insulated Metal	0.60
Wood	0.50
Insulated, nonmetal edge, not exceeding 45% glazing, any glazing double pane	0.35

**TABLE N1101.10.3(3) [R303.1.3(3)]
DEFAULT GLAZED FENESTRATION SHGC AND VT**

	SINGLE GLAZED		DOUBLE GLAZED		GLAZED BLOCK
	Clear	Tinted	Clear	Tinted	
SHGC	0.8	0.7	0.7	0.6	0.6
VT	0.6	0.3	0.6	0.3	0.6

N1101.10.4 (R303.1.4) Insulation product rating. The thermal resistance, *R*-value, of insulation shall be determined in accordance with Part 460 of US-FTC CFR Title 16 in units of h × ft^2 × °F/Btu at a mean temperature of 75°F (24°C).

N1101.10.4.1 (R303.1.4.1) Insulated siding. The thermal resistance, *R*-value, of *insulated siding* shall be determined in accordance with ASTM C1363. Installation for testing shall be in accordance with the manufacturer's instructions.

N1101.10.5 (R303.1.5) Air-impermeable insulation. Insulation having an air permeability not greater than 0.004 cubic feet per minute per square foot [0.002 L/(s × m^2)] under pressure differential of 0.3 inch water gauge (75 Pa) when tested in accordance with ASTM E2178 shall be determined air-impermeable insulation.

N1101.11 (R303.2) Installation. Materials, systems and equipment shall be installed in accordance with the manufacturer's instructions and this code.

N1101.11.1 (R303.2.1) Protection of exposed foundation insulation. Insulation applied to the exterior of *basement walls*, *crawl space* walls and the perimeter of slab-on-grade floors shall have a rigid, opaque and weather-resistant protective covering to prevent the degradation of the insulation's thermal performance. The protective covering shall cover the exposed exterior insulation and extend not less than 6 inches (153 mm) below grade.

N1101.12 (R303.3) Maintenance information. Maintenance instructions shall be furnished for equipment and systems that require preventive maintenance. Required regular maintenance actions shall be clearly stated and incorporated on a readily visible label. The label shall include the title or publication number for the operation and maintenance manual for that particular model and type of product.

N1101.13 (R401.2) Application. Residential buildings shall comply with Section N1101.13.5 and Section N1101.13.1, N1101.13.2, N1101.13.3 or N1101.13.4.

Exception: Additions, *alterations*, repairs and changes of occupancy to existing buildings complying with Section N1109.

N1101.13.1 (R401.2.1) Prescriptive Compliance Option. The Prescriptive Compliance Option requires compliance with Sections N1101 through N1104.

N1101.13.2 (R401.2.2) Total Building Performance Option. The Total Building Performance Compliance Path requires compliance with Section N1105.

N1101.13.3 (R401.2.3) Energy Rating Index Option. The Energy Rating Index (ERI) option requires compliance with Section N1106.

N1101.13.4 (R401.2.4) Tropical Climate Region Option. The Tropical Climate Region Option requires compliance with Section N1107.

N1101.13.5 (R401.2.5) Additional energy efficiency. This section establishes additional requirements applicable to all compliance approaches to achieve additional energy efficiency.

1. For buildings complying with Section N1101.13.1, one of the additional efficiency package options shall be installed according to Section N1108.2.
2. For buildings complying with Section N1101.13.2, the building shall meet one of the following:
   2.1. One of the additional efficiency package options in Section N1108.2 shall be installed without including such measures in the proposed design under Section N1105.
   2.2. The proposed design of the building under Section N1105.3 shall have an annual energy cost that is less than or equal to 95 percent of the annual energy cost of the standard reference design.

3. For buildings complying with the Energy Rating Index alternative Section N1101.13.3, the Energy Rating Index value shall be at least 5 percent less than the Energy Rating Index target specified.

The option selected for compliance shall be identified on the certificate required by Section N1101.14.

N1101.14 (R401.3) Certificate. A permanent certificate shall be completed by the builder or other *approved* party and posted on a wall in the space where the furnace is located, a utility room or an *approved* location inside the *building*. Where located on an electrical panel, the certificate shall not cover or obstruct the visibility of the circuit directory *label*, service disconnect *label* or other required *labels*. The certificate shall indicate the following:

1. The predominant *R*-values of insulation installed in or on ceilings, roofs, walls, foundation components such as slabs, *basement walls*, *crawl space walls* and floors, and ducts outside *conditioned spaces*.
2. *U*-factors of fenestration and the *solar heat gain coefficient* (SHGC) of fenestration. Where there is more than one value for any component of the building envelope, the certificate shall indicate both the value covering the largest area and the area weighted average value if available.
3. The results from any required duct system and building envelope air leakage testing performed on the building.
4. The types, sizes and efficiencies of heating, cooling and service water-heating equipment. Where a gas-fired unvented room heater, electric furnace, or baseboard electric heater is installed in the residence, the certificate shall indicate "gas-fired unvented room heater," "electric furnace" or "baseboard electric heater," as appropriate. An efficiency is not required to be indicated for gas-fired unvented room heaters, electric furnaces and electric baseboard heaters.
5. Where on-site *photovoltaic panel* systems have been installed, the array capacity, inverter efficiency, panel tilt and orientation shall be noted on the certificate.
6. For buildings where an Energy Rating Index score is determined in accordance with Section N1106, the Energy Rating Index score, both with and without any on-site generation, shall be listed on the certificate.
7. The code edition under which the structure was permitted and the compliance path used.

SECTION N1102 (R402) BUILDING THERMAL ENVELOPE

N1102.1 (R402.1) General. The *building thermal envelope* shall comply with the requirements of Sections N1102.1.1 through N1102.1.5.

Exceptions:

1. The following low-energy *buildings*, or portions thereof, separated from the remainder of the building by *building thermal envelope* assemblies complying with this section shall be exempt from the *building thermal envelope* provisions of Section N1102.
   1.1. Those with a peak design rate of energy usage less than 3.4 Btu/h × ft^2 (10.7 W/m^2) or 1.0 watt/ft^2 of floor area for space-conditioning purposes.
   1.2. Those that do not contain *conditioned space*.
2. Log homes designed in accordance with ICC 400.

N1102.1.1 (R402.1.1) Vapor retarder. Wall assemblies in the *building thermal envelope* shall comply with the vapor retarder requirements of Section R702.7.

N1102.1.2 (R402.1.2) Insulation and fenestration criteria. The *building thermal envelope* shall meet the requirements of Table N1102.1.2 based on the *climate zone* specified in Section N1101.7. Assemblies shall have a *U*-factor equal to or less than that specified in Table N1102.1.2. Fenestration shall have a *U*-factor and glazed fenestration SHGC equal to or less than that specified in Table N1102.1.2.

N1102.1.3 (R402.1.3) *R*-value alternative. Assemblies with *R*-value of insulation materials equal to or greater than that specified in Table N1102.1.3 shall be an alternative to the *U*-factor in Table N1102.1.2.

N1102.1.4 (R402.1.4) *R*-value computation. Cavity insulation alone shall be used to determine compliance with the cavity insulation *R*-value requirements in Table N1102.1.3. Where cavity insulation is installed in multiple layers, the *R*-values of the cavity insulation layers shall be summed to determine compliance with the cavity insulation *R*-value requirements. The manufacturer's settled *R*-value shall be used for blown-in insulation. Continuous insulation (ci) alone shall be used to determine compliance with the continuous insulation *R*-value requirements in Table N1102.1.3. Where continuous insulation is installed in multiple layers, the *R*-values of the continuous insulation layers shall be summed to determine compliance with the continuous insulation *R*-value requirements. Cavity insulation *R*-values shall not be used to determine compliance with the continuous insulation *R*-value requirements in Table N1102.1.3. Computed *R*-values shall not include an *R*-value for other building materials or air films. Where *insulated siding* is used for the purpose of complying with the continuous insulation requirements of Table N1102.1.3, the manufacturer's *labeled* *R*-value for *insulated siding* shall be reduced by R-0.6.

N1102.1.5 (R402.1.5) Total UA alternative. Where the total *building thermal envelope* UA, the sum of *U*-factor times assembly area, is less than or equal to the total UA resulting from multiplying the *U*-factors in Table N1102.1.2 by the same assembly area as in the proposed *building*, the *building* shall be considered to be incompliance with Table N1102.1.2. The UA calculation shall be performed using a method consistent with the ASHRAE *Handbook of Fundamentals* and shall include the thermal bridging effects of framing materials. In addition to UA compliance, the SHGC requirements of Table N1102.1.2 and the maximum fenestration *U*-factors of Section N1102.5 shall be met.

N1102.2 (R402.2) Specific insulation requirements. In addition to the requirements of Section N1102.1, insulation shall meet the specific requirements of Sections N1102.2.1 through N1102.2.12.

N1102.2.1 (R402.2.1) Ceilings with attic spaces. Where Section N1102.1.3 requires R-49 insulation in the ceiling or attic, installing R-38 insulation over 100 percent of the ceiling or attic area requiring insulation shall satisfy the requirement for R-49 insulation wherever the full height of uncompressed R-38 insulation extends over the wall top plate at the eaves. Where Section N1102.1.2 requires R-60 insulation in the ceiling, installing R-49 over 100 percent of the ceiling area requiring insulation shall satisfy the requirement for R-60 insulation wherever the full height of uncompressed R-49 insulation extends over the wall top plate at the eaves. This reduction shall not apply to the insulation and fenestration criteria in Section N1102.1.2 and the Total UA alternative in Section N1102.1.5.

N1102.2.2 (R402.2.2) Ceilings without attics. Where Section N1102.1.3 requires insulation *R*-values greater than R-30 in the interstitial space above a ceiling and below the structural roof deck, and the design of the roof/ceiling assembly does not allow sufficient space for the required insulation, the minimum required insulation *R*-value for such roof/ceiling assemblies shall be R-30. Insulation shall extend over the top of the wall plate to the outer edge of such plate and shall not be compressed. This reduction of insulation from the requirements of Section N1102.1.3 shall be limited to 500 square feet (46 m^2) or 20 percent of the total insulated ceiling area, whichever is less. This reduction shall not apply to the Total UA alternative in Section N1102.1.5.

N1102.2.3 (R402.2.3) Eave baffle. For air-permeable insulation in vented attics, a baffle shall be installed adjacent to soffit and eave vents. Baffles shall maintain a net free area opening equal to or greater than the size of the vent. The baffle shall extend over the top of the attic insulation. The baffle shall be permitted to be any solid

TABLE R1102.1.2 (R402.1.2)
MAXIMUM ASSEMBLY *U*-FACTORS[a] AND FENESTRATION REQUIREMENTS

CLIMATE ZONE	FENESTRATION *U*-FACTOR[f]	SKYLIGHT *U*-FACTOR	GLAZED FENESTRATION SHGC[d, e]	CEILING *U*-FACTOR	FRAME WALL *U*-FACTOR	MASS WALL *U*-FACTOR[b]	FLOOR *U*-FACTOR	BASEMENT WALL *U*-FACTOR	CRAWL SPACE WALL *U*-FACTOR
0	0.50	0.75	0.25	0.035	0.084	0.197	0.064	0.360	0.477
1	0.50	0.75	0.25	0.035	0.084	0.197	0.064	0.360	0.477
2	0.40	0.65	0.25	0.026	0.084	0.165	0.064	0.360	0.477
3	0.30	0.55	0.25	0.026	0.060	0.098	0.047	0.091[c]	0.136
4 except Marine	0.30	0.55	0.40	0.024	0.045	0.098	0.047	0.059	0.065
5 and Marine 4	0.30	0.55	NR	0.024	0.045	0.082	0.033	0.050	0.055
6	0.30	0.55	NR	0.024	0.045	0.060	0.033	0.050	0.055
7 and 8	0.30	0.55	NR	0.024	0.045	0.057	0.028	0.050	0.055

For SI: 1 foot = 304.8 mm.

a. Nonfenestration *U*-factors shall be obtained from measurement, calculation or an approved source.

b. Mass walls shall be in accordance with Section R402.2.5. Where more than half the insulation is on the interior, the mass wall *U*-factors shall not exceed 0.17 in Climate Zones 0 and 1, 0.14 in Climate Zone 2, 0.12 in Climate Zone 3, 0.087 in Climate Zone 4 except Marine, 0.065 in Climate Zone 5 and Marine 4, and 0.057 in Climate Zones 6 through 8.

c. In Warm Humid locations as defined by Figure R301.1 and Table R301.1, the basement wall *U*-factor shall not exceed 0.360.

d. The SHGC column applies to all glazed fenestration.

Exception: In Climate Zones 0 through 3, skylights shall be permitted to be excluded from glazed fenestration SHGC requirements provided that the SHGC for such skylights does not exceed 0.30.

e. There are no SHGC requirements in the Marine Zone.

f. A maximum *U*-factor of 0.32 shall apply in Marine Climate Zone 4 and Climate Zones 5 through 8 to vertical fenestration products installed in buildings located either:

1. Above 4,000 feet in elevation above sea level, or
2. In windborne debris regions where protection of openings is required by Section R301.2.1.2.

material. The baffle shall be installed to the outer edge of the *exterior wall* top plate so as to provide maximum space for attic insulation coverage over the top plate. Where soffit venting is not continuous, baffles shall be installed continuously to prevent ventilation air in the eave soffit from bypassing the baffle.

N1102.2.4 (R402.2.4) Access hatches and doors. Access hatches and doors from conditioned to unconditioned spaces such as attics and crawl spaces shall be insulated to the same *R*-value required by Table N1102.1.3 for the wall or ceiling in which they are installed.

Exceptions:

1. Vertical doors providing access from conditioned spaces to unconditioned spaces that comply with the fenestration requirements of Table N1102.1.3 based on the applicable climate zone specified in Chapter 3.
2. Horizontal pull-down, stair-type access hatches in ceiling assemblies that provide access from conditioned to unconditioned spaces in Climate Zones 0 through 4 shall not be required to comply with the insulation level of the surrounding surfaces provided that the hatch meets all of the following:
   2.1. The average *U*-factor of the hatch shall be less than or equal to U-0.10 or have an average insulation *R*-value of R-10 or greater.

TABLE N1102.1.3 (R402.1.3)
INSULATION MINIMUM *R*-VALUES AND FENESTRATION REQUIREMENTS BY COMPONENT[a]

CLIMATE ZONE	FENESTRATION *U*-FACTOR[b, i]	SKYLIGHT[b] *U*-FACTOR	GLAZED FENESTRATION SHGC[b, e]	CEILING *R*-VALUE	WOOD FRAME WALL *R*-VALUE[g]	MASS WALL *R*-VALUE[h]	FLOOR *R*-VALUE	BASEMENT[c, g] WALL *R*-VALUE	SLAB[d] *R*-VALUE & DEPTH	CRAWL SPACE[c, g] WALL *R*-VALUE
0	NR	0.75	0.25	30	13 or 0 + 10	3/4	13	0	0	0
1	NR	0.75	0.25	30	13 or 0 + 10	3/4	13	0	0	0
2	0.40	0.65	0.25	49	13 or 0 + 10	4/6	13	0	0	0
3	0.30	0.55	0.25	49	20 or 13 + 5ci or 0 + 15	8/13	19	5ci or 13[f]	10ci, 2 ft	5ci or 13[f]
4 except Marine	0.30	0.55	0.40	60	20 + 5 or 13 + 10ci or 0 + 15	8/13	19	10ci or 13	10ci, 4 ft	10ci or 13
5 and Marine 4	0.30	0.55	0.40	60	20 + 5 or 13 + 10ci or 0 + 15	13/17	30	15ci or 19 or 13 + 5ci	10ci, 4 ft	15ci or 19 or 13 + 5ci
6	0.30	0.55	NR	60	20 + 5ci or 13 + 10ci or 0 + 20	15/20	30	15ci or 19 or 13 + 5ci	10ci, 4 ft	15ci or 19 or 13 + 5ci
7 and 8	0.30	0.55	NR	60	20 + 5ci or 13 + 10ci or 0 + 20	19/21	38	15ci or 19 or 13 + 5ci	10ci, 4 ft	15ci or 1 9 or 13 + 5ci

For SI: 1 foot = 304.8 mm.

NR = Not Required.

ci = continuous insulation.

a. *R*-values are minimums. *U*-factors and SHGC are maximums. Where insulation is installed in a cavity that is less than the label or design thickness of the insulation, the installed *R*-value of the insulation shall be not less than the *R*-value specified in the table.

b. The fenestration *U*-factor column excludes skylights. The SHGC column applies to all glazed fenestration.

Exception: In Climate Zones 0 through 3, skylights shall be permitted to be excluded from glazed fenestration SHGC requirements provided that the SHGC for such skylights does not exceed 0.30.

c. "5ci or 13" means R-5 continuous insulation (ci) on the interior or exterior surface of the wall or R-13 cavity insulation on the interior side of the wall. "10ci or 13" means R-10 continuous insulation (ci) on the interior or exterior surface of the wall or R-13 cavity insulation on the interior side of the wall. "15ci or 19 or 13 + 5ci" means R-15 continuous insulation (ci) on the interior or exterior surface of the wall; or R-19 cavity insulation on the interior side of the wall; or R-13 cavity insulation on the interior of the wall in addition to R-5 continuous insulation on the interior or exterior surface of the wall.

d. R-5 insulation shall be provided under the full slab area of a heated slab in addition to the required slab edge insulation *R*-value for slabs. as indicated in the table. The slab-edge insulation for heated slabs shall not be required to extend below the slab.

e. There are no SHGC requirements in the Marine Zone.

f. Basement wall insulation shall not be required in Warm Humid locations as defined by Figure N1101.7 and Table N1101.7.

g. The first value is cavity insulation; the second value is continuous insulation. Therefore, as an example, "13 + 5" means R-13 cavity insulation plus R-5 continuous insulation.

h. Mass walls shall be in accordance with Section N1102.2.5. The second *R*-value applies where more than half of the insulation is on the interior of the mass wall.

i. A maximum *U*-factor of 0.32 shall apply in Climate Zones 3 through 8 to vertical fenestration products installed in buildings located either:

1. Above 4,000 feet in elevation, or
2. In windborne debris regions where protection of openings is required by Section R301.2.1.2.

2.2. Not less than 75 percent of the panel area shall have an insulation *R*-value of R-13 or greater.

2.3. The net area of the framed opening shall be less than or equal to 13.5 square feet (1.25 m^2).

2.4. The perimeter of the hatch edge shall be weatherstripped.

The reduction shall not apply to the total UA alternative in Section N1102.1.5.

N1102.2.4.1 (R402.2.4.1) Access hatch and door insulation installation and retention. Vertical or horizontal access hatches and doors from *conditioned spaces* to *unconditioned spaces* such as attics and crawl spaces shall be weatherstripped. Access that prevents damaging or compressing the insulation shall be provided to all equipment. Where loose-fill insulation is installed, a wood-framed or equivalent baffle or retainer, or dam shall be installed to prevent loose-fill insulation from spilling into living spaces, from higher to lower sections of the attic, and from attics covering conditioned spaces to unconditioned spaces. The baffle or retainer shall provide a permanent means of maintaining the installed *R*-value of the loose-fill insulation.

N1102.2.5 (R402.2.5) Mass walls. Mass walls where used as a component of the *building thermal envelope* shall be one of the following:

1. Above-ground walls of concrete block, concrete, insulated concrete form, masonry cavity, brick but not brick veneer, adobe, compressed earth block, rammed earth, solid timber, mass timber or solid logs.
2. Any wall having a heat capacity greater than or equal to 6 Btu/ft^2 × °F (123 kJ/m^2 × K).

N1102.2.6 (R402.2.6) Steel-frame ceilings, walls, and floors. Steel-frame ceilings, walls, and floors shall comply with the insulation requirements of Table N1102.2.6 or the *U*-factor requirements of Table N1102.1.2. The calculation of the *U*-factor for a steel-frame envelope assembly shall use a series-parallel path calculation method.

N1102.2.7 (R402.2.7) Floors. Floor *cavity* insulation shall comply with one of the following:

1. Insulation shall be installed to maintain permanent contact with the underside of the subfloor decking in accordance with manufacturer instructions to maintain required *R*-value or readily fill the available cavity space.
2. Floor framing cavity insulation shall be permitted to be in contact with the top side of sheathing separating the cavity and the unconditioned space below. Insulation shall extend from the bottom to the top of all perimeter floor framing members and the framing members shall be air sealed.
3. A combination of cavity and continuous insulation shall be installed so that the cavity insulation is in contact with the top side of the continuous insulation that is installed on the underside of the floor framing separating the cavity and the unconditioned space below. The combined *R*-value of the cavity and continuous insulation shall equal the required *R*-value for floors. Insulation shall extend from the bottom to the top of all perimeter floor framing members and the framing members shall be air sealed.

TABLE N1102.2.6 (R402.2.6)
STEEL-FRAME CEILING, WALL AND FLOOR INSULATION *R*-VALUES

WOOD FRAME *R*-VALUE REQUIREMENT	COLD-FORMED STEEL-FRAME EQUIVALENT *R*-VALUE[a]
Steel Truss Ceilings[b]	
R-30	R-38 or R-30 + 3 or R-26 + 5
R-38	R-49 or R-38 + 3
R-49	R-38 + 5
Steel Joist Ceilings[b]	
R-30	R-38 in 2 × 4 or 2 × 6 or 2 × 8 R-49 in any framing
R-38	R-49 in 2 × 4 or 2 × 6 or 2 × 8 or 2 × 10
Steel-frame Wall, 16 inches on center	
R-13	R-13 + 4.2 or R-21 + 2.8 or R-0 + 9.3 or R-15 + 3.8 or R-21 + 3.1
R-13 + 5	R-0 + 15 or R-13 + 9 or R-15 + 8.5 or R-19 + 8 or R-21 + 7
R-13 + 10	R-0 + 20 or R-13 + 15 or R-15 + 14 or R-19 + 13 or R-21 + 13
R-20	R-0 + 14.0 or R-13 + 8.9 or R-15 + 8.5 or R-19 + 7.8 or R-21 + 7.5
R-20 + 5	R-13 + 12.7 or R-15 + 12.3 or R-19 + 11.6 or R-21 + 11.3 or R-25 + 10.9
R-21	R-0 + 14.6 or R-13 + 9.5 or R-15 + 9.1 or R-19 + 8.4 or R-21 + 8.1 or R-25 + 7.7
Steel-frame Wall, 24 inches on center	
R-13	R-0 + 9.3 or R-13 + 3.0 or R-15 + 2.4
R-13 + 5	R-0 + 15 or R-13 + 7.5 or R-15 + 7 or R-19 + 6 or R-21 + 6
R-13 + 10	R-0 + 20 or R-13 + 13 or R-15 + 12 or R-19 + 11 or R-21 + 11
R-20	R-0 + 14.0 or R-13 + 7.7 or R-15 + 7.1 or R-19 + 6.3 or R-21 + 5.9
R-20 + 5	R-13 + 11.5 or R-15 + 10.9 or R-19 + 10.1 or R-21 + 9.7 or R-25 + 9.1
R-21	R-0 + 14.6 or R-13 + 8.3 or R-15 + 7.7 or R-19 + 6.9 or R-21 + 6.5 or R-25 + 5.9
Steel Joist Floor	
R-13	R-19 in 2 × 6, or R-19 + 6 in 2 × 8 or 2 × 10
R-19	R-19 + 6 in 2 × 6, or R-19 + 12 in 2 × 8 or 2 × 10

For SI: 1 inch = 25.4 mm.

a. The first value is cavity insulation *R*-value; the second value is continuous insulation *R*-value. Therefore, for example, "R-30 + 3" means R-30 cavity insulation plus R-3 continuous insulation.

b. Insulation exceeding the height of the framing shall cover the framing.

N1102.2.8 (R402.2.8) Basement walls. *Basement walls* shall be insulated in accordance with Table N1102.1.3.

Exception: Basement walls associated with unconditioned basements where all of the following requirements are met:

1. The floor overhead, including the underside stairway stringer leading to the basement, is insulated in accordance with Section N1102.1.3 and applicable provisions of Sections N1102.2 and N1102.2.7.
2. There are no uninsulated duct, domestic hot water or hydronic heating surfaces exposed to the basement.
3. There are no HVAC supply or return diffusers serving the basement.
4. The walls surrounding the stairway and adjacent to conditioned space are insulated in accordance with Section N1102.1.3 and applicable provisions of Section N1102.2.
5. The door(s) leading to the basement from conditioned spaces are insulated in accordance with Section N1102.1.3 and applicable provisions of Section N1102.2, and weatherstripped in accordance with Section N1102.4.
6. The building thermal envelope separating the basement from adjacent conditioned spaces complies with Section N1102.4.

N1102.2.8.1 (R402.2.8.1) Basement wall insulation installation. Where *basement walls* are insulated, the insulation shall be installed from the top of the *basement wall* down to 10 feet (3048 mm) below grade or to the basement floor, whichever is less.

N1102.2.9 (R402.2.9) Slab-on-grade floors. Slab-on-grade floors with a floor surface less than 12 inches (305 mm) below grade shall be insulated in accordance with Table N1102.1.3.

Exception: Slab-edge insulation is not required in jurisdictions designated by the code official as having a very heavy termite infestation.

N1102.2.9.1 (R402.2.9.1) Slab-on-grade floor insulation installation. Where installed, the insulation shall extend downward from the top of the slab on the outside or inside of the foundation wall. Insulation located below grade shall extend the distance provided in Table N1102.1.3 or the distance of the proposed design, as applicable, by any combination of vertical insulation, insulation extending under the slab or insulation extending out from the building. Insulation extending away from the building shall be protected by pavement or by not less than 10 inches (254 mm) of soil. The top edge of the insulation installed between the *exterior wall* and the edge of the interior slab shall be permitted to be cut at a 45-degree (0.79 rad) angle away from the *exterior wall.*

N1102.2.10 (R402.2.10) Crawl space walls. Crawl space walls shall be insulated in accordance with Table N1102.1.3.

Exception: Crawl space walls associated with a crawl space that is vented to the outdoors and the floor overhead is insulated in accordance with Table N1102.1.3 and Section N1102.2.7.

N1102.2.10.1 (R402.2.10.1) Crawl space wall insulation installation. Where crawl space wall insulation is installed, it shall be permanently fastened to the wall and shall extend downward from the floor to the finished grade elevation and then vertically or horizontally for not less than an additional 24 inches (610 mm). Exposed earth in unvented crawl space foundations shall be covered with a continuous Class I vapor retarder in accordance with this code. Joints of the vapor retarder shall overlap by 6 inches (153 mm) and be sealed or taped. The edges of the vapor retarder shall extend not less than 6 inches (153 mm) up the stem walls and shall be attached to the stem walls.

N1102.2.11 (R402.2.11) Masonry veneer. Insulation shall not be required on the horizontal portion of a foundation that supports a masonry veneer.

N1102.2.12 (R402.2.12) Sunroom and heated garage insulation. *Sunrooms* enclosing *conditioned space* and heated garages shall meet the insulation requirements of this code.

Exception: For *sunrooms* and heated garages provided with *thermal isolation*, and enclosing *conditioned space*, the following exceptions to the insulation *requirements* of this code shall apply:

1. The minimum ceiling insulation *R*-values shall be R-19 in *Climate Zones* 0 through 4 and R-24 in Climate Zones 5 through 8.
2. The minimum wall insulation *R*-value shall be R-13 in all climate zones. Walls separating a *sunroom* or heated garage with *thermal isolation* from *conditioned space* shall comply with the *building thermal envelope* requirements of this code.

N1102.3 (R402.3) Fenestration. In addition to the requirements of Section N1102, fenestration shall comply with Sections N1102.3.1 through N1102.3.5.

N1102.3.1 (R402.3.1) *U*-factor. An area-weighted average of fenestration products shall be permitted to satisfy the *U*-factor requirements.

N1102.3.2 (R402.3.2) Glazed fenestration SHGC. An area-weighted average of fenestration products more than 50-percent glazed shall be permitted to satisfy the SHGC requirements.

Dynamic glazing shall be permitted to satisfy the SHGC requirements of Table N1102.1.2 provided that the ratio of the higher to lower *labeled* SHGC is greater than or equal to 2.4, and the dynamic glazing is automatically controlled to modulate the amount of solar gain into the space in multiple steps. Dynamic glazing shall be considered separately from other fenestration, and area-weighted averaging with other fenestration that is not dynamic glazing shall be prohibited.

Exception: Dynamic glazing shall not be required to comply with this section where both the lower and higher *labeled* SHGC comply with the requirements of Table N1102.1.2.

N1102.3.3 (R402.3.3) Glazed fenestration exemption. Not greater than 15 square feet (1.4 m^2) of glazed fenestration per *dwelling unit* shall be exempt from the *U*-factor and SHGC requirements in Section N1102.1.2. This exemption shall not apply to the Total UA alternative in Section N1102.1.5.

N1102.3.4 (R402.3.4) Opaque door exemption. One side-hinged opaque door assembly not greater than 24 square feet (2.22 m^2) in area shall be exempt from the *U*-factor requirement in Section N1102.1.2. This exemption shall not apply to the and the Total UA alternative in Section N1102.1.5.

N1102.3.5 (R402.3.5) Sunroom and heated garage fenestration. *Sunrooms* and heated garages enclosing *conditioned space* shall comply with the fenestration requirements of this code.

Exception: In Climate Zones 2 through 8, for *sunrooms* and heated garages with *thermal isolation* and enclosing *conditioned space*, the fenestration *U*-factor shall not exceed 0.45 and the skylight *U*-factor shall not exceed 0.70.

New fenestration separating a sunroom or heated garages with thermal isolation from conditioned space shall comply with the building thermal envelope requirements of this code.

N1102.4 (R402.4) Air leakage. The *building thermal envelope* shall be constructed to limit air leakage in accordance with the requirements of Sections N1102.4.1 through N1102.4.5.

N1102.4.1 (R402.4.1) Building thermal envelope. The *building thermal envelope* shall comply with Sections N1102.4.1.1 through N1102.4.1.3. The sealing methods between dissimilar materials shall allow for differential expansion and contraction.

N1102.4.1.1 (R402.4.1.1) Installation. The components of the *building thermal envelope* as indicated in Table N1102.4.1.1 shall be installed in accordance with the manufacturer's instructions and the criteria indicated in Table N1102.4.1.1, as applicable to the method of construction. Where required by the *code official*, an *approved* third party shall inspect all components and verify compliance.

N1102.4.1.2 (R402.4.1.2) Testing. The *building* or *dwelling unit* shall be tested for air leakage. The maximum air leakage rate for any *building* or *dwelling unit* under any compliance path shall not exceed 5.0 air changes per hour or 0.28 cubic feet per minute (CFM) per square foot [0.0079 m^3/(s × m^2)] of dwelling unit enclosure area. Testing shall be conducted in accordance with ANSI/RESNET/ICC 380, ASTM E779 or ASTM E1827 and reported at a pressure of 0.2 inch w.g. (50 Pascals). Where required by the *code official*, testing shall be conducted by an *approved* third party. A written report of the results of the test shall be signed by the party conducting the test and provided to the *code official*. Testing shall be performed at any time after creation of all penetrations of the *building thermal envelope* have been sealed.

Exception: For heated, attached private garages and heated, detached private garages accessory to one- and two-family dwellings and townhouses not more than three stories above *grade plane* in height, building envelope tightness and insulation installation shall be considered acceptable where the items in Table N1102.4.1.1, applicable to the method of construction, are field verified. Where required by the code official, an *approved* third party independent from the installer shall inspect both air barrier and insulation installation criteria. Heated, attached private garage space and heated, detached private garage space shall be thermally isolated from all other *conditioned spaces* in accordance with Sections N1102.2.12 and N1102.3.5, as applicable.

During testing:

1. Exterior windows and doors, fireplace and stove doors shall be closed, but not sealed, beyond the intended weatherstripping or other infiltration control measures.
2. Dampers including exhaust, intake, makeup air, backdraft and flue dampers shall be closed, but not sealed beyond intended infiltration control measures.
3. Interior doors, where installed at the time of the test, shall be open.
4. Exterior or interior terminations for continuous ventilation systems shall be sealed.
5. Heating and cooling systems, where installed at the time of the test, shall be turned off.
6. Supply and return registers, where installed at the time of the test, shall be fully open.

Exception: When testing individual *dwelling units*, an air leakage rate not exceeding 0.30 cubic feet per minute per square foot [0.008 m^3/(s × m^2)] of the *dwelling unit* enclosure area, tested in accordance with ANSI/RESNET/ICC 380, ASTM E779 or ASTM E1827 and reported at a pressure of 0.2 inch water gauge (50 Pa), shall be permitted in all climate zones for:

1. Attached single- and multiple-family building *dwelling units*.
2. Buildings or *dwelling units* that are 1,500 square feet (139.4 m^2) or smaller.

Mechanical *ventilation* shall be provided in accordance with Section M1505 of this code or Section

403.3.2 of the *International Mechanical Code*, as applicable, or with other *approved* means of *ventilation*.

N1102.4.1.3 (R402.4.1.3) Leakage rate. Where complying with Section N1101.13.1, the building or *dwelling unit* shall have an air leakage rate not exceeding 5.0 air changes per hour in Climate Zones 0, 1 and 2, and 3.0 air changes per hour in Climate Zones 3 through 8, when tested in accordance with Section N1102.4.1.2.

TABLE N1102.4.1.1 (R402.4.1.1)
AIR BARRIER, AIR SEALING AND INSULATION INSTALLATION[a]

COMPONENT	AIR BARRIER CRITERIA	INSULATION INSTALLATION CRITERIA
General requirements	A continuous air barrier shall be installed in the building envelope. Breaks or joints in the air barrier shall be sealed.	Air-permeable insulation shall not be used as a sealing material.
Ceiling/attic	The air barrier in any dropped ceiling or soffit shall be aligned with the insulation and any gaps in the air barrier shall be sealed. Access openings, drop-down stairs or knee wall doors to unconditioned attic spaces shall be sealed.	The insulation in any dropped ceiling/soffit shall be aligned with the air barrier.
Walls	The junction of the foundation and sill plate shall be sealed. The junction of the top plate and the top of exterior walls shall be sealed. Knee walls shall be sealed.	Cavities within corners and headers of frame walls shall be insulated by completely filling the cavity with a material having a thermal resistance, *R*-value, of not less than R-3 per inch. Exterior thermal envelope insulation for framed walls shall be installed in substantial contact and continuous alignment with the air barrier.
Windows, skylights and doors	The space between framing and skylights, and the jambs of windows and doors, shall be sealed.	—
Rim joists	Rim joists shall include an exterior air barrier.[b] The junctions of the rim board to the sill plate and the rim board and the subfloor shall be air sealed.	Rim joists shall be insulated so that the insulation maintains permanent contact with the exterior rim board.[b]
Floors, including cantilevered floors and floors above garages	The air barrier shall be installed at any exposed edge of insulation.	Floor framing cavity insulation shall be installed to maintain permanent contact with the underside of subfloor decking. Alternatively, floor framing cavity insulation shall be in contact with the top side of sheathing, or continuous insulation installed on the underside of floor framing and extending from the bottom to the top of all perimeter floor framing members.
Basement crawl space, and slab foundations	Exposed earth in unvented crawl spaces shall be covered with a Class I vapor retarder/air barrier in accordance with Section R402.2.10. Penetrations through concrete foundation walls and slabs shall be air sealed. Class 1 vapor retarders shall not be used as an air barrier on below-grade walls and shall be installed in accordance with Section R702.7.	Crawl space insulation, where provided instead of floor insulation, shall be installed in accordance with Section R402.2.10. Conditioned basement foundation wall insulation shall be installed in accordance with Section R402.2.8.1. Slab-on-grade floor insulation shall be installed in accordance with Section R402.2.10.
Shafts, penetrations	Duct and flue shafts and other similar penetrations to exterior or unconditioned space shall be sealed to allow for expansion, contraction and mechanical vibration. Utility penetrations of the air barrier shall be caulked, gasketed or otherwise sealed and shall allow for expansion, contraction of materials and mechanical vibration.	Insulation shall be fitted tightly around utilities passing through shafts and penetrations in the building thermal envelope to maintain required *R*-value.
Narrow cavities	Narrow cavities of 1 inch or less that are not able to be insulated shall be air sealed.	Batts to be installed in narrow cavities shall be cut to fit or narrow cavities shall be filled with insulation that on installation readily conforms to the available cavity space.
Garage separation	Air sealing shall be provided between the garage and conditioned spaces.	Insulated portions of the garage separation assembly shall be installed in accordance with Sections R303 and R402.2.7.

(continued)

TABLE N1102.4.1.1 (R402.4.1.1)—continued
AIR BARRIER AND INSULATION INSTALLATION[a]

COMPONENT	AIR BARRIER CRITERIA	INSULATION INSTALLATION CRITERIA
Recessed lighting	Recessed light fixtures installed in the building thermal envelope shall be air sealed in accordance with Section R402.4.5.	Recessed light fixtures installed in the building thermal envelope shall be airtight and IC rated, and shall be buried or surrounded with insulation.
Plumbing, wiring or other obstructions	All holes created by wiring, plumbing or other obstructions in the air barrier assembly shall be air sealed.	Insulation shall be installed to fill the available space and surround wiring, plumbing, or other obstructions, unless the required *R*-value can be met by installing insulation and air barrier systems completely to the exterior side of the obstructions.
Shower/tub on exterior wall	The air barrier installed at exterior walls adjacent to showers and tubs shall separate the wall from the shower or tub.	Exterior walls adjacent to showers and tubs shall be insulated.
Electrical/phone box on exterior walls	The air barrier shall be installed behind electrical and communication boxes. Alternatively, air-sealed boxes shall be installed.	—
HVAC register boots	HVAC supply and return register boots that penetrate building thermal envelope shall be sealed to the subfloor, wall covering or ceiling penetrated by the boot.	—
Concealed sprinklers	Where required to be sealed, concealed fire sprinklers shall only be sealed in a manner that is recommended by the manufacturer. Caulking or other adhesive sealants shall not be used to fill voids between fire sprinkler cover plates and walls or ceilings.	—

For SI: 1 inch = 25.4 mm.

a. Inspection of log walls shall be in accordance with the provisions of ICC 400.

b. Air barrier and insulation full enclosure is not required in unconditioned/ventilated attic spaces and at rim joists.

N1102.4.2 (R402.4.2) Fireplaces. New wood-burning fireplaces shall have tight-fitting flue dampers or doors, and outdoor combustion air Where using tight-fitting doors on factory-built fireplaces *listed* and *labeled* in accordance with UL 127, the doors shall be tested and *listed* for the fireplace.

N1102.4.3 (R402.4.3) Fenestration air leakage. Windows, *skylights* and sliding glass doors shall have an air infiltration rate of not greater than 0.3 cfm per square foot (1.5 L/s/m^2), and for swinging doors not greater than 0.5 cfm per square foot (2.6 L/s/m^2), when tested in accordance with NFRC 400 or AAMA/WDMA/CSA 101/I.S.2/A440 by an accredited, independent laboratory and *listed* and *labeled* by the manufacturer.

Exception: Site-built windows, *skylights* and doors.

N1102.4.4 (R402.4.4) Rooms containing fuel-burning appliances. In Climate Zones 3 through 8, where open-combustion airducts provide combustion air to open combustion fuel-burning appliances, the appliances and combustion air opening shall be located outside the *building thermal envelope* or enclosed in a room that is isolated from inside the thermal envelope. Such rooms shall be sealed and insulated in accordance with the envelope requirements of Table N1102.1.3, where the walls, floors and ceilings shall meet a minimum of the *basement wall* *R*-value requirement. The door into the room shall be fully gasketed and any water lines and ducts in the room insulated in accordance with Section N1103. The combustion air duct shall be insulated where it passes through *conditioned space* to an *R*-value of not less than R-8.

Exceptions:

1. Direct vent appliances with both intake and exhaust pipes installed continuous to the outside.
2. Fireplaces and stoves complying with Sections N1102.4.2 and R1006.

N1102.4.5 (R402.4.5) Recessed lighting. Recessed luminaires installed in the *building thermal envelope* shall be sealed to limit air leakage between conditioned and *unconditioned spaces.* Recessed luminaires shall be IC-rated and *labeled* as having an air leakage rate of not greater than 2.0 cfm (0.944 L/s) when tested in accordance with ASTM E283 at a pressure differential of 1.57 psf (75 Pa). Recessed luminaires shall be sealed with a gasket or caulked between the housing and the interior wall or ceiling covering.

N1102.4.6 (R402.4.6) Electrical and communication outlet boxes (air-sealed boxes). Electrical and communication outlet boxes installed in the building thermal envelope shall be sealed to limit air leakage between conditioned and unconditioned spaces. Electrical and communication outlet boxes shall be tested in accordance with NEMA OS 4, *Requirements for Air-Sealed Boxes for Electrical and Communication Applications*, and shall have an air leakage rate of not greater than 2.0 cubic feet per minute (0.944 L/s) at a pressure differential of 1.57

psf (75 Pa). Electrical and communication outlet boxes shall be marked "NEMA OS 4" or "OS 4" in accordance with NEMA OS 4. Electrical and communication outlet boxes shall be installed per the manufacturer's instructions and with any supplied components required to achieve compliance with NEMA OS 4.

N1102.5 (R402.5) Maximum fenestration U-factor and SHGC. The area-weighted average maximum fenestration *U*-factor permitted using tradeoffs from Section N1102.1.5 or N1105 shall be 0.48 in Climate Zones 4 and 5 and 0.40 in Climate Zones 6 through 8 for vertical fenestration, and 0.75 in Climate Zones 4 through 8 for *skylights*. The area-weighted average maximum fenestration SHGC permitted using tradeoffs from Section N1105 in Climate Zones 0 through 3 shall be 0.40.

Exception: The maximum *U*-factor and solar heat gain coefficient (SHGC) for fenestration shall not be required in *storm shelters* complying with ICC 500.

SECTION N1103 (R403) SYSTEMS

N1103.1 (R403.1) Controls. Not less than one thermostat shall be provided for each separate heating and cooling system.

N1103.1.1 (R403.1.1) Programmable thermostat. The thermostat controlling the primary heating or cooling system of the *dwelling unit* shall be capable of controlling the heating and cooling system on a daily schedule to maintain different temperature set points at different times of the day and different days of the week. This thermostat shall include the capability to set back or temporarily operate the system to maintain *zone* temperatures of not less than 55°F (13°C) to not greater than 85°F (29°C). The thermostat shall be programmed initially by the manufacturer with a heating temperature setpoint of not greater than 70°F (21°C) and a cooling temperature setpoint of not less than 78°F (26°C).

N1103.1.2 (R403.1.2) Heat pump supplementary heat. Heat pumps having supplementary electric-resistance heat shall have controls that, except during defrost, prevent supplemental heat operation when the heat pump compressor can meet the heating load.

N1103.2 (R403.2) Hot water boiler outdoor temperature reset. The manufacturer shall equip each gas, oil and electric boiler (other than a boiler equipped with a tankless domestic water heating coil) with automatic means of adjusting the water temperature supplied by the boiler to ensure incremental change of the inferred heat load will cause an incremental change in the temperature of the water supplied by the boiler. This can be accomplished with outdoor reset, indoor reset or water temperature sensing.

N1103.3 (R403.3) Ducts. Ducts and air handlers shall be installed in accordance with Sections N1103.3.1 through N1103.3.7.

N1103.3.1 (R403.3.1) Ducts located outside conditioned space. Supply and return ducts located outside *conditioned space* shall be insulated to an *R*-value of not less than R-8 for ducts 3 inches (76 mm) in diameter and larger and not less than R-6 for ducts smaller than 3 inches (76 mm) in diameter. Ducts buried beneath a building shall be insulated as required by this section or have an equivalent thermal distribution efficiency. Underground ducts utilizing the thermal distribution efficiency method shall be listed and *labeled* to indicate the *R*-value equivalency.

N1103.3.2 (R403.3.2) Ducts located in conditioned space. For ductwork to be considered inside a *conditioned space*, it shall comply with one of the following:

1. The duct system is located completely within the *continuous air barrier* and within the *building thermal envelope.*
2. Ductwork in ventilated attic spaces is buried within ceiling insulation in accordance with Section N1103.3.3 and all of the following conditions exist:
    - 2.1. The air handler is located completely within the *continuous air barrier* and within the *building thermal envelope.*
    - 2.2. The duct leakage, as measured either by a rough-in test of the ducts or a post-construction total system leakage test to outside the *building thermal envelope* in accordance with Section N1103.3.6, is less than or equal to 1.5 cubic feet per minute (42.5 L/min) per 100 square feet (9.29 m^2) of *conditioned floor area* served by the duct system.
    - 2.3. The ceiling insulation *R*-value installed against and above the insulated duct is greater than or equal to the proposed ceiling insulation *R*-value, less the *R*-value of the insulation on the duct.
3. Ductwork in floor cavities located over unconditioned space shall have the following:
    - 3.1. A *continuous air barrier* installed between unconditioned space and the duct.
    - 3.2. Insulation installed in accordance with Section N1102.2.7.
    - 3.3. A minimum R-19 insulation installed in the cavity width separating the duct from unconditioned space.
4. Ductwork located within *exterior walls* of the *building thermal envelope* shall have the following:
    - 4.1. A *continuous air barrier* installed between unconditioned space and the duct.
    - 4.2. Minimum R-10 insulation installed in the cavity width separating the duct from the outside sheathing.
    - 4.3. The remainder of the cavity insulation fully insulated to the drywall side.

N1103.3.3 (R403.3.3) Ducts buried within ceiling insulation. Where supply and return air ducts are partially or completely buried in ceiling insulation, such ducts shall comply with all of the following:

1. The supply and return duct shall have an insulation *R*-value not less than R-8.
2. At all points along each duct, the sum of the ceiling insulation *R*-values against and above the top of the duct, and against and below the bottom of the duct shall be not less than R-19, excluding the *R*-value of the duct insulation.
3. In Climate Zones 0A, 1A, 2A and 3A, the supply ducts shall be completely buried within ceiling insulation, insulated to an *R*-value of not less than R-13 and in compliance with the vapor retarder requirements of Section M1601.4.6.

   Exception: Sections of the supply duct that are less than 3 feet (914 mm) from the supply outlet shall not be required to comply with these requirements.

N1103.3.3.1 (R403.3.3.1) Effective *R*-value of deeply buried ducts. Where using the Total Building Performance Compliance Option in accordance with Section N1101.13.2, sections of ducts that are installed in accordance with Section N1103.3.3, located directly on or within 5.5 inches (140 mm) of the ceiling, surrounded with blown-in attic insulation having an *R*-value of R-30 or greater and located such that the top of the duct is not less than 3.5 inches (89 mm) below the top of the insulation, shall be considered as having an effective duct insulation *R*-value of R-25.

N1103.3.4 (R403.3.4) Sealing. Ducts, air handlers and filter boxes shall be sealed. Joints and seams shall comply with Section M1601.4.1.

N1103.3.4.1 (R403.3.4.1) Sealed air handler. Air handlers shall have a manufacturer's designation for an air leakage of not greater than 2 percent of the design airflow rate when tested in accordance with ASHRAE 193.

N1103.3.5 (R403.3.5) Duct testing. Ducts shall be pressure tested in accordance with ANSI/RESNET/ICC 380 or ASTM E1554 to determine air leakage by one of the following methods:

1. Rough-in test: Total leakage shall be measured with a pressure differential of 0.1 inch w.g. (25 Pa) across the system, including the manufacturer's air handler enclosure if installed at the time of the test. Registers shall be taped or otherwise sealed during the test.
2. Postconstruction test: Total leakage shall be measured with a pressure differential of 0.1 inch w.g. (25 Pa) across the entire system, including the manufacturer's air handler enclosure. Registers shall be taped or otherwise sealed during the test.

Exception: A duct air-leakage test shall not be required for ducts serving heating, cooling or ventilation systems that are not integrated with ducts serving heating or cooling systems.

N1103.3.6 (R403.3.6) Duct leakage. The total leakage of the ducts, where measured in accordance with Section N1103.3.5, shall be as follows:

1. Rough-in test: The total leakage shall be less than or equal to 4.0 cubic feet per minute (113.3 L/min) per 100 square feet (9.29 m^2) of *conditioned floor area* where the air handler is installed at the time of the test. Where the air handler is not installed at the time of the test, the total leakage shall be less than or equal to 3.0 cubic feet per minute (85 L/min) per 100 square feet (9.29 m^2) of *conditioned floor area.*
2. Postconstruction test: Total leakage shall be less than or equal to 4.0 cubic feet per minute (113.3 L/min) per 100 square feet (9.29 m^2) of conditioned floor area.
3. Test for ducts within thermal envelope: Where all ducts and air handlers are located entirely within the *building thermal envelope*, total leakage shall be less than or equal to 8.0 cubic feet per minute (226.6 L/min) per 100 square feet (9.29 m^2) of *conditioned floor area.*

N1103.3.7 (R403.3.7) Building cavities. *Building* framing cavities shall not be used as ducts or plenums.

N1103.4 (R403.4) Mechanical system piping insulation. Mechanical system piping capable of carrying fluids greater than 105°F (41°C) or less than 55°F (13°C) shall be insulated to an *R*-value of not less than R-3.

N1103.4.1 (R403.4.1) Protection of piping insulation. Piping insulation exposed to weather shall be protected from damage, including that caused by sunlight, moisture, equipment maintenance and wind. The protection shall provide shielding from solar radiation that can cause degradation of the material. Adhesive tape shall be prohibited.

N1103.5 (R403.5) Service hot water systems. Energy conservation measures for service hot water systems shall be in accordance with Sections N1103.5.1 through N1103.5.3.

N1103.5.1 Heated water circulation and temperature maintenance systems. Heated water circulation systems shall be in accordance with Section N1103.5.1.1. Heat trace temperature maintenance systems shall be in accordance with Section N1103.5.1.2. Automatic controls, temperature sensors and pumps shall be in a location with access. Manual controls shall be in a location with *ready access.*

N1103.5.1.1 (R403.5.1.1) Circulation systems. Heated water circulation systems shall be provided with a circulation pump. The system return pipe shall be a dedicated return pipe or a cold water supply pipe. Gravity and thermosyphon circulation systems shall be prohibited. Controls for circulating hot water system pumps shall automatically turn off the pump when the water in the circulation loop is at the desired temperature and when there is no demand for hot water. The controls shall limit the temperature of the water entering the cold water piping to not greater than 104°F (40°C).

N1103.5.1.1.1 (R403.5.1.1.1) Demand recirculation water systems. Where installed, *demand recirculation water systems* shall have controls that start the pump upon receiving a signal from the action of a user of a fixture or appliance, sensing the presence of a user of a fixture or sensing the flow of hot or tempered water to a fixture fitting or appliance.

N1103.5.1.2 Heat trace systems. Electric heat trace systems shall comply with IEEE 515.1 or UL 515. Controls for such systems shall automatically adjust the energy input to the heat tracing to maintain the desired water temperature in the piping in accordance with the times when heated water is used in the occupancy.

N1103.5.2 (R403.5.2) Hot water pipe insulation. Insulation for service hot water piping with a thermal resistance, *R*-value, of not less than R-3 shall be applied to the following:

1. Piping $^{3}/_{4}$ inch (19 mm) and larger in nominal diameter located inside the *conditioned space.*
2. Piping serving more than one *dwelling unit.*
3. Piping located outside the *conditioned space.*
4. Piping from the water heater to a distribution manifold.
5. Piping located under a floor slab.
6. Buried piping.
7. Supply and return piping in circulation and recirculation systems other than cold water pipe return demand recirculation systems.

N1103.5.3 (R403.5.3) Drain water heat recovery units. Where installed, drain water heat recovery units shall comply with CSA B55.2. Drain water heat recovery units shall be tested in accordance with CSA B55.1. Potable water-side pressure loss of drain water heat recovery units shall be less than 3 psi (20.7 kPa) for individual units connected to one or two showers. Potable water-side pressure loss of drain water heat recovery units shall be less than 2 psi (13.8 kPa) for individual units connected to three or more showers.

N1103.6 (R403.6) Mechanical ventilation. *Buildings* and *dwelling units* shall be provided with mechanical *ventilation* that complies with the requirements of Section M1505 or with other *approved* means of *ventilation.* Outdoor air intakes and exhausts shall have automatic or gravity dampers that close when the *ventilation* system is not operating.

N1103.6.1 (R403.6.1) Heat or energy recovery ventilation. *Dwelling units* shall be provided with a heat recovery or energy recovery ventilation system in Climate Zones 7 and 8. The system shall be balanced with a minimum sensible heat recovery efficiency of 65 percent at 32°F (0°C) at a flow greater than or equal to the design airflow.

N1103.6.2 (R403.6.2) Whole-dwelling mechanical ventilation system fan efficacy. Fans used to provide whole-dwelling mechanical ventilation shall meet the efficacy requirements of Table N1103.6.2 at one or more rating points. Fans shall be tested in accordance with HVI 916 and listed. The airflow shall be reported in the product listing or on the label. Fan efficacy shall be reported in the product listing or shall be derived from the input power and airflow values reported in the product listing or on the label. Fan efficacy for fully ducted HRV, ERC, balanced and in-line fans shall be determined at a static pressure of not less than 0.2 inch water column (49.82 Pa). Fan efficacy for ducted range hoods, bathroom, and utility room fans shall be determined at a static pressure of not less than 0.1 inch water column (24.91 Pa).

TABLE N1103.6.2 (R403.6.2)
WHOLE-DWELLING MECHANICAL VENTILATION SYSTEM FAN EFFICACY[a]

FAN LOCATION	AIRFLOW RATE MINIMUM (CFM)	MINIMUM EFFICACY (CFM/WATT)
HRV, ERV	Any	1.2 cfm/watt
In-line supply or exhaust fan	Any	3.8 cfm/watt
Other exhaust fan	< 90	2.8 cfm/watt
Other exhaust fan	≥ 90	3.5 cfm/watt
Air-handler that is integrated to tested and listed HVAC equipment	Any	1.2 cfm/watt

For SI: 1 cubic foot per minute = 28.3 L/min.
a. Design outdoor airflow rate/watts of fan used.

N1103.6.3 (R403.6.3) Testing. Mechanical ventilation systems shall be tested and verified to provide the minimum ventilation flow rates required by Section N1103.6. Testing shall be performed according to the ventilation *equipment* manufacturer's instructions, or by using a flow hood or box, flow grid, or other airflow measuring device at the mechanical ventilation fan's inlet terminals or grilles, outlet terminals or grilles, or in the connected ventilation ducts. Where required by the code official, testing shall be conducted by an *approved* third party. A written report of the results of the test shall be signed by the party conducting the test and provided to the code official.

Exception: Kitchen range hoods that are ducted to the outside with 6-inch (152 mm) or larger duct and not more than one 90-degree (1.57 rad) elbow or equivalent in the duct run.

N1103.7 (R403.7) Equipment sizing and efficiency rating. Heating and cooling *equipment* shall be sized in accordance with ACCA Manual S based on *building* loads calculated in accordance with ACCA Manual J or other *approved* heating and cooling calculation methodologies. New or replacement heating and cooling *equipment* shall have an efficiency rating equal to or greater than the minimum required by federal law for the geographic location where the *equipment* is installed.

N1103.8 (R403.8) Systems serving multiple dwelling units. Systems serving multiple *dwelling units* shall comply with Sections C403 and C404 of the *International Energy Conservation Code*—Commercial Provisions instead of Section N1103.

N1103.9 (R403.9) Snow melt system controls. Snow- and ice-melting systems, supplied through energy service to the building, shall include automatic controls capable of shutting off the system when the pavement temperature is greater than 50°F (10°C) and precipitation is not falling, and an automatic or manual control that will allow shutoff when the outdoor temperature is greater than 40°F (4.8°C).

N1103.10 (R403.10) Energy consumption of pools and spas. The energy consumption of pools and permanent spas shall be controlled by the requirements in Sections N1103.10.1 through N1103.10.3.

N1103.10.1 (R403.10.1) Heaters. The electric power to heaters shall be controlled by an on-off switch that is an integral part of the heater mounted on the exterior of the heater in a location with *ready access*, or external to and within 3 feet (914 mm) of the heater. Operation of such switch shall not change the setting of the heater thermostat. Such switches shall be in addition to a circuit breaker for the power to the heater. Gas-fired heaters shall not be equipped with continuously burning ignition pilots.

N1103.10.2 (R403.10.2) Time switches. Time switches or other control methods that can automatically turn heaters and pump motors off and on according to a preset schedule shall be installed for heaters and pump motors. Heaters and pump motors that have built-in time switches shall be in compliance with this section.

Exceptions:

1. Where public health standards require 24-hour pump operation.
2. Pumps that operate solar- and waste-heat-recovery pool heating systems.

N1103.10.3 (R403.10.3) Covers. Outdoor heated pools and outdoor permanent spas shall be provided with a vapor-retardant cover or other *approved* vapor-retardant means.

Exception: Where more than 75 percent of the energy for heating, computed over an operation season of not **fewer** than 3 calendar months, is from a heat pump or an on-site renewable energy system, covers or other vapor-retardant means shall not be required.

N1103.11 (R403.11) Portable spas. The energy consumption of electric-powered portable spas shall be controlled by the requirements of APSP 14.

N1103.12 (R403.12) Residential pools and permanent residential spas. Where installed, the energy consumption of residential swimming pools and permanent residential spas shall be controlled in accordance with the requirements of APSP 15.

SECTION N1104 (R404) ELECTRICAL POWER AND LIGHTING SYSTEMS

N1104.1 (R404.1) Lighting equipment. All permanently installed lighting fixtures, excluding kitchen appliance lighting fixtures, shall contain only high-efficacy lighting sources.

N1104.1.1 (R404.1.1) Exterior lighting. Connected exterior lighting for Group R-2, R-3 and R-4 buildings shall comply with Section C405.4 of the *International Energy Conservation Code*—Commercial Provisions.

Exceptions:

1. Detached one- and two-family dwellings.
2. Townhouses.
3. Solar-powered lamps not connected to any electrical service.
4. Luminaires controlled by a motion sensor.

N1104.1.2 (R404.1.1) Fuel gas lighting equipment. Fuel gas lighting systems shall not have continuously burning pilot lights.

N1104.2 (R404.2) Interior lighting controls. Permanently installed lighting fixtures shall be controlled with a dimmer, an occupant sensor control or another control that is installed or built into the fixture.

Exception: Lighting controls shall not be required for the following:

1. Bathrooms.
2. Hallways.
3. Exterior lighting fixtures.
4. Lighting designed for safety or security.

N1104.3 (R404.3) Exterior lighting controls. Where the total permanently installed exterior lighting power is greater than 30 watts, the permanently installed exterior lighting shall comply with the following:

1. Lighting shall be controlled by a manual on and off switch that permits automatic shut-off actions.

   Exception: Lighting serving multiple *dwelling units.*

2. Lighting shall be automatically shut off when daylight is present and satisfies the lighting needs.
3. Controls that override automatic shut-off actions shall not be allowed unless the override automatically returns automatic control to its normal operation within 24 hours.

SECTION N1105 (R405)
TOTAL BUILDING PERFORMANCE

N1105.1 (R405.1) Scope. This section establishes criteria for compliance using total building performance analysis. Such analysis shall include heating, cooling, mechanical ventilation and service water-heating energy only.

N1105.2 (R405.2) Performance-based compliance. Compliance based on total building performance requires that a *proposed design* meets all of the following:

1. The requirements of the sections indicated within Table N1105.2.
2. The building thermal envelope greater than or equal to levels of efficiency and solar heat gain coefficients in Table R402.1.1 or R402.1.3 of the 2009 *International Energy Conservation Code*.
3. An annual energy cost that is less than or equal to the annual energy cost of the *standard reference design*. Energy prices shall be taken from a source *approved* by the *code official*, such as the Department of Energy, Energy Information Administration's State Energy Data System Prices and Expenditures reports. *Code officials* shall be permitted to require time-of-use pricing in energy cost calculations.

   Exception: The energy use based on source energy expressed in Btu or Btu per square foot of conditioned floor area shall be permitted to be substituted for the energy cost. The source energy multiplier for electricity shall be 3.16. The source energy multiplier for fuels other than electricity shall be 1.1.

N1105.3 (R405.3) Documentation. Documentation of the software used for the performance design and the parameters for the *building* shall be in accordance with Sections N1105.3.1 through N1105.3.2.2.

N1105.3.1 (R405.3.1) Compliance software tools. Documentation verifying that the methods and accuracy of the compliance software tools conform to the provisions of this section shall be provided to the *code official*.

N1105.3.2 (R405.3.2) Compliance report. Compliance software tools shall generate a report that documents that the *proposed design* complies with Section N1105.3. A compliance report on the *proposed design* shall be

**TABLE N1105.2 (R405.2)
REQUIREMENTS FOR TOTAL BUILDING PERFORMANCE**

SECTION[a]	TITLE
General	
N1101.13.5	Additional energy efficiency
N1101.14	Certificate
Building Thermal Envelope	
N1102.1.1	Vapor retarder
N1102.2.3	Eave baffle
N1102.2.4.1	Access hatches and doors
N1102.2.10.1	Crawl space wall insulation installation
N1102.4.1.1	Installation
N1102.4.1.2	Testing
N1102.5	Maximum fenestration *U*-factor and SHGC
Mechanical	
N1103.1	Controls
N1103.3, including N1103.3.1, except Sections N1103.3.2, N1103.3.3 and N1103.3.6	Ducts
N1103.4	Mechanical system piping insulation
N1103.5.1	Heated water circulation and temperature maintenance systems
N1103.5.3	Drain water heat recovery units
N1103.6	Mechanical ventilation
N1103.7	Equipment sizing and efficiency rating
N1103.8	Systems serving multiple dwelling units
N1103.9	Snow melt system controls
N1103.10	Energy consumption of pools and spas
N1103.11	Portable spas
N1103.12	Residential pools and permanent residential spas
Electrical Power and Lighting Systems	
N1104.1	Lighting equipment
N1104.2	Interior lighting controls

a. Reference to a code section includes all the relative subsections except as indicated in the table.

submitted with the application for the building permit. Upon completion of the building, a confirmed compliance report based on the confirmed condition of the building shall be submitted to the *code official* before a certificate of occupancy is issued.

Compliance reports shall include information in accordance with Sections N1105.3.1 and N1105.3.2.2.

N1105.3.2.1 (R405.3.2.1) Compliance report for permit application. A compliance report submitted with the application for building permit shall include the following:

1. Building street address, or other *building site* identification.
2. The name of the individual performing the analysis and generating the compliance report.
3. The name and version of the compliance software tool.
4. Documentation of all inputs entered into the software used to produce the results for the reference design and/or the rated home.
5. A certificate indicating that the proposed design complies with Section N1105.3. The certificate shall document the building components' energy specifications that are included in the calculation, including component-level insulation *R*-values or *U*-factors; duct system and building envelope air leakage testing assumptions; and the type and rated efficiencies of proposed heating, cooling, mechanical ventilation and service water-heating equipment to be installed. If on-site renewable energy systems will be installed, the certificate shall report the type and production size of the proposed system.
6. When a site-specific report is not generated, the proposed design shall be based on the worst-case orientation and configuration of the rated home.

N1105.3.2.2 (R405.3.2.2) Compliance report for certificate of occupancy. A compliance report submitted for obtaining the certificate of occupancy shall include the following:

1. Building street address, or other building site identification.
2. Declaration of the total building performance path on the title page of the energy report and the title page of the building plans.
3. A statement, bearing the name of the individual performing the analysis and generating the report, indicating that the as-built building complies with Section N1105.3.
4. The name and version of the compliance software tool.
5. A site-specific energy analysis report that is in compliance with Section N1105.3.
6. A final confirmed certificate indicating compliance based on inspection, and a statement indicating that the confirmed rated design of the built home complies with Section N1105.3. The certificate shall report the energy features that were confirmed to be in the home, including component-level insulation *R*-values or *U*-factors; results from any required duct system and building envelope air leakage testing; and the type and rated efficiencies of the heating, cooling, mechanical ventilation and service water heating equipment installed.
7. Where on-site renewable energy systems have been installed, the certificate shall report the type and production size of the installed system.

N1105.4 (R405.4) Calculation procedure. Calculations of the performance design shall be in accordance with Sections N1105.4.1 and N1105.4.2.

N1105.4.1 (R405.4.1) General. Except as specified by this section, the *standard reference design* and *proposed design* shall be configured and analyzed using identical methods and techniques.

N1105.4.2 (R405.4.2) Residence specifications. The *standard reference design* and *proposed design* shall be configured and analyzed as specified by Table N1105.4.2(1). Table N1105.4.2(1) shall include, by reference, all notes contained in Table N1102.1.3.

N1105.5 (R405.5) Calculation software tools. Calculation software, where used, shall be in accordance with Sections N1105.5.1 through N1105.5.3.

N1105.5.1 (R405.5.1) Minimum capabilities. Calculation procedures used to comply with this section shall be software tools capable of calculating the annual energy consumption of all building elements that differ between the *standard reference design* and the *proposed design* and shall include the following capabilities:

1. Computer generation of the *standard reference design* using only the input for the *proposed design*. The calculation procedure shall not allow the user to directly modify the building component characteristics of the *standard reference design*.
2. Calculation of whole-building (as a single *zone*) sizing for the heating and cooling *equipment* in the *standard reference design* residence in accordance with Section N1103.6.
3. Calculations that account for the effects of indoor and outdoor temperatures and part-load ratios on the performance of heating, ventilating and air-conditioning *equipment* based on climate and *equipment* sizing.
4. Printed *code official* inspection checklist listing each of the *proposed design* component characteristics from Table N1105.4.2(1) determined by the analysis to provide compliance, along with their respective performance ratings such as *R*-value, *U*-factor, SHGC, HSPF, AFUE, SEER and EF.

TABLE N1105.4.2(1) [R405.4.2(1)]
SPECIFICATIONS FOR THE STANDARD REFERENCE AND PROPOSED DESIGNS

BUILDING COMPONENT	STANDARD REFERENCE DESIGN	PROPOSED DESIGN
Above-grade walls	Type: mass where the proposed wall is a mass wall; otherwise wood frame.	As proposed
	Gross area: same as proposed.	As proposed
	U-factor: as specified in Table N1102.1.2.	As proposed
	Solar absorptance = 0.75.	As proposed
	Emittance = 0.90.	As proposed
Basement and crawl space walls	Type: same as proposed.	As proposed
	Gross area: same as proposed.	As proposed
	U-factor: as specified in Table N1102.1.2, with the insulation layer on the interior side of the walls.	As proposed
Above-grade floors	Type: wood frame.	As proposed
	Gross area: same as proposed.	As proposed
	U-factor: as specified in Table N1102.1.2.	As proposed
Ceilings	Type: wood frame.	As proposed
	Gross area: same as proposed.	As proposed
	U-factor: as specified in Table N1102.1.2.	As proposed
Roofs	Type: composition shingle on wood sheathing.	As proposed
	Gross area: same as proposed.	As proposed
	Solar absorptance = 0.75.	As proposed
	Emittance = 0.90.	As proposed
Attics	Type: vented with an aperture of 1 ft^2 per 300 ft^2 of ceiling area.	As proposed
Foundations	Type: same as proposed.	As proposed
	Foundation wall area above and below grade and soil characteristics: same as proposed.	As proposed
Opaque doors	Area: 40 ft^2.	As proposed
	Orientation: North.	As proposed
	U-factor: same as fenestration as specified in Table N1102.1.2.	As proposed
Vertical fenestration other than opaque doors	Total area[h] = (a) The proposed glazing area, where the proposed glazing area is less than 15 percent of the conditioned floor area. (b) 15 percent of the conditioned floor area, where the proposed glazing area is 15 percent or more of the conditioned floor area.	As proposed
	Orientation: equally distributed to four cardinal compass orientations (N, E, S & W).	As proposed
	U-factor: as specified in Table N1102.1.2.	As proposed
	SHGC: as specified in Table N1102.1.2 except for climate zones without an SHGC requirement, the SHGC shall be equal to 0.40.	As proposed
	Interior shade fraction: 0.92 – (0.21 × SHGC for the standard reference design).	Interior shade fraction: 0.92 – (0.21 × SHGC as proposed)
	External shading: none	As proposed
Skylights	None	As proposed
Thermally isolated sunrooms	None	As proposed

(continued)

TABLE N1105.4.2(1) [R405.4.2(1)]—continued
SPECIFICATIONS FOR THE STANDARD REFERENCE AND PROPOSED DESIGNS

BUILDING COMPONENT	STANDARD REFERENCE DESIGN	PROPOSED DESIGN
Air exchange rate	The air leakage rate at a pressure of 0.2 inch w.g. (50 Pa) shall be Climate Zones 0 through 2: 5.0 air changes per hour. Climate Zones 3 through 8: 3.0 air changes per hour.	The measured air exchange rate.[a]
	The mechanical ventilation rate shall be in addition to the air leakage rate and shall be the same as in the proposed design, but not greater than $0.01 \times CFA + 7.5 \times (N_{br} + 1)$ where: CFA = conditioned floor area, ft^2. N_{br} = number of bedrooms. The mechanical ventilation system type shall be the same as in the proposed design. Energy recovery shall not be assumed for mechanical ventilation.	The mechanical ventilation rate[b] shall be in addition to the air leakage rate and shall be as proposed.
Mechanical ventilation	Where mechanical ventilation is not specified in the proposed design: None Where mechanical ventilation is specified in the proposed design, the annual vent fan energy use, in units of kWh/yr, shall equal $(1/e_f) \times [0.0876 \times CFA + 65.7 \times (N_{br}+1)]$ where: e_f = the minimum exhaust fan efficacy, as specified in Table N1103.6.2, corresponding to the system type at a flow rate of $0.01 \times CFA + 7.5 \times (N_{br}+1)$ CFA = conditioned floor area, ft^2. N_{br} = number of bedrooms.	As proposed
Internal gains	IGain, in units of Btu/day per dwelling unit, shall equal $17,900 + 23.8 \times CFA + 4,104 \times N_{br}$ where: CFA = conditioned floor area, ft^2. N_{br} = number of bedrooms.	Same as standard reference design.
Internal mass	Internal mass for furniture and contents: 8 pounds per square foot of floor area.	Same as standard reference design, plus any additional mass specifically designed as a thermal storage element[c] but not integral to the building envelope or structure.
Structural mass	For masonry floor slabs: 80 percent of floor area covered by R-2 carpet and pad, and 20 percent of floor directly exposed to room air.	As proposed
	For masonry basement walls: as proposed, but with insulation as specified in Table N1102.1.3, located on the interior side of the walls.	As proposed
	For other walls, ceilings, floors, and interior walls: wood frame construction.	As proposed
Heating systems[d, e]	For other than electric heating without a heat pump: as proposed. Where the proposed design utilizes electric heating without a heat pump, the standard reference design shall be an air source heat pump meeting the requirements of Section C403 of the IECC—Commercial Provisions. Capacity: sized in accordance with Section N1103.7.	As proposed
Cooling systems[d, f]	As proposed. Capacity: sized in accordance with Section N1103.7.	As proposed

(continued)

TABLE N1105.4.2(1) [R405.4.2(1)]—continued
SPECIFICATIONS FOR THE STANDARD REFERENCE AND PROPOSED DESIGNS

<table>
<tr><th>BUILDING COMPONENT</th><th>STANDARD REFERENCE DESIGN</th><th colspan="3">PROPOSED DESIGN</th></tr>
<tr><td rowspan="7">Service water heating[d, g]</td><td rowspan="7">As proposed.
Use, in units of gal/day = 30 + (10 × N_{br})
where:
N_{br} = number of bedrooms.</td><td colspan="3">As proposed
Use, in units of gal/day = 25.5 + (8.5 × N_{br}) × (1 – $HWDS$)
where:
N_{br} = number of bedrooms.
$HWDS$ = factor for the compactness of the hot water distribution system.</td></tr>
<tr><td colspan="2">Compactness ratio[i] factor</td><td>HWDS</td></tr>
<tr><td>1 story</td><td>2 or more stories</td><td></td></tr>
<tr><td>$> 60\%$</td><td>$> 30\%$</td><td>0</td></tr>
<tr><td>$> 30\%$ to $\leq 60\%$</td><td>$> 15\%$ to $\leq 30\%$</td><td>0.05</td></tr>
<tr><td>$> 15\%$ to $\leq 30\%$</td><td>$> 7.5\%$ to $\leq 15\%$</td><td>0.10</td></tr>
<tr><td>$< 15\%$</td><td>$< 7.5\%$</td><td>0.15</td></tr>
<tr><td>Thermal distribution systems</td><td>Duct insulation: in accordance with Section N1103.3.1.
A thermal distribution system efficiency (DSE) of 0.88 shall be applied to both the heating and cooling system efficiencies for all systems other than tested duct systems.
Duct location: same as proposed design.
Exception: For nonducted heating and cooling systems that do not have a fan, the standard reference design thermal distribution system efficiency (DSE) shall be 1. For tested duct systems, the leakage rate shall be 4 cfm (113.3 L/min) per 100 ft^2 (9.29 m^2) of conditioned floor area at a pressure of differential of 0.1 inch w.g. (25 Pa).</td><td colspan="3">Duct insulation: as proposed.
As tested or, where not tested, as specified in Table N1105.4.2(2).</td></tr>
<tr><td>Thermostat</td><td>Type: Manual, cooling temperature setpoint = 75°F;
Heating temperature setpoint = 72°F.</td><td colspan="3">Same as standard reference design.</td></tr>
<tr><td>Dehumidistat</td><td>Where a mechanical ventilation system with latent heat recovery is not specified in the proposed design:
None.
Where the proposed design utilizes a mechanical ventilation system with latent heat recovery:
Dehumidistat type: manual, setpoint = 60% relative humidity.
Dehumidifier: whole-dwelling with integrated energy factor = 1.77 liters/kWh.</td><td colspan="3">Same as standard reference design.</td></tr>
</table>

(continued)

TABLE N1105.4.2(1) [R405.4.2(1)]—continued
SPECIFICATIONS FOR THE STANDARD REFERENCE AND PROPOSED DESIGNS

For SI: 1 square foot = 0.93 m^2, 1 British thermal unit = 1055 J, 1 pound per square foot = 4.88 kg/m^2, 1 gallon (US) = 3.785 L, °C = (°F – 32)/1.8, 1 degree = 0.79 rad.

a. Where required by the code official, testing shall be conducted by an approved party. Hourly calculations as specified in the ASHRAE *Handbook of Fundamentals,* or the equivalent, shall be used to determine the energy loads resulting from infiltration.

b. The combined air exchange rate for infiltration and mechanical ventilation shall be determined in accordance with Equation 43 of 2001 ASHRAE *Handbook of Fundamentals,* page 26.24 and the "Whole-house Ventilation" provisions of 2001 ASHRAE *Handbook of Fundamentals,* page 26.19 for intermittent mechanical ventilation.

c. Thermal storage element shall mean a component that is not part of the floors, walls or ceilings that is part of a passive solar system, and that provides thermal storage such as enclosed water columns, rock beds, or phase-change containers. A thermal storage element shall be in the same room as fenestration that faces within 15 degrees (0.26 rad) of true south, or shall be connected to such a room with pipes or ducts that allow the element to be actively charged.

d. For a proposed design with multiple heating, cooling or water heating systems using different fuel types, the applicable standard reference design system capacities and fuel types shall be weighted in accordance with their respective loads as calculated by accepted engineering practice for each equipment and fuel type present.

e. For a proposed design without a proposed heating system, a heating system having the prevailing federal minimum efficiency shall be assumed for both the standard reference design and proposed design.

f. For a proposed design home without a proposed cooling system, an electric air conditioner having the prevailing federal minimum efficiency shall be assumed for both the standard reference design and the proposed design.

g. For a proposed design with a nonstorage-type water heater, a 40-gallon storage-type water heater having the prevailing federal minimum energy factor for the same fuel as the predominant heating fuel type shall be assumed. For a proposed design without a proposed water heater, a 40-gallon storage-type water heater having the prevailing federal minimum efficiency for the same fuel as the predominant heating fuel type shall be assumed for both the proposed design and standard reference design.

h. For residences with conditioned basements, R-2 and R-4 residences, and for townhouses, the following formula shall be used to determine glazing area:

$AF = A_s \times FA \times F$

where:

AF = Total glazing area.

A_s = Standard reference design total glazing area.

FA = (Above-grade thermal boundary gross wall area)/(above-grade boundary wall area + 0.5 × below-grade boundary wall area).

F = (above-grade thermal boundary wall area)/(above-grade thermal boundary wall area + common wall area) or 0.56, whichever is greater.

and where:

Thermal boundary wall is any wall that separates conditioned space from unconditioned space or ambient conditions.

Above-grade thermal boundary wall is any thermal boundary wall component not in contact with soil.

Below-grade boundary wall is any thermal boundary wall in soil contact.

Common wall area is the area of walls shared with an adjoining dwelling unit.

i. The factor for the compactness of the hot water distribution system is the ratio of the area of the rectangle that bounds the source of hot water and the fixtures that it serves (the "hot water rectangle") divided by the floor area of the dwelling.

1. Sources of hot water include water heaters, or in multiple-family buildings with central water heating systems, circulation loops or electric heat traced pipes.
2. The hot water rectangle shall include the source of hot water and the points of termination of all hot water fixture supply piping.
3. The hot water rectangle shall be shown on the floor plans and the area shall be computed to the nearest square foot.
4. Where there is more than one water heater and each water heater serves different plumbing fixtures and appliances, it is permissible to establish a separate hot water rectangle for each hot water distribution system and add the area of these rectangles together to determine the compactness ratio.
5. The basement or attic shall be counted as a story when it contains the water heater.
6. Compliance shall be demonstrated by providing a drawing on the plans that shows the hot water distribution system rectangle(s), comparing the area of the rectangle(s) to the area of the dwelling and identifying the appropriate compactness ratio and *HWDS* factor.

TABLE N1105.4.2(2) [R405.4.2(2)]
DEFAULT DISTRIBUTION SYSTEM EFFICIENCIES FOR PROPOSED DESIGNS[a]

DISTRIBUTION SYSTEM CONFIGURATION AND CONDITION	FORCED AIR SYSTEMS	HYDRONIC SYSTEMS[b]
Distribution system components located in unconditioned space	—	0.95
Untested distribution systems entirely located in conditioned space[c]	0.88	1
"Ductless" systems[d]	1	—

a. Default values this table are for untested distribution systems, which must still meet minimum requirements for duct system insulation.
b. Hydronic systems shall mean those systems that distribute heating and cooling energy directly to individual spaces using liquids pumped through closed-loop piping and that do not depend on ducted, forced airflow to maintain space temperatures.
c. Entire system in conditioned space shall mean that no component of the distribution system, including the air handler unit, is located outside of the conditioned space.
d. Ductless systems shall be allowed to have forced airflow across a coil but shall not have any ducted airflow external to the manufacturer's air handler enclosure.

N1105.5.2 (R405.5.2) Specific approval. Performance analysis tools meeting the applicable provisions of Section N1105 shall be permitted to be *approved.* Tools are permitted to be *approved* based on meeting a specified threshold for a *jurisdiction.* The *code official* shall be permitted to approve such tools for a specified application or limited scope.

N1105.5.3 (R405.5.3) Input values. When calculations require input values not specified by Sections N1102, N1103, N1104 and N1105, those input values shall be taken from an *approved* source.

SECTION N1106 (R406) ENERGY RATING INDEX COMPLIANCE ALTERNATIVE

N1106.1 (R406.1) Scope. This section establishes criteria for compliance using an Energy Rating Index (ERI) analysis.

N1106.2 (R406.2) ERI compliance. Compliance based on the Energy Rating Index (ERI) requires that the rated design meet all of the following:

1. The requirements of the sections indicated within Table N1106.2.
2. Maximum ERI of Table N1106.5.

N1106.3 (R406.3) Building thermal envelope. Building and portions thereof shall comply with Section N1106.3.1 or N1106.3.2.

N1106.3.1 (R406.3.1) On-site renewables are not included. Where on-site renewable energy is not included for compliance using the ERI analysis of Section N1106.4, the proposed total building thermal envelope UA, which is sum of *U*-factor times assembly area, shall be less than or equal to the building thermal envelope UA using the prescriptive *U*-factors from Table N1102.1.2 multiplied by 1.15 in accordance with Equation 11-4. The area-weighted maximum fenestration SHGC permitted in Climate Zones 0 through 3 shall be 0.30.

$$UA_{\text{Proposed design}} = 1.15 \times UA_{\text{Prescriptive reference design}}$$

(Equation 11-4)

N1106.3.2 (R406.3.2) On-site renewables are included. Where on-site renewable energy is included for compliance using the ERI analysis of Section N1106.4, the *building thermal envelope* shall be greater than or equal to the levels of efficiency and SHGC in Table N1102.1.2, or Table R402.1.4 of the 2015 *International Energy Conservation Code.*

N1106.4 (R406.4) Energy Rating Index. The Energy Rating Index (ERI) shall be determined in accordance with RESNET/ICC 301 except that the *ERI reference design* ventilation rate shall be in accordance with Equation 11-5.

$$\text{Ventilation rate, CFM} = (0.01 \times \text{total square foot area of house}) + [7.5 \times (\text{number of bedrooms} + 1)]$$

(Equation 11-5)

Energy used to recharge or refuel a vehicle used for transportation on roads that are not on the *building site* shall not be included in the *ERI reference design* or the *rated design.* For compliance purposes, any reduction in energy use of the rated design associated with on-site renewable energy shall not exceed 5 percent of the total energy use.

N1106.5 (R406.5) ERI-based compliance. Compliance based on an ERI analysis requires that the *rated proposed design* and confirmed built dwelling be shown to have an ERI less than or equal to the appropriate value indicated in Table N1106.5 when compared to the *ERI reference design.*

TABLE N1106.5 (R406.5)
MAXIMUM ENERGY RATING INDEX

CLIMATE ZONE	ENERGY RATING INDEX
0–1	52
2	52
3	51
4	54
5	55
6	54
7	53
8	53

N1106.6 (R406.6) Verification by approved agency. Verification of compliance with Section N1106 as outlined in Sections N1106.4 and N1106.6 shall be completed by an *approved* third party. Verification of compliance with Section N1106.2 shall be completed by the authority having jurisdiction or an *approved* third-party inspection agency in accordance with Section R105.4.

TABLE N1106.2 (R406.2)
REQUIREMENTS FOR ENERGY RATING INDEX

SECTION[a]	TITLE
General	
N1101.13.5	Additional efficiency packages
N1101.14	Certificate
Building Thermal Envelope	
N1102.1.1	Vapor retarder
N1102.2.3	Eave baffle
N1102.2.4.1	Access hatches and doors
N1102.2.10.1	Crawl space wall insulation installation
N1102.4.1.1	Installation
N1102.4.1.2	Testing
Mechanical	
N1103.1	Controls
N1103.3 except Sections N1103.3.2, N1103.3.3 and N1103.3.6	Ducts
N1103.4	Mechanical system piping insulation
N1103.5.1	Heated water circulation and temperature maintenance systems
N1103.5.3	Drain water heat recovery units
N1103.6	Mechanical ventilation
N1103.7	Equipment sizing and efficiency rating
N1103.8	Systems serving multiple dwelling units
N1103.9	Snow melt system controls
N1103.10	Energy consumption of pools and spas
N1103.11	Portable spas
N1103.12	Residential pools and permanent residential spas
Electrical Power and Lighting Systems	
N1104.1	Lighting equipment
N1104.2	Interior lighting controls
N1106.3	Building thermal envelope

a. Reference to a code section includes all of the relative subsections except as indicated in the table.

N1106.7 (R406.7) Documentation. Documentation of the software used to determine the ERI and the parameters for the *residential building* shall be in accordance with Sections N1106.7.1 through N1106.7.4.

N1106.7.1 (R406.7.1) Compliance software tools. Software tools used for determining ERI shall be *Approved* Software Rating Tools in accordance with RESNET/ICC 301.

N1106.7.2 (R406.7.2) Compliance report. Compliance software tools shall generate a report that documents that the home and the ERI score of the *rated design* comply with Sections N1106.2, N1106.3 and N1106.4. Compliance documentation shall be created for the proposed design and shall be submitted with the application for the building permit. Confirmed compliance documents of the built *dwelling unit* shall be created and submitted to the code official for review before a certificate of occupancy is issued. Compliance reports shall include information in accordance with Sections N1106.7.2.1 and N1106.7.2.2.

N1106.7.2.1 (R406.7.2.1) Proposed compliance report for permit application. Compliance reports submitted with the application for a building permit shall include the following:

1. Building street address, or other *building site* identification.
2. Declaration of ERI on the title page and on the building plans.
3. The name of the individual performing the analysis and generating the compliance report.
4. The name and version of the compliance software tool.
5. Documentation of all inputs entered into the software used to produce the results for the reference design and/or the rated home.
6. A certificate indicating that the proposed design has an ERI less than or equal to the appropriate score indicated in Table N1106.5 when compared to the ERI reference design.

The certificate shall document the building component energy specifications that are included in the calculation, including: component level insulation *R*-values or *U*-factors; assumed duct system and building envelope air leakage testing results; and the type and rated efficiencies of proposed heating, cooling, mechanical ventilation and service water-heating equipment to be installed. If on-site renewable energy systems will be installed, the certificate shall report the type and production size of the proposed system.

7. When a site-specific report is not generated, the proposed design shall be based on the worst-case orientation and configuration of the rated home.

N1106.7.2.2 (R406.7.2.2) Confirmed compliance report for a certificate of occupancy. A confirmed compliance report submitted for obtaining the certificate of occupancy shall be made site and address specific and include the following:

1. Building street address or other *building site* identification.
2. Declaration of ERI on the title page and on the building plans.
3. The name of the individual performing the analysis and generating the report.
4. The name and version of the compliance software tool.
5. Documentation of all inputs entered into the software used to produce the results for the reference design and/or the rated home.
6. A final confirmed certificate indicating that the confirmed rated design of the built home complies with Sections N1106.2 and N1106.4. The certificate shall report the energy features that were confirmed to be in the home, including: component-level insulation *R*-values or *U*-factors; results from any required duct system and building envelope air leakage testing; and the type and rated efficiencies of the heating, cooling, mechanical ventilation, and service water-heating equipment installed. Where on-site renewable energy systems have been installed on or in the home, the certificate shall report the type and production size of the installed system.

N1106.7.3 (R406.7.3) Renewable energy certificate (REC) documentation. Where on-site renewable energy is included in the calculation of an ERI, one of the following forms of documentation shall be provided to the code official:

1. Substantiation that the RECs associated with the on-site renewable energy are owned by, or retired on behalf of, the homeowner.
2. A contract that conveys to the homeowner the RECs associated with the on-site renewable energy, or conveys to the homeowner an equivalent quantity of RECs associated with other renewable energy.

N1106.7.4 (R406.7.4) Additional documentation. The *code official* shall be permitted to require the following documents:

1. Documentation of the building component characteristics of the *ERI reference design.*
2. A certification signed by the builder providing the building component characteristics of the *rated design.*
3. Documentation of the actual values used in the software calculations for the *rated design.*

N1106.7.5 (R406.7.5) Specific approval. Performance analysis tools meeting the applicable subsections of Section N1106 shall be *approved.* Documentation demonstrating the approval of performance analysis tools in accordance with Section N1106.7.1 shall be provided.

N1106.7.6 (R406.7.6) Input values. Where calculations require input values not specified by Sections N1102, N1103, N1104 and N1105, those input values shall be taken from RESNET/ICC 301.

SECTION N1107 (R407) TROPICAL CLIMATE REGION COMPLIANCE PATH

N1107.1 (R407.1) Scope. This section establishes alternative criteria for residential buildings in the tropical region at elevations less than 2,400 feet (731.5 m) above sea level. **

N1107.2 (R407.2) Tropical climate region. Compliance with this section requires the following:

1. Not more than one-half of the occupied space is air conditioned.
2. The occupied space is not heated.
3. Solar, wind or other renewable energy source supplies not less than 80 percent of the energy for service water heating.
4. Glazing in conditioned spaces has a solar heat gain coefficient (SHGC) of less than or equal to 0.40, or has an overhang with a projection factor equal to or greater than 0.30.
5. Permanently installed lighting is in accordance with Section N1104.
6. The exterior roof surface complies with one of the options in Table C402.3 of the International Energy Conservation Code or the roof or ceiling has insulation with an *R*-value of R-15 or greater. Where attics are present, attics above the insulation are vented and attics below the insulation are unvented.
7. Roof surfaces have a slope of not less than $^1/_4$ unit vertical in 12 units horizontal (21-percent slope). The finished roof does not have water accumulation areas.
8. Operable fenestration provides a ventilation area of not less than 14 percent of the floor area in each

room. Alternatively, equivalent ventilation is provided by a ventilation fan.

9. Bedrooms with exterior walls facing two different directions have operable fenestration on exterior walls facing two directions.
10. Interior doors to bedrooms are capable of being secured in the open position.
11. A ceiling fan or ceiling fan rough-in is provided for bedrooms and the largest space that is not used as a bedroom.

SECTION N1108 (R408) ADDITIONAL EFFICIENCY PACKAGE OPTIONS

N1108.1 (R408.1) Scope. This section establishes additional efficiency package options to achieve additional energy efficiency in accordance with Section N1101.13.5.

N1108.2 (R408.2) Additional efficiency package options. Additional efficiency package options for compliance with Section N1101.13.5 are set forth in Sections N1108.2.1 through N1108.2.5.

N1108.2.1 (R408.2.1) Enhanced envelope performance option. The total *building thermal envelope* UA, the sum of *U*-factor times assembly area, shall be less than or equal to 95 percent of the total UA resulting from multiplying the *U*-factors in Table N1102.1.2 by the same assembly area as in the proposed building. The UA calculation shall be performed in accordance with Section N1102.1.5. The area-weighted average SHGC of all glazed fenestration shall be less than or equal to 95 percent of the maximum glazed fenestration SHGC in Table N1102.1.2.

N1108.2.2 (R408.2.2) More efficient HVAC equipment performance option. Heating and cooling *equipment* shall meet one of the following efficiencies:

1. Greater than or equal to 95 AFUE natural gas furnace and 16 SEER air conditioner.
2. Greater than or equal to 10 HSPF/16 SEER air source heat pump.
3. Greater than or equal to 3.5 COP ground source heat pump. For multiple cooling systems, all systems shall meet or exceed the minimum efficiency requirements in this section and shall be sized to serve 100 percent of the cooling design load. For multiple heating systems, all systems shall meet or exceed the minimum efficiency requirements in this section and shall be sized to serve 100 percent of the heating design load.

N1108.2.3 (R408.2.3) Reduced energy use in service water-heating option. The hot water system shall meet one of the following efficiencies:

1. Greater than or equal to 82 EF fossil fuel service water-heating system.
2. Greater than or equal to 2.0 EF electric service water-heating system.
3. Greater than or equal to 0.4 solar fraction solar water-heating system.

N1108.2.4 (R408.2.4) More efficient duct thermal distribution system option. The thermal distribution system shall meet one of the following efficiencies:

1. 100 percent of ducts and air handlers located entirely within the building thermal envelope.
2. 100 percent of ductless thermal distribution system or hydronic thermal distribution system located completely inside the building thermal envelope.
3. 100 percent of duct thermal distribution system located in conditioned space as defined by Section N1103.3.2.

N1108.2.5 (R408.2.5) Improved air sealing and efficient ventilation system option. The measured air leakage rate shall be less than or equal to 3.0 ACH50, with either an Energy Recovery Ventilator (ERV) or Heat Recovery Ventilator (HRV) installed. Minimum HRV and ERV requirements, measured at the lowest tested net supply airflow, shall be greater than or equal to 75 percent Sensible Recovery Efficiency (SRE), less than or equal to 1.1 cubic feet per minute per watt (0.03 m^3/min/watt) and shall not use recirculation as a defrost strategy. In addition, the ERV shall be greater than or equal to 50 percent Latent Recovery/Moisture Transfer (LRMT).

SECTION N1109 (R501) EXISTING BUILDINGS—GENERAL

N1109.1 (R501.1) Scope. The provisions of Sections N1109 through N1113 shall control the *alteration, repair, addition* and change of occupancy of existing *buildings* and structures.

N1109.1.1 (R501.1.1) General. Except as specified in this chapter, this code shall not be used to require the removal, *alteration* or abandonment of, nor prevent the continued use and maintenance of, an existing *building* or *building* system lawfully in existence at the time of adoption of this code. Unaltered portions of the existing *building*, or *building* supply system shall not be required to comply with this code.

N1109.2 (R501.2) Compliance. *Additions, alterations, repairs* or changes of occupancy to, or relocation of, an existing *building, building* system or portion thereof shall comply with Section N1110, N1111, N1112 or N1113, respectively, in this code. Changes where unconditioned space is changed to *conditioned space* shall comply with Section N1110.

N1109.3 (R501.3) Maintenance. *Buildings* and structures, and parts thereof, shall be maintained in a safe and sanitary condition. Devices and systems that are required by this code shall be maintained in compliance with the code edition under which installed. The *owner* or the owner's authorized agent shall be responsible for the maintenance of *buildings* and structures. The requirements of this chapter shall not provide the basis for removal or abrogation of energy conservation, fire protection and safety systems and devices in existing structures.

N1109.4 (R501.4) Compliance. *Alterations, repairs, additions* and changes of occupancy to, or relocation of, existing *buildings* and structures shall comply with the provisions for *alterations, repairs, additions* and changes of occupancy or relocation, respectively, in this code.

N1109.5 (R501.5) New and replacement materials. Except as otherwise required or permitted by this code, materials permitted by the applicable code for new construction shall be used. Like materials shall be permitted for *repairs*, provided that hazards to life, health or property are not created. Hazardous materials shall not be used where the code for new construction would not allow their use in *buildings* of similar occupancy, purpose and location.

N1109.6 (R501.6) Historic buildings. Provisions of this chapter relating to the construction, *repair, alteration*, restoration and movement of structures, and change of occupancy shall not be mandatory for *historic buildings* provided that a report has been submitted to the *building official* and signed by the *owner*, a *registered design professional*, or a representative of the State Historic Preservation Office or the historic preservation authority having *jurisdiction*, demonstrating that compliance with that provision would threaten, degrade or destroy the historic form, fabric or function of the *building*.

SECTION N1110 (R502)
ADDITIONS

N1110.1 (R502.1) General. *Additions* to an existing *building, building* system or portion thereof shall conform to the provisions of this chapter as they relate to new construction without requiring the unaltered portion of the existing *building* or *building* system to comply with this chapter. *Additions* shall not create an unsafe or hazardous condition or overload existing *building* systems. An *addition* shall be deemed to comply with this chapter where the *addition* alone complies, where the existing *building* and *addition* comply with this chapter as a single *building*, or where the *building* with the *addition* does not use more energy than the existing *building*. *Additions* shall be in accordance with Section N1110.1.1 or N1110.1.2.

N1110.2 (R502.2) Change in space conditioning. Any unconditioned or low energy space that is altered to become *conditioned space* shall be required to be brought into full compliance with this chapter.

Exceptions:

1. Where the simulated performance option in Section N1105 is used to comply with this section, the annual energy cost of the *proposed design* is permitted to be 110 percent of the annual energy cost otherwise allowed by Section N1105.3.
2. Where the Total UA, as determined in Section N1102.1.5, of the existing *building* and the *addition*, and any *alterations* that are part of the project, is less than or equal to the Total UA generated for the existing *building*.
3. Where complying in accordance with Section N1105 and the annual energy cost or energy use of the *addition* and the existing *building*, and any *alterations* that are part of the project, is less than or equal to the annual energy cost of the existing *building*. The *addition* and any *alterations* that are part of the project shall comply with Section N1105 in its entirety.

N1110.3 (R502.3) Prescriptive compliance. *Additions* shall comply with Sections N1110.3.1 through N1110.3.4.

N1110.3.1 (R502.3.1) Building envelope. New *building* envelope assemblies that are part of the *addition* shall comply with Sections N1102.1, N1102.2, N1102.3.1 through N1102.3.5, and N1102.4.

Exception: New envelope assemblies are exempt from the requirements of Section N1102.4.1.2.

N1110.3.2 (R502.3.2) Heating and cooling systems. HVAC ducts newly installed as part of an *addition* shall comply with Section N1103.

Exception: Where ducts from an existing heating and cooling system are extended to an *addition*.

N1110.3.3 (R502.3.3) Service hot water systems. New service hot water systems that are part of the *addition* shall comply with Section N1103.5.

N1110.3.4 (R502.3.4) Lighting. New lighting systems that are part of the *addition* shall comply with Section N1104.1.

SECTION N1111 (R503)
ALTERATIONS

N1111.1 (R503.1) General. *Alterations* to any *building* or structure shall comply with the requirements of the code for new construction, without requiring the unaltered portions of the existing building or building system to comply with this chapter. *Alterations* shall be such that the existing *building* or structure is not less conforming with the provisions of this chapter than the existing building or structure was prior to the *alteration*.

Alterations shall not create an unsafe or hazardous condition or overload existing *building* systems. *Alterations* shall be such that the existing *building* or structure does not use more energy than the existing *building* or structure prior to the *alteration*. *Alterations* to existing *buildings* shall comply with Sections N1111.1.1 through N1111.1.4.

N1111.1.1 (R503.1.1) Building envelope. *Building* envelope assemblies that are part of the *alteration* shall comply with Section N1102.1.2 or N1102.1.4, Sections N1102.2.1 through N1102.2.12, N1102.3.1, N1102.3.2, N1102.4.3 and N1102.4.5.

Exception: The following *alterations* shall not be required to comply with the requirements for new construction provided that the energy use of the *building* is not increased:

1. Storm windows installed over existing fenestration.

2. Existing ceiling, wall or floor cavities exposed during construction provided that these cavities are filled with insulation.
3. Construction where the existing roof, wall or floor cavity is not exposed.
4. *Roof recover.*
5. Roofs without insulation in the cavity and where the sheathing or insulation is exposed during *reroofing* shall be insulated either above or below the sheathing.
6. Surface-applied window film installed on existing single-pane fenestration assemblies to reduce solar heat gain provided that the code does not require the glazing or fenestration assembly to be replaced.

N1111.1.1.1 (R503.1.1.1) Replacement fenestration. Where some or all of an existing fenestration unit is replaced with a new fenestration product, including sash and glazing, the replacement fenestration unit shall meet the applicable requirements for *U*-factor and SHGC as specified in Table N1102.1.3. Where more than one replacement fenestration unit is to be installed, an area-weighted average of the *U*-factor, SHGC or both of all replacement fenestration units shall be an alternative that can be used to show compliance.

N1111.1.2 (R503.1.2) Heating and cooling systems. HVAC ducts newly installed as part of an *alteration* shall comply with Section N1103.

Exception: Where ducts from an existing heating and cooling system are extended to an *addition.*

N1111.1.3 (R503.1.3) Service hot water systems. New service hot water systems that are part of the *alteration* shall comply with Section N1103.5.

N1111.1.4 (R503.1.4) Lighting. New lighting systems that are part of the *alteration* shall comply with Section N1104.1.

Exception: *Alterations* that replace less than 10 percent of the luminaires in a space, provided that such *alterations* do not increase the installed interior lighting power.

SECTION N1112 (R504) REPAIRS

N1112.1 (R504.1) General. *Buildings,* structures and parts thereof shall be repaired in compliance with Section N1109.3 and this section. Work on nondamaged components necessary for the required repair of damaged components shall be considered to be part of the *repair* and shall not be subject to the requirements for *alterations* in this chapter. Routine maintenance required by Section N1109.3, ordinary *repairs* exempt from *permit,* and abatement of wear due to normal service conditions shall not be subject to the requirements for *repairs* in this section.

N1112.2 (R504.2) Application. For the purposes of this code, the following shall be considered to be *repairs*:

1. Glass-only replacements in an existing sash and frame.
2. Roof *repairs.*
3. *Repairs* where only the bulb, ballast or both within the existing luminaires in a space are replaced provided that the replacement does not increase the installed interior lighting power.

SECTION N1113 (R505) CHANGE OF OCCUPANCY OR USE

N1113.1 (R505.1) General. Any space that is converted to a *dwelling unit* or portion thereof from another use or occupancy shall comply with this chapter.

Exception: Where the simulated performance option in Section N1105 is used to comply with this section, the annual energy cost of the *proposed design* is permitted to be 110 percent of the annual energy cost allowed by Section N1105.2.

N1113.1.1 (R505.1.1) Unconditioned space. Any unconditioned or low-energy space that is altered to become a *conditioned space* shall comply with Section N1108.

Part V—Mechanical

CHAPTER 12
MECHANICAL ADMINISTRATION

User notes:

About this chapter: *Chapter 12 supplements Chapter 1 and establishes the scope of coverage for Chapters 13 through 24. The applicability of code provisions to existing mechanical systems and appliances is established herein.*

Code development reminder: *Code change proposals to this chapter will be considered by the IRC—Plumbing/Mechanical Code Development Committee during the 2021 (Group A) Code Development Cycle.*

SECTION M1201
GENERAL

M1201.1 Scope. The provisions of Chapters 12 through 24 shall regulate the design, installation, maintenance, *alteration* and inspection of mechanical systems that are permanently installed and used to control environmental conditions within buildings. These chapters shall also regulate those mechanical systems, system components, *equipment* and *appliances* specifically addressed in this code.

M1201.2 Application. In addition to the general administration requirements of Chapter 1, the administrative provisions of this chapter shall apply to the mechanical requirements of Chapters 13 through 24.

SECTION M1202
EXISTING MECHANICAL SYSTEMS

M1202.1 Additions, alterations or repairs. *Additions, alterations,* renovations or repairs to a mechanical system shall conform to the requirements for a new mechanical system without requiring the existing mechanical system to comply with all of the requirements of this code. *Additions, alterations* or repairs shall not cause an existing mechanical system to become unsafe, hazardous or overloaded. Minor *additions, alterations* or repairs to existing mechanical systems shall meet the provisions for new construction, unless such work is done in the same manner and arrangement as was in the existing system, is not hazardous, and is *approved.*

M1202.2 Existing installations. Except as otherwise provided for in this code, a provision in this code shall not require the removal, *alteration* or abandonment of, nor prevent the continued use and maintenance of, an existing mechanical system lawfully in existence at the time of the adoption of this code.

M1202.3 Maintenance. Mechanical systems, both existing and new, and parts thereof shall be maintained in proper operating condition in accordance with the original design and in a safe and sanitary condition. Devices or safeguards that are required by this code shall be maintained in compliance with the code edition under which such devices and safeguards were installed. The *owner* or the owner's designated agent shall be responsible for maintenance of the mechanical systems. To determine compliance with this provision, the *building official* shall have the authority to require a mechanical system to be reinspected.

CHAPTER 13

GENERAL MECHANICAL SYSTEM REQUIREMENTS

User notes:

About this chapter: *Chapter 13 contains general requirements that apply broadly and that would not be at home in other chapters that address specific subject matter. Coverage includes: Testing and certification of materials, installation requirements, listing and labeling, access to appliances, clearances to combustibles, and protection of mechanical systems and the building structure.*

Code development reminder: *Code change proposals to this chapter will be considered by the IRC—Plumbing/Mechanical Code Development Committee during the 2021 (Group A) Code Development Cycle.*

SECTION M1301
GENERAL

M1301.1 Scope. The provisions of this chapter shall govern the installation of mechanical systems not specifically covered in other chapters applicable to mechanical systems. Installations of mechanical *appliances, equipment* and systems not addressed by this code shall comply with the applicable provisions of the *International Fuel Gas Code* and the *International Mechanical Code*.

M1301.1.1 Flood-resistant installation. In flood hazard areas as established by Table R301.2(1), mechanical *appliances, equipment* and systems shall be located or installed in accordance with Section R322.1.6.

M1301.2 Identification. Each length of pipe and tubing and each pipe fitting utilized in a mechanical system shall bear the identification of the manufacturer.

M1301.3 Installation of materials. Materials shall be installed in strict accordance with the standards under which the materials are accepted and *approved*. In the absence of such installation procedures, the manufacturer's instructions shall be followed. Where the requirements of referenced standards or manufacturer's instructions do not conform to minimum provisions of this code, the provisions of this code shall apply.

M1301.4 Plastic pipe, fittings and components. Plastic pipe, fittings and components shall be third-party certified as conforming to NSF 14.

M1301.5 Third-party testing and certification. Piping, tubing and fittings shall comply with the applicable referenced standards, specifications and performance criteria of this code and shall be identified in accordance with Section M1301.2. Piping, tubing and fittings shall either be tested by an *approved* third-party testing agency or certified by an *approved* third-party certification agency.

SECTION M1302
APPROVAL

M1302.1 Listed and labeled. *Appliances* regulated by this code shall be *listed* and *labeled* for the application in which they are installed and used, unless otherwise *approved* in accordance with Section R104.11.

SECTION M1303
LABELING OF APPLIANCES

M1303.1 Label information. A permanent factory-applied nameplate(s) shall be affixed to *appliances* on which shall appear, in legible lettering, the manufacturer's name or trademark, the model number, a serial number and the seal or *mark* of the testing agency. A *label* also shall include the following:

1. Electrical *appliances*. Electrical rating in volts, amperes and motor phase; identification of individual electrical components in volts, amperes or watts and motor phase; and in Btu/h (W) output and required clearances.
2. Absorption units. Hourly rating in Btu/h (W), minimum hourly rating for units having step or automatic modulating controls, type of fuel, type of refrigerant, cooling capacity in Btu/h (W) and required clearances.
3. Fuel-burning units. Hourly rating in Btu/h (W), type of fuel approved for use with the *appliance* and required clearances.
4. Electric comfort-heating *appliances*. The electric rating in volts, amperes and phase; Btu/h (W) output rating; individual marking for each electrical component in amperes or watts, volts and phase; and required clearances from combustibles.
5. Maintenance instructions. Required regular maintenance actions and title or publication number for the operation and maintenance manual for that particular model and type of product.

SECTION M1304
TYPE OF FUEL

M1304.1 Fuel types. Fuel-fired *appliances* shall be designed for use with the type of fuel to which they will be connected and the altitude at which they are installed. *Appliances* that comprise parts of the building mechanical system shall not be converted for the use of a different fuel, except where *approved* and converted in accordance with the manufacturer's instructions. The fuel input rate shall not be increased or decreased beyond the limit rating for the altitude at which the *appliance* is installed.

SECTION M1305
APPLIANCE ACCESS

M1305.1 Appliance access for inspection service, repair and replacement. *Appliances* shall be located to allow for access for inspection, service, repair and replacement without removing permanent construction, other *appliances*, or any other piping or ducts not connected to the *appliance* being inspected, serviced, repaired or replaced. A level working space not less than 30 inches deep and 30 inches wide (762 mm by 762 mm) shall be provided in front of the control side to service an *appliance*.

M1305.1.1 Appliances in rooms. *Appliances* installed in a compartment, alcove, *basement* or similar space shall be accessed by an opening or door and an unobstructed passageway measuring not less than 24 inches (610 mm) wide and large enough to allow removal of the largest *appliance* in the space, provided there is a level service space of not less than 30 inches (762 mm) deep and the height of the *appliance*, but not less than 30 inches (762 mm), at the front or service side of the *appliance* with the door open.

M1305.1.2 Appliances in attics. *Attics* containing *appliances* shall be provided with an opening and a clear and unobstructed passageway large enough to allow removal of the largest *appliance*, but not less than 30 inches (762 mm) high and 22 inches (559 mm) wide and not more than 20 feet (6096 mm) long measured along the centerline of the passageway from the opening to the *appliance*. The passageway shall have continuous solid flooring in accordance with Chapter 5 not less than 24 inches (610 mm) wide. A level service space not less than 30 inches (762 mm) deep and 30 inches (762 mm) wide shall be present along all sides of the *appliance* where access is required. The clear access opening dimensions shall be not less than of 20 inches by 30 inches (508 mm by 762 mm), and large enough to allow removal of the largest *appliance*.

Exceptions:

1. The passageway and level service space are not required where the *appliance* can be serviced and removed through the required opening.
2. Where the passageway is unobstructed and not less than 6 feet (1829 mm) high and 22 inches (559 mm) wide for its entire length, the passageway shall be not more than 50 feet (15 250 mm) long.

M1305.1.2.1 Electrical requirements. A luminaire controlled by a switch located at the required passageway opening and a receptacle outlet shall be installed at or near the *appliance* location in accordance with Chapter 39. Exposed lamps shall be protected from damage by location or lamp guards.

M1305.1.3 Appliances under floors. Underfloor spaces containing *appliances* shall be provided with an unobstructed passageway large enough to remove the largest *appliance*, but not less than 30 inches (762 mm) high and 22 inches (559 mm) wide, nor more than 20 feet (6096 mm) long measured along the centerline of the passageway from the opening to the *appliance*. A level service space not less than 30 inches (762 mm) deep and 30 inches (762 mm) wide shall be present at the front or service side of the *appliance*. If the depth of the passageway or the service space exceeds 12 inches (305 mm) below the adjoining grade, the walls of the passageway shall be lined with concrete or masonry extending 4 inches (102 mm) above the adjoining grade in accordance with Chapter 4. The rough-framed access opening dimensions shall be not less than 22 inches by 30 inches (559 mm by 762 mm), and large enough to remove the largest *appliance*.

Exceptions:

1. The passageway is not required where the level service space is present when the access is open, and the *appliance* can be serviced and removed through the required opening.
2. Where the passageway is unobstructed and not less than 6 feet (1829 mm) high and 22 inches (559 mm) wide for its entire length, the passageway shall not be limited in length.

M1305.1.3.1 Ground clearance. *Equipment* and *appliances* supported from the ground shall be level and firmly supported on a concrete slab or other *approved* material extending not less than 3 inches (76 mm) above the adjoining ground. Such support shall be in accordance with the manufacturer's installation instructions. *Appliances* suspended from the floor shall have a clearance of not less than 6 inches (152 mm) from the ground.

M1305.1.3.2 Pit locations. Appliances installed in pits or excavations shall not come in direct contact with the surrounding soil and shall be installed not less than 3 inches (76 mm) above the pit floor. The sides of the pit or excavation shall be held back not less than 12 inches (305 mm) from the *appliance*. Where the depth exceeds 12 inches (305 mm) below adjoining grade, the walls of the pit or excavation shall be lined with concrete or masonry. Such concrete or masonry shall extend not less than 4 inches (102 mm) above adjoining grade and shall have sufficient lateral load-bearing capacity to resist collapse. Excavation on the control side of the *appliance* shall extend horizontally not less than 30 inches (762 mm). The *appliance* shall be protected from flooding in an *approved* manner.

M1305.1.3.3 Electrical requirements. A luminaire controlled by a switch located at the required passageway opening and a receptacle outlet shall be installed at or near the *appliance* location in accordance with Chapter 39. Exposed lamps shall be protected from damage by location or lamp guards.

SECTION M1306 CLEARANCES FROM COMBUSTIBLE CONSTRUCTION

M1306.1 Appliance clearance. *Appliances* shall be installed with the clearances from unprotected *combustible materials* as indicated on the *appliance label* and in the manufacturer's installation instructions.

M1306.2 Clearance reduction. The reduction of required clearances to combustible assemblies or *combustible materials* shall be based on Section M1306.2.1 or M1306.2.2.

M1306.2.1 Labeled assemblies. The allowable clearance shall be based on an approved reduced clearance protective assembly that is *listed* and *labeled* in accordance with UL 1618.

M1306.2.2 Reduction table. Reduction of clearances shall be in accordance with the *appliance* manufacturer's instructions and Table M1306.2. Forms of protection with ventilated airspace shall conform to the following requirements:

1. Not less than 1-inch (25 mm) airspace shall be provided between the protection and combustible wall surface.
2. Air circulation shall be provided by having edges of the wall protection open not less than 1 inch (25 mm).
3. If the wall protection is mounted on a single flat wall away from corners, air circulation shall be provided by having the bottom and top edges, or the side and top edges not less than 1 inch (25 mm).
4. Wall protection covering two walls in a corner shall be open at the bottom and top edges not less than 1 inch (25 mm).

M1306.2.3 Solid-fuel appliances. Table M1306.2 shall not be used to reduce the clearance required for solid-fuel *appliances listed* for installation with minimum clearances of 12 inches (305 mm) or less. For *appliances listed* for installation with minimum clearances greater than 12 inches (305 mm), Table M1306.2 shall not be used to reduce the clearance to less than 12 inches (305 mm).

SECTION M1307 APPLIANCE INSTALLATION

M1307.1 General. Installation of *appliances* shall conform to the conditions of their *listing* and *label* and the manufacturer's instructions. The manufacturer's operating and installation instructions shall remain attached to the *appliance.*

M1307.2 Anchorage of appliances. *Appliances* designed to be fixed in position shall be fastened or anchored in an *approved* manner. In Seismic Design Categories D_0, D_1 and D_2, and in townhouses in Seismic Design Category C, water heaters and thermal storage units shall be anchored or strapped to resist horizontal displacement caused by earthquake motion in accordance with one of the following:

1. Anchorage and strapping shall be designed to resist a horizontal force equal to one-third of the operating weight of the water heater storage tank, acting in any horizontal direction. Strapping shall be at points within the upper one-third and lower one-third of the *appliance's* vertical dimensions. At the lower point, the strapping shall maintain a minimum distance of 4 inches (102 mm) above the controls.
2. The anchorage strapping shall be in accordance with the appliance manufacturer's recommendations.

M1307.3 Elevation of ignition source. *Appliances* having an *ignition source* shall be elevated such that the source of ignition is not less than 18 inches (457 mm) above the floor in garages. For the purpose of this section, rooms or spaces that are not part of the *living space* of a *dwelling unit* and that communicate with a private garage through openings shall be considered to be part of the garage.

Exception: Elevation of the *ignition source* is not required for *appliances* that are *listed* as flammable-vapor-ignition resistant.

M1307.3.1 Protection from impact. *Appliances* shall not be installed in a location subject to vehicle damage except where protected by *approved* barriers.

M1307.4 Hydrogen-generating and refueling operations. *Ventilation* shall be required in accordance with Section M1307.4.1, M1307.4.2 or M1307.4.3 in private garages that contain hydrogen-generating *appliances* or refueling systems. For the purpose of this section, rooms or spaces that are not part of the *living space* of a *dwelling unit* and that communicate directly with a private garage through openings shall be considered to be part of the private garage.

M1307.4.1 Natural ventilation. Indoor locations intended for hydrogen-generating or refueling operations shall be limited to a maximum floor area of 850 square feet (79 m^2) and shall communicate with the outdoors in accordance with Sections M1307.4.1.1 and M1307.4.1.2. The maximum rated output capacity of hydrogen-generating *appliances* shall not exceed 4 standard cubic feet per minute (1.9 L/s) of hydrogen for each 250 square feet (23 m^2) of floor area in such spaces. The minimum cross-sectional dimension of air openings shall be 3 inches (76 mm). Where ducts are used, they shall be of the same cross-sectional area as the free area of the openings to which they connect. In those locations, *equipment* and *appliances* having an *ignition source* shall be located so that the source of ignition is not within 12 inches (305 mm) of the ceiling.

TABLE M1306.2
REDUCTION OF CLEARANCES WITH SPECIFIED FORMS OF PROTECTION[a, c, d, e, f, g, h, i, j, k, l]

TYPE OF PROTECTION APPLIED TO AND COVERING ALL SURFACES OF COMBUSTIBLE MATERIAL WITHIN THE DISTANCE SPECIFIED AS THE REQUIRED CLEARANCE WITH NO PROTECTION (See Figures M1306.1 and M1306.2)	WHERE THE REQUIRED CLEARANCE WITHOUT PROTECTION FROM APPLIANCE, VENT CONNECTOR, OR SINGLE-WALL METAL PIPE IS:									
	36 inches		18 inches		12 inches		9 inches		6 inches	
	Allowable clearances with specified protection (Inches)[b]									
	Use Column 1 for clearances above an appliance or horizontal connector. Use column 2 for clearances from an appliance, vertical connector and single-wall metal pipe.									
	Above column 1	Sides and rear column 2	Above column 1	Sides and rear column 2	Above column 1	Sides and rear column 2	Above column 1	Sides and rear column 2	Above column 1	Sides and rear column 2
$3^1/_2$-inch-thick masonry wall without ventilated airspace	—	24	—	12	—	9	—	6	—	5
$^1/_2$-inch insulation board over 1-inch glass fiber or mineral wool batts	24	18	12	9	9	6	6	5	4	3
Galvanized sheet steel having a minimum thickness of 0.0236-inch (No. 24 gage) over 1-inch glass fiber or mineral wool batts reinforced with wire or rear face with a ventilated airspace	18	12	9	6	6	4	5	3	3	3
$3^1/_2$-inch-thick masonry wall with ventilated airspace	—	12	—	6	—	6	—	6	—	6
Galvanized sheet steel having a minimum thickness of 0.0236-inch (No. 24 gage) with a ventilated airspace 1 inch off the combustible assembly	18	12	9	6	6	4	5	3	3	2
$^1/_2$-inch-thick insulation board with ventilated airspace	18	12	9	6	6	4	5	3	3	3
Galvanized sheet steel having a minimum thickness of 0.0236-inch (No. 24 gage) with ventilated airspace over 24 gage sheet steel with a ventilated space	18	12	9	6	6	4	5	3	3	3
1-inch glass fiber or mineral wool batts sandwiched between two sheets of galvanized sheet steel having a minimum thickness of 0.0236-inch (No. 24 gage) with a ventilated airspace	18	12	9	6	6	4	5	3	3	3

For SI: 1 inch = 25.4 mm, 1 pound per cubic foot = 16.019 kg/m^3, °C = [(°F) – 32/1.8], 1 Btu/(h × ft^2 × °F/in.) = 0.001442299 (W/cm^2 × °C/cm).

a. Reduction of clearances from combustible materials shall not interfere with combustion air, draft hood clearance and relief, and accessibility of servicing.

b. Clearances shall be measured from the surface of the heat-producing appliance or equipment to the outer surface of the combustible material or combustible assembly.

c. Spacers and ties shall be of noncombustible material. Spacers and ties shall not be used directly opposite appliance or connector.

d. Where all clearance reduction systems use a ventilated airspace, adequate provision for air circulation shall be provided as described (see Figures M1306.1 and M1306.2).

e. There shall be not less than 1 inch between clearance reduction systems and combustible walls and ceilings for reduction systems using ventilated airspace.

f. If a wall protector is mounted on a single flat wall away from corners, adequate air circulation shall be permitted to be provided by leaving only the bottom and top edges or only the side and top edges open with not less than a 1-inch air gap.

g. Mineral wool and glass fiber batts (blanket or board) shall have a minimum density of 8 pounds per cubic foot and a minimum melting point of 1,500°F.

h. Insulation material used as part of a clearance reduction system shall have a thermal conductivity of 1.0 Btu inch per square foot per hour °F or less. Insulation board shall be formed of noncombustible material.

i. There shall be not less than 1 inch between the appliance and the protector. The clearance between the appliance and the combustible surface shall not be reduced below that allowed in this table.

j. All clearances and thicknesses are minimum; larger clearances and thicknesses are acceptable.

k. Listed single-wall connectors shall be permitted to be installed in accordance with the terms of their listing and the manufacturer's instructions.

l. For limitations on clearance reduction for solid-fuel-burning appliances, see Section M1306.2.3.

CONSTRUCTION USING COMBUSTIBLE MATERIAL, PLASTERED OR UNPLASTERED
A
B
C
C
SHEET METAL OR OTHER PROTECTION
EQUIPMENT OR VENT CONNECTOR

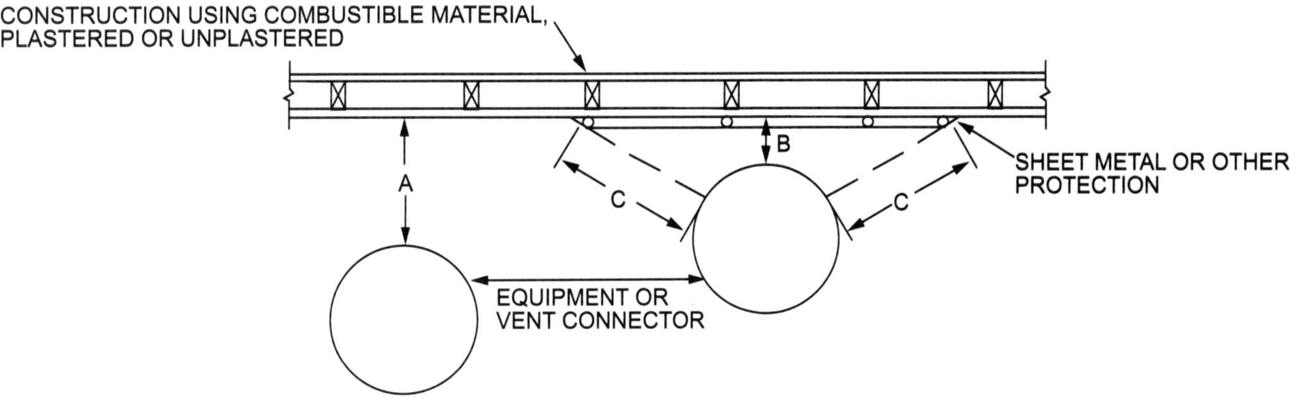

Note: "A" equals the required clearance with no protection. "B" equals the reduced clearance permitted in accordance with Table M1306.2. The protection applied to the construction using combustible material shall extend far enough in each direction to make "C" equal to "A."

FIGURE M1306.1
REDUCED CLEARANCE DIAGRAM

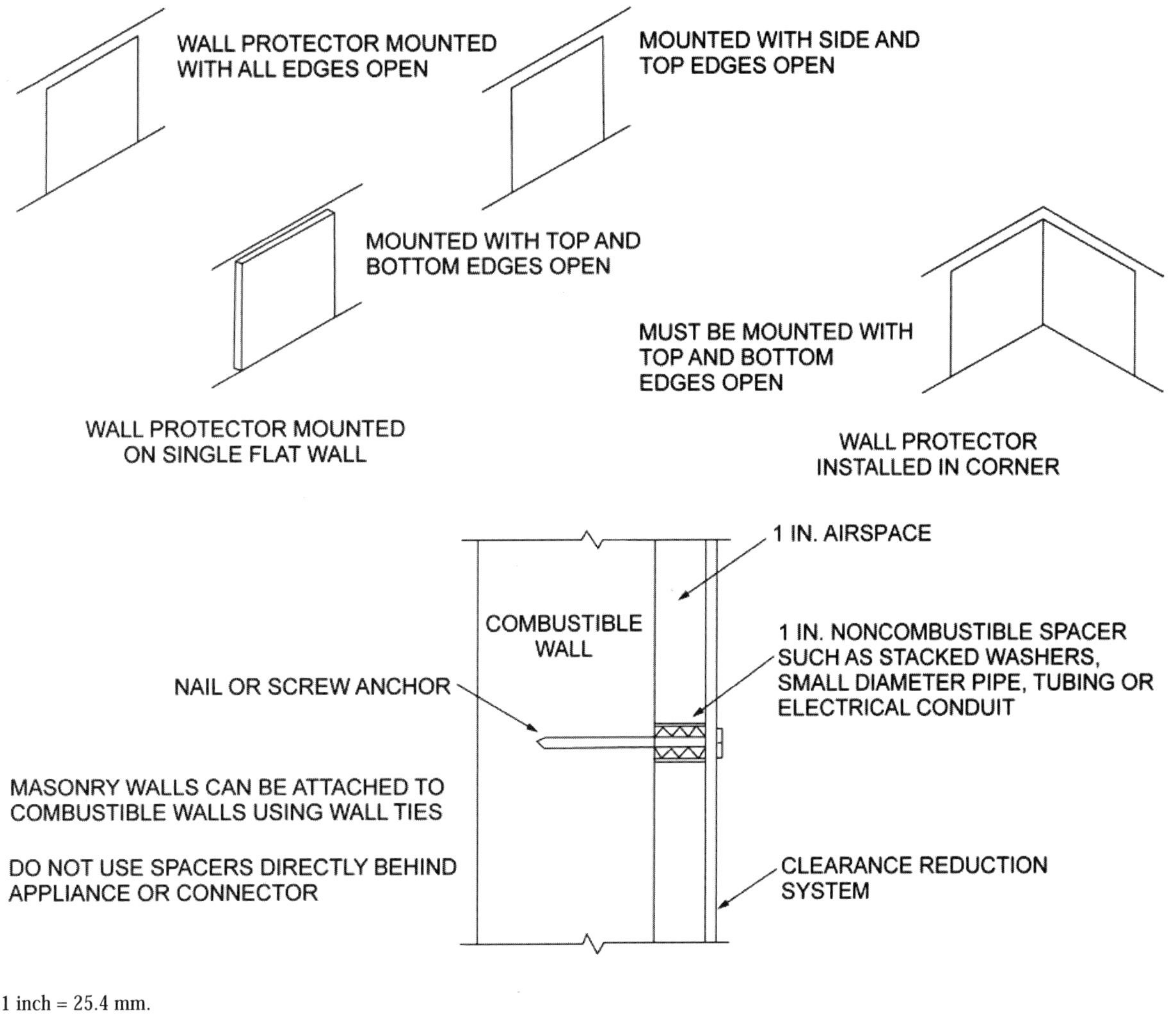

WALL PROTECTOR MOUNTED WITH ALL EDGES OPEN
MOUNTED WITH SIDE AND TOP EDGES OPEN
MOUNTED WITH TOP AND BOTTOM EDGES OPEN
MUST BE MOUNTED WITH TOP AND BOTTOM EDGES OPEN
WALL PROTECTOR MOUNTED ON SINGLE FLAT WALL
WALL PROTECTOR INSTALLED IN CORNER
1 IN. AIRSPACE
COMBUSTIBLE WALL
1 IN. NONCOMBUSTIBLE SPACER SUCH AS STACKED WASHERS, SMALL DIAMETER PIPE, TUBING OR ELECTRICAL CONDUIT
NAIL OR SCREW ANCHOR
MASONRY WALLS CAN BE ATTACHED TO COMBUSTIBLE WALLS USING WALL TIES
DO NOT USE SPACERS DIRECTLY BEHIND APPLIANCE OR CONNECTOR
CLEARANCE REDUCTION SYSTEM

For SI:1 inch = 25.4 mm.

FIGURE M1306.2
WALL PROTECTOR CLEARANCE REDUCTION SYSTEM

M1307.4.1.1 Two openings. Two permanent openings shall be constructed within the garage. The upper opening shall be located entirely within 12 inches (305 mm) of the ceiling of the garage. The lower opening shall be located entirely within 12 inches (305 mm) of the floor of the garage. Both openings shall be constructed in the same exterior wall. The openings shall communicate directly with the outdoors and shall have a minimum free area of $^1/_2$ square foot per 1,000 cubic feet (1.7 m^2/1000 m^3) of garage volume.

M1307.4.1.2 Louvers and grilles. In calculating free area required by Section M1307.4.1, the required size of openings shall be based on the net free area of each opening. If the free area through a design of louver or grille is known, it shall be used in calculating the size opening required to provide the free area specified. If the design and free area are not known, it shall be assumed that wood louvers will have a 25-percent free area and metal louvers and grilles will have a 75-percent free area. Louvers and grilles shall be fixed in the open position.

M1307.4.2 Mechanical ventilation. Indoor locations intended for hydrogen-generating or refueling operations shall be ventilated in accordance with Section 502.16 of the *International Mechanical Code*. In these locations, *equipment* and *appliances* having an *ignition source* shall be located so that the source of ignition is below the mechanical *ventilation* outlet(s).

M1307.4.3 Specially engineered installations. As an alternative to the provisions of Sections M1307.4.1 and M1307.4.2, the necessary supply of air for *ventilation* and dilution of flammable gases shall be provided by an *approved* engineered system.

M1307.5 Electrical appliances. Electrical *appliances* shall be installed in accordance with Chapters 14, 15, 19, 20 and 34 through 43.

M1307.6 Plumbing connections. Potable water and drainage system connections to *equipment* and *appliances* regulated by this code shall be in accordance with Chapters 29 and 30.

M1307.7 Prohibited support. Gypsum board shall not be used as a support base under an *appliance*.

SECTION M1308 MECHANICAL SYSTEMS INSTALLATION

M1308.1 Drilling and notching. Wood-framed structural members shall be drilled, notched or altered in accordance with the provisions of Sections R502.8, R602.6, R602.6.1 and R802.7. Holes in load-bearing members of cold-formed steel *light-frame construction* shall be permitted only in accordance with Sections R505.2.6, R603.2.6 and R804.2.6. In accordance with the provisions of Sections R505.3.5, R603.3.4 and R804.3.3, cutting and notching of flanges and lips of load-bearing members of cold-formed steel light frame construction shall not be permitted. Structural insulated panels (SIPs) shall be drilled and notched or altered in accordance with the provisions of Section R610.7.

M1308.2 Protection against physical damage. Where piping will be concealed within *light-frame construction* assemblies, the piping shall be protected against penetration by fasteners in accordance with Sections M1308.2.1 through M1308.2.3.

Exception: Cast-iron piping and galvanized steel piping shall not be required to be protected.

M1308.2.1 Piping through bored holes or notches. Where *piping* is installed through holes or notches in framing members and is located less than 1$^1/_2$ inches (38 mm) from the framing member face to which wall, ceiling or floor membranes will be attached, the pipe shall be protected by shield plates that cover the width of the pipe and the framing member and that extend 2 inches (51 mm) to each side of the framing member. Where the framing member that the piping passes through is a bottom plate, bottom track, top plate or top track, the shield plates shall cover the framing member and extend 2 inches (51 mm) above the bottom framing member and 2 inches (51 mm) below the top framing member.

M1308.2.2 Piping in other locations. Where piping is located within a framing member and is less than 1$^1/_2$ inches (38 mm) from the framing member face to which wall, ceiling or floor membranes will be attached, the piping shall be protected by shield plates that cover the width and length of the piping. Where piping is located outside of a framing member and is located less than 1$^1/_2$ inches (38 mm) from the nearest edge of the face of the framing member to which the membrane will be attached, the piping shall be protected by shield plates that cover the width and length of the piping.

M1308.2.3 Shield plates. Shield plates shall be of steel material having a thickness of not less than 0.0575 inch (1.463 mm) (No. 16 gage).

CHAPTER 14

HEATING AND COOLING EQUIPMENT AND APPLIANCES

User notes:

About this chapter: *Chapter 14 addresses the indoor environmental control systems and appliances typically found in dwelling units. Coverage includes general requirements for equipment and appliance sizing, condensate disposal, access and support, and specific coverage for more than a dozen different types of space conditioning equipment and appliances common to dwelling units.*

Code development reminder: *Code change proposals to this chapter will be considered by the IRC—Plumbing/Mechanical Code Development Committee during the 2021 (Group A) Code Development Cycle.*

SECTION M1401 GENERAL

M1401.1 Installation. Heating and cooling *equipment* and *appliances* shall be installed in accordance with the manufacturer's instructions and the requirements of this code.

M1401.2 Access. Heating and cooling *equipment* and *appliances* shall be located with respect to building construction and other *equipment* and *appliances* to permit maintenance, servicing and replacement. Clearances shall be maintained to permit cleaning of heating and cooling surfaces; replacement of filters, blowers, motors, controls and vent connections; lubrication of moving parts; and adjustments.

Exception: Access shall not be required for ducts, piping, or other components *approved* for concealment.

M1401.3 Equipment and appliance sizing. Heating and cooling *equipment* and *appliances* shall be sized in accordance with ACCA Manual S or other *approved* sizing methodologies based on building loads calculated in accordance with ACCA Manual J or other *approved* heating and cooling calculation methodologies.

Exception: Heating and cooling *equipment* and *appliance* sizing shall not be limited to the capacities determined in accordance with ACCA Manual S where either of the following conditions applies:

1. The specified *equipment* or *appliance* utilizes multistage technology or variable refrigerant flow technology and the loads calculated in accordance with the *approved* heating and cooling calculation methodology are within the range of the manufacturer's published capacities for that *equipment* or *appliance*.
2. The specified *equipment* or *appliance* manufacturer's published capacities cannot satisfy both the total and sensible heat gains calculated in accordance with the *approved* heating and cooling calculation methodology and the next larger standard size unit is specified.

M1401.4 Outdoor installations. *Equipment* and *appliances* installed outdoors shall be *listed* and *labeled* for outdoor installation. Supports and foundations shall prevent excessive vibration, settlement or movement of the *equipment*. Supports and foundations shall be in accordance with Section M1305.1.3.1.

M1401.5 Flood hazard. In flood hazard areas as established by Table R301.2, heating and cooling *equipment* and *appliances* shall be located or installed in accordance with Section R322.1.6.

SECTION M1402 CENTRAL FURNACES

M1402.1 General. Oil-fired central furnaces shall conform to ANSI/UL 727. Electric furnaces shall conform to UL 1995 or UL/CSA/ANCE 60335-2-40.

M1402.2 Clearances. Clearances shall be provided in accordance with the *listing* and the manufacturer's installation instructions.

M1402.3 Combustion air. *Combustion air* shall be supplied in accordance with Chapter 17. *Combustion air* openings shall be unobstructed for a distance of not less than 6 inches (152 mm) in front of the openings.

SECTION M1403 HEAT PUMP EQUIPMENT

M1403.1 Heat pumps. Electric heat pumps shall be *listed* and *labeled* in accordance with UL 1995 or UL/CSA/ANCE 60335-2-40.

SECTION M1404 REFRIGERATION COOLING EQUIPMENT

M1404.1 Compliance. Refrigeration cooling *equipment* shall comply with Section M1411.

SECTION M1405 BASEBOARD CONVECTORS

M1405.1 General. Electric baseboard convectors shall be installed in accordance with the manufacturer's instructions and Chapters 34 through 43. Electric baseboard heaters shall be *listed* and *labeled* in accordance with UL 1042.

SECTION M1406 RADIANT HEATING SYSTEMS

M1406.1 General. Electric radiant heating systems shall be installed in accordance with the manufacturer's instructions and Chapters 34 through 43 and shall be *listed* for the application.

M1406.2 Clearances. Clearances for radiant heating panels or elements to any wiring, outlet boxes and junction boxes used for installing electrical devices or mounting luminaires shall comply with Chapters 34 through 43.

M1406.3 Installation of radiant panels. Radiant panels installed on wood framing shall conform to the following requirements:

1. Heating panels shall be installed parallel to framing members and secured to the surface of framing members or mounted between framing members.
2. Mechanical fasteners shall penetrate only the unheated portions provided for this purpose. Panels shall not be fastened at any point closer than $^{1}/_{4}$ inch (6.4 mm) to an element. Other methods of attachment of the panels shall be in accordance with the panel manufacturer's instructions.
3. Unless *listed* and *labeled* for field cutting, heating panels shall be installed as complete units.

M1406.4 Installation in concrete or masonry. Radiant heating systems installed in concrete or masonry shall conform to the following requirements:

1. Radiant heating systems shall be identified as being suitable for the installation, and shall be secured in place as specified in the manufacturer's installation instructions.
2. Radiant heating panels or radiant heating panel sets shall not be installed where they bridge expansion joints unless protected from expansion and contraction.

M1406.5 Finish surfaces. Finish materials installed over radiant heating panels or systems shall be installed in accordance with the manufacturer's instructions. Surfaces shall be secured so that nails or other fastenings do not pierce the radiant heating elements.

SECTION M1407 DUCT HEATERS

M1407.1 General. Electric duct heaters shall be installed in accordance with the manufacturer's instructions and Chapters 34 through 43. Electric duct heaters shall comply with UL 1996.

M1407.2 Installation. Electric duct heaters shall be installed so that they will not create a fire hazard. Class 1 ducts, duct coverings and linings shall be interrupted at each heater to provide the clearances specified in the manufacturer's installation instructions. Such interruptions are not required for duct heaters *listed* and *labeled* for zero clearance to *combustible materials*. Insulation installed in the immediate area of each heater shall be classified for the maximum temperature produced on the duct surface.

M1407.3 Installation with heat pumps and air conditioners. Duct heaters located within 4 feet (1219 mm) of a heat pump or air conditioner shall be *listed* and *labeled* for such installations. The heat pump or air conditioner shall additionally be *listed* and *labeled* for such duct heater installations.

M1407.4 Access. Duct heaters shall be located to allow access for servicing, and clearance shall be maintained to permit adjustment, servicing and replacement of controls and heating elements.

M1407.5 Fan interlock. The fan circuit shall be provided with an interlock to prevent heater operation when the fan is not operating.

SECTION M1408 VENTED FLOOR FURNACES

M1408.1 General. Oil-fired vented *floor furnaces* shall comply with UL 729 and shall be installed in accordance with their *listing*, the manufacturer's instructions and the requirements of this code.

M1408.2 Clearances. Vented *floor furnaces* shall be installed in accordance with their listing and the manufacturer's instructions.

M1408.3 Location. Location of *floor furnaces* shall conform to the following requirements:

1. Floor registers of *floor furnaces* shall be installed not less than 6 inches (152 mm) from a wall.
2. Wall registers of *floor furnaces* shall be installed not less than 6 inches (152 mm) from the adjoining wall at inside corners.
3. The furnace register shall be located not less than 12 inches (305 mm) from doors in any position, draperies or similar combustible objects.
4. The furnace register shall be located not less than 5 feet (1524 mm) below any projecting *combustible materials*.
5. The *floor furnace* burner assembly shall not project into an occupied under-floor area.
6. The *floor furnaces* shall not be installed in concrete floor construction built on grade.
7. The *floor furnaces* shall not be installed where a door can swing within 12 inches (305 mm) of the grille opening.

M1408.4 Access. An opening in the foundation not less than 18 inches by 24 inches (457 mm by 610 mm), or a trap door not less than 22 inches by 30 inches (559 mm by 762 mm) shall be provided for access to a *floor furnace*. The opening and passageway shall be large enough to allow replacement of any part of the *equipment*.

M1408.5 Installation. *Floor furnace* installations shall conform to the following requirements:

1. Thermostats controlling *floor furnaces* shall be located in the room in which the register of the *floor furnace* is located.
2. *Floor furnaces* shall be supported independently of the furnace floor register.
3. *Floor furnaces* shall be installed not closer than 6 inches (152 mm) to the ground. The minimum clearance shall be 2 inches (51 mm), where the lower 6 inches (152 mm) of the furnace is sealed to prevent water entry.
4. Where excavation is required for a *floor furnace* installation, the excavation shall extend 30 inches (762 mm) beyond the control side of the *floor furnace* and 12 inches (305 mm) beyond the remaining sides. Excavations shall slope outward from the perimeter of the base of the excavation to the surrounding *grade* at an angle not exceeding 45 degrees (0.79 rad) from horizontal.
5. *Floor furnaces* shall not be supported from the ground.

SECTION M1409 VENTED WALL FURNACES

M1409.1 General. Oil-fired vented wall furnaces shall comply with UL 730 and shall be installed in accordance with their *listing*, the manufacturer's instructions and the requirements of this code.

M1409.2 Location. The location of vented wall furnaces shall conform to the following requirements:

1. Vented wall furnaces shall be located where they will not cause a fire hazard to walls, floors, combustible furnishings or doors. Vented wall furnaces installed between bathrooms and adjoining rooms shall not circulate air from bathrooms to other parts of the building.
2. Vented wall furnaces shall not be located where a door can swing within 12 inches (305 mm) of the furnace air inlet or outlet measured at right angles to the opening. Doorstops or door closers shall not be installed to obtain this clearance.

M1409.3 Installation. Vented wall furnace installations shall conform to the following requirements:

1. Required wall thicknesses shall be in accordance with the manufacturer's installation instructions.
2. Ducts shall not be attached to a wall furnace. Casing extensions or boots shall be installed only where listed as part of a *listed* and *labeled appliance*.
3. A manual shutoff valve shall be installed ahead of all controls.

M1409.4 Access. Vented wall furnaces shall be provided with access for cleaning of heating surfaces; removal of burners; replacement of sections, motors, controls, filters and other working parts; and for adjustments and lubrication of parts requiring such attention. Panels, grilles and access doors that must be removed for normal servicing operations shall not be attached to the building construction.

SECTION M1410 VENTED ROOM HEATERS

M1410.1 General. Vented room heaters shall be tested in accordance with ASTM E1509 for pellet-fuel burning, UL 896 for oil-fired or UL 1482 for solid fuel-fired and installed in accordance with their *listing*, the manufacturer's installation instructions and the requirements of this code.

M1410.2 Floor mounting. Room heaters shall be installed on noncombustible floors or *approved* assemblies constructed of *noncombustible materials* that extend not less than 18 inches (457 mm) beyond the *appliance* on all sides.

Exceptions:

1. *Listed* room heaters shall be installed on noncombustible floors, assemblies constructed of *noncombustible materials* or floor protectors *listed* and *labeled* in accordance with UL 1618. The materials and dimensions shall be in accordance with the *appliance* manufacturer's instructions.
2. Room heaters *listed* for installation on combustible floors without floor protection shall be installed in accordance with the *appliance* manufacturer's instructions.

SECTION M1411 HEATING AND COOLING EQUIPMENT

M1411.1 Approved refrigerants. Refrigerants used in direct refrigerating systems shall conform to the applicable provisions of ANSI/ASHRAE 34.

M1411.2 Refrigeration coils in warm-air furnaces. Where a cooling coil is located in the supply plenum of a warm-air furnace, the furnace blower shall be rated at not less than 0.5-inch water column (124 Pa) static pressure unless the furnace is *listed* and *labeled* for use with a cooling coil. Cooling coils shall not be located upstream from heat exchangers unless *listed* and *labeled* for such use. Conversion of existing furnaces for use with cooling coils shall be permitted provided that the furnace will operate within the temperature rise specified for the furnace.

M1411.3 Condensate disposal. Condensate from cooling coils and evaporators shall be conveyed from the drain pan outlet to an *approved* place of disposal. Such piping shall maintain a minimum horizontal slope in the direction of discharge of not less than $^1/_8$ unit vertical in 12 units horizontal (1-percent slope). Condensate shall not discharge into a street, alley or other area where it would cause a nuisance.

M1411.3.1 Auxiliary and secondary drain systems. In addition to the requirements of Section M1411.3, a secondary drain or auxiliary drain pan shall be required for each cooling or evaporator coil where damage to any building components will occur as a result of overflow from the *equipment* drain pan or stoppage in the condensate drain piping. Such piping shall maintain a minimum horizontal slope in the direction of discharge of not less than $^{1}/_{8}$ unit vertical in 12 units horizontal (1-percent slope). Drain piping shall be not less than $^{3}/_{4}$-inch (19 mm) nominal pipe size. One of the following methods shall be used:

1. An auxiliary drain pan with a separate drain shall be installed under the coils on which condensation will occur. The auxiliary pan drain shall discharge to a conspicuous point of disposal to alert occupants in the event of a stoppage of the primary drain. The pan shall have a minimum depth of 1.5 inches (38 mm), shall be not less than 3 inches (76 mm) larger than the unit or the coil dimensions in width and length and shall be constructed of corrosion-resistant material. Galvanized sheet steel pans shall have a minimum thickness of not less than 0.0236-inch (0.6010 mm) (No. 24 Gage). Nonmetallic pans shall have a minimum thickness of not less than 0.0625 inch (1.6 mm).
2. A separate overflow drain line shall be connected to the drain pan installed with the *equipment.* This overflow drain shall discharge to a conspicuous point of disposal to alert occupants in the event of a stoppage of the primary drain. The overflow drain line shall connect to the drain pan at a higher level than the primary drain connection.
3. An auxiliary drain pan without a separate drain line shall be installed under the coils on which condensation will occur. This pan shall be equipped with a water level detection device conforming to UL 508 that will shut off the *equipment* served prior to overflow of the pan. The pan shall be equipped with a fitting to allow for drainage. The auxiliary drain pan shall be constructed in accordance with Item 1 of this section.
4. A water-level detection device conforming to UL 508 shall be installed that will shut off the *equipment* served in the event that the primary drain is blocked. The device shall be installed in the primary drain line, the overflow drain line or the *equipment*-supplied drain pan, located at a point higher than the primary drain line connection and below the overflow rim of such pan.

M1411.3.1.1 Water-level monitoring devices. On down-flow units and other coils that do not have secondary drain or provisions to install a secondary or auxiliary drain pan, a water-level monitoring device shall be installed inside the primary drain pan. This device shall shut off the *equipment* served in the event that the primary drain becomes restricted. Devices shall not be installed in the drain line.

M1411.3.1.2 Appliance, equipment and insulation in pans. Where *appliances, equipment* or insulation are subject to water damage when auxiliary drain pans fill, that portion of the *appliance, equipment* and insulation shall be installed above the rim of the pan. Supports located inside of the pan to support the *appliance* or *equipment* shall be water resistant and *approved.*

M1411.3.2 Drain pipe materials and sizes. Components of the condensate disposal system shall be ABS, cast iron, copper, cross-linked polyethylene, CPVC, galvanized steel, PE-RT, polyethylene, polypropylene or PVC pipe or tubing. Components shall be selected for the pressure and temperature rating of the installation. Joints and connections shall be made in accordance with the applicable provisions of Chapter 30. Condensate waste and drain line size shall be not less than $^{3}/_{4}$-inch (19 mm) nominal diameter from the drain pan connection to the place of condensate disposal. Where the drain pipes from more than one unit are manifolded together for condensate drainage, the pipe or tubing shall be sized in accordance with an *approved* method.

M1411.3.3 Drain line maintenance. Condensate drain lines shall be configured to permit the clearing of blockages and performance of maintenance without requiring the drain line to be cut.

M1411.3.4 Appliances, equipment and insulation in pans. Where *appliances, equipment* or insulation are subject to water damage when auxiliary drain pans fill, those portions of the *appliances, equipment* and insulation shall be installed above the flood level rim of the pan. Supports located inside of the pan to support the *appliance* or *equipment* shall be water resistant and *approved.*

M1411.4 Condensate pumps. Condensate pumps located in uninhabitable spaces, such as attics and crawl spaces, shall be connected to the *appliance* or *equipment* served such that when the pump fails, the *appliance* or *equipment* will be prevented from operating. Pumps shall be installed in accordance with the manufacturer's instructions.

M1411.5 Auxiliary drain pan. Category IV condensing *appliances* shall have an auxiliary drain pan where damage to any building component will occur as a result of stoppage in the condensate drainage system. These pans shall be installed in accordance with the applicable provisions of Section M1411.3.

Exception: Fuel-fired *appliances* that automatically shut down operation in the event of a stoppage in the condensate drainage system.

M1411.6 Insulation of refrigerant piping. Piping and fittings for refrigerant vapor (suction) lines shall be insulated with insulation having a thermal resistivity of not less than R-3 and having external surface permeance not exceeding 0.05 perm [2.87 ng/(s $\times$ m^2 $\times$ Pa)] when tested in accordance with ASTM E96.

M1411.6.1 Refrigerant line insulation protection. Refrigerant piping insulation shall be protected in accordance with Section N1103.4.1.

M1411.7 Location and protection of refrigerant piping. Refrigerant piping installed within $1^{1}/_{2}$ inches (38 mm) of the underside of *roof decks* shall be protected from damage caused by nails and other fasteners.

M1411.8 Support of refrigerant piping. Refrigerant piping and tubing shall be securely fastened to a permanent support within 6 feet (1829 mm) of the condensing unit.

M1411.9 Locking access port caps. Refrigerant circuit access ports located outdoors shall be fitted with locking-type tamper-resistant caps or shall be otherwise secured to prevent unauthorized access.

SECTION M1412 ABSORPTION COOLING EQUIPMENT

M1412.1 Approval of equipment. Absorption systems shall be installed in accordance with the manufacturer's instructions. Absorption *equipment* shall comply with UL 1995 or UL/CSA/ANCE 60335-2-40.

M1412.2 Condensate disposal. Condensate from the cooling coil shall be disposed of as provided in Section M1411.3.

M1412.3 Insulation of piping. Refrigerant piping, brine piping and fittings within a building shall be insulated to prevent condensation from forming on piping.

M1412.4 Pressure-relief protection. Absorption systems shall be protected by a pressure-relief device. Discharge from the pressure-relief device shall be located where it will not create a hazard to persons or property.

SECTION M1413 EVAPORATIVE COOLING EQUIPMENT

M1413.1 General. Evaporative cooling *equipment* and *appliances* shall comply with UL 1995 or UL/CSA/ANCE 60335-2-40 and shall be installed:

1. In accordance with the manufacturer's instructions.
2. On level platforms in accordance with Section M1305.1.3.1.
3. So that openings in exterior walls are flashed in accordance with Section R703.4.
4. So as to protect the potable water supply in accordance with Section P2902.
5. So that air intake opening locations are in accordance with Section R303.5.1.

SECTION M1414 FIREPLACE STOVES

M1414.1 General. Fireplace stoves shall be *listed, labeled* and installed in accordance with the terms of the listing. Fireplace stoves shall be tested in accordance with UL 737.

M1414.2 Hearth extensions. Hearth extensions for fireplace stoves shall be installed in accordance with the *listing* of the fireplace stove. The supporting structure for a hearth extension for a fireplace stove shall be at the same level as the supporting structure for the fireplace unit. The hearth extension shall be readily distinguishable from the surrounding floor area.

SECTION M1415 MASONRY HEATERS

M1415.1 General. *Masonry heaters* shall be constructed in accordance with Section R1002.

CHAPTER 15
EXHAUST SYSTEMS

User notes:

About this chapter: *Chapter 15 is specific to exhaust systems related to clothes dryers, domestic cooking, toilet rooms, bathrooms and whole-house ventilation systems. Included are requirements for exhaust discharge locations, protection of exhaust ducts from damage, exhaust duct construction, duct length limits, and exhaust termination clearances. This chapter contains prohibitions for exhaust recirculation and discharge locations and addresses the design of whole-house ventilation systems required by Chapter 3.*

Code development reminder: *Code change proposals to this chapter will be considered by the IRC—Plumbing/Mechanical Code Development Committee during the 2021 (Group A) Code Development Cycle.*

SECTION M1501 GENERAL

M1501.1 Outdoor discharge. The air removed by every mechanical exhaust system shall be discharged to the outdoors in accordance with Section M1504.3. Air shall not be exhausted into an attic, soffit, ridge vent or *crawl space.*

> **Exception:** Whole-house *ventilation*-type attic fans that discharge into the attic space of *dwelling units* having private *attics* shall be permitted.

SECTION M1502 CLOTHES DRYER EXHAUST

M1502.1 General. Clothes dryers shall be exhausted in accordance with the manufacturer's instructions.

M1502.2 Independent exhaust systems. Dryer exhaust systems shall be independent of all other systems and shall convey the moisture to the outdoors.

> **Exception:** This section shall not apply to *listed* and *labeled* condensing (ductless) clothes dryers.

M1502.3 Duct termination. Exhaust ducts shall terminate on the outside of the building. Exhaust duct terminations shall be in accordance with the dryer manufacturer's installation instructions. If the manufacturer's instructions do not specify a termination location, the exhaust duct shall terminate not less than 3 feet (914 mm) in any direction from openings into buildings, including openings in ventilated soffits. Exhaust duct terminations shall be equipped with a backdraft damper. Screens shall not be installed at the duct termination.

M1502.3.1 Exhaust termination outlet and passageway size. The passageway of dryer exhaust duct terminals shall be undiminished in size and shall provide an open area of not less than 12.5 square inches (8065 mm^2).

M1502.4 Dryer exhaust ducts. Dryer exhaust ducts shall conform to the requirements of Sections M1502.4.1 through M1502.4.8.

M1502.4.1 Material and size. Exhaust ducts shall have a smooth interior finish and shall be constructed of metal not less than 0.0157 inch (0.3950 mm) in thickness (No. 28 gage). The duct shall be 4 inches (102 mm) nominal in diameter.

M1502.4.2 Duct installation. Exhaust ducts shall be supported at intervals not to exceed 12 feet (3658 mm) and shall be secured in place. The insert end of the duct shall extend into the adjoining duct or fitting in the direction of airflow. Exhaust duct joints shall be sealed in accordance with Section M1601.4.1 and shall be mechanically fastened. Ducts shall not be joined with screws or similar fasteners that protrude more than $^1/_8$ inch (3.2 mm) into the inside of the duct. Where dryer exhaust ducts are enclosed in wall or ceiling cavities, such cavities shall allow the installation of the duct without deformation.

M1502.4.3 Transition duct. Transition ducts used to connect the dryer to the exhaust *duct system* shall be a single length that is *listed* and *labeled* in accordance with UL 2158A. Transition ducts shall be not greater than 8 feet (2438 mm) in length. Transition ducts shall not be concealed within construction.

M1502.4.4 Dryer exhaust duct power ventilators. Domestic dryer exhaust duct power ventilators shall conform to UL 705 for use in dryer exhaust duct systems. The dryer exhaust duct power ventilator shall be installed in accordance with the manufacturer's instructions.

M1502.4.5 Booster fans prohibited. Domestic booster fans shall not be installed in dryer exhaust systems.

M1502.4.6 Duct length. The maximum allowable exhaust duct length shall be determined by one of the methods specified in Sections M1502.4.6.1 through M1502.4.6.3.

M1502.4.6.1 Specified length. The maximum length of the exhaust duct shall be 35 feet (10 668 mm) from the connection to the transition duct from the dryer to the outlet terminal. Where fittings are used, the maximum length of the exhaust duct shall be reduced in accordance with Table M1502.4.6.1. The maximum length of the exhaust duct does not include the transition duct.

TABLE M1502.4.6.1
DRYER EXHAUST DUCT FITTING EQUIVALENT LENGTH

DRYER EXHAUST DUCT FITTING TYPE	EQUIVALENT LENGTH
4-inch radius mitered 45-degree elbow	2 feet 6 inches
4-inch radius mitered 90-degree elbow	5 feet
6-inch radius smooth 45-degree elbow	1 foot
6-inch radius smooth 90-degree elbow	1 foot 9 inches
8-inch radius smooth 45-degree elbow	1 foot
8-inch radius smooth 90-degree elbow	1 foot 7 inches
10-inch radius smooth 45-degree elbow	9 inches
10-inch radius smooth 90-degree elbow	1 foot 6 inches

For SI: 1 inch = 25.4 mm, 1 foot = 304.8 mm, 1 degree = 0.0175 rad.

M1502.4.6.2 Manufacturer's instructions. The size and maximum length of the exhaust duct shall be determined by the dryer manufacturer's installation instructions. The code official shall be provided with a copy of the installation instructions for the make and model of the dryer at the concealment inspection. In the absence of fitting equivalent length calculations from the clothes dryer manufacturer, Table M1502.4.6.1 shall be used.

M1502.4.6.3 Dryer exhaust duct power ventilator. The maximum length of the exhaust duct shall be determined in accordance with the manufacturer's instructions for the dryer exhaust duct power ventilator.

M1502.4.7 Length identification. Where the exhaust duct equivalent length exceeds 35 feet (10 668 mm), the equivalent length of the exhaust duct shall be identified on a permanent *label* or tag. The *label* or tag shall be located within 6 feet (1829 mm) of the exhaust duct connection.

M1502.4.8 Exhaust duct required. Where space for a clothes dryer is provided, an exhaust *duct system* shall be installed. Where the clothes dryer is not installed at the time of occupancy the exhaust duct shall be capped or plugged in the space in which it originates and identified and marked "future use."

Exception: Where a *listed* condensing clothes dryer is installed prior to occupancy of the structure.

M1502.5 Protection required. Protective shield plates shall be placed where nails or screws from finish or other work are likely to penetrate the clothes dryer exhaust duct. Shield plates shall be placed on the finished face of framing members where there is less than $1^1/_4$ inches (32 mm) between the duct and the finished face of the framing member. Protective shield plates shall be constructed of steel, shall have a minimum thickness of 0.062 inch (1.6 mm) and shall extend not less than 2 inches (51 mm) above sole plates and below top plates.

SECTION M1503
DOMESTIC COOKING EXHAUST EQUIPMENT

M1503.1 General. Domestic cooking exhaust equipment shall comply with the requirements of this section.

M1503.2 Domestic cooking exhaust. Where domestic cooking exhaust equipment is provided, it shall comply with one of the following:

1. The fan for overhead range hoods and downdraft exhaust equipment not integral with the cooking *appliance* shall be *listed* and *labeled* in accordance with UL 507.
2. Overhead range hoods and downdraft exhaust equipment with integral fans shall comply with UL 507.
3. Domestic cooking *appliances* with integral downdraft exhaust equipment shall be *listed* and *labeled* in accordance with ANSI Z21.1 or UL 858.
4. Microwave ovens with integral exhaust for installation over the cooking surface shall be *listed* and *labeled* in accordance with UL 923.

M1503.2.1 Open-top broiler exhaust. Domestic open-top broiler units shall be provided with a metal exhaust hood having a thickness of not less than 0.0157 inch (0.3950 mm) (No. 28 gage). Such hoods shall be installed with a clearance of not less than $^1/_4$ inch (6.4 mm) between the hood and the underside of *combustible material* and cabinets. A clearance of not less than 24 inches (610 mm) shall be maintained between the cooking surface and *combustible material* and cabinets. The hood width shall be not less than the width of the broiler unit and shall extend over the entire unit.

Exception: Broiler units that incorporate an integral exhaust system, and that are *listed* and *labeled* for use without an exhaust hood, shall not be required to have an exhaust hood.

M1503.3 Exhaust discharge. Domestic cooking exhaust equipment shall discharge to the outdoors through a duct. The duct shall have a smooth interior surface, shall be airtight, shall be equipped with a backdraft damper and shall be independent of all other exhaust systems. Ducts serving domestic cooking exhaust equipment shall not terminate in an attic or *crawl space* or areas inside the building.

Exception: Where installed in accordance with the manufacturer's instructions, and where mechanical or natural *ventilation* is otherwise provided, *listed* and *labeled* ductless range hoods shall not be required to discharge to the outdoors.

M1503.4 Duct material. Ducts serving domestic cooking exhaust equipment shall be constructed of galvanized steel, stainless steel or copper.

Exception: Ducts for domestic kitchen cooking *appliances* equipped with down-draft exhaust systems shall be permitted to be constructed of schedule 40 PVC pipe and fittings provided that the installation complies with all of the following:

1. The duct is installed under a concrete slab poured on grade.
2. The underfloor trench in which the duct is installed is completely backfilled with sand or gravel.
3. The PVC duct extends not more than 1 inch (25 mm) above the indoor concrete floor surface.
4. The PVC duct extends not more than 1 inch (25 mm) above grade outside of the building.
5. The PVC ducts are solvent cemented.

M1503.5 Kitchen exhaust rates. Where domestic kitchen cooking *appliances* are equipped with ducted range hoods or down-draft exhaust systems, the fans shall be sized in accordance with Section M1505.4.4.

M1503.6 Makeup air required. Where one or more gas, liquid or solid fuel-burning *appliance* that is neither direct-vent nor uses a mechanical draft venting system is located within a dwelling unit's air barrier, each exhaust system capable of exhausting in excess of 400 cubic feet per minute (0.19 m^3/s) shall be mechanically or passively provided with makeup air at a rate approximately equal to the exhaust air rate. Such makeup air systems shall be equipped with not fewer than one damper complying with Section M1503.6.2.

Exception: Makeup air is not required for exhaust systems installed for the exclusive purpose of space cooling and intended to be operated only when windows or other air inlets are open.

M1503.6.1 Location. Kitchen exhaust makeup air shall be discharged into the same room in which the exhaust system is located or into rooms or *duct systems* that communicate through one or more permanent openings with the room in which such exhaust system is located. Such permanent openings shall have a net cross-sectional area not less than the required area of the makeup air supply openings.

M1503.6.2 Makeup air dampers. Where makeup air is required by Section M1503.6, makeup air dampers shall comply with this section. Each damper shall be a gravity damper or an electrically operated damper that automatically opens when the exhaust system operates. Dampers shall be located to allow access for inspection, service, repair and replacement without removing permanent construction or any other ducts not connected to the damper being inspected, serviced, repaired or replaced. Gravity or barometric dampers shall not be used in passive makeup air systems except where the dampers are rated to provide the design makeup airflow at a pressure differential of 0.01 in. w.c. (3 Pa) or less.

SECTION M1504 EXHAUST DUCTS AND EXHAUST OPENINGS

M1504.1 Duct construction. Where exhaust duct construction is not specified in this chapter, construction shall comply with Chapter 16.

M1504.2 Duct length. The length of exhaust and supply ducts used with ventilating *equipment* shall not exceed the lengths determined in accordance with Table M1504.2.

Exception: Duct length shall not be limited where the *duct system* complies with the manufacturer's design criteria or where the flow rate of the installed ventilating *equipment* is verified by the installer or *approved* third party using a flow hood, flow grid or other airflow measuring device.

M1504.3 Exhaust openings. Air exhaust openings shall terminate as follows:

1. Not less than 3 feet (914 mm) from property lines.
2. Not less than 3 feet (914 mm) from gravity air intake openings, operable windows and doors.
3. Not less than 10 feet (3048 mm) from mechanical air intake openings except where the exhaust opening is located not less than 3 feet (914 mm) above the air intake opening. Openings shall comply with Sections R303.5.2 and R303.6.

SECTION M1505 MECHANICAL VENTILATION

M1505.1 General. Where local exhaust or whole-house mechanical *ventilation* is provided, the ventilation system shall be designed in accordance with this section.

M1505.2 Recirculation of air. Exhaust air from bathrooms and toilet rooms shall not be recirculated within a residence or circulated to another *dwelling unit* and shall be exhausted directly to the outdoors. Exhaust air from bathrooms, toilet rooms and kitchens shall not discharge into an attic, *crawl space* or other areas inside the building. This section shall not prohibit the installation of ductless range hoods in accordance with the exception to Section M1503.3.

M1505.3 Exhaust equipment. Exhaust fans and whole-house mechanical ventilation fans shall be *listed* and *labeled* as providing the minimum required airflow in accordance with ANSI/AMCA 210-ANSI/ASHRAE 51.

M1505.4 Whole-house mechanical ventilation system. Whole-house mechanical ventilation systems shall be designed in accordance with Sections M1505.4.1 through M1505.4.4.

TABLE M1504.2
DUCT LENGTH

DUCT TYPE	FLEX DUCT								SMOOTH-WALL DUCT							
Fan airflow rating (CFM @ 0.25 inch wc[a])	50	80	100	125	150	200	250	300	50	80	100	125	150	200	250	300
Diameter[b] (inches)	Maximum length[c, d, e] (feet)															
3	X	X	X	X	X	X	X	X	5	X	X	X	X	X	X	X
4	56	4	X	X	X	X	X	X	114	31	10	X	X	X	X	X
5	NL	81	42	16	2	X	X	X	NL	152	91	51	28	4	X	X
6	NL	NL	158	91	55	18	1	X	NL	NL	NL	168	112	53	25	9
7	NL	NL	NL	NL	161	78	40	19	NL	NL	NL	NL	NL	148	88	54
8 and above	NL	NL	NL	NL	NL	189	111	69	NL	NL	NL	NL	NL	NL	198	133

For SI: 1 inch = 25.4 mm, 1 foot = 304.8 mm.
a. Fan airflow rating shall be in accordance with ANSI/AMCA 210-ANSI/ASHRAE 51.
b. For noncircular ducts, calculate the diameter as four times the cross-sectional area divided by the perimeter.
c. This table assumes that elbows are not used. Fifteen feet of allowable duct length shall be deducted for each elbow installed in the duct run.
d. NL = no limit on duct length of this size.
e. X = not allowed. Any length of duct of this size with assumed turns and fittings will exceed the rated pressure drop.

M1505.4.1 System design. The whole-house ventilation system shall consist of one or more supply or exhaust fans, or a combination of such, and associated ducts and controls. Local exhaust or supply fans are permitted to serve as such a system. Outdoor air ducts connected to the return side of an air handler shall be considered as providing supply ventilation.

M1505.4.2 System controls. The whole-house mechanical ventilation system shall be provided with controls that enable manual override. Controls shall include text or a symbol indicating their function.

M1505.4.3 Mechanical ventilation rate. The whole-house mechanical ventilation system shall provide outdoor air at a continuous rate not less than that determined in accordance with Table M1505.4.3(1) or not less than that determined by Equation 15-1.

Ventilation rate in cubic feet per minute = (0.01 × total square foot area of house) + [7.5 × (number of bedrooms + 1)]

(Equation 15-1)

Exceptions:

1. Ventilation rate credit. The minimum mechanical ventilation rate determined in accordance with Table M1505.4.3(1) or Equation 15-1 shall be reduced by 30 percent, provided that both of the following conditions apply:
   1.1. A ducted system supplies ventilation air directly to each bedroom and to one or more of the following rooms:
      1.1.1. Living room.
      1.1.2. Dining room.
      1.1.3. Kitchen.

TABLE M1505.4.3(1)
CONTINUOUS WHOLE-HOUSE MECHANICAL VENTILATION SYSTEM AIRFLOW RATE REQUIREMENTS

DWELLING UNIT FLOOR AREA (square feet)	NUMBER OF BEDROOMS				
	0–1	2–3	4–5	6–7	> 7
	Airflow in CFM				
< 1,500	30	45	60	75	90
1,501–3,000	45	60	75	90	105
3,001–4,500	60	75	90	105	120
4,501–6,000	75	90	105	120	135
6,001–7,500	90	105	120	135	150
> 7,500	105	120	135	150	165

For SI: 1 square foot = 0.0929 m^2, 1 cubic foot per minute = 0.0004719 m^3/s.

TABLE M1505.4.3(2)
INTERMITTENT WHOLE-HOUSE MECHANICAL VENTILATION RATE FACTORS[a, b]

RUN-TIME PERCENTAGE IN EACH 4-HOUR SEGMENT	25%	33%	50%	66%	75%	100%
Factor[a]	4	3	2	1.5	1.3	1.0

a. For ventilation system run-time values between those given, the factors are permitted to be determined by interpolation.
b. Extrapolation beyond the table is prohibited.

1.2. The whole-house ventilation system is a balanced ventilation system.

2. Programmed intermittent operation. The whole-house mechanical ventilation system is permitted to operate intermittently where the system has controls that enable operation for not less than 25 percent of each 4-hour segment and the ventilation rate prescribed in Table M1505.4.3(1), by Equation 15-1 or by Exception 1 is multiplied by the factor determined in accordance with Table M1505.4.3(2).

M1505.4.4 Local exhaust rates. *Local exhaust* systems shall be designed to have the capacity to exhaust the minimum airflow rate determined in accordance with Table M1505.4.4.

TABLE M1505.4.4
MINIMUM REQUIRED LOCAL EXHAUST RATES FOR ONE- AND TWO-FAMILY DWELLINGS

AREA TO BE EXHAUSTED	EXHAUST RATES[a]
Kitchens	100 cfm intermittent or 25 cfm continuous
Bathrooms-Toilet Rooms	Mechanical exhaust capacity of 50 cfm intermittent or 20 cfm continuous

For SI: 1 cubic foot per minute = 0.0004719 m^3/s,
1 inch water column = 0.2488 kPa.

a. The listed exhaust rate for bathrooms-toilet rooms shall equal or exceed the exhaust rate at a minimum static pressure of 0.25 inch water column in accordance with Section M1505.3.

CHAPTER 16
DUCT SYSTEMS

User notes:

About this chapter: Chapter 16 addresses duct construction for HVAC and most exhaust systems. This chapter covers duct materials, duct construction, duct installation, duct insulation properties, duct sealing, above-ground and underground ducts, return air intake locations and air plenums.

Code development reminder: Code change proposals to this chapter will be considered by the IRC—Plumbing/Mechanical Code Development Committee during the 2021 (Group A) Code Development Cycle.

SECTION M1601 DUCT CONSTRUCTION

M1601.1 Duct design. *Duct systems* serving heating, cooling and *ventilation equipment* shall be installed in accordance with the provisions of this section and ACCA *Manual D*, the *appliance* manufacturer's installation instructions or other *approved* methods.

M1601.1.1 Above-ground duct systems. Above-ground *duct systems* shall conform to the following:

1. *Equipment* connected to *duct systems* shall be designed to limit discharge air temperature to not greater than 250°F (121°C).
2. Factory-made ducts shall be *listed* and *labeled* in accordance with UL 181 and installed in accordance with the manufacturer's instructions.
3. Fibrous glass duct construction shall conform to the SMACNA *Fibrous Glass Duct Construction Standards* or NAIMA *Fibrous Glass Duct Construction Standards.*
4. Field-fabricated and shop-fabricated metal and flexible duct constructions shall conform to the SMACNA *HVAC Duct Construction Standards—Metal and Flexible* except as allowed by Table M1601.1.1. Galvanized steel shall conform to ASTM A653.
5. The use of gypsum products to construct return air ducts or plenums is permitted, provided that the air temperature does not exceed 125°F (52°C) and exposed surfaces are not subject to condensation.
6. *Duct systems* shall be constructed of materials having a flame spread index of not greater than 200.
7. Stud wall cavities and the spaces between solid floor joists to be used as air plenums shall comply with the following conditions:
   7.1. These cavities or spaces shall not be used as a plenum for supply air.
   7.2. These cavities or spaces shall not be part of a required fire-resistance-rated assembly.
   7.3. Stud wall cavities shall not convey air from more than one floor level.
   7.4. Stud wall cavities and joist-space plenums shall be isolated from adjacent concealed spaces by tight-fitting fireblocking in accordance with Section R302.11. Fireblocking materials used for isolation shall comply with Section R302.11.1.
   7.5. Stud wall cavities in the outside walls of building envelope assemblies shall not be utilized as air plenums.
   7.6. Building cavities used as plenums shall be sealed.
8. Volume dampers, equipment and other means of supply, return and exhaust air adjustment used in system balancing shall be provided with access.

M1601.1.2 Underground duct systems. Underground *duct systems* shall be constructed of *approved* concrete, clay, metal or plastic. The maximum design temperature for systems utilizing plastic duct and fittings shall be 150°F (66°C). Metal ducts shall be protected from corrosion in an *approved* manner or shall be completely encased in concrete not less than 2 inches (51 mm) thick. Nonmetallic ducts shall be installed in accordance with the manufacturer's instructions. Plastic pipe and fitting materials shall conform to cell classification 12454-B of ASTM D1248 or ASTM D1784 and external loading properties of ASTM D2412. Ducts shall slope to a drainage point that has access. Ducts shall be sealed, secured and tested prior to encasing the ducts in concrete or direct burial. Duct tightness shall be verified as required by Section N1103.3. Metallic ducts having an *approved* protective coating and nonmetallic ducts shall be installed in accordance with the manufacturer's instructions.

M1601.2 Vibration isolators. Vibration isolators installed between mechanical equipment and metal ducts shall be fabricated from *approved* materials and shall not exceed 10 inches (254 mm) in length.

M1601.3 Duct insulation materials. Duct insulation materials shall conform to the following requirements:

1. Duct coverings and linings, including adhesives where used, shall have a flame spread index not higher than 25, and a *smoke-developed index* not over 50 when tested in accordance with ASTM E84

or UL 723, using the specimen preparation and mounting procedures of ASTM E2231.

Exception: Spray application of polyurethane foam to the exterior of ducts in *attics* and *crawl spaces* shall be permitted subject to all of the following:

1. The flame spread index is not greater than 25 and the *smoke-developed index* is not greater than 450 at the specified installed thickness.
2. The foam plastic is protected in accordance with the ignition barrier requirements of Sections R316.5.3 and R316.5.4.
3. The foam plastic complies with the requirements of Section R316.

2. Duct coverings and linings shall not flame, glow, smolder or smoke when tested in accordance with ASTM C411 at the temperature to which they are exposed in service. The test temperature shall not fall below 250°F (121°C). Coverings and linings shall be *listed* and *labeled.*

3. External reflective duct insulation shall be legibly printed or identified at intervals not greater than 36 inches (914 mm) with the name of the manufacturer, the product *R*-value at the specified installed thickness and the flame spread and smoke-developed indices. The installed thickness of the external duct insulation shall include the enclosed airspace(s). The product *R*-value for external reflective duct insulation shall be determined in accordance with ASTM C1668.

4. External duct insulation and factory-insulated flexible ducts shall be legibly printed or identified at intervals not longer than 36 inches (914 mm) with the name of the manufacturer, the thermal resistance *R*-value at the specified installed thickness and the flame spread and smoke-developed indices of the composite materials. Spray polyurethane foam manufacturers shall provide the same product information and properties, at the nominal installed thickness, to the customer in writing at the time of foam application. Nonreflective duct insulation product *R*-values shall be based on insulation only, excluding air films, vapor retarders or other duct components, and shall be based on tested C-values at 75°F (24°C) mean temperature at the installed thickness, in accordance with recognized industry procedures. The installed thickness of duct insulation used to determine its *R*-value shall be determined as follows:

   4.1. For duct board, duct liner and factory-made rigid ducts not normally subjected to compression, the nominal insulation thickness shall be used.

   4.2. For ductwrap, the installed thickness shall be assumed to be 75 percent (25-percent compression) of nominal thickness.

   4.3. For factory-made flexible air ducts, The installed thickness shall be determined by dividing the difference between the actual

TABLE M1601.1.1
DUCT CONSTRUCTION MINIMUM SHEET METAL THICKNESS FOR SINGLE DWELLING UNITS[a]

ROUND DUCT DIAMETER (inches)	STATIC PRESSURE			
	1/2-inch water gauge		1-inch water gauge	
	Thickness (inches)		Thickness (inches)	
	Galvanized	Aluminum	Galvanized	Aluminum
< 12	0.013	0.018	0.013	0.018
12 to 14	0.013	0.018	0.016	0.023
15 to 17	0.016	0.023	0.019	0.027
18	0.016	0.023	0.024	0.034
19 to 20	0.019	0.027	0.024	0.034
RECTANGULAR DUCT DIMENSION (inches)	STATIC PRESSURE			
	1/2-inch water gauge		1-inch water gauge	
	Thickness (inches)		Thickness (inches)	
	Galvanized	Aluminum	Galvanized	Aluminum
≤ 8	0.013	0.018	0.013	0.018
9 to 10	0.013	0.018	0.016	0.023
11 to 12	0.016	0.023	0.019	0.027
13 to 16	0.019	0.027	0.019	0.027
17 to 18	0.019	0.027	0.024	0.034
19 to 20	0.024	0.034	0.024	0.034

For SI: 1 inch = 25.4 mm, 1 inch water gage = 249 Pa.

a. Ductwork that exceeds 20 inches by dimension or exceeds a pressure of 1 inch water gauge shall be constructed in accordance with SMACNA *HVAC Duct Construction Standards—Metal and Flexible.*

outside diameter and nominal inside diameter by two.

4.4. For spray polyurethane foam, the aged *R*-value per inch measured in accordance with recognized industry standards shall be provided to the customer in writing at the time of foam application. In addition, the total *R*-value for the nominal application thickness shall be provided.

M1601.4 Installation. Duct installation shall comply with Sections M1601.4.1 through M1601.4.10.

M1601.4.1 Joints, seams and connections. Longitudinal and transverse joints, seams and connections in metallic and nonmetallic ducts shall be constructed as specified in SMACNA *HVAC Duct Construction Standards—Metal and Flexible* and NAIMA *Fibrous Glass Duct Construction Standards.* Joints, longitudinal and transverse seams, and connections in ductwork shall be securely fastened and sealed with welds, gaskets, mastics (adhesives), mastic-plus-embedded-fabric systems, liquid sealants or tapes. Tapes and mastics used to seal fibrous glass ductwork shall be *listed* and *labeled* in accordance with UL 181A and shall be marked "181A-P" for pressure-sensitive tape, "181 A-M" for mastic or "181 A-H" for heat-sensitive tape.

Tapes and mastics used to seal metallic and flexible air ducts and flexible air connectors shall comply with UL 181B and shall be marked "181 B-FX" for pressure-sensitive tape or "181 BM" for mastic. Duct connections to flanges of air distribution system *equipment* shall be sealed and mechanically fastened. Mechanical fasteners for use with flexible nonmetallic air ducts shall comply with UL 181B and shall be marked 181B-C. Crimp joints for round metallic ducts shall have a contact lap of not less than 1 inch (25 mm) and shall be mechanically fastened by means of not less than three sheet-metal screws or rivets equally spaced around the joint.

Closure systems used to seal all ductwork shall be installed in accordance with the manufacturers' instructions.

Exceptions:

1. Spray polyurethane foam shall be permitted to be applied without additional joint seals.
2. Where a duct connection is made that is partially without access, three screws or rivets shall be equally spaced on the exposed portion of the joint so as to prevent a hinge effect.
3. For ducts having a static pressure classification of less than 2 inches of water column (500 Pa), additional closure systems shall not be required for continuously welded joints and seams and locking-type joints and seams. This exception shall not apply to snap-lock and button-lock type joints and seams that are located outside of *conditioned spaces.*

M1601.4.2 Duct lap. Crimp joints for round and oval metal ducts shall be lapped not less than 1 inch (25 mm) and the male end of the duct shall extend into the adjoining duct in the direction of airflow.

M1601.4.3 Plastic duct joints. Joints between plastic ducts and plastic fittings shall be made in accordance with the manufacturer's installation instructions.

M1601.4.4 Support. Factory-made ducts *listed* in accordance with UL 181 shall be supported in accordance with the manufacturer's installation instructions. Field- and shop-fabricated fibrous glass ducts shall be supported in accordance with the SMACNA *Fibrous Glass Duct Construction Standards* or the NAIMA *Fibrous Glass Duct Construction Standards.* Field- and shop-fabricated metal and flexible ducts shall be supported in accordance with the SMACNA *HVAC Duct Construction Standards—Metal and Flexible.*

M1601.4.5 Fireblocking. Duct installations shall be fireblocked in accordance with Section R302.11.

M1601.4.6 Duct insulation. Duct insulation shall be installed in accordance with the following requirements:

1. A vapor retarder having a permeance of not greater than 0.05 perm [2.87 ng/(s $\times$ m^2 $\times$ Pa)] in accordance with ASTM E96, or aluminum foil with a thickness of not less than 2 mils (0.05 mm), shall be installed on the exterior of insulation on cooling supply ducts that pass through unconditioned spaces conducive to condensation except where the insulation is spray polyurethane foam with a water vapor permeance of not greater than 3 perms per inch [1722 ng/(s $\times$ m^2 $\times$ Pa)] at the installed thickness.
2. Outdoor *duct systems* shall be protected against the elements.
3. Duct coverings shall not penetrate a fireblocked wall or floor.

M1601.4.7 Factory-made air ducts. Factory-made air ducts shall not be installed in or on the ground, in tile or metal pipe, or within masonry or concrete.

M1601.4.8 Duct separation. Ducts shall be installed with not less than 4 inches (102 mm) separation from earth except where they meet the requirements of Section M1601.1.2.

M1601.4.9 Ducts located in garages. Ducts in garages shall comply with the requirements of Section R302.5.2.

M1601.4.10 Flood hazard areas. In flood hazard areas as established by Table R301.2, *duct systems* shall be located or installed in accordance with Section R322.1.6.

M1601.5 Under-floor plenums. Under-floor plenums shall be prohibited in new structures. Modification or repairs to under-floor plenums in existing structures shall conform to the requirements of this section.

M1601.5.1 General. The space shall be cleaned of loose *combustible materials* and scrap, and shall be tightly enclosed. The ground surface of the space shall be covered with a moisture barrier having a thickness of not less than 4 mils (0.1 mm). Plumbing waste cleanouts shall not be located within the space.

Exception: Plumbing waste cleanouts shall be permitted to be located in unvented *crawl spaces* that receive *conditioned air* in accordance with Section R408.3.

M1601.5.2 Materials. The under-floor space, including the sidewall insulation, shall be formed by materials having flame spread index values not greater than 200 when tested in accordance with ASTM E84 or UL 723.

M1601.5.3 Furnace connections. A duct shall extend from the furnace supply outlet to not less than 6 inches (152 mm) below the combustible framing. This duct shall comply with the provisions of Section M1601.1. A noncombustible receptacle shall be installed below any floor opening into the plenum in accordance with the following requirements:

1. The receptacle shall be securely suspended from the floor members and shall be not more than 18 inches (457 mm) below the floor opening.
2. The area of the receptacle shall extend 3 inches (76 mm) beyond the opening on all sides.
3. The perimeter of the receptacle shall have a vertical lip not less than 1 inch (25 mm) in height at the open sides.

M1601.5.4 Access. Access to an under-floor plenum shall be provided through an opening in the floor with minimum dimensions of 18 inches by 24 inches (457 mm by 610 mm).

M1601.5.5 Furnace controls. The furnace shall be equipped with an automatic control that will start the air-circulating fan when the air in the furnace bonnet reaches a temperature not higher than 150°F (66°C). The furnace shall additionally be equipped with an *approved* automatic control that limits the outlet air temperature to 200°F (93°C).

M1601.6 Independent garage HVAC systems. Furnaces and air-handling systems that supply air to living spaces shall not supply air to or return air from a garage.

SECTION M1602
RETURN AIR

M1602.1 Outdoor air openings. Outdoor intake openings shall be located in accordance with Section R303.5.1. Opening protection shall be in accordance with Section R303.6

M1602.2 Return air openings. Return air openings for heating, *ventilation* and air-conditioning systems shall comply with all of the following:

1. Openings shall not be located less than 10 feet (3048 mm) measured in any direction from an open combustion chamber or draft hood of another *appliance* located in the same room or space.
2. The amount of return air taken from any room or space shall be not greater than the flow rate of supply air delivered to such room or space.
3. Return and transfer openings shall be sized in accordance with the *appliance* or *equipment* manufacturer's installation instructions, Manual D or the design of the *registered design professional.*
4. Return air shall not be taken from a closet, bathroom, toilet room, kitchen, garage, mechanical room, boiler room, furnace room or unconditioned attic.

   Exceptions:

   1. Taking return air from a kitchen is not prohibited where such return air openings serve the kitchen only, and are located not less than 10 feet (3048 mm) from the cooking *appliances.*
   2. Dedicated forced-air systems serving only the garage shall not be prohibited from obtaining return air from the garage.
5. For other than dedicated HVAC systems, return air shall not be taken from indoor swimming pool enclosures and associated deck areas except where the air in such spaces is dehumidified,
6. Taking return air from an unconditioned *crawl space* shall not be accomplished through a direct connection to the return side of a forced-air furnace. Transfer openings in the *crawl space* enclosure shall not be prohibited.
7. Return air from one *dwelling unit* shall not be discharged into another *dwelling unit.*

CHAPTER 17
COMBUSTION AIR

User notes:

About this chapter: *Chapter 17 applies only to oil-fired and solid fuel-fired appliances. Chapter 24 applies to combustion air for gas-fired appliances.*

Code development reminder: *Code change proposals to this chapter will be considered by the IRC—Plumbing/Mechanical Code Development Committee during the 2021 (Group A) Code Development Cycle.*

SECTION M1701
GENERAL

M1701.1 Scope. Solid fuel-burning *appliances* shall be provided with *combustion air* in accordance with the *appliance* manufacturer's installation instructions. Oil-fired *appliances* shall be provided with *combustion air* in accordance with NFPA 31. The methods of providing *combustion air* in this chapter do not apply to fireplaces, fireplace stoves and direct-vent *appliances.* The requirements for combustion and dilution air for gas-fired *appliances* shall be in accordance with Chapter 24.

M1701.2 Opening location. In flood hazard areas as established in Table R301.2, *combustion air* openings shall be located at or above the elevation required in Section R322.2.1 or R322.3.2.

CHAPTER 18

CHIMNEYS AND VENTS

User notes:

About this chapter: *Chapter 18 addresses chimneys and vents that serve oil- and solid fuel-fired appliances, including wood pellet appliances. Gas-fired appliances are vented in accordance with Chapter 24. Chapter 10 addresses chimneys for fireplaces and masonry and factory-built chimneys in general. Note that chimneys and vents are distinct.*

Code development reminder: *Code change proposals to this chapter will be considered by the IRC—Plumbing/Mechanical Code Development Committee during the 2021 (Group A) Code Development Cycle.*

SECTION M1801 GENERAL

M1801.1 Venting required. Fuel-burning *appliances* shall be vented to the outdoors in accordance with their *listing* and *label* and manufacturer's installation instructions except *appliances* listed and *labeled* for unvented use. Venting systems shall consist of *approved* chimneys or vents, or venting assemblies that are integral parts of *labeled appliances*. Gas-fired *appliances* shall be vented in accordance with Chapter 24.

M1801.2 Draft requirements. A venting system shall satisfy the draft requirements of the *appliance* in accordance with the manufacturer's installation instructions, and shall be constructed and installed to develop a positive flow to convey combustion products to the outside atmosphere.

M1801.3 Existing chimneys and vents. Where an *appliance* is permanently disconnected from an existing chimney or vent, or where an *appliance* is connected to an existing chimney or vent during the process of a new installation, the chimney or vent shall comply with Sections M1801.3.1 through M1801.3.4.

M1801.3.1 Size. The chimney or vent shall be resized as necessary to control flue gas condensation in the interior of the chimney or vent and to provide the *appliance*, or *appliances* served, with the required draft. For the venting of oil-fired *appliances* to masonry chimneys, the resizing shall be done in accordance with NFPA 31.

M1801.3.2 Flue passageways. The flue gas passageway shall be free of obstructions and combustible deposits and shall be cleaned if previously used for venting a solid or liquid fuel-burning *appliance* or fireplace. The flue liner, chimney inner wall or vent inner wall shall be continuous and free of cracks, gaps, perforations, or other damage or deterioration that would allow the escape of combustion products, including gases, moisture and creosote.

M1801.3.3 Cleanout. Masonry chimneys shall be provided with a cleanout opening complying with Section R1003.17.

M1801.3.4 Clearances. Chimneys and vents shall have airspace clearance to combustibles in accordance with this code and the chimney or vent manufacturer's installation instructions.

Exception: Masonry chimneys equipped with a chimney lining system tested and *listed* for installation in chimneys in contact with combustibles in accordance with UL 1777, and installed in accordance with the manufacturer's instructions, shall not be required to have a clearance between *combustible materials* and exterior surfaces of the *masonry chimney*. Noncombustible firestopping shall be provided in accordance with this code.

M1801.4 Space around lining. The space surrounding a flue lining system or other vent installed within a *masonry chimney* shall not be used to vent any other *appliance*. This shall not prevent the installation of a separate flue lining in accordance with the manufacturer's installation instructions and this code.

M1801.5 Mechanical draft systems. A *mechanical draft system* shall be used only with *appliances listed* and *labeled* for such use. Provisions shall be made to prevent the flow of fuel to the *equipment* when the draft system is not operating. Forced draft systems and portions of induced draft systems under positive pressure during operation shall be designed and installed to prevent leakage of flue gases into a building.

M1801.6 Direct-vent appliances. Direct-vent *appliances* shall be installed in accordance with the manufacturer's instructions.

M1801.7 Support. Venting systems shall be adequately supported for the weight of the material used.

M1801.8 Duct penetrations. Chimneys, vents and vent connectors shall not extend into or through supply and return air ducts or plenums.

M1801.9 Fireblocking. Vent and chimney installations shall be fireblocked in accordance with Section R602.8.

M1801.10 Unused openings. Unused openings in any venting system shall be closed or capped.

M1801.11 Multiple-appliance venting systems. Two or more *listed* and *labeled appliances* connected to a common natural draft venting system shall comply with the following requirements:

1. *Appliances* that are connected to common venting systems shall be located on the same floor of the *dwelling*.

   Exception: Engineered systems as provided for in Section G2427.

2. Inlets to common venting systems shall be offset such that no portion of an inlet is opposite another inlet.
3. Connectors serving *appliances* operating under a natural draft shall not be connected to any portion of a *mechanical draft system* operating under positive pressure.

M1801.12 Multiple solid fuel prohibited. A solid fuel-burning *appliance* or fireplace shall not connect to a chimney passageway venting another *appliance*.

SECTION M1802 VENT COMPONENTS

M1802.1 Draft hoods. Draft hoods shall be located in the same room or space as the *combustion air* openings for the *appliances*.

M1802.2 Vent dampers. Vent dampers shall comply with Sections M1802.2.1 and M1802.2.2.

M1802.2.1 Manually operated. Manually operated dampers shall not be installed except in connectors or chimneys serving solid fuel-burning *appliances*.

M1802.2.2 Automatically operated. Automatically operated dampers shall conform to UL 17 and be installed in accordance with the terms of their *listing* and *label*. The installation shall prevent firing of the burner when the damper is not opened to a safe position.

M1802.3 Draft regulators. Draft regulators shall be provided for oil-fired *appliances* that must be connected to a chimney. Draft regulators provided for solid fuel-burning *appliances* to reduce draft intensity shall be installed and set in accordance with the manufacturer's installation instructions.

M1802.3.1 Location. Where required, draft regulators shall be installed in the same room or enclosure as the *appliance* so that a difference in pressure will not exist between the air at the regulator and the *combustion air* supply.

M1802.4 Blocked vent switch. Oil-fired *appliances* shall be equipped with a device that will stop burner operation in the event that the venting system is obstructed. Such device shall have a manual reset and shall be installed in accordance with the manufacturer's instructions.

SECTION M1803 CHIMNEY AND VENT CONNECTORS

M1803.1 General. Connectors shall be used to connect fuel-burning *appliances* to a vertical chimney or vent except where the chimney or vent is attached directly to the *appliance*.

M1803.2 Connectors for oil and solid fuel-burning appliances. Connectors for oil and solid fuel-burning *appliances* shall be constructed of *factory-built chimney* material, Type L vent material or single-wall metal pipe having resistance to corrosion and heat and thickness not less than that of galvanized steel as specified in Table M1803.2.

TABLE M1803.2
THICKNESS FOR SINGLE-WALL METAL PIPE CONNECTORS

DIAMETER OF CONNECTOR (inches)	GALVANIZED SHEET METAL GAGE NUMBER	MINIMUM THICKNESS (inch)
Less than 6	26	0.019
6 to 10	24	0.024
Over 10 through 16	22	0.029

For SI: 1 inch = 25.4 mm.

M1803.3 Installation. Vent and chimney connectors shall be installed in accordance with the manufacturer's instructions and within the space where the *appliance* is located. *Appliances* shall be located as close as practical to the vent or chimney. Connectors shall be as short and straight as possible and installed with a slope of not less than $^1/_4$ inch (6 mm) rise per foot of run. Connectors shall be securely supported and joints shall be fastened with sheet metal screws or rivets. Devices that obstruct the flow of flue gases shall not be installed in a connector unless *listed* and *labeled* or *approved* for such installation.

M1803.3.1 Floor, ceiling and wall penetrations. A chimney connector or vent connector shall not pass through any floor or ceiling. A chimney connector or vent connector shall not pass through a wall or partition unless the connector is *listed* and *labeled* for wall pass-through, or is routed through a device *listed* and *labeled* for wall pass-through and is installed in accordance with the conditions of its *listing* and *label*. Connectors for oil-fired *appliances listed* and *labeled* for Type L vents, passing through walls or partitions shall be in accordance with the following:

1. Type L vent material for oil *appliances* shall be installed with not less than *listed* and *labeled* clearances to *combustible material*.
2. Single-wall metal pipe shall be *guarded* by a ventilated metal thimble not less than 4 inches (102 mm) larger in diameter than the vent connector. Not less than 6 inches (152 mm) of clearance shall be maintained between the thimble and combustibles.

M1803.3.2 Length. The horizontal run of an uninsulated connector to a natural draft chimney shall not exceed 75 percent of the height of the vertical portion of the chimney above the connector. The horizontal run of a *listed* connector to a natural draft chimney shall not exceed 100 percent of the height of the vertical portion of the chimney above the connector.

M1803.3.3 Size. A connector shall not be smaller than the flue collar of the *appliance*.

Exception: Where installed in accordance with the *appliance* manufacturer's instructions.

M1803.3.4 Clearance. Connectors shall be installed with clearance to combustibles as set forth in Table M1803.3.4. Reduced clearances to *combustible materials* shall be in accordance with Table M1306.2 and Figure M1306.1.

**TABLE M1803.3.4
CHIMNEY AND VENT CONNECTOR
CLEARANCES TO COMBUSTIBLE MATERIALS[a]**

TYPE OF CONNECTOR	MINIMUM CLEARANCE (inches)
Single-wall metal pipe connectors:	
Oil and solid-fuel appliances	18
Oil appliances listed for use with Type L vents	9
Type L vent piping connectors:	
Oil and solid-fuel appliances	9
Oil appliances listed for use with Type L vents	3[b]

For SI: 1 inch = 25.4 mm.

a. These minimum clearances apply to unlisted single-wall chimney and vent connectors. Reduction of required clearances is permitted as in Table M1306.2.

b. Where listed Type L vent piping is used, the clearance shall be in accordance with the vent listing.

M1803.3.5 Access. The entire length of a connector shall allow access for inspection, cleaning and replacement.

M1803.4 Connection to fireplace flue. Connection of *appliances* to chimney flues serving fireplaces shall comply with Sections M1803.4.1 through M1803.4.4.

M1803.4.1 Closure and accessibility. A noncombustible seal shall be provided below the point of connection to prevent entry of room air into the flue. Means shall be provided for access to the flue for inspection and cleaning.

M1803.4.2 Connection to factory-built fireplace flue. A different *appliance* shall not be connected to a flue serving a factory-built fireplace unless the *appliance* is specifically *listed* for such an installation. The connection shall be made in compliance with the *appliance* manufacturer's instructions.

M1803.4.3 Connection to masonry fireplace flue. A connector shall extend from the *appliance* to the flue serving a masonry fireplace to convey the flue gases directly into the flue. The connector shall be provided with access or shall be removable for inspection and cleaning of both the connector and the flue. *Listed* direct-connection devices shall be installed in accordance with their *listing*.

M1803.4.4 Size of flue. The size of the fireplace flue shall be in accordance with Section M1805.3.1.

SECTION M1804
VENTS

M1804.1 Type of vent required. *Appliances* shall be provided with a *listed* and *labeled* venting system as set forth in Table M1804.1.

**TABLE M1804.1
VENT SELECTION CHART**

VENT TYPES	APPLIANCE TYPES
Type L oil vents	Oil-burning appliances listed and labeled for venting with Type L vents
Pellet vents	Pellet fuel-burning appliances listed and labeled for use with pellet vents

M1804.2 Termination. Vent termination shall comply with Sections M1804.2.1 through M1804.2.6.

M1804.2.1 Through the roof. Vents passing through a roof shall extend through flashing and terminate in accordance with the manufacturer's installation requirements.

M1804.2.2 Decorative shrouds. Decorative shrouds shall not be installed at the termination of vents except where the shrouds are *listed* and *labeled* for use with the specific venting system and are installed in accordance with the manufacturer's instructions.

M1804.2.3 Natural draft appliances. Vents for natural draft *appliances* shall terminate not less than 5 feet (1524 mm) above the highest connected *appliance* outlet, and natural draft gas vents serving wall furnaces shall terminate at an elevation not less than 12 feet (3658 mm) above the bottom of the furnace.

M1804.2.4 Type L vent. Type L venting systems shall conform to UL 641 and shall terminate with a *listed* and *labeled* cap in accordance with the vent manufacturer's installation instructions not less than 2 feet (610 mm) above the roof and not less than 2 feet (610 mm) above any portion of the building within 10 feet (3048 mm).

M1804.2.5 Direct vent terminations. Vent terminals for direct-vent *appliances* shall be installed in accordance with the manufacturer's instructions.

M1804.2.6 Mechanical draft systems. *Mechanical draft systems* shall comply with UL 378 and shall be installed in accordance with their *listing*, the manufacturer's instructions and, except for direct-vent *appliances*, the following requirements:

1. The vent terminal shall be located not less than 3 feet (914 mm) above a forced air inlet located within 10 feet (3048 mm).
2. The vent terminal shall be located not less than 4 feet (1219 mm) below, 4 feet (1219 mm) horizontally from, or 1 foot (305 mm) above any door, window or gravity air inlet into a *dwelling*.

3. The vent termination point shall be located not closer than 3 feet (914 mm) to an interior corner formed by two walls perpendicular to each other.
4. The bottom of the vent terminal shall be located not less than 12 inches (305 mm) above finished ground level.
5. The vent termination shall not be mounted directly above or within 3 feet (914 mm) horizontally of an oil tank vent or gas meter.
6. Power exhauster terminations shall be located not less than 10 feet (3048 mm) from *lot lines* and adjacent buildings.
7. The discharge shall be directed away from the building.

M1804.3 Installation. Type L and pellet vents shall be installed in accordance with the terms of their *listing* and *label* and the manufacturer's instructions.

M1804.3.1 Size of single-appliance venting systems. An individual vent for a single *appliance* shall have a cross-sectional area equal to or greater than the area of the connector to the *appliance*, but not less than 7 square inches (4515 mm^2) except where the vent is an integral part of a *listed* and *labeled appliance*.

M1804.4 Door swing. Appliance and *equipment* vent terminals shall be located such that doors cannot swing within 12 inches (305 mm) horizontally of the vent terminals. Door stops or closers shall not be installed to obtain this clearance.

SECTION M1805 MASONRY AND FACTORY-BUILT CHIMNEYS

M1805.1 General. Masonry and factory-built chimneys shall be built and installed in accordance with Sections R1003 and R1005, respectively. Flue lining for masonry chimneys shall comply with Section R1003.11.

M1805.2 Masonry chimney connection. A chimney connector shall enter a *masonry chimney* not less than 6 inches (152 mm) above the bottom of the chimney. Where it is not possible to locate the connector entry not less than 6 inches (152 mm) above the bottom of the chimney flue, a cleanout shall be provided by installing a capped tee in the connector next to the chimney. A connector entering a *masonry chimney* shall extend through, but not beyond, the wall and shall be flush with the inner face of the liner. Connectors, or thimbles where used, shall be firmly cemented into the masonry.

M1805.3 Size of chimney flues. The effective area of a natural draft chimney flue for one *appliance* shall be not less than the area of the connector to the *appliance*. The area of chimney flues connected to more than one *appliance* shall be not less than the area of the largest connector plus 50 percent of the areas of additional chimney connectors.

Exception: Chimney flues serving oil-fired *appliances* sized in accordance with NFPA 31.

M1805.3.1 Size of chimney flue for solid-fuel appliance. Except where otherwise specified in the manufacturer's installation instructions, the cross-sectional area of a flue connected to a solid fuel-burning *appliance* shall be not less than the area of the flue collar or connector, and not larger than three times the area of the flue collar.

CHAPTER 19

SPECIAL APPLIANCES, EQUIPMENT AND SYSTEMS

User notes:

About this chapter: *Chapter 19 is specific to appliances and systems that are not related to HVAC, including cooking appliances, sauna heaters, fuel cells and hydrogen systems. Chapter 24 also applies to cooking appliances and sauna heaters.*

Code development reminder: *Code change proposals to this chapter will be considered by the IRC—Plumbing/Mechanical Code Development Committee during the 2021 (Group A) Code Development Cycle.*

SECTION M1901 RANGES AND OVENS

M1901.1 Clearances. Freestanding or built-in ranges shall have a vertical clearance above the cooking top of not less than 30 inches (762 mm) to unprotected *combustible material.* Reduced clearances are permitted in accordance with the *listing* and *labeling* of the range hoods or ovens with integral exhaust.

M1901.2 Cooking appliances. Cooking *appliances* shall be *listed* and *labeled* for household use and shall be installed in accordance with the manufacturer's instructions. The installation shall not interfere with *combustion air* or access for operation and servicing. Electric cooking *appliances* shall comply with UL 858 or UL 1026. Solid-fuel-fired fireplace stoves shall comply with UL 737. Microwave ovens shall comply with UL 923.

SECTION M1902 SAUNA HEATERS

M1902.1 Locations and protection. Sauna heaters shall be protected from accidental contact by persons with a guard of material having a low thermal conductivity, such as wood. The guard shall not have a substantial effect on the transfer of heat from the heater to the room.

M1902.2 Installation. Sauna heaters shall be installed in accordance with the manufacturer's instructions. Sauna heaters shall comply with UL 875.

M1902.3 Combustion air. *Combustion air* and venting for a nondirect vent-type heater shall be provided in accordance with Chapters 17 and 18, respectively.

M1902.4 Controls. Sauna heaters shall be equipped with a thermostat that will limit room temperature to not greater than 194°F (90°C). Where the thermostat is not an integral part of the heater, the heat-sensing element shall be located within 6 inches (152 mm) of the ceiling.

SECTION M1903 STATIONARY FUEL CELL POWER PLANTS

M1903.1 General. Stationary fuel cell power plants having a power output not exceeding 1,000 kW shall comply with ANSI/CSA America FC 1 and shall be installed in accordance with the manufacturer's instructions and NFPA 853.

SECTION M1904 GASEOUS HYDROGEN SYSTEMS

M1904.1 Installation. Gaseous hydrogen systems shall be installed in accordance with the applicable requirements of Sections M1307.4 and M1903.1, the *International Building Code,* the *International Fire Code* and the *International Fuel Gas Code.*

CHAPTER 20

BOILERS AND WATER HEATERS

User notes:

About this chapter: *Chapter 20 is specific to boilers and water heaters. The provisions of this chapter apply to appliances generally without regard to the energy source. Gas-fired boilers and water heaters are also addressed in Chapter 24; therefore, Chapters 20 and 24 both apply to such appliances.*

Code development reminder: *Code change proposals to this chapter will be considered by the IRC—Plumbing/Mechanical Code Development Committee during the 2021 (Group A) Code Development Cycle.*

SECTION M2001 BOILERS

M2001.1 Installation. In addition to the requirements of this code, the installation of boilers shall conform to the manufacturer's instructions. The manufacturer's rating data, the nameplate and operating instructions of a permanent type shall be attached to the boiler. Boilers shall have their controls set, adjusted and tested by the installer. A complete control diagram together with complete boiler operating instructions shall be furnished by the installer. Solid and liquid fuel-burning boilers shall be provided with *combustion air* as required by Chapter 17.

M2001.1.1 Standards. Packaged oil-fired boilers shall be *listed* and *labeled* in accordance with UL 726. Packaged electric boilers shall be *listed* and *labeled* in accordance with UL 834. Solid fuel-fired boilers shall be *listed* and *labeled* in accordance with UL 2523. Boilers shall be designed, constructed and certified in accordance with the *ASME Boiler and Pressure Vessel Code*, Section I or IV. Controls and safety devices for boilers with fuel input ratings of 12,500,000 Btu/hr (3663 kW) or less shall meet the requirements of ASME CSD-1. Gas-fired boilers shall conform to the requirements listed in Chapter 24.

M2001.2 Clearance. Boilers shall be installed in accordance with their *listing* and *label*.

M2001.3 Valves. Every boiler or modular boiler shall have a shutoff valve in the supply and return piping. For multiple boiler or multiple modular boiler installations, each boiler or modular boiler shall have individual shutoff valves in the supply and return piping.

Exception: Shutoff valves are not required in a system having a single low-pressure steam boiler.

M2001.4 Flood-resistant installation. In flood hazard areas established in Table R301.2, boilers, water heaters and their control systems shall be located or installed in accordance with Section R322.1.6.

SECTION M2002 OPERATING AND SAFETY CONTROLS

M2002.1 Safety controls. Electrical and mechanical operating and safety controls for boilers shall be *listed* and *labeled*.

M2002.2 Hot water boiler gauges. Every hot water boiler shall have a pressure gauge and a temperature gauge, or combination pressure and temperature gauge. The gauges shall indicate the temperature and pressure within the normal range of the system's operation.

M2002.3 Steam boiler gauges. Every steam boiler shall have a water-gauge glass and a pressure gauge. The pressure gauge shall indicate the pressure within the normal range of the system's operation. The gauge glass shall be installed so that the midpoint is at the normal water level.

M2002.4 Pressure relief valve. Boilers shall be equipped with pressure relief valves with minimum rated capacities for the equipment served. Pressure relief valves shall be set at the maximum rating of the boiler. Discharge shall be piped to drains by gravity to within 18 inches (457 mm) of the floor or to an open receptor.

M2002.5 Boiler low-water cutoff. Steam and hot water boilers shall be protected with a low-water cutoff control.

Exception: A low-water cutoff is not required for coil-type and water-tube-type boilers that require forced circulation of water through the boiler and that are protected with a flow-sensing control.

M2002.6 Operation. Low-water cutoff controls and flow-sensing controls required by Section M2002.5 shall automatically stop the combustion operation of the *appliance* when the water level drops below the lowest safe water level as established by the manufacturer or when the water circulation flow is less than that required for safe operation of the *appliance*, respectively.

SECTION M2003 EXPANSION TANKS

M2003.1 General. Hot water boilers shall be provided with expansion tanks. Nonpressurized expansion tanks shall be securely fastened to the structure or boiler and supported to carry twice the weight of the tank filled with water. Provisions shall be made for draining nonpressurized tanks without emptying the system.

M2003.1.1 Pressurized expansion tanks. Pressurized expansion tanks shall be consistent with the volume and capacity of the system. Tanks shall be capable of withstanding a hydrostatic test pressure of two and one-half times the allowable working pressure of the system.

M2003.2 Minimum capacity. The minimum capacity of expansion tanks shall be determined from Table M2003.2.

TABLE M2003.2
EXPANSION TANK MINIMUM CAPACITY[a]
FOR FORCED HOT-WATER SYSTEMS

SYSTEM VOLUME[b] (gallons)	PRESSURIZED DIAPHRAGM TYPE	NONPRESSURIZED TYPE
10	1.0	1.5
20	1.5	3.0
30	2.5	4.5
40	3.0	6.0
50	4.0	7.5
60	5.0	9.0
70	6.0	10.5
80	6.5	12.0
90	7.5	13.5
100	8.0	15.0

For SI: 1 gallon = 3.785 L, 1 pound per square inch gauge = 6.895 kPa, °C = [(°F) – 32]/1.8.

a. Based on average water temperature of 195°F, fill pressure of 12 psig and an operating pressure of not greater than 30 psig.

b. System volume includes volume of water in boiler, convectors and piping, not including the expansion tank.

SECTION M2004 WATER HEATERS USED FOR SPACE HEATING

M2004.1 General. Water heaters used to supply both potable hot water and hot water for space heating shall be installed in accordance with this chapter, Chapter 24, Chapter 28 and the manufacturer's instructions.

SECTION M2005 WATER HEATERS

M2005.1 General. Water heaters shall be installed in accordance with Chapter 28, the manufacturer's instructions and the requirements of this code. Water heaters installed in an attic shall comply with the requirements of Section M1305.1.2. Gas-fired water heaters shall comply with the requirements in Chapter 24. Domestic electric water heaters shall comply with UL 174. Oiled-fired water heaters shall comply with UL 732. Solar thermal water heating systems shall comply with Chapter 23 and ICC 900/SRCC 300. Solid fuel-fired water heaters shall comply with UL 2523.

M2005.2 Prohibited locations. Fuel-fired water heaters shall not be installed in a room used as a storage closet. Water heaters located in a bedroom or bathroom shall be installed in a sealed enclosure so that *combustion air* will not be taken from the living space. Installation of direct-vent water heaters within an enclosure is not required.

M2005.2.1 Water heater access. Access to water heaters that are located in an *attic* or underfloor *crawl space* is permitted to be through a closet located in a sleeping room or bathroom where *ventilation* of those spaces is in accordance with this code.

M2005.3 Electric water heaters. Electric water heaters shall be installed in accordance with the applicable provisions of Chapters 34 through 43.

M2005.4 Supplemental water-heating devices. Potable water-heating devices that use refrigerant-to-water heat exchangers shall be *approved* and installed in accordance with the manufacturer's instructions.

SECTION M2006 POOL HEATERS

M2006.1 General. Pool and spa heaters shall be installed in accordance with the manufacturer's installation instructions. Oil-fired pool heaters shall comply with UL 726. Electric pool and spa heaters shall comply with UL 1261. Pool and spa heat pump water heaters shall comply with UL 1995, UL/CSA/ANCE 60335-2-40 or CSA C22.2 No. 236.

Exception: Portable residential spas and portable residential exercise spas shall comply with UL 1563 or CSA C22.2 No. 218.1.

M2006.2 Clearances. The clearances shall not interfere with *combustion air*, draft hood or flue terminal relief, or accessibility for servicing.

M2006.3 Bypass valves. Where an integral bypass system is not provided as a part of the pool heater, a bypass line and valve shall be installed between the inlet and outlet piping for use in adjusting the flow of water through the heater.

CHAPTER 21
HYDRONIC PIPING

User notes:

About this chapter: *Chapter 21 is specific to hydronic piping, which includes steam, hot water and ground-source heat-pump system loop piping. This chapter addresses piping materials, joining methods, support, protection of the structure, testing, protection of potable water and general installation requirements.*

Code development reminder: *Code change proposals to this chapter will be considered by the IRC—Plumbing/Mechanical Code Development Committee during the 2021 (Group A) Code Development Cycle.*

SECTION M2101
HYDRONIC PIPING SYSTEMS INSTALLATION

M2101.1 General. Hydronic piping shall conform to Table M2101.1. *Approved* piping, valves, fittings and connections shall be installed in accordance with the manufacturer's instructions. Pipe and fittings shall be rated for use at the operating temperature and pressure of the hydronic system. Used pipe, fittings, valves or other materials shall be free of foreign materials.

M2101.2 System drain down. Hydronic piping systems shall be installed to permit draining of the system. Where the system drains to the plumbing drainage system, the installation shall conform to the requirements of Chapters 25 through 32 of this code.

Exception: The buried portions of systems embedded underground or under floors.

M2101.3 Protection of potable water. The potable water system shall be protected from backflow in accordance with the provisions listed in Section P2902.

M2101.4 Pipe penetrations. Openings through concrete or masonry building elements shall be sleeved.

M2101.5 Contact with building material. A hydronic piping system shall not be in direct contact with any building material that causes the piping material to degrade or corrode.

M2101.6 Drilling and notching. Wood-framed structural members shall be drilled, notched or altered in accordance with the provisions of Sections R502.8, R602.6, R602.6.1 and R802.7. Holes in load-bearing members of cold-formed steel *light-frame construction* shall be permitted only in accordance with Sections R505.2.6, R603.2.6 and R804.2.6. In accordance with the provisions of Sections R505.3.5, R603.3.4 and R804.3.3, cutting and notching of flanges and lips of load-bearing members of cold-formed steel *light-frame construction* shall not be permitted. Structural insulated panels (SIPs) shall be drilled and notched or altered in accordance with the provisions of Section R610.7.

M2101.7 Prohibited tee applications. Fluid in the supply side of a hydronic system shall not enter a tee fitting through the branch opening.

M2101.8 Expansion, contraction and settlement. Piping shall be installed so that piping, connections and *equipment* shall not be subjected to excessive strains or stresses. Provisions shall be made to compensate for expansion, contraction, shrinkage and structural settlement.

M2101.9 Piping support. Hangers and supports shall be of material of sufficient strength to support the piping, and shall be fabricated from materials compatible with the piping material. Piping shall be supported at intervals not exceeding the spacing specified in Table M2101.9.

TABLE M2101.9
HANGER SPACING INTERVALS

PIPING MATERIAL	MAXIMUM HORIZONTAL SPACING (feet)	MAXIMUM VERTICAL SPACING (feet)
ABS	4	10[a]
CPVC ≤ 1-inch pipe or tubing	3	5[a]
CPVC ≥ $1^1/_4$ inches	4	10[a]
Copper or copper-alloy pipe	12	10
Copper or copper-alloy tubing	6	10
PB pipe or tubing	2.67	4
PE pipe or tubing	2.67	4
PE-RT ≤ 1 inch	2.67	10[a]
PE-RT ≥ $1^1/_4$ inches	4	10[a]
PEX tubing ≤ 1 inch	2.67	4
PEX tubing ≥ $1^1/_4$ inches	4	10[a]
PP < 1-inch pipe or tubing	2.67	4
PP > $1^1/_4$ inches	4	10[a]
PVC	4	10[a]
Steel pipe	12	15
Steel tubing	8	10

For SI: 1 inch = 25.4 mm, 1 foot = 304.8 mm.

a. For sizes 2 inches and smaller, a guide shall be installed midway between required vertical supports. Such guides shall prevent pipe movement in a direction perpendicular to the axis of the pipe.

TABLE M2101.1
HYDRONIC PIPING AND FITTING MATERIALS

MATERIAL	USE CODE[a]	STANDARD[b]	JOINTS	NOTES
Acrylonitrile butadiene styrene (ABS) plastic pipe	1, 5	ASTM D1527, ASTM F2806, ASTM F2969	Solvent cement joints	—
Chlorinated poly (vinyl chloride) (CPVC) pipe and tubing	1, 2, 3	ASTM D2846	Solvent cement joints, compression joints and threaded adapters	—
Copper and copper-alloy pipe	1	ASTM B42, ASTM B43, ASTM B302	Brazed, soldered and mechanical fittings threaded, welded and flanged	—
Copper and copper-alloy tubing (Type K, L or M)	1, 2	ASME B16.51, ASTM B75, ASTM B88, ASTM B135, ASTM B251, ASTM B306	Brazed, soldered, press-connected and flared mechanical fittings	Joints embedded in concrete shall be brazed
Cross-linked polyethylene (PEX)	1, 2, 3	ASTM F876; ASTM F3253	(See PEX fittings)	Install in accordance with manufacturer's instructions
Cross linked polyethylene/aluminum/cross-linked polyethylene (PEX-AL-PEX) pressure pipe	1, 2	ASTM F1281 or CAN/CSA B137.10	Mechanical, crimp/insert	Install in accordance with manufacturer's instructions
PEX fittings	—	ASTM F877, ASTM F1807, ASTM F1960, ASTM F2098, ASTM F2159, ASTM F2735, ASTM F3253	Copper crimp/insert fittings, cold expansion fittings, stainless steel clamp, insert fittings	Install in accordance with manufacturer's instructions
Polybutylene (PB) pipe and tubing	1, 2, 3	ASTM D3309	Heat-fusion, crimp/insert and compression	Joints in concrete shall be heat-fused
Polyethylene/aluminum/polyethylene (PE-AL-PE) pressure pipe	1, 2, 3	ASTM F1282, CSA B137.9	Mechanical, crimp/insert	—
Polypropylene (PP)	1, 2, 3	ISO 15874, ASTM F2389	Heat-fusion joints, mechanical fittings, threaded adapters, compression joints	—
Raised temperature polyethylene (PE-RT)	1, 2, 3	ASTM F2623, ASTM F2769, CSA B137.18	Copper crimp/insert fitting, stainless steel clamp, insert fittings	—
Raised temperature polyethylene (PE-RT) fittings	1, 2, 3	ASTM D3261, ASTM F1807, ASTM F2098, ASTM F2159, ASTM F2735, ASTM F2769, CSA B137.18	Copper crimp/insert fitting, stainless steel clamp, insert fittings	—
Steel pipe	1, 2	ASTM A53, ASTM A106	Brazed, welded, threaded, flanged and mechanical fittings	Joints in concrete shall be welded. Galvanized pipe shall not be welded or brazed.
Steel tubing	1	ASTM A254	Mechanical fittings, welded	—

For SI: °C = [(°F) − 32]/1.8.

a. Use code:
1. Above ground.
2. Embedded in radiant systems.
3. Temperatures below 180°F only.
4. Low-temperature (below 130°F) applications only.
5. Temperatures below 160°F only.

b. Standards as listed in Chapter 44.

M2101.10 Tests. Hydronic piping systems shall be tested hydrostatically at a pressure of one and one-half times the maximum system design pressure, but not less than 100 pounds per square inch (689 kPa). The duration of each test shall be not less than 15 minutes.

Exception: For PEX piping systems, testing with a compressed gas shall be an alternative to hydrostatic testing where compressed air or other gas pressure testing is specifically authorized by all of the manufacturers' instructions for the PEX pipe and fittings products installed at the time the system is being tested, and compressed air or other gas testing is not otherwise prohibited by applicable codes, laws, or regulations outside of this code.

M2101.11 Used materials. Used pipe, fittings, valves and other materials shall not be reused in hydronic systems.

M2101.12 Material rating. Pipe and tubing shall be rated for the operating temperature and pressure of the system. Fittings shall be suitable for the pressure applications and recommended by the manufacturer for use with the pipe and tubing material installed. Where used underground, materials shall be suitable for burial.

M2101.13 Joints and connections. Joints and connections shall be of an *approved* type. Joints and connections shall be tight for the pressure of the system. Joints used underground shall be *approved* for such applications.

M2101.13.1 Joints between different piping materials. Joints between different piping materials shall be made with *approved* transition fittings.

M2101.14 Preparation of pipe ends. Pipe shall be cut square and shall be free of burrs and obstructions. Pipe ends shall have full-bore openings and shall be prepared in accordance with the pipe manufacturer's instructions.

M2101.15 Joint preparation and installation. Where required by Sections M2101.16 through M2101.18, the preparation and installation of mechanical and thermoplastic-welded joints shall comply with Sections M2101.15.1 and M2101.15.2.

M2101.15.1 Mechanical joints. Mechanical joints shall be installed in accordance with the manufacturer's instructions.

M2101.15.2 Thermoplastic-welded joints. Joint surfaces for thermoplastic-welded joints shall be cleaned by an *approved* procedure. Joints shall be welded in accordance with the manufacturer's instructions.

M2101.16 CPVC plastic pipe. Joints between CPVC plastic pipe or fittings shall be solvent cemented in accordance with Section P2906.9.1.2. Threaded joints between fittings and CPVC plastic pipe shall be in accordance with Section M2101.16.1.

M2101.16.1 Threaded joints. Threads shall conform to ASME B1.20.1 The pipe shall be Schedule 80, 40 or heavier plastic pipe and shall be threaded with dies specifically designed for plastic pipe. Thread lubricant, pipe-joint compound or tape shall be applied on the male threads only and shall be *approved* for application on the piping material.

M2101.17 Cross-linked polyethylene (PEX) plastic tubing. Joints between cross-linked polyethylene plastic tubing and fittings shall comply with Sections M2101.17.1 and M2101.17.2. Mechanical joints shall comply with Section M2101.15.1.

M2101.17.1 Compression-type fittings. Where compression-type fittings include inserts and ferrules or O-rings, the fittings shall be installed without omitting the inserts and ferrules or O-rings.

M2101.17.2 Plastic-to-metal. Solder joints in a metal pipe shall not occur within 18 inches (457 mm) of a transition from such metal pipe to plastic pipe or tubing.

M2101.18 Polyethylene plastic pipe and tubing. Joints between polyethylene plastic pipe and tubing or fittings for systems shall be heat-fusion joints complying with Section M2101.18.1, electrofusion joints complying with Section M2101.18.2, or stab-type insertion joints complying with Section M2101.18.3.

M2101.18.1 Heat-fusion joints. Joints shall be of the socket-fusion, saddle-fusion or butt-fusion type, and joined in accordance with ASTM D2657. Joint surfaces shall be clean and free from moisture. Joint surfaces shall be heated to melt temperatures and joined. The joint shall remain undisturbed until cool. Fittings shall be manufactured in accordance with ASTM D2683 or ASTM D3261.

M2101.18.2 Electrofusion joints. Joints shall be of the electrofusion type. Joint surfaces shall be clean and free from moisture, and scoured to expose virgin resin. Joint surfaces shall be heated to melt temperatures for the period of time specified by the manufacturer. The joint shall remain undisturbed until cool. Fittings shall be manufactured in accordance with ASTM F1055.

M2101.18.3 Stab-type insert fittings. Joint surfaces shall be clean and free from moisture. Pipe ends shall be chamfered and inserted into the fittings to full depth. Fittings shall be manufactured in accordance with ASTM F1924.

M2101.19 Polypropylene (PP) plastic. Joints between PP plastic pipe and fittings shall comply with Sections M2101.19.1 and M2101.19.2.

M2101.19.1 Heat-fusion joints. Heat-fusion joints for polypropylene (PP) pipe and tubing joints shall be installed with socket-type heat-fused polypropylene fittings, electrofusion polypropylene fittings or by butt fusion. Joint surfaces shall be clean and free from moisture. The joint shall remain undisturbed until cool. Joints shall be made in accordance with ASTM F2389.

M2101.19.2 Mechanical and compression sleeve joints. Mechanical and compression sleeve joints shall be installed in accordance with the manufacturer's instructions.

M2101.20 Raised temperature polyethylene (PE-RT) plastic tubing. Joints between raised temperature polyethylene tubing and fittings shall comply with Sections M2101.20.1 through M2101.20.4. Mechanical joints shall comply with Section M2101.15.1.

M2101.20.1 Compression-type fittings. Where compression-type fittings include inserts and ferrules or O-rings, the fittings shall be installed without omitting the inserts and ferrules or O-rings.

M2101.20.2 PE-RT-to-metal connections. Solder joints in a metal pipe shall not occur within 18 inches (457 mm) of a transition from such metal pipe to PE-RT pipe or tubing.

M2101.20.3 Heat-fusion joints. Heat-fusion joints shall be of the socket-fusion, saddle-fusion or butt-fusion type, and shall be joined in accordance with ASTM D2657. Joint surfaces shall be clean and free from moisture. Joint surfaces shall be heated to melt temperatures and joined. The joint shall remain undisturbed until cool. Fittings shall be manufactured in accordance with ASTM D2683 or ASTM D3261.

M2101.20.4 Electrofusion joints. Joints shall be of the electrofusion type. Joint surfaces shall be clean and free from moisture and scoured to expose virgin resin. Joint surfaces shall be heated to melt temperatures for the period of time specified by the manufacturer and joined. The joint shall remain undisturbed until cool. Fittings shall be manufactured in accordance with ASTM F1055.

M2101.21 PVC plastic pipe. Joints between PVC plastic pipe or fittings shall be solvent cemented in accordance with Section P2906.9.1.4. Threaded joints between fittings and PVC plastic pipe shall be in accordance with Section M2101.16.1.

M2101.22 Shutoff valves. Shutoff valves shall be installed in ground-source loop piping systems in the locations indicated in Sections M2101.22.1 through M2101.22.6.

M2101.22.1 Heat exchangers. Shutoff valves shall be installed on the supply and return sides of a heat exchanger.

Exception: Shutoff valves shall not be required where heat exchangers are integral with a boiler or are a component of a manufacturer's boiler and heat exchanger packaged unit and are capable of being isolated from the hydronic system by the supply and return valves required by Section M2001.3.

M2101.22.2 Central systems. Shutoff valves shall be installed on the building supply and return of a central utility system.

M2101.22.3 Pressure vessels. Shutoff valves shall be installed on the connection to any pressure vessel.

M2101.22.4 Pressure-reducing valves. Shutoff valves shall be installed on both sides of a pressure-reducing valve.

M2101.22.5 Equipment and appliances. Shutoff valves shall be installed on connections to mechanical equipment and *appliances*. This requirement does not apply to components of ground-source loop systems such as pumps, air separators, metering devices, and similar equipment.

M2101.22.6 Expansion tanks. Shutoff valves shall be installed at connections to nondiaphragm-type expansion tanks.

M2101.23 Reduced pressure. A pressure relief valve shall be installed on the low-pressure side of a hydronic piping system that has been reduced in pressure. The relief valve shall be set at the maximum pressure of the system design. The valve shall be installed in accordance with Section M2002.

M2101.24 Installation. Piping, valves, fittings and connections shall be installed in accordance with the manufacturer's instructions.

M2101.25 Protection of potable water. Where hydronic systems have a connection to a potable water supply, the potable water system shall be protected from backflow in accordance with Section P2902.

M2101.26 Pipe penetrations. Openings for pipe penetrations in walls, floors and ceilings shall be larger than the penetrating pipe. Openings through concrete or masonry building elements shall be sleeved. The annular space surrounding pipe penetrations shall be protected in accordance with Section P2606.1.

M2101.27 Clearance from combustibles. A pipe in a piping system having an exterior surface temperature exceeding 250°F (121°C) shall have a clearance of not less than 1 inch (25 mm) from *combustible materials*.

M2101.28 Contact with building material. A piping system shall not be in direct contact with building materials that cause the piping or fitting material to degrade or corrode, or that interfere with the operation of the system.

M2101.29 Strains and stresses. Piping shall be installed so as to prevent detrimental strains and stresses in the pipe. Provisions shall be made to protect piping from damage resulting from expansion, contraction and structural settlement. Piping shall be installed so as to avoid structural stresses or strains within building components.

M2101.29.1 Flood hazard. Piping located in a flood hazard area shall be capable of resisting hydrostatic and hydrodynamic loads and stresses, including the effects of buoyancy, during the occurrence of flooding to the design flood elevation.

M2101.30 Chemical compatibility. Antifreeze and other materials used in the system shall be chemically compatible with the pipe, tubing, fittings and mechanical systems.

M2101.31 Makeup water. The transfer fluid shall be compatible with the makeup water supplied to the system.

SECTION M2102
BASEBOARD CONVECTORS

M2102.1 General. Baseboard convectors shall be installed in accordance with the manufacturer's instructions. Convectors shall be supported independently of the hydronic piping.

SECTION M2103
FLOOR HEATING SYSTEMS

M2103.1 Piping materials. Piping for embedment in concrete or gypsum materials shall be standard-weight steel pipe, copper and copper-alloy pipe and tubing, cross-linked polyethylene/aluminum/cross-linked polyethylene (PEX-AL-PEX) pressure pipe, chlorinated polyvinyl chloride (CPVC), cross-linked polyethylene (PEX) tubing, polyethylene of raised temperature (PE-RT) or polypropylene (PP) with a rating of not less than 80 pounds per square inch at 180°F (552 kPa at 82°C).

M2103.2 Thermal barrier required. Radiant floor heating systems shall have a thermal barrier in accordance with Sections M2103.2.1 and M2103.2.2. Insulation *R*-values for slab-on-grade and suspended floor installations shall be in accordance with Chapter 11.

Exception: Insulation shall not be required in engineered systems where it can be demonstrated that the insulation will decrease the efficiency or have a negative effect on the installation.

M2103.2.1 Thermal break required. A thermal break consisting of asphalt expansion joint materials or similar insulating materials shall be provided at a point where a heated slab meets a foundation wall or other conductive slab.

M2103.2.2 Thermal barrier material marking. Insulating materials used in thermal barriers shall be installed so that the manufacturer's *R*-value mark is readily observable upon inspection.

M2103.3 Piping joints. Copper and copper-alloy systems shall be soldered, brazed, or press connected. Soldering shall be in accordance with ASTM B828. Fluxes for soldering shall be in accordance with ASTM B813. Brazing fluxes shall be in accordance with AWS A5.31. Press-connect joints shall be in accordance with ASME B16.51. Piping joints that are embedded shall be installed in accordance with the following requirements:

1. Steel pipe joints shall be welded.
2. Copper tubing shall be joined by brazing complying with Section P3003.6.1.
3. Polybutylene pipe and tubing joints shall be installed with socket-type heat-fused polybutylene fittings.
4. CPVC tubing shall be joined using solvent cement joints.
5. Polypropylene pipe and tubing joints shall be installed with socket-type heat-fused polypropylene fittings.
6. Cross-linked polyethylene (PEX) tubing shall be joined using cold expansion, insert or compression fittings.
7. Raised temperature polyethylene (PE-RT) tubing shall be joined using insert or compression fittings.

M2103.4 Testing. Piping or tubing to be embedded shall be tested by applying a hydrostatic pressure of not less than 100 psi (690 kPa). The pressure shall be maintained for 30 minutes, during which the joints shall be visually inspected for leaks.

SECTION M2104
LOW TEMPERATURE PIPING

M2104.1 Piping materials. Low temperature piping for embedment in concrete or gypsum materials shall be as indicated in Table M2101.1.

M2104.2 Piping joints. Piping joints that are embedded, other than those in Section M2103.3, shall comply with the following requirements:

1. Cross-linked polyethylene (PEX) tubing shall be installed in accordance with the manufacturer's instructions.
2. Polyethylene tubing shall be installed with heat-fusion joints.
3. Polypropylene (PP) tubing shall be installed in accordance with the manufacturer's instructions.
4. Raised temperature polyethylene (PE-RT) shall be installed in accordance with the manufacturer's instructions.

M2104.3 Raised temperature polyethylene (PE-RT) plastic tubing. Joints between raised temperature polyethylene tubing and fittings shall conform to Sections M2104.3.1 through M2104.3.3. Mechanical joints shall be installed in accordance with the manufacturer's instructions.

M2104.3.1 Compression-type fittings. Where compression-type fittings include inserts and ferrules or O-rings, the fittings shall be installed without omitting such inserts and ferrules or O-rings.

M2104.3.2 PE-RT-to-metal connections. Solder joints in a metal pipe shall not occur within 18 inches (457 mm) of a transition from such metal pipe to PE-RT pipe.

M2104.3.3 PE-RT insert fittings. PE-RT insert fittings shall be installed in accordance with the manufacturer's instructions.

M2104.4 Polyethylene/aluminum/polyethylene (PE-AL-PE) pressure pipe. Joints between polyethylene/aluminum/polyethylene pressure pipe and fittings shall conform to Sections M2104.4.1 and M2104.4.2. Mechanical joints shall be installed in accordance with the manufacturer's instructions.

M2104.4.1 Compression-type fittings. Where compression-type fittings include inserts and ferrules or O-rings, the fittings shall be installed without omitting such inserts and ferrules or O-rings.

M2104.4.2 PE-AL-PE-to-metal connections. Solder joints in a metal pipe shall not occur within 18 inches (457 mm) of a transition from such metal pipe to PE-AL-PE pipe.

SECTION M2105 GROUND-SOURCE HEAT-PUMP SYSTEM LOOP PIPING

M2105.1 Plastic ground-source heat-pump loop piping. Plastic piping and tubing material used in water-based ground-source heat-pump ground-loop systems shall conform to the standards specified in this section.

M2105.2 Used materials. Reused pipe, fittings, valves, and other materials shall not be used in ground-source heat-pump loop systems.

M2105.3 Material rating. Pipe and tubing shall be rated for the operating temperature and pressure of the ground-source heat-pump loop system. Fittings shall be suitable for the pressure applications and recommended by the manufacturer for installation with the pipe and tubing material installed. Where used underground, materials shall be suitable for burial.

M2105.4 Piping and tubing materials standards. Ground-source heat-pump ground-loop pipe and tubing shall conform to the standards listed in Table M2105.4.

M2105.5 Fittings. Ground-source heat-pump pipe fittings shall be *approved* for installation with the piping materials to be installed, shall conform to the standards listed in Table M2105.5 and, where installed underground, shall be suitable for burial.

M2105.6 Joints and connections. Joints and connections shall be of an *approved* type. Joints and connections shall be tight for the pressure of the ground-source loop system. Joints used underground shall be *approved* for such applications.

M2105.6.1 Joints between different piping materials. Joints between different piping materials shall be made with *approved* transition fittings.

M2105.7 Preparation of pipe ends. Pipe shall be cut square and shall be free of burrs and obstructions. Pipe ends shall have full-bore openings and shall be prepared in accordance with the pipe manufacturer's instructions.

M2105.8 Joint preparation and installation. Where required by Sections M2105.9 through M2105.11, the preparation and installation of mechanical and thermoplastic-welded joints shall comply with Sections M2105.8.1 and M2105.8.2.

M2105.8.1 Mechanical joints. Mechanical joints shall be installed in accordance with the manufacturer's instructions.

TABLE M2105.4 GROUND-SOURCE LOOP PIPE

MATERIAL	STANDARD
Chlorinated polyvinyl chloride (CPVC)	ASTM D2846; ASTM F437; ASTM F438; ASTM F439; ASTM F441; ASTM F442; CSA B137.6
Cross-linked polyethylene (PEX)	ASTM F876; CSA B137.5; ANSI/CSA/IGSHPA C448; NSF 358-3
High-density polyethylene (HDPE)	ASTM D2737; ASTM D3035; ASTM F714; AWWA C901; CSA B137.1; ANSI/CSA/IGSHPA C448; NSF 358-1
Polyethylene/aluminum/polyethylene (PE-AL-PE) pressure pipe	ASTM F1282; AWWA C903; CSA B137.9
Polypropylene (PP-R)	ASTM F2389; CSA B137.11; NSF 358-2
Polyvinyl chloride (PVC)	ASTM D1785; ASTM D2241; CSA B137.3
Raised temperature polyethylene (PE-RT)	ASTM F2623; ASTM F2769; CSA B137.18; ANSI/CSA/IGSHPA C448; NSF 358-4

TABLE M2105.5 GROUND-SOURCE LOOP PIPE FITTINGS

PIPE MATERIAL	STANDARD
Chlorinated polyvinyl chloride (CPVC)	ASTM D2846; ASTM F437; ASTM F438; ASTM F439; ASTM F1970; CSA B137.6
Cross-linked polyethylene (PEX)	ASTM F877; ASTM F1807; ASTM F1960; ASTM F2080; ASTM F2159; ASTM F2434; CSA B137.5; ANSI/CSA/IGSHPA C448; NSF 358-3
High-density polyethylene (HDPE)	ASTM D2683; ASTM D3261; ASTM F1055; CSA B137.1; ANSI/CSA/IGSHPA C448; NSF 358-1
Polyethylene/aluminum/polyethylene (PE-AL-PE)	ASTM F1282; ASTM F2434; CSA B137.9
Polypropylene (PP-R)	ASTM F2389; CSA B137.11; NSF 358-2
Polyvinyl chloride (PVC)	ASTM D2464; ASTM D2466; ASTM D2467; ASTM F1970; CSA B137.2; CSA B137.3
Raised temperature polyethylene (PE-RT)	ASTM D2683; ASTM D3261; ASTM F1055; ASTM F1807; ASTM F2098; ASTM F2159; ASTM F2735; ASTM F2769; CSA B137.1; CSA B137.18; ANSI/CSA/IGSHPA C448; NSF 358-4

M2105.8.2 Thermoplastic-welded joints. Joint surfaces for thermoplastic-welded joints shall be cleaned by an *approved* procedure. Joints shall be welded in accordance with the manufacturer's instructions.

M2105.9 CPVC plastic pipe. Joints between CPVC plastic pipe or fittings shall be solvent-cemented in accordance with Section P2906.9.1.2. Threaded joints between fittings and CPVC plastic pipe shall be in accordance with Section M2105.9.1.

M2105.9.1 Threaded joints. Threads shall conform to ASME B1.20.1. The pipe shall be Schedule 80 or heavier plastic pipe and shall be threaded with dies specifically designed for plastic pipe. Thread lubricant, pipe-joint compound or tape shall be applied on the male threads only and shall be *approved* for application on the piping material.

M2105.10 Cross-linked polyethylene (PEX) plastic tubing. Joints between cross-linked polyethylene plastic tubing and fittings shall comply with Sections M2105.10.1 and M2105.10.2. Mechanical joints shall comply with Section M2105.8.1.

M2105.10.1 Compression-type fittings. Where compression-type fittings include inserts and ferrules or O-rings, the fittings shall be installed without omitting the inserts and ferrules or O-rings.

M2105.10.2 Plastic-to-metal connections. Solder joints in a metal pipe shall not occur within 18 inches (457 mm) of a transition from such metal pipe to plastic pipe or tubing.

M2105.11 Polyethylene plastic pipe and tubing. Joints between polyethylene plastic pipe and tubing or fittings for ground-source heat-pump loop systems shall be heat-fusion joints complying with Section M2105.11.1, electrofusion joints complying with Section M2105.11.2, or stab-type insertion joints complying with Section M2105.11.3.

M2105.11.1 Heat-fusion joints. Joints shall be of the socket-fusion, saddle-fusion or butt-fusion type, and joined in accordance with ASTM D2657. Joint surfaces shall be clean and free from moisture. Joint surfaces shall be heated to melt temperatures and joined. The joint shall remain undisturbed until cool. Fittings shall be manufactured in accordance with ASTM D2683 or ASTM D3261.

M2105.11.2 Electrofusion joints. Joints shall be of the electrofusion type. Joint surfaces shall be clean and free from moisture, and scoured to expose virgin resin. Joint surfaces shall be heated to melt temperatures for the period of time specified by the manufacturer. The joint shall remain undisturbed until cool. Fittings shall be manufactured in accordance with ASTM F1055.

M2105.11.3 Stab-type insert fittings. Joint surfaces shall be clean and free from moisture. Pipe ends shall be chamfered and inserted into the fittings to full depth. Fittings shall be manufactured in accordance with ASTM F1924.

M2105.12 Polypropylene (PP) plastic. Joints between PP plastic pipe and fittings shall comply with Sections M2105.12.1 and M2105.12.2.

M2105.12.1 Heat-fusion joints. Heat-fusion joints for polypropylene (PP) pipe and tubing joints shall be installed with socket-type heat-fused polypropylene fittings, electrofusion polypropylene fittings or by butt fusion. Joint surfaces shall be clean and free from moisture. The joint shall remain undisturbed until cool. Joints shall be made in accordance with ASTM F2389.

M2105.12.2 Mechanical and compression sleeve joints. Mechanical and compression sleeve joints shall be installed in accordance with the manufacturer's instructions.

M2105.13 Raised temperature polyethylene (PE-RT) plastic tubing. Joints between raised temperature polyethylene tubing and fittings shall comply with Sections M2105.13.1 through M2105.13.4. Mechanical joints shall comply with Section M2105.8.1.

M2105.13.1 Compression-type fittings. Where compression-type fittings include inserts and ferrules or O-rings, the fittings shall be installed without omitting the inserts and ferrules or O-rings.

M2105.13.2 PE-RT-to-metal connections. Solder joints in a metal pipe shall not occur within 18 inches (457 mm) of a transition from such metal pipe to PE-RT pipe or tubing.

M2105.13.3 Heat-fusion joints. Heat-fusion joints shall be of the socket-fusion, saddle-fusion or butt-fusion type, and shall be joined in accordance with ASTM D2657. Joint surfaces shall be clean and free from moisture. Joint surfaces shall be heated to melt temperatures and joined. The joint shall remain undisturbed until cool. Fittings shall be manufactured in accordance with ASTM D2683 or ASTM D3261.

M2105.13.4 Electrofusion joints. Joints shall be of the electrofusion type. Joint surfaces shall be clean and free from moisture and scoured to expose virgin resin. Joint surfaces shall be heated to melt temperatures for the period of time specified by the manufacturer and joined. The joint shall remain undisturbed until cool. Fittings shall be manufactured in accordance with ASTM F1055.

M2105.14 PVC plastic pipe. Joints between PVC plastic pipe or fittings shall be solvent-cemented in accordance with Section P2906.9.1.4. Threaded joints between fittings and PVC plastic pipe shall be in accordance with Section M2105.9.1.

M2105.15 Shutoff valves. Shutoff valves shall be installed in ground-source loop piping systems in the locations indicated in Sections M2105.15.1 through M2105.15.6.

M2105.15.1 Heat exchangers. Shutoff valves shall be installed on the supply and return side of a heat exchanger.

Exception: Shutoff valves shall not be required where heat exchangers are integral with a boiler or are a component of a manufacturer's boiler and heat exchanger packaged unit and are capable of being isolated from the hydronic system by the supply and return valves required by Section M2001.3.

M2105.15.2 Central systems. Shutoff valves shall be installed on the building supply and return of a central utility system.

M2105.15.3 Pressure vessels. Shutoff valves shall be installed on the connection to any pressure vessel.

M2105.15.4 Pressure-reducing valves. Shutoff valves shall be installed on both sides of a pressure-reducing valve.

M2105.15.5 Equipment and appliances. Shutoff valves shall be installed on connections to mechanical equipment and *appliances*. This requirement does not apply to components of ground-source loop systems such as pumps, air separators, metering devices, and similar equipment.

M2105.15.6 Expansion tanks. Shutoff valves shall be installed at connections to nondiaphragm-type expansion tanks.

M2105.16 Reduced pressure. A pressure relief valve shall be installed on the low-pressure side of a hydronic piping system that has been reduced in pressure. The relief valve shall be set at the maximum pressure of the system design. The valve shall be installed in accordance with Section M2002.

M2105.17 Installation. Piping, valves, fittings, and connections shall be installed in accordance with the manufacturer's instructions.

M2105.18 Protection of potable water. Where ground-source heat-pump ground-loop systems have a connection to a potable water supply, the potable water system shall be protected from backflow in accordance with Section P2902.

M2105.19 Pipe penetrations. Openings for pipe penetrations in walls, floors and ceilings shall be larger than the penetrating pipe. Openings through concrete or masonry building elements shall be sleeved. The annular space surrounding pipe penetrations shall be protected in accordance with Section P2606.1.

M2105.20 Clearance from combustibles. A pipe in a ground-source heat pump piping system having an exterior surface temperature exceeding 250°F (121°C) shall have a clearance of not less than 1 inch (25 mm) from *combustible materials*.

M2105.21 Contact with building material. A ground-source heat-pump ground-loop piping system shall not be in direct contact with building materials that cause the piping or fitting material to degrade or corrode, or that interfere with the operation of the system.

M2105.22 Strains and stresses. Piping shall be installed so as to prevent detrimental strains and stresses in the pipe. Provisions shall be made to protect piping from damage resulting from expansion, contraction and structural settlement. Piping shall be installed so as to avoid structural stresses or strains within building components.

M2105.22.1 Flood hazard. Piping located in a flood hazard area shall be capable of resisting hydrostatic and hydrodynamic loads and stresses, including the effects of buoyancy, during the occurrence of flooding to the *design flood elevation*.

M2105.23 Pipe support. Pipe shall be supported in accordance with Section M2101.9.

M2105.24 Velocities. Ground-source heat-pump ground-loop systems shall be designed so that the flow velocities do not exceed the maximum flow velocity recommended by the pipe and fittings manufacturer. Flow velocities shall be controlled to reduce the possibility of water hammer.

M2105.25 Labeling and marking. Ground-source heat-pump ground-loop system piping shall be marked with tape, metal tags or other methods where it enters a building. The marking shall state the following words: "GROUND-SOURCE HEAT-PUMP LOOP SYSTEM." The marking shall indicate if antifreeze is used in the system and shall indicate the chemicals by name and concentration.

M2105.26 Chemical compatibility. Antifreeze and other materials used in the system shall be chemically compatible with the pipe, tubing, fittings and mechanical systems.

M2105.27 Makeup water. The transfer fluid shall be compatible with the makeup water supplied to the system.

M2105.28 Testing. Before connection header trenches are backfilled, the assembled loop system shall be pressure tested with water at 100 psi (689 kPa) for 15 minutes without observed leaks. Flow and pressure loss testing shall be performed and the actual flow rates and pressure drops shall be compared to the calculated design values. If actual flow rate or pressure drop values differ from calculated design values by more than 10 percent, the cause shall be identified and corrective action taken.

M2105.29 Embedded piping. Ground-source heat-pump ground-loop piping to be embedded in concrete shall be pressure tested prior to pouring concrete. During pouring, the pipe shall be maintained at the proposed operating pressure.

CHAPTER 22

SPECIAL PIPING AND STORAGE SYSTEMS

User notes:

About this chapter: *Chapter 22 addresses fuel oil piping and storage related to oil-fired heating appliances. Materials, joining methods, tanks, pumps, valves and installation of such are covered.*

Code development reminder: *Code change proposals to this chapter will be considered by the IRC—Plumbing/Mechanical Code Development Committee during the 2021 (Group A) Code Development Cycle.*

SECTION M2201 OIL TANKS

M2201.1 Materials. Supply tanks shall be *listed* and *labeled* and shall conform to UL 58 for underground tanks and UL 80 for indoor tanks.

M2201.2 Above-ground tanks. The maximum amount of fuel oil stored above ground or inside of a building shall be 660 gallons (2498 L). The supply tank shall be supported on rigid noncombustible supports to prevent settling or shifting.

> **Exception:** The storage of fuel oil, used for space or water heating, above ground or inside buildings in quantities exceeding 660 gallons (2498 L) shall comply with NFPA 31.

M2201.2.1 Tanks within buildings. Supply tanks for use inside of buildings shall be of such size and shape to permit installation and removal from *dwellings* as whole units. Supply tanks larger than 10 gallons (38 L) shall be placed not less than 5 feet (1524 mm) from any fire or flame either within or external to any fuel-burning *appliance*.

M2201.2.2 Outdoor above-ground tanks. Tanks installed outdoors, above ground shall be not less than 5 feet (1524 mm) from an adjoining property line. Such tanks shall be suitably protected from the weather and from physical damage.

M2201.3 Underground tanks. Excavations for underground tanks shall not undermine the foundations of existing structures. The clearance from the tank to the nearest wall of a *basement*, pit or property line shall be not less than 1 foot (305 mm). Tanks shall be set on and surrounded with noncorrosive inert materials such as clean earth, sand or gravel well-tamped in place. Tanks shall be covered with not less than 1 foot (305 mm) of earth. Corrosion protection shall be provided in accordance with Section M2203.7.

M2201.4 Multiple tanks. Cross connection of two supply tanks shall be permitted in accordance with Section M2203.6.

M2201.5 Oil gauges. Inside tanks shall be provided with a device to indicate when the oil in the tank has reached a predetermined safe level. Glass gauges or a gauge subject to breakage that could result in the escape of oil from the tank shall not be used. Liquid-level indicating gauges shall comply with UL 180.

M2201.6 Flood-resistant installation. In flood hazard areas as established by Table R301.2, tanks shall be installed in accordance with Section R322.2.4 or R322.3.10.

M2201.7 Tanks abandoned or removed. Outdoor above-grade fill piping shall be removed when tanks are abandoned or removed. Tank abandonment and removal shall be in accordance with the *International Fire Code*.

SECTION M2202 OIL PIPING, FITTING AND CONNECTIONS

M2202.1 Materials. Piping shall consist of steel pipe, copper and copper-alloy pipe and tubing, steel tubing conforming to ASTM A539, or stainless steel tubing conforming to ASTM A254 or ASTM A269. Aluminum tubing shall not be used between the fuel-oil tank and the burner units.

M2202.2 Joints and fittings. Piping shall be connected with fittings compatible with the piping material. Cast-iron fittings shall not be used for oil piping. Unions requiring gaskets or packings, right or left couplings, and sweat fittings employing solder having a melting point less than 1,000°F (538°C) shall not be used for oil piping. Threaded joints and connections shall be made tight with a lubricant or pipe thread compound.

M2202.3 Flexible connectors. Flexible metallic hoses shall be *listed* and *labeled* in accordance with UL 536 and shall be installed in accordance with their *listing* and *labeling* and the manufacturer's installation instructions. Connectors made from *combustible materials* shall not be used inside of buildings or above ground outside of buildings.

SECTION M2203 INSTALLATION

M2203.1 General. Piping shall be installed in a manner to avoid placing stresses on the piping, and to accommodate expansion and contraction of the piping system.

M2203.2 Supply piping. Supply piping used in the installation of oil burners and *appliances* shall be not smaller than $^3/_8$-inch (9 mm) pipe or $^3/_8$-inch (9 mm) outside diameter tubing. Copper tubing and fittings shall be Type L or heavier.

M2203.3 Fill piping. Fill piping shall terminate outside of buildings at a point not less than 2 feet (610 mm) from any building opening at the same or lower level. Fill openings shall be equipped with a tight metal cover.

M2203.4 Vent piping. Vent piping shall be not smaller than $1^1/_4$-inch (32 mm) pipe. Vent piping shall be laid to drain toward the tank without sags or traps in which the liquid can collect. Vent pipes shall not be cross connected with fill pipes, lines from burners or overflow lines from auxiliary tanks. The lower end of a vent pipe shall enter the tank through the top and shall extend into the tank not more than 1 inch (25 mm).

M2203.5 Vent termination. Vent piping shall terminate outside of buildings at a point not less than 2 feet (610 mm), measured vertically or horizontally, from any building opening. Outer ends of vent piping shall terminate in a weatherproof cap or fitting having an unobstructed area equal to or greater than the cross-sectional area of the vent pipe, and shall be located sufficiently above the ground to avoid being obstructed by snow and ice.

M2203.6 Cross connection of tanks. Cross connection of two supply tanks, not exceeding 660 gallons (2498 L) aggregate capacity, with gravity flow from one tank to another, shall be acceptable providing that the two tanks are on the same horizontal plane.

M2203.7 Corrosion protection. Underground tanks and buried piping shall be protected by corrosion-resistant coatings or special alloys or fiberglass-reinforced plastic.

SECTION M2204
OIL PUMPS AND VALVES

M2204.1 Pumps. Oil pumps shall be positive displacement types that automatically shut off the oil supply when stopped. Automatic pumps shall be *listed* and *labeled* in accordance with UL 343 and shall be installed in accordance with their *listing*.

M2204.2 Shutoff valves. A manual shutoff valve shall be installed between the oil supply tank and the burner. Such valve shall be provided with *ready access*. Where the shutoff valve is installed in the discharge line of an oil pump, a pressure relief valve shall be incorporated to bypass or return surplus oil. Valves shall comply with UL 842.

M2204.3 Maximum pressure. Pressure at the oil supply inlet to an *appliance* shall be not greater than 3 pounds per square inch (20.7 kPa).

M2204.4 Relief valves. Fuel-oil lines incorporating heaters shall be provided with relief valves that will discharge to a return line when excess pressure exists.

CHAPTER 23

SOLAR THERMAL ENERGY SYSTEMS

User notes:

About this chapter: *Chapter 23 is specific to thermal solar systems and equipment. Solar voltaic systems are not addressed in this chapter. This chapter covers solar collectors, system design, safety devices, relief valves, freeze protection, expansion tanks, signage, labeling, heat transfer fluids, protection of potable water and potable water heating.*

Code development reminder: *Code change proposals to this chapter will be considered by the IRC—Plumbing/Mechanical Code Development Committee during the 2021 (Group A) Code Development Cycle.*

SECTION M2301
SOLAR THERMAL ENERGY SYSTEMS

M2301.1 General. This section provides for the design, construction, installation, *alteration* and *repair* of equipment and systems using solar thermal energy to provide space heating or cooling, hot water heating and swimming pool heating.

M2301.2 Design and installation. The design and installation of solar thermal energy systems shall comply with Sections M2301.2.1 through M2301.2.13.

M2301.2.1 Access. Access shall be provided to solar energy equipment for maintenance. Solar systems and appurtenances shall not obstruct or interfere with the operation of any doors, windows or other building components requiring operation or access. Roof-mounted solar thermal equipment shall not obstruct or interfere with the operation of roof-mounted equipment, *appliances*, chimneys, plumbing vents, roof hatches, smoke vents, skylights and other roof penetrations and openings.

M2301.2.2 Collectors and panels. Solar collectors and panels shall comply with Sections M2301.2.2.1 and M2301.2.2.2.

M2301.2.2.1 Roof-mounted collectors. The roof shall be constructed to support the loads imposed by roof-mounted solar collectors. Roof-mounted solar collectors that serve as a roof covering shall conform to the requirements for roof coverings in Chapter 9 of this code. Where mounted on or above the roof coverings, the collectors and supporting structure shall be constructed of *noncombustible materials* or fire-retardant-treated wood equivalent to that required for the roof construction.

M2301.2.2.2 Collector sensors. Collector sensor installation, sensor location and the protection of exposed sensor wires from degradation shall be in accordance with ICC 900/SRCC 300.

M2301.2.3 Pressure and temperature relief valves and system components. System components containing fluids shall be protected with temperature and pressure relief valves or pressure relief valves. Relief devices shall be installed in sections of the system so that a section cannot be valved off or isolated from a relief device. Direct systems and the potable water portion of indirect systems shall be equipped with a relief valve in accordance with Section P2804. For indirect systems, pressure relief valves in solar loops shall comply with ICC 900/SRCC 300. System components shall have a working pressure rating of not less than the setting of the pressure relief device.

M2301.2.4 Vacuum relief. System components that might be subjected to a vacuum during operation or shutdown shall be designed to withstand such a vacuum or shall be protected with vacuum relief valves.

M2301.2.5 Piping insulation. Piping shall be insulated in accordance with the requirements of Chapter 11. Exterior insulation shall be protected from ultraviolet degradation. The entire solar loop shall be insulated. Where split-style insulation is used, the seam shall be sealed. Fittings shall be fully insulated.

Exceptions:

1. Those portions of the piping that are used to help prevent the system from overheating shall not be required to be insulated.
2. Those portions of piping that are exposed to solar radiation, made of the same material as the solar collector absorber plate and are covered in the same manner as the solar collector absorber, or that are used to collect additional solar energy, shall not be required to be insulated.
3. Piping in thermal solar systems using unglazed solar collectors to heat a swimming pool shall not be required to be insulated.

M2301.2.6 Protection from freezing. System components shall be protected from damage resulting from freezing of heat-transfer liquids at the winter design temperature provided in Table R301.2. Freeze protection shall be provided in accordance with ICC 900/SRCC 300. Drain-back systems shall be installed in compliance with Section M2301.2.6.1. Systems utilizing freeze-protection valves shall comply with Section M2301.2.6.2.

Exception: Where the 97.5-percent winter design temperature is greater than or equal to 48°F (9°C).

M2301.2.6.1 Drain-back systems. Drain-back systems shall be designed and installed to allow for manual gravity draining of fluids from areas subject to freezing to locations not subject to freezing, and air filling of the components and piping. Such piping and components shall maintain a horizontal slope in the direction of flow of not less than $^1/_4$ unit vertical in 12 units horizontal (2-percent slope). Piping and components subject to manual gravity draining shall permit subsequent air filling upon drainage and air venting upon refilling.

M2301.2.6.2 Freeze-protection valves. Freeze-protection valves shall discharge in a manner that does not create a hazard or structural damage.

M2301.2.7 Storage tank sensors. Storage tank sensors shall comply with ICC 900/SRCC 300.

M2301.2.8 Expansion tanks. Expansion tanks in *solar energy systems* shall be installed in accordance with Section M2003 in solar collector loops that contain pressurized heat transfer fluid. Where expansion tanks are used, the system shall be designed in accordance with ICC 900/SRCC 300 to provide an expansion tank that is sized to withstand the maximum operating pressure of the system.

Exception: Expansion tanks shall not be required in the collector loop of *drain-back systems.*

M2301.2.9 Roof and wall penetrations. Roof and wall penetrations shall be flashed and sealed in accordance with Chapter 9 to prevent entry of water, rodents and insects.

M2301.2.10 Description and warning labels. Solar thermal systems shall comply with description *label* and warning *label* requirements of Section M2301.2.11.2 and ICC 900/SRCC 300.

M2301.2.11 Solar loop. Solar loops shall be in accordance with Sections M2301.2.11.1 and M2301.2.11.2.

M2301.2.11.1 Solar loop isolation. Valves shall be installed to allow the solar loop to be isolated from the remainder of the system.

M2301.2.11.2 Drain and fill valve labels and caps. Drain and fill valves shall be *labeled* with a description and warning that identifies the fluid in the solar loop and a warning that the fluid might be discharged at high temperature and pressure. Drain caps shall be installed at drain and fill valves.

M2301.2.12 Maximum temperature limitation. Systems shall be equipped with means to limit the maximum water temperature of the system fluid entering or exchanging heat with any pressurized vessel inside the *dwelling* to 180°F (82°C). This protection is in addition to the required temperature and pressure relief valves required by Section M2301.2.3.

M2301.2.13 Thermal storage unit seismic bracing. In Seismic Design Categories D_0, D_1 and D_2 and in townhouses in Seismic Design Category C, thermal storage units shall be anchored in accordance with Section M1307.2.

M2301.3 Labeling. *Labeling* shall comply with Sections M2301.3.1 and M2301.3.2.

M2301.3.1 Collectors and panels. Solar thermal collectors and panels shall be *listed* and *labeled* in accordance with ICC 901/SRCC 100. Factory-built collectors shall bear a *label* indicating the manufacturer's name, model number and serial number.

M2301.3.2 Thermal storage units. Pressurized water storage tanks shall bear a *label* indicating the manufacturer's name and address, model number, serial number, storage unit maximum and minimum allowable operating temperatures and storage unit maximum and minimum allowable operating pressures. The *label* shall clarify that these specifications apply only to the water storage tanks.

M2301.4 Heat transfer gases or liquids and heat exchangers. *Essentially toxic transfer fluids*, ethylene glycol, flammable gases and flammable liquids shall not be used as heat transfer fluids. Heat transfer gases and liquids shall be rated to withstand the system's maximum design temperature under operating conditions without degradation. Heat exchangers used in solar thermal systems shall comply with Section P2902.5.2 and ICC 900/SRCC 300.

Heat transfer fluids shall be in accordance with ICC 900/SRCC 300. The flash point of the heat transfer fluids utilized in solar thermal systems shall be not less than 50°F (28°C) above the design maximum nonoperating or no-flow temperature attained by the fluid in the collector.

M2301.5 Backflow protection. Connections from the potable water supply to solar systems shall comply with Section P2902.5.5.

M2301.6 Filtering. Air provided to *occupied spaces* that passes through thermal mass storage systems by mechanical means shall be filtered for particulates at the outlet of the thermal mass storage system.

M2301.7 Solar thermal systems for heating potable water. Where a solar thermal system heats potable water to supply a potable hot water distribution system, the solar thermal system shall be in accordance with Sections M2301.7.1, M2301.7.2 and P2902.5.5.

M2301.7.1 Indirect systems. Heat exchangers that are components of indirect solar thermal heating systems shall comply with Section P2902.5.2.

M2301.7.2 Direct systems. Where potable water is directly heated by a solar thermal system, the pipe, fittings, valves and other components that are in contact with the potable water in the solar heating system shall comply with the requirements of Chapter 29.

Part VI—Fuel Gas

CHAPTER 24
FUEL GAS

The text of this chapter is extracted from the 2021 edition of the International Fuel Gas Code *and has been modified where necessary to conform to the scope of application of the* International Residential Code for One- and Two-Family Dwellings. *The section numbers appearing in parentheses after each section number are the section numbers of the corresponding text in the* International Fuel Gas Code.

User notes:

About this chapter: *Chapter 24 addresses fuel gas piping, appliances, combustion air, appliance venting and specific appliances, among other subjects. Note that Chapter 24 includes definitions that are unique to this chapter. The text of this chapter is identical to that of the* International Fuel Gas Code®, *except that this chapter contains coverage only for that which is typically found in residential occupancies, consistent with the scope of this code.*

Code development reminder: *Code change proposals to this chapter will be considered by the IRC—Plumbing/Mechanical Code Development Committee during the 2021 (Group A) Code Development Cycle.*

SECTION G2401 (101) GENERAL

G2401.1 (101.2) Application. This chapter covers those fuel gas *piping systems*, fuel-gas *appliances* and related accessories, *venting systems* and *combustion air* configurations most commonly encountered in the construction of one- and two-family dwellings and structures regulated by this code.

Coverage of *piping systems* shall extend from the *point of delivery* to the outlet of the *appliance* shutoff *valves* (see definition of "*Point of delivery*"). *Piping systems* requirements shall include design, materials, components, fabrication, assembly, installation, testing, inspection, operation and maintenance. Requirements for gas *appliances* and related accessories shall include installation, combustion and ventilation air and venting and connections to *piping systems*.

The omission from this chapter of any material or method of installation provided for in the *International Fuel Gas Code* shall not be construed as prohibiting the use of such material or method of installation. Fuel-gas *piping systems*, fuel-gas *appliances* and related accessories, *venting systems* and *combustion air* configurations not specifically covered in these chapters shall comply with the applicable provisions of the *International Fuel Gas Code*.

Gaseous hydrogen systems shall be regulated by Chapter 7 of the *International Fuel Gas Code*.

This chapter shall not apply to the following:

1. Liquefied natural gas (LNG) installations.
2. Temporary LP-*gas piping* for buildings under construction or renovation that is not to become part of the permanent *piping system*.
3. Except as provided in Section G2412.1.1, gas *piping*, meters, gas pressure regulators, and other appurtenances used by the serving gas supplier in the distribution of gas, other than undiluted LP-gas.
4. Portable LP-gas *appliances* and equipment of all types that is not connected to a fixed fuel *piping system*.
5. Portable fuel cell *appliances* that are neither connected to a fixed *piping system* nor interconnected to a power grid.
6. Installation of hydrogen gas, LP-gas and compressed natural gas (CNG) systems on vehicles.

SECTION G2402 (201) GENERAL

G2402.1 (201.1) Scope. Unless otherwise expressly stated, the following words and terms shall, for the purposes of this chapter, have the meanings indicated in this chapter.

G2402.2 (201.2) Interchangeability. Words used in the present tense include the future; words in the masculine gender include the feminine and neuter; the singular number includes the plural and the plural, the singular.

G2402.3 (201.3) Terms defined in other codes. Where terms are not defined in this code and are defined in the *International Building Code, International Fire Code, International Mechanical Code, International Fuel Gas Code* or *International Plumbing Code*, such terms shall have meanings ascribed to them as in those *codes*.

SECTION G2403 (202) GENERAL DEFINITIONS

ACCESS (TO). That which enables a device, *appliance* or equipment to be reached by ready *access* or by a means that first requires the removal or movement of a panel, door or similar obstruction (see also "*Ready access*").

AIR, EXHAUST. Air being removed from any space or piece of *equipment* or *appliance* and conveyed directly to the atmosphere by means of openings or ducts.

AIR, MAKEUP. Any combination of outdoor and transfer air intended to replace exhaust air and exfiltration.

AIR CONDITIONER, GAS-FIRED. A gas-burning, automatically operated *appliance* for supplying cooled air, dehumidified air, or both, or chilled liquid.

AIR CONDITIONING. The treatment of air so as to control simultaneously the temperature, humidity, cleanness and distribution of the air to meet the requirements of a *conditioned space.*

AIR-HANDLING UNIT. A blower or fan used for the purpose of distributing supply air to a room, space or area.

ALTERATION. A change in a system that involves an extension, addition or change to the arrangement, type or purpose of the original installation.

ANODELESS RISER. A transition assembly in which plastic *piping* is installed and terminated above ground outside of a building.

APPLIANCE. Any apparatus or device that utilizes a fuel or a raw material as a fuel to produce light, heat, power, refrigeration or air conditioning. Also, an apparatus that compresses fuel gases.

APPLIANCE, AUTOMATICALLY CONTROLLED. *Appliances* equipped with an automatic *burner* ignition and safety shutoff device and other automatic devices, that accomplish complete turn-on and shutoff of the gas to the *main burner* or *burners*, and graduate the gas supply to the *burner* or *burners*, but do not affect complete shutoff of the gas.

APPLIANCE, FAN-ASSISTED COMBUSTION. An *appliance* equipped with an integral mechanical means to either draw or force products of combustion through the combustion chamber or heat exchanger.

APPLIANCE, UNVENTED. An *appliance* designed or installed in such a manner that the products of combustion are not conveyed by a vent or *chimney* directly to the outside atmosphere.

APPLIANCE, VENTED. An *appliance* designed and installed in such a manner that all of the products of combustion are conveyed directly from the *appliance* to the outside atmosphere through an *approved chimney* or vent system.

APPROVED. Acceptable to the *code official.*

APPROVED AGENCY. An established and recognized agency that is regularly engaged in conducting tests, furnishing inspection services or furnishing certification, where such agency has been *approved* by the *code official.*

ATMOSPHERIC PRESSURE. The pressure of the weight of air and water vapor on the surface of the earth, approximately 14.7 pounds per square inch (psia) (101 kPa absolute) at sea level.

AUTOMATIC IGNITION. Ignition of gas at the *burner(s)* when the gas controlling device is turned on, including reignition if the flames on the *burner(s)* have been extinguished by means other than by the closing of the gas controlling device.

BAROMETRIC DRAFT REGULATOR. A balanced *damper* device attached to a *chimney*, vent *connector*, breeching or flue gas manifold to protect combustion *appliances* by controlling *chimney draft.* A double-acting *barometric draft regulator* is one whose balancing *damper* is free to move in either direction to protect combustion *appliances* from both excessive *draft* and backdraft.

BOILER, LOW-PRESSURE. A self-contained *appliance* for supplying steam or hot water.

Hot water heating boiler. A boiler in which no steam is generated, from which hot water is circulated for heating purposes and then returned to the boiler, and that operates at water pressures not exceeding 160 pounds per square inch gauge (psig) (1100 kPa gauge) and at water temperatures not exceeding 250°F (121°C) at or near the boiler outlet.

Hot water supply boiler. A boiler, completely filled with water, which furnishes hot water to be used externally to itself, and that operates at water pressures not exceeding 160 psig (1100 kPa gauge) and at water temperatures not exceeding 250°F (121°C) at or near the boiler outlet.

Steam heating boiler. A boiler in which steam is generated and that operates at a steam pressure not exceeding 15 psig (100 kPa gauge).

BONDING JUMPER. A conductor installed to electrically connect metallic gas *piping* to the grounding electrode system.

BRAZING. A metal-joining process wherein coalescence is produced by the use of a nonferrous filler metal having a melting point above 1,000°F (538°C), but lower than that of the base metal being joined. The filler material is distributed between the closely fitted surfaces of the joint by capillary action.

BTU. Abbreviation for British thermal unit, which is the quantity of heat required to raise the temperature of 1 pound (454 g) of water 1°F (0.56°C) (1 *Btu* = 1055 J).

BURNER. A device for the final conveyance of the gas, or a mixture of gas and air, to the combustion zone.

Induced-draft. A *burner* that depends on *draft* induced by a fan that is an integral part of the *appliance* and is located downstream from the *burner.*

Power. A *burner* in which gas, air or both are supplied at pressures exceeding, for gas, the line pressure, and for air, atmospheric pressure, with this added pressure being applied at the *burner.*

CHIMNEY. A primarily vertical structure containing one or more flues, for the purpose of carrying gaseous products of *combustion* and air from an *appliance* to the outside atmosphere.

Factory-built chimney. A *listed* and *labeled* chimney composed of factory-made components, assembled in the field in accordance with manufacturer's instructions and the conditions of the listing.

Masonry chimney. A field-constructed chimney composed of solid masonry units, bricks, stones or concrete.

CLEARANCE. The minimum distance through air measured between the heat-producing surface of the mechanical *appliance,* device or *equipment* and the surface of the *combustible material* or assembly.

CLOTHES DRYER. An *appliance* used to dry wet laundry by means of heated air.

Type 1. Factory-built package, multiple production. Primarily used in the family living environment. Usually the smallest unit physically and in function output.

CODE. These regulations, subsequent amendments thereto, or any emergency rule or regulation that the administrative authority having *jurisdiction* has lawfully adopted.

CODE OFFICIAL. The officer or other designated authority charged with the administration and enforcement of this code, or a duly authorized representative.

COMBUSTIBLE ASSEMBLY. Wall, floor, ceiling or other assembly constructed of one or more component materials that are not defined as noncombustible.

COMBUSTIBLE MATERIAL. Any material not defined as noncombustible.

COMBUSTION. In the context of this code, refers to the rapid oxidation of fuel accompanied by the production of heat or heat and light.

COMBUSTION AIR. Air necessary for complete combustion of a fuel, including theoretical air and excess air.

COMBUSTION CHAMBER. The portion of an *appliance* within which combustion occurs.

COMBUSTION PRODUCTS. Constituents resulting from the combustion of a fuel with the oxygen of the air, including the inert gases, but excluding excess air.

CONCEALED LOCATION. A location that cannot be accessed without damaging permanent parts of the building structure or finish surface. Spaces above, below or behind readily removable panels or doors shall not be considered as concealed.

CONCEALED PIPING. *Piping* that is located in a *concealed location* (see "*Concealed location*").

CONDENSATE. The liquid that condenses from a gas (including flue gas) caused by a reduction in temperature or increase in pressure.

CONNECTOR, APPLIANCE (Fuel). Rigid metallic *pipe* and fittings, semirigid metallic *tubing* and fittings or a *listed* and *labeled* device that connects an *appliance* to the *gas piping system.*

CONNECTOR, CHIMNEY OR VENT. The *pipe* that connects an *appliance* to a chimney or vent.

CONTROL. A manual or automatic device designed to regulate the gas, air, water or electrical supply to, or operation of, a mechanical system.

CONVERSION BURNER. A unit consisting of a *burner* and its *controls* for installation in an *appliance* originally utilizing another fuel.

COPPER ALLOY. A homogeneous mixture of not less than two metals where not less than 50 percent of the finished metal is copper.

CUBIC FOOT. The amount of gas that occupies 1 cubic foot (0.02832 m^3) when at a temperature of 60°F (16°C), saturated with water vapor and under a pressure equivalent to that of 30 inches of mercury (101 kPa).

DAMPER. A manually or automatically controlled device to regulate *draft* or the rate of flow of air or combustion gases.

DECORATIVE APPLIANCE, VENTED. A *vented appliance* wherein the primary function lies in the aesthetic effect of the flames.

DECORATIVE APPLIANCES FOR INSTALLATION IN VENTED FIREPLACES. A *vented appliance* designed for installation within the fire chamber of a vented *fireplace,* wherein the primary function lies in the aesthetic effect of the flames.

DEMAND. The maximum amount of gas input required per unit of time, usually expressed in cubic feet per hour, or *Btu*/h (1 *Btu*/h = 0.2931 W).

DESIGN FLOOD ELEVATION. The elevation of the "design flood," including wave height, relative to the datum specified on the community's legally designated flood hazard map. In areas designated as Zone AO, the *design flood elevation* shall be the elevation of the highest existing grade of the *building's* perimeter plus the depth number (in feet) specified on the flood hazard map. In areas designated as Zone AO where a depth number is not specified on the map, the depth number shall be taken as being equal to 2 feet (610 mm).

DILUTION AIR. Air that is introduced into a *draft hood* and is mixed with the *flue gases.*

DIRECT-VENT APPLIANCES. *Appliances* that are constructed and installed so that all air for combustion is derived directly from the outside atmosphere and all *flue gases* are discharged directly to the outside atmosphere.

DRAFT. The pressure difference existing between the *appliance* or any component part and the atmosphere, that causes a continuous flow of air and products of combustion through the gas passages of the *appliance* to the atmosphere.

Mechanical or induced draft. The pressure difference created by the action of a fan, blower or ejector that is located between the *appliance* and the chimney or vent termination.

Natural draft. The pressure difference created by a vent or chimney because of its height, and the temperature difference between the *flue gases* and the atmosphere.

DRAFT HOOD. A nonadjustable device built into an *appliance*, or made as part of the vent *connector* from an *appliance*, that is designed to: provide for ready escape of the *flue gases* from the *appliance* in the event of no *draft*, backdraft, or stoppage beyond the *draft hood*; prevent a backdraft from entering the *appliance*; and neutralize the effect of stack action of the chimney or gas vent upon operation of the *appliance*.

DRAFT REGULATOR. A device that functions to maintain a desired *draft* in the *appliance* by automatically reducing the *draft* to the desired value.

DRIP. The container placed at a low point in a system of *piping* to collect *condensate* and from which the *condensate* is removable.

DUCT FURNACE. A warm-air *furnace* normally installed in an air distribution duct to supply warm air for heating. This definition shall apply only to a warm-air heating *appliance* that depends for air circulation on a blower not furnished as part of the *furnace*.

DWELLING UNIT. A single unit providing complete, independent living facilities for one or more persons, including permanent provisions for living, sleeping, eating, cooking and sanitation.

EQUIPMENT. Apparatus and devices other than *appliances*.

EXCESS FLOW VALVE (EFV). A valve designed to activate when the fuel gas passing through it exceeds a prescribed flow rate.

EXTERIOR MASONRY CHIMNEYS. Masonry chimneys exposed to the outdoors on one or more sides below the roof line.

FIREPLACE. A fire chamber and hearth constructed of *noncombustible material* for use with solid fuels and provided with a chimney.

Factory-built fireplace. A *fireplace* composed of *listed* factory-built components assembled in accordance with the terms of listing to form the completed *fireplace*.

Masonry fireplace. A hearth and fire chamber of solid masonry units such as bricks, stones, *listed* masonry units or reinforced concrete, provided with a suitable chimney.

FLAME SAFEGUARD. A device that will automatically shut off the fuel supply to a *main burner* or group of *burners* when the means of ignition of such *burners* becomes inoperative, and when flame failure occurs on the *burner* or group of *burners*.

FLASHBACK ARRESTOR CHECK VALVE. A device that will prevent the backflow of one gas into the supply system of another gas and prevent the passage of flame into the gas supply system.

FLOOD HAZARD AREA. The greater of the following two areas:

1. The area within a floodplain subject to a 1 percent or greater chance of flooding in any given year.
2. This area designated as a *flood hazard area* on a community's flood hazard map, or otherwise legally designated.

FLOOR FURNACE. A completely self-contained *furnace* suspended from the floor of the space being heated, taking air for combustion from outside such space and with means for observing flames and lighting the *appliance* from such space.

FLUE, APPLIANCE. The passage(s) within an *appliance* through which *combustion products* pass from the *combustion chamber* of the *appliance* to the *draft hood* inlet opening on an *appliance* equipped with a *draft hood* or to the outlet of the *appliance* on an *appliance* not equipped with a *draft hood*.

FLUE COLLAR. That portion of an *appliance* designed for the attachment of a *draft hood*, *vent connector* or venting system.

FLUE GASES. Products of combustion plus excess air in *appliance flues* or heat exchangers.

FLUE LINER (LINING). A system or material used to form the inside surface of a flue in a *chimney* or vent, for the purpose of protecting the surrounding structure from the effects of *combustion products* and for conveying *combustion products* without leakage to the atmosphere.

FUEL GAS. A natural gas, manufactured gas, *liquefied petroleum gas* or mixtures of these gases.

FURNACE. A completely self-contained heating unit that is designed to supply heated air to spaces remote from or adjacent to the *appliance* location.

FURNACE, CENTRAL. A self-contained *appliance* for heating air by transfer of heat of *combustion* through metal to the air, and designed to supply heated air through ducts to spaces remote from or adjacent to the *appliance* location.

FURNACE PLENUM. An air compartment or chamber to which one or more ducts are connected and that forms part of an air distribution system.

GAS CONVENIENCE OUTLET. A permanently mounted, manually operated device that provides the means for connecting an *appliance* to, and disconnecting an *appliance* from, the supply *piping*. The device includes an integral, manually operated valve with a nondisplaceable valve member and is designed so that disconnection of an *appliance* only occurs when the manually operated valve is in the closed position.

GAS PIPING. An installation of pipe, valves or fittings installed on a premises or in a building and utilized to convey fuel gas.

HAZARDOUS LOCATION. Any location considered to be a fire hazard for flammable vapors, dust, combustible fibers or other highly combustible substances. The location is not necessarily categorized in the *International Building Code* as a high-hazard use group classification.

HOUSE PIPING. See *"Piping system."*

IGNITION PILOT. A *pilot* that operates during the lighting cycle and discontinues during *main burner* operation.

IGNITION SOURCE. A flame spark or hot surface capable of igniting flammable vapors or fumes. Such sources include *appliance burners, burner* ignitors and electrical switching devices.

INFRARED RADIANT HEATER. A heater that directs a substantial amount of its energy output in the form of infrared radiant energy into the area to be heated. Such heaters are of either the vented or unvented type.

JOINT, FLARED. A metal-to-metal compression joint in which a conical spread is made on the end of a tube that is compressed by a flare nut against a mating flare.

JOINT, MECHANICAL. A general form of gastight joints obtained by the joining of metal parts through a positive-holding mechanical construction, such as a press-connect joint, flanged joint, threaded joint, flared joint or compression joint.

JOINT, PLASTIC ADHESIVE. A joint made in thermoset plastic *piping* by the use of an adhesive substance that forms a continuous bond between the mating surfaces without dissolving either one of them.

LABELED. Equipment, materials or products to which have been affixed a *label*, seal, symbol or other identifying *mark* of a nationally recognized testing laboratory, inspection agency or other organization concerned with product evaluation that maintains periodic inspection of the production of the above-*labeled* items and whose labeling indicates either that the *equipment*, material or product meets identified standards or has been tested and found suitable for a specified purpose.

LEAK CHECK. An operation performed on a gas *piping system* to verify that the system does not leak.

LIQUEFIED PETROLEUM GAS or LPG (LP-GAS). *Liquefied petroleum gas* composed predominately of propane, propylene, butanes or butylenes, or mixtures thereof that is gaseous under normal atmospheric conditions, but is capable of being liquefied under moderate pressure at normal temperatures.

LISTED. Equipment, materials, products or services included in a list published by an organization acceptable to the code official and concerned with evaluation of products or services that maintains periodic inspection of production of *listed equipment* or materials or periodic evaluation of services and whose listing states either that the *equipment*, material, product or service meets identified standards or has been tested and found suitable for a specified purpose.

LIVING SPACE. Space within a *dwelling unit* utilized for living, sleeping, eating, cooking, bathing, washing and sanitation purposes.

LOG LIGHTER. A manually operated solid-fuel ignition *appliance* for installation in a vented solid fuel-burning *fireplace*.

MAIN BURNER. A device or group of devices essentially forming an integral unit for the final conveyance of gas or a mixture of gas and air to the combustion zone, and on which combustion takes place to accomplish the function for which the *appliance* is designed.

METER. The instrument installed to measure the volume of gas delivered through it.

MODULATING. Modulating or throttling is the action of a *control* from its maximum to minimum position in either predetermined steps or increments of movement as caused by its actuating medium.

NONCOMBUSTIBLE MATERIALS. Materials that, where tested in accordance with ASTM E136, have not fewer than three of four specimens tested meeting all of the following criteria:

1. The recorded temperature of the surface and interior thermocouples shall not at any time during the test rise more than 54°F (30°C) above the furnace temperature at the beginning of the test.
2. There shall not be flaming from the specimen after the first 30 seconds.
3. If the weight loss of the specimen during testing exceeds 50 percent, the recorded temperature of the surface and interior thermocouples shall not at any time during the test rise above the furnace air temperature at the beginning of the test, and there shall not be flaming of the specimen.

OFFSET (VENT). A combination of *approved* bends that make two changes in direction bringing one section of the vent out of line, but into a line parallel with the other section.

OUTLET. The point at which a gas-fired *appliance* connects to the gas *piping system*.

OXYGEN DEPLETION SAFETY SHUTOFF SYSTEM (ODS). A system designed to act to shut off the gas supply to the main and *pilot burners* if the oxygen in the surrounding atmosphere is reduced below a predetermined level.

PILOT. A small flame that is utilized to ignite the gas at the *main burner* or *burners*.

PIPING. Where used in this code, *"piping"* refers to either *pipe* or *tubing*, or both.

Pipe. A rigid conduit of iron, steel, copper, copper-alloy or plastic.

Tubing. Semirigid conduit of copper, copper-alloy, aluminum, plastic or steel.

PIPING SYSTEM. The fuel *piping*, valves and fittings from the outlet of the *point of delivery* to the outlets of the *appliance* shutoff valves.

PLASTIC, THERMOPLASTIC. A plastic that is capable of being repeatedly softened by increase of temperature and hardened by decrease of temperature.

POINT OF DELIVERY. For natural gas systems, the *point of delivery* is the outlet of the service meter assembly or the outlet of the service regulator or service shutoff valve where a meter is not provided. Where a system shutoff valve is provided after the outlet of the service meter assembly, such valve shall be considered to be downstream of the *point of delivery*. For undiluted liquefied petroleum gas systems, the *point of delivery* shall be considered to be the outlet of the service pressure regulator, exclusive of line gas regulators, in the system.

PRESS-CONNECT JOINT. A permanent mechanical joint incorporating an elastomeric seal or an elastomeric seal and corrosion-resistant grip or bite ring. The joint is made with a pressing tool and jaw or ring approved by the fitting manufacturer.

PRESSURE DROP. The loss in pressure due to friction or obstruction in pipes, valves, fittings, *regulators* and *burners*.

PRESSURE TEST. An operation performed to verify the gastight integrity of *gas piping* following its installation or modification.

PURGE. To free a gas conduit of air or gas, or a mixture of gas and air.

READY ACCESS (TO). That which enables a device, *appliance* or *equipment* to be directly reached, without requiring the removal or movement of any panel, door or similar obstruction. (See "*Access.*")

REGULATOR. A device for controlling and maintaining a uniform gas supply pressure, either pounds-to-inches water column (MP regulator) or inches-to-inches water column (*appliance regulator*).

REGULATOR, GAS APPLIANCE. A *pressure regulator* for controlling pressure to the manifold of the gas *appliance*.

REGULATOR, LINE GAS PRESSURE. A device placed in a gas line between the *service pressure regulator* and the *appliance* for controlling, maintaining or reducing the pressure in that portion of the *piping system* downstream of the device.

REGULATOR, MEDIUM-PRESSURE (MP Regulator). A line *pressure regulator* that reduces gas pressure from the range of greater than 0.5 psig (3.4 kPa) and less than or equal to 5 psig (34.5 kPa) to a lower pressure.

REGULATOR, MONITORING. A pressure regulator set in series with another pressure regulator for the purpose of preventing an overpressure in the downstream piping system.

REGULATOR, PRESSURE. A device placed in a gas line for reducing, controlling and maintaining the pressure in that portion of the *piping system* downstream of the device.

REGULATOR, SERVICE PRESSURE. For natural gas systems, a device installed by the serving gas supplier to reduce and limit the service line pressure to delivery pressure. For undiluted liquefied petroleum gas systems, the regulator located upstream from all line gas pressure regulators, where installed, and downstream from any first stage or a high pressure regulator in the system.

RELIEF OPENING. The opening provided in a *draft hood* to permit the ready escape to the atmosphere of the flue products from the *draft hood* in the event of no *draft*, backdraft or stoppage beyond the *draft hood*, and to permit air into the *draft hood* in the event of a strong chimney updraft.

RELIEF VALVE (DEVICE). A safety valve designed to forestall the development of a dangerous condition by relieving either pressure, temperature or vacuum in the hot water supply system.

RELIEF VALVE, PRESSURE. An *automatic valve* that opens and closes a relief vent, depending on whether the pressure is above or below a predetermined value.

RELIEF VALVE, TEMPERATURE.

Manual reset type. A valve that automatically opens a *relief* vent at a predetermined temperature and that must be manually returned to the closed position.

Reseating or self-closing type. An *automatic valve* that opens and closes a relief vent, depending on whether the temperature is above or below a predetermined value.

RELIEF VALVE, VACUUM. A valve that automatically opens and closes a vent for relieving a vacuum within the hot water supply system, depending on whether the vacuum is above or below a predetermined value.

RISER, GAS. A vertical *pipe* supplying fuel gas.

ROOM HEATER, UNVENTED. See "*Unvented room heater.*"

ROOM HEATER, VENTED. A free-standing heating unit used for direct heating of the space in and adjacent to that in which the unit is located. (See "*Vented room heater.*")

SAFETY SHUTOFF DEVICE. See "*Flame safeguard.*"

SERVICE METER ASSEMBLY. The meter, valve, regulator, piping, fittings and equipment installed by the service gas supplier before the *point of delivery*.

SHAFT. An enclosed space extending through one or more stories of a building, connecting vertical openings in successive floors, or floors and the roof.

SPECIFIC GRAVITY. As applied to gas, *specific gravity* is the ratio of the weight of a given volume to that of the same volume of air, both measured under the same condition.

SYSTEM SHUTOFF. A valve installed after the *point of delivery* to shut off the entire piping system.

THERMOSTAT. (See types that follow.)

Electric switch type. A device that senses changes in temperature and controls electrically, by means of separate components, the flow of gas to the *burner(s)* to maintain selected temperatures.

Integral gas valve type. An automatic device, actuated by temperature changes, designed to control the gas supply to the burner(s) in order to maintain temperatures between predetermined limits, and in which the thermal actuating element is an integral part of the device.

1. Graduating thermostat. A thermostat in which the motion of the valve is approximately in direct

proportion to the effective motion of the thermal element induced by temperature change.

2. Snap-acting thermostat. A thermostat in which the thermostatic valve travels instantly from the closed to the open position, and vice versa.

THIRD-PARTY CERTIFICATION AGENCY. An *approved* agency operating a product or material certification system that incorporates initial product testing, assessment and surveillance of a manufacturer's quality control system.

THIRD-PARTY CERTIFIED. Certification obtained by the manufacturer indicating that the function and performance characteristics of a product or material have been determined by testing and ongoing surveillance by an *approved* third-party certification agency. Assertion of certification is in the form of identification in accordance with the requirements of the third-party certification agency.

THIRD-PARTY TESTED. Procedure by which an *approved* testing laboratory provides documentation that a product, material or system conforms to specified requirements.

TOILET, GAS-FIRED. A packaged and completely assembled *appliance* containing a toilet that incinerates refuse instead of flushing it away with water.

TRANSITION FITTINGS, PLASTIC TO STEEL. An adapter for joining plastic *pipe* to steel *pipe*. The purpose of this fitting is to provide a permanent, pressure-tight connection between two materials that cannot be joined directly one to another.

UNIT HEATER. A self-contained, automatically controlled, vented, fuel-gas-burning, space-heating *appliance*, intended for installation in the space to be heated without the use of ducts, and having integral means for circulation of air.

UNVENTED ROOM HEATER. An unvented heating *appliance* designed for stationary installation and utilized to provide comfort heating. Such *appliances* provide radiant heat or convection heat by gravity or fan circulation directly from the heater and do not utilize ducts.

VALVE. A device used in *piping* to control the gas supply to any section of a system of *piping* or to an *appliance*.

Appliance shutoff. A *valve* located in the *piping system*, used to isolate individual *appliances* for purposes such as service or replacement.

Automatic. An automatic or semiautomatic device consisting essentially of a *valve* and an operator that control the gas supply to the *burner(s)* during operation of an *appliance*. The operator shall be actuated by application of gas pressure on a flexible diaphragm, by electrical means, by mechanical means or by other *approved* means.

Automatic gas shutoff. A *valve* used in conjunction with an automatic gas shutoff device to shut off the gas supply to a water-heating system. It shall be constructed integrally with the gas shutoff device or shall be a separate assembly.

Individual main burner. A *valve* that controls the gas supply to an individual *main burner*.

Main burner control. A *valve* that controls the gas supply to the *main burner* manifold.

Manual main gas-control. A manually operated *valve* in the gas line for the purpose of completely turning on or shutting off the gas supply to the *appliance*, except to *pilot* or pilots that are provided with independent shutoff.

Manual reset. An automatic shutoff valve installed in the gas supply *piping* and set to shut off when unsafe conditions occur. The device remains closed until manually reopened.

Service shutoff. A valve, installed by the serving gas supplier between the source of supply and the *point of delivery*, to shut off the entire *piping system*.

VENT. A *pipe* or other conduit composed of factory-made components, containing a passageway for conveying *combustion products* and air to the atmosphere, *listed* and *labeled* for use with a specific type or class of *appliance*.

Special gas vent. A vent *listed* and *labeled* for use with *listed* Category II, III and IV gas *appliances*.

Type B vent. A vent *listed* and *labeled* for use with *appliances* with *draft hoods* and other Category I *appliances* that are *listed* for use with Type B vents.

Type BW vent. A vent *listed* and *labeled* for use with wall *furnaces*.

Type L vent. A vent *listed* and *labeled* for use with *appliances* that are *listed* for use with Type L or Type B vents.

VENT CONNECTOR. See "*Connector, Chimney or Vent.*"

VENT PIPING.

Breather. *Piping* run from a pressure-regulating device to the outdoors, designed to provide a reference to *atmospheric pressure*. If the device incorporates an integral pressure *relief* mechanism, a breather vent can also serve as a *relief* vent.

Relief. *Piping* run from a pressure-regulating or pressure-limiting device to the outdoors, designed to provide for the safe venting of gas in the event of excessive pressure in the *gas piping system*.

VENTED APPLIANCE CATEGORIES. *Appliances* that are categorized for the purpose of vent selection are classified into the following four categories:

Category I. An *appliance* that operates with a nonpositive vent static pressure and with a vent gas temperature that avoids excessive *condensate* production in the vent.

Category II. An *appliance* that operates with a nonpositive *vent* static pressure and with a vent gas temperature that is capable of causing excessive *condensate* production in the vent.

Category III. An *appliance* that operates with a positive vent static pressure and with a vent gas temperature that avoids excessive *condensate* production in the vent.

Category IV. An *appliance* that operates with a positive vent static pressure and with a vent gas temperature that is capable of causing excessive *condensate* production in the vent.

VENTED ROOM HEATER. A vented self-contained, free-standing, nonrecessed *appliance* for furnishing warm air to the space in which it is installed, directly from the heater without duct connections.

VENTED WALL FURNACE. A self-contained vented *appliance* complete with grilles or equivalent, designed for incorporation in or permanent attachment to the structure of a building, mobile home or travel trailer, and furnishing heated air circulated by gravity or by a fan directly into the space to be heated through openings in the casing. This definition shall exclude *floor furnaces, unit heaters* and *central furnaces* as herein defined.

VENTING SYSTEM. A continuous open passageway from the *flue collar* or *draft hood* of an *appliance* to the outdoor atmosphere for the purpose of removing flue or vent gases. A venting system is usually composed of a vent or a chimney and *vent connector*, if used, assembled to form the open passageway.

WALL HEATER, UNVENTED TYPE. A room heater of the type designed for insertion in or attachment to a wall or partition. Such heater does not incorporate concealed venting arrangements in its construction and discharges all products of *combustion* through the front into the room being heated.

WATER HEATER. Any heating *appliance* or *equipment* that heats potable water and supplies such water to the potable hot water distribution system.

SECTION G2404 (301) GENERAL

G2404.1 (301.1) Scope. This section shall govern the approval and installation of all *equipment* and *appliances* that comprise parts of the installations regulated by this code in accordance with Section G2401.

G2404.2 (301.1.1) Other fuels. The requirements for *combustion* and *dilution air* for gas-fired *appliances* shall be governed by Section G2407. The requirements for *combustion* and *dilution air* for *appliances* operating with fuels other than fuel gas shall be regulated by Chapter 17.

G2404.3 (301.3) Listed and labeled. *Appliances* regulated by this code shall be *listed* and *labeled* for the application in which they are used unless otherwise *approved* in accordance with Section R104.11. The approval of unlisted *appliances* in accordance with Section R104.11 shall be based on *approved* engineering evaluation.

G2404.4 (301.8) Vibration isolation. Where means for isolation of vibration of an *appliance* is installed, an *approved* means for support and restraint of that *appliance* shall be provided.

G2404.5 (301.9) Repair. Defective material or parts shall be replaced or repaired in such a manner so as to preserve the original approval or listing.

G2404.6 (301.10) Wind resistance. *Appliances* and supports that are exposed to wind shall be designed and installed to resist the wind pressures determined in accordance with this code.

G2404.7 (301.11) Flood hazard. For structures located in flood hazard areas, the *appliance*, equipment and system installations regulated by this code shall be located at or above the elevation required by Section R322 for utilities and attendant equipment.

> **Exception:** The *appliance*, equipment and system installations regulated by this code are permitted to be located below the elevation required by Section R322 for utilities and attendant equipment provided that they are designed and installed to prevent water from entering or accumulating within the components and to resist hydrostatic and hydrodynamic loads and stresses, including the effects of buoyancy, during the occurrence of flooding to such elevation.

G2404.8 (301.12) Seismic resistance. Where earthquake loads are applicable in accordance with this code, the supports shall be designed and installed for the seismic forces in accordance with this code.

G2404.9 (301.14) Rodentproofing. Buildings or structures and the walls enclosing habitable or occupiable rooms and spaces in which persons live, sleep or work, or in which feed, food or foodstuffs are stored, prepared, processed, served or sold, shall be constructed to protect against the entry of rodents.

G2404.10 (307.5) Auxiliary drain pan. Category IV condensing *appliances* shall be provided with an auxiliary drain pan where damage to any building component will occur as a result of stoppage in the *condensate* drainage system. Such pan shall be installed in accordance with the applicable provisions of Section M1411.

> **Exception:** An auxiliary drain pan shall not be required for *appliances* that automatically shut down operation in the event of a stoppage in the *condensate* drainage system.

G2404.11 (307.6) Condensate pumps. Condensate pumps located in uninhabitable spaces, such as attics and crawl spaces, shall be connected to the *appliance* or *equipment* served such that when the pump fails, the *appliance* or *equipment* will be prevented from operating. Pumps shall be installed in accordance with the manufacturer's instructions.

SECTION G2405 (302) STRUCTURAL SAFETY

G2405.1 (302.1) Structural safety. The building shall not be weakened by the installation of any gas *piping*. In the process of installing or repairing any gas *piping*, the finished floors, walls, ceilings, tile work or any other part of the building or premises that is required to be changed or replaced shall be left in a safe structural condition in accordance with the requirements of this code.

G2405.2 (302.4) Alterations to trusses. Truss members and components shall not be cut, drilled, notched, spliced or otherwise altered in any way without the written concurrence and approval of a *registered design professional. Alterations* resulting in the addition of loads to any member, such as HVAC equipment and water heaters, shall not be permitted without verification that the truss is capable of supporting such additional loading.

G2405.3 (302.3.1) Engineered wood products. Cuts, notches and holes bored in trusses, *structural composite lumber*, structural glued-laminated members and I-joists are prohibited except where permitted by the manufacturer's recommendations or where the effects of such *alterations* are specifically considered in the design of the member by a *registered design professional*.

SECTION G2406 (303)
APPLIANCE LOCATION

G2406.1 (303.1) General. *Appliances* shall be located as required by this section, specific requirements elsewhere in this code and the conditions of the *equipment* and *appliance* listing.

G2406.2 (303.3) Prohibited locations. *Appliances* shall not be located in sleeping rooms, bathrooms, toilet rooms, storage closets or surgical rooms, or in a space that opens only into such rooms or spaces, except where the installation complies with one of the following:

1. The *appliance* is a direct-vent *appliance* installed in accordance with the conditions of the listing and the manufacturer's instructions.
2. *Vented room heaters*, wall *furnaces*, vented decorative *appliances*, vented gas *fireplaces*, vented gas *fireplace* heaters and decorative *appliances* for installation in vented solid fuel-burning *fireplaces* are installed in rooms that meet the required volume criteria of Section G2407.5.
3. A single wall-mounted *unvented room heater* is installed in a bathroom and such *unvented room heater* is equipped as specified in Section G2445.6 and has an input rating not greater than 6,000 *Btu*/h (1.76 kW). The bathroom shall meet the required volume criteria of Section G2407.5.
4. A single wall-mounted *unvented room heater* is installed in a bedroom and such *unvented room heater* is equipped as specified in Section G2445.6 and has an input rating not greater than 10,000 *Btu*/h (2.93 kW). The bedroom shall meet the required volume criteria of Section G2407.5.
5. The *appliance* is installed in a room or space that opens only into a bedroom or bathroom, and such room or space is used for no other purpose and is provided with a solid weather-stripped door equipped with an *approved* self-closing device. *Combustion air* shall be taken directly from the outdoors in accordance with Section G2407.6.
6. A clothes dryer is installed in a residential bathroom or toilet room having a permanent opening with an area of not less than 100 square inches (0.06 m^2) that communicates with a space outside of a sleeping room, bathroom, toilet room or storage closet.

G2406.3 (303.6) Outdoor locations. *Appliances* installed in outdoor locations shall be either *listed* for outdoor installation or provided with protection from outdoor environmental factors that influence the operability, durability and safety of the *appliance*.

SECTION G2407 (304)
COMBUSTION, VENTILATION AND DILUTION AIR

G2407.1 (304.1) General. Air for *combustion*, *ventilation* and dilution of *flue gases* for *appliances* installed in buildings shall be provided by application of one of the methods prescribed in Sections G2407.5 through G2407.9. Where the requirements of Section G2407.5 are not met, outdoor air shall be introduced in accordance with one of the methods prescribed in Sections G2407.6 through G2407.9. *Direct-vent appliances*, gas *appliances* of other than *natural draft* design, vented gas *appliances* not designated as Category I and appliances equipped with power burners, shall be provided with *combustion*, ventilation and *dilution air* in accordance with the *appliance* manufacturer's instructions.

Exception: *Type 1 clothes dryers* that are provided with *makeup air* in accordance with Section G2439.5.

G2407.2 (304.2) Appliance location. *Appliances* shall be located so as not to interfere with proper circulation of *combustion*, *ventilation* and *dilution air*.

G2407.3 (304.3) Draft hood/regulator location. Where used, a *draft hood* or a *barometric draft regulator* shall be installed in the same room or enclosure as the *appliance* served to prevent any difference in pressure between the hood or regulator and the *combustion air* supply.

G2407.4 (304.4) Makeup air provisions. Where exhaust fans, *clothes dryers* and kitchen ventilation systems interfere with the operation of *appliances*, *makeup air* shall be provided.

G2407.5 (304.5) Indoor combustion air. The required volume of indoor air shall be determined in accordance with Section G2407.5.1 or G2407.5.2, except that where the air infiltration rate is known to be less than 0.40 air changes per hour (ACH), Section G2407.5.2 shall be used. The total required volume shall be the sum of the required volume calculated for all *appliances* located within the space. Rooms communicating directly with the space in which the *appliances* are installed through openings not furnished with doors, and through *combustion air* openings sized and located in accordance with Section G2407.5.3, are considered to be part of the required volume.

G2407.5.1 (304.5.1) Standard method. The minimum required volume shall be 50 cubic feet per 1,000 *Btu*/h (4.8 m^3/kW) of the *appliance* input rating.

G2407.5.2 (304.5.2) Known air-infiltration-rate method. Where the air infiltration rate of a structure is known, the minimum required volume shall be determined as follows:

For *appliances* other than fan-assisted, calculate volume using Equation 24-1.

$$Required\ Volume_{other} \geq \frac{21\text{ft}^3}{ACH}\left(\frac{I_{other}}{1{,}000\ \text{Btu/h}}\right)$$

(Equation 24-1)

For fan-assisted *appliances*, calculate volume using Equation 24-2.

$$Required\ Volume_{fan} \geq \frac{15\text{ft}^3}{ACH}\left(\frac{I_{fan}}{1{,}000\ \text{Btu/h}}\right)$$

(Equation 24-2)

where:

I_{other} = All *appliances* other than fan assisted (input in *Btu*/h).

I_{fan} = Fan-assisted *appliance* (input in *Btu*/h).

ACH = Air change per hour (percent of volume of space exchanged per hour, expressed as a decimal).

For purposes of this calculation, an infiltration rate greater than 0.60 *ACH* shall not be used in Equations 24-1 and 24-2.

G2407.5.3 (304.5.3) Indoor opening size and location. Openings used to connect indoor spaces shall be sized and located in accordance with Sections G2407.5.3.1 and G2407.5.3.2 (see Figure G2407.5.3).

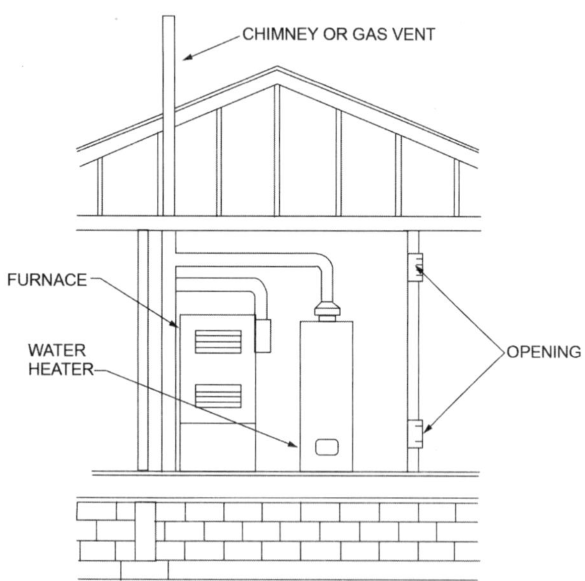

FIGURE G2407.5.3 (304.5.3)
ALL AIR FROM INSIDE THE BUILDING
(see Section G2407.5.3)

G2407.5.3.1 (304.5.3.1) Combining spaces on the same story. Where combining spaces on the same story, each opening shall have a minimum free area of 1 square inch per 1,000 *Btu*/h (2,200 mm²/kW) of the total input rating of all *appliances* in the space, but not less than 100 square inches (0.06 m²). One permanent opening shall commence within 12 inches (305 mm) of the top and one permanent opening shall commence within 12 inches (305 mm) of the bottom of the enclosure. The minimum dimension of air openings shall be not less than 3 inches (76 mm).

G2407.5.3.2 (304.5.3.2) Combining spaces in different stories. The volumes of spaces in different stories shall be considered to be communicating spaces where such spaces are connected by one or more permanent openings in doors or floors having a total minimum free area of 2 square inches per 1,000 *Btu*/h (4402 mm²/kW) of total input rating of all *appliances*.

G2407.6 (304.6) Outdoor combustion air. Outdoor *combustion* air shall be provided through opening(s) to the outdoors in accordance with Section G2407.6.1 or G2407.6.2. The minimum dimension of air openings shall be not less than 3 inches (76 mm).

G2407.6.1 (304.6.1) Two-permanent-openings method. Two permanent openings, one commencing within 12 inches (305 mm) of the top and one commencing within 12 inches (305 mm) of the bottom of the enclosure, shall be provided. The openings shall communicate directly or by ducts with the outdoors or spaces that freely communicate with the outdoors.

Where directly communicating with the outdoors, or where communicating with the outdoors through vertical ducts, each opening shall have a minimum free area of 1 square inch per 4,000 *Btu*/h (550 mm²/kW) of total input rating of all *appliances* in the enclosure [see Figures G2407.6.1(1) and G2407.6.1(2)].

Where communicating with the outdoors through horizontal ducts, each opening shall have a minimum free area of not less than 1 square inch per 2,000 *Btu*/h (1100 mm²/kW) of total input rating of all *appliances* in the enclosure [see Figure G2407.6.1(3)].

G2407.6.2 (304.6.2) One-permanent-opening method. One permanent opening, commencing within 12 inches (305 mm) of the top of the enclosure, shall be provided. The *appliance* shall have *clearances* of not less than 1 inch (25 mm) from the sides and back and 6 inches (152 mm) from the front of the *appliance*. The opening shall directly communicate with the outdoors or through a vertical or horizontal duct to the outdoors, or spaces that freely communicate with the outdoors (see Figure G2407.6.2) and shall have a minimum free area of 1 square inch per 3,000 *Btu*/h (734 mm²/kW) of the total input rating of all *appliances* located in the enclosure and not less than the sum of the areas of all *vent connectors* in the space.

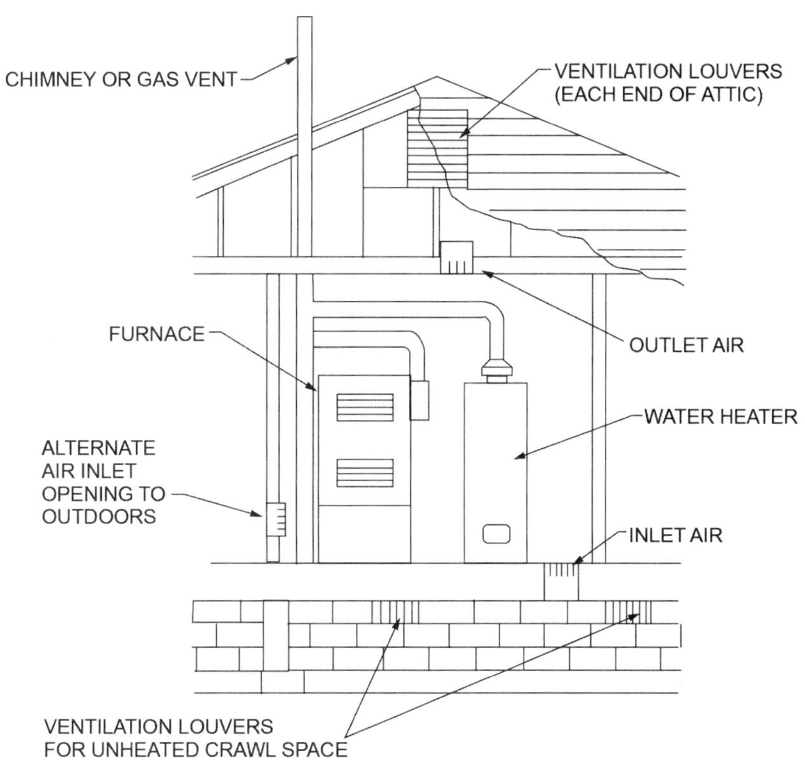

FIGURE G2407.6.1(1) [304.6.1(1)]
ALL AIR FROM OUTDOORS—INLET AIR FROM VENTILATED CRAWL SPACE AND OUTLET AIR TO VENTILATED ATTIC
(see Section G2407.6.1)

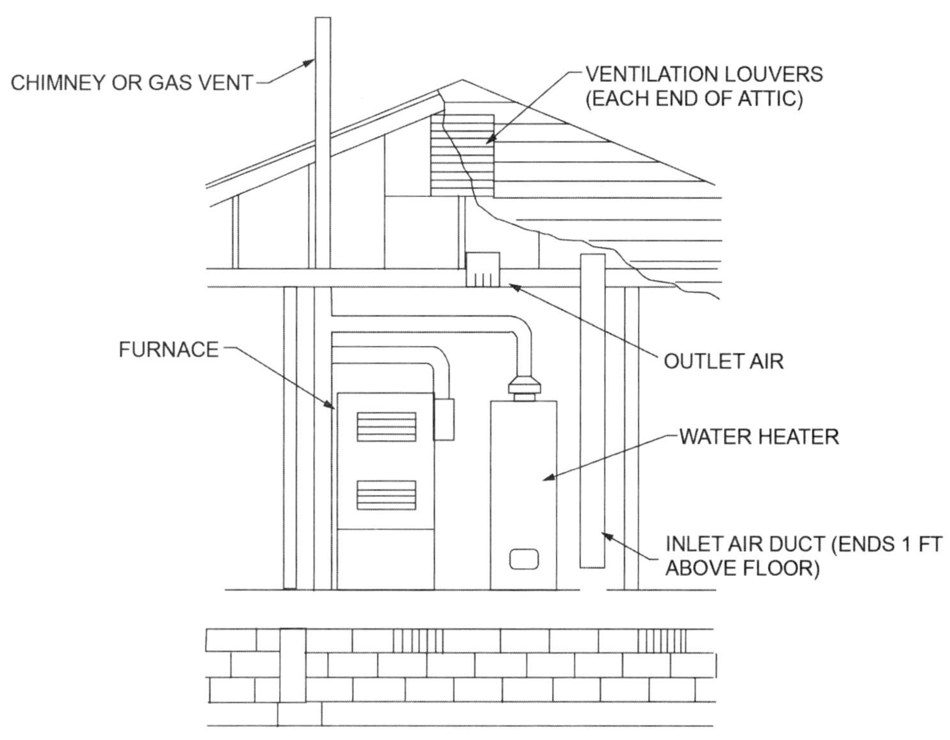

For SI: 1 foot = 304.8 mm.

FIGURE G2407.6.1(2) [304.6.1(2)]
ALL AIR FROM OUTDOORS THROUGH VENTILATED ATTIC
(see Section G2407.6.1)

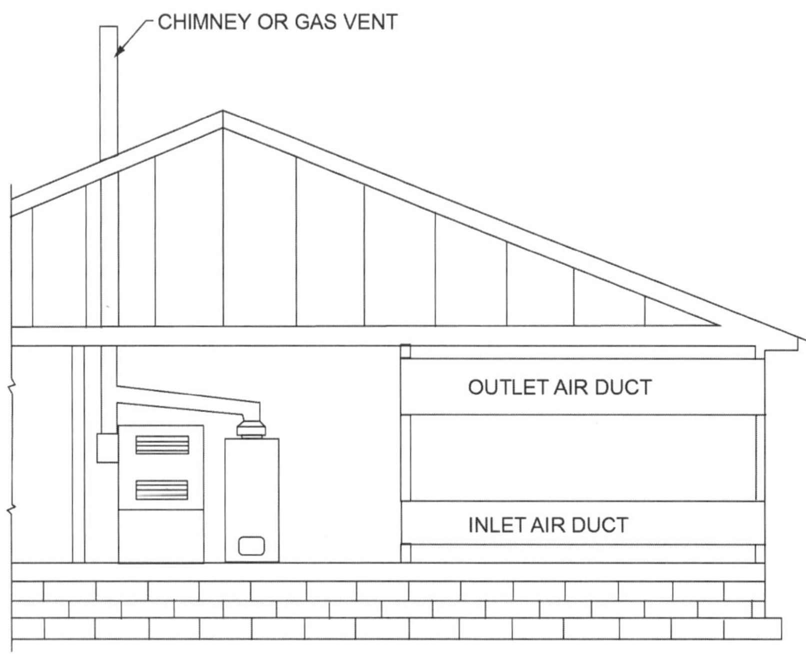

FIGURE G2407.6.1(3) [304.6.1(3)]
ALL AIR FROM OUTDOORS (see Section G2407.6.1)

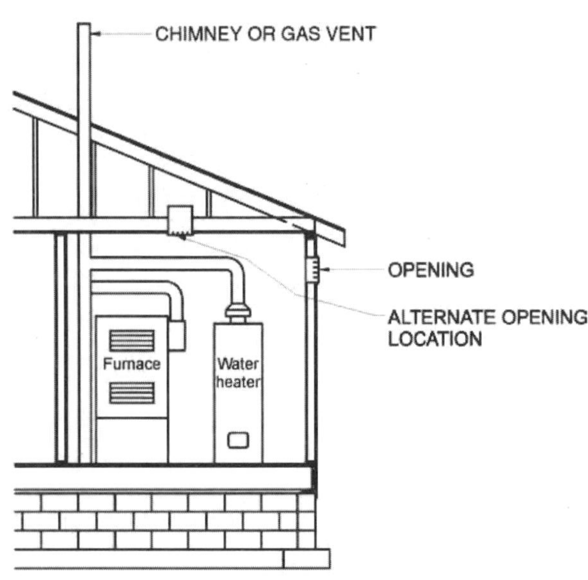

FIGURE G2407.6.2 (304.6.2)
SINGLE COMBUSTION AIR OPENING,
ALL AIR FROM OUTDOORS
(see Section G2407.6.2)

G2407.7 (304.7) Combination indoor and outdoor combustion air. The use of a combination of indoor and outdoor *combustion air* shall be in accordance with Sections G2407.7.1 through G2407.7.3.

G2407.7.1 (304.7.1) Indoor openings. Where used, openings connecting the interior spaces shall comply with Section G2407.5.3.

G2407.7.2 (304.7.2) Outdoor opening location. Outdoor opening(s) shall be located in accordance with Section G2407.6.

G2407.7.3 (304.7.3) Outdoor opening(s) size. The outdoor opening(s) size shall be calculated in accordance with the following:

1. The ratio of interior spaces shall be the available volume of all communicating spaces divided by the required volume.
2. The outdoor size reduction factor shall be one minus the ratio of interior spaces.
3. The minimum size of outdoor opening(s) shall be the full size of outdoor opening(s) calculated in accordance with Section G2407.6, multiplied by the reduction factor. The minimum dimension of air openings shall be not less than 3 inches (76 mm).

G2407.8 (304.8) Engineered installations. Engineered *combustion air* installations shall provide an adequate supply of *combustion*, *ventilation* and *dilution air* determined using *approved* engineering methods.

G2407.9 (304.9) Mechanical combustion air supply. Where all *combustion air* is provided by a mechanical air supply system, the *combustion air* shall be supplied from the outdoors at a rate not less than 0.35 cubic feet per minute per 1,000 *Btu*/h (0.034 m^3/min per kW) of total input rating of all *appliances* located within the space.

G2407.9.1 (304.9.1) Makeup air. Where exhaust fans are installed, *makeup air* shall be provided to replace the exhausted air.

G2407.9.2 (304.9.2) Appliance interlock. Each of the *appliances* served shall be interlocked with the mechanical air supply system to prevent *main burner* operation when the mechanical air supply system is not in operation.

G2407.9.3 (304.9.3) Combined combustion air and ventilation air system. Where *combustion air* is provided by the building's mechanical ventilation system, the system shall provide the specified *combustion air* rate in addition to the required ventilation air.

G2407.10 (304.10) Louvers and grilles. The required size of openings for *combustion, ventilation* and *dilution air* shall be based on the net free area of each opening. Where the free area through a design of louver, grille or screen is known, it shall be used in calculating the size opening required to provide the free area specified. Where the design and free area of louvers and grilles are not known, it shall be assumed that wood louvers will have 25-percent free area and metal louvers and grilles will have 75-percent free area. Screens shall have a mesh size not smaller than $^1/_4$ inch (6.4 mm). Nonmotorized louvers and grilles shall be fixed in the open position. Motorized louvers shall be interlocked with the *appliance* so that they are proven to be in the full open position prior to *main burner* ignition and during *main burner* operation. Means shall be provided to prevent the *main burner* from igniting if the louvers fail to open during *burner* start-up and to shut down the *main burner* if the louvers close during operation.

G2407.11 (304.11) Combustion air ducts. *Combustion air* ducts shall comply with all of the following:

1. Ducts shall be constructed of galvanized steel complying with Chapter 16 or of a material having equivalent corrosion resistance, strength and rigidity.

   Exception: Within dwellings units, unobstructed stud and joist spaces shall not be prohibited from conveying *combustion air*, provided that not more than one required fireblock is removed.

2. Ducts shall terminate in an unobstructed space allowing free movement of *combustion air* to the *appliances.*
3. Ducts shall serve a single enclosure.
4. Ducts shall not serve both upper and lower *combustion air* openings where both such openings are used. The separation between ducts serving upper and lower *combustion air* openings shall be maintained to the source of *combustion air.*
5. Ducts shall not be screened where terminating in an attic space.
6. Horizontal upper *combustion air* ducts shall not slope downward toward the source of *combustion air.*
7. The remaining space surrounding a *chimney* liner, gas vent, special gas vent or plastic *piping* installed within a masonry, metal or factory-built *chimney* shall not be used to supply *combustion air.*

   Exception: Direct-vent gas-fired *appliances* designed for installation in a solid fuel-burning *fireplace* where installed in accordance with the manufacturer's instructions.

8. *Combustion air* intake openings located on the exterior of a building shall have the lowest side of such openings located not less than 12 inches (305 mm) vertically from the adjoining finished ground level.

G2407.12 (304.12) Protection from fumes and gases. Where corrosive or flammable process fumes or gases, other than products of *combustion,* are present, means for the disposal of such fumes or gases shall be provided. Such fumes or gases include carbon monoxide, hydrogen sulfide, ammonia, chlorine and halogenated hydrocarbons.

In barbershops, beauty shops and other facilities where chemicals that generate corrosive or flammable products, such as aerosol sprays, are routinely used, nondirect vent-type *appliances* shall be located in a mechanical room separated or partitioned off from other areas with provisions for *combustion air* and *dilution air* from the outdoors. *Direct-vent appliances* shall be installed in accordance with the *appliance* manufacturer's instructions.

SECTION G2408 (305) INSTALLATION

G2408.1 (305.1) General. *Equipment* and *appliances* shall be installed as required by the terms of their approval, in accordance with the conditions of listing, the manufacturer's instructions and this code. Manufacturer's installation instructions shall be available on the job site at the time of inspection. Where a code provision is less restrictive than the conditions of the listing of the *equipment* or *appliance* or the manufacturer's installation instructions, the conditions of the listing and the manufacturer's installation instructions shall apply.

Unlisted *appliances approved* in accordance with Section G2404.3 shall be limited to uses recommended by the manufacturer and shall be installed in accordance with the manufacturer's instructions, the provisions of this code and the requirements determined by the *code official.*

G2408.2 (305.3) Elevation of ignition source. *Equipment* and *appliances* having an *ignition source* shall be elevated such that the source of ignition is not less than 18 inches (457 mm) above the floor in *hazardous locations* and public garages, private garages, repair garages, motor fuel-dispensing facilities and parking garages. For the purpose of this section, rooms or spaces that are not part of the *living space* of a *dwelling unit* and that communicate directly with a private garage through openings shall be considered to be part of the private garage.

Exception: Elevation of the *ignition source* is not required for *appliances* that are *listed* as flammable-vapor-ignition resistant.

G2408.2.1 (305.3.1) Installation in residential garages. In residential garages where *appliances* are installed in a separate, enclosed space having access only from outside of the garage, such *appliances* shall be permitted to be installed at floor level, provided that the required *combustion air* is taken from the exterior of the garage.

G2408.3 (305.5) Private garages. *Appliances* located in private garages shall be installed with a minimum *clearance* of 6 feet (1829 mm) above the floor.

Exception: The requirements of this section shall not apply where the *appliances* are protected from motor vehicle impact and installed in accordance with Section G2408.2.

G2408.4 (305.7) Clearances from grade. *Equipment* and *appliances* installed at grade level shall be supported on a level concrete slab or other *approved* material extending not less than 3 inches (76 mm) above adjoining grade or shall be suspended not less than 6 inches (152 mm) above adjoining grade. Such supports shall be installed in accordance with the manufacturer's instructions.

G2408.5 (305.8) Clearances to combustible construction. Heat-producing *equipment* and *appliances* shall be installed to maintain the required clearances to combustible construction as specified in the listing and manufacturer's instructions. Such *clearances* shall be reduced only in accordance with Section G2409. *Clearances* to combustibles shall include such considerations as door swing, drawer pull, overhead projections or shelving and window swing. Devices, such as door stops or limits and closers, shall not be used to provide the required *clearances.*

G2408.6 (305.12) Avoid strain on gas piping. *Appliances* shall be supported and connected to the *piping* so as not to exert undue strain on the connections.

SECTION G2409 (308) CLEARANCE REDUCTION

G2409.1 (308.1) Scope. This section shall govern the reduction in required clearances to *combustible materials,* including gypsum board, and *combustible assemblies* for chimneys, vents, *appliances,* devices and equipment. Clearance requirements for air-conditioning equipment and central heating boilers and furnaces shall comply with Sections G2409.3 and G2409.4.

G2409.2 (308.2) Reduction table. The allowable *clearance* reduction shall be based on one of the methods specified in Table G2409.2 or shall utilize a reduced *clearance* protective assembly *listed* and *labeled* in accordance with UL 1618. Where required clearances are not listed in Table G2409.2, the reduced clearances shall be determined by linear interpolation between the distances listed in the table. Reduced clearances shall not be derived by extrapolation below the range of the table. The reduction of the required clearances to combustibles for *listed* and *labeled appliances* and *equipment* shall be in accordance with the requirements of this section, except that such clearances shall not be reduced where reduction is specifically prohibited by the terms of the *appliance* or equipment listing [see Figures G2409.2(1) through 2409.2(3)].

G2409.3 (308.3) Clearances for indoor air-conditioning appliances. *Clearance* requirements for indoor air-conditioning *appliances* shall comply with Sections G2409.3.1 through G2409.3.4.

G2409.3.1 (308.3.1) Appliances clearances. Air-conditioning *appliances* shall be installed with clearances in accordance with the manufacturer's instructions.

G2409.3.2 (308.3.2) Clearance reduction. Air-conditioning *appliances* shall be permitted to be installed with reduced clearances to *combustible material,* provided that the *combustible material* or *appliance* is protected as described in Table G2409.2 and such reduction is allowed by the manufacturer's instructions.

G2409.3.3 (308.3.3) Plenum clearances. Where the *furnace plenum* is adjacent to plaster on metal lath or *noncombustible material* attached to *combustible material,* the *clearance* shall be measured to the surface of the plaster or other noncombustible finish where the *clearance* specified is 2 inches (51 mm) or less.

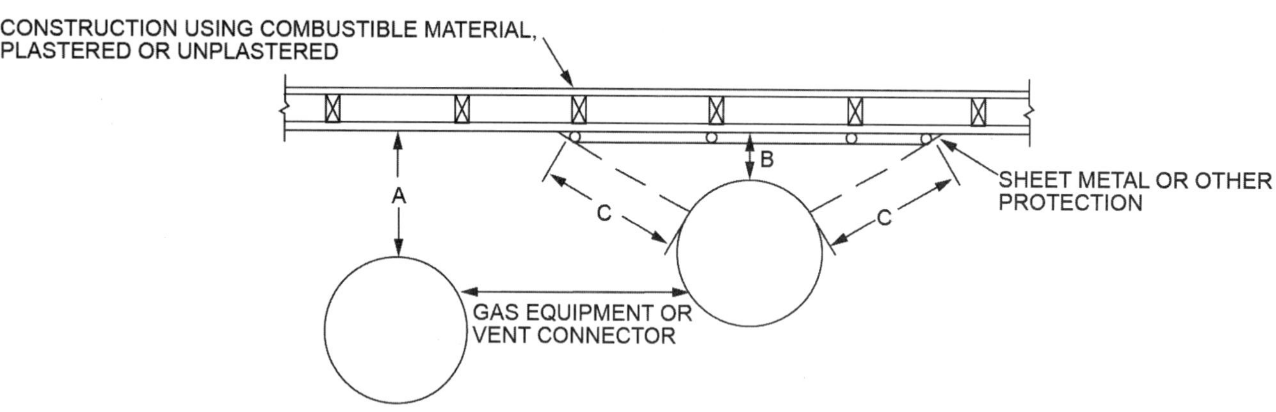

NOTES:

A = The clearance without protection.

B = The reduced clearance permitted in accordance with Table G2409.2. The protection applied to the construction using combustible material shall extend far enough in each direction to make "C" equal to "A."

FIGURE G2409.2(1) [308.2(1)]
EXTENT OF PROTECTION NECESSARY TO REDUCE CLEARANCES FROM GAS EQUIPMENT OR VENT CONNECTORS

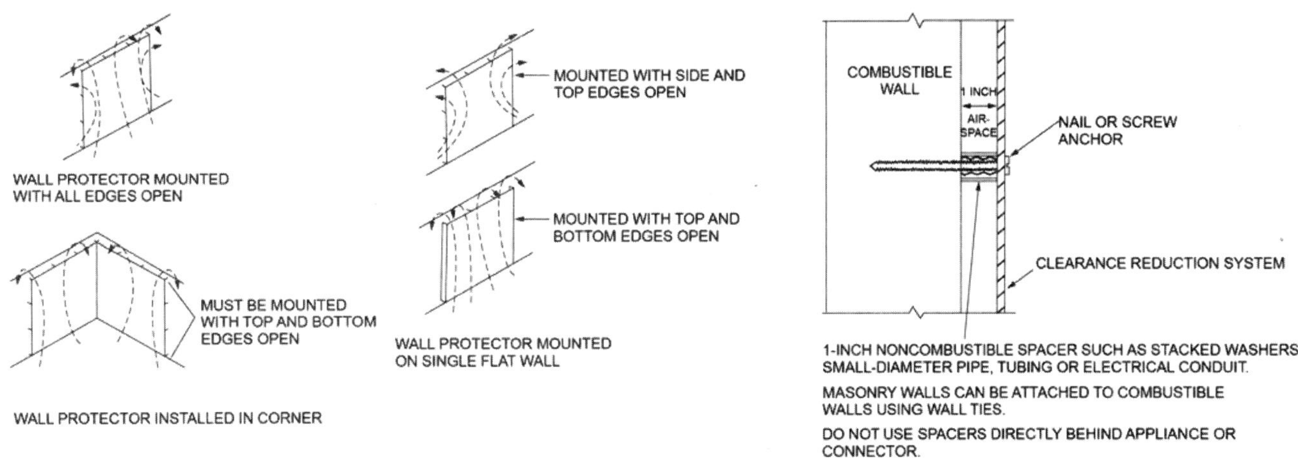

For SI: 1 inch = 25.4 mm.

FIGURE G2409.2(2) [308.2(2)]
WALL PROTECTOR CLEARANCE REDUCTION SYSTEM

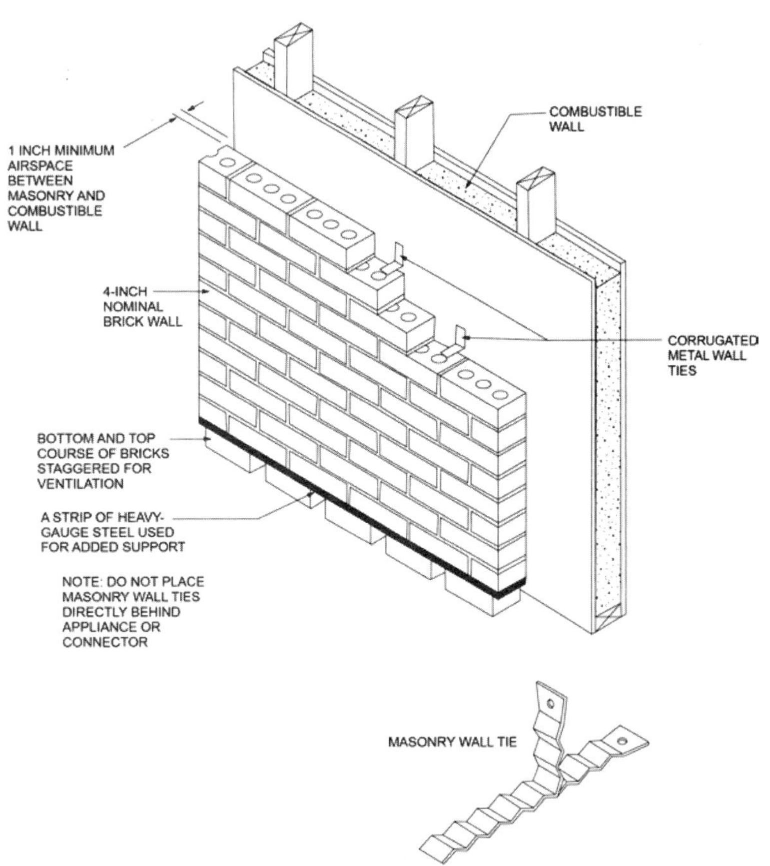

For SI: 1 inch = 25.4 mm.

FIGURE G2409.2(3) [308.2(3)]
MASONRY CLEARANCE REDUCTION SYSTEM

TABLE G2409.2 (308.2)
REDUCTION OF CLEARANCES WITH SPECIFIED FORMS OF PROTECTION [a through k]

TYPE OF PROTECTION APPLIED TO AND COVERING ALL SURFACES OF COMBUSTIBLE MATERIAL WITHIN THE DISTANCE SPECIFIED AS THE REQUIRED CLEARANCE WITH NO PROTECTION [see Figures G2409.2(1), G2409.2(2), and G2409.2(3)]	WHERE THE REQUIRED CLEARANCE WITH NO PROTECTION FROM APPLIANCE, VENT CONNECTOR, OR SINGLE-WALL METAL PIPE IS: (inches)									
	36		18		12		9		6	
	Allowable clearances with specified protection (inches)									
	Use Column 1 for clearances above appliance or horizontal connector. Use Column 2 for clearances from appliance, vertical connector and single-wall metal pipe.									
	Above Col. 1	Sides and rear Col. 2	Above Col. 1	Sides and rear Col. 2	Above Col. 1	Sides and rear Col. 2	Above Col. 1	Sides and rear Col. 2	Above Col. 1	Sides and rear Col. 2
1. $3^1/_2$-inch-thick masonry wall without ventilated airspace	—	24	—	12	—	9	—	6	—	5
2. $^1/_2$-inch insulation board over 1-inch glass fiber or mineral wool batts	24	18	12	9	9	6	6	5	4	3
3. 0.024-inch (nominal 24 gage) sheet metal over 1-inch glass fiber or mineral wool batts reinforced with wire on rear face with ventilated airspace	18	12	9	6	6	4	5	3	3	3
4. $3^1/_2$-inch-thick masonry wall with ventilated airspace	—	12	—	6	—	6	—	6	—	6
5. 0.024-inch (nominal 24 gage) sheet metal with ventilated airspace	18	12	9	6	6	4	5	3	3	2
6. $^1/_2$-inch-thick insulation board with ventilated airspace	18	12	9	6	6	4	5	3	3	3
7. 0.024-inch (nominal 24 gage) sheet metal with ventilated airspace over 0.024-inch (nominal 24 gage) sheet metal with ventilated airspace	18	12	9	6	6	4	5	3	3	3
8. 1-inch glass fiber or mineral wool batts sandwiched between two sheets 0.024-inch (nominal 24 gage) sheet metal with ventilated airspace	18	12	9	6	6	4	5	3	3	3

For SI: 1 inch = 25.4 mm, °C = [(°F – 32)/1.8], 1 pound per cubic foot = 16.02 kg/m^3, 1 Btu per inch per square foot per hour per °F = 0.144 W/m^2 × K.

a. Reduction of clearances from combustible materials shall not interfere with combustion air, draft hood clearance and relief, and accessibility of servicing.
b. Clearances shall be measured from the outer surface of the combustible material to the nearest point on the surface of the appliance, disregarding any intervening protection applied to the combustible material.
c. Spacers and ties shall be of noncombustible material. A spacer or tie shall not be used directly opposite an appliance or connector.
d. For all clearance reduction systems using a ventilated airspace, adequate provision for air circulation shall be provided as described [see Figures G2409.2(2) and G2409.2(3)].
e. There shall be not less than 1 inch between clearance reduction systems and combustible walls and ceilings for reduction systems using ventilated airspace.
f. Where a wall protector is mounted on a single flat wall away from corners, it shall have a minimum 1-inch air gap. To provide air circulation, the bottom and top edges, or only the side and top edges, or all edges shall be left open.
g. Mineral wool batts (blanket or board) shall have a density of 8 pounds per cubic foot and a minimum melting point of 1500°F.
h. Insulation material used as part of a clearance reduction system shall have a thermal conductivity of 1.0 Btu per inch per square foot per hour per °F or less.
i. There shall be not less than 1 inch between the appliance and the protector. In no case shall the clearance between the appliance and the combustible surface be reduced below that allowed in this table.
j. Clearances and thicknesses are minimum; larger clearances and thicknesses are acceptable.
k. Listed single-wall connectors shall be installed in accordance with the manufacturer's instructions.

G2409.3.4 (308.3.4) Clearance from supply ducts. Supply air ducts connecting to *listed* central heating furnaces shall have the same minimum clearance to combustibles as required for the furnace supply plenum for a distance of not less than 3 feet (914 mm) from the supply plenum. Clearance is not required beyond the 3-foot (914 mm) distance.

G2409.4 (308.4) Central-heating boilers and furnaces. *Clearance* requirements for central-heating boilers and *furnaces* shall comply with Sections G2409.4.1 through G2409.4.5. The *clearance* to these *appliances* shall not interfere with *combustion air, draft hood clearance* and relief, and accessibility for servicing.

G2409.4.1 (308.4.1) Appliances clearances. Central-heating furnaces and low-pressure boilers shall be installed with clearances in accordance with the manufacturer's instructions.

G2409.4.2 (308.4.2) Clearance reduction. Central-heating furnaces and low-pressure boilers shall be permitted to be installed with reduced clearances to *combustible material* provided that the *combustible material* or *appli-*

ance is protected as described in Table G2409.2 and such reduction is allowed by the manufacturer's instructions.

G2409.4.3 (308.4.4) Plenum clearances. Where the *furnace plenum* is adjacent to plaster on metal lath or *noncombustible material* attached to *combustible material*, the *clearance* shall be measured to the surface of the plaster or other noncombustible finish where the *clearance* specified is 2 inches (51 mm) or less.

G2409.4.4 (308.4.5) Clearance from supply ducts. Supply air ducts connecting to *listed* central heating furnaces shall have the same minimum clearance to combustibles as required for the furnace supply plenum for a distance of not less than 3 feet (914 mm) from the supply plenum. Clearance is not required beyond the 3-foot (914 mm) distance.

G2409.4.5 (308.4.3) Clearance for servicing appliances. Front *clearance* shall be sufficient for servicing the *burner* and the *furnace* or boiler.

SECTION G2410 (309) ELECTRICAL

G2410.1 (309.1) Grounding. *Gas piping* shall not be used as a grounding electrode.

G2410.2 (309.2) Connections. Electrical connections between *appliances* and the building wiring, including the grounding of the *appliances*, shall conform to Chapters 34 through 43.

SECTION G2411 (310) ELECTRICAL BONDING

G2411.1 (310.1) Pipe and tubing other than CSST. Each above-ground portion of a *gas piping system* other than corrugated stainless steel tubing (CSST) that is likely to become energized shall be electrically continuous and bonded to an effective ground-fault current path. *Gas piping* other than CSST shall be considered to be bonded where it is connected to an *appliance* that is connected to the equipment grounding conductor of the circuit that supplies that *appliance*.

G2411.2 (310.2) CSST. This section applies to corrugated stainless steel tubing (CSST) that is not *listed* with an arc-resistant jacket or coating system in accordance with ANSI LC1/CSA 6.26. CSST *gas piping systems* and piping systems containing one or more segments of CSST shall be electrically continuous and bonded to the electrical service grounding electrode system or, where provided, the lightning protection grounding electrode system.

G2411.2.1 (310.2.1) Point of connection. The bonding jumper shall connect to a metallic pipe, pipe fitting or CSST fitting.

G2411.2.2 (310.2.2) Size and material of jumper. The bonding jumper shall be not smaller than 6 AWG copper wire or equivalent.

G2411.2.3 (310.2.3) Bonding jumper length. The length of the bonding jumper between the connection to a *gas piping system* and the connection to a grounding electrode system shall not exceed 75 feet (22 860 mm). Any additional grounding electrodes installed to meet this requirement shall be bonded to the electrical service grounding electrode system or, where provided, the lightning protection grounding electrode system.

G2411.2.4 (310.2.4) Bonding connections. Bonding connections shall be in accordance with NFPA 70.

G2411.2.5 (310.2.5) Connection devices. Devices used for making the bonding connections shall be *listed* for the application in accordance with UL 467.

G2411.3 (310.3) Arc-resistant CSST. This section applies to corrugated stainless steel tubing (CSST) that is *listed* with an arc-resistant jacket or coating system in accordance with ANSI LC1/CSA 6.26. The CSST shall be electrically continuous and bonded to an effective ground fault current path. Where any CSST component of a *piping system* does not have an arc-resistant jacket or coating system, the bonding requirements of Section G2411.2 shall apply. Arc-resistant-jacketed CSST shall be considered to be bonded where it is connected to an *appliance* that is connected to the *appliance* grounding conductor of the circuit that supplies that *appliance*.

SECTION G2412 (401) GENERAL

G2412.1 (401.1) Scope. This section shall govern the design, installation, modification and maintenance of *piping systems*. The applicability of this code to *piping systems* extends from the *point of delivery* to the connections with the *appliances* and includes the design, materials, components, fabrication, assembly, installation, testing, inspection, operation and maintenance of such *piping systems*.

G2412.1.1 (401.1.1) Utility piping systems located within buildings. Utility service *piping* located within buildings shall be installed in accordance with the structural safety and fire protection provisions of this code.

G2412.2 (401.2) Liquefied petroleum gas storage. The storage system for *liquefied petroleum gas* shall be designed and installed in accordance with the *International Fire Code* and NFPA 58.

G2412.3 (401.3) Modifications to existing systems. In modifying or adding to existing *piping systems*, sizes shall be maintained in accordance with this chapter.

G2412.4 (401.4) Additional appliances. Where an additional *appliance* is to be served, the existing *piping* shall be checked to determine if it has adequate capacity for all *appliances* served. If inadequate, the existing system shall be enlarged as required or separate *piping* of adequate capacity shall be provided.

G2412.5 (401.5) Identification. For other than steel *pipe* and CSST, exposed *piping* shall be identified by a yellow *label* marked "Gas" in black letters. The marking shall be spaced at intervals not exceeding 5 feet (1524 mm). The marking shall not be required on piping located in the same room as the *appliance* served. CSST shall be identified as required by ANSI LC1/CSA 6.26.

G2412.6 (401.6) Interconnections. Where two or more *meters* are installed on the same premises but supply separate consumers, the *piping systems* shall not be interconnected on the outlet side of the *meters.*

G2412.7 (401.7) Piping meter identification. *Piping* from multiple *meter* installations shall be marked with an *approved* permanent identification by the installer so that the *piping system* supplied by each *meter* is readily identifiable.

G2412.8 (401.8) Minimum sizes. Pipe utilized for the installation, extension and *alteration* of any *piping system* shall be sized to supply the full number of outlets for the intended purpose and shall be sized in accordance with Section G2413.

G2412.9 (401.9) Identification. Each length of pipe and tubing and each pipe fitting, utilized in a fuel gas system, shall bear the identification of the manufacturer.

Exceptions:

1. Steel pipe sections that are 2 feet (610 mm) and less in length and are cut from longer sections of pipe.
2. Steel pipe fittings 2 inches and less in size.
3. Where identification is provided on the product packaging or crating.
4. Where other *approved* documentation is provided.

G2412.10 (401.10) Piping materials standards. Piping, tubing and fittings shall be manufactured to the applicable referenced standards, specifications and performance criteria listed in Section G2414 and shall be identified in accordance with Section G2412.9.

SECTION G2413 (402) PIPE SIZING

G2413.1 (402.1) General considerations. *Piping systems* shall be of such size and so installed as to provide a supply of gas sufficient to meet the maximum *demand* and supply gas to each *appliance* inlet at not less than the minimum supply pressure required by the *appliance.*

G2413.2 (402.2) Maximum gas demand. The volumetric flow rate of gas to be provided shall be the sum of the maximum input of the *appliances* served.

The total connected hourly load shall be used as the basis for pipe sizing, assuming that all *appliances* could be operating at full capacity simultaneously. Where a diversity of load can be established, pipe sizing shall be permitted to be based on such loads.

The volumetric flow rate of gas to be provided shall be adjusted for altitude where the installation is above 2,000 feet (610 m) in elevation.

G2413.3 (402.3) Sizing. *Gas piping* shall be sized in accordance with one of the following:

1. *Pipe* sizing tables or sizing equations in accordance with Section G2413.4 or G2413.5, as applicable.
2. The sizing tables included in a *listed piping* system's manufacturer's installation instructions.
3. *Approved engineering methods.*

G2413.4 (402.4) Sizing tables and equations. This section applies to piping materials other than noncorrugated stainless steel tubing. Where Tables G2413.4(1) through G2413.4(21) are used to size *piping* or *tubing*, the *pipe* length shall be determined in accordance with Section G2413.4.1, G2413.4.2 or G2413.4.3.

Where Equations 24-3 and 24-4 are used to size *piping* or *tubing*, the *pipe* or *tubing* shall have smooth inside walls and the pipe length shall be determined in accordance with Section G2413.4.1, G2413.4.2 or G2413.4.3.

1. Low-pressure gas equation [less than $1^1/_2$ pounds per square inch (psi) (10.3 kPa)]:

$$D = \frac{Q^{0.381}}{19.17\left(\frac{\Delta H}{C_r \times L}\right)^{0.206}} \qquad \textbf{(Equation 24-3)}$$

2. High-pressure gas equation [$1^1/_2$ psi (10.3 kPa) and above]:

$$D = \frac{Q^{0.381}}{18.93\left[\frac{(P_1^2 - P_2^2) \times Y}{C_r \times L}\right]^{0.206}} \qquad \textbf{(Equation 24-4)}$$

where:

C_r = Value determined by Table G2413.4.

D = Inside diameter of pipe, inches (mm).

Q = Input rate appliance(s), cubic feet per hour at 60°F (16°C) and 30-inch mercury column.

P_1 = Upstream pressure, psia (P_1 + 14.7).

P_2 = Downstream pressure, psia (P_2 + 14.7).

L = Equivalent length of pipe, feet.

Y = Value determined by Table G2413.4.

ΔH = Pressure drop, inch water column (27.7-inch water column = 1 psi).

TABLE G2413.4 (402.4)
C_r AND Y VALUES FOR NATURAL GAS AND UNDILUTED PROPANE AT STANDARD CONDITIONS

GAS	EQUATION FACTORS	
	C_r	Y
Natural gas	0.6094	0.9992
Undiluted propane	1.2462	0.9910

For SI: 1 cubic foot = 0.028 m^3, 1 foot = 305 mm,
1-inch water column = 0.249 kPa,
1 pound per square inch = 6.895 kPa,
1 British thermal unit per hour = 0.293 W.

G2413.4.1 (402.4.1) Longest length method. The *pipe* size of each section of *gas piping* shall be determined using the longest length of *piping* from the *point of delivery* to the most remote *outlet* and the load of the section.

TABLE G2413.4(1) [402.4(2)] SCHEDULE 40 METALLIC PIPE

Gas	Natural
Inlet Pressure	Less than 2 psi
Pressure Drop	0.5 in. w.c.
Specific Gravity	0.60

	PIPE SIZE (inches)													
Nominal	1/2	3/4	1	1 1/4	1 1/2	2	2 1/2	3	4	5	6	8	10	12
Actual ID	0.622	0.824	1.049	1.380	1.610	2.067	2.469	3.068	4.026	5.047	6.065	7.981	10.020	11.938
Length (ft)	Capacity in Cubic Feet of Gas per Hour													
10	172	360	678	1,390	2,090	4,020	6,400	11,300	23,100	41,800	67,600	139,000	252,000	399,000
20	118	247	466	957	1,430	2,760	4,400	7,780	15,900	28,700	46,500	95,500	173,000	275,000
30	95	199	374	768	1,150	2,220	3,530	6,250	12,700	23,000	37,300	76,700	139,000	220,000
40	81	170	320	657	985	1,900	3,020	5,350	10,900	19,700	31,900	65,600	119,000	189,000
50	72	151	284	583	873	1,680	2,680	4,740	9,660	17,500	28,300	58,200	106,000	167,000
60	65	137	257	528	791	1,520	2,430	4,290	8,760	15,800	25,600	52,700	95,700	152,000
70	60	126	237	486	728	1,400	2,230	3,950	8,050	14,600	23,600	48,500	88,100	139,000
80	56	117	220	452	677	1,300	2,080	3,670	7,490	13,600	22,000	45,100	81,900	130,000
90	52	110	207	424	635	1,220	1,950	3,450	7,030	12,700	20,600	42,300	76,900	122,000
100	50	104	195	400	600	1,160	1,840	3,260	6,640	12,000	19,500	40,000	72,600	115,000
125	44	92	173	355	532	1,020	1,630	2,890	5,890	10,600	17,200	35,400	64,300	102,000
150	40	83	157	322	482	928	1,480	2,610	5,330	9,650	15,600	32,100	58,300	92,300
175	37	77	144	296	443	854	1,360	2,410	4,910	8,880	14,400	29,500	53,600	84,900
200	34	71	134	275	412	794	1,270	2,240	4,560	8,260	13,400	27,500	49,900	79,000
250	30	63	119	244	366	704	1,120	1,980	4,050	7,320	11,900	24,300	44,200	70,000
300	27	57	108	221	331	638	1,020	1,800	3,670	6,630	10,700	22,100	40,100	63,400
350	25	53	99	203	305	587	935	1,650	3,370	6,100	9,880	20,300	36,900	58,400
400	23	49	92	189	283	546	870	1,540	3,140	5,680	9,190	18,900	34,300	54,300
450	22	46	86	177	266	512	816	1,440	2,940	5,330	8,620	17,700	32,200	50,900
500	21	43	82	168	251	484	771	1,360	2,780	5,030	8,150	16,700	30,400	48,100
550	20	41	78	159	239	459	732	1,290	2,640	4,780	7,740	15,900	28,900	45,700
600	19	39	74	152	228	438	699	1,240	2,520	4,560	7,380	15,200	27,500	43,600
650	18	38	71	145	218	420	669	1,180	2,410	4,360	7,070	14,500	26,400	41,800
700	17	36	68	140	209	403	643	1,140	2,320	4,190	6,790	14,000	25,300	40,100
750	17	35	66	135	202	389	619	1,090	2,230	4,040	6,540	13,400	24,400	38,600
800	16	34	63	130	195	375	598	1,060	2,160	3,900	6,320	13,000	23,600	37,300
850	16	33	61	126	189	363	579	1,020	2,090	3,780	6,110	12,600	22,800	36,100
900	15	32	59	122	183	352	561	992	2,020	3,660	5,930	12,200	22,100	35,000
950	15	31	58	118	178	342	545	963	1,960	3,550	5,760	11,800	21,500	34,000
1,000	14	30	56	115	173	333	530	937	1,910	3,460	5,600	11,500	20,900	33,100
1,100	14	28	53	109	164	316	503	890	1,810	3,280	5,320	10,900	19,800	31,400
1,200	13	27	51	104	156	301	480	849	1,730	3,130	5,070	10,400	18,900	30,000
1,300	12	26	49	100	150	289	460	813	1,660	3,000	4,860	9,980	18,100	28,700
1,400	12	25	47	96	144	277	442	781	1,590	2,880	4,670	9,590	17,400	27,600
1,500	11	24	45	93	139	267	426	752	1,530	2,780	4,500	9,240	16,800	26,600
1,600	11	23	44	89	134	258	411	727	1,480	2,680	4,340	8,920	16,200	25,600
1,700	11	22	42	86	130	250	398	703	1,430	2,590	4,200	8,630	15,700	24,800
1,800	10	22	41	84	126	242	386	682	1,390	2,520	4,070	8,370	15,200	24,100
1,900	10	21	40	81	122	235	375	662	1,350	2,440	3,960	8,130	14,800	23,400
2,000	NA	20	39	79	119	229	364	644	1,310	2,380	3,850	7,910	14,400	22,700

For SI: 1 inch = 25.4 mm, 1 foot = 304.8 mm, 1 pound per square inch = 6.895 kPa, 1-inch water column = 0.2488 kPa, 1 British thermal unit per hour = 0.2931 W, 1 cubic foot per hour = 0.0283 m^3/h, 1 degree = 0.01745 rad.

Notes:

1. NA means a flow of less than 10 cfh.
2. Table entries have been rounded to three significant digits.

TABLE G2413.4(2) [402.4(5)]
SCHEDULE 40 METALLIC PIPE

Gas	Natural
Inlet Pressure	2.0 psi
Pressure Drop	1.0 psi
Specific Gravity	0.60

	PIPE SIZE (inches)								
Nominal	$^{1}/_{2}$	$^{3}/_{4}$	1	$1^{1}/_{4}$	$1^{1}/_{2}$	2	$2^{1}/_{2}$	3	4
Actual ID	0.622	0.824	1.049	1.380	1.610	2.067	2.469	3.068	4.026
Length (ft)	Capacity in Cubic Feet of Gas per Hour								
10	1,510	3,040	5,560	11,400	17,100	32,900	52,500	92,800	189,000
20	1,070	2,150	3,930	8,070	12,100	23,300	37,100	65,600	134,000
30	869	1,760	3,210	6,590	9,880	19,000	30,300	53,600	109,000
40	753	1,520	2,780	5,710	8,550	16,500	26,300	46,400	94,700
50	673	1,360	2,490	5,110	7,650	14,700	23,500	41,500	84,700
60	615	1,240	2,270	4,660	6,980	13,500	21,400	37,900	77,300
70	569	1,150	2,100	4,320	6,470	12,500	19,900	35,100	71,600
80	532	1,080	1,970	4,040	6,050	11,700	18,600	32,800	67,000
90	502	1,010	1,850	3,810	5,700	11,000	17,500	30,900	63,100
100	462	934	1,710	3,510	5,260	10,100	16,100	28,500	58,200
125	414	836	1,530	3,140	4,700	9,060	14,400	25,500	52,100
150	372	751	1,370	2,820	4,220	8,130	13,000	22,900	46,700
175	344	695	1,270	2,601	3,910	7,530	12,000	21,200	43,300
200	318	642	1,170	2,410	3,610	6,960	11,100	19,600	40,000
250	279	583	1,040	2,140	3,210	6,180	9,850	17,400	35,500
300	253	528	945	1,940	2,910	5,600	8,920	15,800	32,200
350	232	486	869	1,790	2,670	5,150	8,210	14,500	29,600
400	216	452	809	1,660	2,490	4,790	7,640	13,500	27,500
450	203	424	759	1,560	2,330	4,500	7,170	12,700	25,800
500	192	401	717	1,470	2,210	4,250	6,770	12,000	24,400
550	182	381	681	1,400	2,090	4,030	6,430	11,400	23,200
600	174	363	650	1,330	2,000	3,850	6,130	10,800	22,100
650	166	348	622	1,280	1,910	3,680	5,870	10,400	21,200
700	160	334	598	1,230	1,840	3,540	5,640	9,970	20,300
750	154	322	576	1,180	1,770	3,410	5,440	9,610	19,600
800	149	311	556	1,140	1,710	3,290	5,250	9,280	18,900
850	144	301	538	1,100	1,650	3,190	5,080	8,980	18,300
900	139	292	522	1,070	1,600	3,090	4,930	8,710	17,800
950	135	283	507	1,040	1,560	3,000	4,780	8,460	17,200
1,000	132	275	493	1,010	1,520	2,920	4,650	8,220	16,800
1,100	125	262	468	960	1,440	2,770	4,420	7,810	15,900
1,200	119	250	446	917	1,370	2,640	4,220	7,450	15,200
1,300	114	239	427	878	1,320	2,530	4,040	7,140	14,600
1,400	110	230	411	843	1,260	2,430	3,880	6,860	14,000
1,500	106	221	396	812	1,220	2,340	3,740	6,600	13,500
1,600	102	214	382	784	1,180	2,260	3,610	6,380	13,000
1,700	99	207	370	759	1,140	2,190	3,490	6,170	12,600
1,800	96	200	358	736	1,100	2,120	3,390	5,980	12,200
1,900	93	195	348	715	1,070	2,060	3,290	5,810	11,900
2,000	91	189	339	695	1,040	2,010	3,200	5,650	11,500

For SI: 1 inch = 25.4 mm, 1 foot = 304.8 mm, 1 pound per square inch = 6.895 kPa, 1-inch water column = 0.2488 kPa, 1 British thermal unit per hour = 0.2931 W, 1 cubic foot per hour = 0.0283 m^3/h, 1 degree = 0.01745 rad.

Note: Table entries have been rounded to three significant digits.

TABLE G2413.4(3) [402.4(9)]
SEMIRIGID COPPER TUBING

Gas	Natural
Inlet Pressure	Less than 2 psi
Pressure Drop	0.5 in. w.c.
Specific Gravity	0.60

		TUBE SIZE (inches)								
Nominal	K & L	1/4	3/8	1/2	5/8	3/4	1	1 1/4	1 1/2	2
	ACR	3/8	1/2	5/8	3/4	7/8	1 1/8	1 3/8	—	—
Outside		0.375	0.500	0.625	0.750	0.875	1.125	1.375	1.625	2.125
Inside		0.305	0.402	0.527	0.652	0.745	0.995	1.245	1.481	1.959
Length (ft)		Capacity in Cubic Feet of Gas per Hour								
10		27	55	111	195	276	590	1,060	1,680	3,490
20		18	38	77	134	190	406	730	1,150	2,400
30		15	30	61	107	152	326	586	925	1,930
40		13	26	53	92	131	279	502	791	1,650
50		11	23	47	82	116	247	445	701	1,460
60		10	21	42	74	105	224	403	635	1,320
70		NA	19	39	68	96	206	371	585	1,220
80		NA	18	36	63	90	192	345	544	1,130
90		NA	17	34	59	84	180	324	510	1,060
100		NA	16	32	56	79	170	306	482	1,000
125		NA	14	28	50	70	151	271	427	890
150		NA	13	26	45	64	136	245	387	806
175		NA	12	24	41	59	125	226	356	742
200		NA	11	22	39	55	117	210	331	690
250		NA	NA	20	34	48	103	186	294	612
300		NA	NA	18	31	44	94	169	266	554
350		NA	NA	16	28	40	86	155	245	510
400		NA	NA	15	26	38	80	144	228	474
450		NA	NA	14	25	35	75	135	214	445
500		NA	NA	13	23	33	71	128	202	420
550		NA	NA	13	22	32	68	122	192	399
600		NA	NA	12	21	30	64	116	183	381
650		NA	NA	12	20	29	62	111	175	365
700		NA	NA	11	20	28	59	107	168	350
750		NA	NA	11	19	27	57	103	162	338
800		NA	NA	10	18	26	55	99	156	326
850		NA	NA	10	18	25	53	96	151	315
900		NA	NA	NA	17	24	52	93	147	306
950		NA	NA	NA	17	24	50	90	143	297
1,000		NA	NA	NA	16	23	49	88	139	289
1,100		NA	NA	NA	15	22	46	84	132	274
1,200		NA	NA	NA	15	21	44	80	126	262
1,300		NA	NA	NA	14	20	42	76	120	251
1,400		NA	NA	NA	13	19	41	73	116	241
1,500		NA	NA	NA	13	18	39	71	111	232
1,600		NA	NA	NA	13	18	38	68	108	224
1,700		NA	NA	NA	12	17	37	66	104	217
1,800		NA	NA	NA	12	17	36	64	101	210
1,900		NA	NA	NA	11	16	35	62	98	204
2,000		NA	NA	NA	11	16	34	60	95	199

For SI: 1 inch = 25.4 mm, 1 foot = 304.8 mm, 1 pound per square inch = 6.895 kPa, 1-inch water column = 0.2488 kPa, 1 British thermal unit per hour = 0.2931 W, 1 cubic foot per hour = 0.0283 m^3/h, 1 degree = 0.01745 rad.

Notes:

1. Table capacities are based on Type K copper tubing inside diameter (shown), which has the smallest inside diameter of the copper tubing products.
2. NA means a flow of less than 10 cfh.
3. Table entries have been rounded to three significant digits

TABLE G2413.4(4) [402.4(12)] SEMIRIGID COPPER TUBING

Gas	Natural
Inlet Pressure	2.0 psi
Pressure Drop	1.0 psi
Specific Gravity	0.60

		TUBE SIZE (inches)								
Nominal	K & L	1/4	3/8	1/2	5/8	3/4	1	1 1/4	1 1/2	2
	ACR	3/8	1/2	5/8	3/4	7/8	1 1/8	1 3/8	—	—
Outside		0.375	0.500	0.625	0.750	0.875	1.125	1.375	1.625	2.125
Inside		0.305	0.402	0.527	0.652	0.745	0.995	1.245	1.481	1.959
Length (ft)		Capacity in Cubic Feet of Gas per Hour								
10		245	506	1,030	1,800	2,550	5,450	9,820	15,500	32,200
20		169	348	708	1,240	1,760	3,750	6,750	10,600	22,200
30		135	279	568	993	1,410	3,010	5,420	8,550	17,800
40		116	239	486	850	1,210	2,580	4,640	7,310	15,200
50		103	212	431	754	1,070	2,280	4,110	6,480	13,500
60		93	192	391	683	969	2,070	3,730	5,870	12,200
70		86	177	359	628	891	1,900	3,430	5,400	11,300
80		80	164	334	584	829	1,770	3,190	5,030	10,500
90		75	154	314	548	778	1,660	2,990	4,720	9,820
100		71	146	296	518	735	1,570	2,830	4,450	9,280
125		63	129	263	459	651	1,390	2,500	3,950	8,220
150		57	117	238	416	590	1,260	2,270	3,580	7,450
175		52	108	219	383	543	1,160	2,090	3,290	6,850
200		49	100	204	356	505	1,080	1,940	3,060	6,380
250		43	89	181	315	448	956	1,720	2,710	5,650
300		39	80	164	286	406	866	1,560	2,460	5,120
350		36	74	150	263	373	797	1,430	2,260	4,710
400		33	69	140	245	347	741	1,330	2,100	4,380
450		31	65	131	230	326	696	1,250	1,970	4,110
500		30	61	124	217	308	657	1,180	1,870	3,880
550		28	58	118	206	292	624	1,120	1,770	3,690
600		27	55	112	196	279	595	1,070	1,690	3,520
650		26	53	108	188	267	570	1,030	1,620	3,370
700		25	51	103	181	256	548	986	1,550	3,240
750		24	49	100	174	247	528	950	1,500	3,120
800		23	47	96	168	239	510	917	1,450	3,010
850		22	46	93	163	231	493	888	1,400	2,920
900		22	44	90	158	224	478	861	1,360	2,830
950		21	43	88	153	217	464	836	1,320	2,740
1,000		20	42	85	149	211	452	813	1,280	2,670
1,100		19	40	81	142	201	429	772	1,220	2,540
1,200		18	38	77	135	192	409	737	1,160	2,420
1,300		18	36	74	129	183	392	705	1,110	2,320
1,400		17	35	71	124	176	376	678	1,070	2,230
1,500		16	34	68	120	170	363	653	1,030	2,140
1,600		16	33	66	116	164	350	630	994	2,070
1,700		15	31	64	112	159	339	610	962	2,000
1,800		15	30	62	108	154	329	592	933	1,940
1,900		14	30	60	105	149	319	575	906	1,890
2,000		14	29	59	102	145	310	559	881	1,830

For SI: 1 inch = 25.4 mm, 1 foot = 304.8 mm, 1 pound per square inch = 6.895 kPa, 1-inch water column = 0.2488 kPa, 1 British thermal unit per hour = 0.2931 W, 1 cubic foot per hour = 0.0283 m^3/h, 1 degree = 0.01745 rad.

Notes:

1. Table capacities are based on Type K copper tubing inside diameter (shown), which has the smallest inside diameter of the copper tubing products.
2. Table entries have been rounded to three significant digits.

TABLE G2413.4(5) [402.4(15)] CORRUGATED STAINLESS STEEL TUBING (CSST)

Gas	Natural
Inlet Pressure	Less than 2 psi
Pressure Drop	0.5 in. w.c.
Specific Gravity	0.60

	TUBE SIZE (EHD)													
Flow Designation	13	15	18	19	23	25	30	31	37	39	46	48	60	62
Length (ft)	Capacity in Cubic Feet of Gas per Hour													
5	46	63	115	134	225	270	471	546	895	1,037	1,790	2,070	3,660	4,140
10	32	44	82	95	161	192	330	383	639	746	1,260	1,470	2,600	2,930
15	25	35	66	77	132	157	267	310	524	615	1,030	1,200	2,140	2,400
20	22	31	58	67	116	137	231	269	456	536	888	1,050	1,850	2,080
25	19	27	52	60	104	122	206	240	409	482	793	936	1,660	1,860
30	18	25	47	55	96	112	188	218	374	442	723	856	1,520	1,700
40	15	21	41	47	83	97	162	188	325	386	625	742	1,320	1,470
50	13	19	37	42	75	87	144	168	292	347	559	665	1,180	1,320
60	12	17	34	38	68	80	131	153	267	318	509	608	1,080	1,200
70	11	16	31	36	63	74	121	141	248	295	471	563	1,000	1,110
80	10	15	29	33	60	69	113	132	232	277	440	527	940	1,040
90	10	14	28	32	57	65	107	125	219	262	415	498	887	983
100	9	13	26	30	54	62	101	118	208	249	393	472	843	933
150	7	10	20	23	42	48	78	91	171	205	320	387	691	762
200	6	9	18	21	38	44	71	82	148	179	277	336	600	661
250	5	8	16	19	34	39	63	74	133	161	247	301	538	591
300	5	7	15	17	32	36	57	67	95	148	226	275	492	540

For SI: 1 inch = 25.4 mm, 1 foot = 304.8 mm, 1 pound per square inch = 6.895 kPa, 1-inch water column = 0.2488 kPa, 1 British thermal unit per hour = 0.2931 W, 1 cubic foot per hour = 0.0283 m^3/h, 1 degree = 0.01745 rad.

Notes:

1. Table includes losses for four 90-degree bends and two end fittings. Tubing runs with larger numbers of bends or fittings shall be increased by an equivalent length of tubing to the following equation: $L = 1.3n$, where L is additional length (feet) of tubing and n is the number of additional fittings or bends.
2. EHD—Equivalent Hydraulic Diameter, which is a measure of the relative hydraulic efficiency between different tubing sizes. The greater the value of EHD, the greater the gas capacity of the tubing.
3. Table entries have been rounded to three significant digits.

TABLE G2413.4(6) [402.4(18)] CORRUGATED STAINLESS STEEL TUBING (CSST)

Gas	Natural
Inlet Pressure	2.0 psi
Pressure Drop	1.0 psi
Specific Gravity	0.60

	TUBE SIZE (EHD)													
Flow Designation	13	15	18	19	23	25	30	31	37	39	46	48	60	62
Length (ft)	Capacity in Cubic Feet of Gas Per Hour													
10	270	353	587	700	1,100	1,370	2,590	2,990	4,510	5,037	9,600	10,700	18,600	21,600
25	166	220	374	444	709	876	1,620	1,870	2,890	3,258	6,040	6,780	11,900	13,700
30	151	200	342	405	650	801	1,480	1,700	2,640	2,987	5,510	6,200	10,900	12,500
40	129	172	297	351	567	696	1,270	1,470	2,300	2,605	4,760	5,380	9,440	10,900
50	115	154	266	314	510	624	1,140	1,310	2,060	2,343	4,260	4,820	8,470	9,720
75	93	124	218	257	420	512	922	1,070	1,690	1,932	3,470	3,950	6,940	7,940
80	89	120	211	249	407	496	892	1,030	1,640	1,874	3,360	3,820	6,730	7,690
100	79	107	189	222	366	445	795	920	1,470	1,685	3,000	3,420	6,030	6,880
150	64	87	155	182	302	364	646	748	1,210	1,389	2,440	2,800	4,940	5,620
200	55	75	135	157	263	317	557	645	1,050	1,212	2,110	2,430	4,290	4,870
250	49	67	121	141	236	284	497	576	941	1,090	1,890	2,180	3,850	4,360
300	44	61	110	129	217	260	453	525	862	999	1,720	1,990	3,520	3,980
400	38	52	96	111	189	225	390	453	749	871	1,490	1,730	3,060	3,450
500	34	46	86	100	170	202	348	404	552	783	1,330	1,550	2,740	3,090

For SI: 1 inch = 25.4 mm, 1 foot = 304.8 mm, 1 pound per square inch = 6.895 kPa, 1-inch water column = 0.2488 kPa, 1 British thermal unit per hour = 0.2931 W, 1 cubic foot per hour = 0.0283 m^3/h, 1 degree = 0.01745 rad.

Notes:

1. Table does not include effect of pressure drop across the line regulator. Where regulator loss exceeds $^3/_4$ psi, DO NOT USE THIS TABLE. Consult with the regulator manufacturer for pressure drops and capacity factors. Pressure drops across a regulator can vary with flow rate.
2. CAUTION: Capacities shown in the table might exceed maximum capacity for a selected regulator. Consult with the regulator or tubing manufacturer for guidance.
3. Table includes losses for four 90-degree bends and two end fittings. Tubing runs with larger numbers of bends or fittings shall be increased by an equivalent length of tubing to the following equation: $L = 1.3n$ where L is additional length (feet) of tubing and n is the number of additional fittings or bends.
4. EHD—Equivalent Hydraulic Diameter, which is a measure of the relative hydraulic efficiency between different tubing sizes. The greater the value of EHD, the greater the gas capacity of the tubing.
5. Table entries have been rounded to three significant digits.

TABLE G2413.4(7) [402.4(21)]
POLYETHYLENE PLASTIC PIPE

Gas	Natural
Inlet Pressure	Less than 2 psi
Pressure Drop	0.5 in. w.c.
Specific Gravity	0.60

PIPE SIZE (inches)						
Nominal OD	$^1/_2$	$^3/_4$	1	$1^1/_4$	$1^1/_2$	2
Designation	SDR 9	SDR 11	SDR 11	SDR 10	SDR 11	SDR 11
Actual ID	0.660	0.860	1.077	1.328	1.554	1.943
Length (ft)	Capacity in Cubic Feet of Gas per Hour					
10	201	403	726	1,260	1,900	3,410
20	138	277	499	865	1,310	2,350
30	111	222	401	695	1,050	1,880
40	95	190	343	594	898	1,610
50	84	169	304	527	796	1,430
60	76	153	276	477	721	1,300
70	70	140	254	439	663	1,190
80	65	131	236	409	617	1,110
90	61	123	221	383	579	1,040
100	58	116	209	362	547	983
125	51	103	185	321	485	871
150	46	93	168	291	439	789
175	43	86	154	268	404	726
200	40	80	144	249	376	675
250	35	71	127	221	333	598
300	32	64	115	200	302	542
350	29	59	106	184	278	499
400	27	55	99	171	258	464
450	26	51	93	160	242	435
500	24	48	88	152	229	411

For SI: 1 inch = 25.4 mm, 1 foot = 304.8 mm, 1 pound per square inch = 6.895 kPa, 1-inch water column = 0.2488 kPa, 1 British thermal unit per hour = 0.2931 W, 1 cubic foot per hour = 0.0283 m^3/h, 1 degree = 0.01745 rad.

Note: Table entries have been rounded to three significant digits.

TABLE G2413.4(8) [402.4(22)]
POLYETHYLENE PLASTIC PIPE

Gas	Natural
Inlet Pressure	2.0 psi
Pressure Drop	1.0 psi
Specific Gravity	0.60

	PIPE SIZE (inches)					
Nominal OD	1/2	3/4	1	1 1/4	1 1/2	2
Designation	SDR 9	SDR 11	SDR 11	SDR 10	SDR 11	SDR 11
Actual ID	0.660	0.860	1.077	1.328	1.554	1.943
Length (ft)	Capacity in Cubic Feet of Gas per Hour					
10	1,860	3,720	6,710	11,600	17,600	31,600
20	1,280	2,560	4,610	7,990	12,100	21,700
30	1,030	2,050	3,710	6,420	9,690	17,400
40	878	1,760	3,170	5,490	8,300	14,900
50	778	1,560	2,810	4,870	7,350	13,200
60	705	1,410	2,550	4,410	6,660	12,000
70	649	1,300	2,340	4,060	6,130	11,000
80	603	1,210	2,180	3,780	5,700	10,200
90	566	1,130	2,050	3,540	5,350	9,610
100	535	1,070	1,930	3,350	5,050	9,080
125	474	949	1,710	2,970	4,480	8,050
150	429	860	1,550	2,690	4,060	7,290
175	395	791	1,430	2,470	3,730	6,710
200	368	736	1,330	2,300	3,470	6,240
250	326	652	1,180	2,040	3,080	5,530
300	295	591	1,070	1,850	2,790	5,010
350	272	544	981	1,700	2,570	4,610
400	253	506	913	1,580	2,390	4,290
450	237	475	856	1,480	2,240	4,020
500	224	448	809	1,400	2,120	3,800
550	213	426	768	1,330	2,010	3,610
600	203	406	733	1,270	1,920	3,440
650	194	389	702	1,220	1,840	3,300
700	187	374	674	1,170	1,760	3,170
750	180	360	649	1,130	1,700	3,050
800	174	348	627	1,090	1,640	2,950
850	168	336	607	1,050	1,590	2,850
900	163	326	588	1,020	1,540	2,770
950	158	317	572	990	1,500	2,690
1,000	154	308	556	963	1,450	2,610
1,100	146	293	528	915	1,380	2,480
1,200	139	279	504	873	1,320	2,370
1,300	134	267	482	836	1,260	2,270
1,400	128	257	463	803	1,210	2,180
1,500	124	247	446	773	1,170	2,100
1,600	119	239	431	747	1,130	2,030
1,700	115	231	417	723	1,090	1,960
1,800	112	224	404	701	1,060	1,900
1,900	109	218	393	680	1,030	1,850
2,000	106	212	382	662	1,000	1,800

For SI: 1 inch = 25.4 mm, 1 foot = 304.8 mm, 1 pound per square inch = 6.895 kPa, 1-inch water column = 0.2488 kPa, 1 British thermal unit per hour = 0.2931 W, 1 cubic foot per hour = 0.0283 m^3/h, 1 degree = 0.01745 rad.

Note: Table entries have been rounded to three significant digits.

TABLE G2413.4(9) [402.4(25)]
SCHEDULE 40 METALLIC PIPE

Gas	Undiluted Propane
Inlet Pressure	10.0 psi
Pressure Drop	1.0 psi
Specific Gravity	1.50

INTENDED USE: PIPE SIZING BETWEEN FIRST STAGE (high-pressure regulator) AND SECOND STAGE (low-pressure regulator)									
PIPE SIZE (inches)									
Nominal	$^1/_2$	$^3/_4$	1	$1^1/_4$	$1^1/_2$	2	$2^1/_2$	3	4
Actual ID	0.622	0.824	1.049	1.380	1.610	2.067	2.469	3.068	4.026
Length (ft)	Capacity in Thousands of Btu per Hour								
10	3,320	6,950	13,100	26,900	40,300	77,600	124,000	219,000	446,000
20	2,280	4,780	9,000	18,500	27,700	53,300	85,000	150,000	306,000
30	1,830	3,840	7,220	14,800	22,200	42,800	68,200	121,000	246,000
40	1,570	3,280	6,180	12,700	19,000	36,600	58,400	103,000	211,000
50	1,390	2,910	5,480	11,300	16,900	32,500	51,700	91,500	187,000
60	1,260	2,640	4,970	10,200	15,300	29,400	46,900	82,900	169,000
70	1,160	2,430	4,570	9,380	14,100	27,100	43,100	76,300	156,000
80	1,080	2,260	4,250	8,730	13,100	25,200	40,100	70,900	145,000
90	1,010	2,120	3,990	8,190	12,300	23,600	37,700	66,600	136,000
100	956	2,000	3,770	7,730	11,600	22,300	35,600	62,900	128,000
125	848	1,770	3,340	6,850	10,300	19,800	31,500	55,700	114,000
150	768	1,610	3,020	6,210	9,300	17,900	28,600	50,500	103,000
175	706	1,480	2,780	5,710	8,560	16,500	26,300	46,500	94,700
200	657	1,370	2,590	5,320	7,960	15,300	24,400	43,200	88,100
250	582	1,220	2,290	4,710	7,060	13,600	21,700	38,300	78,100
300	528	1,100	2,080	4,270	6,400	12,300	19,600	34,700	70,800
350	486	1,020	1,910	3,930	5,880	11,300	18,100	31,900	65,100
400	452	945	1,780	3,650	5,470	10,500	16,800	29,700	60,600
450	424	886	1,670	3,430	5,140	9,890	15,800	27,900	56,800
500	400	837	1,580	3,240	4,850	9,340	14,900	26,300	53,700
550	380	795	1,500	3,070	4,610	8,870	14,100	25,000	51,000
600	363	759	1,430	2,930	4,400	8,460	13,500	23,900	48,600
650	347	726	1,370	2,810	4,210	8,110	12,900	22,800	46,600
700	334	698	1,310	2,700	4,040	7,790	12,400	21,900	44,800
750	321	672	1,270	2,600	3,900	7,500	12,000	21,100	43,100
800	310	649	1,220	2,510	3,760	7,240	11,500	20,400	41,600
850	300	628	1,180	2,430	3,640	7,010	11,200	19,800	40,300
900	291	609	1,150	2,360	3,530	6,800	10,800	19,200	39,100
950	283	592	1,110	2,290	3,430	6,600	10,500	18,600	37,900
1,000	275	575	1,080	2,230	3,330	6,420	10,200	18,100	36,900
1,100	261	546	1,030	2,110	3,170	6,100	9,720	17,200	35,000
1,200	249	521	982	2,020	3,020	5,820	9,270	16,400	33,400
1,300	239	499	940	1,930	2,890	5,570	8,880	15,700	32,000
1,400	229	480	903	1,850	2,780	5,350	8,530	15,100	30,800
1,500	221	462	870	1,790	2,680	5,160	8,220	14,500	29,600
1,600	213	446	840	1,730	2,590	4,980	7,940	14,000	28,600
1,700	206	432	813	1,670	2,500	4,820	7,680	13,600	27,700
1,800	200	419	789	1,620	2,430	4,670	7,450	13,200	26,900
1,900	194	407	766	1,570	2,360	4,540	7,230	12,800	26,100
2,000	189	395	745	1,530	2,290	4,410	7,030	12,400	25,400

For SI: 1 inch = 25.4 mm, 1 foot = 304.8 mm, 1 pound per square inch = 6.895 kPa, 1-inch water column = 0.2488 kPa,
1 British thermal unit per hour = 0.2931 W, 1 cubic foot per hour = 0.0283 m^3/h, 1 degree = 0.01745 rad.

Note: Table entries have been rounded to three significant digits.

TABLE G2413.4(10) [402.4(26)]
SCHEDULE 40 METALLIC PIPE

Gas	Undiluted Propane
Inlet Pressure	10.0 psi
Pressure Drop	3.0 psi
Specific Gravity	1.50

INTENDED USE: PIPE SIZING BETWEEN FIRST STAGE (high-pressure regulator) AND SECOND STAGE (low-pressure regulator)									
PIPE SIZE (inches)									
Nominal	1/2	3/4	1	1 1/4	1 1/2	2	2 1/2	3	4
Actual ID	0.622	0.824	1.049	1.380	1.610	2.067	2.469	3.068	4.026
Length (ft)	Capacity in Thousands of Btu per Hour								
10	5,890	12,300	23,200	47,600	71,300	137,000	219,000	387,000	789,000
20	4,050	8,460	15,900	32,700	49,000	94,400	150,000	266,000	543,000
30	3,250	6,790	12,800	26,300	39,400	75,800	121,000	214,000	436,000
40	2,780	5,810	11,000	22,500	33,700	64,900	103,000	183,000	373,000
50	2,460	5,150	9,710	19,900	29,900	57,500	91,600	162,000	330,000
60	2,230	4,670	8,790	18,100	27,100	52,100	83,000	147,000	299,000
70	2,050	4,300	8,090	16,600	24,900	47,900	76,400	135,000	275,000
80	1,910	4,000	7,530	15,500	23,200	44,600	71,100	126,000	256,000
90	1,790	3,750	7,060	14,500	21,700	41,800	66,700	118,000	240,000
100	1,690	3,540	6,670	13,700	20,500	39,500	63,000	111,000	227,000
125	1,500	3,140	5,910	12,100	18,200	35,000	55,800	98,700	201,000
150	1,360	2,840	5,360	11,000	16,500	31,700	50,600	89,400	182,000
175	1,250	2,620	4,930	10,100	15,200	29,200	46,500	82,300	167,800
200	1,160	2,430	4,580	9,410	14,100	27,200	43,300	76,500	156,100
250	1,030	2,160	4,060	8,340	12,500	24,100	38,400	67,800	138,400
300	935	1,950	3,680	7,560	11,300	21,800	34,800	61,500	125,400
350	860	1,800	3,390	6,950	10,400	20,100	32,000	56,500	115,300
400	800	1,670	3,150	6,470	9,690	18,700	29,800	52,600	107,300
450	751	1,570	2,960	6,070	9,090	17,500	27,900	49,400	100,700
500	709	1,480	2,790	5,730	8,590	16,500	26,400	46,600	95,100
550	673	1,410	2,650	5,450	8,160	15,700	25,000	44,300	90,300
600	642	1,340	2,530	5,200	7,780	15,000	23,900	42,200	86,200
650	615	1,290	2,420	4,980	7,450	14,400	22,900	40,500	82,500
700	591	1,240	2,330	4,780	7,160	13,800	22,000	38,900	79,300
750	569	1,190	2,240	4,600	6,900	13,300	21,200	37,400	76,400
800	550	1,150	2,170	4,450	6,660	12,800	20,500	36,200	73,700
850	532	1,110	2,100	4,300	6,450	12,400	19,800	35,000	71,400
900	516	1,080	2,030	4,170	6,250	12,000	19,200	33,900	69,200
950	501	1,050	1,970	4,050	6,070	11,700	18,600	32,900	67,200
1,000	487	1,020	1,920	3,940	5,900	11,400	18,100	32,000	65,400
1,100	463	968	1,820	3,740	5,610	10,800	17,200	30,400	62,100
1,200	442	923	1,740	3,570	5,350	10,300	16,400	29,000	59,200
1,300	423	884	1,670	3,420	5,120	9,870	15,700	27,800	56,700
1,400	406	849	1,600	3,280	4,920	9,480	15,100	26,700	54,500
1,500	391	818	1,540	3,160	4,740	9,130	14,600	25,700	52,500
1,600	378	790	1,490	3,060	4,580	8,820	14,100	24,800	50,700
1,700	366	765	1,440	2,960	4,430	8,530	13,600	24,000	49,000
1,800	355	741	1,400	2,870	4,300	8,270	13,200	23,300	47,600
1,900	344	720	1,360	2,780	4,170	8,040	12,800	22,600	46,200
2,000	335	700	1,320	2,710	4,060	7,820	12,500	22,000	44,900

For SI: 1 inch = 25.4 mm, 1 foot = 304.8 mm, 1 pound per square inch = 6.895 kPa, 1-inch water column = 0.2488 kPa,
1 British thermal unit per hour = 0.2931 W, 1 cubic foot per hour = 0.0283 m^3/h, 1 degree = 0.01745 rad.

Note: Table entries have been rounded to three significant digits.

TABLE G2413.4(11) [402.4(27)]
SCHEDULE 40 METALLIC PIPE

Gas	Undiluted Propane
Inlet Pressure	2.0 psi
Pressure Drop	1.0 psi
Specific Gravity	1.50

INTENDED USE: PIPE SIZING BETWEEN 2 PSIG SERVICE AND LINE PRESSURE REGULATOR									
	PIPE SIZE (inches)								
Nominal	1/2	3/4	1	1 1/4	1 1/2	2	2 1/2	3	4
Actual ID	0.622	0.824	1.049	1.380	1.610	2.067	2.469	3.068	4.026
Length (ft)	Capacity in Thousands of Btu per Hour								
10	2,680	5,590	10,500	21,600	32,400	62,400	99,500	176,000	359,000
20	1,840	3,850	7,240	14,900	22,300	42,900	68,400	121,000	247,000
30	1,480	3,090	5,820	11,900	17,900	34,500	54,900	97,100	198,000
40	1,260	2,640	4,980	10,200	15,300	29,500	47,000	83,100	170,000
50	1,120	2,340	4,410	9,060	13,600	26,100	41,700	73,700	150,000
60	1,010	2,120	4,000	8,210	12,300	23,700	37,700	66,700	136,000
70	934	1,950	3,680	7,550	11,300	21,800	34,700	61,400	125,000
80	869	1,820	3,420	7,020	10,500	20,300	32,300	57,100	116,000
90	815	1,700	3,210	6,590	9,880	19,000	30,300	53,600	109,000
100	770	1,610	3,030	6,230	9,330	18,000	28,600	50,600	103,000
125	682	1,430	2,690	5,520	8,270	15,900	25,400	44,900	91,500
150	618	1,290	2,440	5,000	7,490	14,400	23,000	40,700	82,900
175	569	1,190	2,240	4,600	6,890	13,300	21,200	37,400	76,300
200	529	1,110	2,080	4,280	6,410	12,300	19,700	34,800	71,000
250	469	981	1,850	3,790	5,680	10,900	17,400	30,800	62,900
300	425	889	1,670	3,440	5,150	9,920	15,800	27,900	57,000
350	391	817	1,540	3,160	4,740	9,120	14,500	25,700	52,400
400	364	760	1,430	2,940	4,410	8,490	13,500	23,900	48,800
450	341	714	1,340	2,760	4,130	7,960	12,700	22,400	45,800
500	322	674	1,270	2,610	3,910	7,520	12,000	21,200	43,200
550	306	640	1,210	2,480	3,710	7,140	11,400	20,100	41,100
600	292	611	1,150	2,360	3,540	6,820	10,900	19,200	39,200
650	280	585	1,100	2,260	3,390	6,530	10,400	18,400	37,500
700	269	562	1,060	2,170	3,260	6,270	9,990	17,700	36,000
750	259	541	1,020	2,090	3,140	6,040	9,630	17,000	34,700
800	250	523	985	2,020	3,030	5,830	9,300	16,400	33,500
850	242	506	953	1,960	2,930	5,640	9,000	15,900	32,400
900	235	490	924	1,900	2,840	5,470	8,720	15,400	31,500
950	228	476	897	1,840	2,760	5,310	8,470	15,000	30,500
1,000	222	463	873	1,790	2,680	5,170	8,240	14,600	29,700
1,100	210	440	829	1,700	2,550	4,910	7,830	13,800	28,200
1,200	201	420	791	1,620	2,430	4,680	7,470	13,200	26,900
1,300	192	402	757	1,550	2,330	4,490	7,150	12,600	25,800
1,400	185	386	727	1,490	2,240	4,310	6,870	12,100	24,800
1,500	178	372	701	1,440	2,160	4,150	6,620	11,700	23,900
1,600	172	359	677	1,390	2,080	4,010	6,390	11,300	23,000
1,700	166	348	655	1,340	2,010	3,880	6,180	10,900	22,300
1,800	161	337	635	1,300	1,950	3,760	6,000	10,600	21,600
1,900	157	327	617	1,270	1,900	3,650	5,820	10,300	21,000
2,000	152	318	600	1,230	1,840	3,550	5,660	10,000	20,400

For SI: 1 inch = 25.4 mm, 1 foot = 304.8 mm, 1 pound per square inch = 6.895 kPa, 1-inch water column = 0.2488 kPa, 1 British thermal unit per hour = 0.2931 W, 1 cubic foot per hour = 0.0283 m^3/h, 1 degree = 0.01745 rad.

Note: Table entries have been rounded to three significant digits.

TABLE G2413.4(12) [402.4(28)] SCHEDULE 40 METALLIC PIPE

Gas	Undiluted Propane
Inlet Pressure	11.0 in. w.c.
Pressure Drop	0.5 in. w.c.
Specific Gravity	1.50

INTENDED USE: PIPE SIZING BETWEEN SINGLE- OR SECOND-STAGE (low pressure) REGULATOR AND APPLIANCE									
	PIPE SIZE (inches)								
Nominal	1/2	3/4	1	1 1/4	1 1/2	2	2 1/2	3	4
Actual ID	0.622	0.824	1.049	1.380	1.610	2.067	2.469	3.068	4.026
Length (ft)	Capacity in Thousands of Btu per Hour								
10	291	608	1,150	2,350	3,520	6,790	10,800	19,100	39,000
20	200	418	787	1,620	2,420	4,660	7,430	13,100	26,800
30	160	336	632	1,300	1,940	3,750	5,970	10,600	21,500
40	137	287	541	1,110	1,660	3,210	5,110	9,030	18,400
50	122	255	480	985	1,480	2,840	4,530	8,000	16,300
60	110	231	434	892	1,340	2,570	4,100	7,250	14,800
80	101	212	400	821	1,230	2,370	3,770	6,670	13,600
100	94	197	372	763	1,140	2,200	3,510	6,210	12,700
125	89	185	349	716	1,070	2,070	3,290	5,820	11,900
150	84	175	330	677	1,010	1,950	3,110	5,500	11,200
175	74	155	292	600	899	1,730	2,760	4,880	9,950
200	67	140	265	543	814	1,570	2,500	4,420	9,010
250	62	129	243	500	749	1,440	2,300	4,060	8,290
300	58	120	227	465	697	1,340	2,140	3,780	7,710
350	51	107	201	412	618	1,190	1,900	3,350	6,840
400	46	97	182	373	560	1,080	1,720	3,040	6,190
450	42	89	167	344	515	991	1,580	2,790	5,700
500	40	83	156	320	479	922	1,470	2,600	5,300
550	37	78	146	300	449	865	1,380	2,440	4,970
600	35	73	138	283	424	817	1,300	2,300	4,700
650	33	70	131	269	403	776	1,240	2,190	4,460
700	32	66	125	257	385	741	1,180	2,090	4,260
750	30	64	120	246	368	709	1,130	2,000	4,080
800	29	61	115	236	354	681	1,090	1,920	3,920
850	28	59	111	227	341	656	1,050	1,850	3,770
900	27	57	107	220	329	634	1,010	1,790	3,640
950	26	55	104	213	319	613	978	1,730	3,530
1,000	25	53	100	206	309	595	948	1,680	3,420
1,100	25	52	97	200	300	578	921	1,630	3,320
1,200	24	50	95	195	292	562	895	1,580	3,230
1,300	23	48	90	185	277	534	850	1,500	3,070
1,400	22	46	86	176	264	509	811	1,430	2,930
1,500	21	44	82	169	253	487	777	1,370	2,800
1,200	24	50	95	195	292	562	895	1,580	3,230
1,300	23	48	90	185	277	534	850	1,500	3,070
1,400	22	46	86	176	264	509	811	1,430	2,930
1,500	21	44	82	169	253	487	777	1,370	2,800
1,600	20	42	79	162	243	468	746	1,320	2,690
1,700	19	40	76	156	234	451	719	1,270	2,590
1,800	19	39	74	151	226	436	694	1,230	2,500
1,900	18	38	71	146	219	422	672	1,190	2,420
2,000	18	37	69	142	212	409	652	1,150	2,350

For SI: 1 inch = 25.4 mm, 1 foot = 304.8 mm, 1 pound per square inch = 6.895 kPa, 1-inch water column = 0.2488 kPa, 1 British thermal unit per hour = 0.2931 W, 1 cubic foot per hour = 0.0283 m^3/h, 1 degree = 0.01745 rad.

Note: Table entries have been rounded to three significant digits.

TABLE G2413.4(13) [402.4(29)]
SEMIRIGID COPPER TUBING

Gas	Undiluted Propane
Inlet Pressure	10.0 psi
Pressure Drop	1.0 psi
Specific Gravity	1.50

INTENDED USE: SIZING BETWEEN FIRST STAGE (high-pressure regulator) AND SECOND STAGE (low-pressure regulator)										
TUBE SIZE (inches)										
Nominal	K & L	$^1/_4$	$^3/_8$	$^1/_2$	$^5/_8$	$^3/_4$	1	$1^1/_4$	$1^1/_2$	2
	ACR	$^3/_8$	$^1/_2$	$^5/_8$	$^3/_4$	$^7/_8$	$1^1/_8$	$1^3/_8$	—	—
Outside		0.375	0.500	0.625	0.750	0.875	1.125	1.375	1.625	2.125
Inside		0.305	0.402	0.527	0.652	0.745	0.995	1.245	1.481	1.959
Length (ft)		Capacity in Thousands of Btu per Hour								
10		513	1,060	2,150	3,760	5,330	11,400	20,500	32,300	67,400
20		352	727	1,480	2,580	3,670	7,830	14,100	22,200	46,300
30		283	584	1,190	2,080	2,940	6,290	11,300	17,900	37,200
40		242	500	1,020	1,780	2,520	5,380	9,690	15,300	31,800
50		215	443	901	1,570	2,230	4,770	8,590	13,500	28,200
60		194	401	816	1,430	2,020	4,320	7,780	12,300	25,600
70		179	369	751	1,310	1,860	3,980	7,160	11,300	23,500
80		166	343	699	1,220	1,730	3,700	6,660	10,500	21,900
90		156	322	655	1,150	1,630	3,470	6,250	9,850	20,500
100		147	304	619	1,080	1,540	3,280	5,900	9,310	19,400
125		131	270	549	959	1,360	2,910	5,230	8,250	17,200
150		118	244	497	869	1,230	2,630	4,740	7,470	15,600
175		109	225	457	799	1,130	2,420	4,360	6,880	14,300
200		101	209	426	744	1,060	2,250	4,060	6,400	13,300
250		90	185	377	659	935	2,000	3,600	5,670	11,800
300		81	168	342	597	847	1,810	3,260	5,140	10,700
350		75	155	314	549	779	1,660	3,000	4,730	9,840
400		70	144	292	511	725	1,550	2,790	4,400	9,160
450		65	135	274	480	680	1,450	2,620	4,130	8,590
500		62	127	259	453	643	1,370	2,470	3,900	8,120
550		59	121	246	430	610	1,300	2,350	3,700	7,710
600		56	115	235	410	582	1,240	2,240	3,530	7,350
650		54	111	225	393	558	1,190	2,140	3,380	7,040
700		51	106	216	378	536	1,140	2,060	3,250	6,770
750		50	102	208	364	516	1,100	1,980	3,130	6,520
800		48	99	201	351	498	1,060	1,920	3,020	6,290
850		46	96	195	340	482	1,030	1,850	2,920	6,090
900		45	93	189	330	468	1,000	1,800	2,840	5,910
950		44	90	183	320	454	970	1,750	2,750	5,730
1,000		42	88	178	311	442	944	1,700	2,680	5,580
1,100		40	83	169	296	420	896	1,610	2,540	5,300
1,200		38	79	161	282	400	855	1,540	2,430	5,050
1,300		37	76	155	270	383	819	1,470	2,320	4,840
1,400		35	73	148	260	368	787	1,420	2,230	4,650
1,500		34	70	143	250	355	758	1,360	2,150	4,480
1,600		33	68	138	241	343	732	1,320	2,080	4,330
1,700		32	66	134	234	331	708	1,270	2,010	4,190
1,800		31	64	130	227	321	687	1,240	1,950	4,060
1,900		30	62	126	220	312	667	1,200	1,890	3,940
2,000		29	60	122	214	304	648	1,170	1,840	3,830

For SI: 1 inch = 25.4 mm, 1 foot = 304.8 mm, 1 pound per square inch = 6.895 kPa, 1-inch water column = 0.2488 kPa, 1 British thermal unit per hour = 0.2931 W, 1 cubic foot per hour = 0.0283 m^3/h, 1 degree = 0.01745 rad.

Notes:

1. Table capacities are based on Type K copper tubing inside diameter (shown), which has the smallest inside diameter of the copper tubing products.
2. Table entries have been rounded to three significant digits.

TABLE G2413.4(14) [402.4(30)]
SEMIRIGID COPPER TUBING

Gas	Undiluted Propane
Inlet Pressure	11.0 in. w.c.
Pressure Drop	0.5 in. w.c.
Specific Gravity	1.50

INTENDED USE: SIZING BETWEEN SINGLE- OR SECOND-STAGE (low-pressure regulator) AND APPLIANCE										
		TUBE SIZE (inches)								
Nominal	K & L	$^1/_4$	$^3/_8$	$^1/_2$	$^5/_8$	$^3/_4$	1	$1^1/_4$	$1^1/_2$	2
	ACR	$^3/_8$	$^1/_2$	$^5/_8$	$^3/_4$	$^7/_8$	$1^1/_8$	$1^3/_8$	—	—
Outside		0.375	0.500	0.625	0.750	0.875	1.125	1.375	1.625	2.125
Inside		0.305	0.402	0.527	0.652	0.745	0.995	1.245	1.481	1.959
Length (ft)		Capacity in Thousands of Btu per Hour								
10		45	93	188	329	467	997	1,800	2,830	5,890
20		31	64	129	226	321	685	1,230	1,950	4,050
30		25	51	104	182	258	550	991	1,560	3,250
40		21	44	89	155	220	471	848	1,340	2,780
50		19	39	79	138	195	417	752	1,180	2,470
60		17	35	71	125	177	378	681	1,070	2,240
70		16	32	66	115	163	348	626	988	2,060
80		15	30	61	107	152	324	583	919	1,910
90		14	28	57	100	142	304	547	862	1,800
100		13	27	54	95	134	287	517	814	1,700
125		11	24	48	84	119	254	458	722	1,500
150		10	21	44	76	108	230	415	654	1,360
175		NA	20	40	70	99	212	382	602	1,250
200		NA	18	37	65	92	197	355	560	1,170
250		NA	16	33	58	82	175	315	496	1,030
300		NA	15	30	52	74	158	285	449	936
350		NA	14	28	48	68	146	262	414	861
400		NA	13	26	45	63	136	244	385	801
450		NA	12	24	42	60	127	229	361	752
500		NA	11	23	40	56	120	216	341	710
550		NA	11	22	38	53	114	205	324	674
600		NA	10	21	36	51	109	196	309	643
650		NA	NA	20	34	49	104	188	296	616
700		NA	NA	19	33	47	100	180	284	592
750		NA	NA	18	32	45	96	174	274	570
800		NA	NA	18	31	44	93	168	264	551
850		NA	NA	17	30	42	90	162	256	533
900		NA	NA	17	29	41	87	157	248	517
950		NA	NA	16	28	40	85	153	241	502
1,000		NA	NA	16	27	39	83	149	234	488
1,100		NA	NA	15	26	37	78	141	223	464
1,200		NA	NA	14	25	35	75	135	212	442
1,300		NA	NA	14	24	34	72	129	203	423
1,400		NA	NA	13	23	32	69	124	195	407
1,500		NA	NA	13	22	31	66	119	188	392
1,600		NA	NA	12	21	30	64	115	182	378
1,700		NA	NA	12	20	29	62	112	176	366
1,800		NA	NA	11	20	28	60	108	170	355
1,900		NA	NA	11	19	27	58	105	166	345
2,000		NA	NA	11	19	27	57	102	161	335

For SI: 1 inch = 25.4 mm, 1 foot = 304.8 mm, 1 pound per square inch = 6.895 kPa, 1-inch water column = 0.2488 kPa, 1 British thermal unit per hour = 0.2931 W, 1 cubic foot per hour = 0.0283 m^3/h, 1 degree = 0.01745 rad.

Notes:

1. Table capacities are based on Type K copper tubing inside diameter (shown), which has the smallest inside diameter of the copper tubing products.
2. NA means a flow of less than 10,000 Btu/hr.
3. Table entries have been rounded to three significant digits.

TABLE G2413.4(15) [402.4(31)]
SEMIRIGID COPPER TUBING

Gas	Undiluted Propane
Inlet Pressure	2.0 psi
Pressure Drop	1.0 psi
Specific Gravity	1.50

INTENDED USE: TUBE SIZING BETWEEN 2 PSIG SERVICE AND LINE PRESSURE REGULATOR										
TUBE SIZE (inches)										
Nominal	K & L	$^{1}/_{4}$	$^{3}/_{8}$	$^{1}/_{2}$	$^{5}/_{8}$	$^{3}/_{4}$	1	$1^{1}/_{4}$	$1^{1}/_{2}$	2
	ACR	$^{3}/_{8}$	$^{1}/_{2}$	$^{5}/_{8}$	$^{3}/_{4}$	$^{7}/_{8}$	$1^{1}/_{8}$	$1^{3}/_{8}$	—	—
Outside		0.375	0.500	0.625	0.750	0.875	1.125	1.375	1.625	2.125
Inside		0.305	0.402	0.527	0.652	0.745	0.995	1.245	1.481	1.959
Length (ft)		Capacity in Thousands of Btu per Hour								
10		413	852	1,730	3,030	4,300	9,170	16,500	26,000	54,200
20		284	585	1,190	2,080	2,950	6,310	11,400	17,900	37,300
30		228	470	956	1,670	2,370	5,060	9,120	14,400	29,900
40		195	402	818	1,430	2,030	4,330	7,800	12,300	25,600
50		173	356	725	1,270	1,800	3,840	6,920	10,900	22,700
60		157	323	657	1,150	1,630	3,480	6,270	9,880	20,600
70		144	297	605	1,060	1,500	3,200	5,760	9,090	18,900
80		134	276	562	983	1,390	2,980	5,360	8,450	17,600
90		126	259	528	922	1,310	2,790	5,030	7,930	16,500
100		119	245	498	871	1,240	2,640	4,750	7,490	15,600
125		105	217	442	772	1,100	2,340	4,210	6,640	13,800
150		95	197	400	700	992	2,120	3,820	6,020	12,500
175		88	181	368	644	913	1,950	3,510	5,540	11,500
200		82	168	343	599	849	1,810	3,270	5,150	10,700
250		72	149	304	531	753	1,610	2,900	4,560	9,510
300		66	135	275	481	682	1,460	2,620	4,140	8,610
350		60	124	253	442	628	1,340	2,410	3,800	7,920
400		56	116	235	411	584	1,250	2,250	3,540	7,370
450		53	109	221	386	548	1,170	2,110	3,320	6,920
500		50	103	209	365	517	1,110	1,990	3,140	6,530
550		47	97	198	346	491	1,050	1,890	2,980	6,210
600		45	93	189	330	469	1,000	1,800	2,840	5,920
650		43	89	181	316	449	959	1,730	2,720	5,670
700		41	86	174	304	431	921	1,660	2,620	5,450
750		40	82	168	293	415	888	1,600	2,520	5,250
800		39	80	162	283	401	857	1,540	2,430	5,070
850		37	77	157	274	388	829	1,490	2,350	4,900
900		36	75	152	265	376	804	1,450	2,280	4,750
950		35	72	147	258	366	781	1,410	2,220	4,620
1,000		34	71	143	251	356	760	1,370	2,160	4,490
1,100		32	67	136	238	338	721	1,300	2,050	4,270
1,200		31	64	130	227	322	688	1,240	1,950	4,070
1,300		30	61	124	217	309	659	1,190	1,870	3,900
1,400		28	59	120	209	296	633	1,140	1,800	3,740
1,500		27	57	115	201	286	610	1,100	1,730	3,610
1,600		26	55	111	194	276	589	1,060	1,670	3,480
1,700		26	53	108	188	267	570	1,030	1,620	3,370
1,800		25	51	104	182	259	553	1,000	1,570	3,270
1,900		24	50	101	177	251	537	966	1,520	3,170
2,000		23	48	99	172	244	522	940	1,480	3,090

For SI: 1 inch = 25.4 mm, 1 foot = 304.8 mm, 1 pound per square inch = 6.895 kPa, 1-inch water column = 0.2488 kPa,
1 British thermal unit per hour = 0.2931 W, 1 cubic foot per hour = 0.0283 m^3/h, 1 degree = 0.01745 rad.

Notes:

1. Table capacities are based on Type K copper tubing inside diameter (shown), which has the smallest inside diameter of the copper tubing products.
2. Table entries have been rounded to three significant digits.

TABLE G2413.4(16) [402.4(32)] CORRUGATED STAINLESS STEEL TUBING (CSST)

Gas	Undiluted Propane
Inlet Pressure	11.0 in. w.c.
Pressure Drop	0.5 in. w.c.
Specific Gravity	1.50

INTENDED USE: SIZING BETWEEN SINGLE OR SECOND STAGE (Low Pressure) REGULATOR AND THE APPLIANCE SHUTOFF VALVE														
	TUBE SIZE (EHD)													
Flow Designation	13	15	18	19	23	25	30	31	37	39	46	48	60	62
Length (ft)	Capacity in Thousands of Btu per Hour													
5	72	99	181	211	355	426	744	863	1,420	1,638	2,830	3,270	5,780	6,550
10	50	69	129	150	254	303	521	605	971	1,179	1,990	2,320	4,110	4,640
15	39	55	104	121	208	248	422	490	775	972	1,620	1,900	3,370	3,790
20	34	49	91	106	183	216	365	425	661	847	1,400	1,650	2,930	3,290
25	30	42	82	94	164	192	325	379	583	762	1,250	1,480	2,630	2,940
30	28	39	74	87	151	177	297	344	528	698	1,140	1,350	2,400	2,680
40	23	33	64	74	131	153	256	297	449	610	988	1,170	2,090	2,330
50	20	30	58	66	118	137	227	265	397	548	884	1,050	1,870	2,080
60	19	26	53	60	107	126	207	241	359	502	805	961	1,710	1,900
70	17	25	49	57	99	117	191	222	330	466	745	890	1,590	1,760
80	15	23	45	52	94	109	178	208	307	438	696	833	1,490	1,650
90	15	22	44	50	90	102	169	197	286	414	656	787	1,400	1,550
100	14	20	41	47	85	98	159	186	270	393	621	746	1,330	1,480
150	11	15	31	36	66	75	123	143	217	324	506	611	1,090	1,210
200	9	14	28	33	60	69	112	129	183	283	438	531	948	1,050
250	8	12	25	30	53	61	99	117	163	254	390	476	850	934
300	8	11	23	26	50	57	90	107	147	234	357	434	777	854

For SI: 1 inch = 25.4 mm, 1 foot = 304.8 mm, 1 pound per square inch = 6.895 kPa, 1-inch water column = 0.2488 kPa, 1 British thermal unit per hour = 0.2931 W, 1 cubic foot per hour = 0.0283 m^3/h, 1 degree = 0.01745 rad.

Notes:

1. Table includes losses for four 90-degree bends and two end fittings. Tubing runs with larger numbers of bends or fittings shall be increased by an equivalent length of tubing to the following equation: $L = 1.3n$ where L is additional length (feet) of tubing and n is the number of additional fittings or bends.
2. EHD—Equivalent Hydraulic Diameter, which is a measure of the relative hydraulic efficiency between different tubing sizes. The greater the value of EHD, the greater the gas capacity of the tubing.
3. Table entries have been rounded to three significant digits.

TABLE G2413.4(17) [402.4(33)]
CORRUGATED STAINLESS STEEL TUBING (CSST)

Gas	Undiluted Propane
Inlet Pressure	2.0 psi
Pressure Drop	1.0 psi
Specific Gravity	1.50

INTENDED USE: SIZING BETWEEN 2 PSI SERVICE AND THE LINE PRESSURE REGULATOR														
TUBE SIZE (EHD)														
Flow Designation	13	15	18	19	23	25	30	31	37	39	46	48	60	62
Length (ft)	Capacity in Thousands of Btu per Hour													
10	426	558	927	1,110	1,740	2,170	4,100	4,720	7,130	7,958	15,200	16,800	29,400	34,200
25	262	347	591	701	1,120	1,380	2,560	2,950	4,560	5,147	9,550	10,700	18,800	21,700
30	238	316	540	640	1,030	1,270	2,330	2,690	4,180	4,719	8,710	9,790	17,200	19,800
40	203	271	469	554	896	1,100	2,010	2,320	3,630	4,116	7,530	8,500	14,900	17,200
50	181	243	420	496	806	986	1,790	2,070	3,260	3,702	6,730	7,610	13,400	15,400
75	147	196	344	406	663	809	1,460	1,690	2,680	3,053	5,480	6,230	11,000	12,600
80	140	189	333	393	643	768	1,410	1,630	2,590	2,961	5,300	6,040	10,600	12,200
100	124	169	298	350	578	703	1,260	1,450	2,330	2,662	4,740	5,410	9,530	10,900
150	101	137	245	287	477	575	1,020	1,180	1,910	2,195	3,860	4,430	7,810	8,890
200	86	118	213	248	415	501	880	1,020	1,660	1,915	3,340	3,840	6,780	7,710
250	77	105	191	222	373	448	785	910	1,490	1,722	2,980	3,440	6,080	6,900
300	69	96	173	203	343	411	716	829	1,360	1,578	2,720	3,150	5,560	6,300
400	60	82	151	175	298	355	616	716	1,160	1,376	2,350	2,730	4,830	5,460
500	53	72	135	158	268	319	550	638	1,030	1,237	2,100	2,450	4,330	4,880

For SI: 1 inch = 25.4 mm, 1 foot = 304.8 mm, 1 pound per square inch = 6.895 kPa, 1-inch water column = 0.2488 kPa,
1 British thermal unit per hour = 0.2931 W, 1 cubic foot per hour = 0.0283 m^3/h, 1 degree = 0.01745 rad.

Notes:

1. Table does not include effect of pressure drop across the line regulator. Where regulator loss exceeds $^1/_2$ psi (based on 13 in. w.c. outlet pressure), DO NOT USE THIS TABLE. Consult with the regulator manufacturer for pressure drops and capacity factors. Pressure drops across a regulator can vary with flow rate.
2. CAUTION: Capacities shown in the table might exceed maximum capacity for a selected regulator. Consult with the regulator or tubing manufacturer for guidance.
3. Table includes losses for four 90-degree bends and two end fittings. Tubing runs with larger numbers of bends or fittings shall be increased by an equivalent length of tubing to the following equation: $L = 1.3n$ where L is additional length (feet) of tubing and n is the number of additional fittings or bends.
4. EHD—Equivalent Hydraulic Diameter, which is a measure of the relative hydraulic efficiency between different tubing sizes. The greater the value of EHD, the greater the gas capacity of the tubing.
5. Table entries have been rounded to three significant digits.

TABLE G2413.4(18) [402.4(34)]
CORRUGATED STAINLESS STEEL TUBING (CSST)

Gas	Undiluted Propane
Inlet Pressure	5.0 psi
Pressure Drop	3.5 psi
Specific Gravity	1.50

	TUBE SIZE (EHD)													
Flow Designation	13	15	18	19	23	25	30	31	37	39	46	48	60	62
Length (ft)	Capacity in Thousands of Btu per Hour													
10	826	1,070	1,710	2,060	3,150	4,000	7,830	8,950	13,100	14,441	28,600	31,200	54,400	63,800
25	509	664	1,090	1,310	2,040	2,550	4,860	5,600	8,400	9,339	18,000	19,900	34,700	40,400
30	461	603	999	1,190	1,870	2,340	4,430	5,100	7,680	8,564	16,400	18,200	31,700	36,900
40	396	520	867	1,030	1,630	2,030	3,820	4,400	6,680	7,469	14,200	15,800	27,600	32,000
50	352	463	777	926	1,460	1,820	3,410	3,930	5,990	6,717	12,700	14,100	24,700	28,600
75	284	376	637	757	1,210	1,490	2,770	3,190	4,920	5,539	10,300	11,600	20,300	23,400
80	275	363	618	731	1,170	1,450	2,680	3,090	4,770	5,372	9,990	11,200	19,600	22,700
100	243	324	553	656	1,050	1,300	2,390	2,760	4,280	4,830	8,930	10,000	17,600	20,300
150	196	262	453	535	866	1,060	1,940	2,240	3,510	3,983	7,270	8,210	14,400	16,600
200	169	226	393	464	755	923	1,680	1,930	3,050	3,474	6,290	7,130	12,500	14,400
250	150	202	352	415	679	828	1,490	1,730	2,740	3,124	5,620	6,390	11,200	12,900
300	136	183	322	379	622	757	1,360	1,570	2,510	2,865	5,120	5,840	10,300	11,700
400	117	158	279	328	542	657	1,170	1,360	2,180	2,498	4,430	5,070	8,920	10,200
500	104	140	251	294	488	589	1,050	1,210	1,950	2,247	3,960	4,540	8,000	9,110

For SI: 1 inch = 25.4 mm, 1 foot = 304.8 mm, 1 pound per square inch = 6.895 kPa, 1-inch water column = 0.2488 kPa, 1 British thermal unit per hour = 0.2931 W, 1 cubic foot per hour = 0.0283 m^3/h, 1 degree = 0.01745 rad.

Notes:

1. Table does not include effect of pressure drop across line regulator. Where regulator loss exceeds 1 psi, DO NOT USE THIS TABLE. Consult with the regulator manufacturer for pressure drops and capacity factors. Pressure drop across regulator can vary with the flow rate.
2. CAUTION: Capacities shown in the table might exceed maximum capacity of selected regulator. Consult with the tubing manufacturer for guidance.
3. Table includes losses for four 90-degree bends and two end fittings. Tubing runs with larger numbers of bends or fittings shall be increased by an equivalent length of tubing to the following equation: $L = 1.3n$ where L is additional length (feet) of tubing and n is the number of additional fittings or bends.
4. EHD—Equivalent Hydraulic Diameter, which is a measure of the relative hydraulic efficiency between different tubing sizes. The greater the value of EHD, the greater the gas capacity of the tubing.
5. Table entries have been rounded to three significant digits.

TABLE G2413.4(19) [402.4(35)] POLYETHYLENE PLASTIC PIPE

Gas	Undiluted Propane
Inlet Pressure	11.0 in. w.c.
Pressure Drop	0.5 in. w.c.
Specific Gravity	1.50

INTENDED USE: PE PIPE SIZING BETWEEN INTEGRAL 2-STAGE REGULATOR AT TANK OR SECOND STAGE (low-pressure regulator) AND BUILDING						
PIPE SIZE (inches)						
Nominal OD	1/2	3/4	1	1 1/4	1 1/2	2
Designation	SDR 9	SDR 11	SDR 11	SDR 10	SDR 11	SDR 11
Actual ID	0.660	0.860	1.077	1.328	1.554	1.943
Length (ft)	Capacity in Thousands of Btu per Hour					
10	340	680	1,230	2,130	3,210	5,770
20	233	468	844	1,460	2,210	3,970
30	187	375	677	1,170	1,770	3,180
40	160	321	580	1,000	1,520	2,730
50	142	285	514	890	1,340	2,420
60	129	258	466	807	1,220	2,190
70	119	237	428	742	1,120	2,010
80	110	221	398	690	1,040	1,870
90	103	207	374	648	978	1,760
100	98	196	353	612	924	1,660
125	87	173	313	542	819	1,470
150	78	157	284	491	742	1,330
175	72	145	261	452	683	1,230
200	67	135	243	420	635	1,140
250	60	119	215	373	563	1,010
300	54	108	195	338	510	916
350	50	99	179	311	469	843
400	46	92	167	289	436	784
450	43	87	157	271	409	736
500	41	82	148	256	387	695

For SI: 1 inch = 25.4 mm, 1 foot = 304.8 mm, 1 pound per square inch = 6.895 kPa, 1-inch water column = 0.2488 kPa, 1 British thermal unit per hour = 0.2931 W, 1 cubic foot per hour = 0.0283 m^3/h, 1 degree = 0.01745 rad.

Note: Table entries have been rounded to three significant digits.

TABLE G2413.4(20) [402.4(36)]
POLYETHYLENE PLASTIC PIPE

Gas	Undiluted Propane
Inlet Pressure	2.0 psi
Pressure Drop	1.0 psi
Specific Gravity	1.50

INTENDED USE: PE PIPE SIZING BETWEEN 2 PSIG SERVICE REGULATOR AND LINE PRESSURE REGULATOR						
PIPE SIZE (inches)						
Nominal OD	1/2	3/4	1	1 1/4	1 1/2	2
Designation	SDR 9	SDR 11	SDR 11	SDR 10	SDR 11	SDR 11
Actual ID	0.660	0.860	1.077	1.328	1.554	1.943
Length (ft)	Capacity in Thousands of Btu per Hour					
10	3,130	6,260	11,300	19,600	29,500	53,100
20	2,150	4,300	7,760	13,400	20,300	36,500
30	1,730	3,450	6,230	10,800	16,300	29,300
40	1,480	2,960	5,330	9,240	14,000	25,100
50	1,310	2,620	4,730	8,190	12,400	22,200
60	1,190	2,370	4,280	7,420	11,200	20,100
70	1,090	2,180	3,940	6,830	10,300	18,500
80	1,010	2,030	3,670	6,350	9,590	17,200
90	952	1,910	3,440	5,960	9,000	16,200
100	899	1,800	3,250	5,630	8,500	15,300
125	797	1,600	2,880	4,990	7,530	13,500
150	722	1,450	2,610	4,520	6,830	12,300
175	664	1,330	2,400	4,160	6,280	11,300
200	618	1,240	2,230	3,870	5,840	10,500
250	548	1,100	1,980	3,430	5,180	9,300
300	496	994	1,790	3,110	4,690	8,430
350	457	914	1,650	2,860	4,320	7,760
400	425	851	1,530	2,660	4,020	7,220
450	399	798	1,440	2,500	3,770	6,770
500	377	754	1,360	2,360	3,560	6,390
550	358	716	1,290	2,240	3,380	6,070
600	341	683	1,230	2,140	3,220	5,790
650	327	654	1,180	2,040	3,090	5,550
700	314	628	1,130	1,960	2,970	5,330
750	302	605	1,090	1,890	2,860	5,140
800	292	585	1,050	1,830	2,760	4,960
850	283	566	1,020	1,770	2,670	4,800
900	274	549	990	1,710	2,590	4,650
950	266	533	961	1,670	2,520	4,520
1,000	259	518	935	1,620	2,450	4,400
1,100	246	492	888	1,540	2,320	4,170
1,200	234	470	847	1,470	2,220	3,980
1,300	225	450	811	1,410	2,120	3,810
1,400	216	432	779	1,350	2,040	3,660
1,500	208	416	751	1,300	1,960	3,530
1,600	201	402	725	1,260	1,900	3,410
1,700	194	389	702	1,220	1,840	3,300
1,800	188	377	680	1,180	1,780	3,200
1,900	183	366	661	1,140	1,730	3,110
2,000	178	356	643	1,110	1,680	3,020

For SI: 1 inch = 25.4 mm, 1 foot = 304.8 mm, 1 pound per square inch = 6.895 kPa, 1-inch water column = 0.2488 kPa, 1 British thermal unit per hour = 0.2931 W, 1 cubic foot per hour = 0.0283 m^3/h, 1 degree = 0.01745 rad.

Note: Table entries have been rounded to three significant digits.

TABLE G2413.4(21) [402.4(37)] POLYETHYLENE PLASTIC TUBING

Gas	Undiluted Propane
Inlet Pressure	11.0 in. w.c.
Pressure Drop	0.5 in. w.c.
Specific Gravity	1.50

INTENDED USE: PE PIPE SIZING BETWEEN INTEGRAL 2-STAGE REGULATOR AT TANK OR SECOND STAGE (low-pressure regulator) AND BUILDING		
Plastic Tubing Size (CTS) (inch)		
Nominal OD	1/2	1
Designation	SDR 7	SDR 11
Actual ID	0.445	0.927
Length (ft)	Capacity in Cubic Feet of Gas per Hour	
10	121	828
20	83	569
30	67	457
40	57	391
50	51	347
60	46	314
70	42	289
80	39	269
90	37	252
100	35	238
125	31	211
150	28	191
175	26	176
200	24	164
225	22	154
250	21	145
275	20	138
300	19	132
350	18	121
400	16	113
450	15	106
500	15	100

For SI: 1 inch = 25.4 mm, 1 foot = 304.8 mm, 1 pound per square inch = 6.895 kPa, 1-inch water column = 0.2488 kPa, 1 British thermal unit per hour = 0.2931 W, 1 cubic foot per hour = 0.0283 m^3/h, 1 degree = 0.01745 rad.

Note: Table entries have been rounded to three significant digits.

G2413.4.2 (402.4.2) Branch length method. *Pipe* shall be sized as follows:

1. *Pipe* size of each section of the longest *pipe* run from the *point of delivery* to the most remote *outlet* shall be determined using the longest run of *piping* and the load of the section.
2. The *pipe* size of each section of branch *piping* not previously sized shall be determined using the length of *piping* from the *point of delivery* to the most remote *outlet* in each branch and the load of the section.

G2413.4.3 (402.4.3) Hybrid pressure. The *pipe* size for each section of higher pressure *gas piping* shall be determined using the longest length of *piping* from the *point of delivery* to the most remote line *pressure regulator*. The *pipe* size from the line *pressure regulator* to each *outlet* shall be determined using the length of *piping* from the *regulator* to the most remote outlet served by the *regulator*.

G2413.5 (402.5) Noncorrugated stainless steel tubing. Noncorrugated stainless steel tubing shall be sized in accordance with Equations 24-3 and 24-4 of Section 2413.4 in conjunction with Section 2413.4.1, 2413.4.2 or 2413.4.3.

G2413.6 (402.6) Allowable pressure drop. The design pressure loss in any *piping system* under maximum demand, from the *point of delivery* to the inlet connection of all *appliances* served, shall be such that the supply pressure at each *appliance* inlet is greater than or equal to the minimum pressure required by the *appliance*.

G2413.7 (402.7) Maximum operating pressure. The maximum design operating pressure for *piping systems* located inside buildings shall not exceed 5 pounds per square inch gauge (psig) (34 kPa gauge) except where one or more of the following conditions are met:

1. The *piping* joints are welded or brazed.
2. The *piping* is joined by fittings *listed* to ANSI LC4/CSA 6.32 and installed in accordance with the manufacturer's instructions.
3. The *piping* joints are flanged and pipe-to-flange connections are made by welding or brazing.
4. The *piping* is located in a ventilated chase or otherwise enclosed for protection against accidental gas accumulation.
5. The *piping* is a temporary installation for buildings under construction.

G2413.7.1 (402.7.1) Operation below -5°F (-21°C). LP-gas systems designed to operate below -5°F (-21°C) or with butane or a propane-butane mix shall be designed to either accommodate liquid LP-gas or prevent LP-gas vapor from condensing into a liquid.

SECTION G2414 (403) PIPING MATERIALS

G2414.1 (403.1) General. Materials used for *piping systems* shall comply with the requirements of this chapter or shall be *approved*.

G2414.2 (403.2) Used materials. *Pipe*, fittings, *valves* or other materials shall not be used again unless they are free from foreign materials and have been ascertained to be adequate for the service intended.

G2414.3 (403.3) Metallic pipe. Metallic *pipe* shall comply with Sections G2414.3.1 and G2414.3.2.

G2414.3.1 (403.3.1) Cast iron. Cast-iron *pipe* shall not be used.

G2414.3.2 (403.3.2) Steel. Steel, stainless steel and wrought-iron *pipe* shall not be lighter than Schedule 10 and shall comply with the dimensional standards of ASME B36.10M and one of the following standards:

1. ASTM A53/A53M.
2. ASTM A106.
3. ASTM A312.

G2414.4 (403.4) Metallic tubing. *Tubing* shall not be used with gases corrosive to the tubing material.

G2414.4.1 (403.4.1) Steel tubing. Steel *tubing* shall comply with ASTM A254.

G2414.4.2 (403.4.2) Stainless steel. Stainless steel *tubing* shall comply with ASTM A268 or ASTM A269.

G2414.4.3 (403.4.3) Copper or copper-alloy tubing. Copper *tubing* shall comply with Standard Type K or L of ASTM B88 or ASTM B280.

Copper and copper-alloy *tubing* shall not be used if the gas contains more than an average of 0.3 grains of hydrogen sulfide per 100 standard cubic feet of gas (0.7 milligrams per 100 liters).

G2414.4.4 (403.4.5) Corrugated stainless steel tubing. Corrugated stainless steel *tubing* shall be *listed* in accordance with ANSI LC1/CSA 6.26.

G2414.5 (403.6) Plastic pipe, tubing and fittings. Polyethylene plastic pipe, *tubing* and fittings used to supply fuel gas shall conform to ASTM D2513. Such pipe shall be marked "Gas" and "ASTM D2513."

Polyamide pipe, *tubing* and fittings shall be identified and conform to ASTM F2945. Such pipe shall be marked "Gas" and "ASTM F2945."

Polyvinyl chloride (PVC) and chlorinated polyvinyl chloride (CPVC) plastic pipe, *tubing* and fittings shall not be used to supply fuel gas.

G2414.5.1 (403.5.1) Anodeless risers. Plastic pipe, *tubing* and anodeless risers shall comply with the following:

1. Factory-assembled anodeless risers shall be recommended by the manufacturer for the gas used and shall be leak tested by the manufacturer in accordance with written procedures.
2. Service head adapters and field-assembled anodeless risers incorporating service head adapters shall be recommended by the manufacturer for the gas used, and shall be designed and certified to meet the requirements of Category I of ASTM D2513, and US Department of Transportation, Code of Federal Regulations, Title 49 CFR, Part 192.281(e). The manufacturer shall provide the user with qualified installation instructions as prescribed by the US Department of Transportation, Code of Federal Regulations, Title 49 CFR, Part 192.283(b).

G2414.5.2 (403.5.2) LP-gas systems. The use of plastic pipe, *tubing* and fittings in undiluted liquefied petroleum gas *piping* systems shall be in accordance with NFPA 58.

G2414.5.3 (403.5.3) Regulator vent piping. Plastic pipe and fittings used to connect *regulator* vents to remote vent terminations shall be of PVC conforming to ANSI/UL 651. PVC vent *piping* shall not be installed indoors.

G2414.6 (403.6) Workmanship and defects. *Pipe, tubing* and fittings shall be clear and free from cutting burrs and defects in structure or threading, and shall be thoroughly brushed, and chip and scale blown.

Defects in *pipe, tubing* and fittings shall not be repaired. Defective *pipe, tubing* and fittings shall be replaced. (See Section G2417.1.2.)

G2414.7 (403.7) Protective coating. Where in contact with material or atmosphere exerting a corrosive action, metallic *piping* and fittings coated with a corrosion-resistant material shall be used. External or internal coatings or linings used on *piping* or components shall not be considered as adding strength.

G2414.8 (403.8) Metallic pipe threads. Metallic *pipe* and fitting threads shall be taper *pipe* threads and shall comply with ASME B1.20.1.

G2414.8.1 (403.8.1) Damaged threads. *Pipe* with threads that are stripped, chipped, corroded or otherwise damaged shall not be used. Where a weld opens during the operation of cutting or threading, that portion of the *pipe* shall not be used.

G2414.8.2 (403.8.2) Number of threads. Field threading of metallic *pipe* shall be in accordance with Table G2414.8.2.

TABLE G2414.8.2 (403.8.2)
SPECIFICATIONS FOR THREADING METALLIC PIPE

IRON PIPE SIZE (inches)	APPROXIMATE LENGTH OF THREADED PORTION (inches)	APPROXIMATE NO. OF THREADS TO BE CUT
$^1/_2$	$^3/_4$	10
$^3/_4$	$^3/_4$	10
1	$^7/_8$	10
$1^1/_4$	1	11
$1^1/_2$	1	11

For SI: 1 inch = 25.4 mm.

G2414.8.3 (403.8.3) Threaded joint sealing. Threaded joints shall be made using a thread joint sealing material. Thread joint sealing materials shall be nonhardening and shall be resistant to the chemical constituents of the gases to be conveyed through the *piping*. Thread joint sealing materials shall be compatible with the pipe and fitting materials on which the sealing materials are used.

G2414.9 (403.9) Metallic piping joints and fittings. The type of *piping* joint used shall be suitable for the pressure-temperature conditions and shall be selected giving consideration to joint tightness and mechanical strength under the service conditions. The joint shall be able to sustain the maximum end force caused by the internal pressure and any additional forces caused by temperature expansion or contraction, vibration, fatigue, or to the weight of the *pipe* and its contents.

G2414.9.1 (403.9.1) Pipe joints. Schedule 40 and heavier *pipe* joints shall be threaded, flanged, brazed, welded or assembled with press-connect fittings *listed* in accordance with ANSI LC4/CSA 6.32. Pipe lighter than Schedule 40 shall be connected using press-connect fittings, flanges, brazing or welding. Where nonferrous *pipe* is brazed, the *brazing* materials shall have a melting point in excess of 1,000°F (538°C). *Brazing* alloys shall not contain more than 0.05-percent phosphorus.

G2414.9.2 (403.9.2) Copper tubing joints. Copper *tubing* joints shall be assembled with *approved gas tubing* fittings, shall be brazed with a material having a melting point in excess of 1,000°F (538°C) or assembled with press-connect fittings *listed* in accordance with ANSI LC4/CSA 6.32. *Brazing alloys* shall not contain more than 0.05-percent phosphorus.

G2414.9.3 (403.9.3) Stainless steel tubing joints. Stainless steel *tubing* joints shall be welded, assembled with *approved tubing* fittings, brazed with a material having a melting point in excess of 1,000°F (538°C), or assembled with press-connect fittings *listed* in accordance with ANSI LC4/CSA 6.32.

G2414.9.4 (403.10.4) Flared joints. *Flared joints* shall be used only in systems constructed from nonferrous *pipe* and *tubing* where experience or tests have demonstrated that the joint is suitable for the conditions and where provisions are made in the design to prevent separation of the joints.

G2414.9.5 (403.9.5) Metallic fittings. Metallic fittings shall comply with the following:

1. Fittings used with steel, stainless steel or wrought-iron *pipe* shall be steel, stainless steel, copper alloy, malleable iron or cast iron.
2. Fittings used with copper or copper alloy *pipe* shall be copper or copper alloy.
3. Cast-iron bushings shall be prohibited.
4. Special fittings. Fittings such as couplings, proprietary-type joints, saddle tees, gland-type compression fittings, and flared, flareless and compression-type *tubing* fittings shall be: used within the fitting manufacturer's pressure-temperature recommendations; used within the service conditions anticipated with respect to vibration, fatigue, thermal expansion and contraction; and shall be *approved.*
5. Where pipe fittings are drilled and tapped in the field, the operation shall be in accordance with all of the following:
   - 5.1. The operation shall be performed on systems having operating pressures of 5 psi (34.5 kPa) or less.
   - 5.2. The operation shall be performed by the gas supplier or the gas supplier's designated representative.
   - 5.3. The drilling and tapping operation shall be performed in accordance with written procedures prepared by the gas supplier.
   - 5.4. The fittings shall be located outdoors.
   - 5.5. The tapped fitting assembly shall be inspected and proven to be free of leakage.

G2414.10 (403.10) Plastic piping, joints and fittings. Plastic *pipe, tubing* and fittings shall be joined in accordance with the manufacturers' instructions. Such joints shall comply with the following:

1. The joints shall be designed and installed so that the longitudinal pull-out resistance of the joint will be greater than or equal to the tensile strength of the plastic *piping* material.
2. Heat-fusion joints shall be made in accordance with qualified procedures that have been established and proven by test to produce gastight joints as strong as or stronger than the *pipe* or *tubing* being joined. Joints shall be made with the joining method recommended by the *pipe* manufacturer. Polyethylene heat fusion fittings shall be marked "ASTM D2513." Polyamide heat fusion fittings shall be marked "ASTM F2945."
3. Where compression-type *mechanical joints* are used, the gasket material in the fitting shall be compatible with the plastic *piping* and with the gas distributed by the system. An internal tubular rigid stiffener shall be used in conjunction with the fitting. The stiffener shall be flush with the end of the *pipe* or *tubing* and shall extend to or beyond the outside end of the compression fitting when installed. The stiffener shall be free of rough or sharp edges and shall not be a force-fit in the plastic. Split tubular stiffeners shall not be used.
4. Plastic *piping* joints and fittings for use in *liquefied petroleum gas piping systems* shall be in accordance with NFPA 58.

SECTION G2415 (404)
PIPING SYSTEM INSTALLATION

G2415.1 (404.1) Installation of materials. Materials used shall be installed in strict accordance with the standards under which the materials are accepted and *approved.* In the absence of such installation procedures, the manufacturer's instructions shall be followed. Where the requirements of referenced standards or manufacturer's instructions do not conform to minimum provisions of this code, the provisions of this code shall apply.

G2415.2 (404.2) CSST. CSST piping systems shall be installed in accordance with the terms of their approval, the conditions of listing, the manufacturer's instructions and this code.

G2415.3 (404.3) Prohibited locations. *Piping* shall not be installed in or through a ducted supply, return or exhaust, or a clothes chute, *chimney* or gas vent, dumbwaiter or elevator shaft. *Piping* installed downstream of the *point of delivery* shall not extend through any townhouse unit other than the unit served by such *piping.*

G2415.4 (404.4) Piping in solid partitions and walls. *Concealed piping* shall not be located in solid partitions and solid walls, unless installed in a chase or casing.

G2415.5 (404.5) Fittings in concealed locations. Fittings installed in concealed locations shall be limited to the following types:

1. Threaded elbows, tees, couplings, plugs and caps.
2. Brazed fittings.
3. Welded fittings.
4. Fittings *listed* to ANSI LC1/CSA 6.26 or ANSI LC4/CSA 6.32.

G2415.6 (404.6) Underground penetrations prohibited. Gas *piping* shall not penetrate building foundation walls at any point below grade. Gas *piping* shall enter and exit a building at a point above grade and the annular space between the *pipe* and the wall shall be sealed.

G2415.7 (404.7) Protection against physical damage. Where *piping* will be concealed within *light-frame construction* assemblies, the *piping* shall be protected against penetration by fasteners in accordance with Sections G2415.7.1 through G2415.7.3.

Exception: Black steel *piping* and galvanized steel *piping* shall not be required to be protected.

G2415.7.1 (404.7.1) Piping through bored holes or notches. Where *piping* is installed through holes or notches in framing members and the *piping* is located less than $1^1/_2$ inches (38 mm) from the framing member face to which wall, ceiling or floor membranes will be attached, the pipe shall be protected by shield plates that cover the width of the pipe and the framing member and that extend not less than 4 inches (102 mm) to each side of the framing member. Where the framing member that the *piping* passes through is a bottom plate, bottom track, top plate or top track, the shield plates shall cover the framing member and extend not less than 4 inches (102 mm) above the bottom framing member and not less than 4 inches (102 mm) below the top framing member.

G2415.7.2 (404.7.2) Piping installed in other locations. Where the *piping* is located within a framing member and is less than $1^1/_2$ inches (38 mm) from the framing member face to which wall, ceiling or floor membranes will be attached, the *piping* shall be protected by shield plates that cover the width and length of the *piping.* Where the *piping* is located outside of a framing member and is located less than $1^1/_2$ inches (38 mm) from the nearest edge of the face of the framing member to which the membrane will be attached, the *piping* shall be protected by shield plates that cover the width and length of the *piping.*

2415.7.3 (404.7.3) Shield plates. Shield plates shall be of steel material having a thickness of not less than 0.0575 inch (1.463 mm) (No. 16 gage).

G2415.8 (404.8) Piping in solid floors. *Piping* in solid floors shall be laid in channels in the floor and covered in a manner that will allow access to the *piping* with a minimum amount of damage to the building. Where such *piping* is subject to exposure to excessive moisture or corrosive substances, the *piping* shall be protected in an *approved* manner. As an alternative to installation in channels, the *piping* shall be installed in a conduit of Schedule 40 steel, wrought iron, PVC or ABS pipe in accordance with Section G2415.8.1 or G2415.8.2.

G2415.8.1 (404.8.1) Conduit with one end terminating outdoors. The conduit shall extend into an occupiable portion of the building and, at the point where the conduit terminates in the building, the space between the conduit and the *gas piping* shall be sealed to prevent the possible entrance of any gas leakage. The conduit shall extend not less than 2 inches (51 mm) beyond the point where the *pipe* emerges from the floor. If the end sealing is capable of withstanding the full pressure of the gas *pipe,* the conduit shall be designed for the same pressure as the *pipe.* Such conduit shall extend not less than 4 inches (102 mm) outside the building, shall be vented above grade to the outdoors and shall be installed so as to prevent the entrance of water and insects.

G2415.8.2 (404.8.2) Conduit with both ends terminating indoors. Where the conduit originates and terminates within the same building, the conduit shall originate and terminate in an accessible portion of the building and shall not be sealed. The conduit shall extend not less than 2 inches (51 mm) beyond the point where the pipe emerges from the floor.

G2415.9 (404.9) Above-ground piping outdoors. *Piping* installed outdoors shall be elevated not less than $3^1/_2$ inches (89 mm) above ground and where installed across roof surfaces, shall be elevated not less than $3^1/_2$ inches (89 mm) above the roof surface. *Piping* installed above ground, outdoors, and installed across the surface of roofs shall be securely supported and located where it will be protected from physical damage. Where passing through an outside wall, the *piping* shall be protected against corrosion by coating or wrapping with an inert material. Where *piping* is encased in a protective pipe sleeve, the annular space between the *piping* and the sleeve shall be sealed.

G2415.10 (404.10) Isolation. Metallic *piping* and metallic *tubing* that conveys *fuel gas* from an LP-gas storage container shall be provided with an *approved* dielectric fitting to electrically isolate the underground portion of the pipe or tube from the above-ground portion that enters a building. Such dielectric fitting shall be installed above ground, outdoors.

G2415.11 (404.11) Protection against corrosion. Steel pipe or *tubing* exposed to corrosive action, such as soil condition or moisture, shall be protected in accordance with Sections G2415.11.1 through G2415.11.4.

G2415.11.1 (404.11.1) Galvanizing. Zinc coating shall not be deemed adequate protection for underground gas piping.

G2415.11.2 (404.11.2) Protection methods. Underground piping shall comply with one or more of the following:

1. The piping shall be made of corrosion-resistant material that is suitable for the environment in which it will be installed.
2. Pipe shall have a factory-applied, electrically insulating coating. Fittings and joints between sections of coated pipe shall be coated in accordance with the coating manufacturer's instructions.
3. The piping shall have a cathodic protection system installed and the system shall be monitored and maintained in accordance with an *approved* program.

G2415.11.3 (404.11.3) Dissimilar metals. Where dissimilar metals are joined underground, an insulating coupling or fitting shall be used.

G2415.11.4 (404.11.4) Protection of risers. Steel risers connected to plastic piping shall be cathodically protected by means of a welded anode, except where such risers are anodeless risers.

G2415.12 (404.12) Minimum burial depth. Underground *piping systems* shall be installed a minimum depth of 12 inches (305 mm) below grade, except as provided for in Section G2415.12.1.

G2415.12.1 (404.12.1) Individual outdoor appliances. Individual lines to outdoor lights, grills and other *appliances* shall be installed not less than 8 inches (203 mm) below finished grade, provided that such installation is *approved* and is installed in locations not susceptible to physical damage.

G2415.13 (404.13) Trenches. The trench shall be graded so that the pipe has a firm, substantially continuous bearing on the bottom of the trench.

G2415.14 (404.14) Piping underground beneath buildings. *Piping* installed underground beneath buildings is prohibited except where the *piping* is encased in a conduit of wrought iron, plastic pipe, steel pipe, a piping or encasement system *listed* for installation beneath buildings, or other *approved* conduit material designed to withstand the superimposed loads. The conduit shall be protected from corrosion in accordance with Section G2415.11 and shall be installed in accordance with Section G2415.14.1 or G2415.14.2.

G2415.14.1 (404.14.1) Conduit with one end terminating outdoors. The conduit shall extend into an occupiable portion of the building and, at the point where the conduit terminates in the building, the space between the conduit and the *gas piping* shall be sealed to prevent the possible entrance of any gas leakage. The conduit shall extend not less than 2 inches (51 mm) beyond the point where the *pipe* emerges from the floor. Where the end sealing is capable of withstanding the full pressure of the gas pipe, the conduit shall be designed for the same pressure as the pipe. Such conduit shall extend not less than 4 inches (102 mm) outside the building, shall be vented above grade to the outdoors and shall be installed so as to prevent the entrance of water and insects.

G2415.14.2 (404.14.2) Conduit with both ends terminating indoors. Where the conduit originates and terminates within the same building, the conduit shall originate and terminate in an accessible portion of the building and shall not be sealed. The conduit shall extend not less than 2 inches (51 mm) beyond the point where the pipe emerges from the floor.

G2415.15 (404.15) Outlet closures. Gas *outlets* that do not connect to *appliances* shall be capped gastight.

Exception: *Listed* and *labeled* flush-mounted-type quick-disconnect devices and *listed* and *labeled gas convenience outlets* shall be installed in accordance with the manufacturer's instructions.

G2415.16 (404.16) Location of outlets. The unthreaded portion of *piping outlets* shall extend not less than 1 inch (25 mm) through finished ceilings and walls and where extending through floors or outdoor patios and slabs, shall be not less than 2 inches (51 mm) above them. The *outlet* fitting or *piping* shall be securely supported. *Outlets* shall not be placed behind doors. *Outlets* shall be located in the room or space where the *appliance* is installed.

Exception: *Listed* and *labeled* flush-mounted-type quick-disconnect devices and *listed* and *labeled gas convenience outlets* shall be installed in accordance with the manufacturer's instructions.

G2415.17 (404.17) Plastic pipe. The installation of plastic *pipe* shall comply with Sections G2415.17.1 through G2415.17.3.

G2415.17.1 (404.17.1) Limitations. Plastic pipe shall be installed outdoors underground only. Plastic pipe shall not be used within or under any building or slab or be operated at pressures greater than 100 psig (689 kPa) for natural gas or 30 psig (207 kPa) for LP-gas.

Exceptions:

1. Plastic pipe shall be permitted to terminate above ground outside of buildings where installed in premanufactured *anodeless risers* or service head adapter risers that are installed in accordance with the manufacturer's instructions.
2. Plastic pipe shall be permitted to terminate with a wall head adapter within buildings where the plastic pipe is inserted in a *piping* material for *fuel gas* use in buildings.
3. Plastic pipe shall be permitted under outdoor patio, walkway and driveway slabs provided that the burial depth complies with Section G2415.12.

G2415.17.2 (404.17.2) Connections. Connections made outdoors and underground between metallic and plastic *piping* shall be made only with transition fittings conforming to ASTM D2513 Category I or ASTM F1973.

G2415.17.3 (404.17.3) Tracer. A yellow-insulated copper tracer wire or other *approved* conductor, or a product specifically designed for that purpose, shall be installed adjacent to underground nonmetallic *piping*. Access shall be provided to the tracer wire or the tracer wire shall terminate above ground at each end of the nonmetallic *piping*. The tracer wire size shall be not less than 18 AWG and the insulation type shall be suitable for direct burial.

G2415.18 (404.18) Pipe debris removal. The interior of *piping* shall be clear of debris. The use of a flammable or combustible gas to clean or remove debris from a *piping system* shall be prohibited.

G2415.19 (404.19) Prohibited devices. A device shall not be placed inside the *piping* or fittings that will reduce the cross-sectional area or otherwise obstruct the free flow of gas.

Exceptions:

1. *Approved* gas filters.

2. An *approved* fitting or device where the *gas piping system* has been sized to accommodate the pressure drop of the fitting or device.

G2415.20 (404.20) Testing of piping. Before any system of *piping* is put in service or concealed, it shall be tested to ensure that it is gastight. Testing, inspection and purging of *piping systems* shall comply with Section G2417.

SECTION G2416 (405) PIPING BENDS AND CHANGES IN DIRECTION

G2416.1 (405.1) General. Changes in direction of pipe shall be permitted to be made by the use of fittings, factory bends or field bends.

G2416.2 (405.2) Metallic pipe. Metallic pipe bends shall comply with the following:

1. Bends shall be made only with bending tools and procedures intended for that purpose.
2. Bends shall be smooth and free from buckling, cracks or other evidence of mechanical damage.
3. The longitudinal weld of the pipe shall be near the neutral axis of the bend.
4. Pipe shall not be bent through an arc of more than 90 degrees (1.6 rad).
5. The inside radius of a bend shall be not less than six times the outside diameter of the pipe.

G2416.3 (405.3) Plastic pipe. Plastic pipe bends shall comply with the following:

1. The pipe shall not be damaged and the internal diameter of the pipe shall not be effectively reduced.
2. Joints shall not be located in pipe bends.
3. The radius of the inner curve of such bends shall be not less than 25 times the inside diameter of the pipe.
4. Where the *piping* manufacturer specifies the use of special bending tools or procedures, such tools or procedures shall be used.

SECTION G2417 (406) INSPECTION, TESTING AND PURGING

G2417.1 (406.1) General. Prior to acceptance and initial operation, all *piping* installations shall be visually inspected and pressure tested to determine that the materials, design, fabrication and installation practices comply with the requirements of this code.

G2417.1.1 (406.1.1) Inspections. Inspection shall consist of visual examination, during or after manufacture, fabrication, assembly or *pressure tests.*

G2417.1.2 (406.1.2) Repairs and additions. In the event repairs or additions are made after the *pressure test,* the affected *piping* shall be tested.

Minor repairs and additions are not required to be *pressure tested* provided that the work is inspected and connections are tested with a noncorrosive leak-detecting fluid or other *approved* leak-detecting methods.

G2417.1.3 (406.1.3) New branches. Where new branches are installed to new *appliances,* only the newly installed branches shall be required to be *pressure tested.* Connections between the new *piping* and the existing *piping* shall be tested with a noncorrosive leak-detecting fluid or other *approved* leak-detecting methods.

G2417.1.4 (406.1.4) Section testing. A *piping system* shall be permitted to be tested as a complete unit or in sections. A *valve* in a line shall not be used as a bulkhead between gas in one section of the *piping system* and test medium in an adjacent section, except where a double block and bleed valve system is installed. A valve shall not be subjected to the test pressure unless it can be determined that the valve, including the valve-closing mechanism, is designed to safely withstand the test pressure.

G2417.1.5 (406.1.5) Regulators and valve assemblies. *Regulator* and valve assemblies fabricated independently of the *piping system* in which they are to be installed shall be permitted to be tested with inert gas or air at the time of fabrication.

G2417.1.6 (406.1.6) Pipe clearing. Prior to testing, the interior of the pipe shall be cleared of all foreign material.

G2417.2 (406.2) Test medium. The test medium shall be air, nitrogen, carbon dioxide or an inert gas. Oxygen shall not be used as a test medium.

G2417.3 (406.3) Test preparation. *Pipe* joints, including welds, shall be left exposed for examination during the test.

Exception: Covered or *concealed pipe* end joints that have been previously tested in accordance with this code.

G2417.3.1 (406.3.1) Expansion joints. Expansion joints shall be provided with temporary restraints, if required, for the additional thrust load under test.

G2417.3.2 (406.3.2) Appliance and equipment isolation. *Appliances* and *equipment* that are not to be included in the test shall be either disconnected from the *piping* or isolated by blanks, blind flanges or caps.

G2417.3.3 (406.3.3) Appliance and equipment disconnection. Where the *piping system* is connected to *appliances* or *equipment* designed for operating pressures of less than the test pressure, such *appliances* or *equipment* shall be isolated from the *piping system* by disconnecting them and capping the *outlet(s).*

G2417.3.4 (406.3.4) Valve isolation. Where the *piping system* is connected to *appliances* or *equipment* designed for operating pressures equal to or greater than the test pressure, such *appliances* or *equipment* shall be isolated from the *piping system* by closing the individual *appliance* or *equipment* shutoff valve(s).

G2417.3.5 (406.3.5) Testing precautions. Testing of *piping* systems shall be performed in a manner that protects the safety of employees and the public during the test.

G2417.4 (406.4) Test pressure measurement. Test pressure shall be measured with a manometer or with a pressure-measuring device designed and calibrated to read, record, or indicate a pressure loss caused by leakage during the *pressure test* period. The source of pressure shall be isolated before the *pressure tests* are made. Mechanical gauges used to measure test pressures shall have a range such that the highest end of the scale is not greater than five times the test pressure.

G2417.4.1 (406.4.1) Test pressure. The test pressure to be used shall be not less than $1^1/_2$ times the proposed maximum working pressure, but not less than 3 psig (20 kPa gauge), irrespective of design pressure. Where the test pressure exceeds 125 psig (862 kPa gauge), the test pressure shall not exceed a value that produces a hoop stress in the *piping* greater than 50 percent of the specified minimum yield strength of the pipe.

G2417.4.2 (406.4.2) Test duration. The test duration shall be not less than 10 minutes.

G2417.5 (406.5) Detection of leaks and defects. The *piping system* shall withstand the test pressure specified without showing any evidence of leakage or other defects. Any reduction of test pressures as indicated by pressure gauges shall be deemed to indicate the presence of a leak unless such reduction can be readily attributed to some other cause.

G2417.5.1 (406.5.1) Detection methods. The leakage shall be located by means of an *approved* gas detector, a noncorrosive leak detection fluid or other *approved* leak detection methods.

G2417.5.2 (406.5.2) Corrections. Where leakage or other defects are located, the affected portion of the *piping system* shall be repaired or replaced and retested.

G2417.6 (406.6) Piping system and equipment leakage check. Leakage checking of systems and *equipment* shall be in accordance with Sections G2417.6.1 through G2417.6.4.

G2417.6.1 (406.6.1) Test gases. Leak checks using fuel gas shall be permitted in *piping systems* that have been pressure tested in accordance with Section G2417.

G2417.6.2 (406.6.2) Before turning gas on. During the process of turning gas on into a system of new *gas piping*, the entire system shall be inspected to determine that there are no open fittings or ends and that all *valves* at unused outlets are closed and plugged or capped.

G2417.6.3 (406.6.3) Leak check. Immediately after the gas is turned on into a new system or into a system that has been initially restored after an interruption of service, the *piping system* shall be checked for leakage. Where leakage is indicated, the gas supply shall be shut off until the necessary repairs have been made.

G2417.6.4 (406.6.4) Placing appliances and equipment in operation. *Appliances* and *equipment* shall not be placed in operation until after the *piping system* has been checked for leakage in accordance with Section G2417.6.3, the *piping system* has been purged in accordance with Section G2417.7 and the connections to the *appliances* have been checked for leakage.

G2417.7 (406.7) Purging. The purging of *piping* shall be in accordance with Sections G2417.7.1 through 2417.7.3.

G2417.7.1 (406.7.1) Piping systems required to be purged outdoors. The purging of *piping systems* shall be in accordance with the provisions of Sections G2417.7.1.1 through G2417.7.1.4 where the *piping system* meets either of the following:

1. The design operating gas pressure is greater than 2 psig (13.79 kPa).
2. The *piping* being purged contains one or more sections of pipe or tubing meeting the size and length criteria of Table G2417.7.1.1.

G2417.7.1.1 (406.7.1.1) Removal from service. Where existing *gas piping* is opened, the section that is opened shall be isolated from the gas supply and the line pressure vented in accordance with Section G2417.7.1.3. Where *gas piping* meeting the criteria of Table G2417.7.1.1 is removed from service, the residual fuel gas in the *piping* shall be displaced with an inert gas.

TABLE G2417.7.1.1 (406.7.1.1)
SIZE AND LENGTH OF PIPING

NOMINAL PIPE SIZE (inches)[a]	LENGTH OF PIPING (feet)
$\geq 2^1/_2 < 3$	> 50
$\geq 3 < 4$	> 30
$\geq 4 < 6$	> 15
$\geq 6 < 8$	> 10
≥ 8	Any length

For SI: 1 inch = 25.4 mm, 1 foot = 304.8 mm.

a. CSST EHD size of 62 is equivalent to nominal 2-inch pipe or tubing size.

G2417.7.1.2 (406.7.1.2) Placing in operation. Where *gas piping* containing air and meeting the criteria of Table G2417.7.1.1 is placed in operation, the air in the *piping* shall first be displaced with an inert gas. The inert gas shall then be displaced with fuel gas in accordance with Section G2417.7.1.3.

G2417.7.1.3 (406.7.1.3) Outdoor discharge of purged gases. The open end of a *piping* system being pressure vented or purged shall discharge directly to an outdoor location. Purging operations shall comply with all of the following requirements:

1. The point of discharge shall be controlled with a shutoff valve.
2. The point of discharge shall be located not less than 10 feet (3048 mm) from sources of ignition, not less than 10 feet (3048 mm) from building openings and not less than 25 feet (7620 mm) from mechanical air intake openings.
3. During discharge, the open point of discharge shall be continuously attended and monitored with a combustible gas indicator that complies with Section G2417.7.1.4.
4. Purging operations introducing fuel gas shall be stopped when 90 percent fuel gas by volume is detected within the pipe.

5. Persons not involved in the purging operations shall be evacuated from all areas within 10 feet (3048 mm) of the point of discharge.

G2417.7.1.4 (406.7.1.4) Combustible gas indicator. Combustible gas indicators shall be *listed* and shall be calibrated in accordance with the manufacturer's instructions. Combustible gas indicators shall numerically display a volume scale from zero percent to 100 percent in 1-percent or smaller increments.

G2417.7.2 (406.7.2) Piping systems allowed to be purged indoors or outdoors. The purging of *piping systems* shall be in accordance with the provisions of Section G2417.7.2.1 where the *piping system* meets both of the following:

1. The design operating gas pressure is 2 psig (13.79 kPa) or less.
2. The *piping* being purged is constructed entirely from pipe or tubing not meeting the size and length criteria of Table G2417.7.1.1.

G2417.7.2.1 (406.7.2.1) Purging procedure. The *piping system* shall be purged in accordance with one or more of the following:

1. The *piping* shall be purged with fuel gas and shall discharge to the outdoors.
2. The *piping* shall be purged with fuel gas and shall discharge to the indoors or outdoors through an *appliance* burner not located in a combustion chamber. Such burner shall be provided with a continuous source of ignition.
3. The *piping* shall be purged with fuel gas and shall discharge to the indoors or outdoors through a burner that has a continuous source of ignition and that is designed for such purpose.
4. The *piping* shall be purged with fuel gas that is discharged to the indoors or outdoors, and the point of discharge shall be monitored with a *listed* combustible gas detector in accordance with Section G2417.7.2.2. Purging shall be stopped when fuel gas is detected.
5. The *piping* shall be purged by the gas supplier in accordance with written procedures.

G2417.7.2.2 (406.7.2.2) Combustible gas detector. Combustible gas detectors shall be *listed* and shall be calibrated or tested in accordance with the manufacturer's instructions. Combustible gas detectors shall be capable of indicating the presence of fuel gas.

G2417.7.3 (406.7.3) Purging appliances and equipment. After the *piping system* has been placed in operation, *appliances* and *equipment* shall be purged before being placed into operation.

SECTION G2418 (407) PIPING SUPPORT

G2418.1 (407.1) General. *Piping* shall be provided with support in accordance with Section G2418.2.

G2418.2 (407.2) Design and installation. *Piping* shall be supported with metal pipe hooks, metal pipe straps, metal bands, metal brackets, metal hangers or building structural components suitable for the size of *piping*, of adequate strength and quality, and located at intervals so as to prevent or damp out excessive vibration. *Piping* shall be anchored to prevent undue strains on connected *appliances* and shall not be supported by other *piping*. Pipe hangers and supports shall conform to the requirements of MSS SP-58 and shall be spaced in accordance with Section G2424. Supports, hangers and anchors shall be installed so as not to interfere with the free expansion and contraction of the *piping* between anchors. The components of the supporting *equipment* shall be designed and installed so that they will not be disengaged by movement of the supported *piping*.

SECTION G2419 (408) DRIPS AND SLOPED PIPING

G2419.1 (408.1) Slopes. *Piping* for other than dry gas conditions shall be sloped not less than $^{1}/_{4}$ inch in 15 feet (6.3 mm in 4572 mm) to prevent traps.

G2419.2 (408.2) Drips. Where wet gas exists, a *drip* shall be provided at any point in the line of pipe where *condensate* could collect. A *drip* shall be provided at the outlet of the *meter* and shall be installed so as to constitute a trap wherein an accumulation of *condensate* will shut off the flow of gas before the *condensate* will run back into the *meter*.

G2419.3 (408.3) Location of drips. *Drips* shall be provided with *ready access* to permit cleaning or emptying. A *drip* shall not be located where the *condensate* is subject to freezing.

G2419.4 (408.4) Sediment trap. Where a sediment trap is not incorporated as part of the *appliance*, a sediment trap shall be installed downstream of the *appliance* shutoff valve as close to the inlet of the *appliance* as practical. The sediment trap shall be either a tee fitting having a capped nipple of any length installed vertically in the bottommost opening of the tee as illustrated in Figure G2419.4 or other device *approved* as an effective sediment trap. Illuminating *appliances*, ranges, clothes dryers, decorative vented *appliances* for installation in vented fireplaces, gas fireplaces and outdoor grills need not be so equipped.

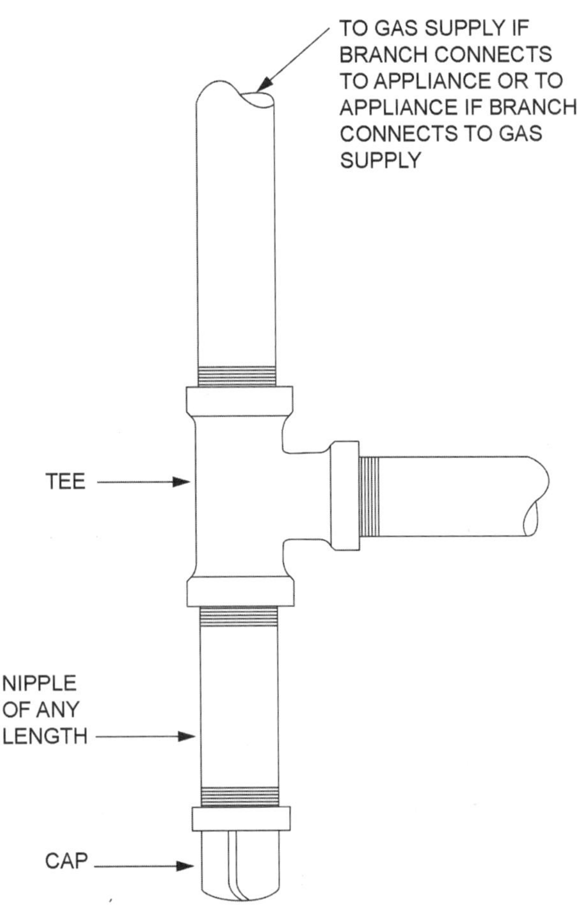

FIGURE G2419.4 (408.4)
METHOD OF INSTALLING A TEE FITTING SEDIMENT TRAP

SECTION G2420 (409) SHUTOFF VALVES

G2420.1 (409.1) General. *Piping systems* shall be provided with shutoff valves in accordance with this section.

G2420.1.1 (409.1.1) Valve approval. Shutoff valves shall be of an *approved* type; shall be constructed of materials compatible with the *piping*; and shall comply with the standard that is applicable for the pressure and application, in accordance with Table G2420.1.1.

G2420.1.2 (409.1.2) Prohibited locations. Shutoff valves shall be prohibited in *concealed locations* and *furnace plenums*.

G2420.1.3 (409.1.3) Access to shutoff valves. Shutoff *valves* shall be located in places so as to provide access for operation and shall be installed so as to be protected from damage.

G2420.2 (409.2) Meter valve. Every *meter* shall be equipped with a shutoff valve located on the supply side of the *meter*.

G2420.3 (409.3.2) Individual buildings. In a common system serving more than one building, shutoff valves shall be installed outdoors at each building.

G2420.4 (409.4) MP regulator valves. A *listed* shutoff valve shall be installed immediately ahead of each MP *regulator*.

G2420.5 (409.5) Appliance shutoff valve. Each *appliance* shall be provided with a shutoff valve in accordance with Section G2420.5.1, G2420.5.2 or G2420.5.3.

G2420.5.1 (409.5.1) Located within same room. The shutoff valve shall be located in the same room as the *appliance*. The shutoff valve shall be within 6 feet (1829 mm) of the *appliance*, and shall be installed upstream of the union, connector or quick disconnect device it serves. Such shutoff *valves* shall be provided with *access*. Shutoff valves serving movable *appliances*, such as cooking *appliances* and clothes dryers, shall be considered to be provided with access where installed behind such *appliances*. *Appliance shutoff valves* located in the firebox of a *fireplace* shall be installed in accordance with the *appliance* manufacturer's instructions.

G2420.5.2 (409.5.2) Vented decorative appliances and room heaters. Shutoff valves for vented decorative *appliances*, room heaters and decorative *appliances* for installation in vented *fireplaces* shall be permitted to be installed in an area remote from the *appliances* where such valves are provided with *ready access*. Such *valves* shall be permanently identified and shall not serve another *appliance*. The *piping* from the shutoff valve to within 6 feet (1829 mm) of the *appliance* shall be designed, sized and installed in accordance with Sections G2412 through G2419.

TABLE G2420.1.1 (409.1.1)
MANUAL GAS VALVE STANDARDS

VALVE STANDARDS	APPLIANCE SHUTOFF VALVE APPLICATION UP TO $^1/_2$ psig PRESSURE	OTHER VALVE APPLICATIONS			
		UP TO $^1/_2$ psig PRESSURE	UP TO 2 psig PRESSURE	UP TO 5 psig PRESSURE	UP TO 125 psig PRESSURE
ANSI Z21.15/CSA 9.1	X	—	—	—	—
ASME B16.44	X	X	X[a]	X[b]	—
ASME B16.33	X	X	X	X	X

For SI: 1 pound per square inch gauge = 6.895 kPa.
a. If labeled 2G.
b. If labeled 5G.

G2420.5.3 (409.5.3) Located at manifold. Where the *appliance* shutoff valve is installed at a manifold, such shutoff valve shall be located within 50 feet (15 240 mm) of the *appliance* served and shall be readily accessible and permanently identified. The *piping* from the manifold to within 6 feet (1829 mm) of the *appliance* shall be designed, sized and installed in accordance with Sections G2412 through G2419.

G2420.6 (409.7) Shutoff valves in tubing systems. Shutoff valves installed in *tubing* systems shall be rigidly and securely supported independently of the *tubing*.

SECTION G2421 (410) FLOW CONTROLS

G2421.1 (410.1) Pressure regulators. A line *pressure regulator* shall be installed where the *appliance* is designed to operate at a lower pressure than the supply pressure. *Line gas pressure regulators* shall be *listed* as complying with ANSI Z21.80. *Access* shall be provided to *pressure regulators*. *Pressure regulators* shall be protected from physical damage. *Regulators* installed on the exterior of the building shall be *approved* for outdoor installation.

G2421.2 (410.2) MP regulators. MP *pressure regulators* shall comply with the following:

1. The MP *regulator* shall be *approved* and shall be suitable for the inlet and outlet gas pressures for the application.
2. The MP *regulator* shall maintain a reduced outlet pressure under lock-up (no-flow) conditions.
3. The capacity of the MP *regulator*, determined by published ratings of its manufacturer, shall be adequate to supply the *appliances* served.
4. The MP *pressure regulator* shall be provided with *access*. Where located indoors, the *regulator* shall be vented to the outdoors or shall be equipped with a leak-limiting device, in either case complying with Section G2421.3.
5. A tee fitting with one opening capped or plugged shall be installed between the MP *regulator* and its upstream shutoff valve. Such tee fitting shall be positioned to allow connection of a pressure-measuring instrument and to serve as a sediment trap.
6. A tee fitting with one opening capped or plugged shall be installed not less than 10 pipe diameters downstream of the MP *regulator* outlet. Such tee fitting shall be positioned to allow connection of a pressure-measuring instrument. The tee fitting is not required where the MP regulator serves an *appliance* that has a pressure test port on the gas control inlet side and the appliance is located in the same room as the MP regulator.
7. Where connected to rigid *piping*, a union shall be installed within 1 foot (304 mm) of either side of the MP *regulator*.

G2421.3 (410.3) Venting of regulators. *Pressure regulators* that require a vent shall be vented directly to the outdoors. The vent shall be designed to prevent the entry of insects, water and foreign objects.

Exception: A vent to the outdoors is not required for *regulators* equipped with and *labeled* for utilization with an *approved* vent-limiting device installed in accordance with the manufacturer's instructions.

G2421.3.1 (410.3.1) Vent piping. Vent *piping* for relief vents and breather vents shall be constructed of materials allowed for *gas piping* in accordance with Section G2414. Vent *piping* shall be not smaller than the vent connection on the pressure-regulating device. Vent *piping* serving relief vents and combination relief and breather vents shall be run independently to the outdoors and shall serve only a single device vent. Vent *piping* serving only breather vents is permitted to be connected in a manifold arrangement where sized in accordance with an *approved* design that minimizes backpressure in the event of diaphragm rupture. *Regulator* vent *piping* shall not exceed the length specified in the *regulator* manufacturer's instructions.

G2421.4 (410.4) Excess flow valves. Where automatic *excess flow valves* are installed, they shall be *listed* in accordance with ANSI Z21.93/CSA 6.30 and shall be sized and installed in accordance with the manufacturer's instructions.

G2421.5 (410.5) Flashback arrestor check valve. Where fuel gas is used with oxygen in any hot work operation, a *listed* protective device that serves as a combination flashback arrestor and backflow check valve shall be installed at an *approved* location on both the fuel gas and oxygen supply lines. Where the pressure of the piped fuel gas supply is insufficient to ensure such safe operation, *approved* equipment shall be installed between the gas meter and the *appliance* that increases pressure to the level required for such safe operation.

SECTION G2422 (411) APPLIANCE CONNECTIONS

G2422.1 (411.1) Connecting appliances. *Appliances* shall be connected to the *piping system* by one of the following:

1. Rigid metallic pipe and fittings.
2. Corrugated stainless steel *tubing* (CSST) where installed in accordance with the manufacturer's instructions.
3. *Listed* and *labeled appliance connectors* in compliance with ANSI Z21.24/CSA 6.10 and installed in accordance with the manufacturer's instructions and located entirely in the same room as the *appliance*.
4. *Listed* and *labeled* quick-disconnect devices in compliance with ANSI Z21.41/CSA 6.9 used in conjunction with *listed* and *labeled appliance connectors*.

5. *Listed* and *labeled* convenience outlets in compliance with ANSI Z21.90/CSA 6.24 used in conjunction with *listed* and *labeled appliance connectors.*
6. *Listed* and *labeled* outdoor *appliance connectors* in compliance with ANSI Z21.75/CSA 6.27 and installed in accordance with the manufacturer's instructions.
7. *Listed* outdoor gas hose connectors in compliance with ANSI Z21.54 used to connect portable outdoor *appliances.* The gas hose connection shall be made only in the outdoor area where the *appliance* is used, and shall be to the gas *piping* supply at an *appliance* shutoff valve, a *listed* quick-disconnect device or *listed* gas convenience outlet.

G2422.1.1 (411.1.2) Protection from damage. Connectors and *tubing* shall be installed so as to be protected against physical damage.

G2422.1.2 (411.1.3) Connector installation. *Appliance* fuel connectors shall be installed in accordance with the manufacturer's instructions and Sections G2422.1.2.1 through G2422.1.2.4.

G2422.1.2.1 (411.1.3.1) Maximum length. Connectors shall have an overall length not to exceed 6 feet (1829 mm). Measurement shall be made along the centerline of the connector. Only one connector shall be used for each *appliance.*

Exception: Rigid metallic *piping* used to connect an *appliance* to the *piping system* shall be permitted to have a total length greater than 6 feet (1829 mm), provided that the connecting pipe is sized as part of the *piping system* in accordance with Section G2413 and the location of the *appliance* shutoff valve complies with Section G2420.5.

G2422.1.2.2 (411.1.3.2) Minimum size. Connectors shall have the capacity for the total *demand* of the connected *appliance.*

G2422.1.2.3 (411.1.3.3) Prohibited locations and penetrations. Connectors shall not be concealed within, or extended through, walls, floors, partitions, ceilings or *appliance* housings.

Exceptions:

1. Connectors constructed of materials allowed for *piping systems* in accordance with Section G2414 shall be permitted to pass through walls, floors, partitions and ceilings where installed in accordance with Section G2420.5.2 or G2420.5.3.
2. Rigid steel pipe connectors shall be permitted to extend through openings in *appliance* housings.
3. *Fireplace* inserts that are factory equipped with grommets, sleeves or other means of protection in accordance with the listing of the *appliance.*
4. Semirigid *tubing* and *listed* connectors shall be permitted to extend through an opening in an *appliance* housing, cabinet or casing where the tubing or connector is protected against damage.

G2422.1.2.4 (411.1.3.4) Shutoff valve. A shutoff valve not less than the nominal size of the connector shall be installed ahead of the connector in accordance with Section G2420.5.

G2422.1.3 (411.1.5) Connection of gas engine-powered air conditioners. Internal combustion engines shall not be rigidly connected to the gas supply *piping.*

G2422.1.4 (411.1.6) Unions. A union fitting shall be provided for *appliances* connected by rigid metallic pipe. Such unions shall be accessible and located within 6 feet (1829 mm) of the *appliance.*

G2422.1.5 (411.1.4) Movable appliances. Where *appliances* are equipped with casters or are otherwise subject to periodic movement or relocation for purposes such as routine cleaning and maintenance, such *appliances* shall be connected to the supply system *piping* by means of an *appliance connector listed* as complying with ANSI Z21.69/CSA 6.16 or by means of Item 1 of Section G2422.1. Such flexible connectors shall be installed and protected against physical damage in accordance with the manufacturer's instructions.

G2422.2 (411.3) Suspended low-intensity infrared tube heaters. Suspended low-intensity infrared tube heaters shall be connected to the building *piping system* with a connector *listed* for the application complying with ANSI Z21.24/CSA 6.10. The connector shall be installed as specified by the tube heater manufacturer's instructions.

SECTION G2423 (413) COMPRESSED NATURAL GAS MOTOR VEHICLE FUEL-DISPENSING FACILITIES

G2423.1 (413.1) General. Motor fuel-dispensing facilities for CNG fuel shall be in accordance with Section 413 of the *International Fuel Gas Code.*

SECTION G2424 (415) PIPING SUPPORT INTERVALS

G2424.1 (415.1) Interval of support. *Piping* shall be supported at intervals not exceeding the spacing specified in Table G2424.1. Spacing of supports for CSST shall be in accordance with the CSST manufacturer's instructions.

TABLE G2424.1 (415.1)
SUPPORT OF PIPING

STEEL PIPE, NOMINAL SIZE OF PIPE (inches)	SPACING OF SUPPORTS (feet)	NOMINAL SIZE OF TUBING SMOOTH-WALL (inch O.D.)	SPACING OF SUPPORTS (feet)
$^1/_2$	6	$^1/_2$	4
$^3/_4$ or 1	8	$^5/_8$ or $^3/_4$	6
$1^1/_4$ or larger (horizontal)	10	$^7/_8$ or 1 (horizontal)	8
$1^1/_4$ or larger (vertical)	Every floor level	1 or larger (vertical)	Every floor level

For SI: 1 inch = 25.4 mm, 1 foot = 304.8 mm.

SECTION G2425 (501) GENERAL

G2425.1 (501.1) Scope. This section shall govern the installation, maintenance, repair and approval of factory-built *chimneys, chimney* liners, vents and connectors and the utilization of masonry chimneys serving gas-fired *appliances.*

G2425.2 (501.2) General. Every *appliance* shall discharge the products of combustion to the outdoors, except for *appliances* exempted by Section G2425.8.

G2425.3 (501.3) Masonry chimneys. *Masonry chimneys* shall be constructed in accordance with Section G2427.5 and Chapter 10.

G2425.4 (501.4) Minimum size of chimney or vent. *Chimneys* and vents shall be sized in accordance with Sections G2427 and G2428.

G2425.5 (501.5) Abandoned inlet openings. Abandoned inlet openings in *chimneys* and vents shall be closed by an *approved* method.

G2425.6 (501.6) Positive pressure. Where an *appliance* equipped with a mechanical forced *draft* system creates a positive pressure in the venting system, the venting system shall be designed for positive pressure applications.

G2425.7 (501.7) Connection to fireplace. Connection of *appliances* to *chimney* flues serving *fireplaces* shall be in accordance with Sections G2425.7.1 through G2425.7.3.

G2425.7.1 (501.7.1) Closure and access. A noncombustible seal shall be provided below the point of connection to prevent entry of room air into the flue. Means shall be provided for *access* to the flue for inspection and cleaning.

G2425.7.2 (501.7.2) Connection to factory-built fireplace flue. An *appliance* shall not be connected to a flue serving a *factory-built fireplace* unless the *appliance* is specifically *listed* for such installation. The connection shall be made in accordance with the *appliance* manufacturer's installation instructions.

G2425.7.3 (501.7.3) Connection to masonry fireplace flue. A connector shall extend from the *appliance* to the flue serving a *masonry fireplace* such that the *flue gases* are exhausted directly into the flue. The connector shall be accessible or removable for inspection and cleaning of both the connector and the flue. *Listed* direct connection devices shall be installed in accordance with their listing.

G2425.8 (501.8) Appliances not required to be vented. The following *appliances* shall not be required to be vented:

1. Ranges.
2. Built-in domestic cooking units *listed* and marked for optional venting.
3. Hot plates and laundry stoves.
4. *Type 1 clothes dryers* (*Type 1 clothes dryers* shall be exhausted in accordance with the requirements of Section G2439).
5. Refrigerators.
6. Counter *appliances.*
7. Room heaters *listed* for unvented use.

Where the *appliances* listed in Items 5 through 7 are installed so that the aggregate input rating exceeds 20 Btu per hour per cubic foot (207 W/m^3) of volume of the room or space in which such *appliances* are installed, one or more shall be provided with venting *systems* or other *approved* means for conveying the *vent gases* to the outdoor atmosphere so that the aggregate input rating of the remaining *unvented appliances* does not exceed 20 Btu per hour per cubic foot (207 W/m^3). Where the room or space in which the *appliance* is installed is directly connected to another room or space by a doorway, archway or other opening of comparable size that cannot be closed, the volume of such adjacent room or space shall be permitted to be included in the calculations.

G2425.9 (501.9) Chimney entrance. Connectors shall connect to a *masonry chimney* flue at a point not less than 12 inches (305 mm) above the lowest portion of the interior of the *chimney* flue.

G2425.10 (501.10) Connections to exhauster. *Appliance* connections to a *chimney* or vent equipped with a power exhauster shall be made on the inlet side of the exhauster. Joints on the positive pressure side of the exhauster shall be sealed to prevent flue-gas leakage as specified by the manufacturer's installation instructions for the exhauster.

G2425.11 (501.11) Masonry chimneys. *Masonry chimneys* utilized to vent *appliances* shall be located, constructed and sized as specified in the manufacturer's installation instructions for the *appliances* being vented and Section G2427.

G2425.12 (501.12) Residential and low-heat appliances flue lining systems. *Flue lining* systems for use with residential-type and low-heat *appliances* shall be limited to the following:

1. Clay *flue lining* complying with the requirements of ASTM C315 or equivalent. Clay *flue lining* shall be installed in accordance with Chapter 10.
2. *Listed chimney* lining systems complying with UL 1777.
3. Other *approved* materials that will resist, without cracking, softening or corrosion, *flue gases* and *condensate* at temperatures up to 1,800°F (982°C).

G2425.13 (501.13) Category I appliance flue lining systems. *Flue lining* systems for use with Category I *appliances* shall be limited to the following:

1. *Flue lining* systems complying with Section G2425.12.
2. *Chimney* lining systems *listed* and *labeled* for use with gas *appliances* with *draft hoods* and other Category I gas *appliances listed* and *labeled* for use with Type B vents.

G2425.14 (501.14) Category II, III and IV appliance venting systems. The design, sizing and installation of vents for Category II, III and IV *appliances* shall be in accordance with the *appliance* manufacturer's instructions.

G2425.15 (501.15) Existing chimneys and vents. Where an *appliance* is permanently disconnected from an existing *chimney* or vent, or where an *appliance* is connected to an existing *chimney* or vent during the process of a new installation, the *chimney* or vent shall comply with Sections G2425.15.1 through G2425.15.4.

G2425.15.1 (501.15.1) Size. The *chimney* or vent shall be resized as necessary to control flue gas condensation in the interior of the *chimney* or vent and to provide the *appliance* or *appliances* served with the required *draft*. For Category I *appliances*, the resizing shall be in accordance with Section G2426.

G2425.15.2 (501.15.2) Flue passageways. The flue gas passageway shall be free of obstructions and combustible deposits and shall be cleaned if previously used for venting a solid or liquid fuel-burning *appliance* or *fireplace*. The *flue liner*, *chimney* inner wall or vent inner wall shall be continuous and shall be free of cracks, gaps, perforations, or other damage or deterioration that would allow the escape of *combustion products*, including gases, moisture and creosote.

G2425.15.3 (501.15.3) Cleanout. *Masonry chimney* flues shall be provided with a cleanout opening having a minimum height of 6 inches (152 mm). The upper edge of the opening shall be located not less than 6 inches (152 mm) below the lowest *chimney* inlet opening. The cleanout shall be provided with a tight-fitting, noncombustible cover.

G2425.15.4 (501.15.4) Clearances. *Chimneys* and vents shall have airspace *clearance* to combustibles in accordance with Chapter 10 and the *chimney* or vent manufacturer's installation instructions.

Exception: *Masonry chimneys* without the required airspace *clearances* shall be permitted to be used if lined or relined with a *chimney* lining system *listed* for use in *chimneys* with reduced *clearances* in accordance with UL 1777. The *chimney clearance* shall be not less than permitted by the terms of the *chimney* liner listing and the manufacturer's instructions.

G2425.15.4.1 (501.15.4.1) Fireblocking. Noncombustible fireblocking shall be provided in accordance with Chapter 10.

SECTION G2426 (502) VENTS

G2426.1 (502.1) General. Vents, except as provided in Section G2427.7, shall be *listed* and *labeled*. Type B and BW vents shall be tested in accordance with UL 441. Type L vents shall be tested in accordance with UL 641. Vents for Category II and III *appliances* shall be tested in accordance with UL 1738. Plastic vents for Category IV *appliances* shall not be required to be *listed* and *labeled* where such vents are as specified by the *appliance* manufacturer and are installed in accordance with the *appliance* manufacturer's instructions.

G2426.2 (502.2) Connectors required. Connectors shall be used to connect *appliances* to the vertical *chimney* or vent, except where the *chimney* or vent is attached directly to the *appliance*. Vent *connector* size, material, construction and installation shall be in accordance with Section G2427.

G2426.3 (502.3) Vent application. The application of vents shall be in accordance with Table G2427.4.

G2426.4 (502.4) Insulation shield. Where vents pass through insulated assemblies, an insulation shield constructed of steel having a minimum thickness of 0.0187 inch (0.4712 mm) (No. 26 gage) shall be installed to provide *clearance* between the vent and the insulation material. The *clearance* shall be not less than the *clearance* to combustibles specified by the vent manufacturer's installation instructions. Where vents pass through attic space, the shield shall terminate not less than 2 inches (51 mm) above the insulation materials and shall be secured in place to prevent displacement. Insulation shields provided as part of a *listed* vent system shall be installed in accordance with the manufacturer's instructions.

G2426.5 (502.5) Installation. Vent systems shall be sized, installed and terminated in accordance with the vent and *appliance* manufacturer's installation instructions and Section G2427.

G2426.6 (502.6) Support of vents. All portions of vents shall be adequately supported for the design and weight of the materials employed.

G2426.7 (502.7) Protection against physical damage. In *concealed locations*, where a vent is installed through holes or notches in studs, joists, rafters or similar members less than $1^1/_2$ inches (38 mm) from the nearest edge of the member, the vent shall be protected by shield plates. Protective steel shield plates having a minimum thickness of 0.0575-inch (1.463 mm) (No. 16 gage) shall cover the area of the vent where the member is notched or bored and shall extend not less than 4 inches (102 mm) above sole plates, below top plates and to each side of a stud, joist or rafter.

G2426.7.1 (502.7.1) Door swing. Appliance and equipment vent terminals shall be located such that doors cannot swing within 12 inches (305 mm) horizontally of the vent terminal. Door stops or closures shall not be installed to obtain this clearance.

SECTION G2427 (503) VENTING OF APPLIANCES

G2427.1 (503.1) General. The venting of *appliances* shall be in accordance with Sections G2427.2 through G2427.16.

G2427.2 (503.2) Venting systems required. Except as permitted in Sections G2425.8, G2427.2.1 and G2427.2.2, all *appliances* shall be connected to *venting systems*.

G2427.2.1 (503.2.3) Direct-vent appliances. *Listed direct-vent appliances* shall be installed in accordance with the manufacturer's instructions. Through-the-wall vent terminations for listed direct-vent *appliances* shall be in accordance with Section G2427.8.

G2427.2.2 (503.2.4) Appliances with integral vents. *Appliances* incorporating integral venting means shall be installed in accordance with Section G2427.8, Items 1 and 2.

G2427.3 (503.3) Design and construction. *Venting systems* shall be designed and constructed so as to convey all flue and *vent gases* to the outdoors.

G2427.3.1 (503.3.1) Appliance draft requirements. A *venting system* shall satisfy the *draft* requirements of the *appliance* in accordance with the manufacturer's instructions.

G2427.3.2 (503.3.2) Design and construction. *Appliances* required to be vented shall be connected to a *venting system* designed and installed in accordance with the provisions of Sections G2427.4 through G2427.16.

G2427.3.3 (503.3.3) Mechanical draft systems. Mechanical *draft* systems shall comply with the following:

1. Mechanical *draft* systems shall be *listed* in accordance with UL 378 and shall be installed in accordance with the manufacturer's instructions for both the *appliance* and the mechanical *draft* system.
2. *Appliances* requiring venting shall be permitted to be vented by means of mechanical *draft* systems of either forced or induced *draft* design.
3. Forced *draft* systems and all portions of induced *draft* systems under positive pressure during operation shall be designed and installed so as to prevent leakage of flue or *vent gases* into a building.
4. *Vent connectors* serving *appliances* vented by natural *draft* shall not be connected into any portion of mechanical *draft* systems operating under positive pressure.
5. Where a mechanical *draft* system is employed, provisions shall be made to prevent the flow of gas to the *main burners* when the *draft* system is not performing so as to satisfy the operating requirements of the *appliance* for safe performance.

G2427.3.4 (503.3.5) Air ducts and furnace plenums. *Venting systems* shall not extend into or pass through any fabricated air duct or *furnace plenum*.

G2427.3.5 (503.3.6) Above-ceiling air-handling spaces. Where a *venting system* passes through an above-ceiling air-handling space or other nonducted portion of an air-handling system, the *venting system* shall conform to one of the following requirements:

1. The *venting system* shall be a *listed* special gas vent; other *venting system* serving a Category III or Category IV *appliance*; or other positive pressure vent, with joints sealed in accordance with the *appliance* or vent manufacturer's instructions.
2. The *venting system* shall be installed such that fittings and joints between sections are not installed in the above-ceiling space.
3. The *venting system* shall be installed in a conduit or enclosure with sealed joints separating the interior of the conduit or enclosure from the ceiling space.

G2427.4 (503.4) Type of venting system to be used. The type of *venting system* to be used shall be in accordance with Table G2427.4.

TABLE G2427.4 (503.4)
TYPE OF VENTING SYSTEM TO BE USED

APPLIANCES	TYPE OF VENTING SYSTEM
Listed Category I appliances Listed appliances equipped with draft hood Appliances listed for use with Type B gas vent	Type B gas vent (Section G2427.6) Chimney (Section G2427.5) Single-wall metal pipe (Section G2427.7) Listed chimney lining system for gas venting (Section G2427.5.2) Special gas vent listed for these appliances (Section G2427.4.2)
Listed vented wall furnaces	Type B-W gas vent (Sections G2427.6, G2436)
Category II, Category III and Category IV appliances	As specified or furnished by manufacturers of listed appliances (Sections G2427.4.1, G2427.4.2)
Unlisted appliances	Chimney (Section G2427.5)
Decorative appliances in vented fireplaces	Chimney
Direct-vent appliances	See Section G2427.2.1
Appliances with integral vent	See Section G2427.2.2

G2427.4.1 (503.4.1) Plastic piping. Where plastic piping is used to vent an *appliance*, the *appliance* shall be *listed* for use with such venting materials and the *appliance* manufacturer's installation instructions shall identify the specific plastic piping material. The plastic pipe venting materials shall be *labeled* in accordance with the product standards specified by the *appliance* manufacturer or shall be *listed* in accordance with UL 1738.

G2427.4.1.1 (503.4.1.1) Plastic vent joints. Plastic *pipe* and fittings used to vent *appliances* shall be installed in accordance with the *appliance* manufacturer's instructions. Plastic pipe venting materials *listed* and *labeled* in accordance with UL 1738 shall be installed in accordance with the vent manufacturer's instructions. Where a primer is required, it shall be of a contrasting color.

G2427.4.2 (503.4.2) Special gas vent. Special gas *vent* shall be *listed* and *labeled* in accordance with UL 1738 and installed in accordance with the special gas *vent* manufacturer's instructions.

G2427.5 (503.5) Masonry, metal and factory-built chimneys. Masonry, metal and factory-built *chimneys* shall comply with Sections G2427.5.1 through G2427.5.10.

G2427.5.1 (503.5.1) Factory-built chimneys. Factory-built *chimneys* shall be *listed* in accordance with UL 103. Factory-built *chimneys* used to vent *appliances* that operate at a positive vent pressure shall be *listed* for such application.

G2427.5.2 (503.5.3) Masonry chimneys. Masonry *chimneys* shall be built and installed in accordance with NFPA 211 and shall be lined with an *approved* clay *flue lining*, a *chimney* lining system *listed* and *labeled* in accordance with UL 1777 or other *approved* material that will resist corrosion, erosion, softening or cracking from vent gases at temperatures up to 1,800°F (982°C).

Exception: Masonry *chimney* flues serving *listed* gas *appliances* with *draft hoods*, Category I *appliances* and other gas *appliances listed* for use with Type B vents shall be permitted to be lined with a *chimney* lining system specifically *listed* for use only with such *appliances*. The liner shall be installed in accordance with the liner manufacturer's instructions. A permanent identifying *label* shall be attached at the point where the connection is to be made to the liner. The *label* shall read: "This *chimney* liner is for *appliances* that burn gas only. Do not connect to solid or liquid fuel-burning *appliances* or incinerators."

G2427.5.3 (503.5.4) Chimney termination. *Chimneys* for residential-type or low-heat *appliances* shall extend not less than 3 feet (914 mm) above the highest point where they pass through a roof of a building and not less than 2 feet (610 mm) higher than any portion of a building within a horizontal distance of 10 feet (3048 mm). *Chimneys* for medium-heat *appliances* shall extend not less than 10 feet (3048 mm) higher than any portion of any building within 25 feet (7620 mm). *Chimneys* shall extend not less than 5 feet (1524 mm) above the highest connected *appliance draft hood* outlet or *flue collar*. Decorative shrouds shall not be installed at the termination of factory-built *chimneys* except where such shrouds are *listed* and *labeled* for use with the specific factory-built *chimney* system and are installed in accordance with the manufacturer's instructions.

G2427.5.4 (503.5.5) Size of chimneys. The effective area of a *chimney* venting system serving *listed appliances* with *draft hoods*, Category I *appliances*, and other *appliances* listed for use with Type B vents shall be determined in accordance with one of the following methods:

1. The provisions of Section G2428.
2. The effective areas of the vent connector and chimney flue of a venting system serving a single *appliance* with a *draft hood* shall be not less than the area of the *appliance flue collar* or *draft hood* outlet, nor greater than seven times the *draft hood* outlet area.
3. The effective area of a chimney flue or a venting system serving two *appliances* with *draft hoods*.
4. *Chimney venting systems* using mechanical *draft* shall be sized in accordance with *approved* engineering methods.
5. Other *approved* engineering methods.

G2427.5.5 (503.5.6) Inspection of chimneys. Before replacing an existing *appliance* or connecting a vent *connector* to a *chimney*, the *chimney* passageway shall be examined to ascertain that it is clear and free of obstructions and it shall be cleaned if previously used for venting solid or liquid fuel-burning *appliances* or *fireplaces*.

G2427.5.5.1 (503.5.6.1) Chimney lining. *Chimneys* shall be lined in accordance with NFPA 211.

G2427.5.5.2 (503.5.6.2) Cleanouts. Cleanouts shall be examined and where they do not remain tightly closed when not in use, they shall be repaired or replaced.

G2427.5.5.3 (503.5.6.3) Unsafe chimneys. Where inspection reveals that an existing *chimney* is not safe for the intended application, it shall be repaired, rebuilt, lined, relined or replaced with a vent or *chimney* to conform to NFPA 211 and it shall be suitable for the *appliances* to be vented.

G2427.5.6 (503.5.7) Chimneys serving appliances burning other fuels. *Chimneys* serving *appliances* burning other fuels shall comply with Sections G2427.5.6.1 through G2427.5.6.4.

G2427.5.6.1 (503.5.7.1) Solid fuel-burning appliances. An *appliance* shall not be connected to a *chimney* flue serving a separate *appliance* designed to burn solid fuel.

G2427.5.6.2 (503.5.7.2) Liquid fuel-burning appliances. Where one *chimney* flue serves gas *appliances* and liquid fuel-burning *appliances*, the *appliances* shall be connected through separate openings or shall be connected through a single opening where joined by a suitable fitting located as close as practical to the *chimney.* Where two or more openings are provided into one *chimney* flue, they shall be at different levels. Where the *appliances* are automatically controlled, they shall be equipped with *safety shutoff devices.*

G2427.5.6.3 (503.5.7.3) Combination gas- and solid fuel-burning appliances. A combination gas- and solid fuel-burning *appliance* shall be permitted to be connected to a single *chimney* flue where equipped with a manual reset device to shut off gas to the *main burner* in the event of sustained backdraft or flue gas spillage. The *chimney* flue shall be sized to properly vent the *appliance.*

G2427.5.6.4 (503.5.7.4) Combination gas- and oil fuel-burning appliances. Where a single chimney flue serves a *listed* combination gas- and oil fuel-burning *appliance,* such flue shall be sized in accordance with the *appliance* manufacturer's instructions.

G2427.5.7 (503.5.8) Support of chimneys. All portions of *chimneys* shall be supported for the design and weight of the materials employed. Factory-built *chimneys* shall be supported and spaced in accordance with the manufacturer's installation instructions.

G2427.5.8 (503.5.9) Cleanouts. Where a *chimney* that formerly carried flue products from liquid or solid fuel-burning *appliances* is used with an *appliance* using *fuel gas,* an accessible cleanout shall be provided. The cleanout shall have a tight-fitting cover and be installed so its upper edge is not less than 6 inches (152 mm) below the lower edge of the lowest *chimney* inlet opening.

G2427.5.9 (503.5.10) Space surrounding lining or vent. The remaining space surrounding a *chimney* liner, gas vent, special gas *vent* or plastic *piping* installed within a masonry *chimney* flue shall not be used to vent another *appliance.* The insertion of another liner or vent within the *chimney* as provided in this code and the liner or vent manufacturer's instructions shall not be prohibited.

The remaining space surrounding a *chimney* liner, gas vent, special gas vent or plastic *piping* installed within a masonry, metal or factory-built *chimney* shall not be used to supply *combustion air.* Such space shall not be prohibited from supplying *combustion air* to *direct-vent appliances* designed for installation in a solid fuel-burning *fireplace* and installed in accordance with the manufacturer's instructions.

G2427.5.10 (503.5.11) Insulation shield. Where a factory-built chimney passes through insulated assemblies, an insulation shield constructed of steel having a thickness of not less than 0.0187 inch (0.475 mm) (nominal 26 gage) shall be installed to provide clearance between the chimney and the insulation material. The clearance shall be not less than the clearance to combustibles specified by the chimney manufacturer's installation instructions. Where chimneys pass through attic space, the shield shall terminate not less than 2 inches (51 mm) above the installation materials and shall be secured in place to prevent displacement.

G2427.6 (503.6) Gas vents. Gas vents shall comply with Sections G2427.6.1 through G2427.6.12. (See Section G2403, General Definitions.)

G2427.6.1 (503.6.1) Materials. Type B and BW gas vents shall be *listed* in accordance with UL 441. Vents for *listed* combination gas- and oil-burning *appliances* shall be *listed* in accordance with UL 641.

G2427.6.2 (503.6.2) Installation, general. Gas vents shall be installed in accordance with the manufacturer's instructions.

G2427.6.3 (503.6.3) Type B-W vent capacity. A Type B-W gas vent shall have a listed capacity not less than that of the *listed vented wall furnace* to which it is connected.

G2427.6.4 (503.6.5) Gas vent terminations. A gas vent shall terminate in accordance with one of the following:

1. Gas vents that are 12 inches (305 mm) or less in size and located not less than 8 feet (2438 mm) from a vertical wall or similar obstruction shall terminate above the roof in accordance with Figure G2427.6.4.
2. Gas vents that are over 12 inches (305 mm) in size or are located less than 8 feet (2438 mm) from a vertical wall or similar obstruction shall terminate not less than 2 feet (610 mm) above the highest point where they pass through the roof and not less than 2 feet (610 mm) above any portion of a building within 10 feet (3048 mm) horizontally.
3. As provided for direct-vent systems in Section G2427.2.1.
4. As provided for *appliances* with integral vents in Section G2427.2.2.
5. As provided for mechanical *draft* systems in Section G2427.3.3.

G2427.6.4.1 (503.6.5.1) Decorative shrouds. Decorative shrouds shall not be installed at the termination of gas vents except where such shrouds are *listed* for use with the specific gas venting system and are installed in accordance with manufacturer's instructions.

G2427.6.5 (503.6.6) Minimum height. A Type B or L gas vent shall terminate not less than 5 feet (1524 mm) in vertical height above the highest connected *appliance draft hood* or *flue collar.* A Type B-W gas vent shall terminate not less than 12 feet (3658 mm) in vertical height above the bottom of the wall *furnace.*

G2427.6.6 (503.6.7) Roof terminations. Gas vents shall extend through the roof flashing, roof jack or roof thimble and terminate with a *listed* cap or *listed roof assembly.*

G2427.6.7 (503.6.8) Forced air inlets. Gas vents shall terminate not less than 3 feet (914 mm) above any forced air inlet located within 10 feet (3048 mm).

G2427.6.8 (503.6.9) Exterior wall penetrations. A gas *vent* extending through an exterior wall shall not terminate adjacent to the wall or below eaves or parapets, except as provided in Sections G2427.2.1 and G2427.3.3.

G2427.6.9 (503.6.10) Size of gas vents. *Venting systems* shall be sized and constructed in accordance with Sections G2427.6.9.1 through G2427.6.9.4 and the *appliance* manufacturer's installation instructions.

G2427.6.9.1 (503.6.10.1) Category I appliances. The sizing of *natural draft venting systems* serving one or more *listed appliances* equipped with a *draft hood* or *appliances listed* for use with Type B gas vent, installed in a single story of a building, shall be in accordance with one of the following methods:

1. The provisions of Section G2428.
2. For sizing an individual gas vent for a single, *draft hood*-equipped *appliance*, the effective area of the vent *connector* and the gas vent shall be not less than the area of the *appliance draft hood* outlet, nor greater than seven times the *draft hood* outlet area.
3. For sizing a gas vent connected to two *appliances* with *draft hoods*, the effective area of the vent shall be not less than the area of the larger *draft hood* outlet plus 50 percent of the area of the smaller *draft hood* outlet, nor greater than seven times the smaller *draft hood* outlet area.
4. *Approved* engineering methods.

G2427.6.9.2 (503.6.10.2) Vent offsets. Type B and L vents sized in accordance with Item 2 or 3 of Section G2427.6.8.1 shall extend in a generally vertical direction with offsets not exceeding 45 degrees (0.79 rad), except that a vent system having not more than one 60-degree (1.04 rad) *offset* shall be permitted. Any angle

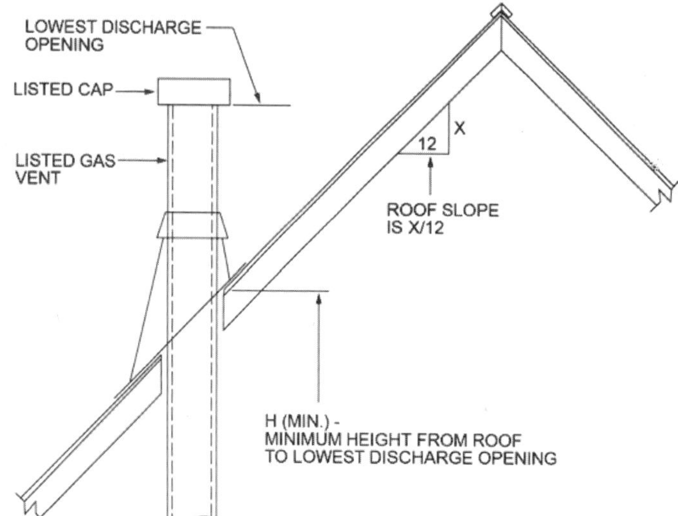

ROOF SLOPE	H (minimum) ft
Flat to 6/12	1.0
Over 6/12 to 7/12	1.25
Over 7/12 to 8/12	1.5
Over 8/12 to 9/12	2.0
Over 9/12 to 10/12	2.5
Over 10/12 to 11/12	3.25
Over 11/12 to 12/12	4.0
Over 12/12 to 14/12	5.0
Over 14/12 to 16/12	6.0
Over 16/12 to 18/12	7.0
Over 18/12 to 20/12	7.5
Over 20/12 to 21/12	8.0

For SI: 1 inch = 25.4 mm, 1 foot = 304.8 mm.

FIGURE G2427.6.4 (503.6.5)
TERMINATION LOCATIONS FOR GAS VENTS WITH LISTED CAPS 12 INCHES OR LESS IN SIZE NOT LESS THAN 8 FEET FROM A VERTICAL WALL

greater than 45 degrees (0.79 rad) from the vertical is considered horizontal. The total horizontal distance of a vent plus the horizontal vent *connector* serving *draft hood*-equipped *appliances* shall be not greater than 75 percent of the vertical height of the vent.

G2427.6.9.3 (503.6.10.3) Category II, III and IV appliances. The sizing of gas vents for Category II, III and IV *appliances* shall be in accordance with the *appliance* manufacturer's instructions. The sizing of plastic pipe that is specified by the *appliance* manufacturer as a venting material for Category II, III and IV *appliances*, shall be in accordance with the manufacturer's instructions.

G2427.6.9.4 (503.6.10.4) Mechanical draft. *Chimney venting systems* using mechanical *draft* shall be sized in accordance with *approved* engineering methods.

G2427.6.10 (503.6.12) Support of gas vents. Gas vents shall be supported and spaced in accordance with the manufacturer's installation instructions.

G2427.6.11 (503.6.13) Marking. In those localities where solid and liquid fuels are used extensively, gas vents shall be permanently identified by a *label* attached to the wall or ceiling at a point where the *vent connector* enters the gas vent. The determination of where such localities exist shall be made by the *code official*. The *label* shall read:

"This gas vent is for *appliances* that burn gas. Do not connect to solid or liquid fuel-burning *appliances* or incinerators."

G2427.6.12 (503.6.14) Fastener penetrations. Screws, rivets and other fasteners shall not penetrate the inner wall of double-wall gas vents, except at the transition from an *appliance draft hood* outlet, a *flue collar* or a single-wall metal connector to a double-wall vent.

G2427.7 (503.7) Single-wall metal pipe. Single-wall metal *pipe* vents shall comply with Sections G2427.7.1 through G2427.7.13.

G2427.7.1 (503.7.1) Construction. Single-wall metal pipe shall be constructed of galvanized sheet steel not less than 0.0304 inch (0.7 mm) thick, or other *approved*, noncombustible, corrosion-resistant material.

G2427.7.2 (503.7.2) Cold climate. Uninsulated single-wall metal pipe shall not be used outdoors for venting *appliances* in regions where the 99-percent winter design temperature is below 32°F (0°C).

G2427.7.3 (503.7.3) Termination. Single-wall metal pipe shall terminate not less than 5 feet (1524 mm) in vertical height above the highest connected *appliance draft hood* outlet or *flue collar*. Single-wall metal pipe shall extend not less than 2 feet (610 mm) above the highest point where it passes through a roof of a building and not less than 2 feet (610 mm) higher than any portion of a building within a horizontal distance of 10 feet (3048 mm). An *approved* cap or *roof assembly* shall be attached to the terminus of a single-wall metal pipe.

G2427.7.4 (503.7.4) Limitations of use. Single-wall metal pipe shall be used only for runs directly from the space in which the *appliance* is located through the roof or exterior wall to the outdoor atmosphere.

G2427.7.5 (503.7.5) Roof penetrations. A pipe passing through a roof shall extend without interruption through the roof flashing, roof jack or roof thimble. Where a single-wall metal pipe passes through a roof constructed of *combustible material*, a noncombustible, nonventilating thimble shall be used at the point of passage. The thimble shall extend not less than 18 inches (457 mm) above and 6 inches (152 mm) below the roof with the annular space open at the bottom and closed only at the top. The thimble shall be sized in accordance with Section G2427.7.7.

G2427.7.6 (503.7.6) Installation. Single-wall metal pipe shall not originate in any unoccupied attic or concealed-space and shall not pass through any attic, inside wall, concealed space, or floor. The installation of a single-wall metal pipe through an exterior combustible wall shall comply with Section G2427.7.7.

G2427.7.7 (503.7.7) Single-wall penetrations of combustible walls. A single-wall metal pipe shall not pass through a combustible exterior wall unless guarded at the point of passage by a ventilated metal thimble not smaller than the following:

1. For *listed appliances* with *draft hoods* and *appliances listed* for use with Type B gas vents, the thimble shall be not less than 4 inches (102 mm) larger in diameter than the metal pipe. Where there is a run of not less than 6 feet (1829 mm) of metal pipe in the open between the *draft hood* outlet and the thimble, the thimble shall be permitted to be not less than 2 inches (51 mm) larger in diameter than the metal pipe.
2. For unlisted *appliances* having *draft hoods*, the thimble shall be not less than 6 inches (152 mm) larger in diameter than the metal pipe.
3. For residential and low-heat *appliances*, the thimble shall be not less than 12 inches (305 mm) larger in diameter than the metal pipe.

Exception: In lieu of thimble protection, all *combustible material* in the wall shall be removed a sufficient distance from the metal pipe to provide the specified *clearance* from such metal pipe to *combustible material*. Any material used to close up such opening shall be noncombustible.

G2427.7.8 (503.7.8) Clearances. Minimum *clearances* from single-wall metal pipe to *combustible material* shall be in accordance with Table G2427.10.5. The *clearance* from single-wall metal pipe to *combustible material* shall be permitted to be reduced where the *combustible material* is protected as specified for *vent connectors* in Table G2409.2.

G2427.7.9 (503.7.9) Size of single-wall metal pipe. A venting system constructed of single-wall metal pipe shall be sized in accordance with one of the following methods and the *appliance* manufacturer's instructions:

1. For a *draft hood*-equipped *appliance*, in accordance with Section G2428.
2. For a venting system for a single *appliance* with a *draft hood*, the areas of the connector and the pipe each shall be not less than the area of the *appliance flue collar* or *draft hood* outlet, whichever is smaller. The vent area shall be not greater than seven times the *draft hood* outlet area.
3. *Approved* engineering methods.

G2427.7.10 (503.7.10) Pipe geometry. Any shaped single-wall metal pipe shall be permitted to be used, provided that its equivalent effective area is equal to the effective area of the round pipe for which it is substituted, and provided that the minimum internal dimension of the pipe is not less than 2 inches (51 mm).

G2427.7.11 (503.7.11) Termination capacity. The vent cap or a *roof assembly* shall have a venting capacity of not less than that of the pipe to which it is attached.

G2427.7.12 (503.7.12) Support of single-wall metal pipe. All portions of single-wall metal pipe shall be supported for the design and weight of the material employed.

G2427.7.13 (503.7.13) Marking. Single-wall metal pipe shall comply with the marking provisions of Section G2427.6.11.

G2427.8 (503.8) Venting system terminal clearances. The clearances for through-the-wall direct-vent and nondirect-vent terminals shall be in accordance with Figure G2427.8 and Table G2427.8.

Exception: The clearances in Table G2427.8 shall not apply to the *combustion air* intake of a direct-vent *appliance*.

G2427.9 (503.9) Condensation drainage. Provisions shall be made to collect and dispose of *condensate* from *venting systems* serving Category II and IV *appliances* and noncategorized condensing *appliances*. Drains for *condensate* shall be installed in accordance with the *appliance* and vent manufacturer's instructions.

G2427.10 (503.10) Vent connectors for Category I appliances. Vent *connectors* for Category I *appliances* shall comply with Sections G2427.10.1 through G2427.10.14.

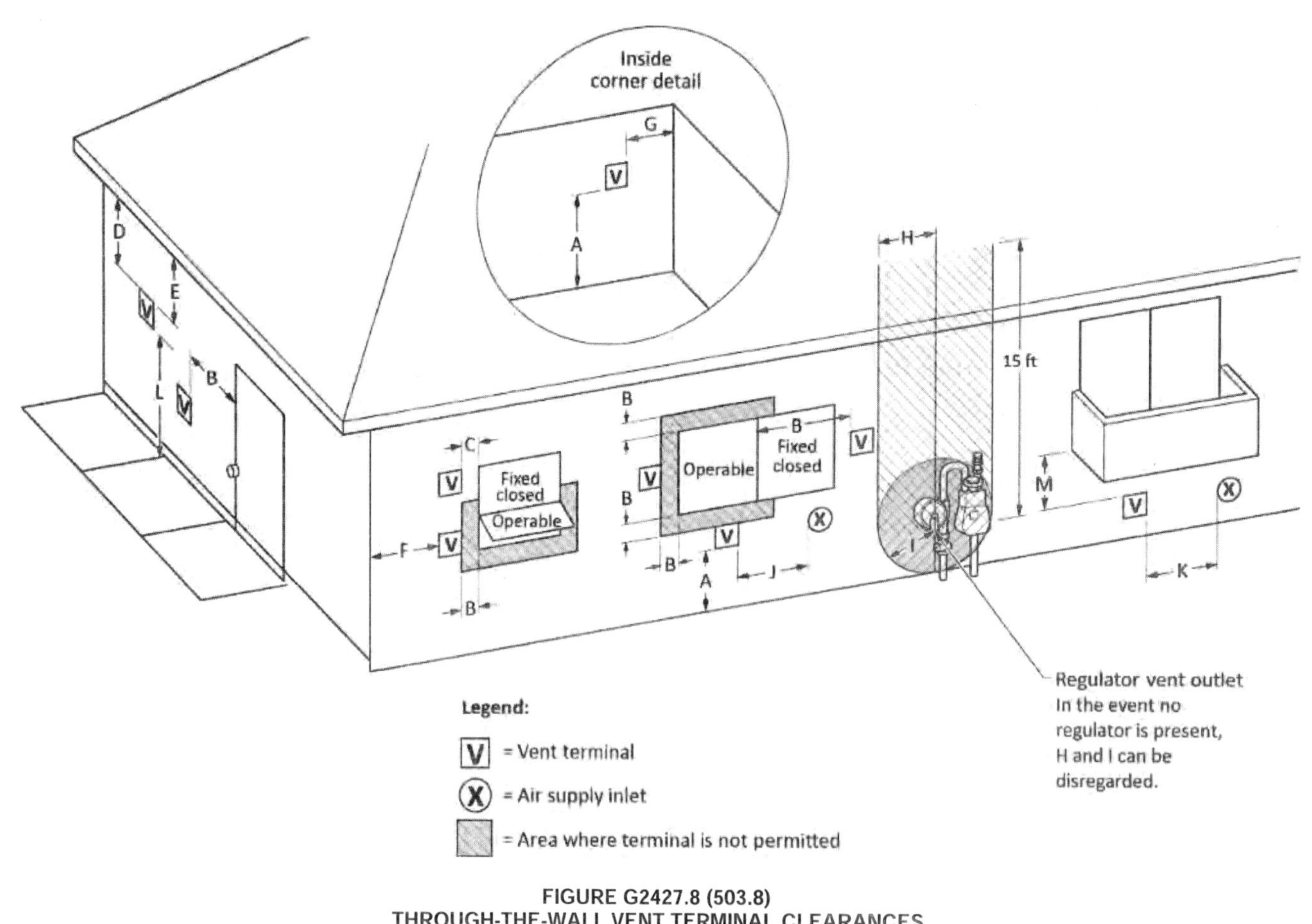

FIGURE G2427.8 (503.8)
THROUGH-THE-WALL VENT TERMINAL CLEARANCES

TABLE G2427.8 (503.8)
THROUGH-THE-WALL VENT TERMINAL CLEARANCES

FIGURE CLEARANCE	CLEARANCE LOCATION	MINIMUM CLEARANCES FOR DIRECT-VENT TERMINALS	MINIMUM CLEARANCES FOR NONDIRECT-VENT TERMINALS
A	Clearance above finished grade level, veranda, porch, deck or balcony	12 inches	
B	Clearance to window or door that is openable	6 inches: Appliances ≤ 10,000 Btu/hr 9 inches: Appliances > 10,000 Btu/hr ≤ 50,000 Btu/hr 12 inches: Appliances > 50,000 Btu/hr ≤ 150,000 Btu/hr Appliances > 150,000 Btu/hr, in accordance with the appliance manufacturer's instructions and not less than the clearances specified for nondirect-vent terminals in Row B	4 feet below or to side of opening or 1 foot above opening
C	Clearance to nonopenable window	None unless otherwise specified by the appliance manufacturer	
D	Vertical clearance to ventilated soffit located above the terminal within a horizontal distance of 2 feet from the centerline of the terminal	None unless otherwise specified by the appliance manufacturer	
E	Clearance to unventilated soffit	None unless otherwise specified by the appliance manufacturer	
F	Clearance to outside corner of building	None unless otherwise specified by the appliance manufacturer	
G	Clearance to inside corner of building	None unless otherwise specified by the appliance manufacturer	
H	Clearance to each side of centerline extended above regulator vent outlet	3 feet up to a height of 15 feet above the regulator vent outlet	
I	Clearance to service regulator vent outlet in all directions	3 feet for gas pressures up to 2 psi; 10 feet for gas pressures above 2 psi	
J	Clearance to nonmechanical air supply inlet to building and the combustion air inlet to any other appliance	Same clearance as specified for Row B	
K	Clearance to a mechanical air supply inlet	10 feet horizontally from inlet or 3 feet above inlet	
L	Clearance above paved sidewalk or paved driveway located on public property	7 feet and shall not be located above public walkways or other areas where condensate or vapor can cause a nuisance or hazard	
M	Clearance to underside of veranda, porch deck or balcony	12 inches where the area beneath the veranda, porch deck or balcony is open on not less than two sides. The vent terminal is prohibited in this location where only one side is open.	

For SI: 1 inch = 25.4 mm, 1 foot = 304.8 mm, 1 pound per square inch = 6.895 kPa, 1 Btu/h = 0.293 W.

G2427.10.1 (503.10.1) Where required. A vent *connector* shall be used to connect an *appliance* to a gas vent, *chimney* or single-wall metal pipe, except where the gas vent, *chimney* or single-wall metal pipe is directly connected to the *appliance*.

G2427.10.2 (503.10.2) Materials. *Vent connectors* shall be constructed in accordance with Sections G2427.10.2.1 through G2427.10.2.4.

G2427.10.2.1 (503.10.2.1) General. A *vent connector* shall be made of noncombustible corrosion-resistant material capable of withstanding the vent gas temperature produced by the *appliance* and of sufficient thickness to withstand physical damage.

G2427.10.2.2 (503.10.2.2) Vent connectors located in unconditioned areas. Where the *vent connector* used for an *appliance* having a *draft hood* or a Category I *appliance* is located in or passes through attics, crawl spaces or other unconditioned spaces, that portion of the *vent connector* shall be *listed* Type B, Type L or *listed* vent material having equivalent insulation properties.

Exception: Single-wall metal pipe located within the exterior walls of the building in areas having a local 99-percent winter design temperature of 5°F (-15°C) or higher shall be permitted to be used in unconditioned spaces other than attics and crawl spaces.

G2427.10.2.3 (503.10.2.3) Residential-type appliance connectors. Where *vent connectors* for residential-type *appliances* are not installed in attics or other unconditioned spaces, connectors for *listed appliances* having *draft hoods, appliances* having *draft hoods* and equipped with *listed conversion burners* and Category I *appliances* shall be one of the following:

1. Type B or L vent material.
2. Galvanized sheet steel not less than 0.018 inch (0.46 mm) thick.
3. Aluminum (1100 or 3003 alloy or equivalent) sheet not less than 0.027 inch (0.69 mm) thick.
4. Stainless steel sheet not less than 0.012 inch (0.31 mm) thick.
5. Smooth interior wall metal pipe having resistance to heat and corrosion equal to or greater than that of Item 2, 3 or 4.
6. A *listed* vent *connector.*

Vent connectors shall not be covered with insulation.

Exception: *Listed* insulated *vent connectors* shall be installed in accordance with the manufacturer's instructions.

G2427.10.2.4 (503.10.2.4) Low-heat appliance. A *vent connector* for a nonresidential, low-heat *appliance* shall be a factory-built *chimney* section or steel *pipe* having resistance to heat and corrosion equivalent to that for the appropriate galvanized pipe as specified in Table G2427.10.2.4. Factory-built *chimney* sections shall be joined together in accordance with the *chimney* manufacturer's instructions.

**TABLE G2427.10.2.4 (503.10.2.4)
MINIMUM THICKNESS FOR GALVANIZED STEEL VENT CONNECTORS FOR LOW-HEAT APPLIANCES**

DIAMETER OF CONNECTOR (inches)	MINIMUM THICKNESS (inch)
Less than 6	0.019
6 to less than 10	0.023
10 to 12 inclusive	0.029
14 to 16 inclusive	0.034
Over 16	0.056

For SI: 1 inch = 25.4 mm.

G2427.10.3 (503.10.3) Size of vent connector. *Vent connectors* shall be sized in accordance with Sections G2427.10.3.1 through G2427.3.5.

G2427.10.3.1 (503.10.3.1) Single draft hood and fan-assisted. A *vent connector* for an *appliance* with a single *draft hood* or for a Category I fan-assisted *combustion* system *appliance* shall be sized and installed in accordance with Section G2428 or *approved* engineering methods.

G2427.10.3.2 (503.10.3.2) Multiple draft hood. Where a single *appliance* having more than one *draft hood* outlet or *flue collar* is installed, the manifold shall be constructed according to the instructions of the *appliance* manufacturer. Where there are no instructions, the manifold shall be designed and constructed in accordance with *approved* engineering methods. As an alternate method, the effective area of the manifold shall equal the combined area of the *flue collars* or *draft hood* outlets and the *vent connectors* shall have a rise of not less than 12 inches (305 mm).

G2427.10.3.3 (503.10.3.3) Multiple appliances. Where two or more *appliances* are connected to a common *vent* or *chimney*, each *vent connector* shall be sized in accordance with Section G2428 or *approved* engineering methods.

As an alternative method applicable only where all of the *appliances* are *draft hood* equipped, each *vent connector* shall have an effective area not less than the area of the *draft hood* outlet of the *appliance* to which it is connected.

G2427.10.3.4 (503.10.3.4) Common connector/manifold. Where two or more *appliances* are vented through a common *vent connector* or vent manifold, the common *vent connector* or vent manifold shall be located at the highest level consistent with available headroom and the required *clearance* to *combustible materials* and shall be sized in accordance with Section G2428 or *approved* engineering methods.

As an alternate method applicable only where there are two *draft hood*-equipped *appliances*, the effective area of the common *vent connector* or vent manifold and all junction fittings shall be not less than the area of the larger *vent connector* plus 50 percent of the area of the smaller *flue collar* outlet.

G2427.10.3.5 (503.10.3.5) Size increase. Where the size of a *vent connector* is increased to overcome installation limitations and obtain connector capacity equal to the *appliance* input, the size increase shall be made at the *appliance draft hood* outlet.

G2427.10.4 (503.10.4) Two or more appliances connected to a single vent or chimney. Where two or more *vent connectors* enter a common vent, *chimney* flue, or single-wall metal pipe, the smaller connector shall enter at the highest level consistent with the available headroom or *clearance* to *combustible material. Vent connectors* serving Category I *appliances* shall not be connected to any portion of a *mechanical draft* system operating under positive static pressure, such as those serving Category III or IV *appliances.*

G2427.10.4.1 (503.10.4.1) Two or more openings. Where two or more openings are provided into one *chimney* flue or vent, the openings shall be at different levels, or the connectors shall be attached to the vertical portion of the *chimney* or vent at an angle of 45 degrees (0.79 rad) or less relative to the vertical.

G2427.10.5 (503.10.5) Clearance. Minimum *clearances* from *vent connectors* to *combustible material* shall be in accordance with Table G2427.10.5.

Exception: The *clearance* between a *vent connector* and *combustible material* shall be permitted to be reduced where the *combustible material* is protected as specified for *vent connectors* in Table G2409.2.

G2427.10.6 (503.10.6) Joints. Joints between sections of connector piping and connections to *flue collars* and *draft hood* outlets shall be fastened by one of the following methods:

1. Sheet metal screws.
2. *Vent connectors* of *listed* vent material assembled and connected to *flue collars* or *draft hood* outlets in accordance with the manufacturer's instructions.
3. Other *approved* means.

G2427.10.7 (503.10.7) Connector junctions. Where *vent connectors* are joined together, the connection shall be made with a tee or wye fitting manufactured for the purpose.

G2427.10.8 (503.10.8) Slope. A *vent connector* shall be installed without dips or sags and shall slope upward toward the vent or *chimney* not less than $^1/_4$ inch per foot (21 mm/m).

Exception: *Vent connectors* attached to a mechanical *draft* system installed in accordance with the *appliance* and *draft* system manufacturers' instructions.

G2427.10.9 (503.10.9) Length of vent connector. The maximum horizontal length of a single-wall connector shall be 75 percent of the height of the *chimney* or vent except for engineered systems. The maximum horizontal length of a Type B double-wall connector shall be 100 percent of the height of the *chimney* or vent except for engineered systems.

G2427.10.10 (503.10.10) Support. A *vent connector* shall be supported for the design and weight of the material employed to maintain *clearances* and prevent physical damage and separation of joints.

G2427.10.11 (503.10.11) Chimney connection. Where entering a flue in a masonry or metal *chimney*, the *vent connector* shall be installed above the extreme bottom to avoid stoppage. Where a thimble or slip joint is used to facilitate removal of the connector, the connector shall be firmly attached to or inserted into the thimble or slip joint to prevent the connector from falling out. Means shall be employed to prevent the connector from entering so far as to restrict the space between its end and the opposite wall of the *chimney* flue (see Section G2425.9).

G2427.10.12 (503.10.12) Inspection. The entire length of a *vent connector* shall be provided with *ready access* for inspection, cleaning and replacement.

G2427.10.13 (503.10.13) Fireplaces. A *vent connector* shall not be connected to a *chimney* flue serving a *fireplace* unless the *fireplace* flue opening is permanently sealed.

G2427.10.14 (503.10.14) Passage through ceilings, floors or walls. Single-wall metal pipe connectors shall not pass through any wall, floor or ceiling except as permitted by Section G2427.7.4.

G2427.11 (503.11) Vent connectors for Category II, III and IV appliances. *Vent connectors* for Category II, III and IV *appliances* shall be as specified for the *venting systems* in accordance with Section G2427.4.

G2427.12 (503.12) Draft hoods and draft controls. The installation of *draft hoods* and draft controls shall comply with Sections G2427.12.1 through G2427.12.7.

G2427.12.1 (503.12.1) Appliances requiring draft hoods. *Vented appliances* shall be installed with *draft hoods.*

Exception: Dual oven-type combination ranges; *direct-vent appliances*; fan-assisted *combustion* system *appliances*; *appliances* requiring *chimney draft* for operation; single firebox boilers equipped with *conversion burners* with inputs greater than 400,000 *Btu* per hour (117 kW); *appliances* equipped with blast, power or pressure *burners* that are not *listed* for use with *draft hoods*; and *appliances* designed for forced venting.

G2427.12.2 (503.12.2) Installation. A *draft hood* supplied with or forming a part of a *listed vented appliance* shall be installed without *alteration*, exactly as furnished and specified by the *appliance* manufacturer.

TABLE G2427.10.5 (503.10.5)
CLEARANCES FOR CONNECTORS[a]

APPLIANCE	MINIMUM DISTANCE FROM COMBUSTIBLE MATERIAL			
	Listed Type B gas vent material	Listed Type L vent material	Single-wall metal pipe	Factory-built chimney sections
Listed appliances with draft hoods and appliances listed for use with Type B gas vents	As listed	As listed	6 inches	As listed
Residential boilers and furnaces with listed gas conversion burner and with draft hood	6 inches	6 inches	9 inches	As listed
Residential appliances listed for use with Type L vents	Not permitted	As listed	9 inches	As listed
Listed gas-fired toilets	Not permitted	As listed	As listed	As listed
Unlisted residential appliances with draft hood	Not permitted	6 inches	9 inches	As listed
Residential and low-heat appliances other than above	Not permitted	9 inches	18 inches	As listed
Medium-heat appliances	Not permitted	Not permitted	36 inches	As listed

For SI: 1 inch = 25.4 mm.

a. These clearances shall apply unless the manufacturer's installation instructions for a listed appliance or connector specify different clearances, in which case the listed clearances shall apply.

G2427.12.2.1 (503.12.2.1) Draft hood required. If a *draft hood* is not supplied by the *appliance* manufacturer where one is required, a *draft hood* shall be installed, shall be of a *listed* or *approved* type and, in the absence of other instructions, shall be of the same size as the *appliance flue collar*. Where a *draft hood* is required with a *conversion burner*, it shall be of a *listed* or *approved* type.

G2427.12.3 (503.12.3) Draft control devices. Where a *draft control* device is part of the *appliance* or is supplied by the *appliance* manufacturer, it shall be installed in accordance with the manufacturer's instructions. In the absence of manufacturer's instructions, the device shall be attached to the *flue collar* of the *appliance* or as near to the *appliance* as practical.

G2427.12.4 (503.12.4) Additional devices. *Appliances* requiring a controlled *chimney draft* shall be permitted to be equipped with a *listed* double-acting barometric-*draft regulator* installed and adjusted in accordance with the manufacturer's instructions.

G2427.12.5 (503.12.5) Location. *Draft hoods* and *barometric draft regulators* shall be installed in the same room or enclosure as the *appliance* in such a manner as to prevent any difference in pressure between the hood or *regulator* and the *combustion air* supply.

G2427.12.6 (503.12.6) Positioning. *Draft hoods* and *draft regulators* shall be installed in the position for which they were designed with reference to the horizontal and vertical planes and shall be located so that the *relief opening* is not obstructed by any part of the *appliance* or adjacent construction. The *appliance* and its *draft hood* shall be located so that the *relief opening* is accessible for checking *vent* operation.

G2427.12.7 (503.12.7) Clearance. A *draft hood* shall be located so its *relief opening* is not less than 6 inches (152 mm) from any surface except that of the *appliance* it serves and the venting system to which the *draft hood* is connected. Where a greater or lesser *clearance* is indicated on the *appliance label*, the *clearance* shall be not less than that specified on the *label*. Such *clearances* shall not be reduced.

G2427.13 (503.13) Manually operated dampers. A manually operated *damper* shall not be placed in the vent *connector* for any *appliance*. Fixed baffles and balancing baffles shall not be classified as manually operated *dampers*.

G2427.13.1 (503.13.1) Balancing baffles. Balancing baffles shall be *listed* in accordance with UL 378 and shall be mechanically locked in the desired position before placing the *appliance* in operation.

G2427.14 (503.14) Automatically operated vent dampers. An automatically operated vent damper shall be *listed*.

G2427.15 (503.15) Obstructions. Devices that retard the flow of *vent gases* shall not be installed in a *vent connector*, *chimney* or vent. The following shall not be considered as obstructions:

1. *Draft regulators* and safety *controls* specifically *listed* for installation in *venting systems* and installed in accordance with the manufacturer's instructions.
2. *Approved draft regulators* and safety *controls* that are designed and installed in accordance with *approved* engineering methods.
3. *Listed* heat reclaimers and automatically operated vent dampers installed in accordance with the manufacturer's instructions.
4. *Approved* economizers, heat reclaimers and recuperators installed in *venting systems* of *appliances* not required to be equipped with *draft hoods*, provided that the *appliance* manufacturer's instructions cover the installation of such a device in the venting system and performance in accordance with Sections G2427.3 and G2427.3.1 is obtained.
5. Vent dampers serving *listed* appliances installed in accordance with Sections G2428.2.1 and G2428.3.1 or *approved* engineering methods.

G2427.16 (503.16) (IFGS) Outside wall penetrations. Where vents, including those for *direct-vent appliances*, penetrate outside walls of buildings, the annular spaces around such penetrations shall be permanently sealed using *approved* materials to prevent entry of *combustion products* into the building.

SECTION G2428 (504) SIZING OF CATEGORY I APPLIANCE VENTING SYSTEMS

G2428.1 (504.1) Definitions. The following definitions apply to the tables in this section:

APPLIANCE CATEGORIZED VENT DIAMETER/AREA. The minimum vent area/diameter permissible for Category I appliances to maintain a nonpositive vent static pressure when tested in accordance with nationally recognized standards.

FAN + FAN. The maximum combined appliance input rating of two or more Category I fan-assisted appliances attached to the common vent.

FAN Max. The maximum input rating of a Category I fan-assisted appliance attached to a vent or connector.

FAN Min. The minimum input rating of a Category I fan-assisted appliance attached to a vent or connector.

FAN + NAT. The maximum combined appliance input rating of one or more Category I fan-assisted appliances and one or more Category I draft hood-equipped appliances attached to the common vent.

FAN-ASSISTED COMBUSTION SYSTEM. An appliance equipped with an integral mechanical means to either draw or force products of combustion through the combustion chamber or heat exchanger.

NA. Vent configuration is not allowed due to potential for condensate formation or pressurization of the venting system, or not applicable due to physical or geometric restraints.

NAT Max. The maximum input rating of a Category I draft hood-equipped appliance attached to a vent or connector.

NAT + NAT. The maximum combined appliance input rating of two or more Category I draft hood-equipped appliances attached to the common vent.

G2428.2 (504.2) Application of single-appliance vent Tables G2428.2(1) and G2428.2(2). The application of Tables G2428.2(1) and G2428.2(2) shall be subject to the requirements of Sections G2428.2.1 through G2428.2.17.

G2428.2.1 (504.2.1) Vent obstructions. These venting tables shall not be used where obstructions, as described in Section G2427.15, are installed in the venting system. The installation of vents serving *listed appliances* with vent dampers shall be in accordance with the *appliance* manufacturer's instructions or in accordance with the following:

1. The maximum capacity of the vent system shall be determined using the "NAT Max" column.
2. The minimum capacity shall be determined as if the *appliance* were a fan-assisted *appliance*, using the "FAN Min" column to determine the minimum capacity of the vent system. Where the corresponding "FAN Min" is "NA," the vent configuration shall not be permitted and an alternative venting configuration shall be utilized.

G2428.2.2 (504.2.2) Minimum size. Where the vent size determined from the tables is smaller than the *appliance draft hood outlet* or *flue collar*, the smaller size shall be permitted to be used provided that all of the following requirements are met:

1. The total vent height (*H*) is not less than 10 feet (3048 mm).
2. Vents for *appliance draft hood* outlets or *flue collars* 12 inches (305 mm) in diameter or smaller are not reduced more than one table size.
3. Vents for *appliance draft hood* outlets or *flue collars* larger than 12 inches (305 mm) in diameter are not reduced more than two table sizes.
4. The maximum capacity listed in the tables for a fan-assisted *appliance* is reduced by 10 percent (0.90 × maximum table capacity).
5. The *draft hood* outlet is greater than 4 inches (102 mm) in diameter. Do not connect a 3-inch-diameter (76 mm) vent to a 4-inch-diameter (102 mm) *draft hood* outlet. This provision shall not apply to fan-assisted *appliances*.

G2428.2.3 (504.2.3) Vent offsets. Single-*appliance* venting configurations with zero (0) lateral lengths in Tables G2428.2(1) and G2428.2(2) shall not have elbows in the *venting system*. Single-*appliance* venting configurations with lateral lengths include two 90-degree (1.57 rad) elbows. For each additional elbow up to and including 45 degrees (0.79 rad), the maximum capacity listed in the venting tables shall be reduced by 5 percent. For each additional elbow greater than 45 degrees (0.79 rad) up to and including 90 degrees (1.57 rad), the maximum capacity listed in the venting tables shall be reduced by 10 percent. Where multiple *offsets* occur in a vent, the total lateral length of all *offsets* combined shall not exceed that specified in Tables G2428.2(1) and G2428.2(2).

G2428.2.4 (504.2.4) Zero lateral. Zero (0) lateral (L) shall apply only to a straight vertical vent attached to a top outlet *draft hood* or *flue collar*.

G2428.2.5 (504.2.5) High-altitude installations. Sea-level input ratings shall be used when determining maximum capacity for high-altitude installation. Actual input, derated for altitude, shall be used for determining minimum capacity for high-altitude installation.

G2428.2.6 (504.2.6) Multiple input rate appliances. For *appliances* with more than one input rate, the minimum vent capacity (FAN Min) determined from the tables shall be less than the lowest *appliance* input rating, and the maximum vent capacity (FAN Max/NAT Max) determined from the tables shall be greater than the highest *appliance* rating input.

G2428.2.7 (504.2.7) Liner system sizing and connections. *Listed* corrugated metallic *chimney* liner systems in masonry *chimneys* shall be sized by using Table G2428.2(1) or G2428.2(2) for Type B vents with the maximum capacity reduced by 20 percent (0.80 × maximum capacity) and the minimum capacity as shown in Table G2428.2(1) or G2428.2(2). Corrugated metallic liner systems installed with bends or offsets shall have their maximum capacity further reduced in accordance with Section G2428.2.3. The 20-percent reduction for corrugated metallic *chimney* liner systems includes an allowance for one long-radius 90-degree (1.57 rad) turn at the bottom of the liner.

Connections between *chimney* liners and *listed* double-wall connectors shall be made with *listed* adapters designed for such purpose.

G2428.2.8 (504.2.8) Vent area and diameter. Where the vertical vent has a larger diameter than the *vent connector*, the vertical vent diameter shall be used to determine the minimum vent capacity, and the connector diameter shall be used to determine the maximum vent capacity. The flow area of the vertical vent shall not exceed seven times the flow area of the *listed appliance* categorized vent area, *flue collar* area, or *draft hood* outlet area unless designed in accordance with *approved* engineering methods.

TABLE G2428.2(1) [504.2(1)] TYPE B DOUBLE-WALL GAS VENT

Number of Appliances	Single
Appliance Type	Category I
Appliance Vent Connection	Connected directly to vent

HEIGHT (*H*) (feet)	LATERAL (*L*) (feet)	VENT DIAMETER (*D*)—inches																				
		3			4			5			6			7			8			9		
		APPLIANCE INPUT RATING IN THOUSANDS OF BTU/H																				
		FAN		NAT	FAN		NAT	FAN		NAT	FAN		NAT	FAN		NAT	FAN		NAT	FAN		NAT
		Min	Max	Max	Min	Max	Max	Min	Max	Max	Min	Max	Max	Min	Max	Max	Min	Max	Max	Min	Max	Max
6	0	0	78	46	0	152	86	0	251	141	0	375	205	0	524	285	0	698	370	0	897	470
	2	13	51	36	18	97	67	27	157	105	32	232	157	44	321	217	53	425	285	63	543	370
	4	21	49	34	30	94	64	39	153	103	50	227	153	66	316	211	79	419	279	93	536	362
	6	25	46	32	36	91	61	47	149	100	59	223	149	78	310	205	93	413	273	110	530	354
8	0	0	84	50	0	165	94	0	276	155	0	415	235	0	583	320	0	780	415	0	1,006	537
	2	12	57	40	16	109	75	25	178	120	28	263	180	42	365	247	50	483	322	60	619	418
	5	23	53	38	32	103	71	42	171	115	53	255	173	70	356	237	83	473	313	99	607	407
	8	28	49	35	39	98	66	51	164	109	64	247	165	84	347	227	99	463	303	117	596	396
10	0	0	88	53	0	175	100	0	295	166	0	447	255	0	631	345	0	847	450	0	1,096	585
	2	12	61	42	17	118	81	23	194	129	26	289	195	40	402	273	48	533	355	57	684	457
	5	23	57	40	32	113	77	41	187	124	52	280	188	68	392	263	81	522	346	95	671	446
	10	30	51	36	41	104	70	54	176	115	67	267	175	88	376	245	104	504	330	122	651	427
15	0	0	94	58	0	191	112	0	327	187	0	502	285	0	716	390	0	970	525	0	1,263	682
	2	11	69	48	15	136	93	20	226	150	22	339	225	38	475	316	45	633	414	53	815	544
	5	22	65	45	30	130	87	39	219	142	49	330	217	64	463	300	76	620	403	90	800	529
	10	29	59	41	40	121	82	51	206	135	64	315	208	84	445	288	99	600	386	116	777	507
	15	35	53	37	48	112	76	61	195	128	76	301	198	98	429	275	115	580	373	134	755	491
20	0	0	97	61	0	202	119	0	349	202	0	540	307	0	776	430	0	1,057	575	0	1,384	752
	2	10	75	51	14	149	100	18	250	166	20	377	249	33	531	346	41	711	470	50	917	612
	5	21	71	48	29	143	96	38	242	160	47	367	241	62	519	337	73	697	460	86	902	599
	10	28	64	44	38	133	89	50	229	150	62	351	228	81	499	321	95	675	443	112	877	576
	15	34	58	40	46	124	84	59	217	142	73	337	217	94	481	308	111	654	427	129	853	557
	20	48	52	35	55	116	78	69	206	134	84	322	206	107	464	295	125	634	410	145	830	537
30	0	0	100	64	0	213	128	0	374	220	0	587	336	0	853	475	0	1,173	650	0	1,548	855
	2	9	81	56	13	166	112	14	283	185	18	432	280	27	613	394	33	826	535	42	1,072	700
	5	21	77	54	28	160	108	36	275	176	45	421	273	58	600	385	69	811	524	82	1,055	688
	10	27	70	50	37	150	102	48	262	171	59	405	261	77	580	371	91	788	507	107	1,028	668
	15	33	64	NA	44	141	96	57	249	163	70	389	249	90	560	357	105	765	490	124	1,002	648
	20	56	58	NA	53	132	90	66	237	154	80	374	237	102	542	343	119	743	473	139	977	628
	30	NA	NA	NA	73	113	NA	88	214	NA	104	346	219	131	507	321	149	702	444	171	929	594
50	0	0	101	67	0	216	134	0	397	232	0	633	363	0	932	518	0	1,297	708	0	1,730	952
	2	8	86	61	11	183	122	14	320	206	15	497	314	22	715	445	26	975	615	33	1,276	813
	5	20	82	NA	27	177	119	35	312	200	43	487	308	55	702	438	65	960	605	77	1,259	798
	10	26	76	NA	35	168	114	45	299	190	56	471	298	73	681	426	86	935	589	101	1,230	773
	15	59	70	NA	42	158	NA	54	287	180	66	455	288	85	662	413	100	911	572	117	1,203	747
	20	NA	NA	NA	50	149	NA	63	275	169	76	440	278	97	642	401	113	888	556	131	1,176	722
	30	NA	NA	NA	69	131	NA	84	250	NA	99	410	259	123	605	376	141	844	522	161	1,125	670

For SI: 1 inch = 25.4 mm, 1 foot = 304.8 mm, 1 British thermal unit per hour = 0.2931 W.

TABLE G2428.2(2) [504.2(2)]
TYPE B DOUBLE-WALL GAS VENT

Number of Appliances	Single
Appliance Type	Category I
Appliance Vent Connection	Single-wall metal connector

HEIGHT (*H*) (feet)	LATERAL (*L*) (feet)	VENT DIAMETER (*D*)—inches																									
		3			4			5			6			7			8			9			10			12		
		APPLIANCE INPUT RATING IN THOUSANDS OF BTU/H																										
		FAN		NAT	FAN		NAT	FAN		NAT	FAN		NAT	FAN		NAT	FAN		NAT	FAN		NAT	FAN		NAT	FAN		NAT
		Min	Max	Max	Min	Max	Max	Min	Max	Max	Min	Max	Max	Min	Max	Max	Min	Max	Max	Min	Max	Max	Min	Max	Max	Min	Max	Max
6	0	38	77	45	59	151	85	85	249	140	126	373	204	165	522	284	211	695	369	267	894	469	371	1,118	569	537	1,639	849
	2	39	51	36	60	96	66	85	156	104	123	231	156	159	320	213	201	423	284	251	541	368	347	673	453	498	979	648
	4	NA	NA	33	74	92	63	102	152	102	146	225	152	187	313	208	237	416	277	295	533	360	409	664	443	584	971	638
	6	NA	NA	31	83	89	60	114	147	99	163	220	148	207	307	203	263	409	271	327	526	352	449	656	433	638	962	627
8	0	37	83	50	58	164	93	83	273	154	123	412	234	161	580	319	206	777	414	258	1,002	536	360	1,257	658	521	1,852	967
	2	39	56	39	59	108	75	83	176	119	121	261	179	155	363	246	197	482	321	246	617	417	339	768	513	486	1,120	743
	5	NA	NA	37	77	102	69	107	168	114	151	252	171	193	352	235	245	470	311	305	604	404	418	754	500	598	1,104	730
	8	NA	NA	33	90	95	64	122	161	107	175	243	163	223	342	225	280	458	300	344	591	392	470	740	486	665	1,089	715
10	0	37	87	53	57	174	99	82	293	165	120	444	254	158	628	344	202	844	449	253	1,093	584	351	1,373	718	507	2,031	1,057
	2	39	61	41	59	117	80	82	193	128	119	287	194	153	400	272	193	531	354	242	681	456	332	849	559	475	1,242	848
	5	52	56	39	76	111	76	105	185	122	148	277	186	190	388	261	241	518	344	299	667	443	409	834	544	584	1,224	825
	10	NA	NA	34	97	100	68	132	171	112	188	261	171	237	369	241	296	497	325	363	643	423	492	808	520	688	1,194	788
15	0	36	93	57	56	190	111	80	325	186	116	499	283	153	713	388	195	966	523	244	1,259	681	336	1,591	838	488	2,374	1,237
	2	38	69	47	57	136	93	80	225	149	115	337	224	148	473	314	187	631	413	232	812	543	319	1,015	673	457	1,491	983
	5	51	63	44	75	128	86	102	216	140	144	326	217	182	459	298	231	616	400	287	795	526	392	997	657	562	1,469	963
	10	NA	NA	39	95	116	79	128	201	131	182	308	203	228	438	284	284	592	381	349	768	501	470	966	628	664	1,433	928
	15	NA	NA	NA	NA	NA	72	158	186	124	220	290	192	272	418	269	334	568	367	404	742	484	540	937	601	750	1,399	894
20	0	35	96	60	54	200	118	78	346	201	114	537	306	149	772	428	190	1,053	573	238	1,379	750	326	1,751	927	473	2,631	1,346
	2	37	74	50	56	148	99	78	248	165	113	375	248	144	528	344	182	708	468	227	914	611	309	1,146	754	443	1,689	1,098
	5	50	68	47	73	140	94	100	239	158	141	363	239	178	514	334	224	692	457	279	896	596	381	1,126	734	547	1,665	1,074
	10	NA	NA	41	93	129	86	125	223	146	177	344	224	222	491	316	277	666	437	339	866	570	457	1,092	702	646	1,626	1,037
	15	NA	NA	NA	NA	NA	80	155	208	136	216	325	210	264	469	301	325	640	419	393	838	549	526	1,060	677	730	1,587	1,005
	20	NA	NA	NA	NA	NA	NA	186	192	126	254	306	196	309	448	285	374	616	400	448	810	526	592	1,028	651	808	1,550	973
30	0	34	99	63	53	211	127	76	372	219	110	584	334	144	849	472	184	1,168	647	229	1,542	852	312	1,971	1,056	454	2,996	1,545
	2	37	80	56	55	164	111	76	281	183	109	429	279	139	610	392	175	823	533	219	1,069	698	296	1,346	863	424	1,999	1,308
	5	49	74	52	72	157	106	98	271	173	136	417	271	171	595	382	215	806	521	269	1,049	684	366	1,324	846	524	1,971	1,283
	10	NA	NA	NA	91	144	98	122	255	168	171	397	257	213	570	367	265	777	501	327	1,017	662	440	1,287	821	620	1,927	1,234
	15	NA	NA	NA	115	131	NA	151	239	157	208	377	242	255	547	349	312	750	481	379	985	638	507	1,251	794	702	1,884	1,205
	20	NA	NA	NA	NA	NA	NA	181	223	NA	246	357	228	298	524	333	360	723	461	433	955	615	570	1,216	768	780	1,841	1,166
	30	NA	NA	NA	NA	NA	NA	NA	NA	NA	NA	NA	NA	389	477	305	461	670	426	541	895	574	704	1,147	720	937	1,759	1,101

(continued)

TABLE G2428.2(2) [504.2(2)]—continued
TYPE B DOUBLE-WALL GAS VENT

Number of Appliances	Single
Appliance Type	Category I
Appliance Vent Connection	Single-wall metal connector

HEIGHT (*H*) (feet)	LATERAL (*L*) (feet)	VENT DIAMETER (*D*)—inches																										
		3			4			5			6			7			8			9			10			12		
		APPLIANCE INPUT RATING IN THOUSANDS OF BTU/H																										
		FAN		NAT	FAN		NAT	FAN		NAT	FAN		NAT	FAN		NAT	FAN		NAT	FAN		NAT	FAN		NAT	FAN		NAT
		Min	Max	Max	Min	Max	Max	Min	Max	Max	Min	Max	Max	Min	Max	Max	Min	Max	Max	Min	Max	Max	Min	Max	Max	Min	Max	Max
50	0	33	99	66	51	213	133	73	394	230	105	629	361	138	928	515	176	1,292	704	220	1,724	948	295	2,223	1,189	428	3,432	1,818
	2	36	84	61	53	181	121	73	318	205	104	495	312	133	712	443	168	971	613	209	1,273	811	280	1,615	1,007	401	2,426	1,509
	5	48	80	NA	70	174	117	94	308	198	131	482	305	164	696	435	204	953	602	257	1,252	795	347	1,591	991	496	2,396	1,490
	10	NA	NA	NA	89	160	NA	118	292	186	162	461	292	203	671	420	253	923	583	313	1,217	765	418	1,551	963	589	2,347	1,455
	15	NA	NA	NA	112	148	NA	145	275	174	199	441	280	244	646	405	299	894	562	363	1,183	736	481	1,512	934	668	2,299	1,421
	20	NA	NA	NA	NA	NA	NA	176	257	NA	236	420	267	285	622	389	345	866	543	415	1,150	708	544	1,473	906	741	2,251	1,387
	30	NA	NA	NA	NA	NA	NA	NA	NA	NA	315	376	NA	373	573	NA	442	809	502	521	1,086	649	674	1,399	848	892	2,159	1,318

For SI: 1 inch = 25.4 mm, 1 foot = 304.8 mm, 1 British thermal unit per hour = 0.2931 W.

G2428.2.9 (504.2.9) Chimney and vent locations. Tables G2428.2(1) and G2428.2(2) shall be used only for chimneys and vents not exposed to the outdoors below the roof line. A Type B vent or *listed* chimney lining system passing through an unused *masonry chimney* flue shall not be considered to be exposed to the outdoors. Where vents extend outdoors above the roof more than 5 feet (1524 mm) higher than required by Figure G2427.6.4, and where vents terminate in accordance with Section G2427.6.4, Item 2, the outdoor portion of the vent shall be enclosed as required by this section for vents not considered to be exposed to the outdoors or such venting system shall be engineered. A Type B vent shall not be considered to be exposed to the outdoors where it passes through an unventilated enclosure or chase insulated to a value of not less than R-8.

G2428.2.10 (504.2.10) Corrugated vent connector size. Corrugated *vent connectors* shall be not smaller than the *listed appliance* categorized *vent* diameter, *flue collar* diameter, or *draft hood* outlet diameter.

G2428.2.11 (504.2.11) Vent connector size limitation. *Vent connectors* shall not be increased in size more than two sizes greater than the *listed appliance* categorized vent diameter, *flue collar* diameter or *draft hood* outlet diameter.

G2428.2.12 (504.2.12) Component commingling. In a single run of vent or *vent connector*, different diameters and types of vent and connector components shall be permitted to be used, provided that all such sizes and types are permitted by the tables.

G2428.2.13 (504.2.13) Draft hood conversion accessories. *Draft hood* conversion accessories for use with *masonry chimneys* venting *listed* Category I fan-assisted *appliances* shall be *listed* and installed in accordance with the manufacturer's instructions for such *listed* accessories.

G2428.2.14 (504.2.14) Table interpolation. Interpolation shall be permitted in calculating capacities for vent dimensions that fall between the table entries.

G2428.2.15 (504.2.15) Extrapolation prohibited. Extrapolation beyond the table entries shall not be permitted.

G2428.2.16 (504.2.16) Engineering calculations. Where a *vent* height is less than 6 feet (1829 mm) or greater than shown in the tables, an engineering method shall be used to calculate the *vent* capacity.

G2428.2.17 (504.2.17) Height entries. Where the actual height of a vent falls between entries in the height column of the applicable table in Tables G2428.2(1) and G2428.2(2), either interpolation shall be used or the lower *appliance* input rating shown in the table entries shall be used for FAN Max and NAT Max column values and the higher *appliance* input rating shall be used for the FAN Min column values.

G2428.3 (504.3) Application of multiple appliance vent Tables G2428.3(1) through G2428.3(4). The application of Tables G2428.3(1) through G2428.3(4) shall be subject to the requirements of Sections G2428.3.1 through G2428.3.24.

G2428.3.1 (504.3.1) Vent obstructions. These venting tables shall not be used where obstructions, as described in Section G2427.15, are installed in the venting system. The installation of vents serving *listed appliances* with vent dampers shall be in accordance with the *appliance* manufacturer's instructions or in accordance with the following:

1. The maximum capacity of the *vent connector* shall be determined using the NAT Max column.
2. The maximum capacity of the vertical vent or *chimney* shall be determined using the FAN + NAT column where the second *appliance* is a fan-assisted *appliance*, or the NAT + NAT column where the second *appliance* is equipped with a *draft hood.*
3. The minimum capacity shall be determined as if the *appliance* were a fan-assisted *appliance.*
   3.1. The minimum capacity of the *vent connector* shall be determined using the FAN Min column.
   3.2. The FAN + FAN column shall be used where the second *appliance* is a fan-assisted *appliance*, and the FAN + NAT column shall be used where the second *appliance* is equipped with a *draft hood*, to determine whether the vertical vent or *chimney* configuration is not permitted (NA). Where the vent configuration is NA, the vent configuration shall not be permitted and an alternative venting configuration shall be utilized.

G2428.3.2 (504.3.2) Connector length limit. The *vent connector* shall be routed to the vent utilizing the shortest possible route. Except as provided in Section G2428.3.3, the maximum *vent connector* horizontal length shall be $1^1/_2$ feet for each inch (18 mm per mm) of connector diameter as shown in Table G2428.3.2.

TABLE G2428.3.2 (504.3.2)
MAXIMUM VENT CONNECTOR LENGTH

CONNECTOR DIAMETER (inches)	CONNECTOR MAXIMUM HORIZONTAL LENGTH (feet)
3	$4^1/_2$
4	6
5	$7^1/_2$
6	9
7	$10^1/_2$
8	12
9	$13^1/_2$

For SI: 1 inch = 25.4 mm, 1 foot = 304.8 mm.

TABLE G2428.3(1) [504.3(1)]
TYPE B DOUBLE-WALL VENT

Number of Appliances	Two or more
Appliances Type	Category I
Appliances Vent Connection	Type B double-wall connector

VENT CONNECTOR CAPACITY																									
VENT HEIGHT (*H*) (feet)	CONNECTOR RISE (*R*) (feet)	TYPE B DOUBLE-WALL VENT AND CONNECTOR DIAMETER (*D*)—inches																							
		3			4			5			6			7			8			9			10		
		APPLIANCE INPUT RATING LIMITS IN THOUSANDS OF BTU/H																							
		FAN		NAT	FAN		NAT	FAN		NAT	FAN		NAT	FAN		NAT	FAN		NAT	FAN		NAT	FAN		NAT
		Min	Max	Max	Min	Max	Max	Min	Max	Max	Min	Max	Max	Min	Max	Max	Min	Max	Max	Min	Max	Max	Min	Max	Max
6	1	22	37	26	35	66	46	46	106	72	58	164	104	77	225	142	92	296	185	109	376	237	128	466	289
	2	23	41	31	37	75	55	48	121	86	60	183	124	79	253	168	95	333	220	112	424	282	131	526	345
	3	24	44	35	38	81	62	49	132	96	62	199	139	82	275	189	97	363	248	114	463	317	134	575	386
8	1	22	40	27	35	72	48	49	114	76	64	176	109	84	243	148	100	320	194	118	408	248	138	507	303
	2	23	44	32	36	80	57	51	128	90	66	195	129	86	269	175	103	356	230	121	454	294	141	564	358
	3	24	47	36	37	87	64	53	139	101	67	210	145	88	290	198	105	384	258	123	492	330	143	612	402
10	1	22	43	28	34	78	50	49	123	78	65	189	113	89	257	154	106	341	200	125	436	257	146	542	314
	2	23	47	33	36	86	59	51	136	93	67	206	134	91	282	182	109	374	238	128	479	305	149	596	372
	3	24	50	37	37	92	67	52	146	104	69	220	150	94	303	205	111	402	268	131	515	342	152	642	417
15	1	21	50	30	33	89	53	47	142	83	64	220	120	88	298	163	110	389	214	134	493	273	162	609	333
	2	22	53	35	35	96	63	49	153	99	66	235	142	91	320	193	112	419	253	137	532	323	165	658	394
	3	24	55	40	36	102	71	51	163	111	68	248	160	93	339	218	115	445	286	140	565	365	167	700	444
20	1	21	54	31	33	99	56	46	157	87	62	246	125	86	334	171	107	436	224	131	552	285	158	681	347
	2	22	57	37	34	105	66	48	167	104	64	259	149	89	354	202	110	463	265	134	587	339	161	725	414
	3	23	60	42	35	110	74	50	176	116	66	271	168	91	371	228	113	486	300	137	618	383	164	764	466
30	1	20	62	33	31	113	59	45	181	93	60	288	134	83	391	182	103	512	238	125	649	305	151	802	372
	2	21	64	39	33	118	70	47	190	110	62	299	158	85	408	215	105	535	282	129	679	360	155	840	439
	3	22	66	44	34	123	79	48	198	124	64	309	178	88	423	242	108	555	317	132	706	405	158	874	494

COMMON VENT CAPACITY																					
VENT HEIGHT (*H*) (feet)	TYPE B DOUBLE-WALL COMMON VENT DIAMETER (*D*)—inches																				
	4			5			6			7			8			9			10		
	COMBINED APPLIANCE INPUT RATING IN THOUSANDS OF BTU/H																				
	FAN + FAN	FAN + NAT	NAT + NAT	FAN + FAN	FAN + NAT	NAT + NAT	FAN + FAN	FAN + NAT	NAT + NAT	FAN + FAN	FAN + NAT	NAT + NAT	FAN + FAN	FAN + NAT	NAT + NAT	FAN + FAN	FAN + NAT	NAT + NAT	FAN + FAN	FAN + NAT	NAT + NAT
6	92	81	65	140	116	103	204	161	147	309	248	200	404	314	260	547	434	335	672	520	410
8	101	90	73	155	129	114	224	178	163	339	275	223	444	348	290	602	480	378	740	577	465
10	110	97	79	169	141	124	243	194	178	367	299	242	477	377	315	649	522	405	800	627	495
15	125	112	91	195	164	144	283	228	206	427	352	280	556	444	365	753	612	465	924	733	565
20	136	123	102	215	183	160	314	255	229	475	394	310	621	499	405	842	688	523	1,035	826	640
30	152	138	118	244	210	185	361	297	266	547	459	360	720	585	470	979	808	605	1,209	975	740
50	167	153	134	279	244	214	421	353	310	641	547	423	854	706	550	1,164	977	705	1,451	1,188	860

For SI: 1 inch = 25.4 mm, 1 foot = 304.8 mm, 1 British thermal unit per hour = 0.2931 W.

TABLE G2428.3(2) [504.3(2)]
TYPE B DOUBLE-WALL VENT

Number of Appliances	Two or more
Appliances Type	Category I
Appliances Vent Connection	Single-wall metal connector

VENT CONNECTOR CAPACITY																									
VENT HEIGHT (*H*) (feet)	CONNECTOR RISE (*R*) (feet)	SINGLE-WALL METAL VENT CONNECTOR DIAMETER (*D*)—inches																							
		3			4			5			6			7			8			9			10		
		APPLIANCE INPUT RATING LIMITS IN THOUSANDS OF BTU/H																							
		FAN		NAT	FAN		NAT	FAN		NAT	FAN		NAT	FAN		NAT	FAN		NAT	FAN		NAT	FAN		NAT
		Min	Max	Max	Min	Max	Max	Min	Max	Max	Min	Max	Max	Min	Max	Max	Min	Max	Max	Min	Max	Max	Min	Max	Max
6	1	NA	NA	26	NA	NA	46	NA	NA	71	NA	NA	102	207	223	140	262	293	183	325	373	234	447	463	286
	2	NA	NA	31	NA	NA	55	NA	NA	85	168	182	123	215	251	167	271	331	219	334	422	281	458	524	344
	3	NA	NA	34	NA	NA	62	121	131	95	175	198	138	222	273	188	279	361	247	344	462	316	468	574	385
8	1	NA	NA	27	NA	NA	48	NA	NA	75	NA	NA	106	226	240	145	285	316	191	352	403	244	481	502	299
	2	NA	NA	32	NA	NA	57	125	126	89	184	193	127	234	266	173	293	353	228	360	450	292	492	560	355
	3	NA	NA	35	NA	NA	64	130	138	100	191	208	144	241	287	197	302	381	256	370	489	328	501	609	400
10	1	NA	NA	28	NA	NA	50	119	121	77	182	186	110	240	253	150	302	335	196	372	429	252	506	534	308
	2	NA	NA	33	84	85	59	124	134	91	189	203	132	248	278	183	311	369	235	381	473	302	517	589	368
	3	NA	NA	36	89	91	67	129	144	102	197	217	148	257	299	203	320	398	265	391	511	339	528	637	413
15	1	NA	NA	29	79	87	52	116	138	81	177	214	116	238	291	158	312	380	208	397	482	266	556	596	324
	2	NA	NA	34	83	94	62	121	150	97	185	230	138	246	314	189	321	411	248	407	522	317	568	646	387
	3	NA	NA	39	87	100	70	127	160	109	193	243	157	255	333	215	331	438	281	418	557	360	579	690	437
20	1	49	56	30	78	97	54	115	152	84	175	238	120	233	325	165	306	425	217	390	538	276	546	664	336
	2	52	59	36	82	103	64	120	163	101	182	252	144	243	346	197	317	453	259	400	574	331	558	709	403
	3	55	62	40	87	107	72	125	172	113	190	264	164	252	363	223	326	476	294	412	607	375	570	750	457
30	1	47	60	31	77	110	57	112	175	89	169	278	129	226	380	175	296	497	230	378	630	294	528	779	358
	2	51	62	37	81	115	67	117	185	106	177	290	152	236	397	208	307	521	274	389	662	349	541	819	425
	3	54	64	42	85	119	76	122	193	120	185	300	172	244	412	235	316	542	309	400	690	394	555	855	482

COMMON VENT CAPACITY																					
VENT HEIGHT (*H*) (feet)	TYPE B DOUBLE-WALL COMMON VENT DIAMETER (*D*)—inches																				
	4			5			6			7			8			9			10		
	COMBINED APPLIANCE INPUT RATING IN THOUSANDS OF BTU/H																				
	FAN + FAN	FAN + NAT	NAT + NAT	FAN + FAN	FAN + NAT	NAT + NAT	FAN + FAN	FAN + NAT	NAT + NAT	FAN + FAN	FAN + NAT	NAT + NAT	FAN + FAN	FAN + NAT	NAT + NAT	FAN +FAN	FAN +NAT	NAT +NAT	FAN +FAN	FAN +NAT	NAT +NAT
6	NA	78	64	NA	113	99	200	158	144	304	244	196	398	310	257	541	429	332	665	515	407
8	NA	87	71	NA	126	111	218	173	159	331	269	218	436	342	285	592	473	373	730	569	460
10	NA	94	76	163	137	120	237	189	174	357	292	236	467	369	309	638	512	398	787	617	487
15	121	108	88	189	159	140	275	221	200	416	343	274	544	434	357	738	599	456	905	718	553
20	131	118	98	208	177	156	305	247	223	463	383	302	606	487	395	824	673	512	1,013	808	626
30	145	132	113	236	202	180	350	286	257	533	446	349	703	570	459	958	790	593	1,183	952	723
50	159	145	128	268	233	208	406	337	296	622	529	410	833	686	535	1,139	954	689	1,418	1,157	838

For SI: 1 inch = 25.4 mm, 1 foot = 304.8 mm, 1 British thermal unit per hour = 0.2931 W.

TABLE G2428.3(3) [504.3(3)] MASONRY CHIMNEY

Number of Appliances	Two or more
Appliances Type	Category I
Appliances Vent Connection	Type B double-wall connector

VENT CONNECTOR CAPACITY																									
VENT HEIGHT (*H*) (feet)	CONNECTOR RISE (*R*) (feet)	TYPE B DOUBLE-WALL VENT CONNECTOR DIAMETER (*D*)—inches																							
		3			4			5			6			7			8			9			10		
		APPLIANCE INPUT RATING LIMITS IN THOUSANDS OF BTU/H																							
		FAN		NAT	FAN		NAT	FAN		NAT	FAN		NAT	FAN		NAT	FAN		NAT	FAN		NAT	FAN		NAT
		Min	Max	Max	Min	Max	Max	Min	Max	Max	Min	Max	Max	Min	Max	Max	Min	Max	Max	Min	Max	Max	Min	Max	Max
6	1	24	33	21	39	62	40	52	106	67	65	194	101	87	274	141	104	370	201	124	479	253	145	599	319
	2	26	43	28	41	79	52	53	133	85	67	230	124	89	324	173	107	436	232	127	562	300	148	694	378
	3	27	49	34	42	92	61	55	155	97	69	262	143	91	369	203	109	491	270	129	633	349	151	795	439
8	1	24	39	22	39	72	41	55	117	69	71	213	105	94	304	148	113	414	210	134	539	267	156	682	335
	2	26	47	29	40	87	53	57	140	86	73	246	127	97	350	179	116	473	240	137	615	311	160	776	394
	3	27	52	34	42	97	62	59	159	98	75	269	145	99	383	206	119	517	276	139	672	358	163	848	452
10	1	24	42	22	38	80	42	55	130	71	74	232	108	101	324	153	120	444	216	142	582	277	165	739	348
	2	26	50	29	40	93	54	57	153	87	76	261	129	103	366	184	123	498	247	145	652	321	168	825	407
	3	27	55	35	41	105	63	58	170	100	78	284	148	106	397	209	126	540	281	147	705	366	171	893	463
15	1	24	48	23	38	93	44	54	154	74	72	277	114	100	384	164	125	511	229	153	658	297	184	824	375
	2	25	55	31	39	105	55	56	174	89	74	299	134	103	419	192	128	558	260	156	718	339	187	900	432
	3	26	59	35	41	115	64	57	189	102	76	319	153	105	448	215	131	597	292	159	760	382	190	960	486
20	1	24	52	24	37	102	46	53	172	77	71	313	119	98	437	173	123	584	239	150	752	312	180	943	397
	2	25	58	31	39	114	56	55	190	91	73	335	138	101	467	199	126	625	270	153	805	354	184	1,011	452
	3	26	63	35	40	123	65	57	204	104	75	353	157	104	493	222	129	661	301	156	851	396	187	1,067	505

COMMON VENT CAPACITY																								
VENT HEIGHT (*H*) (feet)	MINIMUM INTERNAL AREA OF MASONRY CHIMNEY FLUE (square inches)																							
	12			19			28			38			50			63			78			113		
	COMBINED APPLIANCE INPUT RATING IN THOUSANDS OF BTU/H																							
	FAN + FAN	FAN + NAT	NAT + NAT	FAN + FAN	FAN + NAT	NAT + NAT	FAN + FAN	FAN + NAT	NAT + NAT	FAN + FAN	FAN + NAT	NAT + NAT	FAN + FAN	FAN + NAT	NAT + NAT	FAN + FAN	FAN + NAT	NAT + NAT	FAN + FAN	FAN + NAT	NAT + NAT	FAN + FAN	FAN + NAT	NAT + NAT
6	NA	74	25	NA	119	46	NA	178	71	NA	257	103	NA	351	143	NA	458	188	NA	582	246	1,041	853	NA
8	NA	80	28	NA	130	53	NA	193	82	NA	279	119	NA	384	163	NA	501	218	724	636	278	1,144	937	408
10	NA	84	31	NA	138	56	NA	207	90	NA	299	131	NA	409	177	606	538	236	776	686	302	1,226	1,010	454
15	NA	NA	36	NA	152	67	NA	233	106	NA	334	152	523	467	212	682	611	283	874	781	365	1,374	1,156	546
20	NA	NA	41	NA	NA	75	NA	250	122	NA	368	172	565	508	243	742	668	325	955	858	419	1,513	1,286	648
30	NA	NA	NA	NA	NA	NA	NA	270	137	NA	404	198	615	564	278	816	747	381	1,062	969	496	1,702	1,473	749
50	NA	NA	NA	NA	NA	NA	NA	NA	NA	NA	NA	NA	NA	620	328	879	831	461	1,165	1,089	606	1,905	1,692	922

For SI: 1 inch = 25.4 mm, 1 square inch = 645.16 mm^2, 1 foot = 304.8 mm, 1 British thermal unit per hour = 0.2931 W.

TABLE G2428.3(4) [504.3(4)] MASONRY CHIMNEY

Number of Appliances	Two or more
Appliances Type	Category I
Appliances Vent Connection	Single-wall connector

VENT CONNECTOR CAPACITY

SINGLE-WALL METAL VENT CONNECTOR DIAMETER (*D*)—inches; APPLIANCE INPUT RATING LIMITS IN THOUSANDS OF BTU/H

VENT HEIGHT (*H*) (feet)	CONNECTOR RISE (*R*) (feet)	3			4			5			6			7			8			9			10		
		FAN		NAT	FAN		NAT	FAN		NAT	FAN		NAT	FAN		NAT	FAN		NAT	FAN		NAT	FAN		NAT
		Min	Max	Max	Min	Max	Max	Min	Max	Max	Min	Max	Max	Min	Max	Max	Min	Max	Max	Min	Max	Max	Min	Max	Max
6	1	NA	NA	21	NA	NA	39	NA	NA	66	179	191	100	231	271	140	292	366	200	362	474	252	499	594	316
	2	NA	NA	28	NA	NA	52	NA	NA	84	186	227	123	239	321	172	301	432	231	373	557	299	509	696	376
	3	NA	NA	34	NA	NA	61	134	153	97	193	258	142	247	365	202	309	491	269	381	634	348	519	793	437
8	1	NA	NA	21	NA	NA	40	NA	NA	68	195	208	103	250	298	146	313	407	207	387	530	263	529	672	331
	2	NA	NA	28	NA	NA	52	137	139	85	202	240	125	258	343	177	323	465	238	397	607	309	540	766	391
	3	NA	NA	34	NA	NA	62	143	156	98	210	264	145	266	376	205	332	509	274	407	663	356	551	838	450
10	1	NA	NA	22	NA	NA	41	130	151	70	202	225	106	267	316	151	333	434	213	410	571	273	558	727	343
	2	NA	NA	29	NA	NA	53	136	150	86	210	255	128	276	358	181	343	489	244	420	640	317	569	813	403
	3	NA	NA	34	97	102	62	143	166	99	217	277	147	284	389	207	352	530	279	430	694	363	580	880	459
15	1	NA	NA	23	NA	NA	43	129	151	73	199	271	112	268	376	161	349	502	225	445	646	291	623	808	366
	2	NA	NA	30	92	103	54	135	170	88	207	295	132	277	411	189	359	548	256	456	706	334	634	884	424
	3	NA	NA	34	96	112	63	141	185	101	215	315	151	286	439	213	368	586	289	466	755	378	646	945	479
20	1	NA	NA	23	87	99	45	128	167	76	197	303	117	265	425	169	345	569	235	439	734	306	614	921	347
	2	NA	NA	30	91	111	55	134	185	90	205	325	136	274	455	195	355	610	266	450	787	348	627	986	443
	3	NA	NA	35	96	119	64	140	199	103	213	343	154	282	481	219	365	644	298	461	831	391	639	1,042	496

COMMON VENT CAPACITY

MINIMUM INTERNAL AREA OF MASONRY CHIMNEY FLUE (square inches); COMBINED APPLIANCE INPUT RATING IN THOUSANDS OF BTU/H

VENT HEIGHT (*H*) (feet)	12			19			28			38			50			63			78			113		
	FAN + FAN	FAN + NAT	NAT + NAT	FAN + FAN	FAN + NAT	NAT + NAT	FAN + FAN	FAN + NAT	NAT + NAT	FAN + FAN	FAN + NAT	NAT + NAT	FAN + FAN	FAN + NAT	NAT + NAT	FAN + FAN	FAN + NAT	NAT + NAT	FAN + FAN	FAN + NAT	NAT + NAT	FAN + FAN	FAN + NAT	NAT + NAT
6	NA	NA	25	NA	118	45	NA	176	71	NA	255	102	NA	348	142	NA	455	187	NA	579	245	NA	846	NA
8	NA	NA	28	NA	128	52	NA	190	81	NA	276	118	NA	380	162	NA	497	217	NA	633	277	1,136	928	405
10	NA	NA	31	NA	136	56	NA	205	89	NA	295	129	NA	405	175	NA	532	234	171	680	300	1,216	1,000	450
15	NA	NA	36	NA	NA	66	NA	230	105	NA	335	150	NA	400	210	677	602	280	866	772	360	1,359	1,139	540
20	NA	NA	NA	NA	NA	74	NA	247	120	NA	362	170	NA	503	240	765	661	321	947	849	415	1,495	1,264	640
30	NA	NA	NA	NA	NA	NA	NA	NA	135	NA	398	195	NA	558	275	808	739	377	1,052	957	490	1,682	1,447	740
50	NA	NA	NA	NA	NA	NA	NA	NA	NA	NA	NA	NA	NA	612	325	NA	821	456	1,152	1,076	600	1,879	1,672	910

For SI: 1 inch = 25.4 mm, 1 square inch = 645.16 mm^2, 1 foot = 304.8 mm, 1 British thermal unit per hour = 0.2931 W.

G2428.3.3 (504.3.3) Connectors with longer lengths. Connectors with longer horizontal lengths than those listed in Section G2428.3.2 are permitted under the following conditions:

1. The maximum capacity (FAN Max or NAT Max) of the *vent connector* shall be reduced 10 percent for each additional multiple of the length allowed by Section G2428.3.2. For example, the maximum length listed in Table G2428.3.2 for a 4-inch (102 mm) connector is 6 feet (1829 mm). With a connector length greater than 6 feet (1829 mm) but not exceeding 12 feet (3658 mm), the maximum capacity must be reduced by 10 percent (0.90 × maximum vent *connector* capacity). With a connector length greater than 12 feet (3658 mm), but not exceeding 18 feet (5486 mm), the maximum capacity must be reduced by 20 percent (0.80 × maximum vent capacity).
2. For a connector serving a fan-assisted *appliance*, the minimum capacity (FAN Min) of the connector shall be determined by referring to the corresponding single-*appliance* table. For Type B double-wall connectors, Table G2428.2(1) shall be used. For single-wall connectors, Table G2428.2(2) shall be used. The height (H) and lateral (L) shall be measured according to the procedures for a single-*appliance* vent, as if the other *appliances* were not present.

G2428.3.4 (504.3.4) Vent connector manifold. Where the *vent connectors* are combined prior to entering the vertical portion of the common vent to form a common vent manifold, the size of the common vent manifold and the common vent shall be determined by applying a 10-percent reduction (0.90 × maximum common vent capacity) to the common vent capacity part of the common vent tables. The length of the common *vent connector* manifold (L_m) shall not exceed $1^1/_2$ feet for each inch (18 mm per mm) of common *vent connector* manifold diameter (D).

G2428.3.5 (504.3.5) Common vertical vent offset. Where the common vertical vent is *offset*, the maximum capacity of the common vent shall be reduced in accordance with Section G2428.3.6. The horizontal length of the common vent *offset* (L_o) shall not exceed $1^1/_2$ feet for each inch (18 mm per mm) of common vent diameter (D). Where multiple *offsets* occur in a common vent, the total horizontal length of all *offsets* combined shall not exceed $1^1/_2$ feet for each inch (18 mm per mm) of the common vent diameter (D).

G2428.3.6 (504.3.6) Elbows in vents. For each elbow up to and including 45 degrees (0.79 rad) in the common vent, the maximum common vent capacity listed in the venting tables shall be reduced by 5 percent. For each elbow greater than 45 degrees (0.79 rad) up to and including 90 degrees (1.57 rad), the maximum common vent capacity listed in the venting tables shall be reduced by 10 percent.

G2428.3.7 (504.3.7) Elbows in connectors. The *vent connector* capacities listed in the common vent sizing tables include allowance for two 90-degree (1.57 rad) elbows. For each additional elbow up to and including 45 degrees (0.79 rad), the maximum *vent connector* capacity listed in the venting tables shall be reduced by 5 percent. For each elbow greater than 45 degrees (0.79 rad) up to and including 90 degrees (1.57 rad), the maximum *vent connector* capacity listed in the venting tables shall be reduced by 10 percent.

G2428.3.8 (504.3.8) Common vent minimum size. The cross-sectional area of the common vent shall be equal to or greater than the cross-sectional area of the largest connector.

G2428.3.9 (504.3.9) Common vent fittings. At the point where tee or wye fittings connect to a common vent, the opening size of the fitting shall be equal to the size of the common vent. Such fittings shall not be prohibited from having reduced-size openings at the point of connection of *appliance vent connectors.*

G2428.3.9.1 (504.3.9.1) Tee and wye fittings. Tee and wye fittings connected to a common gas vent shall be considered to be part of the common gas vent and shall be constructed of materials consistent with that of the common gas vent.

G2428.3.10 (504.3.10) High-altitude installations. Sea-level input ratings shall be used when determining maximum capacity for high-altitude installation. Actual input, derated for altitude, shall be used for determining minimum capacity for high-altitude installation.

G2428.3.11 (504.3.11) Connector rise measurement. Connector rise (R) for each *appliance connector* shall be measured from the *draft hood* outlet or *flue collar* to the centerline where the vent gas streams come together.

G2428.3.12 (504.3.12) Vent height measurement. For multiple *appliances* all located on one floor, available total height (H) shall be measured from the highest *draft hood* outlet or *flue collar* up to the level of the outlet of the common vent.

G2428.3.13 (504.3.17) Vertical vent maximum size. Where two or more *appliances* are connected to a vertical vent or *chimney*, the flow area of the largest section of vertical vent or *chimney* shall not exceed seven times the smallest *listed appliance* categorized vent areas, *flue collar* area, or *draft hood* outlet area unless designed in accordance with *approved* engineering methods.

G2428.3.14 (504.3.18) Multiple input rate appliances. The minimum *vent connector* capacity (FAN Min) for *appliances* with more than one input rate shall be determined from the tables and shall be less than the lowest *appliance* input rating. The maximum *vent connector* capacity (FAN Max or NAT Max) for *appliances* with more than one input rate shall be determined from the tables and shall be greater than the highest *appliance* input rating.

G2428.3.15 (504.3.19) Liner system sizing and connections. *Listed*, corrugated metallic *chimney* liner systems in masonry *chimneys* shall be sized by using Table G2428.3(1) or G2428.3(1) for Type B vents, with the maximum capacity reduced by 20 percent (0.80 × maximum capacity) and the minimum capacity as shown in Table G2428.3(1) or G2428.3(1). Corrugated metallic liner systems installed with bends or offsets shall have their maximum capacity further reduced in accordance with Sections G2428.3.5 and G2428.3.6. The 20-percent reduction for corrugated metallic *chimney* liner systems includes an allowance for one long-radius 90-degree (1.57 rad) turn at the bottom of the liner. Where double-wall connectors are required, tee and wye fittings used to connect to the common vent *chimney* liner shall be *listed* double-wall fittings. Connections between *chimney* liners and *listed* double-wall fittings shall be made with *listed* adapter fittings designed for such purpose.

G2428.3.16 (504.3.20) Chimney and vent location. Tables G2428.3(1), G2428.3(2), G2428.3(3) and G2428.3(4) shall be used only for chimneys and vents not exposed to the outdoors below the roof line. A Type B vent or *listed* chimney lining system passing through an unused *masonry chimney* flue shall not be considered to be exposed to the outdoors. Where vents extend outdoors above the roof more than 5 feet (1524 mm) higher than required by Figure G2427.6.4 and where vents terminate in accordance with Section G2427.6.4, Item 2, the outdoor portion of the vent shall be enclosed as required by this section for vents not considered to be exposed to the outdoors or such venting system shall be engineered. A Type B vent shall not be considered to be exposed to the outdoors where it passes through an unventilated enclosure or chase insulated to a value of not less than R-8.

G2428.3.17 (504.3.21) Connector maximum and minimum size. *Vent connectors* shall not be increased in size more than two sizes greater than the *listed appliance* categorized vent diameter, *flue collar* diameter or *draft hood* outlet diameter. *Vent connectors* for *draft hood*-equipped *appliances* shall not be smaller than the *draft hood* outlet diameter. Where a *vent connector* size(s) determined from the tables for a fan-assisted *appliance(s)* is smaller than the *flue collar* diameter, the use of the smaller size(s) shall be permitted provided that the installation complies with all of the following conditions:

1. *Vent connectors* for fan-assisted *appliance flue collars* 12 inches (305 mm) in diameter or smaller are not reduced by more than one table size [for example, 12 inches to 10 inches (305 mm to 254 mm) is a one-size reduction] and those larger than 12 inches (305 mm) in diameter are not reduced more than two table sizes [for example, 24 inches to 20 inches (610 mm to 508 mm) is a two-size reduction].
2. The fan-assisted *appliance(s)* is common vented with a *draft hood*-equipped *appliance(s)*.
3. The vent *connector* has a smooth interior wall.

G2428.3.18 (504.3.22) Component commingling. Combinations of pipe sizes and combinations of single-wall and double-wall metal pipe shall be allowed within any connector run(s) or within the common vent, provided that all of the appropriate tables permit all of the desired sizes and types of pipe, as if they were used for the entire length of the subject connector or vent. Where single-wall and Type B double-wall metal pipes are used for *vent connectors* within the same venting system, the common vent must be sized using Table G2428.3(2) or G2428.3(4), as appropriate.

G2428.3.19 (504.3.23) Draft hood conversion accessories. *Draft hood* conversion accessories for use with *masonry chimneys* venting *listed* Category I fan-assisted *appliances* shall be *listed* and installed in accordance with the manufacturer's instructions for such *listed* accessories.

G2428.3.20 (504.3.24) Multiple sizes permitted. Where a table permits more than one diameter of pipe to be used for a connector or vent, all of the permitted sizes shall be permitted to be used.

G2428.3.21 (504.3.25) Table interpolation. Interpolation shall be permitted in calculating capacities for vent dimensions that fall between table entries.

G2428.3.22 (504.3.26) Extrapolation prohibited. Extrapolation beyond the table entries shall not be permitted.

G2428.3.23 (504.3.27) Engineering calculations. For vent heights less than 6 feet (1829 mm) and greater than shown in the tables, engineering methods shall be used to calculate vent capacities.

G2428.3.24 (504.3.28) Height entries. Where the actual height of a vent falls between entries in the height column of the applicable table in Tables G2428.3(1) through G2428.3(4), either interpolation shall be used or the lower *appliance* input rating shown in the table shall be used for FAN Max and NAT Max column values and the higher *appliance* input rating shall be used for the FAN Min column values.

SECTION G2429 (505) DIRECT-VENT, INTEGRAL VENT, MECHANICAL VENT AND VENTILATION/EXHAUST HOOD VENTING

G2429.1 (505.1) General. The installation of direct-vent and integral vent *appliances* shall be in accordance with Section G2427. Mechanical *venting systems* shall be designed and installed in accordance with Section G2427.

SECTION G2430 (506) FACTORY-BUILT CHIMNEYS

G2430.1 (506.1) Listing. Factory-built *chimneys* for building heating *appliances* producing *flue gases* having a temperature not greater than 1,000°F (538°C), measured at the entrance to the *chimney*, shall be *listed* and *labeled* in accordance with UL 103 and shall be installed and terminated in accordance with the manufacturer's instructions.

G2430.2 (506.2) Support. Where factory-built *chimneys* are supported by structural members, such as joists and rafters, such members shall be designed to support the additional load.

SECTION G2431 (601) GENERAL

G2431.1 (601.1) Scope. Sections G2432 through G2453 shall govern the approval, design, installation, construction, maintenance, *alteration* and *repair* of the *appliances* and *equipment* specifically identified herein.

SECTION G2432 (602) DECORATIVE APPLIANCES FOR INSTALLATION IN FIREPLACES

G2432.1 (602.1) General. Decorative *appliances* for installation in *approved* solid fuel-burning *fireplaces* shall be *listed* in accordance with ANSI Z21.60/CSA 6.26 and shall be installed in accordance with the manufacturer's instructions. Manually lighted natural gas decorative *appliances* shall be *listed* in accordance with ANSI Z21.84.

G2432.2 (602.2) Flame safeguard device. Decorative *appliances* for installation in *approved* solid fuel-burning *fireplaces*, with the exception of those *listed* in accordance with ANSI Z21.84, shall utilize a direct ignition device, an ignitor or a *pilot* flame to ignite the fuel at the *main burner*, and shall be equipped with a *flame safeguard* device. The *flame safeguard* device shall automatically shut off the fuel supply to a *main burner* or group of *burners* when the means of ignition of such *burners* becomes inoperative.

G2432.3 (602.3) Prohibited installations. Decorative *appliances* for installation in *fireplaces* shall not be installed where prohibited by Section G2406.2.

SECTION G2433 (603) LOG LIGHTERS

G2433.1 (603.1) General. Log lighters shall be *listed* in accordance with CSA 8 and shall be installed in accordance with the manufacturer's instructions.

SECTION G2434 (604) VENTED GAS FIREPLACES (DECORATIVE APPLIANCES)

G2434.1 (604.1) General. Vented gas *fireplaces* shall be *listed* in accordance with ANSI Z21.50/CSA 2.22, shall be installed in accordance with the manufacturer's instructions and shall be designed and equipped as specified in Section G2432.2.

G2434.2 (604.2) Access. Panels, grilles and access doors that are required to be removed for normal servicing operations shall not be attached to the building.

SECTION G2435 (605) VENTED GAS FIREPLACE HEATERS

G2435.1 (605.1) General. *Vented* gas *fireplace* heaters shall be installed in accordance with the manufacturer's instructions, shall be *listed* in accordance with Z21.88/CSA 2.33 and shall be designed and equipped as specified in Section G2432.2.

SECTION G2436 (608) VENTED WALL FURNACES

G2436.1 (608.1) General. *Vented wall furnaces* shall be *listed* in accordance with ANSI Z21.86/CSA 2.32 and shall be installed in accordance with the manufacturer's instructions.

G2436.2 (608.2) Venting. *Vented wall furnaces* shall be vented in accordance with Section G2427.

G2436.3 (608.3) Location. *Vented wall furnaces* shall be located so as not to cause a fire hazard to walls, floors, combustible furnishings or doors. *Vented wall furnaces* installed between bathrooms and adjoining rooms shall not circulate air from bathrooms to other parts of the building.

G2436.4 (608.4) Door swing. *Vented wall furnaces* shall be located so that a door cannot swing within 12 inches (305 mm) of an air inlet or air outlet of such *furnace* measured at right angles to the opening. Doorstops or door closers shall not be installed to obtain this *clearance*.

G2436.5 (608.5) Ducts prohibited. Ducts shall not be attached to wall *furnaces*. Casing extension boots shall not be installed unless *listed* as part of the *appliance*.

G2436.6 (608.6) Access. *Vented wall furnaces* shall be provided with access for cleaning of heating surfaces, removal of *burners*, replacement of sections, motors, *controls*, filters and other working parts, and for adjustments and lubrication of parts requiring such attention. Panels, grilles and access doors that are required to be removed for normal servicing operations shall not be attached to the building construction.

SECTION G2437 (609) FLOOR FURNACES

G2437.1 (609.1) General. *Floor furnaces* shall be *listed* in accordance with ANSI Z21.86/CSA 2.32 and shall be installed in accordance with the manufacturer's instructions.

G2437.2 (609.2) Placement. The following provisions apply to *floor furnaces*:

1. Floors. *Floor furnaces* shall not be installed in the floor of any doorway, stairway landing, aisle or passageway of any enclosure, public or private, or in an exitway from any such room or space.
2. Walls and corners. The register of a *floor furnace* with a horizontal warm air outlet shall not be placed closer than 6 inches (152 mm) to the nearest wall. A distance of not less than 18 inches (457 mm) from two adjoining sides of the *floor furnace* register to

walls shall be provided to eliminate the necessity of occupants walking over the warm-air discharge. The remaining sides shall be permitted to be placed not closer than 6 inches (152 mm) to a wall. Wall-register models shall not be placed closer than 6 inches (152 mm) to a corner.

3. Draperies. The *furnace* shall be placed so that a door, drapery, or similar object cannot be nearer than 12 inches (305 mm) to any portion of the register of the *furnace*.

4. Floor construction. *Floor furnaces* shall not be installed in concrete floor construction built on grade.

5. *Thermostat*. The controlling *thermostat* for a *floor furnace* shall be located within the same room or space as the *floor furnace* or shall be located in an adjacent room or space that is permanently open to the room or space containing the *floor furnace*.

G2437.3 (609.3) Bracing. The floor around the *furnace* shall be braced and headed with a support framework designed in accordance with Chapter 5.

G2437.4 (609.4) Clearance. The lowest portion of the *floor furnace* shall have not less than a 6-inch (152 mm) *clearance* from the grade level; except where the lower 6-inch (152 mm) portion of the *floor furnace* is sealed by the manufacturer to prevent entrance of water, the minimum *clearance* shall be not less than 2 inches (51 mm). Where such *clearances* cannot be provided, the ground below and to the sides shall be excavated to form a pit under the *furnace* so that the required *clearance* is provided beneath the lowest portion of the *furnace*. A 12-inch (305 mm) minimum clearance shall be provided on all sides except the *control* side, which shall have an 18-inch (457 mm) minimum *clearance*.

G2437.5 (609.5) First-floor installation. Where the basement story level below the floor in which a *floor furnace* is installed is utilized as *habitable space*, such *floor furnaces* shall be enclosed as specified in Section G2437.6 and shall project into a nonhabitable space.

G2437.6 (609.6) Upper-floor installations. *Floor furnaces* installed in upper stories of buildings shall project below into nonhabitable space and shall be separated from the nonhabitable space by an enclosure constructed of *noncombustible materials*. The *floor furnace* shall be provided with access, *clearance* to all sides and bottom of not less than 6 inches (152 mm) and *combustion air* in accordance with Section G2407.

SECTION G2438 (613) CLOTHES DRYERS

G2438.1 (613.1) General. *Clothes dryers* shall be *listed* in accordance with ANSI Z21.5.1/CSA 7.1 and shall be installed in accordance with the manufacturer's instructions.

SECTION G2439 (614) CLOTHES DRYER EXHAUST

G2439.1 (614.1) Installation. *Clothes dryers* shall be exhausted in accordance with the manufacturer's instructions. Dryer exhaust systems shall be independent of all other systems and shall convey the moisture and any products of *combustion* to the outside of the building.

G2439.2 (614.2) Duct penetrations. Ducts that exhaust *clothes dryers* shall not penetrate or be located within any fireblocking, draftstopping or any wall, floor/ceiling or other assembly required by this code to be fire-resistance rated, unless such duct is constructed of galvanized steel or aluminum of the thickness specified in the mechanical provisions of this code and the fire-resistance rating is maintained in accordance with this code. Fire dampers shall not be installed in *clothes dryer* exhaust duct systems.

G2439.3 (614.4) Exhaust installation. Exhaust ducts for *clothes dryers* shall terminate on the outside of the building and shall be equipped with a backdraft *damper*. Screens shall not be installed at the duct termination. Ducts shall not be connected or installed with sheet metal screws or other fasteners that will obstruct the flow. *Clothes dryer* exhaust ducts shall not be connected to a *vent connector*, vent or *chimney*. *Clothes dryer* exhaust ducts shall not extend into or through ducts or plenums. Clothes dryer exhaust ducts shall be sealed in accordance with Section M1601.4.1.

G2439.3.1 (614.4.1) Exhaust termination outlet and passageway. The passageway of dryer exhaust duct terminals shall be undiminished in size and shall provide an open area of not less than 12.5 square inches (8065 mm^2).

G2439.4 (614.5) Dryer exhaust duct power ventilators. Domestic dryer exhaust duct power ventilators shall be *listed* and *labeled* to UL 705 for use in dryer exhaust duct systems. The dryer exhaust duct power ventilator shall be installed in accordance with the manufacturer's instructions.

G2439.5 (614.7) Makeup air. Installations exhausting more than 200 cfm (0.09 m^3/s) shall be provided with *makeup air*.

G2439.5.1 (614.7.1) Closet installation. Where a closet is designed for the installation of a *clothes dryer*, an opening having an area of not less than 100 square inches (645 mm^2) for *makeup air* shall be provided in the closet enclosure, or *makeup air* shall be provided by other *approved* means.

G2439.6 (614.8) Protection required. Protective shield plates shall be placed where nails or screws from finish or other work are likely to penetrate the *clothes dryer* exhaust duct. Shield plates shall be placed on the finished face of all framing members where there is less than $1^1/_4$ inches (32 mm) between the duct and the finished face of the framing member. Protective shield plates shall be constructed of steel, shall have a minimum thickness of 0.062 inch (1.6 mm) and shall extend not less than 2 inches (51 mm) above sole plates and below top plates.

G2439.7 (614.9) Domestic clothes dryer exhaust ducts. Exhaust ducts for domestic *clothes dryers* shall conform to the requirements of Sections G2439.7.1 through G2439.7.6.

G2439.7.1 (614.9.1) Material and size. Exhaust ducts shall have a smooth interior finish and shall be constructed of metal not less than 0.016 inch (0.4 mm) in thickness. The exhaust duct size shall be 4 inches (102 mm) nominal in diameter.

G2439.7.2 (614.9.2) Duct installation. Exhaust ducts shall be supported at 4-foot (1219 mm) intervals and secured in place. The insert end of the duct shall extend into the adjoining duct or fitting in the direction of airflow. Ducts shall not be joined with screws or similar fasteners that protrude more than $^1/_8$ inch (3.2 mm) into the inside of the duct. Where dryer exhaust ducts are enclosed in wall or ceiling cavities, such cavities shall allow the installation of the duct without deformation.

G2439.7.3 (614.9.3) Transition ducts. Transition ducts used to connect the dryer to the exhaust *duct system* shall be a single length that is *listed* and *labeled* in accordance with UL 2158A. Transition ducts shall be not more than 8 feet (2438 mm) in length and shall not be concealed within construction.

G2439.7.4 (614.9.4) Duct length. The maximum allowable exhaust duct length shall be determined by one of the methods specified in Sections G2439.7.4.1 through G2439.7.4.3.

G2439.7.4.1 (614.9.4.1) Specified length. The maximum length of the exhaust duct shall be 35 feet (10 668 mm) from the connection to the transition duct from the dryer to the outlet terminal. Where fittings are used, the maximum length of the exhaust duct shall be reduced in accordance with Table G2439.7.4.1.

TABLE G2439.7.4.1 (TABLE 614.9.4.1)
DRYER EXHAUST DUCT FITTING EQUIVALENT LENGTH

DRYER EXHAUST DUCT FITTING TYPE	EQUIVALENT LENGTH
4-inch radius mitered 45-degree elbow	2 feet 6 inches
4-inch radius mitered 90-degree elbow	5 feet
6-inch radius smooth 45-degree elbow	1 foot
6-inch radius smooth 90-degree elbow	1 foot 9 inches
8-inch radius smooth 45-degree elbow	1 foot
8-inch radius smooth 90-degree elbow	1 foot 7 inches
10-inch radius smooth 45-degree elbow	9 inches
10-inch radius smooth 90-degree elbow	1 foot 6 inches

For SI: 1 inch = 25.4 mm, 1 foot = 304.8 mm, 1 degree = 0.0175 rad.

G2439.7.4.2 (614.9.4.2) Manufacturer's instructions. The maximum length of the exhaust duct shall be determined by the dryer manufacturer's installation instructions. The *code official* shall be provided with a copy of the installation instructions for the make and model of the dryer. Where the exhaust duct is to be concealed, the installation instructions shall be provided to the *code official* prior to the concealment inspection. In the absence of fitting equivalent length calculations from the clothes dryer manufacturer, Table G2439.7.4.1 shall be utilized.

G2439.7.4.3 (614.9.4.3) Dryer exhaust duct power ventilator length. The maximum length of the exhaust duct shall be determined by the dryer exhaust duct power ventilator manufacturer's installation instructions.

G2439.7.5 (614.9.5) Length identification. Where the exhaust duct equivalent length exceeds 35 feet (10 668 mm), the equivalent length of the exhaust duct shall be identified on a permanent *label* or tag. The *label* or tag shall be located within 6 feet (1829 mm) of the exhaust duct connection

G2439.7.6 (614.9.6) Exhaust duct required. Where space for a *clothes dryer* is provided, an exhaust *duct system* shall be installed.

Where the *clothes dryer* is not installed at the time of occupancy, the exhaust duct shall be capped at the location of the future dryer.

Exception: Where a *listed* condensing *clothes dryer* is installed prior to occupancy of the structure.

SECTION G2440 (615) SAUNA HEATERS

G2440.1 (615.1) General. Sauna heaters shall be installed in accordance with the manufacturer's instructions.

G2440.2 (615.2) Location and protection. Sauna heaters shall be located so as to minimize the possibility of accidental contact by a person in the room.

G2440.2.1 (615.2.1) Guards. Sauna heaters shall be protected from accidental contact by an *approved* guard or barrier of material having a low coefficient of thermal conductivity. The guard shall not substantially affect the transfer of heat from the heater to the room.

G2440.3 (615.3) Access. Panels, grilles and access doors that are required to be removed for normal servicing operations, shall not be attached to the building.

G2440.4 (615.4) Combustion and dilution air intakes. Sauna heaters of other than the direct-vent type shall be installed with the *draft hood* and *combustion air* intake located outside the sauna room. Where the *combustion air* inlet and the *draft hood* are in a dressing room adjacent to the sauna room, there shall be provisions to prevent physically blocking the *combustion air* inlet and the *draft hood* inlet, and to prevent physical contact with the *draft hood* and vent assembly, or warning notices shall be posted to avoid such contact. Any warning notice shall be easily readable, shall contrast with its background and the wording shall be in letters not less than $^1/_4$ inch (6.4 mm) high.

G2440.5 (615.5) Combustion and ventilation air. *Combustion air* shall not be taken from inside the sauna room. *Combustion* and ventilation air for a sauna heater not of the direct-vent type shall be provided to the area in which the *combustion air* inlet and *draft hood* are located in accordance with Section G2407.

G2440.6 (615.6) Heat and time controls. Sauna heaters shall be equipped with a *thermostat* that will limit room temperature to 194°F (90°C). If the *thermostat* is not an integral part of the sauna heater, the heat-sensing element shall be located within 6 inches (152 mm) of the ceiling. If the heat-sensing element is a capillary tube and bulb, the assembly shall be attached to the wall or other support, and shall be protected against physical damage.

G2440.6.1 (615.6.1) Timers. A timer, if provided to *control main burner* operation, shall have a maximum operating time of 1 hour. The *control* for the timer shall be located outside the sauna room.

G2440.7 (615.7) Sauna room. A ventilation opening into the sauna room shall be provided. The opening shall be not less than 4 inches by 8 inches (102 mm by 203 mm) located near the top of the door into the sauna room.

SECTION G2441 (617) POOL AND SPA HEATERS

G2441.1 (617.1) General. Pool and spa heaters shall be *listed* in accordance with ANSI Z21.56/CSA 4.7 and shall be installed in accordance with the manufacturer's instructions.

SECTION G2442 (618) FORCED-AIR WARM-AIR FURNACES

G2442.1 (618.1) General. Forced-air warm-air *furnaces* shall be *listed* in accordance with ANSI Z21.47/CSA 2.3 or UL 795 and shall be installed in accordance with the manufacturer's instructions.

G2442.2 (618.2) Dampers. Volume dampers shall not be placed in the air inlet to a *furnace* in a manner that will reduce the required air to the *furnace*.

G2442.3 (618.3) Prohibited sources. Outdoor or return air for forced-air heating and cooling systems shall not be taken from the following locations:

1. Closer than 10 feet (3048 mm) from an *appliance* vent outlet, a vent opening from a plumbing drainage system or the discharge outlet of an exhaust fan, unless the outlet is 3 feet (914 mm) above the outside air inlet.
2. Where there is the presence of objectionable odors, fumes or flammable vapors; or where located less than 10 feet (3048 mm) above the surface of any abutting *public way* or driveway; or where located at grade level by a sidewalk, street, alley or driveway.
3. A hazardous or insanitary location or a refrigeration machinery room as defined in the *International Mechanical Code*.
4. A room or space, the volume of which is less than 25 percent of the entire volume served by such system. Where connected by a permanent opening having an area sized in accordance with this code, adjoining rooms or spaces shall be considered to be a single room or space for the purpose of determining the volume of such rooms or spaces.

   Exception: The minimum volume requirement shall not apply where the amount of return air taken from a room or space is less than or equal to the amount of supply air delivered to such room or space.
5. A room or space containing an *appliance* where such a room or space serves as the sole source of return air.

   Exceptions: This shall not apply where:
   1. The *appliance* is a direct-vent *appliance* or an *appliance* not requiring a vent in accordance with Section G2425.8.
   2. The room or space complies with the following requirements:
      - 2.1. The return air shall be taken from a room or space having a volume exceeding 1 cubic foot for each 10 Btu/h (9.6 L/W) of combined input rating of all fuel-burning *appliances* therein.
      - 2.2. The volume of supply air discharged back into the same space shall be approximately equal to the volume of return air taken from the space.
      - 2.3. Return-air inlets shall not be located within 10 feet (3048 mm) of a *draft hood* in the same room or space or the combustion chamber of any atmospheric burner *appliance* in the same room or space.
   3. Rooms or spaces containing solid fuel-burning *appliances*, provided that return-air inlets are located not less than 10 feet (3048 mm) from the firebox of such appliances.
6. A closet, bathroom, toilet room, kitchen, garage, boiler room, furnace room or unconditioned attic.

   Exceptions:
   1. Where return air intakes are located not less than 10 feet (3048 mm) from cooking *appliances* and serve only the kitchen area, taking return air from a kitchen area shall not be prohibited.
   2. Dedicated forced-air systems serving only a garage shall not be prohibited from obtaining return air from the garage.
7. A *crawl space* by means of direct connection to the return side of a forced-air system. Transfer openings in the *crawl space* enclosure shall not be prohibited.

G2442.4 (618.4) Screen. Required outdoor air inlets shall be covered with a screen having $^{1}/_{4}$-inch (6.4 mm) openings.

G2442.5 (618.5) Return-air limitation. Return air from one *dwelling unit* shall not be discharged into another *dwelling unit*.

G2442.6 (618.6) Furnace plenums and air ducts. Where a *furnace* is installed so that supply ducts carry air circulated by the *furnace* to areas outside of the space containing the *furnace*, the return air shall be handled by a duct(s) sealed to the *furnace* casing and terminating outside of the space containing the *furnace*.

SECTION G2443 (619) CONVERSION BURNERS

G2443.1 (619.1) Conversion burners. The installation of *conversion burners* shall conform to ANSI Z21.8.

SECTION G2444 (620) UNIT HEATERS

G2444.1 (620.1) General. *Unit heaters* shall be *listed* in accordance with ANSI Z83.8/CSA 2.6 and shall be installed in accordance with the manufacturer's instructions.

G2444.2 (620.2) Support. Suspended-type *unit heaters* shall be supported by elements that are designed and constructed to accommodate the weight and dynamic loads. Hangers and brackets shall be of *noncombustible material*.

G2444.3 (620.3) Ductwork. Ducts shall not be connected to a unit heater unless the heater is *listed* for such installation.

G2444.4 (620.4) Clearance. Suspended-type *unit heaters* shall be installed with *clearances* to *combustible materials* of not less than 18 inches (457 mm) at the sides, 12 inches (305 mm) at the bottom and 6 inches (152 mm) above the top where the unit heater has an internal *draft hood* or 1 inch (25 mm) above the top of the sloping side of the vertical *draft hood*.

Floor-mounted-type *unit heaters* shall be installed with *clearances* to *combustible materials* at the back and one side only of not less than 6 inches (152 mm). Where the *flue gases* are vented horizontally, the 6-inch (152 mm) *clearance* shall be measured from the *draft hood* or *vent* instead of the rear wall of the unit heater. Floor-mounted-type *unit heaters* shall not be installed on combustible floors unless *listed* for such installation.

Clearances for servicing all *unit heaters* shall be in accordance with the manufacturer's installation instructions.

Exception: *Unit heaters listed* for reduced *clearance* shall be permitted to be installed with such *clearances* in accordance with their listing and the manufacturer's instructions.

SECTION G2445 (621) UNVENTED ROOM HEATERS

G2445.1 (621.1) General. *Unvented room heaters* shall be *listed* in accordance with ANSI Z21.11.2 and shall be installed in accordance with the conditions of the listing and the manufacturer's instructions.

G2445.2 (621.2) Prohibited use. One or more *unvented room heaters* shall not be used as the sole source of comfort heating in a *dwelling unit*.

G2445.3 (621.3) Input rating. *Unvented room heaters* shall not have an input rating in excess of 40,000 *Btu*/h (11.7 kW).

G2445.4 (621.4) Prohibited locations. The location of *unvented room heaters* shall comply with Section G2406.2.

G2445.5 (621.5) Room or space volume. The aggregate input rating of all *unvented appliances* installed in a room or space shall not exceed 20 *Btu*/h per *cubic foot* (207 W/m^3) of volume of such room or space. Where the room or space in which the *appliances* are installed is directly connected to another room or space by a doorway, archway or other opening of comparable size that cannot be closed, the volume of such adjacent room or space shall be permitted to be included in the calculations.

G2445.6 (621.6) Oxygen-depletion safety system. *Unvented room heaters* shall be equipped with an oxygen-depletion-sensitive safety shutoff system. The system shall shut off the gas supply to the main and *pilot burners* when the oxygen in the surrounding atmosphere is depleted to the percent concentration specified by the manufacturer, but not lower than 18 percent. The system shall not incorporate field adjustment means capable of changing the set point at which the system acts to shut off the gas supply to the room heater.

G2445.7 (621.7) Unvented decorative room heaters. An unvented decorative room heater shall not be installed in a *factory-built fireplace* unless the *fireplace* system has been specifically tested, *listed* and *labeled* for such use in accordance with UL 127.

G2445.7.1 (621.7.1) Ventless firebox enclosures. Ventless firebox enclosures used with unvented decorative room heaters shall be *listed* as complying with ANSI Z21.91.

SECTION G2446 (622) VENTED ROOM HEATERS

G2446.1 (622.1) General. *Vented room heaters* shall be *listed* in accordance with ANSI Z21.86/CSA 2.32, shall be designed and equipped as specified in Section G2432.2 and shall be installed in accordance with the manufacturer's instructions.

SECTION G2447 (623) COOKING APPLIANCES

G2447.1 (623.1) Cooking appliances. Cooking *appliances* that are designed for permanent installation, including ranges, ovens, stoves, broilers, grills, fryers, griddles, hot plates and barbecues, shall be *listed* in accordance with ANSI Z21.1 or ANSI Z21.58/CSA 1.6 and shall be installed in accordance with the manufacturer's instructions.

G2447.2 (623.2) Prohibited location. Cooking *appliances* designed, tested, *listed* and *labeled* for use in commercial occupancies shall not be installed within *dwelling units* or within any area where domestic cooking operations occur.

Exception: *Appliances* that are also *listed* as domestic cooking *appliances*.

G2447.3 (623.3) Domestic appliances. Cooking *appliances* installed within *dwelling units* and within areas where domestic cooking operations occur shall be *listed* and *labeled* as household-type *appliances* for domestic use.

G2447.4 (623.4) Range installation. Ranges installed on combustible floors shall be set on their own bases or legs and shall be installed with *clearances* of not less than that shown on the *label.*

G2447.5 (623.7) Vertical clearance above cooking top. Household cooking *appliances* shall have a vertical *clearance* above the cooking top of not less than 30 inches (760 mm) to *combustible material* and metal cabinets. A minimum *clearance* of 24 inches (610 mm) is permitted where one of the following is installed:

1. The underside of the *combustible material* or metal cabinet above the cooking top is protected with not less than $^{1}/_{4}$-inch (6.4 mm) insulating millboard covered with sheet metal not less than 0.0122 inch (0.3 mm) thick.
2. A metal ventilating hood constructed of sheet metal not less than 0.0122 inch (0.3 mm) thick is installed above the cooking top with a *clearance* of not less than $^{1}/_{4}$ inch (6.4 mm) between the hood and the underside of the *combustible material* or metal cabinet. The hood shall have a width not less than the width of the *appliance* and shall be centered over the *appliance.*
3. A *listed* cooking *appliance* or microwave oven is installed over a *listed* cooking *appliance* and in compliance with the terms of the manufacturer's installation instructions for the upper *appliance.*

SECTION G2448 (624) WATER HEATERS

G2448.1 (624.1) General. Water heaters shall be *listed* in accordance with ANSI Z21.10.1/CSA 4.1 or ANSI Z21.10.3/CSA 4.3 and shall be installed in accordance with the manufacturer's instructions.

G2448.1.1 (624.1.1) Installation requirements. The requirements for *water heaters* relative to sizing, *relief valves,* drain pans and scald protection shall be in accordance with this code.

G2448.2 (624.2) Water heaters utilized for space heating. *Water heaters* utilized both to supply potable hot water and provide hot water for space-heating applications shall be *listed* and *labeled* for such applications by the manufacturer and shall be installed in accordance with the manufacturer's instructions and this code.

SECTION G2449 (627) AIR-CONDITIONING APPLIANCES

G2449.1 (627.1) General. Gas-fired air-conditioning *appliances* shall be *listed* in accordance with ANSI Z21.40.1/CSA 2.91 or ANSI Z21.40.2/CSA 2.92 and shall be installed in accordance with the manufacturer's instructions.

G2449.2 (627.2) Independent piping. *Gas piping* serving heating *appliances* shall be permitted to also serve cooling *appliances* where such heating and cooling *appliances* cannot be operated simultaneously (see Section G2413).

G2449.3 (627.3) Connection of gas-engine-powered air conditioners. To protect against the effects of normal vibration in service, gas engines shall not be rigidly connected to the gas supply *piping.*

G2449.4 (627.6) Installation. Air-conditioning *appliances* shall be installed in accordance with the manufacturer's instructions. Unless the *appliance* is *listed* for installation on a combustible surface such as a floor or roof, or unless the surface is protected in an *approved* manner, the *appliance* shall be installed on a surface of noncombustible construction with *noncombustible material* and surface finish, and *combustible material* shall not be against the underside thereof.

SECTION G2450 (628) ILLUMINATING APPLIANCES

G2450.1 (628.1) General. Illuminating *appliances* shall be *listed* in accordance with ANSI Z21.42 and shall be installed in accordance with the manufacturer's instructions.

G2450.2 (628.2) Mounting on buildings. Illuminating *appliances* designed for wall or ceiling mounting shall be securely attached to substantial structures in such a manner that they are not dependent on the *gas piping* for support.

G2450.3 (628.3) Mounting on posts. Illuminating *appliances* designed for post mounting shall be securely and rigidly attached to a post. Posts shall be rigidly mounted. The strength and rigidity of posts greater than 3 feet (914 mm) in height shall be at least equivalent to that of a $2^{1}/_{2}$-inch-diameter (64 mm) post constructed of 0.064-inch-thick (1.6 mm) steel or a 1-inch (25 mm) Schedule 40 steel *pipe.* Posts 3 feet (914 mm) or less in height shall not be smaller than a $^{3}/_{4}$-inch (19.1 mm) Schedule 40 steel *pipe.* Drain openings shall be provided near the base of posts where there is a possibility of water collecting inside them.

G2450.4 (628.4) Appliance pressure regulators. Where an *appliance pressure regulator* is not supplied with an illuminating *appliance* and the service line is not equipped with a *service pressure regulator,* an *appliance pressure regulator* shall be installed in the line to the illuminating *appliance.* For multiple installations, one *regulator* of adequate capacity shall be permitted to serve more than one illuminating *appliance.*

SECTION G2451 (630) INFRARED RADIANT HEATERS

G2451.1 (630.1) General. Infrared radiant heaters shall be *listed* in accordance with ANSI Z83.19 or Z83.20 and shall be installed in accordance with the manufacturer's instructions.

G2451.2 (630.2) Support. *Infrared radiant heaters* shall be fixed in a position independent of gas and electric supply lines. Hangers and brackets shall be of *noncombustible material.*

SECTION G2452 (631) BOILERS

G2452.1 (631.1) Standards. Boilers shall be *listed* in accordance with the requirements of ANSI Z21.13/CSA 4.9 or UL 795. If applicable, the boiler shall be designed and constructed in accordance with the requirements of ASME CSD-1 and as applicable, the *ASME Boiler and Pressure Vessel Code,* Sections I, II, IV, V and IX and NFPA 85.

G2452.2 (631.2) Installation. In addition to the requirements of this code, the installation of boilers shall be in accordance with the manufacturer's instructions. Operating instructions of a permanent type shall be attached to the boiler. Boilers shall have all *controls* set, adjusted and tested by the installer. A complete *control* diagram together with complete boiler operating instructions shall be furnished by the installer. The manufacturer's rating data and the nameplate shall be attached to the boiler.

G2452.3 (631.3) Clearance to combustible material. *Clearances* to *combustible materials* shall be in accordance with Section G2409.4.

SECTION G2453 (635) OUTDOOR DECORATIVE APPLIANCES

G2453.1 (635.1) General. Permanently fixed-in-place outdoor decorative *appliances* shall be *listed* in accordance with ANSI Z21.97 and shall be installed in accordance with the manufacturer's instructions.

Part VII—Plumbing

CHAPTER 25
PLUMBING ADMINISTRATION

User notes:

About this chapter: *Chapter 25 covers regulations for existing plumbing installations and testing of new or repaired systems. These general requirements can be superseded by more specific requirements for certain applications in Chapters 26 through 33.*

Code development reminder: *Code change proposals to this chapter will be considered by the IRC—Mechanical/Plumbing Code Development Committee during the 2021 (Group A) Code Development Cycle.*

SECTION P2501
GENERAL

P2501.1 Scope. The provisions of this chapter shall establish the general administrative requirements applicable to plumbing systems and inspection requirements of this code.

P2501.2 Application. In addition to the general administration requirements of Chapter 1, the administrative provisions of this chapter shall apply to the plumbing requirements of Chapters 25 through 33.

SECTION P2502
EXISTING PLUMBING SYSTEMS

P2502.1 Existing building sewers and building drains. Where the entire sanitary drainage system of an existing building is replaced, existing *building drains* under concrete slabs and existing *building sewers* that will serve the new system shall be internally examined to verify that the piping is sloping in the correct direction, is not broken, is not obstructed and is sized for the drainage load of the new plumbing drainage system to be installed.

P2502.2 Additions, alterations or repairs. Additions, *alterations*, renovations or repairs to any plumbing system shall conform to that required for a new plumbing system without requiring the existing plumbing system to comply with the requirements of this code. Additions, *alterations* or repairs shall not cause an existing system to become unsafe, insanitary or overloaded.

Minor additions, *alterations*, renovations and repairs to existing plumbing systems shall be permitted in the same manner and arrangement as in the existing system, provided that such repairs or replacement are not hazardous and are *approved*.

SECTION P2503
INSPECTION AND TESTS

P2503.1 Inspection required. New plumbing work and parts of existing systems affected by new work or *alterations* shall be inspected by the *building official* to ensure compliance with the requirements of this code.

P2503.2 Concealment. A plumbing or drainage system, or part thereof, shall not be covered, concealed or put into use until it has been tested, inspected and *approved* by the *building official*.

P2503.3 Responsibility of permittee. Test equipment, materials and labor shall be furnished by the permittee.

P2503.4 Building sewer testing. The *building sewer* shall be tested by insertion of a test plug at the point of connection with the public sewer, filling the *building sewer* with water and pressurizing the sewer to not less than a 10-foot (3048 mm) head of water. The test pressure shall not decrease during a period of not less than 15 minutes. The *building sewer* shall be watertight at all points.

A forced sewer test shall consist of pressurizing the piping to a pressure of not less than 5 psi (34.5 kPa) greater than the pump rating and maintaining such pressure for not less than 15 minutes. The forced sewer shall be watertight at all points.

P2503.5 Drain, waste and vent systems testing. Rough-in and finished plumbing installations of drain, waste and vent systems shall be tested in accordance with Sections P2503.5.1 and P2503.5.2.

P2503.5.1 Rough plumbing. DWV systems shall be tested on completion of the rough piping installation by water, by air for piping systems other than plastic, or by a vacuum of air for plastic piping systems, without evidence of leakage. The test shall be applied to the drainage system in its entirety or in sections after rough-in piping has been installed, as follows:

1. Water test. Each section shall be filled with water to a point not less than 10 feet (3048 mm) above the highest fitting connection in that section, or to the highest point in the completed system. Water shall be held in the section under test for a period of 15 minutes. The system shall prove leak free by visual inspection.

2. Air test. The portion under test shall be maintained at a gauge pressure of 5 pounds per square inch (psi) (34 kPa) or 10 inches of mercury column (34 kPa). This pressure shall be held without introduction of additional air for a period of 15 minutes.
3. Vacuum test. The portion under test shall be evacuated of air by a vacuum-type pump to achieve a uniform gauge pressure of -5 pounds per square inch or a negative 10 inches of mercury column (-34 kPa). This pressure shall be held without the removal of additional air for a period of 15 minutes.

P2503.5.2 Finished plumbing. After the plumbing fixtures have been set and their traps filled with water, their connections shall be tested and proved gastight or watertight as follows:

1. Watertightness. Each fixture shall be filled and then drained. Traps and fixture connections shall be proven watertight by visual inspection.
2. Gastightness. Where required by the local administrative authority, a final test for gastightness of the DWV system shall be made by the smoke or peppermint test as follows:
    2.1. Smoke test. Introduce a pungent, thick smoke into the system. When the smoke appears at vent terminals, such terminals shall be sealed and a pressure equivalent to a 1-inch water column (249 Pa) shall be applied and maintained for a test period of not less than 15 minutes.
    2.2. Peppermint test. Introduce 2 ounces (59 mL) of oil of peppermint into the system. Add 10 quarts (9464 mL) of hot water and seal the vent terminals. The odor of peppermint shall not be detected at any trap or other point in the system.

P2503.6 Shower liner test. Where shower floors and receptors are made watertight by the application of materials required by Section P2709.2, the completed liner installation shall be tested. The pipe from the shower drain shall be plugged watertight for the test. The floor and receptor area shall be filled with potable water to a depth of not less than 2 inches (51 mm) measured at the threshold. Where a threshold of not less than 2 inches (51 mm) in height does not exist, a temporary threshold shall be constructed to retain the test water in the lined floor or receptor area to a level not less than 2 inches (51 mm) in depth measured at the threshold. The water shall be retained for a test period of not less than 15 minutes and there shall not be evidence of leakage.

P2503.7 Water-supply system testing. Upon completion of the water-supply system or a section of it, the system or portion completed shall be tested and proved tight under a water pressure of not less than the working pressure of the system or, for piping systems other than plastic, by an air test of not less than 50 psi (345 kPa). This pressure shall be held for not less than 15 minutes. The water used for tests shall be obtained from a potable water source.

Exception: For PEX piping systems, testing with a compressed gas shall be an alternative to hydrostatic testing where compressed air or other gas pressure testing is specifically authorized by the manufacturer's instructions for the PEX pipe and fittings products installed at the time the system is being tested, and compressed air or other gas testing is not otherwise prohibited by applicable codes, laws or regulations outside of this code.

P2503.8 Inspection and testing of backflow prevention devices. Inspection and testing of backflow prevention devices shall comply with Sections P2503.8.1 and P2503.8.2.

P2503.8.1 Inspections. Inspections shall be made of backflow prevention assemblies to determine whether they are operable.

P2503.8.2 Testing. Reduced pressure principle, double check, double check detector and pressure vacuum breaker backflow preventer assemblies shall be tested at the time of installation, immediately after repairs or relocation and every year thereafter.

P2503.9 Test gauges. Gauges used for testing shall be as follows:

1. Tests requiring a pressure of 10 psi or less shall utilize a testing gauge having increments of 0.10 psi (0.69 kPa) or less.
2. Tests requiring a pressure higher than 10 psi (0.69 kPa) but less than or equal to 100 psi (690 kPa) shall use a testing gauge having increments of 1 psi (6.9 kPa) or less.
3. Tests requiring a pressure higher than 100 psi (690 kPa) shall use a testing gauge having increments of 2 psi (14 kPa) or less.

CHAPTER 26
GENERAL PLUMBING REQUIREMENTS

User notes:

About this chapter: *Chapter 26 contains general requirements that could apply to Chapters 26 through 33. Placing such requirements in only one location eliminates code development coordination issues associated with the same requirement in multiple locations. This chapter covers liquid waste disposal requirements, special installation provisions for flood hazard areas and requirements for third-party certification of materials and products that are required to comply with a referenced standard.*

Code development reminder: *Code change proposals to this chapter will be considered by the IRC—Mechanical/Plumbing Code Development Committee during the 2021 (Group A) Code Development Cycle.*

SECTION P2601 GENERAL

P2601.1 Scope. The provisions of this chapter shall govern the installation of plumbing not specifically covered in other chapters applicable to plumbing systems. The installation of plumbing, *appliances, equipment* and systems not addressed by this code shall comply with the applicable provisions of the *International Plumbing Code.*

P2601.2 Connections to drainage system. Plumbing fixtures, drains, appurtenances and *appliances* used to receive or discharge liquid wastes or sewage shall be directly connected to the sanitary drainage system of the building or premises, in accordance with the requirements of this code. This section shall not be construed to prevent indirect waste connections where required by the code.

> **Exception:** Bathtubs, showers, lavatories, clothes washers and laundry trays shall not be required to discharge to the sanitary drainage system where such fixtures discharge to systems complying with Sections P2910 and P2911.

P2601.3 Flood hazard areas. In flood hazard areas as established by Table R301.2, plumbing fixtures, drains, and *appliances* shall be located or installed in accordance with Section R322.1.6.

SECTION P2602 INDIVIDUAL WATER SUPPLY AND SEWAGE DISPOSAL

P2602.1 General. The water-distribution system of any building or premises where plumbing fixtures are installed shall be connected to a public water supply. Where a public water-supply system is not available, or connection to the supply is not feasible, an individual water supply shall be provided. Individual water supplies shall be constructed and installed in accordance with the applicable state and local laws. Where such laws do not address the requirements set forth in NGWA-01, individual water supplies shall comply with NGWA-01 for those requirements not addressed by state and local laws.

Sanitary drainage piping from plumbing fixtures in buildings and sanitary drainage piping systems from premises shall be connected to a public sewer. Where a public sewer is not available, the sanitary drainage piping and systems shall be connected to a private sewage disposal system in compliance with state or local requirements. Where state or local requirements do not exist for private sewage disposal systems, the sanitary drainage piping and systems shall be connected to an *approved* private sewage disposal system that is in accordance with the *International Private Sewage Disposal Code.*

> **Exception:** Sanitary drainage piping and systems that convey only the discharge from bathtubs, showers, lavatories, clothes washers and laundry trays shall not be required to connect to a public sewer or to a private sewage disposal system provided that the piping or systems are connected to a system in accordance with Section P2910 or P2911.

P2602.2 Flood-resistant installation. In flood hazard areas as established by Table R301.2:

1. Water supply systems shall be designed and constructed to prevent infiltration of floodwaters.
2. Pipes for sewage disposal systems shall be designed and constructed to prevent infiltration of floodwaters into the systems and discharges from the systems into floodwaters.

SECTION P2603 STRUCTURAL AND PIPING PROTECTION

P2603.1 General. In the process of installing or repairing any part of a plumbing and drainage installation, the finished floors, walls, ceilings, tile work or any other part of the building or premises that must be changed or replaced shall be left in a safe structural condition in accordance with the requirements of the building portion of this code.

P2603.2 Drilling and notching. Wood-framed structural members shall not be drilled, notched or altered in any manner except as provided in Sections R502.8, R602.6, R802.7 and R802.7.1. Holes in load-bearing members of cold-formed steel *light-frame construction* shall be made only in accordance with Sections R505.2.6, R603.2.6 and R804.2.6. In accordance with the provisions in Sections R505.3.5, R603.3.3 and R804.3.4, cutting and notching of flanges and lips of load-bearing members of cold-formed steel *light-frame construction* shall be prohibited. Structural insulated panels (SIPs) shall be drilled and notched or altered in accordance with the provisions of Section R610.7.

P2603.2.1 Protection against physical damage. In concealed locations, where piping, other than cast-iron or galvanized steel, is installed through holes or notches in studs, joists, rafters or similar members less than $1^1/_4$ inches (31.8 mm) from the nearest edge of the member, the pipe shall be protected by steel shield plates. Such shield plates shall have a thickness of not less than 0.0575 inch (1.463 mm) (No. 16 Gage). Such plates shall cover the area of the pipe where the member is notched or bored, and shall extend not less than 2 inches (51 mm) above sole plates and below top plates.

P2603.3 Protection against corrosion. Metallic piping, except for cast iron, ductile iron and galvanized steel, shall not be placed in direct contact with steel framing members, concrete or masonry. Metallic piping shall not be placed in direct contact with corrosive soil. Where sheathing is used to prevent direct contact, the sheathing material thickness shall be not less than 0.008 inch (8 mil) (0.203 mm) and shall be made of plastic. Where sheathing protects piping that penetrates concrete or masonry walls or floors, the sheathing shall be installed in a manner that allows movement of the piping within the sheathing.

P2603.4 Pipes through foundation walls. A pipe that passes through a foundation wall shall be provided with a relieving arch, or a pipe sleeve shall be built into the foundation wall. The sleeve shall be two pipe sizes greater than the pipe passing through the wall.

P2603.5 Freezing. In localities having a winter design temperature of 32°F (0°C) or lower as shown in Table R301.2 of this code, a water, soil or waste pipe shall not be installed outside of a building, in exterior walls, in *attics* or crawl spaces, or in any other place subjected to freezing temperature unless adequate provision is made to protect it from freezing by insulation or heat or both. Water service pipe shall be installed not less than 12 inches (305 mm) deep and not less than 6 inches (152 mm) below the frost line.

P2603.5.1 Sewer depth. *Building sewers* that connect to private sewage disposal systems shall be not less than [NUMBER] inches (mm) below finished grade at the point of septic tank connection. *Building sewers* shall be not less than [NUMBER] inches (mm) below grade.

SECTION P2604 TRENCHING AND BACKFILLING

P2604.1 Trenching and bedding. Where trenches are excavated such that the bottom of the trench forms the bed for the pipe, solid and continuous load-bearing support shall be provided between joints. Where over-excavated, the trench shall be backfilled to the proper grade with compacted earth, sand, fine gravel or similar granular material. Piping shall not be supported on rocks or blocks at any point. Rocky or unstable soil shall be over-excavated by two or more pipe diameters and brought to the proper grade with suitable compacted granular material.

P2604.2 Water service and building sewer in same trench. Where the water service piping and *building sewer* piping is installed in same trench, the installation shall be in accordance with Section P2906.4.1.

P2604.3 Backfilling. Backfill shall be free from discarded construction material and debris. Backfill shall be free from rocks, broken concrete and frozen chunks until the pipe is covered by not less than 12 inches (305 mm) of tamped earth. Backfill shall be placed evenly on both sides of the pipe and tamped to retain proper alignment. Loose earth shall be carefully placed in the trench in 6-inch (152 mm) layers and tamped in place.

P2604.4 Protection of footings. Trenching installed parallel to footings and walls shall not extend into the bearing plane of a footing or wall. The upper boundary of the bearing plane is a line that extends downward, at an angle of 45 degrees (0.79 rad) from horizontal, from the outside bottom edge of the footing or wall.

SECTION P2605 SUPPORT

P2605.1 General. Piping shall be supported in accordance with the following:

1. Piping shall be supported to ensure alignment and prevent sagging, and allow movement associated with the expansion and contraction of the piping system.
2. Piping in the ground shall be laid on a firm bed for its entire length, except where support is otherwise provided.
3. Hangers and *anchors* shall be of sufficient strength to maintain their proportional share of the weight of pipe and contents and of sufficient width to prevent distortion to the pipe. Hangers and strapping shall be of *approved* material that will not promote galvanic action.
4. Where horizontal pipes 4 inches (102 mm) and larger convey drainage or waste, and where a pipe fitting changes the flow direction greater than 45 degrees (0.79 rad), rigid bracing or other rigid support arrangements shall be installed to resist movement of the upstream pipe in the direction of flow. A change of flow direction into a vertical pipe shall not require the upstream pipe to be braced.
5. Piping shall be supported at distances not to exceed those indicated in Table P2605.1.

SECTION P2606 PENETRATIONS

P2606.1 Sealing of annular spaces. The annular space between the outside of a pipe and the inside of a pipe sleeve or between the outside of a pipe and an opening in a building envelope wall, floor, or ceiling assembly penetrated by a pipe shall be sealed with caulking material or foam sealant or closed with a gasketing system. The caulking material, foam sealant or gasketing system shall be designed for the conditions at the penetration location and shall be compatible with the pipe, sleeve and building materials in contact with the sealing materials. Annular spaces created by pipes penetrating fire-resistance-rated assemblies or membranes of such assemblies shall be sealed or closed in accordance with the building portion of this code.

SECTION P2607 WATERPROOFING OF OPENINGS

P2607.1 Pipes penetrating roofs. Where a pipe penetrates a roof, a flashing of lead, copper, galvanized steel or an *approved* elastomeric material shall be installed in a manner that prevents water entry into the building. Counterflashing into the opening of pipe serving as a vent terminal shall not reduce the required internal cross-sectional area of the vent pipe to less than the internal cross-sectional area of one pipe size smaller.

P2607.2 Pipes penetrating exterior walls. Where a pipe penetrates an exterior wall, a waterproof seal shall be made on the exterior of the wall by one of the following methods:

1. A waterproof sealant applied at the joint between the wall and the pipe.
2. A flashing of an *approved* elastomeric material.

SECTION P2608 WORKMANSHIP

P2608.1 General. Valves, pipes and fittings shall be installed in correct relationship to the direction of the flow. Burred ends shall be reamed to the full bore of the pipe.

SECTION P2609 MATERIALS EVALUATION AND LISTING

P2609.1 Identification. Each length of pipe and each pipe fitting, trap, fixture, material and device utilized in a plumbing system shall bear the identification of the manufacturer and any markings required by the applicable referenced standards. Nipples created from the cutting and threading of *approved* pipe shall not be required to be identified.

> **Exception:** Where the manufacturer identification cannot be marked on pipe fittings and pipe nipples because of the small size of such fittings, the identification shall be printed on the item packaging or on documentation provided with the item.

P2609.2 Installation of materials. Materials used shall be installed in strict accordance with the standards under which the materials are accepted and *approved.* In the absence of such installation procedures, the manufacturer's instructions

TABLE P2605.1 PIPING SUPPORT

PIPING MATERIAL	MAXIMUM HORIZONTAL SPACING (feet)	MAXIMUM VERTICAL SPACING (feet)
ABS pipe	4	10[b]
Aluminum tubing	10	15
Cast-iron pipe	5[a]	15
Copper or copper-alloy pipe	12	10
Copper or copper-alloy tubing ($1^1/_4$ inches in diameter and smaller)	6	10
Copper or copper-alloy tubing ($1^1/_2$ inches in diameter and larger)	10	10
Cross-linked polyethylene (PEX) pipe, 1 inch and smaller	2.67 (32 inches)	10[b]
Cross-linked polyethylene (PEX) pipe, $1^1/_4$ inches and larger	4	10[b]
Cross-linked polyethylene/aluminum/cross-linked polyethylene (PEX-AL-PEX) pipe	2.67 (32 inches)	4[b]
CPVC pipe or tubing (1 inch in diameter and smaller)	3	10[b]
CPVC pipe or tubing ($1^1/_4$ inches in diameter and larger)	4	10[b]
Lead pipe	Continuous	4
PB pipe or tubing	2.67 (32 inches)	4
Polyethylene of raised temperature (PE-RT) pipe, 1 inch and smaller	2.67 (32 inches)	10[b]
Polyethylene of raised temperature (PE-RT) pipe, $1^1/_4$ inches and larger	4	10[b]
Polypropylene (PP) pipe or tubing (1 inch and smaller)	2.67 (32 inches)	10[b]
Polypropylene (PP) pipe or tubing ($1^1/_4$ inches and larger)	4	10[b]
PVC pipe	4	10[b]
Stainless steel drainage systems	10	10[b]
Steel pipe	12	15

For SI: 1 inch = 25.4 mm, 1 foot = 304.8 mm.

a. The maximum horizontal spacing of cast-iron pipe hangers shall be increased to 10 feet where 10-foot lengths of pipe are installed.

b. For sizes 2 inches and smaller, a guide shall be installed midway between required vertical supports. Such guides shall prevent pipe movement in a direction perpendicular to the axis of the pipe.

shall be followed. Where the requirements of referenced standards or manufacturer's instructions do not conform to the minimum provisions of this code, the provisions of this code shall apply.

P2609.3 Plastic pipe, fittings and components. Plastic pipe, fittings and components shall be third-party certified as conforming to NSF 14.

P2609.4 Third-party certification. Plumbing products and materials required by the code to be in compliance with a referenced standard shall be *listed* by a third-party certification agency as complying with the referenced standards. Products and materials shall be identified in accordance with Section P2609.1.

P2609.5 Water supply systems. Water service pipes, water distribution pipes and the necessary connecting pipes, fittings, control valves, faucets and appurtenances used to dispense water intended for human ingestion shall be evaluated and *listed* as conforming to the requirements of NSF 61.

CHAPTER 27
PLUMBING FIXTURES

User notes:

About this code: *Because fixture design and quality are paramount to ensure that plumbing fixtures operate properly, Chapter 27 specifies numerous product and material standards for plumbing fixtures. Regulations for locating plumbing fixtures and constructing field-built shower receptors are provided.*

Code development reminder: *Code change proposals to this chapter will be considered by the IRC—Mechanical/Plumbing Code Development Committee during the 2021 (Group A) Code Development Cycle.*

SECTION P2701
FIXTURES, FAUCETS AND FIXTURE FITTINGS

P2701.1 Quality of fixtures. Plumbing fixtures, faucets and fixture fittings shall have smooth impervious surfaces, shall be free from defects, shall not have concealed fouling surfaces, and shall conform to the standards indicated in Table P2701.1 and elsewhere in this code.

SECTION P2702
FIXTURE ACCESSORIES

P2702.1 Plumbing fixtures. Plumbing fixtures, other than water closets, shall be provided with *approved* strainers.

Exception: Hub drains receiving only clear water waste and standpipes shall not require strainers.

P2702.2 Waste fittings. Waste fittings shall conform to ASME A112.18.2/CSA B125.2, ASTM F409 or shall be made from pipe and pipe fittings complying with any of the standards indicated in Tables P3002.1(1) and P3002.3.

P2702.3 Plastic tubular fittings. Plastic tubular fittings shall conform to ASTM F409 as indicated in Table P2701.1.

P2702.4 Carriers for wall-hung water closets. Carriers for wall-hung water closets shall conform to ASME A112.6.2.

SECTION P2703
TAIL PIECES

P2703.1 Minimum size. Fixture tail pieces shall be not less than $1^1/_2$ inches (38 mm) in diameter for sinks, dishwashers, laundry tubs, bathtubs and similar fixtures, and not less than $1^1/_4$ inches (32 mm) in diameter for bidets, lavatories and similar fixtures.

SECTION P2704
SLIP-JOINT CONNECTIONS

P2704.1 Slip joints. Slip-joint connections shall be installed only for tubular waste piping and only between the trap outlet of a fixture and the connection to the drainage piping. Slip-joint connections shall be made with an *approved* elastomeric sealing gasket. Slip-joint connections shall be accessible. Such access shall provide an opening that is not less than 12 inches (305 mm) in its smallest dimension.

SECTION P2705
INSTALLATION

P2705.1 General. The installation of fixtures shall conform to the following:

1. Floor-outlet or floor-mounted fixtures shall be secured to the drainage connection and to the floor, where so designed, by screws, bolts, washers, nuts and similar fasteners of copper, copper alloy or other corrosion-resistant material.
2. Wall-hung fixtures shall be rigidly supported so that strain is not transmitted to the plumbing system.
3. Where fixtures come in contact with walls and floors, the contact area shall be watertight.
4. Plumbing fixtures shall be usable.
5. Water closets, lavatories and bidets. A water closet, lavatory or bidet shall not be set closer than 15 inches (381 mm) from its center to any side wall, partition or vanity or closer than 30 inches (762 mm) center-to-center between adjacent fixtures. There shall be a clearance of not less than 21 inches (533 mm) in front of a water closet, lavatory or bidet to any wall, fixture or door.
6. The location of piping, fixtures or equipment shall not interfere with the operation of windows or doors.
7. In flood hazard areas as established by Table R301.2, plumbing fixtures shall be located or installed in accordance with Section R322.1.6.
8. Integral fixture-fitting mounting surfaces on manufactured plumbing fixtures or plumbing fixtures constructed on site, shall meet the design requirements of ASME A112.19.2/CSA B45.1 or ASME A112.19.3/CSA B45.4.

TABLE P2701.1
PLUMBING FIXTURES, FAUCETS AND FIXTURE FITTINGS

MATERIAL	STANDARD
Air gap fittings for use with plumbing fixtures, appliances and appurtenances	ASME A112.1.3
Bathtub/whirlpool pressure-sealed doors	ASME A112.19.15
Diverters for faucets with hose spray, anti-syphon type, residential application	ASME A112.18.1/CSA B125.1
Enameled cast-iron plumbing fixtures	ASME A112.19.1/CSA B45.2
Floor drains	ASME A112.6.3
Framing-affixed supports for off-the-floor water closets with concealed tanks	ASME A112.6.2
Hose connection vacuum breaker	ASSE 1052
Hot water dispensers, household storage type, electrical	ASSE 1023
Household disposers	ASSE 1008
Hydraulic performance for water closets and urinals	ASME A112.19.2/CSA B45.1
Individual automatic compensating valves for individual fixture fittings	ASME A112.18.1/CSA B125.1
Individual shower control valves anti-scald	ASSE 1016/ASME A112.1016/CSA B125.16
Macerating toilet systems and related components	ASME A112.3.4/CSA B45.9
Nonvitreous ceramic plumbing fixtures	ASME A112.19.2/CSA B45.1
Plastic bathtub units	CSA B45.5/IAPMO Z124; ASME A112.19.2/ CSA B45.1
Plastic lavatories	CSA B45.5/IAPMO Z124
Plastic shower receptors and shower stalls	CSA B45.5/IAPMO Z124
Plastic sinks	CSA B45.5/IAPMO Z124
Plastic water closet bowls and tanks	CSA B45.5/IAPMO Z124
Plumbing fixture fittings	ASME A112.18.1/CSA B125.1
Plumbing fixture waste fittings	ASME A112.18.2/CSA B125.2; ASTM F409
Porcelain-enameled formed steel plumbing fixtures	ASME A112.19.1/CSA B45.2
Pressurized flushing devices for plumbing fixtures	ASSE 1016/ASME 112.1016/CSA B125.16; CSA B125.3
Specification for copper sheet and strip for building construction	ASTM B370
Stainless steel plumbing fixtures	ASME A112.19.3/CSA B45.4
Suction fittings for use in whirlpool bathtub appliances	ASME A112.19.7/CSA B45.10
Temperature-actuated, flow reduction valves to individual fixture fittings	ASSE 1062
Thermoplastic accessible and replaceable plastic tube and tubular fittings	ASTM F409
Trench drains	ASME A112.6.3
Trim for water closet bowls, tanks and urinals	ASME A112.19.5/CSA B45.15
Vacuum breaker wall hydrant-frost-resistant, automatic-draining type	ASSE 1019
Vitreous china plumbing fixtures	ASME A112.19.2/CSA B45.1
Wall-mounted and pedestal-mounted, adjustable and pivoting lavatory and sink carrier systems	ASME A112.19.12
Water closet flush tank fill valves	ASSE 1002/ASME A112.1002/CSA B125.12; CSA B125.3
Whirlpool bathtub appliances	ASME A112.19.7/CSA B45.10

SECTION P2706 WASTE RECEPTORS

P2706.1 General. For other than hub drains that receive only clear-water waste and standpipes, a removable strainer or basket shall cover the waste outlet of waste receptors. Waste receptors shall not be installed in concealed spaces. Waste receptors shall not be installed in plenums, attics, crawl spaces or interstitial spaces above ceilings and below floors. Waste receptors shall be *readily accessible*.

P2706.1.1 Hub drains. Hub drains shall be in the form of a hub or a pipe that extends not less than 1 inch (25.4 mm) above a water-impervious floor.

P2706.1.2 Standpipes. Standpipes shall extend not less than 18 inches (457 mm) and not greater than 42 inches (1067 mm) above the trap weir.

P2706.1.2.1 Laundry tray connection to standpipe. Where a laundry tray waste line connects into a standpipe for an automatic clothes washer drain, the standpipe shall extend not less than 30 inches (762 mm) above the standpipe trap weir and shall extend above the flood level rim of the laundry tray. The outlet of the laundry tray shall not be greater than 30 inches (762 mm) horizontally from the standpipe trap.

P2706.2 Prohibited waste receptors. Plumbing fixtures that are used for washing or bathing shall not be used to receive the discharge of indirect waste piping.

Exceptions:

1. A kitchen sink trap is acceptable for use as a receptor for a dishwasher.
2. A laundry tray is acceptable for use as a receptor for a clothes washing machine.

SECTION P2707 DIRECTIONAL FITTINGS

P2707.1 Directional fitting required. *Approved* directional-type branch fittings shall be installed in fixture tailpieces receiving the discharge from food-waste disposer units or dishwashers.

SECTION P2708 SHOWERS

P2708.1 General. Shower compartments shall have not less than 900 square inches (0.6 m^2) of interior cross-sectional area. Shower compartments shall be not less than 30 inches (762 mm) in minimum dimension measured from the finished interior dimension of the shower compartment, exclusive of fixture valves, shower heads, soap dishes, and safety grab bars or rails. The minimum required area and dimension shall be measured from the finished interior dimension at a height equal to the top of the threshold and at a point tangent to its centerline and shall be continued to a height of not less than 70 inches (1778 mm) above the shower drain outlet. Hinged shower doors shall open outward. The wall area above built-in tubs having installed shower heads and in shower compartments shall be constructed in accordance with Section R702.4. Such walls shall form a watertight joint with each other and with either the tub, receptor or shower floor.

Exceptions:

1. Fold-down seats shall be permitted in the shower, provided that the required 900-square-inch (0.6 m^2) dimension is maintained when the seat is in the folded-up position.
2. Shower compartments having not less than 25 inches (635 mm) in minimum dimension measured from the finished interior dimension of the compartment provided that the shower compartment has a cross-sectional area of not less than 1,300 square inches (0.838 m^2).

P2708.1.1 Access. The shower compartment access and egress opening shall have a clear and unobstructed finished width of not less than 22 inches (559 mm).

P2708.2 Shower drain. Shower drains shall have an outlet size of not less than $1^1/_2$ inches (38 mm) in diameter.

P2708.2.1 Waste fittings. Waste fittings shall conform to ASME A112.18.2/CSA B125.2.

P2708.3 Water supply riser. Water supply risers from the shower valve to the shower head outlet, whether exposed or concealed, shall be attached to the structure using support devices designed for use with the specific piping material or fittings anchored with screws.

P2708.4 Shower control valves. Individual shower and tub/shower combination valves shall be balanced-pressure, thermostatic or combination balanced-pressure/thermostatic valves that conform to the requirements of ASSE 1016/ASME 112.1016/CSA B125.16 or ASME A112.18.1/CSA B125.1. Shower control valves shall be rated for the flow rate of the installed shower head. Such valves shall be installed at the point of use. Shower and tub/shower combination valves required by this section shall be equipped with a means to limit the maximum setting of the valve to 120°F (49°C), which shall be field adjusted in accordance with the manufacturer's instructions to provide water at a temperature not to exceed 120°F (49°C). In-line thermostatic valves shall not be utilized for compliance with this section.

P2708.5 Hand showers. Hand-held showers shall conform to ASME A112.18.1/CSA B125.1. Hand-held showers shall provide backflow protection in accordance with ASME A112.18.1/CSA B125.1 or shall be protected against backflow by a device complying with ASME A112.18.3.

SECTION P2709 SHOWER RECEPTORS

P2709.1 Construction. Where a shower receptor has a finished curb threshold, it shall be not less than 1 inch (25.4 mm) below the sides and back of the receptor. The curb shall be not less than 2 inches (51 mm) and not more than 9 inches (229 mm) deep when measured from the top of the curb to the top of the drain. The finished floor shall slope uniformly toward the drain not less than $^1/_4$ unit vertical in 12 units horizontal (2-percent slope) nor more than $^1/_2$ unit vertical per 12 units horizontal (4-percent slope) and floor drains shall be flanged to provide a watertight joint in the floor.

P2709.2 Lining required. The adjoining walls and floor framing enclosing on-site built-up shower receptors shall be lined with one of the following materials:

1. Sheet lead.
2. Sheet copper.
3. Plastic liner material that complies with ASTM D4068 or ASTM D4551.
4. Hot mopping in accordance with Section P2709.2.3.
5. Sheet-applied load-bearing, bonded waterproof membranes that comply with ANSI A118.10.

The lining material shall extend not less than 2 inches (51 mm) beyond or around the rough jambs and not less than 2 inches (51 mm) above finished thresholds. Sheet-applied load bearing, bonded waterproof membranes shall be applied in accordance with the manufacturer's instructions.

P2709.2.1 PVC sheets. Plasticized polyvinyl chloride (PVC) sheet shall meet the requirements of ASTM D4551. Sheets shall be joined by solvent welding in accordance with the manufacturer's instructions.

P2709.2.2 Chlorinated polyethylene (CPE) sheets. Nonplasticized chlorinated polyethylene sheet shall meet the requirements of ASTM D4068. The liner shall be joined in accordance with the manufacturer's instructions.

P2709.2.3 Hot-mopping. Shower receptors lined by hot mopping shall be built-up with not less than three layers of standard grade Type 15 asphalt-impregnated roofing felt. The bottom layer shall be fitted to the formed subbase and each succeeding layer thoroughly hot-mopped to that below. Corners shall be carefully fitted and shall be made strong and watertight by folding or lapping, and each corner shall be reinforced with suitable webbing hot-mopped in place. Folds, laps and reinforcing webbing shall extend not less than 4 inches (102 mm) in all directions from the corner and webbing shall be of *approved* type and mesh, producing a tensile strength of not less than 50 pounds per inch (893 kg/m) in either direction.

P2709.2.4 Liquid-type, trowel-applied, load-bearing, bonded waterproof materials. Liquid-type, trowel-applied, load-bearing, bonded waterproof materials shall meet the requirements of ANSI A118.10 and shall be applied in accordance with the manufacturer's instructions.

P2709.3 Installation. Lining materials shall be sloped $^1/_4$ unit vertical in 12 units horizontal (2-percent slope) to weep holes in the subdrain by means of a smooth, solidly formed subbase, shall be properly recessed and fastened to *approved* backing so as not to occupy the space required for the wall covering, and shall not be nailed or perforated at any point less than 1 inch (25.4 mm) above the finished threshold.

P2709.3.1 Materials. Lead and copper linings shall be insulated from conducting substances other than the connecting drain by 15-pound (6.80 kg) asphalt felt or its equivalent. Sheet lead liners shall weigh not less than 4 pounds per square foot (19.5 kg/m^2). Sheet copper liners shall weigh not less than 12 ounces per square foot (3.7 kg/m^2). Joints in lead and copper pans or liners shall be burned or silver brazed, respectively. Joints in plastic liner materials shall be joined in accordance with the manufacturer's instructions.

P2709.4 Receptor drains. An *approved* flanged drain shall be installed with shower subpans or linings. The flange shall be placed flush with the subbase and be equipped with a clamping ring or other device to make a watertight connection between the lining and the drain. The flange shall have weep holes into the drain.

P2709.4.1 Waste fittings. Flanged drains shall conform to ASME A112.18.2/CSA B125.2.

SECTION P2710
SHOWER WALLS

P2710.1 Bathtub and shower spaces. Walls in shower compartments and walls above bathtubs that have a wall-mounted showerhead shall be finished in accordance with Section R307.2.

SECTION P2711
LAVATORIES

P2711.1 Approval. Lavatories shall conform to ASME A112.19.1/CSA B45.2, ASME A112.19.2/CSA B45.1, ASME A112.19.3/CSA B45.4 or CSA B45.5/IAPMO Z124.

P2711.2 Cultured marble lavatories. Cultured marble vanity tops with an integral lavatory shall conform to CSA B45.5/IAPMO Z124.

P2711.3 Lavatory waste outlets. Lavatories shall have waste outlets not less than $1^1/_4$ inch (32 mm) in diameter. A strainer, pop-up stopper, crossbar or other device shall be provided to restrict the clear opening of the waste outlet.

P2711.4 Movable lavatory systems. Movable lavatory systems shall comply with ASME A112.19.12.

SECTION P2712
WATER CLOSETS

P2712.1 Approval. Water closets shall conform to the water consumption requirements of Section P2903.2 and shall conform to ASME A112.19.2/CSA B45.1, ASME A112.19.3/CSA B45.4 or CSA B45.5/IAPMO Z124. Water closets shall conform to the hydraulic performance requirements of ASME A112.19.2/CSA B45.1. Water closet tanks shall conform to ASME A112.19.2/CSA B45.1, ASME A112.19.3/CSA B45.4 or CSA B45.5/IAPMO Z124. Water closets that have an invisible seal and unventilated space or walls that are not thoroughly washed at each discharge shall be prohibited. Water closets that allow backflow of the contents of the bowl into the flush tank shall be prohibited. Water closets equipped with a dual flushing device shall comply with ASME A112.19.14.

P2712.2 Flushing devices required. Water closets shall be provided with a flush tank, flushometer tank or flushometer valve designed and installed to supply water in sufficient quantity and flow to flush the contents of the fixture, to cleanse the fixture and refill the fixture trap in accordance with ASME A112.19.2/CSA B45.1.

P2712.3 Water supply for flushing devices. An adequate quantity of water shall be provided to flush and clean the fixture served. The water supply to flushing devices equipped for manual flushing shall be controlled by a float valve or other automatic device designed to refill the tank after each discharge and to completely shut off the water flow to the tank when the tank is filled to operational capacity. Provision shall be made to automatically supply water to the fixture so as to refill the trap after each flushing.

P2712.4 Flush valves in flush tanks. Flush valve seats in tanks for flushing water closets shall be not less than 1 inch (25.4 mm) above the flood-level rim of the bowl connected thereto, except an *approved* water closet and flush tank combination designed so that when the tank is flushed and the fixture is clogged or partially clogged, the flush valve will close tightly so that water will not spill continuously over the rim of the bowl or backflow from the bowl to the tank.

P2712.5 Overflows in flush tanks. Flush tanks shall be provided with overflows discharging to the water closet connected thereto and such overflow shall be of sufficient size to prevent flooding the tank at the maximum rate at which the tanks are supplied with water according to the manufacturer's design conditions.

P2712.6 Access. Parts in a flush tank shall be accessible for repair and replacement.

P2712.7 Water closet seats. Water closets shall be equipped with seats of smooth, nonabsorbent material and shall be properly sized for the water closet bowl type.

P2712.8 Flush tank lining. Sheet copper used for flush tank linings shall have a weight of not less than 10 ounces per square foot (3 kg/m^2).

P2712.9 Electro-hydraulic water closets. Electro-hydraulic water closets shall conform to ASME A112.19.2/CSA B45.1.

SECTION P2713 BATHTUBS

P2713.1 Bathtub waste outlets and overflows. Bathtubs shall be equipped with a waste outlet that is not less than $1^1/_2$ inches (38 mm) in diameter. The waste outlet shall be equipped with a watertight stopper. Where an overflow is installed, the overflow shall be not less than $1^1/_2$ inches (38 mm) in diameter.

P2713.2 Bathtub enclosures. Doors within a bathtub enclosure shall conform to ASME A112.19.15.

P2713.3 Bathtub and whirlpool bathtub valves. Bathtubs and whirlpool bathtub valves shall have or be supplied by a water-temperature-limiting device that conforms to ASSE 1070/ASME A112.1070/CSA B125.70, except where such valves are combination tub/shower valves in accordance with Section P2708.4. The water-temperature-limiting device required by this section shall be equipped with a means to limit the maximum setting of the device to 120°F (49°C), and, where adjustable, shall be field adjusted in accordance with the manufacturer's instructions to provide hot water at a temperature not to exceed 120°F (49°C). Access shall be provided to water-temperature-limiting devices that conform to ASSE 10705/ASME A112.1070/CSA B125.70.

> **Exception:** Access is not required for nonadjustable water-temperature-limiting devices that conform to ASSE 1070/ASME A112.1070/CSA B125.70 and are integral with a fixture fitting, provided that the fixture fitting itself can be accessed for replacement.

SECTION P2714 SINKS

P2714.1 Sink waste outlets. Sinks shall be provided with waste outlets not less than $1^1/_2$ inches (38 mm) in diameter. A strainer, crossbar or other device shall be provided to restrict the clear opening of the waste outlet.

P2714.2 Movable sink systems. Movable sink systems shall comply with ASME A112.19.12.

SECTION P2715 LAUNDRY TUBS

P2715.1 Laundry tub waste outlet. Each compartment of a laundry tub shall be provided with a waste outlet not less than $1^1/_2$ inches (38 mm) in diameter. A strainer or crossbar shall restrict the clear opening of the waste outlet.

SECTION P2716 FOOD-WASTE DISPOSER

P2716.1 Food-waste disposer waste outlets. Food-waste disposers shall be connected to a drain of not less than $1^1/_2$ inches (38 mm) in diameter.

P2716.2 Water supply required. A sink equipped with a food-waste disposer shall be provided with a faucet.

SECTION P2717 DISHWASHING MACHINES

P2717.1 Protection of water supply. The water supply to a dishwasher shall be protected against backflow by an *air gap* complying with ASME A112.1.3 or A112.1.2 that is installed integrally within the machine or a backflow preventer in accordance with Section P2902.

P2717.2 Sink and dishwasher. The combined discharge from a dishwasher and a one- or two-compartment sink, with or without a food-waste disposer, shall be served by a trap of not less than $1^1/_2$ inches (38 mm) in outside diameter. The dishwasher discharge pipe or tubing shall rise to the underside of the counter and be fastened or otherwise held in that position before connecting to the head of the food-waste disposer or to a wye fitting in the sink tailpiece.

SECTION P2718
CLOTHES WASHING MACHINE

P2718.1 Waste connection. The discharge from a clothes washing machine shall be through an *air break*.

SECTION P2719
FLOOR DRAINS

P2719.1 Floor drains. Floor drains shall have waste outlets not less than 2 inches (51 mm) in diameter and a removable strainer. Floor drains shall be constructed so that the drain can be cleaned. Access shall be provided to the drain inlet. Floor drains shall not be located under or have their access restricted by permanently installed appliances.

SECTION P2720
WHIRLPOOL BATHTUBS

P2720.1 Access to pump. Access shall be provided to circulation pumps in accordance with the fixture or pump manufacturer's installation instructions. Where the manufacturer's instructions do not specify the location and minimum size of field-fabricated access openings, an opening of not less than 12 inches by 12 inches (305 mm by 305 mm) shall be installed for access to the circulation pump. Where pumps are located more than 2 feet (610 mm) from the access opening, an opening of not less than 18 inches by 18 inches (457 mm by 457 mm) shall be installed. A door or panel shall be permitted to close the opening. The access opening shall be unobstructed and be of the size necessary to permit the removal and replacement of the circulation pump.

P2720.2 Piping drainage. The circulation pump shall be accessibly located above the crown weir of the trap. The pump drain line shall be properly graded to ensure minimum water retention in the volute after fixture use. The circulation piping shall be installed to be self-draining.

P2720.3 Leak testing. Leak testing and pump operation shall be performed in accordance with the manufacturer's instructions.

P2720.4 Manufacturer's instructions. The product shall be installed in accordance with the manufacturer's instructions.

SECTION P2721
BIDET INSTALLATIONS

P2721.1 Water supply. The bidet shall be equipped with either an air-gap-type or vacuum-breaker-type fixture supply fitting.

P2721.2 Bidet water temperature. The discharge water temperature from a bidet fitting shall be limited to not greater than 110°F (43°C) by a water-temperature-limiting device conforming to ASSE 1070/ASME A112.1070/CSA B125.70.

SECTION P2722
FIXTURE FITTING

P2722.1 General. Fixture supply valves and faucets shall comply with ASME A112.18.1/CSA B125.1 as indicated in Table P2701.1. Faucets and fixture fittings that supply drinking water for human ingestion shall conform to the requirements of NSF 61, Section 9. Flexible water connectors shall conform to the requirements of Section P2906.7.

P2722.2 Hot water. Fixture fittings supplied with both hot and cold water shall be installed and adjusted so that the left-hand side of the water temperature control represents the flow of hot water when facing the outlet.

> **Exception:** Shower and tub/shower mixing valves conforming to ASSE 1016/ASME 112.1016/CSA B125.16, where the water temperature control corresponds to the markings on the device.

P2722.3 Hose-connected outlets. Faucets and fixture fittings with hose-connected outlets shall conform to ASME A112.18.3 or ASME A112.18.1/CSA B125.1.

P2722.4 Individual pressure-balancing in-line valves for individual fixture fittings. Individual pressure-balancing in-line valves for individual fixture fittings shall comply with ASSE 1066. Such valves shall be installed in an accessible location and shall not be used as a substitute for the balanced pressure, thermostatic or combination shower valves required in Section P2708.4.

P2722.5 Water closet personal hygiene devices. Personal hygiene devices integral to water closets or water closet seats shall conform to ASME A112.4.2/CSA B45.16.

SECTION P2723
MACERATING TOILET SYSTEMS

P2723.1 General. Macerating toilet systems shall be installed in accordance with manufacturer's instructions.

P2723.2 Drain. The size of the drain from the macerating toilet system shall be not less than $^3/_4$ inch (19 mm) in diameter.

SECTION P2724
SPECIALTY TEMPERATURE CONTROL DEVICES AND VALVES

P2724.1 Temperature-actuated mixing valves. Temperature-actuated mixing valves, which are installed to reduce water temperatures to defined limits, shall comply with ASSE 1017. Such valves shall be installed at the hot water source.

P2724.2 Temperature-actuated, flow-reduction devices for individual fixtures. Temperature-actuated, flow-reduction devices, where installed for individual fixture fittings, shall conform to ASSE 1062. Such valves shall not be used as a substitute for the balanced pressure, thermostatic or combination shower valves required for showers in Section P2708.4.

SECTION P2725 NONLIQUID SATURATED TREATMENT SYSTEMS

P2725.1 General. Materials, design, construction and performance of nonliquid saturated treatment systems shall comply with NSF 41.

CHAPTER 28

WATER HEATERS

User notes:

About this chapter: *Chapter 28 contains regulations concerning the safety of water heating units and hot water storage tanks. Heated potable water is needed for plumbing fixtures that are associated with washing, bathing and kitchen activities. Heated water is commonly stored in pressurized storage tanks that must be protected against explosion by pressure and temperature relief valves specified in this chapter. This chapter also covers the access requirements to water heaters and hot water storage tanks to allow for the maintenance and replacement of that equipment.*

Code development reminder: *Code change proposals to this chapter will be considered by the IRC—Mechanical/Plumbing Code Development Committee during the 2021 (Group A) Code Development Cycle.*

SECTION P2801 GENERAL

P2801.1 Required. *Hot water* shall be supplied to plumbing fixtures and plumbing *appliances* intended for bathing, washing or culinary purposes.

P2801.2 Drain valves. Drain valves for emptying shall be installed at the bottom of each tank-type water heater and hot water storage tank. The drain valve inlet shall be not less than $^{3}/_{4}$-inch (19.1 mm) nominal iron pipe size and the outlet shall be provided with a male hose thread.

P2801.3 Installation. Water heaters shall be installed in accordance with this chapter and Chapters 20 and 24.

P2801.4 Location. Water heaters and storage tanks shall be installed in accordance with Section M1305 and shall be located and connected to provide access for observation, maintenance, servicing and replacement.

P2801.5 Prohibited locations. Water heaters shall be located in accordance with Chapter 20.

P2801.6 Required pan. Where a storage tank-type water heater or a hot water storage tank is installed in a location where water leakage from the tank will cause damage, the tank shall be installed in a pan constructed of one of the following:

1. Galvanized steel or aluminum of not less than 0.0236 inch (0.6010 mm) in thickness.
2. Plastic not less than 0.036 inch (0.9 mm) in thickness.
3. Other *approved* materials.

A plastic pan beneath a gas-fired water heater shall be constructed of material having a flame spread index of 25 or less and a *smoke-developed index* of 450 or less when tested in accordance with ASTM E84 or UL 723.

P2801.6.1 Pan size and drain. The pan shall be not less than $1^{1}/_{2}$ inches (38 mm) deep and shall be of sufficient size and shape to receive dripping or condensate from the tank or water heater. The pan shall be drained by an indirect waste pipe of not less than $^{3}/_{4}$ inch (19 mm) diameter. Piping for safety pan drains shall be of those materials indicated in Table P2906.5.

Where a pan drain was not previously installed, a pan drain shall not be required for a replacement water heater installation.

P2801.6.2 Pan drain termination. The pan drain shall extend full-size and terminate over a suitably located indirect waste receptor or shall extend to the exterior of the building and terminate not less than 6 inches (152 mm) and not more than 24 inches (610 mm) above the adjacent ground surface.

P2801.7 Water heaters installed in garages. Water heaters having an *ignition source* shall be elevated such that the source of ignition is not less than 18 inches (457 mm) above the garage floor.

Exception: Elevation of the *ignition source* is not required for *appliances* that are *listed* as flammable vapor ignition-resistant.

P2801.8 Water heater seismic bracing. In Seismic Design Categories D_0, D_1 and D_2 and townhouses in Seismic Design Category C, water heaters shall be anchored or strapped in the upper one-third and in the lower one-third of the *appliance* to resist a horizontal force equal to one-third of the operating weight of the water heater, acting in any horizontal direction, or in accordance with the *appliance* manufacturer's recommendations.

SECTION P2802 SOLAR WATER HEATING SYSTEMS

P2802.1 Water temperature control. Where heated water is discharged from a solar thermal system to a hot water distribution system, a temperature-actuated mixing valve complying with ASSE 1017 shall be installed to temper the water to a temperature of not greater than 140°F (60°C). Solar thermal systems supplying hot water for both space heating and domestic uses shall comply with Section P2803.2. A temperature-indicating device shall be installed to indicate the temperature of the water discharged from the outlet of the mixing valve. The temperature-actuated mixing valve required by this section shall not be a substitute for water-temperature limiting devices required by Chapter 27 for specific fixtures.

P2802.2 Isolation valves. Isolation valves in accordance with Section P2903.9.2 shall be provided on the cold water feed to the water heater. Isolation valves and associated piping shall be provided to bypass solar storage tanks where the system contains multiple storage tanks.

SECTION P2803 WATER HEATERS USED FOR SPACE HEATING

P2803.1 Protection of potable water. Piping and components connected to a water heater for space heating applications shall be suitable for use with potable water in accordance with Chapter 29. Water heaters that will be used to supply potable water shall not be connected to a heating system or components previously used with nonpotable-water heating *appliances*. Chemicals for boiler treatment shall not be introduced into the water heater.

P2803.2 Temperature control. Where a combination water heater-space heating system requires water for space heating at temperatures exceeding 140°F (60°C), a temperature-actuated mixing valve complying with ASSE 1017 shall be installed to temper the water to a temperature of not greater than 140°F (60°C) for domestic uses.

SECTION P2804 RELIEF VALVES

P2804.1 Relief valves required. *Appliances* and equipment used for heating water or storing hot water shall be protected by one of the following:

1. A separate pressure-relief valve and a separate temperature-relief valve.
2. A combination pressure-and-temperature relief valve.

P2804.2 Rating. Relief valves shall have a minimum rated capacity for the equipment served and shall conform to ANSI Z21.22.

P2804.3 Pressure relief valves. Pressure relief valves shall have a relief rating adequate to meet the pressure conditions for the *appliances* or equipment protected. In tanks, they shall be installed directly into a tank tapping or in a water line close to the tank. They shall be set to open at not less than 25 psi (172 kPa) above the system pressure and not greater than 150 psi (1034 kPa). The relief-valve setting shall not exceed the rated working pressure of the tank.

P2804.4 Temperature relief valves. Temperature relief valves shall have a relief rating compatible with the temperature conditions of the *appliances* or equipment protected. The valves shall be installed such that the temperature-sensing element monitors the water within the top 6 inches (152 mm) of the tank. The valve shall be set to open at a temperature of not greater than 210°F (99°C).

P2804.5 Combination pressure-and-temperature relief valves. Combination-pressure-and-temperature relief valves shall comply with the requirements for separate pressure and temperature relief valves.

P2804.6 Installation of relief valves. A check or shutoff valve shall not be installed in any of the following locations:

1. Between a relief valve and the termination point of the relief valve discharge pipe.
2. Between a relief valve and a tank.
3. Between a relief valve and heating *appliances* or equipment.

P2804.6.1 Requirements for discharge pipe. The discharge piping serving a pressure relief valve, temperature relief valve or combination valve shall:

1. Not be directly connected to the drainage system.
2. Discharge through an *air gap* located in the same room as the water heater.
3. Not be smaller than the diameter of the outlet of the valve served and shall discharge full size to the *air gap*.
4. Serve a single relief device and shall not connect to piping serving any other relief device or equipment.
5. Discharge to the floor, to the pan serving the water heater or storage tank, to a waste receptor or to the outdoors.
6. Discharge in a manner that does not cause personal injury or structural damage.
7. Discharge to a termination point that is readily observable by the building occupants.
8. Not be trapped.
9. Be installed to flow by gravity.
10. Terminate not more than 6 inches (152 mm) and not less than two times the discharge pipe diameter above the floor or waste receptor flood level rim.
11. Not have a threaded connection at the end of the piping.
12. Not have valves or tee fittings.
13. Be constructed of those materials indicated in Section P2906.5 or materials tested, rated and *approved* for such use in accordance with ASME A112.4.1.
14. Be one nominal size larger than the size of the relief-valve outlet, where the relief-valve discharge piping is installed with insert fittings. The outlet end of such tubing shall be fastened in place.

P2804.7 Vacuum relief valve. Bottom fed tank-type water heaters and bottom fed tanks connected to water heaters shall have a vacuum relief valve installed that complies with ANSI Z21.22.

CHAPTER 29
WATER SUPPLY AND DISTRIBUTION

User notes:

About this chapter: *Many plumbing fixtures require a supply of potable water. Other fixtures could be supplied with nonpotable water such as reclaimed water. Chapter 29 covers the requirements for water distribution piping systems to and within buildings. The regulations include the types of materials and the connection methods for such systems. This chapter regulates the assemblies, devices and methods that are used for the prevention of backflow of contaminated or polluted water into the potable water system. Also contained in this chapter are the design requirements for the installation of fire sprinkler systems, as such systems are connected to the potable water supply for the building. Storm water and some liquid waste from a building can be a source of nonpotable water that can used to reduce the volume of potable water supplied to the building. This chapter provides the requirements for storage, treatment and distribution of this resource. This chapter also regulates the piping systems for reclaimed water supplied by a wastewater treatment facility.*

Code development reminder: *Code change proposals to this chapter will be considered by the IRC—Mechanical/Plumbing Code Development Committee during the 2021 (Group A) Code Development Cycle.*

SECTION P2901
GENERAL

P2901.1 Potable water required. Potable water shall be supplied to plumbing fixtures and plumbing *appliances* except where treated rainwater, treated graywater or municipal reclaimed water is supplied to water closets, urinals and trap primers. The requirements of this section shall not be construed to require signage for water closets and urinals.

P2901.2 Identification of nonpotable water systems. Where *nonpotable* water systems are installed, the piping conveying the nonpotable water shall be identified either by color marking, metal tags or tape in accordance with Sections P2901.2.1 through P2901.2.2.3.

P2901.2.1 Signage required. Nonpotable water outlets such as hose connections, open-ended pipes and faucets shall be identified with signage that reads as follows: "Nonpotable water is utilized for [application name]. CAUTION: NONPOTABLE WATER. DO NOT DRINK." The words shall be legibly and indelibly printed on a tag or sign constructed of corrosion-resistant waterproof material or shall be indelibly printed on the fixture. The letters of the words shall be not less than 0.5 inches (12.7 mm) in height and in colors in contrast to the background on which they are applied. In addition to the required wordage, the pictograph shown in Figure P2901.2.1 shall appear on the required signage.

P2901.2.2 Distribution pipe labeling and marking. Nonpotable distribution piping shall be: purple in color and embossed or integrally stamped or marked with the words, "CAUTION: NONPOTABLE WATER. DO NOT DRINK"; or installed with a purple identification tape or wrap. Pipe identification shall include the contents of the piping system and an arrow indicating the direction of flow. Hazardous piping systems shall contain information addressing the nature of the hazard. Pipe identification shall be repeated at intervals not exceeding 25 feet (7620 mm) and at each point where the piping passes through a wall, floor or roof. Lettering shall be readily observable within the room or space where the piping is located.

FIGURE P2901.2.1
PICTOGRAPH—DO NOT DRINK

P2901.2.2.1 Color. The color of the pipe identification shall be discernable and consistent throughout the building. The color purple shall be used to identify reclaimed water, rainwater and graywater distribution systems.

P2901.2.2.2 Lettering size. The size of the background color field and lettering shall comply with Table P2901.2.2.2.

TABLE P2901.2.2.2
SIZE OF PIPE IDENTIFICATION

PIPE DIAMETER (inches)	LENGTH OF BACKGROUND COLOR FIELD (inches)	SIZE OF LETTERS (inches)
$^3/_4$ to $1^1/_4$	8	0.5
$1^1/_2$ to 2	8	0.75
$2^1/_2$ to 6	12	1.25
8 to 10	2	2.5
Over 10	32	3.5

For SI: 1 inch = 25.4 mm.

P2901.2.2.3 Identification tape. Where used, identification tape shall be not less than 3 inches (76 mm) wide and have white or black lettering on a purple field stating, "CAUTION: NONPOTABLE WATER—DO NOT DRINK." Identification tape shall be installed on top of nonpotable rainwater distribution pipes and fastened not greater than every 10 feet (3048 mm) to each pipe length, and run continuously the entire length of the pipe.

SECTION P2902 PROTECTION OF POTABLE WATER SUPPLY

P2902.1 General. A potable water supply system shall be designed and installed as to prevent contamination from nonpotable liquids, solids or gases being introduced into the potable water supply. Connections shall not be made to a potable water supply in a manner that could contaminate the water supply or provide a cross connection between the supply and a source of contamination except where *approved* backflow prevention assemblies, backflow prevention devices or other means or methods are installed to protect the potable water supply. Cross connections between an individual water supply and a potable public water supply shall be prohibited.

P2902.2 Plumbing fixtures. The supply lines and fittings for every plumbing fixture shall be installed so as to prevent backflow. Plumbing fixture fittings shall provide backflow protection in accordance with ASME A112.18.1/CSA B125.1.

P2902.3 Backflow protection. A means of protection against backflow shall be provided in accordance with Sections P2902.3.1 through P2902.3.7. Backflow prevention applications shall conform to Table P2902.3, except as specifically stated in Sections P2902.4 through P2902.5.5.

P2902.3.1 Air gaps. *Air gaps* shall comply with ASME A112.1.2 and *air gap* fittings shall comply with ASME A112.1.3. An *air gap* shall be measured vertically from the lowest end of a water outlet to the flood level rim of the fixture or receptor into which the water outlets discharges to the floor. The required *air gap* shall be not less than twice the diameter of the effective opening of the outlet and not less than the values specified in Table P2902.3.1.

P2902.3.2 Atmospheric-type vacuum breakers. Atmospheric-type vacuum breakers shall conform to ASSE 1001 or CSA B64.1.1. Hose-connection vacuum breakers shall conform to ASSE 1011, ASSE 1019, ASSE 1035, ASSE 1052, CSA B64.2, CSA B64.2.1, CSA B64.2.1.1, CSA B64.2.2 or CSA B64.7. Both types of vacuum breakers shall be installed with the outlet continuously open to the atmosphere. The critical level of the atmospheric vacuum breaker shall be set at not less than 6 inches (152 mm) above the highest elevation of downstream piping and the flood level rim of the fixture or device.

P2902.3.3 Backflow preventer with intermediate atmospheric vent. Backflow preventers with intermediate atmospheric vents shall conform to ASSE 1012, ASSE 1081 or CSA B64.3. These devices shall be permitted to be installed where subject to continuous pressure conditions. These devices shall be prohibited as a means of protection where any hazardous chemical additives are introduced downstream of the device. The relief opening shall discharge by *air gap* and shall be prevented from being submerged.

P2902.3.4 Pressure vacuum breaker assemblies. Pressure vacuum breaker assemblies shall conform to ASSE 1020 or CSA B64.1.2. Spill-resistant vacuum breaker assemblies shall comply with ASSE 1056. These assemblies are designed for installation under continuous pressure conditions where the critical level is installed at the required height. The critical level of a pressure vacuum breaker and a spill-resistant vacuum breaker assembly shall be set at not less than 12 inches (304 mm) above the highest elevation of downstream piping and the flood level rim of the fixture or device. Pressure vacuum breaker assemblies shall not be installed in locations where spillage could cause damage to the structure.

P2902.3.5 Reduced pressure principle backflow prevention assemblies. Reduced pressure principle backflow prevention assemblies and reduced pressure principle fire protection backflow prevention assemblies shall conform to ASSE 1013, AWWA C511, CSA B64.4 or CSA B64.4.1. Reduced pressure detector fire protection backflow prevention assemblies shall conform to ASSE 1047. These devices shall be permitted to be installed where subject to continuous pressure conditions. The relief opening shall discharge by *air gap* and shall be prevented from being submerged.

P2902.3.6 Double-check backflow prevention assemblies. Double-check backflow prevention assemblies shall conform to ASSE 1015, AWWA C510, CSA B64.5 or CSA B64.5.1. Double-check detector fire protection backflow prevention assemblies shall conform to ASSE 1048. These assemblies shall be capable of operating under continuous pressure conditions.

P2902.3.7 Dual check backflow preventer. Dual check backflow preventers shall conform with ASSE 1024 or CSA B64.6.

P2902.4 Protection of potable water outlets. Potable water openings and outlets shall be protected by an *air gap*, a reduced pressure principle backflow prevention assembly, an atmospheric vent, an atmospheric-type vacuum breaker, a pressure-type vacuum breaker assembly or a hose connection backflow preventer.

P2902.4.1 Fill valves. Flush tanks shall be equipped with an antisiphon fill valve conforming to ASSE 1002/ASME A112.1002/CSA B125.12 or CSA B125.3. The critical level of the fill valve shall be located not less than 1 inch (25 mm) above the top of the flush tank overflow pipe.

TABLE P2902.3 APPLICATION FOR BACKFLOW PREVENTERS

DEVICE	DEGREE OF HAZARD[a]	APPLICATION[b]	APPLICABLE STANDARDS
Backflow Prevention Assemblies			
Double-check backflow prevention assembly and double-check fire protection backflow prevention assembly	Low hazard	Backpressure or backsiphonage sizes $^3/_8$″ – 16″	ASSE 1015, AWWA C510, CSA B64.5, CSA B64.5.1
Double-check detector fire protection backflow prevention assemblies	Low hazard	Backpressure or backsiphonage sizes 2″ – 16″	ASSE 1048
Pressure vacuum breaker assembly	High or low hazard	Backsiphonage only sizes $^1/_2$″ – 2″	ASSE 1020, CSA B64.1.2
Reduced pressure principle backflow prevention assembly and reduced pressure principle fire protection backflow prevention assembly	High or low hazard	Backpressure or backsiphonage sizes $^3/_8$″ – 16″	ASSE 1013, AWWA C511, CSA B64.4, CSA B64.4.1
Reduced pressure detector fire protection backflow prevention assemblies	High or low hazard	Backsiphonage or backpressure (automatic sprinkler systems)	ASSE 1047
Spill-resistant vacuum breaker	High or low hazard	Backsiphonage only sizes $^1/_4$″ – 2″	ASSE 1056, CSA B64.1.3
Backflow Preventer Plumbing Devices			
Antisiphon-type fill valves for gravity water closet flush tanks	High hazard	Backsiphonage only	ASSE 1002/ASME A112.1002/CSA B125.12, CSA B125.3
Backflow preventer with intermediate atmospheric vents	Low hazard	Backpressure or backsiphonage sizes $^1/_4$″ – $^3/_8$″	ASSE 1012, CSA B64.3
Backflow preventer with intermediate atmospheric vent and pressure-reducing valve	Low hazard	Backpressure or backsiphonage sizes $^1/_4$″ – $^3/_8$″	ASSE 1081
Dual-check-valve-type backflow preventers	Low hazard	Backpressure or backsiphonage sizes $^1/_4$″ – 1″	ASSE 1024, CSA B64.6
Hose-connection backflow preventer	High or low hazard	Low head backpressure, rated working pressure backpressure or backsiphonage sizes $^1/_2$″ – 1″	ASSE 1052, CSA B64.2.1.1
Hose-connection vacuum breaker	High or low hazard	Low head backpressure or backsiphonage sizes $^1/_2$″, $^3/_4$″, 1″	ASSE 1011, CSA B64.2, CSA B64.2.1
Laboratory faucet backflow preventer	High or low hazard	Low head backpressure and backsiphonage	ASSE 1035, CSA B64.7
Pipe-applied atmospheric-type vacuum breaker	High or low hazard	Backsiphonage only sizes $^1/_4$″ – 4″	ASSE 1001, CSA B64.1.1
Vacuum breaker wall hydrants, frost-resistant, automatic-draining type	High or low hazard	Low head backpressure or back siphonage sizes $^3/_4$″ – 1″	ASSE 1019, CSA B64.2.2
Other Means or Methods			
Air gap	High or low hazard	Backsiphonage only	ASME A112.1.2
Air gap fittings for use with plumbing fixtures, appliances and appurtenances	High or low hazard	Backsiphonage or backpressure	ASME A112.1.3

For SI: 1 inch = 25.4 mm.

a. Low hazard—See "*Pollution*" (Section R202). High hazard—See "*Contamination*" (Section R202).

b. See "*Backpressure*" (Section R202). See "*Backpressure, Low Head*" (Section R202). See "*Backsiphonage*" (Section R202).

TABLE P2902.3.1 MINIMUM AIR GAPS

FIXTURE	MINIMUM AIR GAP	
	Away from a wall[a] (inches)	Close to a wall (inches)
Effective openings greater than 1 inch	Two times the diameter of the effective opening	Three times the diameter of the effective opening
Lavatories and other fixtures with effective openings not greater than $^1/_2$ inch in diameter	1	1.5
Over-rim bath fillers and other fixtures with effective openings not greater than 1 inch in diameter	2	3
Sink, laundry trays, gooseneck back faucets and other fixtures with effective openings not greater than $^3/_4$ inch in diameter	1.5	2.5

For SI: 1 inch = 25.4 mm.

a. Applicable where walls or obstructions are spaced from the nearest inside edge of the spout opening a distance greater than three times the diameter of the effective opening for a single wall, or a distance greater than four times the diameter of the effective opening for two intersecting walls.

P2902.4.2 Deck-mounted and integral vacuum breakers. *Approved* deck-mounted or equipment-mounted vacuum breakers and faucets with integral atmospheric vacuum breakers or spill-resistant vacuum breaker assemblies shall be installed in accordance with the manufacturer's instructions and the requirements for labeling. The critical level of the breakers and assemblies shall be located at not less than 1 inch (25 mm) above the *flood level rim.*

P2902.4.3 Hose connection. Sillcocks, hose bibbs, wall hydrants and other openings with a hose connection shall be protected by an atmospheric-type or pressure-type vacuum breaker, a pressure vacuum-breaker assembly or a permanently attached hose connection vacuum breaker.

Exceptions:

1. This section shall not apply to water heater and boiler drain valves that are provided with hose connection threads and that are intended only for tank or vessel draining.
2. This section shall not apply to water supply valves intended for connection of clothes washing machines where backflow prevention is otherwise provided or is integral with the machine.

P2902.5 Protection of potable water connections. Connections to the potable water shall conform to Sections P2902.5.1 through P2902.5.5.

P2902.5.1 Connections to boilers. Where chemicals will not be introduced into a boiler, the potable water supply to the boiler shall be protected from the boiler by a backflow preventer with an intermediate atmospheric vent complying with ASSE 1012 or CSA B64.3. Where chemicals will be introduced into a boiler, the potable water supply to the boiler shall be protected from the boiler by an *air gap* or a reduced pressure principle backflow prevention assembly complying with ASSE 1013, AWWA C511 or CSA B64.4.

P2902.5.2 Heat exchangers. Heat exchangers using an essentially toxic transfer fluid shall be separated from the potable water by double-wall construction. An *air gap* open to the atmosphere shall be provided between the two walls. Single-wall construction heat exchangers shall be used only where an *essentially nontoxic transfer fluid* is utilized.

P2902.5.3 Lawn irrigation systems. The potable water supply to lawn irrigation systems shall be protected against backflow by an atmospheric vacuum breaker, a pressure vacuum-breaker assembly or a reduced pressure principle backflow prevention assembly. Valves shall not be installed downstream from an atmospheric vacuum breaker. Where chemicals are introduced into the system, the potable water supply shall be protected against backflow by a reduced pressure principle backflow prevention assembly.

P2902.5.4 Connections to automatic fire sprinkler systems. The potable water supply to automatic fire sprinkler systems shall be protected against backflow by a double-check backflow prevention assembly, a double-check fire protection backflow prevention assembly, a reduced pressure principle backflow prevention assembly or a reduced pressure principle fire protection backflow prevention assembly.

Exception: Where sprinkler systems are installed in accordance with Section P2904.1, backflow protection for the water supply system shall not be required.

P2902.5.4.1 Additives or nonpotable source. Where systems contain chemical additives or antifreeze, or where systems are connected to a nonpotable secondary water supply, the potable water supply shall be protected against backflow by a reduced pressure principle backflow prevention assembly or a reduced pressure principle fire protection backflow prevention assembly. Where chemical additives or antifreeze is added to only a portion of an automatic fire sprinkler or standpipe system, the reduced pressure principle fire protection backflow preventer shall be permitted to be located so as to isolate that portion of the system.

P2902.5.5 Solar thermal systems. Where a solar thermal system heats potable water to supply a potable hot water distribution or any other type of heating system, the solar thermal system shall be in accordance with Section P2902.5.5.1, P2902.5.5.2 or P2902.5.5.3 as applicable.

P2902.5.5.1 Indirect systems. Water supplies of any type shall not be connected to the solar heating loop of an indirect solar thermal hot water heating system. This requirement shall not prohibit the presence of inlets or outlets on the solar heating loop for the purposes of servicing the fluid in the solar heating loop.

P2902.5.5.2 Direct systems for potable water distribution systems. Where a solar thermal system directly heats potable water for a potable water distribution system, the pipe, fittings, valves and other components that are in contact with the potable water in the system shall comply with the requirements of Chapter 29.

P2902.5.5.3 Direct systems for other than potable water distribution systems. Where a solar thermal system directly heats water for a system other than a potable water distribution system, a potable water supply connected to such system shall be protected by a backflow preventer with an intermediate atmospheric vent complying with ASSE 1012. Where a solar thermal system directly heats chemically treated water for a system other than a potable water distribution system, a potable water supply connected to such system shall be protected by a reduced pressure principle backflow prevention assembly complying with ASSE 1013.

P2902.6 Location of backflow preventers. Access shall be provided to backflow preventers as specified by the manufacturer's installation instructions.

P2902.6.1 Outdoor enclosures for backflow prevention devices. Outdoor enclosures for backflow prevention devices shall comply with ASSE 1060.

P2902.6.2 Protection of backflow preventers. Backflow preventers shall not be located in areas subject to freezing except where they can be removed by means of unions, or are protected by heat, insulation or both.

P2902.6.3 Relief port piping. The indirect waste receptor and drainage piping shall be sized to drain the maximum discharge flow rate from the relief port as published by the backflow preventer manufacturer. The termination of the piping from the relief port or air gap fitting of the backflow preventer shall discharge to an *approved* indirect waste receptor or to the outdoors where it will not cause damage or create a nuisance.

SECTION P2903 WATER SUPPLY SYSTEM

P2903.1 Water supply system design criteria. The water service and water distribution systems shall be designed and sized for peak demand using values shown in Table P2903.1.

P2903.2 Maximum flow and water consumption. The maximum water consumption flow rates and quantities for plumbing fixtures and fixture fittings shall be in accordance with Table P2903.2.

P2903.3 Minimum pressure. Where the water pressure supplied by the public water main or an individual water supply system is insufficient to provide for the minimum pressures and quantities for the plumbing fixtures in the building, the pressure shall be increased by means of an elevated water tank, a hydropneumatic pressure booster system or a water pressure booster pump.

P2903.3.1 Pumps handling drinking water. Pumps intended to supply drinking water shall conform to NSF 61.

P2903.3.2 Maximum pressure. The static water pressure shall be not greater than 80 psi (551 kPa). Where the main pressure exceeds 80 psi (551 kPa), an *approved* pressure-reducing valve conforming to ASSE 1003 or CSA B356 shall be installed on the domestic water branch main or riser at the connection to the water service pipe.

P2903.4 Thermal expansion control. A means for controlling increased pressure caused by thermal expansion shall be installed where required in accordance with Sections P2903.4.1 and P2903.4.2.

P2903.4.1 Pressure-reducing valve. For water service system sizes up to and including 2 inches (51 mm), a device for controlling pressure shall be installed where,

TABLE P2903.1
REQUIRED CAPACITIES AT POINT OF OUTLET DISCHARGE

FIXTURE SUPPLY OUTLET SERVING	FLOW RATE (gpm)	FLOW PRESSURE (psi)
Bathtub, balanced-pressure, thermostatic or combination balanced-pressure/thermostatic mixing valve	4	20
Bidet, thermostatic mixing valve	2	20
Dishwasher	2.75	8
Laundry tray	4	8
Lavatory	0.8	8
Shower, balanced-pressure, thermostatic or combination balanced-pressure/thermostatic mixing valve	2.5[a]	20
Sillcock, hose bibb	5	8
Sink	1.75	8
Water closet, flushometer tank	1.6	20
Water closet, tank, close coupled	3	20
Water closet, tank, one-piece	6	20

For SI: 1 pound per square inch = 6.895 kPa, 1 gallon per minute = 3.785 L/m.

a. Where the shower mixing valve manufacturer indicates a lower flow rating for the mixing valve, the lower value shall be applied.

TABLE P2903.2
MAXIMUM FLOW RATES AND CONSUMPTION FOR PLUMBING FIXTURES AND FIXTURE FITTINGS[b]

PLUMBING FIXTURE OR FIXTURE FITTING	MAXIMUM FLOW RATE OR QUANTITY
Lavatory faucet	2.2 gpm at 60 psi
Shower head[a]	2.5 gpm at 80 psi
Sink faucet	2.2 gpm at 60 psi
Water closet	1.6 gallons per flushing cycle

For SI: 1 gallon per minute = 3.785 L/m, 1 pound per square inch = 6.895 kPa.

a. A hand-held shower spray shall be considered to be a shower head.

b. Consumption tolerances shall be determined from referenced standards.

because of thermal expansion, the pressure on the downstream side of a pressure-reducing valve exceeds the pressure-reducing valve setting.

P2903.4.2 Backflow prevention device or check valve. Where a backflow prevention device, check valve or other device is installed on a water supply system using storage water heating equipment such that thermal expansion causes an increase in pressure, a device for controlling pressure shall be installed.

P2903.5 Water hammer. The flow velocity of the water distribution system shall be controlled to reduce the possibility of water hammer. A water-hammer arrestor shall be installed where quick-closing valves are utilized. Water-hammer arrestors shall be installed in accordance with the manufacturer's instructions. Water-hammer arrestors shall conform to ASSE 1010.

P2903.6 Determining water supply fixture units. Supply loads in the building water distribution system shall be determined by total load on the pipe being sized, in terms of water supply fixture units (w.s.f.u.), as shown in Table P2903.6, and gallon per minute (gpm) flow rates [see Table P2903.6(1)]. For fixtures not listed, choose a w.s.f.u. value of a fixture with similar flow characteristics.

P2903.7 Size of water-service mains, branch mains and risers. The size of the water service pipe shall be not less than $^3/_4$ inch (19 mm) diameter. The size of water service mains, branch mains and risers shall be determined from the water supply demand [gpm (L/m)], available water pressure [psi (kPa)] and friction loss caused by the water meter and *developed length* of pipe [feet (m)], including *equivalent length* of fittings. The size of each water distribution system shall be determined according to design methods conforming to acceptable engineering practice, such as those methods in Appendix AP and shall be *approved* by the *building official.*

P2903.8 Gridded and parallel water distribution systems. Hot water and cold water manifolds installed with parallel-connected individual distribution lines and cold water manifolds installed with gridded distribution lines to each fixture or fixture fitting shall be designed in accordance with Sections P2903.8.1 through P2903.8.5. Gridded systems for hot water distribution systems shall be prohibited.

P2903.8.1 Sizing of manifolds. Manifolds shall be sized in accordance with Table P2903.8.1. Total gallons per minute is the demand for all outlets.

TABLE P2903.8.1
MANIFOLD SIZING[a]

PLASTIC		METALLIC	
Nominal Size ID (inches)	Maximum[b] gpm	Nominal Size ID (inches)	Maximum[b] gpm
$^3/_4$	17	$^3/_4$	11
1	29	1	20
$1^1/_4$	46	$1^1/_4$	31
$1^1/_2$	66	$1^1/_2$	44

For SI: 1 inch = 25.4 mm, 1 gallon per minute = 3.785 L/m, 1 foot per second = 0.3048 m/s.

a. See Table P2903.6 for w.s.f.u and Table P2903.6(1) for gallon-per-minute (gpm) flow rates.

b. Based on velocity limitation: plastic, 12 feet per second; metal, 8 feet per second.

P2903.8.2 Minimum size. Where the *developed length* of the distribution line is 60 feet (18 288 mm) or less, and the available pressure at the meter is not less than 40 pounds per square inch (276 kPa), the size of individual distribution lines shall be not less than $^3/_8$ inch (10 mm) diameter. Certain fixtures such as one-piece water closets and whirlpool bathtubs shall require a larger size where specified by the manufacturer. Where a water heater is fed from the end of a cold water manifold, the manifold shall be one size larger than the water heater feed.

TABLE P2903.6
WATER SUPPLY FIXTURE UNIT VALUES FOR VARIOUS PLUMBING FIXTURES AND FIXTURE GROUPS

TYPE OF FIXTURES OR GROUP OF FIXTURES	WATER-SUPPLY FIXTURE-UNIT VALUE (w.s.f.u.)		
	Hot	Cold	Combined
Bathtub (with/without overhead shower head)	1.0	1.0	1.4
Clothes washer	1.0	1.0	1.4
Dishwasher	1.4	—	1.4
Full-bath group with bathtub (with/without shower head) or shower stall	1.5	2.7	3.6
Half-bath group (water closet and lavatory)	0.5	2.5	2.6
Hose bibb (sillcock)[a]	—	2.5	2.5
Kitchen group (dishwasher and sink with or without food-waste disposer)	1.9	1.0	2.5
Kitchen sink	1.0	1.0	1.4
Laundry group (clothes washer standpipe and laundry tub)	1.8	1.8	2.5
Laundry tub	1.0	1.0	1.4
Lavatory	0.5	0.5	0.7
Shower stall	1.0	1.0	1.4
Water closet (tank type)	—	2.2	2.2

For SI: 1 gallon per minute = 3.785 L/m.

a. The fixture-unit value 2.5 assumes a flow demand of 2.5 gallons per minute, such as for an individual lawn sprinkler device. If a hose bibb or sill cock will be required to furnish a greater flow, the equivalent fixture-unit value may be obtained from this table or Table P2903.6(1).

TABLE P2903.6(1)
CONVERSIONS FROM WATER SUPPLY FIXTURE UNIT TO GALLON PER MINUTE FLOW RATES

SUPPLY SYSTEMS PREDOMINANTLY FOR FLUSH TANKS			SUPPLY SYSTEMS PREDOMINANTLY FOR FLUSHOMETER VALVES		
Load	Demand		Load	Demand	
(Water supply fixture units)	(Gallons per minute)	(Cubic feet per minute)	(Water supply fixture units)	(Gallons per minute)	(Cubic feet per minute)
1	3.0	0.04104	—	—	—
2	5.0	0.0684	—	—	—
3	6.5	0.86892	—	—	—
4	8.0	1.06944	—	—	—
5	9.4	1.256592	5	15.0	2.0052
6	10.7	1.430376	6	17.4	2.326032
7	11.8	1.577424	7	19.8	2.646364
8	12.8	1.711104	8	22.2	2.967696
9	13.7	1.831416	9	24.6	3.288528
10	14.6	1.951728	10	27.0	3.60936
11	15.4	2.058672	11	27.8	3.716304
12	16.0	2.13888	12	28.6	3.823248
13	16.5	2.20572	13	29.4	3.930192
14	17.0	2.27256	14	30.2	4.037136
15	17.5	2.3394	15	31.0	4.14408
16	18.0	2.90624	16	31.8	4.241024
17	18.4	2.459712	17	32.6	4.357968
18	18.8	2.513184	18	33.4	4.464912
19	19.2	2.566656	19	34.2	4.571856
20	19.6	2.620128	20	35.0	4.6788
25	21.5	2.87412	25	38.0	5.07984
30	23.3	3.114744	30	42.0	5.61356
35	24.9	3.328632	35	44.0	5.88192
40	26.3	3.515784	40	46.0	6.14928
45	27.7	3.702936	45	48.0	6.41664
50	29.1	3.890088	50	50.0	6.684

For SI: 1 gallon per minute = 3.785 L/m, 1 cubic foot per minute = 0.4719 L/s.

P2903.8.3 Support and protection. Plastic piping bundles shall be secured in accordance with the manufacturer's instructions and supported in accordance with Section P2605. Bundles that have a change in direction equal to or greater than 45 degrees (0.79 rad) shall be protected from chafing at the point of contact with framing members by sleeving or wrapping.

P2903.8.4 Valving. Fixture valves, when installed, shall be located either at the fixture or at the manifold. Valves installed at the manifold shall be *labeled* indicating the fixture served.

P2903.8.5 Hose bibb bleed. A *readily accessible* air bleed shall be installed in hose bibb supplies at the manifold or at the hose bibb exit point.

P2903.9 Valves. Valves shall be installed in accordance with Sections P2903.9.1 through P2903.9.5.

P2903.9.1 Service valve. Each *dwelling unit* shall be provided with an accessible main shutoff valve near the entrance of the water service. The valve shall be of a full-open type having nominal restriction to flow, with provision for drainage such as a bleed orifice or installation of a separate drain valve. Additionally, the water service shall be valved at the curb or *lot line* in accordance with local requirements.

P2903.9.2 Water heater valve. A *readily accessible* full-open valve shall be installed in the cold-water supply pipe to each water heater at or near the water heater.

P2903.9.3 Fixture valves and access. Shutoff valves shall be required on each fixture supply pipe to each plumbing *appliance* and to each plumbing fixture other than bathtubs and showers. Valves serving individual plumbing fixtures, *plumbing appliances,* risers and branches shall be accessible.

P2903.9.4 Valve requirements. Valves shall be compatible with the type of piping material installed in the system. Valves shall conform to one of the standards indicated in Table P2903.9.4 or shall be *approved.* Valves intended to supply drinking water shall meet the requirements of NSF 61.

P2903.9.5 Valves and outlets prohibited below grade. Potable water outlets and combination stop-and-waste valves shall not be installed underground or below grade. Freezeproof yard hydrants that drain the riser into the ground are considered to be stop-and-waste valves.

> **Exception:** Installation of freezeproof yard hydrants that drain the riser into the ground shall be permitted if the potable water supply to such hydrants is protected upstream of the hydrants in accordance with Section P2902 and the hydrants are permanently identified as nonpotable outlets by *approved* signage that reads, "CAUTION, NONPOTABLE WATER. DO NOT DRINK."

P2903.10 Hose bibb. Hose bibbs subject to freezing, including the "frostproof" type, shall be equipped with an accessible stop-and-waste-type valve inside the building so that they can be controlled and drained during cold periods.

> **Exception:** Frostproof hose bibbs installed such that the stem extends through the building insulation into an open heated or *semiconditioned space* need not be separately valved (see Figure P2903.10).

P2903.11 Drain water heat recovery units. Drain water heat recovery units shall be in accordance with Section N1103.5.4.

SECTION P2904
DWELLING UNIT FIRE SPRINKLER SYSTEMS

P2904.1 General. The design and installation of residential fire sprinkler systems shall be in accordance with NFPA 13D or Section P2904, which shall be considered to be equivalent to NFPA 13D. Partial residential sprinkler systems shall be permitted to be installed only in buildings not required to be equipped with a residential sprinkler system. Section P2904 shall apply to stand-alone and multipurpose wet-pipe sprinkler systems that do not include the use of antifreeze. A multipurpose fire sprinkler system shall provide domestic water to both fire sprinklers and plumbing fixtures. A stand-alone sprinkler system shall be separate and independent from the water distribution system. A backflow preventer shall not be required to separate a sprinkler system from the water distribution system, provided that the sprinkler system complies with all of the following:

1. The system complies with NFPA 13D or Section P2904.
2. The piping material complies with Section P2906.
3. The system does not contain antifreeze.
4. The system does not have a fire department connection.

TABLE P2903.9.4
VALVES

MATERIAL	STANDARD
Chlorinated polyvinyl chloride (CPVC) plastic	ASME A112.4.14, ASME A112.18.1/CSA B125.1, ASTM F1970, CSA B125.3, MSS SP-122
Copper or copper alloy	ASME A112.4.14, ASME A112.18.1/CSA B125.1, ASME B16.34, CSA B125.3, MSS SP-67, MSS SP-80, MSS SP-110, MSS SP-139
Gray and ductile iron	ASTM A126, AWWA C500, AWWA C504, AWWA C507, MSS SP-42, MSS SP-67, MSS SP-70, MSS SP-71, MSS SP-72, MSS SP-78
Cross-linked polyethylene (PEX) plastic	ASME A112.4.14, ASME A112.18.1/CSA B125.1, CSA B125.3, NSF 359
Polypropylene (PP) plastic	ASME A112.4.14, ASTM F2389
Polyvinyl chloride (PVC) plastic	ASME A112.4.14, ASTM F1970, MSS SP-122

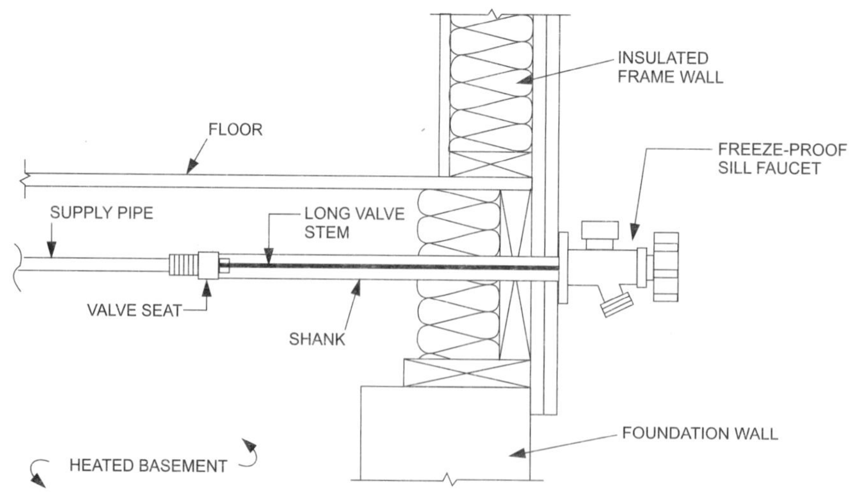

FIGURE P2903.10
TYPICAL FROSTPROOF HOSE BIBB INSTALLATION NOT REQUIRING SEPARATE VALVE

P2904.1.1 Required sprinkler locations. Sprinklers shall be installed to protect all areas of a *dwelling unit.*

Exceptions:

1. *Attics*, crawl spaces and normally unoccupied concealed spaces that do not contain fuel-fired *appliances* do not require sprinklers. In *attics*, crawl spaces and normally unoccupied concealed spaces that contain fuel-fired equipment, a sprinkler shall be installed above the equipment; however, sprinklers shall not be required in the remainder of the space.
2. Clothes closets, linen closets and pantries not exceeding 24 square feet (2.2 m^2) in area, with the smallest dimension not greater than 3 feet (915 mm) and having wall and ceiling surfaces of gypsum board.
3. Bathrooms not more than 55 square feet (5.1 m^2) in area.
4. Garages; carports; exterior porches; unheated entry areas, such as mud rooms, that are adjacent to an exterior door; and similar areas.

P2904.2 Sprinklers. Sprinklers shall be new *listed* residential sprinklers and shall be installed in accordance with the sprinkler manufacturer's instructions.

P2904.2.1 Temperature rating and separation from heat sources. Except as provided for in Section P2904.2.2, sprinklers shall have a temperature rating of not less than 135°F (57°C) and not more than 225°F (107°C). Sprinklers shall be separated from heat sources as required by the sprinkler manufacturer's installation instructions.

P2904.2.2 Intermediate temperature sprinklers. Sprinklers shall have an intermediate temperature rating not less than 175°F (79°C) and not more than 225°F (107°C) where installed in the following locations:

1. Directly under skylights, where the sprinkler is exposed to direct sunlight.
2. In *attics.*
3. In concealed spaces located directly beneath a roof.
4. Within the distance to a heat source as specified in Table P2904.2.2.

P2904.2.3 Freezing areas. Piping shall be protected from freezing as required by Section P2603.5 or by using one of the following:

1. A dry-pipe automatic sprinkler system that is listed for residential occupancy applications.
2. Dry-sidewall or dry-pendent sprinklers extending from a nonfreezing area into a freezing area.

P2904.2.4 Sprinkler coverage. Sprinkler coverage requirements and sprinkler obstruction requirements shall be in accordance with Sections P2904.2.4.1 and P2904.2.4.2.

P2904.2.4.1 Coverage area limit. The area of coverage of a single sprinkler shall not exceed 400 square feet (37 m^2) and shall be based on the sprinkler *listing* and the sprinkler manufacturer's installation instructions.

P2904.2.4.2 Obstructions to coverage. Sprinkler discharge shall not be blocked by obstructions unless additional sprinklers are installed to protect the obstructed area. Additional sprinklers shall not be required where the sprinkler separation from obstructions complies with either the minimum distance indicated in Figure P2904.2.4.2 or the minimum distances specified in the sprinkler manufacturer's instructions where the manufacturer's instructions permit a lesser distance.

TABLE P2904.2.2
LOCATIONS WHERE INTERMEDIATE TEMPERATURE SPRINKLERS ARE REQUIRED

HEAT SOURCE	RANGE OF DISTANCE FROM HEAT SOURCE WITHIN WHICH INTERMEDIATE TEMPERATURE SPRINKLERS ARE REQUIRED[a, b] (inches)
Coal and wood burning stove	12 to 42
Fireplace, front of recessed fireplace	36 to 60
Fireplace, side of open or recessed fireplace	12 to 36
Front of wall-mounted warm-air register	18 to 36
Heating duct, not insulated	9 to 18
Hot water pipe, not insulated	6 to 12
Kitchen range top	9 to 18
Luminaire up to 250 watts	3 to 6
Luminaire 250 watts up to 499 watts	6 to 12
Oven	9 to 18
Side of ceiling or wall warm-air register	12 to 24
Vent connector or chimney connector	9 to 18
Water heater, furnace or boiler	3 to 6

For SI: 1 inch = 25.4 mm.

a. Sprinklers shall not be located at distances less than the minimum table distance unless the sprinkler listing allows a lesser distance.

b. Distances shall be measured in a straight line from the nearest edge of the heat source to the nearest edge of the sprinkler.

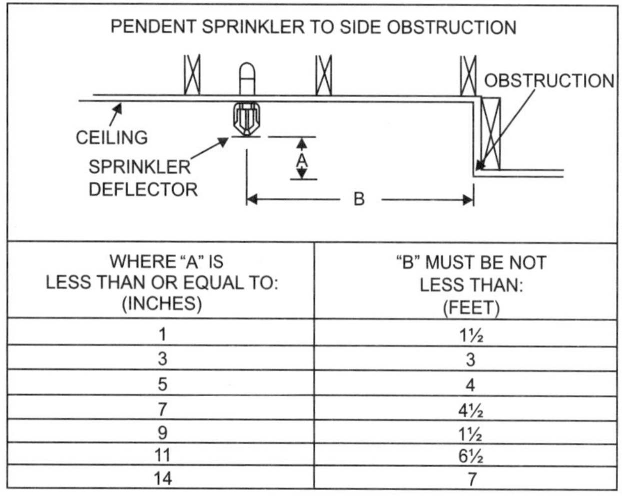

WHERE "A" IS LESS THAN OR EQUAL TO: (INCHES)	"B" MUST BE NOT LESS THAN: (FEET)
1	1½
3	3
5	4
7	4½
9	1½
11	6½
14	7

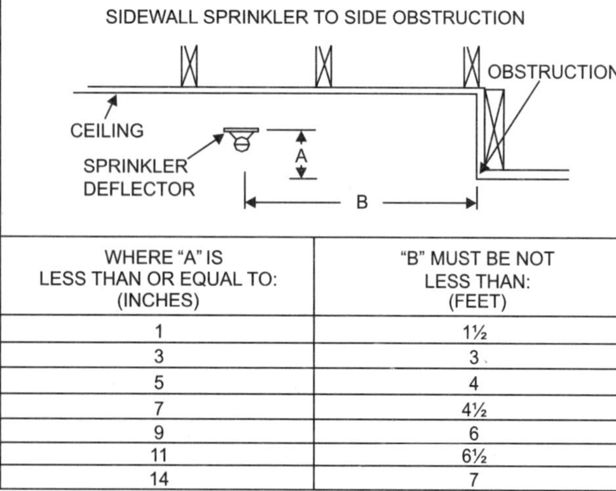

WHERE "A" IS LESS THAN OR EQUAL TO: (INCHES)	"B" MUST BE NOT LESS THAN: (FEET)
1	1½
3	3
5	4
7	4½
9	6
11	6½
14	7

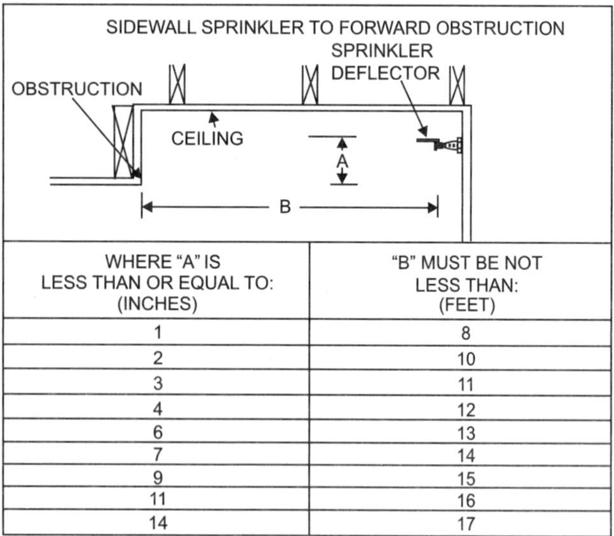

WHERE "A" IS LESS THAN OR EQUAL TO: (INCHES)	"B" MUST BE NOT LESS THAN: (FEET)
1	8
2	10
3	11
4	12
6	13
7	14
9	15
11	16
14	17

For SI: 1 inch = 25.4 mm, 1 foot = 304.8 mm.

FIGURE P2904.2.4.2
MINIMUM ALLOWABLE DISTANCE BETWEEN SPRINKLER AND OBSTRUCTION

P2904.2.4.2.1 Additional requirements for pendent sprinklers. Pendent sprinklers within 3 feet (915 mm) of the center of a ceiling fan, surface-mounted ceiling luminaire or similar object shall be considered to be obstructed, and additional sprinklers shall be installed.

P2904.2.4.2.2 Additional requirements for sidewall sprinklers. Sidewall sprinklers within 5 feet (1524 mm) of the center of a ceiling fan, surface-mounted ceiling luminaire or similar object shall be considered to be obstructed, and additional sprinklers shall be installed.

P2904.2.5 Sprinkler installation on systems assembled with solvent cement. The solvent cementing of threaded adapter fittings shall be completed and threaded adapters for sprinklers shall be verified as being clear of excess cement prior to the installation of sprinklers on systems assembled with solvent cement.

P2904.2.6 Sprinkler modifications prohibited. Painting, caulking or modifying of sprinklers shall be prohibited. Sprinklers that have been painted, caulked, modified or damaged shall be replaced with new sprinklers.

P2904.3 Sprinkler piping system. Sprinkler piping shall be supported in accordance with requirements for cold water distribution piping. Sprinkler piping shall comply with the requirements for cold water distribution piping. For multipurpose piping systems, the sprinkler piping shall connect to and be a part of the cold water distribution piping system.

Exception: For plastic piping, it shall be permissible to follow the manufacturer's installation instructions.

P2904.3.1 Nonmetallic pipe and tubing. Nonmetallic pipe and tubing, such as CPVC, PEX, and PE-RT shall be *listed* for use in residential fire sprinkler systems.

P2904.3.1.1 Nonmetallic pipe protection. Nonmetallic pipe and tubing systems shall be protected from exposure to the living space by a layer of not less than $^{3}/_{8}$-inch-thick (9.5 mm) gypsum wallboard, $^{1}/_{2}$-inch-thick (13 mm) plywood, or other material having a 15-minute fire rating.

Exceptions:

1. Pipe protection shall not be required in areas that do not require protection with sprinklers as specified in Section P2904.1.1.
2. Pipe protection shall not be required where exposed piping is permitted by the pipe *listing.*

P2904.3.2 Shutoff valves prohibited. With the exception of shutoff valves for the entire water distribution system or a single master control valve for the automatic sprinkler system that is locked in the open position, valves shall not be installed in any location where the valve would isolate piping serving one or more sprinklers.

P2904.3.3 Single dwelling limit. Piping beyond the service valve located at the beginning of the water distribution system shall not serve more than one *dwelling.*

P2904.3.4 Drain. A means to drain the sprinkler system shall be provided on the system side of the water distribution shutoff valve.

P2904.4 Determining system design flow. The flow for sizing the sprinkler piping system shall be based on Sections P2904.4.1 and P2904.4.2.

P2904.4.1 Determining required flow rate for each sprinkler. The minimum required flow for each sprinkler shall be determined using the sprinkler manufacturer's published data for the specific sprinkler model based on all of the following:

1. The area of coverage.
2. The ceiling configuration, in accordance with Sections P2904.4.1.1 through P2904.4.1.3.
3. The temperature rating.
4. Any additional conditions specified by the sprinkler manufacturer.

P2904.4.1.1 Ceiling configurations. Manufacturer's published flow rates for sprinklers tested under a ceiling 8 feet (2438 mm) in height, in accordance with the sprinkler listing, shall be used for the following ceiling configurations, provided that the ceiling surface does not have significant irregularities, lumps or indentations and is continuous in a single plane.

1. Ceilings that are horizontal or that have a slope not exceeding 8 units vertical in 12 units horizontal (67 percent), without beams, provided that the ceiling height, measured to the highest point, does not exceed 24 feet (7315 mm) above the floor. Where the slope exceeds 2 units vertical in 12 units horizontal (17 percent), the highest sprinkler installed along the sloped portion of a ceiling shall be positioned above all communicating openings connecting the sloped ceiling compartment with an adjacent space.
2. Ceilings that are horizontal or that have a slope not exceeding 8 units vertical in 12 units horizontal (67 percent), with beams, provided that the ceiling height, measured to the highest point, does not exceed 24 feet (7315 mm) above the floor. Beams shall not exceed 14 inches (350 mm) in depth, and pendent sprinklers shall be installed under the beams as described at the end of this section. The compartment containing the beamed ceiling shall not exceed 600 square feet (56 m^2) in area. Where the slope does not exceed 2 units vertical in 12 units horizontal (17 percent), the highest sprinkler in the compartment shall be above all communicating openings connecting the compartment with an adjacent space. Where the slope exceeds 2 units vertical in 12 units horizontal (17 percent), the highest sprinkler installed along the sloped portion of a ceiling shall be positioned above all communicating openings connecting the sloped ceiling compartment with an adjacent space.
3. Ceilings that have a slope exceeding 2 units vertical in 12 units horizontal (17 percent) but not exceeding 8 units vertical in 12 units horizontal (67 percent), with beams of any depth, provided that the ceiling height, measured to the highest point, does not exceed 24 feet (7315 mm) above the floor. Sidewall or pendent sprinklers shall be installed in each pocket formed by beams. The compartment containing the sloped, beamed ceiling shall not exceed 600 square feet (56 m^2) in area.

Pendent, recessed pendent and flush-type pendent sprinklers installed directly under a beam having a maximum depth of 14 inches (356 mm) shall have the sprinkler deflector located not less than 1 inch (25 mm) or more than 2 inches (51 mm) below the bottom of the beam. Pendent sprinklers installed adjacent to the bottom of a beam having a maximum depth of 14 inches (356 mm) shall be positioned such that the vertical centerline of the sprinkler is not more than 2 inches (51 mm) from the edge of the beam, with the sprinkler deflector located not less than 1 inch (25 mm) or more than 2 inches (51 mm) below the bottom of the beam. Pendent sprinklers shall also be permitted to be installed less than 1 inch (25 mm) below the bottom of a beam where in accordance with manufacturer's instructions for installation of flush sprinklers.

P2904.4.1.2 Ceiling configurations with special sprinkler listings. For ceiling configurations not specified in Section 2904.4.1.1, the manufacturer's published flow rate for sprinklers that have been listed for protection of such configurations shall be used.

P2904.4.1.3 Other ceiling configurations. For ceiling configurations not addressed by Section P2904.4.1.1 or P2904.4.1.2, the flow rate shall be subject to approval by the building official.

P2904.4.2 System design flow rate. The design flow rate for the system shall be based on the following:

1. The design flow rate for a room having only one sprinkler shall be the flow rate required for that sprinkler, as determined by Section P2904.4.1.
2. The design flow rate for a room having two or more sprinklers shall be determined by identifying the sprinkler in that room with the highest required flow rate, based on Section P2904.4.1, and multiplying that flow rate by 2.
3. Where the sprinkler manufacturer specifies different criteria for ceiling configurations that are not smooth, flat and horizontal, the required flow rate for that room shall comply with the sprinkler manufacturer's instructions.
4. The design flow rate for the sprinkler system shall be the flow required by the room with the largest flow rate, based on Items 1, 2 and 3.

5. For the purpose of this section, it shall be permissible to reduce the design flow rate for a room by subdividing the space into two or more rooms, where each room is evaluated separately with respect to the required design flow rate. Each room shall be bounded by walls and a ceiling. Openings in walls shall have a lintel not less than 8 inches (203 mm) in depth and each lintel shall form a solid barrier between the ceiling and the top of the opening.

P2904.5 Water supply. The water supply shall provide not less than the required design flow rate for sprinklers in accordance with Section P2904.4.2 at a pressure not less than that used to comply with Section P2904.6.

P2904.5.1 Water supply from individual sources. Where a *dwelling unit* water supply is from a tank system, a private well system or a combination of these, the available water supply shall be based on the minimum pressure control setting for the pump.

P2904.5.2 Required capacity. The water supply shall have the capacity to provide the required design flow rate for sprinklers for a period of time as follows:

1. Seven minutes for *dwelling units* one *story* in height and less than 2,000 square feet (186 m^2) in area.
2. Ten minutes for *dwelling units* two or more stories in height or equal to or greater than 2,000 square feet (186 m^2) in area.

Where a well system, a water supply tank system or a combination thereof is used, any combination of well capacity and tank storage shall be permitted to meet the capacity requirement.

P2904.6 Pipe sizing. The piping to sprinklers shall be sized for the flow required by Section P2904.4.2. The flow required to supply the plumbing fixtures shall not be required to be added to the sprinkler design flow.

P2904.6.1 Method of sizing pipe. Piping supplying sprinklers shall be sized using the prescriptive method in Section P2904.6.2 or by hydraulic calculation in accordance with NFPA 13D. The minimum pipe size from the water supply source to any sprinkler shall be $^3/_4$ inch (19 mm) nominal. Threaded adapter fittings at the point where sprinklers are attached to the piping shall be not less than $^1/_2$ inch (13 mm) nominal.

P2904.6.2 Prescriptive pipe sizing method. Pipe shall be sized by determining the available pressure to offset friction loss in piping and identifying a piping material, diameter and length using the equation in Section P2904.6.2.1 and the procedure in Section P2904.6.2.2.

P2904.6.2.1 Available pressure equation. The pressure available to offset friction loss in the interior piping system (P_t) shall be determined in accordance with the Equation 29-1.

$$P_t = P_{sup} - PL_{svc} - PL_m - PL_d - PL_e - P_{sp}$$

(Equation 29-1)

where:

P_t = Pressure used in applying Tables P2904.6.2(4) through P2904.6.2(9).

P_{sup} = Pressure available from the water supply source.

PL_{svc} = Pressure loss in the water service pipe. [Table P2904.6.2(1)]

PL_m = Pressure loss in the water meter. [Table P2904.6.2(2)]

PL_d = Pressure loss from devices other than the water meter.

PL_e = Pressure loss associated with changes in elevation. [Table P2904.6.2(3)]

P_{sp} = Maximum pressure required by a sprinkler.

P2904.6.2.2 Calculation procedure. Determination of the required size for water distribution piping shall be in accordance with the following procedure:

Step 1—Determine P_{sup}

Obtain the static supply pressure that will be available from the water main from the water purveyor, or for an individual source, the available supply pressure shall be in accordance with Section P2904.5.1.

Step 2—Determine PL_{svc}

Use Table P2904.6.2(1) to determine the pressure loss in the water service pipe based on the selected size of the water service.

Step 3—Determine PL_m

Use Table P2904.6.2(2) to determine the pressure loss from the water meter, based on the selected water meter size.

Step 4—Determine PL_d

Determine the pressure loss from devices other than the water meter installed in the piping system supplying sprinklers, such as pressure-reducing valves, backflow preventers, water softeners or water filters. Device pressure losses shall be based on the device manufacturer's specifications. The flow rate used to determine pressure loss shall be the rate from Section P2904.4.2, except that 5 gpm (0.3 L/s) shall be added where the device is installed in a water service pipe that supplies more than one *dwelling*. As an alternative to deducting pressure loss for a device, an automatic bypass valve shall be installed to divert flow around the device when a sprinkler activates.

Step 5—Determine PL_e

Use Table P2904.6.2(3) to determine the pressure loss associated with changes in elevation. The elevation used in applying the table shall be the difference between the elevation where the water source pressure was measured and the elevation of the highest sprinkler.

Step 6—Determine P_{sp}

Determine the maximum pressure required by any individual sprinkler based on the flow rate from Section P2904.4.1. The required pressure is provided in the sprinkler manufacturer's published data for the specific sprinkler model based on the selected flow rate.

Step 7—Calculate P_t

Using Equation 29-1, calculate the pressure available to offset friction loss in water-distribution piping between the service valve and the sprinklers.

Step 8—Determine the maximum allowable pipe length

Use Tables P2904.6.2(4) through P2904.6.2(9) to select a material and size for water distribution piping. The piping material and size shall be acceptable if the *developed length* of pipe between the service valve and the most remote sprinkler does not exceed the maximum allowable length specified by the applicable table. Interpolation of P_t between the tabular values shall be permitted.

The maximum allowable length of piping in Tables P2904.6.2(4) through P2904.6.2(9) incorporates an adjustment for pipe fittings. Additional consideration of friction losses associated with pipe fittings shall not be required.

P2904.7 Instructions and signs. An owner's manual for the fire sprinkler system shall be provided to the *owner.* A sign or valve tag shall be installed at the main shutoff valve to the water distribution system stating, "Warning, the water system for this home supplies fire sprinklers that require certain flows and pressures to fight a fire. Devices that restrict the flow or decrease the pressure or automatically shut off the water to the fire sprinkler system, such as water softeners, filtration systems and automatic shutoff valves, shall not be added to this system without a review of the fire sprinkler system by a fire protection specialist. Do not remove this sign."

P2904.8 Inspections. The water distribution system shall be inspected in accordance with Sections P2904.8.1 and P2904.8.2.

TABLE P2904.6.2(1)
WATER SERVICE PRESSURE LOSS (PL_{svc})[a, b]

FLOW RATE[c] (gpm)	$^3/_4$-INCH WATER SERVICE PRESSURE LOSS (psi)				1-INCH WATER SERVICE PRESSURE LOSS (psi)				$1^1/_4$-INCH WATER SERVICE PRESSURE LOSS (psi)			
	Length of water service pipe (feet)				Length of water service pipe (feet)				Length of water service pipe (feet)			
	40 or less	41 to 75	76 to 100	101 to 150	40 or less	41 to 75	76 to 100	101 to 150	40 or less	41 to 75	76 to 100	101 to 150
8	5.1	8.7	11.8	17.4	1.5	2.5	3.4	5.1	0.6	1.0	1.3	1.9
10	7.7	13.1	17.8	26.3	2.3	3.8	5.2	7.7	0.8	1.4	2.0	2.9
12	10.8	18.4	24.9	NP	3.2	5.4	7.3	10.7	1.2	2.0	2.7	4.0
14	14.4	24.5	NP	NP	4.2	7.1	9.6	14.3	1.6	2.7	3.6	5.4
16	18.4	NP	NP	NP	5.4	9.1	12.4	18.3	2.0	3.4	4.7	6.9
18	22.9	NP	NP	NP	6.7	11.4	15.4	22.7	2.5	4.3	5.8	8.6
20	27.8	NP	NP	NP	8.1	13.8	18.7	27.6	3.1	5.2	7.0	10.4
22	NP	NP	NP	NP	9.7	16.5	22.3	NP	3.7	6.2	8.4	12.4
24	NP	NP	NP	NP	11.4	19.3	26.2	NP	4.3	7.3	9.9	14.6
26	NP	NP	NP	NP	13.2	22.4	NP	NP	5.0	8.5	11.4	16.9
28	NP	NP	NP	NP	15.1	25.7	NP	NP	5.7	9.7	13.1	19.4
30	NP	NP	NP	NP	17.2	NP	NP	NP	6.5	11.0	14.9	22.0
32	NP	NP	NP	NP	19.4	NP	NP	NP	7.3	12.4	16.8	24.8
34	NP	NP	NP	NP	21.7	NP	NP	NP	8.2	13.9	18.8	NP
36	NP	NP	NP	NP	24.1	NP	NP	NP	9.1	15.4	20.9	NP

For SI: 1 inch = 25.4 mm, 1 foot = 304.8 mm, 1 gallon per minute = 0.063 L/s, 1 pound per square inch = 6.895 kPa.

NP = Not Permitted. Pressure loss exceeds reasonable limits.

a. Values are applicable for underground piping materials listed in Table P2906.4 and are based on an SDR of 11 and a Hazen Williams C Factor of 150.

b. Values include the following length allowances for fittings: 25-percent length increase for actual lengths up to 100 feet and 15-percent length increase for actual lengths over 100 feet.

c. Flow rate from Section P2904.4.2. Add 5 gpm to the flow rate required by Section P2904.4.2 where the water service pipe supplies more than one dwelling.

TABLE P2904.6.2(2)
MINIMUM WATER METER PRESSURE LOSS (PL_m)[a]

FLOW RATE (gallons per minute, gpm)[b]	5/8-INCH METER PRESSURE LOSS (pounds per square inch, psi)	3/4-INCH METER PRESSURE LESS (pounds per square inch, psi)	1-INCH METER PRESSURE LOSS (pounds per square inch, psi)
8	3	3	1
10	3	3	1
12	4	3	1
14	6	5	1
16	7	6	1
18	9	7	2
20	11	9	2
23	14	11	3
26	18	14	3
31	26	22	4
39	38	35	6
52	NP	NP	10

For SI: 1 inch = 25.4 mm, 1 pound per square inch = 6.895 kPa, 1 gallon per minute = 0.063 L/s.
NP = Not permitted unless the actual water meter pressure loss is known.
a. Table P2904.6.2(2) establishes conservative values for water meter pressure loss or installations where the water meter loss is unknown. Where the actual water meter pressure loss is published and available from the meter manufacturer, PL_m shall be the published pressure loss for the selected meter.
b. Flow rate from Section P2904.4.2. Add 5 gpm to the flow rate required by Section P2904.4.2 where the water service pipe supplies more than one dwelling.

TABLE P2904.6.2(3)
ELEVATION LOSS (PL_e)

ELEVATION (feet)	PRESSURE LOSS (psi)
5	2.2
10	4.4
15	6.5
20	8.7
25	10.9
30	13
35	15.2
40	17.4

For SI: 1 foot = 304.8 mm, 1 pound per square inch = 6.895 kPa.

TABLE P2904.6.2(4)
ALLOWABLE PIPE LENGTH FOR 3/4-INCH TYPE M COPPER WATER TUBING

SPRINKLER FLOW RATE[a] (gpm)	WATER DISTRIBUTION SIZE (inch)	AVAILABLE PRESSURE—P_t (psi)									
		15	20	25	30	35	40	45	50	55	60
		Allowable length of pipe from service valve to farthest sprinkler (feet)									
8	3/4	217	289	361	434	506	578	650	723	795	867
9	3/4	174	232	291	349	407	465	523	581	639	697
10	3/4	143	191	239	287	335	383	430	478	526	574
11	3/4	120	160	200	241	281	321	361	401	441	481
12	3/4	102	137	171	205	239	273	307	341	375	410
13	3/4	88	118	147	177	206	235	265	294	324	353
14	3/4	77	103	128	154	180	205	231	257	282	308
15	3/4	68	90	113	136	158	181	203	226	248	271
16	3/4	60	80	100	120	140	160	180	200	220	241
17	3/4	54	72	90	108	125	143	161	179	197	215
18	3/4	48	64	81	97	113	129	145	161	177	193
19	3/4	44	58	73	88	102	117	131	146	160	175
20	3/4	40	53	66	80	93	106	119	133	146	159
21	3/4	36	48	61	73	85	97	109	121	133	145
22	3/4	33	44	56	67	78	89	100	111	122	133
23	3/4	31	41	51	61	72	82	92	102	113	123
24	3/4	28	38	47	57	66	76	85	95	104	114
25	3/4	26	35	44	53	61	70	79	88	97	105
26	3/4	24	33	41	49	57	65	73	82	90	98
27	3/4	23	30	38	46	53	61	69	76	84	91
28	3/4	21	28	36	43	50	57	64	71	78	85
29	3/4	20	27	33	40	47	53	60	67	73	80
30	3/4	19	25	31	38	44	50	56	63	69	75
31	3/4	18	24	29	35	41	47	53	59	65	71
32	3/4	17	22	28	33	39	44	50	56	61	67
33	3/4	16	21	26	32	37	42	47	53	58	63
34	3/4	NP	20	25	30	35	40	45	50	55	60
35	3/4	NP	19	24	28	33	38	42	47	52	57
36	3/4	NP	18	22	27	31	36	40	45	49	54
37	3/4	NP	17	21	26	30	34	38	43	47	51
38	3/4	NP	16	20	24	28	32	36	40	45	49
39	3/4	NP	15	19	23	27	31	35	39	42	46
40	3/4	NP	NP	18	22	26	29	33	37	40	44

For SI: 1 inch = 25.4 mm, 1 foot = 304.8 mm, 1 pound per square inch = 6.895 kPa, 1 gallon per minute = 0.963 L/s.
NP = Not Permitted.
a. Flow rate from Section P2904.4.2.

TABLE P2904.6.2(5)
ALLOWABLE PIPE LENGTH FOR 1-INCH TYPE M COPPER WATER TUBING

SPRINKLER FLOW RATE[a] (gpm)	WATER DISTRIBUTION SIZE (inch)	AVAILABLE PRESSURE—P_t (psi)									
		15	20	25	30	35	40	45	50	55	60
		Allowable length of pipe from service valve to farthest sprinkler (feet)									
8	1	806	1,075	1,343	1,612	1,881	2,149	2,418	2,687	2,955	3,224
9	1	648	864	1,080	1,296	1,512	1,728	1,945	2,161	2,377	2,593
10	1	533	711	889	1,067	1,245	1,422	1,600	1,778	1,956	2,134
11	1	447	586	745	894	1,043	1,192	1,341	1,491	1,640	1,789
12	1	381	508	634	761	888	1,015	1,142	1,269	1,396	1,523
13	1	328	438	547	657	766	875	985	1,094	1,204	1,313
14	1	286	382	477	572	668	763	859	954	1,049	1,145
15	1	252	336	420	504	588	672	756	840	924	1,008
16	1	224	298	373	447	522	596	671	745	820	894
17	1	200	266	333	400	466	533	600	666	733	799
18	1	180	240	300	360	420	479	539	599	659	719
19	1	163	217	271	325	380	434	488	542	597	651
20	1	148	197	247	296	345	395	444	493	543	592
21	1	135	180	225	270	315	360	406	451	496	541
22	1	124	165	207	248	289	331	372	413	455	496
23	1	114	152	190	228	267	305	343	381	419	457
24	1	106	141	176	211	246	282	317	352	387	422
25	1	98	131	163	196	228	261	294	326	359	392
26	1	91	121	152	182	212	243	273	304	334	364
27	1	85	113	142	170	198	226	255	283	311	340
28	1	79	106	132	159	185	212	238	265	291	318
29	1	74	99	124	149	174	198	223	248	273	298
30	1	70	93	116	140	163	186	210	233	256	280
31	1	66	88	110	132	153	175	197	219	241	263
32	1	62	83	103	124	145	165	186	207	227	248
33	1	59	78	98	117	137	156	176	195	215	234
34	1	55	74	92	111	129	148	166	185	203	222
35	1	53	70	88	105	123	140	158	175	193	210
36	1	50	66	83	100	116	133	150	166	183	199
37	1	47	63	79	95	111	126	142	158	174	190
38	1	45	60	75	90	105	120	135	150	165	181
39	1	43	57	72	86	100	115	129	143	158	172
40	1	41	55	68	82	96	109	123	137	150	164

For SI: 1 inch = 25.4 mm, 1 foot = 304.8 mm, 1 pound per square inch = 6.895 kPa, 1 gallon per minute = 0.963 L/s.
a. Flow rate from Section P2904.4.2.

TABLE P2904.6.2(6)
ALLOWABLE PIPE LENGTH FOR 3/4-INCH CPVC PIPE

SPRINKLER FLOW RATE[a] (gpm)	WATER DISTRIBUTION SIZE (inch)	AVAILABLE PRESSURE—P_t (psi)									
		15	20	25	30	35	40	45	50	55	60
		Allowable length of pipe from service valve to farthest sprinkler (feet)									
8	3/4	348	465	581	697	813	929	1,045	1,161	1,278	1,394
9	3/4	280	374	467	560	654	747	841	934	1,027	1,121
10	3/4	231	307	384	461	538	615	692	769	845	922
11	3/4	193	258	322	387	451	515	580	644	709	773
12	3/4	165	219	274	329	384	439	494	549	603	658
13	3/4	142	189	237	284	331	378	426	473	520	568
14	3/4	124	165	206	247	289	330	371	412	454	495
15	3/4	109	145	182	218	254	290	327	363	399	436
16	3/4	97	129	161	193	226	258	290	322	354	387
17	3/4	86	115	144	173	202	230	259	288	317	346
18	3/4	78	104	130	155	181	207	233	259	285	311
19	3/4	70	94	117	141	164	188	211	234	258	281
20	3/4	64	85	107	128	149	171	192	213	235	256
21	3/4	58	78	97	117	136	156	175	195	214	234
22	3/4	54	71	89	107	125	143	161	179	197	214
23	3/4	49	66	82	99	115	132	148	165	181	198
24	3/4	46	61	76	91	107	122	137	152	167	183
25	3/4	42	56	71	85	99	113	127	141	155	169
26	3/4	39	52	66	79	92	105	118	131	144	157
27	3/4	37	49	61	73	86	98	110	122	135	147
28	3/4	34	46	57	69	80	92	103	114	126	137
29	3/4	32	43	54	64	75	86	96	107	118	129
30	3/4	30	40	50	60	70	81	91	101	111	121
31	3/4	28	38	47	57	66	76	85	95	104	114
32	3/4	27	36	45	54	63	71	80	89	98	107
33	3/4	25	34	42	51	59	68	76	84	93	101
34	3/4	24	32	40	48	56	64	72	80	88	96
35	3/4	23	30	38	45	53	61	68	76	83	91
36	3/4	22	29	36	43	50	57	65	72	79	86
37	3/4	20	27	34	41	48	55	61	68	75	82
38	3/4	20	26	33	39	46	52	59	65	72	78
39	3/4	19	25	31	37	43	50	56	62	68	74
40	3/4	18	24	30	35	41	47	53	59	65	71

For SI: 1 inch = 25.4 mm, 1 foot = 304.8 mm, 1 pound per square inch = 6.895 kPa, 1 gallon per minute = 0.963 L/s.
a. Flow rate from Section P2904.4.2.

TABLE P2904.6.2(7)
ALLOWABLE PIPE LENGTH FOR 1-INCH CPVC PIPE

SPRINKLER FLOW RATE[a] (gpm)	WATER DISTRIBUTION SIZE (inch)	AVAILABLE PRESSURE—P_t (psi)									
		15	20	25	30	35	40	45	50	55	60
		Allowable length of pipe from service valve to farthest sprinkler (feet)									
8	1	1,049	1,398	1,748	2,098	2,447	2,797	3,146	3,496	3,845	4,195
9	1	843	1,125	1,406	1,687	1,968	2,249	2,530	2,811	3,093	3,374
10	1	694	925	1,157	1,388	1,619	1,851	2,082	2,314	2,545	2,776
11	1	582	776	970	1,164	1,358	1,552	1,746	1,940	2,133	2,327
12	1	495	660	826	991	1,156	1,321	1,486	1,651	1,816	1,981
13	1	427	570	712	854	997	1,139	1,281	1,424	1,566	1,709
14	1	372	497	621	745	869	993	1,117	1,241	1,366	1,490
15	1	328	437	546	656	765	874	983	1,093	1,202	1,311
16	1	291	388	485	582	679	776	873	970	1,067	1,164
17	1	260	347	433	520	607	693	780	867	954	1,040
18	1	234	312	390	468	546	624	702	780	858	936
19	1	212	282	353	423	494	565	635	706	776	847
20	1	193	257	321	385	449	513	578	642	706	770
21	1	176	235	293	352	410	469	528	586	645	704
22	1	161	215	269	323	377	430	484	538	592	646
23	1	149	198	248	297	347	396	446	496	545	595
24	1	137	183	229	275	321	366	412	458	504	550
25	1	127	170	212	255	297	340	382	425	467	510
26	1	118	158	197	237	276	316	355	395	434	474
27	1	111	147	184	221	258	295	332	368	405	442
28	1	103	138	172	207	241	275	310	344	379	413
29	1	97	129	161	194	226	258	290	323	355	387
30	1	91	121	152	182	212	242	273	303	333	364
31	1	86	114	143	171	200	228	257	285	314	342
32	1	81	108	134	161	188	215	242	269	296	323
33	1	76	102	127	152	178	203	229	254	280	305
34	1	72	96	120	144	168	192	216	240	265	289
35	1	68	91	114	137	160	182	205	228	251	273
36	1	65	87	108	130	151	173	195	216	238	260
37	1	62	82	103	123	144	165	185	206	226	247
38	1	59	78	98	117	137	157	176	196	215	235
39	1	56	75	93	112	131	149	168	187	205	224
40	1	53	71	89	107	125	142	160	178	196	214

For SI: 1 inch = 25.4 mm, 1 foot = 304.8 mm, 1 pound per square inch = 6.895 kPa, 1 gallon per minute = 0.963 L/s.

a. Flow rate from Section P2904.4.2.

TABLE P2904.6.2(8) ALLOWABLE PIPE LENGTH FOR 3/4-INCH PEX AND PE-RT TUBING

SPRINKLER FLOW RATE[a] (gpm)	WATER DISTRIBUTION SIZE (inch)	AVAILABLE PRESSURE—P_t (psi)									
		15	20	25	30	35	40	45	50	55	60
		Allowable length of pipe from service valve to farthest sprinkler (feet)									
8	3/4	93	123	154	185	216	247	278	309	339	370
9	3/4	74	99	124	149	174	199	223	248	273	298
10	3/4	61	82	102	123	143	163	184	204	225	245
11	3/4	51	68	86	103	120	137	154	171	188	205
12	3/4	44	58	73	87	102	117	131	146	160	175
13	3/4	38	50	63	75	88	101	113	126	138	151
14	3/4	33	44	55	66	77	88	99	110	121	132
15	3/4	29	39	48	58	68	77	87	96	106	116
16	3/4	26	34	43	51	60	68	77	86	94	103
17	3/4	23	31	38	46	54	61	69	77	84	92
18	3/4	21	28	34	41	48	55	62	69	76	83
19	3/4	19	25	31	37	44	50	56	62	69	75
20	3/4	17	23	28	34	40	45	51	57	62	68
21	3/4	16	21	26	31	36	41	47	52	57	62
22	3/4	NP	19	24	28	33	38	43	47	52	57
23	3/4	NP	17	22	26	31	35	39	44	48	52
24	3/4	NP	16	20	24	28	32	36	40	44	49
25	3/4	NP	NP	19	22	26	30	34	37	41	45
26	3/4	NP	NP	17	21	24	28	31	35	38	42
27	3/4	NP	NP	16	20	23	26	29	33	36	39
28	3/4	NP	NP	15	18	21	24	27	30	33	36
29	3/4	NP	NP	NP	17	20	23	26	28	31	34
30	3/4	NP	NP	NP	16	19	21	24	27	29	32
31	3/4	NP	NP	NP	15	18	20	23	25	28	30
32	3/4	NP	NP	NP	NP	17	19	21	24	26	28
33	3/4	NP	NP	NP	NP	16	18	20	22	25	27
34	3/4	NP	NP	NP	NP	NP	17	19	21	23	25
35	3/4	NP	NP	NP	NP	NP	16	18	20	22	24
36	3/4	NP	NP	NP	NP	NP	15	17	19	21	23
37	3/4	NP	NP	NP	NP	NP	NP	16	18	20	22
38	3/4	NP	NP	NP	NP	NP	NP	16	17	19	21
39	3/4	NP	NP	NP	NP	NP	NP	NP	16	18	20
40	3/4	NP	NP	NP	NP	NP	NP	NP	16	17	19

For SI: 1 inch = 25.4 mm, 1 foot = 304.8 mm, 1 pound per square inch = 6.895 kPa, 1 gallon per minute = 0.963 L/s.
NP = Not Permitted.
a. Flow rate from Section P2904.4.2.

TABLE P2904.6.2(9)
ALLOWABLE PIPE LENGTH FOR 1-INCH PEX AND PE-RT TUBING

SPRINKLER FLOW RATE[a] (gpm)	WATER DISTRIBUTION SIZE (inch)	AVAILABLE PRESSURE—P_t (psi)									
		15	20	25	30	35	40	45	50	55	60
		Allowable length of pipe from service valve to farthest sprinkler (feet)									
8	1	314	418	523	628	732	837	941	1,046	1,151	1,255
9	1	252	336	421	505	589	673	757	841	925	1,009
10	1	208	277	346	415	485	554	623	692	761	831
11	1	174	232	290	348	406	464	522	580	638	696
12	1	148	198	247	296	346	395	445	494	543	593
13	1	128	170	213	256	298	341	383	426	469	511
14	1	111	149	186	223	260	297	334	371	409	446
15	1	98	131	163	196	229	262	294	327	360	392
16	1	87	116	145	174	203	232	261	290	319	348
17	1	78	104	130	156	182	208	233	259	285	311
18	1	70	93	117	140	163	187	210	233	257	280
19	1	63	84	106	127	148	169	190	211	232	253
20	1	58	77	96	115	134	154	173	192	211	230
21	1	53	70	88	105	123	140	158	175	193	211
22	1	48	64	80	97	113	129	145	161	177	193
23	1	44	59	74	89	104	119	133	148	163	178
24	1	41	55	69	82	96	110	123	137	151	164
25	1	38	51	64	76	89	102	114	127	140	152
26	1	35	47	59	71	83	95	106	118	130	142
27	1	33	44	55	66	77	88	99	110	121	132
28	1	31	41	52	62	72	82	93	103	113	124
29	1	29	39	48	58	68	77	87	97	106	116
30	1	27	36	45	54	63	73	82	91	100	109
31	1	26	34	43	51	60	68	77	85	94	102
32	1	24	32	40	48	56	64	72	80	89	97
33	1	23	30	38	46	53	61	68	76	84	91
34	1	22	29	36	43	50	58	65	72	79	86
35	1	20	27	34	41	48	55	61	68	75	82
36	1	19	26	32	39	45	52	58	65	71	78
37	1	18	25	31	37	43	49	55	62	68	74
38	1	18	23	29	35	41	47	53	59	64	70
39	1	17	22	28	33	39	45	50	56	61	67
40	1	16	21	27	32	37	43	48	53	59	64

For SI: 1 inch = 25.4 mm, 1 foot = 304.8 mm, 1 pound per square inch = 6.895 kPa, 1 gallon per minute = 0.963 L/s.

a. Flow rate from Section P2904.4.2.

P2904.8.1 Preconcealment inspection. The following items shall be verified prior to the concealment of any sprinkler system piping:

1. Sprinklers are installed in all areas as required by Section P2904.1.1.
2. Where sprinkler water spray patterns are obstructed by construction features, luminaires or ceiling fans, additional sprinklers are installed as required by Section P2904.2.4.2.
3. Sprinklers are the correct temperature rating and are installed at or beyond the required separation distances from heat sources as required by Sections P2904.2.1 and P2904.2.2.
4. The pipe size equals or exceeds the size used in applying Tables P2904.6.2(4) through P2904.6.2(9) or, if the piping system was hydraulically calculated in accordance with Section P2904.6.1, the size used in the hydraulic calculation.
5. The pipe length does not exceed the length permitted by Tables P2904.6.2(4) through P2904.6.2(9) or, if the piping system was hydraulically calculated in accordance with Section P2904.6.1, pipe lengths and fittings do not exceed those used in the hydraulic calculation.
6. Nonmetallic piping that conveys water to sprinklers is *listed* for use with fire sprinklers.
7. Piping is supported in accordance with the pipe manufacturer's and sprinkler manufacturer's installation instructions.
8. The piping system is tested in accordance with Section P2503.7.

P2904.8.2 Final inspection. The following items shall be verified upon completion of the system:

1. Sprinklers are not painted, damaged or otherwise hindered from operation.
2. Where a pump is required to provide water to the system, the pump starts automatically upon system water demand.
3. Pressure-reducing valves, water softeners, water filters or other impairments to water flow that were not part of the original design have not been installed.
4. The sign or valve tag required by Section P2904.7 is installed and the owner's manual for the system is present.

SECTION P2905 HEATED WATER DISTRIBUTION SYSTEMS

P2905.1 Heated water circulation systems and heat trace systems. Circulation systems and heat trace systems that are installed to bring heated water in close proximity to one or more fixtures shall meet the requirements of Section N1103.5.2.

P2905.2 Demand recirculation systems. *Demand recirculation water systems* shall be in accordance with Section N1103.5.2.1.1.

P2905.3 Hot water supply to fixtures. The *developed length* of hot water piping, from the source of the hot water to the fixtures that require hot water, shall not exceed 100 feet (30 480 mm). Water heaters and recirculating system piping shall be considered to be sources of hot water.

SECTION P2906 MATERIALS, JOINTS AND CONNECTIONS

P2906.1 Soil and groundwater. The installation of water service pipe, water distribution pipe, fittings, valves, appurtenances and gaskets shall be prohibited in soil and groundwater that is contaminated with solvents, fuels, organic compounds or other detrimental materials that cause permeation, corrosion, degradation or structural failure of the water service or water distribution piping material.

P2906.1.1 Investigation required. Where detrimental conditions are suspected by or brought to the attention of the *building official*, a chemical analysis of the soil and groundwater conditions shall be required to ascertain the acceptability of the water service material for the specific installation.

P2906.1.2 Detrimental condition. Where a detrimental condition exists, *approved* alternate materials or alternate routing shall be required.

P2906.2 Lead content. The lead content in pipe and fittings used in the water supply system shall be not greater than 8 percent.

P2906.2.1 Lead content of drinking water pipe and fittings. Pipe, pipe fittings, joints, valves, faucets and fixture fittings utilized to supply water for drinking or cooking purposes shall comply with NSF 372 and shall have a weighted average lead content of 0.25-percent lead or less.

P2906.3 Polyethylene plastic piping installation. Polyethylene pipe shall be cut square using a cutter designed for plastic pipe. Except where joined by heat fusion, pipe ends shall be chamfered to remove sharp edges. Pipe that has been kinked shall not be installed. For bends, the installed radius of pipe curvature shall be greater than 30 pipe diameters or the coil radius where bending with the coil. Coiled pipe shall not be bent beyond straight. Bends within 10 pipe diameters of any fitting or valve shall be prohibited. Joints between polyethylene plastic pipe and fittings shall comply with Section P2906.3.1 or P2906.3.2.

P2906.3.1 Heat-fusion joints. Joint surfaces shall be clean and free from moisture. Joint surfaces shall be heated to melting temperature and joined. The joint shall be undisturbed until cool. Joints shall be made in accordance with ASTM D2657.

P2906.3.2 Mechanical joints. Mechanical joints shall be installed in accordance with the manufacturer's instructions.

P2906.4 Water service pipe. Water service pipe shall conform to NSF 61 and shall conform to one of the standards indicated in Table P2906.4. Water service pipe or tubing, installed underground and outside of the structure, shall have a working pressure rating of not less than 160 pounds per square inch at 73°F (1103 kPa at 23°C). Where the water pressure exceeds 160 pounds per square inch (1103 kPa), piping material shall have a rated working pressure equal to or greater than the highest available pressure. Water service piping materials not third-party certified for water distribution shall terminate at or before the full open valve located at the entrance to the structure. Ductile iron water service piping shall be cement mortar lined in accordance with AWWA C104/A21.4.

P2906.4.1 Separation of water service and building sewer. Trenching, pipe installation and backfilling shall be in accordance with Section P2604. Where water service piping is located in the same trench with the *building sewer*, such sewer shall be constructed of materials listed in Table P3002.1(2). Where the *building sewer* piping is not constructed of materials indicated in Table P3002.1(2), the water service pipe and the *building sewer* shall be horizontally separated by not less than 5 feet (1524 mm) of undisturbed or compacted earth. The required separation distance shall not apply where a water service pipe crosses a sewer pipe, provided that the water service is sleeved to a point not less than 5 feet (1524 mm) horizontally from the sewer pipe centerline on both sides of such crossing. The sleeve shall be of pipe materials indicated in Table P2906.4, P3002.1(2) or P3002.2. The required separation distance shall not apply where the bottom of the water service pipe that is located within 5 feet (1524 mm) of the sewer is not less than 12 inches (305 mm) above the highest point of the top of the *building sewer.*

P2906.5 Water distribution pipe. Water distribution piping within *dwelling units* shall conform to NSF 61 and shall conform to one of the standards indicated in Table P2906.5. Water distribution pipe and tubing shall have a pressure rating of not less than 100 psi at 180°F (689 kPa at 82°C).

P2906.6 Fittings. Pipe fittings shall be *approved* for installation with the piping material installed and shall comply with the applicable standards indicated in Table P2906.6. Pipe fittings used in water supply systems shall comply with NSF 61.

P2906.6.1 Saddle tap fittings. The use of saddle tap fittings and combination saddle tap and valve fittings shall be prohibited.

P2906.7 Flexible water connectors. Flexible water connectors, exposed to continuous pressure, shall conform to ASME A112.18.6/CSA B125.6. Access shall be provided to flexible water connectors.

P2906.8 Joint and connection tightness. Joints and connections in the plumbing system shall be gastight and watertight for the intended use or required test pressure.

P2906.9 Plastic pipe joints. Joints in plastic piping shall be made with *approved* fittings by solvent cementing, heat fusion, corrosion-resistant metal clamps with insert fittings or compression connections. Flared joints for polyethylene pipe shall be permitted in accordance with Section P2906.10.1.

TABLE P2906.4
WATER SERVICE PIPE

MATERIAL	STANDARD
Acrylonitrile butadiene styrene (ABS) plastic pipe	ASTM D1527; ASTM D2282
Chlorinated polyvinyl chloride (CPVC) plastic pipe	ASTM D2846; ASTM F441; ASTM F442/F442M; CSA B137.6
Chlorinated polyvinyl chloride/aluminum/chlorinated polyvinyl chloride (CPVC/AL/CPVC) plastic pipe	ASTM F2855
Copper or copper-alloy pipe	ASTM B42; ASTM B43; ASTM B302
Copper or copper-alloy tubing (Type K, WK, L, WL, M or WM)	ASTM B75/B75M; ASTM B88; ASTM B251; ASTM B447
Cross-linked polyethylene/aluminum/cross-linked polyethylene (PEX-AL-PEX) pipe	ASTM F1281; ASTM F2262; CSA B137.10
Cross-linked polyethylene/aluminum/high-density polyethylene (PEX-AL-HDPE) pipe	ASTM F1986
Cross-linked polyethylene (PEX) plastic tubing	ASTM F876; AWWA C904; CSA 137.5
Ductile iron water pipe	AWWA C115/A21.15; AWWA C151/A21.51
Galvanized steel pipe	ASTM A53
Polyethylene/aluminum/polyethylene (PE-AL-PE) pipe	ASTM F1282; CSA B137.9
Polyethylene (PE) plastic pipe	ASTM D2104; ASTM D2239; AWWA C901; CSA 137.1
Polyethylene (PE) plastic tubing	ASTM D2737; AWWA C901; CSA 137.1
Polyethylene of raised temperature (PE-RT) plastic tubing	ASTM F2769; CSA B137.18
Polypropylene (PP) plastic tubing	ASTM F2389; CSA B137.11
Polyvinyl chloride (PVC) plastic pipe	ASTM D1785; ASTM D2241; ASTM D2672; CSA B137.3
Stainless steel (Type 304/304L) pipe	ASTM A312; ASTM A778
Stainless steel (Type 316/316L) pipe	ASTM A312; ASTM A778

TABLE P2906.5
WATER DISTRIBUTION PIPE

MATERIAL	STANDARD
Chlorinated polyvinyl chloride (CPVC) plastic pipe and tubing	ASTM D2846; ASTM F441; ASTM F442/F442M; CSA B137.6
Chlorinated polyvinyl chloride/aluminum/chlorinated polyvinyl chloride (CPVC/AL/CPVC) plastic pipe	ASTM F2855
Copper or copper-alloy pipe	ASTM B42; ASTM B43; ASTM B302
Copper or copper-alloy tubing (Type K, WK, L, WL, M or WM)	ASTM B75/B75M; ASTM B88; ASTM B251; ASTM B447
Cross-linked polyethylene (PEX) plastic tubing	ASTM F876; CSA B137.5
Cross-linked polyethylene/aluminum/cross-linked polyethylene (PEX-AL-PEX) pipe	ASTM F1281; ASTM F2262; CSA B137.10
Cross-linked polyethylene/aluminum/high-density polyethylene (PEX-AL-HDPE) pipe	ASTM F1986
Galvanized steel pipe	ASTM A53
Polyethylene/aluminum/polyethylene (PE-AL-PE) composite pipe	ASTM F1282
Polyethylene of raised temperature (PE-RT) plastic tubing	ASTM F2769; CSA B137.18
Polypropylene (PP) plastic pipe or tubing	ASTM F2389; CSA B137.11
Stainless steel (Type 304/304L) pipe	ASTM A312; ASTM A778

TABLE P2906.6
PIPE FITTINGS

MATERIAL	STANDARD
Acrylonitrile butadiene styrene (ABS) plastic	ASTM D2468
Cast iron	ASME B16.4
Chlorinated polyvinyl chloride (CPVC) plastic	ASSE 1061; ASTM D2846; ASTM F437; ASTM F438; ASTM F439; CSA B137.6
Copper or copper alloy	ASME B16.15; ASME B16.18; ASME B16.22; ASME B16.26; ASME B16.51; ASSE 1061; ASTM F3226
Cross-linked polyethylene/aluminum/high-density polyethylene (PEX-AL-HDPE)	ASTM F1986
Fittings for cross-linked polyethylene (PEX) plastic tubing	ASSE 1061; ASTM F877; ASTM F1807; ASTM F1960; ASTM F2080; ASTM F2098; ASTM F2159; ASTM F2434; ASTM F2735; CSA B137.5
Gray iron and ductile iron	AWWA C110/A21.10; AWWA C153/A21.53
Malleable iron	ASME B16.3
Insert fittings for polyethylene/aluminum/polyethylene (PE-AL-PE) and cross-linked polyethylene/aluminum/cross-linked polyethylene (PEX-AL-PEX)	ASTM F1281; ASTM F1282; ASTM F1974; CSA B137.9; CSA B137.10
Polyethylene (PE) plastic	ASTM D2609; CSA B137.1
Fittings for polyethylene of raised temperature (PE-RT) plastic tubing	ASSE 1061; ASTM D2683; ASTM D3261; ASTM F1055; ASTM F1807; ASTM F2098; ASTM F2159; ASTM F2735; ASTM F2769; CSA B137.18
Polypropylene (PP) plastic pipe or tubing	ASTM F2389; CSA B137.11
Polyvinyl chloride (PVC) plastic	ASTM D2464; ASTM D2466; ASTM D2467; CSA B137.2; CSA B137.3
Stainless steel (Type 304/304L) pipe	ASTM A312; ASTM A778
Stainless steel (Type 316/316L) pipe	ASTM A312; ASTM A778
Steel	ASME B16.9; ASME B16.11; ASME B16.28

P2906.9.1 Solvent cementing. Solvent-cemented joints shall comply with Sections P2906.9.1.1 through P2906.9.1.4.

P2906.9.1.1 ABS plastic pipe. Solvent cement for ABS plastic pipe conforming to ASTM D2235 shall be applied to all joint surfaces.

P2906.9.1.2 CPVC plastic pipe. Joint surfaces shall be clean and free from moisture. Joints shall be made in accordance with the pipe, fitting or solvent cement manufacturer's installation instructions. Where such instructions require a primer to be used, an *approved* primer shall be applied, and a solvent cement, orange in color and conforming to ASTM F493, shall be applied to joint surfaces. Where such instructions allow for a one-step solvent cement, yellow or red in color and conforming to ASTM F493, to be used, the joint surfaces shall not require application of a primer before the solvent cement is applied. The joint shall be made while the cement is wet, and in accordance with ASTM D2846 or ASTM F493. Solvent cement joints shall be permitted above or below ground.

P2906.9.1.3 CPVC/AL/CPVC pipe. Joint surfaces shall be clean and free from moisture, and an *approved* primer shall be applied. Solvent cement, orange in color and conforming to ASTM F493, shall be applied to all joint surfaces. The joint shall be made while the cement is wet, and in accordance with ASTM D2846 or ASTM F493. Solvent-cemented joints shall be installed above or below ground.

Exception: A primer shall not be required where all of the following conditions apply:

1. The solvent cement used is third-party certified as conforming to ASTM F493.
2. The solvent cement used is yellow in color.
3. The solvent cement is used only for joining $^1/_2$-inch (12.7 mm) through 1-inch (25 mm) diameter CPVC/AL/CPVC pipe and CPVC fittings.
4. The CPVC fittings are manufactured in accordance with ASTM D2846.

P2906.9.1.4 PVC plastic pipe. A purple primer, or other *approved* primer that conforms to ASTM F656 shall be applied to PVC solvent-cemented joints. Solvent cement for PVC plastic pipe conforming to ASTM D2564 shall be applied to all joint surfaces.

P2906.10 Cross-linked polyethylene plastic (PEX). Joints between cross-linked polyethylene plastic tubing or fittings shall comply with Section P2906.9.10.1 or Section P2906.9.10.2.

P2906.10.1 Flared joints. Flared pipe ends shall be made by a tool designed for that operation.

P2906.10.2 Mechanical joints. Mechanical joints shall be installed in accordance with the manufacturer's instructions. Fittings for cross-linked polyethylene (PEX) plastic tubing shall comply with the applicable standards indicated in Table P2906.6 and shall be installed in accordance with the manufacturer's instructions. PEX tubing shall be factory marked with the applicable standards for the fittings that the PEX manufacturer specifies for use with the tubing.

P2906.11 Polypropylene (PP) plastic. Joints between polypropylene plastic pipe and fittings shall comply with Section P2906.11.1 or P2906.11.2.

P2906.11.1 Heat-fusion joints. Heat fusion joints for polypropylene pipe and tubing joints shall be installed with socket-type heat-fused polypropylene fittings, butt fusion polypropylene fittings or electrofusion polypropylene fittings. Joint surfaces shall be clean and free from moisture. The joint shall be undisturbed until cool. Joints shall be made in accordance with ASTM F2389.

P2906.11.2 Mechanical and compression sleeve joints. Mechanical and compression sleeve joints shall be installed in accordance with the manufacturer's instructions.

P2906.12 Cross-linked polyethylene/aluminum/cross-linked polyethylene. Joints between polyethylene/aluminum/polyethylene (PE-AL-PE) and cross-linked polyethylene/aluminum/cross-linked polyethylene (PEX-AL-PEX) pipe and fittings shall comply with Section P2906.12.1.

P2906.12.1 Mechanical joints. Mechanical joints shall be installed in accordance with the manufacturer's instructions. Fittings for PE-AL-PE and PEX-AL-PEX as described in ASTM F1281, ASTM F1282, ASTM F1974, CSA B137.9 and CSA B137.10 shall be installed in accordance with the manufacturer's instructions.

P2906.13 Stainless steel. Joints between stainless steel pipe and fittings shall comply with Section P2906.13.1 or P2906.13.2.

P2906.13.1 Mechanical joints. Mechanical joints shall be installed in accordance with the manufacturer's instructions.

P2906.13.2 Welded joints. Joint surfaces shall be cleaned. The joint shall be welded autogenously or with an *approved* filler metal in accordance with ASTM A312.

P2906.14 Threaded pipe joints. Threaded joints shall conform to American National Taper Pipe Thread specifications. Pipe ends shall be deburred and chips removed. Pipe joint compound shall be used only on male threads.

P2906.15 Soldered and brazed joints. Soldered joints in copper and copper alloy tubing shall be made with fittings *approved* for water piping and shall conform to ASTM B828. Surfaces to be soldered shall be cleaned bright. Fluxes for soldering shall be in accordance with ASTM B813. Brazing fluxes shall be in accordance with AWS A5.31M/A5.31. Solders and fluxes used in potable water-supply systems shall have a lead content of not greater than 0.2 percent. Solder and flux joining pipe or fittings intended to supply drinking water shall conform to NSF 61.

P2906.16 Flared joints. Flared joints in water tubing shall be made with *approved* fittings. The tubing shall be reamed and then expanded with a flaring tool.

P2906.17 Above-ground joints. Joints within the building between copper pipe or CPVC tubing, in any combination with compatible outside diameters, shall be permitted to be made with the use of *approved* push-in mechanical fittings of a pressure-lock design.

P2906.18 Joints between different materials. Joints between different piping materials shall be made in accordance with Section P2906.18.1, P2906.18.2, P2906.18.3 or P2906.18.4, or with a mechanical joint of the compression or mechanical sealing type having an elastomeric seal conforming to ASTM D1869 or ASTM F477. Joints shall be installed in accordance with the manufacturer's instructions.

P2906.18.1 Copper or copper-alloy tubing to galvanized steel pipe. Joints between copper or copper-alloy tubing and galvanized steel pipe shall be made with a copper alloy fitting or dielectric fitting. The copper tubing shall be joined to the fitting in an *approved* manner, and the fitting shall be screwed to the threaded pipe.

P2906.18.2 Joint between PVC water service and CPVC water distribution. Where a PVC water service pipe connects to a CPVC pipe at the beginning of a water distribution system, the transition shall be by a mechanical fitting, an *approved* adapter fitting, a transition fitting or by a single, solvent-cemented transition joint. A single, solvent-cemented transition joint shall be in compliance with ASTM F493 and the pipe, fitting and solvent cement manufacturers' instructions. Solvent cement joint surfaces shall be clean, free from moisture and prepared with an *approved* primer. Solvent cement conforming to ASTM F493 shall be applied to the joint surfaces and the joint assembled while the cement is wet.

P2906.18.3 Plastic pipe or tubing to other piping material. Joints between different types of plastic pipe or between plastic pipe and other piping material shall be made with an *approved* adapter fitting.

P2906.18.4 Stainless steel. Joints between stainless steel and different piping materials shall be made with a mechanical joint of the compression or mechanical-sealing type or a dielectric fitting.

P2906.19 Press-connected joints. Press-connected joints shall conform to one of the standards indicated in Table P2906.6. Press-type mechanical joints in copper tubing shall be made in accordance with the manufacturer's instructions. Cut tube ends shall be reamed to the full inside diameter of the tube end. Joint surfaces shall be cleaned. The tube shall be fully inserted into the press-connected fitting. Press-connected joints shall be pressed with a tool certified by the manufacturer.

P2906.20 Polyethylene of raised temperature plastic. Joints between polyethylene of raised temperature plastic tubing and fittings shall be in accordance with Sections P2906.20.1, P2906.20.2 and P2906.20.3.

P2906.20.1 Mechanical joints. Mechanical joints shall be installed in accordance with the manufacturer's instructions. Fittings for polyethylene of raised temperature plastic tubing shall comply with the applicable standards indicated in Table P2906.6 and shall be installed in accordance with the manufacturer's instructions. Polyethylene of raised temperature plastic tubing shall be factory marked with the applicable standards for the fittings that the manufacturer of the tubing specifies for use with the tubing.

P2906.20.2 Heat fusion joints. Joints shall be of the socket-fusion, saddle-fusion, or butt-fusion type, and shall be joined in accordance with ASTM D2657. Joint surfaces shall be clean and free of moisture. Joint surfaces shall be heated to melt temperatures and joined. The joint shall remain undisturbed until cool. Fittings shall be manufactured in accordance with ASTM D2683 or ASTM D3261.

P2906.20.3 Electrofusion joints. Joints shall be of the electrofusion type. Joint surfaces shall be clean and free of moisture and scoured to expose virgin resin. Joint surfaces shall be heated to melt temperatures for a period of time specified by the manufacturer and joined. The joint shall remain undisturbed until cool. Fittings shall be manufactured in accordance with ASTM F1055.

P2906.21 Push-fit fitting joints. Push-fit fittings shall be used only on copper-tube-size outside diameter dimensioned CPVC, PEX, PE-RT and copper tubing. Push-fit fittings shall conform to ASSE 1061 and shall be installed in accordance with the manufacturer's instructions.

SECTION P2907 CHANGES IN DIRECTION

P2907.1 Bends. Changes in direction in copper tubing shall be permitted to be made with bends having a radius of not less than four diameters of the tube, provided that such bends are made by use of forming equipment that does not deform or create loss in cross-sectional area of the tube.

SECTION P2908 SUPPORT

P2908.1 General. Pipe and tubing support shall conform to Section P2605.

SECTION P2909 DRINKING WATER TREATMENT UNITS

P2909.1 Design. Drinking water treatment units shall meet the requirements of NSF42, NSF 44, NSF 53, NSF 62 or CSA B483.1.

P2909.2 Reverse osmosis drinking water treatment units. Point-of-use reverse osmosis drinking water treatment units, designed for residential use, shall meet the requirements of CSA B483.1 or NSF 58. Waste or discharge from reverse osmosis drinking water treatment units shall enter the drainage system through an *air gap* or an *air gap* device that meets the requirements of NSF 58.

P2909.3 Connection tubing. The tubing to and from drinking water treatment units shall be of a size and material as recommended by the manufacturer. The tubing shall comply with NSF 14, NSF 42, NSF 44, NSF 53, NSF 58 or NSF 61.

SECTION P2910 NONPOTABLE WATER SYSTEMS

P2910.1 Scope. The provisions of this section shall govern the materials, design, construction and installation of systems for the collection, storage, treatment and distribution of nonpotable water. The use and application of nonpotable water shall comply with laws, rules and ordinances applicable in the *jurisdiction.*

P2910.2 Water quality. Nonpotable water for each end use application shall meet the minimum water quality requirements as established for the intended application by the laws, rules and ordinances applicable in the *jurisdiction.* Where nonpotable water from different sources is combined in a system, the system shall comply with the most stringent requirements of this code applicable to such sources.

P2910.2.1 Residual disinfectants. Where chlorine is used for disinfection, the nonpotable water shall contain not more than 4 ppm (4 mg/L) of chloramines or free chlorine. Where ozone is used for disinfection, the nonpotable water shall not contain gas bubbles having elevated levels of ozone at the point of use.

Exception: Reclaimed water sources shall not be required to comply with the requirements of this section.

P2910.2.2 Filtration required. Nonpotable water utilized for water closet and urinal flushing applications shall be filtered by a 100 micron or finer filter.

Exception: Reclaimed water sources shall not be required to comply with the requirements of this section.

P2910.3 Signage required. Nonpotable water outlets such as hose connections, open-ended pipes and faucets shall be identified at the point of use for each outlet with signage that reads, "Nonpotable water is utilized for [application name]. CAUTION: NONPOTABLE WATER. DO NOT DRINK." The words shall be legibly and indelibly printed on a tag or sign constructed of corrosion-resistant, waterproof material or shall be indelibly printed on the fixture. The letters of the words shall be not less than 0.5 inches (12.7 mm) in height and in colors contrasting the background on which they are applied. In addition to the required wordage, the pictograph shown in Figure P2910.3 shall appear on the signage required by this section.

P2910.4 Permits. *Permits* shall be required for the construction, installation, *alteration* and *repair* of nonpotable water systems. *Construction documents,* engineering calculations, diagrams and other such data pertaining to the nonpotable water system shall be submitted with each *permit* application.

P2910.5 Potable water connections. Where a potable system is connected to a nonpotable water system, the potable water supply shall be protected against backflow in accordance with Section P2902.

P2910.6 Approved components and materials. Piping, plumbing components and materials used in collection and conveyance systems shall be manufactured of material *approved* for the intended application and compatible with any disinfection and treatment systems used.

P2910.7 Insect and vermin control. The system shall be protected to prevent the entrance of insects and vermin into storage tanks and piping systems. Screen materials shall be compatible with contacting system components and shall not accelerate the corrosion of system components.

P2910.8 Freeze protection. Where sustained freezing temperatures occur, provisions shall be made to keep storage tanks and the related piping from freezing.

P2910.9 Nonpotable water storage tanks. Nonpotable water storage tanks shall comply with Sections P2910.9.1 through P2910.9.11.

P2910.9.1 Sizing. The holding capacity of the storage tank shall be sized in accordance with the anticipated demand.

P2910.9.2 Location. Storage tanks shall be installed above or below grade. Above-grade storage tanks shall be protected from direct sunlight and shall be constructed using opaque, UV-resistant materials such as, but not limited to, heavily tinted plastic, lined metal, concrete and wood; or painted to prevent algae growth; or shall have specially constructed sun barriers including, but not limited to, installation in garages, crawl spaces or sheds. Storage tanks and their manholes shall not be located directly under any soil piping, waste piping or any source of contamination.

P2910.9.3 Materials. Where collected on site, water shall be collected in an *approved* tank constructed of durable, nonabsorbent and corrosion-resistant materials. The storage tank shall be constructed of materials compatible with any disinfection systems used to treat water upstream of the tank and with any systems used to maintain water quality within the tank. Wooden storage tanks that are not equipped with a makeup water source shall be provided with a flexible liner.

FIGURE P2910.3
PICTOGRAPH—DO NOT DRINK

P2910.9.4 Foundation and supports. Storage tanks shall be supported on a firm base capable of withstanding the weight of the storage tank when filled to capacity. Storage tanks shall be supported in accordance with this code.

P2910.9.4.1 Ballast. Where the soil can become saturated, an underground storage tank shall be ballasted or otherwise secured to prevent the tank from floating out of the ground when empty. The combined weight of the tank and hold-down ballast shall meet or exceed the buoyancy force of the tank. Where the installation requires a foundation, the foundation shall be flat and shall be designed to support the storage tank weight when full, consistent with the bearing capability of adjacent soil.

P2910.9.4.2 Structural support. Where installed below grade, storage tank installations shall be designed to withstand earth and surface structural loads without damage and with minimal deformation when empty or filled with water.

P2910.9.5 Makeup water. Where an uninterrupted nonpotable water supply is required for the intended application, potable or reclaimed water shall be provided as a source of makeup water for the storage tank. The makeup water supply shall be protected against backflow by means of an *air gap* not less than 4 inches (102 mm) above the overflow or an *approved* backflow device in accordance with Section P2902. A full-open valve located on the makeup water supply line to the storage tank shall be provided. Inlets to the storage tank shall be controlled by fill valves or other automatic supply valves installed to prevent the tank from overflowing and to prevent the water level from dropping below a predetermined point. Where makeup water is provided, the water level shall be prohibited from dropping below the source water inlet or the intake of any attached pump.

P2910.9.5.1 Inlet control valve alarm. Makeup water systems shall be fitted with a warning mechanism that alerts the user to a failure of the inlet control valve to close correctly. The alarm shall activate before the water within the storage tank begins to discharge into the overflow system.

P2910.9.6 Overflow. The storage tank shall be equipped with an overflow pipe having a diameter not less than that shown in Table P2910.9.6. The overflow outlet shall discharge at a point not less than 6 inches (152 mm) above the roof or roof drain; floor or floor drain; or over an open water-supplied fixture. The overflow outlet shall be covered with a corrosion-resistant screen of not less than 16 by 20 mesh per inch (630 by 787 mesh per m) and by $^1/_4$-inch (6.4 mm) hardware cloth or shall terminate in a horizontal angle seat check valve. Drainage from overflow pipes shall be directed to prevent freezing on roof walks. The overflow drain shall not be equipped with a shutoff valve. Not less than one cleanout shall be provided on each overflow pipe in accordance with Section P3005.2.

TABLE P2910.9.6
SIZE OF DRAIN PIPES FOR WATER TANKS

TANK CAPACITY (gallons)	DRAIN PIPE (inches)
Up to 750	1
751 to 1,500	$1^1/_2$
1,501 to 3,000	2
3,001 to 5,000	$2^1/_2$
5,001 to 7,500	3
Over 7,500	4

For SI: 1 gallon = 3.875 liters, 1 inch = 25.4 mm.

P2910.9.7 Access. Not less than one access opening shall be provided to allow inspection and cleaning of the tank interior. Access openings shall have an *approved* locking device or other *approved* method of securing access. Below-grade storage tanks, located outside of the building, shall be provided with a manhole either not less than 24 inches (610 mm) square or with an inside diameter not less than 24 inches (610 mm). Manholes shall extend not less than 4 inches (102 mm) above ground or shall be designed to prevent water infiltration. Finished grade shall be sloped away from the manhole to divert surface water. Manhole covers shall be secured to prevent unauthorized access. Service ports in manhole covers shall be not less than 8 inches (203 mm) in diameter and shall be not less than 4 inches (102 mm) above the finished grade level. The service port shall be secured to prevent unauthorized access.

Exception: Storage tanks under 800 gallons (3028 L) in volume installed below grade shall not be required to be equipped with a manhole, but shall have a service port not less than 8 inches (203 mm) in diameter.

P2910.9.8 Venting. Storage tanks shall be provided with a vent sized in accordance with Chapter 31 and based on the aggregate diameter of all tank influent pipes. The reservoir vent shall not be connected to sanitary drainage system vents. Vents shall be protected from contamination by means of an *approved* cap or a U-bend installed with the opening directed downward. Vent outlets shall extend not less than 4 inches (102 mm) above grade, or as necessary to prevent surface water from entering the storage tank. Vent openings shall be protected against the entrance of vermin and insects in accordance with the requirements of Section P2910.7.

P2910.9.9 Drain. A drain shall be located at the lowest point of the storage tank. The tank drain pipe shall discharge as required for overflow pipes and shall not be smaller in size than specified in Table P2910.9.6. Not less than one cleanout shall be provided on each drain pipe in accordance with Section P3005.2.

P2910.10 Marking and signage. Each nonpotable water storage tank shall be *labeled* with its rated capacity. The contents of storage tanks shall be identified with the words, "CAUTION: NONPOTABLE WATER. DO NOT DRINK."

Where an opening is provided that could allow the entry of personnel, the opening shall be marked with the words, "DANGER—CONFINED SPACE." Markings shall be indelibly printed on the tank, or on a tag or sign constructed of corrosion-resistant waterproof material that is mounted on the tank. The letters of the words shall be not less than 0.5 inches (12.7 mm) in height and shall be of a color in contrast with the background on which they are applied.

P2910.11 Storage tank tests. Storage tanks shall be tested in accordance with the following:

1. Storage tanks shall be filled with water to the overflow line prior to and during inspection. Seams and joints shall be left exposed and the tank shall remain watertight without leakage for a period of 24 hours.
2. After 24 hours, supplemental water shall be introduced for a period of 15 minutes to verify proper drainage of the overflow system and leaks do not exist.
3. Following a successful test of the overflow, the water level in the tank shall be reduced to a level that is 2 inches (51 mm) below the makeup water trigger point by using the tank drain. The tank drain shall be observed for proper operation. The makeup water system shall be observed for proper operation, and successful automatic shutoff of the system at the refill threshold shall be verified. Water shall not be drained from the overflow at any time during the refill test.

P2910.12 System abandonment. If the *owner* of an on-site nonpotable water reuse system or rainwater collection and conveyance system elects to cease use of or fails to properly maintain such system, the system shall be abandoned and shall comply with the following:

1. System piping connecting to a utility-provided water system shall be removed or disabled.
2. The distribution piping system shall be replaced with an *approved* potable water supply piping system. Where an existing potable water pipe system is already in place, the fixtures shall be connected to the existing system.
3. The storage tank shall be secured from accidental access by sealing or locking tank inlets and access points, or filled with sand or equivalent.

P2910.13 Separation requirements for nonpotable water piping. Nonpotable water collection and distribution piping and reclaimed water piping shall be separated from the *building sewer* and potable water piping underground by 5 feet (1524 mm) of undisturbed or compacted earth. Nonpotable water collection and distribution piping shall not be located in, under or above cesspools, septic tanks, septic tank drainage fields or seepage pits. Buried nonpotable water piping shall comply with the requirements of Section P2604.

Exceptions:

1. The required separation distance shall not apply where the bottom of the nonpotable water pipe within 5 feet (1524 mm) of the sewer is not less than 12 inches (305 mm) above the top of the highest point of the sewer and the pipe materials conforms to Table P3002.2.
2. The required separation distance shall not apply where the bottom of the potable water service pipe within 5 feet (1524 mm) of the nonpotable water pipe is not less than 12 inches (305 mm) above the top of the highest point of the nonpotable water pipe and the pipe materials comply with the requirements of Table P2906.5.
3. The required separation distance shall not apply where a nonpotable water pipe is located in the same trench with a *building sewer* that is constructed of materials that comply with the requirements of Table P3002.2.
4. The required separation distance shall not apply where a nonpotable water pipe crosses a sewer pipe provided that the nonpotable water pipe is sleeved to not less than 5 feet (1524 mm) horizontally from the sewer pipe centerline on both sides of such crossing, with pipe materials that comply with Table P3002.2.
5. The required separation distance shall not apply where a potable water service pipe crosses a nonpotable water pipe, provided that the potable water service pipe is sleeved for a distance of not less than 5 feet (1524 mm) horizontally from the centerline of the nonpotable pipe on both sides of such crossing, with pipe materials that comply with Table P3002.2.
6. The required separation distance shall not apply to irrigation piping located outside of a building and downstream of the backflow preventer where nonpotable water is used for outdoor applications.

P2910.14 Outdoor outlet access. Sillcocks, hose bibbs, wall hydrants, yard hydrants and other outdoor outlets supplied by nonpotable water shall be located in a locked vault or shall be operable only by means of a removable key.

SECTION P2911
ON-SITE NONPOTABLE WATER REUSE SYSTEMS

P2911.1 General. The provisions of this section shall govern the construction, installation, *alteration* and *repair* of on-site nonpotable water reuse systems for the collection, storage, treatment and distribution of on-site sources of nonpotable water as permitted by the *jurisdiction*.

P2911.2 Sources. On-site nonpotable water reuse systems shall collect waste discharge only from the following sources: bathtubs, showers, lavatories, clothes washers and laundry trays. Water from other *approved* nonpotable sources including swimming pool backwash operations, air conditioner condensate, rainwater, foundation drain water, fluid cooler discharge water and fire pump test water shall be permitted to be collected for reuse by on-site nonpotable water reuse systems, as *approved* by the *building official* and as appropriate for the intended application.

P2911.2.1 Prohibited sources. Reverse osmosis system reject water, water softener backwash water, kitchen sink wastewater, dishwasher wastewater and wastewater containing urine or fecal matter shall not be collected for reuse within an on-site nonpotable water reuse system.

P2911.3 Traps. Traps serving fixtures and devices discharging wastewater to on-site nonpotable water reuse systems shall comply with the Section P3201.2.

P2911.4 Collection pipe. On-site nonpotable water reuse systems shall utilize drainage piping *approved* for use within plumbing drainage systems to collect and convey untreated water for reuse. Vent piping *approved* for use within plumbing venting systems shall be utilized for vents within the graywater system. Collection and vent piping materials shall comply with Section P3002.

P2911.4.1 Installation. Collection piping conveying untreated water for reuse shall be installed in accordance with Section P3005.

P2911.4.2 Joints. Collection piping conveying untreated water for reuse shall utilize joints *approved* for use with the distribution piping and appropriate for the intended applications as specified in Section P3002.

P2911.4.3 Size. Collection piping conveying untreated water for reuse shall be sized in accordance with drainage sizing requirements specified in Section P3005.4.

P2911.4.4 Marking. Additional marking of collection piping conveying untreated water for reuse shall not be required beyond that required for sanitary drainage, waste and vent piping by Chapter 30.

P2911.5 Filtration. Untreated water collected for reuse shall be filtered as required for the intended end use. Filters shall be accessible for inspection and maintenance. Filters shall utilize a pressure gauge or other *approved* method to provide indication when a filter requires servicing or replacement. Filters shall be installed with shutoff valves immediately upstream and downstream to allow for isolation during maintenance.

P2911.6 Disinfection. Nonpotable water collected on site for reuse shall be disinfected, treated or both to provide the quality of water needed for the intended end-use application. Where the intended end-use application does not have requirements for the quality of water, disinfection and treatment of water collected on site for reuse shall not be required. Nonpotable water collected on site containing untreated graywater shall be retained in collection reservoirs for not more than 24 hours.

P2911.6.1 Graywater used for fixture flushing. Graywater used for flushing water closets and urinals shall be disinfected and treated by an on-site water reuse treatment system complying with NSF 350.

P2911.7 Storage tanks. Storage tanks utilized in on-site nonpotable water reuse systems shall comply with Section P2910.9 and Sections P2911.7.1 through P2911.7.3.

P2911.7.1 Location. Storage tanks shall be located with a minimum horizontal distance between various elements as indicated in Table P2911.7.1.

**TABLE P2911.7.1
LOCATION OF NONPOTABLE
WATER REUSE STORAGE TANKS**

ELEMENT	MINIMUM HORIZONTAL DISTANCE FROM STORAGE TANK (feet)
Critical root zone (CRZ) of protected trees	2
Lot line adjoining private lots	5
Public water main	10
Seepage pits	5
Septic tanks	5
Streams and lakes	50
Water service	5
Water wells	50

For SI: 1 foot = 304.8 mm

P2911.7.2 Inlets. Storage tank inlets shall be designed to introduce water into the tank with minimum turbulence, and shall be located and designed to avoid agitating the contents of the storage tank.

P2911.7.3 Outlets. Outlets shall be located not less than 4 inches (102 mm) above the bottom of the storage tank, and shall not skim water from the surface.

P2911.8 Valves. Valves shall be supplied on on-site nonpotable water reuse systems in accordance with Sections P2911.8.1 and P2911.8.2.

P2911.8.1 Bypass valve. One three-way diverter valve certified to NSF 50 or other *approved* device shall be installed on collection piping upstream of each storage tank, or drainfield, as applicable, to divert untreated on-site reuse sources to the sanitary sewer to allow servicing and inspection of the system. Bypass valves shall be installed downstream of fixture traps and vent connections. Bypass valves shall be *labeled* to indicate the direction of flow, connection and storage tank or drainfield connection. Bypass valves shall be installed in accessible locations. Two shutoff valves shall not be installed to serve as a bypass valve.

P2911.8.2 Backwater valve. Backwater valves shall be installed on each overflow and tank drain pipe. Backwater valves shall be in accordance with Section P3008.

P2911.9 Pumping and control system. Mechanical equipment including pumps, valves and filters shall be accessible and removable in order to perform repair, maintenance and cleaning. The minimum flow rate and flow pressure delivered by the pumping system shall be appropriate for the application and in accordance with Section P2903.

P2911.10 Water pressure-reducing valve or regulator. Where the water pressure supplied by the pumping system exceeds 80 psi (552 kPa) static, a pressure-reducing valve shall be installed to reduce the pressure in the nonpotable water distribution system piping to 80 psi (552 kPa) static or less. Pressure-reducing valves shall be specified and installed in accordance with Section P2903.3.2.

P2911.11 Distribution pipe. Distribution piping utilized in on-site nonpotable water reuse systems shall comply with Sections PP2911.11.1 through P2911.11.3.

Exception: Irrigation piping located outside of the building and downstream of a backflow preventer.

P2911.11.1 Materials, joints and connections. Distribution piping shall conform to the standards and requirements specified in Section P2906 for nonpotable water.

P2911.11.2 Design. On-site nonpotable water reuse distribution piping systems shall be designed and sized in accordance with Section P2903 for the intended application.

P2911.11.3 Marking. On-site nonpotable water distribution piping labeling and marking shall comply with Section P2901.2

P2911.12 Tests and inspections. Tests and inspections shall be performed in accordance with Sections P2911.12.1 through P2911.12.6.

P2911.12.1 Collection pipe and vent test. Drain, waste and vent piping used for on-site water reuse systems shall be tested in accordance with Section P2503.

P2911.12.2 Storage tank test. Storage tanks shall be tested in accordance with Section P2910.11.

P2911.12.3 Water supply system test. The testing of makeup water supply piping and distribution piping shall be conducted in accordance with Section P2503.7.

P2911.12.4 Inspection and testing of backflow prevention assemblies. The testing of backflow preventers and backwater valves shall be conducted in accordance with Section P2503.8.

P2911.12.5 Inspection of vermin and insect protection. Inlets and vents to the system shall be inspected to verify that each is protected to prevent the entrance of insects and vermin into the storage tank and piping systems in accordance with Section P2910.7.

P2911.12.6 Water quality test. The quality of the water for the intended application shall be verified at the point of use in accordance with the requirements of the *jurisdiction.*

P2911.13 Operation and maintenance manuals. Operation and maintenance materials shall be supplied with nonpotable on-site water reuse systems in accordance with Sections P2911.13.1 through P2911.13.4.

P2911.13.1 Manual. A detailed operations and maintenance manual shall be supplied in hard-copy form for each system.

P2911.13.2 Schematics. The manual shall include a detailed system schematic, the location of system components and a list of system components that includes the manufacturers and model numbers of the components.

P2911.13.3 Maintenance procedures. The manual shall provide a schedule and procedures for system components requiring periodic maintenance. Consumable parts including filters shall be noted along with part numbers.

P2911.13.4 Operations procedures. The manual shall include system startup and shutdown procedures. The manual shall include detailed operating procedures for the system.

SECTION P2912 NONPOTABLE RAINWATER COLLECTION AND DISTRIBUTION SYSTEMS

P2912.1 General. The provisions of this section shall govern the construction, installation, *alteration* and repair of rainwater collection and conveyance systems for the collection, storage, treatment and distribution of rainwater for nonpotable applications. For nonpotable rainwater systems, the provisions of CSA B805/ICC 805 shall be an alternative for regulating the materials, design, construction and installation of systems for rainwater collection, storage, treatment and distribution of nonpotable water. The use and application of nonpotable water shall comply with laws, rules and ordinances applicable in the *jurisdiction.*

P2912.2 Collection surface. Rainwater shall be collected only from above-ground impervious roofing surfaces constructed from *approved* materials. Collection of water from vehicular parking or pedestrian walkway surfaces shall be prohibited except where the water is used exclusively for landscape irrigation. Overflow and bleed-off pipes from roof-mounted *appliances* including, but not limited to, evaporative coolers, water heaters and solar water heaters shall not discharge onto rainwater collection surfaces.

P2912.3 Debris excluders. Downspouts and leaders shall be connected to a roof washer and shall be equipped with a debris excluder or equivalent device to prevent the contamination of collected rainwater with leaves, sticks, pine needles and similar material. Debris excluders and equivalent devices shall be self-cleaning.

P2912.4 Roof washer. An amount of rainwater shall be diverted at the beginning of each rain event, and not allowed to enter the storage tank, to wash accumulated debris from the collection surface. The amount of rainfall to be diverted shall be field adjustable as necessary to minimize storage tank water contamination. The roof washer shall not rely on manually operated valves or devices, and shall operate automatically. Diverted rainwater shall not be drained to the roof surface, and shall be discharged in a manner consistent with the stormwater runoff requirements of the *jurisdiction.* Roof washers shall be accessible for maintenance and service.

P2912.5 Roof gutters and downspouts. Gutters and downspouts shall be constructed of materials that are compatible with the collection surface and the rainwater quality for the desired end use. Joints shall be watertight.

P2912.5.1 Slope. Roof gutters, leaders and rainwater collection piping shall slope continuously toward collection inlets and shall be free of leaks. Gutters and downspouts shall have a slope of not less than $^{1}/_{8}$ inch per foot (10.4 mm/m) along their entire length. Gutters and downspouts shall be installed so that water does not pool at any point.

P2912.5.2 Cleanouts. Cleanouts shall be provided in the water conveyance system to allow access to filters, flushes, pipes and downspouts.

P2912.6 Drainage. Water drained from the roof washer or debris excluder shall not be drained to the sanitary sewer. Such water shall be diverted from the storage tank and shall discharge to a location that will not cause erosion or damage to property. Roof washers and debris excluders shall be provided with an automatic means of self-draining between rain events and shall not drain onto roof surfaces.

P2912.7 Collection pipe. Rainwater collection and conveyance systems shall utilize drainage piping *approved* for use within plumbing drainage systems to collect and convey captured rainwater. Vent piping *approved* for use within plumbing venting systems shall be utilized for vents within the rainwater system. Collection and vent piping materials shall comply with Section P3002.

P2912.7.1 Installation. Collection piping conveying captured rainwater shall be installed in accordance with Section P3005.3.

P2912.7.2 Joints. Collection piping conveying captured rainwater shall utilize joints *approved* for use with the distribution piping and appropriate for the intended applications as specified in Section P3003.

P2912.7.3 Size. Collection piping conveying captured rainwater shall be sized in accordance with drainage-sizing requirements specified in Section P3005.4.

P2912.7.4 Marking. Additional marking of collection piping conveying captured rainwater for reuse shall not be required beyond that required for sanitary drainage, waste, and vent piping by Chapter 30.

P2912.8 Filtration. Collected rainwater shall be filtered as required for the intended end use. Filters shall be accessible for inspection and maintenance. Filters shall utilize a pressure gauge or other *approved* method to provide indication when a filter requires servicing or replacement. Filters shall be installed with shutoff valves installed immediately upstream and downstream to allow for isolation during maintenance.

P2912.9 Disinfection. Where the intended application for rainwater requires disinfection or other treatment or both, it shall be disinfected as needed to ensure that the required water quality is delivered at the point of use.

P2912.10 Storage tanks. Storage tanks utilized in nonpotable rainwater collection and conveyance systems shall comply with Section P2910.9 and Sections P2912.10.1 through P2912.10.3.

P2912.10.1 Location. Storage tanks shall be located with a minimum horizontal distance between various elements as indicated in Table P2912.10.1.

TABLE P2912.10.1
LOCATION OF RAINWATER STORAGE TANKS

ELEMENT	MINIMUM HORIZONTAL DISTANCE FROM STORAGE TANK (feet)
Critical root zone (CRZ) of protected trees	2
Lot line adjoining private lots	5
Seepage pits	5
Septic tanks	5

For SI: 1 foot = 304.8 mm

P2912.10.2 Inlets. Storage tank inlets shall be designed to introduce collected rainwater into the tank with minimum turbulence, and shall be located and designed to avoid agitating the contents of the storage tank.

P2912.10.3 Outlets. Outlets shall be located not less than 4 inches (102 mm) above the bottom of the storage tank and shall not skim water from the surface.

P2912.11 Valves. Valves shall be supplied on rainwater collection and conveyance systems in accordance with Sections P2912.11.1 and P2912.11.2.

P2912.11.1 Influent diversion. A means shall be provided to divert storage tank influent to allow for maintenance and repair of the storage tank system.

P2912.11.2 Backwater valve. Backwater valves shall be installed on each overflow and tank drain pipe. Backwater valves shall be in accordance with Section P3008.

P2912.12 Pumping and control system. Mechanical equipment including pumps, valves and filters shall be easily accessible and removable in order to perform repair, maintenance and cleaning. The minimum flow rate and flow pressure delivered by the pumping system shall appropriate for the application and in accordance with Section P2903.

P2912.13 Water pressure-reducing valve or regulator. Where the water pressure supplied by the pumping system exceeds 80 psi (552 kPa) static, a pressure-reducing valve shall be installed to reduce the pressure in the rainwater distribution system piping to 80 psi (552 kPa) static or less. Pressure-reducing valves shall be specified and installed in accordance with Section P2903.3.2.

P2912.14 Distribution pipe. Distribution piping utilized in rainwater collection and conveyance systems shall comply with Sections P2912.14.1 through P2912.14.3.

Exception: Irrigation piping located outside of the building and downstream of a backflow preventer.

P2912.14.1 Materials, joints and connections. Distribution piping shall conform to the standards and requirements specified in Section P2906 for nonpotable water.

P2912.14.2 Design. Distribution piping systems shall be designed and sized in accordance with the Section P2903 for the intended application.

P2912.14.3 Labeling and marking. Nonpotable rainwater distribution piping labeling and marking shall comply with Section P2901.2.

P2912.15 Tests and inspections. Tests and inspections shall be performed in accordance with Sections P2912.15.1 through P2912.15.8.

P2912.15.1 Roof gutter inspection and test. Roof gutters shall be inspected to verify that the installation and slope is in accordance with Section P2912.5.1. Gutters shall be tested by pouring not less than 1 gallon of water (3.8 L) into the end of the gutter opposite the collection point. The gutter being tested shall not leak and shall not retain standing water.

P2912.15.2 Roofwasher test. Roofwashers shall be tested by introducing water into the gutters. Proper diversion of the first quantity of water in accordance with the requirements of Section P2912.4 shall be verified.

P2912.15.3 Collection pipe and vent test. Drain, waste and vent piping used for rainwater collection and conveyance systems shall be tested in accordance with Section P2503.

P2912.15.4 Storage tank test. Storage tanks shall be tested in accordance with the Section P2910.11.

P2912.15.5 Water supply system test. The testing of makeup water supply piping and distribution piping shall be conducted in accordance with Section P2503.7.

P2912.15.6 Inspection and testing of backflow prevention assemblies. The testing of backflow preventers and backwater valves shall be conducted in accordance with Section P2503.8.

P2912.15.7 Inspection of vermin and insect protection. Inlets and vents to the system shall be inspected to verify that each is protected to prevent the entrance of insects and vermin into the storage tank and piping systems in accordance with Section P2910.7.

P2912.15.8 Water quality test. The quality of the water for the intended application shall be verified at the point of use in accordance with the requirements of the *jurisdiction*.

P2912.16 Operation and maintenance manuals. Operation and maintenance manuals shall be supplied with rainwater collection and conveyance systems in accordance with Sections P2912.16.1 through P2912.16.4.

P2912.16.1 Manual. A detailed operations and maintenance manual shall be supplied in hard-copy form for each system.

P2912.16.2 Schematics. The manual shall include a detailed system schematic, the location of system components and a list of system components that includes the manufacturers and model numbers of the components.

P2912.16.3 Maintenance procedures. The manual shall provide a maintenance schedule and procedures for system components requiring periodic maintenance. Consumable parts, including filters, shall be noted along with part numbers.

P2912.16.4 Operations procedures. The manual shall include system startup and shutdown procedures, and detailed operating procedures.

SECTION P2913 RECLAIMED WATER SYSTEMS

P2913.1 General. The provisions of this section shall govern the construction, installation, *alteration* and *repair* of systems supplying nonpotable reclaimed water.

P2913.2 Water pressure-reducing valve or regulator. Where the reclaimed water pressure supplied to the building exceeds 80 psi (552 kPa) static, a pressure-reducing valve shall be installed to reduce the pressure in the reclaimed water distribution system piping to 80 psi (552 kPa) static or less. Pressure-reducing valves shall be specified and installed in accordance with Section P2903.3.2

P2913.3 Reclaimed water systems. The design of the reclaimed water systems shall conform to accepted engineering practice.

P2913.3.1 Distribution pipe. Distribution piping shall comply with Sections P2913.3.1.1 through P2913.3.1.3.

Exception: Irrigation piping located outside of the building and downstream of a backflow preventer.

P2913.3.1.1 Materials, joints and connections. Distribution piping conveying reclaimed water shall conform to standards and requirements specified in Section P2906 for nonpotable water.

P2913.3.1.2 Design. Distribution piping systems shall be designed and sized in accordance with Section P2903 for the intended application.

P2913.3.1.3 Labeling and marking. Nonpotable rainwater distribution piping labeling and marking shall comply with Section P2901.2.

P2913.4 Tests and inspections. Tests and inspections shall be performed in accordance with Sections P2913.4.1 and P2913.4.2.

P2913.4.1 Water supply system test. The testing of makeup water supply piping and reclaimed water distribution piping shall be conducted in accordance with Section P2503.7.

P2913.4.2 Inspection and testing of backflow prevention assemblies. The testing of backflow preventers shall be conducted in accordance with Section P2503.8.

CHAPTER 30

SANITARY DRAINAGE

User notes:

About this chapter: *Chapter 30 regulates methods and piping systems that remove water that has been used for purposes such as flushing water closets, bathing, culinary activities and equipment discharges. The types of materials, drainage fitting and connection methods for these systems, beginning at the receiving fixtures and ending at the point of disposal for the liquid waste, are covered. A design method for a gravity flow system of vertical and horizontal piping is provided based on the probability of flows from specific fixtures.*

Code development reminder: *Code change proposals to this chapter will be considered by the IRC—Mechanical/Plumbing Code Development Committee during the 2021 (Group A) Code Development Cycle.*

SECTION P3001 GENERAL

P3001.1 Scope. The provisions of this chapter shall govern the materials, design, construction and installation of sanitary drainage systems. Plumbing materials shall conform to the requirements of this chapter. The drainage, waste and vent (DWV) system shall consist of piping for conveying wastes from plumbing fixtures, *appliances* and appurtenances, including fixture traps; above-grade drainage piping; below-grade drains within the building, such as a *building drain*; below- and above-grade venting systems; and piping to the public sewer or private septic system.

P3001.2 Protection from freezing. Portions of the above-grade DWV system, other than vent terminals, shall not be located outside of a building, in *attics* or crawl spaces, concealed in outside walls, or in any other place subjected to freezing temperatures unless adequate provision is made to protect them from freezing by insulation or heat or both, except in localities having a winter design temperature greater than 32°F (0°C) (ASHRAE 97.5 percent column, winter, see Chapter 3).

P3001.3 Flood-resistant installation. In flood hazard areas as established by Table R301.2, drainage, waste and vent systems shall be located and installed to prevent infiltration of floodwaters into the systems and discharges from the systems into floodwaters.

SECTION P3002 MATERIALS

P3002.1 Piping within buildings. Drain, waste and vent (DWV) piping in buildings shall be as indicated in Tables P3002.1(1) and P3002.1(2) except that galvanized wrought-iron or galvanized steel pipe shall not be used underground and shall be maintained not less than 6 inches (152 mm) above ground. Allowance shall be made for the thermal expansion and contraction of plastic piping.

P3002.2 Building sewer. *Building sewer* piping shall be as indicated in Table P3002.2 Forced main sewer piping shall conform to one of the standards for ABS plastic pipe, copper or copper-alloy tubing, PVC plastic pipe or pressure-rated pipe indicated in Table P3002.2.

P3002.2.1 Building sewer pipe near the water service. The proximity of a *building sewer* to a water service shall comply with Section P2906.1.4.

P3002.3 Fittings. Pipe fittings shall be *approved* for installation with the piping material installed and shall comply with the applicable standards indicated in Table P3002.3.

**TABLE P3002.1(1)
ABOVE-GROUND DRAINAGE AND VENT PIPE**

MATERIAL	STANDARD
Acrylonitrile butadiene styrene (ABS) plastic pipe in IPS diameters, including Schedule 40, DR 22 (PS 200) and DR 24 (PS 140); with a solid, cellular core or composite wall	ASTM D2661; ASTM D2680; ASTM F628; ASTM F1488; CSA B181.1
Cast-iron pipe	ASTM A74; ASTM A888; CISPI 301
Copper or copper-alloy pipe	ASTM B42; ASTM B43; ASTM B302
Copper or copper-alloy tubing (Type K, L, M or DWV)	ASTM B75/B75M; ASTM B88; ASTM B251/B251M; ASTM B306
Galvanized steel pipe	ASTM A53/A53M
Polyolefin pipe	CSA B181.3
Polyvinyl chloride (PVC) plastic pipe in IPS diameters, including Schedule 40, DR 22 (PS 200) and DR 24 (PS 140); with a solid, cellular core or composite wall	ASTM D2665; ASTM F891; ASTM F1488; CSA B181.2
Polyvinyl chloride (PVC) plastic pipe with a 3.25-inch O.D. and a solid, cellular core or composite wall	ASTM D2949; ASTM F1488
Stainless steel drainage systems, Types 304 and 316L	ASME A112.3.1

For SI: 1 inch = 25.4 mm.

TABLE P3002.1(2)
UNDERGROUND BUILDING DRAINAGE AND VENT PIPE

PIPE	STANDARD
Acrylonitrile butadiene styrene (ABS) plastic pipe in IPS diameters, including Schedule 40, DR 22 (PS 200) and DR 24 (PS 140); with a solid, cellular core or composite wall	ASTM D2661; ASTM F628; ASTM F1488; CSA B181.1
Cast-iron pipe	ASTM A74; ASTM A888; CISPI 301
Copper or copper-alloy tubing (Type K, L, M or DWV)	ASTM B75/B75M; ASTM B88; ASTM B251; ASTM B306
Polyethylene (PE) plastic pipe (SDR-PR)	ASTM F714
Polyolefin pipe	ASTM F714; ASTM F1412; CSA B181.3
Polyvinyl chloride (PVC) plastic pipe in IPS diameters, including Schedule 40, DR 22 (PS 200) and DR 24 (PS 140); with a solid, cellular core or composite wall	ASTM D2665; ASTM F891; ASTM F1488; CSA B181.2
Polyvinyl chloride (PVC) plastic pipe with a 3.25-inch O.D. and a solid, cellular core or composite wall	ASTM D2949; ASTM F1488
Stainless steel drainage systems, Type 316L	ASME A112.3.1

For SI: 1 inch = 25.4 mm.

TABLE P3002.2
BUILDING SEWER PIPE

MATERIAL	STANDARD
Acrylonitrile butadiene styrene (ABS) plastic pipe in IPS diameters, including Schedule 40, DR 22 (PS 200) and DR 24 (PS 140); with a solid, cellular core or composite wall	ASTM D2661; ASTM F628; ASTM F1488
Acrylonitrile butadiene styrene (ABS) plastic pipe in sewer and drain diameters, including SDR 42 (PS 20), PS35, SDR 35 (PS 45), PS50, PS100, PS140, SDR 23.5 (PS 150) and PS200; with a solid, cellular core or composite wall	ASTM D2751; ASTM F1488
Polyvinyl chloride (PVC) plastic pipe in sewer and drain diameters, including PS 25, SDR 41 (PS 28), PS 35, SDR 35 (PS 46), PS 50, PS 100, SDR 26 (PS 115), PS140 and PS 200; with a solid, cellular core or composite wall	ASTM D3034; ASTM F891; ASTM F1488; CSA B182.2; CSA B182.4
Cast-iron pipe	ASTM A74; ASTM A888; CISPI 301
Concrete pipe	ASTM C14; ASTM C76; CSA A257.1; CSA A257.2
Copper or copper-alloy tubing (Type K or L)	ASTM B75/B75M; ASTM B88; ASTM B251/B251M
Polyethylene (PE) plastic pipe (SDR-PR)	ASTM F714
Polyolefin pipe	ASTM F1412; CSA B181.3
Polyvinyl chloride (PVC) plastic pipe in IPS diameters, including Schedule 40, DR 22 (PS 200) and DR 24 (PS 140); with solid, cellular core or composite wall	ASTM D2665; ASTM D2949; ASTM D3034; ASTM F1412; CSA B182.2; CSA B182.4
Polyvinyl chloride (PVC) plastic pipe with a 3.25-inch O.D. and a solid, cellular core or composite wall	ASTM D2949, ASTM F1488
Stainless steel drainage systems, Types 304 and 316L	ASME A112.3.1
Vitrified clay pipe	ASTM C425; ASTM C700

For SI: 1 inch = 25.4 mm.

TABLE P3002.3
PIPE FITTINGS

PIPE MATERIAL	FITTING STANDARD
Acrylonitrile butadiene styrene (ABS) plastic pipe in IPS diameters	ASME A112.4.4; ASTM D2661; ASTM D3311; ASTM F628; CSA B181.1
Acrylonotrile butadiene styrene (ABS) plastic pipe in sewer and drain diameters	ASTM D2751
Cast-iron	ASME B16.4; ASME B16.12; ASTM A74; ASTM A888; CISPI 301
Copper or copper alloy	ASME B16.15; ASME B16.18; ASME B16.22; ASME B16.23; ASME B16.26; ASME B16.29
Gray iron and ductile iron	AWWA C110/A21.10
Polyethylene	ASTM D2683
Polyolefin	ASTM F1412; CSA B181.3
Polyvinyl chloride (PVC) plastic in IPS diameters	ASME A112.4.4; ASTM D2665; ASTM D3311; ASTM F1866
Polyvinyl chloride (PVC) plastic pipe in sewer and drain diameters	ASTM D3034
Polyvinyl chloride (PVC) plastic pipe with a 3.25-inch O.D.	ASTM D2949
PVC fabricated fittings	ASTM F1866
Stainless steel drainage systems, Types 304 and 316L	ASME A112.3.1
Vitrified clay	ASTM C700

For SI: 1 inch = 25.4 mm.

P3002.3.1 Drainage. Drainage fittings shall have a smooth interior waterway of the same diameter as the piping served. Fittings shall conform to the type of pipe used. Drainage fittings shall not have ledges, shoulders or reductions that can retard or obstruct drainage flow in the piping. Threaded drainage pipe fittings shall be of the recessed drainage type, black or galvanized. Drainage fittings shall be designed to maintain $^1/_4$ unit vertical in 12 units horizontal (2-percent slope) grade. This section shall not be applicable to tubular waste fittings used to convey vertical flow upstream of the trap seal liquid level of a fixture trap.

P3002.4 Other materials. Sheet lead, lead bends, lead traps and sheet copper shall comply with Sections P3002.4.1 through P3002.4.3.

P3002.4.1 Sheet lead. Sheet lead shall weigh not less than indicated for the following applications:

1. Flashing of vent terminals, 3 psf (15 kg/m^2).
2. Prefabricated flashing for vent pipes, $2^1/_2$ psf (12 kg/m^2).

P3002.4.2 Lead bends and traps. Lead bends and lead traps shall be not less than $^1/_8$-inch (3 mm) wall thickness.

P3002.4.3 Sheet copper. Sheet copper shall weigh not less than indicated for the following applications:

1. General use, 12 ounces per square feet (4 kg/m^2).
2. Flashing for vent pipes, 8 ounces per square feet (2.5 kg/m^2).

SECTION P3003 JOINTS AND CONNECTIONS

P3003.1 Tightness. Joints and connections in the DWV system shall be gastight and watertight for the intended use or pressure required by test.

P3003.1.1 Threaded joints, general. Pipe and fitting threads shall be tapered.

P3003.2 Prohibited joints. Running threads and bands shall not be used in the drainage system. Drainage and vent piping shall not be drilled, tapped, burned or welded.

The following types of joints and connections shall be prohibited:

1. Cement or concrete.
2. Mastic or hot-pour bituminous joints.
3. Joints made with fittings not *approved* for the specific installation.
4. Joints between different diameter pipes made with elastomeric rolling O-rings.
5. Solvent-cement joints between different types of plastic pipe except where provided for in Section P3003.13.4.
6. Saddle-type fittings.

P3003.3 ABS plastic. Joints between ABS plastic pipe or fittings shall comply with Sections P3003.3.1 through P3003.3.4.

P3003.3.1 Mechanical joints. Mechanical joints on drainage pipes shall be made with an elastomeric seal conforming to ASTM C1173, ASTM D3212 or CSA B602. Mechanical joints shall be installed only in underground systems unless otherwise *approved.* Joints shall be installed in accordance with the manufacturer's instructions.

P3003.3.2 Solvent cementing. Joint surfaces shall be clean and free from moisture. Solvent cement that conforms to ASTM D2235 or CSA B181.1 shall be applied to joint surfaces. The joint shall be made while the cement is wet. Joints shall be made in accordance with ASTM D2235, ASTM D2661, ASTM F628 or CSA B181.1. Solvent-cement joints shall be permitted above or below ground.

P3003.3.3 Threaded joints. Threads shall conform to ASME B1.20.1. Schedule 80 or heavier pipe shall be permitted to be threaded with dies specifically designed for plastic pipe. *Approved* thread lubricant or tape shall be applied on the male threads only.

P3003.3.4 Push-fit fitting joints. Push-fit DWV fittings shall be *listed* and *labeled* to ASME A112.4.4 and shall be installed in accordance with the manufacturer's instructions.

P3003.4 Cast iron. Joints between cast-iron pipe or fittings shall comply with Sections P3003.4.1 through P3003.4.3.

P3003.4.1 Caulked joints. Joints for hub and spigot pipe shall be firmly packed with oakum or hemp. Molten lead shall be poured in one operation to a depth of not less than 1 inch (25 mm). The lead shall not recede more than $^1/_8$ inch (3 mm) below the rim of the hub and shall be caulked tight. Paint, varnish or other coatings shall not be permitted on the jointing material until after the joint has been tested and *approved.* Lead shall be run in one pouring and shall be caulked tight.

P3003.4.2 Compression gasket joints. Compression gaskets for hub and spigot pipe and fittings shall conform to ASTM C564. Gaskets shall be compressed when the pipe is fully inserted.

P3003.4.3 Mechanical joint coupling. Mechanical joint couplings for hubless pipe and fittings shall consist of an elastomeric sealing sleeve and a metallic shield that comply with CISPI 310, ASTM C1277 or ASTM C1540. The elastomeric sealing sleeve shall conform to ASTM C564 or CSA B602 and shall have a center stop. Mechanical joint couplings shall be installed in accordance with the manufacturer's instructions.

P3003.5 Concrete joints. Joints between concrete pipe and fittings shall be made with an elastomeric seal conforming to ASTM C443, ASTM C1173, CSA A257.3 or CSA B602.

P3003.6 Copper and copper-alloy pipe and tubing. Joints between copper or copper-alloy pipe tubing or fittings shall comply with Sections P3003.6.1 through P3003.6.4.

P3003.6.1 Brazed joints. All joint surfaces shall be cleaned. An *approved* flux shall be applied where required. Brazing materials shall have a melting point in excess of 1,000°F (538°C). Brazing alloys filler metal shall be in accordance with AWS A5.8.

P3003.6.2 Mechanical joints. Mechanical joints shall be installed in accordance with the manufacturer's instructions.

P3003.6.3 Soldered joints. Copper and copper-alloy joints shall be soldered in accordance with ASTM B828. Cut tube ends shall be reamed to the full inside diameter of the tube end. Joint surfaces shall be cleaned. Fluxes for soldering shall be in accordance with ASTM B813 and shall become noncorrosive and nontoxic after soldering. The joint shall be soldered with a solder conforming to ASTM B32.

P3003.6.4 Threaded joints. Threads shall conform to ASME B1.20.1. Pipe-joint compound or tape shall be applied on the male threads only.

P3003.7 Steel. Joints between galvanized steel pipe or fittings shall comply with Sections P3003.7.1 and P3003.7.2.

P3003.7.1 Threaded joints. Threads shall conform to ASME B1.20.1. Pipe-joint compound or tape shall be applied on the male threads only.

P3003.7.2 Mechanical joints. Joints shall be made with an *approved* elastomeric seal. Mechanical joints shall be installed in accordance with the manufacturer's instructions.

P3003.8 Lead. Joints between lead pipe or fittings shall comply with Sections P3003.8.1 and P3003.8.2.

P3003.8.1 Burned. Burned joints shall be uniformly fused together into one continuous piece. The thickness of the joint shall be not less than the thickness of the lead being joined. The filler metal shall be of the same material as the pipe.

P3003.8.2 Wiped. Joints shall be fully wiped, with an exposed surface on each side of the joint not less than $^3/_4$ inch (19 mm). The joint shall be not less than $^3/_8$ inch (9.5 mm) thick at the thickest point.

P3003.9 PVC plastic. Joints between PVC plastic pipe or fittings shall comply with Sections P3003.9.1 through P3003.9.4.

P3003.9.1 Mechanical joints. Mechanical joints on drainage pipe shall be made with an elastomeric seal conforming to ASTM C1173, ASTM D3212 or CSA B602. Mechanical joints shall not be installed in above-ground systems, unless otherwise *approved.* Joints shall be installed in accordance with the manufacturer's instructions.

P3003.9.2 Solvent cementing. Joint surfaces shall be clean and free from moisture. A purple primer, or other *approved* primer, that conforms to ASTM F656 shall be applied. Solvent cement not purple in color and conforming to ASTM D2564, CSA B137.3 or CSA B181.2 shall be applied to all joint surfaces. The joint shall be made while the cement is wet, and shall be in accordance with ASTM D2855. Solvent-cement joints shall be installed above or below ground.

Exception: A primer shall not be required where all of the following conditions apply:

1. The solvent cement used is third-party certified as conforming to ASTM D2564.
2. The solvent cement is used only for joining PVC drain, waste and vent pipe and fittings in nonpressure applications in sizes up to and including 4 inches (102 mm) in diameter

P3003.9.3 Threaded joints. Threads shall conform to ASME B1.20.1. Schedule 80 or heavier pipe shall be permitted to be threaded with dies specifically designed for plastic pipe. *Approved* thread lubricant or tape shall be applied on the male threads only.

P3003.9.4 Push-fit joints. Push-fit joints shall conform to ASME A112.4.4 and shall be installed in accordance with the manufacturer's instructions.

P3003.10 Vitrified clay. Joints between vitrified clay pipe or fittings shall be made with an elastomeric seal conforming to ASTM C425, ASTM C1173 or CSA B602.

P3003.11 Polyolefin plastic. Joints between polyolefin plastic pipe and fittings shall comply with Sections P3003.11.1 and P3003.11.2.

P3003.11.1 Heat-fusion joints. Heat-fusion joints for polyolefin pipe and tubing joints shall be installed with socket-type heat-fused polyolefin fittings or electrofusion polyolefin fittings. Joint surfaces shall be clean and free from moisture. The joint shall be undisturbed until cool. Joints shall be made in accordance with ASTM F1412 or CSA B181.3.

P3003.11.2 Mechanical and compression sleeve joints. Mechanical and compression sleeve joints shall be installed in accordance with the manufacturer's instructions.

P3003.12 Polyethylene plastic pipe. Joints between polyethylene plastic pipe and fittings shall be underground and shall comply with Section P3003.12.1 or P3003.12.2.

P3003.12.1 Heat fusion joints. Joint surfaces shall be clean and free from moisture. Joint surfaces shall be cut, heated to melting temperature and joined using tools specifically designed for the operation. Joints shall be undisturbed until cool. Joints shall be made in accordance with ASTM D2657 and the manufacturer's instructions.

P3003.12.2 Mechanical joints. Mechanical joints in drainage piping shall be made with an elastomeric seal conforming to ASTM C1173, ASTM D3212 or CSA B602. Mechanical joints shall be installed in accordance with the manufacturer's instructions.

P3003.13 Joints between different materials. Joints between different piping materials shall be made with a mechanical joint of the compression or mechanical-sealing type conforming to ASTM C1173, ASTM C1460 or ASTM C1461. Connectors and adapters shall be *approved* for the application and such joints shall have an elastomeric seal conforming to ASTM C425, ASTM C443, ASTM C564, ASTM C1440, ASTM D1869, ASTM F477, CSA A257.3 or CSA B602, or as required in Sections P3003.13.1 through P3003.13.6. Joints between glass pipe and other types of materials shall be made with adapters having a TFE seal. Joints shall be installed in accordance with the manufacturer's instructions.

P3003.13.1 Copper pipe or tubing to cast-iron hub pipe. Joints between copper pipe or tubing and cast-iron hub pipe shall be made with a copper-alloy ferrule or compression joint. The copper pipe or tubing shall be soldered to the ferrule in an *approved* manner, and the ferrule shall be joined to the cast-iron hub by a caulked joint or a mechanical compression joint.

P3003.13.2 Copper pipe or tubing to galvanized steel pipe. Joints between copper pipe or tubing and galvanized steel pipe shall be made with a copper-alloy or dielectric fitting. The copper tubing shall be soldered to the fitting in an *approved* manner, and the fitting shall be screwed to the threaded pipe.

P3003.13.3 Cast-iron pipe to galvanized steel or copper-alloy pipe. Joints between cast-iron and galvanized steel or copper-alloy pipe shall be made by either caulked or threaded joints or with an *approved* adapter fitting.

P3003.13.4 Plastic pipe or tubing to other piping material. Joints between different types of plastic pipe shall be made with an *approved* adapter fitting or by a solvent-cement joint only where a single joint is made between ABS and PVC pipes at the end of a building drainage pipe and the beginning of a *building sewer* pipe using a solvent cement complying with ASTM D3138. Joints between plastic pipe and other piping material shall be made with an *approved* adapter fitting. Joints between plastic pipe and cast-iron hub pipe shall be made by a caulked joint or a mechanical compression joint.

P3003.13.5 Lead pipe to other piping material. Joints between lead pipe and other piping material shall be made by a wiped joint to a caulking ferrule, soldering nipple, or bushing or shall be made with an *approved* adapter fitting.

P3003.13.6 Stainless steel drainage systems to other materials. Joints between stainless steel drainage systems and other piping materials shall be made with *approved* mechanical couplings.

P3003.14 Joints between drainage piping and water closets. Joints between drainage piping and water closets or similar fixtures shall be made by means of a closet flange or a waste connector and sealing gasket compatible with the drainage system material, securely fastened to a structurally firm base. The joint shall be bolted, with an *approved* gasket flange to fixture connection complying with ASME A112.4.3 or setting compound between the fixture and the closet flange or waste connector and sealing gasket. The waste connector and sealing gasket joint shall comply with the joint-tightness test of ASME A112.4.3 and shall be installed in accordance with the manufacturer's instructions.

SECTION P3004 DETERMINING DRAINAGE FIXTURE UNITS

P3004.1 DWV system load. The load on DWV-system piping shall be computed in terms of drainage fixture unit (d.f.u.) values in accordance with Table P3004.1.

SECTION P3005 DRAINAGE SYSTEM

P3005.1 Drainage fittings and connections. Changes in direction in drainage piping shall be made by the appropriate use of sanitary tees, wyes, sweeps, bends or by a combination of these drainage fittings in accordance with Table P3005.1. Change in direction by combination fittings, heel or side inlets or increasers shall be installed in accordance with Table P3005.1 and Sections P3005.1.1 through P3005.1.4, based on the pattern of flow created by the fitting.

P3005.1.1 Horizontal to vertical (multiple connection fittings). Double fittings such as double sanitary tees and tee-wyes or *approved* multiple connection fittings and back-to-back fixture arrangements that connect two or more branches at the same level shall be permitted as long as directly opposing connections are the same size and the discharge into directly opposing connections is from similar fixture types or fixture groups. Double sanitary tee patterns shall not receive the discharge of back-to-back water closets and fixtures or *appliances* with pumping action discharge.

Exception: Back-to-back water closet connections to double sanitary tee patterns shall be permitted where the horizontal *developed length* between the outlet of

TABLE P3004.1
DRAINAGE FIXTURE UNIT (d.f.u.) VALUES FOR VARIOUS PLUMBING FIXTURES

TYPE OF FIXTURE OR GROUP OF FIXTURES	DRAINAGE FIXTURE UNIT VALUE (d.f.u.)[a]
Bar sink	1
Bathtub (with or without a shower head or whirlpool attachments)	2
Bidet	1
Clothes washer standpipe	2
Dishwasher	2
Floor drain[b]	0
Kitchen sink	2
Lavatory	1
Laundry tub	2
Shower stall	2
Water closet (1.6 gallons per flush)	3
Water closet (greater than 1.6 gallons per flush)	4
Full-bath group with bathtub (with 1.6-gallons-per-flush water closet, and with or without shower head and/or whirlpool attachment on the bathtub or shower stall)	5
Full-bath group with bathtub (water closet greater than 1.6 gallons per flush, and with or without shower head and/or whirlpool attachment on the bathtub or shower stall)	6
Half-bath group (1.6-gallons-per-flush water closet plus lavatory)	4
Half-bath group (water closet greater than 1.6 gallons per flush plus lavatory)	5
Kitchen group (dishwasher and sink with or without food-waste disposer)	2
Laundry group (clothes washer standpipe and laundry tub)	3
Multiple-bath groups[c]:	
1.5 baths	7
2 baths	8
2.5 baths	9
3 baths	10
3.5 baths	11

For SI: 1 gallon = 3.785 L, 1 gallon per minute = 3.785 L/m.

a. For a continuous or semicontinuous flow into a drainage system, such as from a pump or similar device, 1.5 fixture units shall be allowed per gpm of flow. For a fixture not listed, use the highest d.f.u. value for a similar listed fixture.

b. A floor drain itself does not add hydraulic load. Where used as a receptor, the fixture unit value of the fixture discharging into the receptor shall be applicable.

c. Add 2 d.f.u. for each additional full bath.

the water closet and the connection to the double sanitary tee is 18 inches (457 mm) or greater.

P3005.1.2 Heel- or side-inlet quarter bends, drainage. Heel-inlet quarter bends shall be an acceptable means of connection, except where the quarter bends serves a water closet. A low-heel inlet shall not be used as a wet-vented connection. Side-inlet quarter bends shall be an acceptable means of connection for both drainage, wet venting and stack venting arrangements.

P3005.1.3 Heel- or side-inlet quarter bends, venting. Heel-inlet or side-inlet quarter bends, or any arrangement of pipe and fittings producing a similar effect, shall be acceptable as a dry vent where the inlet is placed in a vertical position. The inlet is permitted to be placed in a *horizontal* position only where the entire fitting is part of a dry vent arrangement.

P3005.1.4 Water closet connection between flange and pipe. One-quarter bends 3 inches (76 mm) in diameter shall be acceptable for water closet or similar connections, provided that a 4-inch by 3-inch (102 mm by 76 mm) flange is installed to receive the closet fixture horn. Alternately, a 4-inch by 3-inch (102 mm by 76 mm) elbow shall be acceptable with a 4-inch (102 mm) flange.

P3005.1.5 Provisions for future fixtures. Where drainage has been roughed-in for future fixtures, the drainage unit values of the future fixtures shall be considered in determining the required drain sizes. Such future installations shall be terminated with an accessible permanent plug or cap fitting.

P3005.1.6 Drainage piping size reduction in the direction of flow. The size of the drainage piping shall not be reduced in the direction of the flow. The following shall not be considered a reduction in size in the direction of flow:

1. A 4-inch by 3-inch (102 mm by 76 mm) water closet flange.
2. A water closet bend fitting having a 4-inch (102 mm) inlet and a 3-inch (76 mm) outlet provided that the 4-inch (102 mm) leg of the fitting is upright and below, but not necessarily directly connected to, the water closet flange.
3. An offset closet flange.

P3005.2 Cleanouts required. Cleanouts shall be provided for drainage piping in accordance with Sections P3005.2.1 through P3005.2.11

P3005.2.1 Horizontal drains and building drains. *Horizontal* drainage pipes in buildings shall have cleanouts located at intervals of not more than 100 feet (30 480 mm). *Building drains* shall have cleanouts located at intervals of not more than 100 feet (30 480 mm) except where manholes are used instead of cleanouts, the manholes shall be located at intervals of not more than 400 feet (122 m). The interval length shall be measured from the cleanout or manhole opening, along the *developed length* of the piping to the next drainage fitting providing access for cleaning, the end of the *horizontal* drain or the end of the *building drain.*

Exception: Horizontal *fixture drain* piping serving a nonremovable trap shall not be required to have a cleanout for the section of piping between the trap and the vent connection for such trap.

P3005.2.2 Building sewers. *Building sewers* smaller than 8 inches (203 mm) shall have cleanouts located at intervals of not more than 100 feet (30 480 mm). *Building sewers* 8 inches (203 mm) and larger shall have a manhole located not more than 200 feet (60 960 mm) from the junction of the *building drain* and *building sewer* and at intervals of not more than 400 feet (122 m). The interval length shall be measured from the cleanout or manhole opening, along the *developed length* of the piping to the next drainage fitting providing access for cleaning, a manhole or the end of the *building sewer.*

P3005.2.3 Building drain and building sewer junction. The junction of the *building drain* and the *building sewer* shall be served by a cleanout that is located at the junction or within 10 feet (3048 mm) *developed length* of piping

TABLE P3005.1
FITTINGS FOR CHANGE IN DIRECTION

TYPE OF FITTING PATTERN	CHANGE IN DIRECTION		
	Horizontal to vertical[c]	Vertical to horizontal	Horizontal to horizontal
Sixteenth bend	X	X	X
Eighth bend	X	X	X
Sixth bend	X	X	X
Quarter bend	X	X[a]	X[a]
Short sweep	X	X[a,b]	X[a]
Long sweep	X	X	X
Sanitary tee	X[c]	—	—
Wye	X	X	X
Combination wye and eighth bend	X	X	X

For SI: 1 inch = 25.4 mm.

a. The fittings shall only be permitted for a 2-inch or smaller fixture drain.

b. Three inches and larger.

c. For a limitation on multiple connection fittings, see Section P3005.1.1.

upstream of the junction. For the requirements of this section, removal of a water closet shall not be required to provide cleanout access.

P3005.2.4 Changes of direction. Where a *horizontal* drainage pipe, a *building drain* or a *building sewer* has a change of horizontal direction greater than 45 degrees (0.79 rad), a cleanout shall be installed at the change of direction. Where more than one change of horizontal direction greater than 45 degrees (0.79 rad) occurs within 40 feet (12 192 mm) of *developed length* of piping, the cleanout installed for the first change of direction shall serve as the cleanout for all changes in direction within that 40 feet (12 192 mm) of *developed length* of piping.

P3005.2.5 Cleanout size. Cleanouts shall be the same size as the piping served by the cleanout, except cleanouts for piping larger than 4 inches (102 mm) need not be larger than 4 inches (102 mm).

Exceptions:

1. A removable P-trap with slip- or ground-joint connections can serve as a cleanout for drain piping that is one size larger than the P-trap size.
2. Cleanouts located on stacks can be one size smaller than the stack size.
3. The size of cleanouts for cast-iron piping can be in accordance with the referenced standards for cast iron fittings as indicated in Table P3002.3.

P3005.2.6 Cleanout plugs. Cleanout plugs shall be copper alloy, plastic or other *approved* materials. Cleanout plugs for borosilicate glass piping systems shall be of borosilicate glass. Copper-alloy cleanout plugs shall conform to ASTM A74 and shall be limited for use only on metallic piping systems. Plastic cleanout plugs shall conform to the referenced standards for plastic pipe fittings as indicated in Table P3002.3. Cleanout plugs shall have a raised square head, a countersunk square head or a countersunk slot head. Where a cleanout plug will have a trim cover screw installed into the plug, the plug shall be manufactured with a blind end threaded hole for such purpose.

P3005.2.7 Manholes. Manholes and manhole covers shall be of an *approved* type. Manholes located inside of a building shall have gastight covers that require tools for removal.

P3005.2.8 Installation arrangement. The installation arrangement of a cleanout shall enable cleaning of drainage piping only in the direction of drainage flow.

Exceptions:

1. Test tees serving as cleanouts.
2. A two-way cleanout installation that is *approved* for meeting the requirements of Section P3005.2.3.

P3005.2.9 Required clearance. Cleanouts for 6-inch (153 mm) and smaller piping shall be provided with a clearance of not less than 18 inches (457 mm) from, and perpendicular to, the face of the opening to any obstruction. Cleanouts for 8-inch (203 mm) and larger piping shall be provided with a clearance of not less than 36 inches (914 mm) from, and perpendicular to, the face of the opening to any obstruction.

P3005.2.10 Cleanout access. Required cleanouts shall not be installed in concealed locations. For the purposes of this section, concealed locations include, but are not limited to, the inside of plenums, within walls, within floor/ceiling assemblies, below grade and in crawl spaces where the height from the *crawl space* floor to the nearest obstruction along the path from the *crawl space* opening to the cleanout location is less than 24 inches (610 mm). Cleanouts with openings at a finished wall shall have the face of the opening located within $1^1/_2$ inches (38 mm) of the finished wall surface. Cleanouts located below grade shall be extended to grade level so that the top of the cleanout plug is at or above grade. A cleanout installed in a floor or walkway that will not have a trim cover installed shall have a counter-sunk plug installed so the top surface of the plug is flush with the finished surface of the floor or walkway.

P3005.2.10.1 Cleanout equivalent. A fixture trap or a fixture with an integral trap, removable without altering the concealed piping, shall be acceptable as a cleanout equivalent.

P3005.2.10.2 Cleanout plug trim covers. Trim covers and access doors for cleanout plugs shall be designed for such purposes. Trim cover fasteners that thread into cleanout plugs shall be corrosion resistant. Cleanout plugs shall not be covered with mortar, plaster or any other permanent material.

P3005.2.10.3 Floor cleanout assemblies. Where it is necessary to protect a cleanout plug from the loads of vehicular traffic, cleanout assemblies in accordance with ASME A112.36.2M shall be installed.

P3005.2.11 Prohibited use. The use of a threaded cleanout opening to add a fixture or extend piping shall be prohibited except where another cleanout of equal size is installed with the required access and clearance.

P3005.3 Horizontal drainage piping slope. *Horizontal* drainage piping shall be installed in uniform alignment at uniform slopes not less than $^1/_4$ unit vertical in 12 units horizontal (2-percent slope) for $2^1/_2$-inch (64 mm) diameter and less, and not less than $^1/_8$ unit vertical in 12 units horizontal (1-percent slope) for diameters of 3 inches (76 mm) or more.

P3005.4 Drain pipe sizing. Drain pipes shall be sized according to drainage fixture unit (d.f.u.) loads. The size of the drainage piping shall not be reduced in size in the direction of flow. The following general procedure is permitted to be used:

1. Draw an isometric layout or riser diagram denoting fixtures on the layout.
2. Assign d.f.u. values to each fixture group plus individual fixtures using Table P3004.1.
3. Starting with the top floor or most remote fixtures, work downstream toward the *building drain* accumulating d.f.u. values for fixture groups plus individual

fixtures for each branch. Where multiple bath groups are being added, use the reduced d.f.u. values in Table P3004.1, which take into account probability factors of simultaneous use.

4. Size branches and stacks by equating the assigned d.f.u. values to pipe sizes shown in Table P3005.4.1.
5. Determine the pipe diameter and slope of the *building drain* and *building sewer* based on the accumulated d.f.u. values, using Table P3005.4.2.

P3005.4.1 Branch and stack sizing. Branches and stacks shall be sized in accordance with Table P3005.4.1. Below grade drain pipes shall be not less than $1^1/_2$ inches (38 mm) in diameter. Drain stacks shall be not smaller than the largest horizontal branch connected.

Exceptions:

1. A 4-inch by 3-inch (102 mm by 76 mm) closet bend or flange.
2. A 4-inch (102 mm) closet bend connected to a 3-inch (76 mm) stack tee shall not be prohibited.

TABLE P3005.4.1
MAXIMUM FIXTURE UNITS ALLOWED TO BE CONNECTED TO BRANCHES AND STACKS

NOMINAL PIPE SIZE (inches)	ANY HORIZONTAL FIXTURE BRANCH	ANY ONE VERTICAL STACK OR DRAIN
$1^1/_4$ [a, b]	—	—
$1^1/_2$ [b]	3	4
2[b]	6	10
$2^1/_2$ [b]	12	20
3	20	48
4	160	240

For SI: 1 inch = 25.4 mm.

a. $1^1/_4$-inch pipe size limited to a single-fixture drain. See Table P3201.7.

b. Water closets prohibited.

P3005.4.2 Building drain and sewer size and slope. Pipe sizes and slope shall be determined from Table P3005.4.2 on the basis of drainage load in fixture units (d.f.u.) computed from Table P3004.1.

TABLE P3005.4.2
MAXIMUM NUMBER OF FIXTURE UNITS ALLOWED TO BE CONNECTED TO THE BUILDING DRAIN, BUILDING DRAIN BRANCHES OR THE BUILDING SEWER

DIAMETER OF PIPE (inches)	SLOPE PER FOOT		
	$^1/_8$ inch	$^1/_4$ inch	$^1/_2$ inch
$1^1/_2$ [a, b]	—	Note a	Note a
2[b]	—	21	27
$2^1/_2$ [b]	—	24	31
3	36	42	50
4	180	216	250

For SI: 1 inch = 25.4 mm, 1 foot = 304.8 mm.

a. $1^1/_2$-inch pipe size limited to a building drain branch serving not more than two waste fixtures, or not more than one waste fixture if serving a pumped discharge fixture or food waste disposer discharge.

b. No water closets.

P3005.5 Connections to offsets and bases of stacks. Horizontal branches shall connect to the bases of stacks at a point located not less than 10 times the diameter of the drainage stack downstream from the stack. Horizontal branches shall connect to horizontal stack offsets at a point located not less than 10 times the diameter of the drainage stack downstream from the upper stack.

SECTION P3006
SIZING OF DRAIN PIPE OFFSETS

P3006.1 Vertical offsets. An offset in a vertical drain, with a change of direction of 45 degrees (0.79 rad) or less from the vertical, shall be sized as a straight vertical drain.

P3006.2 Horizontal offsets above the lowest branch. A stack with an offset of more than 45 degrees (0.79 rad) from the vertical shall be sized as follows:

1. The portion of the stack above the offset shall be sized as for a regular stack based on the total number of fixture units above the offset.
2. The offset shall be sized as for a *building drain* in accordance with Table P3005.4.2.
3. The portion of the stack below the offset shall be sized as for the offset or based on the total number of fixture units on the entire stack, whichever is larger.

P3006.3 Horizontal offsets below the lowest branch. In soil or waste stacks below the lowest horizontal branch, a change in diameter shall not be required if the offset is made at an angle not greater than 45 degrees (0.79 rad) from the vertical. If an offset greater than 45 degrees (0.79 rad) from the vertical is made, the offset and stack below it shall be sized as a *building drain* in accordance with Table P3005.4.2.

SECTION P3007
SUMPS AND EJECTORS

P3007.1 Building subdrains. Building subdrains that cannot be discharged to the sewer by gravity flow shall be discharged into a tightly covered and vented sump from which the liquid shall be lifted and discharged into the building gravity drainage system by automatic pumping equipment or other *approved* method. In other than existing structures, the sump shall not receive drainage from any piping within the building capable of being discharged by gravity to the *building sewer.*

P3007.2 Valves required. A check valve and a full open valve located on the discharge side of the check valve shall be installed in the pump or ejector discharge piping between the pump or ejector and the gravity drainage system. Access shall be provided to such valves. Such valves shall be located above the sump cover required by Section P3007.3.2 or, where the discharge pipe from the ejector is below grade, the valves shall be accessibly located outside the sump below grade in an access pit with a removable access cover.

P3007.3 Sump design. The sump pump, sump and discharge piping shall conform to the requirements of Sections P3007.3.1 through P3007.3.5.

P3007.3.1 Sump pump. The sump pump capacity and head shall be appropriate to anticipated use requirements.

P3007.3.2 Sump. The sump shall be not less than 18 inches (457 mm) in diameter and 24 inches (610 mm) deep, unless otherwise *approved.* The sump shall be accessible and located so that drainage flows into the sump by gravity. The sump shall be constructed of tile, concrete, steel, plastic or other *approved* materials. The sump bottom shall be solid and provide permanent support for the pump. The sump shall be fitted with a gastight removable cover that is installed not more than 2 inches (51 mm) below grade or floor level. The cover shall be adequate to support anticipated loads in the area of use. The sump shall be vented in accordance with Chapter 31.

P3007.3.3 Discharge pipe and fittings. Discharge pipe and fittings serving sump pumps and ejectors shall be constructed of materials in accordance with Sections P3007.3.3.1 and P3007.3.3.2.

P3007.3.3.1 Materials. Pipe and fitting materials shall be constructed of copper alloy, copper, CPVC, ductile iron, PE, or PVC.

P3007.3.3.2 Ratings. Pipe and fittings shall be rated for the maximum system operating pressure and temperature. Pipe fitting materials shall be compatible with the pipe material. Where pipe and fittings are buried in the earth, they shall be suitable for burial.

P3007.3.4 Maximum effluent level. The effluent level control shall be adjusted and maintained to at all times prevent the effluent in the sump from rising to within 2 inches (51 mm) of the invert of the gravity drain inlet into the sump.

P3007.3.5 Ejector connection to the drainage system. Pumps connected to the drainage system shall connect to a *building sewer, building drain,* soil stack, waste stack or *horizontal* branch drain. Where the discharge line connects into *horizontal* drainage piping, the connection shall be made through a wye fitting into the top of the drainage piping and such wye fitting shall be located not less than 10 pipe diameters from the base of any soil stack, waste stack or *fixture drain.*

P3007.4 Sewage pumps and sewage ejectors. A sewage pump or sewage ejector shall automatically discharge the contents of the sump to the building drainage system.

P3007.5 Macerating toilet systems and pumped waste systems. Macerating toilet systems and pumped waste systems shall comply with ASME A112.3.4/CSA B45.9 and shall be installed in accordance with the manufacturer's instructions.

P3007.6 Capacity. Sewage pumps and sewage ejectors shall have the capacity and head for the application requirements. Pumps and ejectors that receive the discharge of water closets shall be capable of handling spherical solids with a diameter of up to and including 2 inches (51 mm). Other pumps or ejectors shall be capable of handling spherical solids with a diameter of up to and including $^1/_2$ inch (13 mm). The minimum capacity of a pump or ejector based on the diameter of the discharge pipe shall be in accordance with Table 3007.6.

Exceptions:

1. Grinder pumps or grinder ejectors that receive the discharge of water closets shall have a discharge opening of not less than $1^1/_4$ inches (32 mm).
2. Macerating toilet assemblies that serve single water closets shall have a discharge opening of not less than $^3/_4$ inch (19 mm).

TABLE 3007.6
MINIMUM CAPACITY OF SEWAGE PUMP OR SEWAGE EJECTOR

DIAMETER OF THE DISCHARGE PIPE (inches)	CAPACITY OF PUMP OR EJECTOR (gpm)
2	21
$2^1/_2$	30
3	46

For SI: 1 inch = 25.4 mm, 1 gallon per minute = 3.785 L/m.

SECTION P3008 BACKWATER VALVES

P3008.1 Where required. Where the flood level rims of plumbing fixtures are below the elevation of the manhole cover of the next upstream manhole in the public sewer, the fixtures shall be protected by a backwater valve installed in the *building drain,* branch of the *building drain* or *horizontal* branch serving such fixtures.

P3008.2 Allowable installations. Where plumbing fixtures are installed on a floor with a finished floor elevation above the elevation of the manhole cover of the next upstream manhole in the public sewer, and a backwater valve is installed in the *building drain* or horizontal branch serving such fixtures, the backwater valve shall be of the normally open type.

Exception: Normally closed backwater valve installations for existing buildings shall not be prohibited.

P3008.3 Material. Backwater valves shall comply with ASME A112.14.1, CSA B181.1 or CSA B181.2.

P3008.4 Location. Backwater valves shall be installed so that access is provided to the working parts.

SECTION P3009 GRAYWATER SOIL ABSORPTION SYSTEMS

P3009.1 Scope. The provisions of this section shall govern the materials, design, construction and installation of subsurface graywater soil absorption systems connected to nonpotable water from on-site water reuse systems.

P3009.2 Materials. Above-ground drain, waste and vent piping for subsurface graywater soil absorption systems shall conform to one of the standards indicated in Table P3002.1(1). Subsurface graywater soil absorption, underground building drainage and vent pipe shall conform to one of the standards indicated in Table P3002.1(2).

P3009.3 Tests. Drain, waste and vent piping for subsurface graywater soil absorption systems shall be tested in accordance with Section P2503.

P3009.4 Inspections. Subsurface graywater soil absorption systems shall be inspected in accordance with Section R109.

P3009.5 Disinfection. Disinfection shall not be required for on-site nonpotable reuse water for subsurface graywater soil absorption systems.

P3009.6 Coloring. On-site nonpotable reuse water used for subsurface graywater soil absorption systems shall not be required to be dyed.

P3009.7 Sizing. The system shall be sized in accordance with the sum of the output of all water sources connected to the graywater soil absorption system. Where graywater collection piping is connected to subsurface graywater soil absorption systems, graywater output shall be calculated according to the gallons-per-day-per-occupant (liters per day per occupant) number based on the type of fixtures connected. The graywater discharge shall be calculated by the following equation:

$$C = A \times B \qquad \textbf{(Equation 30-1)}$$

where:

A = Number of occupants:

Number of occupants shall be determined by the actual number of occupants, but not less than two occupants for one bedroom and one occupant for each additional bedroom.

B = Estimated flow demands for each occupant:

25 gallons (94.6 L) per day per occupant for showers, bathtubs and lavatories and 15 gallons (56.7 L) per day per occupant for clothes washers or laundry trays.

C = Estimated graywater discharge based on the total number of occupants.

P3009.8 Percolation tests. The permeability of the soil in the proposed absorption system shall be determined by percolation tests or permeability evaluation.

P3009.8.1 Percolation tests and procedures. Not less than three percolation tests in each system area shall be conducted. The holes shall be spaced uniformly in relation to the bottom depth of the proposed absorption system. More percolation tests shall be made where necessary, depending on system design.

P3009.8.1.1 Percolation test hole. The test hole shall be dug or bored. The test hole shall have vertical sides and a horizontal dimension of 4 inches to 8 inches (102 mm to 203 mm). The bottom and sides of the hole shall be scratched with a sharp-pointed instrument to expose the natural soil. Loose material shall be removed from the hole and the bottom shall be covered with 2 inches (51 mm) of gravel or coarse sand.

P3009.8.1.2 Test procedure, sandy soils. The hole shall be filled with clear water to not less than 12 inches (305 mm) above the bottom of the hole for tests in sandy soils. The time for this amount of water to seep away shall be determined, and this procedure shall be repeated if the water from the second filling of the hole seeps away in 10 minutes or less. The test shall proceed as follows: Water shall be added to a point not more than 6 inches (152 mm) above the gravel or coarse sand. Thereupon, from a fixed reference point, water levels shall be measured at 10-minute intervals for a period of 1 hour. Where 6 inches (152 mm) of water seeps away in less than 10 minutes, a shorter interval between measurements shall be used. The water depth shall not exceed 6 inches (152 mm). Where 6 inches (152 mm) of water seeps away in less than 2 minutes, the test shall be stopped and a rate of less than 3 minutes per inch (7.2 s/mm) shall be reported. The final water level drop shall be used to calculate the percolation rate. Soils not meeting these requirements shall be tested in accordance with Section P3009.8.1.3.

P3009.8.1.3 Test procedure, other soils. The hole shall be filled with clear water, and a minimum water depth of 12 inches (305 mm) shall be maintained above the bottom of the hole for a 4-hour period by refilling whenever necessary or by use of an automatic siphon. Water remaining in the hole after 4 hours shall not be removed. Thereafter, the soil shall be allowed to swell not less than 16 hours or more than 30 hours. Immediately after the soil swelling period, the measurements for determining the percolation rate shall be made as follows: any soil sloughed into the hole shall be removed and the water level shall be adjusted to 6 inches (152 mm) above the gravel or coarse sand. Thereupon, from a fixed reference point, the water level shall be measured at 30-minute intervals for a period of 4 hours, unless two successive water level drops do not vary by more than $^{1}/_{16}$ inch (1.59 mm). Not less than three water level drops shall be observed and recorded. The hole shall be filled with clear water to a point not more than 6 inches (152 mm) above the gravel or coarse sand whenever it becomes nearly empty. Adjustments of the water level shall not be made during the three measurement periods except to the limits of the last measured water level drop. When the first 6 inches (152 mm) of water seeps away in less than 30 minutes, the time interval between measurements shall be 10 minutes and the test run for 1 hour. The water depth shall not exceed 5 inches (127 mm) at any time during the measurement period. The drop that occurs during the final measurement period shall be used in calculating the percolation rate.

P3009.8.1.4 Mechanical test equipment. Mechanical percolation test equipment shall be of an *approved* type.

P3009.8.2 Permeability evaluation. Soil shall be evaluated for estimated percolation based on structure and texture in accordance with accepted soil evaluation practices. Borings shall be made in accordance with Section P3009.8.1.1 for evaluating the soil.

P3009.9 Subsurface graywater soil absorption system site location. The surface grade of soil absorption systems shall be located at a point lower than the surface grade of any water well or reservoir on the same or adjoining lot. Where this is not possible, the site shall be located so surface water drainage from the site is not directed toward a well or reservoir. The soil absorption system shall be located with a minimum horizontal distance between various elements as indicated in Table P3009.9. Private sewage disposal systems in compacted areas, such as parking lots and driveways, are prohibited. Surface water shall be diverted away from any soil absorption site on the same or neighboring lots.

**TABLE P3009.9
LOCATION OF SUBSURFACE IRRIGATION SYSTEM**

ELEMENT	MINIMUM HORIZONTAL DISTANCE	
	STORAGE TANK (feet)	ABSORPTION FIELD (feet)
Buildings	5	2
Lot line adjoining private property	5	5
Public water main	10	10
Seepage pits	5	5
Septic tanks	0	5
Streams and lakes	50	50
Water service	5	5
Water wells	50	100

For SI: 1 foot = 304.8 mm.

P3009.10 Installation. Absorption systems shall be installed in accordance with Sections P3009.10.1 through P3009.11.1.

P3009.10.1 Absorption area. The total absorption area required shall be computed from the estimated daily graywater discharge and the design-loading rate based on the percolation rate for the site. The required absorption area equals the estimated graywater discharge divided by the design loading rate from Table P3009.10.1.

**TABLE P3009.10.1
DESIGN LOADING RATE**

PERCOLATION RATE (minutes per inch)	DESIGN LOADING FACTOR (gallons per square foot per day)
0 to less than 10	1.2
10 to less than 30	0.8
30 to less than 45	0.72
45 to 60	0.4

For SI: 1 minute per inch = min/25.4 mm,
1 gallon per square foot = 40.7 L/m^2.

P3009.10.2 Seepage trench excavations. Seepage trench excavations shall be not less than 1 foot (304 mm) in width and not greater than 5 feet (1524 mm) in width. Trench excavations shall be spaced not less than 2 feet (610 mm) apart. The soil absorption area of a seepage trench shall be computed by using the bottom of the trench area (width) multiplied by the length of pipe. Individual seepage trenches shall be not greater than 100 feet (30 480 mm) in *developed length*.

P3009.10.3 Seepage bed excavations. Seepage bed excavations shall be not less than 5 feet (1524 mm) in width and have more than one distribution pipe. The absorption area of a seepage bed shall be computed by using the bottom of the trench area. Distribution piping in a seepage bed shall be uniformly spaced not greater than 5 feet (1524 mm) and not less than 3 feet (914 mm) apart, and greater than 3 feet (914 mm) and not less than 1 foot (305 mm) from the sidewall or headwall.

P3009.10.4 Excavation and construction. The bottom of a trench or bed excavation shall be level. Seepage trenches or beds shall not be excavated where the soil is so wet that such material rolled between the hands forms a soil wire. Smeared or compacted soil surfaces in the sidewalls or bottom of seepage trench or bed excavations shall be scarified to the depth of smearing or compaction and the loose material removed. Where rain falls on an open excavation, the soil shall be left until sufficiently dry so a soil wire will not form when soil from the excavation bottom is rolled between the hands. The bottom area shall then be scarified and loose material removed.

P3009.10.5 Aggregate and backfill. Not less than 6 inches (150 mm) in depth of aggregate ranging in size from $^1/_2$ to $2^1/_2$ inches (12.7 mm to 64 mm) shall be laid into the trench below the distribution piping elevation. The aggregate shall be evenly distributed not less than 2 inches (51 mm) in depth over the top of the distribution pipe. The aggregate shall be covered with *approved* synthetic materials or 9 inches (229 mm) of uncompacted marsh hay or straw. Building paper shall not be used to cover the aggregate. Not less than 9 inches (229 mm) of soil backfill shall be provided above the covering.

P3009.11 Distribution piping. Distribution piping shall be not less than 3 inches (76 mm) in diameter. Materials shall comply with Table P3009.11. The top of the distribution pipe shall be not less than 8 inches (203 mm) below the original surface. The slope of the distribution pipes shall be not less than 2 inches (51 mm) and not greater than 4 inches (102 mm) per 100 feet (30 480 mm).

**TABLE P3009.11
DISTRIBUTION PIPE**

MATERIAL	STANDARD
Polyethylene (PE) plastic pipe	ASTM F405
Polyvinyl chloride (PVC) plastic pipe	ASTM D2729
Polyvinyl chloride (PVC) plastic pipe with a 3.5-inch O.D. and solid, cellular core or composite wall	ASTM F1488

For SI: 1 inch = 25.4 mm.

P3009.11.1 Joints. Joints in distribution pipe shall be made in accordance with Section P3003.

SECTION P3010
REPLACEMENT OF UNDERGROUND BUILDING SEWERS AND BUILDING DRAINS BY PIPE BURSTING METHODS

P3010.1 General. This section shall govern the replacement of existing *buildingPP sewer* and *building drain* piping by pipe-bursting methods.

P3010.2 Applicability. The replacement of *building sewer* and *building drain* piping by pipe bursting methods shall be limited to gravity drainage piping of sizes 6 inches (150 mm) and smaller. The replacement piping shall be of the same nominal size as the existing piping.

P3010.3 Preinstallation inspection. The existing piping sections to be replaced shall be inspected internally by a recorded video camera survey. The survey shall include notations of the position of cleanouts and the depth of connections to the existing piping.

P3010.4 Pipe. The replacement pipe shall be made of a high-density polyethylene (HDPE) and shall be in compliance with ASTM F714.

P3010.5 Pipe fittings. Pipe fittings to be connected to the replacement pipe shall be made of high-density polyethylene (HDPE) and shall be in compliance with ASTM D2683.

P3010.6 Cleanouts. Where the existing *building sewer* or *building drain* did not have cleanouts meeting the requirements of this code, cleanout fittings shall be installed as required by this code.

P3010.7 Post-installation inspection. The completed replacement piping section shall be inspected internally by a recorded video camera survey. The video survey shall be reviewed and *approved* by the *building official* prior to pressure testing of the replacement piping system.

P3010.8 Pressure testing. The replacement piping system and the connections to the replacement piping shall be tested in accordance with Section P2503.4.

SECTION P3011
RELINING OF BUILDING SEWERS AND BUILDING DRAINS

P3011.1 General. This section shall govern the relining of existing *building sewer* and building drainage piping.

P3011.2 Applicability. The relining of existing *building sewer* piping and building drainage piping shall be limited to gravity drainage piping 4 inches (102 mm) in diameter and larger. The relined piping shall be of the same nominal size as the existing piping.

P3011.3 Preinstallation requirements. Prior to commencement of the relining installation, the existing piping sections to be relined shall be descaled and cleaned. After the cleaning process has occurred and water has been flushed through the system, the piping shall be inspected internally by a recorded video camera survey.

P3011.3.1 Preinstallation recorded video camera survey. The video survey shall include verification of the project address location. The video shall include notations of the cleanout and fitting locations, and the approximate depth of the existing piping. The video shall also include notations of the length of piping at intervals not greater than 25 feet (7620 mm).

P3011.4 Permitting. Prior to issuing a permit for relining, the building official shall review and evaluate the preinstallation recorded video camera survey to determine whether the piping system is able to be relined in accordance with the proposed lining system manufacturer's installation requirements and applicable referenced standards.

P3011.5 Prohibited applications. Where the preinstallation recorded video camera survey reveals that piping systems are not installed correctly, or defects exist, relining shall not be permitted. The defective portions of piping shall be exposed and repaired with pipe and fittings in accordance with this code. Defects shall include, but are not limited to, backslope or insufficient slope, complete pipe wall deterioration or complete separations such as from tree root invasion or improper support.

P3011.6 Relining materials. The relining materials shall be manufactured in compliance with applicable standards and certified as required in Section P2609. Fold-and-form pipe reline materials shall be manufactured in compliance with ASTM F1504 or ASTM F1871.

P3011.7 Installation. The installation of relining materials shall be performed in accordance with the manufacturer's installation instructions, applicable referenced standards and this code.

P3011.7.1 Material data report. The installer shall record the data as required by the relining material manufacturer and applicable standards. The recorded data shall include, but is not limited to, the location of the project, relining material type, amount of product installed and conditions of the installation. A copy of the data report shall be provided to the building official prior to final approval.

P3011.8 Post-installation recorded video camera survey. The completed relined piping system shall be inspected internally by a recorded video camera survey after the system has been flushed and flow-tested with water. The video survey shall be submitted to the building official prior to finalization of the permit. The video survey shall be reviewed and evaluated to provide verification that no defects exist. Any defects identified shall be repaired and replaced in accordance with this code.

P3011.9 Certification. Certification shall be provided in writing to the building official, from the permit holder, that the relining materials have been installed in accordance with the manufacturer's installation instructions, the applicable standards and this code.

P3011.10 Approval. Upon verification of compliance with the requirements of Sections P3011.1 through P3011.9, the building official shall approve the installation.

CHAPTER 31
VENTS

User notes:

About this chapter: *Chapter 31 regulates connection locations, various venting system arrangements and the sizing of piping for vent systems. The proper operation of a gravity flow drainage system (Chapter 30) depends on maintaining an air path throughout the system to prevent waste and odor "blow back" into fixtures and siphoning of the trap seal in fixture traps (Chapter 32).*

Code development reminder: *Code change proposals to this chapter will be considered by the IRC—Mechanical/Plumbing Code Development Committee during the 2021 (Group A) Code Development Cycle.*

SECTION P3101
VENT SYSTEMS

P3101.1 General. This chapter shall govern the selection and installation of piping, tubing and fittings for vent systems. This chapter shall control the minimum diameter of vent pipes, circuit vents, branch vents and individual vents, and the size and length of vents and various aspects of vent stacks and stack vents. Additionally, this chapter regulates vent grades and connections, height above fixtures and relief vents for stacks and fixture traps, and the venting of sumps and sewers.

P3101.2 Trap seal protection. The plumbing system shall be provided with a system of vent piping that will allow the admission or emission of air so that the liquid seal of any fixture trap shall not be subjected to a pressure differential of more than 1 inch of water column (249 Pa).

P3101.2.1 Venting required. Every *trap* and trapped fixture shall be vented in accordance with one of the venting methods specified in this chapter.

P3101.3 Use limitations. The plumbing vent system shall not be used for purposes other than the venting of the plumbing system.

P3101.4 Extension outside a structure. In climates where the 97.5-percent value for outside design temperature is 0°F (-18°C) or less (ASHRAE 97.5-percent column, winter, see Chapter 3), vent pipes installed on the exterior of the structure shall be protected against freezing by insulation, heat or both. Vent terminals shall be protected from frost closure in accordance with Section P3103.2.

P3101.5 Flood resistance. In flood hazard areas as established by Table R301.2, vents shall be located at or above the elevation required in Section R322.1 (flood hazard areas including A Zones) or R322.2 (coastal high-hazard areas including V Zones).

SECTION P3102
VENT STACKS AND STACK VENTS

P3102.1 Required vent extension. The vent system serving each *building drain* shall have not less than one vent pipe that extends to the outdoors.

P3102.2 Installation. The required vent shall be a dry vent that connects to the *building drain* or an extension of a drain that connects to the *building drain.* Such vent shall not be an island fixture vent as permitted by Section P3112.

P3102.3 Size. The required vent shall be sized in accordance with Section P3113.1 based on the required size of the *building drain.*

SECTION P3103
VENT TERMINALS

P3103.1 Vent pipes terminating outdoors. Vent pipes terminating outdoors shall be extended to the outdoors through the roof or a sidewall of the building in accordance with one of the methods identified in Sections P3103.1.1 through P3103.1.4.

P3103.1.1 Roof extension. Open vent pipes that extend through a roof that do not meet the conditions of Section P3103.1.2 or P3103.1.3 shall terminate not less than 6 inches (150 mm) above the roof or 6 inches (150 mm) above the anticipated snow accumulation, whichever is greater.

P3103.1.2 Roof used for recreational purposes. Where a roof is to be used for assembly, as a promenade, observation deck or sunbathing deck, or for similar purposes, open vent pipes shall terminate not less than 7 feet (2134 mm) above the roof.

P3103.1.3 Roof extension covered. Where an open vent pipe terminates above a sloped roof and is covered by either a roof-mounted panel (such as a solar collector or *photovoltaic panel* mounted over the vent opening) or a roof element (such as an architectural feature or a decorative shroud), the vent pipe shall terminate not less than 2 inches (51 mm) above the roof surface. Such roof elements shall be designed to prevent the adverse effects of snow accumulation and wind on the function of the vent. The placement of a panel over a vent pipe and the design of a roof element covering the vent pipe shall provide for an open area for the vent pipe to the outdoors that is not less than the area of the pipe, as calculated from the inside diameter of the pipe. Such vent terminals shall be protected by a method that prevents birds and rodents from entering or blocking the vent pipe opening.

P3103.1.4 Sidewall vent terminal. Vent terminals extending through the wall shall terminate not less than

10 feet (3048 mm) from a *lot line* and not less than 10 feet (3048 mm) above the highest grade elevation within 10 feet (3048 mm) in any direction horizontally of the vent terminal. Vent pipes shall not terminate under the overhang of a structure where the overhang includes soffit vents. Such vent terminals shall be protected by a method that prevents birds and rodents from entering or blocking the vent pipe opening and that does not reduce the open area of the vent pipe.

P3103.2 Frost closure. Where the 97.5-percent value for outside design temperature is 0°F (-18°C) or less, vent extensions through a roof or wall shall be not less than 3 inches (76 mm) in diameter. Any increase in the size of the vent shall be made not less than 1 foot (304.8 mm) inside the thermal envelope of the building.

P3103.3 Flashings and sealing. The juncture of each vent pipe with the roof line shall be made watertight by an *approved* flashing. Vent extensions in walls and soffits shall be made weathertight by caulking.

P3103.4 Prohibited use. A vent terminal shall not be used for any purpose other than a vent terminal.

P3103.5 Location of vent terminal. An open vent terminal from a drainage system shall not be located less than 4 feet (1219 mm) directly beneath any door, openable window, or other air intake opening of the building or of an adjacent building, nor shall any such vent terminal be within 10 feet (3048 mm) horizontally of such an opening unless it is not less than 3 feet (914 mm) above the top of such opening.

SECTION P3104 VENT CONNECTIONS AND GRADES

P3104.1 Connection. Individual branch and circuit vents shall connect to a vent stack, stack vent or extend to the open air.

Exception: Individual, branch and circuit vents shall be permitted to terminate at an *air admittance valve* in accordance with Section P3114.

P3104.2 Grade. Vent and branch vent pipes shall be graded, connected and supported to allow moisture and condensate to drain back to the soil or waste pipe by gravity.

P3104.3 Vent connection to drainage system. A dry vent connecting to a horizontal drain shall connect above the centerline of the horizontal drain pipe.

P3104.4 Vertical rise of vent. A dry vent shall rise vertically to not less than 6 inches (152 mm) above the flood level rim of the highest trap or trapped fixture being vented.

P3104.5 Height above fixtures. A connection between a vent pipe and a vent stack or stack vent shall be made not less than 6 inches (152 mm) above the flood level rim of the highest fixture served by the vent. *Horizontal* vent pipes forming branch vents shall be not less than 6 inches (152 mm) above the flood level rim of the highest fixture served.

P3104.6 Vent for future fixtures. Where the drainage piping has been roughed-in for future fixtures, a rough-in connection for a vent, not less than one-half the diameter of the drain, shall be installed. The vent rough-in shall connect to the vent system or shall be vented by other means as provided in this chapter. The connection shall be identified to indicate that the connection is a vent.

SECTION P3105 FIXTURE VENTS

P3105.1 Distance of trap from vent. Each fixture trap shall have a protecting vent located so that the slope and the *developed length* in the *fixture drain* from the trap weir to the vent fitting are within the limits indicated in Table P3105.1.

Exception: The *developed length* of the *fixture drain* from the trap weir to the vent fitting for self-siphoning fixtures, such as water closets, shall not be limited.

TABLE P3105.1 MAXIMUM DISTANCE OF FIXTURE TRAP FROM VENT

SIZE OF TRAP (inches)	SLOPE (inch per foot)	DISTANCE FROM TRAP (feet)
$1^1/_4$	$^1/_4$	5
$1^1/_2$	$^1/_4$	6
2	$^1/_4$	8
3	$^1/_8$	12
4	$^1/_8$	16

For SI: 1 inch = 25.4 mm, 1 foot = 304.8 mm, 1 inch per foot = 83.3 mm/m.

P3105.2 Fixture drains. The total fall in a *fixture drain* resulting from pipe slope shall not exceed one pipe diameter, nor shall the vent pipe connection to a *fixture drain*, except for water closets, be below the weir of the trap.

P3105.3 Crown vent prohibited. A vent shall not be installed within two pipe diameters of the trap weir.

SECTION P3106 INDIVIDUAL VENT

P3106.1 Individual vent permitted. Each trap and trapped fixture shall be permitted to be provided with an individual vent. The individual vent shall connect to the *fixture drain* of the trap or trapped fixture being vented.

SECTION P3107 COMMON VENT

P3107.1 Individual vent as common vent. An individual vent shall be permitted to vent two traps or trapped fixtures as a common vent. The traps or trapped fixtures being common vented shall be located on the same floor level.

P3107.2 Connection at the same level. Where the *fixture drains* being common vented connect at the same level, the vent connection shall be at the interconnection of the *fixture drains* or downstream of the interconnection.

P3107.3 Connection at different levels. Where the *fixture drains* connect at different levels, the vent shall connect as a vertical extension of the vertical drain. The vertical drain pipe connecting the two *fixture drains* shall be considered to be the vent for the lower *fixture drain*, and shall be sized in accordance with Table P3107.3. The upper fixture shall not be a water closet.

TABLE P3107.3
COMMON VENT SIZES

PIPE SIZE (inches)	MAXIMUM DISCHARGE FROM UPPER FIXTURE DRAIN (d.f.u.)
$1^1/_2$	1
2	4
$2^1/_2$ to 3	6

For SI: 1 inch = 25.4 mm.

SECTION P3108
WET VENTING

P3108.1 Horizontal wet vent permitted. Any combination of fixtures within two *bathroom groups* located on the same floor level shall be permitted to be vented by a horizontal wet vent. The wet vent shall be considered to be the vent for the fixtures and shall extend from the connection of the dry vent along the direction of the flow in the drain pipe to the most downstream *fixture drain* connection. Each *fixture drain* shall connect horizontally to the horizontal branch being wet vented or shall have a dry vent. Each wet-vented *fixture drain* shall connect independently to the horizontal wet vent. Only the fixtures within the *bathroom groups* shall connect to the wet-vented horizontal branch drain. Any additional fixtures shall discharge downstream of the horizontal wet vent.

P3108.2 Dry vent connection. The required dry-vent connection for wet-vented systems shall comply with Sections P3108.2.1 and P3108.2.2.

P3108.2.1 Horizontal wet vent. The dry-vent connection for a horizontal wet-vent system shall be an individual vent or a common vent for any *bathroom group* fixture, except an emergency floor drain. Where the dry vent connects to a water closet *fixture drain*, the drain shall connect horizontally to the horizontal wet-vent system. Not more than one wet-vented *fixture drain* shall discharge upstream of the dry-vented *fixture drain* connection.

P3108.2.2 Vertical wet vent. The dry-vent connection for a vertical wet-vent system shall be an individual vent or common vent for the most upstream *fixture drain*.

P3108.3 Size. Horizontal and vertical wet vents shall be not less than the size as specified in Table P3108.3, based on the fixture unit discharge to the wet vent. The dry vent serving the wet vent shall be sized based on the largest required diameter of pipe within the wet-vent system served by the dry vent.

TABLE P3108.3
WET VENT SIZE

WET VENT PIPE SIZE (inches)	FIXTURE UNIT LOAD (d.f.u.)
$1^1/_2$	1
2	4
$2^1/_2$	6
3	12
4	32

For SI: 1 inch = 25.4 mm.

P3108.4 Vertical wet vent permitted. A combination of fixtures located on the same floor level shall be permitted to be vented by a vertical wet vent. The vertical wet vent shall be considered to be the vent for the fixtures and shall extend from the connection of the dry vent down to the lowest *fixture drain* connection. Each wet-vented fixture shall connect independently to the vertical wet vent. All water closet drains shall connect at the same elevation. Other *fixture drains* shall connect above or at the same elevation as the water closet *fixture drains*. The dry-vent connection to the vertical wet vent shall be an individual or common vent serving one or two fixtures.

P3108.5 Trap weir to wet-vent distances. The maximum *developed length* of wet-vented *fixture drains* shall comply with Table P3105.1.

SECTION P3109
WASTE STACK VENT

P3109.1 Waste stack vent permitted. A waste stack shall be considered to be a vent for all of the fixtures discharging to the stack where installed in accordance with the requirements of this section.

P3109.2 Stack installation. The waste stack shall be vertical, and both horizontal and vertical offsets shall be prohibited between the lowest *fixture drain* connection and the highest *fixture drain* connection to the stack. Every *fixture drain* shall connect separately to the waste stack. The stack shall not receive the discharge of water closets or urinals.

P3109.3 Stack vent. A stack vent shall be installed for the waste stack. The size of the stack vent shall be not less than the size of the waste stack. Offsets shall be permitted in the stack vent and shall be located not less than 6 inches (152 mm) above the flood level of the highest fixture, and shall be in accordance with Section P3104.5. The stack vent shall be permitted to connect with other stack vents and vent stacks in accordance with Section P3113.3.

P3109.4 Waste stack size. The waste stack shall be sized based on the total discharge to the stack and the discharge within a *branch interval* in accordance with Table P3109.4. The waste stack shall be the same size throughout the length of the waste stack.

TABLE P3109.4
WASTE STACK VENT SIZE

STACK SIZE (inches)	MAXIMUM NUMBER OF FIXTURE UNITS (d.f.u.)	
	Total discharge into one branch interval	Total discharge for stack
$1^1/_2$	1	2
2	2	4
$2^1/_2$	No limit	8
3	No limit	24
4	No limit	50

For SI: 1 inch = 25.4 mm.

SECTION P3110
CIRCUIT VENTING

P3110.1 Circuit vent permitted. Not greater than eight fixtures connected to a horizontal branch drain shall be permitted to be circuit vented. Each *fixture drain* shall connect horizontally to the horizontal branch being circuit vented. The horizontal branch drain shall be classified as a vent from the most downstream *fixture drain* connection to the most upstream *fixture drain* connection to the horizontal branch.

P3110.2 Vent connection. The circuit vent connection shall be located between the two most upstream *fixture drains.* The vent shall connect to the horizontal branch and shall be installed in accordance with Section P3104. The circuit vent pipe shall not receive the discharge of any soil or waste.

P3110.3 Slope and size of horizontal branch. The slope of the vent section of the horizontal branch drain shall be not greater than 1 unit vertical in 12 units horizontal (8-percent slope). The entire length of the vent section of the horizontal branch drain shall be sized for the total drainage discharge to the branch in accordance with Table P3005.4.1.

P3110.4 Additional fixtures. Fixtures, other than circuit-vented fixtures, shall be permitted to discharge to the horizontal branch drain. Such fixtures shall be located on the same floor as the circuit-vented fixtures and shall be either individually or common vented.

SECTION P3111
COMBINATION WASTE AND VENT SYSTEM

P3111.1 Type of fixtures. A combination waste and vent system shall only serve floor drains, sinks, lavatories and drinking fountains. A combination waste and vent system shall be considered to be the vent for those fixtures. The *developed length* of a *fixture drain* to the combination waste and vent system piping shall not exceed the limitations of Table P3105.1.

P3111.1.1 Single-fixture systems. A horizontal *fixture drain* shall be considered to be a combination waste and vent system provided that the *fixture drain* size complies with Table P3111.3.

P3111.2 Installation. The only vertical pipe of a combination waste and vent system shall be the connection between a *fixture drain* and a horizontal combination waste and vent pipe. The length of the vertical pipe shall be not greater than 8 feet (2438 mm).

P3111.2.1 Slope. The slope of horizontal combination waste and vent piping shall be not greater than $^1/_2$ unit vertical in 12 units horizontal (4-percent slope) and shall be not less than that indicated in Section P3005.2.

P3111.2.2 Vent connection. A combination waste and vent system shall be provided with a dry vent connected at any point within the system, or the system shall connect to a horizontal drain or *building drain* that serves vented fixtures located on the same floor. Combination waste and vent systems connecting to *building drains* receiving only the discharge from one or more stacks shall be provided with a dry vent. The dry vent connected to the combination waste and vent pipe shall extend vertically to a point not less than 6 inches (152 mm) above the flood level rim of the highest fixture being vented by the combination waste and vent system before horizontal offsets in the dry vent piping are allowed.

P3111.2.3 Vent size. The dry vent connected to the combination waste and vent system shall be sized for the total drainage fixture unit load in accordance with Section P3111.1.

P3111.3 Size and length. The size of a combination drain and vent piping shall be not less than that specified in Table P3111.3. The horizontal length of a combination drain and vent system shall be unlimited.

TABLE P3111.3
SIZE OF COMBINATION WASTE AND VENT PIPE

DIAMETER PIPE (inches)	MAXIMUM NUMBER OF FIXTURE UNITS (d.f.u.)	
	Connecting to a horizontal branch or stack	Connecting to a building drain or building subdrain
2	3	4
$2^1/_2$	6	26
3	12	31
4	20	50

For SI: 1 inch = 25.4 mm.

SECTION P3112
ISLAND FIXTURE VENTING

P3112.1 Limitation. Island fixture venting shall not be permitted for fixtures other than sinks and lavatories. Kitchen sinks with a dishwasher waste connection, a food waste disposer, or both, in combination with the kitchen sink waste, shall be permitted to be vented in accordance with this section.

P3112.2 Vent connection. The island fixture vent shall connect to the *fixture drain* as required for an individual or common vent. The vent shall rise vertically to above the drainage outlet of the fixture being vented before offsetting horizontally or vertically downward. The vent or branch vent for multiple island fixture vents shall extend not less than 6 inches (152 mm) above the highest island fixture being vented before connecting to the outside vent terminal.

P3112.3 Vent installation below the fixture flood level rim. The vent located below the flood level rim of the fixture being vented shall be installed as required for drainage piping in accordance with Chapter 30, except for sizing. The vent shall be sized in accordance with Section P3113.1. The lowest point of the island fixture vent shall connect full size to the drainage system. The connection shall be to a vertical drain pipe or to the top half of a horizontal drain pipe. Cleanouts shall be provided in the island fixture vent to permit rodding of all vent piping located below the flood level rim of the fixtures. Rodding in both directions shall be permitted through a cleanout.

SECTION P3113 VENT PIPE SIZING

P3113.1 Size of vents. The required diameter of individual vents, branch vents, circuit vents, vent stacks and stack vents shall be not less than one-half the required diameter of the drain served. The required size of the drain shall be determined in accordance with Chapter 30. Vent pipes shall be not less than $1^{1}/_{4}$ inches (32 mm) in diameter. Vents exceeding 40 feet (12 192 mm) in *developed length* shall be increased by one nominal pipe size for the entire *developed length* of the vent pipe.

P3113.2 Developed length. The *developed length* of individual, branch and circuit vents shall be measured from the farthest point of vent connection to the drainage system, to the point of connection to the vent stack, stack vent or termination outside of the building.

P3113.3 Branch vents. Where branch vents are connected to a common branch vent, the common branch vent shall be sized in accordance with this section, based on the size of the common horizontal drainage branch that is or would be required to serve the total drainage fixture unit (d.f.u.) load being vented.

P3113.4 Sump vents. Sump vent sizes shall be determined in accordance with Sections P3113.4.1 and P3113.4.2.

P3113.4.1 Sewage pumps and sewage ejectors other than pneumatic. Drainage piping below sewer level shall be vented in the same manner as that of a gravity system. Building sump vent sizes for sumps with sewage pumps or sewage ejectors, other than pneumatic, shall be determined in accordance with Table P3113.4.1.

P3113.4.2 Pneumatic sewage ejectors. The air pressure relief pipe from a pneumatic sewage ejector shall be connected to an independent vent stack terminating as required for vent extensions through the roof. The relief pipe shall be sized to relieve air pressure inside the ejector to atmospheric pressure, but shall be not less than $1^{1}/_{4}$ inches (32 mm) in size.

SECTION P3114 AIR ADMITTANCE VALVES

P3114.1 General. Vent systems using *air admittance valves* shall comply with this section. Individual and branch-type air admittance valves shall conform to ASSE 1051. Stack-type air admittance valves shall conform to ASSE 1050.

P3114.2 Installation. The valves shall be installed in accordance with the requirements of this section and the manufacturer's instructions. *Air admittance valves* shall be installed after the DWV testing required by Section P2503.5.1 or P2503.5.2 has been performed.

P3114.3 Where permitted. Individual vents, branch vents, circuit vents and stack vents shall be permitted to terminate with a connection to an *air admittance valve*. Individual and branch-type air admittance valves shall vent only fixtures that are on the same floor level and connect to a horizontal branch drain.

P3114.4 Location. Individual and branch *air admittance valves* shall be located not less than 4 inches (102 mm) above the horizontal branch drain or *fixture drain* being vented. Stack-type air admittance valves shall be located not less than 6 inches (152 mm) above the flood level rim of the highest fixture being vented. The *air admittance valve* shall be located within the maximum *developed length* permitted for the vent. The *air admittance valve* shall be installed not less than 6 inches (152 mm) above insulation materials where installed in *attics*.

P3114.5 Access and ventilation. Access shall be provided to *air admittance valves*. Such valves shall be installed in a location that allows air to enter the valve.

**TABLE P3113.4.1
SIZE AND LENGTH OF SUMP VENTS**

DISCHARGE CAPACITY OF PUMP (gpm)	MAXIMUM DEVELOPED LENGTH OF VENT (feet)[a]				
	Diameter of vent (inches)				
	$1^{1}/_{4}$	$1^{1}/_{2}$	2	$2^{1}/_{2}$	3
10	No limit[b]	No limit	No limit	No limit	No limit
20	270	No limit	No limit	No limit	No limit
40	72	160	No limit	No limit	No limit
60	31	75	270	No limit	No limit

For SI: 1 inch = 25.4 mm, 1 foot = 304.8 mm, 1 gallon per minute (gpm) = 3.785 L/m.

a. Developed length plus an appropriate allowance for entrance losses and friction caused by fittings, changes in direction and diameter. Suggested allowances shall be obtained from NBS Monograph 31 or other approved sources. An allowance of 50 percent of the developed length shall be assumed if a more precise value is not available.

b. Actual values greater than 500 feet.

P3114.6 Size. The *air admittance valve* shall be rated for the size of the vent to which the valve is connected.

P3114.7 Vent required. Within each plumbing system, not less than one stack vent or a vent stack shall extend outdoors to the open air.

P3114.8 Prohibited installations. *Air admittance valves* shall not be used to vent sumps or tanks except where the vent system for the sump or tank has been designed by an engineer. *Air admittance valves* shall not be installed on outdoor vent terminals for the sole purpose of reducing clearances to gravity or mechanical air intakes.

CHAPTER 32
TRAPS

User notes:

About this chapter: *Chapter 32 regulates the design of fixture traps, methods for preventing evaporation of trap seals in traps, and the required locations for interceptors and separators. The trap seal of a trap is an essential feature of a drainage system to prevent odors from the drainage piping from entering the building. The discharge of various processes such as cooking and laundry creates the need for equipment to retain detrimental greases and solids from entering the drainage systems.*

Code development reminder: *Code change proposals to this chapter will be considered by the IRC—Mechanical/Plumbing Code Development Committee during the 2021 (Group A) Code Development Cycle.*

SECTION P3201
FIXTURE TRAPS

P3201.1 Design of traps. Traps shall be of standard design, shall have smooth uniform internal waterways, shall be self-cleaning and shall not have interior partitions except where integral with the fixture. Traps shall be constructed of lead, cast iron, copper or copper alloy or *approved* plastic. Copper or copper-alloy traps shall be not less than No. 20 gage (0.8 mm) thickness. Solid connections, slip joints and couplings shall be permitted to be used on the trap inlet, trap outlet, or within the trap seal. Traps having slip-joint connections shall comply with Section P2704.1.

P3201.2 Trap seals. Each fixture trap shall have a liquid seal of not less than 2 inches (51 mm) and not more than 4 inches (102 mm).

P3201.2.1 Trap seal protection. Traps seals of emergency floor drain traps and traps subject to evaporation shall be protected by one of the methods in Sections P3201.2.1.1 through P3201.2.1.4.

P3201.2.1.1 Potable water-supplied trap seal primer valve. A potable water-supplied trap seal primer valve shall supply water to the trap. Water-supplied trap seal primer valves shall conform to ASSE 1018. The discharge pipe from the trap seal primer valve shall connect to the trap above the trap seal on the inlet side of the trap.

P3201.2.1.2 Reclaimed or graywater-supplied trap seal primer valve. A reclaimed or graywater-supplied trap seal primer valve shall supply water to the trap. Water-supplied trap seal primer valves shall conform to ASSE 1018. The quality of reclaimed or graywater supplied to trap seal primer valves shall be in accordance with the requirements of the manufacturer of the trap seal primer valve. The discharge pipe from the trap seal primer valve shall connect to the trap above the trap seal on the inlet side of the trap.

P3201.2.1.3 Wastewater-supplied trap primer device. A wastewater-supplied trap primer device shall supply water to the trap. Wastewater-supplied trap primer devices shall conform to ASSE 1044. The discharge pipe from the trap seal primer device shall connect to the trap above the trap seal on the inlet side of the trap.

P3201.2.1.4 Barrier-type trap seal protection device. A barrier-type trap seal protection device shall protect the floor drain trap seal from evaporation. Barrier-type floor drain trap seal protection devices shall conform to ASSE 1072. The devices shall be installed in accordance with the manufacturer's instructions.

P3201.3 Trap setting and protection. Traps shall be set level with respect to their water seals and shall be protected from freezing. Trap seals shall be protected from siphonage, aspiration or back pressure by an *approved* system of venting (see Section P3101).

P3201.4 Building traps. Building traps shall be prohibited.

P3201.5 Prohibited trap designs. The following types of traps are prohibited:

1. Bell traps.
2. Separate fixture traps with interior partitions, except those lavatory traps made of plastic, stainless steel or other corrosion-resistant material.
3. "S" traps.
4. Drum traps.
5. Trap designs with moving parts.

P3201.6 Number of fixtures per trap. Each plumbing fixture shall be separately trapped by a water seal trap. The vertical distance from the fixture outlet to the trap weir shall not exceed 24 inches (610 mm) and the horizontal distance shall not exceed 30 inches (762 mm) measured from the centerline of the fixture outlet to the centerline of the inlet of the trap. The height of a clothes washer standpipe above a trap shall conform to Section P2706.1.2. Fixtures shall not be double trapped.

Exceptions:

1. Fixtures that have integral traps.
2. A single trap shall be permitted to serve two or three like fixtures limited to kitchen sinks, laundry tubs and lavatories. Such fixtures shall be adjacent to each other and located in the same room with a continuous waste arrangement. The trap shall be installed at the center fixture where three fixtures are installed. Common trapped fixture outlets shall be not more than 30 inches (762 mm) apart.

3. Connection of a laundry tray waste line into a standpipe for the automatic clothes-washer drain shall be permitted in accordance with Section P2706.1.2.1.

P3201.7 Size of fixture traps. Trap sizes for plumbing fixtures shall be as indicated in Table P3201.7. Where the tailpiece of a plumbing fixture is larger than that indicated in Table P3201.7, the trap size shall be the same nominal size as the fixture tailpiece. A trap shall not be larger than the drainage pipe into which the trap discharges.

TABLE P3201.7
SIZE OF TRAPS FOR PLUMBING FIXTURES

PLUMBING FIXTURE	TRAP SIZE MINIMUM (inches)
Bathtub (with or without shower head and/or whirlpool attachments)	$1^1/_2$
Bidet	$1^1/_4$
Clothes washer standpipe	2
Dishwasher (on separate trap)	$1^1/_2$
Floor drain	2
Kitchen sink (one or two traps, with or without dishwasher and food waste disposer)	$1^1/_2$
Laundry tub (one or more compartments)	$1^1/_2$
Lavatory	$1^1/_4$
Shower (based on the total flow rate through showerheads and body sprays) Flow rate:	
5.7 gpm and less	$1^1/_2$
More than 5.7 gpm up to 12.3 gpm	2
More than 12.3 gpm up to 25.8 gpm	3
More than 25.8 gpm up to 55.6 gpm	4

For SI: 1 inch = 25.4 mm, 1 gallon per minute = 3.785 L/m.

CHAPTER 33

STORM DRAINAGE

User notes:

About this chapter: *Chapter 33 regulates methods and systems that control storm water from a building, such as from subsoil drainage systems and rainfall on roof surfaces. Regulations for sumps and pumping systems for subsoil drainage systems are provided in this chapter.*

Code development reminder: *Code change proposals to this chapter will be considered by the IRC—Mechanical/Plumbing Code Development Committee during the 2021 (Group A) Code Development Cycle.*

SECTION P3301 GENERAL

P3301.1 Scope. The provisions of this chapter shall govern the materials, design, construction and installation of storm drainage.

SECTION P3302 SUBSOIL DRAINS

P3302.1 Subsoil drains. *Subsoil drains* shall be open-jointed, horizontally split or perforated pipe conforming to one of the standards indicated in Table P3302.1. Such drains shall be not less than 4 inches (102 mm) in diameter. Where the building is subject to backwater, the *subsoil drain* shall be protected by an accessibly located backwater valve. *Subsoil drains* shall discharge to a trapped area drain, sump, dry well or *approved* location above ground. The subsoil sump shall not be required to have either a gastight cover or a vent. The sump and pumping system shall comply with Section P3303.

SECTION P3303 SUMPS AND PUMPING SYSTEMS

P3303.1 Pumping system. The sump pump, sump and discharge piping shall conform to Sections P3303.1.1 through P3303.1.4.

P3303.1.1 Pump capacity and head. The sump pump shall be of a capacity and head appropriate to anticipated use requirements.

P3303.1.2 Sump pit. The sump shall be not less than 18 inches (457 mm) in diameter and 24 inches (610 mm) deep, unless otherwise *approved.* The sump shall be *accessible* and located so that all drainage flows into the sump by gravity. The sump shall be constructed of tile, steel, plastic, cast iron, concrete or other *approved* material, with a removable cover adequate to support anticipated loads in the area of use. The sump floor shall be solid and provide permanent support for the pump.

P3303.1.3 Electrical. Electrical outlets shall meet the requirements of Chapters 34 through 43.

P3303.1.4 Piping. Discharge piping shall meet the requirements of Sections P3002.1, P3002.2, P3002.3 and P3003. Discharge piping shall include an *accessible* full-flow check valve. Pipe and fittings shall be the same size as, or larger than, the pump discharge tapping.

TABLE P3302.1
SUBSOIL DRAIN PIPE

MATERIAL	STANDARD
Cast-iron pipe	ASTM A74; ASTM A888; CISPI 301
Polyethylene (PE) plastic pipe	ASTM F405; CSA B182.1; CSA B182.6; CSA B182.8
Polyvinyl chloride (PVC) plastic pipe (type sewer pipe, SDR 35, PS25, PS50 or PS100)	ASTM D2729; ASTM D3034; ASTM F891; CSA B182.2; CSA B182.4
Stainless steel drainage systems, Type 316L	ASME A112.3.1
Vitrified clay pipe	ASTM C4; ASTM C700

Part VIII—Electrical

CHAPTER 34

GENERAL REQUIREMENTS

 The section numbers appearing in parentheses or brackets after IRC text are the section numbers of the corresponding text in the *National Electrical Code* (NFPA 70).

IMPORTANT NOTICE AND DISCLAIMER CONCERNING THE NEC AND THIS ELECTRICAL PART.

ICC user note:

About this chapter: *Chapter 34 contains broadly applicable requirements including provisions for the protection of the structural elements of a building, inspection of work, general installation and conductor identification. This chapter requires that all electrical system components be listed and labeled by an approved agency. The electrical provisions of this code are identical to the intent of the NEC provisions except that this code requires all electrical system components be listed and labeled. The code does not contain unique electrical requirements. A dwelling built to the code will have electrical systems identical to those required by the respective edition of the NEC. This code addresses only those electrical systems that are common to dwelling construction, and the NEC is referenced for any subject not addressed in the code.*

SECTION E3401 GENERAL

E3401.1 Applicability. The provisions of Chapters 34 through 43 shall establish the general scope of the electrical system and equipment requirements of this code. Chapters 34 through 43 cover those wiring methods and materials most commonly encountered in the construction of one- and two-family dwellings and structures regulated by this code. Other wiring methods, materials and subject matter covered in NFPA 70 are also allowed by this code.

E3401.2 Scope. Chapters 34 through 43 shall cover the installation of electrical systems, equipment and components indoors and outdoors that are within the scope of this code, including services, power distribution systems, fixtures, appliances, devices and appurtenances. Services within the scope of this code shall be limited to 120/240-volt, 0- to 400-ampere, single-phase systems. These chapters specifically cover the equipment, fixtures, appliances, wiring methods and materials that are most commonly used in the construction or *alteration* of one- and two-family dwellings and *accessory structures* regulated by this code. The omission from these chapters of any material or method of construction provided for in the referenced standard NFPA 70 shall not be construed as prohibiting the use of such material or method of construction. Electrical systems, equipment or components not specifically covered in these chapters shall comply with the applicable provisions of NFPA 70.

E3401.3 Not covered. Chapters 34 through 43 do not cover the following:

1. Installations, including associated lighting, under the exclusive control of communications utilities and electric utilities.
2. Services over 400 amperes.

E3401.4 Additions and alterations. Any addition or *alteration* to an existing electrical system shall be made in conformity to the provisions of Chapters 34 through 43. Where additions subject portions of existing systems to loads exceeding those permitted herein, such portions shall be made to comply with Chapters 34 through 43.

SECTION E3402 BUILDING STRUCTURE PROTECTION

E3402.1 Drilling and notching. Wood-framed structural members shall not be drilled, notched or altered in any manner except as provided for in this code.

E3402.2 Penetrations of fire-resistance-rated assemblies. Electrical installations in hollow spaces, vertical shafts and *ventilation* or air-handling ducts shall be made so that the possible spread of fire or products of combustion will not be substantially increased. Electrical penetrations into or through fire-resistance-rated walls, partitions, floors or ceilings shall be protected by *approved* methods to maintain the fire-resistance rating of the element penetrated. Penetrations of fire-resistance-rated walls shall be limited as specified in Section R302.4. (300.21)

E3402.3 Penetrations of firestops and draftstops. Penetrations through fire blocking and draftstopping shall be protected in an *approved* manner to maintain the integrity of the element penetrated.

SECTION E3403 INSPECTION AND APPROVAL

E3403.1 Approval. Electrical materials, components and equipment shall be approved. (110.2)

E3403.2 Inspection required. New electrical work and parts of existing systems affected by new work or *alterations* shall be inspected by the building official to ensure compliance with the requirements of Chapters 34 through 43.

E3403.3 Listing and labeling. Electrical materials, components, devices, fixtures and equipment shall be *listed* for the application, shall bear the *label* of an *approved* agency and shall be installed, and used, or both, in accordance with any instructions included in the listing and labeling. [110.3(B)]

SECTION E3404 GENERAL EQUIPMENT REQUIREMENTS

E3404.1 Voltages. Throughout Chapters 34 through 43, the voltage considered shall be that at which the circuit operates. The voltage rating of electrical equipment shall be not less than the nominal voltage of the circuit to which it is connected. (110.4)

E3404.2 Interrupting rating. Equipment intended to interrupt current at fault levels shall have a minimum interrupting rating of 10,000 amperes at the nominal circuit voltage. Equipment intended to interrupt current at levels other than fault levels shall have an interrupting rating at nominal circuit voltage of not less than the current that must be interrupted. (110.9)

E3404.3 Circuit characteristics. The overcurrent protective devices, total impedance, equipment short-circuit current ratings and other characteristics of the circuit to be protected shall be so selected and coordinated as to permit the circuit protective devices that are used to clear a fault to do so without extensive damage to the electrical equipment of the circuit. This fault shall be assumed to be either between two or more of the circuit conductors or between any circuit conductor and the equipment grounding conductors permitted in Section E3908.9. *Listed* equipment applied in accordance with its listing shall be considered to meet the requirements of this section. (110.10)

E3404.4 Enclosure types. Enclosures, other than surrounding fences or walls, of panelboards, meter sockets, enclosed switches, transfer switches, circuit breakers, pullout switches and motor controllers, rated not over 600 volts nominal and intended for such locations, shall be marked with an enclosure-type number as shown in Table E3404.4.

Table E3404.4 shall be used for selecting these enclosures for use in specific locations other than hazardous (classified) locations. The enclosures are not intended to protect against conditions such as condensation, icing, corrosion, or contamination that might occur within the enclosure or enter through the raceway or unsealed openings. (110.28)

E3404.5 Protection of equipment. Equipment not identified for outdoor use and equipment identified only for indoor use, such as "dry locations," "indoor use only" "damp locations," or enclosure Type 1, 2, 5, 12, 12K and 13, shall be protected against damage from the weather during construction. (110.11)

E3404.6 Unused openings. Unused openings, other than those intended for the operation of equipment, those intended for mounting purposes, and those permitted as part of the design for *listed* equipment, shall be closed to afford protection substantially equivalent to the wall of the equipment. Where metallic plugs or plates are used with nonmetallic enclosures they shall be recessed at least $^1/_4$ inch (6.4 mm) from the outer surface of the enclosure. [110.12(A)]

E3404.7 Integrity of electrical equipment. Internal parts of electrical equipment, including busbars, wiring terminals, insulators and other surfaces, shall not be damaged or contaminated by foreign materials such as paint, plaster, cleaners or abrasives, and corrosive residues. There shall not be any damaged parts that might adversely affect safe operation or mechanical strength of the equipment such as parts that are broken; bent; cut; deteriorated by corrosion, chemical action, or overheating. Foreign debris shall be removed from equipment. [110.12(B)]

E3404.8 Mounting. Electrical equipment shall be firmly secured to the surface on which it is mounted. Wooden plugs driven into masonry, concrete, plaster, or similar materials shall not be used. [110.13(A)]

E3404.9 Energized parts guarded against accidental contact. *Approved* enclosures shall guard energized parts that are operating at 50 volts or more against accidental contact. [110.27(A)]

E3404.10 Prevent physical damage. In locations where electrical equipment is likely to be exposed to physical damage, enclosures or guards shall be so arranged and of such strength as to prevent such damage. [110.27(B)]

E3404.11 Equipment identification.

1. The manufacturer's name, trademark or other descriptive marking by which the organization responsible for the product can be identified shall be placed on all electric equipment. Other markings shall be provided that indicate voltage, current, wattage or other ratings as specified elsewhere in Chapters 34 through 43. The marking shall have the durability to withstand the environment involved. [110.21(A)(1)]
2. Reconditioned equipment shall be marked with the names, or trademark, or other descriptive marking by which the organization responsible for reconditioning the electrical equipment can be identified, along with the date of the reconditioning. Reconditioned equipment shall be identified as "reconditioned" and the original listing *mark* removed. Approval of the reconditioned equipment shall not be based solely on the equipment's original listing. [110.21(A)(2)]

E3404.12 Field-applied hazard markings. Where caution, warning, or danger signs or *labels* are required by this code, the *labels* shall meet the following requirements:

1. The marking shall warn of the hazard using effective words, colors, symbols or any combination thereof.

TABLE E3404.4 (Table 110.28)
ENCLOSURE SELECTION

PROVIDES A DEGREE OF PROTECTION AGAINST THE FOLLOWING ENVIRONMENTAL CONDITIONS	FOR OUTDOOR USE									
	Enclosure type number									
	3	3R	3S	3X	3RX	3SX	4	4X	6	6P
Incidental contact with the enclosed equipment	X	X	X	X	X	X	X	X	X	X
Rain, snow and sleet	X	X	X	X	X	X	X	X	X	X
Sleet[a]	—	—	X	—	—	X	—	—	—	—
Windblown dust	X	—	X	X	—	X	X	X	X	X
Hosedown	—	—	—	—	—	—	X	X	X	X
Corrosive agents	—	—	—	X	X	X	—	X	—	X
Temporary submersion	—	—	—	—	—	—	—	—	X	X
Prolonged submersion	—	—	—	—	—	—	—	—	—	X
PROVIDES A DEGREE OF PROTECTION AGAINST THE FOLLOWING ENVIRONMENTAL CONDITIONS	FOR INDOOR USE									
	Enclosure type number									
	1	2	4	4X	5	6	6P	12	12K	13
Incidental contact with the enclosed equipment	X	X	X	X	X	X	X	X	X	X
Falling dirt	X	X	X	X	X	X	X	X	X	X
Falling liquids and light splashing	—	X	X	X	X	X	X	X	X	X
Circulating dust, lint, fibers and flyings	—	—	X	X	—	X	X	X	X	X
Settling airborne dust, lint, fibers and flyings	—	—	X	X	X	X	X	X	X	X
Hosedown and splashing water	—	—	X	X	—	X	X	—	—	—
Oil and coolant seepage	—	—	—	—	—	—	—	X	X	X
Oil or coolant spraying and splashing	—	—	—	—	—	—	—	—	—	X
Corrosive agents	—	—	—	X	—	—	X	—	—	—
Temporary submersion	—	—	—	—	—	X	X	—	—	—
Prolonged submersion	—	—	—	—	—	—	X	—	—	—

a. Mechanism shall be operable when ice covered.

Note 1: The term "raintight" is typically used in conjunction with Enclosure Types 3, 3S, 3SX, 3X, 4, 4X, 6 and 6P. The term "rainproof" is typically used in conjunction with Enclosure Types 3R and 3RX. The term "watertight" is typically used in conjunction with Enclosure Types 4, 4X, 6 and 6P. The term "driptight" is typically used in conjunction with Enclosure Types 2, 5, 12, 12K and 13. The term "dusttight" is typically used in conjunction with Enclosure Types 3, 3S, 3SX, 3X, 5, 12, 12K and 13.

Note 2: Ingress protection (IP) ratings are found in ANSI/NEMA 60529, *Degrees of Protection Provided by Enclosures.* IP ratings are not a substitute for enclosure-type ratings.

2. *Labels* shall be permanently affixed to the equipment or wiring method.
3. *Labels* shall not be handwritten except for portions of *labels* or markings that are variable, or that could be subject to changes. *Labels* shall be legible.
4. *Labels* shall be of sufficient durability to withstand the environment involved. [110.21(B)]

E3404.13 Identification of disconnecting means. Each disconnecting means shall be legibly marked to indicate its purpose, except where located and arranged so that the purpose is evident. The marking shall have the durability to withstand the environment involved. [110.22(A)]

SECTION E3405 EQUIPMENT LOCATION AND CLEARANCES

E3405.1 Working space and clearances. Access and working space shall be provided and maintained around all electrical equipment to permit ready and safe operation and maintenance of such equipment in accordance with this section and Figure E3405.1. (110.26)

E3405.2 Working clearances for energized equipment and panelboards. Except as otherwise specified in Chapters 34 through 43, the dimension of the working space in the direction of access to panelboards and live parts of other equipment likely to require examination, adjustment, servicing or maintenance while energized shall be not less than 36 inches (914 mm) in depth. Distances shall be measured from the energized parts where such parts are exposed or from the enclosure front or opening where such parts are enclosed. In addition to the 36-inch dimension (914 mm), the work space shall not be less than 30 inches (762 mm) wide in front of the electrical equipment and not less than the width of such equipment. The work space shall be clear and shall extend from the floor or platform to a height of 6.5 feet (1981 mm) or the height of the equipment, whichever is greater. In all cases, the work space shall allow at least a 90-degree (1.57 rad) opening of equipment doors or hinged panels. Equipment or support structures, such as concrete pads, associated with the electrical installation located above or below the electrical equipment shall be permitted to extend not more than 6 inches (152 mm) beyond the front of the electrical equipment.

Where such equipment is required by installation instruction or function to be located in a space with limited access, all of the following shall apply:

1. Where the equipment is installed above a lay-in ceiling, there shall be an opening not smaller than 22 inches by 22 inches (559 mm by 559 mm), or in a *crawl space*, there shall be an accessible opening not smaller than 22 inches by 30 inches (559 mm by 762 mm).
2. The width of the working space shall be the width of the equipment enclosure or not less than 30 inches (762 mm), whichever is greater.
3. Enclosure doors and hinged panels shall be capable of opening not less than 90 degrees.
4. The space in front of the enclosure shall comply with the depth requirements of Table 110.26(A)(1) of NFPA 70. The maximum height of the working space shall be the height necessary to install the equipment in the limited space. A horizontal ceiling structural member or access panel shall be permitted in this space. [110.26(A) (1), (2), (3), (4)]

Exceptions:

1. In existing dwelling units, service equipment and panelboards that are not rated in excess of 200 amperes shall be permitted in spaces where the height of the working space is less than 6.5 feet (1981 mm). [110.26(A)(3) Exception No. 1]
2. Meters that are installed in meter sockets shall be permitted to extend beyond the other equipment. Meter sockets shall not be exempt from the requirements of this section. [110.26(A)(3) Exception No. 2]

E3405.3 Indoor dedicated panelboard space. The indoor space equal to the width and depth of the panelboard and extending from the floor to a height of 6 feet (1829 mm) above the panelboard, or to the structural ceiling, whichever is lower, shall be dedicated to the electrical installation. Piping, ducts, leak protection apparatus and other equipment foreign to the electrical installation shall not be installed in such dedicated space. The area above the dedicated space shall be permitted to contain foreign systems, provided that protection is installed to avoid damage to the electrical equipment from condensation, leaks and breaks in such foreign systems (see Figure E3405.1).

Exception: Suspended ceilings with removable panels shall be permitted within the 6-foot (1829 mm) dedicated space.

E3405.4 Outdoor dedicated panelboard space. The outdoor space equal to the width and depth of the panelboard, and extending from grade to a height of 6 feet (1829 mm) above the panelboard, shall be dedicated to the electrical installation. Piping and other equipment foreign to the electrical installation shall not be located in this zone.

E3405.5 Location of working spaces and equipment. Required working space shall not be designated for storage. Panelboards and overcurrent protection devices shall not be located in clothes closets, in bathrooms, or over the steps of a stairway. [110.26(B), 240.24(D), (E), (F)]

E3405.6 Access and entrance to working space. Access shall be provided to the required working space. [110.26(C)(1)]

E3405.7 Illumination. Artificial illumination shall be provided for all working spaces for service equipment and panelboards installed indoors and all illumination shall not be controlled by automatic means only. Additional lighting outlets shall not be required where the work space is illuminated by an adjacent light source or as permitted by Exception 1 of Section E3903.2 for switched receptacles. [110.26(D)]

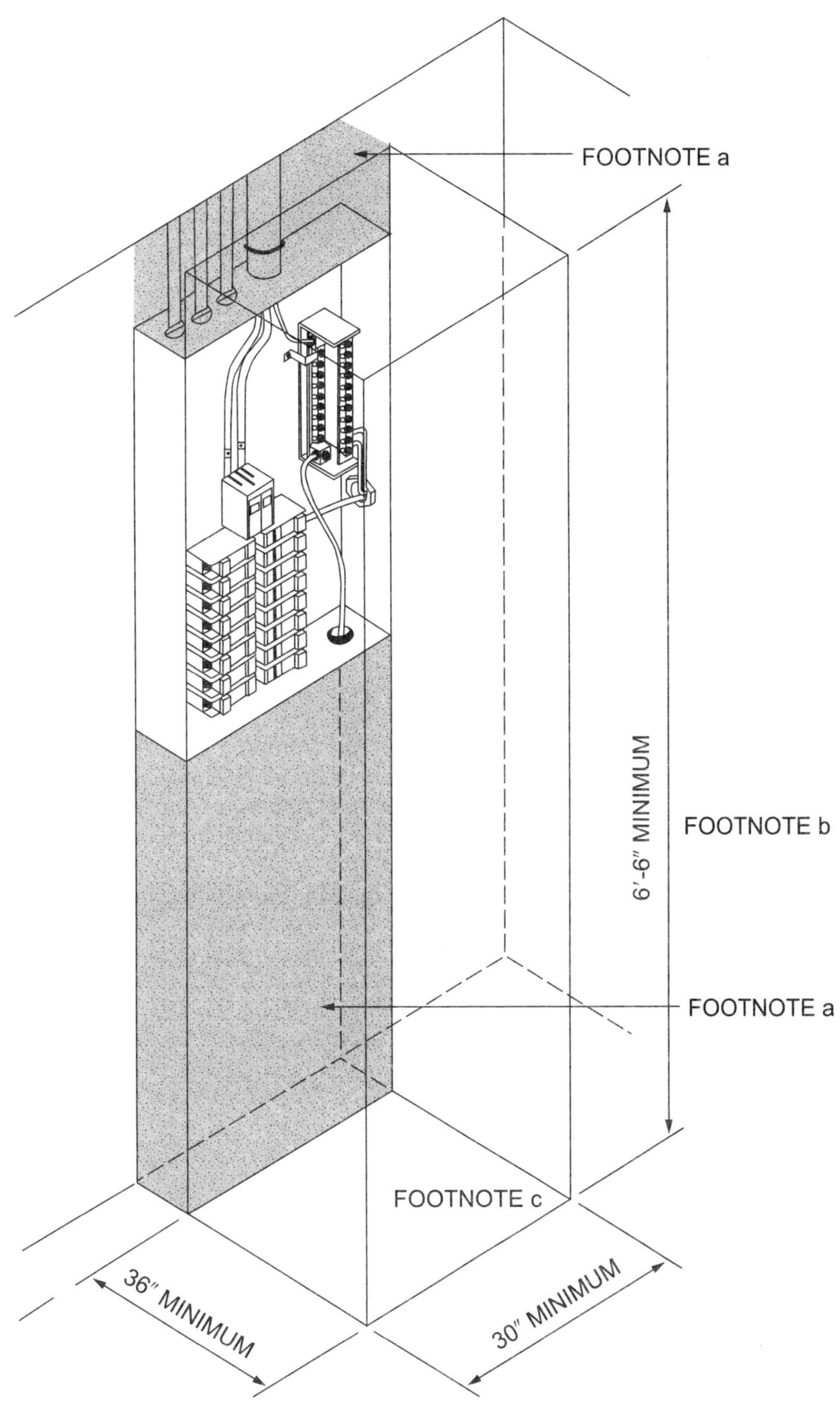

For SI: 1 inch = 25.4 mm, 1 foot = 304.8 mm.

a. Equipment, piping and ducts foreign to the electrical installation shall not be placed in the shaded areas extending from the floor to a height of 6 feet above the panelboard enclosure or to the structural ceiling, whichever is lower.

b. The working space shall be clear and unobstructed from the floor to a height of 6.5 feet or the height of the equipment, whichever is greater.

c. The working space shall not be designated for storage.

d. Panelboards, service equipment and similar enclosures shall not be located in bathrooms, toilet rooms, clothes closets or over the steps of a stairway.

e. Such work spaces shall be provided with artificial lighting where located indoors and shall not be controlled by automatic means only.

FIGURE E3405.1[a, b, c, d, e]
WORKING SPACE AND CLEARANCES

SECTION E3406 ELECTRICAL CONDUCTORS AND CONNECTIONS

E3406.1 General. This section provides general requirements for conductors, connections and splices. These requirements do not apply to conductors that form an integral part of equipment, such as motors, appliances and similar equipment, or to conductors specifically provided for elsewhere in Chapters 34 through 43. (310.1)

E3406.2 Conductor material. Conductors used to conduct current shall be of copper, aluminum or copper-aluminum except as otherwise provided in Chapters 34 through 43. Where the conductor material is not specified, the material and the sizes given in these chapters shall apply to copper conductors. Where other materials are used, the conductor sizes shall be changed accordingly. (110.5)

E3406.3 Minimum size of conductors. The minimum size of conductors for feeders and branch circuits shall be 14 AWG copper and 12 AWG aluminum or copper-clad aluminum. The minimum size of service conductors shall be as specified in Chapter 36. The minimum size of Class 2 remote control, signaling and power-limited circuits conductors shall be as specified in Chapter 43. [310.3(A)]

E3406.4 Stranded conductors. Where installed in raceways, conductors 8 AWG and larger shall be stranded. A solid 8 AWG conductor shall be permitted to be installed in a raceway only to meet the requirements of Sections E3610.2 and E4204. [310.3(C)]

E3406.5 Individual conductor insulation. Except where otherwise permitted in Sections E3605.1 and E3908.10, and E4303, current-carrying conductors shall be insulated. Insulated conductors shall have insulation types identified as RHH, RHW, RHW-2, THHN, THHW, THW, THW-2, THWN, THWN-2, TW, UF, USE, USE-2, XHHW or XHHW-2. Insulation types shall be *approved* for the application. [310.10(B), (C), 310.4]

E3406.6 Conductors in parallel. Circuit conductors that are connected in parallel shall be limited to sizes 1/0 AWG and larger. Conductors in parallel shall: be of the same length; consist of the same conductor material; be the same circular mil area and have the same insulation type. Conductors in parallel shall be terminated in the same manner. Where run in separate raceways or cables, the raceway or cables shall have the same physical characteristics. Where conductors are in separate raceways or cables, the same number of conductors shall be used in each raceway or cable. [310.10(G)]

E3406.7 Conductors of the same circuit. All conductors of the same circuit and, where used, the grounded conductor and all equipment grounding conductors and bonding conductors shall be contained within the same raceway, cable or cord. [300.3(B)]

E3406.8 Aluminum and copper connections. Terminals and splicing connectors shall be identified for the material of the conductors joined. Conductors of dissimilar metals shall not be joined in a terminal or splicing connector where physical contact occurs between dissimilar conductors such as copper and aluminum, copper and copper-clad aluminum, or aluminum and copper-clad aluminum, except where the device is *listed* for the purpose and conditions of application. Materials such as inhibitors and compounds shall be suitable for the application and shall be of a type that will not adversely affect the conductors, installation or equipment. (110.14)

E3406.9 Fine stranded conductors. Connectors and terminals for conductors that are more finely stranded than Class B and Class C stranding, as shown in Table E3406.9, shall

TABLE E3406.9 (Chapter 9, Table 10) CONDUCTOR STRANDING[c]

CONDUCTOR SIZE		NUMBER OF STRANDS		
		Copper		Aluminum
AWG or kcmil	mm²	Class B	Class C	Class B
24-30	0.20-0.05	Note a	—	—
22	0.32	7	—	—
20	0.52	10	—	—
18	0.82	16	—	—
16	1.3	26	—	—
14-2	2.1-33.6	7	19	7[b]
1-4/0	42.4-107	19	37	19
250-500	127-253	37	61	37
600-1000	304-508	61	91	61
1250-1500	635-759	91	127	91
1750-2000	886-1016	127	271	127

a. Number of strands varies.

b. Aluminum 14 AWG (2.1 mm²) is not available.

c. With the permission of Underwriters Laboratories, Inc., this material is reproduced from UL Standard 486A-486B, *Wire Connectors,* which is copyrighted by Underwriters Laboratories, Inc., Northbrook, Illinois. While use of this material has been authorized, UL shall not be responsible for the manner in which the information is presented, nor for any interpretations thereof.

be identified for the specific conductor class or classes. (110.14)

E3406.10 Terminals. Connection of conductors to terminal parts shall be made without damaging the conductors and shall be made by means of pressure connectors, including set-screw type, by means of splices to flexible leads, or for conductor sizes of 10 AWG and smaller, by means of wire binding screws or studs and nuts having upturned lugs or the equivalent. Terminals for more than one conductor and terminals for connecting aluminum conductors shall be identified for the application. [110.14(A)]

E3406.11 Splices. Conductors shall be spliced or joined with splicing devices *listed* for the purpose. Splices and joints and the free ends of conductors shall be covered with an insulation equivalent to that of the conductors or with an insulating device *listed* for the purpose. Wire connectors or splicing means installed on conductors for direct burial shall be *listed* for such use. [110.14(B)]

E3406.11.1 Continuity. Conductors in raceways shall be continuous between outlets, boxes, and devices and shall be without splices or taps in the raceway.

Exception: Splices shall be permitted within surface-mounted raceways that have a removable cover. [300.13(A)]

E3406.11.2 Device connections. The continuity of a grounded conductor in multiwire branch circuits shall not be dependent on connection to devices such as receptacles and lampholders. The arrangement of grounding connections shall be such that the disconnection or the removal of a receptacle, luminaire or other device fed from the box does not interfere with or interrupt the grounding continuity. [300.13(B)]

E3406.11.3 Length of conductor for splice or termination. Where conductors are to be spliced, terminated or connected to fixtures or devices, a minimum length of 6 inches (152 mm) of free conductor shall be provided at each outlet, junction or switch point. The required length shall be measured from the point in the box where the conductor emerges from its raceway or cable sheath. Where the opening to an outlet, junction or switch point is less than 8 inches (200 mm) in any dimension, each conductor shall be long enough to extend at least 3 inches (75 mm) outside of such opening. (300.14)

E3406.12 Installation. Tightening torque values for terminal connections shall be as indicated on equipment or in installation instructions provided by the manufacturer. An *approved* means shall be used to achieve the indicated torque value. [110.14 (D)]

E3406.13 Grounded conductor continuity. The continuity of a grounded conductor shall not depend on connection to a metallic enclosure, raceway or cable armor. [200.2(B)]

E3406.14 Connection of grounding and bonding equipment. The connection of equipment grounding conductors, grounding electrode conductors and bonding jumpers shall be in accordance with Sections E3406.14.1 and E3406.14.2.

E3406.14.1 Permitted methods. Equipment grounding conductors, grounding electrode conductors, and bonding jumpers shall be connected by one or more of the following means:

1. *Listed* pressure connectors.
2. Terminal bars.
3. Pressure connectors *listed* as grounding and bonding equipment.
4. Exothermic welding process.
5. Machine screw-type fasteners that engage not less than two threads or are secured with a nut.
6. Thread-forming machine screws that engage not less than two threads in the enclosure.
7. Connections that are part of a *listed* assembly.
8. Other *listed* means. [250.8 (A)]

E3406.14.2 Methods not permitted. Connection devices or fittings that depend solely on solder shall not be used. [250.8 (B)]

SECTION E3407 CONDUCTOR AND TERMINAL IDENTIFICATION

E3407.1 Grounded conductors. Insulated grounded conductors of sizes 6 AWG or smaller shall be identified by a continuous white or gray outer finish or by three continuous white or gray stripes on other than green insulation along the entire length of the conductors. Conductors of sizes 4 AWG or larger shall be identified either by a continuous white or gray outer finish or by three continuous white or gray stripes on other than green insulation along its entire length or at the time of installation by a distinctive white or gray marking at its terminations. This marking shall encircle the conductor or insulation. [200.6(A) & (B)]

E3407.2 Equipment grounding conductors. Equipment grounding conductors of sizes 6 AWG and smaller shall be identified by a continuous green color or a continuous green color with one or more yellow stripes on the insulation or covering, except where bare. Conductors with insulation or individual covering that is green, green with one or more yellow stripes, or otherwise identified as permitted by this section shall not be used for ungrounded or grounded circuit conductors. (250.119)

Equipment grounding conductors 4 AWG and larger AWG that are not identified as required for conductors of sizes 6 AWG and smaller shall, at the time of installation, be permanently identified as an equipment grounding conductor at each end and at every point where the conductor is accessible, except where such conductors are bare.

The required identification for conductors 4 AWG and larger shall encircle the conductor and shall be accomplished by one of the following:

1. Stripping the insulation or covering from the entire exposed length.
2. Coloring the exposed insulation or covering green at the termination.

3. Marking the exposed insulation or covering with green tape or green adhesive *labels* at the termination. [250.119(A)]

Exceptions:

1. Conductors 4 AWG and larger shall not be required to be identified in conduit bodies that do not contain splices or unused hubs. [250.119(A)(1) Exception]
2. Power-limited, Class 2 or Class 3 circuit cables containing only circuits operating at less than 50 volts shall be permitted to use a conductor with green insulation for other than equipment grounding purposes. [250.119 Exception No. 1]

E3407.3 Ungrounded conductors. Insulation on the ungrounded conductors shall be a continuous color other than white, gray and green. [310.6(C)]

Exception: An insulated conductor that is part of a cable or flexible cord assembly and that has a white or gray finish or a finish marking with three continuous white or gray stripes shall be permitted to be used as an ungrounded conductor where it is permanently reidentified to indicate its use as an ungrounded conductor by marking tape, painting, or other effective means at all terminations and at each location where the conductor is visible and accessible. Identification shall encircle the insulation and shall be a color other than white, gray, and green. [200.7(C)(1), (2)]

Where used for single-pole, 3-way or 4-way switch loops, the reidentified conductor with white or gray insulation or three continuous white or gray stripes shall be used only for the supply to the switch, not as a return conductor from the switch to the outlet. [200.7(C)(1)]

E3407.4 Identification of terminals. Terminals for attachment to conductors shall be identified in accordance with Sections E3407.4.1 and E3407.4.2.

E3407.4.1 Device terminals. All devices excluding panelboards, provided with terminals for the attachment of conductors and intended for connection to more than one side of the circuit shall have terminals properly marked for identification, except where the terminal intended to be connected to the grounded conductor is clearly evident. [200.10(A)]

Exception: Terminal identification shall not be required for devices that have a normal current rating of over 30 amperes, other than polarized attachment caps and polarized receptacles for attachment caps as required in Section E3407.4.2. [200.10(A) Exception]

E3407.4.2 Receptacles, plugs and connectors. Receptacles, polarized attachment plugs and cord connectors for plugs and polarized plugs shall have the terminal intended for connection to the grounded (white) conductor identified. Identification shall be by a metal or metal coating substantially white or silver in color or by the word "white" or the letter "W" located adjacent to the identified terminal. Where the terminal is not visible, the conductor entrance hole for the connection shall be colored white or marked with the word "white" or the letter "W." [200.10(B)]

CHAPTER 35

ELECTRICAL DEFINITIONS

ICC user note:

About this chapter: *The electrical trade, like other construction trades, has its vernacular. If people do not understand the language of the code text, they will have difficulty interpreting that text. Many words have a unique meaning in the context of a particular code and may be defined differently in a dictionary.*

Because much code text meaning depends on the definitions of the terms used in the text, Chapter 35 lists the definitions of terms that do not have everyday, common, universally accepted or dictionary-defined meanings. Chapter 35 is provided as the key to understanding the electrical text.

SECTION E3501 GENERAL

E3501.1 Scope. This chapter contains definitions that shall apply only to the electrical requirements of Chapters 34 through 43. Unless otherwise expressly stated, the following terms shall, for the purpose of this code, have the meanings indicated in this chapter. Words used in the present tense include the future; the singular number includes the plural and the plural the singular. Where terms are not defined in this section and are defined in Section R202 of this code, such terms shall have the meanings ascribed to them in that section. Where terms are not defined in these sections, they shall have their ordinarily accepted meanings or such as the context implies.

ACCESSIBLE.

As applied to equipment. Capable of being reached for operation, renewal and inspection.

As applied to wiring methods. Capable of being removed or exposed without damaging the building structure or finish, or not permanently closed in by the structure or finish of the building.

ACCESSIBLE, READILY. Capable of being reached quickly for operation, renewal or inspections, without requiring those to whom *ready access* is requisite to take actions such as to use tools, other than keys, to climb over or under, to remove obstacles or to resort to portable ladders, etc.

AMPACITY. The maximum current in amperes that a conductor can carry continuously under the conditions of use without exceeding its temperature rating.

APPLIANCE. Utilization equipment, normally built in standardized sizes or types, that is installed or connected as a unit to perform one or more functions such as clothes washing, air conditioning, food mixing, deep frying, etc.

APPROVED. Acceptable to the authority having *jurisdiction.*

ARC-FAULT CIRCUIT INTERRUPTER. A device intended to provide protection from the effects of arc-faults by recognizing characteristics unique to arcing and by functioning to de-energize the circuit when an arc-fault is detected.

ATTACHMENT PLUG (PLUG CAP) (PLUG). A device that, by insertion into a receptacle, establishes connection between the conductors of the attached flexible cord and the conductors connected permanently to the receptacle.

AUTOMATIC. Performing a function without the necessity of human intervention.

BATHROOM. An area, including a sink (basin), with one or more of the following: a toilet, a urinal, a tub, a shower, a bidet, or similar plumbing fixture.

BONDED (BONDING). Connected to establish electrical continuity and conductivity.

BONDING CONDUCTOR OR JUMPER. A reliable conductor to ensure the required electrical conductivity between metal parts required to be electrically connected.

BONDING JUMPER (EQUIPMENT). The connection between two or more portions of the equipment grounding conductor.

BONDING JUMPER, MAIN. The connection between the grounded circuit conductor and the equipment grounding conductor, or the supply-side bonding jumper, or both, at the service.

BONDING JUMPER, SUPPLY-SIDE. A conductor installed on the supply side of a service or within a service equipment enclosure(s) that ensures the required electrical conductivity between metal parts required to be electrically connected.

BRANCH CIRCUIT. The circuit conductors between the final overcurrent device protecting the circuit and the outlet(s).

BRANCH CIRCUIT, APPLIANCE. A branch circuit that supplies energy to one or more outlets to which appliances are to be connected, and that has no permanently connected luminaires that are not a part of an appliance.

BRANCH CIRCUIT, GENERAL PURPOSE. A branch circuit that supplies two or more receptacle outlets or outlets for lighting and appliances.

BRANCH CIRCUIT, INDIVIDUAL. A branch circuit that supplies only one utilization equipment.

BRANCH CIRCUIT, MULTIWIRE. A branch circuit consisting of two or more ungrounded conductors having voltage difference between them, and a grounded conductor having equal voltage difference between it and each ungrounded conductor of the circuit, and that is connected to the neutral or grounded conductor of the system.

CABINET. An enclosure designed either for surface or flush mounting and provided with a frame, mat or trim in which a swinging door or doors are or may be hung.

CIRCUIT BREAKER. A device designed to open and close a circuit by nonautomatic means and to open the circuit automatically on a predetermined overcurrent without damage to itself when properly applied within its rating.

CLOTHES CLOSET. A nonhabitable room or space intended primarily for storage of garments and apparel.

CONCEALED. Rendered inaccessible by the structure or finish of the building.

CONDUCTOR.

Bare. A conductor having no covering or electrical insulation whatsoever.

Covered. A conductor encased within material of composition or thickness that is not recognized by this code as electrical insulation.

Insulated. A conductor encased within material of composition and thickness that is recognized by this code as electrical insulation.

CONDUIT BODY. A separate portion of a conduit or tubing system that provides access through a removable cover(s) to the interior of the system at a junction of two or more sections of the system or at a terminal point of the system. Boxes such as FS and FD or larger cast or sheet metal boxes are not classified as conduit bodies.

CONNECTOR, PRESSURE (SOLDERLESS). A device that establishes a connection between two or more conductors or between one or more conductors and a terminal by means of mechanical pressure and without the use of solder.

CONTINUOUS LOAD. A load where the maximum current is expected to continue for 3 hours or more.

COOKING UNIT, COUNTER-MOUNTED. A cooking appliance designed for mounting in or on a counter and consisting of one or more heating elements, internal wiring and built-in or separately mountable controls.

COPPER-CLAD ALUMINUM CONDUCTORS. Conductors drawn from a copper-clad aluminum rod with the copper metallurgically bonded to an aluminum core. The copper forms a minimum of 10 percent of the cross-sectional area of a solid conductor or each strand of a stranded conductor.

CUTOUT BOX. An enclosure designed for surface mounting and having swinging doors or covers secured directly to and telescoping with the walls of the box proper (see "*Cabinet*").

DEAD FRONT. Without live parts exposed to a person on the operating side of the equipment.

DEMAND FACTOR. The ratio of the maximum demand of a system, or part of a system, to the total connected load of a system or the part of the system under consideration.

DEVICE. A unit of an electrical system that carries or controls electrical energy as its principal function.

DISCONNECTING MEANS. A device, or group of devices, or other means by which the conductors of a circuit can be disconnected from their source of supply.

DWELLING.

Dwelling unit. A single unit, providing complete and independent living facilities for one or more persons, including permanent provisions for living, sleeping, cooking and sanitation.

One-family dwelling. A building consisting solely of one dwelling unit.

Two-family dwelling. A building consisting solely of two dwelling units.

EFFECTIVE GROUND-FAULT CURRENT PATH. An intentionally constructed, low-impedance electrically conductive path designed and intended to carry current under ground-fault conditions from the point of a ground fault on a wiring system to the electrical supply source and that facilitates the operation of the overcurrent protective device or ground-fault detectors.

ENCLOSED. Surrounded by a case, housing, fence or walls that will prevent persons from accidentally contacting energized parts.

ENCLOSURE. The case or housing of apparatus, or the fence or walls surrounding an installation, to prevent personnel from accidentally contacting energized parts or to protect the equipment from physical damage.

ENERGIZED. Electrically connected to, or is, a source of voltage.

EQUIPMENT. A general term including material, fittings, devices, appliances, luminaires, apparatus, machinery and the like used as a part of, or in connection with, an electrical installation.

EXPOSED.

As applied to live parts. Capable of being inadvertently touched or approached nearer than a safe distance by a person.

As applied to wiring methods. On or attached to the surface or behind panels designed to allow access.

EXTERNALLY OPERABLE. Capable of being operated without exposing the operator to contact with live parts.

FEEDER. All circuit conductors between the service equipment, or the source of a separately derived system, or other power supply source and the final branch-circuit overcurrent device.

FITTING. An accessory such as a locknut, bushing or other part of a wiring system that is intended primarily to perform a mechanical rather than an electrical function.

GROUND. The earth.

GROUNDED (GROUNDING). Connected (connecting) to ground or to a conductive body that extends the ground connection.

GROUNDED, EFFECTIVELY. Intentionally connected to earth through a ground connection or connections of sufficiently low impedance and having sufficient current-carrying capacity to prevent the buildup of voltages that may result in undue hazards to connected equipment or to persons.

GROUNDED CONDUCTOR. A system or circuit conductor that is intentionally grounded.

GROUND-FAULT CIRCUIT INTERRUPTER. A device intended for the protection of personnel that functions to de-energize a circuit or portion thereof within an established period of time when a ground-fault current exceeds the value for a Class A device.

GROUND-FAULT CURRENT PATH. An electrically conductive path from the point of a ground fault on a wiring system through normally noncurrent-carrying conductors, grounded conductors, equipment, or the earth to the electrical supply source.

Examples of ground-fault current paths are any combination of equipment grounding conductors, metallic raceways, metallic cable sheaths, electrical equipment, and any other electrically conductive material such as metal, water, and gas piping; steel framing members; stucco mesh; metal ducting; reinforcing steel; shields of communications cables; and the earth itself.

GROUNDING CONDUCTOR, EQUIPMENT (EGC). A conductive path that is part of an effective ground-fault current path and connects normally noncurrent-carrying metal parts of equipment together and to the system grounded conductor, the grounding electrode conductor or both.

GROUNDING ELECTRODE. A conducting object through which a direct connection to earth is established.

GROUNDING ELECTRODE CONDUCTOR. A conductor used to connect the system grounded conductor or the equipment to a grounding electrode or to a point on the grounding electrode system.

GUARDED. Covered, shielded, fenced, enclosed or otherwise protected by means of suitable covers, casings, barriers, rails, screens, mats or platforms to remove the likelihood of approach or contact by persons or objects to a point of danger.

HABITABLE ROOM. A room in a building for living, sleeping, eating or cooking, but excluding bathrooms, toilet rooms, closets, hallways, storage or utility spaces, and similar areas.

IDENTIFIED. (As applied to equipment.) Recognizable as suitable for the specific purpose, function, use, environment, application, etc., where described in a particular code requirement.

IN SIGHT FROM (Within sight from, within sight). Where this code specifies that one piece of equipment shall be "in sight from," "within sight from," "within sight of," or similarly stated from/of another piece of equipment, the specified equipment must be visible and not more than 50 feet (15.2 m) distant from the other.

INTERRUPTING RATING. The highest current at rated voltage that a device is identified to interrupt under standard test conditions.

INTERSYSTEM BONDING TERMINATION. A device that provides a means for connecting intersystem bonding conductors for communications systems to the grounding electrode system.

ISOLATED. (As applied to location.) Not readily accessible to persons unless special means for access are used.

KITCHEN. An area with a sink and permanent provisions for food preparation and cooking.

LABELED. Equipment or materials to which has been attached a *label*, symbol or other identifying *mark* of an organization acceptable to the authority having *jurisdiction* and concerned with product evaluation that maintains periodic inspection of production of *labeled* equipment or materials and by whose labeling the manufacturer indicates compliance with appropriate standards or performance in a specified manner.

LAUNDRY AREA. An area containing or designed to contain a laundry tray, clothes washer or clothes dryer.

LIGHTING OUTLET. An outlet intended for the direct connection of a lampholder or luminaire.

LIGHTING TRACK (Track Lighting). A manufactured assembly designed to support and energize luminaires that are capable of being readily repositioned on the track. Its length can be altered by the addition or subtraction of sections of track.

LISTED. Equipment, materials or services included in a list published by an organization that is acceptable to the authority having *jurisdiction* and concerned with evaluation of products or services, that maintains periodic inspection of production of *listed* equipment or materials or periodic evaluation of services, and whose listing states either that the equipment, material or services meets identified standards or has been tested and found suitable for a specified purpose.

LIVE PARTS. Energized conductive components.

LOCATION, DAMP. Location protected from weather and not subject to saturation with water or other liquids but subject to moderate degrees of moisture.

LOCATION, DRY. A location not normally subject to dampness or wetness. A location classified as dry may be temporarily subject to dampness or wetness, as in the case of a building under construction.

LOCATION, WET. Installations underground or in concrete slabs or masonry in direct contact with the earth and locations subject to saturation with water or other liquids, such as vehicle-washing areas, and locations exposed to weather.

LUMINAIRE. A complete lighting unit consisting of a light source such as a lamp or lamps together with the parts designed to position the light source and connect it to the power supply. A luminaire can include parts to protect the light source or the ballast or to distribute the light. A lampholder itself is not a luminaire.

MULTIOUTLET ASSEMBLY. A type of surface, or flush, or freestanding raceway; designed to hold conductors and receptacles, assembled in the field or at the factory.

NEUTRAL CONDUCTOR. The conductor connected to the neutral point of a system that is intended to carry current under normal conditions.

NEUTRAL POINT. The common point on a wye connection in a polyphase system or midpoint on a single-phase, 3-wire system, or midpoint of a single-phase portion of a 3-phase delta system, or a midpoint of a 3-wire, direct-current system.

OUTLET. A point on the wiring system at which current is taken to supply utilization equipment.

OVERCURRENT. Any current in excess of the rated current of equipment or the ampacity of a conductor. Such current might result from overload, short circuit or ground fault.

OVERLOAD. Operation of equipment in excess of normal, full-load rating, or of a conductor in excess of rated ampacity that, when it persists for a sufficient length of time, would cause damage or dangerous overheating. A fault, such as a short circuit or ground fault, is not an overload.

PANELBOARD. A single panel or group of panel units designed for assembly in the form of a single panel, including buses and automatic overcurrent devices, and equipped with or without switches for the control of light, heat or power circuits, designed to be placed in a cabinet or cutout box placed in or against a wall, partition or other support and accessible only from the front.

PLENUM. A compartment or chamber to which one or more air ducts are connected and that forms part of the air distribution system.

POWER OUTLET. An enclosed assembly that may include receptacles, circuit breakers, fuseholders, fused switches, buses and watt-hour meter mounting means, intended to supply and control power to mobile homes, recreational vehicles or boats, or to serve as a means for distributing power required to operate mobile or temporarily installed equipment.

PREMISES WIRING (SYSTEM). Interior and exterior wiring, including power, lighting, control and signal circuit wiring together with all of their associated hardware, fittings and wiring devices, both permanently and temporarily installed. This includes wiring from the service point or power source to the outlets and wiring from and including the power source to the outlets where there is no service point. Such wiring does not include wiring internal to appliances, luminaires, motors, controllers, and similar equipment.

QUALIFIED PERSON. One who has the skills and knowledge related to the construction and operation of the electrical equipment and installations and has received safety training to recognize and avoid the hazards involved.

RACEWAY. An enclosed channel expressly for holding wires, cables, or busbars, with additional functions as permitted in this code.

RAINTIGHT. Constructed or protected so that exposure to a beating rain will not result in the entrance of water under specified test conditions.

RAINPROOF. Constructed, protected or treated so as to prevent rain from interfering with the successful operation of the apparatus under specified test conditions.

RECEPTACLE. A contact device installed at the outlet for the connection of an attachment plug, or for the direct connection of electrical utilization equipment designed to mate with the corresponding contact device. A single receptacle is a single contact device with no other contact device on the same yoke or strap. A multiple receptacle is two or more contact devices on the same yoke or strap.

RECEPTACLE OUTLET. An outlet where one or more receptacles are installed.

RECONDITIONED. Electromechanical systems, equipment, apparatus or components that are restored to operating conditions. This process differs from the normal servicing of equipment that remains within a facility, or replacement of *listed* equipment on a one-to-one basis.

SERVICE. The conductors and equipment connecting the serving utility to the wiring system of the premises served.

SERVICE CABLE. Service conductors made up in the form of a cable.

SERVICE CONDUCTORS. The conductors from the service point to the service disconnecting means.

SERVICE CONDUCTORS, OVERHEAD. The overhead conductors between the service point and the first point of connection to the service-entrance conductors at the building or other structure.

SERVICE CONDUCTORS, UNDERGROUND. The underground conductors between the service point and the first point of connection to the service-entrance conductors in a terminal box, meter, or other enclosure, inside or outside of the building wall.

SERVICE DROP. The overhead conductors between the serving utility and the service point.

SERVICE EQUIPMENT. The necessary equipment, consisting of a circuit breaker(s) or switch(es) and fuse(s), and their accessories, connected to the serving utility and intended to constitute the main control and disconnect of the serving utility.

SERVICE LATERAL. The underground service conductors between the electric utility supply system and the service point.

SERVICE POINT. The point of connection between the facilities of the serving utility and the premises wiring.

SERVICE-ENTRANCE CONDUCTORS, OVERHEAD SYSTEM. The service conductors between the terminals of the service equipment and a point usually outside of the building, clear of building walls, where joined by tap or splice to the service drop or overhead service conductors.

SERVICE-ENTRANCE CONDUCTORS, UNDERGROUND SYSTEM. The service conductors between the terminals of the service equipment and the point of connection to the service lateral or underground service conductors.

STRUCTURE. That which is built or constructed, other than equipment.

SWITCHES.

General-use snap switch. A form of general-use switch constructed so that it can be installed in device boxes or on box covers or otherwise used in conjunction with wiring systems recognized by this code.

General-use switch. A switch intended for use in general distribution and branch circuits. It is rated in amperes and is capable of interrupting its rated current at its rated voltage.

Isolating switch. A switch intended for isolating an electric circuit from the source of power. It has no interrupting rating and is intended to be operated only after the circuit has been opened by some other means.

Motor-circuit switch. A switch, rated in horsepower that is capable of interrupting the maximum operating overload current of a motor of the same horsepower rating as the switch at the rated voltage.

UNGROUNDED. Not connected to ground or to a conductive body that extends the ground connection.

UTILIZATION EQUIPMENT. Equipment that utilizes electric energy for electronic, electromechanical, chemical, heating, lighting or similar purposes.

VENTILATED. Provided with a means to permit circulation of air sufficient to remove an excess of heat, fumes or vapors.

VOLTAGE (OF A CIRCUIT). The greatest root-mean-square (rms) (effective) difference of potential between any two conductors of the circuit concerned.

VOLTAGE, NOMINAL. A nominal value assigned to a circuit or system for the purpose of conveniently designating its voltage class (e.g., 120/240). The actual voltage at which a circuit operates can vary from the nominal within a range that permits satisfactory operation of equipment.

VOLTAGE TO GROUND. For grounded circuits, the voltage between the given conductor and that point or conductor of the circuit that is grounded. For ungrounded circuits, the greatest voltage between the given conductor and any other conductor of the circuit.

WATERTIGHT. Constructed so that moisture will not enter the enclosure under specified test conditions.

WEATHERPROOF. Constructed or protected so that exposure to the weather will not interfere with successful operation.

CHAPTER 36
SERVICES

ICC user note:

About this chapter: *Chapter 36 is the first of the logical order of chapters that mimics the normal sequence of dwelling construction. The first step in dwelling wiring is typically the sizing, design and installation of the service that is the source of power for the building. This chapter addresses the sizing of services, service conductor sizing and installation, system grounding and bonding, overcurrent protection, disconnecting means and the grounding electrode system.*

This chapter requires services to be properly sized to serve the load. This is intended to prevent overloading and to provide the utility expected by the occupants. This chapter is also intended to protect occupants and the building from fire and protect the occupants from electrical shock hazards associated with service conductors and equipment.

SECTION E3601 GENERAL SERVICES

E3601.1 Scope. This chapter covers service conductors and equipment for the control and protection of services and their installation requirements. (230.1)

E3601.2 Number of services. One- and two-family dwellings shall be supplied by only one service. (230.2)

E3601.3 One building or other structure not to be supplied through another. Service conductors supplying a building or other structure shall not pass through the interior of another building or other structure. (230.3)

E3601.4 Other conductors in raceway or cable. Conductors other than service conductors shall not be installed in the same service raceway or service cable in which the service conductors are installed. (230.7)

Exceptions:

1. Grounding electrode conductors or supply side bonding jumpers or conductors shall be permitted within service raceways.
2. Load management control conductors having overcurrent protection shall be permitted within service raceways.

E3601.5 Raceway seal. Where a service raceway enters from an underground distribution system, it shall be sealed in accordance with Section E3803.6. (230.8)

E3601.6 Service disconnect required. Means shall be provided to disconnect all conductors in a building or other structure from the service entrance conductors. (230.70)

E3601.6.1 Marking of service equipment and disconnects. Service disconnects shall be permanently marked as a service disconnect. [230.70(B)]

E3601.6.2 Service disconnect location. The service disconnecting means shall be installed at a readily accessible location either outside of a building or inside nearest the point of entrance of the service conductors. Service disconnecting means shall not be installed in bathrooms. Each occupant shall have access to the disconnect serving the dwelling unit in which they reside. [230.70(A)(1)(2), 230.72(C)]

E3601.7 Maximum number of disconnects. Each service shall have only one disconnecting means unless installed using one or more of the methods specified in Sections E3601.7.1 through E3601.7.3. In all cases, the maximum number of disconnecting means for any service shall not exceed six and the multiple service disconnecting means shall be grouped.

E3601.7.1 Separate enclosures. A service with two to six separate enclosures with a single main service disconnecting means in each enclosure shall be permitted.

E3601.7.2 Panelboards. A service with two to six separate panelboards with a single main service disconnecting means in each panelboard shall be permitted.

E3601.7.3 Metering centers. A service with two to six disconnecting means in separate compartments of a metering center shall be permitted. [230.71 (B), 230.72 (A)]

Exception: Disconnecting means installed as part of *listed* equipment and used solely for the following shall not be considered a service disconnecting means:

1. Power monitoring equipment.
2. Surge-protective device(s).
3. Power-operable service disconnecting means. [230.71 (A)]

E3601.8 Emergency disconnects. For one- and two-family dwelling units, all service conductors shall terminate in disconnecting means having a short-circuit current rating equal to or greater than the available fault current, installed in a readily accessible outdoor location. If more than one disconnect is provided, they shall be grouped. Each disconnect shall be one of the following:

1. Service disconnects marked as follows: EMERGENCY DISCONNECT, SERVICE DISCONNECT.
2. Meter disconnect switches that have a short-circuit current rating equal to or greater than the available fault current and all metal housings and service enclosures are grounded in accordance with Section E3908.7 and bonded in accordance with Section 3609. A meter disconnect switch shall be capable of interrupting the load served and shall be marked as

follows: EMERGENCY DISCONNECT, METER DISCONNECT, NOT SERVICE EQUIPMENT.

3. Other *listed* disconnect switches or circuit breakers on the supply side of each service disconnect that are suitable for use as service equipment and marked as follows: EMERGENCY DISCONNECT, NOT SERVICE EQUIPMENT.

Markings shall comply with Section E3404.12. [230.82 (3), 230.85]

SECTION E3602 SERVICE SIZE AND RATING

E3602.1 Ampacity of ungrounded conductors. Ungrounded service conductors shall have an ampacity of not less than the load served. For one-family dwellings, the ampacity of the ungrounded conductors shall be not less than 100 amperes, 3 wire. For all other installations, the ampacity of the ungrounded conductors shall be not less than 60 amperes. [230.42(B), 230.79(C) & (D)]

E3602.2 Service load. The minimum load for ungrounded service conductors and service devices that serve 100 percent of the dwelling unit load shall be computed in accordance with Table E3602.2. Ungrounded service conductors and service devices that serve less than 100 percent of the dwelling unit load shall be computed as required for feeders in accordance with Chapter 37. [220.82(A)]

E3602.2.1 Services under 100 amperes. Services that are not required to be 100 amperes shall be sized in accordance with Chapter 37. [230.42(A), (B), and (C)].

E3602.3 Rating of service disconnect. The combined rating of all individual service disconnects serving a single dwelling unit shall be not less than the load determined from Table E3602.2 and shall be not less than as specified in Section E3602.1. (230.79 & 230.80)

E3602.4 Voltage rating. Systems shall be three-wire, 120/240-volt, single-phase with a grounded neutral. [220.82(A)]

SECTION E3603 SERVICE, FEEDER AND GROUNDING ELECTRODE CONDUCTOR SIZING

E3603.1 Grounded and ungrounded service conductor size. Service and feeder conductors supplied by a single-phase, 120/240-volt system shall be sized in accordance with Sections E3603.1.1 through E3603.1.5 and Table E3705.1.

E3603.1.1 Ungrounded service conductors. For a service rated at 100 through 400 amperes, the service conductors supplying the entire load associated with a one-family dwelling, or the service conductors supplying the entire load associated with an individual dwelling unit in a two-family dwelling, shall have an ampacity of not less than 83 percent of the service rating. The service rating is based on the standard ampere ratings in Section E3705.6. If no adjustment or correction factors are required, Table E3603.1.1 shall be permitted to be applied.

TABLE E3602.2 [220.82(B) & (C)]
MINIMUM SERVICE LOAD CALCULATION

LOADS AND PROCEDURE
3 volt-amperes per square foot of floor area for general lighting and general use receptacle outlets.
Plus
1,500 volt-amperes multiplied by total number of 20-ampere-rated small appliance and laundry circuits.
Plus
The nameplate volt-ampere rating of all fastened-in-place, permanently connected or dedicated circuit-supplied appliances such as ranges, ovens, cooking units, clothes dryers not connected to the laundry branch circuit and water heaters.
Apply the following demand factors to the above subtotal:
The minimum subtotal for the loads above shall be 100 percent of the first 10,000 volt-amperes of the sum of the above loads plus 40 percent of any portion of the sum that is in excess of 10,000 volt-amperes.
Plus the largest of the following:
One hundred percent of the nameplate rating(s) of the air-conditioning and cooling equipment.
One hundred percent of the nameplate rating(s) of the heat pump where a heat pump is used without any supplemental electric heating.
One hundred percent of the nameplate rating of the electric thermal storage and other heating systems where the usual load is expected to be continuous at the full nameplate value. Systems qualifying under this selection shall not be figured under any other category in this table.
One hundred percent of nameplate rating of the heat pump compressor and 65 percent of the supplemental electric heating load for central electric space-heating systems. If the heat pump compressor is prevented from operating at the same time as the supplementary heat, the compressor load does not need to be added to the supplementary heat load for the total central electric space-heating load.
Sixty-five percent of nameplate rating(s) of electric space-heating units if less than four separately controlled units.
Forty percent of nameplate rating(s) of electric space-heating units of four or more separately controlled units.
The minimum total load in amperes shall be the volt-ampere sum calculated above divided by 240 volts.

TABLE E3603.1.1 (Table 310.12)
SINGLE-PHASE DWELLING SERVICES AND FEEDERS

SERVICE OR FEEDER RATING (amperes)	CONDUCTOR (AWG or kcmil)	
	Copper	Aluminum or Copper-clad Aluminum
100	4	2
110	3	1
125	2	1/0
150	1	2/0
175	1/0	3/0
200	2/0	4/0
225	3/0	250
250	4/0	300
300	250	350
350	350	500
400	400	600

E3603.1.2 Ungrounded feeder conductors. For a feeder rated at 100 through 400 amperes, the feeder conductors supplying the entire load associated with a one-family dwelling, or the feeder conductors supplying the entire load associated with an individual dwelling unit in a two-family dwelling, shall have an ampacity of not less than 83 percent of the feeder rating. The feeder rating is based on the standard ampere ratings in Section E3705.6. If no adjustment or correction factors are required, Table E3603.1.1 shall be permitted to be applied.

E3603.1.3 Feeder size relative to service size. A feeder for an individual dwelling unit shall not be required to have an ampacity greater than that specified in Sections E3603.1.1 and E3603.1.2.

E3603.1.4 Grounded conductors. The grounded conductor ampacity shall be not less than the maximum unbalance of the load and the size of the grounded conductor shall be not smaller than the required minimum grounding electrode conductor size specified in Table E3603.4. [310.12]

E3603.1.5 Adjustment/correction factors. Where correction or adjustment factors are required by Section E3705.2 or E3705.3, they shall be permitted to be applied to the ampacity associated with the temperature rating of the conductor.

E3603.2 Ungrounded service conductors for accessory buildings and structures. Ungrounded conductors for other than dwelling units shall have an ampacity of not less than 60 amperes and shall be sized as required for feeders in Chapter 37. [230.79(D)]

Exceptions:

1. For limited loads of a single branch circuit, the service conductors shall have an ampacity of not less than 15 amperes. [230.79(A)]
2. For loads consisting of not more than two two-wire branch circuits, the service conductors shall have an ampacity of not less than 30 amperes. [230.79(B)]

E3603.3 Overload protection. Each ungrounded service conductor shall have overload protection. (230.90)

E3603.3.1 Ungrounded conductor. Overload protection shall be provided by an overcurrent device installed in series with each ungrounded service conductor. The overcurrent device shall have a rating or setting not higher than the allowable service or feeder rating specified in Section E3603.1. A set of fuses shall be considered to be all of the fuses required to protect all of the ungrounded conductors of a circuit. Single pole circuit breakers, grouped in accordance with Section E3601.7, shall be considered as one protective device. [230.90(A)]

Exception: Two to six circuit breakers or sets of fuses shall be permitted as the overcurrent device to provide the overload protection. The sum of the ratings of the circuit breakers or fuses shall be permitted to exceed the ampacity of the service conductors, provided that the calculated load does not exceed the ampacity of the service conductors. [230.90(A) Exception No. 3]

E3603.3.2 Not in grounded conductor. Overcurrent devices shall not be connected in series with a grounded service conductor except where a circuit breaker is used that simultaneously opens all conductors of the circuit. [230.90(B)]

E3603.3.3 Location. The service overcurrent device shall be an integral part of the service disconnecting means or shall be located immediately adjacent thereto. Where fuses are used as the service overcurrent device, the disconnecting means shall be located on the supply side of the fuses. (230.91)

E3603.4 Grounding electrode conductor size. The grounding electrode conductors shall be sized based on the size of the service entrance conductors as required in Table E3603.4. (250.66)

TABLE E3603.4
GROUNDING ELECTRODE CONDUCTOR SIZE[a, b, c, d, e, f]

SIZE OF LARGEST UNGROUNDED SERVICE-ENTRANCE CONDUCTOR OR EQUIVALENT AREA FOR PARALLEL CONDUCTORS (AWG/kcmil)		SIZE OF GROUNDING ELECTRODE CONDUCTOR (AWG/kcmil)	
Copper	Aluminum or copper-clad aluminum	Copper	Aluminum or copper-clad aluminum
2 or smaller	1/0 or smaller	8	6
1 or 1/0	2/0 or 3/0	6	4
2/0 or 3/0	4/0 or 250	4	2
Over 3/0 through 350	Over 250 through 500	2	1/0
Over 350 through 600	Over 500 through 900	1/0	3/0

a. If multiple sets of service-entrance conductors connect directly to a service drop, set of overhead service conductors, set of underground service conductors, or service lateral, the equivalent size of the largest service-entrance conductor shall be determined by the largest sum of the areas of the corresponding conductors of each set. (Table 250.66)

b. Where there are no service-entrance conductors, the grounding electrode conductor size shall be determined by the equivalent size of the largest service-entrance conductor required for the load to be served. (Table 250.66)

(continued)

TABLE E3603.4—continued

c. Where protected by a ferrous metal raceway, grounding electrode conductors shall be electrically bonded to the ferrous metal raceway at both ends. [250.64(E)(1)]

d. An 8 AWG grounding electrode conductor shall be protected with rigid metal conduit, intermediate metal conduit, rigid polyvinyl chloride (Type PVC) nonmetallic conduit, rigid thermosetting resin (Type RTRC) nonmetallic conduit, electrical metallic tubing or cable armor. [250.64(B)]

e. Where not protected, 6 AWG grounding electrode conductor shall closely follow a structural surface for physical protection. The supports shall be spaced not more than 24 inches on center and shall be within 12 inches of any enclosure or termination. [250.64(B)]

f. Where the grounding electrode conductor or bonding jumper connected to a single or multiple rod, pipe, or plate electrode(s) or any combination thereof, as described in Section E3608.3, does not extend on to other types of electrodes that require a larger size of conductor, the grounding electrode conductor shall not be required to be larger than 6 AWG copper or 4 AWG aluminum. Where the grounding electrode conductor or bonding jumper connected to a single or multiple concrete-encased electrode(s), as described in Section E3608.1.2, does not extend on to other types of electrodes that require a larger size of conductor, the grounding electrode conductor shall not be required to be larger than 4 AWG copper conductor. [250.66(A) and (B)]

E3603.5 Temperature limitations. Except where the equipment is marked otherwise, conductor ampacities used in determining equipment termination provisions shall be based on Table E3705.1. [110.14(C)(1)]

SECTION E3604 OVERHEAD SERVICE AND SERVICE-ENTRANCE CONDUCTOR INSTALLATION

E3604.1 Clearances on buildings. Open conductors and multiconductor cables without an overall outer jacket shall have a clearance of not less than 3 feet (914 mm) from the sides of doors, porches, decks, stairs, ladders, fire escapes and balconies, and from the sides and bottom of windows that open. See Figure E3604.1. [230.9(A)]

E3604.2 Vertical clearances. Overhead service conductors shall not have *ready access* and shall comply with Sections E3604.2.1 and E3604.2.2. (230.24)

E3604.2.1 Above roofs. Conductors shall have a vertical clearance of not less than 8 feet (2438 mm) above the roof surface. The vertical clearance above the roof level shall be maintained for a distance of not less than 3 feet (914 mm) in all directions from the edge of the roof. See Figure E3604.2.1. [230.24(A)]

Exceptions:

1. Conductors above a roof surface subject to pedestrian traffic shall have a vertical clearance from the roof surface in accordance with Section E3604.2.2. [230.24(A) Exception No. 1]
2. Where the roof has a slope of 4 inches (102 mm) in 12 inches (305 mm), or greater, the minimum clearance shall be 3 feet (914 mm). [230.24(A) Exception No. 2]
3. The minimum clearance above only the overhanging portion of the roof shall not be less than 18 inches (457 mm) where not more than 6 feet (1829 mm) of overhead service conductor length passes over 4 feet (1219 mm) or less of roof surface measured horizontally and such conductors are terminated at a through-the-roof raceway or *approved* support. [230.24(A) Exception No. 3]

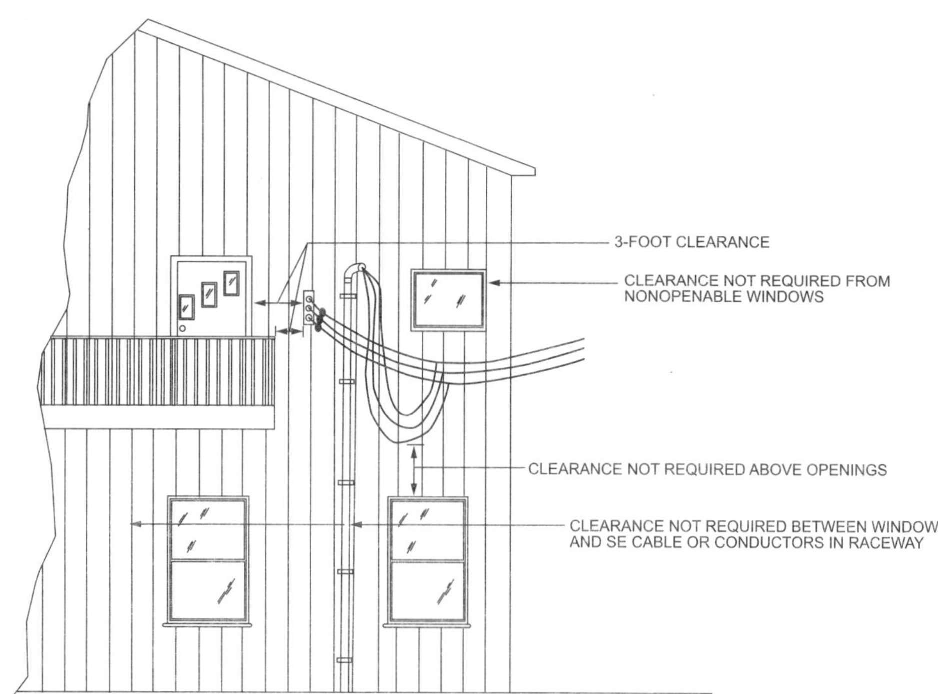

For SI: 1 foot = 304.8 mm.

FIGURE E3604.1 CLEARANCES FROM BUILDING OPENINGS

4. The requirement for maintaining the vertical clearance for a distance of 3 feet (914 mm) from the edge of the roof shall not apply to the final conductor span where the service drop is attached to the side of a building. [230.24(A) Exception No. 4]
5. Where the voltage between conductors does not exceed 300 and the roof area is guarded or isolated, a reduction in clearance to 3 feet (914 mm) shall be permitted. [230.24(A) Exception No. 5]

E3604.2.2 Vertical clearance from grade. Overhead service conductors shall have the following minimum clearances from final grade:

1. For conductors supported on and cabled together with a grounded bare messenger wire, the minimum vertical clearance shall be 10 feet (3048 mm) at the electric service entrance to buildings, at the lowest point of the drip loop of the building electric entrance, and above areas or sidewalks accessed by pedestrians only. Such clearance shall be measured from final grade or other accessible surfaces.
2. Twelve feet (3658 mm)—over residential property and driveways.
3. Eighteen feet (5486 mm)—over public streets, alleys, roads or parking areas subject to truck traffic. [(230.24(B)(1), (2), and (4)]

E3604.3 Point of attachment. The point of attachment of the overhead service conductors to a building or other structure shall provide the minimum clearances as specified in Sections E3604.1 through E3604.2.2. The point of attachment shall be not less than 10 feet (3048 mm) above finished grade. (230.26)

E3604.4 Means of attachment. Multiconductor cables used for overhead service conductors shall be attached to buildings or other structures by fittings *approved* for the purpose. (230.27)

E3604.5 Service masts as supports. A service mast used for the support of service-drop or overhead service conductors shall comply with Sections E3604.5.1 and E3604.5.2. Only power service drop or overhead service conductors shall be attached to a service mast.

E3604.5.1 Strength. The service mast shall be of adequate strength or shall be supported by braces or guy wires to safely withstand the strain imposed by the service-drop or overhead service conductors. Hubs intended for use with a conduit that serves as a service mast shall be identified for use with service-entrance equipment.

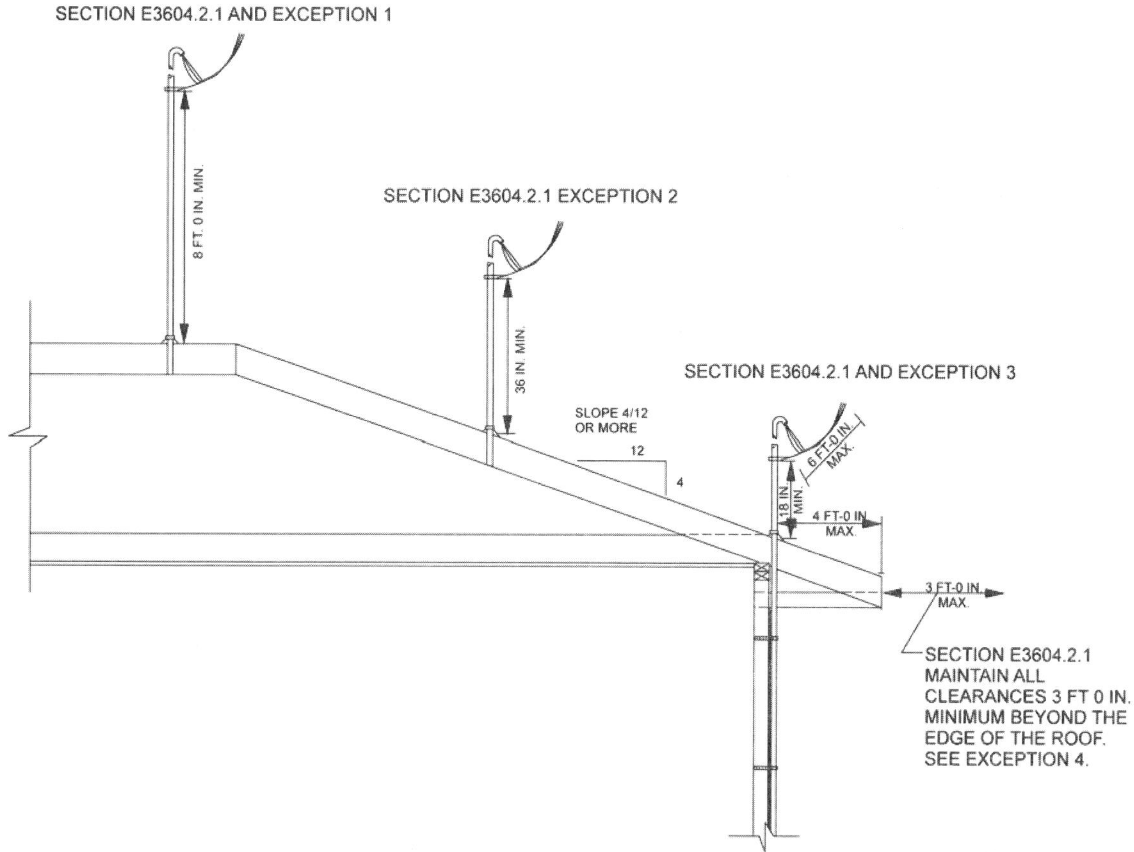

For SI: 1 inch = 25.4 mm, 1 foot = 304.8 mm.

**FIGURE E3604.2.1
CLEARANCES FROM ROOFS**

E3604.5.2 Attachment. Service-drop or overhead service conductors shall not be attached to a service mast at a point between a coupling and a weatherhead or the end of the conduit, where the coupling is located above the last point of securement of the building or other structure or is located above the building or other structure. [230.28(A) & (B)]

E3604.6 Supports over buildings. Service conductors passing over a roof shall be securely supported by a substantial structure. For a grounded system, where the substantial structure is metal, it shall be bonded by means of a bonding jumper and *listed* connector to the grounded overhead service conductor. Where practicable, such supports shall be independent of the building. (230.29)

SECTION E3605 SERVICE-ENTRANCE CONDUCTORS

E3605.1 Insulation of service-entrance conductors. Service-entrance conductors entering or on the exterior of buildings or other structures shall be insulated in accordance with Section E3406.5. (230.41)

Exceptions:

1. A copper grounded conductor shall not be required to be insulated where it is:
   - 1.1. In a raceway or part of a service cable assembly,
   - 1.2. Directly buried in soil of suitable condition, or
   - 1.3. Part of a cable assembly *listed* for direct burial without regard to soil conditions.
2. An aluminum or copper-clad aluminum grounded conductor shall not be required to be insulated where part of a cable or where identified for direct burial or utilization in underground raceways. (230.41 Exception)

E3605.2 Wiring methods for services. Service-entrance wiring methods shall be installed in accordance with the applicable requirements in Chapter 38. (230.43)

E3605.3 Spliced conductors. Service-entrance conductors shall be permitted to be spliced or tapped. Splices shall be made in enclosures or, if directly buried, with *listed* underground splice kits. Conductor splices shall be made in accordance with Chapters 34, 37, 38 and 39. Power distribution blocks, pressure connectors, and devices for splices and taps shall be *listed.* Power distribution blocks installed on service conductors shall be marked "suitable for use on the line side of the service equipment" or equivalent. (230.33, 230.46)

E3605.4 Protection of underground service entrance conductors. Underground service-entrance conductors shall be protected against physical damage in accordance with Chapter 38. (230.32)

E3605.5 Protection of all other service cables. Aboveground service-entrance cables, where subject to physical damage, shall be protected by one or more of the following: rigid metal conduit, intermediate metal conduit, Schedule 80 PVC conduit, electrical metallic tubing, reinforced thermosetting resin conduit or other *approved* means. [230.50(1)]

E3605.6 Locations exposed to direct sunlight. Insulated conductors and cables used where exposed to direct rays of the sun shall comply with one of the following:

1. The conductors and cables shall be *listed,* or *listed* and marked, as being sunlight resistant.
2. The conductors and cables are covered with insulating material, such as tape or sleeving, that is *listed,* or *listed* and marked, as being sunlight resistant. [310.10(D)]

E3605.7 Mounting supports. Service-entrance cables shall be supported by straps or other *approved* means within 12 inches (305 mm) of every service head, gooseneck or connection to a raceway or enclosure and at intervals not exceeding 30 inches (762 mm). [230.51(A)]

E3605.8 Raceways to drain. Where exposed to the weather, raceways enclosing service-entrance conductors shall be *listed* or *approved* for use in wet locations and arranged to drain. Where embedded in masonry, raceways shall be arranged to drain. (230.53)

E3605.9 Overhead service locations. Connections at service heads shall be in accordance with Sections E3605.9.1 through E3605.9.7. (230.54)

E3605.9.1 Raintight service head. Service raceways shall be equipped with a service head at the point of connection to service-drop or overhead conductors. The service head shall be *listed* for use in wet locations. [230.54(A)]

E3605.9.2 Service cable, service head or gooseneck. Service-entrance cable shall be equipped with a service head or shall be formed into a gooseneck in an approved manner. The service head shall be *listed* for use in wet locations. [230.54(B)]

E3605.9.3 Service-head location. Service heads on raceways or service-entrance cables, and goosenecks in service-entrance cables, shall be located above the point of attachment of the service-drop or overhead service conductors to the building or other structure. [230.54(C)]

Exception: Where it is impracticable to locate the service head or gooseneck above the point of attachment, the service head or gooseneck location shall be not more than 24 inches (610 mm) from the point of attachment. [230.54(C) Exception]

E3605.9.4 Separately bushed openings. Service heads shall have conductors of different potential brought out through separately bushed openings. [230.54(E)]

E3605.9.5 Drip loops. Drip loops shall be formed on individual conductors. To prevent the entrance of moisture, service-entrance conductors shall be connected to the service-drop or overhead conductors either below the level of the service head or below the level of the termination of the service-entrance cable sheath. [230.54(F)]

E3605.9.6 Conductor arrangement. Service-entrance and overhead service conductors shall be arranged so that water will not enter service raceways or equipment. [230.54(G)]

E3605.9.7 Secured. Service-entrance cables shall be held securely in place. [230.54(D)]

SECTION E3606 SERVICE EQUIPMENT—GENERAL

E3606.1 Service equipment enclosures. Energized parts of service equipment shall be enclosed. (230.62)

E3606.2 Working space. The working space in the vicinity of service equipment shall be not less than that specified in Chapter 34. (110.26)

E3606.3 Available short-circuit current. Service equipment shall be suitable for the maximum fault current available at its supply terminals, but not less than 10,000 amperes. (110.9)

E3606.4 Marking. Service equipment shall be marked to identify it as being suitable for use as service equipment. Service equipment shall be listed or field labeled. Individual meter socket enclosures shall not be considered as service equipment but shall be listed and rated for the voltage and ampacity of the service. [230.66 (A) & (B)]

Exception: Meter sockets supplied by and under the exclusive control of an electric utility shall not be required to be listed. (230.66 Exception)

E3606.5 Surge protection. All services supplying one- and two-family dwelling units shall be provided with a surge-protective device (SPD) installed in accordance with Sections E3606.5.1 through E3606.5.3.

E3606.5.1 Location. The SPD shall be an integral part of the service equipment or shall be located immediately adjacent thereto.

Exception: The SPD shall not be required to be located in the service equipment if located at each next-level distribution equipment downstream toward the load.

E3606.5.2 Type. The SPD shall be a Type 1 or Type 2 SPD.

E3606.5.3 Replacement. Where service equipment is replaced, all of the requirements of this section shall apply. [230.67]

SECTION E3607 SYSTEM GROUNDING

E3607.1 System service ground. The premises wiring system shall be grounded at the service with a grounding electrode conductor connected to a grounding electrode system as required by this code. Grounding electrode conductors shall be sized in accordance with Table E3603.4. [250.20(B)(1) and 250.24(A)]

E3607.2 Location of grounding electrode conductor connection. The grounding electrode conductor shall be connected to the grounded service conductor at any accessible point from the load end of the overhead service conductors, service drop, underground service conductors, or service lateral to and including the terminal or bus to which the grounded service conductor is connected at the service disconnecting means. A grounding connection shall not be made to any grounded circuit conductor on the load side of the service disconnecting means, except as provided in Section E3607.3.2. [250.24(A)(1) and (A)(5)]

E3607.3 Buildings or structures supplied by feeder(s) or branch circuit(s). A building or structure supplied by a feeder(s) or branch circuit(s) shall have a grounding electrode system and grounding electrode conductor installed in accordance with Section E3608. Where there is no existing grounding electrode, the grounding electrode(s) required in Section E3608 shall be installed. [250.32(A)]

Exception: A grounding electrode shall not be required where only one branch circuit, including a multiwire branch circuit, supplies the building or structure and the branch circuit includes an equipment grounding conductor for grounding the noncurrent-carrying parts of all equipment. For the purposes of this section, a multiwire branch circuit shall be considered as a single branch circuit. [250.32(A) Exception]

E3607.3.1 Equipment grounding conductor. An equipment grounding conductor as described in Section E3908 shall be run with the supply conductors and connected to the building or structure disconnecting means and to the grounding electrode(s). The equipment grounding conductor shall be used for grounding or bonding of equipment, structures or frames required to be grounded or bonded. The equipment grounding conductor shall be sized in accordance with Section E3908.13. Any installed grounded conductor shall not be connected to the equipment grounding conductor or to the grounding electrode(s). [250.32(B) and Table 250.122]

E3607.3.2 Grounded conductor, existing premises. For installations made in compliance with previous editions of this code that permitted such connection and where an equipment grounding conductor is not run with the supply conductors to the building or structure, there are no continuous metallic paths bonded to the grounding system in both buildings or structures involved, and ground-fault protection of equipment has not been installed on the supply side of the feeder(s), the grounded conductor run with the supply to the buildings or structure shall be connected to the building or structure disconnecting means and to the grounding electrode(s) and shall be used for grounding or bonding of equipment, structures, or frames required to be grounded or bonded. Where used for grounding in accordance with this provision, the grounded conductor shall be not smaller than the larger of:

1. That required by Section E3704.3.
2. That required by Section E3908.13. [250.32(B)(1) Exception]

E3607.4 Grounding electrode conductor. A grounding electrode conductor shall be used to connect the equipment grounding conductors, the service equipment enclosures, and the grounded service conductor to the grounding electrode(s). This conductor shall be sized in accordance with Table E3603.4. [250.24(D)]

E3607.5 Main bonding jumper. An unspliced main bonding jumper shall be used to connect the equipment grounding conductor(s) and the service-disconnect enclosure to the grounded conductor of the system within the enclosure for each service disconnect. [250.24(B)]

E3607.6 Common grounding electrode. Where an AC system is connected to a grounding electrode in or at a building or structure, the same electrode shall be used to ground conductor enclosures and equipment in or on that building or structure. Where separate services, feeders or branch circuits supply a building and are required to be connected to a grounding electrode(s), the same grounding electrode(s) shall be used. Two or more grounding electrodes that are effectively bonded together shall be considered as a single grounding electrode system. (250.58)

SECTION E3608 GROUNDING ELECTRODE SYSTEM

E3608.1 Grounding electrode system. All electrodes specified in Sections E3608.1.1, E3608.1.2, E3608.1.3, E3608.1.4 E3608.1.5 and E3608.1.6 that are present at each building or structure served shall be bonded together to form the grounding electrode system. Where none of these electrodes are present, one or more of the electrodes specified in Sections E3608.1.3, E3608.1.4, E3608.1.5 and E3608.1.6 shall be installed and used. (250.50)

Exception: Concrete-encased electrodes of existing buildings or structures shall not be required to be part of the grounding electrode system where the steel reinforcing bars or rods are not accessible for use without disturbing the concrete. (250.50 Exception)

E3608.1.1 Metal underground water pipe. A metal underground water pipe that is in direct contact with the earth for 10 feet (3048 mm) or more, including any well casing effectively bonded to the pipe and that is electrically continuous, or made electrically continuous by bonding around insulating joints or insulating pipe to the points of connection of the grounding electrode conductor and the bonding conductors, shall be considered as a grounding electrode (see Section E3608.1). [250.52(A)(1)]

E3608.1.1.1 Interior metal water piping. Interior metal water piping located more than 5 feet (1524 mm) from the point of entrance into the building shall not be used as a conductor to interconnect electrodes of the grounding electrode system. [250.68(C)(1)]

E3608.1.1.2 Installation. Continuity of the grounding path or the bonding connection to interior piping shall not rely on water meters, filtering devices and similar equipment. A metal underground water pipe shall be supplemented by an additional electrode of a type specified in Sections E3608.1.2 through E3608.1.6. The supplemental electrode shall be bonded to the grounding electrode conductor, the grounded service-entrance conductor, a nonflexible grounded service raceway, any grounded service enclosure or to the equipment grounding conductor provided in accordance with Section E3607.3.1. Where the supplemental electrode is a rod, pipe or plate electrode in accordance with Section E3608.1.4 or E3608.1.5, it shall comply with Section E3608.4.

Where the supplemental electrode is a rod, pipe or plate electrode in accordance with Section E3608.1.4 or E3608.1.5, that portion of the bonding jumper that is the sole connection to the supplemental grounding electrode shall not be required to be larger than 6 AWG copper or 4 AWG aluminum wire. [250.53(D) and (E)]

E3608.1.2 Concrete-encased electrode. A concrete-encased electrode consisting of not less than 20 feet (6096 mm) of either of the following shall be considered as a grounding electrode:

1. One or more bare or zinc-galvanized or other electrically conductive coated steel reinforcing bars or rods not less than $^1/_2$ inch (13 mm) in diameter, installed in one continuous 20-foot (6096 mm) length, or if in multiple pieces connected together by the usual steel tie wires, exothermic welding, welding, or other effective means to create a 20-foot (6096 mm) or greater length.
2. A bare copper conductor not smaller than 4 AWG.

Metallic components shall be encased by at least 2 inches (51 mm) of concrete and shall be located horizontally within that portion of a concrete foundation or footing that is in direct contact with the earth or within vertical foundations or structural components or members that are in direct contact with the earth.

Where multiple concrete-encased electrodes are present at a building or structure, only one shall be required to be bonded into the grounding electrode system. [250.52(A)(3)]

E3608.1.3 Ground rings. A ground ring encircling the building or structure, in direct contact with the earth at a depth below the earth's surface of not less than 30 inches (762 mm), consisting of at least 20 feet (6096 mm) of bare copper conductor not smaller than 2 AWG shall be considered as a grounding electrode. [250.52(A)(4)]

E3608.1.4 Rod and pipe electrodes. Rod and pipe electrodes not less than 8 feet (2438 mm) in length and consisting of the following materials shall be considered as a grounding electrode:

1. Grounding electrodes of pipe or conduit shall not be smaller than trade size $^3/_4$ (metric designator 21) and, where of iron or steel, shall have the outer surface galvanized or otherwise metal-coated for corrosion protection.
2. Rod-type grounding electrodes of stainless steel and copper or zinc-coated steel shall be at least $^5/_8$ inch (15.9 mm) in diameter unless *listed.* [250.52(A)(5)]

E3608.1.4.1 Installation. The rod and pipe electrodes shall be installed such that at least 8 feet (2438 mm) of length is in contact with the soil. They shall be driven to a depth of not less than 8 feet (2438 mm) except that, where rock bottom is encountered, electrodes shall be driven at an oblique angle not to exceed 45 degrees (0.79 rad) from the vertical or shall be buried in a trench that is at least 30 inches (762 mm) deep. The upper end of the electrodes shall be flush with or below ground level except where the above-ground end and the grounding electrode conductor attachment are protected against physical damage. [250.53(G)]

E3608.1.5 Plate electrodes. A plate electrode that exposes not less than 2 square feet (0.186 m^2) of surface to exterior soil shall be considered as a grounding electrode. Electrodes of bare or electrically conductive coated iron or steel plates shall be not less than $^1/_4$ inch (6.4 mm) in thickness. Solid, uncoated electrodes of nonferrous metal shall be not less than 0.06 inch (1.5 mm) in thickness. Plate electrodes shall be installed not less than 30 inches (762 mm) below the surface of the earth. [250.52(A)(7)]

E3608.1.6 Other electrodes. In addition to the grounding electrodes specified in Sections E3608.1.1 through E3608.1.5, other *listed* grounding electrodes shall be permitted. [250.52(A)(6)]

E3608.2 Bonding jumper. The bonding jumper(s) used to connect the grounding electrodes together to form the grounding electrode system shall be installed in accordance with Sections E3610.2, and E3610.3, shall be sized in accordance with Section E3603.4, and shall be connected in the manner specified in Section E3611.1. Rebar shall not be used as a conductor to interconnect the electrodes of grounding electrode systems. [250.53(C)]

E3608.3 Rod, pipe and plate electrode requirements. Where practicable, rod, pipe and plate electrodes shall be embedded below permanent moisture level. Such electrodes shall be free from nonconductive coatings such as paint or enamel. Where more than one such electrode is used, each electrode of one grounding system shall be not less than 6 feet (1829 mm) from any other electrode of another grounding system. Two or more grounding electrodes that are effectively bonded together shall be considered as a single grounding electrode system. That portion of a bonding jumper that is the sole connection to a rod, pipe or plate electrode shall not be required to be larger than 6 AWG copper or 4 AWG aluminum wire. [250.53(A)(1), 250.53(B), 250.53(E)]

E3608.4 Supplemental electrode required. A single rod, pipe or plate electrode shall be supplemented by an additional electrode of a type specified in Sections E3608.1.2 through E3608.1.6. The supplemental electrode shall be bonded to one of the following:

1. A rod, pipe or plate electrode.
2. A grounding electrode conductor.
3. A grounded service-entrance conductor.
4. A nonflexible grounded service raceway.
5. A grounded service enclosure.

Where multiple rod, pipe or plate electrodes are installed to meet the requirements of this section, they shall be not less than 6 feet (1829 mm) apart. [250.53(A)(2) and (A)(3)]

Exception: Where a single rod, pipe or plate grounding electrode has a resistance to earth of 25 ohms or less, the supplemental electrode shall not be required. [250.53(A)(2) Exception]

E3608.5 Aluminum electrodes. Aluminum electrodes shall not be permitted. [250.52(B)(2)]

E3608.6 Metal underground gas piping system. A metal underground gas piping system shall not be used as a grounding electrode. [250.52(B)(1)]

E3608.7 Pool, spa and hot tub structures and structural reinforcing steel. The structures and structural reinforcing steel described in Section E4204.2, Items 1 and 2, shall not be used as a grounding electrode. [250.52 (B)(3)]

SECTION E3609
BONDING

E3609.1 General. Bonding shall be provided where necessary to ensure electrical continuity and the capacity to conduct safely any fault current likely to be imposed. (250.90)

E3609.2 Bonding of equipment for services. The noncurrent-carrying metal parts of the following equipment shall be effectively bonded together:

1. Raceways or service cable armor or sheath that enclose, contain, or support service conductors.
2. Service enclosures containing service conductors, including meter fittings, and boxes, interposed in the service raceway or armor. [250.92(A)]

E3609.3 Bonding for communications systems. Communications system bonding terminations shall be connected in accordance with Section E3609.3.1 or E3609.3.2. (250.94)

E3609.3.1 Intersystem bonding termination device. An intersystem bonding termination (IBT) for connecting intersystem bonding conductors shall be provided external to enclosures at the service equipment or metering equipment enclosure and at the disconnecting means for any additional buildings or structures. An IBT shall comply with all of the following:

1. It shall be accessible for connection and inspection.
2. It shall consist of a set of terminals with the capacity for connection of not less than three intersystem bonding conductors.
3. It shall not interfere with opening of the enclosure for a service, building or structure disconnecting means, or metering equipment.
4. Where located at the service equipment, it shall be securely mounted and electrically connected to an enclosure for the service equipment, to the meter enclosure, or to an exposed nonflexible metallic

service raceway, or shall be mounted at one of these enclosures and connected to the enclosure or to the grounding electrode conductor with a 6 AWG or larger copper conductor.

5. Where located at the disconnecting means for a building or structure, it shall be securely mounted and electrically connected to the metallic enclosure for the building or structure disconnecting means, or shall be mounted at the disconnecting means and connected to the metallic enclosure or to the grounding electrode conductor with a 6 AWG or larger copper conductor.
6. It shall be *listed* as grounding and bonding equipment. [250.94(A)]

Exception: Means for connecting intersystem bonding conductors are not required where communications systems are not likely to be used.

E3609.3.2 Other means. An aluminum or copper busbar not less than $^{1}/_{4}$ inch thick by 2 inches wide (6.4 mm by 51 mm) and of sufficient length to accommodate not fewer than three terminations for communications systems in addition to other connections shall be provided. The busbar shall be securely fastened and shall be installed in an accessible location. Connections shall be made by a *listed* connector. Where aluminum busbars are used, the installation shall comply with Section E3610.2.

Exception: Means for connecting intersystem bonding conductors are not required where communications systems are not likely to be used. [250.94(B)]

E3609.4 Method of bonding at the service. Bonding jumpers meeting the requirements of this chapter shall be used around impaired connections, such as reducing washers or oversized, concentric, or eccentric knockouts. Standard locknuts or bushings shall not be the only means for the bonding required by this section but shall be permitted to be installed to make mechanical connections of raceways. Electrical continuity at service equipment, service raceways and service conductor enclosures shall be ensured by one or more of the methods specified in Sections E3609.4.1 through E3609.4.4.

E3609.4.1 Grounded service conductor. Equipment shall be bonded to the grounded service conductor in a manner provided in this code.

E3609.4.2 Threaded connections. Equipment shall be bonded by connections using threaded couplings or *listed* threaded hubs on enclosures. Such connections shall be made wrench tight.

E3609.4.3 Threadless couplings and connectors. Equipment shall be bonded by threadless couplings and connectors for metal raceways and metal-clad cables. Such couplings and connectors shall be made wrench tight. Standard locknuts or bushings shall not be used for the bonding required by this section.

E3609.4.4 Other devices. Equipment shall be bonded by other *listed* devices, such as bonding-type locknuts, bushings and bushings with bonding jumpers. [250.92(B)]

E3609.5 Sizing supply-side bonding jumper and main bonding jumper. The bonding jumper shall not be smaller than the sizes shown in Table E3603.4 for grounding electrode conductors. Where the service-entrance conductors are paralleled in two or more raceways or cables, and an individual supply-side bonding jumper is used for bonding these raceways or cables, the supply-side bonding jumper for each raceway or cable shall be selected from Table E3603.4 based on the size of the ungrounded supply conductors in each raceway or cable. A single supply-side bonding jumper installed for bonding two or more raceways or cables shall be sized in accordance with Table E3603.4 based on the largest set of parallel ungrounded supply conductors. [250.102(C)]

E3609.6 Metal water piping bonding. The metal water piping system shall be bonded to the service equipment enclosure, the grounded conductor at the service, the grounding electrode conductor where of sufficient size, or to the one or more grounding electrodes used. The bonding jumper shall be sized in accordance with Table E3603.4. The points of attachment of the bonding jumper(s) shall be accessible. [250.104(A) and 250.104(A)(1)]

E3609.7 Bonding other metal piping. Where installed in or attached to a building or structure, metal piping systems, including gas piping, capable of becoming energized shall be bonded to the service equipment enclosure, the grounded conductor at the service, the grounding electrode conductor where of sufficient size, or to the one or more grounding electrodes used. The bonding conductor(s) or jumper(s) shall be sized in accordance with Table E3908.13 and equipment grounding conductors shall be sized in accordance with Table E3908.13 using the rating of the circuit capable of energizing the piping. The equipment grounding conductor for the circuit that is capable of energizing the piping shall be permitted to serve as the bonding means. The points of attachment of the bonding jumper(s) shall be accessible. [250.104(B)]

SECTION E3610 GROUNDING ELECTRODE CONDUCTORS

E3610.1 Continuous. The grounding electrode conductor shall be installed in one continuous length without splices or joints and shall run to any convenient grounding electrode available in the grounding electrode system where the other electrode(s), if any, are connected by bonding jumpers in accordance with Section E3608.2, or to one or more grounding electrode(s) individually. The grounding electrode conductor shall be sized for the largest grounding electrode conductor required among all of the electrodes connected to it. [250.64(C)]

Exception: Splicing of the grounding electrode conductor by irreversible compression-type connectors *listed* as grounding and bonding equipment or by the exothermic welding process shall not be prohibited. [250.64(C)(1)]

E3610.2 Securing and protection against physical damage. Where exposed, a grounding electrode conductor or its enclosure shall be securely fastened to the surface on which it is carried. Grounding electrode conductors shall be permitted to be installed on or through framing members. A 6 AWG or larger copper or aluminum grounding electrode conductor not exposed to physical damage shall be permitted to be run along the surface of the building construction without metal covering or protection. A 6 AWG or larger copper or aluminum grounding electrode exposed to physical damage shall be in rigid metal conduit, intermediate metal conduit, rigid polyvinyl chloride (PVC), nonmetallic conduit, reinforced thermosetting resin (RTRC-XW) nonmetallic conduit, electrical metallic tubing or cable armor. Grounding electrode conductors smaller than 6 AWG shall be in rigid metal conduit, intermediate metal conduit, rigid polyvinyl chloride (PVC) nonmetallic conduit, reinforced thermosetting resin (RTRC-XW) nonmetallic conduit, electrical metallic tubing or cable armor. Grounding electrode conductors and grounding electrode bonding jumpers in contact with the earth shall not be required to comply with Section E3803, but shall be buried or otherwise protected if subject to physical damage. [250.64(B)]

Bare aluminum or copper-clad aluminum grounding electrode conductors shall comply with the following:

1. Bare or covered conductors without an extruded polymeric covering shall not be installed where subject to corrosive conditions or be installed in direct contact with concrete.
2. Terminations made within outdoor enclosures that are *listed* and identified for the environment shall be permitted within 18 inches (457 mm) of the bottom of the enclosure.
3. Aluminum or copper-clad aluminum conductors external to buildings or equipment enclosures shall not be terminated within 18 inches (457 mm) of the earth. [250.64(A)]

E3610.3 Raceways and enclosures for grounding electrode conductors. Ferrous metal raceways, enclosures and cable armor for grounding electrode conductors shall be electrically continuous from the point of attachment to cabinets or equipment to the grounding electrode, and shall be securely fastened to the ground clamp or fitting. Ferrous metal raceways, enclosures and cable armor shall be bonded at each end of the raceway or enclosure to the grounding electrode or to the grounding electrode conductor to create an electrically parallel path. Nonferrous metal raceways, enclosures and cable armor shall not be required to be electrically continuous. Bonding methods in compliance with Section E3609.4 for installations at service equipment locations and with Sections E3609.4.2 through E3609.4.4 for other than service equipment locations shall apply at each end and to all intervening ferrous raceways, boxes, and enclosures between the cabinets or equipment and the grounding electrode. The bonding jumper for a grounding electrode conductor(s), raceway(s), enclosure(s) or cable armor shall be the same size or larger than the largest enclosed grounding electrode conductor.

Where a raceway is used as protection for a grounding conductor, the installation shall comply with the requirements of Chapter 38. [250.64(E)(1), (2), (3), (4)]

E3610.4 Prohibited use. An equipment grounding conductor shall not be used as a grounding electrode conductor. (250.121)

Exception: A wire-type equipment grounding conductor shall be permitted to serve as both an equipment grounding conductor and a grounding electrode conductor where installed in accordance with the applicable requirements for both the equipment grounding conductor and the grounding electrode conductor in Chapters 36 and 39. Where used as a grounding electrode conductor, the wire-type equipment grounding conductor shall be installed and arranged in a manner that will prevent objectionable current. [250.121 Exception, 250.6(A)]

SECTION E3611 GROUNDING ELECTRODE CONDUCTOR CONNECTION TO THE GROUNDING ELECTRODES

E3611.1 Methods of grounding conductor connection to electrodes. The grounding or bonding conductor shall be connected to the grounding electrode by exothermic welding, *listed* lugs, *listed* pressure connectors, *listed* clamps or other *listed* means. Connections depending on solder shall not be used. Ground clamps shall be *listed* for the materials of the grounding electrode and the grounding electrode conductor and, where used on pipe, rod or other buried electrodes, shall also be *listed* for direct soil burial or concrete encasement. Not more than one conductor shall be connected to the grounding electrode by a single clamp or fitting unless the clamp or fitting is *listed* for multiple conductors. One of the methods indicated in the following items shall be used:

1. A pipe fitting, pipe plug or other approved device screwed into a pipe or pipe fitting.
2. A *listed* bolted clamp of cast bronze or brass, or plain or malleable iron.
3. For indoor communications purposes only, a *listed* sheet metal strap-type ground clamp having a rigid metal base that seats on the electrode and having a strap of such material and dimensions that it is not likely to stretch during or after installation.
4. Other equally substantial approved means. (250.70)

E3611.2 Accessibility. All mechanical elements used to terminate a grounding electrode conductor or bonding jumper to the grounding electrodes that are not buried or concrete encased shall be accessible. [250.68(A) and 250.68(A) Exception]

E3611.3 Effective grounding path. The connection of the grounding electrode conductor or bonding jumper shall be made in a manner that will ensure a permanent and effective grounding path. Where necessary to ensure effective grounding for a metal piping system used as a grounding electrode, effective bonding shall be provided around insulated joints and sections and around any equipment that is likely to be disconnected for repairs or replacement. Bonding jumpers shall be of sufficient length to permit removal of such equipment while retaining the integrity of the grounding path. [250.68(B)]

E3611.4 Interior metal water piping. Where grounding electrode conductors and bonding jumpers are connected to interior metal water piping as a means to extend the grounding electrode conductor connection to an electrode(s), such piping shall be located not more than 5 feet (1524 mm) from the point of entry into the building.

Where interior metal water piping is used as a conductor to interconnect electrodes that are part of the grounding electrode system, such piping shall be located not more than 5 feet (1524 mm) from the point of entry into the building. [250.68(C)(1)]

E3611.5 Rebar-type concrete-encased electrode. A rebar-type concrete-encased electrode installed in accordance with Section E3608.1.2 with an additional rebar section extended from its location within the concrete foundation or footing to an accessible location that is not subject to corrosion shall be permitted for connection of grounding electrode conductors and bonding jumpers in accordance with the following:

1. The additional rebar section shall be continuous with the grounding electrode rebar or shall be connected to the grounding electrode rebar and connected together by the usual steel tie wires, exothermic welding, welding or other effective means.
2. The rebar extension shall not be exposed to contact with the earth without corrosion protection.
3. The rebar shall not be used as a conductor to interconnect the electrodes of grounding electrode systems. [250.68 (C) (3)]

E3611.6 Protection of ground clamps and fittings. Ground clamps or other fittings shall be approved for applications without protection or shall be protected from physical damage by installing them where they are not likely to be damaged or by enclosing them in metal, wood or equivalent protective coverings. (250.10)

E3611.7 Clean surfaces. Nonconductive coatings (such as paint, enamel and lacquer) on equipment to be grounded or bonded shall be removed from threads and other contact surfaces to ensure good electrical continuity or shall be connected by fittings that make such removal unnecessary. (250.12)

CHAPTER 37

BRANCH CIRCUIT AND FEEDER REQUIREMENTS

ICC user note:

About this chapter: *Chapter 37 addresses the sizing of conductors for feeders and branch circuits, specifies the required branch circuits, provides for overcurrent protection of such conductors, specifies limitations for branch circuit loading, and addresses panel board ratings and protection. Design and installation processes move directly from the service to feeders and branch circuits.*

SECTION E3701 GENERAL

E3701.1 Scope. This chapter covers branch circuits and feeders and specifies the minimum required branch circuits, the allowable loads and the required overcurrent protection for branch circuits and feeders that serve less than 100 percent of the total dwelling unit load. Feeder circuits that serve 100 percent of the dwelling unit load shall be sized in accordance with the procedures in Chapter 36. [310.12(B)]

E3701.2 Branch-circuit and feeder ampacity. Branch-circuit and feeder conductors shall have ampacities not less than the maximum load to be served. Where a branch circuit or a feeder supplies continuous loads or any combination of continuous and noncontinuous loads, the minimum branch-circuit or feeder conductor size shall not be less the larger of Section E3701.2.1 or E3701.2.2 and shall comply with Section E3705.4.

E3701.2.1 Continuous load. Where a branch circuit or feeder supplies continuous loads or any combination of continuous and noncontinuous loads, the minimum branch-circuit or feeder conductor size shall have an ampacity not less than the noncontinuous load plus 125 percent of the continuous load in accordance with Table E3705.1.

E3701.2.2 Ampacity adjustment or correction. The minimum branch-circuit or feeder conductor size shall have an ampacity not less than the maximum load to be served after the application of any adjustment or correction factors in accordance with Tables E3705.1, E3705.2 and E3705.3. [210.19(A)(1) and 215.2(A)(1)]

Exception: The grounded conductors of feeders that are not connected to an overcurrent device shall be permitted to be sized at 100 percent of the continuous and noncontinuous load. [215.1(A)(1) Exception No. 3]

E3701.3 Selection of ampacity. Where more than one calculated or tabulated ampacity could apply for a given circuit length, the lowest value shall be used. [310.14(A)(2)]

Exception: Where different ampacities apply to portions of a circuit, the higher ampacity shall be permitted to be used where the total portion(s) of the circuit with the lower ampacity does not exceed the lesser of 10 feet (3048 mm) or 10 percent of the total circuit. [310.14(A)(2) Exception]

E3701.4 Branch circuits with more than one receptacle. Conductors of branch circuits supplying more than one receptacle for cord-and-plug-connected portable loads shall have ampacities of not less than the rating of the branch circuit. [210.19(A)(2)]

E3701.5 Multiwire branch circuits. All conductors for multiwire branch circuits shall originate from the same panelboard or similar distribution equipment. Except where all ungrounded conductors are opened simultaneously by the branch-circuit overcurrent device, multiwire branch circuits shall supply only line-to-neutral loads or only one appliance. [210.4(A) and 210.4(C)]

E3701.5.1 Disconnecting means. Each multiwire branch circuit shall be provided with a means that will simultaneously disconnect all ungrounded conductors at the point where the branch circuit originates. [210.4(B)]

E3701.5.2 Grouping. The ungrounded and grounded circuit conductors of each multiwire branch circuit shall be grouped by wire markers, cable ties or similar means in at least one location within the panelboard or other point of origination. [200.4(B), 210.4(D)]

Exception: Grouping shall not be required where the circuit conductors enter from a cable or raceway unique to the circuit, thereby making the grouping obvious, or where the conductors pass through a box or conduit body without a loop as described in Section E3905.12.2.1 or without a splice or termination. [200.4(B) Exceptions 1 and 2]

SECTION E3702 BRANCH CIRCUIT RATINGS

E3702.1 Branch-circuit voltage limitations. The voltage ratings of branch circuits that supply luminaires or receptacles for cord-and-plug-connected loads of up to 1,400 volt-amperes or of less than $^{1}/_{4}$ horsepower (0.186 kW) shall be limited to a maximum rating of 120 volts, nominal, between conductors.

Branch circuits that supply cord-and-plug-connected or permanently connected utilization equipment and appliances rated at over 1,440 volt-amperes or $^{1}/_{4}$ horsepower (0.186 kW) and greater shall be rated at 120 volts or 240 volts, nominal. [210.6(A), (B), and (C)]

E3702.2 Branch-circuit ampere rating. Branch circuits shall be rated in accordance with the maximum allowable ampere rating or setting of the overcurrent protection device. The rating for other than individual branch circuits shall be 15, 20, 30, 40 and 50 amperes. Where conductors of higher ampacity are used, the ampere rating or setting of the specified over-current device shall determine the circuit rating. (210.18)

E3702.3 Fifteen- and 20-ampere branch circuits. A 15- or 20-ampere branch circuit shall be permitted to supply lighting units, or other utilization equipment, or a combination of both. The rating of any one cord-and-plug-connected utilization equipment not fastened in place shall not exceed 80 percent of the branch-circuit ampere rating. The total rating of utilization equipment fastened in place, other than luminaires, shall not exceed 50 percent of the branch-circuit ampere rating where lighting units, cord-and-plug-connected utilization equipment not fastened in place, or both, are also supplied. [210.23(A)(1) and (2)]

E3702.4 Thirty-ampere branch circuits. A 30-ampere branch circuit shall be permitted to supply fixed utilization equipment. A rating of any one cord-and-plug-connected utilization equipment shall not exceed 80 percent of the branch-circuit ampere rating. [210.23(B)]

E3702.5 Branch circuits serving multiple loads or outlets. General-purpose branch circuits shall supply lighting outlets, appliances, equipment or receptacle outlets, and combinations of such. Multioutlet branch circuits serving lighting or receptacles shall be limited to a maximum branch-circuit rating of 20 amperes. [210.23(A), (B), and (C)]

E3702.6 Branch circuits serving a single motor. Branch-circuit conductors supplying a single motor shall have an ampacity not less than 125 percent of the motor full-load current rating. [430.22(A)]

E3702.7 Branch circuits serving motor-operated and combination loads. For circuits supplying loads consisting of motor-operated utilization equipment that is fastened in place and that has a motor larger than $^1/_8$ horsepower (0.093 kW) in combination with other loads, the total calculated load shall be based on 125 percent of the largest motor load plus the sum of the other loads. [220.18(A)]

E3702.8 Branch-circuit inductive and LED lighting loads. For circuits supplying luminaires having ballasts or LED drivers, the calculated load shall be based on the total ampere ratings of such units and not on the total watts of the lamps. [220.18(B)]

E3702.9 Branch-circuit load for ranges and cooking appliances. It shall be permissible to calculate the branch-circuit load for one range in accordance with Table E3704.2(2). The branch-circuit load for one wall-mounted oven or one counter-mounted cooking unit shall be the nameplate rating of the appliance. The branch-circuit load for a counter-mounted cooking unit and not more than two wall-mounted ovens all supplied from a single branch circuit and located in the same room shall be calculated by adding the nameplate ratings of the individual appliances and treating the total as equivalent to one range. (220.55 Note 4)

E3702.9.1 Minimum branch circuit for ranges. Ranges with a rating of 8.75 kVA or more shall be supplied by a branch circuit having a minimum rating of 40 amperes. [210.19(A)(3)]

E3702.10 Branch circuits serving heating loads. Branch circuits supplying space heating and water heating equipment shall be sized in accordance with the following:

1. The branch circuit conductors and overcurrent protective device supplying fixed storage-type water heaters having a capacity of 120 gallons (450 L) or less shall be sized not less than 125 percent of the water heater rating. [422.13]
2. The branch circuit conductors supplying electric space-heating equipment and any associated motors shall be sized not less than 125 percent of the equipment load. Branch circuits supplying two or more outlets for fixed electric space-heating equipment shall be rated not over 30 amperes. [424.3(A) & (B)]

E3702.11 Branch circuits for air-conditioning and heat pump equipment. The ampacity of the conductors supplying multimotor and combination load equipment shall be not less than the minimum circuit ampacity marked on the equipment. The branch-circuit overcurrent device rating shall be the size and type marked on the appliance. [440.4(B), 440.35]

E3702.12 Branch circuits serving room air conditioners. A room air conditioner shall be considered as a single motor unit in determining its branch-circuit requirements where all the following conditions are met:

1. It is cord- and attachment plug-connected.
2. The rating is not more than 40 amperes and 250 volts; single phase.
3. Total rated-load current is shown on the room air-conditioner nameplate rather than individual motor currents.
4. The rating of the branch-circuit short-circuit and ground-fault protective device does not exceed the ampacity of the branch-circuit conductors, or the rating of the branch-circuit conductors, or the rating of the receptacle, whichever is less. [440.62(A)]

E3702.12.1 Where no other loads are supplied. The total marked rating of a cord- and attachment plug-connected room air conditioner shall not exceed 80 percent of the rating of a branch circuit where no other appliances are also supplied. [440.62(B)]

E3702.12.2 Where lighting units or other appliances are also supplied. The total marked rating of a cord- and attachment plug-connected room air conditioner shall not exceed 50 percent of the rating of a branch circuit where lighting or other appliances are also supplied. Where the circuitry is interlocked to prevent simultaneous operation of the room air conditioner and energization of other outlets on the same branch circuit, a cord- and attachment plug-connected room air conditioner shall not exceed 80 percent of the branch-circuit rating. [440.62(C)]

E3702.13 Electric vehicle branch circuit. Outlets installed for the purpose of charging electric vehicles shall be supplied by an individual branch circuit. Each circuit shall not supply other outlets. Electric vehicle charging loads shall be considered to be continuous loads. [625.40 and 625.42]

E3702.14 Branch-circuit requirement—summary. The requirements for circuits having two or more outlets, or receptacles, other than the receptacle circuits of Sections E3703.2, E3703.3 and E3703.4, are summarized in Table E3702.14. Branch circuits in dwelling units shall supply only loads within that dwelling unit or loads associated only with that dwelling unit. Branch circuits installed for the purpose of lighting, central alarm, signal, communications or other purposes for public or common areas of a two-family dwelling shall not be supplied from equipment that supplies an individual dwelling unit. (210.24 and 210.25)

TABLE E3702.14 (Table 210.24)
BRANCH-CIRCUIT REQUIREMENTS—SUMMARY[a, b]

	CIRCUIT RATING		
	15 amp	20 amp	30 amp
Conductors: Minimum size (AWG) circuit conductors	14	12	10
Maximum overcurrent-protection device rating Ampere rating	15	20	30
Outlet devices: Lampholders permitted Receptacle rating (amperes)	Any type 15 maximum	Any type 15 or 20	NA 30
Maximum load (amperes)	15	20	30

a. These gages are for copper conductors.
b. NA = Not Allowed.

SECTION E3703 REQUIRED BRANCH CIRCUITS

E3703.1 Branch circuits for central heating. Central heating equipment other than fixed electric space heating shall be supplied by an individual branch circuit. Permanently connected air-conditioning equipment, and auxiliary equipment directly associated with the central heating equipment such as pumps, motorized valves, humidifiers and electrostatic air cleaners, shall not be prohibited from connecting to the same branch circuit as the central heating equipment. (422.12 and 422.12 Exceptions No. 1 and No. 2)

E3703.2 Kitchen and dining area receptacles. A minimum of two 20-ampere-rated branch circuits shall be provided for all receptacle outlets specified by Section E3901.3. [210.11(C)(1)]

Exception: Additional receptacle outlets for specific appliances shall be permitted to be supplied from an individual branch circuit rated 15 amperes or greater. [210.52(B)(1) Exception No. 2]

E3703.3 Laundry circuit. A minimum of one 20-ampere-rated branch circuit shall be provided for receptacle outlets required by Section E3901.8 and shall serve only receptacle outlets located in the laundry area. [210.11(C)(2)]

E3703.4 Bathroom branch circuits. A minimum of one 20-ampere branch circuit shall be provided to supply bathroom receptacle outlet(s) required by Section E3901.6 and any countertop or similar work surface receptacle outlets. Such circuits shall have no other outlets. [210.11(C)(3)]

Exception: Where the 20-ampere circuit supplies a single bathroom, outlets for other equipment within the same bathroom shall be permitted to be supplied in accordance with Section E3702. [210.11(C)(3) Exception]

E3703.5 Garage branch circuits. In addition to the number of branch circuits required by other parts of this section, not less than one 120-volt, 20-ampere branch circuit shall be installed to supply receptacle outlets required by Section E3901.9 in attached garages and in detached garages with electric power. This circuit shall not have other outlets.

Exception: This circuit shall be permitted to supply readily accessible outdoor receptacle outlets.

E3703.6 Number of branch circuits. The minimum number of branch circuits shall be determined from the total calculated load and the size or rating of the circuits used. The number of circuits shall be sufficient to supply the load served. In no case shall the load on any circuit exceed the maximum specified by Section E3702. [210.11(A)]

E3703.7 Branch-circuit load proportioning. Where the branch-circuit load is calculated on a volt-amperes-per-square-foot (m^2) basis, the wiring system, up to and including the branch-circuit panelboard(s), shall have the capacity to serve not less than the calculated load. This load shall be evenly proportioned among multioutlet branch circuits within the panelboard(s). Branch-circuit overcurrent devices and circuits shall only be required to be installed to serve the connected load. [210.11(B)]

SECTION E3704 FEEDER REQUIREMENTS

E3704.1 Conductor size. Feeder conductors that do not serve 100 percent of the dwelling unit load and branch-circuit conductors shall be of a size sufficient to carry the load as determined by this chapter. Feeder conductors shall not be required to be larger than the service-entrance conductors that supply the dwelling unit. The load for feeder conductors that serve as the main power feeder to a dwelling unit shall be determined as specified in Chapter 36 for services. [310.12(B) & (C)]

E3704.2 Feeder loads. The minimum load in volt-amperes shall be calculated in accordance with the load calculation procedure prescribed in Table E3704.2(1). The associated table demand factors shall be applied to the actual load to determine the minimum load for feeders. (220.40)

TABLE E3704.2(1)
[220.14(J), Table 220.42, 220.50, 220.51, 220.52, 220.53, 220.54, 220.55, and 220.60]
FEEDER LOAD CALCULATION

LOAD CALCULATION PROCEDURE	APPLIED DEMAND FACTOR
Lighting and receptacles: A unit load of not less than 3 VA per square foot of total floor area shall constitute the lighting and 120-volt, 15- and 20-ampere general use receptacle load. 1,500 VA shall be added for each 20-ampere branch circuit serving receptacles in the kitchen, dining room, pantry, breakfast area and laundry area.	100 percent of first 3,000 VA or less and 35 percent of that in excess of 3,000 VA.
Plus	
Appliances and motors: The nameplate rating load of all fastened-in-place appliances $^{1}/_{4}$ horsepower or greater, or 500 watts or greater, that are fastened in place, other than dryers, ranges, air-conditioning and space-heating equipment.	100 percent of load for three or less appliances. 75 percent of load for four or more appliances.
Plus	
Fixed motors: Full-load current of motors plus 25 percent of the full load current of the largest motor.	
Plus	
Electric clothes dryer: The dryer load shall be 5,000 VA for each dryer circuit or the nameplate rating load of each dryer, whichever is greater.	
Plus	
Cooking appliances: The nameplate rating of ranges, wall-mounted ovens, counter-mounted cooking units and other cooking appliances rated in excess of 1.75 kVA shall be summed.	Demand factors shall be as allowed by Table E3704.2(2).
Plus the largest of either the heating or cooling load	
Largest of the following two selections: 1. 100 percent of the nameplate rating(s) of the air conditioning and cooling, including heat pump compressors. 2. 100 percent of the fixed electric space heating.	

For SI: 1 square foot = 0.0929 m^2.

TABLE E3704.2(2) (220.55 and Table 220.55)
DEMAND LOADS FOR ELECTRIC RANGES, WALL-MOUNTED OVENS, COUNTER-MOUNTED COOKING UNITS AND OTHER COOKING APPLIANCES OVER $1^{3}/_{4}$ kVA RATING[a, b]

NUMBER OF APPLIANCES	MAXIMUM DEMAND[b, c]	DEMAND FACTORS (percent)[d]	
	Column A maximum 12 kVA rating	Column B less than $3^{1}/_{2}$ kVA rating	Column C $3^{1}/_{2}$ to $8^{3}/_{4}$ kVA rating
1	8 kVA	80	80
2	11 kVA	75	65

a. Column A shall be used in all cases except as provided for in Footnote d.

b. For ranges all having the same rating and individually rated more than 12 kVA but not more than 27 kVA, the maximum demand in Column A shall be increased 5 percent for each additional kVA of rating or major fraction thereof by which the rating of individual ranges exceeds 12 kVA.

c. For ranges of unequal ratings and individually rated more than 8.75 kVA, but none exceeding 27 kVA, an average value of rating shall be computed by adding together the ratings of all ranges to obtain the total connected load (using 12 kVA for any ranges rated less than 12 kVA) and dividing by the total number of ranges; and then the maximum demand in Column A shall be increased 5 percent for each kVA or major fraction thereof by which this average value exceeds 12 kVA.

d. Over 1.75 kVA through 8.75 kVA. As an alternative to the method provided in Column A, the nameplate ratings of all ranges rated more than 1.75 kVA but not more than 8.75 kVA shall be added and the sum shall be multiplied by the demand factor specified in Column B or C for the given number of appliances.

E3704.3 Feeder neutral load. The feeder neutral load shall be the maximum unbalance of the load determined in accordance with this chapter. The maximum unbalanced load shall be the maximum net calculated load between the neutral and any one ungrounded conductor. For a feeder or service supplying electric ranges, wall-mounted ovens, counter-mounted cooking units and electric dryers, the maximum unbalanced load shall be considered as 70 percent of the load on the ungrounded conductors. [220.61(A) and (B)]

E3704.4 Lighting and general use receptacle load. A unit load of not less than 3 volt-amperes shall constitute the minimum lighting and general use receptacle load for each square foot of floor area (33 VA for each square meter of floor area). The floor area for each floor shall be calculated from the outside dimensions of the building. The calculated floor area shall not include open porches, garages, or unused or unfinished spaces not adaptable for future use. [220.11 and 220.14(J)]

E3704.5 Ampacity and calculated loads. The calculated load of a feeder shall be not less than the sum of the loads on the branch circuits supplied, as determined by Section E3704, after any applicable demand factors permitted by Section E3704 have been applied. (220.40)

E3704.6 Equipment grounding conductor. Where a feeder supplies branch circuits in which equipment grounding conductors are required, the feeder shall include or provide an equipment grounding conductor that is one or more or a combination of the types specified in Section E3908.9, to which the equipment grounding conductors of the branch circuits shall be connected. Where the feeder supplies a separate building or structure, the requirements of Section E3607.3.1 shall apply. (215.6)

SECTION E3705 CONDUCTOR SIZING AND OVERCURRENT PROTECTION

E3705.1 General. Ampacities for conductors shall be determined based in accordance with Table E3705.1 and Sections E3705.2 and E3705.3. [310.14(A)]

E3705.2 Correction factor for ambient temperatures. For ambient temperatures other than 30°C (86°F), multiply the allowable ampacities specified in Table E3705.1 by the appropriate correction factor shown in Table E3705.2. [310.15(B)(1)]

E3705.3 Adjustment factor for conductor proximity. Where the number of current-carrying conductors in a raceway or cable exceeds three, or where single conductors or multiconductor cables are stacked or bundled for distances greater than 24 inches (610 mm) without maintaining spacing and are not installed in raceways, the allowable ampacity of each conductor shall be reduced as shown in Table E3705.3. [310.15(C)(1)]

Exceptions:

1. Adjustment factors shall not apply to conductors in raceways having a length not exceeding 24 inches (610 mm). [310.15(C)(1)(b)]

TABLE E3705.1
ALLOWABLE AMPACITIES

CONDUCTOR SIZE	CONDUCTOR TEMPERATURE RATING						CONDUCTOR SIZE
	60°C	75°C	90°C	60°C	75°C	90°C	
AWG kcmil	Types TW, UF	Types RHW, THHW, THW, THWN, USE, XHHW	Types RHW-2, THHN, THHW, THW-2, THWN-2, XHHW, XHHW-2, USE-2	Types TW, UF	Types RHW, THHW, THW, THWN, USE, XHHW	Types RHW-2, THHN, THHW, THW-2, THWN-2, XHHW, XHHW-2, USE-2	AWG kcmil
	Copper			Aluminum or copper-clad aluminum			
14[a]	15	20	25	—	—	—	—
12[a]	20	25	30	15	20	25	12[a]
10[a]	30	35	40	25	30	35	10[a]
8	40	50	55	35	40	45	8
6	55	65	75	40	50	55	6
4	70	85	95	55	65	75	4
3	85	100	115	65	75	85	3
2	95	115	130	75	90	100	2
1	110	130	145	85	100	115	1
1/0	125	150	170	100	120	135	1/0
2/0	145	175	195	115	135	150	2/0
3/0	165	200	225	130	155	175	3/0
4/0	195	230	260	150	180	205	4/0
250	215	255	290	170	205	230	250
300	240	285	320	195	230	260	300
350	260	310	350	210	250	280	350
400	280	335	380	225	270	305	400
500	320	380	430	260	310	350	500
600	350	420	475	285	340	385	600
700	385	460	520	315	375	425	700
750	400	475	535	320	385	435	750
800	410	490	555	330	395	445	800
900	435	520	585	355	425	480	900

For SI: °C = [(°F) – 32]/1.8.

a. See Table E3705.5.3 for conductor overcurrent protection limitations.

2. Adjustment factors shall not apply to underground conductors entering or leaving an outdoor trench if those conductors have physical protection in the form of rigid metal conduit, intermediate metal conduit, or rigid nonmetallic conduit having a length not exceeding 10 feet (3048 mm) and the number of conductors does not exceed four. [310.15(C)(1)(c)]
3. Adjustment factors shall not apply to type AC cable or to type MC cable without an overall outer jacket meeting all of the following conditions:
   3.1. Each cable has not more than three current-carrying conductors.
   3.2. The conductors are 12 AWG copper.
   3.3. Not more than 20 current-carrying conductors are bundled, stacked or supported on bridle rings. [310.15(C)(1)(d)]
4. An adjustment factor of 60 percent shall be applied to Type AC cable and Type MC cable where all of the following conditions apply:
   4.1. The cables do not have an overall outer jacket.
   4.2. The number of current-carrying conductors exceeds 20.
   4.3. The cables are stacked or bundled longer than 24 inches (607 mm) without spacing being maintained. [310.15(B)(3)(4) Exception]

TABLE E3705.3 [310.15(C)(1)]
CONDUCTOR PROXIMITY ADJUSTMENT FACTORS

NUMBER OF CURRENT-CARRYING CONDUCTORS IN CABLE OR RACEWAY	PERCENT OF VALUES IN TABLE E3705.1
4–6	80
7–9	70
10–20	50
21–30	45
31–40	40
41 and above	35

E3705.4 Temperature limitations. The temperature rating associated with the ampacity of a conductor shall be so selected and coordinated to not exceed the lowest temperature rating of any connected termination, conductor or device. Conductors with temperature ratings higher than specified for terminations shall be permitted to be used for ampacity adjustment, correction, or both. Except where the equipment is marked otherwise, conductor ampacities used in determining equipment termination provisions shall be based on Table E3705.1. [110.14(C)]

E3705.4.1 Conductors rated 60°C. Except where the equipment is marked otherwise, termination provisions of equipment for circuits rated 100 amperes or less, or marked for 14 AWG through 1 AWG conductors, shall be used only for one of the following:

1. Conductors rated 60°C (140°F);
2. Conductors with higher temperature ratings, provided that the ampacity of such conductors is determined based on the 60°C (140°F) ampacity of the conductor size used;

TABLE E3705.2 [310.15(B)(1)]
AMBIENT TEMPERATURE CORRECTION FACTORS

For ambient temperatures other than 30°C (86°F), multiply the allowable ampacities specified in the ampacity tables by the appropriate correction factor shown below.

Ambient Temperature (°C)	Temperature Rating of Conductor			Ambient Temperature (°F)
	60°C	75°C	90°C	
10 or less	1.29	1.20	1.15	50 or less
11–15	1.22	1.15	1.12	51–59
16–20	1.15	1.11	1.08	60–68
21–25	1.08	1.05	1.04	69–77
26–30	1.00	1.00	1.00	78–86
31–35	0.91	0.94	0.96	87–95
36–40	0.82	0.88	0.91	96–104
41–45	0.71	0.82	0.87	105–113
46–50	0.58	0.75	0.82	114–122
51–55	0.41	0.67	0.76	123–131
56–60	—	0.58	0.71	132–140
61–65	—	0.47	0.65	141–149
66–70	—	0.33	0.58	150–158
71–75	—	—	0.50	159–167
76–80	—	—	0.41	168–176
81–85	—	—	0.29	177–185

For SI: °C = [(°F) – 32]/1.8.

3. Conductors with higher temperature ratings where the equipment is *listed* and identified for use with such conductors; or
4. For motors marked with design letters B, C, or D conductors having an insulation rating of 75°C (167°F) or higher shall be permitted to be used provided that the ampacity of such conductors does not exceed the 75°C (167°F) ampacity. [110.14(C)(1)(a)]

E3705.4.2 Conductors rated 75°C. Termination provisions of equipment for circuits rated over 100 amperes, or marked for conductors larger than 1 AWG, shall be used only for:

1. Conductors rated 75°C (167°F).
2. Conductors with higher temperature ratings provided that the ampacity of such conductors does not exceed the 75°C (167°F) ampacity of the conductor size used, or provided that the equipment is *listed* and identified for use with such conductors. [110.14(C)(1)(b)]

E3705.4.3 Separately installed pressure connectors. Separately installed pressure connectors shall be used with conductors at the ampacities not exceeding the ampacity at the *listed* and identified temperature rating of the connector. [110.14(C)(2)]

E3705.4.4 Conductors of Type NM cable. Conductors in NM cable assemblies shall be rated at 90°C (194°F). Types NM and NMC cable identified by the markings NM-B and NMC-B meet this requirement. The allowable ampacity of Types NM and NMC cable shall not exceed that of 60°C (140°F) rated conductors and shall comply with Section E3705.1 and Table E3705.5.3. The 90°C (194°F) rating shall be permitted to be used for ampacity adjustment and calculations provided that the final corrected or adjusted ampacity does not exceed that for a 60°C (140°F) rated conductor. Where more than two NM cables containing two or more current-carrying conductors are installed, without maintaining spacing between the cables, through the same opening in wood framing that is to be sealed with thermal insulation, caulk or sealing foam, the allowable ampacity of each conductor shall be adjusted in accordance with Table E3705.3 and the provisions of Section E3701.3, Exception, shall not apply. Where more than two NM cables containing two or more current-carrying conductors are installed in contact with thermal insulation without maintaining spacing between cables, the allowable ampacity of each conductor shall be adjusted in accordance with Table E3705.3. (334.80 and 334.112)

E3705.4.5 Conductors of Type SE cable. Where used as a branch circuit or feeder wiring method within the interior of a building and installed in thermal insulation, the ampacity of the conductors in Type SE cable assemblies with ungrounded conductor sizes 10 AWG and smaller shall be in accordance with the 60°C (140°F) conductor temperature rating. The maximum conductor temperature rating shall be permitted to be used for ampacity adjustment and correction purposes, provided that the final derated ampacity does not exceed that for a 60°C (140°F) rated conductor. [338.10(B)(4)(a)]

E3705.5 Overcurrent protection required. All ungrounded branch-circuit and feeder conductors shall be protected against overcurrent by an overcurrent device installed at the point where the conductors receive their supply. Overcurrent devices shall not be connected in series with a grounded conductor. Overcurrent protection and allowable loads for branch circuits and for feeders that do not supply the entire load associated with a one-family dwelling or the entire load associated with an individual dwelling unit in a two-family dwelling shall be in accordance with this chapter.

Branch-circuit conductors and equipment shall be protected by overcurrent protective devices having a rating or setting not exceeding the allowable ampacity specified in Table E3705.1 and Sections E3705.2, E3705.3 and E3705.4 except where otherwise permitted or required in Sections E3705.5.1 through E3705.5.3. [240.4, 240.21, and 310.12(B)]

E3705.5.1 Cords. Cords shall be protected in accordance with Section E3909.2. [240.5(B)]

E3705.5.2 Overcurrent devices of the next higher rating. The next higher standard overcurrent device rating, above the ampacity of the conductors being protected, shall be permitted to be used, provided that all of the following conditions are met:

1. The conductors being protected are not part of a branch circuit supplying more than one receptacle for cord- and plug-connected portable loads.
2. The ampacity of conductors does not correspond with the standard ampere rating of a fuse or a circuit breaker without overload trip adjustments above its rating (but that shall be permitted to have other trip or rating adjustments).
3. The next higher standard device rating does not exceed 400 amperes. [240.4(B)]

E3705.5.3 Small conductors. Except as specifically permitted by Section E3705.5.4, the rating of overcurrent protection devices shall not exceed the ratings shown in Table E3705.5.3 for the conductors specified therein. [240.4(D)]

**TABLE E3705.5.3 [240.4(D)]
OVERCURRENT-PROTECTION RATING**

COPPER		ALUMINUM OR COPPER-CLAD ALUMINUM	
Size (AWG)	Maximum overcurrent-protection-device rating[a] (amps)	Size (AWG)	Maximum overcurrent-protection-device rating[a] (amps)
14	15	12	15
12	20	10	25
10	30	8	30

a. The maximum overcurrent-protection-device rating shall not exceed the conductor allowable ampacity determined by the application of the correction and adjustment factors in accordance with Sections E3705.2 and E3705.3.

E3705.5.4 Air-conditioning and heat pump equipment. Air-conditioning and heat pump equipment circuit conductors shall be permitted to be protected against overcurrent in accordance with Section E3702.11. [240.4(G)]

E3705.6 Fuses and fixed trip circuit breakers. The standard ampere ratings for fuses and inverse time circuit breakers shall be considered to be as shown in Table E3705.6. (240.6)

**TABLE E3705.6 (Table 240.6)
STANDARD APMPERE RATING
FOR FUSES AND INVERSE TIME CIRCUIT BREAKERS**

STANDARD AMPERE RATINGS				
15	20	25	30	35
40	45	50	60	70
80	90	100	110	125
150	175	200	225	250
300	350	400	—	—

E3705.7 Location of overcurrent devices in or on premises. Circuit breakers and switches containing fuses shall:

1. Be readily accessible. [240.24(A)]
2. Not be located where they will be exposed to physical damage. [240.24(C)]
3. Not be located where they will be in the vicinity of easily ignitable material such as in clothes closets. [240.24(D)]
4. Not be located in bathrooms. [240.24(E)]
5. Not be located over steps of a *stairway*. [240.24(F)]
6. Be installed so that the center of the grip of the operating handle of the switch or circuit breaker, when in its highest position, is not more than 6 feet 7 inches (2007 mm) above the floor or working platform. [240.24(A)]

Exceptions:

1. This section shall not apply to supplementary overcurrent protection that is integral to utilization equipment. [240.24(A)(2)]
2. Overcurrent devices installed adjacent to the utilization equipment that they supply shall be permitted to be accessible by portable means. [240.24(A)(4)]

E3705.8 Ready access for occupants. Each occupant shall have ready access to all overcurrent devices protecting the conductors supplying that occupancy. [240.24(B)]

E3705.9 Enclosures for overcurrent devices. Overcurrent devices shall be enclosed in cabinets, cutout boxes, or equipment assemblies. The operating handle of a circuit breaker shall be permitted to be accessible without opening a door or cover. [240.30(A) and (B)]

SECTION E3706
PANELBOARDS

E3706.1 Panelboard rating. All panelboards shall have a rating not less than that of the minimum service or feeder capacity required for the calculated load. (408.30)

E3706.2 Panelboard circuit identification. All circuits and circuit modifications shall be legibly identified as to their clear, evident, and specific purpose or use. The identification shall include an approved degree of detail that allows each circuit to be distinguished from all others. Spare positions that contain unused overcurrent devices or switches shall be described accordingly. The identification shall be included in a circuit directory located on the face, inside of or in an approved location adjacent to the panel door. Circuits shall not be described in a manner that depends on transient conditions of occupancy. [408.4(A)]

E3706.3 Panelboard overcurrent protection. In addition to the requirement of Section E3706.1, a panelboard shall be protected by an overcurrent protective device having a rating not greater than that of the panelboard. Such overcurrent protective device shall be located within or at any point on the supply side of the panelboard. (408.36)

E3706.4 Grounded conductor terminations. Each grounded conductor shall terminate within the panelboard on an individual terminal that is not also used for another conductor, except that grounded conductors of circuits with parallel conductors shall be permitted to terminate on a single terminal where the terminal is identified for connection of more than one conductor. (408.41 and 408.41 Exception)

E3706.5 Back-fed devices. Plug-in-type overcurrent protection devices or plug-in-type main lug assemblies that are back-fed and used to terminate field-installed ungrounded supply conductors shall be secured in place by an additional fastener that requires other than a pull to release the device from the mounting means on the panel. [408.36(D)]

CHAPTER 38

WIRING METHODS

ICC user note:

About this chapter: *Chapter 38 provides installation details for the wiring methods commonly found in dwelling unit construction, and it dictates where and under what conditions specific wiring methods can be used.*

SECTION E3801
GENERAL REQUIREMENTS

E3801.1 Scope. This chapter covers the wiring methods for services, feeders and branch circuits for electrical power and distribution. (300.1)

E3801.2 Allowable wiring methods. The allowable wiring methods for electrical installations shall be those *listed* in Table E3801.2. Single conductors shall be used only where part of one of the recognized wiring methods *listed* in Table E3801.2. As used in this code, abbreviations of the wiring-method types shall be as indicated in Table E3801.2. [110.8, 300.3(A)]

TABLE E3801.2
ALLOWABLE WIRING METHODS

ALLOWABLE WIRING METHOD	DESIGNATED ABBREVIATION
Armored cable	AC
Electrical metallic tubing	EMT
Electrical nonmetallic tubing	ENT
Flexible metal conduit	FMC
Intermediate metal conduit	IMC
Liquidtight flexible conduit	LFC
Metal-clad cable	MC
Nonmetallic sheathed cable	NM
Rigid metallic conduit	RMC
Rigid polyvinyl chloride conduit (Type PVC)	RNC
Reinforced thermosetting resin conduit (Type RTRC)	RTRC
Service entrance cable	SE
Surface raceways	SR
Underground feeder cable	UF
Underground service cable	USE

E3801.3 Circuit conductors. All conductors of a circuit, including equipment grounding conductors and bonding conductors, shall be contained in the same raceway, trench, cable or cord. [300.3(B)]

E3801.4 Wiring method applications. Wiring methods shall be applied in accordance with Table E3801.4. (Chapter 3 and 300.2)

SECTION E3802
ABOVE-GROUND INSTALLATION REQUIREMENTS

E3802.1 Installation and support requirements. Wiring methods shall be installed and supported in accordance with Table E3802.1. (Chapter 3 and 300.11)

E3802.2 Cables in accessible attics. Cables in attics or roof spaces provided with access shall be installed as specified in Sections E3802.2.1 and E3802.2.2. (320.3 and 334.23)

E3802.2.1 Across structural members. Where run across the top of floor joists, or run within 7 feet (2134 mm) of floor or floor joists across the face of rafters or studding, in attics and roof spaces that are provided with access, the cable shall be protected by substantial guard strips that are at least as high as the cable. Where such spaces are not provided with access by permanent stairs or ladders, protection shall only be required within 6 feet (1829 mm) of the nearest edge of the attic entrance. [320.23(A) and 334.23]

E3802.2.2 Cable installed through or parallel to framing members. Where cables are installed through or parallel to the sides of rafters, studs or floor joists, guard strips and running boards shall not be required, and the installation shall comply with Table E3802.1. [300.4(D), 320.23(B) and 334.23]

E3802.3 Exposed cable. In exposed work, except as provided for in Sections E3802.2 and E3802.4, cable assemblies shall be installed as specified in Sections E3802.3.1 and E3802.3.2. (320.15, 330.15 and 334.15)

E3802.3.1 Surface installation. Cables shall closely follow the surface of the building finish or running boards. [334.15(A)]

E3802.3.2 Protection from physical damage. Where subject to physical damage, cables shall be protected by rigid metal conduit, intermediate metal conduit, electrical metallic tubing, Schedule 80 PVC conduit, RTRC-XW or other approved means. Where passing through a floor, the cable shall be enclosed in rigid metal conduit, intermediate metal conduit, electrical metallic tubing, Schedule 80 PVC conduit RTRC-XW or other approved means extending not less than 6 inches (152 mm) above the floor. [334.15(B)]

TABLE E3801.4 (Chapter 3 and 300.2)
ALLOWABLE APPLICATIONS FOR WIRING METHODS[a, b, c, d, e, f, g, h, i, j, k]

ALLOWABLE APPLICATIONS (application allowed where marked with an "A")	AC	EMT	ENT	FMC	IMC RMC RNC RTRC	LFC[a, g]	MC	NM	SR	SE	UF	USE
Services	—	A	A[h]	A[i]	A	A[i]	A	—	—	A	—	A
Feeders	A	A	A	A	A	A	A	A	—	A[b]	A	A[b]
Branch circuits	A	A	A	A	A	A	A	A	A	A[c]	A	—
Inside a building	A	A	A	A	A	A	A	A	A	A	A	—
Wet locations exposed to sunlight	—	A	A[h]	—	A	A	A	—	—	A	A[e]	A[e]
Damp locations	—	A	A	A[d]	A	A	A	—	—	A	A	A
Embedded in noncinder concrete in dry location	—	A	A	—	A	A[j]	—	—	—	—	—	—
In noncinder concrete in contact with grade	—	A[f]	A	—	A[f]	A[j]	—	—	—	—	—	—
Embedded in plaster not exposed to dampness	A	A	A	A	A	A	A	—	—	A	A	—
Embedded in masonry	—	A	A	—	A[f]	A	A	—	—	—	—	—
In masonry voids and cells exposed to dampness or below grade line	—	A[f]	A	A[d]	A[f]	A	A	—	—	A	A	—
Fished in masonry voids	A	—	—	A	—	A	A	A	—	A	A	—
In masonry voids and cells not exposed to dampness	A	A	A	A	A	A	A	A	—	A	A	—
Run exposed	A	A	A	A	A	A	A	A	A	A	A	—
Run exposed and subject to physical damage	—	—	—	—	A[g]	—	—	—	—	—	—	—
For direct burial	—	A[f]	—	—	A[f]	A	A[f]	—	—	—	A	A

For SI: 1 foot = 304.8 mm.

a. Liquidtight flexible nonmetallic conduit without integral reinforcement within the conduit wall shall not exceed 6 feet in length.
b. Type USE cable shall not be used inside buildings.
c. The grounded conductor shall be insulated.
d. Conductors shall be a type approved for wet locations and the installation shall prevent water from entering other raceways.
e. Shall be listed as "Sunlight Resistant."
f. Metal raceways shall be protected from corrosion and approved for the application. Aluminum RMC requires approved supplementary corrosion protection.
g. RNC shall be Schedule 80. RTRC shall be RTRC-XW
h. Shall be listed as "Sunlight Resistant" where exposed to the direct rays of the sun.
i. Conduit shall not exceed 6 feet in length.
j. Liquidtight flexible nonmetallic conduit is permitted to be encased in concrete where listed for direct burial and only straight connectors listed for use with LFNC are used.
k. In wet locations under any of the following conditions:
   1. The metallic covering is impervious to moisture.
   2. A lead sheath or moisture-impervious jacket is provided under the metal covering.
   3. The insulated conductors under the metallic covering are listed for use in wet locations and a corrosion-resistant jacket is provided over the metallic sheath.

E3802.3.3 Locations exposed to direct sunlight. Insulated conductors and cables used where exposed to direct rays of the sun shall be *listed* or *listed* and marked, as being "sunlight resistant," or shall be covered with insulating material, such as tape or sleeving, that is *listed* or *listed* and marked as being "sunlight resistant." [310.10(D)]

E3802.4 In unfinished basements and crawl spaces. Where Type NM or SE cable is run at angles with joists in unfinished basements and crawl spaces, cable assemblies containing two or more conductors of sizes 6 AWG and larger and assemblies containing three or more conductors of sizes 8 AWG and larger shall not require additional protection where attached directly to the bottom of the joists. Smaller cables shall be run either through bored holes in joists or on running boards. Type NM or SE cable installed on the wall of an unfinished basement shall be permitted to be installed in a *listed* conduit or tubing or shall be protected in accordance with Table E3802.1. Conduit or tubing shall be provided with a suitable insulating bushing or adapter at the point where the cable enters the raceway. The sheath of the Type NM or SE cable shall extend through the conduit or tubing and into the outlet or device box not less than $^1/_4$ inch (6.4 mm). The cable shall be secured within 12 inches (305 mm) of the point where the cable enters the conduit or tubing. Metal conduit, tubing, and metal outlet boxes shall be connected to an equipment grounding conductor complying with Section E3908.14. [334.15(C)]

E3802.5 Bends. Bends shall be made so as not to damage the wiring method or reduce the internal diameter of raceways.

For Types NM, NMC and SE cable, bends shall be so made, and other handling shall be such that the cable will not be damaged and the radius of the curve of the inner edge of any bend shall be not less than five times the diameter of the cable. (334.24 and 338.24)

E3802.6 Cable-securing means. Cables shall be supported and secured by staples; cable ties *listed* and identified for securement and support; or straps, hangers or similar fittings designed and installed so as not to damage the cable. [320.30(A), 330.30(A), 334.30, 338.10(B)(4), 340.10(4)]

E3802.7 Raceways exposed to different temperatures. Where portions of a raceway or sleeve are known to be subjected to different temperatures and where condensation is known to be a problem, as in cold storage areas of buildings or where passing from the interior to the exterior of a building, the raceway or sleeve shall be sealed with an approved material to prevent the circulation of warm air to a colder section of the raceway or sleeve. Sealants shall be identified for use with the cable insulation, conductor insulation, a bare conductor, a shield or other components. [300.7(A)]

E3802.8 Raceways in wet locations above grade. Where raceways are installed in wet locations above grade, the inte-

TABLE E3802.1 (Chapter 3)
GENERAL INSTALLATION AND SUPPORT REQUIREMENTS FOR WIRING METHODS[a, b, c, d, e, f, g, h, i, j, k]

INSTALLATION REQUIREMENTS (Requirement applicable only to wiring methods marked "A")	AC MC	EMT IMC RMC	ENT	FMC LFC	NM UF	RNC RTRC	SE	SR[a]	USE
Where run parallel with the framing member or furring strip, the wiring shall be not less than $1^1/_4$ inches from the edge of a furring strip or a framing member such as a joist, rafter or stud or shall be physically protected.	A	—	A	A	A	—	A	—	—
Bored holes in framing members for wiring shall be located not less than $1^1/_4$ inches from the edge of the framing member or shall be protected with a minimum 0.0625-inch steel plate or sleeve, a listed steel plate or other physical protection.	A[k]	—	A[k]	A[k]	A[k]	—	A[k]	—	—
Where installed in grooves, to be covered by wallboard, siding, paneling, carpeting, or similar finish, wiring methods shall be protected by 0.0625-inch-thick steel plate, sleeve, or equivalent, a listed steel plate or by not less than $1^1/_4$-inch free space for the full length of the groove in which the cable or raceway is installed.	A	—	A	A	A	—	A	A	A
Securely fastened bushings or grommets shall be provided to protect wiring run through openings in metal framing members.	—	—	A[j]	—	A[j]	—	A[j]	—	—
The maximum number of 90-degree bends shall not exceed four between junction boxes.	—	A	A	A	—	A	—	—	—
Bushings shall be provided where entering a box, fitting or enclosure unless the box or fitting is designed to afford equivalent protection.	A	A	A	A	—	A	—	A	—
Ends of raceways shall be reamed to remove rough edges.	—	A	A	A	—	A	—	A	—
Maximum allowable on center support spacing for the wiring method in feet.	4.5[b, c]	10	3[b]	4.5[b]	4.5[i]	3[d]	2.5[e]	—	2.5
Maximum support distance in inches from box or other terminations.	12[b, f]	36	36	12[b, g]	12[h, i]	36	12	—	—

For SI: 1 inch = 25.4 mm, 1 foot = 304.8 mm, 1 degree = 0.0175 rad.

a. Installed in accordance with listing requirements.

b. Supports not required in accessible ceiling spaces between light fixtures where lengths do not exceed 6 feet.

c. Six feet for MC cable.

d. Five feet for trade sizes $1^1/_4$ through 2 inches, 6 feet for trade sizes $2\,^1/_2$ through 3, 7 feet for trade sizes $3^1/_2$ through 5 inches, 8 feet for trade size 6 inches.

e. Two and one-half feet where used for service or outdoor feeder and 4.5 feet where used for branch circuit or indoor feeder.

f. Twenty-four inches for Type AC cable and 36 inches for interlocking Type MC cable where flexibility is necessary.

g. Where flexibility after installation is necessary, lengths of flexible metal conduit and liquidtight flexible metal conduit measured from the last point where the raceway is securely fastened shall not exceed: 36 inches for trade sizes $^1/_2$ through $1^1/_4$, 48 inches for trade sizes $1^1/_2$ through 2, and 5 feet for trade sizes $2^1/_2$ and larger.

h. Within 8 inches of boxes without cable clamps.

i. Flat cables shall not be stapled on edge.

j. Bushings and grommets shall remain in place and shall be listed for the purpose of cable protection.

k. See Sections R502.8 and R802.7 for additional limitations on the location of bored holes in horizontal framing members.

rior of such raceways shall be considered to be a wet location. Insulated conductors and cables installed in raceways in wet locations above grade shall be *listed* for use in wet locations. (300.9)

SECTION E3803 UNDERGROUND INSTALLATION REQUIREMENTS

E3803.1 Minimum cover requirements. Direct buried cable or raceways shall be installed in accordance with the minimum cover requirements of Table E3803.1. [300.5(A)]

E3803.2 Warning ribbon. Underground service conductors that are not encased in concrete and that are buried 18 inches (457 mm) or more below grade shall have their location identified by a warning ribbon that is placed in the trench not less than 12 inches (305 mm) above the underground installation. [300.5(D)(3)]

E3803.3 Protection from damage. Direct buried conductors and cables emerging from the ground shall be protected by enclosures or raceways extending from the minimum cover distance below grade required by Section E3803.1 to a point at least 8 feet (2438 mm) above finished grade. In no case shall the protection be required to exceed 18 inches

TABLE E3803.1 (Table 300.5)
MINIMUM COVER REQUIREMENTS, BURIAL IN INCHES[a, b, c, d, e]

LOCATION OF WIRING METHOD OR CIRCUIT	TYPE OF WIRING METHOD OR CIRCUIT				
	1 Direct burial cables or conductors	2 Rigid metal conduit or intermediate metal conduit	3 Nonmetallic raceways listed for direct burial without concrete encasement or other approved raceways	4 Residential branch circuits rated 120 volts or less with GFCI protection and maximum overcurrent protection of 20 amperes	5 Circuits for control of irrigation and landscape lighting limited to not more than 30 volts and installed with Type UF or in other identified cable or raceway
All locations not specified below	24	6	18	12	6[f, g]
In trench below 2-inch-thick concrete or equivalent	18	6	12	6	6
Under a building	0 (In raceway only or Type MC identified for direct burial)	0	0	0 (In raceway only or Type MC identified for direct burial)	0 (In raceway only or Type MC identified for direct burial)
Under minimum of 4-inch-thick concrete exterior slab with no vehicular traffic and the slab extending not less than 6 inches beyond the underground installation	18	4	4	6 (Direct burial) 4 (In raceway)	6 (Direct burial) 4 (In raceway)
Under streets, highways, roads, alleys, driveways and parking lots	24	24	24	24	24
One- and two-family dwelling driveways and outdoor parking areas, and used only for dwelling-related purposes	18	18	18	12	18
In solid rock where covered by minimum of 2 inches concrete extending down to rock	2 (In raceway only)	2	2	2 (In raceway only)	2 (In raceway only)

For SI: 1 inch = 25.4 mm.

a. Raceways approved for burial only where encased concrete shall require concrete envelope not less than 2 inches thick.

b. Lesser depths shall be permitted where cables and conductors rise for terminations or splices or where access is otherwise required.

c. Where one of the wiring method types listed in columns 1 to 3 is combined with one of the circuit types in columns 4 and 5, the shallower depth of burial shall be permitted.

d. Where solid rock prevents compliance with the cover depths specified in this table, the wiring shall be installed in metal or nonmetallic raceway permitted for direct burial. The raceways shall be covered by a minimum of 2 inches of concrete extending down to the rock.

e. Cover is defined as the shortest distance in inches (millimeters) measured between a point on the top surface of any direct-buried conductor, cable, conduit or other raceway and the top surface of finished grade, concrete, or similar cover.

f. A lesser depth shall be permitted where specified in the installation instructions of a listed low-voltage lighting system.

g. A depth of 6 inches shall be permitted for pool, spa, and fountain lighting that is installed in a nonmetallic raceway, limited to not more than 30 volts and part of a listed low-voltage lighting system.

(457 mm) below finished grade. Conductors entering a building shall be protected to the point of entrance. Where the enclosure or raceway is subject to physical damage, the conductors shall be installed in electrical metallic tubing, rigid metal conduit, intermediate metal conduit, Schedule 80 PVC conduit, RTRC-XW or the equivalent. [300.5(D)(1), (2), (3)]

E3803.4 Splices and taps. Direct buried conductors or cables shall be permitted to be spliced or tapped without the use of splice boxes. The splices or taps shall be made by approved methods with materials *listed* for the application. [300.5(E)]

E3803.5 Backfill. Backfill containing large rock, paving materials, cinders, large or sharply angular substances, or corrosive material shall not be placed in an excavation where such materials cause damage to raceways, cables or other substructures or prevent adequate compaction of fill or contribute to corrosion of raceways, cables or other substructures. Where necessary to prevent physical damage to the raceway or cable, protection shall be provided in the form of granular or selected material, suitable boards, suitable sleeves or other approved means. [300.5(F)]

E3803.6 Raceway seals. Conduits or raceways shall be sealed or plugged at either or both ends where moisture will enter and contact live parts. Spare or unused raceways shall also be sealed. Sealants shall be identified for the use with the cable insulation, conductor insulation, bare conductor or other components. [300.5(G)]

E3803.7 Bushing. A bushing, or terminal fitting, with an integral bushed opening shall be installed on the end of a conduit or other raceway that terminates underground where the conductors or cables emerge as a direct burial wiring method. A seal incorporating the physical protection characteristics of a bushing shall be considered equivalent to a bushing. [300.5(H)]

E3803.8 Single conductors. All conductors of the same circuit and, where present, the grounded conductor and all equipment grounding conductors shall be installed in the same raceway or shall be installed in close proximity in the same trench. [300.5(I)]

Exception: Conductors shall be permitted to be installed in parallel in raceways, multiconductor cables, and direct-buried single conductor cables. Each raceway or multi-conductor cable shall contain all conductors of the same circuit, including equipment grounding conductors. Each direct-buried single conductor cable shall be located in close proximity in the trench to the other single conductor cables in the same parallel set of conductors in the circuit, including equipment grounding conductors. [300.5(I) Exception 1]

E3803.9 Earth movement. Where direct buried conductors, raceways or cables are subject to movement by settlement or frost, direct buried conductors, raceways or cables shall be arranged to prevent damage to the enclosed conductors or to equipment connected to the raceways. [300.5(J)]

E3803.10 Wet locations. The interior of enclosures or raceways installed underground shall be considered to be a wet location. Insulated conductors and cables installed in such enclosures or raceways in underground installations shall be *listed* for use in wet locations. [300.5(B)]

E3803.11 Under buildings. Underground cable and conductors installed under a building shall be in a raceway. [300.5(C)]

Exception: Type MC Cable shall be permitted under a building without installation in a raceway where the cable is *listed* and identified for direct burial or concrete encasement and one or more of the following applies:

1. The metallic covering is impervious to moisture.
2. A moisture-impervious jacket is provided under the metal covering.
3. The insulated conductors under the metallic covering are *listed* for use in wet locations. [300.5(C) Exception 2]

CHAPTER 39
POWER AND LIGHTING DISTRIBUTION

ICC user note:

About this chapter: *Chapter 39 addresses the "rough-in" stage of construction in which the wiring system is installed and receptacle and lighting outlets placed throughout the dwelling. This chapter covers receptacle outlet spacing, GFCI (ground-fault circuit-interrupter) and AFCI (arc-fault circuit-interrupter) protection, lighting outlet locations, raceway and box fill limitations, box and panel board installation, equipment grounding and flexible cords.*

SECTION E3901
RECEPTACLE OUTLETS

E3901.1 General. Outlets for receptacles rated at 125 volts, 15- and 20-amperes shall be provided in accordance with Sections E3901.2 through E3901.11. Receptacle outlets required by this section shall be in addition to any receptacle that is:

1. Part of a luminaire or appliance;
2. Located within cabinets or cupboards;
3. Controlled by a listed wall-mounted control device in accordance with Section E3903.2, Exception 1; or
4. Located over 5.5 feet (1676 mm) above the floor.

Permanently installed electric baseboard heaters equipped with factory-installed receptacle outlets, or outlets provided as a separate assembly by the baseboard manufacturer shall be permitted as the required outlet or outlets for the wall space utilized by such permanently installed heaters. Such receptacle outlets shall not be connected to the heater circuits. (210.52)

E3901.2 General purpose receptacle distribution. In every kitchen, family room, dining room, living room, parlor, library, den, *sunroom*, bedroom, recreation room, or similar room or area of dwelling units, receptacle outlets shall be installed in accordance with the general provisions specified in Sections E3901.2.1 through E3901.2.4 (see Figure E3901.2).

E3901.2.1 Spacing. Receptacles shall be installed so that no point measured horizontally along the floor line of any wall space is more than 6 feet (1829 mm), from a receptacle outlet. [210.52(A)(1)]

E3901.2.2 Wall space. As used in this section, a wall space shall include the following: [210.52(A)(2)]

1. Any space that is 2 feet (610 mm) or more in width, including space measured around corners, and that is unbroken along the floor line by doorways and similar openings, fireplaces, and fixed cabinets that do not have countertops or similar work surfaces.
2. The space occupied by fixed panels in exterior walls, excluding sliding panels.
3. The space created by fixed room dividers such as railings and freestanding bar-type counters.

E3901.2.3 Floor receptacles. Receptacle outlets in floors shall not be counted as part of the required number of receptacle outlets except where located within 18 inches (457 mm) of the wall. [210.52(A)(3)]

E3901.2.4 Countertop and similar work surface receptacles outlets. Receptacles installed for countertop and similar work surfaces as specified in Section E3901.4 shall not be considered as the receptacle outlets required by Section E3901.2. [210.52(A)(4)]

E3901.3 Small appliance receptacles. In the kitchen, pantry, breakfast room, dining room, or similar area of a dwelling unit, the two or more 20-ampere small-appliance branch circuits required by Section E3703.2, shall serve all wall and floor receptacle outlets covered by Sections

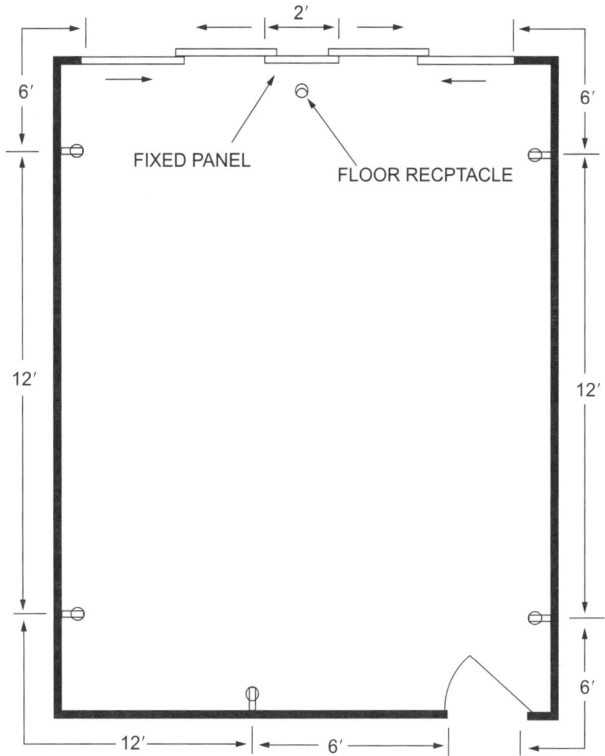

For SI: 1 foot = 304.8 mm.

FIGURE E3901.2
GENERAL USE RECEPTACLE DISTRIBUTION

E3901.2 and E3901.4 and those receptacle outlets provided for refrigeration appliances. [210.52(B)(1)]

Exceptions:

1. In addition to the required receptacles specified by Sections E3901.1 and E3901.2, switched receptacles supplied from a general-purpose 15- or 20-ampere branch circuit as required in Section E3903.2, Exception 1 shall be permitted. [210.52(B)(1) Exception No. 1]
2. In addition to the required receptacles specified by Section E3901.2, a receptacle outlet to serve a specific appliance shall be permitted to be supplied from an individual branch circuit rated at 15 amperes or greater. [210.52(B)(1) Exception No. 2]

E3901.3.1 Other outlets prohibited. The two or more small-appliance branch circuits specified in Section E3901.3 shall serve no other outlets. [210.52(B)(2)]

Exceptions:

1. A receptacle installed solely for the electrical supply to and support of an electric clock in any of the rooms specified in Section E3901.3. [210.52(B)(2) Exception No.1]
2. Receptacles installed to provide power for supplemental equipment and lighting on gas-fired ranges, ovens, and counter-mounted cooking units. [210.52(B)(2) Exception No. 2]

E3901.3.2 Limitations. Receptacles installed in a kitchen to serve countertop surfaces shall be supplied by not less than two small-appliance branch circuits, either or both of which shall also be permitted to supply receptacle outlets in the same kitchen and in other rooms specified in Section E3901.3. Additional small-appliance branch circuits shall be permitted to supply receptacle outlets in the kitchen and other rooms specified in Section E3901.3. A small-appliance branch circuit shall not serve more than one kitchen. [210.52(B)(3)]

E3901.4 Countertop and work surface receptacles. In kitchens pantries, breakfast rooms, dining rooms and similar areas of dwelling units, receptacle outlets for countertop and work surfaces that are 12 inches (305 mm) or wider shall be installed in accordance with Sections E3901.4.1 through E3901.4.3 and shall not be considered as the receptacle outlets required in Section E3901.2.

For the purposes of this section, where using multioutlet assemblies, each 12 inches (305 mm) of multioutlet assembly containing two or more receptacles installed in individual or continuous lengths shall be considered to be one receptacle outlet (see Figure E3901.4). [210.52(C)]

E3901.4.1 Wall spaces. A receptacle outlet shall be installed at each wall countertop and work surface that is 12 inches (305 mm) or wider. Receptacle outlets shall be installed so that no point along the wall line is more than 24 inches (610 mm), measured horizontally, from a receptacle outlet in that space. [210.52(C)(1)]

Exception: Receptacle outlets shall not be required on a wall directly behind a range, counter-mounted cooking unit or sink in the installation described in Figure E3901.4.1. [210.52(C)(1) Exception]

E3901.4.2 Island and peninsular countertops and work spaces. Receptacle outlets shall be installed in accordance with the following: [210.52(C)(2)]

1. At least one receptacle outlet shall be provided for the first 9 square feet (0.84 m^2), or fraction thereof, of the countertop or work surface. A receptacle outlet shall be provided for every additional 18 square feet (1.7 m^2), or fraction thereof, of the countertop or work surface. [210.52(C)(2)(a)]
2. At least one receptacle outlet shall be located within 2 feet (600 mm) of the outer end of a peninsular countertop or work surface. Additional receptacle outlets shall be permitted to be located as determined by the installer, designer or building *owner*. The location of the receptacle outlets shall be in accordance with Section E3901.4.3. [210.52(C)(2)(b)]

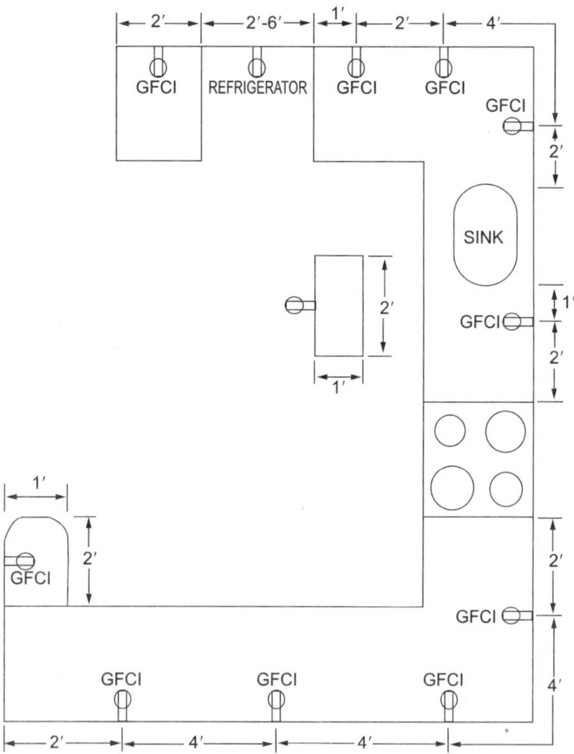

For SI: 1 foot = 304.8 mm.

FIGURE E3901.4
COUNTERTOP RECEPTACLES

A peninsular countertop shall be measured from the connected perpendicular wall. [210.52(C)(2)]

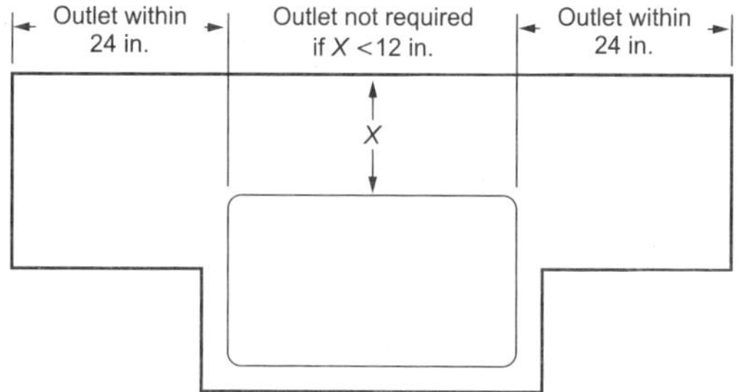

Sink, range or counter-mounted cooking unit extending from face of counter

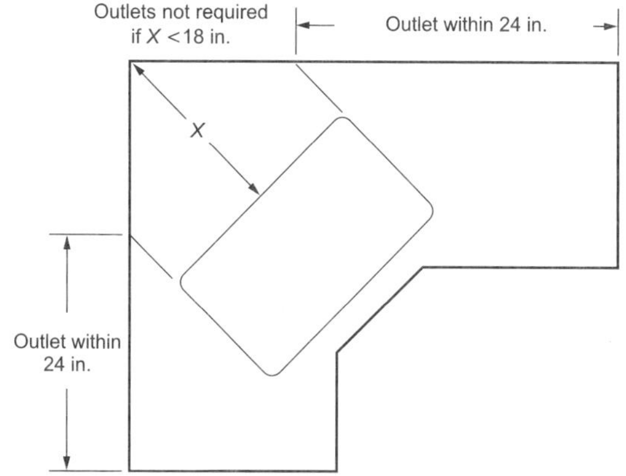

Sink, range or counter-mounted cooking unit mounted in corner

For SI: 1 inch = 25.4 mm.

FIGURE E3901.4.1
DETERMINATION OF AREA BEHIND SINK OR RANGE

E3901.4.3 Receptacle outlet location. Receptacle outlets rendered not readily accessible by appliances fastened in place, appliance garages, sinks, or rangetops as covered in the exception to Section E3901.4.1, or appliances occupying assigned spaces shall not be considered as these required outlets. Required receptacle outlets shall be located in one or more of the following:

1. On or above, but not more than 20 inches (508 mm) above, the countertop or work surface.
2. Receptacle outlet assemblies listed for the use in countertops or work surfaces shall be permitted to be installed in countertops or work surfaces.
3 Not more than 12 inches (305 mm) below the countertop or work surface. Receptacles installed below a countertop or work surface shall not be located where the countertop or work surface extends more than 6 inches (152 mm) beyond its support base. [210.52(C)(3)]

E3901.5 Appliance receptacle outlets. Appliance receptacle outlets installed for specific appliances, such as laundry equipment, shall be installed within 6 feet (1829 mm) of the intended location of the appliance. (210.50(C)]

E3901.6 Bathroom. At least one receptacle outlet shall be installed in bathrooms and such outlet shall be located within 36 inches (914 mm) of the outside edge of each lavatory basin. The receptacle outlet shall be located on a wall or partition that is adjacent to the lavatory basin location, located on the countertop, or installed on the side or face of the basin cabinet. The receptacle shall be located not more than 12 inches (305 mm) below the top of the basin or basin countertop.

Receptacle outlet assemblies installed in countertops shall be *listed* for the application. [210.52(D)]

E3901.7 Outdoor outlets. Not less than one receptacle outlet that is readily accessible from grade level and located not more than 6 feet 6 inches (1981 mm) above grade, shall be installed outdoors at the front and back of each dwelling unit having direct access to grade level. Balconies, decks and porches that are accessible from inside of the dwelling unit shall have at least one receptacle outlet accessible from the balcony, deck or porch. The receptacle shall be located not more than 6 feet 6 inches (1981 mm) above the balcony, deck, or porch surface. [210.52(E)]

E3901.8 Laundry areas. Not less than one receptacle outlet shall be installed in areas designated for the installation of laundry equipment.

E3901.9 Basements, garages and accessory buildings. Not less than one receptacle outlet, in addition to any provided for specific equipment, shall be installed in each separate unfinished portion of a basement; in each vehicle bay not more than 5.5 feet (1676 mm) above the floor in attached garages; in each vehicle bay not more than 5.5 feet (1676 mm) above the floor in detached garages that are provided with electric power and in accessory buildings that are provided with electric power. [210.52(G)(1), (2), and (3)]

E3901.10 Hallways. Hallways of 10 feet (3048 mm) or more in length shall have at least one receptacle outlet. The hall length shall be considered the length measured along the centerline of the hall without passing through a doorway. [210.52(H)]

E3901.11 Foyers. Foyers that are not part of a hallway in accordance with Section E3901.10 and that have an area that is greater than 60 square feet (5.57 m^2) shall have a receptacle(s) located in each wall space that is 3 feet (914 mm) or more in width. Doorways, door-side windows that extend to the floor, and similar openings shall not be considered as wall space. [210.52(H)]

E3901.12 HVAC outlet. A 125-volt, single-phase, 15- or 20-ampere-rated receptacle outlet shall be installed at an accessible location for the servicing of heating, air-conditioning and refrigeration equipment. The receptacle shall be located on the same level and within 25 feet (7620 mm) of the heating, air-conditioning and refrigeration equipment. The receptacle outlet shall not be connected to the load side of the HVAC equipment disconnecting means. (210.63)

Exception: A receptacle outlet shall not be required for the servicing of evaporative coolers. (210.63 Exception)

SECTION E3902 GROUND-FAULT AND ARC-FAULT CIRCUIT-INTERRUPTER PROTECTION

E3902.1 Bathroom receptacles. 125-volt through 250-volt receptacles installed in bathrooms and supplied by single-phase branch circuits rated 150 volts or less to ground shall have ground-fault circuit-interrupter protection for personnel. [210.8(A)(1)]

E3902.2 Garage and accessory building receptacles. 125-volt through 250-volt receptacles installed in garages and grade-level portions of unfinished accessory buildings used for storage or work areas and supplied by single-phase branch circuits rated 150 volts or less to ground shall have ground-fault circuit-interrupter protection for personnel. [210.8(A)(2)]

E3902.3 Outdoor receptacles. 125-volt through 250-volt receptacles installed outdoors and supplied by single-phase branch circuits rated 150 volts or less to ground shall have ground-fault circuit-interrupter protection for personnel. [210.8(A)(3)]

> **Exception:** Receptacles as covered in Section E4101.7. [210.8(A)(3) Exception]

E3902.4 Crawl space receptacles and lighting outlets. Where a *crawl space* is at or below grade level, 125-volt through 250-volt receptacles installed in such spaces and supplied by single-phase branch circuits rated 150 volts or less to ground shall have ground-fault circuit-interrupter protection for personnel. Lighting outlets not exceeding 120 volts shall have ground-fault circuit-interrupter protection. [210.8(A)(4), 2108(E)]

E3902.5 Basement receptacles. 125-volt through 250-volt receptacles installed in basements and supplied by single-phase branch circuits rated 150 volts or less to ground shall have ground-fault circuit-interrupter protection for personnel. [210.8(A)(5)]

> **Exception:** A receptacle supplying only a permanently installed fire alarm or burglar alarm system. A receptacle installed in accordance with this exception shall not be considered as meeting the requirement of Section E3901.9. Receptacles installed in accordance with this exception shall not be considered as meeting the requirement of Section E3901.9. [210.8(A)(5) Exception]

E3902.6 Kitchen receptacles. 125-volt through 250-volt receptacles that serve countertop surfaces and are supplied by single-phase branch circuits rated 150 volts or less to ground shall have ground-fault circuit-interrupter protection for personnel. [210.8(A)(6)]

E3902.7 Sink receptacles. 125-volt through 250-volt receptacles that are located within 6 feet (1829 mm) of the top inside edge of the bowl of the sink and supplied by single-phase branch circuits rated 150 volts or less to ground shall have ground-fault circuit-interrupter protection for personnel. [210.8(A)(7)]

E3902.8 Bathtub or shower stall receptacles. 125-volt through 250-volt receptacles that are located within 6 feet (1829 mm) of the outside edge of a bathtub or shower stall and supplied by single-phase branch circuits rated 150 volts or less to ground shall have ground-fault circuit-interrupter protection for personnel. [210.8(A)(9)]

E3902.9 Laundry areas. 125-volt through 250-volt receptacles installed in laundry areas and supplied by single-phase branch circuits rated 150 volts or less to ground shall have ground-fault circuit-interrupter protection for personnel. [210.8(A)(10)]

E3902.10 Indoor damp and wet locations. 125-volt through 250-volt receptacles installed in indoor damp and wet locations and supplied by single-phase branch circuits rated 150 volts or less to ground shall have ground-fault circuit-interrupter protection for personnel. [210.8(A)(11)]

E3902.11 Kitchen dishwasher branch circuit. Ground-fault circuit-interrupter protection shall be provided for outlets supplied by branch circuits rated 150 volts or less to ground that supply dishwashers in dwelling unit locations. [422.5 (A)]

E3902.12 Boathouse receptacles. 125-volt through 250-volt receptacles installed in boathouses and supplied by single-phase branch circuits rated 150 volts or less to ground shall have ground-fault circuit-interrupter protection for personnel. [210.8(A)(8)]

E3902.13 Boat hoists. Ground-fault circuit-interrupter protection for personnel shall be provided for 240-volt and less outlets that supply boat hoists. [555.9]

E3902.14 Electrically heated floors. Ground-fault circuit-interrupter protection for personnel shall be provided for electric heating cables embedded in concrete or poured masonry floors in bathrooms, kitchens and in hydromassage bathtub, spa and hot tub locations. Heating cables installed under floor coverings shall be provided with ground-fault circuit-interrupter protection for personnel. [424.44(E), 424.45(E)]

E3902.15 Location of ground-fault circuit interrupters. Ground-fault circuit interrupters shall be installed in a readily accessible location. When determining distance from receptacles, the distance shall be measured as the shortest path the supply cord of an appliance connected to the receptacle would follow without piercing a floor, wall, ceiling, or fixed barrier, or the shortest path without passing through a window. [210.8(A)]

E3902.16 Location of arc-fault circuit interrupters. Arc-fault circuit interrupters shall be installed in readily accessible locations.

E3902.17 Arc-fault circuit interrupter protection. Branch circuits that supply 120-volt, single-phase, 15- and 20-ampere outlets installed in kitchens, family rooms, dining rooms, living rooms, parlors, libraries, dens, bedrooms, *sunrooms*, recreations rooms, closets, hallways, laundry areas and similar rooms or areas shall be protected by any of the following: [210.12(A)]

1. A *listed* combination-type arc-fault circuit interrupter, installed to provide protection of the entire branch circuit. [210.12(A)(1)]
2. A *listed* branch/feeder-type AFCI installed at the origin of the branch-circuit in combination with a

listed outlet branch-circuit-type arc-fault circuit interrupter installed at the first outlet box on the branch circuit. The first outlet box in the branch circuit shall be marked to indicate that it is the first outlet of the circuit. [210.12(A)(2)]

3. A *listed* supplemental arc-protection circuit breaker installed at the origin of the branch circuit in combination with a *listed* outlet branch-circuit-type arc-fault circuit-interrupter installed at the first outlet box on the branch circuit where all of the following conditions are met:
   3.1. The branch-circuit wiring shall be continuous from the branch-circuit overcurrent device to the outlet branch-circuit arc-fault circuit interrupter.
   3.2. The maximum length of the branch-circuit wiring from the branch-circuit overcurrent device to the first outlet shall not exceed 50 feet (15.2 m) for 14 AWG conductors and 70 feet (21.3 m) for 12 AWG conductors.
   3.3. The first outlet box on the branch circuit shall be marked to indicate that it is the first outlet on the circuit. [210.12(A)(3)]
4. A *listed* outlet branch-circuit-type arc-fault circuit interrupter installed at the first outlet on the branch circuit in combination with a *listed* branch-circuit overcurrent protective device where all of the following conditions are met:
   4.1. The branch-circuit wiring shall be continuous from the branch-circuit overcurrent device to the outlet branch-circuit arc-fault circuit interrupter.
   4.2. The maximum length of the branch-circuit wiring from the branch-circuit overcurrent device to the first outlet shall not exceed 50 feet (15.2 m) for 14 AWG conductors and 70 feet (21.3 m) for 12 AWG conductors.
   4.3. The first outlet box on the branch circuit shall be marked to indicate that it is the first outlet on the circuit.
   4.4. The combination of the branch-circuit overcurrent device and outlet branch-circuit AFCI shall be identified as meeting the requirements for a system combination-type AFCI and shall be *listed* as such. [210.12(A)(4)]
5. Where metal raceways, metal wireways, metal auxiliary gutters or Type MC or Type AC cable meeting the applicable requirements of Section E3908.9 with metal boxes, metal conduit bodies and metal enclosures are installed for the portion of the branch circuit between the branch-circuit overcurrent device and the first outlet, a *listed* outlet branch-circuit type AFCI installed at the first outlet shall be considered as providing protection for the remaining portion of the branch circuit. [210.12(A)(5)]
6. Where a *listed* metal or nonmetallic conduit or tubing or Type MC cable is encased in not less than 2 inches (50.8 mm) of concrete for the portion of the branch circuit between the branch-circuit overcurrent device and the first outlet, a *listed* outlet branch-circuit-type AFCI installed at the first outlet shall be considered as providing protection for the remaining portion of the branch circuit. [210.12(A)(6)]

Exception: AFCI protection shall not be required for an individual branch circuit supplying a fire alarm system where the branch circuit is installed in a metal raceway, metal auxiliary gutter, steel-armored cable, Type MC or Type AC, meeting the requirements of Section E3908.9, with metal boxes, conduit bodies and enclosures.

E3902.18 Arc-fault circuit-interrupter protection for branch circuit extensions or modifications. Where branch-circuit wiring is modified, replaced, or extended in any of the areas specified in Section E3902.17, the branch circuit shall be protected by one of the following:

1. A combination-type AFCI located at the origin of the branch circuit.
2. An outlet branch-circuit type AFCI located at the first receptacle outlet of the existing branch circuit. [210.12(B)]

Exception: AFCI protection shall not be required where the extension of the existing branch circuit conductors is not more than 6 feet (1.8 m) in length and does not include any additional outlets or devices other than splicing devices. This measurement shall not include the conductors inside an enclosure, cabinet, or junction box. [210.12(B) Exception]

SECTION E3903
LIGHTING OUTLETS

E3903.1 General. Lighting outlets shall be provided in accordance with Sections E3903.2 through E3903.4. [210.70(A)]

E3903.2 Habitable rooms. At least one lighting outlet controlled by a *listed* wall-mounted control device shall be installed in every habitable room, kitchen and bathroom. The wall-mounted control device shall be located near an entrance to the room on a wall. [210.70(A)(1)]

Exceptions:

1. In other than kitchens and bathrooms, one or more receptacles controlled by a *listed* wall-mounted control device shall be considered equivalent to the required lighting outlet. [210.70(A)(1) Exception No. 1]
2. Lighting outlets shall be permitted to be controlled by occupancy sensors that are in addition to *listed* wall-mounted control devices or that are located at a customary wall switch location and equipped with a manual override that will allow the sensor

to function as a wall switch. [210.70(A)(1) Exception No. 2]

E3903.3 Additional locations. At least one lighting outlet controlled by a *listed* wall-mounted control device shall be installed in hallways, *stairways*, attached garages, and detached garages with electric power. At least one lighting outlet controlled by a *listed* wall-mounted control device shall be installed to provide illumination on the exterior side of each outdoor egress door having grade level access, including outdoor egress doors for attached garages and detached garages with electric power. A vehicle door in a garage shall not be considered as an outdoor egress door.

E3903.3.1 Stairway lighting outlet control. Where one or more lighting outlets are installed for interior *stairways*, there shall be a *listed* wall-mounted control device at each floor level and landing level that includes an entryway to control the lighting outlets where the *stairway* between floor levels has six or more risers. Lighting outlets installed to meet this requirement shall not be controlled by the use of *listed* wall-mounted control devices except where the devices provide the full range of dimming control at each switch location. [210.70(A)(2)]

Exception: In hallways, *stairways*, and at outdoor egress doors, remote, central, or automatic control of lighting shall be permitted. [210.70(A)(2) Exception]

E3903.4 Storage or equipment spaces. In attics, underfloor spaces, utility rooms and basements, at least one lighting outlet shall be installed where these spaces are used for storage or contain equipment requiring servicing. Such lighting outlet shall be controlled by a wall switch or *listed* wall-mounted control device. A point of control shall be provided at each entry that permits access to the attic or underfloor space, utility room, or basement. Where a lighting outlet is installed for equipment requiring service, the lighting outlet shall be installed at or near the equipment requiring servicing. [210.70(C)]

SECTION E3904
GENERAL INSTALLATION REQUIREMENTS

E3904.1 Electrical continuity of metal raceways and enclosures. Metal raceways, cable armor and other metal enclosures for conductors shall be mechanically joined together into a continuous electric conductor and shall be connected to all boxes, fittings and cabinets so as to provide effective electrical continuity. Raceways and cable assemblies shall be mechanically secured to boxes, fittings cabinets and other enclosures. (300.10)

Exception: Short sections of raceway used to provide cable assemblies with support or protection against physical damage. (300.10 Exception No. 1)

E3904.2 Mechanical continuity—raceways and cables. Raceways, cable armors and cable sheaths shall be continuous between cabinets, boxes, fittings or other enclosures or outlets. (300.12)

Exception: Short sections of raceway used to provide cable assemblies with support or protection against physical damage. (300.12 Exception No. 1)

E3904.3 Securing and supporting. Raceways, cable assemblies, boxes, cabinets and fittings shall be securely fastened in place. (300.11)

E3904.3.1 Prohibited means of support. Cable wiring methods shall not be used as a means of support for other cables, raceways and nonelectrical equipment. [300.11(D)]

E3904.4 Raceways as means of support. Raceways shall be used as a means of support for other raceways, cables or nonelectric equipment only under the following conditions:

1. Where the raceway or means of support is identified as a means of support; or
2. Where the raceway contains power supply conductors for electrically controlled equipment and is used to support Class 2 circuit conductors or cables that are solely for the purpose of connection to the control circuits of the equipment served by such raceway; or
3. Where the raceway is used to support boxes or conduit bodies in accordance with Sections E3906.8.4 and E3906.8.5. [300.11(C)]

E3904.5 Raceway installations. Raceways shall be installed complete between outlet, junction or splicing points prior to the installation of conductors. (300.18)

Exception: Short sections of raceways used to contain conductors or cable assemblies for protection from physical damage shall not be required to be installed complete between outlet, junction, or splicing points. (300.18 Exception)

E3904.6 Conduit and tubing fill. The maximum number of conductors installed in conduit or tubing shall be in accordance with Tables E3904.6(1) through E3904.6(10). (300.17, Chapter 9, Table 1 and Annex C)

E3904.7 Air handling-stud cavity and joist spaces. Where wiring methods having a nonmetallic covering pass through stud cavities and joist spaces used for air handling, such wiring shall pass through such spaces perpendicular to the long dimension of the spaces. [300.22(C) Exception]

TABLE E3904.6(1) (Annex C, Table C.1)
MAXIMUM NUMBER OF CONDUCTORS IN ELECTRICAL METALLIC TUBING (EMT)[a]

TYPE LETTERS	CONDUCTOR SIZE AWG/kcmil	TRADE SIZES (inches)					
		1/2	3/4	1	1 1/4	1 1/2	2
RHH, RHW, RHW-2	14	4	7	11	20	27	46
	12	3	6	9	17	23	38
	10	2	5	8	13	18	30
	8	1	2	4	7	9	16
	6	1	1	3	5	8	13
	4	1	1	2	4	6	10
	3	1	1	1	4	5	9
	2	1	1	1	3	4	7
	1	0	1	1	1	3	5
	1/0	0	1	1	1	2	4
	2/0	0	1	1	1	2	4
	3/0	0	0	1	1	1	3
	4/0	0	0	1	1	1	3
TW, THHW, THW, THW-2	14	8	15	25	43	58	96
	12	6	11	19	33	45	74
	10	5	8	14	24	33	55
	8	2	5	8	13	18	30
RHH[a], RHW[a], RHW-2[a]	14	6	10	16	28	39	64
	12	4	8	13	23	31	51
	10	3	6	10	18	24	40
	8	1	4	6	10	14	24
RHH[a], RHW[a], RHW-2[a], TW, THW, THHW, THW-2	6	1	3	4	8	11	18
	4	1	1	3	6	8	13
	3	1	1	3	5	7	12
	2	1	1	2	4	6	10
	1	1	1	1	3	4	7
	1/0	0	1	1	2	3	6
	2/0	0	1	1	1	3	5
	3/0	0	1	1	1	2	4
	4/0	0	0	1	1	1	3
THHN, THWN, THWN-2	14	12	22	35	61	84	138
	12	9	16	26	45	61	101
	10	5	10	16	28	38	63
	8	3	6	9	16	22	36
	6	2	4	7	12	16	26
	4	1	2	4	7	10	16
	3	1	1	3	6	8	13
	2	1	1	3	5	7	11
	1	1	1	1	4	5	8
	1/0	1	1	1	3	4	7
	2/0	0	1	1	2	3	6
	3/0	0	1	1	1	3	5
	4/0	0	1	1	1	2	4
XHH, XHHW, XHHW-2	14	8	15	25	43	58	96
	12	6	11	19	33	45	74
	10	5	8	14	24	33	55
	8	2	5	8	13	18	30
	6	1	3	6	10	14	22
	4	1	2	4	7	10	16
	3	1	1	3	6	8	14
	2	1	1	3	5	7	11
	1	1	1	1	4	5	8
	1/0	1	1	1	3	4	7
	2/0	0	1	1	2	3	6
	3/0	0	1	1	1	3	5
	4/0	0	1	1	1	2	4

For SI: 1 inch = 25.4 mm.
a. Types RHW and RHW-2 without outer covering.

TABLE E3904.6(2) (Annex C, Table C.2)
MAXIMUM NUMBER OF CONDUCTORS IN ELECTRICAL NONMETALLIC TUBING (ENT)[a]

TYPE LETTERS	CONDUCTOR SIZE AWG/kcmil	TRADE SIZES (inches)					
		1/2	3/4	1	1 1/4	1 1/2	2
RHH, RHW, RHW-2	14	3	6	10	19	26	43
	12	2	5	9	16	22	36
	10	1	4	7	13	17	29
	8	1	1	3	6	9	15
	6	1	1	3	5	7	12
	4	1	1	2	4	6	9
	3	1	1	1	3	5	8
	2	0	1	1	3	4	7
	1	0	1	1	1	3	5
	1/0	0	0	1	1	2	4
	2/0	0	0	1	1	1	3
	3/0	0	0	1	1	1	3
	4/0	0	0	1	1	1	2
TW, THHW, THW, THW-2	14	7	13	22	40	55	92
	12	5	10	17	31	42	71
	10	4	7	13	23	32	52
	8	1	4	7	13	17	29
RHH[a], RHW[a], RHW-2[a]	14	4	8	15	27	37	61
	12	3	7	12	21	29	49
	10	3	5	9	17	23	38
	8	1	3	5	10	14	23
RHH[a], RHW[a], RHW-2[a], TW, THW, THHW, THW-2	6	1	2	4	7	10	17
	4	1	1	3	5	8	13
	3	1	1	2	5	7	11
	2	1	1	2	4	6	9
	1	0	1	1	3	4	6
	1/0	0	1	1	2	3	5
	2/0	0	1	1	1	3	5
	3/0	0	0	1	1	2	4
	4/0	0	0	1	1	1	3
THHN, THWN, THWN-2	14	10	18	32	58	80	132
	12	7	13	23	42	58	96
	10	4	8	15	26	36	60
	8	2	5	8	15	21	35
	6	1	3	6	11	15	25
	4	1	1	4	7	9	15
	3	1	1	3	5	8	13
	2	1	1	2	5	6	11
THHN, THWN, THWN-2	1	1	1	1	3	5	8
	1/0	0	1	1	3	4	7
	2/0	0	1	1	2	3	5
	3/0	0	1	1	1	3	4
	4/0	0	0	1	1	2	4
XHH, XHHW, XHHW-2	14	7	13	22	40	55	92
	12	5	10	17	31	42	71
	10	4	7	13	23	32	52
	8	1	4	7	13	17	29
	6	1	3	5	9	13	21
	4	1	1	4	7	9	15
	3	1	1	3	6	8	13
	2	1	1	2	5	6	11
	1	1	1	1	3	5	8
	1/0	0	1	1	3	4	7
	2/0	0	1	1	2	3	6
	3/0	0	1	1	1	3	5
	4/0	0	0	1	1	2	4

For SI: 1 inch = 25.4 mm.

a. Types RHW and RHW-2 without outer covering.

TABLE E3904.6(3) (Annex C, Table C.3)
MAXIMUM NUMBER OF CONDUCTORS IN FLEXIBLE METALLIC CONDUIT (FMC)[a]

TYPE LETTERS	CONDUCTOR SIZE AWG/kcmil	TRADE SIZES (inches)					
		1/2	3/4	1	1 1/4	1 1/2	2
RHH, RHW, RHW-2	14	3	7	11	17	25	44
	12	3	6	9	14	21	37
	10	14	5	7	11	17	30
	8	1	2	4	6	9	15
	6	1	1	3	5	7	12
	4	1	1	2	4	5	10
	3	1	1	1	3	5	7
	2	1	1	1	3	4	7
	1	0	1	1	1	2	5
	1/0	0	1	1	1	2	4
	2/0	0	1	1	1	1	3
	3/0	0	0	1	1	1	3
TW, THHW, THW, THW-2	14	9	15	23	36	53	94
	12	7	11	18	28	41	72
	10	5	8	13	21	30	54
	8	3	5	7	11	17	30
RHH[a], RHW[a], RHW-2[a]	14	6	10	15	24	35	62
	12	5	8	12	19	28	50
	10	4	6	10	15	22	39
	8	1	4	6	9	13	23
RHH[a], RHW[a], RHW-2[a], TW, THW, THHW, THW-2	6	1	3	4	7	10	18
	4	1	1	3	5	7	13
	3	1	1	3	4	6	11
	2	1	1	2	4	5	10
	1	1	1	1	2	4	7
	1/0	0	1	1	1	3	6
	2/0	0	1	1	1	3	5
	3/0	0	1	1	1	2	4
	4/0	0	0	1	1	1	3
	4/0	0	0	1	1	1	2
THHN, THWN, THWN-2	14	13	22	33	52	76	134
	12	9	16	24	38	56	98
	10	6	10	15	24	35	62
	8	3	6	9	14	20	35
THHN, THWN, THWN-2	6	2	4	6	10	14	25
	4	1	2	4	6	9	16
	3	1	1	3	5	7	13
	2	1	1	3	4	6	11
	1	1	1	1	3	4	8
	1/0	1	1	1	2	4	7
	2/0	0	1	1	1	3	6
	3/0	0	1	1	1	2	5
	4/0	0	1	1	1	1	4
XHH, XHHW, XHHW-2	14	9	15	23	36	53	94
	12	7	11	18	28	41	72
	10	5	8	13	21	30	54
	8	3	5	7	11	17	30
	6	1	3	5	8	12	22
	4	1	2	4	6	9	16
	3	1	1	3	5	7	13
	2	1	1	3	4	6	11
	1	1	1	1	3	5	8
	1/0	1	1	1	2	4	7
	2/0	0	1	1	2	3	6
	3/0	0	1	1	1	3	5
	4/0	0	1	1	1	2	4

For SI: 1 inch = 25.4 mm.

a. Types RHW and RHW-2 without outer covering.

TABLE E3904.6(4) (Annex C, Table C.4) MAXIMUM NUMBER OF CONDUCTORS IN INTERMEDIATE METALLIC CONDUIT (IMC)[a]

TYPE LETTERS	CONDUCTOR SIZE AWG/kcmil	TRADE SIZES (inches)					
		1/2	3/4	1	1 1/4	1 1/2	2
RHH, RHW, RHW-2	14	4	8	13	22	30	49
	12	4	6	11	18	25	41
	10	3	5	8	15	20	33
	8	1	3	4	8	10	17
	6	1	1	3	6	8	14
	4	1	1	3	5	6	11
	3	1	1	2	4	6	9
	2	1	1	1	3	5	8
	1	0	1	1	2	3	5
	1/0	0	1	1	1	3	4
	2/0	0	1	1	1	2	4
	3/0	0	0	1	1	1	3
	4/0	0	0	1	1	1	3
TW, THHW, THW, THW-2	14	10	17	27	47	64	104
	12	7	13	21	36	49	80
	10	5	9	15	27	36	59
	8	3	5	8	15	20	33
RHH[a], RHW[a], RHW-2[a]	14	6	11	18	31	42	69
	12	5	9	14	25	34	56
	10	4	7	11	19	26	43
	8	2	4	7	12	16	26
RHH[a], RHW[a], RHW-2[a], TW, THW, THHW, THW-2	6	1	3	5	9	12	20
	4	1	2	4	6	9	15
	3	1	1	3	6	8	13
	2	1	1	3	5	6	11
	1	1	1	1	3	4	7
	1/0	1	1	1	3	4	6
	2/0	0	1	1	2	3	5
	3/0	0	1	1	1	3	4
	4/0	0	1	1	1	2	4
THHN, THWN, THWN-2	14	14	24	39	68	91	149
	12	10	17	29	49	67	109
	10	6	11	18	31	42	68
	8	3	6	10	18	24	39
	6	2	4	7	13	17	28
	4	1	3	4	8	10	17
	3	1	2	4	6	9	15
	2	1	1	3	5	7	12
	1	1	1	2	4	5	9
	1/0	1	1	1	3	4	8
	2/0	1	1	1	3	4	6
THHN, THWN, THWN-2	3/0	0	1	1	2	3	5
	2/0	0	1	1	1	2	4
XHH, XHHW, XHHW-2	14	10	17	27	47	64	104
	12	7	13	21	36	49	80
	10	5	9	15	27	36	59
	8	3	5	8	15	20	33
	6	1	4	6	11	15	24
	4	1	3	4	8	11	18
	3	1	2	4	7	9	15
	2	1	1	3	5	7	12
	1	1	1	2	4	5	9
	1/0	1	1	1	3	5	8
	2/0	1	1	1	3	4	6
	3/0	0	1	1	2	3	5
	4/0	0	1	1	1	2	4

For SI: 1 inch = 25.4 mm.

a. Types RHW and RHW-2 without outer covering.

TABLE E3904.6(5) (Annex C, Table C.5)
MAXIMUM NUMBER OF CONDUCTORS IN LIQUID-TIGHT FLEXIBLE NONMETALLIC CONDUIT (FNMC-B)[a]

TYPE LETTERS	CONDUCTOR SIZE AWG/kcmil	TRADE SIZES (inches)						
		3/8	1/2	3/4	1	1 1/4	1 1/2	2
RHH, RHW, RHW-2	14	2	4	7	12	21	27	44
	12	1	3	6	10	17	22	36
	10	1	3	5	8	14	18	29
	8	1	1	2	4	7	9	1
	6	1	1	1	3	6	7	12
	4	0	1	1	2	4	6	9
	3	0	1	1	1	4	5	8
	2	0	1	1	1	3	4	7
	1	0	0	1	1	1	3	5
	1/0	0	0	1	1	1	2	4
	2/0	0	0	1	1	1	1	3
	3/0	0	0	0	1	1	1	3
	4/0	0	0	0	1	1	1	2
TW, THHW, THW, THW-2	14	5	9	15	25	44	57	93
	12	4	7	12	19	33	43	71
	10	3	5	9	14	25	32	53
	8	1	3	5	8	14	18	29
RHH[a], RHW[a], RGW-2[a]	14	3	6	10	16	29	38	62
	12	3	5	8	13	23	30	50
	10	1	3	6	10	18	23	39
	8	1	1	4	6	11	14	23
RHH[a], RHW[a], RHW-2[a],TW, THW, THHW, THW-2	6	1	1	3	5	8	11	18
	4	1	1	1	3	6	8	13
	3	1	1	1	3	5	7	11
	2	0	1	1	2	4	6	9
	1	0	1	1	1	3	4	7
	1/0	0	0	1	1	2	3	6
	2/0	0	0	1	1	2	3	5
	3/0	0	0	1	1	1	2	4
	4/0	0	0	0	1	1	1	3
THHN, THWN, THWN-2	14	8	13	22	36	63	81	133
	12	5	9	16	26	46	59	97
	10	3	6	10	16	29	37	61
	8	1	3	6	9	16	21	35
THHN, THWN, THWN-2	6	1	2	4	7	12	15	25
	4	1	1	2	4	7	9	15
	3	1	1	1	3	6	8	13
	2	1	1	1	3	5	7	11
	1	0	1	1	1	4	5	8
	1/0	0	1	1	1	3	4	7
	2/0	0	0	1	1	2	3	6
	3/0	0	0	1	1	1	3	5
	4/0	0	0	1	1	1	2	4
XHH, XHHW, XHHW-2	14	5	9	15	25	44	57	93
	12	4	7	12	19	33	43	71
	10	3	5	9	14	25	32	53
	8	1	3	5	8	14	18	29
	6	1	1	3	6	10	13	22
	4	1	1	2	4	7	9	16
	3	1	1	1	3	6	8	13
	2	1	1	1	3	5	7	11
	1	0	1	1	1	4	5	8
	1/0	0	1	1	1	3	4	7
	2/0	0	0	1	1	2	3	6
	3/0	0	0	1	1	1	3	5
	4/0	0	0	1	1	1	2	4

For SI: 1 inch = 25.4 mm.
a. Types RHW and RHW-2 without outer covering.

TABLE E3904.6(6) (Annex C, Table C.6)
MAXIMUM NUMBER OF CONDUCTORS IN LIQUID-TIGHT FLEXIBLE NONMETALLIC CONDUIT (FNMC-A)[a]

TYPE LETTERS	CONDUCTOR SIZE AWG/kcmil	TRADE SIZES (inches)						
		3/8	1/2	3/4	1	1 1/4	1 1/2	2
RHH, RHW, RHW-2	14	2	4	7	11	20	27	45
	12	1	3	6	9	17	23	38
	10	1	3	5	8	13	18	30
	8	1	1	2	4	7	9	16
	6	1	1	1	3	5	7	13
	4	0	1	1	2	4	6	10
	3	0	1	1	1	4	5	8
	2	0	1	1	1	3	4	7
	1	0	0	1	1	1	3	5
	1/0	0	0	1	1	1	2	4
	2/0	0	0	1	1	1	1	4
	3/0	0	0	0	1	1	1	3
	4/0	0	0	0	1	1	1	3
TW, THHW, THW, THW-2	14	5	9	15	24	43	58	96
	12	4	7	12	19	33	44	74
	10	3	5	9	14	24	33	55
	8	1	3	5	8	13	18	30
RHH[a], RHW[a], RHW-2[a]	14	3	6	10	16	28	38	64
	12	3	4	8	13	23	31	51
	10	1	3	6	10	18	24	40
	8	1	1	4	6	10	14	24
RHH[a], RHW[a], RHW-2[a], TW, THW, THHW, THW-2	6	1	1	3	4	8	11	18
	4	1	1	1	3	6	8	13
	3	1	1	1	3	5	7	11
	2	0	1	1	2	4	6	10
	1	0	1	1	1	3	4	7
RHH[a], RHW[a], RHW-2[a], TW, THW, THHW, THW-2	1/0	0	0	1	1	2	3	6
	2/0	0	0	1	1	1	3	5
	3/0	0	0	1	1	1	2	4
	4/0	0	0	0	1	1	1	3
THHN, THWN, THWN-2	14	8	13	22	35	62	83	137
	12	5	9	16	25	45	60	100
	10	3	6	10	16	28	38	63
	8	1	3	6	9	16	22	36
	6	1	2	4	6	12	16	26
	4	1	1	2	4	7	9	16
	3	1	1	1	3	6	8	13
	2	1	1	1	3	5	7	11
	1	0	1	1	1	4	5	8
	1/0	0	1	1	1	3	4	7
	2/0	0	0	1	1	2	3	6
	3/0	0	0	1	1	1	3	5
	4/0	0	0	1	1	1	2	4
XHH, XHHW, XHHW-2	14	5	9	15	24	43	58	96
	12	4	7	12	19	33	44	74
	10	3	5	9	14	24	33	55
	8	1	3	5	8	13	18	30
	6	1	1	3	5	10	13	22
	4	1	1	2	4	7	10	16
	3	1	1	1	3	6	8	14
	2	1	1	1	3	5	7	11
	1	0	1	1	1	4	5	8
XHH, XHHW, XHHW-2	1/0	0	1	1	1	3	4	7
	2/0	0	0	1	1	2	3	6
	3/0	0	0	1	1	1	3	5
	4/0	0	0	1	1	1	2	4

For SI: 1 inch = 25.4 mm.

a. Types RHW and RHW-2 without outer covering.

TABLE E3904.6(7) (Annex C, Table C.7)
MAXIMUM NUMBER OF CONDUCTORS IN LIQUID-TIGHT FLEXIBLE METAL CONDUIT (LFMC)[a]

TYPE LETTERS	CONDUCTOR SIZE AWG/kcmil	TRADE SIZES (inches)					
		1/2	3/4	1	1 1/4	1 1/2	2
RHH, RHW, RHW-2	14	4	7	12	21	27	44
	12	3	6	10	17	22	36
	10	3	5	8	14	18	29
	8	1	2	4	7	9	15
	6	1	1	3	6	7	12
	4	1	1	2	4	6	9
	3	1	1	1	4	5	8
	2	1	1	1	3	4	7
	1	0	1	1	1	3	5
	1/0	0	1	1	1	2	4
	2/0	0	1	1	1	1	3
	3/0	0	0	1	1	1	3
	4/0	0	0	1	1	1	2
TW, THHW, THW, THW-2	14	9	15	25	44	57	93
	12	7	12	19	33	43	71
	10	5	9	14	25	32	53
	8	3	5	8	14	18	29
RHH[a], RHW[a], RHW-2[a], THHW, THW, THW-2	14	6	10	16	29	38	62
	12	5	8	13	23	30	50
	10	3	6	10	18	23	39
	8	1	4	6	11	14	23
RHH[a], RHW[a], RHW-2[a], TW, THW, THHW, THW-2	6	1	3	5	8	11	18
	4	1	1	3	6	8	13
	3	1	1	3	5	7	11
	2	1	1	2	4	6	9
	1	1	1	1	3	4	7
	1/0	0	1	1	2	3	6
	2/0	0	1	1	2	3	5
	3/0	0	1	1	1	2	4
	4/0	0	0	1	1	1	3
THHN, THWN, THWN-2	14	13	22	36	63	81	133
	12	9	16	26	46	59	97
	10	6	10	16	29	37	61
	8	3	6	9	16	21	35
	6	2	4	7	12	15	25
	4	1	2	4	7	9	15
	3	1	1	3	6	8	13
	2	1	1	3	5	7	11
	1	1	1	1	4	5	8
	1/0	1	1	1	3	4	7
	2/0	0	1	1	2	3	6
	3/0	0	1	1	1	3	5
	4/0	0	1	1	1	2	4
XHH, XHHW, XHHW-2	14	9	15	25	44	57	93
	12	7	12	19	33	43	71
	10	5	9	14	25	32	53
	8	3	5	8	14	18	29
XHH, XHHW, XHHW-2	6	1	3	6	10	13	22
	4	1	2	4	7	9	16
	3	1	1	3	6	8	13
	2	1	1	3	5	7	11
	1	1	1	1	4	5	8
	1/0	1	1	1	3	4	7
	2/0	0	1	1	2	3	6
	3/0	0	1	1	1	3	5
	4/0	0	1	1	1	2	4

For SI: 1 inch = 25.4 mm.

a. Types RHW and RHW-2 without outer covering.

TABLE E3904.6(8) (Annex C, Table C.8)
MAXIMUM NUMBER OF CONDUCTORS IN RIGID METAL CONDUIT (RMC)[a]

TYPE LETTERS	CONDUCTOR SIZE AWG/kcmil	TRADE SIZES (inches)					
		1/2	3/4	1	1 1/4	1 1/2	2
RHH, RHW, RHW-2	14	4	7	12	21	28	46
	12	3	6	10	17	23	38
	10	3	5	8	14	19	31
	8	1	2	4	7	10	16
	6	1	1	3	6	8	13
	4	1	1	2	4	6	10
	3	1	1	2	4	5	9
	2	1	1	1	3	4	7
	1	0	1	1	1	3	5
	1/0	0	1	1	1	2	4
	2/0	0	1	1	1	2	4
	3/0	0	0	1	1	1	3
	4/0	0	0	1	1	1	3
TW, THHW, THW, THW-2	14	9	15	25	44	59	98
	12	7	12	19	33	45	75
	10	5	9	14	25	34	56
	8	3	5	8	14	19	31
RHH[a], RHW[a], RHW-2[a]	14	6	10	17	29	39	65
	12	5	8	13	23	32	52
	10	3	6	10	18	25	41
	8	1	4	6	11	15	24
RHH[a], RHW[a], RHW-2[a], TW, THW, THHW, THW-2	6	1	3	5	8	11	18
	4	1	1	3	6	8	14
	3	1	1	3	5	7	12
	2	1	1	2	4	6	10
	1	1	1	1	3	4	7
	1/0	0	1	1	2	3	6
	2/0	0	1	1	2	3	5
	3/0	0	1	1	1	2	4
	4/0	0	0	1	1	1	3
THHN, THWN, THWN-2	14	13	22	36	63	85	140
	12	9	16	26	46	62	102
	10	6	10	17	29	39	64
	8	3	6	9	16	22	37
	6	2	4	7	12	16	27
	4	1	2	4	7	10	16
	3	1	1	3	6	8	14
	2	1	1	3	5	7	11
	1	1	1	1	4	5	8
THHN, THWN, THWN-2	1/0	1	1	1	3	4	7
	2/0	0	1	1	2	3	6
	3/0	0	1	1	1	3	5
	4/0	0	1	1	1	2	4
XHH, XHHW, XHHW-2	14	9	15	25	44	59	98
	12	7	12	19	33	45	75
	10	5	9	14	25	34	56
	8	3	5	8	14	19	31
	6	1	3	6	10	14	23
	4	1	2	4	7	10	16
	3	1	1	3	6	8	14
	2	1	1	3	5	7	12
	1	1	1	1	4	5	9
	1/0	1	1	1	3	4	7
	2/0	0	1	1	2	3	6
	3/0	0	1	1	1	3	5
	4/0	0	1	1	1	2	4

For SI: 1 inch = 25.4 mm.

a. Types RHW and RHW-2 without outer covering.

TABLE E3904.6(9) (Annex C, Table C.9)
MAXIMUM NUMBER OF CONDUCTORS IN RIGID PVC CONDUIT, SCHEDULE 80 (PVC-80)[a]

TYPE LETTERS	CONDUCTOR SIZE AWG/kcmil	TRADE SIZES (inches)					
		1/2	3/4	1	1 1/4	1 1/2	2
RHH, RHW, RHW-2	14	3	5	9	17	23	39
	12	2	4	7	14	19	32
	10	1	3	6	11	15	26
	8	1	1	3	6	8	13
	6	1	1	2	4	6	11
	4	1	1	1	3	5	8
	3	0	1	1	3	4	7
	2	0	1	1	3	4	6
	1	0	1	1	1	2	4
	1/0	0	0	1	1	1	3
	2/0	0	0	1	1	1	3
	3/0	0	0	1	1	1	3
	4/0	0	0	0	1	1	2
TW, THHW, THW, THW-2	14	6	11	20	35	49	82
	12	5	9	15	27	38	63
	10	3	6	11	20	28	47
	8	1	3	6	11	15	26
RHH[a], RHW[a], RHW-2[a]	14	4	8	13	23	32	55
	12	3	6	10	19	26	44
	10	2	5	8	15	20	34
	8	1	3	5	9	12	20
RHH[a], RHW[a], RHW-2[a], TW, THW, THHW, THW-2	6	1	1	3	7	9	16
	4	1	1	3	5	7	12
	3	1	1	2	4	6	10
	2	1	1	1	3	5	8
	1	0	1	1	2	3	6
	1/0	0	1	1	1	3	5
RHH[a], RHW[a], RHW-2[a], TW, THW, THHW, THW-2	2/0	0	1	1	1	2	4
	3/0	0	0	1	1	1	3
	4/0	0	0	1	1	1	3
THHN, THWN, THWN-2	14	9	17	28	51	70	118
	12	6	12	20	37	51	86
	10	4	7	13	23	32	54
	8	2	4	7	13	18	31
	6	1	3	5	9	13	22
	4	1	1	3	6	8	14
	3	1	1	3	5	7	12
	2	1	1	2	4	6	10
	1	0	1	1	3	4	7
	1/0	0	1	1	2	3	6
	2/0	0	1	1	1	3	5
	3/0	0	1	1	1	2	4
	4/0	0	0	1	1	1	3
XHH, XHHW, XHHW-2	14	6	11	20	35	49	82
	12	5	9	15	27	38	63
	10	3	6	11	20	28	47
	8	1	3	6	11	15	26
	6	1	2	4	8	11	19
XHH, XHHW, XHHW-2	14	6	11	20	35	49	82
	12	5	9	15	27	38	63
	10	3	6	11	20	28	47
	8	1	3	6	11	15	26
	6	1	2	4	8	11	19
	4	1	1	3	6	8	14
	3	1	1	3	5	7	12
	2	1	1	2	4	6	10
	1	0	1	1	3	4	7
	1/0	0	1	1	2	3	6
	2/0	0	1	1	1	3	5
	3/0	0	1	1	1	2	4
	4/0	0	0	1	1	1	3

For SI: 1 inch = 25.4 mm.

a. Types RHW and RHW-2 without outer covering.

TABLE E3904.6(10) (Annex C, Table C.10)
MAXIMUM NUMBER OF CONDUCTORS IN RIGID PVC CONDUIT SCHEDULE 40 (PVC-40)[a]

TYPE LETTERS	CONDUCTOR SIZE AWG/kcmil	TRADE SIZES (inches)					
		1/2	3/4	1	1 1/4	1 1/2	2
RHH, RHW, RHW-2	14	4	7	11	20	27	45
	12	3	5	9	16	22	37
	10	2	4	7	13	18	30
	8	1	2	4	7	9	15
	6	1	1	3	5	7	12
	4	1	1	2	4	6	10
	3	1	1	1	4	5	8
	2	1	1	1	3	4	7
	1	0	1	1	1	3	5
	1/0	0	1	1	1	2	4
	2/0	0	0	1	1	1	3
	3/0	0	0	1	1	1	3
	4/0	0	0	1	1	1	2
TW, THHW, THW, THW-2	14	8	14	24	42	57	94
	12	6	11	18	32	44	72
	10	4	8	13	24	32	54
	8	2	4	7	13	18	30
RHH[a], RHW[a], RHW-2[a]	14	5	9	16	28	38	63
	12	4	8	13	22	30	50
	10	3	6	10	17	24	39
	8	1	3	6	10	14	23
RHH[a], RHW[a], RHW-2[a], TW, THW, THHW, THW-2	6	1	2	4	8	11	18
	4	1	1	3	6	8	13
	3	1	1	3	5	7	11
	2	1	1	2	4	6	10
	1	0	1	1	3	4	7
	1/0	0	1	1	2	3	6
	2/0	0	1	1	1	3	5
	3/0	0	1	1	1	2	4
	4/0	0	0	1	1	1	3
THHN, THWN, THWN-2	14	11	21	34	60	82	135
	12	8	15	25	43	59	99
	10	5	9	15	27	37	62
	8	3	5	9	16	21	36
THHN, THWN, THWN-2	6	1	4	6	11	15	26
	4	1	2	4	7	9	16
	3	1	1	3	6	8	13
	2	1	1	3	5	7	11
	1	1	1	1	3	5	8
	1/0	1	1	1	3	4	7
	2/0	0	1	1	2	3	6
	3/0	0	1	1	1	3	5
	4/0	0	1	1	1	2	4
XHH, XHHW, XHHW-2	14	8	14	24	42	57	94
	12	6	11	18	32	44	72
	10	4	8	13	24	32	54
	8	2	4	7	13	18	30
	6	1	3	5	10	13	22
	4	1	2	4	7	9	16
	3	1	1	3	6	8	13
	2	1	1	3	5	7	11
	1	1	1	1	3	5	8
	1/0	1	1	1	3	4	7
	2/0	0	1	1	2	3	6
	3/0	0	1	1	1	3	5
	4/0	0	1	1	1	2	4

For SI: 1 inch = 25.4 mm.
a. Types RHW and RHW-2 without outer covering.

SECTION E3905 BOXES, CONDUIT BODIES AND FITTINGS

E3905.1 Box, conduit body or fitting—where required. A box or conduit body shall be installed at each conductor splice point, outlet, switch point, junction point and pull point except as otherwise permitted in Sections E3905.1.1 through E3905.1.6.

Fittings and connectors shall be used only with the specific wiring methods for which they are designed and *listed.* (300.15)

E3905.1.1 Equipment. An integral junction box or wiring compartment that is part of *listed* equipment shall be permitted to serve as a box or conduit body. [300.15(B)]

E3905.1.2 Protection. A box or conduit body shall not be required where cables enter or exit from conduit or tubing that is used to provide cable support or protection against physical damage. A fitting shall be provided on the end(s) of the conduit or tubing to protect the cable from abrasion. [300.15(C)]

E3905.1.3 Integral enclosure. A wiring device with integral enclosure identified for the use, having brackets that securely fasten the device to walls or ceilings of conventional on-site frame construction, for use with nonmetallic-sheathed cable, shall be permitted in lieu of a box or conduit body. [300.15(E)]

E3905.1.4 Fitting. A fitting identified for the use shall be permitted in lieu of a box or conduit body where such fitting is accessible after installation and does not contain spliced or terminated conductors. [300.15(F)]

E3905.1.5 Buried conductors. Splices and taps in buried conductors and cables shall not be required to be enclosed in a box or conduit body where installed in accordance with Section E3803.4.

E3905.1.6 Luminaires. Where a luminaire is *listed* to be used as a raceway, a box or conduit body shall not be required for wiring installed therein. [300.15(J)]

E3905.2 Metal boxes. Metal boxes shall be grounded and bonded in accordance with Section E3908. (314.4)

E3905.3 Nonmetallic boxes. Nonmetallic boxes shall be used only with nonmetallic raceways, cabled wiring methods with entirely nonmetallic sheaths, flexible cords and nonmetallic raceways. (314.3)

Exceptions:

1. Where internal bonding means are provided between all entries, nonmetallic boxes shall be permitted to be used with metal raceways and metal-armored cables. (314.3 Exception No. 1)
2. Where integral bonding means with a provision for attaching an equipment grounding jumper inside the box are provided between all threaded entries in nonmetallic boxes *listed* for the purpose, nonmetallic boxes shall be permitted to be used with metal raceways and metal-armored cables. (314.3 Exception No. 2)

E3905.3.1 Nonmetallic-sheathed cable entering boxes. Where nonmetallic-sheathed cable or other cable assemblies with nonmetallic sheaths are used, the cable assembly, including the sheath, shall extend into the box not less than $^1/_4$ inch (6.4 mm) through a cable knockout opening. The cable sheath shall extend beyond the cable clamp. Wiring methods shall be secured to the boxes. [314.17(B)(2)]

Exception: Where nonmetallic-sheathed cable is used with boxes not larger than a nominal size of $2^1/_4$ inches by 4 inches (57 mm by 102 mm) mounted in walls or ceilings, and where the cable is fastened within 8 inches (203 mm) of the box measured along the sheath, and where the sheath extends through a cable knockout not less than $^1/_4$ inch (6.4 mm), securing the cable to the box shall not be required. [314.17(B)(2) Exception]

E3905.3.2 Conductor rating. Nonmetallic boxes shall be suitable for the lowest temperature-rated conductor entering the box. [314.17(B)(4)]

E3905.4 Minimum depth of boxes for outlets, devices, and utilization equipment. Outlet and device boxes shall have an approved depth to allow equipment installed within them to be mounted properly and without the likelihood of damage to conductors within the box. (314.24)

E3905.4.1 Outlet boxes without enclosed devices or utilization equipment. Outlet boxes that do not enclose devices or utilization equipment shall have an internal depth of not less than $^1/_2$ inch (12.7 mm). [314.24(A)]

E3905.4.2 Utilization equipment. Outlet and device boxes that enclose devices or utilization equipment shall have a minimum internal depth that accommodates the rearward projection of the equipment and the size of the conductors that supply the equipment. The internal depth shall include that of any extension boxes, plaster rings, or raised covers. The internal depth shall comply with all of the applicable provisions that follow: [314.24(B)]

1. Large equipment. Boxes that enclose devices or utilization equipment that projects more than $1^7/_8$ inches (48 mm) rearward from the mounting plane of the box shall have a depth that is not less than the depth of the equipment plus $^1/_4$ inch (6.4 mm). [314.24(B)(1)]
2. Conductors larger than 4 AWG. Boxes that enclose devices or utilization equipment supplied by conductors larger than 4 AWG shall be identified for their specific function. Devices or utilization equipment supplied by conductors larger than 4 AWG shall be permitted to be mounted on or in junction and pull boxes larger than 100 cubic inches (1650 cm^3) if the spacing at the terminals meets the requirements of Section E3907.9. [314.24(B)(2) and Exception]
3. Conductors 8, 6, or 4 AWG. Boxes that enclose devices or utilization equipment supplied by 8, 6, or 4 AWG conductors shall have an internal depth

that is not less than $2^1/_{16}$ inches (52.4 mm). [314.24(B)(3)]

4. Conductors 12 or 10 AWG. Boxes that enclose devices or utilization equipment supplied by 12 or 10 AWG conductors shall have an internal depth that is not less than $1^3/_{16}$ inches (30.2 mm). Where the equipment projects rearward from the mounting plane of the box by more than 1 inch (25.4 mm), the box shall have a depth that is not less than that of the equipment plus $^1/_4$ inch (6.4 mm). [314.24(B)(4)]
5. Conductors 14 AWG and smaller. Boxes that enclose devices or utilization equipment supplied by 14 AWG or smaller conductors shall have a depth that is not less than $^{15}/_{16}$ inch (23.8 mm). [314.24(B)(5)]

Exception: Utilization equipment that is listed to be installed with specified boxes shall be permitted. [314.24(B)(1)–(5) Exception]

E3905.5 Boxes enclosing flush-mounted devices or flush equipment. Boxes enclosing flush-mounted devices or flush equipment shall be of such design that the devices or equipment are completely enclosed at the back and all sides and shall provide support for the devices or equipment. Screws for supporting the box shall not be used for attachment of the device or equipment contained therein. (314.19)

E3905.6 Boxes at luminaire outlets. Outlet boxes used at luminaire or lampholder outlets shall be designed for the support of luminaires and lampholders and shall be installed as required by Section E3904.3. [314.27(A)]

E3905.6.1 Vertical surface outlets. Boxes used at luminaire or lampholder outlets in or on a vertical surface shall be identified and marked on the interior of the box to indicate the maximum weight of the luminaire or lamp holder that is permitted to be supported by the box if other than 50 pounds (22.7 kg). [314.27(A)(1)]

Exception: A vertically mounted luminaire or lampholder weighing not more than 6 pounds (2.7 kg) shall be permitted to be supported on other boxes or plaster rings that are secured to other boxes, provided that the luminaire or its supporting yoke is secured to the box with not fewer than two No. 6 or larger screws. [314.27(A)(1) Exception]

E3905.6.2 Ceiling outlets. For outlets used exclusively for lighting, the box shall be designed or installed so that a luminaire or lampholder can be attached. Such boxes shall be capable of supporting a luminaire weighing up to 50 pounds (22.7 kg). A luminaire that weighs more than 50 pounds (22.7 kg) shall be supported independently of the outlet box, unless the outlet box is *listed* for not less than the weight to be supported. The interior of the box shall be marked by the manufacturer to indicate the maximum weight that the box is permitted to support. [314.27(A)(2)]

E3905.6.3 Separable attachment fittings. Outlet boxes required in Section E3905.6 shall be permitted to support *listed* locking support and mounting receptacles used in combination with compatible attachment fittings. The combination shall be identified for the support of equipment within the weight and mounting orientation limits of the listing. Where the supporting receptacle is installed within a box, it shall be included in the fill calculation covered in Section E3905.12.1.

E3905.7 Floor boxes. Where outlet boxes for receptacles are installed in the floor, such boxes shall be *listed* specifically for that application. [314.27(B)]

E3905.8 Boxes at fan outlets. Outlet boxes and outlet box systems used as the sole support of ceiling-suspended fans (paddle) shall be marked by their manufacturer as suitable for this purpose and shall not support ceiling-suspended fans (paddle) that weigh more than 70 pounds (31.8 kg). For outlet boxes and outlet box systems designed to support ceiling-suspended fans (paddle) that weigh more than 35 pounds (15.9 kg), the required marking shall include the maximum weight to be supported.

Outlet boxes mounted in the ceilings of habitable rooms in a location acceptable for the installation of a ceiling-suspended (paddle) fan shall comply with one of the following:

1. *Listed* for sole support of ceiling-suspended (paddle) fans.
2. An outlet box complying with the applicable requirements of Section E3905.6 and providing access to structural framing capable of supporting of a ceiling-suspended (paddle) fan bracket or equipment. [314.27(C)]

E3905.9 Utilization equipment. Boxes used for the support of utilization equipment other than ceiling-suspended (paddle) fans shall meet the requirements of Sections E3905.6.1 and E3905.6.2 for the support of a luminaire that is the same size and weight. [314.27(D)]

Exception: Utilization equipment weighing not more than 6 pounds (2.7 kg) shall be permitted to be supported on other boxes or plaster rings that are secured to other boxes, provided that the equipment or its supporting yoke is secured to the box with not fewer than two No. 6 or larger screws. [314.27(D) Exception]

E3905.10 Conduit bodies and junction, pull and outlet boxes to be accessible. Conduit bodies and junction, pull and outlet boxes shall be installed so that the wiring therein can be accessed without removing any part of the building or structure or, in underground circuits, without excavating sidewalks, paving, earth or other substance used to establish the finished grade. [314.29(A) & (B)]

Exception: Boxes covered by gravel, light aggregate or noncohesive granulated soil shall be *listed* for the application, and the box locations shall be effectively identified and access shall be provided for excavation. [314.29(B) Exception]

E3905.11 Damp or wet locations. In damp or wet locations, boxes, conduit bodies and fittings shall be placed or equipped so as to prevent moisture from entering or accumulating within the box, conduit body or fitting. Boxes, conduit bodies and fittings installed in wet locations shall be *listed* for use in wet locations.

Where drainage openings are installed in the field in boxes or conduit bodies *listed* for use in damp or wet locations, such openings shall be approved, not smaller than $^1/_8$ inch (3 mm) and not larger than $^1/_4$ inch (6.4 mm). For *listed* drain fittings, larger openings are permitted where installed in the field in accordance with the manufacturer's instructions. (314.15)

E3905.12 Number of conductors in outlet, device, and junction boxes, and conduit bodies. Boxes and conduit bodies shall be of an approved size to provide free space for all enclosed conductors. In no case shall the volume of the box, as calculated in Section E3905.12.1, be less than the box fill calculation as calculated in Section E3905.12.2. The minimum volume for conduit bodies shall be as calculated in Section E3905.12.3. The provisions of this section shall not apply to terminal housings supplied with motors or generators. (314.16)

E3905.12.1 Box volume calculations. The volume of a wiring enclosure (box) shall be the total volume of the assembled sections, and, where used, the space provided by plaster rings, domed covers, extension rings, etc., that are marked with their volume in cubic inches or are made from boxes the dimensions of which are listed in Table E3905.12.1. Where a box is provided with one or more securely installed barriers, the volume shall be apportioned to each of the resulting spaces. Each barrier, if not marked with its volume, shall be considered to take up to 0.5 cubic inch (8.2 cm^3) if metal, and 1.0 cubic inch (16.4 cm^3) if nonmetallic. [314.16(A)]

E3905.12.1.1 Standard boxes. The volumes of standard boxes that are not marked with a cubic-inch capacity shall be as given in Table E3905.12.1. [314.16(A)(1)]

E3905.12.1.2 Other boxes. Boxes 100 cubic inches (1640 cm^3) or less, other than those described in Table E3905.12.1, and nonmetallic boxes shall be durably and legibly marked by the manufacturer with their cubic-inch capacity. Boxes described in Table E3905.12.1 that have a larger cubic inch capacity than is designated in the table shall be permitted to have their cubic-inch capacity marked as required by this section. [314.16(A)(2)]

E3905.12.2 Box fill calculations. The volumes in Section E3905.12.2.1 through Section E3905.12.2.5, as applicable, shall be added together. No allowance shall be required for small fittings such as locknuts and bushings. Each space within a box installed with a barrier shall be calculated separately. [314.16(B)]

E3905.12.2.1 Conductor fill. Each conductor that originates outside the box and terminates or is spliced within the box shall be counted once, and each conductor that passes through the box without splice or

TABLE E3905.12.1 [Table 314.16(A)]
MAXIMUM NUMBER OF CONDUCTORS IN METAL BOXES[a]

BOX DIMENSIONS (inches trade size and type)	MAXIMUM CAPACITY (cubic inches)	MAXIMUM NUMBER OF CONDUCTORS[a]						
		18 Awg	16 Awg	14 Awg	12 Awg	10 Awg	8 Awg	6 Awg
4 × 1$^1/_4$ round or octagonal	12.5	8	7	6	5	5	4	2
4 × 1$^1/_2$ round or octagonal	15.5	10	8	7	6	6	5	3
4 × 2$^1/_8$ round or octagonal	21.5	14	12	10	9	8	7	4
4 × 1$^1/_4$ square	18.0	12	10	9	8	7	6	3
4 × 1$^1/_2$ square	21.0	14	12	10	9	8	7	4
4 × 2$^1/_8$ square	30.3	20	17	15	13	12	10	6
4$^{11}/_{16}$ × 1$^1/_4$ square	25.5	17	14	12	11	10	8	5
4$^{11}/_{16}$ × 1$^1/_2$ square	29.5	19	16	14	13	11	9	5
4$^{11}/_{16}$ × 2$^1/_8$ square	42.0	28	24	21	18	16	14	8
3 × 2 × 1$^1/_2$ device	7.5	5	4	3	3	3	2	1
3 × 2 × 2 device	10.0	6	5	5	4	4	3	2
3 × 2 × 2$^1/_4$ device	10.5	7	6	5	4	4	3	2
3 × 2 × 2$^1/_2$ device	12.5	8	7	6	5	5	4	2
3 × 2 × 2$^3/_4$ device	14.0	9	8	7	6	5	4	2
3 × 2 × 3$^1/_2$ device	18.0	12	10	9	8	7	6	3
4 × 2$^1/_8$ × 1$^1/_2$ device	10.3	6	5	5	4	4	3	2
4 × 2$^1/_8$ × 1$^7/_8$ device	13.0	8	7	6	5	5	4	2
4 × 2$^1/_8$ × 2$^1/_8$ device	14.5	9	8	7	6	5	4	2
3$^3/_4$ × 2 × 2$^1/_2$ masonry box/gang	14.0	9	8	7	6	5	4	2
3$^3/_4$ × 2 × 3$^1/_2$ masonry box/gang	21.0	14	12	10	9	8	7	4

For SI: 1 inch = 25.4 mm, 1 cubic inch = 16.4 cm^3.

a. Where volume allowances are not required by Sections E3905.12.2.2 through E3905.12.2.5.

termination shall be counted once. Each loop or coil of unbroken conductor having a length equal to or greater than twice that required for free conductors by Section E3406.11.3, shall be counted twice. The conductor fill, in cubic inches, shall be computed using Table E3905.12.2.1. A conductor, no part of which leaves the box, shall not be counted. [314.16(B)(1)]

Exception: An equipment grounding conductor or not more than four fixture wires smaller than No. 14, or both, shall be permitted to be omitted from the calculations where such conductors enter a box from a domed fixture or similar canopy and terminate within that box. [314.16(B)(1) Exception]

TABLE E3905.12.2.1 [Table 314.16(B)]
VOLUME ALLOWANCE REQUIRED PER CONDUCTOR

SIZE OF CONDUCTOR	FREE SPACE WITHIN BOX FOR EACH CONDUCTOR (cubic inches)
18 AWG	1.50
16 AWG	1.75
14 AWG	2.00
12 AWG	2.25
10 AWG	2.50
8 AWG	3.00
6 AWG	5.00

For SI: 1 cubic inch = 16.4 cm^3.

E3905.12.2.2 Clamp fill. Where one or more internal cable clamps, whether factory or field supplied, are present in the box, a single volume allowance in accordance with Table E3905.12.2.1 shall be made based on the largest conductor present in the box. An allowance shall not be required for a cable connector having its clamping mechanism outside of the box.

A clamp assembly that incorporates a cable termination for the cable conductors shall be *listed* and marked for use with specific nonmetallic boxes. Conductors that originate within the clamp assembly shall be included in conductor fill calculations provided in Section E3905.12.2.1 as though they entered from outside of the box. The clamp assembly shall not require a fill allowance, but, the volume of the portion of the assembly that remains within the box after installation shall be excluded from the box volume as marked in accordance with Section E3905.12.1.2. [314.16(B)(2)]

E3905.12.2.3 Support fittings fill. Where one or more fixture studs or hickeys are present in the box, a single volume allowance in accordance with Table E3905.12.2.1 shall be made for each type of fitting based on the largest conductor present in the box. [314.16(B)(3)]

E3905.12.2.4 Device or equipment fill. For each yoke or strap containing one or more devices or equipment, a double volume allowance in accordance with Table E3905.12.2.1 shall be made for each yoke or strap based on the largest conductor connected to a device(s) or equipment supported by that yoke or strap. For a device or utilization equipment that is wider than a single 2-inch (51 mm) device box as described in Table E3905.12.1, a double volume allowance shall be made for each ganged portion required for mounting of the device or equipment. [314.16(B)(4)]

E3905.12.2.5 Equipment grounding conductor fill. Where up to four equipment grounding conductors or equipment bonding jumpers enter a box, a single volume allowance in accordance with Table E3905.12.2.1 shall be made based on the largest equipment grounding conductor or equipment bonding jumper entering the box. A one-quarter volume allowance shall be made for each additional equipment grounding conductor or equipment bonding jumper that enters the box, based on the largest equipment grounding conductor or equipment bonding conductor. [314.16(B)(5)]

E3905.12.3 Conduit bodies. Conduit bodies enclosing 6 AWG conductors or smaller, other than short-radius conduit bodies, shall have a cross-sectional area not less than twice the cross-sectional area of the largest conduit or tubing to which they can be attached. The maximum number of conductors permitted shall be the maximum number permitted by Section E3904.6 for the conduit to which it is attached. [314.16(C)(1)]

E3905.12.3.1 Splices, taps or devices. Only those conduit bodies that are durably and legibly marked by the manufacturer with their cubic inch capacity shall be permitted to contain splices, taps or devices. The maximum number of conductors shall be calculated using the same procedure for similar conductors in other than standard boxes. [314.16(C)(2)]

E3905.12.3.2 Short-radius conduit bodies. Conduit bodies such as capped elbows and service-entrance elbows that enclose conductors 6 AWG or smaller and that are only intended to enable the installation of the raceway and the contained conductors, shall not contain splices, taps, or devices and shall be of sufficient size to provide free space for all conductors enclosed in the conduit body. [314.16(C)(3)]

SECTION E3906
INSTALLATION OF BOXES, CONDUIT BODIES AND FITTINGS

E3906.1 Conductors entering boxes, conduit bodies or fittings. Conductors entering boxes, conduit bodies or fittings shall be protected from abrasion. (314.17)

E3906.1.1 Insulated fittings. Where raceways contain 4 AWG or larger insulated circuit conductors and these conductors enter a cabinet, box enclosure, or raceway, the conductors shall be protected in accordance with any of the following: [300.4(G)]

1. An identified fitting providing a smoothly rounded insulating surface. Conduit bushings constructed wholly of insulating material shall not be used to secure a fitting or raceway.
2. A *listed* metal fitting that has smoothly rounded edges.
3. Separation from the fitting or raceway using an identified insulating material that is securely fastened in place.
4. Threaded hubs or bosses that are an integral part of a cabinet, box, enclosure, or raceway providing a smoothly rounded or flared entry for conductors.

Insulating fittings or insulating material shall have a temperature rating not less than the insulation temperature rating of the installed conductors. [300.4(G)]

E3906.2 Openings. Openings through which conductors enter shall be closed in a manner identified for the application. [314.17(A)]

E3906.3 Metal boxes and conduit bodies. Where raceway is installed with metal boxes, or conduit bodies, the raceway shall be secured to such boxes and conduit bodies. [314.17(B)(1)]

E3906.4 Unused openings. Unused openings other than those intended for the operation of equipment, those intended for mounting purposes, or those permitted as part of the design for *listed* equipment, shall be closed to afford protection substantially equivalent to that of the wall of the equipment. Metal plugs or plates used with nonmetallic boxes or conduit bodies shall be recessed at least $^1/_4$ inch (6.4 mm) from the outer surface of the box or conduit body. [110.12(A)]

E3906.5 Flush-mounted installations. Installations within or behind a surface of concrete, tile, gypsum, plaster or other *noncombustible material*, including boxes employing a flush-type cover or faceplate shall be made so that the front edge of the box, plaster ring, extension ring, or *listed* extender will be set back from the finished surface not more than $^1/_4$ inch (6.4 mm). Installations within a surface of wood or other *combustible material*, boxes, plaster rings, extension rings and *listed* extenders shall extend to the finished surface or project therefrom. (314.20)

E3906.6 Noncombustible surfaces. Openings in noncombustible surfaces that accommodate boxes employing a flush-type cover or faceplate shall be made so that there are no gaps or open spaces greater than $^1/_8$ inch (3.2 mm) around the edge of the box. (314.21)

E3906.7 Surface extensions. Surface extensions shall be made by mounting and mechanically securing an extension ring over the box. (314.22)

Exception: A surface extension shall be permitted to be made from the cover of a flush-mounted box where the cover is designed so it is unlikely to fall off, or be removed if its securing means becomes loose. The wiring method shall be flexible for an approved length that permits removal of the cover and provides access to the box interior and shall be arranged so that any bonding or grounding continuity is independent of the connection between the box and cover. (314.22 Exception)

E3906.8 Supports. Boxes and enclosures shall be supported in accordance with one or more of the provisions in Sections E3906.8.1 through E3906.8.6. (314.23)

E3906.8.1 Surface mounting. An enclosure mounted on a building or other surface shall be rigidly and securely fastened in place. If the surface does not provide rigid and secure support, additional support in accordance with other provisions of Section E3906.8 shall be provided. [314.23(A)]

E3906.8.2 Structural mounting. An enclosure supported from a structural member or from grade shall be rigidly supported either directly, or by using a metal, polymeric or wood brace. [314.23(B)]

E3906.8.2.1 Nails and screws. Nails and screws, where used as a fastening means, shall secure boxes by using brackets on the outside of the enclosure, or by using mounting holes in the back or one or more sides of the enclosure, or they shall pass through the interior within $^1/_4$ inch (6.4 mm) of the back or ends of the enclosure. Screws shall not be permitted to pass through the box except where exposed threads in the box are protected by an approved means to avoid abrasion of conductor insulation. Mounting holes made in the field shall be field approved. [314.23(B)(1)]

E3906.8.2.2 Braces. Metal braces shall be protected against corrosion and formed from metal that is not less than 0.020 inch (0.508 mm) thick uncoated. Wood braces shall have a cross section not less than nominal 1 inch by 2 inches (25.4 mm by 51 mm). Wood braces in wet locations shall be treated for the conditions. Polymeric braces shall be identified as being suitable for the use. [314.23(B)(2)]

E3906.8.3 Mounting in finished surfaces. An enclosure mounted in a finished surface shall be rigidly secured there to by clamps, anchors, or fittings identified for the application. [314.23(C)]

E3906.8.4 Raceway supported enclosures without devices or fixtures. An enclosure that does not contain a device(s), other than splicing devices, or support a luminaire, lampholder or other equipment, and that is supported by entering raceways shall not exceed 100 cubic inches (1640 cm^3) in size. The enclosure shall have threaded entries or identified hubs. The enclosure shall be supported by two or more conduits threaded wrenchtight into the enclosure or hubs. Each conduit shall be secured within 3 feet (914 mm) of the enclosure, or within 18 inches (457 mm) of the enclosure if all entries are on the same side of the enclosure. [314.23(E)]

Exception: Rigid metal, intermediate metal, or rigid polyvinyl chloride nonmetallic conduit or electrical metallic tubing shall be permitted to support a conduit body of any size, provided that the conduit body is not larger in trade size than the largest trade size of the supporting conduit or electrical metallic tubing. [314.23(E) Exception]

E3906.8.5 Raceway supported enclosures, with devices or luminaire. An enclosure that contains a device(s), other than splicing devices, or supports a luminaire, lampholder or other equipment and is supported by entering raceways shall not exceed 100 cubic inches (1640 cm^3) in size. The enclosure shall have threaded entries or identified hubs. The enclosure shall be supported by two or more conduits threaded wrench-tight into the enclosure or hubs. Each conduit shall be secured within 18 inches (457 mm) of the enclosure. [314.23(F)]

Exceptions:

1. Rigid metal or intermediate metal conduit shall be permitted to support a conduit body of any size, provided that the conduit bodies are not larger in trade size than the largest trade size of the supporting conduit. [314.23(F) Exception No. 1]
2. An unbroken length(s) of rigid or intermediate metal conduit shall be permitted to support a box used for luminaire or lampholder support, or to support a wiring enclosure that is an integral part of a luminaire and used in lieu of a box in accordance with Section E3905.1.1, where all of the following conditions are met:
   - 2.1. The conduit is securely fastened at a point so that the length of conduit beyond the last point of conduit support does not exceed 3 feet (914 mm).
   - 2.2. The unbroken conduit length before the last point of conduit support is 12 inches (305 mm) or greater, and that portion of the conduit is securely fastened at some point not less than 12 inches (305 mm) from its last point of support.
   - 2.3. Where accessible to unqualified persons, the luminaire or lampholder, measured to its lowest point, is not less than 8 feet (2438 mm) above grade or standing area and at least 3 feet (914 mm) measured horizontally to the 8-foot (2438 mm) elevation from windows, doors, porches, fire escapes, or similar locations.
   - 2.4. A luminaire supported by a single conduit does not exceed 12 inches (305 mm) in any direction from the point of conduit entry.
   - 2.5. The weight supported by any single conduit does not exceed 20 pounds (9.1 kg).
   - 2.6. At the luminaire or lampholder end, the conduit(s) is threaded wrenchtight into the box, conduit body, or integral wiring enclosure, or into hubs identified for the purpose. Where a box or conduit body is used for support, the luminaire shall be secured directly to the box or conduit body, or through a threaded conduit nipple not over 3 inches (76 mm) long. [314.23(F) Exception No. 2]

E3906.8.6 Enclosures in concrete or masonry. An enclosure supported by embedment shall be identified as being suitably protected from corrosion and shall be securely embedded in concrete or masonry. [314.23(G)]

E3906.9 Covers and canopies. In completed installations, each box shall have a cover, faceplate, lampholder or luminaire canopy.

Screws used for the purpose of attaching covers or other equipment to the box shall be either machine screws matching the thread gauge or size that is integral to the box or shall be in accordance with the manufacturer's instructions. (314.25)

E3906.10 Covers and plates. Covers and plates shall be nonmetallic or metal. Metal covers and plates shall be grounded. [314.25(A)]

E3906.11 Exposed combustible finish. Combustible wall or ceiling finish exposed between the edge of a fixture canopy or pan and the outlet box shall be covered with *noncombustible material* where required by Section E4004.2. [314.25(B)]

SECTION E3907
CABINETS AND PANELBOARDS

E3907.1 Space within switch and overcurrent device enclosures. The wiring space within enclosures for switches and overcurrent devices shall be permitted for other wiring and equipment in accordance with Sections E3907.1.1 and E3907.1.2. [312.8]

E3907.1.1 Splices, taps and feed-through conductors. Where the wiring space of enclosures for switches or overcurrent device contains conductors feeding through, spliced, or tapping off to other enclosures, switches, or overcurrent devices, the following conditions shall be met:

1. The total area of all conductors installed at any cross section of the wiring space shall not exceed 40 percent of the cross-sectional area of that space.
2. The total area of all conductors, splices and taps installed at any cross section of the wiring space shall not exceed 75 percent of the cross-sectional area of that space.
3. A warning label shall be applied to the enclosure that identifies the closest disconnecting means for any feed-through conductors. [312.8(A)]

E3907.1.2 Power monitoring or energy management equipment. Where the wiring space of enclosures for switches or overcurrent devices contains power monitoring or energy management equipment, all of the following conditions shall be met:

1. The power monitoring or energy management equipment shall be identified as a field installable accessory as part of the *listed* equipment, or shall be a *listed* kit evaluated for field installation in switch or overcurrent device enclosures.
2. The total area of all conductors, splices, taps and equipment at any cross section of the wiring space shall not exceed 75 percent of the cross-sectional area of that space.
3. Conductors used exclusively for control or instrumentation circuits shall comply with one of the following:
   - 3.1. Conductors are installed as Class 1 circuits in accordance with Section 725.49 of NFPA 70.
   - 3.2. Conductors smaller than 18 AWG, but not smaller than 22 AWG for a single conductor or 26 AWG for a multiconductor cable, shall be permitted to be used where the conductors and cable assemblies meet all of the following conditions:
     - 3.2.1. Are enclosed within raceways or routed along one or more walls of the enclosure and secured at intervals that do not exceed 10 inches (250 mm).
     - 3.2.2. Are secured within 10 inches (250 mm) of terminations.
     - 3.2.3. Are secured to prevent contact with current-carrying components within the enclosure.
     - 3.2.4. Are rated for the system voltage and not less than 600 volts.
     - 3.2.5. Have a minimum insulation temperature rating of 194°F (90°C).

E3907.2 Damp and wet locations. In damp or wet locations, cabinets and panelboards of the surface type shall be placed or equipped so as to prevent moisture or water from entering and accumulating within the cabinet, and shall be mounted to provide an airspace not less than $^1/_4$ inch (6.4 mm) between the enclosure and the wall or other supporting surface. Cabinets installed in wet locations shall be weatherproof. For enclosures in wet locations, raceways and cables entering above the level of uninsulated live parts shall be installed with fittings *listed* for wet locations. (312.2)

Exception: Nonmetallic enclosures installed on concrete, masonry, tile, or similar surfaces shall not be required to be installed with an airspace between the enclosure and the wall or supporting surface. (312.2 Exception)

E3907.3 Position in wall. In walls of concrete, tile or other *noncombustible material,* cabinets and panelboards shall be installed so that the front edge of the cabinet will not set back of the finished surface more than $^1/_4$ inch (6.4 mm). In walls constructed of wood or other *combustible material,* cabinets shall be flush with the finished surface or shall project therefrom. (312.3)

E3907.4 Repairing noncombustible surfaces. Noncombustible surfaces that are broken or incomplete shall be repaired so that there will not be gaps or open spaces greater than $^1/_8$ inch (3.2 mm) at the edge of the cabinet or cutout box employing a flush-type cover. (312.4)

E3907.5 Unused openings. Unused openings, other than those intended for the operation of equipment, those intended for mounting purposes, and those permitted as part of the design for *listed* equipment, shall be closed to afford protection substantially equivalent to that of the wall of the equipment. Metal plugs and plates used with nonmetallic cabinets shall be recessed at least $^1/_4$ inch (6.4 mm) from the outer surface. Unused openings for circuit breakers and switches shall be closed using identified closures, or other approved means that provide protection substantially equivalent to the wall of the enclosure. (110.12(A)]

E3907.6 Conductors entering cabinets. Conductors entering cabinets and panelboards shall be protected from abrasion and shall comply with Section E3906.1.1. (312.5)

E3907.7 Openings to be closed. Openings through which conductors enter cabinets, panelboards and meter sockets shall be closed in an approved manner. [312.5(A)]

E3907.8 Cables. Where cables are used, each cable shall be secured to the cabinet, panelboard, cutout box or meter socket enclosure. [312.5(C)]

Exception: Cables with entirely nonmetallic sheaths shall be permitted to enter the top of a surface-mounted enclosure through one or more sections of rigid raceway not less than 18 inches (457 mm) nor more than 10 feet (3048 mm) in length, provided all the following conditions are met:

1. Each cable is fastened within 12 inches (305 mm), measured along the sheath, of the outer end of the raceway.
2. The raceway extends directly above the enclosure and does not penetrate a structural ceiling.
3. A fitting is provided on each end of the raceway to protect the cable(s) from abrasion and the fittings remain accessible after installation.
4. The raceway is sealed or plugged at the outer end using approved means so as to prevent access to the enclosure through the raceway.
5. The cable sheath is continuous through the raceway and extends into the enclosure beyond the fitting not less than $^1/_4$ inch (6.4 mm).
6. The raceway is fastened at its outer end and at other points in accordance with Section E3802.1.
7. The allowable cable fill for conduit or tubing shall not exceed that permitted by Table E3907.8 and shall be considered as a complete conduit or tubing system. A multiconductor cable having two

or more conductors shall be treated as a single conductor for calculating the percentage of conduit fill area. For cables that have elliptical cross sections, the cross-sectional area calculation shall be based on the major diameter of the ellipse as a circle diameter. [312.5(C) Exception]

**TABLE E3907.8 (Chapter 9, Table 1)
PERCENT OF CROSS SECTION OF
CONDUIT AND TUBING FOR CONDUCTORS**

NUMBER OF CONDUCTORS	MAXIMUM PERCENT OF CONDUIT AND TUBING AREA FILLED BY CONDUCTORS
1	53
2	31
Over 2	40

E3907.9 Wire-bending space within an enclosure containing a panelboard. Wire-bending space within an enclosure containing a panelboard shall comply with the requirements of Sections E3907.9.1 through E3907.9.3.

E3907.9.1 Top and bottom wire-bending space. The top and bottom wire-bending space for a panelboard enclosure shall be sized in accordance with Table E3907.9.1(1) based on the largest conductor entering or leaving the enclosure. [408.55 (A)]

Exceptions:

1. For a panelboard rated at 225 amperes or less and designed to contain not more than 42 overcurrent devices, either the top or bottom wire-bending space shall be permitted to be sized in accordance with Table E3907.9.1(2). For the purposes of this exception, a 2-pole or a 3-pole circuit breaker shall be considered as two or three overcurrent devices, respectively. [408.55(A) Exception No. 1]
2. For any panelboard, either the top or bottom wire-bending space shall be permitted to be sized in accordance with Table E3907.9.1(2) where the wire-bending space on at least one side is sized in accordance with Table

**TABLE E3907.9.1(1) [Table 312.6(B)]
MINIMUM WIRE-BENDING SPACE AT TERMINALS (see Note 1)**

WIRE SIZE (AWG or kcmil)		WIRES PER TERMINAL			
		One (see Note 2)		Two	
All other conductors	Compact stranded AA-8000 aluminum alloy conductors (see Note 3)	inches	mm	inches	mm
14-10	12-8	Not specified	Not specified	—	—
8	6	$1^1/_2$	38.1	—	—
6	4	2	50.8	—	—
4	2	3	76.2	—	—
3	1	3	76.2	—	—
2	1/0	$3^1/_2$	88.9	—	—
1	2/0	$4^1/_2$	114	—	—
1/0	3/0	$5^1/_2$	140	$5^1/_2$	140
2/0	4/0	6	152	6	152
3/0	250	$6^1/_2$[a]	165[a]	$6^1/_2$[a]	165[a]
4/0	300	7[b]	178[b]	$7^1/_2$[c]	190[c]
250	350	$8^1/_2$[d]	216[d]	$8^1/_2$[d]	229[d]
300	400	10[e]	254[e]	10[d]	254[d]
350	500	12[e]	305[e]	12[c]	305[c]
400	600	13[e]	330[e]	13[c]	330[c]
500	700-750	14[e]	356[e]	14[c]	356[c]
600	800-900	15[e]	381[e]	16[c]	406[c]
700	1,000	16[e]	406[e]	18[c]	457[c]

Notes:

1. Bending space at terminals shall be measured in a straight line from the end of the lug or wire connector in a direction perpendicular to the enclosure wall.
2. For removable and lay-in wire terminals intended for only one wire, bending space shall be permitted to be reduced by the following number of inches (millimeters):
   a. $^1/_2$ inch (12.7 mm)
   b. 1 inch (25.4 mm)
   c. $1^1/_2$ inches (38.1 mm)
   d. 2 inches (50.8 mm)
   e. 3 inches (76.2 mm)
3. This column shall be permitted to determine the required wire-bending space for compact stranded aluminum conductors in sizes up to 1,000 kcmil and manufactured using AA-8000 series electrical grade aluminum alloy conductor material.

E3907.9.1(1) based on the largest conductor to be terminated in any side wire-bending space. [408.55(A) Exception No. 2]

3. Where the panelboard is designed and constructed for wiring using only a single 90-degree bend for each conductor, including the grounded circuit conductor, and the wiring diagram indicates and specifies the method of wiring that must be used, the top and bottom wire-bending space shall be permitted to be sized in accordance with Table E3907.9.1(2). [408.55(A) Exception No. 3]
4. Where there are no conductors terminated in that space, either the top or the bottom wire-bending space, shall be permitted to be sized in accordance with Table E3907.9.1(2). [408.55(A) Exception No. 4]

E3907.9.2 Side wire-bending space. Side wire-bending space shall be in accordance with Table E3907.9.1(2) based on the largest conductor to be terminated in that space. [408.55(B)]

E3907.9.3 Back wire-bending space. The distance between the center of the rear entry and the nearest termination for the entering conductors shall be not less than the distance given in Table E3907.9.1(1). Where a raceway or cable entry is in the wall of the enclosure, opposite a removable cover, the distance from that wall to the cover shall be permitted to comply with the distance required in Table E3907.9.1(2). [408.55 (C)]

SECTION E3908 GROUNDING AND BONDING

E3908.1 Metal enclosures. Metal enclosures of conductors, devices and equipment shall be connected to the equipment grounding conductor. (250.86)

Exceptions:

1. Short sections of metal enclosures or raceways used to provide cable assemblies with support or protection against physical damage. (250.86 Exception No. 2)
2. Metal components that are installed in an underground installation of rigid nonmetallic conduit and are isolated from possible contact by a minimum cover of 18 inches (457 mm) to any part of the metal components or that are isolated from possible contact by encasement in not less than 2 inches (51 mm) of concrete. (250.86 Exception No. 3)

E3908.2 Equipment fastened in place or connected by permanent wiring methods (fixed). Exposed, normally noncurrent-carrying metal parts of fixed equipment supplied by or enclosing conductors or components that are likely to become energized shall be connected to the equipment

TABLE E3907.9.1(2) [Table 312.6(A)]
MINIMUM WIRE-BENDING SPACE AT TERMINALS AND MINIMUM WIDTH OF WIRING GUTTERS (see Note 1)

ALL OTHER CONDUCTORS	COMPACT STRANDED AA-8000 ALUMINUM ALLOY CONDUCTORS (see Note 2)	WIRES PER TERMINAL			
		One		Two	
		inches	mm	inches	mm
14-10	12-8	Not specified	Not specified	—	—
8-6	6-4	$1^1/_2$	38.1	—	—
4-3	2-1	2	50.8	—	—
2	1/0	$2^1/_2$	63.5	—	—
1	2/0	3	76.2	—	—
1/0-2/0	3/0-4/0	$3^1/_2$	88.9	5	127
3/0-4/0	250-300	4	102	6	152
250	350	$4^1/_2$	114	6	152
300-350	400-500	5	127	8	203
400-500	600-750	6	152	8	203
600-700	800-1000	8	203	10	254
750-900	—	8	203	12	305
1000-1250	—	10	254	—	—
1500-2000	—	12	305	—	—

Notes:

1. Bending space at terminals shall be measured in a straight line from the end of the lug or wire connector (in the direction that the wire leaves the terminal) to the wall, barrier, or obstruction.
2. This column shall be permitted to be used to determine the minimum wire-bending space for compact stranded aluminum conductors in sizes up to 1000 kcmil and manufactured using AA-8000 series electrical grade aluminum alloy conductor material in accordance with 310.106(B). The minimum width of the wire gutter space shall be determined using the all other conductors value in this table.

grounding conductor where any of the following conditions apply:

1. Where within 8 feet (2438 mm) vertically or 5 feet (1524 mm) horizontally of earth or grounded metal objects and subject to contact by persons;
2. Where located in a wet or damp location and not isolated; or
3. Where in electrical contact with metal. (250.110)

E3908.3 Specific equipment fastened in place (fixed) or connected by permanent wiring methods. Exposed, normally noncurrent-carrying metal parts of the following equipment and enclosures shall be connected to an equipment grounding conductor:

1. Luminaires as provided in Chapter 40. [250.112(J)]
2. Motor-operated water pumps, including submersible types. Where a submersible pump is used in a metal well casing, the well casing shall be connected to the pump circuit equipment grounding conductor. [250.112(L) & (M)]

E3908.4 Effective ground-fault current path. Electrical equipment and wiring and other electrically conductive material likely to become energized shall be installed in a manner that creates a low-impedance circuit facilitating the operation of the overcurrent device or ground detector for high-impedance grounded systems. Such circuit shall be capable of safely carrying the maximum ground-fault current likely to be imposed on it from any point on the wiring system where a ground fault might occur to the electrical supply source. [250.(A)(5)]

E3908.5 Earth as a ground-fault current path. The earth shall not be considered as an effective ground-fault current path. [250.4(A)(5)]

E3908.6 Line-side grounded (neutral) conductor. A grounded circuit conductor shall be permitted to be connected to noncurrent-carrying metal parts of equipment, raceways, and other enclosures at any of the following locations:

1. On the supply side or within the enclosure of the ac service disconnecting means.
2. On the supply side or within the enclosure of the main disconnecting means for separate buildings as provided in Section E3607.3.2. [250.142 (A) (1) & (2)]

E3908.7 Load-side grounded (neutral) conductor. A grounded conductor shall not be connected to normally noncurrent-carrying metal parts of equipment, to equipment grounding conductor(s), or be reconnected to ground on the load side of the service disconnecting means. [250.24(A)(5)]

E3908.8 Load-side equipment. A grounded circuit conductor shall not be used for grounding noncurrent-carrying metal parts of equipment on the load side of the service disconnecting means. [250.142(B)]

E3908.9 Types of equipment grounding conductors. The equipment grounding conductor run with or enclosing the circuit conductors shall be one or more or a combination of the following:

1. A copper, aluminum or copper-clad aluminum conductor. This conductor shall be solid or stranded; insulated, covered or bare; and in the form of a wire or a busbar of any shape. [250.118(1)]
2. Rigid metal conduit. [250.118(2)]
3. Intermediate metal conduit. [250.118(3)]
4. Electrical metallic tubing. [250.118(4)]
5. Armor of Type AC cable in accordance with Section E3908.4. [250.118(8)]
6. Type MC cable that provides an effective ground-fault current path in accordance with one or more of the following:
   6.1. It contains an insulated or uninsulated equipment grounding conductor in compliance with Item 1 of this section.
   6.2. The combined metallic sheath and uninsulated equipment grounding/bonding conductor of interlocked metal tape-type MC cable that is *listed* and identified as an equipment grounding conductor.
   6.3. The metallic sheath or the combined metallic sheath and equipment grounding conductors of the smooth or corrugated tube-type MC cable that is *listed* and identified as an equipment grounding conductor. [250.118(10)]
7. Other electrically continuous metal raceways and auxiliary gutters. [250.118(13)]
8. Surface metal raceways *listed* for grounding. [250.118(14)]

E3908.9.1 Flexible metal conduit. Flexible metal conduit shall be permitted as an equipment grounding conductor where all of the following conditions are met:

1. The conduit is terminated in *listed* fittings.
2. The circuit conductors contained in the conduit are protected by overcurrent devices rated at 20 amperes or less.
3. The size of the conduit does not exceed trade size $1^1/_4$.
4. The combined length of flexible metal conduit and flexible metallic tubing and liquid-tight flexible metal conduit in the same ground return path does not exceed 6 feet (1829 mm).

If used to connect equipment where flexibility is necessary to minimize the transmission of vibration from equipment or to provide flexibility for equipment that requires movement after installation, an equipment grounding conductor shall be installed. [250.118(5)]

E3908.9.2 Liquid-tight flexible metal conduit. Liquid-tight flexible metal conduit shall be permitted as an equipment grounding conductor where all of the following conditions are met:

1. The conduit is terminated in *listed* fittings.
2. For trade sizes $^3/_8$ through $^1/_2$ (metric designator 12 through 16), the circuit conductors contained in the conduit are protected by overcurrent devices rated at 20 amperes or less.
3. For trade sizes $^3/_4$ through $1^1/_4$ (metric designator 21 through 35), the circuit conductors contained in the conduit are protected by overcurrent devices rated at not more than 60 amperes and there is no flexible metal conduit, flexible metallic tubing, or liquid-tight flexible metal conduit in trade sizes $^3/_8$ inch or $^1/_2$ inch (9.5 mm through 12.7 mm) in the ground fault current path.
4. The combined length of flexible metal conduit and flexible metallic tubing and liquid-tight flexible metal conduit in the same ground return path does not exceed 6 feet (1829 mm).

If used to connect equipment where flexibility is necessary to minimize the transmission of vibration from equipment or to provide flexibility for equipment that requires movement after installation, an equipment grounding conductor shall be installed. [250.118(6)]

E3908.9.3 Nonmetallic sheathed cable (Type NM). In addition to the insulated conductors, the cable shall have an insulated, covered, or bare equipment grounding conductor. Equipment grounding conductors shall be sized in accordance with Table E3908.13. (334.108)

E3908.10 Equipment fastened in place or connected by permanent wiring methods. Noncurrent-carrying metal parts of equipment, raceways and other enclosures, where required to be grounded, shall be grounded by one of the following methods: (250.134)

1. By any of the equipment grounding conductors permitted by Sections E3908.9 through E3908.9.3. [250.134(A)]
2. By an equipment grounding conductor contained within the same raceway, cable or cord, or otherwise run with the circuit conductors. Equipment grounding conductors shall be identified in accordance with Section E3407.2. [250.134(B)]

E3908.11 Methods of equipment grounding. Fixtures and equipment shall be considered grounded where mechanically connected to an equipment grounding conductor as specified in Sections E3908.9 through E3908.9.3. Wire type equipment grounding conductors shall be sized in accordance with Section E3908.13. (250 Part VII)

E3908.12 Equipment grounding conductor installation. Where an equipment grounding conductor consists of a raceway, cable armor or cable sheath or where such conductor is a wire within a raceway or cable, it shall be installed in accordance with the provisions of this chapter and Chapters 34 and 38 using fittings for joints and terminations approved for installation with the type of raceway or cable used. All connections, joints and fittings shall be made tight using suitable tools. [250.120(A)]

E3908.12.1 Aluminum and copper-clad aluminum equipment grounding conductors. Equipment grounding conductors of bare, covered, or insulated aluminum or copper-clad aluminum shall comply with the following:

1. Unless part of a suitable Chapter 38 cable wiring method, bare or covered conductors shall not be installed where subject to corrosive conditions or be installed in direct contact with concrete, masonry or the earth.
2. Terminations made within outdoor enclosures that are *listed* and identified for the environment shall be permitted within 18 inches (457 mm) of the bottom of the enclosure.
3. Aluminum or copper-clad aluminum conductors external to buildings or enclosures shall not be terminated within 18 inches (457 mm) of the earth, unless terminated within a *listed* wire connector system. [250.120(B)]

E3908.13 Equipment grounding conductor size. Copper, aluminum and copper-clad aluminum equipment grounding conductors of the wire type shall be not smaller than shown in Table E3908.13, but they shall not be required to be larger than the circuit conductors supplying the equipment. Where a raceway or a cable armor or sheath is used as the equip-

TABLE E3908.13 (Table 250.122)
EQUIPMENT GROUNDING CONDUCTOR SIZING

RATING OR SETTING OF AUTOMATIC OVERCURRENT DEVICE IN CIRCUIT AHEAD OF EQUIPMENT, CONDUIT, ETC., NOT EXCEEDING THE FOLLOWING RATINGS (amperes)	MINIMUM SIZE	
	Copper wire No. (AWG)	Aluminum or copper-clad aluminum wire No. (AWG)
15	14	12
20	12	10
60	10	8
100	8	6
200	6	4
300	4	2
400	3	1

ment grounding conductor, as provided in Section E3908.9, it shall comply with Section E3908.4. Where ungrounded conductors are increased in size for any reason other than as required in Section E3705.2 or E3705.3, wire-type equipment grounding conductors shall be increased in size proportionally according to the circular mil area of the ungrounded conductors. [250.122(A) and (B)]

E3908.13.1 Multiple circuits. A single equipment grounding conductor shall be permitted to be installed for multiple circuits that are installed in the same raceway, cable or trench. It shall be sized in accordance with Table E3908.13 for the largest overcurrent device protecting circuit conductors in the raceway, cable or trench. [250.122(C)]

E3908.14 Continuity of equipment grounding conductors and attachment to boxes. Where circuit conductors are spliced within a box or terminated on equipment within or supported by a box, all wire-type equipment grounding conductors associated with any of those circuit conductors shall be connected within the box or to the box with devices suitable for the use in accordance with Section E3406.13.1. Connections depending solely on solder shall not be used. Splices shall be made in accordance with Section E3406.10 except that insulation shall not be required. The arrangement of grounding connections shall be such that the disconnection or removal of a receptacle, luminaire or other device fed from the box will not interfere with or interrupt the grounding continuity. [250.8(B), 250.148(A) and (B)]

E3908.15 Connecting receptacle grounding terminal to an equipment grounding conductor. An equipment bonding jumper, sized in accordance with Table E3908.13 based on the rating of the overcurrent device protecting the circuit conductors, shall be used to connect the grounding terminal of a grounding-type receptacle to a metal box that is connected to an equipment grounding conductor except as permitted by of the following: (250.146)

1. Surface-mounted box. Where a metal box is mounted on the surface, the direct metal-to-metal contact between the device yoke or strap to the box shall be permitted to provide the required effective ground fault current path. At least one of the insulating washers shall be removed from receptacles that do not have a contact yoke or device to ensure direct metal-to-metal contact. Direct metal-to-metal contact for providing continuity applies to cover-mounted receptacles where the box and cover combination are *listed* as providing satisfactory continuity between the box and the receptacle. A *listed* exposed work cover shall be considered to be the grounding and bonding means where the device is attached to the cover with at least two fasteners that are permanent, such as a rivet or have a thread locking or screw locking means and where the cover mounting holes are located on a flat non-raised portion of the cover. [250.146(A)]
2. Contact devices or yokes. Contact devices or yokes designed and *listed* for the purpose shall be permitted in conjunction with the supporting screws to establish equipment bonding between the device yoke and flush-type boxes. [250.146(B)]
3. Floor boxes. Floor boxes designed for and *listed* as providing satisfactory continuity between the box and the device shall be permitted. [250.146(C)]

E3908.16 Metal boxes. A connection used for no other purpose shall be made between the metal box and the equipment grounding conductor(s) in accordance with Section E3406.14.1. [250.148(C)]

E3908.17 Nonmetallic boxes. One or more equipment grounding conductors brought into a nonmetallic outlet box shall be arranged to allow connection of any fittings or devices installed in that box to an equipment grounding conductor. [250.148(D)]

E3908.18 Clean surfaces. Nonconductive coatings such as paint, lacquer and enamel on equipment to be grounded or bonded shall be removed from threads and other contact surfaces to ensure good electrical continuity or the equipment shall be connected by means of fittings designed so as to make such removal unnecessary. (250.12)

E3908.19 Bonding other enclosures. Metal raceways, cable armor, cable sheath, enclosures, frames, fittings and other metal noncurrent-carrying parts that serve as equipment grounding conductors, with or without the use of supplementary equipment grounding conductors, shall be effectively bonded where necessary to ensure electrical continuity and the capacity to conduct safely any fault current likely to be imposed on them. Any nonconductive paint, enamel and similar coating shall be removed at threads, contact points and contact surfaces, or connections shall be made by means of fittings designed so as to make such removal unnecessary. [250.96(A)]

E3908.20 Size of equipment bonding jumper on load side of an overcurrent device. The equipment bonding jumper on the load side of an overcurrent devices shall be sized, as a minimum, in accordance with Table E3908.13, but shall not be required to be larger than the circuit conductors supplying the equipment. An equipment bonding conductor shall be not smaller than No. 14 AWG.

A single common continuous equipment bonding jumper shall be permitted to connect two or more raceways or cables where the bonding jumper is sized in accordance with Table E3908.13 for the largest overcurrent device supplying circuits therein. [250.102(D) and 250.122]

E3908.21 Installation equipment bonding jumper. Bonding jumpers or conductors and equipment bonding jumpers shall be installed either inside or outside of a raceway or an enclosure in accordance with Sections E3908.21.1 and E3908.21.2. [250.102(E)]

E3908.21.1 Inside raceway or enclosure. Where installed inside a raceway or enclosure, equipment bonding jumpers and bonding jumpers or conductors shall comply with the requirements of Sections E3407.2 and E3908.14. [250.102(E)(1)]

E3908.21.2 Outside a raceway or an enclosure. Where installed outside of a raceway or enclosure, the length of the bonding jumper or conductor or equipment bonding jumper shall not exceed 6 feet (1829 mm) and shall be routed with the raceway or enclosure. [250.102(E)(2)]

Equipment bonding jumpers and supply-side bonding jumpers installed for bonding grounding electrodes and installed at outdoor pole locations for the purpose of bonding or grounding isolated sections of metal raceways or elbows installed in exposed risers of metal conduit or other metal raceway, shall not be limited in length and shall not be required to be routed with a raceway or enclosure. [250.102(E)(2) Exception]

E3908.21.3 Protection. Bonding jumpers or conductors and equipment bonding jumpers shall be installed in accordance with Section E3610.2. [250.102(E)(3)]

SECTION E3909 FLEXIBLE CORDS AND FLEXIBLE CABLES

E3909.1 Where permitted. Flexible cords and flexible cables shall be used only for the connection of appliances where the fastening means and mechanical connections of such appliances are designed to permit ready removal for maintenance, repair or frequent interchange and the appliance is *listed* for flexible cord connection. Flexible cords, flexible cables, cord sets and power supply cords shall not be installed as a substitute for the fixed wiring of a structure; shall not be run through holes in walls, structural ceilings, suspended ceilings, dropped ceilings or floors; shall not be concealed behind walls, floors, ceilings or located above suspended or dropped ceilings; and shall not be attached to building surfaces. (400.10 and 400.12)

E3909.2 Loading and protection. The ampere load of flexible cords serving fixed appliances shall be in accordance with Table E3909.2. This table shall be used in conjunction with applicable end use product standards to ensure selection of the proper size and type. Where flexible cord is approved for and used with a specific *listed* appliance, it shall be considered to be protected where applied within the appliance listing requirements. [240.4, 240.5(A), 240.5(B)(1), 400.5, and 400.16]

**TABLE E3909.2 [Table 400.5(A)(1)]
MAXIMUM AMPERE LOAD FOR FLEXIBLE CORDS**

CORD SIZE (AWG)	CORD TYPES S, SE, SEO, SJ, SJE, SJEO, SJO, SJOO, SJT, SJTO, SJTOO, SO, SOO, SRD, SRDE, SRDT, ST, STD, SV, SVO, SVOO, SVTO, SVTOO	
	Maximum ampere load	
	Three current-carrying conductors	Two current-carrying conductors
18	7	10
16	10	13
14	15	18
12	20	25

E3909.3 Splices. Flexible cord shall be used only in continuous lengths without splices or taps. (400.13)

E3909.4 Attachment plugs. Where used in accordance with Section E3909.1, each flexible cord shall be equipped with an attachment plug and shall be energized from a receptacle outlet. [400.10(B)]

CHAPTER 40
DEVICES AND LUMINAIRES

ICC user note:

About this chapter: Chapter 40 addresses the "trim-out" (final) stage of construction in which devices and fixtures are installed and connected to the installed wiring system.

This chapter covers receptacle ratings and installation, lighting fixture installation, construction and location, and grounding of devices and fixtures.

SECTION E4001 SWITCHES

E4001.1 Rating and application of snap switches. Switches shall be *listed* and used within their ratings and shall control only the following loads:

1. Resistive and inductive loads not exceeding the ampere rating of the switch at the voltage involved.
2. Tungsten-filament lamp loads not exceeding the ampere rating of the switch at 120 volts.
3. Motor loads not exceeding 80 percent of the ampere rating of the switch at its rated voltage.
4. Electric discharge lamp loads not exceeding the marked ampere and voltage rating of the switch.
5. Electronic ballasts, self-ballasted lamps, compact fluorescent lamps, and LED lamp loads with their associated drivers, not exceeding 20 amperes and not exceeding the ampere rating of the switch at the voltage applied. [404.14(A)]

E4001.2 CO/ALR snap switches. Snap switches rated 20 amperes or less directly connected to aluminum conductors shall be marked CO/ALR. [404.14(C)]

E4001.3 Indicating. General-use and motor-circuit switches and circuit breakers shall clearly indicate in a location that is visible when accessing the external operating means whether they are in the open OFF or closed ON position. Where single-throw switches or circuit breaker handles are operated vertically rather than rotationally or horizontally, the up position of the handle shall be the closed (on) position. [404.7]

E4001.4 Time switches and similar devices. Time switches and similar devices shall be of the enclosed type or shall be mounted in cabinets or boxes or equipment enclosures. A barrier shall be used around energized parts to prevent operator exposure when making manual adjustments or switching. (404.5)

E4001.5 Grounding of enclosures. Metal enclosures for switches or circuit breakers shall be connected to an equipment grounding conductor. Metal enclosures for switches or circuit breakers used as service equipment shall comply with the provisions of Section E3609.4. Where nonmetallic enclosures are used with metal raceways or metal-armored cables, provisions shall be made for connecting the equipment grounding conductor.

Nonmetallic boxes for switches shall be installed with a wiring method that provides or includes an equipment grounding conductor. (404.12)

E4001.6 Access. Switches and circuit breakers used as switches shall be located to allow operation from a readily accessible location. Such devices shall be installed so that the center of the grip of the operating handle of the switch or circuit breaker, when in its highest position, will not be more than 6 feet 7 inches (2007 mm) above the floor or working platform. [404.8(A)]

Exception: This section shall not apply to switches and circuit breakers that are accessible by portable means and are installed adjacent to the motors, appliances and other equipment that they supply. [404.8(A) Exception]

E4001.7 Damp or wet locations. A surface mounted switch or circuit breaker located in a damp or wet location or outside of a building shall be enclosed in a weatherproof enclosure or cabinet. A flush-mounted switch or circuit breaker in a damp or wet location shall be equipped with a weatherproof cover. Switches shall not be installed within wet locations in tub or shower spaces unless installed as part of a *listed* tub or shower assembly. [404.4(A), (B), and (C)]

E4001.8 Grounded conductors. Switches or circuit breakers shall not disconnect the grounded conductor of a circuit except where the switch or circuit breaker simultaneously disconnects all conductors of the circuit. [404.2(B)]

E4001.9 Switch connections. Three- and four-way switches shall be wired so that all switching occurs only in the ungrounded circuit conductor. Color coding of switch connection conductors shall comply with Section E3407.3. Where in metal raceways or metal-jacketed cables, wiring between switches and outlets shall be in accordance with Section E3406.7. [404.2(A)]

Exception: Switch loops do not require a grounded conductor. [404.2(A) Exception]

E4001.10 Box mounted. General-use snap switches, dimmers and control switches mounted in boxes that are recessed from the finished wall surfaces as covered in Section E3906.5 shall be installed so that the extension plaster ears are seated against the surface of the wall. Flush-type devices mounted in boxes that are flush with the finished wall surface or project therefrom shall be installed so that the mounting yoke or strap of the device is seated against the box.

Screws used for the purpose of attaching a device to a box shall be of the type provided with a *listed* device, or shall be machine screws having 32 threads per inch or part of *listed* assemblies or systems, in accordance with the manufacturer's instructions. [404.10(B)]

E4001.11 Snap switch faceplates. Faceplates provided for snap switches, dimmers and control devices mounted in boxes and other enclosures shall be installed so as to completely cover the opening and, where the switch is flush mounted, seat against the finished surface. [404.9(A)]

E4001.11.1 Faceplate grounding. Snap switches, including dimmer and similar control switches, shall be connected to an equipment grounding conductor and shall provide a means to connect metal faceplates to the equipment grounding conductor, whether or not a metal faceplate is installed. Metal faceplates shall be bonded to the equipment grounding conductor. Snap switches, dimmers, control switches and metal faceplates shall be connected to an equipment grounding conductor using either of the following methods:

1. The switch is mounted with metal screws to a metal box or metal cover that is connected to an equipment grounding conductor or to a nonmetallic box with integral means for connecting to an equipment grounding conductor.
2. An equipment grounding conductor or equipment bonding jumper is connected to an equipment grounding termination of the snap switch. [404.9(B)]

Exceptions:

1. Where a means to connect to an equipment grounding conductor does not exist within the snap-switch enclosure or where the wiring method does not include or provide an equipment grounding conductor, a snap switch without a grounding connection to an equipment grounding conductor shall be permitted for replacement purposes only. A snap switch wired under the provisions of this exception and located within 8 feet (2438 mm) vertically or 5 feet (1524 mm) horizontally of ground or exposed grounded metal objects, shall be provided with a faceplate of nonconducting *noncombustible material* with nonmetallic attachment screws, except where the switch-mounting strap or yoke is nonmetallic or the circuit is protected by a ground-fault circuit interrupter. [404.9(B) Exception No.1]
2. Listed kits or listed assemblies shall not be required to be connected to an equipment grounding conductor if all of the following conditions apply:
   - 2.1. The device is provided with a nonmetallic faceplate and the device is designed such that no metallic faceplate replaces the one provided.
   - 2.2. The device does not have mounting means to accept other configurations of faceplates.
   - 2.3. The device is equipped with a nonmetallic yoke.
   - 2.4. All parts of the device that are accessible after installation of the faceplate are manufactured of nonmetallic materials. [404.9(B) Exception No. 2]
3. Connection to an equipment grounding conductor shall not be required for snap switches that have an integral nonmetallic enclosure complying with Section E3905.1.3. [404.9(B) Exception No. 3]

E4001.12 Dimmer and electronic control switches. General-use dimmer switches shall be used only to control permanently installed incandescent luminaires (lighting fixtures) except where *listed* for the control of other loads and installed accordingly. Other electronic control switches, such as timing switches and occupancy sensors, shall be used to control permanently connected loads. They shall be marked by their manufacturer with their current and voltage ratings at the voltage approved. [404.14(E)]

E4001.13 Multipole snap switches. A multipole, general-use snap switch shall not be fed from more than a single circuit unless it is *listed* and marked as a two-circuit or three-circuit switch. [404.8(C)]

E4001.14 Cord-and-plug-connected loads. Where snap switches are used to control cord-and-plug-connected equipment on a general-purpose branch circuit, each snap switch controlling receptacle outlets or cord connectors that are supplied by permanently connected cord pendants shall be rated at not less than the rating of the maximum permitted ampere rating or setting of the overcurrent device protecting the receptacles or cord connectors, as provided in Sections E4002.1.1 and E4002.1.2. [404.14(F)]

E4001.15 Switches controlling lighting loads. The grounded circuit conductor for the controlled lighting circuit shall be installed at the location where switches control lighting loads that are supplied by a grounded general-purpose branch circuit serving bathrooms, hallways, *stairways*, and habitable rooms or occupiable spaces as defined in the code. Where multiple switch locations control the same lighting load such that the entire floor area of the room or space is visible from the single or combined switch locations, the grounded conductor shall be required only at one location. A grounded conductor shall not be required to be installed at lighting switch locations under any of the following conditions:

1. Where conductors enter the box enclosing the switch through a raceway, provided that the raceway is large enough for all contained conductors, including a grounded conductor.
2. Where the box enclosing the switch is accessible for the installation of an additional or replacement cable without removing finish materials.

3. Where snap switches with integral enclosures comply with Section E3905.1.3.
4. Where lighting in the area is controlled by automatic means.
5. Where a switch controls a receptacle load. [404.2(C)]

Effective January 1, 2020, the grounded conductor shall be extended to any switch location as necessary and shall be connected to switching devices that require line-to-neutral voltage to operate the electronics of the switch in the standby mode.

The requirement for connection to switching devices shall not apply to replacement or retrofit switches installed in locations prior to the adoption of Section E4001.15 and where the grounded conductor cannot be extended without removing finish materials. The number of electronic lighting control switches on a branch circuit shall not exceed five, and the number connected to any feeder on the load side of a system or main bonding jumper shall not exceed 25.

SECTION E4002 RECEPTACLES

E4002.1 Rating and type. Receptacles and cord connectors shall be rated at not less than 15 amperes, 125 volts, or 15 amperes, 250 volts, and shall not be a lampholder type. Receptacles shall be rated in accordance with this section. [406.3(B)]

E4002.1.1 Single receptacle. A single receptacle installed on an individual branch circuit shall have an ampere rating not less than that of the branch circuit. [210.21(B)(1)]

E4002.1.2 Two or more receptacles. Where connected to a branch circuit supplying two or more receptacles or outlets, receptacles shall conform to the values listed in Table E4002.1.2. [210.21(B)(3)]

**TABLE E4002.1.2 [Table 210.21(B)(3)]
RECEPTACLE RATINGS FOR
VARIOUS SIZE MULTIOUTLET CIRCUITS**

CIRCUIT RATING (amperes)	RECEPTACLE RATING (amperes)
15	15
20	15 or 20
30	30
40	40 or 50
50	50

E4002.2 Grounding type. Receptacles installed on 15- and 20-ampere-rated branch circuits shall be of the grounding type. [406.4(A)]

E4002.3 CO/ALR receptacles. Receptacles rated at 20 amperes or less and directly connected to aluminum conductors shall be marked CO/ALR. [406.3(C)]

E4002.4 Faceplates. Metal face plates shall be grounded. [406.6(B)]

E4002.5 Position of receptacle faces. After installation, receptacle faces shall be flush with or project from face plates of insulating material and shall project a minimum of 0.015 inch (0.381 mm) from metal face plates. Faceplates shall be installed so as to completely cover the opening and seat against the mounting surface.

Receptacle faceplates mounted inside of a box having a recess-mounted receptacle shall effectively close the opening and seat against the mounting surface. [406.5(D), 406.6]

Exception: *Listed* kits or assemblies encompassing receptacles and nonmetallic faceplates that cover the receptacle face, where the plate cannot be installed on any other receptacle, shall be permitted. [406.5(D) Exception]

E4002.6 Receptacles mounted in boxes. Receptacles mounted in boxes that are set back from the finished wall surface as permitted by Section E3906.5 shall be installed so that the mounting yoke or strap of the receptacle is held rigidly at the finished surface of the wall. Screws used for the purpose of attaching receptacles to a box shall be of the type provided with a *listed* receptacle, or shall be machine screws having 32 threads per inch or part of *listed* assemblies or systems, in accordance with the manufacturer's instructions. Receptacles mounted in boxes that are flush with the wall surface or project therefrom shall be so installed that the mounting yoke or strap is seated against the box or raised cover. [406.5(A) and (B)]

E4002.7 Receptacles mounted on covers. Receptacles mounted to and supported by a cover shall be held rigidly against the cover by more than one screw or shall be a device assembly or box cover *listed* and identified for securing by a single screw. [406.5(C)]

E4002.8 Damp locations. A receptacle installed outdoors in a location protected from the weather or in other damp locations shall have an enclosure for the receptacle that is weatherproof when the receptacle cover(s) is closed and an attachment plug cap is not inserted. An installation suitable for wet locations shall also be considered suitable for damp locations. A receptacle shall be considered to be in a location protected from the weather where located under roofed open porches, canopies and similar structures and not subject to rain or water runoff. Fifteen- and 20-ampere, 125- and 250-volt nonlocking receptacles installed in damp locations shall be *listed* a weather-resistant type. [406.9(A)]

E4002.9 Fifteen- and 20-ampere receptacles in wet locations. Where installed in a wet location, 15- and 20-ampere, 125- and 250-volt receptacles shall have an enclosure that is weatherproof whether or not the attachment plug cap is inserted. An outlet box hood installed for this purpose shall be *listed* and identified as "extra-duty."

Exception: 15- and 20-ampere, 125- through 250-volt receptacles installed in a wet location and subject to routine high-pressure spray washing need not have an enclosure that is weatherproof when the attachment plug is inserted.

Fifteen- and 20-ampere, 125- and 250-volt nonlocking receptacles installed in wet locations shall be *listed* and so identified as the weather-resistant type. [406.9(B)(1)]

E4002.10 Other receptacles in wet locations. Where a receptacle other than a 15- or 20-amp, 125- or 250-volt receptacle is installed in a wet location and where the product intended to be plugged into it is not attended while in use, the receptacle shall have an enclosure that is weatherproof both when the attachment plug cap is inserted and when it is removed. Where such receptacle is installed in a wet location and where the product intended to be plugged into it will be attended while in use, the receptacle shall have an enclosure that is weatherproof when the attachment plug cap is removed. [406.9(B)(2)]

E4002.11 Bathtub and shower space. Receptacles shall not be installed within a zone measured 3 feet (90 mm) horizontally and 8 feet (2438 mm) vertically from the top of the bathtub rim or shower stall threshold. The identified zone is all-encompassing and shall include the space directly over the tub or shower stall.

Exception: In bathrooms with less than the required zone, the receptacle(s) shall be permitted to be installed opposite the bathtub rim or shower stall threshold on the farthest wall in the room. [406.9(C)]

E4002.12 Flush mounting with faceplate. In damp or wet locations, the enclosure for a receptacle installed in an outlet box flush-mounted in a finished surface shall be made weatherproof by means of a weatherproof faceplate assembly that provides a watertight connection between the plate and the finished surface. [406.9(E)]

E4002.13 Exposed terminals. Receptacles shall be enclosed so that live wiring terminals are not exposed to contact. [406.5(I)]

E4002.14 Tamper-resistant receptacles. In areas specified in Section E3901.1, 15- and 20-ampere, 125- and 250-volt nonlocking-type receptacles shall be *listed* tamper-resistant receptacles. [406.12]

Exception: Receptacles in the following locations shall not be required to be tamper resistant:

1. Receptacles located more than 5.5 feet (1676 mm) above the floor.
2. Receptacles that are part of a luminaire or appliance.
3. A single receptacle for a single appliance or a duplex receptacle for two appliances where such receptacles are located in spaces dedicated for the appliances served and, under conditions of normal use, the appliances are not easily moved from one place to another. The appliances shall be cord-and-plug-connected to such receptacles in accordance with Section E3909.4. [406.12 Exception]

E4002.15 Receptacles in countertops. Receptacle assemblies for installation in countertop surfaces shall be *listed* for countertop applications. Receptacle assemblies and GFCI receptacle assemblies installed in work surfaces shall be *listed* for work surface or countertop applications. [406.5 (E) and (F)]

E4002.16 Receptacle position. Receptacles shall not be installed in a face-up position in or on countertops surfaces or work surfaces except where the receptacles are *listed* for countertop or work surface applications. [406.5 (G)]

SECTION E4003 LUMINAIRES

E4003.1 Energized parts. Luminaires, lampholders, and lamps shall not have energized parts normally exposed to contact. (410.5)

E4003.2 Luminaires near combustible material. Luminaires shall be installed or equipped with shades or guards so that *combustible material* will not be subjected to temperatures in excess of 90°C (194°F). (410.11)

E4003.3 Exposed conductive parts. Exposed conductive parts that are accessible to unqualified persons a shall be connected to an equipment grounding conductor or be separated from all live parts and other conducting surfaces by a *listed* system of double insulation. Small isolated parts, such as mounting screws, clips and decorative bands on glass spaced at least $1^1/_2$ inches (38 mm) from lamp terminals, shall not require connection to an equipment grounding conductor. Portable luminaires with a polarized attachment plug shall not require connection to an equipment grounding conductor. (410.42)

E4003.4 Screw-shell type. Lampholders of the screw-shell type shall be installed for use as lampholders only. Where supplied by a circuit having a grounded conductor, the grounded conductor shall be connected to the screw shell. (410.90)

E4003.5 Recessed incandescent luminaires. Recessed incandescent luminaires shall have thermal protection and shall be *listed* as thermally protected. [410.115(B)]

Exceptions:

1. Thermal protection shall not be required in recessed luminaires *listed* for the purpose and installed in poured concrete. [410.115(B) Exception No. 1]
2. Thermal protection shall not be required in recessed luminaires having design, construction, and thermal performance characteristics equivalent to that of thermally protected luminaires, and such luminaires are identified as inherently protected. [410.115(B) Exception No. 2]

E4003.6 Thermal protection. The ballast of a fluorescent luminaire installed indoors shall have integral thermal protection. Replacement ballasts shall also have thermal protection integral with the ballast. A simple reactance ballast in a fluorescent luminaire with straight tubular lamps shall not be required to be thermally protected. [410.130(E)(1), (2)]

E4003.7 High-intensity discharge luminaires. Recessed high-intensity luminaires designed to be installed in wall or ceiling cavities shall have thermal protection and be identified as thermally protected. Thermal protection shall not be required in recessed high-intensity luminaires having design, construction and thermal performance characteristics equivalent to that of thermally protected luminaires, and such luminaires are identified as inherently protected. Thermal protection shall not be required in recessed high-intensity discharge luminaires installed in and identified for use in poured concrete. A recessed remote ballast for a high-intensity discharge luminaire shall have thermal protection that is integral with the ballast and shall be identified as thermally protected. [410.130(F)(1), (2), (3), and (4)]

E4003.8 Metal halide lamp containment. Luminaires that use a metal halide lamp other than a thick-glass parabolic reflector lamp (PAR) shall be provided with a containment barrier that encloses the lamp, or shall be provided with a physical means that allows the use of only a lamp that is Type O. [410.130(F)(5)]

E4003.9 Wet or damp locations. Luminaires installed in wet or damp locations shall be installed so that water cannot enter or accumulate in wiring compartments, lampholders or other electrical parts. All luminaires installed in wet locations shall be marked "SUITABLE FOR WET LOCATIONS." All luminaires installed in damp locations shall be marked "SUITABLE FOR WET LOCATIONS" or "SUITABLE FOR DAMP LOCATIONS." (410.10)

E4003.10 Lampholders in wet or damp locations. Lampholders installed in wet locations shall be *listed* for use in wet locations. Lampholders installed in damp locations shall be *listed* for damp locations or shall be *listed* for wet locations. (410.96)

E4003.11 Bathtub and shower areas. Cord-connected luminaires, chain-, cable-, or cord-suspended-luminaires, lighting track, pendants, and ceiling-suspended (paddle) fans shall not have any parts located within a zone measured 3 feet (914 mm) horizontally and 8 feet (2438 mm) vertically from the top of a bathtub rim or shower stall threshold. This zone is all encompassing and includes the space directly over the tub or shower. Luminaires within the actual outside dimension of the bathtub or shower to a height of 8 feet (2438 mm) vertically from the top of the bathtub rim or shower threshold shall be marked for damp locations and where subject to shower spray, shall be marked for wet locations. [410.10(D)]

E4003.12 Luminaires in clothes-closet storage space. For the purposes of this section, storage space shall be defined as a volume bounded by the sides and back closet walls and planes extending from the closet floor vertically to a height of 6 feet (1829 mm) or the highest clothes-hanging rod and parallel to the walls at a horizontal distance of 24 inches (610 mm) from the sides and back of the closet walls respectively, and continuing vertically to the closet ceiling parallel to the walls at a horizontal distance of 12 inches (305 mm) or the width of the shelf, whichever is greater. For a closet that permits access to both sides of a hanging rod, the storage space shall include the volume below the highest rod extending 12 inches (305 mm) on either side of the rod on a plane horizontal to the floor extending the entire length of the rod (see Figure E4003.12). (410.2)

The types of luminaires installed in clothes closets shall be limited to surface-mounted or recessed incandescent or LED luminaires with completely enclosed light sources, surface-mounted or recessed fluorescent luminaires, and surface-mounted fluorescent or LED luminaires identified as suitable for installation within the closet storage area. Incandescent luminaires with open or partially enclosed lamps and pendant luminaires or lamp-holders shall be prohibited. The minimum clearance between luminaires installed in clothes closets and the nearest point of a closet storage area shall be as follows: [410.16(A) and (B)]

1. Surface-mounted incandescent or LED luminaires with a completely enclosed light source shall be installed on the wall above the door or on the ceiling, provided that there is a minimum clearance of 12 inches (305 mm) between the fixture and the nearest point of a storage space.
2. Surface-mounted fluorescent luminaires shall be installed on the wall above the door or on the ceiling, provided that there is a minimum clearance of 6 inches (152 mm).
3. Recessed incandescent luminaires or LED luminaires with a completely enclosed light source shall be installed in the wall or the ceiling provided that there is a minimum clearance of 6 inches (152 mm).
4. Recessed fluorescent luminaires shall be installed in the wall or on the ceiling provided that there is a minimum clearance of 6 inches (152 mm) between the fixture and the nearest point of a storage space.
5. Surface-mounted fluorescent or LED luminaires shall be permitted to be installed within the closet storage space where identified for this use. [410.16(C)]

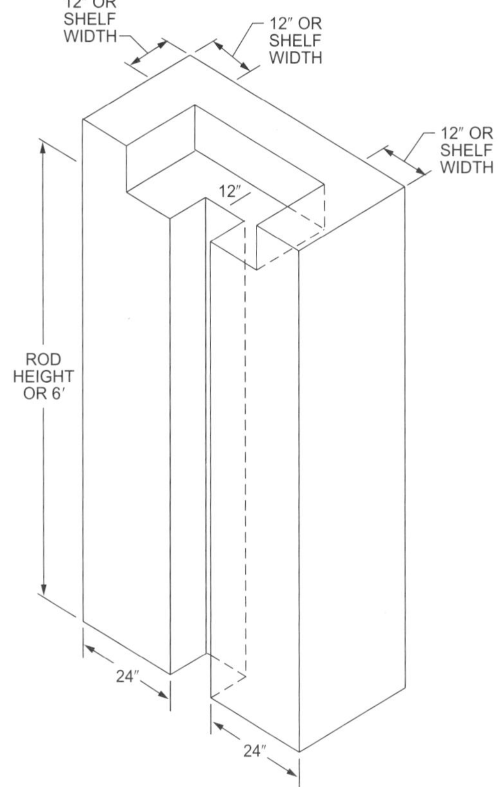

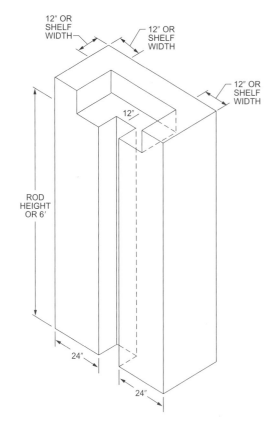

For SI: 1 inch = 25.4 mm, 1 foot = 304.8 mm.

FIGURE E4003.12 CLOSET STORAGE SPACE

E4003.13 Luminaire wiring—general. Wiring on or within luminaires shall be neatly arranged and shall not be exposed to physical damage. Excess wiring shall be avoided. Conductors shall be arranged so that they are not subjected to temperatures above those for which the conductors are rated. (410.48)

E4003.13.1 Polarization of luminaires. Luminaires shall be wired so that the screw shells of lampholders will be connected to the same luminaire or circuit conductor or terminal. The grounded conductor shall be connected to the screw shell. (410.50)

E4003.13.2 Luminaires as raceways. Luminaires shall not be used as raceways for circuit conductors except where such luminaires are *listed* and marked for use as a raceway or are identified for through-wiring.

Luminaires designed for end-to-end connection to form a continuous assembly, and luminaires connected together by recognized wiring methods, shall not be required to be *listed* as a raceway where they contain the conductors of one 2-wire branch circuit or one multiwire branch circuit and such conductors supply the connected luminaires. One additional 2-wire branch circuit that separately supplies one or more of the connected luminaires shall also be permitted. [410.64(A), (B), and (C)]

E4003.14 Reconditioned equipment. Luminaires, lampholders and retrofit kits shall not be permitted to be reconditioned. If a retrofit kit is installed in a luminaire in accordance with the installation instructions, the retrofitted luminaire shall not be considered reconditioned.

SECTION E4004 LUMINAIRE INSTALLATION

E4004.1 Outlet box covers. In a completed installation, each outlet box shall be provided with a cover except where covered by means of a luminaire canopy, lampholder, receptacle that covers the box or is provided with a faceplate, or similar device.(410.22)

E4004.2 Combustible material at outlet boxes. Combustible wall or ceiling finish exposed between the inside edge of a luminaire canopy or pan and the outlet box and having a surface area of 180 square inches (1160 mm^2) or more shall be covered with a *noncombustible material.* (410.23)

E4004.3 Access. Luminaires shall be installed so that the connections between the luminaire conductors and the circuit conductors can be accessed without requiring the disconnection of any part of the wiring. Luminaires that are connected by attachment plugs and receptacles meet the requirement of this section. (410.8)

E4004.4 Supports. Luminaires and lampholders shall be securely supported. A luminaire that weighs more than 6 pounds (2.72 kg) or exceeds 16 inches (406 mm) in any dimension shall not be supported by the screw shell of a lampholder. [410.30(A)]

E4004.5 Means of support. Luminaires shall be permitted to be supported by outlet boxes or fittings installed as required by Sections E3905, and E3906. Outlet boxes complying with Section E3906.12 shall be considered lighting outlets as required by Section E3903. [410.36(A)]

E4004.6 Exposed components. Luminaires having exposed ballasts, transformers, LED drivers or power supplies shall be installed so that such ballasts, transformers, LED drivers or power supplies are not in contact with *combustible material* unless *listed* for such condition. [410.136(A)]

E4004.7 Combustible low-density cellulose fiberboard. Where a surface-mounted luminaire containing a ballast, transformer, LED driver or power supply is installed on combustible low-density cellulose fiberboard, the luminaire shall be marked for this purpose or it shall be spaced not less than $1^1/_2$ inches (38 mm) from the surface of the fiberboard. Where such luminaires are partially or wholly recessed, the provisions of Sections E4004.8 and E4004.9 shall apply. [410.136(B)]

E4004.8 Recessed luminaire clearance. A recessed luminaire that is not identified for contact with insulation shall have all recessed parts spaced at least $^1/_2$ inch (12.7 mm) from *combustible materials.* The points of support and the finish trim parts at the opening in the ceiling, wall or other finished surface shall be permitted to be in contact with *combustible materials.* A recessed luminaire that is identified for contact with insulation, Type IC, shall be permitted to be in contact with *combustible materials* at recessed parts, points of support, and portions passing through the building structure and at finish trim parts at the opening in the ceiling or wall. [410.116(A)(1) and (A)(2)]

E4004.9 Recessed luminaire installation. Thermal insulation shall not be installed above a recessed luminaire or within 3 inches (76 mm) of the recessed luminaire's enclosure, wiring compartment, ballast, transformer, LED driver or power supply except where such luminaire is identified for contact with insulation, Type IC. [410.116(B)]

SECTION E4005 TRACK LIGHTING

E4005.1 Installation. Lighting track shall be permanently installed and permanently connected to a branch circuit having a rating not more than that of the track. [410.151(A) and (B)]

E4005.2 Fittings. Fittings identified for use on lighting track shall be designed specifically for the track on which they are to be installed. Fittings shall be securely fastened to the track, shall maintain polarization and connection to the equipment grounding conductor, and shall be designed to be suspended directly from the track. Only lighting track fittings shall be installed on lighting track. Lighting track fittings shall not be equipped with general-purpose receptacles. [410.151(A) and (B)]

E4005.3 Connected load. The connected load on lighting track shall not exceed the rating of the track. Lighting track shall be supplied by a branch circuit having a rating not greater than that of the track. [410.151(B)]

E4005.4 Prohibited locations. Lighting track shall not be installed in the following locations:

1. Where likely to be subjected to physical damage.
2. In wet or damp locations.
3. Where subject to corrosive vapors.
4. In storage battery rooms.
5. In hazardous (classified) locations.
6. Where concealed.
7. Where extended through walls or partitions.
8. Less than 5 feet (1524 mm) above the finished floor except where protected from physical damage or the track operates at less than 30 volts rms open-circuit voltage.
9. Where prohibited by Section E4003.11. [410.151(C)]

E4005.5 Fastening. Lighting track shall be securely mounted so that each fastening will be suitable for supporting the maximum weight of luminaires that can be installed. Except where identified for supports at greater intervals, a single section 4 feet (1219 mm) or shorter in length shall have two supports and, where installed in a continuous row, each individual section of not more than 4 feet (1219 mm) in length shall have one additional support. (410.154)

E4005.6 Grounding. Lighting track shall be grounded in accordance with Chapter 39, and the track sections shall be securely coupled to maintain continuity of the circuitry, polarization and grounding throughout. [410.155(B)]

CHAPTER 41
APPLIANCE INSTALLATION

ICC user note:

About this chapter: *Chapter 41 covers appliance installation, which is typically the final stage of construction after all wiring, devices and fixtures are installed. This chapter covers flexible cords, overcurrent protection, disconnecting means and installation provisions.*

SECTION E4101
GENERAL

E4101.1 Scope. This section covers installation requirements for appliances and fixed heating equipment. (422.1 and 424.1)

E4101.2 Installation. Appliances and equipment shall be installed in accordance with the manufacturer's installation instructions. Electrically heated appliances and equipment shall be installed with the required clearances to *combustible materials.* [110.3(B) and 422.17]

E4101.3 Flexible cords. Cord-and-plug-connected appliances shall use cords suitable for the environment and physical conditions likely to be encountered. Flexible cords shall be used only where the appliance is *listed* to be connected with a flexible cord. The cord shall be identified as suitable in the installation instructions of the appliance manufacturer. Receptacles for cord-and-plug-connected appliances shall be accessible and shall be located to avoid physical damage to the flexible cord. Except for a *listed* appliance marked to indicate that it is protected by a system of double-insulation, the flexible cord supplying an appliance shall terminate in a grounding-type attachment plug. The cord lengths specified for built-in dishwashers and trash compactors shall be measured from the face of the attachment plug to the plane of the rear of the appliance. A receptacle for a cord-and-plug-connected range hood shall be supplied by an individual branch circuit. A receptacle for a built-in dishwasher shall be located in a space adjacent to the space occupied by the dishwasher, and where the flexible cord passes through an opening, it shall be protected against damage by a bushing, grommet or other approved means. Specific appliances have additional requirements as specified in Table E4101.3 (see Section E3909). [422.16(B)(1), (B)(2), (B)(4)]

TABLE E4101.3
FLEXIBLE CORD LENGTH

APPLIANCE	MINIMUM CORD LENGTH (inches)	MAXIMUM CORD LENGTH (inches)
Electrically operated in-sink waste disposal	18	36
Built-in dishwasher	36	78
Trash compactor	36	48
Range hoods	18	48

For SI: 1 inch = 25.4 mm.

E4101.4 Overcurrent protection. Each appliance shall be protected against overcurrent in accordance with the rating of the appliance and its listing. [110.3(B), 422.11(A)]

E4101.4.1 Single nonmotor-operated appliance. The overcurrent protection for a branch circuit that supplies a single nonmotor-operated appliance shall not exceed that marked on the appliance. Where the overcurrent protection rating is not marked and the appliance is rated at over 13.3 amperes, the overcurrent protection shall not exceed 150 percent of the appliance rated current. Where 150 percent of the appliance rating does not correspond to a standard overcurrent device ampere rating, the next higher standard rating shall be permitted. Where the overcurrent protection rating is not marked and the appliance is rated at 13.3 amperes or less, the overcurrent protection shall not exceed 20 amperes. [422.11(E)]

E4101.5 Disconnecting means. Each appliance shall be provided with a means to disconnect all ungrounded supply conductors. For fixed electric space-heating equipment, means shall be provided to disconnect the heater and any motor controller(s) and supplementary overcurrent-protective devices. Switches and circuit breakers used as a disconnecting means shall be of the indicating type. Disconnecting means shall be as set forth in Table E4101.5. (422.30, 422.35, and 424.19)

E4101.6 Support of ceiling-suspended paddle fans. Ceiling-suspended fans (paddle) shall be supported independently of an outlet box; by a *listed* outlet box or outlet box system identified for the use and installed in accordance with Section E3905.8; or by a *listed* outlet box system, a *listed* locking support and mounting receptacle, and a compatible factory installed attachment fitting designed for support, identified for the use and installed in accordance with Section E3905.6.3. (422.18)

E4101.7 Snow-melting and deicing equipment protection. Outdoor receptacles that are not readily accessible and are supplied from a dedicated branch circuit for electric snow-melting or deicing equipment shall be permitted to be installed without ground-fault circuit-interrupter protection for personnel. However, ground-fault protection of equipment shall be provided for fixed outdoor electric deicing and snow-melting equipment. [210.8(A)(3) Exception, 426.28]

E4101.8 Lockable disconnecting means. Where a disconnecting means is required to be lockable, it shall be capable of being locked in the open position. The provisions for locking shall remain in place with or without the lock installed.

Exception: Locking provisions for a cord-and-plug connection shall not be required to remain in place without the lock installed. [110.25]

TABLE E4101.5
[422.31(A), (B), and (C); 422.33; 422.34; 422.35; 424.19; 424.20; and 440.14]
DISCONNECTING MEANS

DESCRIPTION	ALLOWED DISCONNECTING MEANS
Permanently connected appliance rated at not over 300 volt-amperes or $^1/_8$ horsepower.	Branch-circuit overcurrent device where the switch or circuit breaker is within sight of the appliance or is capable of being locked in the open position in compliance with Section E4101.8.
Permanently connected appliances rated in excess of 300 volt-amperes.	Branch circuit switch or circuit breaker located within sight of the appliance or such devices in any location that are capable of being locked in the open position in compliance with Section E4101.8.
Motor-operated appliances rated over $^1/_8$ horsepower.	For permanently connected motor-operated appliances with motors rated over $^1/_8$ horsepower, the disconnecting means shall be within sight from the appliance or it shall be capable of being locked in the open position in compliance with Section E4101.8. The disconnecting means shall be one of the following types: a listed motor-circuit switch rated in horsepower, a listed molded case circuit breaker, a listed molded case switch, a listed manual motor controller additionally marked "Suitable as Motor Disconnect" where installed between the final motor branch-circuit short-circuit protective device and the motor. For stationary motors rated at 2 hp or less and 300 volts or less, the disconnecting means shall be permitted to be one of the following devices: 1. A general-use switch having an ampere rating not less than twice the full-load current rating of the motor. 2. On AC circuits, a general-use snap switch suitable only for use on AC, not general-use AC–DC snap switches, where the motor full-load current rating is not more than 80 percent of the ampere rating of the switch. 3. A listed manual motor controller having a horsepower rating not less than the rating of the motor and marked "Suitable as Motor Disconnect." The disconnecting means shall have an ampere rating not less than 115 percent of the full-load current rating of the motor except that a listed unfused motor-circuit switch having a horsepower rating not less than the motor horsepower shall be permitted to have an ampere rating less than 115 percent of the full-load current rating of the motor. **Exception:** Where an appliance of more than $^1/_8$ hp is provided with a unit switch with a marked-off position that is a part of the appliance and disconnects all ungrounded conductors, such unit switch shall be permitted as the disconnecting means and the switch or circuit breaker serving as the other disconnecting means shall be permitted to be not within sight from the appliance.
Appliances listed for cord-and-plug connection.	A separable connector or attachment plug and receptacle provided with access.

(continued)

TABLE E4101.5—continued
[422.31(A), (B), and (C); 422.33; 422.34; 422.35; 424.19; 424.20; and 440.14]
DISCONNECTING MEANS

DESCRIPTION	ALLOWED DISCONNECTING MEANS
Permanently installed heating equipment with motors rated at not over $^1/_8$ horsepower with supplementary overcurrent protection.	Disconnect, on the supply side of fuses, within sight from the supplementary overcurrent device, and within sight of the heating equipment or, in any location, where the disconnecting means is capable of being locked in the open position in compliance with Section E4101.8.
Heating equipment containing motors rated over $^1/_8$ horsepower with supplementary overcurrent protection.	Disconnect permitted to serve as required disconnect for both the heating equipment and the controller where, on the supply side of fuses, and within sight from the supplementary overcurrent devices, if the disconnecting means is also within sight from the controller, or is capable of being locked in the open position in compliance with Section E4101.8 and simultaneously disconnects the heater, motor controller(s) and supplementary overcurrent protective devices from all ungrounded conductors. The disconnecting means shall have an ampere rating not less than 125 percent of the total load of the motors and the heaters.
Heating equipment containing no motor rated over $^1/_8$ horsepower without supplementary overcurrent protection.	Branch-circuit switch or circuit breaker where within sight from the heating equipment or capable of being locked in the open position in compliance with Section E4101.8 and simultaneously disconnects the heater and motor controller(s) from all ungrounded conductors. The provision for locking or adding a lock to the disconnecting means shall be installed on or at the switch or circuit breaker used as the disconnecting means and shall remain in place with or without the lock installed. The disconnecting means shall have an ampere rating not less than 125 percent of the total load of the motors and the heaters.
Heating equipment containing motors rated over $^1/_8$ horsepower without supplementary overcurrent protection.	Disconnecting means within sight from motor controller or is capable of being locked in the open position in compliance with Section E4101.8 and simultaneously disconnects the heater, motor controller(s) and supplementary overcurrent protective devices from all ungrounded conductors. The disconnecting means shall have an ampere rating not less than 125 percent of the total load of the motors and the heaters.
Air-conditioning condensing units and heat pump units.	A readily accessible disconnect within sight from unit as the only allowable means.[a]
Appliances and fixed heating equipment with unit switches having a marked OFF position.	Unit switch where an additional individual switch or circuit breaker serves as the other required disconnecting means.
Thermostatically controlled fixed heating equipment.	Thermostats with a marked OFF position that directly open all ungrounded conductors, which when manually placed in the OFF position are designed so that the circuit cannot be energized automatically and that are located within sight of the equipment controlled.

For SI: 1 horsepower = 0.746 kW.

a. The disconnecting means shall be permitted to be installed on or within the unit. It shall not be located on panels designed to allow access to the unit or located so as to obscure the air-conditioning equipment nameplate(s).

CHAPTER 42
SWIMMING POOLS

ICC user note:

About this chapter: Chapter 42 addresses all aspects of wiring, fixtures, motors and electrical accessories for swimming pools, wading pools, hot tubs, spas and hydromassage bathtubs.

This chapter focuses on protection of occupants from electrical shock. The dangers of using electricity around water, wet surfaces, grounded surfaces and plumbing are well known, and this chapter is intended to minimize or eliminate those hazards.

SECTION E4201 GENERAL

E4201.1 Scope. The provisions of this chapter shall apply to the construction and installation of electric wiring and equipment associated with all swimming pools, wading pools, decorative pools, hot tubs and spas, and hydromassage bathtubs, whether permanently installed or storable, and shall apply to metallic auxiliary equipment, such as pumps, filters and similar equipment. Sections E4202 through E4206 provide general rules for permanent pools, spas and hot tubs. Section E4207 provides specific rules for storable pools and storable/portable spas and hot tubs. Section E4208 provides specific rules for spas and hot tubs. Section E4209 provides specific rules for hydromassage bathtubs. (680.1)

E4201.2 Definitions. The definitions in this section shall apply only within this chapter. (680.2)

CORD-AND-PLUG-CONNECTED LIGHTING ASSEMBLY. A lighting assembly consisting of a cord-and-plug-connected transformer and a luminaire intended for installation in the wall of a spa, hot tub, or storable pool.

CORROSIVE ENVIRONMENT. Areas where pool sanitation chemicals are stored, handled, or dispensed, and confined areas under decks adjacent to such areas, as well as areas with circulation pumps, automatic chlorinators, filters, open areas under decks adjacent to or abutting the pool structure, and similar locations.

DRY-NICHE LUMINAIRE. A luminaire intended for installation in the floor or wall of a pool or spa in a niche that is sealed against the entry of water.

FORMING SHELL. A structure designed to support a wet-niche luminaire assembly and intended for mounting in a pool structure.

HYDROMASSAGE BATHTUB. A permanently installed bathtub equipped with a recirculating piping system, pump, and associated equipment. It is designed so it can accept, circulate and discharge water upon each use.

LOW-VOLTAGE CONTACT LIMIT. A voltage not exceeding the following values:

1. 15 volts (RMS) for sinusoidal ac.
2. 21.2 volts peak for nonsinusoidal ac.
3. 30 volts for continuous dc.
4. 12.4 volts peak for dc that is interrupted at a rate of 10 to 200 Hz.

MAXIMUM WATER LEVEL. The highest level that water can reach before it spills out.

NO-NICHE LUMINAIRE. A luminaire intended for installation above or below the water without a niche.

PACKAGED SPA OR HOT TUB EQUIPMENT ASSEMBLY. A factory-fabricated unit consisting of water-circulating, heating and control equipment mounted on a common base, intended to operate a spa or hot tub. Equipment may include pumps, air blowers, heaters, luminaires, controls and sanitizer generators.

PERMANENTLY INSTALLED SWIMMING, WADING, IMMERSION AND THERAPEUTIC POOLS. Those that are constructed in the ground or partially in the ground, and all others capable of holding water with a depth greater than 42 inches (1067 mm), and all pools installed inside of a building, regardless of water depth, whether or not served by electrical circuits of any nature.

POOL. Manufactured or field-constructed equipment designed to contain water on a permanent or semipermanent basis and used for swimming, wading, immersion, or therapeutic purposes.

POOL COVER, ELECTRICALLY OPERATED. Motor-driven equipment designed to cover and uncover the water surface of a pool by means of a flexible sheet or rigid frame.

SELF-CONTAINED SPA OR HOT TUB. A factory-fabricated unit consisting of a spa or hot tub vessel with all water-circulating, heating and control equipment integral to the unit. Equipment may include pumps, air blowers, heaters, luminaires, controls and sanitizer generators.

SPA OR HOT TUB. A hydromassage pool, or tub for recreational or therapeutic use, not located in health care facilities, designed for immersion of users, and usually having a filter, heater, and motor-driven blower. They are installed indoors or outdoors, on the ground or supporting structure, or in the ground or supporting structure. Generally, a spa or hot tub is not designed or intended to have its contents drained or discharged after each use.

STORABLE SWIMMING, WADING OR IMMERSION POOLS; OR STORABLE/PORTABLE SPAS AND HOT TUBS. Swimming, wading, or immersion pools that are intended to be stored when not in use, that are constructed on or above the ground and that are capable of holding water with a maximum depth of 42 inches (1067 mm), or a pool, spa, or hot tub that is constructed on or above the ground with nonmetallic, molded polymeric walls or inflatable fabric walls regardless of dimension.

THROUGH-WALL LIGHTING ASSEMBLY. A lighting assembly intended for installation above grade, on or through the wall of a pool, consisting of two interconnected groups of components separated by the pool wall.

WET-NICHE LUMINAIRE. A luminaire intended for installation in a forming shell mounted in a pool structure where the luminaire will be completely surrounded by water.

SECTION E4202 WIRING METHODS FOR POOLS, SPAS, HOT TUBS AND HYDROMASSAGE BATHTUBS

E4202.1 General. Wiring methods used in conjunction with permanently installed swimming pools, spas or hot tubs that are installed in corrosive environments described in Section E4202.2 shall comply with Table E4202.1, Sections E4202.2 and E4205 and Chapter 38 except as otherwise stated in this section. Wiring methods used in conjunction with permanently installed swimming pools, spas or hot tubs that are not installed in corrosive environments shall comply with Chapter 38. Storable swimming pools shall comply with Section E4207.

Hydromassage bathtubs shall comply with Section E4209. [680.7; 680.14; 680.21(A); 680.23(B) and (F); 680.25(A); 680.42; 680.43; and 680.70]

E4202.2 Wiring methods in corrosive environment. Wiring methods in a corrosive environment shall be *listed* and identified for use in such areas. Rigid metal conduit, intermediate metal conduit, rigid polyvinyl chloride conduit and reinforced thermosetting resin conduit shall be considered to be resistant to the corrosive environment.

E4202.3 Flexible cords. Flexible cords used in conjunction with a pool, spa, hot tub or hydromassage bathtub shall be installed in accordance with the following:

1. For other than underwater luminaires, fixed or stationary equipment shall be permitted to be connected with a flexible cord and plug to facilitate removal or disconnection for maintenance or repair. For other than storable pools, the flexible cord shall not exceed 3 feet (914 mm) in length. Cords that supply swimming pool equipment shall have a copper equipment grounding conductor not smaller than 12 AWG and shall terminate in a grounding-type attachment plug. [680.8(A), (B), and (C); 680.21(A)(3)]
2. Other than *listed* low-voltage lighting systems not requiring grounding, wet-niche luminaires that are supplied by a flexible cord or cable shall have all exposed noncurrent-carrying metal parts connected to an insulated copper equipment grounding conductor that is an integral part of the cord or cable. This equipment grounding conductor shall be connected to a grounding terminal in the supply junction box, transformer enclosure or other enclosure. The equipment grounding conductor shall be not smaller than the supply conductors and not smaller than 16 AWG. [680.23(B)(3)]

TABLE E4202.1[a]
PERMITTED WIRING METHODS IN CORROSIVE ENVIRONMENTS

WIRING LOCATION OR PURPOSE (Application allowed where marked with an "A")	IMC[b], RMC[b], RNC[e]	LFMC	LFNMC	MC[g]	FLEX CORD
Panelboard(s) that supply pool equipment: from service equipment to panelboard	A[f]	—	A	—	—
Wet-niche and no-niche luminaires: from branch circuit OCPD to deck or junction box	A	—	A	—	—
Wet-niche and no-niche luminaires: from deck or junction box to forming shell	A[j]	—	A	—	A[d]
Dry niche: from branch circuit OCPD to luminaires	A	—	A	—	—
Pool-associated motors: from branch circuit OCPD to motor[h]	A	A[c]	A[c]	A	A[d]
Packaged or self-contained outdoor spas and hot tubs with underwater luminaire: from branch circuit OCPD to spa or hot tub	A	A	A	—	A[d]
Packaged or self-contained outdoor spas and hot tubs without underwater luminaire: from branch circuit OCPD to spa or hot tub	A	A	A	—	A[d]
Indoor spas and hot tubs, and other pool, spa or hot tub associated equipment: from branch circuit OCPD to equipment	A	A	A	—	A[d]
Connection at pool lighting transformers or power supplies	A	A[i]	A	—	—

For SI: 1 foot = 304.8 mm.

a. For all wiring methods, see Section E4205 for equipment grounding conductor requirements.
b. See Section E4202.2 for use of metal conduits in corrosive environments.
c. Limited to where necessary to employ flexible connections at or adjacent to a pool motor.
d. Flexible cord shall be installed in accordance with Section E4202.3.
e. Nonmetallic conduit shall be rigid polyvinyl chloride conduit Type PVC or reinforced thermosetting resin conduit Type RTRC.
f. Aluminum conduits shall not be permitted in the pool area where subject to corrosion.
g. Where installed as direct burial cable or in wet locations, Type MC cable shall be listed and identified for the location.
h. See Section E4202.4 for listed, double-insulated pool pump motors.
i. Limited to use in individual lengths not to exceed 6 feet. The total length of all individual runs of LFMC shall not exceed 10 feet.
j. Metal conduit shall be constructed of brass or other approved corrosion-resistant metal.

3. A *listed* packaged spa or hot tub installed outdoors that is GFCI protected shall be permitted to be cord-and-plug-connected provided that such cord does not exceed 15 feet (4572 mm) in length. [680.42(A)(2)]
4. A *listed* packaged spa or hot tub rated at 20 amperes or less and installed indoors shall be permitted to be cord-and-plug-connected to facilitate maintenance and repair. (680.43 Exception No. 1)
5. For other than underwater and storable pool lighting luminaire, the requirements of Item 1 shall apply to any cord-equipped luminaire that is located within 16 feet (4877 mm) radially from any point on the water surface. [680.22(B)(5)]

E4202.4 Double-insulated pool pumps. A *listed* cord-and-plug-connected pool pump incorporating an approved system of double insulation that provides a means for grounding only the internal and nonaccessible, noncurrent-carrying metal parts of the pump shall be connected to any wiring method recognized in Chapter 38 that is suitable for the location. Where the equipment grounding conductor of the motor circuit is connected to the equipotential bonding means in accordance with Section E4204.2, Item 6.1, the branch circuit wiring shall comply with Sections E4202.1 and E4205.5. [680.21(B)]

SECTION E4203 EQUIPMENT LOCATION AND CLEARANCES

E4203.1 Receptacle outlets. Receptacles outlets shall be installed and located in accordance with Sections E4203.1.1 through E4203.1.7. In determining the dimensions in this section addressing receptacle spacings, the distance to be measured shall be the shortest path the supply cord of an appliance connected to the receptacle would follow without piercing a floor, wall, ceiling, doorway with hinged or sliding door, window opening, or other effective permanent barrier. [680.22(A)(6)]

E4203.1.1 Location. Receptacles that provide power for water-pump motors or other loads directly related to the circulation and sanitation system shall be located at least 6 feet (1829 mm) from the inside walls of pools, outdoor spas and hot tubs. These receptacles shall have GFCI protection and be of the grounding type. [680.22(A)(2)].

E4203.1.2 Other receptacles. Other receptacles on the property shall be located not less than 6 feet (1829 mm) from the inside walls of pools and outdoor spas and hot tubs. [680.22 (A)(3)]

E4203.1.3 Where required. No less than one 125-volt, 15- or 20-ampere receptacle supplied by a general-purpose branch circuit shall be located not less than 6 feet (1829 mm) from and not more than 20 feet (6096 mm) from the inside wall of permanently installed pools and outdoor spas and hot tubs. This receptacle shall be located not more than 6 feet, 6 inches (1981 mm) above the floor, platform or grade level serving the pool, spa or hot tub. [680.22(A)(1)]

E4203.1.4 GFCI protection. All 15- and 20-ampere, single phase, 125-volt receptacles located within 20 feet (6096 mm) of the inside walls of pools and outdoor spas and hot tubs shall be protected by a Class A ground-fault circuit interrupter. Outlets supplying all pool motors on branch circuits rated at 150 volts or less to ground, and 60 amperes or less, single- or 3-phase, shall be provided with Class A ground-fault circuit-interrupter protection. [680.21(C) and 680.22(A)(4)]

E4203.1.5 Indoor locations. Receptacles shall be located not less than 6 feet (1829 mm) measured horizontally from the inside walls of indoor spas and hot tubs. A minimum of one 125-volt, 15- or 20-ampere receptacle on a general-purpose branch circuit shall be located between 6 feet (1829 mm) and 10 feet (3048 mm) from the inside walls of indoor spas or hot tubs. [680.43(A) and 680.43(A)(1)]

E4203.1.6 Indoor GFCI protection. Receptacles rated 125 volts and 30 amperes or less and located within 10 feet (3048 mm) of the inside walls of a spa or hot tub installed indoors shall be protected by ground-fault circuit interrupters. [680.43(A)(2)].

E4203.1.7 Pool equipment room. At least one GFCI-protected 125-volt, 15- or 20-ampere receptacle on a general-purpose circuit shall be located within a pool equipment room, and all other receptacles supplied by branch circuits rated 150 volts or less to ground within a pool equipment room shall be GFCI protected.

E4203.2 Switching devices. Switching devices shall be located at least 5 feet (1524 mm) horizontally from the inside walls of pools, spas and hot tubs unless separated from the pool, spa or hot tub by a solid fence, wall, or other permanent barrier that provides at least a 5-foot (1524 mm) reach distance. Alternatively, a switch that is *listed* as being acceptable for use within 5 feet (1524 mm) shall be permitted. Switching devices located in a room or area containing a hydromassage bathtub shall be located in accordance with the general requirements for installing equipment in bathrooms. [680.22(C); 680.43(C); and 680.72]

E4203.3 Disconnecting means. One or more means to simultaneously disconnect all ungrounded conductors for all utilization equipment other than lighting shall be provided. Each of such means shall be readily accessible and within sight from the equipment it serves and shall be located at least 5 feet (1524 mm) horizontally from the inside walls of a pool, spa, or hot tub unless separated from the open water by a permanently installed barrier that provides a 5-foot (1524 mm) or greater reach path. This horizontal distance shall be measured from the water's edge along the shortest path required to reach the disconnect. (680.13)

E4203.4 Luminaires, equipment and ceiling fans. Lighting outlets, luminaires, equipment and ceiling-suspended paddle fans shall be installed and located in accordance with Sections E4203.4.1 through E4203.4.7. [680.22(B)]

E4203.4.1 Outdoor location. In outdoor pool, outdoor spas and outdoor hot tubs areas, luminaires, lighting outlets, and ceiling-suspended paddle fans shall not be installed over the pool or over the area extending 5 feet (1524 mm) horizontally from the inside walls of a pool except where no part of the luminaire or ceiling-suspended paddle fan is less than 12 feet (3658 mm) above the maximum water level. [680.22(B)(1)]

E4203.4.2 Indoor locations. In indoor pool areas, the limitations of Section E4203.4.1 shall apply except where the luminaires, lighting outlets and ceiling-suspended paddle fans comply with all of the following conditions:

1. The luminaires are of a totally enclosed type.
2. Ceiling-suspended paddle fans are identified for use beneath ceiling structures such as porches and patios.
3. A ground-fault circuit interrupter is installed in the branch circuit supplying the luminaires or ceiling-suspended paddle fans.
4. The distance from the bottom of the luminaire or ceiling-suspended paddle fan to the maximum water level is not less than 7 feet, 6 inches (2286 mm). [680.22(B)(2)]

E4203.4.3 Low-voltage luminaires. *Listed* low-voltage luminaires not requiring grounding, not exceeding the low-voltage contact limit, and supplied by *listed* transformers or power supplies that comply with Section E4206.1 shall be permitted to be located less than 5 feet (1.5 m) from the inside walls of the pool. [680.22(B)(6)]

E4203.4.4 Existing lighting outlets and luminaires. Existing lighting outlets and luminaires that are located within 5 feet (1524 mm) horizontally from the inside walls of pools and outdoor spas and hot tubs shall be permitted to be located not less than 5 feet (1524 mm) vertically above the maximum water level, provided that such luminaires and outlets are rigidly attached to the existing structure and are protected by a ground-fault circuit interrupter. [680.22(B)(3)]

E4203.4.5 Indoor spas and hot tubs.

1. Luminaires, lighting outlets and ceiling-suspended paddle fans located over the spa or hot tub or within 5 feet (1524 mm) from the inside walls of the spa or hot tub shall be not less than 7 feet, 6 inches (2286 mm) above the maximum water level and shall be protected by a ground-fault circuit interrupter. [680.43(B)(1)(b)]

   Luminaires, lighting outlets and ceiling-suspended paddle fans that are located 12 feet (3658 mm) or more above the maximum water level shall not require ground-fault circuit-interrupter protection. [680.43(B)(1)(a)]

2. Luminaires protected by a ground-fault circuit interrupter and complying with Item 2.1 or 2.2 shall be permitted to be installed less than 7 feet, 6 inches (2286 mm) over a spa or hot tub.

   2.1. Recessed luminaires shall have a glass or plastic lens and nonmetallic or electrically isolated metal trim, and shall be suitable for use in damp locations.

   2.2. Surface-mounted luminaires shall have a glass or plastic globe and a nonmetallic body or a metallic body isolated from contact. Such luminaires shall be suitable for use in damp locations. [680.43(B)(1)(c)(1) and (2)]

E4203.4.6 GFCI protection in adjacent areas. Luminaires, lighting outlets and ceiling-suspended paddle fans that are installed in the area extending between 5 feet (1524 mm) and 10 feet (3048 mm) from the inside walls of pools and outdoor spas and hot tubs shall be protected by ground-fault circuit interrupters except where such luminaires, lighting outlets and ceiling-suspended paddle fans are installed not less than 5 feet (1524 mm) above the maximum water level and are rigidly attached to the structure. [680.22(B)(4)]

E4203.4.7 Low-voltage gas-fired luminaires, decorative fireplaces, fire pits and similar equipment. *Listed* low-voltage gas-fired luminaires, decorative fireplaces, fire pits and similar equipment that use low-voltage ignitors that do not require grounding, and that are supplied by *listed* transformers or power supplies that comply with Section E4206.1 with outputs that do not exceed the low-voltage contact limit, shall be permitted to be located less than 5 feet (1524 mm) from the inside walls of the pool. Metallic equipment shall be bonded in accordance with the requirements in Section E4204.2. Transformers and power supplies supplying this type of equipment shall be installed in accordance with the requirements of Section E4206.9.1. Metallic gas piping shall be bonded in accordance with the requirements of Sections E3609.7 and 4204.2, Item 7. [680.22 (B)(7)]

E4203.5 Other outlets. Other outlets such as for remote control, signaling, fire alarm and communications shall be not less than 10 feet (3048 mm) from the inside walls of the pool. Measurements shall be determined in accordance with Section E4203.1. [680.22(D)]

E4203.6 Other equipment. Other equipment with ratings exceeding the low-voltage contact limit shall be located at least 5 feet (1524 mm) horizontally from the inside walls of a pool unless separated from the pool by a solid fence, wall or other permanent barrier. [680.22(E)]

E4203.7 Overhead conductor clearances. Except where installed with the clearances specified in Table E4203.7, the following parts of pools and outdoor spas and hot tubs shall not be placed under existing service-drop conductors, overhead service conductor, or any other open overhead wiring; nor shall such wiring be installed above the following:

1. Pools and the areas extending not less than 10 feet (3048 mm) horizontally from the inside of the walls of the pool.
2. Diving structures and the areas extending not less than 10 feet (3048 mm) horizontally from the outer edge of such structures.

3. Observation stands, towers, and platforms and the areas extending not less than 10 feet (3048 mm) horizontally from the outer edge of such structures.

Overhead conductors of network-powered broadband communications systems shall comply with the provisions in Table E4203.7 for conductors operating at 0 to 750 volts to ground.

Utility-owned, -operated and -maintained communications conductors, community antenna system coaxial cables and the supporting messengers shall be permitted at a height of not less than 10 feet (3048 mm) above swimming and wading pools, diving structures, and observation stands, towers, and platforms. [680.9(A), (B), and (C)]

E4203.8 Underground wiring. Underground wiring within 5 feet (1524 mm) horizontally from the inside wall of the pool shall be permitted. Underground wiring shall not be installed under the pool except where this wiring is necessary to supply pool equipment permitted by this chapter. Underground wiring shall be installed in rigid metal conduit, intermediate metal conduit, rigid polyvinyl chloride conduit, reinforced thermosetting resin conduit or Type MC cable, suitable for the conditions subject to that location. The minimum cover depth shall be in accordance with Table E3803.1. (680.11)

SECTION E4204 EQUIPOTENTIAL BONDING

E4204.1 Performance. The equipotential bonding required by this section shall be installed to reduce voltage gradients in the prescribed areas of permanently installed swimming pools and spas and hot tubs other than the storable/portable type.

E4204.2 Bonded parts. The parts of pools, spas, and hot tubs specified in Items 1 through 7 shall be bonded together using insulated, covered or bare solid copper conductors not smaller than 8 AWG or using rigid metal conduit of brass or other identified corrosion-resistant metal. An 8 AWG or larger solid copper bonding conductor provided to reduce voltage gradients in the pool, spa, or hot tub area shall not be required to be extended or attached to remote panelboards, service equipment, or electrodes. Connections shall be made by exothermic welding, by *listed* pressure connectors or clamps that are *labeled* as being suitable for the purpose and that are made of stainless steel, brass, copper or copper alloy, machine screw-type fasteners that engage not less than two threads or are secured with a nut, thread-forming machine screws that engage not less than two-threads, or terminal bars. Connection devices or fittings that depend solely on solder shall not be used. Sheet metal screws shall not be used to connect bonding conductors or connection devices: [680.26(B)]

1. Conductive pool shells. Bonding to conductive pool shells shall be provided as specified in Item 1.1 or 1.2. Cast-in-place concrete, pneumatically applied or sprayed concrete, and concrete block with painted or plastered coatings shall be considered to be conductive materials because of their water permeability and porosity. Vinyl liners and fiberglass composite shells shall be considered to be nonconductive materials. Reconstructed pool shells shall also meet the requirements of this section.
   - 1.1. Structural reinforcing steel. Unencapsulated structural reinforcing steel shall be bonded together by steel tie wires or the equivalent. Where structural reinforcing steel is encapsulated in a nonconductive compound, a copper conductor grid shall be installed in accordance with Item 1.2.
   - 1.2. Copper conductor grid. A copper conductor grid shall be provided and shall comply with Items 1.2.1 through 1.2.4.
     - 1.2.1. It shall be constructed of minimum 8 AWG bare solid copper conductors bonded to each other at all points of crossing.
     - 1.2.2. It shall conform to the contour of the pool.
     - 1.2.3. It shall be arranged in a 12-inch (305 mm) by 12-inch (305 mm) network of conductors in a uniformly spaced perpendicular grid pattern with a tolerance of 4 inches (102 mm).
     - 1.2.4. It shall be secured within or under the pool not more than 6 inches (152 mm) from the outer contour of the pool shell. [680.26(B)(1)]

TABLE E4203.7 [Table 680.8(A)]
OVERHEAD CONDUCTOR CLEARANCES

	INSULATED SUPPLY OR SERVICE DROP CABLES, 0-750 VOLTS TO GROUND, SUPPORTED ON AND CABLED TOGETHER WITH AN EFFECTIVELY GROUNDED BARE MESSENGER OR EFFECTIVELY GROUNDED NEUTRAL CONDUCTOR (feet)	ALL OTHER SUPPLY OR SERVICE DROP CONDUCTORS (feet)	
		Voltage to ground	
		0 to 15 kV	Greater than 15 to 50 kV
A. Clearance in any direction to the water level, edge of water surface, base of diving platform, or permanently anchored raft	22.5	25	27
B. Clearance in any direction to the diving platform	14.5	17	18

For SI: 1 foot = 304.8 mm.

2. Perimeter surfaces. The perimeter surface to be bonded shall be considered to extend for 3 feet (914 mm) horizontally beyond the inside walls of the pool and shall include unpaved surfaces, poured concrete surfaces and other types of paving. Perimeter surfaces that are separated from the pool by a permanent wall or building 5 feet (1524 mm) or more in height shall require equipotential bonding only on the pool side of the permanent wall or building. Bonding to perimeter surfaces shall be provided as specified in Item 2.1 or 2.2 and shall be attached to the pool, spa, or hot tub reinforcing steel or copper conductor grid at a minimum of four points uniformly spaced around the perimeter of the pool, spa, or hot tub. For nonconductive pool shells, bonding at four points shall not be required.

   Exceptions:

   1. Equipotential bonding of perimeter surfaces shall not be required for spas and hot tubs where all of the following conditions apply:
      1.1. The spa or hot tub is *listed* as a self-contained spa for above-ground use.
      1.2. The spa or hot tub is not identified as suitable only for indoor use.
      1.3. The installation is in accordance with the manufacturer's instructions and is located on or above grade.
      1.4. The top rim of the spa or hot tub is not less than 28 inches (711 mm) above all perimeter surfaces that are within 30 inches (762 mm), measured horizontally from the spa or hot tub. The height of nonconductive external steps for entry to or exit from the self-contained spa is not used to reduce or increase this rim height measurement.
   2. The equipotential bonding requirements for perimeter surfaces shall not apply to a *listed* self-contained spa or hot tub located indoors and installed above a finished floor.
      2.1. Structural reinforcing steel. Structural reinforcing steel shall be bonded in accordance with Item 1.1.
      2.2. Copper ring. Where structural reinforcing steel is not available or is encapsulated in a nonconductive compound, a copper conductor(s) shall be used in accordance with Items 2.2.1 through 2.2.5:
         2.2.1. At least one minimum 8 AWG bare solid copper conductor shall be provided.
         2.2.2. The conductors shall follow the contour of the perimeter surface.
         2.2.3. Only *listed* splicing devices or exothermic welding shall be permitted.
         2.2.4. The required conductor shall be 18 to 24 inches (457 to 610 mm) from the inside walls of the pool.
         2.2.5. The required conductor shall be secured within or under the perimeter surface 4 to 6 inches (102 mm to 152 mm) below the subgrade. [680.26(B)(2)]
      2.3. The copper grid shall follow the contour of the perimeter surface extending 3 feet (914 mm) horizontally beyond the inside walls of the pool.
         2.3.1. Only *listed* splicing devices or exothermic welding shall be permitted.
         2.3.2. The required conductor shall be secured within or under the perimeter surface 4 to 6 inches (102 to 152 mm) below the subgrade. [680.26(B)(2)(c)]

3. Metallic components. All metallic parts of the pool structure, including reinforcing metal not addressed in Item 1.1, shall be bonded. Where reinforcing steel is encapsulated with a nonconductive compound, the reinforcing steel shall not be required to be bonded. [680.26(B)(3)]
4. Underwater lighting. All metal forming shells and mounting brackets of no-niche luminaires shall be bonded. [680.26(B)(4)]

   Exception: *Listed* low-voltage lighting systems with nonmetallic forming shells shall not require bonding. [680.26(B)(4) Exception]

5. Metal fittings. All metal fittings within or attached to the pool structure shall be bonded. Isolated parts that are not over 4 inches (102 mm) in any dimension and do not penetrate into the pool structure more than 1 inch (25.4 mm) shall not require bonding. Metallic pool cover anchors intended for insertion in a concrete or masonry deck surface, 1 inch (25 mm) or less in any dimension and 2 inches (51 mm) or less in length, and metallic pool cover anchors intended for insertion in a wood or composite deck surface, 2 inches (51 mm) or less in any flange dimension and 2 inches (51 mm) or less in length, shall not require bonding. [680.26(B)(5)]

6. Electrical equipment. Metal parts of electrical equipment associated with the pool water circulating system, including pump motors and metal parts of equipment associated with pool covers, including electric motors, shall be bonded. [680.26(B)(6)]

   Exception: Metal parts of listed equipment incorporating an approved system of double insulation shall not be bonded. [680.26(B)(6) Exception]

   6.1. Double-insulated water pump motors. Where a double-insulated water pump motor is installed under the provisions of this item, a solid 8 AWG copper conductor of sufficient length to make a bonding connection to a replacement motor shall be extended from the swimming pool equipotential bonding means to an accessible point in the vicinity of the pool pump motor. Where there is no connection between the swimming pool equipotential bonding means and the equipment grounding system for the premises, this bonding conductor shall be connected to the equipment grounding conductor of the motor circuit. [680.26(B)(6)(a)]

   6.2. Pool water heaters. For pool water heaters rated at more than 50 amperes and having specific instructions regarding bonding and grounding, only those parts designated to be bonded shall be bonded and only those parts designated to be grounded shall be grounded. [680.26(B)(6)(b)]

7. All fixed metal parts including, but not limited to, metal-sheathed cables and raceways, metal piping, metal awnings, metal fences and metal door and window frames. [680.26(B)(7)]

   Exceptions:

   1. Those separated from the pool by a permanent barrier that prevents contact by a person shall not be required to be bonded. [680.26(B)(7) Exception No. 1]
   2. Those greater than 5 feet (1524 mm) horizontally from the inside walls of the pool shall not be required to be bonded. [680.26(B)(7) Exception No. 2]
   3. Those greater than 12 feet (3658 mm) measured vertically above the maximum water level of the pool, or as measured vertically above any observation stands, towers, or platforms, or any diving structures, shall not be required to be bonded. [680.26(B)(7) Exception No. 3]

E4204.3 Pool water. Where none of the bonded parts are in direct connection with the pool water, the pool water shall be in direct contact with an approved corrosion-resistant conductive surface that exposes not less than 9 square inches (5800 mm^2) of surface area to the pool water at all times. The conductive surface shall be located where it is not exposed to physical damage or dislodgement during usual pool activities, and it shall be bonded in accordance with Section E4204.2.

E4204.4 Bonding of outdoor hot tubs and spas. Outdoor hot tubs and spas shall comply with the bonding requirements of Sections E4204.1 through E4204.3. Bonding by metal-to-metal mounting on a common frame or base shall be permitted. The metal bands or hoops used to secure wooden staves shall not be required to be bonded as required in Section E4204.2. [680.42 and 680.42(B)]

E4204.5 Bonding of indoor hot tubs and spas. The following parts of indoor hot tubs and spas shall be bonded together:

1. All metal fittings within or attached to the hot tub or spa structure. [680.43(D)(1)]
2. Metal parts of electrical equipment associated with the hot tub or spa water circulating system, including pump motors unless part of a *listed* self-contained spa or hot tub. [680.43(D)(2)]
3. Metal raceway and metal piping that are within 5 feet (1524 mm) of the inside walls of the hot tub or spa and that are not separated from the spa or hot tub by a permanent barrier. [680.43(D)(3)]
4. All metal surfaces that are within 5 feet (1524 mm) of the inside walls of the hot tub or spa and that are not separated from the hot tub or spa area by a permanent barrier. [680.43(D)(4)]

   Exception: Small conductive surfaces not likely to become energized, such as air and water jets and drain fittings, where not connected to metallic piping, towel bars, mirror frames, and similar nonelectrical equipment, shall not be required to be bonded. [680.43(D)(4) Exception]

5. Noncurrent-carrying metal parts of electrical devices and controls that are not associated with the hot tubs or spas and that are located less than 5 feet (1524 mm) from such units. [680.43(D)(5)]

E4204.5.1 Methods. All metal parts associated with the hot tub or spa shall be bonded by any of the following methods:

1. The interconnection of threaded metal piping and fittings. [680.43(E)(1)]
2. Metal-to-metal mounting on a common frame or base. [680.43(E)(2)]
3. The provision of an insulated, covered or bare solid copper bonding jumper not smaller than 8 AWG. It shall not be the intent to require that the 8 AWG or larger solid copper bonding conductor be extended or attached to any remote panelboard, service equipment, or any electrode, but only that it shall be employed to eliminate voltage gradients in the hot tub or spa area as prescribed. [680.43(E)(3)]

E4204.5.2 Connections. Connections to bonded parts shall be made in accordance with Section E3406.14.1.

SECTION E4205 BONDING AND GROUNDING

E4205.1 Equipment to be bonded and grounded. The following equipment shall be bonded and grounded:

1. Through-wall lighting assemblies and underwater luminaires other than those low-voltage lighting products *listed* for the application without an equipment grounding conductor.
2. All electrical equipment located within 5 feet (1524 mm) of the inside wall of the pool, spa or hot tub.
3. All electrical equipment associated with the recirculating system of the pool, spa or hot tub.
4. Junction boxes.
5. Transformer and power supply enclosures.
6. Ground-fault circuit interrupters.
7. Panelboards that are not part of the service equipment and that supply any electrical equipment associated with the pool, spa or hot tub. (680.6)

E4205.2 Luminaires and related equipment. Where branch-circuit wiring on the supply side of enclosures and junction boxes connected to conduits run to underwater luminaires are installed in corrosive environments as described in Section E4202.2, the wiring method of that portion of the branch circuit shall be as required in Section E4202.2 or shall be liquid-tight flexible nonmetallic conduit (LFNMC). Where not installed in corrosive environments, branch circuits shall comply with Chapter 38. Wiring methods shall contain an insulated copper equipment grounding conductor sized in accordance with Table E3809.13 but not smaller than 12 AWG. The equipment grounding conductor between the wiring chamber of the secondary winding of a transformer and a junction box shall be sized in accordance with the transformer secondary overcurrent protection provided.

The insulated copper equipment grounding conductor shall be connected to all through-wall lighting assemblies, wet-niche, dry-niche, or no-niche luminaires other than *listed* low-voltage luminaires not requiring grounding. The junction box, transformer enclosure, or other enclosure in the supply circuit to a wet-niche or no-niche luminaire and the field-wiring chamber of a dry-niche luminaire shall be grounded to the equipment grounding terminal of the panelboard. The equipment grounding terminal shall be directly connected to the panelboard enclosure. The equipment grounding conductor shall be installed without joint or splice. [680.23(F)(1), (F)(2) and 680.23(F)(2) Exception]

Exceptions:

1. Where more than one underwater luminaire is supplied by the same branch circuit, the equipment grounding conductor, installed between the junction boxes, transformer enclosures, or other enclosures in the supply circuit to wet-niche luminaires, or between the field-wiring compartments of dry-niche luminaires, shall be permitted to be terminated on grounding terminals. [680.23(F)(2)(a)]
2. Where an underwater luminaire is supplied from a transformer, ground-fault circuit interrupter, clock-operated switch, or a manual snap switch that is located between the panelboard and a junction box connected to the conduit that extends directly to the underwater luminaire, the equipment grounding conductor shall be permitted to terminate on grounding terminals on the transformer, ground-fault circuit interrupter, clock-operated switch enclosure, or an outlet box used to enclose a snap switch. [680.23(F)(2)(b)]

E4205.3 Nonmetallic conduit. Where a nonmetallic conduit is installed between a forming shell and a junction box, transformer enclosure, or other enclosure, an 8 AWG insulated copper bonding jumper shall be installed in this conduit except where a *listed* low-voltage lighting system not requiring grounding is used. The bonding jumper shall be terminated in the forming shell, junction box or transformer enclosure, or ground-fault circuit-interrupter enclosure. The termination of the 8 AWG bonding jumper in the forming shell shall be covered with, or encapsulated in, a *listed* potting compound to protect such connection from the possible deteriorating effect of pool water. [680.23(B)(2)(b)]

E4205.4 Flexible cords. Other than *listed* low-voltage lighting systems not requiring grounding, wet-niche luminaires that are supplied by a flexible cord or cable shall have all exposed noncurrent-carrying metal parts connected to an insulated copper equipment grounding conductor that is an integral part of the cord or cable. This equipment grounding conductor shall be connected to a grounding terminal in the supply junction box, transformer enclosure, or other enclosure. The equipment grounding conductor shall not be smaller than the supply conductors and not smaller than 16 AWG. [680.23(B)(3)]

E4205.5 Pool motors. Wiring methods installed in the corrosive environment described in Section E4202.2 shall comply with Section E4202.2 or shall be Type MC cable *listed* for that location. Wiring methods installed in corrosive environments described in Section E4202.2 shall contain an insulated copper equipment conductor sized in accordance with Table E3908.12 but not smaller than 12 AWG.

Where installed in noncorrosive environments, branch circuit wiring methods shall comply with Chapter 38. [680.21(A)(1)].

E4205.6 Feeders. These provisions shall apply to any feeder on the supply side of panelboards supplying branch circuits for pool equipment covered in this chapter and on the load side of the service equipment. Where feeders are installed in corrosive environments as described in Section E4202.2, the wiring method of that portion of the feeder shall comply with Section E4202.2 or shall be liquid-tight flexible nonmetallic conduit (LFNMC). Wiring methods installed in corrosive environments as described in Section E4202.2 shall contain an insulated copper equipment grounding conductor sized in accordance with Table E3908.12, but not smaller than 12 AWG.

Where installed in noncorrosive environments, feeder wiring methods shall comply with Chapter 38. [680.25(A)].

E4205.7 Cord-connected equipment. Where fixed or stationary equipment is connected with a flexible cord to facilitate removal or disconnection for maintenance, repair, or storage, as provided in Section E4202.3, the equipment grounding conductors shall be connected to a fixed metal part of the assembly. The removable part shall be mounted on or bonded to the fixed metal part. (680.8)

E4205.8 Other equipment. Other electrical equipment shall be grounded in accordance with Section E3908. (Article 250, Parts V, VI, and VII; and 680.6)

E4205.9 Bonding and equipment grounding terminals. Terminals used for equipment grounding and bonding shall be identified for use in wet and corrosive environments. Field-installed terminals in a damp, wet or corrosive environment shall be composed of copper, copper alloy or stainless steel and shall be *listed* for direct burial use. (680.7)

SECTION E4206 EQUIPMENT INSTALLATION

E4206.1 Transformers and power supplies. Transformers and power supplies used for the supply of underwater luminaires, together with the transformer or power supply enclosure, shall be *listed*, labeled and identified for swimming pool and spa use. The transformer or power supply shall incorporate either a transformer of the isolated-winding type with an ungrounded secondary that has a grounded metal barrier between the primary and secondary windings, or a transformer that incorporates an approved system of double insulation between the primary and secondary windings. [680.23(A)(2)]

E4206.2 Ground-fault circuit interrupters. Ground-fault circuit interrupters (GFCIs) shall be self-contained units, circuit-breaker types, receptacle types or other *listed* types. The GFCI requirements in this chapter, unless otherwise noted, are in addition to the requirements in Section E3902. (680.5)

E4206.3 Wiring on load side of ground-fault circuit interrupters and transformers. For other than grounding conductors, conductors installed on the load side of a ground-fault circuit interrupter or transformer used to comply with the provisions of Section E4206.4, shall not occupy raceways, boxes, or enclosures containing other conductors except where the other conductors are protected by ground-fault circuit-interrupters or are grounding conductors. Supply conductors to a feed-through type ground-fault circuit interrupter shall be permitted in the same enclosure. Ground-fault circuit interrupters shall be permitted in a panelboard that contains circuits protected by other than ground-fault circuit interrupters. [680.23(F)(3)]

E4206.4 Underwater luminaires. The design of an underwater luminaire supplied from a branch circuit, either directly or by way of a transformer or power supply meeting the requirements of Section E4206.1, shall be such that, where the fixture is properly installed without a ground-fault circuit interrupter, there is no shock hazard with any likely combination of fault conditions during normal use (not relamping). In addition, ground-fault circuit-interrupter protection for personnel shall be installed in the branch circuit supplying luminaires operating at voltages greater than the low-voltage contact limit to protect personnel performing lamping, relamping or servicing. The installation of the ground-fault circuit interrupter shall be such that there is no shock hazard with any likely fault-condition combination that involves a person in a conductive path from any ungrounded part of the branch circuit or the luminaire to ground. Compliance with this requirement shall be obtained by the use of a *listed* underwater luminaire and by installation of a *listed* ground-fault circuit-interrupter in the branch circuit or a *listed* transformer or power supply for luminaires operating at more than the low-voltage contact limit. Luminaires that depend on submersion for safe operation shall be inherently protected against the hazards of overheating when not submerged. [680.23(A)(1), (A)(3), (A)(7) and (A)(8)]

E4206.4.1 Maximum voltage. Luminaires shall not be installed for operation on supply circuits over 150 volts between conductors. [680.23(A)(4)]

E4206.4.2 Luminaire location. Luminaires mounted in walls shall be installed with the top of the fixture lens not less than 18 inches (457 mm) below the normal water level of the pool, except where the luminaire is *listed* and identified for use at a depth of not less than 4 inches (102 mm) below the normal water level of the pool. A luminaire facing upward shall have the lens adequately guarded to prevent contact by any person or shall be *listed* for use without a guard. [680.23(A)(5) and (A)(6)]

E4206.5 Wet-niche luminaires. Forming shells shall be installed for the mounting of all wet-niche underwater luminaires and shall be equipped with provisions for conduit entries. Conduit shall extend from the forming shell to a suitable junction box or other enclosure located as provided in Section E4206.9. Metal parts of the luminaire and forming shell in contact with the pool water shall be of brass or other approved corrosion-resistant metal. [680.23(B)(1)]

The end of flexible-cord jackets and flexible-cord conductor terminations within a luminaire shall be covered with, or encapsulated in, a suitable potting compound to prevent the entry of water into the luminaire through the cord or its conductors. If present, the connection of the equipment grounding conductor within a luminaire shall be similarly treated to protect such connection from the deteriorating effect of pool water in the event of water entry into the luminaire. [680.23(B)(4)]

Luminaires shall be bonded to and secured to the forming shell by a positive locking device that ensures a low-resistance contact and requires a tool to remove the luminaire from the forming shell. [680.23(B)(5)]

E4206.5.1 Servicing. All wet-niche luminaires shall be removable from the water for inspection, relamping, or other maintenance. The forming shell location and length of cord in the forming shell shall permit personnel to place the removed luminaire on the deck or other dry location for such maintenance. The luminaire maintenance location shall be accessible without entering or going into the pool water. In spa locations where wet-niche luminaires are installed low in the foot well of the spa, the luminaire shall only be required to reach the bench location, where the spa can be drained to make the bench location dry. [680.23(B)(6)]

E4206.6 Dry-niche luminaires. Dry-niche luminaires shall have provisions for drainage of water. Other than *listed* low-voltage luminaires not requiring grounding, a dry-niche luminaire shall have means for accommodating one equipment grounding conductor for each conduit entry. Junction boxes shall not be required but, if used, shall not be required to be elevated or located as specified in Section E4206.9 if the luminaire is specifically identified for the purpose. [680.23(C)(1) and (C)(2)]

E4206.7 No-niche luminaires. No-niche luminaires shall be *listed* for the purpose and shall be installed in accordance with the requirements of Section E4206.5. Where connection to a forming shell is specified, the connection shall be to the mounting bracket. [680.23(D)]

E4206.8 Through-wall lighting assembly. A through-wall lighting assembly shall be equipped with a threaded entry or hub, or a nonmetallic hub, for the purpose of accommodating the termination of the supply conduit. A through-wall lighting assembly shall meet the construction requirements of Section E4205.4 and be installed in accordance with the requirements of Section E4206.5 Where connection to a forming shell is specified, the connection shall be to the conduit termination point. [680.23(E)]

E4206.9 Junction boxes and enclosures for transformers or ground-fault circuit interrupters. Junction boxes for underwater luminaires and enclosures for transformers and ground-fault circuit interrupters that supply underwater luminaires shall comply with Sections E4206.9.1 through E4206.9.5. [680.24(A)]

E4206.9.1 Junction boxes. A junction box connected to a conduit that extends directly to a forming shell or mounting bracket of a no-niche luminaire shall be:

1. *Listed* as a swimming pool junction box; [680.24(A)(1)]
2. Equipped with threaded entries or hubs or a nonmetallic hub; [680.24(A)(1)(1)]
3. Constructed of copper, brass, suitable plastic, or other approved corrosion-resistant material; [680.24(A)(1)(2)]
4. Provided with electrical continuity between every connected metal conduit and the grounding terminals by means of copper, brass, or other approved corrosion-resistant metal that is integral with the box; and [680.24(A)(1)(3)]
5. Located not less than 4 inches (102 mm), measured from the inside of the bottom of the box, above the ground level, or pool deck, or not less than 8 inches (203 mm) above the maximum pool water level, whichever provides the greatest elevation, and shall be located not less than 4 feet (1219 mm) from the inside wall of the pool, unless separated from the pool by a solid fence, wall or other permanent barrier. Where used on a lighting system operating at the low-voltage contact limit or less, a flush deck box shall be permitted provided that an approved potting compound is used to fill the box to prevent the entrance of moisture; and the flush deck box is located not less than 4 feet (1219 mm) from the inside wall of the pool. [680.24(A)(2)]

E4206.9.2 Other enclosures. An enclosure for a transformer, ground-fault circuit interrupter or a similar device connected to a conduit that extends directly to a forming shell or mounting bracket of a no-niche luminaire shall be:

1. *Listed* and *labeled* for the purpose, comprised of copper, brass, suitable plastic, or other approved corrosion-resistant material; [680.24(B)(1), 680.24(B)(1)(2)]
2. Equipped with threaded entries or hubs or a nonmetallic hub; [680.24(B)(1)(1)]
3. Provided with an approved seal, such as duct seal at the conduit connection, that prevents circulation of air between the conduit and the enclosures; [680.24(B)(1)(3)]
4. Provided with electrical continuity between every connected metal conduit and the grounding terminals by means of copper, brass or other approved corrosion-resistant metal that is integral with the enclosures; and [680.24(B)(1)(4)]
5. Located not less than 4 inches (102 mm), measured from the inside bottom of the enclosure, above the ground level or pool deck, or not less than 8 inches (203 mm) above the maximum pool water level, whichever provides the greater elevation, and shall be located not less than 4 feet (1219 mm) from the inside wall of the pool, except where separated from the pool by a solid fence, wall or other permanent barrier. [680.24(B)(2)]

E4206.9.3 Protection of junction boxes and enclosures. Junction boxes and enclosures mounted above the grade of the finished walkway around the pool shall not be located in the walkway unless afforded additional protection, such as by location under diving boards or adjacent to fixed structures. [680.24(C)]

E4206.9.4 Grounding terminals. Junction boxes, transformer and power supply enclosures, and ground-fault circuit-interrupter enclosures connected to a conduit that extends directly to a forming shell or mounting bracket of a no-niche luminaire shall be provided with grounding terminals in a quantity not less than the number of conduit entries plus one. [680.24(D)]

E4206.9.5 Strain relief. The termination of a flexible cord of an underwater luminaire within a junction box, transformer or power supply enclosure, ground-fault circuit interrupter, or other enclosure shall be provided with a strain relief. [680.24(E)]

E4206.10 Underwater audio equipment. Underwater audio equipment shall be identified as underwater audio equipment. [680.27(A)]

E4206.10.1 Speakers. Each speaker shall be mounted in an approved metal forming shell, the front of which is enclosed by a captive metal screen, or equivalent, that is bonded to and secured to the forming shell by a positive locking device that ensures a low-resistance contact and requires a tool to open for installation or servicing of the speaker. The forming shell shall be installed in a recess in the wall or floor of the pool. [680.27(A)(1)]

E4206.10.2 Wiring methods. Rigid metal conduit of brass or other identified corrosion-resistant metal, rigid polyvinyl chloride conduit, rigid thermosetting resin conduit or liquid-tight flexible nonmetallic conduit (LFNC-B) shall extend from the forming shell to a suitable junction box or other enclosure as provided in Section E4206.9. Where rigid nonmetallic conduit or liquid-tight flexible nonmetallic conduit is used, an 8 AWG solid or stranded insulated copper bonding jumper shall be installed in this conduit with provisions for terminating in the forming shell and the junction box. The termination of the 8 AWG bonding jumper in the forming shell shall be covered with, or encapsulated in, a suitable potting compound to protect such connection from the possible deteriorating effect of pool water. [680.27(A)(2)]

E4206.10.3 Forming shell and metal screen. The forming shell and metal screen shall be of brass or other approved corrosion-resistant metal. Forming shells shall include provisions for terminating an 8 AWG copper conductor. [680.27(A)(3)]

E4206.11 Electrically operated pool covers. The electric motors, controllers, and wiring for pool covers shall be located not less than 5 feet (1524 mm) from the inside wall of the pool except where separated from the pool by a wall, cover, or other permanent barrier. Electric motors installed below grade level shall be of the totally enclosed type. The electric motor and controller shall be connected to a branch circuit protected by a ground-fault circuit interrupter. The device that controls the operation of the motor for an electrically operated pool cover shall be located so that the device operator has full view of the pool. [680.27(B)(1) and (B)(2)]

Exceptions:

1. Motors that are part of *listed* systems with ratings not exceeding the low-voltage contact limit and that are supplied by *listed* transformers or power supplies that comply with Section E4206.1 shall be permitted to be located less than 5 feet (1524 mm) from the inside walls of the pool. [680.27(B)(1) Exception]
2. Motors that are part of *listed* systems with ratings not exceeding the low-voltage contact limit and that are supplied by *listed* transformers or power supplies that comply with Section E4206.1 shall not be required to be connected to a branch circuit protected by a ground fault circuit-interrupter. [680.27 and (B)(2) Exception]

E4206.12 Electric pool water heaters. Electric pool water heaters shall have the heating elements subdivided into loads not exceeding 48 amperes and protected at not more than 60 amperes. The ampacity of the branch-circuit conductors and the rating or setting of overcurrent protective devices shall be not less than 125 percent of the total nameplate load rating. (680.10)

E4206.13 Pool area heating. The provisions of Sections E4206.13.1 through E4206.13.3 shall apply to all pool deck areas, including a covered pool, where electrically operated comfort heating units are installed within 20 feet (6096 mm) of the inside wall of the pool. [680.27(C)]

E4206.13.1 Unit heaters. Unit heaters shall be rigidly mounted to the structure and shall be of the totally enclosed or guarded types. Unit heaters shall not be mounted over the pool or within the area extending 5 feet (1524 mm) horizontally from the inside walls of a pool. [680.27(C)(1)]

E4206.13.2 Permanently wired radiant heaters. Electric radiant heaters shall be suitably guarded and securely fastened to their mounting devices. Heaters shall not be installed over a pool or within the area extending 5 feet (1524 mm) horizontally from the inside walls of the pool and shall be mounted not less than 12 feet (3658 mm) vertically above the pool deck. [680.27(C)(2)]

E4206.13.3 Radiant heating cables prohibited. Radiant heating cables embedded in or below the deck shall be prohibited. [680.27(C)(3)]

SECTION E4207 STORABLE SWIMMING POOLS, STORABLE SPAS AND STORABLE HOT TUBS

E4207.1 Pumps. A cord-and-plug-connected pool filter pump for use with storable pools shall incorporate an approved system of double insulation or its equivalent and shall be provided with means for the termination of an equipment grounding conductor for the connection to the internal and nonaccessible noncurrent-carrying metal parts of the pump.

The means for grounding shall be an equipment grounding conductor run with the power-supply conductors in a flexible cord that is properly terminated in a grounding-type attachment plug having a fixed grounding contact. Cord-and-plug-connected pool filter pumps shall be provided with a ground-fault circuit-interrupter that is an integral part of the attachment plug or located in the power supply cord within 12 inches (305 mm) of the attachment plug. (680.31)

E4207.2 Ground-fault circuit interrupters required. Electrical equipment, including power-supply cords, used with storable pools shall be protected by ground-fault circuit interrupters. 125-volt, 15- and 20-ampere receptacles located within 20 feet (6096 mm) of the inside walls of a storable pool, storable spa, or storable hot tub shall be protected by a ground-fault circuit interrupter. In determining these dimensions, the distance to be measured shall be the shortest path that the supply cord of an appliance connected to the receptacle would follow without passing through a floor, wall, ceiling, doorway with hinged or sliding door, window opening, or other effective permanent barrier. (680.32)

E4207.3 Luminaires. Luminaires for storable pools, storable spas, and storable hot tubs shall not have exposed metal parts and shall be *listed* for the purpose as an assembly. In addition, luminaires for storable pools shall comply with the requirements of Section E4207.3.1 or E4207.3.2. (680.33)

E4207.3.1 Within the low-voltage contact limit. A luminaire installed in or on the wall of a storable pool shall be part of a cord-and-plug-connected lighting assembly. The assembly shall:

1. Have a luminaire lamp that is suitable for the use at the supplied voltage;
2. Have an impact-resistant polymeric lens, luminaire body, and transformer enclosure;
3. Have a transformer meeting the requirements of Section E4206.1 with a primary rating not over 150 volts; and
4. Have no exposed metal parts. [680.33(A)]

E4207.3.2 Over the low-voltage contact limit but not over 150 volts. A lighting assembly without a transformer or power supply, and with the luminaire lamp(s) operating at over the low-voltage contact limit, but not over 150 volts, shall be permitted to be cord-and-plug-connected where the assembly is *listed* as an assembly for the purpose and complies with all of the following:

1. It has an impact-resistant polymeric lens and luminaire body.
2. A ground-fault circuit interrupter with open neutral conductor protection is provided as an integral part of the assembly.
3. The luminaire lamp is permanently connected to the ground-fault circuit interrupter with open-neutral protection.
4. It complies with the requirements of Section E4206.4.
5. It has no exposed metal parts. [680.33(B)]

E4207.4 Receptacle locations. Receptacles shall be located not less than 6 feet (1829 mm) from the inside walls of a storable pool, storable spa or storable hot tub. In determining these dimensions, the distance to be measured shall be the shortest path that the supply cord of an appliance connected to the receptacle would follow without passing through a floor, wall, ceiling, doorway with hinged or sliding door, window opening, or other effective permanent barrier. (680.34)

E4207.5 Clearances. Overhead conductor installations shall comply with Section E4203.7 and underground conductor installations shall comply with Section E4203.8. [680.30]

E4207.6 Disconnecting means. Disconnecting means for storable pools and storable/portable spas and hot tubs shall comply with Section E4203.3. [680.30]

E4207.7 Ground-fault circuit interrupters. Ground-fault circuit interrupters shall comply with Section E4206.2. [680.30]

E4207.8 Bonding and grounding of equipment. Equipment shall be bonded and grounded as required by Section E4205.1. [680.30]

E4207.9 Pool water heaters. Electric pool water heaters shall comply with Section E4206.12.

SECTION E4208
SPAS AND HOT TUBS

E4208.1 Ground-fault circuit interrupters. The outlet(s) that supplies a self-contained spa or hot tub, or a packaged spa or hot tub equipment assembly, or a field-assembled spa or hot tub shall be protected by a ground-fault circuit interrupter. (680.44)

A *listed* self-contained unit or *listed* packaged equipment assembly marked to indicate that integral ground-fault circuit-interrupter protection is provided for all electrical parts within the unit or assembly, including pumps, air blowers, heaters, lights, controls, sanitizer generators and wiring, shall not require that the outlet supply be protected by a ground-fault circuit-interrupter. [680.44(A)]

E4208.2 Electric water heaters. Electric spa and hot tub water heaters shall be *listed* and shall have the heating elements subdivided into loads not exceeding 48 amperes and protected at not more than 60 amperes. The ampacity of the branch-circuit conductors, and the rating or setting of overcurrent protective devices, shall be not less than 125 percent of the total nameplate load rating. (680.10)

E4208.3 Underwater audio equipment. Underwater audio equipment used with spas and hot tubs shall comply with the provisions of Section E4206.10. [680.43(G)]

E4208.4 Emergency switch for spas and hot tubs. A clearly *labeled* emergency shutoff or control switch for the purpose of stopping the motor(s) that provides power to the recirculation system and jet system shall be installed at a point that is readily accessible to the users, adjacent to and within sight of the spa or hot tub and not less than 5 feet (1524 mm) away from the spa or hot tub. This requirement shall not apply to one-family dwellings. (680.41)

SECTION E4209
HYDROMASSAGE BATHTUBS

E4209.1 General. Installations of hydromassage bathtubs shall be required to comply only with Section E4209. The branch circuit wiring method(s) supplying a hydromassage bathtub shall comply with Chapter 38.

E4209.2 Ground-fault circuit interrupters. Hydromassage bathtubs and their associated electrical components shall be supplied by an individual branch circuit(s) and protected by a readily accessible ground-fault circuit interrupter. All 125-volt, single-phase receptacles not exceeding 30 amperes and located within 6 feet (1829 mm) measured horizontally of the inside walls of a hydromassage tub shall be protected by a ground-fault circuit interrupter(s). (680.71)

E4209.3 Other electric equipment. Luminaires, switches, receptacles and other electrical equipment located in the same room, and not directly associated with a hydromassage bathtub, shall be installed in accordance with the requirements of this code relative to the installation of electrical equipment in bathrooms. (680.72)

E4209.4 Accessibility. Hydromassage bathtub electrical equipment shall be accessible without damaging the building structure or building finish.

Where the hydromassage bathtub is cord- and plug-connected with the supply receptacle accessible only through a service access opening, the receptacle shall be installed so that its face is within direct view and not more than 12 inches (305 mm) from the plane of the opening. (680.73)

E4209.5 Bonded parts. The following parts shall be bonded together:

1. Metal fittings within or attached to the tub structure that are in contact with the circulating water.
2. Metal parts of electrical equipment associated with the tub water circulating system, including the pump and blower motors.
3. Metal-sheathed cables, metal raceways and metal piping that are within 5 feet (1524 mm) of the inside walls of the tub and that are not separated from the tub area by a permanent barrier.
4. Exposed metal surfaces that are within 5 feet (1524 mm) of the inside walls of the tub and not separated from the tub area by a permanent barrier.
5. Noncurrent-carrying metal parts of electrical devices and controls that are not associated with the hydromassage tubs and that are located within 5 feet (1524 mm) from such units. [680.74(A)]

Exceptions:

1. Double-insulated motors and blowers shall not be bonded.
2. Small conductive surfaces not likely to become energized, such as air and water jets, supply valve assemblies and drain fittings not connected to metal piping, and towel bars, mirror frames and similar nonelectric equipment not connected to metal framing shall not be required to be bonded. [680.74(A) Exceptions 1 and 2]

E4209.6 Method of bonding. Metal parts required to be bonded by this section shall be bonded together using a solid copper bonding jumper, insulated, covered or bare, not smaller than 8 AWG. The bonding jumper(s) shall be required for equipotential bonding in the area of the hydromassage bathtub and shall not be required to be extended or attached to any remote panelboard, service equipment, or electrode. In all installations, a bonding jumper long enough to terminate on a replacement nondouble-insulated pump or blower motor shall be provided and shall be terminated to the equipment grounding conductor of the branch circuit of the motor where a double-insulated circulating pump or blower motor is used. [680.74(B)]

CHAPTER 43

CLASS 2 REMOTE-CONTROL, SIGNALING AND POWER-LIMITED CIRCUITS

ICC user note:

About this chapter: *Chapter 43 covers low-voltage power-limited circuits such as alarm, doorbell, remote control and signaling circuits.*

SECTION E4301 GENERAL

E4301.1 Scope. This chapter contains requirements for power supplies and wiring methods associated with Class 2 remote-control, signaling, and power-limited circuits that are not an integral part of a device or appliance. Other classes of remote-control, signaling and power-limited conductors shall comply with Article 725 of NFPA 70. (725.1)

E4301.2 Definitions.

CLASS 2 CIRCUIT. That portion of the wiring system between the load side of a Class 2 power source and the connected equipment. Due to its power limitations, a Class 2 circuit considers safety from a fire initiation standpoint and provides acceptable protection from electric shock. (725.2)

REMOTE-CONTROL CIRCUIT. Any electrical circuit that controls any other circuit through a relay or an equivalent device. (Article 100)

SIGNALING CIRCUIT. Any electrical circuit that energizes signaling equipment. (Article 100)

SECTION E4302 POWER SOURCES

E4302.1 Power sources for Class 2 circuits. The power source for a Class 2 circuit shall be one of the following:

1. A *listed* Class 2 transformer.
2. A *listed* Class 2 power supply.
3. Other *listed* equipment marked to identify the Class 2 power source.
4. *Listed* audio/video information technology (computer) communications and industrial equipment limited power circuits.
5. A battery source or battery source system that is *listed* and identified as Class 2. [725.121(A)]

E4302.2 Interconnection of power sources. A Class 2 power source shall not have its output connections paralleled or otherwise interconnected with another Class 2 power source except where *listed* for such interconnection. [725.121(B)]

SECTION E4303 WIRING METHODS

E4303.1 Wiring methods on supply side of Class 2 power source. Conductors and equipment on the supply side of the power source shall be installed in accordance with the appropriate requirements of Chapters 34 through 41. Transformers or other devices supplied from electric light or power circuits shall be protected by an over-current device rated at not over 20 amperes. The input leads of a transformer or other power source supplying Class 2 circuits shall be permitted to be smaller than 14 AWG, if not over 12 inches (305 mm) long and if the conductor insulation is rated at not less than 600 volts. In no case shall such leads be smaller than 18 AWG. (725.127 and 725.127 Exception)

E4303.2 Wiring methods and materials on load side of the Class 2 power source. Class 2 cables installed as wiring within buildings shall be *listed* as being resistant to the spread of fire and *listed* as meeting the criteria specified in Sections E4303.2.1 through E4303.2.4. Cables shall be marked in accordance with Section E4303.2.5. Cable substitutions as described in Table E4303.2 and wiring methods covered in Chapter 38 shall also be permitted. [725.130(B); 725.133; 725.135(A), (C), (G) and (M); 725.154; Table 725.154; [Figure 725.154(A); and 725.179]

TABLE E4303.2 CABLE USES AND PERMITTED SUBSTITUTIONS [Figure 725.154(A)]

CABLE TYPE	USE	PERMITTED SUBSTITUTIONS[a]
CL2P	Class 2 Plenum Cable	CMP, CL3P
CL2R	Class 2 Plenum Cable	CMP, CL3P, CL2P, CMR, CL3R
CL2	Class 2 Cable	CMP, CL3P, CL2P, CMR, CL3R, CL2R CMG, CM, CL3
CL2X	Class 2 Cable, Limited Use	CMP, CL3P CL2P, CMR, CL3R, CL2R, CMG, CM, CL3, CL2, CMX, CL3X

a. For identification of cables other than Class 2 cables, see NFPA 70.

E4303.2.1 Type CL2P cables. Cables installed in ducts, plenums and other spaces for environmental air shall be Type CL2P cables *listed* as suitable for the use and *listed* as having adequate fire-resistant and low smoke-producing characteristics. [725.179(A)]

E4303.2.2 Type CL2 cables. Cables for general-purpose use, shall be *listed* as resistant to the spread of fire and *listed* for the use. [725.179 (C)]

E4303.2.3 Type CL2X cables. Type CL2X limited-use cable shall be *listed* as suitable for use in dwellings and raceways and shall be *listed* as resistant to flame spread. [725.179 (D)]

E4303.2.4 Type CL2R cables. Cables installed in a vertical run in a shaft or installed from floor to floor shall be *listed* as suitable for use in a vertical run in a shaft or from floor to floor and shall be *listed* as having fire-resistant characteristics capable of preventing fire from being conveyed from floor to floor. [725.179(B)]

Exception: CL2X and CL3X cables with a diameter of less than $^1/_4$ inch (6.4 mm) and CL2P, CL3P, CL2R, CL3R, CL2 and CL3 cables shall be permitted in risers in one- and two-family dwelling units. [725.135(G)]

E4303.2.5 Marking. Cables shall be marked in accordance with Table E4303.2.5. Voltage ratings shall not be marked on cables. Cables shall be permitted to be surface marked to indicate special characteristics of the cable materials. [(725.179(J), (K)]

Table E4303.2.5 [Table 725.179(J)]
CABLE MARKING

CABLE MARKING	TYPE
CL2P	Class 2 plenum cable
CL2R	Class 2 riser cable
CL2	Class 2 cable
CL2X	Class 2 cable, limited use

SECTION E4304
INSTALLATION REQUIREMENTS

E4304.1 Separation from other conductors. In cables, compartments, enclosures, outlet boxes, device boxes and raceways, conductors of Class 2 circuits shall not be placed in any cable, compartment, enclosure, outlet box, device box, raceway or similar fitting with conductors of electric light, power, Class 1 and nonpower-limited fire alarm circuits. (725.136)

Exceptions:

1. Where the conductors of the electric light, power, Class 1 and nonpower-limited fire alarm circuits are separated by a barrier from the Class 2 circuits. In enclosures, Class 2 circuits shall be permitted to be installed in a raceway within the enclosure to separate them from Class 1, electric light, power and nonpower-limited fire alarm circuits. [725.136(B), (C)]
2. Class 2 conductors in compartments, enclosures, device boxes, outlet boxes and similar fittings where electric light, power, Class 1 or nonpower-limited fire alarm circuit conductors are introduced solely to connect to the equipment connected to the Class 2 circuits. The electric light, power, Class 1 and nonpower-limited fire alarm circuit conductors shall be routed to maintain a minimum of $^1/_4$ inch (6.4 mm) separation from the conductors and cables of the Class 2 circuits; or the electric light power, Class 1 and nonpower-limited fire alarm circuit conductors operate at 150 volts or less to ground and the Class 2 circuits are installed using Types CL3, CL3R, or CL3P or permitted substitute cables, and provided that these Class 3 cable conductors extending beyond their jacket are separated by a minimum of $^1/_4$ inch (6.4 mm) or by a nonconductive sleeve or nonconductive barrier from all other conductors. [725.136(D)]

E4304.2 Other applications. Conductors of Class 2 circuits shall be separated by not less than 2 inches (51 mm) from conductors of any electric light, power, Class 1 or nonpower-limited fire alarm circuits except where one of the following conditions is met:

1. All of the electric light, power, Class 1 and nonpower-limited fire alarm circuit conductors are in raceways or in metal-sheathed, metal-clad, nonmetallic-sheathed or Type UF cables.
2. All of the Class 2 circuit conductors are in raceways or in metal-sheathed, metal-clad, nonmetallic-sheathed or Type UF cables. [725.136(I)]

E4304.3 Class 2 circuits with communications circuits. Conductors of one or more Class 2 circuits shall be permitted in the same cable with conductors of communications circuits, provided that the cable is a *listed* communications cable that shall be installed in accordance with the requirements of Part V of Article 805 of NFPA 70. The cables shall be *listed* as communications cables. [725.139(D)(1)]

Cables constructed of individually *listed* Class 2 and communications cables under a common jacket shall be permitted to be classified as communications cables. The fire-resistance rating of the composite cable shall be determined by the performance of the composite cable. [725.139 (D)(2)]

E4304.4 Class 2 cables with other circuit cables. Jacketed cables of Class 2 circuits shall be permitted in the same enclosure or raceway with jacketed cables of any of the following:

1. Power-limited fire alarm systems in compliance with Article 760 of NFPA 70.
2. Nonconductive and conductive optical fiber cables in compliance with Article 770 of NFPA 70.
3. Communications circuits in compliance with Article 805 of NFPA 70.

4. Community antenna television and radio distribution systems in compliance with Article 820 of NFPA 70.
5. Low-power, network-powered broadband communications in compliance with Article 830 of NFPA 70. [725.139(E)]

E4304.5 Installation of conductors and cables. Cables and conductors installed exposed on the surface of ceilings and sidewalls shall be supported by the building structure in such a manner that they will not be damaged by normal building use. Such cables shall be supported by straps, staples, hangers, cable ties or similar fittings designed so as to not damage the cable. Nonmetallic cable ties and other nonmetallic accessories used to secure and support cables located in stud cavity and joist space plenums shall be *listed* as having low smoke and heat release properties. The installation shall comply with Table E3802.1 regarding cables run parallel with framing members and furring strips. The installation of wires and cables shall not prevent access to equipment nor prevent removal of panels, including suspended ceiling panels. Raceways shall not be used as a means of support for Class 2 circuit conductors, except where the supporting raceway contains conductors supplying power to the functionally associated equipment controlled by the Class 2 conductors. [300.11 (C) (2), 300.22 (C) (1) and 725.24]

Part IX—Referenced Standards

CHAPTER 44
REFERENCED STANDARDS

User notes:

About this chapter: The one- and two-family dwelling code contains numerous references to standards promulgated by other organizations that are used to provide requirements for materials, products and methods of construction. Chapter 44 contains a comprehensive list of all standards that are referenced in this code. These standards, in essence, are part of this code to the extent of the reference to the standard.

This chapter lists the standards that are referenced in various sections of this document. The standards are listed herein by the promulgating agency of the standard, the standard identification, the effective date and title, and the section or sections of this document that reference the standard. The application of the referenced standards shall be as specified in Section R102.4.

AAMA

American Architectural Manufacturers Association
1900 E. Golf Road, Suite 1250
Schaumburg, IL 60173

450—20: Voluntary Performance Rating Method for Mulled Fenestration Assemblies
R609.8

506—16: Voluntary Specifications for Hurricane Impact and Cycle Testing of Fenestration Products
R609.6.1

711—20: Voluntary Specification for Self-adhering Flashing Used for Installation of Exterior Wall Fenestration Products
R703.4

712—14: Voluntary Specification for Mechanically Attached Flexible Flashing
R703.4

714—20: Voluntary Specification for Liquid Applied Flashing Used to Create a Water-resistive Seal around Exterior Wall Openings in Buildings
R703.4

AAMA/NSA 2100—20: Specifications for Sunrooms
R301.2.1.1.1

AAMA/WDMA/CSA 101/I.S.2/A440—17: North American Fenestration Standards/Specifications for Windows, Doors and Skylights
N1102.4.3, R308.6.9, R609.3

ACCA

Air Conditioning Contractors of America
1330 Braddock Place, Suite 350
Alexandria, VA 22314

ANSI/ACCA 1 Manual D—2016: Residential Duct Systems
Table R301.2(1), M1601.1, M1602.2

ANSI/ACCA 2 Manual J—2016: Residential Load Calculation
N1103.7, M1401.3

ANSI/ACCA 3 Manual S—2014: Residential Equipment Selection
N1103.7, M1401.3

ACI

American Concrete Institute
38800 Country Club Drive
Farmington Hills, MI 48331

318—19: Building Code Requirements for Structural Concrete
R402.2, Table R404.1.2(2), Table R404.1.2(5), Table R404.1.2(6), Table R404.1.2(7), Table R404.1.2(8), R404.1.3, R404.1.3.1, R404.1.3.3, R404.1.3.4, R404.1.4.2, R404.5.1, R608.1, R608.1.1, R608.1.2, R608.2, R608.5.1, R608.6.1, R608.8.2, R608.9.2, R608.9.3

332—20: Residential Code Requirements for Structural Concrete
R402.2, R403.1, R404.1.3, R404.1.3.4, R404.1.4.2, R506.1

AISI

American Iron and Steel Institute
25 Massachusetts Avenue, NW Suite 800
Washington, DC 20001

AISI S100—16 (2020) w/S2—20: North American Specification for the Design of Cold-Formed Steel Structural Members, 2016 Edition (Reaffirmed 2020), with Supplement 2, 2020 Edition
R608.9.2, R608.9.3

AISI S220—20: North American Standard for Cold-Formed Steel Nonstructural Framing, 2020
R702.3.3

AISI S230—18: Standard for Cold-Formed Steel Framing—Prescriptive Method for One- and Two-Family Dwellings, 2018
R301.1.1, R301.2.1.1, R301.2.2.7, R301.2.2.8, R603.6, R603.9.4.1, R603.9.4.2, Figure 608.9(11), R608.9.2, R608.9.3, R608.10

AISI S240—20: North American Standard for Cold-Formed Steel Structural Framing, 2020
R505.1.3, R603.6, R702.3.3, R804.3.6

AMCA

Air Movement and Control Association International
30 West University Drive
Arlington Heights, IL 60004

ANSI/AMCA 210-ANSI/ASHRAE 51—16: Laboratory Methods of Testing Fans for Aerodynamic Performance Rating
Table M1504.2, M1505.3

ANCE

Association of Standardization and Certification
Av. Lázaro Cárdenas No. 869
Fraccion 3
Col. Nva. Industrial Vallejo
Deleg. Gustavo A. Madero
Mexico, D.F.

NMX-J-521/2-40-ANCE—2014/CAN/CSA-22.2 No. 60335-2-40—12/UL 60335-2-40: Safety of Household and Similar Electric Appliances, Part 2-40: Particular Requirements for Heat Pumps, Air-Conditioners and Dehumidifiers
M1403.1, M1412.1, M1413.1

ANSI

American National Standards Institute
25 West 43rd Street, 4th Floor
New York, NY 10036

A108.1A—17: Installation of Ceramic Tile in the Wet-set Method, with Portland Cement Mortar
R702.4.1

A108.1B—2017: Installation of Ceramic Tile, Quarry Tile on a Cured Portland Cement Mortar Setting Bed with Dry-set or Latex Portland Mortar
R702.4.1

A108.4—09: Installation of Ceramic Tile with Organic Adhesives or Water-Cleanable Tile-setting Epoxy Adhesive
R702.4.1

A108.5—19: Installation of Ceramic Tile with Dry-set Portland Cement Mortar or Latex Portland Cement Mortar
R702.4.1

ANSI—continued

A108.6—19: Installation of Ceramic Tile with Chemical-resistant, Water-cleanable Tile-setting and -grouting Epoxy
R702.4.1

A108.11—10: Interior Installation of Cementitious Backer Units
R702.4.1

A118.1—18: American National Standard Specifications for Dry-set Portland Cement Mortar
R702.4.1

A118.3—20: American National Standard Specifications for Chemical-resistant, Water-cleanable Tile-setting and -grouting Epoxy, and Water-cleanable Tile-setting Epoxy Adhesive
R702.4.1

A118.4—18: American National Standard Specifications for Modified Dry-Set Cement Mortar
R606.2.11

A118.10—14: Specification for Load-bearing, Bonded, Waterproof Membranes for Thin-set Ceramic Tile and Dimension Stone Installation
P2709.2, P2709.2.4

A136.1—19: American National Standard Specifications for Organic Adhesives for Installation of Ceramic Tile
R702.4.1

A137.1—19: American National Standard Specifications for Ceramic Tile
R702.4.1

ANSI 117—2020: Standard Specifications for Structural Glued Laminated Timber of Softwood Species
R502.1.3, R602.1.3, R802.1.3

ANSI/CSA FC 1—2014: Fuel Cell Technologies—Part 3-100: Stationary Fuel Cell Power Systems—Safety
M1903.1

LC1/CSA 6.26—2016: Fuel Gas Piping Systems Using Corrugated Stainless Steel Tubing (CSST)
G2411.3, G2414.4.4, G2415.5

LC4/CSA 6.32—12: Press-connect Metallic Fittings for Use in Fuel Gas Distribution Systems
G2414.9.1, G2414.9.2, G2414.9.3, G2415.5

Z21.1—2016: Household Cooking Gas Appliances
M1503.2, G2447.1

Z21.5.1/CSA 7.1—2017: Gas Clothes Dryers—Volume I—Type I Clothes Dryers
G2438.1

Z21.8—94(R2012): Installation of Domestic Gas Conversion Burners
G2443.1

Z21.10.1/CSA 4.1—2012: Gas Water Heaters—Volume I—Storage Water Heaters with Input Ratings of 75,000 Btu per hour or Less
G2448.1

Z21.10.3/CSA 4.3—2017: Gas Water Heaters—Volume III—Storage Water Heaters with Input Ratings above 75,000 Btu per hour, Circulating and Instantaneous
G2448.1

Z21.11.2—2016: Gas-fired Room Heaters—Volume II—Unvented Room Heaters
G2445.1

Z21.13/CSA 4.9—2017: Gas-fired Low-pressure Steam and Hot Water Boilers
G2452.1

Z21.15/CSA 9.1—09(R2014): Manually Operated Gas Valves for Appliances, Appliance Connector Valves and Hose End Valves
Table G2420.1.1

Z21.22—99 (R2003): Relief Valves for Hot Water Supply Systems—with Addenda Z21.22a—2000 (R2003) and 21.22b—2001 (R2003)
P2804.2, P2804.7

Z21.24/CSA 6.10—2015: Connectors for Gas Appliances
G2422.1, G2422.2

Z21.40.1/CSA 2.91—96(2017): Gas-fired, Heat-activated Air-conditioning and Heat Pump Appliances
G2449.1

Z21.40.2/CSA 2.92—96(R2017): Gas-fired Work Activated Air-conditioning and Heat Pump Appliances (Internal Combustion)
G2449.1

ANSI—continued

Z21.41/CSA 6.9—2014: Quick Disconnect Devices for Use with Gas Fuel Appliances
G2422.1

Z21.42—2013: Gas-fired Illuminating Appliances
G2450.1

Z21.47/CSA 2.3—2016: Gas-fired Central Furnaces
G2442.1

Z21.50/CSA 2.22—2016: Vented Decorative Gas Fireplaces
G2434.1

Z21.54—2014: Gas Hose Connectors for Portable Outdoor Gas-fired Appliances
G2422.1

Z21.56/CSA 4.7—17: Gas-fired Pool Heaters
G2441.1

Z21.58—95/CSA 1.6—2015: Outdoor Cooking Gas Appliances
G2447.1

Z21.60/CSA 2.26—2017: Decorative Gas Appliances for Installation in Solid Fuel-burning Fireplaces
G2432.1

Z21.69/CSA 6.16—2015: Connectors for Movable Gas Appliances
G2422.1.5

Z21.75/CSA 6.27—2016: Connectors for Outdoor Gas Appliances and Manufactured Homes
G2422.1

Z21.80/CSA 6.22—11(R2016): Line Pressure Regulators
G2421.1

Z21.84—2017: Manually Lighted, Natural Gas Decorative Gas Appliances for Installation in Solid Fuel-burning Appliances
G2432.1, G2432.2

Z21.86/CSA 2.32—2016: Vented Gas-fired Space Heating Appliances
G2436.1, G2437.1, G2446.1

Z21.88/CSA 2.33—16: Vented Gas Fireplace Heaters
G2435.1

Z21.90/CSA 6.24-2015: Gas Convenience Outlets and Optional Enclosures
G2422.1

Z21.91—2017: Ventless Firebox Enclosures for Gas-fired Unvented Decorative Room Heaters
G2445.7.1

Z21.93/CSA 6.30—2017: Excess Flow Valves for Natural Gas and Propane Gas with Pressures up to 5 psig
G2421.4

Z21.97—2014: Outdoor Decorative Gas Appliances
G2453.1

Z83.6—90 (R1998): Gas-fired Infrared Heaters
G2451.1

Z83.8/CSA 2.6—2016: Gas Unit Heaters, Gas Packaged Heaters, Gas Utility Heaters, and Gas-fired Duct Furnaces
G2444.1

Z83.19—2009 (R2014): Gas-fired High-intensity Infrared Heaters
G2451.1

Z83.20—08: Gas-fired Low-intensity Infrared Heaters Outdoor Decorative Appliances
G2451.1

Z97.1—2014: Safety Glazing Materials Used in Buildings—Safety Performance Specifications and Methods of Test
R308.1.1, R308.3.1

APA

APA—The Engineered Wood Association
7011 South 19th Street
Tacoma, WA 98466

ANSI/A190.1—2017: Structural Glued-laminated Timber
R502.1.3, R602.1.3, R802.1.2

APA—continued

ANSI/APA PRG 320—2019: Standard for Performance-rated Cross Laminated Timber
R502.1.6, R602.1.6, R802.1.6

ANSI/APA PRP 210—2019: Standard for Performance-rated Engineered Wood Siding
R604.1, Table R703.3(1), R703.3.4

ANSI/APA PRR 410—2016: Standard for Performance-rated Engineered Wood Rim Boards
R502.1.7, R602.1.7, R802.1.7

ANSI/APA PRS 610.1—2018: Standard for Performance-Rated Structural Insulated Panels in Wall Applications
R602.1.11, R610.3, R610.4

APA E30—19: Engineered Wood Construction Guide
Table R503.2.1.1(1), R503.2.2, R803.2.2, R803.2.3

APSP

Pool & Hot Tub Alliance (formerly the Association of Pool & Spa Professionals)
211 Eisenhower Avenue, Suite 500
Alexander, VA 22314

ANSI/APSP/ICC 14—2019: American National Standard for Portable Electric Spa Energy Efficiency
N1103.11

ANSI/APSP/ICC 15a—2011: American National Standard for Residential Swimming Pool and Spas—Includes Addenda A Approved January 9, 2013
N1103.12

ASCE/SEI

American Society of Civil Engineers
Structural Engineering Institute
Reston, VA 20191-4400

7—16 with Supplement 1: Minimum Design Loads and Associated Criteria for Buildings and Other Structures
Table R608.7(1A), Table R608.7(1B), Table R608.7(1C), R301.2.1.1, R301.2.1.2, R301.2.1.2.1, R301.2.1.5, R301.2.1.5.1, Table R608.6(1), Table R608.6(2), Table R608.6(3), Table R608.6(4), R608.9.2, R608.9.3, R609.2, R609.6.2

24—14: Flood-resistant Design and Construction
R301.2.4, R301.2.4.1, R322.1, R322.1.1, R322.1.6, R322.1.9, R322.2.2, R322.3.3

32—01: Design and Construction of Frost-protected Shallow Foundations
R403.1.4.1

ASHRAE

ASHRAE
180 Technology Parkway NW
Peachtree Corners, GA 30092

ASHRAE 34—2019: Designation and Safety Classification of Refrigerants
M1411.1

ASHRAE 193—2010(RA 2014): Method of Test for Determining Airtightness of HVAC Equipment
N1103.3.4.1

ASHRAE—2001: 2001 ASHRAE Handbook of Fundamentals
Table N1105.4.2(1)

ASHRAE—2021: ASHRAE Handbook of Fundamentals
N1102.1.5, P3001.2, P3101.4

ASME

American Society of Mechanical Engineers
Two Park Avenue
New York, NY 10016-5990

A18.1—2020: Safety Standard for Platforms and Stairway Chair Lifts
R321.2

A112.1.2—2012(R2022): Air Gaps in Plumbing Systems (For Plumbing Fixtures and Water Connected Receptors)
P2717.1, Table P2902.3, P2902.3.1

ASME—continued

A112.1.3—2000 (Reaffirmed 2020): Air Gap Fittings for Use with Plumbing Fixtures, Appliances and Appurtenances
Table P2701.1, P2717.1, Table P2902.3, P2902.3.1

A112.3.1—2007(R2022): Stainless Steel Drainage Systems for Sanitary, DWV, Storm and Vacuum Applications Above and Below Ground
Table P3002.1(1), Table P3002.1(2), Table P3002.2, Table P3002.3, Table P3302.1

A112.3.4—2020/CSA B45.9—20: Macerating Toilet Systems and Related Components
Table P2701.1, P3007.5

A112.4.1—2019: Water Heater Relief Valve Drain Tubes
P2804.6.1

A112.4.3—1999(R2020): Plastic Fittings for Connecting Water Closets to the Sanitary Drainage System
P3003.14

A112.4.4—2017: Plastic Push-Fit Drain, Waste, and Vent (DWV) Fittings
Table P3002.3, P3003.9.4

A112.4.14—2004(R2019): Manually Operated Valves for Use in Plumbing Systems
Table P2903.9.4

A112.6.2—2022: Framing-affixed Supports for Off-the-floor Water Closets with Concealed Tanks
Table P2701.1, P2702.4

A112.6.3—2019: Floor and Trench Drains
Table P2701.1

A112.14.1—03(2022): Backwater Valves
P3008.3

A112.18.1—2020/CSA B125.1—2020: Plumbing Supply Fittings
Table P2701.1, P2708.5, P2722.1, P2722.3, P2902.2, Table P2903.9.4

A112.18.2—2019/CSA B125.2—2019: Plumbing Waste Fittings
Table P2701.1, P2702.2

A112.18.3M—2002(R2020): Performance Requirements for Backflow Protection Devices and Systems in Plumbing Fixture Fittings
P2708.5, P2722.3

A112.18.6—2021/CSA B125.6—21: Flexible Water Connectors
P2906.7

A112.19.1—2020/CSA B45.2—2020: Enameled Cast-iron and Enameled Steel Plumbing Fixtures
Table P2701.1, P2711.1

A112.19.2—2020/CSA B45.1—2020: Ceramic Plumbing Fixtures
Table P2701.1, P2705.1, P2711.1, P2712.1, P2712.2, P2712.9

A112.19.3—2021/CSA B45.4—08(R2021): Stainless Steel Plumbing Fixtures
Table P2701.1, P2705.1, P2711.1, P2712.1

A112.19.5—2021/CSA B45.15—2021: Flush Valves and Spuds for Water-closets, Urinals and Tanks
Table P2701.1

A112.19.7—2012/CSA B45.10—2012(R2021): Hydromassage Bathtub Systems
Table P2701.1

A112.19.12—2019: Wall-mounted and Pedestal-mounted, Adjustable, Elevating, Tilting, and Pivoting Lavatory and Sink, and Shampoo Bowl Carrier Systems and Drain Waste Systems
Table P2701.1, P2711.4, P2714.2

A112.19.14—2013(R2018): Six-Liter Water Closets Equipped with Dual Flushing Device
P2712.1

A112.19.15—2012(R2017): Bathtub/Whirlpool Bathtubs with Pressure-sealed Doors
Table P2701.1, P2713.2

A112.36.2M—1991(R2017): Cleanouts
P3005.2.10.2

ASME A17.1—2019/CSA B44—2019: Safety Code for Elevators and Escalators
R321.1

ASME A112.4.2—2020/CSA B45.16—20: Water-closet Personal Hygiene Devices
P2722.5

ASME—continued

ASSE 1002—2020/ASME A112.1002—2020/CSA B125.12—20: Anti-Siphon Fill Valves
Table P2701.1, Table P2902.3, P2902.4.1

ASSE 1016—2020/ASME 112.1016—2020/CSA B125.16—2020: Performance Requirements for Automatic Compensating Valves for Individual Showers and Tub/Shower Combinations
Table P2701.1, P2708.4, P2722.2

ASSE 1070—2020/ASME A112.1070—2020: Performance Requirements for Water-temperature-limiting Devices
P2713.3, P2721.2

B1.20.1—2019: Pipe Threads, General-purpose (Inch)
G2414.9, P3003.3.3, P3003.6.4, P3003.7.1, P3003.9.3

B16.3—2021: Malleable-iron-threaded Fittings, 150 and 300
Table P2906.6

B16.4—2021: Gray-iron-threaded Fittings
Table P2906.6, Table P3002.3

B16.9—2018: Factory-made, Wrought-steel Buttwelding Fittings
Table P2906.6

B16.11—2021: Forged Fittings, Socket-welding and Threaded
Table P2906.6

B16.12—2009(R2019): Cast-iron-threaded Drainage Fittings
Table P3002.3

B16.15—2018: Cast Alloy Threaded Fittings: Classes 125 and 250
Table P2906.6, Table P3002.3

B16.18—2018: Cast-copper-alloy Solder Joint Pressure Fittings
Table P2906.6, Table P3002.3

B16.22—2018: Wrought-copper and Copper-alloy Solder Joint Pressure Fittings
Table P2906.6, Table P3002.3

B16.23—2021: Cast-copper-alloy Solder Joint Drainage Fittings (DWV)
Table P3002.3

B16.26—2018: Cast-copper-alloy Fittings for Flared Copper Tubes
Table P2906.6, Table P3002.3

B16.28—1994: Wrought-steel Buttwelding Short Radius Elbows and Returns
Table P2906.6

B16.29—2017: Wrought-copper and Wrought-copper-alloy Solder Joint Drainage Fittings (DWV)
Table P3002.3

B16.33—2012(R2017): Manually Operated Metallic Gas Valves for Use in Gas Piping Systems up to 125 psig (Sizes $^1/_2$ through 2)
Table G2420.1.1

B16.34—2020: Valves—Flanged, Threaded and Welding End
Table P2903.9.4

B16.44—2012(R2017): Manually Operated Metallic Gas Valves for Use in Above-ground Piping Systems up to 5 psi
Table G2420.1.1

B16.51—2018: Copper and Copper Alloy Press-Connect Pressure Fittings
Table M2101.1, M2103.3, Table P2906.6

B36.10M—2018: Welded and Seamless Wrought-steel Pipe
G2414.4.2

BPVC—2019: ASME Boiler and Pressure Vessel Code (Sections I, II, IV, V, VI and VIII)
M2001.1.1, G2452.1

CSD-1—2021: Controls and Safety Devices for Automatically Fired Boilers
M2001.1.1, G2452.1

ASSE

ASSE International
18927 Hickory Creek Drive, Suite 220
Mokena, IL 60448

1001—2017: Performance Requirements for Atmospheric-type Vacuum Breakers
Table P2902.3, P2902.3.2

1003—2011: Performance Requirements for Water-pressure-reducing Valves for Domestic Water Distribution Systems
P2903.3.1

1008—2006: Performance Requirements for Plumbing Aspects of Residential Food Waste Disposer Units
Table P2701.1

1010—2004: Performance Requirements for Water Hammer Arresters
P2903.5

1011—2016: Performance Requirements for Hose Connection Vacuum Breakers
Table P2902.3, P2902.3.2

1012—2009: Performance Requirements for Backflow Preventers with Intermediate Atmospheric Vent
Table P2902.3, P2902.3.3, P2902.5.1, P2902.5.5.3

1013—2017: Performance Requirements for Reduced Pressure Principle Backflow Preventers and Reduced Pressure Principle Fire Protection Backflow Preventers
Table P2902.3, P2902.3.5, P2902.5.1, P2902.5.5.3

1015—2017: Performance Requirements for Double Check Backflow Prevention Assemblies and Double Check Fire Protection Backflow Prevention Assemblies
Table P2902.3, P2902.3.6

1017—2009: Performance Requirements for Temperature-actuated Mixing Valves for Hot Water Distribution Systems
P2724.1, P2802.1, P2803.2

1018—2001: Performance Requirements for Trap Seal Primer Valves; Potable Water Supplied
P3201.2.1.1, P3201.2.1.2

1019—2011(R2016): Performance Requirements for Freeze-resistant, Wall Hydrants, Vacuum Breaker, Draining Types
Table P2701.1, Table P2902.3, P2902.3.2

1020—2004: Performance Requirements for Pressure Vacuum Breaker Assembly
Table P2902.3, P2902.3.4

1023—1979: Performance Requirements for Hot Water Dispensers, Household-storage-type—Electrical
Table P2701.1

1024—2017: Performance Requirements for Dual Check Backflow Preventers, Anti-siphon-type, Residential Applications
Table P2902.3, P2902.3.7

1035—2008: Performance Requirements for Laboratory Faucet Backflow Preventers
Table P2902.3, P2902.3.2

1044—2015: Performance Requirements for Trap Seal Primer Devices Drainage Types and Electronic Design Types
P3201.2.1.3

1047—2011: Performance Requirements for Reduced Pressure Detector Fire Protection Backflow Prevention Assemblies
Table P2902.3, P2902.3.5

1048—2011: Performance Requirements for Double Check Detector Fire Protection Backflow Prevention Assemblies
Table P2902.3, P2902.3.6

1050—2009: Performance Requirements for Stack Air Admittance Valves for Sanitary Drainage Systems
P3114.1

1051—2009: Performance Requirements for Individual and Branch-type Air Admittance Valves for Plumbing Drainage Systems
P3114.1

1052—2016: Performance Requirements for Hose Connection Backflow Preventers
Table P2701.1, Table P2902.3, P2902.3.2

1056—2013: Performance Requirements for Spill-resistant Vacuum Breakers
Table P2902.3, P2902.3.4

1060—2016: Performance Requirements for Outdoor Enclosures for Fluid-conveying Components
P2902.6.1

1061—2015: Performance Requirements for Push Fit Fittings
Table P2906.6, P2906.21

ASSE—continued

1062—2017: Performance Requirements for Temperature-actuated, Flow Reduction (TAFR) Valves for Individual Supply Fittings
Table P2701.1, P2724.2

1066—1997: Performance Requirements for Individual Pressure Balancing In-line Valves for Individual Fixture Fittings
P2722.4

1072—2007: Performance Requirements for Barrier-type Floor Drain Trap Seal Protection Devices
P3201.2.1.4

1081—2014: Performance Requirements for Backflow Preventers with Integral Pressure Reducing Boiler Feed Valve and Intermediate Atmospheric Vent Style for Domestic and Light Commercial Water Distribution Systems
Table P2902.3, P2902.3.3

ASSE 1002—2020/ASME A112.1002—2020/CSA B125.12—20: Anti-Siphon Fill Valves
Table P2701.1, Table P2902.3, P2902.4.1

ASSE 1016—2020/ASME 112.1016—2020/CSA B125.16—2020: Performance Requirements for Automatic Compensating Valves for Individual Showers and Tub/Shower Combinations
Table P2701.1, P2708.4, P2722.2

ASSE 1037—2015/ASME A112.1037—2015/CSA B125.37—15: Performance Requirements for Pressurized Flushing Devices for Plumbing Fixtures
Table P2701.1

ASSE 1070—2020/ASME A112.1070—2020/ CSA B125.70—20: Performance Requirements for Water-temperature-limiting Devices
P2713.3, P2721.2, P2724.1

ASTM

ASTM International
100 Barr Harbor Drive, P.O. Box C700
West Conshohocken, PA 19428

A36/A36M—14: Specification for Carbon Structural Steel
R608.5.2.2

A53/A53M—2018: Specification for Pipe, Steel, Black and Hot-dipped, Zinc-coated Welded and Seamless
R407.3, Table M2101.1, G2414.4.2, Table P2906.4, Table P2906.5, Table P3002.1(1)

A74—2017: Specification for Cast-iron Soil Pipe and Fittings
Table P3002.1(1), Table P3002.1(2), Table P3002.2, Table P3002.3, P3005.2.6, Table P3302.1

A106/A106M—2018: Specification for Seamless Carbon Steel Pipe for High-temperature Service
Table M2101.1, G2414.4.2

A123/A123M—2017: Standard Specification for Zinc (Hot-Dip Galvanized) Coatings on Iron and Steel Products
Table 507.2.3

A126—04(2014): Gray Iron Castings for Valves, Flanges and Pipe Fittings
Table P2903.9.4

A153/A153M—2016A: Specification for Zinc Coating (Hot Dip) on Iron and Steel Hardware
R317.3, Table 507.2.3, Table R606.3.4.1, R703.6.3, R905.7.5, R905.8.6

A167—99(2009): Specification for Stainless and Heat-resisting Chromium-nickel Steel Plate, Sheet and Strip
Table R606.3.4.1

A240/A240M—17: Standard Specification for Chromium and Chromium-nickel Stainless Steel Plate, Sheet and Strip for Pressure Vessels and for General Applications
Table R905.10.3(1)

A254—2010(2018): Specification for Copper Brazed Steel Tubing
Table M2101.1, G2414.5.1

A268/A268M—2010(16): Standard Specification for Seamless and Welded Ferritic and Martensitic Stainless Steel Tubing for General Service
G2414.5.2

A269/A269M—2015A: Standard Specification for Seamless and Welded Austenitic Stainless Steel Tubing for General Service
G2414.5.2

A307—2014E1: Specification for Carbon Steel Bolts and Studs, 60,000 psi Tensile Strength
Table R507.2.3, R608.5.2.2

ASTM—continued

A312—2018: Specification for Seamless, Welded and Heavily Cold Worked Austenitic Stainless Steel Pipes
Table P2906.4, Table P2906.5, Table P2906.6, P2906.13.2

A463/A463M—15: Standard Specification for Steel Sheet, Aluminum-coated by the Hot-dip Process
Table R905.10.3(2)

A539—99: Specification for Electric-resistance-welded Coiled Steel Tubing for Gas and Fuel Oil Lines
M2202.1

A563—15: Standard Specification for Carbon and Alloy Steel Nuts
Table R507.2.3

A615/A615M—2015aE1: Specification for Deformed and Plain Carbon-steel Bars for Concrete Reinforcement
R402.3.1, R403.1.3.5.1, R404.1.3.3.7.1, R608.5.2.1

A641/A641M—09a(2014): Specification for Zinc-coated (Galvanized) Carbon Steel Wire
Table R606.3.4.1

A653—2017: Specification for Steel Sheet, Zinc-coated (Galvanized) or Zinc-iron Alloy-coated (Galvannealed) by the Hot-dip Process
R317.3.1, R505.2.2, Table R507.2.3, R603.2.2, Table R606.3.4.1, R608.5.2.3, R804.2.2, R804.2.3, Table R905.10.3(1), Table R905.10.3(2), M1601.1.1

A706M—2016: Specification for Low-alloy Steel Deformed and Plain Bars for Concrete Reinforcement
R402.3.1, R403.1.3.5.1, R404.1.3.3.7.1, R608.5.2.1

A755M—2016E1: Specification for Steel Sheet, Metallic Coated by the Hot-dip Process and Prepainted by the Coil-coating Process for Exterior Exposed Building Products
Table R905.10.3(2)

A778M—2016: Specification for Welded Unannealed Austenitic Stainless Steel Tubular Products
Table P2906.4, Table P2906.5, Table P2906.6

A792/A792M—10(2015): Specification for Steel Sheet, 55% Aluminum-zinc Alloy-coated by the Hot-dip Process
R505.2.2, R603.2.2, R608.5.2.3, R804.2.2, Table 905.10.3(2)

A875/A875M—13: Specification for Steel Sheet, Zinc-5%, Aluminum Alloy-coated by the Hot-dip Process
R608.5.2.3, Table R905.10.3(2)

A888—2018: Specification for Hubless Cast Iron Soil Pipe and Fittings for Sanitary and Storm Drain, Waste and Vent Piping Application
Table P3002.1(1), Table P3002.1(2), Table P3002.2, Table P3002.3, Table P3302.1

A924M—2017A: Standard Specification for General Requirements for Steel Sheet, Metallic-coated by the Hot-dip Process
Table R905.10.3(1)

A996M—2016: Specifications for Rail-steel and Axle-steel Deformed Bars for Concrete Reinforcement
R403.1.3.5.1, Table R404.1.2(9), R404.1.3.3.7.1, R608.5.2.1, Table R608.5.4(2)

A1003/A1003M—15: Standard Specification for Steel Sheet, Carbon, Metallic and Nonmetallic-coated for Cold-formed Framing Members
R505.2.1, R505.2.2, R603.2.1, R603.2.2, R804.2.1, R804.2.2

B32—08(2014): Specification for Solder Metal
P3003.6.3

B42—2015A: Specification for Seamless Copper Pipe, Standard Sizes
Table M2101.1, Table P2906.4, Table P2906.5, Table P3002.1(1)

B43—15: Specification for Seamless Red Brass Pipe, Standard Sizes
Table M2101.1, Table P2906.4, Table P2906.5, Table P3002.1(1)

B75/B75M—11: Specification for Seamless Copper Tube
Table M2101.1, Table P2906.4, Table P2906.5, Table P3002.1(1), Table P3002.1(2), Table P3002.2

B88—2016: Specification for Seamless Copper Water Tube
Table M2101.1, G2414.5.2, Table P2906.4, Table P2906.5, Table P3002.1(1), Table P3002.1(2), Table P3002.2

B101—12: Specification for Lead-coated Copper Sheet and Strip for Building Construction
Table R905.2.8.2, Table R905.10.3(1)

B135M—2017: Specification for Seamless Brass Tube
Table M2101.1

ASTM—continued

B209—14: Specification for Aluminum and Aluminum-alloy Sheet and Plate
Table 905.10.3(1)

B251/B251M—2017: Specification for General Requirements for Wrought Seamless Copper and Copper-alloy Tube
Table M2101.1, Table P2906.4, Table P2906.5, Table P3002.1(1), Table P3002.1(2), Table P3002.2

B280—13: Standard Specification for Seamless Copper Tube for Air-Conditioning and Refrigeration Field Service
G2414.4.3

B302—2017: Specification for Threadless Copper Pipe, Standard Sizes
Table M2101.1, Table P2906.4, Table P2906.5, Table P3002.1(1)

B306—13: Specification for Copper Drainage Tube (DWV)
Table M2101.1, Table P3002.1(1), Table P3002.1(2)

B370—12: Specification for Copper Sheet and Strip for Building Construction
Table R905.2.8.2, Table R905.10.3(1), Table P2701.1

B447—12a: Specification for Welded Copper Tube
Table P2906.4, Table P2906.5

B695—2004(2016): Standard Specification for Coatings of Zinc Mechanically Deposited on Iron and Steel
R317.3.1, R317.3.3, Table R507.2.3

B813—2016: Specification for Liquid and Paste Fluxes for Soldering Applications of Copper and Copper Alloy Tube
Table M2101.1, M2103.3, P2906.15, P3003.6.3

B828—2016: Practice for Making Capillary Joints by Soldering of Copper and Copper Alloy Tube and Fittings
M2103.3, P2906.15, P3003.6.3

C4—2004(2018): Specification for Clay Drain Tile and Perforated Clay Drain Tile
Table P3302.1

C5—2018: Specification for Quicklime for Structural Purposes
R702.2.1

C14—15a: Specification for Non-reinforced Concrete Sewer, Storm Drain and Culvert Pipe
Table P3002.2

C22/C22M—2015: Specification for Gypsum
R702.2.1, R702.3.1

C27—1998(2018): Specification for Standard Classification of Fireclay and High-alumina Refractory Brick
R1001.5

C28/C28M—10(2015): Specification for Gypsum Plasters
R702.2.1

C33/C33M—2018: Specification for Concrete Aggregates
R403.4.1

C34—2017: Specification for Structural Clay Load-bearing Wall Tile
Table R301.2(1), R606.2.2

C35/C35M—(2014): Specification for Inorganic Aggregates for Use in Gypsum Plaster
R702.2.1

C55—2017: Specification for Concrete Building Brick
R202, Table R301.2(1), R606.2.1

C56—2013(2017): Standard Specification for Structural Clay Nonloadbearing Tile
R606.2.2

C59/C59M—00(2015): Specification for Gypsum Casting Plaster and Molding Plaster
R702.2.1

C61/C61M—00(2015): Specification for Gypsum Keene's Cement
R702.2.1

C62—2017: Standard Specification for Building Brick (Solid Masonry Units Made from Clay or Shale)
R202, Table R301.2(1), R606.2.2

C73—2017: Specification for Calcium Silicate Face Brick (Sand Lime Brick)
R202, Table R301.2(1), R606.2.1

C76—2018A: Specification for Reinforced Concrete Culvert, Storm Drain and Sewer Pipe
Table P3002.2

ASTM—continued

C90—2016A: Specification for Load-bearing Concrete Masonry Units
Table R301.2(1), 606.2.1

C91M—2018A: Specification for Masonry Cement
R702.2.2, R703.7.2

C94/C94M—2017A: Standard Specification for Ready-mixed Concrete
R404.1.3.3.2, R608.5.1.2

C126—2017: Standard Specification for Ceramic Glazed Structural Clay Facing Tile, Facing Brick, and Solid Masonry Units
R606.2.2

C129—2017: Specification for Nonload-bearing Concrete Masonry Units
Table R301.2(1)

C143/C143M—15A: Test Method for Slump of Hydraulic Cement Concrete
R404.1.3.3.4, R608.5.1.4

C145—85: Specification for Solid Load-bearing Concrete Masonry Units
R202, Table R301.2(1)

C150/C150M—2018: Specification for Portland Cement
R608.5.1.1, R702.7.2

C199—1984(2016): Test Method for Pier Test for Refractory Mortar
R1001.5, R1001.8, R1003.12

C207—2018: Specification for Hydrated Lime for Masonry Purposes
Table R606.2.8

C208—2012(2017)E1: Specification for Cellulosic Fiber Insulating Board
R602.1.10, Table R602.3(1), Table R906.2

C212—2017: Standard Specification for Structural Clay Facing Tile
R606.2.2

C216—2017A: Specification for Facing Brick (Solid Masonry Units Made from Clay or Shale)
R202, Table R301.2(1), R606.2.2

C270—14A: Specification for Mortar for Unit Masonry
R606.2.8, Table R606.2.8, R606.2.11

C315—2007(2016): Specification for Clay Flue Liners and Chimney Pots
R1001.8, R1003.11.1, Table R1003.14(1), G2425.12

C406/C406M—2015: Specifications for Roofing Slate
R905.6.4

C411—2017: Test Method for Hot-surface Performance of High-temperature Thermal Insulation
M1601.3

C425—2004(2018): Specification for Compression Joints for Vitrified Clay Pipe and Fittings
Table P3002.2, P3003.10, P3003.13

C443—2012(2017): Specification for Joints for Concrete Pipe and Manholes, Using Rubber Gaskets
P3003.5, P3003.13

C475M—2017: Specification for Joint Compound and Joint Tape for Finishing Gypsum Wallboard
R702.3.1

C476—2018: Specification for Grout for Masonry
R606.2.12

C503M—2015: Standard Specification for Marble Dimension Stone
R606.2.4

C514—04(2014): Specification for Nails for the Application of Gypsum Wallboard
R702.3.1

C552—2017E1: Standard Specification for Cellular Glass Thermal Insulation
Table R906.2

C557—2003(2017): Specification for Adhesives for Fastening Gypsum Wallboard to Wood Framing
R702.3.1.1

C564—14: Specification for Rubber Gaskets for Cast Iron Soil Pipe and Fittings
P3003.4.2, P3003.4.3, P3003.13

ASTM—continued

C568M—2015: Standard Specification for Limestone Dimension Stone
R606.2.4

C578—2018: Specification for Rigid, Cellular Polystyrene Thermal Insulation
R316.8, R403.3, Table 703.8.4(2), Table R703.15.1, Table R703.15.2, Table R703.16.1, Table R703.16.2, Table R906.2

C587—2004(2018): Specification for Gypsum Veneer Plaster
R702.2.1

C595/C595M—2018: Specification for Blended Hydraulic Cements
R608.5.1.1, R702.2.2, R703.7.2

C615/C615M—2018E1: Standard Specification for Granite Dimension Stone
R606.2.4

C616/C616M—2015: Standard Specification for Quartz-based Dimension Stone
R606.2.4

C629/C629M—2015: Standard Specification for Slate Dimension Stone
R606.2.4

C631—09(2014): Specification for Bonding Compounds for Interior Gypsum Plastering
R702.2.1

C645—2018: Specification for Nonstructural Steel Framing Members
R702.3.3

C652—2017A: Specification for Hollow Brick (Hollow Masonry Units Made from Clay or Shale)
R202, Table R301.2(1), R606.2.2

C685/C685M—2017: Specification for Concrete Made by Volumetric Batching and Continuous Mixing
R404.1.3.3.2, R608.5.1.2

C700—2018: Specification for Vitrified Clay Pipe, Extra Strength, Standard Strength and Perforated
Table P3002.2, Table P3002.3, Table P3302.1

C726—2017: Standard Specification for Mineral Wool Roof Insulation Board
Table R906.2

C728—2017A: Standard Specification for Perlite Thermal Insulation Board
Table R906.2

C744—2016: Standard Specification for Prefaced Concrete and Calcium Silicate Masonry Units
R606.2.1

C836/C836M—2018: Specification for High Solids Content, Cold Liquid-applied Elastomeric Waterproofing Membrane for Use with Separate Wearing Course
R905.15.2

C841—2003(2018): Standard Specification for Installation of Interior Lathing and Furring
R702.2.1

C842—05(2015): Standard Specification for Application of Interior Gypsum Plaster
R702.2.1

C843—2017: Specification for Application of Gypsum Veneer Plaster
R702.2.1

C844—2015: Specification for Application of Gypsum Base to Receive Gypsum Veneer Plaster
R702.2.1

C847—2018: Specification for Metal Lath
R702.2.1, R702.2.2

C887—13: Specification for Packaged, Dry, Combined Materials for Surface Bonding Mortar
R406.1, R606.2.9

C897—15: Specification for Aggregate for Job-mixed Portland Cement-based Plasters
R702.2.2

C920—2018: Standard Specification for Elastomeric Joint Sealants
R406.4.1

C926—2018B: Specification for Application of Portland Cement-based Plaster
R702.2.2, R702.2.2.1, R703.7, R703.7.2, R703.7.2.1, R703.7.4

ASTM—continued

C933—2018: Specification for Welded Wire Lath
R702.2.1, R702.2.2

C946—2018: Standard Practice for Construction of Dry-Stacked, Surface-Bonded Walls
R606.2.9

C954—2018: Specification for Steel Drill Screws for the Application of Gypsum Panel Products or Metal Plaster Bases to Steel Studs from 0.033 in (0.84 mm) or to 0.112 in. (2.84 mm) in Thickness
R505.2.5, R603.2.5, R702.3.5.1, R804.2.5

C957/C957M—2017: Specification for High-solids Content, Cold Liquid-applied Elastomeric Waterproofing Membrane for Use with Integral Wearing Surface
R905.15.2

C1002—2018: Specification for Steel Self-piercing Tapping Screws for the Application of Gypsum Panel Products or Metal Plaster Bases to Wood Studs or Steel Studs
R702.3.1, R702.3.5.1

C1029—15: Specification for Spray-applied Rigid Cellular Polyurethane Thermal Insulation
R905.14.2

C1032—2018: Specification for Woven Wire Plaster Base
R702.2.1, R702.2.2

C1047—14a: Specification for Accessories for Gypsum Wallboard and Gypsum Veneer Base
R702.2.1, R702.2.2, R702.3.1

C1063—2018B: Specification for Installation of Lathing and Furring to Receive Interior and Exterior Portland Cement-based Plaster
R702.2.2, R703.7, R703.7.1

C1088—2018: Standard Specification for Thin Veneer Brick Units Made from Clay or Shale
R606.2.2

C1107/C1107M—2017: Standard Specification for Packaged Dry, Hydraulic-cement Grout (Nonshrink)
R402.3.1

C1116/C116M—10(2015): Standard Specification for Fiber-reinforced Concrete and Shotcrete
R402.3.1

C1157—11/C1157M—2017: Standard Performance Specification for Hydraulic Cement
R608.5.1.1, R703.7.2

C1167—2011(2017): Specification for Clay Roof Tiles
R905.3.4

C1173—2018: Specification for Flexible Transition Couplings for Underground Piping Systems
P3003.3.1, P3003.5, P3003.9.1, P3003.10, P3003.12.2, P3003.13

C1177/C1177M—2017: Specification for Glass Mat Gypsum Substrate for Use as Sheathing
R702.3.1, Table 906.2

C1178/C1178M—2018: Specification for Glass Mat Water-resistant Gypsum Backing Panel
R702.3.1, R702.3.7, Table R702.4.2

C1186—2008(2016): Specification for Flat Fiber Cement Sheets
R703.10.1, R703.10.2

C1261—2013(2017)E1: Specification for Firebox Brick for Residential Fireplaces
R1001.5, R1001.8

C1277—2018: Specification for Shielded Couplings Joining Hubless Cast Iron Soil Pipe and Fittings
P3003.4.3

C1278/C1278M—2017: Specification for Fiber-reinforced Gypsum Panels
R702.3.1, R702.3.7, Table R702.4.2, Table R906.2

C1280—18: Standard Specification for Application of Exterior Gypsum Panel Products for Use as Sheathing
R602.3(1)

C1283—2015: Practice for Installing Clay Flue Lining
R1003.9.1, R1003.12

C1288—2017: Standard Specification for Discrete Nonasbestos Fiber-cement Interior Substrate Sheets
Table R503.2.1.1(1), Table R503.2.1.1(2), Table 602.3(2), Table R702.4.2

ASTM—continued

C1289—2018: Standard Specification for Faced Rigid Cellular Polyisocyanurate Thermal Insulation Board
R316.8, Table R703.15.1, Table R703.15.2, Table R703.16.1, Table R703.16.2, Table R906.2

C1325—2018: Standard Specification for Nonasbestos Fiber-mat Reinforced Cement Interior Substrate Sheets Backer Units
Table R702.4.2

C1328/C1328M—12: Specification for Plastic (Stucco) Cement
R702.2.2, R703.7.2

C1363—11: The Standard Test Method for Thermal Performance of Building Materials and Envelope Assemblies by Means of a Hot Box Apparatus
N1101.10.4.1

C1364—2017: Standard Specification for Architectural Cast Stone
R606.2.5

C1396/C1396M—2017: Specification for Gypsum Board
Table R602.3(1), R702.2.1, R702.2.2, R702.3.1, R702.3.7

C1405—2016: Standard Specification for Glazed Brick (Single Fired, Brick Units)
R606.2.2

C1440—2017: Specification for Thermoplastic Elastomeric (TPE) Gasket Materials for Drain, Waste and Vent (DWV), Sewer, Sanitary and Storm Plumbing Systems
P3003.13

C1460—2017: Specification for Shielded Transition Couplings for Use with Dissimilar DWV Pipe and Fittings Above Ground
P3003.13

C1461—2008(2017): Specification for Mechanical Couplings Using Thermoplastic Elastomeric (TPE) Gaskets for Joining Drain, Waste and Vent (DWV) Sewer, Sanitary and Storm Plumbing Systems for Above and Below Ground Use
P3003.13

C1492—2003(2016): Specification for Concrete Roof Tile
R905.3.5

C1513—2018: Standard Specification for Steel Tapping Screws for Cold-formed Steel Framing Connections
R505.2.5, R603.2.5, R702.3.5.1, Table R703.3(2), Table R703.16.1, Table R703.16.2, R804.2.5

C1540—2018: Specification for Heavy Duty Shielded Couplings Joining Hubless Cast-iron Soil Pipe and Fittings
P3003.4.3

C1634—2017: Standard Specification for Concrete Facing Brick
R606.2.1

C1658/C1658M—2018: Standard Specification for Glass Mat Gypsum Panels
R702.3.1

C1668—13a: Standard Specification for Externally Applied Reflective Insulation Systems on Rigid Duct in Heating, Ventilation, and Air Conditioning (HVAC) Systems
M1601.3

C1670/1670M—2018: Standard Specification for Adhered Manufactured Stone Masonry Veneer Units
R606.2.6

C1691—2011(2017): Standard Specification for Unreinforced Autoclaved Aerated Concrete (AAC) Masonry Units
R606.2.3

C1693—2011(2017): Standard Specification for Autoclaved Aerated Concrete (AAC)
R606.2.3

C1766—2015: Standard Specification for Factory-Laminated Gypsum Panel Products
R702.3.1

D41/D41M—2011(2016): Specification for Asphalt Primer Used in Roofing, Dampproofing and Waterproofing
Table R905.9.2, Table R905.11.2

D43/D43M—2000(20118): Specification for Coal Tar Primer Used in Roofing, Dampproofing and Waterproofing
Table R905.9.2

D226/D226M—2017: Specification for Asphalt-saturated (Organic Felt) Used in Roofing and Waterproofing
R703.2, R905.1.1, Table R905.1.1(1), R905.8.4, Table R905.9.2

D227/D227M—2003(2018): Specification for Coal Tar Saturated (Organic Felt) Used in Roofing and Waterproofing
Table R905.9.2

ASTM—continued

D312/D312M—2016M: Specification for Asphalt Used in Roofing
Table R905.9.2

D422—63(2007)E2: Test Method for Particle-size Analysis of Soils
R403.1.8.1

D449/D449M—03(2014)E1: Specification for Asphalt Used in Dampproofing and Waterproofing
R406.2

D450/D450M—2017(2018): Specification for Coal-tar Pitch Used in Roofing, Dampproofing and Waterproofing
Table R905.9.2

D1227—13: Specification for Emulsified Asphalt Used as a Protective Coating for Roofing
Table R905.9.2, Table R905.11.2, R905.15.2

D1248—2016: Specification for Polyethylene Plastics Extrusion Materials for Wire and Cable
M1601.1.2

D1527—99(2005): Specification for Acrylonitrile-butadiene-styrene (ABS) Plastic Pipe, Schedules 40 and 80
Table P2906.4

D1693—15: Test Method for Environmental Stress-cracking of Ethylene Plastics
Table M2101.1

D1784—11: Standard Specification for Rigid Poly (Vinyl Chloride) (PVC) Compounds and Chlorinated Poly (Vinyl Chloride) (CPVC) Compounds
M1601.1.2

D1785—15E1: Specification for Poly (Vinyl Chloride) (PVC) Plastic Pipe, Schedules 40, 80 and 120
Table P2906.4

D1863/D1863M—2005(2018): Specification for Mineral Aggregate Used in Built-up Roofs
Table R905.9.2

D1869—15: Specification for Rubber Rings for Fiber-Reinforced Cement Pipe
P2906.18, P3003.13

D1970/D1970M—2017A: Specification for Self-adhering Polymer Modified Bitumen Sheet Materials Used as Steep Roofing Underlayment for Ice Dam Protection
R905.1.1, R905.2.8.2, R905.11.2.1

D2104—03: Specification for Polyethylene (PE) Plastic Pipe, Schedule 40
Table P2906.4

D2178/D2178M—15A: Specification for Asphalt Glass Felt Used in Roofing and Waterproofing
Table R905.9.2

D2235—2004(2016): Specification for Solvent Cement for Acrylonitrile-butadiene-styrene (ABS) Plastic Pipe and Fittings
P2906.9.1.1, P3003.3.2

D2239—12A: Specification for Polyethylene (PE) Plastic Pipe (SIDR-PR) Based on Controlled Inside Diameter
Table P2906.4

D2241—15: Specification for Poly (Vinyl Chloride) (PVC) Pressure-rated Pipe (SDR-Series)
Table P2906.4

D2282—99(2005): Specification for Acrylonitrile-butadiene-styrene (ABS) Plastic Pipe (SDR-PR)
Table P2906.4

D2412—2011(2018): Test Method for Determination of External Loading Characteristics of Plastic Pipe by Parallel-plate Loading
M1601.1.2

D2447—03: Specification for Polyethylene (PE) Plastic Pipe Schedules 40 and 80, Based on Outside Diameter
Table M2101.1

D2464—15: Specification for Threaded Poly (Vinyl Chloride) (PVC) Plastic Pipe Fittings, Schedule 80
Table P2906.6

D2466—20: 17: Specification for Poly (Vinyl Chloride) (PVC) Plastic Pipe Fittings, Schedule 40
Table P2906.6

D2467—15: Specification for Poly (Vinyl Chloride) (PVC) Plastic Pipe Fittings, Schedule 80
Table P2906.6

D2468—96a: Specification for Acrylonitrile-butadiene-styrene (ABS) Plastic Pipe Fittings, Schedule 40
Table P2906.6

ASTM—continued

D2513—2018A: Specification for Gas Pressure Pipe, Tubing and Fittings
Table M2101.1, G2414.6, G2414.6.1, G2414.11, G2415.17.2

D2564—2012(2018): Specification for Solvent Cements for Poly (Vinyl Chloride) (PVC) Plastic Piping Systems
P2906.9.1.4, P3003.9.2

D2609—15: Specification for Plastic Insert Fittings for Polyethylene (PE) Plastic Pipe
Table P2906.6

D2626/D2626M—04 (2012)e1: Specification for Asphalt-saturated and Coated Organic Felt Base Sheet Used in Roofing
Table R905.1.1(1), Table R905.9.2

D2657—2007(2015): Standard Practice for Heat Fusion-joining of Polyolefin Pipe Fittings
M2105.11.1, P2906.3.1, P2906.20.2, P3003.12.1

D2661—14E1: Specification for Acrylonitrile-butadiene-styrene (ABS) Schedule 40 Plastic Drain, Waste, and Vent Pipe and Fittings
Table P3002.1(1), Table P3002.1(2), Table P3002.2, Table P3002.3, P3003.3.2

D2665—14: Specification for Poly (Vinyl Chloride) (PVC) Plastic Drain, Waste and Vent Pipe and Fittings
Table P3002.1(1), Table P3002.1(2), Table P3002.2, Table P3002.3

D2672—14: Specification for Joints for IPS PVC Pipe Using Solvent Cement
Table P2906.4

D2680—01(2014): Standard Specification for Acrylonitrile-Butadiene-Styrene (ABS) and Poly(Vinyl Chloride) (PVC) Composite Sewer Piping
Table P3002.2

D2683—14: Specification for Socket-type Polyethylene Fittings for Outside Diameter-controlled Polyethylene Pipe and Tubing
Table M2105.5, M2105.11.1, P2906.20.2, P3002.3, P3010.5

D2729—2017: Specification for Poly (Vinyl Chloride) (PVC) Sewer Pipe and Fittings
P3009.11, Table P3302.1

D2737—2012A: Specification for Polyethylene (PE) Plastic Tubing
Table P2906.4

D2751—05: Specification for Acrylonitrile-butadiene-styrene (ABS) Sewer Pipe and Fittings
Table P3002.2, Table P3002.3

D2822/D2822M—2005(2011): Specification for Asphalt Roof Cement, Asbestos Containing
Table R905.9.2

D2823/D2823M—05(2011)e1: Specification for Asphalt Roof Coatings, Asbestos Containing
Table R905.9.2

D2824/D2824M—2018: Specification for Aluminum-pigmented Asphalt Roof Coatings, Nonfibered, Asbestos Fibered and Fibered without Asbestos
Table R905.9.2, Table R905.11.2

D2846/D2846M—2017BE1: Specification for Chlorinated Poly (Vinyl Chloride) (CPVC) Plastic Hot- and Cold-water Distribution Systems
Table M2101.1, Table P2906.4, Table P2906.5, Table P2906.6, P2906.9.1.2, P2906.9.1.3

D2855—2015: Standard Practice for Making Solvent-cemented Joints with Poly (Vinyl Chloride) (PVC) Pipe and Fittings
P3003.9.2

D2898—2010(2017): Test Methods for Accelerated Weathering of Fire-retardant-treated Wood for Fire Testing
R802.1.5.4, R802.1.5.8

D2949—10: Specification for 3.25-in. Outside Diameter Poly (Vinyl Chloride) (PVC) Plastic Drain, Waste and Vent Pipe and Fittings
Table P3002.1(1), Table P3002.1(2), Table P3002.2, Table P3002.3

D3019/D3019—2017: Specification for Lap Cement Used with Asphalt Roll Roofing, Nonfibered, Asbestos Fibered and Nonasbestos Fibered
Table R905.9.2, Table R905.11.2

D3034—2016: Specification for Type PSM Poly (Vinyl Chloride) (PVC) Sewer Pipe and Fittings
Table P3002.2, Table P3002.3

D3035—15: Specification for Polyethylene (PE) Plastic Pipe (DR-PR) Based on Controlled Outside Diameter
Table M2105.4

ASTM—continued

D3138—04(2011): Standard Specification for Solvent Cements for Transition Joints Between Acrylonitrile-Butadiene-Styrene (ABS) and Poly (Vinyl Chloride) (PVC) Non-Pressure Piping Components
P3003.13.4

D3161/D3161M—2016A: Test Method for Wind-Resistance of Steep Slope Roofing Products (Fan Induced Method)
R905.2.4.1, Table R905.2.4.1, R905.16.6

D3201/D3201M—2013: Test Method for Hygroscopic Properties of Fire-retardant Wood and Wood-base Products
R802.1.5.9

D3212—07(2013): Specification for Joints for Drain and Sewer Plastic Pipes Using Flexible Elastomeric Seals
P3003.3.1, P3003.9.1, P3003.12.2

D3261—2016: Specification for Butt Heat Fusion Polyethylene (PE) Plastic Fittings for Polyethylene (PE) Plastic Pipe and Tubing
Table M2101.1, Table M2105.5, M2105.11.1, M2105.13.3, P2906.20.2

D3309—96a(2002): Specification for Polybutylene (PB) Plastic Hot- and Cold-water Distribution System
Table M2101.1

D3311—2017: Specification for Drain, Waste and Vent (DWV) Plastic Fittings Patterns
P3002.3

D3350—14: Specification for Polyethylene Plastic Pipe and Fitting Materials
Table M2101.1

D3462/D3462M—10A: Specification for Asphalt Shingles Made from Glass Felt and Surfaced with Mineral Granules
R905.2.4

D3468/D3468M—99(2013)E1: Specification for Liquid-applied Neoprene and Chlorosulfanated Polyethylene Used in Roofing and Waterproofing
R905.15.2

D3679—2017: Specification for Rigid Poly (Vinyl Chloride) (PVC) Siding
R703.11

D3737—2018E1: Practice for Establishing Allowable Properties for Structural Glued Laminated Timber (Glulam)
R502.1.3, R602.1.3, R802.1.2

D3747—79(2007): Specification for Emulsified Asphalt Adhesive for Adhering Roof Insulation
Table R905.9.2, Table R905.11.2

D3909/D3909M—14: Specification for Asphalt Roll Roofing (Glass Felt) Surfaced with Mineral Granules
R905.2.8.2, R905.5.4, Table R905.9.2

D4022/D4022M—2007(2012)e1: Specification for Coal Tar Roof Cement, Asbestos Containing
Table R905.9.2

D4068—2017: Specification for Chlorinated Polyethylene (CPE) Sheeting for Concealed Water Containment Membrane
P2709.2, P2709.2.2

D4318—2017E1: Test Methods for Liquid Limit, Plastic Limit and Plasticity Index of Soils
R403.1.8.1

D4434/D4434M—2015: Specification for Poly (Vinyl Chloride) Sheet Roofing
R905.13.2

D4479/D4479M—2007(2018): Specification for Asphalt Roof Coatings—asbestos-free
Table R905.9.2

D4551—2017: Specification for Poly (Vinyl Chloride) (PVC) Plastic Flexible Concealed Water-containment Membrane
P2709.2, P2709.2.1

D4586/D4586M—2007(2018): Specification for Asphalt Roof Cement—asbestos-free
Table R905.9.2

D4601/D4601M—04(2012)e1: Specification for Asphalt-coated Glass Fiber Base Sheet Used in Roofing
Table R905.9.2, R905.11.2.1

D4637/D4637M—2015: Specification for EPDM Sheet Used in Single-ply Roof Membrane
R905.12.2

D4829—11: Test Method for Expansion Index of Soils
R403.1.8.1

D4869/D4869M—2016A: Specification for Asphalt-saturated (Organic Felt) Underlayment Used in Steep Slope Roofing
R905.1.1, Table R905.1.1(1)

ASTM—continued

D4897/D4897M—2016: Specification for Asphalt Coated Glass-fiber Venting Base Sheet Used in Roofing
Table R905.9.2

D4990—1997a(2013): Specification for Coal Tar Glass Felt Used in Roofing and Waterproofing
Table R905.9.2

D5019—07a: Specification for Reinforced Nonvulcanized Polymeric Sheet Used in Roofing Membrane
R905.12.2

D5055—2016: Specification for Establishing and Monitoring Structural Capacities of Prefabricated Wood I-joists
R502.1.2, R802.1.8

D5456—2018: Standard Specification for Evaluation of Structural Composite Lumber Products
R502.1.5, R602.1.5, R802.1.4

D5516—2018: Test Method for Evaluating the Flexural Properties of Fire-retardant-treated Softwood Plywood Exposed to the Elevated Temperatures
R802.1.5.6

D5643/D5643M—2006(2018): Specification for Coal Tar Roof Cement Asbestos-free
Table R905.9.2

D5664—2017: Test Methods for Evaluating the Effects of Fire-retardant Treatments and Elevated Temperatures on Strength Properties of Fire-retardant-treated Lumber
R802.1.5.7

D5665/D5665M—99a(2014)E1: Specification for Thermoplastic Fabrics Used in Cold-applied Roofing and Waterproofing
Table R905.9.2

D5726—98(2013): Specification for Thermoplastic Fabrics Used in Hot-applied Roofing and Waterproofing
Table R905.9.2

D6083/D6083M—2018: Specification for Liquid-applied Acrylic Coating Used in Roofing
Table R905.9.2, Table R905.11.2, Table R905.14.3, R905.15.2

D6162/D6162M—2016: Specification for Styrene Butadiene Styrene (SBS) Modified Bituminous Sheet Materials Using a Combination of Polyester and Glass Fiber Reinforcements
Table R905.11.2

D6163/D6163M—2016: Specification for Styrene Butadiene Styrene (SBS) Modified Bituminous Sheet Materials Using Glass Fiber Reinforcements
Table R905.11.2

D6164/D6164M—2016: Specification for Styrene Butadiene Styrene (SBS) Modified Bituminous Sheet Materials Using Polyester Reinforcements
Table R905.11.2

D6222/D6222M—2016: Specification for Atactic Polypropylene (APP) Modified Bituminous Sheet Materials Using Polyester Reinforcements
Table R905.11.2

D6223/D6223M—2016: Specification for Atactic Polypropylene (APP) Modified Bituminous Sheet Materials Using a Combination of Polyester and Glass Fiber Reinforcement
Table R905.11.2

D6298—2016: Specification for Fiberglass-reinforced Styrene Butadiene Styrene (SBS) Modified Bituminous Sheets with a Factory Applied Metal Surface
Table R905.11.2

D6305—08(2015)E1: Practice for Calculating Bending Strength Design Adjustment Factors for Fire-retardant-treated Plywood Roof Sheathing
R802.1.5.6

D6380/D6380—2003(2018): Standard Specification for Asphalt Roll Roofing (Organic Felt)
Table R905.1.1(1), R905.2.8.2, R905.5.4

D6464—2003A(2017): Standard Specification for Expandable Foam Adhesives for Fastening Gypsum Wallboard to Wood Framing
R702.3.1.1

D6694/D6694M—2015: Standard Specification for Liquid-applied Silicone Coating Used in Spray Polyurethane Foam Roofing Systems
Table R905.14.3, R905.15.2

D6754/D6754M—2015: Standard Specification for Ketone-ethylene-ester-based Sheet Roofing
R905.13.2

ASTM—continued

D6757/D6757M—2018: Specification for Underlayment Felt Containing Inorganic Fibers Used with Steep Slope Roofing
R905.1.1, Table R905.1.1(1)

D6841—2016: Standard Practice for Calculating Design Value Treatment Adjustment Factors for Fire-retardant-treated Lumber
R802.1.5.7

D6878/D6878M—2017: Standard Specification for Thermoplastic-polyolefin-based Sheet Roofing
R905.13.2

D6947/D6947M—2016: Standard Specification for Liquid Applied Moisture Cured Polyurethane Coating Used in Spray Polyurethane Foam Roofing System
Table R905.14.3, R905.15.2

D7032—2017: Standard Specification for Establishing Performance Ratings for Wood-plastic Composite Deck Boards and Guardrail Systems (Guards or Handrails)
R507.2.2, R507.2.2.1, 507.2.2.3, 507.2.2.4

D7158—D7158M—2019: Standard Test Method for Wind Resistance of Asphalt Shingles (Uplift Force/Uplift Resistance Method)
R905.2.4.1, Table R905.2.4.1

D7254—2017: Standard Specification for Polypropylene (PP) siding
Table R703.3(1), R703.14

D7425/D7425M—13: Standard Specification for Spray Polyurethane Foam Used for Roofing Application
R905.14.2

D7672—2014E1: Standard Specification for Evaluating Structural Capacities of Rim Board Products and Assemblies
R502.1.7, R602.1.7, R802.1.7

D7793—2017: Standard Specification for Insulated Vinyl Siding
Table R703.3(1), R703.13

E84—2018B: Standard Test Method for Surface Burning Characteristics of Building Materials
R202, R302.9.3, R302.9.4, R302.10.1, R302.10.2, R316.3, R316.5.9, R316.5.11, R507.2.2.2, R703.14.3, R802.1.5, M1601.3, M1601.5.2, P2801.6

E96/E96M—2016: Test Method for Water Vapor Transmission of Materials
R202, Table R806.5, M1411.6, M1601.4.6

E108—2017: Test Methods for Fire Tests of Roof Coverings
R302.2.4, R902.1

E119—2018B: Test Methods for Fire Tests of Building Construction and Materials
Table R302.1(1), Table R302.1(2), R302.2.1, R302.2.2, R302.3, R302.4.1, R302.11.1, R606.2.2

E136—2019: Test Method for Assessing Combustibility of Materials Using a Vertical Tube Furnace at 750°C
R202, R302.11

E283—2004(2012): Test Method for Determining the Rate of Air Leakage through Exterior Windows, Curtain Walls and Doors Under Specified Pressure Differences across the Specimen
R202, N1102.4.5

E330/E330M—14: Test Method for Structural Performance of Exterior Windows, Curtain Walls and Doors by Uniform Static Air Pressure Difference
R609.4, R609.5, R609.6.2, R703.1.2

E331—2000(2016): Test Method for Water Penetration of Exterior Windows, Skylights, Doors and Curtain Walls by Uniform Static Air Pressure Difference
R703.1.1

E779—2010(2018): Standard Test Method for Determining Air Leakage Rate by Fan Pressurization
N1102.4.1.2

E814—2013A(2017): Standard Test Method for Fire Tests of Penetration Firestop Systems
R302.4.1.2

E970—2017: Standard Test Method for Critical Radiant Flux of Exposed Attic Floor Insulation Using a Radiant Heat Energy Source
R302.10.5

E1509—2012(2017): Standard Specification for Room Heaters, Pellet Fuel-burning Type
M1410.1

ASTM—continued

E1554/E1554 M—13(2018): Standard Test Methods for Determining Air Leakage of Air Distribution Systems by Fan Pressurization

N1103.3.5

E1602—2003(20117): Guide for Construction of Solid Fuel Burning Masonry Heaters

R1002.2

E1745—17: Standard Specification for Plastic Water Vapor Retarders Used in Contact with Soil or Granular Fill under Concrete Slabs

R506.2.3

E1827—2011(2017): Standard Test Methods for Determining Airtightness of Building Using an Orifice Blower Door

N1102.4.1.2

E1886—2013A: Test Method for Performance Impact Protective Systems Impacted by Missile(s) and Exposed to Cyclic Pressure Differentials

R301.2.1.2, R609.6.1, R609.6.2, Table R703.11.2

E1996—2017: Standard Specification for Performance of Exterior Windows, Curtain Walls, Doors and Impact Protective Systems Impacted by Windborne Debris in Hurricanes

R301.2.1.2, R301.2.1.2.1, R609.6.1, R609.6.2

E2178—2013: Standard Test Method for Air Permeance of Building Materials

R202

E2231—2018: Standard Practice for Specimen Preparation and Mounting of Pipe and Duct Insulation Materials to Assess Surface Burning Characteristics

M1601.3

E2273—2018: Standard Test Method for Determining the Drainage Efficiency of Exterior Insulation and Finish Systems (EIFS) Clad Wall Assemblies

R703.9.2

E2556/E2556M—10: Standard Specification for Vapor Permeable Flexible Sheet Water-Resistive Barriers Intended for Mechanical Attachment

R703.2

E2568—2017A: Standard Specification for PB Exterior Insulation and Finish Systems

R703.9.1, R703.9.2

E2570/E2570M—07(2014)E1: Standard Test Methods for Evaluating Water-resistive Barrier (WRB) Coatings Used Under Exterior Insulation and Finish Systems (EIFS) or EIFS with Drainage

R703.9.2

E2634—2018: Standard Specification for Flat Wall Insulating Concrete Form (ICF) Systems

R404.1.3.3.6.1, R608.4.4

E2925—17: Standard Specification for Manufactured Polymeric Drainage and Ventilation Materials Used to Provide a Rainscreen Function

R703.7.3.2

F405—05: Specification for Corrugated Polyethylene (PE) Pipe and Fittings

Table P3009.11, Table P3302.1

F409—2017: Specification for Thermoplastic Accessible and Replaceable Plastic Tube and Tubular Fittings

Table P2701.1, P2702.2, P2702.3

F437—15: Specification for Threaded Chlorinated Poly (Vinyl Chloride) (CPVC) Plastic Pipe Fittings, Schedule 80

Table P2906.6

F438—2017: Specification for Socket-type Chlorinated Poly (Vinyl Chloride) (CPVC) Plastic Pipe Fittings, Schedule 40

Table P2906.6

F439—13: Specification for Chlorinated Poly (Vinyl Chloride) (CPVC) Plastic Pipe Fittings, Schedule 80

Table P2906.6

F441/F441M—15: Specification for Chlorinated Poly (Vinyl Chloride) (CPVC) Plastic Pipe, Schedules 40 and 80

Table P2906.4, Table P2906.5

F442/F442M—13E1: Specification for Chlorinated Poly (Vinyl Chloride) (CPVC) Plastic Pipe (SDR-PR)

Table P2906.4, Table P2906.5

F477—14: Specification for Elastomeric Seals (Gaskets) for Joining Plastic Pipe

P2906.18, P3003.13

ASTM—continued

F493—14: Specification for Solvent Cements for Chlorinated Poly (Vinyl Chloride) (CPVC) Plastic Pipe and Fittings
P2906.9.1.2, P2906.9.1.3, P2906.18.2

F628—2012E2: Specification for Acrylonitrile-butadiene-styrene (ABS) Schedule 40 Plastic Drain, Waste and Vent Pipe with a Cellular Core
Table P3002.1(1), Table P3002.1(2), Table P3002.2, Table P3002.3, P3003.3.2

F656—2015: Specification for Primers for Use in Solvent Cement Joints of Poly (Vinyl Chloride) (PVC) Plastic Pipe and Fittings
P2906.9.1.4, P3003.9.2

F714—13: Specification for Polyethylene (PE) Plastic Pipe (SDR-PR) Based on Outside Diameter
Table P3002.1(2), Table P3002.2, P3010.4

F844—07a(2013): Standard Specification for Washers, Steel, Plain (Flat), Unhardened for General Use
Table R507.2.3

F876—2017: Specification for Cross-linked Polyethylene (PEX) Tubing
Table M2101.1, Table P2906.4, Table P2906.5

F877—2018A: Specification for Cross-linked Polyethylene (PEX) Plastic Hot- and Cold-water Distribution Systems
Table M2101.1, Table P2906.6

F891—2016: Specification for Coextruded Poly (Vinyl Chloride) (PVC) Plastic Pipe with a Cellular Core
Table P3002.1(1), Table P3002.1(2), Table P3002.2, Table P3302.1

F1055—2016A: Specification for Electrofusion Type Polyethylene Fittings for Outside Diameter Controlled Polyethylene and Crosslinked Polyethylene Pipe and Tubing
Table M2105.5, M2105.11.2, P2906.20.2

F1281—2017: Specification for Cross-linked Polyethylene/Aluminum/Cross-linked Polyethylene (PEX-AL-PEX) Pressure Pipe
Table M2101.1, P2506.12.1, Table P2906.4, Table P2906.5, Table P2906.6

F1282—2017: Specification for Polyethylene/Aluminum/Polyethylene (PE-AL-PE) Composite Pressure Pipe
Table M2101.1, Table P2906.4, Table P2906.5, Table P2906.6, P2906.12.1

F1412—2016: Specification for Polyolefin Pipe and Fittings for Corrosive Waste Drainage
Table P3002.1(2), Table P3002.2, Table P3002.3, P3003.11.1

F1488—14E1: Specification for Coextruded Composite Pipe
Table P3002.1(1), Table P3002.1(2), Table P3002.2, Table P3009.11

F1504—2014: Standard Specification for Folded Poly (Vinyl Chloride) (PVC) for Existing Sewer and Conduit Rehabilitation
P3011.4

F1554—2018: Specification for Anchor Bolts, Steel, 36, 55 and 105-ksi Yield Strength
R608.5.2.2

F1667—2018: Specification for Driven Fasteners, Nails, Spikes and Staples
R317.3, Table R507.2.3, Table R602.3(1), R703.3.3, R703.6.3, Table R703.15.1, Table R703.15.2, R905.2.5

F1807—2018: Specification for Metal Insert Fittings Utilizing a Copper Crimp Ring for SDR9 Cross-linked Polyethylene (PEX) Tubing and SDR9 Polyethylene of Raised Temperature (PE-RT) Tubing
Table M2101.1, Table P2906.6

F1866—2018: Specification for Poly (Vinyl Chloride) (PVC) Plastic Schedule 40 Drainage and DWV Fabricated Fittings
Table P3002.3

F1871—2011: Standard Specification for Folded/Formed Poly (Vinyl Chloride) Pipe Type A for Existing Sewer and Conduit Rehabilitation
P3011.4

F1924—12: Standard Specification for Plastic Mechanical Fittings for Use on Outside Diameter Controlled Polyethylene Gas Distribution Pipe and Tubing
M2105.11.1

F1960—2018: Specification for Cold Expansion Fittings with PEX Reinforcing Rings for Use with Cross-linked Polyethylene (PEX) Tubing
Table M2101.1, Table P2906.6

F1970—12E1: Standard Specification for Special Engineered Fittings, Appurtenances or Valves for Use in Poly (Vinyl Chloride) (PVC) or Chlorinated Poly (Vinyl Chloride) (CPVC) Systems
M2105.5, Table 2903.9.4

ASTM—continued

F1973—2013(2018): Standard Specification for Factory Assembled Anodeless Risers and Transition Fittings in Polyethylene (PE) and Polyamide 11 (PA 11) Fuel Gas Distribution Systems
G2415.17.2

F1974—09(2015): Specification for Metal Insert Fittings for Polyethylene/Aluminum/Polyethylene and Cross-linked Polyethylene/Aluminum/Cross-linked Polyethylene Composite Pressure Pipe
Table P2906.6, P2906.12.1

F1986—2001(2011): Multilayer Pipe Type 2, Compression Joints for Hot and Cold Drinking Water Systems
Table P2906.4, Table P2906.5, Table P2906.6

F2080—2016: Specification for Cold-expansion Fittings with Metal Compression-sleeves for Cross-linked Polyethylene (PEX) Pipe
Table P2906.6

F2090—17: Specification for Window Fall Prevention Devices with Emergency Escape (Egress) Release Mechanisms
R310.1.1, R312.2.1, R312.2.2

F2098—2015: Standard Specification for Stainless Steel Clamps for Securing SDR9 Cross-linked Polyethylene (PEX) Tubing to Metal Insert and Plastic Insert Fittings
Table M2101.1, Table P2906.6

F2159—2018: Standard Specification for Plastic Insert Fittings Utilizing a Copper Crimp Ring for SDR9 Cross-linked Polyethylene (PEX) Tubing and SDR9 Polyethylene of Raised Temperature (PE-RT) Tubing
Table P2906.6

F2262—09: Standard Specification for Cross-linked Polyethylene/Aluminum/Cross-linked Polyethylene Tubing OD Controlled SDR9
Table P2906.4, Table P2906.5

F2389—2017A: Standard for Pressure-rated Polypropylene (PP) Piping Systems
Table P2906.4, Table P2906.5, Table P2906.6, P2906.11.1

F2434—14: Standard Specification for Metal Insert Fittings Utilizing a Copper Crimp Ring for Polyethylene/Aluminum/Cross-linked Polyethylene (PEX-AL-PEX) Tubing
Table P2906.6

F2623—14: Standard Specification for Polyethylene of Raised Temperature (PE-RT) SDRG Tubing
Table M2101.1

F2735—2009(2016): Standard Specification for Plastic Insert Fittings for SDR9 Cross-linked Polyethylene (PEX) and Polyethylene of Raised Temperature (PE-RT) Tubing
Table M2101.1, Table P2906.6

F2769—2018: Polyethylene or Raised Temperature (PE-RT) Plastic Hot and Cold-Water Tubing and Distribution Systems
Table M2101.1, Table P2906.4, Table P2906.5, Table P2906.6

F2806—10(2015): Standard Specification for Acrylonitrile-butadiene-styrene (ABS) Plastic Pipe (Metric SDR-PR)
Table M2101.1

F2855—12: Standard Specification for Chlorinated Poly (Vinyl Chloride)/Aluminum/Chlorinated Poly (Vinyl Chloride) (CPVC AL CPVC) Composite Pressure Tubing
Table P2906.4, Table P2906.5

F2945—2018: Standard Specification for Polyamide 11 Gas Pressure Pipe, Tubing and Fittings
G2414.6

F2969—12: Standard Specification for Acrylonitrile-butadiene-styrene (ABS) IPS Dimensioned Pressure Pipe
Table M2101.1

F3226/F3226M—16: Standard Specification for Metallic Press-Connect Fittings for Piping and Tubing Systems
Table P2906.6

F3253—2017: Standard Specification for Crosslinked Polyethylene (PEX) Tubing with Oxygen Barrier for Hot- and Cold-Water Hydronic Distribution Systems
Table M2101.1

AWC

American Wood Council
222 Catoctin Circle SE, Suite 201
Leesburg, VA 20175

ANSI/AWC NDS—2018: National Design Specification (NDS) for Wood Construction—with 2018 Supplement
R404.2.2, R502.2, Table R503.1, R507.2.1, R602.3, R608.9.2, R608.9.3, Table R703.15.1, Table R703.15.2, R802.2

ANSI/AWC PWF—2021: Permanent Wood Foundation Design Specification
R317.3.2, R401.1, R404.2.3

ANSI/AWC WFCM—2018: Wood Frame Construction Manual for One- and Two-family Dwellings
R301.1.1, R301.2.1.1, R602.10.8.2, Figure R608.9(9), R608.9.2, R608.9.3, R608.10

AWC STJR—2021: Span Tables for Joists and Rafters
R502.3, R802.4.1, R802.5.1

AWPA

American Wood Protection Association
P.O. Box 361784
Birmingham, AL 35236-1784

C1—03: All Timber Products—Preservative Treatment by Pressure Processes
R902.2

M4—15: Standard for the Care of Preservative-treated Wood Products
R317.1.1, R318.1.2

U1—20: USE CATEGORY SYSTEM: User Specification for Treated Wood Except Commodity Specification H
R317.1, R402.1.2, R504.3, R703.6.3, R905.7.5, Table R905.8.5, R905.8.6

AWS

American Welding Society
8669 NW 36 Street, #130
Miami, FL 33166

A5.8M/A5.8—2011—AMD1: Specifications for Filler Metals for Brazing and Braze Welding
P3003.6.1

ANSI/AWS A5.31M/A5.31—2012: Specification for Fluxes for Brazing and Braze Welding Edition: 2nd
M2103.3, M2202.2, P2906.15

AWWA

American Water Works Association
6666 West Quincy Avenue
Denver, CO 80235

C104/A21.4—16: Cement-mortar Lining for Ductile-iron Pipe and Fittings
P2906.4

C110/A21.10—12: Ductile-iron and Gray-iron Fittings
Table P2906.6, P3002.3

C115/A21.15—11: Flanged Ductile-iron Pipe with Ductile-iron or Gray-iron Threaded Flanges
Table P2906.4

C151/A21.51—17: Ductile-iron Pipe, Centrifugally Cast, for Water
Table P2906.4

C153/A21.53—11: Ductile-iron Compact Fittings for Water Service
Table P2906.6

C500—09: Standard for Metal-seated Gate Valves for Water Supply Service
Table P2903.9.4

C504—15: Standard for Rubber-seated Butterfly Valves
Table P2903.9.4

C507—15: Standard for Ball Valves, 6 In. Through 60 In. (150 mm through 1,500 mm)
Table P2903.9.4

C510—07: Double Check Valve Backflow Prevention Assembly
Table P2902.3, P2902.3.6

AWWA—continued

C511—17: Reduced-pressure Principle Backflow Prevention Assembly
Table P2902.3, P2902.3.5, P2902.5.1

C901—16: Polyethylene (PE) Pressure Pipe and Tubing $^1/_2$ in. (13 mm) through 3 in. (76 mm) for Water Service
P2906.4

C903—16: Polyethylene-aluminum-polyethylene (PE-AL-PE) Composite Pressure Pipe, 12 mm ($^1/_2$ in.) through 50 mm (2 in.), for Water Service
Table M2105.4

C904—16: Cross-linked Polyethylene (PEX) Pressure Tubing, $^1/_2$ in. (13 mm) through 3 in. (76 mm) for Water Service
P2906.4

CEN

European Committee for Standardization (EN)
Rue de la Science 23
Brussels, Belgium B - 1040

EN 15250—2007: Slow Heat Release Appliances Fired by Solid Fuel Requirements and Test Methods
R1002.2

CISPI

Cast Iron Soil Pipe Institute
2401 Fieldcrest Drive
Mundelein, IL 60060

301—18: Standard Specification for Hubless Cast Iron Soil Pipe and Fittings for Sanitary and Storm Drain, Waste and Vent Piping Applications
Table P3002.1(1), Table P3002.1(2), Table P3002.2, Table P3002.3, Table P3302.1

310—18: Standard Specification for Coupling for Use in Connection with Hubless Cast Iron Soil Pipe and Fittings for Sanitary and Storm Drain, Waste and Vent Piping Applications
P3003.4.3

CPA

Composite Panel Association
19465 Deerfield Avenue, Suite 306
Leesburg, VA 20176

A208.1—2016: Particleboard
R503.3.1, R602.1.9, R605.1

ANSI A135.4—2012: Basic Hardboard
Table R602.3(2)

ANSI A135.5—2012: Prefinished Hardboard Paneling
R702.5

ANSI A135.6—2012: Engineered Wood Siding
R703.5

ANSI A135.7—2012: Engineered Wood Trim
R703.5

CPSC

Consumer Product Safety Commission
4330 East-West Highway
Bethesda, MD 20814

16 CFR, Part 1201—(2002): Safety Standard for Architectural Glazing
R308.1.1, R308.3.1, Table R308.3.1(1)

16 CFR, Part 1209—(2002): Interim Safety Standard for Cellulose Insulation
R302.10.3

16 CFR, Part 1404—(2002): Cellulose Insulation
R302.10.3

CSA

CSA Group
8501 East Pleasant Valley Road
Cleveland, OH 44131-5516

A112.18.6—2021/CSA B125.6—2021: Flexible Water Connectors
P2906.7

A112.19.5—2017/CSA B45.15—2017: Flush Valves and Spuds for Water-closets, Urinals and Tanks
Table P2701.1

A112.19.7—2012/CSA B45.10—2012(R2021): Hydromassage Bathtub Systems
Table P2701.1

A257.2—14: Reinforced Circular Concrete Culvert, Storm Drain, Sewer Pipe and Fittings
Table P3002.2, P3003.13

A257.3—14: Joints for Circular Concrete Sewer and Culvert Pipe, Manhole Sections and Fittings Using Rubber Gaskets
P3003.5, P3003.13

AAMA/WDMA/CSA 101/I.S.2/A440—17: North American Fenestration Standard/Specification for Windows, Doors and Unit Skylights
R308.6.9, R609.3, N1102.4.3

ANSI/CSA FC I—2014: Fuel Cell Technologies—Part 3-100; Stationary fuel cell power systems—Safety
M1903.1

ANSI/CSA/IGSHPA C448 Series—16: Design and Installation of Ground Source Heat Pump Systems for Commercial and Residential Buildings
Table M2105.4, Table M2105.5

ASME A17.1/CSA B44—2019: Safety Code for Elevators and Escalators
R321.1

ASME A112.3.4—2013/CSA B45.9—18: Macerating Toilet Systems and Related Components
Table P2701.1, P3007.5

ASME A112.4.2—2015/CSA B45.16—15: Water-closet Personal Hygiene Device
P2722.5

ASME A112.18.1—2018/CSA B125.1—2018: Plumbing Supply Fittings
Table P2701.1, P2708.4, P2708.5, P2722.1, P2722.3, P2902.2, Table P2903.9.4

ASME A112.18.2—2019/CSA B125.2—2019: Plumbing Waste Fittings
Table P2701.1, P2702.2

ASME A112.19.1—2018/CSA B45.2—18: Enameled Cast-iron and Enameled Steel Plumbing Fixtures
Table P2701.1, P2711.1

ASME A112.19.2—2018/CSA B45.1—18: Ceramic Plumbing Fixtures
Table P2701.1, P2705.1, P2711.1, P2712.1, P2712.2, P2712.9

ASME A112.19.3—2017/CSA B45.4—2017: Stainless Steel Plumbing Fixtures
Table P2701.1, P2705.1, P2711.1, P2712.1

ASSE 1002—2020/ASME A112.1002—2020/CSA B125.12—20: Anti-Siphon Fill Valves for Water Closet Tanks
Table P2701.1, Table P2902.3, P2902.4.1

ASSE 1016—2017/ASME 112.1016—2017/CSA B125.16—2017: Performance Requirements for Automatic Compensating Valves for Individual Showers and Tub/Shower Combinations
Table P2701.1, P2708.4, P2722.2

ASSE 1070—2015/ASME A112.1070—2015/CSA B125.70—15: Performance Requirements for Water-temperature-limiting Devices
P2713.3, P2721.2, P2724.1

B44—2019: Safety Code for Elevators and Escalators
R321.1

B55.1—2015: Test Method for Measuring Efficiency and Pressure Loss of Drain Water Heat Recovery Units
N1103.5.4

B55.2—2015: Drain Water Heat Recovery Units
N1103.5.4

CSA—continued

B64.1.1—11(R2016): Vacuum Breakers, Atmospheric Type (AVB)
Table P2902.3, P2902.3.2

B64.1.2—11(R2016): Pressure Vacuum Breakers (PVB)
Table P2902.3, P2902.3.4

B64.1.3—11(R2016): Spill Resistant Pressure Vacuum Breakers (SRPVB)
Table P2902.3

B64.2—11(R2016): Vacuum Breakers, Hose Connection Type (HCVB)
Table P2902.3, P2902.3.2

B64.2.1—11(R2016): Hose Connection Vacuum Breakers (HCVB) with Manual Draining Feature
Table P2902.3, P2902.3.2

B64.2.1.1—11(2016): Hose Connection Dual Check Vacuum Breakers (HCDVB)
Table P2902.3, P2902.3.2

B64.2.2—11(2016): Vacuum Breakers, Hose Connection Type (HCVB) with Automatic Draining Feature
Table P2902.3, P2902.3.2

B64.3—11(2016): Dual Check Backflow Preventers with Atmospheric Port (DCAP)
Table P2902.3, P2902.3.2, P2902.5.1

B64.4—11(2016): Backflow Preventers, Reduced Pressure Principle Type (RP)
Table P2902.3, P2902.3.5

B64.4.1—11(2016): Reduced Pressure Principle for Fire Sprinklers (RPF)
Table P2902.3, P2902.3.5

B64.5—11(2016): Double Check Backflow Preventers (DCVA)
Table P2902.3, P2902.3.6

B64.5.1—11(2016): Double Check Valve Backflow Preventers, Type for Fire Systems (DCVAF)
Table P2902.3, P2902.3.6

B64.6—11(2016): Dual Check Valve Backflow Preventers (DuC)
Table P2902.3, P2902.3.7

B64.7—11(2016): Laboratory Faucet Vacuum Breakers (LFVB)
Table P2902.3, P2902.3.2

B125.3—18: Plumbing Fittings
Table P2701.1, P2713.3, P2721.2, Table P2902.3, P2902.4.1, Table P2903.9.4

B137.1—17: Polyethylene (PE) Pipe, Tubing and Fittings for Cold Water Pressure Services
Table P2906.4, Table P2906.6

B137.2—17: Polyvinylchloride PVC Injection-moulded Gasketed Fittings for Pressure Applications
Table P2906.6

B137.3—17: Rigid Poly (Vinyl Chloride) (PVC) Pipe for Pressure Applications
Table P2906.4, Table P2906.6, P3003.9.2

B137.5—17: Cross-linked Polyethylene (PEX) Tubing Systems for Pressure Applications
Table P2906.4, Table P2906.5, Table P2906.6

B137.6—17: Chlorinated polyvinylchloride CPVC Pipe, Tubing and Fittings For Hot- and Cold-water Distribution Systems
Table P2906.4, Table P2906.5, Table 2906.6

B137.9—17: Polyethylene/Aluminum/Polyethylene (PE-AL-PE) Composite Pressure Pipe Systems
Table M2101.1, Table P2906.4

B137.10—17: Cross-linked Polyethylene/Aluminum/Cross-linked Polyethylene (PE-AL-PE) Composite Pressure Pipe Systems
Table M2101.1, Table P2906.4, Table P2906.5, Table P2906.6, P2906.12.1

B137.11—17: Polypropylene (PP-R) Pipe and Fittings for Pressure Applications
Table P2906.4, Table P2906.5, Table P2906.6

B137.18—17: Polyethylene of Raised Temperature (PE-RT) Tubing Systems for Pressure Applications
Table M2101.1, Table M2105.4, Table M2105.5, Table P2906.4, Table P2906.5, Table P2906.6

B181.1—18: Acrylonitrile-butadiene-styrene (ABS) Drain, Waste and Vent Pipe and Pipe Fittings
Table P3002.1(1), Table P3002.1(2), Table P3002.3, P3003.3.2

B181.2—18: Polyvinylchloride (PVC) and chlorinated polyvinylchloride (CPVC) Drain, Waste and Vent Pipe and Pipe Fittings
Table P3002.1(1), Table P3002.1(2), P3003.9.2, P3008.3

CSA—continued

B181.3—18: Polyolefin and polyvinylidene (PVDF) Laboratory Drainage Systems
Table P3002.1(1), Table P3002.1(2), Table P3002.2, Table P3002.3, P3003.11.1

B182.1—18: Plastic Drain and Sewer Pipe and Pipe Fittings
Table P3302.1

B182.2—18: PSM Type polyvinylchloride (PVC) Sewer Pipe and Fittings
Table P3002.2, Table P3302.1

B182.4—18: Profile polyvinylchloride (PVC) Sewer Pipe & Fittings
Table P3002.2, Table P3302.1

B182.6—18: Profile Polyethylene (PE) Sewer Pipe and Fittings for leak-proof Sewer Applications
Table P3302.1

B182.8—18: Profile Polyethylene (PE) Storm Sewer and Drainage Pipe and Fittings
Table P3302.1

B356—10(R2015): Water Pressure Reducing Valves for Domestic Water Supply Systems
P2903.3.1

B483.1—07(R2017): Drinking Water Treatment Systems
P2909.1, P2909.2

B602—16: Mechanical Couplings for Drain, Waste and Vent Pipe and Sewer Pipe
P3003.3.1, P3003.4.3, P3003.5, P3003.9.1, P3003.10, P3003.12.2, P3003.13

C22.2 No. 218.1—13(R2017): Spas, Hot Tubs and Associated Equipment
M2006.1

C22.2 No. 236—15: Heating and Cooling Equipment
M2006.1

CAN/CSA/C22.2 No. 60335-2-40—2012: Safety of Household and Similar Electrical Appliances, Part 2-40: Particular Requirements for Electrical Heat Pumps, Air-Conditioners and Dehumidifiers
M1403.1, M1412.1, M1413.1

CSA 8—93: Requirements for Gas Fired Log Lighters for Wood Burning Fireplaces
G2433.1

CSA A257.1—2014: Non-reinforced Circular Concrete Culvert, Storm Drain, Sewer Pipe and Fittings
Table P3002.2

CSA B45.5—2017/IAPMO Z124—2017 with Errata dated August 2017: Plastic Plumbing Fixtures
Table P2701.1, P2711.1, P2711.2, P2712.1

CSA B805—18/ICC 805—18: Rainwater Harvesting Systems
P2912.1

CSA O325—16: Construction Sheathing
R503.2.1, R602.1.8, R604.1, R803.2.1

O437-Series—93(R2011): Standards on OSB and Waferboard
R503.2.1, R602.1.8, R604.1, R803.2.1

CSSB

Cedar Shake & Shingle Bureau
P.O. Box 1178
Sumas, WA 98295-1178

CSSB—97: Grading and Packing Rules for Western Red Cedar Shakes and Western Red Shingles of the Cedar Shake and Shingle Bureau
R702.6, R703.6

DASMA

Door & Access Systems Manufacturers Association International
1300 Sumner Avenue
Cleveland, OH 44115-2851

105—2017: Test Method for Thermal Transmittance and Air Infiltration of Garage Doors and Rolling Doors
N1101.10.3

DASMA—continued

108—2017: Standard Method for Testing Garage Doors, Rolling Doors and Flexible Doors; Determination of Structural Performance Under Uniform Static Air Pressure Difference
R609.4

115—2017: Standard Method for Testing Sectional Garage Doors, Rolling Doors and Flexible Doors: Determination of Structural Performance Under Missile Impact and Cyclic Wind Pressure
R301.2.1.2

DHA

Decorative Hardwoods Association (formerly HPVA)
42777 Trade West Drive
Sterling, Virginia 20166

ANSI/HPVA HP-1—2016: American National Standard for Hardwood and Decorative Plywood
R702.5

DOC

United States Department of Commerce
1401 Constitution Avenue, NW
Washington, DC 20230

PS 1—19: Structural Plywood
R404.2.1, Table R404.2.3, R503.2.1, R602.1.8, R604.1, R803.2.1

PS 2—18: Performance Standard for Wood Structural Panels
R404.2.1, Table R404.2.3, R503.2.1, R602.1.8, R604.1, R803.2.1

PS 20—05: American Softwood Lumber Standard
R404.2.1, R502.1.1, R602.1.1, R802.1.1

DOTn

U.S. Department of Transportation
1200 New Jersey Avenue SE
Washington, DC 20590

49 CFR, Parts 192.281(e) & 192.283 (b) (2009): Transportation of Natural and Other Gas by Pipeline: Minimum Federal Safety Standards
G2414.5.1

FEMA

Federal Emergency Management Agency
500 C Street SW
Washington, DC 20472

FEMA TB-2—08: Flood Damage-resistant Materials Requirements
R322.1.8

FEMA TB-11—01: Crawlspace Construction for Buildings Located in Special Flood Hazard Area
R408.7

FM

FM Approvals
Headquarters Office
Norwood, MA 02062

4450—(1989): Approval Standard for Class 1 Insulated Steel Deck Roofs—with Supplements through July 1992
R906.1

4474—2011: American National Standard for Evaluating the Simulated Wind Uplift Resistance of Roof Assemblies Using Static Positive and/or Negative Differential Pressures
R905.4.4.1

ANSI/FM 4880—(2017): American National Standard for Evaluating the Fire Performance of Insulated Building Panel Assemblies and Interior Finish Materials
R316.6

GA

Gypsum Association
6525 Belcrest Road, Suite 480
Hyattsville, MD 20782

GA-253—2018: Application of Gypsum Sheathing
Table R602.3(1)

HVI

Home Ventilating Institute
1740 Dell Range Blvd., Suite H, PMB 450
Cheyenne, WY 82009

916—18: Airflow Test Procedure
N1103.6.2

IAPMO

IAPMO Group
4755 E. Philadelphia Street
Ontario, CA 91761-USA

CSA B45.5—17/IAPMO Z124—2017: Plastic Plumbing Fixtures
Table P2701.1, P2711.1, P2711.2, P2712.1

ICC

International Code Council, Inc.
500 New Jersey Avenue NW
6th Floor
Washington, DC 20001

ANSI/RESNET/ICC 301—2019: Standard for the Calculation and Labeling of the Energy Performance of Low-rise Dwelling and Sleeping Units Using the Energy Rating Index
N1106.4

ANSI/RESNET/ICC 380—2019: Standard for Testing Airtightness of Building, Dwelling Unit and Sleeping Unit Enclosures; Airtightness of Heating and Cooling Air Distribution Systems and Airflow of Mechanical Ventilation Systems
N1102.4.1.2

IBC—21: International Building Code®
R320.1.1, R101.2, R202, R301.1.1, R301.1.3, R301.2.1.1, R301.2.2.1.1, R301.2.2.1.2, R301.3, Table R302.1(1), Table R302.1(2), R302.2.1, R302.2.2, R302.3, R308.5, R320.1, R321.3, R403.1.8, Table R602.10.3(3), Table R606.12.2.1, R609.2, R802.1.5.4, R905.10.3, G2402.3

ICC 400—17: Standard on the Design and Construction of Log Structures
R301.1.1, R502.1.4, R602.1.4, R703.1, R802.1.3, N1102.1, Table N1102.4.1.1

ICC 500—2020: ICC/NSSA Standard for the Design and Construction of Storm Shelters
R323.1

ICC 600—2020: Standard for Residential Construction in High-wind Regions
R301.2.1.1

ICC 900/SRCC 300—2020: Solar Thermal System Standard
M2301.2.2.2, M2301.2.3, M2301.2.6, M2301.2.7, M2301.2.8, M2301.2.10, M2301.4

ICC 901/SRCC 100—2020: Solar Thermal Collector Standard
M2301.3.1

ICC/ANSI A117.1—17: Accessible and Usable Buildings and Facilities
R321.3

IEBC—21: International Existing Building Code®
R110.2

IECC—06: International Energy Conservation Code®
N1101.6

IECC—21: International Energy Conservation Code®
N1101.1, N1103.8, Table N1105.4.2(1)

IFC—21: International Fire Code®
R102.7, R324.2, M2201.7, G2402.3, G2412.2

ICC—continued

IFGC—21: International Fuel Gas Code®
G2401.1, G2402.3, G2423.1

IMC—21: International Mechanical Code®
G2402.3

IPC—21: International Plumbing Code®
Table R301.2(1), R903.4.1, G2402.3, P2601.1

IPMC—21: International Property Maintenance Code®
R102.7

IPSDC—21: International Private Sewage Disposal Code®
R322.1.7

ISPSC—21: International Swimming Pool and Spa Code®
R327.1

IEEE

Institute of Electrical and Electronic Engineers, Inc.
3 Park Avenue, 17th Floor
New York, NY 10016-5997

515.1—2012: IEEE Standard for the Testing, Design, Installation and Maintenance of Electrical Resistance Trace Heating for Commercial Applications
N1103.5.2.2

ISO

International Organization for Standardization
Chemin de Blandonnet 8
Geneva, Switzerland CP 401 - 1214

8336—2009: Fibre-cement Flat Sheets-product Specification and Test Methods
Table R503.2.1.1(1), Table R503.2.1.1(2), Table R602.3(2), Table R702.4.2, R703.10.1, R703.10.2

15874—2002: Polypropylene Plastic Piping Systems for Hot and Cold Water Installations
Table M2101.1

MSS

Manufacturers Standardization Society of the Valve and Fittings Industry
127 Park Street, NE
Vienna, VA 22180

SP-42—2013: Corrosion Resistant Gate, Globe, Angle and Check Valves with Flanged and Butt Weld Ends (Glasses 150, 300 & 600)
Table P2903.9.4

SP-58—2018: Pipe Hangers and Supports—Materials, Design, Manufacture, Selection, Application and Installation
G2418.2

SP-67—2017: Butterfly Valves
Table P2903.9.4

SP-70—2011: Gray Iron Gate Valves, Flanged and Threaded Ends
Table P2903.9.4

SP-71—2018: Gray Iron Swing Check Valves, Flanged and Threaded Ends
Table P2903.9.4

SP-72—2010a: Ball Valves with Flanged or Butt-Welding Ends for General Service
P2903.9.4

SP-78—2011: Cast Iron Plug Valves, Flanged and Threaded Ends
Table P2903.9.4

SP-80—2013: Bronze Gate, Globe, Angle and Check Valves
Table P2903.9.4

SP-110—2010: Ball Valves, Threaded, Socket Welded, Solder Joint, Grooved and Flared Ends
Table P2903.9.4

MSS—continued

SP-122—2017: Plastic Industrial Ball Valves
Table P2903.9.4

SP-139—2014: Copper Alloy Gate, Globe, Angle, and Check Valves for Low Pressure/ Low Temperature Plumbing Applications
Table P2903.9.4

NAIMA

North American Insulation Manufacturers Association
11 Canal Center Plaza, Suite 101
Alexandria, VA 22314

AH 116—09: Fibrous Glass Duct Construction Standards, Fifth Edition
M1601.1.1

NEMA

National Electrical Manufacturers Association
12300 17th Street North No. 900
Arlington, VA 22209

OS 4—2016: Requirements for Air-Sealed Boxes for Electrical and Communication Applications
N1102.4.6

NFPA

National Fire Protection Association
1 Batterymarch Park
Quincy, MA 02169-7471

13D—19: Standard for the Installation of Sprinkler Systems in One- and Two-family Dwellings and Manufactured Homes
R313.1.1, R313.2.1, R324.6.2.1, P2904.1, P2904.6.1

13R—19: Standard for the Installation of Sprinkler Systems in Low-Rise Residential Occupancies
R325.5

31—20: Standard for the Installation of Oil-burning Equipment
M1701.1, M1801.3.1, M1805.3, M2201.2

58—20: Liquefied Petroleum Gas Code
G2412.2, G2414.5.2

70—20: National Electrical Code
R107.3, R324.3, R328.6, E3401.1, E3401.2, E4301.1, Table E4303.2, E4304.3, E4304.4

72—19: National Fire Alarm and Signaling Code
R314.1, R314.7.1

85—19: Boiler and Combustion Systems Hazards Code
G2452.1

211—19: Standard for Chimneys, Fireplaces, Vents and Solid Fuel Burning Appliances
R1002.5, G2427.5.5.1

259—18: Standard for Test Method for Potential Heat of Building Materials
R316.5.7, R316.5.8

275—17: Standard Method of Fire Tests for the Evaluation of Thermal Barriers
R316.4

276—15: Standard Method of Fire Tests for Determining the Heat Release Rate of Roofing Assemblies with Combustible Above-deck Roofing Components
R906.1

286—19: Standard Methods of Fire Tests for Evaluating Contribution of Wall and Ceiling Interior Finish to Room Fire Growth
R302.9.4, R316.6

501—17: Standard on Manufactured Housing
R202

720—15: Standard for the Installation of Carbon Monoxide (CO) Detectors and Warning Equipment
R315.7.1, R315.7.2

853—20: Standard on the Installation of Stationary Fuel Cell Power Systems
M1903.1

NFRC

National Fenestration Rating Council, Inc.
6305 Ivy Lane, Suite 140
Greenbelt, MD 20770

100—2020: Procedure for Determining Fenestration Products *U*-Factors
N1101.10.3

200—2020: Procedure for Determining Fenestration Product Solar Heat Gain Coefficients and Visible Transmittance at Normal Incidence
N1101.10.3

400—2020: Procedure for Determining Fenestration Product Air Leakage
N1102.4.3

NGWA

National Ground Water Association
601 Dempsey Road
Westerville, OH 43081

ANSI/NGWA 01—14: Water Well Construction Standard
P2602.1

NSF

NSF International
789 N. Dixboro Road
Ann Arbor, MI 48105

14—2017: Plastics Piping System Components and Related Materials
M1301.4, P2609.3, P2909.3

41—2016: Nonliquid Saturated Treatment Systems (Composting Toilets)
P2725.1

42—2017: Drinking Water Treatment Units—Aesthetic Effects
P2909.1, P2909.3

44—2017: Residential Cation Exchange Water Softeners
P2909.1, P2909.3

50—2017: Equipment for Swimming Pools, Spas, Hot Tubs and Other Recreational Water Facilities
P2911.8.1

53—2017: Drinking Water Treatment Units—Health Effects
P2909.1, P2909.3

58—2017: Reverse Osmosis Drinking Water Treatment Systems
P2909.2, P2909.3

61—2018: Drinking Water System Components—Health Effects
P2609.5, P2722.1, P2903.9.4, P2906.4, P2906.5, P2906.6, P2909.3

62—2017: Drinking Water Distillation Systems
P2909.1

350—2017a: Onsite Residential and Commercial Water Reuse Treatment Systems
P2911.6.1

358-1—2017: Polyethylene Pipe and Fittings for Water-based Ground Source "Geothermal" Heat Pump Systems
M2105.4, M2105.5

358-2—2017: Polypropylene Pipe and Fittings for Water-based Ground Source "Geothermal" Heat Pump Systems
Table M2105.4, M2105.5

358-3—2016: Cross-linked Polyethylene (PEX) Pipe and Fittings for Water-based Ground-source (Geothermal) Heat Pump Systems
Table M2105.4, Table M2105.5

358-4—2017: Polyethylene of Raised Temperature (PE-RT) Pipe and Fittings for Water-based Ground-source (Geothermal) Heat Pump Systems
Table 2105.4, Table 2105.5

359—2011(R2016): Valves for Crosslinked Polyethylene (PEX) Water Distribution Tubing Systems
Table P2903.9.4

372—2016: Drinking Water Systems Components—Lead Content
P2906.2.1

PCA

Portland Cement Association
5420 Old Orchard Road
Skokie, IL 60077

100—17: Prescriptive Design of Exterior Concrete Walls for One- and Two-family Dwellings (Pub. No. PCA 100.3)
R301.2.2.5, R404.1.3, R404.1.3.2.1, R404.1.3.2.2, R404.1.3.4, R404.1.4.2, R608.1, R608.2, R608.5.1, R608.9.2, R608.9.3

SBCA

Structural Building Components Association
6300 Enterprise Lane
Madison, WI 53719

ANSI/FS100—12(R2018): Standard Requirements for Wind Pressure Resistance of Foam Plastic Insulating Sheathing Used in Exterior Wall Covering Assemblies
R316.8

BCSI—2018: Building Component Safety Information Guide to Good Practice for Handling, Installing, Restraining & Bracing of Metal Plate Connected Wood Trusses
R502.11.2, R802.10.3

CFS-BCSI—(updated June 2016): Cold-formed Steel Building Component Safety Information (CFSBCSI) Guide to Good Practice for Handling, Installing & Bracing of Cold-formed Steel Trusses
R505.1.3, R804.3.6

SMACNA

Sheet Metal & Air Conditioning Contractors National Assoc. Inc.
4021 Lafayette Center Road
Chantilly, VA 22021

SMACNA—10: Fibrous Glass Duct Construction Standards 7th edition
M1601.1.1, M1601.4.1

SMACNA/ANSI—2016: HVAC Duct Construction Standards—Metal and Flexible 4th Edition (ANSI)
M1601.4.1

TMS

The Masonry Society
105 South Sunset Street, Suite Q
Longmont, CO 80501

402—2016: Building Code Requirements for Masonry Structures
R404.1.2, R606.1, R606.1.1, R606.12.1, R606.12.2.3.1, R606.12.3.1, R703.12

403—2017: Direct Design Handbook for Masonry Structures
R606.1, R606.1.1, R606.12.1, R606.12.3.1

404—2016: Standard for the Design of Architectural Cast Stone
R606.1

602—2016: Specification for Masonry Structures
R606.2.10, R606.2.13, R703.12

TPI

Truss Plate Institute
2670 Crain Highway, Suite 203
Waldorf, MD 20601

TPI 1—2014: National Design Standard for Metal Plate Connected Wood Truss Construction
R502.11.1, R802.10.2

UL

UL LLC
333 Pfingsten Road
Northbrook, IL 60062

17—2008: Vent or Chimney Connector Dampers for Oil-fired Appliances—with revisions through September 2013
M1802.2.2

UL—continued

55A—2004: Materials for Built-up Roof Coverings
R905.9.2

58—2018: Steel Underground Tanks for Flammable and Combustible Liquids
M2201.1

80—2007: Steel Tanks for Oil-burner Fuel—with revisions through January 2014
M2201.1

103—2010: Factory-built Chimneys for Residential Type and Building Heating Appliances—with revisions through March 2017
R202, R1005.3, G2430.1

127—2011: Factory-built Fireplaces—with revisions through July 2016
R1001.11, R1004.1, R1004.4, R1004.5, R1005.4, N1102.4.2, G2445.7

174—2004: Household Electric Storage Tank Water Heaters—with revisions through December 2016
M2005.1

180—2012: Liquid-level Indicating Gauges for Oil Burner Fuels and Other Combustible Liquids—with revisions through May 2017
M2201.5

181—2005: Factory-made Air Ducts and Air Connectors—with revisions through April 2017
M1601.1.1, M1601.4.1

181A—2013: Closure Systems for Use with Rigid Air Ducts and Air Connectors—with revisions through March 2017
M1601.2, M1601.4.1

181B—2013: Closure Systems for Use with Flexible Air Ducts and Air Connectors—with revisions through March 2017
M1601.4.1

217—2015: Smoke Alarms—with revisions through November 2016
R314.1.1, R315.1.1

263—2011: Fire Test of Building Construction and Materials—with revisions through March 2018
Table R302.1(2), R302.2, R302.2.1, R302.2.2, R302.4.1, R302.11.1, R606.2.2

268—2016: Smoke Detectors for Fire Alarm Systems—with revisions through July 2016
R314.7.1, R314.7.4, R315.7.4

325—2017: Door, Drapery, Gate, Louver and Window Operations and Systems
R309.4

343—2017: Pumps for Oil-burning Appliances
M2204.1

378—06: Draft Equipment—with revisions through September 2013
M1804.2.6, G2427.3.3

441—16: Gas Vents—with revisions through July 2016
G2426.1, G2427.6.1

467—13: Grounding and Bonding Equipment
G2411.2.5

507—2017: Electric Fans—with revisions through August 2018
M1503.2

508—2018: Industrial Control Equipment
M1411.3.1

515—11: Electrical Resistance Heat Tracing for Commercial and Industrial Applications Including Revisions through July 2015
N1103.5.1.2

536—2014: Flexible Metallic Hose
M2202.3

580—2006: Test for Uplift Resistance of Roof Assemblies—with Revisions through October 2013
R905.4.4.1

641—2010: Type L, Low-temperature Venting Systems—with revisions through April 2018
R202, R1003.11.5, M1804.2.4, G2426.1, G2427.6.1

651—2011: Schedule 40, Type EB and A Rigid PVC Conduit and Fittings—with revisions through June 2016
G2414.5.3

705—2017: Power Ventilators—with revisions through October 2018
M1502.4.4

UL—continued

723—2018: Standard for Test for Surface Burning Characteristics of Building Materials
R202, R302.9.3, R302.9.4, R302.10.1, R302.10.2, R316.3, R316.5.9, R316.5.11, R507.2.2.2, R703.14.3, R802.1.5, M1601.3, M1601.5.2, P2801.6

726—95: Oil-fired Boiler Assemblies—with revisions through October 2013
M2001.1.1, M2006.1

727—2018: Oil-fired Central Furnaces
M1402.1

729—2003: Oil-fired Floor Furnaces
M1408.1

730—03: Oil-fired Wall Furnaces—with revisions through November 2016
M1409.1

732—2018: Oil-fired Storage Tank Water Heaters—with revisions through August 2018
M2005.1

737—2011: Fireplaces Stoves
M1414.1, M1901.2

790—04: Standard Test Methods for Fire Tests of Roof Coverings—with revisions through October 2018
R302.2.4, R902.1

795—2016: Commercial-industrial Gas Heating Equipment
G2442.1, G2452.1

834—2004: Heating, Water Supply and Power Boilers—Electric—with revisions through September 2018
M2001.1.1

842—2015: Valves for Flammable Fluids—with revisions through May 2015
M2204.2

858—2014: Household Electric Ranges—with revisions through June 2018
M1503.2, M1901.2

875—09: Electric Dry-bath Heaters—with revisions through September 2017
M1902.2

896—1993: Oil-burning Stoves—with revisions through November 2016
M1410.1

907—94: Fireplace Accessories—with revisions through November 2014
R1001.13

923—2013: Microwave Cooking Appliances—with revisions through July 2017
M1503.2, M1504.1, M1901.2

959—2010: Medium Heat Appliance Factory-built Chimneys—with revisions through June 2014
R1005.6

1026—2012: Electric Household Cooking and Food Serving Appliances—with revisions through July 2018
M1901.2

1040—1996: Fire Test of Insulated Wall Construction—with revisions through April 2017
R316.6

1042—2009: Electric Baseboard Heating Equipment—with revisions through December 2016
M1405.1

1256—02: Fire Test of Roof Deck Construction—with revisions through August 2018
R906.1

1261—2016: Electric Water Heaters for Pools and Tubs—with revisions through September 2017
M2006.1

1479—2015: Fire Tests of Through-Penetration Firestops
R302.4.1.2

1482—2011: Solid-Fuel-type Room Heaters—with revisions through August 2015
R1002.2, R1002.5, M1410.1

1563—2009: Electric Spas, Equipment Assemblies, and Associated Equipment—with revisions through October 2017
M2006.1

UL—continued

1618—2015: Wall Protectors, Floor Protectors, and Hearth Extensions—with revisions through January 2018
R1004.2, M1410.2

1693—2010: Electric Radiant Heating Panels and Heating Panel Sets—with revisions through October 2011
M1406.1

1703—2002: Flat-plate Photovoltaic Modules and Panels—with revisions through September 2018
R324.3.1, R902.4, R905.16.4

1715—97: Fire Test of Interior Finish Material—with revisions through April 2017
R316.6

1738—2010: Venting Systems for Gas-burning Appliances, Categories II, III and IV
G2426.1, G2427.4.1, G2427.4.1.1, G2427.4.2

1741—2010: Inverters, Converters, Controllers and Interconnection System Equipment with Distributed Energy Resources—with revisions through February 2018
R324.3.1, R328.6

1777—07: Chimney Liners—with revisions through April 2014
R1003.11.1, R1003.18, M1801.3.4, G2425.12, G2425.15.4, G2427.5.1, G2427.5.2

1897—15: Uplift Tests for Roof Covering Systems
R905.4.4.1

1995—2015: Heating and Cooling Equipment—with revisions through August 2018
M1402.1, M1403.1, M1407.1, M1412.1, M1413.1, M2006.1

1996—2009: Electric Duct Heaters—with revisions through July 2016
M1402.1, M1407.1

2034—2017: Standard for Single- and Multiple-station Carbon Monoxide Alarms—with revisions through September 2018
R314.1.1, R315.1.1

2075—2013: Gas and Vapor Detectors and Sensors—with revisions through December 2017
R314.7.4, R315.7.1, R315.7.4

2158A—2013: Outline of Investigation for Clothes Dryer Transition Duct—with revisions through April 2017
M1502.4.3, G2439.7.3

2200—2012: Stationary Engine Generator Assemblies—with revisions through October 2015
R329.1

2523—2009: Standard for Solid Fuel-fired Hydronic Heating Appliances, Water Heaters and Boilers—with revisions through March 2018
M2001.1.1, M2005.1

2703—2014: Mounting Systems, Mounting Devices, Clamping/Retention Devices and Ground Lugs for Use with Flat-Plate Photovoltaic Modules and Panels—with revisions through December 2019
R902.4

7103—19: Outline of Investigation for Building-Integrated Photovoltaic Roof Covering
R902.3, R905.16.4, Table 905.16.6, R905.17.5

9540—2016: Standard for Energy Storage Systems and Equipment
R328.2, R328.6

61730-1—2017: Photovoltaic (PV) Module Safety Qualification—Part 1: Requirements for Construction
R324.3.1 , R905.16.4 , 905.17.5

61730-2—2017: Photovoltaic (PV) Module Safety Qualification—Part 2: Requirements for Testing
R324.3.1, R905.16.4, R905.17.5

UL/CSA/ANCE 60335-2-40—2012: Standard for Household and Similar Electrical Appliances, Part 2: Particular Requirements for Motor-compressors
M1402.1, M1403.1, M1412.1, M1413.1, M2006.1

ULC

ULC
13775 Commerce Parkway
Richmond, BC V6V 2V4

CAN/ULC S 102.2—2018: Standard Method of Test for Surface Burning Characteristics of Building Materials and Assemblies
R302.10.1, R302.10.2

US-FTC

United States-Federal Trade Commission
600 Pennsylvania Avenue NW
Washington, DC 20580

CFR Title 16(2015): R-value Rule
N1101.10.4

WDMA

Window and Door Manufacturers Association
2025 M Street NW, Suite 800
Washington, DC 20036-3309

AAMA/WDMA/CSA 101/I.S.2/A440—17: North American Fenestration Standard/Specifications for Windows, Doors and Skylights
R308.6.9, R609.3, N1102.4.3

I.S. 11—16: Industry Standard Analytical Method for Design Pressure (DP) Ratings of Fenestration Products
R308.6.9.1, R609.3.1

WMA

World Millwork Alliance (formerly Association of Millwork Distributors Standards AMD)
10047 Robert Trent Parkway
New Port Richey, FL 34655-4649

ANSI WMA 100—2018: Standard Method of Determining Structural Performance Ratings of Side-Hinged Exterior Door Systems and Procedures for Component Substitution
R609.3

APPENDIX AA

SIZING AND CAPACITIES OF GAS PIPING

This appendix is an excerpt from the 2021 International Fuel Gas Code® *informative Appendix A. Table references in the text, other than AA tables, are as numbered in the* International Fuel Gas Code *(IFGC). For related table references in this code, you can find the IFGC table number in brackets adjacent to the table number in Chapter 24 of this code. For example, Table 402.4(2) in the* IFGC *is related to Table G2413.4(1) [402.4(2)] in this code.*

User note:

About this appendix: *Appendix AA provides commentary, guidance and examples for sizing of gas piping systems.*

SECTION AA101 GENERAL PIPING CONSIDERATIONS

The first goal of determining the pipe sizing for a fuel gas *piping* system is to make sure that there is sufficient gas pressure at the inlet to each *appliance*. The majority of systems are residential and the *appliances* will all have the same, or nearly the same, requirement for minimum gas pressure at the *appliance* inlet. This pressure will be about 5-inch water column (w.c.) (1.25 kPa), which is enough for proper operation of the *appliance* regulator to deliver about 3.5-inches water column (w.c.) (875 kPa) to the burner itself. The pressure drop in the *piping* is subtracted from the source delivery pressure to verify that the minimum is available at the *appliance*.

There are other systems, however, where the required inlet pressure to the different appliances is quite varied. In such cases, the greatest inlet pressure required must be satisfied, as well as the farthest *appliance*, which is almost always the critical *appliance* in small systems.

There is an additional requirement to be observed besides the capacity of the system at 100-percent flow. That requirement is that at minimum flow, the pressure at the inlet to any *appliance* does not exceed the pressure rating of the *appliance* regulator. This would seldom be of concern in small systems if the source pressure is $^1/_2$ psi (14-inch w.c.) (3.5 kPa) or less but it should be verified for systems with greater gas pressure at the point of supply.

To determine the size of *piping* used in a gas *piping* system, the following factors must be considered:

1. Allowable loss in pressure from *point of delivery* to *appliance*.
2. Maximum gas demand.
3. Length of *piping* and number of fittings.
4. Specific gravity of the gas.
5. Diversity factor.

For any gas *piping* system or special *appliance*, or for conditions other than those covered by the tables provided in this code such as longer runs, greater gas demands or greater pressure drops, the size of each gas *piping* system should be determined by standard engineering practices acceptable to the code official.

SECTION AA102 DESCRIPTION OF TABLES

AA102.1 General. The quantity of gas to be provided at each *outlet* should be determined, whenever possible, directly from the manufacturer's gas input Btu/h rating of the *appliance* that will be installed. In case the ratings of the *appliances* to be installed are not known, Table R402.2 shows the approximate consumption (in Btu per hour) of certain types of typical household *appliances*.

To obtain the cubic feet per hour of gas required, divide the total Btu/h input of all *appliances* by the average Btu heating value per cubic feet of the gas. The average Btu per cubic feet of the gas in the area of the installation can be obtained from the serving gas supplier.

AA102.2 Low pressure natural gas tables. Capacities for gas at low pressure [less than 2.0 psig (13.8 kPa gauge)] in cubic feet per hour of 0.60 specific gravity gas for different sizes and lengths are shown in Tables 402.4(1) and 402.4(2) for iron pipe or equivalent rigid pipe; in Tables 402.4(8) through 402.4(11) for smooth wall semirigid tubing; and in Tables 402.4(15) through 402.4(17) for corrugated stainless steel tubing. Tables 402.4(1) and 402.4(6) are based on a pressure drop of 0.3-inch w.c. (75 Pa), whereas Tables 402.4(2), 402.4(9) and 402.4(15) are based on a pressure drop of 0.5- inch w.c. (125 Pa). Tables 402.4(3), 402.4(4), 402.4(10), 402.4(11), 402.4(16) and 402.4(17) are special low-pressure applications based on pressure drops greater than 0.5-inch w.c. (125 Pa). In using these tables, an allowance (in equivalent length of pipe) should be considered for any *piping* run with four or more fittings (see Table AA102.2).

TABLE AA102.2
EQUIVALENT LENGTHS OF PIPE FITTINGS AND VALVES

		SCREWED FITTINGS[1]				90° WELDING ELBOWS AND SMOOTH BENDS[2]					
		45°/Ell	90°/Ell	180° close return bends	Tee	R/d = 1	R/d = 1 1/3	R/d = 2	R/d = 4	R/d = 6	R/d = 8
k factor =		0.42	0.90	2.00	1.80	0.48	0.36	0.27	0.21	0.27	0.36
L/d' ratio[4] *n* =		14	30	67	60	16	12	9	7	9	12
Nominal pipe size, inches	Inside diameter *d*, inches, Schedule 40[6]	*L* = Equivalent Length in Feet of Schedule 40 (Standard-weight) Straight Pipe[6]									
1/2	0.622	0.73	1.55	3.47	3.10	0.83	0.62	0.47	0.36	0.47	0.62
3/4	0.824	0.96	2.06	4.60	4.12	1.10	0.82	0.62	0.48	0.62	0.82
1	1.049	1.22	2.62	5.82	5.24	1.40	1.05	0.79	0.61	0.79	1.05
1 1/4	1.380	1.61	3.45	7.66	6.90	1.84	1.38	1.03	0.81	1.03	1.38
1 1/2	1.610	1.88	4.02	8.95	8.04	2.14	1.61	1.21	0.94	1.21	1.61
2	2.067	2.41	5.17	11.5	10.3	2.76	2.07	1.55	1.21	1.55	2.07
2 1/2	2.469	2.88	6.16	13.7	12.3	3.29	2.47	1.85	1.44	1.85	2.47
3	3.068	3.58	7.67	17.1	15.3	4.09	3.07	2.30	1.79	2.30	3.07
4	4.026	4.70	10.1	22.4	20.2	5.37	4.03	3.02	2.35	3.02	4.03
5	5.047	5.88	12.6	28.0	25.2	6.72	5.05	3.78	2.94	3.78	5.05
6	6.065	7.07	15.2	33.8	30.4	8.09	6.07	4.55	3.54	4.55	6.07
8	7.981	9.31	20.0	44.6	40.0	10.6	7.98	5.98	4.65	5.98	7.98
10	10.02	11.7	25.0	55.7	50.0	13.3	10.0	7.51	5.85	7.51	10.0
12	11.94	13.9	29.8	66.3	59.6	15.9	11.9	8.95	6.96	8.95	11.9
14	13.13	15.3	32.8	73.0	65.6	17.5	13.1	9.85	7.65	9.85	13.1
16	15.00	17.5	37.5	83.5	75.0	20.0	15.0	11.2	8.75	11.2	15.0
18	16.88	19.7	42.1	93.8	84.2	22.5	16.9	12.7	9.85	12.7	16.9
20	18.81	22.0	47.0	105.0	94.0	25.1	18.8	14.1	11.0	14.1	18.8
24	22.63	26.4	56.6	126.0	113.0	30.2	22.6	17.0	13.2	17.0	22.6

(continued)

TABLE AA102.2—continued
EQUIVALENT LENGTHS OF PIPE FITTINGS AND VALVES

		MITER ELBOWS[3] (No. of miters)					WELDING TEES		VALVES (screwed, flanged, or welded)			
		1-45°	1-60°	1-90°	2-90°[5]	3-90°[5]	Forged	Miter[3]	Gate	Globe	Angle	Swing Check
k factor =		0.45	0.90	1.80	0.60	0.45	1.35	1.80	0.21	10	5.0	2.5
L/d' ratio[4] n =		15	30	60	20	15	45	60	7	333	167	83
Nominal pipe size, inches	Inside diameter d, inches, Schedule 40[6]	L = Equivalent Length in Feet of Schedule 40 (Standard-weight) Straight Pipe[6]										
$^{1}/_{2}$	0.622	0.78	1.55	3.10	1.04	0.78	2.33	3.10	0.36	17.3	8.65	4.32
$^{3}/_{4}$	0.824	1.03	2.06	4.12	1.37	1.03	3.09	4.12	0.48	22.9	11.4	5.72
1	1.049	1.31	2.62	5.24	1.75	1.31	3.93	5.24	0.61	29.1	14.6	7.27
$1^{1}/_{4}$	1.380	1.72	3.45	6.90	2.30	1.72	5.17	6.90	0.81	38.3	19.1	9.58
$1^{1}/_{2}$	1.610	2.01	4.02	8.04	2.68	2.01	6.04	8.04	0.94	44.7	22.4	11.2
2	2.067	2.58	5.17	10.3	3.45	2.58	7.75	10.3	1.21	57.4	28.7	14.4
$2^{1}/_{2}$	2.469	3.08	6.16	12.3	4.11	3.08	9.25	12.3	1.44	68.5	34.3	17.1
3	3.068	3.84	7.67	15.3	5.11	3.84	11.5	15.3	1.79	85.2	42.6	21.3
4	4.026	5.04	10.1	20.2	6.71	5.04	15.1	20.2	2.35	112.0	56.0	28.0
5	5.047	6.30	12.6	25.2	8.40	6.30	18.9	25.2	2.94	140.0	70.0	35.0
6	6.065	7.58	15.2	30.4	10.1	7.58	22.8	30.4	3.54	168.0	84.1	42.1
8	7.981	9.97	20.0	40.0	13.3	9.97	29.9	40.0	4.65	222.0	111.0	55.5
10	10.02	12.5	25.0	50.0	16.7	12.5	37.6	50.0	5.85	278.0	139.0	69.5
12	11.94	14.9	29.8	59.6	19.9	14.9	44.8	59.6	6.96	332.0	166.0	83.0
14	13.13	16.4	32.8	65.6	21.9	16.4	49.2	65.6	7.65	364.0	182.0	91.0
16	15.00	18.8	37.5	75.0	25.0	18.8	56.2	75.0	8.75	417.0	208.0	104.0
18	16.88	21.1	42.1	84.2	28.1	21.1	63.2	84.2	9.85	469.0	234.0	117.0
20	18.81	23.5	47.0	94.0	31.4	23.5	70.6	94.0	11.0	522.0	261.0	131.0
24	22.63	28.3	56.6	113.0	37.8	28.3	85.0	113.0	13.2	629.0	314.0	157.0

For SI: 1 inch = 25.4 mm, 1 foot = 304.8 mm, 1 degree = 0.01745 rad.

Note: Values for welded fittings are for conditions where bore is not obstructed by weld spatter or backing rings. If appreciably obstructed, use values for "Screwed Fittings."

1. Flanged fittings have three-fourths the resistance of screwed elbows and tees.
2. Tabular figures give the extra resistance due to curvature alone to which should be added the full length of travel.
3. Small size socket-welding fittings are equivalent to miter elbows and miter tees.
4. Equivalent resistance in number of diameters of straight pipe computed for a value of (f – 0.0075) from the relation (n – $k/4f$).
5. For condition of minimum resistance where the centerline length of each miter is between d and $2^{1}/_{2}d$.
6. For pipe having other inside diameters, the equivalent resistance can be computed from the n values.

Source: Crocker, S. *Piping Handbook*, 4th ed., Table XIV, pp. 100–101. Copyright 1945 by McGraw-Hill, Inc. Used by permission of McGraw-Hill Book Company.

AA102.3 Undiluted liquefied petroleum tables. Capacities in thousands of Btu per hour of undiluted liquefied petroleum gases based on a pressure drop of 0.5-inch w.c. (125 Pa) for different sizes and lengths are shown in Table 402.4(28) for iron pipe or equivalent rigid pipe, in Table 402.4(30) for smooth wall semi-rigid tubing, in Table 402.4(32) for corrugated stainless steel tubing, and in Tables 402.4(35) and 402.4(37) for polyethylene plastic pipe and tubing. Tables 402.4(33) and 402.4(34) for corrugated stainless steel tubing and Table 402.4(36) for polyethylene plastic pipe are based on operating pressures greater than $1^1/_2$ pounds per square inch (psi) (3.5 kPa) and pressure drops greater than 0.5-inch w.c. (125 Pa). In using these tables, an allowance (in equivalent length of pipe) should be considered for any *piping* run with four or more fittings [see Table AA102.2.

AA102.4 Natural gas specific gravity. Gas *piping* systems that are to be supplied with gas of a specific gravity of 0.70 or less can be sized directly from the tables provided in this code, unless the code official specifies that a gravity factor be applied. Where the specific gravity of the gas is greater than 0.70, the gravity factor should be applied.

Application of the gravity factor converts the figures given in the tables provided in this code to capacities for another gas of different specific gravity. Such application is accomplished by multiplying the capacities given in the tables by the multipliers shown in Table AA102.4. In case the exact specific gravity does not appear in the table, choose the next higher value specific gravity shown.

AA102.5 Higher pressure natural gas tables. Capacities for gas at pressures 2.0 psig (13.8 kPa) or greater in cubic feet per hour of 0.60 specific gravity gas for different sizes and lengths are shown in Tables 402.4(5) through 402.4(7) for iron pipe or equivalent rigid pipe; Tables 402.4(12) to 402.4(14) for semirigid tubing; Tables 402.4(18) and 402.4(19) for corrugated stainless steel tubing; and Table 402.4(22) for polyethylene plastic pipe.

SECTION AA103 USE OF CAPACITY TABLES

AA103.1 Longest length method. This sizing method is conservative in its approach by applying the maximum operating conditions in the system as the norm for the system and by setting the length of pipe used to size any given part of the *piping* system to the maximum value.

To determine the size of each section of gas *piping* in a system within the range of the capacity tables, proceed as follows (also see sample calculations included in this Appendix):

1. Divide the *piping* system into appropriate segments consistent with the presence of tees, branch lines and main runs. For each segment, determine the gas load (assuming all *appliances* operate simultaneously) and its overall length. An allowance (in equivalent length of pipe) as determined from Table AA102.2 shall be considered for *piping* segments that include four or more fittings.
2. Determine the gas demand of each *appliance* to be attached to the *piping* system. Where Tables 402.4(1) through 402.4(24) are to be used to select the *piping* size, calculate the gas demand in terms of cubic feet per hour for each *piping* system *outlet.* Where Tables 402.4(25) through 402.4(37) are to be used to select the *piping* size, calculate the gas demand in terms of thousands of Btu per hour for each *piping* system *outlet.*
3. Where the *piping* system is for use with other than undiluted liquefied petroleum gases, determine the design system pressure, the allowable loss in pressure (pressure drop), and specific gravity of the gas to be used in the *piping* system.
4. Determine the length of *piping* from the *point of delivery* to the most remote *outlet* in the building/*piping* system.
5. In the appropriate capacity table, select the row showing the measured length or the next longer length if the table does not give the exact length. This is the only length used in determining the size of any section of gas *piping.* If the gravity factor is to be

TABLE AA102.4 MULTIPLIERS TO BE USED WITH TABLES 402.4(1) THROUGH 402.4(22) WHERE THE SPECIFIC GRAVITY OF THE GAS IS OTHER THAN 0.60

SPECIFIC GRAVITY	MULTIPLIER
0.35	1.31
0.40	1.23
0.45	1.16
0.50	1.10
0.55	1.04
0.60	1.00
0.65	0.96
0.70	0.93
0.75	0.90
0.80	0.87
0.85	0.84
0.90	0.82
1.00	0.78
1.10	0.74
1.20	0.71
1.30	0.68
1.40	0.66
1.50	0.63
1.60	0.61
1.70	0.59
1.80	0.58
1.90	0.56
2.00	0.55
2.10	0.54

applied, the values in the selected row of the table are multiplied by the appropriate multiplier from Table AA102.4.

6. Use this horizontal row to locate ALL gas demand figures for this particular system of *piping*.
7. Starting at the most remote *outlet*, find the gas demand for that *outlet* in the horizontal row just selected. If the exact figure of demand is not shown, choose the next larger figure left in the row.
8. Opposite this demand figure, in the first row at the top, the correct size of gas *piping* will be found.
9. Proceed in a similar manner for each *outlet* and each section of gas *piping*. For each section of *piping*, determine the total gas demand supplied by that section.

Where a large number of *piping* components (such as elbows, tees and valves) are installed in a pipe run, additional pressure loss can be accounted for by the use of equivalent lengths. Pressure loss across any *piping* component can be equated to the pressure drop through a length of pipe. The equivalent length of a combination of only four elbows/tees can result in a jump to the next larger length row, resulting in a significant reduction in capacity. The equivalent lengths in feet shown in Table AA102.2 have been computed on a basis that the inside diameter corresponds to that of Schedule 40 (standard-weight) steel pipe, which is close enough for most purposes involving other schedules of pipe. Where a more specific solution for equivalent length is desired, this can be made by multiplying the actual inside diameter of the pipe in inches by $n/12$, or the actual inside diameter in feet by n (n can be read from the table heading). The equivalent length values can be used with reasonable accuracy for copper or brass fittings and bends although the resistance per foot of copper or brass pipe is less than that of steel. For copper or brass valves, however, the equivalent length of pipe should be taken as 45 percent longer than the values in the table, which are for steel pipe.

AA103.2 Branch length method. This sizing method reduces the amount of conservatism built into the traditional Longest Length Method. The longest length as measured from the meter to the farthest remote *appliance* is only used to size the initial parts of the overall *piping* system. The Branch Length Method is applied in the following manner:

1. Determine the gas load for each of the connected *appliances*.
2. Starting from the meter, divide the *piping* system into a number of connected segments, and determine the length and amount of gas that each segment would carry assuming that all *appliances* were operated simultaneously. An allowance (in equivalent length of pipe) as determined from Table AA102.2 should be considered for piping segments that include four or more fittings.
3. Determine the distance from the *outlet* of the gas meter to the *appliance* farthest removed from the meter.
4. Using the longest distance (found in Step 3), size each *piping* segment from the meter to the most remote *appliance outlet*.
5. For each of these *piping* segments, use the longest length and the calculated gas load for all of the connected *appliances* for the segment and begin the sizing process in Steps 6 through 8.
6. Referring to the appropriate sizing table (based on operating conditions and *piping* material), find the longest length distance in the first column or the next larger distance if the exact distance is not listed. The use of alternative operating pressures or pressure drops will require the use of a different sizing table, but will not alter the sizing methodology. In many cases, the use of alternative operating pressures or pressure drops will require the approval of both the code official and the local gas serving utility.
7. Trace across this row until the gas load is found or the closest larger capacity if the exact capacity is not listed.
8. Read up the table column and select the appropriate pipe size in the top row. Repeat Steps 6, 7 and 8 for each pipe segment in the longest run.
9. Size each remaining section of branch *piping* not previously sized by measuring the distance from the gas meter location to the most remote *outlet* in that branch, using the gas load of attached *appliances* and following the procedures of Steps 2 through 8.

AA103.3 Hybrid pressure method. The sizing of a 2 psi (13.8 kPa) gas *piping* system is performed using the traditional Longest Length Method but with modifications. The 2 psi (13.8 kPa) system consists of two independent pressure zones, and each zone is sized separately. The Hybrid Pressure Method is applied as follows:

The sizing of the 2 psi (13.8 kPa) section (from the meter to the line regulator) is as follows:

1. Calculate the gas load (by adding up the name plate ratings) from all connected *appliances*. (In certain circumstances the installed gas load can be increased up to 50 percent to accommodate future addition of *appliances*.) Ensure that the line regulator capacity is adequate for the calculated gas load and that the required pressure drop (across the regulator) for that capacity does not exceed $^3/_4$ psi (5.2 kPa) for a 2 psi (13.8 kPa) system. If the pressure drop across the regulator is too high (for the connected gas load), select a larger regulator.
2. Measure the distance from the meter to the line regulator located inside the building.
3. If there are multiple line regulators, measure the distance from the meter to the regulator farthest removed from the meter.
4. The maximum allowable pressure drop for the 2 psi (13.8 kPa) section is 1 psi (6.9 kPa).
5. Referring to the appropriate sizing table (based on piping material) for 2 psi (13.8 kPa) systems with a 1 psi

(6.9 kPa) pressure drop, find this distance in the first column, or the closest larger distance if the exact distance is not listed.

6. Trace across this row until the gas load is found or the closest larger capacity if the exact capacity is not listed.
7. Read up the table column to the top row and select the appropriate pipe size.
8. If there are multiple regulators in this portion of the *piping* system, each line segment must be sized for its actual gas load, but using the longest length previously determined in steps 2 and 3.

The low pressure section (all *piping* downstream of the line regulator) is sized as follows:

1. Determine the gas load for each of the connected *appliances*.
2. Starting from the line regulator, divide the piping system into a number of connected segments or independent parallel piping segments, and determine the amount of gas that each segment would carry assuming that all *appliances* were operated simultaneously. An allowance (in equivalent length of pipe) as determined from Table AA102.2 should be considered for piping segments that include four or more fittings.
3. For each piping segment, use the actual length or longest length (if there are sub-branchlines) and the calculated gas load for that segment and begin the sizing process as follows:
   a. Referring to the appropriate sizing table (based on operating pressure and piping material), find the longest length distance in the first column or the closest larger distance if the exact distance is not listed. The use of alternative operating pressures or pressure drops will require the use of a different sizing table, but will not alter the sizing methodology. In many cases, the use of alternative operating pressures or pressure drops can require the approval of the code official.
   b. Trace across this row until the *appliance* gas load is found or the closest larger capacity if the exact capacity is not listed.
   c. Read up the table column to the top row and select the appropriate pipe size.
   d. Repeat this process for each segment of the piping system.

AA103.4 Pressure drop per 100 feet method. This sizing method is less conservative than the others, but it allows the designer to immediately see where the largest pressure drop occurs in the system. With this information, modifications can be made to bring the total drop to the critical *appliance* within the limitations that are presented to the designer.

Follow the procedures described in the Longest Length Method for Steps 1 through 4 and 9.

For each *piping* segment, calculate the pressure drop based on pipe size, length as a percentage of 100 feet (30 480 mm) and gas flow. Table AA103.4 shows pressure drop per 100 feet (30 480 mm) for pipe sizes from $^1/_2$ inch (12.7 mm) through 2 inches (51 mm). The sum of pressure drops to the critical *appliance* is subtracted from the supply pressure to verify that sufficient pressure will be available. If not, the layout can be examined to find the high drop section(s) and sizing selections modified.

Note: Other values can be obtained by using the following equation:

$$\text{Desired Value} = MBH \times \sqrt{\frac{\text{Desired Drop}}{\text{Table Drop}}}$$

For example, if it is desired to get flow through $^3/_4$-inch (19.1 mm) pipe at 2 inches/100 feet, multiply the capacity of $^3/_4$-inch (19.1 mm) pipe at 1 inch/100 feet by the square root of the pressure ratio:

$$147\ MBH \times \sqrt{\frac{2''\ \text{w.c.}}{1''\ \text{w.c.}}} = 147 \times 1.414 = 208\ MBH$$

$$(MBH = 1000\ \text{Btu/h})$$

SECTION AA104 USE OF SIZING EQUATIONS

Capacities of smooth wall pipe or tubing can be determined by using the following formulae:

1. High Pressure [1.5 psi (10.3 kPa) and above]:

$$Q = 181.6\sqrt{\frac{D^5 \times (P_1^2 - P_2^2) \times Y}{C_r \times fba \times L}}$$

$$= 2237\ D^{2.623}\left[\frac{(P_1^2 - P_2^2) \times Y}{C_r \times L}\right]^{0.541}$$

TABLE AA103.4
THOUSANDS OF BTU/H (MBH) OF NATURAL GAS PER 100 FEET OF PIPE AT VARIOUS PRESSURE DROPS AND PIPE DIAMETERS

PRESSURE DROP PER 100 FEET IN INCHES W.C.	PIPE SIZES (inch)					
	$^1/_2$	$^3/_4$	1	$1^1/_4$	$1^1/_2$	2
0.2	31	64	121	248	372	716
0.3	38	79	148	304	455	877
0.5	50	104	195	400	600	1,160
1.0	71	147	276	566	848	1,640

For SI: 1 inch = 25.4 mm, 1 foot = 304.8 mm, 1 British thermal unit per hour = 0.293 W.

2. Low Pressure [Less than 1.5 psi (10.3 kPa)]:

$$Q = 187.3\sqrt{\frac{D^5 \times \Delta H}{C_r \times fba \times L}}$$

$$= 2313 D^{2.623}\left(\frac{\Delta H}{C_r \times L}\right)^{0.541}$$

where:

Q = Rate, cubic feet per hour at 60°F and 30-inch mercury column

D = Inside diameter of pipe, in.

P_1 = Upstream pressure, psia

P_2 = Downstream pressure, psia

Y = Superexpansibility factor = 1/supercompressibility factor

C_r = Factor for viscosity, density and temperature*

$= 0.00354\ ST\left(\frac{Z}{S}\right)^{0.152}$

*Note: See Table 402.4 for Y and C_r for natural gas and propane.

S = Specific gravity of gas at 60°F and 30-inch mercury column (0.60 for natural gas, 1.50 for propane), or = 1488μ

T = Absolute temperature, °F or = t + 460

t = Temperature, °F

Z = Viscosity of gas, centipoise (0.012 for natural gas, 0.008 for propane), or = 1488μ

fba = Base friction factor for air at 60°F (CF = 1)

L = Length of pipe, ft

ΔH = Pressure drop, in. w.c. (27.7 in. H_2O = 1 psi)

(For SI, see Section 402.4)

SECTION AA105 PIPE AND TUBE DIAMETERS

Where the internal diameter is determined by the formulas in Section 402.4, Tables AA105.1 and AA105.2 can be used to select the nominal or standard pipe size based on the calculated internal diameter.

SECTION AA106 EXAMPLES OF PIPING SYSTEM DESIGN AND SIZING

AA106.1 Example 1: Longest length method. Determine the required pipe size of each section and *outlet* of the *piping* system shown in Figure AA106.1, with a designated pressure drop of 0.5-inch w.c. (125 Pa) using the Longest Length Method. The gas to be used has 0.60 specific gravity and a heating value of 1,000 Btu/ft^3 (37.5 MJ/m^3).

Solution:

1. Maximum gas demand for *Outlet* A:

$$\frac{\text{Consumption (rating plate input) or Table 402.2 if necessary}}{\text{Btu of gas}} =$$

$$\frac{\text{35,000 Btu per hour rating}}{\text{1,000 Btu per cubic foot}} = \text{35 cubic feet per hour} = \text{35 cfh}$$

Maximum gas demand for *Outlet* B:

$$\frac{\text{Consumption}}{\text{Btu of gas}} = \frac{75{,}000}{1{,}000} = \text{75 cfh}$$

Maximum gas demand for *Outlet* C:

$$\frac{\text{Consumption}}{\text{Btu of gas}} = \frac{35{,}000}{1{,}000} = \text{35 cfh}$$

Maximum gas demand for *Outlet* D:

$$\frac{\text{Consumption}}{\text{Btu of gas}} = \frac{100{,}000}{1{,}000} = \text{100 cfh}$$

2. The length of pipe from the *point of delivery* to the most remote *Outlet* (A) is 60 feet (18 288 mm). This is the only distance used.
3. Using the row marked 60 feet (18 288 mm) in Table 402.4(2):
   a. *Outlet* A, supplying 35 cfh (0.99 m^3/hr), requires $^1/_2$-inch pipe.
   b. *Outlet* B, supplying 75 cfh (2.12 m^3/hr), requires $^3/_4$-inch pipe.
   c. Section 1, supplying *Outlets* A and B, or 110 cfh (3.11 m^3/hr), requires $^3/_4$-inch pipe.
   d. Section 2, supplying *Outlets* C and D, or 135 cfh (3.82 m^3/hr), requires $^3/_4$-inch pipe.
   e. Section 3, supplying *Outlets* A, B, C and D, or 245 cfh (6.94 m^3/hr), requires 1-inch pipe.
4. If a different gravity factor is applied to this example, the values in the row marked 60 feet (18 288 mm) of Table 402.4(2) would be multiplied by the appropriate multiplier from Table AA102.4 and the resulting cubic feet per hour values would be used to size the *piping*.

TABLE AA105.1
SCHEDULE 40 STEEL PIPE STANDARD SIZES

NOMINAL SIZE (inch)	INTERNAL DIAMETER (inch)
$^1/_4$	0.364
$^3/_8$	0.493
$^1/_2$	0.622
$^3/_4$	0.824
1	1.049
$1^1/_4$	1.380
$1^1/_2$	1.610
2	2.067
$2^1/_2$	2.469
3	3.068
$3^1/_2$	3.548
4	4.026

For SI: 1 inch = 25.4 mm.

TABLE AA105.2
COPPER TUBE STANDARD SIZES

TUBE TYPE	NOMINAL OR STANDARD SIZE (inches)	INTERNAL DIAMETER (inches)
K	$^1/_4$	0.305
L	$^1/_4$	0.315
ACR (D)	$^3/_8$	0.315
ACR (A)	$^3/_8$	0.311
K	$^3/_8$	0.402
L	$^3/_8$	0.430
ACR (D)	$^1/_2$	0.430
ACR (A)	$^1/_2$	0.436
K	$^1/_2$	0.527
L	$^1/_2$	0.545
ACR (D)	$^5/_8$	0.545
ACR (A)	$^5/_8$	0.555
K	$^5/_8$	0.652
L	$^5/_8$	0.666
ACR (D)	$^3/_4$	0.666
ACR (A)	$^3/_4$	0.680
K	$^3/_4$	0.745
L	$^3/_4$	0.785
ACR	$^7/_8$	0.785
K	1	0.995
L	1	1.025
ACR	$1^1/_8$	1.025
K	$1^1/_4$	1.245
L	$1^1/_4$	1.265
ACR	$1^3/_8$	1.265
K	$1^1/_2$	1.481
L	$1^1/_2$	1.505
ACR	$1^5/_8$	1.505
K	2	1.959
L	2	1.985
ACR	$2^1/_8$	1.985
K	$2^1/_2$	2.435
L	$2^1/_2$	2.465
ACR	$2^5/_8$	2.465
K	3	2.907
L	3	2.945
ACR	$3^1/_8$	2.945

For SI: 1 inch = 25.4 mm.

AA106.2 Example 2: Hybrid or dual pressure systems. Determine the required CSST size of each section of the *piping* system shown in Figure AA106.2, with a designated pressure drop of 1 psi (6.9 kPa) for the 2 psi (13.8 kPa) section and 3-inch w.c. (0.75 kPa) pressure drop for the 13-inch w.c. (2.49 kPa) section. The gas to be used has 0.60 specific gravity and a heating value of 1,000 Btu/ft^3 (37.5 MJ/ m^3).

Solution:

1. Size 2 psi (13.8 kPa) line using Table 402.4(18).
2. Size 10-inch w.c. (2.5 kPa) lines using Table 402.4(16).
3. Using the following, determine if sizing tables can be used.
   a. Total gas load shown in Figure AA106.2 equals 110 cfh (3.11 m^3/hr).
   b. Determine pressure drop across regulator [see notes in Table 402.4(18)].
   c. If pressure drop across regulator exceeds $^3/_4$ psig (5.2 kPa), Table 402.4(18) cannot be used. Note: If pressure drop exceeds $^3/_4$ psi (5.2 kPa), then a larger regulator must be selected or an alternative sizing method must be used.

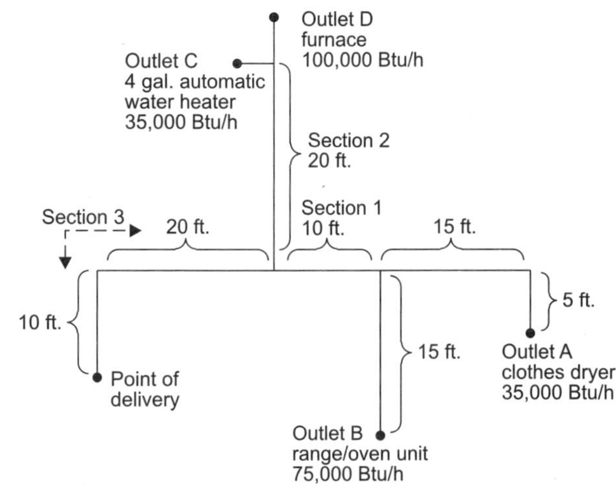

For SI: 1 foot = 304.8 mm, 1 gallon = 3.79 liters, 1 Btu = 1055 J.

FIGURE AA106.1
PIPING PLAN SHOWING A STEEL PIPING SYSTEM

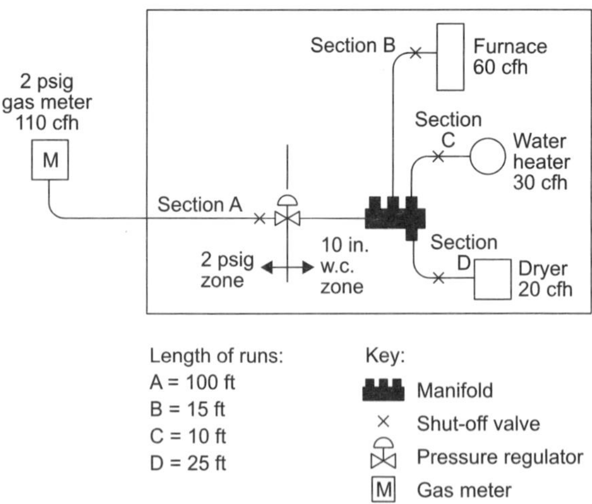

For SI: 1 foot = 304.8 mm, 1 cubic foot per hour = 0.028 m^3/hr.

FIGURE AA106.2
PIPING PLAN SHOWING A CSST SYSTEM

d. Pressure drop across the line regulator [for 110 cfh (3.11 m^3/hr)] is 4-inch w.c. (0.99 kPa) based on manufacturer's performance data.

e. Assume the CSST manufacturer has tubing sizes or EHDs of 13, 18, 23 and 30.

4. Section A [2 psi (13.8 kPa) zone]
   a. Distance from meter to regulator = 100 feet (30 480 mm).
   b. Total load supplied by A = 110 cfh (3.11 m^3/hr) (furnace + water heater + dryer).
   c. Table 402.4(18) shows that EHD size 18 should be used.

   Note: It is not unusual to oversize the supply line by 25 to 50 percent of the as-installed load. EHD size 18 has a capacity of 189 cfh (5.35 m^3/hr).

5. Section B (low pressure zone)
   a. Distance from regulator to furnace is 15 feet (4572 mm).
   b. Load is 60 cfh (1.70 m^3/hr).
   c. Table 402.4(16) shows that EHD size 13 should be used.

6. Section C (low pressure zone)
   a. Distance from regulator to water heater is 10 feet (3048 mm).
   b. Load is 30 cfh (0.85 m^3/hr).
   c. Table 402.4(16) shows that EHD size 13 should be used.

7. Section D (low pressure zone)
   a. Distance from regulator to dryer is 25 feet (7620 mm).
   b. Load is 20 cfh (0.57 m^3/hr).
   c. Table 402.4(16) shows that EHD size 13 should be used.

AA106.3 Example 3: Branch length method. Determine the required semirigid copper tubing size of each section of the *piping* system shown in Figure AA106.3, with a designated pressure drop of 1-inch w.c. (250 Pa) (using the Branch Length Method). The gas to be used has 0.60 specific gravity and a heating value of 1,000 Btu/ft^3 (37.5 MJ/m^3).

Solution:

1. Section A
   a. The length of tubing from the *point of delivery* to the most remote *appliance* is 50 feet (15 240 mm), A + C.
   b. Use this longest length to size Sections A and C.
   c. Using the row marked 50 feet (15 240 mm) in Table 402.4(10), Section A, supplying 220 cfh (6.2 m^3/hr) for four *appliances* requires 1-inch tubing.

2. Section B
   a. The length of tubing from the *point of delivery* to the range/oven at the end of Section B is 30 feet (9144 mm), A + B.
   b. Use this branch length to size Section B only.
   c. Using the row marked 30 feet (9144 mm) in Table 402.4(10), Section B, supplying 75 cfh (2.12 m^3/hr) for the range/oven requires $^1/_2$-inch tubing.

3. Section C
   a. The length of tubing from the *point of delivery* to the dryer at the end of Section C is 50 feet (15 240 mm), A + C.
   b. Use this branch length to size Section C.
   c. Using the row marked 50 feet (15 240 mm) in Table 402.4(10), Section C, supplying 30 cfh (0.85 m^3/hr) for the dryer requires $^3/_8$-inch tubing.

4. Section D
   a. The length of tubing from the *point of delivery* to the water heater at the end of Section D is 30 feet (9144 mm), A + D.
   b. Use this branch length to size Section D only.
   c. Using the row marked 30 feet (9144 mm) in Table 402.4(10), Section D, supplying 35 cfh (0.99 m^3/hr) for the water heater requires $^3/_8$-inch tubing.

5. Section E
   a. The length of tubing from the *point of delivery* to the furnace at the end of Section E is 30 feet (9144 mm), A + E.
   b. Use this branch length to size Section E only.

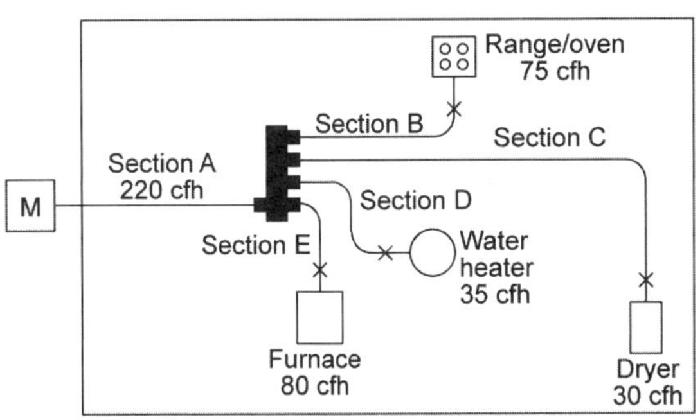

For SI: 1 foot = 304.8 mm, 1 cubic foot per hour = 0.028 m^3/hr.

FIGURE AA106.3
PIPING PLAN SHOWING A COPPER TUBING SYSTEM

c. Using the row marked 30 feet (9144 mm) in Table 402.4(10), Section E, supplying 80 cfh (2.26 m^3/hr) for the furnace requires $^1/_2$-inch tubing.

AA106.4 Example 4: Modification to existing piping system. Determine the required CSST size for Section G (retrofit application) of the *piping* system shown in Figure AA106.4, with a designated pressure drop of 0.5-inch w.c. (125 Pa) using the branch length method. The gas to be used has 0.60 specific gravity and a heating value of 1,000 Btu/ft^3 (37.5 MJ/m^3).

Solution:

1. The length of pipe and CSST from the *point of delivery* to the retrofit *appliance* (barbecue) at the end of Section G is 40 feet (12 192 mm), A + B + G.
2. Use this branch length to size Section G.
3. Assume the CSST manufacturer has tubing sizes or EHDs of 13, 18, 23 and 30.
4. Using the row marked 40 feet (12 192 mm) in Table 402.4(15), Section G, supplying 40 cfh (1.13 m^3/hr) for the barbecue requires EHD 18 CSST.
5. The sizing of Sections A, B, F and E must be checked to ensure adequate gas carrying capacity since an *appliance* has been added to the *piping* system (see Section AA106.1 for details).

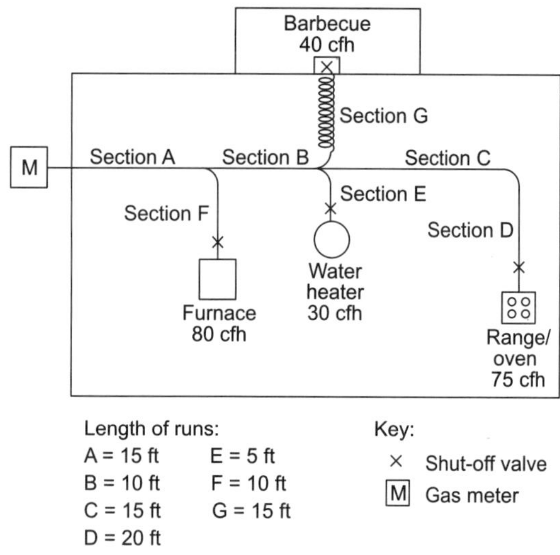

For SI: 1 foot = 304.8 mm, 1 cubic foot per hour = 0.028 m^3/hr.

FIGURE AA106.4
PIPING PLAN SHOWING A
MODIFICATION TO EXISTING PIPING SYSTEM

AA106.5 Example 5: Calculating pressure drops due to temperature changes. A test *piping* system is installed on a warm autumn afternoon when the temperature is 70°F (21°C). In accordance with local custom, the new *piping* system is subjected to an air pressure test at 20 psig (138 kPa). Overnight, the temperature drops and when the inspector shows up first thing in the morning the temperature is 40°F (4°C).

If the volume of the *piping* system is unchanged, then the formula based on Boyle's and Charles' law for determining the new pressure at a reduced temperature is as follows:

$$\frac{T_1}{T_2} = \frac{P_1}{P_2}$$

where:

T_1 = Initial temperature, absolute (T_1 + 459)

T_2 = Final temperature, absolute (T_2 + 459)

P_1 = Initial pressure, psia (P_1 + 14.7)

P_2 = Final pressure, psia (P_2 + 14.7)

$$\frac{(70+459)}{(40+459)} = \frac{(20+14.7)}{(P_2+14.7)}$$

$$\frac{529}{499} = \frac{34.7}{(P_2+14.7)}$$

$$(P_2+14.7) \times \frac{529}{499} = 34.7$$

$$(P_2+14.7) \times \frac{34.7}{1.060}$$

P_2 = 32.7 - 14.7

P_2 = 18 psig

Therefore, the gauge could be expected to register 18 psig (124 kPa) when the ambient temperature is 40°F (4°C).

AA106.6 Example 6: Pressure drop per 100 feet of pipe method. Using the layout shown in Figure AA106.1 and ΔH = pressure drop, in w.c. (27.7 in. H_2O = 1 psi), proceed as follows:

1. Length to A = 20 feet, with 35,000 Btu/hr.

   For $^1/_2$-inch pipe, $\Delta H = {}^{20\text{ feet}}/_{100\text{ feet}}$ × 0.3 inch w.c. = 0.06 in w.c.

2. Length to B = 15 feet, with 75,000 Btu/hr.

   For $^3/_4$-inch pipe, $\Delta H = {}^{15\text{ feet}}/_{100\text{ feet}}$ × 0.3 inch w.c. = 0.045 in w.c.

3. Section 1 = 10 feet, with 110,000 Btu/hr. Here there is a choice:

   For 1-inch pipe: $\Delta H = {}^{10\text{ feet}}/_{100\text{ feet}}$ × 0.2 inch w.c. = 0.02 in w.c.

   For $^3/_4$-inch pipe: $\Delta H = {}^{10\text{ feet}}/_{100\text{ feet}}$ × [0.5 inch w.c. + ${}^{(110{,}000\text{ Btu/hr}-104{,}000\text{ Btu/hr})}/_{(147{,}000\text{ Btu/hr}-104{,}000\text{ Btu/hr})}$ × (1.0 inches w.c. - 0.5 inch w.c.)] = 0.1 × 0.57 inch w.c. ≈ 0.06 inch w.c.

   Note that the pressure drop between 104,000 Btu/hr and 147,000 Btu/hr has been interpolated as 110,000 Btu/hr.

4. Section 2 = 20 feet, with 135,000 Btu/hr. Here there is a choice:

   For 1-inch pipe: $\Delta H = {}^{20\text{ feet}}/_{100\text{ feet}} \times$ [0.2 inch w.c. + ${}^{(14{,}000\text{ Btu/hr})}/_{(27{,}000\text{ Btu/hr})} \times$ 0.1 inch w.c.] = 0.05 inch w.c.

   For $^3/_4$-inch pipe: $\Delta H = {}^{20\text{ feet}}/_{100\text{ feet}} \times$ 1.0 inch w.c. = 0.2 inch w.c.

   Note that the pressure drop between 121,000 Btu/hr and 148,000 Btu/hr has been interpolated as 135,000 Btu/hr, but interpolation for the ¾-inch pipe (trivial for 104,000 Btu/hr to 147,000 Btu/hr) was not used.

5. Section 3 = 30 feet, with 245,000 Btu/hr. Here there is a choice:

   For 1-inch pipe: $\Delta H = {}^{30\text{ feet}}/_{100\text{ feet}} \times$ 1.0 inches w.c. = 0.3 inch w.c.

   For $1^1/_4$ -inch pipe: $\Delta H = {}^{30\text{ feet}}/_{100\text{ feet}} \times$ 0.2 inch w.c. = 0.06 inch w.c.

   Note that interpolation for these options is ignored since the table values are close to the 245,000 Btu/hr carried by that section.

6. The total pressure drop is the sum of the section approaching A, Sections 1 and 3, or either of the following, depending on whether an absolute minimum is needed or the larger drop can be accommodated.

   Minimum pressure drop to farthest *appliance*:

   ΔH = 0.06 inch w.c. + 0.02 inch w.c. + 0.06 inch w.c. = 0.14 inch w.c.

   Larger pressure drop to the farthest *appliance*:

   ΔH = 0.06 inch w.c. + 0.06 inch w.c. + 0.3 inch w.c. = 0.42 inch w.c.

   Notice that Section 2 and the run to B do not enter into this calculation, provided that the appliances have similar input pressure requirements.

   For SI units: 1 Btu/hr = 0.293 W, 1 cubic foot = 0.028 m^3, 1 foot = 0.305 m, 1 inch w.c. = 249 Pa.

APPENDIX AB

SIZING OF VENTING SYSTEMS SERVING APPLIANCES EQUIPPED WITH DRAFT HOODS, CATEGORY I APPLIANCES AND APPLIANCES LISTED FOR USE WITH TYPE B VENTS

This appendix is an excerpt from the 2021 International Fuel Gas Code® *informative Appendix B. Section and table references in the text, other than AB sections and tables, are as numbered in the* International Fuel Gas Code *(IFGC). For related table references in this code, you can find the IFGC table number in brackets adjacent to the table number in Chapter 24 of this code. For example, Table 504.2(2) in the IFGC is related to Table G2428.2(2) [504.2(2)] in this code.*

User note:

About this appendix: *Appendix AB provides commentary, guidance and examples for the design of venting systems for the types of appliances that vent by natural draft and have draft hoods or are listed as Category I or are listed for use with Type B vents.*

SECTION AB101 EXAMPLES USING SINGLE-APPLIANCE VENTING TABLES

AB101.1 Example 1: Single draft hood-equipped appliance. An installer has a 120,000 British thermal unit (Btu) per hour input *appliance* with a 5-inch-diameter draft hood outlet that needs to be vented into a 10-foot-high Type B vent system. What size vent should be used assuming:

1. A 5-foot lateral single-wall metal vent connector is used with two 90-degree elbows, or
2. A 5-foot lateral single-wall metal vent connector is used with three 90-degree elbows in the vent system.

Solution:

Table 504.2(2) of the *International Fuel Gas Code* should be used to solve this problem because single-wall metal vent connectors are being used with a Type B vent.

1. Read down the first column in Table 504.2(2) of the *International Fuel Gas Code* until the row associated with a 10-foot height and 5-foot lateral is found. Read across this row until a vent capacity greater than 120,000 Btu per hour is located in the shaded columns labeled "NAT Max" for draft hood-equipped appliances. In this case, a 5-inch-diameter vent has a capacity of 122,000 Btu per hour and can be used for this application.
2. If three 90-degree elbows are used in the vent system, then the maximum vent capacity listed in the tables must be reduced by 10 percent (see Section 504.2.3 of the *International Fuel Gas Code* for single-appliance vents). This implies that the 5-inch-diameter vent has an adjusted capacity of only 110,000 Btu per hour. In this case, the vent system must be increased to 6 inches in diameter (see the following calculations).

   122,000(0.90) = 110,000 for 5-inch vent

   From Table 504.2(2) of the *International Fuel Gas Code*, select 6-inch vent.

   186,000(0.90) = 167,000; This is greater than the required 120,000. Therefore, use a 6-inch vent and connector where three elbows are used.

See Figure AB101.1 for an example.

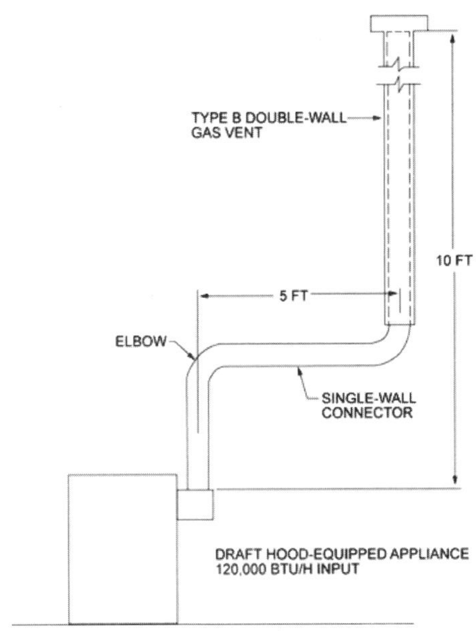

For SI: 1 foot = 304.8 mm, 1 British thermal unit per hour = 0.2931 W.

FIGURE AB101.1 EXAMPLE 1: SINGLE DRAFT HOOD-EQUIPPED APPLIANCE

AB101.2 Example 2: Single fan-assisted appliance. An installer has an 80,000 Btu per hour input fan-assisted *appliance* that must be installed using 10 feet of lateral connector attached to a 30-foot-high Type B vent. Two 90-degree elbows are needed for the installation. Can a single-wall metal vent connector be used for this application?

Solution:

Table 504.2(2) of the *International Fuel Gas Code* refers to the use of single-wall metal vent connectors with Type B vent. In the first column find the row associated with a 30-foot height and a 10-foot lateral. Read across this row, looking at the FAN Min and FAN Max columns, to find that a 3-inch-diameter single-wall metal vent connector is not recommended. Moving to the next larger size single wall connector (4 inches), note that a 4-inch-diameter single-wall metal connector has a recommended minimum vent capacity of 91,000 Btu per hour and a recommended maximum vent capacity of 144,000 Btu per hour. The 80,000 Btu per hour fan-assisted *appliance* is outside this range, so the conclusion is that a single-wall metal vent connector cannot be used to vent this *appliance* using 10 feet of lateral for the connector.

However, if the 80,000 Btu per hour input *appliance* could be moved to within 5 feet of the vertical vent, then a 4-inch single-wall metal connector could be used to vent the *appliance.* Table 504.2(2) of the *International Fuel Gas Code* shows the acceptable range of vent capacities for a 4-inch vent with 5 feet of lateral to be between 72,000 Btu per hour and 157,000 Btu per hour.

If the *appliance* cannot be moved closer to the vertical vent, then Type B vent could be used as the connector material. In this case, Table 504.2(1) of the *International Fuel Gas Code* shows that for a 30-foot-high vent with 10 feet of lateral, the acceptable range of vent capacities for a 4-inch-diameter vent attached to a fan-assisted *appliance* is between 37,000 Btu per hour and 150,000 Btu per hour.

See Figure AB101.2 for an example.

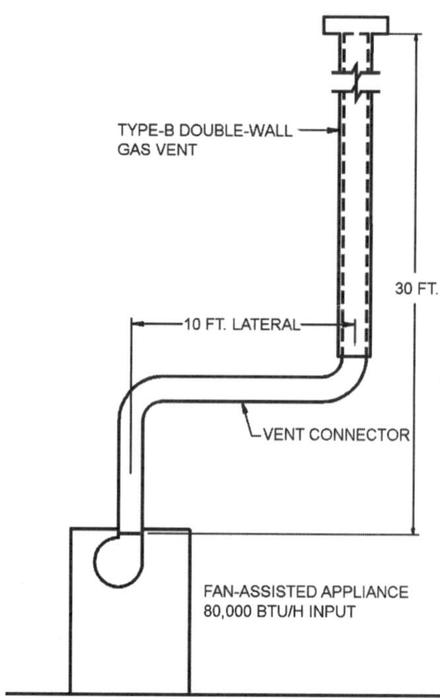

For SI: 1 foot = 304.8 mm, 1 British thermal unit per hour = 0.2931 W.

FIGURE AB101.2
EXAMPLE 2: SINGLE FAN-ASSISTED APPLIANCE

AB101.3 Example 3: Interpolating between table values. An installer has an 80,000 Btu per hour input *appliance* with a 4-inch-diameter draft hood outlet that needs to be vented into a 12-foot-high Type B vent. The vent connector has a 5-foot lateral length and is also Type B. Can this *appliance* be vented using a 4-inch-diameter vent?

Solution:

Table 504.2(1) of the *International Fuel Gas Code* is used in the case of an all Type B vent system. However, since there is no entry in Table 504.2(1) of the *International Fuel Gas Code* for a height of 12 feet, interpolation must be used. Read down the 4-inch diameter NAT Max column to the row associated with 10-foot height and 5-foot lateral to find the capacity value of 77,000 Btu per hour. Read further down to the 15-foot height, 5-foot lateral row to find the capacity value of 87,000 Btu per hour. The difference between the 15-foot height capacity value and the 10-foot height capacity value is 10,000 Btu per hour. The capacity for a vent system with a 12-foot height is equal to the capacity for a 10-foot height plus $^2/_5$ of the difference between the 10-foot and 15-foot height values, or 77,000 + $^2/_5$(10,000) = 81,000 Btu per hour. Therefore, a 4-inch-diameter vent can be used in the installation.

AB101.4 Figures. See Figures AB101.4(1) through AB101.4(5) for examples of single-appliance venting.

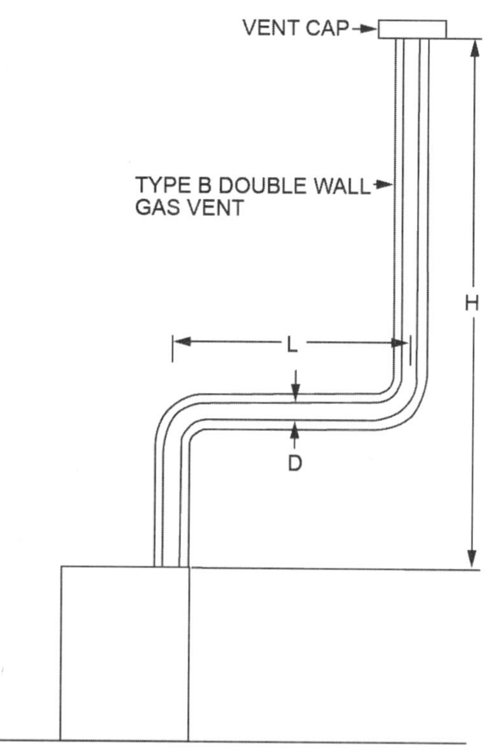

Table 504.2(1) of the *International Fuel Gas Code* is used where sizing Type B double-wall gas vent connected directly to the appliance.

Note: The appliance may be either Category I draft hood-equipped or fan-assisted type.

FIGURE AB101.4(1)
TYPE B DOUBLE-WALL VENT SYSTEM SERVING A SINGLE APPLIANCE WITH A TYPE B DOUBLE-WALL VENT

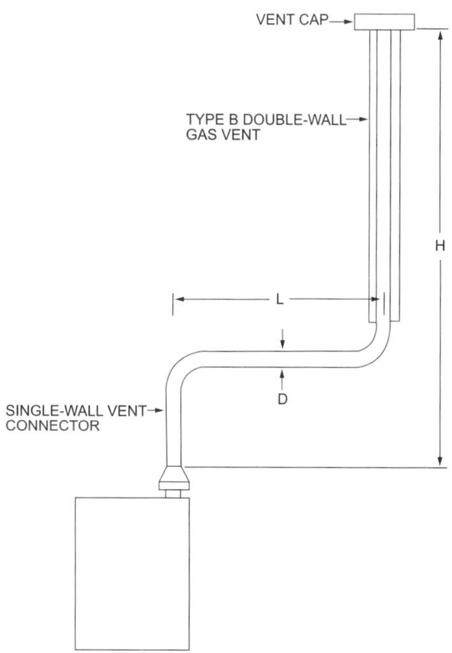

Table 504.2(2) of the *International Fuel Gas Code* is used where sizing a single-wall metal vent connector attached to a Type B double-wall gas vent.

Note: The appliance may be either Category I draft hood-equipped or fan-assisted type.

**FIGURE AB101.4(2)
TYPE B DOUBLE-WALL VENT SYSTEM
SERVING A SINGLE APPLIANCE WITH
A SINGLE-WALL METAL VENT CONNECTOR**

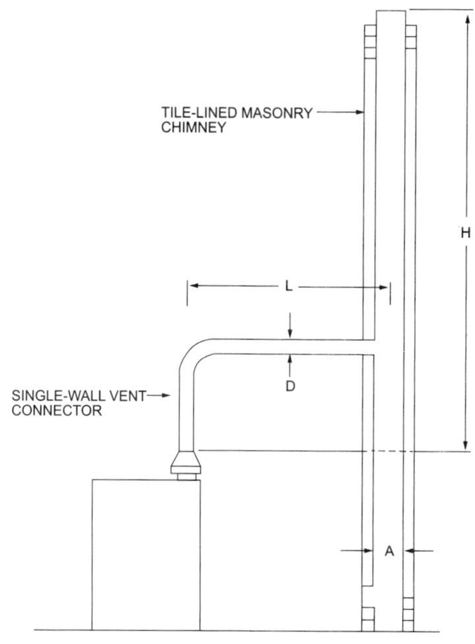

Table 504.2(4) of the *International Fuel Gas Code* is used where sizing a single-wall vent connector attached to a tile-lined masonry chimney.

Note: "A" is the equivalent cross-sectional area of the tile liner.

Note: The appliance can be either Category I draft hood-equipped or fan-assisted type.

**FIGURE AB101.4(4)
VENT SYSTEM SERVING A
SINGLE APPLIANCE USING A MASONRY CHIMNEY
AND A SINGLE-WALL METAL VENT CONNECTOR**

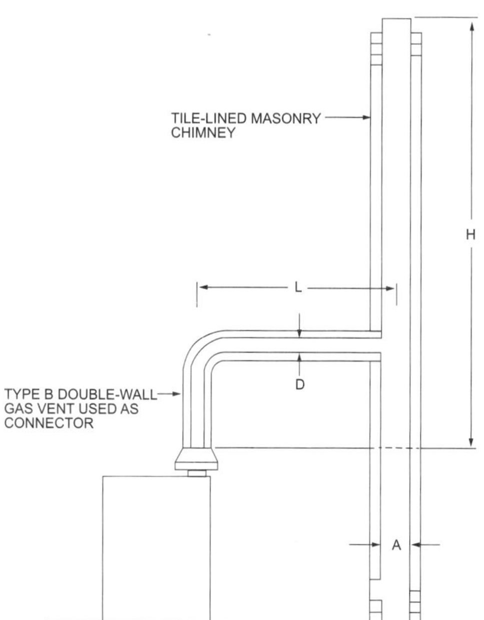

Table 504.2(3) of the *International Fuel Gas Code* is used where sizing a Type B double-wall gas vent connector attached to a tile-lined masonry chimney.

Note: "A" is the equivalent cross-sectional area of the tile liner.

Note: The appliance can be either Category I draft hood-equipped or fan-assisted type.

**FIGURE AB101.4(3)
VENT SYSTEM SERVING A SINGLE
APPLIANCE WITH A MASONRY CHIMNEY OF
TYPE B DOUBLE-WALL VENT CONNECTOR**

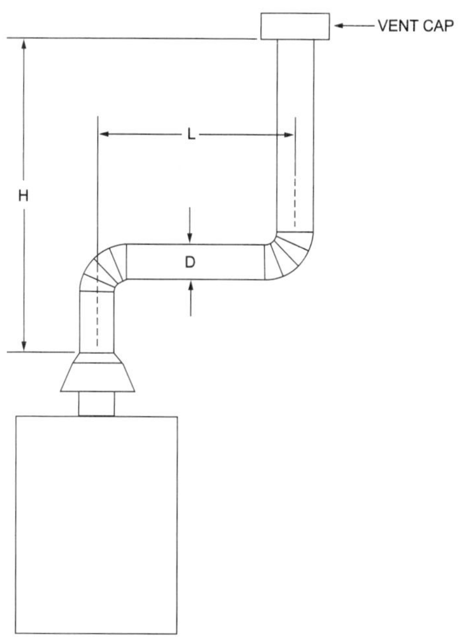

Asbestos cement Type B or single-wall metal vent serving a single draft hood-equipped appliance [see Table 504.2(5) of the *International Fuel Gas Code*].

**FIGURE AB101.4(5)
ASBESTOS CEMENT TYPE B OR
SINGLE-WALL METAL VENT SYSTEM SERVING
A SINGLE DRAFT HOOD-EQUIPPED APPLIANCE**

SECTION AB102 EXAMPLES USING COMMON VENTING TABLES

AB102.1 Example 4: Common venting two draft hood-equipped appliances. A 35,000 Btu per hour water heater is to be common vented with a 150,000 Btu per hour furnace using a common vent with a total height of 30 feet. The connector rise is 2 feet for the water heater with a horizontal length of 4 feet. The connector rise for the furnace is 3 feet with a horizontal length of 8 feet. Assume single-wall metal connectors will be used with Type B vent. What size connectors and combined vent should be used in this installation?

Solution:

Table 504.3(2) of the *International Fuel Gas Code* should be used to size single-wall metal vent connectors attached to Type B vertical vents. In the vent connector capacity portion of Table 504.3(2) of the *International Fuel Gas Code*, find the row associated with a 30-foot vent height. For a 2-foot rise on the vent connector for the water heater, read the shaded columns for draft hood-equipped appliances to find that a 3-inch-diameter vent connector has a capacity of 37,000 Btu per hour. Therefore, a 3-inch single-wall metal vent connector can be used with the water heater. For a draft hood-equipped furnace with a 3-foot rise, read across the appropriate row to find that a 5-inch-diameter vent connector has a maximum capacity of 120,000 Btu per hour (which is too small for the furnace) and a 6-inch-diameter vent connector has a maximum vent capacity of 172,000 Btu per hour. Therefore, a 6-inch-diameter vent connector should be used with the 150,000 Btu per hour furnace. Since both vent connector horizontal lengths are less than the maximum lengths listed in Section 504.3.2 of the *International Fuel Gas Code*, the table values can be used without adjustments.

In the common vent capacity portion of Table 504.3(2) of the *International Fuel Gas Code*, find the row associated with a 30-foot vent height and read over to the NAT + NAT portion of the 6-inch-diameter column to find a maximum combined capacity of 257,000 Btu per hour. Since the two appliances total only 185,000 Btu per hour, a 6-inch common vent can be used.

See Figure AB102.1 for an example.

AB102.2 Example 5a: Common venting a draft hood-equipped water heater with a fan-assisted furnace into a Type B vent. In this case, a 35,000 Btu per hour input draft hood-equipped water heater with a 4-inch-diameter draft hood *outlet*, 2 feet of connector rise, and 4 feet of horizontal length is to be common vented with a 100,000 Btu per hour fan-assisted furnace with a 4-inch-diameter flue collar, 3 feet of connector rise, and 6 feet of horizontal length. The common vent consists of a 30-foot height of Type B vent. What are the recommended vent diameters for each connector and the common vent? The installer would like to use a single-wall metal vent connector.

Solution: [Table 504.3(2) of the *International Fuel Gas Code*].

Water Heater Vent Connector Diameter. Since the water heater vent connector horizontal length of 4 feet is less than the maximum value listed in Section 504.3.2 of the *International Fuel Gas Code*, the venting table values can be used without adjustments. Using the Vent Connector Capacity portion of Table 504.3(2), read down the Total Vent Height (*H*) column to 30 feet and read across the 2-foot Connector Rise (*R*) row to the first Btu per hour rating in the NAT Max column that is equal to or greater than the water heater input rating. The table shows that a 3-inch vent connector has a maximum input rating of 37,000 Btu per hour. Although this is greater than the water heater input rating, a 3-inch vent connector is prohibited by Section 504.3.21 of the *International Fuel Gas Code*. A 4-inch vent connector has a maximum input rating of not more than 67,000 Btu per hour and is equal to the draft hood *outlet* diameter. A 4-inch vent connector is selected. Since the water heater is equipped with a draft hood, there are no minimum input rating restrictions.

Furnace Vent Connector Diameter. Using the Vent Connector Capacity portion of Table 504.3(2) of the *International Fuel Gas Code*, read down the Total Vent Height (*H*) column to 30 feet and across the 3-foot Connector Rise (*R*) row. Since the furnace has a fan-assisted combustion system, find the first FAN Max column with a Btu per hour rating greater than the furnace input rating. The 4-inch vent connector has a maximum input rating of 119,000 Btu per hour and a minimum input rating of 85,000 Btu per hour. The 100,000 Btu per hour furnace in this example falls within this range, so a 4-inch connector is adequate. Since the furnace vent connector horizontal length of 6 feet does not exceed the maximum value listed in Section 504.3.2 of the *International Fuel Gas Code*, the venting table values can be used without adjustment. If the furnace had an input rating of 80,000 Btu per hour, then a Type B vent connector [see Table 504.3(1) of the *International Fuel Gas Code*] would be needed in order to meet the minimum capacity limit.

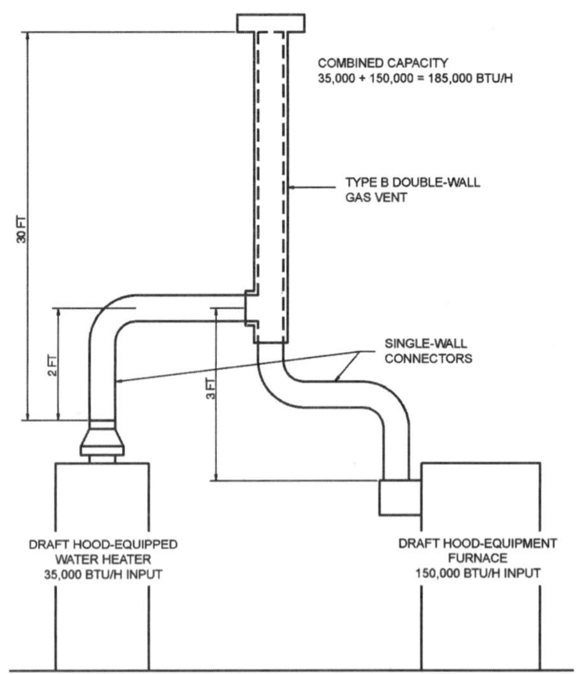

For SI: 1 British thermal unit per hour = 0.2931 W.

FIGURE AB102.1 EXAMPLE 4: COMMON VENTING TWO DRAFT HOOD-EQUIPPED APPLIANCES

Common Vent Diameter. The total input to the common vent is 135,000 Btu per hour. Using the Common Vent Capacity portion of Table 504.3(2) of the *International Fuel Gas Code*, read down the Total Vent Height (*H*) column to 30 feet and across this row to find the smallest vent diameter in the FAN + NAT column that has a Btu per hour rating equal to or greater than 135,000 Btu per hour. The 4-inch common vent has a capacity of 132,000 Btu per hour and the 5-inch common vent has a capacity of 202,000 Btu per hour. Therefore, the 5-inch common vent should be used in this example.

Summary. In this example, the installer can use a 4-inch-diameter, single-wall metal vent connector for the water heater and a 4-inch-diameter, single-wall metal vent connector for the furnace. The common vent should be a 5-inch-diameter Type B vent.

See Figure AB102.2 for an example.

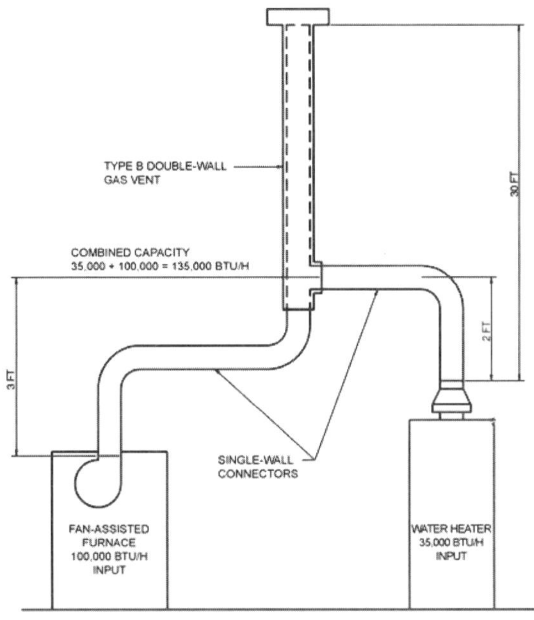

For SI: British thermal unit per hour = 0.2931 W.

FIGURE AB102.2 EXAMPLE 5a: COMMON VENTING A DRAFT HOOD WITH A FAN-ASSISTED FURNACE INTO A TYPE B DOUBLE-WALL COMMON VENT

AB102.3 Example 5b: Common venting into a masonry chimney. In this case, the water heater and fan-assisted furnace of Example 5a are to be common vented into a clay tile-lined *masonry chimney* with a 30-foot height. The chimney is not exposed to the outdoors below the roof line. The internal dimensions of the clay tile liner are nominally 8 inches by 12 inches. Assuming the same vent connector heights, laterals and materials found in Example 5a, what are the recommended vent connector diameters, and is this an acceptable installation?

Solution:

Table 504.3(4) of the *International Fuel Gas Code* is used to size common venting installations involving single-wall connectors into masonry chimneys. Water Heater Vent Connector Diameter. Using Table 504.3(4) of the *International Fuel Gas Code*, Vent Connector Capacity, read down the Total Vent Height (H) column to 30 feet, and read across the 2-foot Connector Rise (R) row to the first Btu-per-hour rating in the NAT Max column that is equal to or greater than the water heater input rating. The table shows that a 3-inch vent connector has a maximum input of only 31,000 Btu per hour while a 4-inch vent connector has a maximum input of 57,000 Btu per hour. A 4-inch vent connector must therefore be used.

Furnace Vent Connector Diameter. Using the Vent Connector Capacity portion of Table 504.3(4) of the *International Fuel Gas Code*, read down the Total Vent Height (H) column to 30 feet and across the 3-foot Connector Rise (R) row. Since the furnace has a fan-assisted combustion system, find the first FAN Max column with a Btu-per-hour rating greater than the furnace input rating. The 4-inch vent connector has a maximum input rating of 127,000 Btu per hour and a minimum input rating of 95,000 Btu per hour. The 100,000 Btu-per-hour furnace in this example falls within this range, so a 4-inch connector is adequate.

Masonry Chimney. From Table AB102.3, the equivalent area for a nominal liner size of 8 inches by 12 inches is 63.6 square inches. Using Table 504.3(4) of the *International Fuel Gas Code*, Common Vent Capacity, read down the FAN + NAT column under the Minimum Internal Area of Chimney value of 63 to the row for 30-foot height to find a capacity value of 739,000 Btu per hour. The combined input rating of the furnace and water heater, 135,000 Btu per hour, is less than the table value, so this is an acceptable installation.

Section 504.3.17 of the *International Fuel Gas Code* requires the common vent area to be not greater than seven times the smallest *listed appliance* categorized vent area, flue collar area or draft hood outlet area. Both *appliances* in this installation have 4-inch-diameter outlets. From Table AB102.3, the equivalent area for an inside diameter of 4 inches is 12.2 square inches. Seven times 12.2 equals 85.4, which is greater than 63.6, so this configuration is acceptable.

AB102.4 Example 5c: Common venting into an exterior masonry chimney. In this case, the water heater and fan-assisted furnace of Examples 5a and 5b are to be common vented into an exterior *masonry chimney*. The chimney height, clay tile liner dimensions, and vent connector heights and laterals are the same as in Example 5b. This system is being installed in Charlotte, North Carolina. Does this exterior *masonry chimney* need to be relined? If so, what corrugated metallic liner size is recommended? What vent connector diameters are recommended?

Solution:

In accordance with Section 504.3.20 of the *International Fuel Gas Code*, Type B vent connectors are required to be used with exterior masonry chimneys. Use Tables 504.3(7a) and (7b) of the *International Fuel Gas Code* to size FAN+NAT common venting installations involving Type-B double wall connectors into exterior masonry chimneys.

TABLE AB102.3
MASONRY CHIMNEY LINER DIMENSIONS WITH CIRCULAR EQUIVALENTS[a]

NOMINAL LINER SIZE (inches)	INSIDE DIMENSIONS OF LINER (inches)	INSIDE DIAMETER OR EQUIVALENT DIAMETER (inches)	EQUIVALENT AREA (square inches)
4 × 8	$2^1/_2 \times 6^1/_2$	4	12.2
		5	19.6
		6	28.3
		7	38.3
8 × 8	$6^3/_4 \times 6^3/_4$	7.4	42.7
		8	50.3
8 × 12	$6^1/_2 \times 10^1/_2$	9	63.6
		10	78.5
12 × 12	$9^3/_4 \times 9^3/_4$	10.4	83.3
		11	95
12 × 16	$9^1/_2 \times 13^1/_2$	11.8	107.5
		12	113.0
		14	153.9
16 × 16	$13^1/_4 \times 13^1/_4$	14.5	162.9
		15	176.7
16 × 20	13 × 17	16.2	206.1
		18	254.4
20 × 20	$16^3/_4 \times 16^3/_4$	18.2	260.2
		20	314.1
20 × 24	$16^1/_2 \times 20^1/_2$	20.1	314.2
		22	380.1
24 × 24	$20^1/_4 \times 20^1/_4$	22.1	380.1
		24	452.3
24 × 28	$20^1/_4 \times 20^1/_4$	24.1	456.2
28 × 28	$24^1/_4 \times 24^1/_4$	26.4	543.3
		27	572.5
30 × 30	$25^1/_2 \times 25^1/_2$	27.9	607
		30	706.8
30 × 36	$25^1/_2 \times 31^1/_2$	30.9	749.9
		33	855.3
36 × 36	$31^1/_2 \times 31^1/_2$	34.4	929.4
		36	1017.9

For SI: 1 inch = 25.4 mm, 1 square inch = 645.16 m^2.

a. Where liner sizes differ dimensionally from those shown in Table AB102.3, equivalent diameters can be determined from published tables for square and rectangular ducts of equivalent carrying capacity or by other engineering methods.

The local 99-percent winter design temperature needed to use Table 504.3(7b) of the *International Fuel Gas Code* can be found in the ASHRAE *Handbook of Fundamentals.* For Charlotte, North Carolina, this design temperature is 19°F.

Chimney Liner Requirement. As in Example 5b, use the 63 square inch Internal Area columns for this size clay tile liner. Read down the 63 square inch column of Table 504.3(7a) of the *International Fuel Gas Code* to the 30-foot height row to find that the combined *appliance* maximum input is 747,000 Btu per hour. The combined input rating of the *appliances* in this installation, 135,000 Btu per hour, is less than the maximum value, so this criterion is satisfied. Table 504.3(7b) of the *International Fuel Gas Code,* at a 19°F design temperature, and at the same vent height and internal area used above, shows that the minimum allowable input rating of a space-heating *appliance* is 470,000 Btu per hour. The furnace input rating of 100,000 Btu per hour is less than this minimum value. So this criterion is not satisfied, and an alternative venting design needs to be used, such as a Type B vent shown in Example 5a or a *listed* chimney liner system shown in the remainder of the example.

In accordance with Section 504.3.19, Table 504.3(1) or Table 504.3(2) of the *International Fuel Gas Code* is used for sizing corrugated metallic liners in masonry chimneys, with the maximum common vent capacities reduced by 20 percent. This example will be continued assuming Type B vent connectors.

Water Heater Vent Connector Diameter. Using Table 504.3(1), Vent Connector Capacity, of the *International Fuel Gas Code* read down the Total Vent Height (*H*) column to 30 feet, and read across the 2-foot Connector Rise (*R*) row to the first Btu/h rating in the NAT Max column that is equal to or greater than the water heater input rating. The table shows that a 3-inch vent connector has a maximum capacity of 39,000 Btu/h. Although this rating is greater than the water heater input rating, a 3-inch vent connector is prohibited by Section 504.3.21 of the *International Fuel Gas Code.* A 4-inch vent connector has a maximum input rating of 70,000 Btu/h and is equal to the draft hood outlet diameter. A 4-inch vent connector is selected.

Furnace Vent Connector Diameter. Using Table 504.3(1), Vent Connector Capacity, of the *International Fuel Gas Code* read down the Vent Height (*H*) column to 30 feet, and read across the 3-foot Connector Rise (*R*) row to the first Btu per hour rating in the FAN Max column that is equal to or greater than the furnace input rating. The 100,000 Btu per hour furnace in this example falls within this range, so a 4-inch connector is adequate.

Chimney Liner Diameter. The total input to the common vent is 135,000 Btu per hour. Using the Common Vent Capacity portion of Table 504.3(1) of the *International Fuel Gas Code,* read down the Vent Height (*H*) column to 30 feet and across this row to find the smallest vent diameter in the FAN+NAT column that has a Btu per hour rating greater than 135,000 Btu per hour. The 4-inch common vent has a capacity of 138,000 Btu per hour. Reducing the maximum capacity by 20 percent (Section 504.3.19 of the *International Fuel Gas Code*) results in a maximum capacity for a 4-inch corrugated liner of 110,000 Btu per hour, less than the total input of 135,000 Btu per hour. So a larger liner is needed. The 5-inch common vent capacity *listed* in Table 504.3(1) of the *International Fuel Gas Code* is 210,000 Btu per hour, and after reducing by 20 percent is 168,000 Btu per hour. Therefore, a 5-inch corrugated metal liner should be used in this example.

Single-Wall Connectors. Once it has been established that relining the chimney is necessary, Type B double-wall vent connectors are not specifically required. This example could be redone using Table 504.3(2) of the *International Fuel Gas Code* for single-wall vent connectors. For this case, the vent connector and liner diameters would be the same as found above with Type B double-wall connectors.

AB102.5 Figures. See Figures AB102.5(1) through AB102.5(9) for examples of common venting. See Figure AB102.5(10) for the 99% Winter Design Temperatures for the Contiguous United States.

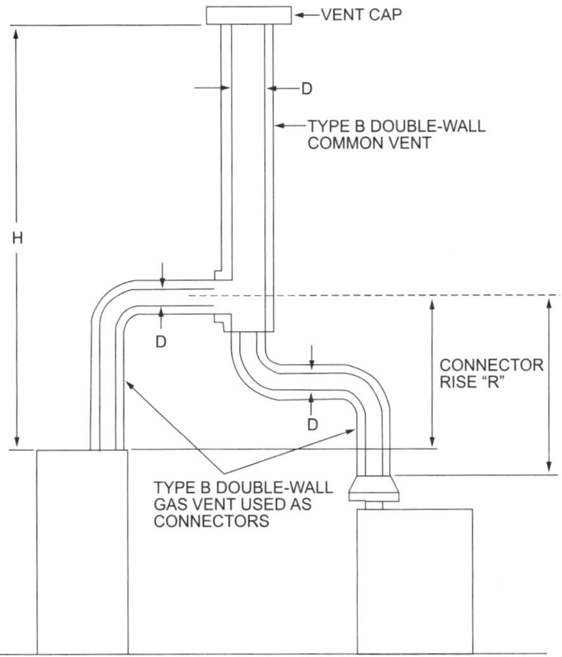

Table 504.3(1) of the *International Fuel Gas Code* is used where sizing Type B double-wall vent connectors attached to a Type B double-wall common vent.

Note: Each appliance can be either Category I draft hood-equipped or fan-assisted type.

**FIGURE AB102.5(1)
VENT SYSTEM SERVING TWO OR MORE
APPLIANCES WITH TYPE B DOUBLE-WALL VENT AND
TYPE B DOUBLE-WALL VENT CONNECTOR**

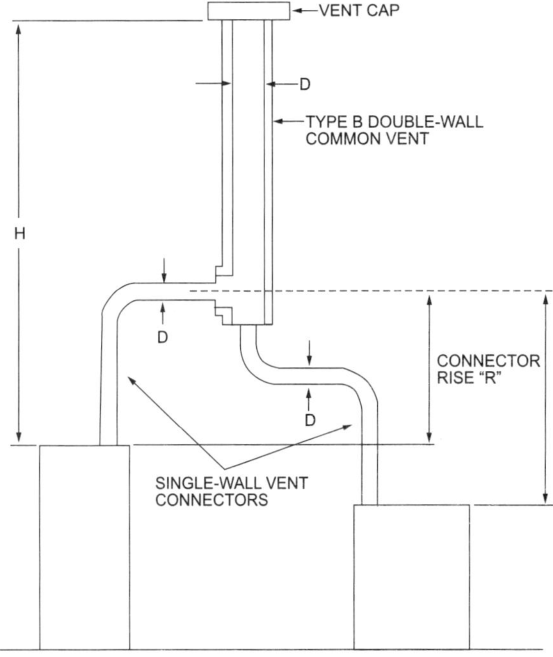

Table 504.3(2) of the *International Fuel Gas Code* is used where sizing single-wall vent connectors attached to a Type B double-wall common vent.

Note: Each appliance can be either Category I draft hood-equipped or fan-assisted type.

**FIGURE AB102.5(2)
VENT SYSTEM SERVING TWO OR MORE
APPLIANCES WITH TYPE B DOUBLE-WALL VENT AND
SINGLE-WALL METAL VENT CONNECTORS**

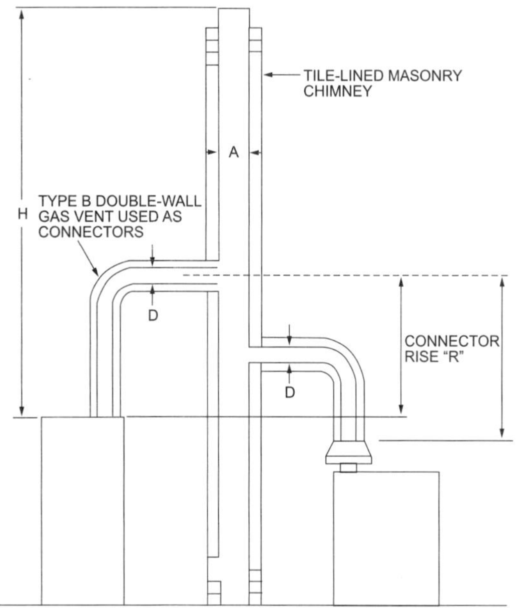

Table 504.3(3) of the *International Fuel Gas Code* is used where sizing Type B double-wall vent connectors attached to a tile-lined masonry chimney.

Note: "A" is the equivalent cross-sectional area of the tile liner.

Note: Each appliance can be either Category I draft hood-equipped or fan-assisted type.

**FIGURE AB102.5(3)
MASONRY CHIMNEY SERVING
TWO OR MORE APPLIANCES WITH
TYPE B DOUBLE-WALL VENT CONNECTOR**

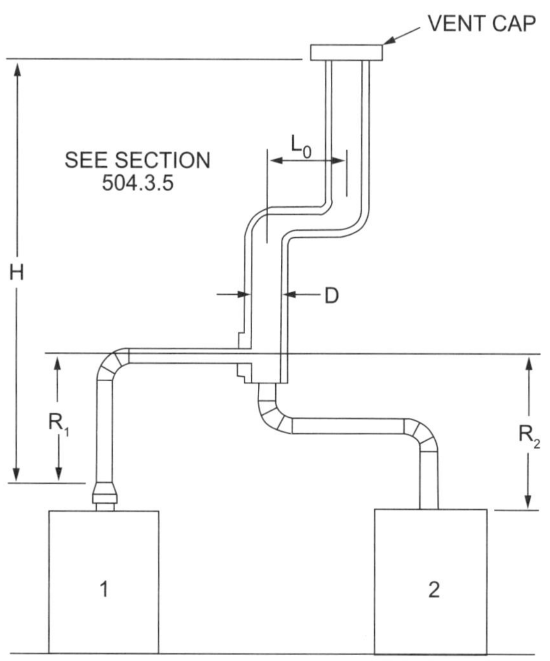

Asbestos cement Type B or single-wall metal pipe vent serving two or more draft hood-equipped appliances [see Table 504.3(5) of the *International Fuel Gas Code*].

**FIGURE AB102.5(5)
ASBESTOS CEMENT TYPE B OR
SINGLE-WALL METAL VENT SYSTEM SERVING TWO OR
MORE DRAFT HOOD-EQUIPPED APPLIANCES**

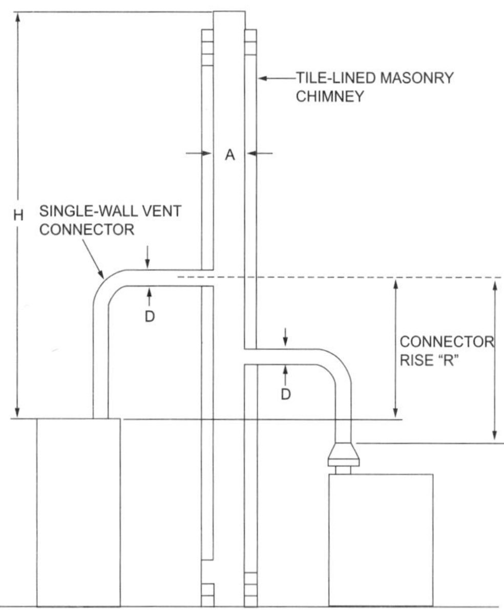

Table 504.3(4) of the *International Fuel Gas Code* is used where sizing single-wall metal vent connectors attached to a tile-lined masonry chimney.

Note: "A" is the equivalent cross-sectional area of the tile liner.

Note: Each appliance can be either Category I draft hood-equipped or fan-assisted type.

**FIGURE AB102.5(4)
MASONRY CHIMNEY SERVING TWO OR MORE APPLIANCES
WITH SINGLE-WALL METAL VENT CONNECTORS**

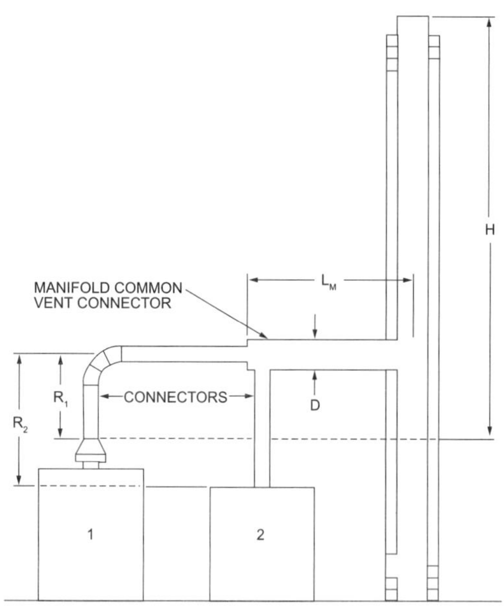

Example: Manifolded Common Vent Connector L_M shall be no greater than 18 times the common vent connector manifold inside diameter; i.e., a 4-inch (102 mm) inside diameter common vent connector manifold shall not exceed 72 inches (1829 mm) in length (see Section 504.3.4 of the *International Fuel Gas Code*).

Note: This is an illustration of a typical manifolded vent connector. Different appliance, vent connector or common vent types are possible. Consult Section 502.3 of the *International Fuel Gas Code*.

**FIGURE AB102.5(6)
USE OF MANIFOLD COMMON VENT CONNECTOR**

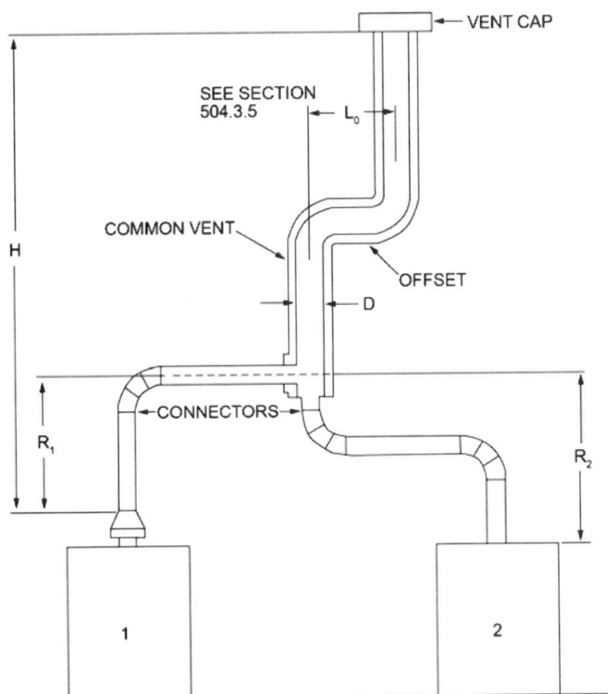

Example: Offset Common Vent

Note: This is an illustration of a typical offset vent. Different appliance, vent connector, or vent types are possible. Consult Sections 504.2 and 504.3 of the *International Fuel Gas Code.*

FIGURE AB102.5(7)
USE OF OFFSET COMMON VENTS WITH SINGLE-WALL METAL VENT CONNECTORS

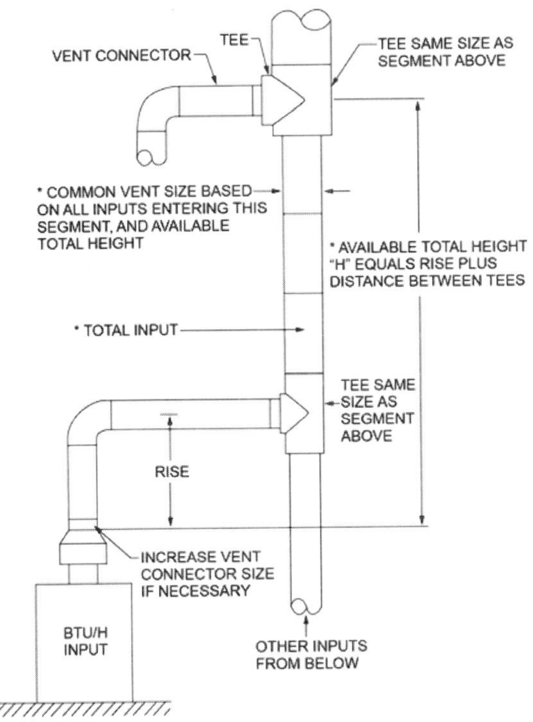

FIGURE AB102.5(8)
MULTISTORY GAS VENT DESIGN PROCEDURE FOR EACH SEGMENT OF SYSTEM

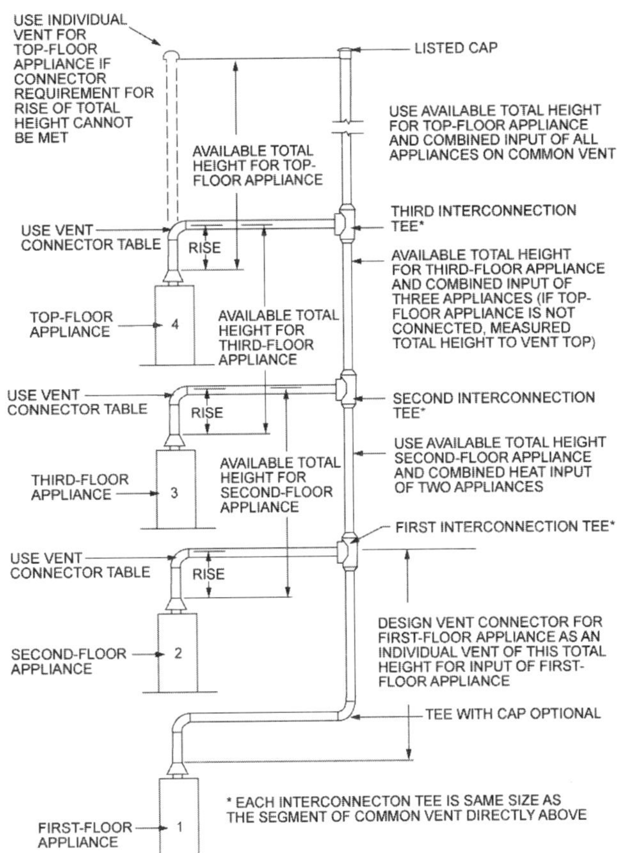

Principles of design of multistory vents using vent connector and common vent design tables (see Sections 504.3.11 through 504.3.17 of the *International Fuel Gas Code*).

FIGURE AB102.5(9)
MULTISTORY VENT SYSTEMS

FIGURE AB102.5(10)
99% WINTER DESIGN TEMPERATURES FOR THE CONTIGUOUS UNITED STATES

APPENDIX AC
EXIT TERMINALS OF MECHANICAL DRAFT AND DIRECT-VENT VENTING SYSTEMS

This appendix is informative and is not part of the code. This appendix is an excerpt from the 2018 International Fuel Gas Code®, *coordinated with the section numbering of the* International Residential Code.

User note:

About this appendix: *Appendix AC provides a graphic depiction of the venting terminal location requirements of the code.*

SECTION AC101 GENERAL

AC101.1 Exit terminal locations. Location requirements of exit terminals of mechanical draft and direct vent venting systems are provided in Figure AC101.1.

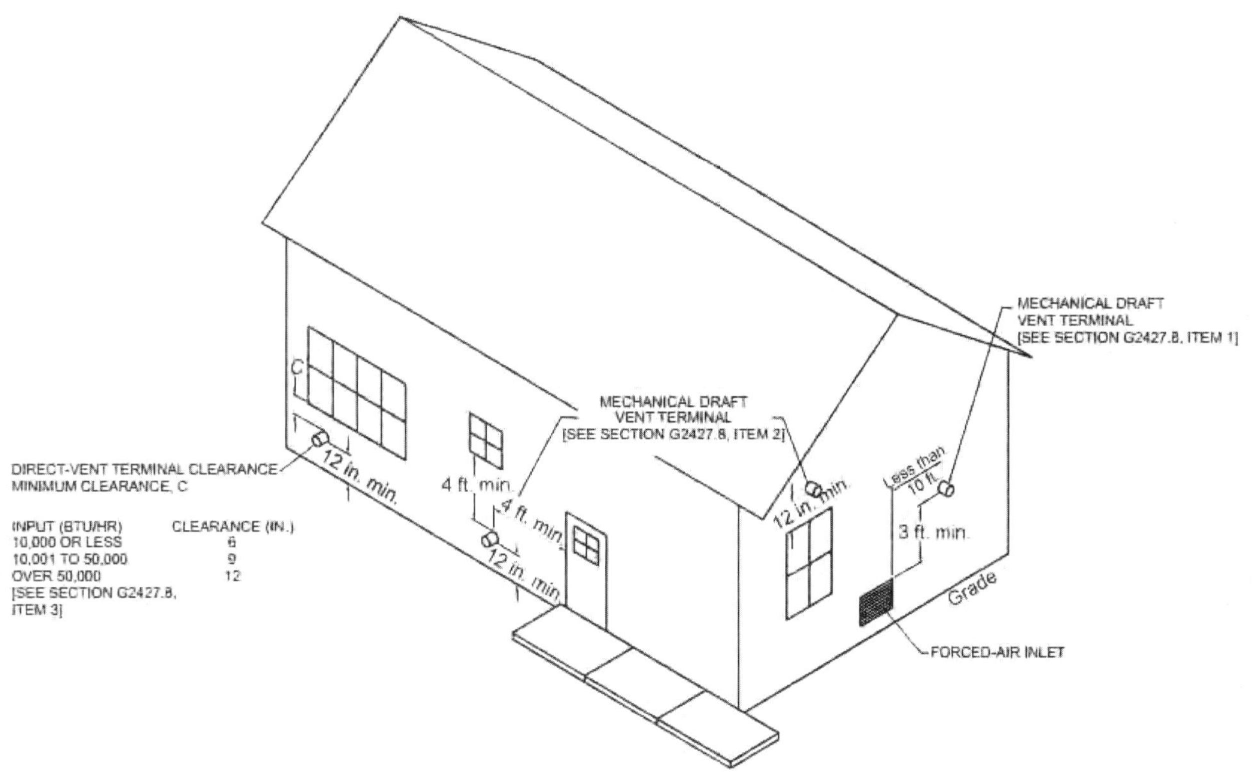

For SI: 1 inch = 25.4 mm, 1 foot = 304.8 mm, 1 British thermal unit per hour = 0.2931 W.

FIGURE AC101.1
EXIT TERMINALS OF MECHANICAL DRAFT AND DIRECT-VENT VENTING SYSTEMS

APPENDIX AD
RECOMMENDED PROCEDURE FOR SAFETY INSPECTION OF AN EXISTING APPLIANCE INSTALLATION

This appendix is an excerpt from the 2021 International Fuel Gas Code® *informative Appendix D, coordinated with the section numbering of the* International Residential Code.

User note:

About this appendix: Appendix AD provides procedures for testing and inspecting existing gas appliance installations for safe operation.

SECTION AD101
GENERAL

The following procedure is intended as a guide to aid in determining that an appliance is properly installed and is in a safe condition for continued use. Where a gas supplier performs an inspection, their written procedures should be followed.

AD101.1 Application. This procedure is intended for existing residential installations of a furnace, boiler, room heater, water heater, cooking appliance, fireplace appliance and clothes dryer. This procedure should be performed prior to any attempt to modify the appliance installation or building envelope.

AD101.2 Weatherization programs. Before a building envelope is to be modified as part of a weatherization program, the existing appliance installation should be inspected in accordance with these procedures. After all unsafe conditions are repaired, and immediately after the weatherization is complete, the appliance inspections in Section AD105.2 are to be repeated.

AD101.3 Inspection procedure. The safety of the building occupant and inspector are to be determined as the first step as described in Section AD102. Only after the ambient environment is found to be safe should inspections of gas piping and appliances be undertaken. It is recommended that all inspections described in Sections AD103, AD104, and AD106, where the appliance is in the off mode, be completed and any unsafe conditions repaired or corrected before continuing with inspections of an operating appliance described in Sections AD105 and AD106.

AD101.4 Manufacturer instructions. Where available, the manufacturer's installation and operating instructions for the installed appliances should be used as part of these inspection procedures to determine if it is installed correctly and is operating properly.

AD101.5 Instruments. The inspection procedures include measuring for fuel gas and carbon monoxide (CO) and will require the use of a combustible gas detector (CGD) and a CO detector. It is recommended that both types of detectors be *listed.* Prior to any inspection, the detectors should be calibrated or tested in accordance with the manufacturer's instructions. In addition, it is recommended that the detectors have the following minimum specifications.

1. Gas Detector: The CGD should be capable of indicating the presence of the type of fuel gas for which it is to be used, for example, natural gas or propane. The combustible gas detector should be capable of the following:
   a. *PPM:* Numeric display with a parts per million (ppm) scale from 1ppm to 900 ppm in 1 ppm increments.
   b. *LEL:* Numeric display with a percent lower explosive limit (% LEL) scale from 0 percent to 100 percent in 1 percent increments.
   c. *Audio:* An audio sound feature to locate leaks.
2. CO Detector: The CO detector should be capable of the following functions and have a numeric display scale as follows:
   a. *PPM:* For measuring ambient room and appliance emissions a display scale in parts per million (ppm) from 0 to 1,000 ppm in 1 ppm increments.
   b. *Alarm:* A sound alarm function where hazardous levels of ambient CO is found (see AD102 for alarm levels)
   c. *Air Free:* Capable of converting CO measurements to an air free level in ppm. Where a CO detector is used without an air free conversion function, the CO air free can be calculated in accordance with Note 3 in Table AD106.

SECTION AD102
OCCUPANT AND INSPECTOR SAFETY

Prior to entering a building, the inspector should have both a combustible gas detector (CGD) and CO detector turned on, calibrated, and operating. Immediately upon entering the building, a sample of the ambient atmosphere should be taken. Based on CGD and CO detector readings, the inspector should take the following actions:

1. The CO detector indicates a carbon monoxide level of 70 ppm or greater.[1] The inspector should immedi-

ately notify the occupant of the need for themselves and any building occupant to evacuate; the inspector shall immediately evacuate and call 911.

2. Where the CO detector indicates a reading between 30 ppm and 70 ppm.[1] The inspector should advise the occupant that high CO levels have been found and recommend that all possible sources of CO should be turned off immediately and windows and doors opened. Where it appears that the source of CO is a permanently installed appliance, advise the occupant to keep the appliance off and have the appliance serviced by a qualified servicing agent.
3. Where CO detector indicates CO below 30 ppm[1] the inspection can continue.
4. The CGD indicates a combustible gas level of 20-percent LEL or greater. The inspector should immediately notify the occupant of the need for themselves and any building occupant to evacuate; the inspector shall immediately evacuate and call 911.
5. The CGD indicates a combustible gas level below 20-percent LEL, the inspection can continue.

If during the inspection process it is determined a condition exists that could result in unsafe appliance operation, shut off the appliance and advise the *owner* of the unsafe condition. Where a gas leak is found that could result in an unsafe condition, advise the owner of the unsafe condition and call the gas supplier to turn off the gas supply. The inspector should not continue a safety inspection on an operating appliance, venting system, and piping system until repairs have been made.

SECTION AD103
GAS PIPING AND CONNECTIONS INSPECTIONS

1. *Leak Checks.* Conduct a test for gas leakage using either a noncorrosive leak detection solution or a CGD confirmed with a leak detection solution.

   The preferred method for leak checking is by use of gas leak detection solution applied to all joints. This method provides a reliable visual indication of significant leaks.

   The use of a CGD in its audio sensing mode can quickly locate suspect leaks but can be overly sensitive indicating insignificant and false leaks. Suspect leaks found through the use of a CGD should be confirmed using a leak detection solution.

   Where gas leakage is confirmed, the *owner* should be notified that repairs must be made. The inspection should include the following components:

   a. Gas piping fittings located within the appliance space.
   b. Appliance connector fittings.
   c. Appliance gas valve/regulator housing and connections.
2. *Appliance Connector.* Verify that the appliance connection type is compliant with Section G2422. Inspect flexible appliance connections to determine if they are free of cracks, corrosion and signs of damage. Verify that there are no uncoated brass connectors. Where connectors are determined to be unsafe or where an uncoated brass connector is found, the appliance shutoff valve should be placed in the off position and the *owner* notified that the connector must be replaced.
3. *Piping Support.* Inspect piping to determine that it is adequately supported, that there is no undue stress on the piping, and if there are any improperly capped pipe openings.
4. *Bonding.* Verify that the electrical bonding of gas piping is compliant with Section G2411.

SECTION AD104
INSPECTIONS TO BE PERFORMED WITH THE APPLIANCE NOT OPERATING

The following safety inspection procedures are performed on appliances that are not operating. These inspections are applicable to all appliance installations.

1. *Preparing for Inspection.* Shut off all gas and electrical power to the appliances located in the same room being inspected. For gas supply, use the shutoff valve in the supply line or at the manifold serving each appliance. For electrical power, place the circuit breaker in the off position or remove the fuse that serves each appliance. A lock type device or tag should be installed on each gas shutoff valve and at the electrical panel to indicate that the service has been shut off for inspection purposes.
2. *Vent System Size and Installation.* Verify that the existing venting system size and installation are compliant with Chapter 5 of the *International Fuel Gas Code.* The size and installation of venting systems for other than natural draft and Category I appliances should be in compliance with the manufacturer's installation instructions. Inspect the venting system to determine that it is free of blockage, restriction, leakage, corrosion, and other deficiencies that could cause an unsafe condition. Inspect masonry chimneys to determine if they are lined. Inspect plastic venting system to determine that it is free of sagging and it is sloped in an upward direction to the outdoor vent termination.
3. *Combustion Air Supply.* Inspect provisions for *combustion air* as follows:

   a. *Nondirect-vent Appliances.* Determine that non-direct vent appliance installations are compliant with the *combustion air* requirements in Section G2407. Inspect any interior and exterior *combustion air* openings and any connected *combustion air* ducts to determine that there is no blockage, restriction, corrosion

[1] US Consumer Product Safety Commission, *Responding to Residential Carbon Monoxide Incidents, Guidelines For Fire and Other Emergency Response Personnel*, Approved 7/23/02

or damage. Inspect to determine that the upper horizontal *combustion air* duct is not sloped in a downward direction toward the air supply source.

b. *Direct Vent Appliances.* Verify that the *combustion air* supply ducts and pipes are securely fastened to direct vent appliance and determine that there are no separations, blockage, restriction, corrosion or other damage. Determine that the combustion air source is located in the outdoors or to areas that freely communicate to the outdoors.

c. *Unvented Appliances.* Verify that the total input of all unvented room heaters and gas-fired refrigerators installed in the same room or rooms that freely communicate with each other does not exceed 20 Btu/hr/ft^3.

4. *Flooded Appliances.* Inspect for flood damage to the appliance. Signs of flooding include a visible water submerge line on the appliance housing, excessive surface or component rust, deposited debris on internal components, and mildew-like odor. Inform the owner that any part of the appliance control system and any appliance gas control that has been under water must be replaced. Flood-damaged plumbing, heating, cooling and electrical appliances should be replaced.

5. *Flammable Vapors.* Inspect the room/space where the appliance is installed to determine if the area is free of the storage of gasoline or any flammable products such as oil-based solvents, varnishes or adhesives. Where the appliance is installed where flammable products will be stored or used, such as a garage, verify that the appliance burner(s) is not less than 18 inches above the floor unless the appliance is *listed* as flammable vapor ignition resistant.

6. *Clearances to Combustibles.* Inspect the immediate location where the appliance is installed to determine if the area is free of rags, paper or other combustibles. Verify that the appliance and venting system are compliant with clearances to combustible building components in accordance with Sections G2408.5, G2425.15.4, G2426.5, G2427.6.2, G2427.10.5 and other applicable sections of Section G2427.

7. *Appliance Components.* Inspect internal components by removing access panels or other components for the following:

   a. Inspect burners and crossovers for blockage and corrosion. The presence of soot, debris, and signs of excessive heating are potential indicators of incomplete combustion caused by blockage or improper burner adjustments.

   b. Metallic and nonmetallic hoses for signs of cracks, splitting, corrosion, and lose connections.

   c. Signs of improper or incomplete repairs.

   d. Modifications that override controls and safety systems.

   e. Electrical wiring for loose connections; cracks, missing or worn electrical insulation; and indications of excessive heat or electrical shorting. Appliances requiring an external electrical supply should be inspected for proper electrical connection in accordance with the *National Electrical Code.*

8. *Placing Appliances Back in Operation.* Return all inspected appliances and systems to their preexisting state by reinstalling any removed access panels and components. Turn on the gas supply and electricity to each appliance found in safe condition. Proceed to the operating inspections in Sections AD105 through AD106.

SECTION AD105 INSPECTIONS TO BE PERFORMED WITH THE APPLIANCE OPERATING

The following safety inspection procedures are to be performed on appliances that are operating where there are no unsafe conditions or where corrective repairs have been completed.

AD105.1 General appliance operation.

1. *Initial Startup.* Adjust the thermostat or other control device to start the *appliance.* Verify that the *appliance* starts up normally and is operating properly.

   Determine that the pilot(s), where provided, is burning properly and that the main burner ignition is satisfactory, by interrupting and re-establishing the electrical supply to the *appliance* in any convenient manner. If the *appliance* is equipped with a continuous pilot(s), test all pilot safety devices to determine whether they are operating properly by extinguishing the pilot(s) when the main burner(s) is off and determining, after 3 minutes, that the main burner gas does not flow upon a call for heat. If the *appliance* is not provided with a pilot(s), test for proper operation of the ignition system in accordance with the *appliance* manufacturer's lighting and operating instructions.

2. *Flame Appearance.* Visually inspect the flame appearance for proper color and appearance. Visually determine that the main burner gas is burning properly (i.e., without floating, lifting or flashback). Adjust the primary air shutter as required. If the *appliance* is equipped with high and low flame controlling or flame modulation, check for proper main burner operation at low flame.

3. *Appliance Shutdown.* Adjust the thermostat or other control device to shut down the *appliance.* Verify that the appliance shuts off properly.

AD105.2 Test for combustion air and vent drafting for natural draft and Category I appliances. *Combustion air* and vent draft procedures are for natural draft and Category I appliances equipped with a draft hood and connected to a natural draft venting system.

1. *Preparing for Inspection.* Close all exterior building doors and windows and all interior doors between the space in which the *appliance* is located and other spaces of the building that can be closed. Turn on any clothes dryer. Turn on any exhaust fans, such as range hoods and bathroom exhausts, so they will operate at maximum speed. Do not operate a summer exhaust fan. Close fireplace dampers and any fireplace doors.
2. *Placing the Appliance in Operation.* Place the *appliance* being inspected in operation. Adjust the thermostat or control so the *appliance* will operate continuously.
3. *Spillage Test.* Verify that all *appliances* located within the same room are in their standby mode and ready for operation. Follow lighting instructions for each *appliance* as necessary. Test for spillage at the draft hood relief opening as follows:
    a. After 5 minutes of main burner operation, check for spillage using smoke.
    b. Immediately after the first check, turn on all other fuel gas burning *appliances* within the same room so they will operate at their full inputs and repeat the spillage test.
    c. Shut down all *appliances* to their standby mode and wait for 15 minutes.
    d. Repeat the spillage test steps a through c on each *appliance* being inspected.
4. Additional Spillage Tests: Determine if the *appliance* venting is impacted by other door and air handler settings by performing the following tests.
    a. Set initial test condition in accordance with Section AD105.2, Item 1.
    b. Place the appliance(s) being inspected in operation. Adjust the thermostat or control so the appliance(s) will operate continuously.
    c. Open the door between the space in which the appliance(s) is located and the rest of the building. After 5 minutes of main burner operation, check for spillage at each *appliance* using smoke.
    d. Turn on any other central heating or cooling air handler fan that is located outside of the area where the *appliances* are being inspected. After 5 minutes of main burner operation, check for spillage at each *appliance* using smoke. The test should be conducted with the door between the space in which the appliance(s) is located and the rest of the building in the open and in the closed position.
5. Return doors, windows, exhaust fans, fireplace dampers, and any other fuel gas burning *appliance* to their previous conditions of use.
6. If, after completing the spillage test it is believed sufficient *combustion air* is not available, the *owner* should be notified that an alternative *combustion air* source is needed in accordance with Section G2407. Where it is believed that the venting system does not provide adequate natural draft, the *owner* should be notified that alternative vent sizing, design or configuration is needed in accordance with Chapter 24. If spillage occurs, the *owner* should be notified as to its cause, be instructed as to which position of the door (open or closed) would lessen its impact, and that corrective action by a HVAC professional should be taken.

SECTION AD106 APPLIANCE-SPECIFIC INSPECTIONS

The following *appliance*-specific inspections are to be performed as part of a complete inspection. These inspections are performed either with the *appliance* in the off or standby mode (indicated by "OFF") or on an *appliance* that is operating (indicated by "ON"). The CO measurements are to be undertaken only after the *appliance* is determined to be properly venting. The CO detector should be capable of calculating CO emissions in ppm air free.

1. Forced Air Furnaces:
    a. OFF. Verify that an air filter is installed and that it is not excessively blocked with dust.
    b. OFF. Inspect visible portions of the furnace combustion chamber for cracks, ruptures, holes, and corrosion. A heat exchanger leakage test should be conducted.
    c. ON. Verify both the limit control and the fan control are operating properly. Limit control operation can be checked by blocking the circulating air inlet or temporarily disconnecting the electrical supply to the blower motor and determining that the limit control acts to shut off the main burner gas.
    d. ON. Verify that the blower compartment door is properly installed and can be properly resecured if opened. Verify that the blower compartment door safety switch operates properly.
    e. ON. Check for flame disturbance before and after blower comes on which can indicate heat exchanger leaks.
    f. ON. Measure the CO in the vent after 5 minutes of main burner operation. The CO should not exceed threshold in Table AD106.
2. Boilers:
    a. OFF and ON. Inspect for evidence of water leaks around boiler and connected piping.
    b. ON. Verify that the water pumps are in operating condition. Test low water cutoffs, automatic feed controls, pressure and

temperature limit controls, and relief valves in accordance with the manufacturer's recommendations to determine that they are in operating condition.

c. ON. Measure the CO in the vent after 5 minutes of main burner operation. The CO should not exceed threshold in Table AD106.

3. Water Heaters:

a. OFF. Verify that the pressure-temperature relief valve is in operating condition. Water in the heater should be at operating temperature.

b. OFF. Verify that inspection covers, glass, and gaskets are intact and in place on a flammable vapor ignition resistant (FVIR) type water heater.

c. ON. Verify that the thermostat is set in accordance with the manufacturer's operating instructions and measure the water temperature at the closest tub or sink to verify that it is not greater than 120°F.

d. OFF. Where required by the local building code in earthquake prone locations, inspect that the water heater is secured to the wall studs in two locations (high and low) using appropriate metal strapping and bolts.

e. ON. Measure the CO in the vent after 5 minutes of main burner operation. The CO should not exceed threshold in Table AD106.

4. Cooking Appliances

a. OFF. Inspect oven cavity and range-top exhaust vent for blockage with aluminum foil or other materials.

b. OFF. Inspect cook top to verify that it is free from a build-up of grease.

c. ON. Measure the CO above each burner and at the oven exhaust vents after 5 minutes of burner operation. The CO should not exceed threshold in Table AD106.

5. Vented Room Heaters

a. OFF. For built-in room heaters and wall furnaces, inspect that the burner compartment is free of lint and debris.

b. OFF. Inspect that furnishings and combustible building components are not blocking the heater.

c. ON. Measure the CO in the vent after 5 minutes of main burner operation. The CO should not exceed threshold in Table AD106.

6. Vent-Free (unvented) Heaters

a. OFF. Verify that the heater input is not more than 40,000 Btu input, but not more than 10,000 Btu where installed in a bedroom, and 6,000 Btu where installed in a bathroom.

b. OFF. Inspect the ceramic logs provided with gas log type vent free heaters that they are properly located and aligned.

c. OFF. Inspect the heater that it is free of excess lint build-up and debris.

d. OFF. Verify that the oxygen depletion safety shutoff system has not been altered or bypassed.

e. ON. Verify that the main burner shuts down within 3 minutes by extinguishing the pilot light. The test is meant to simulate the operation of the oxygen depletion system (ODS).

f. ON. Measure the CO after 5 minutes of main burner operation. The CO should not exceed threshold in Table AD106.

7. Gas Log Sets and Gas Fireplaces

a. OFF. For gas logs installed in wood burning fireplaces equipped with a damper, verify that the fireplace damper is in a fixed open position.

b. ON. Measure the CO in the firebox (log sets installed in wood burning fireplaces or in the vent (gas fireplace) after 5 minutes of main burner operation. The CO should not exceed threshold in Table AD106.

8. Gas Clothes Dryer

a. OFF. Where installed in a closet, verify that a source of make-up air is provided and inspect that any make-up air openings, louvers, and ducts are free of blockage.

b. OFF. Inspect for excess amounts of lint around the dryer and on dryer components. Inspect that there is a lint trap properly installed and it does not have holes or tears. Verify that it is in a clean condition.

c. OFF. Inspect visible portions of the exhaust duct and connections for loose fittings and connections, blockage, and signs of corrosion. Verify that the duct termination is not blocked and that it terminates in an outdoor location. Verify that only approved metal vent ducting material is installed (plastic and vinyl materials are not approved for gas dryers).

d. ON. Verify mechanical components including drum and blower are operating properly.

e. ON. Operate the clothes dryer and verify that exhaust system is intact and exhaust is exiting the termination.

f. ON. Measure the CO at the exhaust duct or termination after 5 minutes of main burner operation. The CO should not exceed threshold in Table AD106.

TABLE AD106
CO THRESHOLDS

Boilers (all categories)	400 ppm air free
Central Furnace (all categories)	400 ppm[1] air free[2, 3]
Floor Furnace	400 ppm air free
Gravity Furnace	400 ppm air free
Wall Furnace (BIV)	200 ppm air free
Wall Furnace (Direct Vent)	400 ppm air free
Vented Room Heater	200 ppm air free
Vent-Free Room Heater	200 ppm air free
Water Heater	200 ppm air free
Oven/Broiler	225 ppm as measured
Top Burner	25 ppm as measured (per burner)
Clothes Dryer	400 ppm air free
Refrigerator	25 ppm as measured
Gas Log (gas fireplace)	25 ppm as measured in vent
Gas Log (installed in wood burning fireplace)	400 ppm air free in firebox

1. Parts per million.
2. Air-free emission levels are based on a mathematical equation (involving carbon monoxide and oxygen or carbon dioxide readings) to convert an actual diluted flue gas carbon monoxide testing sample to an undiluted air-free flue gas carbon monoxide level utilized in the appliance certification standards. For natural gas or propane, using as-measured CO ppm and O_2 percentage:

$$CO_{AFppm} = \left(\frac{20.9}{20.9 - O_2}\right) \times CO_{ppm}$$

where:

CO_{AFppm} = Carbon monoxide, air-free ppm

CO_{ppm} = As-measured combustion gas carbon monoxide ppm

O_2 = Percentage of oxygen in combustion gas, as a percentage

3. An alternate method of calculating the CO air-free when access to an oxygen meter is not available:

$$CO_{AFppm} = \left(\frac{UCO_2}{CO_2}\right) \times CO$$

where:

UCO_2 = Ultimate concentration of carbon dioxide for the fuel being burned in percent for natural gas (12.2 percent) and propane (14.0 percent)

CO_2 = Measured concentration of carbon dioxide in combustion products in percent

CO = Measured concentration of carbon monoxide in combustion products in percent

APPENDIX AE

MANUFACTURED HOUSING USED AS DWELLINGS

The provisions contained in this appendix are not mandatory unless specifically referenced in the adopting ordinance.

User note:

About this appendix: *Appendix AE regulates the installation, relocation, maintenance and repair of manufactured housing, including mobile homes. It addresses permits, fees, inspections, utility service, location on a lot and foundation systems. This appendix is not intended to regulate the design and construction of those portions of manufactured housing or mobile homes that are above the foundation system except where manufactured housing or mobile homes are moved or altered. Federal standards regulate those portions of manufactured housing and mobile homes that are above the foundation system.*

SECTION AE101 SCOPE

AE101.1 General. These provisions shall be applicable only to a *manufactured home* used as a single *dwelling unit* installed on privately owned (nonrental) lots and shall apply to the following:

1. Construction, *alteration* and *repair* of any foundation system that is necessary to provide for the installation of a *manufactured home* unit.
2. Construction, installation, addition, *alteration, repair* or maintenance of the building service equipment that is necessary for connecting *manufactured homes* to water, fuel, or power supplies and sewage systems.
3. *Alterations, additions* or repairs to existing *manufactured homes.* The construction, *alteration,* moving, demolition, *repair* and use of accessory buildings and structures, and their building service equipment, shall comply with the requirements of the codes adopted by this *jurisdiction.*

These provisions shall not be applicable to the design and construction of *manufactured homes* and shall not be deemed to authorize either modifications or *additions* to *manufactured homes* where otherwise prohibited.

AE101.2 Flood hazard areas. New and replacement manufactured homes to be installed in flood hazard areas as established in Table R301.2 shall meet the applicable requirements of Section R322.

SECTION AE102 APPLICATION TO EXISTING MANUFACTURED HOMES AND BUILDING SERVICE EQUIPMENT

AE102.1 General. *Manufactured homes* and their building service equipment to which *additions, alterations* or repairs are made shall comply with all the requirements of these provisions for new facilities, except as specifically provided in this section.

AE102.2 Additions, alterations or repairs. *Additions* made to a *manufactured home* shall conform to one of the following:

1. Be certified under the National Manufactured Housing Construction and Safety Standards Act of 1974 (42 U.S.C. Section 5401, et seq.).
2. Be designed and constructed to comply with the applicable provisions of the National Manufactured Housing Construction and Safety Standards Act of 1974 (42 U.S.C. Section 5401, et seq.).
3. Be designed and constructed in compliance with the code adopted by this *jurisdiction.*

Additions shall be structurally separated from the *manufactured home.*

Exception: A structural separation need not be provided where structural calculations are provided to justify the omission of such separation.

Alterations or repairs may be made to any *manufactured home* or to its building service equipment without requiring the existing *manufactured home* or its building service equipment to comply with all the requirements of these provisions, provided that the *alteration* or repair conforms to that required for new construction, and provided further that hazard to life, health or safety will not be created by such *additions, alterations* or repairs.

Alterations or repairs to an existing *manufactured home,* which are nonstructural and do not adversely affect any structural member or any part of the building or structure having required fire protection, shall be made with materials equivalent to those of which the *manufactured home* structure is constructed, subject to approval by the *building official.*

Exception: The installation or replacement of glass shall be required for new installations.

Minor *additions, alterations* and repairs to existing building service equipment installations may be made in accordance with the codes in effect at the time the original installation was made, subject to the approval of the *building official*, and provided that such *additions*, *alterations* and repairs will not cause the existing building service equipment to become unsafe, insanitary or overloaded.

AE102.3 Existing installations. Building service equipment lawfully in existence at the time of the adoption of the applicable codes shall have their use, maintenance or repair continued if the use, maintenance or repair is in accordance with the original design and hazard to life, health or property has not been created by such building service equipment.

AE102.4 Existing occupancy. *Manufactured homes* that are in existence at the time of the adoption of these provisions shall have their existing use or occupancy continued if such use or occupancy was legal at the time of the adoption of these provisions, provided that such continued use is not dangerous to life, health and safety.

The use or occupancy of any existing *manufactured home* shall not be changed unless evidence satisfactory to the *building official* is provided to show compliance with all applicable provisions of the codes adopted by this *jurisdiction*. Upon any change in use or occupancy, the *manufactured home* shall cease to be classified as such within the intent of these provisions.

AE102.5 Maintenance. *Manufactured homes* and their building service equipment, existing and new, and all parts thereof, shall be maintained in a safe and sanitary condition. Devices or safeguards that are required by applicable codes or by the Manufactured Home Standards shall be maintained in conformance to the code or standard under which it was installed. The *owner* or the owner's designated agent shall be responsible for the maintenance of *manufactured homes*, accessory buildings, structures and their building service equipment. To determine compliance with this section, the *building official* has the authority to cause any *manufactured home*, accessory building or structure to be reinspected.

AE102.6 Relocation. *Manufactured homes* that are to be relocated within this *jurisdiction* shall comply with these provisions.

SECTION AE103 DEFINITIONS

AE103.1 General. For the purpose of these provisions, certain abbreviations, terms, phrases, words and their derivatives shall be construed as defined or specified herein.

ACCESSORY BUILDING. Any building or structure or portion thereto, located on the same property as a *manufactured home*, which does not qualify as a *manufactured home* as defined herein.

BUILDING SERVICE EQUIPMENT. Refers to the plumbing, mechanical and electrical equipment, including piping, wiring, fixtures and other accessories that provide sanitation, lighting, heating, ventilation, cooling, fire protection and facilities essential for the habitable occupancy of a *manufactured home* or accessory building or structure for its designated use and occupancy.

MANUFACTURED HOME. A structure transportable in one or more sections that, in the traveling mode, is 8 body feet (2438 body mm) or more in width or 40 body feet (12 192 body mm) or more in length or, where erected on site, is 320 or more square feet (30 m^2), and is built on a permanent chassis and designed to be used as a *dwelling* with or without a permanent foundation when connected to the required utilities, and includes the plumbing, heating, air-conditioning and electrical systems contained therein; except that such term shall include any structure that meets all of the requirements of this paragraph, except the size requirements and with respect to which the manufacturer voluntarily files a certification required by the Secretary of the US Department of Housing and Urban Development (HUD) and complies with the standards established under this title.

For mobile homes built prior to June 15, 1976, a *label* certifying compliance with the *Standard for Mobile Homes*, NFPA 501, ANSI 119.1, in effect at the time of manufacture, is required. For the purpose of these provisions, a mobile home shall be considered to be a *manufactured home*.

MANUFACTURED HOME INSTALLATION. Construction that is required for the installation of a *manufactured home*, including the construction of the foundation system, required structural connections thereto and the installation of on-site water, gas, electrical and sewer systems and connections thereto that are necessary for the normal operation of the *manufactured home*.

MANUFACTURED HOME STANDARDS. The *Manufactured Home Construction and Safety Standards* as promulgated by the HUD.

PRIVATELY OWNED (NONRENTAL) LOT. A parcel of real estate outside of a *manufactured home* rental community (park) where the land and the *manufactured home* to be installed thereon are held in common ownership.

SECTION AE104 PERMITS

AE104.1 Initial installation. A *manufactured home* shall not be installed on a foundation system, reinstalled or altered without first obtaining a *permit* from the *building official*. A separate *permit* shall be required for each *manufactured home* installation. Where *approved* by the *building official*, such *permit* may include accessory buildings and structures, and their building service equipment, if the accessory buildings or structures will be constructed in conjunction with the *manufactured home* installation.

AE104.2 Additions, alterations and repairs to a manufactured home. A *permit* shall be obtained to alter, remodel, repair or add accessory buildings or structures to a *manufactured home* subsequent to its initial installation. *Permit* issuance and fees therefor shall be in conformance to the codes applicable to the type of work involved.

An *addition* made to a *manufactured home*, as defined in these provisions, shall comply with these provisions.

AE104.3 Accessory buildings. Except as provided in Section AE104.1, *permits* shall be required for all accessory buildings and structures, and their building service equipment. *Permit* issuance and fees therefor shall be in conformance to the codes applicable to the types of work involved.

AE104.4 Exempted work. A *permit* shall not be required for the types of work specifically exempted by the applicable codes. Exemption from the *permit* requirements of any of said codes shall not be deemed to grant authorization for any work to be done in violation of the provisions of said codes or any other laws or ordinances of this *jurisdiction*.

SECTION AE105 APPLICATION FOR PERMIT

AE105.1 Application. To obtain a *manufactured home* installation *permit*, the applicant shall first file an application, in writing, on a form furnished by the *building official* for that purpose. At the option of the *building official*, every such application shall:

1. Identify and describe the work to be covered by the *permit* for which application is made.
2. Describe the land on which the proposed work is to be done by legal description, street address or similar description that will readily identify and definitely locate the proposed building or work.
3. Indicate the use or occupancy for which the proposed work is intended.
4. Be accompanied by plans, diagrams, computations and specifications, and other data as required in Section AE105.2.
5. Be accompanied by a soil investigation where required by Section AE114.2.
6. State the valuation of any new building or structure; or any *addition*, remodeling or *alteration* to an existing building.
7. Be signed by the permittee, or permittee's authorized agent, who may be required to submit evidence to indicate such authority.
8. Give other data and information where required by the *building official*.

AE105.2 Plans and specifications. Plans, engineering calculations, diagrams and other data as required by the *building official* shall be submitted in not less than two sets with each application for a *permit*. The *building official* has the authority to require plans, computations and specifications to be prepared and designed by an engineer or architect licensed by the state to practice as such.

Where unusual site conditions do not exist, the *building official* has the authority to accept *approved* standard foundation plans and details in conjunction with the manufacturer's *approved* installation instructions without requiring the submittal of engineering calculations.

AE105.3 Information on plans and specifications. Plans and specifications shall be drawn to scale on substantial paper or cloth, and shall be of sufficient clarity to indicate the location, nature and extent of the work proposed and shown in detail that it will conform to these provisions and all relevant laws, ordinances, rules and regulations. The *building official* shall determine what information is required on plans and specifications to ensure compliance.

SECTION AE106 PERMITS ISSUANCE

AE106.1 Issuance. The application, plans and specifications, and other data filed by an applicant for *permit* shall be reviewed by the *building official*. Such plans may be reviewed by other departments of this *jurisdiction* to verify compliance with any applicable laws under their *jurisdiction*. If the *building official* finds that the work described in an application for a *permit*, and the plans, specifications and other data filed therewith conform to the requirements of these provisions and other pertinent codes, laws and ordinances, and that the fees specified in Section AE107 have been paid, the *building official* shall issue a *permit* therefor to the applicant.

Upon issuing a *permit* where plans are required, the *building official* shall endorse in writing or stamp the plans and specifications *APPROVED*. Such *approved* plans and specifications shall not be changed, modified or altered without authorization from the *building official*, and all work shall be done in accordance with the *approved* plans.

AE106.2 Retention of plans. One set of *approved* plans and specifications shall be returned to the applicant and shall be kept on the site of the building or work at all times during which the work authorized thereby is in progress. One set of *approved* plans, specifications and computations shall be retained by the *building official* until final approval of the work.

AE106.3 Validity of permit. The issuance of a *permit* or approval of plans and specifications shall not be construed to be a *permit* for, or an approval of, any violation of any of these provisions or other pertinent codes of any other ordinance of the *jurisdiction*. A *permit* presuming to give authority to violate or cancel these provisions shall not be valid.

The issuance of a *permit* based on plans, specifications and other data shall not prevent the *building official* from thereafter requiring the correction of errors in said plans, specifications and other data, or from preventing building operations being carried on thereunder when in violation of these provisions or of any other ordinances of this *jurisdiction*.

AE106.4 Expiration. Every *permit* issued by the *building official* under these provisions shall expire by limitation and become null and void if the work authorized by such *permit* is not commenced within 180 days from the date of such *permit*, or if the work authorized by such *permit* is suspended or abandoned at any time after the work is commenced for a period of 180 days. Before such work can be recommenced, a new *permit* shall be first obtained, and the fee therefor shall be one-half the amount required for a new *permit* for such work, provided that changes have not been made or will not be made in the original plans and specifications for such work, and provided further that such suspension or abandonment has not exceeded 1 year. In order to renew action on a *permit* after expiration, the permittee shall pay a new full *permit* fee.

Any permittee holding an unexpired *permit* may apply for an extension of the time within which work shall commence under that *permit* where the permittee is unable to commence work within the time required by this section for good and satisfactory reasons. The *building official* has the authority to extend the time for action by the permittee for a period not exceeding 180 days upon written request by the permittee showing that circumstances beyond the control of the permittee have prevented action from being taken. A *permit* shall not be extended more than once.

AE106.5 Suspension or revocation. The *building official* may, in writing, suspend or revoke a *permit* issued under these provisions whenever the *permit* is issued in error or on the basis of incorrect information supplied, or in violation of any ordinance or regulation or any of these provisions.

SECTION AE107 FEES

AE107.1 Permit fees. The fee for each *manufactured home* installation *permit* shall be established by the *building official.*

Where *permit* fees are to be based on the value or valuation of the work to be performed, the determination of value or valuation under these provisions shall be made by the *building official.* The value to be used shall be the total value of all work required for the *manufactured home* installation plus the total value of all work required for the construction of accessory buildings and structures for which the *permit* is issued, as well as all finish work, painting, roofing, electrical, plumbing, heating, air conditioning, elevators, fire-extinguishing systems and any other permanent equipment that is a part of the accessory building or structure. The value of the *manufactured home* itself shall not be included.

AE107.2 Plan review fees. Where plans or other data are required to be submitted by Section AE105.2, a plan review fee shall be paid at the time of submitting plans and specifications for review. Said plan review fee shall be as established by the *building official.* Where plans are incomplete or changed so as to require additional plan review, an additional plan review fee shall be charged at a rate as established by the *building official.*

AE107.3 Other provisions.

AE107.3.1 Expiration of plan review. Applications for which a *permit* has not been issued within 180 days following the date of application shall expire by limitation, and plans and other data submitted for review shall thereafter be returned to the applicant or destroyed by the *building official.* The *building official* has the authority to extend the time for action by the applicant for a period not exceeding 180 days upon request by the applicant showing that circumstances beyond the control of the applicant have prevented action from being taken. An application shall not be extended more than once. In order to renew action on an application after expiration, the applicant shall resubmit plans and pay a new plan review fee.

AE107.3.2 Investigation fees—work without a permit.

AE107.3.2.1 Investigation. Whenever any work for which a *permit* is required by these provisions has been commenced without first obtaining said *permit,* a special investigation shall be made before a *permit* is issued for such work.

AE107.3.2.2 Fee. An investigation fee, in addition to the *permit* fee, shall be collected whether or not a *permit* is then or subsequently issued. The investigation fee shall be equal to the amount of the *permit* fee required. The minimum investigation fee shall be the same as the minimum fee established by the *building official.* The payment of such investigation fee shall not exempt any person from compliance with all other provisions of either these provisions or other pertinent codes or from any penalty prescribed by law.

AE107.3.3 Fee refunds.

AE107.3.3.1 Permit fee erroneously paid or collected. The *building official* has the authority to authorize the refunding of any fee paid hereunder that was erroneously paid or collected.

AE107.3.3.2 Permit fee paid where no work done. The *building official* has the authority to authorize the refunding of not more than 80 percent of the *permit* fee paid where no work has been done under a *permit* issued in accordance with these provisions.

AE107.3.3.3 Plan review fee. The *building official* has the authority to authorize the refunding of not more than 80 percent of the plan review fee paid where an application for a *permit* for which a plan review fee has been paid is withdrawn or canceled before any plan reviewing is done.

The *building official* shall not authorize the refunding of any fee paid, except upon written application by the original permittee not later than 180 days after the date of the fee payment.

SECTION AE108 INSPECTIONS

AE108.1 General. All construction or work for which a *manufactured home* installation *permit* is required shall be subject to inspection by the *building official,* and certain types of construction shall have continuous inspection by special inspectors as specified in Section AE109. The *building official* has the authority to require a survey of the *lot* to verify that the structure is located in accordance with the *approved* plans.

It shall be the duty of the *permit* applicant to cause the work to be accessible and exposed for inspection purposes. Neither the *building official* nor this *jurisdiction* shall be liable for expense entailed in the removal or replacement of any material required to allow inspection.

AE108.2 Inspection requests. It shall be the duty of the person doing the work authorized by a *manufactured home* installation *permit* to notify the *building official* that such work is ready for inspection. The *building official* has the authority to require that every request for inspection be filed not less than one working day before such inspection is desired. Such request shall be in writing or by telephone at the option of the *building official.*

It shall be the duty of the person requesting any inspections required, either by these provisions or other applicable codes, to provide access to and means for proper inspection of such work.

AE108.3 Inspection record card. Work requiring a *manufactured home* installation *permit* shall not be commenced until the *permit* holder or the *permit* holder's agent shall have posted an inspection record card in a conspicuous place on the premises and in such position as to allow the *building official* conveniently to make the required entries thereon regarding inspection of the work. This card shall be maintained in such position by the *permit* holder until final approval has been issued by the *building official.*

AE108.4 Approval required. Work shall not be done on any part of the *manufactured home* installation beyond the point indicated in each successive inspection without first obtaining the approval of the *building official.* Such approval shall be given only after an inspection has been made of each successive step in the construction as indicated by each of the inspections required in Section AE108.5. There shall be a final inspection and approval of the *manufactured home* installation, including connections to its building service equipment, when completed and ready for occupancy or use.

AE108.5 Required inspections.

AE108.5.1 Structural inspections for the manufactured home installation. Reinforcing steel or structural framework of any part of any *manufactured home* foundation system shall not be covered or concealed without first obtaining the approval of the *building official.* The *building official,* upon notification from the *permit* holder or the *permit* holder's agent, shall make the following inspections and shall either approve that portion of the construction as completed or shall notify the *permit* holder or the *permit* holder's agent wherein the same fails to comply with these provisions or other applicable codes:

1. Foundation inspection: To be made after excavations for footings are completed and any required reinforcing steel is in place. For concrete foundations, any required forms shall be in place prior to inspection. Materials for the foundation shall be on the job, except where concrete from a central mixing plant (commonly termed "transit mixed") is to be used, the concrete materials need not be on the job. Where the foundation is to be constructed of *approved* treated wood, requirements for additional framing inspections shall be subject to the evaluation of the *building official.*
2. Concrete slab or under-floor inspection: To be made after all in-slab or under-floor building service equipment, conduit, piping accessories and other ancillary equipment items are in place but before any concrete is poured or the *manufactured home* is installed.
3. Anchorage inspection: To be made after the *manufactured home* has been installed and permanently anchored.

AE108.5.2 Structural inspections for accessory building and structures. Inspections for accessory buildings and structures shall be made as set forth in this code.

AE108.5.3 Building service equipment inspections. Building service equipment that is required as a part of a *manufactured home* installation, including accessory buildings and structures authorized by the same *permit,* shall be inspected by the *building official.* Building service equipment shall be inspected and tested as required by the applicable codes. Such inspections and testing shall be limited to site construction and shall not include building service equipment that is a part of the *manufactured home* itself. No portion of any building service equipment intended to be concealed by any permanent portion of the construction shall be concealed until inspected and *approved.* Building service equipment shall not be connected to a water, fuel or power supply, or sewer system, until authorized by the *building official.*

AE108.5.4 Final inspection. Where finish grading and the *manufactured home* installation, including the installation of all required building service equipment, is completed and the *manufactured home* is ready for occupancy, a final inspection shall be made.

AE108.6 Other inspections. In addition to the called inspections specified in Section AE108.5.4, the *building official* has the authority to make or require other inspections of any construction work to ascertain compliance with these provisions or other codes and laws that are enforced by the code enforcement agency.

SECTION AE109 SPECIAL INSPECTIONS

AE109.1 General. In addition to the inspections required by Section AE108, the *building official* has the authority to require the *owner* to employ a special inspector during construction of specific types of work as described in this code.

SECTION AE110 UTILITY SERVICE

AE110.1 General. Utility service shall not be provided to any building service equipment regulated by these provisions or other applicable codes, and for which a *manufactured home* installation *permit* is required by these provisions, until *approved* by the *building official.*

SECTION AE111 OCCUPANCY CLASSIFICATION

AE111.1 Manufactured homes. A *manufactured home* shall be limited in use to a single *dwelling unit.*

AE111.2 Accessory buildings. Accessory buildings shall be classified as to occupancy by the *building official* as set forth in this code.

SECTION AE112 LOCATION ON PROPERTY

AE112.1 General. *Manufactured homes* and accessory buildings shall be located on the property in accordance with applicable codes and ordinances of this *jurisdiction.*

SECTION AE113 DESIGN

AE113.1 General. A *manufactured home* shall be installed on a foundation system designed and constructed to sustain within the stress limitations specified in this code and all loads specified in this code.

> **Exception:** Where specifically authorized by the *building official,* foundation and anchorage systems that are constructed in accordance with the methods specified in Section AE120 of these provisions, or in the HUD, *Permanent Foundations for Manufactured Housing,* 1984 Edition, Draft, shall be deemed to meet the requirements of this appendix.

AE113.2 Manufacturer's installation instructions. The installation instructions as provided by the manufacturer of the *manufactured home* shall be used to determine permissible points of support for vertical loads and points of attachment for anchorage systems used to resist horizontal and uplift forces.

AE113.3 Rationality. Any system or method of construction to be used shall submit to a rational analysis in accordance with well-established principles of mechanics.

SECTION AE114 FOUNDATION SYSTEMS

AE114.1 General. Foundation systems designed and constructed in accordance with this section shall be considered a permanent installation.

AE114.2 Soil classification. The classification of the soil at each *manufactured home* site shall be determined where required by the *building official.* The *building official* has the authority to require that the determination be made by an engineer or architect licensed by the state to conduct soil investigations.

The classification shall be based on observation and any necessary tests of the materials disclosed by borings or excavations made in appropriate locations. Additional studies may be necessary to evaluate soil strength, the effect of moisture variation on soil-bearing capacity, compressibility and expansiveness.

Where required by the *building official,* the soil classification design-bearing capacity and lateral pressure shall be shown on the plans.

AE114.3 Footings and foundations. Footings and foundations, unless otherwise specifically provided, shall be constructed of materials specified by this code for the intended use and in all cases shall extend below the frost line. Footings of concrete and masonry shall be of solid material. Foundations supporting untreated wood shall extend not less than 8 inches (203 mm) above the adjacent finish *grade.* Footings shall have a minimum depth below finished *grade* of 12 inches (305 mm) unless a greater depth is recommended by a foundation investigation.

Piers and bearing walls shall be supported on masonry or concrete foundations or piles, or other *approved* foundation systems that shall be of sufficient capacity to support all loads.

AE114.4 Foundation design. Where a design is provided, the foundation system shall be designed in accordance with the applicable structural provisions of this code and shall be designed to minimize differential settlement. Where a design is not provided, the minimum foundation requirements shall be as set forth in this code.

AE114.5 Drainage. Provisions shall be made for the control and drainage of surface water away from the *manufactured home.*

AE114.6 Under-floor clearances—ventilation and access. A minimum clearance of 12 inches (305 mm) shall be maintained beneath the lowest member of the floor support framing system. Clearances from the bottom of wood floor joists or perimeter joists shall be as specified in this code.

Under-floor spaces shall be ventilated with openings as specified in this code. If *combustion air* for one or more heat-producing *appliance* is taken from within the under-floor spaces, ventilation shall be adequate for proper *appliance* operation.

Under-floor access openings shall be provided. Such openings shall be not less than 18 inches (457 mm) in any dimension and not less than 3 square feet (0.279 m^2) in area, and shall be located so that any water supply and sewer drain connections located under the *manufactured home* are accessible.

SECTION AE115 SKIRTING AND PERIMETER ENCLOSURES

AE115.1 Skirting and permanent perimeter enclosures. Skirting and permanent perimeter enclosures shall be installed only where specifically required by other laws or ordinances. Skirting shall be of material suitable for exterior exposure and contact with the ground. Permanent perimeter enclosures shall be constructed of materials as required by this code for regular foundation construction.

Skirting shall be installed in accordance with the skirting manufacturer's installation instructions. Skirting shall be adequately secured to ensure stability, minimize vibration and susceptibility to wind damage, and compensate for possible frost heave.

AE115.2 Retaining walls. Where retaining walls are used as a permanent perimeter enclosure, they shall resist the lateral displacements of soil or other materials and shall conform to this code as specified for foundation walls. Retaining walls and foundation walls shall be constructed of *approved* treated wood, concrete, masonry or other *approved* materials or combination of materials as for foundations as specified in this code. Siding materials shall extend below the top of the exterior of the retaining or foundation wall, or the joint between the siding and enclosure wall shall be flashed in accordance with this code.

SECTION AE116 STRUCTURAL ADDITIONS

AE116.1 General. Accessory buildings shall not be structurally supported by or attached to a *manufactured home* unless engineering calculations are submitted to substantiate any proposed structural connection.

> **Exception:** The *building official* has the authority to waive the submission of engineering calculations if it is found that the nature of the work applied for is such that engineering calculations are not necessary to show conformance to these provisions.

SECTION AE117 BUILDING SERVICE EQUIPMENT

AE117.1 General. The installation, *alteration*, *repair*, replacement, addition to or maintenance of the building service equipment within the *manufactured home* shall conform to regulations set forth in the Manufactured Home Standards. Such work that is located outside the *manufactured home* shall comply with the applicable codes adopted by this *jurisdiction*.

SECTION AE118 EXITS

AE118.1 Site development. Exterior *stairways* and *ramps* that provide egress to the *public way* shall comply with the applicable provisions of this code.

AE118.2 Accessory buildings. Every accessory building or portion thereof shall be provided with exits as required by this code.

SECTION AE119 OCCUPANCY, FIRE SAFETY AND ENERGY CONSERVATION STANDARDS

AE119.1 General. *Alterations* made to a *manufactured home* subsequent to its initial installation shall conform to the occupancy, fire safety and energy conservation requirements set forth in the Manufactured Home Standards.

SECTION AE120 SPECIAL REQUIREMENTS FOR FOUNDATION SYSTEMS

AE120.1 General. This section is applicable only where specifically authorized by the *building official*.

SECTION AE121 FOOTINGS AND FOUNDATIONS

AE121.1 General. The capacity of individual load-bearing piers and their footings shall be sufficient to sustain all loads specified in this code within the stress limitations specified in this code. Footings, unless otherwise *approved* by the *building official*, shall be placed level on firm, undisturbed soil or an engineered fill that is free of organic material, such as weeds and grasses. Where used, an engineered fill shall provide a minimum load-bearing capacity of not less than 1,000 pounds per square foot (48 kN/m^2). Continuous footings shall conform to the requirements of this code. Section AE114 of these provisions shall apply to footings and foundations constructed under the provisions of this section.

SECTION AE122 PIER CONSTRUCTION

AE122.1 General. Piers shall be designed and constructed to distribute loads evenly. Multiple-section homes may have concentrated roof loads that will require special consideration. Load-bearing piers shall be constructed utilizing one of the following methods listed. Such piers shall be considered to resist only vertical forces acting in a downward direction. They shall not be considered as providing any resistance to horizontal loads induced by wind or earthquake forces.

1. A prefabricated load-bearing device that is *listed* and *labeled* for the intended use.
2. Mortar shall comply with ASTM C270, Type M, S or N; this may consist of one part Portland cement, one-half part hydrated lime and four parts sand by volume. Lime shall not be used with plastic or waterproof cement.
3. A cast-in-place concrete pier with concrete having specified compressive strength at 28 days of 2,500 pounds per square inch (17 225 kPa).

Alternative materials and methods of construction used for piers shall be designed by an engineer or architect licensed by the state to practice as such.

Caps and leveling spacers may be used for leveling of the *manufactured home*. Spacing of piers shall be as specified in the manufacturer's installation instructions, if available, or by an *approved* designer.

SECTION AE123 HEIGHT OF PIERS

AE123.1 General. Piers constructed as indicated in Section AE122 shall have heights as follows:

1. Except for corner piers, piers 36 inches (914 mm) or less in height shall be constructed of *masonry units*, placed with cores or cells vertically. Piers shall be installed with their long dimension at right angles to the main frame member they support and shall have a minimum cross-sectional area of 128 square inches (82 560 mm^2). Piers shall be capped with minimum 4-inch (102 mm) *solid masonry* units or equivalent.
2. Piers between 36 and 80 inches (914 and 2032 mm) in height and all corner piers greater than 24 inches (610 mm) in height shall be not less than 16 inches by 16 inches (406 mm by 406 mm) consisting of interlocking *masonry units* and shall be fully capped with minimum 4-inch (102 mm) *solid masonry* units or equivalent.
3. Piers greater than 80 inches (2032 mm) in height shall be constructed in accordance with the provisions of Item 2, provided that the piers shall be filled solid with grout and reinforced with four continuous No. 5 bars. One bar shall be placed in each corner cell of hollow masonry unit piers or in each corner of the grouted space of piers constructed of *solid masonry* units.
4. Cast-in-place concrete piers meeting the same size and height limitations of Items 1, 2 and 3 may be substituted for piers constructed of masonry units.

SECTION AE124 ANCHORAGE INSTALLATIONS

AE124.1 Ground anchors. Ground anchors shall be designed and installed to transfer the anchoring loads to the ground. The load-carrying portion of the ground anchors shall be installed to the full depth called for by the manufacturer's installation instructions and shall extend below the established frost line into undisturbed soil.

Manufactured ground anchors shall be *listed* and installed in accordance with the terms of their listing and the anchor manufacturer's instructions, and shall include the means of attachment of ties meeting the requirements of Section AE125. Ground anchor manufacturer's installation instructions shall include the amount of preload required and load capacity in various types of soil. These instructions shall include tensioning adjustments where needed to prevent damage to the *manufactured home*, particularly damage that can be caused by frost heave. Each ground anchor shall be marked with the manufacturer's identification and *listed* model identification number, which shall be visible after installation. Instructions shall accompany each *listed* ground anchor specifying the types of soil for which the anchor is suitable under the requirements of this section.

Each *approved* ground anchor, when installed, shall be capable of resisting an allowable working load not less than 3,150 pounds (14 kN) in the direction of the tie plus a 50-percent overload [4,725 pounds (21 kN) total] without failure. Failure shall be considered to have occurred when the anchor moves more than 2 inches (51 mm) at a load of 4,725 pounds (21 kN) in the direction of the tie installation. Those ground anchors that are designed to be installed so that loads on the anchor are other than direct withdrawal shall be designed and installed to resist an applied design load of 3,150 pounds (14 kN) at 40 to 50 degrees from vertical or within the angle limitations specified by the home manufacturer without displacing the tie end of the anchor more than 4 inches (102 mm) horizontally. Anchors designed for the connection of multiple ties shall be capable of resisting the combined working load and overload consistent with the intent expressed herein.

Where it is proposed to use ground anchors and the *building official* has reason to believe that the soil characteristics at a given site are such as to render the use of ground anchors advisable, or where there is doubt regarding the ability of the ground anchors to obtain their *listed* capacity, the *building official* has the authority to require that a representative field installation be made at the site in question and tested to demonstrate ground-anchor capacity. The *building official* shall approve the test procedures.

AE124.2 Anchoring equipment. Anchoring equipment, where installed as a permanent installation, shall be capable of resisting all loads as specified within these provisions. Where the stabilizing system is designed by an engineer or architect licensed by the state to practice, such alternative designs shall include anchoring equipment capable of withstanding a load equal to 1.5 times the calculated load. Anchoring equipment shall be *listed* and *labeled* as being capable of meeting the requirements of these provisions. Anchors as specified in this code shall be attached to the main frame of the *manufactured home* by an *approved* $^{3}/_{16}$-inch-thick (4.76 mm) slotted steel plate anchoring device. Other anchoring devices or methods meeting the requirements of these provisions shall be subject to the evaluation and *approval* of the *building official*.

Anchoring systems shall be so installed as to be permanent. Anchoring equipment shall be so designed to prevent self-disconnection with no hook ends used.

AE124.3 Resistance to weather deterioration. All anchoring equipment, tension devices and ties shall have a resistance to deterioration as required by this code.

AE124.4 Tensioning devices. Tensioning devices, such as turnbuckles or yoke-type fasteners, shall be ended with clevis or welded eyes.

SECTION AE125 TIES, MATERIALS AND INSTALLATION

AE125.1 General. Steel strapping, cable, chain or other *approved* materials shall be used for ties. Ties shall be fastened to ground anchors and drawn tight with turnbuckles or other adjustable tensioning devices or devices supplied with the ground anchor. Tie materials shall be capable of resisting an allowable working load of 3,150 pounds (14 kN) with not more than 2-percent elongation and shall withstand a 50-percent overload [4,750 pounds (21 kN)]. Ties shall comply with the weathering requirements of Section AE124.3. Ties shall connect the ground anchor and the main structural frame. Ties shall not connect to steel outrigger beams that fasten to and intersect the main structural frame unless specifically stated in the manufacturer's installation instructions. Connection of cable ties to main frame members shall be $^{5}/_{8}$-inch (15.9 mm) closed-eye bolts affixed to the frame member in an *approved* manner. Cable ends shall be secured with not fewer than two U-bolt cable clamps with the "U" portion of the clamp installed on the short (dead) end of the cable to ensure strength equal to that required by this section.

Wood floor support systems shall be fixed to perimeter foundation walls in accordance with provisions of this code. The minimum number of ties required per side shall be sufficient to resist the wind load stated in this code. Ties shall be as evenly spaced as practicable along the length of the *manufactured home* with the distance from each end of the home and the tie nearest that end not exceeding 8 feet (2438 mm). Where continuous straps are provided as vertical ties, such ties shall be positioned at rafters and studs. Where a vertical tie and diagonal tie are located at the same place, such ties connected to a single anchor that is capable of carrying both loads. Multiple-section *manufactured homes* require diagonal ties only. Diagonal ties shall be installed on the exterior main frame and slope to the exterior at an angle of 40 to 50 degrees from the vertical or within the angle limitations specified by the home manufacturer. Vertical ties that are not continuous over the top of the *manufactured home* shall be attached to the main frame.

SECTION AE126 REFERENCED STANDARDS

AE126.1 General. See Table AE126.1 for standards that are referenced in various section of this appendix. Standards are listed by the standard indentification with the effective date, the standard title, and the section or sections of this appendix that reference this standard.

TABLE AE126.1 REFERENCED STANDARDS

STANDARD ACRONYM	STANDARD NAME	SECTIONS HEREIN REFERENCED
ASTM C270—14A	*Specification for Mortar for Unit Masonry*	AE122.1
NFPA 501—17	*Standard on Manufactured Housing*	AE103.1

APPENDIX AF
RADON CONTROL METHODS

The provisions contained in this appendix are not mandatory unless specifically referenced in the adopting ordinance.

User note:

About this appendix: *Appendix AF contains provisions that are intended to mitigate the transfer of radon gases from the soil into dwelling units. Radon is a radioactive gas that has been identified as a cancer-causing agent. Radon comes from the natural breakdown of uranium in soil, rock and water.*

SECTION AF101 SCOPE

AF101.1 General. This appendix contains requirements for new construction in *jurisdictions* where radon-resistant construction is required.

Inclusion of this appendix by *jurisdictions* shall be determined through the use of locally available data or determination of Zone 1 designation in Figure AF101.1 and Table AF101.1.

SECTION AF102 DEFINITIONS

AF102.1 General. For the purpose of these requirements, the terms used shall be defined as follows:

DRAIN TILE LOOP. A continuous length of drain tile or perforated pipe extending around all or part of the internal or external perimeter of a *basement* or *crawl space* footing.

RADON GAS. A naturally occurring, chemically inert, radioactive gas that is not detectable by human senses. As a gas, it can move readily through particles of soil and rock, and can accumulate under the slabs and foundations of homes where it can easily enter into the living space through construction cracks and openings.

SOIL-GAS-RETARDER. A continuous membrane of 6-mil (0.15 mm) polyethylene or other equivalent material used to retard the flow of soil gases into a building.

SUBMEMBRANE DEPRESSURIZATION SYSTEM. A system designed to achieve lower submembrane air pressure relative to *crawl space* air pressure by use of a vent drawing air from beneath the soil-gas-retarder membrane.

SUBSLAB DEPRESSURIZATION SYSTEM (Active). A system designed to achieve lower subslab air pressure relative to indoor air pressure by use of a fan-powered vent drawing air from beneath the slab.

SUBSLAB DEPRESSURIZATION SYSTEM (Passive). A system designed to achieve lower subslab air pressure relative to indoor air pressure by use of a vent pipe routed through the *conditioned space* of a building and connecting the subslab area with outdoor air, thereby relying on the convective flow of air upward in the vent to draw air from beneath the slab.

SECTION AF103 REQUIREMENTS

AF103.1 General. The following construction techniques are intended to resist radon entry and prepare the building for post-construction radon mitigation, if necessary (see Figure AF103.1). These techniques are required in areas where designated by the *jurisdiction.*

AF103.2 Subfloor preparation. A layer of gas-permeable material shall be placed under all concrete slabs and other floor systems that directly contact the ground and are within the walls of the living spaces of the building, to facilitate future installation of a subslab depressurization system, if needed. The gas-permeable layer shall consist of one of the following:

1. A uniform layer of clean aggregate, not less than 4 inches (102 mm) thick. The aggregate shall consist of material that will pass through a 2-inch (51 mm) sieve and be retained by a $^{1}/_{4}$-inch (6.4 mm) sieve.
2. A uniform layer of sand (native or fill), not less than 4 inches (102 mm) thick, overlain by a layer or strips of geotextile drainage matting designed to allow the lateral flow of soil gases.
3. Other materials, systems or floor designs with demonstrated capability to permit depressurization across the entire subfloor area.

AF103.3 Soil-gas-retarder. A minimum 6-mil (0.15 mm) [or 3-mil (0.075 mm) cross-laminated] polyethylene or equivalent flexible sheeting material shall be placed on top of the gas-permeable layer prior to casting the slab or placing the floor assembly to serve as a soil-gas-retarder by bridging any cracks that develop in the slab or floor assembly, and to prevent concrete from entering the void spaces in the aggregate base material. The sheeting shall cover the entire floor area with separate sections of sheeting lapped not less than 12 inches (305 mm). The sheeting shall fit closely around any pipe, wire or other penetrations of the material. Punctures or tears in the material shall be sealed or covered with additional sheeting.

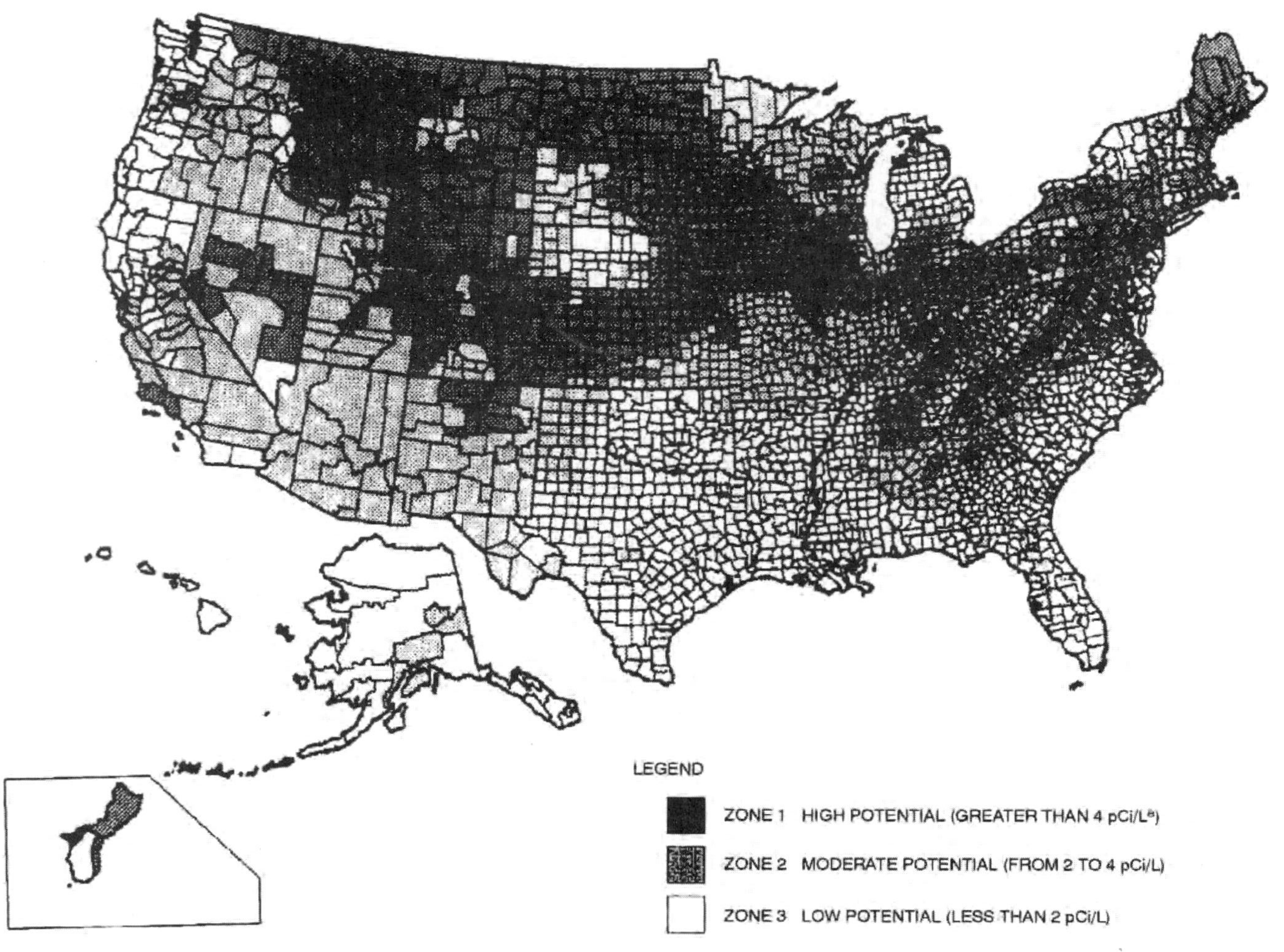

a. pCi/L stands for picocuries per liter of radon gas. The US Environmental Protection Agency (EPA) recommends that homes that measure 4 pCi/L and greater be mitigated.

The EPA and the US Geological Survey have evaluated the radon potential in the United States and have developed a map of radon zones designed to assist building officials in deciding whether radon-resistant features are applicable in new construction.

The map assigns each of the 3,141 counties in the United States to one of three zones based on radon potential. Each zone designation reflects the average short-term radon measurement that can be expected to be measured in a building without the implementation of radon-control methods. The radon zone designation of highest priority is Zone 1. Table AF101.1 lists the Zone 1 counties illustrated on the map. More detailed information can be obtained from state-specific booklets (EPA-401-R-93-021 through 070) available through the State Radon Offices or from the EPA Regional Offices.

FIGURE AF101.1
EPA MAP OF RADON ZONES

TABLE AF101.1
HIGH RADON-POTENTIAL (ZONE 1) COUNTIES[a]

ALABAMA
Calhoun
Clay
Cleburne
Colbert
Coosa
Franklin
Jackson
Lauderdale
Lawrence
Limestone
Madison
Morgan
Talladega
CALIFORNIA
Santa Barbara
Ventura
COLORADO
Adams
Arapahoe
Baca
Bent
Boulder
Chaffee
Cheyenne
Clear Creek
Crowley
Custer
Delta
Denver
Dolores
Douglas
El Paso
Elbert
Fremont
Garfield
Gilpin
Grand
Gunnison
Huerfano
Jackson
Jefferson
Kiowa
Kit Carson
Lake
Larimer
Las Animas
Lincoln
Logan
Mesa
Moffat
Montezuma
Montrose
Morgan
Otero
Ouray
Park
Phillips
Pitkin
Prowers
Pueblo
Rio Blanco
San Miguel
Summit
Teller
Washington
Weld
Yuma
CONNECTICUT
Fairfield
Middlesex
New Haven
New London
GEORGIA
Cobb
De Kalb
Fulton
Gwinnett
IDAHO
Benewah
Blaine
Boise
Bonner
Boundary
Butte
Camas
Clark
Clearwater
Custer
Elmore
Fremont
Gooding
Idaho
Kootenai
Latah
Lemhi
Shoshone
Valley
ILLINOIS
Adams
Boone
Brown
Bureau
Calhoun
Carroll
Cass
Champaign
Coles
De Kalb
De Witt
Douglas
Edgar
Ford
Fulton
Greene
Grundy
Hancock
Henderson
Henry
Iroquois
Jersey
Jo Daviess
Kane
Kendall
Knox
La Salle
Lee
Livingston
Logan
Macon
Marshall
Mason
McDonough
McLean
Menard
Mercer
Morgan
Moultrie
Ogle
Peoria
Piatt
Pike
Putnam
Rock Island
Sangamon
Schuyler
Scott
Stark
Stephenson
Tazewell
Vermilion
Warren
Whiteside
Winnebago
Woodford
INDIANA
Adams
Allen
Bartholomew
Benton
Blackford
Boone
Carroll
Cass
Clark
Clinton
De Kalb
Decatur
Delaware
Elkhart
Fayette
Fountain
Fulton
Grant
Hamilton
Hancock
Harrison
Hendricks
Henry
Howard
Huntington
Jay
Jennings
Johnson
Kosciusko
LaGrange
Lawrence
Madison
Marion
Marshall
Miami
Monroe
Montgomery
Noble
Orange
Putnam
Randolph
Rush
Scott
Shelby
St. Joseph
Steuben
Tippecanoe
Tipton
Union
Vermillion
Wabash
Warren
Washington
Wayne
Wells
White
Whitley
IOWA
All Counties

(continued)

TABLE AF101.1—continued
HIGH RADON-POTENTIAL (ZONE 1) COUNTIES[a]

KANSAS	Pawnee	Monroe	Lenawee	Pennington
Atchison	Phillips	Nelson	St. Joseph	Pipestone
Barton	Pottawatomie	Pendleton	Washtenaw	Polk
Brown	Pratt	Pulaski	**MINNESOTA**	Pope
Cheyenne	Rawlins	Robertson	Becker	Ramsey
Clay	Republic	Russell	Big Stone	Red Lake
Cloud	Rice	Scott	Blue Earth	Redwood
Decatur	Riley	Taylor	Brown	Renville
Dickinson	Rooks	Warren	Carver	Rice
Douglas	Rush	Woodford	Chippewa	Rock
Ellis	Saline	**MAINE**	Clay	Roseau
Ellsworth	Scott	Androscoggin	Cottonwood	Scott
Finney	Sheridan	Aroostook	Dakota	Sherburne
Ford	Sherman	Cumberland	Dodge	Sibley
Geary	Smith	Franklin	Douglas	Stearns
Gove	Stanton	Hancock	Faribault	Steele
Graham	Thomas	Kennebec	Fillmore	Stevens
Grant	Trego	Lincoln	Freeborn	Swift
Gray	Wallace	Oxford	Goodhue	Todd
Greeley	Washington	Penobscot	Grant	Traverse
Hamilton	Wichita	Piscataquis	Hennepin	Wabasha
Haskell	Wyandotte	Somerset	Houston	Wadena
Hodgeman	**KENTUCKY**	York	Hubbard	Waseca
Jackson	Adair	**MARYLAND**	Jackson	Washington
Jewell	Allen	Baltimore	Kanabec	Watonwan
Johnson	Barren	Calvert	Kandiyohi	Wilkin
Kearny	Bourbon	Carroll	Kittson	Winona
Kingman	Boyle	Frederick	Lac Qui Parle	Wright
Kiowa	Bullitt	Harford	Le Sueur	Yellow Medicine
Lane	Casey	Howard	Lincoln	**MISSOURI**
Leavenworth	Clark	Montgomery	Lyon	Andrew
Lincoln	Cumberland	Washington	Mahnomen	Atchison
Logan	Fayette	**MASS.**	Marshall	Buchanan
Marion	Franklin	Essex	Martin	Cass
Marshall	Green	Middlesex	McLeod	Clay
McPherson	Harrison	Worcester	Meeker	Clinton
Meade	Hart	**MICHIGAN**	Mower	Holt
Mitchell	Jefferson	Branch	Murray	Iron
Nemaha	Jessamine	Calhoun	Nicollet	Jackson
Ness	Lincoln	Cass	Nobles	Nodaway
Norton	Marion	Hillsdale	Norman	Platte
Osborne	Mercer	Jackson	Olmsted	
Ottawa	Metcalfe	Kalamazoo	Otter Tail	

(continued)

TABLE AF101.1—continued
HIGH RADON-POTENTIAL (ZONE 1) COUNTIES[a]

MONTANA
Beaverhead
Big Horn
Blaine
Broadwater
Carbon
Carter
Cascade
Chouteau
Custer
Daniels
Dawson
Deer Lodge
Fallon
Fergus
Flathead
Gallatin
Garfield
Glacier
Granite
Hill
Jefferson
Judith Basin
Lake
Lewis and Clark
Madison
McCone
Meagher
Missoula
Park
Phillips
Pondera
Powder River
Powell
Prairie
Ravalli
Richland
Roosevelt
Rosebud
Sanders
Sheridan
Silver Bow
Stillwater
Teton
Toole
Valley
Wibaux
Yellowstone
NEBRASKA
Adams
Boone
Boyd
Burt
Butler
Cass
Cedar
Clay
Colfax
Cuming
Dakota
Dixon
Dodge
Douglas
Fillmore
Franklin
Frontier
Furnas
Gage
Gosper
Greeley
Hamilton
Harlan
Hayes
Hitchcock
Hurston
Jefferson
Johnson
Kearney
Knox
Lancaster
Madison
Nance
Nemaha
Nuckolls
Otoe
Pawnee
Phelps
Pierce
Platte
Polk
Red Willow
Richardson
Saline
Sarpy
Saunders
Seward
Stanton
Thayer
Washington
Wayne
Webster
York
NEVADA
Carson City
Douglas
Eureka
Lander
Lincoln
Lyon
Mineral
Pershing
White Pine
NEW HAMPSHIRE
Carroll
NEW JERSEY
Hunterdon
Mercer
Monmouth
Morris
Somerset
Sussex
Warren
NEW MEXICO
Bernalillo
Colfax
Mora
Rio Arriba
San Miguel
Santa Fe
Taos
NEW YORK
Albany
Allegany
Broome
Cattaraugus
Cayuga
Chautauqua
Chemung
Chenango
Columbia
Cortland
Delaware
Dutchess
Erie
Genesee
Greene
Livingston
Madison
Onondaga
Ontario
Orange
Otsego
Putnam
Rensselaer
Schoharie
Schuyler
Seneca
Steuben
Sullivan
Tioga
Tompkins
Ulster
Washington
Wyoming
Yates
N. CAROLINA
Alleghany
Buncombe
Cherokee
Henderson
Mitchell
Rockingham
Transylvania
Watauga
N. DAKOTA
All Counties
OHIO
Adams
Allen
Ashland
Auglaize
Belmont
Butler
Carroll
Champaign
Clark
Clinton
Columbiana
Coshocton
Crawford
Darke
Delaware
Fairfield
Fayette
Franklin
Greene
Guernsey
Hamilton
Hancock
Hardin
Harrison
Holmes
Huron
Jefferson
Knox
Licking
Logan
Madison
Marion
Mercer
Miami
Montgomery
Morrow
Muskingum
Perry

(continued)

TABLE AF101.1—continued
HIGH RADON-POTENTIAL (ZONE 1) COUNTIES[a]

OHIO—continued
Pickaway
Pike
Preble
Richland
Ross
Seneca
Shelby
Stark
Summit
Tuscarawas
Union
Van Wert
Warren
Wayne
Wyandot
PENNSYLVANIA
Adams
Allegheny
Armstrong
Beaver
Bedford
Berks
Blair
Bradford
Bucks
Butler
Cameron
Carbon
Centre
Chester
Clarion
Clearfield
Clinton
Columbia
Cumberland
Dauphin
Delaware
Franklin
Fulton
Huntingdon
Indiana
Juniata
Lackawanna
Lancaster
Lebanon
Lehigh
Luzerne
Lycoming
Mifflin
Monroe
Montgomery
Montour
Northampton
Northumberland
Perry
Schuylkill
Snyder
Sullivan
Susquehanna
Tioga
Union
Venango
Westmoreland
Wyoming
York
RHODE ISLAND
Kent
Washington
S. CAROLINA
Greenville
S. DAKOTA
Aurora
Beadle
Bon Homme
Brookings
Brown
Brule
Buffalo
Campbell
Charles Mix
Clark
Clay
Codington
Corson
Davison
Day
Deuel
Douglas
Edmunds
Faulk
Grant
Hamlin
Hand
Hanson
Hughes
Hutchinson
Hyde
Jerauld
Kingsbury
Lake
Lincoln
Lyman
Marshall
McCook
McPherson
Miner
Minnehaha
Moody
Perkins
Potter
Roberts
Sanborn
Spink
Stanley
Sully
Turner
Union
Walworth
Yankton
TENNESEE
Anderson
Bedford
Blount
Bradley
Claiborne
Davidson
Giles
Grainger
Greene
Hamblen
Hancock
Hawkins
Hickman
Humphreys
Jackson
Jefferson
Knox
Lawrence
Lewis
Lincoln
Loudon
Marshall
Maury
McMinn
Meigs
Monroe
Moore
Perry
Roane
Rutherford
Smith
Sullivan
Trousdale
Union
Washington
Wayne
Williamson
Wilson
UTAH
Carbon
Duchesne
Grand
Piute
Sanpete
Sevier
Uintah
VIRGINIA
Alleghany
Amelia
Appomattox
Augusta
Bath
Bland
Botetourt
Bristol
Brunswick
Buckingham
Buena Vista
Campbell
Chesterfield
Clarke
Clifton Forge
Covington
Craig
Cumberland
Danville
Dinwiddie
Fairfax
Falls Church
Fluvanna
Frederick
Fredericksburg
Giles
Goochland
Harrisonburg
Henry
Highland
Lee
Lexington
Louisa
Martinsville
Montgomery
Nottoway
Orange
Page
Patrick
Pittsylvania
Powhatan
Pulaski
Radford
Roanoke
Rockbridge
Rockingham
Russell
Salem
Scott
Shenandoah
Smyth
Spotsylvania

(continued)

TABLE AF101.1—continued
HIGH RADON-POTENTIAL (ZONE 1) COUNTIES[a]

VIRGINIA—continued
Stafford
Staunton
Tazewell
Warren
Washington
Waynesboro
Winchester
Wythe
WASHINGTON
Clark
Ferry
Okanogan
Pend Oreille
Skamania
Spokane
Stevens
W. VIRGINIA
Berkeley
Brooke
Grant
Greenbrier
Hampshire
Hancock
Hardy
Jefferson
Marshall
Mercer
Mineral
Monongalia
Monroe
Morgan
Ohio
Pendleton
Pocahontas
Preston
Summers
Wetzel
WISCONSIN
Buffalo
Crawford
Dane
Dodge
Door
Fond du Lac
Grant
Green
Green Lake
Iowa
Jefferson
Lafayette
Langlade
Marathon
Menominee
Pepin
Pierce
Portage
Richland
Rock
Shawano
St. Croix
Vernon
Walworth
Washington
Waukesha
Waupaca
Wood
WYOMING
Albany
Big Horn
Campbell
Carbon
Converse
Crook
Fremont
Goshen
Hot Springs
Johnson
Laramie
Lincoln
Natrona
Niobrara
Park
Sheridan
Sublette
Sweetwater
Teton
Uinta
Washakie

a. The EPA recommends that this county listing be supplemented with other available state and local data to further understand the radon potential of a Zone 1 area.

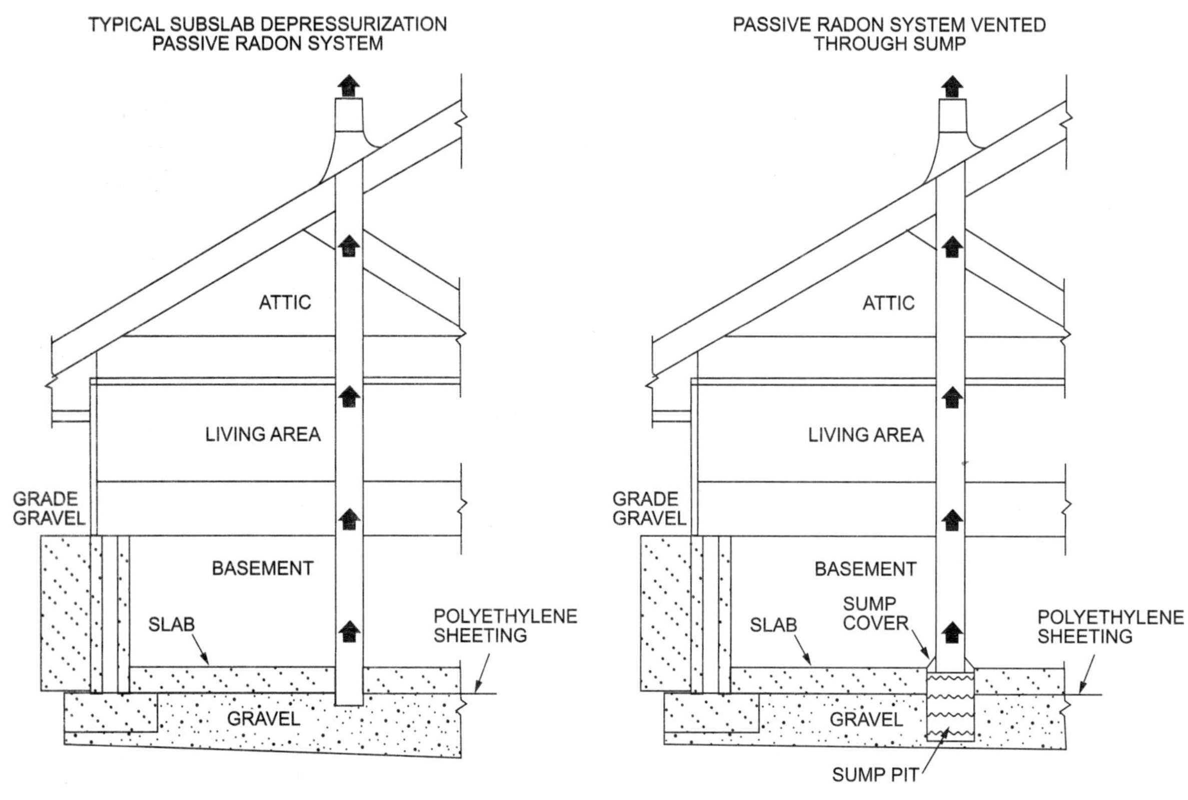

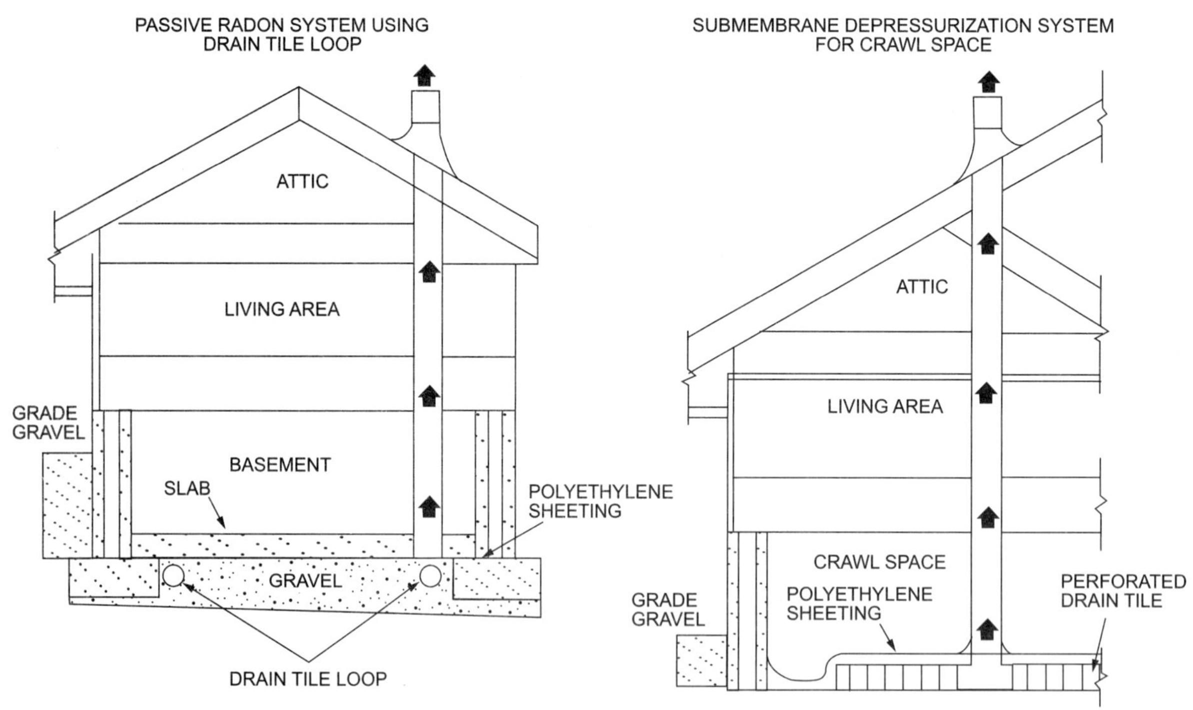

FIGURE AF103.1
RADON-RESISTANT CONSTRUCTION DETAILS FOR FOUR FOUNDATION TYPES

AF103.4 Entry routes. Potential radon entry routes shall be closed in accordance with Sections AF103.4.1 through AF103.4.10.

AF103.4.1 Floor openings. Openings around bathtubs, showers, water closets, pipes, wires or other objects that penetrate concrete slabs, or other floor assemblies, shall be filled with a polyurethane caulk or equivalent sealant applied in accordance with the manufacturer's recommendations.

AF103.4.2 Concrete joints. Control joints, isolation joints, construction joints, and any other joints in concrete slabs or between slabs and foundation walls shall be sealed with a caulk or sealant. Gaps and joints shall be cleared of loose material and filled with polyurethane caulk or other elastomeric sealant applied in accordance with the manufacturer's recommendations.

AF103.4.3 Condensate drains. Condensate drains shall be trapped or routed through nonperforated pipe to daylight.

AF103.4.4 Sumps. Sump pits open to soil or serving as the termination point for subslab or exterior drain tile loops shall be covered with a gasketed or otherwise sealed lid. Sumps used as the suction point in a subslab depressurization system shall have a lid designed to accommodate the vent pipe. Sumps used as a floor drain shall have a lid equipped with a trapped inlet.

AF103.4.5 Foundation walls. Hollow block masonry foundation walls shall be constructed with either a continuous course of *solid masonry*, one course of masonry grouted solid, or a solid concrete beam at or above finished ground surface to prevent the passage of air from the interior of the wall into the living space. Where a brick veneer or other masonry ledge is installed, the course immediately below that ledge shall be sealed. Joints, cracks or other openings around all penetrations of both exterior and interior surfaces of masonry block or wood foundation walls below the ground surface shall be filled with polyurethane caulk or equivalent sealant. Penetrations of concrete walls shall be filled.

AF103.4.6 Dampproofing. The exterior surfaces of portions of concrete and masonry block walls below the ground surface shall be dampproofed in accordance with Section R406.

AF103.4.7 Air-handling units. Air-handling units in crawl spaces shall be sealed to prevent air from being drawn into the unit.

Exception: Units with gasketed seams or units that are otherwise sealed by the manufacturer to prevent leakage.

AF103.4.8 Ducts. Ductwork passing through or beneath a slab shall be of seamless material unless the air-handling system is designed to maintain continuous positive pressure within such ducting. Joints in such ductwork shall be sealed to prevent air leakage.

Ductwork located in crawl spaces shall have seams and joints sealed by closure systems in accordance with Section M1601.4.1.

AF103.4.9 Crawl space floors. Openings around all penetrations through floors above crawl spaces shall be caulked or otherwise filled to prevent air leakage.

AF103.4.10 Crawl space access. Access doors and other openings or penetrations between *basements* and adjoining crawl spaces shall be closed, gasketed or otherwise filled to prevent air leakage.

AF103.5 Passive submembrane depressurization system. In buildings with *crawl space* foundations, the following components of a passive submembrane depressurization system shall be installed during construction.

Exception: Buildings in which an *approved* mechanical *crawl space* ventilation system or other equivalent system is installed.

AF103.5.1 Ventilation. Crawl spaces shall be provided with vents to the exterior of the building. The minimum net area of ventilation openings shall comply with Section R408.1.

AF103.5.2 Soil-gas-retarder. The soil in crawl spaces shall be covered with a continuous layer of minimum 6-mil (0.15 mm) polyethylene soil-gas-retarder. The ground cover shall be lapped not less than 12 inches (305 mm) at joints and shall extend to all foundation walls enclosing the *crawl space* area.

AF103.5.3 Vent pipe. A plumbing tee or other *approved* connection shall be inserted horizontally beneath the sheeting and connected to a 3- or 4-inch-diameter (76 or 102 mm) fitting with a vertical vent pipe installed through the sheeting. The vent pipe shall be extended up through the building floors, and terminate not less than 12 inches (305 mm) above the roof in a location not less than 10 feet (3048 mm) away from any window or other opening into the *conditioned spaces* of the building that is less than 2 feet (610 mm) below the exhaust point, and 10 feet (3048 mm) from any window or other opening in adjoining or adjacent buildings.

AF103.6 Passive subslab depressurization system. In *basement* or slab-on-grade buildings, the following components of a passive subslab depressurization system shall be installed during construction.

AF103.6.1 Vent pipe. A minimum 3-inch-diameter (76 mm) ABS, PVC or equivalent gastight pipe shall be embedded vertically into the subslab aggregate or other permeable material before the slab is cast. A "T" fitting or equivalent method shall be used to ensure that the pipe opening remains within the subslab permeable material. Alternatively, the 3-inch (76 mm) pipe shall be inserted directly into an interior perimeter drain tile loop or through a sealed sump cover where the sump is exposed to the subslab aggregate or connected to it through a drainage system.

The pipe shall be extended up through the building floors, and terminate not less than 12 inches (305 mm) above the surface of the roof in a location not less than 10 feet (3048 mm) away from any window or other opening into the *conditioned spaces* of the building that is less than 2 feet (610 mm) below the exhaust point, and 10 feet

(3048 mm) from any window or other opening in adjoining or adjacent buildings.

AF103.6.2 Multiple vent pipes. In buildings where interior footings or other barriers separate the subslab aggregate or other gas-permeable material, each area shall be fitted with an individual vent pipe. Vent pipes shall connect to a single vent that terminates above the roof or each individual vent pipe shall terminate separately above the roof.

AF103.7 Vent pipe drainage. Components of the radon vent pipe system shall be installed to provide positive drainage to the ground beneath the slab or soil-gas-retarder.

AF103.8 Vent pipe accessibility. Radon vent pipes shall be accessible for future fan installation through an attic or other area outside the *habitable space.*

Exception: The radon vent pipe need not be accessible in an attic space where an *approved* roof-top electrical supply is provided for future use.

AF103.9 Vent pipe identification. Exposed and visible interior radon vent pipes shall be identified with not less than one *label* on each floor and in accessible *attics.* The *label* shall read: "Radon Reduction System."

AF103.10 Combination foundations. Combination *basement/crawl space* or slab-on-grade/*crawl space* foundations shall have separate radon vent pipes installed in each type of foundation area. Each radon vent pipe shall terminate above the roof or shall be connected to a single vent that terminates above the roof.

AF103.11 Building depressurization. Joints in air ducts and plenums in unconditioned spaces shall meet the requirements of Section M1601. Thermal envelope air infiltration requirements shall comply with the energy conservation provisions in Chapter 11. Fireblocking shall meet the requirements contained in Section R302.11.

AF103.12 Power source. To provide for future installation of an active submembrane or subslab depressurization system, an electrical circuit terminated in an *approved* box shall be installed during construction in the attic or other anticipated location of vent pipe fans. An electrical supply shall be accessible in anticipated locations of system failure alarms.

SECTION AF104
TESTING

AF104.1 Testing. Where radon-resistant construction is required, radon testing shall be as specified in Items 1 through 11:

1. Testing shall be performed after the dwelling passes its air tightness test.
2. Testing shall be performed after the radon control system and HVAC installations are complete. The HVAC system shall be operating during the test. Where the radon system has an installed fan, the dwelling shall be tested with the radon fan operating.
3. Testing shall be performed at the lowest occupied floor level, whether or not that space is finished. Spaces that are physically separated and served by different HVAC systems shall be tested separately.
4. Testing shall not be performed in a closet, hallway, *stairway,* laundry room, furnace room, bathroom or kitchen.
5. Testing shall be performed with a commercially available radon test kit or testing shall be performed by an *approved* third party with a continuous radon monitor. Testing with test kits shall include two tests, and the test results shall be averaged. Testing shall be in accordance with this section and the testing laboratory kit manufacturer's instructions.
6. Testing shall be performed with the windows closed. Testing shall be performed with the exterior doors closed, except when being used for entrance or exit. Windows and doors shall be closed for not fewer than 12 hours prior to the testing.
7. Testing shall be performed by the builder, a *registered design professional* or an *approved* third party.
8. Testing shall be conducted over a period of not less than 48 hours or not less that the period specified by the testing device manufacturer, whichever is longer.
9. Written radon test results shall be provided by the test lab or testing party. The final written test report with results less than 4 picocuries per liter (pCi/L) shall be provided to the code official.
10. Where the radon test result is 4 pCi/L or greater, the fan for the radon vent pipe shall be installed as specified in Sections AF103.9 and AF103.12.
11. Where the radon test result is 4 pCi/L or greater, the system shall be modified and retested until the test result is less than 4 pCi/L.

Exception: Testing is not required where the occupied space is located above an unenclosed open space.

APPENDIX AG

PIPING STANDARDS FOR VARIOUS APPLICATIONS

User note:

About this appendix: *Plastic piping is commonly used in the construction of buildings covered by this code. Appendix AG provides a table of the most commonly used standards for plastic piping in a variety of applications. Although some standards in this table are not referenced in the code, the majority of standards indicated are listed in the Referenced Standards chapter of this code.*

SECTION AG101
PLASTIC PIPING STANDARDS

AG101.1 Plastic piping. Table AG101.1 provides a list of plastic piping product standards for various applications.

SECTION AG102
REFERENCED STANDARDS

AG102.1 General. See Table AG102.1 for standards that are referenced in this appendix. Standards are listed by the standard identification with the effective date, standard title, and the table that references the standard.

TABLE AG101.1
PLASTIC PIPING STANDARDS FOR VARIOUS APPLICATIONS[a, b]

APPLICATION	LOCATION	TYPE OF PLASTIC PIPING								
		ABS	CPVC	PE	PE-AL-PE	PE-RT	PEX	PEX-AL-PEX	PP	PVC
Central vacuum	System piping	—	—	—	—	—	—	—	—	ASTM F2158
Foundation drainage	System piping	ASTM F628	—	ASTM F405	—	—	—	—	—	ASTM D2665; ASTM D2729; ASTM D3034
Geothermal ground loop	System piping	—	ASTM D2846; ASTM F441; ASTM F442; ASTM F2855; CSA B137.6	ASTM D2239; ASTM D2737; ASTM D3035	ASTM F1282	ASTM F2623; ASTM F2769; CSA B137.18	ASTM F876; CSA B137.5	ASTM F1281	ASTM F2389; CSA B137.11	ASTM D1785; ASTM D2241; CSA B137.3
	Loop piping	—	—	ASTM D2239; ASTM D2737; ASTM D3035; NSF 358-1	ASTM F1282	ASTM F2623; ASTM F2769; CSA B137.18	ASTM F876; CSA B137.5	—	ASTM F2389; CSA B137.11	—
Graywater	Nonpressure distribution/ collection	ASTM F628	—	ASTM D2239; ASTM D2737; ASTM D3035; ASTM F2306	—	—	—	—	ASTM F2389; CSA B137.11	ASTM D1785; ASTM D2729; ASTM D2949; ASTM D3034; ASTM F891; ASTM F1760; CSA B137.3
	Pressure/ distribution	—	ASTM D2846; ASTM F441; ASTM F442; ASTM F2855; CSA B137.6	ASTM D2239; ASTM D2737; ASTM D3035	ASTM F1282	ASTM F2623; ASTM F2769; CSA B137.18	ASTM F876; CSA B137.5	ASTM F1281	ASTM F2389; CSA B137.11	ASTM D1785; ASTM D2241; CSA B137.3
Radiant cooling	Loop piping	—	ASTM D2846; ASTM F441; ASTM F442; ASTM F2855	ASTM D2239; ASTM D2737; ASTM D3035	ASTM F1282	ASTM F2623; ASTM F2769; CSA B137.18	ASTM F876; CSA B137.5	ASTM F1281	ASTM F2389; CSA B137.11	—

(continued)

TABLE AG101.1—continued
PLASTIC PIPING STANDARDS FOR VARIOUS APPLICATIONS[a, b]

APPLICATION	LOCATION	TYPE OF PLASTIC PIPING								
		ABS	CPVC	PE	PE-AL-PE	PE-RT	PEX	PEX-AL-PEX	PP	PVC
Radiant heating	Loop piping	—	ASTM D2846; ASTM F441; ASTM F442; ASTM F2855	—	ASTM F1282	ASTM F2623; ASTM F2769; CSA B137.18	ASTM F876; CSA B137.5	ASTM F1281	ASTM F2389; CSA B137.11	—
Rainwater harvesting	Nonpressure/ collection	ASTM F628	—	ASTM F1901	—	—	—	—	ASTM F2389; CSA B137.11	ASTM D1785; ASTM D2729; ASTM D2949; ASTM F891; ASTM F1760; CSA B137.3
	Pressure/ distribution	—	ASTM D2846/ D2846M; ASTM F441; ASTM F442; ASTM F2855; CSA B137.6	ASTM D2239 ASTM D2737; ASTM D3035	ASTM F1282	ASTM F2623; ASTM F2769; CSA B137.18	ASTM F876; CSA B137.5	ASTM F1281	ASTM F2389; CSA B137.11	ASTM D1785; ASTM D2241; CSA B137.3
Radon venting	System piping	ASTM F628	—	—	—	—	—	—	—	ASTM D1785; ASTM F891; ASTM F1760
Reclaimed water	Main to building service	—	ASTM D2846/ D2846M; ASTM F441; ASTM F442; ASTM F2855; CSA B137.6	ASTM D3035; AWWA C901; CSA B137.1	ASTM F1282	ASTM F2623; ASTM F2769; CSA B137.18	ASTM F876; AWWA C904; CSA B137.5	—	ASTM F2389; CSA B137.11	ASTM D1785; ASTM D2241; AWWA C905; CSA B137.3
	Pressure/ distribution/ irrigation	—	ASTM D2846; ASTM F441; ASTM F442; ASTM F2855; CSA B137.6	ASTM D2239; ASTM D2737; ASTM D3035	ASTM F1282	ASTM F2623; ASTM F2769; CSA B137.18	ASTM F876; CSA B137.5	ASTM F1281	ASTM F2389; AWWA C900; CSA B137.11	ASTM D1785; ASTM D2241; AWWA C900

(continued)

TABLE AG101.1—continued
PLASTIC PIPING STANDARDS FOR VARIOUS APPLICATIONS[a, b]

APPLICATION	LOCATION	TYPE OF PLASTIC PIPING								
		ABS	CPVC	PE	PE-AL-PE	PE-RT	PEX	PEX-AL-PEX	PP	PVC
Residential fire sprinklers[c]	Sprinkler piping	—	ASTM F441; ASTM F442; CSA B137.6; UL 1821	—	—	ASTM F2769; CSA B137.18	ASTM F876; CSA B137.5; UL 1821	—	ASTM F2389; CSA B137.11	—
Solar heating	Pressure/ distribution	—	ASTM D2846; ASTM F441; ASTM F442; ASTM F2855	—	—	ASTM F2623; ASTM F2769; CSA B137.18	ASTM F876; CSA B137.5	ASTM F1281	ASTM F2389; CSA B137.11	—

a. This table indicates manufacturing standards for plastic piping materials that are suitable for use in the applications indicated. Such applications support green and sustainable building practices. The system designer or the installer of piping shall verify that the piping chosen for an application complies with local codes and the recommendations of the manufacturer of the piping.

b. Fittings applicable for the piping shall be as recommended by the manufacturer of the piping.

c. Piping systems for fire sprinkler applications shall be listed for the application.

TABLE AG102.1
REFERENCED STANDARDS

STANDARD ACRONYM	STANDARD NAME	TABLE HEREIN REFERENCED
ASTM D1785—15E1	*Specification for Poly (Vinyl Chloride) (PVC) Plastic Pipe, Schedules 40, 80 and 120*	Table AG101.1
ASTM D2239—12A	*Specification for Polyethylene (PE) Plastic Pipe (SIDR-PR) Based on Controlled Inside Diameter*	Table AG101.1
ASTM D2241—15	*Specification for Poly (Vinyl Chloride) (PVC) Pressure-rated Pipe (SDR-Series)*	Table AG101.1
ASTM D2665—14	*Specification for Poly (Vinyl Chloride) (PVC) Plastic Drain, Waste and Vent Pipe and Fittings*	Table AG101.1
ASTM D2729—17	*Specification for Poly (Vinyl Chloride) (PVC) Sewer Pipe and Fittings*	Table AG101.1
ASTM D2737—2012A	*Specification for Polyethylene (PE) Plastic Tubing*	Table AG101.1
ASTM D2846/ D2846M—2017BE1	*Specification for Chlorinated Poly (Vinyl Chloride) (CPVC) Plastic Hot- and Cold-water Distribution Systems*	Table AG101.1
ASTM D2949—10	*Specification for 3.25-in. Outside Diameter Poly (Vinyl Chloride) (PVC) Plastic Drain, Waste and Vent Pipe and Fittings*	Table AG101.1
ASTM D3034—2016	*Specification for Type PSM Poly (Vinyl Chloride) (PVC) Sewer Pipe and Fittings*	Table AG101.1
ASTM D3035—15	*Specification for Polyethylene (PE) Plastic Pipe (DR-PR) Based on Controlled Outside Diameter*	Table AG101.1
ASTM F405—05	*Specification for Corrugated Polyethylene (PE) Pipe and Fittings*	Table AG101.1
ASTM F441/F441M—15	*Specification for Chlorinated Poly (Vinyl Chloride) (CPVC) Plastic Pipe, Schedules 40 and 80*	Table AG101.1
ASTM F442/F442M—13E1	*Specification for Chlorinated Poly (Vinyl Chloride) (CPVC) Plastic Pipe (SDR-PR)*	Table AG101.1
ASTM F628—2012E2	*Specification for Acrylonitrile-butadiene-styrene (ABS) Schedule 40 Plastic Drain, Waste and Vent Pipe with a Cellular Core*	Table AG101.1
ASTM F876—2017	*Specification for Cross-linked Polyethylene (PEX) Tubing*	Table AG101.1
ASTM F891—2016	*Specification for Coextruded Poly (Vinyl Chloride) (PVC) Plastic Pipe with a Cellular Core*	Table AG101.1
ASTM F1281—2017	*Specification for Cross-linked Polyethylene/Aluminum/ Cross-linked Polyethylene (PEX-AL-PEX) Pressure Pipe*	Table AG101.1
ASTM F1282—2017	*Specification for Polyethylene/Aluminum/Polyethylene (PE-AL-PE) Composite Pressure Pipe*	Table AG101.1
ASTM F1760—01 (2011)	*Standard Specification for Coextruded Poly (Vinyl Chloride) (PVC) Non-Pressure Plastic Pipe Having Reprocessed-Recycled Content*	Table AG101.1
ASTM F1901—10	*Standard Specification for Polyethylene (PE) Pipe and Fittings for Roof Drain Systems*	Table AG101.1
ASTM F2158—08 (2016)	*Standard for Residential Central-vacuum Tube and Fittings*	Table AG101.1
ASTM F2306/ F2306M—2018	*Standard Specification for 12" to 60" Annular Corrugated Profile-wall Polyethylene (PE) Pipe and Fittings for Gravity Flow Storm Sewer and Sub-surface Drainage Applications*	Table AG101.1
ASTM F2389—2017A	*Standard for Pressure-rated Polypropylene (PP) Piping Systems*	Table AG101.1
ASTM F2623—14	*Standard Specification for Polyethylene of Raised Temperature (PE-RT) SDRG Tubing*	Table AG101.1
ASTM F2769—18	*Polyethylene or Raised Temperature (PE-RT) Plastic Hot and Cold-Water Tubing and Distribution Systems*	Table AG101.1
ASTM F2855—12	*Standard Specification for Chlorinated Poly (Vinyl Chloride)/ Aluminum/Chlorinated Poly (Vinyl Chloride) (CPVC AL CPVC) Composite Pressure Tubing*	Table AG101.1

(continued)

TABLE AG102.1—continued
REFERENCED STANDARDS

STANDARD ACRONYM	STANDARD NAME	TABLE HEREIN REFERENCED
AWWA C900—07	*Polyvinyl chloride (PVC) Pressure Pipe and Fabricated Fittings, 4 in. through 12 in. (350 mm through 1200 mm), for Water Transmission and Distribution*	Table AG101.1
AWWA C901—16	*Polyethylene (PE) Pressure Pipe and Tubing $^1/_2$ in. (13 mm) through 3 in. (76 mm) for Water Service*	Table AG101.1
AWWA C904—16	*Cross-linked Polyethylene (PEX) Pressure Tubing, $^1/_2$ in. (13 mm) through 3 in. (76 mm) for Water Service*	Table AG101.1
AWWA C905—10	*Polyvinyl chloride (PVC) Pressure Pipe and Fabricated Fittings, 14 in. through 48 in. (100 mm through 300 mm)*	Table AG101.1
CSA B137.1—17	*Polyethylene (PE) Pipe, Tubing and Fittings for Cold Water Pressure Services*	Table AG101.1
CSA B137.3—17	*Rigid Poly (Vinyl Chloride) (PVC) Pipe for Pressure Applications*	Table AG101.1
CSA B137.5—17	*Cross-linked Polyethylene (PEX) Tubing Systems for Pressure Applications*	Table AG101.1
CSA B137.6—17	*Chlorinated polyvinylchloride CPVC Pipe, Tubing and Fittings for Hot- and Cold-water Distribution Systems*	Table AG101.1
CSA B137.11—17	*Polypropylene (PP-R) Pipe and Fittings for Pressure Applications*	Table AG101.1
CSA B137.18—17	*Polyethylene of Raised Temperature (PE-RT) Tubing Systems for Pressure Applications*	Table AG101.1
NSF 358-1—2017	*Polyethylene Pipe and Fittings for Water-based Ground Source "Geothermal" Heat Pump Systems*	Table AG101.1
UL 1821—2011	*Standard for Thermoplastic Sprinkler Pipe and Fittings for Fire Protection Service—with revisions through August 2015*	Table AG101.1

APPENDIX AH
PATIO COVERS

The provisions contained in this appendix are not mandatory unless specifically referenced in the adopting ordinance.

User note:

About this appendix: *Appendix AH relaxes certain provisions contained in the body of the code as related to patio covers, including those regarding: permitted uses; exterior wall insect screens; glazing and translucent or transparent plastic; light, ventilation and emergency egress; height; structural design loads; and footings. This appendix also includes provisions that are specifically applicable to hurricane-prone regions.*

SECTION AH101
GENERAL

AH101.1 Scope. Patio covers shall conform to the requirements of Sections AH101 through AH106.

AH101.2 Permitted uses. Patio covers detached from or attached to *dwelling units* shall be used only for recreational, outdoor living purposes, and not as carports, garages, storage rooms or habitable rooms.

SECTION AH102
DEFINITION

AH102.1 General. The following word and term shall, for the purposes of this appendix, have the meaning shown herein.

PATIO COVER. A structure with open or glazed walls that is used for recreational, outdoor living purposes associated with a *dwelling unit.*

SECTION AH103
EXTERIOR WALLS AND OPENINGS

AH103.1 Enclosure walls. Enclosure walls shall be permitted to be of any configuration, provided that the open or glazed area of the longer wall and one additional wall is not less than 65 percent of the area below 6 feet 8 inches (2032 mm) of each wall, measured from the floor. Openings shall be enclosed with any of the following:

1. Insect screening.
2. *Approved* translucent or transparent plastic not more than 0.125 inch (3.2 mm) in thickness.
3. Glass conforming to the provisions of Section R308.
4. Any combination of the foregoing.

AH103.2 Light, ventilation and emergency egress. Exterior openings required for light and ventilation into a patio structure conforming to Section AH101 shall be unenclosed where such openings serve as emergency egress or rescue openings from sleeping rooms. Where such exterior openings serve as an exit from the *dwelling unit,* the patio structure, unless unenclosed, shall be provided with exits conforming to the provisions of Section R311.

SECTION AH104
HEIGHT

AH104.1 Height. Patio covers are limited to one-story structures not exceeding 12 feet (3657 mm) in height.

SECTION AH105
STRUCTURAL PROVISIONS

AH105.1 Design loads. Patio covers shall be designed and constructed to sustain, within the stress limits of this code, all dead loads plus a vertical *live load* of not less than 10 pounds per square foot (0.48 kN/m^2), except that snow loads shall be used where such snow loads exceed this minimum. Such covers shall be designed to resist the minimum wind loads set forth in Section R301.2.1.

AH105.2 Footings. In areas with a frostline depth of zero as specified in Table R301.2, for patio covers supported on a slab-on-*grade* without footings, the slab shall conform to the provisions of Section R506, shall be not less than 3.5 inches (89 mm) thick and the columns shall not support live and dead loads in excess of 750 pounds (3.34 kN) per column.

SECTION AH106
SPECIAL PROVISIONS FOR ALUMINUM SCREEN ENCLOSURES IN HURRICANE-PRONE REGIONS

AH106.1 General. Screen enclosures in *hurricane-prone regions* shall be in accordance with the provisions of this section.

AH106.1.1 Habitable spaces. Screen enclosures shall not be considered *habitable spaces.*

AH106.1.2 Minimum ceiling height. Screen enclosures shall have a ceiling height of not less than 7 feet (2134 mm).

AH106.2 Definition. The following word and term shall, for the purposes of this appendix, have the meaning shown herein.

SCREEN ENCLOSURE. A building or part thereof, in whole or in part self-supporting, and having walls of insect screening, and a roof of insect screening, plastic, aluminum or similar lightweight material.

AH106.3 Screen enclosures. Screen enclosures shall comply with Sections AH106.3.1 and AH106.3.2.

AH106.3.1 Thickness. Actual wall thickness of extruded aluminum members shall be not less than 0.040 inch (1.02 mm).

AH106.3.2 Density. Screen density shall be not more than 20 threads per inch by 20 threads per inch mesh.

AH106.4 Design. The structural design of screen enclosures shall comply with Sections AH106.4.1 through AH106.4.3.

AH106.4.1 Wind load. Structural members supporting screen enclosures shall be designed to support the minimum wind loads given in Tables AH106.4.1(1) and AH106.4.1(2) for the ultimate design wind speed, V_{ult}, determined from Figure AH106.4.1. Where any value is less than 10 pounds per square foot (psf) (0.479 kN/m^2) use 10 pounds per square foot (0.479 kN/m^2).

AH106.4.2 Deflection limit. For members supporting screen surfaces only, the total load deflection shall not exceed *l*/60. Screen surfaces shall be permitted to include not more than 25-percent solid flexible finishes.

AH106.4.3 Roof live load. The roof *live load* shall be not less than 10 psf (0.479 kN/m^2).

AH106.5 Footings. In areas with a frost line depth of zero, screen enclosures supported on a concrete slab-on-*grade* without footings shall conform to the provisions of Section R506, be not less than $3^1/_2$ inches (89 mm) thick and the columns shall not support loads in excess of 750 pounds (3.36 kN) per column.

TABLE AH106.4.1(2)
ADJUSTMENT FACTOR FOR BUILDING HEIGHT AND EXPOSURE

MEAN ROOF HEIGHT (feet)	EXPOSURE		
	B	C	D
15	1.00	1.21	1.47
20	1.00	1.29	1.55
25	1.00	1.35	1.61
30	1.00	1.40	1.66
35	1.05	1.45	1.70
40	1.09	1.49	1.74
45	1.12	1.53	1.78
50	1.16	1.56	1.81
55	1.19	1.59	1.84
60	1.22	1.62	1.87

For SI: 1 foot = 304.8 mm.

TABLE AH106.4.1(1)
DESIGN WIND PRESSURES FOR SCREEN ENCLOSURE FRAMING[a, b, e, f, g, h]

LOAD CASE	WALL	ULTIMATE DESIGN WIND SPEED, V_{ult} (mph)											
		90	95	100	105	110	120	130	140	150	160	170	180
		Exposure Category B Design Pressure (psf)											
A[c]	Windward and leeward walls (flow thru) and windward wall (nonflow thru) *L/W* = 0-1	5	6	6	7	8	9	11	13	14	16	18	21
A[c]	Windward and leeward walls (flow thru) and windward wall (nonflow thru) *L/W* = 2	6	7	7	8	9	11	12	14	16	19	21	24
B[d]	Windward: Nongable roof	7	8	9	10	11	13	15	18	21	23	26	30
B[d]	Windward: Gable roof	10	10	11	13	14	16	19	22	26	29	33	37
	ROOF												
All[e]	Roof-screen	2	2	2	3	3	3	4	4	5	6	7	7
All[e]	Roof-solid	6	6	7	8	8	10	12	13	15	18	20	22

For SI: 1 inch = 25.4 mm, 1 mile per hour = 0.44 m/s, 1 pound per square foot = 0.0479 kPa, 1 foot = 304.8 mm.

a. Design pressure shall be not less than 10 psf in accordance with Section AH106.4.1.

b. Loads are applicable to screen enclosures with a mean roof height of 30 feet or less in Exposure B. For screen enclosures of different heights or exposure, the pressures given shall be adjusted by multiplying the table pressure by the adjustment factor given in Table AH106.4.1(2).

c. For Load Case A flow-thru condition, the pressure given shall be applied simultaneously to both the upwind and downwind screen walls acting in the same direction as the wind. The structure shall be analyzed for wind coming from the opposite direction. For the nonflow-thru condition, the screen enclosure wall shall be analyzed for the load applied acting toward the interior of the enclosure.

d. For Load Case B, the table pressure multiplied by the projected frontal area of the screen enclosure is the total drag force, including drag on screen surfaces parallel to the wind, that must be transmitted to the ground. Use Load Case A for members directly supporting the screen surface perpendicular to the wind. Load Case B loads shall be applied only to structural members that carry wind loads from more than one surface.

e. The roof structure shall be analyzed for the pressure given occurring both upward and downward.

f. Table pressures are MWFRS loads. The design of solid roof panels and their attachments shall be based on component and cladding loads for enclosed or partially enclosed structures as appropriate.

g. Table pressures apply to 20-inch by 20-inch by 0.013-inch mesh screen. For 18-inch by 14-inch by 0.013-inch mesh screen, pressures on screen surfaces shall be permitted to be multiplied by 0.88. For screen densities greater than 20 inches by 20 inches by 0.013 inch, pressures for enclosed buildings shall be used.

h. Linear interpolation shall be permitted.

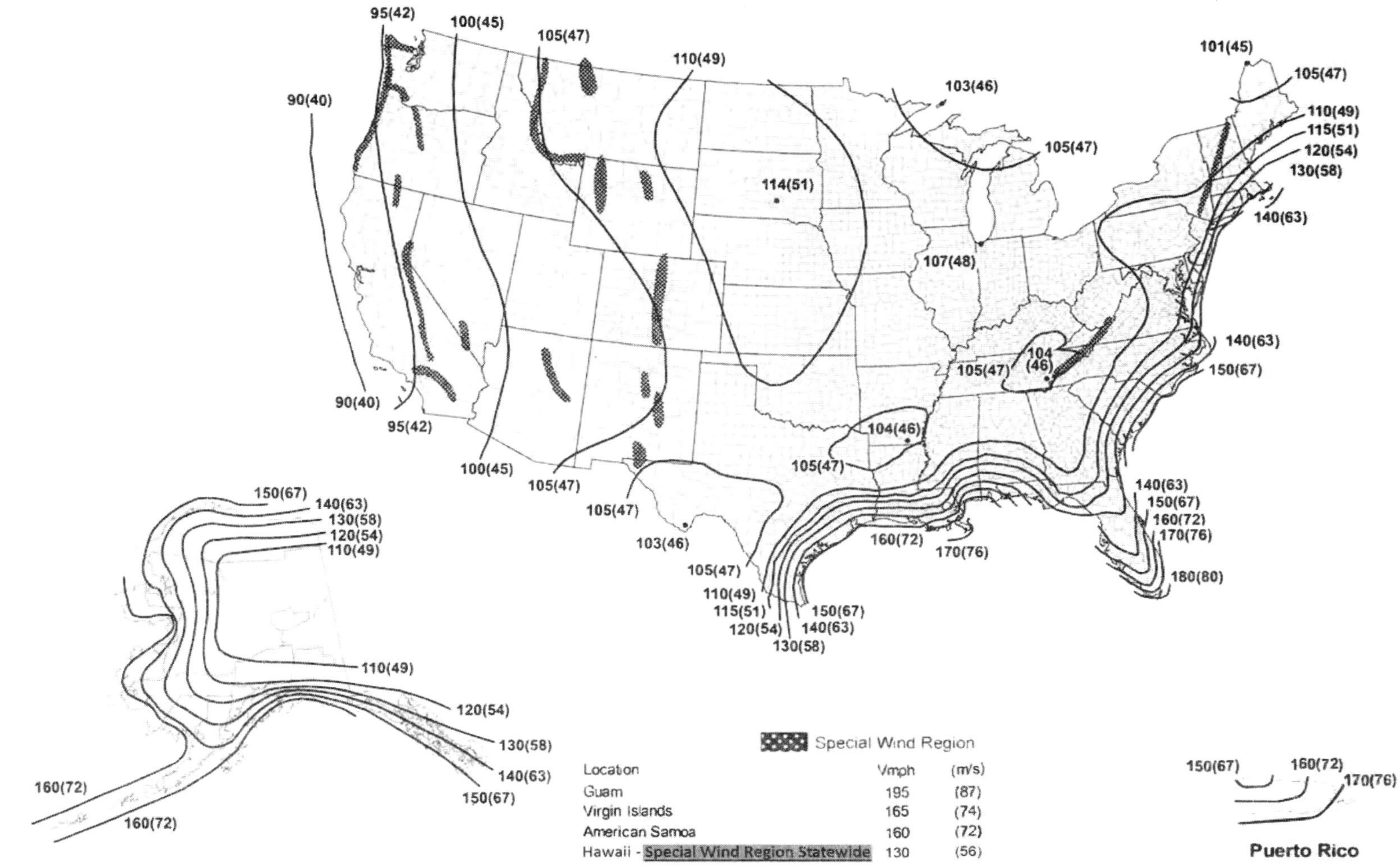

Location	Vmph	(m/s)
Guam	195	(87)
Virgin Islands	165	(74)
American Samoa	160	(72)
Hawaii - Special Wind Region Statewide	130	(56)

For SI: 1 foot = 304.8 mm, 1 mph = 0.447 m/s.

Notes:

1. Values are nominal design 3-second gust wind speeds in miles per hour (m/s) at 33 feet above ground for Exposure C category.
2. Linear interpolation between contours is permitted.
3. Islands and coastal areas outside the last contour shall use the last wind speed contour of the coastal area.
4. Mountainous terrain, gorges, ocean promontories and special wind regions shall be examined for unusual wind conditions.
5. Wind speeds correspond to approximately a 7-percent probability of exceedence in 50 years (Annual Exceedence Probability = 0.00143, MRI = 700 years).

FIGURE AH106.4.1
ULTIMATE DESIGN WIND SPEEDS FOR PATIO COVERS AND SCREEN ENCLOSURES

APPENDIX AI

PRIVATE SEWAGE DISPOSAL

The provisions contained in this appendix are not mandatory unless specifically referenced in the adopting ordinance.

User note:

About this appendix: *Appendix AI has one simple requirement that private sewage disposal systems must be in accordance with the* International Private Sewage Disposal Code®.

SECTION AI101
GENERAL

AI101.1 Scope. Private sewage disposal systems shall conform to the *International Private Sewage Disposal Code.*

APPENDIX AJ

EXISTING BUILDINGS AND STRUCTURES

The provisions contained in this appendix are not mandatory unless specifically referenced in the adopting ordinance.

User note:

About this appendix: *Appendix AJ regulates the repair, renovation alteration and reconstruction of existing buildings that are within the scope of this code. It is intended to encourage the continued safe use of existing buildings and ensure that new work conforms to the intent of the code and that exiting conditions remain at their current level of compliance or are improved.*

SECTION AJ101 PURPOSE AND INTENT

AJ101.1 General. The purpose of these provisions is to encourage the continued use or reuse of legally existing buildings and structures. These provisions are intended to permit work in existing buildings that is consistent with the purpose of this code. Compliance with these provisions shall be deemed to meet the requirements of this code.

AJ101.2 Classification of work. For purposes of this appendix, work in existing buildings shall be classified into the categories of *repair*, renovation, *alteration* and reconstruction. Specific requirements are established for each category of work in these provisions.

AJ101.3 Multiple categories of work. Work of more than one category shall be part of a single work project. Related work permitted within a 12-month period shall be considered to be a single work project. Where a project includes one category of work in one building area and another category of work in a separate and unrelated area of the building, each project area shall comply with the requirements of the respective category of work. Where a project with more than one category of work is performed in the same area or in related areas of the building, the project shall comply with the requirements of the more stringent category of work.

SECTION AJ102 COMPLIANCE

AJ102.1 General. Regardless of the category of work being performed, the work shall not cause the structure to become unsafe or adversely affect the performance of the building; shall not cause an existing mechanical or plumbing system to become unsafe, hazardous, insanitary or overloaded; and unless expressly permitted by these provisions, shall not make the building any less compliant with this code or to any previously *approved* alternative arrangements than it was before the work was undertaken.

AJ102.2 Requirements by category of work. Repairs shall conform to the requirements of Section AJ107. Renovations shall conform to the requirements of Section AJ108. *Alterations* shall conform to the requirements of Section AJ109 and the requirements for renovations. Reconstructions shall conform to the requirements of Section AJ110 and the requirements for *alterations* and renovations.

AJ102.3 Smoke detectors. Regardless of the category of work, smoke detectors shall be provided where required by Section R314.2.2.

AJ102.4 Replacement windows. Regardless of the category of work, where an existing window, including the sash and glazed portion, or safety glazing is replaced, the replacement window or safety glazing shall comply with the requirements of Sections AJ102.4.1 through AJ102.4.4, as applicable.

AJ102.4.1 Energy efficiency. Replacement windows shall comply with the requirements of Chapter 11.

AJ102.4.2 Safety glazing. Replacement glazing in hazardous locations shall comply with the safety glazing requirements of Section R308.

AJ102.4.3 Replacement windows for emergency escape and rescue openings. Where windows are required to provide emergency escape and rescue openings, replacement windows shall be exempt from Sections R310.2 and R310.4 provided that the replacement window meets the following conditions:

1. The replacement window is the manufacturer's largest standard size window that will fit within the existing frame or existing rough opening. The replacement window shall be permitted to be of the same operating style as the existing window or a style that provides for an equal or greater window opening area than the existing window.
2. Where the replacement window is not part of a change of occupancy.

Window opening control devices and fall prevention devices complying with ASTM F2090 shall be permitted for use on windows serving as required emergency escape and rescue openings.

AJ102.4.3.1 Control devices. Emergency escape and rescue openings with window opening control devices or fall prevention devices complying with ASTM F2090, after operation to release the control device allowing the window to fully open, shall not reduce the net clear opening area of the window unit. Emergency escape and rescue openings shall be operational from the inside of the room without the use of keys or tools.

AJ102.4.4 Window control devices. Window opening control devices or fall prevention devices complying with ASTM F2090 shall be installed where an existing window is replaced and where all of the following apply to the replacement window:

1. The window is operable.
2. One of the following applies:
   2.1. The window replacement includes replacement of the sash and the frame.
   2.2. The window replacement includes the sash only when the existing frame remains.
3. The bottom of the clear opening of the window opening is at a height less than 24 inches (610 mm) above the finished floor.
4. The window will permit openings that will allow passage of a 4-inch-diameter (102 mm) sphere where the window is in its largest opened position.
5. The vertical distance from the top of the sill of the window opening to the finished grade or other surface below, on the exterior of the building, is greater than 72 inches (1829 mm).

AJ102.5 Flood hazard areas. Work performed in existing buildings located in a flood hazard area as established by Table R301.2(1) shall be subject to the provisions of Section R105.3.1.1.

AJ102.6 Equivalent alternatives. Work performed in accordance with the *International Existing Building Code* shall be deemed to comply with the provisions of this appendix. These provisions are not intended to prevent the use of any alternative material, alternative design or alternative method of construction not specifically prescribed herein, provided that any alternative has been deemed to be equivalent and its use authorized by the *building official.*

AJ102.7 Other alternatives. Where compliance with these provisions or with this code as required by these provisions is technically infeasible or would impose disproportionate costs because of construction or dimensional difficulties, the building official shall have the authority to accept alternatives. These alternatives include materials, design features and operational features.

AJ102.8 More restrictive requirements. Buildings or systems in compliance with the requirements of this code for new construction shall not be required to comply with any more restrictive requirement of these provisions.

AJ102.9 Features exceeding code requirements. Elements, components and systems of existing buildings with features that exceed the requirements of this code for new construction, and are not otherwise required as part of *approved* alternative arrangements or deemed by the *building official* to be required to balance other building elements not complying with this code for new construction, shall not be prevented by these provisions from being modified as long as they remain in compliance with the applicable requirements for new construction.

SECTION AJ103
PRELIMINARY MEETING

AJ103.1 General. If a building *permit* is required at the request of the prospective *permit* applicant, the *building official* or his or her designee shall meet with the prospective applicant to discuss plans for any proposed work under these provisions prior to the application for the *permit.* The purpose of this preliminary meeting is for the *building official* to gain an understanding of the prospective applicant's intentions for the proposed work, and to determine, together with the prospective applicant, the specific applicability of these provisions.

SECTION AJ104
EVALUATION OF AN EXISTING BUILDING

AJ104.1 General. The *building official* shall have the authority to require an existing building to be investigated and evaluated by a *registered design professional* in the case of proposed reconstruction of any portion of a building. The evaluation shall determine the existence of any potential nonconformities to these provisions, and shall provide a basis for determining the impact of the proposed changes on the performance of the building. The evaluation shall use the following sources of information, as applicable:

1. Available documentation of the existing building.
   1.1. Field surveys.
   1.2. Tests (nondestructive and destructive).
   1.3. Laboratory analysis.

Exception: Detached one- or two-family dwellings that are not irregular buildings under Section R301.2.2.6 and are not undergoing an extensive reconstruction shall not be required to be evaluated.

SECTION AJ105
PERMIT

AJ105.1 Identification of work area. The work area shall be clearly identified on the *permits* issued under these provisions.

SECTION AJ106
DEFINITIONS

AJ106.1 General. For purposes of this appendix, the terms used are defined as follows:

ALTERATION. The reconfiguration of any space; the addition or elimination of any door or window; the reconfiguration or extension of any system; or the installation of any additional equipment.

CATEGORIES OF WORK. The nature and extent of construction work undertaken in an existing building. The categories of work covered in this appendix, listed in increasing order of stringency of requirements, are *repair,* renovation, *alteration* and reconstruction.

DANGEROUS. Where the stresses in any member; the condition of the building, or any of its components or elements or attachments; or other condition that results in an overload exceeding 150 percent of the stress allowed for the member or material in this code.

EQUIPMENT OR FIXTURE. Any plumbing, heating, electrical, ventilating, air-conditioning, refrigerating and fire protection equipment; and elevators, dumb waiters, boilers, pressure vessels, and other mechanical facilities or installations that are related to building services.

MATERIALS AND METHODS REQUIREMENTS. Those requirements in this code that specify material standards; details of installation and connection; joints; penetrations; and continuity of any element, component or system in the building. The required quantity, fire resistance, flame spread, acoustic or thermal performance, or other performance attribute is specifically excluded from materials and methods requirements.

RECONSTRUCTION. The reconfiguration of a space that affects an exit, a renovation or *alteration* where the work area is not permitted to be occupied because existing means-of-egress and fire protection systems, or their equivalent, are not in place or continuously maintained; or there are extensive *alterations* as defined in Section AJ109.3.

REHABILITATION. Any *repair*, renovation, *alteration* or reconstruction work undertaken in an existing building.

RENOVATION. The change, strengthening or addition of load-bearing elements; or the refinishing, replacement, bracing, strengthening, upgrading or extensive repair of existing materials, elements, components, equipment or fixtures. Renovation does not involve reconfiguration of spaces. Interior and exterior painting are not considered refinishing for purposes of this definition, and are not renovation.

REPAIR. The patching, restoration or minor replacement of materials, elements, components, equipment or fixtures for the purposes of maintaining those materials, elements, components, equipment or fixtures in good or sound condition.

WORK AREA. That portion of a building affected by any renovation, *alteration* or reconstruction work as initially intended by the *owner* and indicated as such in the *permit*. Work area excludes other portions of the building where incidental work entailed by the intended work must be performed, and portions of the building where work not initially intended by the *owner* is specifically required by these provisions for a renovation, *alteration* or reconstruction.

SECTION AJ107
REPAIRS

AJ107.1 Materials. Except as otherwise required herein, work shall be done using like materials or materials permitted by this code for new construction.

AJ107.1.1 Hazardous materials. Hazardous materials no longer permitted, such as asbestos and lead-based paint, shall not be used.

AJ107.1.2 Plumbing materials and supplies. The following plumbing materials and supplies shall not be used:

1. All-purpose solvent cement, unless *listed* for the specific application.
2. Flexible traps and tailpieces, unless *listed* for the specific application.
3. Solder having more than 0.2-percent lead in the repair of potable water systems.

AJ107.2 Water closets. Where any water closet is replaced with a newly manufactured water closet, the replacement water closet shall comply with the requirements of Section P2903.2.

AJ107.3 Electrical. Repair or replacement of existing electrical wiring and equipment undergoing repair with like material shall be permitted.

Exceptions:

1. Replacement of electrical receptacles shall comply with the requirements of Chapters 34 through 43.
2. Plug fuses of the Edison-base type shall be used for replacements only where there is not evidence of overfusing or tampering in accordance with the applicable requirements of Chapters 34 through 43.
3. For replacement of nongrounding-type receptacles with grounding-type receptacles and for branch circuits that do not have an equipment grounding conductor in the branch circuitry, the grounding conductor of a grounding-type receptacle outlet shall be permitted to be grounded to any accessible point on the grounding electrode system, or to any accessible point on the grounding electrode conductor, as allowed and described in Chapters 34 through 43.

SECTION AJ108
RENOVATIONS

AJ108.1 Materials and methods. The work shall comply with the materials and methods requirements of this code.

AJ108.2 Door and window dimensions. Minor reductions in the clear opening dimensions of replacement doors and windows that result from the use of different materials shall be allowed, whether or not they are permitted by this code.

AJ108.3 Interior finish. Wood paneling and textile wall coverings used as an interior finish shall comply with the flame spread requirements of Section R302.9.

AJ108.4 Structural. Unreinforced masonry buildings located in Seismic Design Category D_2 or E shall have parapet bracing and wall anchors installed at the roofline whenever a *reroofing permit* is issued. Such parapet bracing and wall anchors shall be of an *approved* design.

SECTION AJ109 ALTERATIONS

AJ109.1 Newly constructed elements. Newly constructed elements, components and systems shall comply with the requirements of this code.

Exceptions:

1. Added openable windows are not required to comply with the light and *ventilation* requirements of Section R303.
2. Newly installed electrical equipment shall comply with the requirements of Section AJ109.5.

AJ109.2 Nonconformities. The work shall not increase the extent of noncompliance with the requirements of Section AJ110, or create nonconformity to those requirements that did not previously exist.

AJ109.3 Extensive alterations. Where the total area of all of the work areas included in an *alteration* exceeds 50 percent of the area of the *dwelling unit*, the work shall be considered to be a reconstruction and shall comply with the requirements of these provisions for reconstruction work.

Exception: Work areas in which the *alteration* work is exclusively plumbing, mechanical or electrical shall not be included in the computation of the total area of all work areas.

AJ109.4 Structural. The minimum design loads for the structure shall be the loads applicable at the time the building was constructed, provided that a dangerous condition is not created. Structural elements that are uncovered during the course of the *alteration* and that are found to be unsound or dangerous shall be made to comply with the applicable requirements of this code.

AJ109.5 Electrical equipment and wiring.

AJ109.5.1 Materials and methods. Newly installed electrical equipment and wiring relating to work done in any work area shall comply with the materials and methods requirements of Chapters 34 through 43.

Exception: Electrical equipment and wiring in newly installed partitions and ceilings shall comply with the applicable requirements of Chapters 34 through 43.

AJ109.5.2 Electrical service. Service to the *dwelling unit* shall be not less than 100 ampere, three-wire capacity and service *equipment* shall be dead front having no live parts exposed that could allow accidental contact. Type "S" fuses shall be installed where fused equipment is used.

Exception: Existing service of 60 ampere, three-wire capacity, and feeders of 30 ampere or larger two- or three-wire capacity shall be accepted if adequate for the electrical load being served.

AJ109.5.3 Additional electrical requirements. Where the work area includes any of the following areas within a *dwelling unit*, the requirements of Sections AJ109.5.3.1 through AJ109.5.3.5 shall apply.

AJ109.5.3.1 Enclosed areas. Enclosed areas other than closets, kitchens, *basements*, garages, hallways, laundry areas and bathrooms shall have not less than two duplex receptacle outlets, or one duplex receptacle outlet and one ceiling- or wall-type lighting outlet.

AJ109.5.3.2 Kitchen and laundry areas. Kitchen areas shall have not less than two duplex receptacle outlets. Laundry areas shall have not less than one duplex receptacle outlet located near the laundry equipment and installed on an independent circuit.

AJ109.5.3.3 Ground-fault circuit interruption. Ground-fault circuit interruption shall be provided on newly installed receptacle outlets if required by Chapters 34 through 43.

AJ109.5.3.4 Lighting outlets. Not less than one lighting outlet shall be provided in every bathroom, hallway, *stairway*, attached garage and detached garage with electric power to illuminate outdoor entrances and exits, and in utility rooms and *basements* where these spaces are used for storage or contain equipment requiring service.

AJ109.5.3.5 Clearance. Clearance for electrical service equipment shall be provided in accordance with Chapters 34 through 43.

AJ109.6 Ventilation. Reconfigured spaces intended for occupancy and spaces converted to habitable or occupiable space in any work area shall be provided with *ventilation* in accordance with Section R303.

AJ109.7 Ceiling height. *Habitable spaces* created in existing *basements* shall have ceiling heights of not less than 6 feet, 8 inches (2032 mm), except that the ceiling height at obstructions shall be not less than 6 feet 4 inches (1930 mm) from the *basement* floor. Existing finished ceiling heights in nonhabitable spaces in *basements* shall not be reduced.

AJ109.8 Stairs.

AJ109.8.1 Stair width. Existing *basement* stairs and *handrails* not otherwise being altered or modified shall be permitted to maintain their current clear width at, above and below existing *handrails*.

AJ109.8.2 Stair headroom. Headroom height on existing *basement* stairs being altered or modified shall not be reduced below the existing *stairway* finished headroom. Existing *basement* stairs not otherwise being altered shall be permitted to maintain the current finished headroom.

AJ109.8.3 Stair landing. Landings serving existing *basement* stairs being altered or modified shall not be reduced below the existing *stairway* landing depth and width. Existing *basement* stairs not otherwise being altered shall be permitted to maintain the current landing depth and width.

SECTION AJ110 RECONSTRUCTION

AJ110.1 Stairways, handrails and guards.

AJ110.1.1 Stairways. *Stairways* within the work area shall be provided with illumination in accordance with Section R303.6.

AJ110.1.2 Handrails. Every required exit *stairway* that has four or more risers, is part of the means of egress for any work area, and is not provided with not fewer than one *handrail*, or in which the existing *handrails* are judged to be in danger of collapsing, shall be provided with *handrails* designed and installed in accordance with Section R311 for the full length of the run of steps on not less than one side.

AJ110.1.3 Guards. Every open portion of a *stair*, landing or balcony that is more than 30 inches (762 mm) above the floor or *grade* below, is part of the egress path for any work area, and does not have *guards*, or in which the existing *guards* are judged to be in danger of collapsing, shall be provided with *guards* designed and installed in accordance with Section R312.

AJ110.2 Wall and ceiling finish. The interior finish of walls and ceilings in any work area shall comply with the requirements of Section R302.9. Existing interior finish materials that do not comply with those requirements shall be removed or shall be treated with an *approved* fire-retardant coating in accordance with the manufacturer's instructions to secure compliance with the requirements of this section.

AJ110.3 Separation walls. Where the work area is in an attached *dwelling unit*, walls separating *dwelling units* that are not continuous from the foundation to the underside of the roof sheathing shall be constructed to provide a continuous fire separation using construction materials consistent with the existing wall or complying with the requirements for new structures. Performance of work shall be required only on the side of the wall of the *dwelling unit* that is part of the work area.

AJ110.4 Ceiling height. *Habitable spaces* created in existing *basements* shall have ceiling heights of not less than 6 feet, 8 inches (2032 mm), except that the ceiling height at obstructions shall be not less than 6 feet 4 inches (1930 mm) from the *basement* floor. Existing finished ceiling heights in nonhabitable spaces in *basements* shall not be reduced.

SECTION AJ111
REFERENCED STANDARDS

AJ111.1 General. See Table AJ111.1 for standards that are referenced in various sections of this appendix. Standards are listed by the standard identification with the effective date, the standard title and the section or sections of this appendix that reference the standard.

TABLE AJ111.1
REFERENCED STANDARDS

STANDARD ACRONYM	STANDARD NAME	SECTION HEREIN REFERENCED
ASTM F2090—17	*Specification for Window Fall Prevention Devices with Emergency Escape (Egress) Release Mechanisms*	AJ102.4.3, AJ102.4.4
IEBC—21	*International Existing Building Code*	AJ102.6

APPENDIX AK
SOUND TRANSMISSION

The provisions contained in this appendix are not mandatory unless specifically referenced in the adopting ordinance.

User note:

About this appendix: *Sound transmission relates directly to the psychological and long-term physical well-being of building occupants. Many human activities cannot be accommodated efficiently or comfortably in various types of building spaces without proper attention to the mitigation of sound transmission from other spaces within the building, or from outside of the building. In Appendix AK, attention is specifically paid to the mitigation of sound transmission between dwelling units and other dwelling units and occupancies.*

SECTION AK101 GENERAL

AK101.1 General. Wall and floor-ceiling assemblies separating *dwelling units*, including those separating adjacent townhouse units, shall provide airborne sound insulation for walls, and both airborne and impact sound insulation for floor-ceiling assemblies.

SECTION AK102 AIRBORNE SOUND

AK102.1 General. Airborne sound insulation for wall and floor-ceiling assemblies shall meet a sound transmission class (STC) rating of 45 where tested in accordance with ASTM E90 or a Normalized Noise Isolation Class (NNIC) rating of 42 where tested in accordance with ASTM E336. Penetrations or openings in construction assemblies for piping; electrical devices; recessed cabinets; bathtubs; soffits; or heating, ventilating or exhaust ducts shall be sealed, lined, insulated or otherwise treated to maintain the required ratings. *Dwelling unit* entrance doors, which share a common space, shall be tight fitting to the frame and sill.

AK102.1.1 Masonry. The sound transmission class of *concrete masonry* and *clay masonry* assemblies shall be calculated in accordance with TMS 0302 or determined through testing in accordance with ASTM E90.

SECTION AK103 STRUCTURAL-BORNE SOUND

AK103.1 General. Floor/ceiling assemblies between *dwelling units*, or between a *dwelling unit* and a public or service area within a structure, shall have an impact insulation class (IIC) rating of not less than 45 when tested in accordance with ASTM E492 or a Normalized Impact Sound Rating (NISR) of 42 where tested in accordance with ASTM E1007.

SECTION AK104 REFERENCED STANDARDS

AK104.1 General. See Table AK104.1 for standards that are referenced in various sections of this appendix. Standards are listed by the standard identification with the effective date, the standard title, and the section or sections of this appendix that references the standard.

TABLE AK104.1 REFERENCED STANDARDS

STANDARD ACRONYM	STANDARD NAME	SECTIONS HEREIN REFERENCED
ASTM E90—09	*Test Method for Laboratory Measurement of Airborne Sound Transmission Loss of Building Partitions and Elements*	AK102.1, AK102.1.1
ASTM E336—17a	*Standard Test Method for Measurement of Airborne Sound Attenuation between Rooms in Buildings*	AK102.1
ASTM E492—09	*Specification for Laboratory Measurement of Impact Sound Transmission through Floor-ceiling Assemblies Using the Tapping Machine*	AK103.1
ASTM E1007—16	*Standard Test Method for Field Measurement of Tapping Machine Impact Sound Transmission Through Floor-Ceiling Assemblies and Associated Support Structures*	AK103.1
TMS 0302—12	*Standard for Determining the Sound Transmission Class Rating for Masonry Walls*	AK102.1.1

APPENDIX AL

PERMIT FEES

The provisions contained in this appendix are not mandatory unless specifically referenced in the adopting ordinance.

User note:

About this appendix: *Appendix AL is intended to provide guidance to building departments in their efforts to set fees for building permits. This appendix provides examples that may be used as a reference when setting fee schedules and are not intended to be literally applied.*

SECTION AL101 GENERAL

AL101.1 Permit fee schedule. *Permit* fees shall be in accordance with Table AL101.1.

TABLE AL101.1 PERMIT FEE SCHEDULE

TOTAL VALUATION	FEE
$1 to $500	$24
$501 to $2,000	$24 for the first $500; plus $3 for each additional $100 or fraction thereof, up to and including $2,000
$2,001 to $40,000	$69 for the first $2,000; plus $11 for each additional $1,000 or fraction thereof, up to and including $40,000
$40,001 to $100,000	$487 for the first $40,000; plus $9 for each additional $1,000 or fraction thereof, up to and including $100,000
$100,000 to $500,000	$1,027 for the first $100,000; plus $7 for each additional $1,000 or fraction thereof, up to and including $500,000
$500,001 to $1,000,000	$3,827 for the first $500,000; plus $5 for each additional $1,000 or fraction thereof, up to and including $1,000,000
$1,000,001 to $5,000,000	$6,327 for the first $1,000,000; plus $3 for each additional $1,000 or fraction thereof, up to and including $5,000,000
$5,000,001 and over	$18,327 for the first $5,000,000; plus $1 for each additional $1,000 or fraction thereof

APPENDIX AM
HOME DAY CARE—R-3 OCCUPANCY

The provisions contained in this appendix are not mandatory unless specifically referenced in the adopting ordinance.

User note:

About this appendix: *Appendix AM is intended to apply to scenarios where day care is provided to more than five children in dwellings that are under the scope of this code. The* International Building Code® *considers such structures R-3 residential occupancies and allows them to be constructed in accordance with this code. Although there are many general provisions in the body of the code that apply to home day care as well as other occupancies, this appendix contains the provisions that are specific to home day care.*

SECTION AM101
GENERAL

AM101.1 General. This appendix shall apply to a home day care operated within a *dwelling*. It is to include buildings and structures occupied by persons of any age who receive custodial care for less than 24 hours by individuals other than parents or guardians or relatives by blood, marriage, or adoption, and in a place other than the home of the person cared for.

SECTION AM102
DEFINITION

AM102.1 General. The following term shall, for the purposes of this appendix, have the meaning shown herein.

EXIT ACCESS. That portion of a means-of-egress system that leads from any occupied point in a building or structure to an exit.

SECTION AM103
MEANS OF EGRESS

AM103.1 Exits required. If the occupant load of the residence is more than nine, including those who are residents, during the time of operation of the day care, two exits are required from the ground-level *story*. Two exits are required from a home day care operated in a *manufactured home*, regardless of the occupant load. Exits shall comply with Section R311.

AM103.1.1 Exit access prohibited. An exit access from the area of day care operation shall not pass through bathrooms, bedrooms, closets, garages, fenced rear yards or similar areas.

Exception: An exit may discharge into a fenced yard if the gate or gates remain unlocked during day care hours. The gates may be locked if there is an area of refuge located within the fenced yard and more than 50 feet (15 240 mm) from the *dwelling*. The area of refuge shall be large enough to allow 5 square feet (0.5 m^2) per occupant.

AM103.1.2 Basements. If the *basement* of a *dwelling* is to be used in the day care operation, two exits are required from the *basement* regardless of the occupant load. One of the exits may pass through the *dwelling* and the other shall lead directly to the exterior of the *dwelling*.

An emergency and escape window used as the second means of egress from a *basement* shall comply with Sections R310 and AM103.1.1.

AM103.1.3 Yards. If the yard is to be used as part of the day care operation it shall be fenced.

AM103.1.3.1 Type of fence and hardware. The fence shall be of durable materials and be not less than 6 feet (1529 mm) tall, completely enclosing the area used for the day care operations. Each opening shall be a gate or door equipped with a self-closing and self-latching device to be installed at not less than 5 feet (1528 mm) above the ground.

Exception: The door of any *dwelling* that forms part of the enclosure need not be equipped with self-closing and self-latching devices.

AM103.1.3.2 Construction of fence. Openings in the fence, wall or enclosure required by this section shall have intermediate rails or an ornamental pattern that do not allow a sphere 4 inches (102 mm) in diameter to pass through. In addition, the following criteria must be met:

1. The maximum vertical clearance between *grade* and the bottom of the fence, wall or enclosure shall be 2 inches (51 mm).
2. Solid walls or enclosures that do not have openings, such as masonry or stone walls, shall not contain indentations or protrusions, except for tooled masonry joints.
3. Maximum mesh size for chain link fences shall be $1^1/_4$ inches (32 mm) square, unless the fence has slats at the top or bottom that reduce the opening to not more than $1^3/_4$ inches (44 mm). The wire shall be not less than 9 gage [0.148 inch (3.8 mm)].

AM103.1.3.3 Decks. Decks that are more than 12 inches (305 mm) above *grade* shall have a *guard* in compliance with Section R312.

AM103.2 Width and height of an exit. The minimum width of a required exit is 36 inches (914 mm) with a net clear width of 32 inches (813 mm). The minimum height of a required exit is 6 feet 8 inches (2032 mm).

AM103.3 Type of lock and latches for exits. Regardless of the occupant load served, exit doors shall be openable from the inside without the use of a key or any special knowledge or effort. Where the occupant load is 10 or less, a night latch, dead bolt or security chain may be used, provided that such devices are openable from the inside without the use of a key or tool, and are mounted at a height not to exceed 48 inches (1219 mm) above the finished floor.

AM103.4 Landings. Landings for *stairways* and doors shall comply with Section R311, except that landings shall be required for the exterior side of a sliding door where a home day care is being operated in a Group R-3 occupancy.

SECTION AM104
SMOKE DETECTION

AM104.1 General. Smoke detectors shall be installed in *dwelling units* used for home day care operations. Detectors shall be installed in accordance with the approved manufacturer's instructions. If the current smoke detection system in the *dwelling* is not in compliance with the currently adopted code for smoke detection, it shall be upgraded to meet the currently adopted code requirements and Section AM103 before day care operations commence.

AM104.2 Power source. Required smoke detectors shall receive their primary power from the building wiring where that wiring is served from a commercial source and shall be equipped with a battery backup. The detector shall emit a signal when the batteries are low. Wiring shall be permanent and without a disconnecting switch other than those required for overcurrent protection. Required smoke detectors shall be interconnected such that if one detector is activated, all detectors are activated.

AM104.3 Location. A detector shall be located in each bedroom and any room that is to be used as a sleeping room, and centrally located in the corridor, hallway or area giving access to each separate sleeping area. Where the *dwelling unit* has more than one *story,* and in *dwellings* with *basements,* a detector shall be installed on each *story* and in the *basement.* In *dwelling units* where a *story* or *basement* is split into two or more levels, the smoke detector shall be installed on the upper level, except that where the lower level contains a sleeping area, a detector shall be installed on each level. Where sleeping rooms are on the upper level, the detector shall be placed at the ceiling of the upper level in close proximity to the stairway. In *dwelling units* where the ceiling height of a room open to the hallway serving the bedrooms or sleeping areas exceeds that of the hallway by 24 inches (610 mm) or more, smoke detectors shall be installed in the hallway and the adjacent room. Detectors shall sound an alarm audible in all sleeping areas of the *dwelling unit* in which they are located.

APPENDIX AN

VENTING METHODS

This appendix is informative and is not part of the code. This appendix provides examples of various venting methods.

User note:

About this appendix: *Venting for plumbing systems is often best understood using diagrams such as isometrics. Appendix AN illustrates a variety of venting methods indicated in Chapter 31 of this code.*

SECTION AN101 VENTING METHODS

AN101.1 General. Venting methods are illustrated in Figures AN101.1(1) through AN101.1(7).

LAV
1 1/2 IN.
1 1/4 IN.
TUB
WC
1 1/2 IN.
2 IN.
3 IN.

A. TYPICAL SINGLE-BATH ARRANGEMENT

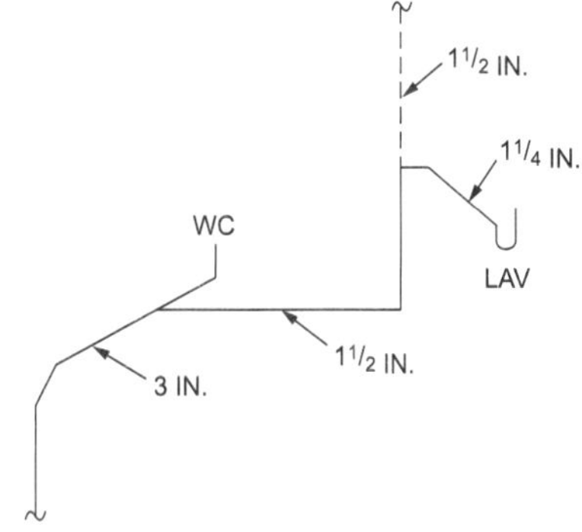

B. TYPICAL POWDER ROOM

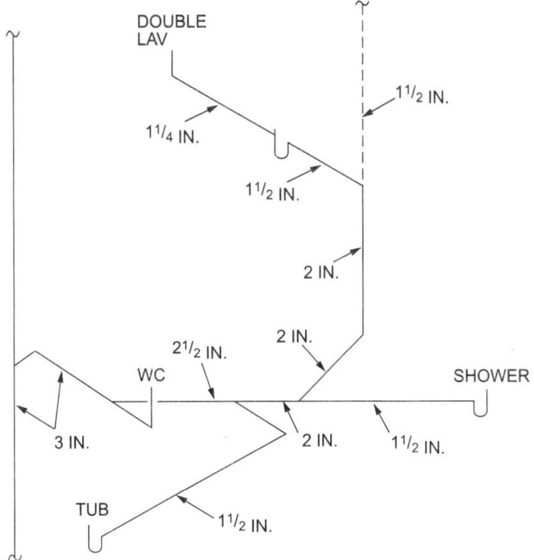

C. MORE ELABORATE SINGLE-BATH ARRANGEMENT

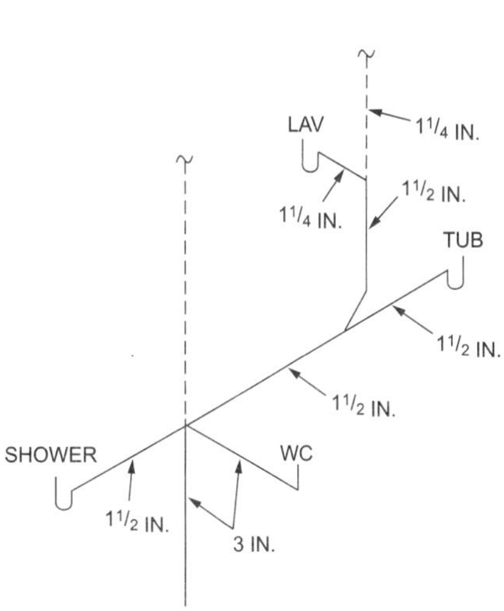

D. COMBINATION WET- AND STACK-VENTING WITH STACK FITTING

For SI: 1 inch = 25.4 mm.

**FIGURE AN101.1(1)
TYPICAL SINGLE-BATH WET-VENT ARRANGEMENTS**

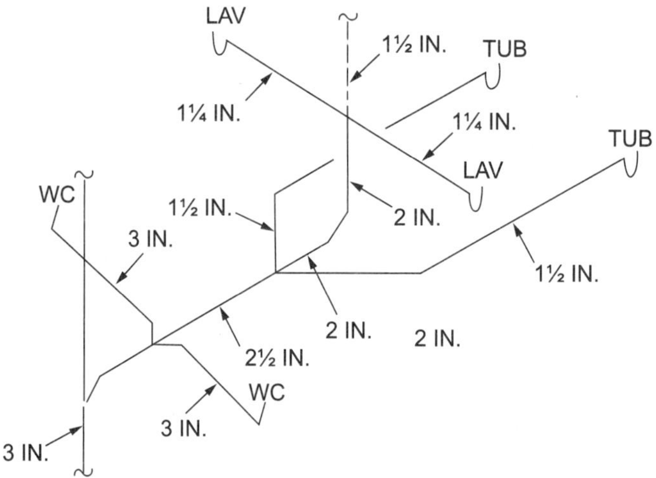

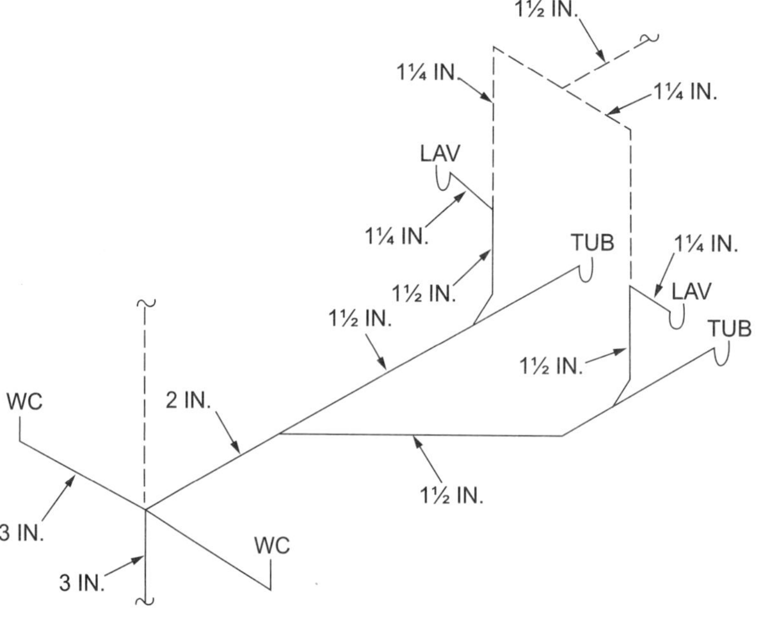

For SI: 1 inch = 25.4 mm.

FIGURE AN101.1(2)
TYPICAL DOUBLE-BATH WET-VENT ARRANGEMENTS

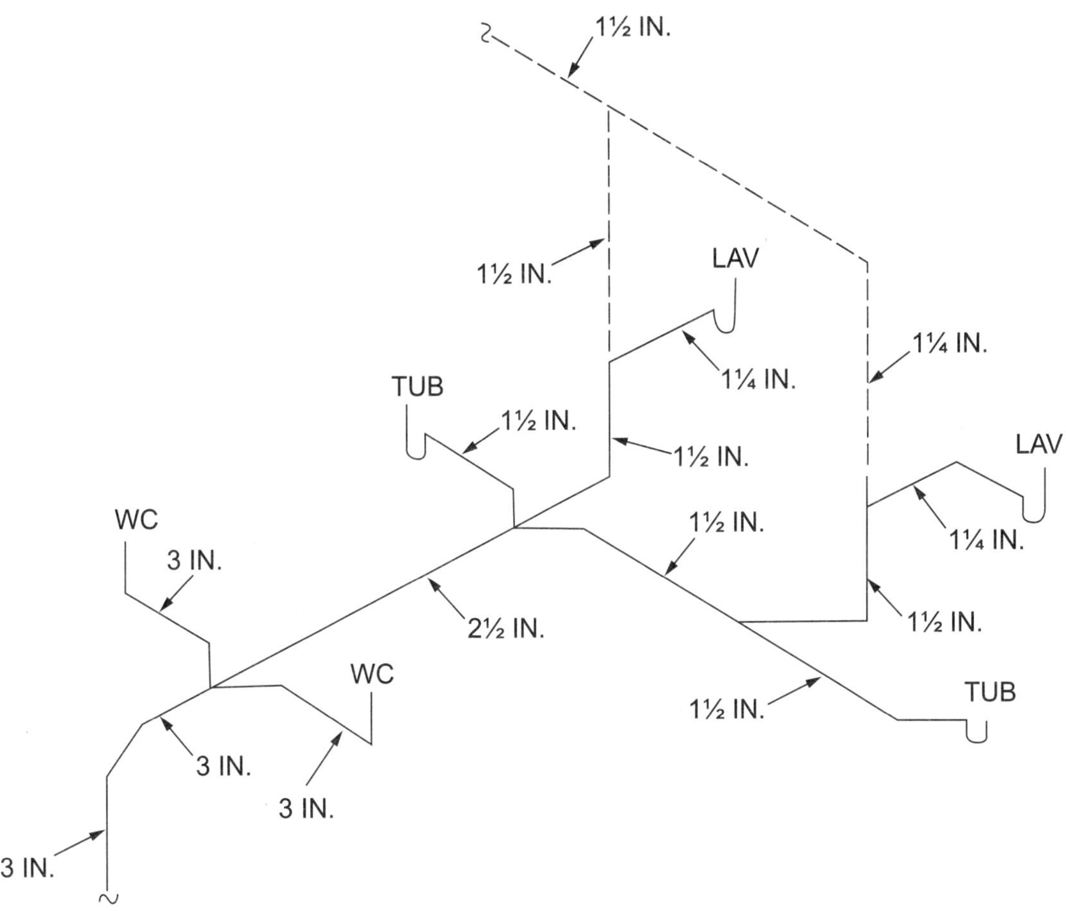

For SI: 1 inch = 25.4 mm.

FIGURE AN101.1(3)
TYPICAL HORIZONTAL WET VENTING

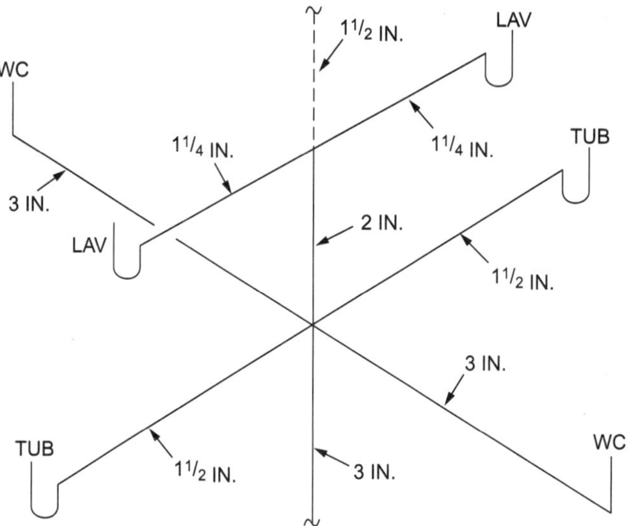

A. VERTICAL WET VENTING

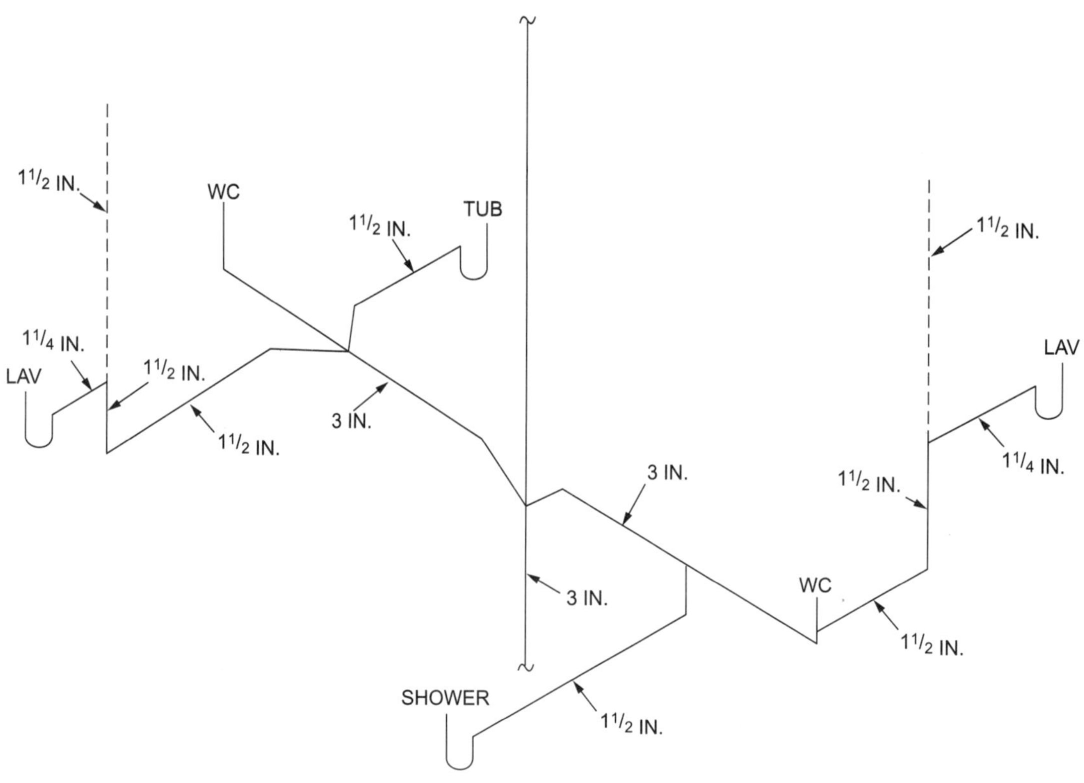

B. HORIZONTAL WET VENTING

For SI: 1 inch = 25.4 mm.

FIGURE AN101.1(4)
TYPICAL METHODS OF WET VENTING

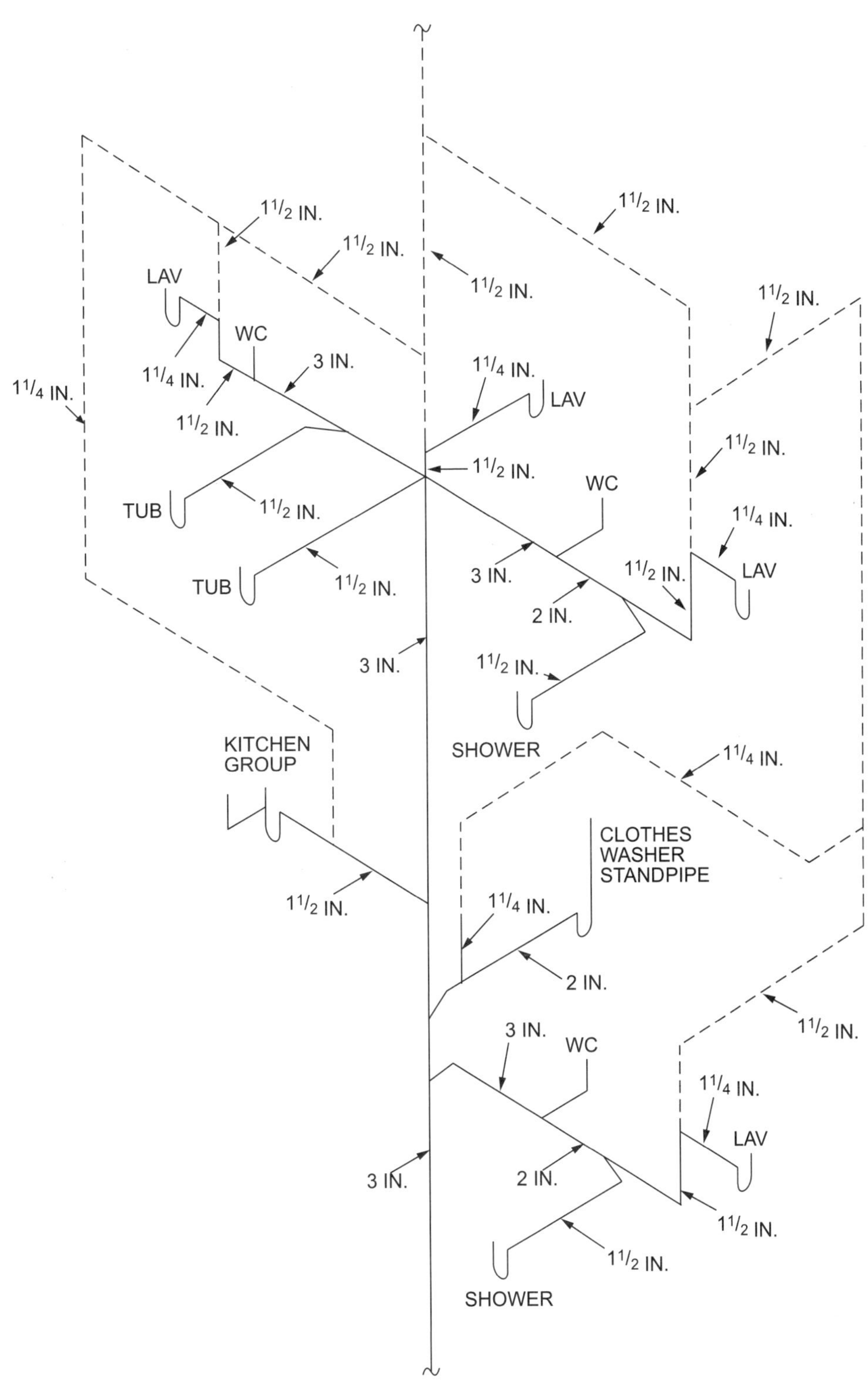

For SI: 1 inch = 25.4 mm.

FIGURE AN101.1(5)
SINGLE-STACK SYSTEM FOR A TWO-STORY DWELLING

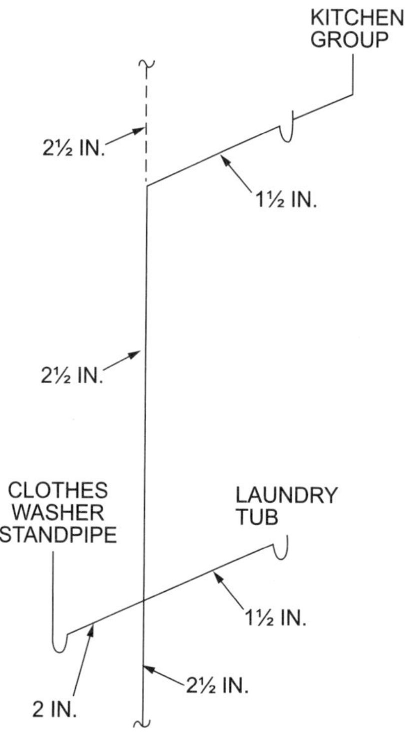

For SI: 1 inch = 25.4 mm.

FIGURE AN101.1(6)
WASTE STACK VENTING

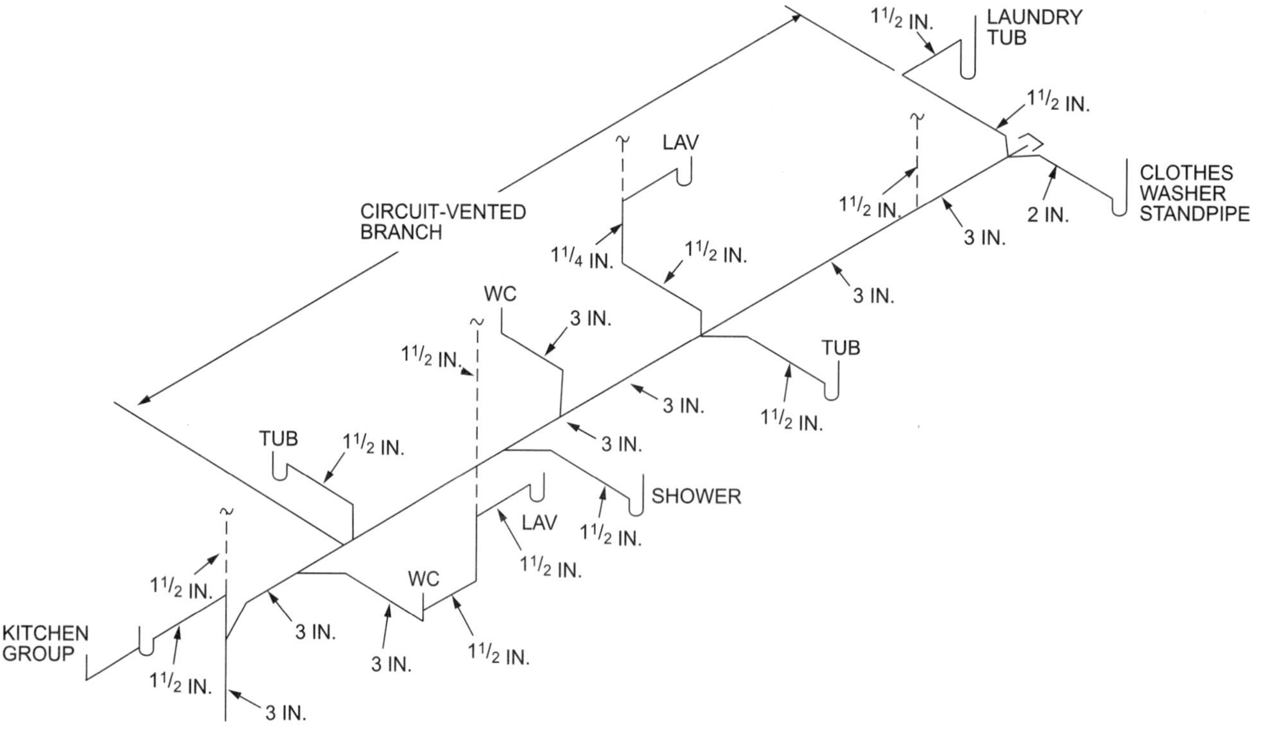

For SI: 1 inch = 25.4 mm.

FIGURE AN101.1(7)
CIRCUIT VENT WITH ADDITIONAL NONCIRCUIT-VENTED BRANCH

APPENDIX AO
AUTOMATIC VEHICULAR GATES

The provisions contained in this appendix are not mandatory unless specifically referenced in the adopting ordinance.

User note:

About this appendix: *Appendix AO provides requirements for automatic vehicular gates, including a definition of and references to standards that regulate such gates.*

SECTION AO101 GENERAL

AO101.1 General. The provisions of this appendix shall control the design and construction of automatic vehicular gates installed on the lot of a one- or two-family dwelling.

SECTION AO102 DEFINITION

AO102.1 General. The following term shall, for the purposes of this appendix, have the meaning shown herein.

VEHICULAR GATE. A gate that is intended for use at a vehicular entrance or exit to the lot of a one- or two-family dwelling, and that is not intended for use by pedestrian traffic.

SECTION AO103 AUTOMATIC VEHICULAR GATES

AO103.1 Vehicular gates intended for automation. Vehicular gates intended for automation shall be designed, constructed and installed to comply with the requirements of ASTM F2200.

AO103.2 Vehicular gate openers. Vehicular gate openers, where provided, shall be *listed* in accordance with UL 325.

SECTION AO104 REFERENCED STANDARDS

AO104.1 General. See Table AO104.1 for standards that are referenced in various sections of this appendix. Standards are listed by the standard identification with the effective date, the standard title, and the section or sections of this appendix that reference the standard.

TABLE AO104.1 REFERENCED STANDARDS

STANDARD ACRONYM	STANDARD NAME	SECTIONS HEREIN REFERENCED
ASTM F2200—14	*Standard Specification for Automated Vehicular Gate Construction*	AO103.1
UL 325—02	*Door, Drapery, Gate, Louver and Window Operations and Systems—with revisions through May 2015*	AO103.2

APPENDIX AP

SIZING OF WATER PIPING SYSTEM

The provisions contained in this appendix are not mandatory unless specifically referenced in the adopting ordinance.

User note:

About this appendix: *Chapter 29 has the basic information to begin sizing of a water service and water distribution piping system. Appendix AP provides several methods that can be used to complete pipe sizing for a building.*

SECTION AP101 GENERAL

AP101.1 Scope.

AP101.1.1 Two procedures. This appendix outlines two procedures for sizing a water piping system (see Sections AP103.3 and AP201.1). The design procedures are based on the minimum static pressure available from the supply source, the head changes in the system caused by friction and elevation, and the rates of flow necessary for operation of various fixtures.

AP101.1.2 Variable conditions. Because of the variable conditions encountered in hydraulic design, it is impractical to specify definite and detailed rules for sizing of the water piping system. Accordingly, other sizing or design methods conforming to good engineering practice standards are acceptable alternatives to those presented herein.

SECTION AP102 INFORMATION REQUIRED

AP102.1 Preliminary. Obtain the necessary information regarding the minimum daily static service pressure in the area where the building is to be located. If the building supply is to be metered, obtain information regarding friction loss relative to the rate of flow for meters in the range of sizes likely to be used. Friction loss data can be obtained from most manufacturers of water meters.

AP102.2 Demand load.

AP102.2.1 Fixture demand. Estimate the supply demand of the building main and the principal branches and risers of the system by totaling the corresponding demand from the applicable part of Table AP103.3(3).

AP102.2.2 Continuous demand. Estimate continuous supply demands, in gallons per minute (gpm) (L/m) such as for lawn sprinklers and air conditioners, and add the sum to the total demand for fixtures. The result is the estimated supply demand for the building supply.

SECTION AP103 SELECTION OF PIPE SIZE

AP103.1 General. Decide from Table P2903.1 what is the desirable minimum residual pressure that should be maintained at the highest fixture in the supply system. If the highest group of fixtures contains flushometer valves, the pressure for the group should be not less than 15 pounds per square inch (psi) (103.4 kPa) flowing. For flush tank supplies, the available pressure should be not less than 8 psi (55.2 kPa) flowing, except blowout action fixtures must not be less than 25 psi (172.4 kPa) flowing.

AP103.2 Pipe sizing.

AP103.2.1 Pipe size selection. Pipe sizes can be selected using the following procedure or by use of other design methods conforming to acceptable engineering practice that are *approved* by the *building official*. The sizes selected must not be less than the minimum required by this code.

AP103.2.2 Pressures and losses. Water pipe sizing procedures are based on a system of pressure requirements and losses, the sum of which must not exceed the minimum pressure available at the supply source. These pressures are as follows:

1. Pressure required at fixture to produce required flow. See Section P2903.1 of this code and Section 604.3 of the *International Plumbing Code.*
2. Static pressure loss or gain (due to head) is computed at 0.433 psi per foot (9.8 kPa/m) of elevation change.

   Example: Assume that the highest fixture supply outlet is 20 feet (6096 mm) above or below the supply source. This produces a static pressure differential of 8.66 psi (59.8 kPa) loss [20 feet by 0.433 psi per foot (2096 mm by 9.8 kPa/m)].
3. Loss through water meter. The friction or pressure loss can be obtained from meter manufacturers.
4. Loss through taps in water main.
5. Loss through special devices, such as filters, softeners, backflow prevention devices and pressure regulators. These values must be obtained from the manufacturer.
6. Loss through valves and fittings. Losses for these items are calculated by converting to the *equivalent length* of piping and adding to the total pipe length.
7. Loss caused by pipe friction can be calculated where the pipe size, pipe length and flow through

the pipe are known. With these three items, the friction loss can be determined. For piping flow charts not included, use manufacturers' tables and velocity recommendations.

8. For the purposes of all examples, the following metric conversions are applicable:

   1 cubic foot per minute = 0.4719 L/s.

   1 square foot = 0.0929 m^2.

   1 degree = 0.0175 rad.

   1 pound per square inch = 6.895 kPa.

   1 inch = 25.4 mm.

   1 foot = 304.8 mm.

   1 gallon per minute = 3.785 L/m.

AP103.3 Segmented loss method. The size of water service mains, branch mains and risers by the segmented loss method, must be determined by knowing the water supply demand [gpm (L/m)], available water pressure [psi (kPa)] and friction loss caused by the water meter and *developed length* of pipe [feet (m)], including the *equivalent length* of fittings. This design procedure is based on the following parameters:

- The calculated friction loss through each length of pipe.
- A system of pressure losses, the sum of which must not exceed the minimum pressure available at the street main or other source of supply.
- Pipe sizing based on estimated peak demand, total pressure losses caused by difference in elevation, equipment, *developed length* and pressure required at the most remote fixture; loss through taps in water main; losses through fittings, filters, backflow prevention devices, valves and pipe friction.

Because of the variable conditions encountered in hydraulic design, it is impractical to specify definite and detailed rules for the sizing of the water piping system. Current sizing methods do not address the differences in the probability of use and flow characteristics of fixtures between types of occupancies. Creating an exact model of predicting the demand for a building is impossible and final studies assessing the impact of water conservation on demand are not yet complete. The following steps are necessary for the segmented loss method.

1. **Preliminary.** Obtain the necessary information regarding the minimum daily static service pressure in the area where the building is to be located. If the building supply is to be metered, obtain information regarding friction loss relative to the rate of flow for meters in the range of sizes to be used. Friction loss data can be obtained from manufacturers of water meters. Enough pressure must be available to overcome all system losses caused by friction and elevation so that plumbing fixtures operate properly. Section 604.6 of the *International Plumbing Code* requires that the water distribution system be designed for the minimum pressure available taking into consideration pressure fluctuations. The lowest pressure must be selected to guarantee a continuous, adequate supply of water. The lowest pressure in the public main usually occurs in the summer because of lawn sprinkling and supplying water for air-conditioning cooling towers. Future demands placed on the public main as a result of large growth or expansion should be considered. The available pressure will decrease as additional loads are placed on the public system.

2. **Demand load.** Estimate the supply demand of the building main and the principal branches and risers of the system by totaling the corresponding demand from the applicable part of Table AP103.3(3). When estimating peak demand, sizing methods typically use water supply fixture units (w.s.f.u.) [see Table AP103.3(2)]. This numerical factor measures the load-producing effect of a single plumbing fixture of a given kind. The use of fixture units can be applied to a single basic probability curve (or table), found in the various sizing methods [see Table AP103.3(3)]. The fixture units are then converted into a gpm (L/m) flow rate for estimating demand.

   2.1. Estimate continuous supply demand in gpm (L/m) such as for lawn sprinklers, air conditioners, etc., and add the sum to the total demand for fixtures. The result is the estimated supply demand for the building supply. Fixture units cannot be applied to constant-use fixtures, such as hose bibbs, lawn sprinklers and air conditioners. These types of fixtures must be assigned the gpm (L/m) value.

3. **Selection of pipe size.** This water pipe sizing procedure is based on a system of pressure requirements and losses, the sum of which must not exceed the minimum pressure available at the supply source. These pressures are as follows:

   3.1. Pressure required at the fixture to produce required flow. See Section P2903.1 of this code and Section 604.3 of the *International Plumbing Code*.

   3.2. Static pressure loss or gain (because of head) is computed at 0.433 psi per foot (9.8 kPa/m) of elevation change.

   3.3. Loss through a water meter. The friction or pressure loss can be obtained from the manufacturer.

   3.4. Loss through taps in water main [see Table AP103.3(4)].

   3.5. Loss through special devices, such as filters, softeners, backflow prevention devices and pressure regulators. These values must be obtained from the manufacturers.

   3.6. Loss through valves and fittings [see Tables AP103.3(5) and AP103.3(6)]. Losses for these items are calculated by converting to the *equivalent length* of piping and adding to the total pipe length.

3.7. Loss caused by pipe friction can be calculated where the pipe size, pipe length and flow through the pipe are known. With these three items, the friction loss can be determined using Figures AP103.3(2) through AP103.3(7). Where using charts, use pipe inside diameters. For piping flow charts not included, use manufacturers' tables and velocity recommendations. Before attempting to size any water supply system, it is necessary to gather preliminary information including available pressure, piping material, select design velocity, elevation differences and *developed length* to the most remote fixture. The water supply system is divided into sections at major changes in elevation or where branches lead to fixture groups. The peak demand must be determined in each part of the hot and cold water supply system. The expected flow through each section is determined in w.s.f.u. and converted to gpm (L/m) flow rate. Sizing methods require determination of the "most hydraulically remote" fixture to compute the pressure loss caused by pipe and fittings. The hydraulically remote fixture represents the most downstream fixture along the circuit of piping requiring the most available pressure to operate properly. Consideration must be given to all pressure demands and losses, such as friction caused by pipe, fittings and equipment; elevation; and the residual pressure required by Table P2903.1. The two most common and frequent complaints about water supply system operation are lack of adequate pressure and noise.

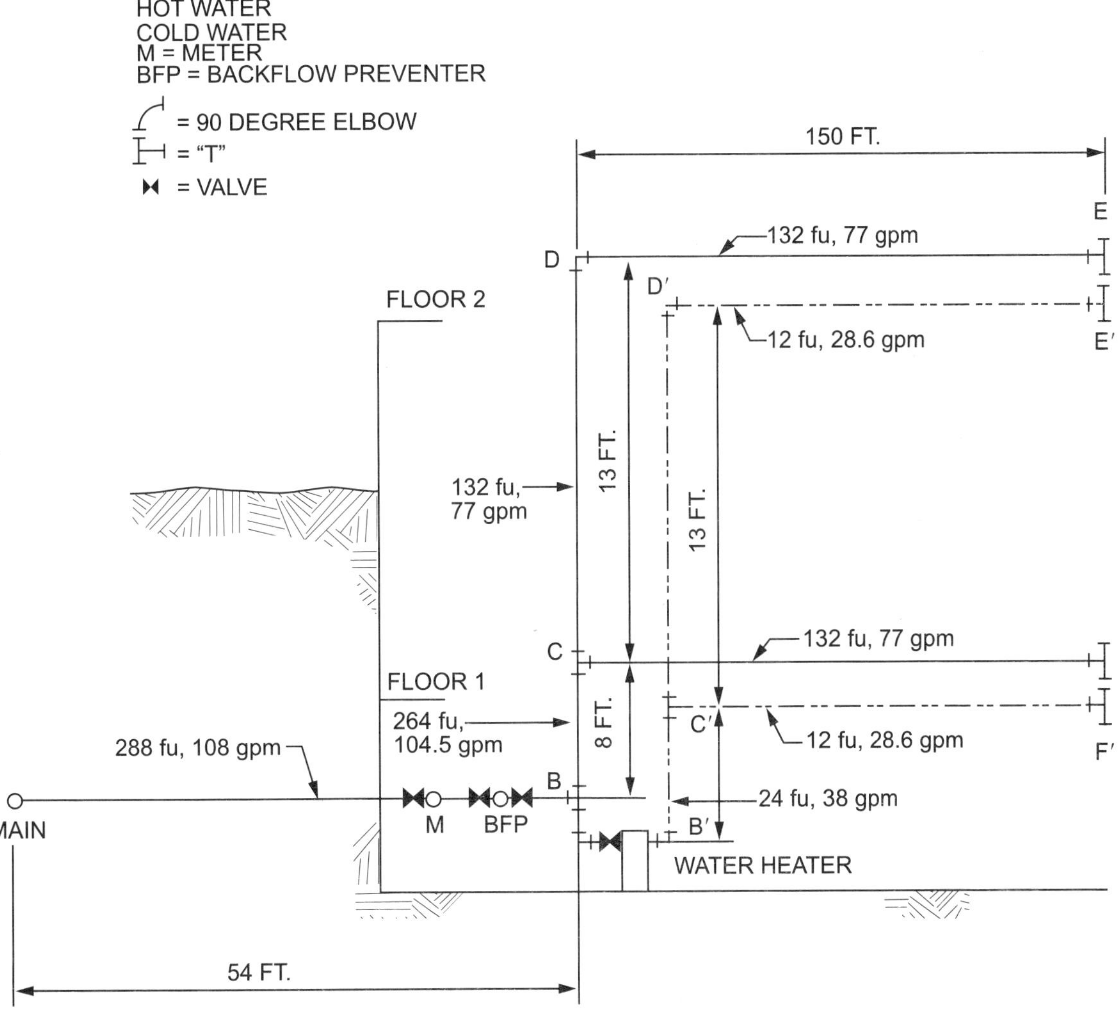

For SI: 1 foot = 304.8 mm, 1 gallon per minute = 3.785 L/m.

FIGURE AP103.3(1)
EXAMPLE—SIZING

TABLE AP103.3(1)
RECOMMENDED TABULAR ARRANGEMENT FOR USE IN SOLVING PIPE SIZING PROBLEMS

COLUMN	1	2		3	4	5	6	7	8	9	10
Line	Description		Pounds per square inch	Gallons per min through section	Length of section (feet)	Trial pipe size (inches)	Equivalent length of fittings and valves (feet)	Total equivalent length [(Col. 4 + Col. 6)/ 100 feet)]	Friction loss per 100 feet of trial size pipe (psi)	Friction loss in equivalent length Column 8 x Column 7 (psi)	Excess pressure over friction losses (psi)
A	Service and cold water distribution piping[a]	Minimum pressure available at main	55.00								
B		Highest pressure required at a fixture (see Table P2903.1)	15.00								
C		Meter loss 2″ meter	11.00								
D		Tap in main loss 2″ tap [see Table AP103.3(4)]	1.61								
E		Static head loss 21 ft × 0.43 psi/ft	9.03								
F		Special fixture loss backflow preventer	9.00	—	—	—	—	—	—	—	—
G		Special fixture loss—Filter	0.00								
H		Special fixture loss—Other	0.00								
I		Total overall losses and requirements (Sum of Lines B through H)	45.64								
J		Pressure available to overcome pipe friction (Line A minus Line I)	9.36								
	Pipe section (from diagram) Cold water distribution piping	A-B	288	108.0	54	$2^1/_2$	15.00	0.69	3.2	2.21	—
		B-C	264	104.5	8	$2^1/_2$	0.5	0.085	3.1	0.26	—
		C-D	132	77.0	13	$2^1/_2$	7.00	0.20	1.9	0.38	—
		C-F[b]	132	77.0	150	$2^1/_2$	12.00	1.62	1.9	3.08	—
		D-E[b]	132	77.0	150	$2^1/_2$	12.00	1.62	1.9	3.08	—
K	Total pipe friction losses (cold)			—	—	—	—	—	—	5.93	—
L	Difference (Line J minus Line K)			—	—	—	—	—	—	—	3.43
	Pipe section (from diagram) Hot water distribution piping	A′B′	288	108.0	54	$2^1/_2$	12.00	0.69	3.3	2.21	—
		B′C′	24	38.0	8	2	7.5	0.16	1.4	0.22	—
		C′D′	12	28.6	13	$1^1/_2$	4.0	0.17	3.2	0.54	—
		C′F′[b]	12	28.6	150	$1^1/_2$	7.00	1.57	3.2	5.02	—
		D′E′[b]	12	28.6	150	$1^1/_2$	7.00	1.57	3.2	5.02	—
K	Total pipe friction losses (hot)			—	—	—	—	—	—	7.99	—
L	Difference (Line J minus Line K)			—	—	—	—	—	—	—	1.37

For SI: 1 inch = 25.4 mm, 1 foot = 304.8 mm, 1 pound per square inch = 6.895 kPa, 1 gallon per minute = 3.785 L/m.

a. To be considered as pressure gain for fixtures below main (to consider separately, omit from "I" and add to "J").

b. To consider separately, in Line K use Section C-F only if greater loss than the loss in Section D-E.

TABLE AP103.3(2)
LOAD VALUES ASSIGNED TO FIXTURES[a]

FIXTURE	OCCUPANCY	TYPE OF SUPPLY CONTROL	LOAD VALUES, IN WATER SUPPLY FIXTURE UNITS (w.s.f.u.)		
			Cold	Hot	Total
Bathroom group	Private	Flush tank	2.7	1.5	3.6
Bathroom group	Private	Flushometer valve	6.0	3.0	8.0
Bathtub	Private	Faucet	1.0	1.0	1.4
Bathtub	Public	Faucet	3.0	3.0	4.0
Bidet	Private	Faucet	1.5	1.5	2.0
Combination fixture	Private	Faucet	2.25	2.25	3.0
Dishwashing machine	Private	Automatic	—	1.4	1.4
Drinking fountain	Offices, etc.	$^{3}/_{8}$″ valve	0.25	—	0.25
Kitchen sink	Private	Faucet	1.0	1.0	1.4
Kitchen sink	Hotel, restaurant	Faucet	3.0	3.0	4.0
Laundry trays (1 to 3)	Private	Faucet	1.0	1.0	1.4
Lavatory	Private	Faucet	0.5	0.5	0.7
Lavatory	Public	Faucet	1.5	1.5	2.0
Service sink	Offices, etc.	Faucet	2.25	2.25	3.0
Shower head	Public	Mixing valve	3.0	3.0	4.0
Shower head	Private	Mixing valve	1.0	1.0	1.4
Urinal	Public	1″ flushometer valve	10.0	—	10.0
Urinal	Public	$^{3}/_{4}$″ flushometer valve	5.0	—	5.0
Urinal	Public	Flush tank	3.0	—	3.0
Washing machine (8 lb)	Private	Automatic	1.0	1.0	1.4
Washing machine (8 lb)	Public	Automatic	2.25	2.25	3.0
Washing machine (15 lb)	Public	Automatic	3.0	3.0	4.0
Water closet	Private	Flushometer valve	6.0	—	6.0
Water closet	Private	Flush tank	2.2	—	2.2
Water closet	Public	Flushometer valve	10.0	—	10.0
Water closet	Public	Flush tank	5.0	—	5.0
Water closet	Public or private	Flushometer tank	2.0	—	2.0

For SI: 1 inch = 25.4 mm, 1 pound = 0.454 kg.

a. For fixtures not listed, loads should be assumed by comparing the fixture to one listed using water in similar quantities and at similar rates. The assigned loads for fixtures with both hot and cold water supplies are given for separate hot and cold water loads, and for total load. The separate hot and cold water loads are three-fourths of the total load for the fixture in each case.

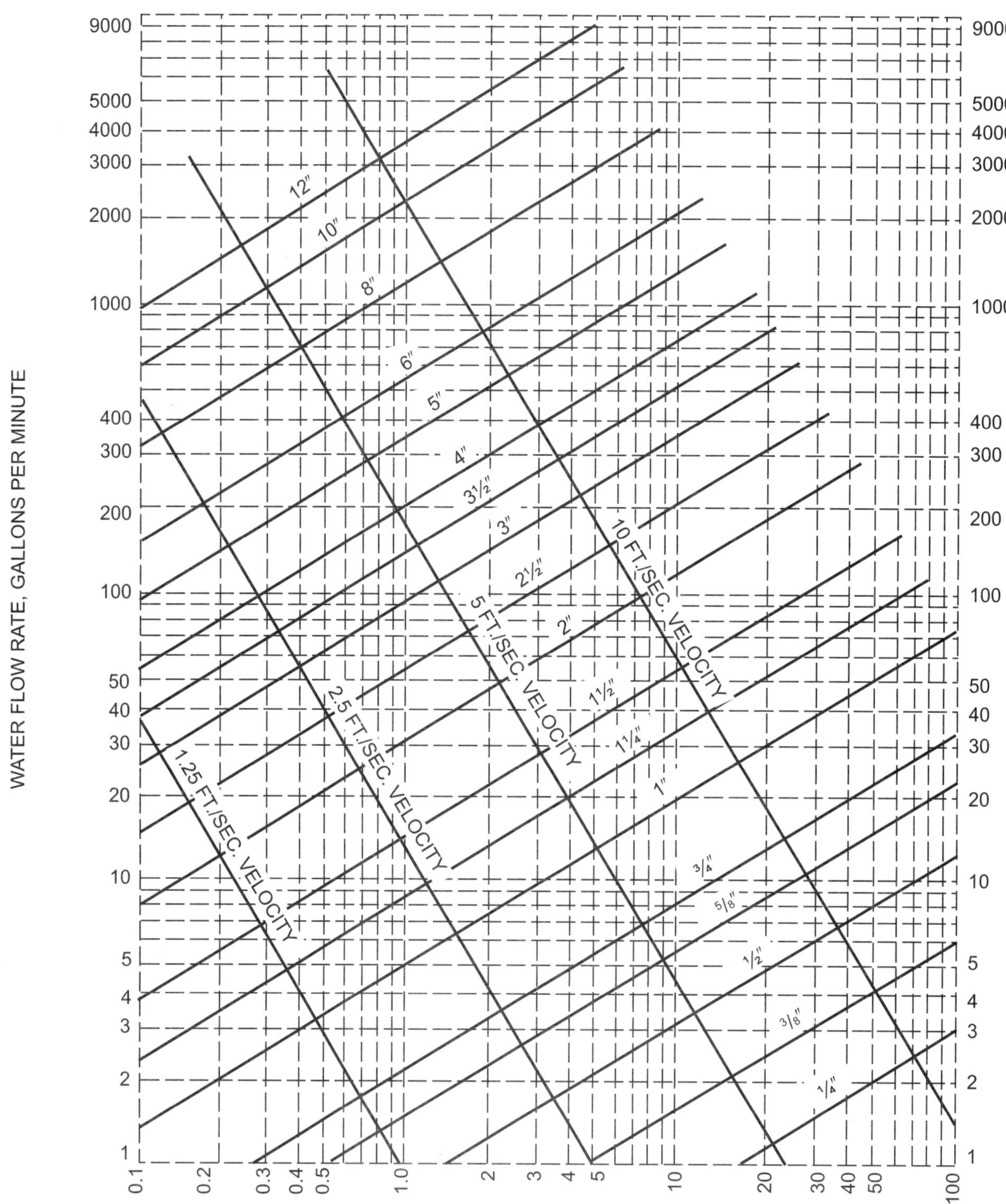

Note: Fluid velocities in excess of 5 to 8 feet per second are not usually recommended.

For SI: 1 inch = 25.4 mm, 1 foot = 304.8 mm, 1 gallon per minute = 3.785 L/m, 1 pound per square inch = 6.895 kPa, 1 foot per second = 0.305 m/s.

a. This figure applies to smooth new copper tubing with recessed (streamline) soldered joints and to the actual sizes of types indicated on the diagram.

FIGURE AP103.3(2)
FRICTION LOSS IN SMOOTH PIPE[a] (TYPE K, ASTM B88 COPPER TUBING)

TABLE AP103.3(3)
TABLE FOR ESTIMATING DEMAND

SUPPLY SYSTEMS PREDOMINANTLY FOR FLUSH TANKS			SUPPLY SYSTEMS PREDOMINANTLY FOR FLUSHOMETERS		
Load	Demand		Load	Demand	
(w.s.f.u.)	(gpm)	(cfm)	(w.s.f.u.)	(gpm)	(cfm)
1	3.0	0.04104	—	—	—
2	5.0	0.0684	—	—	—
3	6.5	0.86892	—	—	—
4	8.0	1.06944	—	—	—
5	9.4	1.256592	5	15.0	2.0052
6	10.7	1.430376	6	17.4	2.326032
7	11.8	1.577424	7	19.8	2.646364
8	12.8	1.711104	8	22.2	2.967696
9	13.7	1.831416	9	24.6	3.288528
10	14.6	1.951728	10	27.0	3.60936
11	15.4	2.058672	11	27.8	3.716304
12	16.0	2.13888	12	28.6	3.823248
13	16.5	2.20572	13	29.4	3.930192
14	17.0	2.27256	14	30.2	4.037136
15	17.5	2.3394	15	31.0	4.14408
16	18.0	2.90624	16	31.8	4.241024
17	18.4	2.459712	17	32.6	4.357968
18	18.8	2.513184	18	33.4	4.464912
19	19.2	2.566656	19	34.2	4.571856
20	19.6	2.620128	20	35.0	4.6788
25	21.5	2.87412	25	38.0	5.07984
30	23.3	3.114744	30	42.0	5.61356
35	24.9	3.328632	35	44.0	5.88192
40	26.3	3.515784	40	46.0	6.14928
45	27.7	3.702936	45	48.0	6.41664
50	29.1	3.890088	50	50.0	6.684
60	32.0	4.27776	60	54.0	7.21872
70	35.0	4.6788	70	58.0	7.75344
80	38.0	5.07984	80	61.2	8.181216
90	41.0	5.48088	90	64.3	8.595624
100	43.5	5.81508	100	67.5	9.0234
120	48.0	6.41664	120	73.0	9.75864
140	52.5	7.0182	140	77.0	10.29336
160	57.0	7.61976	160	81.0	10.82808
180	61.0	8.15448	180	85.5	11.42964
200	65.0	8.6892	200	90.0	12.0312
225	70.0	9.3576	225	95.5	12.76644
250	75.0	10.026	250	101.0	13.50168
275	80.0	10.6944	275	104.5	13.96956
300	85.0	11.3628	300	108.0	14.43744
400	105.0	14.0364	400	127.0	16.97736
500	124.0	16.57632	500	143.0	19.11624

(continued)

TABLE AP103.3(3)—continued
TABLE FOR ESTIMATING DEMAND

SUPPLY SYSTEMS PREDOMINANTLY FOR FLUSH TANKS			SUPPLY SYSTEMS PREDOMINANTLY FOR FLUSHOMETERS		
Load	Demand		Load	Demand	
(w.s.f.u.)	(gpm)	(cfm)	(w.s.f.u.)	(gpm)	(cfm)
750	170.0	22.7256	750	177.0	23.66136
1,000	208.0	27.80544	1,000	208.0	27.80544
1,250	239.0	31.94952	1,250	239.0	31.94952
1,500	269.0	35.95992	1,500	269.0	35.95992
1,750	297.0	39.70296	1,750	297.0	39.70296
2,000	325.0	43.446	2,000	325.0	43.446
2,500	380.0	50.7984	2,500	380.0	50.7984
3,000	433.0	57.88344	3,000	433.0	57.88344
4,000	535.0	70.182	4,000	525.0	70.182
5,000	593.0	79.27224	5,000	593.0	79.27224

For SI: 1 gallon per minute = 3.785 L/m, 1 cubic foot per minute = 0.000471 m^3/s.

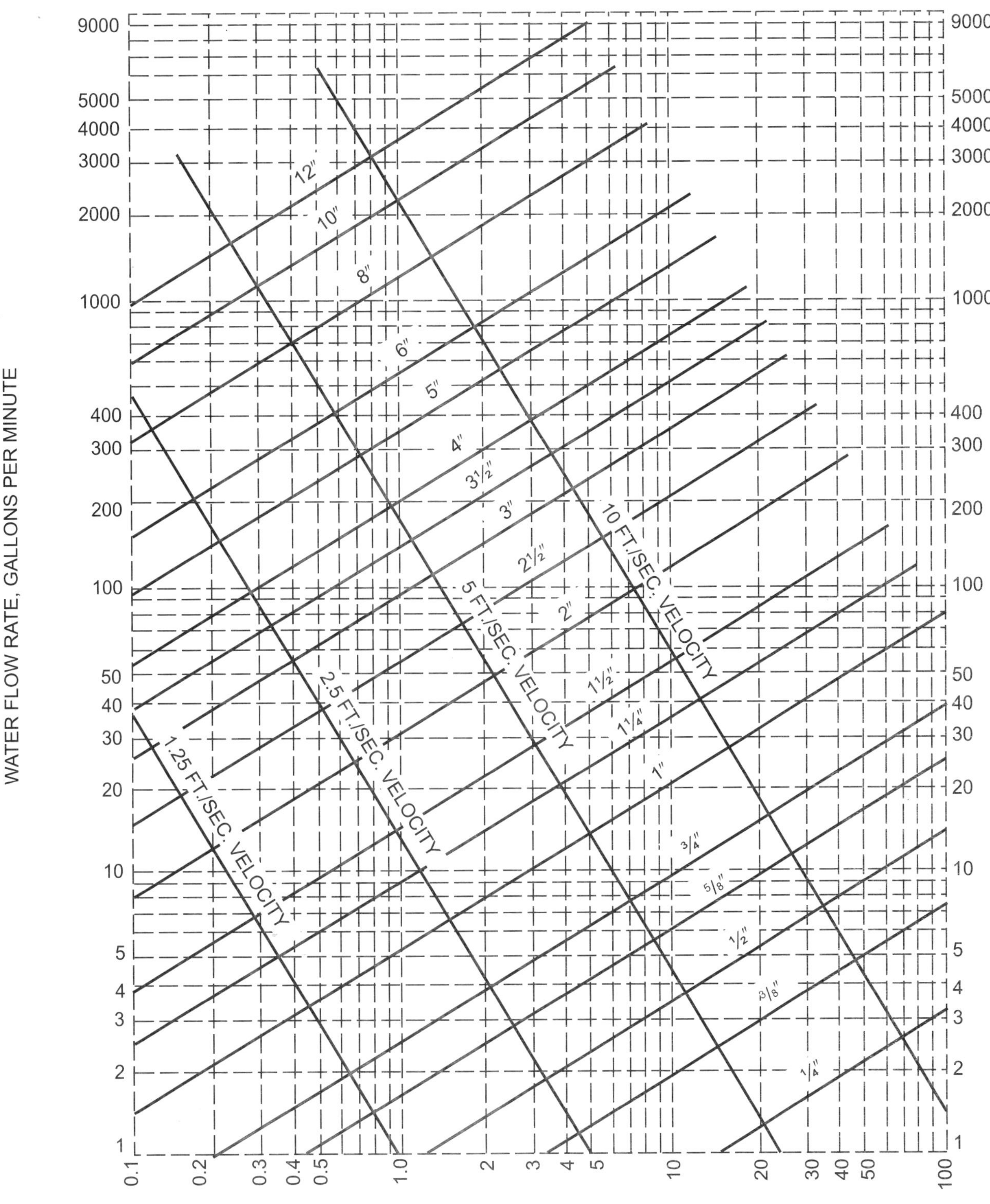

Note: Fluid velocities in excess of 5 to 8 feet per second are not usually recommended.

For SI: 1 inch = 25.4 mm, 1 foot = 304.8 mm, 1 gallon per minute = 3.785 L/m, 1 pound per square inch = 6.895 kPa, 1 foot per second = 0.305 m/s.

a. This figure applies to smooth new copper tubing with recessed (streamline) soldered joints and to the actual sizes of types indicated on the diagram.

FIGURE AP103.3(3)
FRICTION LOSS IN SMOOTH PIPE[a] (TYPE L, ASTM B88 COPPER TUBING)

TABLE AP103.3(4)
LOSS OF PRESSURE THROUGH TAPS AND TEES IN POUNDS PER SQUARE INCH (psi)

GALLONS PER MINUTE	SIZE OF TAP OR TEE (inches)						
	$^5/_8$	$^3/_4$	1	$1^1/_4$	$1^1/_2$	2	3
10	1.35	0.64	0.18	0.08	—	—	—
20	5.38	2.54	0.77	0.31	0.14	—	—
30	12.10	5.72	1.62	0.69	0.33	0.10	—
40	—	10.20	3.07	1.23	0.58	0.18	—
50	—	15.90	4.49	1.92	0.91	0.28	—
60	—	—	6.46	2.76	1.31	0.40	—
70	—	—	8.79	3.76	1.78	0.55	0.10
80	—	—	11.50	4.90	2.32	0.72	0.13
90	—	—	14.50	6.21	2.94	0.91	0.16
100	—	—	17.94	7.67	3.63	1.12	0.21
120	—	—	25.80	11.00	5.23	1.61	0.30
140	—	—	35.20	15.00	7.12	2.20	0.41
150	—	—	—	17.20	8.16	2.52	0.47
160	—	—	—	19.60	9.30	2.92	0.54
180	—	—	—	24.80	11.80	3.62	0.68
200	—	—	—	30.70	14.50	4.48	0.84
225	—	—	—	38.80	18.40	5.60	1.06
250	—	—	—	47.90	22.70	7.00	1.31
275	—	—	—	—	27.40	7.70	1.59
300	—	—	—	—	32.60	10.10	1.88

For SI: 1 inch = 25.4 mm, 1 pound per square inch = 6.895 kPa, 1 gallon per minute = 3.785 L/m.

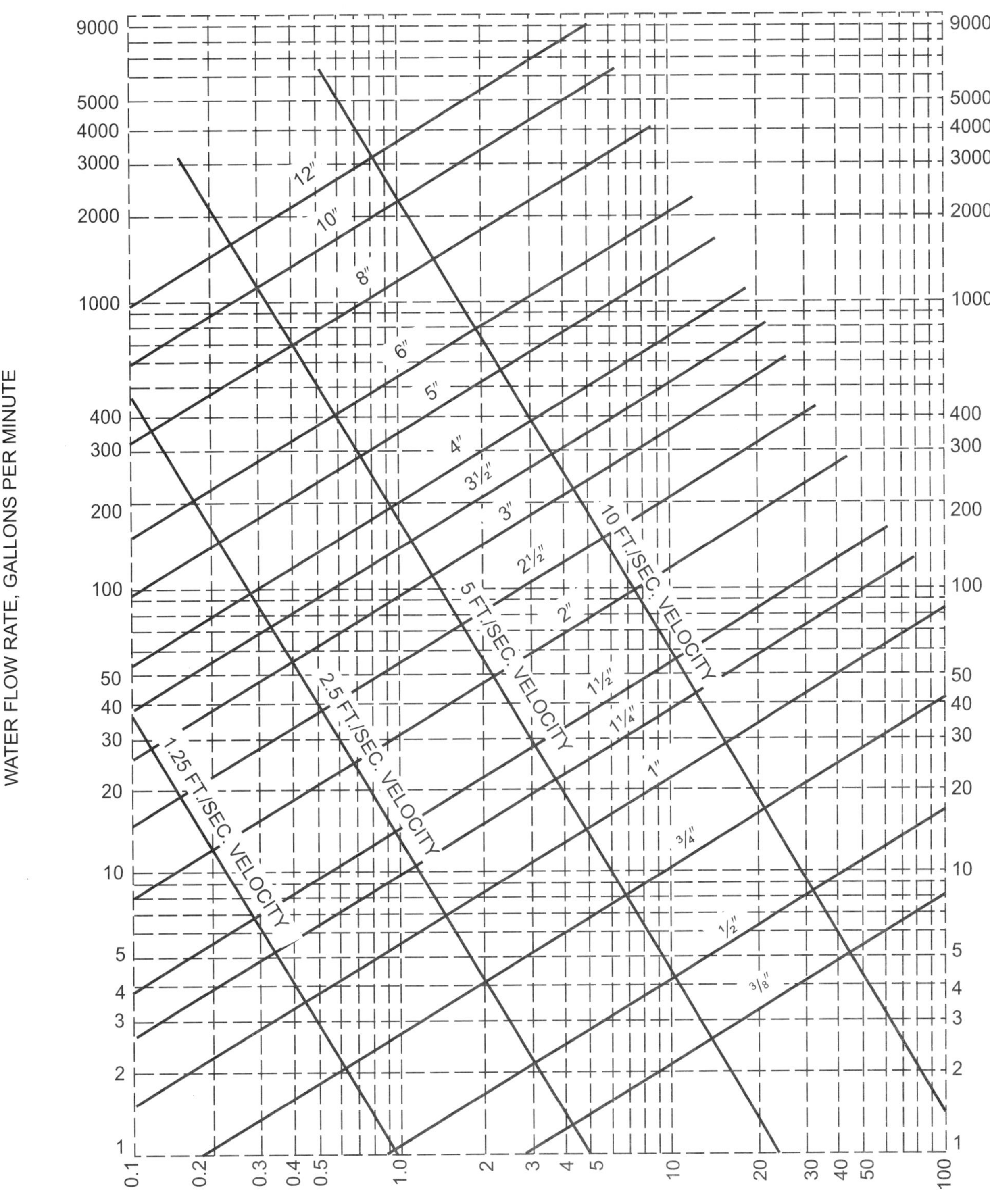

Note: Fluid velocities in excess of 5 to 8 feet per second are not usually recommended.

For SI: 1 inch = 25.4 mm, 1 foot = 304.8 mm, 1 gallon per minute = 3.785 L/m, 1 pound per square inch = 6.895 kPa, 1 foot per second = 0.305 m/s.

a. This figure applies to smooth new copper tubing with recessed (streamline) soldered joints and to the actual sizes of types indicated on the diagram.

FIGURE AP103.3(4)
FRICTION LOSS IN SMOOTH PIPE[a] (TYPE M, ASTM B88 COPPER TUBING)

TABLE AP103.3(5)
ALLOWANCE IN EQUIVALENT LENGTHS OF PIPE FOR FRICTION LOSS IN VALVES AND THREADED FITTINGS (feet)

FITTING OR VALVE	PIPE SIZE (inches)							
	$^1/_2$	$^3/_4$	1	$1^1/_4$	$1^1/_2$	2	$2^1/_2$	3
45-degree elbow	1.2	1.5	1.8	2.4	3.0	4.0	5.0	6.0
90-degree elbow	2.0	2.5	3.0	4.0	5.0	7.0	8.0	10.0
Tee, run	0.6	0.8	0.9	1.2	1.5	2.0	2.5	3.0
Tee, branch	3.0	4.0	5.0	6.0	7.0	10.0	12.0	15.0
Gate valve	0.4	0.5	0.6	0.8	1.0	1.3	1.6	2.0
Balancing valve	0.8	1.1	1.5	1.9	2.2	3.0	3.7	4.5
Plug-type cock	0.8	1.1	1.5	1.9	2.2	3.0	3.7	4.5
Check valve, swing	5.6	8.4	11.2	14.0	16.8	22.4	28.0	33.6
Globe valve	15.0	20.0	25.0	35.0	45.0	55.0	65.0	80.0
Angle valve	8.0	12.0	15.0	18.0	22.0	28.0	34.0	40.0

For SI: 1 inch = 25.4 mm, 1 foot = 304.8 mm, 1 degree = 0.0175 rad.

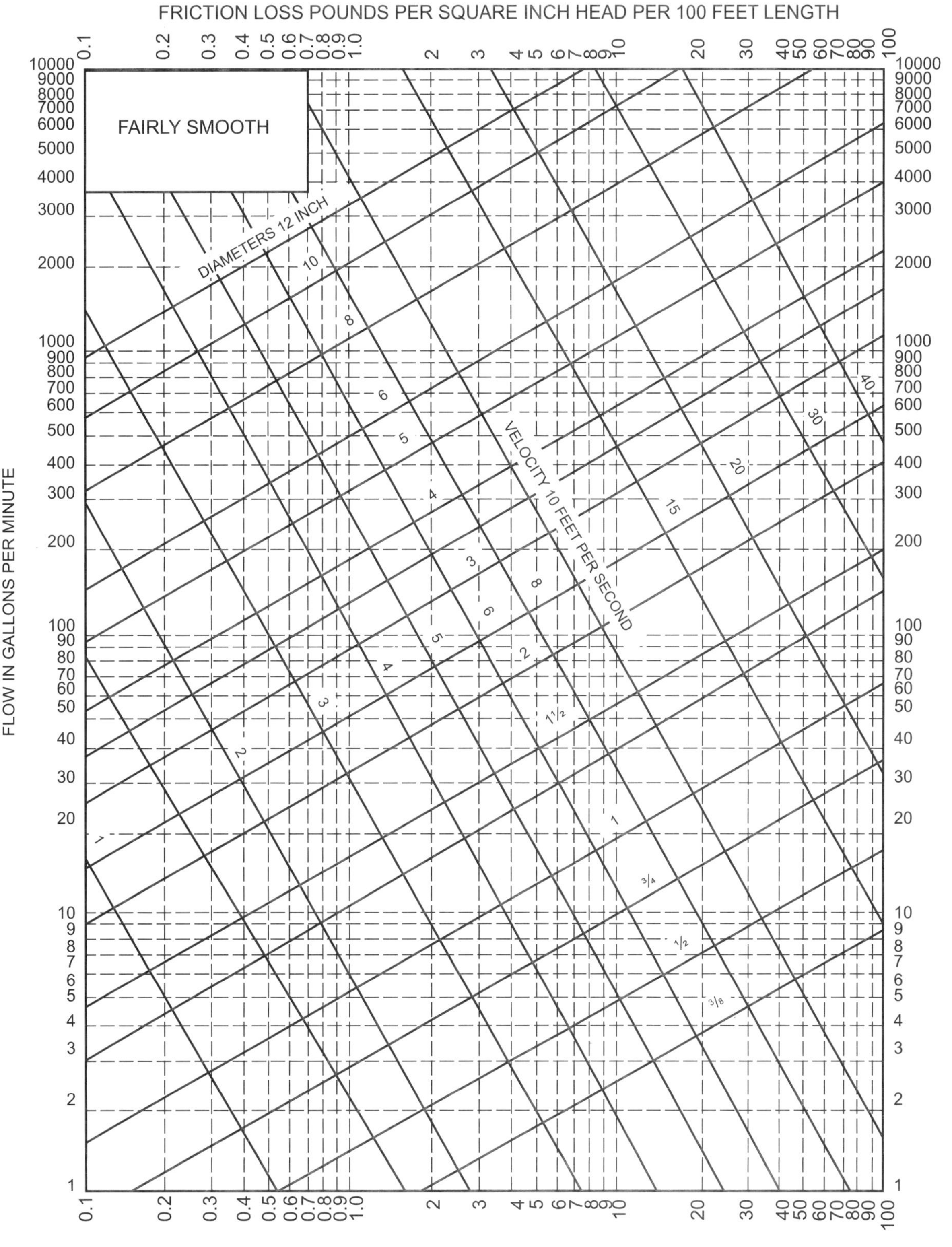

For SI: 1 inch = 25.4 mm, 1 foot = 304.8 mm, 1 gallon per minute = 3.785 L/m, 1 pound per square inch = 6.895 kPa, 1 foot per second = 0.305 m/s.

a. This figure applies to smooth new steel (fairly smooth) pipe and to actual diameters of standard-weight pipe.

FIGURE AP103.3(5)
FRICTION LOSS IN FAIRLY SMOOTH PIPE[a]

TABLE AP103.3(6)
PRESSURE LOSS IN FITTINGS AND VALVES EXPRESSED AS EQUIVALENT LENGTH OF TUBE[a] (feet)

NOMINAL OR STANDARD SIZE (inches)	FITTINGS					VALVES			
	Standard Ell		90-degree Tee		Coupling	Ball	Gate	Butterfly	Check
	90 Degree	45 Degree	Side Branch	Straight Run					
$^3/_8$	0.5	—	1.5	—	—	—	—	—	1.5
$^1/_2$	1	0.5	2	—	—	—	—	—	2
$^5/_8$	1.5	0.5	2	—	—	—	—	—	2.5
$^3/_4$	2	0.5	3	—	—	—	—	—	3
1	2.5	1	4.5	—	—	0.5	—	—	4.5
$1^1/_4$	3	1	5.5	0.5	0.5	0.5	—	—	5.5
$1^1/_2$	4	1.5	7	0.5	0.5	0.5	—	—	6.5
2	5.5	2	9	0.5	0.5	0.5	0.5	7.5	9
$2^1/_2$	7	2.5	12	0.5	0.5	—	1	10	11.5
3	9	3.5	15	1	1	—	1.5	15.5	14.5
$3^1/_2$	9	3.5	14	1	1	—	2	—	12.5
4	12.5	5	21	1	1	—	2	16	18.5
5	16	6	27	1.5	1.5	—	3	11.5	23.5
6	19	7	34	2	2	—	3.5	13.5	26.5
8	29	11	50	3	3	—	5	12.5	39

For SI: 1 inch = 25.4 mm, 1 foot = 304.8 mm, 1 degree = 0.01745 rad.

a. Allowances are for streamlined soldered fittings and recessed threaded fittings. For threaded fittings, double the allowances shown in the table. The equivalent lengths presented in the table are based on a C factor of 150 in the Hazen-Williams friction loss formula. The lengths shown are rounded to the nearest half-foot.

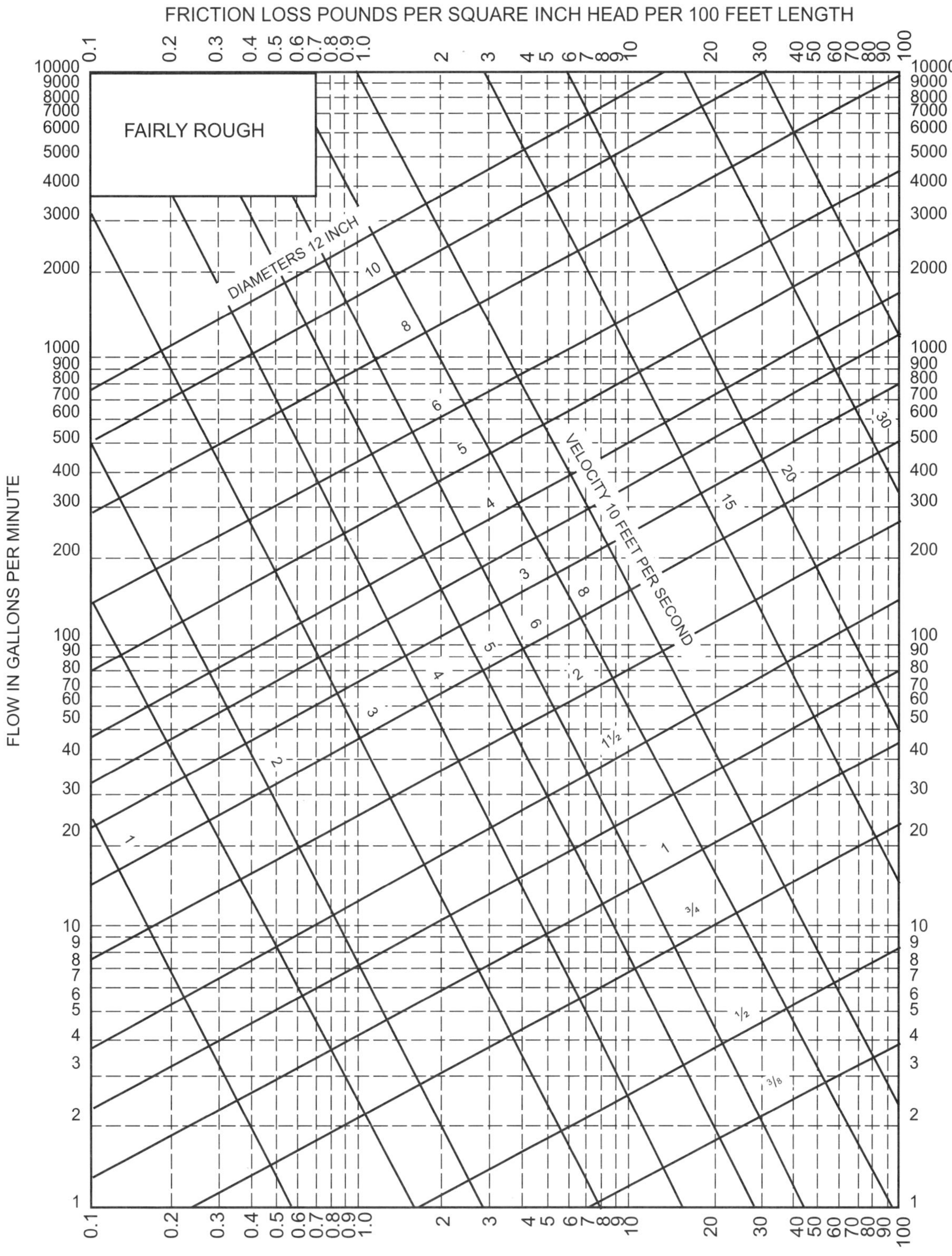

For SI: 1 inch = 25.4 mm, 1 foot = 304.8 mm, 1 gallon per minute = 3.785 L/m, 1 pound per square inch = 6.895 kPa, 1 foot per second = 0.305 m/s.

a. This figure applies to fairly rough pipe and to actual diameters, which, in general, will be less than the actual diameters of the new pipe of the same kind.

FIGURE AP103.3(6)
FRICTION LOSS IN FAIRLY ROUGH PIPE[a]

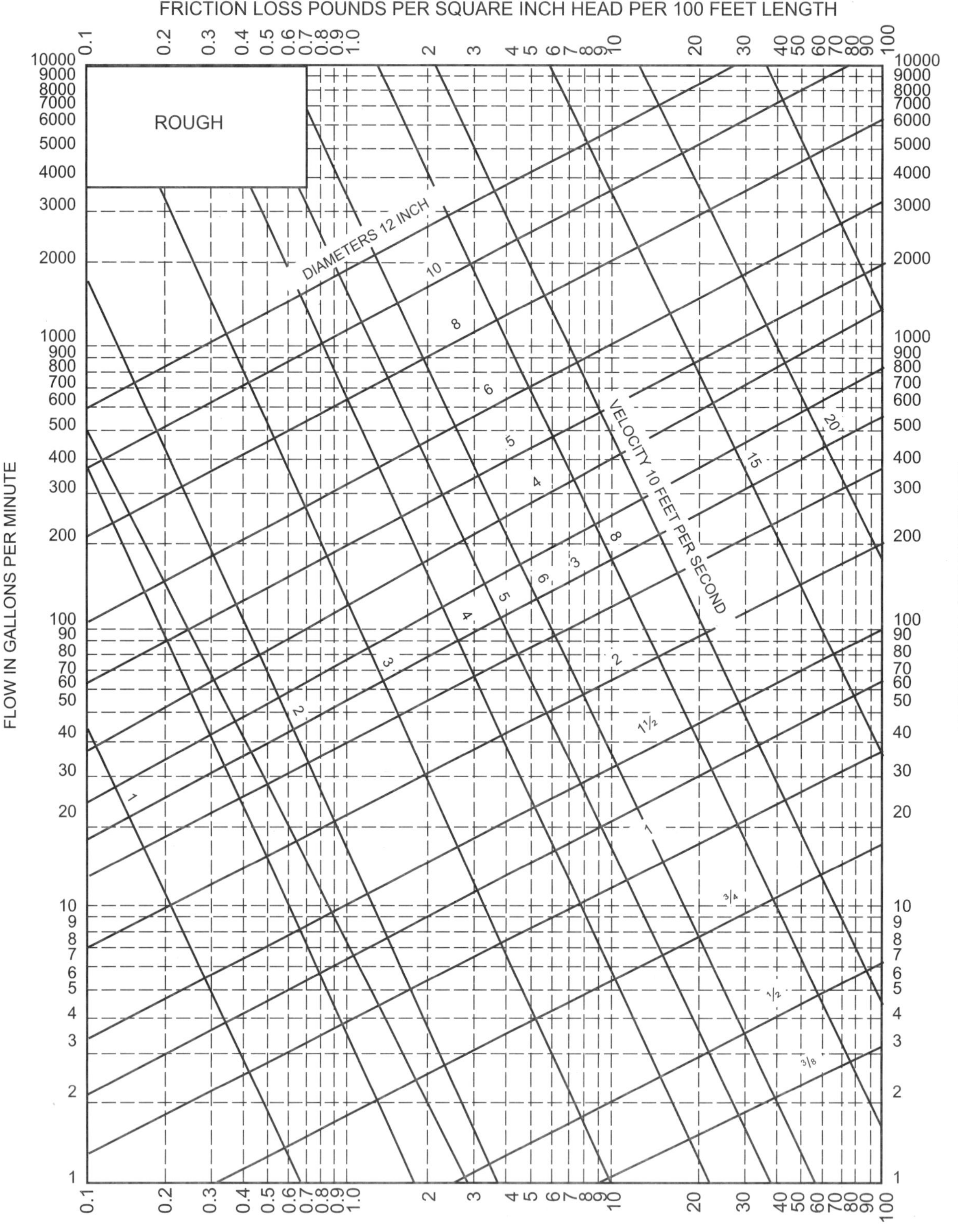

For SI: 1 inch = 25.4 mm, 1 foot = 304.8 mm, 1 gallon per minute = 3.785 L/m, 1 pound per square inch = 6.895 kPa, 1 foot per second = 0.305 m/s.

a. This figure applies to very rough pipe and existing pipe, and to their actual diameters.

FIGURE AP103.3(7)
FRICTION LOSS IN ROUGH PIPE[a]

AP103.3.1 Sample problem. What size Type L copper water pipe, service and distribution will be required to serve a two-story factory building having on each floor, back-to-back, two toilet rooms each equipped with hot and cold water? The highest fixture is 21 feet above the street main, which is tapped with a 2-inch corporation cock at which point the minimum pressure is 55 psi. In the building basement, a 2-inch meter with a pressure drop of not more than 11 psi and 3-inch reduced pressure principle backflow preventer with a pressure drop of not more than 9 psi are to be installed. The system is shown in Figure AP103.3(1). To be determined are the pipe sizes for the service main, and the cold and hot water distribution pipes.

Problem solution: A tabular arrangement such as shown in Table AP103.3(1) should first be constructed. The steps to be followed are indicated by the tabular arrangement itself as they are in sequence, Columns 1 through 10 and Lines A through L.

Step 1

Columns 1 and 2: Divide the system into sections breaking at major changes in elevation or where branches lead to fixture groups. After Point B [see Figure AP103.3(1)], separate consideration will be given to the hot and cold water piping. Enter the sections to be considered in the service and cold water piping in Column 1 of the tabular arrangement. Column 1 of Table AP103.3(1) provides a line-by-line, recommended tabular arrangement for use in solving pipe sizing.

The objective in designing the water supply system is to ensure an adequate water supply and pressure to all fixtures and equipment. Column 2 provides the psi (kPa) to be considered separately from the minimum pressure available at the main. Losses to take into consideration are the following: the differences in elevations between the water supply source and the highest water supply outlet; meter pressure losses; the tap in main loss; special fixture devices, such as water softeners and backflow prevention devices; and the pressure required at the most remote fixture outlet.

The difference in elevation can result in an increase or decrease in available pressure at the main. Where the water supply outlet is located above the source, this results in a loss in the available pressure and is subtracted from the pressure at the water source. Where the highest water supply outlet is located below the water supply source, there will be an increase in pressure that is added to the available pressure of the water source.

Column 3: Using Table AP103.3(3), determine the gpm (L/m) of flow to be expected in each section of the system. These flows range from 28.6 to 108 gpm. Load values for fixtures must be determined as w.s.f.u. and then converted to a gpm rating to determine peak demand. Where calculating peak demands, the w.s.f.u. are added and then converted to the gpm rating. For continuous flow fixtures, such as hose bibbs and lawn sprinkler systems, add the gpm demand to the intermittent demand of fixtures. For example, a total of 120 w.s.f.u. is converted to a demand of 48 gpm. Two hose bibbs × 5 gpm demand = 10 gpm. Total gpm rating = 48.0 gpm + 10 gpm = 58.0 gpm demand.

Step 2

Line A: Enter the minimum pressure available at the main source of supply in Column 2. This is 55 psi (379.2 kPa). The local water authorities generally keep records of pressures at different times of the day and year. The available pressure can be checked from nearby buildings or from fire department hydrant checks.

Line B: Determine from Table P2903.1 the highest pressure required for the fixtures on the system, which is 15 psi (103.4 kPa), to operate a flushometer valve. The most remote fixture outlet is necessary to compute the pressure loss caused by pipe and fittings, and represents the most downstream fixture along the circuit of piping requiring the available pressure to operate properly as indicated by Table P2903.1.

Line C: Determine the pressure loss for the meter size given or assumed. The total water flow from the main through the service as determined in Step 1 will serve to aid in the meter selected. There are three common types of water meters; the pressure losses are determined by the American Water Works Association Standards for displacement type, compound type and turbine type. The maximum pressure loss of such devices takes into consideration the meter size, safe operating capacity [gpm (L/m)] and maximum rates for continuous operations [gpm (L/m)]. Typically, equipment imparts greater pressure losses than piping.

Line D: Select from Table AP103.3(4) and enter the pressure loss for the tap size given or assumed. The loss of pressure through taps and tees in psi (kPa) is based on the total gpm (L/m) flow rate and size of the tap.

Line E: Determine the difference in elevation between the main and source of supply and the highest fixture on the system. Multiply this figure, expressed in feet (mm), by 0.43 psi. Enter the resulting psi (kPa) loss on Line E. The difference in elevation between the water supply source and the highest water supply outlet has a significant impact on the sizing of the water supply system. The difference in elevation usually results in a loss in the available pressure because the water supply outlet is generally located above the water supply source. The loss is caused by the pressure required to lift the water to the outlet. The pressure loss is subtracted from the pressure at the water source. Where the highest water supply outlet is located below the water source, there will be an increase in pressure that is added to the available pressure of the water source.

Lines F, G and H: The pressure losses through filters, backflow prevention devices or other special fixtures must be obtained from the manufacturer or estimated and entered on these lines. Equipment, such as backflow prevention devices, check valves, water softeners, instantaneous, or tankless water heaters, filters and strainers, can impart a much greater pressure loss than the piping. The pressure losses can range from 8 to 30 psi.

Step 3

Line I: The sum of the pressure requirements and losses that affect the overall system (Lines B through H) is entered on this line. Summarizing the steps, all of the system losses are subtracted from the minimum water pressure. The remainder is the pressure available for friction, defined as the energy available to push the water through the pipes to each fixture. This force can be used as an average pressure loss, as long as the pressure available for friction is not exceeded. Saving a certain amount for available water supply pressures as an area incurs growth, or because of the aging of the pipe or equipment added to the system is recommended.

Step 4

Line J: Subtract Line I from Line A. This gives the pressure that remains available from overcoming friction losses in the system. This figure is a guide to the pipe size that is chosen for each section, incorporating the total friction losses to the most remote outlet (measured length is called *developed length*).

Exception: Where the main is above the highest fixture, the resulting psi (kPa) must be considered a pressure gain (static head gain) and omitted from the sums of Lines B through H and added to Line J.

The maximum friction head loss that can be tolerated in the system during peak demand is the difference between the static pressure at the highest and most remote outlet at no-flow conditions and the minimum flow pressure required at that outlet. If the losses are within the required limits, every run of pipe will be within the required friction head loss. Static pressure loss is at the most remote outlet in feet × 0.433 = loss in psi caused by elevation differences.

Step 5

Column 4: Enter the length of each section from the main to the most remote outlet (at Point E). Divide the water supply system into sections breaking at major changes in elevation or where branches lead to fixture groups.

Step 6

Column 5: Where selecting a trial pipe size, the length from the water service or meter to the most remote fixture outlet must be measured to determine the developed length. However, in systems having a flushometer valve or temperature-controlled shower at the topmost floors, the developed length would be from the water meter to the most remote flushometer valve on the system. A rule of thumb is that size will become progressively smaller as the system extends farther from the main source of supply. A trial pipe size can be arrived at by the following formula:

Line J: (Pressure available to overcome pipe friction) × 100/equivalent length of run total developed length to most remote fixture × percentage factor of 1.5 (Note: a percentage factor is used only as an estimate for friction losses imposed for fittings for initial trial pipe size) = psi (average pressure drop per 100 feet of pipe).

For trial pipe size, see Figure AP103.3(3) (Type L copper) based on 2.77 psi and 108 gpm = $2^1/_2$ inches. To determine the equivalent length of run to the most remote outlet, the developed length is determined and added to the friction losses for fittings and valves. The developed lengths of the designated pipe sections are shown in Table AP103.3.1(1).

The equivalent length of the friction loss in fittings and valves must be added to the developed length (most remote outlet). Where the size of fittings and valves is not known, the added friction loss should be approximated. A general rule that has been used is to add 50 percent of the developed length to allow for fittings and valves. For example, the equivalent length of run equals the developed length of run (225 feet × 1.5 = 338 feet). The total equivalent length of run for determining a trial pipe size is 338 feet.

Example: 9.36 (pressure available to overcome pipe friction) × 100/338 (equivalent length of run = 225 × 1.5) = 2.77 psi (average pressure drop per 100 feet of pipe).

TABLE AP103.3.1(1)
SUMMATION OF DEVELOPED PIPING LENGTHS

SEGMENT	LENGTH[a] (feet)
A-B	54
B-C	8
C-D	13
D-E	150

For SI: 1 foot = 304.8 mm.
a. Total developed length = 225 feet.

Step 7

Column 6: Select from Table AP103.3(6) the equivalent lengths for the trial pipe size of fittings and valves on each pipe section. Enter the sum for each section in Column 6. (The number of fittings to be used in this example must be an estimate). The equivalent length of piping is the developed length plus the equivalent lengths of pipe corresponding to the friction head losses for fittings and valves. Where the size of fittings and valves is not known, the added friction head losses must be approximated. An estimate for this example is found in Table AP103.3.1(2).

Step 8

Column 7: Add the figures from Columns 4 and 6, and enter in Column 7. Express the sum in hundreds of feet.

Step 9

Column 8: Select from Figure AP103.3(3) the friction loss per 100 feet of pipe for the gpm flow in a section (Column 3) and trial pipe size (Column 5). Maximum friction head loss per 100 feet is determined on the basis of the total pressure available for friction head loss and the longest equivalent length of run. The selection is based on the gpm demand, uniform friction head loss and maximum design velocity. Where the size indicated by the hydraulic table indicates a velocity in excess of the selected velocity, a size must be selected that produces the required velocity.

Step 10

Column 9: Multiply the figures in Columns 7 and 8 for each section and enter in Column 9.

Total friction loss is determined by multiplying the friction loss per 100 feet for each pipe section in the total developed length by the pressure loss in fittings expressed as equivalent length in feet (mm). Note: Section C-F should be considered in the total pipe friction losses only if greater loss occurs in Section C-F than in pipe Section D-E. Section C-F is not considered in the total developed length. Total friction loss in equivalent length is determined in Table AP103.3.1(3).

Step 11

Line K: Enter the sum of the values in Column 9. The value is the total friction loss in equivalent length for each designated pipe section.

Step 12

Line L: Subtract Line J from Line K and enter in Column 10.

The result should always be a positive or plus figure. If it is not, repeat the operation using Columns 5, 6, 8 and 9 until a balance or near balance is obtained. If the difference between Lines J and K is a high positive number, it is an indication that the pipe sizes are too large and should be reduced, thus saving materials. In such a case, the operations using Columns 5, 6, 8 and 9 should be repeated.

The total friction losses are determined and subtracted from the pressure available to overcome pipe friction for the trial pipe size. This number is critical because it provides a guide to whether the pipe size selected is too large and the process should be repeated to obtain an economically designed system.

Answer: The final figures entered in Column 5 become the design pipe size for the respective sections. Repeating this operation a second time using the same sketch but considering the demand for hot water, it is possible to size the hot water distribution piping. This has been worked up as a part of the overall problem in the tabular arrangement used for sizing the service and water distribution piping. Note that consideration must be given to the pressure losses from the street main to the water heater (Section A-B) in determining the hot water pipe sizes.

SECTION AP201 SELECTION OF PIPE SIZE

AP201.1 Size of water-service mains, branch mains and risers. The minimum size water service pipe shall be $^3/_4$ inch (19.1 mm). The size of water service mains, branch mains and risers shall be determined according to water supply demand [gpm (L/m)], available water pressure [psi (kPa)] and friction loss caused by the water meter and *developed length* of pipe [feet (m)], including the *equivalent length* of fittings. The size of each water distribution system shall be

TABLE AP103.3.1(2)
FITTING PRESSURE LOSSES EXPRESSED IN EQUIVALENT LENGTHS (formerly Table AP.1)

COLD WATER PIPE SECTION	FITTINGS/VALVES	PRESSURE LOSS EXPRESSED AS EQUIVALENT LENGTH OF TUBE (feet)	HOT WATER PIPE SECTION	FITTINGS/VALVES	PRESSURE LOSS EXPRESSED AS EQUIVALENT OF TUBE (feet)
A-B	3 – $2^1/_2$″ Gate valves	3	A-B	3 – $2^1/_2$″ Gate valves	3
	1 – $2^1/_2$″ Side branch tee	12	—	1 – $2^1/_2$″ Side branch tee	12
B-C	1 – $2^1/_2$″ Straight run tee	0.5	B-C	1 – 2″ Straight run tee	7
	—	—	—	1 – 2″ 90-degree ell	0.5
C-F	1 – $2^1/_2$″ Side branch tee	12	C-F	1 – $1^1/_2$″ Side branch tee	7
C-D	1 – $2^1/_2$″ 90-degree ell	7	C-D	1 – $^1/_2$″ 90-degree ell	4
D-E	1 – $2^1/_2$″ Side branch tee	12	D-E	1 – $1^1/_2$″ Side branch tee	7

For SI: 1 inch = 25.4 mm, 1 foot = 304.8 mm, 1 degree = 0.01745 rad.

TABLE AP103.3.1(3)
TOTAL FRICTION LOSS EQUIVALENT PIPING LENGTH (formerly Table AP.2)

PIPE SECTIONS	FRICTION LOSS EQUIVALENT LENGTH (feet)	
	Cold Water	Hot Water
A-B	0.69 × 3.2 = 2.21	0.69 × 3.2 = 2.21
B-C	0.085 × 3.1 = 0.26	0.16 × 1.4 = 0.22
C-D	0.20 × 1.9 = 0.38	0.17 × 3.2 = 0.54
D-E	1.62 × 1.9 = 3.08	1.57 × 3.2 = 5.02
Total pipe friction losses (Line K)	5.93	7.99

For SI: 1 foot = 304.8 mm.

determined according to the procedure outlined in this section or by other design methods conforming to acceptable engineering practice and *approved* by the *building official*:

1. Supply load in the building water distribution system shall be determined by the total load on the pipe being sized, in terms of w.s.f.u., as shown in Table AP103.3(2). For fixtures not *listed*, choose a w.s.f.u. value of a fixture with similar flow characteristics.
2. Obtain the minimum daily static service pressure [psi (kPa)] available (as determined by the local water authority) at the water meter or other source of supply at the installation location. Adjust this minimum daily static pressure [psi (kPa)] for the following conditions:
   - 2.1. Determine the difference in elevation between the source of supply and the highest water supply outlet. Where the highest water supply outlet is located above the source of supply, deduct 0.5 psi (3.4 kPa) for each foot (0.3 m) of difference in elevation. Where the highest water supply outlet is located below the source of supply, add 0.5 psi (3.4 kPa) for each foot (0.3 m) of difference in elevation.
   - 2.2. Where a water pressure-reducing valve is installed in the water distribution system, the minimum daily static water pressure available is 80 percent of the minimum daily static water pressure at the source of supply or the set pressure downstream of the water pressure-reducing valve, whichever is smaller.
   - 2.3. Deduct all pressure losses caused by special equipment, such as a backflow preventer, water filter and water softener. Pressure loss data for each piece of equipment shall be obtained through the manufacturer of the device.
   - 2.4. Deduct the pressure in excess of 8 psi (55 kPa) resulting from the installation of the special plumbing fixture, such as temperature-controlled shower and flushometer tank water closet. Using the resulting minimum available pressure, find the corresponding pressure range in Table AP201.1.
3. The maximum *developed length* for water piping is the actual length of pipe between the source of supply and the most remote fixture, including either hot (through the water heater) or cold water branches multiplied by a factor of 1.2 to compensate for pressure loss through fittings. Select the appropriate column in Table AP201.1 equal to or greater than the calculated maximum *developed length*.
4. To determine the size of the water service pipe, meter and main distribution pipe to the building using the appropriate table, follow down the selected "maximum *developed length*" column to a fixture unit equal to or greater than the total installation demand calculated by using the "combined" w.s.f.u. column of Table AP201.1. Read the water service pipe and meter sizes in the first left-hand column and the main distribution pipe to the building in the second left-hand column on the same row.
5. To determine the size of each water distribution pipe, start at the most remote outlet on each branch (either hot or cold branch) and, working back toward the main distribution pipe to the building, add up the w.s.f.u. demand passing through each segment of the distribution system using the related hot or cold column of Table AP201.1. Knowing demand, the size of each segment shall be read from the second left-hand column of the same table and the maximum *developed length* column selected in Steps 1 and 2, under the same or next smaller size meter row. The size of any branch or main need never be larger than the size of the main distribution pipe to the building established in Step 4.

TABLE AP201.1
MINIMUM SIZE OF WATER METERS, MAINS AND DISTRIBUTION PIPING BASED ON WATER SUPPLY FIXTURE UNIT VALUES (w.s.f.u.)

METER AND SERVICE PIPE (inches)	DISTRIBUTION PIPE (inches)	MAXIMUM DEVELOPMENT LENGTH (feet)									
Pressure Range 30 to 39 psi		40	60	80	100	150	200	250	300	400	500
$^{3}/_{4}$	$^{1}/_{2}$[a]	2.5	2	1.5	1.5	1	1	0.5	0.5	0	0
$^{3}/_{4}$	$^{3}/_{4}$	9.5	7.5	6	5.5	4	3.5	3	2.5	2	1.5
$^{3}/_{4}$	1	32	25	20	16.5	11	9	7.8	6.5	5.5	4.5
1	1	32	32	27	21	13.5	10	8	7	5.5	5
$^{3}/_{4}$	$1^{1}/_{4}$	32	32	32	32	30	24	20	17	13	10.5
1	$1^{1}/_{4}$	80	80	70	61	45	34	27	22	16	12
$1^{1}/_{2}$	$1^{1}/_{4}$	80	80	80	75	54	40	31	25	17.5	13
1	$1^{1}/_{2}$	87	87	87	87	84	73	64	56	45	36
$1^{1}/_{2}$	$1^{1}/_{2}$	151	151	151	151	117	92	79	69	54	43
2	$1^{1}/_{2}$	151	151	151	151	128	99	83	72	56	45
1	2	87	87	87	87	87	87	87	87	87	86
$1^{1}/_{2}$	2	275	275	275	275	258	223	196	174	144	122
2	2	365	365	365	365	318	266	229	201	160	134
2	$2^{1}/_{2}$	533	533	533	533	533	495	448	409	353	311

METER AND SERVICE PIPE (inches)	DISTRIBUTION PIPE (inches)	MAXIMUM DEVELOPMENT LENGTH (feet)									
Pressure Range 40 to 49 psi		40	60	80	100	150	200	250	300	400	500
$^{3}/_{4}$	$^{1}/_{2}$[a]	3	2.5	2	1.5	1.5	1	1	0.5	0.5	0.5
$^{3}/_{4}$	$^{3}/_{4}$	9.5	9.5	8.5	7	5.5	4.5	3.5	3	2.5	2
$^{3}/_{4}$	1	32	32	32	26	18	13.5	10.5	9	7.5	6
1	1	32	32	32	32	21	15	11.5	9.5	7.5	6.5
$^{3}/_{4}$	$1^{1}/_{4}$	32	32	32	32	32	32	32	27	21	16.5
1	$1^{1}/_{4}$	80	80	80	80	65	52	42	35	26	20
$1^{1}/_{2}$	$1^{1}/_{4}$	80	80	80	80	75	59	48	39	28	21
1	$1^{1}/_{2}$	87	87	87	87	87	87	87	78	65	55
$1^{1}/_{2}$	$1^{1}/_{2}$	151	151	151	151	151	130	109	93	75	63
2	$1^{1}/_{2}$	151	151	151	151	151	139	115	98	77	64
1	2	87	87	87	87	87	87	87	87	87	87
$1^{1}/_{2}$	2	275	275	275	275	275	275	264	238	198	169
2	2	365	365	365	365	365	349	304	270	220	185
2	$2^{1}/_{2}$	533	533	533	533	533	533	533	528	456	403

(continued)

TABLE AP201.1—continued
MINIMUM SIZE OF WATER METERS, MAINS AND DISTRIBUTION PIPING BASED ON WATER SUPPLY FIXTURE UNIT VALUES (w.s.f.u.)

METER AND SERVICE PIPE (inches)	DISTRIBUTION PIPE (inches)	MAXIMUM DEVELOPMENT LENGTH (feet)									
Pressure Range 50 to 60 psi		40	60	80	100	150	200	250	300	400	500
$^3/_4$	$^1/_2$[a]	3	3	2.5	2	1.5	1	1	1	0.5	0.5
$^3/_4$	$^3/_4$	9.5	9.5	9.5	8.5	6.5	5	4.5	4	3	2.5
$^3/_4$	1	32	32	32	32	25	18.5	14.5	12	9.5	8
1	1	32	32	32	32	30	22	16.5	13	10	8
$^3/_4$	$1^1/_4$	32	32	32	32	32	32	32	32	29	24
1	$1^1/_4$	80	80	80	80	80	68	57	48	35	28
$1^1/_2$	$1^1/_4$	80	80	80	80	80	75	63	53	39	29
1	$1^1/_2$	87	87	87	87	87	87	87	87	82	70
$1^1/_2$	$1^1/_2$	151	151	151	151	151	151	139	120	94	79
2	$1^1/_2$	151	151	151	151	151	151	146	126	97	81
1	2	87	87	87	87	87	87	87	87	87	87
$1^1/_2$	2	275	275	275	275	275	275	275	275	247	213
2	2	365	365	365	365	365	365	365	329	272	232
2	$2^1/_2$	533	533	533	533	533	533	533	533	533	486

METER AND SERVICE PIPE (inches)	DISTRIBUTION PIPE (inches)	MAXIMUM DEVELOPMENT LENGTH (feet)									
Pressure Range Over 60		40	60	80	100	150	200	250	300	400	500
$^3/_4$	$^1/_2$[a]	3	3	3	2.5	2	1.5	1.5	1	1	0.5
$^3/_4$	$^3/_4$	9.5	9.5	9.5	9.5	7.5	6	5	4.5	3.5	3
$^3/_4$	1	32	32	32	32	32	24	19.5	15.5	11.5	9.5
1	1	32	32	32	32	32	28	28	17	12	9.5
$^3/_4$	$1^1/_4$	32	32	32	32	32	32	32	32	32	30
1	$1^1/_4$	80	80	80	80	80	80	69	60	46	36
$1^1/_2$	$1^1/_4$	80	80	80	80	80	80	76	65	50	38
1	$1^1/_2$	87	87	87	87	87	87	87	87	87	84
$1^1/_2$	$1^1/_2$	151	151	151	151	151	151	151	144	114	94
2	$1^1/_2$	151	151	151	151	151	151	151	151	118	97
1	2	87	87	87	87	87	87	87	87	87	87
$1^1/_2$	2	275	275	275	275	275	275	275	275	275	252
2	2	365	368	368	368	368	368	368	368	318	273
2	$2^1/_2$	533	533	533	533	533	533	533	533	533	533

For SI: 1 inch = 25.4, 1 foot = 304.8 mm, 1 pound per square inch = 6.895 kPa.
a. Minimum size for building supply is a $^3/_4$-inch pipe.

APPENDIX AQ

TINY HOUSES

The provisions contained in this appendix are not mandatory unless specifically referenced in the adopting ordinance.

User note:

About this appendix: *Appendix AQ relaxes various requirements in the body of the code as they apply to houses that are 400 square feet in area or less. Attention is specifically paid to features such as compact stairs, including stair handrails and headroom, ladders, reduced ceiling heights in lofts and guard and emergency escape and rescue opening requirements at lofts.*

SECTION AQ101 GENERAL

AQ101.1 Scope. This appendix shall be applicable to *tiny houses* used as single *dwelling units. Tiny houses* shall comply with this code except as otherwise stated in this appendix.

SECTION AQ102 DEFINITIONS

AQ102.1 General. The following words and terms shall, for the purposes of this appendix, have the meanings shown herein. Refer to Chapter 2 of this code for general definitions.

EGRESS ROOF ACCESS WINDOW. A *skylight* or roof window designed and installed to satisfy the emergency escape and rescue opening requirements of Section R310.2.

LANDING PLATFORM. A landing provided as the top step of a stairway accessing a *loft.*

LOFT. A floor level located more than 30 inches (762 mm) above the main floor, open to the main floor on one or more sides with a ceiling height of less than 6 feet 8 inches (2032 mm) and used as a living or sleeping space.

TINY HOUSE. A *dwelling* that is 400 square feet (37 m^2) or less in floor area excluding *lofts.*

SECTION AQ103 CEILING HEIGHT

AQ103.1 Minimum ceiling height. *Habitable space* and hallways in *tiny houses* shall have a ceiling height of not less than 6 feet 8 inches (2032 mm). Bathrooms, toilet rooms and kitchens shall have a ceiling height of not less than 6 feet 4 inches (1930 mm). Obstructions including, but not limited to, beams, girders, ducts and lighting, shall not extend below these minimum ceiling heights.

Exception: Ceiling heights in *lofts* are permitted to be less than 6 feet 8 inches (2032 mm).

SECTION AQ104 LOFTS

AQ104.1 Minimum loft area and dimensions. *Lofts* used as a sleeping or living space shall meet the minimum area and dimension requirements of Sections AQ104.1.1 through AQ104.1.3.

AQ104.1.1 Minimum area. *Lofts* shall have a floor area of not less than 35 square feet (3.25 m^2).

AQ104.1.2 Minimum horizontal dimensions. *Lofts* shall be not less than 5 feet (1524 mm) in any horizontal dimension.

AQ104.1.3 Height effect on loft area. Portions of a *loft* with a sloped ceiling measuring less than 3 feet (914 mm) from the finished floor to the finished ceiling shall not be considered as contributing to the minimum required area for the loft. See Figure AQ104.1.3.

Exception: Under gable roofs with a minimum slope of 6 units vertical in 12 units horizontal (50-percent slope), portions of a *loft* with a sloped ceiling measuring less than 16 inches (406 mm) from the finished floor to the finished ceiling shall not be considered as contributing to the minimum required area for the *loft.*

AQ104.2 Loft access and egress. The access to and primary egress from *lofts* shall be of any type described in Sections AQ104.2.1 through AQ104.2.5. The *loft* access and egress element along its required minimum width shall meet the *loft* where its ceiling height is not less than 3 feet (914 mm).

AQ104.2.1 Stairways. Stairways accessing *lofts* shall comply with this code or with Sections AQ104.2.1.1 through AQ104.2.1.7.

AQ104.2.1.1 Width. Stairways accessing a *loft* shall not be less than 17 inches (432 mm) in clear width at or above the *handrail.* The width below the *handrail* shall be not less than 20 inches (508 mm).

AQ104.2.1.2 Headroom. The headroom above stairways accessing a *loft* shall be not less than 6 feet 2 inches (1880 mm), as measured vertically, from a sloped line connecting the tread, landing or landing platform *nosings* in the center of their width and vertically from the landing platform along the center of its width.

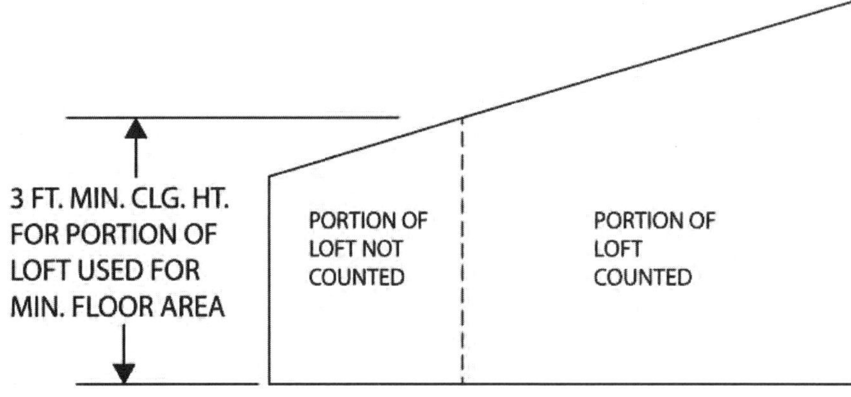

For SI: 1 foot = 304.8 mm.

FIGURE AQ104.1.3
HEIGHT EFFECT ON LOFT AREA

AQ104.2.1.3 Treads and risers. *Risers* for stairs accessing a *loft* shall be not less than 7 inches (178 mm) and not more than 12 inches (305 mm) in height. Tread depth and riser height shall be calculated in accordance with one of the following formulas:

1. The tread depth shall be 20 inches (508 mm) minus four-thirds of the riser height.
2. The riser height shall be 15 inches (381 mm) minus three-fourths of the tread depth.

AQ104.2.1.4 Landings. Intermediate landings and landings at the bottom of stairways shall comply with Section R311.7.6, except that the depth in the direction of travel shall be not less than 24 inches (610 mm).

AQ104.2.1.5 Landing platforms. The top tread and *riser* of stairways accessing *lofts* shall be constructed as a *landing platform* where the *loft* ceiling height is less than 6 feet 2 inches (1880 mm) where the stairway meets the *loft*. The *landing platform* shall be not less than 20 inches (508 mm) in width and in depth measured horizontally from and perpendicular to the *nosing* of the landing platform. The landing platform riser height to the loft floor shall be not less than 16 inches (406 mm) and not greater than 18 inches (457 mm).

AQ104.2.1.6 Handrails. *Handrails* shall comply with Section R311.7.8.

AQ104.2.1.7 Stairway guards. Guards at open sides of stairways, landings and landing platforms shall comply with Section R312.1.

AQ104.2.2 Ladders. Ladders accessing *lofts* shall comply with Sections AQ104.2.1 and AQ104.2.2.2.

AQ104.2.2.1 Size and capacity. Ladders accessing *lofts* shall have a rung width of not less than 12 inches (305 mm), and 10-inch (254 mm) to 14-inch (356 mm) spacing between rungs. Ladders shall be capable of supporting a 300-pound (136 kg) load on any rung. Rung spacing shall be uniform within $^{3}/_{8}$ inch (9.5 mm).

AQ104.2.2.2 Incline. Ladders shall be installed at 70 to 80 degrees from horizontal.

AQ104.2.3 Alternating tread devices. Alternating tread devices accessing *lofts* shall comply with Sections R311.7.11.1 and R311.7.11.2. The clear width at and below the *handrails* shall be not less than 20 inches (508 mm).

AQ104.2.4 Ship's ladders. Ship's ladders accessing *lofts* shall comply with Sections R311.7.12.1 and R311.7.12.2. The clear width at and below *handrails* shall be not less than 20 inches (508 mm).

AQ104.2.5 Loft guards. *Loft* guards shall be located along the open sides of *lofts*. *Loft* guards shall be not less than 36 inches (914 mm) in height or one-half of the clear height to the ceiling, whichever is less. *Loft* guards shall comply with Section R312.1.3 and Table R301.5 for their components.

SECTION AQ105 EMERGENCY ESCAPE AND RESCUE OPENINGS

AQ105.1 General. *Tiny houses* shall meet the requirements of Section R310 for emergency escape and rescue openings.

Exception: *Egress roof access windows* in *lofts* used as sleeping rooms shall be deemed to meet the requirements of Section R310 where installed such that the bottom of the opening is not more than 44 inches (1118 mm) above the *loft* floor, provided the egress roof access window complies with the minimum opening area requirements of Section R310.2.1.

SECTION AQ106 ENERGY CONSERVATION

AQ106.1 Air leakage testing. The air leakage rate for *tiny houses* shall not exceed 0.30 cubic feet per minute at 50 Pascals of pressure per square foot of the *dwelling unit* enclosure area. The air leakage testing shall be in accordance with the testing methods required in Section N1102.4.1.2. The *dwelling unit* enclosure area shall be the sum of the areas of ceilings, floors and walls that separate the conditioned space of a *dwelling unit* from the exterior, its adjacent unconditioned spaces and adjacent *dwelling units.*

AQ106.1.1 Whole-house mechanical ventilation. Where the air leakage rate is in accordance with Section AQ106.1, the *tiny house* shall be provided with whole-house mechanical ventilation in accordance with Section M1505.4.

AQ106.2 Alternative compliance. *Tiny houses* shall be deemed to be in compliance with Chapter 11 of this code and Chapter R4 of the *International Energy Conservation Code*, provided that the following conditions are met:

1. The insulation and fenestration meet the requirements of Table N1102.1.2.
2. The thermal envelope meets the requirements of Section N1102.4.1.1 and Table N1102.4.1.1.
3. Solar, wind or other renewable energy source supplies not less than 90 percent of the energy use for the structure.
4. Solar, wind or other renewable energy source supplies not less than 90 percent of the energy for service water heating.
5. Permanently installed lighting is in accordance with Section N1104.
6. Mechanical ventilation is provided in accordance with Section M1505 and operable fenestration is not used to meet ventilation requirements.

APPENDIX AR

LIGHT STRAW-CLAY CONSTRUCTION

The provisions contained in this appendix are not mandatory unless specifically referenced in the adopting ordinance.

User note:

About this appendix: *While heavier forms of straw-clay construction have been used in various parts of the world for thousands of years, light forms of straw-clay construction began to appear in Europe in 1950 and in the United States in 1990. These lighter forms of straw-clay construction are intended as infill materials in nonload-bearing walls. The advantages of light straw-clay construction, such as regulated by Appendix AR, include thermal performance and low environmental impact.*

SECTION AR101 GENERAL

AR101.1 Scope. This appendix shall govern the use of light straw-clay as a nonbearing building material and wall infill system in *Seismic Design Categories* A and B. Use of light straw-clay in *Seismic Design Categories* C, D_0, D_1 and D_2 shall require an *approved* engineered design by a *registered design professional* in accordance with Section R301.1.3.

SECTION AR102 DEFINITIONS

AR102.1 General. The following words and terms shall, for the purposes of this appendix, have the meanings shown herein. Refer to Chapter 2 for general definitions.

CLAY. Inorganic soil with particle sizes of less than 0.00008 inch (0.002 mm) having the characteristics of high to very high dry strength and medium to high plasticity.

CLAY SLIP. A suspension of clay or clay subsoil particles in water.

CLAY SUBSOIL. Subsoil sourced directly from the earth or refined, containing clay and not more than trace amounts organic matter.

INFILL. Light straw-clay that is placed between the structural and nonstructural members of a building.

LIGHT STRAW-CLAY. A mixture of straw and clay slip compacted and dried to form insulation and plaster substrate between or around structural and nonstructural members in a wall.

NONBEARING. Not bearing the weight of the building other than the weight of the light straw-clay itself and its finish.

STRAW. The dry stems of cereal grains after the seed heads have been removed.

VOID. Any space in a light straw-clay wall wider than $^1/_4$ inch (6 mm), greater than 2 inches (51 mm) in horizontal length and greater than 2 inches (51 mm) in depth.

SECTION AR103 NONBEARING LIGHT STRAW-CLAY CONSTRUCTION

AR103.1 General. Light straw-clay shall be limited to infill between or around structural and nonstructural wall framing members.

AR103.2 Structure. The structure of buildings using light straw-clay shall be in accordance with the *International Residential Code* or shall be in accordance with an *approved* design by a *registered design professional.*

AR103.2.1 Number of stories. Use of light straw-clay infill shall be limited to buildings that are not more than one *story above grade plane.*

Exception: Buildings using light straw-clay infill that are greater than one *story above grade plane* shall be in accordance with an *approved* design by a *registered design professional.*

AR103.2.2 Bracing. Bracing for buildings with light straw-clay infill shall be in accordance with Section R602.10. Walls with light straw-clay infill shall use Method LIB and shall not be sheathed with solid sheathing. Walls without light straw-clay infill shall comply with any bracing method prescribed by this code.

AR103.2.3 Requirements and properties of light straw-clay mixtures. The requirements and properties of light straw-clay mixtures shall be in accordance with Table AR103.2.3.

AR103.2.4 Stabilization of light straw-clay. Light straw-clay shall be stabilized as follows, or shall be in accordance with an *approved* design by a *registered design professional*:

1. Vertical stabilization shall be of structural or nonstructural wood framing in accordance with Figure AR103.2.4(1), AR103.2.4(2) or AR103.2.4(3). Framing members that are both load-bearing and stabilization members shall meet the requirements of Section R602 and this section. Nonstructural stabilization members shall be not more than 32 inches (813 mm) on center.

2. Horizontal stabilization shall be installed at not more than 24 inches (610 mm) on center and in accordance with Figure AR103.2.4(1), AR103.2.4(2) or AR103.2.4(3). Horizontal stabilization shall be of any of the following with the stated minimum dimensions: $^{3}/_{4}$-inch (19.1 mm) bamboo, $^{1}/_{2}$-inch (12.7 mm) fiberglass rod, 1-inch (25 mm) wood dowel or nominal 1-inch by 2-inch (25 mm by 51 mm) wood.

AR103.3 Materials. The materials used in light straw-clay construction shall be in accordance with Sections AR103.3.1 through AR103.3.3.

AR103.3.1 Straw requirements. Straw shall be stems of wheat, rye, oats, rice or barley, and shall be free of visible decay, insects and green plant material.

AR103.3.2 Clay subsoil requirements. Suitability of clay subsoil shall be determined in accordance with Table AR103.2.3.

TABLE AR103.2.3
REQUIREMENTS AND PROPERTIES OF LIGHT STRAW-CLAY MIXTURES[a]

Density (pcf)	Straw (pcf)	Subsoil (pcf)	Water (gal/cf)[b]	Min. % clay in subsoil	Minimum clay: silt ratio	Subsoil testing method[c, d]	Max. wall thickness, inches	*R*-value (hr/F°/cf/BTU/inch)
10	6.7	3.3	1.55	70	3.5:1	A	15	1.80
12	6.7	5.3	1.63	46	1.7:1	A	15	1.72
13	6.7	6.3	1.67	40	1.33:1	A	15	1.69
15	6.7	8.3	1.74	35	0.95:1	A	15	1.63
20	6.7	13.3	1.93	30	0.60:1	A	12	1.48
30	6.7	23.3	2.31	NA	NA	B	12	1.22
40	6.7	33.3	2.70	NA	NA	B	12	1.01
50	6.7	43.3	3.08	NA	NA	B	12	0.84

a. Interpolation permitted. Extrapolation not permitted.
b. Water mixed with subsoil equals clay slip.
c. Subsoil Testing Methods:
   1. Lab test for percent of clay, silt and sand via hydrometer method.
   2. The Figure 2 Ribbon Test and the Figure 3 Ball Test in the Appendix of ASTM E2392/E2392M.
d. Trace amounts of organic materials are acceptable.

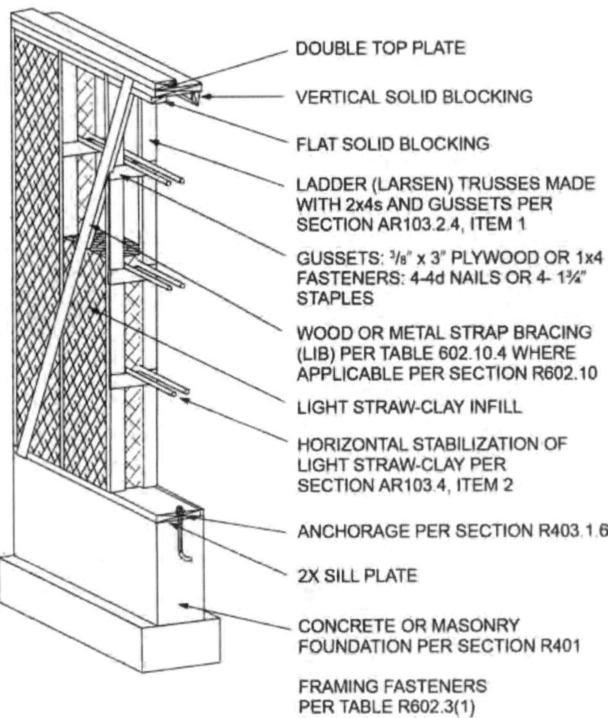

For SI: 1 inch = 25.4 mm.

FIGURE AR103.2.4(1)
LIGHT STRAW-CLAY WALL WITH LARSEN TRUSSES

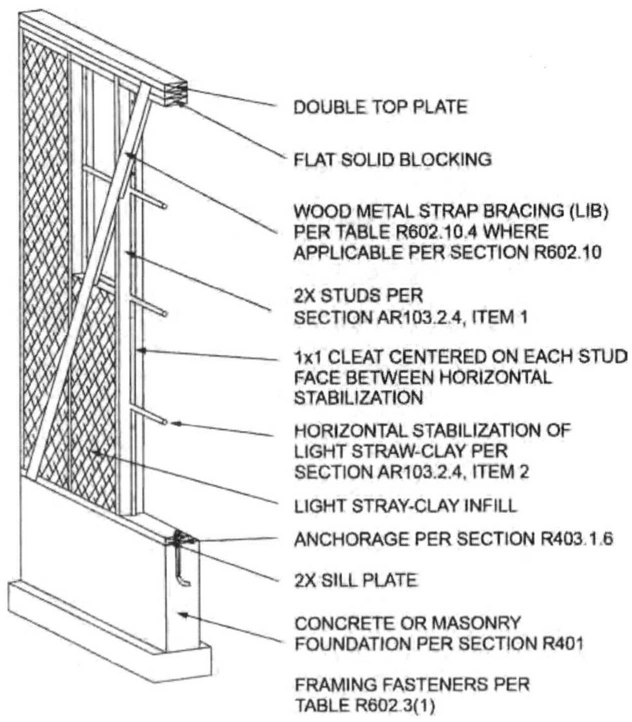

FIGURE AR103.2.4(2)
LIGHT STRAW-CLAY WALL SINGLE STUD WIDTH

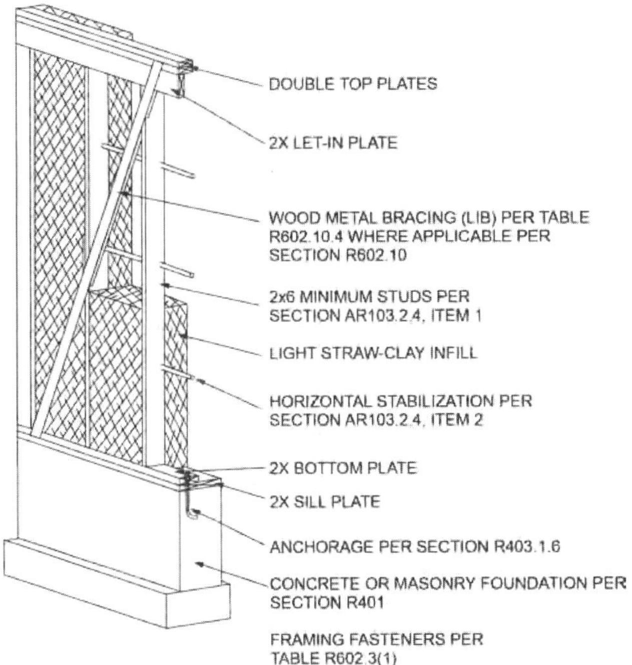

FIGURE AR103.2.4(3)
LIGHT STRAW-CLAY WALL WITH BLIND STUDS

AR103.3.3 Light straw-clay mixture. A light straw-clay mixture shall consist of loose straw mixed and coated with clay slip such that there is not more than 5 percent uncoated straw, and shall be in accordance with Table AR103.2.3.

AR103.4 Wall construction. Light straw-clay wall construction shall be in accordance with the requirements of Sections AR103.4.1 through AR103.4.7.

AR103.4.1 Light straw-clay maximum thickness. The maximum thickness of light straw-clay shall be in accordance with Table AR103.2.3.

AR103.4.2 Distance above grade. Light straw-clay and its exterior finish shall be not less than 8 inches (203 mm) above exterior finished *grade*.

AR103.4.3 Moisture barrier. An *approved* moisture barrier shall separate the bottom of light straw-clay walls from any masonry or concrete foundation or slab that directly supports the walls. Penetrations and joints in the barrier shall be sealed with an *approved* sealant.

AR103.4.4 Contact with wood members. Light straw-clay shall be permitted to be in contact with untreated wood members.

AR103.4.5 Contact with nonwood structural members. Nonwood structural members in contact with light straw-clay shall be resistant to corrosion or shall be coated to prevent corrosion with an *approved* coating.

AR103.4.6 Installation. Light straw-clay shall be installed in accordance with the following:

1. Formwork shall be sufficiently strong to resist bowing where the light straw-clay is compacted into the forms.
2. Light straw-clay shall be uniformly placed into forms and evenly tamped to achieve stable walls free of voids. Light straw-clay shall be placed in lifts of not more than 6 inches (152 mm) and shall be thoroughly tamped before additional material is added.
3. Temporary formwork shall be removed from walls within 24 hours after tamping, and walls shall remain exposed until moisture content is in accordance with Section AR103.5.1. Visible voids shall be filled with light straw-clay or other insulative material prior to plastering.

AR103.4.7 Openings in walls. Openings in walls shall be in accordance with the following:

1. Rough framing for doors and windows shall be fastened to structural members in accordance with the *International Residential Code*. Windows and doors shall be flashed in accordance with the *International Residential Code*.
2. An *approved* moisture barrier shall be installed at window sills in light straw-clay walls prior to installation of windows.

AR103.5 Wall finishes. The interior and exterior surfaces of light straw-clay walls shall be protected with a finish in accordance with Sections AR103.5.1 through AR103.5.5.

AR103.5.1 Dimensional stability of light straw-clay prior to application of plaster finish. Light straw-clay infill having a density of 30 pounds per cubic foot (480.6 kg/m^3) or greater shall be dry to a moisture content of not more than 20 percent at a depth of 4 inches (102 mm), as measured from each side of the wall. Light straw-clay infill having a density of less than 30 pounds per cubic foot (480.6 kg/m^3) shall be sufficiently dry such that the overall shrinkage of the light straw-clay is dimensionally stable.

AR103.5.2 Plaster finish. Exterior plaster shall be clay plasters or lime plasters. Interior plasters shall be clay plasters, lime plasters or gypsum plasters. Plasters shall be permitted to be applied directly to the surface of the light straw-clay walls without reinforcement, except that the juncture of dissimilar substrates shall be in accordance with Section AR103.5.4. Plasters shall have a thickness of not less than $^1/_2$ inch (12.7 mm) and not more than 1 inch (25 mm) and shall be installed in not less than two coats. Rain-exposed clay plasters shall be finished with a lime-based or silicate-mineral coating.

AR103.5.3 Separation of wood and plaster. Where wood framing occurs in light straw-clay walls, such wood surfaces shall be separated from exterior plaster with No.15 asphalt felt, Grade D paper or other *approved* material except where the wood is preservative treated or naturally durable.

Exception: Exterior clay plasters shall not be required to be separated from wood.

AR103.5.4 Bridging across dissimilar substrates. Bridging shall be installed across dissimilar substrates prior to the application of plaster. Acceptable bridging materials include expanded metal lath, woven wire mesh, welded wire mesh, fiberglass mesh, reed matting or burlap. Bridging shall extend not less than 4 inches (102 mm), on both sides of the juncture.

AR103.5.5 Exterior cladding. Exterior cladding shall be spaced not less than $^1/_2$ inch (12.7 mm) from the light straw-clay such that a ventilation space is created to allow for moisture diffusion. Furring strips that create this ventilation space shall be securely fastened to the stabilization members or framing. The cladding shall be fastened to the wood furring strips in accordance with the manufacturer's instructions. Insect screening shall be provided at the top and bottom of the ventilation space.

SECTION AR104 THERMAL PERFORMANCE

AR104.1 Thermal characteristics. Walls with light straw-clay infill of densities of greater than or equal to 20 pounds per cubic foot (480.6 kg/m^3) shall be classified as mass walls in accordance with Section N1102.2.5 (R402.2.5) and shall meet the *R*-value requirements for mass walls in Table N1102.1.3 (R402.1.2). Walls with light straw-clay infill of densities less than 20 pounds per cubic foot (480.6 kg/m^3) shall meet the *R*-value requirements for wood frame walls in Table N1102.1.3 (R402.1.2).

AR104.2 Thermal resistance. Light straw-clay shall be deemed to have a thermal resistance as specified in Table AR103.2.3.

SECTION AR105 REFERENCED STANDARDS

AR105.1 General. See Table AR105.1 for standards that are referenced in various section of this appendix. Standards are listed by the standard identification with the effective date, the standard title, and the section or sections of this appendix that reference this standard.

TABLE AR105.1 REFERENCED STANDARDS

STANDARD ACRONYM	STANDARD NAME	SECTIONS HEREIN REFERENCED
ASTM E2392/E2392M—10	*Standard Guide for Design of Earthen Wall Building Systems*	Table AR103.2.3

APPENDIX AS
STRAWBALE CONSTRUCTION

The provisions contained in this appendix are not mandatory unless specifically referenced in the adopting ordinance.

User note:

About this appendix: *The use of strawbale construction has steadily increased since the 1980s such that there are now buildings of strawbale construction in every state in the United States and in more than 50 countries around the globe. Estimates are that there are over 1,000 buildings of strawbale construction in California alone, including both residential and commercial buildings. Appendix AS provides prescriptive requirements for the construction of exterior and interior walls, both structural and nonstructural, in buildings that are under the scope of this code.*

SECTION AS101 GENERAL

AS101.1 Scope. This appendix provides prescriptive and performance-based requirements for the use of baled *straw* as a building material. Other methods of *strawbale* construction shall be subject to approval in accordance with Section R104.11 of this code. *Buildings* using *strawbale* walls shall comply with this code except as otherwise stated in this appendix.

AS101.2 Strawbale wall systems. *Strawbale* wall systems include those shown in Figure AS101.2 and *approved* variations.

SECTION AS102 DEFINITIONS

AS102.1 Definitions. The following words and terms shall, for the purposes of this appendix, have the meanings shown herein. Refer to Chapter 2 of the *International Residential Code* for general definitions.

BALE. Equivalent to *straw bale.*

BRACED WALL PANEL, STRAWBALE. A *strawbale* wall designed and constructed to resist in-plane shear loads through the interaction of the stacked *straw bales*, the reinforced plaster and its connections to the top plate, sill plates and foundation. The panel's length meets the requirements for the particular wall type and contributes toward the total amount of bracing required along its braced wall line in accordance with Sections AS106.13 and R602.10.1.

CLAY. Inorganic soil with particle sizes less than 0.00008 inch (0.002 mm) having the characteristics of high to very high dry strength and medium to high plasticity, used as the binder of other component materials in clay plaster and straw-clay.

CLAY SLIP. A suspension of clay or *clay subsoil* particles in water.

CLAY SUBSOIL. Subsoil sourced directly from the earth, containing *clay*, sand, silt and not more than trace amounts of organic matter.

FINISH. Completed combination of materials on the interior or exterior faces of stacked *bales.*

FLAKE. An intact section of compressed *straw* removed from an untied *bale.*

LAID FLAT. The orientation of a *bale* with its largest faces horizontal, its longest dimension parallel with the wall plane, its *ties* concealed in the unfinished wall and its *straw* lengths oriented predominantly across the thickness of the wall. See Figure AS102.1.

LOAD-BEARING WALL. A *strawbale* wall that supports more than 100 pounds per linear foot (1459 N/m) of vertical load in addition its own weight.

MESH. An openwork fabric of linked strands of metal, plastic, or natural or synthetic fiber.

NONSTRUCTURAL WALL. Walls other than *load-bearing* walls or *shear walls.*

ON-EDGE. The orientation of a *bale* with its largest faces vertical, its longest dimension parallel with the wall plane, its *ties* on the face of the wall and its *straw* lengths oriented predominantly vertically. See Figure AS102.1.

ON-END. The orientation of a *bale* with its longest dimension vertical. For use in *nonstructural strawbale* walls only. See Figure AS102.1.

PIN. A vertical metal rod, wood dowel or bamboo, driven into the center of stacked *bales*, or placed on opposite surfaces of stacked *bales* and through-tied.

PLASTER. Clay, soil-cement, gypsum, lime, clay-lime, lime-cement or cement plaster, as described in Section AS104.

PRECOMPRESSION. Permanent vertical compression of stacked *bales* before the application of finish.

REINFORCED PLASTER. A *plaster* containing mesh reinforcement.

ROOF-BEARING ASSEMBLY. In load-bearing *strawbale* walls, a structural assembly at the top of the wall that bears and distributes roof loads to the wall.

RUNNING BOND. The placement of *straw bales* such that the head joints in successive courses are offset not less than one-quarter the bale length.

SHEAR WALL. A *strawbale* wall designed and constructed to resist in-plane lateral seismic and wind forces in accordance with Section AS106.13. This term is synonymous with "Braced wall panel."

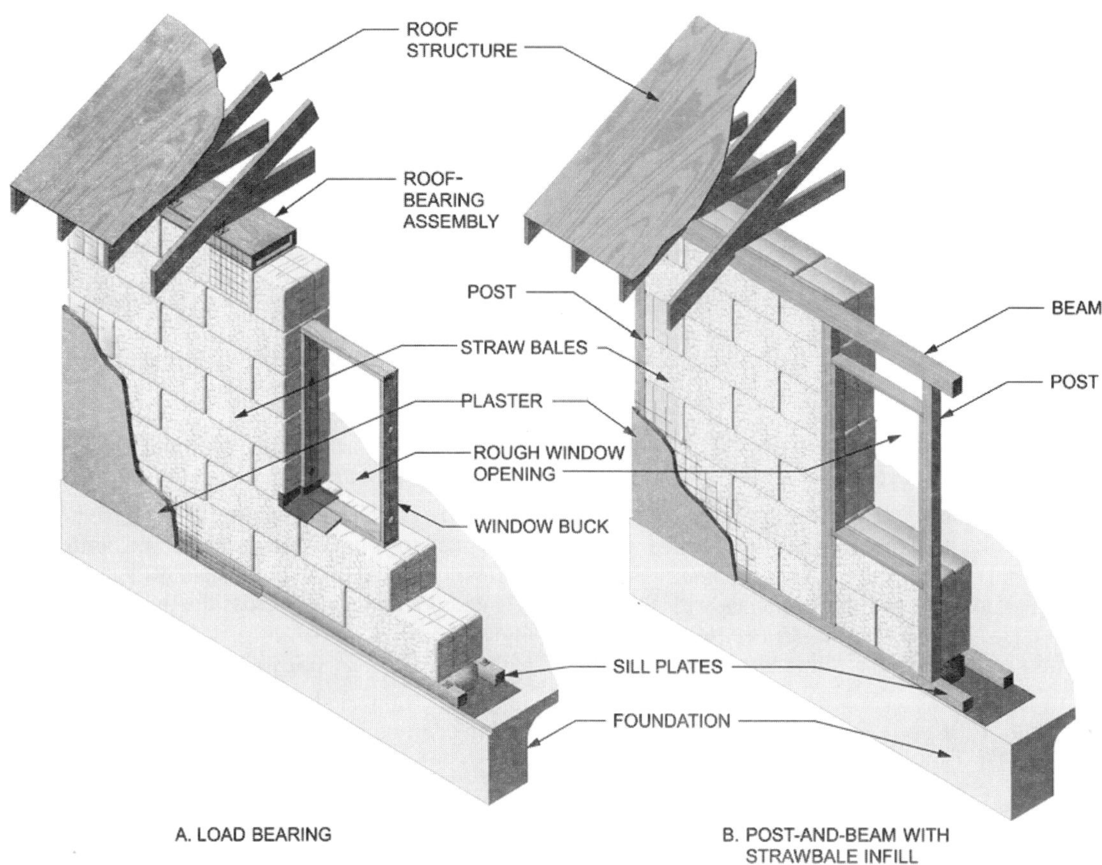

Note: See Figures AS105.1(1) through AS105.1(4) for detailed views and section references. Other strawbale wall systems or variations are permitted as approved.

FIGURE AS101.2
TYPICAL STRAWBALE WALL SYSTEMS

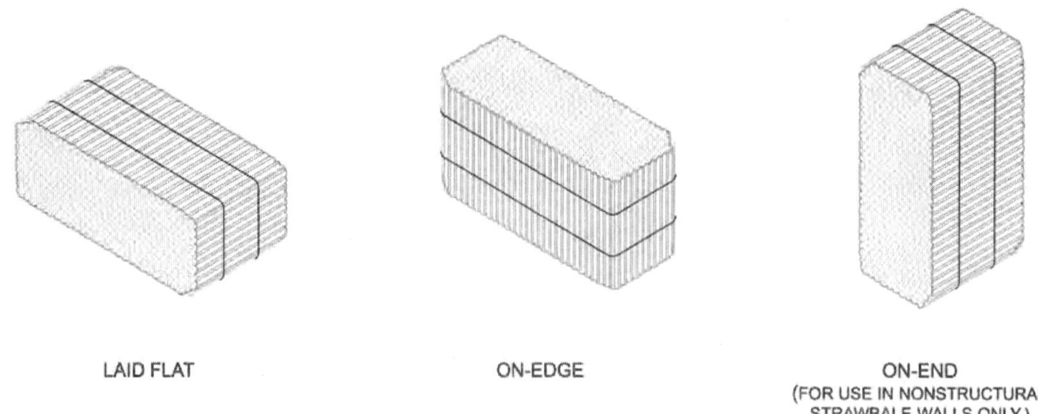

Note: Illustrations also show the predominant direction of the lengths of straw in a typical straw bale. However, some randomness of direction is normal.

FIGURE AS102.1
BALE ORIENTATIONS

SKIN. The compilation of *plaster* and reinforcing, if any, applied to the surface of stacked *bales.*

STACK BOND. The placement of *straw bales* such that head joints in successive courses are vertically aligned.

STRAW. The dry stems of cereal grains after the seed heads have been removed.

STRAW BALE. A rectangular compressed block of *straw,* bound by *ties.*

STRAWBALE. The adjective form of *straw bale.*

STRAW-CLAY. Loose *straw* mixed and coated with *clay* slip.

STRUCTURAL WALL. A wall that meets the definition for a *load-bearing* wall or *shear wall.*

TIE. A synthetic fiber, natural fiber or metal wire used to confine a *straw bale.*

TRUTH WINDOW. An area of a *strawbale* wall left without its finish, to allow view of the *straw* otherwise concealed by its finish.

SECTION AS103 BALES

AS103.1 Shape. *Bales* shall be rectangular in shape, except for partial bales made to fill nonrectangular spaces in accordance with Section AS103.6.

AS103.2 Size. *Bales* shall have a height and thickness of not less than 12 inches (305 mm), except as otherwise permitted or required in this appendix. *Bales* used within a continuous wall shall be of consistent height and thickness to ensure even distribution of loads within the wall system. See Figure AS103.2 for approximate dimensions of common *straw bales.*

AS103.3 Ties. *Bales* shall be confined by synthetic fiber, natural fiber or metal *ties* sufficient to maintain required *bale* density. *Ties* shall be not less than 3 inches (76 mm) and not more than 6 inches (152 mm) from the two untied faces and shall be spaced not more than 12 inches (305 mm) apart. *Bales* with broken *ties* shall be retied with sufficient tension to maintain required bale density.

AS103.4 Moisture content. The moisture content of *bales* at the time of application of the first coat of *plaster* or the installation of another finish shall not exceed 20 percent of the weight of the *bale.* The moisture content of *bales* shall be determined with a moisture meter designed for use with *baled straw* or hay, equipped with a probe of sufficient length to reach the center of the *bale.* Not less than 5 percent and not fewer than 10 bales shall be randomly selected and tested.

AS103.5 Density. *Bales* shall have a dry density of not less than 6.5 pounds per cubic foot (104 kg/cubic meter). The dry density shall be calculated by subtracting the weight of the moisture in pounds (kg) from the actual *bale* weight and dividing by the volume of the *bale* in cubic feet (cubic meters). Not less than 2 percent and not fewer than five *bales* shall be randomly selected and tested on-site.

AS103.6 Partial bales. Partial *bales* made after original fabrication shall be retied with *ties* complying with Section AS103.3.

AS103.7 Types of straw. *Bales* shall be composed of *straw* from wheat, rice, rye, barley or oat. The dry stems of other cereal grains or similar crops shall be acceptable where *approved* by the *building official.* Bales shall not be composed of hay.

AS103.8 Orientation of bales. Straw bales shall be placed *laid flat, on-edge* or *on end* in accordance with this appendix.

SECTION AS104 FINISHES

AS104.1 General. Finishes applied to *strawbale* walls shall comply with this section and with Chapters 3 and 7 unless stated otherwise in this section.

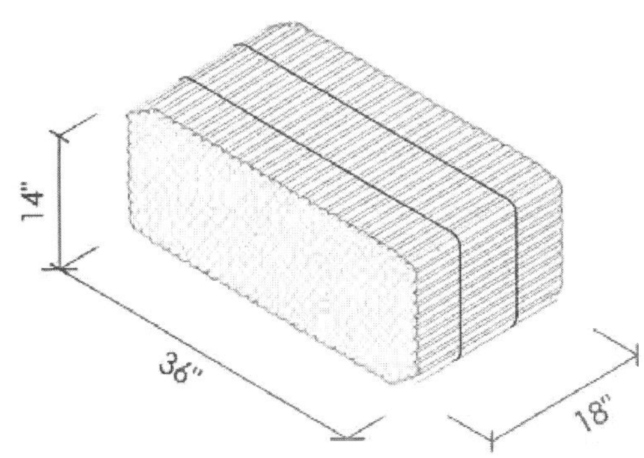

TWO-STRING BALE

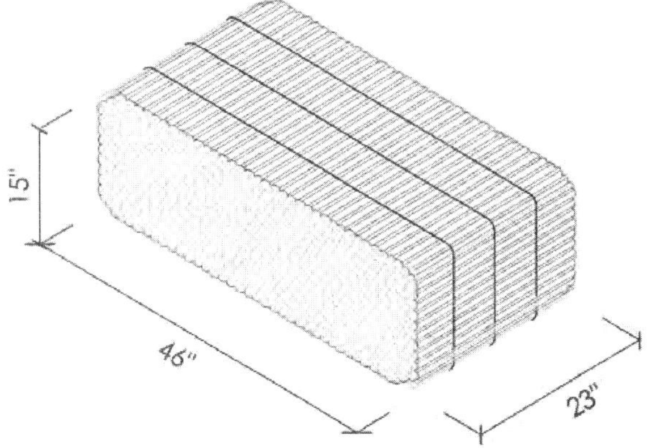

THREE-STRING BALE

For SI: 1 inch = 25.4 mm.

FIGURE AS103.2
APPROXIMATE DIMENSIONS OF COMMON STRAW BALES

AS104.1.1 Exterior wall finishes. Exterior wall finishes shall be plasters in accordance with Section AS104.4, or nonplaster exterior wall coverings in accordance with Section R703 and other finish systems complying with all of the following:

1. With *approved* specifications and details showing the finish system's means of attachment to the wall or its independent support, and a means of draining or evaporating water that penetrates the exterior finish to the exterior.
2. The vapor permeance of the combination of finish materials shall be 5 perms or greater to allow the transpiration of water vapor through the wall.
3. Finish systems with weights greater than 10 or less than or equal to 20 pounds per square foot (> 48.9 and ≤ 97.8 kg/m) of wall area require a factor of 1.2 for minimum total length of braced wall panels in Table AS106.13(3).
4. Finish systems with weights greater than 20 pounds per square foot (97.8 kg/m) of wall area require an engineered design.

AS104.2 Purpose, and where required. *Strawbale* walls shall be finished so as to provide mechanical protection, fire resistance and protection from weather and to restrict the passage of air through the *bales*, in accordance with this appendix and this code. Vertical *strawbale* wall surfaces shall receive a coat of *plaster* not less than $^3/_8$ inch (10 mm) thick, or greater where required elsewhere in this appendix, or shall fit tightly against a solid wall panel or dense-packed cellulose insulation with a density of not less than 3.5 pounds per cubic foot (56 kg/m^3) blown into an adjacent framed wall. The tops of *strawbale* walls shall receive a coat of *plaster* not less than $^3/_8$ inch (10 mm) thick or be tightly covered by gypsum board or a roof-bearing assembly.

Exception: *Truth windows* shall be permitted where a fire-resistance rating is not required. Weather-exposed *truth windows* shall be fitted with a weather-tight cover. Interior truth windows in Climate Zones 5, 6, 7, 8 and Marine 4 shall be fitted with an airtight cover.

AS104.3 Vapor retarders. Class I and II vapor retarders shall not be used on a *strawbale wall*, nor shall any other material be used that has a vapor permeance rating of less than 5 perms, except as permitted or required elsewhere in this appendix.

AS104.4 Plaster. *Plaster* applied to *bales* shall be any type described in this section, and as required or limited in this appendix. *Plaster* thickness shall not exceed 2 inches (51 mm).

AS104.4.1 Plaster and membranes. *Plaster* shall be applied directly to *strawbale* walls to facilitate transpiration of moisture from the *bales*, and to secure a mechanical bond between the *skin* and the *bales*, except where a membrane is allowed or required elsewhere in this appendix.

AS104.4.2 Lath and mesh for plaster. The surface of the *straw bales* functions as lath, and other lath or *mesh* shall not be required, except as required for out-of-plane resistance by Table AS105.4 or for structural walls by Tables AS106.12 and AS106.13(1).

AS104.4.3 Clay plaster. *Clay plaster* shall comply with Sections AS104.4.3.1 through AS104.4.3.6.

AS104.4.3.1 General. *Clay plaster* shall be any plaster having a clay or *clay subsoil* binder. Such plaster shall contain sufficient clay to fully bind the sand or other aggregate, and shall be permitted to contain reinforcing fibers. Acceptable reinforcing fibers include chopped straw, sisal and animal hair.

AS104.4.3.2 Clay subsoil requirements. The suitability of *clay subsoil* shall be determined in accordance with the Figure 2 Ribbon Test and the Figure 3 Ball Test in the appendix of ASTM E2392/E2392M.

AS104.4.3.3 Thickness and coats. *Clay plaster* shall be not less than 1 inch (25 mm) thick, except where required to be thicker for structural walls as described elsewhere in this appendix, and shall be applied in not less than two coats.

AS104.4.3.4 Rain-exposed. *Clay plaster,* where exposed to rain, shall be finished with lime wash, lime plaster, linseed oil or other *approved* erosion-resistant finish.

AS104.4.3.5 Prohibited finish coat. *Plaster* containing Portland cement shall not be permitted as a finish coat over *clay plasters.*

AS104.4.3.6 Plaster additives. Additives shall be permitted to increase *plaster* workability, durability, strength or water resistance.

AS104.4.4 Soil-cement plaster. Soil-cement plaster shall comply with Sections AS104.4.4.1 through AS104.4.4.3.

AS104.4.4.1 General. Soil-cement *plaster* shall be composed of *clay subsoil,* sand and not less than 10 percent and not more than 20 percent Portland cement by volume, and shall be permitted to contain reinforcing fibers.

AS104.4.4.2 Lath and mesh. Soil-cement *plaster* shall use any corrosion-resistant lath or *mesh* permitted by this code, or as required in Section AS106 where used on structural walls.

AS104.4.4.3 Thickness. Soil-cement *plaster* shall be not less than 1 inch (25 mm) thick.

AS104.4.5 Gypsum plaster. Gypsum *plaster* shall comply with Section R702.2.1. Gypsum *plaster* shall be limited to use on interior surfaces of *nonstructural* walls, and as an interior *finish* coat over a structural *plaster* that complies with this appendix.

AS104.4.6 Lime plaster. Lime *plaster* shall comply with Sections AS104.4.6.1 through AS104.4.6.3.

AS104.4.6.1 General. Lime *plaster* is any *plaster* with a binder that is composed of calcium hydroxide (CaOH) including Type N or S hydrated lime, hydraulic lime, natural hydraulic lime or slaked quicklime. Hydrated lime shall comply with ASTM C206. Hydraulic lime shall comply with ASTM C1707.

Natural hydraulic lime shall comply with ASTM C141 and CEN EN 459. Quicklime shall comply with ASTM C5.

AS104.4.6.2 Thickness and coats. Lime *plaster* shall be not less than $^{7}/_{8}$ inch (22 mm) thick, and shall be applied in not less than three coats.

AS104.4.6.3 On structural walls. Lime *plaster* on *strawbale* structural walls in accordance with Table AS106.12 or AS106.13.1) shall use hydraulic or natural hydraulic lime.

AS104.4.7 Clay-lime plaster. Clay-lime plaster shall be composed of refined clay or clay subsoil, sand, and lime, and shall be permitted to contain reinforcing fibers.

AS104.4.8 Cement-lime plaster. Cement-lime plaster shall be *plaster* mixes CL, F or FL, as described in ASTM C926.

AS104.4.9 Cement plaster. Cement *plaster* shall conform to ASTM C926 and shall comply with Sections R703.7.4 and R703.7.5, except that the amount of lime in plaster coats shall be not less than 1 part lime to 4 parts cement to allow a minimum acceptable vapor permeability. The combined thickness of *plaster* coats shall be not more than $1^{1}/_{2}$ inches (38 mm).

SECTION AS105
STRAWBALE WALLS—GENERAL

AS105.1 General. *Strawbale walls* shall be designed and constructed in accordance with this section and with Figures AS105.1(1) through AS105.1(4) or an *approved* alternative design. *Strawbale structural walls* shall be in accordance with the additional requirements of Section AS106.

AS105.2 Building limitations and requirements for use of strawbale nonstructural walls. *Buildings* using *strawbale nonstructural walls* shall be subject to the following limitations and requirements:

1. Number of stories: not more than one, except that two stories shall be allowed with an *approved* engineered design.
2. *Building* height: not more than 25 feet (7620 mm), except that greater heights shall be allowed with an *approved* engineered design.
3. Wall height: in accordance with Table AS105.4.
4. Braced wall panel lengths: in accordance with Section R602.10.3, with the additional requirements that Table R602.10.3(3) shall apply to all *buildings* in *Seismic Design Category* C, and the minimum total length of braced wall panels in Table R602.10.3(3) shall be increased by 60 percent for *buildings* in *Seismic Design Categories* C, D_0, D_1 and D_2.

AS105.3 Sill plates. Sill plates shall be installed in accordance with Figure AS105.1(1) or AS105.1(2). Sill plates shall support and be flush with each face of the *straw bales* above and shall be of naturally durable or preservative-treated wood where required by this code. Sill plates shall be not less than nominal 2 inches by 4 inches (51 mm by 102 mm) with anchoring complying with Section R403.1.6 and the additional requirements of Table AS105.4, where applicable, and Sections AS106.13.2 and AS106.13.3 for strawbale braced wall panels.

AS105.3.1 Exterior sill plate flashing. Exterior sill plates shall receive flashing across the plate to slab or foundation joints.

AS105.4 Out-of-plane resistance methods and unrestrained wall dimension limits. *Strawbale* walls shall employ a method of out-of-plane load resistance in accordance with Table AS105.4, and comply with its associated limits and requirements.

AS105.4.1 Determination of out-of-plane loading. Out-of-plane loading for the use of Table AS105.4 shall be in terms of the ultimate design wind speed and seismic design category as determined in accordance with Sections R301.2.1 and R301.2.2. An engineered design in accordance with Section R301.2.1 shall be required where the building is located in a special wind region or where wind design is required in accordance with Figure R301.2(2) and Section R301.2.1.1, respectively.

AS105.4.2 Pins. *Pins* used for out-of-plane resistance shall comply with the following or shall be in accordance with an *approved* engineered design. *Pins* shall be external, internal or a combination of the two.

1. *Pins* shall be $^{1}/_{2}$-inch-diameter (12.7 mm) steel, $^{3}/_{4}$-inch-diameter (19.1 mm) wood or $^{1}/_{2}$-inch-diameter (12.7 mm) bamboo.
2. External *pins* shall be installed vertically on both sides of the wall at a spacing of not more than 24 inches (610 mm) on center. External *pins* shall have full lateral bearing on the sill plate and the top plate or roof-bearing element, and shall be tightly tied through the wall to an opposing *pin* with *ties* spaced not more than 32 inches (813 mm) apart and not more than 8 inches (203 mm) from each end of the *pins*.
3. Internal *pins* shall be installed vertically within the center third of the *bales*, at spacing of not more than 24 inches (610 mm) and shall extend from top course to bottom course. The bottom course shall be connected to its support and the top course shall be connected to the roof- or floor-bearing member above with *pins* or other *approved* means. Internal *pins* shall be continuous or shall overlap through not less than one *bale* course.

AS105.5 Connection of light-frame walls to strawbale walls. *Light-frame* walls perpendicular to, or at an angle to a *strawbale* wall assembly, shall be fastened to the bottom and top wood members of the *strawbale* wall in accordance with requirements for wood or cold-formed steel *light-frame* walls in this code, or the abutting stud shall be connected to alternating *strawbale* courses with a $^{1}/_{2}$-inch diameter (12.7 mm) steel, $^{3}/_{4}$-inch-diameter (19.1 mm) wood or $^{5}/_{8}$-inch-diameter (15.9 mm) bamboo dowel, with not less than 8-inch (203 mm) penetration.

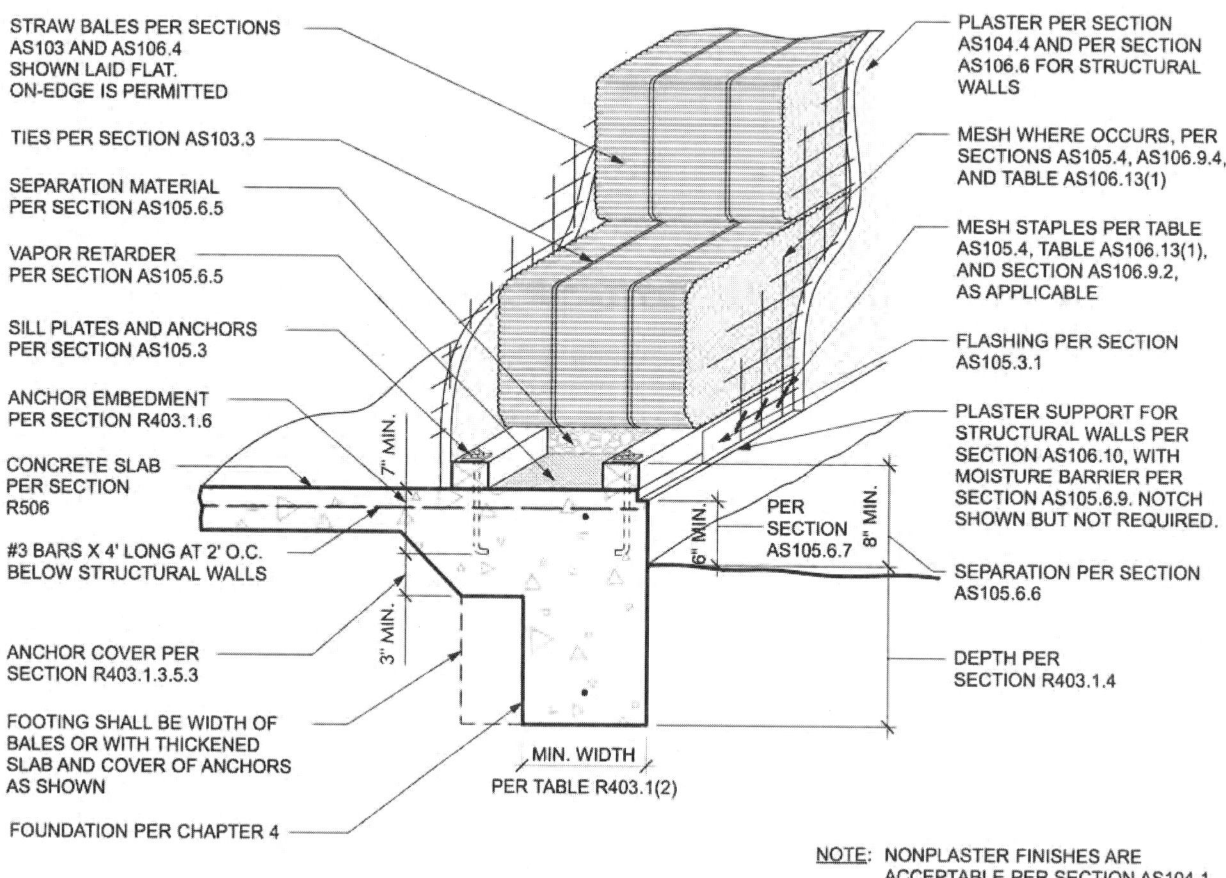
STRAW BALES PER SECTIONS AS103 AND AS106.4 SHOWN LAID FLAT. ON-EDGE IS PERMITTED
TIES PER SECTION AS103.3
SEPARATION MATERIAL PER SECTION AS105.6.5
VAPOR RETARDER PER SECTION AS105.6.5
SILL PLATES AND ANCHORS PER SECTION AS105.3
ANCHOR EMBEDMENT PER SECTION R403.1.6
CONCRETE SLAB PER SECTION R506
#3 BARS X 4' LONG AT 2' O.C. BELOW STRUCTURAL WALLS
ANCHOR COVER PER SECTION R403.1.3.5.3
FOOTING SHALL BE WIDTH OF BALES OR WITH THICKENED SLAB AND COVER OF ANCHORS AS SHOWN
FOUNDATION PER CHAPTER 4
7" MIN.
3" MIN.
6" MIN.
8" MIN.
PER SECTION AS105.6.7
MIN. WIDTH PER TABLE R403.1(2)
PLASTER PER SECTION AS104.4 AND PER SECTION AS106.6 FOR STRUCTURAL WALLS
MESH WHERE OCCURS, PER SECTIONS AS105.4, AS106.9.4, AND TABLE AS106.13(1)
MESH STAPLES PER TABLE AS105.4, TABLE AS106.13(1), AND SECTION AS106.9.2, AS APPLICABLE
FLASHING PER SECTION AS105.3.1
PLASTER SUPPORT FOR STRUCTURAL WALLS PER SECTION AS106.10, WITH MOISTURE BARRIER PER SECTION AS105.6.9. NOTCH SHOWN BUT NOT REQUIRED.
SEPARATION PER SECTION AS105.6.6
DEPTH PER SECTION R403.1.4
NOTE: NONPLASTER FINISHES ARE ACCEPTABLE PER SECTION AS104.1

For SI: 1 inch = 25.4 mm.

FIGURE AS105.1(1)
TYPICAL BASE OF PLASTERED STRAWBALE WALL ON CONCRETE SLAB AND FOOTING

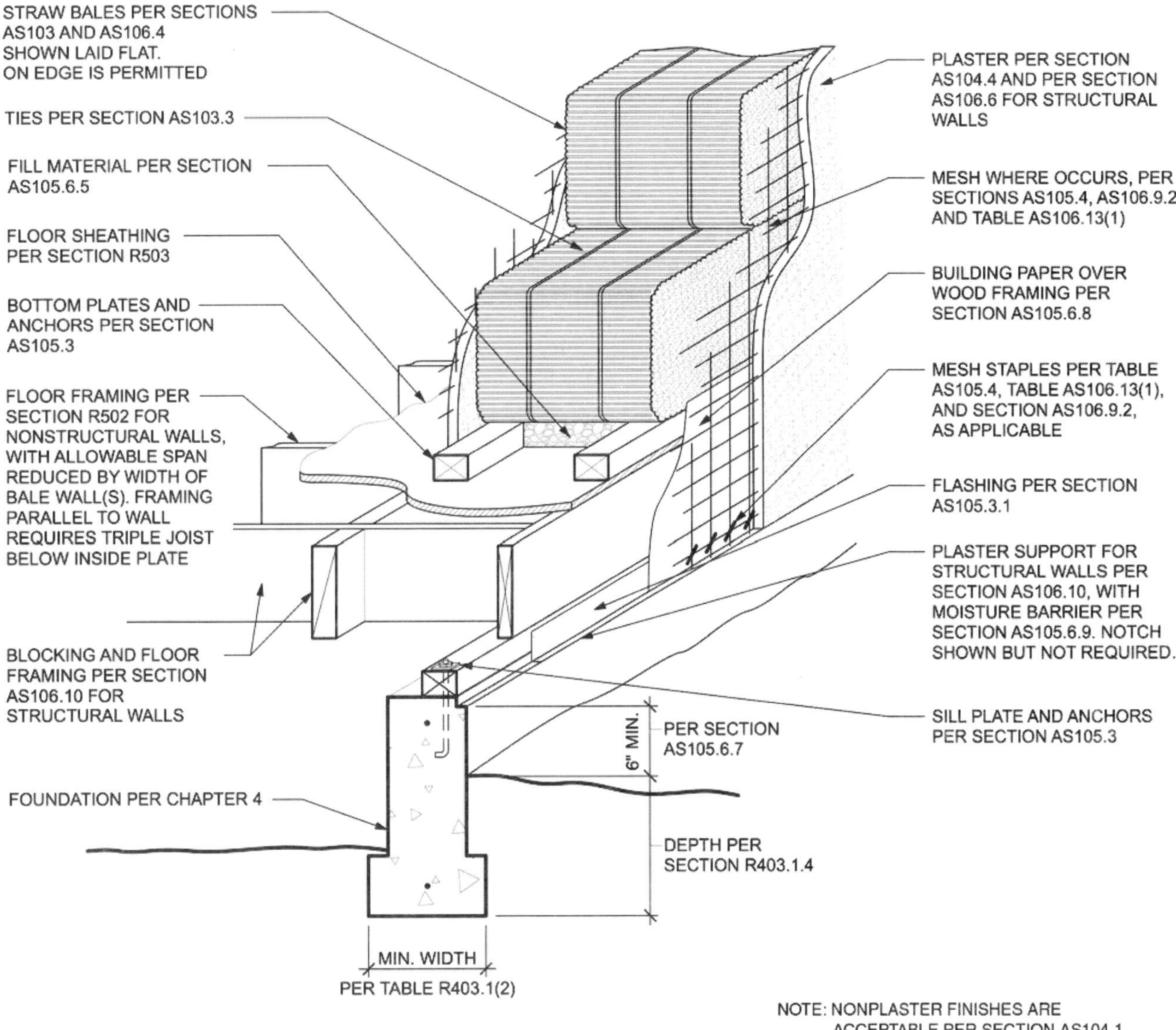
STRAW BALES PER SECTIONS AS103 AND AS106.4 SHOWN LAID FLAT. ON EDGE IS PERMITTED
TIES PER SECTION AS103.3
FILL MATERIAL PER SECTION AS105.6.5
FLOOR SHEATHING PER SECTION R503
BOTTOM PLATES AND ANCHORS PER SECTION AS105.3
FLOOR FRAMING PER SECTION R502 FOR NONSTRUCTURAL WALLS, WITH ALLOWABLE SPAN REDUCED BY WIDTH OF BALE WALL(S). FRAMING PARALLEL TO WALL REQUIRES TRIPLE JOIST BELOW INSIDE PLATE
BLOCKING AND FLOOR FRAMING PER SECTION AS106.10 FOR STRUCTURAL WALLS
FOUNDATION PER CHAPTER 4
PLASTER PER SECTION AS104.4 AND PER SECTION AS106.6 FOR STRUCTURAL WALLS
MESH WHERE OCCURS, PER SECTIONS AS105.4, AS106.9.2 AND TABLE AS106.13(1)
BUILDING PAPER OVER WOOD FRAMING PER SECTION AS105.6.8
MESH STAPLES PER TABLE AS105.4, TABLE AS106.13(1), AND SECTION AS106.9.2, AS APPLICABLE
FLASHING PER SECTION AS105.3.1
PLASTER SUPPORT FOR STRUCTURAL WALLS PER SECTION AS106.10, WITH MOISTURE BARRIER PER SECTION AS105.6.9. NOTCH SHOWN BUT NOT REQUIRED.
SILL PLATE AND ANCHORS PER SECTION AS105.3
6" MIN.
PER SECTION AS105.6.7
DEPTH PER SECTION R403.1.4
MIN. WIDTH
PER TABLE R403.1(2)
NOTE: NONPLASTER FINISHES ARE ACCEPTABLE PER SECTION AS104.1

For SI: 1 inch = 25.4 mm.

FIGURE AS105.1(2)
TYPICAL BASE OF PLASTERED STRAWBALE WALL OVER RAISED FLOOR

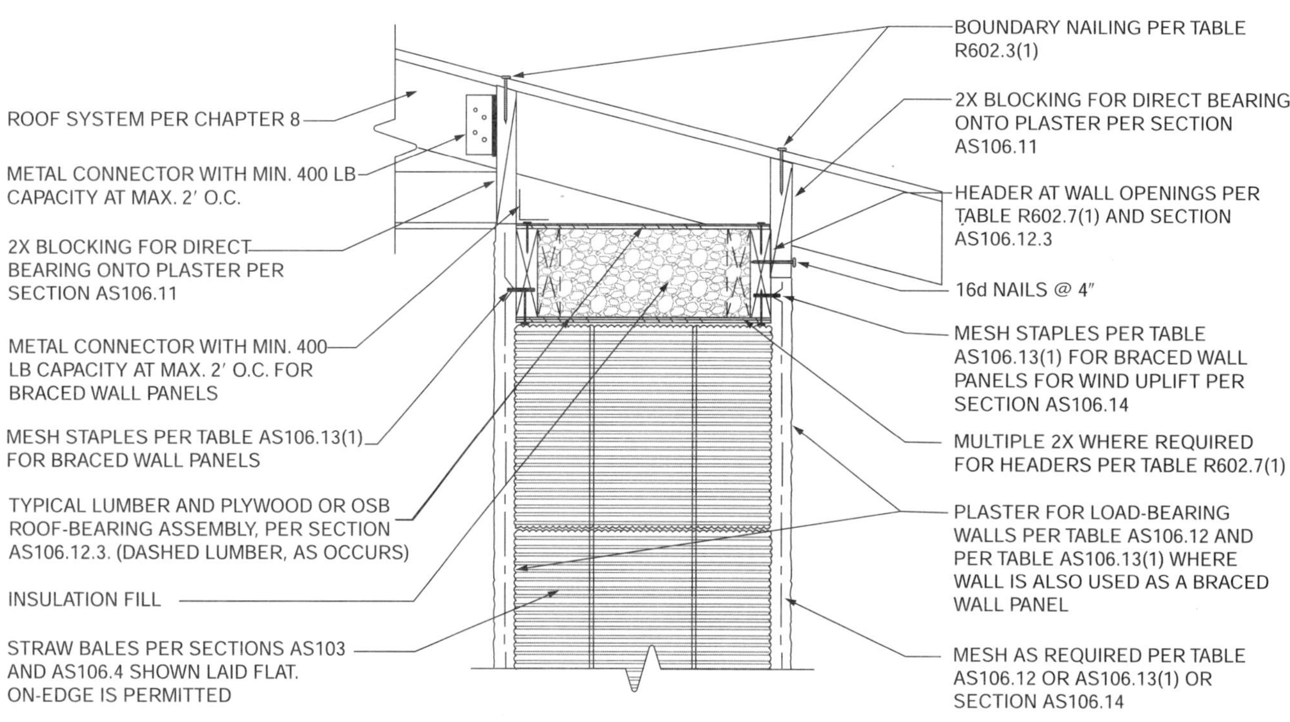
BOUNDARY NAILING PER TABLE R602.3(1)
2X BLOCKING FOR DIRECT BEARING ONTO PLASTER PER SECTION AS106.11
HEADER AT WALL OPENINGS PER TABLE R602.7(1) AND SECTION AS106.12.3
16d NAILS @ 4″
MESH STAPLES PER TABLE AS106.13(1) FOR BRACED WALL PANELS FOR WIND UPLIFT PER SECTION AS106.14
MULTIPLE 2X WHERE REQUIRED FOR HEADERS PER TABLE R602.7(1)
PLASTER FOR LOAD-BEARING WALLS PER TABLE AS106.12 AND PER TABLE AS106.13(1) WHERE WALL IS ALSO USED AS A BRACED WALL PANEL
MESH AS REQUIRED PER TABLE AS106.12 OR AS106.13(1) OR SECTION AS106.14
ROOF SYSTEM PER CHAPTER 8
METAL CONNECTOR WITH MIN. 400 LB CAPACITY AT MAX. 2′ O.C.
2X BLOCKING FOR DIRECT BEARING ONTO PLASTER PER SECTION AS106.11
METAL CONNECTOR WITH MIN. 400 LB CAPACITY AT MAX. 2′ O.C. FOR BRACED WALL PANELS
MESH STAPLES PER TABLE AS106.13(1) FOR BRACED WALL PANELS
TYPICAL LUMBER AND PLYWOOD OR OSB ROOF-BEARING ASSEMBLY, PER SECTION AS106.12.3. (DASHED LUMBER, AS OCCURS)
INSULATION FILL
STRAW BALES PER SECTIONS AS103 AND AS106.4 SHOWN LAID FLAT. ON-EDGE IS PERMITTED

For SI: 1 inch = 25.4 mm, 1 foot = 304.8 mm, 1 pound = 2.2 kg.

FIGURE AS105.1(3)
TYPICAL TOP OF LOAD-BEARING STRAWBALE WALL

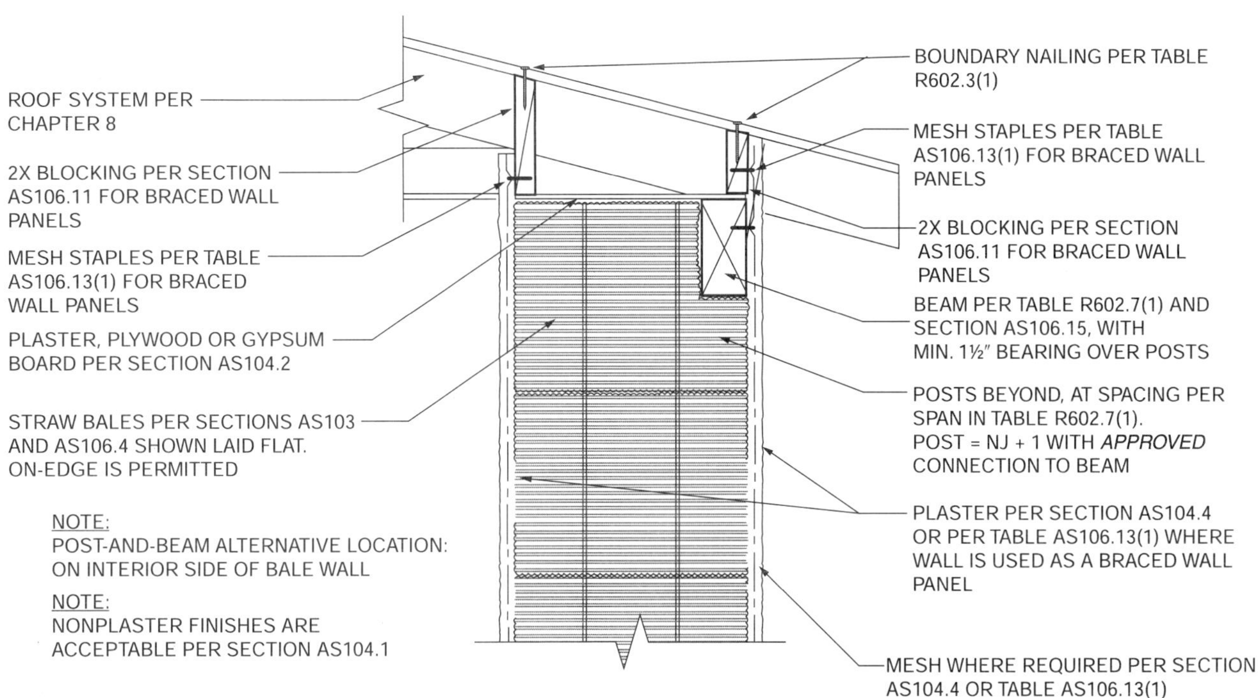

For SI: 1 inch = 25.4 mm.

FIGURE AS105.1(4)
TYPICAL TOP OF POST-AND-BEAM WALL WITH PLASTERED STRAWBALE INFILL

TABLE AS105.4
OUT-OF-PLANE RESISTANCE METHODS AND UNRESTRAINED WALL DIMENSION LIMITS

METHOD OF OUT-OF-PLANE LOAD RESISTANCE[a]	FOR ULTIMATE DESIGN WIND SPEEDS (mph)	FOR SEISMIC DESIGN CATEGORIES	UNRESTRAINED WALL DIMENSIONS, H^b		MESH STAPLE SPACING AT BOUNDARY RESTRAINTS
			Absolute limit in feet	Limit based on bale thickness T^c in feet (mm)	
Nonplaster finish or unreinforced plaster	≤ 130	A, B, C, D_0	$H \le 8$	$H \le 5T$	None required
Pins per Section AS105.4.2	≤ 130	A, B, C, D_0	$H \le 12$	$H \le 8T$	None required
Pins per Section AS105.4.2	≤ 140	A, B, C, D_0, D_1, D_2	$H \le 10$	$H \le 7T$	None required
Reinforced[d] clay plaster	≤ 140	A, B, C, D_0, D_1, D_2	$H \le 10$	$H \le 8T^{0.5}$ ($H \le 140T^{0.5}$)	≤ 6 inches
Reinforced[d] clay plaster	≤ 140	A, B, C, D_0, D_1, D_2	$10 < H \le 12$	$H \le 8T^{0.5}$ ($H \le 140T^{0.5}$)	≤ 4 inches[e]
Reinforced[d] cement, cement-lime, lime or soil-cement plaster	≤ 140	A, B, C, D_0, D_1, D_2	$H \le 10$	$H \le 9T^{0.5}$ ($H \le 157T^{0.5}$)	≤ 6 inches
Reinforced[d] cement, cement-lime, lime or soil-cement plaster	≤ 155	A, B, C, D_0, D_1, D_2	$H \le 12$	$H \le 9T^{0.5}$ ($H \le 157T^{0.5}$)	≤ 4 inches[e]
2×6 load-bearing wood studs[f] at max. 6′ o.c.	≤ 140	A, B, C, D_0, D_1, D_2	$H^g \le 9$	NA	None required
2×6 load-bearing wood studs[f] at max. 4′ o.c.	≤ 140	A, B, C, D_0, D_1, D_2	$H^g \le 10$	NA	None required
2×6 load-bearing wood studs[f] at max. 2′ o.c.	≤ 140	A, B, C, D_0, D_1, D_2	$H^g \le 12$	NA	None required
2×4 load-bearing wood studs[f] at max. 2′ o.c.	≤ 140	A, B, C, D_0, D_1, D_2	$H^g \le 10$	NA	None required
2×6 nonload-bearing wood studs[f] at max. 6′ o.c.	≤ 140	A, B, C, D_0, D_1, D_2	$H^g \le 12$	NA	None required

For SI: 1 inch = 25.4 mm, 1 foot = 304.8 mm, 1 mile per hour = 0.447 m/s.

NA = Not Applicable.

a. Finishes applied to both sides of stacked bales. Where different finishes are used on opposite sides of a wall, the more restrictive requirements shall apply.

b. H = Stacked bale height in feet (mm) between sill plate and top plate or other approved horizontal restraint, or the horizontal distance in feet (mm) between approved vertical restraints. For load-bearing walls, H refers to vertical height only.

c. T = Bale thickness in feet (mm).

d. Plaster reinforcement shall be any mesh allowed in Table AS106.13(1) for the matching plaster type, and with staple spacing in accordance with this table. Mesh shall be installed in accordance with Section AS106.9.

e. Sill plate attachment shall be with $^5/_8$-inch anchor bolts or approved equivalent at not more than 48 inches on center where staple spacing is required to be ≤ 4 inches.

f. Bales shall be attached to the studs by an approved method. Horizontal framing and attachment at top and bottom of studs shall be in accordance with Section R602 or an approved alternative. Table R602.7(1) shall be used to determine the top framing member where load-bearing stud spacing exceeds 24 inches o.c.

g. H is vertical height only.

AS105.6 Moisture control. *Strawbale* walls shall be protected from moisture intrusion and damage in accordance with Sections AS105.6.1 through AS105.6.9.

AS105.6.1 Water-resistant barriers and vapor permeance ratings. *Plastered bale* walls shall be constructed without any membrane barrier between *straw* and *plaster* to facilitate transpiration of moisture from the *bales*, and to secure a structural bond between *straw* and *plaster*, except as permitted or required elsewhere in this appendix. Where a water-resistant barrier is placed behind an exterior finish, it shall have a vapor permeance rating of not less than 5 perms, except as permitted or required elsewhere in this appendix.

AS105.6.2 Vapor retarders. Wall *finishes* shall have an equivalent vapor permeance rating of a Class III vapor retarder on the interior side of exterior *strawbale walls* in Climate Zones 5, 6, 7, 8 and Marine 4, as defined in Chapter 11. *Bales* in walls enclosing showers or steam rooms shall be protected on the interior side by a Class I or Class II vapor retarder.

AS105.6.3 Penetrations in exterior strawbale walls. Penetrations in exterior *strawbale* walls shall be sealed with an *approved* sealant or gasket on the exterior side of the wall in all climate zones, and on the interior side of the wall in Climate Zones 5, 6, 7, 8 and Marine 4, as defined in Chapter 11.

AS105.6.4 Horizontal surfaces. *Bale* walls and other *bale* elements shall be provided with a water-resistant barrier at weather-exposed horizontal surfaces. The water-resistant barrier shall be of a material and installation that will prevent water from entering the wall system. Horizontal surfaces shall include exterior window sills, sills at exterior niches and buttresses. Horizontal surfaces shall be sloped not less than 1 unit vertical in 12 units horizontal (8-percent slope) and shall drain away from *bale* walls and elements. Where the water-resistant barrier is below the finish material, it shall be sloped not less than 1 unit vertical in 12 units horizontal (8-percent slope) and shall drain to the outside surface of the *bale* wall's vertical finish.

AS105.6.5 Separation of bales and concrete. A sheet or liquid-applied Class II *vapor retarder* shall be installed between bales and supporting concrete or masonry. The bales shall be separated from the vapor retarder by not less than $^3/_4$ inch (19.1 mm), and that space shall be filled with an insulating material such as wood or rigid insulation, or a material that allows vapor dispersion such as gravel, or other *approved* insulating or vapor dispersion material. Sill plates shall be installed at this interface in accordance with Section AS105.3. Where bales abut a concrete or masonry wall that retains earth, a Class II vapor retarder shall be provided between such wall and the bales.

AS105.6.6 Separation of bales and earth. *Bales* shall be separated from earth by not less than 8 inches (203 mm).

AS105.6.7 Separation of exterior plaster and earth. Exterior plaster applied to *straw bales* shall be located not less than 6 inches (102 mm) above earth or 3 inches (51 mm) above paved areas.

AS105.6.8 Separation of wood and plaster. Where wood framing or wood sheathing occurs at the exterior face of *strawbale* walls, such wood surfaces shall be separated from exterior plaster with two layers of Grade D paper, No. 15 asphalt felt or other *approved* material in accordance with Section R703.7.3, extending not less than 1 inch (25 mm) past the edges of the framing member.

Exceptions:

1. Where the wood is preservative treated or *naturally durable* and is not greater than $1^1/_2$ inches (38 mm) in width.
2. Clay plaster shall not be required to be separated from untreated wood that is not greater than $1^1/_2$ inches (38 mm) in width.

AS105.6.9 Separation of exterior plaster and foundation. Exterior plaster shall be separated from the building foundation with a moisture barrier.

AS105.7 Inspections. The *building official* shall inspect the following aspects of *strawbale* construction in accordance with Section R109.1:

1. Sill plate anchors, as part of and in accordance with Section R109.1.1.
2. Mesh placement and attachment, where mesh is required by this appendix.
3. *Pins*, where required by and in accordance with Section AS105.4.

AS105.8 Voids and stuffing. Voids between *bales* and between *bales* and framing members shall not exceed 4 inches (102 mm) in width, and such voids shall be tightly stuffed with *flakes*, loose straw or *straw-clay* before application of finish.

SECTION AS106 STRAWBALE WALLS—STRUCTURAL

AS106.1 General. Plastered *strawbale* walls shall be permitted to be used as structural walls in accordance with the prescriptive provisions of this section.

AS106.2 Building limitations and requirements for use of strawbale structural walls. *Buildings* using strawbale *structural* walls shall be subject to the following limitations and requirements:

1. Number of stories: Not more than one, except that two stories shall be allowed with an *approved* engineered design.
2. *Building* height: Not more than 25 feet (7620 mm), except that greater heights shall be allowed with an *approved* engineered design.
3. Wall height: In accordance with Table AS105.4, AS106.13(2) or AS106.13(3) as applicable, whichever is most restrictive.
4. Braced wall panel lengths: The greater of the values determined in accordance with Tables AS106.13(2) and AS106.13(3) for *buildings* using strawbale braced wall panels, or in accordance with Item 4 of Section AS105.2 for *buildings* with *load-bearing strawbale walls* that do not use *strawbale* braced wall panels.

AS106.3 Loads and other limitations. Live and dead loads and other limitations shall be in accordance with Section R301. *Strawbale* wall dead loads shall not exceed 60 psf (2872 N/m^2) per face area of wall.

AS106.4 Foundations. Foundations for plastered *strawbale* walls shall be in accordance with Chapter 4, Figure AS105.1(1), Figure AS105.1(2) or an *approved* engineered design.

AS106.5 Orientation and configuration of bales. *Bales* in *strawbale* structural walls shall be laid flat or on-edge and in a *running bond* or *stack bond*, except that bales in structural walls with unreinforced plasters shall be laid in a *running bond* only.

AS106.6 Plaster on structural walls. Plaster on *load-bearing* walls shall be in accordance with Table AS106.12. Plaster on *shear walls* shall be in accordance with Table AS106.13(1).

AS106.6.1 Compressive strength. For plaster on *strawbale* structural walls, the *building official* is authorized to require a 2-inch (51mm) cube test conforming to ASTM C109 to demonstrate a minimum compressive strength in accordance with Table AS106.6.1.

**TABLE AS106.6.1
MINIMUM COMPRESSIVE STRENGTH
FOR PLASTERS ON STRUCTURAL WALLS**

PLASTER TYPE	MINIMUM COMPRESSIVE STRENGTH (psi)
Clay	100
Soil-cement	1,000
Lime	600
Cement-lime	1,000
Cement	1,400

For SI: 1 pound per square inch = 6894.76 N/m^2.

AS106.7 Straightness of plaster. Plaster on *strawbale* structural walls shall be straight, as a function of the bale wall surfaces they are applied to, in accordance with all of the following:

1. As measured across the face of a *bale*, *straw* bulges shall not protrude more than $^3/_4$ inch (19.1 mm) across 2 feet (610 mm) of its height or length.
2. As measured across the face of a *bale* wall, *straw* bulges shall not protrude from the vertical plane of a *bale* wall more than 2 inches (51 mm) over 8 feet (2438 mm).
3. The vertical faces of adjacent *bales* shall not be offset more than $^3/_8$ inch (9.5 mm).

AS106.8 Plaster and membranes on structural walls. *Strawbale* structural walls shall not have a membrane between straw and plaster, or shall have attachment through the *bale* wall from one plaster skin to the other in accordance with an *approved* engineered design.

AS106.9 Mesh. Mesh in plasters on *strawbale* structural walls, and where required by Table AS105.4, and where used to resist wind uplift in accordance with Section AS106.14, shall be installed in accordance with Sections AS106.9.1 through AS106.9.4.

AS106.9.1 Mesh laps. Mesh required by Table AS105.4 or AS106.12 shall be installed with not less than 4-inch (102 mm) laps. Mesh required by Table AS106.13(1) or in walls designed to resist wind uplift of more than 100 plf (1459 N/m) in accordance with Section AS106.14, shall run continuous vertically from sill plate to the top plate or roof-bearing element, or shall lap not less than 8 inches (203 mm). Horizontal laps in such mesh shall be not less than 4 inches (102 mm).

AS106.9.2 Mesh attachment. Mesh shall be attached with staples to top plates or roof-bearing elements and to sill plates in accordance with all of the following:

1. **Staples.** Staples shall be pneumatically driven, stainless steel or electro-galvanized, 16 gage with $1^1/_2$-inch (38 mm) legs, $^7/_{16}$-inch (11.1 mm) crown; or manually driven, galvanized, 15 gage with 1-inch (25 mm) legs. Other staples shall be as designed by a *registered design professional*. Staples into preservative-treated wood shall be stainless steel.
2. **Staple orientation.** Staples shall be firmly driven diagonally across mesh intersections at the required spacing.
3. **Staple spacing.** Staples shall be spaced not more than 4 inches (102 mm) on center, except where a lesser spacing is required by Table AS106.13(1) or Section AS106.14, as applicable.

AS106.9.3 Steel mesh. Steel mesh shall be galvanized, and shall be separated from preservative-treated wood by Grade D paper, No. 15 roofing felt or other *approved* barrier.

AS106.9.4 Mesh in plaster. Required mesh shall be embedded in the plaster except where staples fasten the mesh to horizontal boundary elements.

AS106.10 Support of plaster skins. Plaster *skins* on *strawbale* structural walls shall be continuously supported along their bottom edge. Acceptable supports include: a concrete or masonry stem wall, a concrete slab-on-grade, a wood-framed floor in accordance with Figure AS105.1(2) and an *approved* engineered design or a steel angle anchored with an *approved* engineered design. A weep screed as described in Section R703.7.2.1 is not an acceptable support.

AS106.11 Transfer of loads to and from plaster skins. Where plastered *strawbale* walls are used to support superimposed vertical loads, such loads shall be transferred to the plaster *skins* by continuous direct bearing in accordance with Figure AS105.1(3) or by an *approved* engineered design. Where plastered *strawbale* walls are used to resist in-plane lateral loads, such loads shall be transferred to the reinforcing mesh from the structural member or assembly above in accordance with Figure AS105.1(3) or AS105.1(4) and to the sill plate in accordance with Figure AS105.1(1) or AS105.1(2) and with Table AS106.13(1).

AS106.12 Load-bearing walls. Bearing capacities for plastered *strawbale* walls used as *load-bearing walls* in one-*story buildings* to support vertical loads imposed in accordance with Section R301 shall be in accordance with Table AS106.12.

AS106.12.1 Precompression of load-bearing strawbale walls. Prior to application of plaster, walls designed to be load-bearing shall be precompressed by a uniform load of not less than 100 plf (1459 N/m).

AS106.12.2 Concentrated loads. Concentrated loads shall be distributed by structural elements capable of distributing the loads to the bearing wall within the allowable bearing capacity listed in Table AS106.12 for the plaster type used.

AS106.12.3 Roof-bearing assembly. Roof-bearing assemblies shall be of nominal 2-inch by 6-inch (51 mm by 152 mm) lumber with $^{15}/_{32}$-inch (12 mm) plywood or OSB panels fastened with 8d nails at 6 inches (152 mm) on center in accordance with Figure AS105.1(3) and Items 1 through 6, or be of an *approved* engineered design.

1. Assembly shall be a box assembly on the top course of *bales*, with the panels horizontal.
2. Assembly shall be the width of the *strawbale* wall and shall comply with Section AS106.11.

3. Discontinuous lumber shall be spliced with a metal strap with not less than a 500-pound (2224 N) allowable wind or seismic load tension capacity. Where the wall line includes a braced wall panel the strap shall have not less than a 2,000-pound (8896 N) capacity.
4. Panel joints shall be blocked.
5. Roof and ceiling framing shall be attached to the roof-bearing assembly in accordance with Table R602.3(1), Items 2 and 6.
6. Where the roof-bearing assembly spans wall openings, it shall comply with Section AS106.12.3.1

AS106.12.3.1 Roof-bearing assembly spanning openings. Roof-bearing assemblies that span openings in *strawbale* walls shall comply with the following at each opening:

1. Lumber on each side of the assembly shall be of the dimensions and quantity required to span each opening in accordance with Table R602.7(1).
2. The required lumber in the assembly shall be supported at each side of the opening by the number of jack studs required by Table R602.7(1), or shall extend beyond the opening on both sides a distance, D, using the following equation:

$$D = S \times R/2 \,/\, (1\text{-}R) \qquad \textbf{(Equation AS-1)}$$

where:

D = Minimum distance (in feet) for required spanning lumber to extend beyond the opening

S = Span in feet

R = B_L/B_C

B_L = Design load on the wall (in pounds per lineal foot) in accordance with Sections R301.4 and R301.6

B_C = Allowable bearing capacity of the wall in accordance with Table AS106.12

AS106.13 Braced wall panels. Plastered *strawbale* walls used as braced wall panels for one-story *buildings* shall be in accordance with Section R602.10 and Tables AS106.13(1), AS106.13(2) and AS106.13(3). Wind design criteria shall be in accordance with Section R301.2.1. Seismic design criteria shall be in accordance with Section R301.2.2. An *approved* engineered design in accordance with Section R301.2.1 shall be required where the building is located in a special wind region or where wind design is required in accordance with Figure R301.2(2) and Section R301.2.1.1, respectively.

AS106.13.1 Bale wall thickness. The thickness of *strawbale* braced wall panels without their plaster shall be not less than 15 inches (381 mm).

AS106.13.2 Sill plates. Sill plates shall be in accordance with Table AS106.13(1).

AS106.13.3 Sill plate fasteners. Sill plates shall be fastened with not less than $^5/_8$-inch-diameter (15.9 mm) steel anchor bolts with 3-inch by 3-inch by $^3/_{16}$-inch (76.2 mm by 76.2 mm by 4.8 mm) steel washers, with not less than 7-inch (177.8 mm) embedment in a concrete or masonry foundation, or shall be an *approved* equivalent, with the spacing shown in Table AS106.13(1). Anchor bolts or other fasteners into framed floors shall be of an *approved* engineered design.

AS106.14 Resistance to wind uplift forces. Plaster mesh in *skins* of *strawbale walls* that resist uplift forces from the *roof assembly*, as determined in accordance with Section R802.11, shall be in accordance with all of the following:

1. Plaster shall be any type and thickness allowed in Section AS104.
2. Mesh shall be any type allowed in Table AS106.13(1), and shall be attached to top plates or roof-bearing elements and to sill plates in accordance with Section AS106.9.2.
3. Sill plates shall be not less than nominal 2-inch by 4-inch (51 mm by 102 mm) with anchoring complying with Section R403.1.6.
4. Mesh attached with staples at 4 inches (51 mm) on center shall be considered to be capable of resisting uplift forces of 100 plf (1459 N/m) for each plaster skin.

TABLE AS106.12
ALLOWABLE SUPERIMPOSED VERTICAL LOADS (LBS/FOOT) FOR PLASTERED LOAD-BEARING STRAWBALE WALLS

WALL DESIGNATION	PLASTER[a] (both sides) Minimum thickness in inches each side	MESH[b]	STAPLES[c]	ALLOWABLE BEARING CAPACITY[d] (plf)
A	Clay $1^1/_2$	None required	None required	400
B	Soil-cement 1	Required	Required	800
C	Lime[e] $^7/_8$	Required	Required	500
D	Cement-lime $^7/_8$	Required	Required	800
E	Cement $^7/_8$	Required	Required	800

For SI: 1 inch = 25.4 mm, 1 pound per foot = 14.5939 N/m.

a. Plasters shall conform to Sections AS104.4.3 through AS104.4.9, AS106.7 and AS106.10.

b. Any metal mesh allowed by this appendix and installed in accordance with Section AS106.9.

c. In accordance with Section AS106.9.2, except as required to transfer roof loads to the plaster skins in accordance with Section AS106.11.

d. For walls with a different plaster on each side, the lower value shall be used. For walls with plaster on only one side, half of the tabular value shall be used.

e. Shall use hydraulic or natural hydraulic lime.

**TABLE AS106.13(1)
PLASTERED STRAWBALE BRACED WALL PANEL TYPES**

WALL DESIGNATION	PLASTER[a] (both sides)		SILL PLATES[b] (nominal size in inches)	ANCHOR BOLT[c] SPACING (inches on center)	MESH[d] (inches)	STAPLE SPACING[e] (inches on center)
	Type	Thickness (minimum in inches each side)				
A1	Clay	1.5	2 × 4	32	None	None
A2	Clay	1.5	2 × 4	32	2 × 2 high-density polypropylene	2
A3	Clay	1.5	2 × 4	32	2 × 2 × 14 gage	4
B	Soil-cement	1	4 × 4	24	2 × 2 × 14 gage	2
C1	Lime[f]	$^7/_8$	2 × 4	32	17-gage woven wire	3
C2	Lime[f]	$^7/_8$	4 × 4	24	2 × 2 × 14 gage	2
D1	Cement-lime	$^7/_8$	4 × 4	32	17-gage woven wire	2
D2	Cement-lime	$^7/_8$	4 × 4	24	2 × 2 × 14 gage	2
E1	Cement	$^7/_8$	4 × 4	32	2 × 2 × 14 gage	2
E2	Cement	1.5	4 × 4	24	2 × 2 × 14 gage	2

For SI: 1 inch = 25.4 mm.

a. Plasters shall comply with Sections AS104.4.3 through AS104.4.9, AS106.7, AS106.8 and AS106.12.

b. Sill plates shall be Douglas fir-larch or southern pine and shall be preservative treated where required by the *International Residential Code.*

c. Anchor bolts shall be in accordance with Section AS106.13.3 at the spacing shown in this table.

d. Installed in accordance with Section AS106.9.

e. Staples shall be in accordance with Section AS106.9.2 at the spacing shown in this table.

f. Shall use hydraulic or natural hydraulic lime.

5. Mesh attached with staples at 2 inches (51 mm) on center shall be considered to be capable of resisting uplift forces of 200 plf (2918 N/m) for each plaster skin.

AS106.15 Post-and-beam with strawbale infill. Post-and-beam with *strawbale* infill systems shall be in accordance with Figure AS105.1(4) and Items 1 through 7, or be of an *approved* engineered design.

1. Beams shall be of the dimensions and number of members in accordance with Table R602.7(1), where the space between posts equals the span in the table.
2. Beam ends shall bear over posts not less than $1^1/_2$ inches (38 mm) or be supported by a framing anchor in accordance with Table R602.7(1).
3. Discontinuous beam ends shall be spliced with a metal strap with not less than 1,000-pound (454 kg) wind or seismic load tension capacity. Where the wall line includes a braced wall panel, the strap shall have not less than a 4,000-pound (1814 kg) capacity.
4. Each post shall equal NJ + 1 in accordance with Table R602.7(1), where the space between posts equals the span in the table.
5. Posts shall be connected to the beam by an *approved* means.
6. Roof and ceiling framing shall be attached to the beam in accordance with Table R602.3(1), Items 2 and 6.
7. Posts shall be supported by the sill plate of the bale wall in accordance with Section AS105.3 or AS106.13.2, with fastening in accordance with Table R602.3(1), Item 16, or shall be supported and fastened at their base by an *approved* means.

SECTION AS107
FIRE RESISTANCE

AS107.1 Fire-resistance rating. *Strawbale* walls shall not be considered to exhibit a fire-resistance rating, except for walls constructed in accordance with Section AS107.1.1 or AS107.1.2. Alternately, fire-resistance ratings of strawbale walls shall be determined in accordance with Section R302.

AS107.1.1 One-hour-rated clay-plastered wall. One-hour fire-resistance-rated nonload-bearing clay plastered *strawbale* walls shall comply with all of the following:

1. *Bales* shall be laid flat or on-edge in a *running bond.*
2. *Bales* shall maintain thickness of not less than 18 inches (457 mm).
3. *Bales* shall have a minimum dry density of 7.25 pounds per cubic foot (116 kg/m^3).
4. Gaps shall be stuffed with *straw-clay.*
5. *Clay* plaster on each side of the wall shall be not less than 1 inch (25 mm) thick and shall be composed of a mixture of 3 parts clay, 2 parts chopped straw and 6 parts sand, or an alternative *approved clay* plaster.
6. Plaster application shall be in accordance with Section AS104.4.3.3 for the number and thickness of coats.

TABLE AS106.13(2)
BRACING REQUIREMENTS FOR STRAWBALE BRACED WALL PANELS BASED ON WIND SPEED

• EXPOSURE CATEGORY B[d] • 25-FOOT MEAN ROOF HEIGHT • 10-FOOT EAVE-TO-RIDGE HEIGHT[d] • 10-FOOT WALL HEIGHT[d] • 2 BRACED WALL LINES[d]			MINIMUM TOTAL LENGTH (FEET) OF STRAWBALE BRACED WALL PANELS REQUIRED ALONG EACH BRACED WALL LINE[a, b, c, d]		
Ultimate design windspeed (mph)	Story location	Braced wall line spacing (feet)	Strawbale braced wall panel[e] A2, A3	Strawbale braced wall panel[e] C1, C2, D1	Strawbale braced wall panel[e] B, D2, E1, E2
≤ 110	One-story building	10	6.4	3.8	3.0
		20	8.5	5.1	4.0
		30	10.2	6.1	4.8
		40	13.3	6.9	5.5
		50	16.3	7.7	6.1
		60	19.4	8.3	6.6
≤ 115	One-story building	10	6.4	3.8	3.0
		20	8.5	5.1	4.0
		30	11.2	6.4	5.1
		40	14.3	7.2	5.7
		50	18.4	8.1	6.5
		60	21.4	8.8	7.0
≤ 120	One-story building	10	7.1	4.3	3.4
		20	9.0	5.4	4.3
		30	12.2	6.6	5.3
		40	16.3	7.7	6.1
		50	19.4	8.3	6.6
		60	23.5	9.2	7.3
≤ 130	One-story building	10	7.1	4.3	3.4
		20	10.2	6.1	4.8
		30	14.3	7.2	5.7
		40	18.4	8.1	6.5
		50	22.4	9.0	7.1
		60	26.5	9.8	7.8
≤ 140	One-story building	10	7.8	4.7	3.7
		20	11.2	6.4	5.1
		30	16.3	7.7	6.1
		40	21.4	8.8	7.0
		50	26.5	9.8	7.8
		60	30.6	11.0	8.3

For SI: 1 foot = 305 mm, 1 mile per hour = 0.447 m/s.

a. Linear interpolation shall be permitted.

b. All braced wall panels shall be without openings and shall have an aspect ratio (H:L) ≤ 2:1.

c. Tabulated minimum total lengths are for braced wall lines using single-braced wall panels with an aspect ratio (H:L) ≤ 2:1, or using multiple braced wall panels with aspect ratios (H:L) ≤ 1:1. For braced wall lines using two or more braced wall panels with an aspect ratio (H:L) > 1:1, the minimum total length shall be multiplied by the largest aspect ratio (H:L) of braced wall panels in that line.

d. Subject to applicable wind adjustment factors associated with "All methods" in Table AS106.13(2)

e. Strawbale braced panel types indicated shall comply with Sections AS106.13.1 through AS106.13.3 and with Table AS106.13(1).

TABLE AS106.13(3)
BRACING REQUIREMENTS FOR STRAWBALE BRACED WALL PANELS BASED ON SEISMIC DESIGN CATEGORY

• SOIL CLASS D[f] • WALL HEIGHT = 10 FEET[d] • 15 PSF ROOF-CEILING DEAD LOAD[d] • BRACED WALL LINE SPACING ≤ 25 FEET[d]			MINIMUM TOTAL LENGTH (FEET) OF STRAWBALE BRACED WALL PANELS REQUIRED ALONG EACH BRACED WALL LINE[a, b, c, d]	
Seismic Design Category	Story location	Braced wall line length (feet)	Strawbale braced wall panel[e] A2, A3[g], C1, C2, D1	Strawbale braced wall panel[e] B, D2, E1, E2
C	One-story building	10	5.7	4.6
		20	8.0	6.5
		30	9.8	7.9
		40	12.9	9.1
		50	16.1	10.4
D_0	One-story building	10	6.0	4.8
		20	8.5	6.8
		30	10.9	8.4
		40	14.5	9.7
		50	18.1	11.7
D_1	One-story building	10	6.3	5.1
		20	9.0	7.2
		30	12.1	8.8
		40	16.1	10.4
		50	20.1	13.0
D_2	One-story building	10	7.1	5.7
		20	10.1	8.1
		30	15.1	9.9
		40	20.1	13.0
		50	25.1	16.3

For SI: 1 inch = 25.4 mm, 1 foot = 305 mm, 1 pound per square foot = 0.0479 kPa.

a. Linear interpolation shall be permitted.

b. Braced wall panels shall be without openings and shall have an aspect ratio (H:L) ≤ 2:1.

c. Tabulated minimum total lengths are for braced wall lines using single braced wall panels with an aspect ratio (H:L) ≤ 2:1, or using multiple braced wall panels with aspect ratios (H:L) ≤ 1:1. For braced wall lines using two or more braced wall panels with an aspect ratio (H:L) > 1:1, the minimum total length shall be multiplied by the largest aspect ratio (H:L) of braced wall panels in that line.

d. Subject to applicable seismic adjustment factors associated with "All methods" in Table R602.10.3(4), except "Wall dead load."

e. Strawbale braced wall panel types indicated shall comply with Sections AS106.13.1 through AS106.13.3 and Table AS106.13(1).

f. Wall bracing lengths are based on a soil site class "D." Interpolation of bracing lengths between S_{ds} values associated with the seismic design categories is allowable where a site-specific S_{ds} value is determined in accordance with Section 1613.3 of the *International Building Code*.

g. Where using wall Type A3, the minimum total length of braced wall panels in this column shall be multiplied by 1.25.

AS107.1.2 Two-hour-rated cement-plastered wall. Two-hour fire-resistance-rated nonload-bearing cement-plastered strawbale walls shall comply with all of the following:

1. Bales shall be laid flat or on-edge in a *running bond.*
2. Bales shall maintain a thickness of not less than 14 inches (356 mm).
3. *Bales* shall have a minimum dry density of 7.25 pounds per cubic foot (116 kg/m^3).
4. Gaps shall be stuffed with *straw-clay.*
5. A single section of $^1/_2$-inch (38 mm) by 17-gage galvanized woven wire mesh shall be attached to wood members with $1^1/_2$-inch (38 mm) staples at 6 inches (152 mm) on center. 9 gage U-pins with not less than 8-inch (203 mm) legs shall be installed at 18 inches (457 mm) on center to fasten the mesh to the *bales.*
6. Cement plaster on each side of the wall shall be not less than 1 inch (25 mm) thick.
7. Plaster application shall be in accordance with Section AS104.4.9 for the number and thickness of coats.

AS107.2 Openings in rated walls. Openings and penetrations in *bale* walls required to have a fire-resistance rating shall satisfy the same requirements for openings and penetrations as prescribed in this code.

AS107.3 Clearance to fireplaces and chimneys. *Strawbale* surfaces adjacent to fireplaces or chimneys shall be finished with not less than $^3/_8$-inch-thick (10 mm) plaster of any type permitted by this appendix. Clearance from the face of such plaster to fireplaces and chimneys shall be maintained as required from fireplaces and chimneys to combustibles in Chapter 10, or as required by manufacturer's instructions, whichever is more restrictive.

SECTION AS108 THERMAL INSULATION

AS108.1 R-value. The unit *R*-value of a *strawbale* wall with bales laid flat is R-1.55 for each inch of *bale* thickness. The unit *R*-value of a *strawbale* wall with *bales* on-edge is R-1.85 for each inch of *bale* thickness.

AS108.2 Compliance with Section R302.10.1. *Straw bales* meet the requirements for insulation materials in Section R302.10.1 for flame spread index and *smoke-developed index* as tested in accordance with ASTM E84.

SECTION AS109 REFERENCED STANDARDS

AS109.1 General. See Table AS109.1 for standards that are referenced in various sections of this appendix. Standards are listed by the standard identification with the effective date, the standard title and the section or sections of this appendix that reference the standard.

TABLE AS109.1 REFERENCED STANDARDS

STANDARD ACRONYM	STANDARD NAME	SECTIONS HEREIN REFERENCED
ASTM C5—10	Standard Specification for Quicklime for Structural Purposes	AS104.4.6.1
ASTM C109/C109M—2015el	*Standard Test Method for Compressive Strength of Hydraulic Cement Mortars*	AS106.6.1
ASTM C141/C141M—14	*Standard Specification for Hydrated Hydraulic Lime for Structural Purposes*	AS104.4.6.1
ASTM C206—14	*Standard Specification for Finishing Hydrated Lime*	AS104.4.6.1
ASTM C926—15B	*Standard Specification for Application of Portland Cement Based Plaster*	AS104.4.8, AS104.4.9
ASTM C1707—11	*Standard Specification for Pozzolanic Hydraulic Lime for Structural Purposes*	AS104.4.6.1
ASTM E2392/ASTM E2392M—10	*Standard Guide for Design of Earthen Wall Building Systems*	AS104.4.3.2
CEN EN 459—2015	*Part 1: Building Lime. Definitions, Specifications and Conformity Criteria; Part 2: Test Methods*	AS104.4.6.1

APPENDIX AT [RE]

SOLAR-READY PROVISIONS—DETACHED ONE- AND TWO-FAMILY DWELLINGS AND TOWNHOUSES

The provisions contained in this appendix are not mandatory unless specifically referenced in the adopting ordinance.

User note:

About this appendix: *Harnessing the heat or radiation from the sun's rays is a method to reduce the energy consumption of a building. Although Appendix AT does not require solar systems to be installed for a building, it does require the space(s) for installing such systems, providing pathways for connections and requiring adequate structural capacity of roof systems to support solar systems.*

Section numbers in parenthesis are those in Appendix RB of the residential provisions of the International Energy Conservation Code®.

SECTION AT101 (RB101) SCOPE

AT101.1 (RB101.1) General. These provisions shall be applicable for new construction where solar-ready provisions are required.

SECTION AT102 (RB102) GENERAL DEFINITION

AT102.1 (RB102.1) General. The following term shall, for the purpose of this appendix, have the meaning shown herein.

SOLAR-READY ZONE. A section or sections of the roof or building overhang designated and reserved for the future installation of a solar photovoltaic or solar thermal system.

SECTION AT103 (RB103) SOLAR-READY ZONE

AT103.1 (RB103.2) General. New detached one- and two-family dwellings, and townhouses with not less than 600 square feet (55.74 m^2) of roof area oriented between 110 degrees and 270 degrees of true north, shall comply with Sections AT103.2 through AT103.10.

Exceptions:

1. New residential buildings with a permanently installed on-site renewable energy system.
2. A building where all areas of the roof that would otherwise meet the requirements of Section AT103 are in full or partial shade for more than 70 percent of daylight hours annually.

AT103.2 (RB103.2) Construction document requirements for solar-ready zone. *Construction documents* shall indicate the solar-ready zone.

AT103.3 (RB103.3) Solar-ready zone area. The total solar-ready zone area shall be not less than 300 square feet (27.87 m^2) exclusive of mandatory access or setback areas as required by the *International Fire Code.* New townhouses three stories or less in height above *grade plane* and with a total floor area less than or equal to 2,000 square feet (185.8 m^2) per dwelling shall have a solar-ready zone area of not less than 150 square feet (13.94 m^2). The solar-ready zone shall be composed of areas not less than 5 feet (1524 mm) in width and not less than 80 square feet (7.44 m^2) exclusive of access or set-back areas as required by the *International Fire Code.*

AT103.4 (RB103.4) Obstructions. Solar-ready zones shall be free from obstructions, including but not limited to vents, chimneys, and roof-mounted equipment.

AT103.5 (RB103.5) Shading. The solar-ready zone shall be set back from any existing or new, permanently affixed object on the building or site that is located south, east or west of the solar zone a distance not less than two times the object's height above the nearest point on the roof surface. Such objects include, but are not limited to, taller portions of the building itself, parapets, chimneys, antennas, signage, rooftop equipment, trees and roof plantings.

AT103.6 (RB103.6) Capped roof penetration sleeve. A capped roof penetration sleeve shall be provided adjacent to a solar-ready zone located on a roof slope of not greater than 1 unit vertical in 12 units horizontal (8-percent slope). The capped roof penetration sleeve shall be sized to accommodate the future photovoltaic system conduit, but shall have an inside diameter of not less than $1^1/_4$ inches (32 mm).

AT103.7 (RB103.7) Roof load documentation. The structural design loads for roof dead load and roof *live load* shall be clearly indicated on the *construction documents.*

AT103.8 (RB103.8) Interconnection pathway. *Construction documents* shall indicate pathways for routing of conduit or plumbing from the solar-ready zone to the electrical service panel or service hot water system.

AT103.9 (RB103.9) Electrical service reserved space. The main electrical service panel shall have a reserved space to allow installation of a dual pole circuit breaker for future solar electric installation and shall be *labeled* "For Future Solar Electric." The reserved space shall be positioned at the opposite (load) end from the input feeder location or main circuit location.

AT103.10 (RB103.10) Construction documentation certificate. A permanent certificate, indicating the solar-ready zone and other requirements of this section, shall be posted near the electrical distribution panel, water heater or other conspicuous location by the builder or *registered design professional.*

APPENDIX AU

COB CONSTRUCTION (MONOLITHIC ADOBE)

This appendix is informative and is not part of the code.

User note:

About this appendix: *Cob construction has been used for thousands of years around the world, notably in England and Northern Europe, the Middle East, West Africa, China and the Southwestern United States. An estimated 20,000 cob homes are still inhabited in the English county of Devon alone, some dating from the 15th century. The term "cob" derives from an Old English word for "lump," since historical structures were often constructed one handful at a time.*

SECTION AU101 GENERAL

AU101.1 Scope. This appendix provides prescriptive and performance-based requirements for the use of *natural cob* as a building material. Buildings using *cob* walls shall comply with this code except as otherwise stated in this appendix.

AU101.2 Intent. In addition to the intent described in Section R101.3, the purpose of this appendix is to establish minimum requirements for cob structures that provide flexibility in the application of certain provisions of the code, to permit the use of site-sourced and local materials, and to permit combinations of historical and modern techniques.

AU101.3 Tests and empirical evidence. Tests for an alternative material, design or method of construction shall be in accordance with Section R104.11.1, and the *building official* shall have the authority to consider evidence of a history of successful use in lieu of testing.

AU101.4 Cob wall systems. *Cob* wall systems include those shown in Figure AU101.4 and *approved* variations.

AU101.5 Definitions. The words and terms in Section AU102 shall, for the purposes of this appendix, have the meanings shown herein. Refer to Chapter 2 for general definitions.

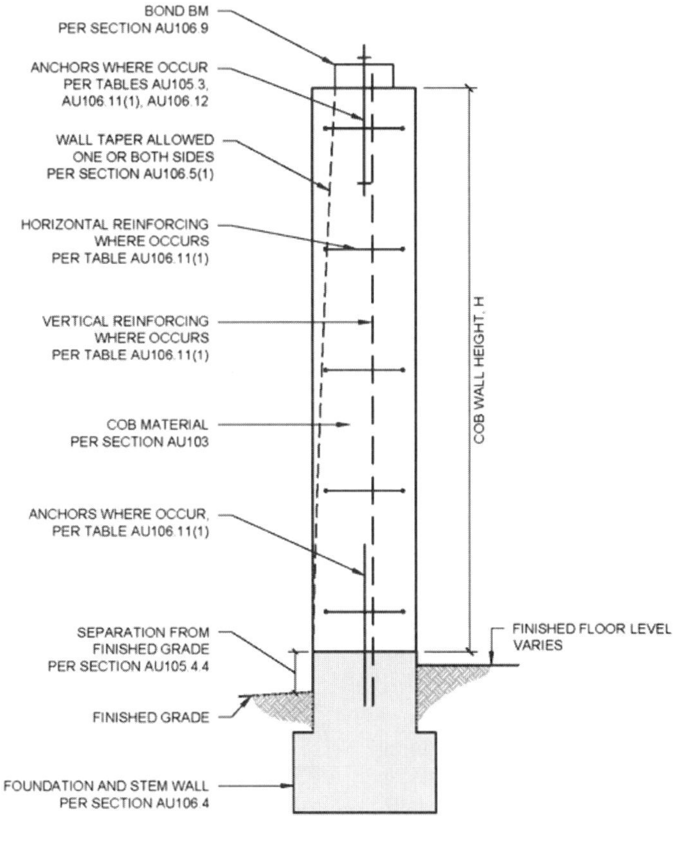

FIGURE AU101.4
TYPICAL COB WALL

SECTION AU102 DEFINITIONS

BRACED WALL PANEL. A *cob* wall designed and constructed to resist in-plane shear loads through the interaction of the cob material, its reinforcing and its connections to its bond beam and foundation. The panel's length meets the requirements for the particular wall type and contributes toward the total amount of bracing required along its braced wall line in accordance with Sections AU106.11 and R602.10.1.

BUTTRESS. A mass set at an angle to or bonded to a wall that it strengthens or supports.

CLAY. Inorganic soil with particle sizes less than 0.00008 inch (0.002 mm) and having the characteristics of high to very high dry strength and medium to high plasticity, used as the binder of other component materials in a mix of *cob* or of clay plaster.

CLAY SUBSOIL. Subsoil sourced directly from the earth, containing clay, sand and silt, and containing not more than trace amounts of organic matter.

COB. A composite building material consisting of refined *clay* or *clay subsoil* wet-mixed with loose straw and sometimes sand. Also known as "*Monolithic adobe.*"

COB CONSTRUCTION. A wall system of layers or lifts of moist *cob* placed to create monolithic walls, typically without formwork.

DRY JOINT. The boundary between a layer of moist *cob* and a previously laid and significantly drier, nonmalleable layer of *cob* that requires wetting to achieve bonding between the layers.

FINISH. Completed combination of materials on the face of a *cob* wall.

LIFT. A layer of installed *cob.*

LOAD-BEARING WALL. A *cob* wall that supports more than 100 pounds per linear foot (1459 N/m) of vertical load in addition to its own weight.

MONOLITHIC ADOBE. See "*Cob.*"

NATURAL COB. *Cob* not containing admixtures such as Portland cement, lime, asphalt emulsion or oil. Synonymous with "*Unstabilized cob.*"

NONSTRUCTURAL WALL. Walls other than *load-bearing walls* or *shear walls.*

PLASTER. Clay, soil-cement, gypsum, lime, clay-lime, cement-lime or cement plaster as described in Section AU104.

SHEAR WALL. A *cob* wall designed and constructed to resist in-plane lateral seismic and wind forces in accordance with Section AU106.11. Synonymous with "*Braced wall panel.*"

STABILIZED. *Cob* or other earthen material containing admixtures, such as Portland cement, lime, asphalt emulsion or oil, that are intended to help limit water absorption, stabilize volume, increase strength and increase durability.

STRAW. The dry stems of cereal grains after the seed heads have been removed.

STRUCTURAL WALL. A wall that meets the definition for a "*Load-bearing wall*" or "*Shear wall.*"

UNSTABILIZED. A *cob* or other earthen material that does not contain admixtures such as Portland cement, lime, asphalt emulsion or oil.

UNSTABILIZED COB. See "*Natural cob.*"

SECTION AU103 MATERIALS, MIXING AND INSTALLATION

AU103.1 Clay subsoil. *Clay subsoil* for a *cob* mix shall be acceptable if the mix it produces meets the requirements of Section AU103.4.

AU103.2 Sand. Sand or other aggregates such as, but not limited to, gravel, pumice and lava rock, when added to *cob* mixes, shall yield a mix that meets the requirements of Section AU103.4.

AU103.3 Straw. *Straw* for *cob* mixes shall be from wheat, rice, rye, barley or oat, or similar reinforcing fibers with similar performance. Before mixing, the straw or other reinforcing fibers shall be dry to the touch and free of visible decay.

AU103.4 Mix proportions. *Cob* mixes shall be of any proportions of refined *clay* or *clay subsoil*, added sand (if any) and straw that produce a dried mix that passes the shrinkage test in accordance with Section AU103.4.1, complies with the compressive strength requirements of Section AU106.6 and complies with the modulus of rupture requirements of Section AU106.7.

AU103.4.1 Shrinkage test for cob mixes. Each proposed *cob* mix of different mix proportions shall be placed moist to completely fill a 24-inch by $3^1/_2$-inch by $3^1/_2$-inch (610 mm by 89 mm by 89 mm) wooden form on a plastic or paper slip sheet and dried to ambient moisture conditions, or oven dried. The total shrinkage of the length shall not exceed 1 inch (25 mm), as measured from the dried edges of the material to the insides of the form. Cracks in the sample greater than $^1/_{16}$ inch (1.5 mm) shall first be closed manually. The shrinkage test shall be shown to the building official for approval before placement of the *cob* mix onto walls

AU103.5 Mixing. The *clay subsoil*, sand and straw for *cob* shall be thoroughly mixed by manual or mechanical means with water sufficient to produce a mix of a plastic consistency capable of bonding of successively placed layers or *lifts.*

AU103.6 Installation. *Cob* shall be installed on the wall in *lifts* of a height that supports itself with minimal slumping.

AU103.7 Dry joints. Each layer of *cob* shall be prevented from drying until the next layer is installed, to ensure bonding of successive layers. The top of each layer shall be kept moist and malleable with one or more of the following methods:

1. Covering with a material that prevents loss of or holds moisture.
2. Covering with a material that shades it from direct sun.
3. Wetting.

Where *dry joints* are unavoidable, the previous layer shall be wetted prior to application of the next layer.

AU103.8 Drying holes. Where holes to facilitate drying are used, such holes shall be of any depth and not exceeding $^3/_4$ inch (19 mm) in diameter on the face of *cob* walls. Drying holes shall not be spaced closer than 10 hole-diameters. Drying holes shall not be placed in *braced wall panels*. The design load on *load-bearing walls* with drying holes shall not exceed 90 percent of the allowable bearing capacity as determined in accordance with Section AU106.8. Drying holes shall be filled with *cob* before final inspection.

AU103.9 Adding roof loads to walls. Roof and ceiling loads shall not be added until walls are sufficiently dry to support them without compressing.

SECTION AU104 FINISHES

AU104.1 General. *Cob* walls shall not require a *finish*, except as required by Section AU104.2. *Finishes* applied to *cob* walls shall comply with this section and Chapters 3 and 7 unless stated otherwise in this section.

AU104.1.1 Interior wall finishes. Where installed, interior wall *finishes* and interior fire protection shall comply with the applicable provisions of Section R302, and shall be *plasters* in accordance with Section AU104.4 or nonplaster wall coverings in accordance with Section R702.

AU104.1.2 Exterior wall finishes. Where installed, exterior wall *finishes* shall be *plasters* in accordance with Section AU104.4, nonplaster exterior wall coverings in accordance with Section R703, or other *finish* systems in accordance with the following:

1. Specifications and details of the *finish* system's means of attachment to the wall or its independent support and means of draining or evaporating water that penetrates the exterior finish shall be provided.
2. The vapor permeance of the combination of *finish* materials shall be 5 perms or greater to allow the transpiration of water vapor from the wall.
3. *Finish* systems with weights greater than 10 pounds per square foot (48.9 kg/m) and less than or equal to20 pounds per square foot (97.8 kg/m) of wall area shall require that the minimum total length of *braced wall panels* in Table AU106.11(3) be multiplied by a factor of 1.2.
4. *Finish* systems with weights greater than 20 pounds per square foot (97.8 kg/m) of wall area shall require an engineered design.

AU104.2 Where required. *Cob* walls exposed to rain due to local climate, building design and wall orientation shall be *finished* or clad to provide protection from excessive erosion.

AU104.3 Vapor retarders. Class I and II vapor retarders shall not be used on *cob* walls, except at *cob* walls surrounding showers or as required or addressed elsewhere in this appendix.

AU104.4 Plaster. *Plaster* applied to *cob* walls shall be any type described in this section. *Plaster* thickness shall not exceed 3 inches (76 mm) on each face except where an *approved* engineered design is provided.

AU104.4.1 Plaster and membranes. *Plaster* shall be applied directly to *cob* walls to facilitate transpiration of moisture from the walls and to secure a mechanical bond between the *plaster* and the *cob*. A membrane shall not be located between the *cob* wall and the *plaster*.

AU104.4.2 Plaster lath. The surface of *cob* walls shall be permitted to function as lath for *plaster*, with no other lath required. Metal, plastic, and natural fiber lath shall be permitted to be used to limit *plaster* cracking, increase the *plaster* bond to the wall, or to bridge dissimilar materials.

AU104.4.3 Clay plaster. *Clay plaster* shall comply with Sections AU104.4.3.1 and AU104.4.3.2.

AU104.4.3.1 General. *Clay plaster* shall be any *plaster* having a *clay* or *clay subsoil* binder. Such *plaster* shall contain sufficient clay to fully bind the sand or other aggregate and any reinforcing fibers. Reinforcing fibers shall be chopped straw, sisal, hemp, animal hair or other similar approved fibers.

AU104.4.3.2 Clay subsoil requirements. The suitability of *clay subsoil* shall be determined in accordance with the Figure 2 Ribbon Test and the Figure 3 Ball Test in the appendix of ASTM E2392/E2392M.

AU104.4.4 Soil-cement plaster. Soil-cement *plaster* shall be composed of *clay subsoil*, sand, not more than 7 percent Portland cement by volume and, where provided, reinforcing fibers.

AU104.4.5 Gypsum plaster. Gypsum *plaster* shall comply with Section R702.2.1 and shall be limited to interior use.

AU104.4.6 Lime plaster. Lime *plaster* is any *plaster* with a binder composed of calcium hydroxide including Type N or S hydrated lime, hydraulic lime, natural hydraulic lime or slaked quicklime. Hydrated lime shall comply with ASTM C206. Hydraulic lime shall comply with ASTM C1707. Natural hydraulic lime shall comply with ASTM C141 and EN 459. Quicklime shall comply with ASTM C5.

AU104.4.7 Clay-lime plaster. Clay-lime *plaster* shall be composed of refined *clay* or *clay subsoil*, sand, lime and, where provided, reinforcing fibers.

AU104.4.8 Cement-lime plaster. Cement-lime *plaster* shall be plaster mix types CL, F or FL, as described in ASTM C926.

AU104.4.9 Cement plaster. Cement *plaster* shall have not less than 1 part lime to 4 parts cement and be not thicker than $1^1/_2$ inches (38 mm), to ensure minimum acceptable vapor permeability

SECTION AU105 COB WALLS—GENERAL

AU105.1 General. *Cob* walls shall be designed and constructed in accordance with this section and Figure AU101.4 or an *approved* alternative design. In addition to the general requirements for c*ob* walls in this section, c*ob structural walls* shall comply with Section AU106.

AU105.2 Building limitations and requirements for cob wall construction. *Cob* walls shall be subject to the following limitations and requirements:

1. Number of stories: not more than one.
2. Building height: not more than 20 feet (6096 mm).
3. *Seismic design categories*: limited to use in *Seismic Design Categories* A, B and C, except where an approved engineered design is provided.
4. Wall height: in accordance with Table AU105.3, and with Table AU106.11(1) for *braced wall panels.*
5. Wall thickness, excluding *finish,* shall be not less than 10 inches (254 mm), not greater than 24 inches (610 mm) at the top two-thirds, not limited at the bottom third and, for structural walls, shall comply with Section AU106.2, Item 2. Wall taper is permitted in accordance with Section AU106.5, Item 1.
6. Interior *cob* walls shall require an *approved* engineered design that accounts for the seismic load of the interior *cob* walls, except in Seismic Design Category A for walls with a height to thickness ratio less than or equal to 6.

AU105.3 Out-of-plane resistance methods and unrestrained wall height limits. *Cob* walls shall employ a method of out-of-plane load resistance in accordance with Table AU105.3, and comply with its associated height limits and requirements.

AU105.3.1 Determination of out-of-plane loading. Out-of-plane loading for the use of Table AU105.3 shall be in accordance with the ultimate design wind speed and seismic design category requirements of Sections R301.2.1 and R301.2.2, respectively. An *approved* engineered design shall be required where the building is located in a special wind region or where wind design is required in accordance with Figure R301.2.1.1.

AU105.3.2 Bond beams for nonstructural walls. Nonstructural *cob* walls shall be provided with a bond beam at the top of the wall that complies with Section AU106.9, except for requirements relating to roof and/or ceiling loads or *braced wall panels.*

AU105.3.3 Lintels in nonstructural walls. Door, window and other openings in nonstructural *cob* walls shall require a lintel in accordance with Section AU106.10, except for requirements relating to roof and/or ceiling loads or *braced wall panels.*

AU105.3.4 Reinforcing at wall openings. Reinforcing shall be installed at window, door, and similar wall openings and penetrations greater than 2 feet (610 mm) in width in accordance with Sections AU105.3.4.1 through AU105.3.4.3. Surface voids deeper than 25 percent of the wall thickness shall be considered an opening.

AU105.3.4.1 Opening size limit. Openings shall not exceed 6 feet (1829 mm) in width, and the height of the *cob* wall below openings shall not exceed 6 feet (1829 mm) above the top of the foundation.

AU105.3.4.2 Horizontal reinforcing. Two-inch by 2-inch (51 mm by 51 mm) 14-gage galvanized steel mesh shall be embedded 4 inches (102 mm) in the *cob* above the rough opening and below the rough opening for windows, and shall extend 12 inches (305 mm) beyond the sides of the opening. Walls below rough window openings greater than 4 feet 6 inches (1372 mm) in height shall be provided with additional horizontal reinforcing at midheight.

AU105.3.4.3 Vertical reinforcing. Full-height $^5/_8$-inch (16 mm) threaded rod shall be installed 4 inches (102 mm) from each side of the opening, centered in the thickness of the *cob* wall. The threaded rods shall be embedded 7 inches (178 mm) in the foundation, and 4 inches (102 mm) in concrete bond beams or shall penetrate through wood bond beams and be secured with a nut and washer. The threaded rods shall be embedded in concrete lintels or pass through a drilled hole in wood lintels.

AU105.3.5 Minimum length of cob walls. Sections of *cob* walls between openings shall be not less than 2 feet, 6 inches (762 mm) in length. Wall sections less than 4 feet (1219 mm) and not less than 2 feet, 6 inches (762 mm) in length shall contain vertical reinforcing in accordance with Section AU105.3.4.3.

AU105.4 Moisture control. *Cob* walls shall be protected from moisture intrusion and damage in accordance with Sections AU105.4.1 through AU105.4.5.

AU105.4.1 Water-resistant barriers and vapor permeance. *Cob* walls shall be constructed without a membrane barrier between the *cob* wall and *plaster* to facilitate transpiration of water vapor from the wall, and to secure a mechanical bond between the *cob* and *plaster*, except as otherwise required elsewhere in this appendix. Where a water-resistant barrier is placed behind an exterior *finish*, it shall be considered part of the *finish* system and shall comply with Item 2 of Section AU104.1.2 for the combined vapor permeance rating.

AU105.4.2 Horizontal surfaces. *Cob* walls and other *cob* elements shall be provided with a water-resistant barrier at weather-exposed horizontal surfaces. The water-resistant barrier shall be of a material and installation that will prevent erosion and prevent water from entering the wall system. Horizontal surfaces, including exterior window sills, sills at exterior niches and exterior buttresses, shall be sloped not less than 1 unit vertical in 12 units horizontal to drain away from *cob* walls or other *cob* elements.

AU105.4.3 Separation of cob and foundation. A liquid-applied or bituminous Class II vapor retarder shall be installed between *cob* and supporting concrete or masonry.

> **Exception:** Where local climate, site conditions and foundation design limit ground moisture migration into the base of the *cob* wall, including but not limited to the use of a moisture barrier or capillary break between the supporting concrete or masonry and the earth.

AU105.4.4 Separation of cob and finished grade. *Cob* shall be not less than 8 inches (203 mm) above finished *grade*.

> **Exception:** The minimum separation shall be 4 inches (102 mm) in dry climate zones as defined in Section N1101.7.2, and shall be 2 inches (51mm) on walls that are not weather exposed.

AU105.4.5 Installation of windows and doors. Windows and doors shall be installed in accordance with the manufacturer's instructions to a wooden frame of not less than nominal 2-inch by 4-inch (51 mm by 102 mm) wood members anchored into the *cob* wall with 16d galvanized nails half-driven at a maximum 6-inch (152 mm) spacing, with the protruding half embedded in the *cob*. The wood frame shall be embedded not less than $1^1/_2$ inches (38 mm) in the *cob* and shall be set in from each face of the wall not less than 3 inches (76 mm). Alternative window and door installation methods shall be capable of resisting the wind loads in Table R301.2.1(1).

TABLE AU105.3
OUT-OF-PLANE RESISTANCE METHODS AND UNRESTRAINED WALL HEIGHT LIMITS

WALL TYPE[a, g, h] AND METHOD OF OUT-OF-PLANE LOAD RESISTANCE	FOR ULTIMATE DESIGN WIND SPEEDS (mph)	FOR SEISMIC DESIGN CATEGORIES	UNRESTRAINED COB WALL HEIGHT $H^{b, c, h}$		TOP ANCHOR[e] SPACING (inches)	TENSION TIE[f] SPACING (inches)
			Absolute Limit (feet)	Limit Based on Wall Thickness T^d (feet)		
Wall 1[i]: no anchors, no steel wall reinforcing	≤ 110	A	$H \leq 8$	$H \leq 6T$	None	48
Wall 2: top anchors[j], continuous vertical 6″ × 6″ × 6″ gage steel mesh in center of wall embedded in foundation 12 inches	≤ 140	A, B, C	$H \leq 8$	$H \leq 8T$	12	24
Wall A[i]: top anchors, no vertical steel reinforcing	≤ 120	A, B	$H \leq 8$	$H \leq 6T$	12	48
Wall B[i]: top and bottom anchors, no vertical steel reinforcing	≤ 130	A, B	$H \leq 8$	$H \leq 6T$	12	48
Wall C: top and bottom anchors, continuous vertical threaded rod at 4 feet on center embedded in foundation and connected to bond beam	≤ 140	A, B, C	$H \leq 8$	$H \leq 8T$	12	24
Wall D: continuous vertical threaded rod at 1 foot on center embedded in foundation and connected to bond beam	≤ 140	A, B, C	$H \leq 8$	$H \leq 8T$	N/A	24
Wall E: top anchors, continuous vertical 6″ × 6″ × 6″ gage steel mesh 2 inches from each face of wall embedded in foundation	≤ 140	A, B, C	$H \leq 8$	$H \leq 8T$	12	24

For SI: 1 inch = 25.4 mm, 1 foot = 304.8 mm, 1 mile per hour = 0.447 m/s.

N/A = Not Applicable

a. See Table AU106.11(1) for reinforcing and anchorage specifications for wall Types A, B, C, D and E.
b. *H* = height of the cob portion of the wall only. See Figure AU101.4. The maximum *H* is the absolute limit or the limit based on wall thickness, whichever is more restrictive.
c. Bond beams or other horizontal restraints are capable of separating a wall into more than one unrestrained wall height with an approved engineered design.
d. *T* = Cob wall thickness (in feet) at its minimum, without plaster.
e. $^5/_8$-inch threaded rod anchors at prescribed spacing with 12-inch embedment in cob, full embedment in concrete bond beams or full penetration in wood bond beam with a nut and washer.
f. Attach rafters to bond beam with 4-inch by 3-inch by 3-inch by 18 gage tension tie angles at prescribed spacing. See Figure AU106.9.5. Where rafters are attached to tension ties, roof sheathing shall be edge nailed.
g. All walls shall be tested for compressive strength in accordance with Section AU106.6.
h. For curved walls with an arc length to radius ratio of 1.5:1 or greater, the *H/T* factor shall be increased by 1, and the absolute height limit by 1 foot.
i. Wall type requires a modulus of rupture test in accordance with Section AU106.7.
j. See wall Type A in Table AU106.11(1) for top anchor requirements.

Windows and doors in *cob* walls shall be installed so as to mitigate the passage of air or moisture into or through the wall system. Window sills shall comply with Section AU105.4.2.

AU105.5 Inspections. In addition to ensuring compliance with Section R109.1, the building official shall inspect the following aspects of *cob* construction:

1. Anchors and vertical and horizontal reinforcing in *cob* walls, where required in accordance with Tables AU105.3 and AU106.11(1) and Sections AU105.3.4 through AU105.3.5.
2. Reinforcing in any concrete bond beams or lintels, in accordance with Section AU106.9.2 and Table AU106.10.

SECTION AU106
COB WALLS—STRUCTURAL

AU106.1 General. *Cob* structural walls shall be in accordance with the prescriptive provisions of this section. Designs or portions of designs not complying with this section shall require an *approved* engineered design.

AU106.2 Requirements for cob structural walls. In addition to the requirements of Section AU105.2, *cob* structural walls shall be subject to the following:

1. Wall height: shall be in accordance with Table AU105.3 for load-bearing *cob* walls or Table AU106.11(1) for *cob braced wall panels*, as applicable and most restrictive.
2. Wall thickness: shall be in accordance with Sections AU105.2, Item 5 and Section AU106.8.1 for load-bearing *cob* walls or Table AU106.11(1) for *cob braced wall panels*, as applicable and most restrictive.
3. *Braced wall panel* lengths: for buildings using *cob braced wall panels*, the greater of the values determined in accordance with Table AU106.11(2) for wind loads and Table AU106.11(3), AU106.11(4) or AU106.11(5) for seismic loads shall be used.

AU106.3 Loads and other limitations. Live and dead loads and other limitations shall be in accordance with Section R301, except that the dead load for *cob* walls shall be determined by Equation AU-1.

$$CW_{DL} = (H \times T_{avg} \times D) \qquad \textbf{(Equation AU-1)}$$

CW_{DL} = *Cob* wall dead load (in pounds per lineal foot of wall).

H = Height of *cob* portion of wall (in feet).

T_{avg} = Average thickness of wall (in feet).

D = Density of *cob* = 110 (in pounds per cubic foot), unless a lesser value at equilibrium moisture content is demonstrated to the building official.

AU106.4 Foundations. Foundations for *cob* walls shall be in accordance with Chapter 4. The width of foundations for *cob* walls shall be not less than the width of the *cob* at its base, excluding *finish*.

AU106.5 Wall taper, straightness and surface voids for cob walls. *Cob* walls shall be in accordance with the following:

1. *Cob* structural and *nonstructural walls* shall be vertical or shall taper from bottom to top with the wall thickness in accordance with Section AU105.2, Item 5 and the wall height in accordance with Section AU105.2, Item 4.
2. *Cob* structural and *nonstructural walls* shall be straight or curved. Curved *braced wall panels* shall be in accordance with Sections AU106.11.2 and AU106.11.3.
3. Niches and other surface voids in *load-bearing walls* are limited to 12 inches (305 mm) in width and height and 25 percent of the wall thickness, and shall be located in the top two-thirds of the wall. Surface voids that exceed these limits shall be considered wall openings, and shall receive a lintel in accordance with Section AU106.10 and be reinforced in accordance with Section AU105.3.4. Surface voids are prohibited in *braced wall panels*.

AU106.6 Compressive strength of cob structural and nonstructural walls. All *cob* walls shall have a minimum compressive strength of 60 psi (414 kPa). *Cob* in walls used as *braced wall panels* shall have a minimum compressive strength of 85 psi (586 kPa).

AU106.6.1 Demonstration of compressive strength. The compressive strength of the *cob* mix to be used in structural walls and *nonstructural walls* as required in Section AU106.6 shall be demonstrated to the building official before the placement of *cob* onto walls, with compressive strength tests and an associated report by an *approved* laboratory or with an *approved* on-site test as follows:

1. Five samples of the proposed *cob* mix shall be placed moist to completely fill a 4-inch by 4-inch by 4-inch (102 mm by 102 mm by 102 mm) form and dried to ambient moisture conditions.
2. Samples shall not be oven dried.
3. Any opposite faces shall be faced with plaster of paris if needed to achieve smooth, parallel faces, after which the sample shall reach ambient moisture conditions before testing.
4. The horizontal cross section of the dried sample as tested, and the maximum applied load at failure shall be used to calculate the sample's compressive strength.
5. The fourth-lowest value shall be used to determine the mix's compressive strength.

AU106.7 Modulus of rupture of cob structural walls. *Cob* in walls used as *braced wall panels* shall have a minimum modulus of rupture of 50 pounds per square inch (345 kPa).

AU106.7.1 Demonstration of modulus of rupture. The modulus of rupture of *cob* used in structural walls shall be demonstrated to the building official before the placement of *cob* onto walls, with modulus of rupture tests and an associated report by an *approved* laboratory or with an *approved* on-site test as follows:

1. Five samples of the proposed *cob* mix shall be placed moist to completely fill a 6-inch by 6-inch by 12-inch (152 mm by 152 mm by 305 mm) form and dried to indoor ambient moisture conditions.
2. Samples shall not be oven dried.
3. Each sample shall be tested with the 12-inch (305 mm) dimension horizontal.
4. The fourth-lowest value shall be used to determine if the mix meets the minimum required modulus of rupture.

AU106.8 Bearing capacity. The allowable bearing capacity for *cob load-bearing walls* supporting vertical roof and/or ceiling loads imposed in accordance with Section R301 shall be determined by Equation AU-2.

$$BC = 144\ (C \times T_{min})/3 - (H \times T_{avg} \times D) \qquad \textbf{(Equation AU-2)}$$

BC = Allowable bearing capacity of wall (in pounds per lineal foot of wall).

C = Compressive strength (in psi) as determined in accordance with Section AU106.6.

T_{min} = Thickness of wall (in feet) at its minimum.

H = Height of *cob* portion of wall (in feet).

T_{avg} = Average thickness of wall (in feet).

D = Density of *cob* = 110 (in pounds per cubic foot), unless a lesser value at equilibrium moisture content is demonstrated.

AU106.8.1 Support of uniform loads. Uniform roof and/or ceiling loads shall be supported by *cob load-bearing walls* not exceeding their allowable bearing capacity, as demonstrated in accordance with Equation AU-3.

$$BL \leq BC \qquad \textbf{(Equation AU-3)}$$

BL = Design load on the wall (in pounds per lineal foot) determined in accordance with Sections R301.4 and R301.6.

BC = Allowable bearing capacity of wall (in pounds per lineal foot of wall) determined in accordance with Section AU106.8.

AU106.8.2 Support of concentrated loads. Concentrated roof and ceiling loads shall be distributed by structural elements capable of distributing the loads to the *cob load-bearing wall* and within its allowable bearing capacity as determined in accordance with Section AU106.8. Concentrated loads over lintels or over bond beams spanning openings shall require an *approved* engineered design.

AU106.9 Bond beams. *Cob* structural walls shall require a bond beam at the top of the wall in accordance with Section AU106.9.1, AU106.9.2 or AU106.9.3, and shall be anchored to the *cob* below in accordance with Tables AU105.3, AU106.11(1) and AU106.12 as applicable and most restrictive. Bond beams spanning openings shall be in accordance with Section AU106.9.4.

AU106.9.1 Wood bond beams. Wood bond beams shall be not less than nominal 4 inches high by 8 inches wide and shall comply with Sections AU106.9.1.1 through AU106.9.1.3.

AU106.9.1.1 Wood species and grade. Wood bond beams shall be of a species with an extreme fiber in bending (F_b) of not less than 850 psi (5.9 MPa), a modulus of elasticity (E) of not less than 1,300,000 psi (8964 MPa), and No. 2 grade or better. Composite lumber bond beams shall have an F_b of not less than 850 psi (5.9 MPa), and an E of not less than 1,300,000 psi (8964 MPa).

AU106.9.1.2 Discontinuity. Discontinuous wood bond beams shall be spliced on top with a metal strap with not less than the allowable wind or seismic load tension capacity in accordance with the following, whichever is more restrictive:

1. For *seismic design categories*: A, 2,500 pounds (11 kN); B, 4,500 pounds (20 kN); C, 6,000 pounds (26.7 kN).
2. For braced wall line lengths, when wind governs: 10 feet, 2,500 pounds (11 kN); 20 feet, 3,400 pounds (15.1 kN); 30 feet, 5,000 pounds (22.2 kN).

AU106.9.1.3 Corners and curved walls. Wood bond beams at corners and discontinuities atop curved walls shall be connected across their exterior faces with a metal strap with a capacity of not less than that determined in accordance with Section AU106.9.1.2.

AU106.9.2 Concrete bond beams. Concrete bond beams shall be not less than 6 inches (152 mm) high by 8 inches (305 mm) wide. Concrete bond beams shall be reinforced with two No. 4 bars, 2 inches (51 mm) clear from the bottom and 2 inches (51 mm) clear from the sides. Lap splices shall comply with Table R608.5.4(1). Reinforcing at corners shall be in accordance with the horizontal reinforcing requirements in Section R608.6.4. The concrete shall have a compressive strength of not less than 2,500 pounds per square inch (17.2 MPa) at 28 days.

AU106.9.3 Other bond beams. Bond beams of other materials, including earthen materials, require an *approved* engineered design.

AU106.9.4 Bond beams spanning openings. Bond beams that support uniform roof and/or ceiling loads and span openings in *cob* walls shall be in accordance with Table AU106.10. Bond beams shall be continuous across the opening and not less than 1 foot (305 mm) beyond each side of the opening.

AU106.9.5 Connection of roof framing to bond beams. Roof and ceiling framing shall be attached to bond beams in accordance with Table R602.3(1), Items 2 and 6, and Figure AU106.9.5. Roof sheathing shall be attached to roof framing in accordance with Figure AU106.9.5. A minimum nominal 2-inch by 6-inch (51 mm by 152 mm) wood plate shall be installed on concrete bond beams with $^5/_8$-inch (16 mm) diameter anchor bolts with 5-inch

(127 mm) embedment at 2 feet (610 mm) on center to allow the required fastening of roof and ceiling framing, including tension ties and straps.

AU106.9.6 Bond beams and connections at gable and shed roof end walls. Bond beams and connections at end walls of buildings with gable or shed roofs shall comply with Figure AU106.9.6 and the following:

1. End walls shall not exceed 20 feet (6096 mm) in length.
2. Bond beams shall be continuous and straight for the entire wall line.
3. Wood bond beams shall comply with the following:
   - 3.1. Not less than nominal 4 inches by 8 inches (102 mm by 203 mm) where wind design governs in accordance with Table AU106.11(2) and where seismic design governs in accordance with Table AU106.11(3), AU106.11(4) or AU106.11(5) for wall lengths less than or equal to 20 feet (6096 mm) in Seismic Design Category A or wall lengths less than or equal to 10 feet (3048 mm) in Seismic Design Categories B and C.
   - 3.2. Not less than nominal 4 inches by 10 inches (102 mm by 254 mm) for wall lengths less than or equal to 20 feet (6096 mm) in Seismic Design Category B.
   - 3.3. Not less than nominal 6 inches by 12 inches (152 mm by 305 mm) or 4 inches by 16 inches (102 mm by 406 mm) for wall lengths less than or equal to 20 feet (6096 mm) in Seismic Design Category C.
   - 3.4. Corners shall be connected in accordance with Section AU106.9.3.
4. Concrete bond beams when used shall be in accordance with Section AU106.9.2 in Seismic Design Categories A, B and C and for ultimate design wind speeds less than or equal to 140 mph (63.6 m/s).
5. Walls between the bond beam and roof shall be of wood-framed construction in accordance with Section R602. The ratio of its greatest height to its length shall not exceed 1:2. The wall shall not contain openings.

AU106.10 Lintels. Door, window and other openings in load-bearing *cob* walls shall be provided with a lintel of wood or concrete in accordance with Table AU106.10.

AU106.11 Cob braced wall panels. *Cob braced wall panels* shall be in accordance with Section R602.10 and Tables AU106.11(1), AU106.11(2), AU106.11(3), AU106.11(4) and AU106.11(5). Wind design criteria shall be in accordance with Section R301.2.1. Seismic design criteria shall be in accordance with Section R301.2.2. An *approved* engineered design shall be required in accordance with Section R301.2.1 where the building is located in a special wind region or where wind design is required in accordance with Figure R301.2.1.1.

AU106.11.1 Nonorthogonal braced wall panels. *Braced wall panels* at an angle to the orthogonal *braced wall lines* shall be considered to contribute to the minimum

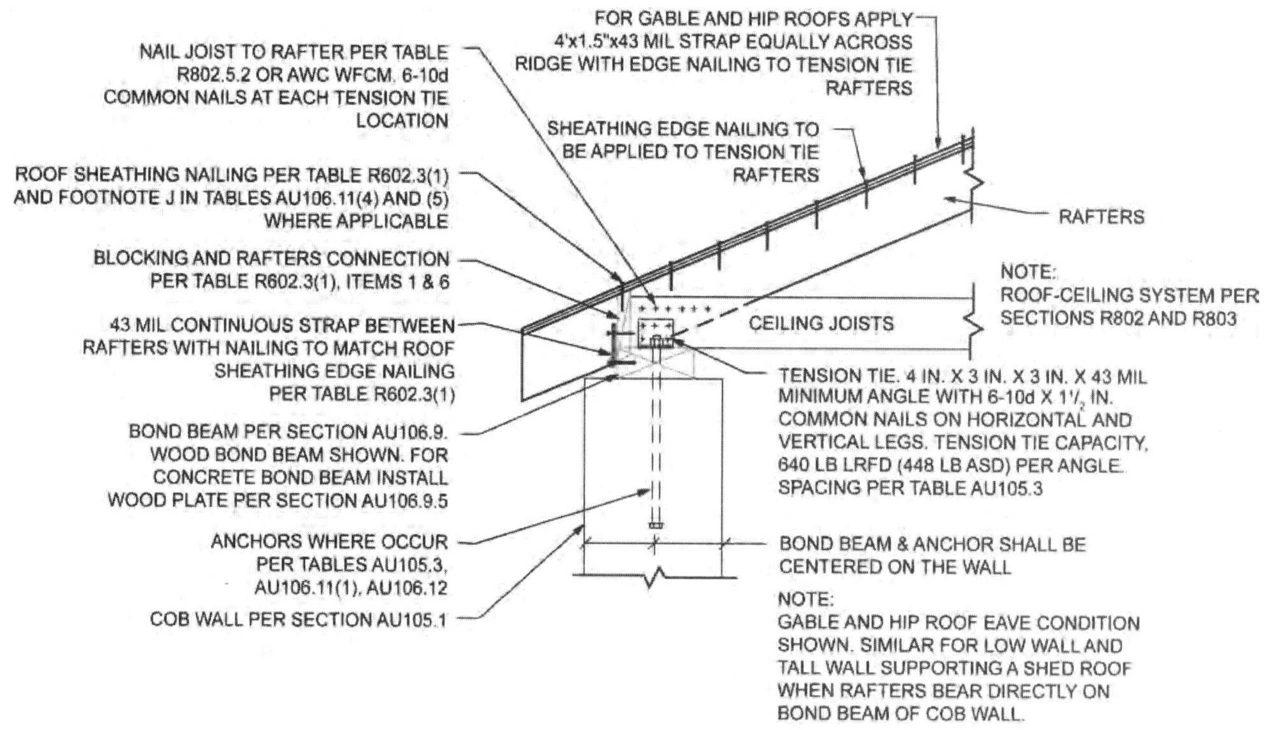

FIGURE AU106.9.5
CONNECTION OF ROOF FRAMING TO BOND BEAMS

total braced wall lengths in Tables AU106.11(2), AU106.11(3), AU106.11(4) and AU106.11(5), as follows:

1. A *braced wall panel* not more than 45 degrees and greater than 30 degrees to an adjacent orthogonal braced wall line shall contribute 50 percent of its length to that line.
2. A *braced wall panel* not more than 30 degrees to an orthogonal braced wall line shall contribute 65 percent of its length to that line.
3. A *braced wall panel* greater than 45 degrees and not more than 60 degrees to an orthogonal braced wall line shall contribute 35 percent of its length to that line.
4. The angle of a curved *braced wall panel* to a braced wall line shall be determined with the chord of that section of wall, connecting the end points of the arc at the center of the wall.

AU106.11.2 Braced wall lines for buildings with curved walls. Buildings with curved *cob* walls shall contain two *braced wall lines* in two orthogonal directions. The spacing of the *braced wall lines* for wind design in Table AU106.11(2) and the spacing and length of the *braced wall lines* for seismic design in Tables AU106.11(3), AU106.11(4) and AU106.11(5) shall be the maximum widths of the building in the two orthogonal directions.

AU106.11.3 Radius, thickness and length of curved braced wall panels. *Cob* curved *braced wall panels* shall have an inside radius of not less than 5 feet (1524 mm), shall be of the thickness required in Table AU106.11(1) and of the length determined in accordance with Section AU106.11. The length of the curved wall shall be considered to be the length of the arc at the center of the wall, in accordance with Figure AU106.11.3 and determined by Equation AU-4.

$$ARC_C = 0.0175\ R_C \times A \qquad \textbf{(Equation AU-4)}$$

ARC_C = Length of arc at center of wall (in feet).

R_C = Radius at center of wall = $R_i + 0.5T$ (in feet).

R_i = Inside radius of wall (in feet).

T = Thickness of wall without *finish* (in feet).

A = Angle of extent of *braced wall panel* from the center of the arc (in degrees).

AU106.12 Resistance to wind uplift forces. *Cob* walls that resist uplift forces from the *roof assembly*, as determined in accordance with Section R802.11, shall be in accordance with Table AU106.12.

AU106.13 Post-and-beam with cob infill. Post-and-beam with *cob* infill wall systems shall be in accordance with an *approved* engineered design.

AU106.14 Buttresses. *Cob buttresses* that are intended to provide out-of-plane wall bracing or additional capacity for *braced wall panels* shall be in accordance with an *approved* engineered design.

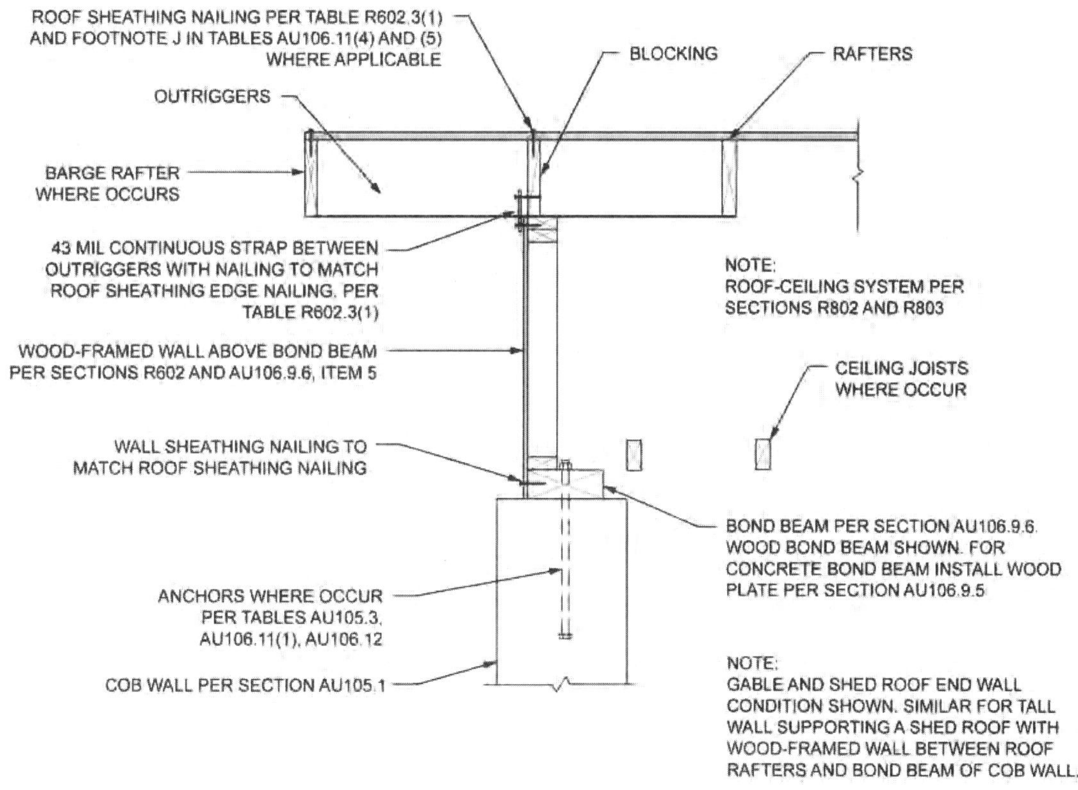

FIGURE AU106.9.6
CONNECTIONS AT GABLE AND SHED ROOF END WALLS

TABLE AU106.10
LINTELS AND BOND BEAMS SPANNING OPENINGS

GROUND SNOW LOAD ≤ 30 PSF			WOOD: • F_b ≥ 850 psi • E ≥ 1,300,000 psi • No. 2 Grade or better • Oriented flat • 1 piece or 2 equal-width pieces • Extend 1 foot beyond opening sides		CONCRETE: • 2500 psi compressive strength • Height = 6 inches • Extend 1 foot beyond opening sides • Reinforcement two No. 4 bars[a] • 2 inches clear from bottom • 2 inches clear from sides[a]	
Building Width (feet)	Cob above Lintel (feet)	Total Cob Wall and Plaster Thickness (inches)	Size of Wood Lintel or Bond Beam—*H* × *W* (nominal inches)		Width of Concrete Lintel or Bond Beam (inches)	
			For Span ≤ 4 ft	For Span ≤ 6 ft	For Span ≤ 6 ft	For Span ≤ 8 ft
10	0	≤ 27	4 × 8	4 × 8	8	8
10	1	15	4 × 12	4 × 12	12	12
10	1	19	4 × 16	4 × 16	16	16
10	1	27	4 × 24	4 × 24	24	24
10	2	15	4 × 12	6 × 12	12	12
10	2	19	4 × 16	6 × 16	16	16
10	2	27	4 × 24	4 × 24	24	24
20	0	≤ 27	4 × 8	6 × 8	8	8
20	1	15	4 × 12	6 × 12	12	12
20	1	19	4 × 16	6 × 16	16	16
20	1	27	4 × 24	4 × 24	24	24
20	2	15	4 × 12	6 × 12	12	NP
20	2	19	4 × 16	6 × 16	16	NP
20	2	27	4 × 24	6 × 24	24	NP
30	0	≤ 27	4 × 8	6 × 8	8	NP
30	1	15	4 × 12	6 × 12	12	NP
30	1	19	4 × 13	6 × 16	16	NP
30	1	27	4 × 24	6 × 24	24	NP
30	2	15	4 × 12	6 × 12	12	NP
30	2	19	4 × 16	6 × 16	16	NP
30	2	27	4 × 24	6 × 24	24	NP

For SI: 1 inch = 25.4 mm, 1 foot = 304.8 mm, 1 pound per square inch = 6.895 kPA.

NP = Not Permitted.

a. Concrete bond beams spanning openings, and lintels greater than 16 inches in width, shall have an additional No. 4 bar in the center of their width.

**TABLE AU106.11(1)
COB BRACED WALL PANEL TYPES**

WALL TYPE[a] DESIGNATION	ANCHORS TO FOUNDATION[b]	ANCHORS TO BOND BEAM[c]	VERTICAL STEEL REINFORCING[b, c]	HORIZONTAL STEEL REINFORCING	MAXIMUM HEIGHT H^d (in feet)	MAXIMUM ASPECT RATIO (H:L)
A	none	$^5/_8$″ threaded rod @ 12″; 4″ from wall ends; 12″ embedment in cob	none	none	7[e]	1:1
B	#5 bar @ 12″; 16″ embedment in cob	$^5/_8$″ threaded rod @ 12″; 4″ from wall ends; 16″ embedment in cob; 2″ × 2″ × $^1/_4$″ washer and nut at cob end	none	2″ × 2″ × 14 gage welded wire mesh[f] @ 18″; 6″ from foundation and bond beam	7[e]	1:1
C	#5 bar @ 12″; 16″ embedment in cob	$^5/_8$″ threaded rod @ 12″; 16″ embedment in cob	$^5/_8$″ threaded rod; 4″ from each end of braced wall panel; continuous from foundation to bond beam	2″ × 2″ × 14 gage welded wire mesh[f] @ 18″; 6″ from foundation and bond beam	7[e]	2:1
D	(see vertical steel reinforcing)	(see vertical steel reinforcing)	$^5/_8$″ threaded rod; 4″ from each end of braced wall panel and @ 12″; continuous from foundation to bond beam	2″ × 2″ × 14 gage welded wire mesh[f] @ 18″; 6″ from foundation and bond beam	7[e]	2:1
E	6″ × 6″× 6 gage welded wire mesh; 12″ embedment in foundation	$^5/_8$″ threaded rod @ 12″; 4″ from wall ends; 12″ embedment in cob	6″ × 6″ × 6 gage welded wire mesh; 2″ from each wall face	none	7.5	1:1

For SI: 1 inch = 25.4 mm, 1 foot = 304.8 mm.

a. Braced wall panel Types A, B, C and D shall be not less than 16 inches thick. Braced wall panel Type E shall be not less than 12 inches thick. All braced wall panels shall be not greater than 24 inches thick.

b. Not less than 8-inch embedment into foundation, unless otherwise stated.

c. Not less than 4-inch embedment into concrete bond beams. Full penetration through wood bond beam, secured with nut and washer.

d. H = height of the cob portion of the wall only. See Figure AU101.4.

e. Maximum height shall be 8 feet when wall thickness is increased to 18 inches.

f. Galvanized mesh.

TABLE AU106.11(2)
BRACING REQUIREMENTS FOR COB BRACED WALL PANELS BASED ON WIND SPEED

• EXPOSURE CATEGORY B[d] • 25-FOOT MEAN ROOF HEIGHT • 10-FOOT EAVE-TO-RIDGE HEIGHT[d] • 10-FOOT WALL HEIGHT[d] • 2 BRACED WALL LINES[d]			MINIMUM TOTAL LENGTH (FEET) OF COB BRACED WALL PANELS REQUIRED ALONG EACH BRACED WALL LINE[a, b, c, d]			
Ultimate Design Wind Speed (mph)	Story Location	Braced Wall Line Spacing (feet)	Cob Braced Wall Panel[e] A; (aspect ratio $H:L \leq 1:1$)	Cob Braced Wall Panel[e] B; (aspect ratio $H:L \leq 1:1$)	Cob Braced Wall Panel[e] C, D; (aspect ratio $H:L \leq 2:1$)	Cob Braced Wall Panel[e] E; (aspect ratio $H:L \leq 1:1$)
≤ 110	One-story building	10	6.0	6.0	3.7	6.0
≤ 110	One-story building	20	7.9	7.4	7.4	6.0
≤ 110	One-story building	30	11.8	11.0	11.0	6.9
≤ 115	One-story building	10	6.0	6.0	4.1	6.0
≤ 115	One-story building	20	8.7	8.1	8.1	6.0
≤ 115	One-story building	30	13.0	12.1	12.1	7.6
≤ 120	One-story building	10	6.0	6.0	4.4	6.0
≤ 120	One-story building	20	9.4	8.8	8.8	6.0
≤ 120	One-story building	30	14.1	13.1	13.1	8.3
≤ 130	One-story building	10	6.0	6.0	5.1	6.0
≤ 130	One-story building	20	11.0	10.3	10.3	6.5
≤ 130	One-story building	30	16.5	15.4	15.4	9.7
≤ 140	One-story building	10	6.0	6.0	5.9	6.0
≤ 140	One-story building	20	12.7	11.9	11.9	7.5
≤ 140	One-story building	30	19.1	17.8	17.8	11.2

For SI: 1 foot = 304.8 mm, 1 mile per hour = 0.447 m/s.

a. Linear interpolation shall be permitted.

b. Braced wall panels shall be without openings.

c. Braced wall panel Types A, B and E shall have an aspect ratio $(H:L) \leq 1:1$. Braced wall panel Types C and D shall have an aspect ratio $(H:L) \leq 2:1$.

d. Subject to applicable wind adjustment factors associated with Items 1 and 2 of Table R602.10.3(2).

e. Cob braced wall panel types indicated shall comply with Section AU106.11 and Table AU106.11(1).

TABLE AU106.11(3)
BRACING REQUIREMENTS FOR COB BRACED WALL PANELS BASED ON SEISMIC DESIGN CATEGORY A

• SOIL CLASS D[f] • TOTAL WALL HEIGHT = 10 FEET (INCLUDING STEM WALL AND BOND BEAM) • COB WALL HEIGHT PER TABLE AU106.11(1) • 15 PSF ROOF-CEILING DEAD LOAD[d] • STORY LOCATION: ONE-STORY BUILDING • SEISMIC DESIGN CATEGORY A • 1.5″ PLASTER THICKNESS EACH SIDE[g]				MINIMUM TOTAL LENGTH (FEET) OF COB BRACED WALL PANELS REQUIRED ALONG EACH BRACED WALL LINE[a, b, c, d, e]		
Braced Wall Line Spacing (feet)	Braced Wall Line Length (feet)	Min. Braced Wall Line % Openings	Min. Perpendicular Braced Wall Line % Openings	Cob Braced Wall Panel[e] A, B	Cob Braced Wall Panel[e] C, D	Cob Braced Wall Panel[e] E
10	30	0	0	—	3.4	6.0
20	20	0	0	—	3.5	6.0
20	30	0	0	—	4.5	6.0
30	30	0	0	—	5.6	6.0

For SI: 1 inch = 25.4 mm, 1 foot = 304.8 mm, 1 pound per square foot = 0.0479 kPa.

a. Interpolation is not permitted.

b. Braced wall panels shall be without openings.

c. Braced wall panel Types A, B and E shall have an aspect ratio $(H:L) \leq 1:1$. Braced wall panel Types C and D shall have an aspect ratio $(H:L) \leq 2:1$.

d. Subject to applicable seismic adjustment factors associated with Item 5 in Table R602.10.3(4).

e. Cob braced wall panel types indicated shall comply with Section AU106.11 and Table AU106.11(1).

f. Wall bracing lengths are based on a soil site class D. Interpolation of bracing lengths between S_{DS} values associated with the seismic design categories is allowable where a site-specific S_{DS} value is determined in accordance with Section 1613 of the *International Building Code*.

g. For total plaster thickness between 3 inches and 6 inches, the minimum total length of braced wall panels shall be multiplied by 1.2.

TABLE AU106.11(4)
BRACING REQUIREMENTS FOR COB BRACED WALL PANELS BASED ON SEISMIC DESIGN CATEGORY B

• SOIL CLASS D[f] • TOTAL WALL HEIGHT = 10 FEET (INCLUDING STEM WALL AND BOND BEAM) • COB WALL HEIGHT PER TABLE AU106.11(1) • 15 PSF ROOF-CEILING DEAD LOAD[d] • STORY LOCATION: ONE-STORY BUILDING • SESIMIC DESIGN CATEGORY B • 1.5″ PLASTER THICKNESS EACH SIDE[g]				MINIMUM TOTAL LENGTH (FEET) OF COB BRACED WALL PANELS REQUIRED ALONG EACH BRACED WALL LINE[a, b, c, d, e]		
Braced Wall Line Spacing (feet)	Braced Wall Line Length (feet)	Min. Braced Wall Line % Openings	Min. Perpendicular Braced Wall Lines % Openings	Cob Braced Wall Panel[e] A, B	Cob Braced Wall Panel[e] C, D	Cob Braced Wall Panel[e] E
10	10	0	0	6.0	3.2	6.0
10	20	0	0	6.0	4.9	6.0
10	20	50	0	6.0	3.5	6.0
10	30	0	0	7.1	6.6	6.0
10	30	50	0	6.0	4.5	6.0
20	10	0	0	6.0[h]	4.9[h]	6.0
20	10	0	50	6.0	3.5	6.0
20	10	50	0	NP	4.2	NP
20	10	50	50	NP	3.0	NP
20	20	0	0	7.4	6.9	6.0
20	20	0	50	6.0	5.5	6.0
20	20	50	0	6.0	5.5	6.0
20	20	50	50	6.0	4.1	6.0
20	30	0	0	9.4	8.8	6.0
20	30	0	50	7.9	7.4	6.0
20	30	50	0	7.2	6.7	6.0
20	30	50	50	6.0	5.3	6.0
30	10	0	0	7.1	6.6	6.0
30	20	0	0	9.4	8.8	6.0
30	20	0	50	7.2	6.7	6.0
30	20	50	0	7.9	7.4	6.0
30	30	0	0	11.8	11.0	6.0
30	30	0	50	9.5	8.9	6.0
30	30	50	0	9.5	8.9	6.0
30	30	50	50	7.3	6.8	6.0

For SI: 1 inch = 25.4 mm, 1 foot = 304.8 mm, 1 pound per square foot = 0.0479 kPa.

NP = Not Permitted.

a. Interpolation is not permitted.

b. Braced wall panels shall be without openings.

c. Braced wall panel Types A, B and E shall have an aspect ratio $(H{:}L) \leq 1{:}1$. Braced wall panel Types C and D shall have an aspect ratio $(H{:}L) \leq 2{:}1$.

d. Subject to applicable seismic adjustment factors associated with Item 5 in Table R602.10.3(4).

e. Cob braced panel types indicated shall comply with Section AU106.11 and Table AU106.11(1).

f. Wall bracing lengths are based on a soil site class D. Interpolation of bracing lengths between S_{DS} values associated with the seismic design categories is allowable where a site-specific S_{DS} value is determined in accordance with Section 1613 of the *International Building Code*.

g. For total plaster thicknesses 3 inches to 6 inches, the minimum total length of braced wall panels shall be multiplied by 1.2.

h. Total plaster thicknesses shall be not greater than 3 inches. Substitute $^{15}/_{32}$″ roof sheathing and 10d at 6″ edge nailing for requirements in Table R602.3(1).

TABLE AU106.11(5)
BRACING REQUIREMENTS FOR COB BRACED WALL PANELS BASED ON SEISMIC DESIGN CATEGORY C

• SOIL CLASS D[f] • TOTAL WALL HEIGHT = 10 FEET (INCLUDING STEM WALL AND BOND BEAM) • COB WALL HEIGHT PER TABLE AU106.11(1) • 15 PSF ROOF-CEILING DEAD LOAD[d] • STORY LOCATION: ONE-STORY BUILDING • SESIMIC DESIGN CATEGORY C • 1.5″ PLASTER THICKNESS EACH SIDE[g]				MINIMUM TOTAL LENGTH (FEET) OF COB BRACED WALL PANELS REQUIRED ALONG EACH BRACED WALL LINE[a, b, c, d, e]		
Braced Wall Line Spacing (feet)	Braced Wall Line Length (feet)	Min. Braced Wall Line % Openings	Min. Perpendicular Braced Wall Lines % Openings	Cob Braced Wall Panel[e] A, B	Cob Braced Wall Panel[e] C, D	Cob Braced Wall Panel[e] E
10	10	0	0	8.3[h]	7.8[h]	6.0
10	10	0	50	6.5	6.1	6.0
10	10	25	0	7.4[h]	6.9[h]	6.0
10	10	50	50	NP	4.4	6.0
10	15	0	0	10.6	9.9	6.0
10	15	0	50	8.7	8.2	6.0
10	15	50	0	NP	7.3	6.0
10	15	50	50	6.0	5.6	6.0
10	20	0	0	12.8	11.9	6.0
10	20	0	50	11.0	10.2	6.0
10	20	50	0	9.1	8.5	6.0
10	20	50	50	7.3	6.8	6.0
15	10	25	0	NP	NP	6.0[h]
15	10	0	50	7.8	7.3	6.0
15	10	50	0	NP	NP	NP
15	10	50	50	NP	NP	NP
15	15	0	0	12.9	12.1	6.0
15	15	0	50	10.2	9.5	6.0
15	15	50	0	NP	NP	6.0
15	15	50	50	7.5	7.0	6.0
15	20	0	0	15.3	14.3	6.0
15	20	0	50	12.6	11.7	6.0
15	20	50	0	NP	NP	6.0
15	20	50	50	8.9	8.3	6.0
20	10	25	0	NP	NP	NP
20	10	0	50	9.1	8.5	6.0
20	10	50	0	NP	NP	NP
20	10	50	50	NP	NP	NP
20	15	0	0	NP	14.3[h]	6.0[h]
20	15	0	50	11.7[h]	10.9[h]	6.0
20	15	50	0	NP	NP	6.0[h]
20	15	50	50	NP	NP	6.0
20	20	0	0	17.8	16.7	6.9
20	20	0	50	14.2	13.3	6.0
20	20	50	0	NP	NP	6.0
20	20	50	50	NP	9.9	6.0

For SI: 1 inch = 25.4 mm, 1 foot = 304.8 mm, 1 pound per square foot = 0.0479 kPa.
NP = Not Permitted.

a. Interpolation is not permitted.
b. Braced wall panels shall be without openings.
c. Braced wall panel Types A, B and E shall have an aspect ratio $(H{:}L) \leq 1{:}1$. Braced wall panel Types C and D shall have an aspect ratio $(H{:}L) \leq 2{:}1$.
d. Subject to applicable seismic adjustment factors associated with Item 5 in Table R602.10.3(4).
e. Cob braced panel types indicated shall comply with Section AU106.11 and Table AU106.11(1).
f. Wall bracing lengths are based on a soil site class D. Interpolation of bracing lengths between S_{DS} values associated with the seismic design categories is allowable where a site-specific S_{DS} value is determined in accordance with Section 1613 of the *International Building Code*.
g. For total plaster thicknesses 3″ to 6″, multiply the minimum total length of braced wall panels by 1.2.
h. Total plaster thickness shall not be greater than 3 inches. Substitute $^{15}/_{32}$″ roof sheathing and 10d at 6″ edge nailing for requirements in Table R602.3(1).

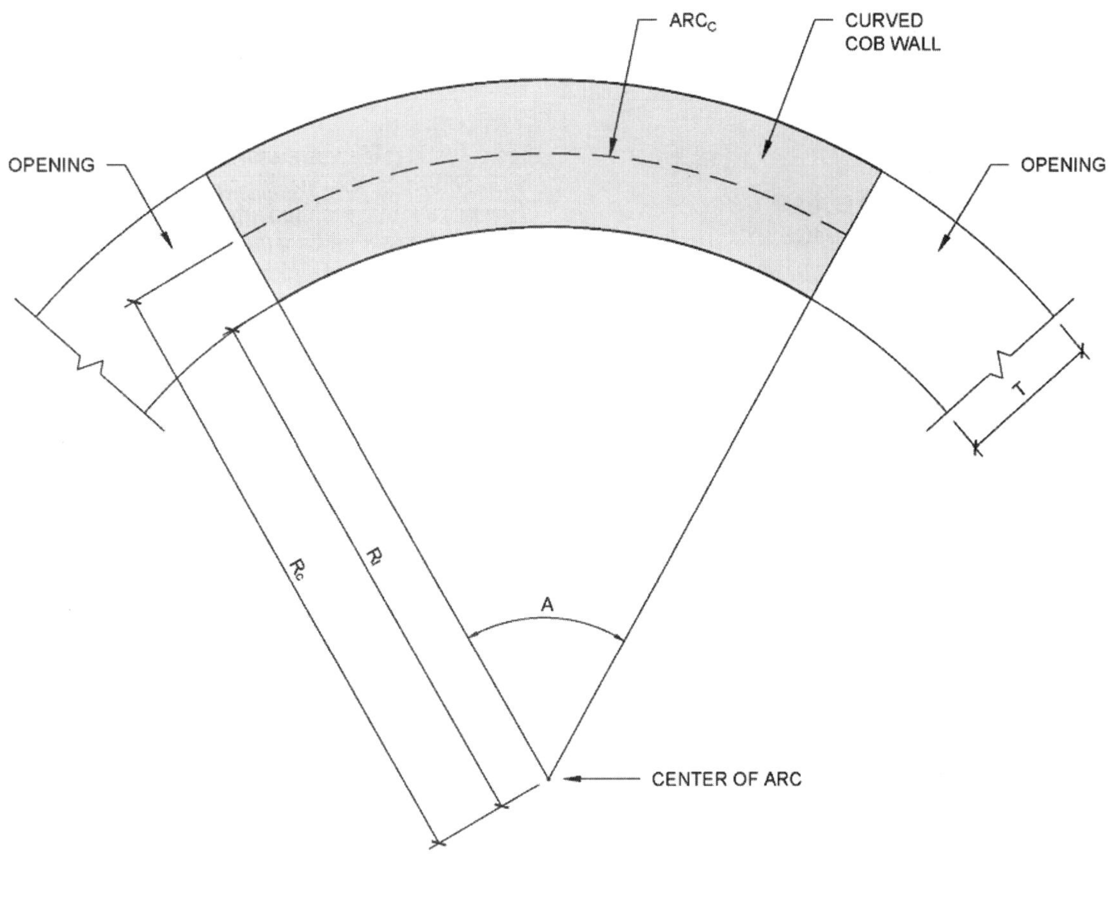

FIGURE AU106.11.3
CURVED BRACED WALL PANEL

TABLE AU106.12
ANCHORAGE OF BOND BEAMS FOR WIND UPLIFT

ANCHORS: • $^5/_8$" ALL THREAD AT 12" O.C.[a, b] • 2" × 2" × $^1/_4$" WASHERS AND NUT AT END IN COB • 4" EMBEDMENT IN CONCRETE BOND BEAMS • FULL PENETRATION THROUGH WOOD BOND BEAMS WITH 2" × 2" × $^1/_4$" WASHER AND NUT			
WIND UPLIFT FORCE FROM TABLE R802.11 (PLF)	ANCHORAGE DEPTH IN INCHES, PER WALL WIDTH AND WIND UPLIFT FORCE		
	≤ 12" Wall Width[c]	≤ 16" Wall Width[c]	≤ 24" wall width[c]
< 75	16	12	12
< 100	24	16	12
< 150	48 o.c. continuous from foundation to bond beam[d]	24	16
< 200	48 o.c. continuous from foundation to bond beam[d]	48 o.c. continuous from foundation to bond beam[d]	24

For SI: 1 inch = 25.4 mm.

a. For wood bond beams a maximum of 6 inches from bond beam ends.

b. For minimum 6-inch by 8-inch concrete bond beams, at 18″ o.c. for wind uplift forces less than 75 pounds per linear foot, and at 16″ o.c. for wind uplift forces less than 100 pounds per linear foot.

c. Excluding finishes.

d. With 7-inch embedment in foundation, 4-inch embedment in concrete bond beam or full penetration through wood bond beam with 2-inch by 2-inch by $^1/_4$-inch washer and nut.

SECTION AU107
COB FLOORS

AU107.1 Cob floors. *Cob* floors supported by *grade* shall be in accordance with an *approved* specification. Straw shall not be required in the material mix.

SECTION AU108
FIRE RESISTANCE

AU108.1 Fire-resistance rating. *Cob* walls are not fire-resistance rated.

AU108.2 Clearance to fireplaces and chimneys. *Cob* walls or other *cob* surfaces shall not require clearance to fireplaces and chimneys, except where clearance to noncombustibles is required by the manufacturer's instructions.

SECTION AU109
THERMAL PERFORMANCE

AU109.1 Thermal characteristics. *Cob* walls shall be classified as mass walls in accordance with Section N1102.2.5 and shall meet the *R*-value requirements for mass walls in Table N1102.1.3.

AU109.2 Thermal resistance. The unit *R*-value for *cob* walls with a density of 110 pounds per cubic foot (1762 kg/m^3) shall be R-0.22 per inch of *cob* thickness. Walls that vary in thickness along their height or length shall use the average thickness of the wall to determine its *R*-value. The thermal resistance values of air films and finish materials or additional insulation shall be added to the *cob* wall's thermal resistance value to determine the *R*-value of the wall assembly.

AU109.3 Additional insulation. Where insulating materials are added to the face of a *cob* wall, the combination of additional insulation and any associated connecting, weather-resisting or protective materials shall comply with Section AU104.1.2, Items 1–4.

SECTION AU110
REFERENCED STANDARDS

AU110.1 General. See Table AU110.1 for standards that are referenced in various sections of this appendix. Standards are listed by the standard identification with the effective date, the standard title and the section or sections of this appendix that reference the standard.

TABLE AU110.1
REFERENCED STANDARDS

STANDARD ACRONYM	STANDARD NAME	SECTIONS HEREIN REFERENCED
ASTM C5—10	*Standard Specification for Quicklime for Structural Purposes*	AU104.4.6
ASTM C141/C141M—14	*Standard Specification for Hydrated Hydraulic Lime for Structural Purposes*	AU104.4.6
ASTM C206—14	*Standard Specification for Finishing Hydrated Lime*	AU104.4.6
ASTM C926	*Specification for Appliance of Portland Cement-based Plaster*	AU104.4.8
ASTM C1707—11	*Standard Specification for Pozzolanic Hydraulic Lime for Structural Purposes*	AU104.4.6
ASTM E2392/E2392M—10	*Standard Guide for Design of Earthen Wall Building Systems*	AU104.4.3.2
ASTM BS1, ASTM BS EN 459—2015	Part 1: *Building Lime. Definitions, Specifications and Conformity Criteria*; Part 2: *Test Methods*	AU104.4.6

APPENDIX AV

BOARD OF APPEALS

The provisions contained in this appendix are not mandatory unless specifically referenced in the adopting ordinance.

User Note:

About this appendix: Appendix AV provides criteria for Board of Appeals members. Also provided are procedures by which the Board of Appeals should conduct its business.

SECTION AV101 GENERAL

AV101.1 Scope. A board of appeals shall be established within the *jurisdiction* for the purpose of hearing applications for modification of the requirements of this code pursuant to the provisions of Section R112. The board shall be established and operated in accordance with this section, and shall be authorized to hear evidence from appellants and the *building official* pertaining to the application and intent of this code for the purpose of issuing orders pursuant to these provisions.

AV101.2 Application for appeal. Any person shall have the right to appeal a decision of the *building official* to the board. An application for appeal shall be based on a claim that the intent of this code or the rules legally adopted hereunder have been incorrectly interpreted, the provisions of this code do not fully apply or an equally good or better form of construction is proposed. The application shall be filed on a form obtained from the *building official* within 20 days after the notice was served.

AV101.2.1 Limitation of authority. The board shall not have authority to waive requirements of this code or interpret the administration of this code.

AV101.2.2 Stays of enforcement. Appeals of notice and orders, other than Imminent Danger notices, shall stay the enforcement of the notice and order until the appeal is heard by the board.

AV101.3 Membership of board. The board shall consist of five voting members appointed by the chief appointing authority of the *jurisdiction.* Each member shall serve for [INSERT NUMBER OF YEARS] years or until a successor has been appointed. The board member's terms shall be staggered at intervals, so as to provide continuity. The *building official* shall be an ex officio member of said board but shall not vote on any matter before the board.

AV101.3.1 Qualifications. The board shall consist of five individuals, who are qualified by experience and training to pass on matters pertaining to building construction and are not employees of the *jurisdiction.*

AV101.3.2 Alternate members. The chief appointing authority is authorized to appoint two alternate members who shall be called by the board chairperson to hear appeals during the absence or disqualification of a member. Alternate members shall possess the qualifications required for board membership, and shall be appointed for the same term or until a successor has been appointed.

AV101.3.3 Vacancies. Vacancies shall be filled for an unexpired term in the same manner in which original appointments are required to be made.

AV101.3.4 Chairperson. The board shall annually select one of its members to serve as chairperson.

AV101.3.5 Secretary. The chief appointing authority shall designate a qualified clerk to serve as secretary to the board. The secretary shall file a detailed record of all proceedings which shall set forth the reasons for the board's decision, the vote of each member, the absence of a member and any failure of a member to vote.

AV101.3.6 Conflict of interest. A member with any personal, professional or financial interest in a matter before the board shall declare such interest and refrain from participating in discussions, deliberations and voting on such matters.

AV101.3.7 Compensation of members. Compensation of members shall be determined by law.

AV101.3.8 Removal from the board. A member shall be removed from the board prior to the end of their term only for cause. Any member with continued absence from regular meeting of the board may be removed at the discretion of the chief appointing authority.

AV101.4 Rules and procedures. The board shall establish policies and procedures necessary to carry out its duties consistent with the provisions of this code and applicable state law. The procedures shall not require compliance with strict rules of evidence, but shall mandate that only relevant information be presented.

AV101.5 Notice of meeting. The board shall meet upon notice from the chairperson, within 10 days of the filing of an appeal or at stated periodic intervals.

AV101.5.1 Open hearing. All hearings before the board shall be open to the public. The appellant, the appellant's representative, the *building official* and any person whose interests are affected shall be given an opportunity to be heard.

AV101.5.2 Quorum. Three members of the board shall constitute a quorum.

AV101.5.3 Postponed hearing. When five members are not present to hear an appeal, either the appellant or the appellant's representative shall have the right to request a postponement of the hearing.

AV101.6 Legal counsel. The *jurisdiction* shall furnish legal counsel to the board to provide members with general legal advice concerning matters before them for consideration. Members shall be represented by legal counsel at the *jurisdiction*'s expense in all matters arising from service within the scope of their duties.

AV101.7 Board decision. The board shall only modify or reverse the decision of the *building official* by a concurring vote of three or more members.

AV101.7.1 Resolution. The decision of the board shall be by resolution. Every decision shall be promptly filed in writing in the office of the *building official* within 3 days and shall be open to the public for inspection. A certified copy shall be furnished to the appellant or the appellant's representative and to the *building official.*

AV101.7.2 Administration. The *building official* shall take immediate action in accordance with the decision of the board.

AV101.8 Court review. Any person, whether or not a previous party of the appeal, shall have the right to apply to the appropriate court for a writ of certiorari to correct errors of law. Application for review shall be made in the manner and time required by law following the filing of the decision in the office of the chief administrative officer.

APPENDIX AW

3D-PRINTED BUILDING CONSTRUCTION

The provisions contained in this appendix are not mandatory unless specifically referenced in the adopting ordinance.

User Note:

About this appendix: *Appendix AW provides for the design, construction and inspection of 3D building construction. UL 3401 was developed to evaluate critical aspects of this construction process to result in consistent 3D-printed building techniques that comply with a level of safety and performance equivalent to legacy construction techniques currently in the code.*

SECTION AW101 GENERAL

AW101.1 Scope. Buildings, structures and building elements fabricated in whole or in part using 3D-printed construction techniques shall be designed, constructed and inspected in accordance with the provisions contained in this appendix and other applicable requirements in this code.

AW101.2 Definitions. The words and terms in Section AW102 shall, for the purposes of this appendix, have the meanings shown herein. Refer to Chapter 2 of this code for general definitions.

SECTION AW102 DEFINITIONS

3D-PRINTED BUILDING CONSTRUCTION. A process for fabricating buildings, structures and building elements from 3D model data using automated equipment that deposits construction material in a layer-upon-layer fashion.

ADDITIVE MANUFACTURING MATERIALS. Materials used by the 3D printer to produce the building structure or system components of the building.

FABRICATION PROCESS. Preparation of the job site and construction material, the deposition, curing, finishing, insertion of components and other methods used to construct building elements such as walls, partitions, *roof assemblies* and structural components, and the means used to connect assemblies together.

PRODUCTION EQUIPMENT. The equipment, including the 3D printer, its settings, nozzles and other accessories used in the fabrication process.

SYSTEM COMPONENTS. Devices, equipment and *appliances* that are installed in the building elements as part of the wiring, plumbing, HVAC and other systems. These include, but are not limited to, electrical outlet boxes, conduit, wiring, piping, tubing and HVAC ducts, each of which is covered by a product standard or installation code requirement.

SECTION AW103 BUILDING DESIGN

AW103.1 Design organization. 3D-printed buildings, structures and building elements shall be designed by an organization certified in accordance with UL 3401 by an *approved* agency and approved by the *building official* in accordance with this section.

AW103.2 Design approval. The structural design, *construction documents* and UL 3401 report of findings shall be submitted for review and approval in accordance with Section 104.11.

SECTION AW104 BUILDING CONSTRUCTION

AW104.1 Construction. 3D-printed buildings, structures and building elements shall be constructed in accordance with this section.

AW104.2 Construction method. The building construction method, consisting of the manufacturer's production equipment and fabrication process, shall be in accordance with the UL 3401 report of findings. The unique identifier of the construction method used shall match the identifier in the UL 3401 report of findings.

AW104.3 Additive manufacturing materials. Only the *listed* additive manufacturing materials identified in the UL 3401 report of findings shall be used to fabricate the building structure or system components. Containers of the additive manufacturing materials shall be *labeled.*

AW104.4 Depositing of manufacturing materials. Manufacturing materials shall only be deposited where ambient temperature and environmental conditions at the job site are within limits specified in the UL 3401 report of findings. The maximum number of layers permitted, specified curing time and any surface preparation or finishing shall be performed as specified in the UL 3401 report of findings.

SECTION AW105 SPECIAL INSPECTIONS

AW105.1 Initial inspection. An initial inspection of the production equipment, including 3D printer, and the fabrication process shall be performed after the production equipment is located on site and before building fabrication has begun. The inspection shall be conducted by representatives of the approved agency that evaluated the fabrication process for compliance with UL 3401. The inspection shall verify that the fabrication process, including production equipment, 3D-printing parameters and additive manufacturing materials, are in accordance with the UL 3401 report of findings and the proprietary information in the UL 3401 detailed report of findings.

Exception: Where *approved* by the *building official*, inspections of the production equipment, including 3D printer, and the fabrication process used in a single housing tract shall be conducted on the first building to be constructed, and on a selected number of subsequent buildings, where the same equipment, equipment operators and fabrication process are used on all buildings. The number of inspections to be performed shall be determined by the *building official.*

SECTION AW106 REFERENCED STANDARDS

AW106.1 General. See Table AW106.1 for standards that are referenced in various sections of this appendix. Standards are listed by the standard identification with the effective date, the standard title and the section or sections of this appendix that reference the standard.

TABLE AW106.1 REFERENCED STANDARDS

STANDARD ACRONYM	STANDARD NAME	SECTIONS HEREIN REFERENCED
UL 3401—19	*Outline of Investigation for 3D Printed Building Construction*	AW103.2, AW104.2, AW104.3, AW104.4, AW105.1

INDEX

A

B

G

H

I

J

K

L

M

N

O

P

R

S

T

U

V

W

Y

RESOURCE A

RECOMMENDED PRACTICES FOR REMOTE VIRTUAL INSPECTIONS (RVI)

User note:

About this resource: *The typical process of inspecting projects by inspectors driving to job sites and performing on-site inspections has certain challenges that impacts timeliness and resource efficiencies both for building construction and safety industry and regulating jurisdictions. The time spent driving to job sites, particularly in larger cities with busy traffic patterns, takes up a substantial part of the day, reducing the number of inspections possible to complete and creating a backlog of requested inspections.*

To address some of the challenges, many jurisdictions have implemented remote virtual inspections (RVI) for more routine and simpler inspections such as water heater replacements or other similar items. RVI is an alternative to on-site inspections using a video call with the inspector. With advances in technology and availability of sophisticated smart phones and tablets, RVI has become more common, and some jurisdictions plan to implement it for more complicated and larger inspection items or projects.

To assist the building construction industry and member jurisdictions in adoption of an RVI program, in May 2020, the International Code Council® (ICC®) published the Recommended Practices for Remote Virtual Inspections (RVI). *This publication offers a comprehensive framework for both local jurisdictions and building industry professionals that desire to implement a remote inspection program.*

Recommended Practices for Remote Virtual Inspections (RVI)

Recommended Practices for Remote Virtual Inspections (RVI)

Second Printing: September 2020

ISBN: 978-1-952468-23-0

The information contained in this document is believed to be accurate; however, it is being provided for informational purposes only and is intended for use only as a reference for recommended practices. Publication of this document by the ICC should not be construed as the ICC engaging in or rendering engineering, legal or other professional services. Use of the information contained in this publication should not be considered by the user as a substitute for the advice of a registered professional engineer, attorney or other professional. If such advice is required, it should be sought through the services of a registered professional engineer, licensed attorney or other professional. This Recommended Practices document does not intend to overrule national building codes or local practices, but to provide a reference for minimum levels of quality to safeguard health, property and general welfare.

PRINTED IN THE USA

Table of Contents

Preface

Technological advances have created enormous possibilities in all aspects of life, including the building construction and safety industry. Digital and online tools for building design, construction and administrative functions, such as permit application, plan review, inspection and commissioning, have drastically increased the efficiency and accuracy of achieving safe and resilient communities. Local, state and national governments have taken advantage of advancing technologies and have incorporated various levels of digitization into their processes in order to save time and reduce costs. Examples of such efforts include online offering of permit applications, payment of permit fees, submittal of plans and digital plan review.

The speed of adoption and implementation of technology, however, varies by geographic region and depends on a number of factors, including the availability of financial resources and the infrastructure needed to support the technology. Many Authorities Having Jurisdiction (AHJs) have implemented technology at various levels with good success and have embraced greater reliance on digitization as time goes by.

The 2020 global coronavirus pandemic created an impetus in speeding the implementation of modern technologies and taking advantage of new ideas in a much shorter time frame. The spread of COVID-19 and the closing of most businesses and social activities in many parts of the world to create social distancing resulted in many sectors of the economy searching to find new solutions for conducting business.

Many AHJs needed to come up with solutions to perform all aspects of codes and standards administration from remote locations and/or home offices. One such solution using available technology is Remote Virtual Inspections (RVI).

RVI is a method of inspection that allows the needed inspections to proceed in a timely manner by the owner or contractor located on the jobsite and the inspector or inspection teams performing the inspection remotely. While this practice gained good acceptance and implementation during the weeks and months of COVID-19 social distancing, its advantages are so great that it will likely become a popular and routine tool for the foreseeable future.

The advantages and opportunities created by RVI locally, nationally and globally are enormous, allowing those with technical expertise in their specific subjects to offer their services across the globe. Building code specialists, inspectors and consultants will be able to provide services and consulting from far distances and to help building safety and resiliency anywhere needed at the local, national or global level.

Recommended Practices for Remote Virtual Inspections (RVI) was developed based on study, research, and discussions related to items that should be considered and addressed for an effective and consistent RVI program and to assist AHJs in implementing the readily available technologies in the adoption and implementation of their own RVI program.

ICC welcomes your comments and feedback to improve future editions of this Recommended Practices publication. Submit feedback at www.iccsafe.org/RVI.

About the International Code Council®

The International Code Council is a nonprofit association that provides a wide range of building safety solutions including product evaluation, accreditation, certification, codification and training. It develops model codes and standards used worldwide to construct safe, sustainable, affordable and resilient structures. The mission of the Code Council is to provide the highest quality codes, standards, products and services for all concerned with the safety and performance of the built environment. ICC Evaluation Service (ICC-ES) is the industry leader in performing technical evaluations for code compliance fostering safe and sustainable design and construction.

Washington DC Headquarters:
500 New Jersey Avenue, NW, 6th Floor
Washington, DC 20001-2070

Regional Offices:

- Eastern Regional Office (BIR)
- Central Regional Office (CH)
- Western Regional Office (LA)
- Distribution Center (Lenexa, KS)

888-ICC-SAFE (888-422-7233)
www.iccsafe.org

Family of Solutions:

1.0 Introduction

Hand-held devices such as smartphones and tablets have capabilities for real time, online communication of videos and photos. Use of advanced tools and technologies, combined with the power of such hand-held devices, has made it possible for anyone to observe the construction activities of a jobsite from any location, near or thousands of miles away. Using Remote Virtual Inspection (RVI) allows construction projects to continue without impediment and allows the Authority Having Jurisdiction (AHJ) to continue to provide the vital services needed for construction of safe buildings.

Purpose and Scope

The purpose and scope of these Recommended Practices is to provide guidance to the Authority Having Jurisdiction (AHJ) when implementing a Remote Virtual Inspection (RVI) program as well as to the construction industry user. This document specifically addresses implementation and administration of RVI. These procedures are organized in a fashion that can be readily implemented by the AHJ as part of their inspection procedures. This document also provides recommended practices to construction industry professionals submitting to an RVI.

Until recently, Remote Virtual Inspections have been conducted only by a few AHJs at varying levels. As a result, there has not been a standardized program that addresses how to prepare for, conduct and participate in these types of inspections.

2.0 Definitions and Acronyms

1. **RVI: Remote Virtual Inspection:** Remote Virtual Inspection, also known as RVI, is a form of visual inspection which uses visual or electronic aids to allow an inspector or team of inspectors to observe products and/or materials from a distance because the objects are inaccessible or are in dangerous environments, or whereby circumstances or conditions prevent an in-person inspection.
2. **AHJ: Authority Having Jurisdiction.**

3.0 Remote Virtual Inspection Process

Remote Virtual Inspections (RVI) may provide benefits to AHJs and customers alike. In certain circumstances, an RVI may provide a better quality inspection with an increase in efficiency and cost savings. It will increase the efficiency of the inspection process utilizing modern technology. Depending on the location and complexity of a project, some limitations may impact its use. In cases where an RVI is not suitable or technology fails to provide sufficient visual clarity (i.e., poor/no service or Wi-Fi, poor lighting, etc.), an onsite inspection may be required. Subject to local approval, the AHJ may choose to use an approved third-party inspection agency or utilize staff inspectors. Where Wi-Fi and/or cellular reception are poor or not available, some AHJs may consider allowing the contractor to provide an acceptable electronic documentation of the area that needs an inspection for review by the assigned inspector or team of inspectors.

A clear understanding of the RVI requirements and communication throughout the process by both parties is paramount to the completion of a successful inspection. The inspector will check all aspects of the permitted construction project to the adopted codes and other applicable laws and regulations no differently than if it were an onsite inspection. Identification of the project jobsite location, posted address and its location within the building will be a critical part of the process.

The applicable Codes and Standards to be used for RVI are the same as the adopted codes and referenced standards of the AHJ. The implementation of the RVI is intended to achieve the same results as the typical in-person site inspection by applying the provisions of adopted codes such as the IBC®, IRC®, IPC®, IFC® and other applicable and adopted International Codes.

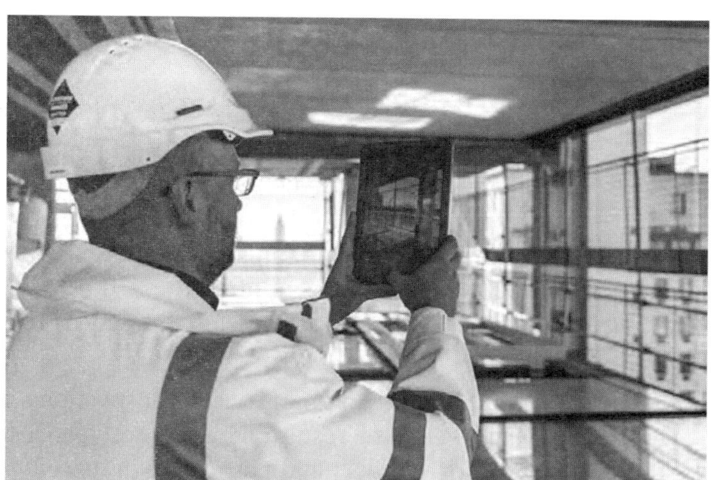

Customer: Requesting a Remote Virtual Inspection

1. Ensure there is an active permit issued or certificate application filed or obtain the appropriate one prior to attempting to schedule an inspection for the project in question.
2. Electronically sign a notice indicating that the permit holder of record or representative:
   - 2.1. Consents to the use of the remote inspections.
   - 2.2. Is responsible for their own safety during the remote inspection.
   - 2.3. Allows the complete use of the videos and photos of the remote inspection by the AHJ.
   - 2.4. Certifies they are making available the site and inspection items truthfully and to the best of their ability.
   - 2.5. Is responsible for compliance with all codes and standards applicable to the project.
   - 2.6. Acknowledges that participation in the remote inspection program is voluntary (if not a mandatory program within the AHJ's jurisdiction).
   - 2.7. Acknowledges that the decision to perform an RVI is at the sole discretion of the AHJ.

3. Prior to contacting the department to schedule the inspection, confirm that the minimum criteria for a remote inspection are met. See Appendix A for examples of qualified inspection activities.
    3.1. Note that some types of inspections may be too complex or otherwise not compatible for remote inspections.
4. Call or go online to schedule an appointment for inspection with the AHJ.
5. Must be at least 18 years old or with an adult to perform the virtual inspection.
6. When scheduling the inspection, provide the address, permit number, and type and number of requested inspections.

AHJ: Scheduling Remote Virtual Inspection

1. Schedule Inspection Time.
    1.1. All remote inspections scheduled should be requested by the customer a minimum of one business day prior to the desired inspection date.
    1.2. Schedule inspection either online or by telephone.
    1.3. Schedule sufficient time for the type of inspection requested.
    1.4. AHJ to send an inspection confirmation email or text to the customer with the date, approximate time of RVI and name of inspector.
    1.5. Send notice of customer consent and acknowledgment for electronic signature. Must be returned by customer prior to inspection scheduled time.
2. Time slots for inspections.
    2.1. Anticipated length of inspections per type (i.e., water heater installation, HVAC replacement, etc.) needs to be established.
    2.2. Each customer will be given an approximate time window for inspection.
3. Post the earliest available time for remote inspections and the latest time of the day a remote inspection may be scheduled Monday through Friday or other days selected by the AHJ.
4. Schedule after-hours or emergency inspections on a case-by-case basis.
5. Determine the types of inspections allowed for remote inspections. See Appendix A for examples of qualified inspection activities.
    5.1. All inspections may qualify for an RVI, depending on the AHJ's resources and policies.
6. Determine which type of videotelephony is available for use and is compatible with the AHJ's permitting software and videotelephony equipment.
    6.1. Videotelephony platform examples: FaceTime, Google Duo, Zoom, WhatsApp, Skype, Tango, WebEx, Microsoft Teams, GoToMeeting, etc.

Customer: Prepare for Remote Virtual Inspection

1. Prior to the inspection, ensure that:
   1.1. The jobsite is safe at all times for the individual(s) using the device during the remote inspection including health safety.
   1.2. The device (smartphone, tablet, drone, etc.) is fully charged and has a suitably charged additional power supply (battery pack).
   1.3. The use of a noise-canceling headset is recommended.
   1.4. The jobsite has high-speed Wi-Fi connectivity or minimum 4G cellular service with a strong signal.
   1.5. The necessary tools based on type of inspection are readily available.
      1.5.1. For example, carry a flashlight, tape measure, level, step ladder (for close ups of ceiling), GFCI tester, etc. An extending pole for the video device, such as selfie pole, may be very helpful in taking the smartphone or other video device closer to the point of inspection in various places such as very high ceilings.
2. Have approved plans, permit card, and other necessary construction documents available onsite.
3. Make sure good lighting is available and clear the area of any unnecessary objects.
4. All features applicable to the required inspection must be visible at the time of the remote inspection. These features must be captured sufficiently and clearly for the inspector to evaluate.
5. If at any point the inspector believes that the remote inspection process is not allowing them to properly assess compliance, they may require that a site inspection be conducted at a future date or instruct the customer to make different arrangements.
   5.1. In areas within the jobsite where there is no Wi-Fi or cell service, at the sole discretion of the inspector, the contractor may be allowed to provide video and/or photographic documentation of the item(s) to be inspected for review by the authorized inspector at a later time.
6. The onsite inspection may be conducted by an approved third-party inspection agency or by the AHJ's inspection staff.

Customer: Prepare to Receive Remote Virtual Inspection Call

1. Ensure that the lens and screen of any device being used to capture images or video has been cleaned. Dust, grit, smudges, etc., might interfere with the image quality and distorting the inspector's view.
2. To minimize interruptions during the RVI and to ensure that the video feed will be uninterrupted, make sure that all notifications are turned off in the Settings of the mobile device used for the RVI. Should the video be interrupted, the inspection could be delayed or have to be rescheduled.
3. Be prepared to answer the inspector's call at any time during the scheduled timeframe. Be cooperative and closely follow the inspector's instructions.

4. As each site and inspection is different, allot the proper amount of time for the type of inspection and accessibility of the site.
5. Carefully follow the inspector's instructions for where to direct the device and for covering the site. Do not rush the inspector but allow him or her adequate time to conduct the RVI to his or her satisfaction.
6. As much as possible, minimize background noise as that can interfere with communication with the inspector.

What to Expect During the Inspection

1. Begin inspection at the street view looking at the structure with the address or other required jobsite identification in the video display.
   1.1. Inspector may also verify location through GPS/Geotagging where the service is available.
2. Follow the directions of the inspector with respect to the order and direction of inspection.
3. As the inspection progresses, write down any items that the inspector finds that need to be corrected. Be sure the notes are detailed and ask questions of or seek clarification from the inspector at the time of the RVI.
4. If provided a permit card, do not write on it. During the next in-person visit, the inspector should update it then.
5. In most cases, the inspector will relay the results of the inspection before the end of the RVI of passing, failing or not ready for inspection.
6. Do not cover any work needing corrections until corrections are verified by reinspection. Reinspection fees may apply in accordance with the AHJ's policies.
7. Note: At a minimum, there must be an adult of the required legal age on site who will represent the owner/representative during the entire duration of the RVI.
8. The owner/representative must be able to verbally communicate with the remote inspector at all times during the inspection.

Inspection Results

1. Results of the inspection will be entered into the AHJ's permit database as soon as practicable after the RVI is completed. It is important to note that the inspection was completed using the RVI process.
2. Where an approval tag for utility connections is required, the AHJ should work directly with the utility company.
3. Following the inspection:
   3.1. Inspection comments will be available on the AHJ's website, within the AHJ's normal timelines, indicating passing or failing with the list of corrections when applicable.
   3.2. In addition, the inspector may email the inspection information upon request to the customer as soon as inspection information is available.

3.3. The inspector will determine whether additional fee(s) for reinspection is required.

4. Scheduling a reinspection or the next inspection needed is based on availability of time slots.
5. The authorized inspector may provide an option for the owner/representative to submit electronic documentation that a deficiency or deficiencies have been corrected.
6. It is incumbent on the owner/representative to provide the address and permit number on all submitted correspondence or communications.

Maintaining Records of Inspections

Required inspection records, including, but not limited to, correction notices, electronic media, recordings or photo documentation, shall be maintained in accordance with the AHJ's policy, laws, regulations, and applicable codes, and may be subject to disclosure.

4.0 Training and Communication

Training and effective communication of processes, procedures and requirements are essential and a critical part to the success of any program. This program is no different as it lends itself to new technology, new programs, and methods that are in many cases, new to the building construction and safety industry. Therefore, training of the AHJ's staff as well as the building industry on the various programs and procedures will save time and money and make the administrative and enforcement process a positive experience with minimal confusion. Training also leads to better communications between an AHJ and its customers.

Staff Training

1. Ensure all staff are trained in the appropriate areas of responsibility.
2. Permit Technicians:
   2.1. Review of approved permit applications relative to RVI requirements.
   2.2. Required departmental approvals are complete.
   2.3. Fee collection process.
   2.4. Required documents for the project (plans, calculations, etc.).
3. Remote Inspection Staff:
   3.1. Inspection software and hardware.
   3.2. Remote inspection procedures.
   3.3. Types of platforms used (Facetime, Skype, etc.).
   3.4. Reinspection fee procedures.
   3.5. Recording inspection results in permit tracking system.

Customer/Applicant

1. Ensure the owner and representative are trained in their areas of responsibility.
2. Permit applicant:
    2.1. Knowledge of the AHJ's departmental approvals required for the project.
    2.2. Knowledge of the AHJ's RVI protocol.
    2.3. Ensuring project meets RVI protocol.
    2.4. Ensure that the project is ready for the RVI at the scheduled time.
    2.5. Comply with the inspector's direction.
3. Owner/Contractor/Subcontractor:
    3.1. Requesting remote inspection process.
    3.2. Knowledge of remote inspections procedures.
    3.3. Platform required (Facetime, Skype, Google Duo, etc.).
    3.4. Jobsite communication requirements (Wi-Fi, 4G, etc.).
    3.5. Communication skills.

Additional Considerations

1. Adopt basic online security practices. Consult with your IT department for guidance.
2. Consult with your legal counsel to ensure compliance with all federal, state and local requirements related to your RVI program. For example, you may want to consult counsel to find out whether a homeowner's release is needed to conduct an RVI.
3. Ensure that all staff have access to the codes and standards that are applicable to what they are inspecting. The Code Council's Digital Codes Library (https://codes.iccsafe.org/) offers online access to all ICC model codes and standards and most state codes.
4. Document lessons learned to improve your RVI program and to support potential long-term establishment of virtual inspection processes.

5.0 Appendix A (Examples of Potential Activities)

The following are a few examples of construction activities that may be considered to be included in a RVI Program. This list is not all-inclusive. The determination of whether an inspection can be conducted remotely is at the sole discretion of the AHJ.

- Plumbing system repairs or fixture replacements.
- Construction trailer installations.
- Swimming pool excavations.
- Gas line repairs or gas utility clearance.
- Electric utility clearances.
- HVAC direct replacement or repair.
- Minor residential electrical.
- Miscellaneous repair/exterior repair or upgrades (stucco, windows, etc.).
- Re-roofing/roof covering replacement.
- Water heater or water softener direct replacement.
- New residential plumbing rough-in.
- New residential rough framing inspections.
- Residential rooftop-mounted photovoltaic panel systems.
- HUD manufactured home installation verification.
- Any other inspection approved by the AHJ.

KEYS TO SUCCESS FOR REMOTE VIRTUAL INSPECTIONS

The Value of Communication

WHAT ARE REMOTE VIRTUAL INSPECTIONS (RVI)?

Remote Virtual Inspections, also known as RVI, are a form of inspections which use visual or electronic aids to allow an inspector or team of inspectors to observe certain types of construction, products and/or materials from a distance.

RVI are a solution to help inspectors observe construction and objects that might be inaccessible or in dangerous environments, or whereby circumstances or conditions prevent an in-person inspection.

RVI BENEFITS

- Construction projects can continue without impediment
- Building professionals can continue providing services with minimal health risk during pandemics such as COVID-19
- Authorities Having Jurisdiction (AHJs), testing agencies, manufacturers, laboratories, home builders and contractors are able to provide the vital services needed on all levels for the construction of safe buildings
- Inspectors can continue providing services remotely while saving time and money
- Safe and resilient construction projects can continue to grow and thrive anywhere needed at the local, national or global level

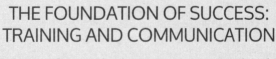

THE FOUNDATION OF SUCCESS: TRAINING AND COMMUNICATION

With the use of new technology, new programs, and methods, AHJs, testing agencies, manufacturers, laboratories, home builders and contractors can implement effective training of these programs to their staff to create a foundation for successful Remote Virtual Inspections.

A clear understanding of RVI requirements and communication throughout the process by all parties involved is paramount to the completion of successful inspections.

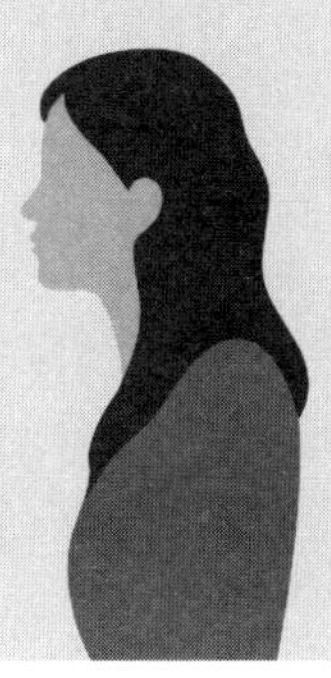

THE KEY STEPS TO A REMOTE VIRTUAL INSPECTION
1
Scheduling
The customer works with the AHJ to schedule remote virtual inspections and ensure both parties have the required documents and equipment necessary
2
Customer Preparation
The customer prepares the jobsite and devices as well as minimizes disruptions for remote virtual inspection, ready to cooperate with the inspector when on the call
3
Jobsite Identification
The inspector makes sure the jobsite is identified on the video display and begins inspection, keeping close communication with the customer
4
Inspection Results
Results are entered into the AHJ's permit database as soon as RVI is completed and communicated with the customer
5
Maintaining Records of Inspections
Required inspection records are maintained in accordance with the AHJ's policies, laws, regulations, and applicable codes, and may be subject to disclosure

LEARN MORE

Read the full version of our Recommended Practices for Virtual Remote Inspections here.

Recommended Practices for Remote Virtual Inspections (RVI)

Recommended Practices for Remote Virtual Inspections (RVI) is the most complete source of information on remote inspections. RVI is an alternative to on-site inspections using a video call on a 4G or WiFi telephony (smartphone, tablet, etc.) in order to interact with the inspector. It is a comprehensive tool for local jurisdictions and the building industry alike that desire to implement a remote inspection program.

This publication covers the RVI process, inspection scheduling, preparation, what the owner/contractor should expect, training and communications, and recording and maintaining records. While all types of inspections may not be suitable for RVI, a list of potential construction activities suitable for remote inspections is provided.

RVI also lends itself to connect seamlessly as part of an overall online program that will allow jurisdictions to provide complete services to the public utilizing the latest technology. Online permitting and electronic plan review, together with remote virtual inspections, can provide a complete program that keeps the construction industry moving while providing a healthy environment for all participants.

ISBN 978-1-952468-23-0

9 781952 468230

Item No. 7072S1